Scientific Basis for Nuclear Waste Management XX

MATERIALS RESEARCH SOCIETY
SYMPOSIUM PROCEEDINGS VOLUME 465

Scientific Basis for Nuclear Waste Management XX

Symposium held December 2–6, 1996, Boston, Massachusetts, U.S.A

EDITORS:

Walter J. Gray
Pacific Northwest National Laboratory
Richland, Washington, U.S.A.

Ines R. Triay
Los Alamos National Laboratory
Los Alamos, New Mexico, U.S.A.

PITTSBURGH, PENNSYLVANIA

A major portion of the financial support for this symposium was provided by the U.S. Department of Energy (DOE). The opinions, findings, conclusions, and recommendations expressed herein are those of the author(s) and do not necessarily reflect the view of the DOE.

Single article reprints from this publication are available through
University Microfilms Inc., 300 North Zeeb Road, Ann Arbor, Michigan 48106

CODEN: MRSPDH

Published by:

Materials Research Society
9800 McKnight Road
Pittsburgh, Pennsylvania   15237
Telephone (412) 367-3003
Fax (412) 367-4373
Website: http://www.mrs.org/

Library of Congress Cataloging in Publication Data

Scientific basis for nuclear waste management XX ; symposium held December
    2–6, 1996, Boston, Massachusetts / editors, Walter J. Gray, Ines R. Triay
        p. cm—(Materials Research Society symposium proceedings,
        ISSN 0275-0012; v. 465)
        Includes bibliographical references and index.
        ISBN 1-55899-369-X
        I. Gray, Walter J. II. Triay, Ines R. III. Title: Scientific basis for nuclear
    waste management twenty.

Manufactured in the United States of America

CONTENTS

PART I: OPENING ADDRESS

PART II: GLASS FORMULATIONS AND PROPERTIES

*Invited Paper

PART III: <u>GLASS/WATER INTERACTIONS</u>

*Invited Paper

PART VII: <u>WASTE PROCESSING AND TREATMENT</u>

PART VIII: <u>RADIATION EFFECTS IN CERAMICS, GLASSES, AND NUCLEAR WASTE MATERIALS</u>

PART IX: <u>WASTE PACKAGE MATERIALS</u>

*Invited Paper

*Invited Paper

PART XIII: <u>REPOSITORY BACKFILL</u>

*Invited Paper

PART XIV: <u>PERFORMANCE ASSESSMENT</u>

*Invited Paper

PART XVII: <u>CHERNOBYL-RELATED WASTE DISPOSAL ISSUES</u>

*Invited Paper

PREFACE

The 20th Materials Research Society symposium on "Scientific Basis for Nuclear Waste Management" included 13 oral and 15 poster sessions during which 166 papers were presented by scientists from 16 countries.

One highlight of the symposium was the opening talk by Rustum Roy, who was instrumental in establishing the first symposium entitled "Science Underlying Radioactive Waste Management," in 1978. Professor Roy summarized his views of the past 19 years of progress in this field. Two symposia were held in 1982, one in June in Berlin and the other in November in Boston, which accounts for the difference between the number of years and number of symposia.

A second highlight was the participation by several Russian and Ukrainian scientists who authored many papers on nuclear waste disposal aspects of the Chernobyl Unit 4 reactor that exploded in April 1986.

A panel discussion on the role of performance assessment for radioactive waste management was held on Wednesday afternoon. A summary of that discussion is published at the end of Part XIV: *Performance Assessment* of this volume.

Three sessions were conducted jointly with concurrent MRS symposia, as follows:

An oral session entitled *Radiation Effects in Ceramics, Glasses, and Nuclear Waste Media* was jointly conducted with Symposium B, "Microstructure Evolution During Irradiation." Four papers from that session are published in this volume. Other papers will appear in the published proceedings of Symposium B, edited by Ian M. Robertson, Gary S. Was, Linn W. Hobbs, and Tomas Diaz de la Rubia.

An oral session entitled *Glass Formulations and Properties for Nuclear Waste Management* was jointly conducted with Symposium T, "Glasses and Glass Formers: Current Issues." All twelve papers from that session are published in this volume. Related papers will appear in the published proceedings of Symposium T, edited by C. Angell, Takeshi Egami, John Kieffer, Kia Ngai, and Uli Nienhaus.

A poster session entitled *Cements in Radioactive Waste Management* was jointly conducted with Symposium HH, "Structure-Property Relationships in Hardened Cement Paste and Composites." Five papers from that session are published in this volume. No published proceedings for Symposium HH is planned.

Walter J. Gray
Ines R. Triay

March 1997

ACKNOWLEDGMENTS

The organizers of the symposium wish to acknowledge the vital role played by the many individuals and organizations who contributed to the success of the symposium.

The session chairs listed below, besides chairing the actual sessions, spent much time and energy reviewing the 166 papers published in this volume. Many of the session chairs also served on the program committee that was responsible for selecting the papers to be presented and organizing them into appropriate sessions.

Glass Formulations and Properties
- W. Lutze, University of New Mexico
- A.S. Aloy, V.G. Khlopin Radium Institute
- N.E. Bibler, Westinghouse Savannah River Company
- P. Van Iseghem, SCK/CEN

Glass/Water Interactions
- B.P. McGrail, Pacific Northwest National Laboratory
- Y. Inagaki, Kyushu University
- J.K. Bates, Argonne National Laboratory
- A. Jouan, CEA/VALRHO - Marcoule

Cements in Radioactive Waste Management
- D.A. Knecht, Lockheed Martin Idaho Technologies Company
- F.P. Glasser, University of Aberdeen

Ceramics and Crystalline Waste Forms
- R.C. Ewing, University of New Mexico
- E.R. Vance, Australian Nuclear Science and Technology Organization
- D.J. Wronkiewicz, Argonne National Laboratory
- D.A. Knecht, Lockheed Martin Idaho Technologies Company
- B.E. Burakov, V.G. Khlopin Radium Institute
- M.L. Balmer, Pacific Northwest National Laboratory
- K.P. Hart, Australian Nuclear Science and Technology Organization

Spent Nuclear Fuel
- J.C. Tait, AECL Whiteshell Nuclear Research Establishment
- J. de Pablo, Universitat Politècnica Catalunya
- M.R. Louthan, Westinghouse Savannah River Company
- L.O. Werme, Swedish Nuclear Fuel & Waste Management Company

Radiation Effects in Ceramics, Glasses, and Nuclear Waste Media
- J.K. Bates, Argonne National Laboratory
- L.W. Hobbs, Massachusetts Institute of Technology

Waste Package Materials
- M.R. Louthan, Westinghouse Savannah River Laboratory
- R. Fujisawa, Mitsubishi Materials Corporation

Radionuclide Solubility and Speciation
A. Simmons, Lawrence Berkeley National Laboratory
C. Degueldre, University of Geneva

Radionuclide Sorption
A. Simmons, Lawrence Berkeley National Laboratory
C. Degueldre, University of Geneva

Radionuclide Transport
A. Meijer, GXC Inc.
S. Stroes-Gascoyne, AECL Whiteshell Nuclear Research Establishment

Repository Backfill
W.E. Glassley, Lawrence Livermore National Laboratory
I. Neretnieks, Royal Institute of Technology/Stockholm

Performance Assessment
M.L. Wilson, Sandia National Laboratories
M.J. Apted, QuantiSci, Inc.

Natural Analogues
S.S. Levy, Los Alamos National Laboratory
P. Sellin, Swedish Nuclear Fuel & Waste Management Company

Excess Plutonium Dispositioning
R.A. Van Konynenburg, Lawrence Livermore National Laboratory
E.B. Anderson, V.G. Khlopin Radium Institute

Chernobyl-Related Waste Disposal Issues
R.C. Ewing, University of New Mexico
B.E. Burakov, V.G. Khlopin Radium Institute
V.M. Oversby, VMO Konsult, Stockholm

The U. S. Department of Energy, Argonne National Laboratory, Los Alamos National Laboratory, and Power Reactor and Nuclear Fuel Development Corporation (Japan) provided financial assistance for the symposium. This funding helped to support administrative and editorial functions, program committee activities, and the distribution of dozens of copies of these proceedings to participating individuals and organizations. It also provided travel expenses for some of the invited speakers as well as many of our Russian and Ukrainian scientists, who otherwise would not have been able to participate.

Of special recognition in helping to secure financial support, our thanks go to Dieter Knecht, Lockheed Martin Idaho Technologies Company.

The organizers thank the management of Los Alamos National Laboratory and Pacific Northwest National Laboratory for support of staff and facilities that contributed to the symposium.

The Materials Research Society staff provided excellent guidance and assistance in organizing the symposium.

Special thanks for administrative support are due to Marie Jimenez, Pacific Northwest National Laboratory, whose dedicated efforts prior to, during, and following the symposium were a major contribution to the success of the symposium and the publication of this volume of proceedings.

Editorial assistance was provided by Margaret Knecht, Lockheed Martin Idaho Technologies Company. Her many hours spent editing papers before, during, and after the symposium were greatly appreciated by the organizers and individual authors.

MATERIALS RESEARCH SOCIETY SYMPOSIUM PROCEEDINGS

Prior Materials Research Society Symposium Proceedings available by contacting Materials Research Society

Part I

Opening Address

SCIENCE UNDERLYING RADIOACTIVE WASTE MANAGEMENT: STATUS AND NEEDS—TWENTY YEARS LATER

RUSTUM ROY
Materials Research Laboratory, The Pennsylvania State University, University Park, PA 16802

ABSTRACT

The setting of the technical subsystem within the overall socio-political-economic-technical radwaste <u>system</u> will be described and the highly interactive nature of the larger setting emphasized. It will be shown that because of the dominance of the socio-political subsystem, the importance of the technical subsystem is overshadowed. Moreover the key issue in the technical subsystem, whether to put more reliance on the <u>immobilization</u> (via the waste package or engineeered barriers) or the <u>isolation</u> (via the geology) has stayed tilted toward the latter since 1978, when we organized the first symposium on radwaste science (at the Materials Research Society meeting). Now that the isolation strategy is stymied, the opportunity arises again for the <u>materials</u> community to make a compelling case for the waste package's true significance.

This review is made mainly from the perspective of one laboratory in one country, the U.S. What is remarkable about the state of research in the technical subsystem is how little the big picture of the *science* or the <u>technology</u> has changed after some billions spent on R/D. The borosilicate glass (almost unchanged) is still the establishment's choice of reference waste form for HLW. Cost, however, is finally forcing cement encapsulated forms to be given a second look. Mineral-modeled ceramics have received a great deal of scientific attention but remain esoteric to managers. It will be shown that in the author's opinion an enormous amount of detailed science has been done but most of it is unlikely to prove to be of any relevance or use. The policy implications for future R/D are discussed.

INTRODUCTION

Penn State's Materials Research Laboratory (MRL) is probably the only academic institution which has been actively involved in studying the science behind the immobilization of radwaste for 25 years. In a retrospective look across that time interval at where the nation and world stands I can do no better than quote Omar Khayyam in his Rubaiyat:

> *"Myself when young did eagerly frequent,*
> *Doctor and saint, and heard great argument,*
> *About it and about: but evermore*
> *Came out by the same door as in I went."*

Twenty years ago the first efforts at public scientific meetings on radwaste were initiated by us. And in 1978 with DOE's Ray Walton's help Greg McCarthy and I organized the first open symposium, ever on the topic at the Materials Research Society (MRS) meeting. This also was used by me to significantly broaden the society scope away from electronic materials. That event triggered a very substantial increase in research effort, first in the U.S. and then all over the world. The surge in R/D lasted only about 5 years, and then it leveled off for a decade and is slowly declining with less and less science and more engineering.

Quantification of Scientific Effort

One of the ways to describe the science underlying a field is to attempt to quantify <u>how</u> much science has been done. I have done this by searching the <u>Chemical Abstract</u> database under various keywords to be able to report what is, in fact, in the open scientific literature. Figures 1(a)–(k) present the <u>number</u> of papers published in the open literature as a function of time under the explicit keywords shown on each figure. (Note that the vertical scales are very different.) From these data, one can draw three conclusions:

Mat. Res. Soc. Symp. Proc. Vol. 465 © 1997 Materials Research Society

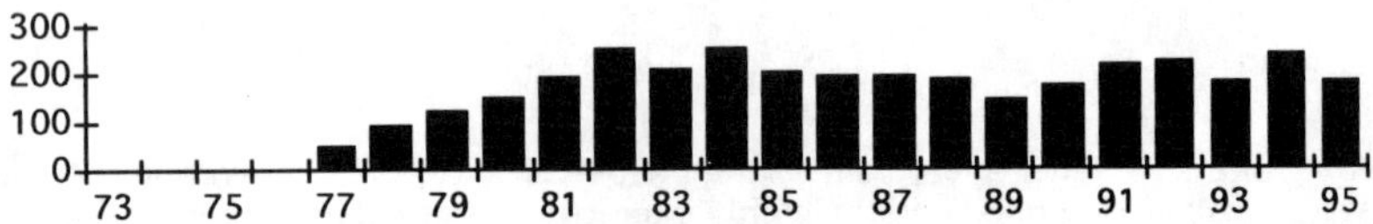

Key Words: Nuclear Waste (Number of Citations in <u>Chemical Abstracts</u> = 3688)

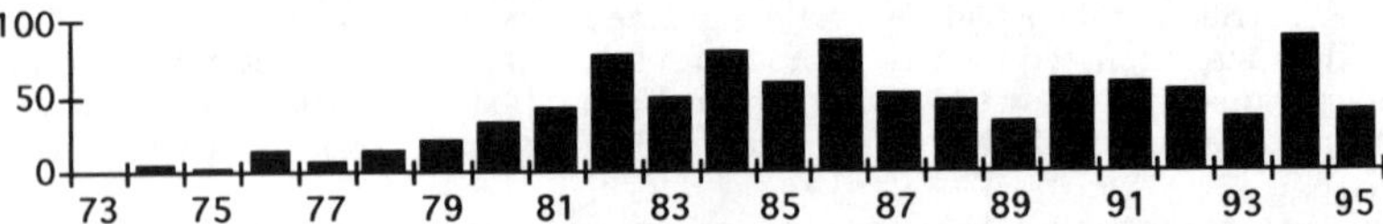

Key Words: Nuclear Waste Glass (Number of Citations in <u>Chemical Abstracts</u> = 960)

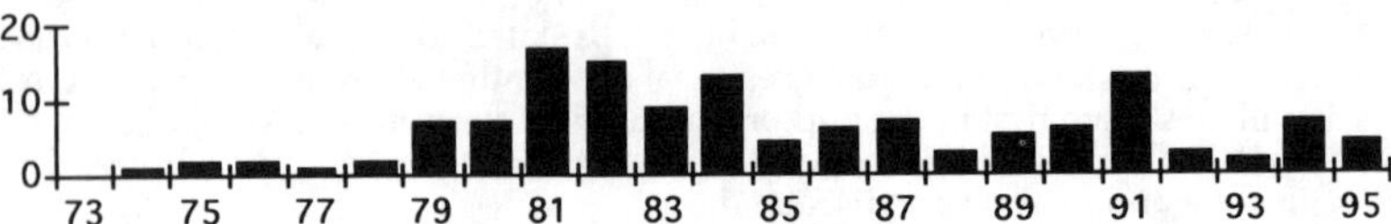

Key Words: Nuclear Waste Ceramics (Number of Citations in <u>Chemical Abstracts</u> = 136)

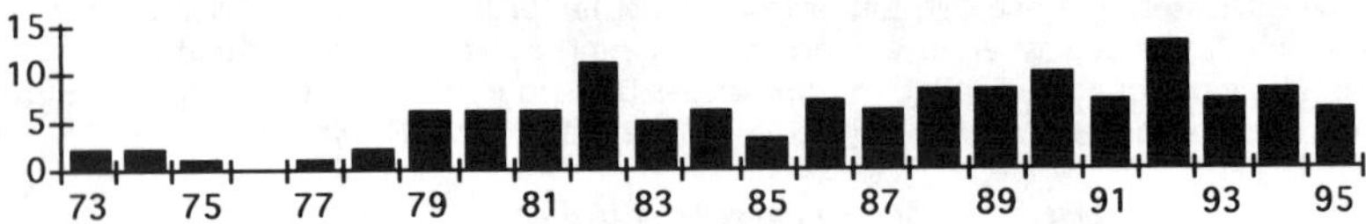

Key Words: Nuclear Waste Cement (Number of Citations in <u>Chemical Abstracts</u> = 131)

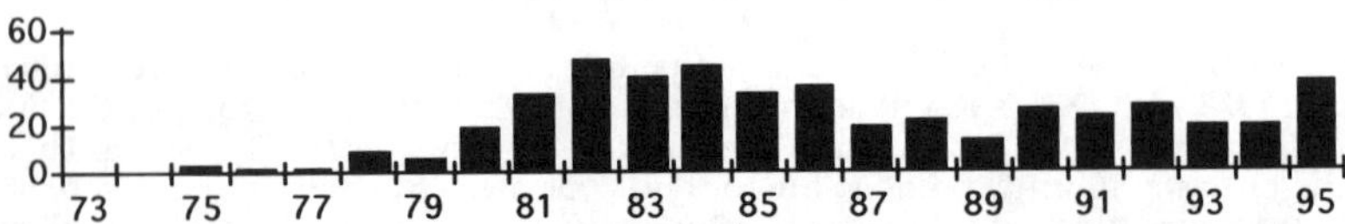

Key Words: Nuclear Waste Leaching (Number of Citations in <u>Chemical Abstracts</u> = 470)

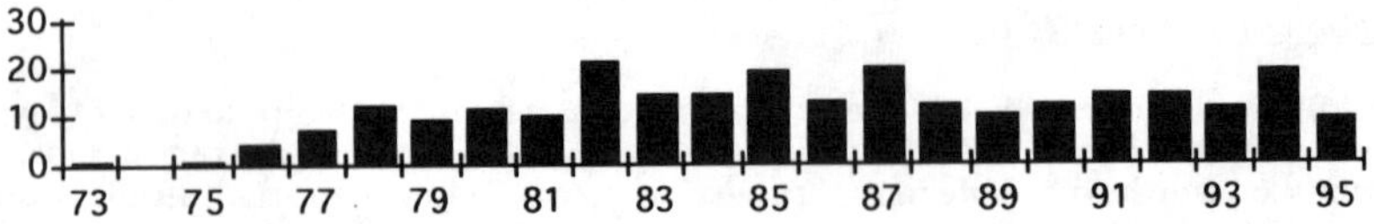

Key Words: Nuclear Waste Containers (Number of Citations in <u>Chemical Abstracts</u> = 257)

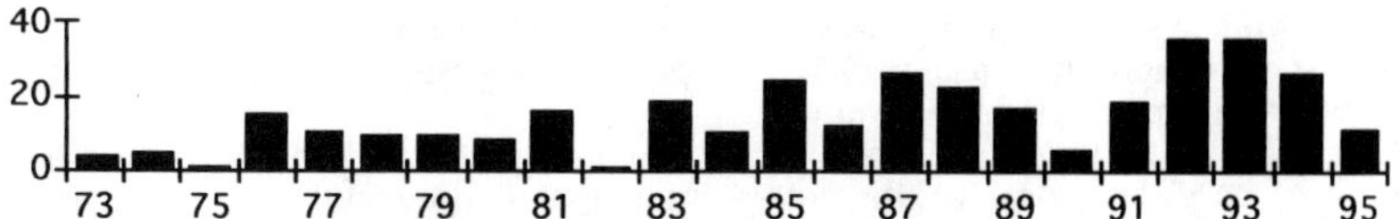

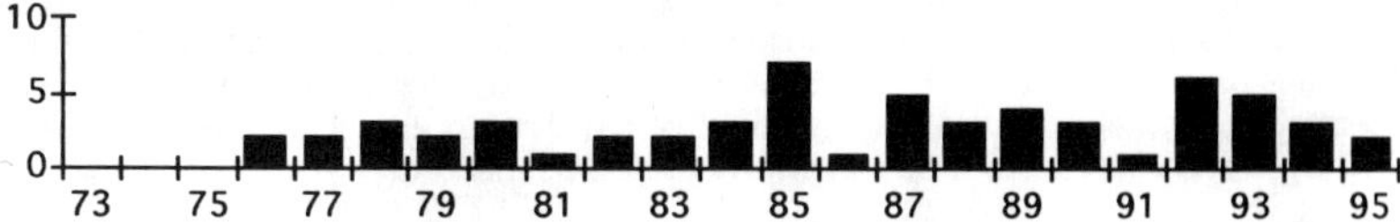

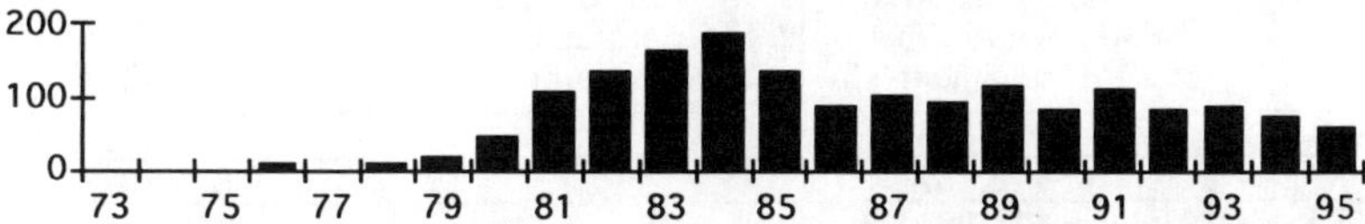

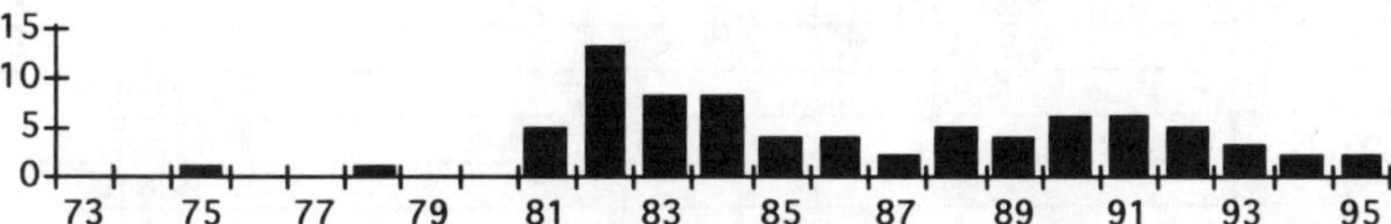

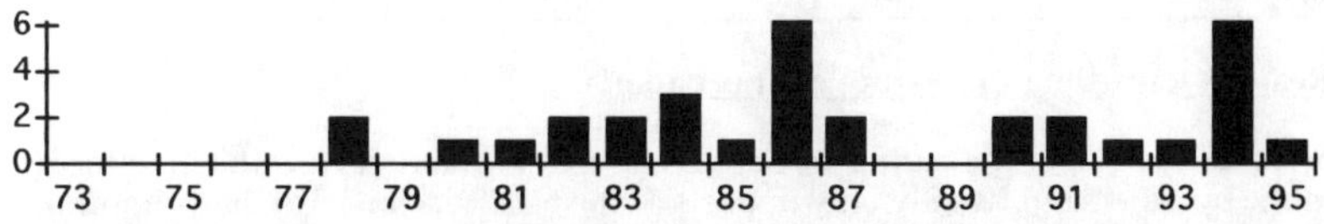

Figure 1. Number of abstracts in Chem. Abstracts as a measure of the scientific activity in RWM as a function of time, 1973-1995.

1. Starting just over 20 years ago, there was a major increase in research activity ('76–'81). However with the arrival of the Reagan administration, and the accompanying policies in U.S. D.O.E. this *increase* terminated and after a modest decline, a lower level has been maintained.
2. There are distinct differences in activity by area of research. Thus it is clearly demonstrated that the nuclear waste form was the focus of attention for the late seventies, peaked in the early eighties, and declined rapidly thereafter.

3. While glass as a waste form has received the lion's share of research activity (960 papers), the alternatives focus on ceramics and cement (especially the latter) had had a not inconsiderable effort of some 271 papers.
4. Containers have had almost as much attention as waste form (T = 257).
5. The last element of the waste package—the overpack—has only 33 papers but this may be due to keyword misidentifications.

Another way to quantify the "science," or research activity is to study the resources devoted to research. Table 1 presents the data—unfortunately only on the U.S. program. Again the data are misleading because although these figures are from the U.S. Secretary of Energy in November 1996, funding is provided from so many other channels for radwaste research they are not counted in DOE's own accounting system as part of the total. The research publications certainly do not correlate—even with a 2-3 year lag—with the dramatic decline in funding from '87-'91. If one includes the budgets of the U.S.-N.R.C., the construction costs for the Savannah River glass plant and Yucca Mountain construction, one can estimate that a total of a few billion dollars has been spent on the research development and engineering of radioactive waste. And a retrospective look at both the science produced and effort expended can only conclude that Omar Khayyam's 700-year-old observation has been repeated.

Table 1. U.S. Department of Energy Defense and Civilian Waste Management Research (dollars in thousands). (Office of the Secretary, DOE, Nov. 1996).

	Office of Civilian Radioactive Waste Management	Office of Environmental Management	Total DOE
FY 1984	$14,180.	$0.	$14,180.
FY 1985	$26,387.	$0.	$26,387.
FY 1986	$15,991.	$0.	$15,991.
FY 1987	$6,500.	$0.	$6,500.
FY 1988	$5,000.	$0.	$5,000.
FY 1989	$2,498.	$0.	$2,498.
FY 1990	$1,000.	$0.	$1,000.
FY 1991	$698.	$96,804.	$97,502.
FY 1992	$4,700.	$106,738.	$111,438.
FY 1993	$4,583.	$135,057.	$139,640.
FY 1994	$646.	$136,002.	$136,648.
FY 1995	$684.	$131,617.	$132,301.
FY 1996	$0.	$146,293.	$146,293.
FY 1997	$0.	$166,873.	$166,873.

Note for Readers Regarding the Nature of This Paper

In the rest of this paper the author makes no effort to formally review the literature. That has been done by others but usually only in parts of the total system. Also the voluminous periodic Proceedings volumes of M.R.S.'s waste management and other meetings of societies provide the reader with ready access to the details of such research.

Instead, the author has chosen to present only the big picture of the entire RWM system. Moreover, he has decided to present it from the viewpoint of one senior investigator and one university laboratory, albeit a rather significant one. This has obvious drawbacks (see above) but it does make for a consistent retrospective viewpoint. The most serious deficiency of this approach is that it is necessarily limited largely to the U.S. situation and views, with its special problems dominated by defense waste and reprocessed wastes in general, which would appear to distort many of the claims—if they were assumed to apply to all wastes, and the world situation.

<u>The Radwaste Management (RWM) System</u>

In 1976 one of the key points I was making to scientists and policy makers and the public through major TV shows on BBC, Walter Cronkite, Danish TV, etc., was that the radwaste field was a highly interactive <u>system</u>. I used Figure 2 to provide a simple image showing the major components of the system. I made the point forcefully then, and repeat it here, that the <u>technological</u> subsystem cannot function in isolation. It must be developed in parallel with developments in the other subsystems of legislation and public acceptance. U.S. policy makers consistently ignored this posture and refused to invest even minuscule amounts of money in education of the public. Hence progress on the public's understanding was zero or negative and all the attention to technology has come to naught. My position is unchanged—the U.S. will <u>never</u> have an implemented waste disposal system without 20 years of prior, consistent, <u>education by schools, colleges and not-for-profits</u> (<u>not</u> directly by the Department of Energy or NRC, which should provide the modest funds needed).

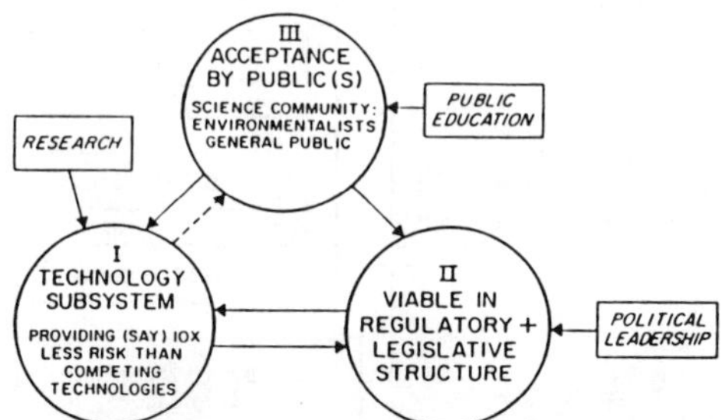

Figure 2. The total system for radioactive waste management.

<u>The Technical Subsystem</u>

Figure 3 shows the detailed breakdown of the technical element of the national RWM system. The two key sub-sub-elements are those labeled <u>Immobilization</u> (in the waste form) and <u>Isolation</u> (in a suitable geological environment). In the early seventies [including my 1978 paper (1) and later book (2)] I argued the case for placing much greater reliance on the waste form than on the geology. Alvin Weinberg, founding Director of Oak Ridge, took up the cause and argued that one will <u>never</u> be able to make a credible case against <u>some</u> geological event in 10,000 years. Hence the engineers should place much greater reliance on that which can be engineered—the waste package. Since the 1950s in Britain the scientifically simple idea of melting the waste oxides into a "glass"—with literally zero research—had become the "reference scenario." We argued the case, that at the then (1980) reference case of a 500°C centerline temperature, all such glasses would begin to crystallize in a few hours in the presence of water. Our two-page paper [McCarthy et al., 1979 (3)] in <u>Nature</u> proving that case, had a profound effect worldwide. Confronted with such evidence the case for examining alternative waste forms had become unstoppable. Two results ensued:
   1. Under the Carter Administration a genuine program of alternative waste form research was begun. Albeit it was organized and managed by an organization and individuals who favored and had predetermined a complex, expensive use of glass as <u>the</u> waste form. The results are obvious from Figure 1.
   2. The re-engineering of the repository was easily carried out by diluting the radio-ion content of the glass. Of course this multiplied several fold both the transportation and the repository cost—both, naturally, perceived as <u>advantages</u> to a GOCO contractor with access to unlimited public funds.

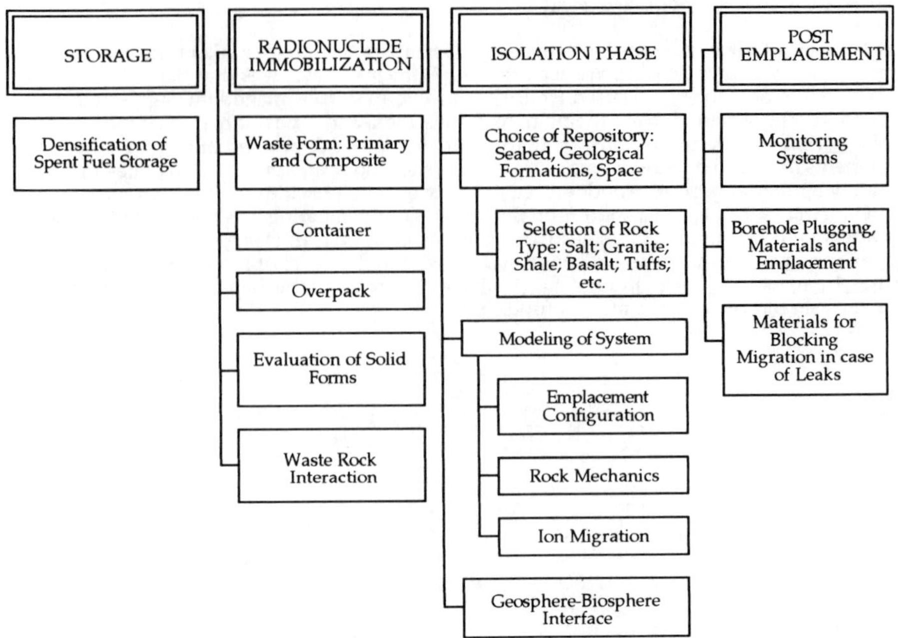

Figure 3. The technical "system" for nuclear waste isolation.

By the mid-eighties, the Reagan Administration had decided to move to ultimate disposal as soon as possible, research was slowed, and alternative waste forms became passé. For much of the last 20 years radwaste management research has focused almost exclusively on "tuning" the technological system and within that on the geologic isolation subsystem (see Figs. 2 and 3 to put this in perspective). After a false start on parallel development of several locations, much of the effort has been on both developing and justifying the Yucca Mountain, NV (and WIPP site in NM) ultimate disposal locations.

Both these locations are perceived to be geologically justifiable on "scientific" grounds. But on system grounds they were not sufficiently justified either to the scientific community nor to the public. Nor was there any public education or debate for and against the more complelling "make minimum changes" argument: Leave waste largely on sites where large amounts have been generated, and where total site cleanup is utterly impractical. In any case Savannah River and Hanford became obvious candidates then. NRC committees had urged the DOE to consider Savannah River in the early 80s. In Britain Sellafield is moving precisely in this logical direction. Finally the most obvious dedicated sites are in the Nevada Test Site, where nearly 1,000 separate final "radioactive waste disposal" sites already exist, in the separate holes created for each weapons test. Recently Daryl Siemer of INEL (4) has worked out a complete scenario for solidifying in concrete waste forms the entire inventory of the INEL lab, transporting, and disposing of the inventory in these already created permanent disposal sites in Nevada.

THE 20-YEAR RETROSPECTIVE

<u>The System</u>

The style of the following section is novel. It tries to capture in figures a summary of the key points and provides relatively modest amounts of text. Similarly with references, only a few key relevant ones are included here. A pamphlet with 400 of our own works arranged by topic is available to anyone who requests it.

Figure 4 summarizes the distinctive feature of the last 20 years—the neglect of the system aspect. This point has already been made above.

A second key feature has been the very large expenditures on committee studies and reports. Table 2 simply shows the titles of the most recent reports. And among these I quote (Fig. 5) only the Executive Summary of the most recently aptly titled—especially relevant to this paper—"Barriers to Science."

This reinforces my own conclusions presented earlier based on intimate study on NAS committees and as a bench researcher: (a) alternative waste forms have never been considered seriously; (b) little new work has been done on waste forms for 15 years; (c) the system was run for the benefit of the local site contractors, not for the nation or the environment.

Neglect of Context

- End of Golden Age of Science: July 17, 1945 (first atom bomb exploded)–October 21, 1993 (Supercollider terminated).
- Failure by scientists to accept the reality that science does not drive any system: Technology, public acceptance or applications drive in industry and (soon) the public sector.
- Radwaste "science" has been "decoration," and employment security for tens of thousands. Zero incentive to solve the overall problem, while doing good science.
- Interminable sequence of reports. Employment for NRC and committee members.

Figure 4. The context in which RWM R/D is done was ignored.

Table 2. Titles of 1995-96 Studies and Reports on Radwaste Management.

- The Waste Isolation Pilot Plant
- Barriers to Science
- The Hanford Tanks
- Review of new York State Low-level Radioactive Waste Siting Process

Barriers to Science
National Research Council • Washington, DC • 1996
Executive Summary

The Committee on Remediation of Buried and Tank Wastes has spent several years studying various scientific and technical issues related to the environmental remediation of the defense radioactive waste sites managed by the Department of Energy. A variety of problems have been noted: (1) planning that is driven by existing organizational structures rather than problems to be solved; 2) commitments that are made without adequately considering technical feasibility, cost, and schedule; 3) an inability to look at more than one alternative at a time; 4) priorities that are driven by narrow interpretations of regulations rather than the regulations' purpose of protecting public health and the environment; 5) the production of documents as an end in itself, rather than as a means to achieve a goal; 6) a lack of organizational coordination; and 7) a "not-invented-here" syndrome at individual sites. The committee observes a common pattern of behavior in these problems: What happens is driven too often by the internal needs of the organizations charged with the remediation work rather than by the overall goal of environmental remediation. Efforts to remedy this situation must involve not only the Department of Energy, but also external stakeholders who have influenced its ways of doing business, including Congress, involved states, and the public.

From the text:

As a result, nearly a decade of research and development has been lost with no substantive change in waste form technology for the past 15 years. What happens is driven too often by the internal needs of the organizations charged with the remediation work rather than by the overall goal of environmental remediation.

Figure 5. A succinct retrospective on the status of radwaste science.

<u>Waste Form Research: Retrospective</u>

Since the author and his colleagues (McCarthy, D. M. Roy, B. E. Scheetz, M. R. Silsbee, W. B. White) have focused their research almost exclusively on waste form design, processing, and evaluation the rest of this review will deal only with this work.

Table 3 summarizes the history of the Penn State-Materials Research Laboratory involvement in RWM research. This lab, under McCarthy and Roy, conceived and championed the mineral-modeled ceramic waste form (of which the Australian "Synroc" was a subset, designed later). We also designed the materials for various modified concrete waste forms and Oak Ridge designed the processes, and D. M. Roy and B. E. Scheetz have continued to specialize for the last decade on cement and concrete's role in many aspects of RWM.

Table 3. Twenty-five Year History of Penn State Involvement in Radwaste Science.

- We recognized our scientific specialties could help in a national problem. No agency initiative for years.
- 1969-70. G. McCarthy, new assistant professor, experienced in rare earth compounds working with me on gel route to glass encapsulation.
- 1971-72. Proposal made to NSF and funded by I. Warshaw under Research Initiation Grants on hot pressing routes to encapsulation of radwaste in glass. $16,000 April '71-March '73.
- 1973. First Penn State paper: G. McCarthy, Quartz Matrix Isolation of Radioactive Wastes, *J. Mat.Sci.* 8, 1358 (1973).
- 1973. Visit to Penn State-MRL by Klingsberg (NAS), J. McElroy and J. Mendel (DOE). R. Roy presents Russian Doll model of redundant protection in waste form, and the obvious advantage of ceramics over glass. D. Roy presents hot pressed cement. AEC sold on it. McCarthy and D. Roy get first AEC grant (October 1973); continued by ERDA, then DOE.
- 1974. Tailored ceramic waste form designed; "supercalcine concept." McCarthy and Davidson start Penn State theme "crystal chemistry and phase formation in fired nuclear waste," *Bull. Am. Ceram. Soc.* 54, 783 (1975).
- 1975. Key concept of most stable minerals in nature as indicators for target ceramics. (Resistant) Mineral modelled waste forms designed.
- 1975-76. Many presentations and publications. Penn State still has pretty much of a monopoly on this field.
- Ted Ringwood (old friend and colleague) asks for briefing and set of reports and papers. "Artificial minerals" becomes synthetic rock, "Synroc."
- 1977. Normal base-broadening: Roy approaches Ray Walton to support a symposium on radwaste at MRS.
- 1978. Key paper appears changing reference scenario of *all nations* by demonsrating that glass could not possibly survive 500°C centerline temperature: McCarthy et al., "Interaction between nuclear waste and surrounding rock in geological disposal," *Nature* 273, 216 (1978).
- 1978. November. First MRS Radwaste Symposium; 572 pages. (Roy, Chair; McCarthy, Program Chair.)
- *1978-85. Extensive work by team on mineral modelled ceramic waste forms: Waste-rock interaction; overpack.*
- *1978-present. Continuous work on cement-based waste forms.*
- *1985-present. NZP as unique, ceramic waste form.*

Some 300 papers have been published on RWM by Penn State-MRL faculty, available upon request.

Figure 6 summarizes this author's perspective on what has occurred in the past 20 years in the science and technology of RWM.

In Figures 7–9, there are presented the major developments in glass and ceramic waste forms. The glass summary is perhaps most revealing in that the basic borosilicate glass (not selected after any real research in the seventies) is basically unchanged after some hundreds of millions of dollars of research. The single major innovation by Harrison and Pope on the sol-gel aluminosilicate glass has never been followed up. In ceramics (Fig. 8), the story is only different in that it started 20 years later, but McCarthy et al. (5) formulations have basically stood the test of time. And in Fig. 9, we see that the Synroc subset of ceramic waste forms has also stayed <u>pretty much</u> the same (6). Among the ceramic waste form the serendipitous "NZP" discovery is rather different, both as being a potential single phase ceramic form for PW-4 waste (7) and for its applicability to Pu (Fig. 10). Finally, the low temperature ceramics which can also serve as nano-encapsulants for Cs and Sr (with spectacular K_d's in the thousands to tens of thousands

---

SCIENCE

1. Additions to thermodynamic databases. Modeling of reactions, hydrology, thermal stresses.
2. Waste form. NZP appears serendipitously. Cement steady progress in Europe.

METASCIENCE

Risk assessment   } Emphasized and
Performance assessment   } advanced

EDUCATION OF PUBLIC:
Totally ignored

TECHNOLOGY
All solutions force fitted to glass technology.

Figure 6. 1976-96 Status of RWM technical sub-system.

	Process	Composition
1965-79. Fingal's Cave	In can melter	Borosilicate glass
1980. Marcoule	Calcine + frit, in tank	Borosilicate
1975–?. Mol-Julich	Platinum bucket → beads in lead	Phosphate
1980-81. Westinghouse R/D Lab	Alkoxide sol-gel in can	Aluminosilicate (rejected)

1990.   • Savannah River
      • NFS
      • Proposed for other sites
	Tank	Borosilicate (with minor addition of alumina)

Figure 7. Glass waste form research shows how little change has occurred.

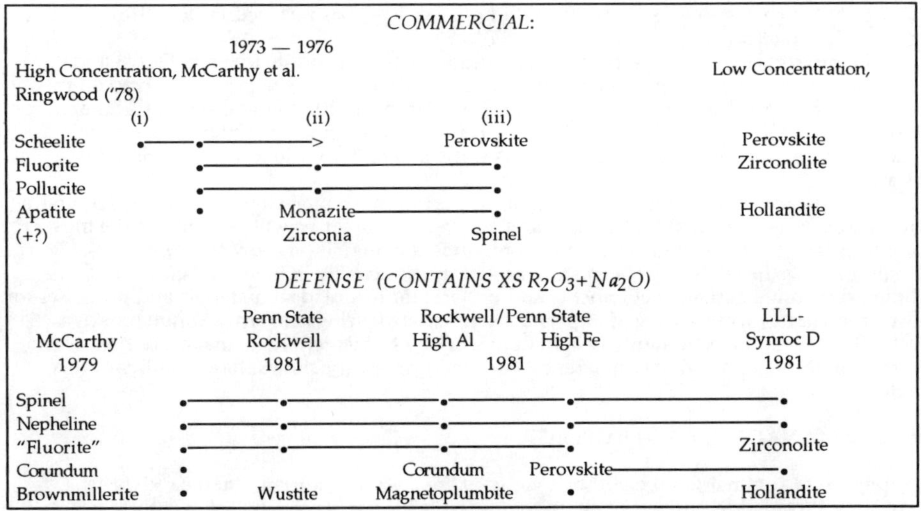

Figure 8. Compositions of tailored ceramic waste forms with time, showing only minor changes.

Synroc C (Perovskite, zirconolite, hollandite, rutile) for Purex HLW			
Synroc-D (SRP waste)	Synroc-F (spent fuel)	Actinide-partitioned wastes	Weapons Pu
Spinel, Nepheline, Zirconolite, Perovskite	Pyrochlore, Uraninite, Hollandite, Perovskite	Zirconolite-rich (+ minor Syn. C)	Zirconolite-rich Synroc

Figure 9. Tailored ceramics II: Minor changes in ceramic phases chosen for Synroc (data supplied by E. R. Vance).

1. Studied detailed crystal chemistry of "NZP" = $NaZr_2P_3O_{12}$, originally studied as fast-ion conductor, and our discovery of zero expansion ceramics in this family.
2. Noted new compound as possible radiophase "$CsZr_2P_3O_{12}$." Reported by Vance and Roy.
3. Then shown that NZP structure can accommodate virtually all ions of Purex up to 10%, up to 70% in NZP+monazite.
4. Ideal host for Pu. Russians far along in using it.

Figure 10. Serendipitous Finding of Only New Ceramic Waste Form.

1966	1976	1980	1985	1990	1991
"Shale," attapulgite	"Zeolites" Fe, Mn, Al-oxyhydroxides	Tailored mixtures of clinoptilolite + mordenite + phillipsite	"Thermo Bottle" *Cold* Cs and Sr saturated minerals in overpack NZP Family	New superclathrates Na-4 mica for Sr K-depleted phlogopite for Cs	Crystalline silicotitanates

Figure 11. New minerals and synthetic materials for low-temperature fixation or overpack.

of mL/g) in cement encapsulated forms are listed in Fig. 11. Relevant references are grouped under Reference 8. These same materials can obviously also serve as overpack materials. The concept of the <u>cold</u> stable Cs- or Sr-saturated overpack (9) has received far too little attention in remediation efforts.

The progress in ceramic waste forms, including Synroc, is summarized in Figs. 8 and 9.

The only new ceramic phase with the unique property that a single phase can accommodate all the reprocess waste ions is "NZP," $NaZr_2P_3O_{12}$ (see Fig. 10). Mineral synthesis tailored for adsorption and low temperature waste forms is the one area in which unique <u>new</u> ceramic phases such as K^+-depleted and Na-4 micas were created for radio-ion incorporation (see Fig. 11 and Ref. 8).

Although cement has received minimal support from the major agencies in the U.S., on the world scene the realization that cement and concrete waste forms will certainly be the most significant single family of nuclear waste containment materials has now slowly sunk in. Figure 12 summarizes the evolution of cement waste forms principally in the U.S. Figure 13 summarizes some of the developments and projects for future ideal materials and processes for using cement in a wide variety of high and intermediate level waste forms. Summaries by D. M. Roy (10) and recent highly imaginative work by D. Siemer (11) to make a cement waste form from INEL waste which can later be vitrified <u>if necessary</u> shows the versatility of this material.

WASTE MANAGEMENT—PROSPECT

Figure 14 summarizes the author's views of "real world solutions" to the radwaste problems. Worldwide, costs—which up to now have been ignored—will slowly force action. It is increasingly clear that a planned slowdown of all timetables is in everybody's interest. Above ground storage can buy 50-75 years within which widespread public education and modest scale research can be conducted. Both of these are the <u>sine qua non</u> of a healthy policy. As one part of the research, the re-examination of leaving waste on-site at all major facilities will have to be considered <u>de novo</u>. A tilt back toward reliance on the immobilization (=waste form) over isolation (geology) is also long overdue. And among the waste form options which this author believes will be most effective, <u>low temperature</u> tailored phases and <u>low temperature</u> processes are sure to win out, because they are both more stable and safer to process.

<1970	1973-80
U.S. Practice	New Materials at Penn State
• Hydrofracture ORNL Grouting (shale repository)	• Low-temperature Ceramic/Cement Waste Form
	• Warm Pressing
70-76	New Processing at ORNL
IAEA Tests developed Low- and Intermediate-level Waste Solidified Shale and Clay Additives; ORNL	• FUETAP (modified hydrothermal)

Figure 12. Cement waste forms evolution.

1980-90	1987-96	Future Ideal for HLW
• Penn State Cement Encapsulated Waste Forms	• Improved Thermodynamic Calculations; U.K., Glasser et al.	• Better for Acid and Basic Solutions (denitrate with sugar)
• Al, Cs-substituted tobermorites tobermorites	Atkinson et al.; Harwell, NIREX "Saltstone"; SRL/Penn State	• Adsorb Cs and Sr in super clathrates
• Composite Waste Forms (incl. CsZP, etc.)	• Better Models for Phase Relations, Stability, Solubility incl. blended cements Glasser et al.	• Low Water Solids Ratio
• Slag, Fly Ash, High-silica Cements		• Additives of Slag or Silica, to maximize compatibility with host rock
• Thermodynamic Stability – Preliminary	• Better Characterization	• In Can Solidification
• Ancient Analogs	• Intermediate-level Cement Waste Solidification, Scale-up; BNFL Sellafield	• Low (<200°C) Temperature Reaction and Dehydration
• Cement Favorable for TRU		
• Canadian United Cask	• Alkali Activated (including Zeolitic) Cements	• Intermediate Wastes Self-heat
	• "Hydro-ceramic" and Vitrifiable Cement Waste Form INEL/Penn State	

Figure 13. Cement Waste Forms Evolution (continued.)

Acknowledgments

The author is grateful to his colleagues D. M. Roy, B. E. Scheetz, E. R. Vance, and Ray Walton of ANL-Washington and Daryl Siemer of INEL for sending data and holding many discussions on this topic.

Main Driver:	Cost effectiveness.
Main Strategy:	Do minimum necessary. Do not commit to anything. Announce policy of 50-year moratorium on decision.
Actions:	• Relieve pressure on spent fuel storage. Centralized (partly) spent fuel storage. • Store above ground in concrete casks for 50-75 years. • Start education and long-term systems all-options research. • Shift focus from isolation to immobilization, i.e. rely on insolubility of waste package rather than reliability of geology. • Consider (a) concrete canisters in Nevada Test Site for INEL wastes (then other defense wastes); (b) concrete in Pacific seabed for fuel pins.

Figure 14. Future of high level radwaste management.

SOME PROJECTIONS: WASTE FORM INNOVATION

Materials

- NZP will be followed up by non-U.S. whatever U.S. does.
- Sub-optimum cement with <1% of R/D funds wins in all objective data comparisons.
- Tailored low-temperature ceramics. Super clathrates for Cs, Sr. Incorporate in lower water-solids special cements.
- Supercalcine rediscovered for ICPP. Can go to either cement or glass. Siemer (11) data impressive.

Processing

- Low temperature advantages are insuperable in energy and materials saving; safer.
- Utilize addition of ultrasonic and microwave fluxes.

Figure 15. Immobilization must get equal time (=Money).

REFERENCES

1. R. Roy, "Science Underlying RWM: Status and Needs," <u>Scientific Basis for Nuclear Waste Management</u>, G. J. McCarthy (ed.), Plenum, NY (1979).
2. R. Roy, <u>Radioactive Waste Disposal; The Waste Package,</u> Pergamon Press,NY (1982).
3. G. J. McCarthy et al., "Interactions Between Nuclear Waste and Surrounding Rock," <u>Nature</u> 272: 216 (1978).
4. D. Siemer, E. J. Bonanno et al., "The Disposal of Orphan Wastes Using the Greater Confinement Disposal Facility," <u>Waste Management '91</u>, Vol. 1, Post and Wacked (eds.), pp. 861-868 (1991).
5. G. McCarthy and M. T. Davidson, "Ceramic Nuclear Waste Forms," <u>Bull. Am. Ceram. Soc.</u> 54:782 (1975).
6. See for details E. R. Vance et al., this volume, p. ___, p. ___.
7. R. Roy, E. R. Vance, and J. Alamo, "[NZP] A New Radiophase for Ceramic Nuclear Waste Form," <u>mat. Res. Bull</u>. 17:585 (1982).
8. <u>Hydroxylated Ceramic Waste Forms and the Absurdity of Leach Tests</u>, Proc., Intl. Seminar on High Level Waste Solidification, R. Udoj and E. Merz (eds), pp. 576-602 (1982). See also W. J. Paulus, S. Komarneni, and R. Roy, "Bulk Synthesis and Selective Exchange of Strontium Ions in $Na_4Mg_6Al_4Si_4O_{20}F_4$ Mica," <u>Nature</u> 357:571-373 (1992); S. Komarneni and R. Roy, "A Cesium Selective Ion Sieve made by Topotactic Leaching of Pholgopite Mica," <u>Science</u> 239:1286-1288 (1988).
9. R. Roy, <u>Reverse Thermodynamic Chemical Barrier for Nuclear Waste</u>, U.S. Patent 4,430,256 (1982); S. Komarneni and R. Roy, "Hydrothermal Transformations in Candidate Overpack Materials,"<u>Nucl. Tech.</u> 54:118-122 (1981).
10. D. M. Roy, "Cementitious Materials in Nuclear Waste Management," 9th Intl. Congress on the Chemistry of Cement, Congress Reports <u>VI</u>, 88-113 (NCB,New Delhi, India 110049, 1992).
11. D. Siemer, B. E. Scheetz and M. L. Gougar, "Hot Isostatic Press (HIP) Vitrification of Radwaste Concretes," <u>Scientific Basis for Nuclear Waste Management XIX</u>, 412:403-410 (1995).

Part II

Glass Formulations and Properties

ISOTHERMAL CRYSTALLIZATION KINETICS IN SIMULATED HIGH-LEVEL NUCLEAR WASTE GLASS

JOHN D. VIENNA, PAVEL HRMA, and DONALD E. SMITH
Pacific Northwest National Laboratory, Richland, WA 99352

ABSTRACT

Crystallization kinetics of a simulated high-level waste (HLW) glass were measured and modelled. Kinetics of acmite growth in the standard HW39-4 glass were measured using the isothermal method. A time-temperature-transformation (TTT) diagram was generated from these data. Classical glass-crystal transformation kinetic models were empirically applied to the crystallization data. These models adequately describe the kinetics of crystallization in complex HLW glasses (i.e., RSquared = 0.908). An approach to measurement, fitting, and use of TTT diagrams for prediction of crystallinity in a HLW glass canister is proposed.

INTRODUCTION

Many examples of crystallization kinetic models are available in the literature for simple glass systems.[1-15] These studies range from purely empirical to theoretical. Few such studies are available for complex glasses such as nuclear waste glasses. These glasses typically contain over 30 chemical components and inclusions such as insoluble noble metals. Studies of crystallization in nuclear waste glass generally do not provide methodology for kinetic modelling or prediction of crystalline phase fraction as a function of thermal history.[16-21] This study demonstrates that crystallization in a complex glass exhibits the general temperature and time dependencies of a simple glass.

Glass-crystal transformation follows a two-step process: crystal nucleation and growth. Johnson and Mehl have modeled the transformation rate in glass as a function of time at a constant temperature.[1] Their theory held well for experimental data with low crystal volume fractions. Later Avrami modified the basic model to account for soft and hard crystal impingement.[2-4] The Avrami equation has later been simplified to analytically describe the transformation extent (ξ),

$$\xi(t) = \frac{X}{X_e} = 1 - Exp\left[-(kt)^n\right] , \tag{1}$$

where X is the volume fraction transformed, the subscript e denotes the equilibrium value, k is a compilation of time-independent constants, t is the time, and n is the time exponent or Avrami number. The Avrami number is related to the mechanism and dimensionality of crystallization. Table I summarizes the phenomena and associated n value for crystallization.

Table I. Time exponent values (n) for various crystallization conditions[5]

	Crystallization Mechanism	
	Diffusion	Interface
Constant nucleation rate		
3-dimensional growth	5/2	4
2-dimensional growth	3/2	3
1-dimensional growth	1/2	2
Constant particle number		
3-dimensional growth	3/2	3
2-dimensional growth	2/2	2
1-dimensional growth	1/2	1
Surface nucleation	1/2	1

Mat. Res. Soc. Symp. Proc. Vol. 465 © 1997 Materials Research Society

Practically, nucleation and crystal growth kinetics can be combined into a single time-independent kinetic constant (k in Equation 1), which can be assumed to vary with temperature in an Arrhenian manner:

$$k = k_0 \, \mathrm{Exp}\left[-\frac{B_k}{T} \right],$$

(2)

where k_0 is the preexponential, T is absolute temperature, and B_k is a characteristic temperature related to an apparent activation energy. Strict similarity in the temperature dependence of nucleation and growth rates is needed for superposition of these parameters into k. Equation 2 is compatible with this requirement and is commonly used as a first approximation. This treatment shows reasonably good agreement with experimental data for simple glass crystallization.[5]

Crystallization in glass can be characterized by several means including isothermal time-temperature-transformation (TTT) diagrams, non-isothermal TTT diagrams (i.e., constant or logarithmic cooling rate diagrams), or crystallinity generated by a standard heat treatment (i.e., the temperature history expected of HLW canister cooling). Theoretically, the volume fraction of crystals produced using any of these methods can be described using the Avrami equation discussed below.

EXPERIMENTAL PROCEDURES

To test the applicability of these models to the prediction of crystallinity in nuclear waste glass, a reference simulated high-level waste glass (HW39-4) was used. Table II shows the nominal composition of this glass. The glass was fabricated using oxide, carbonate, and boric acid precursors, melted in a covered Pt crucible at 1150°C for 1 h and quenched on a steel plate. The glass was crushed and remelted for 1 h, and poured into ($30 \times 1 \times 0.5$ cm) Pt/Au boats for heat treatment. The glass was heat treated in a linear temperature gradient from the glass transition temperature (T_g) (determined dilatometrically to be 490°C) to the liquidus temperature (T_L) (reported[22] to be 870°C) for 4, 16, 24, 48, 72, and 96 h. The sample was then annealed at 490°C for 2 h and sectioned for optical microscopy. The 4 h sample showed no crystallization and will not be considered again. Five samples were heat treated at 670°C in an isothermal furnace for 16, 24, 48, 96, and 192 h and one sample at 770°C for 48 h to confirm gradient furnace results.

Table II. Composition of HW39-4 glass (as batched oxide basis)

Oxide	Mass %	Mol %	Oxide	Mass %	Mol %	Oxide	Mass %	Mol %
Al_2O_3	2.31	1.47	La_2O_3	0.66	0.13	Rb_2O	0.05	0.02
B_2O_3	10.53	9.83	Li_2O	3.75	8.16	Rh_2O_3	0.05	0.01
BaO	0.11	0.05	MgO	0.84	1.36	RuO_2	0.16	0.08
CaO	0.83	0.96	MnO_2	0.16	0.12	SiO_2	53.53	57.92
CdO	0.80	0.41	MoO_3	0.32	0.14	Sm_2O_3	0.05	0.01
CeO_2	0.16	0.06	Na_2O	11.25	11.80	SO_3	0.29	0.24
Cr_2O_3	0.13	0.06	Nd_2O_3	1.31	0.25	SrO	0.11	0.07
Cs_2O	0.16	0.04	NiO	0.61	0.53	Y_2O_3	0.05	0.01
CuO	0.16	0.13	P_2O_5	0.11	0.05	ZrO_2	3.85	2.03
F	0.32	1.10	PdO	0.05	0.03	Total	100.01	100.00
Fe_2O_3	7.19	2.93	Pr_6O_{11}	0.11	0.01			

Reflected light microscopy (Olympus, PGM3) was used (at 1000 and 500×) in conjunction with image analysis (Leco, L2001) to measure the volume fraction transformed in each sample. Ten random measurements were taken and averaged for each sample. Gradient furnace samples were examined at 10°C intervals. X-ray diffraction and scanning electron microscopy with energy dispersive spectroscopy (SEM/EDS) were performed on selected samples for confirmation of crystal structure and composition and to ensure that 1000× magnification was sufficient for characterization.

RESULTS AND MODELING

The crystalline phase found in each sample has been identified as clinopyroxene — $M_1M_2[Si_2O_6]$, where M_1 represents Na and Ca, and M_2 represents Fe, Mg, Ni, and Cr. The SEM/EDS analysis has identified acmite ($NaFe[Si_2O_6]$) over most of the temperature range of interest (525 to 740°C). Crystallization of spinel — $(Ni,Fe,Mn)(Fe,Cr)_2O_4$ was also seen but not quantified in this glass over the same temperature range.

Figure 1 shows the crystal volume fraction from isothermal heat treatments. The apparent overlap of crystallinity curves plus the additional crystallization "shoulder" seen at high temperatures is attributed to surface tension driven convective flow at high end of the temperature gradient. This conclusion has been confirmed by visual examination of the sample, which showed a convective role in this area and by a uniform temperature data point at 770°C, which showed no crystallization after 48 h. These data have not been considered in the analysis of crystallization kinetics for the acmite phase. Samples heat treated in an isothermal furnace at 670°C for 16, 24, 48, 96, and 192 h are in agreement with gradient furnace samples.

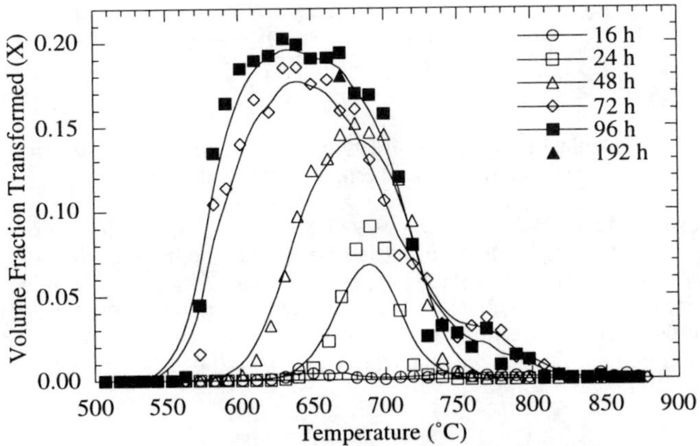

Figure 1. Volume fraction of clinopyroxene in HW39-4 glass as a function of temperature for 16, 24, 48, 72, and 96 h isothermal heat treatments

The X_e value was estimated by extrapolation of X vs. ln(t) curves at each temperature past 96 h. A 192 h sample was mesured at 670°C, the crystal fraction in the sample was slightly below that for the 96 h sample demonstrating that equilibrium had been reached. The equilibrium volume fraction of crystals is related to temperature according to:

$$\frac{X_e}{X_{e0}} = 1 - \mathrm{Exp}\left[-B_{Xe}\left(\frac{1}{T} - \frac{1}{T_L}\right)\right], \tag{3}$$

where the equilibrium factor (B_{Xe}) and T_L are calculated from a plot of $\ln(1-X_e)$ vs. $1/T$, and X_{e0} is the equilibrium volume fraction of crystal at room temperature. The values of 44,302 K for B_{Xe}

and 1,004 K for T_L are taken from the slope and intercept of the $\ln(1-X_e)$ vs. $1/T$ line, and X_{e0} was determined to be 0.203 by extrapolation of the X_e data to T=0 in Equation 3. Figure 2 shows X_e as a function of temperature overlayed on the experimental data. A thermodynamic model (FACT™) has also been used to predict X_e of acmite in HW39-4 glass. The thermodynamic calculation of B_{Xe} (48,380 K) and X_{e0} (0.2) match closely those measured, however, the calculated T_L (1,148 K) is higher.

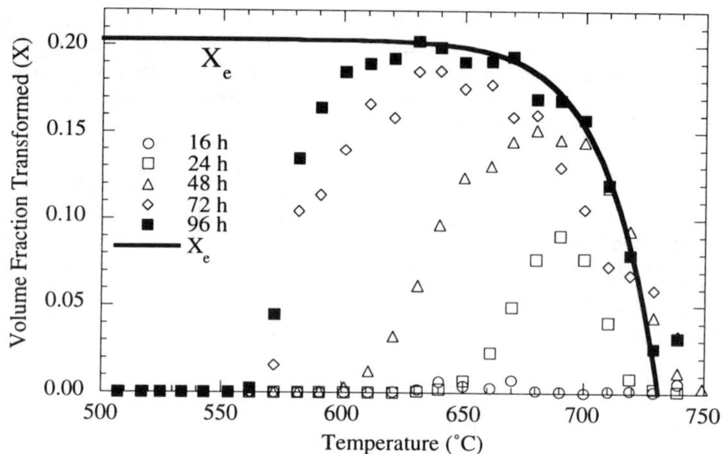

Figure 2. Calculated equilibrium fraction transformed (X_e) as a function of temperature with isothermal crystallization data

Many glasses including HW39-4 show a very distinct induction time (σ) for crystallization. This induction time is estimated by the extrapolated intercept of the linear portion of ξ vs. ln t curves as described by Weinberg.[23] Effectively, σ is a combined effect of the induction time for nucleation and for growth. It was shown by Weinberg that this induction time is not uniquely determined and is dependent on experimental conditions.[23] In the case of HW39-4, σ is Arrhenian in nature (as seen in Figure 3) with a preexponential σ_0 of 0.01206 s and a temperature coefficient B_σ of 14,595 K:

$$\sigma = \sigma_0 \text{Exp}\left(\frac{B_\sigma}{T}\right).$$

(4)

The relationship between σ and t is given by

$$t = t' - \sigma$$

(5)

where t' is the time each sample is exposed to heat treatment temperature and t is the effective time used in Equation 2.

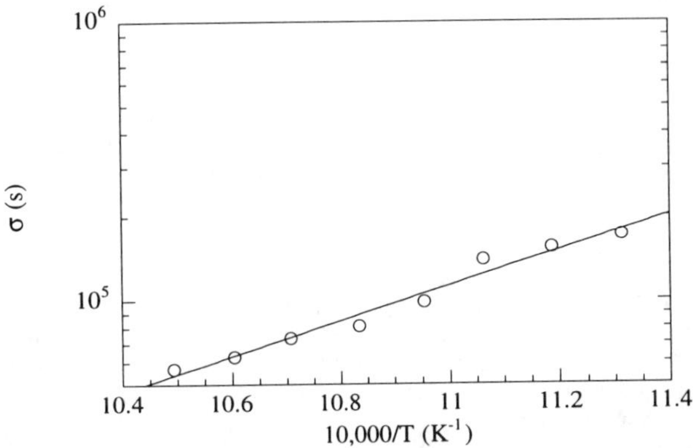

Figure 3. Induction times as a function of inverse temperature

The Avrami number (n) from Equations 1, 2, and 5 is calculated from the slope of $\ln(-\ln(1-\xi))$ vs. $\ln(t)$ plots at constant temperature. Using this method, an average value of 2.52 was calculated for n. Assuming three-dimensional diffusion-controlled growth with constant nucleation, the corresponding theoretical n value is 2.5 (this value is used in the calculations below).

The kinetic constant (k) from Equation 2 is calculated from plots of $\ln(-\ln(1-\xi))$ vs $1/T$ at constant times. From this treatment, nB_k is given as the slope of each curve. Using 2.5 as n, b was found to be 5848.9 K, and k_0 was fitted by least squares of X predicted vs. X measured for all experimental data to be 4.92×10^{-3} s^{-1}.

Figure 4 shows a comparison of predicted and measured crystal volume fraction using Equations 1 and 2. The high temperature shoulder has been removed.

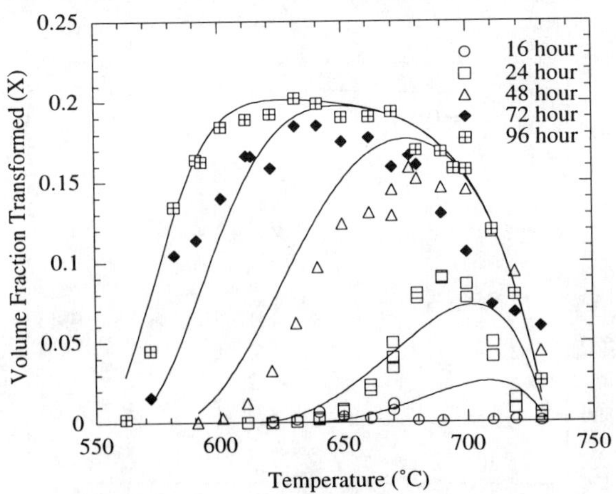

Figure 4. Measured crystal volume fraction and calculated crystallinity
curves asa function of temperature

Theoretically, the volume fraction of crystals in a glass canister can be predicted by solving

Equation 1 with the parameters fitted to isothermal data by numerical solution of

$$\dot{X} = \frac{dX}{dt} = n(X_e - X)k^n t^{n-1}. \tag{6}$$

However, it is not clear how the induction time measured by isothermal testing will affect the prediction of canister crystallinity. This will be addressed in an subsequent study.

SUMMARY AND CONCLUSIONS

An empirical approach is demonstrated for the modeling of crystallization in very complex glasses. Piece-wise fitting of parameters was found to better predict experimental data when compared to full-scale non-linear regression of all empirical coefficients. This approach gives the ability to thoroughly examine data for consistency and use judgement in weighting experimental data points. Kinetic parameters fit to both isothermal and non-isothermal crystallization of acmite from HW39-4 are presented in Table III.

Table III. Parameters fit for crystallization of acmite from HW39-4 glass

Parameter	n	B_{Xe}	T_L	X_{e0}	B_σ	σ_0	B_k	k_0
Equation	1, 2	3	3	3	4	4	2, 3	2
Values	2.5	44,302	1,004	0.203	14,595	0.01206	5,849	0.00492
Units		K	K		K	s	K	s^{-1}

Figure 5 shows the TTT diagram for acmite in HW39-4 glass. This diagram can be used to estimate the amount of crystallinity in a sample heat treated isothermally for a known time. A numerical method is proposed for use of isothermal crystallization kinetics data for the calculation of crysallinity as a function of cooling history and likewise location within the canister. It is not clear how the measured induction time, which is known to be dependent upon experimental conditions, will affect the extension of this data to prediction of crystallinity in a HLW glass canister.

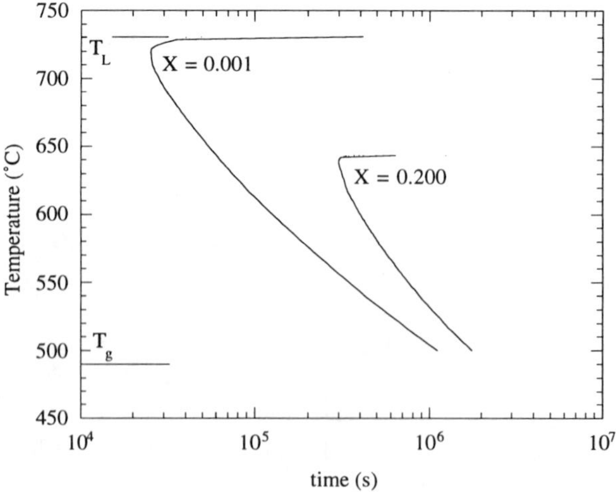

Figure 5. Calculated time-temperature-transformation diagram

ACKNOWLEDGEMENTS

The authors would like to acknowledge F. Mo and J. Coleman for SEM and EDS analysis, D. S. Kim for insightful suggestions, P. Brushaber for independent overcheck of data, and M. J. Schweiger for laboratory assistance. The Pacific Northwest National Laboratory is operated for the U. S. Department of Energy by Battelle under contract DE-AC06-76RLO 1830.

REFERENCES

1. W. A. Johnson and R. F. Mehl, "Reaction Kinetics in Processes of Nucleation and Growth," *Trans. Am. Inst. Min. Met. Eng.* 135, 416-443 (1939).
2. M. Avrami, "Kinetics of Phase Change. I: General Theory," *J. Chem. Phys.* 7, 1103-1112 (1939).
3. M. Avrami, "Kinetics of Phase Change. II: Transformation-Time Relations for Random Distribution of Nuclei," *J. Chem. Phys.* 8, 212-24 (1940).
4. M. Avrami, "Granulation, Phase Change, and Microstructure: Kinetics of Phase Change. III," *J. Chem. Phys.* 9, 177-84 (1941).
5. D. R. MacFarlane and M. Fragoulis, "Theory of Devitrification in Multicomponent Glass Forming Systems Under Diffusion Control," *Physics and Chemistry of Glasses*, Vol. 27, [6], 228-34 (1986).
6. D. R. Uhlmann, "Crystal Growth in Glass Forming Systems: a Ten Year Perspective," in Advances in Ceramics - Volume 4: Nucleation and Crystallization in Glasses, eds. J. H. Simmons et al. ACS Westerville, OH (1981)
7. D. R. Uhlmann and E. V. Uhlmann, "Crystal Growth and Melting in Glass Forming Systems, a View from 1992," in Ceramic Transactions - Vol. 30: Nucleation and Crystallization in Liquids and Glasses, ed. M. C. Weinberg, ACS Westerville, OH (1991).
8. H. Scholze, Glass: Nature, Structure and Properties. 1st English Edition. Springer - Verlag, NY (1990).
9. D. R. Uhlmann and H. Yinnon, "The Formation of Glass," in Glass Science and Technology, ed. D.R. Uhlmann and N. J. Kreidl. Academic Press, NY (1983).
10. W. Hinz, "Nucleation and Growth," Glass 1977, XIth International Congress on Glass, J. Gotz ed., North Holland Publishing Co., NY (1977).
11. N. H. Christensen, A. R. Cooper, and B. S. Rawal, "Kinetics of Dendritic Precipitaion of Crystobalite from a Potassium Silicate Melt," *J. Am. Ceram. Soc.*, Vol. 56, [11] 557-61 (1973).
12. R. A. Grange and J. M. Kiefer, "Transformation of Austenite on Continuous Cooling and Its Relation to Transformation at Constant Temperature," Transactions of the ASM, Cleveland, OH (1940).
13. H. Yinnon and D. R. Uhlmann, "Applications of Thermoanalytical Techniques to the Study of Crystallization Kinetics in Glass-Forming Liquids, Part 1: Theory," *J. Non-Cryst. Sol.* Vol. 54, 253-75 (1983).
14. M. C. Weinberg, "On the Analysis of Non-isothermal Thermoanalytic Crystallization Experiments," *J. Non-Cryst. Sol.* Vol. 127, 151-8 (1991).
15. M. C. Weinberg, "The Use of Site Saturation and Arrhenius Assumptions in the Interpretation of Non-Isothermal DTA/DSC Crystallization Experiments," *Thermochemica Acta*, Vol. 194, 93-107 (1992).
16. C. M. Jantzen and D. F. Bickford, "Leaching of Devitrified Glass Containing Simulated SRP Nuclear Waste," Scientific Basis for Nuclear Waste Management, VIII, C. M. Jantzen et al. eds, Materials Research Society, Pittsburg, PA (1985).
17. D. B. Spilman, L. L. Hench, and D. E. Clark, "Devitrification and Subsequent Effects on The Leach Behavior of Simulated Borosilicate Nuclear Waste Glass," Nuclear and Chemical Waste Management, Vol. 6, 107-119 (1986).
18. J. T. Dalton, K. A. Boult, H. E. Chamberlain, and J. A. C. Marples, "The Influence Of Metal Oxides on the Leach Rate and Crystallization Behavior of Waste Glasses," Procedings of the International Seminar on Chemistry and Process Engineering for High-Level Liquid

Waste Solidification, R. Odoj and E. Merz eds., Nuclear Research Est., Julich (1981).

19. R. P. Turcotte, J. W. Wald, and R. P. May, "Divitrification of Nuclear Waste Glasses," Scientific Basis for Nuclear Waste Management, Vol. 2, C. J. M. Northrup ed., 141-6, Materials Research Society, Pittsburg, PA (1980).

20. C. M. Jantzen, D. F. Bickford, D. G. Karraker, and G. G. Wicks, "Time-Temperature-Transformation Kinetics in SRL Glass," Advances in Ceramics - Nuclear Waste Management, Vol. 8, W. A. Ross and G. G. Wicks eds., 30-8, American Ceramic Soc., Westerville, OH (1984).

21. D. F. Bickford and C. M. Jantzen, "Devitrification Behavior of SRL Defense Waste Glass," Scientific Basis for Nuclear Waste Management, VII., G. L. McVay ed., 557-66, Materials Research Society, Pittsburg, PA (1984).

22. P. Hrma, G. F. Piepel, M. J. Schweiger, D. E. Smith, D. S. Kim, P. E. Redgate, J. D. Vienna, C. A. LoPresti, D. B. Simpson, D. K. Peeler, and M. H. Langowski, Property / Composition Relationships for Hanford High-Level Waste Glasses Melting at 1150°C, PNL-10359, vol 1 and 2, Pacific Northwest Laboratory (1994).

23. M. C. Weinberg, "Induction Time for Crystal Growth," J. Non-Cryst. Sol. Vol 170, 300-302 (1994)..

A DIRECT, SINGLE-STEP PLASMA ARC-VITREOUS CERAMIC PROCESS FOR STABILIZING SPENT NUCLEAR FUELS, SLUDGES, and ASSOCIATED WASTES

Xiangdong Feng*, Robert E. Einziger*, and Richard C. Eschenbach**
*Pacific Northwest National Laboratory, Richland, WA 99352, x_feng@pnl.gov
**Retech, A Division of M4 Environmental, Ukiah, CA 95482

ABSTRACT

A single-step plasma arc-vitreous ceramic (PAVC) process is described for converting spent nuclear fuel (SNF), SNF sludges, and associated wastes into a vitreous ceramic waste form. This proposed technology is built on extensive experience of nuclear waste form development and nuclear waste treatment using the commercially available plasma arc centrifugal (PAC) system. SNF elements will be loaded directly into a PAC furnace with minimum additives and converted into vitreous ceramics with up to 90 wt% waste loading. The vitreous ceramic waste form should meet the functional requirements for borosilicate glasses for permanent disposal in a geologic repository and for interim storage. Criticality safety would be ensured through the use of "batch" modes, and controlling the amount of fuel processed in one batch. The minimum requirements on SNF characterization and pretreatment, the one-step process, and minimum secondary waste generation may reduce treatment duration, radiation exposure, and treatment cost.

INTRODUCTION

The U.S. Department of Energy (DOE) owns approximately 2700 metric tons of heavy metals in spent nuclear fuels (SNF), stored mainly in water-filled pools at 29 DOE facilities around the country [1]. Some of these spent fuels exist in small quantities along with different matrix materials, cladding materials, and overall general characteristics. Many of the spent nuclear fuel elements are damaged and generate sludges that form from the eroded metal cladding, deteriorating concrete, crumbled SNF, and dirt in the storage basins. This sludge also contains plutonium, uranium, and fission products. A cost-effective technical strategy is needed for interim management and ultimate disposition of the SNF and associated debris and sludge wastes. The cost of characterizing these waste streams at the level needed for conditioning treatment or to develop waste forms for the different types of SNF would be prohibitive.

A plasma arc-vitreous ceramic (PAVC) process (see Figure 1) for converting many types of spent nuclear fuels (i.e., metallic, oxide, carbide, hydride, sodium-bearing, aluminum, and zircaloy) and the associated wastes (contaminated sludge, soil, metal fragments, metal containers, and concrete grits) into vitreous ceramic (VC) final waste forms is being developed at Pacific Northwest National Laboratory (PNNL). Vitreous ceramics are composed of crystalline phases embedded in a silicate glass matrix and can be produced in a direct, single-step melter. Crystalline phases such as zirconolite, perovskite, and zircon in the VC can incorporate more than 30 wt% uranium and other fissile elements in their lattices and are extremely corrosion-resistant [2-6]. The durable residual glassy matrix should immobilize other hazardous elements of the SNF wastes that are not incorporated in the crystalline phases.

The most advantageous features of the PAVC process are 1) its simplicity (i.e., SNF in assemblies and associated wastes such as debris and sludges are loaded directly into a PAC furnace and melted into waste form in a single step), 2) high waste loading (i.e., the vitreous ceramics produced in the PAVC process have very high waste loading up to 90% [4] and the

Mat. Res. Soc. Symp. Proc. Vol. 465 © 1997 Materials Research Society

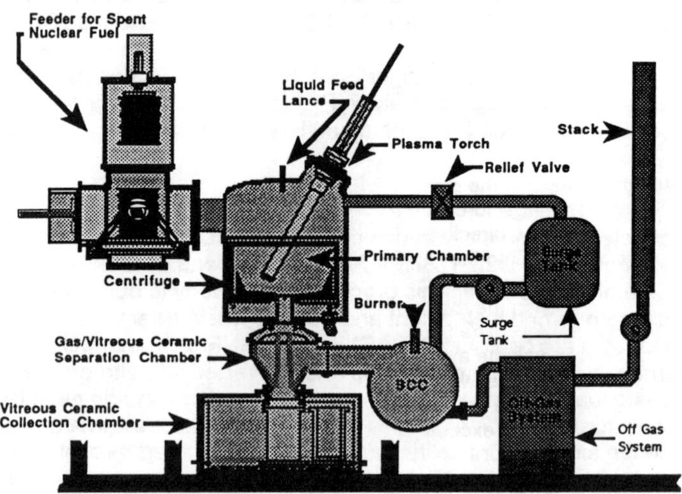

Figure 1. A Schematic of the Plasma Arc-Vitreous Ceramic (PAVC) Process for Treating Spent Nuclear Fuel

final waste form volume is small), 3) reduced SNF characterization (i.e., SNF characterization can be extremely expensive, and there are fewer requirements on fuel characterization before a PAVC processing than alternative technologies), 4) minimal pretreatment required (i.e., the handling and pretreatment of SNF are difficult and expensive, however, little pretreatment in the form of characterization, sorting, size reduction, and separation is needed for the PAVC processing), and 5) the resulting durable and safe waste form (i.e., the vitreous ceramics produced are criticality safe, durable, and suitable for long term storage and disposal). A simpler process is preferred for remote operation in a high-radiation environment. Smaller waste volume means less processing time, less canister handling, less waste form to store, less to transport, and less to deposit into the repository. During the life cycle of treatment and disposal processes, PAVC results in reduced radiation exposure to personnel and lower life-cycle costs.

DESCRIPTION

A schematic of a PAC system is provided in Figure 1. The major components of a PAC system are the feeder, the primary chamber with rotating tub, the plasma torch for melting and homogenizing, an oxygen lance for oxidizing the metal fuel, an optional secondary combustion chamber with after-burner, a vitreous ceramic collection chamber, and an off-gas treatment system.

The 37-liter capacity of the molten waste forms in a 2-ft diameter rotating (centrifugal) tub of a PAC furnace corresponds to about 140 kg of vitreous ceramics. Before a batch melting cycle begins, quantities of contaminated soils/other wastes, additives such as SiO_2, CaO and oxides of neutron absorbers (if necessary), and depleted uranium (if necessary) are placed in the rotating tub for initiating the arc. The rotating tub spins on a vertical axis at 15 to 80 rpm.

Plasma is generated by passing an electric current through a gas (such as helium, air, argon-oxygen, or nitrogen), which heats the gas to a very high temperature (>6,000°C) and ionizes it. The plasma torch creates the high-temperature plasma by injecting plasma-forming gas into the gap between the hollow-torch electrode and the torch nozzle, then initiating and maintaining a direct current electrical discharge between the torch electrode and the rotating tub. The torch can be positioned in the rotating tub with a three-axis, hydraulically assisted mechanism. The plasma torch in a 2-ft-diameter PAC furnace usually operates at 150-175 KW.

The starting materials in the rotating tub are heated to 1600°C or higher, and melted into a molten pool. The centrifugal force keeps the melted materials in the rotating tub. The SNF element would be then loaded directly into the fuel chamber and sliced into small pieces (less than 6 inches) before dropping into the rotating tub. The small pieces of the SNF would be melted, oxidized, and uniformly mixed with the molten pool materials under centrifugal force and under the torch gas depression in the rotating tub. The associated wastes can also be fed simultaneously into the tub. The amount of SNF fuels to be fed into each melt batch is determined by the total volume available in the rotating tub, the requirement on criticality, the compositions of the associated wastes and additives, and the fuel compositions. Two pinhole cameras allow viewing of the arc and chamber. A burst disk is installed on the primary chamber to relieve pressure, should it exceed 15 psig.

After all the fuels are fed into the furnace, another 30-60 minutes may be needed to oxidize the fuels completely and homogenize the melt. The speed of the rotating tub is reduced, and the molten vitreous ceramics pour through an axial hole at the bottom of the rotating tub, directly into the canister under gravity. Following the pour, the tub will be inspected for radioactivity to ensure there is no accumulation of highly enriched uranium (HEU). The PAC furnace is then ready for another batch.

The batch operation offers advantages for criticality control and may limit the output of the process. However, the dynamic agitation of the melt in a PAVC process achieved from the vigorous depression of the plasma gas and centrifugal action significantly reduce the residence time of each melt to a few hours, resulting in a higher output. The high waste loading of vitreous ceramic significantly reduces the final waste form volume in comparison of, for example, a glass waste form option for stabilizing aluminium-SNF. Furthermore, the PAC system is quite compact and multiple system can also be used to increase the output.

MINIMUM SNF CHARACTERIZATION

The PAVC process requires minimal SNF characterization in terms of process necessity when compared to alternative treatment methods although it may have the same amount of requirements on characterization in terms of product assurance. The vitreous ceramic waste form has the flexibility to accommodate feed composition changes within large ranges [4, 6]. The elemental composition information needed for waste-form formulation can usually be obtained through information on the types and the history of fuel production and service. If such information is not available, then estimates through X-ray fluorescence or other types of simple nondestructive methods to provide elemental compositions with +/- 5-10 wt% accuracies are sufficient. If the service history cannot provide HEU and actinide contents accurate enough for criticality control, then an adequate analysis of these elements will be required, similar to other treatment technologies. For these reasons, accurate determinations of hydrides, water, organics, and surface areas, or physical conditions of the fuels are minimized.

MINIMUM PRETREATMENT

Minimum pretreatment is required for the PAVC process because SNF in assemblies, debris, and sludges would be loaded directly into a PAC system feeder without prior size reduction, separation of cladding from fuel, or any other pretreatment. The fuel assembly and large debris would be sliced into small pieces by a plasma torch and are dropped into the centrifugal tub of the furnace.

OFF-GAS TREATMENT

The PAVC process is similar to any other high-temperature process in which glasses or ceramic waste forms are produced, in that particulates and volatile and/or semi-volatile elements such as Cs, Tc, Se, Ru, and I will escape from the melts. An off-gas treatment system is necessary to capture these particulates and volatile/semi-volatile elements. Extensive work has been performed during the development of the plasma-arc process for treating mixed and low-level wastes where many more volatile elements are presented [2-3]. These developed off-gas systems have been shown to be in full compliance with Environmental Protection Agency emission requirements. A dry off-gas system modified from the existing systems is expected to be used for the SNF treatment so all the used materials and filters in the operation can be recycled back into the melter to minimize secondary wastes or the used filters (baghouse, glass filter, packed bed absorber and HEPA filters) can be recycled back to the furnace to be made into a special vitreous ceramic composition that has been shown to retain over 99% of the Cs_2O in a melt at 1550°C [4]. A certain amount of secondary waste streams related to the volatile radionuclides are expected to be generated and need further treatment as with any other high temperature processes.

CRITICALITY CONTROL

Criticality safety is addressed in both the processing and in the products. The criticality safety during processing would be ensured through equipment design and operation in batch mode. The system has a rotating tub of 2 ft or less in diameter (the geometric limit will be determined through criticality considerations) and only critically safe amounts of HEU (about 5.51 kg of ^{235}U per batch of 144 kg of vitreous ceramic melt) and Pu would be allowed in each batch of melting. The melted vitreous ceramics are completely discharged and the rotating tub will be checked for possible accumulation of HEU and other fissile materials before the next batch of a melt begins. Further R&D is needed to assure the complete dissolution of Pu, HfO_2, and Gd_2O_3 under the melt conditions. Pu is expected to be melted, completely oxidized, and mixed with the rest of the melt and with neutron absorbers because of the high plasma temperature at the presence of oxygen, and the vigorous mixing of plasma torch gas depression and centrifugal action. The exact amount of the neutron absorbers needed can only be assessed through future R&D in understanding the local lattice structure of the absorbers in the waste form and the long-term leaching behavior with respect to Pu and HEU.

For the example of treating aluminum SNF, there would be a maximum of 144 kg of vitreous ceramics during each melt, among which there is a maximum of 3.823% ^{235}U while the total uranium content is about 21% (i.e., at 18.2% enrichment). Preliminary criticality control calculations indicated no criticality concerns under operating conditions even when cooling water entered the rotating tub and soaked the tub with water. Parameters regarding each batch of a melt are shown below:

Total Weight of Vitreous Ceramics	144 kg
Total Weight of Al-SNF	71 kg
Total Additives and associated wastes (SiO2, MgO, DU, Gd, Hf, and others)	35 kg
Total U metal	30 kg
Total ^{235}U	5.5 kg

The criticality safety in the product will be ensured through controlling the enrichment in the product and by relying on the capability of vitreous ceramic to incorporate neutron absorbers into the same lattice structure occupied by uranium and fissile materials.

Criticality can occur with HEU metal in sufficient quantities, but it cannot occur when the HEU is dissolved in the vitreous ceramics at an appropriate concentration, such as below 4%, while the total uranium content can be as high as 21%. These limits, based on criticality calculations, govern the final product compositions. In treating Al SNF, for example, the vitreous ceramic could be poured into canisters half the size of the defense waste processing facility (DWPF) canisters. This results in a total of 1046 kg of vitreous ceramics with an HEU content of 40 kg. The canister can then be inserted in the middle of the waste package with four DWPF canisters of borosilicate glasses so the total HEU content in the waste package will not exceed the criticality limit of 42 kg at 18.2% enrichment.

The incorporation of neutron absorbers, such as Gd or Hf, in the vitreous ceramics can be used to achieve high HEU loading in the waste form. Gd and Hf have been shown to have chemical behaviors similar to uranium in vitreous ceramics, and they occupy the same lattice spaces as uranium and plutonium.

WASTE FORMS

Vitreous ceramics promote the formation of stable and low-solubility (in water) crystalline phases (e.g., zirconolite, perovskite, zircon, and spinel) that are embedded in a glass matrix rich in network formers such as SiO_2 and Al_2O_3 [4, 6]. Vitreous ceramic waste form can 1) incorporate large contents of uranium (more than 30 wt%), plutonium (about 15 wt%), other fission products, and neutron absorbers such as Gd, Hf, and other rare earths in the lattices; 2) incorporate a large amount of Al, Fe, Ni, Cr, Zr, and Ti oxides to accommodate the large metallic content of the SNF; 3) use little or no network-forming elements, such as silicon, so these latter elements are enriched in the residual glassy matrix; 4) maintain durability and insolubility in water because these crystals duplicate those that have existed in nature for billions of years; and 5) bind tightly to the glassy matrix to maintain the physical integrity and mechanical strength of the waste form. The residual glassy matrixes in vitreous ceramics are enriched in network formers and deficient in alkalis and are therefore also more durable than high-level waste borosilicate glasses. Furthermore, the residual glassy matrix can immobilize hazardous elements that cannot be incorporated in the crystals and offers much higher flexibility than pure crystalline waste forms, such as synroc and zircon. The targeted crystals to be formed in each melt depend on the types of fuel and associated wastes uses as feedstock. The crystals such as zircon have been shown to have high radiation stability [5] although the radiation stability of the vitreous cermics need more investigation.

Recent work [2-11] indicates that vitreous ceramic waste forms are chemically and physically similar to borosilicate glass waste forms, however, better chemical durability and higher waste loadings were also achieved. Vitreous ceramic waste forms have been increasingly recognized as a viable final waste form by the waste-management efforts both within and outside the U.S. This waste form has been studied by the DOE's Minimum Additive Waste Stabilization Program [6, 9, 12] for mixed waste; by WINCO of Idaho Falls for calcined high-level wastes [13]; by MSE, Inc. [14], SAIC [15], and EG&G [16] for mixed and buried wastes; by Switzerland [17] for spent fuel; by Russia [18, 19] for low- and intermediate-level waste; by Australia [3], and other international efforts [2] for various wastes.

The product quality will be controlled by assessing mass balance of each melt, on-line nondestructive examination of the homogeneity and radioactivity of the product, and the waste form property modeling efforts. However, waste form work using actual SNF has not been performed. Further R&D is needed to varify the waste form work under PAVC conditions using actual wastes.

RELIABILITY AND SAFETY OF THE PAC SYSTEM

Reliability and safety are very important considerations in selecting any technology for SNF treatment. The PAC furnace of the PAVC process has a long history of industrial uses and both pilot and fully scale experiences on a variety of waste treatment.

Plasma systems have been used for melting metals since 1980. To date, there are about 25 units in industrial operation. Teledyne-Allvac at Monroe, North Carolina, has a four-torch system that makes rotating-grade titanium ingots on a 24-h-day, 6-day-week schedule. Operational reliability has been excellent, with a total torch time of thousands of hours. Electrodes are replaced between ingots at intervals of 250 to 300 h.

In waste treatment, Retech has an 8-ft PAC furnace in Muttenz, Switzerland, in production operation for treating medical and industrial wastes. Electrode life is 400 h. In Butte, Montana, a 100-h reliability test was performed with a 6-ft PAC furnace as a proof-of-process test in the period prior to award of the Pit 9 remediation program at the Idaho National Engineering Laboratory [20]. The PAC system has also been demonstrated for treating radioactive wastes [21].

A series of eight tests was successfully conducted at MSE, Inc. in a 6-ft PAC furnace between February and June 1993 to evaluate the PAC system for use in demilitarization [22]. Twenty different small-caliber and hand-held pyrotechnic, smoke, and dye items from the demilitarization stockpile were treated in a 6-ft PAC without requiring disassembly before demilitarization. The conclusion from these tests is that PAC technology is suitable for processing completely assembled, small-caliber and hand-held pyrotechnic, smoke, and dye ordnance. No problems related to torch operation, localized equipment overheating, or vitreous ceramic consistency were encountered during the introduction of ordnance into the PAC system. These results suggest that PAC system can handle the possible pyrophorocity of the uranium metal during processing SNF when the same caution is taken as for processing ordnance.

The existing commercial PAC systems need to be modified into remote operational systems and substantial development can be based on existing work on systems that treating low-level nuclear wastes and on the previous efforts in remoterize the plasma system at PNNL and Retech.

FLEXIBILITY OF PAVC PROCESS

The PAVC process can handle various feed materials, such as metal, oxides, carbides, hydrides, and other forms without pretreatment. When this technology is developed for treating one type of SNF, it can most likely be also extended to treat many other types of fuels, such as zircaloy fuels, with minimum further development.

In particular, DOE's Office of Environmental Management (EM) is in possession of a large quantity of plutonium residuals of varying chemical composition, and diverse spent-fuel types with different matrix material, cladding material and overall general characteristics that exist in small quantities. The cost of characterization of these waste streams at the level needed for most waste forms and development of different waste forms for each type of waste would be prohibitive. The PAVC process has the potential to be expanded to handle the Pu residuals and odd amounts of spent fuel with little characterization and pretreatment; this extension is expected to require minimal development.

The PAVC technology could also potentially handle highly radioactive metal wastes, such as the replaced melters from DWPF, from the PAVC process, and any other operations. These used melters could be sliced and fed directly into a PAC system to produce durable vitreous ceramics for geologic disposal.

SUMMARY

The PAVC process for stabilizing various types of SNF and associated wastes is proposed based on extensive experience of nuclear waste form development and low-level nuclear waste treatment using the commercially available PAC system. SNF elements could be loaded directly into a PAC system and converted into vitreous ceramics. An effective off-gas treatment system for the plasma-arc process has been developed to treat volatile and semi-volatile materials for low-level/mixed nuclear waste treatment. The loaded filters and other wastes captured in the off-gas treatment are recycled back into the furnace to minimize secondary wastes. The existing off-gas system has to be tested/modified for treating SNF. The vitreous ceramic waste form is similar to borosilicate nuclear waste glass, and it will meet the requirements specified for borosilicate glasses in the "Waste Acceptance System Requirements Document" for permanent disposal in a geologic repository and for interim storage. Criticality safety would be ensured through the use of "batch" modes (processing one batch at a time), and controlling the amount of fuel processed in one batch. Minimum product volume would be achieved because high-density vitreous ceramic is produced almost entirely from SNF and associated wastes with minimum additives and minimum secondary wastes. The minimum requirements on SNF characterization and pretreatment, the one-step process, and minimum secondary waste generation reduce treatment duration, reduce radiation exposure, and reduce treatment cost. The PAVC technology is flexible and should be applicable directly to treat many other types of DOE fuels besides aluminum HEU SNF. Necessary R&D using actual SNF and associated wastes is needed to addressing the issues related to off-gas treatment, criticality control, remote operation, and process control.

ACKNOWLEDGMENT

The authors thank the following individuals who have contributed their time and efforts to make this document possible: D. A. Lamar, J. Shuen, J. M. Perez, T. D. Cooper, P. K. Shen, R. K. Womack, and C. G. Whitworth. Work for this project was supported by the U.S. Department of Energy under contract DE-AC06-76RLO 1830.

REFERENCES

1. *DOE-Owned Spent Nuclear Fuel Technology Integration Plan*, DOE/SNF/PP-002, Revision 1, May 1996.

2. Lutze W. and R.C. Ewing.   1988. *Radioactive Waste Forms for the Future*, North-Holland, Amsterdam.

3. Vance, E. R., P.J. Angel, B.D. Begg and R.A. Day, 1994.   "Zirconolite-Rich Titanate Ceramics for High-level Actinide Wastes," *Sci. Bas. Nucl. Waste Manag. XVII*, 333, 293, Philadelphia, Pennsylvania.

4. Feng, X., W.K. Hahn, W. Gong, L. Wang, and R.C. Ewing, "Minimum Additive Waste Stabilization Using Vitreous Ceramics," Pacific Northwest National Laboratory, Report #PNL-10826, September 1995, Richland, WA.

5. Ewing, R.C., W. Lutze, and W.J. Weber, "Zircon: A host-phase for the disposal of weapons Plutonium," J. Mater. Res. 10(2), 243(1995)

6. Feng, X. 1994. "Development of Vitreous Ceramic as Final Waste Forms for Mixed Wastes," *Proc. American Chemical Society* "Emerging Technologies in Hazardous Waste Management VI," Atlanta, Georgia.

7. MSE, Inc. 1994. "Test Results of Screening High-Metal-Content Wastes in a Plasma Centrifugal Furnace Under the Minimum Additive Waste Stabilization Program," Report PTP-3, Butte, Montana.

8. Hassel, G. R., J. A. Batdorf, R.M. Geimer, G.L. Leatherman, J.M. Wilson, W.P. Wolf, A. Wollerman. 1991. "Evaluation of the Test Results from the Plasma Hearth Process," Science Applications International Corporation, SAIC-94/1095, Idaho Falls, Idaho.

9. Feng, X., G. Ordaz, and P. Krumrine. 1994a. "Glassy Slag - A Complementary Waste Form to Homogenous Glass for the Implementation of MAWS in Treating DOE Low-Level/Mixed Wastes," *Proc. Spectrum '94*, Atlanta, Georgia.

10. Feng, X., D.J. Wronkiewicz, J.K. Bates, N.R. Brown, E.C. Buck, N.L. Dietz, M. Gong, and J.W. Emery, 1994b. "Vitreous ceramic for Minimum Additive Waste Stabilization, Interim Program Report, May 1993 - February 1994," Argonne National Laboratory Report # ANL-94/24, Argonne, Illinois.

11. Feng, X., et al. 1994c. "Comparison of Glassy Slag Waste Forms produced in Laboratory Crucibles and in a Pilot Scale Plasma Furnace," *Proc. American Chemical Society* "Emerging Technologies in Hazardous Waste Management VI," Atlanta, Georgia.

12. U. S. Department of Energy, Office of Technology Development. 1994. "Minimum Additive Waste Stabilization (MAWS)," Technology Summary DOE/EM-124P. Washington, DC.

13. Vinjamuri, K. 1994. "Solid-Based Glass-Ceramic Waste Forms for Immobilization of the Fluorine/Sodium Calcined High-Level Waste Stored at the Idaho Chemical Plant," *Proc. Spectrum'94*, Atlanta, Georgia.

14. Battleson, D. 1994. "Latest Developments of Plasma Technology at the CDIF," *Proc. Spectrum '94,* Atlanta, Georgia.

15. Geimer, R., C. Dwight, and G. McClellan. 1994. "The Plasma Hearth Process Demonstration Project for Mixed Waste Treatment," *Pro. Spectrum'94*, Atlanta, Georgia.

16. Quapp, W.J., T.L. Eddy, G.A. Reimann, R.L. Miller. 1993. "Feasibility of Vitrification of Spent Nuclear Fuel Using Iron-Enriched Basalt," *Proc. Emerging Tech. Hazardous Waste Manag.*, Washington, DC.

17. Hoffelner, W., A. Chrubasik, R.C. Eschenbach, M.R. Funfschilling, and B. Pellaud. 1992. "Plasma Technology for Rapid Oxidation, Melting and Vitrification of Low/Medium Radioactive Waste," *Nuclear Eng. Intern.*,

18. Dmitriyev, S.A., S.V. Stefanovsky, I.A. Knyazev, F.A. Lifanov. 1995. "Characterization of Slag Product from Plasma Furnace for Unsorted Solid Radioactive Waste Treatment," *Sci. Basis Nucl. Waste Manag. XVIII,* 1323, Philadelphia, Pennsylvania.

19. Sobolev, I. A., S. V. Stefanovsky, F. A. Lifanov. 1995. "Synthetic Melted Rock-Type Waste Forms," *Sci. Bas. Nucl. Manag. XVIII, 833,* Philadelphia, Pennsylvania.

20. Rivers, T.J., and J.W. Ruffner, "Duration Testing of the Plasma Arc Centrifugal Treatment System," at the 1994 Incineration Conference.

21. Yeast, T.F., R.C. Eschenbach, G.D. Pierce, M.B. Arndt, and J. Ginsburg, "Volatility Studies in A Rorating Furnace," 1994 Winter Meeting of the American Nuclear Society, Washington D.C., November 13-17, 1994

22. MSE, Inc. 1993. "Demilitarization of Pyrotechnic Ordnance Using A Plasma Arc Furnace," Report #PCF-D021R1, Butte, Montana.

An Electrodeless Melter for Vitrification of Nuclear Waste

J. P. FREIDBERG, A. J. SHAJII, K. W. WENZEL, and J. R. LIERZER
MIT Nuclear Engineering Department and Plasma Science and Fusion Center,
NW16, 77 Massachusetts Avenue, Cambridge, MA 02139 USA, freidberg@psfc.mit.edu

Abstract

This paper describes a new concept for a high-temperature, electrodeless melter for vitrifying radioactive wastes. Based on the principles of induction heating, it circumvents a number of difficulties associated with existing technology. The melter can operate at higher temperatures (1500-2000°C vs 1150°C), allowing for a higher quality, more durable glass which reduces the long-term leaching rate. Higher processing temperatures also enable conversion from borosilicate to high-silica glass which can accommodate 2 to 3 times as much radioactive waste, potentially halving the ultimate required long-term disposal space. Finally, with high temperatures, conversion of nuclear waste into ceramics can also be considered. This too leads to higher waste loading and the reduction of repository space. The melter is toroidal, linked by an iron core transformer that allows efficient electrical operation even at 60 Hz. One-dimensional electrical and thermal analyses are presented.

I. Introduction

We have considered an improved method for vitrifying high-level radioactive waste (HLW). The improvement consists of a practical, economical idea for a high-temperature electrodeless glass melter based on the principles of induction heating. Although HLW represents the primary waste stream, there are several possible additional applications. These include the formation of ceramics (as opposed to glass) and the treatment of mixed nuclear waste with the goal of obtaining higher waste loadings. The electrodeless melter can also contribute to the ongoing national evaluation of induction heaters versus plasma arc melters for the treatment of mixed wastes. The new concept accomplishes its goal by an innovative spin-off from magnetic fusion research, which is elegant in its engineering simplicity. To appreciate the advantages of the new melter, as well as to understand the motivation and gains from the electrodeless induction melter, one must first recognize the magnitude of the problems involved and the difficulties associated with the existing technology.

I.A. High level nuclear waste in the USA

Nuclear fuel reprocessing has created large amounts of radioactive waste. An astounding 400,000 m^3 of high-level waste (HLW) from defense activities is stored in the U.S. alone [1,2]. The international consesus is to immobilize this waste in borosilicate glass, because it is stable, tough, nearly insoluble in water, and can be formed at relatively low temperatures in existing melters.[3] High silica glass has also been considered for HLW disposal, because it is more durable and can handle higher waste loading. However, high-silica glasses require higher temperature melters than are currently available. Typical melters process on the order of 100 kg/hr of glass or less [4]. Therefore, it would require about 100 melters, operating continuously almost 14 years to process all the existing U.S. defense HLW.

I.B. Current status of glass melters

Two main types of melters are used for vitrification of HLW. The first is the Joule-heated melter that relies on metal (usually inconel) electrodes in contact with the waste to drive an electric current through the melt. The other type, developed in France, is a high-frequency induction melter, which heats the waste indirectly by heating a metal chamber holding the waste. A fundamental limit to both systems is that they must operate at temperatures below 1150°C because of intense corrosion of the metals at higher temperatures.

Industrial-scale Joule heaters are operating in the U.S. at the Defense Waste Processing Facility and the West Valley Demonstration Project. Industrial-scale induction-heated waste processing facilities are also in operation in Marcoule and La Hague, France. In these facilities, liquid HLW is calcined then vitrified in an inconel chamber. The metal chamber is heated inductively to 1100°C by 100 kW of AC power at a frequency of 10 kHz. Up to 18-25

Mat. Res. Soc. Symp. Proc. Vol. 465 © 1997 Materials Research Society

kg/hr of glass can be produced in these plants [4]. Both Russia and France have developed cold-crucible induction melters, which exhibit some of the same advantages listed here, but must operate at high frequency ($f \geq 10$ kHz) [5,6].

I.C. Advantages of an electrodeless melter

Electrodeless melters have substantial advantages. First, the melter could operate at higher temperatures (1500-2000°C) since it is not limited by electrode erosion. This would enable a conversion from borosilicate to high-silica glass which can accommodate 2 to 3 times as much radioactive waste (i.e. 40-50% vs 15-25%). The net result is a potential doubling of nuclear waste throughput for a given size melter and a halving of the ultimate required long term storage space. Second, glass produced at higher temperatures is generally of a higher quality, exhibiting less porosity and greater leach resistance. Third, without electrodes, which require periodic replacement, the reliability and maintainability of the induction melter should be superior. A related issue is that induction systems that couple to the melt provide nearly uniform heating.

I.D. An induction-heated electrodeless melter

These considerations have led us to a high temperature electrodeless glass melter based on the principles of induction heating. Induction heating is a mature technology, widely applied in the metals processing industries. Even so, the transition from liquid metal applications to glass melting is not straightforward, because molten glass has a far higher electrical resistivity and viscosity than a typical liquid metal. Aluminum and stainless steel have resistivities of 4×10^{-8} and 72×10^{-8} Ω-m, respectively, whereas molten glass has a resistivity of 4×10^{-3} Ω-m. This large difference makes inductive heating an inherently less efficient energy conversion scheme often requiring 100 times more high frequency (10 kHz) reactive power than the heating power delivered to the melt. It is for this reason that induction furnaces for glass melting have received only small attention in the research community.

Our design overcomes these difficulties by switching from a "standard" cylindrical configuration to a toroidal geometry, in which an iron core transformer can be used to greatly enhance the inductive coupling coefficient, essentially raising its value to unity. The transformer, in combination with a larger melter volume (10 m^3 vs 1 m^3) reduces the operating frequency to a more economical industrial range (50-1000 Hz), increases the throughput per melter (from 100 to 300 kg/hr) and may well reduce the unit cost per mass of nuclear waste treated because of economies of scale. Also, even with the larger volume the total power requirements, on the order of 300 kW, remain well within practical limits.

II. Technical Description of the New Induction Melter
II.A. The electromagnetic configuration

A schematic of the new melter is shown in Fig. 1. The melting chamber is a torus, and a coil surrounds the outer leg of an iron core to act as the primary of a transformer while the glass melt serves as the secondary. There are two crucial features to the design. First, the primary of the transformer is linked by an iron core. This core makes the coupling coefficient between primary and secondary effectively unity even in the presence of a large volume of furnace bricks. The electrical conductivity of the bricks must be much less than that of the glass melt, a condition easily satisfied in practice. The second feature is to take advantage of economies of scale and consider larger volume melters than are currently in use. Typically, volumes on the order of 10 m^3 are of interest as opposed to the existing 0.1-3 m^3.

These two features dramatically improve the electrical performance of induction melters. At the optimum operating frequency the reactive power can be made substantially smaller than the dissipated power. Equally important, the optimum frequency is reduced from the 10 kHz range to approximately 1 kHz. Each of these represents a much more economical and robust operating regime. For a small increase in either operating or capital costs, one can drive the melter at 60 Hz, thereby eliminating the need for high-frequency power supplies. This would be a substantial technological simplification and cost saving. There are two additional benefits to the configuration shown in Fig. 1. First, it produces a relatively uniform power deposition. This results in small temperature gradients and no reliance on thermal or external stirring for temperature equilibration. (Initial 2D calculations show less than 20% temperature variation across the melt for the reference case described below.) Second, with the coils on the outside leg, remote maintenance is much easier. It may actually be possible, with a slightly larger iron core, to locate the coils outside the radiation shield, thereby allowing hands-on coil maintenance.

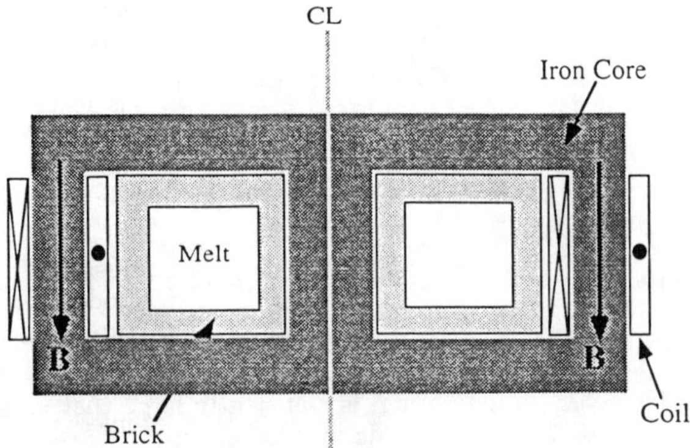

Fig. 1: Schematic of the proposed melter.

II.B. One-dimensional electrical analysis of the melter

Electrical properties of the melter are quantified by means of a 1-D model using the geometry in Fig. 2. Consider sinusoidal steady-state operation at a frequency $\omega = 2\pi f$. The relative permeability of the iron core is denoted by μ_r and for simplicity the resistivity of the melt, η_2, is assumed uniform in space. In the regime of interest, η_2 remains sufficiently high that the generalized Ohm's law for the melt, $\vec{E} + \vec{v} \times \vec{B} = \eta_2 \vec{J}$, is accurately approximated by $\vec{E} = \eta_2 \vec{J}$. All quantities are assumed to depend only upon r.

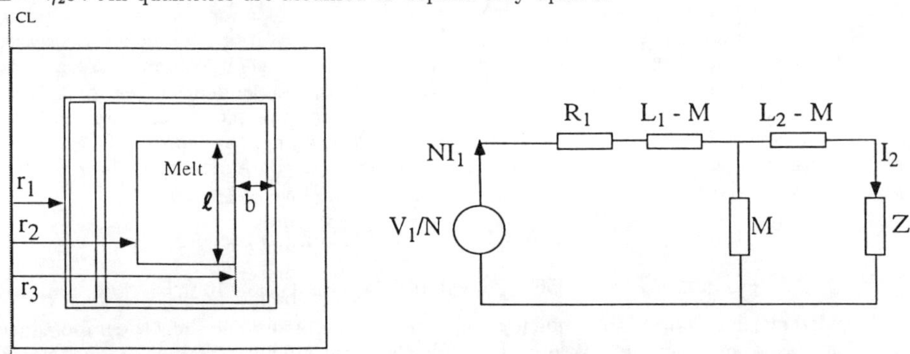

Fig. 2: Melter geometry

Fig. 3: Equivalent circuit of the melter

Application of Maxwell's equations leads to the well known equivalent circuit for a transformer as shown in Fig. 3. Subscripts "1" and "2" denote primary and secondary, respectively. The quantity Z is a complex, frequency dependent portion of the secondary impedance. The frequency dependence is a consequence of the fact that the melt must be treated as a distributed current source since, in the regime of interest, the skin depth is comparable to the chamber width. The value of Z used in these calculations has been obtained by analytically solving the magnetic diffusion equation in the melt region.

A great deal of insight is obtained by considering the limit $\mu_r >> 1$ and approximating Z by its low frequency, large skin depth value. The circuit elements are then

$$R_1 = 0, \tag{1}$$

$$L_1 - M = 0, \tag{2}$$

$$M = \pi \mu_0 \mu_r r_1^2 / l_i, \tag{3}$$

$$L_2 - M \equiv L = \pi \mu_0 (r_2^2 - r_1^2)/l, \tag{4}$$

$$Z \approx R_2 = 2\pi \eta_2 / l \ln(r_3/r_2), \tag{5}$$

where l_i is the total path length around the iron core and N is the number of turns in the primary. Also, note that $M >> L$ for large μ_r and $M << L$ for $\mu_r = 1$.

A critical feature of the melter design is that there is an optimum frequency for maximizing performance, which is defined as that frequency at which the ratio of dissipated power to reactive power for a specified secondary current (required for heating) is maximized. At very low frequencies the reactive impedances are small compared to R_2, but the mutual inductance essentially shorts out the secondary thereby allowing only a small fraction of the total current to flow in the melt. At very high frequencies the reactive impedances are large compared to R_2. In this case the mutual inductance is effectively an open circuit. However, the reactive impedance associated with L is sufficiently large that only a small fraction of the voltage appears across R_2 again leading to a low dissipated power. A simple calculation shows that the optimum frequency and corresponding power ratio are given by

$$\omega = (R_2/L)\sqrt{\alpha/(1+\alpha)}, \tag{6}$$

$$P_{diss}/P_{reac} = 0.5\sqrt{1/\alpha(1+\alpha)}, \tag{7}$$

where

$$\alpha \equiv \frac{L}{M} = \left(\frac{r_2^2}{r_1^2 - 1}\right)\left(\frac{l_i}{\mu_r l}\right). \tag{8}$$

The important scaling predictions are as follows. The optimum frequency decreases as the glass resistivity decreases and relative permeability and melt volume increase. At the optimum, the ratio of dissipated to reactive power is independent of resistivity and increases with increasing relative permeability. These scaling predictions are confirmed by the numerical solution of the circuit equations (Fig. 4). Shown here are curves of P_{diss}/P_{reac} versus the frequency for various values of μ_r, η_2, and volume. Parameters for the reference configuration are given by:

$$l = 1.0 \text{ m}; \quad l_i = 11.5 \text{ m}; \quad b = 1.0 \text{ m}; \quad r_1 = 0.75 \text{ m}; \quad r_2 = 1.75 \text{ m}; \tag{9}$$

$$r_3 = 2.5 \text{ m}; \quad \mu_r = 3500; \quad \eta_2 = 0.035 \ \Omega - \text{m}; \quad \text{Vol} = 10 \text{ m}^3 \tag{10}$$

The existence of an optimum frequency is apparent. Figs. 4a and 4c show the importance of an iron core and larger melt volume for high electrical efficiency at low frequency. A detailed comparison of critical parameters is given in Table 1 for four cases: (1) the reference case, (2) reference geometry with an air core transformer, (3) an iron core transformer with a reduced volume of 1 m^3, and (4) reference case at a frequency of 60 Hz.

For the reference case the optimum frequency is approximately 1 kHz, a reasonably low value, and the ratio of dissipated to reactive power is 3.5 thereby completely eliminating the problem of low electrical efficiency. Operation is degraded when the iron core is eliminated, leading to unacceptably low efficiencies for realistic applications. This is because the parameter $\alpha \equiv L/M$ increases from the value 0.015 (with iron) to 51 (with air). Note that the

Table 1: Parameters for the four melter configurations.

	Case 1 (reference)	Case 2 $\mu_r = 1$	Case 3 Vol=1 m^3	Case 4 f=60 Hz
f_{opt} (Hz)	1000	7500	6600	60
P_{diss}/P_{reac}	3.5	0.0072	3.8	0.41

optimum frequency increases to 7.5 kHz and that P_{diss}/P_{reac} reduces to 0.0072. To assess the desirability of a more compact system the volume has been decreased from 10 to 1 m^3 by scaling r$_3$ and l_i while holding r$_2$ constant and for fixed brick and iron core thickness. In this case the high electrical efficiency is preserved but the optimum frequency remains high at 6.6 kHz. Smaller systems can be efficient but require higher frequency. They are thus somewhat less desirable than larger systems from the electrical point of view, but remain acceptable if an iron core is used. Finally, in the reference case one has the opportunity for an enormous gain in electrical simplicity and economy by operating slightly off optimum at 60 Hz. In particular, no high frequency power supplies would be required, and it is generally recognized that the price of delivering power declines at lower frequency. At this low frequency the reactive power is approximately 2.5 times larger than the power dissipated in the melt, far superior to the factor of 140 at 7.5 kHz for the air core system. The factor 2.5 is sufficiently modest that one could consider simply purchasing the excess reactive power directly from the regional utility. Alternatively, for a single, up-front cost one could install on-site reactive power compensation equipment (essentially a capacitor bank). We conclude from this analysis that an inductively heated glass melter is feasible from an electrical point of view.

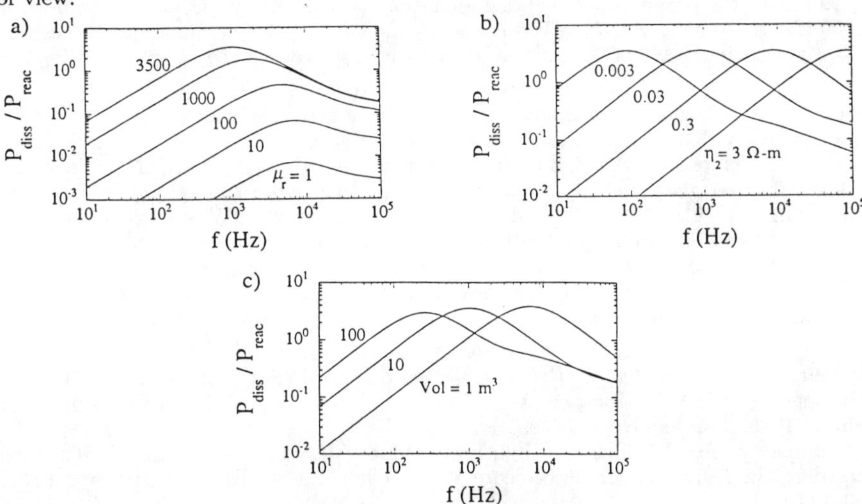

Fig. 4: Ratio of P_{diss} to P_{reac} versus frequency for various a) μ_r, b) η_2, and c) Volume

II.C. Simple thermal analysis of the melter

The purpose of the thermal analysis is to determine the magnitude of the Joule heating ($I_2^2 R_2$) required to balance heat losses to the surroundings and maintain the melt at 1500°C during steady-state operation. A simple control volume analysis of the melter shows that in steady-state, heat losses are due to the following three different mechanisms:

1. Conduction heat losses to air at the outer brick interface.
2. Radiation heat losses at the outer surface of the brick.

3. Convective heat loss from inlet of cold glass and exit of molten glass.

By properly choosing the thickness of the bricks (approximately 0.8 m thick), their outside surface can be maintained at relatively low temperatures (T = 65°C). At this value, radiation heat losses are negligible. In addition, heat conduction losses at the brick/air interface are much smaller than the heat required to raise the temperature of the cold glass entering the melt to 1500°C. That is, item 3 above is the dominant heat loss mechanism in the system. Thus, in steady state the following relation is satisfied:

$$I_2^2 R_2 = \dot{m}\Delta H, \tag{11}$$

where I_2 (A) is the melt current, R_2 (Ω) is the melt resistance, $\dot{m}$ (kg/s) is the throughput of the system, and ΔH is the energy density (J/kg) required to raise the cold (unprocessed) glass to the melt temperature. For glass alone ΔH is given by $\Delta H = C\Delta T$ where C is the specific heat of the glass (J/kg) and the temperature rise is given by $\Delta T \approx 1500°C$. To raise the glass temperature alone one requires 420 kW-hr/tonne. If in addition to the glass, water or other solutes are present, an additional energy must be supplied for the boiling off of such substances. For the purpose of this analysis, it is assumed that as much as 300-600 kW-hr/tonne of energy is required to boil any solute present in the system. Thus, the total power per throughput for the melter is approximately ΔH=700-1000 kW-hr/tonne.

In order to determine the melt current one now requires the resistance of the melt (determined from the electrodynamics calculations to be $R_2 = 0.6\ \Omega$) as well as the throughput $\dot{m}$. Consider existing large electrode melters in the US which have a typical throughput of 100 kg/hr for a melt volume of approximately 3 m^3. By scaling the throughput with the melt volume (10 m^3 for the proposed melter) one finds that $\dot{m}$ =300 kg/hr. Using this value and ΔH= 1000 kW-hr/tonne yields $I_2 = 700$ A and P_{diss}=300 kW. This is the required current to maintain the temperature of the melt at 1500°C during steady-state operation.

Summary and Conclusions

A description and 1-D analysis has been presented of a new electrodeless melter for the vitrification of HLW. The combination of a toroidal iron-core transformer and larger than usual melt volume eliminates the high electrical inefficiencies (i.e. large P_{reac}/P_{diss}) inherent in inductively heated glass melters. For a volume of 10 m^3 the throughput is expected to be approximately $\dot{m} \approx 300$ kg/hr, which requires $P_{diss} \approx 300$ kW and corresponds to a normalized power per throughput of 1 MW-hr/tonne. At the theoretically optimized frequency of 1000 Hz the reactive power is only 86 kVA. Even off optimum, at the highly desirable frequency of 60 Hz, the reactive power is 730 kVA, which could be readily handled by a reasonable, one-time initial cost for reactive power compensation equipment.

References

1. R. L. Murray, *Understanding Radioactive Waste*, Battelle Press, Columbus, OH (1994).
2. Estimating the Cold War Mortgage, U.S. Department of Energy Office of Environmental Management 1995. Baseline Report (1995).
3. J. C. Cunnane and J. M. Allison, "High Level Waste Glass Compendium; What it Tells Us Concerning the Durability of Borosilicate Waste Glass", Mat. Res. Soc. Symp. Proc. Vol. **333** 3 (1994).
4. W. Lutze and R. C. Ewing *Radioactive Waste Forms for the Future*, North-Holland, Amsterdam (1988).
5. I. A. Sobolev, F. A. Lifanov, S. A. Dmitriyev, *et al.*, "Vitrifciation of Radioactive Wastes by Coreless Induction Melting in Cold Crucible", Intl. Nuclear and Hazardous Waste Mgmt. Conf., Atlanta (1994).
6. A. Jouan, J.-P. Moncouyoux, S. Merlin, and P. Roux, "Multiple Applications of Cold-Crucible Melting", Radwaste Magazine, **3** #2, p. 81 (1996).

FUNDAMENTAL CHEMISTRY AND MATERIALS SCIENCE OF AMERICIUM IN SELECTED IMMOBILIZATION GLASSES

R. G. HAIRE *, N. A. Stump **
*Oak Ridge National Laboratory, P. O. Box 2008, Oak Ridge, TN 37831-6375, rgh@ornl.gov
**Dept. Physical Sciences, Winston-Salem State University, 601 Martin Luther King, Jr. Dr., Winston-Salem, NC 27110

ABSTRACT

We have pursued some of the fundamental chemistry and materials science of americium in three glass matrices, two being high-temperature ($850°$ and $1400°C$ melting points) silicate-based glasses and the third a sol-gel glass. Optical spectroscopy was the principal investigating tool in the studies. One aspect of this work was to determine the oxidation state exhibited by americium in these matrices, as well as factors that control and/or may alter this state. We have noted a correlation between the oxidation state of the f-elements in the two high-temperature glasses with their high-temperature oxide chemistries. One exception was americium: although americium dioxide is the stable oxide encountered in air, when this dioxide was incorporated into the high-temperature glasses, only trivalent americium was found in the products. When trivalent americium was used to prepare the sol-gel glasses at ambient temperature, and after these products were heated in air to $800°C$, again only trivalent americium was observed. Potential explanations for the unexpected behavior of americium is offered in the context of its basic chemistry. Experimental spectra, spectroscopic assignments and other pertinent data obtained in the studies are discussed.

INTRODUCTION

The science of immobilization materials for nuclear waste has been an active topic for decades and many studies on this subject have been performed and reported. These have ranged from examining physical properties of the host materials to generating modeling approaches for projecting the performance of materials. The latter depend heavily on valid, fundamental information about the materials. One of our objectives has been to generate information about the fundamental chemistry and materials science of 4f- and 5f-elements in three selected glass matrices. These have included two high-temperature ($850°$ and $1400°$ melting points), silicate-based glasses and glasses prepared via an aqueous sol-gel route. Optical spectroscopy (absorption, fluorescence, Raman) was the principal investigating tool used in the studies.

One aspect of this work was the determination of the oxidation states exhibited by these f-elements in these matrices, as well as the factors that may control and/or alter these states. With the high-temperature silicate-based glasses, we noted a correlation existed between the oxidation state of the f-elements in the glasses with their high-temperature oxide chemistries. The correlation seems plausible given the silica based network encountered in the glasses, and that the products were formed and maintained in air. One exception to this general correlation was observed with americium, where a trivalent rather than a tetravalent oxidation state was observed. Oxidation states reported in the literature for americium range from two through seven. Although trivalent americium is frequently encountered in acidic solutions, when americium salts or oxides are calcined in air up to $1200°C$, the oxide formed is nominally americium dioxide. In contrast, after heating americium in air in these three silicon-oxygen, glass matrices, only trivalent americium was observed in the products. As trivalent americium was used in preparing the sol-

Mat. Res. Soc. Symp. Proc. Vol. 465 © 1997 Materials Research Society

gel glasses and americium dioxide was dissolved in the glasses prepared at high temperature, this required the reduction of americium in two glasses and avoidance of oxidation in the third glass. The lanthanide homolog of americium, europium, both in general and in these three glass matrices, is encountered in its trivalent oxidation state. Thus, americium behaves similarly to europium in the glass products studied but behaves differently from europium in its oxide formation in air. Americium's behavior in the high-temperature glasses mimics the behaviors of all of the lanthanide elements, except for cerium.

Many glass matrices prepared with multi-component f-elements to their maximum solubility limit are not suitable for spectroscopic studies, frequently being opaque and/or very dark (e.g., especially glasses containing cerium, neodymium and plutonium). To aid our spectroscopic studies, our approach was to initiate these studies using a single f-element component (e.g., americium) and at concentrations best suited for spectroscopic analyses (typically 5-20 weight %, depending upon the element). In this studies of americium reported here, both absorption and fluorescence spectral measurements were employed.

Some of the main scientific questions about immobilizing glasses deal with the stability of the materials formed, mobility/diffusion characteristics, and/or the leaching of the f-elements contained in the matrices [1]. The oxidation states of these elements in such hosts is of primary importance in this regard, affecting many of the physicochemical properties of the solid phase, the susceptibility of the material to leaching from the matrix if contacted by aqueous media, and the nature of the species that may be solubilized. There are serious efforts to model the behavior and performance of such isolation matrices, especially given that their performance must be evaluated for time frames which exceed experimental capabilities. The oxidation states of the elements immobilized in such matrices is therefore of critical importance and must be one of the major considerations in modeling and evaluations.

The behavior of americium in the glass hosts described here addresses both its fundamental and technological science. Reported here are the results of studies aimed at examining the oxidation state exhibited by americium in three glass hosts prepared at ambient and elevated temperatures. Potential explanations for the behavior of americium are offered in terms of its basic chemistry and the nature of the americium-oxygen system.

EXPERIMENTAL

<u>Materials</u>

The americium (Am-243; $t_{1/2}$ = 7.4 x 10^3 y) employed in these studies was obtained as a product of the High Flux Isotope Reactor located at Oak Ridge National Laboratory and made available through the BES program of DOE. Aqueous solutions of americium (III) in 0.5 M nitric acid were obtained by diluting with water the eluant obtained from a cation ion-exchange column during the purification procedure [2]. Americium dioxide was prepared from this stock solution by precipitating its oxalate salt and calcining it in air at 1000°C.

Two high-temperature glass formulations were used in these studies. One was a higher melting material (1400°C) with the following nominal weight percentages: SiO_2 (30 %); B_2O_3 (6%); BaO (3%); Al_2O_3 (13%); PbO (10%) and f-element oxide (nominally 38%). To allow a variation of the americium content while maintaining a constant f-element content, a lanthanum oxide-americium oxide ratio was used in preparing these particular glasses. The components without the americium were first melted to form a stock glass: cooled; ground to a fine powder; and the latter then ground with the desired quantity of americium dioxide. This final mixture was then heated in platinum boats to 1450°C, maintained at this temperature until dissolution was complete (4-8 hours) and then slowly cooled to room temperature. The products were obtained in the form of small, transparent glass "jewels", light pink in color.

The second glass formulation was a lower melting glass (850°C) with the following nominal weight percentages: SiO_2 (50%); B_2O_3 (18%) Na_2O (24%); CaO (3%) and AmO_2 (5%). The components without the americium were melted and after cooling to a solid, ground to a fine powder. The americium glasses were then prepared by melting an appropriate quantity of this glass formulation in a platinum container at 900°C and then adding the solid americium dioxide to the molten glass. After complete dissolution, the molten material was slowly cooled to room temperature. The final products were similar in appearance to the higher-temperature products described above.

The sol-gel glasses were prepared by mixing 0.1 M Am (III) solutions (0.5 M nitric acid) with equal volumes of methanol and tetramethylorthosilicate in small glass cylinders with a restricted top to depress volatilization of the volatile components. After 24-48 hours, clear, solid "glass" (solid gel; pink in color) products were formed, which were removed as a single piece from the cylinders. To remove remaining volatile components, it is necessary to heat these products to elevated temperatures.

Spectroscopy

Portions of the each glass product were taken and sealed in glass tubes with optically flat sides designed in-house for our spectroscopic measurements. A Jobin Yvon/Instruments SA model Ramanor HG.2S spectrometer (resolution 0.5 cm^{-1} at 514 nm) was employed in conjunction with a microscopic attachment (Nachet, model NS-400). For the absorption studies, a xenon-arc lamp was used and the dispersed light measured via a photon counting system (cooled photomultipler; multichannel analyzer) interfaced with a PC computer using Galactic Industries, "Spectra-Calc" software. Fluorescence measurements were made using many of these components but employing a 5 watt argon laser (at reduced power levels to avoid excessive heating of the samples) as the excitation source.

RESULTS

Absorption Spectra

The spectral region of the present microscopic instrumentation was limited to 400 - 850 nm. Due to the irregular shaped specimens of the americium glass, absorption extinction coefficients were not determined. The main purpose of the spectral examinations was to determine qualitatively the oxidation state(s) of the americium in the glasses.

An absorption spectrum obtained from the americium glass prepared by dissolving americium dioxide in the molten host at 1450°C is shown in figure 1A. Absorption spectra for these americium glasses show unequivocally that the americium in the glass exhibits an oxidation of three. This solid state absorption spectrum matches well with solution spectra reported for Am (III) ions in different solution media [3,4], in the solid state [5] and the free ion energy levels for the Am^{3+} ion [6]. The major absorption maxima in the absorption spectrum were at 509, 784, and 834 nm (see Table I), with the main absorption being at 509 nm (well-established absorption for trivalent americium). This peak has a definitive shoulder at a slightly longer wavelength. Absorptions assigned to other oxidation states of americium were not observed.

The americium glass prepared at 1450°C also displayed emission peaks when excited by 457.9 nm radiation (Figure 2). Four emission peaks were observed: a weak emission at 11990 cm^{-1} or

834 nm; a very strong emission at 14,431 cm^{-1} or 693 nm; a medium emission at 16587 cm^{-1} or 603 nm; and another weak emission at 19,153 cm^{-1} or 522 nm (see Table I). These emissions are also characteristic of americium having a trivalent oxidation state.

The absorption spectrum obtained from the second glass (lower melting, 850°C, soda borosilicate) prepared at elevated temperatures is not shown. These products gave nearly identical absorption and emission spectra to those shown in figures 1a and 2 for the higher melting glass product. The shoulder at about 515-520 nm was even more pronounced for this product. Again, the spectra show that only trivalent americium was present in the glass.

The two remaining spectra in figure 1 (1b and 1c) were obtained from glasses prepared via a sol-gel technique. The spectrum in figure 1b was obtained from the glass following solidification at ambient temperatures, whereas the spectrum in figure 1c was obtained at ambient temperatures after the glass had been heated to 800°C. Emission spectra obtained from these two products were essentially the same as shown in figure 2. It was concluded again that both products contained only americium in its plus three oxidation state. The main difference noted between these two spectra was that the spectrum in figure 1b was "simpler", essentially having absorptions only at 509 nm and 834 nm.

The absorption spectrum for the sol-gel glass prepared at ambient temperature (1b) is interesting in that there is only a hint of a shoulder at 515 nm and a single broad peak at 834 nm. It bears the best resemblance to the solution spectrum reported for dilute perchloric acid solutions of trivalent americium [3], which was reported to have a very sharp absorption (no shoulder) at 505 nm and a smaller, slightly broadened absorption peak at 810 nm. With greater complexation in aqueous media (e.g., changing to more concentrated nitric acid media) broadening and/or the splitting of these two absorption peaks occurs, producing shoulders or peaks at 515 nm and 780 nm [3]. The spectra for the two high-temperature americium glasses and the sol-gel glass heated to 800°C all show these "shoulders". We believe that these changes (broadening /splitting) reflect either or both a "complexation effect" and/or the effect of a stronger crystal field encountered in the glass matrix; crystal fields in the solid matrix may be stronger than those in the aqueous media. A more detailed examination of the absorption and emission spectra, the specific spectroscopic assignments, and interpretations of bonding/crystal field effects will be covered in a subsequent paper. The main conclusion of importance from the absorption and emission spectra in the present paper is that in the three glasses, americium was present only in its trivalent oxidation state.

TABLE I.   Absorption Data for the Three Americium Glasses

Preparation	Absorption maxima, nm	Emission, nm: 457.9 nm excitation
AmO$_2$ in 1450°C glass	509 major	833 weak (not shown)
and	520 shoulder	692 strong
Am (III) in sol-gel (800°C)	784 broad, medium	581 weak +
	834 " "	522 weak
Am(III) sol-gel 25°C	505; sharp; 825 broad	similar to the above

On the oxidation State of Americium in the Glasses

Given that the glass matrix consists essentially of a silicon-oxygen network, and that the glasses are prepared/heated in air, it would seem that some correlation should exist between the

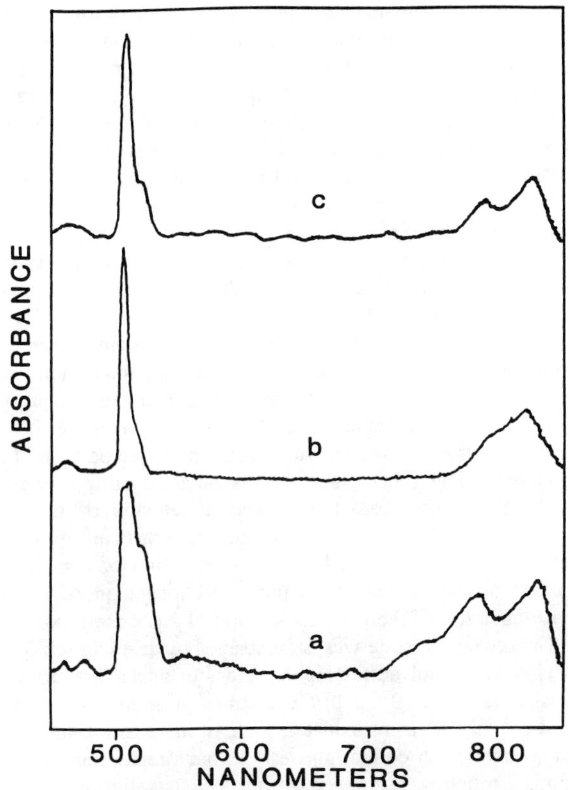

Figure 1: Solid State Absorption Spectra For Am In Glasses;
(a) 1450°C Glass; (b) Sol-Gel Glass; (c) Sol-Gel Glass
After Heating to 800°C.

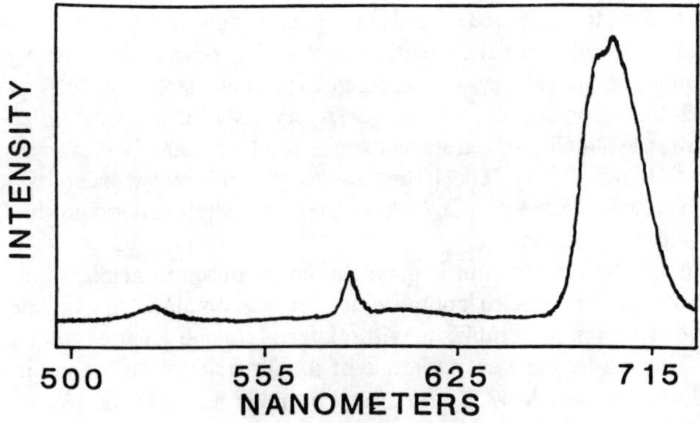

Figure 2: Emission Spectrum From Am in the 1450°C Glass;
Excitation With 457.9 nm Radiation.

oxidation states acquired in oxides in air and in the glass products formed in air. Indeed, in our studies of the oxidation states of the 4f- and 5f-elements acquired in glasses (all the 4f elements; and the 5f-elements, Np-Es), it would appear that only americium and uranium deviate in this regard. In such glasses, the oxidation state of americium is plus three and for uranium an oxidation of four, five or six (depending of the oxygen fugacity) based on reports by others [5-8]. The expected oxides at these temperatures should consist of tetravalent americium and uranium. Trivalent and tetravalent ions are generally accepted as being incorporated into the silicate matrices and mono- and divalent ions as network modifiers, but this difference does not provide a rational for the observed behaviors of these two elements. For glasses prepared between 600 and 1000°C, glasses would be expected to contain tetravalent americium and uranium based on the oxides that are stable at these temperatures in air. As we have not studied uranium glasses, additional discussions here will not include them.

As the first glass we examined was prepared at 1450°C with americium dioxide, it was first thought that thermal reduction of the dioxide before or during dissolution in the molten glass may have been responsible for the generation of trivalent americium in the product. The loss of oxygen from americium dioxide is known to occur at elevated temperatures [9]. The concept was that upon solidification, which occurred at sufficiently high temperature (e.g., 1400°C), re-oxidation upon subsequent cooling may not have been achieved. However, also obtaining trivalent americium in the glasses prepared at 850°C and the retention of trivalent americium in the sol-gel glasses after heating up to 800°C in air, both suggested this concept was not sufficient. In the latter case, it would have been expected that decomposition of residual nitrate at elevated temperatures (e.g., about 600°C) should have promoted formation of tetravalent americium. Further, microscopic examination of the glass dissolution of americium dioxide at 850°C showed that dissolution of the americium dioxide was accompanied by an evolution of gas, believed to be oxygen, as observed when we dissolved the higher oxides of praseodymium and terbium (Pr_6O_{11} and Tb_4O_7) in this glass matrix at 850°C. In the cases of praseodymium and terbium, the glass products also contained only the trivalent oxidation state of these lanthanides.

Several studies have examined cerium glasses. In our preparations and in other studies of cerium glasses, cerium is frequently found in its tetravalent oxidation state, although mixtures of trivalent and tetravalent cerium have also been reported. The equilibria between trivalent and tetravalent states can be shifted by the experimental parameters. One report [10] suggests that the tetravalent state is favored by greater basicity of the glass, a higher concentration of cerium, and by the oxygen activity or potential in the glass (a function of the glass composition that is difficult to measure quantitatively); temperature may also be an important factor (e.g., > 1400°C).

Comparing oxidation couples for the f-elements can serve as an approximate guide for which oxidation state (or multiple states) may be expected for the element when it is placed in a glass host. Although such oxidation couples (or oxygen potentials) for glass media are not available, some guide in the order of stability of tetravalent states can be obtained by examining general III-IV couples for the f-elements [11]. The latter suggest the following order of stability for the tetravalent oxidation states: Np > Pu > Ce > Am >Tb > Pr, where americium has only a slightly lower indicated stability than cerium.

It is known that trivalent americium is generally encountered in acidic solutions and higher oxidation states of americium require complexation or are obtained in alkaline media, where stabilization is afforded by anion complexes with oxygen. In the americium-oxygen solid state system, stability is achieved by the fluorite lattice of the dioxide. With the cubic sesquioxide of americium (body center cubic, α-Mn_2O_3 structure) care must be exercised to avoid its oxidation by rapid insertion of oxygen into the lattice vacancies to form the dioxide. Thus, a loss of stability afforded by the fluorite lattice (e.g., during dissolution of it into a glass matrix) can alter the stability of americium's tetravalent state in the glass matrix, even though it would seem the silicate's oxygen network should also stabilize the tetravalent state.

When f-element oxides are put into a glass matrix, they may dissolve or be merely dispersed in the glass. When only dispersed, a true homogeneous glass is not formed. The product may not appear transparent and X-ray diffraction analysis can yield weak diffraction patterns. We believe the three types of americium glasses discussed here contained fully dissolved americium. All were transparent under microscopic examination and did not yield a X-ray diffraction pattern. Thus, when the americium dioxide dissolved in the two high temperature glass preparations, the stability of the fluorite lattice was lost, oxygen was evolved, and the resulting trivalent americium was the oxidation state stable in the glass matrix.

In the sol-gel glasses, the trivalent americium was dispersed homogeneously together with residual solvents and anions in a solid silicon-oxygen and/or hydroxyl network. During aging at ambient temperatures, and subsequent step-wise heating up to $800^{\circ}C$ in air, the trivalent state of americium remained sufficiently stable to be retained in the changing glass matrix and during nitrate decomposition. In separate experiments with plutonium in identical sol-gel preparations [12], trivalent plutonium could be incorporated and maintained into the solidified sol-gel matrix at ambient temperatures. However, the plutonium in these products was sensitive to oxidation and formed tetravalent plutonium. Heating to $100^{\circ}C$ or above rapidly caused the trivalent plutonium to be oxidized totally to its tetravalent state. The tetravalent state is also the state exhibited by plutonium in glasses formed at high temperatures in air by dissolving plutonium dioxide in glass matrices. Thus, it is the thermodynamic stability (e.g., pseudo-oxidation potentials or oxygen potentials) of the f-element under the particular conditions which determine the oxidation state(s) that will be exhibited by the f-elements in the different glasses. This state may change when stabilization is afforded by other conditions (e.g., the fluorite lattice in the case of americium).

CONCLUSION

The goal in this study was to ascertain fundamental information regarding the nature of americium in glasses. One specific question regarded the oxidation state that americium would acquire in the glasses, as it is expected to have important implications in modeling considerations, diffusion processes, long-term mobility and the leachibility of americium in such glass hosts. In short, its chemistry will be a function of the oxidation state. In the work described here, it is clear that trivalent americium was obtained in the glasses prepared by high-temperature dissolution's of americium dioxide. Further, when trivalent americium was incorporated into sol-gel glasses at ambient temperature, and then heated to various temperatures up to $800^{\circ}C$ in air, there was no evidence to suggest oxidation of the americium and trivalent americium was observed in the resulting products when cooled to $25^{\circ}C$. This was in contrast to the behavior reported for trivalent plutonium under identical conditions and to the behavior encountered when common salts of trivalent americium are calcined in air. Apparently, the silicon-oxygen network provided by the glasses, even though basic in nature, does not offer sufficient stability for tetravalent americium.

The retention of a trivalent state for americium reflects the thermodynamic stability of this oxidation state in this particular environment, which can be contemplated by comparing relative pseudo-oxidation states or oxygen potentials of americium and other f-elements. It is very likely that the additional stability afforded by the cubic, fluorite lattice of americium dioxide plays an important role in the formation of tetravalent americium in air atmospheres below $1200^{\circ}C$. Indeed, americium tetrafluoride decomposes at lower temperatures than does the dioxide, indicating a lower stability of tetravalent americium in this fluoride matrix. Principally, americium is in a pivotal position in the actinide series, where the transplutonium elements are becoming more lanthanide-like and the trivalent oxidation state is becoming the preferred oxidation state.

The results of this study are that americium will acquire a trivalent oxidation state in glass media, for the particular glasses examined and those with similar properties. However, it must be recognized that in other formulations (different basic nature, oxidants, complexants, etc.), and with the presence of other elements or materials (e.g., CrO_3, peroxides, etc.) that may influence the americium's redox couples, it may be possible to acquire higher oxidation states, in part or total, for americium.

ACKNOWLEDGMENT

Research sponsored by the Division of Chemical Sciences, Office of Basic Energy Sciences, US Dept. of Energy, under contract DE-ACO5-96OR22464 with Oak Ridge National Laboratory, managed by Lockheed Martin Energy Research Corp.

REFERENCES

[1] R. C. Ewing and W. Lutze, MRS Bulletin, Volume XIX, number 12, Dec., 1994, pp.16-17.

[2] L. J. King, J. E. Bigelow and E. D. Collins, in Transplutonium Elements-Production and Recovery, edited by James D. Navratil and Wallace W. Schulz (ACS, Washington, DC, 1981), pp.133-145; and references therein.

[3] W. W. Schulz, The Chemistry of Americium, ERDA Critical Review Series, Report TID-26971 (Technical Information Center, Oak Ridge, TN, 1976), pp. 72-77.

[4] W. W. Schulz and Robert A. Penneman, in The Chemistry of the Actinide Elements, 2nd ed., edited by J. J. Katz, G.T. Seaborg and L. R. Morss (Chapman and Hall, New York, 1986),pp. 926-930.

[5] D. G. Karraker, J. Am. Ceram. Soc., **65**, 53 (1982).

[6] W. T. Carnall and H. M. Crosswhite, Report ANL-84-90, USDOE Report , 1985.

[7] I. Poirot, M. Beauvy, M. Boge, D. Bonnisseau, A. Blaise, J. M. Fournier and P. G. Therond, J. Less-Common Metals, **121**,656 (1986).

[8] H. D. Schreiber, J. Less-Common Metals, **91**,129 (1983).

[9] T. D. Chikalla and L. Eyring, J. Inorg. Nucl. Chem., **29**, 2281(1967).

[10] A. Paul and R. W. Douglas, Physics and Chem. of glasses, **6**,212 (1965).

[11] R. G. Haire and L. Eyring, in Handbook on the Physics and Chemistry of Rare Earths, Vol. 18, Lanthanides/Actinides: Chemistry, edited by K.A. Gschneidner, Jr., L. Eyring, G. R. Choppin and G. H. Lander, (North-Holland, New York, 1994) pp. 481-482.

[12] N. Stump, R. G. Haire and S. Dai, Proceedings, Dec., 1996, MRS Symposium II (to be published).

SPECTROSCOPIC INVESTIGATIONS OF NEPTUNIUM'S AND PLUTONIUM'S OXIDATION STATES IN SOL-GEL GLASSES AS A FUNCTION OF INITIAL VALENCE AND THERMAL HISTORY

N. A. STUMP[*], R. G. HAIRE[**], S. DAI[***]
[*]Department of Physical Sciences, Winston-Salem State University, 601 Martin Luther King Jr. Drive, Winston-Salem, NC 27110
[**]Oak Ridge National Laboratory, P.O. Box 2008, Oak Ridge, TN 37831-6375
[***]Oak Ridge National Laboratory, P.O. Box 2008, Oak Ridge, TN 37831-6181

ABSTRACT

Several oxidation states of neptunium and plutonium, Pu(III), Pu(IV), Pu(VI), Np(IV), Np(V), and Np(VI), were studied in glasses prepared by a sol-gel technology. The oxidation state of these actinides in the sol-gel product was examined by absorption spectroscopy after solidification, aging, and thermal treatment. The oxidation state of the actinides in the starting solutions was essentially maintained through the solidification process of the silica matrix. However, during densification and removal of residual solvents at elevated temperatures, both actinides converted eventually to their tetravalent states while in the different sol-gel products. This finding is in accord with reports that tetravalent states of plutonium and neptunium are acquired in glass products prepared by dissolution of the actinide in molten glasses. Comparisons between room temperature spectra obtained from neptunium and plutonium in heated sol-gel products and from molten glass products showed subtle differences that can be related to the metal ion's environments.

INTRODUCTION

The incorporation of lanthanide and actinide ions into molten glasses at high temperature normally results in the formation of reproducible, characteristic, and stable oxidation states which is the most stable under the given preoperative conditions. Strongly reactive conditions (e.g. oxidizing or reducing atmospheres, use of chemical additives, acidity/basicity of the glass, etc.) can potentially affect the oxidation state obtained, when different oxidation states of similar stability exist for the element.[1] When incorporating a lanthanide (e.g. for optical components or laser applications)[2] or actinide (e.g. for waste disposal)[3] in a glass, it may be of fundamental or technological interest to incorporate the element with an oxidation state other than that which is most stable in a glass. One method to accomplish this is to use sol-gel technology, which can avoid high-temperature synthesis steps.

The sol-gel method for preparing a silica matrix allows greater control of the chemistry during the formation of the solidified matrix.[4] The simplicity and ambient temperature aspects of the synthesis permit incorporation of different oxidation states of the metal ion into the glass. To form a silica matrix, free of residual solvents, the sol-gel product must go through a densification process following solidification, which is normally achieved by controlled heating of it to elevated temperatures. The retention and/or the change of the element's oxidation state in the glass during this process then becomes important.

Mat. Res. Soc. Symp. Proc. Vol. 465 © 1997 Materials Research Society

In this investigation, several oxidation states of plutonium and neptunium were individually incorporated into solid sol-gel matrices and the resulting glasses examined by absorption spectroscopy. This spectral analysis provides a convenient method of investigating the valance states of these actinides as a function of experimental parameters.[5,6] In addition, since the actinide ions exhibit characteristic spectral features reflecting their electronic nature, bonding, coordination, etc., spectral data can also be used to ascertain important fundamental aspects about the element in the matrices.

Discussed here are the oxidation state behaviors of plutonium and neptunium in glasses prepared via sol-gel techniques. The glasses were examined after the solidification process, aging, and thermal treatment by absorption spectroscopy. These results from the sol-gel studies are discussed in terms of oxidation state behavior displayed and the nature of the absorption spectra for samples prepared by the sol-gel route and via molten glasses.

EXPERIMENTAL

The synthesis of the different glasses employed Np-237 and Pu-242 which had been purified by standard ion exchange techniques as starting materials. The specific oxidation state of each actinide in the starting solutions were obtained through the used of well known oxidants or reductants. The sol-gel glasses were synthesized by mixing the actinide containing solution in dilute HCl or HNO_3 with equal volumes of methanol and tetramethylorthosilicate in small glass cylinders with open but restricted tops (to control solvent evaporation). After 24-48 hours, a clear, glass-like product was formed. The solid samples were readily removed from the cylinders and portions placed in glass tubes designed in-house for spectroscopic measurements. The coloration of each glass product reflected the oxidation state of the actinide present. Heat treatment of the sol-gel produced samples was accomplished using a conventional resistance furnace. These sol-gel samples were not observed to melt; the maximum temperature of the thermal annealing studies was 800°C.

The light source for absorbance measurements consisted of a xenon-arc lamp (Oriel, model number: 66181). The light was aimed through the sample, which was placed on the stage of a microscope attachment for the monochromator (Nachet model number NS-400). The double-meter monochromator (Jobin Yvon/Instruments SA model number Ramanor HG.2S spectrophotometer) has a resolution of 0.5 cm^{-1} at 514.5 nm. The light dispersed by the monochromator was detected by a photon counting system. This system included a cooled photomultiplier tube (Hamamatsu, model number: R636) and multichannel analyzer (Nicolet, model number: 1170) interfaced with a PC using "Spectra Calc" software (Galactic Industries, version 2.12).

RESULTS

This work is based on the investigation of the oxidation states displayed by neptunium and plutonium in sol-gel glasses in relation to the individual element's fundamental science. The tenet of this paper was the determination of the oxidation states of plutonium and neptunium which could be incorporated into a glass matrix through the use of a sol-gel method of preparation. Each actinide ion studied in this matrix, before and after thermal treatment, is discussed individually.

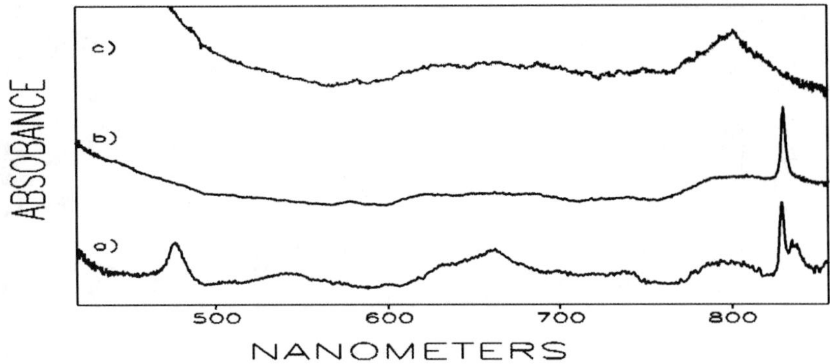

Figure 1.   Absorption spectra exhibited by sol-gel glass synthesized with Pu(VI) following synthesis (a); following heating to 100°C (b); and following heating to 800°C (c).

<u>Plutonium(VI)</u>

When Pu(VI) was used to form a sol-gel glass, the hexavalent state of plutonium was retained in the glass, as seen from its absorption spectrum (Figure 1a). Characteristic absorption near 830nm can be assigned to electronic transitions of Pu(VI).[5] The two features at 830nm and 836nm both result from Pu(VI). The splitting of this feature can be correlated with multiple complexes of Pu(VI).[7] The absorption spectrum (Figure 1a) also exhibits a series of absorption maxima with wavelengths of 476nm, 545nm, 660nm, 740nm, and 800nm, which are all characteristic of Pu(IV).[5] Thus, while Pu(VI) was incorporated in the sol-gel glass matrix, part of the plutonium was found to be tetravalent after solidification of the glass.

A portion of this product was heated to 100°C and then cooled to room temperature. The heating step improved the quality of the absorption spectrum (less noise, better peak definition), possibly due to the evaporation of residual solvent materials. In the absorption spectrum obtained from the sample (Figure 1b), the main feature at 830nm is retained, while the feature at 836nm has disappeared. The disappearance of this peak may reflect a change in environment of the plutonium ion brought about by the loss of volatile coordinating solvents. The absorption features associated with Pu(IV) seen in figure 1a appear different from those observed in figure 1b. The strong, characteristic feature of Pu(IV) at 476nm has disappeared and weak absorption maxima are apparent at 740nm, 800nm, and between 600 and 700nm. These changes can be associated with multiple environments of the Pu(IV).[8,9]

Another portion of the sol-gel sample was heated up to 800°C. The absorption spectrum for this material is shown in Figure 1c. In this spectrum the absence of absorption features in the 830nm region indicates that reduction of Pu(VI) has occurred. The remaining spectral features in figure 1c can be matched with those in figure 1b, which can be associated with Pu(IV).[8,9] The absorption spectrum of this sample following heating to 800°C indicates

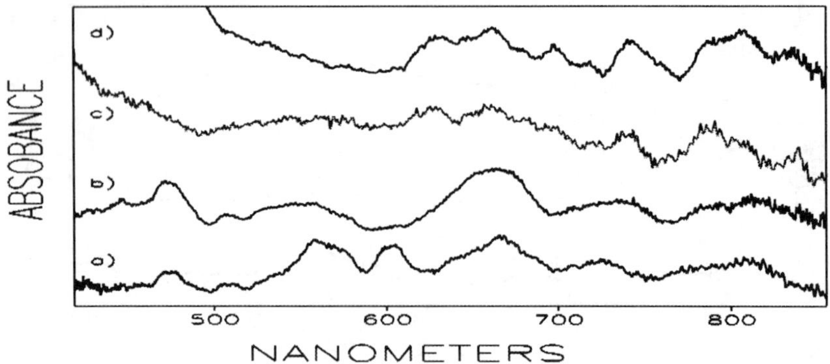

Figure 2.   Absorption spectra of sol-gel glass with Pu(III) as synthesized (a); following 12 hr storage in humid environment (b); heated to 100°C (c); heated to 800°C (d).

that conversion of Pu(VI) to Pu(IV) was complete.

<u>Plutonium(III)</u>

When an aqueous sample of plutonium(III) was used to prepare the sol-gel glass, a blue-violet glass (the blue coloration is characteristic of Pu(III)) was formed. The oxidation state of the plutonium showed variable stability during storage of the sol-gel glass. The material appeared stable (up to one year) when stored in a flowing-air glove box environment at room temperatures. The glass's coloration changes from blue (Pu(III)) to green (characteristic of Pu(IV)) after storage in certain laboratory conditions. It is believed that environmental moisture plays a part in the loss of Pu(III) during "aging" or storage. Storage of the sol-gel product under vacuum permitted complete retention of the trivalent state.

The absorption spectrum of the Pu(III) glass shortly after the solidification process is shown in figure 2a. The spectrum is comprised of features arising from Pu(III) (absorption maxima at wavelengths of 570, 600, 670, and 800nm). Some feature characteristic of Pu(IV) were also observed including maxima at wavelengths of 476, 560, 660, 740, and 800nm.[5]

After storage of the sample in humid air, the violet-blue glass turned greenish brown in color and exhibited the absorption spectrum shown in Figure 2b. In this spectrum the features associated with Pu(III) are no longer apparent, while those associated with Pu(IV) have become pronounced. Thus while Pu(III) can be incorporated into the sol-gel glass, care must be exercised to maintain this oxidation state, even at ambient temperatures.

A small portion of the sol-gel glass was heated 100°C. Upon heating the color of the glass changed rapidly, indicating the loss of Pu(III). This was confirmed by the new absorption spectrum exhibited by the sample (Figure 2c). The absorption spectrum showed maxima between 600 and 700nm, and maxima near 740 and 800nm. These maxima correspond to Pu(IV) as discussed previously.[8,9] Thus, the Pu(III) in the initial material was unstable with respect to being oxidized to Pu(IV) during heating at temperatures as low as 100°C.

Another portion of the initial Pu(III) containing sol-gel sample was heated to 800°C. The absorption spectrum recorded for this sample is shown in Figure 2d. The absorption maxima seen in this spectrum are essentially identical to those exhibited by this glass after the 100°C heat treatment. This indicates that Pu(IV) remains stable at temperatures up to 800°C.

Neptunium(VI)

Incorporation of neptunium(VI) into a sol-gel matrix was also attempted. The absorption spectrum of the solidified glass (Figure 3) exhibits a broad feature near 555nm, which may be assigned to Np(VI).[6] Two features characteristic of Np(V) can also be identified in the absorption spectrum (maxima at 470 and 620nm).[6] Three features at 687, 726, and 771nm can be correlated with absorption maxima characteristic of Np(IV). It appears some Np(VI) was retained through the gelation process of the glass, but Np(V) and Np(IV) were also found in the product.

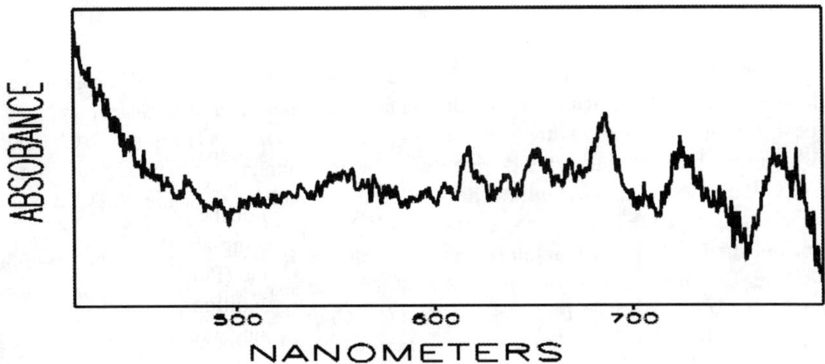

Figure 3.   Absorption spectra exhibited by sol-gel glass synthesized with Np(VI) following synthesis.

Neptunium(V)

Neptunium(V) was incorporated into a sol-gel matrix using this technique. The absorption spectrum exhibited by this glass product is displayed in figure 4a. Absorption maxima at 620 and 470nm indicate that the Np(V) was retained[6] through the formation of the glass matrix. A weak, broad absorbance maxima located at 555nm may imply the presence of a small amount of Np(VI).[6] Three features characteristic of Np(IV) at 687, 726, and 771nm are also apparent in this spectrum. The amounts of Np(VI) and Np(IV) in this are quite small and the primary oxidation state in this glass is Np(V). This implies a greater stability of the pentavalent state through the gelation process of sol-gel matrix.

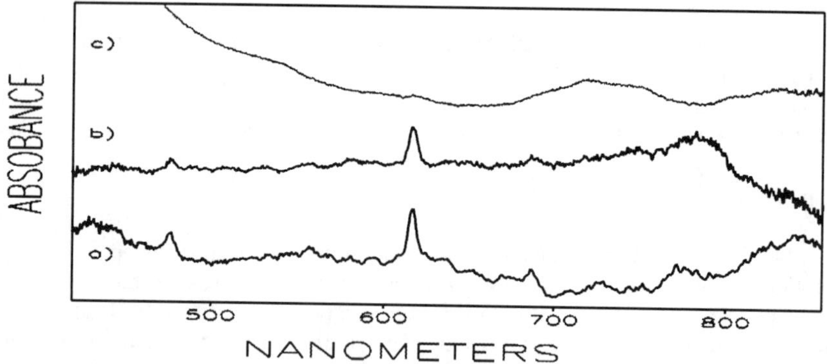

Figure 4.   Absorption spectra exhibited by sol-gel glass synthesized with Np(V)
following synthesis (a); following heating to 100°C (b); and following heating
to 800°C (c).

A separate portion of the Np(V) glass was heated to 100°C. The absorption spectrum
(figure 4b) shows that the primary oxidation state of neptunium in this material is Np(V). The
disappearance of the broad feature at 550nm suggests that no Np(VI) remains. Changes in the
Np(IV) features indicate that some difference in the environment surrounding the Neptunium
ion affects the absorption spectrum of this ion. It is concluded that while Np(V) is stable at
temperatures up to 100°C in the sol-gel matrix, Np(VI) is not. Further, the absorption
spectrum of the Np(IV) that is retained, reflects a change in environment of the neptunium ion.

After heating a portion of this glass to 800°C, the absorption spectrum of the product
(Figure 4c) now primarily exhibits spectral features characteristic of Np(IV). The spectrum
exhibits the same Np(IV) profile noted in the sample heated to 100°C. The very broad features
at 550 and 725nm indicate that the majority of the neptunium is tetravalent.[10,11] Thus, Np(IV)
is the most stable oxidation state at 800°C for the sol-gel product, followed closely by Np(V).

High Temperature Borosilicate Glass Matrix

The preparation of actinide-doped borosilicate matrices from molten glasses results in
the formation of tetravalent oxidation states for both elements[12]. This correlates well with the
observations made in this report that tetravalent forms of both neptunium and plutonium result
in the sol-gel matrix upon heating of the sample to 800°C. Comparison of the absorption
spectra obtained from the sol-gel products with that seen from the borosilicate glasses showed
a difference between the final products. The glasses prepared by the sol-gel technique and
heated to 800°C exhibited absorption spectra that were characteristic of a different environment
around the metal ion. In contrast, the glasses prepared at high temperatures have been
reported to exhibit spectra that are more closely approximated by the spectra seen from
aqueous tetravalent ions of the two elements[12], and by the unheated sol-gel samples. The full

significance of these observations has not been established and may result from several factors, including the glass compositions, complexing components present, oxygen bridging, interactions between the metal ions, etc. An in-depth analysis of these differences will be pursued in future work.

TABLE I

Summary of Plutonium and Neptunium Oxidation States
Observed in the Studied Sol-Gel Glasses

	Initial Actinide Oxidation State			
	Pu(VI)	Pu(III)	Np(VI)	Np(V)
Oxidation State Following Solidification	Pu(VI), Pu(IV)	Pu(III)[a], Pu(IV)	Np(IV), Np(V), Np(VI)	Np(V) [Np(IV),Np(VI)][b]
Oxidation State Following 100°C Heat Treatment	Pu(VI), Pu(IV)	Pu(IV)	-----	Np(V), Np(IV)
Oxidation State Following 800°C Heat Treatment	Pu(IV)	Pu(IV)	-----	Np(IV) [Np(V)][b]

[a] The observance of Pu(III) was dependant on the storage of the sample.
[b] These include minor constituents of the total actinide content.

CONCLUSIONS

Through the use of sol-gel technology, it is possible to generate glass matrices containing different oxidation states of plutonium and neptunium. The majority of these states were retained through the solidification process of the sol-gel preparation. However, the states proved unstable when subjected to heat treatments. At these temperatures the plutonium and neptunium readily converted to their tetravalent state. A summary of the actinide oxidation states observed in the various sol-gel samples is given in Table I.

Regarding the stability of the various oxidation states available for these actinides in the glasses studied, the tetravalent state is clearly the most stable. Np(V) and Pu(VI) both exhibited moderate stability, while Np(VI) and Pu(III) showed lower stability in the sol-gel matrix. The observation of the formation of the tetravalent states for both elements in the heated sol-gel products is in agreement with the results reported for plutonium and neptunium containing glasses prepared by high-temperature syntheses.[12] Comparison of the absorption spectra exhibited by the sol-gel glasses that had been heated to elevated temperature to those characteristic of the elements in borosilicate glasses suggest that differences exist in the local environments provided by the different matrices. These differences are thought to result from variations in the environment of the metal ion.

ACKNOWLEDGMENTS

This research was sponsored by the Division of Chemical Sciences, Office of Basic Energy Sciences, U.S. Department of Energy under contract DE-AC05-96OR22464 with Lockheed Martin Energy Research Corp.

REFERENCES

1. H. L. Smith & A. J. Cohen, Phys. Chem. Glasses **4**, p.173 (1963).

2. I. M. Thomas, S. A. Payne, & G. D. Wilke, J. Non-Cryst. Sol. **151**, p.183 (1992).

3. N. Jacquet-Francillon, R. Bonniaud, & C. Sombret, Radiochim. Acta **25**, p.231 (1978).

4. R. Reisfeld and C. K. Jorgenson, editors, Chemistry, Spectroscopy, and Applications of Sol-Gel Glasses, Structure and Bonding, Vol. 77 (Springer, Berlin, 1992); and L. L. Hench and J. K. West, editors, Chemical Processing of Advanced Materials (Wiley, New York, 1992), p.823.

5. D. Cohen, J. Inorg. Nucl. Chem. **18**, p.211 (1961).

6. P. G. Hagan and J. M. Cleveland, J. Inorg. Nucl. Chem. **28**, p.2905 (1966).

7. R. E. Connick, M. Kasha, W. H. Mcvey, and G. E. Sheline, in The Transuranium Elements, edited by G. T. Seaborg, J. J. Katz, and W. M. Manning, McGraw-Hill Book Company, Inc., New York, 1949, p.559-601.

8. D. A. Costanzo, R. E. Biggers, and J. T. Bell, J. Inorg. Nucl. Chem. **35**, p.609 (1973).

9. J. C. Hindman, in The Transuranium Elements, edited by G. T. Seaborg, J. J. Katz, and W. M. Manning, McGraw-Hill Book Company, Inc., New York, 1949, p.370-386.

10. R. K. Sjoblom and J. C. Hindman, J. Amer. Chem. Soc. **73**, p.1744 (1951).

11. H. Escure, D. Gourisse, and J. Lucas, J. Inorg. Nucl. Chem. **33**, p.1871 (1971).

12. P. G. Eller, G. D. Jarvinen, J. D. Purson, R. A. Penneman, R. R. Ryan, F. W. Lytle, and R. B. Greegor, Radiochim. Acta **39**, p.17 (1985).22

USE OF INDUCTION MELTER WITH A COLD CRUCIBLE (CCIM) FOR HLLW AND PLUTONIUM IMMOBILIZATION

V.V. KUSHNIKOV, Yu. I. MATYUNIN, T.V. SMELOVA, and A.V. DEMIN
MINATOM RF, SSC RF VNIINM, Rogov 5, Moscow, 123060, Russia

ABSTRACT

CCIM technology allows syntheses of various materials over a wide range of compositions. This provides a means of applying the CCIM to the solidification of wastes obtained by different technologies of the spent nuclear fuel reprocessing, including toxic wastes containing heavy metals and those arising from fractionation. Solid compounds produced by this method meet both the requirements for their subsequent disposal in deep geological formations and spent fuel standards.

On the basis of investigations into the REE and Uranium vitrification, the possibilities of CCIM immobilizing weapons Plutonium in preparation of glass compositions are considered.

INTRODUCTION

In connection with considerable reduction in military programs and discarding of a large number of nuclear charges, the problem on safe methods of the handling of released weapons plutonium has been put on the agenda. The development of nuclear power engineering based on thermal and fast reactors in combination with the subsequent waste fuel regeneration and involvement of both secondary uranium and plutonium into the fuel cycle opens up prospects for nuclear power engineering of becoming primarily self-sufficient. This would result in highly efficient utilization of both natural and artificial power sources.

The implementation of the Russian concept of a closed nuclear cycle using surplus weapons plutonium as a fuel source for nuclear power stations calls for close examination of the handling of radioactive wastes produced by reprocessing of plutonium-containing materials as well as of uses of articles that are not subject to reprocessing but require direct, ecologically safe immobilization and disposal. In a conceptual aspect, plutonium-containing wastes can be regarded as belonging to liquid high-level wastes (HLLW), so that the techniques of their immobilization are similar to those used in the HLLW immobilization.

In Russia, the problem concerning the disposal of radioactive waste formed during the nuclear fuel regeneration is resolved through the incorporation of radioactive wastes into solid, fixed forms, which reduce delocalization of radionuclides to a minimum, as well as through the interim controlled storage in a specially designed storage site and the final burial into geologic formations.

The sources of HLLW formation are the tailing water solutions left after uranium and plutonium extraction upon regeneration of the waste nuclear fuel. In the world practice, there are a number of HLLW immobilization technologies [1-4] that have been implemented in industrial facilities and proved their viability over a long period of time. In Russia, a one-stage process of HLLW solidification in the direct electric heating furnace with the formation of phosphate glass (PG) has been developed and implemented on an industrial scale at the PO "Mayak". [5]

Mat. Res. Soc. Symp. Proc. Vol. 465 © 1997 Materials Research Society

EXPERIMENT

Pilot Plant with CCIM

Research into solidification technology for various model high level wastes resulted in creating a pilot facility (at the PO "Mayak" vitrification site) for a two-stage process of the high level waste immobilization in phosphate and borosilicate glasses, as well as in mineral-like materials. The two-stage pilot facility is arranged in a hot chamber and is designed for development of following:

• the technology of high level waste solidification from both the simulated and actual, industrial waste solutions;

• the connectors and mechanical assemblies operating in a remote-controlled mode for dismounting the equipment;

• the transportation-processing scheme for the removal of containers with solidified wastes and equipment.

Figure 1 shows the concepted design of a two-stage pilot facility for HLLW immobilization. A fluxed solution of high level wastes continuously feeds to an inlet of a direct-flow evaporator (DFE). The evaporator is combined with a separator. The initial solution is successively heated and evaporated (with a partial distillation of nitric acid) in the direct-flow evaporator. After separating the vapor-gas phase and molten salt in the separator, the melt flows by gravity into the CCIM. Further dehydration, denitration, and melting of the product proceed in the CCIM with the formation of the glass.The CCIM is a two-region crucible made up of the water-cooled tube sections mounted on a common water collector (Fig. 2). The crucible is surrounded by a water-cooled inductor that is connected to a high-frequency generator.

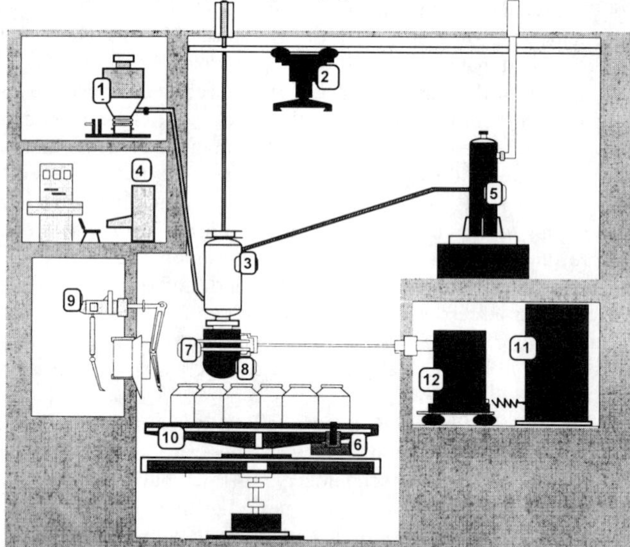

Fig. 1. Concepted design of a two-stage pilot facility for HLLW immobilization
1. Flux batcher. 2. Crane. 3. Direct-flow evaporator. 4. Control board.5 Bubbler-cooler. 6. Weighing device.
7. Inductor. 8. Cold crucible. 9. Duplicating manipulator. 10. Carousel conveyer. 11. HF-generator.
12. Capacitor bank.

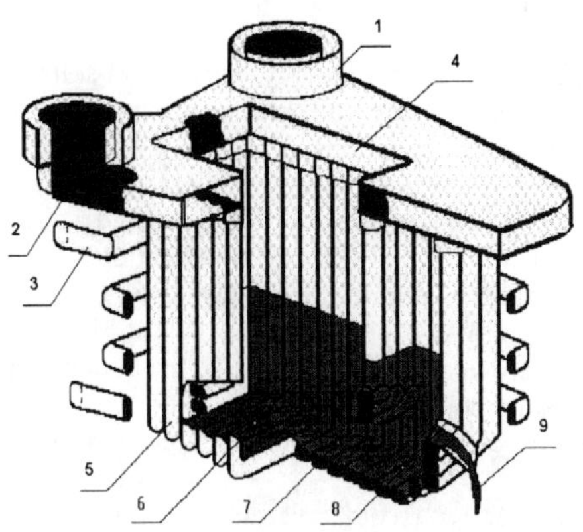

Fig. 2. Induction melter with cold crucible (CCIM)
1. Fusible connector. 2. Water for pipe section cooling. 3. Inductor.
4. Melter inlet for material. 5. Pipe sections of crucible. 6. Glass melt. 7. Partition.
8. Outlet for melt. 9. Stream of melt.

Characteristics of two-region cold crucible for pilot facility at the «Mayak» vitrification site are given below:

– nominal capacity of the bath, dm^3	– 22.6	– output powre of high-frequency generator, kW	– 160
–melt mirror area, dm^2:		– power consumption of high-frequency generator, kW	– 240
melting region	– 6.7		
processing region	– 1.8		
–maximum operating temperature, ^{o}C	– 1500	– operating frequency, MHz	– 1.76
–glass production rate, kg/h	– 18		

Induction melting in a cold crucible opens up certain prospects for the reprocessing of radioactive and toxic wastes and exhibits the following unique features unavailable with melters of other types:

• the possibility for variations in melting temperatures over a wide range;

• the lack of restrictions on composition of the reprocessing wastes and a wide variety of compositions of solidified products (from glassy to mineral-like materials);

• high intensity of convective mixing of a melt, which markedly increases the production rate and does not require the preliminary mixing of waste components with fluxing admixtures;

• reasonable overall dimensions, i.e., the possibility of providing the remote-controlled installation and dismounting of the equipment and also its removal for storage and reprocessing by using the transportation-processing scheme for the removal of vitrified wastes;

• the formation of a skull layer at water-cooled walls of a crucible in the course of the melting process, thus protecting the crucible against a corrosive action of the melt;

• the relation between the parameters of a generator and the operating conditions of the bath with a melt enables to automate the process by using indications of the standard generator instruments;

• the capability of stopping and restarting-up the melter in the case when unforeseen situations occur in any unit of the technological scheme.

<u>Vitrification of HLLW components and Uranium in CCIM as compared with simple-stage furnace</u>

At the initial stage after improving the technology with model solutions, the developed two-stage pilot facility (PO "Mayak") is scheduled to produce and investigate high-level phosphate, borosilicate glasses and mineral-like compositions.

It is known that steady running of melters depends to a large extent on the homogeneity of the glass melt. However, some waste components exhibit a limited solubility in phosphate and borosilicate compositions used in the HLLW immobilization. Disperse particles of the heterogeneous phases formed from platinum group elements (Ru, Rh, and Pd), rare-earth elements (La, Ce, Nd, and Sm), corrosion products, and Pu [6, 7] can precipitate in the melts. Under certain conditions, these particles may accumulate in the bottom part of a melter and affect the properties (viscosity and electric conductivity) of a melt [8].

Table 1 presents the solubility of selected HLLW components in phosphate and borosilicate glasses [9].

The use of an induction melter with cold crucible in the HLLW immobilization process provides a means of synthesizing high-quality glasses with the properties similar to those of glasses prepared by the traditional techniques [10]. However, the CCIM offers an important advantage when utilized in the reprocessing of HLLW components containing nuclides in amounts exceeding their solubility in molten glasses.

A vigorous, electrodynamic thermoconvective stirring of the melt brings about a uniform distribution of the resultant disperse particles over the melter volume. This results in steady running of the CCIM and stabilizes the properties of the end heterogeneous material. Furthermore, investigations into the behavior of rare-earth elements (REEs) and uranium in these melters can be considered as a simulation of the solidification of transuranium elements (including also plutonium).

A. Initial Compositions of Glasses Studied

Table 2 lists the compositions of the initial phosphate and borosilicate matrices used in this study.

The glass compositions were synthesized in a laboratory-scale facility with cold crucible (d=60 mm, h=150 mm, frequency=1.76 MHz) . Analogous materials were concurrently prepared in muffle furnaces (alumina crucible: d=40 mm, h=60 mm).

Phosphate and borosilicate glasses were synthesized for 1 h at temperatures of 1200 - 1250 and 1250 - 1300°C, respectively. The materials synthesized are entirely homogeneous and X-ray amorphous. The results of chemical analyses also indicate a uniform distribution of rare-earth elements in the glasses obtained (Table 3). For the glasses synthesized, the chemical durability, which is determined by the leaching rates of Na^+, Cs^+, and Sr^{2+} (MCC-1*)*, lies within the ranges $(0.6 - 8.7) \times 10^{-5}$ and $(1.3 - 7.9) \times 10^{-6}$ $g \times cm^{-2} \times day^{-1}$ for the first and tenth days, respectively.

The previous investigations [9,10] into the behavior of rare-earth elements in phosphate glasses synthesized in muffle furnaces demonstrated a limited solubility of REEs in this matrix. The tolerable concentrations of a REE mixture in the phosphate systems such as $Na_2O-Al_2O_3-P_2O_5-Ln_xO_y$ (glass A) and $Na_2O-Al_2O_3-P_2O_5-(HLLW)_xO_y-Ln_xO_y$ (glass B) are determined by the equation:

$$S = \Sigma a_i \times n_i,$$

where a_i is the tolerable concentration of a lanthanide in the glass, and n_i is the fraction of this lanthanide in a REE mixture.

Table 1

Solubility of HLLW Components in Molten Phosphate and Borosilicate Glasses (PG and BG) and Their Approximate Content in Actual Products

Nuclide	Solubility (ppm) in PG, 1000°C	Content (ppm) of elements in PG (5 wt % HLLW$_{ox}$)	Solubility (ppm) in BG, 1000°C	Content (ppm) of elements in BG (5 wt % HLLW$_{ox}$)
Ru	20-60	3000	25-40	6000
Rh	20-60	1500	25-40	3000
Pd	300-600	2200	1500-2000	4500
Ag	2600 *)	500	5000 *)	1000
Te	1000 *)	250	1000 *)	500
Tc	~ 10	4500	-	9000
Zr	7000 *)	5500	16000 *)	11000
Mo	7000 *)	5500	3000-12000	11000
Fe	~ 5000	10000	50000 *)	20000
Cr	500-2000	1000	10000 *)	2000
Ni	500-2000	1500	15000 *)	3000
La	11000-14000	1500	50000 *)	3000
Ce	12000-16000	5000	10000-30000	10000
Nd	20000-24000	7000	50000 *)	14000
Sm	28000-32000	2500	50000 *)	5000
U	50000 *)	100	50000 *)	200
Pu	2000 *)	10	~ 2000	20

*) *Max. content of element in prepared glasses. The investigations with large contents of elements were not carried out.*

Table 2

Compositions of phosphate and borosilicate glasses

Oxide	Oxide content, wt %	
	Phosphate glass	Borosilicate glass
Na$_2$O	23.40	21.80
Cs$_2$O *)	0.36	0.36
K$_2$O *)	0.24	0.24
ΣMe$_2$O = 24.0		
Al$_2$O$_3$	18.60	5.40
Fe$_2$O$_3$ *)	0.08	3.38 **)
Cr$_2$O$_3$ *)	0.02	0.02
NiO *)	0.01	0.01
MnO *)	0.03	0.03
CaO *)	0.02	3.32 **)
SrO *)	0.16	0.16
BaO *)	0.16	0.16
ZrO$_2$ *)	0.47	0.47
MoO$_3$ *)	0.31	0.31
La$_2$O$_3$ *)	0.14	0.14
Ce$_2$O$_3$ *)	0.51	0.51
Nd$_2$O$_3$ *)	0.43	0.43
Sm$_2$O$_3$ *)	0.06	0.06
ΣLn$_2$O$_3$ = 1.14		
P$_2$O$_5$	55.00	–
B$_2$O$_3$	–	13.10
SiO$_2$	–	50.10

*) *components of HLLW$_{ox}$*
**) *3.30 wt % Fe$_2$O$_3$ and CaO contents in matrix of glass*

For systems A and B, the above equation has the form:

$$S_A = 1.5n_{(La)} + 2.0n_{(Ce)} + 2.5n_{(Nd)} + 3.7n_{(Sm)}$$
$$S_B = 1.7n_{(La)} + 2.3n_{(Ce)} + 2.6n_{(Nd)} + 3.8n_{(Sm)}.$$

At the ratio La : Ce : Nd : Sm = 0.16 : 0.17 : 0.49 : 0.18 (the ratio of REEs in actual wastes), the calculated values of the total REE solubility are as follows: S_A = 2.45 and S_B = 2.61. The experiments show that the resulting materials retain their homogeneity when the content of REE oxides in phosphate glasses is less than or equal to the calculated S_A and S_B values.

In the case when the REE solubility is exceeded (by 0.2 - 0.3%) in phosphate glasses, disperse crystalline particles of the rare-earth element orthophosphates $LnPO_4$ (where Ln = La, Ce, Nd, and Sm) are formed in the melts. Upon prolonged treatment of a melt, the precipitate of crystalline phase, whose properties differ from those of the major portion of melt, accumulates at the bottom of a melter.

Behavior of Excess REE in Phosphate Glass Melts

When glass is prepared in the CCIM, the resulting disperse particles can be uniformly distributed throughout the bulk of glass melt due to its vigorous stirring. These particles then do not affect substantially the properties of the end material. The behavior of rare-earth elements during the synthesis of phosphate glasses by the CCIM method was studied with the ternary Na_2O-Al_2O_3-P_2O_5 system involving different amounts of REEs.

The glass compositions were synthesized in a laboratory-scale facility with cold crucible (d=60 mm, h=150 mm, frequency=1.76 MHz) and in muffle furnaces (alumina crucible: d=40 mm, h=60 mm).

Phosphate glasses were synthesized by two techniques: (1) the melting of preliminary prepared glass frit containing 3 and 10 wt % REE oxides in the ratio corresponding to an actual content of the above nuclides in the HLLW, and (2) the melting of glasses with a small REE content and the addition of highly enriched calcinates to obtain the end materials involving 3 and 10 wt % REE oxides.

The glasses were synthesized at a melt temperature of 1200^oC for 1 h. The glass samples taken from the melt after 15, 30, and 45 min were homogeneous transparent materials. The glass blocks produced upon cooling of the melt in a crucible were transparent and showed inclusions of the crystallized regions (with a predominant size from 1 to 5 mm across).

Analysis of the resultant materials shows the same REE content both in the crystallized regions and in the surrounding homogeneous glass. It can be assumed that the crystallized regions in a glass block are regions where the glass solution (melt) supersaturated with rare-earth elements crystallizes at the crystalline nuclei. Upon the spontaneous cooling of glass melt in a crucible, virtually all microscopic species can serve as crystalline nuclei (Tables 4 - 6). It is well known [11] that rare-earth elements significantly enhance a tendency for glasses to crystallize. Actually, we observed similar regions of the crystallized glass around the $LnPO_4$ crystals.

Glasses of analogous compositions synthesized by the traditional technique in muffle furnaces in the same temperature-time schedules are opaque glasses containing a disperse crystalline phase, namely, rare-earth element orthophosphates. The sedimentation of disperse particles and formation of the bottom sediment substantially enriched with rare-earth elements are observed in the absence of convective stirring after the heat-treatment of a melt for 1 h at 1200^oC .

The studies of the REE behavior in phosphate glasses synthesized by the CCIM method demonstrated a possibility of synthesizing the phosphate glass materials with a high content of rare-earth elements without technological problems.

Table 3

Distribution of REE over phosphate and borosilicate glass blocks
manufactured in the CCIM *)

Material	Sampling	REE content, wt %			
type	conditions	La$_2$O$_3$	CeO$_2$	Nd$_2$O$_3$	Sm$_2$O$_3$
Phosphate glass	30 min **)	0.15	0.44	0.48	0.04
	60 min **)	0.14	0.44	0.48	0.05
	top ***)	0.15	0.45	0.48	0.05
	bulk ***)	0.15	0.45	0.46	0.05
	bottom ***)	0.14	0.45	0.48	0.05
Borosilicate glass	30 min **)	0.12	0.42	0.36	0.04
	60 min **)	0.12	0.42	0.38	0.04
	top ***)	0.12	0.42	0.38	0.04
	bulk ***)	0.12	0.43	0.37	0.04
	bottom ***)	0.12	0.43	0.38	0.04

*) *Data were obtained by atomic-absorption spectroscopy after dissolution of glasses in the HNO$_3$ + HCl mixture.*

**) *The glass samples taken after 30 and 60 min. holding of a molten glass in the crucible.*

***) *Glass samples were taken from different regions of the resultant glass block after cooling of the melt in a crucible.*

Table 4

Distribution of REE over the height of phosphate glass blocks prepared in the CCIM and muffle
furnace upon addition of calcinate with a high REE content to the glass melt (1200°C, 1 h)*)

Composition, preparation technique	Sampling conditions	Content of components, wt %				Note
		La$_2$O$_3$	CeO$_2$	Nd$_2$O$_3$	Sm$_2$O$_3$	
Σ REE 1.5% + 6.0% CCIM, 5.28 MHz	15 min **)	0.31	1.14	1.19	0.15	Homogeneous
	30 min **)	0.51	1.46	1.27	0.20	Homogeneous
	45 min **)	0.40	1.33	1.11	0.14	Homogeneous
	top ***)	0.36	1.25	1.15	0.17	Inclusions of crystallized glass
	bulk ***)	0.38	1.39	1.17	0.17	
	bottom ***)	0.47	1.44	1.12	0.17	
Calculated content		0.37	1.34	1.13	0.16	
Σ REE 1.5% + 6.0% Muffle furnace	top ***)	0.31	1.15	1.11	0.15	Opacified
	bottom ***)	0.42	1.41	1.14	0.18	Opacified
Σ REE 1.5% + 27.0% CCIM, 5.28 MHz	15 min **)	1.33	4.71	3.25	0.43	Homogeneous
	30 min **)	1.14	4.13	3.25	0.38	Homogeneous
	45 min **)	1.19	4.63	3.21	0.37	Homogeneous
	top ***)	1.28	3.64	3.32	0.37	Inclusions of crystallized glass
	bulk ***)	1.25	4.49	3.15	0.41	
	bottom ***)	1.38	4.76	2.15	0.28	
Calculated content		1.23	4.47	3.77	0.53	
Σ REE 1.5% + 27.0% Muffle furnace	top ***)	0.35	1.05	0.94	0.12	Opacified
	bottom ***)	2.49	7.46	5.24	0.75	Opacified

*) *Data were obtained by X-ray fluorescence and radiometric analyses.*

**) *The glass samples taken after 15, 30 and 45 min holding of a molten glass in the crucible.*

***) *Glass samples were taken from different regions of the resultant glass block after cooling of the melt in a crucible.*

Table 5

Distribution of REE over the height of phosphate glass blocks prepared in the CCIM and muffle furnace upon melting of phosphate glasses with the REE contents of 3 and 10% (1200°C, 1 h)*[)]

Composition, preparation technique	Sampling conditions	Content of components, wt %				Note
		La_2O_3	CeO_2	Nd_2O_3	Sm_2O_3	
Σ REE 3.0% CCIM, 5.28 MHz	15 min **[)]	0.34	1.51	1.15	0.16	Homogeneous
	30 min **[)]	0.34	1.30	1.22	0.13	Homogeneous
	45 min **[)]	0.32	1.37	1.07	0.19	Homogeneous
	top ***[)]	0.33	1.30	1.18	0.14	Homogeneous
	bulk ***[)]	0.38	1.29	1.18	0.22	Homogeneous
	bottom ***[)]	0.38	1.33	1.19	0.18	Homogeneous
	skull ***[)]	0.39	1.43	0.89	0.10	Opacified
Calculated content		0.37	1.34	1.13	0.16	
Σ REE 3.0% Muffle furnace	top ***[)]	0.34	1.40	0.89	0.15	Opacified
	bottom ***[)]	0.57	2.06	1.71	0.21	Opacified
Σ REE 10.0% CCIM, 5.28 MHz	15 min **[)]	1.24	4.76	3.97	0.61	Homogeneous
	30 min **[)]	1.17	4.61	3.92	0.55	Homogeneous
	45 min **[)]	1.43	4.89	3.74	0.54	Inclusions
	top ***[)]	1.22	4.39	3.72	0.50	of
	bulk ***[)]	1.38	4.76	3.77	0.61	crystallized
	bottom ***[)]	1.23	4.35	3.71	0.54	glass
	skull ***[)]	1.24	4.23	3.11	0.57	Opacified
Calculated content		1.23	4.47	3.77	0.53	
Σ REE 10.0% Muffle furnace	top ***[)]	0.35	1.05	0.94	0.12	Opacified
	bottom ***[)]	2.49	7.46	5.24	0.75	Opacified

*[)] *Data were obtained by X-ray fluorescence analysis.*

**[)] *The glass samples taken after 15, 30 and 45 min holding of a molten glass in the crucible.*

***[)] *Glass samples were taken from different regions of the resultant glass block after cooling of the melt in a crucible.*

Table 6

Analysis of the crystallized phases in phosphate glasses synthesized by the CCIM technique upon addition of highly enriched calcinate to the melt surface *[)]

Composition	Content of components, wt %				Note
	La_2O_3	CeO_2	La_2O_3	Sm_2O_3	
Σ REE 1.5% + 6.0%	0.74	2.68	2.26	0.32	Calculated content in calcinate
	0.37	1.34	1.13	0.16	Calculated content in glass
	0.41	1.47	1.09	0.13	Phase, dark center
	0.37	1.34	1.38	0.21	Phase, light center
Σ REE 1.5% + 27.0%	3.32	12.08	10.18	1.42	Calculated content in calcinate
	1.23	4.47	3.77	0.53	Calculated content in glass
	1.05	4.15	1.60	0.33	Phase at the surface
	1.12	4.78	3.06	0.38	First phase at the bottom
	1.29	4.80	2.65	0.34	Second phase at the bottom

*[)] *Some scatter in the data of X-ray powder diffraction analysis is due to the porous structure of phases upon calibration against a homogeneous glass.*

Table 7

Composition of a glass frit used in syntheses of the uranium-containing borosilicate glasses by the CCIM method

Material type	Content of components, wt %						
	Na_2O	K_2O	Al_2O_3	B_2O_3	SiO_2	CaO	Fe_2O_3
GP-91-F	19.38	3.01	5.68	13.36	51.71	3.42	3.42

Behavior of Excess Uranium in Borosilicate Glass Melts

The behavior of uranium in borosilicate glass synthesized in the CCIM was studied by the technique of gradual saturation of a glass melt with uranium up to a level of 35 - 40 wt %. Table 7 presents the composition of preliminary prepared borosilicate glass frit used as the initial glass.

The glass compositions were synthesized in a laboratory-scale facility with cold crucible (d=100 mm, h=250 mm, frequency=5.28 MHz).

In the course of experiment, a mixture of glass frit with uranium oxide (taken in a specified ratio) was fed portionwise onto a glass mirror at a temperature of $1000^{\circ}C$. The glass sampling (the discharge of a melt portion) was performed within 1.5 h after every addition of a mixture, and 11 blocks (200-300 g) of uranium-containing borosilicate glass were obtained in this fashion. After cooling of the melt in a crucible, additional glass samples were taken from different regions of the resultant glass block (3 kg) in order to carry out further investigations.

A series of uranium-containing borosilicate glasses were concurrently synthesized by the melting of a mixture of the glass frit and uranium oxide in a muffle furnace (alundum crucible; $1200^{\circ}C$; heat treatment of a melt, up to 8 h).

The disperse particles and their conglomerates were found in the bulk of glasses that were synthesized in a muffle furnace for 2 h and incorporated more than 7 wt % uranium. As the duration of heat treatment (at $1200^{\circ}C$) of the melts containing more than 7 wt % uranium oxide increased to 8 h, the undissolved uranium oxide precipitated at the bottom of a crucible.

Therefore, upon the melting of a glass frit and uranium oxide under steady-state conditions, the solubility of uranium oxide is limited to 5 - 7 wt %. An excess amount of undissolved uranium oxide accumulates at the bottom of a glass block.

The study on glasses obtained during the immobilization of uranium oxide in borosilicate glass by the CCIM method demonstrates that materials synthesized are X-ray amorphous, completely homogeneous, and free from any disperse and crystalline species. This indicates that borosilicate glasses synthesized in the CCIM incorporate uranium only in the dissolved form (the content of dissolved uranium in the synthesized glasses is as much as 28 - 30 wt %).

We consider in greater detail the glass block obtained upon cooling of the melt directly in a cold crucible.

The glass samples taken from the top and middle parts of glass block are X-ray amorphous and homogeneous without any inclusions. Regions of a material that differs in color and structure from the bulk of glass are observed on the surface of a glass block removed from a cold crucible. X-ray powder diffraction analysis of these regions shows the presence of crystalline sodium uranate $Na_2U_2O_7$ and trace amounts of $U_{1-x}O_2$ uraninite.

A layer of the precipitate, which is inhomogeneous in density and consists of disperse particles with predominantly uncertain shape and also of their conglomerates, is revealed in the bottom part of glass block above the skull layer. The transparent dark-brown glass is observed above the precipitate.

A glassy material in the form of a cone with disperse, needle-shaped particles and their conglomerates is found in the middle of the bottom part of a glass block above the skull layer and the precipitate.

X-ray powder diffraction analysis of the glass samples taken from the above-described regions indicates the presence of the only crystalline phase, namely, $U_{1-x}O_2$ uraninite.

CONCLUSIONS

The use of CCIM for vitrification of the simulative HLLW provides a way of synthesixing high-performance glasses with properties similar to those of glasses prepared in a crucuble in a muffle furnace.

The solubility of rare-earth elements in phosphate glasses perpared in ceramic melters is limited to 2-3 wt. %. The application of CCIM makes it possible to obtain phosphate glass-like materials with uniform distribution of rare-earth element oxides (up to 10 wt. %).

The investigation into the behavior of uranium in borosilicate glass prepared with the CCIM technology demonstrates a feasibility of synthesizing the X-ray amorphous, homogeneous borosilicate glasses incorporating up to 25 - 28 wt % uranium, which is 4-5 times larger than that in glasses obtained by the traditional (melting in muffle furnace) technique.

REFERENCES

1. I.A. Sobolev, F.A. Lifanov, S.A. Dmitriyev, S.V. Stefanovsky, A.P. Kobelev, Vitrification of Radioactive Wates by Coreless Induction Melting in Cold Crucible, Proceedings, SPECTRUM'94, Vol.3, p.2250.

2. J.P. Moncouyoux, R. Boen, M. Puyou, A.Jouan, New Vitrification Techniques, RECORD'91, 3-rd Int.Conf. Nucl. Fuel Reprocess. And Waste Manag., Sendai, April 14-18, 1991, p. 307, v.1, Tokyo, 1991

3. W.Lutze, K.D. Closs, G.Tittel, P. Brennecke, W. Kutz, German Program for Vitrified HLW and Spent Fuel Management, ICEM'93 Proceedings of the 1993 International Conference on Nuclear Waste Management and Environmental Remediation, Prague, Chech Republic, September 5-11, Proceeding V.1 p.75.

4. M. Kitamura, K.Miwa, M. Ayabe, Vitrification Process development and Design for Large Scale Plant, ICEM'95 Proceedings of the Fifth International Conference on Radioactive Waste Management and Environmental Remediation, Berlin, Germany, September 3-7, Proceeding V.2, p.385.

5. E.G. Drozhko, A.P. Suslov, V.I. Fetisov, Y.G. Glagolenko, G.M. Medvedev, V.I. Osnovin, E.G. Dzekun, Basic Directions and Problems of Radioactive Waste Management Program in the «Mayak» Production Association, Chelyabinsk, Russia, ICEM'93 Proceedings of the 1993 International Conference on Nuclear Waste Management and Environmental Remediation, Prague, Chech Republic, September 5-11, Proceeding V.2 p.17.

6. Demin A.V., Matyunin Yu.I., Fedorova M.I., Noble Metal (NM) Behavior During Simulated HLLW Vitrification in Induction Melter with Cold Crucible, Fifth International Conference on Radioactive Waste Management and Environmental Remediation, ICEM'95 Berlin, Germany, September 3-7, Proceeding Vol. 1, p. 443.

7. Yu. I. Matyunin, A.V. Demin, V.V. Koushnikov, M.I. Fedorova, Investigation of uranium and plutonium in phosphate and borosilicate materials, meant for liguid HLW vitrification, International Conference "Actinide-93", Santa-Fe, NM, USA, 19-23 September, 1993.

8. E.Evest, H.Wiese, High-level liquid waste vitrification with the Pamela-plant in Belgium, IAEA-CN-48/177, 1987, 20p

9. A.V. Demin, Yu.I. Matyunin, A.S. Polyakov, M.I. Fedorova, Study of high-level waste component behaviour during solidification to produce phosphte and borosilicate vitreous materials, 1993 International Conference on Nuclear Waste Management and Invironmental Remediation, Praque, Chech Republic, 5-11 September, 1993, v. 1, p. 435.

10. V.V. Kushnikov, Yu.I. Matyunin, T.V. Smelova, Use of CCIM for plutonium Immobilization, Proc. Embed. Top. Meeting on DOE Spend Nuclear Fuel & Fissible Material Manag., Reno, Nevada, June 16-20, 1996, p. 192.

11. A.A. Appen, Khimiya stekla, Khimiya, 1974, p. 224-229 (Russia)

THE KINETICS OF SPINEL CRYSTALLIZATION FROM A HIGH-LEVEL WASTE GLASS

J.G. REYNOLDS* and P. HRMA**
* Division of Soils, University of Idaho, Moscow, Idaho 83843, ryno9522@uidaho.edu
** Pacific Northwest National Laboratory,[1] Richland, Washington 99352, pr_hrma@pnl.gov

ABSTRACT

The kinetics of spinel crystallization from a molten high-iron simulated high-level nuclear waste glass was studied using isothermal heat treatments. Optical microscopy with image analysis was used to measure volume fraction of spinel as a function of heat treatment time and temperature. The Johnson-Mehl-Avrami equation was fitted to data to determine kinetic coefficients for spinel crystallization. The liquidus temperature and Avrami number are $T_L = 1337\text{K}$ and $n = 1.5$.

INTRODUCTION

Spinel commonly precipitates from molten high-level nuclear waste (HLW) glasses. With increasing concentration of Cr, Ni, and Fe in the waste, spinel precipitates at increasingly higher temperatures [1] and may cause an undesirable accumulation of sludge in the waste glass melter. Since spinel sludge can obstruct the melt flow, spinel precipitation kinetics is needed for waste loading optimization. Without an undesirable effect, spinel precipitates in most HLW glasses during cooling. The information about the content of crystalline phases in the waste glass is a qualification requirement [2]. Consequently, an understanding of spinel formation kinetics in the waste glass is also useful for waste form characterization.

Kinetic studies of spinel precipitation from simulated nuclear waste glasses have been conducted by Turcotte et al. [3], Bickford et al. [4], Marra et al. [5], and Simpson et al.[6]. Without modeling spinel crystallization kinetics analytically, these authors generated time-temperature-transformation (TTT) diagrams and described the sequence in which crystals of different phases occurred as the heat treatment temperature decreased.

The purpose of this study was to fit isothermal kinetic and thermodynamic parameters of spinel precipitation from a simulated HLW glass using the Johnson-Mehl-Avrami equation [7]

$$\frac{C}{C_o} = 1 - \exp\left[-\left(\frac{t}{\tau}\right)^n\right] \tag{1}$$

where C is the crystalline phase volume fraction, C_o its value at equilibrium, t time, τ time constant, and n Avrami number. Both C_o and τ are functions of temperature:

$$\frac{C_o}{C_{\max}} = 1 - \exp\left[-B_L\left(\frac{1}{T} - \frac{1}{T_L}\right)\right] \tag{2}$$

$$\tau = \tau_O \exp\left(\frac{B_\tau}{T}\right) \tag{3}$$

where B_L, B_τ, $C_{\max}$, and τ_o are constants and T_L is the liquidus temperature. Though spinel is a solid solution and its precipitation may involve complex nucleation and growth mechanisms, the results of this study show that Eqs. (1-3) may describe spinel formation with sufficient accuracy for prediction purposes.

---

[1] Pacific Northwest National Laboratory is operated for the US Department of Energy by Battelle under Contract DE-ACO6-76RLO 1830.

Mat. Res. Soc. Symp. Proc. Vol. 465 © 1997 Materials Research Society

EXPERIMENTAL

The simulated nuclear waste glass (Table I) was designed to maximize the waste-to-additive ratio for a blended Hanford Site waste [8]. The batched chemicals (mostly oxides and carbonates) were mixed in an agate milling chamber for 4 min. The batch was melted in a platinum-10%-rhodium crucible in a Deltech Dt-31 furnace for 1 h at 1150°C, quenched on a stainless steel plate, and milled in a tungsten carbide chamber for 4 min., obtaining glass powder of 1 to 10 μm grain size.

Table I. The composition of the simulated high level nuclear waste glass

Component	Mass %	Component	Mass %	Component	Mass %
Al_2O_3	8.33	K_2O	0.14	P_2O_5	3.02
B_2O_3	5.00	La_2O_3	0.27	PbO	0.23
Bi_2O_3	1.25	Li_2O	1.86	RuO_2*	0.03
CaO	1.32	MgO	0.05	SiO_2	39.60
CdO	0.06	MnO*	0.95	SO_3	0.22
CeO_2	1.75	MoO_3	0.05	SrO	0.27
Cr_2O_3*	0.29	Na_2O	16.21	WO_3	0.11
F	0.36	Nd_2O_3	5.60	ZrO_2	4.54
Fe_2O_3*	7.05	NiO*	1.45		

* Component of spinel

Spinel crystallization was induced by constant-temperature heat treatment in both gradient temperature and uniform temperature furnaces. The gradient temperature furnace was suitable for measurement at $T < 800°C$. At higher temperatures, evaporation of alkali borates creates a surface tension gradient that drives within the melt a convective cell, thus dragging crystals away from their original positions. Therefore, the uniform temperature furnace was used for $T > 800°C$.

For the uniform temperature furnace, the powdered glass was remelted at 1150°C for 1 h, quenched, and crushed to 1-5 mm-sized pieces. The glass was put into ~1 ml square platinum-5%-gold containers and heat treated at a constant temperature within the range of 800 to 1150°C for 1 to 96 h in a Deltech Dt-31-HW furnace. The samples were annealed at 505°C in a Thermolyne 2000 furnace for 2 h, and allowed to cool to room temperature inside the furnace. Annealed samples were thin sectioned and polished.

Glass for the gradient furnace was remelted in 250×10×10 mm platinum-5%-gold boats in a Lindberg furnace (type 51442) and air quenched. Glass-filled boats were placed in a linear temperature gradient furnace (Marshall by Thermcraft) spanning a temperature interval from 550 to 795°C (1°C/mm gradient) for 1, 3, 8, 24, 72, and 96 h. The samples were annealed at 505°C for 2 h in the Lindberg furnace, and allowed to cool to room temperature. The glass was taken from the boats, mounted in epoxy, thin sectioned and polished.

Spinel was identified in the thin sections by optical microscopy (Olympus PMG 3). Spinel area fraction was obtained with an image analyzer (Leco, 2001) attached to the microscope, using reflected light. The gradient furnace samples were measured every 20 mm (20°C) between 640 and 780°C. A total of 20 measurement for the uniform temperature furnace and 15 for the gradient furnace were taken at random locations on the sample for each time and temperature evaluated. The average area fraction of spinel as a function of the number of measurements (N) converged on a single average by N = 15. Because of the isometric character of spinel crystals, volume and area fractions were assumed to be equal. Below 640°C, spinel crystals were too small (<1 μm) for accurate resolution with an optical microscope. In addition to spinel, cerium zirconium oxide precipitated between 800 and 1050°C in small quantities, mostly by surface crystallization on the original glass particles. These crystals were difficult to distinguish from spinel with the image analyzer. A sodium-calcium-rare-earth phosphate phase was identified by scanning electron

microscopy with energy dispersive analysis spectroscopy (SEM/EDS) in glass heat treated at
≤841°C. This phase was easily distinguishable, but could affect spinel detection in some samples.
The data from glass samples treated in the gradient and uniform temperature furnaces at $T < 800°C$
were found in a good agreement with each other. The composition of spinel was examined with an
ElectroScan model E3 SEM, by EDS Link EXL, in samples heat treated for 24 h.

R ESULTS

The volume fraction of spinel as a function of time and temperature is plotted in Fig. 1. For
$T > 650°C$ and $t \geq 24$ h, the fraction of spinel no longer changes with the heat treatment time.
Hence, the mixture effectively reached phase equilibrium in 24 h at all but the lowest heat treatment
temperatures. This is rapid crystallization compared to clinopyroxene [9].
The $t \geq 24$ h values were fitted by Eq. (2). Nonlinear least squares regression yielded
$T_L = 1337$ K, $B_L = 3783$ K, and $C_{max} = 0.03707$. The liquidus temperature is somewhat higher
than 1050°C, which is considered a safe minimum for the current Joule-heated melters. However,
the melt will not contain more than a fraction of one percent of spinel while in the melter. In the
canister, less than 3 vol% spinel can be expected to form. The outliers at $t \geq 72$ h and $T > 800°C$,
probably caused by precipitation of crystals other than spinel (phosphate), were not used in fitting
Eq. (2) to data.

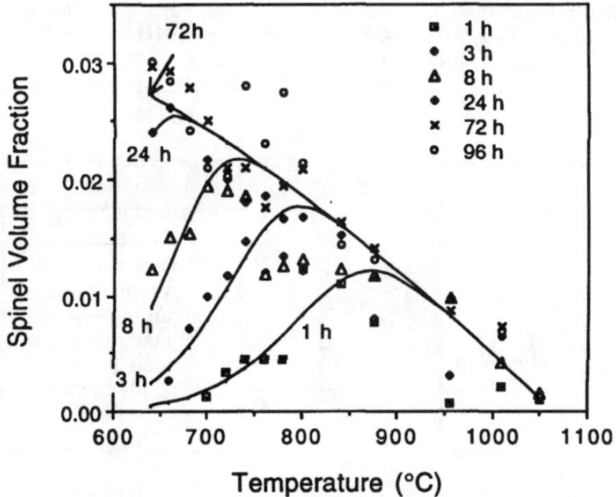

Fig. 1. Volume fractions of spinel versus temperature and time of isothermal heat treatment. The
lines were fitted using Eqs. (1-3).

Using calculated C_o values, $\ln[-\ln(1 - C / C_o)]$ was plotted against $\ln t$ for $t \leq 24$ h data for
different values of temperature. By Eq. (1) and linear regression, a value of 1.5 was obtained for n
from these plots. An $n = 1.5$ corresponds to that theoretically predicted for constant crystal number
density and three-dimensional crystal growth [10]. With $n = 1.5$, Eq. (1) was solved for $\ln(\tau)$ for
each (C, t) pair. Fitting Eq. (3) to these values by linear regression analysis resulted in
$B_\tau = 14250$ K, and $\tau_o = 6.311 \times 10^{-3}$ s. The lines in Fig. 1 represent Eq. (1) with the coefficients
obtained. Fig. 2 displays crystallization kinetics in alternative forms, $C = f_T(t)$ and $T = f_C(t)$,
representing the isothermal evolution of crystallinity and the TTT diagram, respectively. The
curves were obtained from Eqs. (1-3), using the above numerical values of coefficients.

Along with trace amounts of ruthenium, the dominant metals found in spinel by SEM/EDS were iron, nickel, chromium and manganese. Spinel progressively became enriched in chromium and impoverished in nickel with increased temperature, but the composition was essentially temperature independent below ~760°C. Unlike clinopyroxene [9], spinel does not exhibited a shoulder on the $C = f_t(T)$ curve at temperatures close to T_L from compositional changes associated with the heat treatment temperature. According to Bickford and Jantzen [4], spinel nucleates on precipitated or undissolved RuO_2 particles. According to Capobianco and Drake [11], RuO_2 is a common minor component in spinel, which appears to be the case in the present study.

DISCUSSION

The Avrami number value of 1.5 suggests that spinel nucleation occurred virtually instantaneously and did not control precipitation kinetics. The crystals grew without an induction period. Although new crystals were observed for at least 3 h at $T \leq 800°$C, a short time considering the more sluggish crystal growth below 800°C, it was not clear if at $t \leq 3h$ new crystals were nucleated or if they were too small to be seen. Therefore, a change of Avrami number with temperature was not considered. The short-time data for samples treated above 800°C indicate a possible delay in crystallization (see Fig. 1), but since crystal count did not change with time, delayed nucleation is ruled out; more likely, the low fraction of crystals made the volume fraction measurement less accurate.

Crystallization kinetics could also be affected by the changing composition of spinel and by the glass redox condition. Increasing the Fe^{2+}/Fe^{3+} ratio in the glass has been shown to increase both the rate [6] and extent [4, 12] of spinel crystallization. Redox was not measured in this glass, but for typical high-iron HLW glasses processed at 1150°C, Fe^{2+}/Fe^{3+} is approximately 0.03 [13].

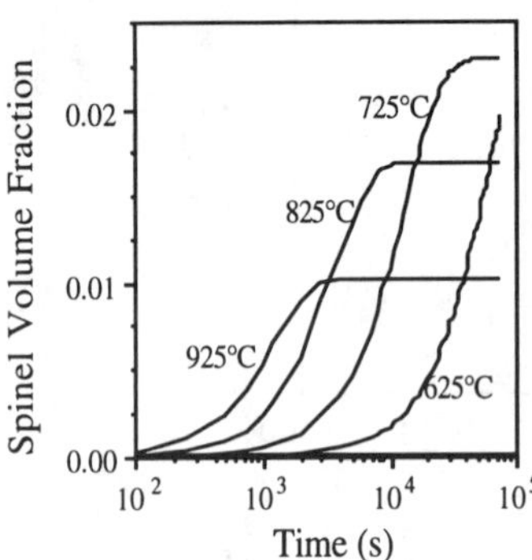

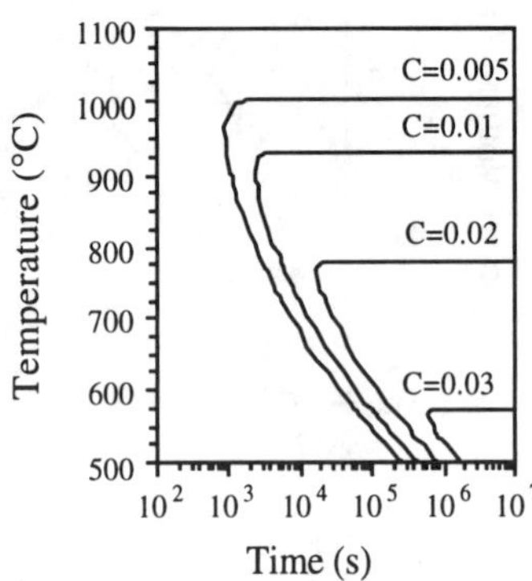

Fig. 2. Time evolution of isothermal spinel crystallization (left) and the time-temperature-transformation diagram (right).

Probably because of the changing composition of the crystalline phase, Eq. (1) does not seem to provide the best fit to measured C_o values, as is evident in Fig. 1. However, a wide data scatter discouraged experimentation with more complex functional forms. Nevertheless, this did not preclude a reasonable fit for crystallization kinetics, as demonstrated in Fig. 1.

The kinetic coefficients for spinel precipitation from isothermal heat treatment may be applied to simple nonisothermal processes, such as cooling. Such an application is promising for the prediction of spatial distribution of spinel crystals in large monoliths of glass, which is of interest for waste form characterization. Spinel precipitation in a monolith is controlled by the time evolution of the temperature field during cooling after the glass has been poured into stainless steel canisters. Spinel distribution can be obtained by solving the differential equation

$$\frac{dC}{dt} = n(C_o - C)\frac{t^{n-1}}{\tau^n} \tag{4}$$

of which Eq. (1) is a solution for $T = const$. However, no attempt has so far been made to perform such calculation.

CONCLUSIONS

The kinetics of spinel crystallization in a simulated nuclear waste glass have fitted to the Johnson-Mehl-Avrami equation. Spinel grows rapidly from pre-existing nuclei. The kinetic and thermodynamic parameters are $T_L = 1337\,\text{K}$, $B_L = 3783$ K, $C_{max} = 0.03707$, $n = 1.5$, $\tau_o = 6.311 \times 10^{-3}$ s, and $B_\tau = 14250\,\text{K}$. Because spinel crystals grow without a measurable induction period, it is expected that these parameters derived from an isothermal study are sufficient for predicting spinel crystallization for simple nonisothermal temperature histories—a task for characterization of nuclear waste glass monoliths. Although compositional changes may complicate the nonisothermal crystallization kinetics, it seems from the results of this study that their effect is insignificant.

Acknowledgments. The authors would like to thank J. Coleman for SEM/EDS analysis, J.D. Vienna for insightful support, and M.J. Schweiger, D.E. Smith, and R.C. Reed for laboratory advice. J.G. Reynolds is grateful to AWU (Associated Western Universities, Inc.) for a fellowship appointment at PNNL.

REFERENCES

1. M. Mika, M.J. Schweiger, J.D. Vienna, and P. Hrma (this volume).
2. U.S. Department of Energy, Waste Acceptance Product Specifications, February, 1993.
3. R.P. Turcotte, J.W. Wald, and R.P. May, in: *Scientific Basis for Nuclear Waste Management*, Vol. 2, edited by C.J.M. Northrup (Plenum Press, New York. 1980) p. 141.
4. D.F. Bickford and C.M. Jantzen, in: *Scientific Basis for Nuclear Waste Management VII*, Vol. 26, edited by G.L. McVay (North Holland, New York, 1984) p. 557.
5. S.L. Marra, M.K. Andrews, and C.A. Cicero, in: *Ceram. Trans.*, Vol. 39, eds. G.B. Mellinger (American Ceramic Society, Ohio, 1993) p. 283.
6. J.C. Simpson, D. Oksoy, T.C. Cleveland, L.D. Pye and V. Jain, in: *Ceram. Trans.*, Vol. 45, eds. D. Bickford, S. Bates, V. Jain, and G. Smith (American Ceramic Society, Ohio, 1994) p. 377.
7. M. Avrami, *J. Chem. Phys.* 7 (1939) 1103; 8 (1940) 212; 9 (1941) 177.
8. D.S. Kim, P. Hrma, D.A. Lamar, and M.L. Elliott, in: *Ceram. Trans.*, Vol. 45, eds. D.F. Bickford, S. Bates, V. Jain, and G. Smith (American Ceramic Society, Ohio, 1994) p. 39.
9. J.D. Vienna, P. Hrma, and D.E. Smith (this volume).
10. D.R. Macfarlane and M. Fragoulis, *Phys. Chem. Glass.* 27 (1986) 228.
11. C.J. Capobianco and M.J. Drake, *Geochim. Cosmochim. Acta* 54 (1990) 869.
12. V. Jain and S.M. Barnes, in: *Ceram. Trans.*, Vol. 23, eds. G.G. Wicks, D. F. Bickford, and L.R. Bunnell (American Ceramic Society, Ohio, 1991) p. 251.
13. P. Hrma, G.F. Piepel, M.J. Schweiger, D. E. Smith, D.S. Kim, P.E. Redgate, J.D. Vienna, C.A. LoPresti, D.B. Simpson, D.K. Peeler, and M.H. Langowski, "Property/Composition Relationships for Hanford High-Level Waste Glasses Melting at 1150°C," PNL-10359, Vol. 1 and 2, Pacific Northwest National Laboratory, Richland, Washington (1994).

LIQUIDUS TEMPERATURE OF SPINEL PRECIPITATING HIGH-LEVEL WASTE GLASSES

M. MIKA, M.J. SCHWEIGER, J. D. VIENNA, P. HRMA
Pacific Northwest National Laboratory, Richland, Washington 99352

ABSTRACT

The liquidus temperature (T_L) often limits the loading of high-level waste in glass through the constraint that T_L must be at least $100\,^{\circ}C$ below the temperature at which the glass viscosity is 5 Pa·s. In this study, values of T_L for spinel primary crystalline phase were measured as a function of glass composition. The test glasses were based on high-iron Hanford Site tank wastes. All studied glasses precipitated spinel $(Ni,Fe,Mn)(Cr,Fe)_2O_4$ as the primary crystalline phase. T_L was increased by additions of Cr_2O_3, NiO, Al_2O_3, Fe_2O_3, MgO, and MnO; while Li_2O, Na_2O, B_2O_3, and SiO_2 had a negative effect. Empirical mixture models were fitted to data.

INTRODUCTION

If crystalline phases precipitate in a continuous fed melter for high-level waste (HLW) glass, serious processing problems can ensue, such as sludge formation on the melter bottom or even melter electrode shorting if the phases are electrically conductive [1]. Spinel (Fe,Ni,Mn) $(Fe,Cr)_2O_4$ precipitates as the primary crystalline phase in HLW glasses that contain high concentrations of Fe_2O_3, Cr_2O_3, and NiO (West Valley, Savannah River, and most Hanford HLW glasses). Accumulation of spinel crystals at the melter bottom or in the overflow can cause clogging of glass melt flow. Settling of crystals follows Stokes' law, but will be affected by convective flow, by melt viscosity (which changes as the particle moves through the temperature field), and by particle agglomeration [2]. Removal of the sludge from the melter bottom is extremely difficult; however, sludge accumulation can be minimized by construction of melters with inclined bottoms and bottom drains. For the current melters, a constraint has been imposed on the waste glass that T_L be at least $100\,^{\circ}C$ below the melting temperature (T_M), defined as the temperature at which the glass viscosity is 5 Pa·s. This constraint limits the waste loading in glass for most Hanford HLW. Therefore, to maximize waste loading and optimize waste pretreatment and blending, it is important to know the effect of glass composition on the liquidus temperature. The purpose of this work is to measure T_L of spinel as a function of glass composition and to fit an empirical mixture model to T_L data. The model can be used in glass formulation to maximize the waste loading while meeting the T_L constraint.

THEORY

The dependence of T_L on chemical composition of glass can be approximated by empirical mixture models. The validity of such a model is usually restricted to a composition range confined within a single primary phase field. The general forms of the first and second-order empirical mixture models for liquidus temperature are [3,4]

$$T_L = \sum_{i=1}^{n} b_i g_i \qquad (\text{first-order}) \qquad (1)$$

Mat. Res. Soc. Symp. Proc. Vol. 465 © 1997 Materials Research Society

$$T_L = \sum_{i=1}^{n} b_i g_i + \sum_{i=1}^{n-1} \sum_{j \geq i}^{n} b_{ij} g_i g_j \qquad \text{(second-order)} \qquad (2)$$

where g_i is the i-th component mass fraction, b_i and b_{ij} are the i-th and j-th component coefficients, and n is the number of components. To obtain coefficient values, Eqs.(1) and (2) were fitted to test data by least-squares regression. A first-order model was fitted to data from this study (SP). For fitting of a second-order model, SP data and data from the Composition Variation Study (CVS) (67 glasses) [3,5] were used. The SP composition region was a subregion of the CVS composition region. A first-order model for liquidus temperature of spinel precipitating glasses was previously reported by Kim and Hrma [6].

EXPERIMENTAL

Test matrix design

To study the composition effects on liquidus temperature of simulated Hanford HLW glasses, a test matrix (Table 1) was designed containing 33 glasses with 10 components (Al_2O_3, B_2O_3, Cr_2O_3, Fe_2O_3, Li_2O, MgO, MnO, Na_2O, NiO, SiO_2) and Others (Ag_2O, BaO, CaO, CdO, CeO_2, Cl, Co_2O_3, CuO, F, K_2O, La_2O_3, MoO_3, Nd_2O_3, P_2O_5, PbO, Rh_2O_3, RuO_2, Sb_2O_3, SeO_2, SO_3, SrO, TiO_2, ZnO, ZrO_2). The test matrix was derived from the baseline glass containing 47 wt% water-washed Hanford nominal waste blend [7]. The baseline composition was altered, one-component-at-a-time, while maintaining the same relative proportions of the remaining components. The mass fractions of components were varied within the spinel primary phase field and the composition envelope that has been estimated for HLW glasses (Table 2). One duplicate glass (the baseline composition) was added to the test matrix (SP-Li-3) to assess measurement variability.

Glass preparation and liquidus temperature measurement

The batched chemicals (oxides, boric acid, and carbonates) were mixed in an agate milling chamber for 5 minutes. The batch was melted for 1 h in a platinum-10%-rhodium crucible covered with a Pt lid in a Deltech DT-31 furnace. The T_M was the temperature at which the calculated glass viscosity was 5 Pa·s. T_M was calculated using a first order model [8]. The model accuracy was within 20°C. After melting, the glass was quenched on a stainless steel plate, milled in a tungsten carbide chamber for 8 min., remelted for 1 h, quenched, and broken into pieces.

Liquidus temperatures were measured using uniform temperature furnaces. Approximately 2.5 g glass samples were placed into square platinum-5%-gold containers with lids and heat treated at a constant temperature for 48 h in a DT-31 furnace. Residual crystals were present in the glass. They were rounded at 24 hours and completely dissolved within 48 hours. Four to 10 samples were treated at different temperatures for each glass. Samples were first examined under a low-power metallurgical microscope using reflected light to determine if crystals were present in the sample. For further examination, selected samples were thin sectioned and polished. Crystals were identified in the thin sections by optical microscopy (Olympus PMG 3) using transmitted light. The difference in temperature between heat-treated glass samples with crystals and samples without crystals was narrowed down to 10°C or less.

Table 1. Glass compositions (in mass fractions)

Oxide	Al_2O_3	B_2O_3	Cr_2O_3	Fe_2O_3	Li_2O	MgO	MnO	Na_2O	NiO	SiO_2	Others
SP-1	**0.0800**	**0.0700**	**0.0022**	**0.1250**	**0.0300**	**0.0060**	**0.0036**	**0.1573**	**0.0052**	**0.4600**	**0.0607**
SP-Al-1	**0.0400**	0.0730	0.0023	0.1304	0.0313	0.0063	0.0038	0.1641	0.0054	0.4800	0.0633
SP-Al-2	**0.1200**	0.0670	0.0021	0.1196	0.0287	0.0057	0.0034	0.1505	0.0050	0.4400	0.0581
SP-Al-3	**0.1600**	0.0639	0.0020	0.1141	0.0274	0.0055	0.0033	0.1436	0.0047	0.4200	0.0554
SP-B-1	0.0860	**0.0000**	0.0024	0.1344	0.0323	0.0065	0.0039	0.1691	0.0056	0.4946	0.0653
SP-B-2	0.0852	**0.0100**	0.0023	0.1331	0.0319	0.0064	0.0038	0.1674	0.0055	0.4897	0.0646
SP-B-3	0.0826	**0.0400**	0.0023	0.1290	0.0310	0.0062	0.0037	0.1624	0.0054	0.4748	0.0627
SP-B-4	0.0757	**0.1200**	0.0021	0.1183	0.0284	0.0057	0.0034	0.1488	0.0049	0.4353	0.0574
SP-Cr-1	0.0802	0.0702	**0.0000**	0.1253	0.0301	0.0060	0.0036	0.1576	0.0052	0.4610	0.0608
SP-Cr-2	0.0798	0.0698	**0.0050**	0.1246	0.0299	0.0060	0.0036	0.1569	0.0052	0.4587	0.0605
SP-Cr-3	0.0795	0.0696	**0.0080**	0.1243	0.0298	0.0060	0.0036	0.1564	0.0052	0.4573	0.0603
SP-Cr-4	0.0792	0.0693	**0.0120**	0.1238	0.0297	0.0059	0.0036	0.1558	0.0051	0.4555	0.0601
SP-Fe-1	0.0859	0.0752	0.0024	**0.0600**	0.0322	0.0064	0.0039	0.1690	0.0056	0.4942	0.0652
SP-Fe-2	0.0832	0.0728	0.0023	**0.0900**	0.0312	0.0062	0.0037	0.1636	0.0054	0.4784	0.0631
SP-Fe-3	0.0777	0.0680	0.0021	**0.1500**	0.0291	0.0058	0.0035	0.1528	0.0051	0.4469	0.0590
SP-Li-1	0.0825	0.0722	0.0023	0.1289	**0.0000**	0.0062	0.0037	0.1622	0.0054	0.4742	0.0626
SP-Li-2	0.0816	0.0714	0.0022	0.1276	**0.0100**	0.0061	0.0037	0.1605	0.0053	0.4695	0.0620
SP-Li-3	0.0800	0.0700	0.0022	0.1250	**0.0300**	0.0060	0.0036	0.1573	0.0052	0.4600	0.0607
SP-Mg-1	0.0797	0.0697	0.0022	0.1245	0.0299	**0.0100**	0.0036	0.1567	0.0052	0.4581	0.0605
SP-Mg-2	0.0789	0.0690	0.0022	0.1232	0.0296	**0.0200**	0.0035	0.1551	0.0051	0.4535	0.0598
SP-Mg-3	0.0757	0.0662	0.0021	0.1182	0.0284	**0.0600**	0.0034	0.1488	0.0049	0.4350	0.0574
SP-Mn-1	0.0803	0.0703	0.0022	0.1255	0.0301	0.0060	**0.0000**	0.1579	0.0052	0.4617	0.0609
SP-Mn-2	0.0795	0.0696	0.0022	0.1242	0.0298	0.0060	**0.0100**	0.1563	0.0052	0.4570	0.0603
SP-Mn-3	0.0771	0.0674	0.0021	0.1204	0.0289	0.0058	**0.0400**	0.1516	0.0050	0.4432	0.0585
SP-Na-1	0.0873	0.0764	0.0024	0.1365	0.0328	0.0066	0.0039	**0.0800**	0.0057	0.5022	0.0663
SP-Na-2	0.0835	0.0731	0.0023	0.1305	0.0313	0.0063	0.0038	**0.1200**	0.0054	0.4804	0.0634
SP-Na-3	0.0759	0.0665	0.0021	0.1187	0.0285	0.0057	0.0034	**0.2000**	0.0049	0.4367	0.0576
SP-Ni-1	0.0804	0.0704	0.0022	0.1257	0.0302	0.0060	0.0036	0.1581	**0.0000**	0.4624	0.0610
SP-Ni-2	0.0796	0.0697	0.0022	0.1244	0.0299	0.0060	0.0036	0.1565	**0.0100**	0.4578	0.0604
SP-Ni-3	0.0780	0.0683	0.0021	0.1219	0.0293	0.0059	0.0035	0.1534	**0.0300**	0.4485	0.0592
SP-Si-1	0.0919	0.0804	0.0025	0.1435	0.0344	0.0069	0.0041	0.1806	0.0060	**0.3800**	0.0697
SP-Si-2	0.0711	0.0622	0.0020	0.1111	0.0267	0.0053	0.0032	0.1398	0.0046	**0.5200**	0.0540
SP-Si-3	0.0593	0.0519	0.0016	0.0926	0.0222	0.0044	0.0027	0.1165	0.0039	**0.6000**	0.0450

Analysis of spinel crystals

Spinel crystals were qualitatively and quantitatively analyzed using Scanning Electron Microscopy (SEM)/ Energy Dispersive Analysis Spectroscopy (EDS). Polished glass samples were examined with an ElectroScan Environmental SEM equipped with a Link eXL x-ray microanalysis system. Spinel crystals were analyzed quantitatively (ZAF corrected) in the spot mode using the Link virtual standards pack (an element profile library and an element reference

Table 2.   Composition regions for the first-order model and second-order models A and B

Variable	Composition regions in mass fractions	
	first-order model	second-order model A and B
SiO_2	0.380 - 0.600	0.370 - 0.601
B_2O_3	0.000 - 0.120	0.000 - 0.200
Na_2O	0.080 - 0.200	0.050 - 0.221
Li_2O	0.000 - 0.030	0.000 - 0.072
CaO		0.000 - 0.100
MgO	0.004 - 0.060	0.000 - 0.080
Fe_2O_3	0.060 - 0.150	0.005 - 0.174
Al_2O_3	0.040 - 0.150	0.000 - 0.160
ZrO_2		0.000 - 0.080
Cr_2O_3	0.000 - 0.012	0.000 - 0.012
MnO	0.000 - 0.040	0.000 - 0.040
NiO	0.000 - 0.030	0.000 - 0.030
Others	0.045 - 0.070	0.001 - 0.246

Table 3.   Composition of spinel crystals

Element	Spinel in SP-Si-1 glass at 1101°C, mol%		Spine in SP-1 (baseline) glass at 997°C, mol%		Spinel in SP-Si-3 glass at 1010°C, mol%	
Cr	6.1	6.1	13.6	16.6	47.6	45.1
Fe	69.8	72.1	59.7	57.3	33.4	35.6
Ni	21.4	19.7	24.8	24.8	17.0	18.0
Mn	2.7	2.1	1.9	1.3	2.0	1.3
Sum	100	100	100	100	100	100

file), which allows analysis to be carried out in a standardless manner. Elements other than Si, Cr, Ni, and Fe were not used for quantitation. For reference, polished 316 stainless steel was also analyzed.

RESULTS AND DISCUSSION

Rod-shaped crystals were found in glasses SP-Ni-1, SP-Fe-1, SP-Si-2, SP-Si-3, and SP-Na-3. These crystals were rich in Ru, probably RuO_2. Disregarding this minor phase, the primary crystalline phase identified in all glasses was spinel.

The dominant metals found in spinel by SEM/EDS were Fe, Ni, Cr, and Mn. Table 3 shows the chemical composition of spinel crystals precipitating in three glasses with different content of SiO_2. Two spinel crystals were analyzed in each glass. The average composition of spinel

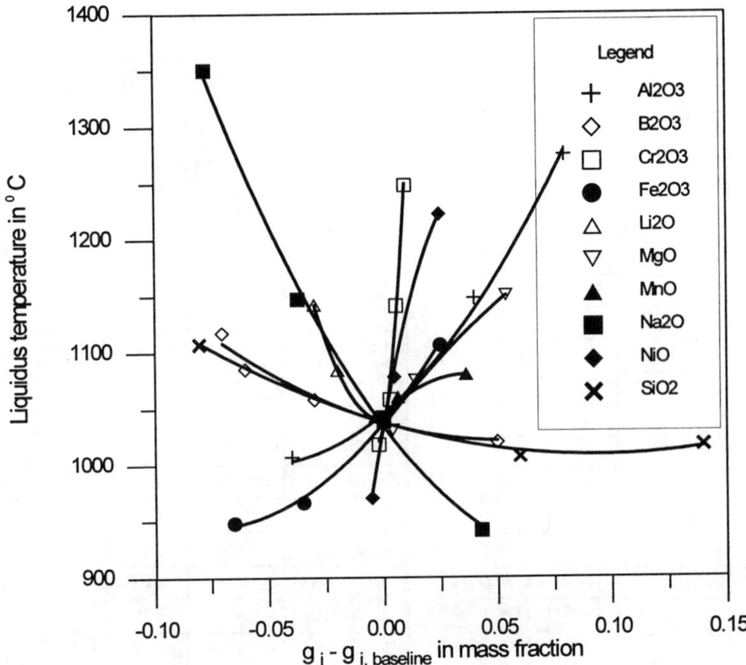

$g_i - g_{i,\,baseline}$ in mass fraction

Figure 1. Component effects on T_L

crystals, expressed as solid solutions of oxides, was

$(Ni_{0.62}Mn_{0.07}Fe_{0.31})O\cdot(Cr_{0.09}Fe_{0.91})_2O_3$	in	SP-Si-1	(38 wt% SiO_2)
$(Ni_{0.74}Mn_{0.05}Fe_{0.21})O\cdot(Cr_{0.23}Fe_{0.77})_2O_3$	in	SP-1	(46 wt% SiO_2)
$(Ni_{0.53}Mn_{0.05}Fe_{0.42})O\cdot(Cr_{0.69}Fe_{0.31})_2O_3$	in	SP-Si-3	(60 wt% SiO_2)

Glass composition comparison shows that Cr concentration in spinel increases and Fe concentration decreases with increasing SiO_2 content in glass (while other components in glass decrease). This suggests that the solubility of Cr_2O_3 in glass decreases with increasing glass acidity.

Measured T_L for 33 glasses are plotted in Figure 1 and listed in Table 3. The difference in liquidus temperature of two independently prepared glasses of the baseline composition (SP-1 and SP-Li-3) was 3 °C. As Figure 1 shows, the largest positive effect on T_L is exhibited by Cr_2O_3, followed by NiO, Al_2O_3, Fe_2O_3, MgO, and MnO; T_L is decreased strongly by Na_2O and Li_2O and slightly by B_2O_3 and SiO_2.

The first-order model was fitted to data with R^2=0.94. The component coefficients are presented in Table 5. The difference between measured and predicted T_L (Table 4) was within 24 °C for most glasses (25) and within 54 °C for all glasses (33). As can be seen in Figure 1, the dependence of T_L on composition is non-linear, which causes the deviation between the

Table 4. Measured and predicted T_L in °C

Glass	$T_L(M)$ °C	$T_L(1)$ °C	$T_L(2A)$ °C	$T_L(2B)$ °C
SP-1	1039	1051	1047	1051
SP-Al-1	1007	953	950	962
SP-Al-2	1148	1149	1147	1139
SP-Al-3	1275	1247	1248	1225
SP-B-1	1117	1101	1093	1097
SP-B-2	1085	1094	1086	1091
SP-B-3	1058	1072	1067	1072
SP-B-4	1020	1016	1014	1016
SP-Cr-1	1018	1012	1007	1013
SP-Cr-2	1058	1101	1099	1100
SP-Cr-3	1141	1155	1154	1151
SP-Cr-4	1248	1226	1227	1218
SP-Fe-1	948	933	940	896
SP-Fe-2	966	987	990	960
SP-Fe-3	1106	1097	1088	1127
SP-Li-1	1142	1129	1125	1134
SP-Li-2	1084	1103	1096	1106
SP-Li-3	1042	1051	1047	1051
SP-Mg-1	1031	1058	1055	1059
SP-Mg-2	1075	1076	1074	1080
SP-Mg-3	1150	1148	1149	1159
SP-Mn-1	1043	1048	1049	1051
SP-Mn-2	1060	1056	1047	1053
SP-Mn-3	1080	1081	1082	1065
SP-Na-1	1350	1315	1324	1271
SP-Na-2	1147	1179	1167	1158
SP-Na-3	941	905	942	927
SP-Ni-1	970	1014	970	963
SP-Ni-2	1078	1086	1106	1116
SP-Ni-3	1222	1230	1219	1214
SP-Si-1	1107	1083	1099	1087
SP-Si-2	1007	1027	1025	1035
SP-Si-3	1017	995	1018	1028

Table 5. T_L component coefficients for spinel primary crystalline phase in °C

Variable	Component coefficients, °C		
	1	2A	2B
SiO_2	834	1373	1239
B_2O_3	395	263	-47
Na_2O	-1826	-5174	-1848
Li_2O	-1470	10318	-2090
CaO		3445	-26206
MgO	2827	2760	2602
Fe_2O_3	2644	2313	8943
Al_2O_3	3307	3824	2623
ZrO_2		11365	3991
Cr_2O_3	18864	19182	-11466
MnO	1870	35697	7265
NiO	8210	5913	23903
Others	4419	374	583
$Na_2O \times Na_2O$		10436	
$Li_2O \times SiO_2$		-25193	
$MnO \times Al_2O_3$		-187582	
$MnO \times ZrO_2$		-1099613	
$NiO \times Na_2O$		69430	
$NiO \times NiO$		-250962	-352543
$CaO \times SiO_2$			52152
$Fe_2O_3 \times SiO_2$			-10855
$Fe_2O_3 \times CaO$			-32059
$Fe_2O_3 \times ZrO_2$			-29474
$Cr_2O_3 \times CaO$			2930960
$NiO \times MnO$			-1253900
R^2	0.94	0.97	0.93

$T_L(M)$...measured T_L
$T_L(1)$....T_L predicted by the first-order model
$T_L(2A)$..T_L predicted by the second-order model A
$T_L(2B)$..T_L predicted by the second-order model B
1........first-order model
2A.....second-order model A
2B.....second-order model B

measured and predicted values.

The constraint that T_L be at least 100°C below the melting temperature limits the waste loading (W) in HLW glass. It is possible to minimize T_L and thus maximize W by an optimized glass formulation. Glass composition, waste loading, and the compositions of waste and glassforming additives are related through the mass balance equation [9]

$$g_i = W w_i + (1 - W) a_i \qquad (3)$$

where w_i and a_i are the i-th component fractions in waste and additives. Substituting for g_i from Eq.(3) in Eq.(1), the following is obtained

$$W = \frac{T_L - \sum_{i=1}^{n} b_i a_i}{\sum_{i=1}^{n} b_i (w_i - a_i)} \qquad (4)$$

It is desirable to predict W within a 1 wt% accuracy, for which the first-order model for T_L with coefficients listed in Table 5 is clearly insufficient. To improve T_L predictions second-order models were developed using SP and CVS data. CVS glasses were measured both in a gradient furnace (31 glasses) and in uniform temperature furnaces (CVS-UT) (36 glasses). The second-order model A (R^2=0.97) presented in Table 5 was fitted to SP and CVS-UT data. SP and all CVS data were used for calculation of the second-order model B (R^2=0.93) (Table 5). The predicted values from both second-order models (Table 4) are not significantly better than values predicted from the first-order model. Therefore, additional experimental data using a statistically designed test matrix with many-components-at-a-time variations are needed for better T_L prediction.

CONCLUSIONS

Spinel precipitated as the primary crystalline phase in all glasses studied (if traces of crystalline RuO_2 are omitted). Analyzed spinel crystals contained Fe, Ni, Cr, and Mn and their chemical composition was dependent on glass composition. T_L increased with increasing additions of Cr_2O_3, NiO, Al_2O_3, Fe_2O_3, MgO, and MnO and decreased by Na_2O, Li_2O, B_2O_3, and, SiO_2 in the order of impact. This dependence was non-linear. The first- and second-order empirical mixture models were fitted to SP and CVS data. Because the one-component-at-a-time experimental design was used in this study (SP), it was impossible to determine the effect of component interactions on T_L from SP data. To determine interactions, a statistically designed test matrix with many-components-at-a-time variations is needed. Though the SP data set was augmented by CVS data, the second-order terms are uncertain because of the lack of an overall statistical design. A statistically designed study is necessary for more accurate predictions required for waste loading maximization.

ACKNOWLEDGMENTS

This work was sponsored by the U.S. Department of Energy under contract DE-AC06-76RLO 1830. The authors thank J.S.Young for SEM analysis. M.Mika is grateful to Associated Western Universities, Inc., for his fellowship appointment at PNNL.

REFERENCES

1. L.R. Bunnell, <u>Laboratory Work in Support of West Valley Glass Development</u>, PNL-6539, Pacific Northwest Laboratory, Richland, Washington (1988).

2. W.T. Cobb and P. Hrma in <u>Nuclear Waste Management IV</u>, edited by G.G. Wicks, D.F. Bickford, and L.R. Bunnell (Am. Ceram. Soc. Proc., Ceram. Trans. **23**, Westerville, Ohio, 1991) pp. 233-237.

3. P. Hrma, G.F. Piepel, M.J. Schweiger, D.E. Smith, D.S. Kim, P.E. Redgate, J.D. Vienna, C.A. LoPresti, D.B. Simpson, D.K. Peeler, and M.H. Langowski, <u>Property/Composition Relationships for Hanford High-Level Waste Glasses Melting at 1150°C</u>, PNL-10359, UC-72, Pacific Northwest National Laboratory, Richland, Washington (1994).

4. J.A. Cornell, <u>Experiments With Mixtures: Designs, Models, and the Analysis of Mixture Data</u>, 2nd ed. (John Wiley and Sons, New York, 1990).

5. J.D. Vienna, P. Hrma, M.J. Schweiger, M.H. Langowski, P.E. Redgate, D.S. Kim, G.F. Piepel, D.E. Smith, C.Y. Chang, D.E. Rinehart, S.E. Palmer, and H. Li, <u>Effect of Composition and Temperature on the Properties of High-Level Waste (HLW) Glass Melting Above 1200°C</u>, PNNL-10987, Pacific Northwest National Laboratory, Richland, Washington (1996).

6. D.S. Kim and P. Hrma in <u>Environmental and Waste Management Issues in the Ceramic Industry II</u>, edited by D. Bickford, S. Bates, V. Jain, and G. Smith (Am. Ceram. Soc. Proc., Ceram. Trans. **45**, Westerville, Ohio, 1994) pp. 327-337.

7. U.S. Department of Energy, <u>TWRS Privatization Request for Proposals</u>, NO. DE-RP06-96RL13308, Richland, Washington (1996).

8. P. Hrma, J.D. Vienna, and M.J. Schweiger, <u>Liquidus Temperature Limited Waste Loading Maximization for Vitrified HLW</u> (to be published in Am. Ceram. Soc. Proc., Ceram. Trans. 1996).

9. P. Hrma and A.W. Bailey in <u>Fifth International Conference on Radioactive Waste Management and Environmental Remediation</u>, edited by S. Slate, F. Feizollahi, and J. Creer (Berlin, Germany, 1995) pp. 447-451.

LIQUIDUS TEMPERATURE MODEL FOR HANFORD HIGH-LEVEL WASTE GLASSES WITH HIGH CONCENTRATIONS OF ZIRCONIA

J.V. CRUM, M.J. SCHWEIGER, P. HRMA, and J.D. VIENNA
Pacific Northwest National Laboratory, Richland WA 99352

ABSTRACT

A study was conducted on glasses based on a simulated transuranic waste with high concentrations of ZrO_2 and Bi_2O_3 to determine the compositional dependence of primary crystalline phases and liquidus temperature (T_L). Starting from a baseline composition, glasses were formulated by changing mass fractions of Al_2O_3, B_2O_3, Bi_2O_3, CeO_2, Li_2O, Na_2O, P_2O_5, SiO_2, and ZrO_2, one at a time, while keeping the remaining components in the same relative proportions as in the baseline glass. Liquidus temperature was measured by heat treating glass samples for 24 h in a uniform temperature furnace. The primary crystalline phase in the baseline glass and the majority of the glasses was zircon ($ZrSiO_4$). A change in the concentration of certain components (Al_2O_3, ZrO_2, Li_2O, B_2O_3, and SiO_2) changed the primary phase to baddeleyite (ZrO_2), while cerium oxide (CeO_2) precipitated from glasses with more than 3 wt% CeO_2. Zircon T_L was strongly increased by Al_2O_3, ZrO_2 and CeO_2, and slightly by P_2O_5 and SiO_2; decreased strongly by Li_2O and Na_2O and moderately by B_2O_3. A first-order model was constructed for T_L as a function of composition for zircon primary crystalline phase glass.

INTRODUCTION

Most Hanford high-level wastes (HLW) have high concentrations of refractory components, that increase the T_L of borosilicate HLW glasses. Melter performance is the major concern for these glasses because of the tendency of refractory constituents, such as Cr_2O_3 and ZrO_2, to precipitate crystals in the melter. Crystallization could lead to formation of a sludge and crystal growth in the glass discharge region of the melter.

To avoid crystallization of solid phases in the melter, T_L of glass is usually required to be at least 100°C below the operating temperature. Most HLW glasses are prone to precipitation of spinel during glass processing due to high concentrations of Fe, Mn, Ni, and Cr. However, transuranic (TRU) waste glasses precipitate zirconium-containing phases because of the high concentration of zirconia in the waste from cladding removal operations.[1] Based on the results from the composition variation study (CVS),[2] glasses with high concentrations of zirconia have a potential to precipitate zirconium silicate (zircon, $ZrSiO_4$); zirconium oxide (baddeleyite, ZrO_2); or sodium-zirconium silicate ($Na_2ZrSi_2O_7$) at temperatures above 1050°C.

A first-order mixture model was fitted by Kim and Hrma[3] to T_L as a function of glass composition for all zirconia-containing phases taken together. The small number of T_L data available from the CVS did not allow fitting of a higher-order mixture model to zirconium-containing phases or generation of models for single primary phase fields. The present work was undertaken to enlarge the T_L database for zirconium-containing primary crystalline phases. Glasses were produced by a systematic variation of a baseline composition that precipitated zircon as the primary phase. A database of T_L values of 36 glass compositions was generated. A first-order model was fitted to measured T_L data in which zircon was the observed primary phase.

EXPERIMENTAL

Test Matrix Design

A test matrix (Table I) was designed to measure the T_L of simulated TRU glasses around a baseline (central) glass by altering its composition, one component at a time, while keeping the remaining components in the same relative proportions. The baseline glass was formulated to melt at 1150°C with 40 wt% water washed TRU waste blend from Hanford's tank farms. Components expected to influence T_L (Al_2O_3, B_2O_3, Bi_2O_3, CeO_2, Li_2O, Na_2O, P_2O_5, SiO_2, and ZrO_2) were

79

varied to the extent of a typical Hanford TRU glass composition envelope. The remaining components in the baseline glass (Ag_2O, BaO, CaO, Co_2O_3, Cr_2O_3, CuO, F, Fe_2O_3, K_2O, La_2O_3, MgO, MnO, NiO, PbO, SrO, and ZnO) were grouped into one component called Others (Oth), which was not varied in this study.

Table I. Glass Compositions for TRU Test Matrix in Mass Fraction

	Al_2O_3	B_2O_3	Bi_2O_3	CeO_2	Li_2O	Na_2O	P_2O_5	SiO_2	ZrO_2	Oth	(a)	(b)	(c)
TRU-BL-1	0.023	0.081	0.054	0.000	0.046	0.114	0.009	0.468	0.107	0.098	ZS	1012	1016
TRU-Al-1	**0.060**	0.078	0.052	0.000	0.044	0.110	0.008	0.450	0.103	0.095	Z	1196	
TRU-Al-4	**0.075**	0.077	0.051	0.000	0.043	0.108	0.008	0.443	0.101	0.093	Z	1308	
TRU-Al-6	**0.040**	0.080	0.053	0.000	0.045	0.112	0.009	0.459	0.105	0.097	ZS	1069	1081
TRU-B-1	0.025	**0.000**	0.059	0.000	0.050	0.124	0.009	0.509	0.116	0.107	Z	1155	
TRU-B-2	0.025	**0.020**	0.058	0.000	0.049	0.122	0.009	0.499	0.114	0.105	ZS	1031	1041
TRU-B-3	0.024	**0.050**	0.056	0.000	0.047	0.118	0.009	0.484	0.111	0.102	ZS	1030	1029
TRU-B-4	0.022	**0.120**	0.052	0.000	0.044	0.109	0.008	0.448	0.102	0.094	ZS	992	1000
TRU-B-5	0.025	**0.010**	0.058	0.000	0.049	0.123	0.009	0.504	0.115	0.106	Z	1116	
TRU-Bi-1	0.024	0.086	**0.000**	0.000	0.048	0.120	0.009	0.494	0.113	0.104	ZS	1015	1014
TRU-Bi-2	0.024	0.084	**0.025**	0.000	0.047	0.117	0.009	0.482	0.110	0.101	ZS	1017	1015
TRU-Bi-3	0.022	0.080	**0.075**	0.000	0.045	0.111	0.009	0.457	0.105	0.096	ZS	1009	1016
TRU-Bi-4	0.022	0.078	**0.100**	0.000	0.044	0.108	0.008	0.445	0.102	0.094	ZS	1023	1017
TRU-Ce-1	0.023	0.080	0.053	**0.015**	0.045	0.112	0.009	0.461	0.105	0.097	ZS	1019	1040
TRU-Ce-2	0.022	0.079	0.053	**0.030**	0.044	0.111	0.008	0.454	0.104	0.095	ZS	1062	1065
TRU-Ce-3	0.022	0.078	0.052	**0.045**	0.044	0.109	0.008	0.447	0.102	0.094	C	1120	
TRU-Ce-4	0.022	0.077	0.051	**0.060**	0.043	0.107	0.008	0.440	0.101	0.092	C	1178	
TRU-Li-1	0.024	0.085	0.057	0.000	**0.000**	0.119	0.009	0.490	0.112	0.103	Z	1271	
TRU-Li-2	0.024	0.084	0.056	0.000	**0.015**	0.118	0.009	0.483	0.110	0.102	ZS	1155	1134
TRU-Li-3	0.023	0.083	0.055	0.000	**0.030**	0.116	0.009	0.475	0.109	0.100	ZS	1074	1077
TRU-Li-4	0.023	0.080	0.053	0.000	**0.060**	0.112	0.009	0.461	0.105	0.097	ZS	982	961
TRU-Na-1	0.025	0.088	0.058	0.000	0.049	**0.045**	0.009	0.504	0.115	0.106	ZS	1350	
TRU-Na-2	0.024	0.086	0.057	0.000	0.048	**0.070**	0.009	0.491	0.112	0.103	ZS	1223	1231
TRU-Na-3	0.023	0.083	0.055	0.000	0.047	**0.095**	0.009	0.478	0.109	0.100	ZS	1110	1108
TRU-Na-4	0.023	0.080	0.053	0.000	0.045	**0.130**	0.009	0.459	0.105	0.097	ZS	930	937
TRU-P-1	0.023	0.081	0.054	0.000	0.045	0.113	**0.020**	0.462	0.106	0.097	ZS	1042	1030
TRU-P-2	0.022	0.080	0.053	0.000	0.045	0.111	**0.030**	0.458	0.105	0.096	ZS	1052	1043
TRU-P-3	0.022	0.079	0.052	0.000	0.044	0.110	**0.040**	0.453	0.104	0.095	ZS	1057	1056
TRU-P-4	0.022	0.078	0.052	0.000	0.044	0.109	**0.050**	0.448	0.103	0.094	ZS	1070	1069
TRU-Si-1	0.025	0.090	0.060	0.000	0.051	0.126	0.010	**0.410**	0.119	0.109	ZS	1010	1011
TRU-Si-2	0.021	0.073	0.049	0.000	0.041	0.103	0.008	**0.520**	0.096	0.089	ZS	1022	1019
TRU-Si-4	0.019	0.066	0.044	0.000	0.037	0.092	0.007	**0.570**	0.086	0.079	ZS	1022	1023
TRU-Si-5	0.026	0.093	0.062	0.000	0.053	0.131	0.010	**0.390**	0.123	0.113	Z	1112	
TRU-Zr-1	0.024	0.084	0.056	0.000	0.047	0.117	0.009	0.482	**0.080**	0.101	ZS	937	926
TRU-Zr-2	0.023	0.080	0.053	0.000	0.045	0.112	0.009	0.461	**0.120**	0.097	ZS	1074	1059
TRU-Zr-3	0.022	0.078	0.052	0.000	0.044	0.110	0.008	0.450	**0.140**	0.095	Z	1182	

(a) Primary crystalline phase: ZS = zircon ($ZrSiO_4$), Z = Baddeleyite (ZrO_2), C = Cerium Oxide (CeO_2)
(b) Measured T_L, °C
(c) Predicted T_L by first-order model for zircon, °C

Glass Melting

Glasses were batched using oxides, carbonates, sodium metaphosphate, sodium fluoride, and boric acid. Prior to melting, each batch was homogenized using an agate mill for 6 min. Glasses were melted for 1 h in a platinum crucible covered with a lid at an estimated temperature (T_5) at which viscosity was 5 Pa·s (see appendix).[2] The viscosity was measured for the baseline glass, the measured T_5 (1148°C) was 8°C lower than the predicted. This accuracy was deemed acceptable. After melting, the glass was quenched on a stainless steel plate. Glass samples were examined for crystallinity or undisolved particles using an optical microscope. To ensure homogeneity, each glass was placed in a tungsten carbide mill and crushed for 6 min, and then remelted at the same temperature for 1 h. If particulates were found in the glass sample, following the first or second melt, it was remelted at an increased temperature.

Liquidus Temperature

Approximately 2.5 g glass samples were loaded into platinum-10%-rhodium rectangular crucibles (12.5 mm × 12.5 mm × 8 mm) with lids and heat treated in a uniform temperature furnace for 24 h. Following the heat treatments, samples were first examined under a low magnification (70×) metallurgical microscope to determine if crystals were present in the glass. Selected samples were thin sectioned and examined under an optical microscope (100× to 1000×) using transmitted and reflected light. X-ray diffraction and scanning electron microscopy (SEM) with electron dispersion spectroscopy (EDS) were used to help identify the crystalline phases. The difference in temperature between heat treated glass samples with and without crystals was narrowed down to 5°C. Hence, T_L was determined within the 5°C interval.

Empirical Models

The baseline glass was composed of M=25 components consisting of oxides and fluorine. The glasses were formulated as N-component mixtures, where N=10, in which nine components were varied by the experimental design (Table I). The tenth component (Others) was a mixture of the remaining M+1-N=16 components.

Within a single primary phase field, the relationship between T_L and glass composition can be satisfactorily described using second-order mixture models. However, the T_L dependence on composition within the limited region of the zircon primary phase field tested was nearly linear and could be represented by the first-order model. Thus,

$$T_L = \sum_{i=1}^{N} b_i g_i \tag{1}$$

where b_i is the i-th component coefficient, g_i is the i-th component mass fraction, and N is the number of components independently varied in the data set. First-order mixture models were previously applied to T_L by Babcock[4] for commercial glasses. Kim and Hrma[3] and Hrma, Piepel et al.[2] fitted first-order models to T_L data of HLW glasses with spinel, clinopyroxene, and zirconium-containing primary phases. The temperature was limited by the gradient furnace to $T_L \leq$ 1118°C and could be affected by convection at temperatures as low as 800°C. The model for spinel has been updated by Mika, Schweiger et al.[5]

The experimental design of this study allowed data for each component to be treated as an outcome of a series of pseudo-binary mixtures of the i-th component, and the remainder of the baseline glass, taken as one component. Thus,

$$T_{L(b)i} = T_{L(o)i} + m_i g_i \qquad (i=1,2,3,......,N) \tag{2}$$

where, $T_{L\,(b)i}$ is the liquidus temperature of the i-th binary mixture, $T_{L(O)i}$ is the intercept, and m_i is the i-th component effect.

Component coefficients can be obtained from component effects using the relation

$$b_i = T_{L(b)i} + (1 - g_i)m_i \qquad (i=1,2,3,......,N) \tag{3}$$

The b_i coefficients determined from Eq. (3) are generally slightly different from those obtained by fitting Eq. (1) to all data using multiple regression.

In the experimental design used in this study, no measurements were performed for pseudo-binary mixtures with the Others component as a variable. To determine the coefficient for the Others component, Eq. (1) was applied to the baseline in the form

$$b_{others} = (T_{L(c)} - \sum_{i=1}^{N-1} b_i g_{(c)i}) / g_{(c)others} \tag{4}$$

where the subscript *(c)* denotes the baseline (center point). The baseline liquidus temperature,. $T_{L(c)}$, was obtained as a weighted average:

$$T_{L(c)} = (\sum_{i=1}^{N-1} g_{(c)i})^{-1} (\sum_{i=1}^{N-1} g_{(c)i} T_{L(c)i})$$ (5)

where $g_{(c)i}$ is the i-th component mass fraction in the baseline glass, and $T_{L(c)i}$ is the calculated T_L for the baseline glass using Eq. (2).

Ternary Sub-Mixture

Following Li et al.[6], who successfully identified glasses prone to nepheline crystallization using the Na_2O-Al_2O_3-SiO_2 sub-mixture, an attempt was made to display the primary phases identified in TRU and CVS glasses on the Na_2O-ZrO_2-SiO_2 sub-mixture in order to establish a correlation between these primary phases and the sub-mixture composition.

RESULTS AND DISCUSSIONS

Primary Crystalline Phases and Measured Liquidus Temperatures

All primary phases of TRU glasses were either zircon, baddeleyite, or cerium oxide. The primary phases and measured temperatures are listed in Table I. Liquidus temperature versus the change in composition from the baseline glass, for all primary phases, is shown in Figure 1. The primary crystalline phase in the baseline glass and its compositional neighborhood (the majority of the test glasses) was zircon ($ZrSiO_4$). In glasses where mass fractions of $ZrO_2 > 0.12$, $Al_2O_3 > 0.04$, $Li_2O = 0.00$, $SiO_2 < 0.41$, and $B_2O_3 < 0.02$, the primary phase was baddeleyite. Cerium oxide was the primary phase at > 0.03 CeO_2.

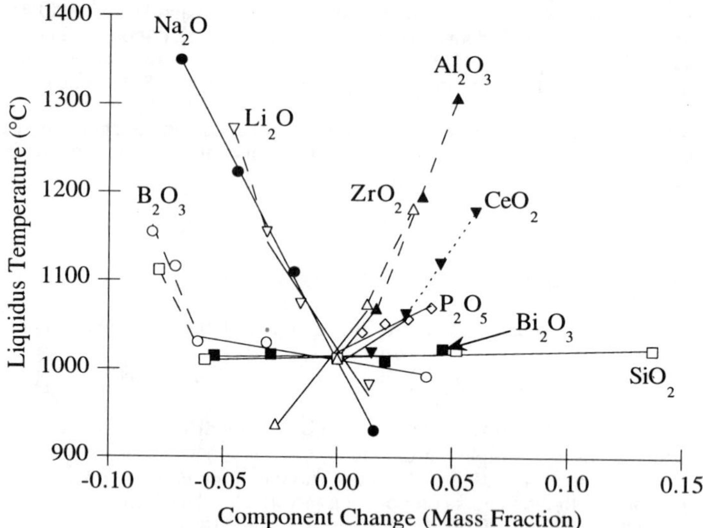

Figure 1. Liquidus temperature versus change in composition; the composition region extends over three primary phases: zircon (solid line), baddeleyite (dashed line), and cerium oxide (dotted line)

Glasses with a primary phase of baddeleyite or cerium oxide showed a steeper dependence of T_L on component concentrations than zircon primary phase glasses. Increasing mass fractions of Al_2O_3 had the strongest effect at raising T_L within the zircon and baddeleyite primary phase fields, followed by ZrO_2, CeO_2 and P_2O_5 in descending order. Conversely, increasing mass fractions of Na_2O or Li_2O lowered the T_L strongly. B_2O_3 strongly lowered T_L at concentrations < 2 wt% (within baddeleyite field). With B_2O_3 > 2 wt%, the effect was moderate on lowering T_L (within zircon field). Increasing concentration of Bi_2O_3 or SiO_2 > 41wt% showed essentially no effect on T_L. However, for SiO_2 < 41wt% (within baddeleyite field), decreasing the concentration of SiO_2 strongly increased T_L.

Displaying primary phases of the TRU and CVS glasses on the Na_2O-ZrO_2-SiO_2 sub-mixture field, as seen in Figure 2, shows that different primary phases tend to occupy different areas. Li et al.[6] found that the occurrence of nepheline in slowly cooled glasses agreed almost exactly with the nepheline liquidus surface of the Na_2O-Al_2O_3-SiO_2 ternary mixture. No such coincidence exists for zirconium-containing phases (see the inset for Figure 2). In addition, the boundaries between the phase fields are not sharp. For example, two glasses within the zircon area have non-zirconium primary phases. However, zircon precipitates in them as the secondary phase. Also, the rare earth zirconate and sodium-zirconium silicate phases overlap each other, the primary phase being dependent on rare earth concentration.

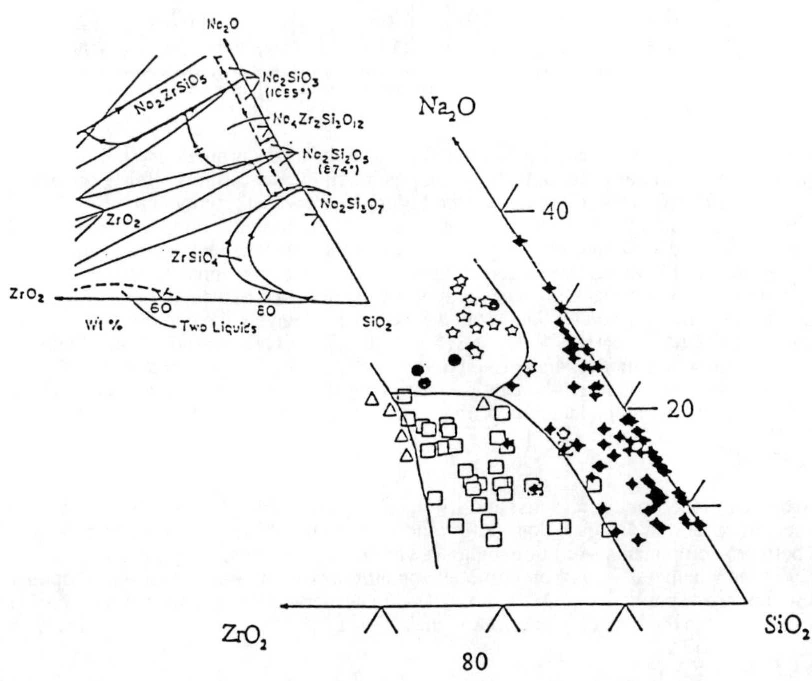

□ Zircon, △ Baddeleyite, ● Sodium-Zirconium Silicate,
☆ Rare Earth Zirconate, ◆ Non-Zirconium Phases

Figure 2. TRU and CVS glasses plotted on the Na_2O-SiO_2-ZrO_2 sub-mixture. Inset shows a corresponding region of the ternary phase diagram.

Liquidus Temperature Coefficients for Zircon Phase Glasses

The present TRU data show an approximately linear relationship for T_L versus composition within the zircon field (Figure 1). Therefore, a first-order model was fitted to T_L data for glasses with only zircon primary phase . The coefficients and composition regions are shown in Table II. Each coefficient was calculated individually from the slopes and intercepts determined by single-component regression using Eqs. (2) and (3). The Others component coefficient was obtained by Eq. (4). Others strongly increases T_L which indicates a presence of oxides that have a strong influence. Predicted values of T_L by the first-order model for TRU glasses with zircon primary crystalline phase are shown in Table I. The first-order model predicted T_L within 9°C of the measured values for most glasses, and within 21°C for all glasses. This accuracy is deemed sufficient for glass formulation optimization.

Table II. Liquidus Temperature Coefficients (in °C) and Composition Region* (in mass fractions) for Zircon Phase Glasses (R^2 = 0.988)

	b_i	g_i,min	g_i,max			b_i	g_i,min	g_i,max
Al_2O_3	4280	0.020	<u>0.040</u>	Na_2O	-3336	0.045	0.130
B_2O_3	632	<u>0.020</u>	0.120	P_2O_5	2295	0.008	0.050
Bi_2O_3	1044	0.000	0.100	SiO_2	1055	<u>0.410</u>	0.570
CeO_2	2673	0.000	<u>0.030</u>	ZrO_2	3995	0.08	<u>0.12</u>
Li_2O	-2660	<u>0.015</u>	0.060	Others	3769	0.079	0.109

*Underlined composition limits are phase boundaries observed during the study, all other composition limits are the highest/lowest concentrations tested.

Several glasses in the test matrix precipitated baddeleyite as the primary crystalline phase. Similar to glasses that precipitate zircon, baddeleyite precipitating glasses show a fairly linear relationship between T_L and glass composition (see Figure 1). However, component effects on T_L in the baddeleyite region are much steeper than the zircon region. Liquidus temperature models for the baddeleyite phase field are important for glass formulation for melters that operate at high temperature (>1150°C). However, the present data are insufficient to design a T_L model for baddeleyite primary phase glasses. Only five components were varied to the extent that the primary phase changed from zircon (at the baseline glass) to baddeleyite; the component coefficients are 7753°C Al_2O_3, 6404°C ZrO_2, -6818°C Li_2O, -5540°C B_2O_3, and -4386°C SiO_2. Alternatively, a second-order model can be fitted to the zircon and baddeleyite phases taken together, based on the component trends seen in Figure 1. A thorough statistical analysis of these and other T_L data for zircon and baddeleyite primary phase fields will be published elsewhere.[7]

CONCLUSIONS

A first-order model for zircon satisfactorily fits T_L data for TRU glasses with zircon primary phase. However, the database does not provide sufficient information regarding interactions between components. Additional glasses varying several components at a time are needed to determine whether a second-order model containing cross-product terms for component interactions will improve predictions. Also, more data are needed for fitting a model relating T_L to glass composition within the baddeleyite primary phase field.

ACKNOWLEDGMENTS

This work was done for the U.S. Department of Energy under contract DE-AC06-76RLO 1830. The authors thank Jim Young for use of the SEM to determine crystalline phases of the waste glasses. Jarrod Crum would like to thank Associated Western Universities, Inc., for his fellowship appointment at PNNL.

REFERENCES

1. S.L. Lambert and D.S. Kim, *Tank Waste Remediation System High-Level Waste Feed Processability Assessment Report* , WHC-SP-1143, Westinghouse Hanford Company, Richland, Washington, 1994.

2. P. Hrma, G.F. Piepel, M.J. Schweiger, D.E. Smith, D.S. Kim, P.E. Redgate, J.D. Vienna, C.A. LoPresti, D.B. Simpson, D.K. Peeler, and M.H. Langowski, *Property / Composition Relationships for Hanford High-Level Waste Glasses Melting at 1150°C* , PNL-10359, Vol. 1 and 2, Pacific Northwest National Laboratory, Richland, Washington (1994).

3. D.S. Kim and P. Hrma, *Ceram. Trans.* 45, 327-337 (1994).

4. C.L. Babcock, *Silicate Glass Technology Methods* (John Wiley & Sons, New York, 1977) p. 225-236.

5. M. Mika, M.J. Schweiger, J. D. Vienna, and P. Hrma, Mater. Res. Soc. Proc. (this vol.).

6. H. Li, J.D. Vienna, P. Hrma, D.E. Smith, and M.J. Schweiger, Mater. Res. Soc. Proc. (this vol.).

7. Q. Rao, G.F. Piepel, P. Hrma, and J.V. Crum, *Liquidus Temperatures of HLW Glasses with Zirconium-Containing Primary Crystalline Phases*, PNNL-SA-28192, Rev.1, Pacific Northwest National Laboratory, Richland, Washington (1996).

8. P. Hrma, G.F. Piepel, P.E. Redgate, D.E. Smith, M.J. Schweiger, J.D. Vienna, and D.S. Kim, *Ceram. Trans.* 61, 505-513 (1995).

APPENDIX: First-Order Model for Melting Temperature

The temperature at which the viscosity is 5 Pa·s (T_5) was estimated using a first-order mixture equation

$$T_L = \sum_{i=1}^{N} T_{Mi} g_i \qquad (A1)$$

where T_{5i} is the i-th component coefficient (Table 1A).

Table IA. Component coefficients for first-order T_5 model in °C

Comp.	T_{5i}	Comp.	T_{5i}	Comp.	T_{5i}	Comp.	T_{5i}
Ag_2O	1797	CeO_2	1797	La_2O_3	1797	P_2O_5	3307
Al_2O_3	2331	Cr_2O_3	1797	Li_2O	-3364	PbO	100
B_2O_3	-86	CuO	1797	MgO	22	SiO_2	2181
BaO	1797	F^-	-200	MnO	1797	SrO	20
Bi_2O_3	200	Fe_2O_3	958	Na_2O	-385	ZnO	1797
CaO	-194	K_2O	-300	NiO	1797	ZrO_2	1892

Alternatively, T_5 can be obtained by first- and second-order models for viscosity as a function of temperature and composition.[2,8]

WASTE VITRIFICATION: PREDICTION OF ACCEPTABLE COMPOSITIONS IN A LIME-SODA-SILICA GLASS-FORMING SYSTEM

T. M. GILLIAM[*] and C. M. JANTZEN[**]
[*]Oak Ridge National Laboratory, Oak Ridge, TN 37831, tmg@ornl.gov
[**]Westinghouse Savannah River Technology Center, Aiken, SC 29802

ABSTRACT

A model is presented based upon calculated bridging oxygens which allows the prediction of the region of acceptable glass compositions for a lime-soda-silica glass-forming system containing mixed waste. The model can be used to guide glass formulation studies (e.g., treatability studies) or assess the applicability of vitrification to candidate waste streams.

INTRODUCTION

Vitrification technology is used around the world for treatment of radioactive waste. Increasingly, it is being considered as the future baseline stabilization technology on U.S. Department of Energy (DOE) sites for the treatment of mixed wastes[1,2]. At the present time, vitrification of mixed waste cannot be considered a mature commercially available technology. The expectation that it can become so, rests primarily on two assumptions: (1) glass making is a well-established technology/industry and the equipment and experience associated with it can be applied to the developing mixed waste vitrification technology; and (2) waste vitrification technology development and experience in the high-level waste arena will be applicable to low-level mixed wastes as well. This rationale is similar to that which led to the birth of cement-based waste-form technology, where it was assumed that equipment and experience from the construction and concrete industries could be applied to waste solidification and where technology development efforts concentrated initially on low-level radioactive waste and next progressed to hazardous and then to mixed waste. Based upon experience with the evolution of cement-based waste-form technology, it is reasonable to assume that vitrification can follow suit and that supporting its development is warranted.

Cement-based waste-form chemistry involves not only physical encapsulation but also chemical interaction of some waste constituents within the solid matrix and chemical reactions within the contained pore structure. The complexity of this chemistry and the lack of understanding concerning it have led to some unexpected problems in the application of the technology to some wastes across the DOE Complex. This, in turn, has led to increased emphasis on developing and defining an alternative waste-form technology such as vitrification. However, glass formation involves both the chemical interaction (i.e., reaction) of some waste constituents in formation of the glass matrix and the dispersion of waste constituents in the fluid glass, leading to their atomistic bonding (and possibly encapsulation, in some cases) by the glass matrix upon cooling. This complex chemistry, like that of cement-based waste forms, is not fully understood, particularly for mixed (i.e., low-level radioactive and chemically hazardous) waste vitrification, in which numerous constituents are present that may affect this chemistry. For this reason, the chemical behavior of mixed waste in a glass matrix cannot be quantitatively predicted.

Mat. Res. Soc. Symp. Proc. Vol. 465 © 1997 Materials Research Society

GLASS FAMILIES OF POTENTIAL INTEREST

There are numerous glass families that have been developed and evaluated on a commercial scale. Examples of these families include soda-lime-silicate (SLS), borosilicate, lead silicate, aluminosilicate, halide, borate, phosphate, sulfide, chalcogenide, chalcohalide, oxyhalide, oxynitride, and oxycarbide glasses. Calcium is a major constituent (i.e., >1 wt %) in the majority of mixed waste sludges located throughout the DOE Complex. Consequently, the study presented herein focused on the SLS system. The SLS system has been evaluated previously for numerous mixed waste sludges[2] and has the advantage of utilizing the calcium content of the wastes as a needed additive.

Historically, the SLS system has used the three-component operating diagram with units of weight percent to illustrate regions of acceptable glass formulations. However, it is well known that the chemistry of any system occurs on a mole basis rather than a weight basis. The primary reason that units of weight percent can successfully illustrate regions within the SLS system is that the molecular weights of the three components are so similar. The molecular weights of Na_2O, CaO, and SiO_2 are 62, 56, and 60 grams/mole, respectively. Thus, a plot of composition in weight percent would accurately (to within a few percent error) reflect composition in mole percent. Obviously, as more components are introduced into the system, this comparison is no longer as accurate as the simple three-component system. This further emphasizes the need to better understand the effects of additional components on the basic system.

<u>The Basic System</u>

The phase diagrams for the basic SLS system, those of the related ternary systems that contain common major waste constituents (e.g., alumina [Al_2O_3]), and the accompanying literature permit useful comparisons to be made between the conditions of composition and temperature for equilibrium systems and those used to vitrify mixed waste. These comparisons lead, in turn, to an understanding of the fundamental chemical issues that operate in the processing of these wastes.

The structures of silicate glasses are based on the polymerizing tendency of SiO_4 tetrahedra, which derives, in part, from the bonding properties of Si(IV) and the small ionic size of silicon relative to that of oxygen. For the present purposes, the bonding properties are deemed to rely primarily on the electronic polarity of the Si-O bond and the coordination number of oxygen atoms that surround the silicon atoms. Based on electronegativity values[3], the Si-O bond is approximately 88% ionic, indicating that the bond exhibits both ionic and covalent properties.

The 12%-covalent character helps to maintain the tetrahedra in the melt, making them the basic "building blocks" in silicate melts and glasses. If crystalline silicates precipitate from the melt, the high ionicity leads to the well-known variety of crystallographic packing arrangements for Si^{4+} and O^{2-}.

These concepts indicate that both the melt and its corresponding glassy state can be considered as differing only moderately from the crystalline silicates that possess similar chemical composition. Consequently, an improved understanding of silicate waste glasses can be gained from knowledge of the glass melt, the corresponding crystalline metal silicates, and the solidified glasses. Analogously, the structural properties of glasses that contain elements such as aluminum and boron would respectively resemble the corresponding crystalline aluminosilicates and borosilicates.

One molecular restriction that applies to the glassy and crystalline states of silicate materials and to their melts is particularly noteworthy. Neighboring tetrahedra that are linked to one another can share only corners; sides and faces of tetrahedra cannot be shared. Hence, two tetrahedra are linked to each other by the sharing of an oxygen atom. This arrangement means that a tetrahedral unit may be linked to two, three, or four other tetrahedra. Two such links per silicon yield, locally and on the average if they predominate, a somewhat "linear" polymer or chain of SiO_4 tetrahedra. Analogously, three bridging oxygens per silicon produce two-dimensional arrays that resemble cross-linked polymers, and four would generate a three-dimensional network.

Clearly, the oxygen-to-silicon ratio is an important parameter for the polymeric and network structures in silicate melts, their corresponding glasses, and the formation of the various crystalline phases that can precipitate during processing. This ratio is 2:1 for unary SiO_2 systems. Thus, crystalline SiO_2 (e.g., quartz) can be understood as composed of tetrahedral arrangements of close-packed ions; this situation explains the relatively high density of quartz. The ionic radii of silicon and oxygen are 0.041 nm and 0.138 nm, respectively. Hence, the corresponding size ratio for Si:O of 0.34:1 leads to the establishment or retention of the SiO_4 tetrahedra in the crystalline state: 2.65 kg m^{-3}, as compared with that of the high-temperature crystalline forms of SiO_2—tridymite (2.26 kg m^{-3}), cristobalite (2.32 kg m^{-3}), and amorphous silica (2.2 kg m^{-3}).

These considerations suggest that the establishment of SiO_4 tetrahedra in a melt requires an additional source of oxygen in order to achieve the oxygen-to-silicon ratio needed to stabilize the 2-dimensional and 3-dimensional networks. Normally, alkali and alkaline oxides are used to provide the needed oxygen while maintaining charge balance between positive and negative ions. Oxides of other metals, such as the transition metals and some of those in Groups IIIB and IVB, can also be used. This function underscores the variety of silicate minerals and the prevalence of complex compositions for tailored glasses and ceramics.

<u>Minimum Bridging Oxygens</u>

Recognizing that the number of bridging oxygens (oxygens which can bond covalently to produce the glass structure) is a critical parameter (above and beyond the simple oxygen-to-silica ratio) suggests a methodology by which one of the boundaries to the range of acceptable compositions can be predicted. The methodology is based upon Stevels,[4] theory in which he calculated the mean possible number of bridging oxygens per SiO_2 based upon the oxygen-to-silica ratio. A minimum of one bridging oxygen is necessary for a plausible glass structure to form[5]: that is, at calculated bridging oxygens of less than one-ionic bonding is favored over covalent bonding. Consequently, there is, in effect, too much oxygen to promote the formation of the basic tetrahedral structure necessary for glass to form. Thus, the region of acceptable glass compositions would be predicted to be bounded by

$$F = 8 - 2(O/Si) \qquad (1)$$

where

F = the mean possible number of bridging oxygens and has a value ≥ 1,
O = moles of oxygen within the glass melt,
Si = moles of silicon within the glass melt.

PREDICTING ACCEPTABLE COMPOSITION REGIONS

A treatability study of vitrification of West End Treatment Facility (WETF) sludge has been published previously[2,6]. For this paper, the data set generated in the study is used solely for the purpose of applying and validating the bridging oxygen model. Consequently, only a brief description of the waste and treatability study is described herein.

Waste Composition[2,6]

Historically tanks have been used for settling and storage of sludge at the Oak Ridge Reservation Y-12 Plant WETF and West Tank Farm. The sludge is composed primarily of WETF process metal hydroxide precipitate (copper, nickel, zinc, and aluminum), biodenitrification sludges (mainly $CaCO_3$) and ferric hydroxide with trace metal hydroxides. Depleted uranium at an average of approximately 0.42% ^{235}U vs 0.72% in natural uranium, is the primary radioisotope of concern. In addition to being low-level radioactive, the waste carries the following Resource Conservation and Recovery Act (RCRA) codes: F001, F002, F006, and F007. Treatability samples were obtained from Tanks 7 and 13, dried, and then homogenized. Characterization data for the two samples, designated as DH-TS02 and DH-TS03, on an oxide basis are presented in Table I.

Table I. Normalized composition of DH-TS02
and DH-TS03 on an oxide basis

Oxide form	DH-TS02 composition (wt %)	DH-TS03 composition (wt %)
Ag_2O	0.0029	0.0044
Al_2O_3	17.2095	16.1895
As_2O_3	0.00000	0.0010
B_2O_3	0.0486	0.1481
BaO	0.1084	0.0538
BeO	0.0048	0.0011
CaO	71.90000	59.20000
CdO	0.0101	0.0081
Cr_2O_3	0.1284	0.1189
Fe_2O_3	2.1246	10.6486
K_2O	0.1534	0.2923
M_gO	3.1317	2.6808
MnO_2	0.0680	0.0532
Na_2O	3.3953	8.2294
NiO	0.2672	0.2056
PbO_2	0.0459	0.1300
SiO_2	0.1278	0.1470
TiO_2	0.0354	0.0397
TI_2O	0.00000	0.0141
UO_2	1.0665	1.5619
ZnO	0.0621	0.1126
ZrO_2	0.1328	0.1359
TOTAL	100.0000	100.0000

With the waste contribution defined, the glass formulation or recipe of interest was selected. It should be noted that for the WETF study, the waste was the sole source of alkaline earths (e.g., CaO). Additives were SiO_2 and alkali (Na_2O and Li_2O, which was partially substituted for Na_2O). The ingredients were weighed, combined, and mixed. The material was then placed in a 99.8% pure alpha-Al_2O_3 crucible with a loose-fitting lid. The crucible and contents were placed in a high-temperature furnace to achieve melting. The furnace was programmed to ramp to the desired melt temperature at 300°C/h and hold at the melt temperature for 4 h, after which, the fluid glass was poured into a stainless steel pan and allowed to cool to ambient temperature. The resulting solidified glass product was then subjected to various types of characterization and analysis.

<u>Operating Diagram</u>

The resulting WETF data, shown in Figure 1, are presented in the traditional ternary diagram form described previously (i.e., alkalis, alkaline earths and glass formers). Line A-A

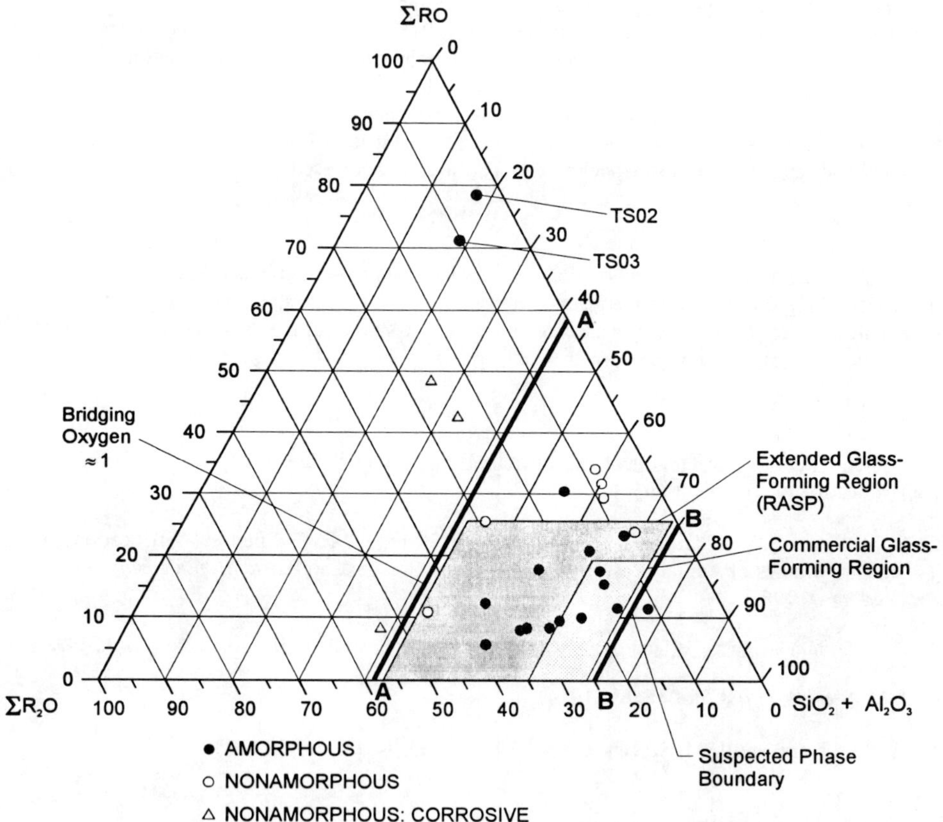

Figure 1. Operating diagram for SLS glass containing WETF sludge

represents calculated bridging oxygens of ≈ 1. Thus, line A-A is one of the boundaries of acceptable glass compositions and, in effect, represents the maximum allowable oxygen content in the melt for this application. Compositions to the right of line A-A are those with calculated F values greater than one. The second boundary for acceptable glass compositions, also shown in Figure 1, is line B-B. Line B-B is particularly noteworthy, in that it represents a suspected phase-separation boundary based upon literature data for the $CaO-Na_2O-SiO_2$ system[7]: that is, compositions beyond this boundary that are "deficient" in alkali and alkaline earth tend to form unary rather than ternary phases. Compositions to the right of line B-B would be expected to produce glass products characterized by phase separation and/or contained crystalline material. The region bounded by lines A-A and B-B represents the predicted region of acceptable compositions.

It should be noted that the use of the bridging oxygen calculations is not unique to the SLS system and should be equally applicable to other silicate systems such as borosilicate glass. However, the second boundary (line B-B) is unique to the SLS system and is not directly applicable to other silicate systems.

As shown in Figure 1, the WETF data compare favorably with the predicted region. Also presented on the operating diagram are acceptable composition regions for commercial SLS glass and the Reactive Additive Stabilization Process (RASP) for Hazardous and Mixed Waste Vitrification[8]. Both lie within the predicted region. Significantly, the predicted region is dependent on the waste composition. As such, it can be used to predict changes in the region of acceptable compositions as the waste or its composition changes. It should be noted, however, that the model is based solely on the bulk chemistry of the glass. As such, constraints like melt viscosity and regulatory performance requirements are not addressed.

CONCLUSIONS

A model using calculated bridging oxygens has been presented which enables the prediction of regions of acceptable glass compositions. The model can be used as a guide to determine glass formulations for evaluation in treatability studies or to assess the applicability of vitrification to a candidate waste stream.

REFERENCES

1. Strategic Plan for the Treatment of Appendix B Wastes, DOE/OR-1083, Rev. 0, U.S. Department of Energy, February 12, 1993.

2. C. M. Jantzen, W. D. Hudson, T. M. Gilliam, A. Bleier, and R. D. Spence, "Vitrification Treatability Studies of Actual Waste Water Treatment Sludges," in *Waste Management '95 Proceedings*, 1995.

3. J. A. Dean, ed., Lange's Handbook of Chemistry, 14th Ed., McGraw-Hill, Inc., New York, 1992.

4. J. M. Stevels, *Glass Ind.*, **35**, 69 (1954).

5. H. J. L. Trapp and J. M. Stevels, *Glastechn. Ber.*, **32**, **K**, VI., 32-52 (1959)

6. T. M. Gilliam, C. M. Jantzen, A. Bleier, S. P. Cooper, C. H. Mattus, C. K. McKeown, M. M. Roe, R. D. Spence, T. Szczygiel, and D. R. Trotter, <u>Vitrification of Wastewater Treatment Sludge. Volume II: Crucible-Scale Formulation Development Studies with Actual West End Treatment Facility Sludge</u>, ORNL/TM-13331, Oak Ridge National Laboratory, Oak Ridge, TN, 1996.

7. E. M. Levin, C. R. Robbins, and H. F. McMurdie, <u>Phase Diagrams for Ceramists,</u> The American Ceramic Society, 1964.

8. C. M. Jantzen, J. B. Pickett, and W. G. Ramsey, "Reactive Additive Stabilization Process for Hazardous and Mixed Waste Vitrification," in *Proceedings of the Second International Symposium on Mixed Waste,* Baltimore, MD, edited by A. A. Moghissi, R. K. Blauvelt, G. A. Benda, and N. E. Rotermich, American Society of Mechanical Engineers, 1993, pp. 4.2.1-4.2.13.

IMMOBILIZATION AND RECOVERY OF THORIUM, A NEPTUNIUM SURROGATE, USING PHASE–SEPARATED GLASSES

T.F. MEAKER, D. KARRAKER, M. TOSTEN, J.M. PAREIZS, W.G. RAMSEY
Savannah River Laboratory, Savannah River Site, Aiken, SC, 29808

ABSTRACT

The Savannah River Site has the majority of the United States' supply of neptunium currently stored in an acid solution in one of their canyon facilities. A program is being developed that could be utilized to ship this material, as glass, to Oak Ridge National Laboratory where the Np could be leached from the glass, purified by ion exchange and made into target material for the production of Pu-238. Ion exchange purification dictates no material be in the leachate making the isolation of the Np difficult. We have developed a process using thorium as a surrogate for Np that could immobilize the Np into a soda borosilicate glass for shipment. To achieve recovery of the Np, the glass can be phase separated prior to leaching with nitric acid. Phase separation would produces a Np-rich sodium-borate phase and a Si-rich phase similar to a Vycor® glass. The nitric acid selectively attacks the sodium-borate phase allowing high Np recovery in a solution that contains only sodium and boron. These can be easily separated from Np by ion exchange. Essentially all of the silicon which would interfere with ion exchange by precipitation is retained in the Vycor®-type phase. This technology may also be applied to other actinides stored in relatively pure solutions.

This paper will report the optimization of variables for maximizing Th (a Np surrogate) recovery while minimizing Si release. Th solubility in glass, heat treatment conditions and leaching parameters will be discussed. Transmission Electron Microscopy (TEM) with energy dispersive spectroscopy (EDS) data will be included to show phase separation after heat treatment.

INTRODUCTION

The Savannah River Site (SRS) presently has neptunium stored in H-canyon. Oak Ridge National Laboratory (ORNL) has a possible need in the future for the neptunium. Neptunium is used as target material in the production of Pu-238. Pu-238 may be needed as a energy source (heat) for future NASA missions. A glass composition designed for the easy recovery of neptunium may be necessary to transport it to ORNL. ORNL requires easy recovery of the Np-237 in order to process efficiently. Tedious chemical separations (i.e. from lanthanides) would adversely impact production schedules. Furthermore, RCRA materials (elements characterized as being hazardous) should not be part of the glass composition to avoid a mixed waste after separation processes. Also, high silica concentrations in the feed could lead to precipitation causing a pressure build-up during separation techniques resulting in dangerous operating conditions.

To meet these objectives, a glass composition resembling a vycor® composition used in industry is being evaluated as a possible solution to the transportation/separation problems. This glass composition would contain only SiO_2, Na_2O, B_2O_3 and NpO_2. In industry, the glass is heat treated causing a glass-in-glass phase separation between the silica and soda-borate phases. Leaching the glass in concentrated nitric acid removes the soda-borate phase leaving a high silica glass intact.[1] It has been shown that transition metals in the glass matrix will separate from the silica phase and remain in the soda-borate phase.[2] Using thorium as a surrogate for neptunium, this glass composition is being evaluated for ease of recovery while not introducing silica into the leachate and subsequent separation processes.

Mat. Res. Soc. Symp. Proc. Vol. 465 © 1997 Materials Research Society

Thorium was chosen as the surrogate because of its similarity to Np with respect to presumed valence state in glass, crystal formation, density, chemistry and ionic radius.

EXPERIMENTAL

Glass Fabrication

Glass compositions were batched with reagent chemicals in 100 mL platinum crucibles. Table 1 shows the two glass compositions used in this study. The melt was ramped to temperature (1250° C) and held for approximately four hours. The crucible was periodically removed from the furnace and stirred with a platinum stir bar. The glass was cast into graphite molds and annealed at 450° C for approximately one hour.

Table 1. Composition of two vycor glasses with thorium loading.

Oxide	Mid10	TG15
SiO2	55.1	55.6
B2O3	26.9	22.7
Na2O	8.0	6.7
ThO2	10.0	15.0

Heat Treatments

The processed glass was fractured with a hammer. Small portions (~5 grams) of the glass were put in 30 mL platinum crucibles and placed in a preheated muffle furnace to induce phase separation. Temperatures ranged from 500° to 700° C and time was varied from 1 to 20 hours.

Phase Separation Characterization

A few small chips of the glass after different heat treatments were examined with transmission electron microscopy (TEM), X-ray diffraction (XRD) and scanning electron microscopy (SEM). TEM was able to show in great detail whether the heat treated samples had glass-in-glass phase separation and/or crystallization.

Dissolution

The 5 gram glass sample of a crystalline heat treated glass, phase separated amorphous glass and a non-heat treated glass were ground in a Tekmar grinder and sieved. Six different size fractions of each glass were collected ranging in size between >40 mesh [equivalent to ~1620 mm^2 surface area] to <200 mesh [equivalent to ~21400 mm^2 surface area]. Approximately 0.5 grams of each size fraction was contacted for one hour with 10 mL of concentrated nitric acid at 110° C in a sealed teflon vessel in a drying oven.

Another 5 gram sample of the phase separated amorphous glass was ground and sieved collecting only the >40 mesh and 40-60 mesh size fractions [SA = 1620 and 3200 mm^2, respectively]. These sizes were contacted with 10 mL of concentrated nitric acid at 110° C for times ranging from 15 minutes to 8 hours.

After the appropriate time, the samples were removed from the oven. A 5 mL aliquot was carefully removed from the vessel so as not to remove any of the remaining glass fragments. This was diluted with 15 mL DI water. The samples were then analyzed by inductively coupled plasma emission spectroscopy (ICP-AES) for Th, Si, Na and B.

RESULTS

<u>Phase Separation Characterization Results</u>

TEM/EDAX was able to identify thorium- and silicon-rich phases and determined if they were amorphous or crystalline in the heat treated glasses. Figures 1 and 4 show the glass samples containing 10 and 15 weight percent ThO_2 with a heat treatment of 650° C for 10 hours. These were used in the dissolution experiments that varied the surface area of the sample while time and dilution volume were kept constant. Notice the 10 weight percent sample had a phase separated amorphous structure with large silicon-rich droplets and a continuous (appears as the mottled region) thorium-rich phase. Figures 2 and 3 show the EDAX spectrum for the different regions of figure 1. Other heat treatments resulted in plate-like crystallization throughout the glass sample. In figure 4, amorphous phase separation similar to figure 1 is present as well as crystal formation. Figure 5 shows the EDAX spectrum of one of the crystal forms. Notice that it is also thorium rich (as is the mottled region). Possible explanations include precipitation of ThO2 or ThSiO4. However, this crystal may be crystabolite formation that has formed with some Th. This has been observed before with HLW glass at Hanford[3].

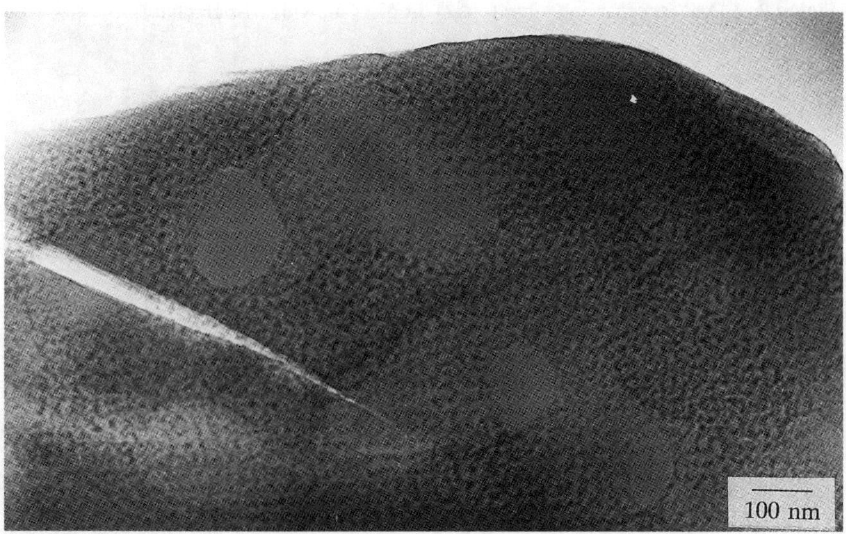

Figure 1. Phase separated amorphous glass showing silicon rich droplets and a mottled thorium rich phase of a 10 weight percent ThO_2 glass heat treated at 650° C for 10 hours.

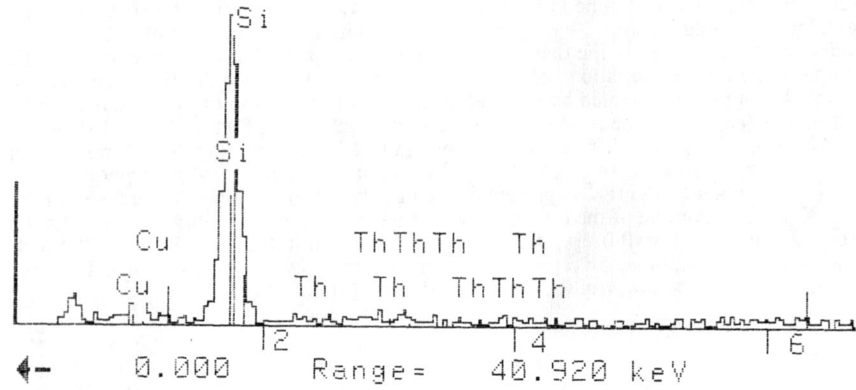

Figure 2. EDAX spectrum of large (clear) droplets showing Si-rich phase.

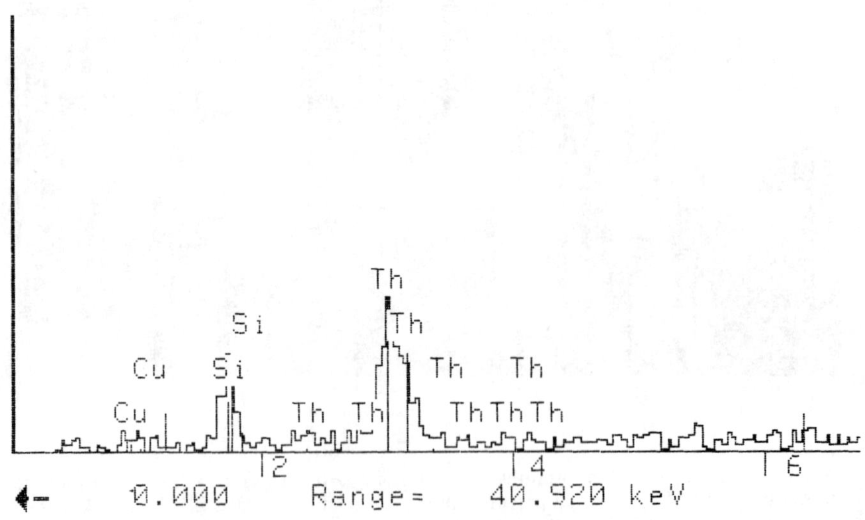

Figure 3. EDAX spectrum of continuous (mottled) region showing Th-rich phase.

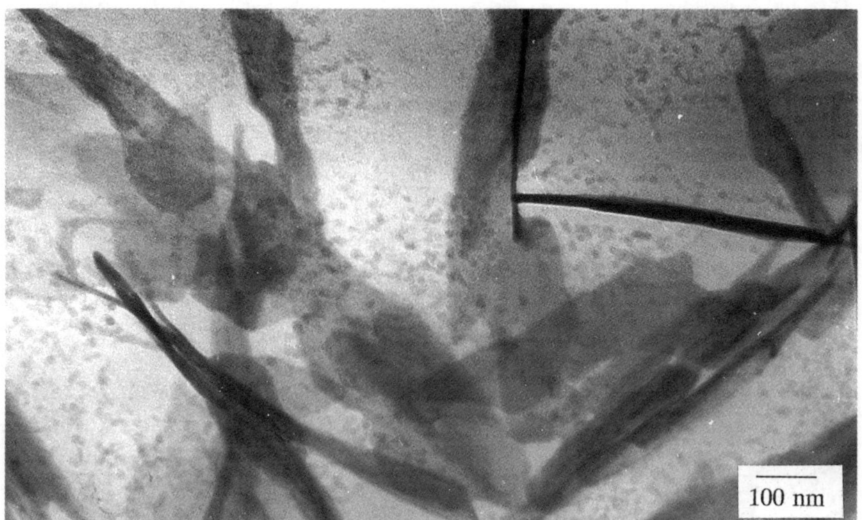

Figure 4. Phase separated glass with crystal formation showing silicon rich droplets, a mottled thorium rich phase and Th-rich crystals of a 15 weight percent ThO_2 glass heat treated at 650° C for 10 hours.

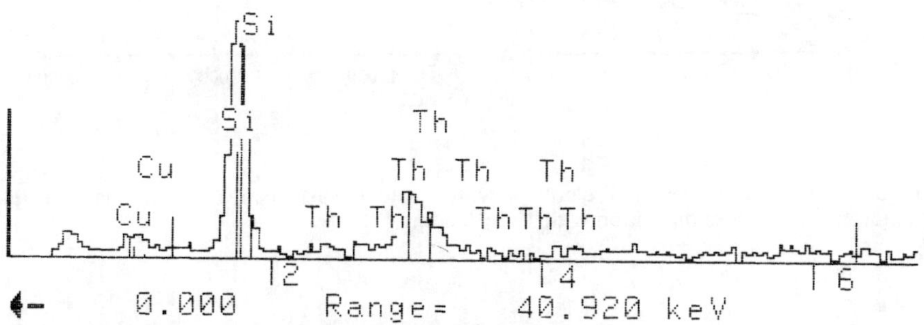

Figure 5. EDAX spectrum of the plate-like crystals seen in figure 4.

Figure 6 shows the recovery rates of all elements of a non-heat treated, amorphous glass. Note the linear relationship between surface area and percent recovery for all elements. Silicon release levels are in the part-per-million range. Figure 7 shows a comparison of thorium recovery levels of two glasses with 10 and 15 weight percent ThO_2 loadings with a heat treatment at 650° C for 10 hours (to induce phase separation) and two non-heat treated glasses with the same loadings. The heat treated glass with the lower loading was crystal free (as seen in the above TEM micrographs). It had a Th recovery rate with respect to surface area far exceeding that of the crystallized sample. The crystallized sample had a recovery rate similar to both non-heat treated glasses. Figure 8 shows the silicon release in parts-per-million of the same glasses with respect to surface area is linear. The heat treated glasses show low levels of silica in the leachate from the larger mesh size fractions. These values are similar to non-heat treated glasses. However, as the particles decrease in size, the silicon concentration in the leachate increases linearly. This shows that recovery strategies need to target large particles to maximize recovery of actinide and minimize unwanted silica.

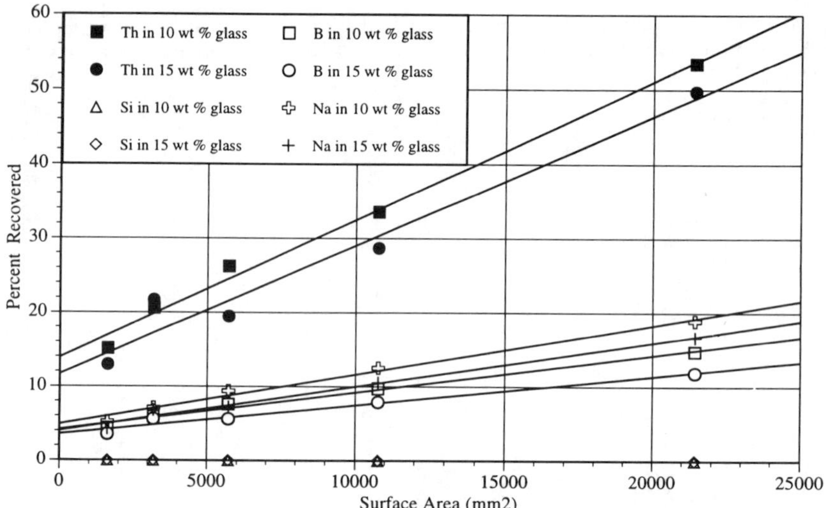

Figure 6. Recovery rates of all elements with respect to surface area in a non-heat treated (amorphous) vycor composition with ThO_2 loading.

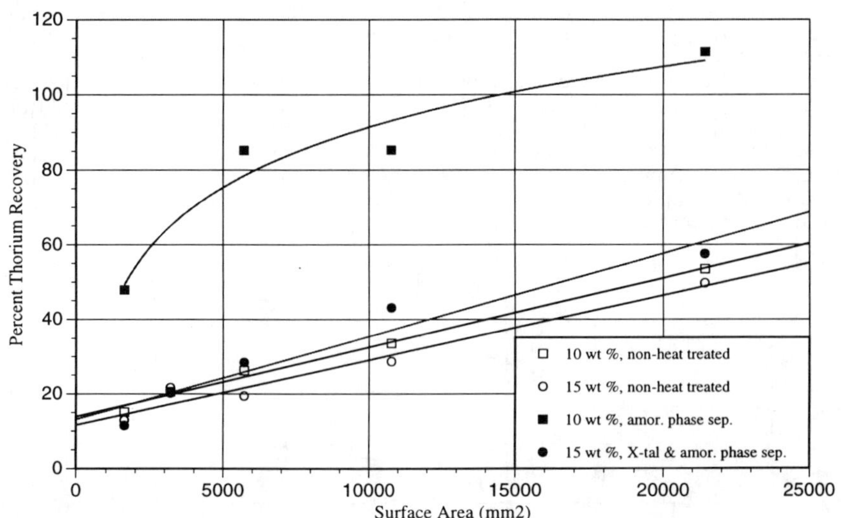

Figure 7. Th recovery rates with respect to surface area of heat treated and non-heat treated glasses with 10 and 15 weight percent ThO_2 loadings.

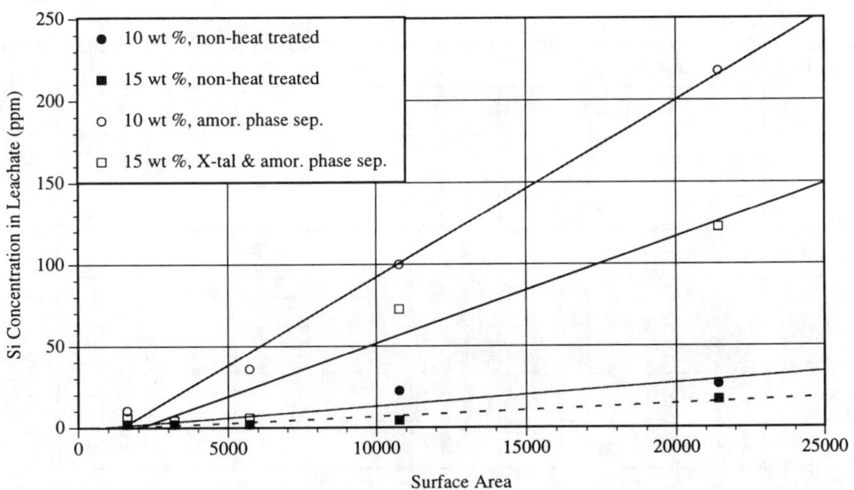

Figure 8. Silicon levels with respect to surface area of heat treated and non-heat treated glasses with 10 and 15 weight percent ThO_2 loadings.

Figure 9 shows Th recovery rates with respect to time of two heat treated glasses with the same temperature and different time and one non-heat treated glass. All glasses had the same ThO2 loading (10 weight percent oxide) and the same composition. The surface area to volume of solids ratio was kept constant. Contact time was varied between 0.25 and 8.0 hours. TEM detected very little crystal formation in either heat treatment, however, the heat treatment of 7 hours had fewer crystals. Over 90 percent of all the Th was recovered from the 40 Mesh size fraction of the 7 hour, 650° C heat treatment after four hours. This is considerably better than the 10 hour heat treatment (~70 %) and significantly better than the non-heat treated sample. The silicon concentration of the same dissolution is depicted in figure 10. All samples contained 2 to 4 ppm silicon in the leachate except the non-heat treated sample after 8 hours in HNO_3 (38 ppm).

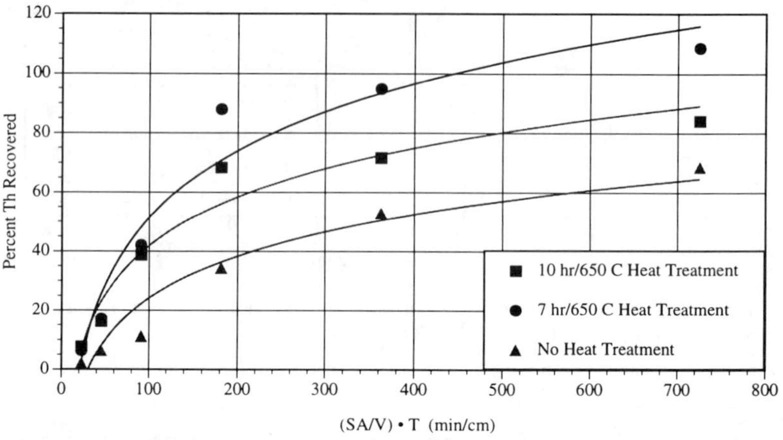

Figure 9. Th recovery rates with respect to time of two different heat treatments and one non-heat treated sample.

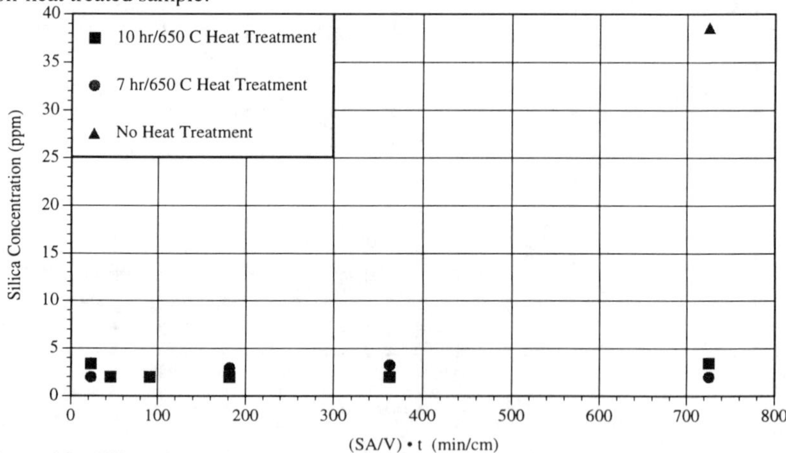

Figure 10. Silicon levels in leachate of 40 mesh size fraction contacted for 0.25 to 8 hours of two different heat treated and one non-heat treated sample.

CONCLUSIONS

The total amount of crystal formation in a heat treated vycor-type glass can be controlled with time, temperature and loading. Heat treatments at lower temperatures and for less time inhibit crystal formation while still allowing significant phase separation. If the Thorium loading exceeds 10 weight percent oxide, crystal formation during heat treatment may not be avoided. The total amount of crystal growth has a direct affect on thorium leachability. An increase in crystal formation limits the Th recovery significantly. High thorium loaded glasses (15 weight percent) with heat treatments (increased crystal formation) leach at approximately the same rate as non-heat treated glasses.

A phase separated (amorphous) glass has been produced using thorium as a surrogate for neptunium. Two different homogeneous vycor compositions targeting 10 and 15 weight percent thorium oxide have been processed, heat treated and leached with concentrated nitric acid at 110° C. Thorium recovery rates have been shown to be considerably better when the glass has been heat treated inducing phase separation that is relatively crystal free. Non-heat treated and crystalline (due to heat treatment) glasses have similar Th recovery rates with respect to surface area. Phase separated amorphous samples were found to have significantly higher thorium concentrations in the leachate compared to non-heat treated and crystalline glasses for all mesh sizes. All glasses had increased thorium concentration in the leachate as surface area increased.

Thorium leach rates of greater than 90 percent have been achieved by contacting a large mesh size, heat treated glass with nitric acid at 110° C for 4 hours. The heat treatment was for 7 hours at 650° C. The glass contained very few crystals after the heat treatment. Thorium leach rates for a slightly more crystallized sample was ~20 percent less. Non-heat treated glasses show ~40 percent less recovery compared to the sample with very few crystals.

Non-heat treated glasses exhibited very low silica releases. Typical levels are at or below 5 ppm of silica dissolved in solution for large mesh sizes and greater than 15 ppm for small mesh sizes. Large glass fragments of phase separated (amorphous) samples had a silica release also below 5 ppm. However, glasses containing amorphous phase separation and crystalline phases did show a ten fold (10X) increase in total silica dissolved as the surface area increased.

ACKNOWLEDGMENTS

The authors would like to acknowlege the laboratory work performed by S. Keel, the TEM micrographs produced by C. Foreman, and the analytical expertise of A. Jurgensen, D. Messimer, M. Summer, J. Durden, M. Cooley, and P. Filpus-Luyckx.

REFERENCES

1. Thomas H. Elmer, "Porous and Reconstructed Glasses," <u>Engineered Materials Handbook, Volume 4,</u> ASM International, Materials Park, OH, 44073, pp. 427-432.

2. M.B. Volf, <u>Chemical Approach to Glass,</u> Elsevier, New York, 1984.

3. D.K. Peeler and P.R. Hrma, "Compositional/Durability Relationship in Simulated Waste Glasses," American Chemical Society, I&EC Division, Emerging Technologies in Hazardous Waste Management VII (1995).

OPTIMIZATION OF LANTHANIDE BOROSILICATE FRIT COMPOSITIONS FOR THE IMMOBILIZATION OF ACTINIDES USING A PLACKETT-BURMAN/SIMPLEX ALGORITHM DESIGN

M.G. MESKO*, T.F. MEAKER**, W.G. RAMSEY**, J.C. MARRA**, D.K. PEELER**
*University of Missouri-Rolla, Rolla, MO 65401 mesko@umr.edu
**Westinghouse Savannah River Co., Aiken SC 29808

ABSTRACT

Immobilization by vitrification is one potential disposition option for a portion of the United States' excess plutonium inventory. Research has been performed at the Savannah River Site (SRS) to determine the optimum composition of a lanthanide borosilicate frit for the vitrification of plutonium using a Plackett-Burman design and simplex algorithm as a statistical tool. This technique uses various response variables to rank and optimize a composition. The variables used in this study correspond to homogeneity, durability, actinide solubility and devitrification after heat-treatment.

The optimized frit composition was determined using a constant ThO_2 loading of 20 wt%. No noticeable trends were followed with respect to the individual components which may indicate a relatively robust system able to accommodate variations in the feed.

Batches containing various loadings of ThO_2 were melted to determine if actinide solubility was improved in the optimized composition compared to that of a similar lanthanide borosilicate glass. No noticeable improvement in ThO_2 solubility was realized as a result of using this optimization technique.

INTRODUCTION

The end of the Cold War, and subsequent reduction of the nuclear weapons arsenal, created a surplus of plutonium in the United States. Many options have been proposed to handle the excess plutonium, including long term storage in vaults and use as a mixed oxide fuel. These options may appear attractive in the short term, however, they are costly and may not be desirable in the future. One attractive option is to immobilize at least a portion of the excess plutonium in a chemically durable glass.

Past studies have been performed on the immobilization of plutonium in a lanthanide borosilicate (LaBS) glass [1,2]. LaBS glass can accommodate up to 10 wt% Pu along with rare earth elements, such as Gd, which may be used as neutron absorbers. The durability of the LaBS glass is approximately 50 times better than the high level waste glasses currently produced at the Defense Waste Processing Facility. The current study differs from past studies in that the frit developed does not contain hazardous barium or lead which many of the other frits contain [2].

From preliminary work focusing on the glass forming limits of a borosilicate glass containing various lanthanide and actinide contents, one composition, SRS-LaBS (Table I), was selected as a base composition for the current study. The preliminary study used less formal optimization experiments to determine the limits of glass formation. The purpose of preforming the Plackett-Burman/simplex algorithm technique was to improve upon the SRS-LaBS composition.

Th was used as a surrogate for Pu in the current study, although a similar study using Pu is being planned. A constant waste loading of 20 wt% ThO_2, which was acheived in the SRS-LaBS glass, was used when determining the optimum base glass composition. After the optimum glass composition was determined, various actinide loadings were used to determine if ThO_2 solubility was improved compared to that of the SRS-LaBS glass.

EXPERIMENT

Glass Formation

Twelve frit compositions were determined using a Plackett-Burman matrix. The high and low values were based on the SRS-LaBS composition and are given in Table I.

Mat. Res. Soc. Symp. Proc. Vol. 465 © 1997 Materials Research Society

Table I: Base composition (wt%) and high and low values used in the Plackett-Burman matrix.

Oxide	SRS-LaBS Composition	Low (-)	High (+)
SiO_2	26.4	*	*
B_2O_3	10.6	6.4	14.8
Al_2O_3	19.5	14.5	23.4
SrO	2.3	0.5	3.9
ZrO_2	1.2	0.0	2.6
Ln_2O_3	20	15	30
ThO_2	20	20	20

*SiO_2 was varied as the other frit components varied to equal 100%

The high and low values were entered into a Plackett-Burman matrix [3] and normalized to 100%. The composition of the initial twelve glasses is given in Table II.

Table II: Batch compositions (wt%) determined by Plackett-Burman matrix.

Composition	SiO_2	B_2O_3	Al_2O_3	SrO	ZrO_2	Ln_2O_3	ThO_2
1	23.1	6.4	18.0	0.5	2.0	30.0	20.0
2	26.8	8.3	23.4	3.9	2.6	15.0	20.0
3	26.1	6.4	14.5	3.0	0	30.0	20.0
4	22.9	14.8	23.4	3.9	0	15.0	20.0
5	19.2	11.3	14.5	3.0	2.0	30.0	20.0
6	26.2	14.8	23.4	0.6	0	15.0	20.0
7	18.2	11.3	18.0	0.5	2.0	30.0	20.0
8	22.6	6.4	18.0	3.0	0	30.0	20.0
9	34.6	8.3	18.9	0.6	2.6	15.0	20.0
10	37.2	8.3	18.9	0.6	0	15.0	20.0
11	23.6	11.4	14.5	0.5	0	30.0	20.0
12	24.9	14.8	18.8	3.9	2.6	15.0	20.0

Reagent grade materials were weighed, added to a platinum crucible, and combined thoroughly prior to melting. The compositions were melted at 1475°C for four hours and stirred twice during melting with a platinum rod to aid in homogenization. The melts were quenched in the crucible by placing the crucible containing the melt in a shallow pan of water. This was done so any insoluble ThO_2 could be detected as a layer on the bottom of the melt. The melting procedure remained constant such that the only variable was the composition of the base glass.

The optimization technique employed requires that the four lowest ranked compositions (ranking described below) be replaced with four new compositions in a process refered to as a simplex algorithm. The new compositions are determined by subtracting each of the four lowest ranked compositions from twice the average of the eight remaining compositions. The simplex algorithm continues until twelve compositions of approximately equal ranking are obtained or there is no longer an improvement in the glass. The current study required a total of 2 simplex algorithms (8 additional compositions) to take place before there was no longer an improvement in the glass. The compositions are listed in Table III.

Table III: Batch compositions (wt%) determined using the simplex algorithm.

Comp. #	SiO_2	B_2O_3	Al_2O_3	SrO	ZrO_2	Ln_2O_3	ThO_2
13	19.6	5.2	13.8	4.8	2.8	33.8	20.0
14	22.4	8.6	22.7	4.9	2.8	18.8	20.0
15	11.2	11.7	18.3	4.8	0.2	33.8	20.0
16	8.6	11.7	18.3	4.8	2.8	33.8	20.0
17	30.4	5.4	24.9	4.7	2.4	12.2	20.0
18	34.4	11.6	25.6	0.4	0.0	8.4	20.0
19	42.8	5.1	21.1	0.4	2.2	8.4	20.0
20	45.4	5.1	21.1	0.4	0.0	8.4	20.0

<u>Ranking</u>

Four response variables were used to evaluate and rank the quenched material. These variables correspond to homogeneity of the material, actinide solubility in the glass, devitrification tendency, and chemical durability.

The first variable, homogeneity, was determined by visual inspection and X-Ray Diffraction (XRD). Glasses were scored a '1' if homogeneous, '0.75' if a small amount of surface crystallization was present, and '0.5' if there was visible (opalescent) amorphous phase separation. A ThO_2 layer on the bottom of the glass was not counted against the homogeneity of the glass as ThO_2 solubility was taken as a separate variable. Compositions found to be crystalline throughout were eliminated from the remainder of the study.

The actinide solubility in the glass was determined using Inductively Coupled Plasma-Mass Spectroscopy (ICP-MS). If present, the undissolved (ThO_2) layer was separated from the homogeneous glass prior to chemical analysis. Glasses were ranked by dividing the ThO_2 content in the glassy phase by the highest ThO_2 content observed.

Devitrification tendency was determined by heat-treatment of the glass following a profile expected to occur during the can-in-can demonstration [4]. The homogeneity of the heat-treated material was determined by visual inspection and XRD. A score of '1' was given if the glass remained homogeneous, '0.75' if slightly phase separated, '0.5' if the top layer was phase separated and '0' if phase separation occurred throughout the glass.

The final variable, chemical durability of the heat-treated material, was determined using the Product Consistency Test (PCT) [5]. The Approved Reference Material (ARM-1) [6] was used to compare the relative durability of the compositions. Scores were computed on a linear scale between '0' for glasses as durable as ARM-1 and '1' for glasses 10 times or more durable than ARM-1.

RESULTS

Seven of the twenty compositions melted were found to contain crystalline material throughout and were therefore dropped from the study. Tables IV to VII give the results and scores for the remaining 13 compositions. The majority of the compositions (10 out of 13) in this study formed a homogeneous or nearly homogeneous glass (small insoluble ThO_2 layer). These compositions are listed in **bold type** in Table IV.

Table IV: Homogeneity of quenched melt and corresponding score.

Composition	Result	Score	Composition	Result	Score
1	**glassy**	**1.0**	**11**	**glassy**	**1.0**
2	**glassy**	**1.0**	12	cryst.	dropped
3	surf. cryst.	0.75	**13**	**glassy**	**1.0**
4	**glassy**	**1.0**	**14**	**glassy**	**1.0**
5	cryst.	dropped	**15**	**glassy**	**1.0**
6	phase sep.	0.5	**16**	**glassy**	**1.0**
7	cryst.	dropped	17	cryst.	dropped
8	**glassy**	**1.0**	18	cryst.	dropped
9	**glassy**	**1.0**	19	cryst.	dropped
10	surf. cryst.	0.75	20	cryst.	dropped

surf. cryst. = surface crystallized
cryst. = crystallized throughout
phase sep. = phase separated
glassy = homogeneous or containing small ThO_2 layer on bottom

All but three of the 13 compositions contained over 19 wt% ThO_2 in the glassy phase based on ICP-MS (Table V). A small amount of insoluble ThO_2 could be dissolved into the glass if melted for longer times or at higher temperatures.

Table V: ThO_2 content in the glass determined by ICP-MS and corresponding score.

Composition	ThO_2 content	Score
1	19.90	0.92
2	20.15	0.94
3	21.52	1.00
4	19.84	0.92
6	19.89	0.92
8	19.87	0.92
9	20.41	0.95
10	19.73	0.92
11	19.39	0.90
13	15.66	0.73
14	17.84	0.83
15	19.95	0.93
16	18.93	0.88

After heat-treatment, three of the 13 glasses exhibited phase separation throughout. Seven of the 13 compositions were found to be homogeneous. Of the 10 compositions found initially to be homogeneous (given a score of '1'), six remained homogeneous after heat-treatment.

Table VI: Homogeneity score before and after heat treatment.

Composition	Before Heat-Treatment	After Heat-Treatment
1	1.0	1.0
2	1.0	1.0
3	0.75	1.0
4	1.0	0.50
6	0.50	0.0
8	1.0	1.0
9	1.0	0.50
10	0.75	0.0
11	1.0	1.0
13	1.0	0.0
14	1.0	1.0
15	1.0	1.0
16	1.0	0.75

All of the glasses had a chemical durability better than the ARM-1 glass. In fact, all but three of the glasses were given a rating of '1' which meant their chemical durability was at least 10 times better than the ARM-1 glass (Table VII).

Table VII: Relative durability of the glasses compared with the ARM-1 glass and corresponding score.

Composition	Normalized Si Release	Normalized B Release	Score
1	0.012	0.020	1.0
2	0.012	0.018	1.0
3	0.136	0.035	0.80
4	0.020	0.020	1.0
6	0.016	0.013	1.0
8	0.011	0.018	1.0
9	0.009	0.015	1.0
10	0.027	0.019	1.0
11	0.003	0.086	0.97
13	0.004	0.056	1.0
14	0.018	0.047	1.0
15	0.011	0.024	1.0
16	0.083	0.043	0.90
ARM-1	0.294	0.560	----

Table VIII shows the scores for the compositions with respect to the four response variables along with a total score. The glasses with the 8 highest total scores are in **bold type**.

Table VIII: Individual scores of the 13 glasses with respect to the four response variables and total score.

Composition	Homogeneity (10%)	Solubility (30%)	Homogeneity after Heat-Treatment (25%)	Durability (35%)	Total
1	**1.0**	**0.92**	**1.0**	**1.0**	**0.98**
2	**1.0**	**0.94**	**1.0**	**1.0**	**0.98**
3	**0.75**	**1.00**	**1.0**	**0.80**	**0.91**
4	1.0	0.92	0.50	1.0	0.85
6	0.50	0.92	0.0	1.0	0.68
8	**1.0**	**0.92**	**1.0**	**1.0**	**0.98**
9	**1.0**	**0.95**	**0.50**	**1.0**	**0.86**
10	0.75	0.92	0.0	1.0	0.70
11	**1.0**	**0.90**	**1.0**	**0.97**	**0.96**
13	1.0	0.73	0.0	1.0	0.67
14	**1.0**	**0.83**	**1.0**	**1.0**	**0.95**
15	**1.0**	**0.93**	**1.0**	**1.0**	**0.98**
16	1.0	0.88	0.75	0.90	0.87

The eight highest ranked compositions were averaged to determine an "optimized" composition. Table IX gives the optimized frit composition based on the 20 compositions melted.

Table IX: Optimized frit composition based on the eight highest ranked compositions.

Oxide	wt%
SiO_2	17.4
B_2O_3	9.9
Al_2O_3	17.8
SrO	3.6
ZrO_2	0.8
Ln_2O_3	30.0
ThO_2	20.0

To determine if the optimized frit composition had an improved ThO_2 solubility, compared to the SRS-LaBS glass, four compositions containing various ThO_2 and Ln_2O_3

contents were melted. The frit components (SiO_2, B_2O_3, Al_2O_3, SrO, and ZrO_2) were normalized to 100% and batched with 25 and 30 wt% Ln_2O_3 and 20 and 25 wt% ThO_2. (Table X).

Table X: Batch composition of glasses containing varying ThO_2 and Ln_2O_3 contents.

comp.	SiO_2	B_2O_3	Al_2O_3	SrO	ZrO_2	Ln_2O_3	ThO_2
a	19.1	10.9	19.6	4.0	0.9	25	20
b	17.4	9.9	17.8	3.6	0.8	30	20
c	17.4	9.9	17.8	3.6	0.8	25	25
d	15.7	8.9	16.0	3.2	0.7	30	25

Compositions a and b, containing 20 wt% ThO_2, produced glasses that were completely amorphous. A small insoluble layer of ThO_2 was present in compositions c and d containing 25 wt% ThO_2. The SRS-LaBS composition could form a homogeneous glass with 25 wt% ThO_2 and 25 wt% Ln_2O_3. The data indicates that with regard to ThO_2 solubility, the LaBS frit composition may not have been optimized.

CONCLUSIONS

This study used a Plackett-Burman/simplex algorithm technique to determine an optimum LaBS frit composition for the immobilization of plutonium. Twenty compositions were melted and ranked on the basis of four response variables. These variables refer to homogeneity, actinide solubility, devitrification tendency, and chemical durability. There were no obvious trends in composition as a function of glass forming ability which may indicate a relatively robust system able to accommodate variations in feed.

The optimized frit composition was used to determine if actinide solubility was improved compared to other LaBS glasses. Homogeneous glasses were melted containing 20 wt% ThO_2, but at 25 wt% ThO_2, an insoluble layer was present. This shows that actinide solubility was not improved over the baseline composition using this optimization technique.

REFERENCES

[1] J. K. Bates, A. J. G. Ellison, J. W. Emery, and J. C. Hoh, Glass as a Waste Form for the Immobilization of Plutonium, in Scientific Basis for Nuclear Waste Management XIX, edited by W. M. Murphy and D. A. Knecht, (Mater. Res. Soc. Proc., Pittsburgh, PA, 1995), p.57.

[2] N. E. Bibler, W. G. Ramsey, T. F. Meaker, and J. M. Pareizs, Durabilities and Microstructures of Radioactive Glasses for Immobilization of Excess Actinides at the Savannah River Site, in Scientific Basis for Nuclear Waste Management XIX, edited by W. M. Murphy and D. A. Knecht, (Mater. Res. Soc. Proc., Pittsburgh, PA, 1995), p. 65.

[3] "Strategy of Experimentation", E. I. duPont de Nemours & Co., Wilmington, DE, 1988.

[4] Immobilization Program Peer Review, Lawrence Livermore National Laboratory, April 24-25, 1996.

[5] ASTM C 1285-94

[6] Savannah River Technology Center, Aiken, SC

AMERICIUM/CURIUM EXTRACTION FROM
A LANTHANIDE BOROSILICATE GLASS

T. S. RUDISILL, J. M. PAREIZS, and W. G. RAMSEY
Westinghouse Savannah River Company

ABSTRACT

A solution containing kilogram quantities of highly radioactive isotopes of americium and curium (Am/Cm) and lanthanide fission products is currently stored in a process tank at the Department of Energy's Savannah River Site (SRS). This tank and its vital support systems are old, subject to deterioration, and prone to possible leakage. For this reason, a program has been initiated to stabilize this material as a lanthanide borosilicate (LBS) glass.[1] The Am/Cm has commercial value and is desired for use by the heavy isotope programs at the Oak Ridge National Laboratory (ORNL).

A recovery flowsheet was demonstrated using a curium-containing glass to extract the Am/Cm from the glass matrix. The procedure involved grinding the glass to less than 200 mesh and dissolving in concentrated nitric acid at 110°C. Under these conditions, the dissolution was essentially 100% after 2 hours except for the insoluble silicon. Using a nonradioactive surrogate, the expected glass dissolution rate during Am/Cm recovery was bracketed by using both static and agitated conditions. The measured rates, 0.0082 and 0.040 g/hr·cm^2, were used to develop a predictive model for the time required to dissolve a spherical glass particle in terms of the glass density, particle size, and measured rate. The calculated dissolution time was in agreement with the experimental observation that the curium glass dissolution was complete in less than 2 hr.

INTRODUCTION

Approximately 15,000 l of solution containing isotopes of Am/Cm are currently stored in the F-Canyon facility at the SRS. These isotopes were recovered during plutonium-242 production campaigns in the mid and late 1970's. The continued storage of this solution was identified as an item of urgent concern in the Defense Nuclear Facility Safety Board's Recommendation 94-1. Currently there are no existing SRS facilities which can be used to stabilize this material for safe long-term storage or transport to the heavy isotope programs at ORNL. An analysis of several alternatives has resulted in the recommendation to stabilize the Am/Cm in a LBS glass. The Multi-Purpose Processing Facility in F-Canyon will be used for the vitrification process. Pretreatment operations will be performed in canyon vessels to separate the actinides and lanthanides from other metal impurities before subsequent vitrification.

Once the Am/Cm solution is stabilized, the glass can be shipped to the Isotope Production and Distribution Program at ORNL for californium-252 production and for use by the transplutonium research community. These potential uses will require the dissolution and recovery of the Am/Cm isotopes. For this reason, feasibility experiments were conducted to demonstrate the dissolution of the glass and solubilization of the Am/Cm. The dissolution flowsheet was based on the existing ORNL procedure for the recovery of Am/Cm from transplutonium production targets. The targets are typically dissolved in nitric acid using a stainless steel dissolver in the ORNL hot cells. The dissolver can be sparged with air to provide a limited amount of agitation.

Mat. Res. Soc. Symp. Proc. Vol. 465 © 1997 Materials Research Society

EXPERIMENT

The feasibility of recovering Am/Cm from a LBS glass was demonstrated using a curium-containing glass prepared to simulate the glass composition proposed for stabilizing the solution stored at the SRS. The dissolution rate of a nonradioactive surrogate was also measured using static and agitated conditions to bracket the dissolution rate which would be expected in a dissolver at ORNL. The design for each experiment is described below.

Curium Extraction Test

The radioactive glass used in the curium extraction test was prepared to simulate the target (baseline) composition proposed for use in stabilizing the Am/Cm solution stored in F-Canyon. A comparison of the simulated and target compositions is shown in Table I.

Table I Composition of Curium and Target Am/Cm Glasses

Oxide	Curium Glass (wt%)	Target Composition (wt%)
SiO_2	27.50	28.70
B_2O_3	5.02	5.97
Al_2O_3	4.10	3.92
BaO	5.41	5.85
PbO	10.50	15.82
La_2O_3	8.85	9.08
CeO_2	5.62	5.34
Pr_2O_3	6.71	5.08
Nd_2O_3	13.30	10.77
Sm_2O_3	2.90	2.52
Eu_2O_3	0.56	0.50
Gd_2O_3	1.38	1.25
Cm_2O_3	0.65	1.08
Other Actinides[†]	--	4.12

† Other Actinides in the glass include AmO_2, PuO_2, and UO_3.

The curium extraction test was performed by placing 0.25 g of -200 mesh ($\leq$74 μm) glass into each of six Teflon™ vessels containing 10 ml of concentrated (15.7 M) nitric acid. The vessels were sealed, weighed, and placed in an oven at 110°C. After 2 hr, 2 vessels were removed for analysis, followed by 2 more at 4 and 8 hr. After cooling each vessel was opened and the contents transferred to a 250 ml flask for dilution with 0.01 M nitric acid. Samples were then analyzed by induction-coupled plasma emission (ICP-ES) and gamma spectroscopy.

Glass Dissolution Rate

The dissolution rate of a 38 wt% LBS glass was determined by measuring the mass and surface area during the dissolution as functions of time. The composition of the glass is shown in Table II.

Table II Composition of 38 wt% LBS Glass

Oxide	Composition (wt%)
La_2O_3	15
CeO_2	7
Nd_2O_3	16
SiO_2	29
Al_2O_3	17
PbO	9
BaO	4
B_2O_3	3

Approximately 30 g of glass were initially melted and cast into a 2 cm diameter graphite crucible. The bottom and top of the glass were cut and polished to produce a right circular cylinder. The glass sample was placed in 8 M nitric acid at nominally 110°C. Periodically, the sample was removed from the acid, rinsed in distilled water, and dried. The sample mass, radius, and height were then measured as a function of the dissolution time. This procedure was initially performed using an air sparge for agitation and with removal of the silica layer from the surface of the glass prior to the measurements. The rate measured under these conditions corresponded to the most rapid dissolution which would be expected in the ORNL dissolver. This procedure was then repeated using the same glass sample with no agitation in the dissolver or removal of silica from the surface of the glass. These conditions corresponded to the slowest expected rate thereby bracketing the actual dissolution rate in the ORNL dissolver.

RESULTS

<u>Curium Extraction Test</u>

Results from the curium extraction test are presented in Table III as the percent of each element recovered from the glass as a function of heating time. The percentages were calculated from the concentrations of the respective elements in the final extraction solutions and the concentrations of the elements in the original glass. The analyzed composition of the glass was used for this calculation except for praseodymium which could not be analyzed due to spectral interferences. The quantity of praseodymium added to the glass batch was used for this calculation. The values shown in Table III are the average of two samples at each extraction time. The results for the duplicate samples were in excellent agreement. Relative standard deviations were nominally ±5%. From these results, it can be seen that all elements excluding silicon can be extracted from the glass in 2 hr or less if the glass is ground to less than 200 mesh. The slightly greater than 100% recoveries are likely due to a low bias in the analyzed composition of the glass. Elemental analysis for the curium glass was performed following a sodium peroxide fusion and nitric acid dissolution. This technique only accounted for approximately 93% of the glass in the original sample (see Table I) which would inflate the recoveries given in Table III. The silicon in the glass which remained essentially insoluble, can be removed from the solution by filtration.

Table III Elemental Recovery from a Curium-Containing Glass

Element	Recovery - 2 hr (%)	Recovery - 4 hr (%)	Recovery - 8 hr (%)
Boron	96.3	97.1	98.9
Aluminum	98.6	101.2	105.7
Barium	99.2	104.1	107.6
Lead	100.9	103.6	106.4
Lanthanum	107.2	107.4	108.6
Cerium	102.7	101.0	102.5
Neodymium	103.2	103.3	104.5
Samarium	103.9	103.5	104.6
Europium	106.2	106.1	107.2
Gadolinium	104.6	103.2	105.0
Curium	102.8	93.7	106.5

<u>Glass Dissolution Rate</u>

Since the actual geometry and conditions used at ORNL for transplutonium target dissolution could not be exactly duplicated in the laboratory, rate measurement experiments were designed to bracket the expected rate. Rapid and slow dissolution rates of the 38 wt% LBS glass were determined by plotting the ratio of the glass mass to surface area as a function of time and fitting least squares lines to the data (see Figures 1 and 2). The measured rates (-slope of the lines), at 95% confidence levels, were 0.040 ± 0.002 and 0.0082 ± 0.0002 g/hr·cm^2.

The more aggressive conditions achieved during rapid dissolution were established by sparging with air and removing the silica layer which formed on the surface of the glass. The exposed surface area enhanced the rate by removing any barrier silica presented for diffusion of acid to the surface of the glass. Energy dispersive x-ray analysis confirmed the residue was pure silica and that all other components of the surrogate glass dissolved in 8 M nitric acid. During the static dissolution test, care was taken not to remove the silica layer; however, silica did flake from the surface and built-up on the bottom of the dissolver. As the silica cracked and flaked from the surface, there was continual renewal of the layer as the dissolution proceeded. Although a silica layer formed on the surface of the glass and was either removed or flaked from the glass cylinder, right circular geometry was maintained throughout both experiments.

Once the glass sample was completely dissolved, the solution and residual solids were separated by filtration. The volume of the filtrate was measured and samples analyzed by ICP-ES for elemental concentrations. The residual solids were washed with 0.5 M nitric acid to redissolve precipitated lead and barium nitrates. The solution was filtered and samples submitted for elemental analysis by ICP-ES. Based on the solution analyses, the recovery of each lanthanide was 100% with an overall elemental recovery (excluding silicon) of approximately 95%. The silicon concentration in the dissolver solution was approximately 20 mg/l.

<u>Glass Dissolution Model</u>

A predictive equation for the time required to dissolve a glass particle of spherical geometry was derived from the slope ($-K_R$) of the line (equation 1) through the data on Figures 1 and 2.

Figure 1 Rapid Glass Dissolution

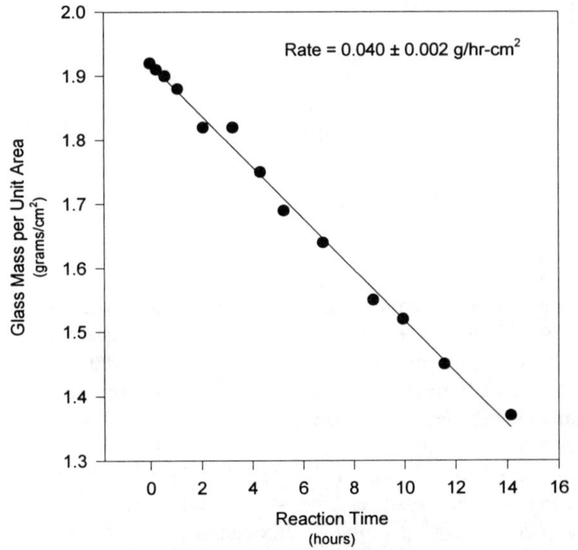

Figure 2 Slow Glass Dissolution

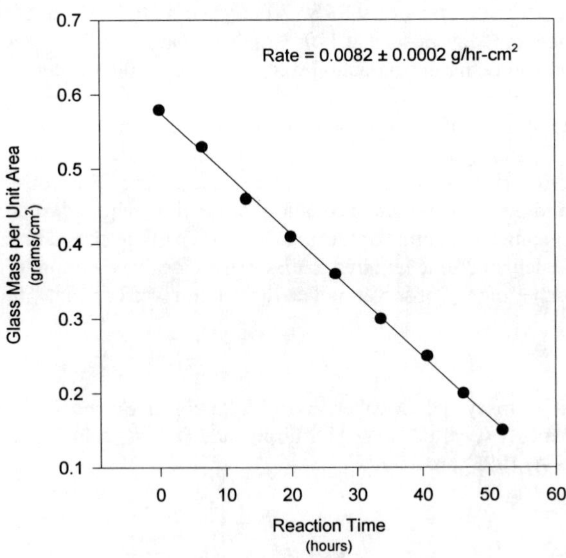

Substituting for the mass (M) and surface area (S_A) in terms of the particle diameter (D_0) and density (ρ_g), followed by integration, yields an equation for the time (t) required for complete dissolution of the glass particle (equation 2).

$$\frac{d\left(\dfrac{M}{S_A}\right)}{dt} = -K_R \tag{1}$$

$$t = \frac{\rho_g D_0}{6K_R} \tag{2}$$

Assuming the curium-containing glass particles were spherical with an equivalent diameter of 74 μm and a density of 4 g/cm^3, the time required for complete dissolution can be calculated from equation (2) as 0.6 hr. This value is consistent with the observation that complete dissolution was achieved in 2 hr or less. By assuming spherical geometry and a maximum particle size, equation (2) will overestimate the time required to dissolve the curium glass. The range of particle sizes with an equivalent diameters less than 74 μm increases the surface area and the rate of dissolution when compared to the idealized case. The higher molarity of nitric acid used to dissolve the glass also enhanced the rate and reduced the dissolution time when compared to the time predicted by equation (2), although this effect has not been quantified.

CONCLUSIONS

The feasibility of recovering Am/Cm from a LBS glass was demonstrated using a curium-containing glass prepared to simulate the composition proposed for stabilizing the Am/Cm solution stored in the F-Canyon facility at SRS. After grinding to less than 200 mesh, the glass was dissolved in concentrated nitric acid at 110 °C in less than 2 hr. Elemental recoveries for each glass component following the extraction were essentially 100% except for silicon which remained mostly insoluble.

The expected rate of dissolution, when the Am/Cm is recovered, was bracketed by measuring the dissolution rate of a surrogate glass under both static and agitated conditions. In each case, the ratio of the mass to surface area was linear with respect to time. The constant rate of dissolution was used to derive a predictive equation for the time required to dissolve a glass particle of spherical geometry in terms of the glass density, particle size, and experimentally measured rate. The calculated time required to dissolve an idealized curium glass particle was in agreement with the experimental observation that dissolution was complete in less than 2 hr.

REFERENCES

1. N.E. Bibler, W.G. Ramsey, T.F. Meaker, and J.M. Pareizs in Scientific Basis for Nuclear Waste Management XIX, edited by W.M. Murphy and D.A. Knecht (Mater. Res. Soc. Proc. **412**, Pittsburgh, PA, 1995) pp. 65-72.

PROCESSING AND CHARACTERIZATION OF GLASS COMPOSITES FOR ASH IMMOBILIZATION

I.A. SOBOLEV, M.I. OJOVAN, O.K. KARLINA, S.V. STEFANOVSKY, V.E. GALTSEV
Moscow Scientific and Industrial Association "Radon", The 7-th Rostovsky Lane, 2/14, Moscow, 119121, Russia, Tel. (7 095)248 1680, fax (7 095)248 1941, E-mail: Oj@nporadon.msk.ru

ABSTRACT

Induction melting by using a cold crucible is a suitable technology for the immobilisation of ash residues after incineration of solid radioactive waste. We investigated the possibility of using glass composites produced by stirring the ash into meltedglass. Glass composites containing 15 - 40 wt. % of ash were obtained in both laboratory and bench scale units. Infrared spectroscopy, electronic paramagnetic resonance, X-ray diffractometry, TEM and SEM analyses were applied in order to characterise the structure of the glass composites obtained. The glass composites consisted of a relatively homogeneous glass matrix with embedded polycrystalline aggregates. The fraction of aggregates increases when the fraction of ash rises. The isothermal curing of composites at 1100 °C leads to dissolution of the ash components into the melt as well as to their inclusion into the glass structure, according to the analysis of the spectra obtained.

INTRODUCTION

Glass composite materials are glass matrices with dispersed small particles of another phase which do not dissolve into the glass melt. The insoluble phase can be, for example, ash. Glass composites were proposed to resolve the problem of phase separation during the vitrification of waste containing large amounts of glass phobic materials such as sulphates and chlorides. As it is composed of waste constituents, the insoluble phase is radioactive and therefore it is usually defined as a radiophase. However, the glass (host) matrix also can contain some portion of radionuclides distributed homogeneously through the matrix [1].

The durability of glass composites can be of the same magnitude and even higher than that of homogeneous glass. Glass composites can be produced by different methods. An induction melter supplied with an agitator was used as an appropriate apparatus in this work. Glass composites will be used at the SIA "Radon" vitrification plant for the vitrification of low and intermediate level waste that contains chemical components incompatible with glass. In this plant, sulphates will be vitrified by forming glass composites at concentrations up to 30 vol.%, using borosilicate glass [2].

Induction melting in a cold crucible is also being investigated at SIA "Radon" for the vitrification of ash residues produced by the incineration of solid radioactive waste[3]. Glass composites can be used also in this case. In the current work the possibility of using glass composites produced by stirring the ash into a glass melt was investigated. Glass composites containing 15 - 40 wt.% of the ash were obtained in both laboratory and bench scale units. Different analytical methods were applied in order to investigate the structure of the glass composites obtained.

EXPERIMENTS

The matrix glass used for preparation of the glass composites (Table I) was developed at SIA "Radon" [1,2]. This glass is chemically durable and has suitable properties in the molten state (viscosity, electric resistivity, surface tension) for production in the cold crucible. Glass composites were produced by using a laboratory unit as well as a bench scale experimental unit based on the cold crucible (see Fig. 1). Both units were equipped with a stainless steel mixer (stirrer). Crushed matrix glass was fed into the crucible and heated to melting at $1100\,^{\circ}C$; then

Mat. Res. Soc. Symp. Proc. Vol. 465 © 1997 Materials Research Society

the simulated ash residue was added to the melt while stirring. The stirrer rotational velocity was $10\ \mathrm{s}^{-1}$.

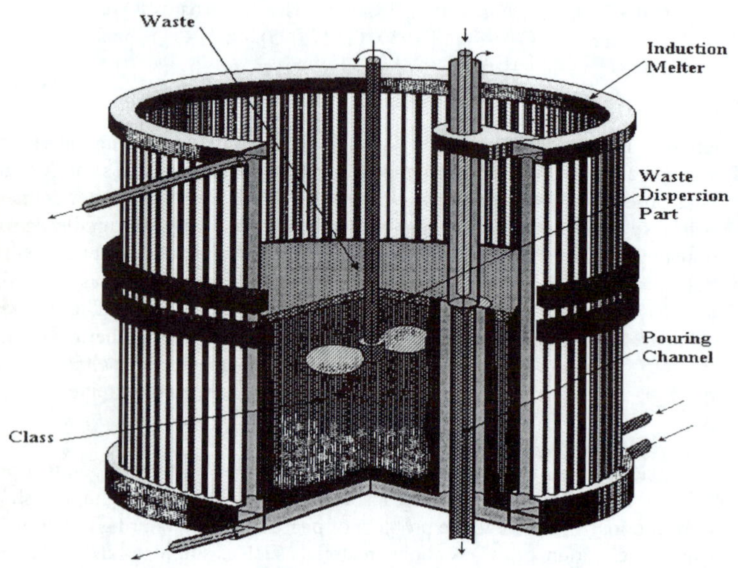

Fig. 1. Cold crucible for the production of glass composite materials

Table I. Chemical composition of matrix glass and ash residue (wt. %)

Material	Na_2O	K_2O	CaO	MgO	Al_2O_3	FeO	Fe_2O_3	B_2O_3	SiO_2	P_2O_5
Glass	20	-	17	-	3	-	3	8	49	-
Ash residue	6	9	14*	5	15	10	-	-	30	11*

*These components were in the form of $Ca_3(PO_4)_2$

The time of stirring was 5 minutes. The ash residue content in the samples and the melt curing time at $1100\ ^{o}C$ after stirring (mixing) were varied as shown in Table II.

Table II. Characteristics of glass composites

Spectrum number	Ash residue content, wt. %	Curing time after mixing at $1100\ ^{o}C$, min
1	15	-
2	25	-
3	40	-
4	40	10
5	40	15

The glass composites produced were investigated by infra red (IR) spectroscopy (IKS-29 spectrophotometer, using powders of the composites pressed with KBr into pellets), electron paramagnetic resonance (Bruker ESP-300 radiospectrometer, using doses of 50 kGr at 77K and 295 K for for non-irradiated and irradiated samples), X-ray diffraction (DRON-3M diffractometer, using the CuK_α radiation), transmission and scanning (TEM and SEM) electron microscopy (UEMV-100 and REMMA-202 instruments, respectively). Chemical durability was investigated by using the basic IAEA leaching technique.

RESULTS AND DISCUSSION

The results of TEM and SEM show that the glass composites consisted of a relatively homogeneous glass matrix with embedded polycrystalline aggregates up to hundreds of microns in size. The fraction of aggregates increases when the ash fraction increases. Simultaneously, this fraction decreases when the time that the composites are held at a constant high temperature increases. X-ray phase analysis shows that the glass composites contain quartz as the main phase and a small fraction of β-vitlockite. If the melt is subjected to longer isothermal curing, the intensity of the diffraction peaks decreases along with an increase in the diffusion halo at small angles of X-ray scattering (Fig.2).

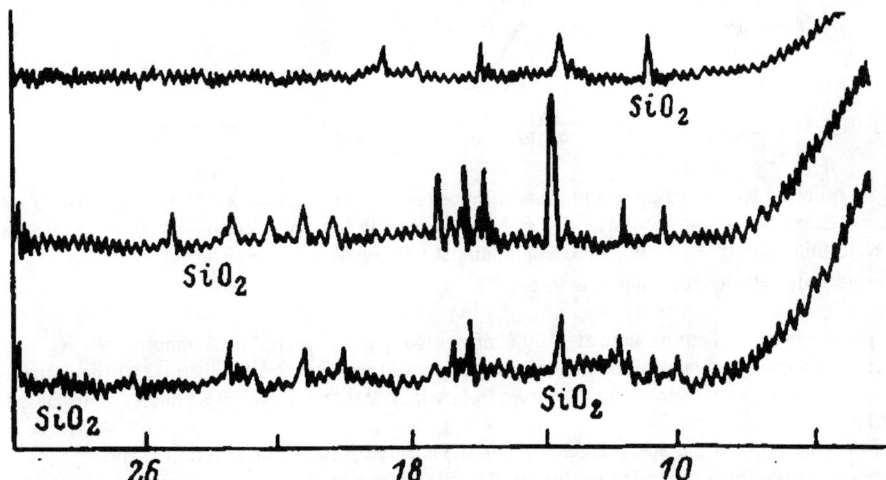

Fig. 2. XRD patterns for glass composites without isothermal exposure [1 (top) and 2(middle)] and after 15 minutes exposure at 1100 °C after stirring (3, bottom). The ash residue content (wt. %) for 1 was 15; for 2 and 3, 40. The vertical axis is I/I_0, in relative units. The horizontal axis is degrees.

In the glass composites which have no additional curing at 1100°C, the crystalline phase fraction increases when the ash residue concentration increases. This can be explained by the dissolution of the crystalline phases of the ash residue in the glassmelt during high temperature curing. The IR spectrum of the matrix glass consists of bands in the 1300 - 1500 cm^{-1} region, with maxima at 1400, 1200, and 850 - 1150 cm^{-1} having a slightly expressed doublet character; and in the 680-750 cm^{-1} region with maxima at 720 and 400-550 cm^{-1}.

With ash residue content increasing up to 40% in the glass composites, the following changes in the infrared spectra can be seen: a band at 1580-1590 cm^{-1} appears and its intensity grows, while the band at 1200-1210 cm^{-1} gradually vanishes, and the maximum of the 850-1150 cm^{-1} band drifts to the high-frequency zone (from 1020 to 1045 cm^{-1}). Also the bands at 790-800, 775, 590-595, 570 and 555 cm^{-1} appear and becomes more intense (Fig. 2, curves 1-3).

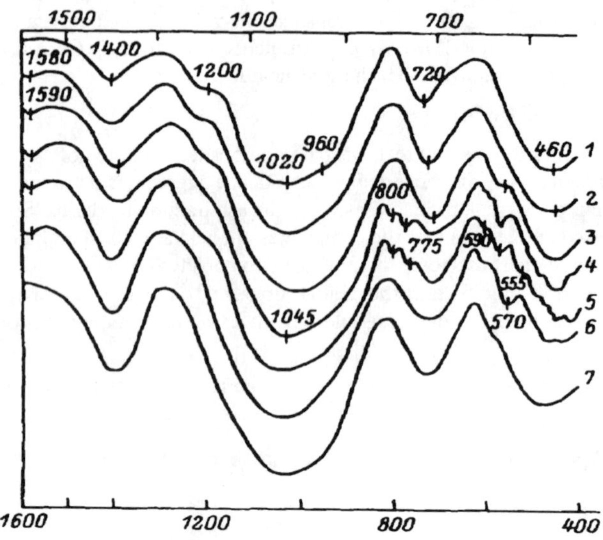

Fig. 3. Infrared spectra of glass and glass composites. 1 - matrix glass, 2 - 15 wt. % of ash, 3 - 25 wt. % of ash, 4 - 40 wt. % of ash, 5 - 40 wt. % of ash after 10 min. curing, 6 - 40 wt. % of ash after 15 min. curing, 7 - 40 wt. % of ash in almost homogeneous glass. The vertical axis is I/I$_0$, in relative units. The horizontal axis is ν, cm^{-1}

When the isothermal curing time at 1100 C after mixing increases up to 15 minutes, all the additional bands gradually vanish, and the infrared spectrum acquires the features of the initial spectrum of the glass matrix (Fig. 3, curves 3-5) with just a few differences which are discussed below.

The interpretation of the observed changes in the infrared spectra as a function of the composition and heat treatment history of the specimens is as follows. The base structure of the initial matrix glass is a boron - silica - oxygen network, consisting of SiO_4 and BO_4 tetrahedra (bands centered at 1150 and 850 cm^{-1} respectively [4]).

When the ash residue content is increased, a number of additional bands appear due to the vibrations of the Al-O$^-$, Fe-O$^-$ and P-O$^-$ bonds in the fragments of the ash structure that transfer to the glass phase. In particular, the melt enrichment with silica from the ash residue shifts the 850-1150 cm^{-1} band maximum to the higher wavenumber region. The disappearance of the 1200 cm$^-$ band can be connected to the change of boron co-ordination from three- to four-fold one. The appearance of numerous relatively narrow bands in the 500-800 cm^{-1} region is a result of the presence of some crystalline phases in the glass composites, which is confirmed by other methods. Among these bands, the one at 1580-1590 cm^{-1} is likely to relate to the P=O bond vibrations; those at 790-800, 570, and 555 cm^{-1}, to the valency and deformation vibrations of P-O$^-$ bonds; and the rest, to the Al-O$^-$ and Fe-O$^-$ bond vibrations. Isothermal curing of the melt leads to ash residue dissolution, making the IR spectrum of the glass composites very similar to that of the matrix glass. However, some differences still exist, for example, the presence of the intense band at 570 cm^{-1} and the shift of the 850-1150 cm^{-1} band maximum to higher wave numbers, resulting from the higher fraction of Si-O-Si bridging bonds and the accumulation of P-O groups in the glass.

The EPR spectra of matrix glass before and after γ-irradiation are shown in Fig.4 (curves 1 and 2). These spectra are characterised by the presence of two intense lines with g-tensor about 4.27 and 2.00, due to Fe^{3+} ions ($3d^5$) in the tetrahedral and octahedral positions, respectively, and a weak additional response on the background of the intense Fe^{3+} response in the vicinity of g_e, apparent only in the irradiated glass.

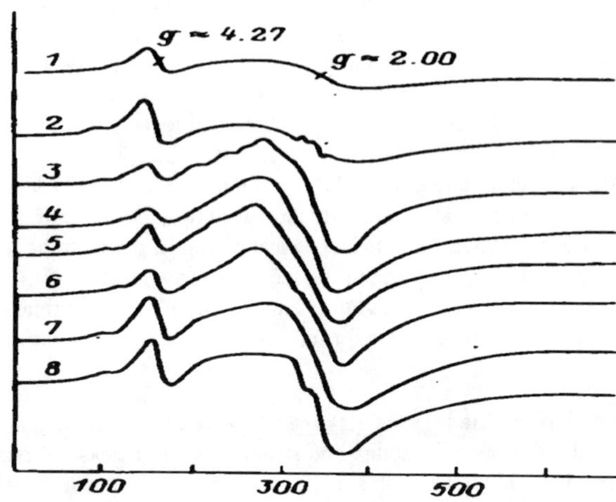

Fig. 4. Electronic paramagnetic spectra of glass composites: 1, 3, 5, and 7 - non irradiated samples; 2, 4, 6, and 8 - irradiated samples; 1 and 2- matrix glass; 3 and 4 - glass composites #1 (15 wt. % ash); 5 and 6 - glass composite #3 (40 wt. % ash); 7 and 8 - glass composite #5 (40 wt. % ash, 15 min. curing at 1100^{0}C after mixing). The vertical axis is dA/dH, in relative units. The horizontal axis is H, mTl.

Computer processing of the irradiated glass spectrum allows us to extract this additional response, which is due to the partly resolved hyperfine structure of ^{11}B (I=3/2) with g=2.011, conventionally assigned to paramagnetic centres (PMC) with four-fold coordinated boron (a hole trapped on a non-bridging oxygen ion of BO_4 tetrahedron [4]). A similar signal can be extracted in the spectrum of the glass composites containing 15 wt. % of ash residue, with the help of the $3r^d$ order line spectrum approximation. However, in this case the line form is more complicated, thus denoting a superposition of another signal from a new PMC.

When the ash content in the glass composites is increased up to 40 wt. %, the spectrum of non irradiated glass reveals additional signals in the low field region (Fig. 4), which become indistinguishable from the spectrum of irradiated glass.

The EPR spectra of glass composites 4 and 5 (Table II) before and after irradiation become more and more similar to the spectrum of the initial matrix glass as the time of the isothermal melt curing increases (Fig. 4). After computer processing, the spectral fragments for the irradiated glass composites 4 and 5 near g_e are quite different from those of the initial matrix glass and glass composites 1 and 2. In particular, they are characterised by the presence of a response with g=2.006-2.008 against a background of the partially resolved hyperfine structure of the boron-oxygen PMC, and by the presence of a shoulder at g=2.032 - 2.035 which is usually seen in glasses with a high content of three-coordinated boron. This additional single line is likely to relate to an Si-O PMC (a hole trapped on a non-bridging oxygen ion of the SiO_4 tetrahedron). Thus, PMCs with four-fold coordinated boron are mainly formed in the irradiated matrix glass. When up to 40 wt. % of ash residue is added to this glass, at least a part of the SiO_2 from the ash incorporates into the structure of the glass with formation of silicon-oxygen PMCs. During the isothermal curing of the glass composite melt, the melt structure is additionally enriched with SiO_2 and a portion of the boron transforms to the three-co-ordinated state, partially due to the effect of P_2O_5 which also is transferred from the ash residue into the glass.

From the comparative analysis of the IR and EPR spectroscopy results (along with the TEM, SEM and XRD data) it is concluded that the dispersion of ash residue into the glass melt results in the partial dissolution of ash components, thus making a portion of the boron three-fold rather than four-fold coordinated and increasing the anionic constituents in the glass structure. This affects the leach rate of alkaline ions.

Investigation of water durability using the IAEA technique gave a Cs-137 leach rate within the range of $(2 - 4) \bullet 10^{-6}$ $g/cm^2 \bullet day$. The leach rate increased up to $1 \bullet 10^{-5}$ $g/cm^2 \bullet day$ when the ash content was increased up to 40 wt. %. The isothermal curing of the composites at 1100 ^oC gives rise to increasing water durability: the leach rate decreases down to $(6 - 8) \bullet 10^{-6}$ $g/cm^2 \bullet day$. Isothermal curing of the melt leads to a higher chemical durability of the final product (glass composites) due to the dissolution of glass forming oxides in the matrix glass.

CONCLUSIONS

Mechanical stirring of ash residue into the glassmelt followed by isothermal curing produces products with practically the same durability and structure as in the case of conventional vitrification of ash residues. However, ash stirring can increase the productivity of the melter [2]. Without isothermal curing one can obtain glass composites with a good durability which is of the same order of magnitude as that for homogeneous glass.

REFERENCES

1.Ojovan M.I. et al., Glass composite materials - a new form of radioactive waste immobilisation, XVI Intern. Congress on Glass, 1992, Madrid, v.4, pp.315-320.

2.Sobolev I.A. et al., Waste vitrification: Using induction melting and glass composite materials, in Proc. of GLOBAL' 95 Intern. Conf., Versaille, France, v.1, pp.734-740.

3.Sobolev I.A. et al., Experience of SIA "Radon in radioactive waste vitrification, in Proc. of TOPSEAL'95 Intern. Meeting, Stockholm, Sweden, v.2, pp.21-24.

4. Wong J., Angell C.A. Glass Structure by Spectroscopy, (Marcel Dekker Inc., New York, 1976).

THE USE OF LASER MICROPROBE-ICP-MS IN
THE EXAMINATION OF NUCLEAR WASTE GLASSES

S. SHUTTLEWORTH*, J. E. MONTEITH**
* Department of Earth and Planetary Sciences, Washington University, 1 Brookings
  Drive, Campus Box 1169, St Louis, MO 63130, USA
** Research and Technology, BNFL, Sellafield, Seascale, Cumbria, CA20 1PG, UK

ABSTRACT

Vitrification is the most developed method for the immobilisation of highly radioactive waste obtained from reprocessing irradiated nuclear fuel. Highly active wastes continue to be vitrified at Sellafield in the Waste Vitrification Plant (WVP) and the product glass has previously been subjected to rigorous characterisation in order to ensure waste form integrity and stability during interim storage. However, the vitrified waste may eventually be placed in an underground repository and so further studies may be needed to confirm the chemical stability of vitrified wastes in different disposal scenarios. One area of uncertainty is in knowing how the alteration products and surface layers on a corroded glass effect the dissolution process and the solution concentrations of long-lived radionuclides. This study shows how the technique of laser microprobe inductively coupled plasma mass spectrometry (LM-ICP-MS) can be used to provide a direct chemical analysis of vitrified samples, both in their as-cast state and after leach testing. LM-ICP-MS, solutions' nebulisation ICP-MS (to provide chemical analyses of the leachants), light microscopy, scanning electron microscopy and confocal laser microscopy were used to examine the effect of leaching on the surfaces of vitrified samples. Data were obtained for a number of simulated WVP-type product glasses that were subjected to soxhlet leach testing for varying lengths of time. The LM-ICP-MS data clearly show that the technique offers a rapid and relatively accurate method for detecting and monitoring minute changes in the surface layers of glass and in the build-up of alteration products. Hence, LM-ICP-MS could be used to obtain valuable mechanistic information on the degradation of glass by examining radionuclide behaviour in the surface layers of glass as it corrodes over extended time periods.

INTRODUCTION

The chemical analysis of glass matrices is of particular interest to the nuclear industry because glass is used to immobilise radioactive waste. However, a number of uncertainties remain in predicting the long-term performance of vitrified wastes, especially with respect to their interaction with water. The effects of leaching on glasses manifest themselves as alteration layers at the surface [1]. It is helpful to be able to chemically map these alteration layers in order to build a better understanding of secondary phase formation and the mobility of radionuclides as they are leached. One technique which presents the opportunity to obtain chemical and isotopic information about small features within a solid sample is laser ablation inductively coupled plasma mass spectrometry (LA-ICP-MS) [2, 3, 4]. This technique uses a laser to ablate the area of sample surface which is of interest. This material is then transported in the form of particulates in an argon gas stream to the inductively coupled plasma where the particles are ionised. The ionised material is extracted from the plasma through a sampling aperture and the ion beam is collimated and focused into a quadruple mass filter where the ions are separated on the basis of their mass to charge ratios. When the

123

Mat. Res. Soc. Symp. Proc. Vol. 465 © 1997 Materials Research Society

laser beam is focused tightly (laser microprobe-ICP-MS), areas as small as 10 microns in diameter have been successfully probed with reported instrumental detection limits in the ppb range based on a one-minute analysis [5, 6]. Laser ablation ICP-MS has been compared with other analysis techniques (proton microprobe, solution ICP-MS, INAA and XRF) and its results have been found to agree well with the results from other methods [7].

EXPERIMENT

Vitrified samples of simulated Magnox and PWR type wastes were prepared by melting with a mixed alkali borosilicate base glass at 1100°C for 4 hours. The compositions of the vitrified samples, as determined by X-ray fluorescence, are given in Table I. Waste loadings of 25 wt% were achieved. The vitrified samples were divided into flat-sided blocks. A reference 'as-cast' sample was kept for each waste type, whilst the remaining blocks were subjected to durability testing. The durability tests were performed using a dynamic method of leach testing based on the International Standard Soxhlet Leach Test Procedure (ISO/TC85/SC5/WC/N37). The two sample sets were tested at 97°C for 2, 4, 6 and 8 weeks. After leaching, the samples were dried to a constant weight in an oven at ~ 80°C.

The surface topography of the leached samples was examined using light microscopy, scanning electron microscopy (SEM) (using an Electroscan Environmental SEM) and confocal laser microscopy (CLM). The CLM used was a Lasertec Corporation Confocal Laser Scanning Microscope, 1LM21, and it was calibrated using a Vickers hardness indentation standard. The CLM was used to generate three-dimensional reconstructions of the sample surfaces by stacking two-dimensional optical sections collected in series.

The LM-ICP-MS was performed using a Fisons Elemental PQ2+ inductively coupled plasma mass spectrometer and a laser microprobe sampling system built in-house. The laser microprobe featured a Spectron 1J (free-running output) Nd: YAG laser, frequency quadrupled to produce output at 266 nm (ultra-violet) and running in single transverse mode. The laser was Q-switched to produce a pulse length of 9 ns, and the output energy was variable in the range of 1-10 mJ. Up to 20 pulses were used. The laser beam was targeted onto the surface of the sample using a modified microscope and sample viewing was facilitated using a camera/video arrangement. The laser beam was focused onto the sample which was housed in a glass ablation cell using a reflecting objective microscope lens. Material was ablated at the surface of the sample when the laser was fired and attenuation of the laser energy was found to produce laser craters of less than 10 microns diameter. Laser sampling occurs at the surface and this can be checked by subjecting the laser craters to analysis by CLM. Bulk analysis can be performed by defocusing the laser beam and utilising a spot diameter of several hundred microns. A stream of argon gas was passed through the sample cell so that ablated particulate material could be carried in this gas stream to the ICP of the mass spectrometer for ionisation and subsequent analysis. The following gas flow rates were used: cool 12.5, auxiliary 0.5 and carrier 1.1 l/min. Figure 1 shows a schematic of the laser beam delivery system.

In LM-ICP-MS it is difficult to produce quantitative data for every solid matrix analysed owing to the shortage and lack of variety in traceable standards. The problem is compounded by the complexity of the ablation process which results in the possibility of vastly differing laser-sample coupling characteristics for different matrices and for different laser parameters. One possible method of calibration is semi-quantitative data processing [1, 7] based on an instrument response curve generated from a standard NIST glass and the element response of

TABLE I. Compositions of the Magnox and PWR vitrified samples (as determined by XRF)

Elemental %		
	Magnox	**PWR**
SiO_2	46.2	47.1
Al_2O_3	5.46	0.10
Fe_2O_3	2.37	0.60
MgO	5.53	<0.1
Na_2O	7.85	7.97
P_2O_5	0.25	0.13
Cr_2O_3	0.41	0.16
ZrO_2	1.75	3.39
HfO_2	0.02	0.04
ZnO	-	0.02
BaO	0.56	1.15
NiO	0.24	0.10
SrO	0.33	0.66
Mn_3O_4	-	0.01
B_2O_3	16.7*	16.6*
Li_2O	3.81*	3.75*
RuO_2	1.00*	1.62*
MoO_3	1.78	3.19
Cs_2O	1.06	1.10
Nd_2O_3	1.63	3.13
Sm_2O_3	0.32	0.55
CeO_2	0.98	1.71
La_2O_3	0.43	0.77
Pr_6O_{11}	0.45	0.92
Y_2O_3	0.21	0.36
Gd_2O_3	-	3.45
Rb_2O	0.12	0.23
TeO_2	0.14	0.21
SO_3 (in the bead after fusion)	~0.15	~0.05

*** By wet methods**

an internal standard in the material to be analysed. The ICP-MS instrumentation was tuned and optimised using the laser ablated signal for lanthanum in NIST 610 glass. An instrument response curve (Figure 2) was generated based on the major and trace elements in the standard glass.

Information for the selection of an internal standard was obtained by subjecting the 'as-cast' samples to bulk analysis using XRF. This allowed an instrument response value to be generated for the internal standard. In both the Magnox and PWR vitrified samples, lanthanum was selected as the internal standard.

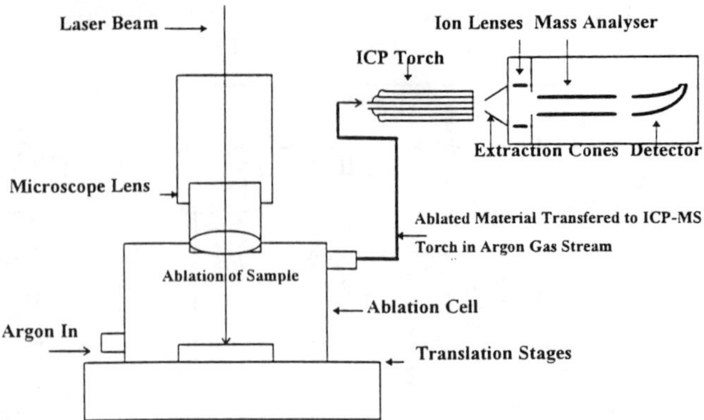

Figure 1. Schematic of the LM-ICP-MS laser delivery system.

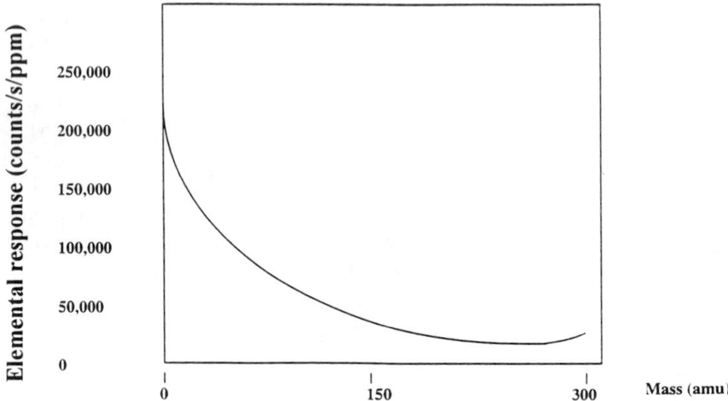

Figure 2. Instrument response curve for NIST 610 glass.

RESULTS

It has been well documented previously that the dissolution of vitrified wastes is dependent on many different parameters eg leachate composition and pH, flow rate and temperature etc and that leaching mechanisms can also change with increasing temperature and as a function of time [8, 9, 10]. Thus, as the glass matrix corrodes, a wide variety of secondary products can form, depending on the temperature and the leachate. The leach tests that were performed, in this study, on the Magnox and PWR waste type glasses were fairly aggressive and involved accelerated extraction under dynamic conditions; the results are given in Table II. This particular type of leach test is not representative of repository conditions but does allow comparisons to be made between different vitrified samples and it is designed to produce a high enough concentration of waste elements for accurate analysis within a reasonably short timescale. However, the results of this study do provide further confirmation of the complex nature of the alteration processes that occur on the surfaces of the vitrified samples as they are leached over time. The light microscopy photographs and

TABLE II. Elemental leach rates for the Magnox and PWR vitrified samples

	Magnox (25 wt%) leach rates ($g/cm^2/day$)				PWR (25 wt%) leach rates ($g/cm^2/day$)			
Weeks	2	4	6	8	2	4	6	8
Si	8.92×10^{-4}	8.59×10^{-4}	5.03×10^{-4}	4.34×10^{-4}	1.50×10^{-3}	6.46×10^{-4}	4.58×10^{-4}	5.38×10^{-4}
B	7.32×10^{-4}	7.68×10^{-4}	7.16×10^{-4}		1.56×10^{-3}	1.57×10^{-3}	1.72×10^{-3}	1.42×10^{-3}
Na	8.62×10^{-4}	1.09×10^{-3}	5.23×10^{-4}	4.93×10^{-4}	2.76×10^{-3}	1.37×10^{-3}	1.38×10^{-3}	1.21×10^{-3}
Li	5.68×10^{-4}	5.58×10^{-4}	3.76×10^{-4}	4.05×10^{-4}	1.04×10^{-3}	1.03×10^{-3}	9.52×10^{-4}	8.02×10^{-4}
Cs	3.46×10^{-4}	3.70×10^{-4}	2.86×10^{-4}	2.88×10^{-4}	8.00×10^{-4}	9.17×10^{-4}	9.78×10^{-4}	9.01×10^{-4}
Mo	5.16×10^{-4}	6.08×10^{-4}	5.16×10^{-4}	5.28×10^{-4}	8.98×10^{-4}	1.15×10^{-3}	1.12×10^{-3}	1.06×10^{-3}
Al	3.28×10^{-4}	3.66×10^{-4}	2.73×10^{-4}	3.19×10^{-4}	-	-	-	-
Fe			2.07×10^{-6}		2.43×10^{-5}		9.27×10^{-6}	
Cr	5.51×10^{-5}	6.22×10^{-5}	4.29×10^{-5}	7.21×10^{-5}	2.65×10^{-5}	1.80×10^{-5}	3.23×10^{-5}	3.84×10^{-5}
Ni	2.05×10^{-4}	2.47×10^{-4}	3.36×10^{-4}	2.02×10^{-4}	4.07×10^{-4}	1.63×10^{-4}	1.97×10^{-4}	4.12×10^{-5}
Gd	-	-	-	-	1.51×10^{-4}	8.76×10^{-5}	6.43×10^{-5}	4.13×10^{-5}
Ba	2.54×10^{-4}	1.80×10^{-4}	1.52×10^{-4}	1.89×10^{-4}	6.18×10^{-4}	1.45×10^{-4}	5.66×10^{-4}	5.91×10^{-4}
Mg	2.02×10^{-4}	2.62×10^{-4}	1.30×10^{-4}	2.28×10^{-4}	-	-	-	-
La	1.52×10^{-6}	1.92×10^{-6}	9.34×10^{-7}	8.93×10^{-7}	1.55×10^{-6}	2.80×10^{-6}	5.93×10^{-7}	5.49×10^{-7}
Ce	6.99×10^{-7}	8.83×10^{-7}	4.29×10^{-7}	4.78×10^{-7}	3.50×10^{-7}	2.34×10^{-6}	1.33×10^{-7}	8.24×10^{-8}
Nd	6.79×10^{-7}	8.83×10^{-7}	2.70×10^{-7}	5.46×10^{-7}	7.02×10^{-7}	2.98×10^{-6}	1.59×10^{-7}	8.95×10^{-8}

SEM micrographs clearly showed how the modified layers progressively changed with leaching time along the sides of the vitrified samples. The as-cast specimens had reasonably smooth surfaces with diffuse surface features but as the leaching process proceeded, banded rippled features grew on the surfaces of the glasses and often, after about four weeks leaching, craters would appear, as if a significant event had occurred in the leaching process, eg. that some replacement of the network modifiers by an ion exchange mechanism, had taken place. The surface topography reconstructions, produced by CLM, also showed how the modified surface layers changed with progressive corrosion. The height calibrations on the vertical axes for the vitrified sample reconstructions suggested that the depth of the alterations at the sample surface increase for the first four weeks of leaching, decrease at six weeks and then increase again, as if the corrosion mechanism involves surface swelling followed by spalling in the gel layer.

The LM-ICP-MS results clearly showed (Table III) that some elements preferentially appear at the surface of corroded glasses eg Li and B because they leach from the reaction zone relatively quickly as the glass matrix breaks down. It is also evident from the results that the less soluble elements such as Zr and Y tend to exhibit far more consistent behaviour and remain concentrated in the gel layer in similar proportions throughout the leaching process.

For the Magnox vitrified samples, there were few distinct trends in elemental surface concentrations with leaching, probably owing to their inherent chemical durability. However, a few elements did appear to behave in a similar way to each other, eg Fe and Mo, where the concentration dropped after two weeks and then steadily increased as leaching proceeded. Reasonably constant levels of Y, Zr, Nd, Ce and Sn were measured in the surfaces of the Magnox samples, and some elements were detected which were impurities in the chemicals used to make the waste simulants: Sn, Hf, Pb and Rh.

The results for the PWR vitrified samples were slightly different to those of the Magnox samples and more obvious patterns of behaviour were apparent. This is probably because PWR vitrified samples (with no additives) are inherently less durable than vitrified Magnox wastes. Similar behaviour was exhibited by Mg, Cr, Fe, Ni and Mn whereby the concentration dropped after two weeks leaching and then steadily increased as corrosion progressed ie similar to the behaviour of Fe and Mo in the Magnox wastes. In vitrified wastes, Fe, Ni and Cr usually localise as spinels and thus they would be expected to behave in a similar manner to each other during leaching processes. Cs, Mo and Rb also acted in a similar way to each other; their concentrations decreased stepwise for four weeks of leaching and then increased again. Yet again, the more insoluble components - Y, Pr, Ce, Zr and Sr - had more consistent concentrations (with a decrease at four weeks leaching) for all the PWR samples. The chemical impurities detected in the PWR samples were Zn, Pb, Hf, Sn, Cd, Mg and Al .

LM-ICP-MS data were also collected for different alteration layers across the surfaces of the leached glasses, however, unfortunately, space does not permit a full appraisal of these results.

CONCLUSIONS

SEM and CLM were used to identify the effects of leaching on the surfaces of vitrified wastes; whilst the LM-ICP-MS was used to measure elemental compositions in the alteration layers of the surfaces. The observations that were made supported the findings of many previous studies [eg 9, 11] ie that as a glass matrix corrodes, a wide variety of secondary

Table III. LM-ICP-MS results: mean elemental compositions for as-cast and leached Magnox and PWR vitrified samples

	MAGNOX Mean compositions in ppm's					PWR Average compositions in ppm's				
	As-cast	2 weeks	4 weeks	6 weeks	8 weeks	As-cast	2 weeks	4 weeks	6 weeks	8 weeks
Lithium	5510	814	8308	5725	6300	6671	9186	3339	5173	5066
Boron	35643	32270	21660	15080	21277	38317	41187	19163	36230	24680
Magnesium	18597	12512	13817	12820	11659	1348	171	687	1587	1990
Aluminium	17107	4675	7718	7536	7589	3251	2330	1639	4361	4868
Chromium	1259	968	1117	1952	1334	1881	443	966	2087	2242
Manganese						216	71	136	328	266
Iron	14563	6922	8944	27180	31740	35607	4296	17832	50750	54330
Nickel	1752	843	4209	2641	1700	8401	701	4230	10150	12280
Zinc	156	72	131	435	390	665	237	365	905	645
Rubidium	476	334	367	322	339	1204	846	645	1138	1001
Strontium	2402	1597	1822	1515	1429	5088	4343	2588	4297	3770
Yttrium	1323	1058	1277	1269	1354	2990	2352	1540	2792	2650
Zirconium	9444	6964	9617	10087	9490	25687	18753	12888	24620	23860
Molybdenum	6543	4467	6738	13279	15827	16723	10810	8409	13000	10380
Ruthenium	3899	3501	2928	5968	4183	9567	6708	4834	12390	7740
Cadmium						328	253	219	312	217
Tin	123	108	94	108	101	211	196	166	146	210
Caesium	3516	3011	2854	2156	2316	3842	3580	1988	3264	2798
Barium	4588	3130	3295	2762	2616	7506	8327	3822	6361	5601
Cerium	7638	6635	5979	5561	6122	11177	12180	5658	9660	10370
Praeseodymium	3727	3401	2822	2531	2622	5928	7119	3035	5188	5841
Neodymium	18487	15465	13960	13650	13887					
Samarium	3423	2870	2394	3557	2352	4384	5944	2265	3456	4474
Europium	387	335	274	258	260					
Gadolinium	234	303	201	252	222	28710	36693	14434	27160	27640
Hafnium	1035	818	595	640	730	1177	2022	678	1086	1127
Rhenium	124	16	2	14	15					
Lead	53	136	23	56	66	50	29	129	44	34

products can form and that elemental migration can take place during leaching. Hence, the results show that LM-ICP-MS can be successfully used to examine glass surfaces. The technique offers a rapid and relatively accurate method for detecting and monitoring minute changes in the surface composition of vitrified waste forms which have been subjected to corrosion testing. Hence, LM-ICP-MS looks to be particularly promising for the examination of radionuclide behaviour in vitrified waste forms because very small samples and a rapid analysis time can be used. LM-ICP-MS with its very low limits of detection could also be used for analysing glasses which have been subjected to leach tests which are more representative of repository conditions and hence be used to generate more realistic data for corrosion models.

ACKNOWLEDGEMENTS

S Shuttleworth would like to thank British Nuclear Fuels for sponsoring this post-doctural research project. P. Lythgoe (Department of Earth Sciences, University of Manchester) is thanked for his help in the design and building of the laser microprobe sampling system. C. Ramshaw (BNFL) is also thanked for producing the vitrified samples and running the soxhlet tests.

REFERENCES

1.   S. Shuttleworth and J. E. Monteith, JAAS. In preparation.

2.   A. L. Gray, Analyst, **110**, p. 551-556 (1985).

3.   P. Arrowsmith, Anal. Chem., **59**, p. 1437-1444 (1987).

4.   I. Abell, D. Gregson and S. Shuttleworth, Laser Ablation ICP-MS Analysis of Ceramic Materials, The Physics and Chemistry of Carbides, Nitrides and Borides. NATO ASI Series. Series E: Applied Sciences (1990), **185**, p. 121-130.

5.   B. J. Fryer, S. E. Jackson and H. P. Longerich, The Canadian Mineralogist, **33**, p. 303-312 (1995).

6.   H. P. Longerich, S. E. Jackson, B. J. Fryer and D. F. Strong, Geosci. Can., **20**, p. 21-27 (1993).

7.   M. D. Norman, N. J. Pearson, A. Sharma and W. L. Griffin. Geostandards Newsletter, **20 (2)**, p. 247-261 (1996).

8.   N. A. Chapman and I. G. McKinley, The Geological Disposal of Nuclear Waste, John Wiley & Sons, Chichester, 1987, p. 81-84.

9.   W. Lutze in Radioactive Waste Forms for the Future, edited by W. Lutze and R. C. Ewing, North-Holland Physics Publishing, The Netherlands, 1988, p. 93-115.

10.   L. Werme, I. K. Bjorner, G. Bart, H. U. Zwicky, B. Grambow, W. Lutze, R. C. Ewing and C. Magrabi, J. Mater. Res., **5(5)**, p. 1130-1146 (1990) and references cited therein.

11.   A. R. Hall, A. Hough and J. A. C. Marples in Scientific Basis for Nuclear Waste Management, (Mater. Res. Soc. Sypm. Proc.), 1982, **5**, p. 83-92 .

DIRECT CONVERSION OF HALOGEN-CONTAINING WASTES TO BOROSILICATE GLASS

C. W. FORSBERG, E. C. BEAHM, and J. C. RUDOLPH
Oak Ridge National Laboratory, Oak Ridge, TN 37831, forsbergcw@ornl.gov

ABSTRACT

Glass has become a preferred waste form worldwide for radioactive wastes; however, there are limitations. Halogen-containing wastes can not be converted to glass because halogens (chlorides, fluorides, etc.) form poor-quality waste glasses. Furthermore, halides in glass melters often form second phases that create operating problems. A new waste vitrification process, the Glass Material Oxidation and Dissolution System (GMODS), removes these limitations by converting halogen-containing wastes into borosilicate glass and a secondary, clean, sodium-halide stream.

INTRODUCTION

Major difficulties exist in the processing and disposal of halogen-containing radioactive wastes such as halogen-containing plastics, incinerator ash from the processing of halogen-containing wastes, plutonium residues [1], Molten Salt Reactor Experiment (MSRE) wastes, and electrometallurgical spent nuclear fuel process wastes. Halides, such as fluorides, chlorides, and iodides, generally make poor-quality waste forms. The analogy used in waste management is that good waste forms (silica, titanates, etc.) for radioactive materials can be found at any ocean beach. Materials that dissolve in water (halides etc.) make poor storage forms.

Glass has become a preferred waste form worldwide for immobilizing radioactive wastes, but like other waste forms, halides in glass result in poor-quality glasses. Furthermore, traditional glass-making operations have encountered major operational difficulties with halides in the melters. With current technology, halide-containing wastes must first be processed to remove the halogens before the wastes can be converted to glass. This process is an expensive and a complex step.

A new waste vitrification process, the Glass Material Oxidation and Dissolution System (GMODS), has been invented [2,3] to convert halogen-containing wastes to borosilicate glass and a secondary sodium-halide stream. The process also (1) converts metals, ceramics, and amorphous solids to glass and (2) oxidizes organics and then converts the residue to glass.

GMODS PROCESS DESCRIPTION

GMODS converts wastes into glass within a melter or melters. The process (Fig. 1) can operate as either a sequential batch or continuous process. The starting conditions are a glass melter filled with molten, lead-borate dissolution glass containing 2+ mol of lead oxide (PbO) per mole of boron oxide (B_2O_3) and initial temperatures of 800 to 1000 °C.

Addition of Feed Materials To The Molten Dissolution Glass (Fig. 1, Sect. 1)

The oxide and amorphous components in the feed dissolve into the molten glass. While metals and organics do not dissolve into conventional molten glasses, the GMODS dissolution glass has

Mat. Res. Soc. Symp. Proc. Vol. 465 © 1997 Materials Research Society

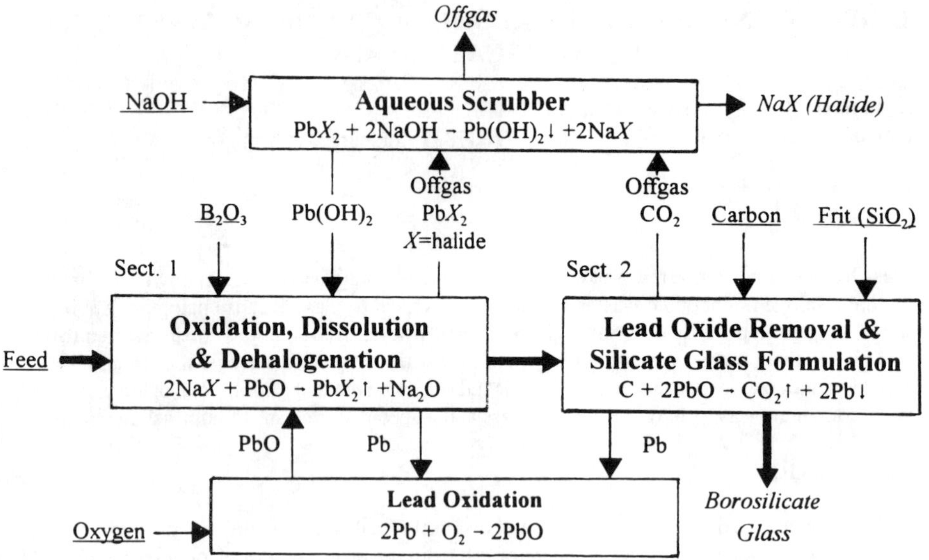

FIGURE 1. GMODS processing of halogen wastes to borosilicate glass.

special properties that permit the processing of these materials *in situ*. The inclusion of the sacrificial oxide—PbO—in the molten glass provides a method to oxidize *in situ* (a) metals [Me] to metal oxides [Me_yO_x] and (b) organics to carbon dioxide (CO_2) gas and steam. These metal oxides dissolve into the glass; carbon oxides (in gaseous form) and steam exit the melter. The B_2O_3 in the melt ensures the rapid dissolution into the glass of (a) protective oxide layers on metal wastes and (b) refractory oxide-like wastes such as those found in some incinerator ashes. The $2PbO:B_2O_3$ ratio in the dissolution glass is chosen to maximize chemical reaction rates and assure high solubility of oxides in the glass. The reaction product, molten lead, separates from the glass and sinks to the bottom of the melter to form a separate layer.

$$yMe + zPbO \rightarrow Me_yO_z + zPb\downarrow \tag{1}$$

$$C + 2PbO \rightarrow CO_2\uparrow + 2Pb\downarrow \tag{2}$$

In the dissolution glass, halogens in the waste react with the PbO to form lead halide compounds. For example, chlorides in the waste form lead chloride ($PbCl_2$), which is volatile at glass melter temperatures and exits to the aqueous sodium hydroxide (NaOH) scrubber. In the scrubber, the $PbCl_2$ reacts with the NaOH to yield insoluble lead hydroxide [$Pb(OH)_2$] and soluble sodium chloride (NaCl). The $Pb(OH)_2$ is recycled back to the melter, in which it decomposes to PbO and steam (H_2O), while the aqueous NaCl stream is cleaned and discharged as a chemical waste.

For halogen removal, there are two primary requirements: (1) low-viscosity glass such as to allow the rapid escape of gaseous lead dihalide compounds from the glass and (2) sufficiently volatile halide compounds. Measurements of glass viscosity in the dissolution glass show that this system has very low viscosities. Thermodynamic analysis and literature data indicate that the lead dihalides are volatile.

Formation of Borosilicate Glass (Fig. 1, Sect. 2)

After dissolution of the wastes, silicon oxide (SiO_2) and other additives are introduced to the glass to produce a high-quality glass. Excess PbO is removed from the glass by adding carbon, which converts the PbO to lead metal and carbon dioxide (CO_2). The lead metal sinks to the bottom of the melter as a separate phase as the CO_2 exits via the off-gas system. The final glass may have some or no PbO depending upon the desired product glass. The product glass is poured from the melter into the canister.

Recycle of Lead Metal

Lead metal from the different chemical reactions is oxidized to PbO with oxygen and recycled back to the beginning of the process. Lead does not leave the system. The oxidation reaction is

$$2Pb + O_2 \rightarrow 2PbO \tag{3}$$

EQUIPMENT

The GMODS equipment configuration depends upon the scale of operation. For throughputs from a few kilograms to several thousand metric tons per year, batch operations are possible with a single melter. In a batch operation, each step is done sequentially: (1) oxidation, dissolution and dehalogenation; (2) PbO removal and frit addition to form borosilicate glass; (3) pouring of product glass into containers while leaving the lead metal at the bottom of the melter; and (4) regeneration of the dissolution glass by oxidation of the lead metal in the melter with O_2 and addition of B_2O_3.

For larger-scale operations, continuous processing equipment would be preferred. In this context, it is recognized the GMODS has similarities to several state-of-the-art lead-smelter processes such as the Queneau-Schuhmann-Lurgi (QSL) continuous lead-smelting process [4]. The QSL furnace has an oxidizing zone (similar to the oxidizing dissolution step in the GMODS process) and a reducing zone (similar to the removal of excess PbO and conversion to borosilicate glass step in the GMODS process).

In the QSL oxidizing zone, oxygen and PbO convert molten slag feed materials (lead sulfide ores and recycle lead battery materials with some organics) to (1) PbO that remains in the slag; (2) lead metal that separates from the slag and sinks to the bottom of the system; and (3) an off-gas stream of sulfur oxides, carbon oxides, and steam. In the QSL reducing step, carbon (coal) is added to the molten glassy slag containing feed impurities and PbO. The PbO is reduced to metal that separates from the slag and sinks to the bottom of the furnace. The glassy slag low in PbO is discharged as a waste. Much (but not all) of the chemistry in the two processes is identical. GMODS is optimized to produce a high-quality glass, whereas QSL is optimized for economic lead recovery from ores. QSL lead smelters have capacities of 60,000 to 75,000 t/year. Such equipment (with modifications) appears suitable for large-scale GMODS operations.

STATUS OF DEVELOPMENT

Thermodynamics

A preliminary thermodynamics study of the GMODS process has been completed [3] for processing metals and oxides. The analysis indicated that noble metals are not oxidized by PbO and will dissolve into the lead. (If noble metals build up in the lead, there are standard lead-industry technologies to recover the noble metals.) Copper and less noble materials will form oxides and enter the glass as an oxide.

An analysis of halides in GMODS has been completed recently. Three criteria indicate the efficacy of the use of GMODS in processing halogen-containing wastes: (1) lead halide compounds form when halogens are introduced into GMODS, (2) lead halide compounds are volatile, and (3) most non-lead-metal halides or oxyhalides are unstable in the presence of PbO. Each of these criteria was investigated. The first criterion is satisfied by showing that reactions of the following type are thermodynamically favorable, where X = F, Cl, Br, I:

$$PbO + B_2O_3 + 2NaX \rightarrow PbX_2\uparrow + Na_2O{:}B_2O_3 \qquad (4)$$

At temperatures from 800 to 1200°C, each of these reactions has an equilibrium constant >1; therefore, lead halides preferentially form. There may be other borates which result in more stable products than the monoborate, but they would enhance the reactions. Metals other than sodium may be introduced into GMODS with the halides, but sodium is a good representative metal.

Table I gives a set of calculated vapor pressures for lead halides. At 1000°C or greater, the vapor pressures of all but lead fluoride (PbF_2) are >1 atm. Although the calculations of the PbF_2 vapor pressure are based on recently assessed data, they are conservative. Measured vapor pressures are significantly higher [5] with a vapor pressure of 100 torr (0.13 atm) at 1077°C and a boiling point of 1292°C. Tests will have to be conducted to determine if the vapor pressure of PbF_2 is high enough for effective fluoride removal without the use of a purge gas through the molten glass. The second criterion for halide processing is met.

Table I. Vapor pressures (atm) of lead halides[a]

Temperature (°C)	Vapor pressures (atm)			
	PbF_2	$PbCl_2$	$PbBr_2$	PbI_2
800	3.8×10^{-4}	1.7×10^{-1}	2.8×10^{-1}	6.9×10^{-1}
900	2.4×10^{-3}	5.9×10^{-1}	9.0×10^{-1}	2.1
1000	1.1×10^{-2}	1.6	2.3	5.2
1100	3.8×10^{-2}	3.7	4.9	$1.1 \times 10^{+1}$
1200	1.1×10^{-1}	7.5	9.2	$2.0 \times 10^{+1}$

[a]Methodology based on Ref.[6] and data from Ref.[7].

Uranium or plutonium halides or oxyhalides were used to assess whether metal oxides or oxyhalides are stable in the presence of PbO. The following reactions were evaluated by thermochemical calculations in the temperature range of 800 – 1200 °C.

$$PbO + UOCl_2 \rightarrow PbCl_2(g)\uparrow + UO_2 \tag{5}$$

$$3PbO + UCl_6(g) \rightarrow 3PbCl_2(g)\uparrow + UO_3 \tag{6}$$

$$2PbO + UF_4(g) \rightarrow 2PbF_2(g)\uparrow + UO_2 \tag{7}$$

$$PbO + UOF_2 \rightarrow PbF_2(g)\uparrow + UO_2 \tag{8}$$

$$PbO + UOF_2 \rightarrow PbF_2\uparrow + UO_2 \tag{9}$$

$$PbO + UO_2F_2 \rightarrow PbF_2(g)\uparrow + UO_3 \tag{10}$$

$$PbO + 2PuOCl \rightarrow PbCl_2(g)\uparrow + Pu_2O_3 \tag{11}$$

$$3PbO + 2PuCl_3 \rightarrow 3PbCl_2(g)\uparrow + Pu_2O_3 \tag{12}$$

In all cases, the thermodynamic equilibrium constants were $>>1$. This finding means that these uranium or plutonium halides or oxyhalides are not stable in the presence of PbO and that, if introduced into the system, they will be converted to more stable oxides with the formation of volatile lead halides. The conclusion at this stage of GMODS development is that the three criteria for treating halogen-containing wastes can be met.

Investigation of Process Steps

Some steps of the GMODS process are new, while other steps are parts of standard industrial processes. Experiments were performed to understand and prove the unique features of GMODS. Literature searches have been conducted to understand those parts of the process that are used in other industrial processes.

The addition of feed materials involves oxidation, dissolution, and mixing of feeds with the molten dissolution glass. Tests demonstrated the dissolution of UO_2, ZrO_2, Al_2O_3, Ce_2O_3, MgO, and other oxides. As expected, the glass melt with high concentrations of B_2O_3 had good dissolution capabilities for oxides. In analytical chemistry, B_2O_3 is the standard chemical reagent for fusion dissolution of unknown oxides because of its capability to dissolve such materials.

Oxidation-dissolution tests demonstrated the oxidation of the following metals and alloys (followed by the dissolution of their oxides into the melt): U, Ce, Zircaloy-2, Al, stainless steel, and other metals. Oxidation-dissolution tests also demonstrated that carbon and graphite are oxidized and that CO_2 is produced.

Limited dehalogenation tests with NaCl demonstrated that lead exits the dissolution glass as $PbCl_2$, thus providing a separation of the chloride from other materials. An Inductively Coupled Plasma Emission Spectrometer (ICP) was used for metals analysis. The experimental apparatus consisted of an alumina crucible which contained the lead borate melt as well as any salt that was added. This was encased in a quartz tube that had entry and exit ports for a carrier gas to enter the system and remove the $PbCl_2$ vapor from above the melt. The off-gas system also had a condenser in the exit gas portion to condense the lead chloride. After exiting the quartz tube the off-gas passed through a sodium hydroxide bubbler as well as a sintered metal filter designed to collect the $PbCl_2$.

In the initial chloride reaction experiment, about 5 wt % of NaCl was added, the temperature of the melt was increased to 900° C, and then the carrier gas (argon) was initiated. An aerosol was detected exiting the system. At this point ($\approx$ 2 h since the carrier gas was initiated), the metal filter clogged, and pressure rose in the system until the glass seal was broken. About 1.9 g of $PbCl_2$ was collected in the off-gas system. However, the efficiency of $PbCl_2$ recovery was low. The low recovery is thought to be a result of several factors including the use of a cool carrier gas that was condensing the $PbCl_2$ chloride vapor back into the melt before the vapor had an opportunity to enter the off-gas system. Additional experiments are planned.

Experimental measurements (Table II) using a rotating platinum cylinder in a platinum cup were made of the viscosity of the dissolution glass with various added materials. Experience in the glass industry indicates that molten glass viscosities should be below 100 cP (about the viscosity of olive oil) for good mixing and creation of homogeneous glasses. This low viscosity is also required for rapid gas release when processing halogen feed materials. Based on our experimental data, the GMODS dissolution glass temperature will need to be between 800 and 1000° C. The final processing temperature (after the addition of the silica) will be above 1000° C because this addition increases glass viscosity.

Table II. Measured viscosities (cP) of lead borate glasses

	Temperature (°C)				
$PbO:B_2O_3$ ratio	**600**	**650**	**700**	**750**	**800**
1:1			260	100	55
2:1	200		45		30
3:1	105	65	45		30
4:1	80		60		25
2:1 with 20 wt % U_3O_8		2000[a]		125	
2:1 with 10 wt % ZrO_2					560[a]

[a] These viscosity measurements have a greater error ($\pm$200 cP) caused by the difficulty with experimental equipment at high viscosities.

Significant industrial experience has occurred in several related technologies that are directly applicable to GMODS. Incinerators and other high-temperature processes with lead and halogens have high losses of lead to off-gas systems because of the volatilization of the lead halides [10]. There is a substantial knowledge about volatile lead halides in high-temperature systems. Addition of glass additives [silicon oxide (SiO_2) etc.] to improve waste-glass quality is identical to operations used for producing many other waste glasses. The addition of carbon to molten glassy slags to recover molten metal from PbO in the slag is used in several lead-smelting processes [9] such as the QSL process. Pouring glass from the furnace followed by solidification is a standard operation in the glass industry. Finally, reoxidation of the lead at the bottom of the melter to PbO by adding O_2 is one of several processes used for producing PbO for batteries and other uses.

DEVELOPMENTAL PERSPECTIVES

The analytical evaluations and laboratory development work have demonstrated each step required for GMODS and have identified equipment, instrumentation, and other components needed for GMODS. A significant development effort, however, will be required to convert GMODS into an industrial technology. This effort will include a better understanding of the process, integration of process steps into a system, and development of equipment. A pilot plant will be required to demonstrate the process.

REFERENCES

1. W. G. Brough and S. T. Boerigter, *Plutonium-Bearing Materials Feed Report for the DOE Fissile Materials Disposition Program Alternatives*, UCRL-IE-120749, Lawrence Livermore National Laboratory, Livermore, California, April 1995.

2. C. W. Forsberg, E. C. Beahm, and G. W. Parker, *Patent Application to the U.S. Office of Trademarks and Patents, U. S. Department of Commerce: The Treatment of Halogen Containing Wastes and Other Waste Materials*, May, 18 1995.

3. C. W. Forsberg, E. C. Beahm, G. W. Parker, J. Rudolph, P. Haas, G. F. Malling, K. R. Elam, L. Ott, *Direct Vitrification of Plutonium-Containing Materials With The Glass Material Oxidation and Dissolution System (GMODS)*, ORNL-6825, Lockheed Martin Energy Systems, Inc., Oak Ridge National Laboratory, Oak Ridge, TN, October 1995.

4. P. E. Queneau and A. Siegmund, JOM, **48** (4), 38 (April 1996).

5 G. Hantke, *Gmelins Handbuch Der Anorganischen Chemie: Blei*, Vol. 47, Verlag Chemie-GMBH, Weinheim, 1969.

6. Outokumpu Research, *HSC Chemistry for Windows*, Outokumpu, Finland, 1992.

7. W. Slough, and G. P. Jones, Chem. **31,** (August 1974).

8. W. P. Linak and J. L. Wendt, Prog. Energ. and Combust. Sci. **19,** p. 145 (1993).

9. M. King, "Lead," in *Encyclopedia of Chemical Technology*, Vol 15, eds. J. I. Kroschwitz and M. Howe-Grant, John Wiley & Sons, New York, 1995, p.69.

OPTIMIZATION OF SAVANNAH RIVER M-AREA MIXED WASTE FOR VITRIFICATION

SABRINA S. FU, HAO GAN, ISABELLE S. MULLER, IAN L. PEGG, AND PEDRO B. MACEDO
Vitreous State Laboratory, The Catholic University of America, Washington, D.C. 20064.

ABSTRACT

Vitrification studies of actual Savannah River M-Area mixed wastes have shown that the limiting factor for high waste loading of this waste stream is its chemical durability as defined by the toxicity characteristics leaching procedure (TCLP). As part of the optimization study of Savannah River M-Area wastes, a number of additives were examined including Na_2O, Li_2O, B_2O_3, ZrO_2, and TiO_2. This paper reports on the effect of varying the boron to total alkali ratio and on the effect of substitutions such as ZrO_2 for waste and TiO_2 for SiO_2 on the chemical durability and processability of M-Area waste glasses.

INTRODUCTION

There are approximately 720,000 gallons of waste water treatment sludge from nickel-plating plants at Savannah River M-Area [1]. These wastes contain 2-5 wt.% uranium oxide on an oxide basis, and are classified as F006 listed waste due to their process history. GTS Duratek of Columbia, Maryland has a contractual requirement with Savannah River to ensure safe treatment and disposal of the M-Area sludge. GTS Duratek has subcontracted the Vitreous State Laboratory (VSL) of the Catholic University of America (CUA) to develop glass formulations for the M-Area sludge that will be processable in a DuraMelter™ and pass the durability requirements set by Savannah River, DOE, and South Carolina state regulators. VSL has worked with GTS Duratek and Savannah River since January 1993 to optimize glass formulations for vitrification of M-Area wastes. Initially studies were first conducted in 1993 to determine the economic feasibility of vitrifying M-Area wastes [2]. Then in the second phase, further crucible studies [3] along with several minimelter runs [4,5] were conducted in 1994 to demonstrate the feasibility of this vitrification project. The third phase of this glass formulation study occurred in 1996 when the various tanks at M-Area were mixed as planned [6]. Each of these three phases of compositional study centered around different waste compositions obtained at each stage of this vitrification project along with different regulatory limits, since the regulations changed with time.

Presently, the requirement for safe disposal is to pass the toxicity characteristic leaching procedure (TCLP) with the following requirements: Less than 0.86 ppm Cr (universal treatment standard), less than 1.8 ppm U (DOE drinking water limit), and less than 2.5 ppm Ni (half of the RCRA limit). In addition to these disposal requirements, the glass formulation needs to produce a processable glass melt at a target processing temperature of 1150°C. A joule-heated DuraMelter™ 5000 with a nominal production rate of 5 metric tons of glass per day [6] has been installed at Savannah River M-Area (construction completed in January 1996) and tested with surrogate feed (April to September of 1996). Actual mixed waste has been fed since October 18th, 1996. DuraMelters™ typically operate optimally with melt viscosities of 10-100 poise and melt conductivities of 0.1-0.6 S/cm. In addition, the glass melt should have a liquidus

Mat. Res. Soc. Symp. Proc. Vol. 465 © 1997 Materials Research Society

temperature below the idling temperature (idling occurs when the melter is not processing feed, at which point the melter temperature is decreased by about 100°C) in order to avoid crystallization events which can cause melter operational problems such as electrical short-circuits and clogs in the draining system. Hence, optimization studies focused on the maximum waste loading achievable with processable viscosity and conductivity, and reasonable liquidus temperature, resulting in a waste glass that passes the contractual requirements for safe disposal, and a target composition which can tolerate compositional variations of the waste stream. This paper reports on some of the interesting findings of this third stage of compositional studies of M-Area wastes; namely, the effect of varying the boron to alkali ratio and the effects of substitutions such as ZrO_2 for waste and TiO_2 for SiO_2.

EXPERIMENT

Glasses were made by combining the appropriate amounts of additives with Savannah River M-Area sludge dried at 110°C. The dried sludge was crushed and sieved to pass mesh #10, combined with various amounts of H_3BO_3, Li_2CO_3, Na_2CO_3, SiO_2, TiO_2, and $ZrSiO_4$, melted at 1150°C in either a clay or quartz crucible for one hour with stirring after half an hour, and then the molten glass poured onto a graphite plate. All of the glasses reported in this study produced homogeneous glass pours. Samples of the glasses were subjected to total acid dissolution followed by direct current plasma atomic emission spectroscopy (DCP-AES) analysis, melt viscosity measurements, melt conductivity measurements, heat treatments, scanning electron microscopy coupled with energy dispersive X-ray spectroscopy (SEM/EDX), and subjected to the Toxicity Characteristics Leaching Procedure (TCLP) test with the leachates analyzed by DCP-AES. The compositional analyses of these glasses typically show better than 90 relative% agreement between target and analyzed values.

The melt viscosity was calculated from measurements of the torque and rotation speed of a calibrated spindle attached to a Brookfield viscometer. Measurements were made over a range of temperatures, usually from 1000-1200°C, and the data interpolated to standard temperatures using the Volger-Fulcher equation: $\ln(\eta)=A/(T-T_o)+B$, where A, B, and T_o are fitting parameters. The equipment was calibrated at room temperature using standard oils of known viscosity, and then checked at 1000°C to 1200°C using a National Institute of Standard and Technology (NIST) standard reference glass (SRM711). The precision and accuracy of the viscosity measurements are estimated to be within ±10%.

The electrical conductivity of the glasses was determined by measuring the resistance of the glass melt as a function of frequency using a calibrated platinum electrode probe. The results were extrapolated to zero frequency to obtain the DC conductivity. These measurements are taken over a temperature range of typically 1000-1200°C, and the data interpolated to standard temperatures using the Arrhenius equation: $\ln(\sigma)=A/T + B$, where A and B are fitting parameters. The equipment was calibrated at room temperature using salt solutions of known concentrations and then checked at 1000-1200°C using the NIST reference standard, SRM1414. Estimated uncertainties in the conductivity measurements are ±10%.

The liquidus temperature is the temperature at which the first signs of crystallization occurs. The liquidus temperature of these glasses was determined from results of many heat treatments. Each treatment consisted of premelting the glass at an appropriate temperature (1150°C) to destroy any pre-existing nuclei, followed by a twenty-hour or longer heat treatment at the prescribed temperature. For practical reasons (melter will not be kept at below 900°C for long periods of time), heat treatments below 900°C were not performed. Hence, if no crystals are produced after 72 hours at $900\,^{\circ}C$, then the liquidus temperature is approximated as below $900\,^{\circ}C$.

The waste glass formed from Savannah River M-Area mixed waste was subjected to the EPA TCLP test [7], and the resulting leachates analyzed by DCP-AES for all constituents of concern plus major glass components of the M-Area waste.

RESULTS AND DISCUSSION

Several crucible studies have been performed for the M-Area waste water treatment sludge due to the large variability of the sludge from batch-to-batch, and due to changes in the Savannah River requirement for chemical durability. Previous investigations [2-5,8] included the minimization of both uranium and nickel release due to the low regulatory limits of both uranium (1.0 ppm) and nickel (0.32 ppm) and the high U and Ni contents of the waste sludge samples at that time, and minimelter runs of actual Savannah River M-Area wastes. Presently, the uranium and nickel release has been increased to 1.8 ppm and 2.5 ppm, respectively. The glasses reported in this paper center around the same consistent waste composition obtained in February 1996 from mixing Tank 8 wastes with the Watts Plating Line solution and half of Tank 10 from the M-Area at Savannah River.

<u>Boron to Alkali Ratio Variation</u>

Table I presents the M-Area waste composition obtained from sludge samples in 1996, both in wt.% and in mole%; in addition, Table I lists the desired processing parameters and the durability requirements of M-Area waste glasses. (Note that the composition presented in Table 1 varied by less than 15 relative% over seven samples taken over a seven month period. However, this does not eliminate the possibility of

Table I. Savannah River M-Area Mixed Waste Composition (dried at $1150\,^{\circ}C$ basis) and Desired Processing Parameters and Required Chemical Durability. Compositions is typical of seven Savannah River M-Area sludge samples from Tank 8 taken over seven months. Variations are less than ±15 relative%.

Sludge Dried at $1150\,^{\circ}C$	wt.%	mole%
Al_2O_3	18.64	12.93
B_2O_3	0.91	0.93
BaO	0.04	0.02
CaO	0.69	0.87
Cr_2O_3	0.03	0.01
Fe_2O_3	0.67	0.30
K_2O	3.05	2.29
Li_2O	0.06	0.14
MgO	0.20	0.36
MnO_2	0.09	0.07
Na_2O	11.88	13.55
NiO	0.34	0.32
P_2O_5	5.69	2.84
SiO_2	55.31	65.11
SrO_2	0.01	0.01
TiO_2	0.10	0.09
U_3O_8	2.30	0.19
ZrO_2	0.00	0.00
Total	100.00	100.00
Viscosity at $1150\,^{\circ}C$ = 10-100 Poise		
Conductivity at $1150\,^{\circ}C$ = 0.1-0.6 S/cm		
Liquidus Temperature = $<1000\,^{\circ}C$		
Savannah River Limits for TCLP Leachate		
U		1.8 ppm
Cr		0.86 ppm
Ni		2.5 ppm

highly variable sludge since there is evidence of some solids at the bottom of these tanks.) The M-Area waste contains considerable amounts of glass formers (SiO_2 and Al_2O_3), and thus requires adding flux to bring the viscosity to the desired processing range at $1150\,^\circ$C. Three additives that behave as fluxes (decrease viscosity) for molten glass are Na_2O, Li_2O, and B_2O_3. However, in addition to effectively decreasing melt viscosity, Na_2O and Li_2O contribute greatly to increasing melt conductivity. Boron oxide, on the other hand, has minimal effect on melt conductivity but has a similar effect on melt viscosity as Na_2O. Conductivity modeling [9] of Savannah River M-Area compositions showed that the total alkali content could vary from 15 to 27 mole% and still be in the acceptable conductivity range for processing in a DuraMelter™. The main concern was the interplay between melt viscosity and chemical durability; decreasing melt viscosity is generally accompanied by a decrease in chemical durability (increase in the leachate concentrations). In order to determine whether the chemical durability could be maximized for lower viscosity glasses, six glasses with 70 to 72 wt.% waste loading (on an oxide basis) was made with various boron to alkali ratio. In two sets of glasses, the total ($Na_2O+B_2O_3$) was kept constant along with all other components, but the ratio of B_2O_3 to Na_2O was varied. Both Set I and Set II show that replacing Na_2O with B_2O_3 can be beneficial, as shown in Table II. Since the major difference between Set I and Set II was the Li_2O content, a sixth glass with intermediate Li_2O content from Set I and Set II was made. The TCLP leachate concentrations of all elements examined correlated well with the $B_2O_3/(B_2O_3+Li_2O+Na_2O)$ molar ratio; a plot of the U in the TCLP leachate as a function of $B_2O_3/(B_2O_3+Li_2O+Na_2O)$ molar ratio is shown as Figure 1. This plot suggests that there is a correlation between the molar ratio of $B_2O_3/(B_2O_3+Li_2O+Na_2O)$ and the uranium content in the TCLP leachate. This correlation is also observed for the other glass constituents examined, as can be seen from the data shown in Table II; however, the constituent of regulatory concern is uranium since that is the only constituent in the TCLP leachates that is greater than one tenth of the regulatory limit. (Note that for four of the glasses, SRM-20, SRM-21, SRM-22, and SRM-18, an additional 0.13 wt.% Cr_2O_3 was added to the crucible melts. This was done because the chromium from new refractories used in melters will initially produce glasses with Cr_2O_3 contents of about 0.1 wt.% Cr_2O_3.)

The correlation between $B_2O_3/(B_2O_3+Li_2O+Na_2O)$ molar ratio and the amount of various constituents leached in the TCLP test suggests that there is an optimal $B_2O_3/(B_2O_3+Li_2O+Na_2O)$ molar ratio which retains uranium and other glass constituents within the glass matrix more effectively. The composition of M-Area waste glasses is predominantly composed of SiO_2, trivalents, and alkalis. A trivalent cation and an alkali cation can substitute for a Si^{4+} cation in a tetrahedral site. Hence, the maximum tetrahedral sites occur when the molar ratio of trivalent oxides to alkali oxides equals one. If the TCLP leachate concentrations of the six glasses shown in Table II and Figure I are plotted as a function of the $(Al_2O_3+B_2O_3)/(Li_2O+K_2O+Na_2O)$ molar ratio, the minimum in TCLP leachate concentration occurs at a $(Al_2O_3+B_2O_3)/(Li_2O+K_2O+Na_2O)$ molar ratio of 1.0. The fact that the same curves (with just different numbers on the x-axis) can be obtained using either $B_2O_3/(B_2O_3+Li_2O+Na_2O)$ molar ratio or $(Al_2O_3+B_2O_3)/(Li_2O+K_2O+Na_2O)$ molar ratio is not surprising since Al_2O_3 and K_2O contents were kept constant for all of these glasses. The fact that a minimum in TCLP leachate concentration occurs at a $(Al_2O_3+B_2O_3)/(Li_2O+K_2O+Na_2O)$ molar ratio of one suggests that the minimum in TCLP leachate concentration shown in Figure I may be due to structural changes which occur with changes in the $B_2O_3/(B_2O_3+Li_2O+Na_2O)$ molar ratio. One indirect method of obtaining structural changes at higher temperatures is by the measurement of melt viscosity. A plot of the natural log of melt viscosity at $1000\,^\circ$C as a function of the $B_2O_3/(B_2O_3+Li_2O+Na_2O)$ ratio is provided in Figure 2. Although this plot does not produce a "universal" curve as seen in

Figure 1. Effect of Varying B2O3/(B2O3+Li2O+Na2O) Ratio on TCLP

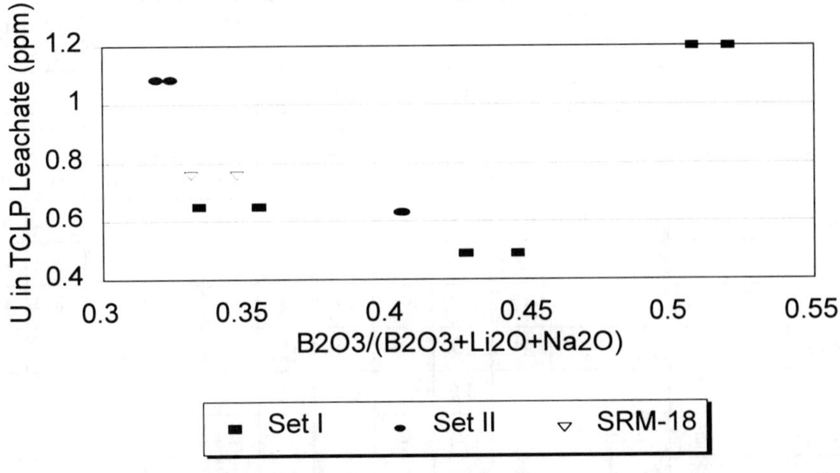

Figure 2. Effect of Varying B2O3/(B2O3+Li2O+Na2O) Ratio on Melt Viscosity

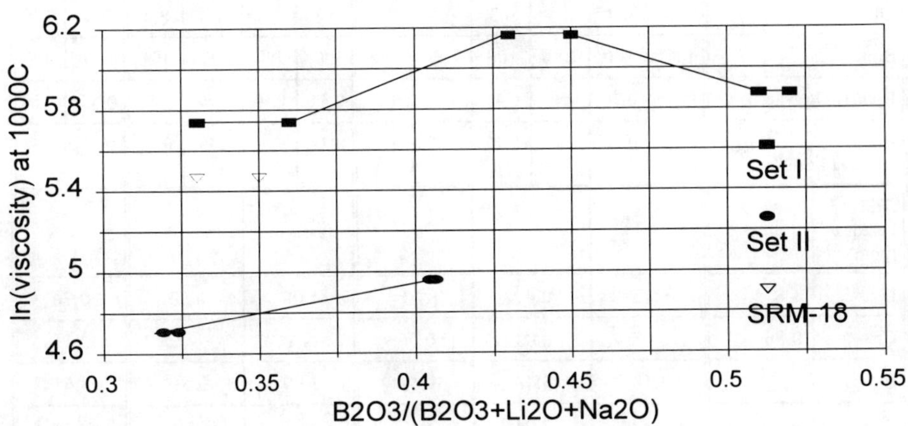

Table II. Effect of Boron Substitution for Sodium on TCLP for Two Different Compositions

	Set I						Set II					
	SRM-20		SRM-21		SRM-22		SRM-12		SRM-26		SRM-18	
Oxides (mole%)	(c)	(a)	(c)	(a)	(c)	(a)	(c)	(a)	(c)	(a)	(c)	(a)
Al_2O_3	9	10	9	10	9	10	9	9	9	10	9	10
B_2O_3	13	13	17	16	21	20	13	13	16	15	13	13
CaO	0.6	0.5	0.6	0.5	0.6	0.5	0.5	0.5	0.6	0.5	0.6	0.6
Fe_2O_3	0.2	0.3	0.2	0.3	0.2	0.3	0.3	0.3	0.2	0.3	0.2	0.3
K_2O	2	2	2	2	2	2	2	2	2	2	2	2
Li_2O	0.1	0.2	0.1	0.1	0.1	0.1	11	11	11	10	4	3.9
Na_2O	27	24	23	20	19	19	17	17	13	13	22	20
NiO	0.2	0.2	0.2	0.2	0.2	0.2	0.2	0.2	0.2	0.2	0.2	0.2
P_2O_5	2	2	2	2	2	2	2	2	2	2	2	2.0
SiO_2	46	48	46	49	46	47	44	45	45	46	46	48
U_3O_8	0.1	0.1	0.1	0.1	0.1	0.1	0.1	0.1	0.1	0.1	0.1	0.1
ZrO_2	0.0	0.0	0.0	0.0	0.0	0.0	1	1	1.1	0.9	0.0	0.0
Others	0.4	0.4	0.4	0.3	0.4	0.3	0.3	0.4	0.4	0.4	0.4	0.4
Sum	100	100	100	100	100	100	100	100	100	100	100	100
$B_2O/(B_2O_3+Na_2O)$	0.33	0.36	0.43	0.45	0.52	0.51	0.44	0.44	0.56	0.55	0.37	0.39
$B_2O_3/(B_2O_3+Li_2O+Na_2O)$	0.33	0.36	0.43	0.45	0.52	0.51	0.32	0.32	0.40	0.41	0.33	0.35

TCLP Leachate (ppm)	Regulatory Limits	SRM-20	SRM-21	SRM-22	SRM-12	SRM-26	SRM-18
U	1.8	0.65	0.49	1.20	1.08	0.63	0.76
Cr	0.86	0.08	0.03	0.06	NA	NA	0.11
B	--	4.48	3.48	9.53	3.99	2.80	3.61
K	--	2.99	2.04	3.47	2.51	1.95	2.46
Si	--	3.28	2.75	2.94	4.25	2.11	2.94
Ni	2.5	0.14	0.10	0.18	0.18	0.10	0.16

c= calculated; a= analyzed

NA= not analyzed

Figure 1, there is a correlation between the melt viscosity and TCLP leachate concentrations within each set of glasses examined. For glasses in Set I, the melt viscosity increases with $B_2O_3/(B_2O_3+Li_2O+Na_2O)$ molar ratio from about 0.35 to 0.45, and then decreases with $B_2O_3/(B_2O_3+Li_2O+Na_2O)$ molar ratio from 0.45 to 0.51. This maximum in melt viscosity for Set I glasses of Figure 2 corresponds to the minimum of uranium TCLP leachate concentrations in Figure 1, at a $B_2O_3/(B_2O_3+Li_2O+Na_2O)$ molar ratio of 0.45, which is at a $(Al_2O_3+B_2O_3)/(Li_2O+K_2O+Na_2O)$ molar ratio of 1.0. Melt viscosity data were taken in the temperature range of 1000-1200°C, and the correlation shown in Figure II for viscosity measurements at 1000°C appears for all other temperatures measured.

Comparing the glasses at a $B_2O_3/(B_2O_3+Li_2O+Na_2O)$ molar ratio of 0.4 to 0.45 in Figures 1 and 2 shows that glasses with similar leach rates but a factor of three difference in melt viscosity can be obtained for similar waste loadings. The ratio of $B_2O_3/(B_2O_3+Li_2O+Na_2O)$ was used in producing Figures 1 and 2 because it is these three constituents that were varied in the Savannah River vitrification studies due to the ability of all three additives to lower the viscosity of the M-Area glass melts to the desired viscosity range. However, since the maximum viscosity and minimum leach rate of these glasses occur at a $(Al_2O_3+B_2O_3)/(Li_2O+K_2O+Na_2O)$ molar ratio of one, it is perhaps the $(Al_2O_3+B_2O_3)/(Li_2O+K_2O+Na_2O)$ molar ratio which is of importance. Future studies on other glass systems will examine the effect of exchanges such as Al_2O_3 for Na_2O as well as determining the universality of the effect of B_2O_3 substitution for Na_2O on TCLP leachate concentrations and melt viscosity.

Effects of ZrO_2-waste and TiO_2-SiO_2 Substitutions

In addition to the B_2O_3-Na_2O substitution studies, the effect of ZrO_2 and TiO_2 replacements for waste (and SiO_2) was briefly examined. A few weight percent of ZrO_2 has been reported to increase both the acid and alkaline durability of glasses significantly. [10] In addition, a few wt.% of ZrO_2 in the glass matrix was found to decrease the corrosion of contact refractories used for vitrification melters. [11] The effect of ZrO_2 on M-Area waste glasses was examined by replacing 2 wt.% of the waste and 1 wt.% of SiO_2 additive with 3 wt.% ZrO_2. The melt conductivity and approximate liquidus temperatures were not affected by this substitution, while the melt viscosity decreased by about 30 relative% and the TCLP leachate concentrations were unaffected within the experimental uncertainties of ± 15 relative%. A similar change in glass properties was observed for the SRM-18 composition when 3 wt.% ZrO_2 additive replaced 3 wt.% of the waste in SRM-18; no detectable changes in TCLP leachate concentrations, with the main change in processing property being a 30 relative% decrease in melt viscosity due to the decrease in SiO_2 and Al_2O_3 that accompanied the decreased waste loading. These two studies of 3 wt.% ZrO_2 additive in place of waste (and SiO_2) showed no beneficial effect on acid durability as measured by TCLP, and the substitutions showed little effect on the processing parameters examined. This is not too surprising given that ZrO_2 was mainly used to replace two of the main constituents of M-Area waste, SiO_2 and Al_2O_3.

Previous studies have shown that a few wt.% of TiO_2 causes the formation of Ti-O-M linkages (M=Na, K, Ca) at ambient temperature and pressure [12]; this suggests a greater stability of alkali in the glass matrix in the presence of titanium, and hence an overall greater stability of the glass matrix. Titanium oxide was used to replace 1 wt.% of the SiO_2 additive; analysis of the TCLP leachates of these two glasses showed no change in leachate concentrations for any of the glass constituents examined (U, Ni, Cr, B, K, and Si). Since no beneficial property changes were detected and since TiO_2 is more expensive than SiO_2, TiO_2 was not used as an additive for Savannah River M-Area waste glasses.

CONCLUSIONS

Examination of Savannah River M-Area waste water treatment sludge shows the waste to consist mainly of silicon, aluminum, and sodium with uranium and nickel as the main contaminants. This listed radioactive waste water treatment sludge can be stabilized into a glass matrix with about 70 wt.% of the glass composed of waste constituents. This represents more than an order in magnitude volume reduction due to a combination of the higher density of glass compared to sludge, the large volatile contents in the sludge, and the high waste loadings achievable for vitrification. Additives which behave as fluxes in molten glass are required in order to decrease the viscosity of the glass melt to optimize melter operations. Boron oxide, lithium oxide, and sodium oxide were chosen as additives, and the ratio of these additives can be varied to reach a minimum in TCLP leachate concentration without adversely affecting desired glass properties, thus providing more durable glasses for remediation without decreasing waste loading.

ACKNOWLEDGEMENTS

The authors would like to acknowledge GTS Duratek for funding of this project. In addition, we thank Gerry DelRosario and his staff for crucible melts, Dr. Shouxiang Hu and his staff for the glass and TCLP leachate analyses, Dr. Shan Tao Lai and his staff for TCLP tests, and Qiuming Yan for heat treatments.

REFERENCES

1. J.B. Pickett, J.C. Musall, and H.L. Martin, Proceedings of the Second International Symposium on Mixed Waste, edited by A.Alan Moghissi, R.K. Blauvelt, G.A. Benda, and N.E. Rothermich, 1993.
2. Proposal to Savannah River from CUA/GTS Duratek, March 1993.
3. S.S. Fu, Characterization Studies in Support of Savannah River M-Area Engineering, Report to GTS Duratek, May 1994.
4. S.S. Fu and K.S. Matlack, Savannah River DuraMelter™ 10 Runs Using VSL Composition, Report to GTS Duratek, October 1994.
5. S.S. Fu, Savannah River DuraMelter™ 10 OTD Runs, Report to GTS Duratek, November 1994.
6. B.W. Bowen and M.M. Brandys, Ceramic Transactions, **61**, 213 (1995).
7. EPA SW-846 Method 1311, Toxicity Characteristic Leaching Procedure.
8. Q. Yan, A.C. Buechele, S. Hu, E. Wang, and S.S. Fu, Ceramic Transactions, **61**, 221 (1995).
9. S.S. Fu, Physics and Chemistry of Glasses, **38**(2), 100 (1997).
10. V. Dimbleby, and W.E.S. Turner, J. Soc. Glass-Technol., **10**, 304 (1926).
11. S. Xing, Y. Lin, R.K. Mohr, and I.L. Pegg, Proceedings to the Materials Research Society Meeting, Fall of 1995.
12. F. Farges, G.E. Brown Jr., A. Navrotsky, H. Gan, and J.J. Rehr, Geochimica et Cosmochimica Acta., **60**, 3039 (1996).

Part III

Glass/Water Interactions

DISSOLUTION RATES OF DWPF GLASSES FROM LONG-TERM PCT

W. L. Ebert and S.-W. Tam
Chemical Technology Division, Argonne National Laboratory, Argonne, IL 60439

ABSTRACT

We have characterized the corrosion behavior of several Defense Waste Processing Facility (DWPF) reference waste glasses by conducting static dissolution tests with crushed glasses. Glass dissolution rates were calculated from measured B concentrations in tests conducted for up to five years. The dissolution rates of all glasses increased significantly after certain alteration phases precipitated. Calculation of the dissolution rates was complicated by the decrease in the available surface area as the glass dissolves. We took the loss of surface area into account by modeling the particles to be spheres, then extracting from the short-term test results the dissolution rate corresponding to a linear decrease in the radius of spherical particles. The measured extent of dissolution in tests conducted for longer times was less than predicted with this linear dissolution model. This indicates that advanced stages of corrosion are affected by another process besides dissolution, which we believe to be associated with a decrease in the precipitation rate of the alteration phases. These results show that the dissolution rate measured soon after the formation of certain alteration phases provides an upper limit for the long-term dissolution rate, and can be used to determine a bounding value for the source term for radionuclide release from waste glasses. The long-term dissolution rates measured in tests at 20,000 m^{-1} at 90°C in tuff groundwater at pH values near 12 are about 0.2, 0.07, and 0.04 g/(m^2•d) for the Environmental Assessment glass and glasses made with SRL 131 and SRL 202 frits, respectively.

INTRODUCTION

We have conducted static dissolution tests to characterize the long-term corrosion behavior of several reference high-level waste glasses for the Defense Waste Processing Facility (DWPF). Many tests were conducted with crushed glass to characterize the dissolution behavior as the solution becomes highly concentrated in glass components. As has been observed with other glasses, after an initially high dissolution rate in dilute leachant solutions, the glass dissolution rate decreases to a very low value as the solution concentrations of glass components increase to apparent saturation values. However, the formation of certain alteration phases has been observed to cause a significant increase in the dissolution rates of many glasses [1-4]. The time required for these phases to form depends on the glass composition, temperature, and glass surface area/solution volume ratio (S/V) of the test, as well as the nucleation kinetics of the precipitated phases. For durable glasses, several years may be required before rate-affecting phases form under a particular set of test conditions, while such phases may form within a few months in similar tests with less durable glasses. The questions we address in this paper are whether the rate expression that has been determined from corrosion behavior prior to formation of rate-affecting alteration phases can also be used to calculate long-term corrosion behavior after they have formed, and whether the measured rate is likely to provide an upper bound to the long-term corrosion rate.

The observed effect of alteration phase formation on the glass dissolution rate is qualitatively consistent with the glass corrosion mechanism used to model long-term corrosion behavior [5-8]. Glass is thermodynamically unstable with respect to a suite of alteration phases; which phases form depends, in part, on the glass composition. Contact of a glass by water provides a kinetically favorable pathway for the transformation of glass to alteration phases via a dissolution-reprecipitation mechanism. As glass dissolves, a quasi-equilibrium between the glass and the solution is approached. This equilibrium is modeled as the hydrolysis of an Si–OSi(OH)$_3$ bond at the surface to release orthosilicic acid into solution. The net dissolution rate of the glass becomes very low as the solution approaches saturation with respect to chalcedony or other surrogates for the glass. The formation of some alteration phases provides a demand for silicon

Mat. Res. Soc. Symp. Proc. Vol. 465 © 1997 Materials Research Society

and affects the equilibrium between the glass and the solution such that the glass dissolution rate increases. Of course, the solution concentrations of other components of the alteration phases will also affect phase formation and the resulting demand for silicon, as given in the equilibrium expression for each alteration phase. For example, the formation of analcime [$NaAlSi_2O_6 \cdot H_2O$] depends on the concentrations of Na, Al, and Si species, as well as the pH and temperature.

The effect of alteration phase formation on the glass dissolution rate will also depend on how fast the phases precipitate. Some Si-bearing phases may form very early in the reaction, but if their precipitation rates are lower than the glass dissolution rate, even under near-saturation conditions, the formation of those phases will not significantly affect the dissolution rate of the glass. For example, clays have been observed to form in tests with many glasses without measurably affecting the dissolution rate [9-11]. After longer reaction times, other phases may form at rates that are much higher than the glass dissolution rate. Formation of these phases may have a pronounced effect on the glass dissolution rate. For example, the formation of analcime has been observed to coincide with an increase in the dissolution rate of some glasses [1-3, 9-13], but may not affect the dissolution rates of others. The effect of alteration phase formation on the glass dissolution rate depends on the relative stabilities of the glass and the alteration phases [8].

The glass dissolution rate prior to the formation of alteration phases can be written as [14]

$$\text{rate} = k_0 10^{\eta pH} e^{-E_a / RT} \left[1 - \left(\frac{Q}{K} \right)^{\sigma} \right] \tag{1}$$

where k_0 is the intrinsic dissolution rate that depends only on the glass composition, η is the coefficient of the pH-dependence, E_a is the activation energy, Q is the ion activity product of the solution, K is the equilibrium constant for the hydrolysis reaction, and σ is the reaction order. The values of η, E_a, K, and σ have been measured to be about 0.4, 80 kJ/mol, $10^{-3.1}$, and 0.1, respectively, for DWPF glasses [15, 16]. In the case that the hydrolysis of an $Si-OSi(OH)_3$ bond at the glass surface is the rate-determining step, Q and K depend only on the activity of orthosilicic acid. The term in brackets is referred to as the affinity term, which may vary between values of 1 in highly dilute solutions to near 0 in nearly saturated solutions. This expression has been found to describe the corrosion behavior of many nuclear waste glasses well prior to the formation of alteration phases. The glass dissolution rate is usually measured experimentally based on the accumulation of B or other highly soluble glass component in the leachate solution. The goal of the present work is to determine if the rate expression in Eq. 1 also describes the corrosion behavior after phases form that increase the dissolution rate, or if additional terms are required.

Alteration phase formation may affect the pH and the value of Q. The other parameters depend on the glass composition. In the limit that the solution chemistry becomes fixed by an equilibrium between the solution and the alteration phases, the value of the affinity term will become constant. If the pH and temperature remain constant, then the glass dissolution rate will also be constant as long as the same suite of alteration phases controls the corrosion behavior. If the formation of alteration phases only affects the value of Q, then Eq. 1 predicts that the long-term dissolution rate will vary with the precipitation rate of the alteration phases. If phases precipitate much faster than the glass can dissolve, then glass dissolution is predicted to proceed at a constant rate. If the precipitation rate decreases, then Q will increase and the glass dissolution rate will decrease. Equation 1 describes glass dissolution, not the precipitation of alteration phases.

We have evaluated the results of long-term Product Consistency Tests (PCTs) to address the use of Eq. 1 to describe corrosion behavior after rate-controlling alteration phases have formed, and to determine if the glass dissolution rate becomes constant at advanced stages of corrosion. One complication of extracting dissolution rates from the results of tests with crushed glass is that the available surface area decreases appreciably as the glass dissolves. We have taken the loss of surface area into account by modeling the particles to be spheres, then extracting the value of a dissolution rate corresponding to a linear decrease in the radius with time. The change in NL(B) corresponding to this constant shrinkage rate was then compared to test results that were adjusted to take into account the decrease in the surface area with the same geometric model.

EXPERIMENTAL

Tests were conducted with crushed DWPF reference glasses of the size fraction -100 +200 mesh and tuff groundwater solutions [2, 3, 9, 10, 17, 18]. The leachant solutions used in the various tests were prepared by reacting groundwater from well J-13 with crushed tuff (<100 mesh) at 90°C for four weeks then passing the solution through a 100-nm filter. The resulting solution is referred to as EJ-13 water. Different batches of water were prepared for each series of tests. The concentrations of the major components are about 35-45 mg/L Si, 45-55 mg/L Na, and 120 mg/L HCO_3^-; the solution pH was near 8 (measured at room temperature). Tests were conducted at glass/leachant mass ratios of 1:10 and 1:1 to attain S/V of about 2000 and 20,000 m^{-1}, respectively. All tests were conducted in convection ovens set at 90°C. At the end of each test, the leachate was analyzed for pH and cation concentrations, and the reacted solids were analyzed. Details regarding phase identification, the disposition of radionuclides, etc. are provided in the above references.

RESULTS AND DISCUSSION

The compositions of the glasses discussed in this paper are given in Table I. The results of static dissolution tests conducted with these glasses are summarized in Table II. The test durations before and after rate-affecting alteration phases formed are listed in Table II. For example, rate-affecting alteration phases were detected in the 364-day test with SRL 202A at 20,000 m^{-1}, but not in the 182-day test. The alteration phases that formed coincident with the increase in the dissolution rate are also listed in Table II. For some glasses, rate-affecting alteration phases had only formed in the test with the longest duration or not enough data were available to determine the rate. In other tests, the glass had completely altered within the test interval in which the rate-affecting phases had formed. While the qualitative behavior observed in those tests is consistent with the effects of alteration phase formation seen in other tests, only lower bounds to the dissolution rates can be extracted. The results of those tests are not evaluated here.

The extent of reaction is commonly expressed in terms of the normalized mass loss, which is calculated by dividing the mass of a glass component in solution by the surface area of glass exposed in the test and by the mass fraction of that component in the glass. The normalized mass loss can be written either in terms of the mass, m_i, or concentration, c_i, of an element in solution

$$NL(i) = \frac{\left(m_i - m_i^o\right)}{\left(S \bullet f_i\right)} = \frac{\left(c_i - c_i^o\right)}{\left([S/V] \bullet f_i\right)} \tag{2}$$

where m_i^o and c_i^o are the mass and concentration of species i in the leachant solution, and f_i is the mass fraction of i in the glass. Note that NL(i) has the units of mass *glass* per area, not mass of i per area, because of the inclusion of the f_i term. Different values of NL(i) will be calculated for different i if glass dissolution is nonstoichiometric or if glass components become incorporated into alteration phases. The rate calculated based on the release of B is usually assumed to provide the best measure of the extent of glass dissolution, since B is highly soluble and is sensitive to the dissolution of the glass matrix.

While the accumulated amount of B is measured directly, the change in the surface area during the test must be estimated analytically. We have calculated NL(B) with the initial surface area and the surface area estimated to remain at the end of the test. The initial surface area was calculated by assuming the glass grains to be spheres having a diameter equal to the arithmetic average of the sieve sizes. For -100 +200 mesh glass, the diameter was assumed to be 112 µm. The remaining surface area at the end of a test was calculated with Eq. 3 [14]:

Table I. Compositions of Glasses

	SRL 131	SRL 202	SRL EA		SRL 131	SRL 202	SRL EA
Al_2O_3	3.27	3.84	3.60	NiO	1.24	0.82	0.53
Am_2O_3	0.0004[a]	0.0004[a]	—	Np_2O_3	0.009[a]	0.009[a]	—
B_2O_3	9.65	7.97	11.16	PbO	—	0.01	—
BaO	0.16	0.22	—	Pu_2O_3	0.009[a]	0.009[a]	—
CaO	0.93	1.20	1.23	SiO_2	43.8	48.9	48.76
Cr_2O_3	0.13	0.08	—	SrO	0.01	0.03	—
CuO	0.02	0.04	—	TcO_2	0.03	0.03	—
Fe_2O_3	12.7[b]	11.4[b]	9.3 b	ThO_2	0.03	0.03	—
K_2O	3.86	3.71	0.04	TiO_2	0.65	0.91	0.65
Li_2O	3.00	4.23	4.21	U_3O_8	2.73	1.93	—
MgO	1.31	1.32	1.79	ZnO	0.02	0.02	0.26
MnO_2	2.43	2.21	1.36	ZrO_2	0.22	0.10	0.48
Na_2O	12.1	8.92	16.88	Total	98.3	98.6	100.25

[a] These radionuclides present in actinide-doped glasses, but not in non-doped glasses.
[b] All iron assumed to be Fe(III).

Table II. Summary of Test Results

Glass	S/V, m^{-1}	Time, days	pH	Alteration Phases[a]	Ref.
SRL 131A	2000	140-280	10.7-11.5	SMEC; ANAL; GYR; WEEK	17
SRL 131A	20,000	98-182	12.1-12.4	SMEC; ANAL; GYR; WEEK	17
SRL 131S[c]	2000	980-1800	11.7-12.0	SMEC; ANAL	2
SRL 131R[b]	2000	>1800[b]	11.9	SMEC	2
SRL 200S[d]	20,000	182-330	11.8-12.3	SMEC; ANAL; CLIN; GYR; WEEK	2
SRL 200R[b]	20,000	>1800[b]	11.7	SMEC	2
SRL 202U	20,000	182-364	11.5-11.7	SMEC; ANAL; GYR; WEEK	17
SRL 202A[c]	2000	1822	11.3	SMEC; ANAL; GYR; WEEK	17
SRL 202A	20,000	182-364	11.3-11.9	SMEC; ANAL; GYR; WEEK	17
SRL EA[d]	2000	313-369	11.8-12.1	SMEC; ANAL; GMEL; GYR; ZEO	18
SRL EA	20,000	<22[e]	12.0-12.4	SMEC; ANAL; GMEL; ZEO	18

[a] SMEC = smectite clay; ANAL = analcime; GYR = gyrolite or other Ca-silicate; WEEK = weeksite; ZEO = Na-Al-silicate phase; GMEL = gmelinite; CLIN = clinoptilolite.
[b] Rate-affecting alteration phases did not form within the longest time tested.
[c] Rate-affecting alteration phases only formed at the longest time tested.
[d] Too few data to extract rate.
[e] Rate-affecting phases formed within shortest time tested.

$$S_t = \left[\frac{3 m_0^{1/3} \left(m_0 - \dfrac{m_B}{f_B} \right)^{2/3}}{(\rho r_0)} \right] \quad (3)$$

where S_t is the surface area remaining at time t, m_0 is the initial mass of glass, m_B is the mass of boron in solution at time t, f_B is the mass fraction of B in the glass, ρ is the density of the glass, and r_0 is the initial radius of the glass spheres. We have calculated S_t with the final surface area to obtain the maximum decrease in surface area over the test duration and to calculate an upper limit for the dissolution rate. The lower limit of the dissolution rate is calculated with the initial surface area. The values of NL(B) for tests with SRL EA, SRL 131A, and SRL 202A glasses conducted with -100 +200 mesh glass at 20,000 m^{-1} and at 90°C that were calculated with the initial surface area and the final surface area are plotted in Fig. 1. The values of NL(B) calculated with the final surface area–NL(B)$_{final}$–are greater than those calculated with the initial surface area–NL(B)$_{initial}$– because the final surface area is less than the initial surface area. The upward curvature of the data is due to the decreasing surface area and does not indicate an increase in the glass dissolution rate.

Analcime and other mineral alteration phases formed within 22 days in tests conducted with SRL EA glass and within 182 days and 364 days in tests with SRL 131A and SRL 202A glasses. Lines in Fig. 1 show the dissolution rates for the three glasses immediately after these phases formed. These lines were drawn through the average of NL(B)$_{initial}$ and NL(B)$_{final}$ in the two tests after the phases had formed for all glasses, except the line for tests with the SRL EA glass. For this glass, the line was drawn through the origin and the average value of NL(B) from the 56-day test. The averages of NL(B)$_{initial}$ and NL(B)$_{final}$ were used because these values better represent the surface area of the glass *during* the test than either the initial or final surface area. The slope of the line for each glass gives what is referred to as the limiting rate. The limiting rates are about 0.22, 0.068, and 0.035 g/(m^2•d) for the SRL EA, SRL 131A, and SRL 202A glasses, respectively. These rates were used to calculate the rate at which the radius of the spherical particles decreases, which is defined as k=dr/dt, by dividing the limiting rates by the density of the glass, which is assumed to be 2.7 g/cm^3 for these three glasses. The rates extracted from the results of tests with several DWPF reference glasses in EJ-13 water at 90°C are summarized in Table III. We emphasize that the glass dissolves at this rate only after rate-affecting alteration phases have formed. An analytical expression relating the mass of glass that has dissolved with the remaining surface area can be written based on the geometry of shrinking spherical particles. In this approximation, the mass of glass that dissolves is simply the density of the glass times the loss of volume of the sphere as the radius decreases over time. The decrease in the volume can then be related to the decrease in surface area through the radius of the sphere. The geometric relationship between the mass of glass dissolved, the remaining surface area, and the dissolution rate can be written as

$$NL(B)(t) = \frac{M(t)}{S(t)} = \left(\frac{\rho kt'}{3}\right)\left\{1 + \frac{2 - \left(\frac{kt'}{R_p}\right)}{\left(1 - \frac{kt'}{R_p}\right)^2}\right\} + X \tag{4}$$

where M(t) is the total mass of glass dissolved through time t, S(t) is the surface area that remains at time t, ρ is the glass density, k is the rate at which the radius of the spheres decrease, and R$_p$ is the radius of the spheres when rate-affecting alteration phases first form. The term X represents glass dissolution that occurred *before* rate-affecting alteration phases formed; its form is identical to the first term, but with a dissolution constant different than k. The contribution of the X term to long-term dissolution is negligible after rate-affecting alteration phases form, and is not further considered here. We emphasize that the first term on the right hand side of Eq. 4 applies to corrosion *after* rate-affecting alteration phases have formed by using t', which is the time after those phases have formed. Since M(t) increases as S(t) decreases, the curve described by Eq. 4 curves upward as time increases. This does not mean that the dissolution rate increases with time.

To better show the common corrosion behavior of these glasses, and because of the small number of tests with any single glass in the presence of alteration phases, the data for several glasses is normalized as follows so data for the three glasses could be presented on a single plot. The time τ that is required for the glass to completely dissolve after alteration phases form is defined as $\tau = R_p/k$. The reaction time is scaled by τ to generate a dimensionless time variable, t$_R$, as t$_R$ = t'/τ; the value of t$_R$ is 1 when the glass is completely dissolved. Similarly, the amount of

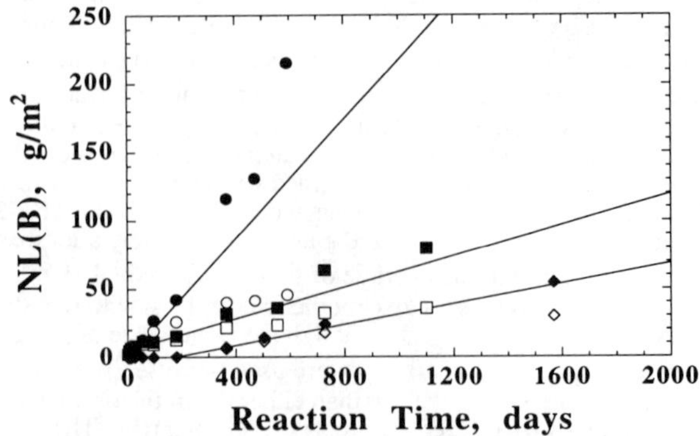

Figure 1.   NL(B) for Tests at 20,000 m^{-1} with SRL EA glass (●), SRL 131A glass (■), and SRL 202A glass (◆). Open symbols calculated with initial surface area, filled symbols calculated with surface area remaining at the end of the test. Lines show the estimated dissolution rate for each glass (see text).

Table III. Dissolution Rates and pH Values Attained After Alteration Phases Form in Tests with Various DWPF Glasses

Glass	S/V, m^{-1}	t_p, d[a]	R_p, μm[b]	k, nm/d
SRL 131A	20,000	98	50	25
SRL 202A	20,000	182	56	13
SRL EA	20,000	0	56	82

[a]Test duration prior to formation of rate-affecting phases.
[b]Estimated grain size at t_p.

glass dissolved per unit area is normalized to a dimensionless quantity M_R, as M_R = $\{M(t)/S(t)\}/\{\rho \cdot R_P\}$. The values of k, t_P and R_P for tests conducted at 20,000 m^{-1} are included in Table III. By substituting these dimensionless quantities, Eq. 4 transforms to

$$M_R = \left(\frac{t_R}{3}\right)\left(1 + \frac{(2 - t_R)}{(1 - t_R)^2}\right) \qquad (5)$$

This gives a universal relationship between the mass measured in solution and the extent to which a glass dissolves if the radius of the glass particles decreases at a constant rate. The plot of M_R vs. t_R is shown in Fig. 2 along with the experimental results that have been transformed into reduced coordinates by dividing NL(B)$_{final}$ by the quantity ρR_P and scaling the test time as t_R = $(t-t_P) \cdot k/R_P$.

Average values of duplicate tests are plotted in Fig. 2 for clarity. After the differences in the values of k and t_P for the different glasses are taken into account, all three glasses show essentially the same behavior. The experimental values agree with the theoretical curve for reduced times less than about 0.3, but data from tests in which the extent of corrosion has progressed further (i.e., those data at t_R>0.3) clearly fall below the curve. Tests with SRL EA glass have the greatest extent of corrosion and show the deviation most clearly, while tests with SRL 202 show only a small deviation at the longest reaction time. Note that the deviation from the curve does *not*

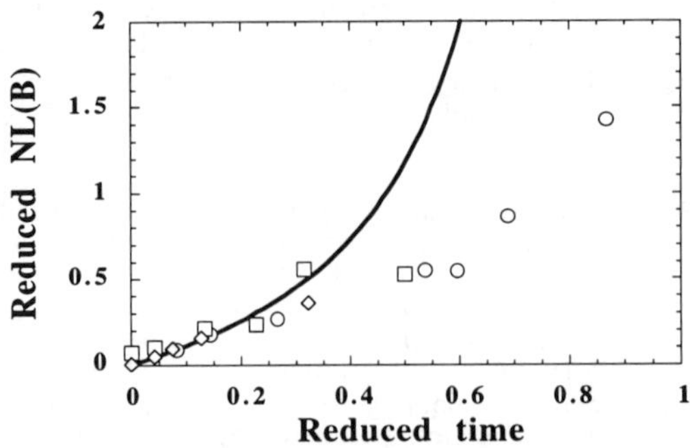

Figure 2. Theoretical Curve and Results of Tests at 20,000 m^{-1} with SRL EA (O), SRL 131A ($\square$), and SRL 202A glass ($\Diamond$), in Dimensionless Reduced Units.

correspond with the formation of the alteration phases, which occurs at $t_R = 0$ for all glasses. Neither is the deviation a result of using the final surface area of the glass in the calculations, since both M_R and the experimental rate are calculated with the final surface area.

The observation that the data follow the curve soon after rate-affecting phases form, while data from more advanced tests do not, indicates that glass corrosion cannot be modeled with a single rate constant or by a single mechanism. This suggests that the glass dissolution rate becomes moderated by another process after extended time periods (i.e., for $t_R > 0.3$) that is not taken into account in Eq. 1. This other process may be associated with the glass, the solution, or the alteration phases. Alteration layers are known to form on the surface of the glass particles as the glass corrodes, and the thickness of the layer increases with the extent of corrosion. However, these layers are very porous and probably ineffective diffusion barriers [2, 3].

In these PCTs, the crushed glass settles at the bottom of the vessel and is covered by a 2-3 cm layer of water in tests at 20,000 m^{-1}. Precipitated phases are generally observed to form as a layer of sediment on top of the glass, which suggests that most precipitates form in the bulk solution and not within the solution between the glass grains. Diffusion of material from the glass to the bulk solution may affect the rate at long times. Because of differences in the compositions of the glass and of the suite of alteration phases, the solution will become depleted of some components needed to form alteration phases over time. This may slow the formation of some phases and limit their abundance, and may result in a slowing of glass dissolution. For example, based on the Na:Al:Si ratios of these glasses, the formation of zeolite alteration phases will be limited by the amount of Al. Hence, the solution concentration of Al will eventually become low enough to limit the formation of these phases.

The impact of other assumptions made in the present analysis on these results must be further evaluated, such as assumption of spherical particles and the assumption that B released from the glass is completely dissolved. Incorporation of small amounts of B into alteration phases would result in an apparent decrease in the dissolution rate. The significant change in the relative amounts of glass and alteration phases that occurs as the test proceeds may also affect the dissolution rate of the glass. Clearly, more work is needed to elucidate phenomena that affect the long-term corrosion behavior of waste glasses.

Regardless of the cause for the difference between the measured and predicted extents of dissolution that are plotted in Fig. 2, that predicted with a linear dissolution rate (the solid curve) does provide an upper limit for the measured extents of dissolution over long test durations. The long-term dissolution rates of these glasses at 90°C in solutions with pH values near 12 are: 0.05 g/(m^2•d) for SRL 131 frit-based glasses, 0.04 g/(m^2•d) for SRL 202 frit-based glasses, and 0.2 g/(m^2•d) for the EA glass. Finally, based on Eq. 2, the glass dissolution rates will depend on the pH. The rates that were extracted in this paper are only relevant at pH values near 12. Tests conducted at lower S/V usually result in leachate solutions with lower pH values than tests at 20,000 m^{-1}. A few very long term tests are in progress at 2000 m^{-1} that may provide dissolution rates at lower pH values that can be compared to those discussed in this paper.

CONCLUSIONS

Corrosion of DWPF glasses in some laboratory tests results in the generation of highly concentrated solutions and the formation of alteration phases with a concomitant increase in the dissolution rate. The decrease in the surface area of the glass particles that occurs when the dissolution rate increases complicates the calculation of the dissolution rate. By modeling the glass particles as spheres, we have shown that dissolution is consistent with a constant decrease in the radius soon after certain phases form but not at long test durations. This may indicate that the precipitation rate of the alteration phases is limiting the dissolution rate of the glass, and that the expression currently used to calculate long-term glass behavior must be modified to accurately model long-term performance. Further testing is required to identify the cause of the effect and to assess its possible impact on the long-term corrosion behavior in a disposal site. Regardless, the present analysis suggests that the rate measured soon after alteration phases form probably provides an upper bound to the long-term dissolution rate.

ACKNOWLEDGMENTS

Work supported by the U.S. Department of Energy, Office of Environmental Management, under contract W-31-109-ENG-38.

REFERENCES

1.   K. Lemmens and P. Van Isegham, Mater. Res. Soc. Symp. Proc. <u>257</u>, 49-56 (1992).
2.   X. Feng, J. K. Bates, E. C. Buck, C. R. Bradley, and M. Gong, Nucl. Technol. <u>104</u>(2), 193-206 (1993).
3.   W. L. Ebert, J. K. Bates, E. C. Buck, and C. R. Bradley, Mater. Res. Soc. Symp. Proc. <u>294</u>, 569-576 (1993).
4.   W.L. Ebert, A.J. Bakel, and N. R. Brown, in Proceedings International Topical Meeting on Nuclear and Hazardous Waste Management, Spectrum '96, Seattle, WA, August 18-23, 1996, pp. 569-575 (1996).
5.   B. Grambow, *Nuclear Waste Glass Dissolution: Mechanism, Model, and Application*, JSS Report 87-02 (1987).
6.   T. Advocat, J.L. Crovisier, B. Fritz, and E. Vernaz, Mater. Res. Soc. Symp. Proc. <u>176</u>, 241-248 (1990).
7.   D. M. Strachan, W. L. Bourcier, and B. P. McGrail, Radioactive Waste Management and Environmental Restoration <u>19</u>, 129-145 (1994).
8.   P. Van Iseghem and B. Grambow, Mater. Res. Soc. Symp. Proc. <u>112</u>, 631-639 (1988).
9.   W. L. Ebert, J. K. Bates, C. R. Bradley, E. C. Buck, N. L. Dietz, and N. R. Brown, Ceram. Trans. <u>39</u>, 333-340 (1993).
10. X. Feng, Mater. Res. Soc. Symp. Proc. <u>333</u>, 55-68 (1994).
11. J. K. Bates, W. L. Ebert, J. J. Mazer, J. P. Bradley, C. R. Bradley, and N. L. Dietz, Mater. Res. Soc. Symp. Proc. <u>212</u>, 77-87 (1991).
12. J. K. Bates and M. J. Steindler, Mater. Res. Soc. Symp. Proc. <u>15</u>, 83-90 (1983).
13. A. J. Bakel, W. L. Ebert, and J. S. Luo, Ceram. Trans. <u>61</u>, 515-522 (1995).
14. B. P. McGrail and D. K. Peeler, *Evaluation of the Single-Pass Flow-Through Test to Support a Low-Activity Waste Specification*, Pacific Northwest Laboratory Report PNL-10746 (1995).
15. W. L. Bourcier, S. A. Carroll, and B. L. Phillips, Mater. Res. Soc. Symp. Proc. <u>333</u>, 507-512 (1994).
16. K. G. Knauss, W. L. Bourcier, K. D. McKeegan, C. I. Merzbacher, S. N. Nguyen, F. J. Ryerson, D. K. Smith, and H. C. Weed, Mater. Res. Soc. Symp. Proc. <u>176</u>, 371-381 (1990).
17. W. L. Ebert, and J. K. Bates, Nuclear Technol. <u>104</u>(3), 372-384 (1993).
18. J. K. Bates, et al., *ANL Technical Support Program for DOE Environmental Restoration and Waste Management. Annual Report October 1993-September 1994*, Argonne National Laboratory Report ANL-95/20 (1995).

SIMULATION OF NATURAL CORROSION BY VAPOR HYDRATION TEST: SEVEN-YEAR RESULTS

J. S. Luo, W. L. Ebert, J. J. Mazer, and J. K. Bates
Chemical Technology Division, Argonne National Laboratory, Argonne, IL 60439

ABSTRACT

We have investigated the alteration behavior of synthetic basalt and SRL 165 borosilicate waste glasses that had been reacted in water vapor at 70°C for time periods up to seven years. The nature and extent of corrosion of glasses have been determined by characterizing the reacted glass surface with optical microscopy, scanning electron microscopy (SEM), transmission electron microscopy (TEM), and energy dispersive x-ray spectroscopy (EDS). Alteration in 70°C laboratory tests was compared to that which occurs at 150-200°C and also with Hawaiian basaltic glasses of 480 to 750 year old subaerially altered in nature. Synthetic basalt and waste glasses, both containing about 50 wt % SiO_2, were found to react with water vapor to form an amorphous hydrated gel that contained small amounts of clay, nearly identical to palagonite layers formed on naturally altered basaltic glass. This result implies that the corrosion reaction in nature can be simulated with a vapor hydration test. These tests also provide a means for measuring the corrosion kinetics, which are difficult to determine by studying natural samples because alteration layers have often spalled off the samples and we have only limited knowledge of the conditions under which alteration occurred.

INTRODUCTION

Various laboratory tests have been developed to study corrosion behavior and to establish chemical models with which to calculate the long-term behavior of glass. In particular, the vapor hydration test developed at Argonne National Laboratory accelerates the glass corrosion process by combining the effects of high surface-area-to-volume ratios (S/V) and high temperatures [1]. However, it is challenging to demonstrate that the vapor hydration test method produces results that can be directly related to the behavior of glass waste forms over time periods of thousands of years. One method to build confidence in the applicability of this test method is to study the corrosion of natural materials that has taken place in natural environments over very long periods [2]. Basalt glass has been chosen for such a comparison study because its silica content is similar to that of a typical waste glass. If it can be demonstrated that the long-term alteration of a glass by natural processes can be simulated in laboratory experiments, then the same test can be used to study the long-term corrosion behavior of waste glasses. Such a natural analogue study will provide confidence that the kinetic models that have been developed based on short-term experiments can be applied to calculate the long-term corrosion of waste glass. In this paper, we compare alteration patterns that occur in naturally altered basalt glasses with those that occur in vapor hydration tests conducted at 70°C.

EXPERIMENTAL

The composition of the synthetic basalt used in the laboratory tests was modeled after a Hawaiian alkali basalt glass (52 wt% SiO_2, 1.7 wt% TiO_2, 12.7 wt% Al_2O_3, 12.7 wt% Fe_2O_3, 0.2 wt% MnO_2, 6.2 wt% MgO, 10.7 wt% CaO, 3.8 wt% Na_2O, and 0.6 wt% K_2O) [3]. The glass was made by melting appropriate amounts of reagent-grade oxides and carbonates at 1600°C, quenching and homogenizing the resulting frit, remelting, and annealing the final product at 550°C for about two hours. Tests were also conducted with a reference nuclear waste glass from Savannah River Laboratory (SRL) 165 black frit for comparison. The SRL 165 glass contains 55.7 wt% SiO_2, 4.3 wt% Al_2O_3, 12.0 wt% Fe_2O_3, 2.9 wt% MnO_2, 0.6 wt% MgO, 1.5 wt% CaO, 10.5 wt% Na_2O, and 0.6 wt% K_2O [3]. Test samples were prepared as wafers approximately 1 cm in diameter and 1 mm thick. Each face of the sample was polished with 600-grit carborundum paper, followed by a final finish with diamond pastes of 6 to 1 μm with silicon oil as

Mat. Res. Soc. Symp. Proc. Vol. 465 © 1997 Materials Research Society

a lubricant. Each sample was thoroughly cleaned and suspended with a Teflon thread from a type 304L stainless steel support bar in the test vessel. Samples were reacted in vapor of 100% relative humidity at 70°C for time periods up to seven years. The details of vapor hydration tests have been described elsewhere [1]. At the end of a test, each sample was examined with optical microscopy, scanning electron microscopy (SEM), transmission electron microscopy (TEM), and energy dispersive x-ray spectroscopy (EDS) to observe the alteration layer and characterize its chemical composition. The details for preparing glass samples for SEM and AEM observation have been described previously [1,3,6]. The crystallinity of precipitates in the alteration layer was also determined with selected area electron diffraction, and crystalline phases were identified by matching the measured composition and diffraction patterns with the reference materials.

For comparison studies, ten samples of volcanic basaltic glass received from two archaeological sites at Oahu, Hawaii, were examined. These samples were located 5 to 85 cm below the ground level in generally wet and cool environments, although the exact humidity and temperature values are not known. The ages of the samples range from 480 to 720 ± 80 years B.P. (Before Present), according to radiocarbon dating of the occupation layer of charcoal [4]. The glasses have been altered by meteoric water and a humid atmosphere in nature. The Hawaiian basalt glasses were analyzed with optical microscopy, SEM, TEM, and EDS.

RESULTS AND DISCUSSION

<u>Natural Weathering of Hawaiian Basalt Glass</u>

All the Hawaiian glass samples examined in our study, regardless of age, exhibit surface alteration with similar features. These features include a brown-colored palagonite layer at the surface (Fig. 1a) and a high density of etching pits in areas from which the palagonite has spalled off the sample (Fig. 1b). The palagonite layers were often not continuous on surfaces (which allowed for the observation of etching pits underneath), and its coverage of the glass surface differed significantly from sample to sample. The spallation of the palagonite layer may have serious implications in determining the kinetics of corrosion on these samples, because the reaction rates are calculated based on the measurement of the thickness of palagonite layers formed on the glass surface [1]. Examination of cross sections shows that the palagonite layers are predominantly amorphous, with various precipitates of (Fe,Ti)-oxides distributed in the layer (see Fig. 2a). Analysis of the amorphous palagonite with EDS shows depletion of Na, Mg, and Ca and enrichment of Fe, Ti, and Al compared to the unreacted glass (see Fig. 2b).

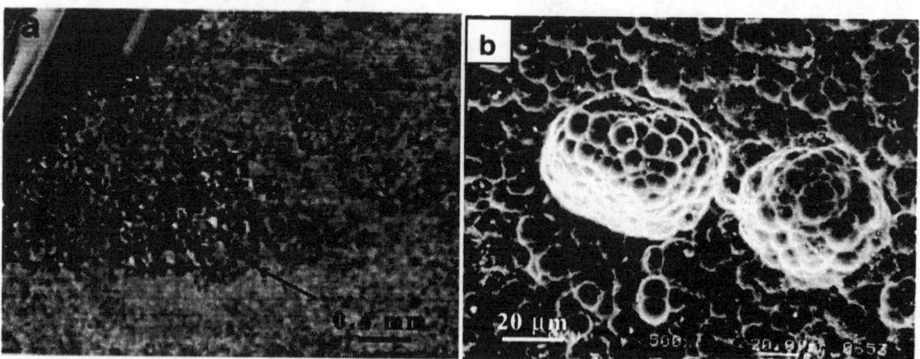

Fig. 1.   (a) Optical image of the palagonite at the surface with evidence of spallation (see arrows) and (b) SEM image of etching pits in areas where the palagonite spalled.

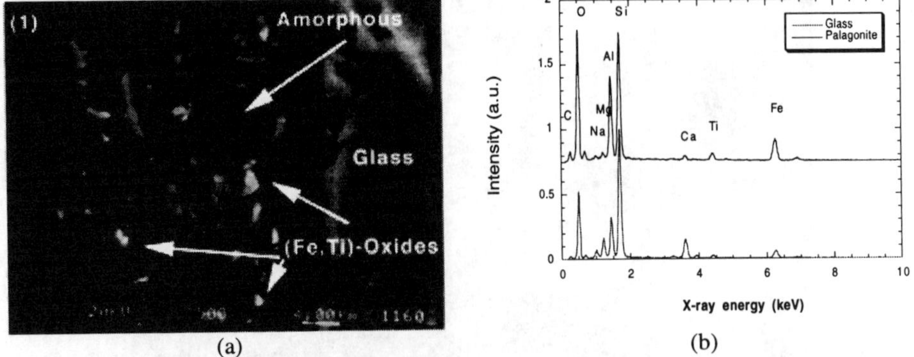

(a)   (b)

Fig. 2.   (a) SEM cross-sectional view of palagonite layer and (b) EDS spectra comparing the glass and palagonite layer. The spectrum was acquired with an ultrathin EDS detector attached to the SEM.

Examination of the palagonite with AEM provides additional information at higher resolution. At high magnification, it can be seen that palagonite is composed of aggregates of small particles (about 10 nm) together with some clay particles (see Fig. 3). The clay was identified with electron diffraction and EDS to be Fe-rich smectite. The presence of nanometer-sized colloids is consistent with a dissolution/precipitation mechanism, by which glass is first dissolved into solution and then insoluble materials precipitate onto the glass surface to form palagonite. The existence of etching pits at the glass surface seems to indicate a dominant dissolution process of glass during its reaction with water. The formation of a crystalline clay phase could result from precipitation from solution or *in situ* conversion of the amorphous phase over time, with dissolution occurring at the layer/glass interface.

<u>Laboratory Hydration of Synthetic Basalt Glass</u>

Tests were performed with synthetic basalt at 70°C in saturated water vapor for up to seven years. A hydrated layer of about 0.5 μm thick was found to have formed on the surface, similar to the palagonite formed on the naturally altered basalt. Examination of the layer at high magnification with TEM shows an amorphous phase plus small amount of clay (Fig. 4a), a feature characteristic of naturally altered basalt. Analysis of the alteration layer with EDS shows depletion of Ca and enrichment of Fe and Ti, compared to the unreacted glass core (Fig. 4b), similar to that observed on naturally altered Hawaiian basalts. The exact location of Ca is unclear at this stage. It is possible that Ca may diffuse to the surface, dissolve and accumulate in a thin layer of water, and then precipitate as Ca-rich phases or be flushed away when in contact with water during the sample preparation. The above results indicate that nearly identical alteration layers are produced in both laboratory-reacted and naturally-occurring basalt glasses and, therefore, the long-term alteration by natural processes can be simulated in laboratory vapor hydration tests.

We also compared alteration at 70°C to that observed during previous tests conducted at 150-240°C [3]. The alteration layer formed at 70°C is similar to the layer formed during the initial reaction at higher temperatures (120-240°C) in short time periods (several days), which exhibits an amorphous microstructure that contains some smectite clays. This suggests that the corrosion of basalt glass may be further accelerated at higher temperatures without altering the fundamental reaction processes, since increasing the reaction temperature resulted in higher reaction rates.

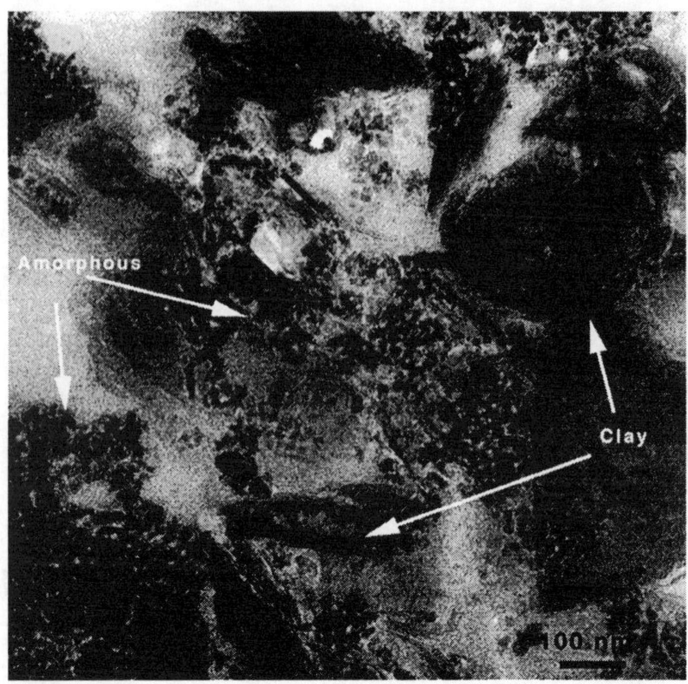

Fig. 3.   Transmission electron microscope image of alteration layer of Hawaiian basalt glass
weathered for 700 years. The layer consists mainly of nanometer-sized amorphous
particles and clay crystals (see arrows).

The fact that the corrosion products in nature can be simulated under accelerated laboratory
conditions suggests that the glass reacts by similar reaction processes. Alternatively, the dominant
corrosion process under a specific reaction condition can be inferred through the study of the
reaction kinetics and may depend on reaction conditions. The kinetic expression for glass
corrosion can be developed based on measurements of layer thickness, and the reaction kinetics of
basalt glass hydrated in vapor hydration tests have been reported as both diffusion controlled and a
linear function of time [1, 5]. However, it is difficult to use data from natural samples to verify a
kinetic expression based on laboratory experiments. Because the alteration layers spalled from the
base glass during wet/dry cycling and we only have limited knowledge of the reaction
environment, the reaction kinetics cannot be established for naturally subaerially altered basalt
samples.

<u>Laboratory Hydration of SRL 165 Waste Glass</u>

Vapor hydration tests were also performed with SRL 165 waste glass at 70°C in saturated
water vapor for seven years. These tests produced a hydrated gel-like alteration layer, which
consisted of an amorphous phase plus small amount of clay (Fig. 5a). This microstructure is
nearly identical to those observed in the basalt glass reacted under the same conditions (see Fig. 4)
or subaerially altered in nature (see Fig. 3). The alteration layer is enriched in Mn, Ni, Al, and Cl
and depleted in Ca in comparison with the unreacted glass core (Fig. 5b). The alteration of
SRL 165 at 70°C is also found to be similar to the initial reaction of this glass in tests conducted at

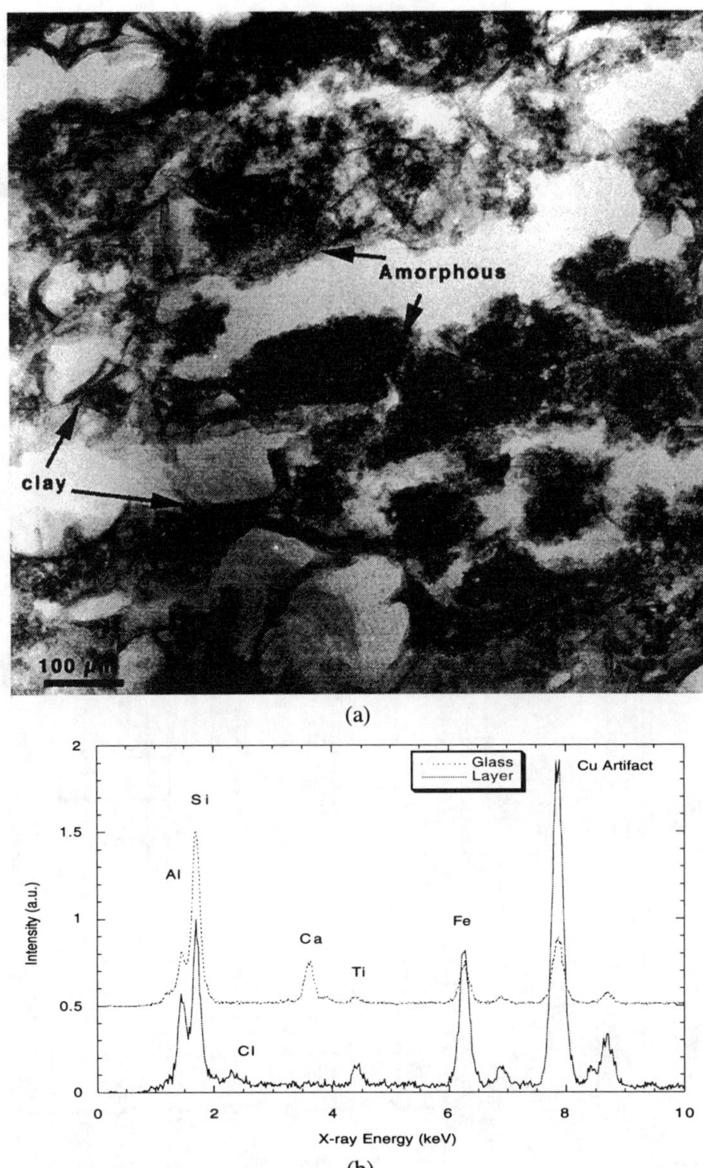

Fig. 4. Analyses with TEM and EDS of synthetic basalt reacted at 70°C in 100% RH for 7 years: (a) TEM image of alteration layer and (b) EDS spectra comparing the glass and amorphous layer (peaks located at energies lower than Al, such as O, Na and Mg, were not detected by the Be-coated Si EDS detector attached to TEM).

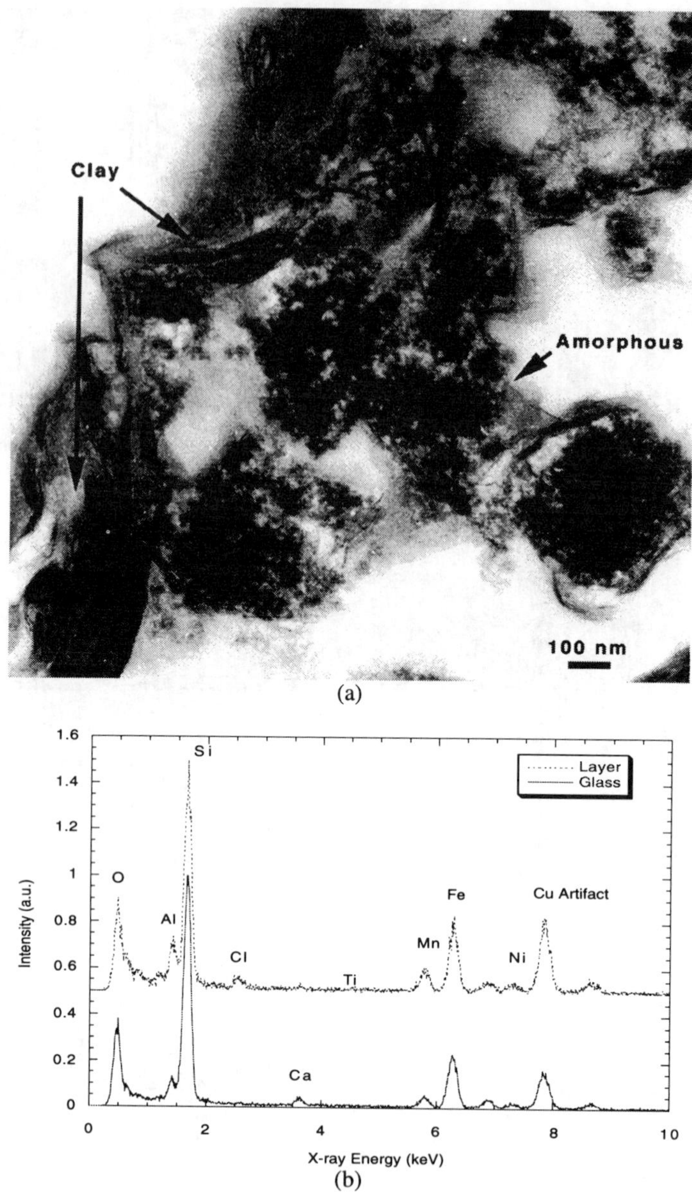

Fig. 5. Analyses with TEM and EDS of SRL 165 waste glass reacted at 70°C in 100% RH for 7 years: (a) TEM image of alteration layer and (b) EDS spectra comparing the glass and amorphous layer.

higher temperatures (120-240°C) in short time periods (7 to 91 days) [3, 6, 7]. At later stages of reaction at high temperatures, other alteration phases precipitate at the surface [1, 3]. The similarity of alteration products between SRL 165 waste glass and basalt glass suggests that both glasses initially corrode by the same reaction processes, at least under vapor hydration test conditions at low temperatures.

CONCLUSIONS

We have examined the alteration products formed on basalt and waste glasses reacted at 70°C in water vapor for up to seven years and compared the laboratory corrosion to subaerial alteration of 450- to 700- year old Hawaiian basalt. By characterizing the alteration products formed in laboratory tests and in nature, we have shown that the long-term subaerial alteration by natural processes can be simulated in vapor hydration tests. We have also shown that the same reaction products form in vapor hydration tests with a waste glass. These findings provide confidence in the application of this test method for studying the long-term corrosion of waste glass.

ACKNOWLEDGMENT

This research was sponsored by the U.S. Department of Energy, Office of Environmental Management, under contract W-31-109-ENG-38. The authors also thank Dr. D. Strachan for critical reading of the manuscript.

REFERENCES

1.   J. K. Bates et al., Nucl. Chem. Waste Manage. 5, (1984) 63.

2.   R. C. Ewing, Scientific Basis for Nuclear Waste Management, edited by G. J. McCarthy, Plenum Press, New York, (1979) 57.

3.   C. D. Byers, R. C. Ewing, and M. J. Jercinovic, Adv. Ceram. 20, (1986) 733.

4.   M. F. Riford, Bishop Museum, private communication, July 2, 1992.

5.   F. E. Diebold and J. K. Bates, Adv. Ceram. 20, (1986) 515.

6.   W. L. Ebert et al., Mater. Res. Soc. Symp. Proc. 176, (1990) 339.

7.   D. J. Wronkiewicz et al,. Geological Society Annual Meeting, Oct. 25-28, 1993, Boston, MA.

SOLUTION-BORNE COLLOIDS FROM DRIP TESTS USING ACTINIDE-DOPED AND FULLY-RADIOACTIVE WASTE GLASSES

Jeffrey A. Fortner, Stephen F. Wolf, Edgar C. Buck, Carol J. Mertz, and John K. Bates.
Argonne National Laboratory, Argonne, IL 60439, USA.

ABSTRACT

Drip tests designed to replicate the synergistic interactions between waste glass, repository groundwater, water vapor, and sensitized 304L stainless steel in the potential Yucca Mountain Repository have been ongoing in our laboratory for over ten years. Results will be presented from three sets of these drip tests: two with actinide-doped glasses, and one with a fully-radioactive glass. Periodic sampling of these tests have revealed trends in actinide release behavior that are consistent with their entrainment in colloidal material when as-cast glass is reacted. Results from vapor hydrated glass show that initially the actinides are completely dissolved in solution, but as the reaction proceeds, the actinides become suspended in solution. Sequential filtering and alpha spectroscopy of colloid-bearing leachate solutions indicate that more than 80% of the plutonium and americium are bound to particles that are captured by a 0.1 μm filter, while less than 10% of the neptunium is stopped by a 0.1 μm filter. Analytical transmission electron microscopy has been used to examine particles from leachate solutions and to identify several actinide-bearing phases which are responsible for the majority of actinide release during glass corrosion.

INTRODUCTION

Drip tests designed to simulate potential conditions in the proposed Yucca Mountain Repository have been in progress with both actinide-doped and fully radioactive glasses. These tests react slowly dripping, tuff-equilibrated groundwater with a monolithic waste glass sample, sensitized 304L stainless steel, and water vapor to determine synergistic interactions possible in a compromised pour canister under unsaturated conditions. The tests using actinide-doped Savannah River Laboratory (SRL) 165 glass, termed the N2 test series, have been ongoing for over 10 years. Tests with a West Valley Demonstration Project former reference glass (ATM-10) have been in progress for 8 years and are termed the N3 test series. A modified drip test with a radioactive sludge-based SRL 200R glass that had been pre-aged, along with its 304L stainless steel holder, in 200°C water vapor has been termed the N4 test series, and has been ongoing for nearly 5 years. During this time, the release of actinides and fission products to solution has been measure d[1-4], and the role of solution-borne colloids in this release has been explored as presented below.

While it has been established that radioactive waste glass can release radionuclides via colloids [1,5], details of the long-term role of colloids in the source term from glass corrosion have not been well documented. Clearly, radionuclides can be sorbed from solution onto aboriginal groundwater colloids (pseudocolloids) or become incorporated into colloidal material that nucleates from dissolved species in solution (radiocolloids). Each of these colloid formation mechanisms would be limited in their potential for transport by the solubility of the radionuclides, and thus are unlikely to substantially increase the release of these elements from the waste package environment. However, as glass reacts it alters to clay and other mineral phases which may spall from the surface, often as colloid-sized particles. This third type, termed primary colloids, can release radionuclides at levels bounded only by the reaction rate of the source glass.

When solution conditions become unstable for colloids, the colloids will flocculate and sediment, attaching themselves either to the test vessel or to the surface of the glass [6]. In the case of drip tests, the low flow results in particles dislodging from the reacted surface. Alternatively, if the ionic strength of the passing water is low, or if the pH is high, alteration phases may be dislodged through increasing particle-particle repulsive forces [6]. At Yucca Mountain, migration may be enhanced if the surface of the fractures in the tuff possess a similar surface electrical potential to the clay colloids, whereas other types of waste-glass-derived colloids may be retarded if the surface charges are different. Therefore, it is important to know the physical properties of the colloids generated due to waste glass reaction. Pseudocolloids and radiocolloids, which can

Mat. Res. Soc. Symp. Proc. Vol. 465 © 1997 Materials Research Society

only marginally increase transport above the solubility limits, may serve to decrease net release by sweeping radionuclides from solution and later plating out onto nearby host rock in the repository. Primary colloids may allow rapid release of radionuclides above solubility limits, which is particularly important for the release behavior of glass. In our laboratory tests, the glasses release nearly all of the long-lived radionuclide content (except neptunium) in the form of colloids. In the experimental work discussed in this paper, we address preliminary efforts to address the formation and characterization of the colloidal material formed in the very near-field, and do not address the potential for the observed species to migrate from the proposed repository. More detailed information regarding the characterization of colloids is required to complete the performance assessment of the repository.

EXPERIMENTAL

<u>Unsaturated (Drip) Test Method</u>

The drip tests, which follow the Unsaturated Test Method, are described in detail elsewhere [2,3], but will be described briefly in order to define a context for the data. The Unsaturated Test Method simulates the potential repository conditions where glass, stainless steel, and water (liquid and vapor) interact. Test parameters can be adjusted to reflect changes in water chemistry and flow rate that may occur.

Each ongoing test series consists of 3 identically-prepared samples and a blank. Every 3.5 days, approximately 3 drops (0.075 mL) of conditioned groundwater from the J-13 well (EJ-13 water) is injected into the airtight vessel where it contacts the top surface of the waste package assemblage (WPA). The WPA consists of a glass monolith contacted on the top and bottom by two perforated retainer plates made from sensitized 304L stainless steel, which are held in place by two wire posts, also made from 304L stainless steel. The entire apparatus is enclosed in a 90°C oven except when samples are taken. Water drips down the sides of the glass and accumulates at the bottom of the WPA. Eventually the water drips from the WPA to the bottom of the test vessel. When tests are sampled (currently at 26-week intervals) the WPA is examined visually to qualitatively ascertain the degree of reaction, including evidence of alteration phase formation and possible spalling of the alteration phases and clay layer. After observation, the WPA is transferred to a fresh test vessel, the test solution removed for analysis, and the just-used vessel is acid-stripped to determine sorbed species.

<u>Analytical Radiochemistry</u>

The sampled solutions from the bottom of the test vessel are analyzed using inductively-coupled plasma mass spectrometry (ICP/MS) plus high-resolution alpha and gamma spectrometry. When sufficient solution is collected, sequential filtering is performed with filters of 1 μm and smaller. Recently, solution has been filtered through a Millipore MC® filter in an ultracentrifuge. The filtered solutions from each test series typically have more than ~80% of the Am and Pu removed by a 0.1 μm filter. The ultracentrifuged filtered samples have had nearly 100% of the Am and Pu removed, with full recovery (no removal) of the Np.

As the glass reacts, material is released from the glass either truly dissolved in solution or as particulate material. The solution is also in contact with the pre-sensitized 304L stainless steel retainer during the reaction process, so the analysis of the solution collected in the bottom of the test vessel represents all the material that is transported from the glass and glass retainer. The solution is analyzed for its constituent parts as described above, but all the material analyzed in the test solution is considered to have been released from the glass/stainless steel assembly.

<u>Analytical Transmission Electron Microscopy</u>

A 5 μL aliquot is prepared for analytical transmission electron microscopy (AEM) by wicking the unfiltered solution through a "holey" (perforated) carbon sample grid. The AEM is carried out with a JEOL 2000FX II transmission electron microscope equipped with two Noran energy dispersive X-ray fluorescence (EDS) detectors (one at high-angle and the other having an ultra-thin window), plus a Gatan model 666 parallel electron energy loss spectrometer (EELS). Identification of crystalline materials with AEM is often possible using electron diffraction. However, the radionuclide content of the colloids and general physical characteristics, e.g., size distribution and charge, cannot be determined with AEM. Additionally, the colloids identified are representative only of the material in the 5 μL sample.

<u>Dynamic Light Scattering</u>

Dynamic light scattering (DLS) is used to probe the dynamics of microscopic particle motion. The principles governing the application of light scattering to study many systems is described elsewhere [7,8]. The technique is used to determine the size and distribution of colloidal-sized particles in the range of one or two nm to several μm. This technique is based on the measurement of fluctuations in scattered light intensity caused by the Brownian motion of the particles in solution. Time scale fluctuations in light intensity are measured as individual scattered photons into time windows that span a range of the scattering signal. The measured autocorrelation function of these data is fit by well-accepted methods to determine the particle size distribution [9-11].

The scattering experiments were carried out on a modified 4700C Photon Correlation Spectrometer from Malvern Instruments, Ltd. The particle sizing data analysis was done using standard methods [7,9]. Light scattering measurements were made on solution extracted from the N4 tests since measurements are taken immediately upon sampling and at time periods thereafter. These tests were just recently sampled, during the development of the DLS method. Laser light scattering system calibration was performed with a series of NIST traceable polystyrene (PS) latex standards (32, 73, and 304 nm) from Duke Scientific Corp. Standard solutions were sampled in filtered deionized water according to certification instructions. The PS data were fit by cumulants and were in excellent agreement with the certified standards. The EJ-13 water contained no DLS-measurable colloids prior to interacting with glass in a test.

RESULTS AND DISCUSSION

An example of the apparent role of colloids in actinide release appears in Fig. 1. Here, the cumulative normalized release (from *unfiltered* test solution) of the elements B, Np, Pu, and Am from one of the three ongoing tests in the N2 series display a behavior that is best explained by a release of particulate material to solution. The soluble elements B and Np were initially released at a substantially greater rate than the relatively insoluble elements Pu and Am. At later times (after ~8 years), the spallation of alteration phases, some of which have incorporated Pu and Am, led to total release of these elements approaching that expected from congruent dissolution of the glass. Sequential filtering has confirmed that more than 80% of the Pu and Am, but very little of the Np, in solution from the N2 series are trapped by a 0.1 μm filter. Some of the potential actinide-bearing alteration phases have been observed using AEM. Many of these particles have small (<1 μm) linear dimensions and high surface areas characteristic of colloids. Examples of these particles, plus others from the N3 and N4 series, are described in detail below.

In both the N2 and N3 tests, the majority of colloidal particles observed by AEM have been either a smectite-type clay or a variety of iron-silicates (Fig. 2). Both clays and iron silicates can sorb actinides, and thus these colloids represent potential transport mechanisms for insoluble elements. As stated previously, more than 80% of the Pu and Am in solution from the N2 and N3 tests appears to be associated with particulate matter that will not pass through a 0.1 μm filter. In the N2 tests, both the clay and iron-silicate colloids are sometimes observed to contain small amounts of uranium. Uranium is also observed on occasion in the clays and iron-silicates from the N3 tests; thorium is generally detected only in an entrained alteration phase such as the thorium calcium orthophosphate mineral *brockite* [1,5,12], and not in the clay itself. Examples of brockite-containing colloids, including evidence for their high transuranic content, have been published elsewhere [1,5]. The aged 200R glass in the N4 drip tests also produced a brockite-like alteration phase rich in rare earth elements (Fig. 3). None of the above colloid particles were observed in wicked solution from the blank tests.

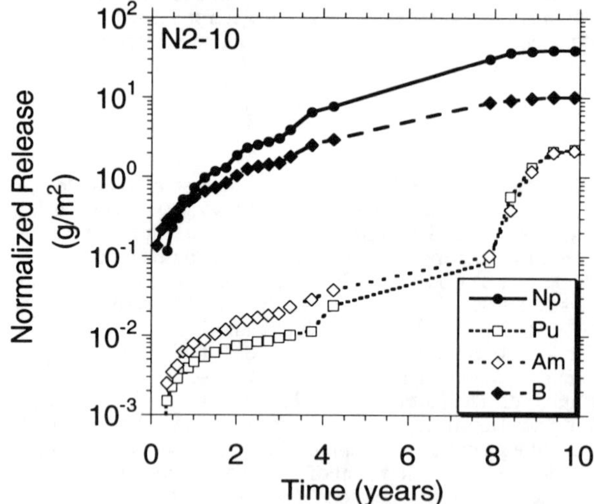

Fig. 1. The cumulative normalized release of the elements B, Np, Pu, and Am from one of the tests in the N2 series. The sudden increase in the release of the insoluble elements Pu and Am shortly after 400 weeks is strong evidence of the role of colloids in the release of these elements from the WPA. Almost no Am or Pu appears in filtered solution, while the Np in filtered solution is essentially the same as that in the unfiltered. The lines are a guide to the eye.

Sparingly soluble or insoluble elements that are not incorporated into the clay structure, or that are present at levels beyond what can be accommodated by clay, will precipitate as alteration phases on the reacted glass and nearby surfaces, such as the stainless steel retainer. As the solution composition evolves with time during glass and steel corrosion, so too will the assemblage of alteration phases evolve. The major constituents in the glass that are not completely dissolved into solution will be incorporated into the clay or will form distinct alteration phases (such as brockite). Excess iron is also present due to the corroding 304L stainless steel.

Once nucleated, the clay layer continues to grow and evolve. Growth occurs as the hydrated glass layer beneath the existing clay dissolves in a nearly congruent manner, and the elements from solution saturate and reprecipitate onto the surface of the clay. Voids are frequently observed between the clay and glass, a result of either disproportionate growth/dissolution rates or accumulating strain due to different materials properties between the glass and the clay layer. The etching of the glass beneath the clay continues in an irregular manner, leaving some points of contact between the clay and glass, holding the layer in place. This clay layer may spall away, however, as stress builds up in the structure and as the glass etches away beneath. This is likely the origin of the clay colloid particles observed in solution from the drip tests. Thus, as the clay becomes detached from the WPA, actinide-bearing alteration phases may be released. This results in the delayed release of Pu and Am as colloidally-suspended particles, as illustrated in the solution data of Fig. 1. However, the transuranic content of the specific phases (as in Figs. 2 and 3) cannot be determined with AEM, and this additional information is required to perform performance assessment calculations which include release from glass in the source term.

The source of primary colloids is the altered surface of the WPA. In order to confirm the characteristics of the source of these colloids, sections of the reacted surface plus alteration phases were removed from the N3-9 and N3-10 glass monolith surfaces during the July 1995 sampling of the N3 tests. These samples, measuring about 5 μm (or less) across, were imbedded in epoxy resin, thin-sectioned to approximately 50 nm thickness using a diamond-knife microtoming process, and placed on a transmission electron microscope grid for AEM examination.

The particles removed from the N3-9 surface were discrete alteration phases, while that removed from the N2-10 was a fragment of the spalling clay layer. The N3-9 particles were found to be an amorphous silica alteration phase, plus a variety of iron silicate and iron silicate hydroxide phases, some of which were identified by electron diffraction as *iron hydroxide* [nominally FeO(OH)], *ferripyrophyllite* [nominally $Fe_2Si_4O_{10}(OH)_2$], and *erlianite* [nominally $Fe_6Si_6O_{15}(OH)_8$]. The N3-10 clay particle turned out to be an interesting composite of smectite clay

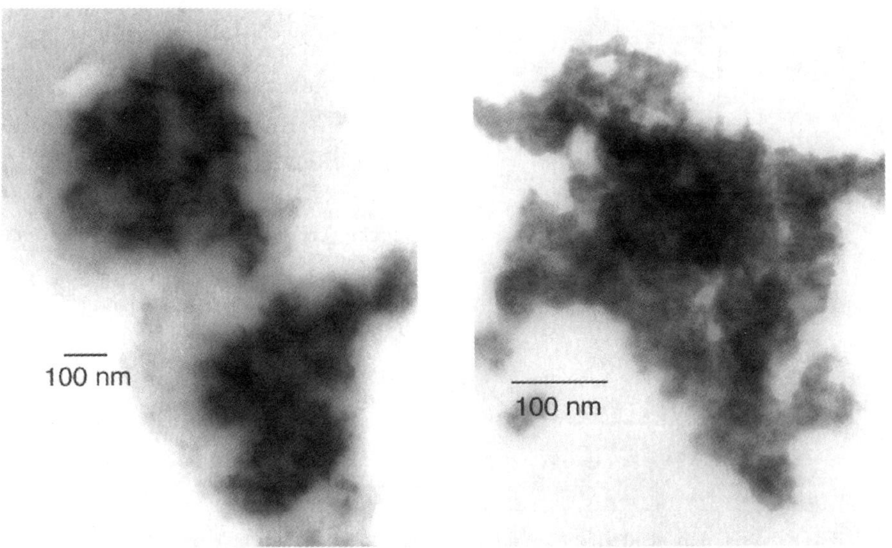

Fig. 2. Typical AEM micrographs of colloid particles extracted from the drip tests. At left are 2 smectite clay colloid particles from the N2-12 Test sampled December 18, 1995. At right is an iron silicate colloid from the N3-9 Test sampled January 11, 1996.

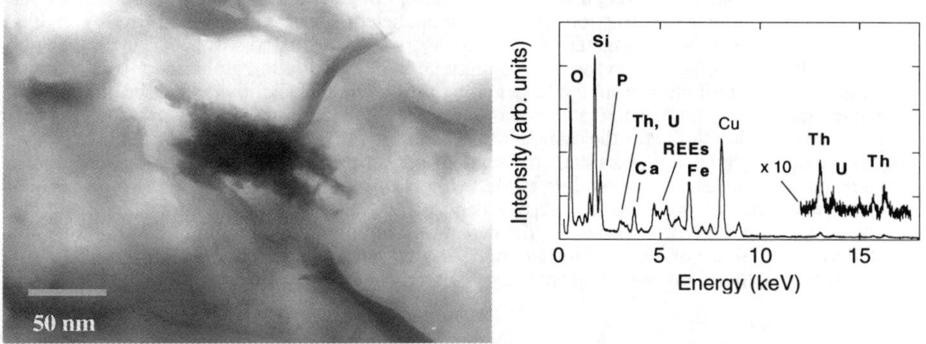

Fig. 3. At left is a TEM image of a rare earth element (REE)-bearing phase from a 200R glass drip test (N4). The EDS analysis of the particle (right) demonstrates that it contains actinide elements.

with copious amounts of *brockite* (nominally, [Th, Ca, REE]PO_4), plus an amorphous thorium-titanium-iron silicate, similar to the mineral thorutite [nominally, (Th, U, Ca)Ti_2O_6]. A transmission electron micrograph of a portion of this N3-10 matrix material appears in Fig. 4. The chemical compositions of the individual phases are characterized by the EDS spectra of Fig. 5, while EELS analysis of REE and actinides in the brockite appears in Fig. 6. The distribution of rare earth and actinides is virtually identical to that observed in tests terminated after 1 year [12].

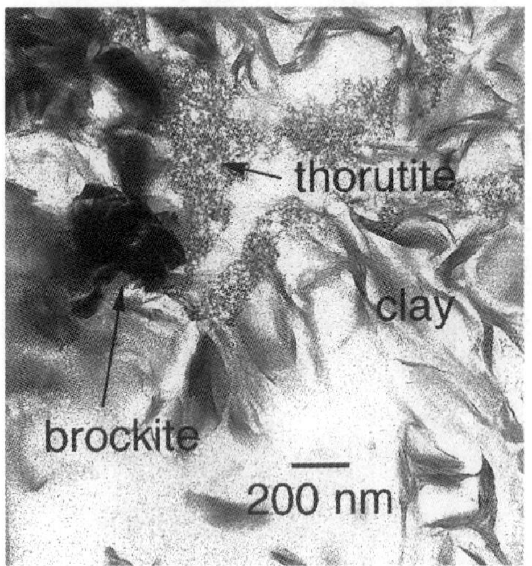

Fig. 4. The clay alteration layer on the N3-10 glass, sampled July 1995, was found to consist of a composite of smectite clay, brockite crystallites, and thorutite, as seen in this transmission electron micrograph. The thorutite forms in the solution between the layer and the dissolving glass and is trapped by the clay layer.

The dynamic light scattering measurements were made on a sample of test solution from the N4-2 test sampled in September 1996, after a 26-week test interval. The sample was held in a temperature-controlled water bath at 25°C during which measurements were made to monitor the stability of the colloids. Representative data from the correlator are shown in Fig. 7. The data indicated a broad distribution or polydisperse sample. Since the cumulants method is limited to narrow distributions with variances of 0.3 or less, the Contin program [9] was used for final data analysis and was able to resolve a bimodal distribution. In general, the largest suspended fraction (>95%) is attributed to a size fraction with a mean diameter of approximately 0.2 μm. A smaller fraction (<5%) with a mean diameter greater than 4 μm was also found suspended in all the samples. These size observations are consistent with both the filtration/alpha spectroscopy and the AEM data. The contribution from the larger size fraction reduced slightly during the ten day measurement period which can be attributed to settling of the larger particles.

The detection limits for the dynamic light scattering measurements were established using the PS standards. In general, particles that scatter light efficiently (e.g., good scattering contrast between the particle and the suspending medium) can be detected quite easily in the ppm concentration range. Particle sizing measurements can be made on concentrations of PS particles as low as 0.2 μg/g(solution). However, the detection limit of the instrument can be severely limited by the presence of dust and foreign material, as well as the scattering contrast of the particles, especially at sub-ppm concentrations. Efforts to quantify the method are ongoing.

CONCLUSIONS

Actinide-bearing colloids are formed during waste glass reaction under potential repository conditions. Candidate materials for actinide entrainment include smectite clay, iron silicates and iron silicate hydroxides. Some alteration phases can incorporate relatively large concentrations of actinides and rare earth elements and become mobilized as colloids when the surface clay is sloughed from the glass surface. The formation of stable, actinide-bearing alteration phases may be of benefit to the repository performance if transport is limited by the hydrology/geology of the site. The ability of these colloids to be transported through the waste package and the repository, and the range of their potential migration cannot be inferred from these tests, and will require separate consideration. However, input to performance assessment of colloidal transport must use a complete physical characterization of the primary colloids that may actually dominate the release of actinides from glass in a scenerio where the steel canisters are breached.

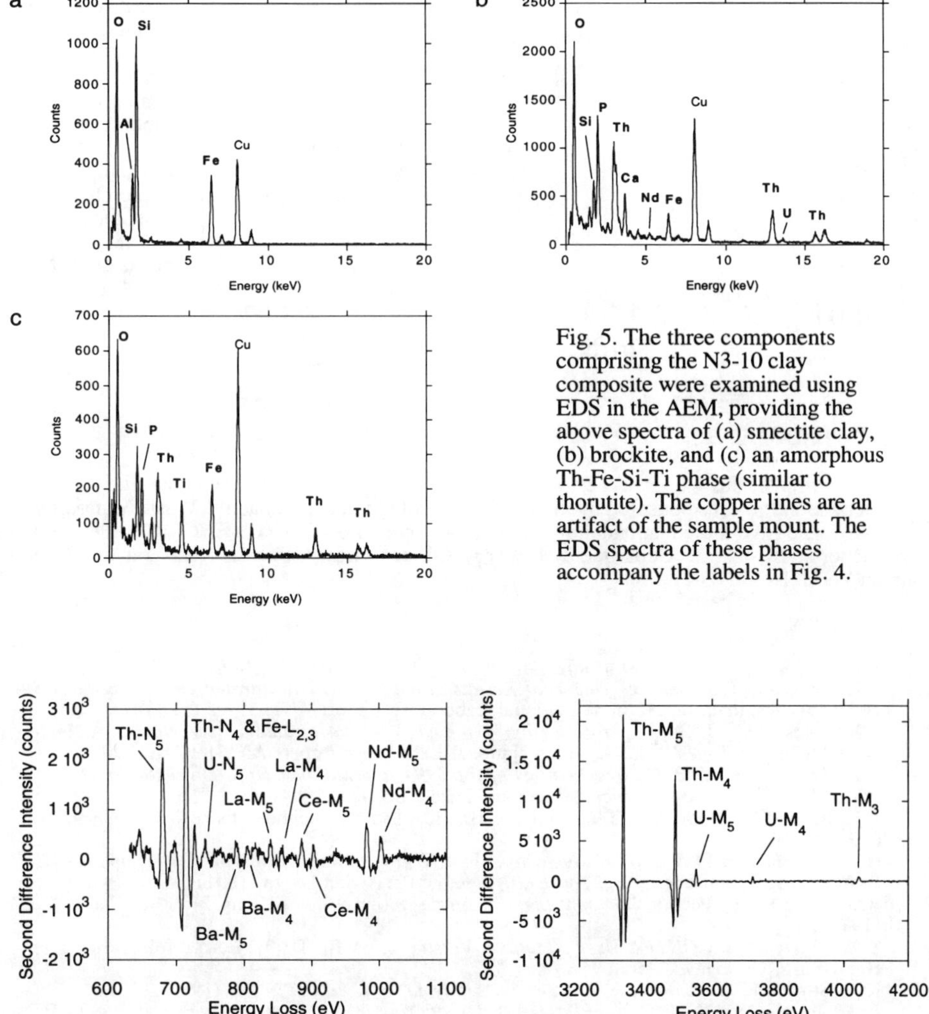

Fig. 5. The three components comprising the N3-10 clay composite were examined using EDS in the AEM, providing the above spectra of (a) smectite clay, (b) brockite, and (c) an amorphous Th-Fe-Si-Ti phase (similar to thorutite). The copper lines are an artifact of the sample mount. The EDS spectra of these phases accompany the labels in Fig. 4.

Fig. 6. The EELS spectra from the brockite indicate nearly identical distributions of REEs and actinides as that observed in brockite particles from tests terminated at times of one year or less [12].

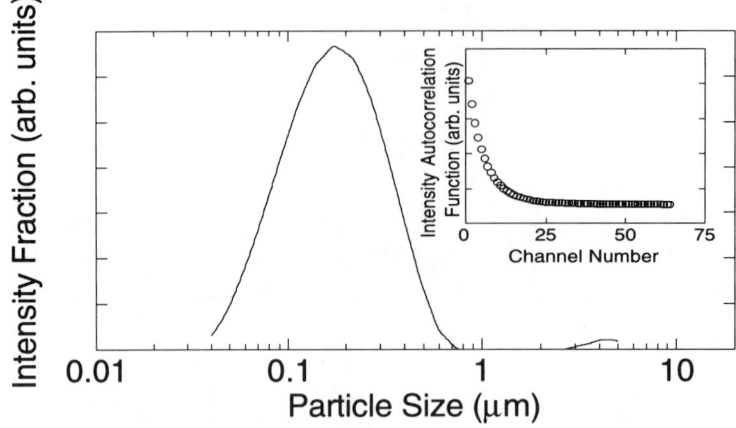

Fig. 7. Particle size distribution (from scattering intensity) with intensity autocorrelation function (inset) from the N4-2 test sampled September 1996. Note the bimodal distribution with peaks at ~0.2 and 4 μm.

ACKNOWLEDGMENTS

A portion of this work was performed under contract to the Yucca Mountain Site Characterization Project, sponsored by the DOE Office of Civilian Radioactive Waste Management at the Lawrence Livermore National Laboratory (contract number W-7405-ENG-48); other portions were supported by the U.S. Department of Energy, Office of Environmental Management, under contract W-31-109-ENG-38.

REFERENCES

1.   J. Fortner and J. K. Bates, *Mat. Res. Soc. Proc.* Vol. **412**, 205-211 (1996).
2.   J. K. Bates and T. J. Gerding, *One-Year Results of the NNWSI Unsaturated Test Procedure: SRL 165 Glass Application*, Argonne National Laboratory Report ANL-85-41 (1986).
3.   J. K. Bates and T. J. Gerding, *Application of the NNWSI Unsaturated Test Method to Actinide Doped SRL 165 Type Glass*, Argonne National Laboratory Report ANL-89/24 (1990).
4.   J. K. Bates and E. C. Buck, *Proceedings of the 1994 International High-Level Radioactive Waste Management Conference*, American Nuclear Society (1994).
5.   J. K. Bates, J. P. Bradley, A. Teetsov, C. R. Bradley, and M. Buchholtz ten Brink, *Science* **256**, 649 (1992).
6.   H. van Olphen, in *Clay Colloid Chemistry;* John Wiley & Sons, New York, 1977, pp. 92-108.
7.   B. Chu, *Laser Light Scattering*, 2nd ed., Academic Press, San Diego (1991).
8.   B. J. Berne and R. Pecora, *Dynamic Light Scattering*, John Wiley & Sons, Inc., New York (1976).
9.   S. W. Provencher, *CONTIN, User's Manual, Version 2*, EMBL-DA07, Euro. Molec. Biol. Lab., Heidelburg, Germany (March 1984).
10. R. S. Stock and W. H. Ray, *J. Polymer Sci.: Polym. Phys. Ed.*, **23**, 1393 (1985).
11. R. M. Johnsen and W. Brown, in *Laser Light Scattering in Biochemistry*, S. E. Harding, D. B. Satelle, and V. A. Bloomfield, Eds., Royal Society of Chemistry, Cambridge, 77 (1992).
12. J. A. Fortner, T. J. Gerding, and J. K. Bates, in *Environmental Issues and Waste Management Technologies in the Ceramic and Nuclear Industries*, Ceramic Trans. Vol. **61**, V. Jain and R. Palmer, eds. pp. 455-462. (1995).

Waste Glass Analytical Bias Correction Using a Reference Standard Glass

Gary L. Smith, Dennis L. Eggett, and Harry D. Smith
Pacific Northwest National Laboratory, P.O. Box 999, Richland, Washington 99352,
gl_smith@pnl.gov, dl_eggett@pnl.gov, hd_smith@pnl.gov

ABSTRACT

Analytical round robins were established to allow laboratories responsible for the analysis of (nuclear) wastes glasses to compare their analytical capabilities and techniques with one another in a non-competitive atmosphere. In addition the quality of analyses using different analytical techniques on the same materials could be compared. Analytical Round Robin 7 had two primary objectives: 1) to evaluate the effect of using the same analytical reference glass as a reference standard for bias correcting analytical results for all the participating laboratories in making the analyses; and 2) to discuss the analytical methods and results from the participating laboratories between the participants for the purpose of evaluating quality and interlaboratory consistency. This paper will primarily address the first objective.

The purpose of an analytical reference material is to bring the analytical results from various laboratories into better agreement by providing them a common reference point. No reference glass is likely to be ideal for each element in the glass being analyzed. Obviously, the closer the reference material is to the unknown, the more confidence there will be in the results because less judgement is called for. Use of the reference glass to bias correct analytical results for a standard set of analyses from several laboratories provides the data needed to determine the effectiveness of an Analytical Reference Glass (ARG) in producing interlaboratory agreement. Concurrently, whether or not the analytical results using ARG-1 are more valid or systematically biased can be evaluated using standard statistical methods. The conclusion is that bias correcting with a standard glass such as ARG-1 for results obtained by the inductively coupled plasma - atomic emission spectroscopy (ICP-AES) analytical technique produces more accurate results.

INTRODUCTION

The Materials Characterization Center (MCC) at Pacific Northwest National Laboratory[a] (PNNL) has been conducting interlaboratory test programs (Nuclear Waste Analytical Round Robins) that focus on analysis of glasses pertinent to nuclear waste repository programs. The program goal is to demonstrate and document achievable levels of performance in the analysis of nuclear waste glass. Chemical analysis data on glass waste forms produced by the nuclear waste community may need to withstand legal challenges for waste material storage qualification and for licensing repositories. Determining the quality of analytical data is important for meeting these challenges, making process operation decisions, and predicting long-term waste behavior. Methods to document the quality and reliability of the analytical data will include selection of appropriate facilities and instrumentation, use of reliable analytical methods/procedures, and application of laboratory quality control and quality assurance procedures.

Round robins give each analytical laboratory an opportunity to evaluate its current analytical capability. This evaluation is done in an absolute sense by comparison against a target composition and in a relative sense by comparison with other laboratories. These round robins are not initiated to foster competition among the participating laboratories, but to instill a spirit of cooperation and learning through interaction at the subsequent workshops, post MCC data reduction and preliminary statistical analysis, held to discuss round robin results. Although each round robin has a slightly different protocol, participating laboratories are always asked to analyze the glasses by using their normal laboratory procedures. This approach enables the participating laboratories to judge how well their laboratory procedures work. Through the interaction at the workshops, many of the participating laboratories have been able to improve their analytical capability.

The results of previous analytical round robins (1 through 6) were summarized and published by Smith and Marschman [1] in response to participants request that the round robin results be

---

[a]Pacific Northwest National Laboratory is operated for the U.S. Department of Energy by Battelle under Contract DE-AC06-76RLO 1830.

Mat. Res. Soc. Symp. Proc. Vol. 465 © 1997 Materials Research Society

published in open literature to inform clients about what constitutes "normal" precision and accuracy when nuclear waste glass components and products are chemically analyzed and to encourage the development and testing of improved analytical equipment and methods. Note that two glass matrix dissolution methods are now also American Society for Testing and Materials (ASTM) approved procedures: "Practice for Dissolution of Silicate or Acid Resistant Matrix Samples" [2] and "Practice for Flux Fusion Sample Dissolution" [3]. Another method involving microwave heating is presently under consideration. All analytical laboratories face similar limitations and problems; this paper will present a summary of Nuclear Waste Analytical Round Robin 7, as was done for analytical round robins (1 through 6), pointing out the level of interlaboratory agreement to be expected in data comparisons so that the data users do not overrate interlaboratory differences.

Researchers who use analytical laboratories need to distinguish between the results that are "routinely achievable and cost-effective" and "best achievable regardless of cost," and when to request and expect these results. Often, because of sample inhomogeneity, field sampling is the limiting step in determining accuracy and intrasample analytical precision. Clients of analytical laboratories need to be aware of this factor and work with the analytical staff to achieve results appropriate for the intended use. It is hoped that this paper will be useful to drafters of waste acceptance criteria, will foster further laboratory interaction to solve existing and future problems, and will publicize analytical methods suitable for the nuclear waste community and encourage their use at additional laboratories.

The inclusion of radioactive elements in the waste glass add particular difficulties to analytical procedures. Additional safeguards and protocols are added which further increase the complexity of the analytical process and require a greater effort to achieve acceptable results.

One of the specific primary objectives of Round Robin 7 was to evaluate the use of a standard reference glass to add consistency and accuracy to nuclear waste glass analyses among the laboratories charged with their analysis by bias correcting to a common standard. To make this evaluation, the participating analytical laboratories were asked to use an analytical reference glass (ARG), ARG-1, and were instructed how to use it as a reference to bias correct their results so that the results from each laboratory would be statistically comparable.

ROUND ROBIN 7 PROTOCOL

Round Robin 7 Protocol

Two glass sample unknowns were used in Round Robin 7, designated Unknown 1 and 2. Unknown 1 is the Environmental Assessment (EA) glass [4]. Unknown 2 is a simulated West Valley Reference 6 glass composition containing uranium and thorium. Both the EA and the West Valley Reference 6 glass were analyzed by Corning, Inc., and the Round Robin 7 participants.

The two different unknown test glasses and a quantity of ARG-1 [5] were sent to each Round Robin 7 participant in the form of a -100 mesh powder. The composition of ARG-1 was included in an attached table. The participants were asked to report concentrations of only those elements listed in the ARG-1 composition table which were listed on the supplied report forms for Round Robin 7 Unknown Glasses 1 & 2.

Round Robin 7 required that each of the two test glasses be analyzed twice. The first analysis of each glass was to be performed using the procedures and reference standard that the participant normally used for such an analysis. The second analysis of both test glasses utilized the analytical reference material (ARG-1) as a reference standard to bias correct the analytical results.

Specific instructions were as follows:

For each glass (ARG-1, Unknown 1, and Unknown 2) make a single dissolution, digestion, fusion, etc. (whatever procedure is used in your individual laboratory). For each glass complete an analysis in triplicate at 0 hours, 4 hours, and 8 hours. This will provide nine analyses for each glass. Analyze ARG-1, Unknown 1, and Unknown 2 as separate samples and report all uncorrected data to the Materials Characterization Center (MCC). Analysis of the glasses should be run in parallel. (The replicate analyses will provide a measure of your precision.) For each replicate, report your average analysis for

each glass based on your customary procedure and the analysis you would report using
ARG-1 as the reference standard (bias correction). In addition to the three replicate
analyses of each unknown glass; determine the reduction state of each unknown glass and
for ARG-1 by measuring the ferrous/ferric ratio (Fe^{+2}/Fe^{+3}) and report these values. The
value given in the ARG-1 composition table is a nominal value and is the average of three
analyses made by Corning. Report the data on the forms provided.

<u>Statistical Analysis of Analytical Data</u>

There were a total of 11 participating laboratories which included: Ames Laboratory, Argonne
National Laboratory, Catholic University of America, Los Alamos National Laboratory, three
different analytical laboratories at Pacific Northwest National Laboratory, three different analytical
laboratories at Westinghouse Savannah River Company, and the U. S. Geological Survey. Six of
the labs submitted two distinct sets of data. Thus, there were 17 resulting sets that were used in
the statistical analysis. The data were separated into sets as follows.

•All Data Set - this contains the data from all 17 data sets.
•Spurious Data Set - three of the data sets had analytical results that were consistently biased.
These data sets were considered to be outliers and were removed from consideration in all data
sets except the All Data Set.
•Reduced Data Set - the All Data Set with the Spurious Data Set removed
•ICP-AES - the data sets from the Reduced Data Set which were the results of ICP-AES
analysis. ICP-AES is the most frequently used analysis method and is the most widely
accepted.
•Other Data Set - the data sets from the Reduced Data Set which were the result of an analysis
method other than ICP-AES.

Statistical analyses were performed on all of the data sets given in Table 1 below.

Table 1. Data Sets from the 11 Participants in Round Robin 7.

Analysis Method	All Data	Reduced Data	Spurious Data	ICP-AES Data	Other Data
ICP - AES	A		S		
ICP - AES	A	R		I	
ICP - AES	A	R		I	
DCP	A	R			O
DCP	A	R			O
ICP - MS	A	R			O
ICP - AES	A	R		I	
ICP - MS	A		S		
Laser Ablation	A		S		
ICP - AES	A	R		I	
ICP - AES	A	R		I	
ICP - MS	A	R			O
ICP - MS	A	R			O
ICP - AES	A	R		I	
ICP - AES	A	R		I	O
XRF	A	R			
ICP - AES	A	R		I	

RESULTS AND DISCUSSION

A main objective of Round Robin 7 was to see if bias correction using an analytical standard (ARG-1 in this case) could improve the results of analyses of unknown samples. Initially it was thought that the bias correction had not helped consistently, but that was because the results had not been completely examined. Subsequently, it was discovered that three of the data sets contained results that were different from the other data sets (identified in the previous section as spurious data sets). When these sets were removed, the results from bias correction improved. Also, further discussions indicated that it might be beneficial to look at the inductively coupled plasma - atomic emission spectroscopy (ICP-AES) results separately (since this is the most widely used, best understood, and most consistent analysis method).

Statistical analyses were performed on all of the data sets described above. The consensus at the Round Robin 7 meeting was that the Joint Bias correction method was unacceptable so only the Individual Bias correction method was applied. Looking at the results for Unknown Glasses 1 and 2, it is observed that, on a regular basis, the bias correction shows a reduction in the lab-to-lab variability and generally little effect in the within-lab variability. The results are summarized in Table 2.

Table 2. A Comparison of the Effectiveness of Bias Correction with the Different Data Sets.

Oxide	[nom] ARG Wt%	[nom] Unk1 Wt%	[nom] Unk2 Wt%	All Data Set	ICP-AES Data Set	Other Data Set	Reduced Data Set	Summary
Al_2O_3	4.73	3.70	6.02	<	<<	=	<	<
B_2O_3	8.67	11.3	12.8	>	<	<	<	<
BaO	0.088	NA	0.14	=	=	=	=	=
CaO	1.43	1.12	0.57	>	<	>>	>	>
Cr_2O_3	0.093	NA	0.17	=	<	=	=	=
CuO	0.004	NA	0.055	>	=	>	>	>
Fe_2O_3	14.0	7.38	11.9	<	<<	>	<	<
K_2O	2.71	0.04	5.13	=	=	=	=	=
Li_2O	3.21	4.26	3.68	=	<	=	=	=
MgO	0.86	1.72	0.92	<	<<	>	<	<
MnO_2	2.31	1.34	1.09	< >	<	= >	< >	<
Na_2O	11.5	16.8	8.12	=	<	<	<	<
NiO	1.05	0.57	0.26	=	<=	=	=	=
P_2O_5	0.22	NA	1.18	<	<	<=	<	<
SiO_2	47.9	48.73	40.9	<	<>	<=	<	<
TiO_2	1.15	0.70	0.80	<	<<	<	<<	<
ZnO	0.02	NA	0.031	<=	<	<<>>	<<=	<
ZrO_2	0.13	0.46	1.30	=	<	<	<	<

Where << indicates a very large reduction in variability, < indicates a reduction in variability, = indicates no reduction in variability, > indicates an increase in variability, and >> indicates a very large increase in variability. When two symbols are given, the first refers to Unknown 1 and the second to Unknown 2. The results indicate that there is a definite reduction in data variability with the bias correction. Those places where an increase in variability is observed are primarily with elements that are present in small amounts (i.e., < 0.5 wt %). From this it is concluded that there is an advantage in bias correcting using a reference glass.

A value that has been calculated in past round robins and used to evaluate progress during the course of the round robins is average relative error. The average relative error is calculated using

$$\text{Ave. Rel. Err. (\%)} = \frac{100}{N}\left[\Sigma\left|\frac{\Delta}{\text{nom}}\right|\right] \tag{1}$$

where [nom] is the nominal concentration in the samples for each element, Δ is the difference between nominal and average analyzed values, and N is the number of elements included. The results of this calculation for each element, for each unknown, and considering all analyses and just ICP-AES analyses are summarized in Tables 3 and 4. The Average Relative Errors summarized from Tables 3 and 4 are as follows:

•Table 3 - EA Glass (Unknown 1)
Participating Round Robin 7 Laboratories
As Analyzed   5.07%
Bias Corrected   6.03%

Laboratories Using ICP-AES in Round Robin 7
As Analyzed   6.56%
Bias Corrected   4.98%

•Table 4 - West Valley Reference Glass #6 (Unknown 2)
Participating Round Robin 7 Laboratories
As Analyzed   7.32%
Bias Corrected   4.94%

Laboratories Using ICP-AES in Round Robin 7
As Analyzed   13.49%
Bias Corrected   3.97%.

"As Analyzed" and "Bias Corrected" are as defined in the Round Robin 7 Protocol section. The results of this analysis also indicate that bias correcting using a reference glass reduces interlaboratory scatter about the nominal composition.

The fact that the average relative error as defined by equation (1) varies from Unknown 1 to Unknown 2 and is a function of analysis technique, suggests that there are a number of interacting factors active in causing analytical results scatter. Two such interacting factors are the closeness in composition of the reference glass to the unknown and the accuracy of the analytical scheme for each element. If the reference standard has a level of concentration of an element that is significantly different from the unknown glass and the technique for analyzing for that element is not very precise, the bias correction may not be affective because of the large potential for scatter and the magnitude of the extrapolation of the bias correction. These factors suggest the need for a grid of analytical reference glasses which provide good range coverage for the elements that are both important in a regulatory sense and difficult to determine in a precise manner. It is anticipated that Round Robin 8 will further resolve the issues by introduction of another analytical reference glass (ARG-2) by demonstrating and quantifying the definition of "normal" precision and accuracy, by continuing to compare results from a variety of analytical methods, by demonstrating and documenting achievable levels of performance of the variety of analytical methods, by developing guidelines and recommendations for analytical practices, and by providing assistance in enhancing quality and interlaboratory consistency of analyses.

Table 3. Average Relative Errors for EA Glass (Unknown 1).

		Environmental Assessment (EA) Glass (Unknown 1)											
		Laboratories Participating in Round Robin 7						Laboratories Using ICP-AES in Round Robin 7					
		As Analyzed			Bias Corrected			As Analyzed			Bias Corrected		
Oxide	Nominal Composition	# Labs	wt%	SD	# Labs	wt%	SD	# Labs	wt%	SD	# Labs	wt%	SD
Al_2O_3	3.70	13	3.46	0.848	13	3.75	0.155	8	3.39	1.076	8	3.75	0.132
B_2O_3	11.30	10	11.26	0.360	10	11.40	0.192	7	11.20	0.324	7	11.40	0.207
BaO	-	5	0.004	0.008	5	0.004	0.007	3	0.007	0.006	3	0.007	0.005
CaO	1.12	14	1.13	0.228	14	1.19	0.447	8	1.08	0.239	8	1.11	0.058
Cr_2O_3	-	6	0.012	0.023	6	0.009	0.016	3	0.015	0.022	3	0.010	0.009
CuO	-	3	0.129	0.374	3	0.211	0.587	-	0.008	-	-	0.001	-
FeO	1.45	-	-	-	-	-	-	-	-	-	-	-	-
Fe_2O_3	7.38	14	8.80	1.299	14	9.10	0.460	8	8.79	1.754	8	9.02	0.172
K_2O	0.04	7	0.062	0.071	7	0.066	0.063	4	0.078	0.081	4	0.084	0.063
La_2O_3	0.42	6	0.376	0.123	-	-	-	3	0.408	0.020	-	-	-
Li_2O	4.26	12	4.29	0.507	12	4.24	0.438	7	4.25	0.319	7	4.30	0.183
MgO	1.72	14	1.72	0.492	14	1.81	0.463	8	1.62	0.431	8	1.75	0.160
MnO_2	1.34	13	1.60	0.240	13	1.64	0.178	7	1.66	0.076	7	1.66	0.019
Na_2O	16.80	10	16.42	2.000	10	16.64	1.892	6	16.44	2.555	6	16.60	2.397
Nd_2O_3	-	1	0.007	-	-	-	-	1	0.007	-	-	-	-
NiO	0.57	10	0.593	0.105	10	0.592	0.110	6	0.581	0.059	6	0.568	0.028
P_2O_5	-	4	0.114	0.404	4	0.079	0.277	-	-	-	-	-	-
SiO_2	48.73	11	47.88	2.111	11	48.40	0.988	7	47.97	2.614	7	48.51	1.129
SrO	-	1	0.005	-	1	0.004	-	1	0.005	-	1	0.004	-
TiO_2	0.70	13	0.694	0.155	13	0.696	0.016	8	0.669	0.184	8	0.694	0.016
ZnO	-	11	0.030	0.018	10	0.021	0.004	6	0.025	0.007	5	0.021	0.004
ZrO_2	0.46	10	0.462	0.039	10	0.444	0.036	6	0.463	0.019	6	0.445	0.039
Sum	99.99		99.048			100.296			98.666			99.934	
Fe^{+2}/Fe^{+3}	0.217	5	0.236	0.033				4	0.239	0.035			

Table 4. Average Relative Errors for West Valley Reference Glass-6 (Unknown 2).

Oxide	Nominal Composition	West Valley Reference Glass-6 (Unknown 2)											
		Laboratories Participating in Round Robin 7						Laboratories Using ICP-AES in Round Robin 7					
		As Analyzed			Bias Corrected			As Analyzed			Bias Corrected		
		# Labs	wt%	SD	# Labs	wt%	SD	# Labs	wt%	SD	# Labs	wt%	SD
Al_2O_3	6.02	9	5.69	1.516	9	6.35	0.380	6	5.53	1.830	6	6.36	0.450
B_2O_3	12.80	7	12.55	0.327	7	12.72	0.282	5	12.51	0.318	5	12.71	0.324
BaO	0.14	8	0.136	0.024	8	0.131	0.023	5	0.137	0.019	5	0.133	0.017
CaO	0.57	10	0.578	0.108	10	0.631	0.158	6	0.567	0.117	6	0.604	0.030
CeO_2	0.17	3	0.167	0.100	-	-	-	2	0.200	0.000	-	-	-
CoO	0.019	-	-	-	-	-	-	-	-	-	-	-	-
Cr_2O_3	0.17	9	0.170	0.027	9	0.161	0.026	5	0.166	0.013	5	0.157	0.009
Cs_2O	0.079	2	0.063	0.033	-	-	-	1	0.050	-	-	-	-
CuO	0.055	8	0.050	0.005	-	-	-	5	0.050	0.003	-	-	-
Fe_2O_3	11.90	10	11.13	1.620	10	11.72	0.922	6	11.15	2.087	6	11.78	0.529
K_2O	5.13	8	4.63	0.646	8	5.08	0.496	5	4.52	0.734	5	5.16	0.581
La_2O_3	0.048	5	0.042	0.020	-	-	-	2	0.042	0.010	-	-	-
Li_2O	3.68	8	3.61	0.414	8	3.56	0.358	5	3.54	0.234	5	3.63	0.111
MgO	0.92	10	0.895	0.200	10	0.954	0.076	6	0.850	0.222	6	0.944	0.044
MnO_2	1.09	9	1.15	0.211	9	1.20	0.323	5	1.14	0.029	5	1.15	0.031
MoO_3	0.039	5	0.042	0.010	-	-	-	2	0.038	0.004	-	-	-
Na_2O	8.12	7	7.89	0.405	7	8.04	0.202	4	7.87	0.405	4	8.04	0.166
Nd_2O_3	0.13	5	0.214	0.192	-	-	-	2	0.322	0.050	-	-	-
NiO	0.26	7	0.259	0.036	7	0.262	0.038	4	0.260	0.037	4	0.264	0.045
P_2O_5	1.18	6	1.06	0.452	6	0.914	0.126	3	0.900	0.480	3	0.918	0.174
Pd	0.031	5	0.168	0.289	-	-	-	2	0.339	0.084	-	-	-
Pr_6O_{11}	0.059	3	0.124	0.255	-	-	-	-	-	-	-	-	-
Rh	0.019	3	0.007	0.001	-	-	-	-	-	-	-	-	-
Ru	0.033	5	0.021	0.017	-	-	-	2	0.029	0.002	-	-	-
SO_3	0.26	-	-	-	-	-	-	-	-	-	-	-	-
SiO_2	40.90	8	39.95	1.397	8	40.88	1.636	5	39.75	0.888	5	41.04	1.713
Sm_2O_3	0.06	3	0.044	0.048	-	-	-	-	-	-	-	-	-
SrO	0.024	7	0.025	0.008	1	0.009	-	4	0.025	0.006	1	0.009	-
ThO_2	3.36	4	2.60	0.736	-	-	-	3	2.80	0.306	-	-	-
TiO_2	0.80	9	0.769	0.188	9	0.800	0.039	6	0.751	0.213	6	0.807	0.025
UO_2	0.56	6	0.740	0.902	-	-	-	1	1.13	-	-	-	-
Y_2O_3	0.042	5	0.034	0.015	-	-	-	2	0.040	0.000	-	-	-
ZnO	0.031	8	0.030	0.008	7	0.026	0.008	5	0.030	0.011	4	0.026	0.006
ZrO_2	1.30	7	1.28	0.195	7	1.24	0.093	4	1.25	0.210	4	1.25	0.123
Sum	99.999		96.118			94.678			95.986			94.982	
Fe^{+2}/Fe^{+3}	0.010	3	0.026	0.025				3	0.026	0.025			

CONCLUSIONS

It is concluded that there is an advantage to be gained in bias correcting using a reference glass. This was one of three conclusions drawn by Smith and Marschman [1] as a necessary condition to improve interlaboratory agreement.

It is recognized that a single reference glass will not appropriately cover the range of compositions of glasses. Certain elements such as silica can be difficult to put in and keep in solution particularly by one digestion regimen. The best analysis results for the particular element are obtained when the reference glass has a similar level of the element as the unknown. Reference glass(es) should be used to evaluate the relative accuracy of different analytical techniques or analytical laboratories (e.g. regulatory issues), to provide a common calibration point for multiple laboratories analyzing the same unknown(s) as for Round Robin 7, to maintain analytical system calibration in laboratories or production facilities analyzing glasses. Cost of additional analyses versus benefit provides the ultimate basis for how much a reference glass is used.

The analytical reference glasses ARG-1 and ARG-2 (West Valley Reference Glass 6) are available from PNNL MCC (Gary L. Smith, 509-372-1957) in quantities for use as analytical standards. The EA Glass is available from Savannah River Laboratory (WSRC) (Carol M. Jantzen, 803-725-2374).

ROUND ROBIN 8

Round Robin 8 is presently in the planning stage but would include radionuclide analyses. Evaluation of the relative merits of the various glass dissolution techniques (sodium peroxide fusion, potassium hydroxide fusion, microwave acid dissolution) as applied to (a) nuclear waste material(s) with and without bias correction is anticipated. Another important objective of Round Robin 8 will be to determine the number and compositions of additional analytical reference glasses needed. Additional reference glasses would complement the present reference glasses focusing on the critical waste glass elements which must be measured but are difficult to quantify.

ACKNOWLEDGMENTS

The authors thank William M. Gerry and J. Leland Daniel of PNNL and Linda P. Adams and Douglas E. Goforth of Corning, Inc., for their extensive work in coordination and data analysis, respectively. In addition, the authors appreciate the Round Robin 7 participants for their efforts in producing with their own funding the analytical data which makes this report possible. Finally, the authors express their gratitude to the U.S. Department of Energy and both West Valley Nuclear Services and Savannah River Laboratory for the financial support of this work.

REFERENCES

1. G. L. Smith and S. C. Marschman, "Nuclear Waste Analytical Round Robins 1-6 Summary Report," in Scientific Basis for Nuclear Waste Management XVII, edited by A. Barkatt and R. A. Van Konynenburg (Mater. Res. Soc. Proc. 333, Pittsburgh, PA 1993), p. 461-472.

2. American Society for Testing and Materials, Standard C1317-95 titled, "Practice for Dissolution of Silicate or Acid Resistant Matrix Samples" ASTM, West Conshohocken, PA.

3. American Society for Testing and Materials, Standard C1342-96 titled, "Practice for Flux Fusion Sample Dissolution" ASTM, West Conshohocken, PA.

4. C. M. Jantzen, N. E. Bibler, D. C. Beam, C. L. Crawford, and M. A. Pickett, "Characterization of the Defence Waste Processing Facility (DWPF) Environmental Assessment (EA) Glass Standard Reference Material (U)," WSRC-TR-92-346, Rev. 1, Westinghouse Savannah River Company, Aiken, SC (1993).

5. G. L. Smith, "Characterization of Analytical Reference Glass-1 (ARG-1)," PNL-8992, Pacific Northwest National Laboratory, Richland, WA (1993).

FURTHER DEVELOPMENT OF THE JANTZEN - PLODINEC MODEL OF GLASS DURABILITY

B. A. Shakhmatkin*, N. M. Vedishcheva*, A. S. Aloy**, M. J. Plodinec***, S. L. Marra***
*   Institute of Silicate Chemistry of the Russian Academy of Sciences, Odoevskogo Str. 24/2, St. Petersburg, 199155, Russia, natalia@termex.spb.su
** Khlopin Radium Institute, 2nd Murinski Av. 28, St. Petersburg, 194021, Russia
*** Westinghouse Savannah River Co., PO Box 616, Aiken, SC 29802

ABSTRACT

The hydration thermodynamic approach to the prediction of glass durability was originally applied to nuclear waste glasses by Jantzen and Plodinec. This approach is useful for control of the production of nuclear waste glasses. However, improvements are necessary if the approach is to be extended to different glasses, particularly those with higher alkali metal concentrations. This is of special significance for vitrification of the salt wastes at Hanford. Various methods for improving the predictive power of the approach have been examined. Combining a more accurate representation of the alkali metal species in the glass with a more rigorous thermodynamic approach is a promising avenue to improved predictive power.

INTRODUCTION

The purpose of vitrifying nuclear waste is to produce a glass form which will hinder the release of radionuclides into groundwater. Clearly, this means that the reactivity of the waste glass with water must be minimized. This can best be accomplished through proper formulation of the waste glass. Proper formulation of waste glasses requires a means to predict the durability of glass from its composition. Jantzen and Plodinec[1] have refined an approach to predict glass durability based on hydration thermodynamics. This approach implies that the logarithm of the extent of reaction in short-term glass-water leach tests should be proportional to a free energy of hydration of the glass, calculated from the composition of the glass. This very simple approach forms the basis for the model being used to control the durability of the glass being produced in the Defense Waste Processing Facility in the US (This paper deals with a predecessor of the model being used for control in the DWPF. Details of that model - and a complete bibliography on this type of model - will appear in a forthcoming publication by C. M. Jantzen.).

However, the approach is based on several approximations, and is very dependent on the set of glass oxide components used to calculate the free energy of hydration. While these factors may not affect the predictive power when the range of glass compositions is restricted, improvements are likely to be necessary if the approach is to be extended to different glasses, particularly those with higher alkali metal concentrations. This is of special significance for vitrification of the salt wastes at Hanford, which result in glasses with ≥ 20 wt% Na_2O. In the following sections, we examine the limits of applicability of the current approach, and some potential modifications which should improve its ability to predict the reactivity of glasses toward water.

DISCUSSION

The limits of applicability of the Jantzen - Plodinec model

The gradient of the chemical potentials ($\Delta\mu$'s) of the components of a glass is the driving force for its dissolution in water. Hence, a rigorous quantitative description of this process can be performed through an analysis of the values of the $\Delta\mu$'s. Since a prediction of the long-term durability of a system does not imply a consideration of its state in every elementary volume at every given moment of time, the above problem can be solved by a consideration of the difference be-

Mat. Res. Soc. Symp. Proc. Vol. 465 © 1997 Materials Research Society

tween the chemical potentials of components in the glass and in solution.

The thermodynamic approach refined by Jantzen and Plodinec[1] is implicitly based on the theory of completely associated solutions. The gross compositions of glasses in various silicate systems are considered in Reference 1 as a combination of certain compounds and the constituent oxides. Thus, sodium silicate glasses are presented as a combination of sodium metasilicate and SiO_2:

$$X_{Na_2O} \cdot Na_2O \cdot \left(1 - X_{Na_2O}\right) SiO_2 \rightarrow X_{Na_2O}\left(Na_2O \cdot SiO_2\right) + \left(1 - 2X_{Na_2O}\right) SiO_2 \quad (1)$$

where X_{Na2O} is the mole fraction of Na_2O in the glass in question. This leads to the following expressions for the chemical potentials of the oxide components. For a given glass, $\Delta\mu_{SiO2}$ and $\Delta\mu_{Na2O}$ are given by:

$$\Delta\mu_{SiO_2} = RT \; \ln \; X^f_{SiO_2} = RT \; \ln \left[\frac{1 - 2X_{Na_2O}}{1 - X_{Na_2O}}\right] \quad (2)$$

where X^f_{SiO2} is the mole fraction of SiO_2 which is not incorporated into sodium metasilicate, and

$$\Delta\mu_{Na_2O} = \Delta G^{\circ}_{Na.Si} + \Delta\mu_{Na.Si} - \Delta\mu_{SiO_2} =$$

$$\Delta G^{\circ}_{Na.Si} + RT \; \ln X_{Na.Si} - \Delta\mu_{SiO_2} = \quad (3)$$

$$\Delta G^{\circ}_{Na.Si} + RT \; \ln\left[\frac{X_{Na_2O}}{1 - X_{Na_2O}}\right] - RT \; \ln\left[\frac{1 - 2X_{Na_2O}}{1 - X_{Na_2O}}\right]$$

where $\Delta G^{\circ}_{Na.Si}$ is the standard Gibbs free energy of formation of sodium metasilicate from the oxides (Na_2O and SiO_2); $\Delta\mu_{Na.Si}$ and $X_{Na.Si}$ are, respectively, the chemical potential and the mole fraction of the $Na_2O \cdot SiO_2$ species in the glass being considered.

As can be inferred from Equations (2) and (3), the composition of a given glass unambiguously determines the chemical potentials of the constituent oxides. However, a limitation of this approach is that restricting the glass to be a combination of metasilicate and SiO_2 does not accurately represent the species actually present in the glass. This is shown in Figure 1, where the numbers of moles of sodium metasilicate and SiO_2 in various glasses determined using the approach in reference 1 are compared with the numbers of moles of a larger variety of species present in sodium silicate glasses according to the model of ideal associated solutions (2,3).

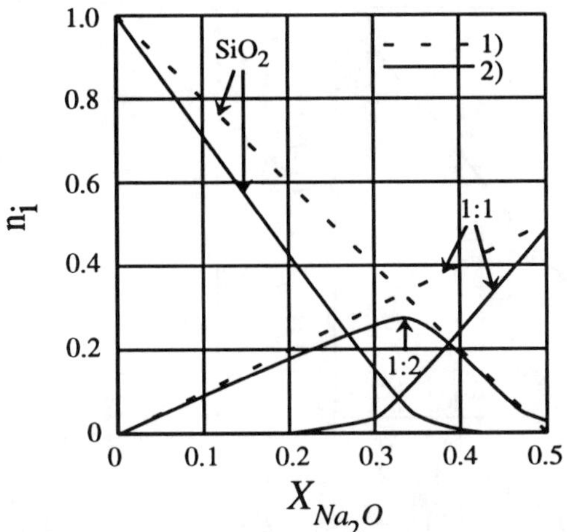

Figure 1. The numbers of moles, n_i, of various species present in sodium silicate glasses. 1 - Jantzen-Plodinec model[1]; 2 - model of ideal associated solutions[2, 3], 1:1 denotes $Na_2O \cdot SiO_2$, 1:2 denotes $Na_2O \cdot 2SiO_2$, and SiO$_2$ denotes silica not incorporated into either meta- or disilicates.

This model considers all chemical processes proceeding in the system in question. Figure 2 shows the chemical potentials of Na_2O and SiO_2 in these glasses calculated using the set of species suggested by Jantzen and Plodinec[1], and using the set of species used by Shakhmatkin[3]. Shakmatkin's values were calculated based on high temperature electrochemical measurements of sodium silicate, as a function of soda content. As the fugure shows, the latter set leads to calculated values of $\Delta\mu_{Na2O}$ and $\Delta\mu_{SiO2}$ which practically coincide with the experimental data.[3] The maximum discrepancy between the data and the more extensive set of species does not exceed 0.5 kcal/mole, while the chemical potentials of Na_2O and SiO_2 modeled on the basis of only silica and metasilicate noticeably deviate from the corresponding experimental values. The observed discrepancy between two series of the modeled chemical potentials of oxides determines the limits of applicability of Jantzen-Plodinec's model. In Figure 3, the extent of reaction of sodium silicate glasses with water (reported as the normalized mass loss of Si from the glass to the leaching solution and denoted "NL$_{Si}$") is compared with the theoretical dependence given in Reference 1. The experimental values were determined by analyzing solutions resulting from exposing samples of the glasses to boiling deionized water. The results show that the experimental and theoretical values of log NL$_{Si}$ are close to each other for compositions with free energies of hydration, ΔG_{hydr}, of approximately -7 kcal/mol. This is the composition region corresponding to the specified durability limit for nuclear waste glasses in the US. Thus, the Jantzen - Plodinec approach is satisfactory for control of production of nuclear waste glass in the US.

In the case of multicomponent systems, the situation is more complicated. As follows from our calculations and the NMR data,[5] in the composition regions of alkali borosilicate systems where $X_{M20}/X_{B203} < 0.5$ (X_{M20} and X_{B203} are the mole fractions of M_2O and B_2O_3 in the glass) alkali silicate groupings, $Na_2O \cdot nSiO_2$, do not form at all and the glasses can be considered as a solution of the components $M_2O \cdot nB_2O_3$ and SiO_2.

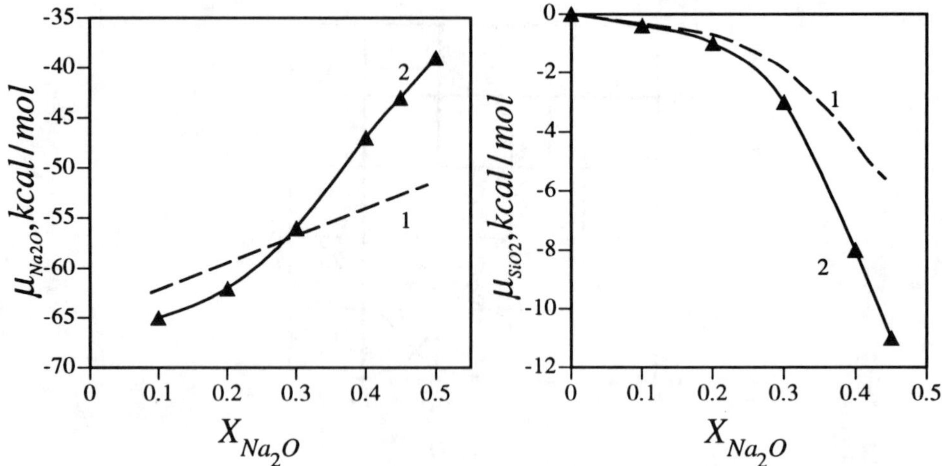

Figure 2. The chemical potentials of Na_2O and SiO_2 in sodium silicate glasses. 1 - Jantzen-Plodinec model [1]; 2 - model of ideal associated solutions [2, 3]; filled triangles - experimental data [3].

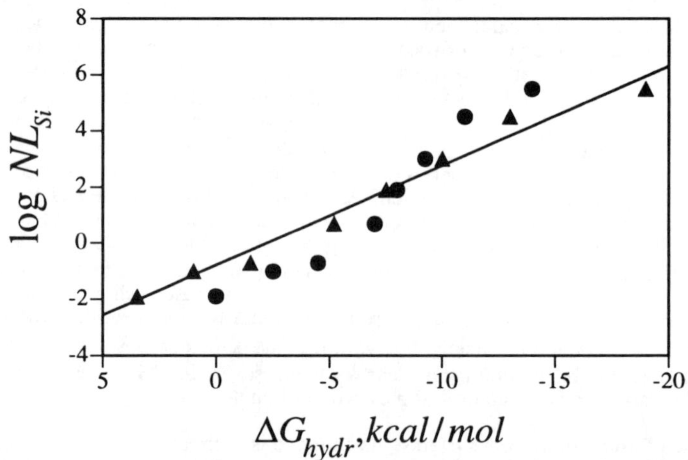

Figure 3. The mass loss (log NL_{Si}) of sodium silicate glasses in contact with water. 1 - theoretical dependence suggested by Jantzen and Plodinec [1]; 2 - experimental data as a function of ΔG_{hydr} calculated using the set of species reported in [1]; 3 - same experimental data as a function of the modified values of ΔG_{hydr} discussed here. Experimental details in reference 4.

Methods for Improving the Jantzen - Plodinec model

Although the Jantzen-Plodinec model cannot be considered completely rigorous, there is no question that it deserves attention due to the simplicity of the approach, and the ease with which it can be applied. There are two possible ways to improve the predictive power of the model. The first one treats the values of ΔG_{hydr} for species present in glass as adjustable parameters. In the case of the sodium silicate system, this is equivalent to the variation of the value $\Delta G^{\circ}_{Na.Si}$ in equation 3 and, hence, to the parallel displacement of the modeled curve for the potential, $\Delta\mu_{Na2O}$. This method is effective if relatively limited composition intervals are considered.

The second way involves the use of a more rigorous approach based on the principles of irreversible thermodynamics together with adequate modeling of the thermodynamic potentials for hydration of glasses. Let us consider in greater detail the driving force of this process, i.e., the change in the free energy of the glass as it hydrates. This can be represented as

$$\Delta G_{hydr} = G_{solution} - G_{glass} = \sum_{solution\ species} n_s \cdot \mu_s - \sum_{glass\ species} n_g \cdot \mu_g =$$

$$\sum_i n_i \cdot \left(v_{i,solution} \cdot \mu^o_{i,solution} - v_{i,glass} \cdot \mu^o_{i,glass} \right) + RT \sum_i n_i \cdot \ln\left(\frac{\prod_i a_{i,solution}^{v_{i,solution}}}{\prod_i a_{i,glass}^{v_{i,glass}}} \right) = \tag{4}$$

$$\Delta G_{hydr}^{J-P} + \Delta G_{hydr}^{conf}$$

where G_{gl} and G_{sol} are the free energies of a given glass and of the solution obtained from its hydration ; n_g and n_s, are, respectively, the numbers of moles of the species present in the glass (in other words, reactants) and those of the products of their hydration; μ_g and μ_s are the chemical potentials of the reactants and those of their hydrated products; $\mu^{\circ}_{i,glass}$ and $\mu^{\circ}_{i,solution}$ are their corresponding standard chemical potentials; $v_{i,glass}$ and $v_{i,solution}$ are the stoichiometric coefficients of the reactants and the products respectively (note that the values n_g and n_s are related to each other through the stoichiometric coefficients); the a_i-values are the activities of the reactants and products; ΔG_{hydr}^{J-P} is the free energy of hydration from the Jantzen - Plodinec model; and ΔG_{hydr}^{conf} is the configurational contribution to the change in the free energy of a given glass composition in the process of hydration.

Thus, as follows from Equation 4, in addition to the standard free energies of hydration of various species present in a glass, which are represented by

$$\sum_i n_i \cdot \left(v_{i,solution} \cdot \mu^o_{i,solution} - v_{i,glass} \cdot \mu^o_{i,glass} \right)$$

the configurational contribution to ΔG_{hydr} has also been taken into account. It should be noted that the contribution from the term ΔG_{hydr}^{conf} increases rapidly in passing from binary to multicomponent systems. For sodium silicate glasses, the introduction of this correction improves the

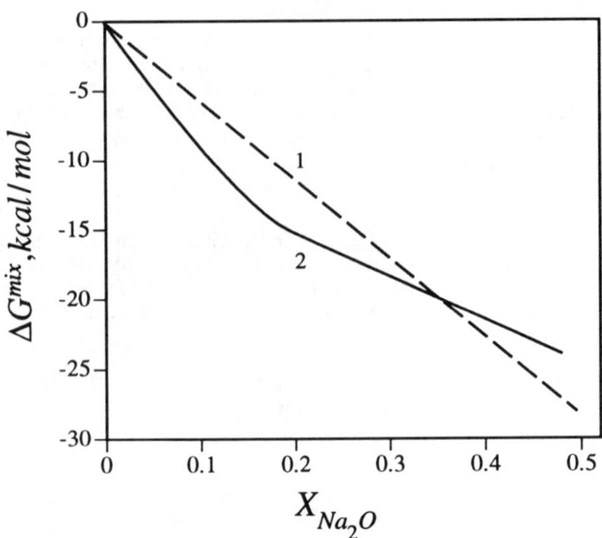

Figure 4. Gibbs free energies of mixing of sodium silicate glasses. Calculations are based on the sets of species reported in: 1 - Reference 1; 2 - Reference 3. Experimental data from Reference 6 are not shown here because they practically coincide with the solid curve 2.

agreement between the experimental free energy of mixing with the corresponding modeled potentials.

A further improvement of the Jantzen - Plodinec model can be achieved through the use of more representative sets of species present in a given glass. In general, for this purpose, a rigorous modeling of the chemical structure of various types of industrial glasses is necessary followed by a generalization of the results obtained (The notion of the chemical structure of glasses implies a relative content of various species in them[2]). For sodium silicate glasses, the results of calculations of ΔG^{mix} based on a more representative set of species[3] are shown in Figure 4. A considerable improvement of the value calculated is obvious.

Figure 3 shows the refined correlation between the values NL_{Si} and the Gibbs free energies of hydration. The correlation has been improved by taking into account the contribution to the potential G_{hydr} from the configurational term and by using a more representative set of species determined using the experimental data reported in Reference 3 for the sodium silicate system. It should be noted that, as follows from our calculations, for glasses containing over 40 mol % Na_2O, the character of solution processes differs from that in the case of low-alkali compositions. The difference is in the fact that the state of equilibrium between a high-alkali glass and its solution cannot be achieved in principle. If this is taken into account and the concentration regions $0 < X_{Na2O} < 0.4$ and $X_{Na2O} > 0.4$ are considered separately, the correlation coefficient of the linear dependence log $NL_{Si} = f(G_{hydr})$ is as high as 0.987, provided that the above-mentioned corrections are used to calculate G_{hydr}.

These corrections have also been estimated for cesium-containing multicomponent glasses. The results obtained have revealed that taking the corrections discussed above into consideration enables a discrepancy between the experimental rate of solution of the glasses and that predicted by the correlation given in Reference 1 to be explained.

CONCLUSIONS

As shown in Figure 3, application of the Jantzen-Plodinec approach does not yield a strongly linear relation between ΔG_{hydr} and the logarithm of the extent of reaction for the wide range of glasses considered here. However, if a more rigorous approach is combined with a better representation of the alkali metal species present in the glass, then the resulting fit is much improved. Further development of the approach along these lines should lead to an even better predictive tool, while hopefully maintaining the simplicity and ease of application of the original approach.

REFERENCES

1. C. M. Jantzen and M. J. Plodinec, **J. Non-Cryst. Solids, 67,** 207 (1984).

2. B. A. Shakhmatkin, N. M. Vedishcheva, M. M. Shultz, and A. C. Wright, **J. Non-Cryst. Solids, 177,** 249 (1994).

3. B. A. Shakhmatkin and M. M. Shultz, **Fiz. Khim. Stekla, 6,** 129 (1980).

4. O. V. Mazurin, M. V. Streltsina and T. P. Shvaiko-Shvaikovskaya, **The Properties of Glasses and Glass-forming Melts,** edited by V. K. Leko and S. V. Nemilov, Nauka, Leningrad, 1973, vol. 1, p. 257.

5. H. Maekawa, T. Maekawa, K. Kawamura, T. Yokakawa, **J. Non-Cryst. Solids, 127,** 53 (1991).

6. E. R. Plante, D. W. Bonnel, J. W. Hastie, in **Advances in the Fusion of Glasses,** (NY State College of Ceramics, Alfred, NY, 1988), p. 26.1-26.18.

THE EFFECT OF HUMIC ACIDS ON THE ELEMENT RELEASE
FROM HIGH LEVEL WASTE GLASS

J. Wei, P. Van Iseghem
SCK•CEN, Boeretang 200, B-2400 Mol, Belgium

ABSTRACT

Eu and Am doped glasses were interacted with synthetic interstitial clay water (SIC) and corresponding reference leachant, humic acids free interstitial solution (IS) to investigate the influence of humic acids on the leaching behaviour of the waste glass. Static leach tests were carried out at 40°C and 90°C. The release of the lanthanide Eu and the actinide Am from the glass was obviously enhanced by the presence of humic acids. The leaching of transition elements, Fe and Ti strongly depends on the humic acids concentration. The leaching of glass matrix components, Al and B was also influenced by the concentrations of humic acids. However, humic acids have little effect on the leaching of glass matrix element Si.

INTRODUCTION

Humic substances are organic polyelectrolyte macromolecules which are the principal organic components of soils and natural waters. Carboxylic and phenolic groups associated with humic acids (HA) have the ability to bind trace metal ions. Considering the disposal of high level nuclear waste forms in deep geological formations, the waste forms may come in contact with groundwaters which contain different concentrations of humic acids depending on the aquifer systems. As a potential repository for the disposal of vitrified waste forms, Boom clay formation has been evaluated extensively in Belgium. The concentration of DOC (dissolved organic carbon) in groundwater related to Boom clay formation is as high as 206 mg/l [1]. The concentration is much higher than that in groundwaters related to other host rocks. To understand the leaching behaviour of waste glasses in the presence of humic acids is important for characterizing the long-term durability of nuclear waste glasses under disposal condition.

Nearly 40 years ago, Ernsberger [2] noted the effects of chelating agents (EDTA and catechol) on the dissolution of soda-lime glass. The dissolution rate of the glass increased since stable complexes of Al, Ca, Mg with EDTA and Si with catechol were formed. Higher solubility of quartz has been observed because silicon was complexed by organic acids [3]. The leach rate of nuclear waste forms is enhanced due to the complexation of Al and Ti with oxalic acids [4]. Several papers have been published on the leaching behaviour of HLW glasses in the presence of humic acids [5-10]. Dran et al [5] found metallic elements (except for Zr) and Si had higher concentrations in the leachate with the increase of HA concentrations. Gin et al [6] discovered the leach rates of transition elements (including Zr), and B in the leachant with high HA concentration (5 g/l)were much higher than those in the absence of humic acids. Moreover, the authors supposed the process by which humic acids form complexes with one or more elements released from the glass could play a kinetically limiting role on the glass corrosion. Gin et al. [7] have investigated the influence of various organic acids or acid salts on the initial dissolution kinetics of the french "R7T7" glass. The monoacid had no effect on the initial glass dissolution rate; conversely, triacid significantly affected glass dissolution. Humic acids are strong catalysts of "R7T7" glass dissolution. The results suggest that the catalytic effect is appreciable only for organic materials with at least two carboxylic acid

Mat. Res. Soc. Symp. Proc. Vol. 465 © 1997 Materials Research Society

functions. Xing et al. [8] pointed out humic acids had no significant effect on the overall glass dissolution rate but it may increase the release of certain elements (Fe and Th) at low temperature. Actinides U and Th were leached readily from the nuclear waste glasses because of the presence of humic acids [6,8], while the effect of humic acids on Pu leaching was not clear because different results were reported [9, 10].

This study attempts to describe the effect of humic acids on the release of the lanthanide Eu, the long-lived actinide Am, transition elements and glass matrix components from HLW glass.

EXPERIMENTAL

PAMELA reference glass SM527 was used in this study. The composition of the glass is listed in Table I. Two kinds of glass were prepared: non-radioactive glass loaded with 1 wt% Eu_2O_3 and radioactive glass doped with 0.0017 wt% [241]Am. The activity of the Am-doped glass is 2.17 MBq/g glass. The glass was melted at 1200°C and annealed at 530°C. The radioactive glass was melted twice, to achieve sufficient homogeneity. The leachant, synthetic interstitial clay water (SIC), whose composition is given in Table II, is a simulate of the groundwater related to Boom clay formation. IS (SIC without humic acids) is selected as the reference leachant. Synthetic clay water with various humic acids concentrations (50, 75, 100, 150, 300, 600, 3000 mg/l) was also used in leach tests. The initial pH values of different leachants are almost the same (about 9.0).

Table I. Composition of glass SM527 (in wt%)

SiO_2	38.75	Fe_2O_3	0.70
B_2O_3	21.70	MnO_2	0.02
Na_2O	8.64	MoO_3	0.05
Li_2O	3.10	ZrO_2	0.05
Al_2O_3	19.96	Nd_2O_3	0.07
CaO	3.87	RuO_2	0.02
MgO	0.14	U_3O_8	0.02
TiO_2	1.55	SO_4	1.35
Cr_2O_3	0.02		

Table II. Composition of the synthetic interstitial clay water (mg/l)

$MgSO_4$	12.0
KCl	20.0
$NaSO_4$	1.50
NaF	8.0
NaCl	44.0
$NaHCO_3$	1250.0
$CaCO_3$	saturated*
humic acids	150.0**

* The solution has to be filtered to remove $CaCO_3$ precipitates.
** Fluka humic acids (product no. 53680)

The procedure adopted for the leach tests was similar to the MCC-3 method [11]. Powdered glass with grain size of 125-250 μm was immersed in 6 ml leachant. The geometrical surface area of the powdered glass is calculated to be 120 cm^2/g. For most leach tests, SA/V is 4000 m^{-1}, but higher

SA/V was applied for some tests. The leach tests were conducted at 40°C and 90°C in N_2 glove box (for radioactive glass) and Ar glove box (for non-radioactive glass). The glass sample was leached in Teflon® containers for durations up to 180 days. At the end of the leach experiments, the leachates were filtered through a filter with pore size of 450 nm and ultrafiltered through a filter with pore size of 1.5 nm to investigate the size distribution of released species. The element concentrations in the leachates (non-radioactive) were analysed by inductively-coupled plasma atomic emission spectroscopy (ICP-AES). The activity of Am was measured by γ-spectrometry.

RESULTS AND DISCUSSION

<u>Eu and Am released from the glass</u>

Table III presents Eu and Am concentrations (mole/l) in leachates after 450 nm filtration. Am concentrations after ultrafiltration are given in brackets, while Eu concentrations after ultrafiltration are usually lower than the detection limit (2.0×10^{-7} M). The Eu concentrations are about three orders of magnitude higher than the Am concentrations after 450 nm filtration. From Table III, it is clear that the release of Eu and Am increases due to the presence of humic acids. Eu concentrations in SIC leachates are about 3-10 times higher than those in IS leachates. Am concentrations in SIC leachates increase at least by a factor of 5 compared to those in IS leachates. The final pH values of SIC and IS leachates have little difference. So the different leach behaviour of Eu and Am in SIC and IS can not be attributed to the pH effect. The difference is caused by the strong complexation of Eu and Am with humic acids in SIC [12-14]. The complexation strength of Eu and Am with humic acids depends on the pH and other properties of the solution. Kim et al [12] obtained a complexation constant for Am(III) of $\log\beta$=6.44 l/mol for the tridentate sites at pH=6.0. However, the complexation constants at pH=9 were determined to be $\log\beta$=13.22-14.05 and $\log\beta$=12-14.3 l/eq H^+ for Eu and Am respectively [13]. Because the average leachate pH is 9.3 in our leach tests, Eu and Am are strongly complexed by humic acids in this pH condition. This leads to higher concentrations of Eu and Am in SIC leachates.

Table III shows that Eu an Am concentrations vary little with time. It implies that the release of Eu and Am is controlled by their solubility though the solubility-controlling phases are not known yet. Vernaz et al. [15] have reported actinide release from the glass in pure water was controlled by the solubility. Complexation with humic acids could result in higher solubilities for Eu and Am. This was supported by measuring the solubility of pure Eu_2O_3 in SIC and IS solutions. Eu_2O_3 solubilities are 4.0×10^{-5} M in SIC solution and 2.2×10^{-6} M in IS solution. The increased solubility in SIC solution certainly results from the complexation of Eu with humic acids. The total activity of Am leached from the glass should be higher than the activity measured in leachate since it is well known that Am is easily adsorbed on container walls. Eu is likely to be absorbed on walls of experimental equipment too [16]. Although Am concentrations are much lower than Eu concentrations in leachates (because of its much lower solubility), the normalized losses for Eu and Am, 10^{-2} g/m² in SIC , are comparable. This means that Eu leach rate is similar to the Am leach rate.

The results in Table III indicate that Eu and Am are predominantly present as colloidal particles in the leachates. Eu colloids in SIC leachates were proved to be Eu-humate colloids [17]. The colloid concentrations of Eu and Am in SIC leachate are much higher than those in IS leachate. This means humic acids play a significant role in colloid formation [17].

Table III. Eu and Am released from the glass (mole/l)

Eu leaching from the glass						
Leachant	Duration(days)					
	30		73		180	
	40°C	90°C	40°C	90°C	40°C	90°C
SIC	5.0×10^{-6}	7.8×10^{-6}	5.3×10^{-6}	9.3×10^{-6}	6.6×10^{-6}	7.2×10^{-6}
IS	5.3×10^{-7}	8.6×10^{-7}	6.6×10^{-7}	1.2×10^{-6}	3.3×10^{-7}	2.6×10^{-6}

Am leaching from the glass						
Leachant	Duration(days)					
	30		73		180	
	40°C	90°C	40°C	90°C	40°C	90°C
SIC	7.2×10^{-9} (1.8×10^{-11})	8.1×10^{-9} (8.5×10^{-11})	5.8×10^{-9} (1.7×10^{-10})	1.1×10^{-8} (8.2×10^{-11})	3.6×10^{-9} (9.8×10^{-10})	8.1×10^{-9} (1.4×10^{-10})
IS	8.5×10^{-10} (6.5×10^{-11})	8.77×10^{-10} (1.4×10^{-10})	9.9×10^{-10} (4.9×10^{-10})	1.9×10^{-9} (5.6×10^{-10})	1.7×10^{-10} (7.4×10^{-11})	7.0×10^{-10} (6.1×10^{-10})

* The leach tests were performed at $SA/V=4000$ m^{-1}.

The relation between HA concentration and Eu release

To further understand the humic acids effect on Eu release, leachants with different HA concentrations (50, 75, 100, 150, 300, 600, 3000 mg/l) were investigated. The leach tests were conducted at $SA/V=4000$ m^{-1} for 30 days. The Eu release in various leachants is plotted in Figure 1 and presented in Table IV. A linear relation between Eu concentrations and HA concentrations can be observed from Figure 1. When leachants contain high HA concentrations(300, 600, 3000 mg/l), the Eu concentrations increased with increasing HA concentrations, but no linear relation is observed. It indicates the release of Eu from the glass strongly depends on the HA concentration in the leachant. Gin et al. [6] found a linear relation between the logarithm of the initial glass dissolution rate and the logarithm of HA concentration. It can be expected that the leaching of the actinide Am will be significantly influenced by HA concentrations in the leachants in a similar way as Eu.

On the other hand the soluble concentration leached for Am shows a different trend than the total amount leached, see Table III. In the presence of humic acids, the soluble concentration leached is smaller than when no humic acids are present. More data are needed to confirm this. This would mean that in case of Am, the presence of humic acids in the contacting solution would be beneficial, because of the decrease of the soluble concentration leached.

Table IV. The relation between HA concentration(mg/l) and element release(concentration in mg/l)

HA concentration	Si	B	Al	Fe	Ti	Eu
300	24.4	248	15.2	3.56	0.25	1.12
600	24.8	264	15.8	6.65	0.33	1.42
3000	24.3	332	17.2	23.3	0.71	2.06

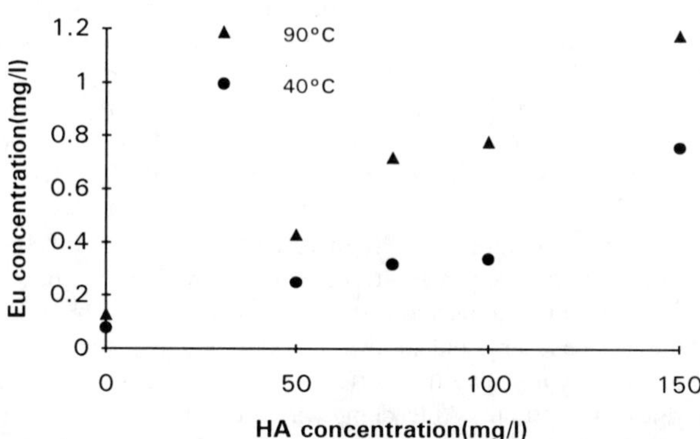

Figure 1. The Eu release as a function of humic acids concentration leaching at 40 and 90°C

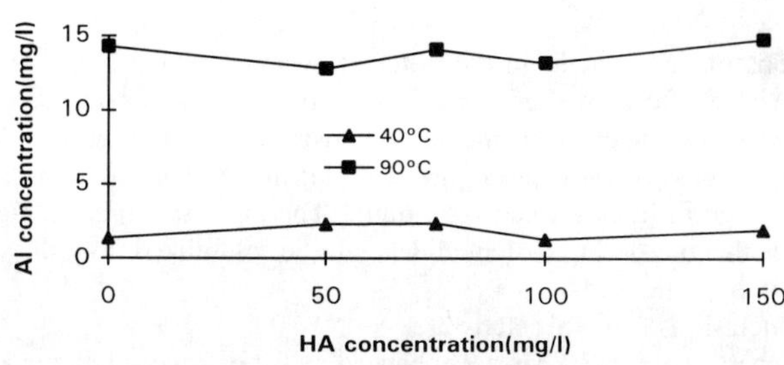

Figure 2. The Al release as a function of humic acids concentration

<u>The effect of humic acids on the leaching of other elements</u>

The relation between HA concentrations and the release of some of the main glass constituents Si, B, Al, Fe,Ti is demonstrated in Table IV and Fig. 2. Humic acids have no effect on the release of Si and Ca, and have minor effect on the release of B and Al, but have strong effect on the release of Fe and Ti. The glass matrix components, Si, B, and Al release from the glass by hydrolysis reaction [22]. The release of Si and Al is limited by their solubilities. B is a very soluble element, and its release could not be controlled by solubility. Fe and Ti are generally considered as immobile elements and retained on the glass surface layer.

Al, Ca and Fe, which have relatively high contents in glass SM527, can be complexed by humic acids [20,21]. There is no information about the complexation of glass matrix elements Si and B with humic acids. It is generally believed that silicic acid and boric acid are formed during the dissolution of waste glasses [22]. The complexation of Fe with humic acids leads to the rapid increase of its concentration in the leachate. Ti concentrations steadily increase with increasing humic acids concentrations. Ti and Fe are transition elements, and their complexation properties may be alike although the complexation of Ti with humic acids has not been reported. The situation for Al and Ca is complicated in this respect. Both elements are known to be complexed by humic acids, but they are also usually involved in secondary phase formation, together with Si and Na [18,19]. The leach tests were performed under saturation conditions ($SA/V=4000$ m^{-1} or higher) in this study, and the secondary phases were probably formed during the leach tests, and precipitated on the glass surface.

We propose the following hypothesis for the increased B release, see Table IV. Because the transition and heavy metal elements, which are usually enriched on the glass surface layer, are complexed by humic acids and become more soluble, B is leached more easily by diffusion process. Analysis of the surface layer could confirm this assumption, if thinner layers would occur. From Fig. 2, it seems humic acids have little effect on the release of Al, while temperature has strong effect on the Al leaching. But the Al leaching was affected when humic acids concentrations were higher (see Table IV). The concentrations of B, Al, Ca, Na, Fe, Zr, Zn, Nd, U, Mo measured in leachates with high HA concentration (5 g/l) are markedly greater than those in the inorganic leachates after 300 days leaching ($SA/V=100$ m^{-1}) [6]. But the leachant pH was imposed at acidic condition in that study[6].

CONCLUSION

Concentrations of humic acids in the order of 10 mg/l, which is the case for natural waters, probably have little effect on the overall dissolution rate of waste glasses, but can result in preferential release of some elements including transition elements, lanthanides and actinides. The glass dissolution is possibly affected significantly if the leachant contains humic acids with high concentrations, such as in clay water (150 mg/l). The information is of interest for performance assessment since the long-living radionuclides will be solubilized from the waste glass due to the complexation with humic acids.

The main conclusions from this study are:
1. The leach rates of Eu and Am are enhanced by the presence of humic acids. A linear relation between humic acids concentration and Eu release is discovered when humic acids concentrations are below 150 mg/l.
2. Under test conditions, humic acids have little effect on the release of glass matrix element Si, but the release of B and Al can be influenced by the presence of humic acids.

Further work is needed to clarify the effect of humic acids on the leaching of long-lived actinides, Pu, Np, Tc, and the overall glass dissolution rate.

ACKNOWLEDGEMENTS

The authors gratefully ackowledge technical support from D.Huys and B.Gielen.

REFERENCES

1. P.N.Henrion, M. Monsecour, A. Fonteyne, M.Put and P.De Regge," Migration of radionuclides in Boom clay", Radioactive Waste Management and the Nuclear Fuel Cycle, Vol 6(3-4), pp. 313-357 (1985)

2. F.M.Ernberger, "Attack of glass by chelating agents" , J. Amer. Ceram. Soc., Vol 42, pp. 373-375 (1959)

3. P.C.Bernnet and D.I.Siegel, "Increased solubility of quartz in water due to complexation by dissolved organic compounds", Nature, Vol 326, pp. 684-686 (1987)

4. Aa.Barkatt, A.Barkatt and W.Sousanpour, "Effects of γ radiation on the leaching kinetics of various nuclear waste-form materials", Nature, Vol 300, pp. 339-341 (1982)

5. J.C.Dran, J. Lombardi, M.C.Magonthier, V.Moulin, J.C. Petit and L.Trotignon ," Leaching of borosilicate glasses by solutions containing humic acids: behaviour of metallic elements", Radiochim. Acta Vol 58/59, pp. 17-20 (1992)

6. S.Gin, N. Godon, J.P. Mestre , E.Y.Vernaz and D.Beaufort," Experimental investigation of aqueous corrosion of R7T7 nuclear glass at 90°C in the presence of humic acid: a kinetic approach", Mat. Res. Soc. Symp. Proc., Vol 333, pp. 565-572 (1994)

7. S.Gin, N. Godon, J.P. Mestre , and E.Y.Vernaz, " Experimental investigation of aqueous corrosion of R7T7 nuclear glass at 90°C in the presence of organic species", Applied Geochemistry Vol 9, pp. 255-269 (1994)

8. Xing Shi-Ben, Keith S.Matlack and Ian L.Pegg," The effect of humic acids on the dissolution of nuclear glasses", Ceram. Trans. Vol 39, pp. 353-365 (1994)

9. Lian Wang, P.Van Iseghem and A. Maes," Plutonium leaching from a reference waste glass in synthtic interstitial claywater", Mat. Res. Soc. Symp. Proc. Vol 294, pp.155-162 (1993)

10. G. Bidoglio, A. De Plano and L. Righetto, " Interactions and transport of plutonium-humic acid particles in groundwater environments", Mat. Res. Soc. Symp. Proc., Vol 127, pp. 823-830 (1989)

11. DOE/TIC 11400, "Nuclear Waste Materials Handbook, Test Methods", PNL, Richland, Washington (1981)

12. J.I.Kim, D.S.Rhee and G.Buckau, "Complexation of Am(III) with humic acids of different orign", Radiochim. Acta, Vol 52/53, pp. 49-55 (1991)

13. A.Maes, J.De Brabandere and A.Cremers, " Complexes of Eu^{3+} and Am^{3+} with Humic Substances", Radiochim Acta Vol 52/53, pp. 41-47 (1991)

14. R.A. Torres and G.R. Choppin," Europium(III) and Americium(III) stability constants with humic acid", Radiochim. Acta, Vol 35, pp. 143-148 (1984)

15. E.Y. Vernaz and N. Godon," Leaching of actinides from nuclear waste glass: french experience", Mat. Res. Soc. Symp. Proc., Vol 257, pp. 37-48 (1992)

16. M.S. Caceci and G.R.Choppin, "An improved technique to minimize cation adsorption in neutral solutions", Radiochim. Acta, Vol 33, pp. 113-114 (1983)

17. Jiang Wei and P. Van Iseghem,"Colloid formation during the interaction of HLW glass with claywater", presented at this symposium

18. T.A.Abrajano, J.K.Bates, A.B.Wodland,"Secondary phas formation during nuclear waste- glass dissolution", Clays and Clay Minerals, Vol 38(5), 537-548 (1990)

19. A.J.G.Ellison, J.J.Mazer, W.L.Ebert,"Effect of glass composition on waste form durability: a critical review", ANL 94/28 (1994)

20. A.Dierckx," Complexation of Europium with Humic Acids, Influence of Cations and Competing Ligands", PhD thesis, Katholieke Universiteit Leuven, 1995

21. V.Moulin, J. Tits and G. Ouzounian,"Actinide speciation in presence of humic substances in natural water conditions", Radiochim. Acta, Vol 58/59, pp. 179-190 (1992)

22. X.Feng, J.C.Cunnane and J.K.Bates,"A literature review of surface alteration effects on waste glass behaviour", Ceram. Trans., Vol 39, pp. 341-352 (1994)

MODELLING GLASS DISSOLUTION IN CLAY WITH ANALYTIC AND STOCHASTIC METHODS

Marc Aertsens, SCK•CEN, Boeretang 200, B–2400 Mol, Belgium.

ABSTRACT

We present a stochastic method to model the dissolution of nuclear glass. Using this method, we solve the diffusion equation in a stochastic way. We do this by giving a large number of particles Brownian displacements. Simultaneously, these particles can participate in other processes, like a chemical reaction or convection.

We apply this method to solve the Pescatore model for the dissolution of nuclear glass in clay. This model combines diffusion of silica in the pore water of the clay with the glass dissolution rate law proposed by Grambow. We use the model for fitting the dissolution data of four glasses in clay slurries (with a high and with a low clay content) and in pure clay. We present the values of the fitting parameters. The solution of the model, obtained by the simulation method, agrees with the analytical solution. We also extend the Pescatore model with a moving boundary, taking into account the receding of the glass surface by corrosion.

INTRODUCTION

In glass dissolution, diffusion is an important process, which may control the long term dissolution behaviour of the glass [1, 2]. During the dissolution process, diffusion occurs in several ways: protons diffuse into the glass to replace alkalis (ion exchange), and silica may diffuse out of the glass through a gel layer [3,4]. If the nuclear glass is embedded in a low permeability porous medium like clay, diffusion of silica in the pore water of the clay may control the dissolution rate [5,6]. Hence, reliable models for glass dissolution should include diffusion. This also means that one needs appropriate methods to solve the mathematics of diffusion. An analytical solution is definitely the best, but it only exists for simple models. A numerical solution, mostly using finite element methods, is always possible. Several computer codes using finite elements exist and work fine. However, a problem with existing packages is that they do not allow to solve all diffusion problems. Since we had a model with boundary conditions that our finite element package cannot handle, we wrote our own code to solve the model. Mainly for the reason of its simplicity, we chose to develop a numerical stochastic code. First, we explain the stochastic method, which is new in the field of glass dissolution. Afterwards we apply this method to a model describing the dissolution of glass surrounded by clay. This model is originally proposed by Pescatore [5], but we extend it by taking into account the receding of the glass surface by corrosion. The reason for this extension is that a fixed boundary was a relevant limitation in a similar model developed by Curti [6]. We formulate the model with the moving boundary and present an analytical solution. Then, we use this model for fitting the experimental data of glass dissolving in pure Boom Clay and in two mixtures of water with Boom Clay. Boom clay is considered in the present work because this formation is studied as a candidate host rock for the geological disposal of high level radioactive waste in Belgium. Finally, we verify the simulation method by comparing the simulation result with the analytically calculated result.

Mat. Res. Soc. Symp. Proc. Vol. 465 © 1997 Materials Research Society

PRESENTATION OF THE STOCHASTIC SIMULATION METHOD

We present a method which leads to a numerical solution of the diffusion equation

$$\frac{\partial C}{\partial t} = D \frac{\partial^2 C}{\partial x^2}$$

(1)

with C the particle concentration and D the diffusion coefficient. Therefore we use the mathematical solution of the diffusion equation in an infinite space. If a particle is at the position $x = 0$ at zero time, the probability that it is at position Δx at time Δt is given by

$$C(\Delta x, \Delta t) = \frac{1}{2\sqrt{\pi D \Delta t}} \exp\left(-\frac{(\Delta x)^2}{4 D \Delta t} \right)$$

(2)

The method consists of taking a large number of particles, with a known initial position. At every time step Δt, each particle receives a random displacement Δx according to the distribution (2). In this way the diffusion is simulated. The exactness of the result depends on the amount of particles. A small number of particles means limited computation time, but also a large statistical error on the result.

The choice of a good time step is crucial for the success of the method as well. It is trivial that the method works best for small time steps, but this is at the expense of large computation times. In general, the method works well as long as the chosen time step does not exceed a certain value. This value depends on the kind of problem.

We also give some examples of how to handle boundary conditions:
- impermeable boundary: use a reflecting boundary condition.
- absorbing boundary: if a particle reaches the boundary, it is taken out of the system.
- flux at the boundary: particles are injected in the system at the boundary.

Advantages of this method are: (1) it is quite easy to program, (2) the programme can easily handle complicated boundary conditions, (3) additional phenomena like particle aggregation, a particle drift velocity or a chemical reaction can be included in an easy way. For instance, for simulating a chemical reaction during diffusion, one gives after every time step Δt to each particle a transition probability between its state in the left side of the chemical reaction and its state in the right side. Of course, these probabilities must be chosen in such a way that the system evolves to chemical equilibrium [7]. A disadvantage of this method is that it is stochastic, which means that a large number of particles is needed. Since nowadays computation power is large and it still increases, we do not consider this as a major problem.

A MODEL FOR THE DISSOLUTION OF GLASS ADJACENT TO CLAY: BASIC EQUATIONS

In the model, we approximate the glass as consisting of silica only. As a result, we neglect all chemical reactions in the glass dissolution process, except silica dissolution and silica sorption on clay. We do not consider diffusion through a glass gel layer neither. When glass dissolves, silica moves from the glass into the neighbouring clay. There, silica diffuses in the pore water of the clay. This process is described by the diffusion equation (1) with C the concentration of silica in the pore water accessible for diffusion and D the diffusion coefficient of silica in the water saturated porous clay. We choose the origin of the x-axis at the initial position of the surface between the glass and the clay. The notation $X(t)$ is used for the position of the surface at any time t $(X(t) < 0)$.

Next, we deduce the boundary condition for the diffusion equation. According to Grambow [3,8], the flux of silica at the surface $X(t)$ is given by

$$J(x=X) = \alpha \left(1 - \frac{C(x=X)}{\gamma} \right) \tag{3}$$

with α the maximum flux and γ a concentration at which the silica dissolution reaction would stop. Both the maximum flux α and the concentration γ depend on pH (e.g. see [6]), but we assume the fluctuations on pH are sufficiently small to approximate α and γ as being constant. Of course, the flux $J(x=X)$ is also given by Fick's law, where we add to the diffusive flux, the flux caused by the receding of the glass surface. This leads to [9]

$$J(x=X) = -\eta R\, D \left(\frac{\partial C}{\partial x} \right)_{x=X} - \eta R\; C(x=X)\, \frac{dX}{dt} \tag{4}$$

with η the diffusion accessible porosity in the porous medium and R the retardation constant. The assumption that dissolved glass is replaced by clay (see (4)), is justified by the self-healing capacity of the plastic clay. Sorption of silica on the negatively charged clay surface, might seem strange at first sight. Indeed, at in situ pH (=8,5) silica diffuses in the clay as neutral $Si(OH)_4$ or anionic $SiO(OH)_3^-$. Neutral or anionic species are normally known as unsorbed species in clay materials. However, the silica sorption might be explained by the formation of surface complexes between silica and aluminium hydroxide groups, present on the clay minerals surface [10]. Silicate might behave in clay like phosphate, which is known to sorb by ligand exchange onto aluminium or iron hydroxide

The total quantity diffused through a unit surface (of the glass/clay interface) at time t is

$$Q(t) = \int_0^t J(x=X,t')dt' \tag{5}$$

and since also $Q(t) = -A\,X(t)$ with $A = C(x<X)$ the concentration of silica in the glass, is

$$J(x=X) = \frac{dQ}{dt} = -A\,\frac{dX}{dt} \tag{6}$$

The three expressions (3), (4) and (6) for the flux of silica $J(X)$ at the surface between the glass and the clay, define the boundary conditions for the diffusion equation (1).

Finally, the initial conditions are

$$C(x>0,\ t=0) = C_0 \quad ; \quad X(t=0) = 0 \tag{7}$$

Neglecting the movement of the wall (mathematically this means taking the limit $A \rightarrow \infty$), this model has already been solved by Pescatore [5].

ANALYTICAL SOLUTION OF THE MODEL

Combining the boundary conditions (3), (4) and (6) leads to

$$\left(\frac{\partial C}{\partial x} \right)_{x=X} = -h\lambda \left(1 - \frac{C(x=X)}{\lambda} + \frac{\eta R\ C(x=X)^2}{A\gamma} \right) \tag{8}$$

with

$$h = \frac{\alpha}{\eta\ R\ D\ \lambda} \quad ; \quad \lambda = \gamma\ \frac{A_*}{A_*+1} \quad ; \quad A_* = \frac{A}{\eta\ R\ \gamma} \tag{9}$$

Assuming that $C(x=X) << (A/(\eta R))+\gamma$, which is true in the real system, it is a good approximation to replace the quadratic term $C(x=X)^2$ by its initial value $C(x=0,t=0)^2$. Next, we solve the diffusion equation (1) with the boundary condition (8) by the Laplace transformation method [9]. This leads to

$$C = C_0 + (\lambda K - C_0) \left(erfc\left(\frac{x-X}{2\sqrt{Dt}} \right) - \exp\!\left(h(x-X) + h^2 Dt \right)\ erfc\left(\frac{x-X}{2\sqrt{Dt}} + h\sqrt{Dt} \right) \right) \tag{10}$$

with

$$K = 1 + \frac{\eta R\ C(x=X,t=0)^2}{A\gamma} \tag{11}$$

Using the same notation as Pescatore, the concentration at the boundary X is

$$C(x=X) = C_0 + (\lambda K - C_0)\left(1 - F(\xi\sqrt{t}) \right) \tag{12}$$

with

$$\xi = h\ \sqrt{D} \quad ; \quad F(\xi\sqrt{t}) = \exp(\xi^2 t)\ erfc\left(\xi\sqrt{t}\right) \tag{13}$$

Using (12) and starting from $C_0 = 0$, the limit of the concentration $C(x,t)$ for large times is $C(x,t\rightarrow\infty) = \lambda < \gamma$. Physically however, the system saturates at the concentration γ. This difference is caused by the approximation to consider $C(x=X)^2$ as a constant. We now try to undo this approximation by replacing K by $1 + (\eta R\ C(x=X)^2/\gamma A)$ in (12). This leads to a quadratic equation for $C(x=X)$. The solution of this equation is

$$C(x=X) = \gamma\ \frac{(A_* + 1) - \sqrt{(A_* + 1)^2 - 4A_*\left(1 - F(\xi\sqrt{t}) \right)^2}}{2\left(1 - F(\xi\sqrt{t}) \right)} \tag{14}$$

Verification shows that the expression (14) is the solution of (8). For $A_* >> 1$, the expression (14) reduces to the solution found by Pescatore [5].

The flux J of particles into the clay is obtained by the substitution of the expression for the concentration $C(X)$, from (14), in the Grambow expression (3). Making this substitution, one observes that the flux J depends on three parameters: the maximum flux α, ξ and A_*.

The total number of particles Q is the integral over time of the flux J (see (5)). It is evident that Q depends on the same three parameters as the flux J. For the limit $A_* \to \infty$, the number of dissolved particles reduces again to the expression found by Pescatore [5]

$$Q(t) = \frac{\alpha}{\xi^2}\left(F(\xi\sqrt{t}) - 1 + \frac{2}{\sqrt{\pi}}\, \xi\sqrt{t} \right) \tag{15}$$

For large times, the function $F(\xi\, t^{0.5})$ goes to zero and the amount $Q(t)$ is approximately equal to (see (9) and (13))

$$Q(t) = \frac{2}{\sqrt{\pi}}\, \frac{\alpha}{\xi}\, \sqrt{t} = \frac{2}{\sqrt{\pi}}\, \eta R\, \gamma\, \sqrt{Dt} \tag{16}$$

Hence, for the amount of dissolved particles $Q(t)$, at large times and $A_* \to \infty$, (1) one observes a square root behaviour and (2) it is possible to fit only one parameter: the ratio α/ξ. This ratio is equal to the product of the bulk concentration of 'glass silica' (silica which originates from the glass, since clay also contains silica) in the clay with the square root of the silica diffusion coefficient in the clay pore water. Note also that in this stage, the dissolution behaviour is independent of the maximum (and initial) leach rate α, and that it is completely determined by the properties of the clay.

FITTING RESULTS

We have fitted the data for the cumulative release Q(t) of silica from the glass for two dissolution experiments by using the model of the previous section. In one experiment, leach tests were executed for two clay to water ratios: 10 g/L (denoted as low clay) and 500 g/L (denoted as high clay). In this experiment, four glasses (SM513, SM527 SON68 and WG124) were exposed to the medium at 90 °C for a surface to volume ratio SA/V of 100 m^{-1}. The normalised mass loss data are based on boron or molybdenum. For the glasses SM513 and SM527, the fit is also done by Pescatore [5]. In the second experiment, three glasses (SON68, SM58 and SAN60) corroded on pure clay at 90°C. Then about 0.5 ml interstitial clay water was mixed with the clay adjacent to the glass samples, to reach a good contact. In this case there is almost no water containing dissolved glass elements, so we fit the total mass loss. Remember that in our model, we assume that the pH remains constant. For about ten measurements taken for each glass during the experiment, the pH fluctuates typically around 9.4 ∓ 0.5 for the low clay and 8.7 ∓ 0.25 for the high clay. The pH was not measured for the pure clay dissolution experiment.

Although the model of the previous section contains three parameters (the maximum flux α, and the parameters ξ and A_*; see eqn. 9 and 13), we could determine from the dissolution data mostly just one parameter (the ratio α/ξ) in an accurate way. For none of the fittings, it was possible to obtain a value for the parameter A_*. A quick estimation of its value leads to $10^3 < A_* < 10^4$ ($A \approx 10^6$ mg/L, $\gamma \approx 10^2$ mg/L, $1 < \eta R < 10$). A numerical study of $Q(t)$ learns that for such

Table 1: Fitting results by applying the Pescatore model to experimental glass dissolution data in pure clay and low clay and high clay mixtures.

Glass type	Clay content	Normalized with respect to	α (g/m^2 day)	α (g/cm^2 sec)	ξ (1/sec$^{0.5}$)	ξ-2 (days)	α/ξ (g/cm^2 sec$^{0.5}$)
SM513	High	boron	14.8	1.7 E-08	6.4 E-03	0.3	2.7 E-06
	Low	boron	7.9	9.2 E-09	6.9 E-03	0.2	1.3 E-06
SM527	High	boron	14.3	1.7 E-08	5.3 E-03	0.4	3.1 E-06
	Low	boron	1.9	2.2 E-09	6.9 E-03	0.2	3.3 E-07
SON68	High	boron	10.4	1.2 E-08	6.9 E-03	0.2	1.7 E-06
	Low	boron	1.2	1.4 E-09	6.8 E-03	0.2	2.0 E-07
WG124	High	molybdenum	14.6	1.7 E-08	4.0 E-03	0.7	4.2 E-06
	Low	molybdenum	1.6	1.9 E-09	6.9 E-03	0.2	2.7 E-07
SON68	Pure	total mass loss	2.1	2.4 E-09	7.3 E-04	22	3.3 E-06
SM58	Pure	total mass loss	1.4	1.6 E-09	1.6 E-04	430	1.0 E-05
SAN60	Pure	total mass loss	2.3	2.6 E-09	6.8 E-03	0.2	3.9 E-07

values the quantity $Q(t)$ is almost insensitive to the exact value of A_*, which means that the influence of a moving boundary is negligible. However, in a similar model, Curti [6] reports that the errors introduced by a fixed boundary are not irrelevant. The reason of the difference may be that our measurement time is not large enough and that our model considers an infinite system instead of a finite system. The fittings also show a close correlation between the two other fit parameters, α and ξ. This means that only the ratio α/ξ can be determined accurately. The reason for this correlation is that the time ξ^{-2}, is in general much smaller than the time of the measurements (see (15-16)).

The fitting results are shown in Table 1, Figure 1 and Figure 2. From Table 1, we note that for the same glass, the ratio α/ξ increases for a higher clay content. Thus, from (16), more silica is dissolved at a higher clay content, which is clearly illustrated for SON68 in Fig. 1. The quality of the fits is good. Comparing our fit results for the glasses SM513 and SM527 with Pescatore's results, we have different values for the individual parameters α and ξ, but approximately the same values for the ratio α/ξ. Since almost all measurement times are larger than ξ^{-2}, this is normal. From Table 1, we also note that for the aluminium rich glass SAN60, the square root release is reached very fast ($\xi^{-2} < 1$ day), while for SM58 this takes much more time ($\xi^{-2} > 1$ year). The value of the third glass, SON68, is three weeks. Because for SAN60, the square root release is reached very fast, this glass dissolves less than SON68 and SM58. One must however remain careful with this conclusion since (1) the parameter ξ cannot be determined accurately (especially when most of the measurement times are larger than ξ^{-2}) and (2) we only have five experimental points for each glass.

VERIFICATION OF THE SIMULATION METHOD

The method works quite well. We have tested it for the Pescatore model (with and without moving boundary) for several parameter values, and in all cases the difference between the analytical result and the simulation result was small. We illustrate this for a simulation of the dissolution of the SON68 glass in pure clay, where we ran the simulation program with the parameter values mentioned in Table 1. During the simulation, we did not allow more than

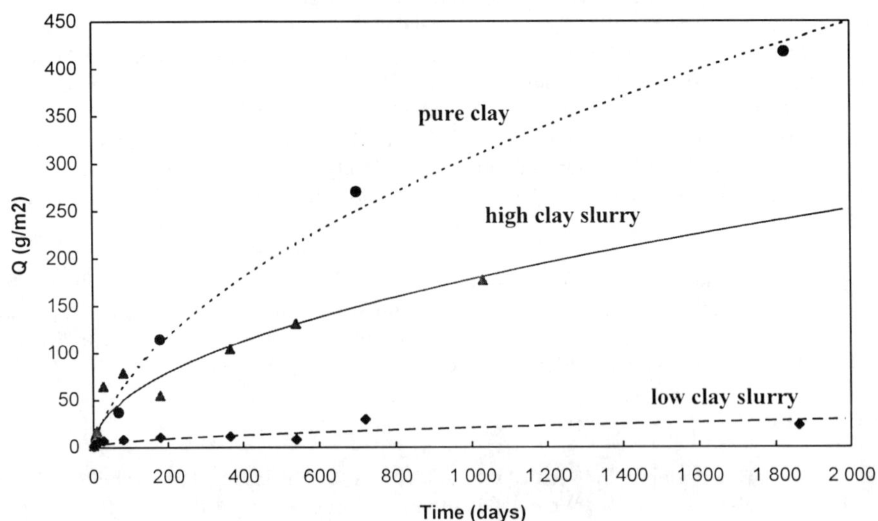

Fig. 1: Experimental data points and theoretical fitting of glass SON68 in pure clay, a high clay mixture and a low clay mixture.

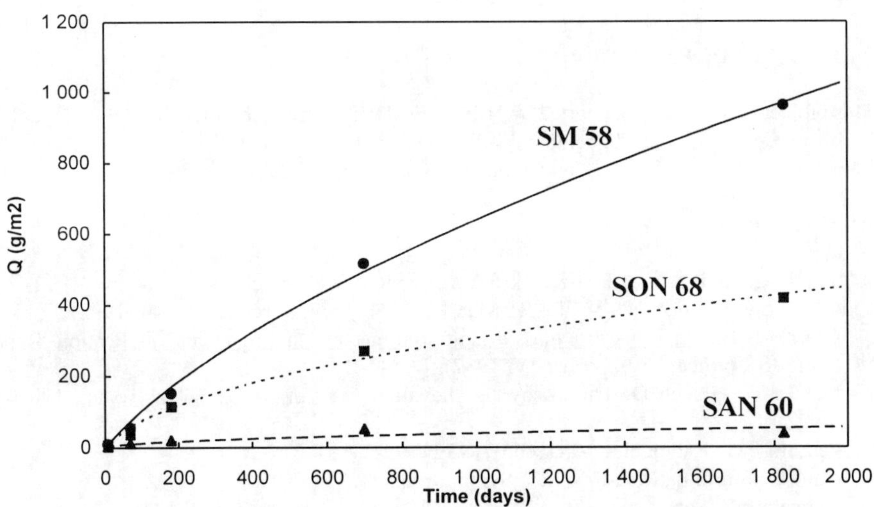

Fig. 2: Experimental data points and theoretical fitting of glasses SON68, SM58 and SAN60 in pure Boom clay.

200000 dissolved particles in the system. The particles dissolve from a surface of size S, which is chosen such that the product S.Q(t) is smaller than but also around 200 000. Since the amount of dissolved particles Q(t) is a monotonously increasing function of time, we decrease the surface S each time the product S.Q(t) becomes larger than 200 000. The time step was chosen small at the start ($\Delta t = 100$ sec) and was slowly increased to $\Delta t \approx 2$ days at the end of the simulation ($\Delta t \approx 20$ years). During this time interval, the difference between the simulation results and the analytical solution changed from around 0.5 % initially to around 1.2 % at the end of the simulation. This shows that for a large interval of time steps, the simulation result is nearly independent of the exact choice of the time step, as it should be.

The simulation method can also be used to solve other models, where diffusion is important. We have implemented it for instance for solving the Grambow model [3], which considers diffusion of silica through the gel layer. Using this method, it is possible to avoid the assumption of a linear concentration profile of dissolved silica in the gel layer.

CONCLUSION

We have developed and verified the validity of a stochastic method for solving diffusion problems in glass dissolution. We extended the Pescatore model for the dissolution of glass adjacent to clay with a moving boundary, but could not determine the parameter introduced by this moving boundary from our experimental data. Even for the original Pescatore model, our data allowed to determine accurately only the ratio between the two model parameters. The experimental data confirm that glass adjacent to clay dissolves according to a square root behaviour. The model can still be improved by taking ion exchange or chemical reaction into account.

ACKNOWLEDGEMENTS

This project is financially supported by NIRAS/ONDRAF under contract CCH090-123 and the European Commission under contract FI4W-CT95-0001. The author thanks P. Van Iseghem, K. Lemmens, P. Lolivier , P. De Cannière and R. Gens for stimulating discussions.

REFERENCES

1. Van Iseghem P., Amaya T.. Suzuki Y., Yamamoto H., J. Nucl. Mat., 190, 269 (1992).
2. Grambow B., Lutze W.. Muller R., Mat. Res. Soc. Symp. Proc., 257, 143 (1992).
3. Grambow B., <u>Nuclear waste glass dissolution: mechanism, model and application</u> (Report to the JSS-project 87-02, phase IV, 1987).
4. Delage F., Ghaleb D., Dussossoy J., Chevallier O., Vernaz E., J. Nucl. Mat., 190, 191 (1992).
5. Pescatore C., Radiochim. Acta, 66/67, 439 (1994).
6. Curti E., Smith P., Mat. Res. Soc. Symp. Proc., 212, 31 (1990).
7. Aertsens M., Van Iseghem P., Mat. Res. Soc. Symp. Proc., 412, p. 271 (1996).
8. Grambow B., Hermansson H., Bjorner I, Werme L., Mat. Res. Soc. Symp. Proc., 50, 187 (1985).
9. Crank J., <u>The mathematics of diffusion</u> (Clarendon Press, Oxford,1975).
10. Tan K., <u>Principles of soil chemistry</u> (Marcel Dekker, New York, Basel, Hong Kong, 1993).

SINTERED GLASSES FOR HIGH-LEVEL WASTES

D.O. RUSSO, N.B. MESSI de BERNASCONI, M.E. STERBA, A.D. HEREDIA, M.
SANFILIPPO, S. PRASTALO and J.C. SBRILLER
CNEA, Centro Atómico Bariloche, RN 8400, ARGENTINA

ABSTRACT

The objective of the work was to evaluate the long-term capacity of sintered glass to retain high-level nuclear wastes (HLW) in near-repository conditions. We have studied the corrosion behavior of waste forms partially devitrified (43 vol.%) in different aqueous media. Devitrified samples were irradiated at doses (γ radiation from a Co^{60} source) ranging from 1.4 x 10^6 Gy to 2.0 x 10^8 Gy, in order to study their aqueous corrosion resistance in simulated underground water. The results show little or no effect of irradiation on the density, microstructure and corrosion resistance. The global dissolution rate was almost constant around a value of $5x10^{-5}$ g. cm^{-2} d^{-1}. Elemental dissolution rates were also unaffected by radiation.

INTRODUCTION

The safe disposal of high-level wastes arising from reprocessing of spent fuel comprises, as a first step, immobilizing them in waste forms with high corrosion resistance to an aqueous medium. Currently, the most developed process on an industrial scale is incorporation of high level wastes into a glass matrix by melting at temperatures from 1400 to 1500 K. One of the main reasons for selecting a glass form material is because of its durability and stability to aqueous corrosion, thermal loading and radiation damage.

In order to evaluate other alternatives, either more advantageous processes or materials, in 1984 we started a research and development program for studying sintering applied to glasses from diverse sources and with different compositions [1, 2]. One of the advantages of the glass sintering process, relative to glass melting, is the lower temperature required (1050 K approximately) which implies less demanding conditions for equipment and a reduction in the evaporation of volatile radionuclides (i.e., Ru, Cs, Mo and Te). Furthermore, crucible erosion that leads to uncontrollable variations in the product composition is avoided and macrosegregation phenomena occurring in molten glass are minimized.

At our laboratories we studied a cold pressing and sintering method (also known as pressureless sintering) and a hot pressing method, to obtain glass samples containing simulated high-level wastes. The last method was also developed up to semi-industrial scale.

In this work we are presenting we try to evaluate how different aqueous media affect the dissolution of a devitrified glass and also the effect of radiation on the corrosion resistance of this glass in contact with underground water.

We studied a devitrified glass because the combination of time and heat generated by the radioactive decay of HLW could produce devitrification of the glass matrix. The main problem with the appearance of crystalline phases is the possibility of segregation of the HLW to interfacial zones where selective water corrosion could occur. Another problematic effect is the interaction between these new crystalline phases and radiation that causes atomic displacements and/or deformation of the crystalline structure, creating stresses in the bulk phase that could also deteriorate the corrosion resistance of the waste form (extensive microcracking can result, increasing the glass surface area exposed and available for radionuclide release).

Mat. Res. Soc. Symp. Proc. Vol. 465 © 1997 Materials Research Society

EXPERIMENTAL PROCEDURES

Materials

In Table I we show the nominal composition and physical characteristics of the aluminoborosilicate glass SG7, formulated at the KfK for the sintering process and manufactured in that country by a private company [3].

The composition of simulated high-level liquid wastes (HLLW) was the one that would be generated from pressurized heavy water reactor (PHWR) fuel as is shown in Table II [4]. It was formulated according to a computer code calculation (ORIGEN2.1) and comprises the wastes generated by the reprocessing of PHWR Argentine reactor fuel where spent UO_2 fuel is discharged from the reactor after reaching a burnup of 7000 MWd/tHM. The discharged fuel has a cooling time of 20 years in interim storage and produces a volume of 250 l HLLW/tHM.

The formulation has included the most abundant actinides and fission products, which were replaced on a molar basis by the corresponding stable elements.

Incorporation of this waste in the glass was done as described in [1]. The waste loading in the sintered samples was 10 wt.%.

Forming process

The samples were produced by the classical cold pressing and sintering technique. We used a cylindrical metal die with a 13 mm internal diameter, lubricated with zinc stearate. The pressure applied was 24 MPa and 3 mm thick discs were obtained. The as-pressed samples were sintered in a tubular electric furnace at 1050 K for 2 h. The sintering atmosphere was air and heating and cooling rates were 200 K/h and 100 K/h, respectively. In this way we got pellets with densities of 96% TD (theoretical density).

Devitrification and irradiation

The devitrification was produced by thermal treatment of the samples at 1050 K for 3000 h. Previous work [1] has demonstrated that this results in 43 volume percent crystallization,

Table I: Chemical composition and some physical characteristics of SG7 glass.

Component	wt %
SiO_2	72.0
Al_2O_3	8.6
B_2O_3	8.3
Na_2O	7.4
CaO	2.7
MgO	1.0
Transition temperature (K)	840
Theoretical density (g/cm^3)	2.42
Median particle size (μm)	7.7

Table II: Composition of the simulated HLLW type PHWR fuel.

Oxide	wt%	Oxide	wt%
SeO_2	0.12	Cs_2O	3.64
Rb_2O	0.60	BaO	3.17
SrO	1.27	La_2O_3	2.39
Y_2O_3	0.90	CeO_2	4.92
ZrO_2	8.16	Pr_6O_{11}	2.33
MoO_3	10.40	Nd_2O_3	7.66
Fe_2O_3	14.40	Sm_2O_3	2.39
NiO	2.60	Gd_2O_3	0.15
Ag_2O	0.17	U_3O_8	28.30
CdO	0.12	Cr_2O_3	2.64
SnO_2	0.10	P_2O_5	1.67
TeO_2	1.17	Co_2O_3	0.79

calculated from x-ray diffraction patterns with the Hermann-Weidinger method [5]. The phases present were quartz, cristobalite and minor diffraction peaks associated with waste compounds (not identified) [1].

The devitrified samples were irradiated in a facility of the Argentine Atomic Commission (CNEA) using a Co^{60} γ-source. The dose rate was 4.34 x 10^4 Gy/h and the total doses were from 1.4 x 10^6 Gy to 2.0 x 10^8 Gy. The latter is equivalent to the total theoretical dose at 20 years, but is one third of the total dose at one million years.

Characterization of the samples

The apparent (geometrical) density of the samples was calculated according to their dimensions as well as by immersion in mercury (non-penetrating liquid). The calculation of open porosity was based on the difference between the apparent density and the density by immersion in water or acetone (penetrating liquid). The density was expressed as percentage of the theoretical density (%TD), calculated from the composition and from stereological measurements.

Microstructural characterization of the samples was performed on ceramographic sections with an optical microscope and a scanning electron microscope (SEM); the distribution of the simulated wastes components in the glass matrix was studied by means of an energy dispersive X-ray analyzer (EDAX) coupled to the SEM.

The devitrification extent was determined by X-ray diffraction analysis of powdered samples, using the method described in ref. [5].

Aqueous corrosion test

The aqueous corrosion behavior of the glass, with and without simulated wastes, was studied using one of the normalized tests recommended by the Material Characterization Center, the MCC-1P [6]. It determines the normalized element mass loss using monolithic samples exposed to aqueous solutions at temperatures lower than 373 K during specific periods of time.

The extent of leaching is expressed by the normalized mass loss rate for element i (NRL_i), which is defined as follows:

$$NRL_i = \frac{m_i}{F_i \cdot S_o \cdot t} \tag{1}$$

where m_i (g) is the mass of the element i in the leachate, F_i (dimensionless) is the mass fraction of the element i in the original sample, S_o (cm^2) is the initial geometrical surface area of the sample and t (day) is the test duration.

In the case of the mass loss measurements, the gravimetric dissolution rates were calculated according to:

$$R = \frac{\Delta m}{S_o \cdot t} \tag{2}$$

where Δm (g) is the sample mass loss after drying at 383 K to constant weight and S_o and t are defined as above.

In order to simulate the repository conditions, apart from the already mentioned devitrification, we made an aqueous solution with a composition similar to that found 500 m deep in a granitic dome of Argentina.

The synthetic water was called ASL and its composition is shown in Table III.

Because different studies with the above mentioned groundwater indicate the possible presence of some basic lead carbonates [7], we prepared another solution with such carbonates (ASL/Pb).

Finally, deionized water (ADI) was used for comparison purposes.

The samples were placed vertically on a Teflon support which was put in a vessel of the same material. Aqueous solution was added in order to have the relation $area_{sample}$ / $volume_{solution}$ equal to 0.1 cm^{-1}, and the vessel was placed in an oven at 363 K for 3, 7, 14 and 28 days.

The concentrations of the elements in the leachate were determined by atomic absorption spectrometry (flame and graphite oven) and inductively coupled plasma spectrometry (ICP).

The measured elements were Si, Ca (as glass matrix elements), Sr, Mo, Fe and Cs (as HLW elements).

We evaluated the corrosion resistance of the devitrified glass in relationship to the irradiation dose and type of aqueous medium.

Table III: Chemical composition of the synthetic laboratory prepared water (ASL)

Element	ASL (mg/l)
Calcium	8.28
Magnesium	2.05
Sodium	80.20
Potassium	0.80
Bicarbonate	199.00
Sulfate	18.80
Chloride	14.50
Fluoride	0.39
pH	8.60

RESULTS

Effect of aqueous medium on non-irradiated samples

As can be seen in Table IV, the dissolution rate (gravimetric and elemental) decreases with time. This can be attributed to a saturation effect. The NRL of Fe is not presented because this element was undetected.

The gravimetric dissolution rate in ADI is around 2 times lower than in ASL, but is almost the same as that in ASL with lead carbonates. We think that the slightly alkaline pH of ASL is responsible for the increase in the dissolution rate. For experiments performed in ASL/Pb, the lower dissolution rate could be explained because PbO tends to concentrate on the glass surface, forming a protective surface layer [8, 9].

From Table IV we can see that the NRL of Si is rather unaffected by the aqueous medium. The high NRL of Ca in ASL and ASL/Pb at a short time, is probably due to an initially selective dissolution effect. At longer times the saturation of the solution and the onset of congruent dissolution make the NRL decrease and approach values similar to those of Si.

In the same Table, it is observed that the more soluble elements Mo and Sr have higher NRLs than Si and Ca. This could be caused by a different leaching process (diffusion through the glass phase and dissolution into the solution [13]). The lower NRL of Sr in ASL/Pb could also be connected with the passivization effect of PbO, but we do not have an explanation for that behavior.

The third soluble element analyzed was Cs and we can see from Table IV that it has the lowest NRL. This effect was also observed by Inagaki [13] and was attributed to a concentration of Cs in the alteration layer.

It appears that surrounding aqueous medium in the repository would not increase the gravimetric corrosion rate more than 2 times, for the case when pure ASL is considered. If there are lead carbonates present, we would not expect any change.

The elemental normalized dissolution rate shows that the aqueous medium affects different elements differently, but all the effects were small (apart from Sr, the differences are lower than a factor of two in the dissolution rate), and we can say that there is not any drastic deterioration of the corrosion resistance of the glass forms under these simulated repository conditions.

Effect of irradiation dose

Studying commercial glasses, Shelby [10] found a gamma radiation-induced densification of borosilicate glasses and related this effect to the boron content. Also Bibler [11] reported a slight ($< 0.05\%$) increase in density of a borosilicate glass produced for immobilization of defense high-level waste.

We do not observe any change in the density of the samples with dose, as is shown in Table V. This could be a result of the lower boron content of our material, the high degree of devitrification of the samples and the remaining closed porosity.

Scanning electron microscopy did not show any difference in the microstructure between irradiated and non-irradiated samples. Due to the nature of the forming process, where temperatures are well below those used in melting technology, the waste particles are only partially dissolved into the glass network.

Table IV: Dissolution rates at different times and in different aqueous media for non irradiated samples

	ADI			
	3 days	7 days	14 days	28 days
Gravimetric	1.85E-05	1.19E-05	9.14E-06	8.70E-06
Si	1.10E-05	1.10E-05	1.17E-05	1.37E-05
Ca	1.38E-04	8.65E-05	2.49E-05	1.93E-05
Cs	1.20E-05	8.73E-06	8.64E-06	8.50E-06
Mo	2.42E-04	8.32E-05	5.15E-05	4.30E-05
Sr	6.08E-04	2.61E-04	1.29E-04	1.05E-04
	ASL			
Gravimetric	3.39E-05	3.80E-05	2.09E-05	2.18E-05
Si	8.80E-05	7.05E-05	3.22E-05	1.39E-05
Ca	1.03E-03	2.43E-04	3.31E-05	5.31E-06
Cs	9.85E-06	5.40E-06	4.18E-06	4.00E-06
Mo	2.91E-04	1.86E-04	1.03E-04	1.01E-04
Sr	7.61E-04	3.83E-04	1.62E-04	1.37E-04
	ASL/Pb			
Gravimetric	1.00E-05	1.02E-05	1.47E-05	9.72E-06
Si	1.10E-05	3.28E-05	1.64E-05	1.20E-05
Ca	1.20E-03	4.56E-04	3.05E-05	3.20E-05
Cs	9.80E-06	2.10E-06	2.10E-06	1.58E-06
Mo	7.00E-04	1.86E-04	1.71E-04	6.08E-05
Sr	1.52E-04	3.38E-05	1.95E-05	1.44E-05

Table V: Behavior of the density of SG7-10% PHWR with irradiation dose

Dose (Gy)	Density	
	(g/cm3)	(%TD)
None	2.460 ± 0.007	96.1 ± 0.3
1.4×10^6	2.463 ± 0.005	96.2 ± 0.2
8.0×10^6	2.462 ± 0.010	96.2 ± 0.4
4.6×10^7	2.455 ± 0.010	95.9 ± 0.4
2.0×10^8	2.457 ± 0.010	96.0 ± 0.4

For this reason, it was possible to distinguish glass particles from simulated waste particles distributed around the glass particles.

Detailed microscopic observation of thin samples did not indicate formation of bubbles in the samples, as was reported by Weber [12]. The porosity is the same in the non-irradiated samples and in the irradiated ones.

In Fig. 1 we can see the X-ray diffraction analysis of devitrified samples irradiated at different doses. The degree of crystallinity and the distribution of crystalline phases were unaffected by the irradiation. The main diffraction peaks correspond to the silica phases (quartz and cristobalite); the smaller peaks belong to waste phases [1].

The dissolution rates of the irradiated samples showed the same saturation effect with time as we mentioned for the unirradiated case.

Fig. 2 describes the behavior of the dissolution rate with dose, for 7 days as an example. We choose this data because the concentration of the elements in the leached medium was high enough for a statistically relevant analysis and because this was a conservative assumption.

The gravimetric dissolution rate at any dose is nearly the same for the irradiated and non-irradiated samples.

The behavior of the dissolution rate of Si is similar to the gravimetric: It increases by a factor of two from the non-irradiated samples to those irradiated with the lowest dose; at higher doses it is almost constant. The case of Cs is similar; the dissolution rate increases one order of magnitude from the non-irradiated to the irradiated state. The behavior of Ca, Sr and Mo is different because there is no large increment in the dissolution rate between irradiated and non-irradiated samples, but the corrosion resistance does show a tendency to increase with applied dose.

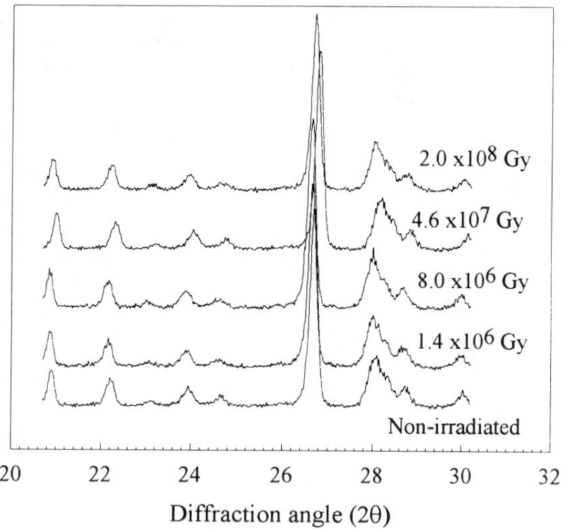

Figure 1: Diffraction patterns of samples irradiated with different gamma doses

These results were qualitatively confirmed by microscopic examination. After leaching, the waste zones were more affected. Further increase in dose did not result in a variation in the elemental dissolution rates.

CONCLUSIONS

Following the research on SG7 glass as a matrix material for the immobilization of HLW of the PHWR type, we have centered our work on the aqueous corrosion behavior of SG7 glass.

The studies to date show that sintered glass forms maintain their corrosion resistance in different lixiviating media, including groundwater.

The dissolution rates obtained with this sintered glass are lower than those obtained with glass formed by melting [14].

We found no adverse influence of the simulated conditions (devitrification, irradiation and aqueous media) on the bulk corrosion resistance or on glass microstructure, but the analyses of the NRLs before and after irradiation are indicating that further studies on the elemental dissolution mechanism are needed.

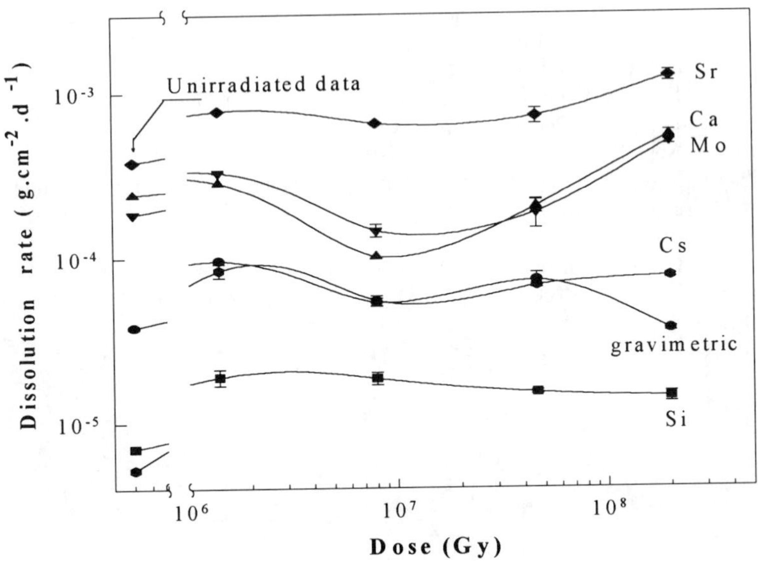

Figure 2: Aqueous corrosion behavior of devitrified samples, irradiated at different gamma doses. Leaching conditions: 363 K, 7 days in ASL.

Finally, we are planning more research on the effect of other types of radiation like electrons and/or alpha particles, including irradiating samples immersed in water to study the combined effect of radiation and aqueous corrosion.

ACKNOWLEDGMENTS

The authors wish to thank to Ing. Silvia Dutrús for her help on SEM analysis. We also thank A. Wiesztort from the Departamento de Investigación Aplicada, Centro Atómico Bariloche and O. Casella from the company Invap S.E. for the chemical analysis of the leachants.

This work was done in the frame of a cooperation agreement with the International Atomic Energy Agency.

REFERENCES

1. M. A. Audero, A. M. Bevilacqua, N. B. Messi de Bernasconi, D. O. Russo and M. E. Sterba, J. Nucl. Mater. **223**, p.151-156 (1995).
2. A. M. Bevilacqua, N. B. Messi de Bernasconi, D. O. Russo, M. A. Audero, M. E. Sterba and A. D. Heredia, J. Nucl. Mater. **229**, p.187-193 (1996).
3. S. Gahlert and G. Ondracek, in Radioactive Waste Forms for the Future, edited by W. Lutze and R.C. Ewing (North-Holland, Amsterdam, 1988) p.161.
4. A. M. Bevilacqua, N. B. Messi de Bernasconi, and M. E. Sterba, Proc. Annual Meeting of the Argentine Association of Nuclear Technology (AATN), Bariloche, 1987.
5. P.H. Hermans and A. Weidinger, Makromol. Chem. **24**, p. 44 (1961).
6. Materials Characterization Center, Materials Characterization Center Test Methods, PNL-3990, Battelle Pacific Northwest Laboratories (1981).
7. C.J. Semino. Final Report of Contract IAEA N°6504/RB.
8. I. Schwartz, M.H. Mintz and N. Shamir, J. Nucl. Mater. **172**, p. 314-318 (1990).
9. C.Q. Buckwalter and L.R. Pederson, J. Am. Ceram. Soc. **65**, p. 431 (1982).
10 J.E. Shelby, J. Appl. Phys. **51**, p. 2561-2565 (1980).
11. N.E. Bibler in Scientific Basis for Nuclear Waste Management, edited by S.V. Topp (North-Holland, New York, 1982) p. 681.
12. W.J. Weber, Nucl. Instr. and Methods in Phys. Res. **B32**, p.471-479 (1988).
13. Y. Inagaki, H. Furuya, K. Idemitsu and S. Yonezawa, J. Nucl. Mater., **208**, p. 27-34 (1994)
14 Hj. Matzke and E. Vernaz, J. Nucl. Mater., **201**, p. 295-309 (1993).

EFFECTS OF REDOX CONDITIONS OF WATER ON Pu AND Cm LEACHING FROM WASTE GLASS

Y.Inagaki, A.Sakai, H.Furuya, K.Idemitsu and T.Arima, Dept. of Nucl. Eng., Kyushu Univ., Fukuoka 812, JAPAN.
T.Banba, T.Maeda, S.Matsumoto and Y.Tamura, Dept. of Environ. Safe. Research, Japan Atomic Energy Research Institute, Tokai-mura, Ibaraki 319-11, JAPAN.

ABSTRACT

Static corrosion tests were performed on the simulated waste glass doped with Pu and Cm (PuO_2; 0.22wt%, Cm_2O_3; 0.09%) in deionized water at 90 °C with S/V ratio of 2600 m^{-1} under oxidizing and reducing conditions, respectively. The corrosion tests under oxidizing conditions were performed in air, while the corrosion tests under reducing conditions were performed in an airtight stainless steel container purged with mixed gas ($Ar+5\%H_2$), where the Eh of the solution was maintained at approximately -0.45 V vs SHE. The results showed that the redox conditions have no remarkable influence on the leaching behavior of Pu and Cm. It was also observed that the Pu and Cm concentrations in the 1.8nm filtered solutions were one or two orders of magnitude lower than those in the 450nm filtered solutions under both redox conditions, which suggests that the insoluble suspended fractions (colloidal particles) with the size from 1.8nm to 450nm are dominant phase for Pu and Cm in the solution under both redox conditions.

INTRODUCTION

Leaching of actinides from high-level nuclear waste glasses is one of the most important phenomena to be evaluated for safety assessments of the disposal system. In recent years, some corrosion tests on actinides-doped waste glasses have been performed, and the leaching behavior of actinides has been investigated[1,2,3,4]. However, most of the previous corrosion tests have been performed under oxidizing conditions, and the leaching behavior of actinides under reducing conditions (predicted in repository environments) has not been well studied. Most of actinides contained in the waste glasses are redox active, and their oxidation states, chemical species and equilibrium solubilities are greatly influenced by redox conditions. Therefore, the leaching behavior of actinides is expected to be greatly affected by redox conditions of water.

The purpose of this study is to understand the effects of redox conditions of water on leaching behavior of actinides from the waste glass. Static corrosion tests were performed on the simulated waste glass doped with Pu and Cm in deionized water at 90 °C with S/V ratio of 2600 m^{-1} under oxidizing and reducing conditions, respectively. The corrosion tests under oxidizing conditions were performed in air, while the corrosion tests under reducing conditions were performed in an airtight stainless steel container purged with mixed gas ($Ar+5\%H_2$), where Eh of the solution was maintained at approximately -0.45 V vs SHE. After the corrosion tests, the solution was filtered through 450nm and 1.8nm filters in order to investigate the distribution of the Pu and Cm particle size fractions, and the solution concentrations of Pu, Cm and other glass constituents were measured by α-spectrometry and ICP-AES. Comparing the experimental results with thermodynamic data from other studies,

Mat. Res. Soc. Symp. Proc. Vol. 465 © 1997 Materials Research Society

the mechanism of the Pu and Cm leaching under reducing conditions are discussed.

EXPERIMENTAL

A powdered simulated waste glass(similar composition to R7T7 glass) doped with Pu and Cm(^{238}PuO$_2$; 0.06wt%, ^{240}PuO$_2$; 0.16wt%,^{244}Cm$_2$O$_3$; 0.09wt%) was used as glass specimen. The grain size of the glass specimen was from 75μm to 150μm, and the specific surface area was 0.083m^2/g measured by BET method. Static corrosion tests on the glass specimen were performed in deionized water at 90℃ under oxidizing and reducing conditions, respectively. The corrosion tests under oxidizing conditions were performed in air. The glass specimen(0.8g) and deionized water(25ml) were placed in a Teflon container with S/V ratio of 2600m^{-1} and kept in a heating furnace at 90℃ for periods of up to 92 days. The pH and Eh of the initial solution(deinonized water) were measured to be 5.6 and +0.50V vs SHE at 25℃, respectively. While, the corrosion tests under reducing conditions were performed in an airtight stainless steel container purged with mixed gas (Ar+5%H$_2$) as shown in Fig.1. In a glove box purged with the mixed gas, the mixed gas was slowly bubbled through deionized water to ensure a O$_2$ and CO$_2$ free environment, and the pH and Eh of the deionized water were measured to be 6.9 and -0.35V vs SHE at 25℃, respectively. The glass specimen(0.8g) and the deionized water with low Eh(25ml) were placed in a Teflon container, and the Teflon container was placed in an airtight stainless steel container in order to maintain the atmosphere for long term. After sealing, the airtight stainless steel container was taken out from the glove box, and kept in a heating furnace at 90℃ for periods of up to 92 days.

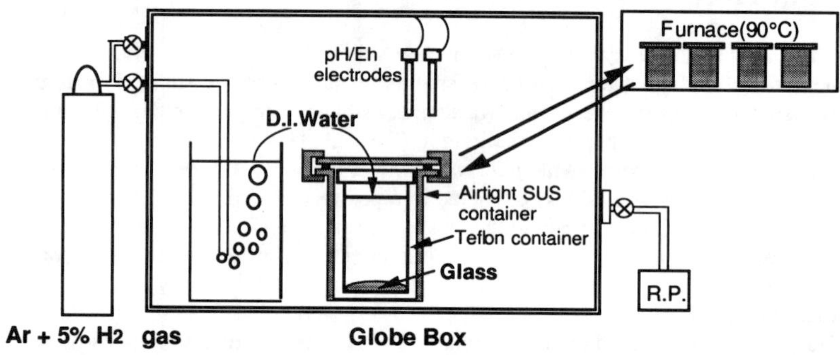

Fig.1. Schematic representative of corrosion tests under reducing conditions.

After the corrosion tests, the Teflon container was cooled to room temperature, and the solution pH and Eh were measured immediately. The solution was filtered through 450nm and 1.8nm filters in order to investigate the distribution of the Pu and Cm particle size fractions, and the solution concentrations of Pu, Cm and other glass constituents were measured by α-spectrometry and ICP-AES. In the case of corrosion tests under reducing conditions, measurement of the pH and Eh, and filtration of the solution were carried out in the glove box.

RESULTS

Fig.2 shows the solution pH and Eh as a function of corrosion time. It can be seen that the solution Eh under reducing conditions was maintained at approximately -0.45V vs SHE during the corrosion tests, which suggests that the solution Eh was controlled by the following reaction[5,6],

$$\frac{1}{2} H_2(gas) = H^+ + e^-. \tag{1}$$

While, the solution pH under both redox conditions decreased slowly with corrosion time after the rapid increase to the pH of 9.5 in a few days, which is expected to be due to the effects of radiolysis of water and gas.

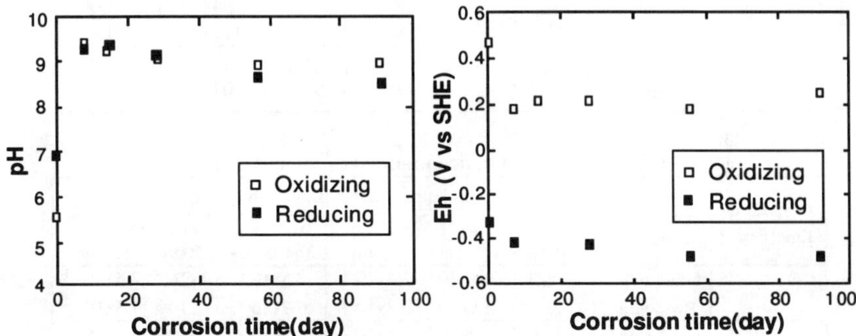

Fig.2. Solution pH and Eh as a function of corrosion time.

The solution concentrations of major glass constituents are listed in table I, and Fig.3 shows the normalized concentrations(NCi) as a function of corrosion time calculated by the following equation[7],

$$NCi = Ci / fi, \tag{2}$$

where, Ci is the solution concentration of element i, and fi is the original fraction of element i in the glass. It can be seen that redox conditions have no remarkable influence on the leaching of major glass constituents. Under both redox conditions, the effects of filtration on the solution concentrations were sufficiently small for Si and soluble elements(B, Li, Na, Mo). While, the effects of filtration were relatively large for Al and Ca, and the concentrations of Al and Ca in 1.8nm filtrates were thirty or forty percent of magnitude lower than those in non-filtered solutions. The present data of NCi for soluble elements from actinides-doped glass are a little higher comparing to the data from non-doped glass obtained in our previous work[6].

The solution concentrations of Pu and Cm are listed in table II, and Fig.4 shows the concentrations as a function of corrosion time. It can be seen that redox conditions also have no remarkable influence on the leaching of Pu and Cm. Under both redox conditions, the effects of filtration were large, and the Pu and Cm concentrations in the 1.8nm filtrates were one or two orders of magnitude lower than those in the 450nm filtrates. While, there was no difference in the Pu and Cm concentrations between non-filtered solutions and 450nm filtrates.

Table I. Solution concentrations of major glass constituents.

Corrosion time[day]	Redox condition	Filtration	Solution concentration[ppm]						
			Si	B	Li	Mo	Na	Ca	Al
7	Oxidizing	no-filtration	45.0	23.3	4.98	6.01	34.1	1.25	0.68
	Reducing		42.7	21.4	4.61	5.76	30.2	1.37	0.59
14	Oxidizing	no-filtration	56.5	35.2	7.57	8.88	49.8	1.31	0.64
	Reducing		53.4	32.6	6.84	8.63	46.1	1.38	0.55
28	Oxidizing	no-filtration	59.3	38.1	8.36	9.34	60.8	1.23	0.66
	Reducing		58.4	36.4	7.85	8.88	59.7	1.16	0.65
56	Oxidizing	no-filtration	67.9	52.7	11.0	11.5	75.8	0.74	0.76
	Reducing		64.9	49.5	10.4	8.68	72.2	1.49	0.72
92	Oxidizing	no-filtration	73.0	73.0	15.1	12.2	94.7	1.02	1.06
	Reducing		70.6	84.2	17.0	10.5	117.0	2.16	1.12
92	Oxidizing	1.8nm	72.7	70.0	14.8	12.0	93.8	0.62	0.73
	Reducing		63.0	75.0	16.0	6.68	104.2	1.72	0.70

Table II. Solution concentrations of Pu and Cm.

Corrosion time[day]	Redox condition	Solution concentration[mol/l]					
		Pu			Cm		
		no-filtration	450nm	1.8nm	no-filtration	450nm	1.8nm
7	Oxidizing	4.0×10^{-8}	4.1×10^{-8}	2.4×10^{-9}	3.5×10^{-9}	3.1×10^{-9}	2.5×10^{-11}
	Reducing	7.6×10^{-8}	5.5×10^{-8}	2.3×10^{-9}	6.9×10^{-9}	6.4×10^{-9}	7.5×10^{-12}
14	Oxidizing	5.1×10^{-8}	3.2×10^{-8}	2.4×10^{-9}	4.8×10^{-9}	4.3×10^{-9}	3.1×10^{-11}
	Reducing	9.5×10^{-8}	1.1×10^{-7}	3.1×10^{-9}	8.6×10^{-9}	7.4×10^{-9}	2.2×10^{-11}
28	Oxidizing	9.8×10^{-8}	1.7×10^{-7}	--	4.8×10^{-9}	3.8×10^{-9}	--
	Reducing	1.3×10^{-7}	1.6×10^{-7}	1.5×10^{-8}	9.9×10^{-9}	7.8×10^{-9}	1.1×10^{-9}
56	Oxidizing	1.4×10^{-7}	1.1×10^{-7}	1.0×10^{-8}	1.2×10^{-8}	8.4×10^{-9}	2.7×10^{-10}
	Reducing	1.5×10^{-7}	1.2×10^{-7}	--	8.8×10^{-9}	7.9×10^{-9}	--
92	Oxidizing	2.5×10^{-7}	1.6×10^{-7}	1.5×10^{-8}	4.3×10^{-8}	9.7×10^{-9}	5.7×10^{-10}
	Reducing	2.3×10^{-7}	2.3×10^{-7}	2.2×10^{-8}	1.4×10^{-8}	1.2×10^{-8}	9.1×10^{-10}

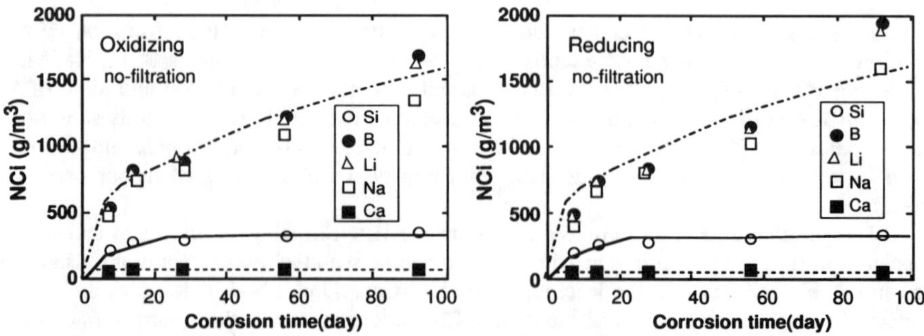

Fig.3. NCi for major glass constituents under oxidizing and reducing conditions.

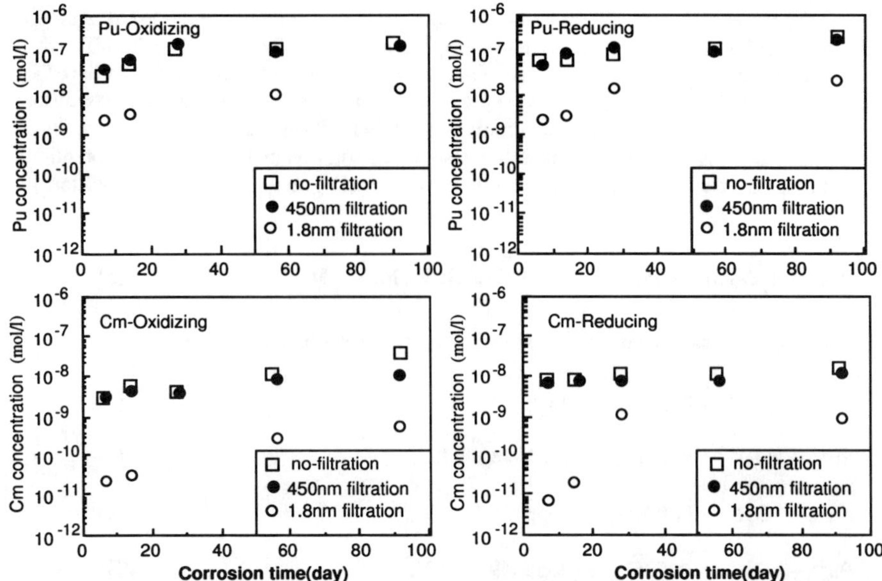

Fig.4. Solution concentrations of Pu and Cm under oxidizing and reducing conditions.

DISCUSSION

The present results showed that the redox conditions have no remarkable influence on the leaching behavior of Pu and Cm, which suggests that dominant oxidation states of the Pu and Cm in the solution under reducing conditions are the same as those under oxidizing conditions. It was also observed that the Pu and Cm concentrations in the 1.8nm filtrates are one or two orders of magnitude lower than those in the 450nm filtrates under both redox conditions. The Pu and Cm concentrations in the 1.8nm filtrates are assumed to correspond to the soluble species controlled by the solubilities, and the difference in Pu and Cm concentrations between the 1.8nm and 450nm filtrates is assumed to correspond to the insoluble suspended fractions(colloidal particles). Therefore, it is suggested that the colloidal particles with the size from 1.8nm to 450nm are dominant phase for Pu and Cm in the solution under both redox conditions. On the other hand, NCi for Pu and Cm are calculated to be $30g/m^3$ and $13g/m^3$ at the maximum, respectively. These values of NCi for Pu and Cm are less than a tenth of the value for Si, which suggests that most of the Pu and Cm remains at the glass surface as the secondary phases under both redox conditions. Similar results of the leaching of actinides can be found in other corrosion tests performed under oxidizing and weakly reducing conditions[1,3,8].

Solubility of Pu in neutral aqueous solutions has been investigated by many researchers, and some thermodynamic data have been reported[9-14]. Comparing the present data on the Pu concentrations with the solubilities calculated by the thermodynamic data, reactions controlling the Pu leaching from glass can be estimated. On the other hand, only a few investigations are known on the solubility of Cm, and the comparison has not been done in the present study. For calculating the Pu solubility, the oxidation states of Pu must be evaluated. From the data of potentials for redox reactions[10], dominant oxidation states of Pu under the present oxidizing conditions (pH=8.8, Eh=+0.25V vs SHE) are expected to be Pu(IV), Pu(V) or Pu(VI), while the dominant oxidation states under the reducing conditions(pH=8.6, Eh=-0.45V vs SHE) are expected to be Pu(III) or Pu(IV).   However, the present results showed that redox conditions have no influence on the solution concentrations of Pu($\approx 10^{-8}$M in the 1.8nm filtrates), which suggests that Pu(IV) is the dominant oxidation state in the solution under both redox conditions. The most feasible reaction controlling the solubility of Pu(IV) under the present conditions[3] is,

$$PuO_2 \cdot xH_2O(am) \leftrightarrow Pu^{4+} + 4OH^- + (x-2)H_2O, \ logK=-56.85[11]. \tag{3}$$

In the solution, Pu^{4+} can make the complexation legands described as follows,

$$Pu^{4+} + OH^- \leftrightarrow Pu(OH)^{3+}, \quad log\beta=13.67[13], \tag{4}$$

$$Pu^{4+} + 2OH^- \leftrightarrow Pu(OH)_2^{2+}, \ log\beta=28.20[13], \tag{5}$$

$$Pu^{4+} + 3OH^- \leftrightarrow Pu(OH)_3^+, \quad log\beta=38.85[13], \tag{6}$$

$$Pu^{4+} + 4OH^- \leftrightarrow Pu(OH)_4^0, \quad log\beta=49.20[13], \tag{7}$$

$$Pu^{4+} + 5OH^- \leftrightarrow Pu(OH)_5^-, \quad log\beta=53.8[13], \tag{8}$$

$$Pu^{4+} + 4OH^- + 2HCO_3^- \leftrightarrow Pu(OH)_2(CO_3)_2^{2-} + 2H_2O, \ log\beta=54.15[14]. \tag{9}$$

The solubility of Pu then can be calculated as the sum of $[Pu^{4+}]$ and the concentrations of complexation legands. Using the equilibrium constants(K) by Rai[11] and the complex stability constants(β) by Nitsche[13] and Yamaguchi[14] described above, the solubility of Pu under the present conditions were calculated. In the calculations the temperature was assumed to be 25℃, because the thermodynamic data can not be found at 90℃.

Table III lists the results of the calculations compared with the measured concentrations of Pu in the 1.8nm filtrates. The concentrations of other hydrolysis species are too small to be taken into account. A relatively good agreement is found between the calculated Pu solubilities and the measured Pu concentrations under both redox conditions. These results suggest that precipitation of $PuO_2 \cdot xH_2O(am)$ is controlling the leaching of soluble species of Pu under both redox conditions, and the dominant soluble species are $Pu(OH)_4^0$ and/or $Pu(OH)_2(CO_3)_2^{2-}$ under oxidizing conditions and $Pu(OH)_4^0$ under reducing conditions. However, the present analysis contains some uncertainties, e.g. no thermodynamic data for Pu solubility at 90℃, widely varying of the thermodynamic data from researcher to researcher even at 25℃ etc., and much more detailed work is necessary. Although Pu(IV) is suggested to be the dominant oxidation state under both redox conditions in the present study, Pu(IV) may be reduced to Pu(III) for longer times, and the corrosion tests for longer times are also necessary.

Table III. Solubility of Pu(IV) calculated by using the equilibrium constants by Rai[11] and the complex stability constants by Nitsche[13] and Yamaguchi[14] compared with the measured concentrations.

Oxidizing conditions (pH=8.8, Eh=+0.25 V vs SHE, $P_{CO2}=10^{-3.5}$)	
Calculated solubility of Pu(IV) (25°C, $PuO_2 \cdot xH_2O$(am) is the solubility control. phase)	**Measured Pu concentration** (90°C, 92days, 1.8nm filtration)
$[Pu(OH)_4^0] = 2.2 \times 10^{-8}$M $[Pu(OH)_5^-] = 5.0 \times 10^{-9}$M $[Pu(OH)_2(CO_3)_2^{2-}] = 1.8 \times 10^{-8}$M	
$[Pu(IV)]_{total} = 4.5 \times 10^{-8}$M	$[Pu]_{total} = 1.5 \times 10^{-8}$M
Reducing conditions (pH=8.6, Eh=-0.45 V vs SHE, $P_{CO2} \approx 0$)	
Calculated solubility of Pu(IV) (25°C, $PuO_2 \cdot xH_2O$(am) is the solubility control. phase)	**Measured Pu concentration** (90°C, 92days, 1.8nm filtration)
$[Pu(OH)_4^0] = 2.2 \times 10^{-8}$M $[Pu(OH)_5^-] = 3.0 \times 10^{-9}$M	
$[Pu(IV)]_{total} = 2.5 \times 10^{-8}$M	$[Pu]_{total} = 2.2 \times 10^{-8}$M

The present results suggest that the colloidal particles with size from 1.8nm to 450nm are dominant phase for Pu and Cm in the solutions under both redox conditions. In addition, the concentrations of the colloidal particles of Pu and Cm was relatively high even in the early stage of glass corrosion and almost saturated after 20days of corrosion time as shown in Fig.4, which tendency is the same as that for other glass constituents such as Si, Ca and Al. While, the concentrations of the soluble species of Pu and Cm(≤1.8nm) were very low in the early stage of glass corrosion and increased slowly with corrosion time. These results suggest that formation of the colloidal particles of Pu and Cm are closely related to the leaching behavior of other glass constituents. For Ca and Al, the concentrations in the 1.8nm filtrates were thirty or forty percent of magnitude lower than those in non-filtered solution as shown in Table I, which suggests that colloidal particles of these elements are present in the solutions. Buck et al.[15] and Fortner et al.[16] showed that clay colloids such as smectite and zeolite are produced as a direct result of the alteration of waste glass itself(primary colloids), and they can sorb or incorporate actinides strongly. Therefore, the colloidal particles of Pu and Cm observed under both redox conditions are expected to be weakly crystalline clay colloids sorbing Pu and Cm which are produced as a direct result of the glass alteration. The future work needs investigation of physical and chemical properties of the collodal particles for the long-term.

CONCLUSIONS

Static corrosion tests were performed on the simulated waste glass doped with Pu and Cm under oxidizing and reducing conditions, respectively. The results showed that redox conditions have no remarkable influence on the leaching behavior of Pu and Cm. It was also shown that colloidal particles with the size from 1.8nm to 450nm are dominant phase for Pu and Cm in the solution under both redox conditions, and the colloidal particles are suggested to be primary colloids. Comparison of the solution concentrations of Pu with the solubility of Pu calculated by use of published thermodynamic data suggests that precipitation of $PuO_2 \cdot xH_2O(am)$ is controlling the leaching of soluble species of Pu under both redox conditions, and the dominant soluble species are $Pu(OH)_4^0$ or $Pu(OH)_2(CO_3)_2^{2-}$ under oxidizing conditions and $Pu(OH)_4^0$ under reducing conditions.

ACKNOWLEDGEMENT

The authors wish to thank O.Kikuchi, M.Kamoshida, H.Yoshinari, K.Kawasaki, H.Umehara and A.Kijima for their helps of the experiments at HOT laboratories.

REFERENCES

[1] E.Y.Vernaz and N.Gordon, Mat.Res.Soc.Proc.Vol.257(1992)37.

[2] K.Lemmens, P.Van Iseghem and L.Wang, Mat.Res.Soc.Proc.Vol.294(1993)147.

[3] L.Wang and P.Van Iseghem, Mat.Res.Soc.Proc.Vol.294(1993)155.

[4] J.A.Fortner and J.K.Bates, Mat.Res.Soc.Proc.Vol.412(1996)205.

[5] J.I.Drever, The Geochemistry of Natural Waters, 2nd ed.(Prentice Hall Inc., Englewood Clif, NJ, 1988)

[6] Y.Inagaki, H.Furuya, K.Idemitsu, T.Maeda and A.Sakai, Mat.Res.Soc.Proc.Vol.353 (1995)23.

[7] Y.Inagaki, A.Ogata, H.Furuya, K.Idemitsu, T.Banba and T.Maeda, Mat.Res.Soc.Proc. Vol.412(1996)257.

[8] J.I.Kim, W.Treiber, Ch.Lierse and P.Offermann, Mat.Res.Soc.Proc.Vol.44 (1985)359.

[9] D.Rai, R.J.Serne and D.A.Moore, Soil Sci.Soc.Am.J.Vol.44(1980)490.

[10] B.Allard, The Solubility of Actinides in Neutral and Basic Solutions in Actinides in Perspective, edited by N.M.Edelstein(Pergamon Press, Oxford,1982).

[11] D.Rai, Radiochimica Acta, 35(1984)97.

[12] J.I.Kim and B.Kanellakopulos, Radiochimica Acta, 48(1989)145.

[13] H.Nitshe, Mat.Res.Soc.Proc.Vol.212(1991)517.

[14] T.Yamaguchi, Y.Sakamoto, T.Ohnuki, Radiochimica Acta, 66/67(1994)9.

[15] E.C.Buck, J.K.Bates et al., Mat.Res.Soc.Proc.Vol.294(1993)199.

[16] J.A.Fortner and J.K.Bates, Mat.Res.Soc.Proc.Vol.412(1996)205.

APPLICABILITY OF V_2O_5-P_2O_5 GLASS SYSTEM FOR LOW-TEMPERATURE VITRIFICATION

T. NISHI*, K. NOSHITA*, T. NAITOH**, T. NAMEKAWA**, K. TAKAHASHI** and M. MATSUDA***
*Power & Industrial Systems R&D Division, Hitachi Ltd., 7-2-1 Omika, Hitachi, Ibaraki 319-12, Japan.
**Hitachi Research Laboratory, Hitachi Ltd., 3-1-1 Saiwai, Hitachi, Ibaraki 317, Japan.
***Hitachi Works, Hitachi Ltd., 3-1-1 Saiwai, Hitachi, Ibaraki 317, Japan.

ABSTRACT

The V_2O_5-P_2O_5 glass system (vanadate glass) has been developed to provide low temperature vitrification of radioactive waste without volatilization of radionuclides. The dissolution behaviour of vanadate glass was determined from the results of the MCC-1 static leach test at 70 ℃ using deionized water, cement-saturated alkaline water, and cement-saturated alkaline water with 3% NaCl. The solubility of vanadate glass was remarkably reduced by the addition of Sb_2O_3 to the glass. It was also lowered in the cement-saturated water due to formation of a precipitation layer. Although the presence of NaCl in the solution accelerated the dissolution of vanadate glass, the addition of alkaline earth metal oxide to the glass gave a dissolution rate three orders lower than non-additive vanadate glass. The results of leach tests using simulated vitrified waste form containing 5mol% of Cs_2O and SrO showed there was a possibility of fixing Cs and Sr into the glass structure.

INTRODUCTION

Borosilicate glass systems have been widely studied for the vitrification of high level radioactive waste regarding a number of points, e.g. vitrification system [1,2], leaching mechanism [3-5], and high waste loading [6]. Although borosilicate glass is superior to other oxide glasses in its chemical and physical durabilities, vitrification using this glass must take place at high temperatures above 1000℃ [5,6], at which volatilization of volatile radionuclides, e.g. cesium or ruthenium, would be inevitable. The following advantages could be expected by reduction of vitrification temperature: controlled volatilization of radionuclides, prevention of cracks due to thermal strain, and extensions of the refractory material lifetime. On the other hand, glasses with a low melting temperature, such as PbO-B_2O_3 glass [7] or chalcogenide glass [8], are not always applicable to waste vitrification, because their solubilities are relatively high.

A low solubility glass with low melting temperature is expected to have an intermediate performance between cement and borosilicate glass, which gives the possibility of application to vitrification of low and intermediate level radioactive waste from a reprocessing plant, e.g. liquid waste and spent adsorbents from the off-gas filter. In this study, the V_2O_5-P_2O_5 glass system [9,10] was selected as a vitrifying agent and basic performance characteristics of a new glass system were evaluated: corrosion rate in water, effect of alkalinity and salt, and leachabilities of Cs and Sr. Additives for the vanadate glass were also investigated to improve the resistance to water and salt.

Mat. Res. Soc. Symp. Proc. Vol. 465 © 1997 Materials Research Society

EXPERIMENTAL

Sample preparation

The vanadate glass specimens were prepared from special grade reagents of V_2O_5 and P_2O_5. Antimony oxide (Sb_2O_3) was also used as a reductant which adjusted the vanadium valency. Mixtures of these oxides were placed in an alumina crucible and heated in an electric furnace for two hours at 1050℃. During this time the melting mixtures were occasionally stirred with an alumina rod. Then the melts were poured into a graphite mold kept at 200℃. The solidified glasses were reheated at 350℃ to remove thermal strain and gradually cooled to room temperature. A specimen of the PbO-B_2O_3 glass system was also prepared by the same procedure to serve as a low softening temperature reference glass.

The simulated waste glass was formed by the addition of Cs_2O and SrO into the base glass (V_2O_5-P_2O_5-Sb_2O_3 glass system) at a temperature of 700-800℃.

Leach test

The water durability of vanadate glass was evaluated by the standard leach test based on the MCC-1 static test [11]. The demolded glass blocks were cut into 10mm cubes with a diamond saw. A glass specimen was then suspended in the leachant which was kept at 70℃. The ratio of the surface area of the glass specimen to the leachant volume (SA/V) was approximately $0.1 cm^{-1}$. All of the leachants were sampled after intervals of a set number of days and the concentrations of leached elements were analyzed by inductively coupled plasma atomic emission spectrometry (ICP-AES).

Normalized elemental mass loss (NL) was calculated by the following equation:

$$NL_i = C_i \cdot V / (f_i \cdot S) \qquad (1),$$

where C_i is the concentration of element i in the leachant, f_i is the original fraction of element i in the glass specimen, V is the volume of leachant, and S is the surface area of the glass specimen.

In this study, three types of leachants were tested, deionized water, cement-saturated alkaline water, which simulated ground water that would leak into the concrete repository, and cement-saturated alkaline water with 3% NaCl, which simulated ground water originating from sea water.

Characterization

The glassification of glass specimens was evaluated by the X-ray powder diffraction (XRD) patterns, which were marked by the absence of diffraction peaks. The molar ratio of tetravalent vanadium (V^{4+}/V_{total}) in the vanadate glass specimens was measured by a redox titration method which is used for vanadium assay in steel [12]. The characteristic temperatures (glass transition temperature, etc.) of the glass specimens were determined by Differential Thermal Analysis (DTA). The surface of the post-leach specimen was characterized by SEM-EDX analysis.

RESULTS AND DISCUSSION

<u>Dissolution behaviour of vanadate glass in water</u>

The original vanadate glass, which is composed of only V_2O_5 and P_2O_5, has a low melting temperature, but high solubility in water, because it tends to form a layered structure due to the properties of pentavalent vanadium. The conversion of the glass structure from a layered one to a three-dimensional network could effectively improve the water resistance of vanadate glass. Therefore, antimony oxide (Sb_2O_3) was selected as an additive which could reduce the vanadium from penta to tetra valency:

$$2V_2O_5 \ + \ Sb_2O_3 \ \rightarrow \ 2V_2O_4 \ + \ Sb_2O_5 \ . \quad (2)$$

Figure 1 shows the relationship between the normalized mass loss of vanadium glass system specimens and the amount of Sb_2O_3 addition. It also shows the change in the molar ratio of tetravalent vanadium (V^{4+}/V_{total}) with the amount of Sb_2O_3 addition. In this leach test, 5mm cube glass specimens, whose molar ratio of V_2O_5 to P_2O_5 was fixed at 7/3, were soaked in deionized water for half an hour at 70℃ . The original vanadate glass without Sb_2O_3 severely corroded in the hot water and was observed to have a stratiform deterioration. The normalized mass loss of glass specimens decreased remarkably with the amount of Sb_2O_3 addition. At a 16mol% addition, it had dropped more than three orders of magnitude. This addition was the limit of glass formation. Since the molar ratio of tetravalent vanadium in the vanadate glass increased with the amount of Sb_2O_3 addition, it was found that the solubility of vanadate glass was closely related to the ratio of crosslinkable tetravalent vanadium.

In order to raise the molar ratio of tetravalent vanadium further, the composition of vanadate glass was optimized by reduction of the V_2O_5/P_2O_5 ratio and the addition of PbO. The optimized composition (V1) of vanadate glass is shown in **Table I.** The molar ratio of tetravalent vanadium was 0.84 and glass transition temperature was approximately 350℃.

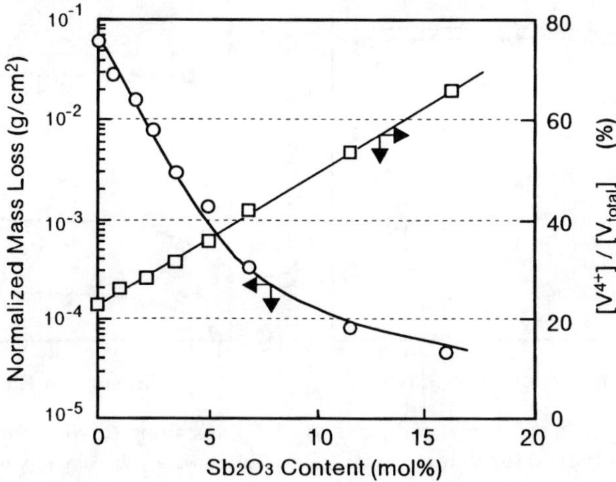

Fig. 1   Influence of Sb₂O₃ addition on solubility and tetravalent vanadium fraction of vanadate glass.

<u>Effects of alkalinity and salt</u>

It has been reported that the glass corrosion rate is affected by alkalinity and salt. In this study, cement-saturated water (pH=12-13) was selected as one of the alkaline leachants in the leach test. Additionally the effect of salt was examined using cement-saturated water containing 3% NaCl. A lead oxide glass (L1), as shown in **Table I**, was included for testing as a reference glass; it had the same characteristic temperatures as those of V1 glass.

The results of the MCC-1 static leach test are shown in **Figure 2**. The normalized mass loss of lead glass, L1, was clearly increased in the alkaline solutions independent of NaCl addition, in comparison with that in deionized water. Since the one-dimensional Pb-O bond is easily dissociated by OH ions, the solubility of lead glass in alkaline water was relatively high. On the other hand, the normalized mass loss of vanadate glass, V1, was somewhat lower in the cement- saturated water than that in deionized water, but it was significantly raised by the presence of NaCl.

Table I   The compositions of two glass specimens

Optimized composition (V1)		Reference composition (L1)	
reagents	composition/wt%	reagents	composition/wt%
V_2O_5	40	PbO	78
P_2O_5	20	B_2O_3	12
Sb_2O_3	25		
PbO	15	others	10

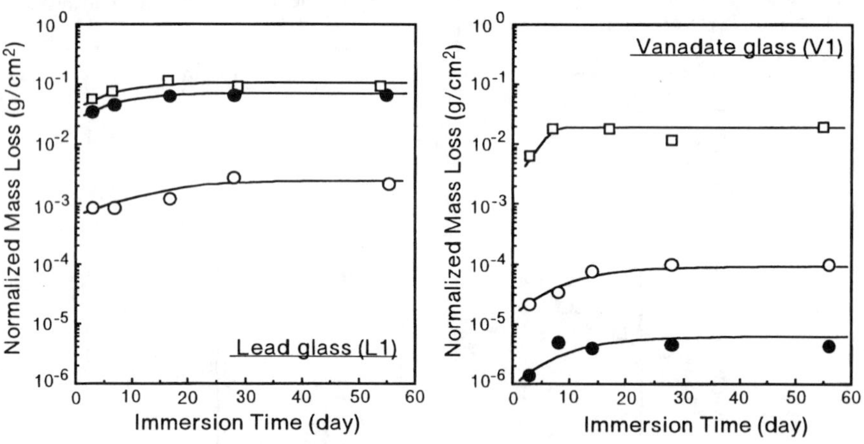

Fig. 2 Changes in the nomalized mass loss of glass samples with immersion time. (MCC-1 static leach test at 70℃, ○; deionized water, ●; cement-saturated water, □; cement-saturated water with 3% NaCl)

The V1 sample leached in cement-saturated water was covered by a white precipitation product, and this was subjected to a surface analysis. As shown in **Figure 3 (a)**, needle-like crystals were observed by SEM on the surface of the post-leach sample. After washing by ultrasonic vibration, fine granular deposits were observed. (**Figure 3 (b)**) These deposits were analyzed by EDX. The results are summarized in **Table II**. Both the needle-like crystals and the fine granular deposits mainly consisted of calcium, silicon and vanadium. It was concluded that the dissolution rate of vanadate glass in cement-saturated alkaline water was lower than that in deionized water, since the precipitation layer, which consisted of insoluble calcium salts, was formed on the glass surface and it limited the dissolution of the glass.

Table II   The elemental composition of deposits on the post-leach V1 sample
(after 56 days' immersion in cement-saturated water)

object	composition/at%						
	V	P	Sb	Pb	Ca	Si	others
bare glass	39.2	34.4	9.7	12.8	-	-	3.9
needle-like crystal	12.3	9.6	-	6.7	46.0	21.8	3.5
granular deposit	8.9	9.6	-	6.8	45.6	35.1	4.0

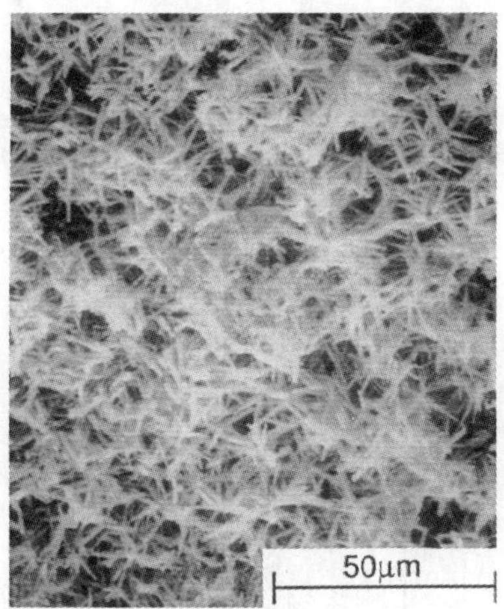

(a) Before ultrasonic cleaning   (b) After ultrasonic cleaning

Fig. 3 SEM photographs of the specimen after leaching in cement-saturated water.   (V1 glass, after 56 days' immersion)

<u>Improved resistance to salt solution</u>

A deterioration mechanism was proposed for the vanadate glass in the cement-saturated alkaline solution with NaCl. It is well known that the vanadium oxide has a negative surface charge in aqueous solution as a result of the following hydration and acid-dissociation reaction [13]:

$$V_2O_5 + 2H_2O \rightarrow H_4V_2O_7 \rightarrow H^+ + H_3V_2O_7^- . \quad (3)$$

Therefore sodium ions tend to permeate into the glass matrix and are adsorbed by negative sites as shown in **Figure 4**. Because the electrostatic force among vanadium oxide layers is neutralized by the sodium ions, the water resistance of vanadate glass can be lowered in the salt solution. This mechanism suggests that the previous placement of divalent cations in the negative sites can not only prevent the permeation of sodium ions, but also reinforce the electrostatic force between vanadium oxide layers.

Enhancement of water resistance to salt solution was examined using additions of alkaline earth metal oxides to the vanadate glass. Four types of vanadate glasses, containing 20mol% of MgO, CaO, SrO and BaO, respectively, were prepared for the leach tests. **Figure 5** shows the normalized mass losses after 3 days' immersion in the cement-saturated water with 3% NaCl. It was found that the additions of an alkaline earth metal to the vanadate glass reduced the corrosion rate in the salt solution, and BaO was most effective for preventing salt corrosion of vanadate glass.

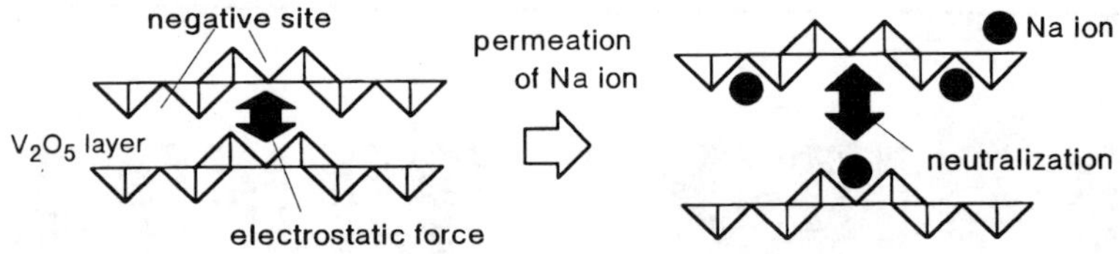

Fig. 4 The proposed detrioration mechanism of vanadate glass in the salt solution.

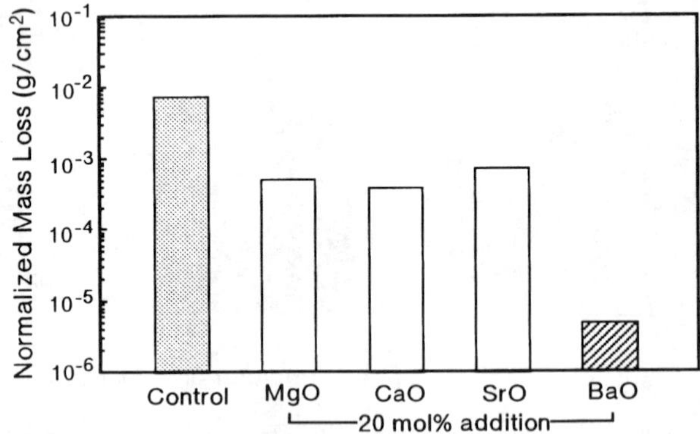

Fig. 5   The effect of alkaline earth metal addition on the solubility of vanadate
glass in the cement-saturated water with 3% NaCl. (after 3 days' immersion)

In order to evaluate the leachability of vanadate glass, non-radioactive tracers, Cs_2O and SrO, were vitrified by the vanadate glass with the optimized composition shown in **Table I** (V1).

Figure 6 shows the changes in X-ray diffraction spectra of the vitrified glass samples with the amount of Cs_2O or SrO. In the case of Cs_2O vitrification, crystallization of vitrified sample was observed above a 10mol% addition. On the other hand, in the SrO addition, homogeneous glass formation was confirmed up to a 20mol% addition. It was found that the vanadate glass was more capable of vitrification of an alkaline earth metal than an alkali metal.

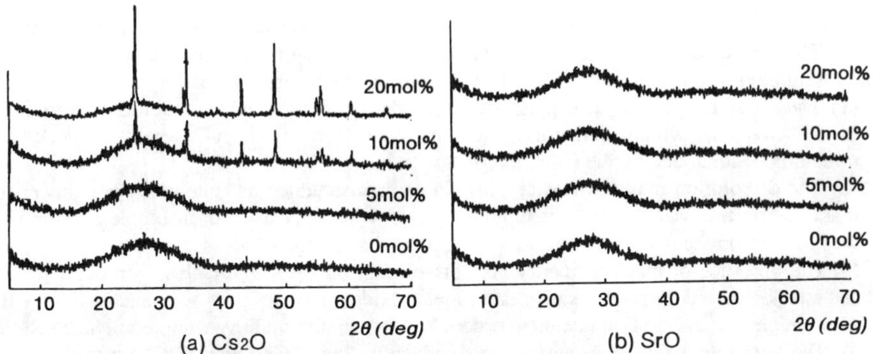

Fig. 6 Changes in X-ray diffraction spectra of vitrified glass with additions of
Cs2O and SrO

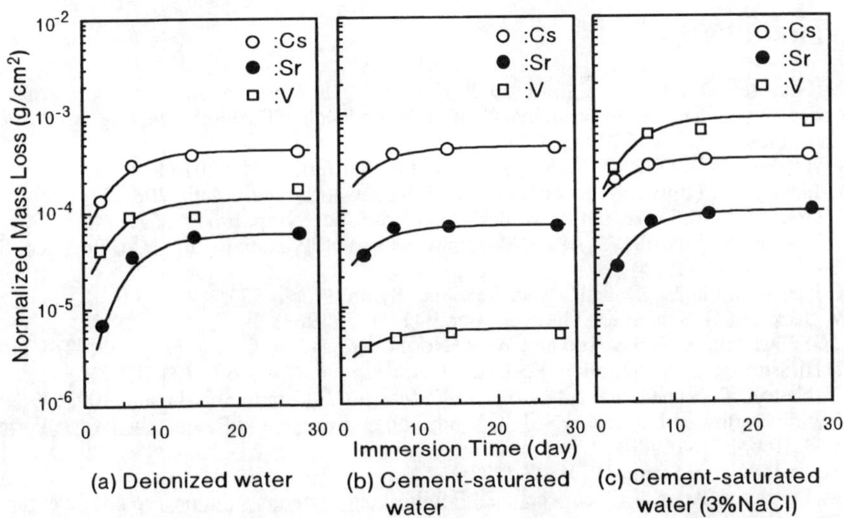

Fig. 7 Changes in the nomalized mass loss of glass sample with immersion time.
(MCC-1 static leach test at 70°C, vitrified 5mol% each of Cs2O and SrO)

The leachability of Cs and Sr from the vitrified vanadate glass containing 5mol% each of Cs_2O and SrO was tested in the three types of leachants. **Figure 7** shows the changes in the normalized mass loss of glass sample with the immersion time. The normalized mass loss obtained from the amount of leached vanadium clearly varied with the type of leachant, but losses obtained from the amount of leached cesium and strontium did not change much with the type. This suggested that the leachabilites of Cs and Sr from the vitrified vanadate glass depended on the solubility of the vanadate unit ($M_x V_2O_5$), which does not change with the type of leachant. The property that vanadate glass can fix the cation in the glass structure was judged suitable for waste vitrification.

CONCLUSIONS

The dissolution behaviour of the V_2O_5-P_2O_5 glass system (vanadate glass) was evaluated by the MCC-1 static leach test at $70°C$ using deionized water, cement-saturated water and cement-saturated water with 3% NaCl. The following results were obtained.

(1) The solubility of vanadate glass was remarkably reduced by the addition of antimony oxide (Sb_2O_3), which converted the glass structure from a layered network to a three-dimensional one by the reduction of V^{5+} to V^{4+}.

(2) The dissolution rate of vanadate glass in cement-saturated alkaline water was lower than that in deionized water, since a precipitation layer, which consisted of insoluble calcium salts, was formed on the glass surface and it limited the dissolution of glass.

(3) The presence of NaCl accelerated the dissolution of V_2O_5-P_2O_5 glass. In order to resist salt attack, the addition of alkaline earth metal oxides into the glass was examined. A BaO addition gave a dissolution rate three orders lower than non-additive sample in NaCl solution.

(4) The vanadate glass could vitrify the simulated wastes, Cs_2O and SrO, up to 5mol% for Cs and 20mol% for Sr, at $700-800°C$. The leachabilities of Cs and Sr from the vitrified vanadate glass did not change much with the type of leachant.

REFERENCES

1. W. Fournier, D. Alexandre, P. Leroy, P. Hugony, C. Sombret, A. Jouan and C. Bernard, Proceedings of The 3rd International Conference on Nuclear Fuel Reprocessing and Waste Management RECOD'91 vol.I, 278 (1991)
2. S. Weisenburger, G. Roth, A. K. Rastogi and H. Seiffert, *ibid.*, 301 (1991)
3. Y. Inagaki, H. Furuya, K. Idemitsu and S. Yonezawa, J. Nucl. Mat., 208, 27 (1994)
4. B. Grambow, W. Lutze and R. Muller, Mat. Res. Soc. Symp. Proc., 257, 143 (1992)
5. H. Yamanaka, J. Nishii, T. Akai, M. Yamashita and H. Wakabayashi, Mat. Res. Soc. Symp. Proc., 353, 95 (1994)
6. K. Kawamura and J. Ohuchi, Mat. Res. Soc. Symp. Proc., 353, 87 (1994)
7. W. Sack and H. Scheidler, Glastech. Ber., 41, 138 (1968)
8. S. S. Flaschen, A. D. Pearson and W. R. Northover, J. Am. Ceram. Soc., 43, 274 (1960)
9. H. Hirashima, K. Nishi and T. Yoshida, J. Am. Ceram. Soc., 66, 704 (1983)
10. T. Naitoh, T. Namekawa, A. Katoh and K. Maeda, J. Ceram. Soc. Japan, 100, 679 (1992)
11. Pacific Northwest Laboratory, MCC Workshop on Leaching of Radioactive Waste Form, PNL Rep.-3318 (1980)
12. Japan Industrial Standard, JIS G-1221
13. M. Pourbaix, Atlas of Electrochemical Equilibria in Aqueous Solutions, p.238, Pergamon Press (1966)

URANIUM SPECIATION IN GLASS CORROSION LAYERS: AN XAFS STUDY

BRUCE M. BIWER#, L. SODERHOLM#, R.B. GREEGOR† and F.W. LYTLE§

#Argonne National Laboratory, Argonne IL 60439, †The Boeing Company, Seattle, WA 98124 and §The EXAFS Company, Pioche, NV 89043

ABSTRACT

Uranium L_3 X-ray absorption data were obtained from two borosilicate glasses, which are considered as models for radioactive wasteforms, both before and after leaching. Surface sensitivity to uranium speciation was attained by a novel application of simultaneous fluorescence and electron-yield detection. Changes in speciation are clearly discernible, from U(VI) in the bulk to $(UO_2)^{2+}$-uranyl in the corrosion layer. The uranium concentrations within the corrosion layer also show variations with leaching times that can be determined from the data.

INTRODUCTION

There has been relatively little work done to determine the valence and coordination environment of metal ions in the surface layers of glasses after leaching. This speciation is of particular concern when the corrosion layers occur on glasses under consideration for radioactive-waste disposal.   An excellent technique for determining metal-ion speciation in non-crystalline media is XAS (X-ray absorption spectroscopy) [1] although it is normally not a surface-sensitive technique, and therefore has not been used extensively to study the speciation of metal ions in glass alteration layers. Surface sensitivity has been achieved in XAS for some uranium-bearing corroded borosilicate glasses by employing glancing-incident detection [2]. This technique relies on the fact that a monochromatic X-ray beam incident on a <u>flat</u> surface has a well-defined critical angle of incidence below which the X-rays are totally reflected. By measuring an absorption spectrum with incident-beam angles below and above this critical angle, it was possible to discern changes that had occurred within about 30Å of the surface (for the U L_3 edge energy). Although surface modifications were observed, because the total reflection geometry requires very flat, polished surfaces, leaching times were very restricted, i.e. limited to 90 minutes or less. This is insufficient time to allow the chemical changes, such as solid-phase formation, that are expected to occur in the surface layer [3,4].

In this manuscript we present a novel combination of fluorescence and electron-yield X-ray absorption fine-structure (XAFS) detection designed to probe the thick surface layer of corroded wasteforms. For the L edges, electron yield relies on the detection of LMM Auger electrons, emitted by the absorbing ion, which for uranium have a range of only about 3000 Å in borosilicate glass. This means that electron yield detection has a sampling depth (defined as $1/e$, the mean free path) such that it provides only information from the altered surface layer of the samples under

Mat. Res. Soc. Symp. Proc. Vol. 465 © 1997 Materials Research Society

study here [5,6]. The more standard fluorescence detection is expected to have a sampling depth of approximately 1000 times that. Because the altered layer is such a small percentage of the total sample probed (less than 10%), the fluorescence signal is representative of only the bulk sample. Therefore, by simultaneously measuring both the fluorescence signal and the electron-yield, we demonstrate herein that the valence and coordination environments of the U present in our samples can be determined in both the bulk, and in the corroded layer to a depth of approximately 3000 Å.

We examined the uranium L X-ray absorption spectra as a function of leach time of two sodium borosilicate glasses that have been considered as models for waste storage, SRL-131 and ATM-1c. The SRL-131 type glass was selected because previous results from leach tests are available [4,7] and have been used in the development of models to predict glass performance [8,9]. The objective here was to define the oxidation state and environment of uranium in the unreacted glasses, and to look for changes at the surface of the glass as they were leached. Although the compositions of these glasses are not expected to be used in final production, these studies may provide information useful to the understanding of uranium speciation in complex glasses.

EXPERIMENTAL SECTION

Sample Preparation

SRL-131 and ATM-1c are two high Fe complex glasses with respective U percentages (as UO_2) of 1.3 and 4.0 %, that have been considered as prototypes for radioactive wasteforms [4,7]. The uranium-doped SRL-131 glass samples used here were synthesized by mixing base SRL-131 frit, as supplied by Savannah River Laboratory in finely powdered form, with U_3O_8 in the appropriate stoichiometry, heating in air in a Pt crucible at 1150°C for 4 hours which completely melted the sample, followed by quenching into water. The sample was reground and melted at 1150°C for 15 min, and then poured into a Pt loaf mold, where it was annealed for 2 hours at 500°C, and then allowed to cool to room temperature. Chemical analyses were confirmed using a combination of acid dissolution/ICP techniques. The ATM-1C glass was used as supplied by Pacific Northwest Laboratory [10].

The sample discs were prepared as previously described [4,7]. Hydrothermal leaching experiments were conducted at 90°C in Teflon® vessels in static tests using a modified MCC-1 testing procedure [11]. SRL 131 glass was leached at surface area-to-volume (S/V) ratios of 10 m^{-1} and 100 m^{-1}. ATM-1c glass samples were leached at S/V ratios of 4 m^{-1} and 10 m^{-1}. Deionized water was used as the reaction leachant in all experiments. Leaching times varied from 7 to 92 days. After each experiment was terminated, the monoliths were rinsed with distilled water, allowed to air dry, and stored in a dry, dust-free container. Preliminary examinations of the altered glass surfaces were performed by both optical and scanning electron microscopy (SEM). In addition to the glass surface, cross-sections of some samples were prepared for SEM analyses of the layer thickness and structure. The corrosion layers were shown to consist of about 0.1 μm of a surface layer of smectite clay overlaying a thin Fe-rich layer covering a 2-10 μm thick dealkalized gel-like phase containing clay precursors. Details of the analyses for similar SRL-131 glass experiments are published elsewhere [4,7]. The electron microscopy of the sectioned surface was analyzed and from these data we conclude that the sampling depth of our electron channel was always within the surface alteration layer, except for the reference-glass samples before leaching.

XAS data collection and analysis

The U L_3 and L_1 spectra were recorded at the Stanford Synchrotron Radiation Laboratory (SSRL) on beam line 4-3. The double crystal monochromator was equipped with Si(220) crystals and was detuned 20% to reject harmonics. The entrance slit to the monochromator, set at 1 mm, defined an energy band width of ~5 eV, which was considerably less than the 8 or 11 eV uncertainty broadening expected for U L_3 or U L_1, respectively [12].

The beam line was equipped with a linear chain of ion chambers to measure the incident beam, I_o, the transmitted beam, I_t, and a reference UO_2 sample, I_r. In addition a combination x-ray fluorescence ion chamber, I_f, and electron yield detector, I_e, was located between I_o and I_t. Note that the e and f channels were measured simultaneously with a detector comprised of an electron collecting chamber separate from and transparent to the fluorescent x-rays [13]. At least two spectra (averaged) were collected for each sample. The L_1 data are not included here for brevity, but their results are consistent with those described for the L_3 analyses.

The data were reduced to normalized x-ray absorption near edge spectra (XANES) and extended x-ray absorption fine structure (EXAFS) files using standard techniques [1]. The data were fit over K from 2-12 $Å^{-1}$ from a back transform over R from 0.5-2.5 Å. Fitting two separate bond distances required 7 parameters. Inclusion of a 3rd bond distance would be the absolute limit (ten variables) for this range of data. The reference compounds were either separated into two-shell entities by Fourier filtering (UO_2) or by averaging the different coordination distances into two groups.

RESULTS AND DISCUSSION

Glass-EXAFS

The U L_3 EXAFS, obtained from the SRL-131 and ATM-1c bulk glasses by monitoring the fluorescence yield, are compared with a standard UO_3 spectrum in Figure 1a. These three spectra appear indistinguishable, as confirmed by the fitted parameters listed in Table I. The FT of the EXAFS spectra are shown in Figure 1b. The broad, single shell observed for both the glasses under study is compared with known U coordinations in Table I. The coordination environment of approximately 2 oxygen at 1.85 Å and approximately 4 oxygen at a slightly longer distance of 2.28 Å is similar to the U coordination found in other borosilicate glasses [2,14,15]. The glass data are inconsistent with the distinctive 2 short (1.76 Å) and 6 longer (2.46 Å) bonds expected for the

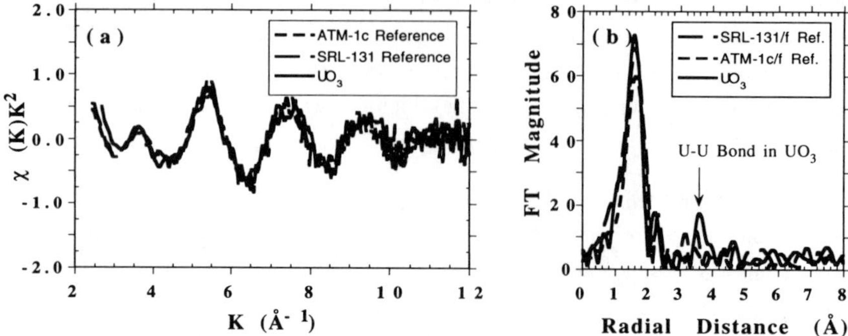

Figure 1. (a) A comparison of the K^3-weighted EXAFS (before phase correction) obtained from glasses with those from UO_3. The Fourier Transform (FT) of these spectra, shown in (b) do not have the two peaks characteristic of uranyl. The peak at ~3.7 Å in UO3 is due to U-U scattering environment for U is also somewhat similar to the metamict betafite [18], although the U-O average bond distance is slightly shorter in our case.

uranyl coordination, as demonstrated for $NaUO_2(CH_3COO)_3$ in Figure 2b. Instead the data are consistent with the pseudo octahedral coordination found for U in UO3 [16,17]. The coordination environment for U is also somewhat similar to the metamict betafite [18], although the U-O average bond distance is slightly shorter in our case.

Not shown is the comparison of the bulk (fluorescence) and surface (e-yield) EXAFS from the samples before leaching, which again are indistinguishable, indicating that the uranium speciation in the bulk is similar to that at the surface for the starting glass. The finding of a similar surface and bulk uranium speciation in the unreacted glass does not contradict the presence of a thin, 50 Å surface hydration layer such as has been previously proposed for samples in contact with air [19]. The EXAFS of the bulk glass remains unchanged for all corroded samples, but the near-surface EXAFS show marked changes with leaching, as demonstrated in Figure 2. Whereas the fit to corrosion-layer spectra looks good, a two-shell model does not adequately represent the data. It appears that approximately 30% of the uranium has converted to a uranyl-type coordination, hence the average environment is very complex, and requires more parameters than are warranted statistically. The finding of some uranyl-type coordination at the surface after leaching is supported by the XANES results.

Glass-XANES

The XANES data, collected in the fluorescence mode, were the same for all the samples measured, as shown in Figure 3a. In addition, the XANES data obtained from the bulk were indistinguishable for the electron-yield data obtained from the near surface for samples before leaching. These spectra show no evidence for the presence of the uranyl moiety, which can be identified by the high energy shoulder on the L3-edge spectra [15,20]. Instead the spectra show a

Table I The parameters obtained by fitting the EXAFS data from the glasses and the standards. N1/R1 are the coordination number and the corresponding distance of the 1st coordination shell, σ^2 are the Debye Waller factors, and the Standard Deviation (S.D.), calculated from the K^3 weighted data, were determined from S.D.=$[S(A_{fit}-A_{data})^2/n]^{1/2}$. The EXAFS data from the corroded SRL-131 glasses are too noisy to fit.

Sample	N1/R1 (Å)	N2/R2 (Å)	$\sigma^2 1/\sigma^2 2$ (Å^2)	S. D. (K^3)
SRL-131 -ref	2.1 O/1.86	4.5 O/2.29	0/.0019	0.3
-corroded	NA	NA	NA	NA
ATM-1c -ref	1.7 O/1.85	3.9 O/2.28	0/.0019	0.36
-corroded	3.2 O/1.84	2.7 O/2.28	0/.0019	0.49
Standards				
-UO$_2$	8 O/2.38	12 U/3.864	.0062/.0052	0.5/0.12
-UO$_3$	2 O/1.865	4 O/2.262	0/.0019	0.78
-NaUO$_2$Ac$_3$	2 O/1.788	6 O/2.450	.0040/.0084	0.28

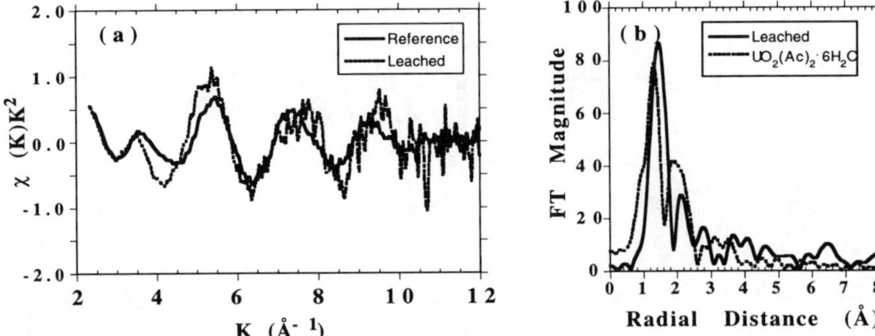

Figure 2. A comparison of the surface (e-yield) data obtained before and after leaching. The FT shows the presence of some uranyl.

edge shift of approximately 2.7 eV to lower energy than expected from the uranyl ion. A shift in the edge to lower energy of this magnitude is often considered evidence for a reduced-metal species, which in this case would be expected to be U(IV). However the presence of a reduced uranium species in our samples is very unlikely based on the synthesis conditions of the glass, the presence of only Fe(III) [21] as evidenced by Mossbauer spectroscopy [22] and the EXAFS data from the same samples (see the comparison of EXAFS functions in Figure 1). A similar edge shift to lower energy has been previously reported in other studies of borosilicate glasses [2,14].

In contrast to the XANES from the bulk, the e-yield data in Figure 3b show marked changes near the surface with leaching times. This result is consistent with previous work, in

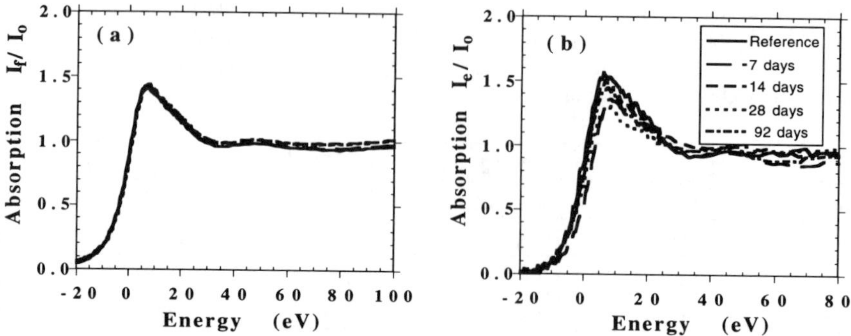

Figure 3. (a) Fluorescence XANES from all the samples appear similar. These data lack the high energy shoulder indicative of the uranyl ion. In addition, the edge energy is lower than expected for uranyl. (b) changes in XANES with leaching times are attributable to increased uranyl ion.

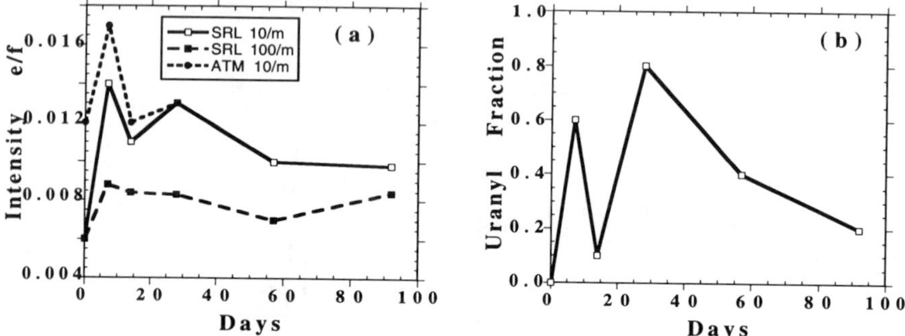

Figure 4. (a) The normalized edge jump intensities, plotted as a function of time, obtained from the electron-yield data for three different samples. These data demonstrate changes are occurring in the corrosion-layer uranium concentration. (b) The fraction of total uranium in the alteration layer that has been converted to a uranyl-like speciation.

which glancing angle X-ray absorption spectroscopy was used to show changes in U speciation at the surface of a glass after as little as 15 minutes exposure to leaching conditions [2]. The evolution of the spectra with leaching times involve changes in both uranium speciation and concentration. The changes in speciation can be modeled with an increase in the percentage of the total uranium present at the surface with uranyl-type coordination. The percentage of uranyl ion can be estimated by taking the initial unreacted XANES pattern, and adding to it a spectrum from the uranyl ion. This method models all of the XANES data well. In the SRL-131 surface layer probed by e-yield, up to 80% of the uranium is in the form of uranyl after 92 days of leaching, whereas approximately 30% of the U in the alteration layer of the ATM-1c glass is present as the uranyl ion after 28 days of leaching.

In addition to the changes in the uranium coordination environment, the XANES data show evidence for changing total uranium concentration in the alteration layer as a function of leach time. The amplitude of the edge jump is proportional to the amount of the absorbing element contributing to the detector signal, therefore it is possible to evaluate changes in concentration within the experimental sampling depth. This is particularly interesting for the corrosion layer, for which changes in metal ion concentrations have been previously reported [2,23]. Using the fluorescence signal from each of the samples as a normalizing factor, the edge jump of the electron-yield signal provides an estimate of the uranium concentration within the corrosion layer. The magnitude of the e/f channel, plotted vs. leach time for all of the leaching sequences, is shown in Figure 4a. For both glasses there was an initial sharp rise for the 7 day samples followed by a gradual decrease for the longer times; however, in all corroded glasses there was significant U enrichment in the near surface. The degree of surface enrichment depended on the S/V ratio (the largest enrichment occurred for the largest volume of leachant) and the duration of the leach test. The initial U surface enrichment in the 4% UO_2 ATM-1c glass was 40%. The initial U surface enrichment in the 1.3% UO_2 SRL-131 glass series was 130% for the 10 m^{-1} and 50% in the 100 m^{-1} series. After 92 days the surface U enrichments of in the now thicker alteration-layers for the same samples were 66% and 30%, respectively.

CONCLUSIONS

Our use of a novel combination of fluorescent x-ray and electron-yield technique has provided unique insight into the speciation of uranium in glass alteration layers prior to dissolution. The x-ray channel probed the bulk and the electron channel probed the surface to a 1/e depth of 3000 Å. This effort extends metal-ion speciation to glasses that have undergone significant surface modification, conditions for which glancing-angle detection is not possible. We found that the speciation of U in two high Fe sodium-borosilicate glasses undergo a significant modification during leaching in water at 90° C, with leach periods from 7-92 days. We demonstrate that the U(VI) converts to a uranyl-like coordination at the surface. Within the hydration layer, the U(VI) was first concentrated and then converted to a uranyl moiety that was subsequently leached from the glass over a period of days. We argue that this conversion is necessary before the uranium dissolves into water as the uranyl, UO_2^{2+} ion. The fraction of uranyl at the surface did not increase in a totally uniform manner with time, as shown in Figure 4b. Instead, there was appreciable scatter in the uranyl fraction. Some of the longer leach period surfaces had a greater fraction of uranyl than others as if the dissolution of uranyl was more rapid in some of the samples than others. This result is consistent with models that suggest that some of the alteration layer spalls from the glass as colloids, although there is no direct evidence from the EXAFS of either U clustering or crystalline phase [24]. These results support the interesting possibility that U dissolution could be retarded if it could be kept from converting to a uranyl-type coordination.

ACKNOWLEDGMENTS

We thank Dr. Mark Antonio for the Mossbauer data, and for helpful discussions. This work was supported by the US-DOE, BES-Chemical Sciences, under contract W-31-109-ENG-38. The efforts of RBG were supported by the U.S. DOE, BES, under Grant No DE-FG06-84ER45121. For user support and synchrotron beamtime, we thank SSRL, which is funded by the Department of Energy.

REFERENCES

1. F.W. Lytle , in *Applications of Synchrotron Radiation,* H. Winick, *et al.,* Editors. 1989, Gordon and Breach, p. 135.
2. G.N. Greaves, N.T. Barrett, G.M. Antonini, F.R. Thornley, B.T.M. Willis, and A. Steel, J. Am. Chem. Soc., **111**, 4313 (1989).
3. T.A. Abrajano, J.K. Bates, A.B. Woodland, J.P. Bradley, and W.L. Bourcier, Clays & Clay Minerals, **38**, 537 (1990).
4. J.J. Mazer, J.K. Bates, B.M. Biwer, and C.R. Bradley, Mat. Res. Soc. Symp. Proc., **257**, 73 (1992).
5. W.T. Elam, J.P. Kirkland, R.A. Neiser, and P.D. Wolf, Phys. Rev. B, **38**, 26 (1988).
6. T. Girardeau, M. Mimault, M. Jaouen, P. Chartier, and G. Tourillon, Phys. Rev. B, **46**, 7144 (1992).
7. B.M. Biwer, J.K. Bates, T.A. Abrajano, and J.P. Bradley, Mat. Res. Soc. Symp. Proc., **176**, 255 (1990).
8. W.L. Bourcier, Mat. Res. Soc. Symp. Proc., **112**, 3 (1991).
9. B. Grambow and D.M. Strachan, Mat. Res. Soc. Symp. Proc., **112**, 713 (1991).
10. J.W. Wald, 1985, Pacific Northwest Laboratory.
11. *MCC-1P Static Leach Test Method, Nuclear Waste Materials Handbook* 1981, Materials Characterization Center.
12. O. Keski-Rahkonen and M.O. Krause, At. Data Nucl. Data Tables, **14**, 139 (1974).
13. The EXAFS Co. Pioche NV 89043.
14. P.G. Eller, G.D. Jarvinen, J.D. Purson, R.A. Penneman, R.R. Ryan, F.W. Lytle, and R.B. Greegor, Radiochim. Acta, **39**, 17 (1985).
15. J. Petiau, G. Calas, D. Petitmaire, A. Bianconi, M. Benfatto, and A. Marcelli, Phys. Rev. B, **34**, 7350 (1986).
16. W.H. Zachariasen, Acta Crystallogr., **1**, 265 (1948).
17. S. Siegel, H.R. Hoekstra, and E. Sherry, Acta Crystallogr., **20**, 292 (1966).
18. R.B. Greegor, F.W. Lytle, B.C. Chakoumakos, G.R. Lumpkin, and R.C. Ewing, Mat. Res. Soc. Symp. Proc., **50**, 387 (1985).
19. L. Hench, D. E.Clark, and E.L. Yen-Bower, Nucl. Chem. Waste Manag., **1**, 59 (1980).
20. E.A. Hudson, J.J. Rehr, and J.J. Bucher, Phys. Rev. B, **22**, 13815 (1995-I).
21. H.D. Schreiber, L.M. Minnix, G.B. Balazs, and B.E. Carpenter, Phys. Chem. Glasses, **25**, 1 (1984).
22. M.R. Antonio and L. Soderholm (unpublished).
23. H.D. Schreiber, G.B. Balazs, and T.N. Solberg, Phys. Chem. Glasses, **26**, 35 (1985).
24. X. Feng, E.C. Buck, C. Mertz, J.K. Bates, J.C. Cunnane, and D.J. Chaiko, Radiochim. Acta, **66/67**, 197 (1994).

COMPARISON OF THE CORROSION BEHAVIOR OF TANK 51 SLUDGE-BASED GLASS AND A NONRADIOACTIVE HOMOLOGUE GLASS

L. Nuñez, W. L. Ebert, S. F. Wolf, and J. K. Bates
Argonne National Laboratory, 9700 South Cass Avenue, Argonne, Illinois 60439

ABSTRACT

We are characterizing the corrosion behavior of the radioactive glass that was made with sludge from Tank 51 at the Defense Waste Processing Facility (DWPF) and a nonradioactive glass having the same composition, except for the absence of radionuclides. Static dissolution tests are being conducted in a tuff groundwater solution at glass surface area/solution volume ratios (S/V) of 2000 and 20,000 m^{-1}. These tests are being conducted to assess the relationship between the behavior of this glass in a 7-day Product Consistency Test and in long-term tests, to assess the effects of radionuclides on the glass corrosion behavior, and to measure the disposition of radionuclides that are released as the radioactive glass corrodes. The radioactive glass reacts slower than the nonradioactive glass through the longest test durations completed to date, which are 140 days for tests at 2000 m^{-1} and about 400 days for tests at 20,000 m^{-1}. This is probably because radiolysis results in lower solution pH values being maintained in tests with the radioactive glass. Rate-affecting alteration phases that had formed within one year in tests with other glasses having compositions similar to the Tank 51 glass have not yet formed in tests with either glass.

INTRODUCTION

We are conducting static dissolution tests with a glass made at the Savannah River Technology Center (SRTC) during a demonstration of the Defense Waste Processing Facility (DWPF) process control for remote vitrification [1]. The glass was made with sludge from Tank 51, SRL 202 frit, and added soda. This glass is similar to waste glasses being made in the current DWPF campaign. Parallel tests are being conducted with a nonradioactive glass made at ANL having the same composition as the radioactive glass, except without the radionuclides. The radioactive and nonradioactive glasses are referred to as 51R and 51S, respectively. The results of these tests provide information pertinent to assessing the long-term corrosion behavior of DWPF glasses, comparing the corrosion behaviors of radioactive and nonradioactive glasses, and characterizing the disposition of radionuclides as the glass corrodes.

An important effect in static tests with radioactive glasses is acidification of the leachate solution through the generation of nitric acid by the radiolysis of moist air [2]. Short-term tests with radioactive glasses and their nonradioactive homologues have generally shown the effects of radiolysis to be minor and the corrosion behaviors of radioactive and nonradioactive homologue glasses to be similar [3]. This is because the pH rise that accompanies the initial dissolution of the glass neutralizes most of the radiolytically produced acid. The slight differences in the reactivities that were observed in those tests can be attributed to minor differences in the compositions of the radioactive and nonradioactive glasses. However, the dissolution rate of a nonradioactive glass increased significantly after certain alteration phases formed in long-term static dissolution tests, while similar phases did not form in tests with a radioactive glass having a similar composition [4]. While similar alteration phases may not have formed in tests with the radioactive glass because of small differences in the glass compositions, the formation of those phases may also have been delayed or prevented by radiolytic effects. The compositions of the 51R and 51S glasses discussed in this paper are much more similar than the sets of radioactive and nonradioactive glasses used in previous tests, and the effects of glass composition and radiolysis on the corrosion behavior can be better distinguished.

Insight into the expected corrosion behavior of the 51R and 51S glasses was gained from tests that were conducted previously with actinide-doped SRL 202 frit-based glasses having similar compositions [5]. Radiolysis did not prevent the formation of rate-affecting alteration phases in

Mat. Res. Soc. Symp. Proc. Vol. 465 © 1997 Materials Research Society

those tests. It was found that the increase in the glass dissolution rate resulting from phase formation did not lead to an increase in the amounts of actinides in the aqueous phase, although the amounts of Pu and Am fixed to the steel reaction vessel increased [6]. The actinide concentrations in the solutions were probably controlled by phases that were sparingly soluble, although the actinide-bearing solid phases were not identified. The effects of radiolysis are expected to be more pronounced in the 51R glass than in that actinide-doped glass.

The solution chemistries of the various radionuclides will affect their distribution between dissolved, colloidal, precipitated, and adsorbed fractions during glass corrosion. The disposition of radionuclides as the glass corrodes in a disposal site is important because it will impact their dispersal. For example, radionuclides that are dissolved or associated with colloidal material may be transported by groundwater flow, while radionuclides that form precipitates or become fixed to other materials, such as steel containers, may become immobilized, depending on the stability of those materials. Several studies have shown the general order of actinide release from high-level waste glasses into solution in laboratory tests to be U, Np > Pu, Am, Cm [7]. The solubilities of these elements will be affected by the pH and Eh of the solution and by the presence of ligands. Therefore, it is important that laboratory tests be conducted with solutions that are relevant to the groundwater chemistry anticipated at the disposal site. The tuff groundwater solution used in the tests discussed in this paper was prepared from the reference groundwater for the Yucca Mountain Site. It contains about 120 mg/L HCO_3^- and has a pH of about 8, and so both the formation of carbonates and hydrolysis will affect the solution chemistries of some actinides. Other species in the groundwater and anions released from the glass, such as F^-, HPO_4^{2-}, and SO_4^{2-}, may also affect the speciation of some actinides.

EXPERIMENTAL

A sample of the 51R glass was obtained from SRTC and the 51S glass was produced at ANL from reagent-grade chemicals. The compositions of the two glasses are given in Table I. The difference in the concentration of each element in the glasses is within the combined analytical uncertainties for analyses of 51R and 51S glasses, which is less than 15% for most components. The biggest difference is in the U contents of the two glasses, which differ by about 11%. Static dissolution tests were conducted similar to the Product Consistency Test (PCT) [8]. Samples of each glass were prepared for testing by crushing and sieving to isolate the -100 +200 mesh fraction (74 -149 μm). The crushed glass was washed with deionized water to remove fines then dried in a 90°C oven. Tests were conducted at glass-to-water mass ratios of 1:10 and 1:1 to achieve glass surface area/solution volume ratios (S/V) of about 2000 and 20,000 m^{-1}, respectively. Most tests were conducted with tuff groundwater from well J-13 at the Nevada Test Site that had been pre-reacted with pulverized tuff at 90°C. The groundwater solution is referred to

Table I. Compositions of the 51S and 51R Glasses, in wt %

Element	51S	51R[a]	Element	51S	51R[a]
Al	2.55	2.56	Ni	0.23	0.23
B	1.99	2.00	P	0.26	0.26
Ca	0.99	0.99	Si	24.4	24.5
Cr	0.30	0.30	Th	0.02	0.02
Fe	9.17	9.21	U	0.86	0.97
K	1.30	1.31	Zn	0.11	0.11
La	0.47	0.47	^{241}Am	—[b]	7.4×10^{-5}
Li	2.09	2.10	^{244}Cm	—	1.1×10^{-5}
Mg	1.24	1.24	^{237}Np	—	4.2×10^{-4}
Mn	0.97	0.98	^{238}Pu	—	1.1×10^{-4}
Na	6.93	6.96	^{239}Pu	—	2.2×10^{-3}

[a]From Ref. [1].
[b]Not present in glass.

as EJ-13. It has a pH of about 8 and its composition, in mg/L, is Al : 1.4, B: 0.23, Ca: 3.2, Fe:0.13, K: 5.9, Li: 0.057, Mg: 0.035, Mn: 0.10, Na: 48, Si: 41.1, F$^-$: 1.8, Cl$^-$: 8.0, NO$_3^-$: 9.9, HPO$_4^{2-}$: <2.0, SO$_4^{2-}$: 20.1, and HCO$_3^-$: 120. Tests with EJ-13 water are being conducted for reaction times from 7 days to longer than 560 days. Tests were also conducted for 7 days at 2000 m^{-1} in deionized water for comparison to similar tests conducted with 51R glass at SRTC. All tests were conducted in Type 304L stainless steel vessels.

Both the leachate solutions and reacted solids were analyzed to characterize the corrosion behavior of each glass. The solutions were analyzed for pH with a combination electrode, anions with ion chromatography, and cations with inductively coupled plasma mass spectrometry (ICP-MS). The radionuclide concentrations in the solutions were measured with ICP-MS and also with alpha and gamma spectroscopy. Aliquots of unfiltered leachate and solution that passed through 1000, 450, 100, 10, and 6 nm filters were analyzed with alpha spectroscopy. An acid-strip solution of the vessel was also analyzed with alpha and gamma spectroscopy.

Some of the reacted glass samples were analyzed with scanning electron microscopy (SEM) and transmission electron microscopy (TEM). Both the SEM and TEM were equipped with energy dispersive X-ray spectrometers (EDS) for compositional analysis. Selected area electron diffraction (SAED) was also performed on some TEM samples.

RESULTS AND DISCUSSION

Tests in Deionized Water —The solution concentrations C(i) and normalized elemental mass losses NL(i) of B, Na, and Si in 7-day PCTs conducted in deionized water at 2000 m^{-1} at ANL and at SRTC are summarized in Table II. The normalized mass loss is calculated by dividing the difference of the measured solution concentration and the background concentration of component i by the S/V of the test and the mass fraction of i in the glass. The results of tests conducted at SRTC were only reported as concentrations, and we have calculated NL(i) values by assuming an S/V of 2000 m^{-1} for those tests.

We first compare the results of tests with 51S and 51R glass conducted at ANL. The measured pH values of tests with 51S glass are consistently higher than those of tests with the 51R glass. Likewise, the values of NL(B) and NL(Na) are higher in tests with 51S glass than in tests with 51R glass. In contrast, the values of NL(Si) are lower in tests with 51S glass than in tests with 51R glass. Dissolution of the 51R glass results in nearly stoichiometric releases of B, Na, and Si, while the 51S glass shows B and Na to be released preferentially to Si.

Next, we compare the results of tests with 51R glass conducted at ANL and SRTC. More 51R glass is measured to have dissolved in tests conducted at SRTC than in tests conducted at ANL based on pH, NL(B), and NL(Na). The values of NL(Si) agree within the overall experimental uncertainty, which is assumed to be about 15%. The differences in the results of tests conducted at SRTC and at ANL may be due to small differences in the compositions of the samples of 51R glass that were tested, differences in the amounts of remaining fines of the crushed glass tested, small differences in the oven temperatures (the PCT procedure specifies a temperature of 90±2°C), etc. For example, slightly different results were obtained from PCTs conducted at SRTC with glass taken from the top and bottom of the pour can [9]. However, the differences in the results are similar to the interlaboratory differences seen previously in tests with radioactive glass [10].

The results of 7-day tests conducted with EA and SRL 202 glasses are included in Table II for comparison. The PCT behavior of the EA glass is the benchmark for high-level waste glasses, and the SRL 202 glass is a reference "blend" glass for the DWPF having a composition similar to that of 51S and 51R that has been tested extensively at ANL [5]. The values for SRL 202 in Table II are from tests in Teflon vessels. The pH values of these tests may have been lowered slightly due to the influx of CO_2, although the tests were conducted in a water bath to minimize that effect. Similar values of NL(B), NL(Na), and NL(Si) are measured in tests with 51R, 51S, and SRL 202 glasses; all are significantly lower than the values measured in tests with the EA glass.

Table II. Results of PCT-A with 51S, 51R, EA, and SRL 202 Glasses

Glass	pH	NC(i), mg/L			NL(i), g/m^2		
		NC(B)	NC(Na)	NC(Si)	NL(B)	NL(Na)	NL(Si)
51S	10.4	14.9	47.3	86.2	0.3	0.3	0.2
51S	10.4	14.6	40.8	87.2	0.3	0.3	0.2
51S	10.3	12.9	40.2	85.7	0.3	0.3	0.2
51R	9.8	9.10	31.0	95.0	0.2	0.2	0.2
51R	9.9	8.80	28.0	94.0	0.2	0.2	0.2
51R	9.9	9.00	27.0	96.0	0.2	0.2	0.2
SRTC 51R[a]	10.3	14.0	44.0	102	0.4	0.3	0.2
SRL 202[b]	1.0	14.0	45.0	94.0	0.3	0.3	0.2
EA[a]	11.9	587	1662	893	8.5	6.7	2.0
EA[c]	11.7	629	1798	999	8.7	7.3	2.2

[a]Average of replicate tests conducted at SRTC [9].
[b]Average of duplicate tests [5].
[c]Average of replicate tests conducted at ANL.

<u>Tests in EJ-13 Water</u> — We now compare the results of tests conducted with EJ-13 water. Tests with 51R and 51S glasses have been completed through 280 and 393 days, respectively; the tests with 51S have progressed to longer times because that glass was available for testing before the 51R glass. The average of the pH values attained in duplicate tests with the 51R and 51S glasses are plotted in Fig. 1. The pH values in tests at 20,000 m^{-1} are consistently higher than those in tests at 2000 m^{-1}. The pH values appear to reach nearly constant values after about 100 days of reaction for tests with both glasses at both S/V.

At a given S/V, the measured pH values for tests with 51R glass are consistently lower than those for corresponding tests with 51S glass. This reflects differences in the dissolution behaviors of the two glasses, the effects of radiolysis, and small compositional differences, including the effects of anions that may be present in the 51R glass but not in the 51S glass. While most waste glasses and sludges are not analyzed for the presence of anions, significant amounts of F$^-$, Cl$^-$, SO$_4^{2-}$, and HPO$_4^{2-}$ have been observed to be released as other sludge-based glass dissolved [4]. Since anion results for the present tests are not yet available, we estimated the effects radiolysis and the presence of anions on the pH. Geochemist's Workbench^{TM} software (GWB) was used to simulate the dissolution of glasses in which the anion concentrations were varied from 0.2 wt% to 2 wt%. These simulation indicated that the presence of these anions had little effect on the pH. Likewise, the small differences in the cation concentrations of the 51S and 51R glasses had a negligible effect. The effects of radiolysis were estimated with the Burns equation [11] using G values from the literature [12]. The results of those calculations indicate that radiolysis accounts for only a small percentage of the measured difference in the final pH. We note that these calculations show that radiolysis causes a negligible change in the amount of N$_2$ in the vessel, even after several years. While the concentrations of anions simulated to be in the leachates must be confirmed with their measured concentrations, it appears that radiolysis and the release of anions do not have a significant direct effect on the pH. Instead, it is likely that the small pH difference due to radiolysis affects the extents of ion exchange and hydrolysis that occurs. This is discussed below.

It is likely that differences in the reactivities of radioactive and simulated glasses that have been attributed to radiolysis or compositional effects in the past may have also been affected by differences in glass melts and sample preparation. This is especially important because of the extent to which nonradioactive homologue glasses have been used as surrogates for radioactive glasses. While further work is needed to understand these differences, we note that the reactivity of the radioactive 51R glass is less than that of the nonradioactive 51S glass, so that, at least for these glasses, the nonradioactive glass is a conservative surrogate.

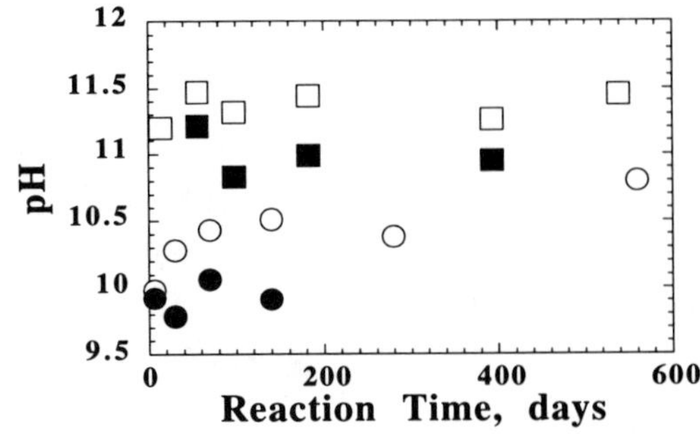

Figure 1. Measured pH vs. Reaction Time for Tests with 51R at 2000 m^{-1} (●) and 20,000 m^{-1} (■), and with 51S at 2000 m^{-1} (○) and 20,000 m^{-1} (□).

The values of NL(B), NL(Na), and NL(Si) for tests conducted with 51S and 51R glasses at 2000 m^{-1} are plotted in Fig. 2a. Uncertainty bars are only shown for the Na results for clarity. The uncertainties in NL(B), NL(Na), and NL(Si) (and other components) are all dominated by the assumed ±10% analytical uncertainty. The values of NL(Na) are significantly higher in tests with 51S glass than in tests with 51R glass, while the values of NL(B) are the same in tests with the two glasses, within experimental uncertainty, as are the values of NL(Si). The lower values of NL(Na) in the tests with the 51R glass are consistent with the lower pH values in those tests. Thus, it appears that less ion exchange to release Na (and also Li and Cs) occurs in tests with 51R glass than in tests with 51S, but the extent of hydrolysis to release B and Si are about the same for both glasses. The lower extent of ion exchange in tests with 51R glass is consistent with the lower final pH in these tests, compared to tests with 51S glass.

The results of tests conducted at 20,000 m^{-1} are shown in Fig. 2b. The average of duplicate test results are shown and the difference between the duplicates is usually within the ±10% analytical uncertainty. As in the tests conducted at 2000 m^{-1}, the values of NL(Na) are higher in tests with 51S glass than in tests with 51R glass over the first 100 days, although the difference is smaller at longer test durations. The values of NL(B) increase with time in tests with 51S glass at a rate of about 0.002 g/(m^2•d), while the values of NL(Na) remain nearly constant.

The values of NL(i) are consistently higher in tests conducted at 2000 m^{-1} than in tests conducted at 20,000 m^{-1} for both glasses. This is because the Si concentrations are higher in tests at 20,000 m^{-1} than in tests at 2000 m^{-1}, and the dissolution rates decrease with increasing concentrations of orthosilicic acid. For example, the Si concentrations measured in tests with 51S glass at 2000 m^{-1} increased from 75 mg/L after 7 days to about 150 mg/L after 282 days, and from about 250 mg/L after 14 days to about 470 mg/L after 393 days in tests with 51S glass at 20,000 m^{-1}. Although a smaller proportion of the Si concentration is orthosilicic acid at the higher pH values attained in tests at 20,000 m^{-1}, speciation calculations show that more orthosilicic acid is present in the 20,000 m^{-1} tests.

The speciation calculation raises another interesting point. Notice in Fig. 2b that the values of NL(Si) for the tests with 51R and 51S glasses are about the same, while the pH values for the tests with 51R glass are lower than for tests with 51S glass. This means that less Si is present in solution as H$_4$SiO$_4$ in tests with 51S glass than with 51R glass. The chemical affinity for corrosion of the 51S glass is, therefore, higher than that for corrosion of the 51R glass, and the dissolution rate should be higher. However, the values of NL(B) become higher in tests with 51R glass than in tests with 51S glass after about 182 days, contrary to what is expected.

Clay Layer— Examination of reacted glass samples with TEM shows the formation of a thin clay layer at the glass surface. Images of 51S and 51R samples reacted at 20,000 m^{-1} for 98 and 182 days, respectively, are shown in Fig. 3. The layers formed on samples of each glass are

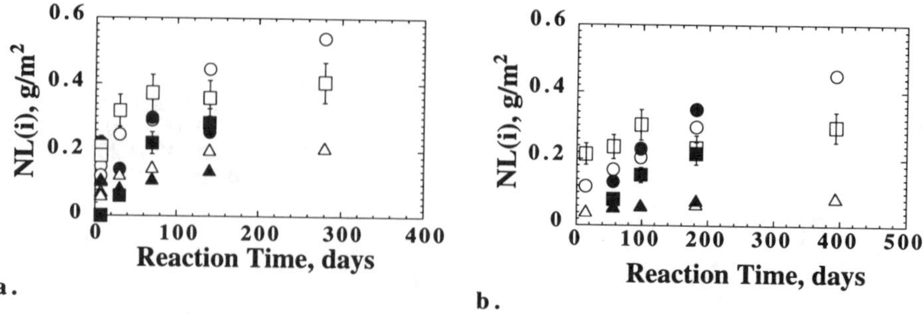

Fig. 2. Normalized Mass Loss vs. Reaction Time for PCTs in EJ-13 at (a) 2000 m⁻¹ and (b) 20,000 m⁻¹: NL(B) (● and ○), NL(Na) (■ and □), and NL(Si) (▲ and △); filled symbols for tests with 51R glass; open symbols for tests with 51S glass.

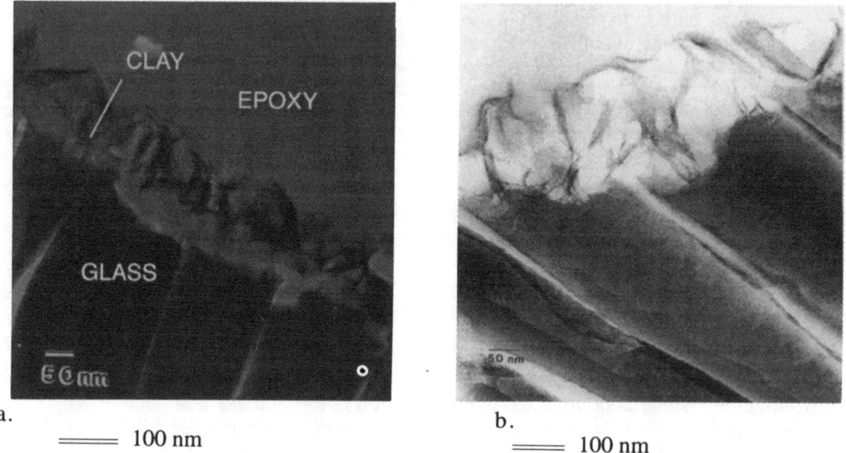

a.
═══ 100 nm

b.
═══ 100 nm

Fig. 3. TEM Photomicrographs of (a) 51S Glass Reacted at 20,000 m⁻¹ for 98 Days, and (b) 51R Glass Reacted at 20,000 m⁻¹ for 182 Days

composed of wisps of clay crystallites, and are less than about 200 nm thick. Although the layers formed in tests completed to date are too thin to be characterized with EDS or electron diffraction, they appear to be identical to the layers of smectite clay that formed during the corrosion of other DWPF glasses [4, 12]. Thicker layers are expected to form in tests conducted for longer durations; these layers will be examined to determine if radionuclides accumulate in the layer.

<u>Radionuclide Disposition</u> —We have measured the amounts of several actinide elements that were dissolved, suspended in the solution, or fixed to the steel reaction vessel. We anticipate measuring the amounts of radionuclides in the clay layer and in alteration phases when samples reacted for longer durations that have thick alteration layers become available. Aliquots of some leachates were sequentially filtered through 1000, 100, 10, and 6 nm filters to measure the size distribution of suspended material. The remaining leachate was filtered with 450 nm filters; that

solution was also analyzed for pH, anions, and other cations. The amounts of Np, Pu, and Cm measured to have passed through the different filter sizes, and the amounts measured in the solution from an acid strip of the vessel are summarized in Table III. Although there is significant scatter in the data, the amounts of all nuclides in all filtrates generally increase with time in tests conducted at 2000 m^{-1}. Most of the Pu is removed from the solution by a 10 nm pore size filter. Quantification of Np with α-spectroscopy is complicated by severe overlap of the Np peak with the tail of the Pu peak. We estimate the uncertainties of the values presented for Np to be on the order of 50%, and the uncertainties in the values of Pu or Cm to be on the order of 25%. The higher uncertainty of the Np results is evidenced by the conflicting results obtained after filtering with 10 and 6 nm filters. Since the measured Np concentrations after filtration with a 10 nm filter are significantly higher than the concentrations measured after the solution was filtered with either 450 or 100 nm filters, these are less reliable than the results of the 6 nm filtration step.

      Most of the Np and Pu that is present in the leachate exists in a form that can be removed by filtering with a 10 nm filter; only a small amount of Np and Pu is dissolved. Tests with the SRL 202 glass showed that colloidal material was present in the leachates of tests run for short times, but that the colloids flocculated and fell out of solution after longer test durations [5]. The decrease in the amounts of Np and Pu in the 1000 and 450 nm filtrates between 70 and 140 days may indicate that flocculation is occurring in these tests, as well. Longer test durations should reveal whether or not the suspended material settles out of solution. The amounts of each nuclide measured to be present in the solution (i.e., either suspended or dissolved) and fixed to the steel vessel are similar.

Table III. Masses of Actinides Measured after Filtration (Np and Pu in ng; Cm in pg)

Time, days	pH	Nuclide	Filter Pore Size					Acid Strip
			1000 nm	450 nm	100 nm	10 nm	6 nm	
			Tests at 2000 m^{-1}					
7	9.8	Np-237	—[a]	20	—	—	—	54
		Pu-239	—	3.8	—	—	—	2.6
		Cm-244	—	0.01	—	—	—	0.00
	9.8	Np-237	—	90	—	—	—	11
		Pu-239	—	8.5	—	—	—	5.7
		Cm-244	—	0.05	—	—	—	0.05
70	10.1	Np-237	600	700	600	1100	60	490
		Pu-239	59	44	22	1.0	1.9	33
		Cm-244	0	0.28	0	0	0.29	0.10
140	10.4	Np-237	1600	300	800	1000	200	2100
		Pu-239	29	22	49	1.0	4.5	59
		Cm-244	0	0	0	0	0	0.03
			Tests at 20,000 m^{-1}					
14	11.2	Np-237	—	0	—	—	—	50
		Pu-239	—	72	—	—	—	3.2
		Cm-244	—	0	—	—	—	0.03
56	10.9	Np-237	—	0	—	—	—	60
		Pu-239	—	0	—	—	—	4.9
		Cm-244	—	0.06	—	—	—	0.06
98	11.0	Np-237	6200	20	—	—	—	40
		Pu-239	72	0.21	—	—	—	11
		Cm-244	0	0.01	—	—	—	0.32

[a] "—" indicates that a particular filtration or analysis was not performed.

CONCLUSIONS

Static dissolution tests are in progress to characterize the long-term corrosion behavior of a glass made with sludge from Tank 51 at the DWPF (51R) and a nonradioactive homologue glass (51S). The compositions of the glasses are essentially identical, except for the radionuclides, so that differences in the corrosion behavior of the two glasses are due to the effects of the radionuclides. Tests completed to date indicate that corrosion of the 51R glass results in slightly lower solution pH values and slightly lower B and Na concentrations than does corrosion of the 51S glass. The difference is probably due to radiolysis effects in the tests with 51R glass, although this remains to be confirmed. Most of the Np and Pu released from the glass is present in the solution as suspended material that can be removed by filtration with a 6 nm filter. Only small amounts of Np, Pu, and Cm are dissolved in the solution. Thin surface layers that are similar in appearance to clays formed during the corrosion of other DWPF glasses have formed during the corrosion of both glasses; these layers were too thin (about 200 nm) to be analyzed to determine if they retained actinides.

ACKNOWLEDGMENTS

J. C. Hoh, J. W. Emery, T. DiSanto, and M. T. Surchik provided technical assistance. Dr. E. C. Buck performed the AEM examinations. This work was supported by the U.S. Department of Energy, Office of Environmental Management, under contract W-31-109-ENG-38.

REFERENCES

1. N. E. Bibler, W. F. Kinard, R. A. Dewberry, and C. J. Coleman, *A Method for the Determination of Waste Acceptance Radionuclides in DWPF Glass and Demonstration of that Method Using SRS Tank 51 Radioactive Sludge and Glass*, Westinghouse Savannah River Company Report WSRC-TR-94-0505 (1994).

2. D. T. Reed and R.A. Van Konynenburg, Mater. Res. Soc. Symp. Proc. 112, 393-404 (1987).

3. D. J. Wronkiewicz, *Effects of Radionuclide Decay on Waste Glass Behavior- A Critical Review*, Argonne National Laboratory Report ANL-93/45 (1993).

4. X. Feng, J. K. Bates, E. C. Buck, C. R. Bradley, and M. Gong, Nucl. Technol. 104, 193-206 (1993).

5. W. L. Ebert and J. K. Bates, Nuclear Technol. 104, 372-384 (1993).

6. W. L. Ebert, J. K. Bates, E. C. Buck, M. Gong, and S. F. Wolf, Ceram. Trans. 45, 231-241 (1994).

7. J. C. Cunnane, ed., *High-Level Waste Borosilicate Glass: A Compendium of Corrosion Characteristics*, U.S. Department of Energy Report DOE-EM-0177 (1994).

8. *Standard Test Methods for Determining Chemical Durability of Nuclear Waste Glasses: The Product Consistency Test (PCT)*, Standard C1285-94, American Society for Testing and Materials, Philadelphia PA.

9. M. K. Andrews and N. E. Bibler, Ceram. Trans. 39, 205-221 (1993).

10. N. E. Bibler and J. K. Bates, Mater. Res. Soc. Symp. Proc. 176, 327-338 (1990).

11. W. G. Burns, A. E. Hughes, J. A. C. Marples, R. S. Nelson, and A. M. Stoneham, Nature 295, 130 (1982).

12. D. T. Reed and D. L. Bowers, Radiochim Acta 51, 119-125 (1990).

13. W.L. Ebert and J.K. Bates, Ceram. Trans. 61, 479-488 (1995).

WASTE GLASS LEACHING AND ALTERATION UNDER CONDITIONS OF OPEN SITE TESTS

I.A. SOBOLEV, M.I. OJOVAN, O.G. BATYUKHNOVA,
N.V. OJOVAN, T.D. SCHERBATOVA
Scientific and Industrial Association "Radon", The 7-th Rostovsky Lane, 2/14, Moscow, 119121, Russia, fax (095)248 1941, E-mail: Oj@nporadon.msk.ru

ABSTRACT

The behaviour of waste glass was investigated under open site disposal conditions. This glass was produced by vitrification of intermediate level radioactive waste from nuclear power plants. Two types of borosilicate glasses were obtained for two different reactor wastes, WWER and RBMK. Leaching and alteration mechanisms are discussed as well as the data processing technique used for these long term tests. The decay of radionuclides was accounted for in order to obtain correct results. The leaching factors obtained can be used for the assesment of radionuclide retention. Discontinuous leaching of Cs-137 has been observed during more than 8 years testing time. The fluctuating leaching rate depends on glass composition. The average leaching rate remains within $(0.4 - 4)$ μg/sq.sm•day.
Alteration of waste glass includes the formation of surface layers and cracks on the glass surface. SEM analysis of glass was used to show these surface layers. The thickness of the layers was determined to be within $2 - 6$ μm. The structure of these layers depends on glass composition and the interfacing environment.

INTRODUCTION

Glass is one of the most useful and widely used materials. One of the many known functions of glass is its use as packaging for safe storage of various substances, including dangerous and toxic ones. Glass is also a suitable matrix for the immobilisation of radioactive waste. The majority of waste components are incorporated into the structure of the glass, and those components which are less compatible with glass may be included in the glass matrix as a dispersion phase. The high chemical and mechanical durability of glass ensures retention of radionuclides over a long period of time. Although initially it was proposed to use glass for the isolation of highly radioactive waste, now vitrification also considered as a promising and economical method for the solidification of intermediate level radioactive waste [1]. The equipment required for intermediate level radioactive waste vitrification is not as complex, and the requirements for the final product are not as strict as those for high level waste. This is due to the relatively short time in which radiotoxic components in the stored intermediate level radioactive waste reach safe levels. This time does not exceed 500-1000 years and as a rule is dependent on the decay of the major radioactive nuclides in the intermediate level radioactive waste: ^{137}Cs, ^{134}Cs, ^{60}Co, and ^{90}Sr. On the other hand, the high durability of glass considerably simplifies the requirements for waste storage conditions and allows a wide use of the most available method of burying waste in near-surface repositories.

Currently there is a large amount of reliable information on the behaviour of vitrified highly radioactive (or simulated) waste (for example, see references 2 and 3). Most of this information also can be used for understanding the nature of the behaviour of intermediate level radioactive glass. However, the peculiarities of content, preparation technology, physical characteristics and storage conditions call for a careful analysis of the behaviour of these materials so that a reliable estimate of their retaining properties is obtained.

Mat. Res. Soc. Symp. Proc. Vol. 465 © 1997 Materials Research Society

The aim of this paper is to analyse leaching and alteration of waste glass during open storage tests conducted over a period of eight to nine years on glass blocks containing intermediate level radioactive waste from nuclear power plants (NPP).

GLASS ALTERATION PROCESSES

The most complete analysis of the glass corrosion process (alteration) during a long storage period was given in references 2 and 3. The authors of reference 3 distinguish seven main glass alteration stages when glass is exposed to water: 1-ion exchange/diffusion, 2-hydrolysis, 3-matrix dissolution, 4-condensation, 5-precipitation, 6-sorption/colloid formation and 7-erosion. Here one shall note the following in connection mainly with the initial period of glass alteration. The point is that in the initial stage (usually the first week or two) surface dissolution takes place and radionuclides are released into the water. This process is described by adding a factor, which disappears with time, to the rate of radionuclide leaching [4]

$$\Delta R = k\, n\, exp(-kt), \tag{1}$$

where k is the rate of dissolution of radionuclide compounds in water and n is the initial concentration of the radionuclides on the glass surface. Although in reality during the time periods assumed for controlling vitrified waste (hundreds or thousands of years) this added factor has a small value, it can be significant in terms of processing experimental data that takes into account the total amount of radionuclides released into the water. This is especially important for waste with intermediate level radioactivity where the main radionuclide is ^{137}Cs.

Besides this, the crack forming process can be separately addressed also. Crack formation can lead to an increase in the area of interface with the environment and even to destruction of glass blocks. On the other hand, being the result of a combination of physical and chemical processes, it can be considered as stage 7.

Currently it is generally accepted that glass alteration leads to the formation of three main layers at the contact surface: a reaction or diffusion zone, a gel layer and a precipitated layer.

The processes described above occur in water, however, they also occur during glass weathering. The term "weathering" is used for defining glass corrosion in unsaturated conditions, i.e. when there is no constant interface with water, damp air (either with or without reaction gases) or steam [5]. For open storage or for disposal in a near surface repository, this term is more appropriate as the interface with water is not constant: during rainfall samples are washed with streams of water, and in dry weather the samples dry up and there is then no interface with water at all. Evaporation of water changes the concentration of ions dissolved in the water that interfaces with the glass, and the pH of the environment also changes. At a pH<7 the dissolution rate of silica increases considerably due to the catalytic action of OH^- at silica-hydrogen connections Si-O-Si [6]. In addition, materials heated by sunshine and then cooled by cold rain water are subjected to mechanical stresses. In winter water gets into cracks and freezes causing additional stresses and crack development. The possibility of using the results of research on ancient glass was considered in reference 7 for the prediction of the long-term durability of waste glass. It has shown that ancient glass subjected to the continuous influence of either a dry or damp environment was less likely to degrade in comparison with glass that was subjected to seasonal water exposure and drying, or freeze-thaw cycles. These results may be considered a fast test for long-term storage. In principle under the conditions of long term open storage we will see processes typical for weathering of rock. So, despite the relatively low radioactivity, we see that the environmental conditions can be similar to those for the storage of highly active waste, and these environmental conditions cause weathering of samples.

LONG TERM TESTS AND DATA PROCESSING

Quantitative models of corrosion have been developed initially for simple homogeneous two component glass matrices. We have the following equation for the quantity of substance leached from the glass:

$$Q = at^{1/2} + bt + n[1 - exp(-kt)], \qquad (2)$$

where t is the exposure time. This equation is based on the mechanisms of diffusion limited ion exchange and matrix dissolution. The last part of this expression differs from that in reference 3 by taking into account the near-surface dissolution of radionuclides according to reference 4.

Constants a and b are determined experimentally and depend on the glass composition, environment pH, temperature, the leaching substances and the storage conditions. The leach rate $R = dQ/dt$ takes into consideration that the initial stage of the surface dissolution is determined by the following formula:

$$R = (a/2t^{1/2}) + b + \Delta R \qquad (3)$$

It is important that for very long intervals of time the leach rate (in an asymptotic regime) tends to some constant value:

$$lim\ R(t \to \infty) = b \qquad (4)$$

This is due to the formation of a gel layer on the surface. The gel layer dissolves slowly, resulting in release of radionuclides into the water. Initially (but later than the surface dissolution stage), radionuclide release is determined by an ion exchange mechanism. The reletive importance of each mechanism (ion exchange and dissolution) depends on the glass characteristics and environmental parameters. Due to the pH drift in the leaching environment, it is possible that the leaching process will nearly always depend on matrix dissolution [3]. Here we shall note the results obtained by Kinoshita and others [8], where for small values of pH the dissolution process for sodium-boron-silicate glass can be described based on percolation theory, allowing the correct prediction of sharp increases in the dissolution rate as some threshold boron concentration limit is exceeded.

It is currently considered that for describing the corrosion process for glasses with a more complex composition the better developed models are the Wallace and Wicks models, the Grambow geochemical approximation and the Jantzen-Plodinec thermodynamic model (see reference 3). The existing models are phenomenological: they require the knowledge of parameters found through experiments, therefore in the description of the process of glass leaching the last word remains with the experiment.

Radionuclide decay corrections have been applied to the leaching data from the long-term tests, when the exposure time approaches or exceeds the half-life of typical radionuclides released from the vitrified waste. During the tests we sample the water interfacing with the glass, determine the radioactivity in each sample and calculate the total quantity of radionuclides released from the glass into the water. These incremental values are added up as the test goes on. Therefore the quantity of radionuclides released into the water is measured, but the quantity in the water is not actually measured. These two amounts would be the same for stable nuclides, but not for radioactive nuclides since the quantity of radionuclides in the water increases due to the release from glass, but decreases due to natural radioactive decay. In the calculation used, the quantity of radionuclides released from the glass into the water Q_{tr} is determined by the integral over time of the radionuclide flux through the glass surface:

$$Q_{tr} = a(\pi/\lambda)^{1/2} \Phi(\sqrt{\lambda t}) + b[1 - exp(-\lambda t)]/\lambda + nk[1 - exp(-(k+\lambda)t]/(k+\lambda), \qquad (5)$$

where λ is the decay constant, and $\Phi(\sqrt{\lambda t})$ is the error integral: $\hat{O}(x) = \dfrac{2}{\sqrt{\pi}} \int\limits_0^x \exp\left(-t^2\right) dt$

For stable nuclides when $\lambda=0$, and for a short exposure period when $\lambda t \to 0$, equation (5) is the same as (2). For the intermediate region where $\lambda t \sim 1$, the differences become considerable. To obtain data that will allow correct interpretation for long term experiments involving radionuclides therefore equation (6) should be used rather than (2). For a very long exposure period when $\lambda t >> 1$, the quantity of radionuclides released into the water for a unit of glass surface Q_{tr}, approaches a constant and maximum possible value:

$$Q_{tr} \to a(\pi/\lambda)^{1/2} + b/\lambda + nk/(k+\lambda) \qquad (10)$$

So, in the end, diffusion and dissolution can lead to the release of some small certain quantity of radionuclides into the water. This amount can be determined by adding up the masses of radionuclides carried away by diffusion from the near surface layer over a time approximately equal to the half-life period; this includes those radionuclides that dissolve along with the glass over this time period and that also dissolve directly from the glass surface during the initial time period. The last formula offers a good opportunity for quantitative assessment of the glass matrix. Actually the release of radionuclides into water from the given glass depends only on glass parameters.

OPEN SITE TESTS

Experimental samples of vitrified radioactive waste were obtained from the SIA "Radon"experimental vitrification plant. The waste glass contained real intermediate level radioactive waste from the Kursk (RBMK reactor) and Kalinin (WWER reactor) NPPs. The chemical composition of these glasses and the glass block characteristics are shown in tables I and II respectively. A glass block (BS-8) with a waste composition and matrix material similar to BS-9, has also been tested for 6 years. The BS-8 block, due to anomalies in technological processing, preparation and storage conditions, had broken into several fragments and was not acceptable for final burial.

Table I. Composition of waste glasses, wt.%.

Compo-nent	SiO_2	Na_2O	CaO	B_2O_3	Na_2SO_4 (SO_3^{2-})	NaCl (Cl^-)	K_2O	Fe_2O_3	Al_2O_3	Other
RBMK waste glass	48.2	16.1	15.5	7.5	1.1	1.2	0.5	1.7	2.5	5.7
WWER waste glass	46.8	24.0	6.2	9.0	0.8	0.9	1.9	1.8	4.3	4.3

Table II.Parameters of vitrified blocks

Specimen	Waste from reactor type	Initial total radioactivity, Bq	Weight, kg	Interface area, cm^2	Testing time, months
BS-9	RBMK	$1.14 \bullet 10^8$	30	3000	108
BS-10	RBMK	$1.14 \bullet 10^8$	30	1120	108
BS-11	WWER	$6.75 \bullet 10^7$	25	380	96
BS-12	WWER	$6.75 \bullet 10^7$	25	380	96

For the open site test, the glass blocks were placed on stainless steel trays (52 X 52 cm), 60 cm above the ground. Atmospheric precipitation that interfaced with the glass blocks was collected from the trays into special containers. Liquid samples were taken from these containers and subjected to radiometric and chemical analysis. Usually water sampling was performed twice a month. The climate parameters for the open testing site are shown in table III.

Table III. Climate parameters for 1988 - 1995 open test site

Parameter	Winter	Spring	Summer	Autumn
Average temperature, ^{o}C	-4.6	8.3	19	5
Average amount of sediment per month, mm	49.2	30.5	61.8	62.9
Average amount of sediment, mm	147.6	91.5	185.4	188.7

The waste glass on the test site has been subjected to solid and liquid atmospheric precipitation, air temperature changes, sunlight, wind, etc. In the summer, major factors affecting the glass are changes in the water ion concentration, e.g., increases resulting from decreased moisture, as well as the changing pH of the environment and the glass component precipitation processes. In the winter, the major factor is the stress in the glass matrix due to water freezing in cracks. In the spring and autumn, freeze-thaw cycles lead to the formation of cracks on the surface of blocks. The intermittant, irregular contact of blocks with water yields an averaged description of the leaching processes [9].

SEM ANALYSIS OF SAMPLES

The microscopic structure of the glass blocks has been analysed using a HITACHI scanning electronic microscope in the secondary electron regime. The samples were prepared by deliberate destruction (chipping) at a depth of up to 1.5 cm. It is noted that the condition of the glass block surface differs from place to place due to the differences in the influences to which it has been subjected. The SEM analysis confirms the existence of surface layers on all examined samples. The outer surface of the glass blocks containing radioactive waste from the RBMK reactor (specimens BS-9 and BS-10) has a "cell" structure with a network of cracks. Such a structure was described early on in studies of leaching from waste glass; the cracks in the alkali depleted surface layer are often called "dehydrated", as they are found on samples dehydrated after leaching [10, 11]. Figure 1 shows a fragment of the interface surface and the bulk matrix structure for glass block BS-9.

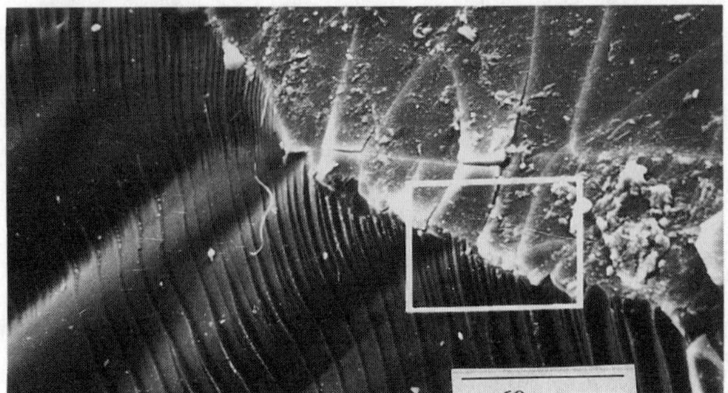

Figure 1. Glass block BS-9, the surface interfacing with the atmosphere.

It is evident that the matrix is in good condition, and there is a layer altered during the glass corrosion process on the glass surface which interfaces with the environment. Figure 2 left and right show the surface layers of samples BS-8 and BS-9 respectively.

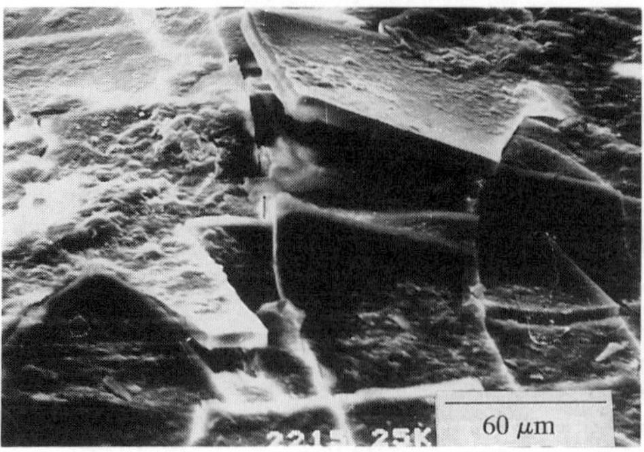
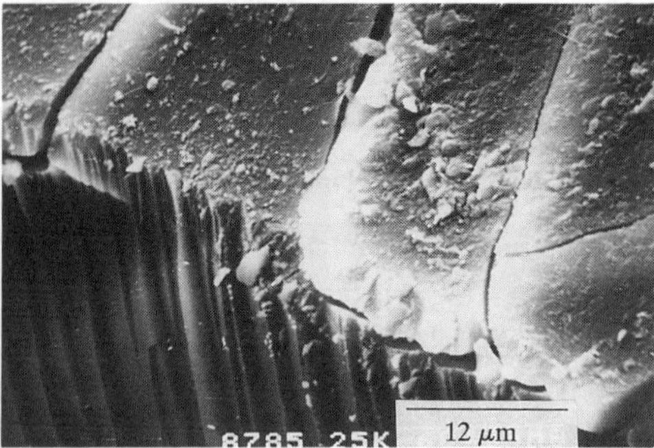

Figure 2. Surface interfacing with the atmosphere:
left-specimen BS-8, right- specimen BS-9.

One can see from Figure 2 (left) that the surface layer, with a thickness of 6 µm, consists of at least two sublayers. The outer sublayer bordering on the environment is ~ 1.5-2 µm thick, and the inner sublayer is ~ 4-4.5 µm thick. Figure 1 of a cross section of the 3 µm surface layer of sample BS-9 does not show such sublayers. However, this could be the result of the conditions used to obtain the desired image contrast. Figure 2 (right) shows a good adherence of the surface layer to the glass. The estimated thickness of the surface layer of block BS-10 is less than 3 µm. The surface layer of sample BS-10 is homogeneous, and we have observed precipitation from secondary phases on the rougher surfaces of BS-8 and BS-9.

Figure 3. Surface deposits on the glass block BS-12.

The surface structure of glass blocks BS-11 and BS-12 containing radioactive waste from the WWER reactor differs from the one described above: it is sponge-like, with pores and other defects. The electron microscopic images of the near-surface areas of these blocks also show the formation of surface layers. Figure 3 shows the image of the block BS-12 surface that interfaced with the environment.

This research confirms that the initial stage of the corrosion process for glass blocks containing intermediate level radioactive waste from the RBMK reactor is ion exchange resulting in the formation of a surface layer of altered glass with a typical network of small cracks on the surface. Moreover the structural X-ray analysis of glass blocks BS-9 and BS-10 shows a narrow quartz line that indicates the beginning of crystallisation. Relatively smooth surface layers on samples BS-8, BS-9 and BS-10 and porous surfaces layers with disturbed surfaces (resulting from the glass dissolution process) on samples BS-11 and BS-12 demonstrate better corrosion resistance of glass blocks containing waste from the RBMK reactor. The deposits on the samples containing waste from the WWER reactor serve as "collectors" for alkali ions [3], and continued weathering may lead to glass hydrolysis, further cracking, and formation of layers on the glass surface.

CS-137 LEACHING

Table IV gives the average leaching factors derived from the experimental leaching curves obtained for the glass blocks over eight years. The actual differential leach rate curve over time would contain leach rate jumps (discontinuities) as well as relatively smoothly varying sections, however the direct data processing yields average leaching factors rather than effective diffusion coefficients [12]. The values of effective diffusion coefficients are almost always considerably less than the values of leaching factors. The leaching factor takes into account the real character of leaching including discontinuities in the leaching rate, therefore it shall be used for long term predictions of radionuclide losses.

Table IV. Leaching parameters for waste glass

Specimen	Leaching rate after 1 year, g/cm^2 day	Leaching rate after 8 years, g/cm^2 day	Leaching factor for 8 yr test time, cm^2/day
BS-9	$8.6 \bullet 10^{-7}$	$3.7 \bullet 10^{-7}$	$1.8 \bullet 10^{-11}$
BS-10	$3.1 \bullet 10^{-6}$	$1.2 \bullet 10^{-6}$	$1.9 \bullet 10^{-10}$
BS-11	$9.8 \bullet 10^{-6}$	$2.5 \bullet 10^{-6}$	$3.6 \bullet 10^{-10}$
BS-12	$2.2 \bullet 10^{-5}$	$4.1 \bullet 10^{-6}$	$8.0 \bullet 10^{-10}$

The different behaviour of the glass blocks contining waste from different sources is explained by the differences in their composition: glass blocks with lower durability contain waste from the WWER reactor, reduced silica content, and a soluble alkali oxide content that is 1.3 - 1.5 times higher than in the glass blocks containing waste from the RBMK reactor.

CONCLUSIONS

The alteration in composition of the surface layer in comparison with the initial material may vary and depend on physical and chemical properties of the glass and of the environment, but for a given glass under the same conditions these alterations are similar. The similar behaviour of glass blocks BS-9 and BS-10 confirms this conclusion, as well as the similar behaviour of glass blocks BS-11 and BS-12. Comparison of the surfaces of glass blocks BS-9, BS-10 and BS-11, BS-12 after exposure to the environment makes it possible to suppose that the first two have ion exchange as the dominant process at this stage of corrosion, whereas in the glass blocks containing waste from WWER reactor, the dominant process is hydrolysis of the lattice. Basically the glass status after prolonged tests remains satisfactory and radionuclide retention is reliable (see table IV). Comparison of the data on layer thickness obtained in this paper with similar parameters from other work allows us to assume that the glass blocks studied here have satisfactory durability.

ACKNOWLEDGEMENTS

Authors wish to thank A.S. Barinov and V.M.Tivansky for discussions and assistance.

REFERENCES

1. Sobolev I.A., Dmitriev S.A., Barinov A.S. *et al.* Management of waste in Russia. HLW, LLW, Mixed Wastes and Environment Restoration in Working Toward a Cleaner Environment. International Conference WM'95, February 26 - March 2, 1995, Tuscon, Arizona, Proceedings on CD-ROM (Abstracts, p.27).

2. Feng X. Surface layer effect on waste glass corrosion. Mat. Res. Soc. Symp. Proc., 1994, Vol.333, pp. 55-68.

3. Clark D.E., Schulz R.L., Wicks G.G., Lodding A.R. Waste glass alteration processes, surface layer evolution and rate limiting steps. Mat. Res. Soc. Symp. Proc. 1994, Vol.333, pp. 107-122.

4. Anisimova L.I., Kulichenko V.V., Krylova N.V. and others. The evaluation of the degree of water contamination during the disposal of vitrified radioactive waste in geological formations. - Atomic Energy (Russia), 1980, Vol.60, #6, pp.413-414.

5. Bates J.K., Buck E.C. Waste glass weathering. Mat. Res. Soc. Symp. Proc., 1994, Vol.333, pp.41-53.

6. El-Shamy T.M., Lewins J., Douglas R.W. The dependence on the pH of the decomposition of glasses by aqueous solutions. Glass Technology, 1972, v.13, pp.81-87.

7. Heimann R.B. Nuclear fuel management and archaeology: are ancient glasses indicator of long-term durability of man made materials? Glass Technology, 1986, v.27, N3, pp.96-101.

8. Kinoshita M., Harada M., Sato Y., Hariguchi Y. Percolation phenomenon for dissolution of sodium borosilicate glasses in aqueous solutions. J. Amer. Ceram. Soc., 1991, v.74, ¹4, pp.783-787.

9. Sobolev I.A., Barinov A.S., Ojovan M.I., Ojovan N.V. Long term natural tests of vitrified radioactive waste. HLW, LLW, Mixed Wastes and Environment Restoration - Working Toward a Cleaner Environment. International Conference WM'96, February 25 - 29, 1996, Tuscon, Arizona, Proceedings on CD-ROM (Abstracts, p.51).

10. Clark D.E. and Hench L.L. Theory of corrosion of alkaliborosilicate glass. Sci. Bas. Nucl. Waste Manag., 1983, v.1, pp.113-124.

11. Van Iseghem P., Timmermans W. and De Batist R. Parametric study of the corrosion behaviour in static disstilled water of simulated European reference high level waste glasses. Mat. Res. Soc. Symp. Proc., 1985, v.44, pp.55-62.

12. Barinov A.S., Ojovan M.I., Ojovan N.V., 7-years leaching tests of NPP vitrified radioactive waste, Mat. Res. Soc. Symp. Proc., 1996, v.412, pp.265-270.

ACCELERATED TESTING OF WASTE FORMS USING A NOVEL PRESSURIZED UNSATURATED FLOW (PUF) METHOD

B. P. MCGRAIL, P. F. MARTIN, AND C. W. LINDENMEIER
Applied Geology and Geochemistry Department, Pacific Northwest National Laboratory,
Richland, Washington 99352, bp_mcgrail@pnl.gov

ABSTRACT

A new experimental technique has been developed to test waste forms and other proposed engineered-barrier materials under unsaturated conditions. Laboratory experiments using the pressurized unsaturated flow (PUF) apparatus have been performed with a Na-Ca-Al borosilicate glass being studied for immobilization of low-activity tank wastes and a reference borosilicate glass for immobilization of high-level wastes (SRL-202). A complex coupling between glass corrosion, secondary phase precipitation, and unsaturated flow behavior was observed. Precipitation of a family of Na-Ca-Al zeolites was also found to cause an acceleration in the reaction rate of the low-activity waste glass. This same effect was observed after approximately 1 year in high solid-to-liquid ratio batch tests but after only 12 days using the PUF method at the same temperature. The onset of secondary phase precipitation was tracked by monitoring changes in volumetric water content and by inline chemical analysis. Finally, elemental release of the major glass components and solution pH were found to differ under unsaturated flow-through conditions as compared with saturated, batch tests. These findings, while preliminary, have important implications for understanding and modeling glass corrosion behavior under unsaturated conditions.

INTRODUCTION

Disposal plans, both in the U.S. and in other countries, for hazardous, low-activity, and high-level radioactive wastes involve subsurface sites located in unsaturated or vadose zones. In the U.S., for example, at the Department of Energy's Hanford Site in Washington State, a shallow subsurface site is planned for over 500,000 MT of glass that will be produced by a private company immobilizing low-activity tank wastes [1]. For waste forms such as high-level waste glass, commercial spent fuel, and specialty glasses or ceramics for immobilization of excess weapons Pu, disposal is planned in the Yucca Mountain (Nevada) repository, which is located above the present water table. However, progress on these disposal actions depends in large part on demonstrating, with scientific credibility, that the long-term public health and safety will be protected. Such performance assessments, like those being done for the low-activity wastes at Hanford [2], rely on models that describe the release of contaminants from the waste form and subsequent transport to the accessible environment.

Current performance assessment models are based on principles of continuum mechanics [3]. If these models are to be improved and tested, experimental methods are needed that can simulate the open-system flow and transport processes relevant to unsaturated disposal systems. Unfortunately, the currently available "unsaturated" test methods have significant drawbacks in this respect. The vapor hydration test (VHT), for example, involves suspending a test specimen in a sealed container, which usually contains just a sufficient mass of water to obtain water vapor

Mat. Res. Soc. Symp. Proc. Vol. 465 © 1997 Materials Research Society

saturation in the internal volume of the container at the test temperature [4]. The VHT only provides information on 1) the types of secondary phases formed under the specific conditions of the VHT and 2) a rough estimate of the corrosion rate as determined from measurements of alteration layer thickness. No data are provided on the solution chemistry in contact with the glass, which is mandatory for modeling. The drip test [5] again calls for suspending the specimen in a sealed container, but, in this case, a small volume of fluid is injected periodically such that several water droplets fall by gravity on the specimen. Although fluid is available for solution analysis, comparison of these data with existing models is problematic. Concentrations of the waste form components and solution pH change over time in the fluid contacting the waste form until the fresh fluid is injected, which then changes the chemistry of the water. Such quasi-static or periodic replenishment experiments cannot be treated adequately with conventional computer codes. Finally, the "drip" scenario does not represent appropriate hydrodynamic conditions for some disposal systems, such as those located in shallow, sandy soils.

Overcoming the limitations of the available test methods for unsaturated conditions, particularly with respect to developing and testing models for performance assessment [3,6] was the motivation behind the test method described in this paper. In an earlier paper, McGrail et al. [7] described the pressurized unsaturated flow (PUF) method developed at Pacific Northwest National Laboratory and presented results from tests with a low-activity waste glass. These results are now extended with new data on a reference borosilicate glass developed at the Savannah River Technology Center for high-level waste.

EXPERIMENTAL METHODS AND MATERIALS

The PUF system was developed with the goals of 1) varying the volumetric water content from saturation to at least 20% of saturation; This design goal was established because it represents an upper bound on the water content of a typical Hanford soil, 2) minimizing the flow rate to increase liquid residence time, and 3) operating at a maximum temperature of 90°C. Although the wastes in the actual site will not necessarily experience the exact moisture levels or elevated temperatures established in the PUF test, the experimental design provides a novel way to study waste form corrosion behavior under unsaturated conditions, subject to open-system flow and transport.

PUF Test Design

The basic test apparatus consists of a column where glass particles (or other material) of a known size and density are compacted to a known bulk density (g/m^3). The remaining void space represents the porosity (ε). Volumetric water content (θ) is the percent volume of water within the total fixed column volume. Thus, when a 20-cm^3 column is packed with glass, leaving a 50% void space (ε), the column is considered fully saturated when 10 ml of water ($\theta = 50\%$) are in the column at any given point in time. Using these same parameters, a column is considered unsaturated when $\theta < \varepsilon$. The rate of water flow through the column is the pore water velocity (U_p, m/s), which is simply the influent flux (m^3/s) divided by the cross-sectional area of the column (m^2) divided by θ. To determine the residence time for an aliquot of water to move through the column, the length (m) of the column is divided by U_p.

Figure 1 is a schematic of the PUF apparatus. The column is fabricated from polyetheretherketone (PEEK), which is chemically inert so that dissolution reactions are not influenced by interaction with the column, nor is the effluent contaminated by contact with column materials. Unlike the previous PUF system [7], a dedicated computer running LabVIEW™ (National Instruments Corporation) software now logs test data to disk from several thermocouples, inline sensors for effluent pH and conductivity, and column weight from an electronic balance to accurately track water mass balance and saturation level. The control program simultaneously controls both column temperature, using a ramp-and-soak algorithm, and a positive-displacement pump for influent delivery capable of flow rates ranging from 0.05 to 100 mL/h.

A porous titanium plate of proprietary design is sealed in the bottom of the column to ensure an adequate pressure

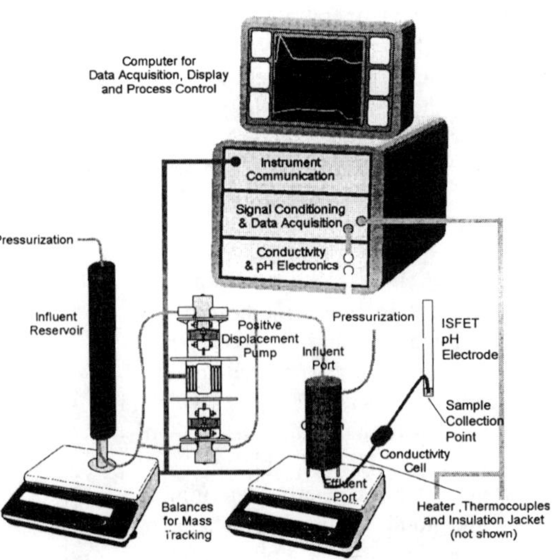

Figure 1. Schematic of PUF Apparatus

differential for the conductance of fluid while operating under unsaturated conditions [8]. Titanium was chosen because it is highly corrosion resistant and has excellent wetting properties. When water saturated, the porous plate allows water but not air to flow through it, as long as the applied pressure differential does not exceed the air entry relief pressure or "bubble pressure" of the plate. Volumetric water contents as low as 8% at 90°C have been achieved with this configuration.

<u>Specific Test Methods and Materials</u>

Tests were performed with 1) a low-activity waste glass formulation, LD6-5412, which contains 12% Al_2O_3, 5% B_2O_3, 4% CaO, 1.5% K_2O, 20% Na_2O, and 55.9% SiO_2 by weight in addition to some other minor components, and 2) a reference high-level waste glass, SRL-202 [9]. The glass was crushed and screened to obtain the -20+70 mesh (210 to 841 μm) size fraction, which was then cleaned according to standard product consistency test (PCT) methods [10]. For each test, the column was filled with the prepared glass, giving an initial porosity of approximately 0.4, and was then vacuum saturated with water at ambient temperature. A temperature controller was programmed to heat the column to 90°C in about 2.5 hour (0.6°C/minute). The column was initially desaturated by injecting air with a gas-tight syringe and the displaced effluent was collected. The influent pump was valved off and the column was vented periodically during the temperature ramp to maintain an internal pressure less than the bubble pressure of the porous plate. After reaching 90°C, the influent valve was opened and influent and effluent flow rates were set equal at 2.0 ml per day. Effluent samples were collected in a receiving vessel, which was periodically drained into tared vials from which samples were extracted for elemental analysis by inductively coupled plasma-atomic emission spectroscopy.

After test termination, the reacted glass material was then carefully scraped out to a series of controlled depths and transferred to tared vials. The filled vials were weighed and then moved to a vacuum oven operating at 35-40°C and dried until a constant weight was obtained. The amount of glass recovered was compared with the amount initially packed into the column. The final volumetric water content in the column was also calculated from these final weights. Samples from the bottom, middle, and top of the column were analyzed for major crystalline phases by powder X-ray diffraction (XRD). Reacted samples are not yet available for analysis from the test with SRL-202 glass, as the experiment is continuing.

RESULTS

Figure 2 shows the time-dependent variation in the water saturation levels and pH for both tests. The computer data logging system was used for the test with the SRL-202 glass, which gave a much higher data density than the test with LD6-5412 glass. Pore saturation values were determined from the total column weight and should be viewed as a global average for the packed bed. Local variation in the water content could occur in the tests, particularly with the onset of secondary phase precipitation, to be discussed below.

For reasons that will not be understood until the reacted glass is examined, the test with SRL-202 glass did not attain the same degree of pore desaturation as in the test with LD6-5412 glass. The difference in water content appears to correlate well with the differences in effluent pH. The low initial water content in the test with LD6-5412 generated a solution pH below 8, but when the moisture content jumped

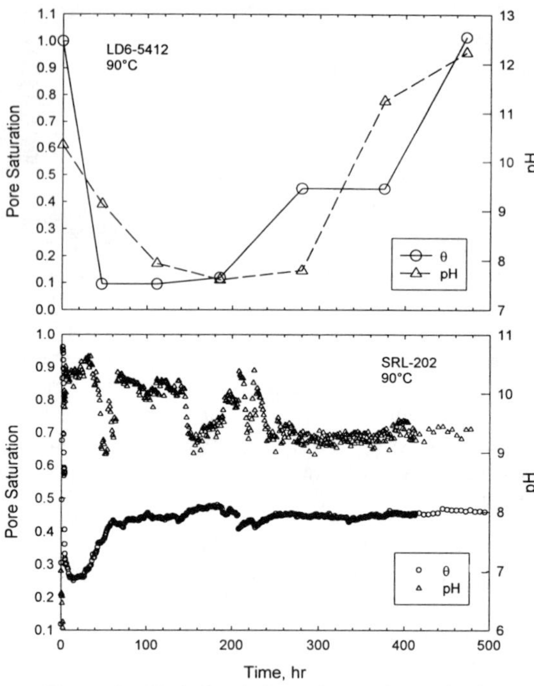

Figure 2. Variation in Pore Saturation and Effluent pH as a Function of Time

to 0.45 and higher, the solution pH increased to above 11. Similarly for the test with SRL-202 glass, the initial desaturation to approximately 25% of pore saturation was matched by the pH drop to about 9, centered at about 50 hours into the test.[a] However, unlike the test with LD6-5412 glass, the column did not continue to desaturate; instead, the water content increased at about 24 hours into the test to a fairly constant level of about 45% of pore saturation. At a fixed flow rate, the higher water content is expected to generate a higher effluent pH because the residence time in the column is longer, which allows a greater extent of solid/liquid reaction, and because it reduces the gas-to-liquid ratio so that less total CO_2 is available to react with the fluid.

---

[a]The pH drop is delayed relative to the water content measurement due to the residence time of fluid in the column.

For the test with LD6-5412 glass, the pH data in Figure 2 provide clear indications of the glass corrosion behavior under unsaturated flow-through conditions. After an initially high reaction rate, the glass reaction rate slows dramatically, which is indicated by the drop in effluent pH values from a high of over 10, to a near-neutral pH of 7.6. However, after 300 hours, a rapid increase in the effluent pH is observed. The change in pH coincides with an increase in the volumetric water content (Figure 2), which indicates that the hydraulic conductivity of the porous glass bed has been reduced. The formation of secondary phases is thought to be the cause of these effects, as will be discussed below. In the test with SRL-202 glass, we suspect that precipitation of a secondary phase, perhaps a clay, is responsible for the increase in water content at about 24 hours into this test. There is also evidence for a second precipitation event at about 210 hours that caused a pH excursion and a minor fluctuation in the moisture content. This speculation will be evaluated once the test is terminated and the reacted solids have been examined.

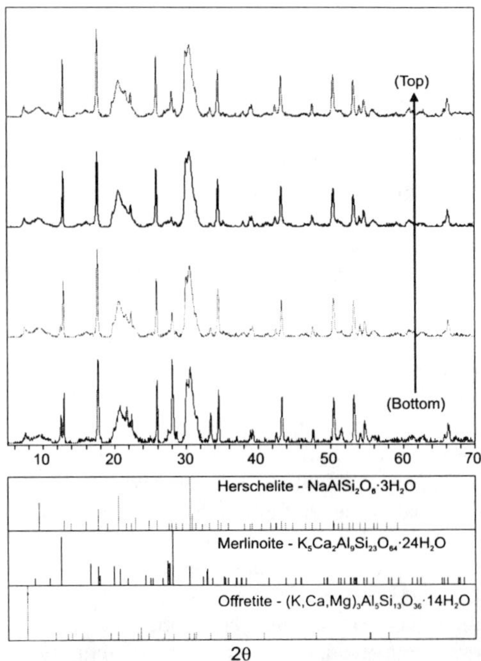

Figure 3. XRD Pattern of Reacted LD6-5412 Glass at Different Locations in PUF Column

In batch PCT tests with LD6-5412 glass, the formation of the zeolite phases phillipsite (KCaAl$_3$Si$_5$O$_{16}$·6H$_2$O) and gobbinsite (Na$_4$CaAl$_6$Si$_{10}$O$_{32}$·12H$_2$O) was found to coincide with the onset of an accelerated rate of reaction [11]. Figure 3 shows the X-ray diffraction analysis of reacted LD6-5412 glass samples taken from the top, middle, and bottom of the PUF column. Approximately 15 to 25 vol% of the sample had been converted into crystalline phases, but unambiguous phase identification was not possible because of the large number of Na-Ca-K-Al silicate phases with only slight differences in diffraction peak positions and intensities. However, structural matches with the zeolites herschelite [(Na,Ca,K)AlSi$_2$O$_6$·3H$_2$O] and merlinoite [(K$_5$Ca$_2$Al$_9$Si$_{23}$O$_{64}$·6H$_2$O] were found in all of the samples, as well as structural matches with a number of sodium aluminum silicate hydrate compounds. The uniformity of diffraction patterns and peak intensities from samples taken from the different column positions suggests a fairly uniform solution chemistry. Simulations of the PUF apparatus using the multiphase flow code STOMP [12] also indicated that a uniform moisture distribution was achieved within a few hours.

Clearly, the rapid rise in effluent pH in the test with LD6-5412 glass (Figure 2) is correlated with the onset of secondary phase precipitation that causes an acceleration of the glass reaction rate. In PCT tests with LD6-5412 glass, at the highest surface area-to-volume ratio (S/V) tested, 20,000 m^{-1}, the accelerated reaction stage did not occur before about 240 days. In the PUF test, the accelerated reaction stage appears to have been reached after only about 12 days, which is a factor of 20 decrease in the time required, even though the specific surface area of the glass used in the test was 10 times smaller than the specific surface area of the glass used in the PCT tests. Precipitation of the zeolite phases, and the concomitant increase in the rate of glass dissolution, are also correlated with increases in volumetric water content. Hydraulic conductivity of unsaturated porous media is a highly non-linear function of porosity, permeability, and water retention characteristics [13]. Relatively small changes in these properties, induced from secondary phase precipitation, could significantly alter hydraulic conductivity, which is manifested by changes in volumetric water content. The ability to monitor, in near-real time, the coupling between secondary phase formation and unsaturated flow hydrodynamics is a unique aspect of the PUF system.

Figure 4 shows the time-dependent normalized concentrations of the major glass components in the effluent samples taken from both tests. The two glasses show distinctly different corrosion behavior under unsaturated flow-through conditions. High solution concentrations are observed at the start of the test with LD6-5412 glass that decline to minimum values between 100 and 200 hours, as did the effluent pH. After 300 hours, solution concentrations of all the glass components, rise dramatically, again corresponding with the changes in volumetric water content (Figure 2). The normalized concentrations show that the release of the glass components is non-stoichiometric, as would be expected in this type of test where secondary phase formation is occurring. However, an unusual feature of the normalized releases is that, before the rate acceleration occurred, B had the highest release rate and the effluent pH was less than 8. In contrast, PCT data on LD6-5412 glass, prior to the rate acceleration, show Na with the highest release and pH values of 11.5 and higher. Unsaturated flow-through conditions establish a fundamentally different physical and chemical environment from the glass/water reaction as compared with water-saturated batch tests.

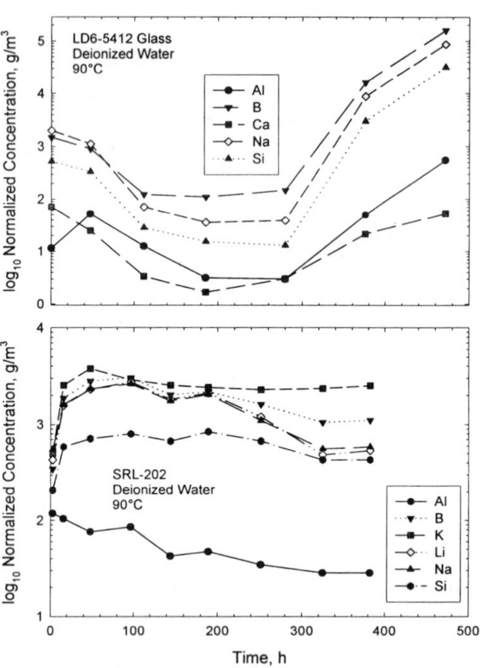

Figure 4. Normalized Concentration of Selected Glass Components as a Function of Time

Unlike the elemental release behavior for LD6-5412 glass, releases of the major glass components for SRL-202 reached peak concentrations at about 100 hours into the test and declined thereafter. The time-dependent variation in the solution concentrations correlates with the variation in moisture content over this same time period (Figure 2). Initial release of B and the alkali metals (Li, Na, K) is congruent. After a suspected precipitation event at approximately 210 hours into the test, differentiation in the elemental release rates is observed. At the last data point available so far , the elemental release rates follow the sequence K>B>Li≈Na≈Si>>Al and solution pH was approximately 9.3, although the K concentrations may not be valid due to leakage of KCl from the inline pH probe. The sequence in the 20,000 m^{-1} PCT results reported by Ebert et al. [9,14], before a rate acceleration was observed, was B>Li>Na≈K>Si and solution pH was 11.0 to 11.4. So far, there is no evidence of a corrosion rate acceleration as was found in the PCTs after 182 days. The PUF experiment with SRL-202 glass will be continued to see if an accelerated reaction period eventually occurs.

CONCLUSION

A second-generation design for a high-temperature pressurized unsaturated flow (PUF) apparatus was described. Experiments can be conducted at temperatures up to 90°C and can operate for extended periods of time. Preliminary results from PUF tests have provided new insights into the corrosion behavior of glasses in unsaturated systems. It was found that the onset of secondary phase precipitation, which has been observed to coincide with accelerated glass reaction rates, can be achieved as much as 20 times faster in the PUF test as compared with standard PCT methods at the same temperature. The PUF test is also unique in that the coupling between secondary phase precipitation, waste form corrosion, and unsaturated flow hydraulics can be tracked in near real-time. Finally, unsaturated flow-through conditions were found to establish significant differences in solution pH and elemental release behavior of the major glass components from a low-activity waste glass (LD6-5412) and a high-level waste glass (SRL-202) as compared with water-saturated tests. These findings raise important questions regarding the applicability of water-saturated batch tests, like the PCT, to evaluate long-term waste form behavior in unsaturated disposal systems. Additional studies are needed to more clearly determine whether the unsaturated flow-through conditions established in the PUF test provide a significantly better representation.

REFERENCES

1. Eiholzer, C. R. 1995. *Disposal Facility Data for the Interim Performance Assessment.* WHC-SD-RPT-159, Rev. 0, Westinghouse Hanford Company, Richland, Washington.

2. Mann, F. M., C. R. Eiholzer, A. H. Lu, P. D. Rittmann, N. W. Kline, Y. Chen, B. P. McGrail, G. F. Williamson, and N. R. Brown. 1996. *Hanford Low-Level Tank Waste Interim Performance Assessment.* WHC-EP-0884, Westinghouse Hanford Company, Richland, Washington.

3. McGrail, B. P., and L. A. Mahoney. 1995. *Selection of a Computer Code for Hanford Low-Level Waste Engineered-System Performance Assessment.* PNL-10830, Pacific Northwest National Laboratory, Richland, Washington.

4. Bates, J. K., L. J. Jardine, and M. J. Steindler. 1982. "Hydration Aging of Nuclear Waste Glass." *Science* **218**:51-54.

5. Bates, J. K., and T. J. Gerding. 1985. *NNWSI Phase II Materials Interaction Test Procedure and Preliminary Results.* ANL-84-81, Argonne National Laboratory, Argonne, Illinois.

6. Chen, Y., B. P. McGrail, and D. W. Engel. 1996. "Source-Term Analysis for Hanford Low-Activity Tank Waste Using the Reaction-Transport Code AREST-CT." In this volume of the *Scientific Basis for Nuclear Waste Management XX*.

7. McGrail, B. P., C. W. Lindenmeier, P. F. Martin, and G. W. Gee. 1996. "The Pressurized Unsaturated Flow (PUF) Test: A New Method for Engineered-Barrier Materials Evaluation." In *Proceedings of the American Ceramic Society,* Indianapolis, Indiana, April 14-18, 1996.

8. Wierenga, P. J., M. H. Young, G. W. Gee, R. G. Hills, C. T. Kincaid, T. J. Nicholson, and R. E. Cady. 1993. *Soil Characterization Methods for Unsaturated Low-Level Waste Sites.* PNL-8480, Pacific Northwest National Laboratory, Richland, Washington.

9. Ebert, W. L., and J. K. Bates. 1993. "A Comparison of Glass Reaction at High and Low Glass Surface/Solution Volume." *Nuc. Tech.* **104**:372-384.

10. ASTM. 1994. *Standard Test Methods for Determining the Chemical Durability of Nuclear Waste Glasses: The Product Consistency Test (PCT).* ASTM C1285-94, Annual Book of ASTM Standards, Philadelphia, Pennsylvania.

11. Bakel? Baker?, A.J., W. L. Ebert, and J. S. Luo. 1995. *Long-Term Performance of Glasses for Hanford Low-Level Waste.* ANL-84526, Argonne National Laboratory, Argonne, Illinois.

12. White, M.D. and M. Oostrom. 1995. *STOMP Subsurface Transport Over Multiple Phases. Theory Guide.* PNL-11217, Pacific Northwest National Laboratory, Richland, Washington.

13. van Genuchten. 1980. "A Closed-Form Equation for Predicting the Hydraulic Conductivity of Unsaturated Soils." *Soil Sci. Soc. Am. J.* **44**:892-898.

14. Ebert, W. L., J. K. Bates, E. C. Buck, and C. R. Bradley. "Accelerated Glass Reaction Under PCT Conditions." In *Scientific Basis for Nuclear Waste Management XVI*, eds. C. G. Interrante and R. T. Pabalan, Materials Research Society, Pittsburgh, Pennsylvannia.

NEPHELINE PRECIPITATION IN HIGH-LEVEL WASTE GLASSES :
COMPOSITIONAL EFFECTS AND IMPACT ON THE WASTE FORM ACCEPTABILITY

H. LI, J.D. VIENNA, P. HRMA, D.E. SMITH, and M.J. SCHWEIGER
Pacific Northwest National Laboratory, Box 999, P8-37, Richland, Washington 99352

ABSTRACT

The impact of crystalline phase precipitation in glass during canister cooling on chemical durability of the waste form limits waste loading in glass, especially for vitrification of certain high-level waste (HLW) streams rich in Na_2O and Al_2O_3. This study investigates compositional effects on nepheline precipitation in simulated Hanford HLW glasses during canister centerline cooling (CCC) heat treatment. It has been demonstrated that the nepheline primary phase field defined by the Na_2O-Al_2O_3-SiO_2 ternary system can be used as an indicator for screening HLW glass compositions that are prone to nepheline formation. Based on the CCC results, the component effects on increasing nepheline precipitation can be approximately ranked as $Al_2O_3 > Na_2O > Li_2O \approx K_2O \approx Fe_2O_3 > B_2O_3 > CaO > SiO_2$. The presence of nepheline in glass is usually detrimental to chemical durability. Using x-ray diffraction data in conjunction with a mass balance and a second-order mixture model for 7-day product consistency test (PCT) normalized B release, the effect of glass crystallization on glass durability can be predicted with an uncertainty less than 50% if the residual glass composition is within the range of the PCT model.

INTRODUCTION

Precipitation of nepheline ($NaAlSiO_4$ or $Na_2O \cdot Al_2O_3 \cdot 2SiO_2$) in glass weakens the network structure of the residual glassy phase by removing three moles of glass former (Al_2O_3 and $2SiO_2$) for every mole of glass modifier (Na_2O). It adversely affects chemical durability of the glass waste form [1-4]. Recently, it was found that for over 200 high-level waste (HLW) glasses subjected to canister centerline cooling (CCC), the glasses that precipitated nepheline were those whose Na_2O-Al_2O_3-SiO_2 (NAS) submixtures were within the nepheline primary phase field (NP field) defined by the Na_2O-Al_2O_3-SiO_2 ternary phase diagram (cf. Figure 1) [5,6]. Therefore, nepheline precipitation in HLW glass during canister cooling limits the loading of HLW streams, which are rich in Na_2O and Al_2O_3.

Effects of glass constituents on nepheline formation in HLW glass have not been fully elucidated in the literature on crystallization kinetics for HLW glasses [1, 7]. This study focused on compositional effects on nepheline precipitation in HLW glasses during CCC treatment. A glass with 57 wt% averaged C106 tank waste [8] was used as a baseline. The component effects were evaluated using the method of single component addition (one component-at-a-time design). The composition ranges of the components studied were (wt%): Na_2O (13.3 - 21.2), Al_2O_3 (6.3 - 19.29), SiO_2 (33.4 - 47.4), B_2O_3 (0 - 20), Li_2O (0 - 8), K_2O (0.1 - 10), CaO (0 - 10), and Fe_2O_3 (5.95 - 12.95).

EXPERIMENTAL

The glass test matrix was designed to change one component-at-a-time in the baseline composition, NP-BL, while keeping the remaining components in constant proportions (Table 1). Chemical reagents (oxides, boric acid, carbonates, and sodium phosphate) were used to prepare the glasses. Glasses were prepared using the procedure described elsewhere [4]. Chemical durability of quenched and CCC-treated glasses was measured using powdered samples (particle size ranged from 75 to 150 μm). The tests followed the procedure of the product consistency test (PCT) [9], namely a ratio of the sample surface area to the volume of water is 2000 m^{-1}; the test was conducted at 90°C for 7 days. Elements released from glass powder were measured using inductively coupled plasma (ICP). Prior to the measurements, the PCT solutions were filtrated through 0.45 μm membrane and then acidified using nitric acid.

Mat. Res. Soc. Symp. Proc. Vol. 465 © 1997 Materials Research Society

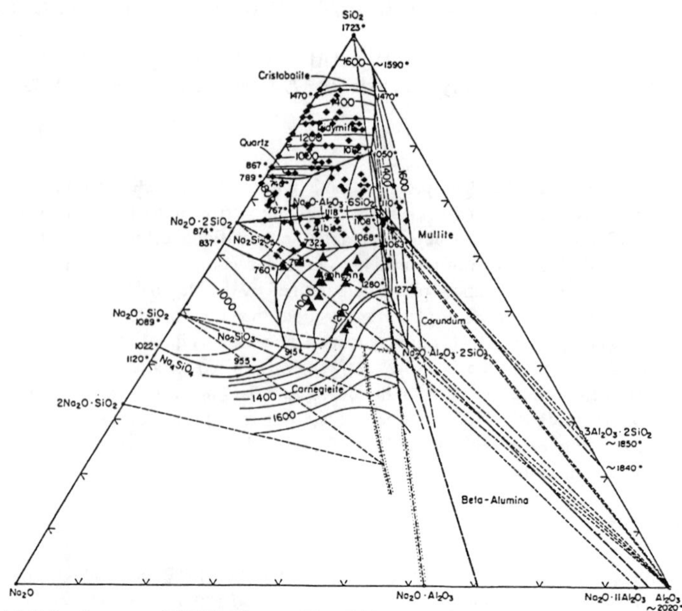

Figure 1. NAS Submixtures of HLW Glasses That Did (▲) and Did Not (◆) Precipitate Nepheline During CCC Treatment [6].

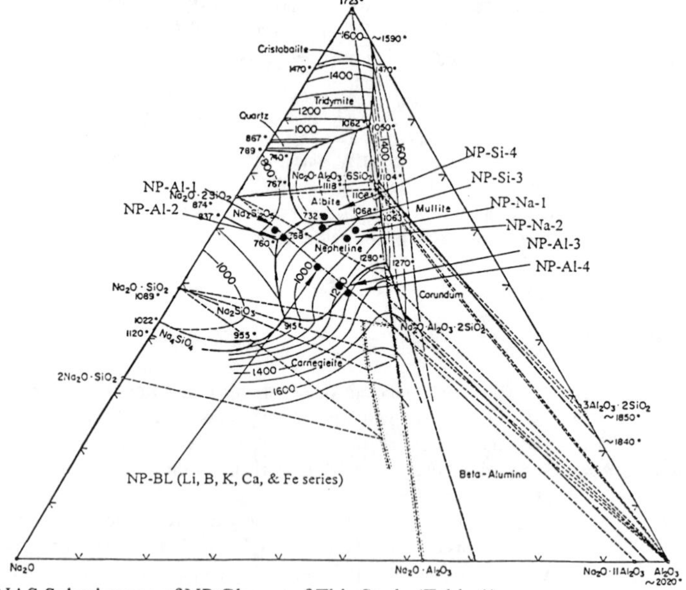

Figure 2. NAS Submixtures of NP Glasses of This Study (Table 1).

Table 1. Nominal Compositions of "Nepheline" (NP) Test Matrix of HLW Glasses

Glass	Components (wt%)										
	SiO$_2$	B$_2$O$_3$	Na$_2$O	Li$_2$O	CaO	MgO	Fe$_2$O$_3$	Al$_2$O$_3$	ZrO$_2$	K$_2$O	Others[a]
NP-BL	38.38	8.00	20.28	4.50	1.12	0.68	9.94	13.76	0.29	0.10	2.86
NP-Si-3	45.31	7.10	18.00	3.99	0.99	0.60	8.82	12.21	0.26	0.09	2.54
NP-Si-4	47.40	6.83	17.31	3.84	0.96	0.58	8.49	11.75	0.25	0.09	2.44
NP-Al-1	41.70	8.69	22.03	4.89	1.22	0.74	10.80	6.30	0.32	0.11	3.11
NP-Al-2	41.06	8.56	21.69	4.81	1.20	0.73	10.63	7.75	0.31	0.11	3.06
NP-Al-3	36.64	7.64	19.36	4.30	1.07	0.65	9.49	17.68	0.28	0.10	2.73
NP-Al-4	35.92	7.49	18.98	4.21	1.05	0.64	9.30	19.29	0.27	0.09	2.68
NP-Na-1	41.74	8.70	13.30	4.89	1.22	0.74	10.81	14.96	0.32	0.11	3.11
NP-Na-2	41.01	8.55	14.81	4.81	1.20	0.73	10.62	14.70	0.31	0.11	3.05
NP-Li-1	40.19	8.38	21.23	0.00	1.17	0.71	10.41	14.41	0.30	0.10	3.00
NP-Li-2	36.98	7.71	19.53	8.00	1.08	0.66	9.58	13.26	0.28	0.10	2.75
NP-B-1	41.72	0.0	22.04	4.89	1.22	0.74	10.80	14.96	0.32	0.11	3.11
NP-B-2	33.38	20.0	17.63	3.91	0.97	0.59	8.64	11.97	0.25	0.09	2.48
NP-K-1	37.27	7.77	19.69	4.37	1.09	0.66	9.65	13.36	0.28	3.00	2.78
NP-K-2	36.12	7.53	19.08	4.23	1.05	0.64	9.35	12.95	0.27	6.00	2.69
NP-Ca-1	38.82	8.09	20.51	4.55	0.00	0.69	10.05	13.92	0.29	0.10	2.89
NP-Ca-2	34.94	7.28	18.46	4.10	10.0	0.62	9.05	12.52	0.26	0.09	2.60
NP-Fe-1	40.12	8.36	21.20	4.70	1.17	0.71	5.95	14.38	0.30	0.10	2.99
NP-Fe-2	39.27	8.19	20.75	4.60	1.15	0.70	7.95	14.08	0.30	0.10	2.93
NP-Fe-3	37.14	7.74	19.62	4.35	1.08	0.66	12.95	13.31	0.28	0.10	2.76

(a) Others are Ag$_2$O, BaO, Cr$_2$O$_3$, MnO, NiO, P$_2$O$_5$, PbO, Sb$_2$O$_3$, SeO$_2$, and TiO$_2$.

Precipitation of nepheline in glass was studied by heat treating glass in a platinum box with a dimension of 25 mm × 25 mm × 25 mm. The particle size was less than 2 mm. Each glass was first brought to its melting temperature and held for 30 min and then cooled according to the selected CCC schedule (Figure 3), i.e., a schedule with a starting temperature, T_S, close to the glass melting temperature. Some glasses were heat treated using two different cooling schedules to examine the effects of cooling on the nepheline precipitation in glass. At the conclusion of each treatment, a polished cross-section of the sample was examined by optical microscopy and scanning electron microscopy/energy dispersive spectrometer (SEM/EDS). Crystalline phases in the CCC-treated glass were identified by powder x-ray diffraction analysis (XRD), including semi-quantitative estimates of the volume fraction of total crystalline phases and the relative concentrations of the phases within the crystallized portion. The experimental errors on the crystallinity estimated by XRD were within 50%.

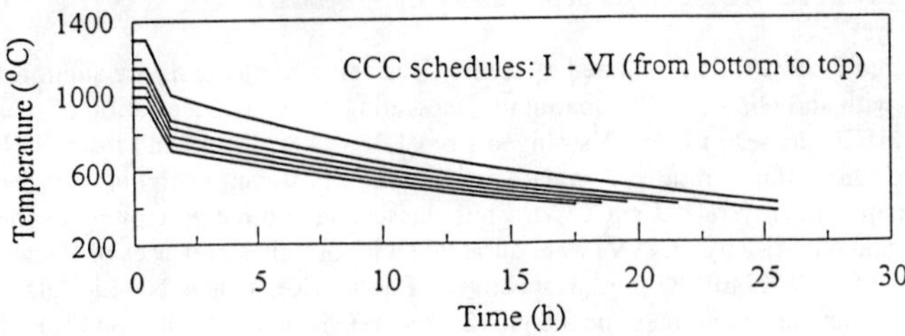

Figure 3. Simulated Canister Centerline Cooling Schedules.

RESULTS AND DISCUSSION

Nepheline Precipitation in Glass During CCC Treatment

Table 2 summarizes the XRD results for the CCC-treated NP glasses. The concentration of nepheline in the glass was calculated using the mass balance equation

$$m_j = f_j [1 + (v_x^{-1} - 1) (\rho_r / \rho_x)]^{-1} \tag{1}$$

where m_j is the j-th crystalline phase concentration in the partially devitrified glass, v_x is the total crystalline phase volume fraction, f_j is the j-th crystalline phase concentration within the crystallized portion, and ρ_r and ρ_x are the densities of the residual glassy phase and the composite of crystallized phases, respectively. Both v_x and f_j were determined by semiquantitative XRD analysis. In Equation (1), ρ_x was determined using the equation:

$$\rho_x = (\sum_{j=1}^{k} f_j / \rho_j)^{-1} \tag{2}$$

where ρ_j is the j-th crystalline phase density. In the calculation of the residual glass composition for the CCC-treated NP glasses, v_x values were adjusted allowing 0.05 wt% Al_2O_3 in the glassy phase whenever Equation (3) yielded a negative Al_2O_3 concentration in the residual glassy phase. This adjustment was made for all glasses with v_x values greater than 0.2. The adjustment varied from 10 to 40% of the reported v_x values.

As Table 2 shows, nepheline formation significantly increased with increasing Al_2O_3, and was significantly less in glasses with lower Na_2O, whereas K_2O had little effect. The effect of SiO_2 was opposite to Al_2O_3 and Na_2O; nepheline precipitation was completely suppressed in NP-Si-3 with higher SiO_2 (by 7 wt%) as compared to the baseline (NP-BL). In the group of nepheline forming oxides, nepheline precipitation increased in the order $Al_2O_3 > Na_2O > K_2O > SiO_2$ (K is commonly found in natural nepheline minerals). In the group of components that do not form nepheline, increasing B_2O_3 and CaO or removing Li_2O from glass suppressed nepheline precipitation. However, the change of Li_2O concentration of the baseline glass from 4.5 to 8.0 wt% seemed to have little effect on the nepheline precipitation. Fe_2O_3 also had little effect on nepheline precipitation. Overall, the component effects on increasing nepheline precipitation can be ranked as $Al_2O_3 > Na_2O > Li_2O \approx K_2O \approx Fe_2O_3 > B_2O_3 > CaO > SiO_2$ (approximated by the change of the calculated nepheline mass fraction in the CCC-treated baseline glass with the addition of 1wt% oxide).

Nepheline precipitation was dependent on the cooling schedule; CCC treatments with lower starting temperatures precipitated more nepheline in the glass (cf. Table 2). As in the previous study [6], glasses with NAS submixtures within the NP field were prone to nepheline formation. However, the NP boundaries of the NP glasses were shifted slightly outside the NP field of the Na_2O-Al_2O_3-SiO_2 ternary.

Chemical Durability of NP Glasses (Quenched and CCC-Treated)

The averaged 7-day PCT normalized releases of the selected elements are summarized in Table 3 for the NP glasses with and without CCC treatment. Depending on the concentration of nepheline crystals in glass, the normalized releases of B and Na ranged from 1.78 to 48.19 g/m² and from 1.17 to 28.01 g/m², respectively. In addition, the normalized release of Cr was significantly higher for glasses with high concentrations of nepheline crystals. The CCC-treated glasses that did not precipitate nepheline (NP-Si-3, NP-Si-4, NP-Al-1, and NP-Al-2 by CCC-V) were durable; their normalized releases of B and Na varied from 0.28 to 1.73 g/m² and from 0.49 to 1.87 g/m², respectively. For the CCC treated NP-Li-2 glass (8 wt% Li_2O), which precipitated a high concentration of nepheline (29.1 wt%), the normalized B and Na releases were not significantly different from the quenched samples. This may be attributed to the precipitation of lithium

Table 2. XRD Analysis of Crystalline Phases and the Calculated Nepheline Concentration in the CCC-Treated NP Glasses

Glass	Treatment	Determined by XRD			Calculated
		v_X (vol fraction)	$f_{Nepheline}$ (wt%)	$f_{Other\ Phases}$ (wt%)	$m_{Nepheline}$ (wt%)
NP-BL	CCC-V	0.4	85	5 Na(Si$_3$Al)O$_8$ (a) 10 Mn$_3$B$_2$O$_{6+x}$ (a)	29.4
	CCC-III	0.4	>95	< 5 Li$_2$SiO$_3$ (a) 0	30.2
NP-Si-3	CCC-V	0.01	0	unkown	0
NP-Si-4	CCC-V	0.01	0	unkown	0
NP-Al-1	CCC-V	0.01	0	100 Ag	0
	CCC-I	0.01	(b)	100 Ag	N/A(b)
NP-Al-2	CCC-V	0.01	0	100 Ag	0
	CCC-I	0.05	100	0	5
NP-Al-3	CCC-V	0.5	85	5 Na(Si$_3$Al)O$_8$ (a) 10 Mn$_3$B$_2$O$_{6+x}$ (a)	37.9
	CCC-IV	0.55	95	5 Na$_3$Bi(PO$_4$)$_2$ (a)	38.8
NP-Al-4	CCC-V	0.5	80	10 Na(Si$_3$Al)O$_8$ (a) 10 Mn$_3$B$_2$O$_{6+x}$ (a)	40.2
NP-Na-1	CCC-V	0.2	75	5 SiO$_2$ 15 spinel 5 Mn$_3$B$_2$O$_{6+x}$ (a)	18.7
NP-Na-2	CCC-V	0.4	90	1 SiO$_2$ < 5 spinel < 5 Mn$_3$B$_2$O$_{6+x}$ (a)	32.3
NP-Li-1	CCC-VI	0.01	45	55 Ag	1
NP-Li-2	CCC-I	0.6	55	35 Li$_2$SiO$_3$ < 5 Li$_8$SiO$_6$ (a) 10 spinel	29.1
NP-B-1	CCC-V	0. 6	10	90 NaAlSiO$_4$	32.8
NP-B-2	CCC-I	0.01	0	100 Ag	0
NP-K-1	CCC-III	0.45	> 95	< 5 SiO$_2$ 1 Li$_2$SiO$_3$	29.3
NP-K-2	CCC-III	0.5	95	5 Li$_2$SiO$_3$ (a)	28.4
NP-Ca-1	CCC-III	0.45	90	10 Li$_2$SiO$_3$ (a)	30.5
NP-Ca-2	CCC-II	0.35	75	25 spinel	4.8
NP-Fe-1	CCC-V	0.35	90	5 Li$_2$SiO$_3$ 5 unknown	31.5
NP-Fe-2	CCC-V	0.3	90	5 Li$_2$SiO$_3$ 5 unknown	28.4
NP-Fe-3	CCC-V	0.45	70	5 Li$_2$SiO$_3$ 15 spinel 10 Fe$_2$O$_3$	29.2

(a) Phase identification was based on the best match available.
(b) Nepheline crystals were detectable using an optical microscope but its concentration was below XRD detection limit.

Table 3. The 7-Day PCT Normalized Elemental Releases (g/m^2) of NP Glasses (the calculated nepheline concentrations, m_{Neph}, in the CCC-treated glasses are included)

Glass	Treatment[a]	m_{Neph} (wt%)	B	Si	Li	Na	K	Ca	Cr	pH
NP-BL	Quenched	0.0	0.54	0.32	0.32	0.78		0.05	0.15	11.55
	CCC-V	29.4	43.24	2.91	12.48	21.01		0.00	6.62	12.47
	CCC-III	30.2	42.38	3.10	13.88	20.61		0.09	4.91	12.43
NP-Si-3	CCC-V	0.0	0.36	0.24	0.34	0.49		0.08	0.16	11.23
NP-Si-4	CCC-V	0.0	0.28	0.21	0.29	0.41		0.08	0.12	11.08
NP-Al-1	CCC-V	0.0	1.64	0.58	1.21	1.79		0.02	0.27	12.01
	CCC-I	N/A	1.73	0.61	1.20	1.87		0.00	0.22	11.97
NP-Al-2	CCC-V	0.0	0.66	0.48	0.98	1.44		0.03	0.28	11.92
	CCC-I	5.0	9.10	2.00	5.32	6.59		0.00	0.66	12.34
NP-Al-3	Quenched	0.0	0.38	0.27	0.27	0.61		0.05	0.17	11.45
	CCC-V	37.9	44.73	0.59	12.19	16.86		0.00	14.43	12.35
	CCC-IV	38.8	46.90	1.30	12.87	17.49		0.00	17.23	12.34
NP-Al-4	Quenched	0.0	0.37	0.26	0.28	0.60		0.05	0.21	11.39
	CCC-V	40.2	44.94	0.93	17.59	15.41		0.00	20.58	12.34
NP-Na-1	CCC-V	18.7	1.60	0.05	0.94	0.40		0.00	0.02	11.43
NP-Na-2	CCC-V	32.3	22.04	0.85	12.11	5.97		0.00	0.01	11.72
NP-Li-1	Quenched	0.0	0.29	0.19	N/A	0.43		0.02	0.05	11.12
	CCC-VI	1.0	0.76	0.20	N/A	0.59		0.03	0.27	10.94
NP-Li-2	Quenched	0.0	1.13	0.51	0.91	1.56		0.06	0.33	12.02
	CCC-I	29.1	1.78	0.11	0.57	1.54		0.02	1.54	12.53
NP-B-1	Quenched	0.0	N/A	0.47	0.75	1.90		0.03	0.41	12.20
	CCC-V	32.8	N/A	1.04	6.84	1.17		0.06	7.67	12.53
NP-B-2	Quenched	0.0	4.27	0.19	3.63	2.85		0.00	0.10	10.98
	CCC-I	0.0	4.35	0.23	3.62	2.94		0.00	0.12	10.83
NP-K-1	Quenched	0.0	1.06	0.45	0.83	1.23	0.55	0.07	0.31	11.84
	CCC-III	29.3	48.19	3.56	12.10	27.06	3.32	0.24	16.80	12.57
NP-K-2	Quenched	0.0	1.76	0.48	1.61	1.81	0.96	0.03	0.33	12.10
	CCC-III	28.4	45.22	3.44	10.96	28.01	11.37	0.18	22.62	12.67
NP-Ca-1	Quenched	0.0	0.81	0.39	0.45	0.98		0.00	0.35	11.70
	CCC-III	30.5	46.68	3.02	8.76	21.22		0.00	0.00	12.45
NP-Ca-2	Quenched	0.0	0.70	0.27	1.07	1.32		0.01	0.16	12.05
	CCC-II	4.8	1.31	0.32	2.48	1.66		0.02	0.33	12.15

(a) See Figure 3 for CCC schedules.

silicates, Li$_2$SiO$_3$ (or Li$_2$O·SiO$_2$), and possibly Li$_8$SiO$_6$ (or 4Li$_2$O·SiO$_2$). According to the empirical model [10], the removal of two moles of Li$_2$O for every mole of SiO$_2$ improves glass durability.

A second-order mixture model [10], in terms of 7-day PCT normalized B release, was used in this study. The model was developed for optimizing Hanford HLW glass formulation. The second-order mixture model is in a form of

$$R_B = \exp\left[\sum_{i=1}^{n} b_i r_i + \sum_{i=1}^{n-1} \sum_{j=i+1}^{n} b_{ij} r_i r_j\right] \quad (3)$$

where R_B is the 7-day PCT normalized B release (g/m^2), b_i and b_{ij} are the first-order and second-order coefficient (g/m^2), respectively, and r_i and r_j are the mass fractions of the i-th and j-th component in the residual glass, respectively. Table 4 lists the b_i and b_{ij} values determined by the model. The residual glass composition, r_i, was determined from the mass balance equation:

$$r_i = (g_i - \sum_{j=1}^{k} c_{ji} m_j) \, / \, (1 - \sum_{j=1}^{k} m_j) \tag{4}$$

where r_i is the i-th component concentration in residual glass, g_i is the i-th component concentration in original glass, and c_{ji} is the i-th component concentration in the j-th crystalline phase.

Using the calculated r_i values and a second-order empirical model [10], the 7-day PCT normalized B releases were predicted and plotted in Figure 4 against the measured values of several CCC-treated glasses (NP-BL, NP-Al-2, NP-Al-3, NP-Al-4, NP-Na-1, NP-Li-2, NP-K-1, NP-K-2, NP-Ca-1, and NP-Ca-2). The predicted values generally agreed with the measured values within an uncertainty less than 50%. However, the normalized B release from the CCC-treated NP-Li-2 glass was over predicted 54 times. This discrepancy may be attributed to the residual glass composition far outside the composition region for which the model was derived. A less likely cause is an error in the XRD analysis: v_X would have to be reduced by ~ 60% to obtain the normalized B release value comparable to the measured. One can conclude that the second-order mixture model for the 7-day PCT normalized B release can be used for assessing glass durability if the nepheline content can be estimated.

Maximizing waste loading for nepheline-limited HLW requires that three models work together: i) a model for nepheline precipitation as a function of glass composition, ii) a model for nepheline crystallization kinetics as a function of glass composition and temperature history, and iii) a model for glass durability. The NAS model for nepheline precipitation and a second-order mixture model for glass durability have been developed and successfully applied to HLW glasses. A model describing nepheline precipitation kinetics is required.

Table 4. Second-Order Mixture Model for 7-day Normalized B Release of Hanford HLW Glasses

Component, i	b_i (g/m^2)	Components, i x j	b_{ij} (g/m^2)
SiO$_2$	- 4.38	Al$_2$O$_3$ × Al$_2$O$_3$	104.22
B$_2$O$_3$	- 3.08	B$_2$O$_3$ × B$_2$O$_3$	76.66
Na$_2$O	21.44	SiO$_2$ × MgO	117.17
Li$_2$O	24.17	Na$_2$O × CaO	- 120.88
CaO	14.02	B$_2$O$_3$ × CaO	- 91.14
MgO	- 49.73	MgO × ZrO$_2$	122.09
Fe$_2$O$_3$	- 1.31		
Al$_2$O$_3$	- 39.58		
ZrO$_2$	- 11.42		
Others	4.21		

(a) The model has a regression coefficient (R^2) 0.9012

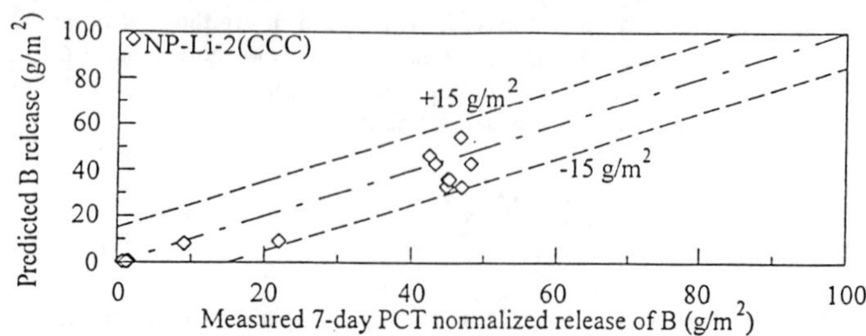

Figure 4. Comparison Between the Measured and Predicted 7-day PCT Normalized B Releases from the CCC-Treated Glasses.

CONCLUSIONS

This study confirmed that the NP field defined by the Na_2O-Al_2O_3-SiO_2 ternary system can be used for screening HLW glasses prone to nepheline formation, though its boundaries for HLW borosilicate glasses are shifted somewhat outside the ternary NP field (additional testing is needed to find their locations). Component effects on increasing nepheline precipitation can be ranked as $Al_2O_3 > Na_2O > Li_2O \approx K_2O \approx Fe_2O_3 > B_2O_3 > CaO > SiO_2$. The presence of nepheline in glass was detrimental for chemical durability. Using a second-order mixture model for 7-day PCT normalized B release, the effect of nepheline on glass durability was predicted with an uncertainty of less than 50% if the residual glass composition is within the range of the PCT model validity. A model for nepheline crystallization kinetics is needed for complete prediction.

ACKNOWLEDGMENT

This work is supported by the U.S. Department of Energy under Contract DE-AC06-76RLO 1830.

REFERENCES

[1] C.M. Jantzen and D.F. Bickford in <u>Scientific Basis for Nuclear Waste Management, VIII,</u>·ed by C.M. Jantzen, J.A.Stone, R.C. Ewing (Mater. Res. Soc. Proc. 44, Pittsburgh, PA, 1980) p. 135-146.

[2] D.S. Kim, D.K. Peeler, and P. Hrma in <u>Ceramic Transaction</u>, Vol. 61, ed. by V. Jain and R. Palmer, (Am. Ceram. Soc. Proc. Westville, OH, 1995) p. 177 - 185,

[3] A. Bailey and P. Hrma, <u>Ceramic Transaction</u>, Vol. 61, ed. by V. Jain and R. Palmer, (Am. Ceram. Soc. Proc. Westville, OH, 1995) p. 549-556.

[4] H. Li, P. Hrma, M.H. Langowski, J. Hlavac, O. Vojtech, V. Krestan, and P. Exnar in <u>Ceramic Transction,</u> (Am. Ceram. Soc. Proc. Westville, OH, 1996) in press.

[5] J.D. Vienna (private communication)

[6] P. Hrma, J.D. Vienna, M.J. Schweiger, D.E. Smith, H. Li, and B. Peng, Report No. 0.2.1.2.4.02A, Pacific Northwest National Laboratory, WA, 1996.

[7] R.P. Turcotte, J.W. Wald, and R.P. May, in <u>Scientific Basis for Nuclear Waste Management,</u> Vol.2, ed. By C. J.M. Northrup, Jr. (Plenum Press, New York, 1980) p.141 - 146.

[8] S. Lambert and G.E. Stegen, and J.D. Vienna, Report No. WHC-SD-WM-TI-768 Rev.0, Westinghouse Hanford Company, Richland, WA, 1996.

[9] C.M. Jantzen and N.E. Bibler, Report No. WSRC-TR-90-539 Rev. 1, Savannah River Laboratory, Aiken, South Carolina, 1990.

[10] P. Hrma, G.F. Piepel, M.J. Schweiger, D.E. Smith, D.S. Kim, P.E. Redgate, J.D. Vienna, C.A. LoPresti, D.B. Simpson, D.K. Peeler and M.H. Langowski, Report No. PNL-10359, Pacific Northwest National Laboratory, Richland, WA, 1994.

COLLOID FORMATION DURING THE INTERACTION OF HLW GLASS WITH INTERSTITIAL CLAY WATER

J. Wei and P. Van Iseghem
SCK•CEN, Boeretang 200, B-2400 Mol, Belgium

ABSTRACT

During the reaction of HLW glass with interstitial clay water at different temperatures with various ratios of glass surface area to solution volume (SA/V) and durations, Eu released from the glass forms predominantly Eu-humate colloids (organic colloids) by a complexation reaction. The size distribution and stability of Eu-humate colloids have been characterized. It is likely that inorganic colloids which are mainly composed of Si, Al and Ca are generated from the corrosion of waste glass by a nucleation reaction.

INTRODUCTION

Colloids are considered to be important in the migration and transport of radionuclides under repository condition. Colloid formation also controls the soluble concentration of the radionuclides in the near field.

Many studies have been carried out on colloid generation and migration in natural systems in the absence of nuclear waste form [1]. Some work on colloid generation and properties has been done in the presence of waste form under repository-relevant conditions [2-10]. Spallation and nucleation have been proposed to be the mechanism of colloid formation during the waste glass reaction [4, 9]. The colloids are identified as clay colloids (inorganic colloids) [4,5,8,9]. For actinides released from the waste glasses, the dominant Pu and Am species in solution are found to be colloidal materials [2-4], whereas Np is considered to be mainly in the ionic form [3,4,10]. The formation of actinide-bearing colloids causes the total concentrations of actinides in leachates to be orders of magnitude higher than the dissolved concentrations [3,4]. The size of Pu, Am and Np colloids is normally < 0.4 μm [2, 3, 7]. The glass matrix elements Si and Al also generate colloidal particles in the leachate [11, 12].

In Belgium, Boom clay is studied as the candidate host rock for geological disposal. The characteristics of Boom clay water are different from the groundwaters collected from other potential disposal sites in Europe [13]. It contains relatively high humic substances (150mg/l). Humic acids can be present in the colloidal form in natural water, and they strongly interact with trivalent and tetravalent metal ions [14-15].

This study attempts to characterize the colloids formed during the reaction of HLW glass with clay water. The results should enhance the knowledge about the radionuclide containing colloids.

EXPERIMENTAL

The PAMELA glass SM527, which is a reference glass for high enriched waste concentrates resulting from the Eurochemic reprocessing activities, was used in this work. Its composition [16] is special compared to that of other HLW reference glasses [17]. It has a particularly high Al_2O_3 content (19.96 wt%). 1 wt% Eu_2O_3 (non-radioactive) is added to the glass, Eu is an analogue of Am.

Mat. Res. Soc. Symp. Proc. Vol. 465 © 1997 Materials Research Society

Different leachants, Synthetic Interstitial Clay Water (SIC), humic acids (HA) free Interstitial Solution (IS) and Real Interstitial Boom Clay Water (RIC) were selected in leach tests. SIC is used to simulate RIC for laboratory research work [16]. Its composition is given in [16]. The composition of IS is the same as SIC but without humic acids. The detail information about Boom clay water can be found in [18]. Boom clay water was sampled from clay layer over 200 m below the surface. pH of the waters is between 8.7 and 9.0. The leach test programme is summarized in Table 1. The type of glass and the leach tests under high SA/V condition were selected because earlier studies showed that under these conditions colloids were readily formed in solution[19].

Table 1. Leach test programme of the study

Temperature(°C)	SA/V(m^{-1})	Leachant	Duration(days)
40, 90, 150	4000, 6500, 10000, 11500	SIC, IS, RIC	30, 73, 180

The high SA/V conditions were achieved by using powdered glass with grain size of 125-250 μm. The test procedure is similar to the MCC-3 test [20]. The powder was fed into 6 ml solution. The Teflon® containers were kept in the ovens at 40°C and 90°C in the Ar glove box under anoxic condition. The leach tests at 150°C were conducted in open air because this temperature is too high for glove box. All tests were performed in duplicate.

For the purpose of comparison, Eu_2O_3 (> 99.99%) was dissolved in SIC and IS at room temperature in the Ar glove box. A stock solution ([Eu(III)]=2.5×10^{-3} M, pH=1.4) was prepared by dissolving Eu_2O_3 in 1N HNO_3 and diluting with distilled water, and the solution was added to SIC and IS. These tests were designed to study the Eu colloid formation during the reaction of Eu_2O_3 and Eu(III) ions with SIC and IS. 0.1 g Eu_2O_3 was added to 10 ml clay waters, or 0.08 g of 2.5×10^{-3} M $Eu(NO_3)_3$ stock solution was added to 25 ml clay waters.

After the tests were terminated, the solutions were filtered or ultrafiltered through a series of filters with pore size ranging from 450 nm to 1.5 nm (Nuclepore and Amicon filters) to characterize the colloid size distribution. The leachates were analysed by inductively-coupled plasma atomic emission spectroscopy (ICP-AES). The HA concentration was determined by spectrophotometer at 280 nm.

RESULTS AND DISCUSSION

<u>Humic acids colloids upon leaching in SIC and RIC</u>

The SIC and RIC leachates were filtered through filters with the following size: 450 nm, 100 nm, 15 nm, 1.5 nm. About 70% and 60% of humic acids is present as colloidal particles in SIC and RIC respectively. About 90% of humic acid colloidal particles are smaller than 15 nm. Kim [21] indicated that about 94% of DOC (mainly humic substances) is smaller than ca. 1 nm in one of the Gorleben groundwaters. Humic acid colloids in SIC and RIC are fairly stable versus time at room temperature, but their stability decreases with the increase of temperature. The solutions of SIC and RIC were kept in the oven for 73 days at 40°C, 90°C and 150°C. At 150°C, most of humic acids are decomposed, and colloid concentration is very low. Below 150°C we did not observe degradation of the humic acids. It has been shown that all proton exchange sites like carboxylic groups and phenolic OH groups are degraded over 150°C [22].

<u>Colloid formation during the corrosion of glass</u>

Table 2 presents Eu concentrations (mg/l) in various leachates. These leach tests were performed at 40°C and 90°C, SA/V=4000 m^{-1}. It is clear that the leaching solutions have an important effect on Eu colloid formation. The leach rate of Eu is the greatest and Eu colloidal species are predominant in SIC. The leach rate of Eu is the smallest and colloid concentration is very low in IS. Eu leaching and colloid formation in RIC is in between SIC and IS. The phenomenon should be attributed to humic acids. The complexation of Eu with humic acids is much stronger in alkaline condition (the average leachate pH=9.3) than in acid condition [23,24]. The enhanced release of certain metallic elements from waste glasses in the presence of humic acids has been reported by other authors[25-27]. Humic acids may enhance the release of some metal elements in waste glasses. As shown by spectrophotometry, humic acids are present mainly as colloidal particles in SIC and RIC. Eu thus forms complexes with humic acids to generate Eu-humate colloids (organic colloids). The formation mechanism of metal-humate colloids has been studied by Choppin [28]. The author pointed out cations can complex with humic acids to form colloidal particles because of the cation charge. This mechanism may be different from that in our study because there are a lot of humic acids colloids in SIC and RIC. The complexing effect of cations is not as important in SICand RIC. Eu colloid concentration is much lower in RIC than that in SIC, although the HA concentration is nearly the same. Henrion [18] has found that the organic colloidal phase in Boom clay water contains large amounts of Ca, Fe and Al. As a result, humic acids in RIC have lower complexation capacity.

Table 2. Eu concentrations (mg/l) leaching from glass SM527

Leachant	Filtration	Duration (days)					
		30		73		180	
		40°C	90°C	40°C	90°C	40°C	90°C
SIC	450 nm	0.76	1.18	0.8	1.41	1.00	1.09
	1.5 nm	< 0.03	< 0.03	0.09	< 0.03	< 0.03	< 0.03
IS	450 nm	0.08	0.13	0.10	0.18	0.05	0.39
	1.5 nm	0.07	< 0.03	0.07	< 0.03	< 0.03	< 0.03
RIC	450 nm	0.20	0.39	0.39	*	0.42	*
	1.5 nm	< 0.03	< 0.03	< 0.03	*	0.04	*

* Not performed

Other experimental observations are that Eu colloid concentration is higher at 90°C than at 40°C and that SA/V has little effect on Eu colloid formation. The concentrations and size distribution of humic acids during the leach tests are almost constant, at both temperatures.

For further understanding the effect of humic acids on Eu colloid formation, Eu_2O_3 and the stock solution $Eu(NO_3)_3$ were added to SIC and IS at room temperature. Table 3 and Table 4 list the Eu concentration (mg/l) in solution. Table 3 indicates more Eu_2O_3 was dissolved and more Eu colloids were formed in SIC due to the presence of humic acids. Table 4 illustrates that free Eu(III) ions will be turned into Eu colloidal particles in SIC, but the change does not occur in IS. It is reasonable to assume that Eu colloids present in SIC solution are Eu-humate colloids, and humic acids play a significant role in the release and colloid formation for Eu. Eu present in Gohy-2227 groundwater is complexed with humic substances and behaves as 'humic colloids'[15]. That result corresponds to our results.

Table 3. The dissolution of Eu_2O_3 in SIC and IS (Eu concentration in mg/l)

Solution	Filtration	Duration(days)	
		14	180
SIC	450 nm	6.13	6.16
	1.5 nm	0.96	0.48
IS	450 nm	0.33	*
	1.5 nm	0.27	*

* Not performed

Table 4. The size distribution of Eu(III) species in SIC and IS**

Solution	Filtration	Eu concentration [mg/l]	% of colloidal fraction
SIC	450 nm	1.21	89.3
	1.5 nm	0.13	
IS	450 nm	0.77	7.8
	1.5 nm	0.71	

** The stock solution was diluted with 50ml SIC and IS. The resulting solutions were left to equilibrate for four days and solution pH was adjusted to 9.30.

The size distribution of Eu-humate colloids was determined by ultrafiltration method. At least 70% of colloids are smaller than 15 nm. This result agrees with the size distribution of humic acids in SIC. Eu concentration in SIC leachate was monitored versus time. The leachate had been kept for 4 months at room temperature. This experiment showed that Eu-humate colloids were quite stable during the 4-month period. Feng [9] reported most of inorganic colloids formed during the waste glass corrosion would settle out of the solution very soon. Hence, organic colloids might be more stable than inorganic colloids in our case.

We observed Si, Al and Ca colloidal particles in the SIC leachate during the leach tests (Table 5). Boron never forms colloids even at very high concentration (1.7 M). As the reacting temperature increased, the colloid inventories of Si, Al and Ca increased also.

Lemmens and Van Iseghem[11] found that Si and Al could form colloidal particles when glass SM527 was contacted with distilled water or SIC at 90°C. Although Al and Ca can form complexes with HA, the complexation of Eu with HA is much stronger [13]. At 150°C, the Eu colloid concentration is quite low because of the decomposition of HA, but there is a large quantity of Al and Ca colloids in the leachate. Si, Al and Ca colloidal particles are probably formed as inorganic colloids by nucleation reactions. As shown in Table 5, the percentage of colloidal fraction of these elements decreases with the leaching duration. Because salt content in leachate increases with the reaction time, colloidal particles agglomerate and lead to eventual sedimentation [5, 9]. Based on X-ray diffraction measurements, these colloidal particles are probably zeolite ($Na_{5.7} Al_{5.7} Si_{10.3} O_{32}$

12H$_2$O). Buck at al. [5] have reported that the colloidal phases formed upon the corrosion of waste glasses are zeolite, heulandite, smectite, dolomite and calcite. For leach tests at 40°C and 90°C, it is possible inorganic colloids as well as organic colloids (Eu-humate colloids) are present in the leachate at the same time. The interaction of two kinds of colloids is not clear. Eu ions are possibly adsorbed by inorganic colloidal materials, SiO_2, Fe_2O_3, Al_2O_3, FeOOH, $Fe(OH)_3$ [29-31]. But the strong interaction of Eu with humic acids could prevent the adsorption of Eu ions to inorganic colloids [29]. Therefore, the sorption of Eu ions on Si, Al and Ca colloids in SIC leachates may not be significant.

Table 5. Dissolved Si, Al and Ca concentrations and percentage of colloids in the leachate[***]

Temperature	Duration (days)	Filtration	Si		Al		Ca	
			C*	%**	C*	%**	C*	%**
90°C	73	450 nm	27.6	38	17.8	40	n.m.	
		1.5 nm	17.1		10.6		n.m.	
	180	450 nm	29.4	18	19.0	16	3.82	38
		1.5 nm	24.0		15.9		2.35	
150°C	30	450 nm	112	48	35.5	54	5.92	51
		1.5 nm	57.9		16.3		2.89	
	73	450 nm	79.9	34	407	49	196	34
		1.5 nm	52.6		205		128	
	180	450 nm	105	22	161	11	42.2	7.8
		1.5 nm	81.9		143		38.9	

C*   Concentration [mg/l]
%** The percentage of colloidal fraction
***   The leach tests were conducted in SIC at SA/V=4000m^{-1}.

<u>Calculation of Eu species in SIC and IS</u>

Eu can form complexes with OH$^-$, CO_3^{2-}, and HA in the solution. The complexation constants are summarized in [13]. Most of complexation constants for Eu(III) with HA were determined at acidic condition to avoid hydrolysis reaction. Maes et al. [23,24] have obtained the complexation constants in alkaline condition (pH=9.0) and found that commercial humic acids (Fluka) has the same complexation capacity with Boom clay humic acids (purified from Boom clay). The complexation constant for Eu with HA in alkaline condition is six orders of magnitude higher than that in acidic condition [23]. Dierckx [32] has provided the experimental evidence of mixed complex formation of Eu(III) with HA and OH$^-$ or CO_3^{2-} ($Eu(OH)_2HA$ and $Eu(CO_3)_2HA$) under in-situ condition. The concentrations of HA and carbonate in SIC were calculated to be 3×10^{-4} eq/l and 1.5×10^{-2} mole/l, respectively. The calculated speciation for Eu(III) is given in Table 6. The calculations were performed for a temperature of 25 °C. The complexation of humic acids with competing cations, Al and Ca, which are present in the SIC leachate, is not taken into account since their complexation constants are not available yet at high pH. Eu mixed complexes are not considered. The calculation shows almost 100% of Eu(III) species are $Eu(CO_3)_2^-$ in IS solution, but Eu(HA) is the predominant species in SIC solution. The results in Table 4 indirectly support these findings. Table 4 suggests Eu(III) forms complexes with humic acids colloids to produce Eu-humate colloids in SIC solution. Eu-carbonate complexes in IS solution usually do not form colloidal phase, as suggested by the low colloidal fraction (Table 4). At this stage of our research we do not calculate the concentration of the various Eu species, because we do not know the concentration of free Eu-ions in the solution.

Table 6. Calculated speciation of Eu(III) in SIC and IS (in mole fraction)

Solution	pH	Eu^{3+}	Eu(HA)	$Eu(CO_3)^+$	$Eu(CO_3)_2^-$	$Eu(OH)_i^{3-i}$
SIC	9.3	6×10^{-11}	0.96	6×10^{-6}	0.04	6×10^{-9}
IS	9.3	2×10^{-9}	/	2×10^{-4}	1.0	2×10^{-7}

* $Eu(OH)_i^{3-i}$, i=1, 2, 3, 4, $\log\beta_{Eu(HA)}$=13.69[23]

CONCLUSION

Eu-humate colloids (organic colloids) and probably inorganic colloids are generated during the interaction of waste glass with claywater. Leachant has a significant effect on Eu colloid formation, and the reacting temperature of glass corrosion can also affect Eu-humate colloid formation. At 150°C, the highest concentration of Si, Al and Ca colloids was observed. Eu-humate colloids were formed through complexation and inorganic colloids were generated by nucleation. A vast majority of the Eu-humate colloids were found to have an average size smaller than 15 nm. Eu-humate colloids were stable for at least 4 months. This study implies that organic colloids of trivalent actinides leached from the waste glasses will likely be formed under repository condition, and trivalent actinides will have a preferential release and have higher solubility due to the presence of humic acids.

For modelling purposes the source term for radionuclide release from glass should include colloidal and dissolved actinides fractions. The dissolved fraction is mobile, and is needed for performance assessment calculations. The colloid inventory may not be necessary for performance assessment calculations, but is important for understanding the mass balance of the soluble inventory.

Finally, it seems that SIC is not a good analogue of the real Boom clay water, because of the different leaching characteristics for Eu in both solutions. The higher leach rate of Eu in SIC could result in overestimating the release of trivalent actinides from nuclear waste glasses in repository condition.

REFERENCES

1. J. I. Kim, MRS Bulletin, XIX(11), pp. 47-56 (1994)

2. J.I. Kim, W. Treiber, Ch. Lierse, and P. Offermann, Mat. Res. Soc. Symp. 44, pp. 359-368 (1985)

3. J. C. Cunnane and J. K. Bates, Ceram. Trans. 23, pp. 65-73 (1991)

4. J. K. Bates, J. P. Bradley, A. Teetsov, C. R. Bradley and M. Buchholtz ten Brink, Science 256, pp. 649-651 (1992)

5. E.C. Buck, J. K. Bates, J. C. Cunnane, W. L. Ebert, X. Feng and D. J. Wronkiewicz, Mat. Res. Soc. Symp. Proc. 287, pp. 199-208 (1993)

6. X. Feng, E. C. Buck, C. Mertz, J. K. Bates, J.C. Cunnane and D. J. Chaiko, Proc. Waste Management'93, Feb. 28 to March 4, Tucson, AZ, Vol. 2, pp. 1015-1024 (1993)

7. L. Wang, P. Van Iseghem and A. Maes, Mat. Res. Soc. Symp. Proc. 294, pp. 155-162 (1993)

8. P. A. Finn, E. C. Buck, M. Gong, J.C. Hoh, J. W. Emery, L. D. Hafenrichter and J.K. Bates, Radiochim. Acta 66/67, pp.181-187 (1994)

9. X. Feng, E. C. Buck, C. Mertz, J. K. Bates, J.C. Cunnane and D. J. Chaiko, Radiochim. Acta 66/67, pp.197-205 (1994)

10. K. P.Hart, B. J. Robinson, T. E. Payne, P. Van Iseghem and K. Lemmens, Mat. Res. Soc. Symp. Proc. 353, pp. 841-845 (1995)

11. K. Lemmens and P. Van Iseghem, Mat. Res. Soc. Symp. Proc. 257, PP. 49-56 (1992)

12. W. L. Ebert, Physics and Chemistry of Glasses, 34(2), pp. 58-67 (1993)

13. A. Dierckx, Complexation of Europium with Humic acidss, Influence of Cations and Competing Ligands. PhD thesis, Katholieke Universiteit Leuven, 1995

14. J. I. Kim, G. Buckau and W. Zhuang, Mat. Res. Soc. Symp. Proc. 84, pp. 747-756 (1987)

15. J. P. L. Dearlove, G. Longworth, M. Ivanovich, J. I. Kim, B. Delakowitz and P. Zeh, Radiochim. Acta 52/53, pp. 83-89 (1991).

16. J. Wei and P. Van Iseghem, The Effect of Humic Acids on the Element Release from High Level Waste Glass. These Proceedings.

17. W. L. Bourcier, Critical Review of Glass Performance Modeling. ANL-94/17 (1994)

18. P. N. Henrion, M. Monsecour, A. Fonteyne, M. Put and P. De Regge, Radioactive Waste Management and the Nuclear Fuel Cycle, 6(3-4), pp. 313-359 (1985)

19. P. Van Iseghem and K. Lemmens, Proceedings of an International Symposium on Geologic Disposal of Spent Fuel, High Level and Alpha Bearing Wastes(1992: Antwerp, Belgium), pp. 209-215 (1993)

20. DOE/TIC 11400, Nuclear Waste Materials Handbook, Test Methods. PNL, Richland, Washington (1981)

21. J. I. Kim, Radiochim. Acta 52/53, pp. 71-81 (1991)

22. M. Schnitzer and J. Hoffmann, Geochimica and Cosmochimica Acta 29, pp. 859-865 (1965)

23. A. Maes, J. De Brabandere and Cremers, Radiochim. Acta 44/45, pp. 51-57 (1988)

24. A. Maes, J. De Brabandere and Cremers, Radiochim. Acta 52/53, pp. 41-47 (1991)

25. J. C. Dran, J. Lombardi, M. C. Magonthier, V. Moulin, J. C. Petit and L. Trotignon, Radiochim. Acta 58/59, pp. 17-20 (1992)

26. S. Gin, N. Godon, J. P. Mestre, E. Y. Vernaz and D. Beaufort, Mat. Res. Soc. Symp. Proc. 333, pp. 565-572 (1994)

27. S. Xing, K. S. Matlack and I. L.Pegg, Ceram. Trans. 39, pp. 353-365 (1994)

28. G. R. Choppin, Radiochim. Acta 44/45, pp. 23-28 (1988)

29. V. Moulin and D. Stammose, Mat. Res. Soc. Symp. Proc.127, pp. 723-727 (1989)

30. L. Righetto, G. Bidoglio, B. Marcandalli and I.R. Bellobono, Radiochim. Acta 44/45, pp. 73-75 (1988)

31. A. Ledin, S. Karlsson, A. Duker and B. Allard, Radiochim. Acta 66/67, pp. 213-220(1994)

32. A. Dierckx, A. Maes and J. Vancluysen, Radiochim. Acta 66/67, pp.149-156 (1994)

PHASE SEPARATION IN SIMULATED PLUTONIUM GLASSES WITH PHOSPHATE AND FLUORINE AND THE EFFECT ON GLASS CORROSION IN WATER

H. Li, J.D. VIENNA, Y.L. CHEN, L.Q. WANG, and J. LIU
Pacific Northwest National Laboratory, Box 999, P8-37, Richland, WA 99352

ABSTRACT

The solubility limit of phosphate in glass was found to decrease as fluorine increased. Amorphous phase separation was found in the glass as a result of interaction between phosphate and fluorine, in which Ce and Gd strongly partitioned. The glasses exhibiting amorphous phase separation showed higher normalized B releases according to the 31-day product consistency test (PCT) compared to glasses without phase separation. Phosphate that leached from glass into water increased the concentrations of Ce and Gd in the 7- and 31-day PCT solutions by more than an order of magnitude. The liquid state ^{31}P-NMR results suggested that phosphate in water interacts with Ce or Gd. Therefore, the observed concentration jump for Ce and Gd in the PCT solution may be attributed to the increases in the solubility limits of Ce and Gd as a result of phosphate complexation with Ce and Gd.

INTRODUCTION

One of the options for immobilizing plutonium-bearing materials is vitrification in borosilicate glass, which has a high tolerance to feed variability. This allows the incorporation of both neutron absorbers and/or proliferation deterrents in the same homogeneous form. Phosphate is present in some plutonium waste streams. An early study [1] showed that the phosphate solubility limit in the glass was affected by glass composition; alkali-containing glasses had higher solubility for phosphate than lanthanide borosilicate glasses without alkalis. A dissolution study of low-level waste glasses showed that the presence of phosphate slightly decreased glass durability [2], which was attributed to localized preferential dissolution [3]. Ce as a simulant for Pu in glass was shown to leach into water at about the same rate as Gd (used as a neutron absorber to address criticality concerns) [1]. In terms of simulating the dissolution behavior of Pu, the use of Ce seems to be justified according to the dissolution studies of borosilicate glasses with Pu, Ce, and Gd in water [4,5]. In a previous study [1], it was also found that the concentrations of Ce and Gd that leached from glass containing phosphate were increased by an order of magnitude as compared to those from glasses without phosphate. In this study, the effect of fluorine on the solubility limit of phosphate (in terms of P_2O_5) in glass was investigated. The effect of phase separation on glass durability, in terms of normalized B release from the PCT, and the effect of phosphate on aqueous concentrations of Ce and Gd that released from the glass in water, were further examined.

EXPERIMENTAL

The compositions, both nominal and measured, of a baseline glass, Pu10S-2, and nominal concentrations of P_2O_5 and F, which were added into the baseline glass, are shown in Table 1 (a&b). All glasses were prepared using chemical reagents described elsewhere [1] and homogenized in a Pt-10%Rh crucible for 2 hr at 1350°C at which the melt viscosity was about 5 Pa·s. Each melt was vigorously stirred using a Pt rod after melting for 1 h. Opalescence in the quenched glasses was visually examined to determine the extent of phase separation. Transmission electron microscopy (TEM) was performed on one glass exhibiting phase separation, which contained both fluorine and phosphate. Glass corrosion in water was evaluated using the product consistency test (PCT) [6]. The PCTs were conducted at 90°C for 7 and 31 days, respectively. The ratio of total surface area of sample powder to volume of water (S/V) was 2000 m^{-1}. The concentrations of elements of interest in the PCT solutions were measured using inductively coupled plasma-atomic emission spectroscopy (ICP-AES). The normalized B release [6] (determined by the ratio of

277

Table 1a. Baseline Composition (wt%) of Simulated Plutonium Glass

Glass	Pu10S-2	
Oxide	nominal	measured
SiO_2	54.45	53.29
Al_2O_3	5.18	4.22
B_2O_3	7.27	7.11
BaO	1.04	0.98
Bi_2O_3	1.05	0.87
CeO_2	6.49	6.11
Cs_2O	1.03	0.87
Gd_2O_3	6.86	6.38
K_2O	2.07	2.13
Li_2O	2.08	(2.08)[a]
Na_2O	10.39	10.67
TiO_2	1.04	1.05
ZrO_2	1.04	0.95
Sum	100.00	96.71

Table 1b. The Baseline Glass with Additions of P_2O_5 and F and the Results of Visual Examination for Glass Phase Separation

Other Glasses	P_2O_5 (wt%)	F (wt%)	Opalescence in Quenched Glass
Pu10S-2-PF1	3.23	3.34	very intense
Pu10S-2-PF2	3.14	2.24	intense
Pu10S-2-PF3	2.17	3.36	mild
Pu10S-2-PF4	1.73	3.31	weak
Pu10S-2-P1 [b,c]	3.30	0.00	not detectable [d]
Pu10S-2-P2 [b,c]	3.90	0.00	intense
Pu10S-2-P3 [b,c]	4.90	0.00	very intense
Pu10S-2-F [b,c]	0.00	3.90	very intense

(a) Li_2O could not be measured by WDXRF and the value in parentheses is based on the nominal concentration. (b) These glasses were previously studied using the same baseline glass, the solubility limits of P_2O_5 and F were 3.3 and 3.4 wt%, respectively [1]. (c) The concentrations of P and F were determined by ICP-AES and ion chromatography, respectively. (d) Both quenched and annealed (24 hr at 600°C) samples were visually examined.

elemental concentration in solution to the product of S/V and mass fraction of the element in the glass) was used to determine the effect of phase separation on glass corrosion susceptibility in water. Liquid ^{31}P nuclear magnetic resonance (NMR) experiments were performed on two PCT solutions with and without Ce and Gd using a chemagnetic spectrometer (300 MHz). The ^{31}P NMR spectra were referenced to the 0 ppm signal from 10% H_3PO_4.

RESULTS AND DISCUSSIONS

The concentrations of P_2O_5 and F in the Pu10S-2-PF glasses were not measured, but were considered to be close to the concentrations as batched. Loss of F from the glass due to evaporation at the glass melting temperature is considered to be small for these glasses because F concentrations (as batched) did not exceed the solubility limit, 3.4 wt% F, for the Pu10S-2 glass [1]. For phosphate in the glass, previous studies showed that its loss due to evaporation was not detectable by ICP regardless of its saturation level in the glass [7,8]. Furthermore, the addition of phosphate in glass was found to decrease glass volatility [9]. Therefore, the nominal concentrations of P_2O_5 in the Pu10S-2-PF glasses are expected to be the same as batched.

Amorphous Phase Separation

Amorphous phase separation was found in all four quenched Pu10S-2-PF glasses that exhibited opalescence. The results of visual inspection are shown in Table 1b. The intensity of glass opalescence was relatively lower for the glass portion contacting the pour plate, suggesting phase separation in glass occurred during glass cooling. The solubility limit of P_2O_5 was 3.3 wt% in the Pu10S-2 without F [1]. At a about constant level of F, the intensity of opalescence in the quenched Pu10S-2-PF1, PF3, and PF4 glasses decreased as the concentration of P_2O_5 decreased, illustrating that the solubility limit of P_2O_5 in the glass was lower with the presence of F.

Figure 1 shows TEM micrographs the Pu10S-2-PF2 (with 2.24 wt% F and 3.14 wt% P_2O_5) glass exhibiting phase separation. A diffusive ring pattern from selected area diffraction (SAD) analysis demonstrated that the separated phase was amorphous (not shown in the figure). The matrix and the separated phase were analyzed by electron energy dispersive x-ray spectroscopy analysis (EDXS) as shown in Figure 2. Comparing with the EDXS result of the glass matrix, the separated phase was shown to have both P and F in addition to Na. Interestingly, Ce and Gd strongly partitioned to the separated phase as well as Ti (the EDXS signals of Si and Al in the separated phase might come from the glass matrix). The EDXS results imply that P and F interact with each other, forming larger structure units that are incompatible with the silicate network.

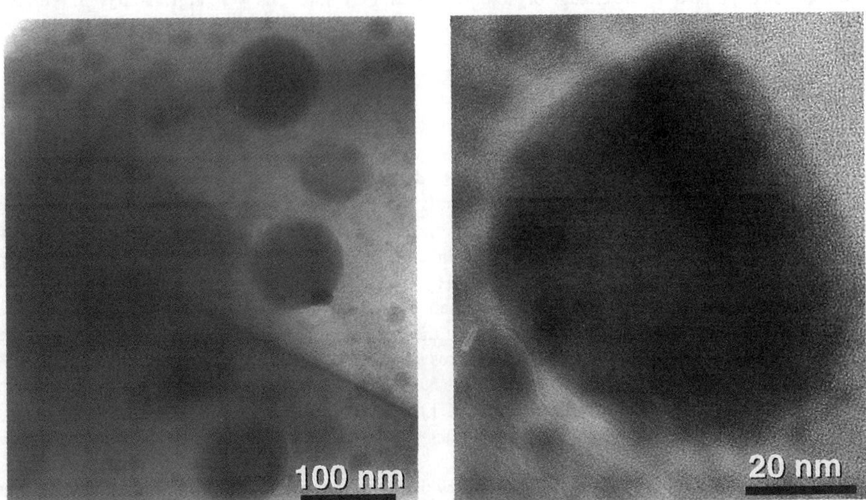

Figure 1. TEM Characterization of Quenched Pu10S-2-PF2 Glass with 2.24 wt% F and 3.14 wt% P_2O_5.

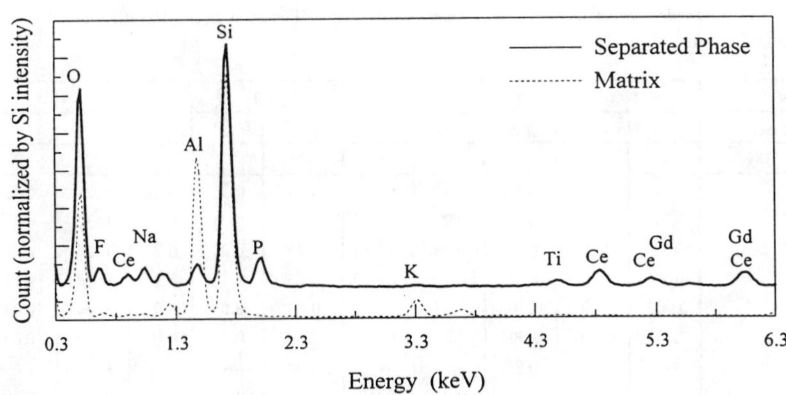

Figure 2. EDXS Spectra of the Glass Matrix and the Separated Phase for Quenched Pu10S-2-PF2 Glass.

Glass Corrosion in Water

Table 2 summarizes the PCT normalized elemental releases for the Pu10S-2 glasses with and without phase separation and solution pH after the PCT. The normalized B release is taken to represent the resistance of the glass network to corrosion by water because its high solubility in water. For all of the Pu10S-2 glasses with and without additions of F and P_2O_5, the 7-day normalized B releases did not change significantly, regardless of the presence of phase separation in the glass. However, the 31-day PCT normalized B releases of the phase separated glasses were significantly higher than the baseline glass, Pu10S-2, (no separated phase). The increase in the 31-day PCT normalized B release in the phase separated glasses varied approximately from 30 to 100%. The observed differences in the normalized B releases between 7 and 31 days could not be explained by the change of solution pH; all PCT solutions were at pH10.0$\pm$0.3. Although EDXS is not capable of detecting B, it is still possible that the separated phase might contain B according to the 31-day PCT results.

In addition, for the Pu10S-2-P2 and Pu10S-2-P3 glasses with higher phosphate concentrations, the 7-day PCT normalized Na releases were significantly higher than the glasses with lower phosphate concentrations. Partitioning of Na in phosphate groups has been extensively studied in silicate glasses [13-15]. High Na releases within a short period of the test (7 days) for the Pu10S-2-P2 and Pu10S-2-P3 glasses exhibiting phase separation is assumed to be a result of selective dissolution of those Na-containing phosphate phase which were preferentially exposed on the surfaces of glass particles during sample crushing. The assumption was indirectly supported by a ^{31}P solid state NMR study on glass particles (a similar type of borosilicate glass) after being corroded in water [16]. In that study, it was found that for the glass with separated Na-phosphate phase, the ^{31}P-NMR signal intensity was substantially reduced for the glass after the 7-day PCT as compared with the intensity observed for the glass prior to the PCT. However, the normalized Na releases of the Pu10S-2-P2, and Pu10S-2-P3 glasses after the 31-day PCT were much lower than those after the 7-day PCT, suggesting the precipitation of the Na-containing phase (scanning electron microscopy was performed on the two glasses after the 31-day PCT). The presence of a precipitated phase was found, yet the chemical analysis of that phase was not conclusive).

The results of the 7- and 31-day PCT also show that the aqueous concentrations of Ce and Gd were increased by one order of magnitude in the glasses with phosphate than those without phosphate, which could not be attributed to the change of solution pH. For the glasses with separated phosphate, the aqueous concentrations of Ce and Gd were even higher. Although the Pu10S-2-PF1 and Pu10S-2-PF2 glasses had

Table 2. PCT Normalized Elemental Releases (g/m^2) from Glasses With and Without Addition of F and P_2O_5

Glass	Pu10S-2		-F	-P1	-P2	-P3		-PF1		-PF2	
Phase Sep.	No		Yes	No	Yes	Yes		Yes		Yes	
Time (days)	7	31	7	7	7	7	31	7	31	7	31
B	0.18	0.19	0.16	0.17	0.16	0.14	0.33	0.18	0.29	0.15	0.20
Li	0.30	0.23	0.21	0.28	0.26	0.22	0.41	0.27	0.31	0.20	0.24
Na	0.26	0.26	0.30	0.33	2.75	2.45	0.37	0.26	0.26	0.21	0.19
K	0.19	0.64	0.15	0.20	0.26	0.15	0.56	0.17	0.24	0.13	0.11
Ce	0.002	0.004	0.001	0.01	0.02	0.04	0.04	0.03	0.04	0.04	0.04
Gd	0.002	0.003	0.001	0.01	0.02	0.03	0.03	0.02	0.02	0.03	0.04
F	----	----	0.20	----	----	----	----	----	0.10	----	0.28
P	----		----	0.07	0.06	0.05	0.04	0.07	0.07	0.05	0.05
pH	10.3	10.4	9.6	10.3	10.2	10.1	10.2	9.7	9.8	9.8	9.9

lower concentrations of P_2O_5 than Pu10S-2-P1, the PCT of the two glasses showed higher aqueous concentrations of Ce and Gd than the latter. The Pu10S-2-F glass also phase separated due to the saturation of fluorine, yet the aqueous concentrations of Ce and Gd were comparable to those for the Pu10S-2 glass without phosphate and phase separation. Therefore, it can be reasoned that the observed increase in the aqueous concentrations of Ce and Gd should be attributed to the amount of phosphate separated in the glass. However, the effect of phosphate, both in the glass and water, on the dissolution of Ce or Gd or their solubility limits in water is still not wellunderstood.

Cerium has two oxidation states, Ce^{4+} and Ce^{3+}, in glass. The Ce^{4+}/Ce^{3+} ratio, 0.57, was determined for the Pu10S-2 glass without phosphate, whereas it was 0.33 ± 0.01 for the Pu10S-2 glasses with 3.3 to 4.9 wt% P_2O_5 [1]. The cerium redox variation may affect the release of Ce from glass into water. However, in this study, the concentration of Ce in the PCT solution is limited by its solubility limit, rather than by the glass network dissolution, because the normalized B release is two orders of magnitude greater than that of Ce. For release of Gd from glass in water, it is also expected that the variation of cerium redox does not affect the glass structure significantly and therefore, the release of Gd should not be affected. These arguments are consistent with a previous result that no effect of Ce^{4+}/Ce^{3+}, varying from 0.32 to 2.04, on the aqueous concentrations of Ce and Gd was found in the 7-day PCT [1]. Therefore, the observed significant increase in the aqueous concentrations of Ce and Gd should be associated with the soluble phosphate group. The TEM/EDXS results shown in Figure 1 and 2 demonstrate that Ce and Gd strongly partitioned to the separated phase rich in P. Hence, it is possible that the higher concentrations of Ce and Gd in the PCT solution were due to the presence of colloids of the separated phosphate phase in the PCT solution. Future study is planned to examine whether (P, Ce, and Gd)-containing colloids exist in the PCT solution.

Another explanation for the observed high concentrations of Ce and Gd in water for glass with phosphate is the formation of P-Ce or P-Gd complex in water. To elucidate chemical environments of P with and without Ce and Gd in water, liquid ^{31}P-NMR experiments were performed on two PCT solutions in this study. The solutions were adjusted to $pH8.5\pm0.1$ for the measurements since the phosphate chemical shift is affected by solution pH. Figure 3 shows the ^{31}P NMR spectra of P chemical shifts for the solutions with and without Ce and Gd. A single P chemical shift at 2.6 ppm was found in the solution without Ce and Gd, while two P environments with the same chemical shift, at 2.5 ppm, were found in the solution with Ce and Gd (the solution from leaching Pu10S-2-P2 glass was used). The difference in chemical shifts of the two sharp P peaks was negligible. The broad P peak found in the solution with Ce and Gd suggested the interaction of P with Ce or Gd, or both. However, whether the interaction that came from the P complexation occurred in water or from the colloids leached from the glass is still unclear.

P complexation of Gd in aqueous solution was studied by Bingler and Byrne [17], which can be described as $Gd^{3+} + HPO_4^{2-} \leftrightarrow GdHPO_4^+$. P complexation of Ce is also expected according to the study by Lee and Byrne [18]. The effect of P complexation with Nd on the solubility of Nd in aqueous solutions was studied for Nd containing borosilicate glasses with and without P by Rai et al [19]. The results suggested that within a pH range between 1 and 9, the solubility of Nd in water was lower for the glass with P than the glass without P. The solubility of Am released form $AmPO_4 \cdot xH_2O$ in water with and without addition of phosphate, PO_4^{3-}, was studied as a function of solution pH by Rai, et al [20]. The overall results suggested that the solubility limits of Am, at pH8 or greater, were higher in a solution with higher concentration of PO_4^{3-}. In summary, the complexation of phosphate with rare earth elements is a complex process, which is affected by solution pH, and types of ligand as well as temperature. Because of the criticality issue for Pu glasses, further studies are planned to investigate the effect of phosphate in glass on the release of Pu and Gd from glass in water.

Glass: 5 wt% P_2O_5

Glass: 3.9 wt% P_2O_5
         6.5 wt% CeO_2
         6.9 wt% Gd_2O_3

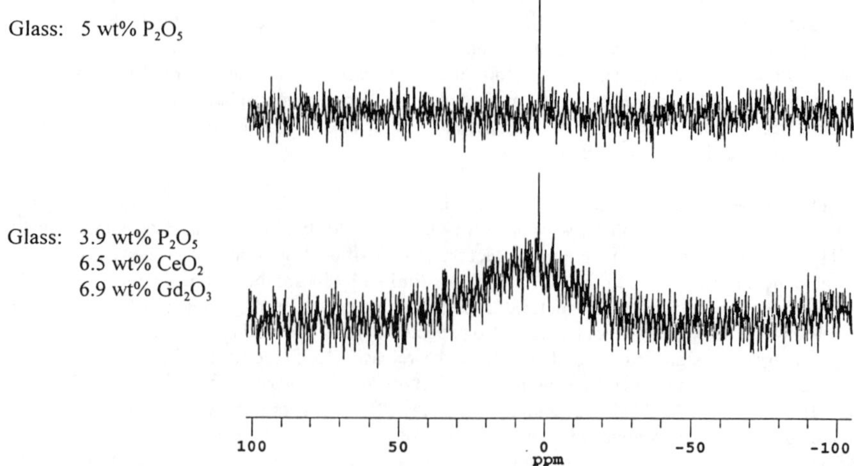

Figure 3. Liquid state ^{31}P NMR spectra for the PCT solutions obtained from leaching phosphate-phase separated glasses with and without Ce and Gd.

CONCLUSIONS

Fluorine was shown to interact with phosphate in glass, which decreased the phosphate solubility in the glass. The interaction of F with P resulted in the separation of amorphous phase containing P, F, Na, Ce, Gd, and Ti. The presence of the separated phase in glass was shown to increase the 31-day PCT normalized B release in water. The concentrations of Ce and Gd in the PCT solutions were increased by an order of magnitude for P-containing glasses and the aqueous concentrations of Ce and Gd were found to be even higher for the glasses with separated phosphate-containing phase.

ACKNOWLEDGMENT

This work is supported by the U.S. Department of Energy under Contract DE-AC06-76RLO 1830. Discussions with Dr. P. Hrma at the Pacific Northwest National Laboratory are appreciated.

REFERENCES

[1]   H. Li, J. D. Vienna, P. Hrma, M.J. Schweiger, D.E. Smith, and M. Gong,
        in <u>Ceramic Transaction</u> (Am. Ceram. Soc. Proc., Westerville, OH, 1996) in press.

[2]   H. Li, J.G. Darab, P.A. Smith, M.J. Schweiger, D.E. Smith, and X. Feng, in
        <u>Nuclear Materials Management,</u> Vol. XXIV (INMM Proc., Northbrook, IL, 1995) p. 460-465.

[3]   H. Li, M. Schweiger, P. Hrma, and X. Feng, in <u>Scientific Basis for Nuclear Waste Management
        XIX,</u> Vol. 412, ed. by W.M. Murphy and D.A. Knecht (Mater. Res. Soc. Proc., Pittsburgh, PA,
        1995) p. 213-220.

[4]   E.Y. Vernaz and N. Godon, in <u>Scientific Basis for Nuclear Waste Management XV</u>, Vol. 257,
        ed. by C.G. Sombret (Mater. Res. Soc. Proc., Pittsburgh, PA, 1992) p. 37-48.

[5]   J.K. Bates, J.W. Emery, J.C. Hoh, and T.R. Johnson, in <u>Scientific Basis for Nuclear Waste Management XIX</u>, Vol. 412, ed. by W.M. Murphy and D.A. Knecht (Mater. Res. Soc. Proc., Pittsburgh, PA, 1995) p. 57-64.

[6]   C.M. Jantzen, and N.E. Bibler, Report No. WSRC-TR-90-539 Rev. 1, Savannah River Laboratory, Aiken, SC, 1990.

[7]   H. Li, J.G. Darab, P.A. Smith, M.J. Schweiger, D.E. Smith, and P. Hrma, in Nuclear Materials Management, Vol. XXIV, (INMM Proc., Northbrook, IL,1995) p. 466-471.

[8]   H. Li, J.D. Vienna, D.K. Peeler, P. Hrma, and M.J. Schweiger, in Proceedings of the Plutonium Stabilization & Immobilization Workshop, CONF-951259, (U.S. Department of Energy, Washington DC, 1995) p. 241-252.

[9]   H. Li, B. Peng, and J.D. Vienna, Pacific Northwest National Laboratory, Richland, WA (unpublished).

[11]   C.W. Ponader and G.E. Brown, Jr., Geochem. Cosmochem. Acta, 53, 2905 (1989).

[12]   F.J. Ryerson and P.C. Hess, Geochem. Cosmochem. Acta, 42, 921 (1978).

[13]   C. Nelson, and D.R. Tallant, Phys. Chem. Glasses, 25, 31 (1984).

[14]   R. Dupree, D. Holland and M.G. Mortuza, Phys. Chem. Glasses, 29, 18 (1988).

[15]   H. Gan, P.C. Hess, and R.J. Kirkpatrick, Geochem. Cosmochem. Acta, 58, 4633 (1994).

[16]   C. Rong, H. Li, P. Hrma, and H. Cho, in <u>Ceramic Transactions</u> (Am. Ceram. Soc. Proc., Westerville, OH, 1996) in press.

[17]   L.S. Bingler and R.H. Byrne, Polyhedron, 8, 1315 (1989).

[18]   J.H. Lee and R.H. Byrne, Geochem. Cosmochem. Acta, 56, 1127 (1992).

[19]   D. Rai, A.R. Felmy, R.W. Fulton, and J.L. Ryan, Radiochimica Acta, 58/59, 9 (1992).

[20]   D. Rai, A.R. Felmy, and R.W. Fulton, Radiochimica Acta, 56, 7 (1992).

Part IV

Cements in Radioactive Waste Management

CHEMISTRY AND PERFORMANCE OF BLENDED CEMENTS AND BACKFILLS FOR USE IN RADIOACTIVE WASTE DISPOSAL

S. Duerden[*], F.P. Glasser[**], K. Goldthorpe[**], J. Pedersen[**], K. Quillin[+], D. Ross[++], S.A. Stronach[**] and M. Tyrer[+++].
[*] Environmental Agency, 2 Marsham Street, London, SW1P 3EB
[**] Department of Chemistry, University of Aberdeen, Old Aberdeen, AB24 3UE Scotland.
[+] Building Research Establishment, Garston, Watford, WD2 7JR, England.
[++] W.S. Atkins, Epsom, Surrey, England.
[+++] Imperial College of Science, Technology and Medicine, London, SW7 2BP, England.

ABSTRACT

The ability of NaCl and $MgSO_4$ to impair the performance of Portland cement, blended cements containing slag and fly ash and of a permeable backfill have been measured. Performance is determined by decrease in pH, changes in mineralogy and loss of physical coherence. Experiments have been made at 25°, 55° and 85°C and extensively backed up by chemical models of cement performance. NaCl, up to 1.5M, has a comparatively slight impact on performance but $MgSO_4$ rapidly and almost quantitatively reacts, lowering system pH's to < 10, conditioned by mixtures of $Mg(OH)_2$ and magnesium silicates with gypsum.

INTRODUCTION

One proposal for the disposal of nuclear waste, currently the focus of much research, is to make use of deep geological disposal at 100-1000m. In this scenario, man made barriers may extensively supplement geological barriers against the return of radionuclides to the biosphere. The man-made barriers include primary containers (steel, concrete), as well as a specially formulated backfill consisting of mixtures of Portland cement, calcium hydroxide and calcium carbonate. Preliminary scenario development suggests that repository temperatures may reach 85°C in the post closure phase while boreholes disclose that the groundwaters percolating such repositories may be strongly saline.

The Environment Agency for England and Wales has commissioned independent research into the future performance of a deep repository. We report here on the man-made cement barriers. The performance of cementitious barriers under these conditions, or similar conditions, is a widespread problem, hence the methodologies used and the conclusions are of general relevance.

Since it is difficult to compress the time factor, a three-strand approach has been taken: experiments on cement powders and blocks exposed to simulate groundwaters; studies of the stability of individual phases in the presence of individual groundwater constituents NaCl, $MgSO_4$ and mixtures of NaCl, $MgSO_4$ - and finally, modelling studies. All three strands of the program used three temperatures, 25°, 55° and 85°C and comparable cement and blended cement compositions.

Many hundreds of experiments were undertaken and in the space of a short publication, it is not practicable to reproduce full details. However, a comprehensive report is expected to become publicly available during the first half of 1997 (1).

EXPERIMENTAL

The complex engineering requirements of encapsulation and repository engineering lead to the use of a range of cement formulations. Table 1 shows details of the formulations which were included in the test program of powders and blocks and which were also used for calculations. For experiments with single cement substances, these conditions were simplified further; three NaCl solutions, 0.5, 1.0 and 1.5M and three $MgSO_4$ solutions, 0.005, 0.01 and 0.05M were used. Additionally, as a worst case scenario, mixed NaCl, $MgSO_4$ solutions were tested.

Mat. Res. Soc. Symp. Proc. Vol. 465 © 1997 Materials Research Society

Single phases, C-S-H (several Ca:Si ratios), ettringite, calcium sulfoaluminate hydrate (AF_m) and $Ca(OH)_2$ were prepared, as reported previously (1).

An accelerated reaction technique has been used to study the reaction chemistry of cements on exposure to high saline groundwaters and elevated temperatures. The technique was based on that used in an earlier programme to study the effects of high sulfate-containing and high carbonate-containing groundwaters on blended Portland cements (2-4). Precured cement powders, prepared using the formulations given in Table 1, were mixed with $100cm^3$ simulated groundwaters (see Table 2) or boiled deionised water and shaken continuously by end-over-end tumbling at temperatures of 25°, 55° and 85°C in sealed bottles for periods of between 7 days and 20 months. A separate sample was used for each test age. At the end of each period the mixes were filtered under nitrogen and the solids dried to constant weight over saturated LiCl solution.

Table 1. Cement Formulations Used in the Present Study

Designation	Notes
Ordinary Portland Cement (OPC)	Similar to an ASTM Type 1
75% Blast Furnace slag - 25% OPC (BFS - OPC)	Ground, glassy granulated iron blast furnace slag.
60% Fly Ash 40% OPC (PFA - OPC)	Fly ash is an ASTM class F (alumino silicate) type.
40.3% OPC, 15.5% $Ca(OH)_2$, 44.2% $CaCO_3$ "backfill"	The $Ca(OH)_2$ and $CaCO_3$ were commercial grades containing 72.6% and 55.1% CaO respectively.

A parallel series of tests were also carried out in which precured cement paste blocks (prepared using a w/c ratio of about 0.8 for all compositions other than the backfill for which the w/c ratio was 0.67) were exposed to groundwater or deionised water for test periods of between 7 days and 18 months at temperatures of 25°, 55° and 85°C. A separate sample was used for each test age. After each test age blocks were dried and inspected visually for signs of degradation.

Reaction products were characterised by X-ray diffraction and analytical electron microscopy. Ettringite was determined quantitatively by X-ray diffractometry using a technique based on that of Crammond (5). Calcite and calcium hydroxide were also determined quantitatively using thermogravimetric analysis. Aqueous phase analyses were performed by appropriate techniques: flame photometry for Na, K and Ca, atomic absorption and colorimetric methods for Al and Si, ion chromatography for oxyanions. Cl was determined using a potentiometric Volhard titration technique. The pH values of all solutions were also determined. The most difficult analytical problems were the characterisation of amorphous, gel-like phases, removing colloids from aqueous solution (usually, 0.45μm filters were used), measuring very high pH's (usually done by titration, to avoid electrode errors) and in avoiding uptake of carbon dioxide in the course of experiments lasting 6 - 20 months.

MODELLING

The use of numerical models is essential in extending our understanding of the behaviour of cementitious systems beyond the spatial and temporal scales possible by experimental observation. The well established thermodynamic equilibrium code PHREEQE (6) has been employed in this work to predict both pore solution chemistry and hydrate phase stability in a range of natural groundwaters. At high ionic strengths, conventional extensions of the Debye-Hückel relationship (such as the Davies equation) offer inadequate means of ionic strength correction. In order to use PHREEQE at high salinity (0.3 - 2 mol dm^{-3}) in mixed electrolyte systems, the code was modified to incorporate the specific interaction theory algorithms (SIT) developed during the recent European Commission CHEMVAL-2 programme (7). This method required re-casting of the available thermodynamic data in order to be consistent with the modelling approach and the resulting database will be published (1).

Table 2. Simulate Groundwater Compositions

	pH	Chemical Contents, ppm					
		Ca	Mg	Na	K	Cl	SO$_4$
A	6.6-7.0	2910	310	76490	490	120840	3390
B	7.8	1050	149	7980	142	15200	1300

The incongruous solubility of the C-S-H solid solution series has been simulated by equilibrium thermodynamics by a number of workers (8-10). These methods divide the solid solution series into discrete regions of specific calcium to silicon ratio and simulate the incongruous dissolution of C-S-H through the co-equilibrium of pairs of congruously soluble salts. The method described by Atkins *et al.* (9) has been adopted in this study and extended to model the solubility behaviour of C-S-H in highly saline systems. The phases used in the present model have the same stoichiometry as true cement hydrates but their dissociation constants have been empirically derived in order to simulate the changing solubility of C-S-H as a function of salinity. This was done by mapping the predicted iso-concentration contours for calcium and silicon in solution onto axes of the apparent log K for each pair of salts. Where a salt is used in more than one model, the mean dissociation constant was chosen, and used to estimate the changes in solubility as a function of salinity. Table 3 (below) shows these derived log K values and the expressions which describe their variation with ionic strength at 25°C.

Table 3. Solubility Relationships Between Structural Components in C-S-H and Salinity.
SIT = Specific Interaction Theory

Ca : Si ratio of CSH in experiments	Stoichiometric phases and dissociation reaction relating to log K$_{diss}$	SIT compatible log K$_{diss}$ (dissociation constants) I = Ionic strength, mol dm^{-3}
0.85	Tobermorite Ca(5) H$_4$SiO$_4$(6) H$_2$O(2) H(-10) Silica H$_4$SiO$_4$(1) H$_2$O(-2)	70.177 + 1.104 log (I) -6.613 - 0.616 log (I)
1.1	Tobermorite Ca (5) H$_4$SiO$_4$(6) H$_2$O(2) H(-10) Afwillite Ca(3) H$_4$SiO$_4$(2) H$_2$O(2) H(-6)	66.502 - 0.373(I) - 0.258 (I)^2 51.508 - 0.015(I) + 0.206 (I)^2
1.1	Tobermorite Ca(5) H$_4$SiO$_4$(6) H$_2$O(2) H(-10) Fictive solid Ca(1) H$_2$SiO$_4$(6)	66.502 - 0.373(I) - 0.258 (I)^2 -8.528 + 0.438(I) - 0.283 (I)^2
1.4	Tobermorite Ca(5) H$_4$SiO$_4$(6) H$_2$O(2) H(-10) Afwillite Ca(3) H$_4$SiO$_4$(2) H$_2$O(2) H(-6)	66.502 - 0.373(I) - 0.258 (I)^2 51.508 - 0.015(I) + 0.206 (I)^2
1.4	Afwillite Ca(3) H$_4$SiO$_4$(2) H$_2$O(2) H(-6) Fictive solid Ca(1) H$_2$SiO$_4$(6)	51.508 - 0.015(I) + 0.206 (I)^2 -8.528 + 0.438(I) - 0.283 (I)^2
1.8	Afwillite Ca(3) H$_4$SiO$_4$(2) H$_2$O(2) H(-6) Portlandite Ca(1) H (-2) H$_2$O(2)	51.508 - 0.015(I) + 0.206 (I)^2 25.5
Portlandite only	Portlandite Ca(1) H (-2) H$_2$O(2)	22.819 - 0.418(I) + 0.091(I)^2

Comparing the observed and predicted pore solution chemistries shows that this approach is valid for simple systems containing synthetic C-S-H minerals in sodium chloride solutions. Calculation has subsequently been used to examine a range of blended cements and a cement-stabilised limestone backfill. Results from the latter are presented here; table 4.

The backfill hydrate mineralogy was predicted using the normative model CEMCHEM (10) which shows that the system will be dominated by calcium carbonate, calcium hydroxide and C-S-H. Following a prolonged thermal pulse in the repository, any ettringite (AF$_t$-SO$_4$) which may have initially formed from the cement, is expected to convert to monosulfate (AF_m), but, owing to the low bulk aluminium content, they constitute a minor phases.

Thermodynamic equilibrium was simulated between the mineral assemblage described above and the two groundwaters present beneath the Sellafield region. These compare directly with the results of the powder-groundwater experiments described above and the most saline case is described here. At 25°C, the pH of the system is buffered by the portlandite component of the C-S-H to around 12.3, with predicted calcium and silicon solution concentration of $7.2x10^{-2}$ and $6.4x10^{-6}$ moles per litre respectively. The equilibrium solution is shown to be oversaturated with respect to a range of AF_m and AF_t type phases. As is to be expected in a saline system, the chloride-bearing phases are the most oversaturated and sequential equilibrium simulations with each phase suggests that both the mono-sulfo and mono-chloro aluminates are expected to precipitate in this system along with minor quantities of Kuzel's salt, $3CaO.Al_2O_3.0.5CaSO_4.0.5CaCl_2.12H_2O$. The first two aluminates were detected experimentally but the latter salt is very close to its phase boundary so the amounts present may be insufficient to give detectable diffractions.

Table 4.   Predicted Mineral Hydrates Expected to Form on Complete Hydration at Elevated Temperature; from ref. (1)

COMPONENT	Percentage by mass of blending component				
OPC	*40.3	28	28	50	50
Lime	15.5	25	10	25	10
Limestone	44.2	47	62	25	40
PHASE	Predicted mole percentage of phase on complete hydration at 85°C				
HT	0.42	0.33	0.33	0.48	0.49
AF_m-SO_4	0.52	0.35	0.37	0.64	0.67
$Ca(OH)_2$	39.50	43.78	25.90	56.11	38.37
C-S-H (Ca:Si = 1.7, Al = 0.5)	14.17	9.74	10.17	17.04	17.84
Hydrogarnet (Si = 0.5)	0.82	0.53	0.58	0.98	1.05
$CaCO_3$	44.58	45.27	62.65	24.76	41.62

* UK Patent Application 9316995.1: Filing date 16 Aug. 1993

RESULTS

<u>Single Cement Substances</u>

$Ca(OH)_2$ is relatively unaffected by the presence of NaCl. Fig. 1 shows the measured solubilities of $Ca(OH)_2$ in NaCl, 0.5 - 1.5M at 25°, 55° and 85°C contrasted with measured values for initially pure H_2O. The NaCl remains essentially constant, so the charge on soluble Ca is effectively balanced by OH^-; hence pH remains high. In general, $Ca(OH)_2$ solubility decreases with increasing temperature but increases with increasing chloride content with the net result that $Ca(OH)_2$ is approximately as soluble in 1.5M NaCl at 85°C as in initially pure water at 25°C.

C-S-H gel is incongruently soluble in water and NaCl solutions. Fig. 2 shows data points for Ca and Si solubilities. Equilibrium is attained rather rapidly at 85°C, probably within a few days or less. If, however, much longer times, up to 1 year, are allowed for reaction the C-S-H gel begins to crystallise. The reaction sequence is shown in Fig. 3; in experiments where little total dissolution occurred i.e. the Ca:Si ratio of the gel was not significantly altered, tobermorite began to crystallise at 25°C in low-ratio gel. However, in no cases, even at 85°C where the kinetics of reaction are more rapid than at 25°C or 55°C, did complete crystallisation occur. Therefore, aqueous pH's tend to be conditioned by gel-like phases and/or $Ca(OH)_2$: significant pH reductions occur only when $Ca(OH)_2$ is absent, at low Ca:Si ratios.

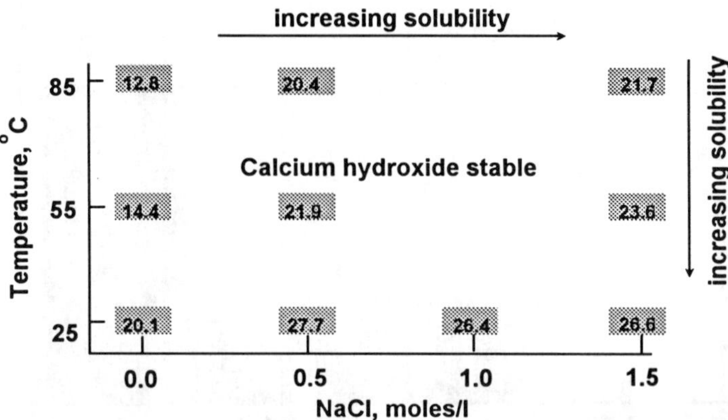

Figure 1.   Summary of solid and aqueous analyses of $Ca(OH)_2$ solubility (mmol/l) in NaCl at 25°, 55° and 85°C

Both $Ca(OH)_2$ and C-S-H react with $MgSO_4$. The $MgSO_4$ concentrations used in these experiments were rather low and, if not renewed, decreased effectively to zero as reaction proceeded. Calculation of the equilibrium threshold for the reaction:

$$Ca(OH)_2 + Mg^{2+}{}_{(aq)} \Leftrightarrow Mg(OH)_2 + Ca^{2+}{}_{(aq)}$$

discloses that the reaction should be displaced towards $Mg(OH)_2$ and $[Mg] \geq 10^{-7}M$ at 25°C, thus corroborating the experimentally observed depletion of the aqueous phases in Mg. This meant that, in practise, the aqueous phase either had to be totally renewed or spiked to maintain [Mg] at the target value: replenishment with spikes of solid $MgSO_4$ was used. The simple base exchange, of Mg for Ca, is inadequate to describe the reaction since sulfate is also removed, but not as effectively as is Mg, by

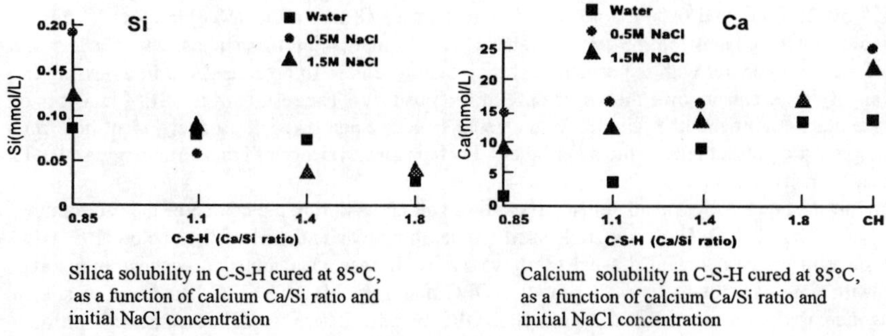

Silica solubility in C-S-H cured at 85°C, as a function of calcium Ca/Si ratio and initial NaCl concentration

Calcium solubility in C-S-H cured at 85°C, as a function of calcium Ca/Si ratio and initial NaCl concentration

Figure 2.   Solubility of calcium silicate hydrate in NaCl, 25°-85°C. Ca, shown in the right hand diagram is much ore soluble than silicon, left hand diagram.

formation of gypsum, $CaSO_4.2H_2O$. When C-S-H is used instead of $Ca(OH)_2$, reaction follows the same progression but silica gel is also present. In summary, therefore, $Ca(OH)_2$ and C-S-H react strongly and rapidly with $MgSO_4$: the products of reaction, $Mg(OH)_2$, $CaSO_4.2H_2O$ and silica gel (from C-S-H), condition aqueous pH's to a maximum of ~ 10.0.

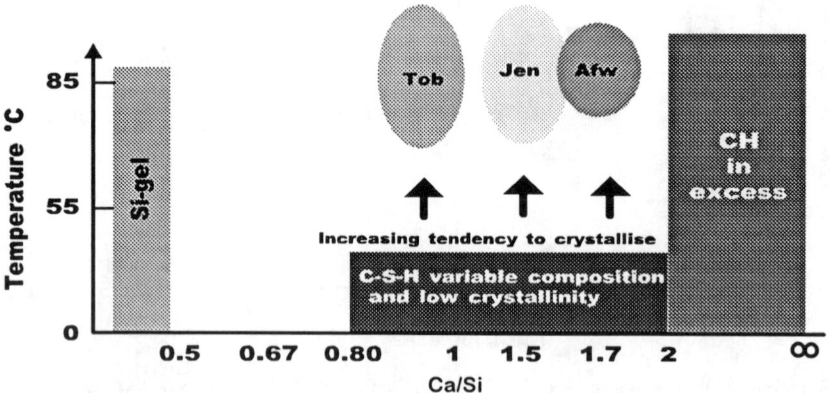

Figure 3. Mode-of-occurrence diagram showing products of crystallisation of calcium silicate hydrate (C-S-H) gels. In compositions at very high or very low Ca:Si ratios, $Ca(OH)_2$ and SiO_2 gel, respectively, also persist.

<u>Cement powder and paste block studies</u>

Phases present in cement powder tests prepared using groundwater A (see Table 2) at the end of the test are given in Table 5. Similar results were obtained using groundwater B although these have not been included here. Many of these phases are commonly formed in cements hydrated under normal conditions. However, the use of accelerated reaction conditions and elevated temperatures led to the formation of several other phases by the end of the test program. C-S-H gel crystallised to form afwillite in OPC and backfill mixes and a tobermorite-like phase in the 75% bfs:25% OPC mix at elevated temperatures. The zeolite minerals thomsonite and philipsite were detected at elevated temperatures in 60% pfa:40% OPC blends. The presence of high saline groundwater led to the formation of a Cl-containing AF_m phase in reference backfill, OPC and 75% bfs:25% OPC mixes. Large amounts of ettringite were also formed in 75% bfs:25 OPC and 60% pfa:40% OPC mixes at 25°C and in the 75% bfs:25 OPC mix at 55°C due to the sulfate content of the groundwater. The formation of these phases will affect the composition of water in contact with the cement and they therefore need to be considered in assessing the likely behaviour of these cements over the service life of a repository. The results of the SEM analyses of the cement paste blocks show that the samples have a zonal structure and reaction products, identified from the powder samples, are present only in the outer layer. There is also evidence of cracks running parallel to the sample faces in many cases.

The pH of the aqueous leachants fell significantly in all cases during the course of the programme, particularly in 60% pfa:40% OPC blends at elevated temperatures where the pH fell to below 10. OPC and reference backfill mixes maintained a pH of above 12 in all cases. Calcium ion concentrations also generally fell with time particularly in 60% pfa:40% OPC and 75% bfs:25% OPC blends. The silicon concentrations increased with time in all 60% pfa:40% OPC blends. These trends are consistent with the formation of crystalline calcium silicate hydrates in the Si and Al-rich formulations. Magnesium ion

Table 5. Solid Phases Formed in Powder Tests Using Groundwater A after 18 months

Type	°C		AF$_t$ (%)	CH (%)	Cc (%)	C-S-H	HG	GH	HT	FS	Gyp	Other
75% bfs: 25% OPC	25	DI	nd			x			x			HC, MS?
		GW	21.8			x				x		
	55	DI	nd			x	x	?	x			
		GW	27.3			x	?			x		
	85	DI	nd			Tob/Afw	x		x			
		GW	nd			x	x			x	x	
60% pfa: 40% OPC	25	DI	nd			x		x				HC, MS?
		GW	48.6			x		x				
	55	DI	nd			x		x				Th?
		GW	nd			x		x		?		Th
	85	DI	nd			x		x				Th, Phi
		GW	nd			x					x	Th
OPC	25	DI	9.2	17		?	x					MC?
		GW	13.8	11		?	x					
	55	DI	12.3	21		Afw	x					MC?
		GW	12.1	18		?	x			x		
	85	DI	nd	24		?	x					MC?
		GW	nd	14		?	?			x	x	
Ref. BF	25	DI	2.4	21	36.6	?	x					MC?
		GW	7.5	14	28.2	?	?			x		
	55	DI	3.1	23	36.6	Afw	?					MC?
		GW	8.2	14	28.9							
	85	DI	nd	20	40.2	?	x					
		GW	nd	13	25.6		?			x		

Key :

Th	Thomsonite	CH	Ca(OH)$_2$	GH	Gehlenite hydrate
Phi	Philipsite	Cc	CaCO$_3$	HT	Hydrotalcite
AF$_t$	Ettringite	HG	Hydrogarnet	MS	Monosulfate
Afw	Afwillite	Tob	Tobermorite	MC	Monocarbonate
Gyp	Gypsum / Anhydrite	HC	Hemicarbonate	FS	Friedel's Salt
GW	simulate ground water	DI	deionised water		
x	present	?	probably present		

concentrations were significantly lower than in the initial (unreacted) groundwater in all cases. Sodium and chloride ion concentrations were comparable with those in the unreacted groundwater. Sulfate levels were lower than those in the unreacted groundwater, especially at temperatures below 55°C. Aqueous sulfate levels generally correlate well with ettringite content of the paste.

The sulfate ion concentrations at   85°C were also lower than in the unreacted groundwater although no obvious sulfate-containing phases could be identified. Sulfate ions may have been taken up by the C-S-H gel.

CONCLUSIONS

The performance of blended Portland cements, exposed to highly saline groundwaters and elevated temperatures, has been studied using experimental and modelling techniques. A method is now available through which thermodynamic calculations of the solubility of C-S-H phases may be undertaken in highly saline systems. The results of the study show that the vault backfill is predicted initially to form calcium monocarboaluminate; as reaction progresses, Friedel's salt and calcium monosulfoaluminate develop; magnesium progressively displaces calcium and the pH decreases.

Studies carried out using powdered cement and cement paste blocks have provided further data concerning the development of the solid and aqueous phases on hydrating blended Portland cements. They have shown that, in addition to the phases formed in hydrating these cements under normal conditions, the long reaction times, aggressive groundwaters and elevated temperatures give rise to additional phases. The composition of the pore solution in contact with backfill is initially strongly buffered by the portlandite and C-S-H phases to a pH in excess of 12. Crystallisation of C-S-H somewhat reduces the high pH buffering reserve although $Ca(OH)_2$ is unaffected. However, magnesium ions exchange with calcium, lowering the pH. Moreover, magnesium salts permeating the pores isolate mineral surfaces from reaction by forming overgrowths, typically of brucite. In this case, a relatively rapid overall lowering of the near field pH may result as phases which might be expected to buffer a high pH become progressively more isolated from physical contact with the permeating aqueous phase. If sulfate is present, gypsum contributes to layer formation. Progressively longer contact times are required to equilibrate fresh portions of simulate groundwaters to high pH even when the pH buffering phases have been only partially reacted. Thus, Mg salts and especially Mg in conjunction with sulfate, are much more deleterious to the longevity of the high pH conditioning induced by cement than is NaCl.

ACKNOWLEDGEMENT

This work has been carried out as part of a collaborative research programme on radioactive waste management carried out at the Building Research Establishment of the UK Department of Environment, the University of Aberdeen, WS Atkins and the Imperial College of Science, Technology and Medicine. The work is supported by the Environment Agency for England and Wales, and the authors thank them for their permission to publish this paper. The results of this work will be used in the formulation of government policy but views expressed in this paper do not necessarily represent government policy.

REFERENCES

1. F.P. Glasser, J. Pedersen, K. Goldthorpe, M. Atkins, M. Tyrer, D. Bennett, D. Ross & K. Quillin, 'The chemistry of blended cement and backfill for radioactive waste disposal'. Rep UKEA (in preparation)
2. K.C. Quillin, S.L. Duerden & A.J. Majumdar, 'Accelerated ageing of blended Portland Cement for use in radioactive waste disposal'. UK DoE report DoE/HMIP/RR/93.023 (1993)
3. K.C. Quillin, S.L. Duerden & A.J. Majumdar, ' Formation of zeolites in OPC - pfa mixtures'. Cem. Conc. Res. 23, 991 (1993)
4. K.C. Quillin, S.L. Duerden & A.J. Majumdar, ' Accelerated ageing of blended OPC cements'. Mat. Res. Soc. Symp. Proc., Vol 333, 341 - 348 (1994)
5. N.J. Crammond, 'Quantitative X-ray diffraction analysis of ettringite, thaumasite and gypsum in concretes and mortars'. Cem. Conc. Res., 15, 431   (1985)
6. D.L. Parkhurst, D.C. Thorstenson & N.L. Plummer, 'PHREEQE - A computer program for geochemical calculations'. Rep. USGS WRI-80-96 (1985)
7. W.E. Falck, D. Read & J.B. Thomas, 'Chemval 2: thermodynamic database'. Final report. European Commission Contract no FI2W/CT92/0122
8. U.R. Berner, 'Modelling the incongruent dissolution of hydrated cement minerals'. Radiochimica Acta 44/45, 387-393 (1988)
9. M. Atkins, F.P. Glasser, L.P. Moroni & J.J. Jack, 'Thermodynamic modelling of blended cements at elevated temperatures'. Rep DoE/HMIP/RR/94.011 (1993)
10. A. Atkinson, J.A. Hearne & C.F. Knights, 'Aqueous chemistry and thermodynamic modelling of $CaO-SiO_2-H_2O$ gels at 80°C'. Rep DoE/HMIP/RR/91/045 (1991)

INFLUENCE OF THE DEGRADED SURFACE LAYER ON THE LONG-TERM BEHAVIOR OF CEMENT PASTES

P. FAUCON*, J.F. JACQUINOT**, F. ADENOT***, J. VIRLET*, J.C. PETIT*
*Service de Chimie Moléculaire, CEA Saclay,
91191 Gif sur Yvette, France.
**Service de Physique de l'Etat Condensé, CEA Saclay,
91191 Gif sur Yvette, France.
***Service d'Entreposage et de Stockage des Déchets, CEA Saclay,
91191 Gif sur Yvette, France

ABSTRACT

Leaching of a cement paste by a demineralized solution results in the progressive dissolution of the zone in contact with the aggressive solution. In the long term, this dissolution will determine the kinetics of degradation of the material. These kinetics will depend principally on the solubility of calcium in the zone in contact with the aggressive solution.

INTRODUCTION

Short-lived, and possibly long-lived, radioactive waste is disposed in concrete containers (casks, repository structures, etc.). It is therefore essential to predict the behaviour of these containers in an aggressive medium. In the case of a deep disposal, water will perhaps enter in contact with the containers. One of the worst-case scenarios will then be leaching of cement paste by flowing water of low ionic content. That is why it is necessary to study this event to evaluate the conditions for a slowest kinetics of degradation.

In a cement paste devoid of alkalis, the hydrates are in equilibrium with an interstitial solution containing about 22 mmol of calcium per litre. Contact with a demineralized solution therefore creates a degraded zone between the core and surface of the leached structure in which there is a decreasing concentration gradient of calcium. This gradient induces the diffusion of calcium into the aggressive solution. A degraded zone therefore forms in the paste and thickens with time (Fig. 1).

In the degraded zone, the decrease in calcium concentration alters the chemical equilibria and provokes dissolution, as well as precipitation of secondary phases [1] which exhibits much faster kinetics than diffusion [2]. Diffusion is therefore not slowed by these phenomena. Models for prediction of the long-term behavior of structures for the disposal of radioactive waste are therefore generally diffusion models [3,4]. Assuming no shrinkage of the material, solution of Fick's equations for a semi-infinite medium indicates that the thickness of the diffusion zone (Fig. 1), e, can be expressed in the following form:

$$e = \sqrt{D_{app}t}$$

where D_{app} is a constant with the same units as a coefficient of diffusion.

The following expression is then obtained for V_1, the rate of degradation in the diffusive regime:

Mat. Res. Soc. Symp. Proc. Vol. 465 © 1997 Materials Research Society

$$V_1 = \frac{\sqrt{D_{app}}}{2\sqrt{t}}$$

V1 will therefore be large in the early stages of degradation, but small for longer times.

The cement paste surface, however, cannot be in equilibrium since it is in contact with a demineralized solution. Below a characteristic calcium concentration in the interstitial solution of the degraded zone, C_{eq} (Fig. 1), the surface layer will be finally attacked by the aggressive solution with a rate V_2 (Fig 1). After a critical time, if V_2 becomes equal to V_1, the diffusive regime will be erroneous.

By studying the degradation of a paste of pure C_3S, we have demonstrated that surface dissolution occurs. We have studied the physical and chemical characteristics of the material in contact with the aggressive solution, and have demonstrated their effects on the kinetics of dissolution of the surface layer in order to evaluate the parameters wichh control V_2.

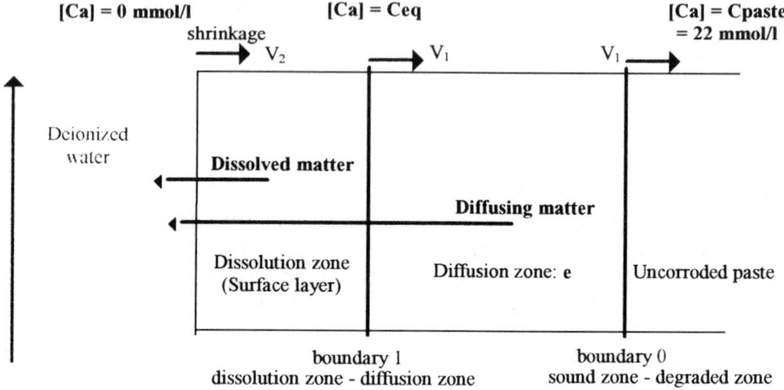

(V_1: speed of advance of diffusion boundary, V_2: speed of advance of dissolution boundary)

Fig. 1. Diagrammatic representation of leaching

EXPERIMENTAL CONDITIONS

A paste of pure C_3S has been synthesized by Ciments Français (Guerville, France). Samples 7 cm in diameter with water/cement ratio of 0.38 were kept for two months in a solution saturated in portlandite, before leaching by a pH 7 solution deionized by recirculation through a mixed bed of ion-exchange resins. Tests were performed under an inert atmosphere (N_2). Sample diameter was measured at different time using a numerical micrometre. After six months of degradation, the altered surface layer (50 to 100 microns thick) was removed and analyzed by X-ray diffraction, [29]Si Nuclear Magnetic Resonance spectroscopy (NMR), and by chemical analysis.

Solid-state NMR experiments were performed on a Bruker MSL 300 NMR spectrometer, in a 7.03 Tesla field, at a ^{29}Si frequency of 59.617 MHz. ^{29}Si MAS-NMR spectra were recorded with a spinning rate of approximately 4 kHz in double-bearing 7-mm ZrO_2 rotors. Spectra were accumulated using single-pulse excitation and high-power 1H decoupling with a 60-kHz rf field. A recycle time of 59 s was used. The number of scans varied from 800 to 2000 for the superficial layer. $[Si(CH_3)_3]_8Si_8O_{20}$ was used as a secondary standard, the major peak being at 11.6 ppm relative to tetramethylsilane (TMS, $Si(CH_3)_4$).

The solubility and kinetics of dissolution of the surface layer in confined medium were also investigated. A 250 mg sample was placed in 15 ml of water. The volume of solid introduced was about 0.11 ml. We monitored the changes in calcium concentration of the solution for over one month in order to study the equilibrium kinetics of the surface layer. These measurements were made at constant temperature (20°C).

EXPERIMENTAL RESULTS

Kinetics of degradation

Figure 2 shows the shrinkage as a function of time for the C_3S paste. This shrinkage indicates that dissolution of the zone in contact with the aggressive solution does occur. Its rate increases with time.

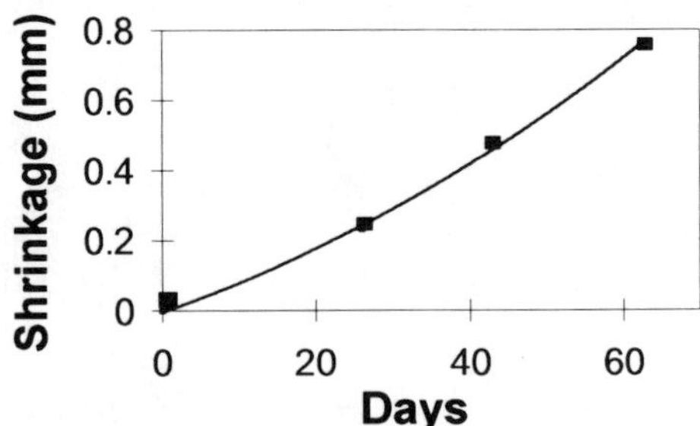

Fig. 2. Shrinkage of the C_3S sample as a function of time

These measurements clearly indicate that the dissolution of the surface, independently of diffusion, should be taken into account in the prediction of the behavior of the cement paste. This dissolution may become substantial in the long term.

X-ray diffraction and chemical analysis.

Figure 3 shows the X-ray diffraction spectrum of the surface layer in contact with the aggressive solution, as well as the molar ratio Ca/Si given by chemical analysis. The spectrum is characteristic of a CSH. The portlandite present in the uncorroded paste dissolved. No other crystalline phase was detected.

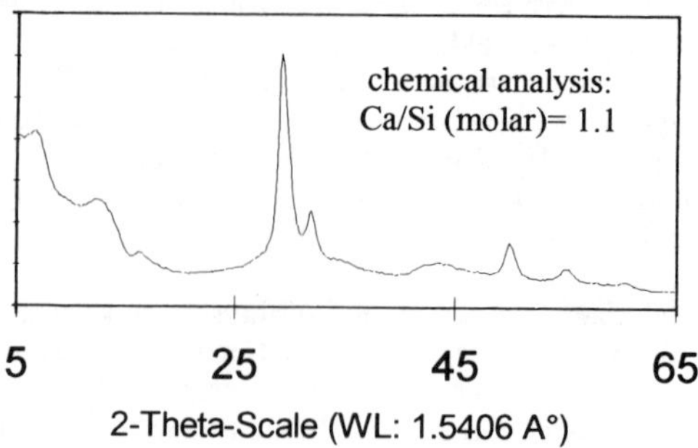

Fig. 3. X-ray diffraction spectrum of the surface layer of the degraded paste.

<u>Silicon NMR</u>

Figure 4 shows the NMR spectra of the silicon of uncorroded paste and of the degraded surface layer.

In the uncorroded zone, the Si-MAS NMR spectrum is characteristic of that of CSH of Ca/Si = 1.7 [5]. The percentages of Q1 (a single silicon for a second neighbor) and Q2 (two silicons for a second neighbor) silicons indicate that the CSH of sound paste comprise short chains of tetrahedra (2 to 5 tetrahedra).

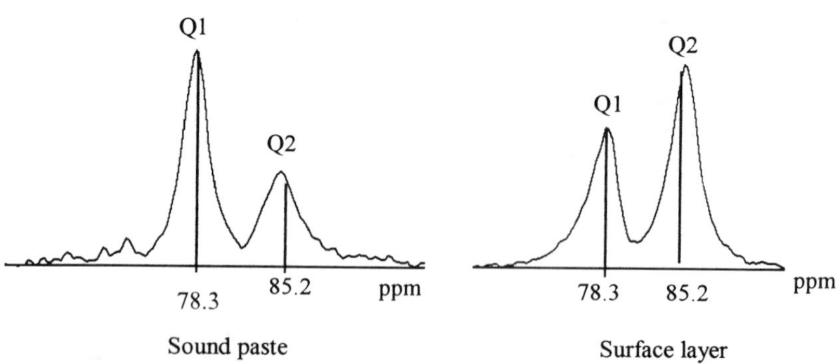

Fig. 4. Si-MAS NMR spectra

In the surface layer, the percentage of Q1 has decreased, and the percentage of Q2 has increased. The chains of tetrahedra of the CSH have therefore lengthened, on average from 2.5 tetrahedra in the sound zone to 5 in the surface layer. **No Q4 or Q3 tetrahedra characteristic of amorphous silica were detected.**

The X-ray diffraction patterns and chemical analysis, as well as the NMR, showed that the surface layer is a homogeneous layer containing a CSH of Ca/Si=1.1.

<u>Reaction kinetics of the surface layer.</u>

In an open system, like that of a sample leached by an aggressive solution, the kinetics of dissolution of a volume of matter (V_S) in a volume of liquid (V_L) can be written as:

$$\frac{\partial S}{\partial t} = a\left(C_{eq} - C\right)^n \tag{1}$$

S: quantity of matter in the unit volume of solid
a: dissolution kinetics constant
n: ordre of reaction

C_{eq}: calcium concentration in solution at equilibrium
C: calcium concentration imposed by the aggressive solution
t: time

$$-\frac{\partial S}{\partial t} = \frac{V_L}{V_S}\frac{\partial C}{\partial t} \tag{2}$$

Equations (1) and (2) can be used to deduce the change in C in the solution as a function of time. If the kinetics are second order, we have:

$$1/(Ceq-C) = k\ t \tag{4}$$

where $k = \dfrac{a\ V_S}{V_L}$

We monitored the changes in calcium concentration when 0.11 ml (V_S) of the surface layer was placed in 15 ml of water (V_L). Figure 5 shows that the kinetics of dissolution are second order.

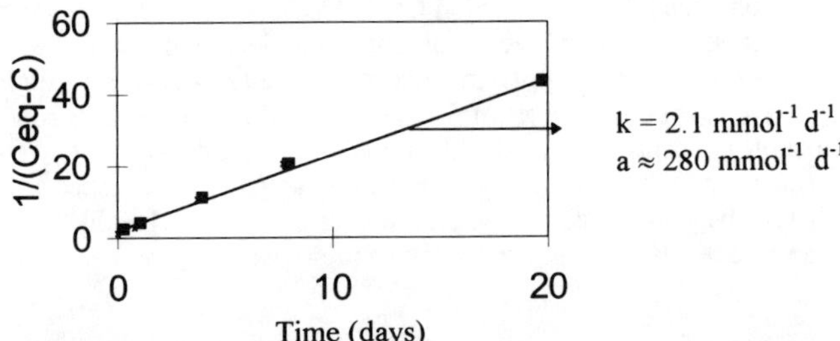

Fig. 5. Change in the ratio $1/(C_{eq} - C)$ as a function of time (C_{eq} is the calcium concentration of the solution after one month).

<u>Solubility of the surface layer</u>

Table 1 gives the solubilities measured after placing 250 mg of the zone in contact with the aggressive solution in 15 ml of water. These solubilities correspond to that of a CSH of Ca/Si = 1.1 [6].

Tab. 1. Solubility of the surface layer

$[Ca]_{eq}$	$[Si]_{eq}$
1.8 mmol/l	0.2 mmol/l

DISCUSSION

Cause of the dissolution of the surface layer

Leaching of a C_3S paste by demineralized water lowers the calcium concentration in the interstitial solution between the core and the surface of the sample. The resulting concentration gradient provokes the dissolution of portlandite and the precipitation of CSH of lower Ca/Si ratio. This mechanism is only possible if there is a local chemical equilibrium, i.e. when the dissociation constant of each CSH is verified:

$$K_S(CSH) = [Ca^{2+}]^X[H_2SiO_4^-]^Y[OH^-]^Z \tag{5}$$

where x, y and z are coefficients that are functions of the Ca/Si ratio of the CSH.

The electrical neutrality imposes a drop in $[OH^-]$ consequent upon the decrease in [Ca] in the degraded layer. Equilibrium is therefore reached when the silicon concentration increases. In the surface layer, leaching will generate a decreasing silicon concentration gradient. The local chemical equilibrium cannot therefore be maintained. The CSH of Ca/Si = 1.1 resulting from dissolution-reprecipitation will then not be in thermodynamic equilibrium and the surface layer will dissolve.

Shrinkage and surface dissolution:

We attempted to model the shrinkage of the interface (see figure 2) from the characteristics of the suspended CSH (Ca/Si = 1.1), which is the material in contact with the aggressive solution. We decomposed the degraded sample into two distinct zones. The first corresponds to a diffusion zone where the local chemical equilibrium is confirmed and which advances at a speed of V_1. The second, or dissolution zone, begins as soon as the calcium concentration in the degraded layer drops below C_{eq} (equilibrium [Ca] of the surface layer). The boundary of this zone advances at a speed of V_2 (Figure 1). Initially, V_2 will be considered negligible compared to V_1. The thickness "e" of this zone will therefore be given by an expression of the type:

$$e = \sqrt{D_{app}t} \tag{6}$$

where D_{app} is a uniform coefficient of diffusion.

In a porous medium where diffusion and dissolution-precipitation occur, Fick's second equation becomes [7]:

$$\frac{\partial C}{\partial t} = D\frac{\partial^2 C}{\partial x^2} - \frac{1}{\phi}\left[\frac{\partial S}{\partial t}\right] \tag{7}$$

C: calcium concentration (in our case) of the interstitial solution at a point in the dissolution zone, expressed in mmol/l

S: calcium concentration of the solid phase at a point in the dissolution zone, expressed in mmol/l of solid.

ϕ: porosity of the dissolution zone
x: distance from the surface (μm)
t: time (s)
D: coefficient of diffusion of calcium in solution in the dissolution zone.

D and ϕ are assumed to be constant in the dissolution zone. This assumption is not strictly true, and will tend to diminish the influence of the dissolution.

Using expression (1), and knowing "a" and "n", we have:

$$\frac{\partial C}{\partial t} = D\frac{\partial^2 C}{\partial x^2} - \frac{a}{\phi}\left(C_{eq} - C\right)^n \tag{8}$$

Given that n=2, this equation can then be solved numerically, taking as initial condition at boundary 1 (Fig. 1):

$$\left(\frac{\partial C}{\partial x}\right) = \frac{(Cpsate - Ceq)}{\sqrt{Dapp\ t}} \tag{9}$$

To solve this equation, we have taken, like Adenot [2], a value of 0.6 for ϕ in the surface layer. The coefficient of diffusion for tritiated water in a thin layer of cement paste of water/cement ratio = 0.5 was chosen for D (3,000,000 μm^2/day). C_{paste} was taken as 22 mmol/l, and n=2.

Figure 6 shows the change in shrinkage and its speed (V_2) as a function of time for C_{eq} = 1.8 mmol/l (C_3S paste). The calculated shrinkage was fully comparable with the experimentally observed shrinkage (Figure 2). The rate of shrinkage increased with time and reached high values. After about one month, V_2 and V_1 were equal. Beyond this critical period, it is therefore impossible to neglect surface dissolution.

These results clearly show that the surface dissolution will control the kinetics of degradation in the long term. Predictive, diffusion models of degradation will be valid only if the critical time after which V_2 becomes equal to V_1 is not reached.

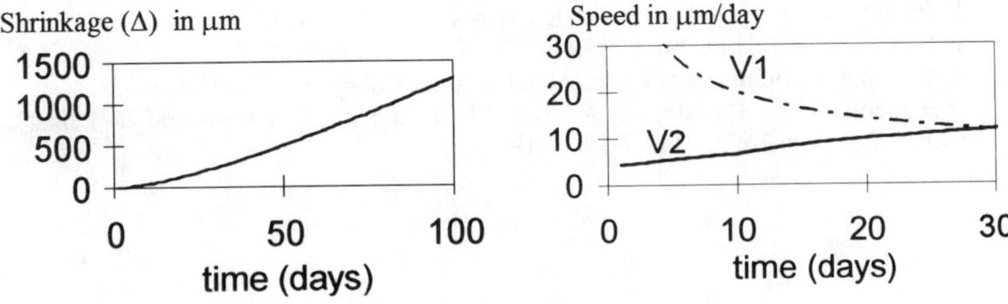

Fig. 6 Calculated shrinkage of the interface and speed of shrinkage (C_{eq} = 1.8 mmol/l)

CONCLUSIONS

The degradation of cement by leaching with water results from the combined action of diffusion and dissolution of surface material. In the short and medium term (t < critical time), degradation results from diffusion of the ionic species in the interstitial solution of the cement paste into the aggressive solution. The kinetics of degradation are then controlled by the kinetics of diffusion in the degraded layer.

In the long term (t > critical time), the rate of surface dissolution exceeds the diffusion rate, and the degradation is then controlled by this dissolution. The properties of the material in contact with the aggressive solution will therefore control the long-term behavior. By lowering its solubility and its coefficient of diffusion, it is possible to reduce the shrinkage caused by this dissolution.

If we want to be sure to avoid dissolution of the cement material in the waste disposal, the critical time must never be reached. This will be the case if deep water that may come in contact with cement has concentrations in Ca and Si higher than the solubilities of the superficial layer.

These conclusions are fundamental for the prediction of the long-term behavior of concrete containers storing radioactive waste. They help to evaluate the concequences of the superficial dissolution which was detected with OPC and slag cement paste too [8], and indicate the key parameters of the durability of the concrete used for disposal structures.

ACKNOWLEDGEMENTS

We would like to express special thanks to B. Félix and F. Pineau of ANDRA (Agence Nationale pour les Déchets Radioactifs), without whom this study would not have been possible.

REFERENCES

1.   P. Faucon, F. Adenot, M. Jorda, M. R., Cabrillac, Mater. Struct. (in press).
2.   F. Adenot, Phd thesis, Université d'Orléans (1992).
3.   F. Adenot, P. Faucon, Mater. Struct. (in press).
4.   B. Gerard, PhD thesis, ENS Cachan-France/ Université Laval-Quebec-Canada (1996).
5.   P. Faucon, PhD thesis, Université de Cergy-Pontoise (1997).
6.   H.F.W. Taylor, J.Chem.Soc.London., 276, p.3682 (1950).
7.   P. Litchner, Geochimica et Cosmochimica Acta, Vol 49, p. 779 (1985).
8.   P. Faucon, P. Le Bescop, F., Adenot, J.F. Jacquinot, F. Pineau and B. Felix, Cem. Concr. Res. Vol 26, No 11, p. 1707 (1996)

PCT Leach Tests of Hot Isostatically Pressed (HIPped) Zeolitic Concretes

D.D. Siemer, LITCO, P.O. Box 1625, Idaho Falls, ID 83415-3485
Della M. Roy, Michael W. Grutzeck, M.L.D. Gougar, and Barry E. Scheetz,
Materials Research Laboratory, The Pennsylvania State University, University Park ,PA 16802.

INTRODUCTION: The logical place for the US Federal government to site a permanent repository for its massive accumulation of cold-war reprocessing radwaste would be at its primary cold-war nuclear device testing reservation, the Nevada Test Site [1]. Regardless of whether it eventually chooses to implement that repository by drilling lateral "drifts" into consolidated rock (e.g. its proposed Yucca Mountain facility) or by augering moderately-deep boreholes into unwelded alluvium beds (e.g. the "Greater Confinement Disposal (GCD) repository implemented at Frenchman Flats in 1984 [2], zeolitic hydroceramic materials would be more stable than would glasses. The thermodynamic rationale for this is that in such regions, soil solutions and soil gasses both tend to "weather" buried natural glasses to zeolitic materials [3]. It has also been demonstrated that the same type of phases form when properly designed cementitious "grouts" are cured under mild hydrothermal conditions [4,5] -- precisely those conditions assumed by DOE's repository modelers for "failed" (i.e. flooded) repositories and under which radwaste-type glasses rapidly decompose to form crystalline zeolitic phases [6,7].

Because the US Federal Government has made a huge investment in radwaste glass-melters at some of its sites, any proposed defense-type waste solidification scheme invoking the manufacture of hydroceramic materials at its other sites should retain a "vitrification" option. Fortunately, recently work indicates that it is simple to convert a steel canister of alkali-activated concrete [8,9,10] with a zeolitic bulk composition into glass-ceramic monoliths that DOE and EPA have agreed to consider as Best Demonstrated Available Technology (BDAT) [11]: first, heat to bake out most of the water, seal and then process in a hot isostatic press (HIP) [12,13]. HIPping conditions needed to "vitrify" concretes are rather mild (e.g. 2-4 hours @ 850°-1000°C & 5-20Kpsi) and the equipment capable of processing 200-liter drums of concrete is commercially available at moderate cost.

Competent implementations of such a "staged" solidification process would require proper waste pretreatment and grout formulation. Raw wastes which contain large amounts of nitrate (or nitrite) salts, free acid, and/or organics (i.e. the majority of DOE's reprocessing radwaste) should be calcined prior to cementitious solidification. Calcination converts such materials into mixtures consisting primarily of sodium oxide (or hydroxide) and sodium aluminate -- the raw materials used in the commercial production of synthetic zeolites. Because zeolites are, in terms of bulk chemistry, hydrated forms of feldspar/feldspathoid minerals -- minerals in which each equivalent of basic components is accompanied by at least one mole of aluminum and at least one of silicon -- grout formulations should contain those elements in approximately the same proportions along with sufficient free hydroxide ions to mobilize silica/alumina during the "curing" process.

The purpose of this particular project was to investigate formulation extremes. Some of the grouts approximated the gross elemental compositions of common minerals -- both in more or less "pure" forms [14] and two-component mixtures one might expect to encounter either in nature or in man-made cementitious materials. [For example, one formulation simulated anorthite ($CaAl_2Si_2O_8$-hydrated equivalent of which is termed gismondine by petrologists and is similar to a synthetic form of zeolite P reported by zeolite and cement chemists) while another approximates an assemblage of 4 parts anorthite and 1 part jadeite ($NaAlSi_2O_6$ -- the zeolitic equivalent is termed analcime)]. Other materials made include a HIPped soda-lime glass formulation, a HIPped radwaste-type concrete loaded with 38% (dry basis) "average" calcined ICPP waste simulant (9) and two HIPed concretes made from simulated incinerator ash originally prepared for a "Mixed Waste Focus Area" (MWFA) project.

Mat. Res. Soc. Symp. Proc. Vol. 465 © 1997 Materials Research Society

The approach described herein invokes the simplest and most cost-effective waste form-making technology consistent with both operational flexibility and top-quality products: solidification with alkali-activated cements[8]. Figure 1 summarizes it. Existing ICPP wastes would first be homogenized and denitrated by "sugar calcining" them together in the facility's existing fluidized bed calciner. The calcined product would then be immediately mixed with water and a mix of cementitious agents chosen so that the overall composition approximates that of common feldspathic/feldspathoid mineral assemblages. The resulting "grouts" would be poured into stainless steel drums/canisters and allowed to cure for a year (or more) under the slightly elevated temperature and atmospheric pressure resulting from the radioactive decay heat in the waste form resulting in reaction conditions between 90° to 150°C at autogeneous pressures. If it should be deemed necessary, an additional processing operation can subsequently be used to drive residual porewater from the cured product (i.e. to FUETAP them). Finished waste forms could either 1) go to long-term storage (repository ?) or 2) again, if it is deemed to be necessary, be converted to glass-ceramics via the HIPping route discussed above.

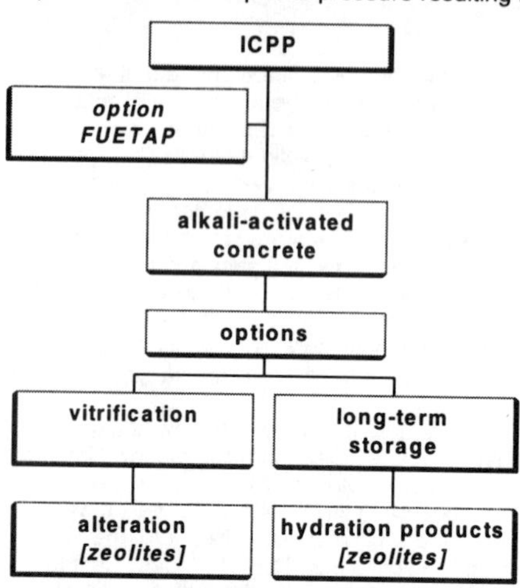

Figure 1. Schematic Flow Sheet for Anticipated Long-Term Behavior of Alkali-Activated, Vitrifiable Concretes with Zeolitic Bulk Compositons.

Zeolite Formation: The zeolitic phases that form in the hydrated un-HIPped hydroceramic materials are members of a family of complex aluminosilicates having a three-dimensional network structure,[15,16] containing relatively large channels and structural openings which are normally occupied by alkali and alkaline-earth cations and loosely held water molecules. Zeolite formation as a component of hydration cementitous nuclear waste systems have been recognized, for example, SRP plant low-level wastes [17] formed by the reaction of alkaline sodium nitrate-rich wastes with fly ash or slag with or without portland cement form gmelinite and other zeolites. It is likely that the degree of zeolitization would have been greater if the nitrate/nitrite "waste" salts had been calcined to basic oxides/hydroxides prior to cementation. In addition, by monitoring leach and pore solution compositions, it has been demonstrated that aluminosilicate-enriched blends of cement and silica fume were much better waste forms for cesium sorption than portland cement because they were forming cesium-containing zeolites such as herschelite and pollucite [18,19,20], as well as calcium silicate hydrate (C-S-H)[5]

There has been extensive research on the synthesis of zeolites from suitable precursors prior to and since the work summarized by Barrer [21]. Additional work[22] generated numerous zeolites by mixing Class F fly ash with NaOH and KOH solutions and with mixtures of fly ash (80%) and portland cement (20%).

Recent work[23] has evaluated formulations containing varying amounts of the Class F fly ash and a simulated ICPP sodium aluminate slurry (SG=1.16) containing 0.5M Al_2O_3 and 3M NaOH. The work represented an attempt to explore the feasibility of a zeolitic waste form from fly ash with simulated calcined ICPP type of waste. A slurry was prepared from a highly reactive calcined alumina, reagent grade NaOH deinonized water. After two months of stirring, the slurry contained soluble sodium aluminate and residual gibbsite. Fly ash was mixed with the slurry (3/2, 1/1, 1/2) and allowed to react at 80°,130° and 150°C for up to 7 days. In all cases, two zeolites formed as more and more fly ash reacted with the slurry. The zeolites have been identified as a gismondine type (NaP1) and a sodalite-type (unnamed 31-1270). The zeolites which form are crystalline and easily differentiated from the starting material fly ash. Micrographs show the glassy portions of the fly ash having reacted and new zeolite phases growing outward from the surface of the fly ash spheres. Sometimes the overall spherical shape of individual fly ash particles are retained.

EXPERIMENTAL: "Stiff" grouts with the nominal compositions listed in Table 1 were mixed and transferred with vibration into molds made from 10-cm length of ~1cm I.D. stainless steel tubing. After being allowed to harden over ambient-temperature water for 16 hours, they were examined to determine the degree of setting that had occurred. The samples were then placed in a glass container of saturated lime water, the bottle capped and further cured at $60^{\circ}C$ for an additional 4 days. After another examination for "setting behavior" the specimens were placed into a "cold" nitrogen-flushed muffle furnace which was rapidly (within ~1 hour) heated to $800^{\circ}C$ and held there for one hour. After the specimens had cooled, headspace air was evacuated and the open ends of the stainless steel specimen tubes were sealed by pinching them off and then welded. These sealed tubes were then HIPped at $1000^{\circ}C$ and 136 MPa for 2 hours.

The Product Consistency Test (PCT) [24] test developed by DOE's Savannah River Site to provide quality control feedback to the operators of the HLW glass melters was used. Basically, the PCT compares hot-water solubilities, at $90^{\circ}C$, of a crushed/sieved (75-150 micron-sized) powdered "unknown" with similarly treated reference material (EA glass). If a sample's gross solubility is not greater than that of the reference material, it is "acceptable". The powders are leached for one week with ten times their mass of static water and the percentage dissolution is determined by measuring the concentrations of the major components in the leachates. For the comparisons in this study, the specific conductance of the leachates was monitored to arrive at the degree of dissolution and not by summing analytical concentrations of individual ions in the leachate; leachates were monitored also as a function of time and not just once after a week's worth of exposure to the leachant; and initial "washout" is measured and included in the cumulative solubility figures (this is especially important for heterogeneous materials such as glass ceramics).

RESULTS AND DISCUSSION: Figure 2 is an exploded diagram of the M_2O-MO-Al_2O_3-SiO_2 system showing the approximate glass forming region of the system and the compositional locations of the samples detailed in Table 1. As can be seen, not all of the formulations in this study were glass forming compositions. The results of the PCT leaching on these vitrified cement specimens are presented in table 2 and figures 3,4 and 5. Two distinct clusters of leach data were measured. The first set had a leachate concentration > 0.1 , Figure 3. On an equivalent basis, the EA glass (DOE's quality control glass) exhibited the largest release of ions into solution over the interval of the test; approximately 10% of the combined alkali and calcium (expressed in milliequivalences/gram) were released during the 7 day test. The next largest release was observed with common soda-lime bottle glass, releasing approximately 5.9% of its available inventory of combined alkali and calcium ions and a calcium aluminate cement (CA) which released approximately 2.5%. Within this set of data, the EA glass exhibited the largest release of ions; both the other three specimens exhibited a more rapid release within the initial 10 hours of the experiment and reached a steady state concentration at least 30 hours before the EA glass.

The rest of the specimens (Figures 4 and 5) released 0.01 to 0.1 meq/g of their alkali/alkaline earth ions into the leachant. These values correspond to roughly 0.1 to 1% of the total inventory.

After being HIPped, all of the specimens formulated to emulate feldspar/feldspathic minerals exhibited chemical durabilities far superior to that of EA glass. All formulations containing an adequate amount of calcia (#'s7-17) were cementitious. Among these, all

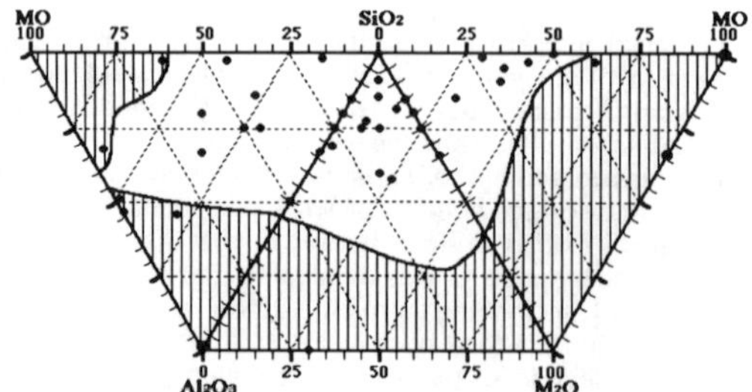

Figure 2. M₂O-MO-Al₂O₃-SiO₂ System Showing the Approximate Glass Forming Area and the Locations of the Study Specimens (mole %)

TABLE 1. NOMINAL COMPOSITIONS OF ALKALI-ACTIVATED, VITRIFIED CEMENTS

I.D. Number	Specimen	Nominal Composition	molar ratios of M₂O, MO, Al₂O₃ & SiO₂
1	EA borosilicate glass		
2	soda-lime glass	$NaCa_{0.5}Al_{0.06}Si_{2.5}O_6$	14-14-1-77
3	natrolite	$Na_2Al_2Si_3O_{10}$	20-0-20-60
4	jadeite	$NaAlSi_2O_6$	17-0-17-66
5	albite	$NaAlSi_3O_8$	12.5-0-12.5-75
6	K-jadeite	$KAlSi_2O_6$	17-0-17-66
7	anorthite	$CaAl_2Si_2O_8$	0-25-25-50
8	anorthite/jadeite	4:1	2-21-24-53
9	monocalcium silicate	$CaSiO_3$	0-50-0-50
10	monocalcium silicate/ jadeite	7:1	3-41-3-53
11	dicalcium silicate/ nepheline	5:1	3-59-3-35
12	CAS	$CaAl_2SiO_6$	0-33.3-33.3-33.3
13	C₂AS₂	$Ca_2Al_2Si_2O_9$	0-40-20-40
14	C₂A	$Ca_2Al_2O_5$	0-67-33-0
15	fly ash/75% MWFA		5-26-12-57
16	fly ash/60% MWFA		5-29-13-53
17	38% ICPP waste/ cement		5.1-36.5-10.1-48.2
18	limit of glass forming	$NaCa_{0.55}Al_{3.05}Si_{4.1}O_{13.8}$	6.1-6.7-37.2-50

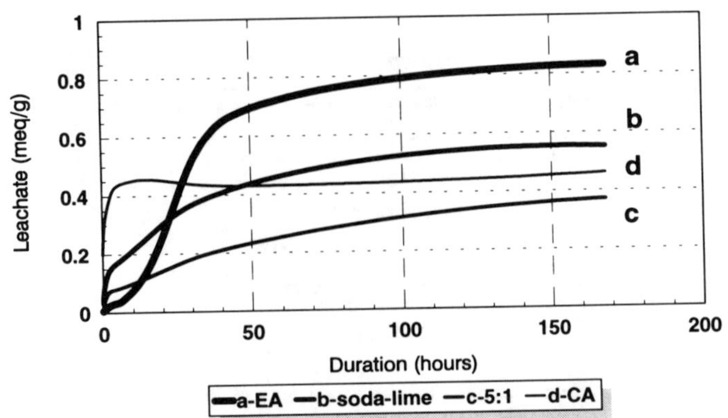

Figure 3. PCT Leach Data for Specimens with Leachates > 0.1 Meq/g

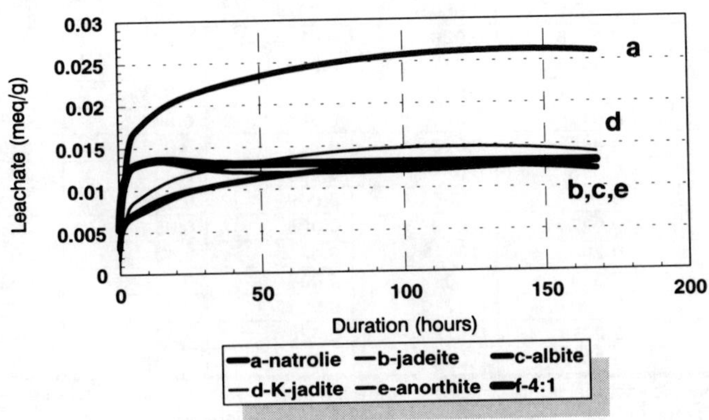

Figure 4. PCT Leach Data for Specimens with Leachates < 0.1 Meq/g.

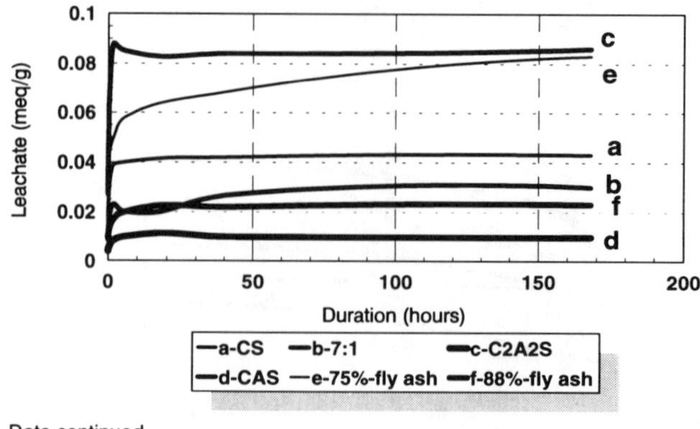

Data continued

Figure 5. PCT Leach Data for Specimens with Leachates <0.1Meq/g, continued

Table 2. PCT Leach Data (meq/g)

I.D. #	%-age of Inventory detected in leachate ^^	washout ^ (meq/g)	cumulative leached @ 1.5hr	cumulative leached @ 6.5hr	cumulative leached @ 44hr	cumulative leached @ 168hr
1	10.0	0.006	0.023	0.038	0.67	0.83
2	6.3	0.032	0.14	0.18	0.42	0.55
3	0.45	0.0066	0.011	0.017	0.023	0.026
4	0.26	0.0027	0.0059	0.0076	0.011	0.013
5	0.32	0.0031	0.0061	0.0071	0.011	0.012
6	0.30	0.0018	0.0063	0.0088	0.013	0.014
7	0.18	0.0054	0.011	0.013	0.012	0.013
8	0.19	0.0054	0.011	0.013	0.013	0.013
9	0.25	0.016	0.39	0.040	0.042	0.043
10	0.20	0.0083	0.023	0.020	0.027	0.030
11	1.85	0.003	0.07	0.085	0.22	0.37
12	0.10	0.004	0.0083	0.0099	0.0098	0.0098
13	0.87	0.027	0.088	0.085	0.084	0.086
14	2.46	0.20	0.37	0.44	0.43	0.46
15	1.22	0.039	0.50	0.058	0.069	0.083
16	0.26	0.0092	0.17	0.020	0.022	0.023
17	0.38	0.0079	0.022	0.025	0.028	0.030
18	0.29	0.0018	0.005	0.0068	0.0079	0.0098

^ the amount of ionic material in the water after ultrasonically-clean the sample powder prior to beginning the leach test – all figures normalized to one gram of sample.

^^ sum of major cationic components (Σ alkali + calcium) in milliequivalents/gram in the original formulation (dry basis). [Water leachates of these sorts of materials generally contain silicon and aluminum as an equivalent amount of anions].

of the examples that also contained an adequate amount of silica HIPped into good vitrified"
products. Of particular note, the doped-fly ash "MWFA" specimens (# 15 & 16), made with a
higher proportion of base-rich cementitious additives, resisted leaching better than did another
specimen (#15) made with lesser amounts of them. This is inconsistent with what free energy of
hydrolysis calculations suggest would be the case. The reason for this is that under mild HIPping
conditions appropriate amounts of bases "fluxes" are required for the otherwise too refractory
matrix-forming components, primarily silica and alumina. Preliminary survey of the stability of
these products was carried out by subjecting powdered samples of specimens #15,16 and 17 to
2.5 days of hydrothermal treatment at 150°C and autogenious pressures. X-ray diffraction and
macroscopic examination of the samples showed no change between the before and after
specimens.

CONCLUSIONS: Provided that it is implemented properly, cementitous solidification of
radwastes represents an attractive alternative to glass-making. The zeolitic mineral
assemblages produced by intrinsically safe, low-temperature cementitious reactions are apt to be
much more stable in any of this nation's likely ultimate repository sites than are radwaste-type
glasses. Another significant advantage is that it is easier and cheaper to accomplish large-scale
solidification with cementitious technologies than it is to attempt to do the same thing with glass
melters -- especially when the operation must be done remotely and the feedstream contains
volatiles. This and previous papers in this series, demonstrate that "good" concretes can be
readily converted to "good" glass-ceramics within hermetically-sealed containers using process
temperatures far below those required to make glasses.

REFERENCES:
1. I.J. Winograd, "Radioactive Waste Disposal in Thick Unsaturated Zones," Science Vol.
   212,pp 1457-1464 (1981).
2. E.J. Bonanno et al., "The Disposal of Orphan Wastes Using the Greater Confinement
   Disposal Facility," Waste management '91, vol. 1, Post & Wacker, Eds. Pp 861-868 (1991).
3. R.C. Surdan and R.A. Sheppard, "Zeolites in Saline, Alkaline Lake Deposits," Proceedings of
   Zeolite '76, International conference on the Occurrence, Properties and Utilization of natural
   Zeolites, L.B. Sand and F.A. Mumpton, Eds., Pergamon Press, pp 145-174 (1976).
4. M. Atkins, F.P. Glasser and J.J. Jack, "Zeolite P in Cements: its Potential for Immobilizing
   Toxic and Radioactive Waste species,", Waste Management, Vol. 151, pp 127-135 (1995).
5. J.L. LaRosa, S. Kwan and M.W. Grutzeck, "Self-generating Zeolite Cement Composites,"
   Mat. Res. Soc., vol. 245, pp 211-216 (1992).
6. W.L. Ebert, K.K. Bates and W.L. Bourcier, "The Hydration of Borosilicate Glass in Liquid
   Water and Steam at 200 C," Waste Management, vol 11, pp 205-221 (1991).
7. "Solidification of High-level Radioactive Wastes, NUREG/CR-895,pp 116-118 (1979).
8. M.L.D. Gougar, M.S. Thesis, The Pennsylvania State University (1997).
9. M.L.D. Gougar, D.D. Siemer and B.E. Scheetz, "Vitrifiable Concrete for Disposal of Spent
   Nuclear Reprocessing Wastes from INEL," Mat. Res. Soc., Vol. 412,pp 395-402 (1996).
10. M.L.D. Gougar, D.D. Siemer and B.E. Scheetz, "Disposal of INEL Spent Nuclear Fuel
    Reprocessing Waste Using a Glass-Forming Cement," Proceedings of the embedded topical
    Meeting on DOE Spent Nuclear Fuel & Fissile material Management, ANS, pp 359-366
    (1996).
11. Federal Register, Vol. 57, No. 101, Tuesday, May 26, 1992, pp 22046-22047 (discusses
    BDAT status of HIPing ICPP-type HLW).
12. D.D. Siemer, "Hot Isostatically Pressed concrete as a Radwaste Form," Amer. Cer. Soc.,
    Symposium on Waste Management, pp 657-664 (1995).
13. D.D. Siemer, Barry E. Scheetz and M.L. Gougar, "Hot Isotatic Press (HIP) Vitrification of
    Radwaste Concretes," Scientific Basis for Nuclear Waste Management XIX, Mat. Res. Soc.,
    Vol. 412, pp 403-410 (1995).

14. In a rigorous sense, none of the grout formulations really represent "pure" mineral phases because many of the materials used in compounding them contain low percentages of elements other than calcium, sodium, potassium, aluminum and silicon. For instance the formulation labeled "anorthite" ($CaAl_2Si_2O_8$) had a composition more along the lines of $CaAl_{1.96}Si_{2.03}Fe_{0.0.47}Mg_{0.06}Na_{0.023}K_{0.009}O_{8.39}$. Of course, in nature, zeolite, feldspar and feldspathoid minerals also contain appreciable amounts of minor and trace elements in their structures.

15. Sand, L.B. and F.A. Mumpton, <u>Natural Zeolites</u>, 546 pp., Pergamon Press, NY (1978).

16. Gottardi, G. and E. Galli, <u>Natural Zeolites</u>, 409 pp., Springer-Verlag, Berlin (1985).

17. Kaushal, S., D.M. Roy, P.H. Licastro and C.A. Langton, "Thermal Properties of Fly Ash Substituted Cement Waste Forms for Disposal of Savannah River Plant Salt, in <u>Fly Ash and Coal Conversion Py-Products: Characterization, Utilization and Disposal II</u>, Mat. Res. Soc., Vol. 165 Eds. G.J. McCarthy, F.P. Glasser and D.M. Roy pp 311-320 (1986).

18. Hoyle, S.L. and M.W. Grutzeck, "Incorporation of Cesium by Hydrating Calcium Aluminosilicates," <u>J. Am. Ceram. Soc. 72</u>, pp.1938-47 (1989).

19. Hoyle, S. and M.W. Grutzeck, "Effect of Pore Solution Composition on Cesium Leachability of Cement-Based Waste Forms," in <u>Scientific Basis for Nuclear Waste Management X</u>, Mat. Res. Soc. Symp. Proc. <u>84</u>, pp. 309-317, J.K. Bates and W.B. Seefeldt (Eds.), Mat. Res. Soc., Pittsburgh, PA (1987).

20. Hoyle, S. and M.W. Grutzeck, "Effects of Phase Composition on the Cesium Leachability of Cement-Based Waste Forms," in <u>Waste Management '86</u>, Proc. Waste Isolation, Tech. Prog. Public Ed. <u>3</u>, pp. 491-496, University of Arizona, Tucson, AZ (1986).

21. Barrer, R.M., "Zeolites and Clay Minerals as Sorbents and Molecular Sieves," Academic Press, New York (1978).

22. LaRosa, J., S. Kwan and M.W. Grutzeck, "Zeolite Formation in Class F Fly Ash Blended Cement Pastes," J. Amer. Ceram. Soc. <u>75</u>, 1574-80 (1992).

23. Grutzeck, M.W. and Siemer, D., "Zeolite-Cement Composite Synthesized from Fly Ash and Sodium Aluminate Waste", J. Amer. Ceram. Soc. (In prearation).

24. C.M. Jantzen et al. ,"Characterization of the Defense Waste Processing Facility (DWPF) Environmental Assessment (EA) glass Standard Reference Material," WSRC-TR-92-346, Rev, June 1, 1993.

CHARACTERIZATION OF SOLIDIFIED RADIOACTIVE WASTE DUE TO THE INCORPORATION OF HIGH-AND LOW-DENSITY POLYETHYLENE GRANULES AND TITANIUM DIOXIDE IN MORTAR MATRICES

Aleksandar Perić
Institute of nuclear sciences "Vinča", P.O.Box 522, 11001 Belgrade, Yugoslavia

ABSTRACT

The rutile form of titanium dioxide (TiO_2) and granules of high density polyethylene (PEHD) and low density polyethylene (PELD) were used to prepare mortar matrices for immobilization of radioactive waste materials containing ^{137}Cs. PELD,PEHD and TiO_2 were added to mortar matrix preparations with the objective of improving physico-chemical characteristics of the radwaste-mortar matrix mixtures, in particular the leach-rate of the immobilized radionuclide. The diameters of the PELD and PEHD used varied from 0.2 to 2.0 mm. One type of PELD and two types of PEHD were used to replace 50 weight percent of stone granules, average diameter of 2 mm, normally used in the matrix, in order to decrease the porosity and density of the mortar matrix and to avoid segregation of the stone particles at the bottom of the immobilized radioactive waste cylindrical form. TiO_2 was also added to the mortar formulation, replacing 5 and 8 weight percent of the total cement weight, for each PEHD and PELD formulation. Cured samples were investigated under temperature stress conditions, where the temperature extremes were: $T_{min} = -20^0C, T_{max} = +70^oC$. Samples were periodically immersed in distilled water at the ambient room temperature, after each freezing and heating treatment. Results of accelerated leaching experiments for these samples and samples prepared exclusively with polyethylenes replacing 100 percents of the stone granules and TiO_2, treated in nonaccelerated leaching experiments, were compared. Even using an accelerated ageing leach test that overestimates ^{137}Cs leach rates, it can be deduced, that radionuclide leach rates from the radioactive waste mortar mixture forms were improved. Leach rates decreased from 5 percent, for the material prepared with stone aggregate, to 3.1 to 4.0 percent, for the materials prepared solely with PEHD, PELD or TiO_2, and to about 3 percents for all six types of the TiO_2-PEHD and TiO_2-PELD mixtures tested.

INTRODUCTION

Using cementation as a technique for immobilizing radioactive waste materials, mortar-radioactive waste mixtures were prepared [1,2,3]. After a curing period in appropriate containers, such solidified radioactive waste-mortar monoliths are planned to be placed in an engineered trench system, adopted as a concept for the final disposal of such waste materials. The hardened mortar matrix presents the first barrier to radionuclides release particularly by leaching, either in normal or accident conditions, in prolonged periods of disposal [4,5,6,7,8,9,10]. The solidified waste forms, hardened either in metal drums or in concrete containers would be placed in trench cells, one above the other, up to four in height [11,12]. Good mechanical and physico-chemical properties of the solidified radioactive waste form are important under normal and accident conditions, e.g., earthquake and flooding. In the investigations presented in this paper, one of the goals was to improve mechanical and physico-chemical properties of the final, immobilized radioactive waste material and to estimate the quality of the new obtained matrix materials with respect to these properties, in a short testing time. Mortar was used as the matrix material to immobilize radioactive waste material, $^{137}CsCl$ solution. In order to optimize results from the investigation, mixtures of high density polyethylene, PEHD, and TiO_2 and low density polyethylene, PELD, and TiO_2, were included in the mortar matrix formulations. PEHD and PELD granules were used in the preparation of the mortar matrices, similar to the preparation of mortar matrices with stone aggregate. PEHD and PELD were used to replace all or a portion of the normally used stone aggregate, in order to decrease porosity and density of the mortar matrix and to avoid the phenomenon of stone aggregate segregation at the bottom of the immobilized radioactive waste cylindric form [13]. Satisfying this goal, we tried to obtain better physico-chemical characteristics of the radioactive waste mortar

Mat. Res. Soc. Symp. Proc. Vol. 465 © 1997 Materials Research Society

mixture, especially mechanical strength and radionuclide leach rates. For these purposes, the rutile form of titanium dioxide (TiO_2) was added in the mortar-PEHD and mortar-PELD mixtures. The quality of the final radioactive waste form is discussed in terms of nuclide leach rates and mechanical strength.

The experimental work was based on previous investigations on mortar matrices, used for the immobilization of radioactive waste solutions, containing ^{137}Cs. These investigations included addition of low density polyethylene (PELD) and high density polyethylene (PEHD) granules to the mortar matrix formulation, where the PELD and PEHD substituted for the stone aggregate normally used in the mortar matrix formulation. These investigations also tried to define the influence of addition of the rutile form of titanium dioxide (TiO_2) to the mortar matrix preparations, with the objective of improving the mechanical and physico-chemical characteristics of the radwaste-mortar matrix mixtures, in particular the radionuclide leach rate. The TiO_2 added to the mortar formulation in these previously performed investigations, replaced the appropriate amount of cement, in the amounts of 1,2,5,8 and 10 weight percent of total cement weight. In the highly basic environment of the mortar (pH above 10), the titanium will form an HTiO-type membrane, that is semi-permeable and selective for cations such as Cs^+ in the pH range above 5.5.

EXPERIMENTAL

Mortar-radioactive waste mixture samples, used for leaching and mechanical strength testing, were made from: cement (PC-20 MPa), stone aggregate (0.2-2 mm), PEHD and PELD granules with diameter varying from 0.2 mm to 2 mm, the rutile form of TiO_2, mixing additives and water. The PEHD and PELD granules had the following properties: density, 0.940-0.955 g/cm^3; dilution index, 0.3-0.4 g/10 min.; breakable temperature, $T_b = -120^0C$ and resistance to an aggressive liquid environment and the influence of mechanical forces, e.g., 30 kp/cm^2 at a temperature of 60^0C.

The water used for mortar preparation was doped with $^{137}CsCl$ solution (pH = 1.2). A water-to-cement ratio, W/C, of 0.36 provided a density of 2.145 g/cm^3 for mortar matrices prepared with stone aggregate and a density of 1.953 to 1.960 g/cm^3 for mortar matrices prepared with PELD, PEHD and TiO_2. The final solidified radwaste form was made with a high water-to-cement ratio (0.36) to allow easy leaching of the immobilized radionuclide, ^{137}Cs. The mixing period for a three-liter mortar batch was ten minutes. Radioactive waste mortar mixtures were poured into plastic molds to harden. After one day, the samples were taken from the molds and cured in an atmosphere of 65 percent relative humidity and T = 20°C for a period of 28 days [4]. Hardened samples all have approximately the same level of bound activity (A_o = 10 kBq per sample). In these investigations, the matrix formulation was not optimized, because of the high water-to-cement ratio, W/C, and absence of radionuclide sorbents in the experimental matrix formulation. Solidified orthocylinder shaped samples, H = 4.5 cm, three for each matrix material formulation, with the exposed surface area completely open to leachant, were used in determining the physico-chemical characteristics of the mortar-radioactive waste mixture forms. In these mixtures, either PELD or PEHD or TiO_2 alone were added. After the curing period, samples were placed inside clear plastic beakers, each containing leachant. The leachant was distilled water. The nonaccelerated ageing leach tests were carried out at a room temperature, T = 20+/-1°C. The leachants were renewed periodically: 5 times in the first month and approximately once per month, thereafter [4,9,10].

Optimization was performed on the basis of the results of the previously performed experiments: TiO_2 powder, 5 and 8 weight percent, was mixed with PELD and each type of PEHD in new matrix formulations, where PELD and PEHD replaced 50 weight percent of stone aggregate [14,15,16,17,18]. Cured samples prepared with PELD, PEHD and TiO_2 were investigated under temperature stress conditions, where the temperature extremes were: $T_{min} = -20^0C$, $T_{max} = +70^0C$. Samples were periodically immersed in leachant, distilled water, after each freezing and heating treatment. This procedure in experimental work was performed in aim to obtain relevant leaching characteristics of the radwaste-mortar mixture form in a relatively short period [13]. The abundance of ^{137}Cs in leachant was determined by gamma-spectrometer. The leach-rate for each matrix formulation was an average value obtained for three measured leachants, from three separate tests. The mechanical strength test were performed using cube shaped samples with a side length of 10 cm. These samples, three for each matrix formulation, were cured under the same conditions as the orthocylinder leach test samples. Mechanical strength test were in a 150 MPa hydraulic press and results for each set of three samples were averaged .

RESULTS AND DISCUSSION

Results of the ^{137}Cs leach-rate measurements for each investigated matrix formulation are presented in Figures 1-3. [137]Cs leach-rate is expressed as percent ratio of the ^{137}Cs cumulative activity leached during experiment versus ^{137}Cs total activity initially immobilized in the sample.

The mechanical strength of the samples prepared with TiO_2 was higher than that of the TiO_2-free samples, and the correlation between the mechanical strength and TiO_2 content appears to be exponential over the composition range explored here [14]. Improvement of the physico-chemical properties of the titanium prepared formulations, is a topic of further investigations. The mechanical strength of the mortar-radioactive waste mixture samples prepared with PELD and PEHD was notable higher then that for the monoliths prepared with the stone aggregate. Though an accelerated ageing leaching test overestimates the ^{137}Cs leach-rate results, compared to those obtained in the nonaccelerated leaching tests, as was shown in previous papers, the radionuclide leach rates for the mortars prepared with 50 weight percent of polyethylenes and the appropriate amounts of TiO_2 were improved [14,15,16,17].

Leach-rates decreased from 5 percent, for the material prepared with stone aggregate, to 3.25 to 4.0 percent, for the materials prepared solely with PEHD, PELD or TiO_2, and to values of about 3 percent for all six types of the TiO_2-PEHD and TiO_2-PELD mixture tested.

Only the rutile form of TiO_2 was observed in the prepared radwaste-mortar mixture samples, using X-ray Fluorescence Spectrometry. Nevertheless, the ^{137}Cs leach-rate for the matrix formulations prepared with TiO_2 was notably lower than the normally prepared (TiO_2 free) samples.

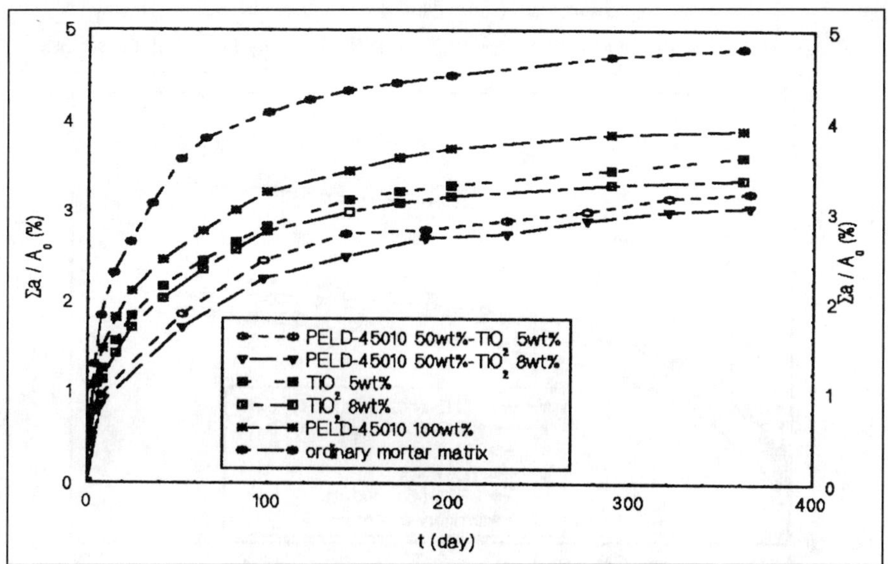

Figure 1. [137]Cs leach-rate for mortar matrix formulations containing PELD-45010 and TiO_2, where:

Σ a (Bq) is ^{137}Cs cumulative activity leached; A_o (Bq) is ^{137}Cs activity initially in the sample.

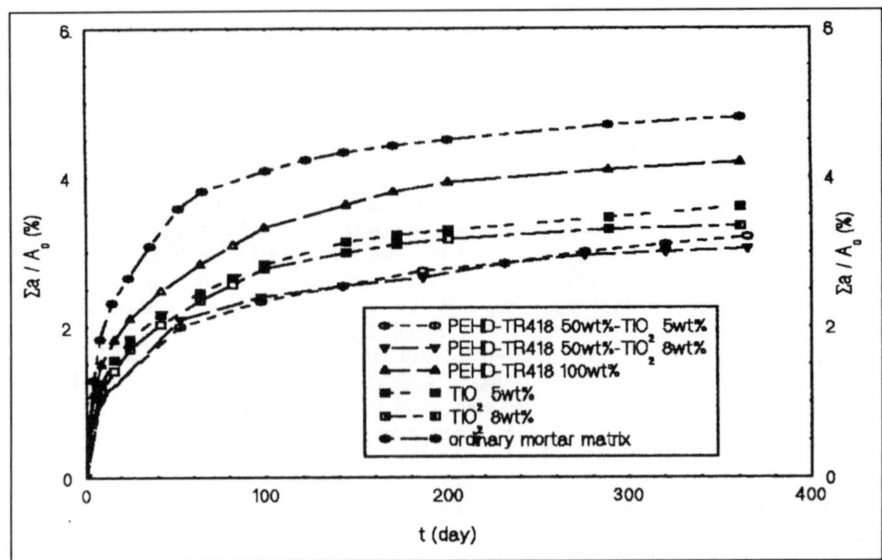

Figure 2. ^{137}Cs leach-rate for mortar matrix formulations containing PEHD-TR 418 and TiO_2, where: Σ a (Bq) is ^{137}Cs cumulative activity leached; A_o (Bq) is ^{137}Cs activity initially in the sample.

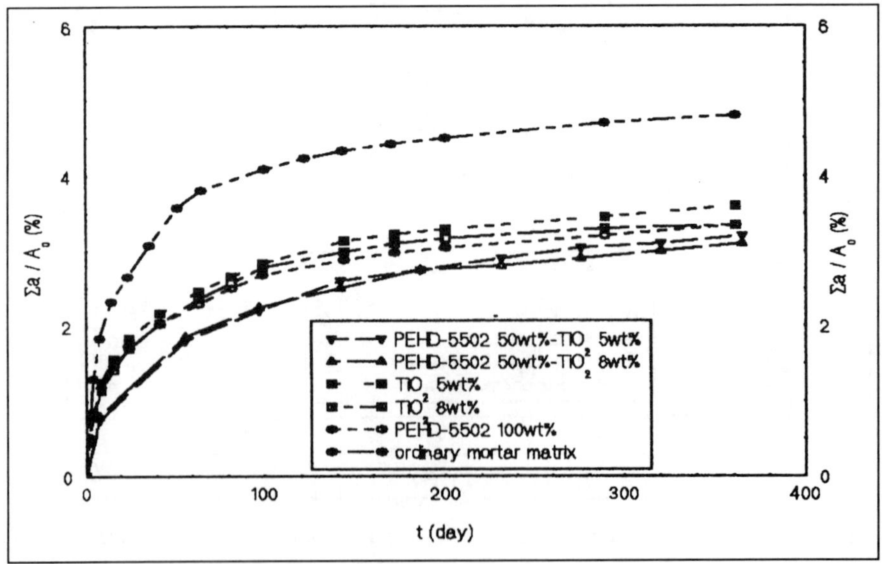

Figure 3. ^{137}Cs leach-rate for mortar matrix formulations containing PEHD-5502 and TiO_2, where: Σ a (Bq) is ^{137}Cs cumulative activity leached; A_o (Bq) is ^{137}Cs activity initially in the sample.

The mechanical strengths of the mortar-radioactive waste mixture samples prepared with PELD and TiO$_2$ or PEHD and TiO$_2$ were notably higher, 15.7 to 21.2 MPa, than the monoliths prepared with PELD and PEHD solely, 14 to 18 MPa, and the ordinary stone aggregate, 11 MPa.

Results of the specified examinations are presented in Table 1. The matrix formulation PEHD-5502 TiO$_2$ 8wt% has shown the best results in mechanical strength examinations, having almost twice the mechanical strength as samples of mortar matrix prepared in the ordinary manner with stone aggregate. The mechanical strengths obtained for samples tested in the accelerated ageing process are slightly lower than the mechanical strengths for samples tested in the nonaccelerated ageing process, up to 2 MPa. A previous investigation indicated that mechanical strength characteristics tend to decrease, as the accelerated ageing process proceeds and the differences between the mechanical characteristics obtained on samples from the accelerated and nonaccelerated ageing processes become more obvious [13].

Table 1.- Mechanical strength resistance for the different mortar matrix formulations.

Type of additive added in matrix formulation	Mechanical strength (MPa) ordinary aged samples one year experimental time	Mechanical strength (MPa) accelerated aged samples equivalent to one year experimental time
PELD-45010	16.25	15.30
PEHD-TR 418	14.65	13.50
PEHD-5502	18.40	17.60
TiO$_2$ 5wt%	12.50	11.55
TiO$_2$ 8wt%	15.00	14.20
PELD-45010 TiO$_2$ 5wt%	17.20	16.50
PELD-45010 TiO$_2$ 8wt%	18.75	18.05
PEHD-TR 418 TiO$_2$ 5wt%	15.70	15.00
PEHD-TR 418 TiO$_2$ 8wt%	16.95	16.20
PEHD-5502 TiO$_2$ 5wt%	20.25	19.45
PEHD-5502 TiO$_2$ 8wt%	21.75	20.85
Stone aggregate	11.2	10.35

Good homogeneity, relatively uniform distribution of granules in the radioactive waste-mortar matrix and a low percent of the void space in the bulk material, that results from the segregation pathways of the stone granules inside the matrix during matrix hardening and increasing the monolith porosity, as well as absence of the phenomenon of particle segregation on the bottom of the matrix material, may results in better physico-chemical and mechanical characteristics of the radioactive waste-mortar matrices prepared with PEHD and PELD.

Parameters obtained in this manner, could be used in the mathematical model and used in calculation of the ^{137}Cs cumulative fraction that could be leached in postulated leaching experiments. Thus, the behaviour of the radioactive waste-mortar matrix mixture in the prolonged disposal period, under the most conservative and undesirable environmental conditions could be predicted.

The fact that the full-scale solidified radioactive waste forms prepared with PEHD and PELD will have up to one fifth lower weight than monoliths prepared with stone granules, underlines the advantages of the mortar matrices prepared with PEHD and PELD granules. When discussing the final disposal of the radioactive waste monolith in the shallow land engineered trench system, in which solidified radioactive waste materials are placed in containers, one above the other, reduction of the total weight of each container is important.

CONCLUSION

Results of the current investigations have shown that there is a possibility of improving the physico-chemical and mechanical characteristics of radioactive waste-mortar mixtures, by use of PEHD and PELD granules and the rutile form of TiO_2 in the mortar matrix, replacing stone aggregate and a certain portion of the cement. It is shown that, using the accelerated ageing of the investigated radioactive waste materials-mortar mixtures, some physico-chemical, as well as mechanical properties of waste forms could be estimated. Accelerated ageing of the investigated samples resulted in increasing the radionuclide leach-rates and decreasing matrix material mechanical strength. Even so, differences between values of the investigated physico-chemical and mechanical characteristics obtained in nonaccelerated and accelerated ageing tests are comparable. Improvement of the physico-chemical characteristics of the radioactive waste-mortar mixtures prepared with PELD and PEHD is a subject for further investigations, including optimization of the weight percentages of the PEHD and PELD granules in the matrix formulation to increase mechanical strength and decrease radionuclide leach rates.

REFERENCES

1. C.A. Mawson, Management of Radioactive Wastes (D.van Nostrand Company Inc., London, 1965).
2. K.Tallberg, P.Aitola, IAEA SM-207/78, 1978.
3. IAEA, Technical Reports Series No.222, 1981.
4. E.Hespe, Atomic Energy Rev. **9**(I), 195 (1971).
5. EC, Second Annual Progress Report of the European Community Programme:1980-1984. Research and Development on Radioactive Waste Management and Storage, Radioactive Waste Management, **8** (1981).
6. A.Bernard, J.C.Nomine, G.Cornec, A.Bonnett, A.Farges, Nuclear and Chemical Waste Management **3**, 161 (1982).
7. S.G.Amarantos, K.G.Papadokostaki, J.H.Petropoulos, CEC Report WAS-305-83-15-GR(B), 1984.
8. C.Y.Hung, Nucl.Chem.Waste Management, **3**, 235 (1982).
9. A.D.Perić, Report on IAEA fellowship, Proposal for the Project "Lyssia",(1991).
10. A.D.Perić, I.B.Plećaš, R.S.Pavlović, S.D.Pavlović in Scientific Basis for Nuclear Waste Management XVI, edited by Aaron Barkatt and Richard A.Van Konynenburg (Material Research Society Proc. **333**, Pittsburg, PA, 1993) pp.377-382.
11. IAEA, Safety Series No.63, 1984.
12. IAEA, "Waste Disposal in Shallow Land Burial", Interregional training course, KfK, Karlsruhe, 1989.
13. A.Perić et al., in Radioactive waste management practices and issues in developing countries, edited by Vladimir Tsyplenkov, (IAEA Proc. **851**, Vienna, Austria 1995) pp.223-227.
14. Aleksandar D.Perić, in Scientific Basis for Nuclear Waste Management XIX, edited by William M. Murphy and Dieter A.Knecht (Mat Res.Soc.Proc. **412**, Pittsburg, PA 1995) pp. 429-433.
15. Aleksandar D.Perić, Ilija B.Plećaš, Radojko S.Pavlović, Environmental International, **22**, 6, (1996).
16. A.D.Perić, I.B.Plećaš and R.S.Pavlović,in International Symposium on The Scientific Basis for Nuclear Waste Management, edited by Tokashi Murakami and Rodney C.Ewing, (Mat.Res.Soc.Proc. **353**, Pittsburgh, PA, 1994) pp.907-911.
17. A.Perić, I.Plećaš, in Resumenes de las ponencias presentadas en la XX Reunion Anual de la Sociedad Nuclear Espanola, (Spanish Nuclear Soc. Proc. **20**, Madrid, Spain, 1994) pp.223-224.
18. Aleksandar Perić, in IRPA'9 International Congress on Radiation Protection, (Inter. Rad.Prot.Ass. Proc. **Vol.II**, Vienna, Austria, 1996) pp.768-770.

EVALUATION OF MICROBIALLY INFLUENCED DEGRADATION AS A METHOD FOR THE DECONTAMINATION OF RADIOACTIVELY CONTAMINATED CONCRETE

R. D. Rogers, M. A. Hamilton, and L. O. Nelson*; J. Benson and M. Green**
*Idaho National Engineering Laboratory, Biotechnology Department, Idaho Falls, ID 83415-2203, rdr2@inel.gov
**British Nuclear Fuels plc., Company Research Laboratory, Preston, UK

ABSTRACT

Because there are literally square kilometers of radioactively contaminated concrete surfaces within the U.S. Department of Energy (DOE) complex, the task (both scope and cost) of decontamination is staggering. Complex-wide cleanup using conventional methodology does not appear to be feasible for every facility because of prioritization, cost, and manual effort required.

We are investigating the feasibility of using microbially influenced degradation (MID) of concrete as a unique, innovative approach for the decontamination of concrete. Currently, work is being conducted to determine the practicality and cost effectiveness of using this environmentally acceptable method for decontamination of large surface concrete structures. Under laboratory conditions, the biodecontamination process has successfully been used to remove 2 mm of the surface of concrete slabs. Subsequently, initial field application data from an ongoing pilot-scale demonstration have shown that an average of 2 mm of surface can be removed from meter-square areas of contaminated concrete. The cost for the process has been estimated as $1.29/m^2$. Methodologies for field application of the process are being developed and will be tested. This paper provides information on the MID process, laboratory evaluation of its use for decontamination, and results from the pilot field application.

INTRODUCTION

Uncoated concrete has been used for the construction of ponds, canals, sumps, and other structures within operating nuclear facilities. Many of these concrete structures have become contaminated over time with various radionuclides. The most frequently occurring radiological contaminants appear to be Cs-137, U-238, and Co-60, followed closely by Sr-90 and tritium.[1] Typically, this contamination is securely fixed on the surface or within the first 1 or 2 mm.

The total area of contaminated concrete within the DOE complex is estimated to be in the range of 73 km^2. The volume of contaminated concrete is estimated at 1.9×10^5 m^3. Because of the immensity of contaminated concrete surfaces within the DOE complex, the task (both scope and cost) of decontamination is staggering. Concrete decontamination needs have been identified as (1) reduction of secondary waste, (2) cost- and schedule-effective technologies, and (3) innovative technologies for floor and wall decontamination.[1] Based on the above information, complex-wide cleanup using conventional methodology does not appear to be feasible for every facility because of prioritization, cost, and the manual effort required.

A practical, cost-effective, environmentally acceptable method for decontamination of large-surface concrete structures could be accomplished through the use of a naturally occurring microbiological process that destroys concrete integrity. Known as microbially influenced degradation (MID) of concrete,[2,3] the process occurs when microorganisms present in the environment produce mineral or organic acids that dissolve or disintegrate the cement matrix. The literature indicates that MID-promoting microorganisms are ubiquitous in the environment and that their mechanisms of attack are consistent with those that have been associated with chemical attack. It seems reasonable that a method could be developed using the same process, under controlled conditions, to decontaminate certain concrete structures.

Data from previous work on the development of a biodegradation test for cement-solidified waste forms[2] demonstrated that the most active of the MID microorganisms were sulfur-oxidizing bacteria, which were therefore selected for development of the biodecontamination process. The

Mat. Res. Soc. Symp. Proc. Vol. 465 © 1997 Materials Research Society

concept of biodecontamination is seen in Figure 1. This paper discusses laboratory evaluation and field prototype testing of a biodecontamination process.

MATERIAL AND METHODS

Thus far in the development cycle, two laboratory studies and a field prototype study have been conducted. The methods, application, and effects of accelerated MID testing on a variety of cement formulations were developed previously.[3] In the first laboratory study, chambers were designed and constructed to expose concrete specimens 10 × 10 cm (100 cm^2 of exposed face) to the effects of accelerated MID. The concrete specimens were made using a commercial mix of portland type II cement and 6.3 mm aggregate. After curing the specimens for 72 hours, the surface (pH 12) was sprayed with 5 mL of an aqueous solution containing 100 ppm cobalt [as $Co(NO_3)_2$] for a total application of 500 µg cobalt per surface. This was done so the specimens could serve as surrogates for concrete contaminated with the radionuclide Co-60.

Two of the Co-60 "contaminated" concrete specimens were used as a treatment and control in this initial laboratory study. Treatment consisted of exposing the specimen to solutions delivered in the form of an intermittent (45-sec/hr), fine spray of *Thiobacillus thiooxidans* lixiviant (10^7 cells/mL). The control was sprayed with a sterile medium solution. Specimens were placed in the exposure chambers at a 45-degree angle to prevent pooling and to promote liquid runoff into a collection reservoir. The collection reservoir was emptied daily, and the quantity of liquid was recorded, with 100 mL saved for later analysis.

The concrete used in the second laboratory study was obtained from a concrete foundation constructed in the mid-1950s. The area being evaluated was the original weathered surface, which had a pH near 10 when moist. To facilitate a dimensional determination of surface loss due to

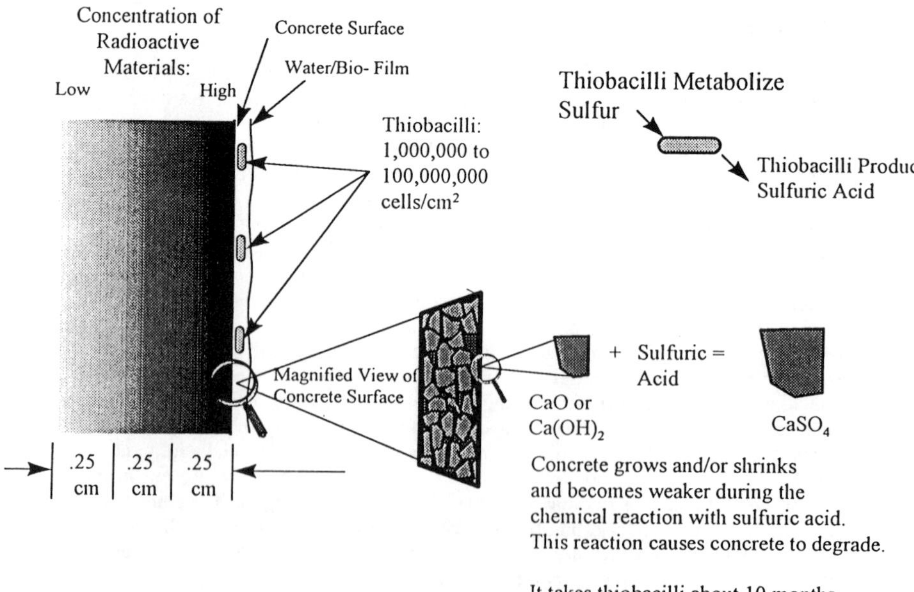

Figure 1. Diagram detailing the conceptualized activity of MID bacteria and how their activity can be used to facilitate the biodecontamination of concrete surfaces.

MID, four cores 2 cm in diameter by 4 cm long were cut out of four quadrants of the specimens. These cores were raised 5 mm above the surface and fixed in place with cement mortar. Epoxy resin used to seal concrete surfaces was applied to the protruding surface of one of the four cores (Figure 2). Since many contaminated concrete surfaces are coated with epoxy resin, this treatment was intended to determine what effect MID would have on the coating.

The study was initiated by spraying the concrete surface with a *T. thiooxidans* monoculture inoculum. The chamber was then flooded with a continuous supply of compressed air amended with a sufficient mixture of 100 ppm H_2S (v/v in N_2) to establish a constant internal concentration of 10 ppm H_2S (v/v). Humidity in the chamber was maintained at 100%.

After successful operation of the laboratory-scale H_2S chamber for a period of 3 months, a field prototype based on its design was fabricated. The location for demonstration of the field prototype was at the Idaho National Engineering Laboratory's Experimental Breeder Reactor I (EBR-I) reactor building. EBR-I was the first power reactor to produce sufficient electricity to light a city. The facility became operational in 1951, and decommissioning was completed in 1964. After a radiological survey, three areas were selected that had fixed contamination ranging from several hundred to a few thousand counts per minute as determined by a hand-held survey meter.

The field demonstration used three larger versions of the air-tight, wall-mounted chamber developed and tested in the laboratory. Chamber sizes were selected to cover areas of contaminated concrete. Two had dimensions of 0.5 m × 0.5 m × 0.05 m, while the third was 1.0 m × 1.0 m × 0.05 m. An attached chamber is seen in Figure 3.

Figure 2. Setup of the laboratory-scale H_2S chamber for evaluating the ability of MID to remove surface layers of environmentally aged concrete (95-633-1-4).

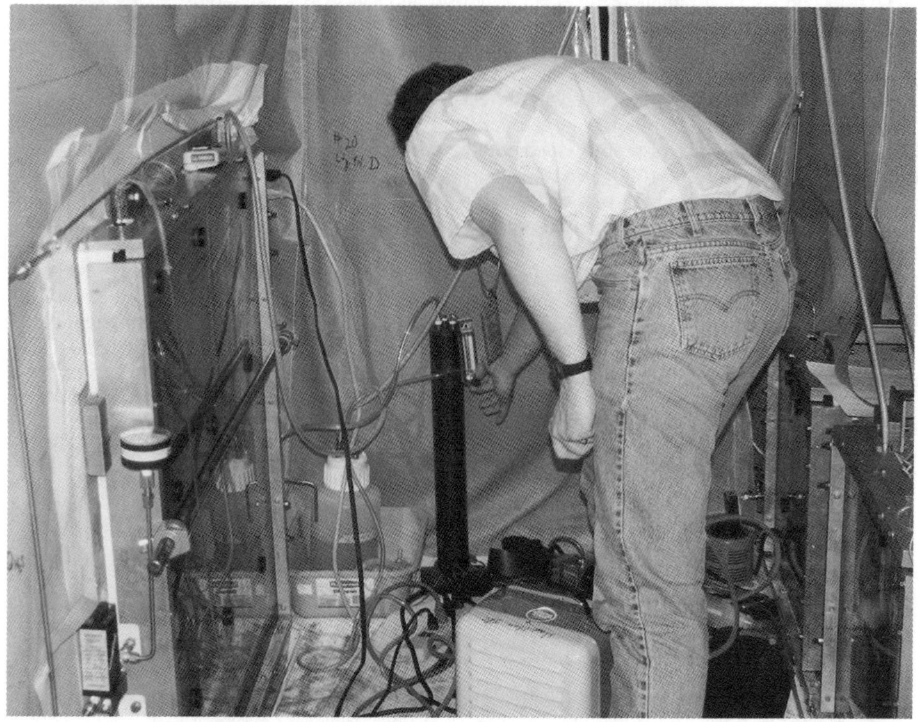

Figure 3. Test setup for evaluation of MID-promoted decontamination of concrete surfaces at the EBR-I field facility (96-114-1-28).

For purposes of designation, the chambers were numbered 1 through 3. Chambers 1 and 3 were designed to evaluate two different sulfur sources (potassium tetrathionate [$K_2S_4O_6$] and hydrogen sulfide [H_2S]), while Chamber 2 was used as an abiotic control. The surfaces encompassed by Chambers 1 and 3 were inoculated with a monoculture of *T. thiooxidans* using the method developed for the laboratory chamber. Valves attached to chamber access ports were used to facilitate sampling of the interior atmosphere. Once per week, sterile cotton swabs were inserted through ports on the chamber cover, and swipes were taken of the concrete surface of each chamber. These samples were used to determine the numbers of viable bacteria growing on the wall surfaces.

RESULTS

Spray Chambers

Chemical analysis showed that over the 10 weeks of this study, nearly 1 g (920 mg) of calcium was removed from the specimen treated with thiobacilli. The rate of loss was nearly linear with time, suggesting that calcium loss would have continued at the same rate had the study been extended. In addition, data showed that at least 100% of the applied cobalt was removed, thus resulting in an almost 100% MID-induced decontamination of the test specimen (Figure 4).

Calcium and cobalt loss from the sterile control was negligible. These data show that the cobalt was fixed to the surface of the concrete.

The potential for decontamination was obvious from the effect that the treatment had on the physical integrity of the concrete surface. The dried surface of the treated specimen had become flaky and was easily removed by scraping. The underlying aggregate was exposed after 3.4 g of the deteriorated surface (an average of 1.2 mm) was removed. This mass, combined with that of the leached calcium, accounted for ~4 g of material being removed by MID decontamination.

<u>Laboratory-Scale H$_2$S Chamber</u>

Depending on the sample location, enumeration showed that a population of 10^4 to 10^6 *T. thiooxidans* cells/cm^2 was established on concrete surfaces within 2 weeks after inoculation. Periodically, the physical condition of the treated concrete surfaces was examined by probing with the blade of a spatula. After 6 months of operation, it was determined by this method that the surfaces had become soft and friable.

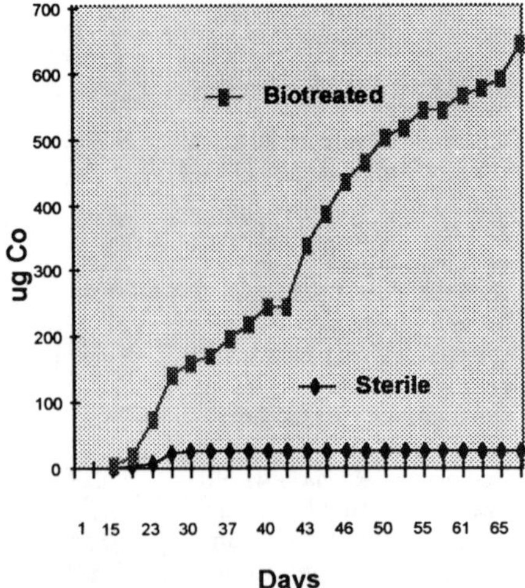

Figure 4. Results of an accelerated "spray chamber" test showing the cumulative removal, over time, of a cobalt contaminant from the surface of concrete.

In general, it was found that there was an overall heterogeneity of surface softness. A 105-cm^2 area was scraped with a spatula blade to remove loosened surface material (between 1 and 3 mm deep). MID activity was also evident on the protruding 2-cm-diameter cores, one of which had the epoxy coating. It was found that 0.05 g of material could be removed from the unprotected core, while 0.29 g of solids could be removed from coated core.

These data confirmed that MID could be initiated on the concrete surface and that the result of this activity was the production of an easily removable, soft, friable surface. Importantly, it was also shown that epoxy coatings can be compromised by MID and that once breached, the resulting space between the coating and concrete surface could promote intense MID activity.

<u>Field Prototype Chambers</u>

Chambers 1 and 2 were set up, tested, and in operation 5 weeks before Chamber 3 became operational. Routine sampling of chamber operation has been ongoing since the initiation of the study. Sufficient numerical data on bacterial numbers and H$_2$S concentrations have been collected to begin the development of two-dimensional representations of these data. It was found that there was a uniform mixing of H$_2$S in the chamber with average concentrations ranging from 55 and 56 ppm (v/v). Others[4] have shown that this H$_2$S concentration is in excess of that required to maintain high numbers of sulfur-oxidizing bacteria on concrete surfaces.

Physical observation of the decontamination process was made after 5 months for the inoculated H$_2$S chamber (Chamber 3) and 7 months for Chamber 1 supplied with K$_2$S$_4$O$_6$ (sampling of the chambers occurred on the same date, but different periods of operation were due to staggered start dates). At the time of sampling, the surface of the contaminated concrete could be observed and samples were retrieved (Figure 5). Probing at various locations on the surfaces showed that softening was occurring. The process of MID had progressed to a point that it was possible to remove loosened, friable concrete from the surface. As much as 1 to 4 mm of the sampled surfaces could be removed. The mass of material obtained from this process was as much as ~1 g/cm^2. Of more importance, however, was that there was a significant reduction in

321

the amount of fixed contamination on the treated surfaces. Surface readings in a scraped area in Chamber 3 were reduced from 600 or 900 cpm to slightly above background. In Chamber 1, the initial contaminant level was reduced from 1,500 cpm to background after surface removal. These findings greatly exceeded our initial expectations. Based on other data,[4] it was assumed that at least 12 months of operation would be required to allow for this significant amount of MID occurrence and decontamination. Operational conditions of the chambers were resumed after this brief examination. Further examination will occur after 12 months of operation.

Figure 5. Surface of concrete at the EBR-I facility after exposure to MID for a period of 5 months. Note the area in which it was possible to remove 3 mm of the surface of the affected concrete (96-567-1-8).

CONCLUSIONS

Prototype applications substantiate that MID can be initiated and managed over a large surface area of concrete walls. In addition, it was shown that managed MID could be used to promote the removal of 2 to 4 mm of the concrete surface. These data serve as a basis to support the concept of biodecontamination.

ACKNOWLEDGMENTS

Funding for this project was from the U.S. Department of Energy, EM Office of Science and Technology, TTP IDDPR06NW0.

REFERENCES

1. K. S. Dickerson, M. J. Wilson-Nichols, and M. I. Morris, *Contaminated Concrete: Occurrence and Emerging Technologies for DOE Decontamination*, DOE/ORO/2034, 1995.

2. R. D. Rogers, M. A. Hamilton, R. H. Veeh, and J. W. McConnell, *Microbial Degradation of Low-Level Radioactive Waste*, NUREG/CR-6188, May 1994.

3. R. D. Rogers, M. A. Hamilton, and J. W. McConnell, *Microbial-Influenced Cement Degradation - Literature Review*, NUREG/CR-5987, March 1993.

4. W. Sand, "Importance of Hydrogen Sulfide, Thiosulfate, and Methylmercaptan for Growth of Thiobacilli During Simulation of Concrete Corrosion," *Appl. and Environ. Microbiol.* 53, pp. 1645–1648, 1987.

Part V

Ceramic and Crystalline Waste Forms

PLUTONIUM AND NEPTUNIUM INCORPORATION IN ZIRCONOLITE

B.D. BEGG, E.R. VANCE, R.A DAY, M. HAMBLEY AND S.D. CONRADSON[*]
Materials Division, ANSTO, PMB 1, Menai, NSW 2234, Australia. email b.begg@ansto.gov.au
[*] Los Alamos National Laboratory, New Mexico, USA

ABSTRACT

The incorporation of Pu and Np in zirconolite ($CaZrTi_2O_7$) has been investigated over a range of redox conditions. Zirconolite formulations designed to favour either trivalent or tetravalent Pu and Np were prepared by limiting the amount of charge compensating additives available to maintain electroneutrality. From near-edge X-ray absorption spectroscopy the Pu valence state was found to vary with the processing atmosphere, from completely tetravalent when fired in air, and located on either the Ca or Zr sites, to trivalent, when substituted on the Ca site after annealing in 3.5% H_2/N_2. Np was predominantly tetravalent over the range of redox conditions examined and was readily incorporated on either of zirconolite's Ca or Zr sites. The charge compensation mechanisms at work in different zirconolites are also discussed.

INTRODUCTION

The long-term immobilisation of actinide-rich waste is an issue of global importance. Currently, strategies are being developed in Japan and Europe for the immobilisation of actinide-rich waste from the reprocessing of spent fuel and subsequent partitioning of the high-level nuclear waste (HLW), whilst in the U.S., efforts are being directed towards the disposition of excess weapons plutonium. Possible options for the immobilisation of these actinide waste streams include vitrification or incorporation in a tailored ceramic such as Synroc.

Synroc is a mineral-analogue based titanate ceramic, consisting of a series of highly durable, mutually compatible phases capable of incorporating HLW elements within their crystal structures. The host mineral phases have successfully demonstrated their ability to retain considerable quantities of naturally occurring radioactive elements (eg. U, Th), in a variety of environments over geological time [1,2]

<u>Immobilisation Strategy</u>

Waste elements are incorporated into the each of the Synroc phases via a number of substitutional solid solution mechanisms. A given waste element is substituted directly for a host matrix element, of a similar ionic size, and where a charge imbalance exists between the waste and the host ions, suitable charge compensation is made to maintain overall charge neutrality. Charge compensation may take the form of an additional ion of appropriate charge substituting on either the same or a separate site, so as to offset the original charge imbalance. In this way, waste ions are chemically bonded into the crystal structure of the durable host Synroc phase.

The key factor in determining the location of a given waste ion within a Synroc phase, is its ionic size. For a waste ion which has a fixed valence, such as Cs^+, the ionic size is fixed and so a suitable host site (Ba in hollandite) of comparable ionic size can be readily selected. However for multivalent waste ions, such as many of the actinides, the ionic size varies depending upon the valence state. Therefore in order to determine the optimal location for many of the actinides within the Synroc mineralogy it is essential to first know their valence state under the selected processing conditions.

There are two main factors that influence the valence state of a given element in a crystalline solid; (a) the oxygen atmosphere under which the material was prepared and (b) crystal-chemical forces derived from the bonding of that element in the selected matrix. The effect of the later factor is clearly illustrated by the chemistry of plutonium, which when heated in air can form tetravalent ions in PuO_2, or trivalent ions in monazite-structured $PuPO_4$ [3].

Mat. Res. Soc. Symp. Proc. Vol. 465 © 1997 Materials Research Society

Zirconolite Design

The primary actinide-bearing Synroc phase is zirconolite ($CaZrTi_2O_7$). Zirconolite is not only capable of incorporating actinides on either its Ca or Zr sites, but is also the most durable of the Synroc phases. The actinides of interest in this study are Pu and Np, both of which exhibit a wide range of possible ionic valence states (from +3 to +6). Under the prevailing Synroc processing conditions however, which are neutral to reducing, the most probable valence alternatives for both Pu and Np are as either trivalent or tetravalent species. Considering their ionic size, the most likely location for trivalent Pu and Np, with ionic radii of 0.100 nm and 0.101 nm respectively, would be on the Ca site (0.100 nm) of zirconolite whilst tetravalent Pu and Np (0.086 nm and 0.087 nm respectively) being of intermediate size, may possibly sit on either the Ca or the Zr site (0.072 nm).

To investigate the incorporation of Pu and Np into zirconolite, a matrix of zirconolite samples was prepared, aimed at producing the alternative Pu and Np valence states (+3 and +4) on both the Ca and Zr sites. This was achieved by replacing part of the designated host site by an equivalent amount of Pu or Np and then ensuring that appropriate charge compensation was present to maintain electroneutrality for the selected valence state on the site of interest. In this way, a single phase zirconolite would result if the sample behaved in accordance with the design. Each sample was then subjected to a range of redox conditions to examine the relative stability of the prevailing valence state.

EXPERIMENTAL

All samples were prepared via a sol-gel route, from a liquid mix of alkoxides and nitrates [4]. The solution was stir-dried prior to calcination in argon at 750°C/1h. Samples were then hot-pressed in a graphite die (1150°C/20MPa) for 2h. They were then divided into three pieces; one was set aside from the original hot-pressing, whilst the remaining two pieces were sintered in air and either nitrogen/argon or 3.5% H_2/N_2 at 1300°C or 1400°C for up to 20 hours. Most of the Pu-zirconolites were prepared with hafnium replacing zirconium on a molar basis (see below).

Samples were prepared for microstructural analysis by polishing to a 1 μm diamond finish. Scanning electron microscopy (SEM) and microanalysis was carried out using a JEOL JSM-6400 scanning electron microscope fitted with a Tracor Northern MICRO-ZII X-ray detector and a Series II TN5502 analysis system. The compositions presented represent an average of at least three separate analyses. X-ray diffraction was carried out with a Siemens D500 instrument.

X-ray absorption spectroscopy (XAS) was performed at the Stanford Synchrotron Radiation Laboratory (SSRL) in fluorescence mode on either the Pu L_{II}-edge or Pu L_{III}-edge and the Np L_{III}-edge. XAS is an element specific tool capable of providing information about the local structure (valence state, coordination, site symmetry) of the absorbing element in a complex matrix. The samples were powdered and triply contained in certified sample holders.

RESULTS AND DISCUSSION

A. PLUTONIUM

To assist with Pu valence determinations in the zirconolites, trivalent and tetravalent Pu valence standards were prepared and characterised. The Pu L_{III}-edge of the tetravalent standard, PuO_2, and trivalent standard $Nd_{0.5}Pu_{0.5}PO_4$, are shown in figure 1. A clear shift in the position of the absorption edge of ~ 3.7 eV may be seen between Pu^{3+} and Pu^{4+}. XRD and SEM confirmed that both standards were phase-pure.

In order to avoid an overlap between the Pu L_{III}-edge and the Zr K-edge in the X-ray absorption spectrum, most of the Pu-bearing zirconolites were prepared with Hf instead of Zr. The complete solid solubility of Hf for Zr in monoclinic zirconolite is well documented [6] and offers numerous potential advantages for inbuilt criticality control within an actinide-bearing Synroc wasteform. The less intense Pu L_{II}-edge was examined in the zirconolite prepared with Zr.

<u>Design: Pu^{3+} on Ca site with Al compensation on the Ti site</u> - Ca$_{0.8}$Pu$_{0.2}$HfTi$_{1.8}$Al$_{0.2}$O$_7$

Figure 2 shows the X-ray absorption near edge (XANES) spectra from the Pu L$_{III}$-edge from the air and H$_2$/N$_2$ reduced samples. A clear shift in the position of the absorption edge ($\sim$ 3.9 eV) may be seen between the two samples. The absolute position of the absorption edge for the air sintered sample corresponds closely to that from the tetravalent Pu standard, PuO$_2$, indicating that the Pu in the air fired zirconolite is present as Pu^{4+}. The $\sim$ -3.9 eV shift in the position of the edge between the air and reduced samples is equivalent to the shift between the Pu^{4+} and Pu^{3+} standards, shown in figure 1, suggesting the trivalent state of the Pu in the zirconolite fired in 3.5% H$_2$/N$_2$. The characterisation of these zirconolites has been summarised in Table I.

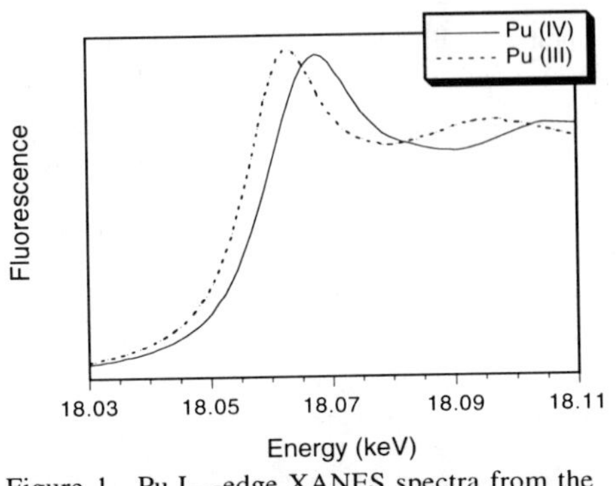

Figure 1. Pu L$_{III}$-edge XANES spectra from the from the trivalent Pu standard Nd$_{0.5}$Pu$_{0.5}$PO$_4$ and the tetravalent standard PuO$_2$.

Figure 2. XANES spectra from the Pu L$_{III}$-edge for Ca$_{0.8}$Pu$_{0.2}$HfTi$_{1.8}$Al$_{0.2}$O$_7$, heated in air and 3.5% H$_2$/N$_2$.

Table I. Ca$_{0.8}$Pu$_{0.2}$HfTi$_{1.8}$Al$_{0.2}$O$_7$ characterisation summary

	3.5% H$_2$/N$_2$	Air
Pu phases	(88%) Ca$_{0.77}$Pu$_{0.13}$Hf$_{1.25}$Ti$_{1.64}$Al$_{0.16}$O$_7$ (Z)	(96%) Ca$_{0.84}$Pu$_{0.17}$Hf$_{1.06}$Ti$_{1.72}$Al$_{0.19}$O$_7$ (Z)
(vol. %)	(10%) Ca$_{0.77}$Pu$_{0.16}$Ti$_{0.90}$Hf$_{0.06}$Al$_{0.04}$O$_3$ (P)	(3%) Hf$_{0.96}$Pu$_{0.04}$TiO$_4$ (HT)
Edge shift[*]	- 3.6 eV	+ 0.3 eV
Valence	$\sim$ 100% Pu (III)	$\sim$ 100% Pu (IV)

[*] Relative to PuO$_2$, Z = zirconolite, P = Perovskite, HT = Hafnium Titanate

Note: Minor non Pu-bearing phases not mentioned in this and subsequent tables include hafnium titanate, rutile and alumina.

Microanalysis of the zirconolite annealed in 3.5% H$_2$/N$_2$ revealed that it only contained $\sim$0.13 formula units of Pu, which was deduced from the overall stoichiometry to be located on the Ca site. The trivalent state of the Pu was compensated by an equivalent amount of Al^{3+} on the Ti^{4+} site. In addition to zirconolite, a small amount of a Pu-bearing perovskite had formed. Significantly, insufficient Al was present on the Ti^{4+} site in the perovskite to compensate for all the trivalent Pu present, indicating that some alternative charge compensating mechanism, such as Ti^{3+} or cation vacancies, must be at work. The presence of Ti^{3+} is not unexpected under the reducing conditions associated with annealing in 3.5% H$_2$/N$_2$ at 1400°C. The existence of Ti^{3+} is known to promote the formation of perovskite at the expense of zirconolite in graphite die hot-pressing [5], and as seen here, perovskite was only present in the sample subjected to the reducing atmosphere heat treatment.

The oxidation of all the Pu^{3+} to Pu^{4+}, associated with the air anneal, resulted in a slight modification to the overall zirconolite formulation. Microanalysis revealed that whilst the amount of Pu in the zirconolite had increased slightly, the zirconolite no longer contained sufficient Al on the Ti^{4+} site to compensate for the tetravalent state of the Pu, which appeared to be on the Ca site from the overall zirconolite stoichiometry. Parallel studies with Ce-doped zirconolites suggest that cation vacancies are the most likely means of compensating for excess positive charge in an oxidising environment [9]. The highly ionised state of cation vacancies in zirconolite, which are either -2 or -4 depending of whether they are Ca or Zr/Ti vacancies, means that only a small fraction of cation vacancies are required to compensate per unit of excess positive charge. This makes direct characterisation of cation vacancies difficult and may be part of the reason, along with the accuracy of the microanalysis, why their presence is not reflected in the composition of the zirconolites.

<u>Design: Pu^{4+} on Ca site with Al compensation on the Ti site</u> - $Ca_{0.8}Pu_{0.2}HfTi_{1.6}Al_{0.4}O_7$

XANES results from the Pu L_{III}-edge, combined with detailed SEM microanalysis, confirmed that Pu^{4+} could be readily stabilised on the Ca site in the air sintered zirconolite (see Table II). However, the zirconolite only contained enough Al to compensate for ~65% Pu^{4+}, pointing to the existence of other charge compensation mechanisms, most likely cation vacancies. Once again, annealing the sample in 3.5% H_2/N_2 stabilised Pu^{3+}, which was still on the Ca site form microanalysis, and resulted in the formation of a small amount of perovskite. Whilst trivalent Pu in the reduced zirconolite was clearly compensated by the presence of an equivalent amount of Al on the Ti^{4+} site, the perovskite only contained enough Al to account for half the trivalent Pu. The presence of a small quantity of Ti^{3+} is the most probable means of compensating for the excess positive charge in the perovskite, formed under a reducing atmosphere.

Table II. $Ca_{0.8}Pu_{0.2}HfTi_{1.6}Al_{0.4}O_7$ characterisation summary

	3.5% H_2/N_2	Air
Pu phases (vol. %)	(77%) $Ca_{0.71}Pu_{0.18}Hf_{1.37}Ti_{1.48}Al_{0.16}O_7$ (Z) (10%) $Ca_{0.74}Pu_{0.18}Ti_{0.84}Hf_{0.06}Al_{0.09}O_3$ (P)	(94%) $Ca_{0.75}Pu_{0.17}Hf_{1.26}Ti_{1.49}Al_{0.28}O_7$ (Z)
Edge shift[*]	- 3.5 eV	+ 0.2 eV
Valence	~ 100% Pu (III)	~ 100% Pu (IV)

[*] Relative to PuO_2, Z = zirconolite, P = perovskite

<u>Design: Plutonium (IV) on Zr site</u> - $CaZr_{0.8}Pu_{0.2}Ti_2O_7$

Despite this zirconolite's design, for tetravalent Pu to be located on the Zr site, XANES showed that all of the Pu was trivalent after being hot-pressed in a graphite die. In addition, SEM and XRD of the hot-pressed material revealed that only 50% of the sample had formed zirconolite, with perovskite making up the balance (see Table III). This major structural reordering is indicative of a problem with the design at this stage of its processing, specifically, the failure of tetravalent Pu to be stabilised under the reducing conditions associated with graphite die hot-pressing. The larger ionic size of Pu^{3+} than that of Pu^{4+} clearly did not permit it to substitute on the deficient Zr site, resulting in significant modifications to the sample's phase assemblage. Indeed, detailed microanalysis of the zirconolite grains that had formed indicated that the trivalent Pu was located on the Ca site. For this to have occurred, the zirconolite had to reformulate itself with less Ca and Ti, in order to make room for the Pu on the Ca site whilst preserving the zirconolite stoichiometry. These elements have subsequently combined, with some Pu, to form the large amount of perovskite present in the sample.

Annealing the hot-pressed sample in air (1300°C/20h) resulted in significant modifications to the overall phase assemblage towards the original design. The proportion of zirconolite increased (~80%), the perovskite was eliminated, and a Pu-rich pyrochlore was formed (~10%). This was consistent with an increase in the proportion of Pu^{4+} (~65%), relative to the graphite hot-pressing (~10%), which enabled the majority of Pu to relocate onto the Zr site of zirconolite as originally planned. Whilst the formation of a second phase was not unexpected, since parallel Ce-analogue studies had indicated that only a maximum of 0.15 formula units of tetravalent ion could be substituted on the Zr site [7], the presence of the Pu-rich pyrochlore was not anticipated. Pyrochlore, which is structurally very similar to zirconolite, is usually only seen, in analogous rare earth zirconolites, when substitutions of greater than 0.5 formula units are made on the Zr site. Its presence here however, combined with the failure of the air anneal to completely oxidise all the Pu^{3+} to Pu^{4+}, probably indicates that the sample has not fully equilibrated.

Table III. $CaZr_{0.8}Pu_{0.2}Ti_2O_7$ characterisation summary

	Graphite		Air	
Pu phases	(49%)	$Ca_{0.72}Pu_{0.23}Zr_{1.10}Ti_{1.94}O_7$ (Z)	(83%)	$Ca_{0.95}Pu_{0.23}Zr_{0.87}Ti_{1.93}O_7$ (Z)
(vol. %)	(48%)	$Ca_{0.81}Pu_{0.19}Ti_{0.99}Zr_{0.01}O_3^{+}$ (P)	(10%)	$Ca_{0.95}Pu_{0.90}Zr_{0.28}Ti_{1.84}O_7$ (Py)
Edge shift*		- 3.4 eV		- 1.3 eV
Valence		~ 100% Pu (III)		~ 65% Pu (IV)

* Relative to PuO_2, Z = zirconolite, P = Perovskite, Py = Pyrochlore

The underlying issue of charge compensation in this zirconolite series has not been resolved. Neither sample contained any charge-compensating Al and yet the majority of trivalent and tetravalent Pu in the graphite, and to a much lesser extent, air preparations were located on divalent Ca sites. As mentioned previously Ti^{3+} is certainly plausible in graphite, and recent work indicates that it should not even be excluded in air [8], although the concentrations required here are large. Investigations using rare-earth simulants suggest cation vacancies are the most probable means of compensating for additional positive charge in zirconolite [9].

<u>Design: Plutonium (IV) on Hf site</u> - $CaHf_{0.9}Pu_{0.1}Ti_2O_7$

In an attempt to promote the formation of a single phase, a second sample was prepared in this zirconolite series in which the amount of Pu substituted on the Zr site was restricted to 0.1 formula unit. The Zr was also replaced on a molar basis by Hf. XANES and SEM revealed that annealing the zirconolite in H_2/N_2 (1400°C/10h) was sufficient to stabilise all the Pu as Pu^{3+} and result in the formation of a small amount of perovskite (~10%, see Table IV). Microanalysis of the zirconolite indicated that the trivalent Pu was located on the Ca site, which was consistent with the previous sample. Annealing this sample in air however, led to the complete oxidation of all the Pu^{3+} to Pu^{4+}, although the Pu remained on the Ca site. The reason why the tetravalent Pu remained on the Ca site in this sample after having been annealed in air may lie in the way in which the sample was designed and prepared. The initial formation of the zirconolite under reducing conditions, associated with graphite die hot-pressing, clearly favours the formation of Pu^{3+} which is to large too substitute on the Hf site and therefore provides a driving force for the zirconolite to reformulate itself in such a manner as to make room for the Pu on the Ca site. Once the zirconolite has formed, with the Pu on the Ca site, the impetus for it to be subsequently restructured when subjected to an oxidising atmosphere for only a finite period of time is reduced, and especially when Pu^{4+} is also capable of being incorporated on the Ca site.

The major driving force against the location of trivalent or tetravalent Pu on the Ca site in this zirconolite series is the absence of charge compensation. This zirconolite composition contains no mechanism to compensate for the presence of anything other than tetravalent Pu on the Hf site. However, parallel investigations of charge compensation mechanisms in rare-earth doped zirconolites suggest that zirconolite can accept a proportion of cation vacancies and/or trivalent titanium to compensate for excess positive charge, which would serve to minimize the resistance to the incorporation of Pu on the Ca site. However, the magnitude of the resistance will increase in proportion to the level of Pu doping. This may well explain why the tetravalent Pu in the current sample, $CaHf_{0.9}Pu_{0.1}Ti_2O_7$, was able to be retained on the Ca site whilst in the more heavily doped zirconolite, $CaZr_{0.8}Pu_{0.2}Ti_2O_7$, the majority of tetravalent Pu was relocated on to the Zr site after being sintered in air. Even so, the zirconolite's ability to self-compensate, in the absence of specific charge compensating additives, only highlights its flexibility as a host for multivalent waste ions, such as the actinides.

Table IV. $CaHf_{0.9}Pu_{0.1}Ti_2O_7$ characterisation summary

	3.5% H_2/N_2		Air	
Pu phases (vol. %)	(88%)	$Ca_{0.84}Pu_{0.10}Hf_{1.13}Ti_{1.86}O_7$ (Z)	(98%)	$Ca_{0.88}Pu_{0.08}Hf_{1.14}Ti_{1.85}O_7$ (Z)
	(10%)	$Ca_{0.85}Pu_{0.09}Ti_{0.96}Hf_{0.05}O_3$ (P)		
Edge shift[*]		- 3.5 eV		0 eV
Valence		~ 100% Pu (III)		~ 100% Pu (IV)

[*] Relative to PuO_2, Z = zirconolite, P = Perovskite

B. NEPTUNIUM

To assist with Np valence determinations in the zirconolites, trivalent and tetravalent Np valence standards were prepared and characterised. The Np L_{III}-edge of the tetravalent standard, NpO_2, and trivalent standard $Nd_{0.9}Np_{0.1}PO_4$, are shown in figure 3. A clear shift in the position of the absorption edge of ~ 3 eV may be seen between Np^{3+} and Np^{4+}.

<u>Design: Neptunium (III) on Ca site</u> - $Ca_{0.8}Np_{0.2}ZrTi_{1.8}Al_{0.2}O_7$

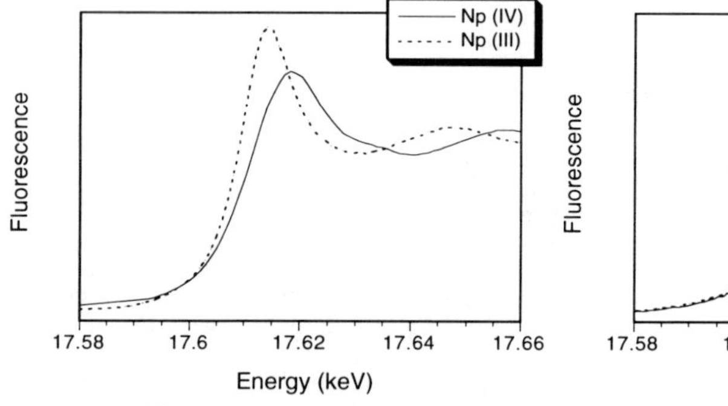

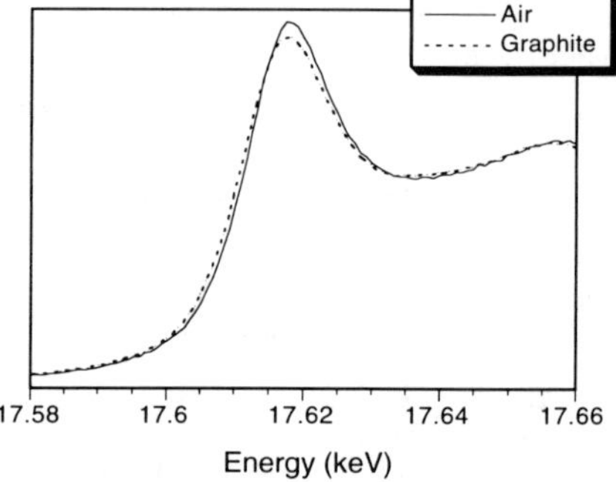

Figure 3. Np L_{III}-edge XANES spectra from the from the trivalent Np standard $Nd_{0.9}Np_{0.1}PO_4$ and the tetravalent standard NpO_2.

Figure 4. XANES spectra from the Np L_{III}-edge for $Ca_{0.8}Np_{0.2}ZrTi_{1.8}Al_{0.2}O_7$, heated in air and graphite.

Figure 4 shows the XANES spectra from the Np L_{III}-edge from the initial graphite hot-press and subsequent air sintered samples. Whilst a slight shift in the position of the absorption edge, of ~ - 0.8 eV, may be seen between the air and graphite samples, the position of the edge from the air-sintered zirconolite is equivalent to that from the tetravalent Np standard, NpO_2. The tetravalent state of the Np after the air anneal was consistent with the result from the parallel Pu sample; however, the Np was less readily reduced when subject to the graphite atmosphere, with only ~ 10% Np^{3+} forming after the graphite die hot-pressing.

Microanalysis from both the graphite hot-pressed and air annealed samples (see Table V) indicated near single-phase zirconolites of composition closely approximating the design, with all the Np located on the Ca site and only sufficient Al on the Ti^{4+} site to compensate for trivalent Np. Previously this had led us to conclude that all the Np in these zirconolites was trivalent [10]. However, the direct valence determinations permitted by XANES has shown that ~ 90% of the Np in the graphite sample and 100% of the Np in the sample annealed in air were tetravalent, testifying to the existence of alternative charge compensation mechanism in these zirconolites. The formation of a near single-phase zirconolite therefore does not necessarily mean that the sample has behaved in accordance with the design.

Table V. $Ca_{0.8}Np_{0.2}ZrTi_{1.8}Al_{0.2}O_7$ characterisation summary

	Graphite	Air
Np phases (vol. %)	(95%) $Ca_{0.79}Np_{0.21}Zr_{0.92}Ti_{1.90}Al_{0.17}O_7$ (Z)	(90%) $Ca_{0.80}Np_{0.21}Zr_{0.99}Ti_{1.79}Al_{0.21}O_7$ (Z) (3%) $Ca_{0.07}Np_{0.05}Zr_{0.83}Ti_{1.07}Al_{0.02}O_4$ [+]
Edge shift [*]	- 0.8 eV	0.0 eV
Valence	~ 90% Np (IV)	~ 100% Np (IV)

[*] Relative to NpO_2, Z = Zirconolite, [+] Zirconium titanate

Design: Neptunium (IV) on Ca site - $Ca_{0.8}Np_{0.2}ZrTi_{1.6}Al_{0.4}O_7$

Microanalysis indicated that this zirconolite series contained a slight excess of Np, see Table VI. Despite this, the samples annealed in air and argon were essentially single phase zirconolite (~98%) and of identical composition. The majority of Np was evidently located on the targeted Ca site, with the excess Np residing on the Zr site. The sample had however picked up a small amount of an iron contaminant which would assist with the charge compensation. XANES indicated that all the Np was in a tetravalent form after both the air and neutral atmosphere heat treatments.

Table VI. $Ca_{0.8}Np_{0.2}ZrTi_{1.6}Al_{0.4}O_7$ characterisation summary

	Argon	Air
Np phases (vol. %)	(98%) $Ca_{0.79}Np_{0.28}Zr_{0.88}Ti_{1.69}Al_{0.30}Fe_{0.06}O_7$ (Z)	(98%) $Ca_{0.80}Np_{0.28}Zr_{0.87}Ti_{1.69}Al_{0.32}Fe_{0.05}O_7$ (Z)
Edge shift [*]	- 0.1 eV	0.0 eV
Valence	~ 100% Np (IV)	~ 100% Np (IV)

[*] Relative to NpO_2, Z = Zirconolite

Design: Neptunium (IV) on Zr site - $CaZr_{0.8}Np_{0.2}Ti_2O_7$

XANES confirmed the relative stability of tetravalent Np in this zirconolite series also annealed in air and a neutral atmosphere (see Table VII). Compositionally, little difference existed between the samples prepared under the different conditions, though both contained two different zirconolites; a regular zirconolite and a Np-rich zirconolite, of the form $CaZr_{1-x}Np_xTi_2O_7$ for x ~ 0.5. The two zirconolites can be easily distinguished by XRD and it is thought that the second zirconolite may contain a hybrid-interlayer structure between zirconolite-2M and pyrochlore [11]. This has been labelled zirconolite-4M. A Pu version of this second zirconolite has also been observed in conjunction with zirconolite and pyrochlore in the more highly doped $CaZr_{0.5}Pu_{0.5}Ti_2O_7$ series.

Table VII. $CaZr_{0.8}Np_{0.2}Ti_2O_7$ characterisation summary

	Nitrogen	Air
Np phases (vol. %)	(78%) $Ca_{0.99}Np_{0.17}Zr_{0.82}Ti_{2.01}O_7$ (Z) (20%) $Ca_{0.94}Np_{0.49}Zr_{0.59}Ti_{1.97}O_7$ (Z')	(78%) $Ca_{1.00}Np_{0.15}Zr_{0.82}Ti_{2.03}O_7$ (Z) (20%) $Ca_{0.97}Np_{0.50}Zr_{0.53}Ti_{2.00}O_7$ (Z')
Edge shift [*]	+ 0.8 eV	+ 0.4 eV
Valence	~ 100% Np (IV)	~ 100% Np (IV)

[*] Relative to NpO_2, Z = zirconolite-2M, Z' = zirconolite-4M

CONCLUSIONS

The Pu valence state in zirconolite (either +3 or +4) is clearly dependent on the prevailing processing atmosphere and the targeted site in zirconolite. Trivalent Pu was readily stabilised during the reducing conditions associated with annealing in 3.5% H_2/N_2, though could be easily oxidised to Pu^{4+} by heating in air. Trivalent and tetravalent Pu may be incorporated on the Ca site of zirconolite, though only tetravalent Pu was small enough to substitute on the Zr site.

Contrary to what we had deduced previously [10], XANES revealed that the Np valence state was predominantly tetravalent, and independent of both the zirconolite's crystal-chemical design and the prevailing oxygen atmosphere. This highlights the fundamental importance of direct valence determinations in the chemical design of zirconolite-rich Synroc for actinide-rich wastes. Tetravalent Np was readily incorporated on either of zirconolite's Ca or Zr sites.

Although explicit solid solution limits were not determined, substitutions of Pu and Np above ~0.15 formula units on the Zr site led to the formation of second phases. Pyrochlore formed in the Pu-zirconolite, $CaZr_{0.8}Pu_{0.2}Ti_2O_7$, whilst a 'second zirconolite' resulted from the parallel Np sample. Investigations of rare-earth analogues of the 'second zirconolite' observed in the Np sample, show it to be zirconolite-4M, which has a structure intermediate between zirconolite and pyrochlore.

Direct valence state determinations permitted by XANES, combined with the detailed zirconolite microanalyses, suggested the existence of alternative charge compensation mechanisms to those in the sample designs. Clearly many of the Pu- and Np-doped zirconolites did not contain sufficient charge-compensating Al to fully neutralise the imposed charge imbalance, pointing to the presence of alternative mechanisms, such as Ti^{3+} or cation vacancies. These possibilities are being investigated by positron annihilation lifetime measurements, high-resolution TEM, and thermal gravimetric analysis on analogous Ce-doped zirconolites and perovskites. The zirconolite's ability to 'self-compensate' in the absence of specific charge compensating additives, highlights its inherent flexibility as a host for multivalent waste ions, such as the actinides.

ACKNOWLEDGEMENTS

The Synchrotron beamtime was provided by the Stanford Synchrotron Radiation Laboratory which is supported by DOE and NIH.

REFERENCES

1. G.R. Lumpkin, K.P. Hart, P.J. McGlinn and T.E. Payne, Radiochimica Acta, **66/67** 469 (1994).
2. G.R. Lumpkin, K.L. Smith, M.G. Blackford, K.P. Hart, P. McGlinn, R. Gieré and C.T. Williams, Proc. 9th Pacific Basin Nuclear Conference, (1994)
3. C.J. Bjorklund, J. Am. Chem. Soc., **79** 6347 (1958).
4. E.R. Vance, C.J. Ball, M.G. Blackford, D.J. Cassidy and K.L. Smith, J. Nucl. Mater.,**175** 55 (1990).
5. B.D. Begg et al., to be published
6. E.R. Vance, K.P. Hart, R.A. Day, B.D. Begg, P.J. Angel, E. Loi, J. Weir and V.M. Oversby, "Excess Pu Disposition in Zirconolite-Rich Synroc", in *Scientific Basis for Nuclear Waste Management XIX*, Ed. W. Murphy and D.A. Knecht, Materials Research Society, Pittsburgh, USA (1996).
7. E.R. Vance, B.D. Begg, R.A. Day and C.J. Ball, "Zirconolite-Rich Ceramics for Actinide Wastes", in *Scientific Basis for Nuclear Waste Management XVIII*, Ed. T. Murakami and R.C. Ewing, Materials Research Society, Pittsburgh, USA, 767 (1995).
8. E.R. Vance, R.A. Day, Z. Zhang, B.D. Begg, C.J. Ball and M.G. Blackford, J. Solid State Chem., **124** 77 (1996)
9. B.D. Begg and E.R. Vance, "The Incorporation of Cerium in Zirconolite", in *Scientific Basis for Nuclear Waste Management XX* (this issue).
10. E.R. Vance, C.J. Ball, K.L. Smith, M.G. Blackford, B.D. Begg and P.J. Angel, J. of Alloys and Compounds, **213/214** 406 (1994).
11. A.A. Coelho, R.W. Cheary and K.L. Smith, (submitted to J. Solid. State Chem.)

THE INCORPORATION OF CERIUM IN ZIRCONOLITE

B.D. BEGG AND E.R. VANCE
Materials Division, ANSTO, PMB 1, Menai, NSW, 2234, Australia. email b.begg@ansto.gov.au

ABSTRACT

Zirconolite ($CaZrTi_2O_7$) is the primary actinide-bearing Synroc phase for the immobilisation of high-level nuclear waste. Using X-ray absorption spectroscopy and microanalysis we have investigated the incorporation of cerium, as a simulant for plutonium, on both zirconolite's Ca and Zr sites under a range of redox conditions. The Ce valence state was found to vary between Ce^{3+} and Ce^{4+} depending on the both the sintering atmosphere and temperature. The existence of alternative charge compensation schemes, predominantly cation vacancies, in addition to those used in the sample design was inferred in many of the zirconolites and will be discussed in detail.

INTRODUCTION

The immobilisation of plutonium is an issue of international concern. The current focus on Pu immobilisation has arisen in part from its presence in reprocessed high-level radioactive waste but more recently as a possible option under consideration for the immobilisation of excess weapons Pu in the U.S.. Synroc is one of a number of options being considered for this task.

Synroc is a multi-phase mineral-analogue ceramic designed to immobilise high-level radioactive waste from the reprocessing of spent nuclear fuel [1]. It consists of three main titanate minerals - zirconolite ($CaZrTi_2O_7$), perovskite ($CaTiO_3$), and hollandite ($BaAl_2Ti_6O_{16}$) in approximately equal proportions along with smaller quantities of rutile TiO_2, and metallic alloy phases.

Waste elements are incorporated into Synroc via a substitutional solid solution mechanism. This requires the direct substitution of a waste element for a host lattice element, of a similar ionic size, in one of the Synroc phases. To ensure that charge neutrality is maintained, where the waste and host lattice elements have a different valence state, suitable charge compensation must be made. This most commonly takes the form of a second element of appropriate valence substituting on either the same or a different site within the selected Synroc phase. The ionic size of the given waste ion therefore determines the possible sites within the Synroc mineralogy into which the waste ion may substitute, whilst the degree of valence mismatch, if any, between the waste and host ions determines the type and amount of charge compensation required.

The incorporation of multivalent waste ions in Synroc is more complex, for associated with an ion's change in valence is a change in its ionic size. Therefore, the range of host sites into which the multivalent ion is capable of substituting, let alone the degree of charge compensation required, will vary depending on the prevailing valence state of the waste ion. The successful incorporation of multivalent ions in Synroc therefore requires an intimate knowledge of their valence states under the given processing environment.

Two main factors influence the valence state of a multivalent ion in a ceramic matrix. They are the partial oxygen pressure in the atmosphere under which the ceramic was processed, and crystal chemical forces associated with the bonding of the multivalent ion into the host matrix. The role of crystal chemical forces is best illustrated in chemistry of Ce and Pu, which are both known to form tetravalent metal oxides and trivalent monazite-structured phosphates when heated in air. This is one reason, along with their comparable ionic sizes (Ce^{4+} = 0.087 nm; Pu^{4+} = 0.086 nm), why Ce is commonly used as a valence simulant for Pu. The incorporation of Ce is also of interest in its own right, as a rare-earth fission product present in reprocessed wastes.

The major rare earth/actinide-bearing Synroc phase is zirconolite. Zirconolite is not only the most durable of the Synroc phases but it is also capable of incorporating rare earth and actinide elements on either its Ca or Zr sites. This paper sets out to investigate the incorporation of Ce in zirconolite by targeting the alternate valence states, +3 and +4, on both the Ca and Zr sites of zirconolite. Individual valence states are targeted by only providing sufficient charge compensation

Mat. Res. Soc. Symp. Proc. Vol. 465 © 1997 Materials Research Society

for the desired valence on the selected site. In this way, a single phase zirconolite should result where the system has behaved in accordance with the design.

Valence determinations were made using X-ray absorption near edge spectroscopy (XANES). XANES is a element-specific, bulk technique that is capable of providing information about the valence state, symmetry and coordination environment of the absorber in its host matrix. A change in the valence state of the absorber gives rise to a shift in the position of the absorption edge, with an increase in edge energy resulting from a corresponding increase in oxidation state (see figure 1). For Ce, in addition to an increase in the energy of the edge associated with the oxidation of Ce^{3+} to Ce^{4+}, a second peak is developed at ~5.734 keV. The presence of this additional peak increases the sensitivity of detecting small amounts of tetravalent Ce in an otherwise trivalent Ce matrix.

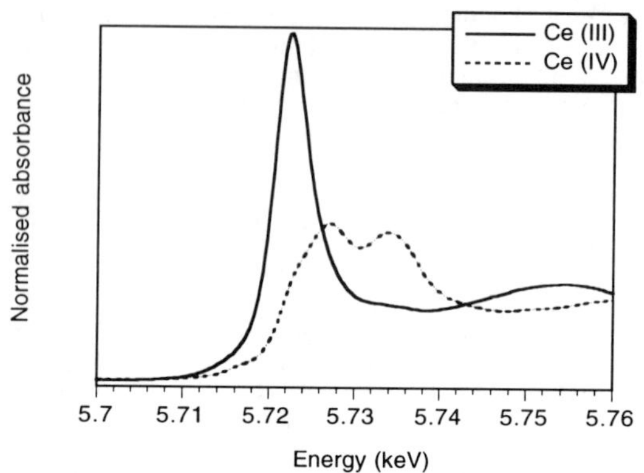

Figure 1. X-ray absorption spectra from the Ce L_{III}-edge showing the increase in the position of the absorption edge and development of a second peak at 5.734 keV associated with a change in Ce valence from Ce^{3+} in $Nd_{0.8}Ce_{0.2}PO_4$ to Ce^{4+} in CeO_2.

EXPERIMENTAL

All samples were prepared from a liquid mix of Ti, Zr and Al alkoxides which was hydrolysed with an aqueous Ca and Ce nitrate solution and stir-dried [2]. The samples were calcined in air at 600°C for 1 hour before being milled to pass through a 120 μm sieve. The calcined powder was then pressed into pellets and sintered at 1200°C for 10 hours in air. These pellets were subsequently reground to < 120 μm, repressed and fired in air for a further 80 hours at 1400°C. Each sample was then powdered and divided into three; one part was set aside and pellets were pressed from the two remaining portions. One was sintered at 1200°C for 12 hours in flowing 3.5% H_2/N_2 and the other sintered in air at 1200°C for 50 hours.

All samples were characterised by scanning electron microscopy (SEM), X-ray diffraction (XRD), XANES and thermogravimetric analysis (TGA). Samples were prepared for microstructural analysis by polishing to a 0.25 μm diamond finish. Scanning electron microscopy (SEM) and microanalysis were carried out using a JEOL JSM-6400 scanning electron microscope fitted with a Tracor Northern MICRO-ZII X-ray detector and a Series II TN5502 analysis system. The compositions presented represent an average of at least three separate analyses. X-ray diffraction was carried out with a Siemens D500 instrument. XANES was performed at the Photon Factory, Tsukuba, Japan, on the Ce L_{III}-edge in transmission at room temperature. The data were collected in 0.25 eV steps. The proportions of trivalent and tetravalent Ce in each sample were estimated by initially fitting the edge, with mixed gaussian/lorentzian peaks, and then calculating the areas of the trivalent and tetravalent components used in the fit. A linear correlation exists between areas of trivalent and tetravalent compenents present in the edge and the proportion of each species in the sample. TGA was carried out in air on ~ 100mg samples using a Setaram TAG 24 thermal analyser.

RESULTS AND DISCUSSION

The results from the various incorporation strategies targeting the alternate Ce valence states (+3 and +4) on both the Ca and Zr sites of zirconolite are presented.

A. Ca site of Zirconolite

<u>Target: Ce^{3+} on Ca site with Al compensation on the Ti site - $Ca_{0.8}Ce_{0.2}ZrTi_{1.8}Al_{0.2}O_7$</u>

Sintering this composition at 1400°C for 80 hours in air led to the formation of a single phase zirconolite. Microanalysis gave a composition of $Ca_{0.80}Ce_{0.21}Zr_{1.12}Ti_{1.68}Al_{0.18}O_7$, which was consistent with the Ce being located on the Ca site, with sufficient Al present on the Ti^{4+} site to charge compensate for trivalent Ce. Annealing this sample further in either air, at 1200°C for 50 hours, or in flowing 3.5% H_2/N_2 at 1200°C for 12 hours, had no significant effect on the phase assembly or the composition of the zirconolite.

The Ce L_{III}-edge XANES spectra taken from these samples after their respective heat treatments are shown in figure 2. The zirconolite initially heated in air at 1400°C for 80 hours contained approximately equal proportions of trivalent and tetravalent Ce. Significantly, equilibrating this sample at a lower temperature in air, 1200°C for 50 hours, resulted in an increase in the amount of tetravalent Ce to ~75%. This may be clearly seen through the shift in the position of the L_{III}-edge towards higher energy and the enhancement of the second peak at 5.734 keV, relative to the 1400°C/80h sinter. The presence of a larger quantity of tetravalent Ce after the lower temperature air anneal reflects the relative instability of high oxidation states (seen here for Ce^{4+}) at elevated temperatures in air. Annealing the initial air-sintered zirconolite in 3.5% H_2/N_2 at 1200°C for 12 hours was sufficient to fully reduce the Ce to Ce^{3+}, as evidenced by the single peak at ~5.723 keV.

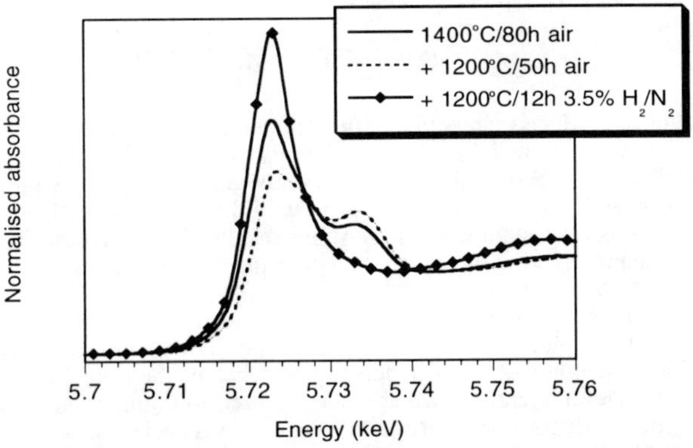

Figure 2. Ce L_{III}-edge X-ray absorption spectra showing the influence of various heat treatments on the Ce valence in $Ca_{0.8}Ce_{0.2}ZrTi_{1.8}Al_{0.2}O_7$.

The ability of this zirconolite formulation to accommodate significant proportions of tetravalent Ce (up to 75%) when fired in air, in the absence of sufficient charge compensating Al on the Ti^{4+} site, poses questions as to the type of charge compensation at work in the air fired zirconolites.

There are three possible means by which the zirconolite may "self-compensate" for an excess of positive charge in the absence of sufficient charge compensating additives. They are trivalent titanium, cation vacancies, or interstitial oxygen. Charge compensation via trivalent Ti requires that a simultaneous charge transfer occurs, such that the additional positive charge resulting from the oxidation of the Ce on the Ca site be offset by the transfer of an equivalent amount of charge away from the Ti, through the reduction of some Ti^{4+} to Ti^{3+}. Whilst the presence of Ti^{3+} might only seem plausible for samples prepared under reducing conditions, recent investigations into charge compensation in the air-sintered $Ca_{1-x}Gd_xTiO_3$ system have indicated that it can not be discounted in samples prepared in air [3]. Although cation vacancies of themselves have no charge, the presence of oxygen ions surrounding the cation vacancy gives it an negative charge that is numerically equivalent to the oxidation state of the absent cation. It is this equivalent negative charge that enables cation vacancies to compensate for additional positive charge. Finally, an excess of positive charge could be counterbalanced through the presence of some additional oxygen. However, for a matrix to be able to accept additional interstitial oxygen atoms without significant structural reordering, which would be energetically unfavourable, the structure would need to contain large spaces where the oxygen could sit (ionic size of O^{2-} = 0.132 nm) in the vicinity of its cations, to avoid electrostatic repulsion from neighbouring oxygen atoms. Materials that exhibit these structural qualities are very unusual, and certainly zirconolite is not one of them. Therefore, of these three alternatives, cation vacancies or trivalent titanium would appear to be the most likely means by which the zirconolite would "self-compensate" for any additional positive charge.

One way of distinguishing between trivalent Ti and cation vacancies in either the reduced or oxidised (1400°C/80h air) zirconolites, depending on the amount of charge compensating Al present, is via TGA measurements. This is possible since the net weight change associated with the oxidation of the Ce^{3+} to Ce^{4+}, by heating the reduced zirconolite to 1400°C in air, will vary depending on whether the sample is being compensated by trivalent Ti or cation vacancies. By definition, the oxidation of trivalent Ce will result in the zirconolite gaining weight through the pick up of oxygen. However, if the oxidation of Ce^{3+} is compensated through the reduction of an equivalent amount of Ti^{4+} to Ti^{3+} then the net change in weight of the zirconolite will be zero, since there has merely been a transfer of charge from the Ti to the Ce. Now if the oxidation of the Ce was compensated through the development of cation vacancies, which requires the presence of the surrounding oxygen ions to compensate for the charge, then the overall mass of the zirconolite would increase in proportion to the level of oxidation of the Ce. This has been illustrated in the following equation (1) for the analogous oxidation of Fe^{2+} to Fe^{3+} in magnesium oxide where cation vacancies ($\square$) are used for charge compensation and the weight gain corresponds to x/2 formula units of oxygen.

$$Mg_{1-x}Fe^{2+}_xO \; \rightarrow \; Mg_{1-x}Fe^{3+}_x\square_{x/2}O_{1+x/2} \tag{1}$$

On heating the reduced zirconolite from the current series to 1400°C in air, TGA revealed a 0.27±0.02% increase in weight, after the initial loss of absorbed species. All the TGA measurements are accurate to ±0.02%. If this weight gain can be solely attributed to the pick up of oxygen associated with the oxidation of Ce, this would correspond to the oxidation of ~60% of the trivalent Ce, which is consistent with the XANES results. This increase in weight would therefore point to cation vacancies as the means of compensating for the tetravalent Ce present in the air fired zirconolites.

The stoichiometric nature of the zirconolite analysis however, does not appear to be consistent with the presence of cation vacancies. The highly ionised nature of cation vacancies in zirconolite, which are either -2 or -4 depending of whether they are Ca or Zr/Ti vacancies, means that only a small fraction of cation vacancies are required to compensate per unit of excess positive charge. This makes direct characterisation of cation vacancies difficult and may be part of the reason, along with the accuracy of the microanalysis, why their presence is not reflected in the composition of the zirconolite.

<u>Target: Ce^{4+} on Ca site with Al compensation on Ti site</u> - $Ca_{0.8}Ce_{0.2}ZrTi_{1.6}Al_{0.4}O_7$

The initial air sinter at 1400°C for 80 hours led to the formation of a zirconolite along with minor quantities of alumina. Microanalysis gave a mean zirconolite composition of $Ca_{0.78}Ce_{0.21}Zr_{1.09}Ti_{1.59}Al_{0.34}O_7$ which would suggest that the Ce is residing on the Ca site, as planned, although there was only sufficient Al present on the Ti^{4+} site to compensate for ~60% of the Ce being present as Ce^{4+}. The lower temperature (1200°C/50h) air sinter did not lead to any significant compositional changes although the reducing 3.5% H_2/N_2 heat treatment stabilised ~5% perovskite, $Ca_{0.79}Ce_{0.20}Ti_{0.88}Al_{0.12}O_3$, in addition to the zirconolite, the composition of which remained essentially constant. Zirconolite is known to be unstable with respect to perovskite when heated under severely reducing conditions [4].

Figure 3 shows the XANES spectra from each of the zirconolites after the various heat treatments. The initial 1400°C/80h air sinter contained ~70% Ce^{4+}, which was consistent with the amount of charge compensating Al present in the zirconolite. Once again the proportion of tetravalent Ce increased, to ~85%, as a result of the lower temperature 1200°C air anneal, whilst the 3.5% H_2/N_2 firing was sufficient to completely reduce the Ce to Ce^{3+}.

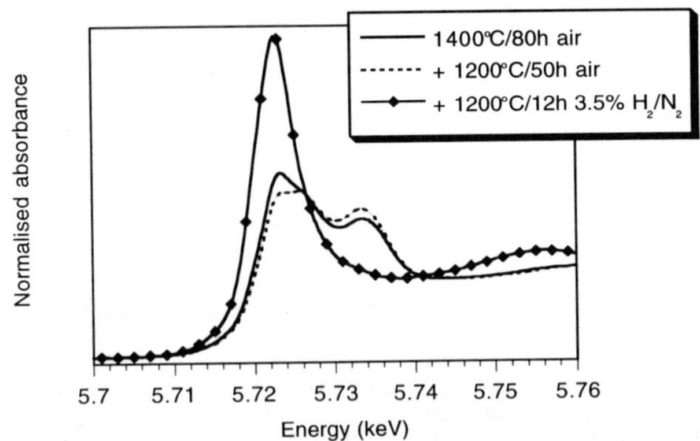

Figure 3. Ce L_{III}-edge X-ray absorption spectra showing the influence of various heat treatments on the Ce valence in $Ca_{0.8}Ce_{0.2}ZrTi_{1.6}Al_{0.4}O_7$.

The zirconolite's capacity to accommodate a significant range of Ce valence states, within an essentially constant composition highlights its ability to employ a wide range of charge compensation mechanisms. Each zirconolite contained enough Al on the Ti^{4+} site to compensate for ~60% of the Ce to exist as Ce^{4+} on the Ca site, which was only appropriate for the initial zirconolite fired in air at 1400°C. This however, represented a severe excess of Al (excess negative charge) for the reduced, 100% Ce^{3+} zirconolite, and a deficiency of Al (excess positive charge) for the zirconolite annealed at 1200°C in air (~85% Ce^{4+}). Whilst cation vacancies could be invoked to compensate for the additional tetravalent Ce in the zirconolite fired in air at 1200°C, in an analogous manner to that described above for Ce^{3+} on Ca zirconolite series, the most probable means of compensating for the excess negative charge associated with the reduced, 100% Ce^{3+} zirconolite, is via oxygen vacancies. To this end, TGA detected a weight gain of ~0.20%, associated with the oxidation of the reduced sample upon heating to 1400°C in air. This would account for the oxidation of ~55% of the Ce^{3+} in the zirconolite, in agreement with the ~70% observed.

<u>Target: Ce^{3+} on Ca site with no compensation - $Ca_{0.9}Ce_{0.1}ZrTi_2O_7$</u>

To further explore the charge compensation and incorporation mechanisms in zirconolite, a formulation was prepared where Ce was targeted on the Ca site without any parallel Al substitution on the Ti^{4+} site for charge compensation. The level of Ce doping was restricted to a 10% substitution on the Ca site to encourage the formation of a single phase. XRD and microanalysis revealed the formation of a single Ce-bearing zirconolite, $Ca_{0.92}Ce_{0.11}Zr_{1.05}Ti_{1.91}O_7$, after each of the three heat treatments. Minor quantities (< 5% total) of other non Ce-bearing phases which were present in some samples included zirconia, rutile and zirconium titanate ($ZrTiO_4$).

The Ce L_{III}-edge XANES spectra from the three zirconolite samples are shown in figure 4. Even in this zirconolite system, without any charge compensating additives, ~65% of the Ce was tetravalent after the initial 1400°C/80 hour sinter, and this increased to ~85% Ce^{4+} after the 1200°C anneal. As for the two previous zirconolite series the Ce was readily reduced to Ce^{3+} after the 3.5% H_2/N_2 heat treatment.

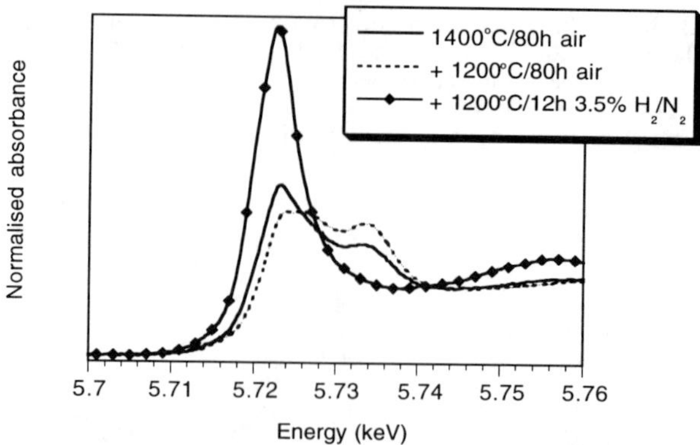

Figure 4. Ce L$_{III}$ -edge X-ray absorption spectra showing the influence of various heat treatments on the Ce valence in $Ca_{0.9}Ce_{0.1}ZrTi_2O_7$.

The capacity for this zirconolite formulation to incorporate a greater proportion of tetravalent Ce on the Ca site than the Al compensated Ce^{3+} zirconolite series and, an equivalent proportion to that of the Al compensated Ce^{4+} zirconolite described earlier, is remarkable. Clearly each of the zirconolites in the current series would appear in the first instance to contain a significant excess of positive charge, which would need to be compensated, most likely through the presence of cation vacancies. TGA yielded a weight gain of ~ 0.14% upon oxidising the reduced zirconolite to 1400°C in air, which would account for the oxidation of ~70% of the trivalent Ce and be consistent with the presence of cation vacancies.

B. Zr site of Zirconolite

Target: Ce^{3+} on Zr site with Ta compensation on Ti site - $CaZr_{0.9}Ce_{0.1}Ti_{1.9}Ta_{0.1}O_7$

Sintering this formulation in air at 1400°C/80h resulted in the formation of two distinct phases, both of which displayed zirconolite-type stoichiometry by microanalysis. The structural similarity of these phases was reflected in conventional powder XRD which indicated that the sample was single phase. Microanalysis from the abundant zirconolite (~90%) gave a mean composition of $Ca_{0.95}Ce_{0.06}Zr_{1.02}Ti_{1.89}Ta_{0.06}O_7$, which would suggest that despite the design, the Ce was located on the Ca site. The other zirconolite (~10%) was considerably richer in Ce and Ta, $Ca_{1.09}Ce_{0.44}Zr_{0.45}Ti_{1.46}Ta_{0.56}O_7$, with all the Ce apparently residing on the Zr site as planned. Annealing this sample further at 1200°C, in both air and H_2/N_2, did not significantly alter the composition of either phase. More detailed XRD is required to structurally characterise the less abundant Ce-rich phase, as previous work [5] has shown the zirconolite solid solution limit for Ce substitutions on the Zr site to be ~15%, well below the amount present in this Ce-rich phase.

The XANES results from this sequence of samples are shown in figure 5. Contrary to the zirconolite's design, the Ce in both samples annealed in air was completely tetravalent. The H_2/N_2 heat treatment was once again sufficient to fully reduce the Ce to Ce^{3+}.

The charge compensation mechanisms at work in this series of zirconolites appear to be quite intricate. Both zirconolites in the reduced, H_2/N_2 annealed sample, appear to contain a slight excess of positive charge, through the combined presence of Ce^{3+} on the Ca site and Ta^{5+} on the Ti site in the abundant zirconolite, and an excess of Ta over that required to compensate for the amount of Ce^{3+} on the Zr site in the Ce-rich zirconolite. Based on the results from the earlier zirconolite compositions, the most probable means of compensating for this excess of positive charge would be via cation vacancies. However, TGA revealed a 0.26% weight gain upon heating this sample to 1400°C in air. This value was greater than the 0.19% required to fully oxidise the Ce, indicating that some additional species, most likely Ti^{4+}, had also been reduced. Therefore, the most likely means of charge compensation within this reduced sample is by a combination of Ti^{3+} and cation vacancies. The lack of significant compositional variation in the oxidised Ce^{4+} zirconolites, after sintering at either 1200°C or 1400°C in air, leave these zirconolites with a additional positive charge. Given the total oxidation of the Ce and Ti, this excess of positive charge must be compensated by cation vacancies.

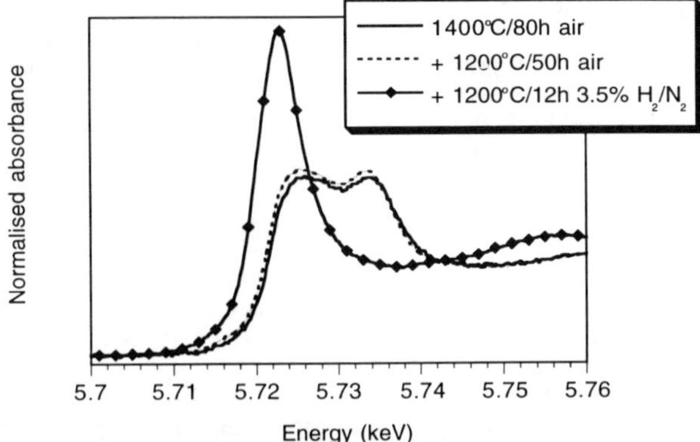

Figure 5. Ce L_{III} -edge X-ray absorption spectra showing the influence of various heat treatments on the Ce valence in $CaZr_{0.9}Ce_{0.1}Ti_{1.9}Ta_{0.1}O_7$.

Target: Ce^{4+} on Zr site - $CaZr_{0.8}Ce_{0.2}Ti_2O_7$

Sintering this zirconolite composition in air at 1400°C for 80 hours led to the development of three separate Ce-bearing phases. Two structurally distinguishable zirconolites were formed; one was the standard 2M polytype (~60%) whilst the other appears to be a hybrid interlayer structure intermediate between zirconolite-2M and pyrochlore. This has been labelled zirconolite-4M (~25%) [6]. The third phase was perovskite (~15%). The compositions of each of the phases are outlined below; zirconolite-2M ($Ca_{0.94}Ce_{0.15}Zr_{1.03}Ti_{1.89}O_7$), zirconolite-4M ($Ca_{0.91}Ce_{0.42}Zr_{0.79}Ti_{1.88}O_7$) and perovskite ($Ca_{0.72}Ce_{0.24}Zr_{0.02}Ti_{1.03}O_3$).

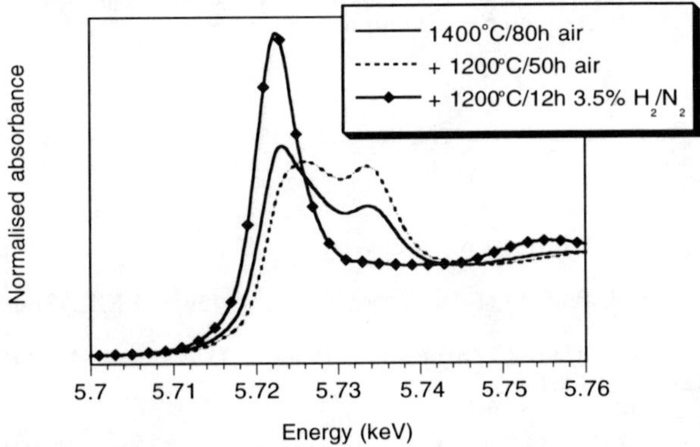

Figure 6. Ce L_{III} -edge X-ray absorption spectra showing the influence of various heat treatments on the Ce valence in $CaZr_{0.8}Ce_{0.2}Ti_2O_7$.

The Ce L_{III}-edge XANES spectra taken from this series of samples annealed under the various environments are shown in figure 6. Whilst the 1400°C/80h air sinter was not sufficient to completely oxidise the Ce ($\sim$65% Ce^{4+}) , this was achieved after the lower 1200°C air heat treatment. Trivalent Ce was formed after the sample was annealed at 1200°C/12h in 3.5% H_2/N_2.

Considering the sample after the 1200°C anneal in air, where the Ce was all tetravalent, each of the phases contain additional positive charge, which has probably been compensated for via cation vacancies. In the case of the reduced sample where the Ce was trivalent, the 2M-zirconolite appears to be compensated through the presence of an equivalent amount of Ce^{3+} on both the Ca^{2+} and Zr^{4+} sites. The 4M-zirconolite however seems to have an excess of negative charge, whilst the perovskite retains a excess of positive charge. Attempts to directly characterise the precise charge compensation mechanisms within an individual phase present in a multiphase sample is at best problematic; a more fruitful approach is to employ the principles obtained from the single phase samples to these more complicated multiphase systems. In light of this, the most probably means of charge compensation in the 4M-zirconolite is via oxygen vacancies whilst trivalent Ti could be expected in the perovskite [3].

CONCLUSIONS

The Ce valence in each of the zirconolites was clearly dependent on both the prevailing oxygen atmosphere and sintering temperature. The initial air sinter at 1400°C/80h resulted in the formation of a significant amount of tetravalent Ce in each of the zirconolites irrespective of the design. The proportion of tetravalent Ce did however vary between the different compositions, reflecting the role of crystal chemical forces. Significantly, annealing each of the zirconolites at a lower temperature in air (1200°C/50h) increased the proportion of tetravalent Ce, highlighting the known relative instability of Ce^{4+} at temperatures $\geq$ 1400°C in air. Annealing the zirconolites at 1200°C/12h in 3.5% H_2/N_2 led to complete reduction of Ce to Ce^{3+} in all instances.

The flexibility of the zirconolite structure to accommodate both trivalent and tetravalent Ce on either its Ca or Zr sites has been clearly demonstrated. However the solid solution limits for substitutions on the Zr site are lower, $\sim$15% for the standard zirconolite-2M structure, than for the Ca site. Substitutions beyond 15% on the Zr site usually result in the formation of the closely related zirconolite-4M, a hybrid between zirconolite-2M and pyrochlore.

In addition, the zirconolite has shown an innate ability to accommodate an in situ change in Ce valence without significant compositional restructuring. This has been facilitated predominantly through the acceptance of cation vacancies and in some instances trivalent Ti and oxygen vacancies. Evidence for these vacancies has been largely circumstantial, as direct characterisation is difficult due to the small quantities of vacancies required for charge compensation and their random distribution. We are proceeding to characterise these materials however with TEM, positron annibilation and high-resolution synchrotron-sourced XRD.

ACKNOWLEDGEMENTS

We thank S. Leung for assistance with the SEM and D.J. Cassidy for the thermal analysis results.

REFERENCES

1. A.E.Ringwood, S.E.Kesson, N.G.Ware, W.Hibberson and A.Major, Nature **278** 219 (1979).
2. E.R. Vance, C.J. Ball, M.G. Blackford, D.J. Cassidy and K.L. Smith, J. Nucl. Mater. **175** 55 (1990).
3. E.R. Vance, R.A. Day, Z. Zhang, B.D. Begg, C.J. Ball and M.G. Blackford, J. Solid State Chem. **124** 77 (1996).
4. B.D. Begg to be published (J. Nucl. Mater).
5. E.R. Vance, B.D. Begg, R.A. Day and C.J. Ball, in <u>Scientific Basis for Nuclear Waste Management XVIII</u>, edited by T. Murakami and R.C. Ewing (Mater. Res. Soc. Proc., Pittsburgh, PA, 1995), pp. 767-74.
6. A.A. Coelho, R.W. Cheary and K.L. Smith, (submitted to J. Solid. State Chem.)

SYNROC DERIVATIVES FOR THE HANFORD WASTE REMEDIATION TASK

E.R. VANCE, K.P. HART, R.A. DAY, M.L. CARTER, M. HAMBLEY, M.G. BLACKFORD
and B.D. BEGG
Australian Nuclear Science and Technology Organisation, Menai, NSW 2234, Australia
fax 61-2-9534-7179, phone 61-2-9717-3019, email erv@nucleus.ansto.gov.au

ABSTRACT

Three wt% each of Cs and Tc were mixed with the standard Synroc precursor and the ceramic was formed by hot-pressing. Attempts were made to incorporate the Tc as either metal or Tc^{4+}, using different redox conditions in processing. Volatile losses of Tc during calcination were $< 0.1\%$ in all cases. Short-term Tc leach rates when the Tc was present as a metal alloy were in the order of 10^{-4} $g/m^2/d$ at $90^{\circ}C$ with frequently changed water, and decreased with increasing leaching time. The valence of the Tc was monitored by X-ray absorption spectroscopy at the drying and calcination stages of the production. The general viability of Synroc/glass composites for immobilising the Hanford HLW sludges is further demonstrated by using further refinements of additive schemes for the inactive "All-blend" formulation and initial studies using the U-containing "All-blend" waste formulation.

INTRODUCTION

The Synroc methodology has two niche applications[1,2] to the US Tank Waste Remediation Task for the Hanford reservation. These are (a) immobilisation of Cs/Sr/Tc/TRU, separated out during cleanup of the waste liquids, and (b) immobilisation of HLW sludges.

Cs/Sr/Tc/TRU can all be readily incorporated[1,3] in the standard Synroc-C wasteform; however because of the higher achievable waste loadings, zirconolite-rich Synroc variants[4,5] are preferred for TRU elements. We have recently reported on the leaching and the metallic nature of Tc in Synroc-C[1], but have not dealt with the incorporation in Synroc of Tc by itself. Nor have we studied Tc volatility losses during processing, although we did prepare Tc^{4+} -substituted perovskite by simply sintering at $1500^{\circ}C$ in an argon atmosphere[1], with the indications being that Tc losses did not exceed 20% of inventory.

The overall strategy is to accommodate Tc in hot-pressed Synroc either in metallic form or as Tc^{4+} ions substituted for Ti ions, depending on the precise application in the Hanford remediation task- notably which ions would accompany Tc.

Here we present (a) volatility loss data on Tc in the calcination stage of Synroc processing; (b) early leaching data on Synroc containing only Cs, Sr and metallic Tc (alloyed with Fe and Ni); (c) X-ray absorption spectroscopy to indicate the Tc speciation in the drying and calcination stages and (d) more detailed electron microscopy of Tc -bearing alloys in Synroc-C.

In our glass/Synroc composite approach [2] to HLW sludges, we aim to prepare mixtures of Synroc phases- perovskite and zirconolite to immobilise REs and TRUs, spinels to immobilise 3d transition group process chemicals and silicate glass to immobilise other fission products, by melting at $\sim 1400^{\circ}C$. We have further optimised some of our earlier work with the (U, Th)-free version of the Feed 2 "All-Blend" waste formulation and have commenced work on the U-bearing "All-Blend" variant.

Mat. Res. Soc. Symp. Proc. Vol. 465 © 1997 Materials Research Society

EXPERIMENTAL

Cs/Tc-bearing Synroc samples were made with standard alkoxide/nitrate aqueous precursors, except for Tc which was introduced as NH_4TcO_4. The precursor composition[3] is as follows (wt%): Al_2O_3 (5.6); BaO (5.4); CaO (11.0); TiO_2 (71.4); ZrO_2 (6.6). Three wt % each of Cs_2O and Tc were also added. In samples in which the desired form of Tc was the metal, extra Fe, Ni and Cr were introduced to alloy with the Tc to try to lower the leach rate of Tc.

Attempts were made to reduce Tc^{7+}, which is volatile at calcination temperatures, to Tc^{4+} by boiling the mixtures in formic acid, using a 5:1 molar ratio of formic acid/(nitrate + pertechnetate). Calcination was carried out at 750°C, using flowing argon or 3.5% H_2/N_2 atmospheres when the Tc state in the final hot-pressed Synroc was targeted as Tc^{4+} or metallic respectively. After dry milling of the calcines to break up agglomerates and to achieve more intimate mixing, the powders were hot-pressed in collapsible metal bellows at 1250°C/20MPa. Ten wt% of Ti metal powder or 10 wt% of NiO powder was generally added before hot-pressing the samples in which the desired state of Tc was metallic or tetravalent respectively.

Samples were first characterised by X-ray diffraction and scanning electron microscopy [4]. Leaching measurements were made at 70°C by a modified MCC-1 method on polished discs, 1 cm in diameter and 2 mm thick. Seven-day PCT tests were also carried out[2]. Transmission electron microscopy was performed using a JEOL JEM 2000FX instrument equipped with a Tracor Northern Si(Li) EDS detector and a Link ISIS microanalyser.

X-ray absorption spectroscopy on the Tc speciation was carried out on finely powdered samples of material corresponding to the drying and calcination stages of the Synroc process. The measurements were performed on the radioactive line 4-2 at the Stanford Synchrotron Radiation Laboratory. Valence standards were dried NH_4TcO_4-bearing materials, together with TcO_2 and Tc formed by calcining NH_4TcO_4 in argon or 3.5%H_2/N_2 respectively at 900°C. The Tc speciation in these latter two standards was confirmed by XRD.

Volatile losses on calcination were studied by placing ~2 g (oxide equivalent) of the dried material in an alumina boat, placing the boat inside a ~60 cm.-long open-ended silica tube, and then inserting the assembly in a gas-tight alumina tube contained in a furnace, such that the silica tube projected at least 30 cm beyond the end of the furnace. The chosen calcination gas flowed through the experimental chamber. A water bubbler was also placed on the downstream side of the furnace. The experimental setup is shown in Fig. 1.

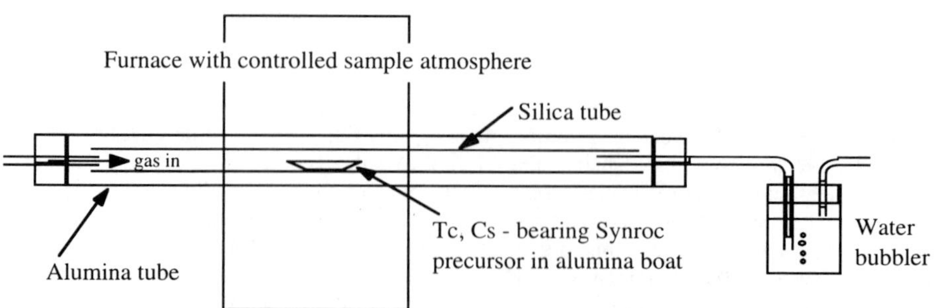

Figure 1. Experimental setup for Tc volatility measurements during calcination.

After calcination the silica tube was closely inspected for deposited material along its length. The tube was then subjected to alkali (1 M NaOH solution) and acid washes (2M HNO$_3$ solution) and the washings were analysed for Tc and Cs, by liquid scintillation counting and atomic absorption spectroscopy respectively. The water in the bubbler was similarly analysed.

Melting of the "All-blend" plus additive mixtures in alumina crucibles was carried out at temperatures of ~1400°C, using flowing atmospheres of argon or air.

RESULTS AND DISCUSSION

Tc and Cs volatility

When formic acid pretreatment was used, only a few drops of condensed steam, and no more than ~ 1 µg of solid material deposited on the inside of the silica tube. However when no formic acid was used and calcination was carried out in argon, a clearly observable deposit of reddish-brown colored material was deposited on the cold end of the tube. This was all removed in the first (alkali) wash.

The results, shown in Table I, represent Tc totals found in the washings plus the water bubbler, and show that Tc losses do not exceed 0.1%, especially under the reducing atmosphere.

Table I. Measured Tc losses by volatilisation

Formic acid pretreatment	Calcination atmosphere	Tc loss (%)
Yes	3.5% H$_2$/N$_2$	1×10^{-5}
Yes	Ar	3×10^{-4}
No	3.5% H$_2$/N$_2$	1×10^{-4}
No	Ar	1×10^{-1}

In each case, the measured Cs loss was ~0.1%, a value in good agreement with previous determinations[3]. Obviously in a real process, losses of Tc etc would occur via dust escape, but the measurements just described show that the potential for intrinsic Tc volatility losses is very small.

Leaching data

Because Tc metal can tarnish in water, it would appear to be necessary in the immobilisation-as-metal approach to alloy the Tc with suitable elements. In a first study, we attempted to produce a Synroc containing 3 wt% each of Cs$_2$O and Tc, in which the Tc was targeted to form an alloy of composition (wt%): Fe(50); Ni(20); Cr(10); Tc(20), comprising 20 wt% of the Synroc sample. The sample was hot-pressed at 1200°C/20 MPa in a graphite die. SEM showed that the Tc alloyed with the Fe and Ni, but that the Cr had entered the ceramic phases, presumably as Cr^{3+}. Mixing at the ~100 µm scale was poor. Even so, the Tc leach rates, given below in Table II, were in order-of-magnitude agreement with those reported[1] for Synroc-C, in which the Tc is variously alloyed with Pd group elements, 3d transition metal corrosion product elements, and Mo and Ag. The leach rates of Cs were also comparable with those in Synroc-C[3]. SEM did not reveal any significant differences between the microstructures before and after leaching.

Table II. Differential leach rates* (g/m^2/d) of Tc and Cs from Synroc sample doped with 3 wt% each of Tc and Cs in modified MCC-1 test at 70°C in polythene containers.

Leach Period, days	Tc	Cs
6	1.5×10^{-4}	0.14
28	1.2×10^{-5}	0.032
56	2.5×10^{-6}	0.014

* Contributions of species deposited on vessel walls were measured after acid stripping, and found to contribute < 1% to totals.

More recent work shows that the addition of Ti metal to the calcine before hot-pressing can allow the Cr to enter the alloy phase, although whether this will lower the Tc leach rate has yet to be determined.

Parallel attempts are being made to produce Tc^{4+} in the hot-pressed samples. The main practical difficulty arises from the fact that a graphite die is used to provide the pressure and the heating via radiofrequency absorption and this creates a very reducing environment. Future work will involve complete sealing of the can prior to hot-pressing.

<u>X-ray absorption spectroscopy of Tc speciation at early stages of processing routes.</u>

Fig 2 shows the near-edge behaviour of the Tc K absorption edge at 21.064 keV in dried and calcined preparations, together with those of the Tc valence standards. In summary, the use of formic acid did not change the Tc valence state from 7+, calcination in argon atmosphere yielded Tc valence states between +4 and +7, and calcination in H$_2$/N$_2$ gave valence states between Tc^0 and Tc^{4+}. More subtle features, such as the pre-edge peaks in the Tc^{7+} spectra at 21.048 keV, do not concern us here. The principal aim of the work was achieved in the sense that we have demonstrated that the very small volatility losses were approximately consistent with the observed average valence states of Tc, and the higher average valence of Tc and Tc loss in the calcine treated in argon as against H$_2$/N$_2$ is consistent with expectations .

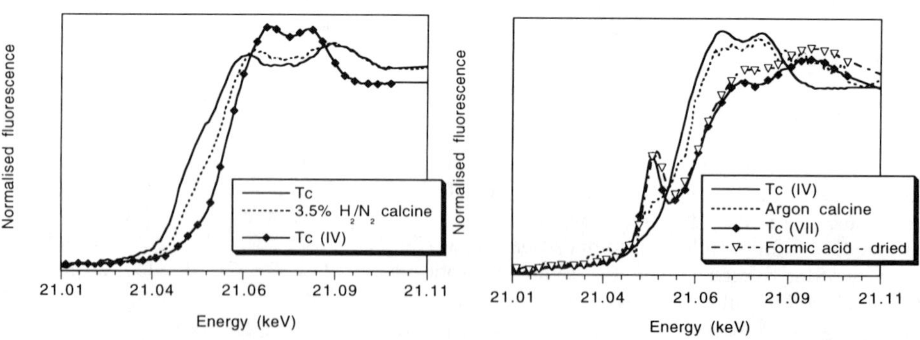

Figure 2   XANES spectra from Tc-bearing preparations after drying and calcination stages of Synroc production, together with spectra from Tc metal, TcO$_2$ (Tc (IV)) and NH$_4$TcO$_4$ (Tc (VII)) standards.

The lowering of Tc volatility by pretreatment with formic acid, even though the Tc valence is unaffected, needs comment. Nitrate was present from the Cs addition. Chemical denitration will take place during the formic acid pretreatment and there will be no oxidising NOx products at the early stages of calcination. Hence there will be less tendency for the Tc to remain in the volatile Tc^{7+} state in the argon calcination atmosphere than if NOx was present.

Further studies of Tc occurrence in Synroc

We have previously reported on SEM studies of the alloy phases in Synroc containing simulated PW-4b waste in which Tc is present, that the Tc concentration in the individual alloy globules was variable. In TEM studies however it was obvious that the globules were polycrystalline and that the Tc was unequally partitioned into the microcrystals constituting each globule. The alloy particles were found to occupy many of the triple points between the ceramic grains and were frequently composed of discrete components with quite different compositions. Occasionally a small amount of porosity and/or glassy material indicated elevated levels of S and Ti, but no Tc. A typical example of an alloy is shown in Fig 3, along with spectra collected from the three discrete components. Table III lists the measured compositions of the three components. In analysing these spectra it was assumed that the alloy density was $10g/cm^3$ and the sample thickness was 200 nm.

Synroc/Glass Composites for Immobilisation of HLW sludges

We have extended our previous studies[2] of the melting route to producing Synroc/glass composites for Hanford HLW sludges. As mentioned earlier, the approach has been to immobilise Na in nepheline, and rare earths/actinides in the Synroc phases zirconolite and perovskite, while allowing iron group transition metal ions and Al to form spinels, with some glass also being produced to incorporate Cs and other radionuclides. Thus we have studied[5] the effects of Al additions, together with Ca, Ti and Si additions as studied previously for the actinide-free "Feed 2" formulation[1]. As expected from the earlier work, the co-additions of 5 or 10 wt% of Al_2O_3 promoted the formation of whitlockite rather than the water-soluble (Na, Ca) phosphate by diverting more Na into additional nepheline. Few significant differences were observed in the either the microstructures or the PCT leaching results of air- and argon-melted samples. The PCT results themselves for the samples melted in argon and with the 5 wt% of Al_2O_3 addition gave rates of <1 $g/m^2/d$ for all elements, including Al, K, Na, Cs, Mo, Na , P and Si. The corresponding results for the 10 wt% of Al_2O_3 co-addition were broadly similar except that the Mo leach rate was 2 $g/m^2/d$.

We have since made a further addition of 5 wt% SiO_2 to the formulation to which the 5 wt% Al_2O_3 co-addition was made, to suppress porosity and cracking via an increased amount of glass. This has been successful, apart from development of a thin foam layer on the surface of the glass melt when an argon (but not an air) atmosphere was used, with the usual melting temperature of $1400^{\circ}C$. This sample consisted of (wt%): HLW (53); CaO (6); Al_2O_3 (9); SiO_2 (22); TiO_2 (10). When melted in an air atmosphere (Fig. 4), this formulation resulted in the desired zirconolite and perovskite as hosts for the REE, as well as Sr in the case of perovksite. The iron-group elements formed spinel. Most of the Na has entered nepheline, while suitable amounts of glass remain to contain Cs, K, Bi, S, and the rest of the Na. Phosphorus has been stabilised in whitlockite, leaving the overall P content of the glass very low. Melting in an argon atmosphere produced a similar phase assemblage except that zirconia formed as well as of zirconolite, and Bi formed an alloy with Ni. The PCT leach rates of the same significant elements (Na, K, Bi, Ca, Ce, Cr, Cs, Fe, La, Mn, Nd, Ni, P, S, Si, Sr and Zr) were all <0.1 $g/m^2/d$, for both melting atmospheres.

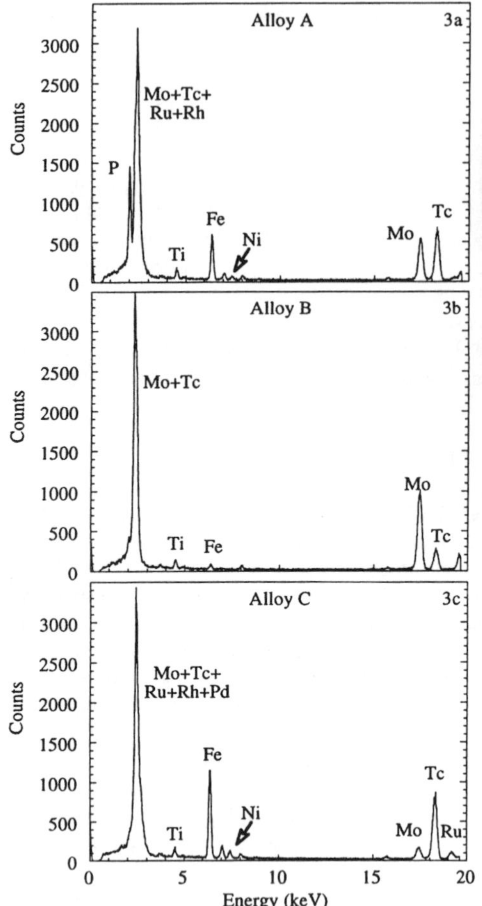

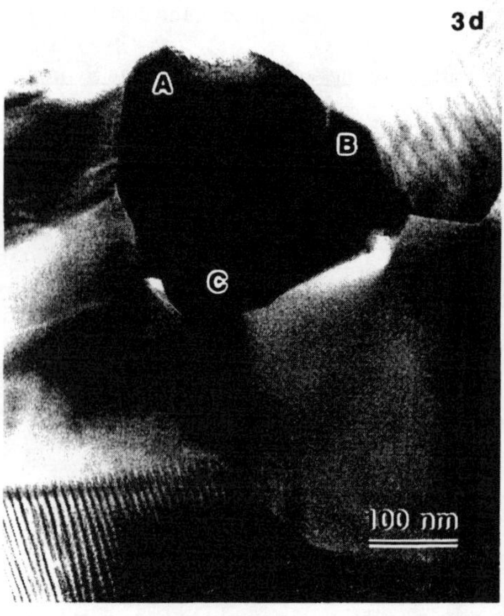

Table III. Composition (wt%) of each alloy component shown in figure 3d.

element	alloy A	alloy B	alloy C
P	11		
Fe	8	1	17
Ni	1		2
Mo	32	77	8
Tc	44	22	54
Ru	1		3
Rh	3		13
Pd			3

Figure 3. a), b) and c) EDS spectra from each of the discrete alloy components identified in the TEM micrograph shown in d).

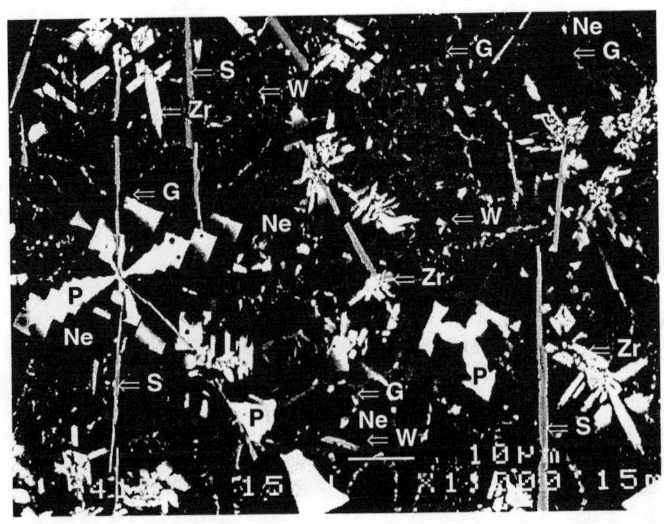

Figure 4   SEM backscattered electron micrograph showing the phases formed in a Synroc/glass composite made from an actinide-free simulant of the Hanford "Feed 2" HLW sludge. Phases present include: zirconolite (Zr), perovskite (P), spinel (S), nepheline (Ne), whitlockite (W), and glass (G).

All-Blend formulation with Zr substituted for U, Th

The composition of the active "All-blend" formulation of HLW sludge is given in Table IV. Initial experiments utilised substitution of Zr for U and Th on a molar basis. The additives were (wt%): CaO (8); TiO_2 (20) and SiO_2 (12). These melted readily at 1350^oC and yielded microstructures broadly similar to those[2] obtained with the inactive "Feed 2".

Table IV. Composition of active "All-blend" HLW sludge simulation.

Component	wt%	Component	wt%	Component	wt%
Al_2O_3	13.0	B_2O_3	0.08	Bi_2O_3	1.95
CaO	2.06	CdO	0.09	CeO_2	2.73
Cl	-	Cr_2O_3	0.45	Cs_2O	0.012
F	0.56	Fe_2O_3	11.0	K_2O	0.22
La_2O_3	0.43	MgO	0.08	MnO_2	1.82
MoO_3	0.07	Na_2O	25.3	Nd_2O_3	0.071
NiO	2.27	P_2O_5	4.71	PbO_2	0.32
SO_3	0.34	SiO_2	10.0	SrO	0.41
WO_3	0.17	ZrO_2	7.08	ThO_2	0.15
U_3O_8	14.3				

The active formulation (Table IV) was then studied, using U substituted for Th, and the same additions of CaO, TiO_2 and SiO_2 as used above. The microstructures for samples melted in air and argon at $1350^{\circ}C$ consisted mainly of glass, nepheline, carnegieite and perovskite, plus for the different atmospheres- zirconolite + $(Na,Ca)_2U_2O_7$ (air) and pyrochlore (argon). However both formulations contained the soluble (Na, Ca) phosphate seen previously[2]. This can be dealt with by adding Al_2O_3(see above) and some extra SiO_2, which will be likely to convert the carnegieite to nepheline[2]. The alkali uranates have poor durability and melting in air is undesirable from that point of view. Further work will therefore concentrate on co-additions of Al_2O_3 and SiO_2 and using argon melting atmospheres.

CONCLUSIONS

Using alloying additions of Fe and Ni, the standard Synroc precursor can be used to incorporate waste Tc as a very leach-resistant metallic species. The incorporation of Tc as Tc^{4+} requires a narrower redox window but offers the potential of very large waste loadings, with all the Ti^{4+} being potentially replaceable by Tc^{4+}. In both approaches, the losses of Tc by volatility upon calcination are very small, <0.1%, with essentially no losses being expected in hot-pressing. X-ray absorption spectroscopy has been used to define Tc valence states at various early stages of Synroc preparation, in which the essential non-crystallinity of the material precludes the use of many other possible techniques. Further information was obtained on the micro-disposition of Tc in Synroc-C. Microstructures for the U-containing "All-blend" Hanford HLW sludge formulation were similar to those obtained previously in the preliminary sets of additives for the inactive versions.

ACKNOWLEDGMENTS

We thank S. D. Conradson for assistance with XAS measurements, E. Loi and J. Weir for the leaching measurements, A. Brownscombe and P. J. Angel for preparation of the Tc-bearing samples, and A. Jostsons for discussions. The synchrotron beamtime was provided by the Stanford Synchrotron Radiation Laboratory which is supported by DOE and NIH.

REFERENCES

1. E. R. Vance, M. L. Carter, R. A. Day and A. Jostsons, A Melting route to Synroc for Hanford HLW Immobilization, in Scientific Basis for Nuclear Waste Management XIX, Eds. W. M. Murphy and D. A. Knecht, Materials Research Society, Pittsburgh, PA, USA, p. 289-96 (1996).
2. K. P. Hart, E. R. Vance, R. A. Day, B. D. Begg, P. J. Angel and A. Jostsons, Immobilization of Separated Tc and Cs/Sr in Synroc, ibid., 281-7.
3. A. E. Ringwood, S. E. Kesson, K. D. Reeve, D. M. Levins and E. J. Ramm, in Radioactive Waste Forms for the Future, Eds. W. Lutze and R. C. Ewing, Elsevier, North-Holland, Amsterdam, Netherlands, pp 233-334 (1988)
4. E. R. Vance, R. A. Day, Z. Zhang, B. D. Begg, C. J. Ball and M. G. Blackford, J. Solid State Chem. 124, 77 (1996)
5. E. R. Vance, M. L. Carter, R. A. Day, B. D. Begg, K. P. Hart and A. Jostsons, Synroc and Synroc-glass Composite Waste Forms for Hanford HLW Immobilization, Proceedings of the Spectrum '96 International Topical Meeting on Nuclear and Hazardous Waste Management, American Nuclear Society, IL, USA, 2072 (1996)

DISSOLUTION OF SYNROC IN DEIONISED WATER AT 150°C

KATHERINE L. SMITH, MICHAEL COLELLA, GORDON J. THOROGOOD, MARK G. BLACKFORD, GREGORY R. LUMPKIN, KAYE P. HART, KATHRYN PRINCE, ELAINE LOI and ADAM JOSTSONS

Materials Division, Australian Nuclear Science and Technology Organisation, Private Mail Bag 1, Menai, NSW 2234, Australia.

ABSTRACT

Synroc containing 20 wt% simulated high level waste (HLW) was subjected to two sets of leach tests at 150°C where the leachant was and was not replaced during the test (replacement and non-replacement testing). The leachant was a KH-phthalate buffered solution (pH 4.2). Samples were characterised before and after leach testing using SEM, AEM and SIMS. Elemental concentrations in leachates were measured using ICP-MS. In concert with the findings of i) a dissolution study of perovskite in a flowing leachant and ii) a previous Synroc dissolution study (wherein Synroc containing 10 wt% simulated HLW was subjected to periodic replacement, leach testing in deionised water at 150°C), the results of this study show that when the leachant replacement frequency is varied from 7 d to the duration of the test, there is no effect on leach rate or leaching mechanisms.

INTRODUCTION

Synroc is a polyphase titanate ceramic designed for the safe disposal of high-level nuclear waste (HLW). It consists primarily of hollandite, zirconolite, perovskite, and rutile-Magnéli phases. Additional minor phases include hibonite, loveringite, calcium aluminium titanate (CAT), and intermetallic phases [1].

Modified MCC-1 type leach testing of Synroc, where the leachant was periodically replaced [2], and flow-through leach testing of perovskite [3] show that Synroc and its constituent phases are highly durable in aqueous fluids. For example, when Smith et al. [4] subjected Synroc containing 10 wt% simulated HLW to leach testing in deionised water at 150°C (wherein the leachant was replaced every 7 d up to 84 and then every 28 d up to 532 d), the average normalised leach rate (based on leach rates of Ca, Cs, Mo, Ba and Sr, which are not significantly subject to reprecipitation) after 336 d was ~10^{-4} g m^{-2} d^{-1} [2,4].

Practical considerations determine the periods between leachant replacement in Synroc leach tests. In short term (<~84 d) Synroc leach tests conducted at 150°C, it is usual to replace the leachant every 7 d, because it takes 8 hrs to cool down a sample and vessel, replace the leachant and reheat the sample and vessel. When leach tests run for longer times (84 < t < 532 d), the time between leachant replacements is increased (to 28 d or longer), to maintain elemental levels above ICP-MS detection limits.

In this study, Synroc containing 20 wt% simulated high level waste (HLW) was subjected to two sets of leach tests where the leachant was and was not replaced during the test in KH-phthalate buffered solution (pH 4.2) at 150°C. The tests were conducted in stainless steel bombs fitted with teflon liners. The results of this study are compared with the results of previous leach tests on Synroc containing 10 wt% simulated HLW in deionised water at 150°C and leach tests on perovskite in flowing leachants at 90°C. The purpose of this work and the comparisons drawn with previous studies is to establish if Synroc's elemental leach rates or leach mechanisms show any variation with the frequency of leachant replacement (when it is varied from 7 d to the duration of the test). In some tests the leachant was replaced after the first day to check for initial release effects.

METHODOLOGY

Synroc containing 20 wt% simulated Purex waste (PW-4b-D) was prepared via the alkoxide route [5]. Polished discs 10 mm in diameter and ~2 mm thick were cut and triplicate sets of discs

Mat. Res. Soc. Symp. Proc. Vol. 465 © 1997 Materials Research Society

were subjected to one of two types of leach tests in potassium-hydrogen-phthalate (KH-phthalate) buffered solution (pH 4.2) at 150°C lasting up to 365 d. In one type of test, the leachant was replaced during the test (Table I shows the replacement schedule) and in the other the leachant was not replaced (replacement and non-replacement testing, RT and NRT, respectively). KH-phthalate buffered solution was used because EQ3NR/6 modelling suggested that at 150°C it would exhibit a pH of ~6.3 which is in the groundwater pH range for proposed HLW repositories. However the buffered solution maintained a pH of 4.2 throughout the tests. The volume of leachant was 20 mL, and the surface area to volume ratio was 0.1 cm^{-1}. The leachates were analysed using inductively coupled plasma mass spectrometry (ICPMS).

Table I. Replacement schedule for leachant in replacement tests.

Total test duration	1 d	7 d	28 d	91 d	182 d	365 d
Leachant replaced at			7 d	1 d	1 d, 91 d.	1 d, 91 d, 182 d.

Secondary ion mass spectrometry (SIMS) data were collected as a function of depth from unleached and leached discs using a Cameca IMS 5f magnetic sector SIMS using a 12.5 kV O^{2+} primary ion beam. The beam was rastered over an area of 250 x 250 μm. An aperture was used to collect positive secondary ions from a ~60 μm diameter region in the centre of the raster area. This configuration eliminated any potential edge effects derived from the sputtered crater.

Secondary phases found on the surfaces of discs after leach tests were investigated using scanning electron microscopy (SEM) and analytical transmission electron micrsocopy (AEM). SEM investigations were performed with a JEOL 6400 operated at 15 kV in standard secondary electron and backscattered electron modes. AEM was performed using either a JEOL 2000FX TEM operated at 200 kV or a VG601 FEG/STEM operated at 100 kV. The JEOL 2000FX was equipped with a Tracor Northern Si(Li) energy dispersive X-ray spectrometer (EDS) with a Be window and a Link ISIS microanalyser. The VG601 was equipped with a Link windowless Si(Li) EDS and a Link ISIS microanalyser. AEM/EDS spectra from the JEOL 2000FX were quantitatively analysed with the ISIS software package TEMQuant, using empirically determined correction factors for Al, Ca, Ti, Fe, Sr, Y, Zr, Ce and Gd and theoretical correction factors calculated by TEMQuant for Mo and Nd. AEM/EDS spectra from the VG601 FEG/STEM were only qualitatively analysed with TEMQuant using theoretical correction factors.

Two types of AEM samples were investigated - replicas and "cross-sectional" specimens. The replication technique [6] was used to extract loosely bound secondary phases from the surface of leached discs onto a replica carbon film. Cross-sectional AEM specimens were prepared by slicing a leached disc across its diameter, glueing two of the flat surfaces together, sectioning the semi-cylindrical block perpendicular to the originally circular leached surface and then preparing ion beam thinned AEM specimens.

RESULTS

Leach data

Figure 1 shows the normalised cumulative mass loss of particular elements from Synroc as a function of leaching time. These data were collated from our present leach tests (RT and NRT) on Synroc containing 20 wt% simulated waste and from leach tests on Synroc containing 10 wt% simulated waste [2,4] wherein the leachant was replaced every 7 d up to 84 d then every 28 d thereafter (frequent replacement testing, FRT).

Cumulative normalised leach data for Ca are shown in Figure 1a. The three curves are parallel within experimental error and level out at ~1 g m^2. Graphs of cumulative normalised leach data for Sr, Al, Ba and Mo are similar to those in Figure 1a and also level out at at ~1 g m^2. Ca and Sr predominantly result from the dissolution of perovskite. Al, Ba and Mo result from the corrosion of aluminates, hollandite and intermetallic alloy particles respectively.

Cumulative normalised leach data for Gd are shown in Figure 1b. The three curves are parallel within experimental error and level out at ~0.1 g m^2. Graphs of cumulative normalised leach data for Y, Nd and Ce are similar to those in Figure 1b and also level out at at ~0.1 g m^2. It is surprising that Ce, Nd, Gd and Y all show similar leach behaviour because Nd and Ce preferentially partition into perovskite (the least resistant major phase) whereas Y and Gd preferentially partition into zirconolite (one of the most resistant major phases). This suggests that the rare earth elements, and in particular Nd and Ce quickly precipitate from the leachate. The presence of Ce and Nd-rich monazite (rare earth phosphate) on replicas in this and a previous study [7] support this suggestion.

<u>SIMS</u>

Figure 2 shows SIMS data collected from a disc leached for 365 d without leachant replacement. Relative to the levels in the unleached bulk of the sample: i) Ca (primarily indicative of perovskite) is depleted to a depth of ~250 nm, ii) Ba and Zr (indicative of hollandite and zirconolite respectively) are possibly depleted to a depth of ~10 nm, and iii) the level of Ti is enhanced to a depth of ~250 nm.

<u>Scanning Electron Microscopy</u>

SEM showed no differences between samples which were leached under different replacement regimes for the same total test duration. In agreement with previous studies in deionised water SEM shows that after leaching for 7 d at 150°C, surficial perovskite grains have been partially replaced by secondary anatase, occasional grains of secondary material are scattered across the other major phases and some alloy particles are highly corroded (Figure 3a). After leaching for 365 d at 150°C, surficial perovskite grains have been covered with anatase, the other major phases are largely covered by material with grain size smaller than that of the secondary anatase (which AEM subsequently showed to be Ti/Fe oxide/hydroxides) and bright pillars (composed of the elements present in the unleached intermetallic alloys) rise out of holes previously filled by intermetallic alloys (Figure 3b).

<u>Analytical Electron Microscopy</u>

Cross-sectional TEM of discs leached for 365 d show that irrespective of leachant replacement regime, perovskite is the only major waste-bearing phase to suffer measurable corrosion (Figure 4). After 365 d, surficial perovskite grains were dissolved to an average depth of 250 nm and polycrystalline plugs of large (~200 Å long) crystals block the holes previously filled by perovskite. Much smaller (~10 Å long) crystals lie on the surfaces of zirconolite, hollandite and rutile/Magneli phases (Figure 4c). Qualitative EDS analyses (performed using the VG601 FEG/TEM) show that the polycrystalline plugs and the small crystals on the surface of zirconolite hollandite and rutile are Ti/Fe oxide/hydroxides. The Ti/Fe ratios of the plugs and the small crystals are ~90/10 and ~50/50 respectively.

Figure 5 shows a typical EDS line scan of a corroded perovskite grain (collected using the VG601 FEG/TEM). The following observations can be made about this line scan in particular and line scans of similar areas in general.
i) The amount of Ti in plug crystals is much greater than that in perovskite.
ii) The dissolution of perovskite is not entirely congruent. There is a thin layer (~40 nm wide) at the edge of the underlying perovskite grain which is depleted in Ca and Sr relative to the bulk of the grain. This layer is not an artifact of electron beam spreading in the specimen as we calculate that the estimated sample thickness of 500 Å would result in a beam diameter of < 10nm.
iii) The amount of Ca, Sr and rare earth elements (REEs) in the bulk of the perovskite grain vary due to original compositional zoning (Figure 5b). As one would expect, the amount of Ca+Sr+REEs stays constant within the bulk of the unaltered perovskite (Figure 5c).

Three phases were found on replicas from the surfaces of discs leached for 365 d in tests where the leachant was not replaced - large (~200 Å long) anatase crystals (~90 wt% Ti / 10 wt% Fe), Ce and Nd-rich monazite (~0.5 μm long) and Fe oxide/hydroxide crystallites (<10 Å). It is likely that the iron in all the secondary phases is due to corrosion of the stainless steel bomb and diffusion of aqueous iron species into the leachate.

DISCUSSION

Comparison of Ca leach data collected in this and previous Synroc leaching studies (Table II) shows that in leach tests lasting approximately one year, the Ca leach rate changes by a factor of less than two, if the frequency of leachant replacement is changed from 7 d to the duration of the

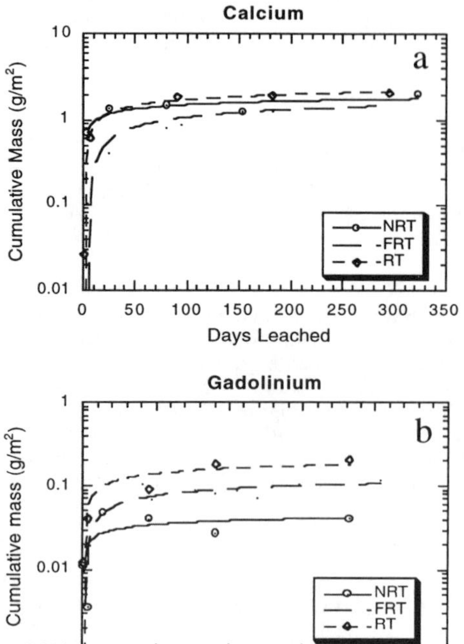

Figure 1. Cumulative normalised mass loss of a) Ca, and b) Gd as a function of leaching time. NRT = non-replacement testing. FRT = frequent replacement testing. RT = replacement testing (Table I). Synroc used in NRT and RT contains 20 wt% simulated HLW and was leached in KH-phthalate buffered solution. Synroc used in FRT contains 10 wt% simulated HLW and was leached in deionised water.

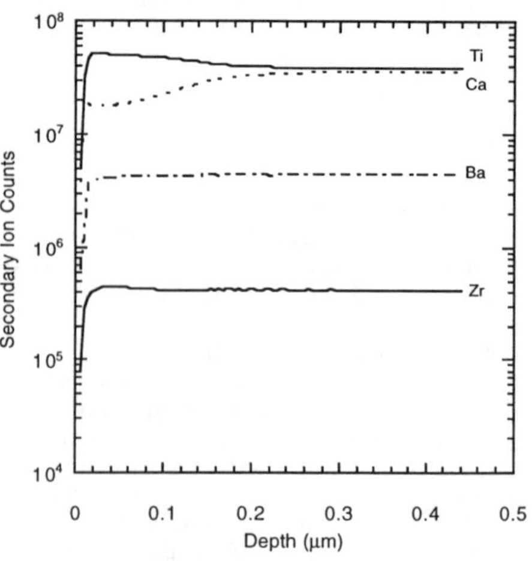

Figure 2. SIMS data from Synroc disc subjected to NRT for 365 d.

Figure 3. SEM images of Synroc discs subjected to NRT for a) 7 d and b) 365 d.

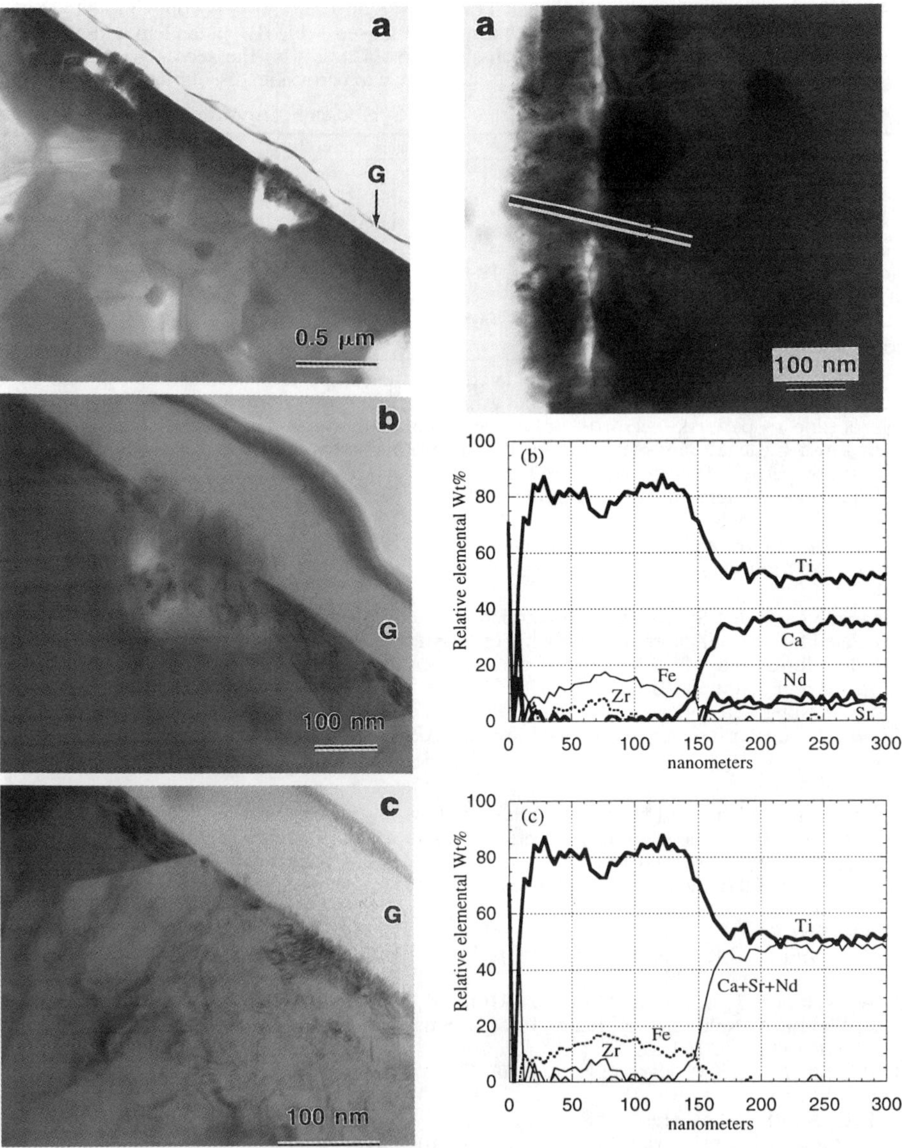

Figure 4. Micrographs showing cross-sections of a disc leached for 365 d. a) General view. b) Corroded perovskite grain plugged with anatase. c) Fine scale secondary phase on rutile (R). G = glue

Figure 5. a) Position of EDS line scan across a corroded perovskite grain in cross-section of disc leached for 365 d. b) & c) Composition along line in a).

test. Furthermore, the normalised Ca leach rate from Synroc at 150°C is comparable to that of perovskite measured in flow through leach tests at 90°C (see Table II). In tandem with the AEM results of this and previous studies which show no Ca in any of the secondary phases, this suggests that the Ca in Synroc leachates is primarily due to perovskite dissolution.

Table II. Normalised Ca leach rates from Synroc and perovskite after ~200 d .

Test type and reference	Leachant	Temperature	Ca leach rate
Replacement testing of Synroc (this study)	KH-phthalate (pH 4.2)	150°C	~1 x 10^{-3} g m^{-2} d^{-1} (close to detection limit)
Frequent replacement testing of Synroc [2,4]	deionised water	150°C	5 x 10^{-3} g m^{-2} d^{-1}
Non-replacement testing of Synroc (this study)	KH-phthalate (pH 4.2)	150°C	7 x 10^{-3} g m^{-2} d^{-1}
Flow through testing of perovskite [3]	buffered solution (pH 6.1)	90°C	4 x 10^{-3} g m^{-2} d^{-1}

Using the perovskite abundance calculated from the compositions of the major phases [7] and data from micrographs of cross-sectional AEM specimens, the normalised Ca mass loss from Synroc due to perovskite corrosion can be calculated. Such a calculation shows that the normalised Ca leach rate measured by ICPMS in this study can be accounted for by the Ca released by perovskite corrosion.

There is good agreement between the SIMS and AEM results. After non-replacement tests lasting 365 d, SIMS shows that Ca is depleted and Ti is enhanced to a depth of ~250 nm from the surface of Synroc discs. This correlates with the AEM observation that surficial perovskite grains have been eaten out to a depth of ~250 nm and that anatase has partially filled the resulting holes.

On the basis that Synroc could be modelled as a uniform mixture of elements, Smith et al.[1] calculated that the bulk normalised leach rate of Synroc at 150°C in deionised water corresponded to a dissolution rate of ~0.1 nm/d. If one assumes that perovskite is the only phase to suffer corrosion and that it constitutes 25 vol% of Synroc, calculations similar to Smith et al.'s suggest that perovskite will corrode at the rate of ~0.4 nm/d which is comparable (only a factor of two different) to the corrosion rate estimated from the AEM data in this study.

CONCLUSION

The results of this and previous studies [2,4] show that when Synroc is leached at 150°C for periods up to 365 d, the leach rate, order and way in which primary Synroc phases are attacked and the speciation of secondary phases are unaffected if the frequency of leachant replacement is changed from 7 d to the duration of the test.

REFERENCES

1. K.L. Smith, G.R. Lumpkin, M.G. Blackford, R.A Day and K.P. Hart (1992), J. Nucl. Mater. 190, 287-294.
2. A. Jostsons, K.L. Smith, M.G. Blackford, K.P. Hart, G.R. Lumpkin, P.J. McGlinn, S. Myhra, A. Netting, D.K. Pham, R. St.C. Smart and P.S. Turner (1990) NERDDP Project No. 1319, pp. 261.
3. P.J. McGlinn, K.P. Hart, E.H. Loi and E.R.Vance (1995), Mater. Res. Soc. Proc., 353, 847-854.
4. K.L. Smith , K.P. Hart, G.R. Lumpkin, P.J. McGlinn, J. Bartlett , P. Lam and M.G. Blackford (1991), Mat. Res. Soc. Symp. Proc., 212, 167-174.
5. A.E. Ringwood, S.E. Kesson, K.D. Reeve, D.M. Levins and E.J. Ramm, in <u>Radioactive Waste Forms for the Future</u>, edited by W. Lutze and R.C. Ewing (North-Holland, Amsterdam, 1988), p. 233.
6. G.R. Lumpkin, K.L. Smith and M.G. Blackford (1991), J. Mater. Res. 6, 2218-2233.
7. G.R. Lumpkin, K.L. Smith and M.G. Blackford (1995), Mat. Res. Soc. Symp. Proc., 353, 855-862. .
8 . G.R. Lumpkin, K.L. Smith and M.G. Blackford (1995), J. Nuc. Mater., 224, 31-42.

ALTERATION OF COLD CRUCIBLE MELTER TITANATE-BASED CERAMICS: COMPARISON WITH HOT-PRESSED TITANATE-BASED CERAMIC

T. ADVOCAT*, G. LETURCQ*, J. LACOMBE*, G. BERGER**, R.A. DAY***, K. HART***, E. VERNAZ* AND A. BONNETIER*

*Commissariat à l'Energie Atomique (CEA) Centre d'Etudes de la Vallée du Rhône, DCC/DRDD, Service de Confinement des Déchets, BP 171, F–30207 Bagnols-sur-Cèze Cedex, France
**Université Paul Sabatier UMR5563, 38 rue des 36 Ponts, F–31400 Toulouse, France
***Austalian Nuclear Science & Technology Organization (ANSTO), Private Mail Bag 1, Menai, NSW 2234, Australia

ABSTRACT

Synroc ceramics were synthesized in an induction-heated cold crucible at laboratory scale (1 kg) from an oxide mixture, and at industrial prototype scale (45 kg) from Synroc previously produced by sintering under load at high temperature. After melting, both materials contained the major phases of Synroc-C. The chemical durability of both melted materials, as determined by static leaching of powder samples in initially pure water at $90°C$ with an SA/V ratio of $20000 \, m^{-1}$, was equivalent to that of conventional hot-pressed Synroc-C. Cerium, used in this investigation to simulate the presence of tri- and tetravalent actinides, was found in steady-state concentrations on the order of 1 ppb (i.e. $NL(Ce) \leq 10^{-6} \, g \cdot m^{-2}$). The concentration in the leachates was independent of the initial CeO_2 content of the Synroc (at least up to 10 wt%); moreover, it is similar to the results obtained with hot-pressed Synroc-C specifically formulated for conditioning long-lived actinides.

INTRODUCTION

Synroc-C is a polyphase, fine-grain ($\leq 1 \, \mu m$) titanate ceramic comprising the following major phases[1]: zirconolite $CaZrTi_2O_7$ (30 wt%), perovskite $CaTiO_3$ (20 wt%), hollandite $Ba(Al,Ti)_2Ti_6O_{16}$ (30 wt%) and titanium oxides Ti_nO_{2n-1} (15 wt%). Additional minor phases consisted mainly of complex oxides, aluminates and metallic alloys (Ru, Mo, Tc, Pd, Fe, Ni, Cr, P). The conventional fabrication process consists in mixing the high-level solution (simulated in this investigation) with a ceramic precursor, then drying, calcining and hot-pressing the mixture for 2 hours at $1200°C$ and 20 MPa under reducing conditions[2]. The material and process were initially developed to condition fission product solutions generated by reprocessing spent fuel from commercial reactors[3]; this type of nuclear waste is now conditioned industrially in borosilicate glass matrices[4]. However, current research on advanced chemical separation processes for the long-lived actinides (Np, Am, Pu, Cm) and the possible option of geological disposal if transmutation is not adopted, have prompted renewed investigation of Synroc as a potential ultimate containment material. The ability of Synroc phases (primarily zirconolite) to incorporate substantial quantities of actinides (e.g. 10% PuO_2), their resistance to aqueous leaching and to significant irradiation doses without any major reduction in their chemical durability has been demonstrated by several studies[5,6,7,8].

The objective to this investigation was to confirm the technical feasibility of producing Synroc by cold crucible melting, and to ensure that the resulting material exhibits the same chemical durability as conventional hot-pressed Synroc-C. Fabricating Synroc from an oxide mixture by melting followed by controlled cooling is not a new idea, and Ringwood obtained this result at about $1330°C$ in 1978[3]. Nevertheless, the development of melting pots capable of withstanding high temperatures, and notably the induction-heated cold crucible melter, permitted the first real demonstration of a controlled melting process by a Russian team in 1993[9]. This process has three significant advantages: the melt forms a solid crust along the cooled crucible wall, effectively protecting it from corrosion by the molten

Mat. Res. Soc. Symp. Proc. Vol. 465 © 1997 Materials Research Society

material; the process is relatively simple and easy to implement; and the "cold cap" of unmelted material at the surface diminishes the process volatility[10]. An industrial-scale induction-heated cold crucible melting process supplied by an external calciner is now under development by the CEA to vitrify fission product solutions, and this development made it possible to conduct melting tests for the present investigation. A prototype cold crucible melter and a laboratory-scale melter were also used.

EXPERIMENTAL METHODS

Fabrication of Synroc by Direct-Induction Cold Crucible Melting

Melting Principle

The technique is illustrated in **Figure 1**. The material to be melted is placed inside the sectorized crucible where it is heated by Joule effect resulting from the electric currents induced by an electromagnetic field. A solidified layer of material forms on contact with the water-cooled crucible wall, allowing refractory materials to be melted at high temperatures without damaging the melter.

A Synroc-C batch synthesized by ANSTO using the conventional sintering method was melted in a prototype facility. A second batch of Synroc was fabricated by melting an oxide mixture in a laboratory-scale melter. The characteristics of the melting facilities are indicated in **Table I**, and the compositions of the Synroc samples in **Table II**.

Test Procedure

The cooling cycle to obtain preferential formation and growth of certain crystalline phases was not optimized for these tests.

• *Prototype-scale test.* A 45 kg batch of Synroc-C (**Table II**), containing 20 wt% simulated PW-4b waste and originally fabricated in 1989 in the ANSTO nonradioactive pilot facility, was

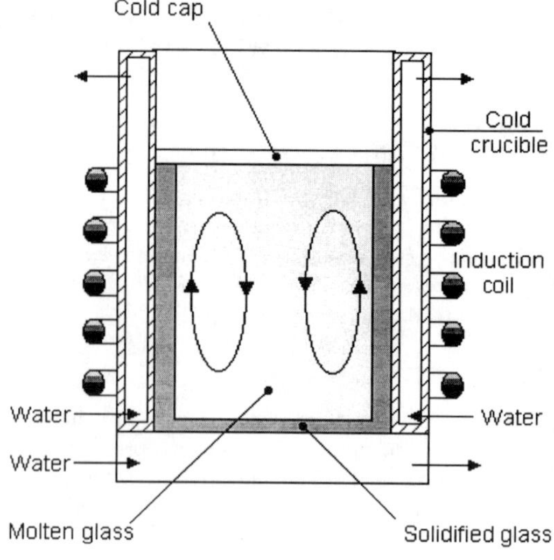

Figure 1. Induction-heated cold crucible schematic

Table II. Calculated compositions (oxide wt%) of hot-pressed or melted Synroc-C and Rocsynmar

Oxide	Synroc-C (20 wt% PW4b) (hot pressed or melted)	Rocsynmar (melted)
TiO_2	58.2	65.3
ZrO_2	7.6	5.6
Al_2O_3	4.2	4.8
BaO	5.3	4.8
CaO	8.6	9.6
SrO	0.7	-
$Cs2O$	2.2	-
La_2O_3	3.0	-
Ce_2O_3	4.2	(CeO_2) 9.9
Nd_2O_3	0.6	-
Y_2O_3	0.4	-
Cr_2O_3	0.2	-
FeO	1.0	-
NiO	0.1	-
MoO_2	3.3	-
P_2O_5	0.4	-
Total	100.0	100.0

Table I. Melting facilities used to fabricate Synroc samples

Type	Crucible diameter (mm)	Crucible height (mm)	Crucible material	Induction generator (kW)	Induction frequency (kHz)	Initial material
Industrial prototype	500	350	Stainless steel	160	300	Hot-pressed Synroc-C (45 kg)
Lab scale	60	250	Stainless steel	20	3000	Oxide mixture (1 kg)

ground into centimeter-sized fragments and placed in the crucible. The material is not an electrical conductor at room temperature, and was therefore preheated by adding about 1 wt% titanium in metallic form. After the induction was initiated, the 45 kg Synroc batch was melted at a temperature exceeding 1600°C for a total of 4 hours. Gas release was very limited throughout the melting process, as a cooler layer or "cold cap" about 5 mm thick formed at the surface of the melt. The induction power input to the cold crucible was then reduced to cool the melted Synroc at a controlled rate of about $350°C \cdot h^{-1}$ from 1600°C to 900°C. This cooling procedure would not be feasible industrially as it would require a melter shutdown; it would be preferable to pour the melt into an insular canister that would ensure a cooling rate compatible with recrystallization of the Synroc phases. After melting, the material formed a monolithic mass about 4 cm thick and 50 cm in diameter.

• *Laboratory-scale test.* The Synroc was fabricated experimentally from elementary oxide powder samples (**Table II**). As in the previous test, induction was initiated by adding about 1 wt% titanium metal. One kilogram of material was completely melted, then cooled by reducing the induction power after one hour. The resulting "Rocsynmar" included cerium oxide to simulate the tri- and tetravalent actinides. After melting, the material was recovered in the form of an ingot about 12 cm high.

Characterization Methods

Scanning electron microscope (SEM) examinations were performed on polished cross sections obtained after embedding the sample in epoxy resin (*JEOL* JSM-6300 SEM with an attached *Noran Instruments* Voyager series III X-ray microanalysis system for melted Synroc-C, and *Cambridge* SEM with an attached *Link* system for Rocsynmar). The mineral phases were also identified with X-ray diffraction on powder samples (*Siemens* D500, 0.01° steps, 2 seconds per step, JCPDS file).

Melted Synroc samples were leached in static mode at 90°C ± 1°C in initially ultrapure water (MillQPlus system) at SA/V ratios of 20000 m^{-1} according to a test protocol described by Vernaz et al.[11] The materials were ground and screened to obtain a grain size between 80 and 150 μm. After cleaning in acetone, the powder sample had a specific area of 0.3477 $m^2 \cdot g^{-1}$ (BET/krypton, 15% uncertainty margin). Melted Synroc powder specimens containing 0.29 g each were placed in contact with 5 ml of ultrapure water in 7 ml *Savillex* PFA containers, which were then sealed in glass flasks maintained in a temperature-regulated oven and stirred continuously. Tests were conducted in triplicate for periods of 7, 15 and 28 days. The Nd, Ti, Ba, Al and Ce concentrations in the leachate were then measured by ICP–MS (*Perkin-Elmer* Elan 5000); the Ca concentration was determined by flamme AES (*Perkin-Elmer* Zeermann 5000).

This leaching protocol is widely used to investigate the long-term behavior of nuclear glass. It is designed to simulate a very high degree of alteration reaction progress in the laboratory by quickly reaching saturation conditions, and to assess the long-term evolution[12].

The leach test results at 20000 m^{-1} are not yet available for hot-pressed Synroc-C; results obtained during a previous test[13] with an SA/V ratio of 10000 m^{-1} are therefore used here for comparison.

CHARACTERIZATION RESULTS

Melted Synroc-C

The X-ray diffraction image (**Figure 2**) shows the presence of the main Synroc phases: zirconolite (JCPDS record 17-0495), rutile (21-1276), perovskite (42-0423) and hollandite (21-1276). Elongated zirconolite and rutile crystals (up to 2 cm long) are visible throughout the sample (**Figure 3**a to 3d ; They are backscattered electron images); nevertheless, most of the zirconolite crystals measured from 100 μm to 1 mm in length. Zirconolite appeared to be the first phase to crystallize. Its composition was relatively uniform: typically $Ca_{0.83}La_{0.02}Ce_{0.06}Nd_{0.02}Y_{0.04}Na_{0.02}Zr_{0.63}Ti_{2.22}Al_{0.15}Fe_{0.02}O_7$.

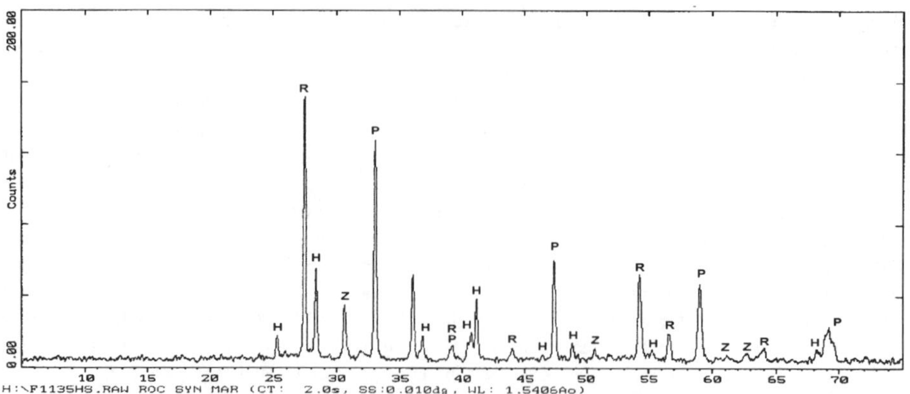

Figure 2. X-ray diffraction image of melted Synroc-C. Z = zirconolite, P = perovskite, H = hollandite, R = rutile

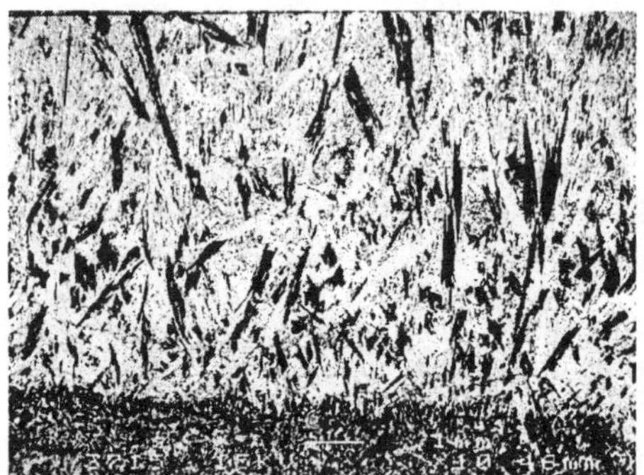

Figure 3a. Melted Synroc-C. Backscattered electron image (×10) of the base of the melt zone adjacent to the fine grained chilled margin. The dark crystals are rutile; others are zirconolite and hollandite.

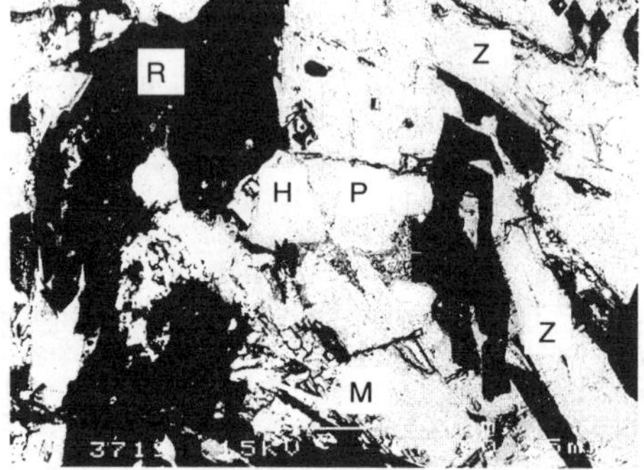

Figure 3b. Melted Synroc-C. Detail view (×200) of interstitial molybdates M (bright) between zirconolite, hollandite, perovskite and rutile crystals.

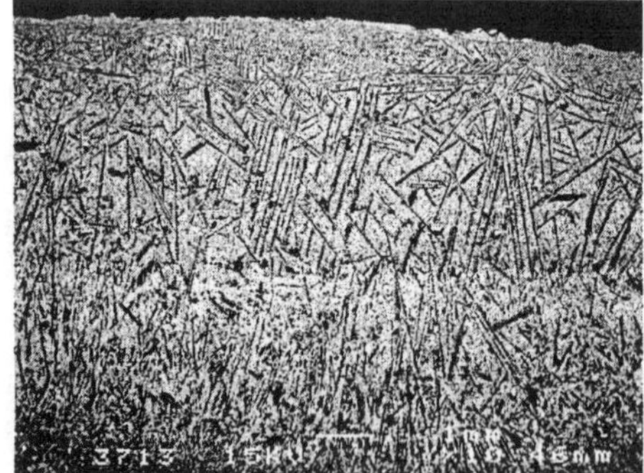

Figure 3c. Melted Synroc-C. (General view). Backscattered electron image (×10) from near surface.

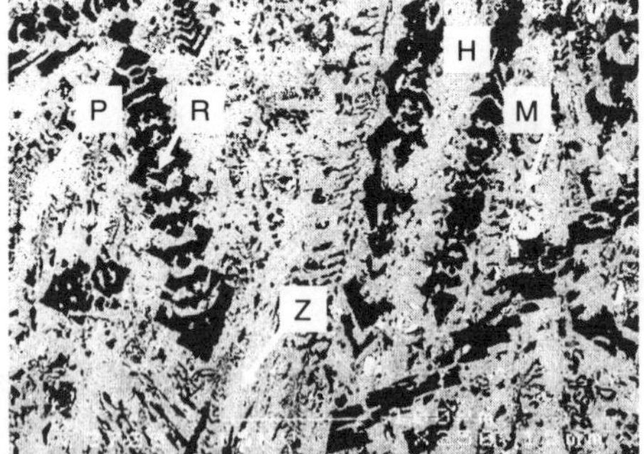

Figure 3d. Melted Synroc-C. Detail view of 3c (×250): rutile crystals with dendritic appearance.

Hollandite crystals (typical sample composition: $Ba_{0.96}Sr_{0.04}Al_{1.84}Fe_{0.21}Ti_{5.94}Zr_{0.02}O_{16}$) were almost as large as the zirconolites and often grew in close association with the zirconolite.

Perovskite (with a typical composition of $Ca_{0.67}Sr_{0.03}La_{0.07}Ce_{0.09}Nd_{0.03}Y_{0.01}Na_{0.01}Ti_{0.98}Al_{0.03}O_3$) revealed composition zoning: the core was richer in rare earth elements. Most of the perovskite crystals were about 100 μm in size in the centers of the melted sample, decreasing to 10 μm in the outer chilled zone near the surface (**Figure 3b** and **Figure 3d**).

Rutile crystals were practically as abundant as zirconolite and hollandite. Although most were highly elongated, some rutile crystals near the upper zone of the sample were characterized by their dendritic appearance (**Figure 3d**). Rutile also showed composition zoning; the core was rich in zirconium oxide (up to 8 wt% ZrO_2) compared with the rim (about 3 wt% ZrO_2).

Minor phases were also observed in the sample (not detected by X-ray diffraction). They included alkali metal and alkaline earth molybdates (phases with very high contrast in **Figure 3b** and **Figure 3d**) with typical compositions of $Ba_{0.89}Sr_{0.09}Ca_{0.07}Mo_{0.98}O_4$ and $Ca_{0.88}Sr_{0.09}Ba_{0.03}Mo_{1.00}O_4$. These were probably the last phases to crystallize, as they were found as interstitial inclusions between zirconolite and hollandite as well as in microcavities and cracks. Phases with a typical composition of $Ca_{8.02}Sr_{0.44}La_{0.23}Ce_{0.31}Nd_{0.12}Na_{0.30}Mg_{0.27}Fe_{0.24}Mo_{0.10}P_{7.03}O_{14}$ (possibly whitlockite, although not identified by X-ray diffraction) were also associated with the molybdates. Another minor phase, sometimes found in the interstices between zirconolite and hollandite crystals or included in the outer perovskite regions, was $Ca_{0.60}Sr_{0.23}La_{0.04}Ce_{0.04}Na_{0.11}Al_{9.50}Ti_{1.33}Fe_{0.59}Ni_{0.11}Mg_{0.31}Zr_{0.04}O_{19}$ (possibly magnetoplumbite), a phase also observed in hot-pressed Synrocs. Very rare phases rich in titanium, measuring from 2 to 5 μm and not identified by X-ray analysis, could be CAT (Ca-Al-Ti) phases.

The entire sample exhibited intergranular porosity on the order of 100 μm, with occasional cavities measuring about 3 mm, containing various minor phases: pale yellowish-brown needles (rutile), greenish crystals (calcium molybdate) and white dendritic barium molybdate crystals.

Rocsynmar

Four types of phases characteristic of hot-pressed Synroc were identified by X-ray diffraction in this sample of melted-oxide Synroc: rutile (JCPDS 21-1276), hollandite (JCPDS 33-0133), perovskite (JCPDS 42-0423) and zirconolite (JCPDS 17-0495). The spectrum was identical with that of the melted Synroc-C (**Figure 2**). SEM examination of polished cross sections revealed:
- dendritic strips up to several millimeters long in which Ti and O (rutile) were detected (black in **Figure 4a** and **Figure 4b**);
- xenomorphous crystals measuring about 50 μm (dark gray in **Figure 4b**) containing Al, Ti, Ba and

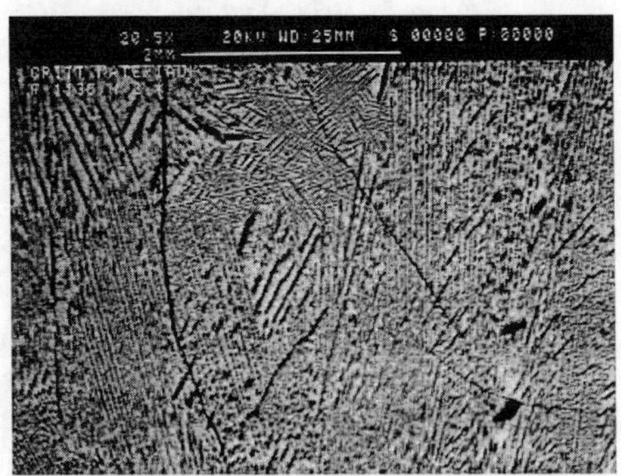

Figure 4a. Rocsynmar.
General view (×20). Backscattered electron image.

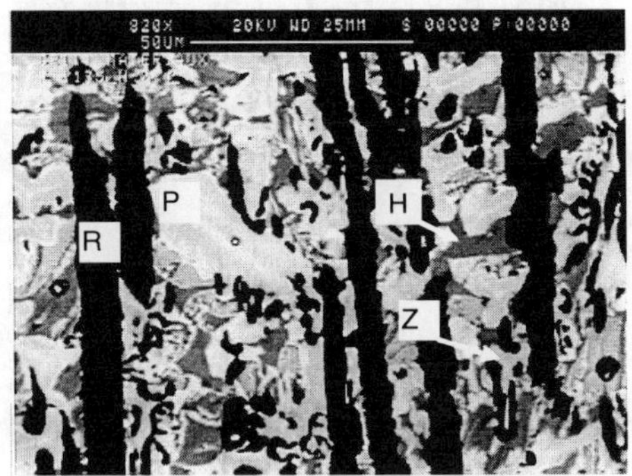

Figure 4b. Rocsynmar. Detail view (×820)
of zirconolite, perovskite, hollandite and rutile crystals.

O (hollandite);

- subautomorphous crystals (dark gray in **Figure 4b**) measuring up to 50 μm, containing Ca, Ce, Ti and O (perovskite) and often exhibiting zoning with exchanges between Ca and Ce;
- star-shaped crystals (light gray in **Figure 4b**) measuring up to 30 μm, containing Ti, Al, Zr, Ce and O (zirconolite).

LEACH TEST RESULTS

The variations in time of the dissolved species concentrations in the leachate are shown in **Figure 5a**, and the normalized elemental mass losses in **Figure 5b** for melted Synroc-C. The Ba concentrations in the leachate remained relatively constant (about 1 ppm) while the Ca concentrations increased in time from about 3 ppm to 5 ppm; the Al concentrations were steady at about 6 ppb. Titanium, the principal Synroc constituent, was found in the leachate at concentrations decreasing from about 20 ppb to 2 ppb over 28 days. The Zr concentrations were relatively constant between about 10 and 1 ppb; the rare earth elements Nd and Ce (simulating the transuranic elements)) were found in concentrations of about 1 ppb. The leachate pH remained constant at approximately 7.

The results obtained for Rocsynmar are shown in **Figure 6a** and **Figure 6b**. The Ba concentrations were slightly lower than for melted Synroc-C, with steady-state values of about 600 ppb. The Al concentration were at all times higher than for melted Synroc-C, diminishing in time from 160 to 50 ppb. The Ti concentrations were again relatively constant at about 2 ppb, while the Zr concentration fluctuated as before between 1 and 10 ppb. The Ce concentration was again about 1 ppb. The pH for all the tests remained steady at about 7.

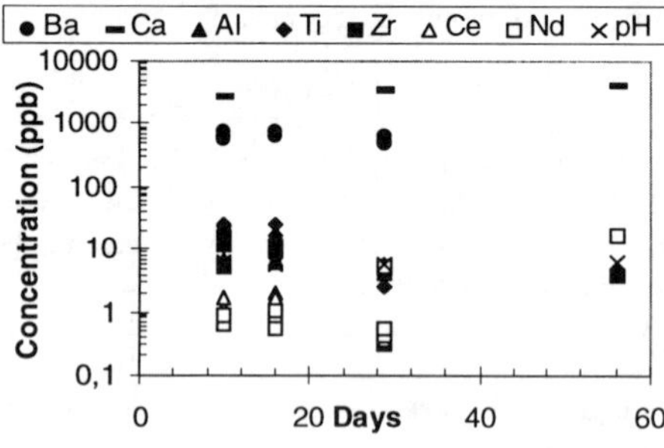

Figure 5a. Melted Synroc-C. Element concentrations (static leach test at 90°C and SA/V ratio of 20000 m⁻¹).

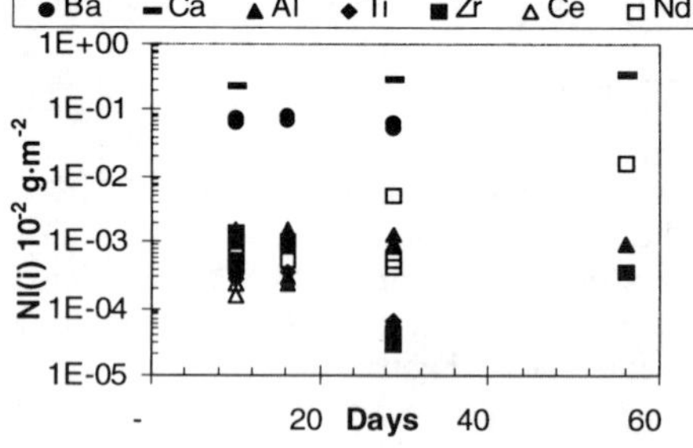

Figure 5b. Melted Synroc-C. Normalized mass loss.

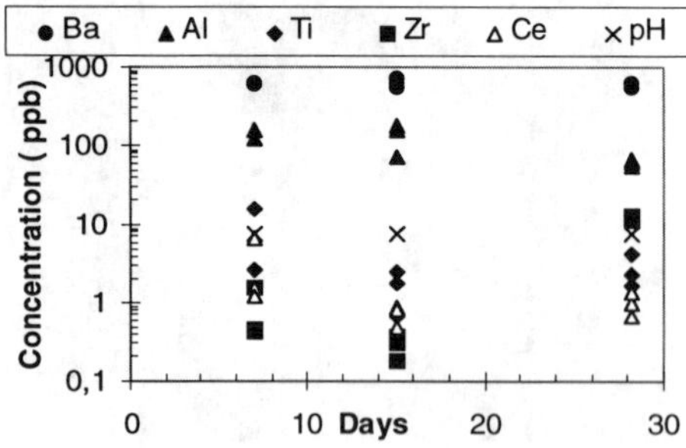

Figure 6a. Rocsynmar. Element concentrations (static leach test at 90°C and SA/V ratio of 20000 m⁻¹).

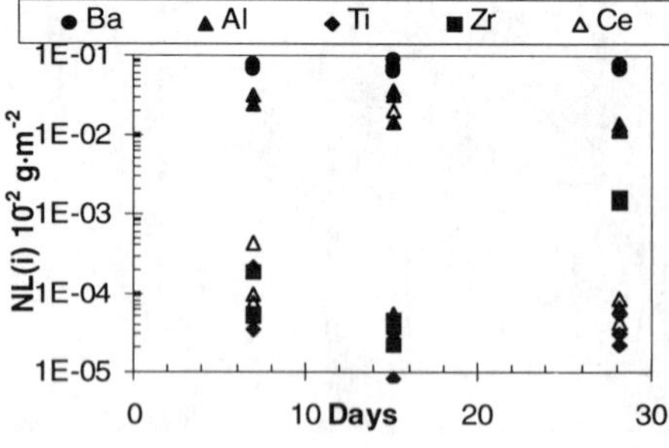

Figure 6b. Rocsynmar. Normalized mass loss.

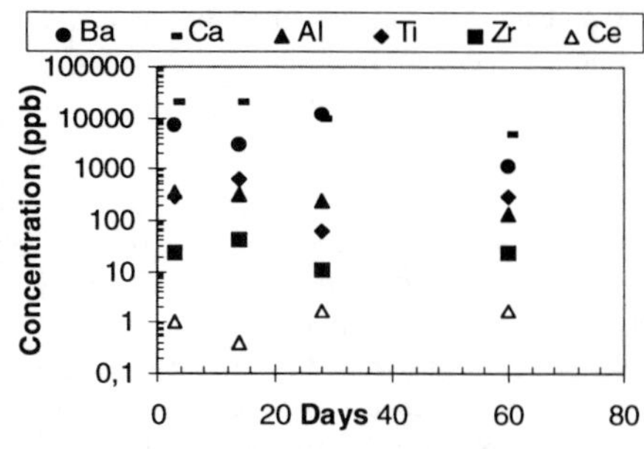

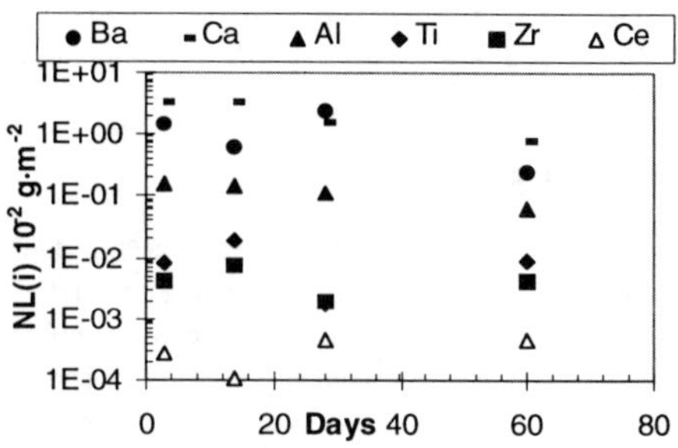

Figure 7a. *Hot-pressed Synroc-C.*
Element concentrations for static leach test[13] at 90°C and SA/V ratio of 10000 m⁻¹.

Figure 7b. *Hot-pressed Synroc-C.*
Normalized mass loss for static leach test[13] at 90°C and SA/V ratio of 10000 m⁻¹.

Leach tests by Van Iseghem *et al.*[13] with hot-pressed Synroc-C (**Figure 7a** and **Figure 7b**) resulted in Ba and Ca concentrations near 10.10^3 ppb and 20.10^3 ppb, respectively; despite a 50% lower SA/V ratio, these values were slightly higher than for the two Synroc compositions obtained by melting (approximately 1.10^3 to 4.10^3 ppb). Similarly, the values observed for titanium were at least 10 times higher than those obtained by leaching the two melted Synrocs. Conversely, the solubilized Al concentrations were comparable to those observed with Rocsynmar (300–100 ppb). The Zr concentrations ranged from 30 to 10 ppb, on the same order of magnitude as for the melted Synrocs (**Figure 5** and **Figure 6**). The cerium concentration remained near 1 ppb, as in the other two tests.

DISCUSSION

Cold crucible melting at prototype scale of Synroc-C containing 20 wt% simulated fission product oxides resulted in the formation of the major phases found in hot-pressed Synroc-C (**Figure 2**), although melting followed by controlled cooling yielded crystals of significantly larger dimensions. However, processing in an oxidizing medium resulted in the formation of molybdate phases containing Ba, Ca, Cs and Sr, preventing their incorporation in the hollandite and metallic alloys (**Figure 3** and **Figure 5**). These results corroborate earlier findings by Russian researchers indicating that if melted Synroc were to be adopted as a conditioning medium for light water reactor fission product solutions, it would have to be produced under reducing conditions.

Cold crucible melting under oxidizing conditions at laboratory scale of an oxide mixture containing 10 wt% CeO_2 to simulate the presence of actinides also resulted in the formation of the standard Synroc host phases. X-ray microanalysis showed that cerium was incorporated in the zirconolite and perovskite, and that the latter exhibited higher Ce concentrations in the core.

When the two ceramic materials were subjected to aqueous leaching, the Ti concentrations in the leachate remained remarkably low, never exceeding about 20 ppb. The Ti concentrations obtained by leaching hot-pressed Synroc-C were 10 times higher on average. These differences may be attributable to the different textures of the materials: hot-pressed Synroc-C is a combination of submicron-sized crystals, while melted Synroc contains crystals several tens of microns in size. The actual reaction surface area accessible to water would thus be greater for hot-pressed Synroc.

Nevertheless, the low titanate solubility ensures low Ce solubilization (found at a steady-state concentration of about 1 ppb in the leachates for all the materials). Without surface examinations of the altered materials, it is impossible to conclude whether the identical secondary phases are responsible for controlling the solubilized Ce concentrations. If calcium and, to a lesser extent, barium (which is mainly incorporated in the hollandite phase) represent Synroc alteration tracers, then the absolute mass of material altered during 28 days of interaction in water is very low in each case: 23×10^{-5} g for melted Synroc, 7×10^{-5} g for Rocsynmar, and 72×10^{-5} g for hot-pressed Synroc.

CONCLUSION

The use of titanate-based ceramics for dedicated conditioning of the rare earth elements (mainly cerium), considered as chemical simulants of the actinides, was found to be feasible both at laboratory scale and at industrial prototype scale by cold crucible melting of the basic component oxides. This process considerably reduces the number and complexity of the steps required to produce hot-pressed Synroc. Although the thermal scenario was not optimized, the zirconolite, perovskite, hollandite and rutile phases crystallized as the melt was cooled. These tests confirmed the results of earlier work by other authors[9] on Synroc-type ceramics with similar concentrations for the major oxides (TiO_2, ZrO_2, CaO, Al_2O_3, BaO) but with lower concentrations of rare earth oxides (mainly CeO_2) simulating the presence of actinides.

Another test objective was to verify the long-term chemical durability of synthetic materials rich in rare earth elements by aqueous leach testing at high SA/V ratios. The preliminary results are encouraging, and confirm the intrinsic low solubility of the titanate phases: titanium, the major component element of the titanate phases incorporating the rare earth elements, reached a maximum concentration of only about 20 ppb. Cerium, simulating the actinides, was found at steady-state concentrations on the order of 1 ppb.

REFERENCES

1. G.R. Lumpkin, K.L. Smith, M.G. Blackford, *Journal of Nuclear Materials*, 224, 31-42 (1995).
2. E.R. Vance, *MRS Bulletin*, Volume XIX, No. 12 (December 1994).
3. A.E. Ringwood, S.E. Kesson, N.G. Ware, W. Hibberson, A. Major, *Nature*, 278, p. 219 (1979).
4. P. Cheron, M. Senoo, C. Fillet et al. *Mat. Res. Soc. Symp. Proc.*, vol. 353, eds. T. Murakami and R.C. Ewing, 767-774 (1995).
5. E.R. Vance, B.D. Begg, R.A. Day, C.J. Ball, *Mat. Res. Soc. Symp. Proc.*, vol. 353, eds. T. Murakami and R.C. Ewing, 767-774 (1995).
6. A. Jostsons, E.R. Vance, D.J. Mercer, V.M. Oversby, *Mat. Res. Soc. Symp. Proc.*, vol. 353, eds. T. Murakami and R.C. Ewing, 775-781 (1995).
7. M.G. Blackford, K.L. Smith, K.P. Hart, *Mat. Res. Soc. Symp. Proc.*, vol. 257, ed. C.G. Sombret, Pittsburgh, PA, 243-249 (1992).
8. A. Hough, J.A.C. Marples, *AEA Technology Report*, AEA-FS-0201 (H), 1993.
9. I.A. Sobolev, S.V. Stefanovsky, F.A. Lifanov, *Radiochemistry* (Russ.), 35, 99 (1993).
10. A. Jouan, J.P. Moncouyoux, S. Merlin, Proceedings of *Waste Management '95*, Feb. 26-March 2, 1995, Tucson, AZ (to be published).
11. E. Vernaz, T. Advocat, J.L. Dussossoy, in *Ceramic Transactions*, vol. 9, Nuclear Waste Management III, ed. G.B. Mellinger, 175-185, (1989).
12. French AFNOR standard X30-407.
13. P. Van Iseghem, W. Jiang, M. Blanchaert, K. Hart, A. Lodding, *Mat. Res. Soc. Symp. Proc.*, vol. 412, edited by W.M. Murphy and D.A. Knecht, *Sci. Bas. for Nucl. Wast. Man.* XIX, Pittsburgh, PA, , 305-312-249, 1992.

STUDY OF MELTED SYNROC DOPED WITH SIMULATED HIGH-LEVEL WASTE

I.A. SOBOLEV*, S.V STEFANOVSKY*, S.V. IOUDINTSEV**,
B.S. NIKONOV**, B.I. OMELIANENKO**, AND A.V. MOKHOV**
* SIA "Radon", 7-th Rostovskii line 2/14, Moscow 119121, Russia, fax: 7-095-919-3194.
** Institute of Geology of Ore Deposits, Staromonetny line 35, Moscow 109017, Russia

ABSTRACT

Preparation and characterization of inductively-melted Synroc containing 20 wt% simulated plant "Mayak" reprocessing waste were performed. The sample bulk composition was as follows, (in wt.%): 55.4 TiO_2; 15.8 ZrO_2; 7.5 CaO; 7.4 BaO; 4.3 Al_2O_3; 2.0 MnO; 1.8 SiO_2; 0.7 Na_2O; 1.9 K_2O; 0.5 Ce_2O_3; 1.0 UO_2; 0.9 NiO; 0.6 Cr_2O_3, and 0.2 FeO. The sample was produced by melting in air at 1550-1600 ^0C under barometric pressure. It is composed of a few crystalline phases and a minor glass phase. Most of the phases (hollandite, zirconolite, perovskite and rutile) are similar to the analogous phases found in the other Synroc formulations. An additional phase with average composition, wt.%: 59.8 TiO_2; 15.6 CaO; 7.0 UO_2; 5.6 ZrO_2; 4.7 MnO; 4.1 Ce_2O_3, and 1.8 Al_2O_3 was found. Some elements (Ba, Si, Ni, K, Na, Fe) were present in the phase in negligible quantities. Its formula $(Ca_{2.65}U_{0.3}Ce_{0.2})(Ti_{7.3}Mn_{0.6}Zr_{0.4}Al_{0.3})O_{20.0}$ is rather close to a rare mineral uhligite - $Ca_3(Ti,Zr,Al)_9O_{20}$. Another possible counterpart of the phase is murataite-like mineral previously described in tailored ceramic designed for Savannah River Plant wastes fixation. This phase as well as zirconolite are the major host for U in the sample. Preliminary data on the material leachability in water at 350 $^\circ C$ and 50 MPa have been obtained. Uranium contents in the solution were about 1 ppb and close to the uranium dioxide solubility in deionized water under the same P-T conditions.

INTRODUCTION

At the present time, aluminum-phosphate glass is used for high level waste (HLW) immobilization in Russia. A major disadvantage of this glass is its low hydrothermal durability [1]. More appropriate wasteforms for safe long-term radionuclides isolation are known to be a crystalline materials, especially Synroc, being highly resistive to hot water attack and radiation damage as well [2]. We have studied a Synroc-type material containing simulated spent fuel reprocessing waste of Production Association (PA) "Mayak" facility. Waste composition includes U and Ce as Pu simulants, corrosion products (Cr, Mn, Ni, Fe) and some fission products (Zr). It did not contain Cs and Sr radionuclides since these elements are assumed to be removed at an earlier stage of waste treatment according to the HLW partitioning concept [3]. The material was produced by inductive melting in a cold crucible described previously [4,5].

EXPERIMENTAL PROCEDURE

A batch for Synroc production was prepared by mixing the crystalline oxides of Ti, Zr, Al, Ba, Ca, U, Ce, Mn, Cr, and Ni. The experimental plant and procedure were described in detail elsewhere [5]. The process temperature was 1550-1600 ^0C. The melt was poured into a metal mold followed by cooling to room temperature at the rate of ~40 $^0C/h$. The material prepared was examined by optical microscopy (POLAM L-213), X-ray diffraction (XRD) using

Mat. Res. Soc. Symp. Proc. Vol. 465 ©1997 Materials Research Society

DRON-4 diffractometer (Cu K_α radiation), scanning electron microscopy (SEM), and electron-probe microanalysis (EPMA) using "JSM-5300 + Link ISIS" analytical system. The dissolution behavior of melted material was also studied (see details below).

MACROSCOPIC AND OPTICAL MICROSCOPY OBSERVATIONS

The ceramic has a dark color and zoned composition. It is composed of three zones: an outer glass-like zone about 10 mm in width, followed by an intermediate fine-crystalline zone (10-15 mm), and finally an inner microcrystalline part about 20 mm in diameter. The structure of the material is mainly determined by the relationships between two major phases - hollandite and zirconolite whose total volume content in the sample reaches 80-85% (Figure 1, a-c).

Zirconolite forms elongate grains with the long-axis dimension decreasing from 0.1-0.5 mm in the central part of the sample to 0.03 mm at the edge. The mineral is often observed as star-like intergrowths of some spear-like grains. In transmitted light zirconolite is slightly green-colored along the *Np* optic axis and pale pink-colored along *Ng*. The other characteristics of the mineral are extinction being close to parallel, negative extension and high interference colors.

Hollandite is observed as grains of elongated or isometric shape that vary in size from 1 mm in the interior of the sample to below 0.2 mm in the peripheral zone. The phase has a red-brownish color whose intensity increases from grain center to its margin. Sometimes the central parts of large grains are green-colored. Hollandite has parallel extinction and positive extension.

Under optical microscopy investigation small amounts of rutile and perovskite were also found. Rutile forms needle-like crystals up to 0.3 mm long with dark-gray color in transmitted light and high interference color. Its distribution in the sample is inhomogeneous with separate spots enriched with this phase. The relative amount of rutile is less than 10%. Perovskite was observed as small pseudo-isometric grains up to 0.1 mm in dimension. It has a pale pink-violet color which together with its isotropy in crossed Nicols allows it to be easily distinguished from the other phases. Its content is near 5%.

XRD AND ELECTRON MICROSCOPY STUDY

XRD and SEM study of the material has shown that it consists of a few crystalline phases and small quantity of residual glass (Figure 1,d, and 2). Four of the phases identified - zirconolite, hollandite, rutile and perovskite - are typical of analogous phases from the other Synroc formulations produced by both hot-pressing and melting. The main peculiarity of the synthetic minerals composition is occurrence of some additional elements (Table I). For example, perovskite is enriched in MnO (up to 2.6%), hollandite is also enriched with MnO (1.6%) as well as K_2O (4.3%), zirconolite has elevated Cr_2O_3 (1.1%), MnO (1.1%), and UO_2 (1.0%) contents. The crystal compositions are varied from grain centers towards their margins (Figure 3), that is caused by a temperature decrease and melt composition variation during the phase crystallisation. For example, the Ni content decreases in hollandite, but K, Mn and Al contents increase toward to grain edges. Outer parts of the zirconolite crystals are enriched with Ti and Ce, but depleted with Zr and Cr. Glass is present in the sample as a minor constituent. Its content is about 5 vol.%. It is located in inter-grain space of crystalline phases and mainly composed of silica, alumina, titania, alkalis, alkaline earths and manganese oxides (Table I). There are also some very rare phases in the material, including zirconia and barium feldspar (celsian).

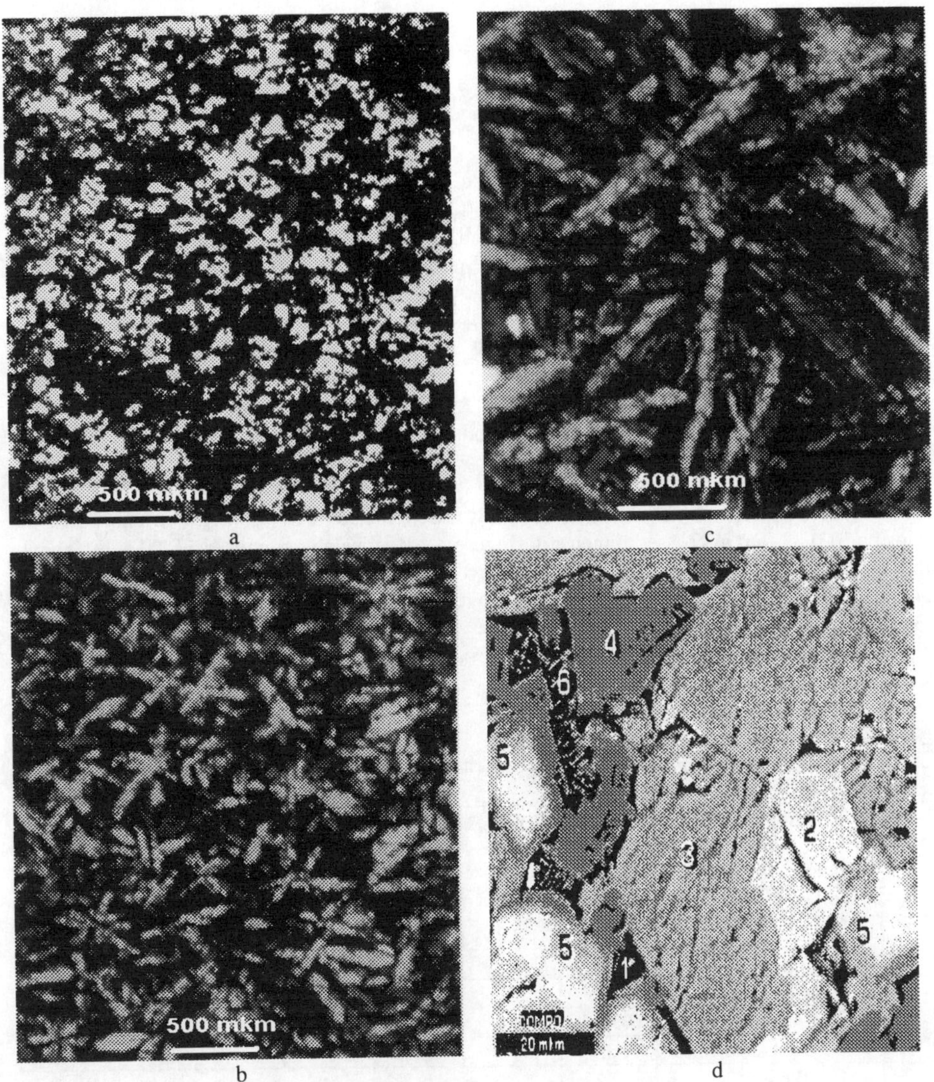

Figure 1. Optical microscopy (a-c) of Synroc containing 20 wt% simulated Mayak HLW and SEM (d) micrographs (backscattered electron image).   NOTE: 1 mkm = 1 μm
  a- rim, b- intermediate part, c- core.
   1. Glass, 2. Zirconolite, 3. Hollandite, 4. Perovskite, 5. X-phase (murataite?, uhligite?), 6. Rutile

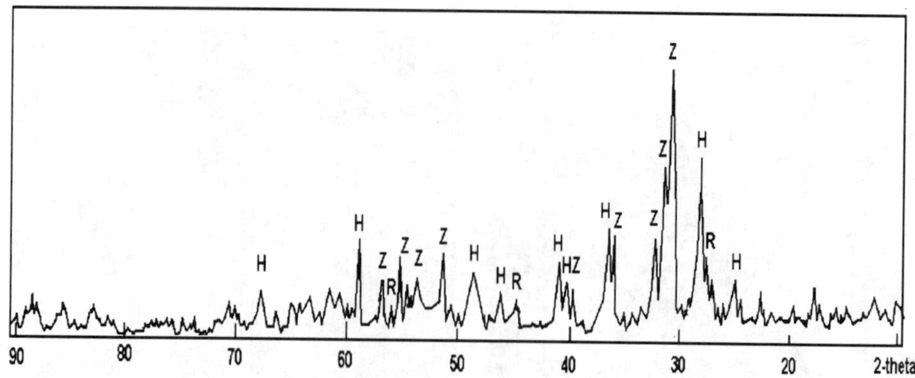

Figure 2. XRD pattern of the melted Synroc with 20 wt% surrogate Mayak waste.
H - Hollandite, R - Rutile, Z - Zirconolite.

Moreover, a specific crystalline X-phase was determined. It has isometric or cubic-shaped grains up to 0.03 mm in crossed dimension (Figure 4). It is mainly composed of Ti, and Ca, but other elements such as U, Mn, Zr, Ce, and Al present as essential additives. Above them small amounts of Na, K, Si, Cr, Ni, Ba and Fe were detected. The chemical composition of this phase is strongly variable (Table II). From the center of grains to their margin U, Zr, Ce, and Ca contents decrease from 12.2 to 0.8% for UO_2, from 6.8 to 3.1% for ZrO_2, from 7.1 to 1.7% for Ce_2O_3, and from 16.6 to 12.9% for CaO.

Table I. Average compositions (oxide wt%) of the phases and the sample determined by SEM method.

Oxide	Sample average	Zirco-nolite	Hollan-dite	Rutile	Perov-skite	X-phase	Glass	Zirco-nia	Barium feld-spar
Na_2O	0.7	-	0.6	-	-	0.1	4.3	-	1.1
Al_2O_3	4.3	1.8	7.2	0.2	0.4	1.8	20.3	0.6	18.5
SiO_2	1.8	0.1	0.2	0.1	0.1	0.3	37.1	0.2	29.0
K_2O	1.9	0.1	4.3	0.1	-	0.1	7.8	-	3.7
CaO	7.5	12.6	0.3	0.4	33.5	15.6	6.5	4.5	6.2
TiO_2	55.4	45.9	66.9	93.4	57.4	59.8	11.1	22.6	13.2
Cr_2O_3	0.6	1.1	0.2	-	0.1	0.1	-	0.8	0.1
MnO	2.0	1.1	1.6	-	2.6	4.7	4.9	0.4	1.3
FeO	0.2	0.1	0.1	-	-	0.1	0.2	-	0.2
NiO	0.9	0.5	2.1	-	-	0.1	-	0.2	0.3
ZrO_2	15.8	34.4	0.5	5.6	0.2	5.6	0.4	69.4	0.2
BaO	7.4	-	15.9	-	1.7	0.6	7.3	-	26.2
Ce_2O_3	0.5	1.3	-	-	3.8	4.1	0.1	0.8	-
UO_2	1.0	1.0	0.1	0.2	0.1	7.0	0.2	0.5	-

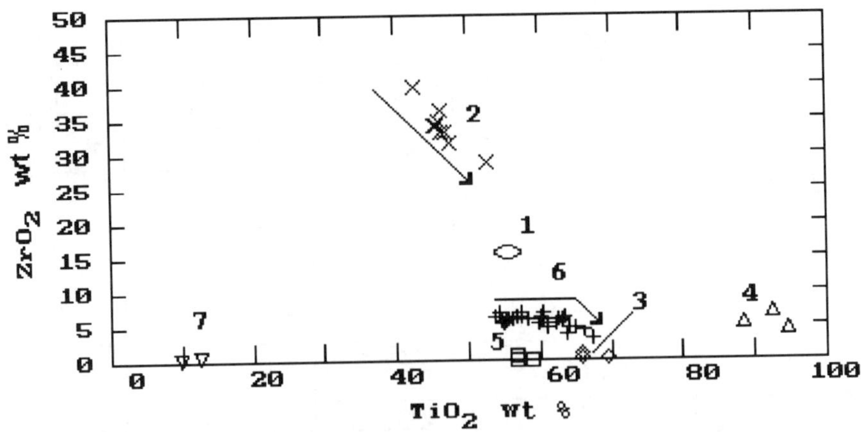

Figure 3. Synroc phase compositions.
1 - bulk, 2 - zirconolite, 3 - hollandite, 4 - rutile, 5 - perovskite, 6 - X-phase, 7 - glass.
Arrows indicate variations of chemical compositions from grains centers to margins.

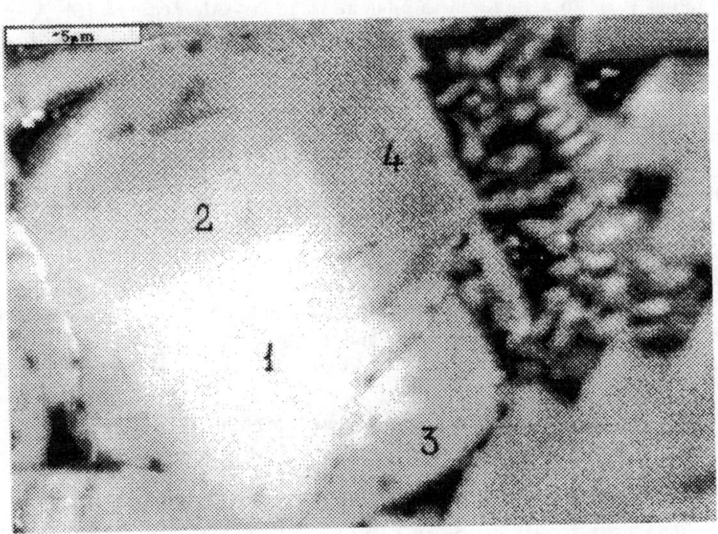

Figure 4. SEM micrograph of X-phase in inductively-melted Synroc, containing 20 wt%
simulated Mayak HLW.
Chemical compositions of zones Nos. 1-4 are given in Table II.

Table II. Chemical composition of different zones in X-phase crystal.

Oxides	1	2	3	4
Al_2O_3	0.8	1.6	3.7	2.8
SiO_2	0.2	0.5	0.2	0.3
K_2O	0.1	0.3	0.7	0.1
CaO	15.6	16.5	13.0	15.9
TiO_2	54.8	56.5	62.9	65.9
Cr_2O_3	0.3	-	0.2	-
MnO	3.6	4.1	5.7	5.6
FeO	0.2	-	0.1	-
NiO	0.2	-	-	0.2
ZrO_2	5.7	6.0	3.8	4.3
BaO	-	3.4	3.7	-
Ce_2O_3	6.0	2.9	2.7	3.7
UO_2	11.9	7.0	2.0	0.8
Total	99.4	98.8	98.7	99.6

Meanwhile, some element contents increase: TiO_2 - from 53.7 to 67.3%, MnO - from 3.3 to 9.6%, and Al_2O_3 from 0.6 to 3.5%. The average phase composition is (in wt.%) 59.8 TiO_2; 15.6 CaO; 7.0 UO_2; 5.6 ZrO_2; 4.7 MnO; 4.1 Ce_2O_3; 1.8 Al_2O_3 which corresponds to the following formula with 20 atoms of oxygen: $(Ca_{2.65}U_{0.3}Ce_{0.2})(Ti_{7.3}Mn_{0.6}Zr_{0.4}Al_{0.3})O_{20}$. The formula is rather close to a rare mineral uhligite $Ca_3(Ti_{5.55}Al_{1.8}Zr_{1.6})O_{20}$ [6]. Another possible counterpart is a murataite-like phase described earlier in a tailored ceramics designed for immobilization of Savannah River Plant radioactive waste [7]. To identify this phase more accurately a research for determination of its unit cell parameters is being conducted.

WASTE ELEMENTS PARTITIONING

Zirconolite and hollandite were found to prevail in the mineral composition of the melted material. Their total content in the sample reaches about 80-85 vol.%. Rutile, perovskite, U-enriched X-phase and glass were present in amounts of about 5% each.

It was established that different phases make various contributions to the total elemental balance (Table III). Ba, Al, K, Na, and Ni are mainly contained by hollandite which contains also about one third of total Fe, Mn and Ti, and relatively small content of Cr. Zirconolite accommodates a major part of Zr, Ca and Cr. Moreover, it contains about half of total U, Ce, Fe

Table III.　Element distribution in the sample (fraction of total, %).

Phase	Na	Al	Si	K	Ca	Ti	Cr	Mn	Fe	Ni	Zr	Ba	Ce	U
Zirconolite	-	18	2	1	65	38	89	30	55	21	94	<1	58	52
Hollandite	47	56	4	78	1	43	9	33	33	78	1	92	1	4
Rutile	-	<1	<1	<1	<1	8	-	<1	-	-	2	-	-	1
Perovskite	-	<1	<1	<1	20	5	1	8	-	-	<1	1	19	1
X-phase	1	2	1	<1	9	5	<1	14	5	1	2	<1	21	41
Glass	52	23	92	20	4	1	-	15	7	-	<1	6	1	1

and Ti of the sample and some Mn, Ni, and Al. The X-phase accomodates a significant part of U,Ce, Mn, and Ca, while perovskite - Ca and Ce. The vitreous phase holds most of Si, Na, and essential part of Al, K, and Mn, while rutile contains only relatively small part of Ti.

These data give rise to the conclusion that the main host for HLW components is zirconolite incorporating corrosion elements (Cr, Fe, Ni), actinides and rare earth elements (U, Ce), and some fission products (Zr). A part of the actinides is also included in the X-phase and perovskite, while hollandite contains corrosion products (Ni, Mn). If Sr and Cs were present in the radioactive waste these elements should be distributed in the sample similar to Ca and K respectively. So we can suppose that Sr will be mainly incorporated into perovskite, while Cs will be distributed between hollandite and the vitreous phase. Zirconia and barium feldspar do not have any visible effect on element partitioning because of their very small amounts in the sample.

Among the radionuclide surrogates there are elements such as U, Cr, and Mn having various valency states. Phases containing the elements in their oxidized forms are characterized by high solubility in hot aqueous solutions. Hence, their presence will lead to elevated leachability of waste elements and should be avoided. The data on the phase compositions indicate clearly that all of these elements are present in their reduced forms: U(IV), Cr(III), and Mn(II). This conclusion is supported by the substitution of U for Zr, Cr for Ti, and Mn for Ca in the mineral structure. Thus, the low valency states of these elements may be stabilized in highly durable minerals, e.g. zirconolite and perovskite without application of a special reducing medium.

The X-phase contains about 40% of the total amount of U and 20% of Ce. The other actinides must obviously be also incorporated in this phase. If their content decreases from the center to the margins as for uranium and cerium it will increase the durability of the phase to hot water attack.

PRELIMINARY DATA ON DURABILITY STUDY OF THE SYNROC MATERIAL

Preliminary research of the sample interaction with hot water was carried out. Static tests were performed at T = 350 °C, P = 50 MPa, oxidizing (in air, $-lgfO_2 = 0.7$ bars) and reducing (slightly below Ni-NiO solid phase buffer, $-lgfO_2 = 32.9-35.1$ bars) conditions. The duration of runs was 18 hours each. A fraction of the powdered material was chosen with particle size ranged between 0.1 and 0.15 mm. The solid to liquid phase weight ratio was equal to be 1:10, and surface area to volume (SA/V) value was of about 1000 m^{-1}. After the runs, contents of U, Ca, Ba, Na, K, and Mn were analyzed in prefiltered solutions (Table IV). As indicated in Table IV the uranium content in solution is very small and does not depend on oxygen partial pressure. The values are similar to uranium dioxide solubility in deionized water under the same parameters. Ba and Mn contents are also very small. Alkali ions content is rather high and may be explained by incongruent dissolution of the glass and crystalline phases.

Table IV. Solution composition (U in ppb, the other elements in ppm) for static runs on leachability study.

Run	lg fO_2	U	Ca	Ba	Mn	Na	K
1	-0.7	1.8	0.73	<0.03	<0.02	58	75
2	-32.9	1.4	0.24	<0.03	<0.02	60	72
3	-34.0	2.2	0.22	<0.03	<0.02	55	70
4	-35.1	0.9	0.27	<0.03	<0.02	58	69

Notes: Data were obtained by A.V.Kudrin in IGEM RAS.

CONCLUSIONS

1. A sample of Synroc-type material with 20 wt.% surrogate PA "Mayak" spent fuel reprocessing waste was produced by inductive melting in the cold crucible. The sample consists mainly of zirconolite and hollandite (about 80-85 vol.%) as well as minor phases of rutile, perovskite, U-enriched mineral and glass ($\leq$ 5 vol.% each). Very small amounts of zirconia and barium feldspar phases were detected.

2. The chemical composition of most of the synthetic minerals is similar to that from the other hot-pressed and melted Synrocs. A specific phase enriched with U has an average composition (in wt%) 59.8 TiO_2; 15.6 CaO; 7.0 UO_2; 5.6 ZrO_2; 4.7 MnO; 4.1 Ce_2O_3, 1.8 Al_2O_3. It contains a large fraction of U, Ce and Mn present in the Synroc sample.

3. An element partitioning study has shown that zirconolite incorporates about 95% of total Zr, 90% of total Cr, and 50-70% each of Ca, Ce, U, and Fe, 40% of total Ti, 20-30% each of Mn, Ni, and Al. Hollandite incorporates up to 70-90% of Ba, K, and Ni, 30-50% each of total Al, Na, Ti, Mn, and Fe, 10% Cr. Rutile may contain up to 10% of total Ti. Perovskite contains up to 20% each of total Ce and Ca. The X-phase incorporates up to 40% U, 10-20% each of Ce, and Mn. Glass concentrates up to 90% of total Si, 50% of total Na, and 10-20% each of total Al, K, and Mn.

4. Experiments on melted Synroc durability to hot water attack also have been started. Earlier results have demonstrated a high hydrothermal stability of the specimen. The uranium content in the co-existing aqueous phase is similar to uranium dioxide solubility in water under the same run conditions.

REFERENCES

1. A.P. Mukhamet-Galeyev, L.O. Magazina, K.A. Levin, N.D. Samotoin, A.V. Zotov and B.I. Omelianenko, in Scientific Basis for Nuclear Waste Management XVIII, edited by T. Murakami and R.C. Ewing (Mat. Res. Soc. Symp. Proc. 353, Pittsburgh, PA, 1995), p. 79-86.

2. A.E. Ringwood, S.E. Kesson and K.D. Reeve, in Radioactive Wastes Forms for the Future, edited by W. Lutze and R.C. Ewing (North Holland, Amsterdam, 1988), p. 233-334.

3. E.G. Dzekun, Y.V. Glagolenko, E.G. Drozhko, et al., in SPECTRUM'96. Proc. International Topical Meeting on Nuclear and Hazardous Waste Management (Seattle, WA, August 1996), Vol.3, p. 2138-2139.

4. I.A. Sobolev, S.V. Stefanovsky, F.A. Lifanov and D.B. Lopukh, in The Proceedings of the V-th International Conference on Radioactive Waste Management and Environmental Remediation, edited by S. Slate, R.Bakerand, G.Benda (Berlin, Germany, 1995), p .419-423.

5. O.A. Knyazev, B.S. Nikonov, B.I. Omelianenko, S.V. Stefanovsky, S.V. Yudintsev, R.A. Day and E.R. Vance, in SPECTRUM'96. Proc. International Topical Meeting on Nuclear and Hazardous Waste Management (Seattle, WA, 1996), Vol.3, p. 2130-2137.

6. A.M.Clark, Hey's Mineral Index: Mineral Species, Varieties and Synopsis (Chapman and Hall, 1993), 852p.

7. W.P. Freeborne and W.B. White, in Advances in Ceramics (Proceedings of the II-nd International Symposium on Ceramics in Nuclear Waste Management, 1983), Vol.8, p. 368-376.

COMPARATIVE STUDY OF SYNROC-C CERAMICS PRODUCED BY HOT-PRESSING AND INDUCTIVE MELTING

I.A. SOBOLEV*, S.V. STEFANOVSKY*, B.I. OMELIANENKO**,
S.V. IOUDINTSEV**, E.R. VANCE***, A. JOSTSONS***
* SIA "Radon", 7th Rostovskii per., 2/14, Moscow 119121 Russia
** Institute of Geology of Ore Deposits, Staromonetnii per., 35, Moscow 109017 Russia
*** ANSTO, New Illawarra Rd, Lucas Heights NSW 2234 PMB 1 Menai, NSW Australia

ABSTRACT

Three Synroc-C samples, containing simulated high level waste were studied. One was produced by the conventional hot-pressing method at ANSTO, Australia, and the others were obtained using cold crucible technology at Radon, Russia. One of the melted samples was prepared using the Australian sol-gel precursor and the second one was obtained from an oxide-nitrate mixture. It was established that the specimens have closely similar mineral compositions, with major hollandite, perovskite, zirconolite, and rutile. Small amounts of hibonite were also found. Unlike the hot-pressed Synroc containing metallic alloy particles, melted Synrocs contain molybdates. An investigation of mineral compositions and elemental distribution in the samples was carried out. Features of hot-pressed and melted ceramics were compared. Unit cell parameters of the Synroc phases were determined and preliminary results on durability of the melted Synroc are presented.

INTRODUCTION

The first Synroc formulation, so-called Synroc A, was produced by melting at ~1300^{0}C [1]. Subsequently the improved Synroc formulations - Synroc B and C with more leaching resistance but higher melting points were developed. Subsolidus hot-pressing at T$\leq$1300^{0}C and P=10-100 MPa was proposed as a basic method of Synroc production [1]. However, attempts to apply a melting route to Synroc were also continued [2]. High melting points of major Synroc minerals required melting process temperatures as high as 1500 ^{0}C. Conventional melters used for nuclear waste vitrification were unsuitable for production of Synroc due to strong corrosion of lining and temperature limitation. Progress in high frequency engineering resulted in development of the cold crucible melting technology which was applied to Synroc melting for the first time by V.I. Vlasov in Russia in the middle of the 1980s [3]. Since 1990 development and testing of the cold crucible technology for Synroc production have been carried out at SIA "Radon" [4]. Most of Synroc formulations developed in Australia were also reproduced by inductive melting in the cold crucible [4-7]. Melted samples were examined by different techniques (XRD, optical microscopy, SEM/TEM, EMPA, etc.) by both Radon specialists separately [4-6] and in co-operation with ANSTO scientists [7].

Currently ANSTO's method of Synroc production is preparation of Synroc precursors by a sol-gel method, followed by calcination and subsolidus hot-pressing at 1150-1200 ^{0}C and 14-21 MPa [8]. It is interesting to try to substitute the cold crucible unit for hot-pressing method and compare process variables and quality of materials produced by different methods. The main aim of this work is comparison of materials produced by three various routes:

- sol-gel preparation of Synroc precursors - calcination - hot-pressing at 1150-1200 ^{0}C and 14-21 MPa;

Mat. Res. Soc. Symp. Proc. Vol. 465 ©1997 Materials Research Society

- sol-gel preparation of Synroc precursor - calcination - cold crucible melting at 0.1 MPa;
- oxide-nitrate mixture - cold crucible melting (0.1 MPa).

EXPERIMENTAL PROCEDURE

Conventionally produced Synroc C specimen as well as a batch of Synroc precursors were passed from ANSTO to SIA "Radon". The oxide-nitrate batch was prepared by mixing of crystalline oxides of Ca, Ba, Fe, Ni, Al, Mo, Ti and Zr as well as nitrates of Cs, Sr, Nd, Gd and Ce.

The experimental procedure for melting of Synroc precursors and oxide-nitrate mixture was the same. The batches were charged into the cold crucible (melt surface area and operation volume were 78.5 cm^2 and 1100 cm^3 respectively) surrounded by a copper inductor and energized from a commercially available high frequency generator VCI 11-60/1.76 (60 kW, 1.76 MHz). Silicon carbide was used as a conductive material to start heating and initiate melting. After starting the melting, the batch was fed into the cold crucible in 100 g portions. A portion of the melt was poured onto a metal plate to be quenched. The residue in the crucible was cooled at the rate of 10 ^{0}C/h by slowly turning down the power and then the solidified block was removed.

The samples were labeled as follows:
- AP - Australian hot-pressed ceramics from the sol-gel calcine route;
- AM - Material prepared by melting of sol-gel calcine Synroc precursors;
- RM - Material prepared by melting of the oxide-nitrate mixture.

Materials prepared were analyzed using an atomic absorption spectrophotometer (S-115, Russian model), emission spectral analysis (ISP-30, Russian model), optical microscopy (POLAM L-213), X-ray diffraction (DRON-4 diffractometer, Cu Kα radiation), scanning electron microscopy (JSM-5300+Link ISIS analytical system), and electron microprobe analysis (EMPA).

To perform a comparative leaching study the Synroc samples were crushed and sieved to separate the fraction with grainsize ranging between 0.10 and 0.15 mm. This powder was placed in the Soxhlet apparatus, heated to 100 ^{0}C and kept at this temperature for one or three days followed by solution renewal by a new portion of deionized water. The powder to water mass ratio was 1:10.

RESULTS AND DISCUSSION

The chemical composition of the melted materials is close to that of the hot-pressed ceramic Synroc C containing 20 wt% of simulated spent fuel reprocessing waste from light water reactors [9] (Table I). However, there is some difference. The samples AP and AM contain numerous components simulating actual high level waste in detail, whereas the Russian sample (RM) contains a simplified waste composition. This is the first reason for a difference in chemical compositions between the samples. The second reason is due to dissolution of a protective glass in the cold crucible, resulting in transfer of sodium, potassium, and silicon oxides to the melted Synrocs. The third reason is partial volatilization of some components, mainly cesium, at high temperatures.

From comparative analysis of X-ray patterns of all the samples (Figure 1), their mineral assemblages are similar. Zirconolite, hollandite, perovskite, and rutile were found as major phases. The sample AP consists of zirconolite, perovskite, hollandite, alloy, and rutile. Hollandite is slightly predominant. Estimated relative amounts of the phases are in good accordance with

Table I. Chemical compositions of the Synroc samples (wt.%).

Oxides	AP	AM, rim	AM, core	RM
SiO_2	2.46	4.60	2.48	1.00
TiO_2	57.90	59.70	60.50	58.30
Al_2O_3	5.47	5.72	5.85	6.77
Fe_2O_3	0.77	0.85	0.81	1.15
MgO	0.10	0.08	0.09	0.02
CoO	-	0.01	0.01	0.25
NiO	0.07	0.05	0.06	0.25
CaO	8.76	9.11	9.20	10.42
SrO	0.23	0.31	0.30	1.17
BaO	4.38	4.77	4.87	4.73
Na_2O	0.15	0.08	0.10	0.35
K_2O	0.15	0.04	0.07	0.03
Cs_2O	0.64	0.65	0.63	0.53
ZrO_2	5.92	6.15	6.76	8.40
REE_2O_3	6.91	5.86	5.22	6.10
$MoO_2 + MoO_3$	4.73	3.42	3.20	2.36

The samples AP and AM also contain Cr_2O_3, P_2O_5, RuO_2, Rh_2O_3, Ag_2O, CdO, TeO_2, etc.

data [9] on Synroc C composition: hollandite (33%), zirconolite (28%), perovskite (19%), rutile (15%), and alloy (~5%).

Diffraction peaks on X-ray patterns of the melted samples are sharper than those of the hot-pressed sample, due to the higher degree of crystallinity and larger crystal sizes in the melted Synrocs. The higher rutile and lower hollandite contents in the sample AM compared to the sample AP with approximately same chemical composition may be caused by the higher oxidizing potential of the melting route. As a result only Ti(IV) exists [7] in the melted specimens, while in hot-pressed ceramics some Ti (III) was also observed [8]. Zirconolite is predominant in the sample RM. Hollandite and perovskite contents are lower, and rutile is minor phase. An increase of zirconolite fraction in the sample is due to features of its chemical composition, in particular, maximum zirconium concentration.

Petrographical studies has shown that the hot-pressed ceramic is black, having a glass-like appearance with conchoidal fracture and unctuous glance. Under optical study in transmitted light, transparent spots of 0.01 to 0.05 mm in size are observed on the basic opaque background (Figure 2,a). These sections in crossed Nicols are presented by microcrystalline aggregate with high interference colors and mosaic extinction. In reflected light they have a higher reflection ability than the basic bulk and weak double reflection. In crossed Nicols strong anisotropy and yellow internal reflections are observed. From SEM data they are formed by rutile with zirconolite admixture. Rare inclusions up to 10 µm in size with isometric form and high reflection ability are observed in the basic bulk. These are isotropic in crossed Nicols. According to EMPA data, these are Ni (70%)-Fe (30%) alloy. Due to the fine grainsize of the hot-pressed ceramics, optical investigation is very complicated.

Investigation with electron microscopy has confirmed inhomogeneous constitution of the hot-pressed ceramics (AP). At low magnification dark colored sections, hundreds of microns across, are clearly observed. In their composition titanium is significantly predominant and their formation is rather from residual titanium particles added to batch in order to reduce oxidizing potential. The basic ceramic bulk is homogenous, and is composed of micron sized grains being

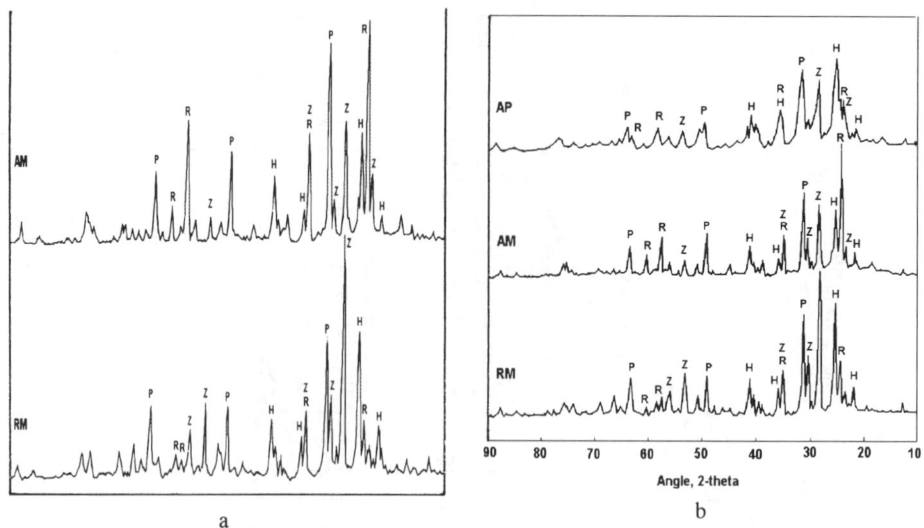

Figure 1. XRD patterns of quenched (a), slowly cooled, and hot-pressed (b) Synroc C samples.

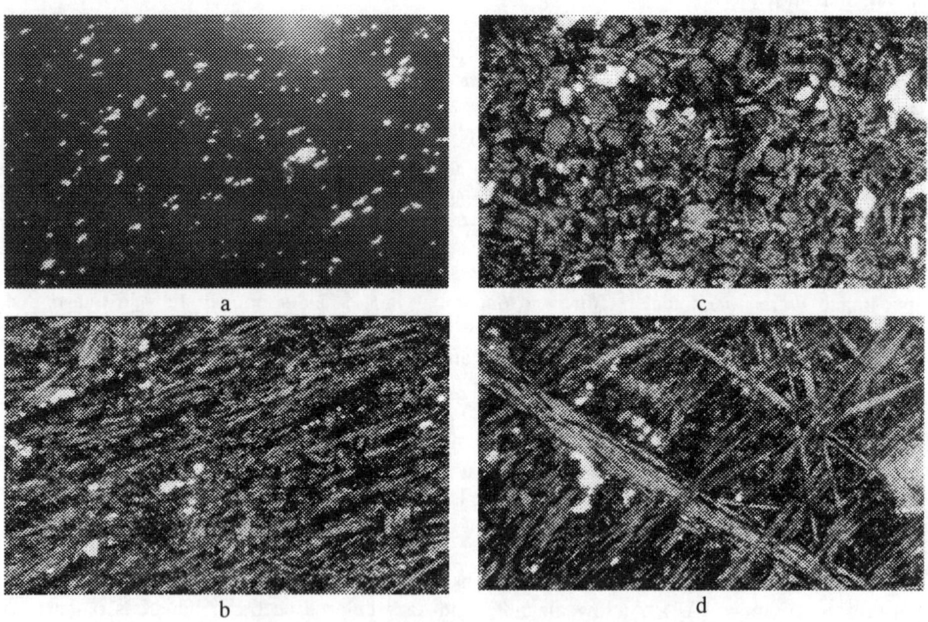

Figure 2. Microphotographs of the Synroc-C samples AP (a), AM (b), and RM (c -rim, d -core). Each photo indicates area of 1.167 x 1.667 mm.

clearly seen due to difference in scattering amplitude in the electron backscattering mode . At high magnification, at least five phases are discernible corresponding to major Synroc minerals - zirconolite, hollandite, perovskite, metallic alloy, and rutile.

The sample AM prepared by melting of Synroc precursors is similar to microcrystalline rock of dark-gray color with conchoidal fracture and weak silky glance. A light grayish-brown glass-like quenching rim of 1 mm in thickness is observed. Cavities partially filled with needle-like hollandite crystals 0.1-0.3 mm in thickness and 1-5 mm in length are located in the central part of the sample AM. From optical microscopy, the sample AM has a fully crystalline structure with grainsize from less than 1 mm to 5 mm. Grainsize in the rim reduces to hundreds of microns. Texture of the sample is determined by hollandite crystals arranged in sub-parallel manner (Figure 2,b) and interspaces filled with thin-needle rutile crystals having here and there a sheaf-like form. Zirconolite occurs as randomly oriented prismatic crystals 0.1-5 mm in length. Perovskite was found to be associated with rutile as grains with irregular shape from tens to hundreds of microns in size. Approximate contents of phases were 30% hollandite, 30% rutile, 20% zirconolite, and 20% perovskite.

Extended needle crystals, green colored along optical axis Ng and yellow-green along Np, and Nm, had high light refraction, very high double refraction, straight extinction, positive extension, and well-defined cleavage with parallel to long axis of grains, and are characteristic of hollandite. Rutile formed a thin needle crystals with pink-brown color, exceptionally high light refraction and double refraction, straight extinction, and positive extension. Zirconolite forms prismatic crystals, almost colorless or very weakly pink-colored, and has straight extinction, negative extension, and very high double refraction and light refraction. Perovskite is pink-brown colored. In crossed Nicols perovskite is clearly discerned due to its isotropy. Structural relationships of minerals in the sample AM indicate hollandite was extra phase with reference to eutectic so it was the first phase crystallized from the melt. The second segregated phase was zirconolite, and rutile and perovskite were formed as the latest .

Melted ceramics from oxide-nitrate mixture (RM) consists of dense, fine-grained material of gray color with weak greenish hue and silky glance. Cavities from fractions of mm to 3 mm in size located predominantly in the sample periphery are occurred. With a microscope the sample appeared to have a fully crystalline constitution with crystal size from 0.1 to 2 mm. Crystal size at the edge zone reduces down to tens of microns (Figure 2,c). The structure of the central part is determined by extended hollandite grains and zirconolite prismatic crystals intergrowths. Isometric perovskite grains and needles of rutile are arranged in interstitials (Figure 2,d). Relative amounts of the minerals are hollandite - 35-40%, zirconolite - 25-30%, perovskite - 30-35%, and rutile - <10%.

Due to relatively large grainsize, the minerals are unambiguously identified on optical properties. Hollandite forms extended grains up to 2 mm in size, with straight extinction, very high light refraction and double refraction, deep yellow color along Np and Nm, and green color along Ng, positive extension, good cleavage. Prismatic zirconolite crystals reach 1 mm in length. Zirconolite presents as both single crystals and cross- or star-like aggregates. At one Nicol it is colorless with straight extinction, negative extension, very high light refraction, and double refraction. Perovskite forms isometric grains, sometimes with weakly appeared faces, hundreds of microns in size, pink-brown colored, and practically isotropic in crossed Nicols. Weak anisotropy is appeared in scattered grain sections. Light refraction of this mineral is very high. The highest idiomorphism degree is characteristic of zirconolite. Perovskite is idiomorphous with reference to hollandite and xenomorphous with reference to zirconolite. It proves the first phase segregated from the melt was zirconolite, the next phases crystallized were perovskite and hollandite, after that all ceramic phases formed simultaneously.

Compositions of the minerals in the hot-pressed and melted samples determined by EMPA are given in Table II. As noted earlier, due to extremely fine grainsizes, analysis of mineral compositions in the sample AP is very complicated. Part of the determinations corresponds to aggregates of zirconolite and hollandite, another part - mixture of rutile, perovskite and zirconolite. A single representative analysis was obtained for rutile.

Table II. Crystal chemical formulae of the Synroc phases.

Ions	Hollandite		Zirconolite		Perovskite		Rutile			Molybdate		Hibo nite
	AM	RM	AM	RM	AM	RM	AP	AM	RM	AM	RM	RM
Na^+	0.01	-	-	0.02	-	0.02	-	-	-	-	-	0.12
Cs^+	0.02	0.04	-	-	-	-	-	-	-	0.01	-	0.01
Ca^{2+}	0.04	0.01	0.77	0.84	0.77	0.78	-	-	-	0.54	0.71	0.42
Sr^{2+}	-	0.03	-	-	0.01	0.02	-	-	-	-	0.12	-
Ba^{2+}	1.07	1.00	0.01	-	-	-	-	-	-	0.27	0.02	-
Mg^{2+}	0.02	-	0.01	-	0.01	0.01	-	-	-	-	-	-
Fe^{2+}	0.23	0.17	0.05	0.02	-	-	-	-	-	0.01	-	0.29
Ni^{2+}	0.04	0.08	-	-	-	-	-	-	-	-	-	0.15
Co^{2+}	0.01	0.07	-	-	-	-	-	-	-	-	-	0.27
Al^{3+}	1.68	1.83	0.23	0.17	0.04	0.02	0.01	0.01	-	0.04	-	8.47
La^{3+}	-	0.01	0.01	-	0.04	0.04	-	-	-	-	-	-
Ce^{3+}	0.06	0.06	0.04	0.02	0.07	0.03	-	-	-	-	-	-
Gd^{3+}	0.01	0.00	0.01	0.04	0.01	0.02	-	-	-	-	-	-
Si^{4+}	0.01	-	-	-	-	-	-	-	-	0.02	-	0.01
Ti^{4+}	5.94	5.89	2.06	2.04	0.98	1.00	0.94	0.94	0.98	0.11	0.01	2.54
Zr^{4+}	0.02	-	0.81	0.84	-	-	0.05	0.05	0.02	-	-	-
Mo^{6+}	-	-	-	-	-	-	-	-	-	0.62	0.71	-
Total	9.16	9.18	3.99	4.01	1.94	1.95	1.00	1.00	1.00	1.62	1.57	12.28
O^{2-}	16.00	16.00	7.00	7.00	3.00	3.00	2.00	2.00	2.00	3.00	3.00	19.00

The mineral composition of the melted materials does not depend on the form of the initial batch. Compositions of analogous phases in the samples prepared from sol-gel or oxide initial batch are close to each other. Some distinctions are due to different bulk composition of the samples. For example Cs, Ni, and Co contents in hollandite of the sample RM are higher, but Fe content is lower compared to the sample AM. In both samples, hollandite differs from theoretical composition by reduced Al and Ti contents.

Zirconolite in the sample AM has lower Ca and Gd contents, but higher concentrations of Fe, Ce, and Al compared to the same phase in the sample RM. They differ from the theoretical composition by the presence of Al and lower Ca and Zr contents, caused by extra components substituting for Ca and Zr. The major difference of perovskite composition from the sample AM compared to the sample PM is extra Ce and lower Gd and Sr contents. Perovskites of both materials differ from theoretical composition by low Ti and Ca contents due to substitution of Al for Ti and TR for Ca. Total TR content in the mineral exceeds 10 wt.%. The main feature of the rutile composition is extra Zr content, especially in the sample AM, which is close to Zr concentration in the same mineral in the hot-pressed ceramic (AP). In the samples of the melted Synroc small amounts of hibonite as well as molybdates were identified.

Unit cell parameters of the melted Synroc minerals (Table III) are close to reference data for the same phases in hot-pressed Synrocs as well as to data for their natural analogs. Results on study of samples solubility in heated solutions are given in Table IV. High concentrations of cesium in solutions after interaction of the melted Synroc samples with hot water are due to presence of Cs molybdates. Leaching rates of other components from melted and hot-pressed Synrocs are comparable. Results in more detail are described in ref. [10].

Table III. Symmetry and unit cell parameters (nm) for minerals in the melted Synroc-C

Lattice parameters	Hollandite	Zirconolite	Perovskite	(Ba,Ca,Sr) molybdate
	Tetragonal	Monoclinic	Rhombic	Tetragonal
a	0.999-1.010	1.259	0.538	0.526
b	0.999-1.010	0.729	0.766	0.526
c	0.287-0.292	1.142	0.546	1.150
β		$100^0 34'$		

Table IV. Elemental concentrations in solutions (in ppm) after leaching of the Synroc samples (fraction 0.1-0.15 mm) at 100^0C.

Test duration, days	Mo	Cs	Na	K	Sr	Ca	Ba	Ti	Zr	Ce
Sample AP										
1	32.2	4.8	0.62	0.54	0.09	2.1	6.0	0.006	0.013	0.07
3	16.5	4.2	1.4	2.2	0.10	0.97	3.5	0.007	0.008	0.04
3	2.9	0.63	0.62	0.38	0.08	2.8	1.0	0.001	0.001	0.004
3	1.4	0.62	0.39	0.23	0.03	1.4	0.77	0.001	0.001	0.003
3	0.70	0.57	0.21	0.14	0.03	1.3	2.1	0.001	0.001	0.003
Sample AM, core										
1	15.6	48.2	0.33	0.46	0.24	2.2	4.3	0.11	0.017	0.07
3	10.6	27.3	1.3	2.2	0.49	3.9	5.3	0.007	0.006	0.04
3	9.9	14.3	0.62	0.38	0.51	3.7	7.0	0.001	0.001	0.008
3	10.9	15.8	0.77	0.42	0.56	4.5	8.2	0.001	0.002	0.01
3	10.8	12.6	0.49	0.48	0.61	4.7	11.9	0.001	0.002	0.01
Sample RM, core										
1	26.0	62.5	1.0	2.0	0.95	2.2	6.6	0.005	0.009	0.04
3	17.0	26.9	0.89	1.5	1.61	2.4	6.3	0.003	0.007	0.02
3	16.4	14.0	1.7	1.6	1.82	4.1	5.6	0.001	0.002	0.01
3	15.7	16.9	2.9	1.9	2.39	5.1	10.6	0.001	0.003	0.01
3	14.0	12.6	1.3	1.3	2.60	6.5	8.5	0.001	0.003	0.009

CONCLUSIONS

The Synroc samples produced by inductive melting in the cold crucible have similar mineral compositions to hot-pressed Synroc. The major phases in these materials are hollandite, zirconolite, perovskite, and rutile. The main differences between them are as follows:

- Molybdenum in the melted Synrocs exists in the form of Ba-Ca and Ca-Sr molybdates rather than metal alloy as in hot-pressed Synrocs; Cs in the melted samples enters molybdates, while in the hot-pressed Synroc this element enters hollandite;
- Grainsizes of phases of the melted Synrocs are 1-2 order of magnitude greater than these in hot-pressed Synroc. The melted Synrocs have higher porosity, reaching 10-15 vol.%.
- Fixation of Cs in the melted Synrocs is substantially lower compared to hot-pressed Synroc due to formation of molybdates under oxidizing melting process conditions. In order to avoid molybdates formation the melting process must be performed under reducing conditions. Leaching of other components (Ti, Zr, TR) is comparable with that in hot-pressed Synroc.

ACKNOWLEDGMENTS

The authors gratefully acknowledge A.A. Gorshkov, A.V. Mokhov, A.V. Sivtsov, I.P. Laputina, A.I. Yakushev, N.V. Korolyova, G.B. Kalentchuk (Institute of Geology of Ore Deposits RAS) for assistance with SEM, EMPA, and chemical analysis of materials. Special thanks are due to A.V. Kudrin for conduction of leaching tests.

REFERENCES

1. A.E. Ringwood, S.E. Kesson, N.G. Ware, W.O. Hibberson and A. Major, Geochem. J. **13**, 141 (1979).
2. H. Pentinghaus, in International Seminar on Chemistry and Process Engineering for High Level Liquid Waste Solidification (Jilich, Germany, 1981, Kernforgchungcentrum, Karlsruhe, 1982), pp. 713-732.
3. V.I. Vlasov, O.L. Kedrovsky, A.S. Nikiforov, A.S. Polyakov and I.Y. Shishtchitz, in Back End of the Nuclear Fuel Cycle: Strategies and Options (Vienna, Austria, IAEA, 1987), pp. 109-117.
4. I.A. Sobolev, F.A. Lifanov, S.A. Dmitriev, S.V. Stefanovsky and A.P. Kobelev, in SPECTRUM'94. Nuclear and Hazardous Waste Management International Topical Meeting (Atlanta, GA, 1994), pp. 2250-2254.
5. I.A. Sobolev, S.V. Stefanovsky and F.A. Lifanov, in Scientific Basis for Nuclear Waste Management XVIII, edited by T. Murakami and R.C. Ewing (Mater. Res. Soc. Symp. Proc. 353, Pittsburgh, PA, 1995) pp. 833-840.
6. F.A.Lifanov, I.A. Sobolev, S.V. Stefanovsky and D.B. Lopukh in ICEM'95. Proceedings of the Fifth International Conference on Radioactive Waste Management and Environmental Remediation, (Berlin, Germany, 1995), American Society of Mechanical Engineers, New York, NY(1995). pp. 413-417.
7. O.A. Knyazev, B.S. Nikonov, B.I. Omelianenko, S.V. Stefanovsky, S.V. Yudintsev, R.A. Day and E.R. Vance, in SPECTRUM'96. Proceedings of the International Topical Meeting on Nuclear and Hazardous Waste Management (Seattle, 18-23 August, 1996), pp. 2130-2137.
8. A.E. Ringwood, S.E. Kesson, K.D. Reeve, D.M. Levins and E.J. Ramm, in Radioactive Waste Forms for the Future, edited by W. Lutze and R.C. Ewing, Elsevier Science Publ. B.V., 1988, pp. 233-334.
9. Ringwood, Miner. Mag., **49** (351), 159, (1985).
10. A.V.Kudrin, S.V. Stefanovsky, B.S. Nikonov, in Scientific Basis for Nuclear Waste Management XX, (these proceedings).

SYNTHESES OF TITANATE-BASED HOSTS
FOR THE IMMOBILIZATION OF Pu(III) AND Am(III)

SHARA S. SHOUP and CARLOS E. BAMBERGER
Chemical and Analytical Sciences Division, Oak Ridge National Laboratory, P.O. Box 2008, Oak Ridge, TN 37831-6110.

ABSTRACT

Syntheses of novel titanate-based waste forms, potentially suitable for immobilizing Pu(III) and Am(III), have been performed. The dititanate compounds $An_2Ti_2O_7$ (An=Pu or Am) were found to have a monoclinic structure. A series of pyrochlore-type solid solutions between $An_2Ti_2O_7$ and the cubic pyrochlore $Ln_2Ti_2O_7$ (Ln=Gd, Er, or Lu) has been established, and limits of solid solubility for $An_2Ti_2O_7$ were determined. These limits were found to increase as the ionic radius of the lanthanide in the host decreased. Strontium-containing titanate compounds $SrAn_2Ti_4O_{12}$ were also prepared and were characterized as exhibiting a perovskite-type structure. Such compounds could serve as waste hosts for the immobilization of not only the actinide but also ^{90}Sr. A 2 month leaching study of $Er_{1.78}Pu_{0.22}Ti_2O_7$ and $SrPu_2Ti_4O_{12}$ by WIPP "A" brine was performed. Very low amounts (< 1 ppm) of Pu(III) were leached. Although the ionic radius of Np(III) is similar to that of Pu(III) and Am(III), analogous Np compounds could not be prepared in any of the titanate systems investigated.

INTRODUCTION

The original motivation behind identifying potential waste forms that would immobilize nuclear waste for thousands of years was the need to safely dispose of the waste generated over the past fifty years. Now an additional motivation further beckons, perhaps more urgently, due to the end of the Cold War. With the agreement to stop producing nuclear weapons and to dismantle many of the existing ones, a decision on what to do with the surplus plutonium from this mass dismantlement is needed. Proposals have been made as to whether the stockpiles should simply be guarded, the plutonium converted for use in energy applications, or diluted in waste forms; the latter two choices would relieve the threat that the plutonium could fall into wrong hands.

Vitrification and incorporation into ceramic materials remain the two most studied modes for immobilizing radioactive elements. Though vitrification appears to lead in large-scale testing, many researchers feel that some inadequate aspects of vitrification, e.g., long-term stability with respect to contact with water, warrant further study of potential ceramic hosts. Synroc, where radioactive elements are distributed into the three minerals hollandite ($BaAl_2Ti_6O_{16}$), perovskite ($CaTiO_3$), and zirconolite ($CaZrTi_2O_7$) [1], has been the most frequently investigated ceramic material, with compounds such as zircon ($ZrSiO_4$) receiving recent attention [2].

Several additional systems of titanate compounds and solid solutions have been identified by the present authors as potential host matrices [3-6]. Initially utilizing lanthanides as surrogates for actinides of comparable ionic radii, cubic pyrochlore solid solutions between monoclinic, light lanthanides $Ln'_2Ti_2O_7$ (Ln'=La-Nd) and cubic pyrochlore, heavy lanthanides $Ln_2Ti_2O_7$ (Ln=Gd, Er, or Lu) have been prepared and solid solubility limits have been established [3,4]. A series of perovskite-type solid solutions ($Sr_{4-x}Ln'_{2x/3}Ti_4O_{12}$) were also

Mat. Res. Soc. Symp. Proc. Vol. 465 © 1997 Materials Research Society

found to form between all compositions of $SrTiO_3$ and $Ln'_2Ti_3O_9$ (Ln'=La-Nd) [5]. This possible waste host could potentially immobilize an actinide together with ^{90}Sr. Having defined the described systems with lanthanides, specifically cerium as a surrogate for plutonium and neodymium for americium, experiments were performed to determine if similar solubility behavior was exhibited by the actinide analogs in the trivalent oxidation state.

Additionally, once compounds potentially suitable for immobilizing radioactive elements were identified, an important consideration was to determine their stability with respect to leaching. Underground salt repositories have been chosen as the most likely storage sites, and the Waste Isolation Pilot Plant (WIPP) in New Mexico is a test facility designed to study this possibility. The composition of a brine that could be present in the event that a breach by water occurred at the site has been determined and named WIPP "A" brine [7]. A 2 month leaching study of two representative plutonium compounds, $Er_{1.78}Pu_{0.22}Ti_2O_7$ and $SrPu_2Ti_4O_{12}$, by WIPP "A" brine has been performed with the results given here and compared to results found for the analogous cerium compounds [6].

EXPERIMENT

<u>Preparation of Compounds and Solid Solutions</u>

The pure compounds and their solid solutions were prepared by several treatments of dry grindings and calcinations to elevated temperatures in air, inert, or reducing atmospheres. The cycles of grinding and calcination were continued until the X-ray diffraction (XRD) patterns of the products did not change. In preparing the compounds $An_2Ti_2O_7$ and $SrAn_2Ti_4O_{12}$, the appropriate reactants $^{242}PuO_2$, $^{241}AmO_2$, TiO_2, and/or $SrCO_3$ (used as a source of SrO) were ground together. TiN was used as a reducing agent to permit the rapid reduction of Pu(IV) to Pu(III) as for example in Eq. (1).

$$2PuO_2 + 0.5TiN + 1.5TiO_2 \rightarrow Pu_2Ti_2O_7 + 0.25N_2 \quad (1)$$

It was not necessary, however, to add the reducing agent to the americium mixtures because elevated temperatures in air were sufficient for the spontaneous reduction of Am(IV) to Am(III). The mixtures were fired between 1200°-$1400^\circ C$ for 70 to 100 hours with 1 to 2 regrindings. The reaction atmosphere for the plutonium compounds was either Ar or $Ar/4\%H_2$ while air was used for mixtures containing americium.

The solid solutions between the heavy lanthanides $Er_2Ti_2O_7$, $Gd_2Ti_2O_7$, and $Lu_2Ti_2O_7$ and $Pu_2Ti_2O_7$ and $Am_2Ti_2O_7$ were synthesized by first forming the lanthanide dititanates. Er_2O_3, Gd_2O_3, or Lu_2O_3 were ground together with TiO_2 and calcined at $1200^\circ C$ in air for 24 hours with 1 regrinding. Then appropriate proportions of the lanthanide and actinide dititanates were ground together, Eq. (2), and calcined at $1425^\circ C$ in $Ar/4\%H_2$ (in the case of plutonium) and 1200°-$1300^\circ C$ in Ar (for americium) for typically 100 hours with up to 3 regrindings.

$$(1-0.5x)Ln_2Ti_2O_7 + (0.5x)An_2Ti_2O_7 \rightarrow Ln_{2-x}An_xTi_2O_7 \quad (2)$$

<u>XRD Characterization</u>

The Debye-Scherrer method was used in obtaining diffraction patterns of the compounds and solid solutions. Cameras of 114 cm diameter and $MoK_{\bar{\alpha}}$ radiation ($\lambda = 0.07107$ nm) were used to identify the products, and lattice parameters were calculated using a modified version of

the LCR-2 least squares program [8]. No internal standard was used because the product
remaining after removing a very small sample for XRD was heated for longer periods to insure
reaction completion.

<u>Leaching Procedures</u>

The compounds $Er_{1.78}Pu_{0.22}Ti_2O_7$ and $SrPu_2Ti_4O_{12}$ were ground with a porcelain mortar
and pestle until the particles could be sieved through a 325 mesh ($\leq 44\mu$m) nylon screen. The
method used to leach the compounds was adapted from the Materials Characterization Center
procedures (MCC-3) [9]. As stated in these procedures, a ratio of 1 g solid/ 10 mL leachant used
in the tests was housed in hermetic Teflon containers that were cleaned accordingly. The
leachant used was WIPP "A" brine, and the test was performed at room temperature. Constant
agitation was achieved by magnetic stirring throughout the test period. The volumes of solution
remaining at the end of the 2 month test period were measured, and the loss of volume was found
to be within acceptable limits of the MCC procedure. Approximately 1 mL of each solution was
filtered through a 0.45 μm pore size syringe filter using a disposable syringe. The leachates were
analyzed by gross alpha counting using a gas-flow proportional counter (Tennelec LB 4000,
Oxford Instrument Co.). The instrument, which had a low alpha background (0.2ppm), was
calibrated using an Am-241 NIST traceable standard. Each sample was plated onto a stainless
steel planchet and counted following the EPA-600/900.0 procedure [10]. The counting time for
each sample was twenty minutes. Alpha pulse height analyses were performed on the same
samples used to obtain the gross alpha counts using an alpha spectrometer (Tennelec TC 257,
Oxford Instrument Co.) and the Genie-ESP Data Acquisition and Processing System (Canberra,
Nuclear Data). The acquisition time for each sample was one hour.

RESULTS

<u>$An_2Ti_2O_7$</u>

The actinide dititanates $Pu_2Ti_2O_7$ and $Am_2Ti_2O_7$ have been prepared and characterized
by XRD as having a monoclinic structure similar to the corresponding light lanthanides [4,11].
$Pu_2Ti_2O_7$ was blue-black in color and maintained the same structure and color after being formed
in either Ar or Ar/4%H_2, confirming that, in both cases, the plutonium was trivalent. The lattice
parameters calculated for the monoclinic $Pu_2Ti_2O_7$ were $a_o = 1.295 \pm 0.008$ nm, $b_o = 0.550 \pm$
0.003 nm, $c_o = 0.772 \pm 0.004$ nm, and $\beta = 98.5 \pm 0.6°$. The formula unit volume calculated from
these parameters was 0.136 ± 0.002 nm^3 and agrees well with 0.137 nm^3 obtained from the
correlation of formula unit volumes of $Ln'_2Ti_2O_7$ (Ln'=La-Nd) and $Ln_2Ti_2O_7$ (Ln=Gd, Er, or Lu)
to ion volumes of Ln'(III) and Ln(III), (CN=6) (Fig. 1). As the coordination number (CN) for the
lanthanide element in the monoclinic dititanates is not well-known, it was elected to arbitrarily
plot formula unit volumes against ion volumes of CN=6 so that the same source of data could be
used for consistency [12]. Plots using other CN values did not significantly affect the
relationship. Brick-red $Am_2Ti_2O_7$ was prepared in air at elevated temperatures without the aid of
the reducing agent necessary with plutonium. The lattice parameters calculated for this
monoclinic dititanate were $a_o = 1.307 \pm 0.003$ nm, $b_o = 0.547 \pm 0.001$ nm, $c_o = 0.769 \pm 0.001$
nm, and $\beta = 98.4 \pm 0.2°$. A formula unit volume having the value 0.136 ± 0.001 nm^3 was
calculated from these parameters and is comparable to the value of 0.135 nm^3 obtained from the
plot in Fig. 1.

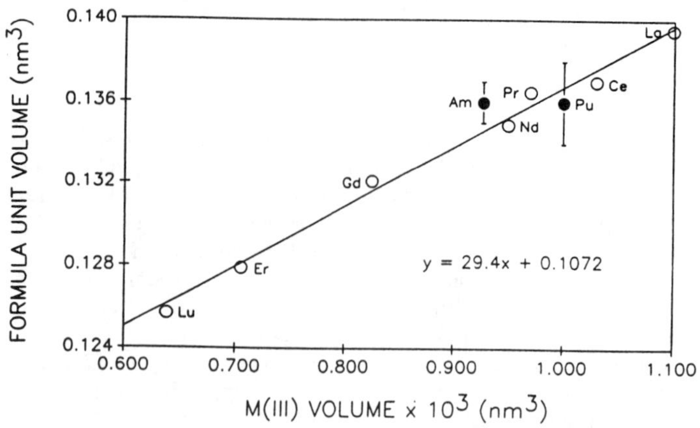

Fig. 1. Correlation of formula unit volumes of $Ln'_2Ti_2O_7$, $Ln_2Ti_2O_7$, $Pu_2Ti_2O_7$, and $Am_2Ti_2O_7$ to ion volumes of Ln'(III), Ln(III), Pu(III), and Am(III), (CN=6).

$Ln_{2-x}An_xTi_2O_7$ Solid Solutions

Cubic pyrochlore solid solutions between the actinide monoclinic dititanates and examples of the cubic pyrochlore, heavy lanthanide dititanates were prepared. $Pu_2Ti_2O_7$ was reacted with $Ln_2Ti_2O_7$ (Ln=Gd, Er, or Lu) while $Am_2Ti_2O_7$ was reacted only with $Er_2Ti_2O_7$ because of the small amounts of americium available. Saturation of the solid solutions was determined by the appearance in the XRD pattern of the monoclinic phase in addition to the cubic pyrochlore solid solution. The monoclinic phase was assumed to be very rich in $An_2Ti_2O_7$, close to pure $An_2Ti_2O_7$. Fig. 2 is a plot of the formula unit volumes of the solid solutions versus the mole percent of the total $Pu_2Ti_2O_7$ added. From the figure it can be seen that as the amount of $Pu_2Ti_2O_7$ added to the solid solution increased, the formula unit volume of the solid solution increased. Once saturation was reached, the volume of the solid solution remained somewhat constant. The saturation point was determined to be the composition half way in between the last composition where only the cubic phase was detected by XRD and the first composition where the cubic and excess monoclinic were present. The slanted and horizontal lines were arbitrarily drawn in Fig. 2 and are not least square fits; their intersection is intended to reflect the estimated saturation point. As the amount of the excess monoclinic phase increased, it became more difficult to calculate the lattice parameter of the cubic phase and thus the volume of the solid solution because of the numerous lines overlapping from both phases. Solid solubility limits of $Pu_2Ti_2O_7$ in $Ln_2Ti_2O_7$ (Ln=Gd, Er, or Lu) were estimated to be 16 ± 4, 22 ± 3, 33 ± 3 mole percent, respectively (Fig. 2). These limits compare well with those found for the analogous cerium compound in the same cubic hosts (19, 22, and 27 mole percent, respectively) [4]. As the ionic radius of the lanthanide ion in the host compound decreased, the amount of $Pu_2Ti_2O_7$ that could be dissolved in the solid solution increased.

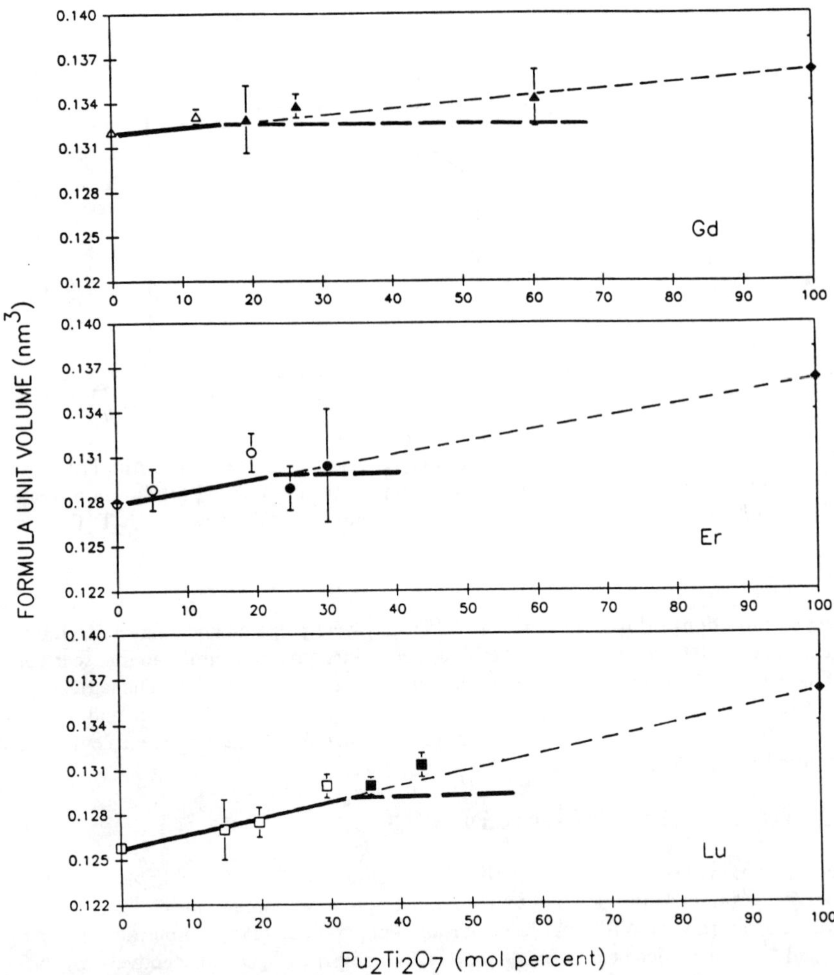

Fig. 2. Measured formula unit volume of $Ln_{2-x}Pu_2Ti_2O_7$ versus mole percent of $Pu_2Ti_2O_7$ in the initial mixture. Sloped lines connect the volume of $Ln_2Ti_2O_7$ with that of $Pu_2Ti_2O_7$, and horizontal lines (filled symbols) indicate saturation of the host matrix with $Pu_2Ti_2O_7$. ($\triangle$Gd, $\bigcirc$Er, $\square$Lu, $\blacklozenge$Pu.)

The solid solubility limit of $Am_2Ti_2O_7$ in $Er_2Ti_2O_7$ was determined similarly to that of $Pu_2Ti_2O_7$ in $Ln_2Ti_2O_7$ described previously; the solubility was estimated to be 59 ± 9 mole percent (Fig. 3) which agrees well with 60.7 mole percent found for $Nd_2Ti_2O_7$ in $Er_2Ti_2O_7$ [3]. In Fig. 3, the formula unit volumes of the unsaturated solid solutions of $Am_2Ti_2O_7$ do not fall on the line connecting the formula unit volumes of the pure end members of the system, unlike in Fig. 2. This may be due to either larger errors affecting the accuracy of the XRD measurements caused by the smaller size of the preparations or, less likely, deviation from Vegard's Law.

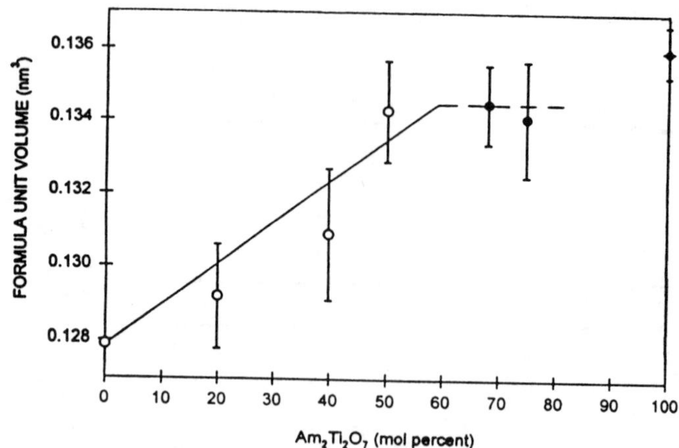

Fig. 3. Measured formula unit volume of $Er_{2-x}Am_2Ti_2O_7$ versus mole percent of $Am_2Ti_2O_7$ in the initial mixture. A sloped line arbitrarily connects the unsaturated solid solutions, and the horizontal line (filled symbols) indicates saturation of $Er_2Ti_2O_7$ with $Am_2Ti_2O_7$.

$SrAn_2Ti_4O_{12}$

The solid solutions $SrPu_2Ti_4O_{12}$ and $SrAm_2Ti_4O_{12}$ were prepared and characterized by XRD as exhibiting the tetragonally distorted (cubic) perovskite structure similar to that seen for $SrLn'_2Ti_4O_{12}$ (Ln'=La-Nd), a representative of the series $Sr_{4-x}Ln_{2x/3}Ti_4O_{12}$ [5]. The lattice parameters calculated for the dark colored $SrPu_2Ti_4O_{12}$ were $a_o = 0.393 \pm 0.001$ nm and $c_o = 0.781 \pm 0.003$ nm, and $a_o = 0.389 \pm 0.001$ nm and $c_o = 0.775 \pm 0.002$ nm were calculated for the brick colored $SrAm_2Ti_4O_{12}$.

Leaching Results for $Er_{1.78}Pu_{0.22}Ti_2O_7$ and $SrPu_2Ti_4O_{12}$

Results from the leaching test indicated that $Er_{1.78}Pu_{0.22}Ti_2O_7$ (9.5 weight % Pu) and $SrPu_2Ti_4O_{12}$ (50.7 weight % Pu) both released 0.2 ± 0.1 ppm of plutonium during the 2 months that they were contacted with WIPP "A" brine at room temperature. (Trace impurities of [241]Am and [241]Am and [239]Pu were detected in $Er_{1.78}Pu_{0.22}Ti_2O_7$ and $SrPu_2Ti_4O_{12}$, respectively, by the pulse height analyses.) These released amounts are on the same order of magnitude as those determined for the analogous cerium compounds [6]. For those compounds, the amounts of cerium leached were below the limit of detection (10 ppm) for the inductively coupled plasma (ICP) atomic emission spectrometer used. The solutions from leached $SrPu_2Ti_4O_{12}$ were not analyzed for strontium; we speculate that the strontium released would be similar to that released from $SrCe_2Ti_4O_{12}$, which is of the same order of magnitude as the solubility of $SrSO_4$ and $SrCO_3$ in water.

CONCLUSIONS

Actinide dititanates $Pu_2Ti_2O_7$ and $Am_2Ti_2O_7$ have been prepared and characterized by XRD to exhibit the monoclinic structure analogous to that displayed by the light lanthanides. Solubility limits of cubic pyrochlore solid solutions formed between these actinide compounds and heavy lanthanide dititanates (Ln=Gd, Er, or Lu) were determined to be 16 ± 4, 22 ± 3, 33 ± 3 mole percent, respectively, for plutonium and 60.7 mole percent for $Am_2Ti_2O_7$ in $Er_2Ti_2O_7$. These limits compare well with those found for cerium and neodymium substantiating the rationale to use them as surrogates for plutonium and americium, respectively. The trend seen for the plutonium solid solutions, similar to that seen in previously studied light lanthanides, is that as the ionic radius of the lanthanide ion in the host compound decreased, the amount of $Pu_2Ti_2O_7$ that can be incorporated into the solid solution increased. Leachability studies in WIPP "A" brine over a 2 month period indicate that the solid solution $Er_{1.78}Pu_{0.22}Ti_2O_7$ might have potential as a durable waste form. Because measurable amounts of strontium were most likely leached from $SrPu_2Ti_4O_{12}$, leaching measurements over longer periods of time are required to determine the effect on the overall ability of the host to immobilize plutonium. Similar leaching results would be expected for analogous americium compounds.

Syntheses of analogous neptunium-containing compounds were attempted but never resulted in the desired single phase product. Other factors besides ionic radii and oxidation state obviously prevail.

This research has established the existence of several titanate-based compounds that can incorporate trivalent plutonium (most of the plutonium-containing titanate compounds discussed in the literature are examples of zirconolites, such as $CaZr_{0.8}Pu_{0.2}Ti_2O_7$ [13] or $CaPuTi_2O_7$ [14], or $PuTi_2O_6$ [15], similar to brannerite, where plutonium is tetravalent) and americium into several different crystal structures that appear to be durable as far as the actinide is concerned in short time leaching studies. Examining the stability of these compounds towards self-irradiation is another aspect that needs to be investigated to determine the true feasibility of these types of trivalent actinide compounds as potential waste and surplus plutonium hosts.

ACKNOWLEDGMENTS

The authors thank Richard G. Haire and C. Leigh Kay, of the Chemical and Analytical Sciences Division, Oak Ridge National Laboratory, for, respectively, obtaining the X-ray diffraction data and providing the gross alpha counting and the alpha pulse height analyses. This research was supported in part by the Division of Materials Sciences, Office of Basic Energy Sciences, U.S. Department of Energy and the Strategic Environmental Research and Development Program (SERDP), U.S. Department of Energy, under Contract No. DE-AC05-96OR24644 with Oak Ridge National Laboratory, managed by Lockheed Martin Energy Research Corporation. S.S.S. is grateful for the support by the Division of Chemical Sciences, Office of Basic Energy Sciences, U.S. Department of Energy, under Grant No. DE-FG05-88ER13865 to the University of Tennessee, Knoxville. This paper was adapted from a dissertation submitted by S.S.S. in partial fulfillment of the requirements for the Ph.D. degree in Chemistry at the University of Tennessee, Knoxville, TN.

REFERENCES

1. A.E. Ringwood, S.E. Kesson, N.G. Ware, W. Hibberson, and A. Major, Nature **278**, p. 219 (1979).

2. B.E. Burakov, E.B. Anderson, V.S. Rovsha, S.V. Ushakov, R.C. Ewing, W. Lutze, and W. J. Weber in <u>Scientific Basis for Nuclear Waste Management XIX</u>, edited by W.M. Murphy and D.A. Knecht (Mater. Res. Soc. Proc. 412, Pittsburgh, PA 1996), p. 33-39.

3. C.E. Bamberger, H.W. Dunn, G.M. Begun, and S.A. Landry, J. Less-Common Met. **109**, p. 209 (1986).

4. C.E. Bamberger, T.J. Haverlock, S.S. Shoup, O.C. Kopp, and N.A. Stump, J. Alloys and Compounds **204**, p. 101 (1994).

5. S.S. Shoup, T.J. Haverlock, and C.E. Bamberger, J. Amer. Ceram. Soc. **78**, p. 1261 (1995).

6. S.S. Shoup, C.E. Bamberger, T.J. Haverlock, and J.R. Peterson, J. Nucl. Mater. in press.

7. R.G. Dosch and A.W. Lynch, Interaction of Radionuclides with Geomedia Associated with the Waste Isolation Pilot Plant Site in New Mexico, Sandia National Laboratory Report SAND78-0297 (1978).

8. D. Williams, A FORTRAN Lattice Constant Refinement Program, Ames Laboratory Report IS-1052 (1964).

9. Materials Characterization Center, Nuclear Waste Materials Handbook Test Methods (Rev. 6), Pacific Northwest Laboratories Report DOE/TIC 11400 (1985).

10. Gross Alpha and Gross Beta Radioactivity in Drinking Water, EPA-600/900.0 (1980).

11. Powder Diffraction File, International Centre for Diffraction Data, Newtown Square, PA.

12. R.D. Shannon, Acta Crystallogr. **A32**, p.751 (1976).

13. E.R. Vance, P.J. Angel, B.D. Begg, and R.A. Day in Scientific Basis for Nuclear Waste Management XVII, edited by R.C. Ewing (Mater. Res. Soc. Proc. 333, Pittsburgh, PA 1994), p. 293-298.

14. F.W. Clinard Jr., L.W. Hobbs, C.C. Land, D.E. Peterson, D.L. Rohr, and R.B. Roof, J. Nucl. Mater. **105**, p. 248 (1982).

15. A. Tabuteau and M. Pagés in The Handbook on the Physics and Chemistry of the Actinides, edited by A.J. Freeman and C. Keller (Elsevier, Amsterdam, Netherlands, 1985), p. 185-241.

16. E.R. Vance, K.P. Hart, R.A. Day, B.D. Begg, P.J. Angel, E. Loi, J. Weir, and V.M. Oversby, in Scientific Basis for Nuclear Waste Management XIX, edited by W.M. Murphy and D.A. Knecht (Mater. Res. Soc. Proc. 412, Pittsburgh, PA 1996), p. 49-55.

PREPARATION OF MONOPHASIC [NZP] RADIOPHASES: POTENTIAL HOST MATRICES FOR THE IMMOBILIZATION OF REPROCESSED COMMERCIAL HIGH-LEVEL WASTES

H. T. HAWKINS*, B. E. SCHEETZ‡ and G. D. GUTHRIE, JR.**
*Nuclear Materials Technology and **Earth and Environmental Sciences Divisions,
Los Alamos National Laboratory, Los Alamos, NM 87545
‡Intercollege Materials Research Laboratory, The Pennsylvania State University,
University Park, PA 16802

ABSTRACT

The compositional flexibility of the sodium zirconium phosphate $(NaZr_2(PO_4)_3)$ structure has been exploited in the design of monophasic radiophases capable of immobilizing the most common cations associated with reprocessed high-level commercial waste streams. Highly crystalline, monophasic members of the $NaZr_2(PO_4)_3$ structural family ([NZP]) have been prepared with conventional processing methods and equipment. These radiophases were tailored to accommodate 10-20 wt % modified PW-4b simulated calcine as single phases isostructural with $NaZr_2(PO_4)_3$. To meet the challenge of designing monophasic materials capable of accommodating the chemical complexity of PW-4b, an ionic substitution scheme based on crystal chemical principles was developed. The radiophases were prepared with inexpensive, inorganic precursors and a solution sol-gel method; these materials were heat treated and/or sintered under a variety of conditions to determine the optimum conditions for single phase [NZP] formation. X-ray powder diffraction provided valuable information that was used to assess the suitability of the ionic substitution model developed in this investigation. The results of this investigation suggest that monophasic [NZP] radiophases capable of accommodating 10-20 wt % modified PW-4b simulated calcine may be continuously processed with conventional ceramic processing methods and equipment. Moreover, the relatively low temperatures involved and the reproducibility of the process make [NZP] radiophases economically attractive hosts for radioactive and heavy metal industrial wastes.

INTRODUCTION

Monophasic members of the sodium zirconium phosphate structural family ([NZP]) possess physical and chemical properties that would permit their application as host matrices for reprocessed commercial high-level wastes (HLW) [1]. [NZP] compounds are resistant to radiation damage [2], thermal expansion [3-4], dissolution, and release of constituents [5-7] over a wide range of experimental conditions. These properties arise from the three-dimensional skeletal structure of [NZP], a hexagonal framework of strongly bonded polyhedra that is capable of accommodating in solid solution the 42 most common cations associated with reprocessed commercial HLW [7].

The prototypical [NZP] structure $(NaZr_2(PO_4)_3)$, which may be represented by the structural formula $M'M''_{\square\square\square}M'''_{\square}A_2(XO_4)_3$, is a three-dimensional network of corner-sharing ZrO_6 octahedra and PO_4 tetrahedra [8]. The basic structural unit of the network, $[Zr_2(PO_4)_3]^-$, consists of two octahedra and three tetrahedra. These structural units are connected to form ribbons parallel to the c-axis of the unit cell; these ribbons are linked by PO_4 tetrahedra normal to the c-axis to generate the three-dimensional network illustrated in Figure 1. The arrangement of $[Zr_2(PO_4)_3]^-$ units in ribbons parallel to the c-axis of the unit cell creates two interconnected interstitial sites, M' and M'', whose occupation determines the symmetry of the [NZP] structure. The M' site, a strongly distorted octahedral site, is located between each $[Zr_2(PO_4)_3]^-$ unit comprising the ribbons; the M'' sites are structural holes that separate the ribbons [9]. The M''' site is a slightly distorted trigonal prism that is often vacant due to strong repulsive forces from adjacent ZrO_6 octahedra [8]. In the prototypical structure, the M' site is occupied by sodium ions (Na^+) whereas the M'' and M''' sites remain vacant. This arrangement maintains the rhombohedral symmetry of the unit cell and the structure belongs to the space group $R\bar{3}c$. Partial occupation of the M'' sites by Na^+ results in a loss of symmetry and conversion of the rhombohedral unit cell to a monoclinic form, C2/c [10]. Atomic parameters for

Mat. Res. Soc. Symp. Proc. Vol. 465 © 1997 Materials Research Society

the rhombohedral $NaZr_2(PO_4)_3$ unit cell and for the monoclinic unit cell of a distorted derivative, $Na_3Zr_2Si_2PO_{12}$, are given in Hong [11].

The three-dimensional skeletal structure of $NaZr_2(PO_4)_3$ offers three distinct crystallographic sites that permit the accommodation of a variety of cations. These sites include the A site (an octahedral site normally occupied by Zr^{4+}) the X site (a tetrahedral site normally occupied by P^{5+}) and two interstitial sites, M' and M", which may or may not be occupied by Na^+ [11]. As the oxidation states of the ions occupying the A and the X sites increase, the stability of the structure and the number of structural variations possible for a particular composition increase; as the oxidation states of the ions occupying A and X decrease, the stability of the skeleton is compromised and the structure collapses. The [NZP] skeleton is strong and flexible, adjusting its dimensions in response to changes in the sizes of its constituents [8]. Members of the [NZP] structural family may be tailored to exhibit negligible thermal expansion, enhanced thermal stability, and super-ionic conductivity through judicious substitution of various ions for those present in the parent composition, $NaZr_2(PO_4)_3$. As a consequence, the ability of [NZP] compounds to incorporate the ionic species of approximately two-thirds of all known elements is well documented in the literature [7]. Partial and complete isovalent, heterovalent, and coupled substitutions have been reported for the crystallographic sites of the $NaZr_2(PO_4)_3$ structure. These substitutions are summarized in Figure 2.

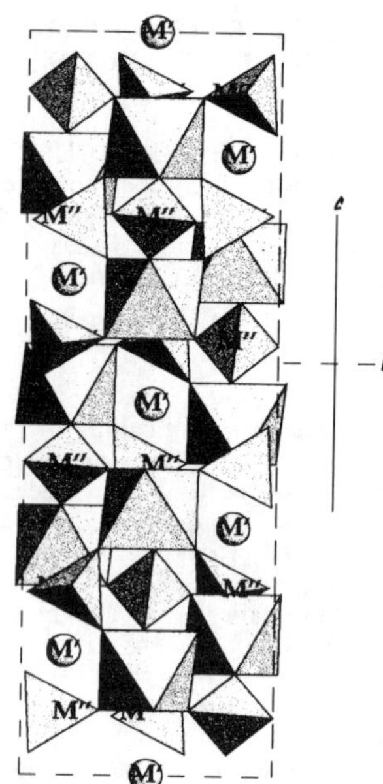

Figure 1. Projection of the $NaZr_2(PO_4)_3$ structure along the c-axis normal to the a-axis of the $R\bar{3}c$ unit cell. The structure is rhombohedral with Na^+ occupying large octahedral cavities (M') between adjacent ZrO_6 and PO_4 polyhedra. M" sites are vacant, but may be occupied by Na^+ upon substitution of a heterovalent cation for Na^+, Zr^{4+}, or P^{5+}. The figure was drawn with ATOMS for WINDOWS (version 3.2 © 1995 Eric Dowty).

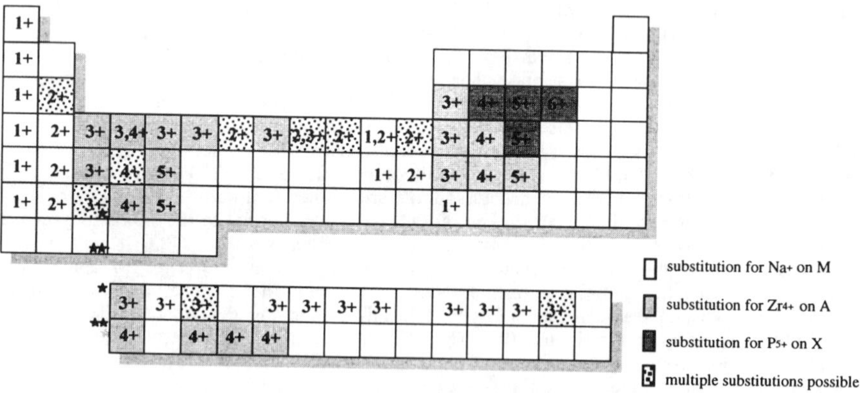

Figure 2. Empirical observations [2, 7-8, 11-13] of cation accommodation on crystallographic sites in the $NaZr_2(PO_4)_3$ structure: $M'M"_{\square\square\square}M'''_{\square}A_2(XO_4)_3$.

The primary objective of this investigation was to exploit the compositional flexibility of the $NaZr_2(PO_4)_3$ structure in the design of monophasic crystalline radiophases that are capable of accommodating in solid solution all of the cations present in PW-4b, a typical commercial reprocessed HLW stream [14]. A secondary objective was to develop processing methods for these materials that could be adapted to an industrial scale. In particular, the radiophases were tailored to accommodate 10-20 wt % modified PW-4b simulated calcine as monophasic compounds isostructural with $NaZr_2(PO_4)_3$.

RADIOPHASE DESIGN

By calculating the radius ratio (Rc/Ra) of each cation present in the PW-4b waste stream and by taking into account the effect of heterovalent substitutions on the electrical neutrality of the $NaZr_2(PO_4)_3$ unit cell, one can predict the most likely site for cation accommodation in the $NaZr_2(PO_4)_3$ structure and develop suitable ionic substitution models; these models may be used to tailor the composition of [NZP] compounds. Values for Rc/Ra

Table 1. MS Model for Ionic Substitution in the $NaZr_2(PO_4)_3$ structure: $M'M''_{\square\square\square}M'''_{\square}A_2(XO_4)_3$

OXIDE	WT %	CATION	Rc (nm)	Rc/Ra*
BaO	4.00	Ba^{2+}	0.135	0.964
Cr_2O_3	0.80	Cr^{3+}	0.062	0.439
Cs_2O	7.30	Cs^{+}	0.167	1.193
Fe_2O_3	3.80	Fe^{3+}	0.065	0.461
MoO_3	13.1	Mo^{6+}	0.041	0.297
NiO	0.30	Ni^{2+}	0.069	0.493
P_2O_5	1.70	P^{5+}	0.017	0.123
PdO	3.70	Pd^{2+}	0.086	0.614
Rb_2O	0.90	Rb^{+}	0.152	1.086
ReO_2	3.30	Re^{4+}	0.063	0.450
RE_2O_3	33.1	Nd^{3+}	0.098	0.702
RuO_2	7.60	Ru^{4+}	0.062	0.443
SrO	2.70	Sr^{2+}	0.118	0.843
TeO_2	1.90	Te^{6+}	0.043	0.312
UO_2	3.20	U^{4+}	0.089	0.636
ZrO_2	12.6	Zr^{4+}	0.072	0.514
SITE	**SUBSTITUTION**		**WT %**	**WT % REQ'D**
M'/M''	Ba, Cs, Rb, Sr Cr, Fe, Ni, Pd, Re		14.90	6.31
A	RE (Nd), Ru, U, Zr		68.40	50.26
X	Mo, P, Te		16.70	43.42

*values determined from Shannon-Prewitt radii [15-16]
**the ionic radius of Nd^{3+} (CN=8) is the same as the weighted average of the other RE^{3+} ionic radii; the essentially identical oxide crystal chemistry of the RE cations which are trivalent under high temperatures and oxidizing conditions allow the substitution of Nd_2O_3 for each of the others (La_2O_3, Pr_2O_3....) [17]

may be calculated for the constituent cations present in a HLW stream, with the ionic radius of each cation, Rc, divided by the ionic radius of oxygen, Ra. Because the values of Rc and Ra vary with coordination number, one must make initial assumptions regarding the coordination of the anions and the cations in a given crystal structure and use appropriate ionic radii to calculate Rc/Ra.

The radius ratios of cations capable of substituting into the octahedrally coordinated A site fall in the range of 0.73-0.41. The radius ratios of cations capable of occupying the tetrahedrally coordinated X site fall in the range of 0.41-0.23. The M' and M'' interstitial sites accept a variety of cations so one must rely on experimental observations, e. g. the data summarized in Figure 2, when assigning cations to these sites. Table 1 lists the cations present in the modified PW-4b simulated calcine and assigns these cations to the three crystallographic sites of the $NaZr_2(PO_4)_3$ structure based on their Rc/Ra values; the model suggests that the electrical neutrality of the structure is achieved through partial occupation of M'' sites by Na^+. A comparison between the amount (wt % oxide) of components provided by the calcine for each crystallographic site and the amount (wt % oxide) required for phase pure $NaZr_2(PO_4)_3$ is also provided in Table 1. The model predicts ionic substitution of cations into the $NaZr_2(PO_4)_3$ structure under oxidizing conditions.

Prior to the preparation of the [NZP] radiophases, the volatility of the calcine with respect to time and temperature was determined with gross gravimetric techniques. Isochronal relationships between average %LOI and temperature are shown in Figure 3. Gray [14] investigated the loss of constituents from a modified PW-4b simulated calcine of similar composition and noted the enhanced volatility of the oxides of molybdenum, ruthenium, tellurium, and cesium after 4-hour heat treatments at 800-1300°C, with tellurium (IV) oxide volatility becoming most significant at 1100°C.

These results and the data shown in Figure 3 suggest that modest temperatures and brief periods of heat treatment are requisite for the minimization of elemental loss during the preparation of [NZP] radiophases tailored to accommodate PW-4b species.

PREPARATION OF RADIOPHASES

An initial step in the preparation of monophasic [NZP] radiophases that are capable of accommodating a maximum volume of modified PW-4b simulated calcine is the selection of appropriate precursors. Because this study is concerned with the processing of [NZP] radiophases with methods, materials, and equipment that may be employed on an industrial scale, the criteria for selecting appropriate precursors are stringent. Precursors must be soluble in H_2O and other polar solvents, mutually compatible, inflammable, and easy to handle in both the laboratory and industrial setting. The precursors must decompose at low temperatures without generating excessive quantities of emissions that would require treatment in a radioactive off-gas facility. With these criteria in mind, sodium phosphate monobasic monohydrate ($NaH_2PO_4 \cdot H_2O$), zirconyl chloride octahydrate ($ZrOCl_2 \cdot 8H_2O$) and orthophosphoric acid (H_3PO_4) were selected as source materials for Na_2O/P_2O_5, ZrO_2 and P_2O_5, respectively.

Solid reaction precursors were intimately mixed in appropriate stoichiometric ratios with pre-determined amounts of modified PW-4b simulated calcine. The batch formulation was adjusted to account for the volatility of the calcine with

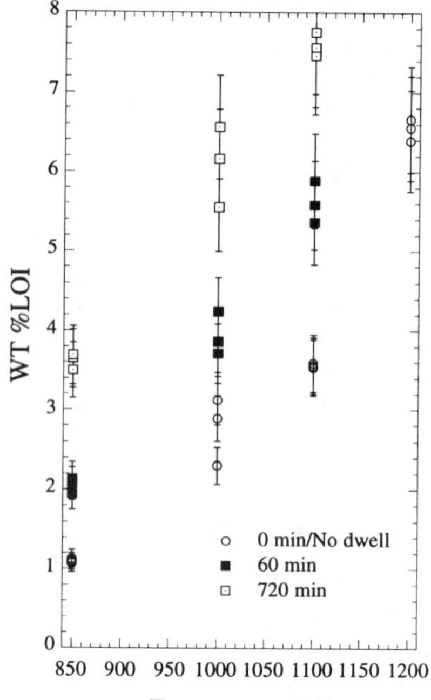

Figure 3. Isochronal relationships between wt % modified PW-4b simulated calcine lost-on-ignition (wt % LOI) and temperature (°C). Data for PW-4b at 1200°C are incomplete due to partial melting of the material.

respect to time and temperature for a specific heat treatment. The powders were ground with an alumina mortar and pestle to increase the homogeneity of the mixtures. Deionized H_2O was added to dissolve soluble components of the mixtures and to improve their rheology. Stoichiometric quantities of H_3PO_4 were added to the mixtures under conditions of constant stirring to generate workable pastes. The pastes were placed in a 100°C drying oven for 60 to 120 minutes. The resulting materials, which were partially free of moisture, were removed from the furnace, ground with an alumina mortar and pestle and placed in a 220°C drying oven for an additional heat treatment of 120 minutes; this heat treatment resulted in complete hydrolysis of the precursors. The dried powders were removed from the drying oven, reground and screened to obtain particle sizes of ≤125 μm. Table 2 lists the compositions of the radiophases prepared in this investigation.

The materials were uniaxially pressed in stainless steel dies under a pressure of 70 MPa and placed in Pt crucibles. Pressed pellets of MS10 and MS20 were heat treated in air to 850-1200°C at a rate of 10°C/min in a programmable furnace. Samples were removed sequentially at 850°C, 1000°C, 1100°C, and 1200°C. Control samples of phase pure $NaZr_2(PO_4)_3$ were processed in an identical manner.

Heat-treated samples were ground with an agate mortar and pestle, packed into the cavities of recessed zero-background slides and analyzed with a Scintag PAD V diffractometer with CuKα radiation. Data were collected from 10-40°2θ at a rate of 1°2θ/min. Standard powder x-ray diffraction (XRD) pattern analyses were performed using Scintag PAD V software and ICDD-JCPDS files; simulated patterns were generated with *POWD10* (version 10), a FORTRAN program that allows one to calculate and index theoretical x-ray diffraction patterns when requisite data, *i. e.* the contents, dimensions, and symmetry of the unit cell, are input into the program [18].

RESULTS AND DISCUSSION

Figures 4 and 5 illustrate the development of [NZP] from the x-ray amorphous MS10 and MS20 precursors with respect to temperature. The reflected x-ray intensity measurements obtained from XRD analysis of each sample (raw data) are plotted on a logarithmic scale against 2θ so that peaks of low intensity are not discounted. The development of [NZP] from the amorphous precursors is rapid. The broad diffraction peaks of MS10 at 850°C, as shown in Figure 4, indicate that the material consists of small crystallites; however, all of the peaks in this pattern are in close agreement with those listed for $NaZr_2(PO_4)_3$ (pattern # 33-1312) in the ICDD-JCPDS files. As the material is subjected to further heat treatment to 1200°C, the diffraction peaks become more intense and sharper, indicating an increase in the sizes of the [NZP]

Table 2 Chemical Compositions of [NZP] Radiophases and $NaZr_2(PO_4)_3$ Control Samples Prepared in this Study (in wt% oxide)

SAMPLE ID	CONTROL	MS10	MS20
PW-4b OXIDES (WT %)			
BaO	0.00	0.40	0.80
Cr_2O_3	0.00	0.08	0.16
Cs_2O	0.00	0.73	1.46
Fe_2O_3	0.00	0.38	0.76
MoO_3	0.00	1.31	2.62
NiO	0.00	0.03	0.06
P_2O_5	0.00	0.17	0.34
PdO	0.00	0.37	0.74
Rb_2O	0.00	0.09	0.18
ReO_2	0.00	0.33	0.66
RE_2O_3*	0.00	3.31	6.62
RuO_2	0.00	0.76	1.52
SrO	0.00	0.27	0.54
TeO_2	0.00	0.19	0.38
UO_2	0.00	0.32	0.64
ZrO_2	0.00	1.26	2.52
% PW-4b	0.00	10.00	20.00
OXIDE ADDITIVES (WT %)			
Na_2O	6.31	4.83	3.34
ZrO_2	50.26	43.42	36.58
P_2O_5	43.42	41.75	40.08
% Additives	100.00	90.00	80.00
TOTAL	100.00	100.00	100.00

*RE_2O_3 was modeled as Nd_2O_3 after McCarthy and Davidson [17]

crystallites. The XRD patterns of MS10 at 1000-1200°C suggest the absence of additional phases, such as Zr_2PO_7 or ZrO_2; the presence of these thermodynamically stable competitive phases would be indicative of poor processing of the material.

The broad diffraction peaks of MS20 at 850°C, as shown in Figure 5, also suggest that the material consists of small crystallites; the peaks in this pattern are in close agreement with those listed for $NaZr_2(PO_4)_3$ in the ICDD-JCPDS files. As the material is subjected to further heat treatment to 1200°C, the [NZP] crystallites increase in size. As a result, the diffraction peaks become more intense and sharper. The XRD pattern of MS20 at 1200°C shows splitting of the peaks at 19.48°2θ and 31.40°2θ which may indicate the formation of an additional [NZP] phase [1]. The XRD patterns of MS20 at each temperature contain peaks of low relative intensity that could not be indexed to Zr_2PO_7, ZrO_2, or to any of the simple oxides present in the modified PW-4b simulated calcine.

The XRD patterns of MS10 and MS20 at 1200°C are compared to those of control samples that were heat treated under identical conditions in Figure 6. The data are plotted on a linear scale to illustrate the high degree of crystallinity of each radiophase. Figure 6 illustrates the close agreement of 2θ values and peak relative intensities (I/I_0) of the control sample (A) and the radiophases MS10 (B) and MS20 (C).

Differences in I/I_0 of the peaks in the XRD patterns of the radiophases from those of the control may be attributed to differences in the scattering factors of Na^+, Zr^{4+}, P^{5+} and the PW-4b cations with which they are replaced in [NZP] radiophases. Although XRD allows one to probe changes in the long-range order and symmetry of a [NZP] radiophase, the technique is less sensitive to assessing changes induced by second order effects which occur at the molecular scale, *i. e.* local distortion and

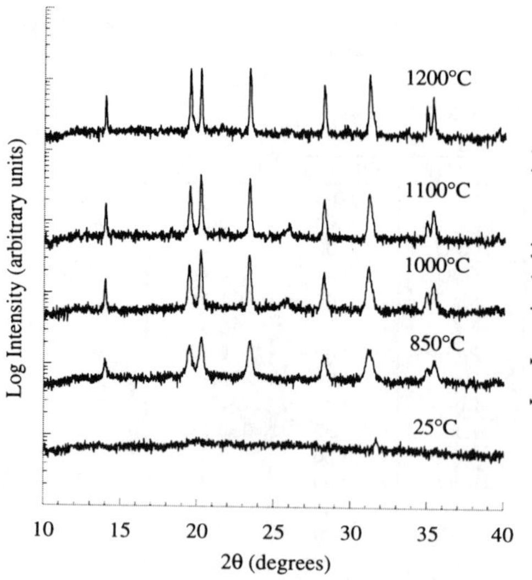

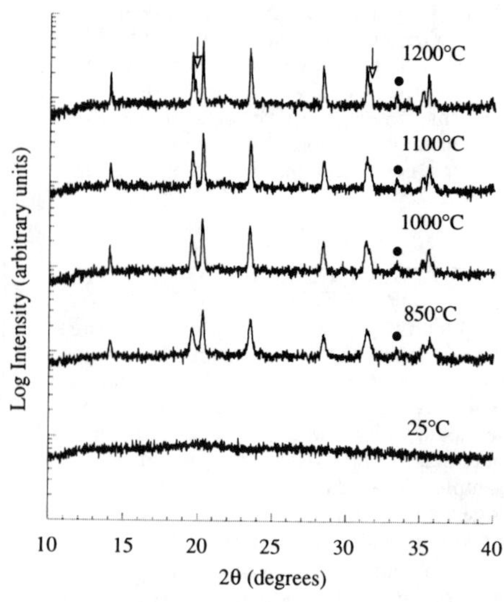

Figure 4. Representative XRD patterns of MS10 prior to heat treatment (25°C) and following heat treatment to: 850°C, 1000°C, 1100°C, and 1200°C. All of the peaks in each pattern are in agreement with those of $NaZr_2(PO_4)_3$ (ICDD-JCPDS pattern #33-1312).

Figure 5. Representative XRD patterns of MS20 prior to heat treatment (25°C) and following heat treatment to: 850°C, 1000°C, 1100°C, and 1200°C. Unlabeled peaks in each pattern closely match those of $NaZr_2(PO_4)_3$. Unknown peaks are labeled as (●). Arrows identify peak splitting which may indicate secondary [NZP] phase formation.

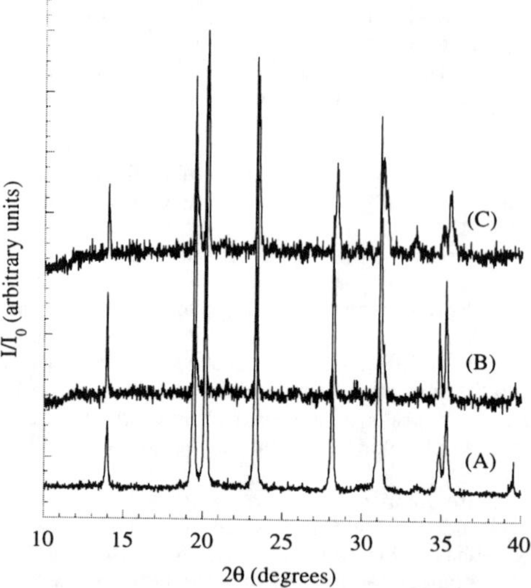

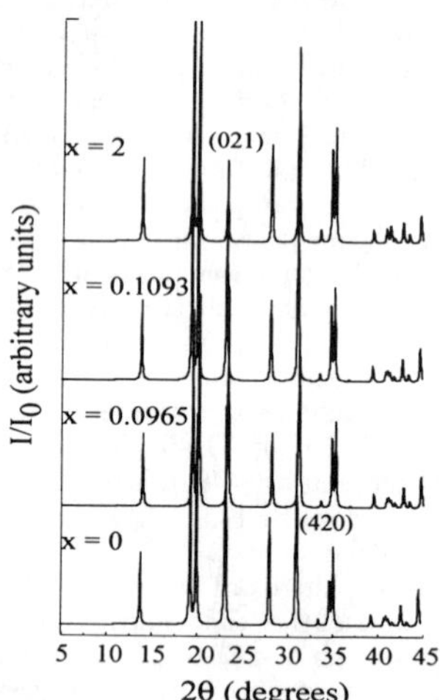

Figure 6. Comparison of 2θ and I/I₀ of $NaZr_2(PO_4)_3$ control sample (A) and MS10 (B) and MS20 (C). Samples were heated to 1200°C at a rate of 10°C/min prior to their removal from the furnace. Note the general agreement of I/I₀ and 2θ values for the radiophases and the control sample.

Figure 7. *POWD10* calculations of Nd^{3+} substitution for Zr^{4+} coupled with M" site occupation by Na^+ in the C2/c unit cell of $NaM''_xNd_xZr_{2-x}(PO_4)_3$. The values x = 0.0965 and x = 0.1093 correspond to the mol fractions of RE^{3+} (modeled as Nd^{3+}) in MS10 and MS20 radiophases, respectively. The values x = 0 and x = 2 represent end member compositions.

cation site preference in an individual [NZP] unit cell. However, one can calculate changes in I/I_0 of the peaks in theoretical XRD patterns of MS10 and MS20 radiophases. The composition of these radiophases is complex and calculations of their theoretical XRD patterns could prove laborious. Due to the fact that RE^{3+} cations (modeled as Nd^{3+}) dominate the composition of PW-4b on both a wt % and mol % basis, the calculation was simplified by modeling changes in the XRD patterns of $NaZr_2(PO_4)_3$ upon coupled Nd^{3+} substitution for Zr^{4+}/Na^+ occupation of M" in the C2/c $NaZr_2(PO_4)_3$ unit cell. The atomic coordinates given in Hong [11] for the $NaZr_2(PO_4)_3$ rhombohedral unit cell defined by hexagonal axes were reindexed on a monoclinic unit cell, C2/c, with the matrix: $a_h = -\frac{1}{2}a_{mon} - \frac{1}{2}b_{mon}$, $b_h = \frac{1}{2}a_{mon} - \frac{1}{2}b_{mon}$, and $c_h = -a_r - 3c_{mon}$ [19].

The calculated patterns given in Figure 7 represent materials with the general formula $NaM''_xNd_xZr_{2-x}(PO_4)_3$, where $x = 0$ for $NaZr_2(PO_4)_3$, $x = 2$ for $Na_3Nd_2(PO_4)_3$, and $x = 0.0965$ and 0.1093 for MS10 and MS20, respectively. These data predict the contribution of Nd^{3+} substitution for Zr^{4+}/Na^+ occupation of M" on differences in I/I_0 in the XRD patterns of $NaZr_2(PO_4)_3$ and the [NZP] radiophases. As the value of x increases, the intensities of certain peaks, i. e. (420), are augmented whereas the intensities of other peaks, i. e. (021), are reduced. Further determination of the effect of site occupancy on the $NaZr_2(PO_4)_3$ structure with *POWD10* will be left to future investigations.

The slight variation in the peak positions in the XRD patterns of the radiophases from those of the control may be attributed to differences in the sizes of Na^+, Zr^{4+}, P^{5+} and the PW-4b cations with which they are replaced in [NZP] radiophases. The following formula may be employed to determine the average ionic radius of PW-4b cations assigned to a particular site S in the NZP structure:

$$\text{Average Ionic Radius (nm)} = \Sigma Rc_{[S]}w_{c[S]} / \Sigma w_{c[S]}$$

where $\Sigma Rc_{[s]}w_{c[s]}$ = sum of the radius of each PW-4b cation assigned to site S multiplied by its amount (wt % PW-4b) and $\Sigma w_{c[S]}$ = wt % of PW-4b cations assigned to site S of the NZP structure. For MS10 radiophases, the average ionic radius of the PW-4b cations assigned to the M' and M" sites of the $NaZr_2(PO_4)_3$ structure is equal to 0.15 nm; this radius is approximately 45% larger than the radius of Na^+ in octahedral coordination with O^{2-}. The average ionic radii of PW-4b cations assigned to the A and X sites are equivalent to 0.08 nm and 0.04 nm, respectively. These radii are approximately 11% larger than octahedrally-coordinated Zr^{4+} and 55% larger than tetrahedrally-coordinated P^{5+}, respectively [15-16]. [NZP] radiophases may be regarded as solid solutions. Therefore, the isovalent and heterovalent substitutions for Na^+, Zr^{4+}, and P^{5+} may cause the lattice parameters of $NaZr_2(PO_4)_3$ to increase or decrease, or in some cases, some parameters to increase and others to decrease [20]. The effect will become more noticeable as the amount of PW-4b cations accommodated by the [NZP] host lattice increases.

CONCLUSIONS

Highly crystalline [NZP] radiophases were tailored to accommodate 10-20 wt % modified simulated PW-4b calcine. To meet the challenge of designing monophasic materials capable of immobilizing all of the chemical species associated with PW-4b, an ionic substitution scheme based on crystal chemical principles was developed. This ionic substitution scheme was used in the preparation of monophasic [NZP] radiophases. Experimental and calculated XRD patterns suggest that MS10 (10 wt % PW-4b) and MS20 (20 wt % PW-4b) radiophases are essentially phase pure and isostructural with $NaZr_2(PO_4)_3$.

The radiophases were processed with inexpensive, inorganic precursors and prepared according to a solution sol-gel method. Pressed pellets of [NZP] precursor powders were heat treated and/or sintered under a variety of temperatures to determine the processing conditions under which the development of single phase [NZP] is favored.

The effect of extended heat treatments on the phase-purity, crystallinity and density of [NZP] radiophases must be rigorously addressed prior to industrial-scale production of these materials. However, the data presented in this investigation suggest that extended heat treatments may not be requisite for the preparation of phase pure [NZP] radiophases. Experimental evidence indicates that phase pure [NZP] radiophases may be continuously processed at temperatures as low as 850°C.

The rapid formation of [NZP] from amorphous, sol-gel derived free powders was demonstrated for MS10 and MS20 precursors. The results of this investigation suggest that single phase [NZP] radiophases may be continuously processed with conventional ceramic processing methods and equipment. Moreover, the relatively low temperatures involved and the reproducibility of the process make [NZP] radiophases economically attractive hosts for radioactive wastes and possibly heavy metal industrial wastes, as well. Specific industrial applications of the material, determination of the effects of site occupancy on the $NaZr_2(PO_4)_3$ structure, characterization of long-term aging effects in actinide-doped [NZP] compounds, identification of secondary and/or amorphous phases, and further refinement of the process are the subject of future studies.

ACKNOWLEDGMENTS

This work was funded by the US Department of Energy Nuclear Materials Stabilization Task Group under the 94-1 Research and Development Program. Work completed at The Pennsylvania State University was supported in part by a grant from Associated Cement Companies, Ltd., of Bombay, India.

REFERENCES

[1]   R. Roy, E. R. Vance, and J. Alamo, Mat. Res. Bull. **17**, 585-589 (1982).

[2]   A. I. Orlova, Yu. F. Volkov, R. F. Melkaya, L. Yu. Masterova, I. A. Kulikov, and V. Alferov, Radiochem. **36** [4], 322-325 (1994).

[3]   D. K. Agrawal and R. Roy, J. Mater. Sci. **20**, 4617-4623 (1985).

[4]   T. Oota and I. Yamai, J. Am. Ceram. Soc. **69**, 1-6 (1986).

[5]   B. E. Scheetz, S. Kormaneni, W. Fajun, L. J. Yang, M. Ollinen, and R. Roy in *Scientific Basis for Nuclear Waste Management VIII*, edited by C. M. Jantzen, J. A. Stone, and R. C. Ewing, (Mater. Res.Soc. Proc. **44**, Pittsburgh, PA, 1985) pp. 903-910.

[6]   A. I. Orlova, V. N. Zyryanov, O. V. Egor'kova, and V. T. Demarin, Radiochem. **38** [1], 20-23 (1996).

[7]   B. E. Scheetz, D. K. Agrawal, E. Breval, and R. Roy, Waste Manage. **14** [6], 489-505 (1994).

[8]   J. Alamo, Solid State Ionics **63-65**, 547-561 (1993).

[9]   R. Roy, D. K. Agrawal, J. Alamo, and R. A. Roy, Mater. Res. Bull. **19**, 471-477 (1984).

[10] H. Kohler and H. Schulz, Mater. Res. Bull. **18**, 589-592 (1983).

[11] H. Y-P. Hong, Mat. Res. Bull. **11,** 173-182 (1976).

[12] M. Alami Talbi, R. Brochu, C. Parent, L. Rabardel, and G. Le Flem, J. Solid State Chem. **110**, 350-355 (1994).

[13] F. Nectoux and A. Tabuteau, Radiochem. and Radioanalyt. Lett. **49** [1], 43-48 (1981).

[14] W. J. Gray, Rad. Waste Manage. **1**, 147-169 (1981).

[15] R. D. Shannon and C. T. Prewitt, Acta Crystall. **B25**, 925-945 (1969).

[16] R. D. Shannon, Acta Crystall. **A32**, 751-767 (1976).

[17] G. J. McCarthy and M. T. Davidson, Ceram. Bull. **54**, 782-786 (1975).

[18] D. K. Smith, M. C. Nichols, and M. E. Zolensky, *POWD 10: A FORTRAN IV Program for Calculating X-ray Powder Diffraction Patterns- Version 10*; Department of Geosciences, The Pennsylvania State University, 1983.

[19] *Mathematical Crystallography (Reviews in Mineralogy Vol. 15)*, M. B. Boisen, Jr. and G. V. Gibbs (The Mineralogical Society of America, Washington, DC, 1985) 56-72.

[20] R. Jenkins in *Modern Powder Diffraction (Reviews in Mineralogy Vol. 20)*, edited by D. L. Bish and J. E. Post (The Mineralogical Society of America, Washington, DC, 1989) 47-69.

SALT-OCCLUDED ZEOLITE WASTE FORMS: CRYSTAL STRUCTURES AND TRANSFORMABILITY

J. W. RICHARDSON, JR.
Intense Pulsed Neutron Source Division, Argonne National Laboratory, Argonne, IL 60439 USA

ABSTRACT

Neutron diffraction studies of salt-occluded zeolite and zeolite/glass composite samples, simulating nuclear waste forms loaded with fission products, have revealed complex structures, with cations assuming the dual roles of charge compensation and occlusion (cluster formation). These clusters roughly fill the 6-8 Å diameter pores of the zeolites. Samples are prepared by equilibrating zeolite-A with complex molten Li, K, Cs, Sr, Ba, Y chloride salts, with compositions representative of anticipated waste systems. Samples prepared using zeolite 4A (which contains exclusively sodium cations) as starting material are observed to transform to sodalite, a denser aluminosilicate framework structure, while those prepared using zeolite 5A (sodium and calcium ions) more readily retain the zeolite-A structure. Because the sodalite framework pores are much smaller than those of zeolite-A, clusters are smaller and more rigorously confined, with a correspondingly lower capacity for waste containment. Details of the sodalite structures resulting from transformation of zeolite-A depend upon the precise composition of the original mixture. The enhanced resistance of salt-occluded zeolites prepared from zeolite 5A to sodalite transformation is thought to be related to differences in the complex chloride clusters present in these zeolite mixtures. Data relating processing conditions to resulting zeolite composition and structure can be used in the selection of processing parameters which lead to optimal waste forms.

INTRODUCTION

This work is carried out in support of a ceramic waste form development project at Argonne National Laboratory for salt waste generated from electrometallurgically conditioned spent nuclear fuel. The ceramic waste form is prepared from hot isistatic pressing of a mixture of salt-occluded zeolite and glass binder. These composite materials were shown to be leach resistant (meeting standard criteria required for waste storage applications). While leach tests are generally satisfactory, however, some variability in performace has been observed.

Powder diffraction analysis of prospective zeolite waste form composites is important for correlating structural features with waste form capacity-controlling factors such as cation segregation, competition between charge balance and occlusion, mixed phase compositions and resistance to phase transformation during processing. This work has emphasized neutron diffraction for a number of reasons. Rietveld profile refinement of zeolite and molecular sieve structures using neutron powder diffraction data is well established [1-4]. In addition, the scattering contrast of important constituent elements in the composites is advantageous. These zeolite composites tend to exhibit strong long-range cation ordering. The strength of this ordering is evidenced by dramatic re-arrangement of oxygens in the zeolite framework to accommodate the long-range ordering. Oxygen is a relatively weak scatterer of X-rays, but an exceptionally strong scatterer of neutrons. Neutron diffraction patterns which are strongly influenced by oxygen scattering, therefore, allow us to follow in detail the accommodation of oxygen to waste salt occlusion. Furthermore, the scattering strength of Li is considerable with neutron diffraction, allowing us to recognize the strong impact Li has on the resulting structures.

Zeolites are very complex materials, even in their simplest chemical forms. Occlusion of simulated waste salts further complicates the structures, by producing strong distortions in AlO_4 and SiO_4 tetrahedra in the zeolite. General statements can, however, be made regarding the relationship between overall composition and resulting structure. This is because cations from the salt (Li in particular) dominate interactions with the zeolite, to induce surprisingly regular long-range structures. Salt-occluded zeolites with relatively high mass fractions of Li (≥ 4 wt%) crystallize in a single phase structure with nearly absolute long-range ordering of cations, largely on the basis of size [5]. As will be discussed in this paper, mixtures prepared from zeolite 4A (with exclusively Na^+ cations) and from zeolite 5A (roughly 60% Ca^{2+}, 40% Na^+) also have structures

Mat. Res. Soc. Symp. Proc. Vol. 465 © 1997 Materials Research Society

involving strong interactions between occluded salt complexes and the zeolite framework, but with different cation ordering schemes.

As part of the waste form processing, salt-occluded zeolites are pressure-bonded with borosilicate glass at an elevated temperature to form the final product. Under many circumstances, this high-temperature processing results in transformation of zeolite-A to the more dense zeolite known as sodalite. The susceptibility of the occluded zeolite to this conversion depends on the starting material; samples prepared with zeolite 5A are less likely to convert. The structures of zeolite 4A and zeolite 5A mixtures have been determined in an effort to identify correlations between cation distribution, etc. and susceptibility to transformation.

EXPERIMENTAL

Salt-occluded zeolite samples are prepared using one of three methods [6-8]: (1) ion exchange, (2) occlusion or (3) a combination of (1) and (2). In method (1), the zeolite is treated with an excess of salt, usually a factor of 10:1 salt:zeolite, and ion exchange predominates. The concentration of fission products in the salt-occluded zeolite is high, usually over 10 wt% and there is little Na, <1 wt%. Most charge compensating ions are Li and K. This type of salt-occluded zeolite is referred to as fully exchanged zeolite. In method (2), excess zeolite is treated with a minimum of salt, and salt occlusion predominates. Typically, there is about 4 times as much zeolite as salt. Thus, the concentration of fission products in the salt-occluded zeolite is about 20% of that in the salt, about 1-3 wt%. The Na concentration is high, typically around 12-13 wt%. This type of salt-occluded zeolite is referred to as blended zeolite. In method (3), both the ion exchange and the occlusion methods are combined. The zeolite is first treated with excess salt, as in an ion exchange process. The zeolite is fully exchanged but contains a significant amount of surface salt. This surface salt is sorbed into new zeolite using the occlusion or second method. The combination method leads to intermediate fission product and Na concentrations.

Shown in Table 1 are the chemical compositions of salt-occluded zeolites used in our neutron powder diffraction studies. These were chosen to be representative of each of the methods described above. Samples identified as BE-4 and BE-5 were prepared using method (1), 3023-E, 4A-Sr, 4A-Y, 5A-Sr and 5A-Y were prepared using method (2), and 12223-D was prepared using method (3).

Table 1. Chemical compositions (wt%) for some of the salt-occluded zeolites studied

Ion	BE-4	BE-5	12223-D	3023-E	4A-Sr	4A-Y	5A-Sr	5A-Y
Li^+	4.00	4.00	1.29	1.72	1.14	1.30	1.25	1.22
Y^{3+}	0.46	1.34	0.01	0.00		1.52		3.01
Na^+	0.60	0.60	7.04	11.20	12.80	13.00	3.49	3.83
Ca^{2+}							7.66	7.45
Sr^{2+}	3.45	1.71	0.06	0.61			3.24	
Ba^{2+}	2.56	3.49	0.19	0.48				
K^+	6.00	6.00	14.60	5.15	4.45	5.00	4.79	4.70
Cs^+	7.75	8.02	0.67	1.38				
Cl^-	16.00	16.00	12.60	12.60	11.30	12.00		

Salt-occluded zeolite samples, prepared as loose powders, were transferred to vanadium sample tubes and placed in the General Purpose Powder Diffractometer (GPPD) [9] at the Intense Pulsed Neutron Source (IPNS) at Argonne National Laboratory. Data collection times were in the range of 12 to 24 hours. Time-of-flight (TOF) neutron diffraction data (Figure 1) are accumulated at a series of fixed scattering angles on the GPPD, in the form of neutron counts as a function of interplanar d-spacings. Data from the highest resolution detectors ($2\Theta=\pm148°$) were analyzed using the Rietveld profile refinement technique [10] modified for TOF data [11]. As will be discussed in the following section, some occluded zeolite samples are single phase, while others are multi-phase.

Rietveld profile refinement provides an identification of space group and unit cell lattice parameters for each phase, along with a quantification of framework distortions and cation and anion site positions and occupancies. Because most populated sites are of relatively high symmetry, far fewer refineable parameters are required to define the structures than the number of distinct diffraction data points (Bragg peak intensities).

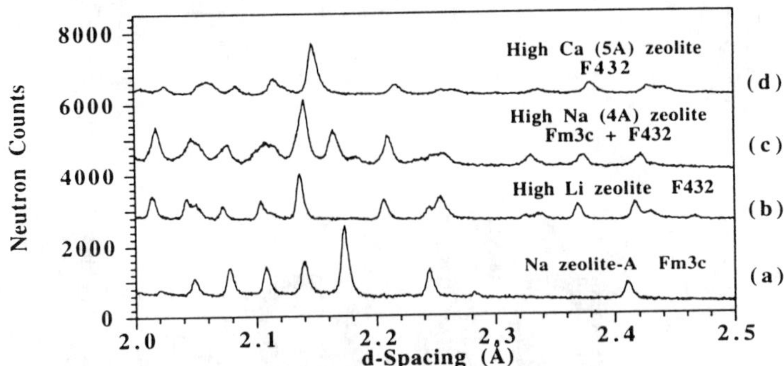

Figure 1. Comparisons of small sections of neutron diffraction patterns of Na zeolite-A and salt-occluded zeolite-A mixtures. These data show that: (a) Na zeolite-A is a simple fcc cubic phase, (b) compositions with high Li content are single phase with a more complex fcc cubic phases, (c) high Na mixtures are two phase; one phase similar to (a) and the other similar to (b), and (d) high Ca mixtures are single phase with a complex fcc cubic phase subtly different than (b). Note that in addition to the introduction of superlattice reflections, occlusion typically contracts the overall lattice, thus shifting reflections to lower d-spacings.

RESULTS

Description of zeolite-A and sodalite structures

The structures of zeolite-A and sodalite are well known, described by numerous crystallographic studies over the past few decades [12-14, and many more]. Both zeolite-A and sodalite are framework structures composed of SiO_4 and AlO_4 tetrahedra linked on corners to form micropores, and contain, as "building units", what are known as sodalite units (see Figure 2). In zeolite-A, isolated sodalite units, referred to as β-cages, are joined through linkages which have the appearance of cubes. Each "cube link" joins two β-cages. Sodalite appears to be assembled by the fusing of sodalite units into a structure made up entirely of sodalite units. A more appropriate description, though, would be to say that the sodalite structure is composed of sodalite units linked at edges, such that the linkages, themselves, are sodalite units. In this representation, each "sodalite link" joins four sodalite units. This picture illustrates a possible mechanism by which zeolite-A can transform to sodalite. If all of the "cube" linkages in zeolite-A are broken, and the dissociated sodalite units re-arrange themselves appropriately, the sodalite structure is generated.

Zeolite-A contains two types of micropores, large α-cages bounded by 4-, 6- and 8-rings, and smaller β-cages bounded by 4- and 6-rings. For reference, a 6-ring, which is designated 6R, is a ring of atoms containing 6 oxygens, 3 silicons, and 3 aluminums (shown as hexagons in Figure 2). Synthetic Na zeolite-A has Na^+ ions sitting in the large α-cages opposite 6-rings (on 6R sites), off-center in 8-rings, and occasionally on 4R sites. The natural form of sodalite [14] has composition $(Na_4Cl)_2 (AlSiO_4)_6$, where chloride ions sit in the centers of sodalite units and sodium ions are on 6R sites, bridging the chloride ion and oxygens in the framework.

Ion-exchanged zeolite-A

In zeolite-A, each $AlSiO_4$ formula unit has a net charge of -1, corresponding to a total charge of -12 for each α-cage. This negative charge must be balanced by extra-framework cations. In Na zeolite-A, this is accomplished with Na^+ ions on 6R, 8R and 4R sites [15]. In cation exchanged forms of zeolite-A, the siting of cations is often more complex [16-20]. Diffraction studies have described the behavior of virtually all of the cations considered here, when they are present alone as ion-exchanged species. In all cases, 6R sites are the most highly populated. Li^+ ions in Li-exchanged zeolite-A (Li-A) are positioned off center in the 6-rings, producing a long range distortion of the framework [work in progress]. In the K^+ [16], Ca^{2+} [17] and Sr^{2+} [18] forms, some 6R ions project into the smaller β-cages. Partially exchanged Cs-A has Cs^+ ions on 6R sites in α- and β-cages and on 8R and 4R sites, with Na^+ ions on 6R sites in α-cages [19]. Commercial 5A zeolites (with Ca^{2+}:Na^+ = 5:2), have Ca^{2+} ions in β-cages on 6R sites and Na^+ ions on 6R sites in α-cages [2].

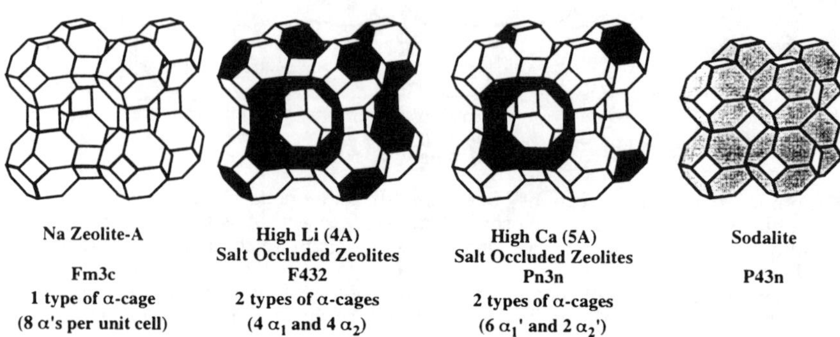

Figure 2. Structural drawings of zeolite-A and sodalite. The vertices of line segments correspond, alternately, to Si and Al atoms, and the mid-point of each line represents an O atom. Zeolite-A is composed of sodalite units (β-cages) joined by "cube linkages", while sodalite is composed of sodalite units joined by "sodalite linkages" (see text for details). Shading of the α-cage surfaces refers to the type of cation clusters present, i.e., predominately Li-K-Cl vs. predominately fission product-Cl.

<u>Salt-occluded zeolites</u>

As described above, salt-occluded zeolites are prepared using a variety of methods. Neutron diffraction studies have shown that the resulting structures (siting of cations) depend both on the method of preparation, and the overall composition of the salt solution used in the preparation. To illustrate this point, the following paragraphs will highlight the characteristics of sample compositions shown in Table 1.

Salt-occluded zeolites prepared using the ion exchange method, which result in mixtures with Li compositions higher than 3 wt%, have a strongly long-range ordered structure [5]. Figure 1 provides direct evidence of the long-range ordering, namely additional superlattice diffraction lines in the neutron diffraction patterns. This ordering is found to be due to the presence of two types of α-cages, $α_1$ with predominately Li^+ on 6R and 4R sites and K^+ on 8R sites, and $α_2$ with predominately large fission product ions (Cs^+ and Ba^{2+}) on 6R sites. The "NaCl-like" ordering of α-cages, whereby α-cages sharing 8-rings are different (i.e., neighboring cages have different colors or shading), is schematically illustrated in Figure 2. The long-range ordering is propagated by very strong distortions of the $AlSiO_4$ framework generated in order for Li^+, with its high charge-to-radius ratio, to closely bond with oxygens in the zeolite. In this model, Li^+ in $α_1$-cages and Cs^+ and Ba^{2+} in $α_2$-cages, in addition to playing the charge balancing role described above, are inter-linked within α-cages through bonds with Cl^- ions. Finally, in the $α_2$-cages, on average only half of the 6R sites are populated by Cs^+ or Ba^{2+}. The remaining 6R sites were found to be populated by Sr^{2+}, Y^{3+} (smaller fission product ions) and Li^+, where the ions project into the smaller β-cages.

Salt-occluded zeolites prepared with Na zeolite-A are two phase mixtures. While it is not readily apparent in Figure 1, where peaks from the two phase are strongly overlapped, Rietveld refinements revealed that two phases are present, and they are qualitatively similar to Na zeolite-A and high Li salt-occluded zeolite, respectively. Although refinements are relatively qualitative, Li-dominated ordering persists in the primary occluded zeolite. In the "Na zeolite-A" phase, in addition to Na^+ and, seemingly, some fission product ions occupying traditional 6R and 8R sites, Cl^- ions link cations in a similar manner to that observed in the high Li zeolites, without producing a superlattice, i.e., there is only one type of α-cage.

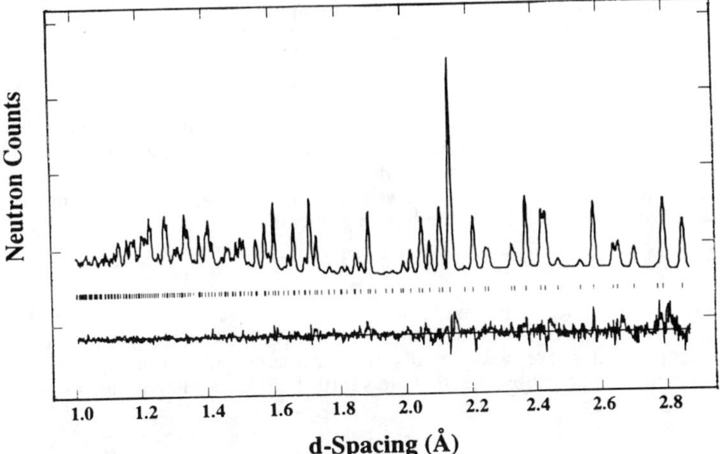

Figure 3. Rietveld profile refinement plot, showing fit to the neutron diffraction pattern of a salt-occluded zeolite, designated Sr-5A, prepared with the occlusion method using zeolite 5A (60% Ca, 40% Na) as starting material. The data points are observed data, the smooth curve through the data is the calculated diffraction spectrum, the residual (obs - calc) is shown at the bottom, and the vertical bars identify d-spacings for reflections.

Samples prepared using the occlusion method with zeolite 5A (Ca^{2+} and Na^+ as charge compensating cations) as starting material have a slightly different structure than those prepared from Na zeolite-A. These samples have single phase, long-range ordered structures, similar to the high Li zeolites. But the ordering pattern is slightly different. Rather than the "NaCl-like" behavior found in Li-rich occluded zeolites, the ordering is more reminiscent of "Cu_3Au-like" ordering in simple intermetallic alloys. Whereas in the "NaCl-like" ordering, 4 out of 8 α-cages per unit cell are of α_1 type and the other 4 of α_2 type, according to this ordering pattern, 6 out of 8 α-cages per unit cell are α_1', and the other 2 are α_2'. Rietveld refinements (see Figure 3 for the profile fit) suggest that in the α_1'-cages, Li^+ and K^+ ions occupy 6R sites, with Cl^- ions linking cations, while the fission products occupy 6R sites in the α_2'-cages, again with Cl^- linkages. The fundamentally different feature in these occluded zeolites is the presence of large amounts of Ca^{2+} ions in β-cages, opposite 6R sites occupied by K^+.

<u>Resistance to sodalite transformation</u>

The postulated mechanism for the transformation of salt-occluded zeolite to sodalite requires that Al-O and Si-O bonds in the linkages between β-cages become de-stabilized. This de-stabilization is probably a consequence of many factors, but must have some relationship to the frequency and type of interactions between the critical oxygens (those whose bonds must be broken) and occluded cations. We have learned from these neutron diffraction analyses that, in Ca-rich salt-occluded zeolites, many Ca^{2+} ions remain in β-cages (i.e., Ca^{2+} is less readily exchanged than Na^+), and that the long-range ordering sequence is different from that of Na-rich occluded zeolites. The degree of de-stabilization is anticipated, then, to be strongly related to the degree of long-range ordering exhibited by the β-cage occluded Ca^{2+} ions.

CONCLUSIONS

Time-of-flight neutron diffraction data from IPNS have provided detailed crystallographic information on the sites and populations of occluded simulated waste ions in salt-occluded zeolites. The unexpected level of detail observed is largely due to the tremendous flexibility of the zeolite-A framework, and its capacity to promote ordering of cations over large length scales. The discovery of strong crystallographic ordering in chloride salt-occluded zeolite-A mixtures may be important when postulating cation migration and transmutation during long-term storage. Finally, it is not absolutely certain how the characteristic cation siting and long-range order interplay to influence the transformability of salt-occluded zeolites in the high temperature bonding environment. Our new understanding of these factors for a variety of compositions at ambient temperature and pressure, though, will be extremely valuable for future studies of these materials under conditions simulating the bonding operation.

ACKNOWLEDGMENTS

The submitted manuscript has been authored by contractors of the U.S. Department of Energy (BES-Materials Sciences) under contract No. W-31-109-ENG-38. Accordingly, the U.S. Government retains a non-exclusive royalty-free license to publish or reproduce the published form of this contribution, or allow others to do so, for U.S. Government purposes.

REFERENCES

1. J. M. Adams and D. A. Haselden, J. Solid State Chem. **47**, 123 (1983).
2. J. M. Adams and D. A. Haselden, J. Solid State Chem. **51**, 83 (1984).
3. J. W. Richardson, Jr., J. V. Smith and J. J. Pluth, J. Phys. Chem. **93**, 8212 (1989).
4. J. W. Richardson, Jr. and E. T. C. Vogt, Zeolites **12**, 13 (1992).
5. J.W. Richardson, Jr., M. A. Lewis and B. R. McCart, Proc. 10th Int. Zeol. Conf. 741 (1994).
6   M. A. Lewis, L. J. Smith and D. F. Fischer, Am. Ceram. Soc. **11**, 2826 (1993).
7. M. A. Lewis, D. F. Fischer and C. D. Murphy, Mat. Reser. Soc. Proc. **333**, 277 (1994).
8. M. A. Lewis, D. F. Fischer and C. D. Murphy in <u>Environmental and Waste Management Issues in the Ceramic Industry II</u>, 277 (1994).
9. J. D. Jorgensen, J. Faber, Jr., J. M. Carpenter, R. K. Crawford, J. R. Hauman, R. L. Hitterman, R. Kleb, G.E. Ostrowski, F.J. Rotella and T.G. Worlton, J. Appl. Cryst. **21**, 321 (1989).
10. H. M. Rietveld, J. Appl. Cryst. **2**, 65 (1969).
11. A. C. Larson and R. B. Von Dreele, GSAS, General Structure Analysis System. Los Alamos National Laboratory Report LAUR86-748 (1986).
12. T. B. Reed and D. W. Breck, J. Am. Chem. Soc. 78 (1956) 5972.
13. V. Gramlich and W. M. Meier, Z. Krist. **133**, 134 (1971).
14. L. Pauling, Z. Krist. **74**, 213 (1930).
15. J. J. Pluth and J. V. Smith, J. Am. Chem. Soc. **102**, 4704 (1980).
16. J. J. Pluth and J. V. Smith, J. Phys. Chem. **83**, 741 (1979).
17. J. J. Pluth and J. V. Smith, J. Am. Chem. Soc. **105**, 2621 (1983).
18. J. J. Pluth and J. V. Smith, J. Am. Chem. Soc. **104**, 6977 (1982).
19. N. H. Heo and K. Seff, J. Am. Chem. Soc. **109**, 7986 (1987).
20. J. M. Newsam, R. H. Jarman and A. J. Jacobson, J. Solid State Chem. **58**, 325 (1985).

PHASE EQUILIBRIA AND ELEMENTS PARTITIONING IN ZIRCONOLITE-RICH REGION OF Ca-Zr-Ti-Al-Gd-Si-O SYSTEM

O.A. KNYAZEV*, S.V. STEFANOVSKY*, S.V. IOUDINTSEV**, B.S. NIKONOV**, B.I. OMELIANENKO**, A.V. MOKHOV**, AND A.I. YAKUSHEV**
* SIA "RADON", 7-th Rostovskii line 2/14, Moscow 119121, Russia
** Institute of Geology of Ore Deposits, Staromonetny line 35, Moscow 109017, Russia

ABSTRACT

Zirconolite-rich ceramics were produced by the cold crucible melting technique in an air atmosphere, at 1550 ± 50 °C and 1 atm. Four samples with overall composition (in wt.%): 4.9-14.3 CaO; 19.0-41.3 ZrO_2; 24.1-42.6 TiO_2; 1.3-11.3 Al_2O_3; 6.8-30.0 Gd_2O_3, and 1.1-8.5 SiO_2 have been studied. Total phases in the ceramics consist of major zirconolite and minor rutile, perovskite, zirconia, aluminium titanate, and glass. The Gd_2O_3 content in zirconolite reaches up to 31.4 wt.% corresponding to the formula: $(Ca_{0.4},Gd_{0.7})Zr_{1.0}(Ti_{1.4},Al_{0.5})O_{7.0}$. The data on the phase composition agree well with coupled Gd incorporation into the mineral structure: $Ca(II) + Ti(IV) = Gd(III) + Al(III)$, and $2Gd(III) = Ca(II) + Zr(IV)$. The highest Gd contents observed in the other phases are 25.4 % for zirconia, 12.6 % in glass, 8.8 % in perovskite, and 1.4 % for rutile. The rest of the elements' distribution in the samples are analyzed.

INTRODUCTION

Zirconolite possesses good isomorphic properties with respect to many components of high level waste (HLW) [1] and the highest radiation stability and hot water durability among the phases of ceramic wasteforms [2,3]. The application of zirconolite-rich ceramics for actinides immobilization was described elsewhere [4,5]. Such ceramics are produced by subsolidus hot-pressing, sintering or pot melting [4-7]. Crystal sizes in the samples obtained are very small. These complicate strongly the investigation of chemical compositions of individual phases and waste elements partitioning between co-existing phases. Using cold crucible inductive melting technique various Synroc-like ceramics may be fabricated [8,9]. Crystal sizes in the specimens are 1-2 orders of magnitude larger than those of the hot-pressed samples. In the present paper recent data on zirconolite-rich ceramics produced through the melting route are discussed.

EXPERIMENTAL PROCEDURE

Four samples were fabricated using the cold crucible melting method formerly developed for high fusible and pure materials production [10] and radioactive waste immobilization [11,12]. It was successfully applied also for rock-type wasteform production [8,9,13-15]. Cold crucible melting is suitable method for zirconolite production due to its high melting point of (~1525°C [15]) which is about 200 degrees superior to that for conventional Synroc-like ceramics.

A starting batch consisted of Ca, Zr, and Ti oxides taken in relative amounts similar to the theoretical composition of zirconolite. Gadolinium was added as an analog of actinides while aluminum was introduced for charge compensation. The experimental procedure was described in detail in our previous work [14]. In order to produce zirconolite-based material the process temperature ranged between 1500 and 1600 °C depending on the batch composition. The

Mat. Res. Soc. Symp. Proc. Vol. 465 © 1997 Materials Research Society

increase of Gd content resulted as a rule in the decrease of process temperature. The melts were poured into steel molds followed by cooling to room temperature at the rate of ~40 ^{0}C/h.

The samples obtained were analyzed with optical microscopy (POLAM L-213), X-ray diffraction (XRD) using DRON-4 diffractometer (Cu K_α radiation) , electron microprobe analysis (EMPA) and scanning electron microscopy using a JSM-5300+Link ISIS analytical system. To study bulk chemical composition of melted material, conventional analytical methods such as spectrophotometry (SF) with S-115, Russian model, and emission spectral analysis (ESA) with ISP-30, Russian model, were also employed.

RESULTS AND DISCUSSION

Four samples with varying elements ratios were produced. Their actual compositions determined by both SEM and chemical analytical routes are given in Table I. The resulting material was found to contain appreciable amounts of silica and minor concentrations of natrium and iron. These components contaminated the samples of materials during melting due to reaction between the Synroc melt and protective putty forming the "skull" [15] which was prepared from zirconia with sodium silicate binder.

Table I. Bulk composition of the samples (wt%) determined by SEM/EMPA and SF/ESA.

Oxides	Z-1		Z-2		Z-3		Z-4	
	SEM/ EMPA	SF/ ESA	SEM/ EMPA	SF/ ESA	SEM/ EMPA	SF/ ESA	SEM/ EMPA	SF/ ESA
Na_2O	n.a.	0.2	n.a.	0.2	n.a.	0.2	n.a.	0.2
Al_2O_3	8.7	6.3	8.8	7.1	11.3	10.3	2.9	1.3
SiO_2	6.0	5.6	5.4	8.5	5.6	5.4	1.1	2.5
CaO	5.3	4.9	5.4	5.6	6.2	6.3	13.2	14.3
TiO_2	26.3	24.1	32.0	29.9	38.2	42.6	33.8	38.3
FeO	0.1	0.2	0.1	0.2	0.1	0.2	0.1	0.1
ZrO_2	23.5	25.4	19.0	21.4	31.0	28.2	41.3	32.6
Gd_2O_3	30.0	n.a.	29.2	27.1	7.6	6.8	7.6	n.a.

n.a. - not analyzed. Z-1, ... Z-4 - sample numbers.

The analytical results obtained by SEM/EMPA and SF/ESA methods are markedly different. Determination of samples compositions by SEM is based on averaging of 4-6 analytical data measured in spots with size from 300x300 to 1000x1000 µm located randomly across the specimens. Due to some microinhomogeneity of the samples, discrepancies between SEM and SF/ESA data appeared. The results obtained by the chemical route provide to characterize more accurately the bulk chemical compositions of the sample studied.

The main phase produced upon cooling the melted material is zirconolite. The material contains from 50 to 80 vol.% zirconolite (Table II). It should be noted that samples Z-1 and Z-2 have two various zirconolite-type phases. The first of them (marked as zirconolite-1) has common zirconolite composition, but the second one (zirconolite-2) is clearly distinguished by an unusually low Zr/Ti ratio (Table III).

Relative amounts of the phases are 4 to 1 in favor of zirconolite-1. In the central part of zirconolite-1 grains, zirconia inclusions occurred always. Maximum content of the latter phase is about 25 vol.% in the sample Z-4. All materials contain also an alumino-silicate vitreous phase located in interstitial regions between the grains. Moreover, the Z-1 sample contains zirconia and glass as additional phases only, while sample Z-2 contains minor rutile, and Z-3 and Z-4 samples contain aluminum titanate and perovskite as minor phases, respectively. The most general sequence of mineral formation corresponds to the following: zirconia - zirconolite - aluminum titanate, perovskite - rutile - glass. Phase relationships for the samples Z-2 and Z-4 are shown on Figure 1.

Table II. Relative contents of the phases, in vol. %.

Phase	Z-1	Z-2	Z-3	Z-4
Zirconolite	80*	75*	65	50
Zirconia	10	10	10	25
Rutile	-	<5	15	5
Perovskite	-	-	-	20
Aluminum titanate	-	-	5	-
Glass	10	10	5	<5

* Sample contains two zirconolite-type phases with different Ti/Zr ratio.

Zirconolite compositions and the mineral formula calculated with 7 atoms of oxygen are given in Table III. Zirconolite-1 has a typical chemical composition for this mineral considering all of possible elemental substitutions. It has a variable composition with the Zr content decreasing and the Ti content increasing from centers towards margins of individual grains.

The Gd content remains almost constant.. The highest Gd_2O_3 content in the zirconolite-1 is about 32 % corresponding to the formula: $(Ca_{0.4},Gd_{0.7})Zr_{1.0}(Ti_{1.4},Al_{0.5})O_{7.0}$. The phase composition is controlled by the following reactions of isomorphic exchange: $Ca(II) + Ti(IV) = Gd(III) + Al(III)$, and $Ca(II) + Zr(IV) = 2Gd(III)$.

Table III. Zirconolite average compositions and formula.

Oxides	Z-1*	Z-1**	Z-2*	Z-2**	Z-3	Z-4
Oxide contents, wt%						
CaO	4.7	4.4	4.8	4.6	7.9	12.9
Gd_2O_3	30.3	35.1	31.4	33.9	19.1	6.9
ZrO_2	32.0	10.5	30.8	10.4	30.3	42.8
TiO_2	26.7	41.8	27.4	42.6	36.3	35.2
Al_2O_3	6.1	7.9	5.5	7.9	6.1	2.0
SiO_2	0.1	-	0.1	0.4	0.2	0.1
FeO	0.1	0.3	-	0.2	0.1	0.1
Crystal chemical formulae						
Ca^{2+}	0.34	0.30	0.35	0.31	0.52	0.84
Gd^{3+}	0.69	0.74	0.72	0.71	0.39	0.14
Zr^{4+}	1.06	0.33	1.03	0.32	0.91	1.26
Ti^{4+}	1.37	2.01	1.42	2.02	1.69	1.60
Al^{3+}	0.49	0.60	0.45	0.59	0.44	0.14
Si^{4+}	0.01	-	0.00	0.02	0.01	0.00
Fe^{2+}	0.01	0.02	-	0.01	0.00	0.00
Total	3.97	4.00	3.97	3.98	3.96	3.98

*Zirconolite-1, **Zirconolite-2. Total - total amount of cations in formulae.

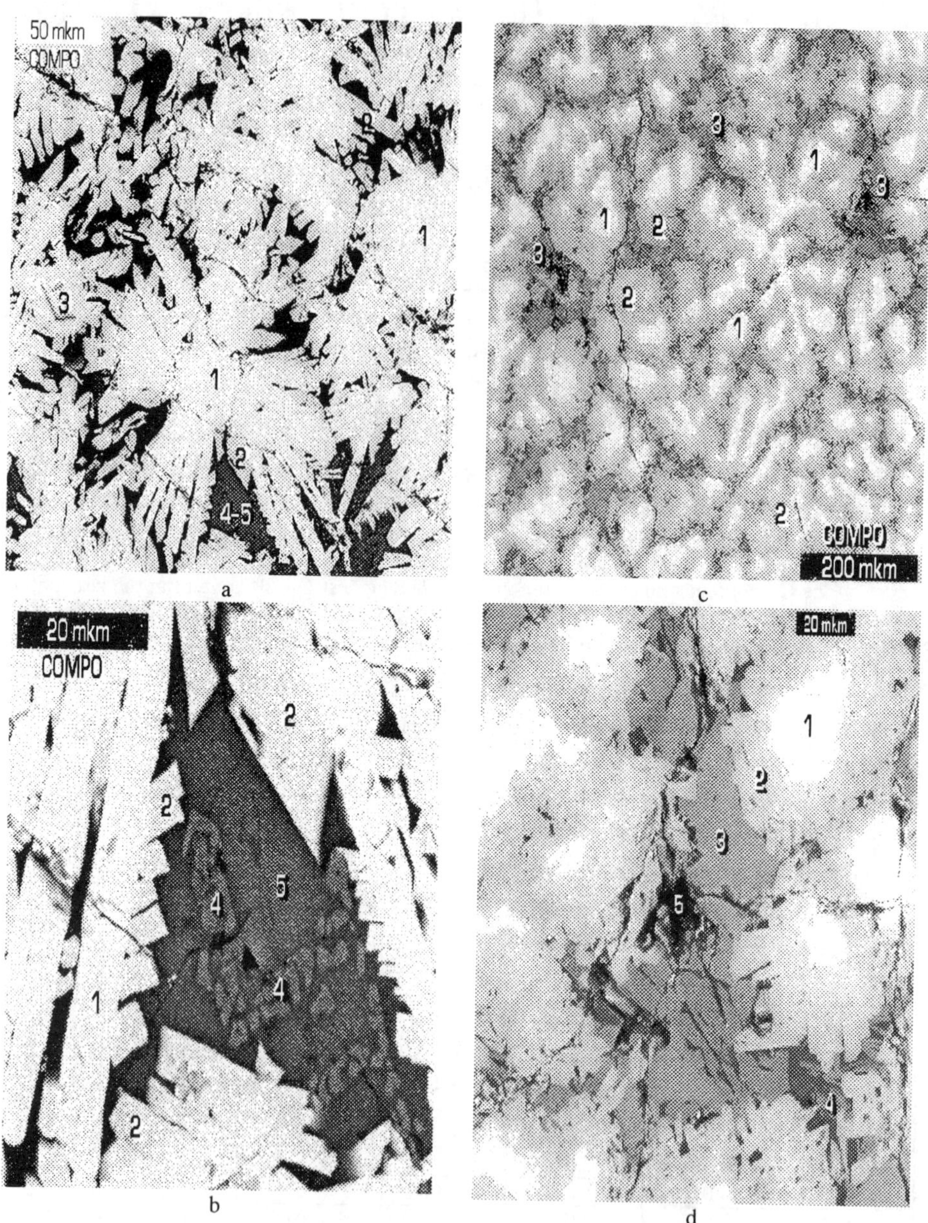

Figure 1. SEM micrographs of Z-2 (a,b) and Z-4 (c,d) samples.   NOTE: 1 mkm = 1 μm
a,b: 1-Zirconolite-1, 2- Zirconolite-2, 3- Zirconia, 4- Rutile, 5- Glass
c,d: 1-Zirconia, 2- Zirconolite, 3- Perovskite, 4- Rutile, 5- Glass.

We did not find any individual gadolinium phase in the samples even containing the highest total Gd concentration. So, the value obtained in the present study is only an estimation for the limit of the element solubility in zirconolite solid solution. The value is in a good agreement with data on the phase isomorphic capacities with respect to rare earth elements published earlier [1]. In the most recent paper [5] the limit value for gadolinium incorporation into zirconolite structure was estimated as 0.7 formula units.

Zirconolite-2 has a small ZrO_2 content, corresponding to about 10 wt.% or 0.3 formula units (Table III). Its composition is characterized by high TiO_2 content and respectively very low Zr/Ti ratio. Al and Gd contents in zirconolite-2 are slightly higher compared to co-existing zirconolite-1. Gd_2O_3 contents reach up to 35 wt.% that corresponds to 0.74 formula unit. Unit cell parameters of this phase are currently being studied in order to characterize the phase more accurately. Besides the common monoclinic zirconolite, some additional zirconolite-structured phases with orthorhombic, and trigonal symmetry were found in hot-pressed ceramics [1,4,5]. For example, variety of probably orthorhombic polymignite-type, or zirconolite-3O as recently recommended [17], was described by E.R.Vance et al. [5].

Zirconia is always found in the central part of the zirconolite-1 grains (Figure 1). It indicates clearly at the earliest phase formation among the other ceramic minerals. Relationships between the host zirconolite and zirconia inclusions give strong evidence for replacement of the zirconia by zirconolite. Zirconia solid solution composes of major Zr with essential additives of Gd, Ti, and sometimes Ca (Table IV). Al contents are varied from 0.3 to 0.4 wt.%, small quantities of Si and Fe may be also present. Gd content in zirconia reaches maximum value of 25 wt.% in sample Z-2 which correlates to the highest Gd content in the sample as a whole. In spite of equal total amounts of Gd in the samples Z-3 and Z-4 the concentration of Gd in zirconia in the first ceramic is a factor of 5 lower than in this phase in the Z-4 sample. Substitution of Gd^{3+}

Table IV. Zirconia and rutile compositions and formulae.

Oxides	Zirconia			Rutile		
	Z-2	Z-3	Z-4	Z-2	Z-3	Z-4
Oxide contents, wt%						
Al_2O_3	0.4	0.3	0.3	0.4	-	0.3
SiO_2	-	0.1	-	0.4	0.2	<0.1
CaO	1.5	0.2	4.5	0.4	0.3	0.4
TiO_2	6.7	9.9	7.4	93.3	86.1	95.5
FeO	0.2	-	-	<0.1	<0.1	0.2
ZrO_2	65.8	87.2	77.5	4.0	13.3	3.6
Gd_2O_3	25.4	2.3	10.3	1.4	<0.1	-
Crystal chemical formulae						
Al^{3+}	0.01	0.01	0.01	0.01	-	0.01
Si^{4+}	-	0.00	-	0.01	0.00	0.00
Ca^{2+}	0.04	0.00	0.10	0.00	0.00	0.01
Ti^{4+}	0.11	0.15	0.11	0.95	0.91	0.97
Fe^{2+}	0.00	-	-	0.00	0.00	0.00
Zr^{4+}	0.72	0.83	0.78	0.03	0.09	0.02
Gd^{3+}	0.19	0.01	0.07	0.01	0.00	-
Total	1.07	1.01	1.07	1.01	1.00	1.01

for Zr^{4+} into zirconia structure results in increasing the sum of cations in the phase formula up to 1.07, compared to the theoretical value of 1.00.

Rutile was identified in three of four ceramic samples. It is mainly composed of Ti but an essential content of Zr also occured (Table IV). The rutile from Z-3 sample has a maximum Zr concentration of about 13 wt.%, and the mineral from materials Z-2 and Z-4 is characterized by similar zirconium contents of ~ 4 wt.%.

Sample Z-3 contains an additional phase of aluminum titanate, forming needle-shaped blue-colored grains. The phase is composed of nearly equal quantities of Al and Ti oxides and minor additives of some other elements - Zr, Si, and Ca. Its average composition is (wt%) Al_2O_3-50.6, TiO_2-46.3, ZrO_2-1.9, SiO_2-0.8, CaO-0.2, FeO-0.1, Gd_2O_3-0.1 which is recalculated to formula: $(Al_{1.84},Si_{0.02})(Ti_{1.07},Zr_{0.03})O_{5.0}$.

Perovskite was found only in the sample Z-4. Ti, Ca, and Gd were the main elements, with Al and Zr present in amounts up to 1 wt% each. Average composition of the phase is (wt.%) Al_2O_3-0.9, SiO_2-0.3, CaO-35.5, TiO_2-53.7, FeO-0.1, ZrO_2-0.7, Gd_2O_3-8.8 corresponding to the formula $(Ca_{0.91},Gd_{0.07})(Ti_{0.96},Al_{0.03},Zr_{0.01})O_{3.0}$. Gd substitutes for Ca in Ca-sites of the perovskite structure, and some Ti atoms exchanged with Al for charge balance: $Ca^{2+} + Ti^{4+} = Gd^{3+} + Al^{3+}$. Taking into account that quantity of Gd in formula is higher than for Al, the additional scheme of replacement Ca by Gd and vacancies may be supposed: $3Ca^{2+} = 2Gd^{3+} +$ vacancy.

In all samples studied, some residual glass was observed. The contents did not exceed 10 volume %. The vitreous phase is enriched in Si, Al, Ca, Ti, but has a very small concentration of Zr (Table V). The highest Na content, relative to the other phases, occurs in the glass. The vitreous phase has a variable Gd content, increasing from <1 wt% in samples Z-3 and Z-4 to 6.5 wt% in Z-1 sample and reaching a maximum (12.6%) in the sample Z-2. Glass compositions were recalculated to a plagioclase formula with 8 atoms of oxygen and a cations sum that is close to 5.0. Unlike plagioclase, the vitreous phase has a significantly higher Ti content. So devitrification of these glasses will probably lead to formation of rutile + plagioclase assemblage. Partial glass devitrification with rutile generation is clearly in the sample Z-2 (Figure 1, b).

Recently, data on the system CaO-TiO_2-ZrO_2 investigation were published [18] where zirconolite, perovskite, rutile and zirconia were described among the phases observed. Moreover a number of additional phases including calzirtite ($Ca_2Zr_5Ti_2O_{16}$), zirconium titanate ($ZrTiO_4$) and two calcium zirconates ($CaZr_4O_9$ and $Ca_6Zr_{19}O_{44}$) were determined. We have not found the latter four phases in the samples we have studied. Instability of these phases in the melted materials is probably explained by a more complicated composition of the system studied.

Table V. Vitreous phase compositions (wt%).

Oxides	Z-1	Z-2	Z-3	Z-4
Na_2O	1.3	not analyzed	1.0	2.4
Al_2O_3	25.3	24.0	29.5	26.6
SiO_2	42.3	40.9	37.2	35.0
CaO	14.1	14.0	16.1	22.4
TiO_2	9.7	8.3	14.4	12.1
FeO	0.1	0.2	0.1	0.4
ZrO_2	0.7	-	1.0	0.7
Gd_2O_3	6.5	12.6	0.7	0.4

CONCLUSIONS

Zirconolite-rich ceramic materials containing up to 80 vol % of the zirconolite were produced by inductive melting in the cold crucible. Two types of zirconolite grains with different compositions occurred. The earlier-formed zirconolite-1 has a composition close to that of hot-pressed ceramics and prevails in the samples. The later zirconolite-2 is characterized by a low value of Zr/Ti ratio, but some Gd and Al enrichment is present. In addition to zirconolite some amounts of zirconia, alumina-silicate glass, rutile, perovskite and aluminum titanate were also found. The additional phase formation depends on the overall sample composition. The vitreous phase has a high silica content. Rutile and aluminum titanate appeared in the sample enriched with Ti and Al. The earliest formation of zirconia is explained by its expanded crystallization field in the system studied. To obtain the highest yield of zirconolite in the ceramics producing through melting route it is necessary to avoid contamination of the system with silica.

In general, melted zirconolite-rich material is similar to those prepared by hot-pressing and sintering routes. This is evidenced from their phase composition, consisting of major zirconolite and minor additional phases - rutile, zirconia, and perovskite. Hot-pressed ceramics often contain two or three zirconolite-structured phases differing in their composition. Moreover, hot-pressed ceramics rich in silica may also contain a vitreous phase [19]. Gd solubility in a zirconolite solid solution is about 0.7 formula units both for hot-pressed and melted ceramics.

The main difference between the hot-pressed and melted ceramics is a larger grain size of melted zirconolite-rich material compared to hot-pressed ceramics. The specific feature of the melted samples is the presence of zirconia. In general, the variations are not so great to strongly affect durability of the melted samples compared to hot-pressed ones. A study of leachability of the melted ceramics has been started.

ACKNOWLEDGMENTS

The authors are grateful to D.B. Lopukh, A.M. Lyubomirov, and A.Y. Petchenkov (University of Electric Engineering, St-Petersburg) for preparation of some samples, and N.V. Koroleva and G.E. Kalenchuk (Institute of Geology of Ore Deposits, Moscow) for assistance in chemical analysis of the samples.

REFERENCES

1. S.E. Kesson, W.G. Sinclair and A.E. Ringwood, Nucl. Chem. Waste Manag. **4**, p. 259 (1983).
2. W.G. Sinclair and A.E. Ringwood, Geochem. J. **15**, p. 229 (1981).
3. G.R. Lumpkin, K.L. Smith and M.G. Blackford, in <u>Scientific Basis for Nuclear Waste Management XVIII</u>, edited by T. Murakami and R.C. Ewing (Mater. Res. Soc. Symp. Proc. 353, Pittsburgh, PA, 1995) p. 855-862.
4. E.R. Vance, B.D. Begg, R.A. Day and C.J. Ball, ibid. p. 767-774.
5. E.R. Vance, A. Jostsons, R.A. Day, C.J. Ball, B.D. Begg and P.G. Angel, in <u>Scientific Basis for Nuclear Waste Management XIX</u>, edited by W.M. Murphy and D.A. Knecht (Mater. Res. Soc. Symp. Proc. 412, Pittsburgh, PA, 1996) p. 41-47.
6. V.M. Oversby and E.R. Vance, in <u>Scientific Basis for Nuclear Waste Management XVIII</u>, edited by T. Murakami and R.C. Ewing (Mater. Res. Soc. Symp. Proc. 353, Pittsburgh, PA, 1995) p. 825-832.

7. E.R. Vance, R.A. Day, M.L. Carter and A. Jostsons, in Scientific Basis for Nuclear Waste Management XIX, edited by W.M. Murphy and D.A. Knecht (Mater. Res. Soc. Symp. Proc. 412, Pittsburgh, PA, 1996) p. 289-295.

8. I.A. Sobolev, F.A. Lifanov, S.A. Dmitriev, S.V. Stefanovsky and A.P. Kobelev, in SPECTRUM'94. Nuclear and Hazardous Waste Management International Topical Meeting (Atlanta, GA, 1994) p. 2250-2254.

9. F.A.Lifanov, I.A. Sobolev, S.V. Stefanovsky and D.B. Lopukh, in ICEM'95. Proceedings of the Fifth International Conference on Radioactive Waste Management and Environmental Remediation, (Berlin, Germany, 1995), American Society of Mechanical Engineers, New York, NY (1995) p. 413-417.

10. Y.B. Petrov, Induction Melting of Oxides, Nauka, Leningrad, 1987.

11. V.I. Vlasov, O.L. Kedrovsky, A.S. Nikiforov, A.S. Polyakov and I. Yu. Shishtchitz, in Back End of the Nuclear Fuel Cycle, (Vienna, IAEA, 1987) p. 109-117.

12. F.A. Lifanov, S.V. Stefanovsky, A.P. Kobelev, V.I. Kornev, A.E. Savkin, O.A. Knyazev, T.N. Lashtchenova, S. Merlin and P. Roux, in Scientific Basis for Nuclear Waste Management XIX, edited by W.M. Murphy and D.A. Knecht (Mater. Res. Soc. Symp. Proc. 412, Pittsburgh, PA, 1996) p. 163-171.

13. I.A. Sobolev, S.V. Stefanovsky and F.A. Lifanov, in Scientific Basis for Nuclear Waste Management XVIII, edited by T. Murakami and R.C. Ewing (Mater. Res. Soc. Symp. Proc. 353, Pittsburgh, PA, 1995) p. 833-840.

14. I.A. Sobolev, F.A. Lifanov, S.V. Stefanovsky and S.A. Dmitriev, in Waste Management'95, HLW, LLW, Mixed Wastes and Environmental Restoration - Working Towards A Cleaner Environment, Abstracts, (Tucson, AZ, WM Symposia Inc., Tucson, AZ 1995), Abstracts, p.19, CD Rom.

15. O.A. Knyazev, B.S. Nikonov, B.I. Omelianenko, S.V. Stefanovsky, S.V. Yudintsev, R.A. Day and E.R. Vance, in SPECTRUM'96. Proceedings of the International Topical Meeting on Nuclear and Hazardous Waste Management (Seattle, 18-23 August, 1996), p. 2130-2137.

16. E.R. Vance, D.J. Cassidy, C.J. Ball and G.J. Thorogood, J. Nucl. Mater. **190**, p. 295 (1992).

17. P. Bayliss, F. Mazzi, R. Munno and T.J.White, Mineral. Mag. **53**, p. 565 (1989).

18. D. Swenson, T.G. Nieh and J.H. Fournelle, in Scientific Basis for Nuclear Waste Management XIX, ed. by W.M. Murphy and D.A. Knecht (Mater. Res. Soc. Symp. Proc. 412, Pittsburgh, PA, 1996) p. 337-344.

19. E.R. Vance, K.P. Hart, R.A. Day, B.D. Begg, P.J. Angel, E. Loi, J. Weir and V.M. Oversby, ibid. p. 49-55.

COMPARISON OF
SODIUM ZIRCONIUM PHOSPHATE-STRUCTURED HLW FORMS
AND SYNROC FOR HIGH-LEVEL NUCLEAR WASTE IMMOBILIZATION

V. N. ZYRYANOV*, E. R. VANCE**
*Chemical Technology Division, Argonne National Laboratory, 9700 South Cass Avenue, Argonne, Illinois 60439-4837, zyryanov@cmt.anl.gov,
**Materials Division, ANSTO, Menai, N.S.W. 2234, Australia.

ABSTRACT

The incorporation of (a) Cs and Sr as; (b) simulated actinides, and (c) simulated Purex waste in sodium zirconium phosphate (NZP) has been studied. The samples were prepared by sintering, by hot pressing and by hot isostatic pressing in metal bellows containers. The short-term chemical durability of the phosphate-based material containing Purex waste was within an order of magnitude of that for Synroc-C, as measured by 7-day MCC-1 tests at 90°C. The dissolution behaviour showed evidence of re-precipitation phenomena, even after times as short as 28 days. Potential for improvement of NZP-based ceramics for HLW management is discussed.

INTRODUCTION

Sodium zirconium phosphate, $NaZr_2(PO_4)_3$ (NZP) is one of the candidate crystalline waste forms for the geological immobilization of high-level nuclear waste. Works [1-8] in the early 1980s identified this as a near-single-phase system for incorporating PW-4b-type Purex waste. At high waste loadings, Cs formed a separate $CsZr_2(PO_4)_3$ phase that, is compatible with NZP and is highly durable in MCC-1 type tests [5]. Some monazite (a very durable mineral[9]) also formed, probably because the solid solubility limit of rare earth elements (REEs) in NZP had been exceeded.

Recently, natural analogs of NZP have been found in the alkali rocks massifs of Russia, Canada[10], Greenland[11], USA[12], and Australia[13]. The NZP structure consists of a three-dimensional network of PO_4 tetrahedra sharing corners with ZrO_6 octahedra, with the interstitial space partly occupied by Na ions [14]. Alkalis can substitute in the Na site. Divalent ions such as Ca, Ba, Fe, Mg, Pb, and Sr substitute for two alkali ions [2]. Rare earth elements are assumed to occupy the Zr site, with charge compensation being necessary for trivalent REEs. Tetravalent Ti substitutes freely on the Zr site. Trivalent Pu substitutes for Zr, forming either $Na_3Pu_2(PO_4)_3$ or $Na_3Pu(PO_4)_2$ depending on the Pu-to-PO_4 ratio [15]. Hexavalent molibdenum is believed to substitute directly on the P site, with appropriate charge compensation on other sites. Scanning electron microscopy (SEM) results indicated that the NZP phase contained all species of simulated waste ions [2], although it was not demonstrated that all of the radwaste entered this phase. Rather than carry out a detailed chemical accounting exercise to establish charge balance, the Zr/Na and P/Na ratios were varied and the phase assemblage was studied by X-ray diffraction (XRD) to establish the ratios giving as close to a single NZP-type phase as possible.

The fabrication of ceramic matrices with the NZP structure utilizes the principle of isomorphous and isodimorphous substitutions similar to those in natural minerals. Formation of a phosphate phase comprising all the major waste components of nuclear fuel cycle including actinides was demonstrated [16]. Alkali neptunium and plutonium phosphates $M^IAn_2(PO_4)_3$ were found to crystallize in several isomorphous modifications. Americium and curium phosphates have structures similar to those of REE phosphates, which isomorphically incorporate for Zr in a matrix. The phosphate NZP matrices may incorporate significant amounts of additional Na[16].

Tetragonal $NaPu_2(PO_4)_3$ was prepared to investigate radiation stability of the NZP matrices [17]. Samples in which the Pu was present as ^{239}Pu showed no perceptible changes during two years of observations in which doses of $9.1x10^{13}$ α–events/mg were sustained, but material made with ^{238}Pu was rendered completely metamict after a dose of $9.3x10^{15}$ α–events/mg. For comparison, zirconolite and perovskite in Synroc become metamict after doses of $4.7x10^{15}$ α–events/mg [17].

Mat. Res. Soc. Symp. Proc. Vol. 465 ©1997 Materials Research Society

Ceramic phosphate matrices with the NZP structure containing simulated waste elements were investigated under hydrothermal conditions. The composition and structure of the ceramics showed no changes after reaction with hot water and aqueous chloride solutions at 200 to 400°C and 600 bars [18].

Russian HLWs are generally high in Na, an element which is not easily accommodated in the Ti-based HLW ceramic Synroc [19]. However Synroc can be reformulated to include Na-bearing phases such as nepheline [20], additional perovskite [21] or freudenbergite [22]. Nepheline and freudenbergite contain around 22 and 9 wt% Na_2O respectively, while NZP contains ~6 wt% of Na_2O.

The aims of present work were as follows: (1) prepare NZP-based waste forms by hot-pressing, rather than by sintering[1-8], and compare leaching and other relevant properties with those of Synroc; (2) investigate separate incorporation of Cs, Sr and actinides as well as PW-4b in NZP; (3) check compatibility of NZP with major Synroc phases to investigate the possibility of hybrid waste forms.

EXPERIMENTAL METHODS

Samples were prepared by methods fairly similar to those used for Synroc. The precursor NZP phase was formed from tetrabutyl zirconate (TBZ) $[Zr(OC_4H_9)_4]$, $NaNO_3$, and 85% H_3PO_4, together with nitric acid-based mixtures of simulated waste, Cs/Sr or RE nitrate solutions. After stir-drying, these mixtures were calcined in air at 700°C. For consolidation, some preliminary work was done by simply pressing pellets and sintering in air at 1000 to 1200°C.

Otherwise, stainless steel (316 SS) bellows were filled with calcined powder, sealed by welding on a lid, and hot uniaxially pressed (HUP) or isostatically pressed (HIP) at approximately 1100°C. In the HUP process, a graphite susceptor is heated by radio-frequency power. With the design of our standard bellows and the small tap density of the precursor, the attainable HUP density was limited because the bellows' walls completely collapsed, thus offering significant resistance to further compaction of the ceramic powder. Though densities of about 92 to 94% of theoretical density were obtained, there was strong layering of the ceramic, which lacked mechanical integrity. These problems were removed by the use of an Eagle HIP, manufactured by International Pressure Systems, Inc., using highly pure argon as the pressure medium.

Archimedes' method was used to measure sample densities and porosities. Powder XRD measurements were made on a Siemens D-500 instrument, using Co radiation. Scanning electron microscopy was performed on a JEOL JXA-480. Leaching experiments were made with 0.1 g of 37-63 μm powders (from which the fines were carefully cleaned by washing in acetone). On the basis of geometrical surface area, the surface area/volume ratio was about 1. The powders were leached for different times, without sampling or replacement of the leach liquid between the beginning and finish of the run. The starting water pH was 5.0 in each case.

RESULTS AND DISCUSSION

I.   <u>Preliminary Studies of Crystallinity by Sintering</u>

The NZP precursors developed essentially full crystallinity during calcination at 700°C. While this favors the formation of desired phases without the occurrence of metastable intermediate phases, some of the chemical driving force to form a dense product during hot-pressing or sintering is reduced. By contrast however, $CsZr_2(PO_4)_3$ did not crystallise until sintering temperatures of about 1000°C were reached. The densities of samples cold-pressed at ~100 Mpa and sintered at 1000 to 1100°C reached approximately 90% but showed visual evidence of layering.

II.   Solid Solubilities of REEs and Mo in NZP

As mentioned above, the REEs and Mo display some solid solubility in NZP, but experimental evidence of solubility limits was lacking. The following target preparations were prepared and sintered in air at 700 to 1100°C:

$$\text{(a) } Na_{(1-x)}Ca_{(x)}Zr_{(2-x)}(Nd/Y)_{(x)}(PO_4)_3 \quad \text{and} \quad \text{(b) } NaZr_{(2-x)}Y_{(x)}P_{(3-x)}Mo_{(x)}O_{12} \tag{1}$$

For the (a) samples, the aim was to substitute the Nd and Y on the Zr site, with Ca substituted on the Na site acting as the charge compensator. Mo was targeted for the P site in the (b) samples, and the charge compensator was Y on the Zr site. Only the NZP phase was observed in the (a) samples after heating at 700°C. Heating at higher temperatures produced additional ZrP_2O_7, presumably due to loss of sodium by evaporation. After heating at 1100°C, no sign of a separate $NdPO_4$ (monazite type) phase was observed in the Nd-doped samples with x = 0.05 and 0.1, but for x >0.2, an additional monazite phase was observed. In the Y-doped samples, similarly heated at 1100°C, a separate YPO_4 (xenotime) phase was observed in addition to NZP and ZrP_2O_7 for x = 0.3 and 0.4, but not at x = 0.1 and 0.2. For the (b) samples, the results were similar to those for the Y-doped (a) samples. The presence of Mo was confirmed by SEM studies of the x = 0.4 sample, showing that Mo volatilisation losses were not serious. However the very fine-grained nature of the powdered samples precluded conclusions about whether the dopant ions were really in solid solution in the NZP phase. To obtain more positive answers, it would be necessary to fire the samples in closed containers and perform TEM studies.

III.   Strategies for Waste Incorporation in NZP

Firstly, Cs and Sr were considered for incorporation in NZP. We employed the actual Cs/Sr ratio in fission product waste, which is approximately 2:1 by weight. The designed total stoichiometry was $Na_{0.5}Cs_{0.2}Sr_{0.15}Zr_2(PO_4)_3$, to give a total waste oxide loading of about 8 wt%. As mentioned above, it was expected that Cs would form a distinct CsZP phase.

Secondly, actinide-rich wastes were considered. Tetravalent actinides would substitute for Zr (see above), but trivalent actinides are another issue. Such actinides were simulated by Gd and Nd. Two strategies were used: (1) Gd and Y were employed as simulations of the waste ions and it was aimed to substitute them in the Zr site, with half of the Y ions in the Na site. The total stoichiometry was $Na_{0.5}Y_{0.5}Gd_{0.5}Y_{0.5}Zr(PO_4)_3$, and the equivalent actinide waste loading was around 40 wt%, assuming a molar substitution of Gd and Y for trivalent actinides. The substitution of the Y in the Na site was quite speculative, but the versatility of the NZP structure is shown by the fact that the Zr^{4+} ions nominally can be completely replaced by a Nb^{5+} with charge compensation provided by a Na ion [23]. (2) Nd was the waste ion simulant used by us and it was substituted on the Zr site; Ca was the charge compensator on the Na site. The equivalent actinide waste loading was about 20 wt%. PW-4b waste (see [2]) was also studied for incorporation in NZP at waste loading of 20 wt%.

IV.   Compatibility Studies

Separate mixtures corresponding to equimolar mixtures of NZP and individual Synroc phases were prepared, dried, calcined in air at 700°C, then ground, pelletised and fired in air at 1100°C. Rutile (TiO_2) + NZP yielded a rutile-zirconia solid solution + NZP; hollandite ($Ba_{1.14}(Al,Ti)_{2.28}Ti_6O_{16}$) + NZP yielded rutile + NZP + unknown phases; perovskite ($CaTiO_3$) + NZP produced rutile + whitlockite ($Ca_3(PO_4)_2$ + NZP, while zirconolite ($CaZrTi_2O_7$) + NZP gave baddeleyite (ZrO_2), rutile + unknown phases.

These results showed incompatibility of NZP and the Synroc phases. Experiments were also undertaken with the Ti analogue of NZP [7], but again compatibility was very limited; Ba and Ca entered the Na site of NZP, and Zr partly replaced the Ti in the Ti analogue of NZP, with residual rutile being formed.

V.   Hot-Pressing Studies

Table I shows the phase assemblages and density/porosity results on various preparations hot isostatically pressed at 1100°C/100MPa for 2 h.

Some SEM micrographs are shown in Figures 1 through 4. In nearly all cases the grain size was less than 2 μm for xenotime and monazite. The NZP matrix in all samples consisted of grains more than 50 μm in size. In samples #48 and 50, a Sr-rich exsolution structure was seen. In sample #50, CsZP formed variably-sized grains coexisting with the NZP matrix.

Table I. Densities and Phases Present in Hot Isostatically Pressed NZP-based Samples

Sample	Composition	Density (g/cm^3)	Porosity (%)	Phases Present
48	$Na_{0.5}Cs_{0.2}Sr_{0.15}Zr_2(PO_4)_3$	3.23	3.6	NZP + ZP
49	$NaZr_2(PO_4)_3$	3.09*	3.6	NZP + tr. ZP
50	NZP+20wt% PW-4b	3.23	1.5	NZP + Mz + ZP + CsZP
51	$Na_{0.5}YZrGd_{0.5}(PO_4)_3$	3.76	3.3	NZP + Xe + ZP
52	$Na_{0.5}Ca_{0.5}Gd_{0.5}Zr_{1.5}(PO_4)_3$	3.43	3.8	NZP + Mz + ZP

*theoretical density = 3.19g/cm^3. Mz-monazite; Xe-xenotime; ZP-ZrP_2O_7; tr.-trace.

Fig. 1. Sample NZP-48.   x300. white regions: separate $CsZr_2(PO_4)_3$ phase; grey: NZP matrix; dark grey spots inside of NZP grains: $(Na,Sr)Zr_2(PO_4)_3$; black: pores. Micron marker: 100 μm

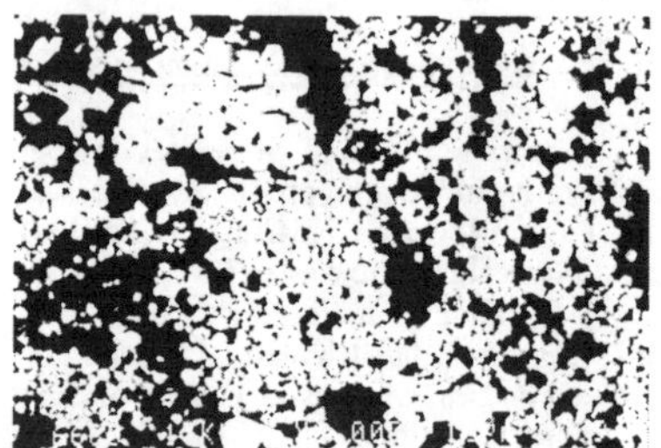

Fig. 2. Sample NZP-51. x2000. white: xenotime; black: NZP. Micron marker: 10 μm

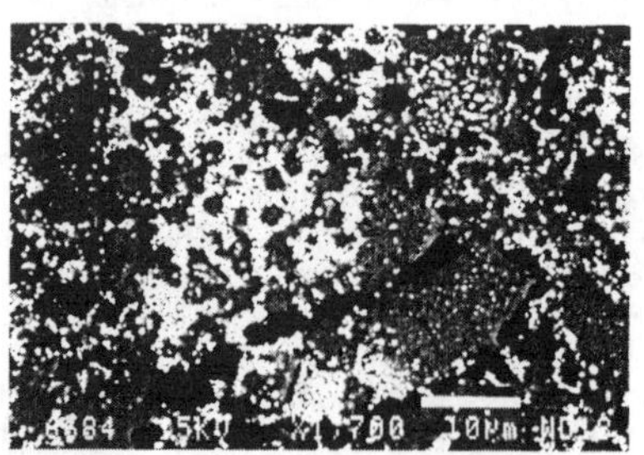

Fig. 3. Sample NZP-52 x1700. white: monazite; grey: NZP; black: $Ca_3(PO_4)_2$. Micron marker: 10 μm

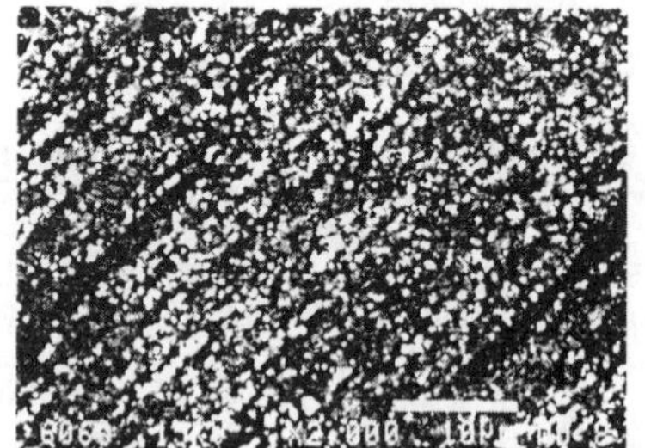

Fig. 4. Sample NZP-50 x2000. white: RE-phases; grey: NZP matrix; black: ZrP_2O_7. Micron marker: 10 μm

The phase assemblages were not unexpected given the previous [1-8], and the results in section II, which showed only limited solid solubilities of REEs and Mo in NZP, although the absence of a CsZP phase in sample 48 indicates some Cs substitution in NZP. Given that the materials densify to ~ 90% of their theoretical value during sintering at around 1000°C, it was puzzling why full densification did not occur in HIPing.

VI.   Leach Testing of HIPed Samples

Only the hot isostatically pressed samples were leach tested. Sections of the samples were crushed to form 37-63 μm diameter powders (see above). A sample of Synroc-C was similarly prepared. Leach results are shown in Tables II and III below.

The 7- and 28-day leach rates of NZP-48 and NZP-49 were quite comparable with values reported essential for Synroc [22]. The corresponding data for the NZPs doped with simulated actinides (samples NZP-51/52) similarly gave results for the matrix elements and the rare earths which were comparable with those reported for Synroc[22]. Data on NZP-50, which contained simulated PW-4b waste, were worse than Synroc by an order of magnitude for the more soluble elements, such as Na, Cs, and Mo. It was further clear from these data, for both the NZP-based materials and Synroc, that the concentrations of some elements in solutions apparently decreased between 7 and 28 days. This observation coupled with the weight loss data showing greater sample weight losses after 7 days than 28 days suggested that re-precipitation of these elements had taken place, although XRD of the 28-day leached powders showed no significant differences from the data obtained from the unleached samples.

The leach rate data on Synroc were a factor 1-10 lower than data obtained on polished samples [19] and showed the usual decrease of leach rate with time for from calcium.

Table II. Solution Leach Rates at 90°C of 37-63 μm Diameter Samples Prepared from Hot Isostatically Pressed NZP-based Samples

Element	NZP-49 wt%	7 days	28 days	NZP-48 wt%	7 days	28 days	NZP-51 wt%	7 days	28 days	NZP-52 wt%	7 days	28 days
Na	4.7	0.12	0.03	2.2	0.012	0.005	2.1	0.01	0.005	2.2	0.008	0.002
Cs	0			5.1	0.02	0.002						
Sr	0			2.5	0.01	0.003						
Ca	0			0						3.8	0.022	0.01
Gd	0			0			13.8	7×10^{-7}	2×10^{-7}			
Y	0			0			16.1	2×10^{-6}	2×10^{-7}			
Nd	0			0						13.7	7×10^{-7}	2×10^{-7}
Zr	3.77	6×10^{-5}	3×10^{-7}	35.2	3×10^{-6}	1×10^{-7}	13.5	2×10^{-6}	2×10^{-7}	26.1	8×10^{-7}	1×10^{-8}
P	18.9	8	0.1	17.9	0.04	0.01	16.9	0.014	0.01	17.7	0.004	0.002
Weight Loss %		18.4	5.3		27.1	5.2		38.8	21.2		31.7	9.4
Final pH		4.76	5.29		4.14	5.28		5.90	4.76		5.25	4.76

Table III. Leaching Results on 37-63 μm Diameter Hot Isostatically Pressed NZP Ceramics with 20 wt% HLW and Synroc-C with 15 wt% HLW

	Leach rate ($g/m^2/day$)						
	NZP-50			Synroc-C			
Element	wt%	7 days	28 days	wt%	7 days	28 days	7 days [18]
Na	3.3	0.8	0.2				
Ca	1.7	0.7	0.2	6.7	0.003	0.006	0.03
Zr	22.8	4×10^{-6}	3×10^{-7}	5.6	1×10^{-6}	4×10^{-7}	
P	19.9	0.4	0.1	0.11	0.3	0.08	
Al	0			2.4	0.0003	0.0001	
Ba	0.72	0.3	0.03	4.8	0.017	0.003	0.03
Ti	0			36.4	2×10^{-6}	6×10^{-7}	
Cr	0.12	0.06	0.003	0.088	0.0007	0.0002	
Fe	0.55	0.1	0.002	0.41	0.013	0.002	
Cs	1.6	1.4	0.1	1.2	0.02	0.005	0.08
Sr	0.47	1.1	0.1	0.41	0.007	0.002	0.03
Y	0.25	0.02	0.002	0.19	0.0001	0.0001	
Ce	2.57	0.0008	5×10^{-5}	1.93	6×10^{-6}	1×10^{-5}	
Nd	2.73	7×10^{-5}	3×10^{-6}	2.05	5×10^{-6}	4×10^{-6}	
Gd	0.65	0.0004	2×10^{-5}				
Ag	2.31	2×10^{-5}	5×10^{-6}	1.73	4×10^{-5}	7×10^{-6}	
Cd	0.04	1.5		0.03	0.04		
Mo	1.80	6.6	2	1.35	0.2	0.05	0.2
Te	0.24	0.2	0.03	0.18	0.001	0.0001	
weight loss %		12.2	14.2		2.3	14.1	
Final pH		5.99	3.52		3.99	5.04	

CONCLUSIONS AND SUGGESTIONS FOR FURTHER WORK

On the basis of the current work, dense NZP-based waste ceramics have chemical durabilities within an order of magnitude of that of Synroc, although the interpretation of the NZP leach results for most of the NZP samples, except NZP-50, appears to involve significant re-precipitation effects even after 28 days in the tests performed here, because the weight losses and some of the solution concentrations in 7-day tests exceeded those in the 28-day tests. Potential for improving of the NZP-based material is seen in the following areas: (1) development of denser precursors; (2) obtaining a more extensive data base on solid solubility of radwaste ions in NZP allowing more careful overall phase design; (3) optimising the hot-pressing sequence, including redox conditions.

ACKNOWLEDGEMENTS

We wish to acknowledge the Department of Industry, Trade and Commerce, Canberra, Australia, for financial support, A. E. Ringwood for helpful discussions, J. Weir for carrying out the leach tests, T. Nicholls for specimen preparation, P. J. Angel and S. S. Moricca for performing the hot-pressings, and S. Leung for assistance with scanning electron microscopy.

REFERENCES

1.   E. R. Vance and T. Adl, in Scientific Basis for Nuclear Waste management,. 3, Ed., S. V. Topp, Plenum, New York, 163-71 (1981).
2.   R. Roy, L. J. Yang, J. Alamo, and E. R. Vance, in Scientific Basis for Nuclear Waste Management VI, Ed., D.G. Brookins, North-Holland, Amsterdam, 15-21 (1983).
3.   L. J. Yang, S. Komarneni, and R. Roy, in Scientific Basis for Nuclear Waste Management VII, Ed., G.L. McVay, North-Holland, Amsterdam, 567-74 (1984).

4. B. E. Scheetz, S. Komarneni, W. Fajun, L. J. Yang, M. Ollinen and R. Roy, in Scientific Basis for Nuclear Waste Management VII, Eds. C. M. Jantzen, J. A. Stone, and R. C. Ewing, Materials Research Society, Pittsburgh, PA, USA, 903-10 (1985).

5. R. Roy, E. R. Vance, and J. Alamo, $CsZr_{2}(PO_4)_3$, Mater. Res. Bull. <u>17</u>, 585-9 (1982).

6. L. J. Yang, S. Komarneni and R. Roy, in Advances in Ceramics. <u>8</u>, Eds., G. G. Wicks and W. A. Ross, American Ceramic Society, Columbus, OH, USA, 368-76 (1984).

7. L. J. Yang, S. Komarneni, and R. Roy, ibid. 255-62.

8. E. R. Vance and F. J. Ahmad, in Scientific Basis for Nuclear Waste Management VII, Ed., D. G. Brookins, North-Holland, Amsterdam, 105-12 (1983).

9. G. J. McCarthy, W. B. White and D. E. Pfoertsch, Mater. Res. Bull., <u>13</u>, 1239-45 (1978).

10. A. M. Mcdonald, G. Y. Chao, and J. D. Grice, Canadians Mineralogist, <u>34</u>, 107-14 (1996).

11. F. Mazzi and L. Ungaretti, Neues Jarbuch fur Mineralogie-Monatshefte, <u>2</u>, 49-66 (1994).

12. M. E. Brownfield, E. E. Foord, S. J. Sutley, and T. Botineely, Am. Mineralogist, <u>78</u>, 653-56 (1993).

13. W. D. Birch, A. Pring, D. J. M. Bevan, and Kharisun, Min. Mag., <u>58</u>, 635-39 (1994).

14. H.Y-P Hong, Mater.Res. Bull <u>11</u>, 173-80 (1976).

15. A. A. Burnaeva, Yu. F. Volkov, A. I. Krjukova, Radiokhimiya <u>36</u>, 289-94 (1994).

16. A. I. Krukova, Yu. F. Volkov, V. N. Zyryanov, G. N. Kazantzev. I. A. Kulikov, S. G. Samoilov, Proceedings of 3rd Annual Conference of Nuclear Society International. Moscow, Sept. 14-18, Management with Radioactive Waste, 673-74, St-Petersburg (1992).

17. A. I. Orlova, G. Yu. Artemjeva, L. Yu. Masterova, V. I. Petkov, A. A. Harlamova, Yu. F. Volkov, V. N. Zyryanov, G. N. Kazantzev, S. G. Samoilov, I. A. Kulikov, and S. V. Stefanovsky, Abstracts of 4th Annual Scientific and Technology Conference of Nuclear Society "Nuclear Energy and Human Safety" NE-93, June 28-July 2, 1993, part 2, 877-78, Nizhniy Novgorod, (1993).

18. A. I. Orlova, V. N. Zyryanov, A. R. Kotelnikov, V. T. Demarin, and E. V. Rakitina, Radiokhimiya <u>6</u>, 120-126 (1993).

19. A. E. Ringwood, S. E. Kesson, N. G. Ware, W. Hibberson and A. Major, Nature (London) <u>278</u>, 219-23 (1979).

20. A. E. Ringwood, S. E. Kesson and N.G. Ware, in Scientific Basis for Nuclear Waste Management Vol. 2., Ed., C.J.M. Northrup, Plenum, New York and London, 265-72 (1980).

21. E. R. Vance and G. J. Thorogood, J.Amer. Ceram. Soc. <u>74</u>, 854-55 (1991).

22. A. E. Ringwood, S. E. Kesson, K. D. Reeve, D. M. Levins, and E. J. Ramm, in Radioactive waste forms for the future, Eds., W.Lutze and R.C. Ewing, Elsevier, New York, 233-34 (1988).

23. I. Yamai and T. Ota, J. Amer. Ceram. Soc., <u>76</u>, 487-91 (1993).

CHEMICAL DURABILITY STUDY OF SYNROC-C CERAMICS PRODUCED BY THROUGH-MELTING METHOD

A.V.KUDRIN*, B.S.NIKONOV* AND S.V.STEFANOVSKY**
*Institute of Geology of Ore Deposits RAS. 109017 Staromonetnii 35, Moscow,
Russia **SIA"Radon" 119121, 7-th Rostovskii per., 2/14, Moscow, Russia

ABSTRACT

An interaction between Synroc-C samples prepared by inductive melting in a cold crucible and deionized water was investigated at room temperature, 100 and 200 °C. The leachability of powdered specimens with grainsize 0.10-0.15 mm was determined by repeated static tests with regular replacement of leachant. The experimental data showed that the most leachable elements are the alkali, alkaline earths and molybdenum. The matrix elements such as titanium, zirconium, and rare earths as well are very leach resistant. Comparison of the calculated leach rates of components with reference data on leaching of alternative materials showed that cesium leachability from the studied specimens are close to borosilicate glass, but leaching behavior of other components is comparable to leachability from Synroc-C prepared by hot-pressing.

INTRODUCTION

Synroc is titanate-based ceramic developed by Ringwood and colleagues at the end of the 1970s [1]. The most conventional method of the Synroc production is hot-pressing (HP) at controlled redox conditions. Numerous leaching tests in distilled water and various solutions have demonstrated the high chemical durability of this ceramics compared to other candidate wasteforms, especially at elevated temperatures.

An alternative method of the Synroc production based on inductive melting in a cold crucible (IMCC) has been proposed by V.I. Vlasov in Russia at the end of the 1980s [2]. The first real Synroc sample whose phase composition was approximately the same as for hot-pressed Synroc was obtained at SIA "Radon" in the beginning of the 1990s [3]. Inductively-melted Synroc has larger crystal sizes, higher porosity, and some additional minor phases [3-5]. The purpose of this work was to study melted Synroc durability in water at elevated temperatures and to compare it with alternative wasteforms.

EXPERIMENTAL

Preparation, Mineralogical and Chemical Composition of Samples

The Synroc-C samples labeled C/1 and C/3 were prepared by IMCC from oxides-nitrate batches at SIA "Radon". Experimental procedure was described in detail elsewhere [5]. C/1 sample was produced by melt pouring onto a metal plate. C/3 sample was cut from the ingot prepared by slow cooling of the Synroc melt in the cold crucible. These samples differ on crystal sizes and porosity. Quenched sample C/1 formed at rapid solidification is a dense microcrystalline aggregate with crystal size ≤ 0.01 mm. Annealed sample C/3 is a porous crystalline aggregate with crystal dimensions ranged between 0.1-0.7 and 1.5 mm. The samples had the same phase and chemical composition (Table I). The mineral assemblage consists of zirconolite (45-50%), hollandite (35-40%), perovskite (10-15%), rutile (~5%) as well as alkali-alkali earth molybdates, and hibonite as minor phases [4,5].

Mat. Res. Soc. Symp. Proc. Vol. 465 © 1997 Materials Research Society

Table I. Chemical compositions (in wt%) of the Synroc-C samples.

Oxides	C/1, C/3	AM	AP
SiO_2	0.99	4.54	2.46
TiO_2	51.34	58.88	62.15
ZrO_2	10.13	6.07	5.92
Al_2O_3	9.24	5.64	5.75
$FeO+Fe_2O_3$	0.89	0.84	0.77
TR_2O_3	4.27*	5.78	4.00
MgO	-	0.08	0.07
NiO	0.70	0.05	0.06
CaO	10.13	8.99	8.70
SrO	1.19	0.31	0.23
BaO	6.65	4.70	4.22
Na_2O	0.40	0.08	0.15
K_2O	0.20	0.04	0.15
Cs_2O	0.60	0.64	0.64
MoO_3	3.28	3.37	4.73
Total	100.00	100.00	100.00

*2.17 Ce_2O_3, 2.10 Nd_2O_3 (analysis was performed at Institute of Geology of Ore Deposits)

Zirconolite and hollandite are the first phases segregated from the melt. These minerals form sheaf-like aggregates. Perovskite occurs as well formed grains in fairly open spaces "interstitial" to the elongate hollandite and zirconolite crystals and has crystallized later than these phases. Rutile and hibonite form small crystals and are also "interstitial" to the zirconolite and hollandite. Rutile tended to grow at the same time and after the perovskite, while hibonite grew slightly earlier. Molybdates were the last phases to form in the crystallization sequence. Alkali earth molybdates are represented as masses of small well formed crystals in microporous areas. Alkali molybdates were found as small crystals within small pores and filling microfractures.

Sample labeled as AM was produced by IMCC at SIA "Radon" from ANSTO calcine prepared using sol-gel process followed by calcination. IMCC procedure was described in [6]. It represents a microcrystalline aggregate of hollandite (~30%), rutile (~30%), zirconolite (~20%), perovskite (~15%) and the same minor phases (~5%) as in C/1 and C/3 samples. Some impurities were also present (Table I). The main difference of this sample from samples C/1 and C/3 is higher TiO_2 and SiO_2 concentrations and lower ZrO_2, Al_2O_3, and BaO contents (Table I).

Sample AP was produced at ANSTO (Australia) using hot-pressing and studied for comparison. It consists of the cryptocrystals of zirconolite (30%), hollandite (30%), perovskite (30%), rutile and CAT phase (15%), and Mo,Ni,Fe - alloy [7]. Nevertheless, based on chemical analysis data on this sample (Table I) it may be concluded the quantitative relationship between these minerals is somewhat different than reported by Ringwood [7].

<u>Leaching procedure</u>

The Synroc-C samples (C/1, C/3, AM, AP) were crushed, and powders were sieved to obtain 0.10-0.15 mm fraction. Each powdered sample (10 g) was washed several times in distilled water (100 cm^3) for 20 minutes to remove the finest particles adhered to the surfaces of the grains. Then each sieve fraction of 0.10-0.15 mm was decanted and dried. The first solution after washing of each powder was filtered and analyzed for the concentrations of Mo, Cs, Na, and K.

The leaching experiments were conducted at 100 and 200 °C in autoclaves fitted with Teflon inserts. Leaching studies were carried out by repeated static tests with regular replacement of leachant after the designed leaching time. The powder specimens (2 g each) were placed into autoclaves (40 cm³) and filled with deionized water (20 cm³). Autoclaves were held in the furnace for the designed period of time followed by water quenching. The leachates were decanted to exclude dissolved component losses through sorption on the filter (the leachates were guite transparent and did not contain suspended solids). After each test the powder specimens were washed twice in distilled water and dried. Inductively coupled plasma spectroscopy was used to analyze the leachates for Mo, Ti, Zr, Ce, and Nd and atomic absorption spectroscopy was used to determine Cs, Na, K, Ca, Ba, and Sr in the leachates.

RESULTS AND DISCUSSION

<u>Leaching at Room Temperature</u>

The data on Mo, Cs, Na, and K concentrations in solutions after the leaching at room temperature (the first solutions after washing of powders for 20 minutes, see the preceding section) are listed in Table II. Leaching behavior of C/1, C/3 and AM specimens is mainly governed by the presence of alkali molybdates. It result from near stoichiometric ratio between molar concentrations of Mo and the sum of alkali which is close to 1:2. The relationships between Cs, Na and K are differ in the leachates from each other, but Cs was the prevailing component always. On the basis of data obtained its amount in molybdates may be estimated as 40-50% of total.

The molybdenum and alkali concentrations in the leachate of specimen AP are one to two orders of magnitude lower then after leaching of IMCC samples (see Table II).

Table II. Leaching from Synroc-C samples at room temperature.

| Sample | Element concentrations in leachates * | | | | | | | | | Cs:Na:K molar ratio |
| | Mo | | Cs | | K | | Na | | (Cs,Na,K) | |
	ppm	logm	ppm	logm	ppm	logm	ppm	logm	logΣm	
C/1	84	-3.06	160	-2.92	7.6	-3.48	11.6	-3.53	-2.74	4:1:1
C/3	70	-3.14	108	-3.05	7.8	-3.47	15.2	-3.41	-2.79	2.5:1:1
AM	24	-3.60	40	-3.52	2.6	-3.94	1.2	-4.50	-3.34	10:4:1
AP	1.3	-4.85	0.48	-5.45	0.82	-4.45	0.46	-4.96	-4.30	1:10:3

* logm - logarithm of the molar concentration in the leachates

<u>Leaching at Elevated Temperatures</u>

Leaching behavior of inductively-melted samples is generally similar at 100 and 200 °C. Figures 1 and 2 show changes in concentrations of the readily leachable elements after repeated leaching tests (the results of the dissolution of the specimens C/1 and C/3 are similar and not indicated). The most leachable components in the first tests of each series are Cs and Mo. These elements were released from alkali molybdates contained in microfractures that were inaccessible for water at room temperature, as well as from other Cs-Mo-containing phases.

Cesium concentration is reduced at the next tests of each series, but molybdenum contents in the leachates remain mostly the same after the second test. Ca, Ba and Sr concentrations also remain almost constant in the repeated tests. Such leaching behavior

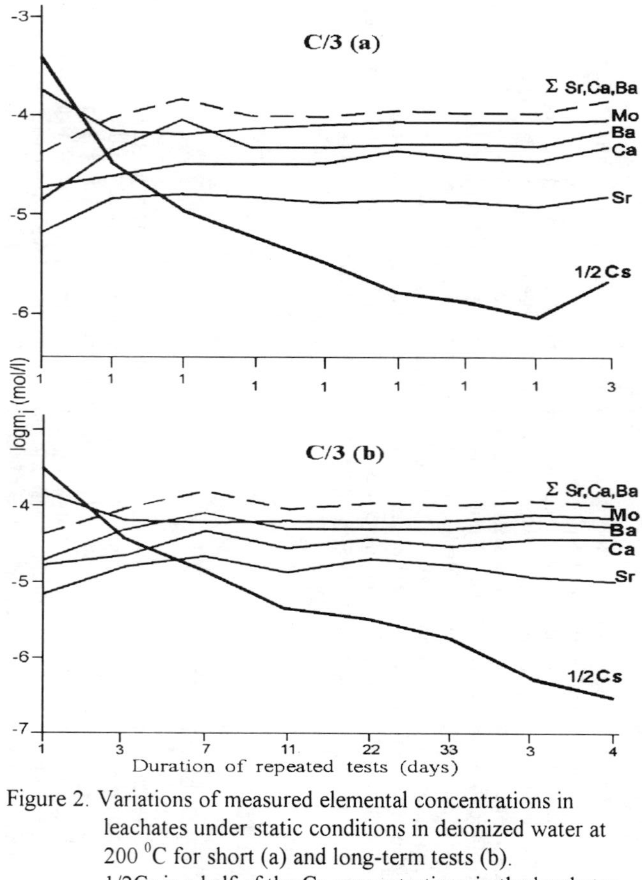

Figure 1. Variations of measured elemental concentrations in leachates under static conditions in deionized water at 100 °C.
1/2Cs is a half of the Cs concentrations in the leachates

Figure 2. Variations of measured elemental concentrations in leachates under static conditions in deionized water at 200 °C for short (a) and long-term tests (b).
1/2Cs is a half of the Cs concentrations in the leachates

of Mo along with Ca, Sr and Ba is controlled by a solubility product of the alkali earth molybdates. The solubility product values of these phases ($\sim 10^{-8}$ at 100 ^{0}C and $\sim 10^{-9}$ at 200 ^{0}C) provide relatively moderate levels of Mo and alkali earths content in solutions (10^{-4} -10^{-5} m). The same levels of these components were maintained in the absence of molybdates at dissolution of the crushed HP Synroc-C samples in deionized water at 90 ^{0}C for 28 days [7,8]. The net concentration of Ca, Ba and Sr in leachates is commonly higher then the Mo concentration that is caused by release of the phases other then alkali earth molybdates. It should be noted also that element concentrations reduce in the order Ba>Ca>Sr at 200 ^{0}C, but at 100 ^{0}C this order is modified as Ca>Ba>Sr.

The leachability of sample AP was determined at 100 ^{0}C only. Concentrations of elements studied are approximately an order of magnitude lower, than in the case of the melted samples.

Ti, Zr, Ce, and Nd concentrations in the leachates are not shown in Figures 1 and 2. These elements have very low leachability and their contents in solutions usually ranged between 0.1 and 0.001 ppm in all tests.

<u>Leach Rates</u>

Figures 3 and 4 demonstrate different leach rates of components at 100 and 200 ^{0}C based on our leaching data as a function of cumulative leaching time (surface area of powders required for calculations were measured by a krypton sorption method). Results of short-term tests at 200 ^{0}C were used for more leachable elements to preclude essential underestimating of the calculated leach rates of the Ca, Ba, Sr and Mo. Reference data on leach rates of components from borosilicate glass and hot-pressed Synroc-C samples are shown for comparison in Figures 3 and 4 [9,10].

Cesium exhibits the highest leach rate at both 100 and 200 ^{0}C. Its leach rates from IMCC Synroc samples are approximately the same as from borosilicate glass and about by two orders of magnitude higher than leach rates from HP Synroc samples. Values of leach rates of alkaline earths and Mo from IMCC Synroc samples are intermediate between borosilicate glass and hot-pressed Synroc at 100 ^{0}C, but these values are close to each other for two types of Synroc at 200 ^{0}C. It can be as a result of both the decrease in alkali earth molybdates solubility (for melted Synroc) and the increase of titanates dissolution (in hot-pressed Synroc).

The main matrix elements - Ti, Zr as well as rare earths demonstrate extremely low leach rates from melted Synroc as in the hot-pressed Synroc.

CONCLUSIONS

Chemical durability study of Synroc-C samples prepared by inductive melting in the cold crucible allows to distinguish three groups of components with different leaching behavior.

1. Cs and Mo (at early stages of leaching). Their yields are determined by the presence of high soluble alkali molybdates containing about a half of total Cs. In the samples studied cesium leachability is close to one from borosilicate glass.

2. Ca, Ba, Sr, Mo. The concentrations of these elements in leachates are controlled predominantly by solubility of alkali earth molybdates. As the solubility product values of these phases are relatively low, the yield of these components from inductively-melted Synroc is comparable to that from hot-pressed Synroc.

3. Ti, Zr, Ce, Nd. These elements present in leachates in very small concentrations at the same level as at leaching of hot-pressed Synroc samples.

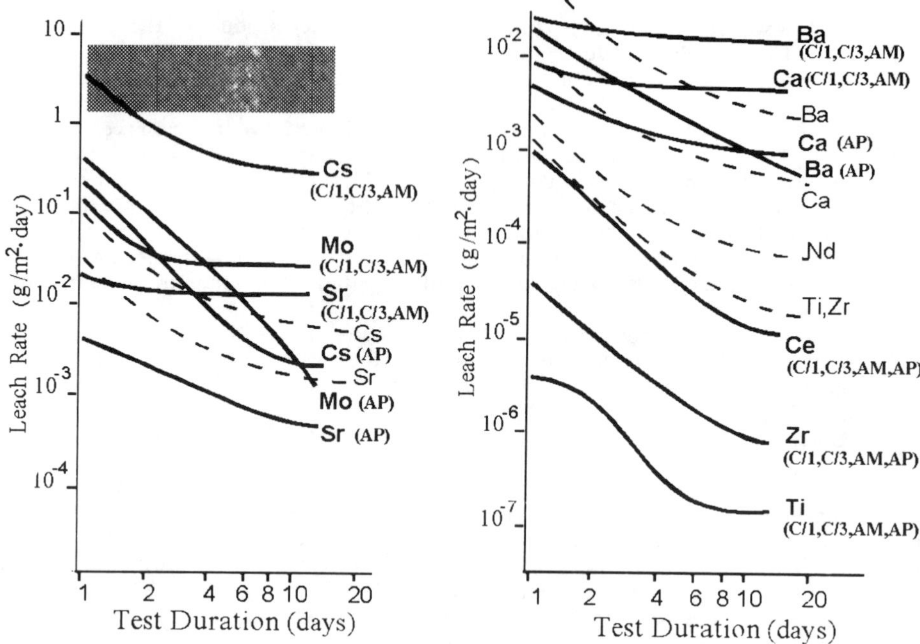

Figure 3. Comparative leaching behavior of Synroc-C prepared by inductive melting (solid curves - this work), hot-pressing (solid curves - this work, dotted curves - ref. [10]), and PNL 76-68 borosilicate glass (dark area - ref. [9]) in water at 100 °C.

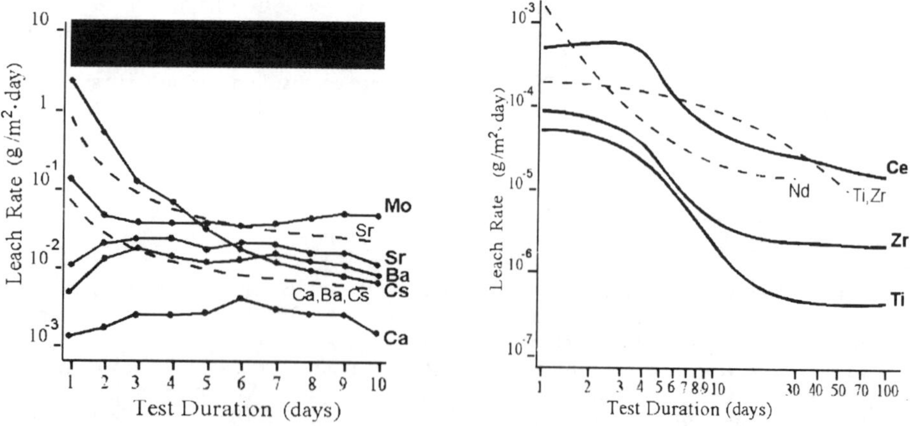

Figure 4. Comparative leaching behavior of Synroc-C prepared by inductive melting (solid curves -this work, sample C/3), hot-pressing (dotted curves - ref.[10]), and PNL 76-68 borosilicate glass (dark area -ref. [9]) in water at 200 °C

Thus, the most undesirable phases in the melted Synroc are alkali molybdates. In order to avoid their formation Synroc melting should be carried out at strongly reducing conditions where the alkali molybdates are thermodynamically unstable or alkali elements must be separated from high level wastes according to high level waste partitioning concept.

ACKNOWLEDGMENTS

Many thanks to Dr. E.R. Vance for the hot-pressed Synroc-C sample used in our study.

REFERENCES

1. A.E.Ringwood, S.E. Kesson, N.G.Ware, W. Hibberson and A. Major, Nature 278, pp 219-223 (1979).
2. V.I. Vlasov, O.L. Kedrovsky, A.S. Nikiforov, A.S.Polyakov, and I. Yu. Shishtchitz, in Back End of the Nuclear Fuel Cycle, Vienna, IAEA, p 109 (1987).
3. I.A. Sobolev, S.V. Stefanovsky, and F.A. Lifanov, Radiochemistry (Russ.) 33, p.99 (1993).
4. S.V. Stefanovsky, S.V. Yudintsev, B.S. Nikonov, and B.I. Omelyanenko, Geoecology (Russ.), 4, pp 58-75 (1996).
5. O.A. Knyazev, B.S. Nikonov, B.I. Omelyanenko, S.V. Stefanovsky, S.V. Yudintsev, R.A. Day, and E.R. Vance, in Nuclear and Hazardous Waste Management International Topical Meeting, Spectrum'96. Seattle. August 18-23, 1996, pp 2130-2137.
6. Nuclear Waste I.A. Sobolev, S.V. Stefanovsky, B.I. Omelyanenko, S.V. Yudintsev, E.R. Vance, and A. Jostsons, Scientific Basis for Management XX (these proceedings).
7. A.E.Ringwood, S.E. Kesson, K.D. Reeve, D.M. Levins and E.J. Ramm, in Radioactive Waste Forms for the Future, edited by W.Lutze and R.C. Ewing (North Holland Publishers, 1988), pp 233-334.
8. P.V. Iseghem, W. Jiang, M. Blanchaert, K. Hart and A. Lodding, in Scientific Basis for Nuclear Waste Management XIX (Mat. Res. Soc. Symp. Proc., 412), p 305 (1995).
9. J.E. Mendel, W.A. Ross, F.P. Roberts, Y. Katayama, J. Westsik, R. Turcotte, J.Wald, and D. Bradley, Characteristics of High-Level Waste Glasses, Annual Report. Battelle Pacific Northwest Lab. Report BWNL-2252, UC-70, pp 1-99 (1977).
10. V.M. Oversby, A.E. Ringwood, Radioactive Waste Management, 2, pp 223-238 (1982).

SYNTHETIC MINERAL-LIKE MATRICES FOR HLLW SOLIDIFICATION: PREPARATION BY INDUCTION MELTER WITH A COLD CRUCIBLE (CCIM)

T.V. SMELOVA, N.V. KRYLOVA, I.N. SHESTOPEROV
MINATOM RF, SSC RF VNIINM, Rogov 5, Moscow, 123060, Russia

ABSTRACT

A technique for solidification of high-level liquid wastes (HLLW) to obtain mineral-like forms through the use of water-cooled melters with direct induction heating of a melt (CCIM) has been developed at the State Scientific Center of the Russian Federation VNIINM (SSC RF VNIINM).

Mineral-like materials of different classes such as pyroxenes, pyrosilicates, garnets of the andradite group, titanosilicates and Synroc D were synthesized by the CCIM method at temperatures of 1250 - 1550°C.

The preliminary X-ray diffraction studies of these materials indicate that the structures synthesized by the CCIM method have compositions that are identical to those of the corresponding natural analogs. The chemical durabilities of the synthesized materials upon their exposure to distilled water at 20°C have been determined.

In addition to the investigation of solidification of liquid HLW, the CCIM method is examined and applied to the synthesis of mineral-like compositions involving simulated waste components (nonsoluble residues, ashes, and pulps) produced by the chemical and metallurgical manufacturing of fissile materials.

INTRODUCTION

A significant improvement in the physicochemical properties of solidified final products together with an increase in the degree of radionuclide incorporation can be achieved with the use of mineral-like materials as waste form matrices. These materials are analogs of natural minerals chosen on the basis of both their geochemical stability and the ability of their crystal lattices to incorporate the cations of the most hazardous long-lived radionuclides in the form of stable solid solutions.

Among such materials are the ceramic material "Supercalcinate" [1], which is a complex of compatible water-nonsoluble crystalline phases that selectively retain radionuclides, and the synthetic stone-like material "Synroc" [2], which is a complex of artificially produced titanate phases, whose lattices firmly retain the elements in radioactive wastes. These materials can be obtained at temperatures above 1200°C. Hence, they are predominantly synthesized by hot pressing, the however application of this technology can cause certain problems connected with the necessities of exactly measuring calcines and inactive additives, their mixing, and the possibility of faines generating dry high-level powders, etc.

The method for manufacturing mineral-like forms of solidified wastes through the use of melters using direct induction heating has been developed at SCC RF VNIINM [3]. The principal features of this cold crucible induction melter (CCIM) process are as follows [4]:

• the absence of direct contact between melt and crucible (due to the presence of a slag lining), which makes it possible to avoid corrosion problems and attain high temperatures in the melting process;

• the possibility of realizing a continuous process;

Mat. Res. Soc. Symp. Proc. Vol. 465 © 1997 Materials Research Society

• the lack of restrictions on the compositions of the reprocessing materials, which makes it possible to considerably extend the range of solidified waste forms.

The synthesis of minerals by melting in CCIMs is similar to the formation of minerals and rocks in the course of various chemical and physicochemical processes occurring in the earths crust due to internal thermal energy (the so-called endogenous magmatic processes).

EXPERIMENT AND RESULTS

The key principle for choosing the mineral-like matrix composition for the incorporation of radionuclides is based on consideration of the liquid HLW macrocomponents to be immobilised in the resulting matrix by the CCIM method. The liquid HLW compositions can be conventionally subdivided into two groups:

(1) liquid HLW in which the total content of alkali and alkaline-earth nuclides together with aluminum comprises (in percentage terms) the major portion of waste oxides;

(2) liquid HLW in which the fission and equipment corrosion products make up the bulk of waste oxides.

Taking into account these wastes, the following mineral-like materials of different classes were chosen to be synthesized by the CCIM method.

(I) Pyroxenes and pyrosilicates.
• Aegirine $NaFe[Si_2O_6]$
• Jadeite $NaAl[Si_2O_6]$
• Aegirine-avgite $NaCaFe[Si_2O_6]$
• Arfvedsonite $Na_3(Mg,Fe)_4(Fe,Al)[Si_4O_{11}]_2[OH,F]_2$
• Ortite $(Ca,Ce)_2(Al,Fe)_3[Si_2O_7][SiO_4]O[O,OH]$
• Tourmaline (sherlite) $(Na,Ca)(Mg,Al)_6[B_3,Al_3,Si_6(O,OH)_{30}]$ is a modification of tourmaline, in which iron replaces multivalent cations.

(II) Garnets of the andradite group.
Andradite contains 33.0% CaO, 31.5% Fe_2O_3, and 36.5% SiO_2. Trivalent elements can substitute for one another. Therefore, three iron-enriched modifications exhibiting intermediate compositions between garnet and orthosilicates (andradite, ferrogarnet I, and ferrogarnet II) were chosen to be synthesized by the CCIM method.

(III) Titanosilicates.
• Sphene $CaTi[SiO_4]O$

(IV) In addition, **Synroc D** was also prepared by melting in a cold crucible. The composition of this material was taken from the data available in the literature and involved 10.0% Al_2O_3, 13.0% CaO, 57.0% TiO_2, 11.0% ZrO_2, and 9.0% BaO.

All the materials were synthesized by the CCIM method at 1250 - 1550°C and investigated under the same conditions. In order to simulate the annealing caused by self-heating, the materials were heat-treated at 650°C for 250 h. To determine the radiation resistance, the materials were exposed to gamma radiation up to 10^{10} rad.

Result and Discussion

Table 1 and Fig. 1 demonstrate the results obtained from short test analyses and illustrate changes in the chemical durability of the studied mineral-like materials. The chemical durability was estimated from the leaching rates of some elements upon exposure of a powdered sample, with a size range of 0.16-0.25 mm, to distilled water at 60°C (using a teflon container in a stainless steel container at the MCC-1 static test) and also from the "depth of destruction". The "depth of destruction" of materials was calculated from the specific electric conductivity of the contact water as measured by the Kolraush method according to the equation:

$$\delta_t = \frac{\lambda_t - \lambda_o}{\alpha\lambda} \times \frac{V}{S} \times 10^{-6},$$

where δ_t is the depth of destruction of a layer at time, A;

λ_o – specific electric conductivity of the initial water, ohm$^{-1}\times$cm^{-1};

λ_t – specific electric conductivity of the solution at a moment in time, ohm$^{-1}\times$cm^{-1};

V – volume of water, ml;

S – surface of a specimen treated, cm^2;

λ – equivalent electric conductivity of a respective base, ohm$^{-1} \times$ cm$^2 \times$ g-eq^{-1};

α – number of gram-eguivalents of the respective oxides in 1 cm^3 of specimen.

Table 1

Rate of leaching from the mineral-like samples
(determined by rapid analysis at a temperature of 60°C for 1 h)

Material type	Initial samples					Heat treatment, 650°C, 250 h; irradiation, 10^{10} rad				
	Leaching rate, g×cm^{-2}×h^{-1} × 10^6				Depth of de-struction	Leaching rate, g×cm^{-2}×h^{-1} × 10^6				Depth of de-struction
	Na$^+$	Cs$^+$	Ca^{2+}	Sr^{2+}	Å	Na$^+$	Cs$^+$	Ca^{2+}	Sr^{2+}	Å
Aegirine	1.9	1.2	–	1.0	100	85	93	–	51	3485
Jadeite	47	0.97	–	0.19	50	9.5	2.4	–	0.96	350
Aegirine-avgite	1.0	0.96	40	1.9	20	4.3	1.0	45	17	668
Arfvedsonite	1.1	4.6	–	1.5	20	2.6	1.1	–	0.45	45
Ortite	–	0.32	0.86	1.6	30	–	0.90	1.1	1.6	115
Sherlite	3.6	0.69	–	2.0	30	8.2	1.7	–	32	5000
Andradite	–	0.21	0.10	0.10	40	–	0.11	0.70	2.7	470
Ferrogarnet I	–	0.85	3.0	0.50	40	–	2.2	0.34	0.30	60
Ferrogarnet II	–		4.5	1.0	40	–	4.0	2.9	0.57	50
Sphene	–	4.6	0.40	1.1	30	–	1.9	0.17	0.88	30
Lovchorrite	22	5.3	13	6.2	400	28	4.1	21	9.6	2000
Synroc	–	78	2.6	0.29	100	–	56	3.0	1.5	155

Rates of Na$^+$ leaching (determined at 20°C) from samples of phosphate and borosilicate glasses are equal to 2.7×10^{-6} and 4.3×10^{-6} g·cm^{-2}·day^{-1}, respectively. After the heat treatment at 650°C for 250 h, the leaching rates of the phosphate and borosilicate glasses are 3.4×10^{-5} and 5.8×10^{-3} g·cm^{-2}·day^{-1}, respectively (MCC-1 test).

These data indicate that crystalline mineral-like materials such as arfvedsonite, sphene, and ferrogarnets I and II are chemically most durable. The leaching rates of individual elements from these materials and depths of destruction remain virtually unchanged upon heat treatment and irradiation.

The lowest chemical durability is observed for the glass materials sherlite, lovchorrite, and egirine. The effect of irradiation and heat treatment on these materials is similar to that of phosphate and borosilicate glasses: the leaching rate of sodium, cesium, and strontium increases by about an order of magnitude.

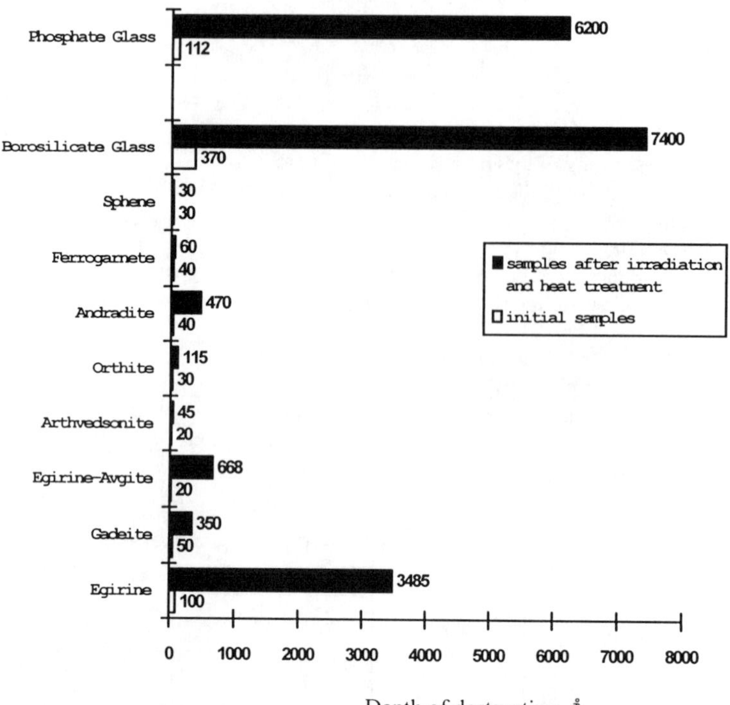

Depth of destruction, Å

Fig. 1 Effect of integral irradiation dose and heat treatment on the depth of destruction (temperature of heat treatment for borosilicate glass and like - mineral materials, 650 ^{0}C; the temperature of heat treatment for phosphate glass, 550^{0}C; heat treatment duration, 250h, irradiation dose, 10^{10} rad)

Glass-crystalline materials such as andradite, jadeite, aegirine-avgite, and Synroc exhibit an intermediate chemical durability.

X-ray diffraction analysis of the synthesized melted mineral-like materials prior to and after the heat treatment and irradiation shows that the majority of compounds, which are visually identified as glass-crystalline, possess an X-ray amorphous structure characterized by the presence of crystalline particles with a size of no more than 1000Å. A crystalline phase uniformly distributed throughout the bulk of the samples was found in the case of ferrogarnets. The composition of this phase is not determined now.

X-ray diffraction showed that sphene features the monoclinic system and interplanar distances identical to those of the natural sphene mineral before treatment. This corroborates the validity of our approach to the synthesis of mineral phases using the CCIM method. The sphene-type materials synthesized by CCIM method have more than 50 % of the crystalline sphene-phase.

Analysis of the results obtained and the chemical compositions of the HLW stored and produced by the currently available technology for the reprocessing of nuclear fuel demonstrates that mineral-like materials of the pyroxene group are the best candidates for solidification of the wastes enriched with alkali elements and aluminum. The mineral-like materials of the ferrogarnet and titanate types are most suitable for solidification of HLW whose basic components are fission products and equipment corrosion products.

Mineral-like materials of the pyroxene and titanosilicate types were prepared without and with cesium, strontium, and rare-earth element oxides (individually and in combination) in amounts from 3 to 5 wt %, using the CCIM method.

Table 2 presents the calculated compositions of the mineral-like materials synthesized by the CCIM method.

Table 2. Calculated compositions of titanosilicates and pyroxenes synthesized by the CCIM method

Material type	Composition	Components (wt %)							
		SiO_2	Na_2O	Fe_2O_3	CaO	TiO_2	Cs_2O	SrO	Ln_2O_3
Titanosilicates (sphene and its modifications)	1	30.6	-	-	28.6	40.8	-	-	-
	2	30.6	-	-	25.6	40.8	3.0	-	-
	3	30.6	-	-	23.6	40.8	-	5.0	-
	4	30.6	-	-	28.6	35.8	-	-	5.0
	5	30.6	-	-	20.6	35.8	3.0	5.0	5.0
Pyroxenes (egirine and its modifications)	1	52.0	13.4	34.6	-	-	-	-	-
	2	52.0	10.4	34.6	-	-	3.0	-	-
	3	52.0	13.4	29.4	-	-	-	5.0	-
	4	52.0	13.4	29.4	-	-	-	-	5.0
	5	52.0	10.4	24.6	-	-	3.0	5.0	5.0

Chemical analysis of the melted materials shows good agreement with the calculated compositions.

These materials are similar in some properties (hardness and density) to natural minerals such as egirine and sphene. However, upon synthesis of egirine by the CCIM method, we succeeded in obtaining not crystalline egirine but its amorphous modification that can also be formed in the earth crust under appropriate conditions.

The preliminary X-ray diffraction studies indicate that the introduction of Sr, Cs, and rare-earth elements (REE) into the base pyroxene matrix brings about a structural change in the parent material: cristobalite (SiO_2) precipitates as a predominant phase and different many-phase crystalline compounds are formed.

X-ray diffraction analyses show that $CaTiSiO_5$ (titanite or sphene) is the predominant crystalline phase in all the samples of titanosilicate mineral-like materials. The incorporation of Sr, Cs, and REE into the matrix composition gives rise to the new multiple-phase formations in small amounts.

The chemical durabilities of the synthesized mineral-like materials of the pyroxene- and titanosilicate-types with the compositions listed in Table 2 were calculated from the rates of leaching of components from the solidified matrices upon their exposure to distilled water at 20°C (Table 3).

Table 3. Rates of leaching from pyroxenes and titanosilicates syntesized by the CCIM method (MCC-1 test)

Material type	Composition (Table 2)	Nuclide	Leaching rate, $g \times cm^{-2} \times day^{-1} \times 10^{6}$		
			1 day	7 day	21 day
Pyroxenes (egirine and its modifications)	1	Na	27	2.7	2.2
		Sr	-	-	-
		Cs	-	-	-
	2	Na	11	0.65	0.35
		Sr	-	-	-
		Cs	4.5	1.5	0.20
	3	Na	50	3.7	0.13
		Sr	44	2.6	0.87
		Cs	-	-	-
	4	Na	1300	7.5	4.4
		Sr	-	-	-
		Cs	-	-	-
		REE	< 7.6	< 1.9	< 1.0
	5	Na	42	5.2	0.19
		Sr	43	12	1.7
		Cs	27	0.90	0.75
		REE	< 7.6	< 1.9	< 1.0
Titanosilicates (sphene and its modifications)	1	Ca	57	23	0.44
		Sr	-	-	-
		Cs	-	-	-
	2	Ca	45	28	0.18
		Sr	-	-	-
		Cs	100	6.6	5.6
	3	Ca	18	13	0.50
		Sr	24	6.0	0.50
		Cs	-	-	-
	4	Ca	100	23	6.8
		Sr	-	-	-
		Cs	-	0	-
		REE	< 6.5	< 1.6	< 0.94
	5	Ca	33	45	0.59
		Sr	16	49	3.4
		Cs	25	15	1.9
		REE	< 6.5	< 1.6	< 0.94

After seven days of exposure to water, the leachability of pure pyroxene materials is equal to 10^{-6}-10^{-7} $g \cdot cm^{-2} \cdot day^{-1}$ (Na^{+}) and the incorporation of Cs, Sr, and REE into the base matrix does not cause considerable deterioration of chemical durability of these materials. However, it should be noted that the fixation of Cs in the pyroxene materials is stronger than that of Na and Sr. In our opinion, this is a consequence of the change in structure of pyroxene material due to the occurrence of cristobalite and its interaction with Cs.

In the first seven days, the leachability of cesium from the pure titanosilicate mineral-like materials (10^{-5}-10^{-6} $g \cdot cm^{-2} \cdot day$-1) was somewhat less than that from pyroxenes. This indirectly confirms the absence of isomorphic incorporation of cesium into these minerals. After three weeks of testing, the leaching rates became equal to 10^{-6}-10^{-7} $g \cdot cm^{-2} \cdot day^{-1}$. The introduction of Sr, Cs, and REE into titanates has little effect leaching.

The leaching rates of the rare-earth elements ($<10^{-6}$ $g \cdot cm^{-2} \cdot day^{-1}$) are indicative of the high degree of their fixation in all the studied matrices.

Therefore, we can conclude that the CCIM method provides a way of obtaining mineral-like materials that correspond in structure to natural minerals. The incorporation of fission-product nuclides into the synthesized matrices brings about changes in the matrix structures. This is characterized by the formation of new crystalline phases, which manifests itself in the X-ray diffraction patterns of these materials and is accompanied by variations in their chemical durability. Currently a detailed investigation of these mineral-like materials synthesized by the CCIM method is being conducted.

In addition to the investigation of HLLW solidification, it is of interest to study applications of the CCIM method to the synthesis of mineral-like compositions involving other simulated wastes (nonsoluble residues, ashes, and pulps) that are produced by the chemical and metallurgical manufacturing of fissile materials. Taking into account the fact that alpha-emitting poorly soluble wastes (PSW) contain all the components in amounts required for the synthesis of glasses and mineral-like compositions, the melting of these wastes is the only reasonable technical solution to their reprocessing. Note that the volume of the initial PSW decreases 5-6 times when they are melted and the alpha radionuclides appear to be chemically bound to the components of the matrix composition.

In this case, the mineral-like materials close in composition to natural garnets of the andradite group are the most suitable as matrices for immobilizing the wastes of chemical and metallurgical manufacturing of fissile materials using the CCIM method.

CONCLUSION

1. The advantages of using a melter with direct induction heating (cold crucible induction melter - CCIM) are demonstrated for solidification of HLW, and the principal features of this process are described.

2. The CCIM method provides a way of obtaining mineral-like materials that correspond in structure to natural minerals.

3. The CCIM synthesis of mineral-like compositions appears also to be applicable to the incorporation of components of waste produced by chemical and metallurgical manufacturing of fissile material.

REFERENCE

1. G.J. McCarthy, <u>High Level Waste Ceramic Material Consideration, Process Simulation and Product Characterization,</u> Nucl. Techn., v.32, 92, 1997.
2. A.E. Ringwood et al, <u>The SYNROC Process A Geochemical Approch to Nuclear Waste Immobilization,</u> Geochem. J., v.13, 141, 1979.
3. V.V. Kushnikov, T.V. Smelova, <u>Induction Melter with Cold Crucible for Immobilizing Long-Life Radionuclides,</u> The First International Environmental Technology Business Action Conference, Moscow-94, report N 262.
4. I.G. Petrov, D.G. Ratnikov, <u>Cold Crucible,</u> Moscow, 1972.

EFFECT OF DIFFERENT GLASS AND ZEOLITE-A COMPOSITIONS ON THE LEACH
RESISTANCE OF CERAMIC WASTE FORMS

M. A. Lewis, M. Hash, and D. Glandorf, Argonne National Laboratory, Chemical Technology
Division, Argonne, IL   60439

ABSTRACT

A ceramic waste form is being developed at Argonne National Laboratory for waste generated
during the electrometallurgical treatment of spent nuclear fuel. The waste is generated when fission
products are removed from the electrolyte, LiCl-KCl eutectic. The ceramic waste form is a
composite, fabricated by hot isostatic pressing a mixture of glass frit and zeolite occluded with fission
products and salt. Past work has shown that the normalized release rate (NRR) is less than 1 g/m^{2}d
for all elements in a Material Characterization Center-Type 1 (MCC-1) leach test run for 28 days in
deionized water at 90°C (363 K). This leach resistance is comparable to that of early Savannah
River glasses. We are investigating how leach resistance is affected by changes in the cationic form
of zeolite and in the glass composition. Composites were made with three forms of zeolite A and six
glasses. We used three-day ASTM C1220-92 (formerly MCC-1) leach tests to screen samples for
development purposes only. The leach test results show that the glass composites of zeolites 5A and
4A retain fission products equally well. The loss of cesium is small, varying from 0.1 to 0.5 wt%,
while the loss of divalent and trivalent fission products is one or more orders of magnitude smaller.
Composites of 5A retain chloride ion better in these short-term screens than 4A and 3A. The more
leach resistant composites were made with durable glasses that were rich in silica and poor in alkaline
earth oxides. The x-ray diffraction (XRD) results show that a salt phase was absent in the leach
resistant composites of 5A and the better glasses but was present in the other composites with poorer
leach performance. Thus, the data show that the absence of a salt phase in a composite's XRD
pattern corresponds to improved leach resistance. The data also suggest that the interactions between
the zeolite and glass depend on the composition of both.

INTRODUCTION

A ceramic waste form is being developed at Argonne National Laboratory for waste generated
during the electrometallurgical treatment of spent nuclear fuel. The waste is generated when fission
products are removed from the electrolyte, LiCl-KCl eutectic, used in the electrorefining step.

Early samples of the ceramic waste form were fabricated by hot uniaxially pressing a mixture
of glass frit and zeolite occluded with fission products and salt [1, 2]. The leach resistance of the
resulting waste form was evaluated with the Material Characterization Center-Type 1 (MCC-1) leach
tests [3]. Most tests were run for 28 days in deionized water at 90°C (363 K), though a few were run
in simulated J-13 well water and brine. Under these test conditions, all elements in the ceramic waste
form had a normalized release rate of less than 1 g/m^{2}d, comparable to that measured in 1982 for
Savannah River borosilicate glass [4].

Recently, the pressing method was changed from uniaxial to isostatic. This change facilitates
fabrication of larger samples. Process parameters were subsequently modified [5]. The temperature
was increased from 700-715°C (973-988 K) to 750°C (1025 K), and the pressure from 6000 psi
(41 Mpa) to 25,000 psi (172 MPa). These conditions were necessary for deformation of the stainless
steel can. The time at temperatures above 600°C (873 K) was minimal to prevent the formation of
other crystalline phases during hot isostatic pressing. These changes prompted a re-examination of
the effect of glass and zeolite compositions on leach resistance.

We are now investigating the leach resistance of ceramic waste forms prepared using three
forms of commercially available zeolite A (designated 5A, 4A, and 3A). Zeolite A (all forms) can be
represented by the formula, $M_{12}Al_{12}Si_{12}O_{48}$, where M represents a univalent charge-compensating
cation [6]. The zeolite A structure contains both alpha and beta cages. The size of the pore
(entrance) to the alpha cage is determined by the number and charge of the charge-compensating
cations. The charge-compensating ions vary in the different forms: 5A contains calcium and sodium;

Mat. Res. Soc. Symp. Proc. Vol. 465 © 1997 Materials Research Society

4A, only sodium; and 3A, potassium and sodium. The number preceding the letter A gives the nominal pore size for the alpha cage, in angstroms. Different forms of zeolite A have different thermal stabilities. We investigated these zeolites to determine if differences in nominal pore size and/or thermal stability affect the leach resistance of their composites. We are also investigating the leach resistance of ceramic waste forms prepared with several glass compositions. Both the binding properties and durability of the glass are important. The glass matrix binds the zeolite powders and reduces the exposed surface area of the zeolite.

All samples were subjected to ASTM C1220-92 leach tests (the American Society Testing and Materials procedure for MCC-1 leach tests) for three days and examined with XRD. These screening tests allowed us to assess potential performance quickly. Leach resistance was evaluated by monitoring the loss of surrogate fission products, chloride, and boron from composite samples during leaching. Obviously, the most important characteristic of a waste form is its ability to retain fission products. The chloride loss is a measure of the chloride ion retention in the ceramic waste form and the boron loss is a measure of the glass durability. Each ceramic waste form sample was examined with XRD. Patterns obtained by this method were compared with those of the starting materials and other phases. The leach data and the XRD data were compared, and correlations were noted.

EXPERIMENTAL

<u>Materials</u>

Waste Salt. The materials used in these tests include simulated waste salt, zeolite A, and glass frit. Two waste salt compositions, given in Table I, were used. Both contained about 10 wt% fission products in KCl-44.2 wt% LiCl eutectic salt. The simulated fission product chlorides were obtained from Aldrich or Johnson-Matthey. They were 99.9 % purity or better, and all were packed under argon in glass ampoules. The LiCl-KCl eutectic salt was obtained from APL Engineered Materials, Inc. (Urbana, IL) and was also packed under an argon atmosphere. The various components were heated at 500°C (773 K) in a furnace well in a dry, argon or helium atmosphere glovebox. After melting, the salt was quenched and then crushed.

Table I. Salt Composition

ID#	Component Salts (wt%)										
	$CsCl$	KCl	$LiCl$	$NaCl$	$BaCl_2$	$SrCl_2$	$CeCl_3$	$LaCl_3$	$NdCl_3$	$PrCl_3$	YCl_3
Salt 1	3.7	45.3	39.6	6.0	1.4	0.6	0.7	1.1	1.0	0	0.1
Salt 2	2.3	48.7	40.7	0.0	1.0	0.4	1.9	0.6	3.8	0.8	0.1

Zeolite. Three commercial forms of zeolite A were used: 5A, 4A, and 3A. They were obtained from the UOP-Molecular Sieves Division (Moorestown, NJ). The nominal compositions of the zeolites are given in Table II.

Table II. Zeolite Compositions

Zeolite Type	Components (Wt%)					
	Al	Ca	K	Na	Si	O
5A	18.1	9.7	NM[a]	4.4	17.9	Remainder
4A	19.0	NM[a]	NM[a]	16.2	19.8	Remainder
3A	16.5	NM[a]	10.1	8.0	16.3	Remainder

[a]NM = not measured.

Glass. Six glass frits were tested. All were aluminoborosilicate glasses. These proprietary glass formulations are designated as G1, G2, G3, G4, G5, and G6, and their compositions are given in

Table III. The glasses were obtained from either Bayer Chemicals (Baltimore, MD) or Corning (Corning, NY).

Table III. Glass Compositions

ID#	Glass Components (wt%)							
	Al_2O_3	B_2O_3	CaO	K_2O	Na_2O	SiO_2	SrO	ZrO_2
G1	8	17	1	1	7	65	0	1
G2	10	14	14	1	6	55	1	0
G3	6	10	11	1	3	61	8	0
G4	3	18	0	9	1	68	0	0
G5	6	11	1	1	7	72	0	0
G6	9	12	0	6	5	68	0	0

Composites. In the following discussion, composites are identified according to zeolite form and glass type. For example, 5A/G1 signifies a composite made with salt-occluded 5A zeolite and glass G1. Except where noted, all composites were prepared from mixtures containing equal weights of salt-occluded zeolite and glass. Two composites also contained an additional 10% dehydrated zeolite. The purpose of this test was to determine if the additional zeolite scavanges non-occluded salt.

Processing

Blending. The salt-occluded zeolite was prepared by mixing dehydrated zeolite and the crushed simulated waste salt in a rotating double-cone blender at 500°C (773 K) [7]. All of the salt-occluded zeolites used in these experiments contained 21 wt% salt. For simplicity in this text we refer to all of the salt-occluded zeolites by the zeolite form used in the blending process, i.e, 5A, 4A, or 3A.

Hot Isostatic Pressing. All composites were hot pressed using a two-step thermal profile that consisted of a temperature increase to 750°C (1023 K) followed by a fast cool down to 600°C (873 K), where the temperature was maintained for one hour. The profiles are discussed in more detail in another paper [5].

Test Method

After hot pressing, cored samples of the composite (ceramic waste form) were leached in deionized water at 90°C (363 K), for three days according to the ASTM C1220-92 procedure. The total mass loss was measured in each test. The leachates were analyzed with a variety of techniques. Chloride ion concentration was measured with a chloride ion selective electrode. Cesium and the rare earth ions were measured with inductively coupled plasma-mass spectroscopy (ICP-MS). The other ions were measured with inductively coupled plasma-atomic emission spectroscopy (ICP-AES). For most leachates, the ICP-AES detection limits were too low for barium (0.02 µg/mL) and strontium (0.005 µg/mL) to be measured. In later tests, ICP-MS was used to measure barium and strontium. In addition, a duplicate sample of each hot-pressed composite was crushed and examined for crystalline phases with XRD.

RESULTS

Leach Tests

The purpose of these experiments was to screen composites made with different zeolites and glasses and to identify the more promising compositions. The most important criterion for the ceramic waste form is that it immobilize fission products. Measurement of fission product loss involves several analyses and is relatively expensive. Therefore, only selected leachates were

435

analyzed for fission products. The fractional loss of cesium is reported in Table IV as a percent. The fractional loss is defined as the ratio of the amount released into the solution relative to the amount initially present. The data are grouped according to zeolite and glass type. The release of fission products was found to depend on valence. Cesium, the only +1 ion, had a higher release than the higher valence fission products. The divalent fission products (barium and strontium) and the trivalent fission products (cerium, lanthanum, neodymium, praseodymium, and yttrium) had very low releases, resulting in a concentration of less than 5 µg/L in the leachate. Cesium releases, by way of contrast, were on the order of 200 µg/L in the leachate. Low releases of cesium and very low releases of the divalent and trivalent ions were common to both 5A and 4A. Typical values for the fraction loss, in percent, for the divalent and trivalent ions are given in Table V for 5A/G1 and 4A/G1.

Table IV.   Summary of Leach Test and XRD Results for Composites of Salt-Occluded Zeolite and Glass

Sample ID	Total Mass Loss (%)	Mass Loss for Cs (%)	Mass Loss for B (%)	Mass Loss for Cl (%)	Phases in XRD Patterns
5A/G1					
HO373	0.12	0.19	0.05	0.59	zeolite
HO424	0.16	0.23	0.05	0.96	zeolite
HO452	0.11	0.23	0.05	0.37	zeolite
HO521	0.12	0.30	0.05	0.45	zeolite
HO624	0.12	0.27	0.04	0.62	zeolite
HO653	0.09	0.46	0.04	0.49	zeolite
HO721	0.07	0.21	0.14	0.32	zeolite
HO741	0.12	0.36	0.09	0.51	zeolite
HO743[a]	0.09	0.26	0.04	0.37	zeolite
5A/G2					
371	0.23	0.22	0.38	0.60	zeolite
453	0.85	NM[b]	0.31	0.89	other + salt
5A/G3, 5A/G4, 5A/G5, 5A/G6					
451 (G3)	0.48	NM[b]	0.68	0.92	other + salt + sodalite
832 (G4)	0.26	0.20	0.51	0.30	zeolite
834 (G5)	0.12	0.25	0.02	0.45	zeolite
835 (G6)	0.17	0.30	ND[c]	0.44	zeolite + other
4A/G1					
423	0.26	0.20	0.10	1.09	zeolite + salt
471[d]	0.36	0.48	0.12	0.88	zeolite + salt
4A/G2					
421	0.44	NM[b]	NM[b]	1.60	zeolite + salt
472[d]	0.43	0.40	0.35	0.85	zeolite + salt + sodalite
3A/G1					
491	0.57	NM	NM	1.67	zeolite + salt

[a]Sample contains 45% glass.

[b]NM = not measured.

[c]ND = not detected.

[d]10% dehydrated zeolite was added to zeolite/glass frit mixture prior to hot pressing.

Table V. Fractional Loss of Divalent and Trivalent Ions from 5A/G1 and 4A/G1

ID #	Zeolite/Glass	Fractional Loss of Higher Valent Ions (%)					
		Ba	Sr	Ce	La	Nd	Y
424	5A/G1	BDL*	BDL*	0.001	0.000	0.001	0.003
423	4A/G1	BDL*	BDL*	0.012	0.001	0.015	0.001

*BDL = below detection limit of 0.02 µG/mL for Ba and 0.005 µG/mL for Sr.

The total mass loss was used as a quality screen. On this basis, three zeolite/glass combinations stand out (5A/G1, 5A/G5, and 5A/G6). All of their mass losses were less than 0.2%. One combination 5A/G1 was studied extensively. Nine samples were prepared using several batches of 5A and several lots of glass frit. The mass losses were consistently low. The average for the 5A/G1 composites was 0.11% with a standard deviation of 0.03%. Two other combinations, 5A/G5 and 5A/G6, also had low mass losses in initial tests. The mass loss for all the other samples shown in Table IV varied from 0.2 to 0.9%, indicating that both zeolite and glass compositions affect leach resistance. For example, G1 composites with 4A and 3A had higher mass losses (0.3 to 0.6%) than with 5A. Likewise, 5A composites with G2 and G3 had higher mass losses (0.2 to 0.9%) than with G1. Possible reasons for the variation in mass losses include differences in the release of chloride and fission products and in the durability of the glass. Fractional losses for chloride, fission products, and boron were examined to gain insight into the leaching mechanisms.

The release of chloride is used to measure chloride dissolution. There are two sources of chloride: one from occluded salt within the zeolite's cages, and the other from "free" salt, which is sometimes seen in XRD patterns of the composites. In general, high chloride losses correlate with high total mass losses (e.g., see 4A/G2 and 3A/G1 composites in Table IV) and, as discussed below, generally track with the presence of "free" salt in the XRD pattern. This suggests chloride loss can be an important leaching mechanism. The data also show that the different forms of zeolite A have different chloride retention properties. The 5A/G1 composites had consistently lower chloride releases than the 4A/G1 and 3A/G1 composites. The 5A/G2 composites also had lower chloride releases than the 4A/G2 composites. Thus, 5A retains salt better than 4A and 3A, for the glasses tested. Additional work was recently started with 5A and other aluminoborosilicate glasses (G3-G6). As shown in Table IV, composites with G4, G5, and G6 had low chloride releases, whereas the composite with G3 had a relatively high release.

Comparison of the compositions of the zeolites, salt, and glasses shows that boron is the only major element present in the glass phase that is not present in the zeolite. Thus, the boron release provides an approximate measure of glass dissolution. The following data confirm that boron loss is a reasonable measure of glass durability. Leach tests of the pure G1 and G2 glasses showed that G1 was more durable than G2. The mass losses measured after a 3-day leach test for pure, hot isostatically pressed glasses were essentially 0% for G1 and 0.24% for G2. The releases from G1 glass of aluminum and silicon were very low, 0.00 and 0.03%, respectively. For G2, the releases of aluminum and silicon were significantly higher, 0.26 and 0.27%, respectively. As shown in Table IV, the fraction of boron leached was lower for G1 composites than for G2 composites, indicating that G1 is more durable in composites than G2. The durability of G3, G4, G5, and G6 has not been as extensively investigated as G1 and G2. The limited data in Table IV show boron releases were very low for G5 and G6, and these glasses appear promising for further testing.

The data in Table IV also show that the boron release was lower for 5A/G1 composites than for 4A/G1 composites. Thus, the form of the zeolite affects the performance of G1 composites.

X-ray Diffraction Results

Sister samples of the leached composites were examined with XRD. These patterns were compared with each other and with the starting materials used in their fabrication. This comparison showed that the composites could be divided into one of three types, according to the predominant crystalline phases. The three types are (1) zeolite, where the pattern of the composite is almost the same as the pattern of the initial zeolite, and there are no additional phases, (2) other, which designates unidentified phases, and (3) a mixture of zeolite, "free" salt, and, in some cases, sodalite. The phases identified in the XRD pattern for each composite are given in Table IV.

Figure 1A is the XRD pattern of 5A with 21 wt% salt. Figure 1B is the XRD pattern of a 5A/G1 composite. All of the XRD patterns for the 5A/G1 composites were similar. The 5A/G5 and 5A/G6 composites were also of this type. (The 5A/G6 composite contained an additional unidentified phase.) In Figure 1B, there is a broad, amorphous peak (between 20-30°) due to the glass component of the composite. Further comparison of Figures 1A and 1B shows that the two patterns are alike in some respects and different in others. For example, the positions of the lines present in both patterns are essentially the same. However, the relative intensities of some lines are very different. One strong line and several weak ones present in Figure 1A are absent in Figure 1B.

Figure 1C is an example of a XRD pattern that consists of "other" phases, and very little, if any, zeolite phase. This can be seen by comparing Figure 1A with Figure 1C, the XRD pattern for a 5A/G2 composite, #453. There is no longer a one-to-one correspondence between the positions of the lines in the XRD patterns for the composite and the zeolite. The XRD pattern of the 5A/G3 composite was also of this type. The "other" phases in these composites are, as yet, unidentified. Easily identified phases such as NaCl or sodalite were present in some of these composites. Samples with these phases are identified in Table IV.

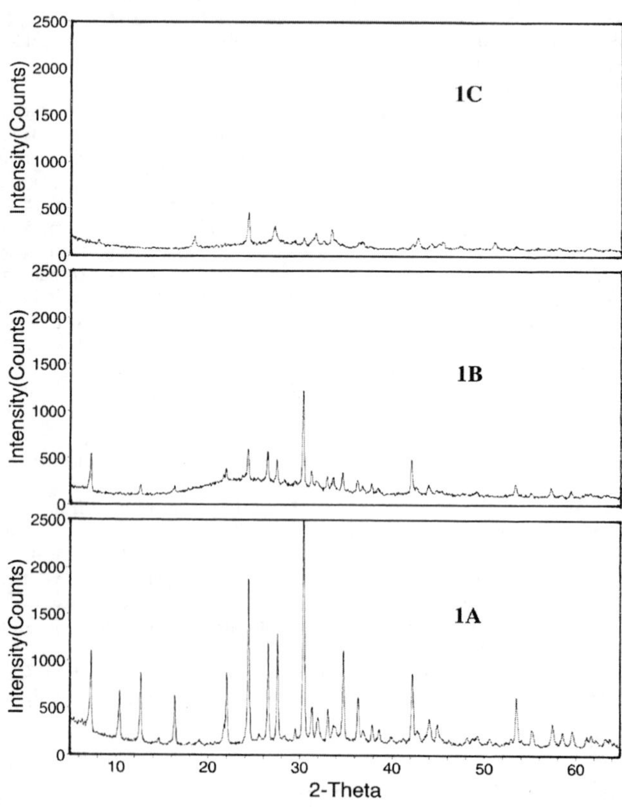

Fig. 1.   X-ray Diffraction Patterns of 5A with 21 wt% salt (A) and
Two Zeolite-Glass Composites, B (5A/G1, #452) and
C (5A/G2, #453)

The data indicate that the zeolite-glass interactions depend on glass composition. For example, three of the composites 5A/G1, 5A/G2, and 5A/G3 (#451, #452, and #453, respectively) had different XRD patterns even though they were made in the same hot pressing run and with the same batch of 5A. Glasses 2 and 3 interacted more strongly with 5A and caused it to transform to other crystalline phases. Glass 1 showed less tendency to effect a transformation of the zeolite.

A comparison of the XRD patterns for 4A and its composites, 4A/G1 and 4A/G2 showed that the patterns of the composites consisted of mixtures of zeolite, NaCl, and, in some, cases, sodalite. The same changes in the zeolite patterns observed in the 5A and 5A/G1 composites were also seen in the 4A samples. There was a one-to-one correspondence in the positions of the zeolite lines but relative intensities of some lines varied. Similar results were seen in the XRD patterns for 3A and the 3A/G1 composite.

Thus, the XRD results suggest that composites made from 5A are less likely to convert to other phases than those made from 4A. We have studied the effect of heating mixtures of 5A/G1 and 4A/G2 powders (prior to hot pressing) at 700°C (973 K). The 4A/G2 mixtures completely converted to sodalite within 4 hours, while the 5A/G1 mixture did not convert to sodalite, even after 24 hours.

The XRD results also show that differences in promoting transformation exist among various glasses. Examination of the compositions in Table III shows that the glasses that react with 5A (G2 and G3) contain more than 10% alkaline earth oxides. Glasses that are less aggressive contain 1% or less alkaline earth oxides.

CONCLUSIONS

Composites comprised of zeolite A containing 21 wt% salt and glass frit were hot isostatically pressed and then leached. Three forms of zeolite A and six glass frits were used. We investigated the effect of compositional changes in the zeolite and the glass by measuring short term leach resistance and crystal structure. The results were the following. First, both zeolite 5A and 4A retain fission products equally well. The release of cesium was small, varying from 0.1 to 0.5 wt%, while the release of divalent and trivalent fission products was one or more orders of magnitude smaller. Second, 5A/G1 composites retained chloride better than 4A/G1 and 3A/G1 composites. The chloride loss was especially high for the 3A/G1 composite. Third, several aluminoborosilicate glasses (G1, G5, and G6) acted as durable glass binders for 5A. The other aluminoborosilicate glasses either reacted with the 5A and transformed its XRD pattern (G2, G3) or were not durable (G4). Finally, test results correlated with the XRD results, i.e., the leach resistance was usually poor when the XRD pattern contained "free" salt, non-zeolitic phases, or sodalite.

Future work will be concerned with measuring the leach resistance for the more promising zeolite-glass composites for longer times. It is important to determine if the trends observed in the 3-day tests are also observed in tests for 28 days or longer. Other planned experiments are concerned with identifying conditions that affect glass reactivity, such as alkaline earth oxide concentration, and identifying other phases that immobilize cesium better.

ACKNOWLEDGEMENT

The authors wish to acknowledge the assistance of A. Essling, D. Huff, E. Huff, P. Johnson, J. Keily, and F. Smith of the Analytical Chemistry Laboratory in developing dissolution methods and analytical techniques, and the helpful comments of J. Harmon of the Editorial staff.

REFERENCES

[1] M. A. Lewis, D. F. Fischer and C. D. Murphy, "Properties of Glass-Bonded Zeolite Monoliths," *Environmental and Waste Management Issues in the Ceramic Industry II*, Am. Ceram. Soc., Westerville, OH, pp. 279-286 (1994).

[2] M. A. Lewis, D. F. Fischer, and C. D. Murphy, "Densification of Salt-Occluded Zeolite A Powders to a Leach Resistant Monolith," Scientific Basis for Nuclear Waste Management, Eds., A. Barkatt and R. Van Konynenburg, Vol. 333, pp. 277-284, Materials Research Society, Pittsburgh, PA (1994)

[3] "Standard Test Method for Static Leaching of Monolithic Waste Forms for Disposal of Radioactive Waste," ASTM C1220-92, American Society for Testing and Materials, Philadelphia, PA (1995)

[4] M. J. Plodenic, G. G. Wicks, and N. E. Bibler, *An Assessment of Savannah River Borosilicate Glass in the Repository Environment*, DP-1629, Savannah River Laboratory, Aiken SC (1982).

[5] M. C. Hash, C. Pereira, M. A. Lewis, R. J. Blaskovitz, V. Zyryanov, and J. P. Ackerman, "Hot Isostatic Pressing of Glass-Zeolite Composites," to be published in *Environmental Issues and Waste Management Technologies in the Ceramic and Nuclear Industries* (1996).

[6] D. W. Breck, *Zeolite Molecular Sieves*, John Wiley & Sons, New York (1974).

[7] C. Pereira, V. N. Zyryanov, M. A. Lewis, and J. P. Ackerman, "Mixing of Zeolite Powders and Molten Salt," to be published in *Environmental Issues and Waste Management Technologies in the Ceramic and Nuclear Industries* (1996).

EVALUATION OF STANDARD DURABILITY TESTS TOWARDS THE QUALIFICATION PROCESS FOR THE GLASS-ZEOLITE CERAMIC WASTE FORM

L. J. SIMPSON and D. J. WRONKIEWICZ
Argonne National Laboratory, 9700 South Cass Avenue, Argonne, IL 60439

ABSTRACT

Glass-bonded zeolite is being developed as a potential ceramic waste form for the disposition of radionuclides associated with the U.S. Department of Energy's (DOE's) spent nuclear fuel conditioning activities. The utility of several standard durability tests [e.g., Materials Characterization Center Test #1 (MCC-1), Product Consistency Test-B (PCT-B), and Vapor Hydration Test (VHT)] was evaluated as a first step in developing methods and criteria that can be applied towards the process of qualifying this material for acceptance into the DOE Civilian Radioactive Waste Management System. The effects of pH, leachant composition, and sample surface-area-to-leachant-volume ratios on the durability test results are discussed, in an attempt to investigate the release mechanisms and other physical and chemical parameters that are important for the acceptance criteria, including the establishment of appropriate test methodologies required for product consistency measurements.

Results from PCT-Bs conducted with 4 μm diameter salt-loaded zeolite powder indicate that a good correlation exists between release rate and ionic size and/or charge for the release behavior of the simulated fission products in deionized water (DIW), EJ-13 groundwater, and brine solutions. Simulated divalent and trivalent fission products [Sr, Ba, and rare earth (RE) ions] were preferentially retained in the zeolite (relative to the singly ionized cations) after tests with the salt-loaded zeolite in DIW. In general, the preferential cation release order for salt-loaded zeolite A in DIW is Li > Na $\geq$ K > Cs > Al > Si > RE > Sr > Ba. Results from PCT-Bs with the salt-loaded zeolite A immersed in high-ionic-strength brines at 90°C indicate a significant increase, relative to DIW tests, in the release rates of the Sr, Ba, and RE ions despite a decrease in the release of the Si and Al ions that make up the framework matrix of the zeolite. An increase in the Mg and Ca concentrations in the reacted zeolites suggests that an ion exchange process may be responsible for this increase.

Vapor hydration and MCC-1 tests were performed with ceramic waste form monoliths of glass-bonded zeolite. The VHTs (temperatures at 120, 150, and 200°C) provided useful information about the effect of glass composition on corrosion rates and alteration phase formation, and about the overall toughness and structural integrity of the ceramic waste form. The MCC-1 test was investigated as an alternative to the PCT for acceptance criteria measurements. The MCC-1 results indicate that corrosion testing with both DIW and high-ionic-strength leachants (that specifically affect the ion exchange behavior of the fission products) are required to fully assess the durability of the ceramic waste form. These preliminary results establish the utility of the MCC-1 test for providing possible acceptance criteria measurements, including elemental release comparisons between the environmental assessment benchmark and the ceramic waste form.

INTRODUCTION

The electrometallurgical treatment program at Argonne National Laboratory is developing conditioning processes for the potential treatment of a diverse inventory of U.S. Department of Energy (DOE) spent nuclear fuel that may not be amenable for direct geologic disposal [1]. These include uranium metal, metal-Na bonded, and Three Mile Island core debris fuels. The electrometallurgical treatment program uses molten anhydrous LiCl-KCl salt in the electrorefining process that accumulates approximately 10 wt% fission products (Sr, Cs, Ba, Ce, La, Nd, and Y). This fuel conditioning generates two waste streams [1]: the residue from the electrorefiner anode and the salt. The anode residue is incorporated in a metal waste form, and fission products that accumulate in the salt are loaded into zeolites through a column ion-exchange process [1,2]. These zeolites, with high concentrations of fission products, are blended with additional zeolites to

Mat. Res. Soc. Symp. Proc. Vol. 465 © 1997 Materials Research Society

absorb any remaining free salt and are hot isostatically pressed with glass to produce a glass-bonded zeolite, termed the "ceramic" waste form. Because the salt with the accumulated fission products is associated with the first stream in the processing of irradiated fuel and may contain transuranic elements, it is considered a high-level waste. This report focuses on the development and evaluation of tests for acceptance of the ceramic waste form.

The development of a technical basis for acceptance of a waste form into the DOE Civilian Radioactive Waste Management System must follow the guidelines presented in the Waste Acceptance System Requirements Document (WASRD, DOE/RW-0351, 1996) for spent nuclear fuel and high-level waste and in the Waste Acceptance Product Specifications (EM-WAPS, 1993) for borosilicate glass [3]. As outlined in the WAPS, the WASRD defines the "borosilicate glass" produced by programs directed from the DOE Office of Environmental Management as a standard high-level waste form. The WASRD specifications define the physical characteristics and consistency required for a waste form. These include chemical composition (elemental and oxide); crystalline phase projections; waste form production control demonstration through a comparison of the waste form with the environmental assessment benchmark (EA) glass (by a product consistency test); structural and compositional phase stability; and hazardous waste accountability.

To be accepted into the DOE Civilian Radioactive Waste Management System, the ceramic waste form must be reviewed and deemed "acceptable" with the identification and development of the appropriate qualification methods and criteria. As an initial step in this process, the present work evaluates the utility of standard durability tests for providing the information needed to qualify the ceramic waste form [3]. The Materials Characterization Center Test #1 (MCC-1, "American Society for Testing and Materials" [ASTM] method C1220-92), Product Consistency Test (PCT) [4], and Vapor Hydration Test (VHT) [5] will be the focus of these initial investigations because of their widespread use in testing borosilicate glass, the ASTM approval of the PCT and MCC-1 test, and their use in qualification of borosilicate glass (e.g., PCT). A broad range of potential repository conditions were incorporated into the testing procedures to determine the bounding parameters appropriate for the corrosion testing of the ceramic waste form and its behavior under accelerated conditions. The effects of pH, leachant composition, sample surface-area-to-leachant-volume ratios, and test temperature on the results from durability tests with the ceramic waste form are discussed. Scanning electron microscopy with electron dispersive spectroscopy, analytical electron microscopy, and X-ray diffraction were used to provide background information about morphology, composition, structure, bonding, and alteration phase formation of both the ceramic waste form and the precursor material of zeolite, blended with simulated waste salt (salt-loaded zeolite).

EXPERIMENTAL

The ceramic waste form group of the electrometallurgical treatment program provided the samples used in these experiments. Equal amounts by weight of salt-loaded zeolite and glass were mixed together in an industrial mill, packed into 304 stainless steel canisters, and hot isostatically pressed to produce the ceramic waste form [7]. The zeolite and glass were uniformly distributed throughout the sample for dimensions greater than 100 μm. Compositional analyses from electron dispersive spectroscopy agreed reasonably well with the values calculated from the initial components of the zeolite and glass mixture (Table I). Electron beam stability of the zeolite during analytical electron microscopy indicated a low intrinsic water content in the ceramic waste form. To minimize pretest leaching, all sample cutting and polishing was performed without the use of any aqueous lubricants. Isopropyl alcohol was used to clean residual polishing material from the specimens. Except for the omission of aqueous lubricants, ASTM protocols were followed in the preparation of samples and in performing the durability tests [ASTM C1220-92, 4, 5].

Table I. Elemental compositions of the salt-loaded zeolite and ceramic waste form.

Elements	Li	B	Na	Al	Si	Cl	K	Ca	Y	Sr	Cs	Ba	La	Ce	Nd	O
SLZ, Wt%	1.36		12.9	15.0	15.6	12.9	5.49		0.01	0.06	0.63	0.21	0.13	0.24	0.39	35.1
CWF, Wt%	0.68	2.46	5.43	9.42	22.1	6.46	3.00	4.27	0.01	0.04	0.32	0.11	0.06	0.12	0.19	45.3

SLZ: Salt-loaded zeolite 4A. CWF: Ceramic waste form.

The evaluation of the durability tests was performed with both the precursor salt-loaded zeolite powders and ceramic waste form monoliths. Initial investigations focused on PCT-B procedures with the salt-loaded zeolite 4A powder that has a mean grain diameter of approximately 4 µm. This work was performed to establish a baseline and investigate the mechanistic behavior of ion exchange and other elemental release properties of the salt-loaded zeolite. This information was required to determine the appropriate test procedures and to help interpret the durability test results with the ceramic waste form.

The evaluation of the VHT and MCC-1 durability tests was performed with monolithic ceramic waste form samples. The utility of these tests for providing useful information related to physical, chemical, and consistency issues of the acceptance criteria were investigated. This evaluation will require the incorporation of a range of potential repository fluid compositions into the PCT-B and MCC-1 test procedures to account for the possible effects of the repository groundwater and leachant variations on the ceramic waste form. The effect of brines on the ceramic waste form may be important when considering disposition in an unsaturated repository where fluids that condense on the waste form surface react with the glass in the sample, creating a concentrated leachate (brine). The composition of such a fluid, with a high solid/liquid ratio, will be controlled by the composition of the solid waste form.

Coupons for the VHTs and MCC-1 tests were "dry" cut from the ceramic waste form samples with a diamond saw. The sample surfaces were polished to 240 grit for the MCC-1 tests and 600 grit for the VHTs (without the use of any lubricant), sonicated for five minutes in isopropyl alcohol to remove any polishing residue, and dried in a 90°C oven for ten minutes. The VHT samples were suspended by a Teflon thread in a stainless steel canister, and placed in high temperature ovens (at 120, 150, and 200°C). Enough DIW was added to the test vessel to provide 100% relative humidity inside the test chamber, but the amount was limited to prevent dripping of accumulated fluid from the samples. This keeps all the dissolution products in contact with the sample surface in a thin film of water, where they precipitate to form alteration phases.

RESULTS AND DISCUSSION

<u>Product Consistency Test-B (PCT-B) with Salt-Loaded Zeolite 4A</u>

In evaluating results from the different durability tests, the relatively small size and interconnected channels of the zeolite framework may be extremely important. Therefore, a detailed understanding of the ion exchange behavior of the salt-loaded zeolite is required. The effects of several leachant compositions on the PCT-B with 4 µm diameter salt-loaded zeolite 4A powders have been examined. The leachants included deionized water (DIW), equilibrated Yucca Mountain groundwater (EJ-13) [12], Ca-Na-K-Cl brine, Mg-Na-K-Cl brine [ASTM C1220-92], and phosphate buffers at different pH values. The incorporation of these fluids, that represent a broad range of leachants to which the waste form may be exposed, into durability test procedures should provide information about the bounding parameters and mechanistic release scenarios appropriate for the corrosion testing of the ceramic waste form and about the behavior of the waste form under accelerated testing conditions.

The PCT-Bs were initiated with the salt-loaded zeolite in DIW to investigate the time dependent release, size-fraction release, and solubility of specific components. Tests were performed on samples that were held at 90°C; for 2 minutes, 1 hour, 3, 7, 28, and 91 days; with a geometric surface area to leachant volume ratio (S/V) of 2000 m^{-1} (Figure 1). Each test used individual material and upon termination the solids and leachates were separated and analyzed. Results from S/V = 2000 m^{-1} and 20,000 m^{-1} PCT-Bs with the salt-loaded zeolite 4A in EJ-13, and ASTM Ca-Na-K-Cl and Mg-Na-K-Cl brines at 90°C for 7, 28, and 91 days were also obtained. Additional S/V = 20,000 m^{-1} tests in DIW were also performed at 90°C for 3, 7, 28, and 91 days.

To better represent the possible ion exchange or preferential ion release processes that may be occurring (where the external geometric surface area is less important and the correlation between the amount released and the total amount of an element in the sample may be more useful) the PCT-B leachate results are reported in fractional release percent [$FR_i\% = 100 \cdot$ (mass of ith element released)/(total mass of the ith element in the sample)]. However, geometric effects are important and must be considered for any meaningful evaluation. The $FR_i\%$ results of these PCT-B experiments indicate that:

1. 20% of the Li and Cl was released within an hour of contact with DIW.
2. 60+% of the Li was released within 3 days with all leachants (Figures 1, 2, and 3).
3. Na and K release occurred at a slower rate, with 20+% released after 3 days (all leachants except brines).
4. 10% of the zeolite Al-Si framework was also released after 3 days in DIW at $2000\ m^{-1}$.
5. An inverse correlation may exist between $FR_i\%$ and ionic radius and/or charge in DIW (Table II).
6. The release rates decreased substantially after 7 days in all leachants.
7. The preferential cation release order at 90°C in all the leachants except the brines is Li > Na $\geq$ K > Cs > Al > Si > RE > Sr > Ba (RE = rare earth).
8. Preferential ion exchange occurred between Ca and Mg species from solution and the Na and K in the zeolite.
9. Substantial increases in RE, Sr, and Ba release were observed in the brines at 90°C, relative to the tests in DIW and EJ-13 leachant (Figures 1 and 2).
10. The RE elements released in the Ca brine leachate formed a RE-Ca-Si-Cl precipitate.

The salt-loaded zeolite powders used in these PCT-Bs have extremely high surface areas that are not representative of the actual material to be disposed and the results are not directly comparable with standardized PCTs conducted with materials having grain sizes between 75 and 150 μm [6]. However, the results from these PCT-Bs do provide information about the relative release rates and mechanisms of the different salt-loaded zeolite components. Table II lists the specific $FR_i\%$ data for leachate ions as a function of ionic charge and radius (7-day PCT-Bs in DIW at 90°C with S/V = 2000 and 20,000 m^{-1}). Elemental release results from these tests are consistent with previously published S/V = 7500 m^{-1} results (Table II) [6], demonstrating a possible correlation between release rate and ionic size and charge of the salt-loaded zeolite constituents, along with possible solubility effects associated with the $FR_i\%$ decrease of some of the elements with increasing S/V.

In most of the tests, the pH of the system is allowed to change in accordance with the reactions taking place. However, to investigate the effect of pH on the durability tests, buffered leachates with defined pH were used. The S/V = 2000 m^{-1} $FR_i\%$ results of PCT-Bs with salt-loaded zeolite 4A in phosphate pH buffered leachants at 90°C for 3 days (Figure 3) indicate that the Si release rate was pH dependent. The $FR_i\%$ of Si increased from 3 to 40% between pH = 9.5 and 3, respectively. Correspondingly, the Ba, Sr, and RE $FR_i\%$ also increased significantly from pH = 9.5 to pH = 3. The Ba, Sr, and RE ions released from the sample passed through a 0.45 μm filters but not through a 0.02 μm filters, suggesting that colloids were forming. Phosphate was found in the zeolite after the tests (as much as 5 wt% at pH = 3), suggesting that phosphate is not an appropriate buffer for this system. The release rates of the singly charged ions (Li, Na, K, Cl, and Cs) were not affected by pH. This result may be due to preferential ion release processes dominating the loss mechanism of these ions.

The above findings indicate that leachant composition, including pH, may significantly affect the evaluation of the results from the durability tests performed on the ceramic waste form. The disproportionately high release of the alkali metals and chloride from the salt-loaded zeolites in the PCT-Bs indicates that preferential ion release (as opposed to bulk dissolution of the entire material) is the primary loss mechanism. Furthermore, the increased release of Ba, Sr, and RE ions in the presence of high-ionic-strength fluids may be important should brine-like fluids contact the waste form in the repository.

<u>MCC-1 Tests with the Ceramic Waste Form</u>

The 3-, 7-, and 28-day MCC-1 tests were performed with the ceramic waste form in DIW, EJ-13, and ASTM Mg-Na-K-Cl brine leachants. Scanning electron microscopy images of the monolith surfaces from the 28-day MCC-1 tests indicate that corrosion pitting increased significantly in the zeolite regions for tests in DIW, compared to tests in EJ-13 or Mg-Na-K-Cl brine leachants. The leachate analyses indicated that the Ba, Sr, and RE ions had very low release rates (at or below detection limits) in DIW and EJ-13 leachants. However, the release of these ions increased substantially in the Mg-Na-K-Cl brine. As evident from the results of the PCT-Bs with the zeolite powders, this increase has been associated with enhanced ion exchange between the zeolite and the high-ionic-strength leachant. These trends suggest that the zeolite in the ceramic waste form should preferentially retain nearly all the contained divalent and trivalent fission products, except in the presence of a brine, where retention of these ions may be more dependent on the leach resistance of the glass matrix.

One of the WAPS criteria for borosilicate glass is a comparison between the release of Li, Na, and boron from the test glass and from the environmental assessment (EA) glass. While the WAPS protocol established the PCT procedure for the comparison, the use of the S/V = 10 m^{-1} MCC-1 test was investigated in these tests as a possible alternative procedure. The 28-day MCC-1 normalized elemental mass loss (NL$_i$, eq. 1) results obtained for the ceramic waste form compare favorably with similar results for the EA glass [9] and a Savannah River Laboratory (SRL) composite glass [10] (Table III and IV). The SRL composite glass was chosen for comparison here because it is similar in composition to the EA glass, it contains Cs (an element not found in the EA glass), and results for brine leachants are reported. Since boron is a highly soluble component in the presence of alteration products, it is often used as an indicator of the glass matrix reaction with the leachate. For the ceramic waste form, boron is present only in the glass and thus is ideally suited to monitor the dissolution of the glass matrix. Similarly, Li is contained only in the zeolite and should provide a useful maximum measure of the alkali metal release. Cesium is contained only in the salt-loaded zeolite and the PCT-Bs with the salt-loaded zeolites indicated that Cs has the greatest release in DIW of all the simulated fission products and should provide a reasonable measure for monitoring maximum radionuclide release via ion exchange with the zeolite crystals. The results listed in Table III indicate that the NL$_i$ of Cs, Si, Li, Na, and boron from the ceramic waste form are lower than either the EA or the SRL composite glass when reacted in DIW. Similarly, the Cs, Si, and boron release for the ceramic waste form are equivalent to the SRL composite glass in brine solutions (Table IV).

$$NL_i = \frac{C_i \, V_s}{f_i \, A} \left(g/m^2 \right) \tag{1}$$

where C_i is the concentration of ith element released to solution, V_s is the volume of leachate, f_i is the fraction of the ith element in the sample, and A is the surface area of the sample. The NL$_i$ is usually associated with the dissolution of the material (primarily a surface effect) and accounts for the external geometric surface area involved.

<u>Vapor Hydration Tests with the Ceramic Waste Form</u>

After termination of several of the tests at set time intervals, examination of the coupons indicated that the VHT samples were significantly affected by differences in structural integrity. Analysis of scanning electron microscopy images of the cross sections of the coupons indicates that severe fracturing occurred (Figure 4a), with more and longer fractures developing in: (1) samples where the zeolite had converted to microsommite [(Ca,Na,K)$_8$(Al,Si)$_{12}$O$_{24}$Cl$_{2.5}$] during the hot isostatic pressing, (2) samples tested for longer times, and (3) samples tested at higher temperatures. Sodium chloride crystals (cubic or dendritic growth features in Figure 4b), KCl crystals, Ca-Al-Si spherical phases (small round bright spots in Figure 4b), and Na-Al-Si-RE "stick" phases (bright "stick" features in Figure 4b) were observed on the surface of the samples after the VHTs. Higher temperatures resulted in an increase in the amount of precipitates observed on the surface. Patches of RE phases were present before and after the VHTs, and were associated with high concentrations of chlorides.

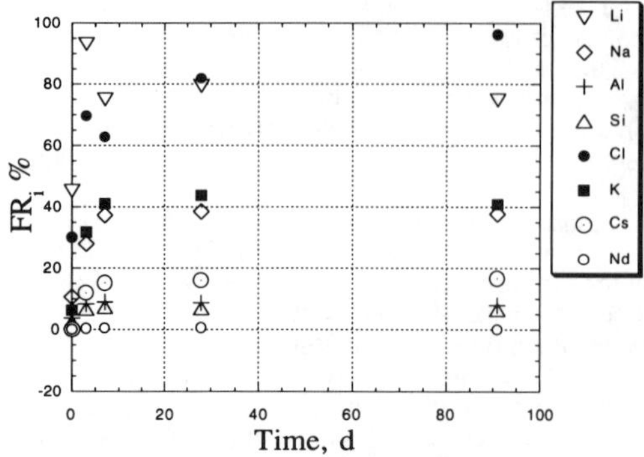

Figure 1. Fractional release of elements from 2000 m⁻¹ PCT-Bs with salt-loaded zeolite 4A in DIW at 90°C (leachant pH = 10 after 91 days).

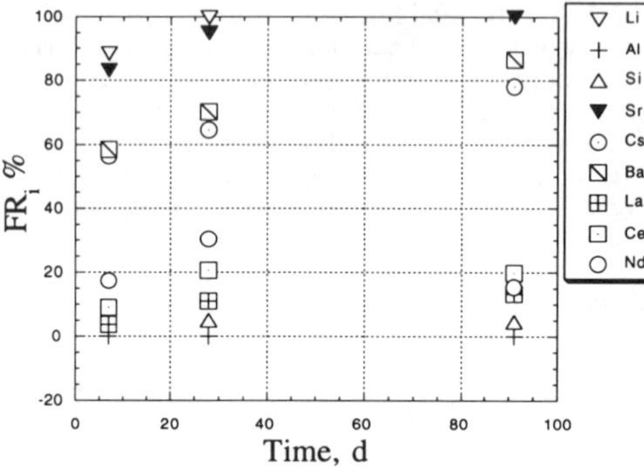

Figure 2. Fractional release of elements from 2000 m⁻¹ PCT-Bs with zeolite 4A in Mg-Na-K-Cl Brine at 90°C (leachant pH = 5.6 after 91 days).

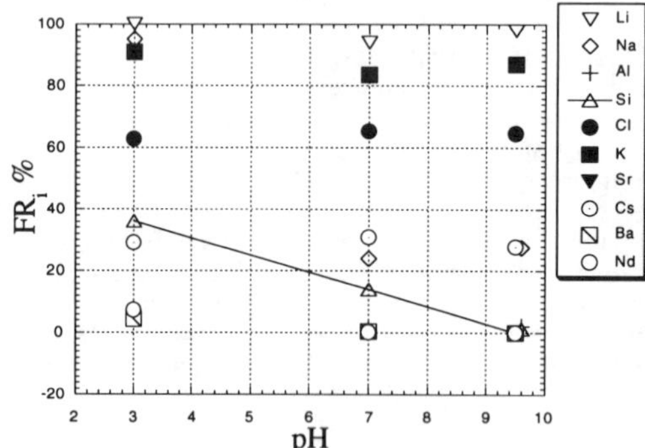

Figure 3. Fractional release of elements from 2000 m⁻¹ PCT-Bs with salt-loaded zeolite 4A in phosphate buffer solutions at 90°C.

Table II. Comparison of $FR_i\%$ in 7 day PCT-B data with ionic charge and radii.

Ions	Ionic Radius (nm)	7 d $FR_i\%$ 2,000 m⁻¹	[6] 7 d $FR_i\%$ 7,500 m⁻¹	7 d $FR_i\%$ 20,000 m⁻¹
Li^+	0.060	75	68	79
Na^+	0.095	39	26	35
K^+	0.133	39	25	34
Cs^+	0.169	15	2.8	10
Sr^{2+}	0.113	0.05		<0.01
Ba^{2+}	0.135	<0.01		<0.01
Nd^{3+}	0.108	0.5		<0.01
Ce^{3+}	0.111	0.4		<0.01
La^{3+}	0.115	0.16		<0.01
Al^{3+}	0.050	8.9	5.3	1.2
Si^{4+}	0.041	7.3	4.7	0.34
Cl^-	0.181	65		84
pH		10.2	10.9	10.9

Table III. Comparison of 28 day NL_i results (g/m²) from 90°C MCC-1 tests in DIW.

Element	Ceramic Waste Form *	Bench-mark EA Glass[9]	SRL Glass Composite[10]
Li	14.3	40.9	na
B	4.5	49.3	22.1
Na	10.9	49.3	na
Si	4.5	34.4	15.4
Cs	15.4	np	38.9

*Leachate pH = 9.

na - test data not available.

np - element is not present in material.

Table IV. Comparison of 28 day NL_i results (g/m²) from 90°C MCC-1 tests in Mg-Na-K-Cl Brines (≈ 25% total dissolved solids).

Element	Ceramic Waste Form *	Bench-mark EA Glass[9]	SRL Glass Composite[10]
B	8.7	na	8.7
Si	7.6	na	5.9
Cs	10.6	na	6.7

*Leachate pH = 5.5.

na - test data not available.

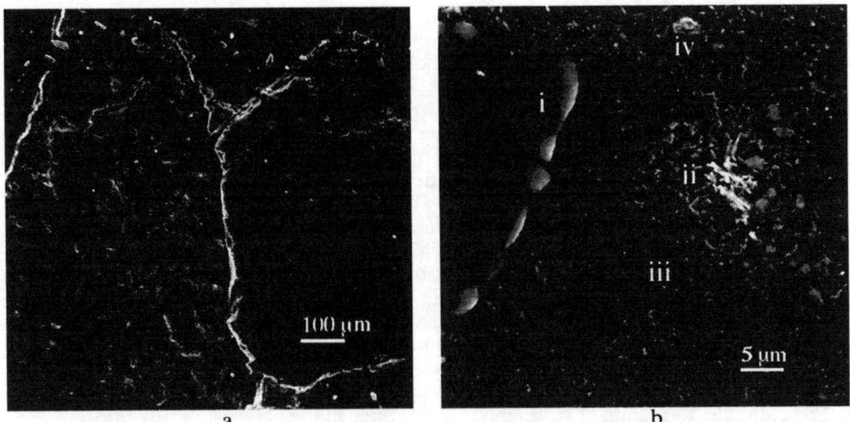

Figure 4. Secondary electron images from a scanning electron microscope. (a) Cross section of a ceramic waste form sample after a 3 day VHT at 200°C, illustrating the significant amount of fracturing that occurred in these samples. A relatively limited amount of chemical alteration occurred at the surface of the sample (upper left corner). (b) Surface of a ceramic waste form sample after a 21-day VHT at 200°C. (i) The large feature on the left is a dendritic growth of NaCl. (ii) The bright, stick-like feature (middle right) contains above average concentrations of RE ions. (iii) The small, gray, square features contain above average concentrations of RE and chloride ions. (iv) A small Ca-Al-Si rich pherical feature (middle top) is also present. Pores due to corrosion are present throughout.

According to preliminary measurements, the rate at which the surface alteration layer forms in these initial VHTs with ceramic waste from monoliths is similar to that for the SRL202 glass [11] reacted under identical conditions. However, quantitative measurement of alteration layer formation was very difficult because of the excessive fracturing of the samples. The VHT provided information about the toughness and structural integrity of the ceramic waste form, but the qualitative nature of layer thickness measurements limits the usefulness of VHT corrosion rates for qualification studies. The VHT can provide information about the ceramic waste form alteration phases that may be useful for the performance assessment modeling efforts required for geologic disposal.

CONCLUSIONS

The utility of the VHT and MCC-1 tests for elucidating the criteria for acceptance of the ceramic waste form into the DOE Civilian Radioactive Waste Management System was investigated. The preliminary results indicate that the MCC-1 test can provide the relevant information required for acceptance of the ceramic waste form with regards to consistency. Additional PCT-B investigations with salt-loaded zeolite 4A powders were performed to determine the release mechanisms and the effect of leachants on the zeolite. The use of DIW as a leachant provides the upper bound for the corrosion of the ceramic waste form where the system pH is neutral or basic. Acidic pH leachants increased the dissolution of the zeolite framework matrix. The use of high-ionic-strength brines increased the release of the alkaline earth and RE elements from the salt-loaded zeolite, and Ca, Mg, and P were preferentially ion exchanged for the zeolite cations Na and K. These results indicate that both DIW, which affects the dissolution of the zeolite framework, and leachants, which specifically affect the ion exchange behavior of the fission products, are required to fully assess the durability of the ceramic waste form. Furthermore, to avoid premature dissolution of alkali metal and chloride ions, the use of water should be limited in any of the sample preparation steps.

As observed in the PCT-Bs with the salt-loaded zeolite, preferential ion release was the dominant loss mechanism. The univalent ions are preferentially released over the divalent and trivalent ions in DIW and EJ-13 leachant. The order of cation release from the salt-loaded zeolites is affected by the leachant solubility of specific ions. Thus, the addition of specific ions to the leachant may have dramatic effects on the ion exchange properties of the zeolite. This was most evident with the salt-loaded zeolite and ceramic waste form samples in the Mg and Ca brines. In general, Sr, Ba, and the RE ions have very low release rates from the zeolites in DIW and low-ionic-strength leachants. However, the high-ionic-strength brines change the equilibrium conditions of the zeolites such that the release of the alkaline earth and the RE ions was dramatically increased (Figure 2), and Mg and Ca were preferentially ion exchanged with the Na and K in the zeolites. Furthermore, Ca and perhaps Mg form precipitates with the RE ions.

The effect of sample glass composition on corrosion rates and alteration phase formation was investigated with VHTs between 120 and 200°C. The VHT provided information about the toughness and structural integrity of the ceramic waste form. However, the qualitative nature of layer thickness measurements limits the usefulness of the VHT corrosion rates for qualification studies. The VHT can provide information about the long-term alteration phases of the ceramic waste form that may be useful for the performance assessment efforts required for geologic disposal. Presently, the formation of stress induced fractures in the hot isostatically pressed samples adversely affects the evaluation of the VHT and may exacerbate the analysis of the MCC-1 tests if fracturing significantly changes the exposed surface areas of these samples during the course of the tests. Fracturing may also affect the results from PCT and other test protocols that are presently being evaluated for use in determining acceptance criteria.

ACKNOWLEDGMENTS

This work was supported by the U.S. Department of Energy, Nuclear Energy Research and Development Program, under Contract No. W-31-109-Eng-38. The insightful and extremely relevant comments from the reviews of this paper by J. K. Bates and D. M. Strachan are greatly appreciated.

REFERENCES

1. C. C. McPheeters, J. P. Ackerman, G. K. Johnson, 1st Seminar on Molten Salts in Nuclear Technologies, Dimitrovgrad, Russia, June 19-23, 1995.
2. A. Dyer, *An Introduction to Zeolite Molecular Sieves* (John Wiley and Sons, NY, 1988).
3. U.S. Department of Energy, Office of Civilian Radiactive Waste Management, *Waste Acceptance System Requirements Document*, Report No. DOE/RW-0351, Rev. 2, 1996.
4. C. M. Jantzen, N. E. Bibler, D. C. Beam, W. G. Ramsey, and B. J. Waters, Westinghouse Savannah River Co. Report No. WSRC-TC-90-539, Rev. 2 (1992).
5. J. K. Bates, M. G. Seitz, and M. J. Steindler, Nucl. Waste Chem. Mgmt. **5**, 63 (1984).
6. M. A. Lewis, D. F. Fischer, and L. J. Smith, J. Am. Ceram. Soc. **76** (11), 2826 (1993).
7. M. A. Lewis, D. F. Fischer, and C. D. Murphy, Ceramic Trans. **45**, 277 (1994).
8. J. W. Sade and G. D. Piepel, Report No. WSRC-TR-90-526, Westinghouse Savannah River Co., Aiken, SC, 1991.
9. P. R. Hrma, G. F. Piepel, M. J. Schweiger, D. E. Smith, D. S. Kim, P. E. Redgate, J. D. Vienna, C. A. LoPresti, D. B. Simpson, D. K. Peeler, and M. H. Langowski, Pacific Northwest Laboratory Report No. 10359, Vol. 2, 1994.
10. M. J. Plodinec, G. G. Wicks, and N. E. Bibler, Savannah River Laboratory Report No. DP-1629 (1982).
11. D. J. Wronkiewicz, L. M. Wang, J. K. Bates, and B. S. Tani, Mat. Res. Soc. Symp. Proc. Vol. **294**, 183 (1993).
12. W. L. Ebert, J. K. Bates, and E. C. Buck, Mat. Res. Soc. Symp. Proc. Vol. **294**, 137 (1993).

THE STRUCTURE AND PROPERTIES OF TWO NEW SILICOTITANATE ZEOLITES

M.L. BALMER*, Y. SU*, I.E. GREY**, A. SANTORO***, R.S.ROTH***, Q HUANG****,
N. HESS* AND B.C. BUNKER*
* Materials and Chemical Sciences Department, Pacific Northwest National Laboratory,
Richland, WA 99352 **CSIRO Division of Minerals, Melbourne, Australia *** Reactor
Division, National Institute of Standards and Technology, Gaithersburg, MD 20899
****University of Maryland, College Park, MD 20742

ABSTRACT

Two new zeolitic crystalline phases with stoichiometry, $CsTiSi_2O_{6.5}$ and $Cs_2TiSi_6O_{15}$,
have been discovered. $CsTiSi_2O_{6.5}$ has a crystal structure isomorphous to the mineral pollucite,
$CsAlSi_2O_6$, with Ti^{+4} replacing Al^{+3}. This replacement requires a mechanism for charge
compensation. A combination of techniques including neutron diffraction, single crystal x-ray
diffraction and x-ray absorption spectroscopy have revealed that eight extra oxygens are present
per unit cell $CsTiSi_2O_{6.5}$ as compared to pollucite. As a result of the extra oxygen, the titanium
coordination geometry is five-fold. Pentacoordinate titanium and tetrahedral silicon form a
network structure with Cs residing in cages formed by the network. The crystal structure of
$Cs_2TiSi_6O_{15}$ is unique, with titanium octahedra and silicon tetrahedra forming an open
framework structure with the Cs residing in large cavities. The largest covalently bonded ring
opening to the Cs cavities in both compounds are smaller than a Cs ion, revealing that the Cs ion
has minimal mobility in the structure. Cesium leach rates for both compounds are lower than or
comparable to borosilicate glass.

INTRODUCTION

Plans are in place to remove, process and permanently dispose of over 100 million gallons
of nuclear waste that is temporarily stored in underground tanks at the Department of Energy's
Hanford site. The baseline process includes separation of the solid and liquid components,
removal of radioactive species such as Cs and Sr from the liquid waste, and disposal of the
resulting high level waste by dissolution in a borosilicate melt to form high level waste (HLW)
glass. A promising ion exchanger for separation of Cs and Sr from tank wastes is the crystalline
silicotitanate (CST) developed at Sandia, Texas A&M and UOP.[1] However, TiO_2 present in
CSTs is only scarcely soluble in borosilicate glass. Analysis of the process flow with the CST
has shown that the total amount of HLW glass produced would increase by 33% as a result of
CST dilution. If these high levels of dilution are required to stabilize these wastes, the resulting
volume of high level borosilicate waste glass can increase disposal costs by billions of dollars. A
significant reduction in waste form volume can be realized if loaded exchangers are thermally
converted to a glass or ceramic waste form rather than diluted in borosilicate glass. Cs-loaded
silicotitanate ion exchangers contain the basic ingredients needed to form a ceramic or glass at high
temperature.

Researchers at PNNL are examining the feasibility of using alternative glass or ceramic
waste forms with compositions similar to the starting loaded ion exchanger for containment of
TiO_2-rich wastes. It was found that several new, unidentified phases precipitate from Cs-loaded

Mat. Res. Soc. Symp. Proc. Vol. 465 © 1997 Materials Research Society

heat treated CSTs.[2] Identification and characterization of all these phases is critical so that thermodynamics and geological evidence can be used to assess long term stability and compatibility with repository conditions. The loaded exchanger represents a complicated chemical mixture containing at least five components, Cs_2O, Na_2O, TiO_2, SiO_2, and a proprietary component. Although the phase selection in the Na_2O-TiO_2-SiO_2 system has been characterized, the Cs_2O-TiO_2-SiO_2 phase diagram and the more complicated four and five component systems have not been investigated. Work to date at PNNL has concentrated on identifying the high temperature phase selection and properties for three key constituents of silicotitanate ion exchange waste; Cs_2O, SiO_2, and TiO_2. As a result of Cs_2O-TiO_2-SiO_2 phase diagram work, two new compounds with zeolitic structures have been identified. The crystal structures, chemical durability and potential of these two new phases for long-term Cs containment are discussed.

EXPERIMENT

Powder samples of $CsTiSi_2O_{6.5}$ were prepared using a sol gel synthesis route. Tetra-isopropyl orthotitanate (TIOT) and tetraethyl orthosilicate (TEOS) were mixed in inert conditions then CsOH, water and ethanol were added dropwise. The hydrolyzed precursor was mixed for a minimum of 15 hours, then dried in air at room temperature. The amorphous, homogeneous precursor was heat treated in air at 800°C for at least one hour to form crystalline $CsTiSi_2O_{6.5}$. Crystalline $Cs_2TiSi_6O_{15}$ powder was made by melting sol gel precursor powders or mixtures of Cs_2CO_3, TiO_2 and SiO_2 at 1200°C, quenching to room temperature, mixing with 40 wt% $CsVO_3$ flux, then heat treating to 790°C for a minimum of 50 hours. Several cycles of grinding and heat treatment were necessary to form a nominally phase pure compound. Residual, unreacted $CsVO_3$ flux was removed by washing three times in warm water at 90°C for 10 minutes followed by centrifuging to separate solids from the liquid. Despite long heat treatment times and several heating and grinding cycles, the final $Cs_2TiSi_6O_{15}$ powder contained approximately 10 wt% of unidentified crystalline impurity compounds.

Single crystals of $CsTiSi_2O_{6.5}$ and $Cs_2TiSi_6O_{15}$ were grown from a powdered sample of $CsTiSi_2O_{6.5}$ and $CsVO_3$ flux. The mixture of $CsTiSi_2O_{6.5}$ and $CsVO_3$ was placed in a Pt tube, sealed at both ends, then heat treated in a vertical tube furnace at the rate of 50°C/hr to 1100°C, held for one hour, then cooled to 650°C at a rate of 1°C/hr. After dissolving the $CsVO_3$ flux in warm water, a mixture of crystals including $CsTiSi_2O_{6.5}$, $Cs_2TiSi_6O_{15}$, rutile and cristobalite were retrieved.

The Cs leach rate was measured using the MCC–1 and the product consistency test method B (PCT) standard leach tests. For MCC-1, heat treated pellets were suspended on a teflon string in de–ionized water contained in a teflon container for the leach test. After the desired time at a constant temperature, a sample of the solution with a known volume was analyzed for the Cs content using atomic absorption spectrometry (AAS). For the PCT test, powder samples were placed in a teflon container in de-ionized water at 90°C for the desired period of time. The leachate solution was analyzed for Cs concentration using AAS.

X-ray single crystal and powder diffraction experiments were performed to elucidate the crystal structure of $CsTiSi_2O_{6.5}$ and $Cs_2TiSi_6O_{15}$. Additionally, neutron diffraction experiments were performed on $CsTiSi_2O_{6.5}$ in order to determine the location of oxygen positions in excess of those predicted by the pollucite crystal structure. Details of the data collection and refinement procedures for neutron powder, single crystal, and powder diffraction are published elsewhere.[3,4]

RESULTS AND DISCUSSION

Structure of $CsTiSi_2O_{6.5}$

Powder x-ray diffraction studies show that $CsTiSi_2O_{6.5}$ is a derivative of the cesium aluminosilicate pollucite, $CsAlSi_2O_6$, with titanium replacing aluminum. The compound crystallizes with the symmetry of space group Ia3d and a lattice parameter a=13.8425. [3,4] Titanium and silicon are disordered on the 48g sites of the cubic space group similar to Si and Al in pollucite. The substitution of Ti^{+4} for Al^{+3} in pollucite requires a mechanism for charge compensation, such as the presence of tri-valent titanium cations, or the incorporation of oxygen in excess of six atoms per formula unit. The presence of Ti^{+3} (resulting in stoichiometry $CsTiSi_2O_6$) is unlikely since the compound is synthesized in air at 800°C, and Ti^{+3} usually only occurs under highly reducing conditions. It is also worth noting that the samples have a white color, typical of Ti^{4+} compounds. In addition, x-ray photoelectron spectroscopy and Ti x-ray absorption near edge structure (XANES) studies on $CsTiSi_2O_{6.5}$ show that Ti is primarily in the Ti^{+4} valence state.[4,5] Evidence of the incorporation of excess oxygen as a charge compensation mechanism could be seen by Ti XANES and Raman spectroscopy which show that Ti is in five fold coordination environment.[5]

Because x-ray diffraction is relatively insensitive to small changes in oxygen composition, neutron diffraction measurements were made using the high resolution powder diffractometer at the National Institute of Standards and Technology reactor to confirm the presence of excess oxygen and to determine their positions. These experiments revealed the presence of eight additional oxygens per unit cell of silicotitanate pollucite (Z=16).[4] The "extra" oxygens are distributed randomly over two partially occupied general sites designated O(2) and O(3). On average the O(2) site is occupied by 0.06 atoms and the O(3) site is occupied by 0.023 atoms. As a consequence of the presence of "extra" oxygen, the average coordination of the central Cs cation becomes thirteen-fold, rather than twelve-fold as in $CsAlSi_2O_6$ pollucite. In addition , when one of the 'extra' sites is occupied by an oxygen, the coordination of the neighboring tetrahedral cations becomes five-fold. In the case where an excess oxygen coordinates a tetrahedral site, it is assumed that the neighboring tetrahedra are occupied by titanium, rather than silicon, based on ionic radii considerations and on experimental XANES results that show Ti(5). Although no ordering on Ti on the 48g site could be detected by single crystal x-ray or neutron diffraction, it is suspected based on stoichiometry (Ti:excess oxygen is 2:1) and on the fact that no Ti(4) could be conclusively identified by XANES or Raman spectroscopy, that randomly distributed pairs of Ti ions share the excess oxygens. The crystal structure of $CsTiSi_2O_{6.5}$ viewed along [111] is shown in Figure 1a, and Figure 1b shows one of the excess O(2) oxygen positions and bond distances with respect to neighboring Ti/Si and Cs. The network structure consists of interconnected rings containing four and six polyhedra. The Cs resides in cages that are bound by two six-rings and that form channels along the non-intersecting three-fold symmetry axis and are joined together in the <110> directions.

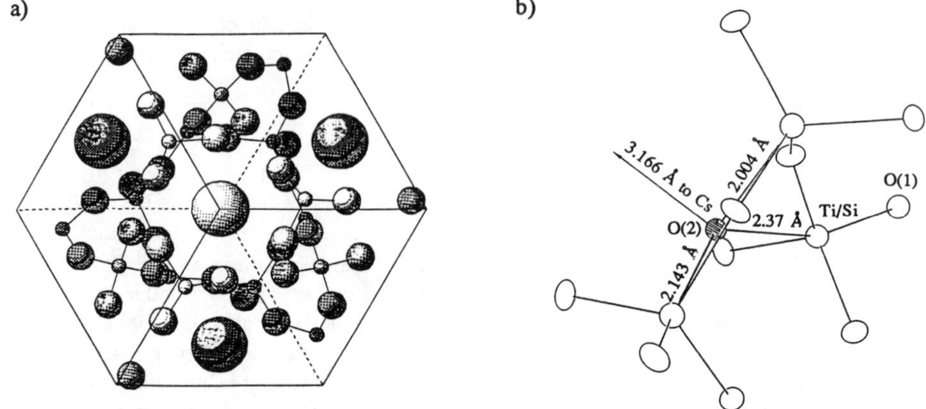

Figure 1. a.) View down the <111> of the Cs cages in $CsTiSi_2O_{6.5}$. Cs cages are formed by rings of four and six Si/Ti polyhedra and Cs ions are coordinated by 13 oxygen ions. b.) Local view of the coordination environment around a tetrahedral site with an O(2) site filled.

Structure of $Cs_2TiSi_6O_{15}$

The compound $Cs_2TiSi_6O_{15}$ has a new crystal structure, unrelated to any known compounds. $Cs_2TiSi_6O_{15}$ crystallizes with monoclinic symmetry and space group C2/c, with a= 13.386 Å, b= 7.423 Å, c = 15.134 Å, β = 107.71° and Z=4.[6] The structure consists of corner shared titanium octahedra and silica tetrahedra that form an open framework. The titanium octahedra are isolated by silica tetrahedra resulting in a local bonding configuration that consists of only of Ti-O-Si and Si-O-Si type bonds.   Similar to pollucite, the Cs ions reside in large cages formed by the tetrahedral and octahedral linkages. The topology of the cage around the Cs can be described by interconnecting rings of polyhedra consisting of one three ring, three five-rings, two six-rings and two eight-rings.[6] A view down the b axis that clearly shows the five and six rings is shown in Figure 2a. It can also be seen in Figure 2a that titanium octahedra are isolated by Si_2O_7 groups and that the Cs ions form rows along the [101].   Figure 2b shows a view of the Cs ions looking down the [101]. The cavities that contain the Cs ions have large apertures normal to the [101] formed by two different type of ellipsoidal 8-rings. In the first type of 8-ring there are two titanium octahedra and six silica tetrahedra and in the second type of 8-ring there are six silica tetrahedra. The average coordination of the central Cs ion is ten.

Chemical Durability

Among the important criteria that a nuclear waste form must satisfy to be suitable for long-term storage are low solubility and low Cs leach rate. The crystal structures of both silicotitanate pollucite and $Cs_2TiSi_6O_{15}$ indicate that the Cs ion will have low mobility in the structure which in turn should result in a low Cs leach rate. The Cs ion in silicotitanate pollucite is in cages where the maximum opening to the cage is defined by a 6-ring. Assuming 2.7Å as the diameter of the oxygens lining the inner peripheries of the 6-ring, the maximum opening size of this ring (in an ideal planar configuration) is 2.7Å, which is considerably smaller than a Cs ion (diameter of at least 3.5 Å). Early investigations of the aluminosilicate pollucite isomorph of

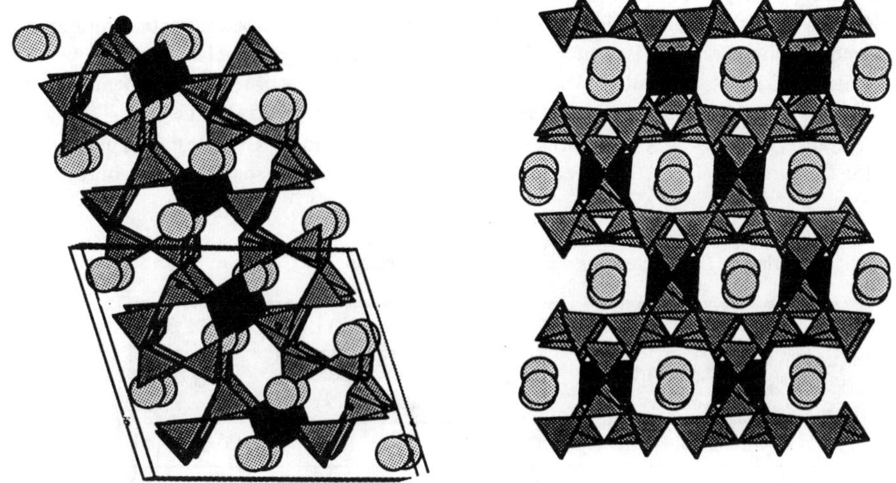

Figure 2. a.) View of the $Cs_2TiSi_6O_{15}$ structure down the b axis showing the 5- and 6-rings of silicon tetrahedra and titanium octahedra. b.) View of $Cs_2TiSi_6O_{15}$ down the <101> showing ellipsoidal rings of eight polyhedra.

$CsTiSi_2O_{6.5}$ show that it has high Cs leach resistance, exhibiting aqueous leach rates from 5.7 x 10^{-4} g m^{-2} d^{-1} to 3 g m^{-2} d^{-1} .[7-9] While incorporation of excess oxygen into $CsTiSi_2O_{6.5}$ as compared to $CsAlSi_2O_6$ influences local bonding configurations, the long range crystal structure and predicted Cs mobility remain largely unchanged.

The largest apertures in the Cs cavity in $Cs_2TiSi_6O_{15}$ are two eight rings formed by eight silica tetrahedra (8T) or by two Ti-octahedra and six tetrahedra (2O + 6T). Based on an oxygen diameter of 2.7Å the opening size of the elliptical 8T ring is 1.8 x 4.2 Å and that of the elliptical 2O + 6T ring is 2.5x4.8 Å.[6] Therefore, the Cs ions can not move freely along the channels, and the compound is expected to have a relatively low Cs leach rate.

Cs leach rate experiments on silicotitanate pollucite and $Cs_2TiSi_6O_{15}$ support crystal structure predictions of low Cs leachability. The equivalent 14 day leach rates normalized to the surface area are 1.8 g m^{-2} d^{-1} for $CsTiSi_2O_{6.5}$ (MCC-1) and 1.47x10^{-5} g m^{-2} d^{-1} for $Cs_2TiSi_6O_{15}$ (PCT) Table 1. The leach rates for silicotitanate pollucite are comparable to hot pressed $CsAlSi_2O_6$ (3 g m^{-2} d^{-1} at 100°C) where a Soxhlet type apparatus and geometric surface area were used to measure leach rate.[9] Values of the normalized leach rate of $CsTiSi_2O_{6.5}$ may be unrealistically high because the geometric surface area used in the MCC-1 static leach test did not reflect the true surface area of the sample. The sample pellet of silicotitanate pollucite was only 60% of theoretical density so that a large amount of surface area in the form of open porosity was not accounted for. The surface area of the powder $Cs_2TiSi_6O_{15}$ used for the PCT test was measured using a He expansion ratio, a much more accurate measurement. The fractional Cs release, $F_R\%$, which reflects the portion of the total Cs released from the sample without normalizing to the surface area is similar for both materials (Table I). It is worth noting that the MCC-1 leach test typically yields conservative durabilities because solution concentrations of elements are comparatively low and leaching continues during the entire leach period. On the other hand, PCT tests typically have high solution concentrations of leached elements which can act to reduce dissolution rates and yield higher durabilities. In the case of $Cs_2TiSi_6O_{15}$ the Cs concentration in the PCT solution was extremely low. PCT and MCC-1 tests are planned on $CsTiSi_2O_{6.5}$ and $Cs_2TiSi_6O_{15}$ respectively so that a direct comparison of the durabilities can be made.

Table I: Cumulative Cs leach rates and fractional Cs release for $CsTiSi_2O_{6.5}$ and $Cs_2TiSi_6O_{15}$.

Time (days)	$CsTiSi_2O_{6.5}$ MCC-1		$Cs_2TiSi_6O_{15}$ PCT	
	Leach Rate (g m^{-2} d^{-1})	Cs $F_R\%$	Leach Rate (g m^{-2} d^{-1})	Cs $F_R\%$
1	4.0	0.41	8.4×10^{-4}	0.91
2	2.73	0.69	2.3×10^{-4}	1.17
3	2.3	0.93	1.1×10^{-4}	1.29
4	-	-	6.8×10^{-5}	1.36
6	1.83	1.5	-	-
7	-	-	3.84×10^{-5}	1.49
9	1.69	2.0	-	-
10	1.66	2.17	-	-
13	1.49	2.57	-	-
14	-	-	1.47×10^{-5}	1.6

The utility of $CsTiSi_2O_{6.5}$ and $Cs_2TiSi_6O_{15}$ as waste forms for Cs containment can be estimated by comparing the Cs leach rates to borosilicate glass; the baseline wasteform for high activity waste. The Cs leach rate using a static leach test similar to MCC-1, of a borosilicate waste glass simulant, R7T7, is 0.16 g m^{-2} d^{-1} after 14 days.[10] The 14 day leach rate of $CsTiSi_2O_{6.5}$ (1.8 g m^{-2} d^{-1}) is higher than the borosilicate simulant, whereas that of $Cs_2TiSi_6O_{15}$ (5×10^{-4} g m^{-2} d^{-1}) is four orders of magnitude lower. These results suggest that $Cs_2TiSi_6O_{15}$ would serve as an excellent wasteform. Although the Cs leach rate of $CsTiSi_2O_{6.5}$ is somewhat higher than borosilicate glass it is within the necessary durability range, and further leach tests on this compound using an exact surface area determination are warranted. It is worth noting that the Cs content and total alkali content of R7T7 (1.42 wt% and 13.26 wt%) are much lower than the Cs content of $CsTiSi_2O_{6.5}$ (42 wt% or 14 mol% Cs_2O) and $Cs_2TiSi_6O_{15}$ (39 wt% or 12.5

mol%) [10] At very high alkali contents, similar to the silicotitanate zeolites, the durability of most silicate glasses is greatly diminished. For example, the leach rate of a silicate glass that contains 15 mol% potassium is 8 g m^{-2} d^{-1} at 85°C. [11] From these results, a silicate glass with 14 mol% Cs_2O (28 mol% Cs) is estimated to exhibit leach rates as high as 15 g m^{-2} d^{-1}.

CONCLUSIONS

Both $CsTiSi_2O_{6.5}$ and $Cs_2TiSi_6O_{15}$ have unique crystal structures where the Cs resides in cages formed by rings of corner sharing silica and titania polyhedra. The largest ring opening size in both structures is smaller than the effective ionic diameter of a Cs ion, indicating that Cs has limited mobility in the structure. The rate of Cs release from $CsTiSi_2O_{6.5}$ is higher than $Cs_2TiSi_6O_{15}$, however, the preliminary leach rate results indicate that both materials may act as suitable waste forms for Cs containment.

ACKNOWLEDGEMENT

The authors would like to gratefully acknowledge Mr. Dave McCready for performing x-ray powder diffraction and Dr. Terrell Vanderah for assistance in fabricating single crystal specimens. Partial support for this work was provided by the Strategic Environmental Research and Development Program (SERDP).

REFERENCES

1. R.G. Dosch, N.E. Brown, D.E. Trudell, R.G. Anthony, D. Gu and C. Thibaud-Erkey, Waste Management 94, February 28, Tucson, AZ, pp. 709-719.
2. Y. Su, M.L. Balmer, and B.C. Bunker, Submitted Materials Research Society Proceedings, Boston, MA (1996).
3. D.E. McCready, M.L. Balmer, and K.D. Keefer, Powder Diffraction, in press.
4. M.L. Balmer, Q. Huang, W. Wong-Ng, R.S. Roth, and A. Santoro J. Sol. State Chem. accepted.
5. N.J. Hess and M.L. Balmer, J. Sol. State Chem., submitted.
6. I.E. Grey, R.S. Roth, and M.L. Balmer, J. Sol. State Chem. submitted.
7. S. Komareni, G.J. McCarthy, and S.A. Gallagher, Inorg. Nucl. Chem. Letters, Vol. 14, pp. 173-177, 1978.
8. E.R. Vance, B.E. Scheetz, M.W. Barnes, and B.J. Bodnar in The Scientific Basis for Nuclear Waste Management, edited by Stephen V. Topp (Elsivier Science Publishing Co. Inc., 1982).
9. K. Yanagisawa, M Nishoka, and N. Yamasaki, J. Nuc. Sci. Tech., 24[1], pp. 51-60, 1987.
10. J.L. Nogues, E.Y. Vernaz, and N. Jacquet-Francillon in in The Scientific Basis for Nuclear Waste Management, Vol. 44 edited by Carol M. Jantzen, John A. Stone, and Rodney C. Ewing, (Materials Research Society, 1984).
11. M.A. Rana, and R.W. Douglas, Phys. And Chem. Of Glasses, Vol. 2, No. 6, pp. 179-195.

EVALUATION OF CESIUM SILICOTITANATES
AS AN ALTERNATIVE WASTE FORM

YALI SU, M. LOU BALMER, and BRUCE C. BUNKER
Pacific Northwest National Laboratory, Richland, WA 99352

ABSTRACT

Silicotitanate ion exchangers are potential materials for the removal of radioactive Cs and Sr from tank wastes. In this paper the viability of direct thermal conversion of Cs-loaded silicotitanates to an acceptable high level waste form has been examined. Results show that in aqueous solutions, the Cs leach rates of crystalline silicotitanates (heat treated at 800°C) are 0.04, 0.18, 0.4 g/m^2day for Cs loadings of 1, 5, and 20 wt%, respectively. Heating the Cs-loaded (up to 20 wt %) silicotitanates at or above 900 °C for 1 hour further reduces the Cs leach rates to approximately zero (beyond the 1ppm detection limits). Moreover, Cs volatilization was found to be ≤ 0.8 wt% at temperatures as high as 1000 °C. These results suggest that thermally converted silicotitanate ion exchangers exhibit excellent chemical durability (comparable to or better than borosilicate glass) and thus, have great potential as an alternative waste form.

INTRODUCTION

Over the past 50 years, nuclear defense activities have produced over 100 million gallons of nuclear wastes that now require safe and permanent disposal[1]. The current proposed disposal procedure is to remove Cs and Sr from the concentrated salt solutions with selective ion exchangers, then treat the solution as low-level waste (LLW). After separation from the salt solutions, the concentrated Cs and Sr are planned to be vitrified, then interim-stored or placed into long-term, geological storage as high level waste (HLW). To implement these procedures, one of the most important steps is to identify the ion exchange material. An acceptable ion exchanger not only must have high selectivity, and chemical, thermal and radiation stabilities, but also must be acceptable to the actual exchange processes, and be able to be converted into a stable waste form for disposal[2]. For the past few years, a number of organic and inorganic ion exchange materials, including resorcinal-formaldehyde resin, aluminosilicate zeolites, zirconium phosphates, and crystalline silicotitanates (CSTs), have been developed and evaluated for the removal of Cs and Sr from the nuclear wastes [2-5]. Recently, studies show that CSTs fulfill many of these requirements[2] and thus, have the potential as ion exchangers for the removal of Cs from nuclear wastes.

The feasibility of converting CSTs into a HLW form is being evaluated. Currently, two options are being considered for processing the Cs-loaded exchangers: 1) elution and concentration, 2) dissolution in borosilicate waste glass melts. Since Cs is retained in CSTs even in acidic solution, this precludes the elution option. Thus, processing the Cs-loaded CSTs requires the direct use of the loaded exchanger as a feed to make the final waste form. In general, this is accomplished by "melt-dissolution" where the loaded exchanger is first mixed with glass frit and/or other radwaste oxides, then dissolved into borosilicate melts to form the HLW glass. For CSTs, this "melt-dissolution" process has been challenged because in current glass

Mat. Res. Soc. Symp. Proc. Vol. 465 © 1997 Materials Research Society

formulations the maximum allowed TiO_2 concentration is only 1 wt%. Assuming ideal column operation can be achieved with the CST, dissolution of the CST into borosilicate glass would increase the total volume of HLW glass to 33% more than the baseline[6]. Researchers at PNNL have identified direct thermal conversion as another option for CST conversion to solid waste[3]. Different from the previous two options, this method directly converts the CSTs into solid waste forms by heating the Cs-loaded CSTs at high temperature with a minimum amount of other additives.

In this paper, we examine the feasibility of direct thermal conversion to produce an alternative waste form. A CST formulation, TAM-5 (synthesized by UOP, initially developed at Sandia and Texas A&M University), has been studied in terms of its Cs volatility during heat treatment and chemical durability after heat treatment. The results show that the thermally converted Cs-loaded silicotitanate ion exchangers have good potential to be an alternative waste form.

EXPERIMENTAL METHODS

TAM-5 ion exchangers were used as received. Four different Cs loadings, 1, 5, 10 and 20 wt% were examined. The Cs loadings (1 to 10 wt%) were accomplished by adding TAM-5 powder to a CsCl solution with the exact Cs concentration, then stirring the solution for 24 hours. Solutions were analyzed by atomic absorption spectrometry (AAS) to assure that all Cs was ion exchanged from solution. The 20 wt% Cs loading was prepared by first adding TAM-5 to a CsCl solution in which the Cs concentration was ten times higher than the Cs exchange capacity. The maximum Cs loading was determined to be 20 wt% by measuring the final Cs concentration in solutions. Cs loaded samples were rinsed with de-ionized water, centrifuged and dried in air. Thermogravimetric analysis (TGA) and differential thermal analysis (DTA) were performed at a rate of 10°C/min on the Cs-loaded samples to determine the temperatures at which decomposition, crystallization, and melting occurred. Heat treatments were performed in air at temperatures ranging from 500 to 1000°C at a rate of 5°C/min. The heat treated samples were analyzed for the volume fraction and structure of crystalline phases using x-ray diffraction (XRD).

The Cs leach rate was measured using the MCC-1 standard static leach test.[7] The Cs-loaded powders were heat treated to 500 °C for one hour then pressed into a pellet. The resulting pellet was further heated at the test temperature for one hour. For the leach test, the heat treated pellet was suspended by a Teflon string in the 90 °C de-ionized water. The leachate solution was analyzed for Cs concentration using (AAS). The geometric surface area of the pellets was used for calculation of the leach rates. Cs volatility measurements were performed in two steps. The Cs-loaded pellet was first heat treated in an enclosed Pt crucible in the hot zone of a furnace. The top of the crucible was extended out from the hot zone and was cooled by convection to allow the volatile components to condense on the inside of the crucible. The volatile components were then collected by washing the crucible in a 0.2 M nitric acid solution. The amount of Cs in the volatile components in the nitric acid solution was determined by AAS.

RESULTS AND DISCUSSION

I. Cs Leach Rate

The leach rates of several Cs loaded TAM–5 samples heat treated at several different temperatures were measured at different time. The results are shown in Figure 1. After heating

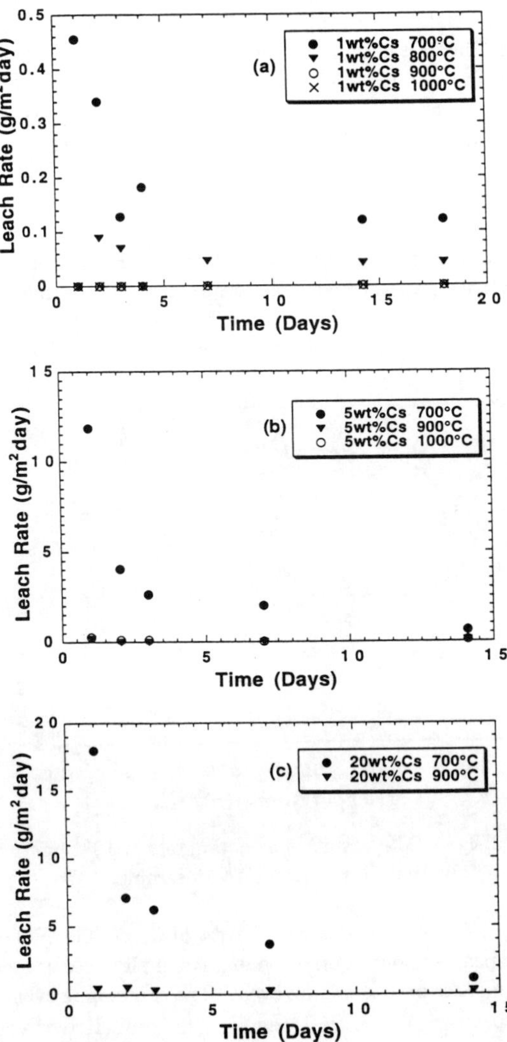

Figure 1. Leach rates measured at 90°C for (a)1wt%, (b) 5 wt%, and (c) 20wt% Cs loaded TAM-5 samples treated at different temperatures.

at 700°C for 1 hour and then leaving the pellet in the deionized water for 24 hours, the 1, 5 and 20wt% Cs loaded samples exhibited high Cs leach rates of 0.45, 12, and 18 g/m²day, respectively. Since the Cs leach rate decays monotonically with time as shown in Fig.1, after 14 days the leach rates reduced from previous values to 0.12, 0.62, and 1.1 g/m²day, respectively. In contrast to these high leach rates, the Cs loaded samples exhibited very low leach rates after heating at higher temperatures. For example, after heating at 800°C for one hour the leach rates of 1, 5, and 20wt% Cs loaded samples became 0.04, 0.18, 0.4 g/m²day, respectively. Thus, increase of the treatment temperature from 700 °C to 800 °C reduces the leach rate by more than an order of magnitude. Increase of temperature to 900°C or higher further reduces the leach rates to approximately zero (beyond the AAS detection limits). These results show that durability highly depends on both the Cs loading and the thermal history. The phase(s) which is (are) responsible for locking in Cs will be discussed in following sections.

II. TGA/DTA results

In addition to the Cs leach rates, melting, crystallization, and weight loss were studied using TGA/DTA to evaluate the phases responsible for the low Cs leach rates. Results showed that samples with different Cs loadings (1 to 10wt%) exhibited similar TGA/DTA patterns, suggesting that different Cs loading did not alter the phase distribution. Figure 2 shows typical

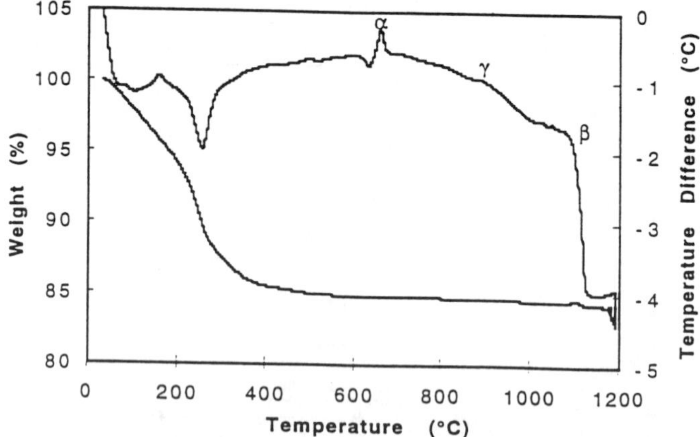

Figure 2. TGA/DTA results at a heating rate of 10°C/min
for 1wt% Cs loaded TAM-5 sample.

TGA/DTA curves of a TAM-5 sample loaded with 1 wt% of Cs. In the DTA curve, an exothermic or endothermic peak without a corresponding weight loss corresponds to the change or formation of a crystalline phase or to a melting event. The exothermic events at 650°C and 850°C (labeled α and γ respectively on Figure 2) represent the formation of a crystalline phase while the endothermic event at 1100 °C labeled β is where melting occurred. Around 250°C, the TGA curve exhibits the highest weight loss rate, suggesting decomposition occurred in that temperature region. The formation of a new crystalline phase at 850°C (labeled γ) corresponds to the significant change in Cs leach rates between 700°C and 900°C. The 20wt% Cs loaded sample exhibits a somewhat different TGA/DTA pattern. In addition to the shifts of endothermic and exothermic peaks in the low temperature region, melting occurs at 1000°C, suggesting that the 20wt% Cs loaded sample has a different phase composition, as will be further discussed in the next section.

III. Phase Evaluation

Cs-Loaded TAM-5 samples have five components, Cs_2O, Na_2O, SiO_2, TiO_2, and a proprietary component. The phase composition of the heat-treated, Cs-loaded TAM-5 samples was analyzed using XRD. Volume percentage of the total crystalline phases was calculated by comparing the total diffracted intensity with the integrated background. Weight percentage of each crystalline phase was determined by the ratio of the 100% peak height to the total diffracted intensity. The crystalline phase, volume fraction and weight percentage of different phases of the heat treated Cs-TAM-5 samples are summarized in Table I.

The phases labeled A–F are new, unknown single phases or mixtures of phases that could not be matched with any existing compounds in the Powder Diffraction File (PDF) database. The phases labeled X and Y are known sodium and sodium titanium-containing phases that also contain a proprietary component. Although phase identification is incomplete, relationships between the phase evolution with temperature and the corresponding leach rates can be identified. As was shown in Section I, the Cs leach rate decreases by more than an order of magnitude between 700 and 900 °C. This temperature range corresponds to the disappearance of phase B and the formation of Phase C, (proprietary phase X remains approximately constant). From this result, it appears that phase C is responsible for binding the Cs and reducing the leach rate. For 20 wt% Cs loadings, additional phase E which does not appear to affect the leach rate, is observed from 700 –900°C. The composition-temperature-phase relationships for the Cs_2O-TiO_2-SiO_2 are currently under investigation and are expected to provide some insights into the phase selection and high durability of heat treated Cs-TAM-5.

IV. Cs Volatility

The Cs weight loss (volatility) of the Cs-TAM-5 was examined at different processing temperatures. Since volatility generally increases with loading, here we only discuss the volatility at the maximum Cs loading (20 wt%) condition. The Cs weight loss as a function of temperature is shown in Figure 3. The results suggest that the Cs losses are extremely low, only 0.6 to 0.8 wt% of the total Cs loading at the expected processing temperatures (800 to 1000 °C). This is comparable to Cs volatility from borosilicate glass for Cs loadings of 1.9 wt% at 900°C, which is well below the melting temperature of borosilicate glass[8]. Cs losses of borosilicate glass can be as high as 70 wt%[9] at the typical processing temperature (1150°C).

V. Comparison of Leach Rates of Cs-TAM-5 with Borosilicate Glass

In order to show that Cs-TAM-5 has the potential to be an alternative waste form, it is important to compare the Cs leach rate of Cs-TAM-5 with those of the current baseline waste

Table I   Phase Composition of Cs-exchanged TAM-5

Cs loading (wt%)	Heat Temperature (°C, 1 hour)	Crystal Vol. Fraction(%)	Weight percentage of Crystalline Phase
1	500	100	Phase A
1	700	100	70% X, 30% Phase B
1	800	100	70% X, 15% Phase B+C, Minor Y
1	900	80	75% X, 10% Phase C, 15% Y
1	1000	70	60% X, 35% Phase C+D, Minor Y
5	500	100	Phase A
5	700	100	70% X, 30% Phase B
5	800	100	70% X, 15% Phase B+C, Minor Y
5	900	80	75% X, 10% Phase C, 15% Y
5	1000	60	70% X, 15% Phase C+D, Minor Y, 5% Rutile
10	500	100	Phase A
10	700	100	70% X, 30% Phase B
10	800	100	70% X, 30% Phase B+C
10	900	55	75% X, 10% Phase C, 15% Y
10	1000	55	55% X, 10% Phase C+D, 35% Rutile
20	500	55	30% X, 70% Phase A
20	700	50	25% X, 75% Phase B+E
20	800	60	45% X, 55% Phase B+C+E
20	900	65	30% X, 55% Phase C+E, 15% Y
20	1000	45	55% X, 20% Phase F, 5% Y, 20% Rutile

Note: Phase A, B, C, D, E, and F are unidentified phases. X and Y are proprietary phases, X and Y phases contain no Cs.

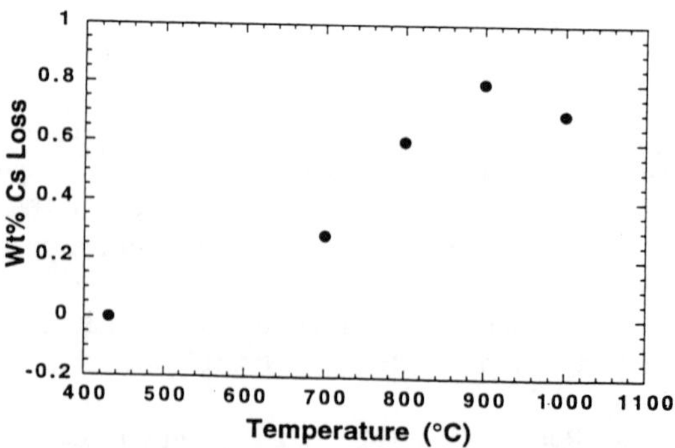

Figure 3. Weight percentage of Cs loss of a Cs-TAM-5 sample (20wt% Cs loading) as a function of treatment temperature.

form-borosilicate glass. However, a direct comparison is difficult since Cs loadings in borosilicate glass are typically much lower than those in Cs-loaded silicotitanates. But if we nonetheless compare the leach rates of borosilicate with those of Cs-TAM-5, we find that the leach rate of borosilicate glass is several times higher than those of the Cs-TAM-5. For example, the Cs leach rates of borosilicate glass simulant (R7T7, Cs_2O content of 1.42 wt% and an overall alkali content of 13.26 wt%) was 0.91 g/m^2day after 1 day of leaching[9]. Under the similar Cs loading, the leach rate of the TAM-5 sample, however, is approximately zero (beyond the 1ppm AAS detection limits) after treated at 900 °C. If borosilicate was loaded with 20 wt% of Cs, the durability of the glasses would be greatly diminished. Furthermore, Rana and Douglas obtained a leach rate of 8 g/m^2day (85 °C) for a silicate glass loaded with 15 mol% potassium[10]. Based on their results, a silicate glass with 20 wt% of Cs (11 mol%) would have a leach rate approximately 6 g/m^2 day, more than two orders of magnitude higher than that of the Cs-TAM-5 treated at 900 °C.

CONCLUSIONS

This study shows that "direct thermal conversion" is a viable processing method to transform Cs loaded silicotitanate ion exchangers into a durable final waste form. Cs-TAM-5 exhibits extremely low Cs leach rates after heat treatment from 800°C to 900°C. The heat treatment is a simple, one step process at temperatures well below the current optimal melting temperature for borosilicate glass. At the optimal processing temperature, the Cs volatility of Cs-TAM-5 is orders of magnitude lower than that of borosilicate glass at its processing temperature (1150°C). The "direct thermal conversion" method concentrates ion exchanger waste resulting in much lower waste volumes than the dissolution-into-borosilicate-melts method. These advantages suggest that direct thermal conversion of Cs-loaded silicotitanate ion exchangers represents a potential alternative to the dissolution of CSTs in borosilicate glass.

ACKNOWLEDGMENTS

The authors would like to thank Mr. D. McCready for performing XRD measurements. Support for this work was provided by the Strategic Environmental Research and Development Program (SERDP). Pacific Northwest National Laboratory is operated for the U. S. Department of Energy by Battelle Memorial Institute.

REFERENCES

1. D. F. Bickford, A. Applewhite-Ramsey, C. M. Jantzen, and K. G. Brown, J. Am. Cer. Soc. **73**, pp. 2896-2902, 1990.
2. B. C. Bunker, "Evaluation of Inorganic Ion Exchangers for Removal of Cs from Tank Wastes", TWRSPP-94-085, Pacific Northwest National Laboratory, Richalnd, WA, September, 1994.
3. M. L. Balmer, and B. C. Bunker, "Inorganic Ion Exchange Evaluation and Design-Silicotitanate Ion Exchange Waste Conversion", PNL-10460 UC-2070 Pacific Northwest National Laboratory, Richland, WA 99352 , March, 1995.

4. R. G. Dosch, N. E. Brown, H.P. Stephens, and R. G. Anthony, Waste Management 93, February 28-March 3, Tucson, Arizona, pp. 1751-1754.

5. E.A. Klavetter, N. E. Brown, D. E. Trudell, R. G. Anthony, D. Gu, and C. Thibaud-Erkey, Waste Management 94, February 28-March 3, Tucson, Arizona, pp. 709-719.

6. WHC Internal Memo, R. A. Kirkbride to D. J. Washenfelder, " Evaluation of the Impact of the Use of Crystalline Silicotitanate Ion Exchanger on High-level Waste Glass Production," 9457566, March, 1995.

7. D. M. Strachan, R. P. Turcotte, and B. O. Barnes, Nuclear Technology, 56, pp. 306-312, 1982.

8. W. J. Gray, Radioactive Waste Management, 12, pp. 147-169, 1980.

9. J. L. Nogues, E. Y. Vernaz, N. Jacquet-Francillon in The Scientific Bases for Nuclear Waste Management, Vol. 44 edited by C. M. Jantzen, J. A. Stone, and R. C. Ewing, (Materials Research Society, 1984).

10. M. A. Rana, and R. W. Douglas, Phys. and Chem. of Glasses, 2, pp. 179-195, 1961.

SPECTROSCOPIC STUDIES OF ALUMINOSILICATE FORMATION
IN TANK WASTE SIMULANTS

YALI SU, LIQIONG WANG, BRUCE C. BUNKER, and CHARLES F. WINDISCH
Pacific Northwest National Laboratory, Richland, WA 99352

ABSTRACT

Aluminosilicates are one of the major class of species controlling the volume of radioactive high-level waste that will be produced from future remediation at Hanford site. Here we present studies of the phases and structures of aluminosilicates as a function of sludge composition using X-ray powder diffraction, solid state ^{27}Al and ^{29}Si NMR, and Raman spectroscopy. The results show that the content of $NaNO_3$ in solution has significant effects on the nature of the insoluble aluminosilicate phases produced. It was found that regardless of the initial Si:Al ratio, nitrate cancrinite was the main phase formed in the solution with pH of 13.5 and 5 M $NaNO_3$. However, at lower $NaNO_3$ concentration with initial Si:Al ratios of 1.1, 2.2, and 11.0 in the solutions, a range of aluminosilicate zeolites was produced with Si:Al ratios of 1.1, 1.3, and 1.5, respectively. Lowering the solution pH appears to promote the formation of amorphous aluminosilicates. The results presented here are important for the prediction of the solubility and dissolution rate of Al in tank wastes.

INTRODUCTION

Aluminum-containing species are the major components controlling the volume of radioactive high level wastes (HLW) that will be produced from hazardous tank waste sludges at Hanford site. The current baseline for tank waste processing calls for washing and leaching of sludges in basic solutions to remove the aluminum prior to glassification. Unfortunately, the effectiveness of Al removal is highly variable. Since sodium and silica are the other two prevalent species existing in the tank wastes, reactions of aluminum with sodium and silica could lead to formation of insoluble aluminosilicates that might affect pretreatment. Since aluminosilicate phases can be highly resistant to dissolution during washing and leaching, they could be retained after pretreatment and impact on the volume of high level waste glass.[1]

Although solutions containing Na, Al, and Si can readily react to form a wide range of aluminosilicate phases, knowledge of the Na:Si:Al ratios in a given sludge composition is insufficient to predict the phases that might be present in waste tanks. Previous studies[2-4], using hydrothermal synthesis at high temperatures (250 to 700°C) and varying ratios of Na:Al:Si via different mixtures of NaOH, SiO_2, and AlOOH, have shown that a number of structures including clays, zeolites, and feldspars could be produced. In addition to the starting Al:Si ratios, phase formation of aluminosilicates also depends on pH, concentrations of each species, temperature, and reaction kinetics. Conditions in previous studies[2-6] also differ from those in waste tanks, which are typically high in salt content and pH. Consequently, understanding the phase formation of aluminosilicate as a function of sludge composition and the factors associated with tank conditions becomes an essential task for waste processing.

In order to properly perform the pretreatment and to predict the solubility and dissolution behavior of Al in sludge, knowledge of structure and composition of aluminosilicates is important. Bunker et al.[1] summarized the dissolution rates of aluminosilicates by extrapolating the existing data to tank processing conditions. The results suggest that, in the

Mat. Res. Soc. Symp. Proc. Vol. 465 © 1997 Materials Research Society

undersaturated solutions, the leaching rates can vary by as much as seven orders of magnitude depending on the phases present in solution and Si:Al ratio in aluminosilicates. In order to relate the relative dissolution rates to the sludge washing and leaching experiments, the removal of Al-containing phases was studied using solution analysis and transmission electron microscopy (TEM)[1]. It was found that, while Gibbsite (dissolution rate is 10^{-7} mole/cm^2•sec) was completely dissolved in the standard washing and leaching procedures, boehmite (dissolution rate is 10^{-9} mole/cm^2•sec) was largely retained under the same procedures. Thus, these two rates set the limits and guidelines for the processing of other waste species. For example, while nepheline (Si:Al ratio is 1, dissolution rate is 10^{-8} mole/cm^2•sec) is likely to be dissolved during the standard washing and leaching, analcime (Si:Al ratio is 2, dissolution rate is 10^{-11} mole/cm^2•sec) and albite (Si:Al ratio is 3, dissolution rate is 10^{-11} mole/cm^2•sec) are insoluble.

In this paper, we present spectroscopic evidence of a range of aluminosilicate phases formed in solutions with the Al:Si ratios resembling those of the tank wastes. The aluminosilicates were characterized using X-ray powder diffraction (XRD), nuclear magnetic resonance (NMR) and Raman spectroscopy. These three techniques provide important and somewhat complementary information that leads to a better understanding of the phases and solubility of aluminosilicates in waste tanks.

EXPERIMENTAL SECTION

The range of Al and Si concentrations for tank wastes is illustrated in Figure 1 along with ranges of solution compositions tested in the first four series of the solutions used in the current experiments. In this study, we choose three tank waste simulants, labeled as A, B, and C with the Si:Al ratios of 1.1, 2.2, and 11 (Figure 1), respectively. This selection represents

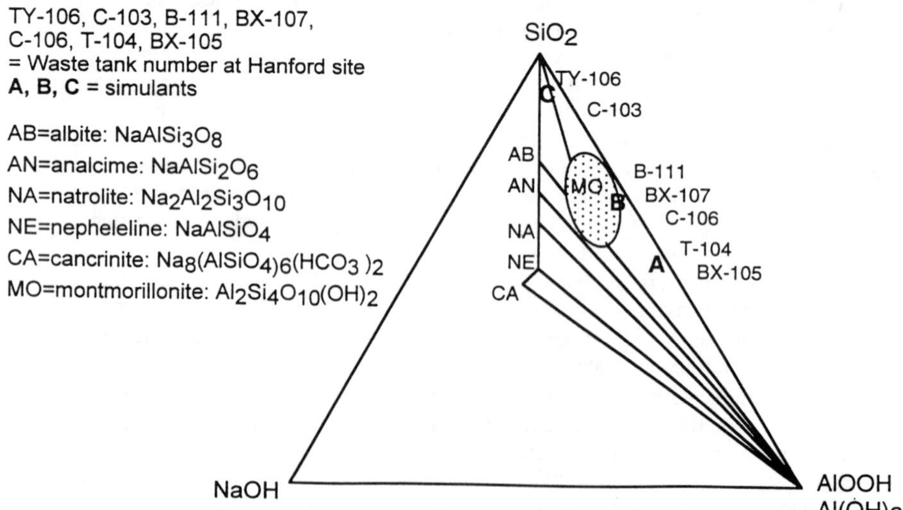

Figure 1. Phase diagram for aluminosilicates in water. Many different aluminosilicate phases can potentially form in tanks vs. composition, pH, temperature, and time. Each phase has different solubility, stability characteristics.

the Si rich tanks. Four sets of solutions with different Si:Al ratios were prepared by varying the relative concentrations of colloidal silica and boehmite (AlOOH) added to the suspension. The pH was further adjusted with NaOH to the desired value. The concentration, pH, and initial Si:Al ratio of the mixtures are listed in Table I. The mixtures were stirred and sonicated for

Table I. Compositions of Initial Solutions

Sample Series	Samples	Al $(10^{-2}$ M)	Si $(10^{-2}$ M)	Na^+ (M)	NO_3^- (M)	pH	Si:Al
1	A-1	5.0	5.33	5	4.47	13.5	1.1
1	B-1	3.33	7.67	5	4.47	13.5	2.2
1	C-1	0.9	10.0	5	4.47	13.5	11.0
2	A-2	5.0	5.33	1	0.5	13.5	1.1
2	B-2	3.33	7.67	1	0.5	13.5	2.2
2	C-2	0.9	10.0	1	0.5	13.5	11.0
3	A-3	5.0	5.33	1	0.965	12.5	1.1
3	B-3	3.33	7.67	1	0.965	12.5	2.2
3	C-3	0.9	10.0	1	0.965	12.5	11.0
4	A-4	200	213	9.7	8.8	13.5	1.1
4	B-4	120	260	9.7	8.8	13.5	2.2
4	C-4	30	330	9.7	8.8	13.5	11.0

homogeneity and then were heated in a 80 °C water bath for a week. After removal from the water bath, samples were centrifuged to separate the solid and liquid phases. The Si and Al contents in the liquid phase were analyzed using inductively coupled plasma emission (ICP). The structures of solid phases were examined using XRD, NMR, and Raman spectroscopy.

X-ray powder diffractometry was performed on a Philips X-ray diffractometer using a Cu Ka source. The Raman spectra were collected on a Spex Triple Raman Spectrometer with the 488 nm line of a Ar^+ laser for excitation. A LN/CCD detector was used with a typical exposure time of 100 s and a slit width of 100 mm. The uncertainty in the Raman peak positions was ±2 cm^{-1}. The ^{29}Si and ^{27}Al NMR spectra were acquired on a Chemagnetic spectrometer using a double resonance probe at frequencies 59.3 and 77.7 MHz for ^{29}Si and ^{27}Al , respectively. ^{29}Si spectra were taken using a single pulse with proton decoupling excitation with a 5 ms pulse, and a 30 s repetition delay. ^{27}Al spectra were collected using a single pulse excitation Bloch-decay method with a 5 ms pulse, and a 0.1 s repetition delay. All the NMR spectra were acquired at 40 ms acquisition times and 50 kHz (Al) and 25 kHz (Si) spectral windows. The number of transients was 500-1000. A Lorentzian line broadening of 24 Hz was used for all spectra. ^{29}Si and ^{27}Al NMR chemical shifts were referenced to tetramethylsilane and 0.1 M $Al(NO_3)_3$ at 0 ppm, respectively.

RESULTS AND DISCUSSIONS

I. Series 1 Samples

X-ray analysis showed that a single cancrinite-type phase was produced from all three solutions of Series 1 (A-1, B-1, and C-1) with nearly 100% yield. Solution analysis using ICP indicated that the amount of Al remaining in solution was at or below ICP detection limits

(20ppm), suggesting that most of the Al was incorporated into the solid phases. The amount of Si remaining in solution corresponded to that expected from the removal of one mole of Si per mole of Al in producing the insoluble solid. Solid state ^{29}Si NMR spectra of the powder exhibited a single peak at –88ppm. This peak is identical to that of cancrinite and is independent of Si:Al ratio in the initial solutions. The relationships between the chemical shift ranges and the number of non-bridging oxygens or the number of AlO_4^- tetrahedra coordination at any given silica tetrahedra have been well established[7-9]. If we use Q_i^j notation [which specifies how many oxygens are bridging (given by superscript j) and how many of those bridging oxygens have the bond configuration of Si-O-Al (given by subscript i)] to describe the bond configuration of Si, all three samples (A-1, B-1, and C-1) are present as Q_4^4 units and the Si:Al ratio in the solid is one regardless the Si:Al ratios in the initial solution.

Based on the known inverse correlation[10-12] between the sizes of the TO_4 rings (T represents Si or Al) present in tectosilicates and the prominent vibrational bands n_s(T–O–T) in 300–600 cm^{-1} region, Raman scattering was used to examine the structure of aluminosilicates. In addition, Raman bands derived from the nitrate ion were also examined to understand the effect of salts in the formation of aluminosilicates. Raman spectra of the series 1 samples [shown in Figure 2(a)] exhibit a strong band at 440 cm^{-1}, corresponding to the n_s(T–O–T) mode that involves six-membered rings. Weak, broad bands at 490 to 510 cm^{-1} are also evident which indicates the co-existance of minor four-membered ring structures. Furthermore, the results show that extensive wash of the sample significantly changes the Raman spectra, as shown in Figure 2. In Figure 2(a), the Raman spectra of the unwashed sample (B-1) are dominated by the three strong Raman bands of sodium nitrate crystaline[13-14] at 1383(m), 1065 (vs), 722 (m) cm^{-1}. Although bands at slightly different frequency such as 716, 1039, 1044, 1052, 1349, and 1380 cm^{-1} are also evident in the spectra, they are much weaker in intensity. This indicates that the unreacted sodium nitrate was a major phase in the unwashed sample.

Extensive washing of the samples in de-ionized water significantly changes the Raman spectra. This is demonstrated in the Raman spectra of the washed sample B-1 shown in figure 2(a). The three strong Raman bands observed in the unwashed sample B-1 disappeared in the washed sample. As a result, Raman bands at slightly different frequencies such as 716, 1039, 1044, 1052, 1349, and 1380 cm^{-1} become more distinct in the spectra. In addition to the changes in intensity, shift and split of the sodium nitrate bands are also evident in the spectra. The disappearance of the unreacted sodium nitrate bands and the increase in intensity of the other bands indicate that the unreacted sodium nitrate phase was dissolved during the washing process, leaving only the sodium nitrate species which interact strongly with the host aluminosilicates. The shift of the Raman bands from ~ 1065 cm^{-1} (free sodium nitrate) to ~ 1040 cm^{-1} region indicates a strong interaction between the remaining nitrate species and aluminosilicates. The splitting of the NO_3^- symmetric stretching band into three bands at 1039, 1044, and 1052 cm^{-1} indicates that NO_3^- is likely to locate at somewhat constrained sites within the aluminosilicate framework with different NO_3–Na–Al interactions. The broadening of those bands must arise from a range of different interactions between the nitrate group and the Al of the aluminosilicate framework at the different constrained sites where it held.

II. Series 2 Samples

Lowering the salt content in solution from 5 M (Series 1) to 1 M (Series 2) has a significant effect on the insoluble aluminosilicate phases produced. Solid state ^{29}Si NMR spectra of the series 2 samples (A-2, B-2, and C-2) in Figure 3 show that in addition to the Q_4^4

peak seen at -88 ppm for Series 1 samples, peaks are evident at -92, -97, and -103 ppm, corresponding to Q_3^4, Q_2^4, and Q_1^4, respectively. Furthermore, the NMR results suggest that instead of a constant Si:Al ratio observed in Series 1 samples, the Si:Al ratio of the three precipitates depends on the Si:Al ratio of the initial solutions. As the Si:Al ratio in solution increases, the population of the Si–O–Si bonds relative to Si–O–Al bonds in the solid phase increases accordingly, as indicated by the change of the Q distribution in the NMR spectra. The relative areas of different Q peaks were used to calculate the Si:Al ratios of solid phases. The Si:Al ratios of the solid phases were found to be 1.1, 1.3, and 1.5, for the initial Si:Al ratios

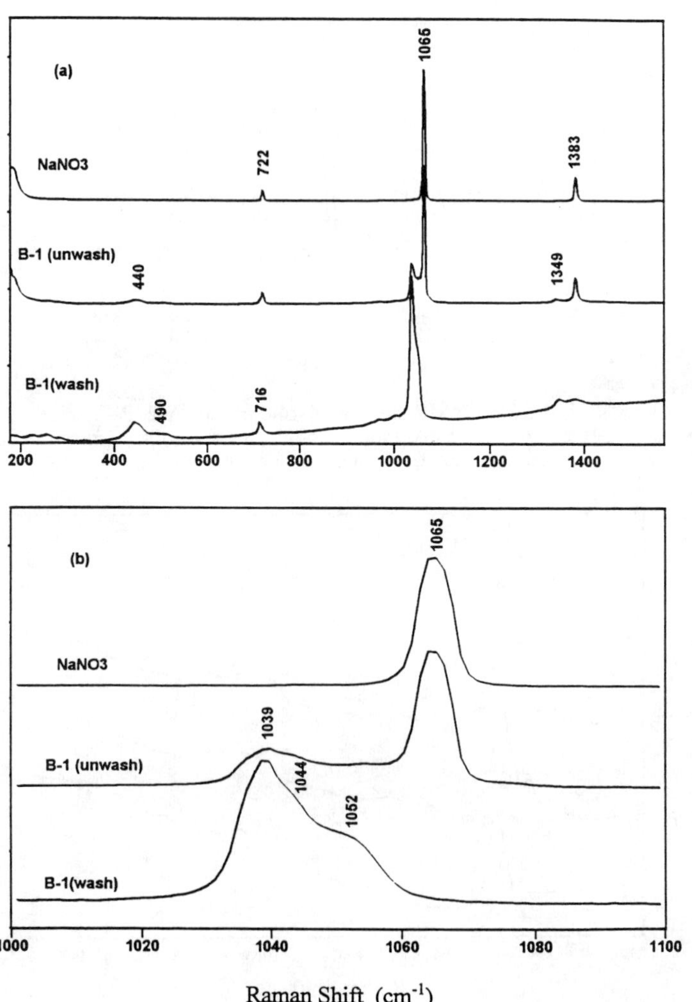

Figure 2. Raman spectra of $NaNO_3$, unwashed sample B-1, and washed sample B-1.

1.1, 2.2, and 11, respectively. Although the variation of the Si:Al ratio of the solid phases seems less significant, it is sufficient to change the initial dissolution rate by more than two orders of magnitude, from 10^{-8} mole/cm^2•sec (Si:Al=1) to $10^{-10.5}$ mole/cm^2•sec (Si:Al=1.5)[1].

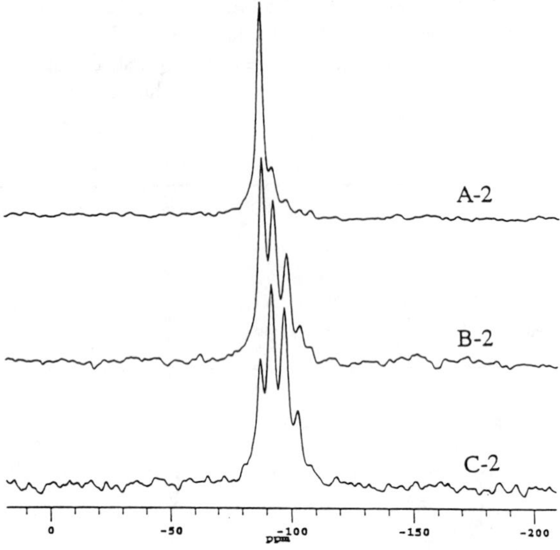

Figure 3. Si NMR spectra of solid phases A-2, B-2, and C-2. Changing the Si:Al ratio in solution changes the atomic structure and composition of aluminosilicate precipitates.

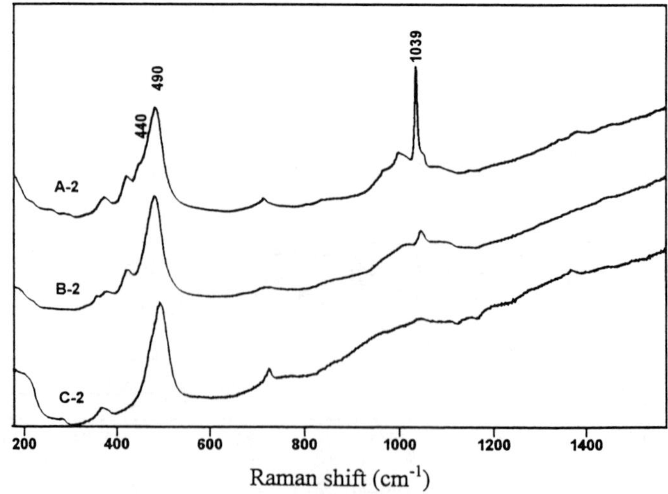

Figure 4. Raman spectra of washed samples A-2, B-2, and C-2.

In addition to the NMR study, Series 2 samples (A-2, B-2, and C-2) were also examined using Raman spectroscopy (Figure 4). A high intensity Raman band near 490 cm^{-1}, which corresponds to n_s(T–O–T) mode involving four-membered rings, is evident in all three samples. However, different from series 1 samples where the six-membered ring dominates the structure, the four-membered ring becomes dominant structure in Series 2 samples. This is demonstrated by the strong Raman band at 490 cm^{-1} (four-membered ring) and weak Raman band at 440 cm^{-1} (six-membered ring). In addition, the Raman spectra also indicates a correlation in intensity between the 440 cm^{-1} shoulder (six-membered ring) and 1040 cm^{-1} band (nitrate). The intensity of both Raman bands decrease accordingly from sample A-2 to C-2, suggesting the nitrate group is likely to be held by the six-memberd ring structure instead of the four-membered ring structure. The difference between Series 1 and 2 samples is likely to be caused by the different sodium nitrate concentration. Sodium nitrate concentration plays a major role in the formation of aluminosilicates. At lower sodium nitrate concentrations, formation of aluminosilicates depends on the initial Si:Al ratios in the solution. A comparison between Series 1 and Series 2 further suggests that the Si:Al ratio in the solid phases tend to be more sensitive to the Si:Al ratio in solution at lower salt content.

III. Series 3 and 4 Samples

Series 3 (A-3, B-3, and C-3) and Series 4 (A-4, B-4, and C-4) solutions did not produce pure-phase crystalline solids. Instead, amorphous gel phases formed initially and then became hard, cement-like materials after the prolonged heat treatment at 80°C. In all cases, consumption of hydroxide via dissolution of the initial silica and boehmite particles lowered the pH to a point where boehmite could no longer dissolve. As a result, solutions contained relatively high concentrations of polymeric species that reacted with each other to form cement rather than to react completely with the dissolved aluminum species. The formation of amorphous phases and incomplete reaction were confirmed by NMR, XRD, and Raman measurements. While ^{29}Si spectra show that the amorphous silicate matrix has a high concentration of non-bridging oxygens, the ^{27}Al NMR spectra show that only a fraction of boehmite is dissolved, and the dissolved Al is incorporated into amorphous aluminosilicate phases with tetrahedral coordination. X-ray analyses suggest that although most of the precipitates are amorphous, there are unreacted boehmite, $NaNO_3$, and cancrinite phases in the precipitates. These unreacted phases are most likely encapsulated by the amorphous cement. The unreacted boehmite was further observed by Raman measurements.

Although a direct comparison of the aluminosilicates formed in the waste tanks and in the simulated solutions is not yet possible, TEM results on real tank sludge suggest that phases similar to the materials reported here are present in tank sludges. Cancrinite is observed in Hanford tank B-111 and possibly in BX-107 sludges. Amorphous cement-like material (Si:Al=2) is seen in BX-107 sludge. An unknown crystalline aluminosilicate (Si:Al=1.7) is also seen in T-104 sludge. Composition analyses indicate that all of the above tanks have lower concentration of free Si and lower pH values than the simulants reported. Based on what has been observed, B-111 sludge is likely to contain cancrinite as the prevalent aluminosilicate. For the other two tanks, the pH is sufficiently low to inhibit gibbsite and boehmite dissolution, resulting in an incomplete reaction. Therefore, the results suggest that the above methodology will be of use in determining the nature of aluminosilicate formation in tank sludges.

CONCLUSIONS

We have investigated the phases and structures of aluminosilicates as a function of tank waste composition using solid state NMR and Raman spectroscopies and XRD methods. Results suggest that $NaNO_3$ has significant effects on the nature of the insoluble aluminosilicate phases produced. At high pH (13.5) and high salt content (5 M of $NaNO_3$), but low initial concentration of Si and Al conditions, nitrate cancrinite is the only phase produced regardless of Si:Al ratio in the initial solutions. At the same pH (13.5), but lower salt concentration (1 M of $NaNO_3$), a range of aluminosilicate zeolites is produced with different Si:Al ratios depending on the Si:Al ratio in the initial solutions. Amorphous cement material forms at lower pH (12.5) with low initial Si and Al concentration ($\sim 10^{-2}$ M) or higher pH (13.5) with high initial Si and Al concentration ($\sim 2M$). The preliminary results also suggest that sodium aluminosilicate might be a good cementitious material.

ACKNOWLEDGMENTS
The authors would like to thank Mr. D. McCready for performing XRD measurement and Ms. Shari X. Li for her assistance with sample preparation. This research was supported by the U. S. Department of Energy under contract DE-AC06-76RLO (1830). Pacific Northwest National Laboratory is operated for the U. S. Department of Energy by Battelle Memorial Institute.

REFERENCES

1. B. C. Bunker, et. al., "Colloidal studies for Solid /Liquid Separations", Chapter 7 in "Tank Waste Treatment Science: Report for the Third Quarter FY 1995", J. P. Lafemina Ed., Pacific Northwest Laboratory, TWRSPP-95-019, June, 1995.
2. L. B. Sand, Rustem Roy, and E. F. Osborn, *Econ. Geol.*, **52**, 169-179, 1957.
3. F. E. Schwochow and G. W. Heinze, Process of zeolite formation in the system in Na_2O-Al_2O_3-SiO_2-H_2O, in Molecular Sieve Zeolites-I, Advances in chemistry series **101**, pp. 102-108, 1971.
4. E. E. Senderov and N. I. Khitarov, Synthesis of thermodynamically stable zeolites in the in Na_2O-Al_2O_3-SiO_2-H_2O system, in Molecular Sieve Zeolites-I, Advances in chemistry series **101**, 1971, pp. 149-154.
5. R. M. Barrer, J. F. Cole, and H. Sticher, J. Chem. Soc. (**A**), pp. 2475-2485, 1968; R. M. Barrer, J. F. Cole, and H. Villiger, J. Chem. Soc. (**A**), pp. 1523-1531, 1970.
6. R. M. Barrer, Hydrothermal chemistry of zeolites, Academic press, 1982.
7. G. Engelhardt and D. Michel, "High-resolution Solid-State NMR of silicates and Zeolites", John Willey & Sons, 1987, and the references therein.
8. M. Magi, E. Lippmaa, A. Samoson, G. Engelhardt, and A. R. Grimmer, J. Phys. Chem., **88**, pp. 1518-22, 1984.
9. E. Lippmaa, M. Magi, A. Samoson, G. Engelhardt, and A. R. Grimmer, J. Am. Chem. Soc., **102**, pp. 4489 , 1980.
10. S. K. Sharma, J. F. Mammone, and F. N. Malcolm, Nature, **292**, pp. 140-141, 1981.
11. S. K. Sharma, J. A. Philpotts, and D. W. Matson, J. Non-Cryst. Solids, **71**, pp. 403-410, 1985.
12. D. W. Matson, S. K. Sharma, and J. A. Philpotts, Amer. Minerl. **71**, pp. 694-704, 1986.
13. I. Nakagawa and J. L. Walter, J. Chem. Phys. 51, pp. 1389, 1969.
14. D. L. Rousseau, R.E. Miller, and G.E. Leroi, J. Chem. Phys. **48**, pp. 3409-3413, 1968.

THE EFFECTS OF SURFACE MODIFICATION ON THE SPECIATION OF METAL IONS INTERCALATED INTO ALUMINOSILICATES

STEPHEN R. WASSERMAN, DANIEL M. GIAQUINTA, STEVEN E. YUCHS AND L. SODERHOLM
Chemistry Division, Argonne National Laboratory, 9700 S. Cass Ave., Argonne IL 60439.

ABSTRACT

The effect of the addition of an organic monolayer to the surface of a clay mineral on the speciation of metal ions intercalated into the clay interlayer is probed by X-ray absorption spectroscopy. The presence of the monolayer changes the surface of the clay from hydrophilic to hydrophobic. It inhibits the interlayer ions from exchanging freely into environmental water and reduces the leach rate of cations out of the clay by approximately a factor of 20. Significant changes are observed when these coated samples are treated under hydrothermal and thermal conditions. Reductions of uranium(VI), in the form of uranyl, and copper(II) ions occur. In addition, the uranium aggregates, forming small particles that appear similar to UO_2. Comparable conglomeration occurs with lead cations and with the reduced copper species.

INTRODUCTION

Clay minerals and zeolites, two types of microporous aluminosilicate, are ion-exchange materials.[1,2] In their most common forms, they have the ability to incorporate external cationic species within their matrices. Because of this property, microporous aluminosilicates have been proposed as storage media for hazardous waste. In this paper we use X-ray absorption spectroscopy (XAS) to examine the structure of cations held within smectite clay minerals and to monitor how modification of the surface of the clay using an organic monolayer affects the coordination of the stored cation. The effects of hydrothermal and thermal processing on the coordination of the ions contained within these systems are also investigated.

Smectite clays consist of sheets of aluminosilicate. The sheets contain a central section of octahedrally coordinated Al^{3+} that is sandwiched between two layers of tetrahedrally coordinated Si^{4+}.[3] Random substitutions, generally of Mg^{2+} for Al^{3+} and Al^{3+} for Si^{4+}, result in a net negative charge in the lattice. This charge is balanced by cations in the interlayer between the sheets. The interlayer also contains water, some of which is complexed to the cations. The interlayer cations can be replaced by ion-exchange techniques with other positively charged species.[4] For this study we have substituted the cations originally present in a clay, usually sodium, calcium, and potassium, with metallic ions, including copper(II), nickel(II), lead(II), and uranyl (UO_2^{2+}). The clay used for these experiments is bentonite, a form of montmorillonite.

The surface of a clay is hydrophilic. This affinity for water, however, also facilitates the leaching of any hazardous species stored within the mineral. In order to isolate the cationic

Mat. Res. Soc. Symp. Proc. Vol. 465 © 1997 Materials Research Society

species within the clay interlayer, we have modified the surfaces of ion-exchanged smectite clays using alkylsilanes of the form $RSiX_3$, $R = C_{18}H_{37}$-, $X = $ -Cl, -OCH$_3$. This process is illustrated schematically in Figure 1. The organosilanes bind covalently to the surface of the clay mineral, creating a polymeric coating that renders the mineral hydrophobic. The presence of this film inhibits the exchange of external water, and accompanying cations, into and out of the interlayer.[5]

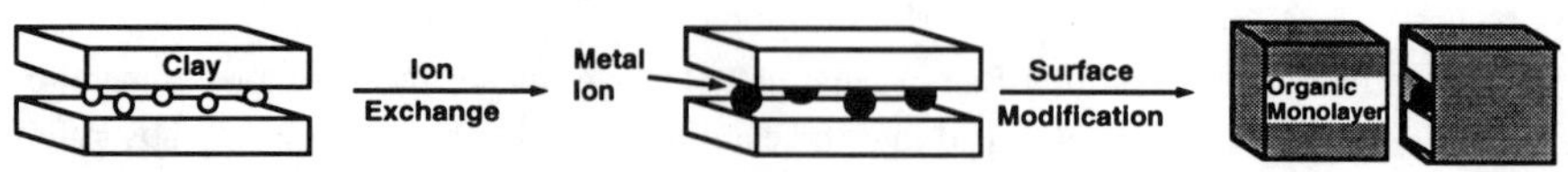

Figure 1. Schematic representation of ion exchange and organic encapsulation of ions within a smectite clay mineral.

EXPERIMENTAL SECTION

The general preparation of ion-exchanged clays has been described elsewhere. The Ca^{2+} form of bentonite was used as received (bentolite L, Southern Clay Products, Gonzales, Texas). The surfaces of the ion-exchanged clays were coated with octadecyltrimethoxysilane, $CH_3(CH_2)_{17}Si(OCH_3)_3$ (OTMS, Aldrich), and octadecyltrichlorosilane, $CH_3(CH_2)_{17}SiCl_3$ (OTCS, Aldrich). The silanes were distilled and stored under dry nitrogen prior to use. The clay (1 g) and silane (1 mL) were added to anhydrous hexane (40 mL). The mixture was stirred for 24 hours at room temperature under a dry nitrogen atmosphere. A catalyst, 3-indolepropionic acid (0.01 g), was added for the reaction of the trimethoxysilane with the clay surface. The samples were collected by centrifugation, washed three times with anhydrous hexane (25 mL) to remove any unreacted silane, and dried at room temperature. The process for the formation of the surface coatings is based on previous work on self-assembled monolayers.[6,7]

The susceptibility of the encapsulated ions to leaching was evaluated by sonicating the clay sample (0.5 g) in a buffered solution of sodium acetate for 2 hours. This solution was prepared by adding glacial acetic acid (5.7 mL) and aqueous sodium hydroxide (64.3 mL of a 1 N solution) to water (1 L). This fluid is the standard extraction fluid for the EPA toxicity characteristic leaching protocol (TCLP). The amounts of extracted ions were determined by ICP-MS.

Hydrothermal processing of the clay samples was performed in a sealed stainless steel Parr high pressure bomb (4746) with a Teflon® insert that contained clay (250 mg) and deionized water (10 mL). The bomb was placed in a Lindberg crucible furnace and heated at 1° C per minute to 200 °C. The bomb was maintained at 200 °C and 15.8 kg/cm^2 (calc) for 20 hours, after which the samples were cooled radiatively to room temperature.

Thermal processing under nitrogen and hydrogen was performed in a specially designed catalyst cell for *in situ* X-ray absorption spectroscopy. The ion-exchanged clay (100 mg) was placed in the sample holder of the cell. The remainder of the volume of the holder was filled with boron nitride to minimize movement of the clay sample. Kapton tape was used as the windows of the cell. The flow rate of nitrogen or hydrogen gas through the cell was approximately 5 mL per minute through a cell volume of 1 cm^3.

X-ray absorption spectra were obtained at the National Synchrotron Light Source (NSLS) on beamlines X23A2 and X10C, and at the Stanford Synchrotron Radiation Laboratory (SSRL) on station 4-3. The beamlines were equipped with either a <220> (SSRL and X10C) or <311> (X23A2) double-crystal Si monochromator. Harmonics were rejected on X10C by use of a mirror and at SSRL by detuning the monochromator to approximately 50% of the maximum X-ray intensity. Fluorescence spectra were collected using a Lytle detector that was purged with Ar gas. An appropriate 3 absorption lengths filter was placed between the sample and the ionization chamber of the detector. Calibration of the edges was maintained through simultaneous acquisition of the transmission spectrum of a reference material. The position of k = 0 Å^{-1} was defined as the maximum in the derivative spectrum from the reference. Analysis of the EXAFS data was performed using theoretical phase shifts and scattering amplitudes from FEFF 3.25.[8]

RESULTS AND DISCUSSION

Ion Coordination. Surface modification of ion-exchanged clays can lead to changes in the local coordination of the ions located within the interlayer. Figure 2 shows the near edge spectra (XANES) and radial structure functions for lead(II) ions in the interlayers of bentonite (Pb-bentonite) before and after addition of organic monolayers formed from octadecyltrichlorosilane (OTCS) and octadecyltrimethoxysilane (OTMS). The clays that result

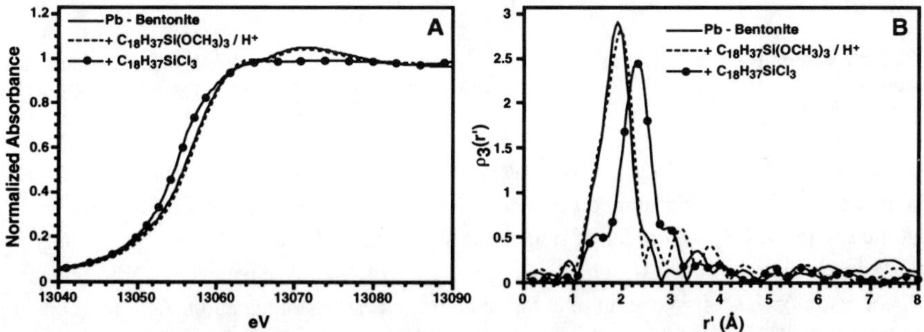

Figure 2. (A) X-ray absorption near edge spectra (XANES) and (B) radial structure functions (not phase corrected) for lead(II) in the interlayer of bentonite: (1) ion-exchanged clay (——), (2) after formation of a monolayer from octadecyltrimethoxysilane (C$_{18}$H$_{37}$Si(OCH$_3$)$_3$, - - - -), and (3) after formation of a monolayer from octadecyltrichlorosilane (CH$_3$(CH$_2$)$_{17}$SiCl$_3$, —●—). The range for the forward Fourier transforms was Δk = 1.7 - 10.0 Å^{-1}.

from addition of these two coatings are hydrophobic. The EXAFS data indicate that, in the original ion-exchanged clay, the coordinating species for the lead ions are oxygen atoms, presumably from water molecules contained within the interlayer of the clay. The XANES and radial distributions after addition of OTMS are similar to those of the original Pb-bentonite. In contrast, when OTCS is used to form the monolayer, the XANES spectrum shifts to lower energy. The bond distance to the atoms in the first coordination sphere in the radial structure function increases by approximately 0.3 Å. The changes in the XANES and EXAFS spectra are consistent with the replacement of the water molecules originally complexed to the Pb(II) ion by chloride anions which are a byproduct from the reaction of the trichlorosilane. Similar results are observed when uranyl or copper(II) ions are placed within bentonite.

This change in the local coordination affects the ability of the clay-monolayer system to retain the interlayer ion. The results of leach tests from ion-exchanged clays before and after surface modification are shown in Table 1. For the four cations examined, the hydrophobic

Table 1. Leachability in mg per g of clay of divalent cations from bentonite before and after surface modification.[a]

Cation	Ex$^+$-bentonite	+ C$_{18}$H$_{37}$Si(OCH$_3$)$_3$	+ C$_{18}$H$_{37}$SiCl$_3$
Ni^{2+}	5.4	0.25	—
Cu^{2+}	8.4	0.24	—
Pb^{2+}	33	1.9	—
UO$_2$$^{2+}$	28	0.95	11

[a]Leach values for the hydrophobic clays are corrected for the mass of organic material contained within the added monolayer. They represent leachability per gram of original unmodified clay.

clays created using OTMS are approximately 20 times more effective in retaining the interior ion than the corresponding unmodified bentonites. In the case of the uranyl cations, the clay modified with OTCS is an order of magnitude worse than the UO$_2$$^{2+}$-bentonite treated with OTMS in keeping the cation within the aluminosilicate matrix. We believe that this behavior is a direct result of the complexation of chloride anions with the stored cation. The reaction of an alkyltrichlorosilane, RSiCl$_3$, with the clay surface and the water adsorbed there, forms siloxane bonds, -Si-O-Si-, and hydrochloric acid, HCl. In the case of divalent cations, the replacement of two or more neutral water molecules with negative chloride anions from the acid results in a complex that is no longer positively charged. These chlorinated species are easily leached from the mineral, since the coulombic forces which normally hold the cation in place are no longer effective. Therefore, both the coulombic and hydrophobic properties of the modified clays are necessary to encapsulate the interlayer cation within these systems.

Hydrothermal and Thermal Processing. In order to simulate the effects of long term storage on these materials, as well as the effects of extreme conditions which may occur in a geologic repository, we have subjected both hydrophobic and hydrophilic clays that contain uranyl (UO_2^{2+}) cations to hydrothermal processing.[9] Figure 3 shows the near edge spectra and radial distributions for uranyl in bentonite before and after such treatment. Figure 3

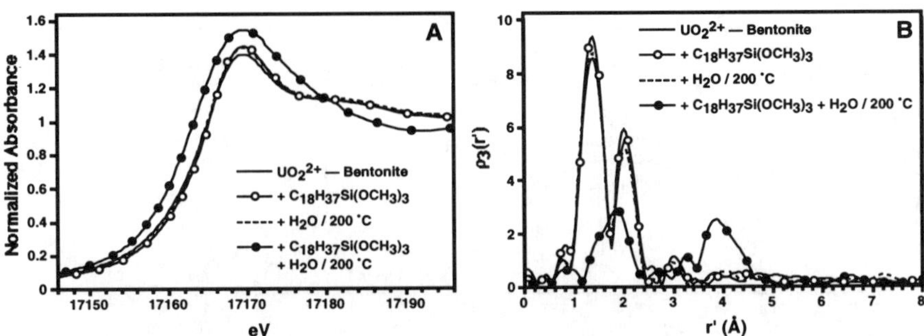

Figure 3. (A) X-ray absorption near edge spectra (XANES) and (B) radial structure functions (not phase corrected) for uranyl (UO_2^{2+}) in the interlayer of bentonite: (1) ion-exchanged clay (——), (2) after hydrothermal processing of the ion-exchanged clay at 200 °C (- - - -), (3) hydrophobic ion-exchanged clay created using octadecyltrimethoxysilane ($C_{18}H_{37}Si(OCH_3)_3$, —O—) and (4) after hydrothermal processing of the hydrophobic uranyl clay (—●—). The range for the forward Fourier transforms was $\Delta k = 1.0$ - 11.0 Å^{-1}.

includes the XAS spectra from a uranyl-bentonite after addition of an organic monolayer formed from OTMS, and after exposure of this hydrophobic clay to hydrothermal conditions. Addition of the silane film to the surface of the clay does not alter the coordination of the uranyl. For the simple uranyl-bentonite, the XANES and radial structures are unchanged after hydrothermal processing. The two peaks in the radial distribution between r' = 1 and 2.5 Å reflect the presence of two axial oxygen atoms, which are directly bound to the uranium, and approximately 5 equatorial oxygen atoms from water molecules which are complexed to the uranyl species.[10] The spectra of the uranyl-clay after application of the organic coating but prior to hydrothermal treatment are identical to those of the uranyl-bentonite. However, when the monolayer is present, exposure to hydrothermal conditions shifts the absorption edge approximately 3 eV to lower energy. This shift is confirmed by a lowering of the energy parameter in the EXAFS fits by 3.5 eV. At the same time, the axial and equatorial features of uranyl have essentially disappeared from the radial structure function, to be replaced by a single peak centered at r' = 1.8 Å. Simultaneously, a uranium-uranium correlation appears at r' = 4 Å. These changes are consistent with the reduction of uranium to the +4 oxidation state, and its aggregation into small, disordered UO_2 particles. The unresolved shoulder in the radial distribution at r' = 1.42 Å may indicate the continued presence of a small amount of unmodified uranyl cations.

We have performed related experiments with Pb-bentonite. The hydrophobic clay that results from the addition of OTMS was heated to 175 °C under nitrogen. The XAS spectra for this clay before and after heating are shown in Figure 4. There is no shift in the energy of the

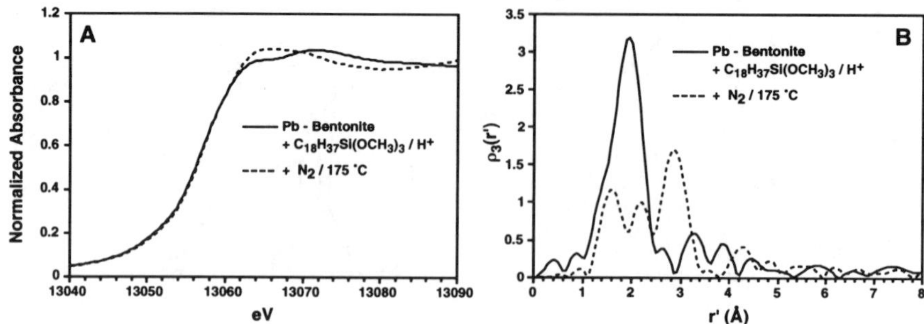

Figure 4. (A) X-ray absorption near edge spectra (XANES) and (B) radial structure functions (not phase corrected) for Pb in the interlayer of bentonite after modification with octadecyltrimethoxysilane ($C_{18}H_{37}Si(OCH_3)_3$): (1) hydrophobic clay (——) and (2) after heating to 175 °C under nitrogen (- - -). The range for the forward Fourier transforms was $\Delta k = 1.7 - 9.0$ Å^{-1}.

absorption edge, indicating that reduction of the lead has not occurred. However, the EXAFS data demonstrate that the lead ions, which are initially separate within the interlayer, coalesce into larger structures. This behavior is similar to that observed when an unmodified Pb-bentonite is exposed to the same conditions.[11]

Differences between thermal and hydrothermal processing do not, apparently, account for the fact that uranyl is reduced while lead is not. Thermal treatment of Cu-bentonite after application of OTMS also results in reduction and aggregation of the cupric ions (Figure 5). The product clay contains a mixture of species. Inspection of the second derivatives of the edge demonstrates that the 1s-4p transitions of both Cu(II) and Cu(I) are present in the near edge spectrum. This fact is confirmed by the continued presence of the 1s-3d transition, although at a lower intensity than that in the original Cu-bentonite. This transition does not appear for cuprous compounds since such species have a filled 3d shell.[12]

The same changes in the XANES and EXAFS of Cu(II) occur when a hydrophilic Cu-bentonite is heated to 180 °C under hydrogen gas. The fact that heating in the presence of an organosilane monolayer or hydrogen results in the same species suggests that there is a reductive source that is the equivalent of hydrogen in the organic systems. This source is either

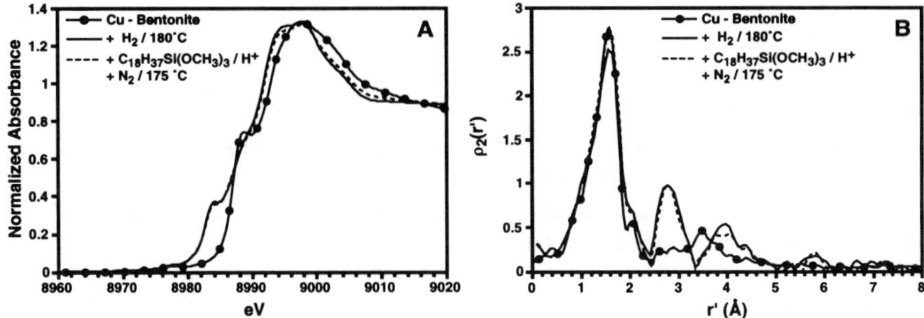

Figure 5. (A) X-ray absorption near edge spectra (XANES) and (B) radial structure functions (not phase corrected) for Cu(II) in the interlayer of bentonite: (1) ion-exchanged clay (—●—), (2) after heating under hydrogen to 180 °C (——) and (3) after thermal processing of the modified Cu-clay created using octadecyltrimethoxysilane ($C_{18}H_{37}Si(OCH_3)_3$, - - -). The range for the forward Fourier transforms was $\Delta k = 1.8 - 13.0$ Å^{-1}.

the silane itself, methanol, which is a by-product of the polymerization of the silane on the surface, or residual amounts of the acid catalyst.

Each of the ions examined aggregates upon hydrothermal or thermal treatment in a surface modified clay. In addition to the ions discussed above, we have found that thorium conglomerates but is not reduced in the modified clays. Whether reduction occurs along with aggregation appears to be a function of the reduction potential of the stored ion. The reduction potentials for Cu^{2+} (to Cu^+), UO_2^{2+} (to UO_2), Pb^{2+} (to Pb(0)), and Th^{4+} (to Th(0)) are 0.16, 0.27, -0.13, and -1.83 V respectively.[13] We have only observed reduction with the former two species, for which reduction results in the formation of a more stable species.

CONCLUSION

These studies have demonstrated that hydrophobic clay minerals can be created by addition of an organic monolayer to the surface of a clay. When such modification is performed using an alkyltrimethoxysilane, the coordination of the cations held within the interlayer is unchanged and leaching of the ions from the matrix is inhibited. If a trichlorosilane, however, is used, the chloride ions formed as a by-product can complex with the interlayer ions. The latter reaction, by lowering the net charge of the metal moiety, apparently reduces the coulombic forces that hold the ions within the clay.

The addition of an organic film to clay minerals also perturbs the behavior of interlayer cations upon thermal and hydrothermal treatment. When a monolayer formed from octadecyltrimethoxysilane is present at the surface, hydrothermal processing results in the reduction and aggregation of the uranyl. Similar aggregation upon heating in nitrogen is observed with lead(II) and copper(II), although reduction of the former does not occur. These examinations of the fate of ions in hydrophobic clays can serve as models for the interactions which occur in nature between ions, clay minerals, and organic compounds.

ACKNOWLEDGMENTS

This research was supported by the Strategic Environmental Research and Development Program of the U. S. Department of Defense (S. R. W. and S. E. Y.) and the Division of Chemical Sciences, Office of Basic Energy Sciences, U. S. Department of Energy (D. M. G. and L. S.) under contract W-31-109-ENG-38.

REFERENCES

1. R. E. Grim, *Clay Mineralogy, 2nd ed.* (McGraw-Hill, New York, 1968).

2. Jule A. Rabo, *Zeolite Chemistry and Catalysis* (American Chemical Society, Washington, D. C., 1976).

3. G. W. Brindley and G. Brown, "Crystal Structures of Clay Minerals and Their X-ray Identification," (Mineralogical Society, London, 1980).

4. D. T. B. Tennakoon, J. M. Thomas, M. J. Tricker, and J. O. Williams, J. Chem. Soc. Dalton **1974**, 2207-2211 (1974).

5. S. R. Wasserman, K. B. Anderson, K. Song, S. E. Yuchs and C. L. Marshall, U. S. Patent Applied For.

6. R. Maoz and J. Sagiv, J. Colloid Interface Sci. **100**, 465-496 (1984).

7. S. R. Wasserman, Y.-T. Tao, and G. M. Whitesides, Langmuir **5**, 1074-1087 (1989).

8. J. Mustre de leon, J. J. Rehr, S. I. Zabinsky, and R. C. Albers, Phys. Rev. **B44**, 4146-4156 (1991).

9. L. L. Hench, D. E. Clark, and E. L. Yen-Bower, Nucl. Chem. Waste Manag. **1**, 59-75 (1980).

10. F. A. Cotton and G. Wilkinson, *Advanced Inorganic Chemistry, 5th ed.* (John Wiley and Sons, New York, 1988).

11. S. R. Wasserman (unpublished).

12. F. W. Lytle, R. B. Greegor, and A. J. Panson, Phys. Rev. B **37** (4), 1550-1562 (1988).

13. A. J. Bard, R. Parsons, and J. Joseph, "Standard Potentials in Aqueous Solutions," (Marcel Dekker, New York, 1985).

THE DEVELOPMENT OF A CERAMIC WASTE FORM FOR IMMOBILISATION OF HIGHLY ACTIVE WASTES FROM RADICAL PUREX REPROCESSING OPERATIONS

Ewan Maddrell
British Nuclear Fuels plc, Sellafield, Seascale, Cumbria, CA20 1PG

ABSTRACT

The development of novel, Radical Purex, reprocessing technologies, leading to fission product waste streams with high levels of inert constituents, may mean that vitrification is no longer the optimum solution for the immobilisation of highly active wastes at the back end of the nuclear fuel cycle. A ceramic phase assemblage is described which uses the inert constituents of the waste stream as a functional component of the waste form, permitting high waste loadings to be achieved and thus enabling waste minimisation considerations to be satisfied. The initial development of this phase assemblage is presented.

INTRODUCTION

Despite the amount of effort that has been devoted to the development of ceramic waste forms, principally Synroc, they have failed to gain widespread acceptance in the nuclear industry because of the proven technology of vitrification. Currently, for the concept of a ceramic waste form to be preferred to vitrification the waste stream must be such that the latter is not viable. One area where this may be the case is in the immobilisation of highly active wastes from Radical Purex reprocessing routes.

The competition between reprocessing and direct disposal at the back end of the nuclear fuel cycle means that reductions in the cost of reprocessing operations must be made for it to remain an attractive option. To this end, a major area of development is in improving the head end processes which dissolve the spent fuel prior to the solvent extraction cycle. A promising candidate is a technique which involves electrochemical dissolution of the entire fuel assembly. For PWR fuel dissolved by this technique, the Zircaloy cladding primarily spalls off as millimetre sized zirconia chunks, which can be treated separately as ILW, although approximately 15 % of the zirconium together with the iron, chromium, nickel and manganese from stainless steel and Inconel components are taken into solution along with the uranium, plutonium and fission products from the spent fuel. After the extraction cycle, the highly active raffinate therefore contains large quantities of first row transition metals in addition to the fission products and minor actinides. An approximate composition of this aqueous raffinate per tonne of uranium (pre-irradiation) for fuel irradiated to 40 Gwd/te U is given in Table I. The fission products shown are a simplification of the actual fission product spectrum but contain elements representative of all those which will occur. Current trends towards fuels with higher burn-up and the advent of MOX fuel mean that the fission product spectrum will differ slightly and a more appropriate waste simulant is now being used. Additionally, the levels of stainless steel and Inconel components within a fuel assembly are being reduced, however, as will be shown, the flexibility of the waste form is sufficient to accommodate such variations in the composition of the waste stream.

In addition to reducing the costs of reprocessing operations, a further important consideration is that of minimising the quantity of waste produced without compromise of the durability of the

Mat. Res. Soc. Symp. Proc. Vol. 465 © 1997 Materials Research Society

TABLE I

COMPOSITION OF SIMULATED HIGHLY ACTIVE WASTE STREAM PRODUCED
FROM A RADICAL PUREX ROUTE USING ELECTROCHEMICAL DISSOLUTION OF
FUEL ASSEMBLIES PER TONNE URANIUM (PRE-IRRADIATION)

INERT COMPONENTS		FISSION PRODUCTS	
Fe_2O_3	63.8 kg	ZrO_2	7.41 kg
ZrO_2	49.8 kg	BaO	2.79 kg
NiO	19.5 kg	SrO	1.51 kg
Cr_2O_3	21.0 kg	RuO_2	3.64 kg
MnO_2	1.9 kg	MoO_3	6.70 kg
SnO_2	0.8 kg	Cs_2O	4.04 kg
MoO_3	0.8 kg	Nd_2O_3	7.24 kg
		Sm_2O_3	1.33 kg
		Ce_2O_3	4.64 kg
		La_2O_3	2.30 kg
		Pr_2O_3	2.26 kg
		Y_2O_3	0.39 kg
		Gd_2O_3	8.76 kg

(added as a neutron poison)

waste form. In the context of Radical Purex, this waste minimisation objective can be applied to comparisons with the total amount of waste produced by current reprocessing technology and also direct disposal. In conventional oxide fuel reprocessing the high level waste consists predominantly of fission products and is vitrified at a waste loading of 20-25 wt %. The high level waste produced by this Radical Purex reprocessing route, however, contains such high quantities of inert material from the fuel assembly that vitrification at the same waste loading would roughly quadruple the final volume of high level waste produced per tonne of fuel reprocessed. Consequently, in order to reduce the final volume of waste a method is required in which the inert constituents of the waste stream form a functional component of the final waste form, and this can be achieved through the use of a suitably tailored multiphase ceramic.

The large quantities of iron and zirconium in the waste stream suggested that two suitable phases for the waste form would be the cubic polytype of zirconia, c-ZrO_2, and the spinel type iron oxide, Fe_3O_4. The former was chosen as a host phase for the rare earth elements within the spectrum of fission products, together with non-recycled actinides, whilst the latter is envisaged as a means of immobilising the iron, manganese and chromium. Both phases have been suggested before in the context of nuclear waste immobilisation.[1] To complete the ceramic phase assemblage a host is required for the larger alkali and alkaline earth elements, principally caesium and strontium, the selected phase being $BaZrO_3$.[2] In principle, caesium accommodation in $BaZrO_3$ is intended to be facilitated by the altervalent substitution of caesium and tantalum on the barium and zirconium sites respectively and the tantalum then mitigates the decay of ^{137}Cs to ^{137}Ba by reduction of Ta^{5+} to Ta^{4+}; in practice, calcination of the waste under an argon/hydrogen atmosphere may prematurely reduce the tantalum to the tetravalent state and caesium accommodation could require simultaneous substitution of lanthanum on the barium site. Strontium replaces barium by direct substitution. In addition to the ceramic phases, processing of the wastes under reducing conditions will lead to the formation of a small quantity of metallic material, principally nickel from the fuel assembly and ruthenium, palladium and silver as fission products. These are expected to take the form of solid solutions based on nickel and silver, however, the waste composition used here is such that only an FCC iron/nickel solid solution has been produced. The fate of molybdenum depends on how completely it has been reduced with accommodation in the magnetite as Mo^{4+} and metallic solutions both possible.

The basic principle of this phase assemblage has already been demonstrated [3] and this paper presents a more detailed study and discussion of two compositional variants. The two samples chosen for study here are designated ZFW-24 and ZFW-28. Both have waste loadings of approximately 65 wt % and immobilise around 1000 kg of fission products per cubic metre of waste form. This latter quantity is seen as useful because it permits comparisons of the total volume of high level waste produced per tonne of fuel reprocessed by differing reprocessing and immobilisation routes. Conventional reprocessing technologies have high level wastes consisting almost solely of fission products and thus lower waste loadings are permissible. The major difference between the two samples is in the varying proportions of the host phases with the intention of establishing the effects of phase composition on leach rates. ZFW-24 was formulated to contain 35 vol % zirconia, 40 vol % magnetite, 20 vol % barium zirconate and 5 vol % nickel. The least durable of these phases is likely to be the barium zirconate which is host for the caesium and strontium. If this phase is interconnecting, it is conceivable that leaching may lead to the creation of percolation pathways within the waste form; if it is at a sufficiently low volume fraction, however, it will cease to be interconnecting and the grains in the bulk will remain encapsulated by the more durable zirconia and magnetite phases, thus minimising leach rates. ZFW-28 was designed to test this hypothesis by reducing the quantity of $BaZrO_3$ produced to an appropriately low volume fraction. For this phase, a target composition of $(Ba_{0.5}Sr_{0.17}Cs_{0.33})(Zr_{0.67}Ta_{0.33})O_3$ was chosen which at these waste loadings gives a nominal volume fraction of 7 %. Vance *et al* [2] have shown that caesium and tantalum substitutions in $BaZrO_3$ up to a composition of $(Ba_{0.3}Cs_{0.7})(Zr_{0.3}Ta_{0.7})O_3$ retain the perovskite structure and recent work in-house suggests that at least 50 % replacement of strontium for barium can also be achieved; consequently it is believed that at this level of substitution the perovskite structure will be stable. To complete the phase assemblage, ZFW-28 was designed to contain 33 vol % magnetite, 55 vol % zirconia and 5 vol % nickel. The quantity of magnetite produced represented the situation in which no additional iron and chromium were added.

EXPERIMENTAL DETAILS

Simulated waste forms have been prepared by hydrolysing appropriate quantities of zirconium propoxide with solutions of the nitrates of the relevant precursor and waste elements. Typically the precursors consisted of yttrium nitrate and zirconia to promote the formation of cubic zirconia; iron, chromium and manganese nitrates to promote the magnetite; barium nitrate to promote the formation of the perovskite type phase and tantalum pentoxide to facilitate charge balancing of the alkali metals. For the purpose of sample production, barium nitrate is troublesome because its low solubility in water means it is difficult to dissolve in the presence of other nitrates. To overcome this, the alkali and alkaline earth nitrates have been dissolved separately in hot water and this solution used for the initial alkoxide hydrolysis, with a concentrated solution of the remaining nitrates being added subsequently. The slurries were homogenised using a high shear mixer and then evaporated to dryness with constant stirring whilst viscosity permitted. The nitrate content of the mixtures meant that high viscosities were reached at relatively low solid contents. The resulting powders were calcined in a flowing Ar - 5% H_2 atmosphere by heating at 5 °C min^{-1} to 950 °C and holding at that temperature for one hour. After calcining the powders were ground in a pestle and mortar and blended with 2 wt % of finely divided iron powder prior to consolidation either by pressureless sintering in a nitrogen atmosphere or hot isostatic pressing. Hot isostatic pressing (HIPping) was conducted at both 1200 °C and 1300 °C under a pressure of 200 MPa for two hours whilst pressureless sintering was for four hours at 1300 °C.

The formulations were designed such that the composition of the zirconia solid solutions was $Zr_{0.75}REE_{0.25}O_{1.875}$. The rare earth element content was made up from the lanthanides present as fission products and further additions of yttria. For equiatomic additions of lanthanum, samarium and yttrium , the above formulation has been shown to be a single phase [4].

Characterisation of the samples has been carried out by X-Ray Diffraction (XRD) using Co Kα radiation and Scanning Electron Microscopy coupled with Energy Dispersive Spectroscopy (SEM/EDS).

<u>RESULTS AND DISCUSSION</u>

X-ray diffraction analysis of the two formulations fabricated with the three conditions described above showed that in all cases the three principal ceramic phases had been produced together with a face centred cubic metallic solid solution. For completeness, the lattice parameters of the ceramic phases are given in Table II, however, there were no significant variations within each variant as a result of the differing fabrication conditions. A range of minor phases were also identified, principally barium molybdate ($BaMoO_4$), wüstite (FeO) and pyrochlore ($REE_2Zr_2O_7$).

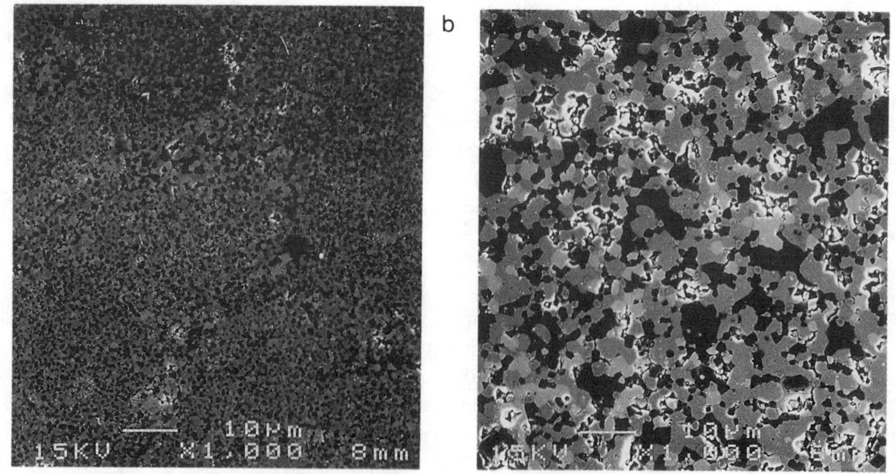

Figures 1 a+b: Secondary electron micrographs of sample ZFW-24 HIPped at 1200 °C and
1300 °C respectively showing residual porosity from agglomerated powders
(1 cm = 10 microns in both figures)

Secondary electron imaging of the sintered samples revealed a high level of porosity indicating that grinding of the powders in a pestle and mortar had not been sufficient to break up the agglomerates that form during drying of the slurries. The samples that had been hot isostatically pressed had lower levels of porosity, Figures 1 a+b, but full densification is still not being achieved due to the formation of hard agglomerates. Bulk densities of samples HIPped at 1200 °C for ZFW-24 and 28 were 5.80 and 5.98 respectively.

Figures 2 a+b are back scattered electron images of ZFW-24 after HIPping at 1200 °C and 1300 °C respectively. Figure 2a reveals the fine, sub-micron, grain structure which is necessary for the long term stability of nuclear waste ceramics whilst Figure 2b enables the contrast between the various phases to be discerned. In Figure 2b, the phases are, in order of increasing brightness, magnetite, wüstite, iron/nickel solid solution, cubic zirconia and barium zirconate. The barium molybdate and pyrochlore showed brightness levels similar to that of that of barium zirconate. The distribution of the wüstite suggested that it was a relic of the iron powder which had been added although it showed traces of manganese from the waste.

Of principal concern in this work is the formation of the barium molybdate phase which indicates that the molybdenum in the waste has not been reduced from the soluble hexavalent state to the preferred tetravalent or metallic states. This is probably due to competition with iron for the hydrogen available for reduction purposes. Its presence implies the existence of the caesium molybdate (Cs_2MoO_4) phase, and SEM examination indeed identified occasional grains of this phase. Figure 3 compares the XRD traces from ZFW-24 produced by each of the three consolidation cycles. The major peak of the barium molybdate phase can be seen to decrease in size with an increase in HIP temperature from 1200 to 1300 °C and also as a result of sintering in nitrogen. This suggests that the molybdenum exists principally as Mo^{6+} after calcining and is largely frozen in this state during HIPping. If the same powder is sintered without being encapsulated the oxygen partial pressure in the furnace appears to be sufficiently low for partial decomposition of MoO_3 to have taken place. Extended heat treatment of the HIPped samples (1300 °C for 24 hours in nitrogen) did not significantly reduce the level of barium molybdate

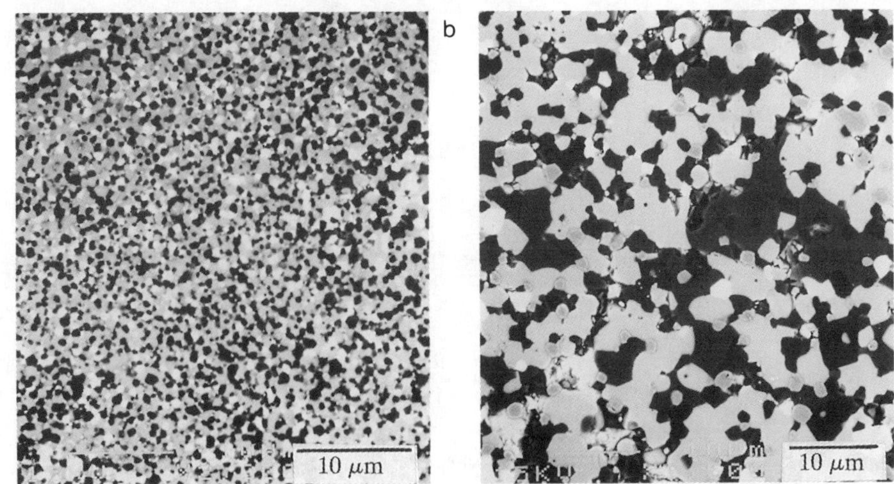

Figures 2 a+b: Backscattered electron images from sample ZFW-24 HIPped at 1200 °C and 1300 °C revealing contrast between phases.

TABLE II

SUMMARY OF LATTICE PARAMETERS FOR THE VARIOUS CERAMIC PHASES
PRODUCED.

	c-ZrO$_2$	BaZrO$_3$	Spinel
ZFW-24 4hrs 1300 °C N$_2$	0.516 nm	0.417 nm	0.842 nm
ZFW-24 2hrs 1300 °C HIP	0.516 nm	0.417 nm	0.844 nm
ZFW-24 2hrs 1200 °C HIP	0.516 nm	0.417 nm	0.842 nm
ZFW-28 4hrs 1300 °C N$_2$	0.516 nm	0.414 nm	0.839 nm
ZFW-28 2hrs 1300 °C HIP	0.516 nm	0.414 nm	0.843 nm
ZFW-28 4hrs 1200 °C HIP	0.516 nm	0.414 nm	0.843 nm

suggesting that it is kinetically stable. Clearly elimination of the Mo^{6+} during calcining is necessary. Thereafter, whether the molybdenum exists in the metallic or tetravalent state is not important because neither will be deleterious to leach resistance.

Such a change may have other consequences for the current waste form concept. Firstly the tantalum added to charge balance the substitution of caesium on the barium site of BaZrO$_3$ may

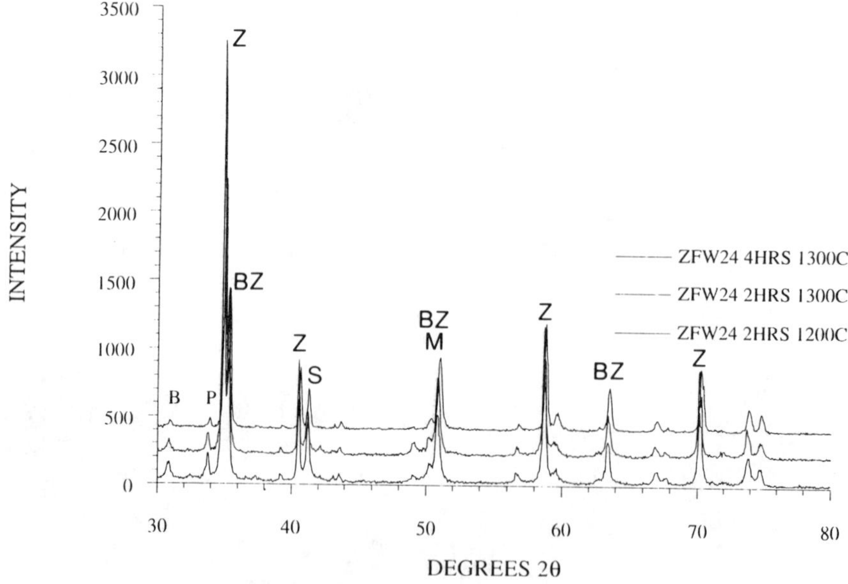

Figure 3: XRD trace from sample ZFW-24 consolidated at various temperatures and times showing gradual elimination of barium molybdate (B) and pyrochlore (P) phases. Principal peaks for zirconia (Z), barium zirconate (BZ), spinel (S) and metal (M) are also indicated.

also be reduced to the tetravalent state. In the current samples, tantalum has been observed within the $BaZrO_3$ grains, however, it may have already been reduced to the tetravalent state and its ability to perform the envisaged charge balancing role cannot be assumed under more reducing conditions. If this turns out to be the case then an alternative charge balancing mechanism will be necessary and the most likely option would appear to be the use of altervalent substitutions with large trivalent ions on the barium site. Additionally, the use of more reducing conditions may lead to the formation of further wüstite above the levels produced from the added iron powder.

The observed pyrochlore also appears to be only a metastable phase, although in no way deleterious to leach resistance. Figure 3 also shows that the principal pyrochlore peak just below $34°$ 2θ diminishes in intensity with increasing temperature and time. The intended composition of the cubic zirconia phase in this sample approximates closely to $Zr_{0.75}(La,Nd,Gd,Y)_{0.25}O_{1.875}$ with the rare earth elements present equiatomically. Preparation of this formulation shows it to be predominantly single phase cubic zirconia, Figure 4. The general composition of zirconias in this work has been targeted as $Zr_{0.75}REE_{0.25}O_{1.875}$ and in all other variants of the phase assemblage yttrium is the most abundant of the trivalent species and assures the formation of a single phase. The situation described here would appear to represent a worst case scenario with regard to incorporation of the rare earth elements into an homogeneous solid solution.

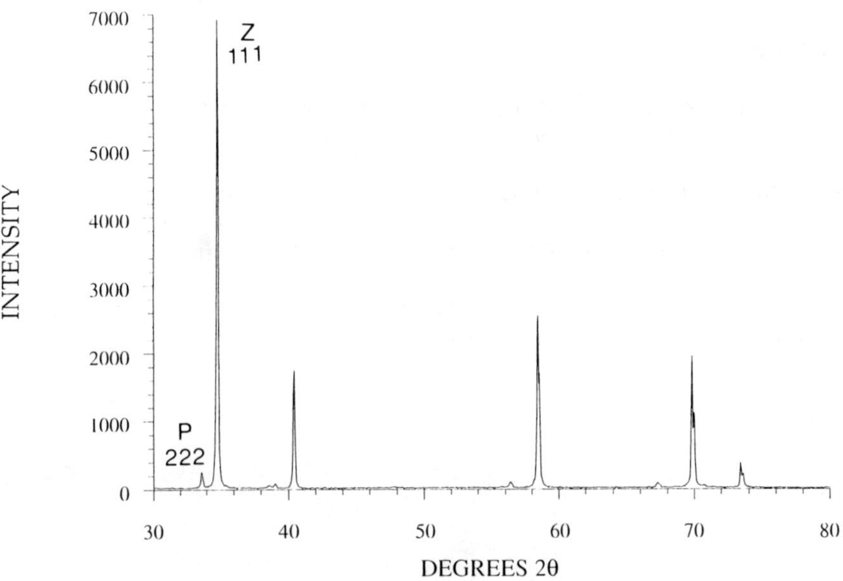

Figure 4: XRD trace from sample formulated to be $Zr_{0.75}(La,Nd,Gd,Y)_{0.25}O_{1.875}$ showing it to be predominantly cubic zirconia with a small trace of pyrochlore. Pyrochlore (222) and zirconia (111) peaks are indicated. Principal peaks are zirconia (111), (200), (220), (311) and (222).

<u>CONCLUSIONS AND FUTURE WORK</u>

The original objective of producing a phase assemblage consisting of three phases, cubic zirconia, magnetite and barium zirconate has been achieved. Additionally, smaller quantities of the non-equilibrium phases barium and caesium molybdate and pyrochlore together with wüstite from the metallic iron oxygen getter. The molybdate phases are of concern because of their deleterious effect on leach rates and have formed because of an inadequate supply of hydrogen during calcination. Elimination of hexavalent molybdenum at the calcining stage is essential, however, it is possible that more aggressive calcining will produce further wüstite and may modify the charge balancing mechanism necessary for the incorporation of alkali metals in the barium zirconate.

The formation of pyrochlore is less problematic because it does not diminish leach resistance and forms only when the zirconia is near to saturation.

The formulations chosen for study here have been deliberately designed to maximise waste loadings in the interests of waste minimisation. In practice, lower waste loadings may be necessary to reduce the thermal output of the waste form but this can be easily achieved by increasing the quantity of the zirconia phase. Additionally, variations in the composition of fuel assemblies, principally in the reduction of stainless steel and Inconel components, can be accommodated simply by a decrease in the volume fraction of the magnetite phase.

<u>REFERENCES</u>

1. Harker AB "Tailored Ceramics" in *Radioactive Waste Forms for the Future,* Eds. W Lutze and RC Ewing (Elsevier Science Publishers BV 1988) p335

2. Vance ER, Roy R, Pepin JG and Agrawal DK, J Mat Sci, **17** p947 (1982)

3. Maddrell ER, to be publshed in *Materials and Nuclear Power* (Institute of Materials, 1996)

4. Maddrell ER, in *Scientific Basis for Nuclear Waste Management XIX*, Eds. WM Murphy and DA Knecht (MRS Symposium Proceedings **412** Pittsburgh PA, 1996)

Part VI

Spent Nuclear Fuel

A KINETIC MODEL FOR THE STABILITY OF THE SPENT FUEL MATRIX UNDER OXIC CONDITIONS: MODEL DEVELOPMENT AGAINST EXPERIMENTAL EVIDENCE

JORDI BRUNO*, E. CERA*, L. DURO*, T.E. ERIKSEN**, P. SELLIN***, K. SPAHIU***
AND L.O. WERME***.

*QuantiSci, Parc Tecnològic del Vallès, 08290 Cerdanyola del Vallès, Barcelona (Spain).
**Dept. of Nuclear Chemistry, Royal Institute of Technology, S 100 44, Stockholm (Sweden).
***Swedish Nuclear Fuel and Waste Management (SKB), Box 5864, S 102 48, Stockholm
(Sweden).

ABSTRACT

A kinetic model recently developed [1] for the radiolytically induced oxidative dissolution of the spent fuel matrix is presented. This is based on experimental studies on the generation and evolution of radiolytic products in a closed system containing fragments of PWR-fuel [2]. The outcome of this model is currently being integrated in the present PA exercise being prepared by SKB. The calibration of the model against various experimental information and its predictive capabilities for the long term performance of the spent fuel matrix are presented.

INTRODUCTION

The chemical stability of a repository is defined by its resistance to radionuclide release to the biosphere. This resistance is primarily dependent on the chemical stability of the spent fuel matrix itself and spent fuel is disposed under chemical conditions that guarantee this stability. These chemical conditions are designed and engineered in order to minimise the potential release of radionuclides and are normally described by two intensive parameters (pH and Eh). It is clear that the redox potential is the most critical one to define the stability of the spent fuel matrix. In this context, it is now well established that the confinement of radionuclides within the matrix is guaranteed if the oxidation state of the UO_2 matrix does not exceed the upper limit of stability of the cubic structure, $UO_{2.33}$, which corresponds to a nominal stoichiometry of U_3O_7.

However, the spent fuel matrix constitutes a dynamic redox system by itself. This is because there is a time-dependent generation of oxidants and reductants at the fuel/water interface due to α, β and γ radiolysis. Under these circumstances, it is problematic to treat the spent fuel/water system as a redox equilibrium and consequently the definition of Eh (or pe) as an intensive parameter becomes questionable. The complexity is further increased if we try to integrate additional redox-sensitive components, like the canister or the pyrite content of the bentonite material, into the system under study.

The role of mineral surfaces in poising the redox state of a solid/water system is a well established phenomenon evolving from the concepts developed for heterogeneous electron transfer [3]. The fact that the UO_2 spent fuel matrix constitutes a semiconductor would indicate that the basic premises for heterogeneous electron transfer reactions are fulfilled. In the repository situation where the Mass to Volume ratio is particularly high, the UO_2 surface is going to play a key role in defining the redox capacities of the system and consequently its redox state.

Experimental evidence on the role of the UO_2 surfaces both as a reductant and on its redox buffering capacity is becoming available. Recently, Eriksen et al. [2], have reported well controlled experiments aimed at obtaining the mass balance of radiolytically produced oxidants and reductants by spent fuel in contact with distilled water. In these experiments, performed in a closed system, a clear deficit was found in the oxidant generation. The authors

491

Mat. Res. Soc. Symp. Proc. Vol. 465 © 1997 Materials Research Society

explained this oxidant deficit by the effect of their uptake by the UO_2 spent fuel surface. Natural uraninite deposits have also brought clear indications of the redox buffering capacity of the UO_2 surface. Redox measurements performed in groundwaters in contact with the uraninite deposit in Palmottu, Finland, indicate that in the reduced zone the redox potential appears to be controlled by the UO_2/U_3O_7 transition [4].

A kinetic model for the stability of the spent fuel matrix under oxic conditions has been recently presented under the name of RDC (ReDuctive Capacity) Model [1], which integrates a mass balance approach with the kinetics of heterogeneous electron transfer and dissolution/precipitation kinetics. The model has proven to reproduce quite satisfactorily experimental data on spent fuel and UO_2 dissolution.

The objectives of this paper are:

-to make a summary presentation of the RDC model

-to present the capabilities of the model to reproduce the behaviour of selected U and Pu dissolution data from leach test experiments.

-to present the results concerning the application of the model to the long-term stability of the matrix, as a part of the PA exercise presently being prepared by SKB.

SUMMARY DESCRIPTION OF THE RDC MODEL

The model has been previously presented in a recent publication [5]. We will shortly describe the main elements of the model.

The reductive capacity of a system according to Scott and Morgan [6] can be described by the following expression:

$$RDC = \sum_i n_i \times [Red] - \sum_j n_j \times [Ox] \tag{1}$$

where:

[Red] and [Ox] are the reductants and the oxidants of the system, and n_i and n_j are the number of electrons transferred in reactions.

The definition of oxidants and reductants requires the establishment of a reference state which is denominated Electron Reference Level (ERL) [6].

In order to describe the redox stability of the spent fuel matrix in a repository, the zero state for the redox capacity, the ERL, has been defined by the oxidation state of the UO_2 matrix before its disruption. This upper limit of oxidation state has been established to be $UO_{2.33}$, which corresponds to the stoichiometry U_3O_7 [7,8].

In Figure 1, we present in a schematic fashion the main processes occurring at the spent fuel matrix/water interface that affect the overall Reducing Capacity of the system.

These can be described by the following set of chemical reactions:

1. The *oxidation of the surface* of spent fuel:

$$> UO_2 \equiv UO_2 + 0.33H_2O \Rightarrow\ > UO_2 \equiv UO_{2.33} + 0.67H^+ + 0.67e^- \tag{2}$$

where "$> UO_2 \equiv UO_2$" notation stands for surface species.

2. The *reduction of the oxidants* radiolytically generated in the system. The main oxidant species produced by α-radiolysis of water are O_2 and H_2O_2, whose reduction reactions can be expressed as follows:

$$O_2 + 4H^+ + 4e^- \Leftrightarrow 2H_2O \tag{3}$$

$$H_2O_2 + 2H^+ + 2e^- \Leftrightarrow 2H_2O \tag{4}$$

3. The *dissolution of the oxidised spent fuel*, i.e., the release of uranium to the solution, described by the following expression:

$$> UO_2 \equiv UO_{2.33} + yH_2O \Rightarrow \; > UO_2 \equiv UO_2 + 0.33U(VI) + xH^+ \tag{5}$$

The U(VI) aqueous speciation will depend on the aqueous chemistry of the system. Thus, in principle two different processes can be proposed at pH in the range 7 to 8 for the dissolution of the oxidised uranium, depending on the predominant U(VI) aqueous speciation:

in distilled water:

$$> UO_2 \equiv UO_{2.33} + 0.33H_2O \Rightarrow > UO_2 \equiv UO_2 + 0.33UO_2(OH)_2(aq) \tag{6}$$

in granitic groundwater:

$$> UO_2 \equiv UO_{2.33} + 0.66HCO_3^- \Rightarrow > UO_2 \equiv UO_2 + 0.33UO_2(CO_3)_2^{2-} + 0.33H_2O \tag{7}$$

In addition the evolution of the aqueous $[U(VI)]_t$ with time is not only controlled by the dissolution from the UO_2 matrix, but also by the precipitation of secondary phases, particularly U(VI) oxyhydroxides ("$UO_2(OH)_2$"). This is given by the following processes, depending on the presence or absence of bicarbonate in the medium:

in distilled water:

$$UO_2(OH)_2(aq) \; \underset{k_d}{\overset{k_p}{\rightleftarrows}} \; \text{``}UO_2(OH)_2\text{''} \; (s) \tag{8}$$

where k_p and k_d stand, respectively, for the rate constants for the precipitation and the dissolution processes.

In granitic groundwater:

$$UO_2(CO_3)_2^{2-} + 2\,H_2O \; \underset{k_d}{\overset{k_p}{\rightleftarrows}} \; \text{``}UO_2(OH)_2\text{''} \; (s) + 2\,HCO_3^- \tag{9}$$

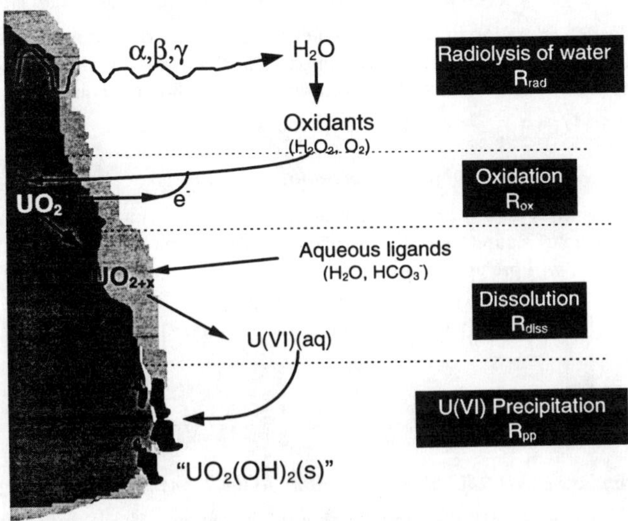

Figure 1. Schematic representation of the main processes occurring at the spent fuel matrix/water interface.

The RDC model takes into consideration the mass balance expressions of the main components involved in the system, oxidant, uranium, oxidised and non-oxidised sites of the surface of the spent fuel as well as their evolution with time.

<u>Oxidant evolution</u>

At any rate, the mass balance equation that would account for the evolution of the oxidant concentration in any hypothetical case, by considering radiolytic generation of oxidants will be expressed by the following equation:

$$[O_2]_t = [O_2]_{rad} - [O_2]_{oxid} \qquad (10)$$

where $[O_2]_{rad}$ stands for the generation of oxidants due to radiolysis of water, and $[O_2]_{oxid}$ represents the consumption of oxidants due to the oxidation of the spent fuel surface. The differential equation accounting for the evolution of $[O_2]$ will be:

$$\frac{d[O_2]_t}{dt} = R_{rad} - 0.166 \times R_{oxid} \qquad (11)$$

<u>Uranium evolution</u>

The evolution of the concentration of uranium in the system is strongly dependent on the hydrological condition of the system. Most of the experiments reported in the literature have been performed under batch conditions and, therefore, the oversaturation of the solution with respect to secondary uranium solid phases may play an important role. Furthermore, the expected flow conditions in a undisturbed spent fuel repository are expected to be *quasistatic,* with very low flow velocities [9]. For this reason, although the kinetic model that we have developed takes into account the possibility of no U(VI) secondary phases oversaturation, it is clear that the conditions in the repository will determine the precipitation of secondary phases.

Neglecting the precipitation of secondary U(VI) solid phases due to oversaturation effects

In this case, the concentration of uranium in the system will increase continuously since no outflow of water from the system occurs. The mass balance equation that will account for the uranium concentration at any time *t* will be expressed by the following equation:

$$[U(VI)]_t = [U(VI)]_{t-1} + R_{diss} \times \Delta t \qquad (12)$$

Precipitation of secondary U(VI) solid phases according to equilibrium

This approach assumes instant precipitation as soon as the concentration of uranium in solution reaches the value of saturation of a U(VI) solid phase (we have considered this phase to be schoepite) the solid will precipitate instantaneously. The set of equations describing this case will be:

$$[U(VI)]_t = \begin{cases} [U(VI)]_{t-1} + R_{diss} \times \Delta t; & \text{for } [U(VI)]_t < [U(VI)]_{eq} \\ \\ [U(VI)]_{eq}; & \text{for } [U(VI)]_t \geq [U(VI)]_{eq} \end{cases} \qquad (13)$$

Precipitation of secondary U(VI) solid phases according to the Transition State Theory

This approach assumes that the precipitation of secondary U(VI) solid phases is governed by kinetics following the well-established Transition State Theory [10], and incorporates a retardation factor which depends on the saturation state of the system with respect to the solid phase that is precipitating.

The mass balance equations accounting for this effect will be different depending on the major dissolved U(VI) species present in solution. As we are mostly interested in the behaviour of the system at pH=7-8 in water with and without bicarbonate, the mass balance equations considered are as follows.

At pH = 7 in the absence of carbonates in solution, distilled water, the speciation of uranium will be dominated by the aqueous complex $UO_2(OH)_2$, according to equation 8. The corresponding mass balance equation by integrating the precipitation process will be:

$$[U(VI)]_t = [U(VI)]_{t-1} + R_{diss} \times \Delta t + k_d \times \left[1 - K_{eq} \times [U(VI)]_t\right] \times \Delta t \qquad (14)$$

where K_{eq} corresponds to the equilibrium constant of the process described in equation 8.

At pH = 8 in the presence of carbonates in solution, granitic groundwater conditions, the speciation of uranium will be dominated by the aqueous complex $UO_2(CO_3)_2^{2-}$, according to equation 9. The corresponding mass balance equation , by including the precipitation process will be:

$$[U(VI)]_t = [U(VI)]_{t-1} + R_{diss} \times \Delta t + k_d \times \left[HCO_3^-\right]^2 \times \left[1 - K_{eq} \times \frac{[U(VI)]_t}{\left[HCO_3^-\right]^2}\right] \times \Delta t \qquad (15)$$

where K_{eq} corresponds to the equilibrium constant of the process described in equation 9.

<u>The evolution of the UO_2 surface sites</u>

- $>UO_2\equiv UO_2$ evolution

$$\left[>UO_2 \equiv UO_2\right]_t = -\left[>UO_2 \equiv UO_2\right]_{ox} + \left[>UO_2 \equiv UO_2\right]_{diss} \qquad (16)$$

$$\left[>UO_2 \equiv UO_2\right]_t = \left[>UO_2 \equiv UO_2\right]_{t-1} - R_{ox} \times \Delta t + R_{diss} \times \Delta t \qquad (17)$$

- $>UO_2\equiv UO_{2.33}$ evolution

$$\left[>UO_2 \equiv UO_{2.33}\right]_t = \left[>UO_2 \equiv UO_{2.33}\right]_{ox} - \left[>UO_2 \equiv UO_{2.33}\right]_{diss} \qquad (18)$$

$$\left[>UO_2 \equiv UO_{2.33}\right]_t = \left[>UO_2 \equiv UO_{2.33}\right]_{t-1} + R_{ox} \times \Delta t - R_{diss} \times \Delta t \qquad (19)$$

The formulation of the model imposes the following constraint on the number of sites:

$$\left[>UO_2 \equiv UO_{2.33}\right]_t + \left[>UO_2 \equiv UO_2\right]_t = \left[>UO_2 \equiv UO_2\right]_0 \qquad (20)$$

where $\left[>UO_2 \equiv UO_2\right]_0$ stands for the initial total number of sites available for the oxidant attack. The density of sites available for oxidation has been extracted from previous data of Casas [11] with a value of $2.94 \cdot 10^{-5}$ moles of sites$\cdot dm^{-2}$.

PARAMETERS OF THE RDC MODEL

Once we have established the main processes and mass balance equations we need to assign values to the key parameters of the model. One of the key elements of the model is the rate of total oxidant generation by water radiolysis. Total oxidant production from radiolysis, R_{rad}, is a function of dose rate, Dose(t), the geometry of the system defined by the surface area to volume ratio, M/V, oxidant consumption by the UO_2 surface, $R_{oxid}(t)$, and composition of the contacting water, in particular the presence or absence of carbonates. Dose rate is a function of time, while the rate of oxidant consumption by the UO_2 surface equals to the rate of oxidation of UO_2 by the main oxidants produced by radiolysis H_2O_2 and $O_2(aq)$. Summarising:

$$R_{rad} = f(Dose(t), M/V, R_{oxid}(t), [HCO_3^-]) \qquad (21)$$

The dose rate, Dose(t), has been calculated by using the inventory of the PWR fuel elements Ringhals DO-7-S14 and a small particle model [12]. The M/V ratio used in the calculations is given in the geometric description of the system considering a hypothetical fuel pellet, contained in the canister, as given in Figure 2.

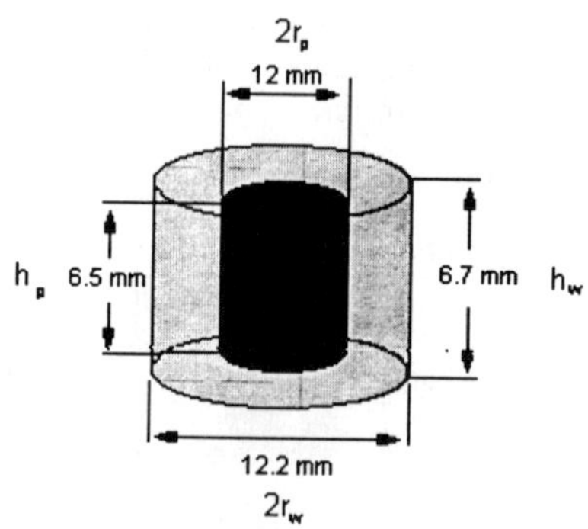

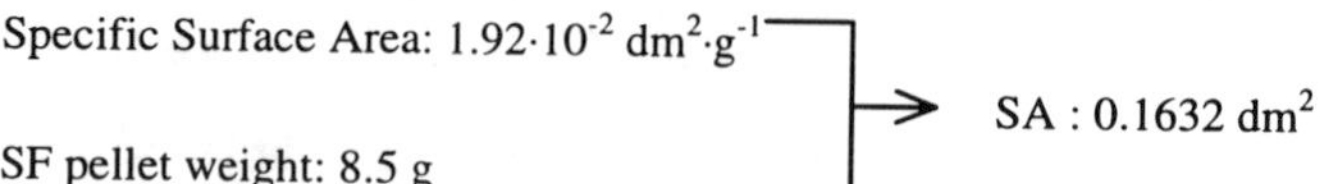

Specific Surface Area: $1.92 \cdot 10^{-2}$ dm$^2 \cdot$ g^{-1}

SF pellet weight: 8.5 g

$\Rightarrow$ SA : 0.1632 dm^2

Figure 2. Geometric description of the system considered.

The bicarbonate concentration considered ranges between 0 (distilled water) and 10 mM (bentonite pore water), with a 2mM concentration for the average composition of the granitic groundwater.

The rate of oxidation of the UO_2 surface by the radiolytically generated oxidants is a critical parameter because it influences the consumption of oxidants, as well as the oxidation of the UO_2 matrix and subsequent dissolution. Therefore, the selection of this datum deserves some discussion:

The dissolution of UO_2 and spent fuel under oxic conditions has been extensively studied. The rates of oxidative dissolution of UO_2, (R^{ox}_{diss}) as summarised by Johnson and Shoesmith (1988) [13], are in the range 10^{-12} to 10^{-10} mole dm^{-2} s^{-1} at 100 kPa oxygen pressure. The lower limit of the range corresponds to the determinations in distilled water, while the upper one corresponds to measurements in carbonate solutions. These values correspond to determinations of the kinetics of UO_2 oxidative dissolution and the two processes (oxidation and dissolution) cannot be decoupled. In this context, Torrero (1995) [14] has determined the rate of oxidation of UO_2 (R_{ox}) as a function of dissolved O_2 by XPS measurements of the surface oxidation. The author has established that the oxidation rate is given by the expression:

$$R_{ox} = k_{ox} \cdot [O_2(aq)] \tag{22}$$

with $k_{ox} = 7 \cdot 10^{-8}$ dm s^{-1}.

Specific measurements of the rate of oxidation of UO_2 by H_2O_2 have not been published to our knowledge. However, Shoesmith and Sunders [15] established the rate of oxidative dissolution of UO_2 by H_2O_2, as given by the expression:

$$R^{ox}_{diss} = k_{H2O2} \cdot [H_2O_2] \tag{23}$$

with $k_{H2O2} = 1.2 \cdot 10^{-10}$ (dm s^{-1})

Giménez [16] also reports a linear dependence of the oxidative dissolution rate on the hydrogen peroxide concentration with a rate constant: $k_{H2O2} = 4.9 \cdot 10^{-10}$ (dm s^{-1}), in reasonable agreement with the data from Shoesmith and Sunders. A third set of experimental data concerning the rate of oxidative UO_2 dissolution in H_2O_2 solutions has been reported by Hiskey [17]. The dissolution studies were performed in ammonium carbonate solutions

(0.005-0.5 mole dm^{-3}) and the author derived an empirical rate equation given by the expression:

$$R^{0x}_{\ diss} = k^{CO3}_{\ H202} \cdot [H_2O_2]^{0.5} \cdot [HCO_3^-]^{0.5} \qquad (24)$$

with a rate constant $k^{CO3}_{\ H202}= 7 \cdot 10^{-7}$ mole $dm^2 s^{-1}$ in the hydrogen peroxide concentration range: 0.001-0.22 mole dm^{-3}.

The oxidative dissolution rates reported by Hiskey [17] are several orders of magnitude higher than the ones reported by Shoesmith and Sunder [15] and Giménez[16]. This is expected from the combined effect of hydrogen peroxide and large carbonate concentrations on the UO_2 dissolution.

Therefore, we have selected an average $R^{0x}_{\ diss}= 3 \cdot 10^{-10} \cdot [H_2O_2]$ by combining Shoesmith and Sunders [15] and Giménez determinations [16]. These values refer to the rates of H_2O_2 - promoted oxidative dissolution of UO_2. The rate of oxidation by H_2O_2 is expected to be higher than in the case of oxygen, therefore we have carried out some sensitivity calculations by using a rate constant which is two orders of magnitude larger ($k_{H2O2}= 3 \cdot 10^{-8}$ dm s^{-1}).

Finally, the dissolution rate of the oxidised surface layer has been extracted from well-controlled flow-through dissolution experiments of the $UO_{2.33}$ surface under nominaly reducing conditions (Pd/Hg) [18]. From this work, we have selected the following value for the rate of dissolution of $UO_{2.33}$:

$$R_{diss} = (1.1 \pm 0.3) \cdot 10^{-12} \cdot [H^+]^{-0.3} \ mol \cdot m^{-2} \cdot s^{-1} \qquad (25)$$

CALIBRATION OF THE RDC MODEL AGAINST SPENT FUEL DISSOLUTION DATA

In order to test the validity of the RDC model we have attempted to model the evolution of U and Pu from selected spent fuel dissolution tests. The data modelled has been previously reported by Forsyth et al. [19]. These authors performed four series of experiments by using as a sample a well-characterised BWR fuel rod irradiated to a burn-up of 42 Mwd/kg U. The dissolution experiments were performed at ambient temperature and by using a synthetic groundwater as solvent. The details of the U calculations can be found in reference [1], while the details for the Pu calculations are described in [20].

Basically, in order to model the evolution of U we have used the same processes and parameters already described. However, the calculations have been performed taking into consideration the specific M/V and geometric arrangements of the dissolution tests. The oxidant evolution with time has been calculated by taking into consideration that these experiments were performed initially in contact with air but in a nominally closed system.

Although radiolytic generation of oxidants was to be expected under these conditions, the expected maximum oxidant concentration was in the level of 10^{-8} mole dm^{-3} of dissolved oxygen. This concentration is negligible compared with the initial dissolved oxygen present in the solution as a result of its contact with air ($2.5 \cdot 10^{-4}$ mole dm^{-3}). Consequently the evolution of oxygen with time has been considered to be given only by the consumption of the initially dissolved oxygen by the oxidation of the spent fuel matrix in the closed system.

The outcome of the calculations carried out is presented in Figure 3, the model has been calculated by considering the precipitation of a secondary uranium solid phase according to the Transition State Theory as described above.

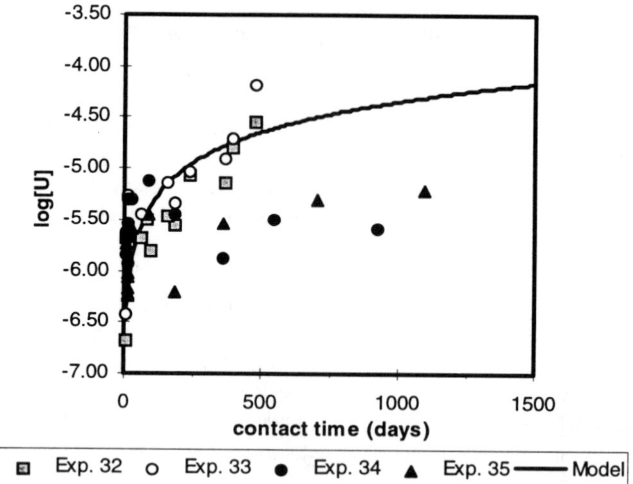

Figure 3. Evolution of uranium dissolved with time (Model). Experimental data reported by Forsyth et al. (1986) [19].The labels correspond to the ones used by the authors in reference 19.

Furthermore, in the case of Pu, we have assumed the following additional processes that control the behaviour of Pu in the leaching experiments:

Pu is present at the fuel surface as a solid solution of Pu(IV) in the UO_2 matrix, symbolised by $Pu_xU_{1-x}O_2$, with x: 0.01-0.02 (1-2% enrichment), depending on the fuel burn-up and history. This will be symbolised as $xPuO_2$ in $UO_2(ss)$. The initial $Pu_xU_{1-x}O_2$ is dissolved at a rate R_0 according to the process:

$$x\ PuO_2 + 2x\ H_2O \Rightarrow x\ Pu(OH)_4(aq) \tag{26}$$

The Pu(IV) dissolution rate will be given by the product of the x fraction of Pu in the matrix, times the oxidative dissolution rate of the UO_2 matrix. This is:

$$R_0(Pu) = x\ R_{ox}(UO_2) = (0.75\pm0.25)\cdot10^{-13}\ \text{mole Pu dm}^{-3}\cdot\text{s}^{-1} \tag{27}$$

Pu is in the IV redox state under radiolytic oxidative conditions and it is only reduced to Pu(III) at redox potentials lower than -150 mV under the granitic groundwater conditions of the experiments. The predominant Pu(IV) species under the conditions of this study is $Pu(OH)_4(aq)$. Furthermore, Pu(IV) has a strong and worrisome tendency to build colloidal dispersions in contact with water. The reaction describing this process can thus be given as:

$$Pu(OH)_4(coll) \underset{k_{-1}}{\overset{k_1}{\rightleftarrows}} Pu(OH)_4(aq) \tag{28}$$

However, these colloids tend to aggregate towards the formation of amorphous Pu(IV) hydroxide, $Pu(OH)_4(am)$, with time in ionic solutions. The equilibrium between the amorphous Pu(IV) hydroxide and the Pu(IV) in solution is given by the reaction:

$$Pu(OH)_4(aq) \underset{k_{-2}}{\overset{k_2}{\rightleftarrows}} Pu(OH)_4(am) \tag{29}$$

The experimental Pu data has been analytically fitted to a function which contains the main kinetic constants describing the evolution of the Pu(IV) concentrations with time.

The expression of the rate of dissolution of the colloids (28) will be given by the following equation:

$$R_1 = \frac{d[Pu]}{dt} = k_1 - k_{-1}\left[Pu(OH)_4(aq)\right] = k_1 - k_{-1}[Pu] \qquad (30)$$

The rate expression for the process expressed in equation 29 will be given by:

$$-R_2 = \frac{d[Pu]}{dt} = k_{-2} - k_2\left[Pu(OH)_4(aq)\right] = k_{-2} - k_2[Pu] \qquad (31)$$

The precipitation of Pu(OH)$_4$(am) (29), will start from the Pu concentration released in the first process (28), therefore, we will integrate the former expression from [Pu]$_0$ to [Pu] and from $t_0 = 0$ to t, where [Pu]$_0$ will be given by the integration of equation 30. The resulting evolution of the Pu concentration with time will be:

$$[Pu] = \frac{k_{-2}}{k_2} - \left\{ \frac{k_{-2}}{k_2} - \frac{k_1}{k_{-1}}\left\{1 - \exp(-k_{-1} \times t)\right\} \right\} \exp\left\{-k_2 \times t\right\} \qquad (32)$$

The goodness of the fitting between the experimental and calculated data can be observed in Figures 4 (a-b).

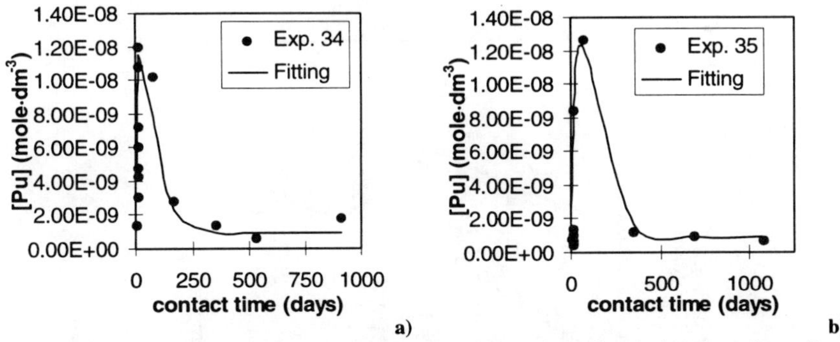

Figure 4. Fit of the model to the experimental data of plutonium dissolution from Spent Fuel. Experimental data reported by Forsyth et al. (1986) [19]. The labels correspond to the ones used by the authors in reference 19.

The best fit to the experimental data was obtained by using the following set of kinetic parameters:

	$\dfrac{k_1}{k_{-1}}$ (K$_1$)	$\dfrac{k_{-2}}{k_2}$ (K$_{so}$)	k_{-1} (d^{-1})	k_2 (d^{-1})
Average	$2.8 \cdot 10^{-8}$	$8.75 \cdot 10^{-10}$	0.046	0.022
St.dev	$0.1 \cdot 10^{-8}$	$0.43 \cdot 10^{-10}$	0.016	0.008

The obtained solubility constant (K$_{s0}$) is in the same range (within one order of magnitude) as the one proposed for the solubility of Pu(OH)$_4$(am) [21], giving support to the chemical meaning of the derived parameters. On the other hand, the reproducibility of the K$_1$ constant for the reaction between colloidal and dissolved Pu(OH)$_4$ would indicate that this process has some degree of reversibility and can be approached from a thermodynamic stand point. The resulting "solubility" constant is 2 orders of magnitude larger than the one corresponding to the amorphous Pu(OH)$_4$ (K$_{s0}$). This would be expected from a phase with much larger interfacial energy.

APPLICATION OF THE RDC MODEL TO A HYPOTHETICAL SPENT FUEL REPOSITORY

Once the model has been stated, and the main processes and parameters have been calibrated, we will explore the consequences for a hypothetical spent fuel repository.

R_{rad}, rate of oxidant generation. In Figure 5 we have plotted the evolution of reductants and oxidants generated by water radiolysis by considering the system specified before (Figure 2). We also show the evolution of the consumed oxygen and hydrogen peroxide by the oxidation of the UO_2 surface. These calculations have been done by using the Macksima Chemist code and a set of 38 relevant reactions, as described by Eriksen et al. [2].The dose calculations have been based on the inventory of the PWR fuel element Ringhals DO-7-S14 with an average burn-up of 40 MWd/kgU used in earlier radiolysis experiments performed by Eriksen and co-workers [2] and a small particle model developed by Jansson et al. [12] The results in Figure 5 refer to a calculation case assuming 2 mM total carbonate and the average value for k_{H2O2} refered before. An increase on k_{H2O2} by two orders of magnitude, results in 60-70 fold increase on the generation rate of H_2.

Therefore, this parameter is of critical importance for the development of the kinetic model for spent fuel oxidative dissolution.

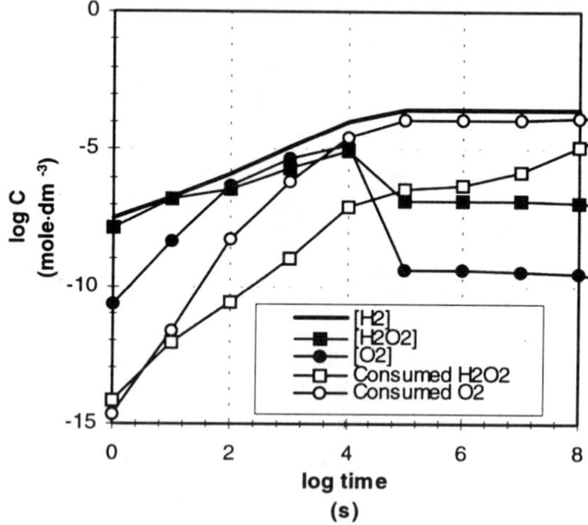

Figure 5. Evolution of reductants and oxidants generated by water radiolysis and consumed oxygen and hydrogen peroxide by the oxidation of the UO_2 surface.

Estimation of the weight loss percentage due to the radiolytic oxidation of spent fuel. Finally, we have applied the RDC model to calculate the fraction of spent fuel weight loss as a function of the radiolytic generation of oxidants. This has been done by integrating the processes and parameters previously described and calculating the evolution of the system with time.

Figure 6 is a plot of the percentage of weight loss of the uranium dioxide pellets, with a geometry as previously described in Figure 2, assuming contacting groundwater with a flow velocity of 0.1 dm^3/year and $[HCO_3^-] = 2 \cdot 10^{-3}$ mole dm^{-3}. The effect of assuming or not secondary $UO_2(OH)_2$ precipitation is relevant for the stability of the system. Furthermore, the assumption of instant or kinetically controlled secondary precipitation has a lower effect.

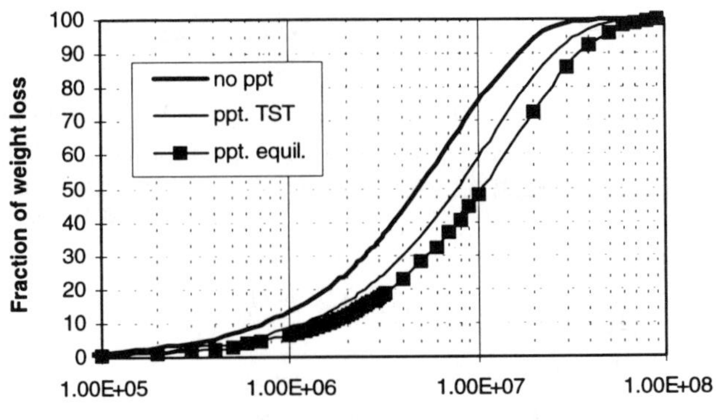

<u>Figure 6</u>. % of release of uranium from SF with and without assuming precipitation of secondary U(VI) solid phases. $[HCO_3^-]$ = 2 mM. Groundwater flow velocity = 0.1 dm^3 per year.

CONCLUSIONS

The UO_2 surface plays an important role in the generation and scavenging of the oxidants produced by water radiolysis.

The combination in the RDC model of electron mass balance equations with the kinetics of oxidant generation by H_2O radiolysis, together with the kinetics of relevant dissolution and precipitation processes help us to understand and model the behaviour of a redox dynamic system like the spent fuel/water interface.

A satisfactory description of the processes involved in the generation and consumption of oxidants at the water/spent fuel interface requires a fundamental understanding of the mechanisms and rates of the UO_2 matrix oxidation by H_2O_2 and O_2(aq). The effect of hydrogen peroxide is critical but our knowledge on the mechanisms of oxidation is still insufficient.

The calibration of the RDC model against early U and Pu dissolution data from spent fuel leaching experiments has proven to be quite encouraging, giving support to the applicability of the model presented.

The application of the model to describe the evolution of the spent fuel matrix under disposal conditions indicates, that the assumption of secondary U(VI) precipitation enhances the stability of the matrix for longer periods. The precipitation of secondary U(VI) solid phases has been identified in some spent fuel laboratory tests [19,22] and evidence for secondary phase formation is substantial in natural system studies [23].

ACKNOWLEDGEMENTS

The financial support of SKB, the Swedish Fuel and Waste Management Co. to this work is gratefully acknowledged. Dr. Joan de Pablo (UPC) has contributed with many discussions and analysis of the UO_2 oxidation data.

REFERENCES

1. J. Bruno, E. Cera, L. Duro, T.E. Eriksen and J. de Pablo, SF-96 QuantiSci Progress Report. Draft version 2.0, July 1996.

2. T.E. Eriksen, U.B. Eklund, L.O. Werme and J. Bruno, J. Nucl. Mater. **227**, 76 (1995).

3. A.F. White and A. Yee, Geochim. Cosmochim. Acta **49**, 1263 (1985).

4. J. Bruno, E. Cera, L. Duro and L. Ahonen, TVO Technical Report, 1996. In press.

5. J. Bruno, E. Cera, L. Duro, T.E. Eriksen and L.O. Werme, J. Nucl. Mater. (1996). In press.

6. M.J. Scott and J.J. Morgan, Energetics and Conservative Properties of Redox Systems, in: Chemical Modeling of Aqueous Systems, Chapter 29 (American Chemical Society, 1990), p. 368.

7. R.J. Finch and R.C. Ewing, J. Nucl. Mater. **190**, 133 (1992).

8. S. Sunder, D.W. Shoesmith, M.B. Bailey, F.W. Stanchell and N.S. McIntyre, J. Electroanal. Chem. **130**, 163 (1981).

9. SKB91, SKB/TR-92-20, 1992.

10. A.C. Lasaga, J.M. Soler, J. Ganor, T.E. Burch and K.L. Nagy, Geochim. Cosmochim. Acta **58**, 2361 (1994).

11. I. Casas, Ph.D. Thesis, Universitat Autònoma de Barcelona, 1989.

12. M. Jansson, M. Jonsson and T.E. Eriksen, SKB Internal Report, 1994.

13. L.H. Johnson, and D.W. Schoesmith, Spent Fuel, in: Radioactive waste forms for the future, edited by W. Lutze and R.C. Ewing (Elsevier Science Publishers, B.V., 1988).

14. M.E. Torrero, E. Baraj, I. Casas, J. Giménez and J. de Pablo, Radiochim. Acta (1995). Submitted.

15. D.W. Shoesmith and S. Sunder, J. Nucl. Mater. **190**, 20 (1992).

16. J. Giménez, Ph.D. Thesis, Universitat de Barcelona, 1996.

17. J.B. Hiskey, The Institution of Mining and Metallurgy 622.775:622.7-349.52, 1980.

18. J. Bruno, I. Casas and I. Puigdomènech, Geochim. Cosmochim. Acta **55**, 647 (1991).

19. R.S. Forsyth, L.O. Werme and J. Bruno, J. Nucl. Mater. **138**, 1 (1986).

20. J. Bruno. and L. Duro, QuantiSci Internal Report, September 1996.

21. I. Puigdomènech and J. Bruno, SKB/TR-91-04, 1991.

22. K. Ollila, Report YJT-95-14, 1995.

23. R.J. Finch. and R.C. Ewing, SKB/TR-91-15, 1991.

DETERMINATION OF Cl IMPURITIES AND ^{36}Cl INSTANT RELEASE FROM USED CANDU FUELS

J.C. Tait[*], R.J.J. Cornett[†], L.A. Chant[†], J. Jirovec[†], J. McConnell[*] and D.L. Wilkin[*]. Atomic Energy of Canada Limited, [*]Whiteshell Laboratories, Pinawa, Manitoba, Canada. [†]Chalk River Laboratories, Chalk River, Ontario, Canada.

ABSTRACT

Chlorine-36 has been identified as a potential source of radiological risk in the disposal of nuclear fuel waste. The radioisotope ^{36}Cl ($t_{½}$ = 3 x 10^5 a) is produced by neutron activation of Cl impurities in UO_2 fuel. The total average Cl impurity level in four unirradiated CANDU UO_2 fuel samples was 2.3 ± 1.1 ppm. ORIGEN-S calculations using a 5 ppm Cl impurity in a CANDU fuel resulted in a ^{36}Cl activity comparable to the activity of ^{129}I and ^{14}C produced in the fuel thus requiring ^{36}Cl to be considered in disposal risk assessments. The "instant release" of ^{36}Cl from the gap and grain boundary regions of the fuel to solution was measured by leaching both clad fuel and fuel samples crushed to grain-sized particles. The ^{36}Cl concentration was measured by Accelerator Mass Spectrometry. The ^{36}Cl releases from fuel samples taken from 8 different fuel bundles ranged from 0.5% to 20.4% of the total ^{36}Cl inventory over a leaching period of 32 days. The ^{36}Cl released was found to correlate with the stable Xe gas release, the fuel burnup and the linear power rating (LPR). For a typical CANDU fuel with an LPR of ~42 kW/m, the "instant release" of ^{36}Cl would be about 5% of the total inventory.

INTRODUCTION

The concept for disposal of used CANDU fuel bundles within corrosion resistant containers, 500 to 1000 m deep in the granite rock of the Canadian Shield, is being investigated in the Canadian Nuclear Fuel Waste Management Program [1]. Groundwaters present in the disposal vault will eventually corrode the containers and contact the UO_2 fuel, resulting in dissolution of the UO_2 fuel and release of radionuclides [2]. Extensive studies on used fuel dissolution have established that radionuclide release is dominated by two processes: the rapid release of a small fraction of the inventory of radionuclides such as ^{137}Cs, ^{129}I and ^{14}C present in the fuel sheath gap and grain-boundary regions; and congruent dissolution from the UO_2 matrix [2]. The model describing radionuclide release thus consists of an "instant release" fraction and a long-term matrix dissolution model describing the release of the remaining radionuclide inventory [3,4].

The "instant release" term has the largest risk impact for dose to humans in the biosphere, because it involves the rapid release of a fraction of the inventory of highly mobile radionuclides such as ^{129}I and ^{14}C. Numerous studies on CANDU fuel have established a strong correlation between the measured gap inventory of ^{137}Cs, ^{129}I and the measured stable Xe fission gas released to the gap region. [5,6,7,8,9]. The prediction of I release is particularly important, since ^{129}I has been shown to be the principal contributor to dose to humans in a postclosure assessment of the Canadian disposal concept [10].

Recently, it became apparent that ^{36}Cl ($t_{1/2}$ = 3 x 10^5 a) produced by neutron activation of naturally abundant ^{35}Cl impurities in the UO_2 and Zircaloy cladding can make a significant contribution to the radiation risk impact and that the contribution to dose from ^{36}Cl would be about forty times lower than that expected from ^{129}I [11]. Analyses on Zr/2.5Nb pressure tubes [12] indicated that typical Cl impurity levels were 1 to 5 ppm (μg/g Zr). This lead to a

Mat. Res. Soc. Symp. Proc. Vol. 465 © 1997 Materials Research Society

speculation that Cl impurities may also be present in the Zircaloy cladding and in the UO_2 fuel pellets at similar levels.

In this work, unirradiated UO_2 fuel pellets have been analysed for total Cl concentration, and irradiated CANDU UO_2 pellets have been analysed for total ^{36}Cl concentration, to provide a basis for the estimation of an upper limit for Cl impurities in UO_2. The "instant release" of ^{36}Cl released to solution over a period of about 30 days has also been measured to support estimates of the potential impact of ^{36}Cl on postclosure assessment studies.

EXPERIMENTAL

Analysis of Chlorine Impurities in Unirradiated UO_2

Chlorine in unirradiated UO_2 samples was analyzed by ion chromatography after its pyrohydrolytic extraction (thermal decomposition in the presence of a V_2O_5 accelerator and high purity oxygen) and subsequent collection of the evolved gaseous products [13]. The UO_2 samples (100 - 500 mg) were selected from unirradiated fuel pellets or batches of UO_2 powder used to manufacture CANDU fuel pellets. UO_2 samples were contained in a nickel boat and pyrolysed in a quartz reaction tube at $1000^{o}C$. The lower limit of detection for the procedure was 0.5 µg Cl/g for a sample size of 500 mg.

Determination of Total ^{36}Cl in Used Fuel

Used CANDU fuel samples (Table I) were analyzed for total ^{36}Cl content using Accelerator Mass Spectrometry (AMS) [14]. Two-cm segments of clad used fuel were added to 100 mL 50% HNO_3 containing 0.7 g NaCl as a carrier. The solution was heated under reflux to $90^{o}C$ for three hours and cooled. A 10 mL aliquot of this solution was diluted to 100 mL with deionized water (DIW) and ammonium molybdophosphate (AMP) was added to remove ^{137}Cs. The solution was filtered, placed in centrifuge tubes and 2 mL 10% $AgNO_3$ was added to precipitate AgCl. The solutions were centrifuged, rinsed with DIW and the solid precipitates redissolved in 100 mL concentrated NH_4OH and submitted for AMS analysis. Sufficient stable Cl was added to the sample containing ^{36}Cl prior to precipitation to provide enough bulk for analysis and to dilute the $^{36}Cl/Cl$ ratio to within an optimal analysis range. Ion Chromatographic analysis of the re-dissolved precipitate indicated 100% recovery of the chloride carrier.

Leaching of Used Fuel Samples

Irradiated CANDU fuels were leached in either DIW or in a borate buffer solution. Approximately 25 g of fuel segments were placed in 100 mL borate buffer (pH 8.5 containing 0.02 g/L KI as an iodine carrier) in a mortar bowl of a motorized mortar and pestle [8]. Grinding reduced the fuel to a powder with >90% of the fuel with a grain size from 5 to 20 µm. The fuel was ground in the borate solution for about 4 hours followed by addition of a few drops of 0.01 mol/L NaOH to maintain a high pH. The solution was allowed to settle and samples were taken using a syringe and filtered sequentially through 0.45 µm and 0.22 µm Millipore syringe filters; AMP was used to remove ^{137}Cs [8]. A 2-cm segment of clad used fuel was also leached (with no grinding) in 100 mL of borate buffer for 7 d and the leachant solution treated similarly to the pulverized fuel leachant. A sample of unirradiated UO_2 was also leached and treated in a similar manner, with and without the AMP extraction procedure. Leachant samples were analyzed for ^{36}Cl using AMS.

Table I Used CANDU Fuel Characteristics

Fuel	Burnup (MWh/kg U)	Linear Power Rating (kW/m)	Fission Gas Release (Measured) (%)
BF21271C OE	327	47	2.75
B12059C OE	269	46	3.44
B12059C INT	222	38	0.09
BJ49093C OE	308	50	7.03
PA13894W OE	202	44	1.28
PA13894W INT	166	36	0.05
PA13894W INN	147	32	0.03
PA19558C OE	221	43	7.8
BG133745 OE	300	50	4.18
PA07993W OE	372	40	4.31
BG00815W INT	134	33	0.17

A second set of leaching experiments was conducted on ~2-cm lengths of clad used fuel samples (~30 g UO_2). The fuels (Table I) were leached in deionized water (DIW) to which 0.7 g NaCl was added as a carrier. Samples (10 mL) were taken at several intervals from 5 to 32 days leaching and these samples were treated as above prior to AMS analysis for ^{36}Cl.

RESULTS

<u>Determination of Cl Impurities in Used Fuel and Unirradiated UO_2</u>

The pyrohydrolytic decomposition analyses were conducted on two separate sources of UO_2 used in the manufacture of CANDU fuel pellets and two unirradiated fuel pellets. Each sample was analysed six to nine times and the results were averaged for each sample. These results indicated Cl impurity levels ranging from 1.6 to 2.6 ppm. An independent analysis was conducted on a fifth unirradiated UO_2 pellet (Teledyne Wah Chang analytical services; Albany, Oregon) using pyrolytic decomposition and coulometry to give a Cl impurity of 3.0 ± 0.1 ppm. The average chlorine impurity level in UO_2 from all these determinations is 2.3 ± 1.1 ppm.

<u>Determination of Total ^{36}Cl in Used CANDU Fuel</u>

Five used CANDU fuel elements were analysed for total ^{36}Cl content (Table II) and the total Cl content was estimated using ORIGEN-S code calculations [15]. Radioisotope generation-depletion calculations were performed on the reference fuel burnup used in performance assessment calculations [10] (190 MWh/kg U, the average CANDU fuel burnup) using an assumed 5 ppm initial Cl impurity concentration. The calculated inventory of ^{36}Cl was 4.22 x 10^5 Bq/kg initial U. The production of ^{36}Cl in the fuel is directly related to the integrated neutron flux, which is directly related to burnup, and thus the quantity of ^{36}Cl produced in any fuel can be estimated using burnup ratios. The measured total ^{36}Cl in the fuels were then compared with the calculated burnup corrected values and the ratio of these values was used to derive the total Cl impurity level in each fuel. The derived Cl impurity levels ranged from 0.3 to 3.4 ppm with an average of 1.5 ± 1.1 ppm.

Table II Analysis of Total ^{36}Cl in Used CANDU Fuels

Fuel	Total ^{36}Cl (Bq)	Weight U (g)	^{36}Cl (meas) (Bq/g)	^{36}Cl (calc)[a] (Bq/g)	Cl Impurity (ppm)
PA13894W OE	2816	30.15	93.4	449	1.04 ± 0,02
PA13894W INN	3320	31.20	106.4	326	1.63 ± 0.06
PA19558C OE	804	15.22	52.8	491	0.54 ± 0.01
PA19558C OE	902	29.62	30.5	491	0.31 ± 0.03
BG13374S OE	4278	15.47	276.5	666	2.08 ± 0.07
BJ49093C OE	6538	14.20	460.4	684	3.37 ± 0.10

a Calculated from ORIGEN-S using 5 ppm Cl and 190 MWh/kg U, and corrected for burnup

Determination of ^{36}Cl Release from Used Fuel

A number of samples for ^{36}Cl analysis were taken from fuel leaching experiments that had been conducted in a borate buffer solution to determine the "instant release" of ^{129}I [8]. The first three fuel samples (Table III) had been ground to a grain size of 5 to 20 µm, which would expose all grain boundaries in the fuel and the total inventory of ^{36}Cl leached should thus be representative of the inventory at the gap and grain boundary regions in these fuels. The average Cl impurity of 2.3 ppm and the fuel burnup was used to calculate the % release for fuels that had not been analyzed for total for ^{36}Cl. The fourth sample in Table III was a clad fuel segment. The % total inventory release was calculated using the derived total ^{36}Cl inventory (Table II). Blank determinations were conducted on an unirradiated UO_2 sample leached in a borate buffer solution for 7 days (Table III) and indicated insignificant ^{36}Cl concentration.

The ^{36}Cl release results from leaching clad used fuels in DIW for 28 and 32 days are presented in Table III along with the results of a blank determination conducted on an unirradiated UO_2 sample. The measured Cl impurity levels (Table II) were used to calculate the % inventory release. For fuels that were not analysed for total ^{36}Cl content, the average measured Cl impurity concentration in unirradiated UO_2 of 2.3 ppm was assumed. Time-dependent measurements indicated that most of the ^{36}Cl release occurred in the first 5 days leaching and that little additional release was observed after 32 days.

DISCUSSION

The close agreement between the measured total Cl in unirradiated fuel (2.3 ppm) and total Cl impurity concentration derived from total ^{36}Cl measurements in used fuels (1.5 ppm) suggests that it is reasonable to assume that Cl impurity levels in used CANDU fuels are less than the 5 ppm concentration assumed in postclosure safety assessment studies [11] and an impurity level of 2.3 ± 1.1 ppm could be assumed for any further assessment studies. Analyses of Light Water Reactor fuel [16] have indicated Cl impurity concentrations at or below the detection limit, a value usually reported as 5 or 10 ppm (µg/g U).

Several techniques were used to measure the "instant release" of ^{36}Cl from used fuel samples. The first three fuels in Table III were fuel segments that were mechanically crushed to reduce the fuel down to grain size and simultaneously leached in a borate buffer solution. This technique was previously used to determine ^{129}I and ^{14}C inventories in both the gap and grain boundary regions of the fuel, the assumption being that if the fuel is reduced to grain-sized particles, all of

Table III ^{36}Cl Released From CANDU Fuels Leached in DIW

Fuel[a]	Time	Weight (g)	Total ^{36}Cl Released (Bq)	Total Cl Impurity (ppm)	Total ^{36}Cl Inventory (Bq)[b]	^{36}Cl Inventory Released (% total)
B21271C OE (P,B)	4 h	22.82	1420	2.30	7620	18.6 ± 0.9
B12059C OE (P,B)	4 h	23.81	460	2.30	6540	7.0 ± 0.4
PA07993W OE (P,B)	4 h	22.83	1380	2.30	8680	15.9 ± 0.8
BJ49093C OE (C,B)	7 d	22.74	1683	3.37[b]	10480	16.1 ± 0.8
B12059C INT (C,D)	32 d	22.78	130	2.30	5170	2.5 ± 0.05
BG00815W INT (C,D)	32 d	21.55	18	2.30	2950	0.6 ± 0.01
PA13894W INT (C,D)	32 d	30.86	32	1.30[b]	3710	0.9 ± 0.01
PA13894W INN (C,D)	32 d	29.34	15	1.63[b]	3120	0.5 ± 0.01
PA13894W INN (C,D)	28 d	36.84	35	1.63[b]	3920	0.9 ± 0.06
PA19558C OE (C,D)	28 d	16.14	32	0.31[b]	492	6.5 ± 0.5
BG13374S OE (C,D)	28 d	17.03	721	2.08[b]	4710	15.3 ± 0.3
BJ49093C OE (C,D)	28 d	17.77	1669	3.37[b]	8180	20.4 ± 0.2
UO$_2$ (B)	7 d	15.40	1.4 ± 0.4			
UO$_2$ (D)	32 d	17.48	1.2 ± 0.4			

a OE outer element from bundle; (P) powdered fuel; (C) fuel leached with cladding intact; (B) borate buffer leachant; (D) deionized water
b Measured Cl impurity (Table II).

the grain-boundaries will be accessible to the leachant solution and will be released to solution [8]. The releases were corrected for U matrix dissolution and thus the measured radionuclide releases should be representative of the gap and grain-boundary inventories [8]. The remainder of the measurements were conducted on fuel fragments and included the fuel cladding (Table III). Release from these fuels should be representative of the gap region plus the contribution from accessible grain boundaries.

Figure 1 shows a clear dependence of the "instant release" (% total inventory) of ^{36}Cl on fuel burnup. An exponential curve fit to this data suggests the "instant release" (IR) can be described by the empirical relationship: $[^{36}Cl]_{IR} = 0.045*exp(0.02*[Burnup])$. At the reference fuel burnup for CANDU fuels of 190 MWh/kg U [10], the average ^{36}Cl release would be about 2%. Since the integrated neutron flux is proportional to burnup, and the release correlates strongly with burnup, it suggests that for high burnup fuels, a significant fraction of the ^{36}Cl may be present in the gap region and readily available for release.

Figure 2 shows the dependence of the "instant release" (% total inventory) of ^{36}Cl as a function of the measured % stable Xe fission gas release (FGR) (Table I). The % Xe FGR was calculated from the measured release of Xe gas from puncture tests on the fuel elements [9]. There is a trend to higher release of ^{36}Cl with increasing FGR. A power curve regression analysis of the "instant release" (IR) of ^{36}Cl and Xe FGR suggests the empirical relationship: $[^{36}Cl]_{IR} = 5.45[Xe]^{0.56}$. Johnson and Joling [17] derived an average total Xe fission gas release to the gap region of 2.2% for Bruce reactor fuels based on the average of all fuels discharged from the reactor core. At this value, the average ^{36}Cl release from Figure 2 would be about 9%. Since FGR is strongly dependent on fuel temperature, it would also be expected that there would

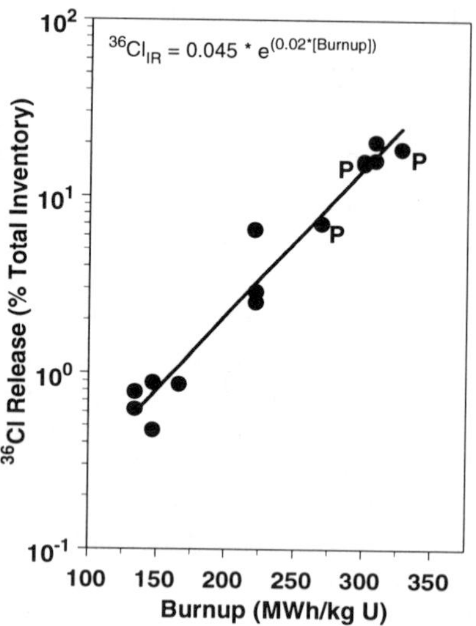

Figure 1 ${}^{36}Cl$ released to solution from clad CANDU fuel (32 d) and from powdered (P) fuel (4 d) at 25°C as a function of fuel burnup.

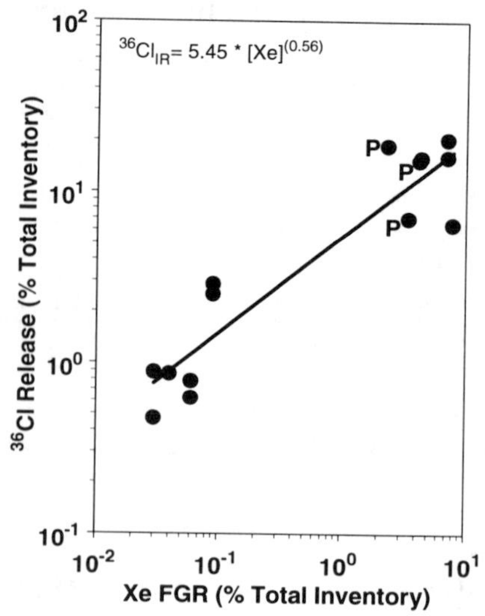

Figure 2 ${}^{36}Cl$ released to solution as in Fig. 1 as a function of measured stable Xe fission gas released to the gap region

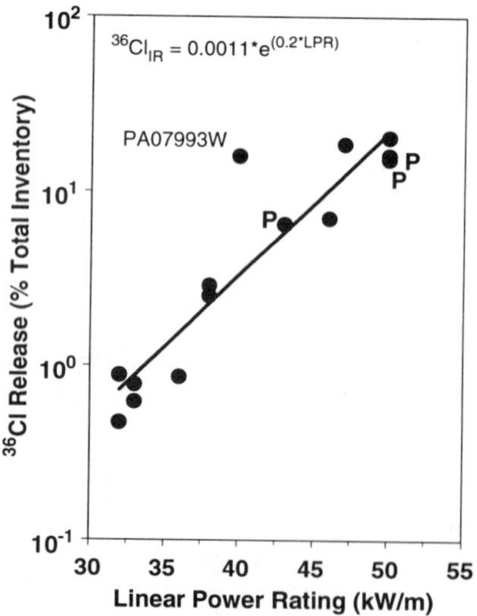

Figure 3 ${}^{36}Cl$ release as in Fig 1 as a function of fuel linear power rating.

be a similar correlation with the LPR (Figure 3). An exponential curve fit to the peak LPR data suggests the "instant release" of ${}^{36}Cl$ can be described by the empirical relationship: $[{}^{36}Cl]_{IR} = 0.0011*\exp(0.2*[LPR])$. This relationship was calculated ignoring the data point from bundle PA07993W as the LPR data for this fuel is suspected to be low. A typical CANDU LPR is about 42 kW/m, which suggests an average "instant release" value for ${}^{36}Cl$ based on LPR of about 5%.

The ${}^{36}Cl$ release from the three powdered fuels (denoted "P" in the Figures) appears to correlate well with the release from clad fuels. If there were significant contributions coming from the grain-boundary regions in the powdered fuels, the ${}^{36}Cl$ releases would be higher and would fall significantly off the trend-line. Typical calculated Xe grain-boundary inventories for these fuels range from about 4 to 13% [8]. The data thus suggest that most of the ${}^{36}Cl$ released is from the gap region and that little resides at grain-boundaries.

The observed release of ^{36}Cl can be compared with the release of ^{129}I from the gap and grain boundary regions. Iodine-129 is generated at U fission sites in the fuel, and because of its incompatibility with the UO_2 matrix, will diffuse to grain-boundaries, where it can accumulate or migrate to cracks and to the fuel/cladding gap region. For the purposes of conservatively modeling release to solution, the total amount of I in the gap and grain boundary regions has been considered to be "instantly" released upon contact with water [8]. However, it is unlikely that chlorine will behave in an analogous manner to I in the fuel.

It is unknown how chlorine impurities are incorporated in the sintered UO_2 matrix, but it is likely that due to the high fabrication temperatures (>1500°C in a H_2 atmosphere), there will be some initial segregation of Cl impurities to grain boundaries and perhaps to the surface of the fuel pellets. Also, the in-reactor centreline temperatures (T_c) in a UO_2 fuel pellet will vary with the average LPR, which, for CANDU fuels, ranges from 20 to 55 kW/m with T_c ranging from 800 to 1700°C [2]. At high centerline temperatures, significant grain growth may occur that will result in the exclusion of species to the gap region. For example, calculations on a fuel with a centerline temperature of 1500°C (LPR ~47 kW/m) [2] show that up to 30% of the fission gas at the center of the fuel could be released; this decreases rapidly, however, to <1% at a fractional radial distance of 0.6.

Because Cl is initially present in the fuel, it is possible that the fraction of the total Cl that is present in cracks and at the fuel-cladding gap may be enhanced compared to I and Xe. Since I and Xe are formed at U fission sites in the fuel, the extent of their migration to the gap region will depend on both their time of formation and their initial location in the fuel at the time of formation (within a grain or near a grain surface). Those fission products formed near grain-boundaries will migrate more rapidly to the gap/grain-boundary region than those formed in the interior of UO_2 grains. In the case of Cl there may be some initial segregation to grain boundaries in the pellet or to the surface of the pellet during fabrication. Migration of Cl (and activated ^{36}Cl) will thus continue throughout the reactor history of the fuel. As a result of high pellet centerline temperatures and fracturing of the pellet during power ramping, Cl could be depleted in the central regions of the fuel and be enhanced at cracks and the fuel/cladding gap region. The observed release results suggest that even for high burnup and high LPR fuels, the majority of the ^{36}Cl remains in the fuel matrix and would be released at a rate comparable with UO_2 matrix dissolution.

An attempt was made to establish a correlation between the "instant release" of ^{36}Cl and ^{129}I releases determined in another study [8]; however, no such correlation was found, suggesting that the release mechanisms are quite different. The correlation between ^{36}Cl and FGR suggests that ^{36}Cl release could be reasonably well predicted from the FGR data provided that the average chlorine impurity concentration can be assumed to be about 2.3 ppm.

Based on the amount of ^{36}Cl released to solution during leaching of various fuels it is recommended that an average "instant release" of 5% total ^{36}Cl inventory be adopted based on the average LPR's for CANDU fuels and that this value also be considered to include release from grain-boundaries. As there is a strong correlation of release with burnup, future postclosure assessment strategies should take into account the distribution of fuel discharge burnups and linear power ratings in the fuels. The "instant release" value for ^{36}Cl (5%) is comparable to the value of 8% used for ^{129}I in postclosure risk assessments [11,17].

SUMMARY AND CONCLUSIONS

In the absence of a large statistical data base for Cl impurity measurements, an average value of 2.3 ± 1.1 ppm initial Cl impurity in CANDU fuel is suggested as the mean concentration.

There is a significant release of ^{36}Cl to the gap region in CANDU fuels. The "instant release" of ^{36}Cl to solution is about 5% of the total inventory, based on average linear power rating and is comparable to the "instant release" value of 8% for ^{129}I used in postclosure safety assessment studies. The release of ^{36}Cl increases significantly with burnup and this factor may need to be taken into account in developing release models for risk assessments for the disposal concept. It appears that most of the ^{36}Cl released originates from the fuel-sheath gap region and that little is present in grain boundaries. The majority of the ^{36}Cl still resides in the fuel matrix and would be released congruently with matrix dissolution.

ACKNOWLEDGMENTS

The Canadian Nuclear Fuel Waste Management Program is jointly funded by AECL and Ontario Hydro under the auspices of the CANDU Owners Group (COG). The authors wish to thank L.H. Johnson, S. Sunder, S. Stroes-Gascoyne and J. Villagran for helpful comments and review of the manuscript.

REFERENCES

1. Atomic Energy of Canada Limited Report, AECL-10711, 1994.
2. L.H. Johnson and D.W. Shoesmith. Spent Fuel. in <u>Radioactive Waste Forms for the Future.</u> W. Lutze and R.C. Ewing (Editors). Elsevier Science Publishers, B.V. 1988.
3. L.H. Johnson, D.M. LeNeveu, D.W. Shoesmith, D.W. Oscarson, M.N. Gray, R.J. Lemire and N.C Garisto. Atomic Energy of Canada Limited Report. AECL-10714, 1994.
4   L.H. Johnson, D.M. LeNeveu, F. king, D.W. Shoesmith, M. Kolar, D.W. Oscarson, S. Sunder, C. Onofrei and J.L. Crosthwaite. Atomic Energy of Canada Limited Report. AECL-11494-2, 1996.
5. S. Stroes-Gascoyne. J. Nucl. Mater. <u>190</u>, p. 87-100. 1992.
6. S. Stroes-Gascoyne, J.C. Tait, N.C. Garisto, R.J. Porth, J.P.M. Ross, G.A. Glowa, and T.R. Barnsdale. in <u>Scientific Basis for Nuclear Waste Management XV</u>, edited by C.G. Sombret. (Mater. Res. Soc. Symp. Proc. 257, Strasbourg, France, 1991) p. 373-380.
7. S. Stroes-Gascoyne, J.C. Tait, R.J. Porth, J.L. McConnell, T.R. Barnsdale, and S. Watson. in <u>Scientific Basis for Nuclear Waste Management XVI</u>, edited by C.G. Interrante and R.T. Pabalan. (Mater. Res. Soc. Symp. Proc. 294, Boston, USA, 1992). p. 41-46.
8. S. Stroes-Gascoyne, S., D.L. Moir, M. Kolar, R.J. Porth, J.L. McConnell and A.H. Kerr. in <u>Scientific Basis for Nuclear Waste Management XVIII</u>, edited by T. Murakami and R.C. Ewing. (Mater. Res. Soc. Symp. Proc. 353, Kyoto, Japan, 1994) p. 625-631.
9. S. Stroes-Gascoyne, L.H. Johnson and D.M. Sellinger. Nucl. Technol. <u>77</u>, p. 320-330, 1987.
10. B.W. Goodwin, D.B. McConnell, T.H. Andres, W.C. Hajas, E.M. LeNeveu, T.W. Melnyk et al. Atomic Energy of Canada Limited Report, AECL-10717, 1994.
11. L.H. Johnson, L.H., B.W. Goodwin, S.C. Sheppard, J.C. Tait, D.M. Wuschke and C.C. Davison. Atomic Energy of Canada Limited Report, AECL-11213, COG-94-527, 1995.
12. I. Aitchison and P.H. Davies. J. Nucl. Mater. <u>203</u>, 206-220, 1993.
13. R.J. Cornett, H.R. Andrews, L. Chant, T. Chaput, Y. Imahori, J. Jirovec, S. Kramer, V.T. Koslowsky, G.M. Milton and J.C.D. Milton. Nucl. Instrum. and Methods in Phys. Research, (in press) 1996.
14. H.R. Andrews, G.C. Ball, R.M. Brown, R.J.J. Cornett, W.G. Davies, B.F. Greiner, Y. Imahori, V.T. Koslowsky, J. McKay, G.M. Milton and J.C.D. Milton. . Nucl. Instrum. and Methods in Phys. Research, B52, p. 243-248 1990.
15. J.C. Tait, I. Gauld and A.H. Kerr. J. Nucl. Materials, <u>223,</u> p. 109-121, 1995.
16   R.J. Guenther, D.E. Blahnik, T.K. Campbell, U.P. Jenquin, J. E. Mendel and C.K. Thornhill. Pacific Northwest Laboratory Report, PNL-5109-106 1988.
17. L.H. Johnson and H.H. Joling. 1984. Atomic Energy of Canada Limited Technical Record TR-00280. 1984.

LEACHING OF USED CANDU FUEL: RESULTS FROM A 19-YEAR LEACH TEST UNDER OXIDIZING CONDITIONS

S. Stroes-Gascoyne, L.H. Johnson, J.C. Tait, J.L. McConnell and R.J. Porth. AECL, Whiteshell Laboratories, Pinawa, Manitoba, Canada, R0E 1L0.

ABSTRACT

A fuel leaching experiment has been in progress since 1977 to study the dissolution behaviour of used CANDU fuel in aerated aqueous solution. The experiment involves exposure of 50-mm clad segments of an outer element of a Pickering fuel bundle (burnup 610 GJ/kg U; linear and peak power ratings 53 and 58 kW/m, respectively), to deionized distilled water (DDH$_2$O, ~2 mg/L carbonate) and tapwater (~50 mg/L carbonate). In 1992, it was observed that the fuel in at least one of the leaching solutions showed some signs of deterioration and, therefore, in 1993, parts of the fuel samples were sacrificed for a detailed analysis of the physical state of the fuel, using SEM and optical microscopy. Leaching results to date show that even after >6900 days only 5 to 7.7% of the total calculated inventory of ^{137}Cs has leached out preferentially and that leach rates suggest a development towards congruent dissolution. Total amounts of ^{137}Cs and ^{90}Sr leached are slightly larger in tapwater than in DDH$_2$O. SEM examinations of leached fuel surface fragments indicate that the fuel surface exposed to DDH$_2$O is covered in a needle-like precipitate. The fuel surface exposed to tapwater shows evidence of leaching but no precipitate, likely because uranium is kept in solution by carbonate. Detailed optical and SEM microscopy examinations on fuel cross sections suggest that grain-boundary dissolution in DDH$_2$O is not prevalent, and in tapwater appears to be limited to the outer ~0.5 mm (pellet/cladding) region of the fuel. Grain boundary attack seems to be limited to microcracks at or near the surface of the fuel. It thus appears that grain-boundary attack occurs only near the fuel pellet surface and is prevalent only in the presence of carbonate in solution.

INTRODUCTION

A leaching experiment was started in 1977 to study the dissolution behaviour of used CANDU fuel in aerated aqueous solution (distilled, deionized H$_2$O (DDH$_2$O) and tapwater, carbonate concentrations ~2 mg/L and ~50 mg/L, respectively) as part of the Canadian Nuclear Fuel Waste Management Program. Results from this experiment were reported previously, in 1980 after the first 3 years of leaching [1,2] and in 1985 after 8 years of leaching [3].

In 1985, the fuel samples were removed from the leaching vessels, and 3-mm-thick sections were sliced from the ends to permit scanning electron microscope (SEM) studies of the leached surfaces. In addition, samples of the precipitate covering the exposed fuel surface were removed with adhesive tape for subsequent examination by SEM/energy dispersive X-ray (EDX) and X-ray diffraction (XRD) analysis [3]. It was concluded that, after 8 years of leaching, the rate of ^{137}Cs release from this fuel remained about two orders of magnitude higher than that of other species (^{90}Sr, ^{238}U, $^{239+240}$P). This was attributed to release of ^{137}Cs from the fuel sheath gap and the grain boundaries. The SEM and surface chemistry analysis showed that the fuel leached in DDH$_2$O was partially covered by a yellow precipitate, composed of UO$_3$.2H$_2$O (schoepite, or a related hydrate) with a Cs-uranate phase as a minor constituent. The fuel leached in tapwater contained relatively litte surface precipitate, likely due to U(VI) remaining in solution as a carbonate complex in tap water [3].

Mat. Res. Soc. Symp. Proc. Vol. 465 © 1997 Materials Research Society

This experiment is continuing to date, with replacement and routine analysis of the leachates every 6 to 9 months. In 1992, it was observed that the fuel in at least one of the leaching solutions (tapwater) showed some signs of deterioration. In 1993 (after 16 years of leaching), parts of the fuel segments were sacrificed for a detailed analysis of the physical state of the fuel. This paper summarizes leaching and dissolution results for ^{137}Cs, ^{90}Sr and [238]U after almost 19 years of leaching and X-ray photoelectron spectroscopy (XPS), optical and SEM microscopy data of the physical state of the UO_2 matrix after 16 years of exposure.

MATERIALS AND METHODS

The experiment has been described in detail elsewhere [1,2]. It involves exposure of 50-mm segments of an outer element of Pickering bundle P15527C (burnup 610 GJ/kg U; linear and peak power ratings 53 and 58 kW/m, respectively), with attached zircaloy cladding, to aerated DDH_2O and tapwater. The latter consists of filtered and slightly chlorinated Winnipeg River water and was chosen in 1977 to approximate the groundwater composition expected to be present in the Lac du Bonnet batholith [1,2], the site of AECL's Underground Research Laboratory. The major difference between the two solutions is the carbonate concentration, ~2 mg/L in DDH_2O and ~50 mg/L in tapwater. The fuel segments (cut dry under air cooling) are leached, in duplicate, in polypropylene bottles containing 100 mL of leachant. After each leaching period (ranging from 1 to 270 days over the course of the years), the fuel segments are transferred to bottles containing fresh leachant. The leachates are acidified in the original bottles (with 5 mL concentrated HNO_3) prior to analysis, to ensure dissolution of particulate material and desorption of radionuclides from the vessel walls prior to radiochemical analysis. Therefore, uranium concentrations measured in the leachates do not reflect equilibrium solution concentrations, but rather the total amounts of uranium released from the fuel segments, including plated-out and particulate material. They do not include any material that may have adsorbed or precipitated on the fuel segments or claddding.

In 1992, small fragments of fuel were pried off the 5-cm-long segments for SEM and XPS analysis (the latter for tapwater only). In 1993, 2-cm-sections were cut from one duplicate each of the fuel segments leached in DDH_2O and tapwater, and examined by optical and SEM microscopy, both in cross-section and longitudinally.

RESULTS

Total fractions released of ^{137}Cs, ^{90}Sr and [238]U in the duplicate leaching experiments are shown in Figure 1 (DDH_2O, denoted as DA and DB) and Figure 2 (tapwater, denoted as TA and TB). Figure 3 shows SEM photographs of the leached fuel surfaces in DDH_2O (DA, DB) and tapwater (TA, TB). Figure 4(a,b) shows 2 locations of a crossection of the fuel segment leached in tapwater (TA). Table 1 gives the degree of surface oxidation as measured by XPS analysis (for tapwater samples (TA,TB) only).

DISCUSSION

Figures 1 and 2 show total fractions leached for the duplicate experiments in DDH_2O and tapwater, respectively, versus time. For experiments DA and TA, data were obtained over 5850

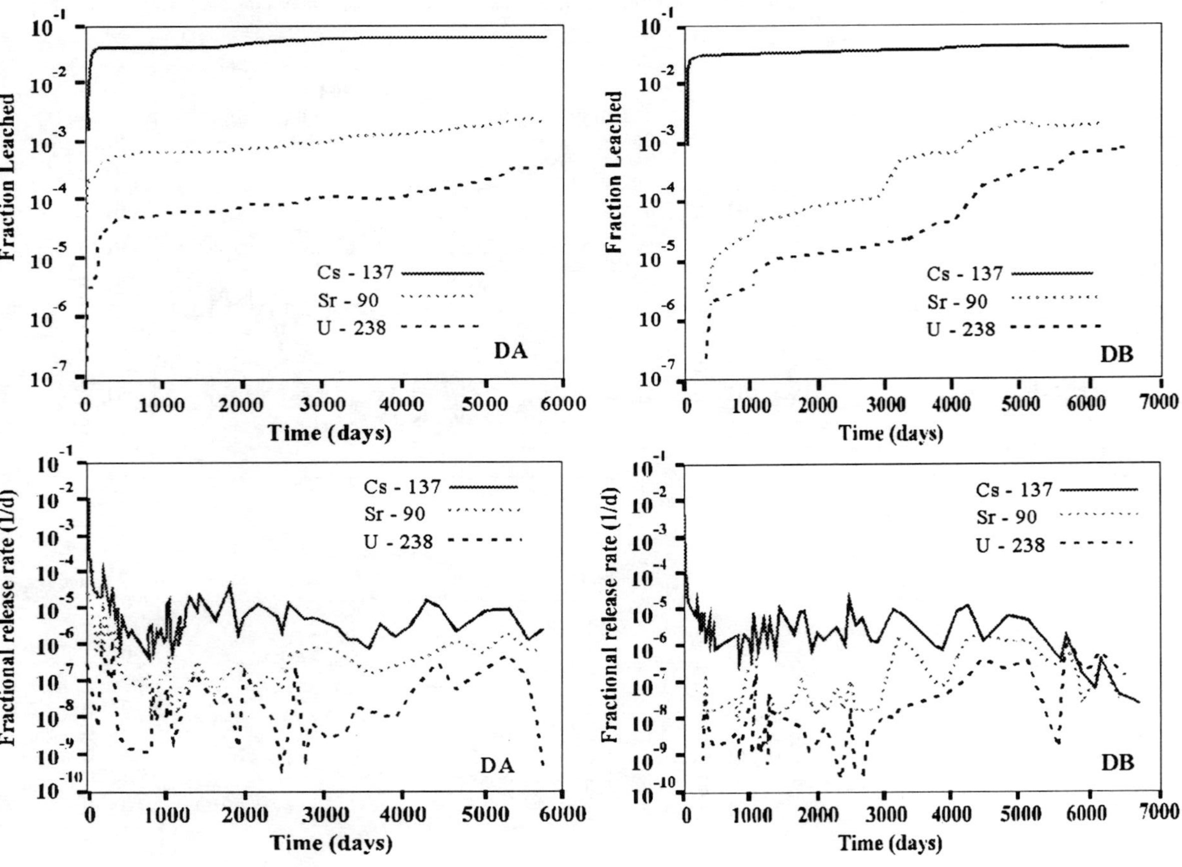

Figure 1: Total fraction leached and fractional release rate as a function of time for ^{137}Cs, ^{90}Sr and ^{238}U in DDH$_2$O(DA,DB)

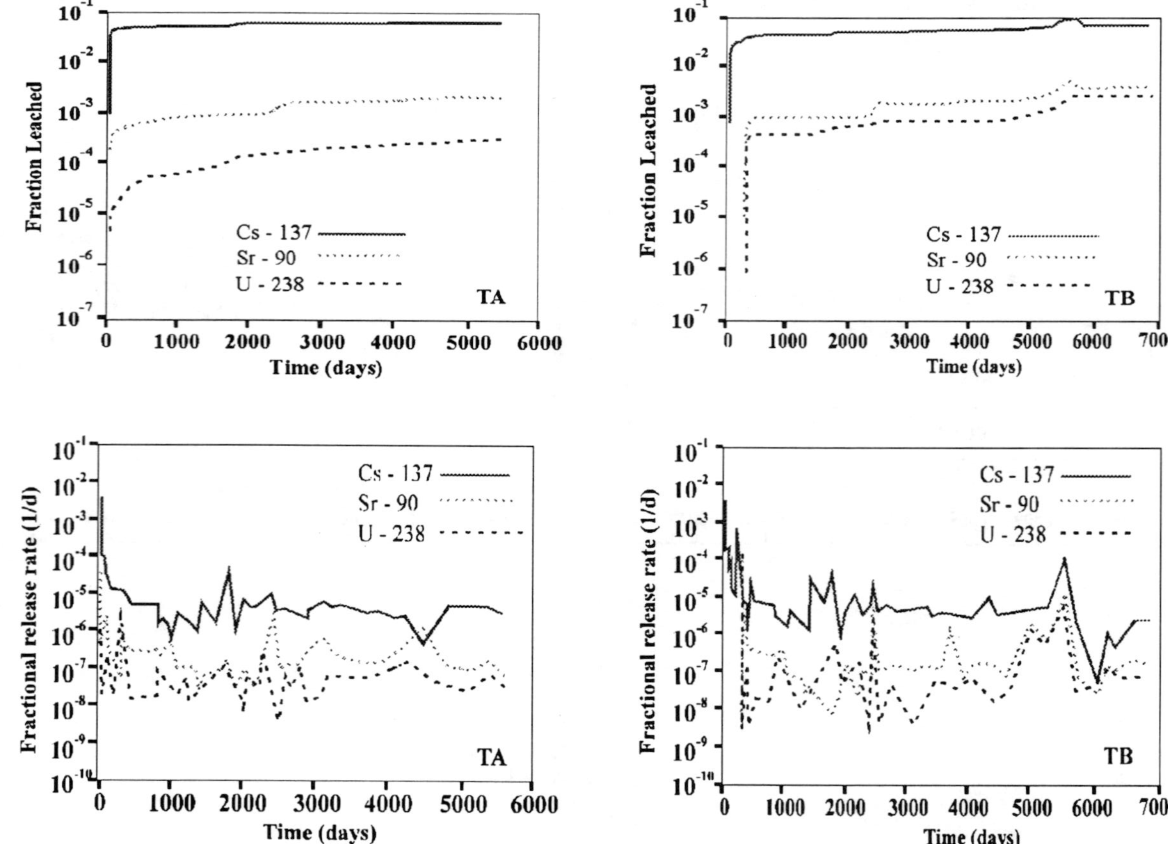

Figure 2: Total fraction leached and fractional release rate as a function of time for ^{137}Cs, ^{90}Sr and ^{238}U in tapwater (TA,TB)

514

d, at which point part of these fuel segments were sacrificed for analysis. For experiments DB and TB, data were obtained for 6920 d. Total fractions of the calculated ^{137}Cs inventory released were 5 to 6.1% in DDH$_2$O, and 5.5 to 8% in tap water after 5850 to 6920 d. Calculated gas release [4] for this fuel was 9.68%, whereas 8.90% was measured in a puncture test on an equivalent fuel element [4]. Assuming an approximately one-to-one relationship between fission gas release and gap inventories for ^{137}Cs [5], these results suggest that, even after >6920 d of leaching, the gap inventory of ^{137}Cs has not entirely leached out.

The total fraction of ^{90}Sr released over the leaching period ranged from 0.22 to 0.24% (of the total ^{90}Sr inventory) in DDH$_2$O and from 0.19 to 0.48% in tapwater. The total fraction of ^{238}U released was ranged from 0.07 to 0.09% in DDH$_2$O and from 0.03 to 0.29% in tapwater.

In one of the tapwater experiments, more ^{137}Cs, ^{90}Sr and ^{238}U leached out than in the duplicate experiments. A visual observation indicated that limited breakup of the fuel in this particular experiment had occurred after about 15 years of exposure. This suggests that the extent of matrix dissolution (i.e., 0.3%) has been sufficient to initiate limited breakup of the fuel. Dissolved uranium is less likely to precipitate onto the fuel segments in tapwater. If, therefore, the total amount of ^{238}U leached may be taken as a measure of matrix dissolution, the results obtained for tapwater indicate preferential dissolution of ^{137}Cs (5.5 to 7.7%) but relatively little preferential dissolution of ^{90}Sr (0.16% to 0.19%). Matrix dissolution cannot be quantified for the DDH$_2$O experiments because U may have precipitated onto the fuel segments, but release of ^{90}Sr appears very similar in DDH$_2$O compared to tapwater.

Fractional release rates in Figures 1 and 2 indicate much higher release rates for ^{137}Cs than for ^{90}Sr and ^{238}U. However, towards the end of the leaching period, there appears to be a tendency for the leach rates to become similar, especially in Figure 1(DB). This may indicate that the easily leachable components have almost disappeared and that the systems are developing towards congruent dissolution. However, a much smaller amount (up to 7.7%) than the calculated combined gap and grain-boundary inventory of ^{137}Cs (>20% for high linear power fuels [5]) has leached out. This suggests that grain-boundary inventories of ^{137}Cs are either overestimated or not accessible to leaching. The former suggestion is consistent with the fact that measured combined gap and grain-boundary inventories of ^{137}Cs are much smaller than the calculated values for high linear power (>44 kW/m) CANDU fuels [6].

Figure 3 shows results of the SEM examinations of the pried-off fuel fragments. Results indicate that the fuel surface exposed to DDH$_2$O is covered in a needle-like precipitate. SEM images of the fuel surface exposed to tapwater show evidence of leaching ('open' and 'round' grain boundaries), but no needle-like precipitate, likely because uranium is kept in solution by carbonate.

Table 1 gives XPS results for the fragments exposed to tap water and indicates surface compositions ranging from UO$_{2.34}$ to UO$_{2.47}$, consistent with an oxidative dissolution process.

Optical and SEM microscopy examinations on (freshly-cut) cross sections of the 2-cm segments were also carried out. The results suggest that evidence for grain-boundary dissolution is greatest in tapwater and appears to be limited to the outer ~0.5 mm (pellet/cladding) region of the fuel (with the highest residual sintering porosity). Grain-boundary attack seems to be limited to microcracks at or near the surface of the fuel (Figure 4a). These microcracks are isolated and occur infrequently in the outer 0.5 mm of the fuel. It appears that attack has occurred to a depth of about 5 to 10 grains in the region of the microcrack. There is no evidence of attack down grain boundaries in the densified mid-pellet regions of the fuel, even where there has been extensive fuel fracturing and microcracks exist (Figure 4b). In DDH$_2$O, there appears to be very little evidence of grain-boundary attack, even in microcracks at the fuel surface. It thus appears that grain-boundary attack only occurs near the fuel-pellet surface and is prevalent only in the

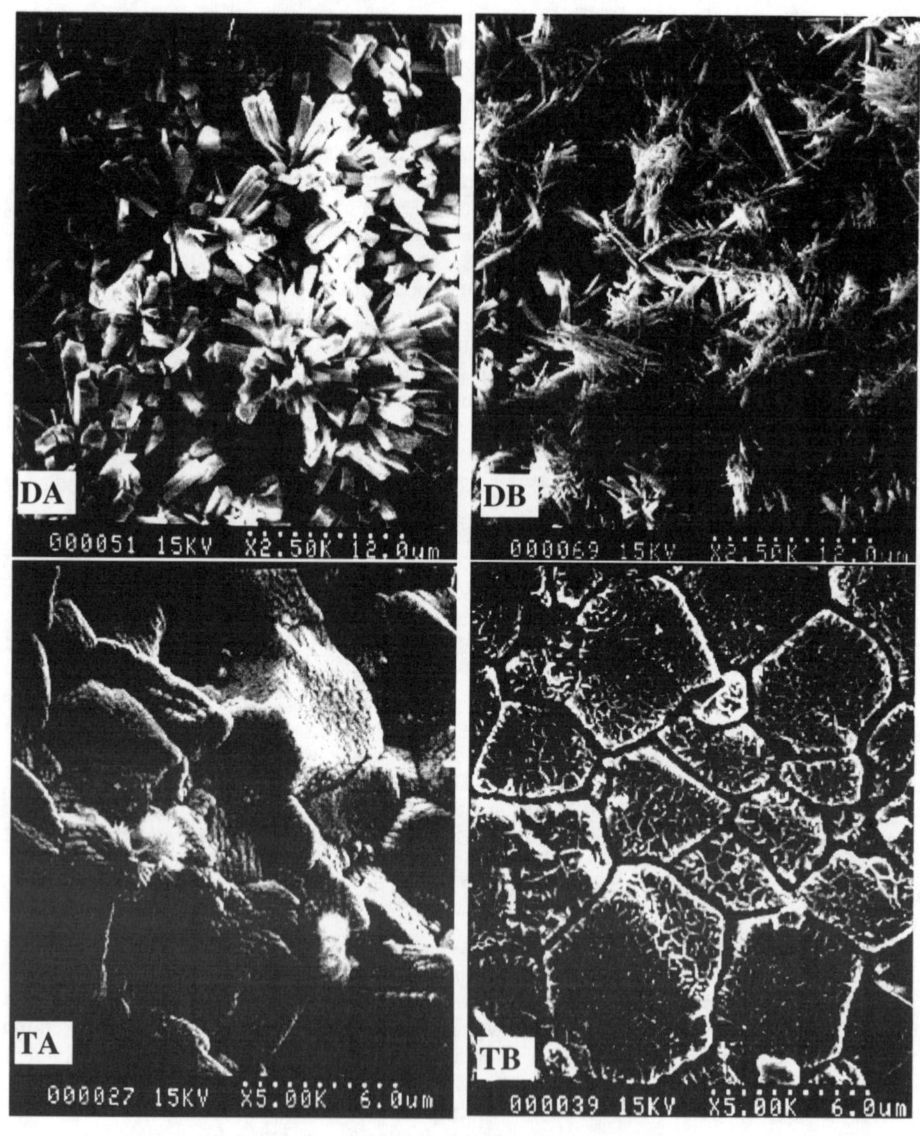

Figure 3:   SEM photographs of the fuel surfaces leached in DDH$_2$O (DA, DB) and in tapwater (TA, TB)

TABLE 1. Degree of surface oxidation of tapwater-leached fuel fragments

Sample	spectra	%U^{6+} [*]	Δ%U^{6+} [**]
TA, original	fair fit + quality	44	41-47
TA, original(R)	fair fit + quality	47	40-54
TA, cleaved	unusable	-	-
TB, original	fair fit + quality	34	30-48
TB, original(R)	fair fit + quality	39	33-48
TB, cleaved	good fit + quality	35	32-40

(R) repeat analysis

[*] Determined by curve-fitting the U4f$_{7/2}$ peak envelope with U^{6+} and U^{4+} components, and expressed as percent U^{6+} [7]

[**] Estimated probable lower and upper limits for percent U^{6+}; these are based on statistical uncertainty, robustness of fit and confidence in the correction for surface charging [7].

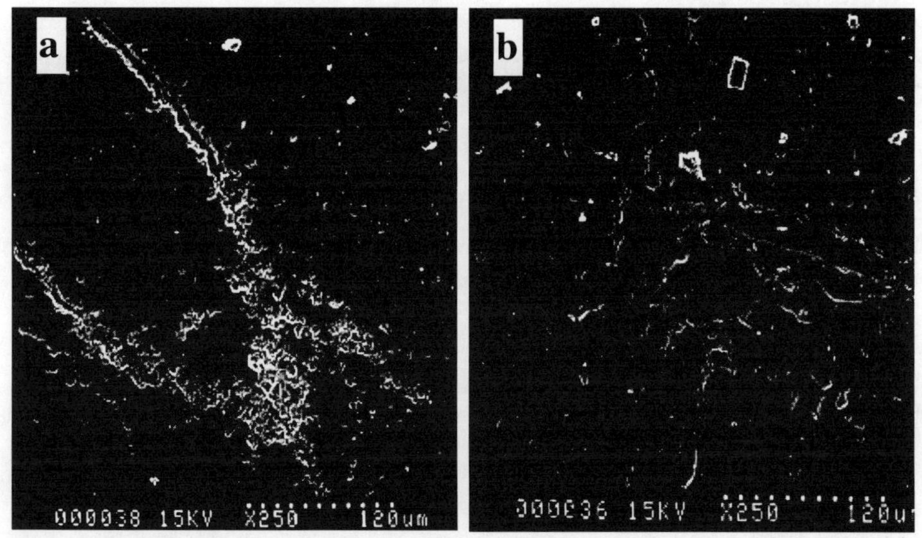

Figure 4:   SEM micrograph of crosssection of CANDU fuel leached in tapwater (TA);
a) Outer pellet region   b) Centre pellet region

presence of carbonate in solution. The lack of secondary phases in the microcracks suggests mass transport of dissolved U out of the attacked region in carbonate solution.

CONCLUSIONS

CANDU fuel segments have been leached for periods of 5850 to 6920 d in both DDH_2O and tapwater. After almost 19 years, only up to 7.7% of the total inventory of ^{137}Cs has leached out and congruent dissolution appears to be developing. These results corroborate the suggestion [6] that calculated grain-boundary inventories of ^{137}Cs in high-linear-power rating CANDU fuels may be overestimated. Results also indicate a slight preferential release of ^{90}Sr. The fuel surface exposed to DDH_2O is covered in a needle-like precipitate, whereas the fuel surface exposed to tapwater shows evidence of leaching but no needle-like precipitate, likely because uranium is kept in solution by carbonate complexation. XPS results indicated surface compositions ranging from $UO_{2.34}$ to $UO_{2.47}$ in tapwater. Grain-boundary dissolution is not evident in DDH_2O and in tapwater appears to be limited to the outer ~0.5 mm (pellet/cladding) region of the fuel. Grain-boundary attack appears to be limited to microcracks at or near the surface of the fuel. These microcracks are isolated and do not account for a large volume of the outer 0.5 mm of the fuel.

ACKNOWLEDGMENTS

This work was funded jointly by AECL and Ontario Hydro. Optical microscopy, SEM and XPS analysis was provided by D.L. Bruneau, R. Behnke and A.M. Duclos. Useful comments on the manuscript were provided by T.T. Vandergraaf and D.W. Shoesmith.

REFERENCES

[1]	Vandergraaf, T.T. 1980. Atomic Energy of Canada Limited Technical Record, TR-100*.

[2]	Vandergraaf, T.T., L.H. Johnson and D.W.P. Lau. 1980. Mat. Res. Soc. Symp. Proc. 2, 335

[3]	Stroes-Gascoyne, S., L.H. Johnson, P.A. Beeley and D.M. Sellinger. 1985. Mat. Res. Soc. Symp. Proc. 50, 317.

[4]	Notley, M.J.F. 1979. Nucl. Technol. 44, 445

[5]	Stroes-Gascoyne, S., L.H. Johnson and D.M. Sellinger. 1987. Nucl. Technol. 77, 320.

[6]	Stroes-Gascoyne, S. 1996. J. Nucl. Mat. (in press).

[7]	Sunder, S., G.D. Boyer and N.H. Miller. 1990. J. Nucl. Mat. 175, 163.

* Internal report, available form SDDO, AECL, Chalk River Laboratories, Chalk River, ON, KOJ1J0

GRAIN BOUNDARY CORROSION AND ALTERATION PHASE FORMATION DURING THE OXIDATIVE DISSOLUTION OF UO_2 PELLETS

David J. Wronkiewicz, Edgar C. Buck, and John K. Bates
Argonne National Laboratory, Argonne, IL 60439-4837, USA, wronkiewicz@cmt.anl.gov

ABSTRACT

The alteration behavior of UO_2 pellets following their reaction under unsaturated drip-test conditions, at 90°C, for time periods of up to 10 years has been examined by solid phase and leachate analyses. Sample reactions were characterized by preferential dissolution of grain boundaries between the original press-sintered UO_2 granules comprising the samples, development of a polygonal network of open channels along the intergrain boundaries, and spallation of surface granules that had undergone severe grain boundary corrosion. The development of a dense mat of alteration phases after two years of reaction trapped loose granules, resulting in reduced rates of particulate uranium release. The paragenetic sequence of alteration phases that formed on the present samples was similar to that observed in surficial weathering zones of natural uraninite (UO_2) deposits, with alkali and alkaline earth uranyl silicates representing the long-term solubility-limiting phases for uranium in both systems.

INTRODUCTION

The dissolution of spent nuclear fuel from commercial power reactors represents the largest potential source of radionuclide release from high-level waste forms in a disposal vault because of the large volumes of spent fuel that need to be disposed and the high concentration of radionuclides contained in the spent fuel rods. Nearly 30,000 metric tons of commercial spent nuclear fuel were in storage at commercial power reactors as of 1995, while future projections suggest that this quantity will more than double by the anticipated opening of the potential Yucca Mountain, Nevada high-level nuclear waste repository in 2010 [1]. Consequently, the dissolution behavior of the spent fuel (irradiated UO_2), and the rates and mechanisms of radionuclide release are critical parameters that will need to be determined before the performance of a nuclear waste repository containing spent fuel can be assessed.

The oxidation of UO_2 leads to the formation of progressively higher oxidation states of uranium and an oxidized surface layer with a composition of UO_{2+x} (where $0 \leq x \leq 1$). This phase is sparingly soluble under relatively reducing conditions, where the oxidation state remains below $UO_{2.33}$. Once the oxidation state reaches $UO_{2.33}$, however, uranium solubility may increase several orders of magnitude, leading to rapid dissolution of the solid [2]. The kinetics of this oxidation step may be further enhanced when the reaction at the sample surface takes place in a thin film of water that is exposed to radiation and/or an oxidizing atmosphere [3-5]. The unsaturated geologic medium at the proposed Yucca Mountain Repository Site represents such an environment, where both an oxidizing atmosphere and limited amounts of liquid water are present. This environment may therefore significantly accelerate the reaction rate of UO_2, and by analogy, spent nuclear fuel in an unsaturated repository setting.

The objective of this program is to evaluate the reaction of UO_2 pellets following their exposure to periodic drops of simulated silicate-bicarbonate groundwater in an air atmosphere at 90°C. The corrosion of the samples was monitored by examining the surficial and cross-sectioned alteration profiles, and changes in the solution composition over time. The UO_2 pellets may represent an effective surrogate material for investigating the behavior of commercial spent fuel because actual spent fuel is composed of ≥ 95 wt% UO_2 [5]. The temperature of the tests is also similar to the boiling point of water at Yucca Mountain (~96°C), and thus, approximates the conditions of the first potential liquid water contact with the waste form at this site. Results from these tests will be used to characterize the (1) corrosion behavior of the pellets under conditions

Mat. Res. Soc. Symp. Proc. Vol. 465 © 1997 Materials Research Society

that simulate an unsaturated setting, (2) formation of alteration phases, and (3) serve as a pilot study for additional tests conducted with spent nuclear fuel [6].

EXPERIMENTAL

The experimental apparatus and materials used to conduct these tests have been described previously [7-9], and thus will only briefly be summarized here. The samples were fabricated and hot-press sintered from a uranium oxide powder with a natural isotopic abundance of uranium, an oxygen/metal ratio of 2.000 ± 0.002, and <70 ppm total contaminants (mostly Cl, Th, and Fe). The pellets were assembled into Zircaloy-4 metal sleeves with their upper and lower surfaces exposed and their sides enclosed by the Zircaloy. The sample assemblies were housed in reaction vessels composed of 304L stainless steel, with the Zircaloy-UO_2 pellet assembly resting on a Teflon® support stand. The test vessels were connected to a leachant injection line and placed in an oven that was maintained at a temperature of $90 \pm 2°C$. Premeasured quantities of EJ-13 silica-bicarbonate simulated groundwater solution [7-9] were injected onto the samples via the injection line at rates of either 0.075 mL every 3.5 days or 0.0375 mL every 7 days.

Periodically, the tests were interrupted to allow for collection of the accumulated leachate. Aliquots were collected for various analyses including anion, pH, carbon, and filtered uranium. After these aliquots were removed, the stainless steel vessel bottom plus the remaining leachate solution were subjected to a 10-minute acid strip in a dilute nitric acid solution. Total uranium release (including the soluble and insoluble fractions) and the concentration of other cation components were determined from this acid strip aliquot. After completion of the sampling process, the vessels and samples were reassembled to begin the next cycle.

Individual tests were periodically terminated to provide samples for studies of solid phase alteration patterns. Five tests were terminated at periods of 1.5, 2.25, 3.5, 3.5, and 8.0 years (78, 117, 183, 183, and 417 weeks, respectively), while the remaining three tests have been allowed to continue into their eleventh year of reaction. The sample surfaces, cross-sectioned samples, and/or individual alteration phases were examined by scanning electron microscopy/energy dispersive X-ray spectrometry (SEM/EDS), optical microscopy, X-ray diffraction, and transmission electron microscopy analysis. Solution aliquots were analyzed by inductively coupled plasma mass spectrometry, Dohrman carbon analysis, and ion chromatography.

RESULTS AND DISCUSSION

Solution Analysis

Uranium release during the first ten years of testing can be broken down into three different periods (Fig. 1) [8,9]; period I, the initial incubation period of low uranium release occurring for approximately the first year of sample reaction; period II, an interval of rapid uranium release, lasting between the first and second years; and period III, an extended time interval of moderate uranium release lasting up to the present 10 years of reaction.

The majority of the uranium release during the period II could be attributed to grain boundary corrosion and spallation of UO_{2+x} particles from the samples. Uranium release during this period exceeded that of thr previous and subsequent intervals combined. An examination of the sample terminated after 1.5-years of reaction revealed the presence of a highly corroded pellet surface and numerous micrometer- to submicrometer-sized anhedral UO_{2+x} particles that were both lying directly on top of the pellets (Fig. 2a) and the alteration phases that had precipitated on this surface. These particles were, however, conspicuously absent from the surfaces of samples during period III reactions. This change occurred concurrently with: (1) the development of a dense mat of alteration phases on the top surfaces of the samples; (2) a depletion in the concentration of alkalis, alkaline earths, and silicon in the leachate solution, relative to EJ-13; and (3) a reduction in total uranium release [8,9]. These patterns indicate that the mat of alteration phases acted as a trap that encapsulated and restricted the migration of spalled UO_{2+x} particles from the sample surface. The release of the UO_{2+x} particles would now be dependent on their complete dissolution or physical release from these encapsulating phases.

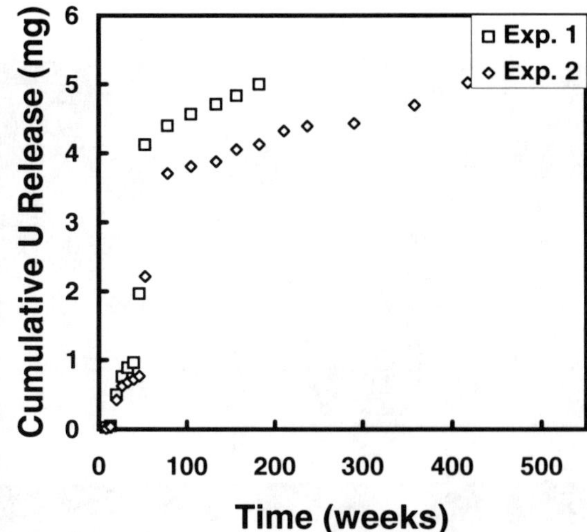

Figure 1. Cumulative release of uranium from experiments 1 and 2. Experiment 1 was terminated after 183 weeks (3.5 years), while experiment 2 was terminated after 417 weeks (8 years) of reaction. Initial pellet weights were 29,522 and 29,166 mg UO_2, respectively.

Calculated uranium release rates for period III reactions are ~0.1 to 0.3 mg/(m^2•day) (normalized for geometric surface area). These rates represent the overall *release* of uranium, but not necessarily the true *dissolution* of uranium from the samples, since the calculated values include the added contribution of particulate uranium release, but do not include dissolved uranium that had been removed from solution through the precipitation of alteration phases.

An analysis of the size-fractioned release patterns during period III indicates that the majority (86 to 97%) of the released uranium was sorbed or precipitated on the walls of the stainless steel test vessel and the Teflon support stand. Between 1 to 12% was present as >5 nm sized particles suspended in the leachate, whereas less than 2% of the total could be attributed to components that passed through a filter with a 5-nm pore-size opening. This latter fraction suggests a uranium concentration of 4 x 10^{-6} M for the leachate collected from the bottom of the test vessel. Electron microscopy and electron diffraction analyses of filtered residues from the eight- to ten-year leachates indicated that needle-shaped grains of uranophane and boltwoodite (Table I) were contributing to the particle release [9].

<u>Solids Analysis</u>

Examinations of the cross-sectioned samples indicate that the boundary regions between the original press-sintered UO_2 granules were preferentially dissolved relative to the cores, resulting in the formation of a polygonal network of open channels (Fig. 2b). These channels generally penetrate into the pellets to a depth of two-to-four grain boundaries (~10 to 20 mm) ahead of the exposed external sample surface. Penetration depths do, however, vary between regions displaying essentially no grain boundary corrosion to regions where penetration has occurred to a depth of approximately ten grains or more.

Spallation of UO_{2+x} surface granules was also noted in regions that had undergone severe grain boundary corrosion. This was especially prevalent on the top sample surfaces, where impact from the water droplets and also possibly flow of water across the surface had aided in the removal of loosened particles, especially during period II reactions. Loss of UO_{2+x} particles from the sample sides and bottoms appeared to be much less prevalent, whereas particle release does not appear to have occurred from the interpellet regions of the samples.

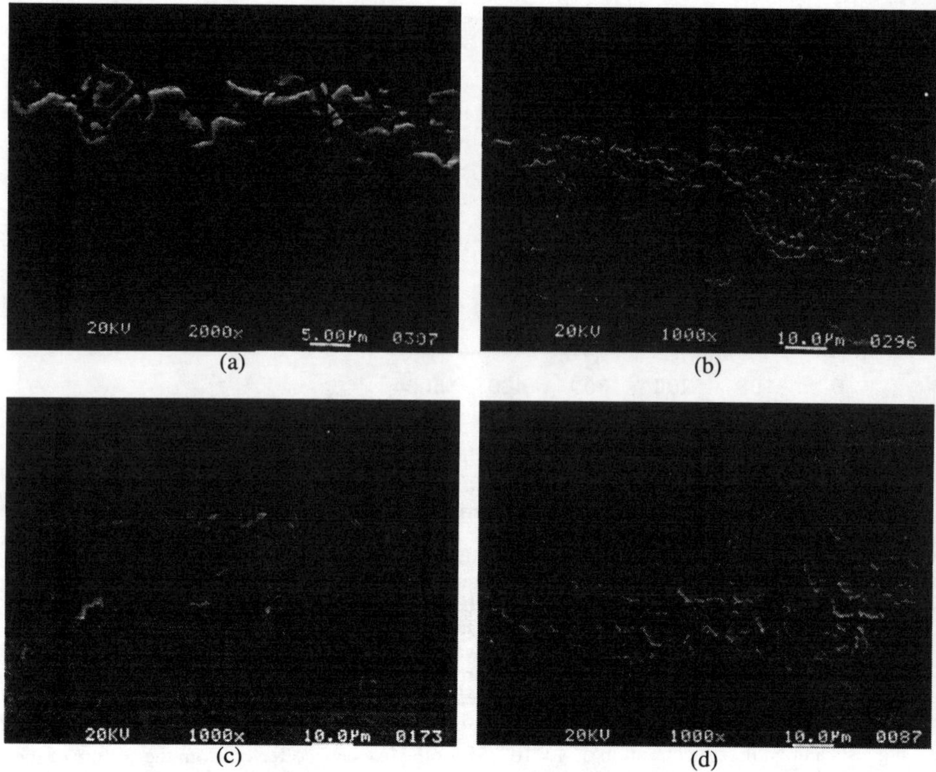

Figure 2. Scanning electron photomicrographs of cross-sectioned top surfaces of the reacted UO_2 samples. The epoxy mounting medium (black) is to the top of each photo, while the unreacted UO_2 is to the bottom. (a) 1.5-year sample displaying grain-boundary corroded surface, loosened particles, and residual pits resulting from the spallation of particles, (b) open porosity resulting from penetrative intergrannular dissolution from the 2.25-year sample. Surface layer (gray) is composed of dehydrated schoepite. Separation of layer from the pellet surface is believed to result from the contraction of the epoxy during drying. Note that the corroded UO_{2+x} surface particles have also been separated along with the alteration layer. (c) development of inner mat of compreignacite and sparse outer uranophane layer on corroded pellet surface of 3.5-year sample, (d) precipitated compreignacite on corroded top surface of 8.0-year sample. Note the continuation of crystal delamination planes from the surface phases into the open porosity region of the sample and the encapsulation of the residual UO_{2+x} surface grains by the alteration phases.

Table I. Summary of UO_2 alteration phases identified on the reacted samples

Uranyl-Oxide Hydrates

Schoepite (meta·schoepite)	$UO_3 \cdot nH_2O$ ($n < 2$)
Dehydrated Schoepite	$UO_3 \cdot (0.8 - 1.0\ H_2O)$
Compreignacite	$(Na,K)_2[(UO_2)_6O_4(OH)_6] \cdot 8\ H_2O$
Becquerelite	$Ca[(UO_2)_6O_4(OH)_6] \cdot 8\ H_2O$

Uranyl Silicate Hydrate

Soddyite	$(UO_2)_2SiO_4 \cdot 2\ H_2O$

Uranyl Alkaline Silicate Hydrates

Uranophane	$Ca(UO_2)(SiO_3)(OH)_2 \cdot 5\ H_2O$
Boltwoodite	$K_2(UO_2)_2(SiO_4)_2(H_3O)_2 \cdot H_2O$
Na-Boltwoodite	$(Na,K)(UO_2)(SiO_4)(H_3O) \cdot H_2O$
Sklodowskite	$Mg(UO_2)_2(SiO_4)_2(H_3O)_2 \cdot 2\ H_2O$

Non-Uranyl Phases

Palygorskite	$(Mg,Al_{0.12-0.66})_5(Si,Al_{0.12-0.66})_8O_{20}(OH)_5 \cdot 4\ H_2O$
Fe-oxides	
Ti-oxides	
Amorphous silica	

The sides, bottoms, and interpellet surfaces of the 3.5- and 8-year samples displayed ample evidence for intergrain boundary corrosion, though only a minimal amount of UO_{2+x} particle spallation appears to have occurred from these regions. On some of the sample bottoms, particle spallation appeared to have occurred in localized clusters, a feature suggesting that the loss of a single grain from the surface influenced the adherence of neighboring grains to the sample.

The cause for the preferential grain boundary dissolution is currently being investigated, but may be related to structural damage to the UO_2 grain-boundaries induced during the hot press-sintering process that was used to produce the pellets, electrochemical potential differences that exist along the grain boundaries, and/or density changes resulting from the partial oxidation of the uranium oxide matrix.

Oxidation from the UO_2 to the $UO_{2.25}$ state results in a slight shrinkage in unit cell dimensions [5]. This shrinkage is known to weaken the grain boundary regions in spent nuclear fuel, inducing a friable nature to the samples where light crushing or polishing caused the fuel to separate into individual grains [10,11]. Detailed backscattered electron-SEM examinations of the UO_2 samples from the present study did not reveal the presence of any density contrasts that could be associated with an increase in the oxidation state of the intergrain boundary regions. This observation suggests that any increase in oxidation state, if it occurred, either affected too small of a region to be detected at the resolution capacity of the SEM (~0.1 mm), or resulted in the rapid dissolution of any uranium that was oxidized. If the latter process occurred, then the rapid dissolution of an oxidized surface layer around the grains would suggest that the oxidation of UO_2 is the rate limiting step for the dissolution of the samples. More detailed examinations of these interpellet regions are currently being conducted to investigate this issue further.

Most of the uranium released to solution during the dissolution of the UO_2 pellets had combined with Si, Na, K, Ca, Mg, and H_2O from the EJ-13 leachant to form alteration phases. These phases precipitated on the test vessel components, sample surfaces, and within the corroded intergrain boundary regions (Figs. 2c and 2d). The end result of this process was the development of a paragenetic sequence of alteration phases, as evident on the top surfaces of the samples, where $UO_2 \Rightarrow$ dehydrated schoepite $\Rightarrow$ compreignacite + becquerelite $\Rightarrow$ soddyite $\Rightarrow$ boltwoodite + uranophane + palygorskite clay (Table I). This trend appears to be controlled by precipitation kinetics and is nearly identical to alteration patterns observed during the weathering of naturally occurring uraninite, such as that that which occurs at the Nopal I uranium deposit in Chihuahua, Mexico [12-14]. This resemblance suggests that the conditions in the present tests have replicated an environment that may be representative of that occurring during uraninite alteration in natural oxidized systems.

The bottom of the eight-year sample displayed a relationship among the crystalline phases that is reminiscent of processes that affected the top surfaces of samples within the first few years of reaction. Dehydrated schoepite, which was the only alteration phase observed on the top surface after 1.5 years and on the bottom surface after 2.25 and 3.5 years, began to display the formation of corrosion pits at eight years [8,9]. The sharp edges of these pits suggest that a period of active dissolution was occurring when the eight-year experiment was terminated. Becquerelite and compreignacite also made their first appearance on the bottom sample surfaces at this time. This pattern is similar to that observed on the top surface after 2.25-years when becquerelite became the predominant phase by displacing the dehydrated schoepite that had formed earlier [8,9]. One may predict then, by analogy with the paragenetic sequence observed on the top sample surface, that the becquerelite that formed on the bottom surface as a replacement for dehydrated schoepite, may eventually itself be replaced by alkali and alkaline earth uranyl silicates.

Solid solution compositional changes were noted in compreignacite crystals with respect to their spatial location on the top sample surface. The molar ratio of K/Na in this phase was highest along the outside edge of the top pellet surface, where nearly pure K phases were found. The K/Na ratio decreased (down to ~0.8) down the sides of the samples, though patterns were more sporadic on the top pellet surface. These trends suggest a preference for potassium ions in the crystalline compreignacite relative to sodium (the K/Na ratio of the EJ-13 leachant is 0.17) as well as a progressive decrease of the potassium activity of the solutions as they migrate across the sample surfaces. Compreignacite was also nearly free of Ca, but did occasionally contain minor amounts of Mg. By contrast, becquerelite was identified only on the top pellet surface, and contained only minor amounts of K and Na, and no Mg. The absence of Mg was unusual in light of the common substitution of Mg for Ca in many crystalline compounds.

A cross-section examination of the samples indicated that the vast majority of the uranium released from the dissolving samples was deposited back on the surface of the UO_2 pellets and Zircaloy cladding as alteration phases (Figs. 2b-d). The quantity of uranium incorporated in these phases was calculated for the eight year sample by determining the volume of precipitated material on the sample surface, proportion of each respective alteration phase present, molar proportion of uranium contained in each phase, and multiplying the calculated volume of each phase present by its respective density. The weight of this sample prior to testing was 29,166 mg (25,709 mg of uranium). Preliminary calculations indicate that ~80 mg of uranium was incorporated into the alteration phases deposited on the sample or Zircaloy surfaces, an amount that far exceeds the 5 mg that was released from the sample-Zircaloy assembly (as recovered in the acid strip component). An additional ~780 mg of uranium remained *in situ* as undissolved UO_{2+x} cores that had undergone some corrosion along their grain boundaries.

CONCLUSIONS

Uranium release from the UO_2 samples was rapid between one and two years. The initial rapid release (period II) could be correlated with an episode of preferential dissolution along the boundaries between the UO_2 pressed grains and spallation of micrometer- to submicrometer-sized UO_{2+x} particles from the sample surfaces. Penetration along the intergrain boundaries typically occurred to a depth of two-to-four grains (~10 to 20 mm) ahead of the exposed external sample surface, but varied from regions with little visible corrosion to those where penetration occurred to a depth of approximately ten grains. Subsequent to this rapid release interval, the development of a mat of alteration phases encapsulated the loosened UO_{2+x}, resulting in a reduction of particle spallation and lower uranium release rates (period III).

Most of the uranium that was released from the samples by dissolution processes eventually precipitated back as alteration phases onto surfaces of the test components (Table I). Uranium release rates during period III, as determined from unfiltered solution aliquots, were generally between 0.1 and 0.3 mg/(m$^2\bullet$day). However, less than 2% of the released uranium occurred in a soluble form (fraction that passed through a 5 nm filter). This latter value correlates to an average filtered uranium solution concentration of 4 x 10^{-6} M.

The sample reaction trends observed in the present tests closely replicated those that occur in natural geologic systems. The experiments display a sequence of phase formation that is characterized by the following paragenetic trend: UO_2 $\Rightarrow$ dehydrated schoepite $\Rightarrow$ compreignacite + becquerelite $\Rightarrow$ soddyite $\Rightarrow$ boltwoodite + uranophane + palygorskite clay. This similarity suggests that the UO_2 in the experiments has reacted by the same mechanism as uraninite in the natural deposits, and thus both the present tests and the natural analogue reactions may be used to simulate the reaction progress expected for uranium at the proposed Yucca Mountain repository. Alkali and alkaline earth uranyl silicates represent the long-term solubility-limiting phases for uranium in these systems.

The overall reaction pathway of UO_2 is controlled by a combination of factors, including oxidative-dissolution, precipitation kinetics, and leachant composition. The results from this study indicate that the alteration rates for UO_2, and by analogy, spent fuel, may be quite rapid in an unsaturated geologic setting. Alteration phases incorporated a large proportion of the uranium that was dissolved from the samples. Such a process may also act as a significant mechanism for retarding the migration of fission product and transuranic elements from spent nuclear fuel. If the effect that the uranyl phases are to have on fission product and transuranic element migration is to be included in performance assessment models, then we will need to have a greater understanding of the compositional variations of these phases, their stability at various temperatures and fluid compositions, and partition coefficient ratios that govern the incorporation of elements into their structures.

ACKNOWLEDGEMENTS

This work is being performed under the guidance of the Yucca Mountain Site Characterization Project/Lawrence Livermore National Laboratory Spent Fuel Scientific Investigation Plan with support for this research has being provided by the U.S. Department of Energy, under contract W-31-109-ENG-38.

REFERENCES

1. Integrated Data Base Report - 1994, U.S. Department of Enery Report DOE/RW-0006, Rev. 11 (1995).
2. J. de Pablo, I. C. J. Giménez, V. Martí, and M. E. Torrero, J. Nucl. Mater. <u>232</u>, 138 (1996).
3. J. Posey-Dowty, E. Axtmann, D. Crerar, M. Borscik, A. Ronk, and W. Woods, Econ. Geol. <u>82</u>, 184 (1987).
4. G. W. McGillivray, D. A. Geeson, and R. C. Greenwood, J. Nucl. Mater. <u>208</u>, 81 (1994).
5. L. H. Johnson and D. W. Shoesmith, in *Radioactive Waste Forms for the Future*, W. Lutze and R. C. Ewing (eds.), Elsevier Science Publishing, Amsterdam, The Netherlands (1988) 635 p.
6. P. A. Finn, J. C. Hoh, S. F. Wolf, M. T. Surchik, E. C. Buck, and J. K. Bates, this volume (1997).
7. D. J. Wronkiewicz, J. K. Bates, T. J. Gerding, E. Veleckis, and B. S. Tani, Argonne National Laboratory Report ANL-91/11 (1991).
8. D. J. Wronkiewicz, J. K. Bates, T. J. Gerding, E. Veleckis, and B. S. Tani, J. Nucl. Mater. <u>190</u>, 107 (1992).
9. D. J. Wronkiewicz, J. K. Bates, S. F. Wolf, and E. C. Buck, J. Nucl. Mater. (in press).
10. W. J. Gray and D. M. Strachan, Mater. Res. Soc. Symp. Proc. <u>212</u>, 205 (1991).

11. R. E. Einziger, L. E. Thomas, H. C. Buchanan, and R. B. Stout, High Level Rad. Waste Mgmt. Proc. Third Inter. Conf. $\underline{2}$, 1449 (1992).
12. B. W. Leslie, E. C. Pearcy, and J. D. Prikryl, Mater. Res. Soc. Symp. Proc. $\underline{294}$, 505 (1993).
13. F. Cesbron, P. Ildefonse, and M-C. Sichere, Mineral. Mag. $\underline{57}$, 301 (1993).
14. W. M. Murphy, Radwaste $\underline{2}$, 44 (1995).

SPENT FUEL REACTION: THE BEHAVIOR OF THE ε-PHASE OVER 3.1 YEARS

P. A. FINN, J. C. HOH, S. F. WOLF, M. T. SURCHIK, E. C. BUCK, and J. K. BATES
Argonne National Laboratory, 9700 South Cass Avenue, Argonne, IL 60439

ABSTRACT

The release fractions of the five elements in the ε-phase (^{99}Tc, ^{97}Mo, Ru, Rh, and Pd) as well as that of ^{238}U are reported for the reaction of two oxide fuels (ATM-103 and ATM-106) in unsaturated tests under oxidizing conditions. The ^{99}Tc release fractions provide a lower limit for the magnitude of the spent fuel reaction. The ^{99}Tc release fractions indicate that a surface reaction might be the rate controlling mechanism for fuel reaction under unsaturated conditions and the oxidant is possibly H_2O_2, a product of alpha radiolysis of water.

INTRODUCTION

A potential site for the U.S. high-level waste repository is in the volcanic tuff beds at Yucca Mountain, Nevada, a hydrologically unsaturated zone. To qualify this site for licensing, information on the corrosion behavior of spent fuel under unsaturated and oxidizing conditions is needed. Performance assessment calculations are needed to license this site and radionuclide release rates are needed as the source term in these calculations. Laboratory testing of spent fuel under unsaturated conditions, i.e., those with limited water, provides the information necessary to determine the magnitude of the potential radionuclide source term at the boundary of the fuel's cladding after the cladding has failed and water as vapor or liquid contacts the fuel.

Tests that simulate the presence of limited water and oxidizing conditions are in progress to evaluate the long-term behavior of commercial spent nuclear fuel at 90°C. In the tests, a thin film of water, which is supplied by saturated water vapor at 90°C, continuously contacts and reacts with the fuel. Water that is dripped on the fuel provides sufficient liquid to transport reacted material beyond the fuel holder.

The purpose of the experiments is to determine the relationship between the rate of fuel alteration and the release rate of different radionuclides under unsaturated conditions. Therefore, the extent of spent fuel alteration, e.g. dissolution, under unsaturated conditions is assessed as is the effect of fuel alteration on radionuclide release. This paper examines the reaction of spent fuel, specifically the ε-phase (Tc, Mo, Ru, Rh, Pd) in high drip rate tests for the first 3.1 years of reaction. The release of ^{99}Tc appears to be a marker for matrix dissolution. The magnitude of the ^{99}Tc interval release fractions as a function of time suggest that the spent fuel reaction is controlled by a surface reaction in which oxidants supplied from alpha radiolysis of water may play an important role.

EXPERIMENTAL

Samples of two pressurized-water-reactor fuels, ATM-103, [1], and ATM-106 [2], with burnups of 30 and 45 MW•d/kg U, respectively, are used in the tests. The grain sizes are 17-20 and 6-16 μm, respectively. The fuel samples, 7-8.5 g each, are fragments with masses in the range of 0.3 to 1.2 g. The average geometric surface area of the fragments is 2.1 x 10^{-4} m^2/g. The measured percent distribution of ^{99}Tc [3] for both fuels in the gap, grain-boundary, and fuel matrix is <0.1, <0.1, and >99.8%, respectively. The reported distribution [4] of ^{99}Tc in ε-phase particles in ATM-103 is as follows. Most ε-phase particles are within the grains and their diameter varies as a function of radial position, being 20 nm diameter at the pellet edge and 50-100 nm diameter in the pellet mid- and center sections. Particles on the grain boundaries are ≤1 μm diameter and are located in the mid- and center sections. For ATM-106 fuel [5], the ε-phase particles within the grains are 20 to 100 nm diameter in the mid- section. Near the pellet center 3 to 10 μm diameter particles were found along the grain boundaries. These particles also contained gas bubbles and secondary precipitates.

527

Mat. Res. Soc. Symp. Proc. Vol. 465 © 1997 Materials Research Society

The test procedures used for the high drip rate tests have been given previously [6]. Groundwater from well J-13 near Yucca Mountain was equilibrated for eighty days at 90°C with crushed core samples of Topopah Spring tuff and is designated EJ-13. Solution compositions were analyzed with inductively coupled plasma-mass spectrometry (ICP-MS) in the single standard scan mode with an indium internal standard. For samples after 2.5 years of reaction, bismuth is used as the internal standard for actinides. This technique provides accurate results ($\pm$15%) for the isotopes in the middle mass range (80-160 atomic mass units), but results for the actinides and the light elements might vary by $\pm$50%. Individual mass concentrations for duplicates of the different samples varied by 0.6-2% for concentrations above 0.5 ppb and by 13-17% below 0.5 ppb. The uncertainties in the ICP-MS data are a maximum of $\pm$50% depending on radionuclide concentration. The assignment of a mass unit to a particular isotope is done by comparison with the isotopic distribution given in results [1,2] from ORIGEN code calculations.

Terminology — The term "interval" refers to the total number of days during which spent fuel is reacted in a given sequential reaction period. The "interval release fraction" for a given radionuclide is defined as the ratio of the mass of radionuclide collected in a given interval, i.e., the sum in the leachate and the acid strip, divided by the amount of the same radionuclide calculated [1,2] to be in the fuel. The material in alteration products, adsorbed on the Zircaloy holder or on the spent fuel is not included. The radionuclide with the largest interval release fractions provides a lower estimate for the spent fuel reaction rate. The interval release fractions are used to compare the releases for various radionuclides for the same interval and for the same test for different intervals. The "cumulative release fraction" is the sum of the interval release fractions.

RESULTS

The fractional releases of ^{99}Tc, ^{97}Mo, and ^{238}U are reported as is the physical appearance of the fuel. Parameters that may control spent fuel reaction and radionuclide release are discussed.

The ε-Phase — Table I provides a summary for successive reaction intervals of the release behavior of the five elements in the ε-phase (Tc, Mo, Ru, Rh, Pd) for tests with ATM-103. The information includes: (1) the released amounts (μg) for the isotope of each element with minimal interference from other elements; (2) the total released amount of each element based on the measured isotope and the element's isotopic distribution; (3) the calculated amount of the ε-phase that reacted based on the ^{99}Tc release and the distribution [1,5] of each element in the ε-phase; and (4) the amount of each element that was retained on the spent fuel based on the difference between the material released (column 2) and that calculated to have reacted (column 3). For ATM-103 at each reaction interval, and for ATM-106, ^{99}Tc was the dominant element released; 10% of the Mo and only trace amounts of the other three elements were detected.

Reaction and dissolution of the ε-phase particles with later incorporation of the non-Tc elements in alteration products is indicated by three additional pieces of data. First, large amounts of Mo were found in a cesium uranyl oxide that was attached to a reacted fuel grain [7]. Second, 1-2 wt% amounts of Ru and Mo were incorporated into a uranyl silicate alteration product, while only ppm of Tc were found. Third, corrosion of ε-phase particles as fuel grains dissolve is evident in Fig. 1 in which individual ε-phase particles <10 nm diameter are shown.

Since the ε-phase particles are homogeneously dispersed in the ATM-103 fuel matrix [1] and both fuels have similar ^{99}Tc release behavior, the use of ^{99}Tc release fractions as an indicator of the minimum extent of the reaction of the fuel matrix is proposed. Observations of ^{99}Tc release behavior in saturated tests were as follows. Forsyth and Werme[8] found that after 2.7 years of reaction under oxidizing conditions ^{99}Tc was the only radionuclide whose release fraction had remained constant. They suggested that ^{99}Tc was oxidized and released as the matrix around the ε-phase particles reacted. Wilson[9] suggested that Tc was oxidized prior to release.

Reaction Progress — Uranium interval release fractions were examined to determine the extent of uranium retention based on ^{99}Tc release fractions as a measure of the extent of fuel reaction. The interval release fractions for ^{99}Tc, ^{97}Mo, and ^{238}U (Table II), and their cumulative release fractions (Table III) indicate that for both fuels, cumulative ^{238}U release fractions were smaller by several orders of magnitude than the ^{99}Tc cumulative release fractions; and, the difference increased as reaction time increased. Alteration products incorporated over 99% of the reacted uranium; a minor amount may have sorbed on the Zircaloy fuel holder.

Table I. Disposition of Elements in ε-Phase for Selected Intervals — ATM-103 High Drip Test

Isotope	Measured[a] Released Element[b] (μg)	Calculated Released Element[c] (μg)	Calculated Amount Reacted (μg)	Element[d] Retained (mass %)
0.3 Years Reaction				
^{99}Tc	20	20	20	—
^{97}Mo	0.9	4	50	93
^{101}Ru	0.02	0.07	50	100
^{103}Rh	0.6	0.6	7	92
^{105}Pd	0.04	0.1	0.5	75
0.8 Years Reaction				
^{99}Tc	10	10	10	—
^{97}Mo	0.05	2	30	94
^{101}Ru	6E-5	2E-4	40	100
^{103}Rh	0.06	0.06	5	99
^{105}Pd	ND[e]	ND	0.3	100
1.6 Years Reaction				
^{99}Tc	40	40	40	—
^{97}Mo	8	30	100	77
^{101}Ru	2E-3	7E-3	200	100
^{103}Rh	0.02	0.02	20	100
^{105}Pd	2E-3	9E-3	1	100
2.1 Years Reaction				
^{99}Tc	5	5	5	—
^{97}Mo	2	10	20	44
^{101}Ru	8E-5	2E-4	20	100
^{103}Rh	7E-3	7E-3	3	100
^{105}Pd	8E-3	0.03	0.2	83
2.5 Years Reaction				
^{99}Tc	10	10	10	—
^{97}Mo	1	6	30	82
^{101}Ru	6E-4	2E-3	30	100
^{103}Rh	0.02	0.02	5	100
^{105}Pd	5E-3	0.02	0.3	94

[a]Measured mass in leachate. Values were rounded to one significant figure.
[b]The isotopic distribution for each element and the mass of the measured isotope was used to determine the total mass released.
[c]For ATM-103, the wt%s in the ε-phase are: [1]: Tc(11.8); Mo(39.9); Ru(42.3); Rh(5.6); Pd(0.4). The released ^{99}Tc was the basis for the reacted amount of a given element.
[d]This is the minimum amount retained and is based on ^{99}Tc and its wt% in the ε-phase.
[e]ND = not detected.

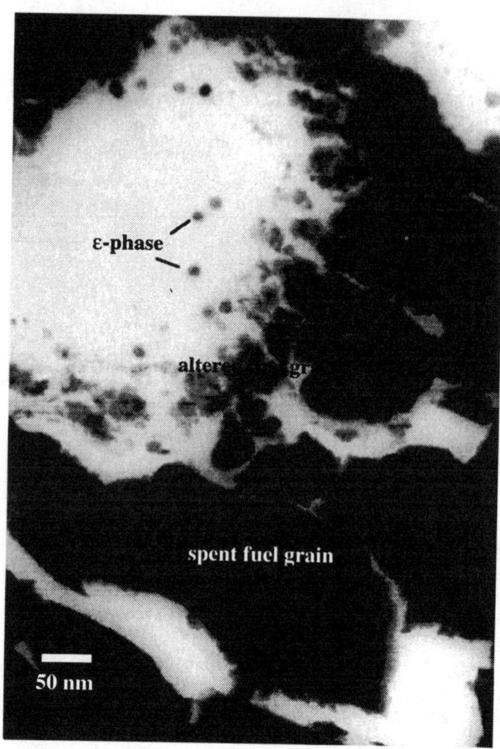

Fig. 1.

Image of corroded pit in altered fuel grain with reacted ε-phase particles at fuel surface.

Table II. Interval Release Fractions — High Drip Rate Tests

Time (yr)	Interval Release Fraction		
	^{99}Tc	^{97}Mo	^{238}U
ATM-103			
0.2	2E-3	1E-5	3E-5
0.3	3E-3	2E-4	2E-5
0.8	2E-3	9E-5	5E-6
1.3	7E-3	2E-4	9E-6
1.6	8E-3	1E-3	2E-5
2.1	1E-3	4E-4	2E-6
2.5	2E-3	3E-4	8E-7
3.1	5E-3	1E-2	3E-6
ATM-106			
0.2	0	0	1E-9
0.3	1E-5	6E-6	2E-5
0.8	1E-4	6E-4	2E-4
1.3	6E-5	9E-6	8E-6
1.6	1E-3	3E-4	1E-6
2.1	4E-3	9E-5	1E-7
2.5	4E-3	9E-5	3E-7
3.1	8E-3	8E-4	3E-7

Table III. Cumulative Release Fractions — High Drip Rate Tests

	^{99}Tc	^{97}Mo	^{238}U
1.6 Year			
ATM-103	2E-2	2E-3	9E-5
ATM-106	2E-3	8E-4	2E-4
2.5 Year			
ATM-103	2E-2	3E-3	9E-5
ATM-106	1E-2	1E-3	2E-4
3.1 Year			
ATM-103	3E-2	1E-2	9E-5
ATM-106	2E-2	2E-3	2E-4

The ^{97}Mo cumulative release fractions increased as a function of reaction time and were also orders of magnitude larger than the ^{238}U cumulative release fractions. No definitive trends can be suggested at this time on the basis of available data.

The ^{99}Tc interval release fractions for both fuels were the same order of magnitude 1 x 10^{-3} to 8 x 10^{-3} after the 1.6 year reaction interval, although earlier the release fractions for the ATM-106 fuel had been smaller, i.e., an induction period had been required for the ATM-106. Since as reaction time increased, the ^{99}Tc interval release fraction was unchanged, this suggests that reaction and dissolution occurs only at the outer grain surface, i.e., the effective surface area for radionuclide release is unchanged although oxidation occurs along the grain boundaries.

Alteration Products — Yellow alteration products were observed on the spent fuel surface after 2.5 years of reaction [7]. Examination of the fuel fragments after 3.7 years of reaction(leachate results are not yet available) indicated that there were two layers of light-yellow alteration products. The outer layer could be easily manipulated with a fine tungsten needle but a thinner layer immediately adjacent to the fuel was denser and adhered tenaciously to the surface. Electron microscopy showed that both layers are uranyl silicates structurally related to uranophane type minerals.

A fragment of reacted ATM-103 fuel was extremely friable and disintegrated under examination producing four smaller pieces and black fuel grains . One of the smaller pieces was reduced to a smear of black-like powder. This behavior indicated that reaction had proceeded down the grain boundaries and into individual grains after less than four years at 90°C. The disintegration of the ATM-103 fragment was similar to fuel behavior in oxidation tests done at 200°C [10] in the presence of water vapor. There, schoepite formed in grain boundaries, but, did not impede further reaction progress throughout the grain-boundary network. Since our tests use a silicate-groundwater, yellow uranyl silicates form on the fuel surface, but our analyses to date, show no reaction products in the grain boundaries.

DISCUSSION

The ε-Phase — The presence of soluble forms of ^{99}Tc and ^{97}Mo in the leachate (pH 7 at room temperature) indicated that the fuel matrix surrounding the ε-phase particles had dissolved exposing them to an oxidizing environment. Kleykamp [11] noted that complete dissolution of the ε-phase occurred with boiling nitric acid but that long reaction times were required, i.e. for particles 0.15 to 1.5 μm in diameter, eight hours was needed for complete dissolution. Thus, ε-phase particles (<1 nm to 100 nm diameter) could react and release ^{99}Tc in each 0.5 year reaction interval, but larger particles (1 μm) might not completely react (see Fig. 1).

Water vapor at 90°C supplies a continuous thin film of water on the spent fuel surface in each test, a film that may be over 50 nm thick. Allen noted that the primary yield of H_2O_2 from alpha radiolysis of water was 1.2 times that from gamma radiolysis of water [12]. The H_2O_2 is produced in dense tracks within the first 30 μm (the alpha particle range) of the fuel surface. Shoesmith et al. [13] noted that for reaction of unirradiated UO_2 at pH 9.5 with an alpha source 30 μm away, the redox potential was 0.11 V versus a saturated calomel electrode, a value similar to that in an equivalent H_2O_2 solution or in a saturated O_2-containing solution. (The corresponding corrosion potential would be -0.14 V.)

The corrosion potential in the unsaturated tests was estimated by considering the Eh required to form soluble ions of Tc, Mo, Ru, Rh, and Pd based on Eh/pH diagrams [14]. For Mo and Tc at pH 7, a minimum Eh of 0 and 0.3 V, respectively, is needed to form TcO_4^- and $HMoO_4^{2-}$, whereas for Pd and Ru, a minimum Eh of 0.5 and 0.6 V, respectively, is needed. For Rh, a soluble form is noted only at pH 4 and an Eh of 0.65 V. If one assumes that these diagrams are appropriate for the conditions of the unsaturated tests, then the oxidation potential at the fuel surface may be as high as 0.6 V since Ru dissolved and was incorporated into alteration products.

An analogue for the behavior of the elements in the ε-phase is the behavior of these elements at the Oklo natural reactor site in Gabon, Africa. Although the natural reactor had an upper temperature of 450°C, out-of-pile experiments showed that the ε-phase can form at temperatures as low as 450°C [15] and thus it was probably present at Oklo. Most of the soluble radionuclides (Tc, Cs, Mo, and Ru) were diminished in the reactor zones with over 50% of the Tc and Ru mass missing. The three release mechanisms suggested [16] were: fission recoil, dissolution of the uraninite grains, and solid-state diffusion. Curtis and Gancarz [17] proposed that oxidative dissolution of the uraninite was caused by water radiolysis in the reactor zones that led to the production of species that oxidized uraninite, dissolved the ε-phase, and permitted transport of soluble species from the reactor zone.

<u>Potential Mechanism in the Spent Fuel Reaction</u> — Grambow et al. [18] found that for unirradiated UO_2 the reaction rate depended only on the concentration of the oxidant and concluded that even under an initially anoxic conditions, the spent fuel dissolution process was oxidative. Therefore, the local steady-state concentration of oxidants at the surface is of key importance and provides an upper limit on reaction.

In the unsaturated tests, there is a steady-state concentration of H_2O_2, as well as other radiolysis products, in the thin film of water at the fuel surface. Reaction with the oxidant occurs at the surface and proceeds down the grain boundaries as shown by the friable nature of the ATM-103 fragment. The alteration products do not form a protective layer that decreases fuel dissolution since the ^{99}Tc interval release fractions were constant over 3.1 years of reaction. Since ppm amounts of ^{99}Tc were present in the alteration phases, the fuel reaction rate is higher than that measured from ^{99}Tc interval release fractions.

If one accepts ^{99}Tc release as a measure of fuel reaction, then it can be used to examine five suggested mechanisms [19]. These are: (1) solubility limited dissolution of alteration products, (2) transport of oxygen to the UO_2 surface, (3) growth of alteration products, (4) the rate of formation of oxidants by radiolysis, and (5) a surface reaction.

For solubility limited dissolution to control rate, radionuclide release has to decrease significantly as a function of time. This might control [238]U release but not its reaction since ^{99}Tc release fractions did not decrease as a function of time.

For oxygen transport to control rate, a diffusion process has to be established. Diffusion does not appear to control rate since the best fit of ^{99}Tc cumulative release was found with a linear dependence on time (see Fig. 2)

For the growth of alteration products to control rate, release should be constant or should increase as the surface area increased, i.e., as grain boundaries opened. For both unsaturated tests, no increase was noted in the ^{99}Tc release fractions as a function of time although the surface area increased, instead, the ^{99}Tc release fractions for ATM-103 were the same order of magnitude during 3.1 years of reaction. Thus, this mechanism does not appear to control the rate.

The rate of formation of oxidants by radiolysis also does not appear to control rate since Grambow et al. noted [18] that even in a reducing environment radiolysis supplied sufficient oxidants to oxidize spent fuel. However, if the rate of a surface reaction is faster than the rate of oxidant formation, this mechanism could be rate controlling.

Since the reaction rate, defined by the ^{99}Tc interval release fractions, remained constant with time a surface reaction appears to be the rate controlling mechanism. Support for control of the rate by a surface reaction was provided by Ericksen et al who observed in tests [20] with spent fuel that nearly 100% of the oxidants from radiolysis was taken up by the fuel surface.

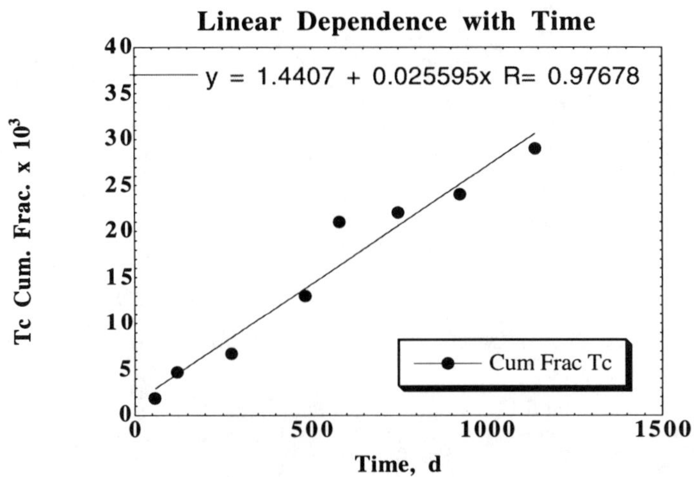

Fig. 2. The linear time dependence of ^{99}Tc cumulative release fractions in the ATM-103 high drip rate test. (Equation is the linear least squares fit.)

CONCLUSIONS

Our conclusions for the high drip rate tests are these. (1) The ^{99}Tc interval release fractions provide a lower limit for the spent fuel reaction. (2) The friability of a spent fuel fragment after less than four years provides support for the use of ^{99}Tc release fractions as a lower limit for reaction. (3) Oxidation of the ε-phase suggests that the oxidizing potential at the fuel surface may be as high as 0.6 V. The oxidant may be H_2O_2, a product of alpha radiolysis of water. (4) The non-variability of the ^{99}Tc release fractions over 3.1 years of reaction provides support for a surface reaction as the rate-controlling mechanism for fuel reaction under unsaturated conditions.

ACKNOWLEDGMENTS

This task was performed under the guidance of the Yucca Mountain Site Characterization Project (YMP) and is part of activity D-20-43 in the YMP/Lawrence Livermore National Laboratory Spent Fuel Scientific Investigation Plan. This work was supported by the U.S. Department of Energy, under contract W-31-109-ENG-38.

REFERENCES

1. R.J. Guenther et al., Characterization of Spent Fuel Approved Testing Material - ATM-103, Pacific Northwest Laboratory Report PNL-5109-103 (1988).
2. R.J. Guenther et al., Characterization of Spent Fuel Approved Testing Material - ATM-106, Pacific Northwest Laboratory Report PNL-5109-106 (1988).
3. W.J. Gray, D.M. Strachan, and C.N. Wilson, in Scientific Basis for Nuclear Waste Management XV, ed. by C.G. Sombret (Mater. Res. Soc. Proc. **257**, Strasbourg, France, 1992), pp. 353-360.

4. L.E. Thomas and R.J. Guenther, in <u>Scientific Basis for Nuclear Waste Management XII</u>, ed. by W. Lutze and R.C. Ewing, (Mater. Res. Soc. Proc. **127**, Berlin, Germany, 1989), pp. 293-300.

5. L.E. Thomas, C.E. Beyer, and L.A. Charlot, J. Nucl. Mater. **188**, 80 (1992).

6. P.A. Finn et al., Radiochimica Acta **66/67**, 189 (1994).

7. P.A. Finn et al., Spent Fuel's Behavior Under Dynamic Drip Tests, Proc. Intl. Conf. on Evaluation of Emerging Nuclear Fuel Cycle Systems, Versailles, France, 240 (1995).

8. R.S. Forsyth and L.O. Werme, Spent Fuel Corrosion and Dissolution, Swedish Nuclear Fuel and Waste Management Report SKB 91-60 (1991).

9. C.N. Wilson, Results from NNWSI Series 3 Spent Fuel Dissolution Tests, Pacific Northwest Laboratory Report PNL-7170 (1990).

10. K.M. Wasywich, et al, Nucl. Tech. **104**, 309 (1993).

11. H. Kleykamp, J. Nucl. Mater. **131**, 221 (1985).

12. A.O. Allen, <u>The Radiation Chemistry of Water and Aqueous Solutions</u>, (Von Nostrand, Princeton 1961).

13. D.W. Shoesmith, et al, in <u>Scientific Basis for Nuclear Waste Management IX</u>, ed. by L.O. Werme (Mater. Res. Soc. Proc. **50**, Stockholm, Sweden, 1985) pp. 309-316.

14. D.G. Brookins, <u>Eh-pH Diagrams for Geochemistry</u>, (Springer-Verlag, Berlin, 1988).

15. J.H. Davies and F.T. Ewart, J. Nucl. Mater. **41,** 143 (1971).

16. L.H. Johnson and D.W. Shoesmith, Spent Fuel, Chap. 11, <u>Radioactive Waste Forms for the Future</u>, ed. by W. Lutze and R.C. Ewing, (Physics Publishing, North Holland, 1988).

17. D.B. Curtis and A.J. Gancarz, Radiolysis in Nature: Evidence from the Oklo Natural Reactors, KBS Technical Report 83-10 (1983).

18. B. Grambow, et al., Chemical Reaction of Fabricated and High Burnup Spent UO_2 Fuel with Saline Brines, Forschungszentrum Karlsruhe Report FZKA 5702 (1996).

19. B. Grambow, et al, in <u>Scientific Basis for Nuclear Waste Management XIII</u>, ed. by V.M. Oversby and P.W. Brown, (Mater. Res. Soc. Proc. **176**, Boston, MA, 1990), pp. 465-474.

20. T. E. Eriksen, et al., J. Nucl. Mater. **227**, 76 (1995).

EFFECT OF TEMPERATURE AND BICARBONATE CONCENTRATION ON THE KINETICS OF UO₂(s) DISSOLUTION UNDER OXIDIZING CONDITIONS

J. de Pablo, I. Casas, J. Giménez, M. Molera and M.E. Torrero
Department of Chemical Engineering, Universitat Politècnica Catalunya
08028 Barcelona. Spain

ABSTRACT

The dissolution rate of unirradiated UO_2 (s) has been studied as a function of hydrogen carbonate concentration at three different temperatures (298.15 K, 313.15 K and 333.15 K) under oxidizing conditions in a continuous flow-through reactor with a thin layer of solid particles (particle size from 100 to 300 µm). From the results of these experiments, two different rate laws have been determined. At high temperature (313.15 K and 333.15 K), we obtained a dissolution rate proportional to hydrogen carbonate concentration while at 298.15 K , the rate almost depends on the square root of the hydrogen carbonate concentration. This indicates a different reaction mechanism depending on temperature which can be related to the oxidation step of the overall process. The apparent activation energy obtained was 41 kJ mol^{-1}.

INTRODUCTION

The dissolution rate of the spent fuel matrix is very important in the performance assessment of the final spent nuclear fuel repository. There have been many investigations to study the dissolution of unirradiated UO_2 and spent fuel in aqueous solutions as a function of several parameters such as pH, carbonate concentration, dose of α-radiolysis, particle size and temperature under both reducing and oxidizing conditions. In this sense, an extensive work has been performed by Gray et al [1,2] to determine the effect of temperature and important water chemistry parameters on the dissolution of both spent fuel and UO_2. However, the conclusions reported in those works are difficult to interpret in terms of oxidation/dissolution mechanisms which are very important to assess the behavior of the matrix barrier in final disposal conditions [3].

Some aspects related to the mechanism of the oxidation/dissolution process under oxidizing conditions, which are expected due to the α-radiolyisis of water, seem to be well established [4,5]. Therefore, the UO_2 surface is oxidized upto $UO_{2.33}$ followed by the detachment of the oxidized U(VI) by acid promoted surface complexation [6] in absence of other complexing agents. In carbonate medium, a similar mechanism has been proposed [5,7] but some discrepancies have been observed when dissolution rates are determined as a function of carbonate concentration. At low carbonate content (0.001 mol L^{-1}), the rate of uraninite dissolution is proportional to total carbonate [8]. At higher concentrations (0.5 mol L^{-1}), the rate depends on the square root of the total carbonate [9]. Recently, the dependence determined by Gray et al. [1] was less than the square root. On the other hand, the effect of the temperature on the dissolution rate has not been extensively studied although apparent activation energy values have been reported in the range of 20 - 60 kJ mol^{-1}. Dissolution rates determined as a function of temperature can allow to elucidate the mechanism of the UO_2 oxidation/dissolution process.

In this work, the dissolution rate of unirradiated UO_2 was determined at oxidizing conditions (air) as a function of carbonate concentration at three different temperatures (298.15 K, 313.15 K and 333.15 K) in a continuous flow-through reactor with a thin layer of solid

Mat. Res. Soc. Symp. Proc. Vol. 465 © 1997 Materials Research Society

particles [1, 10]. This experimental device hinders secondary phase formation and allows to easily determine dissolution rates.

EXPERIMENTAL

We have used unirradiated UO_2 (s) (1.00 g) with a particle size of 100-300 μm obtained by crushing and sieving UO_2 pellets. This solid was washed with ketone before starting the experiments.

The specific surface area measured by the BET method was 0.0113 m^2 g^{-1}.

The ionic strength of the leachant was kept constant at 0.15. Hydrogen carbonate concentration was varied from 10^{-4} to 0.05 mol L^{-1} in NaCl solutions to maintain the ionic strength mentioned above. Oxic conditions were those obtained by equilibrating the leachant solution with air at 25 °C.

Experimental procedure

As it was mentioned above, experiments were carried out in a continuous flow-through thin solid layer reactor which consists in a Flex-column with a jacket to controle the temperature inside the reactor. The column has a cylindrical shape with an interior diameter of 10 mm with two adapters which permit an adjustable height. The exterior diameter of the jacket is of 35 mm. The solid layer thickness was restricted to a maximum of 2 mm.

Flow rate of the leaching solution was mantained in the range of 0.1 to 0.2 L d^{-1} by means of a peristaltic pump. By working in these conditions we ensure that steady state is achieved, determined in a previous work [4]. The pH was continuously monitored by means of an on-line combined glass electrode. Uranium concentration of the outflow solution was analyzed by using a SCINTREX UA-3 Uranium Analyzer [11].

The leaching solution was kept at constant temperature (25 °C) in a water bath in order to maintain the same oxygen concentration in solution in all the experiments. The leaching solution was heated with water at 45 or at 60 °C in the tube before putting in contact with the solid. During this procedure, we did not observe any formation of gas bubbles in the tube so we assume that the oxygen concentration was practically constant. Moreover, the temperature was shifted up and down, showing the reproducibility of the steady state uranium concentrations.

The dissolution rate was calculated from the flow rate (Q in L s^{-1}), the uranium concentration ([U] in mol L^{-1}) at steady state and the total surface area of the solid in the reactor (S in m^2), according to the equation:

$$r \ (\text{mol m}^{-2} \text{ s}^{-1}) = [U] \ Q \ / \ S \qquad (1)$$

RESULTS

Dissolution rates as a function of time at different temperatures are plotted in Figure 1. This figure shows the reproducibility of the methodology used. After shifting up and down the temperature at a constant hydrogen carbonate concentration, we obtained the same dissolution rate. All the dissolution rates obtained in this work have been collected in Table I.

Table I. Dissolution rates as a function of temperature and
hydrogen carbonate concentration

[HCO$_3^-$] (mmol L^{-1})	T (K)	Dissolution rate (mg m^{-2} d^{-1})	Dissolution rate (mol m^{-2} s^{-1})
50	298.15	8.64	$4.20\ 10^{-10}$
50	318.15	116	$5.66\ 10^{-9}$
50	333.15	221	$1.08\ 10^{-8}$
10	298.15	2.20	$1.07\ 10^{-10}$
10	318.15	17.0	$8.27\ 10^{-10}$
10	333.15	59.9	$2.87\ 10^{-9}$
1	298.15	0.84	$4.08\ 10^{-11}$
1	318.15	1.73	$8.43\ 10^{-11}$
1	333.15	2.40	$1.17\ 10^{-10}$
0.1	298.15	0.19	$9.53\ 10^{-12}$
0.1	318.15	0.30	$1.46\ 10^{-11}$
0.1	333.15	0.53	$2.58\ 10^{-11}$

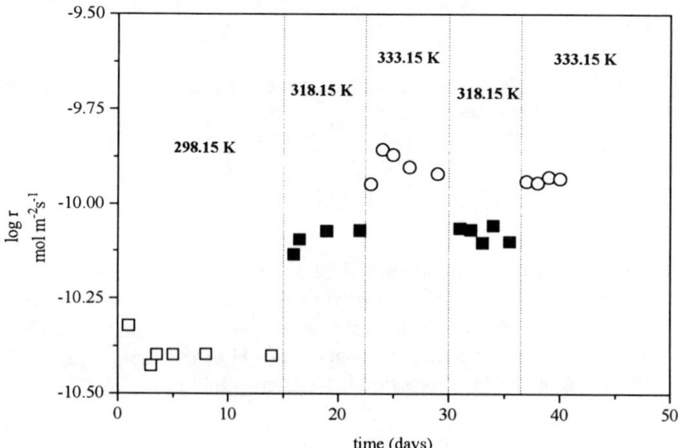

Figure 1. Dissolution rate as a function of time in HCO$_3^-$ 10^{-3} mol dm^{-3}. Dotted lines represent changes in temperature.

At each constant temperature, a general rate equation can be written as:

$$r = k \ [HCO_3^-]^n \qquad (2)$$

The analysis of the results can be made in logarithmic form:

$$\log r = \log k + n \log \ [HCO_3^-] \qquad (3)$$

We have plotted $\log r$ $\underline{vs}$ $\log [HCO_3^-]$ at each temperature in Figure 2. Two different behaviors have been observed depending on temperature. The final form of the rate equation can be written taking into account equation (2) as:

$$(T = 298.15 \ K) \qquad r \ (mol \ m^{-2} \ s^{-1}) = \ 2.0 \ (\pm 0.5) \ 10^{-9} \ [HCO_3^-]^{0.58 \pm 0.05}$$

$$(T = 318.15 \ K) \qquad r \ (mol \ m^{-2} \ s^{-1}) = \ 7.9 \ (\pm 0.3) \ 10^{-8} \ [HCO_3^-]^{0.96 \pm 0.06}$$

$$(T = 333.15 \ K) \qquad r \ (mol \ m^{-2} \ s^{-1}) = \ 2.0 \ (\pm 0.2) \ 10^{-7} \ [HCO_3^-]^{1.0 \pm 0.1}$$

At the lowest temperature the dependency on hydrogen carbonate concentration is a little higher than the square root while at the higher temperatures this dependency is equal to 1. From this result, it seems that the mechanism of the overall process i.e. oxidation/dissolution depends on the temperature.

<u>Activation Energy Calculation</u>

A general form of the rate law for heterogeneous surface reactions taking into account temperature dependence can be written as [12]:

$$r = k_0 \ e^{-E_a / RT} \ a_H^m \prod_i a_i^{n_i} \bullet f(\Delta G) \qquad (4)$$

where k_0 is a rate constant, E_a is the apparent activation energy of the overall reaction, R is the gas constant and T is the temperature. Terms involving activities of species in solution other than H^+, a_i, incorporate other possible effects on the rate. In our case, equation (4) can be rewritten in logarithmic form taking into account that the variation of pH in the range studied 7.6-8.9 does not influence the dissolution rate [6] and reaction is far from equilibrium :

$$\log r = \log k \ - \ \frac{E_a}{RT} + n \log \left[HCO_3^- \right] \qquad (5)$$

Both the reaction order, n, and the apparent activation energy, E_a, were calculated from a multiple linear regression of equation (5) using the least squares method. We have used all the data at 318 and 333 K to evaluate the combined effect of hydrogen carbonate concentration and temperature. The resulting apparent activation energy is 41 ± 8 kJ mol^{-1} and the reaction order is 1.03 ± 0.06, which is in fair agreement with reaction order determined for each temperature in

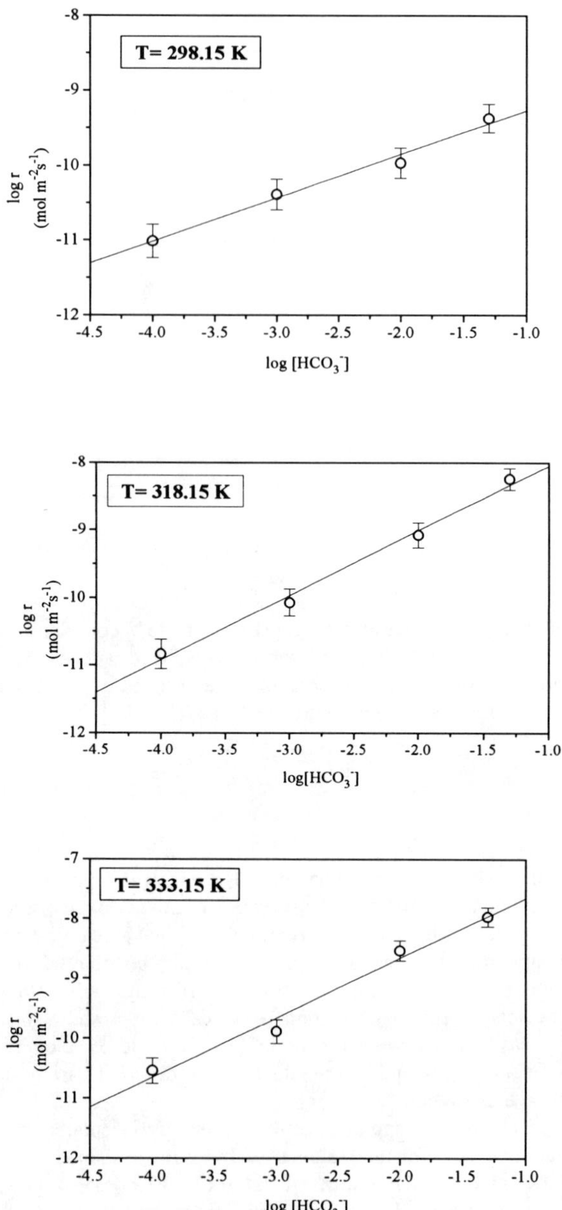

Figure 2. Dissolution rates as a function of bicarbonate concentration at different temperatures.

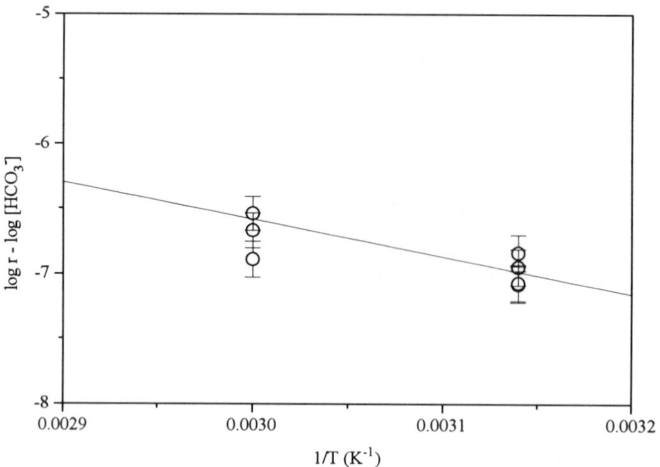

Figure 3. Normalized dissolution rate as a function of 1/T. Only temperatures where the mecahnism is the same are considered.

Figure 2. The errors of the best fit parameters represent the 95% confidence limit calculated from the standard errors of the regression. Figure 3 shows normalized rate as a function of 1/T.

The activation energy value indicates that the processes is surface controlled since activation energies of such process are in the range 40-85 kJ mol^{-1} [13].

DISCUSSION

The results obtained in this work at 298.15 K agree well with those reported by several authors using both spent fuel and unirradiated UO_2 under the same oxidizing conditions. In Figure 4, dissolution rates as a function of hydrogen carbonate concentration are plotted. As it can be observed the reaction order can be approximated to the value obtained in this work.

At higher temperatures, some discrepancies should be pointed out. Gray et al [1,2] obtained at 323.15 and 348.15 K dissolution rates lower than the ones obtained in this work and the reaction order respect to hydrogen carbonate concentration was also lower. One possible explanation is a decrease on oxygen content in [1,2] produced when the temperature was increased. Thus, the final result would be the effect of two different parameters: temperature and dissolved oxygen in the experiment.

As commented above, the apparent activation energy calculated in this work seems to indicate a surface control reaction and agrees with results obtained by other authors [14]. However, some values found in the literature are lower (20-30 kJ mol^{-1}) [2,15,16] although their authors considered this value not indicative of a diffusion-controlled reaction.

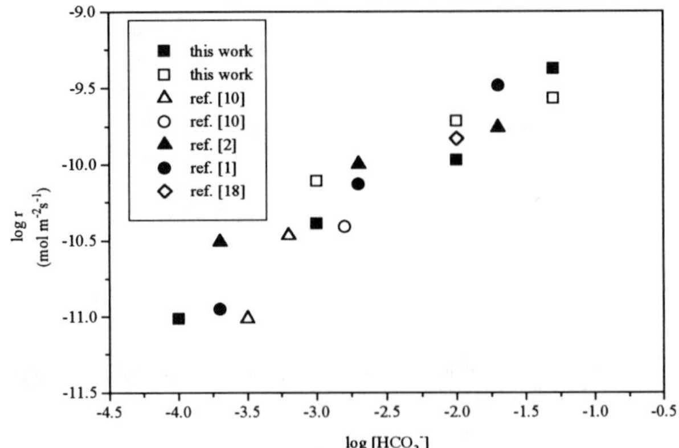

Figure 4. Dissolution rates obtained at 298.15 K as a function of HCO_3^- concentrations.

Finally, the different rate dependence on hydrogen carbonate concentration as a function of temperature can be tentatively explained as follows. The oxidation step could be considered as a surface complexation mechanism, including the precursor complex formation, electron transfer and product release steps [17]. Depending on the adsorption rate of the oxidant or the electron transfer rate, ligand (hydrogen carbonate) would compete with the oxidant molecules (oxygen) for U(IV) surface sites. This competing mechanism promotes the dissolution of any oxidized U(VI) surface complex, leaving an unoxidized solid surface, as observed in the XPS determinations [7] in a very similar way as it has been proposed in the proton-promoted dissolution mechanism of UO_2 [6]. At high temperature, oxidant adsorption and electron transfer is faster than at low temperature and the competition disappears leaving all the oxidized U(VI) sites available for reacting with the hydrogen carbonate. In this case, the overall reaction is controlled by the dissolution step.

Further work is necessary to confirm that temperature favoured more oxidation reactions than dissolution processes.

ACKNOWLEDGEMENTS

Thanks are due to Dr. Lara Duro for valuable discussions. ENRESA (Empresa Nacional de Residuos Radiactivos S.A., Spain) is gratefully acknowledged for its financial support.

REFERENCES

1 W.J Gray and S.A. Steward, <u>High Level Radioactive Waste Management of the Fith International Conference</u> (American Nuclear Society 1994) pp 2602-2608.

2 W.J Gray and S.A. Steward, J. Nucl. Mater. **190**, 46 (1992).

3 J. Bruno et al, this issue.

4 I. Casas, J. Giménez, V. Martí, M.E. Torrero and J. de Pablo, Radiochim. Acta **66/67**, 23 (1994).

5 D.W. Shoesmith, S. Sunder, M.G. Mailey and G.J. Wallace, Corrosion Science **29**, 1115 (1989).

6 M. E. Torrero, E. Baraj, J. de Pablo, J. Giménez and I. Casas, Int. J. Chem. Kinetic (in press).

7 J. Bruno, I. Casas, E. Cera, J. de Pablo, J. Giménez and M.E. Torrero in Scientific Basis for Nuclear Waste Management XVIII, edited by T. Murakami and R.C. Ewing (Mat. Res. Soc. Symp. Ser. **353**, Pittsburg, PA 1995) pp. 601-608.

8 D.E. Grandstaff, Econ. Geol. **71,** 1493 (1976).

9 B. Grambow, Spent Fuel Dissolution and Oxidation. An evaluation of Literature Data, SKB Technical Report 89-13, 1989, pp.1-42.

10 J.Bruno, I. Casas and I. Puigdomenech, Geochim. Cosmochim. Acta **55**, 647 (1991).

11 J.C. Robbins, Can. Inst. Min. Metall. Bull. **71**, 61 (1978).

12 A.C. Lasaga, J.M. Soler, J. Ganor, T.E. Burch and K.L. Nagy, Geochim. Cosmochim. Acta **58**, 2361 (1994).

13 A.C. Lasaga, J. Geophys. Res. **89**, 4009 (1984).

14 J.B. Hiskey, Trans. Inst. Nin. Metall. Sect. C, **88**, C145 (1979).

15 G.F. Thomas and G. Till, Nucl. Chem. Waste Manag. **5**, 141 (1984).

16 L.H. Johnson and H.H. Joling in Scientific Basis for Nuclear Waste Management, vol. 6, edited by S.V. Topp (North-Holland, Amsterdam, 1982) pp 321.

17 A.T. Stone and J.J. Morgan, in Aquatic Surface Chemistry edited by W. Stumm (Wiley Interscience Pub. 1987) pp 247.

18 W.J. Gray, S.A. Steward, J.C. Tait and D.W. Shoesmith, High Level Radioactive Waste Management of the Fith International Conference (American Nuclear Society 1994) pp 2597-2601.

COPRECIPITATION STUDIES FROM SIMFUEL SOLUTIONS IN 5 m NaCl MEDIA

P.Diaz-Arocas, J. Garcia-Serrano
CIEMAT (ITN). Avda. Complutense,22. E-28040 Madrid, Spain

ABSTRACT

Extensive Research is performed in many countries in order to evaluate the spent fuel behaviour under repository conditions. Several aspects as the control of the oxidative spent fuel dissolution by secondary phases formation are not yet clear.

Coprecipitation experiments from SIMFUEL solutions are performed to study if minor elements will influence the formation of secondary phases. Therefore, coprecipitation studies from SIMFUEL solutions aims at identification of stable phases of significant simulated fission products. These experiments provide upper limits for solution concentration and distribution ratios of simulate fission products at several pH values. SIMFUEL pellets, which simulate an irradiated fuel with burnup of 50 GWd/tU were provided by AECL Research Laboratories, Canada. Experiments were carried out by addition of an aliquot of the initial SIMFUEL solution to 5 m NaCl free of carbonates solution. The selected pH was maintained constant during the experiments. The pH range considered was from 5.5 to 9.3. Analyses of the solutions were performed for uranium by Laser fluorescence and for the minor elements by ICP-MS. Solid phases formed at pH 5.5 were dissolved and analysed by ICP-MS. Results of the evolution in solution vs. pH of simulated fission products concentrations are shown in this paper.

INTRODUCTION

In the methods proposed for the spent fuel repository safety, it is common to consider multiple barriers to radionuclide release to the biosphere. The innermost barrier is the limited solubility or dissolution rate of the UO_2 itself and the radionuclides distributed in the fuel matrix.

In laboratory experiments on UO_2 and spent fuel subjected to oxidative dissolution, uranium alteration phases were proposed as UO_2 dissolution controlling phases, and they may control radionuclide release. The influence of minor elements on secondary phases formed is not yet understood.

Coprecipitation studies from SIMFUEL solutions are focussed on the influence of minor elements on solid phases formation. These experiments also provided upper limits for solution concentration of minor elements and distribution ratios of simulated fission products at several pH values.

This paper describes the studies focusing on the understanding and quantification of minor elements coprecipitation with sodium polyuranates formed in high concentrated saline solutions (5 m NaCl).

In order to provide a better picture of the precipitation behaviour of the minor elements of SIMFUEL vs. pH, the results from the experiments performed at pH 5.5, which were previously reported in (1) have also been included in this paper.

A qualitative evaluation of the results obtained is shown in this work. Thermodynamic calculations and more extensive solid phase characterizations are being performed in order to provide a better understanding of the solubility controlling phases (pure phases or solid solution formation).

Mat. Res. Soc. Symp. Proc. Vol. 465 © 1997 Materials Research Society

In a previous work on precipitation of U(VI) phases in NaCl solutions (2, 3) it was shown that Na-polyuranates are the principal stable solid phases in 5 molal NaCl solutions.

Some considerations should be taken into account to understand properly the phases that could control solubility. Na-polyuranates are structurally related to schoepite, and Na^+ ions can easily be exchanged by other cations in the solution (4). In nature, alkali and alkaline-earth uranyl oxides hydrates usually coexist with schoepite (5) due to the ion exchange capacity of these phases. Also it is necessary to take into account (5) that in nature these phases are absent of elements as lanthanides probably due to the persistence of the UO_2^{+2}. Related to other relevant minor elements such as Mo, which commonly coexist in nature with uranium, nevertheless, uranyl molybdates are not yet well known.

The SIMFUEL coprecipitation experiments also provide upper limits of solution concentration of other minor elements of interest as Ru, Rh, Pd. In spent fuel these elements form metallic inclusions with Mo and Tc. In spent fuel dissolution test the metallic inclusions (7) seems to influence the release rate of fission products as Tc.

With this scope and also considering the uncertainties related to the extrapolation of thermodynamic data to 5 molal NaCl solutions, in this paper are shown the results obtained in coprecipitation experiments from SIMFUEL solution.

EXPERIMENTAL PROCEDURE

SIMFUEL pellets, which simulate an irradiated fuel with burnup of 50 GWd/tU were provided by AECL Research Laboratories, Canada. SIMFUEL pellets were dissolved in strong acid solution under oxidising conditions. The obtained SIMFUEL solution was dried and the solid dissolved in 6 N HCl. Table I shows the weight ratios of minor element (EM) and uranium in the initial chloride-SIMFUEL solution.

Experiments were carried out by addition of an aliquot of SIMFUEL solution to 5 m NaCl free of carbonates solution. The pH was maintained constant during the experiments. The considered pH values were 5.5; 5.9; 7.4 and 9.3. The pH of the solution was measured using a Ross electrode. Corrections for liquid junction potential were performed by calibration with solutions of known activity of HCl with reference to the pH convention of Pitzer. The corrected value of pH is given by:

$$pH_{corr} = pH_{meas.} + 0.25$$

Initial uranium concentrations in 5 m NaCl solution were about 1E-3 mol/kg. The experimental procedure was carefully described in (1, 2, 3).

Samples of aqueous suspension were taken during the experiments. Aliquots were filtered through 0.22μm pore size membranes. Experiments were stopped after 50 days; at this time the uranium concentration in solution remained almost constant for 14 days.

Analyses of the solutions were performed for uranium by laser fluorescence and for the minor elements by ICP-MS.

RESULTS AND DISCUSSION

The evolution with time of minor elements concentration of SIMFUEL is shown in Fig. 1 to Fig. 3 for solution at pH 9.3.

Minor elements precipitation ratios are calculated by the ratio between initial and final concentration values in solution. Results are shown in Table II.

Table I. EM/U weight ratio in the initial SIMFUEL solution

Element	EM/U %w	Element	EM/U %w
Sr	2,03 E-03	Pd	2,54 E-03
Y	6,80 E-04	Ba	3,62 E-03
Zr	5,19 E-03	La	2,16 E-03
Mo	5,03 E-03	Ce	8,78 E-03
Ru	6,24 E-04	Nd	9,06 E-03
Rh	7,03 E-04	U	1,00

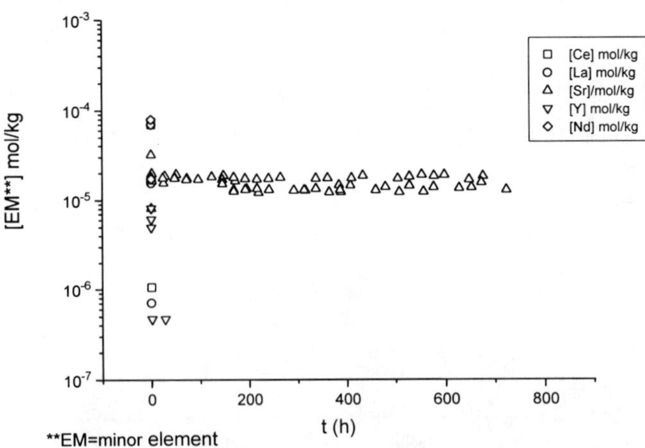

Detection limit (mol/kg) in 5 m NaCl solution of Ce and La is 6E-9, detection limit of Y is 9E-9 and detection limit of Nd is 2E-8

Fig. 1. Ce, La, Y, Sr and Nd concentrations in solution vs. time. pH=9.3, T=25°C, inert atmosphere.

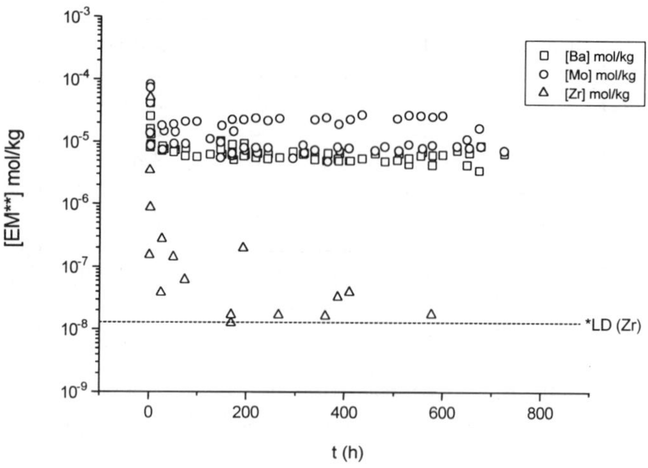

*LD (Zr): detection limit of Zr in 5 m NaCl
**EM:minor element

Fig. 2. Ba, Mo and Zr concentration vs. time at pH=9.3, T=25 °C and inert atmosphere

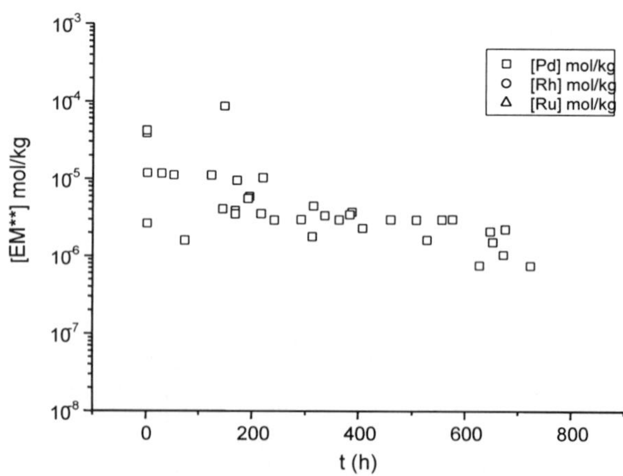

**EM:minor element

Detection Limit (mol/kg) in 5 m NaCl for Rh is 8E-9 and detection limit for Ru is 3E-8. Ru and Rh concentrations in solution were lower than the detection limit

Fig. 3. Pd, Rh, and Ru concentrations in solution vs. time at pH=9.3, T=25°C and inert atmosphere

Table II. Precipitation ratios of minor elements of simfuel at several corrected pH values

pH	Ba	Ce	La	Mo	Nd	Pd	Rh	Ru	Sr	Y	Zr
				Precipitation ratio							
5.5	2.4	1.3	1.4	3.5	1.2	1.2	1.3	2.2	1.5	1.5	120
5.9	2.0	1.8	1.7	12.3	1.8	3.5	1.25	---	1.3	1.3	500
7.4	2.1	6.6	4.9	57.9	8.7	25.7	126	5.6	1	8.1	--
9.3	4.3	--	---	2.8	---	39.8	---	---	1.7	---	2956
9.3	3,7	--	---	2.5	---	---	---	---	1.3	---	--
9.3	2.9	211	29	2.7	---	17.9	---	---	1.3	23.7	--

The precipitation ratios evolution of minor elements of SIMFUEL with pH indicates three different coprecipitation behaviours. Based on the chemical analyses of the solid formed at pH =5.5 (1), we have assumed that precipitation ratios lower than 2 indicate that the minor element considered does not precipitate and precipitation ratios equal or higher than 2 indicate coprecipitation of the minor element.

The Sr precipitation ratio remains constant with pH. The Sr concentration in solution obtained is between 3.5×10^{-5} and 1.5×10^{-5} mol kg^{-1}. Results are consistent with those reported in (1).

The Ba precipitation ratio is higher than the corresponding for Sr, which indicated that this element precipitated. The obtained precipitation ratio evolution with pH is similar to Sr. Only a light increase of the precipitation ratio is observed at pH =9.3. Chemical analyses of the solid formed at pH 5.5 indicated that Ba was in the solid phase (0.17% wt) an Sr was absent (1).

Ce, La, Nd, Rh and Y show a similar behaviour to Sr at pH values lower than 6. However, at pH 7.4 to 9.3 a strong decrease of concentration in solution was observed. Ce concentration in solution for the experiments performed at pH 9.3 were close to the detection limit of the analytical technique used, in the first stage of the precipitation reaction. Therefore, the precipitation ratio obtained is affected by an analytical error higher than 10%. Nevertheless it is sufficient to indicate that Ce is completely precipitated at this pH value. The same situation was found in case of Nd and Y. Detection limits of these elements in 5 m NaCl solutions are for Ce 6E-9 mol/kg, for Y 9E-9 mol/kg, for Nd 2E-8 mol/kg, and for Rh 8E-9 mol/kg.

La simulates itself and also the chemical behaviour of some actinides such as Am and Cm (6). The precipitation ratio obtained at pH 9.3 is high enough to indicate that this element is in the solid phase. The detection limit of this element in 5 m NaCl solutions is 6E-9mol/kg. More work is going to be performed to understand if this minor element forms a solid solution or an independent phase.

Pd remains in solution at pH 5.5 but precipitation ratios increased with pH. Ru precipitation ratios increase with pH. This minor element is in the solid phase at all pH values considered. The detection limit of Ru in 5 m NaCl solution is 3E-8 mol/kg.

Zr concentration in solution strongly decreased in the first stage of the precipitation reaction. Precipitation ratios show that this element is in the solid phase at all pH values tested.

Mo behaves different than the other minor elements considered. From pH 5.5 to 7.4, the precipitation ratios increased with pH but at pH 9.3 the precipitation ratio decreased and it is lower than that obtained at pH 5.5. At pH 7.4, the Mo concentration in solution decreases more than one order of magnitude after 100 hours. In (1) a similar behaviour was found at pH 8.75. From the results the Mo-controlling phase is stable until pH about 9. More work is being performed to identify the solid phase which controls Mo concentration in solution.

CONCLUSIONS

Most of the SIMFUEL minor elements coprecipitated with Na-polyuranates at pH value of 7.4. It is not yet clear if these minor elements in the solid phases are coprecipitated, sorbed or forming independent phases.

The role of Sr as a monitor of the spent fuel matrix dissolution remains unclear. Today, some open questions are related to the dependence of release rate of Sr with redox processes, to the radiolysis influence on Sr release rate and the formation of secondary phases. The results showed in this paper indicate that under the conditions tested there is no objection to use Sr as a monitor of spent fuel matrix dissolution. A comparison of the data showed in this paper and those which will be obtained from coprecipitation experiments performed from spent fuel solutions will provide a better picture about the role of Sr as indicator of spent fuel matrix dissolution.

The Mo-controlling phase is stable until pH about 9. This element shows different precipitation behaviour than the other minor elements contained in SIMFUEL.

Zr completely precipitates at all pH values tested in the first stage of the precipitation reaction.

Thermodynamic calculations and more extensive solid phase characterisations are being performed in order to rationalize the coprecipitation results.

ACKNOWLEDGEMENTS

The authors gratefully acknowledges the help in the laboratory of M.J. Tomas, A. Campo and M.A. Fariñas. ICP-MS analyses were performed by Dr. A. Quejido.

These studies were performed in the frame of the European Community R&D Program 1995-1999 on Nuclear Fission Safety. C.3.2.2: Spent fuel, and under the agreement between CIEMAT (ITN)-ENRESA in the Area of Spent Fuel Behaviour Directly Disposed.

REFERENCES

1. P. Diaz Arocas, J. Garcia-Serrano, J. Quiñones, H. Geckeis and B. Grambow, Radiochimica Acta **74**, 51-58 (1996).
2. P. Diaz Arocas and B. Grambow, Chemistry of uraniumin brines related to the spent fuel disposal in a salt repository (Y), ENRESA Technical Report No. 06/93 (1993).
3. P. Diaz Arocas and B. Grambow, Solid-liquid phase equilibria of U(VI) in NaCl solutions. Geochimica Acta (submitted).
4. B. G. Bilgin, Chim. Acta Turcica **17**, 205-210 (1989).
5. R. Finch and R. Ewing, Uraninite alteration in an oxidizing environment and its relevance to the disposal of spent nuclear fuel. SKB Tech. Report No. 91/15 (1990).
6. P. G. Lucuta, R.A. Verrall, H. J. Matzke and B. J. Palmer, J. Nuclear Matrials **178**, 48-60 (1991)
7. R.S. Forsyth and L.O. Werme, J. Nuclear Materials **190** (1992)

DISSOLUTION OF UNIRRADIATED UO₂ FUEL UNDER OXIDIZING CONDITIONS: EVALUATION OF SOLUBILITY (STEADY STATE) LIMITING FACTORS (EQ3/6)

K. OLLILA
VTT Chemical Technology, Technical Research Center of Finland, PO Box 1404, FIN-02044 VTT, Finland, kaija.ollila@vtt.fi

ABSTRACT

The solubility behavior of unirradiated UO_2 pellets was studied under oxic, air-saturated conditions in deionized water, in $NaHCO_3$ solutions and in two types of synthetic groundwater (25°C). The Allard groundwater represents natural fresh groundwater conditions at great depths in granite bedrock and bentonite groundwater simulates the effects of bentonite on granitic fresh groundwater. The release of uranium was measured during static batch dissolution experiments of long duration (6 years). A comparison was made with the theoretical solubility data calculated with the geochemical code, EQ3/6, in order to evaluate solubility (steady state) limiting factors. Various hypotheses for redox control (redox potential of the bulk solution or redox potential at the surface) were tested in the modeling calculations.

The measured concentrations for uranium at steady-state in deionized water were equal to the solubility of schoepite (pO_2 = 0.2 atm). In $NaHCO_3$ solutions with lower carbonate concentrations (0.98 - 1.96 mmol/l) and in Allard groundwater they were close to the calculated solubilities of U at the U_3O_7/U_3O_8 redox potential. In bentonite groundwater, the results suggest the formation of a secondary phase with a lower solubility. Only uranium oxide with a crystal structure of uraninite ($UO_2 - U_3O_7$) was identified in all waters, when analyzing particulate material in the solutions after contact with UO_2 pellets.

INTRODUCTION

The spent fuel from Finnish nuclear power plants is planned to be disposed of in a repository to be constructed at a depth of about 500 meters in crystalline granitic bedrock. In the concept of the multiple barriers, the barriers for radionuclide release into the environment are the granitic bedrock, the backfill, the copper-steel canister and the fuel itself. The innermost barrier is the limited solubility or dissolution rate of the UO_2-matrix, the main component of the fuel, and of the radionuclides embedded in the UO_2 matrix. Under reducing conditions normally prevailing in deep granitic groundwater, UO_2 (or U_4O_9) is the stable uranium solid. The release of radionuclides from spent fuel is controlled by the solubility of UO_2 (s). However, due to α-radiation on the surface of spent fuel, oxidizing conditions cannot be excluded. Under these conditions, UO_2 (s) is not thermodynamically stable. The oxidative dissolution process is complex including the oxidation of the UO_2 surface to a higher uranium oxide and the possible formation of secondary U(VI) solid phases, depending on the availability of oxygen and other oxidizing agents, and the composition of the groundwater. In this case, the release of radionuclides may be controlled by the overall reaction rate of this process. The release of sparingly soluble radionuclides will depend on their solubilities.

This paper deals with the results obtained from dissolution experiments of unirradiated UO_2 pellets as a function of water composition under oxic air-saturated conditions at 25 °C. The release of uranium was followed during static batch dissolution experiments of long duration (6 years). A comparison of the experimental data was made with the equilibrium solubilities calculated using the geochemical code EQ3/6 in order to evaluate solubility (steady state) limiting factors.

Mat. Res. Soc. Symp. Proc. Vol. 465 © 1997 Materials Research Society

METHODS OF EXPERIMENTS

Materials

All the dissolution experiments of this paper have used unirradiated sintered polycrystalline UO_2 fuel in pellet form. The pellets had an average mass of 9.8 grams and a geometrical surface of $4.5 \cdot 10^{-4}$ m^2. The surfaces of the pellets were not pretreated by polishing prior to the start of the experiments.

Water compositions

The waters included deionized water and sodium bicarbonate solutions with varying bicarbonate contents (0.98-9.83 mmol/l, 60-600 mg/l) as reference waters, and two types of synthetic groundwater, see Table I. Synthetic granitic groundwater by Allard et al [1], so-called Allard groundwater represents natural fresh groundwater conditions at great depths in granitic bedrock. Synthetic bentonite groundwater by Snellman [2] simulates the effects of bentonite on fresh granitic groundwater. The experimental studies by Snellman [2] on the interaction between sodium bentonite and Allard groundwater showed that the most significant effects of bentonite were the increase in the pH and the HCO_3^- , CO_3^{2-} and Na^+ -contents of the water.

Table I.   Composition of the synthetic groundwaters, mmol/l.

	pH	Na^+	K^+	Ca^{2+}	Mg^{2+}	Si^{4+}	HCO_3^-	Cl^-	SO_4^{2-}	F^-	HPO_4^{2-}
Allard groundwater	8.2	2.3	0.1	0.45	0.2	0.13	2.0	2.0	0.1	-	-
Bentonite groundwater	8.4	11.8	0.1	0.45	0.2	0.27	9.8	2.3	0.6	0.4	0.004

Dissolution procedure

The dissolution experiments were performed in batch-without-replenishment fashion. Four UO_2 pellets were immersed in 1 liter of aqueous solution in Schott Duran Flasks of 1 liter total volume. Additionally, a ten-fold ratio of pellet surface area to water volume was applied in one series of tests in Allard groundwater (ten pellets/250 ml of water). Small aliquots (5-10 ml) were taken periodically for further analysis. These aliquots were replaced with fresh water which had similar composition, in order to keep the water volume the same. The amounts of uranium in microfiltered (sample 1, membranes of 0.45 µm pore size) and ultrafiltered (sample 2, membranes of 3 to 4 nm pore size, nominal cut-off value of 50 000 M) samples were measured.

Analytical methods

The uranium contents were analyzed by epithermal neutron activation analysis (INAA). Inductively coupled plasma mass spectrometry (ICP-MS) was used in analyzing the solutions of the last four samplings. The microfiltration of the solutions was carried out using the Millex-HV 0.45 µm filter units of Millipore and the ultrafiltration of the solutions was carried out using the Amicon Ultrafiltration Cells (Amicon Diaflo membranes, XM 50). The contents of anions and cations in the solutions were analyzed by ion chromatography (IC) at the end of the experimental time. The carbonate equilibrium of the solutions was analyzed using the SFS 3005 Standard method for the determination of alkalinity and acidity in water (Metrohm titrator).

The presence of possible secondary phases of uranium on the surface of the pellets or in the solutions was examined with X-ray diffractometry (XRD), Debye-Scherrer camera and X-ray photoelectron spectroscopy (XPS). The XPS analyses were performed in the Universitat Politècnica de Catalunya (UPC) in Spain.

METHODS OF CALCULATIONS

The EQ3/6 package (EQ3NR: versions 3245 R111 and 7.1 R125, EQ6: versions 3245 R104, 7.1 R121) developed by the Lawrence Livermore Laboratory [3] was used to calculate uranium solubilities under experimental conditions. The EQ3/6 code is a theoretical model based on the premise of chemical equilibrium conditions. The EQ3NR code is based both on the charge-balance equation, and on the mass-balance equations for all elements except hydrogen and oxygen. The EQ6 code is based on the mass-balance equations for all elements including oxygen and hydrogen. The redox constraint used was either oxygen fixed gas fugacity, or redox potential was assumed to be determined by the transitions between different UO_{2+x} phases on the surface of UO_2 . The UO_{2+x} phase was considered as a redox buffer, which may have a larger buffering capacity than the solution phase (partial pressure of O_2) . In the EQ3NR simulations, the redox potential was calculated from the mineral equilibrium (e.g. U_3O_7/U_3O_8), which fixes the oxygen fugacity. This also determines the concentration of uranium in solution. The activity coefficients were calculated with the B-dot equation [4].

The thermodynamic databases were the SKBU database (Data 0.3245S02) [5], the NEA database (NEAU, Data0.NEA.R16) [6] and the composite database (COMU, Data0.com.R16) [3]. The SKBU database has been validated for use in the EQ3/6 code package [5].

RESULTS AND DISCUSSION

A general result from the dissolution experiments of unirradiated UO_2 and spent fuel in synthetic groundwater performed in the presence of atmospheric oxygen is that the solution concentration of uranium reaches a constant value after some time period [7,8]. The steady-state solution concentration of uranium appears to be controlled with respect to some kind of oxidized phase. Initial attempts to identify the solubility-controlling phase seemed to indicate that the solution concentration was in close agreement with the solubility of schoepite, $UO_2(OH)_2 \cdot H_2O$ (s) [9]. However, when the refined thermodynamic data for schoepite and for U(VI) hydroxo complexes [10] were used, it was observed that uranium concentrations were about a factor of 10-30 lower than given by the solubility of schoepite [7,10]. Grambow [8] made a comparison of calculated uranium solubilities (EQ3) with Swedish spent fuel dissolution experiments in Allard groundwater under oxic conditions. He used electrochemical arguments to predict solubility at saturation. It was suggested that the controlling factor is the redox potential (corrosion potential) at the UO_2 surface rather than the redox potential of the bulk solution (Eh). The redox potential at the surface may differ substantially from the overall Eh in solution. Two alternative phase boundaries were calculated: U_3O_7/U_3O_8 and U_3O_7/schoepite [8]. The best agreement of calculated uranium solubilities and measured uranium concentrations in spent fuel dissolution experiments in Allard groundwater was obtained when the uranium solubilities were calculated at the U_3O_7/U_3O_8 redox potential.

In the following, the measured solubilities of U in the dissolution experiments of unirradiated UO_2 pellets in deionized water, $NaHCO_3$ solutions and synthetic groundwaters under air-saturated conditions are compared with the theoretical solubilities given by the EQ3/6 modeling. The redox potential was assumed to be controlled by 1) the oxygen partial pressure (pO_2 = 0.2 atm), and the 2) U_3O_7/U_3O_8 or 3) U_3O_7/schoepite boundaries. In the latter two EQ3/6 simulations the mineral equilibrium between different uranium oxides fixes the oxygen partial pressure. This also determines the concentration of dissolved uranium. The redox potential for the latter cases is clearly lower than the redox potential calculated from the partial pressure of O_2.

The measured concentrations of uranium in deionized water as a function of contact time (6 years) are given in Figure 1. The amount of uranium in solution increased slowly and attained a steady state at the concentration of 3-5 · 10^{-6} mol/l, being one order of magnitude higher than the concentrations measured in the dissolution experiments of spent fuel in deionized water by Forsyth and Werme [7].

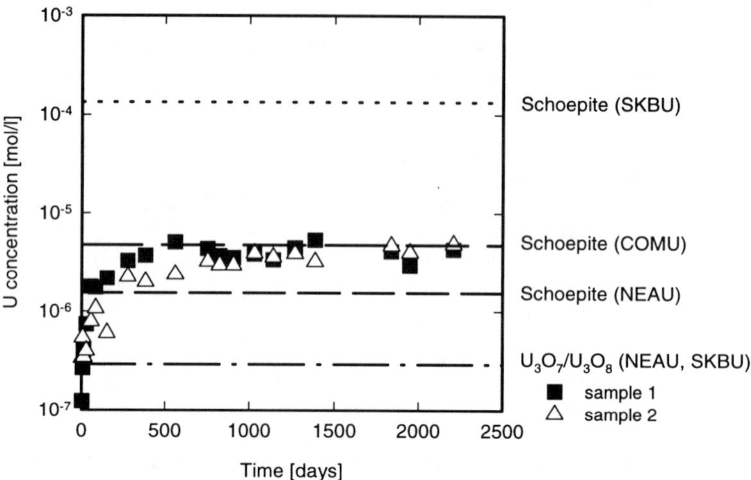

Fig.1   Measured concentrations of uranium in deionized water as a function of contact time, pH= 5.8 (sample 1 and 2: U in micro- and ultrafiltered solutions, respectively). The dashed lines represent the solubilities of schoepite (pO$_2$ = 0.2 atm) and the uranium solubility at the U$_3$O$_7$/U$_3$O$_8$ redox potential. Equilibrium with air was assumed.

The oxidative dissolution mechanism of UO$_2$ in deionized water involves phase alteration, in which UO$_2$ alters to the particulate form of UO$_3$ hydrate, if sufficient oxygen is available. The UO$_3$ hydrates absorb on the surface of UO$_2$ or precipitate elsewhere [11]. The concentration of uranium in solution is not controlled by the solubility of dissolving UO$_2$ (U$_3$O$_7$), but by U(VI) containing solid alteration product. Schoepite or a closely related compound (dehydrated schoepite, UO$_3$ · 2H$_2$O) has been identified in long-term leaching experiments of spent fuel in deionized water [12,13]. Schoepite was also identified, probably as a secondary phase, in the agitated UO$_2$ powder dissolution experiments under oxic (air-saturated) conditions in deionized water [14]. In those experiments, the surface of the UO$_2$ powder was allowed to oxidize in air prior to the start of the dissolution experiments.

Best agreement between the measured data at steady state and the calculations was obtained when the solubility of schoepite was calculated at pO$_2$ = 0.2 atm using the COMU database. The reason for the lower solubility given by the NEAU database is the aqueous species UO$_2$(OH)$_2$(aq) in the COMU database, which has not been included in the NEAU database. The solubility value given by the SKBU database is considerably higher. There is a different value of the solubility product of schoepite in the SKBU and NEAU (COMU) databases (log K = 5.6 and 4.8, respectively). The calculated solubility at the U$_3$O$_7$/U$_3$O$_8$ redox potential is one order of magnitude lower than the measured concentrations at steady state. The detailed results of the EQ3/6 modeling are given elsewhere [15].

The presence of particulate, crystalline material in solution at the end of the experimental time was shown by the XRD and Debye-Scherrer camera analyses of the membrane filter after the filtration of the water phase. It seemed to be uranium oxide with a crystal structure of uraninite (UO_2 - U_3O_7). Schoepite was not observed. The small amount of material retained on the filter made analysis difficult. A new attempt to identify the material on the filter was made with XPS. It was not possible to identify any secondary phase. The XPS determination of the UO_2 pellet surface showed 100 % U(IV).

<u>UO_2 solubility in $NaHCO_3$ solutions</u>

Figure 2 shows the measured concentrations of uranium in $NaHCO_3$ (1.96 mmol HCO_3^-/l) solution as a function of contact time. Uranium attains a steady state at the solubility at the U_3O_7/U_3O_8 redox potential. The solubility of schoepite at pO_2 = 0.2 atm is one order of magnitude higher [15]. Only uranium oxide with a crystal structure of uraninite (UO_2 - U_3O_7) was identified in the material retained on the membrane filter after the filtration of the solution at the end of the dissolution experiment. The XPS determination of the UO_2 pellet surface showed 100 % U(IV). There is a difference between the two databases. The solubilities calculated using the NEAU and COMU databases are increased by the presence of the mixed $(UO_2)_2(CO_3)(OH)_3^-$ complex. This complex was not included in the SKBU database. In another series of dissolution experiments as a function of bicarbonate content (0.98 - 9.83 mmol/l) [15], the uranium concentrations measured after the last dissolution period of 900 days agreed with the uranium solubilities at the U_3O_7/U_3O_8 redox potential at lower carbonate concentrations (0.98 - 1.96 mmol/l), while at higher carbonate concentrations the measured concentrations were lower. These experiments used the batch-with-replenishment method, in which successive dissolution periods were carried out, including water changing between dissolution periods [14,15].

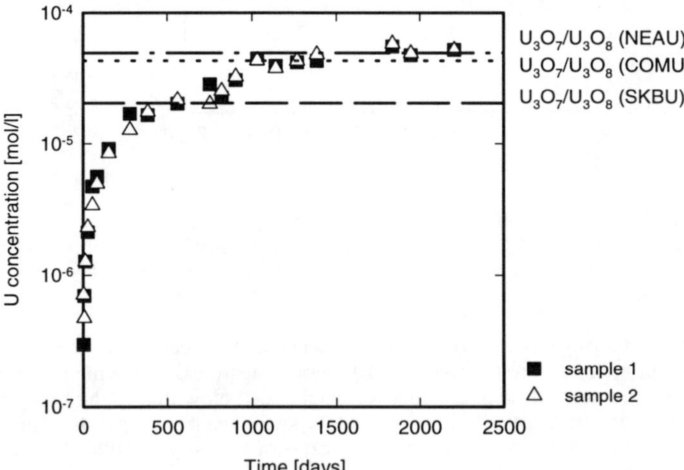

Fig.2   *Measured concentrations of uranium in $NaHCO_3$ solution (1.96 mmol HCO_3^-) as a function of contact time, pH= 8.2 (sample 1 and 2: U in micro- and ultrafiltered solutions, respectively). The dashed lines represent the uranium solubilities at the U_3O_7/U_3O_8 redox potential. Equilibrium with air was assumed.*

<u>UO$_2$ solubilities in synthetic groundwaters</u>

Finally, the measured concentrations of uranium in the synthetic groundwaters are given in Figures 3 and 4. The amount of uranium in Allard groundwater (Figure 3) attains a steady state at a concentration of $1\text{-}2 \cdot 10^{-5}$ mol/l, being in good agreement with the results of the spent fuel dissolution experiments by Forsyth and Werme [7]. In the series with a ten-fold ratio of pellet surface area to water volume [15], uranium leveled off at the same concentration. The steady-state concentration is close to the solubility of uranium at the U_3O_7/U_3O_8 redox potential given by the SKBU database. The solubility given by the COMU database is somewhat higher due to the mixed complex, $(UO_2)_2(CO_3)(OH)_3^-$. The solubility of schoepite is one order of magnitude higher [15]. The COMU data file was used for groundwaters instead of the NEAU data file because of a wider range of minerals, including e.g. calcite. Both Allard groundwater and bentonite water are oversaturated with calcite under air-saturated conditions.

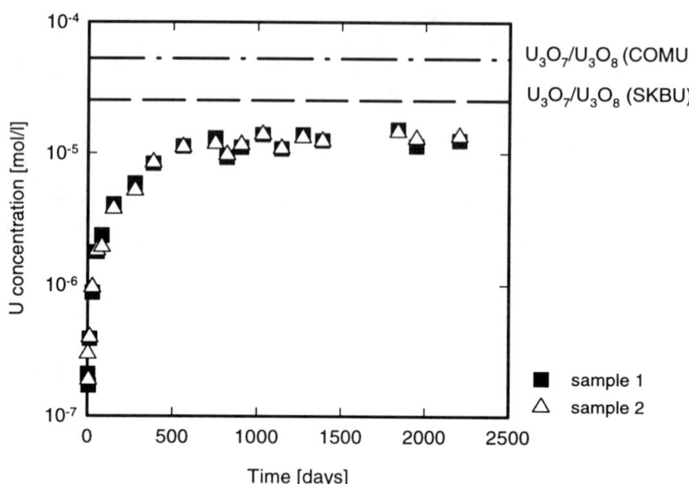

Fig. 3   Measured concentrations of uranium in Allard groundwater as a function of contact time, pH= 8.4 (sample 1 and 2: U in micro- and ultrafiltered solutions, respectively). The dashed lines represent the uranium solubilities at the U_3O_7/U_3O_8 redox potential. Equilibrium with air was assumed.

In bentonite groundwater, a steady state was attained at a concentration of one order of magnitude lower than in Allard groundwater, $1 \cdot 10^{-6}$ mol/l, regardless of the high carbonate content of the water (9.83 mmol/l). It is three orders of magnitude below the solubility of schoepite and the solubilities of uranium at the U_3O_7/U_3O_8 redox potentials. This suggests a different steady-state control, e.g. the formation of a secondary phase with a lower solubility. The uranyl silicates are known to be the most stable long-term phases in oxidizing siliceous groundwater [16]. The solubility of one such phase, haiweeite, $Ca(UO_2)_2(Si_2O_5)_3 \cdot 5\ H_2O$ has been presented for comparison in Figure 4.

The attempts to identify possible secondary phases in the solutions at the end the experimental time were only partly successful because of the small amount of material retained on membrane filters. As in deionized water and the $NaHCO_3$ solution (1.96 mmol HCO_3^-/l), uranium oxide with a crystal structure of uraninite ($UO_2 - U_3O_7$) was identified in both synthetic groundwaters (XRD, Debye-Scherrer camera). Calcite was identified in bentonite groundwater. The composition of the UO_2 pellet surface was, according to the XPS determinations, 92 % U(IV) and 8 % U(VI) after the contact with Allard groundwater, and 100 % U(IV) after the contact with bentonite groundwater. Analyzing of the composition of the solutions after the dissolution experiments did not show any remarkable change in Allard groundwater compared to the situation at the beginning [15]. In bentonite groundwater, the silica (SiO_2) and calcium (magnesium) contents of the solution were lower at the end of the experiment. This refers to the formation of silica-containing secondary uranium phase.

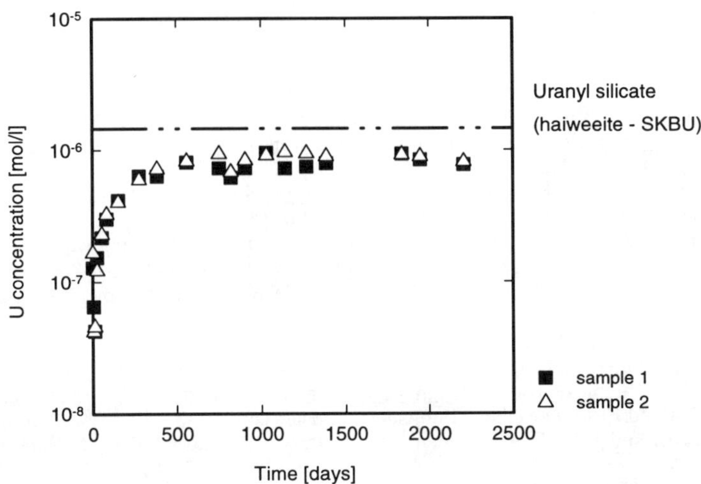

Fig. 4 *Measured concentrations of uranium in bentonite groundwater as a function of contact time, pH= 9.0 (sample 1 and 2: U in micro- and ultrafiltered solutions, respectively). The dashed line represents the solubility of haiweeite (pO_2 = 0.2 atm).*

CONCLUSIONS

Different controlling factors for steady-state concentrations of uranium were observed as a function of water composition under air-saturated conditions. The conclusions are based on the comparison of the measured data in UO_2 dissolution studies (6 years) with the theoretical solubility data (EQ3/6).

In deionized water, the measured data at steady state were equal to the solubility of schoepite at pO_2 = 0.2 atm.

In Allard groundwater and $NaHCO_3$ solutions with low carbonate contents (0.98 - 1.96 mmol/l), the solution concentrations were close to the calculated solubilities of uranium at the U_3O_7/U_3O_8 redox potential.

In bentonite groundwater, the results suggest the formation of a secondary phase with a lower solubility, e. g. uranyl silicate.

Only uranium oxide with a crystal structure of uraninite (UO_2 - U_3O_7) was identified in all the waters at the end of the experimental time.

ACKNOWLEDGEMENTS

Posiva Oy is gratefully acknowledged for financial support. The author takes pleasure in thanking Ms. R. Zilliacus and Dr. R. Rosenberg (VTT Chemical Technology, Molecular Structure and Chemical Analytics) for performing the ICP-MS analyses and Mr. K. Lindqvist (Geological Survey of Finland) for performing the XRD analyses. The author wishes to thank Dr. Joan de Pablo and Dr. E. Torrero (Universitat Politècnica de Catalunya) for the XPS analyses. The contribution of Ms. K. Helosuo also deserves acknowledgement. Thanks are also due to Dr. M. Olin for helping with the EQ3/6 modeling.

REFERENCES

1. B. Allard, S.A. Larson, Y. Albinsson, E.L. Tullborg, M. Karlsson, K. Andersson and B. Torstenfelt, in Near-Field Phenomena in Geologic Repositories for Radioactive Waste, (NEA Workshop Proc., 1981), pp. 93 - 101.
2. M. Snellman, in Third Finnish - German Seminar on Nuclear Waste Management, edited by L. Lamberg (VTT Symposium 87, Espoo 1988), pp. 146 - 168.
3. T.J. Wolery, EQ3/6, A Software Package for Geochemical Modeling of Aqueous Systems (Version 7.0), Lawrence Livermore National Laboratory, UCRL-MA-110662 PT 1-4, 1992.
4. H.C. Helgeson, R.M. Garrels and F.T. Mackenzie, Geochim. Cosmochim. Acta 33, 455 (1969).
5. I. Puigdomenech and J. Bruno, SKB Technical Report 88-21, 1988.
6. I. Grenthe, J. Fuger, R.J.M. Konings, R.J. Lemire, A.B. Muller, C. Nguyen-Trung Cregu and H. Wanner, Chemical Thermodynamics of Uranium, edited by H. Wanner and I. Forest (North-Holland Elsevier Science Publishers, Amsterdam, 1992.
7. R.S. Forsyth and L.O. Werme, Spent fuel corrosion and dissolution, J. Nuclear Materials 190, 3 (1992).
8. B. Grambow, SKB Technical Report 89-13, 1989.
9. J. Bruno, R.S., R.S. Forsyth and L.O. Werme, in Scientific Basis for Nuclear Waste Management VIII, edited by C.M. Jantzen, J.A. Stone and R.C. Ewing (Mater. Res. Soc. Proc., 1984), pp. 413 - 420.
10. J. Bruno and A. Sandino, in Scientific Basis for Nuclear Waste Management XII, edited by W. Lutze and R.C. Ewing (Mater. Res. Soc. Proc., 1989), pp. 871 - 878.
11. D.W. Shoesmith, S. Sunder, M.G. Bailey and G.J. Wallace, in 2nd International Conference on Radioactive Waste Management (Canadien Nuclear Society Proc., 1986), pp. 674 - 679.
12. R.S. Forsyth, U-B. Eklund, O. Mattson and D. Schrire, SKB Technical Report 90-04, 1990.
13. S. Stroes-Gascoyne, L.H. Johnson, P.A. Beeley and D.M. Sellinger, in Scientific Basis for Nuclear Waste Management IX, edited by L.O. Werme (Mater. Res. Soc. Proc., 1986), pp. 317 - 326.
14. K. Ollila and H. Leino-Forsman, Report YJT 93-04,1993.
15. K. Ollila, Report YJT 95-14, 1995.
16. R.J. Finch and R.C. Ewing, SKB Technical Report 89-37, 1989.

COMPARISON AND MODELING OF AQUEOUS DISSOLUTION RATES OF VARIOUS URANIUM OXIDES

S. A. STEWARD and E. T. MONES
Lawrence Livermore National Laboratory, P.O. Box 808, Livermore, CA 94550

ABSTRACT

The purpose of this work has been to measure and model the intrinsic dissolution rates of uranium oxides under a variety of well-controlled conditions that are relevant to a geologic repository. When exposed to air at elevated temperature, spent fuel may form the stable phase U_3O_8. Dehydrated schoepite, $UO_3 \cdot H_2O$, has been shown to exist in drip tests on spent fuel.

Equivalent sets of U_3O_8 and $UO_3 \cdot H_2O$ dissolution experiments allowed a systematic examination of the effects of temperature (25-75°C), pH (8-10) and carbonate (2-200x10^{-4} molar) concentrations at atmospheric oxygen conditions.

Results indicate that $UO_3 \cdot H_2O$ has a much higher dissolution rate (at least ten-fold) than U_3O_8 under the same conditions. The intrinsic dissolution rate of unirradiated U_3O_8 is about twice that of UO_2. Dissolution of both U_3O_8 and $UO_3 \cdot H_2O$ shows a very high sensitivity to carbonate concentration. Present results show a 25 to 50-fold increase in room-temperature $UO_3 \cdot H_2O$ dissolution rates between the highest and lowest carbonate concentrations.

As with the UO_2 dissolution data the classical observed chemical kinetic rate law was used to model the U_3O_8 dissolution rate data. The pH did not have much effect on the models, in agreement with the earlier analysis of the UO_2 and spent fuel dissolution data,. However, carbonate concentration, not temperature, had the strongest effect on the U_3O_8 dissolution rate. The U_3O_8 dissolution activation energy was about 6000 cal/mol, compared with 7300 and 8000 cal/mol for spent fuel and UO_2 respectively.

INTRODUCTION

Understanding the long-term dissolution of spent fuel in groundwater is necessary for its safe disposal in a geological repository. Radionuclides could be released from such a repository by dissolution and transport processes in flowing groundwater. The dissolution of the UO_2 spent fuel matrix is regarded as the rate-limiting step for release of radioactive fission products. Therefore, the intrinsic UO_2 dissolution rate sets an upper limit on the aqueous radionuclide release rate. If the UO_2 in the spent fuel matrix contacts air and is oxidized further, then these dissolution responses also must be measured.

It is commonly assumed that oxidized fuel would dissolve faster than its unoxidized predecessor. The purpose of this and previous work has been to measure the intrinsic dissolution rates of uranium oxides, as well as unoxidized and oxidized spent fuel, under a variety of well-controlled conditions that are relevant to a geological repository and allow for subsequent modeling. When exposed to air at elevated temperature, spent fuel may form the stable phase U_3O_8. A form of the trioxide, dehydrated schoepite, $UO_3 \cdot H_2O$, has been shown to exist in drip tests on spent fuel [1]. The results of essentially identical dissolution experiments performed on depleted U_3O_8 and $UO_3 \cdot H_2O$ will be compared. These are in turn compared with earlier work on spent fuel and UO_2 under similar conditions [2].

Water from wells near Yucca Mountain contain typical aqueous constituents, such as carbonates, sulfates, chlorides, silicates, and calcium. Of the anions commonly found in groundwater, bicarbonate is considered to be the most aggressive towards uranium oxides, forming complexes with the uranyl (UO_2^{+2}) cation. This makes the carbonates good surrogates for all anions in aggressive groundwater. Statistical experimental design was used to plan the set

Mat. Res. Soc. Symp. Proc. Vol. 465 © 1997 Materials Research Society

of U_3O_8 and $UO_3 \cdot H_2O$ dissolution experiments. This approach allows a systematic examination of the effects of temperature, pH and carbonate concentrations on the dissolution rates of these two oxides. It also minimizes the number of experiments required, and provides a robust data set suitable for modeling and comparison with the previously reported UO_2 dissolution data. Because of the already elevated oxidation state, these experiments were run only at 8 ppm dissolved oxygen in the leaching solutions, equivalent to 0.2 atmosphere oxygen in air. The dissolution rates from the design allow a fit to a second-order model in all variables, including interactions between the variables. Additional experiments on UO_2 complete a matrix of dissolution rates measured on all three oxides at the same conditions.

EXPERIMENTAL

As with previous studies [2,3,4], the intrinsic dissolution rates of the uranium oxides were determined by using a single pass flow-through method. Flow rates and specimen size can be controlled with this approach so that the oxides dissolve under conditions that are far from solution saturation (no precipitation of dissolved products). Thus, the dependence of UO_x dissolution kinetics on pH, temperature, oxygen and carbonate/bicarbonate concentrations can be evaluated.

Experiments at three different values of each variable were required, in order to test for nonlinear effects of the three variables on the uranium dissolution rates. The chosen variable ranges were pH's of 8 to 10, temperatures of 25° to 75°C, total carbonate concentrations of 0.2 to 20 millimol/L and 8 ppm dissolved oxygen. The carbonate concentrations bracketed the typical groundwater concentration of 1 millimol/L. The pH range covered a value typical of groundwaters (pH=8) to very alkaline conditions. The dissolved oxygen concentration is the value at atmospheric pressure.

A model that can discriminate nonlinear effects and interactions of the three variables for both oxides has at least 14 terms. A sixteen experiment design is, therefore, the minimum number of experiments required for a numerical regression fit, with extra degrees of freedom to account for experimental variability. For each oxide, a classical three-level, full-factorial experimental design consists of the 27 (3^3) possible combinations of variable settings from the three variables at low, medium and high values. Performing such a large number of 54 experiments was unrealistic. The first 16 experiments listed in Table 1 are a D-optimal design chosen using the RS/Discover computer program from BBN Software [5]. This group represents one of many equivalent designs that could be picked from the candidate set of 54 experiments. For example, if the experiments for the two oxidation states were reversed, an equally good design would result.

The first eight experiments in Table 1, four for each oxide, are a screening design that tests whether each variable has any significant effect on dissolution rate. The D-optimal approach significantly reduced the number of experiments required by classic full- or fractional-factorial designs. These experiments are uniformly distributed over the three-dimensional variable space. The additional eight tests (17-24) in Table 1 were added to the design and represent opposing temperature conditions for those experiments not having a temperature pair in the first 16 runs. These 24 runs include all of the eight possible combinations of the three variables at extreme settings for both oxides, as well as the eight runs with at least one mid-level variable setting needed for non-linear modeling. Run 25 was added so that a dissolution rate for $UO_3 \cdot H_2O$ would be available at the same condition as already obtained for the other two oxides.

Both U_3O_8 and $UO_3 \cdot H_2O$ samples were powders because of the synthetic routes available for each. The U_3O_8 powders were National Bureau of Standards (NBS or NIST) Standard Reference Material (SRM) 750(b). U_3O_8 is the most stable of the uranium oxides and is easily produced by the well known method of heating a uranium compound, UO_2 in this case, to several hundred degrees Celsius in air. The dehydrated schoepite was synthesized via hydrolysis of analytical grade uranyl acetate in a glass distillation apparatus, with continuous additions of water to replace the acetic acid boiled off during the synthesis [6]. Surface areas of both were measured via the traditional BET method using xenon gas. The resulting surface area for the U_3O_8 is 0.18 ± 0.02 m^2/g and 0.31 ± 0.04 m^2/g for the $UO_3 \cdot H_2O$. X-ray Diffraction (XRD) of both materials indicated they were the correct phases. Particle size distributions were also determined by means of

Table 1.   Test Matrix for the UO_{2+x} Dissolution Tests.

RUN	OXIDE	TEMPERATURE (deg C)	CARBONATE (mol/L)	PH Meas./Soln.	DISSOLUTION RATE ($mgU/m^2 \cdot day$)
1	U_3O_8	22.9	0.02	8.6/8.0	19
2	$UO_3 \cdot H_2O$	24.9	0.02	10.0/10.0	~200
3	U_3O_8	74.8	0.02	10.3/10.0	~200
4	$UO_3 \cdot H_2O$	74.7	0.02	8.6/8.0	>1500
5	U_3O_8	22.9	0.0002	8.9/10.0	0.8
6	$UO_3 \cdot H_2O$	24.9	0.0002	7.3/8.0	~100
7	$UO_3 \cdot H_2O$	74.7	0.0002	8.3/10.0	>150
8	U_3O_8	74.8	0.0002	8.1/8.0	~6
9	U_3O_8	22.9	0.02	10.6/10.0	21
10	$UO_3 \cdot H_2O$	24.9	0.0002	7.5/10.0	>100
11	$UO_3 \cdot H_2O$	74.7	0.02	10.0/10.0	>1000
12	U_3O_8	48.0	0.002	8.0/8.0	~10
13	U_3O_8	48.0	0.02	9.0/9.0	>100
14	$UO_3 \cdot H_2O$	25.0	0.002	9.0/9.0	~120
15	$UO_3 \cdot H_2O$	24.9	0.02	8.6/8.0	~700
16	U_3O_8	22.9	0.0002	8.4/9.0	1.3
17	$UO_3 \cdot H_2O$	74.7	0.0002	8.1/8.0	>200
18	U_3O_8	74.8	0.02	8.6/8.0	~150
19	U_3O_8	74.8	0.0002	9.3/10.0	~3
20	U_3O_8	74.8	0.0002	8.1/10.0	~4
21	U_3O_8	21.6	0.002	8.6/8.0	~10
22	U_3O_8	21.6	0.02	9.2/8.0	8.3
23	$UO_3 \cdot H_2O$	75.0	0.002	9.0/ 9.0	>20
24	U_3O_8	22.9	0.002	7.9/8.0	~5
25	$UO_3 \cdot H_2O$	25.0	0.02	9.0/9.0	>1500

sedimentation techniques. The median particle size for the U_3O_8 powder was 2.1 μm with a 25-75 percentile range of 1.0 to 2.8 μm. The median particle size for the $UO_3 \cdot H_2O$ powder was 4.1 μm with a 25-75 percentile range of 2.5 to 5.5 μm. The pore size of the filters used in the sample cells was 0.1 μm. This small pore size, combined with the upward flow of the aqueous solution at a slow velocity of ~3-15 cm/hr, kept the particles within the sample cells.

Test solutions were prepared using analytical-reagent grade chemicals and deionized water. Each solution was continuously sparged with argon gas containing fixed concentrations of oxygen and carbon dioxide to maintain the desired dissolved oxygen concentration and pH of the solution. The test solutions flowed through stainless steel sample cells at rates between 5 and 25 mL/hr. Several times per week, effluent from the cells was collected, acidified to prevent uranium adsorption on the sample vial walls, and analyzed for uranium content using a phosphorescence analyzer. Dissolution rates were calculated from uranium concentrations multiplied by flow rates and divided by surface areas of the test specimens. After steady-state dissolution rates were achieved, the flow rates were occasionally changed to ensure that the observed dissolution rates remained unchanged. Dissolution rates will not be affected by changing flow rates, if the reaction is not solubility- or diffusion-limited.

RESULTS

The 25 measured dissolution rates of U_3O_8 and $UO_3 \cdot H_2O$ at atmospheric oxygen are shown in Table 1, as well as the actual values for the three independent variables, temperature, carbonate concentration and pH. Most experiments lasted about a month. The dissolution rates are the average values after reaching steady-state. The values preceded by an approximation or greater-than sign are estimates, because steady-state could not be reached or the sample was dissolving too rapidly. Rapid sample dissolution was particularly true for the dehydrated schoepite. Table 2 lists the uranium dissolution rates for the three oxides, UO_2 [2], U_3O_8 and $UO_3 \cdot H_2O$ that were measured under atmospheric oxygen conditions. Because the measured values of the independent variables differ between runs, only their nominal values are listed in Table 2. Two new room-temperature UO_2 results were measured at a pH of 10 and 2×10^{-4} and 2×10^{-2} molar total carbonate. These were acquired so that there would be a full set of eight measurements at the extreme conditions (a full-factorial linear experimental design) for each oxide. Results for a spent fuel [2], ATM-103, are listed at equivalent conditions. To facilitate easier comparisons of the dissolution rates and variable effects, the results for the eight experimental conditions at the high and low values of each variable are grouped together as Part 1 of Table 2. They are grouped first by pH, then by carbonate concentration and finally by temperature. Also included are three dissolution rates of UO_2, U_3O_7 and U_3O_8 at a commonly used condition of room-temperature, pH of 8, and 2×10^{-2} molar total carbonate that were estimated from figures in reference [7]. To avoid confusion in comparing trends, the remaining results at intermediate pH and carbonate concentrations are listed separately as Part 2 of Table 2 using the same grouping scheme.

The oxide phase has by far the strongest effect on the uranium dissolution rate. Because U_3O_8 has both U(IV) and U(VI) valence states, its dissolution rates might be expected to be between that of UO_2 and $UO_3 \cdot H_2O$, particularly as carbonate concentrations increase. Indeed, the rate increases significantly in going from UO_2 to U_3O_8 and dramatically from U_3O_8 to $UO_3 \cdot H_2O$. The $UO_3 \cdot H_2O$ dissolution is so rapid that the samples disappear within a few days at the high carbonate levels. With the U_3O_8, unlike UO_2, carbonate affects the dissolution rate to a greater extent than does temperature. Increasing temperature shows its expected effect of enhancing the dissolution rate. The enhancement is particularly strong at the highest carbonate concentration. The data indicate that alkaline pH is the least significant factor in dissolution of spent fuel or any of the uranium oxides under the alkaline conditions of these experiments, although it seems more important in $UO_3 \cdot H_2O$ dissolution. The $UO_3 \cdot H_2O$ dissolution data show strong nonlinearities in dissolution response to all of the variables, pH, temperature and carbonate concentrations. These nonlinearities may only be due to the difficulties in determining appropriate $UO_3 \cdot H_2O$ dissolution rates. A comparison of the leachate and prepared solution pH's of the $UO_3 \cdot H_2O$ experiments in

Table 2, Part 1. Comparison of Dissolution Rates at Boundary Conditions

pH	Carbonate (mol/L)	Oxygen (atm)	Temp °C	Spent Fuel ATM-103 [2]	Dissolution Rates (mgU/m^2·day) UO_2	U_3O_8	UO_3·H_2O
8	0.0002	0.2	25		3.9	~5	~100
8	0.0002	0.2	50		5.4		
8	0.0002	0.2	75	8.6	11	~6	>200
8	0.02	0.2	25	3.5	2.4 (~3)[a]	[U_3O_7 (~3-5)[a]] 19 (~10-15)[a]	~700
8	0.02	0.2	50		38		
8	0.02	0.2	75		54	~150	>1500
10	0.0002	0.2	25	0.63	2.5	0.8	>100
10	0.0002	0.2	50		3.1		
10	0.0002	0.2	75		6.5	~3	>150
10	0.02	0.2	25		20	21	~200
10	0.02	0.2	50		26		
10	0.02	0.2	75	14	77	~200	>1000

a) See reference 7 for estimated rates in parentheses, including U_3O_7.

Table 2, Part 2. Comparison of Dissolution Rates at Intermediate Conditions

pH	Carbonate (mol/L)	Oxygen (atm)	Temp °C	Spent Fuel ATM-103 [2]	Dissolution Rates (mgU/m^2/day) UO_2	U_3O_8	UO_3·H_2O
8	0.002	0.2	25			~10	
8	0.002	0.2	50			~10	
9	0.0002	0.2	25			1.3	
9	0.0002	0.2	75			~4	
9	0.002	0.2	25				~120
9	0.002	0.2	50	6.1	12		
9	0.002	0.2	75		23		>20
9	0.02	0.2	25	2.8	6.7	8.3	>1500
9	0.02	0.2	50			>100	
10	0.002	0.2	25	2.0	9.3		

Table 2 shows that the solution pH sometimes drops upon contact with $UO_3 \cdot H_2O$, particularly at low carbonate concentrations. Nuclear Magnetic Resonance (NMR) analysis of the $UO_3 \cdot H_2O$ shows no residual acetate that might cause the acidity. The $UO_3 \cdot H_2O$ may act as a Brønsted acid by donating a proton, but the actual cause of the acidity is not known. Where the carbonate concentration is low, there is less buffering capacity of the solution.

MODELING

Only the fourteen U_3O_8 dissolution rate data given in Table 2 were modeled. Because the $UO_3 \cdot H_2O$ dissolved so rapidly, their dissolution rates are estimates or minima and not appropriate for modeling. As with the UO_2 dissolution data several approaches to U_3O_8 dissolution modeling are being explored. Again the classical observed chemical kinetic rate law was used and takes the following well-known general form [8]:

$$\text{Rate} = k[A]^a[B]^b[C]^c...\exp(-E_a/RT), \tag{1}$$

This generalized form of the rate law is for homogeneous gas or liquid reaction systems. It does not take into consideration the possibly complex liquid-solid reaction at the UO_x or spent fuel surface. Additional term(s) are needed to account for this element of the reaction, and any radiation effects in the spent fuel, but they are unknown at this time.

Other function forms are being considered. Dissolution of a solid is a thermodynamic nonequilibrium process. An Onsager-type thermodynamic function provides a classical relationship for dissolution rate and is linearly related to the energy change of the solid dissolving into a liquid. This is expected to be descriptive of dissolution response close to thermodynamic equilibrium, particularly those controlled by diffusion kinetics. A form of the Butler-Volmer equation, used in correlation of corrosion and electrochemical rate data, is also being examined. The normal derivation of the Butler-Volmer equations assumes that the electrochemical processes are not at, but near, thermodynamic equilibrium. The Butler-Volmer equations are expected to be descriptive of energy-dependent surface reactions.

Only regression fits of the U_3O_8 dissolution rate data to the classical chemical kinetic rate law are discussed here, for comparison with the previously reported UO_2 and spent fuel models. Model parameters are presented, based on both the leachate pH's used in the UO_2 dissolution models, and the pH's of the original carbonate solutions, before contact with the UO_2 or spent fuel samples, as used previously for the spent fuel data. The pH's of the fresh carbonate leaching solutions are probably more representative of the pH at the sample than the pH of the leachate analysis sample that has been exposed to dissolved CO_2 from the air.

The following equation was fitted using the measured leachate pH's given in Table 1:

U_3O_8 (leachate pH's):
$\log_{10}(DR)\{mgU/m^2 \cdot day\} =$
$7.832 + 0.6910 \cdot \log_{10}[CO_3] + 0.0860 \cdot \log_{10}[H] - 1317/T$
$r^2 = 0.87$.

$$\tag{2}$$

Using the pH's of the prepared carbonate solutions, also given in the same column of the table, we arrive at similar, but perhaps more accurate, coefficients:

U_3O_8 (carbonate soln. pH's):
$\log_{10}(DR)\{mgU/m^2 \cdot day\} =$
$7.951 + 0.6492 \cdot \log_{10}[CO_3] + 0.1065 \cdot \log_{10}[H] - 1333/T$
$r^2 = 0.88$.

$$\tag{3}$$

As with the earlier UO_2 and spent fuel dissolution data, the pH did not have much effect on the model. However, carbonate concentration, not temperature, had the strongest effect on the U_3O_8 dissolution rate. The temperature had half the effect of carbonate concentration on the uranium dissolution rate. The pH was only about one-sixth as effective as carbonate concentration in explaining the changes in U_3O_8 dissolution rates. Leaving out the pH term had a negligible effect on the other coefficients and was absorbed in the constant:

U_3O_8 (carbonate soln. pH's):
$$\log_{10}(DR)\{mgU/m^2{\cdot}day\} =$$
$$6.925 + 0.6486{\cdot}\log_{10}[CO_3] - 1307/T$$
$$r^2=0.86. \tag{4}$$

Temperature and carbonate concentration show significant interaction. The pH shows its importance through interaction with carbonate as well. Additions of cross terms for those interactions to equation 4 improves the fit significantly, with a correlation coefficient of 0.95.

To allow comparisons with this U_3O_8 dissolution data, the previously reported spent fuel and UO_2 atmospheric oxygen models [2] are reproduced in equations 5 and 6. For consistency the UO_2 20% oxygen data were refitted using the fresh carbonate solution pH's and is shown in eq.7. This regression fit had a increased correlation coefficient, compared with the original fit using the leachate pH's (eq. 6). There was a larger change in the coefficients, than with the U_3O_8 results fitted with the two pH sets.

Spent Fuel, ATM-103 (20% oxygen only, carbonate soln. pH's):
$$\log_{10}(DR)\{mgU/m^2{\cdot}day\} =$$
$$7.202 + 0.2260{\cdot}\log_{10}[CO_3] + 0.0905{\cdot}\log_{10}[H] - 1628/T$$
$$r^2=0.95. \tag{5}$$

UO_2 (20% oxygen only, leachate pH's):
$$\log_{10}(DR)\{mgU/m^2{\cdot}day\} =$$
$$4.650 + 0.2742{\cdot}\log_{10}[CO_3] - 0.1868{\cdot}\log_{10}[H] - 1501/T$$
$$r^2=0.79. \tag{6}$$

UO_2 (20% oxygen only, carbonate soln. pH's):
$$\log_{10}(DR)\{mgU/m^2{\cdot}day\} =$$
$$5.828 + 0.3335{\cdot}\log_{10}[CO_3] - 0.1571{\cdot}\log_{10}[H] - 1734/T$$
$$r^2=0.83. \tag{7}$$

CONCLUSIONS

The aim of this work has been the measurement of the intrinsic dissolution rates of uranium oxides under a variety of well-controlled conditions. These experiments are relevant to a geological repository and allow for modeling. After exposure to air at elevated temperature, the stable phase U_3O_8 may form from spent fuel. Formation of dehydrated schoepite, $UO_3{\cdot}H_2O$, has been found in drip tests with spent fuel.

Equivalent sets of U_3O_8 and $UO_3{\cdot}H_2O$ dissolution experiments allowed a systematic examination of the effects of temperature (25-75°C), pH (8-10) and carbonate ($2\text{-}200{\times}10^{-4}$ molar) concentrations at atmospheric oxygen conditions.

Results indicate that $UO_3{\cdot}H_2O$ has a much higher dissolution rate (at least ten-fold) than U_3O_8. Dissolution of both U_3O_8 and $UO_3{\cdot}H_2O$ shows a very high sensitivity to carbonate concentration. Present results show a 25 to 50-fold increase in room-temperature $UO_3{\cdot}H_2O$ dissolution rates between the highest and lowest carbonate concentrations. The intrinsic dissolution rate of unirradiated U_3O_8 is about twice that of UO_2 under similar conditions. At the frequently-studied condition of room temperature, pH of 8, and 2 x 10^{-2} molar total carbonate, this data and the room-temperature dissolution rates of UO_2, U_3O_7, and U_3O_8 estimated from figures in reference [7] show a consistent increase in the intrinsic dissolution rates as the degree of oxidation increases.

As with the UO_2 dissolution data the classical observed chemical kinetic rate law was used to model the U_3O_8 dissolution rate data.   The pH did not have much effect on the models, in concert with the earlier UO_2 and spent fuel dissolution data,. However, carbonate concentration, not temperature, had the strongest effect on the U_3O_8 dissolution rate. The U_3O_8 dissolution activation energy was about 6000 cal/mol, compared with 7300 and 8000 cal/mol for spent fuel and UO_2 respectively.

ACKNOWLEDGMENTS

Work performed under the auspices of the U.S. Department of Energy by Lawrence Livermore National Laboratory under Contract W-7405-ENG-48. This work was supported under activities D-20-53 of the Yucca Mountain Project Spent Fuel Waste form Task (YMP WBS element 1.2.2.4.1) and the AECL/USDOE Cooperative Project sponsored by the DOE Office of Civilian Radioactive Waste Management.

REFERENCES

1.   P.A. Finn et al., <u>Proc. of Topical Meeting on DOE Spent Nuclear Fuel</u>, Salt Lake City, UT, pp. 421-429 (1994).

2.   S. A. Steward and W. J. Gray, <u>Proc. 5th Annual Intl. High-Level Radio. Waste Mgmt. Conf.</u>, **4**, 2602-8 (1994).

3.   W.J. Gray, H.L. Leider and S.A. Steward, J. Nucl. Mater., **190**, 46-52 (1992).

4.   S. A. Steward and H. C. Weed, in <u>Scientific Basis for Nuclear Waste Management XVII</u>, edited by A. Barkatt and R. A. Van Konynenburg (Mater. Res. Soc. Proc. **333**, Pittsburgh, PA, 1994) pp. 409-416.

5.   BBN Software Products Corporation, RS/Discover, Version 2 (1989).

6.   K. H. Gayer and H. Leider, JACS, **77**, 1448 (1955).

7.   W. J. Gray, L. E. Thomas and R. E. Einziger, in <u>Scientific Basis for Nuclear Waste Management XVI</u>, edited by C. G. Interrante and R. T. Pabalan, (Mater. Res. Soc. Proc. **294**, Pittsburgh, PA, 1993) pp. 47-54.

8.   W. Stumm and J. J. Morgan, <u>Aquatic Chemistry: An Introduction Emphasizing Chemical Equilibria In Natural Waters,</u> 2nd ed. (John Wiley and Sons, New York , 1981), Chapter 2.14.

DISSOLUTION STUDIES OF SODDYITE AS A LONG-TERM ANALOGUE OF THE OXIDATIVE ALTERATION OF THE SPENT NUCLEAR FUEL MATRIX

I. Pérez*, I. Casas*, M.E. Torrero*, E. Cera**, L. Duro** and J. Bruno**
* Polytechnic University of Catalunya, Dept. of Chem. Eng., Avd. Diagonal 647, 08028 Barcelona (Spain)
** QuantiSci, Parc Tecnològic del Vallés, 08290 Cerdanyola (Spain)

ABSTRACT

The thermodynamic and kinetic dissolution properties of a synthetically obtained soddyite have been determined at different bicarbonate concentrations. This uranium-silicate is expected to be a secondary solid phase of the oxidative alteration pathway of uranium dioxide in waters with low phosphate content and, consequently, it is likely to constitute one of the long-term uranium solubility limiting solid phases.

The experimental data obtained at the end of the experiments correspond fairly well to the theoretical model calculated with a log K^0_{so} of 3.9 ± 0.7.

On the other hand, the general trend of the total uranium in solution measured in the experiments as a function of time has been fitted by using a kinetic equation obtained from the principle of detailed balancing of the dissolution reaction. In addition, the EQ3/6 code has also been used to model the uranium concentrations as a function of time. In both modeling exercises comparable results were obtained. The dissolution rate, normalized to the total surface area used in the experiments as measured with the BET method, gave an average value of 6.8 (±4.4) 10^{-14} mol $cm^{-2} s^{-1}$.

INTRODUCTION

The performance assessment of HLNW repositories requires the long-term description of the behavior of the waste matrix. This is basically done by using the experience from leaching experiments of actual spent fuel to derive kinetic and thermodynamic models for the dissolution of spent fuel under repository conditions.

However, the time scales of spent nuclear fuel dissolution experiments is of the order of 2 to 10 years, while the performance of a spent fuel repository should be assessed for much longer times (10^5-10^6 years). These time scales can be bridged using appropriate natural phases that give insight into the critical steps for the oxidative alteration of spent fuel in granitic environments.

Recently, much attention has been devoted to the mineralogical and crystallographic studies of the pathways of uraninite alteration and the consequences on the long-term stability of spent fuel [1-4]. In order to better understand the natural complexity, the systematic study of the dissolution behavior of uranium phases is important. Previous studies [5] have demonstrated the complexity of the studies carried out using natural samples. The results have been in some cases of difficult evaluation even considering the careful leaching studies performed as well as the extensive characterizations of the solid phase before and after the dissolution experiments. For this reason, a series of experiments has been started where uranium solid phases have been synthetically obtained under well controlled laboratory conditions.

In the present work, the kinetic and thermodynamic models obtained to describe the dissolution behavior of a synthetic soddyite are presented.

Mat. Res. Soc. Symp. Proc. Vol. 465 © 1997 Materials Research Society

EXPERIMENTAL

Soddyite was synthesized following the procedure proposed by Nguyen et al. [6]. Briefly exposed, this method consists on the following steps: precipitation of soddyite by addition of sodium silicate in an uranyl solution at pH approximatly 4.5, this reaction was left 100 hours at room temperature to proceed. Afterwards the mixture is refluxed at 353 K for 6 hours, and finally, the solid is put in contact with water in a glass Parr bomb at 403 K for 2 weeks. This final step has shown to be crucial to improve the cristallinity of the solid, as it was observed from the XPD characterization. Commercial reagents from Fluka, Merck and Panreac were used.

The characterization of the solid obtained included X-ray Powder Diffraction (XRD), Scanning Electron Microscopy (SEM) and Fourier Transformed Infra Red spectroscopy (FTIR). The XRD measurements were taken with the Cu K α radiation using a typical wavelength of 1.5418 Å on a wide-angle Siemmens D-500 equipped with a quartz primary and graphite secondary monochromator. The database of the powder diffraction standard was the JCPDS (Joint Committee of Powder Diffraction Standard). The Scanning Electron Microscopy (SEM) analysis were performed by using a JEOL 6450, EDX-LINK-LZ5 system. The FTIR absorbance spectra were carried out on a Nicolet 520 FTIR system in the region from 4000 cm^{-1} to 400 cm^{-1}. Each sample was mixed with KBr, ground to a fine powder and pressed into a clear disk. Then, it was directly examined. A He/Ne laser was used as reference to constantly calibrate the scale of frequency of the instrument.

The solid phase characterization studies of the sample were performed before and after the dissolution experiments. The results from these characterizations showed the presence of a soddyite phase of high purity, both at the beginning and at the end of the dissolution experiments.

Because the influence of the total surface area on the rate of dissolution, it was of great importance to determine the specific surface area of the solid phase obtained. The measurements were made by the BET gas adsorption method with a FLOWSORB II 2300 system of Micromeritics$^{\circledR}$. The value obtained was 25.4 ± 0.2 $m^2 g^{-1}$.

Dissolution studies of the synthetic uranyl silicate were performed in order to determine the influence of [HCO_3^-] on the dissolution process. All the studies were made at room temperature. An orbital stirrer was used to keep the solution homogeneously mixed as well as to minimize physical alterations of the solid phase.

Solution composition was selected in order to prevent the formation of secondary solid phases during the leaching of the samples. This selection was based in the so far available database for the uranium-silicon system. Test solutions were prepared with Millipore distilled water. The general composition of the leaching solution was 10^{-3} mol dm^{-3} Na_2SiO_3 and 7.10^{-3} mol dm^{-3} $NaClO_4 \cdot H_2O$. In addition, concentrations of $NaHCO_3$ ranging from 10^{-3} to 2.10^{-2} mol dm^{-3} were used. Perchlorate was used in order to avoid complexation of uranium, except by hydroxyl and carbonate. The solution was kept in contact with air. The pH of the solutions was measured by means of a combined glass electrode.

To allow a relatively short saturation time, the solid surface/solution volume ratio was selected to be somewhat high. This implies the use of a small volume of solution. The initial volume of the test solution was 100 ml with a weight of solid of 0.25 g, which corresponds to a total surface area of 6.35 m^2. This results in a S/V ratio of 63.5 m^2 dm^{-3}.

Total uranium in solution was determined, after filtration through 0.2 μm pore size filters, by using the SCINTREX uranium analyzer.

RESULTS

The release of uranium was followed in all experiments as a function of time. We assumed that the equilibrium was reached when aqueous uranium concentration and pH readings remained constant for several days.Once the equilibrium state was reached, the solution was completely replaced by a fresh one, with the bicarbonate concentration previously set to the desired new value. Table I summarizes the uranium concentration at steady-state for each bicarbonate concentration in solution and the pH values reached. In some cases (1, 2 and 5 mM of total sodium bicarbonate), duplicate experiments were performed, which showed reproducibility of both dissolution rates and uranium saturation levels.

Table I.- Experimental values obtained at the end of the different experiments performed, including the total bicarbonate concentrations, the pH measured at the end of the experiment and the average total uranium concentration in solution, in mol dm^{-3}.

$[HCO_3^-]_{tot}$ (mol dm^{-3})	pH	log $[U]_{tot}$
1.00E-03	8.65±0.25	-4.29±0.05
1.00E-03	8.65±0.17	-4.25±0.01
2.00E-03	8.53±0.18	-4.08±0.01
2.00E-03	8.78±0.30	-4.05±0.07
5.00E-03	8.72±0.03	-3.45±0.04
5.00E-03	8.65±0.04	-3.54±0.03
8.00E-03	9.11±0.09	-3.07±0.002
1.00E-02	8.54±0.07	-2.99±0.03
1.50E-02	8.61±0.17	-2.72±0.03
2.00E-02	8.80±0.14	-2.65±0.01

The behavior of the system showed a fast initial increase of the uranium in solution before the final steady state was reached. The general trend of the total measured uranium concentration in solution is shown in Figures 1.a, 1.b and 1.c (the results are presented in three different figures to allow a better scaling of the experimental data presented).

Soddyite solubility

The aqueous uranium speciation at steady-state was calculated using the HARPHRQ code, and using for uranium the SKBU database [8]. These calculations showed that the $UO_2(CO_3)_3^{4-}$ complex is the dominant aqueous species (more than 90%) at bicarbonate concentrations larger than 5 mM. Hence, the dissolution reaction of soddyite can be written as:

$$(UO_2)_2SiO_4.2H_2O + 6HCO_3^- \Leftrightarrow 2UO_2(CO_3)_3^{4-} + H_4SiO_4 + 2H^+ + 2H_2O \qquad (1)$$

and the corresponding equilibrium constant:

$$K = a_{uranyl\ complex}^2 \cdot a_{silicon} \cdot a_{proton}^2 \cdot a_{bicarbonate}^{-6} \qquad (2)$$

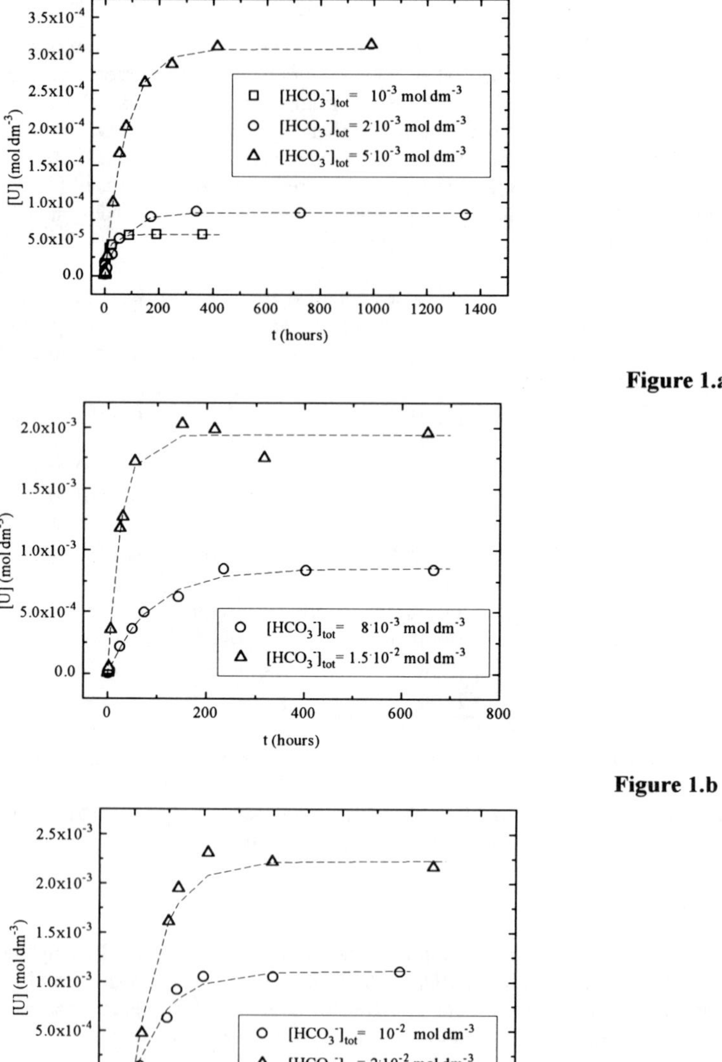

Figure 1.a

Figure 1.b

Figure 1.c

Figures 1.a,b,c.- Total measured uranium in solution as a function of time for different total bicarbonate concentrations. Dashed lines correspond to the dissolution model calculated using equation (5) (see below).

From the total bicarbonate concentration, the silicon in solution (calculated as the initial silicon concentration in solution plus the amount due to dissolution, assuming congruent release with uranium) and the experimental pH, we calculated the equilibrium constant for each experiment using again the HARPHRQ code. These constants were subsequently corrected to zero ionic strength by using the specific ion interaction theory (SIT) [9]. Finally, from the equilibrium constants obtained, K_{S0} values corresponding to the dissolution reaction:

$$(UO_2)_2SiO_4.2H_2O + 4H^+ \Leftrightarrow 2UO_2^{+2} + H_4SiO_4 + 2H_2O \qquad (3)$$

were calculated.

Table II lists values of the ionic strengths of the reaction media obtained from the HARPHRQ code as well as the values of both log $K_{(I=0)}$ and log K_{S0} calculated for each experiment.

Table II.- Calculated equilibrium constant at infinite dilution and the solubility product for each bicarbonate concentration.

$[HCO_3^-]_{tot}$ (mol dm^{-3})	Ionic strength	log K (I_m=0)	log K_{S0}
1.00E-03	0.0087	-12.48	6.30
1.00E-03	0.0087	-12.42	6.36
2.00E-03	0.0095	-13.15	5.63
2.00E-03	0.0096	-13.37	5.41
5.00E-03	0.0129	**-14.18**	**4.60**
5.00E-03	0.0124	**-14.36**	**4.42**
8.00E-03	0.0182	**-14.98**	**3.80**
1.00E-02	0.0196	**-14.54**	**4.24**
1.50E-02	0.0280	**-15.01**	**3.77**
2.00E-02	0.0333	**-16.20**	**2.58**

It can be seen that, at low bicarbonate concentrations, where other complexed uranyl species play an important role, the log K_{S0} values obtained clearly differ from those obtained at higher bicarbonate concentrations. Therefore, only the values of log K_{S0} obtained at bicarbonate concentrations greater than 2 mM were averaged to get the final proposed value of log K_{S0}:

$$\log K_{S0}= 3.9 \pm 0.7 \qquad (4)$$

In Figure 2, we show a comparison of the experimental uranium concentrations and the calculated uranium concentrations obtained by using different values of log K_{S0}. These values include the log K_{S0} calculated for each experiment, as presented in Table II, the average value proposed in this work, and, finally the value reported by Nguyen et al. [6] (log K_{S0}=5.74). The solubility constant recently published in [7] (log K^0_{s0}=6.15) is not included because it would give a model very similar than the one obtained for log K_{s0}=5.74. The model presented in Figure 2 obtained using a log K_{s0}=5.74, clearly gives in all cases higher calculated than experimental uranium concentrations.

The difference on the solubility constant proposed in this work and the previously published in [6, 7] is attributed to the differences in the experimental methodology and conditions. Nguyen et al. and Moll et al. [6, 7] used a single experimental data (at pH=3 in both studies) to extract the solubility product of soddyite, while in our case, a set of data was fitted to calculate K_{S0}. In addition, we used bicarbonate in our test solution. This anion constitutes an important complexing agent for aqueous U(VI), which results in the stabilization of uranium in solution, reducing the possibility of formation of secondary solid phases. Such a secondary solid phase formation has been observed in our laboratory in some experiments performed at neutral to alkaline pH values without bicarbonate in the leaching solution.

On the other hand, the solubility product obtained in this work can also be compared with the value obtained in experiments where a natural uranium phase was used [5]. In that case, a log K_{SO} of 3.0 ± 2.9 was determined, also in the presence of bicarbonate in the test solution.

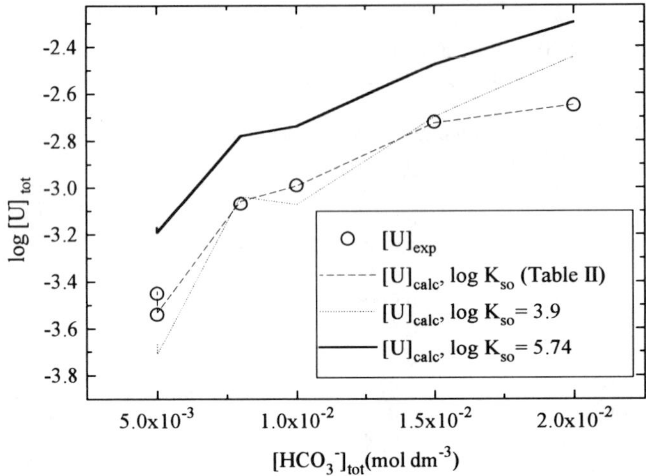

Figure 2.- Experimental solubilities obtained as a function of total bicarbonate concentration as well as calculated solubilities obtained by using different K_{SO} (see legend).

Kinetics of dissolution

Applying the principle of detailed balancing [10] to reaction (1), the net rate of the reaction can be expressed as:

$$r = d[U]/dt = k_1[HCO_3^-]^6 - k_{-1}[UO_2(CO_3)_3^{4-}]^2\,[H_4SiO_4]\,[H^+]^2 \qquad (5)$$

In order to readily solve this equation some approximations were made. First of all, both bicarbonate concentration and pH have been considered to remain constant through the experimental time. This assumption is considered to be reasonable as shown by the experimental measurements. In addition, the silicon released in the dissolution reaction was considered not significant compared to the initial concentration of this element in solution. Hence, a constant silicon concentration of 10^{-3} mol dm^{-3} was used in the modeling exercise. Calculations performed with the HARPHRQ code assuming congruent U-Si dissolution have shown that this approximation can be considered reasonable for most of the experiments performed.

Equation (5) was integrated and fitted to the experimental data. Both forward (k_1) and reverse (k_{-1}) rate constants were determined from the best fit obtained for each experiment. These fittings correspond to the dashed lines shown in Figures 1 a, b and c. As a test of the results obtained, we calculated the equilibrium constant from these rate constants as:

$$K_{eq} = k_1/k_{-1} \qquad (6)$$

From these equilibrium constants, corresponding to reaction (1), we calculated the solubility products corresponding to reaction (3). We obtained an average value of 4.3±0.6, which is in agreement with the solubility constant calculated from the equilibrium data in the preceding section, 3.9±0.7. This result gives confidence to the approximations made in the modeling process.

The experimental results were also modeled by using the EQ3/6 code package. This code uses the following rate equation:

$$R_{diss} = k_{\ diss} (1-10^{SI}) \qquad (7)$$

By comparing equations (7) and (5), we obtain:

$$k_{diss} = k_1 [HCO_3^-]^6 \qquad (8)$$

The comparison between the results obtained from the analytical model with the ones obtained from the EQ3/6 based model are shown in Table III. Only the values that gave the best agreement between the two models are presented.

Table III.- Comparison of the results obtained by applying the principle of detailed balancing to reaction (1) and by using the EQ3/6 program.

$[HCO_3^-]_{tot}$ (mol dm^{-3})	k_{diss} (EQ3/6) (mol cm^{-2} s^{-1})	$k_1[HCO_3^-]^6$ (mol cm^{-2} s^{-1})
8.00E-03	1.23E-14	1.39E-14
1.00E-02	8.23E-14	5.29E-14
1.50E-02	1.23E-13	4.22E-14
2.00E-02	1.23E-13	9.26E-14

Hence, the following average dissolution rate of soddyite is proposed:

$$r_{diss} = 6.8 \ (\pm 4.4) \ 10^{-14} \ mol \ cm^{-2} \ s^{-1} \qquad (9)$$

CONCLUSIONS

The thermodynamic and kinetic properties of soddyite have been determined in the presence of different bicarbonate concentrations, at 25°C. The test solutions were exposed to air throughout the experimental time.

The experimental data obtained at the end of the experiments correspond fairly well to the theoretical model calculated with a log K^0_{so} of 3.9±0.7. This value is approximately 1.5 orders of magnitude lower than the solubility product calculated by Nguyen et al. [6] and Moll et al. [7], while it corresponds fairly well with the value determined from a natural sample [5]. Among the different possibilities that may account for these differences, we consider the fact that the value reported in [6, 7] was extracted from a single experiment at pH=3 in both cases, which may involve a relatively large degree of uncertainty. Also, the presence of bicarbonate in our test solutions can also contribute to prevent the formation of possible secondary solid phases.

On the other hand, the general trend of the total uranium in solution measured in the experiments as a function of time has been fitted by using a kinetic equation obtained from the principle of detailed balancing of the dissolution reaction. In addition, the EQ3/6 code has also been used to model the uranium concentrations as a function of time. Comparable results were obtained from both modeling exercises. The initial dissolution rate, normalized to the total surface area used in the experiments as measured with the BET method, gave an average value of $6.8 \ (\pm 4.4) \ 10^{-14} \ mol \ cm^{-2} \ s^{-1}$.

ACKNOWLEDGMENTS

Montse Marsal and Xavier Alcober are acknowledged for SEM-EDS and XRD characterizations, respectively.
This work has been financially supported by SKB (Swedish Nuclear Fuel and Waste Management Co.).

REFERENCES

[1] Finch R.J. and Ewing R.C. (1989); SKB Technical Report **89-37**.
[2] Finch R.J. and Ewing R.C. (1991); SKB Technical Report **91-15**.
[3] Finch R.J. and Ewing R.C. (1992); J. of Nucl. Mat. **190**, 133-156.
[4] Janeczek J. and Ewing R.C. (1992); J. of Nucl. Mat. **190**, 157-173.
[5] Casas I., Bruno J., Cera E., Finch R.J. and Ewing R.C. (1994); SKB Technical Report **94-16**.
[6] Nguyen S.N., Silva R.J., Weed H.C. and Andrews J.E. Jr. (1992); J. of Chem. Thermodynamics **24**, 359-376.
[7] Moll H., Geipel G., Matz W., Bernhard G. and Nitsche H. (1996); Radiochimica Acta **74**, 3-7
[8] Puigdomènech I. and Bruno J. (1988); SKB Technical Report **88-21**
[9] Grenthe I., Fuger J., Konings R.J.M., Lemire R.J., Muller A.B., Nguyen-Trung C. and Wanner H. (1992a); <u>Chemical Thermodynamics of uranium</u> (eds. Hans Wanner and Isabelle Forest). Elsevier Science Publishers B.V. The Netherlands.
[10] Lasaga A.C., Berner R.A., Fisher G.W., Anderson D.E. and Kirkpatrick R.J. (1983); <u>Kinetics of Geochemical Processes</u>, Mineralogical Society of America, Reviews in Mineralogy Volume 8, Eds. A.C. Lasaga and R.J. Kirkpatrick.

MODELLING OXIDATIVE DISSOLUTION OF SPENT FUEL

IVARS NERETNIEKS
Department of Chemical Engineering and Technology, Royal Institute of Technology
S-100 44 Stockholm, SWEDEN

ABSTRACT

Spent nuclear fuel will, by the radiation, split nearby water into oxidizing and reducing compounds. The reducing compounds are mostly hydrogen that will diffuse away. The remaining oxidizing compounds can oxidize the uranium oxide of the fuel and make it more soluble. The oxidised uranium will dissolve and diffuse away. The nuclides previously incorporated in the spent fuel matrix can then be released and also migrate away from the fuel.

A model is proposed where the produced oxidizing species compete for reaction with the fuel and for escaping out of the system. The chemical reaction rate of oxygen and fuel is taken from literature values based on experiments. The escape rate of oxidants to a receding redox front in the backfill is modelled assuming a redox reaction of oxidizing component and reducing component in the surrounding. The rate of movement of the redox front is determined from the rate of production of oxidants. This is estimated using a previously devised model that has been calibrated to in situ observed radiolysis.

Three cases are modelled. In the first case it is assumed that the reducing compound is insoluble and that the reaction between oxygen and reducing mineral is very fast. In the second case it is assumed that the reducing component has a known solubility and that it can migrate to meet the oxygen and quickly react. In a third case a finite reaction rate is modelled between the oxygen and the reducing species.

The sample calculations show that if the reducing mineral has to be supplied from the backfill a large fraction of the spent fuel could be oxidised. If the corrosion products of a degraded steel canister can supply the reducing species and the redox reaction is fast, very small amounts of the fuel could be oxidised. Literature data indicate that the redox reaction rate may not be so fast that it can be considered instantaneous and then a considerable fraction of the fuel could be oxidised. The model gives a means of exploring which mechanisms and data may be of most importance for radiolytic fuel dissolution, but the realism of the data and the model must be tested further. There is a lack of understanding and data on reaction rates, heterogeneous as well as homogeneous. This is crucial to the results.

BACKGROUND AND INTRODUCTION

The use of Mixed Oxide fuel, MOX, is contemplated in several countries. One of the differences between MOX fuel and Uranium Oxide Fuel, UOF, is that spent MOX will be much more radioactive than UOF and can generate more oxidising products. Oxidation induced dissolution of the spent fuel in a final repository could then possibly become of concern. In this paper a model is proposed that can be used to estimate the rate of radiolysis and the rate of oxidation and dissolution of the fuel. The competing reaction where the oxidants react with reducing components near the fuel is also considered. The reducing components come from the backfill and rock or from the remainder of a degraded steel canister. The model is by necessity based on a simple geometry but uses well established modelling approaches accounting for mass balances, reaction rates and transport by molecular diffusion.

The aim of this study is to gain insight into the most important mechanisms that govern the fuel dissolution and to assess where some major uncertainties in data and assumption are to be found.

CONCEPTUAL MODEL

The setting is simplified to be the following. Spent fuel in a canister is emplaced in a deep geologic repository. The rock surrounding the repository is water saturated and is chemically reducing. This means that there is extremely little oxygen present. Any oxygen or other oxidising

Mat. Res. Soc. Symp. Proc. Vol. 465 © 1997 Materials Research Society

components will readily react with reducing components in the ground water. The ground water is constantly being supplied with the reducing components from minerals in the rock containing these minerals. The most important reducing component is ferrous iron, Fe(II), which is present in percentage quantities in the rock.

The canisters containing the waste are surrounded by compacted bentonite with a very low hydraulic conductivity. Transport in and through the bentonite is dominated by molecular diffusion. Advection plays a very minor role. The bentonite also contains ferrous iron minerals and can thus scavenge any oxidising substance until its supply of ferrous iron minerals is exhausted. Organic matter in the bentonite can also act as a reductant. The remains of a steel canister can also be a major source of reducing components. The redox capacity of bentonite, rock or canister play a very important role in limiting the release of the radionuclides from the near field. This is because as long as the redox conditions are reducing the solubility of the uranium dioxide matrix of the spent fuel and several other radionuclides is very low. The rate of transport out from the fuel's surface will then also be low. If reducing conditions were to prevail at the surface of the fuel, uranium oxide solubility would be very low and the matrix would dissolve at an extremely slow rate.

RADIOLYSIS MEDIATED DISSOLUTION

In oxidising waters the solubility of uranium can be many orders of magnitude larger than under reducing conditions. It is known that the radiation can split water molecules and give rise to a large number of radicals and other products. Some of the species are strongly reducing, others are oxidising. Some of the constituents back react to form water. Some constituents form hydrogen which is a strongly reducing species but which is not very reactive. The reaction kinetics is slow in the absence of catalysts at the temperatures expected in a repository. Hydrogen is, however, a small molecule and diffuses faster than the oxidising constituents. Hydrogen is assumed to leave the system much faster than the oxidising species which therefore are left in the system and can react with other reducing components. Canisters made of steel or copper can be attacked by the oxidants. Corrosion products of a steel canister, i.e., ferrous iron minerals are also effective reducing agents. Such reactions would consume the oxidants if the oxidants could find these materials. If this is the case then there might develop reducing conditions near the fuel surface and dissolution would be slow.

It is assumed here that at least a non-negligible fraction of the produced oxidants are not immediately scavenged by the canister materials. Conditions near the fuel's surface can then become oxidising. The tetravalent uranium U(IV) in UO_2 will be oxidised to hexavalent uranium U(VI) which can form soluble species in water. U(VI) can also form new minerals with different crystal structure than UO_2.

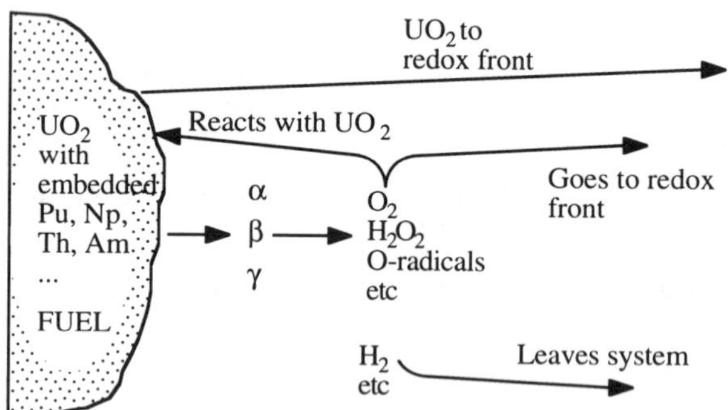

Figure 1. α, β and γ radiolysis generate oxidants which react with the UO$_2$ matrix and release uranium and other nuclides.

In both cases the previously incorporated other nuclides will be released. The rate of release of incorporated nuclides will be proportional to the rate of oxidation of the uranium. Figure 1 shows these processes.

In this model concept it is thus sufficient to determine the rate of production of oxidants by radiolysis to obtain an upper bound on the release rate of radionuclides.

Most of the oxidants are formed very near the fuels surface because the α-particles, which are the main source of radiolytic energy, only reach about 0.04 mm into water. Thus most of the oxidants are formed very near the surface of the fuel and can readily reach it by molecular diffusion. The other reductants in corrosion products and clay backfill will have been exhausted in the immediate vicinity of the surface of the fuel and the oxidant will have to move a longer distance to reach them at the redox front. Thus it would seem that the longer the process has been going on the more readily would the fuel itself be attacked as the redox front recedes. However, the rate of radiolysis decreases with time as the radionuclides decay. These processes are described below

To proceed it is necessary to be able to assess the rate of production of oxidising species and what fraction of the oxidants can oxidise the fuel and other redox sensitive elements in the fuel. To simplify the description it is assumed that the UO_2 is the main reductant and that the other nuclides that are oxidised can be lumped into the UO_2 from a mass balance point of view. It is also assumed that the oxidation products themselves i.e. U(VI) species are oxidising relative to ferrous iron. This is a reasonable assumption because it is known that uranium will precipitate as UO_2 when it encounters ferrous iron minerals and is reduced [1].

BASIC MODEL FOR AN INSOLUBLE REDUCING MINERAL

The oxidising species, be it oxygen, hydrogen peroxide or UO_2^{2+} complexes, diffuse away from the surface outward into the corrosion products and the backfill. Eventually they encounter water that contains ferrous iron or minerals that contain ferrous iron with which they react. If these reactions are fast, which they sometimes are, as can be seen by the sharp redox fronts found in nature around uranium ore bodies, then negligible amounts of oxidants, including uranium, can pass the redox front. The redox front will slowly move outward from the fuel as more and more oxidants are formed by the radiolysis. A pseudo steady state is assumed to develop. This is a reasonable assumption and considerably simplifies the mathematical formulation and solution.

Figure 2 illustrates the movement of the redox front and the concentration profiles of oxidising species and U(VI).

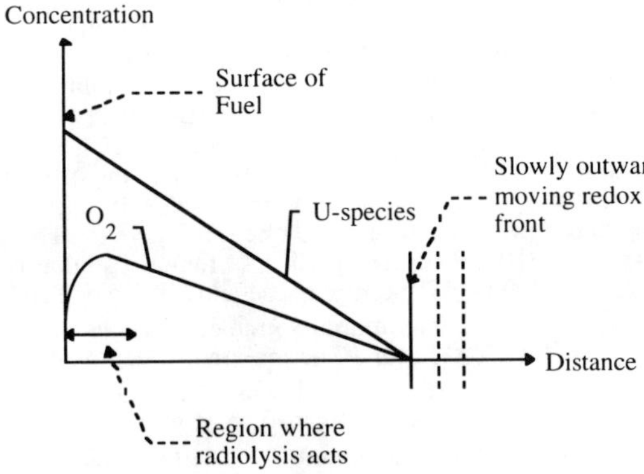

Figure 2   Concentration profiles of U-species and O_2 between fuels surface and redox front. The front slowly recedes to the right. Radiolysis acts only very near the surface.

At this point we make a very important assumption which later will be revisited, namely that *the reducing species is insoluble*. Large organic molecules would fall into this category. This means that the oxidant must move up to the reducing solid mineral before it can react. The reducing species does not dissolve and move to meet the oxidant.

Radiolysis acts only very near the surface of the fuel. The oxidants are modelled as oxygen for simplicity but are mostly in the form of hydrogen peroxide that is more reactive than oxygen. The oxidants formed can either move to the surface of the fuel and react there or move to the redox front. The oxidants that react with the uranium oxide generate some soluble uranium species that also can move out to the redox front and react with the reductants there. This is shown in Figure 2. Some of the oxidised uranium may precipitate at the fuels surface and form new phases. This will happen if the transport rate is not large enough to carry away enough of the uranium to ensure that the concentration does not exceed the solubility limit. This is neglected in the present analysis.

The rate of movement of the front can be directly obtained from the rate of production of oxidising species balanced by the reaction with the reducing components in the surrounding backfill. This has recently been discussed and modelled by Romero [2].

The model is based on a mass balance accounting for the production of oxidants by radiolysis and their reaction with the fuel and other reducing species at the receding redox front. Details of the model, the mathematical formulation and the solutions are given in [3].

SAMPLE CALCULATIONS

In the first example the setting is as follows. 1 tonne of MOX fuel that has been deposited is accessed by water after 1000 years and all the fuel pellets are suddenly wetted by water. The canister has disappeared or corroded in such a way that its chemical and physical properties are the same as the surrounding bentonite backfill. Only 1% of the radiation reaches the water and of this 1 % gives rise to a net production of oxidants. Both these values can be quite realistic [4]. The radiation power is taken from Johnson et al [5] for MOX fuel. The reaction rate of oxygen with the UO_2 is based on the analysis of Bruno et al [6]. The maximum G-value is taken as an average of α, β and γ G-values which do not differ very much [7]. The bentonite clay has 200 equivalents of reductant per m^3. This is equivalent to 11 kg of ferrous iron per m^3 of compacted clay. This is a low but realistic value.

The power generated by radioactive decay is taken from [5]. For MOX fuel the power generated from decay is $p(t) = 57.26 * t^{-0.755}$ W/kg and for UOF fuel the power is 5 times lower. The reaction rate for oxygen with UO_2 is obtained from Gray et al. [8], $k_{UO2} = 2.45 * 10^{-10}$ mol UO_2 s^{-1} m^{-2} (mol $O_2/m^3)^{-0.5}$ This expression implies that the reaction rate is proportional to the oxygen concentration to the power 0.5. Grambow [9] states that at low oxygen concentrations the reaction rate may be a linear function of the oxygen concentration. The square root dependence gives more conservative results for low oxygen concentrations, i.e. higher dissolution rates, and is used in this paper. The data used in the calcualtions are given in Table 1. Other data for the computations are given in [3].

It may be noted that if all the radiation energy from 1 tonne of MOX fuel that reaches the water would generate oxidants, these would in theory be able to oxidise 70 tonnes of uranium oxide over a time period between one thousand and one million years. This also is equivalent to 29 tonnes of Fe(II) oxidised to Fe(III). With the fraction of the energy that is utilised for radiolysis, 0.01, these numbers reduce to 700 kg of uranium oxide and 290 kg of Fe(II) respectively.

The calculations show that the maximum oxygen concentration is 0.02 mol/m^3 at the fuels surface. This will occur after about 2000 years. The maximum fractional dissolution rate of the fuel then is $2.5 \cdot 10^{-5}$/a. After one million years nearly 70% of the MOX fuel could be oxidised in this case. More than 95 % of the oxidising species formed react with the fuel. The maximum uranium concentration at the fuels surface would be about 0.4 mol/m^3. The redox front could move out some meters into the clay after one million years.

Some of the assumptions are not very realistic. The most important assumption is that the canister "disappears" chemically. If an iron canister is used this will have a formidable redox buffering capacity and the redox front will not move very far. This is illustrated by using a concentration of reducing capacity in the surroundings of the fuel that is 100 times higher than that in the clay i.e. 20 000 equ/m^3. This implies that there is about 1100 kg Fe(II) per m^3 accessible corrosion products. The maximum fractional dissolution rate is reduced by about a factor four compared to the previous case. The redox front will move only 1 % of the distance in the previous case i.e. 0.026 m into the canister corrosion products. The cumulative fraction oxidised over one million years is about 50 % or 500 kg.

Table 1.   Other data used in the sample calculations

a	Specific surface of fuel	800	m^2/ m^3
a_m	Mass specific surface (wetted by water)	0.077	m^2/kg
$A_{canister}$	Cross section for transport into clay	10	m^2
d	Diameter of fuel pellet	0.005	m
D	Diffusivity of all species in bentonite*	10^{-10}	m^2/s
f_w	Fraction of radiation energy deposited in water	0.01	-
f_{ox}	Fraction of maximum G-value to produce oxidants	0.01	-
G_{max}	Maximum G-value for oxidising equivalents	3*10^{-7}equ/J	
k_{UO2}		2.45*10^{-10}	
		mol UO$_2$ s^{-1}m^{-2} (mol O$_2$/ m^3)$^{-0.5}$	
M_{UO2}	Molecular weight UO$_2$	0.27	kg/mol
M_{fuel}	Mass of UO$_2$ in a canister	1000	kg
r_{UO2}	Density of uranium dioxide	10 360	kg/m^3

*In the calculations it is assumed that the diffusivity of oxidising species and uranium species is equal.

This means that the fuel oxidation and ensuing release of incorporated nuclides is only slightly influenced by the presence of a steel canister.

The same calculations have been performed for UOF as for MOX fuel. The only difference in data is that the radiation power was decreased by a factor of five. As expected the fractional rate of oxidation was just slightly larger than one fifth of that for MOX fuel. About 150 kg of the fuel would be oxidised after one million years. The redox front would move half a metre.

BASIC MODEL FOR SOLUBLE REDUCING MINERAL

We now revisit the assumption that the reducing species is insoluble and assume instead that it has a solubility of c_{rs}. Figure 3 shows how the oxidant which has a concentration c_{O2} at the surface of the fuel diffuses down gradient to a point x_f where it meets counter diffusing reducing species.   Assuming that the oxidising and reducing species react instantaneously they cannot co-exist. *This is a key assumption and will be revisited later.* This assumption allows an interesting possibility namely that all produced oxidants can be scavenged under certain conditions.

The reducing species diffuses from the point x_{df} in the backfill to the right of which there still is reducing mineral. This dissolves to re-supply the water at the boundary with dissolved reducing species. The concentration profiles are straight because a pseudo stationary state has developed due to the very slow movement of the front at x_{df}.

This case is very interesting because if the supply of reducing species is faster than the generation of oxidising species by radiolysis then all the oxidising species can be scavenged and there will be no oxidation of the uranium oxide of the fuel. This is illustrated by the dotted line in Figure 3.

The solubility of ferrous iron in neutral water is on the order of 0.02 to 0.1 mol/m^3. [10]. It is found that if the reducing compound is to be supplied from the backfill it will not be possible to scavenge all oxygen whereas if the canister corrosion products supply the ferrous iron ions even a solubility of 0.01 mol/m^3 would suffice to scavenge all oxidants. Then there would be no fuel oxidation.

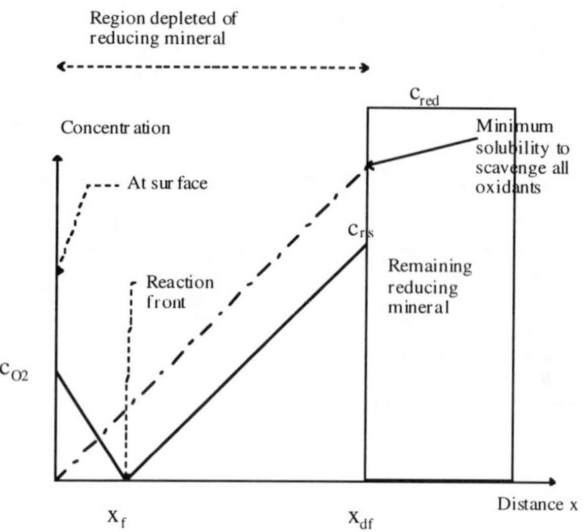

Figure 3   Oxidising species diffuse to the right to meet and react with reducing species. The latter dissolves to its solubility concentration c_{rs} at x_{df}.

REVISITING THE ASSUMPTION THAT THE REACTION RATE BETWEEN OXYGEN AND REDUCING SPECIES IS INSTANTANEOUS.

If the reaction rate is finite and slow the oxygen will not be scavenged fully. Below, the influence of a finite reaction rate is studied. Figure 4 shows the region between the fuels surface and the dissolution front. The concentrations in the non depleted region is also shown.

Concentration

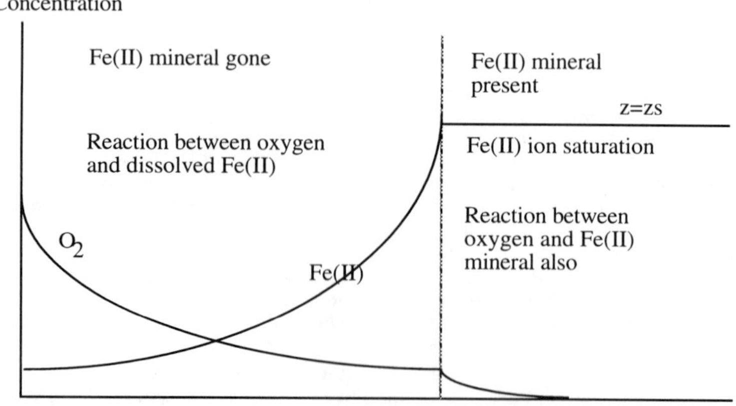

Figure 4.   Illustration of the oxygen and iron concentration

The reaction rate of oxygen and ferrous iron is strongly pH dependent [11]. It is here assumed that pH will be constant and near neutral.

This case can be mathematically formulated accounting for the simultaneous transport of oxidants and reducing species including their reactions in the aqueous phase as well as the reaction and dissolution rates of reducing and oxidizing species [3]. No analytical solution to the coupled differential equations with the boundary conditions has been found so far. However it is possible to study the impact of the reaction rate in the aqueous phase by simplified solutions.

The problem at hand is to assess if the reaction rate is so fast that the earlier solutions based on the assumptions that infinite reaction rate between oxygen and iron is reasonable. For this purpose it is sufficient to reformulate the problem and assume that the Fe(II) concentration is constant everywhere. This implies that the dissolution and transport rates of Fe(II) is very large and will thus show if the reaction rate can be assumed to be "infinitely" large. The dissolved ferrous iron concentration is taken to be 0.1 mol/m^3.

In the sample calculations the maximum oxygen concentration at the surface is found to be 0.002 mol/m^3. This is about 100 times lower than what it would be if there was no reaction with the iron and all produced oxygen would react with the uranium oxide. All oxygen is by no means scavenged as it would be if the reaction was instantaneous. Thousand years after the water comes in contact with the fuel the dissolution rate is on the order of $0.7 \cdot 10^{-5}$/a. This is much slower than for the case with insoluble reducing matter. It is by no means so slow that it can be neglected, however.

The rate coefficient for the oxygen-ferrous iron reaction is 1000 times lower at pH = 5 and 1000 higher at pH = 9. A decrease in pH will thus strongly increase the concentration and with it the oxidation rate of the uranium oxide. However, even a two to three order of magnitude increase in reaction rate is not sufficient to allow the scavenging of all oxidants.

The dissolution of the iron mineral and its transport rate must be large enough to supply the redox reaction with enough iron. It was found earlier (instantaneous reaction rate) that the transport rate is large enough for this when the canister corrosion products supply the reductants but not when they must come from the backfill.

DISCUSSION AND CONCLUSIONS

The model is very simple both conceptually and mathematically. Nevertheless it contains many important mechanisms for fuel dissolution caused by radiolytic action. There are many uncertainties such as simplified assumptions on geometry, uncertainties in data and uncertainties in reaction mechanisms. Provided the core of the model is acceptable the model can be extended to describe different geometries for the fuel and canister. Selecting a more correct geometry may not be very easy. Different assumptions could, however, be tested.

The rate of radiolysis is uncertain due to several reasons. The maximum G-value used is for pure water. It is known that dissolved constituents in water can strongly influence (decrease) the net rate of oxidant production. The data used for utilisation of the radiation energy for net radiolysis are based on field observations where, however, the influence of water chemistry could not be assessed.

The reaction rate coefficient is based on data for oxygen. Hydrogen peroxide is the most common oxidant formed. This is more reactive both with the fuel and the dissolved reductants. At present data are lacking. In addition, the experimental data indicate an uncertainty range of about two orders of magnitude.

It has been assumed that all the oxidised uranium dissolves and also moves out to the redox front. For the release of the other nuclides by congruent dissolution this will not matter because the possibly re-precipitated U(VI) minerals are unlikely to incorporate the other nuclides in the minerals formed. At least it would be difficult to prove this. If secondary uranium minerals precipitate at the fuel's surface, the redox front would not move so fast and the oxygen would be more easily scavenged, leaving less to oxidize the fuel. The same effect would be caused by allowing the diffusion to take place only through a small damage in the canister. Then cross section area for diffusion in relation to the reactive surface would become considerably smaller.

If the reducing mineral dissolves and diffuses towards the fuel to meet the oxygen this will reduce the oxidation rate. With a sufficiently high solubility and reaction rate all oxidant could be scavenged and no fuel oxidation could take place. This is an interesting possibility but it must be studied further. There are some assumptions that may not be entirely acceptable, one crucial assumption is that the reaction of dissolved reducing component and oxidant in solution can be deemed instantaneous. In the sample calculations the reaction rate between oxygen and ferrous iron was used. In real situation the oxidant is mostly hydrogen peroxide which is more reactive. The reductant also may be sulphide and organic matter which also have different rates. Furthermore microbial activity is likely to further the increase the reaction rates. All these factors may lead to a considerable increase in the rate of reaction between oxidant and reductant. Another uncertainty at present is in the dissolution rate of the reducing matter. This must be high enough to supply the water with the reductant that can move towards the oxidant.

ACKNOWLEDGEMENTS

The author gratefully acknowledges the stimulus and financial support from the Swiss National Cooperative for the Storage of Radioactive Waste, Nagra.

REFERENCES

1.   Cross J.E., Haworth A., Neretnieks I., Sharland S.M., Tweed C.J. Modelling of Redox Front and Uranium Movement in a Uranium Mine at Poços de Caldas. Radiochimica Acta 52/53 p 445-451, 1991
2.   Romero L. The near-field transport in a repository for high-level nuclear waste. Ph.D. Thesis, Dept. Chemical Engineering and Technology, Royal Institute of Technology, Stockholm, Sweden, 1995. TRITA-KET R21, ISSN 1104-3466
3.   Neretnieks I. Modelling oxidative dissolution of spent fuel. NAGRA internal report 1996.
4.   Liu Jinsong. Development and test of models in the natural analogue studies of the Cigar Lake uranium deposit. Ph.D. Thesis, Dept. Chemical Engineering and Technology, Royal Institute of Technology, Stockholm, Sweden, 1995. TRITA-KET R24, ISSN 1104-3466
5.   Johnson L.H., Shoesmith D.W., Tit J.C. Release of radionuclides from spent LWR UO_2 and mixed-oxide (MOX) fuel. Nagra Interner Bericht 94-40, AECL research, Whiteshell Laboratories, April 1994.
6.   Bruno J., Cera E., Duro L., Eriksen T.E., Werme L.O. A kinetic model for the stability of spent fuel under oxic conditions, J. Nucl. Mater., In press 1996
7   Choppin G., Rydberg J., Liljenzin J.O. Radiochemistry and Nuclear chemistry, 2nd Ed. Butterworth-Heinemann 1995
8.   Gray W.J., Leider. H.R., Steward S.A. Parametric study of LWR spent fuel dissolution kinetics, J. Nucl. Mater. 190, p 46-52, 1992
9.   Grambow B. Spent fuel dissolution and oxidation. An evaluation of literature data. SKB Technical Report 89-13, 1989
10.   Stumm W., Morgan J.J. Aquatic Chemistry, Wiley & Sons, 1981
11.   Morel M. M., Hering J. J Principles and applications of aquatic chemistry, Wiley, 1993

TRANSURANIUM ELEMENT INCORPORATION INTO THE β–U_3O_8 URANYL SHEET

M.L. MILLER*, P.C. BURNS*[1], R.J. FINCH**[2], AND R.C. EWING*
* Department of Earth & Planetary Sciences, University of New Mexico, Albuquerque, NM 87131, mlm@unm.edu
** Department of Geology, University of Manitoba, Winnipeg, Manitoba R3T 2N2, Canada.

ABSTRACT

Spent nuclear fuel (SNF) is unstable under oxidizing conditions. Although recent studies have determined the paragenetic sequence for uranium phases that result from the corrosion of SNF, there are only limited data on the potential of alteration phases for the incorporation of transuranium elements. The crystal chemical characteristics of transuranic elements (TUE) are to a certain extent similar to uranium; thus TUE incorporation into the sheets of uranyl oxide hydrate structures can be assessed by examination of the structural details of the β-U_3O_8 sheet type.

The sheets of uranyl polyhedra observed in the crystal structure of β-U_3O_8 also occur in the mineral billietite $(Ba[(UO_2)_3O_2(OH)_3]_2(H_2O)_4)$, where they alternate with α-U_3O_8 type sheets. Preliminary crystal structure determinations for the minerals ianthinite, $([U_2^{4+}(UO_2)_4O_6(OH)_4(H_2O)_4](H_2O)_5)$, and "wyartite II" (mineral name not approved by IMA committee on mineral names), $\{CaCO_3\}[U^{4+}(UO_2)_2O_3(OH)_2](H_2O)_4$, indicate that these phases also contain β-U_3O_8 type sheets. The β-U_3O_8 sheet anion topology contains triangular, rhombic, and pentagonal sites in the proportions 2:1:2. In all structures containing β-U_3O_8 type sheets, the triangular sites are vacant. The pentagonal sites are filled with $U^{6+}O_2$ forming pentagonal bipyramids. The rhombic dipyramids filling the rhombic sites contain $U^{6+}O_2$ in billietite, $U^{4+}O_2$ in β-U_3O_8, $U^{4+}(H_2O)_2$ in ianthinite, and $U^{4+}O_3$ in "wyartite-II" (in which one apical anion is replaced by two O atoms forming a shared edge with a carbonate triangle of the interlayer). Interlayer species include: H_2O (billietite, "wyartite II", and ianthinite), Ba^{2+} (billietite) Ca^{2+} ("wyartite II"), and CO_3^{2-} ("wyartite II"); there is no interlayer in β-U_3O_8. The similarity of known TUE coordination polyhedra with those of U suggests that the β-U_3O_8 sheet will accommodate TUE substitution coupled with variations in apical anion configuration and interlayer population providing the required charge balance.

INTRODUCTION

Spent nuclear fuel (SNF) is unstable under the oxidizing conditions present in the proposed repository at Yucca Mountain [1]. Spent nuclear fuel contains approximately ninety-five percent UO_2, one to three percent Pu, and up to four percent other actinides and fission products. Experimental corrosion studies on SNF have determined the paragenetic sequence of phases that will result when groundwater interacts with the fuel rods. Phases that have been observed include ianthinite, schoepite, becquerelite, compreignasite, soddyite, and uranophane [2,3,4,5,6].

---

[1] Current address: Dept. of Geology, University of Illinois at Urbana-Champaign, 245 Natural History Building, 1301 West Greet Street, Urbana, IL 61801

[2] Current address: Argonne National Laboratory, 9700 South Cass Ave., Argonne, IL 60439

Mat. Res. Soc. Symp. Proc. Vol. 465 © 1997 Materials Research Society

The fate of uranium is therefore understood; however, little is known about the disposition of the other actinide elements or the fission products.

Finch and Ewing [7] proposed that ianthinite might be an important host phase for Pu^{4+} incorporation. The potential for accommodation of transuranic element (TUE) within a uranyl sheet structure can be assessed by comparing the range of structural variations exhibited in other structures that are based on the same sheet anion topology with the coordination environments of transuranic elements in the few transuranic element phases for which structural descriptions exist. Further evidence can be obtained by examination of phases that contain sheets of uranium polyhedra that are based on related topologies.

SUMMARY OF COORDINATION ENVIRONMENTS

The ability of a crystal structure to accommodate substitutional impurities is assessed by comparison of the coordination environments of the substituted atoms within the host structure with the coordination environments of the substituting atom observed in crystal structures that normally contain them. The similarity of the substituting atoms coordination to that of the substituted atom must be weighed within the context of the host structures ability to accommodate structural and valence variations. The comparison of uranium coordination with those of transuranic elements in crystal structures known to contain them are summarized in this section; whereas, the ability of the rhombic site in the **R**-chain to accommodate structural and valence variations is demonstrated in the following section.

There are few crystal structures known to contain transuranic elements. Within the phases included in this study, there are seven symmetrically distinct coordination polyhedra containing Th^{4+}, three containing Np^{4+}, three containing Pu^{4+}, five containing Np^{5+}, two containing Pu^{5+}, and one containing Pu^{6+}. Th^{4+} is included in this study because is an important chemical analogue for transuranic elements, and it is present in SNF.

Pentavalent and hexavalent actinide elements tend to form approximately linear actinyl ions which are coordinated to four to six equatorial anions; whereas, tetravalent actinides most commonly occupy more symmetric coordination environments. In the discussion that follows, they are therefore addressed separately.

<u>Tetravalent cations</u>

Comparison of the bond lengths observed for the tetravalent cations in these phases with those of U^{4+} occurring within U^{6+} sheet structures are summarized in Table 1. The mean $<An\text{-}\phi>$ bond distances in Table 1 demonstrate that the mean $An\text{-}\phi$ distances for Th^{4+}, Np^{4+}, and Pu^{4+} are all within one standard deviation of

Table 1. Mean bond distances (nm) and standard deviations for tetravalent species with those of U^{4+} within U^{6+}+ phases. Data for TUEs summarized from [8].

Coordination Number	U^{4+}	Th^{4+}	Np^{4+}	Pu^{4+}
6	0.222(16)	0.237(4)	0.227(9)	0.223(1)
7	0.225(21)			
8		0.241(5)	0.232(2)	0.233(1)
9		0.249(12)		

the mean U-ϕ distance in six-fold coordination. The range of observed U-ϕ distances is 0.1964 to 0.1538 nm. If apical water is excluded, the maximum observed U-ϕ distance is 0.236 nm. It is therefore reasonable to expect that on the basis of cation radius, substitution of these tetravalent species for U^{4+} occupying the rhombic sites of **R**-chains is possible.

<u>Pentavalent and hexavalent cations</u>

Comparison of the mean actinyl bond distances, $<An-\phi_{An}>$, and mean equatorial bond distances, $<An-\phi_{eq}>$, with those of the uranyl ion are tabulated in Tables 2 and 3 respectively. Actinyl ion bond distances are nearly independent of equatorial coordination. Neptunyl bonds are approximately 0.005 nm longer and plutonyl bonds 0.01 nm longer than the corresponding uranyl bonds. In the sheets of uranyl polyhedra that occur within the structures, the oxygen atoms of the uranyl ions extend into the interlayer, so accommodation of the slightly longer actinyl ion bonds is not expected to cause significant structural distortion. The equatorial bonds are sensitive to coordination and to the cation valence. Equatorial bonds to pentavalent neptunium and plutonium are approximately 0.01 nm longer than those to uranyl ions; whereas, equatorial bonds to hexavalent plutonium are approximately 0.01 nm shorter. In every case, equatorial bonds to these transuranic elements are within one standard deviation of those to U^{6+}, and although five-coordinated plutonyl ions have not been reported, it is likely that such a coordination is possible. Furthermore, from the analysis of the coordination environments of these phases, it is probable that both neptunyl and plutonyl ions can substitute for uranyl ions within the sheets of uranyl polyhedra that occur within these phases. The question of whether the crystal structures of phases in the arrowhead group can accommodate variations in cation valence will be addressed in the following section.

Table 2. Mean actinyl ion $<An-O_{An}>$ bond distances (nm). Standard deviations with respect to the last reported decimal place are given in parentheses. Data for TUEs summarized from [8].

Actinyl Ion Bond Distances				
Coordination Number	U^{6+}	Np^{5+}	Pu^{5+}	Pu^{6+}
4	0.179(4)	0.184(0)		
5	0.179(4)	0.186(3)		
6	0.178(3)	0.184(1)	0.194(0)	0.191(0)

Table 3. Mean equatorial $<An-\phi_{\varepsilon\theta}>$ bond distances (nm) to actinyl ions. Standard deviations are given in parentheses. Data for TUEs summarized from [8].

Equatorial Bond Distances to Actinyl Ions				
Coordination Number	U^{6+}	Np^{5+}	Pu^{5+}	Pu^{6+}
4	0.228(5)	0.239(0)		
5	0.239(9)	0.245(4)		
6	0.247(12)	0.256(9)	0.255(0)	0.235(0)

STRUCTURAL DISCRIPTIONS

Of the phases that are known to result form the corrosion of SNF, all but soddyite are sheet structures belonging to the arrowhead group [9], thus they contain similar uranyl sites. Soddyite is a framework and will not be addressed further. Uranyl sheet structures contain uranyl polyhedra (rhombic, pentagonal, and hexagonal bipyramids) that share equatorial edges to form sheets. The sheets may be separated by an interlayer which usually contains water molecules and can also contain monovalent or divalent cations. A useful way with which to describe and compare sheet structures is to reduce the sheets of polyhedra to their underlying sheet anion topologies (SATs).

Sheet anion topologies

A sheet anion topology is derived from a sheet of polyhedra following diagrammed in Figure 1 [10,11]. The topology that results can be divided into a series of chains, each of which is made up of a periodic sequence of polyhedral cation sites. Each site within a topology can be populated with a cation and apical anions as required. In the uranyl phases discussed hear, the rhombic and pentagonal sites are usually populated with approximately linear uranyl ions. Hexagonal sites do not occur in the topologies under consideration here. Triangular sites are vacant in most of the corrosion products of uraninite and SNF; however, in kasolite, the triangular sites are populated with silicon in tetrahedral coordination and the rhombic sites are vacant. The fourth anion of the silicate tetrahedron in kasolite is apical to the sheet and not included in the SAT.

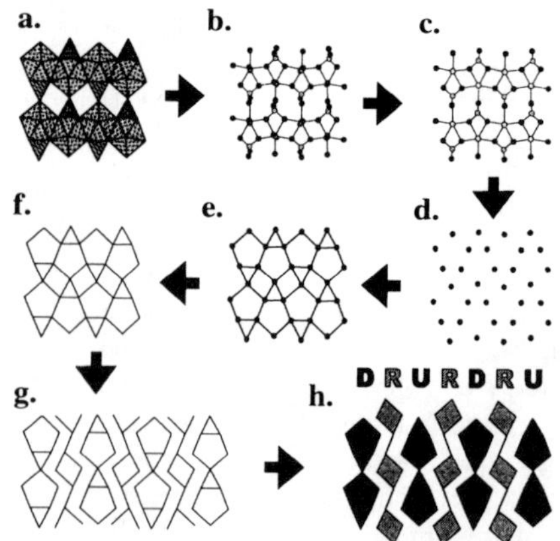

Figure 1. Derivation of the URDR SAT from the sheet of uranyl polyhedra present in the crystal structure of kasolite. a. Polyhedral representation of the sheet. b. Ball and stick representation of same. c. Removal of apical anions. d. Removal of cations. e. Connect anions <0.35 nm. f. Removal of anions yields the topology. g. Identification of chains present in the topology. h. Assignment of chain descriptors.

The arrowhead group

The sheet anion topologies of the phases in the arrowhead group contain arrowhead chains. Arrowhead chains are made up of a structural unit consisting of a pentagonal site sharing an edge with a triangular site, resulting in the arrowhead morphology. The arrowheads share corners to form a chain in which all arrowheads point in the same direction. Within a given sheet arrowhead chains can occur pointing in opposite directions. The chain descriptors **U** and **D** are given to upward and downward pointing arrowhead chains (Figures 1,2). Four other chain types are observed in members of the arrowhead group; only two of which occur in the phases being discussed here: the pentagonal **P**-chain and the rhombic **R**-chain (Figure 2).

The pentagonal **P**-chain is a chain of edge sharing pentagonal sites. Non-adjacent edges are shared such that every other pentagonal site has the same orientation. There are therefore two distinct pentagons, and the chain periodicity of the **P**-chain includes one pentagonal site of each orientation.

The rhombic **R**-chain consists of alternating rhombic sites and edges (Figures 1,2). Edges are not intuitively sites for cation population; however, within the context of a sheet anion topology, an edge can be populated by a cation in tetrahedral coordination if only two anions are shared with other cations of the sheet.

There are thirteen observed topologies in the arrowhead group, representing eighty-three phases. Two topologies representing two phases contain expansion chains that are unique to those phases and are not included in the discussion that follows. All remaining topologies within this group can be constructed from the **UD...** topology by **R**-chain and **P**-chain expansion as shown in Figure 2. All but three topologies, representing ten phases, contain **R**-chains, and five observed topologies contain single **R**-chains. It is the rhombic site of the single **R**-chain, and its ability to adapt to varying cation population that will be the primary focus of this paper.

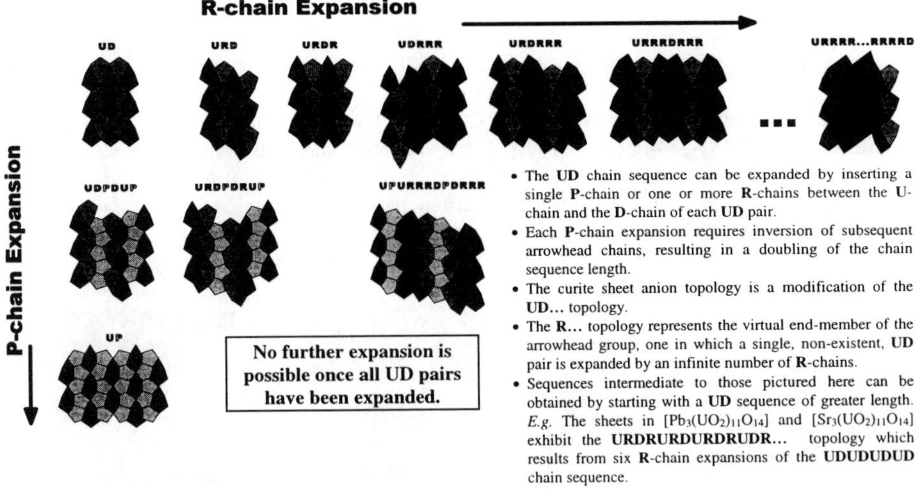

*Figure 2. Topologies resulting from **P**-chain and **R**-chain expansion of the **UD**... sheet anion topology.*

R-CHAIN STRUCTURAL VARIATIONS

The phases known to result from the oxidative corrosion of SNF and natural uraninite include the arrowhead group phases ianthinite, $[U^{4+}_2(U^{6+}O_2)_4O_6(OH)_4(H_2O)_4](H_2O)_5$ [12]; schoepite, $[(U^{6+}O_2)_8O_2(OH)_{12}](H_2O)_{12}$ [13]; becquerelite $Ca[(U^{6+}O_2)_3O_2(OH)_3]_2(H_2O)_8$ [14]; and kasolite, $Pb[(U^{6+}O_2)(SiO_4)](H_2O)$ [15]. These phases exhibit the **UDR...**, **UDPDUP...**, **UP...**, **UP...**, and **UDRD...** topologies, respectively.

The structures of schoepite and becquerelite contains no rhombic sites, so the incorporation of tetravalent transuranic elements is not probable. However, substitution of pentavalent and hexavalent actinyl ions for uranyl ions is possible. Substitution of pentavalent species will require a coupled substitution of OH for O in the sheet.

Rhombic chains occur in both ianthinite and kasolite; however, the rhombic sites are vacant in kasolite so transuranic element substitution is not expected to occur. Furthermore, accumulation of radiogenic lead to levels that will support formation of kasolite from the corrosion of SNF will require extended periods of time. Even then, the triangular sites, being populated with silicon will prevent interstitial substitution of transuranic element cations in the rhombic site without significant structural modification. No arrowhead group phase is known to occur in which both the triangular and rhombic sites are populated.

Ianthinite is the most probable host phase for tetravalent transuranic element substitution. As the preliminary structure refinement of ianthinite indicates that U^{4+} preferentially occupies the rhombic sites of this mineral, no structural modification will be required to account for valence differences if tetravalent transuranic elements substitute for U^{4+}. Bond valence analysis of the ianthinite structure data does indicate that some disorder of U^{4+} into the pentagonal sites occurs, with concomitant occupation of U^{6+} in the rhombic sites, so that it is possible for tetravalent transuranic elements, in addition to the pentavalent and hexavalent species, to occupy the pentagonal sites. Further evidence of this structure's ability to support transuranic element incorporation is provided by examination of other phases with structures based on the **UDR...** sheet anion topology.

Phases exhibiting the **UDR...** topology

In addition to ianthinite, wyartite-II, $\{CaCO_3\}[U^{4+}(U^{6+}O_2)_2O_3(OH)_2](H_2O)_4$; β-U_3O_8 [16]; and billietite, $Ba[(U^{6+}O_2)_3O_2(OH)_3]_2(H_2O)_4$ [11] also contain sheets of uranyl polyhedra that are based on the **UDR...** sheet anion topology (Figure 3). As their structural formulas suggest they are not, however, isostructural. The structural variations exhibited by these phases demonstrate the flexibility of the **UDR...** topology to accommodate structural modifications.

In β-U_3O_8 the apical anions of each sheet are shared with the polyhedra of the adjacent sheets; whereas, in the remaining phases, an interlayer separates each sheet from the adjacent sheets. The interlayers contain: H_2O in ianthinite, H_2O and Ba in billietite, H_2O, Ca, and CO_3 in wyartite-II.

The sheets in each of wyartite-II and β-U_3O_8, are symmetrically related. In contrast, the structures of ianthinite and billietite each contain two alternating symmetrically distinct sheets. Both of the sheets in ianthinite exhibit the **UDR...** topology. Whereas in billietite, sheets with the **UDR...** topology alternate with sheets exhibiting the **UP...** sheet anion topology of becquerelite, protasite, and α-U_3O_8. The interlayer of billietite has a net charge of +2, requiring that the sheet carries a compensating charge of -2. In ianthinite, the sheet is charge neutral. And in wyartite-II, the sheet is neutral if the CO_3 group is considered as part of the interlayer, but it has a charge of -2 if the CO_3 group, which shares an edge with the polyhedra occupying the rhombic site, is considered part of the sheet.

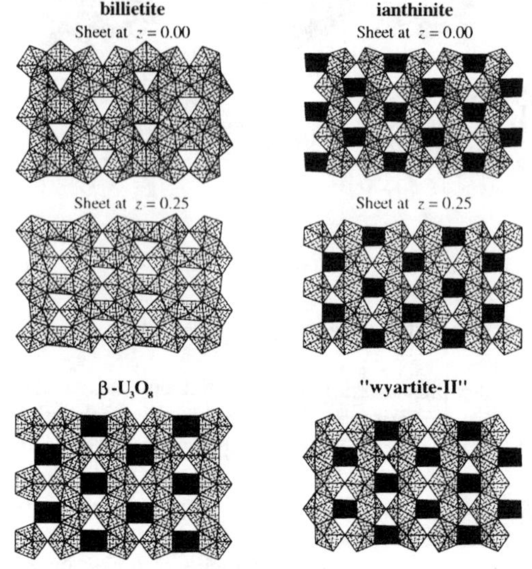

Figure 3. Polyhedral representations of the sheets of uranium polyhedra present in billietite, β-U_3O_8, ianthinite and wyartite-II. Light shading indicates sites populated with U^{6+}; U^{4+} populated sites are indicated with darker shading.

The variations in sheet charge are accomplished with variations in the valence state of the uranium in the rhombic site and with variations in the population of the anion sites (Figure 4). The rhombic sites of wyartite-II, β-U_3O_8, and ianthinite are populated with U^{4+}; whereas in billietite, this site is populated with U^{6+}. Both apical anions and equatorial anions can accommodate charge compensating population variations. The apical anions of the rhombic site are oxygen atoms in billietite and β-U_3O_8, but in ianthinite the apical anion of the rhombic sites are H_2O groups. In wyartite, the single rhombic site has one apical oxygen atom and the other apical position is populated by two oxygen atoms of a carbonate triangle that projects into the interlayer. The U^{4+} is therefore coordinated to seven oxygen atoms. The equatorial anions the of rhombic sites in each sheet in ianthinite are different. The U^{4+} occupying the rhombic U(4) site is coordinated to four equatorial oxygen atoms; whereas, the U^{4+} occupying the rhombic U(2) site of the other sheet is coordinated to two equatorial oxygen atoms and two equatorial hydroxyl groups.

From the structural variability exhibited by the four phases known to be based on the **UDR...** sheet anion topology, it is clear that structures based on this topology can accommodate significant structural modification without disturbing the sheet anion topology. And, as the apical anions

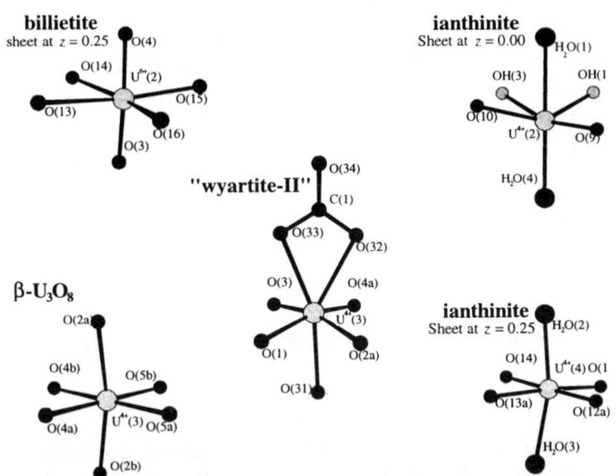

Figure 4. Coordination environments of the cations filling the rhombic sites in phases with UDR... topology.

of the sheet are only weakly bound to the interlayer water groups, it is therefore probable that modifications to individual rhombic sites can occur to accommodate substitution of a tetravalent transuranic element for uranium in the rhombic sites of ianthinite.

Phases exhibiting other arrowhead group topologies

Five of the topologies in the arrowhead group contain single **R**-chains like those in the **UDR...** topology. As these topologies also contain the **URD** (or equivalent **DRU**) chain sequence, examination of phases exhibiting these topologies can provide indication of additional substitution mechanisms that may occur in phases with the **UDR...** topology. Cu, Nb, and Te are observed to occupy the rhombic sites of [(UO$_2$)(CuO$_4$)] [17], K[(UO$_2$)(NbO$_4$)] [18], and [(UO$_2$)(TeO$_3$)] [19], respectively. All of these phases adopt structures that are based on the **URDR...** topology, which is also representative of the sheets in K$_2$[(UO$_2$)$_2$O$_3$] [20] in which the rhombic sites are populated with uranium. Likewise, the pentagonal sites of arrowhead chains are populated with Pb in [Pb$_3$(UO$_2$)$_{11}$O$_{14}$] [21], and Sr in [Sr$_3$(UO$_2$)$_{11}$O$_{14}$] [22], both based on the **URDRURDURDRUDR...** topology; and with Ca in ulrichite, Cu[Ca(UO$_2$)(PO$_4$)$_2$](H$_2$O)$_4$ [23], and with Na in Na$_{5.5}$[(UO$_2$)$_3$(H$_{0.5}$PO$_4$)(PO$_4$)$_3$] [24], both of which are based on the **UDRD...** topology. As these structures contain chain sequence fragments that are the same as those in structures exhibiting the **UDR...** topology, it is reasonable to expect that ianthinite can accommodate a significant amount of impurity cation substitution within the sheets of its structure.

SUMMARY

The structural similarity of the coordination environments of transuranic elements with those of uranium in the phases of the arrowhead group suggests that transuranic element incorporation into the sheets of these phases is likely to occur. Pentavalent and hexavalent transuranic elements will most likely substitute for U^{6+} in pentagonal sites; whereas, tetravalent species are more likely to substitute for uranium (U^{6+} or U^{4+}) in the rhombic sites. Structural modifications required to accommodate local variations in cation charge or size due to cation substitution are within the observed range of structural variations in the structures that contain sheets adopting similar sheet anion topologies. Mechanisms for accommodating cation charge differences include: $O \leftrightarrow OH$ substitution within the sheet anions (equatorial anions), $O \leftrightarrow H_2O$ substitution at the apical anion sites, increased cation coordination at the apical position as observed in wyartite-II ($O \leftrightarrow CO_3$ substitution), and the introduction of a compensating charged species into the interlayer.

ACKNOWLEDGMENTS

This work was supported by the Office of Basic Energy Sciences under Grant No. DE-FG03-95ER14540 awarded to MLM and RCE.

REFERENCES

[1]	W.M. Murphy and R.T. Pabalan, Center for Nucl. Waste Reg. Anal. Rept. **95-014**, 21p. (1995).

[2]	L.H. Johnson and L.O. Werme, *Mater, Res. Soc. Bull.* XIX(12) p. 24 (1994).

[3]	R.S. Forsyth and L.O. Werme, *J. Nucl. Mater.* **190** p. 3 (1992).

[4]	D.J. Wronkiewicz, J.K. Bates, T.J. Gerding, E. Veleckis and B.S. Tani, *J. Nucl. Matr.* **190** p. 107 (1992).

[5]	S. Stroes-Gascoyne, L.J. Johnson, P.A. Beeley, and D.M. Sellinger, In: Scientific Basis for Nuclear Waste Management IX, edited by L.O. Werme (Mater. Res. Soc. Proc. 50, Pittsburgh, PA 1985), p. 317.

[6]	R. Wang and J.B. Katayama, Nucl. *Chem. Waste Management* **3** p. 83 (1982).

[7]	R.J. Finch and R.C. Ewing, In: Scientific Basis for Nuclear Waste Management XII, edited by A. Barkatt and R.A von Konynenburg (Mater. Res. Soc. Proc. 353, Pittsburgh, PA 1993), p. 633.

[8]	P.C. Burns, R.C. Ewing, and M.L. Miller, *J. Nucl. Mater.* (in press).

[9]	M.L. Miller and R.C. Ewing, (in prep)

[10]	M.L. Miller, R.J. Finch, P.C. Burns, and R.C. Ewing, *J. Mater. Res.*, **11**, p. 3048.

[11]	P.C. Burns, M.L. Miller, and R.C. Ewing, *Can. Min.*, **34**, p. 845 (1996).

[12]	P.C. Burns, R..J. Finch, F.C. Hawthorne, M.L. Miller, and R.C. Ewing, In: Scientific Basis for Nuclear Waste Management XX, edited by W.J. Gray and I.R. Triay (Mater. Res. Soc. Proc., this volume, 1996).

[13]	R.J. Finch, M.A. Cooper, F.C. Hawthorne, and R.C. Ewing, *Can. Mineral.* **34** (1996), (in press).

[14]	M.K. Pagoaga, D.E. Appleman, and J.M. Stewart, *Am. Mineral.* **72**, p.1230 (1987).

[15]	A. Rosenzweib and R.R. Ryan, *Cryst. Struct. Commun.* **6**, p.617 (1977).

[16]	B.O. Loopstra, *Acta. Crystallogr.* **B26**, p. 656 (1970).

[17]	P.G. Dickens, G.P. Stuttard, and S. Patat, *J. Mater. Chem.* **3**, p. 339 (1993).

[18]	M. Gasperin, *J. Solid State Chem.* **67**, p.219 (1987).

[19]	G. Meunier and J. Galy, *Acta Crystallog.* **B29**, p. 1251 (1973).

[20]	M.C. Saine, *J. Less-Common Metals* **154**, p. 361 (1989).

[21]	D.J.W. Ijdo, *Acta Crystallogr.* **C49**, p. 654 (1993).

[22]	E.H.P. Cordfunke, P. Van Vlaandersen, M. Onink, and D.J.W. Ijdo, *J. Solid State Chem.* **94**, p. 12 (1991).

[23]	W.D. Birch, W.G. Mumme, and E.R. Segnit, *Aust. Mineral.* **3**, p. 125 (1988).

[24]	Yu.E. Gorbunova, S.A. Linde, A.V. Lavrov, and A.B. Pobedina, *Dokl. Akad.Nauk SSSR* **251**, p. 385 (1980).

Part VII

Waste Processing and Treatment

METHOD OF TREATMENT OF RADIOACTIVE SILTS AND SOILS

A.P. VARLAKOV, I.A. SOBOLEV, A.S. BARINOV, S.A.DMITRIEV,
S.V. KARLIN, V.U. FLIT

Moscow Scientific-Industrial Association RADON , 7 Rostovskij 2/14, Moscow,
119121 , Russia

To solve the problem of silts and soils that are contaminated with radioactive and toxic substances, the following method has been developed at SIA RADON. The material is mixed with limestone and other components, including up to 70 % (mass) of dried residue of liquid radioactive waste. The mixture is heated at 800 to 1000 ºC, shredded, and used to form cement. This cementation process may be used to treat radioactive or other chemical waste.

The work demonstrated that in the case of silts, for example, the product volume is reduced by a factor of 1.5 to 3 compared to the initial silts volume. A fast hardening, durable product is obtained, the quality of which is not inferior to that obtained by using the traditional binders. For some parameters, e.g. , hardening rate and macroelement leaching, the current method is significantly better than the traditional cementation process. It has also been demonstrated that, over a wide range of parameters, when dry residue of liquid radioactive waste is used in the initial mixture, the part of NO_x can be reduced to nitrogen, thereby reducing NO_x emissions.

INTRODUCTION

At present, resolving the problem of silts and soils contaminated with radionuclides or toxic substances is an urgent worldwide need. A number of methods for solving this problem are being studied. Treatment of silts and soils to prevent release and transport of the dangerous substances in the environment, and to allow safe storage of the product, is considered the simplest solution of the problem. For this purpose, cementation method is proposed, i.e. , embedding the radioactive materials in cement block. However , in comparison with the initial volume of material treated, the existing cementation method increases the final product volume [1]. In addition, organic components, including oil products and products from biologic activity, including flora and fauna, destroy the hardened cement matrix.

The Moscow Scientific-Industrial Association (SIA) RADON disposes of mixtures of solid and liquid radioactive waste, containing silts and soils, in cement monoliths in near-surface concrete bunkers [2]. In our experience, the silts and soils form loose layers that include pores that are not penetrated by the cement mortar. A breach of the repository hydro-barrier would result in penetration of surface water into the cement monolith, and washout of the matrix. Hence Moscow SIA RADON is developing methods of conditioning and treating the solid radioactive waste in small containers and barrels to allow homogeneous mixing of the waste in cement monolith. The containers provide an additional isolation barrier during storage and are feasible containers for transporting the treated waste.

Moscow SIA RADON is developing the "Clinker" method to treat radioactive silts and soils [3]. In comparison with traditional cementation, the "Clinker" method reduces the final volume and enhance the strength characteristics of the final product while essentially eliminating problematic organic constituents. The range of applicability of the "Clinker" method is increased by the possibility of treating

Mat. Res. Soc. Symp. Proc. Vol. 465 © 1997 Materials Research Society

materials containing up to 80% (mass) of organic materials, such as turf , flora and fauna decomposition products, and manmade material, including natural materials, such as petroleum products and polymers. In addition, the "Clinker" method does not require expensive waste binders, i.e. , cement.

EXPERIMENT

Radioactive silt or soil that contains organics of natural and manmade origin is mixed with lime and other components, including up to 70% (mass) of dry liquid radioactive waste residue. This mixture is calcined at 800 to 1000 °C. The product is ground to a surface area size of 2500 to 4500 cm^2/g, mixed with water at a water-to-cement ratio not less than 0.25 , and aged to form a solid monolith.

The advantages of the method results from burning the organics in the initial heating stage to form (a) removable volatile compounds and (b) - after further heating and grinding - an inorganic cementitious material that can be mixed with water, and allowed to harden to form a cement block .

In this investigation, the "Clinker" method was compared to the traditional cementation method. Natural silt and soil with the following characteristics were used:

- After natural (gravity) filtration, the water content in the silt was 650 to 700 g/kg.

- After vacuum filtration using a Buchner cone, the water content in silt was 500 to 600 g/kg.

- Losses during calcination of dried silt at 1000 °C were 200 to 400 g/kg.

- Losses during calcination of soil were 100 to 300 g/kg.

- Silt radioactivity in Bq/kg, was: $\Sigma\beta$ Cs^{137} = 1.8 x 10^4 and $\Sigma\alpha$ Pu^{239} = 9,6 x 10^3.

- Soil radioactivity in Bq/kg, was: $\Sigma\beta$ Cs^{137} = 1,9 x 10^5 and $\Sigma\alpha$ Pu^{239} = 5,3 x 10^2.

- The chemical composition of the silt and soil was as given in Table I .

Table I.   Chemical composition of silt and soil.

Material	CaO	SiO$_2$	% (Mass) Al$_2$O$_3$	Fe$_2$O$_3$	Na$_2$O	K$_2$O
Soil	14-18	49-57	14-17	8-10	1-2	1-2
Silt	9-11	59-66	12-15	7-10	1.5-2.0	2.5-3.0

Two pretreatments were compared for silt :

- Natural (gravity) filtration (1).

- Drying at 120 to 130 °C then grinding to a particle size no greater than 0.5 mm (2).

For the "Clinker" method, the required amount of silt or soil was mixed with chemically clean $CaCO_3$ and the other components. The mixture was then calcined at 800 to 1000 °C. Once the desired temperature was achieved, the mixture was held at that temperature for 0.5 to 2 hours (isothermal time lag). The product was ground as previously stated and mixed with water at a water-to-cement ratio of 0.3. Samples for determining the cement block strength were prepared from the resulting cement mortar.

For the traditional cementation method, the required amount of silt or soil was mixed with a cement and water at various a water-to-cement ratio. Samples for determining the cement block strength were prepared from the resulting cement mortar.

RESULTS

Results of the investigation are given in Tables II though IV.

Table II.   Results of silt treated with the traditional cementation method.

Silt Number	Dry Silt Loading in Final Product % (mass)	Water-to-Cement Ratio (mass)	Flowability mm	Compressive Strength, 28th Day of Hardening MPa	Volume Increase Factor*
1	20	0.7	>220	10.56	2.3
1	30	2.22	>220	2.52	1.8
1	35	6.52	>220	0.5	4.1
2	10	0.4	<64	46.5	3.7
2	20	0.55	<64	35.8	2.1
2	30	0.7	<64	8.1	1.5

* Ratio of final product volume to the volume of silt after vacuum filtration.

Table III.   Results of soil treated with the traditional cementation method.

Soil Loading in Final Product % (mass)	Water-to-Cement Ratio (mass)	Flowability mm	Compressive Strength, 28th Day of Hardening MPa	Volume Increase Factor
20	0.7	160	20.5	2.5
30	0.95	140	12.5	1.9
35	1.3	125	9.5	1.7
40	1.5	125	4.5	1.5

Table IV.   Results of silt and soil treatment with the "Clinker" method.

Material	Amount of Dry Material in Initial Mixture % (mass)	Compressive Strength, 28th Day of Hardening MPa	Volume Reduction Factor
soil	50	37	1.2
silt	50	30	1.4
silt	50	28	1.5
soil	60	24	1.4
silt	60	21	1.8
silt	70	19	2.2

As the results show, traditional cementation of silt and soil increases the product volume in comparison with the initial volume of material, the mass loading of the silt and soil and the compressive strength of the cement block product is, as a rule, not high. To improve these characteristics, the silt must be first dried at 100 °C and ground to allow homogeneous distribution of the silts in the cement compound.

Table V compares the results of the traditional cementation method and the "Clinker" method.

Table V. Comparison of results of treatment of silts and soils by the traditional cementation method with the results of treatment by the "Clinker" method.

Method	Compressive Strength MPa	Loading (dry silt or soil) % (mass)	Change in Volume of Wet Silt (Soil) By Treatment	Possible Organic Content in Silt (Soil) % (mass)	Change in Matrix Strength in 1-2 Years
"Clinker"	25 - 50	Up to 80	Reduced by 1.2-6	Up to 80	Hardening
Traditional Cementation	10 - 30	Up to 25	Increased by 1.5-5	Up to 15	Weakening

Measured radionuclide release during the calcination under laboratory conditions was less than 0,5 % of $\Sigma\beta$ for Cs^{137}. No releases were detected based on $\Sigma\alpha$ for Pu^{239}. When treated mixture contained liquid radioactive waste dry residue containing $NaNO_3$, it was noted that calcination of the organic components caused up to 60 to 85 % of the nitric oxides to be reduced to elementary nitrogen.

CONCLUSIONS

The "Clinker" method for treating radioactive silts and soils results in enhanced compressive strength of the cement matrix, reducing the final product volume compared with the volume of material treated. In some cases the volume reduction was a factor of 2 to 5 for silts (soils) with a high organic content (40 to 80 % by mass).

The refinement of the "Clinker" method is not complete. More detailed study of the processes of clinker formation, hydration, syntheses regimes, and cement compound quality is required.

REFERENCES

1. A. S. Nikiforov, V. V. Kulichenko, M. I. Jiharev, The Treatment of Liquid Radioactive Wastes, Energoatomizdat, Moscow, 1985, pp. 130-132.
2. I. A. Sobolev, L. M. Homchik , Management of Radioactive Waste on Centralized Sites, Energoatomizdat, Moscow, 1983, pp. 36-38.
3. I. A. Sobolev, S. A. Dmitriev, A. S. Barinov, A. P. Varlakov, S. V. Karlin, V. G. Savel'ev, A.V. Abakumov, Atom. Energ. 5 , 312 (1995).

RADIATION STABILITY OF CROWN ETHER USED FOR DECONTAMINATION OF RADIONUCLIDES FROM HLW

Vyacheslav M. ABASHKIN, Gennadiy F. EGOROV, Evgeniy FILIPPOV,
Anna K. NARDOVA, and Igor V. MAMAKIN
Russian Research Institute of Chemical Technology, 33 Kashirskoe Ave., Moscow 115230,
vs@khimcon.msk.ru

ABSTRACT

The gamma-ray initiated oxidation of dicyclohexano-18-crown-6 (DCH-6) was studied
in two-phase system under agitation: nitric acid, aqueous solution and organic solvent.

The distribution coefficients of radionuclides, hydrodynamic characteristics (phase
separation and interface surface tension), and the radiolysis products after irradiation at a dose
rate (^{60}Co) of 1.7 Gy/s were determined. The maximum total dose was 3.0×10^5 Gy
(3.0×10^7R).

INTRODUCTION

Crown ethers - stereospecific macrocyclic complexones - exhibited high selectivity for
certain cations and demonstrated good applicability for decontamination of high-level waste
(HLW) from long-lived radionuclides: Strontium-90 and Cesium-137.

Dicyclohexano-18-crown-6 (DCH-6) and its derivatives are the most promising agents in
reprocessing of radwaste solutions for decontamination from Sr [1]. As a rule, the extractant
used for radwaste decontamination is exposed to as much as 50 kGy per extraction-stripping
cycle depending on the extraction regime and equipment. For a cyclic scheme, the total
radiation doses may reach tens of MGy [2]. Usual consequences of radiolysis are:
degradation of the extractant, drop in the recovery efficiency, formation of destruction
products, reduction of the selectivity, and bad phase separation.

Unfortunately, the limited availability of crown ethers and especially a lack of experimental
data for reprocessing of highly irradiated and concentrated aqueous solutions using crown ethers
as extractants have somewhat inhibited the evaluation and recognition of crown ethers as new
agents in the reprocessing of industrial nuclear fuel and radioactive waste.

Nevertheless, the possible application of crown ethers for recovery of radionuclides [Sr, Cs,
transuranium elements (TRU] from real radioactive waste solutions has been reported. The
most promising properties belong to dicyclohexano-18-crown-6 and its derivatives, e.g. solvent
extraction of ^{90}Sr and ^{137}Cs from nitric acid solutions [3,4].

The following properties of the extraction mixture were evaluated during irradiation:
distribution coefficients of radionuclides; their retention in organic phase; hydrodynamic
properties - surface tension, phase demixing rate; and accumulation of radiolysis products.

EXPERIMENTAL

For this study we irradiated an aqueous/organic (1/2 vol.) mixture. The aqueous solution
was 3M HNO_3, and organic phase was 0.1 M solution of DCH-6 in the mixture of fluorinated
alcohol (FA) and isododecyl alcohol (IDA) - 3:2. This two-phase system was gamma-

Mat. Res. Soc. Symp. Proc. Vol. 465 © 1997 Materials Research Society

irradiated by means of ^{60}Co in special stainless steel containers provided with temperature control, continuous shaking (rotation velocity of the shaker was 2500 turns per minute), and intensive bubbling of gaseous oxygen (30-40 ml/min.).

The dose rate, determined by means of a ferrosulfate dosimeter, was 1.7×10^3 W/l (1.7 Gy/s) corrected for the electronic density of organic solution. The maximum total dose of gamma-irradiation was 80 W*h/l (2.88 Gy). The specimens of irradiated organic solution after phase separation were analyzed to determine residual extraction efficiency and hydrodynamic parameters. Both organic and aqueous phase were analyzed to estimate the decay of extractant and composition of radiolysis products, using UV-spectrometry.

RESULTS

EXTRACTION EFFICIENCY

Every time after irradiation and phase separation, the efficiency of the irradiated extractant was tested in extraction of radionuclides (Strontium-85, Cesium-137, Ruthenium-106, and Cerium-144) from model nitric acid (3 M) solution at volume ratio, $W_{org}/W_{aq} = 1/2$. Calculated distribution coefficients ($\alpha = [M]_{org}/[M]_{aq}$) for different irradiation doses are shown in the Figure 1. The extraction efficiency of the DCH-6 solution changed only very slightly with larger absorbed dose. The separation factors of Sr from other radionuclides were quite stable throughout the dose range, except for Ce.

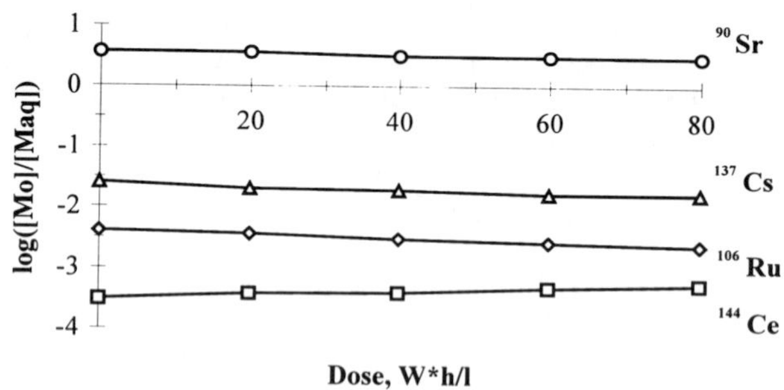

Figure 1. Influence of gamma-irradiation dose on distribution coefficients of radionuclides in two-phase system: 0.1 M DCH-6 + 60% FA + 40% IDA || 3.0 M HNO$_3$ (T = 40°C)

The test extraction was followed by double washing-up of the saturated extractant by diluted nitric acid solution (1-3 g/l) - at $W_{org}/W_{aq} = 3:1$ and T = 20° C, and three-pass stripping by deionized water - $W_{org}/W_{aq} = 2:1$, T = 40°C. Phase mixing and lagging took 5 and 10 minutes, correspondingly.

Table 1 shows the residual percentage of radionuclides in the organic phase for different irradiation doses. The retention factor was calculated as a ratio of the resultant radionuclide concentration to its content in organic solution after test extraction.

Table 1. Retention of radionuclides in the organic phase after different γ -radiation doses

Dose, W*h/l	After washing-up, %				After stripping, %			
	Sr	Cs	Ru	Ce	Sr	Cs	Ru	Ce
0	56	5	3	26	1	<0.1	<0.1	<0.1
20	57	10	5	25	2	<0.1	<0.1	<0.1
40	57	14	7	25	3	<0.1	<0.1	<0.1
60	60	23	11	23	4	<0.1	<0.1	<0.1
80	63	27	19	22	5	<0.1	<0.1	<0.1

The most interesting results were obtained for strontium. Its retention rate ranged from 56% to 63% while washing-up. Water-stripping left 1 to 5% of the strontium in the organic phase, depending on the absorbed dose. However, two additional passes of water back extraction allows to reach residual Sr content less than 0.5% even at higher dose. All traces of Cs, Ru, and Ce were efficiently removed from the organic phase for three passes of water stripping.

Hydrodynamic Effects

The following hydrodynamic characteristics of irradiated DCH-6 solutions were evaluated: interfacial tension (IT), and a rate of demixing (RD). The IT was determined by a stalagmometer using capillary technique. Aqueous nitric acid solutions were used as the dispersive medium and the extractant as the dispersed phase. The hydrophilic surface capillarity prevents a spread of an organic drop on the end of glass capillary. The IT was calculated as:

$$\Sigma = V * \Delta\rho * g / 2\,\pi * r * n,$$

where "v" - volume of "n" drops; $\Delta\rho$ -density difference of organic and aqueous phase; "g" - acceleration of gravity; and "r" - capillary radius. The demixing rate was measured in a special glass mixer with inner diameter of 2.5 cm. The volume of every phase was 10 ml; time of intermixing was 2 minutes.

The effect of irradiation on RD for different aqueous solutions is shown on Figure 2. RD were practically unaffected when the irradiated extractant contacted the concentrated (3 M) nitric acid solution. Perhaps the acidic radiolysis adducts have low interfacial activity. For dilute HNO_3 solution, RD was decreased by irradiation, and there is sharp drop of RD just for the first step of gamma-irradiation (up to 20 W*h/l).

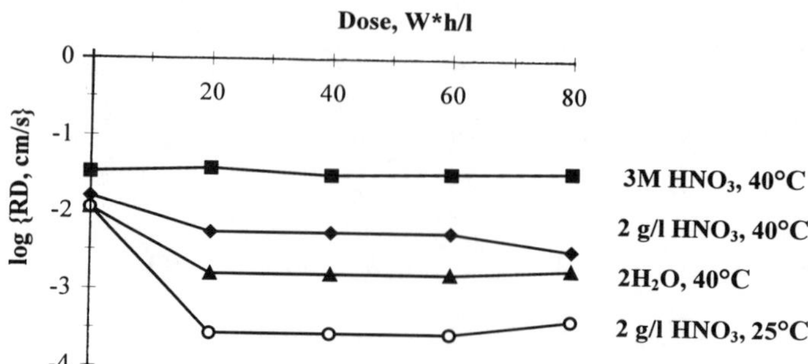

Figure 2. Effect of irradiation on a rate of phase demixing (RD)

It is interesting to note that the temperature decrease, from 40°C to 25°C, causes almost the same depression of RD as irradiation. However, RD values for nonirradated extractant are very close in all cases.

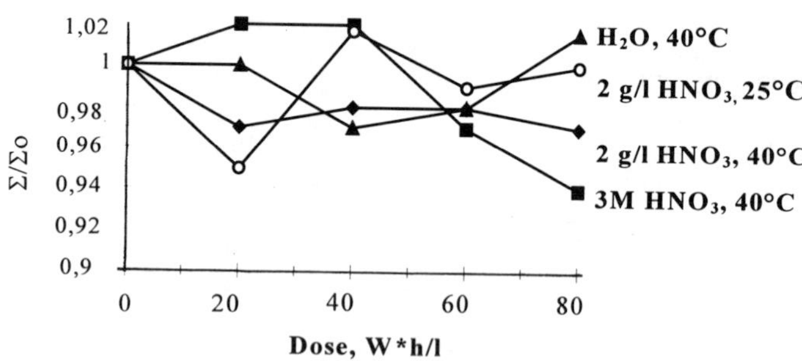

Figure 3. Effect of irradiation on interfacial tension of extractant:
0.1 M DCH-6 + 60% FA + 40% IDA

A comparison of IT as a function of irradiation is given in Figure 3. There is practically the same IT value for all types of aqueous solutions and almost no dependence on the total dose. A lack of simple correlation between RD and IT may be caused by different factors discussed earlier [2].

The decomposition of DCH-6 in the organic solution on irradiation was measured spectrophotometrically using a technique described in [5]. An analysis of infrared spectra of irradiated extractant in the range 1500 through 2000 cm^{-1} revealed a formation of carbonyl derivatives (1730 cm^{-1}), organic nitrates (1640 cm^{-1}) and nitrites (1680 cm^{-1}) produced as a result of oxidation and nitration of the diluent, higher alcohol. Radiation yields for these adducts - measured using appropriate kinetic diagrams are: 13.4 (R=CO), 1.5 (R-ONO_2), and 1.7 (R=NO_2) molecules per 100 eV.

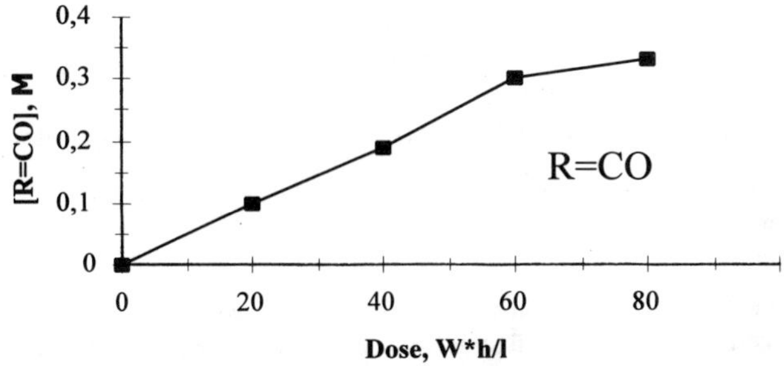

Figure 4. Formation of carbonyl adducts as a result of irradiation of the extractant

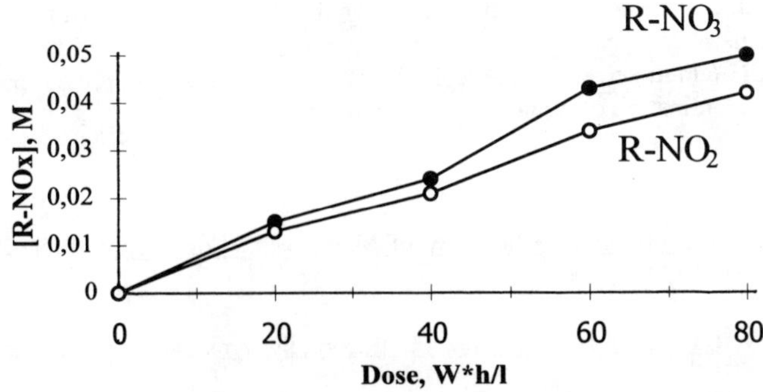

Figure 5. Formation of nitrate and nitrite adducts as a result of irradiation of the extractant

Special attention was paid to an accumulation of fluoride-ions in the aqueous phase as a result of radiolytic decay of the solvent (FA) given in Figure 6. The concentration of F$^-$ increased up to a total dose 60 W*h/l and then slightly decreased with increased dose. A possible reason for the observation of this maximum is the interaction of fluoride-ions with nitric acid and a formation of volatile or extractable compounds. Initial radiation yield of F$^-$

is 5.2 ions per 100 eV, while the highest concentration of the fluoride is 0.14 g*ion per liter - below the permissible corrosion-proof level of the standard extractor.

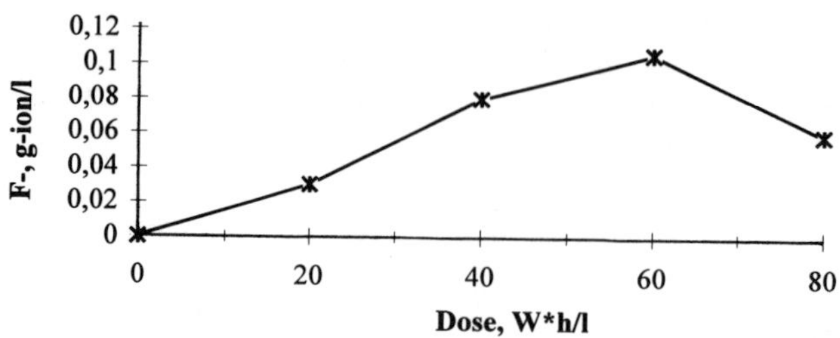

Figure 6. Accumulation of fluoride-ions in aqueous phase as a result of gamma-irradiation

CONCLUSIONS

The subject extractant, a solution of DCH-6 in a mixture of higher and fluorinated alcohols, demonstrated high chemical stability during gamma-irradiation in a two-phase system with nitric acid solution. The distribution coefficient for ^{90}Sr and its separation factors from other radionuclides were high and stable values over the range of absorbed dose studied (at least up to 80 W*h/l).

Irradiation caused a high yield of carbonyl adducts from oxidation of higher alcohol and fluoride-ion - due to the decay of fluorinated alcohol. There was no loss of the crown ether after irradiation.

Thus, DCH-6 in mixed solvent (FA+IDA) may be recommended in the reprocessing of HLW for decontamination of liquid radwaste from Strontium-90.

REFERENCES

1. G.A. Melson, <u>Coordination Chemistry of Macrocyclic Compounds,</u> Plenum Press, New York and London, 1979.

2. G.V. Egorov, <u>Radiochemistry of Solvent Extraction Systems</u> Energoizdat, Moscow, 1986.

3. E.P. Horwitz , M.L. Dietz, and D.E. Fisher , <u>SREX: A new process for the extraction and recovery of Strontium from acidic nuclear waste streams.</u> *Solvent Extr. Ion Exch.*, **8**, 199 (1990).

4. Y. A. Filippov , E.A. Dzekun , A.K. Nardova *et al.* (1992) <u>Application of crown ethers and ferrocyanide based inorganic material for cesium and strontium recovery.</u> In *Proceedings of the Annual Symposium on Waste Management at Tucson AZ*, **2**, 1021(1992).

5. I.V. Pyatnitski <u>Chemistry and Technology of Water</u> (Rus.). Naukova Dumka: Kiev, 1983, **5** , p.224.

NEW SORBENTS AND TECHNOLOGIES FOR LIQUID RADIOACTIVE WASTE MANAGEMENT

D.V. MARININ*, V.A. AVRAMENKO*, V.YU. GLUSHCHENKO*, V.I. SERGIENKO**,
V.A. VASILEVSKIY*, V.V. ZHELEZNOV*
*Laboratory of Sorption Processes, Institute of Chemistry of the Far East Department, Russian
Academy of Sciences, Vladivostok, Russia
**Vice-President, Far East Department, Russian Academy of Sciences

ABSTRACT

The possibilities for using various technologies for liquid radioactive waste (LRW)
management were considered. Technologies based on the application of sorbents highly selective
to radionuclides were shown to have good prospects. Possible ways of increasing selective
sorbent efficiency in cleaning radionuclide-polluted waters of various types were examined. The
results of studying new high-selective sorbents produced in the Institute of Chemistry FEDRAS
are given. The use of these sorbents enables one to treat LRW of various types in a single stage.
The results of testing a sorption filter installation for management of LRW produced during
operation, repair and disposal of nuclear reactors are presented. Use of this installation resolves a
major portion of LRW disposal problems.

INTRODUCTION

Liquid radioactive wastes (LRW) produced as a result of operation, repair and disposal of
nuclear energy installations can be classified in different ways.

First, the source of LRW can be considered as a criterion. In this case one should mention
decontamination LRW; LRW from waste nuclear fuel (WNF) storage basins etc.

All these LRW types differ in chemical and radiochemical composition. The radiochemical
composition changes significantly over time because of decay of short-lived radionuclides. In this
process the fraction of long-lived Cs and Sr isotopes increases continuously. Regarding chemical
composition, LRW are divided into low-salt, salt content up to 20 mg/L (e.g., LRW from WNF
storage basins), and high-salt LRW, salt content higher than 20 mg/L (e.g., decontamination
LRW).

Concerning activity values, LRW are divided into high-active (higher than 10 Ci/L); middle-
active (from 10 down to 10^{-2} Ci/L); low-active (from 10^{-2} down to 10^{-5} Ci/L) and very low-active
(from 10^{-6} down to 10^{-9} Ci/L). LRW of activity lower than 10^{-9} Ci/L are conventionally not
processed but dumped so they will be diluted to below accepted concentration limits.

METHODS AND APPROACHES TO LRW MANAGEMENT

Up to the present, there have existed several directions in the development of LRW
management. The methods that are simplest, most economic, but not as acceptable in ecological
terms are being discontinued: dumping LRW into open water basins (the sea) to achieve dilution
by natural water; pumping them into non-water-penetrable zones of deep geological formations
(economic parameters of this method are only 1.4 times better than those for a traditional process
that includes LRW evaporation [1]); and cleaning in a system of biological ponds that generally
has low decontamination factors regarding basic long-lived radionuclides of interest.

Finally, there are methods of distillation (evaporation) in which the highest extent of LRW
cleaning is possible (4-6 orders of magnitude). However, this method is highly power-consuming

Mat. Res. Soc. Symp. Proc. Vol. 465 © 1997 Materials Research Society

(steam consumption is about 1 ton per 1 m^3 of LRW) and does not allow a high LRW/SRW ratio (index of process ecological safety) to be achieved. Conventional values of this index for distillation vary from 100 up to 200 and are reduced significantly with an increase in LRW salinity.

Sorption-coagulation technologies are quite often used in LRW management practice. Sodium triphosphate, sulphides, iron and especially manganese hydroxides are widely applied for radionuclide co-precipitation. To precipitate cesium radionuclides ferrocyanides of transition metals (mainly nickel and cobalt) are used. Use of co-precipitation in LRW cleaning significantly reduces the process energy consumption, but complex apparatus is required. The LRW/SRW ratio is close to that in distillation.

The electrocoagulation method is conventionally used for preliminary LRW cleaning.

Recently, LRW cleaning has been performed by means of membrane distillation (reverse osmosis). Advantages of this technology include high cleaning coefficients (up to 10^6) at lower energy consumption than for conventional distillation. However, in the case of high salinity LRW, the application of membrane distillation is quite problematic. Besides, this method requires special devices to solidify the salted water formed during the cleaning process.

The ion-exchange technologies for LRW cleaning using ion-exchange resins allow LRW of low salinity to be cleaned quite effectively. However, the cost of high-salinity LRW cleaning on ion-exchange resins is far greater than that for alternative technologies. The use of preliminary electrodialysis increases the efficiency of ion-exchange methods used for LRW cleaning.

At present, sorption technologies are being rapidly developed in LRW management. Such technologies are based on the use of sorbents that are highly selective to radionuclides. Selective sorbents, basically, inorganic ion-exchangers used for LRW cleaning enable rather high coefficients of LRW cleaning (about 10^3 - 10^4) to be obtained using a very simple apparatus.

Most modern technological processes for LRW management include parallel or sequential use of several cleaning technologies, for example:
- coagulation and co-precipitation on MnO_2 ----- ion exchange
- electro-coagulation ----- electrodialysis ----- ion exchange
- distillation ----- ion exchange
- reverse osmosis ----- selective sorption

RESULTS OF SORPTION TECHNOLOGIES APPLICATION IN LRW MANAGEMENT

Laboratory studies of sorbents to be used in sorption technologies

Let us consider in detail some features of sorption methods for LRW cleaning. Those with the most potential use sorbents highly selective to radionuclides. Distribution coefficients, K_d, are conventionally used to characterize sorbents (resins) selectivity and may be determined as:

$$K_d = (C_0 - C_1) \times V / C_1 \times m \qquad (1)$$

where C_0 - initial activity (concentration) of the ion of interest; C_1 - this ion activity (concentration) after contact between sorbent and solution in a batch experiment; V - volume of solution; m - mass of dry sorbent. K_d represents the theoretical volume of solution that can be processed per mass of sorbent (resin) under equilibrium conditions.

During the last few years, the Laboratory of Sorption Processes researchers have been developing a number of new sorbents selective to radionuclides. These are the granulated sorbents ZF and ZP - ferrocyanide and phosphate sorbents on an alumino-silica base and the fiber sorbents FF and FM. Use of these sorbents in sorption filters cleans LRW effectively from the longer-lived radionuclides - cesium-137 and strontium-90. Comparative studies were carried out of the efficiencies of such radionuclide removal from model solutions and real LRW by these sorbents and by those used in LRW management technologies by *Babcock & Wilcox Co., Ltd.*

In laboratory tests the following sorbents were studied:

C-1 - natural zeolite-containing alumino-silica of Chuguevka (Primorsky Territory, Russia) deposit (clinoptilolite).

C-2 - C-1 sorbent coated by iron ferrocyanide (sorbent ZF);

C-3 - fiber sorbent coated by iron ferrocyanide (sorbent FF);

C-4 - C-1 sorbent coated by insoluble phosphates of titanium, zirconium, calcium and magnesium (sorbent ZP);

C-5 - sorbent D-230 *(Babcock & Wilcox Env. Serv.)*;

C-6 - sorbent CN-200-40 (*Babcock & Wilcox Env. Serv.*).

Table I gives typical distribution coefficients, K_d for radionuclides distributed between liquid and solid phase for some ion-exchange resins and selective sorbents used in Russia for radionuclides removal.

Experimental values of K_d (without reference in the Table 1) were determined according to the equation (1) in batch experiments performed in a traditional way: samples containing about 0.1 g of sorbent and 10 mL of solution were shaken in a thermostat for 72 hours that was found to be sufficient time to reach equilibrium conditions. The initial activity was determined in the blank solution. Dry weight of a sorbent was calculated from F-factor ($m = F \times m_{wet}$) determined after drying a sample of sorbent at 105^0C during 24 hours.

The table shows that the selective sorbents have radionuclide distribution coefficients much higher than those of the ion-exchange resins (first two rows). Application of selective sorbents for LRW management allows resolution of one of the most complicated technological problems - cleaning of high-salinity LRW produced during decontamination of nuclear energy installations. Also, it will take a long time to resolve the problem of the ion-exchange resins high cost.

Poor mechanical properties of inorganic ion-exchangers does not allow their use in the most effective filters, which have a stationary sorbent layer. At present, this problem is being solved by the development of composite sorbents where the inorganic ion-exchanger is dispersed into a granule having a porous matrix. Such composite sorbents allow use of sorption technology with a high cleaning coefficient within one stage as a result of forming a stationary sorbent layer in the sorption filter.

Table I. Comparison of cesium-137 and strontium-90 distribution coefficients (K_d, mL/g) for sorbents and ion-exchange resins used in Russia (including those developed by the authors)

Sorbent (resin)	^{137}Cs	^{90}Sr
Dowex 50x8 [2]	2.56	5.12
KU-2x8 [3]	10	175
Seleks KM [3]	1.2×10^3	60
Fezhel	1.5×10^5	-
KG-13 [3]	30	4.3×10^4
ZP	800	1.5×10^4
ZF	5×10^4	-
FF	2×10^6	-

Composite sorbents of the above type are produced on an industrial scale in Russia and abroad (for example, NZhA, Fezhel, D-230, CN-400-20 etc.). At the same time, there are many prospective improvements in the field of composite sorbents in the bench-scale testing. In particular, in the Institute of Chemistry (as presented by the collective authors) new composite selective sorbents ZF, ZP, FF, and FM are being developed which can effectively clean LRW of different salt compositions up to salinity values of 10-35 g/L.

LRW produced by operation, repair and disposal of ships with nuclear energy installations (NEI) are different from LRW produced at other NEI, since in most cases marine LRW are polluted by sea water and, therefore, have high salinity and contain chlorides. Thus, to clean Navy LRW one should use methods that can clean low-salt LRW as effectively as high-salt LRW. As was shown above, distillation is the most reliable method among those used to clean high salinity LRW. However, even in the case of evaporation followed by solidification (using solid heavy oil processing by-products), the LRW/SRW volume ratio for sea water does not exceed 15. Use of selective sorbents with LRW of high salinity allows the efficiency of LRW cleaning to increase significantly.

Sorption properties of the above sorbents were studied for simulated as well as real LRW from storage tanks of Russian Pacific Navy facility "Pinega" and TNT-5 and TNT-27 tankers. An example of the radiochemical and chemical composition of LRW samples used in the experiments is given in Table II.

Table II. Composition of LRW from the TNT-27 tanker

Total activity, Ci/L	**1.4×10^{-6}**
Activity of ^{137}Cs, Ci/L	3.8×10^{-7}
Activity of ^{90}Sr/ ^{90}Y, Ci/L	1.0×10^{-6}
Activity of ^{60}Co, Ci/L	2.0×10^{-8}
Total salt content, mg/L	13 000
Chloride content, mg/L	6 500

The radionuclide sorption was studied by static as well as dynamic methods. The static (batch) experiments were performed as was described above. The radionuclides distribution coefficients, K_d, at different salt content were determined.

Table III. Coefficients of radionuclide distribution on the sorbents studied

Sorbent	K_d ^{137}Cs, mL/g salt content 1.5 g/L	K_d ^{90}Sr, mL/g salt content 1.5 g/L	K_d ^{137}Cs, mL/g salt content 15 g/L	K_d ^{90}Sr, mL/g salt content 15 g/L
C-1	$1.7*10^3$	60	700	5
C-2	$5*10^4$	20	$6*10^3$	-
C-3	$3*10^6$	-	$1.5*10^6$	-
C-4	-	$1.5*10^4$	-	$4.5*10^3$
C-5	$3*10^4$	350		
C-6	40	$1.2*10^3$		

Table IV. LRW decontamination factors for different radionuclides for the sorbents under study in laboratory columns

	^{137}Cs		^{90}Sr	
	DF	**V/Vcol**	**DF**	**V/Vcol**
C-1	5 700	1250	4.2	870
C-2	12 100	8500		
C-3	6 500	1.2×10^5		
C-4			16000	6700
C-5	4 300	4200		
C-6			800	1200

Table IV shows the results of dynamic experiments which were performed in laboratory columns (diameter 10 mm, column height 200 mm, sorbent layer height 70-80 mm, flux 6-10 column volumes/hour). After passing a column solutions were divided into fractions and analyzed. Experiments were performed in a way to simulate real technologies where the above sorbents were supposed to be used. Namely, a process was stopped when the breakthrough point was achieved. As in the real sorption technology developed, where cleaned water is dumped directly into the sea, the breakthrough point was taken at the limit accepted concentration of either radionuclide of interest (1×10^{-9} for ^{137}Cs and 3×10^{-11} for ^{90}Sr). The decontamination factors (DF) were estimated as a ratio between the initial radionuclide concentration C_0 and its average concentration in a whole volume of the solution cleaned C_{av}. The V/V$_{col}$ parameter gives the total number of column volumes of radionuclide-polluted solution that can be cleaned before the breakthrough point.

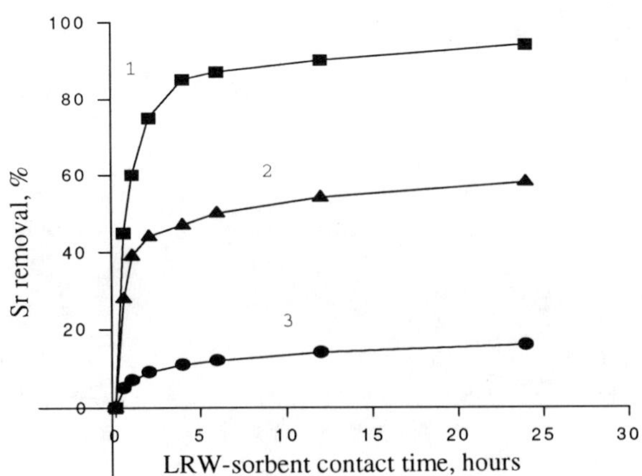

Fig 1. Kinetics of strontium sorption on sorbents: C-1 (3),C-2 (2), C-4 (1).

It can be seen from Tables III and IV that the initial zeolite (sorbent C-1) gives quite good removal of cesium radionuclides but poor removal of strontium and cobalt (not showed). Sorbents modified by ferrocyanides are significantly more effective for cesium removal from solutions of high salinity, and fiber sorbents C-3 appeared to be the most effective among the others. When comparing sorbents developed at the Institute of Chemistry FEDRAS with those used by *Babcock & Wilcox Env. Services* (sorbents C-5 and C-6) from LRW decontamination, one should mention the quite low comparative efficiency of the latter (see Tables III and IV).

Removal of strontium from LRW by sorption is a more complicated problem because of far lower sorbents selectivity in regard to this element. Figs.1 and 2 show kinetic and dynamic sorption curves for LRW using some of the sorbents studied.

The kinetic experiments (Fig. 1) were performed as a series of batch experiments similar to those described above for the distribution coefficients determination: identical samples (each containing 10 mL of solution and 0.1 g of sorbent) were analyzed in different time intervals.

In the dynamic experiments, performed in a similar way as described above for column tests, the comparison was made between the efficiencies of the initial clinoptilolite (C-1) and its ferrocyanide- (C-2) and phosphate-modified (C-4) forms. The results are shown on Fig. 2.

To sum it up, among those under study, C-4 (ZP) appears to be the most effective sorbent to remove strontium from LRW.

As other experiments showed, use of the selective sorbent ZP also resolves the problem of cleaning LRW containing complexed radionuclides - so called LRW TGA (containing a strong complexer - Trilon B). The problem of solutions containing oxalic acid and oxalate was successfully solved in the real sorption installation by a combination of ZP and FM (fiber) sorbents in a composite system.

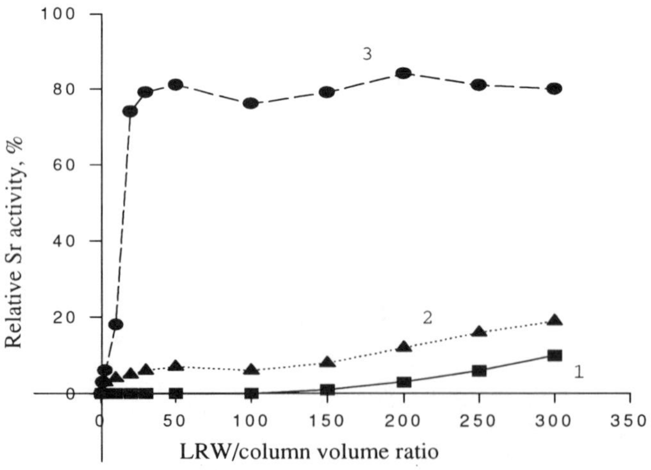

Fig. 2. Dynamics of strontium sorption on sorbents: C-1 (3), C-2 (2), C-4 (1).

<u>Application of the new sorbent technology for LRW management at industrial scale</u>

The work on developing a test industrial installation to clean LRW was conducted at the Far East plant "Zvezda" and the technical tanker "Pinega" during March 1995 through September 1996 and ended with successful tests of the installation and its prototype.

The installation allows to clean different LRW within one stage with very simple control equipment (since it has just three main parts - 1) sorption filter; 2) pumps for LRW feeding/cleaned water removal; 3) analytical unit on the output) and easy maintenance.

During the installation and prototype work about 300 m^3 of LRW of different composition (including decontamination and TGA LRW) were cleaned. Two filtering elements were used which became solid radioactive wastes (without sorbent removal) when expended. They were considered to be expended after the breakthrough point regarding main radionuclides of interest had been achieved. Depending on the LRW type, the filtration was performed sequentially through two filters or in a parallel regime (C-2, C-3, C-4 sorbents were used). In some cases, an additional carbon pre-filter was used to remove organic admixtures and colloids.

The total volume of filtering material in the two filters was 360 dm^3.

The volume of LRW processed in this installation, with decontamination factor about 10^4 without filter change, was not less than 400 m^3.

The installation productivity at LRW salinity 2 g/L was 0.5 m^3/g.

Expended filtering elements were dried, disposed of and replaced by new ones.

CONCLUSIONS

Laboratory studies of radionuclide sorption from model solutions and real LRW showed a principal possibility of one-stage sorption removal of longer-lived radionuclides from LRW solutions of low and high salinity. New sorbents highly selective to longer-lived radionuclides (^{137}Cs, ^{90}Sr) were developed and tested. Use of respective sets of the sorbents would enable effective removal of radionuclides from LRW having a very complex chemical composition (TGA waters, acid and alkali decontamination solutions). Selective sorbents developed in the Institute of Chemistry FEDRAS surpass those used by *Babcock & Wilcox Env. Services* regarding efficiency and allow more effective LRW cleaning technologies to be developed. The results of laboratory studies have been further used to construct an industrial installation for LRW cleaning which showed good results in the management of real LRW solutions.

REFERENCES

1. V.K Bulygin, <u>The experience of cleaning high-salinity oil-containing LRW at Pacific Navy objects by experimental installations "Sharya-04"</u>, Scientific-technical report, Sosnovy Bor, 1995 [in Russian].
2. Yu.Yu. Lurye, <u>Materials for chromatography</u>, Khimia, Moscow, 1972, p.188..
3. V.V. Milyutin, V.M.Gelis, R.A. Penzin <u>Soprtion-selective characteristics of inorganic sorbents and ion-exchange resins in regard to cesium and strontium</u>, Radiokhimia, N 3, p. 76-82 (1993).

PLASMA CHEMICAL PROCESSING OF DEPLETED URANIUM HEXAFLUORIDE

A.V. IVANOV, V.M. ABASHKIN, E.A. FILIPPOV*, Y.N. TUMANOV**
*Russian Research Institute of Chemical Technology, 33 Kashirskoe Avenue, Moscow 115230,
vs@khimcon.msk.ru
**Russian Scientific Center, Kurchatov Institute, Moscow

ABSTRACT

A reprocessing of uranium hexafluoride waste, which was accumulated in the nuclear
industry for many years, is a very vital question now. There are a lot of gas tanks filled with UF_6
that have been exposed for a long time and are posing a serious ecological danger. A plasma
chemical technique is proposed here to transform the uranium hexafluoride depleted with isotope
^{235}U (less than 0.1%) into the safe storage form - powdered uranium oxides. The second product
is HF (70-80% aqueous solution) which may be easily reprocessed by means of the rectification
column up to pure anhydrous hydrogen fluoride.

The designed technological scheme and the basic technical characteristics of the plasma
chemical converter are presented. The proposed technique is efficient, easy controlled and free
of any harmful by-products.

INTRODUCTION

The reprocessing of the uranium hexafluoride wastes that were accumulated for a long
period is a vital question as gas tanks filled with uranium hexafluoride have been exposed for a
long time and are now posing a serious ecological danger.

The objective of this report is a technology eliminating storage of a large quantity of
uranium hexafluoride, depleted in the isotope ^{235}U (less than 0.1%), by conversion into a safe
storage form - uranium oxides. The resolution of this problem also allows fluorine (in the form
of anhydrous hydrogen fluoride) to be returned in the fuel cycle. The proposed technology can
be realized by using an existing industrial operations scheme and by modifying some of the
equipment .

THERMODYNAMICS AND KINETICS OF THE PROCESS

A plasmachemical process for UF_6 conversion in (H-OH)-plasma field is generally described
by the overall reaction (1):

$$UF_6 + 3 \text{ (H-OH)-plasma} \longrightarrow 1/3 \ U_3O_8 + 6 \ HF + 1/6 \ O_2 \tag{1}$$

A detailed picture of thermodynamic and kinetic relations at high temperature was analyzed
to determine proper conditions for the most efficient operation of the process. The following
compounds are formed in the gaseous phase:

$UF_6(g)$, $UF_5(g)$, $UF_4(g)$, $UO_2F_2(g)$, $UO_2(g)$, $F_2(g)$, $O_2(g)$, $HF(g)$, $F(g)$, $H_2(g)$, $H(g)$, $H_2O(g)$, $U(g)$,
$OH(g)$, $O(g)$, and $UO(g)$.

Mat. Res. Soc. Symp. Proc. Vol. 465 ©1997 Materials Research Society

The most significant components of the condensed phase are:

$UF_4(c)$, $UO_2F_2(c)$, UO_3, UO_2, and U_3O_8 .

The chemical composition of the reaction mixture (UF_6 + H_2O) while plasma heating at 50 kPa (0.5 atm) is shown in Figure 1. It is obvious that in the temperature range of 1050 through 1400 K, the initial uranium hexafluoride is transformed almost completely to U_3O_8 (c) and HF(g). Above 1400 K, there is another competitive process - formation of UO_2F_2.

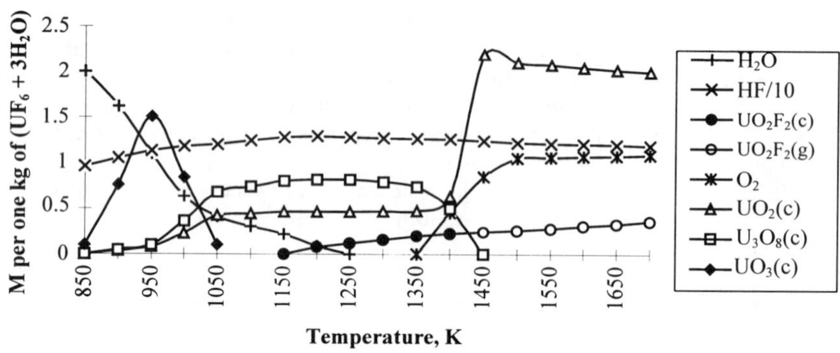

Figure 1. Concentration of principal U- and F- compounds during plasma processing.

Equilibrium analysis of all compounds in the reaction mixture was used to determine the appropriate set of key chemical reactions (ΔH - in kJ/mol ; ΔS - in kJ/mol*K):

$UF_6(g) \rightarrow UF_5(g) + F(g)$; $\Delta H = 278.20$; $\Delta S = 168.30$; (2)

$UF_5(g) \rightarrow UF_4(g) + F(g)$; $\Delta H = 423.06$; $\Delta S = 134.48$; (3)

$UF_6(g) + HOH(g) \rightarrow UOF_4(g) + 2HF(g)$; $\Delta H = 54.83$; $\Delta S = 145.08$; (4)

$UOF_4(g) + HOH(g) \rightarrow UO_2F_2(g) + 2 HF(g)$; $\Delta H = 123.00$; $\Delta S = 138.09$; (5)

$UF_5(g) + 2 HOH(g) \rightarrow UO_2F(g) + 4 HF(g)$; $\Delta H = 342.30$; $\Delta S = 259.68$; (6)

$UF_5(g) + HOH(g) \rightarrow UOF_3(g) + 2 HF(g)$; $\Delta H = 134.40$; $\Delta S = 125.34$; (7)

$UF_4(g) + 2HOH(g) \rightarrow UO_2(g) + 4 HF(g)$; $\Delta H = 518.80$; $\Delta S = 220.50$; (8)

$UO_2F_2(g) + HOH(g) \rightarrow UO_3(g) + 2 HF(g)$; $\Delta H = 250.54$; $\Delta S = 125.62$; (9)

$UF_4(g) + HOH(g) \rightarrow UOF_4(g) + 2 HF(g)$; $\Delta H = 185.86$; $\Delta S = 132.04$; (10)

$UOF_3(g) + 3/2\ HOH(g) + 1/12\ O_2(g) \rightarrow 1/3\ U_3O_8(c) + 3\ HF(g);$ $\quad \Delta H=-138.14;\ \Delta S=-37.95;\ (11)$

$UO_2F(g) + 1/2\ HOH(g) + 1/12\ O_2(g) \rightarrow 1/3\ U_3O_8(c) + HF(g);$ $\quad \Delta H=-346.04;\ \Delta S=-172.30;(12)$

$UOF_2(g) + HOH(g) \rightarrow UO_2(g) + 2\ HF(g);$ $\quad \Delta H= 332.91;\ \Delta S=86.42;\ (13)$

$UO_2F_2(g) \rightarrow UO_2F_2\ (c);$ $\quad \Delta H=-300.00;\ \Delta S=-207.02;(14)$

$UO_2(g) \rightarrow /1/3\ U_3O_8\ (c)+1/6\ O_2(g);$ $\quad \Delta H=-392.36;\ \Delta S=-181.18;(15)$

$UO_2(g) \rightarrow UO_2(c);$ $\quad \Delta H=-605.00;\ \Delta S=-189.28;(16)$

$UF_4(g) \rightarrow UF_4(c);$ $\quad \Delta H=-316.00;\ \Delta S=-211.39;(17)$

$UO_2F_2(c) + HOH(g) \rightarrow 1/3\ U_3O_8 + 2HF(g) +1/6\ O_2(g);$ $\quad \Delta H=157.21;\ \Delta S=151.47;\ (18)$

$UF_4(c) + 2\ HOH(g) \rightarrow UO_2(c) + 4\ HF(g);$ $\quad \Delta H=225.93;\ \Delta S=242.58;\ (19)$

$2\ F(g) \rightarrow F_2(g);$ $\quad \Delta H=158.76;\ \Delta S=242.58;\ (20)$

To determine optimum conditions for complete conversion of UF_6 (by minimizing the contribution of reactions 14 and 18 - condensation of UO_2F_2), the cancellation and approximation procedures were limited to six principal equilibrations: 4, 5, 9, 14, 15 and 18 [2]. The resulting set of kinetic equilibria is the following:

$$d(UF_6)/dt = k_4(UF_6)(HOH); \qquad (21)$$

$$d(UOF_4)/dt = k_4(UF_6)(HOH) - k_5(UOF_4)(HOH); \qquad (22)$$

$$d(UO_2F_2)/dt = k_5(UOF_4)(HOH) - k_9(UO_2F_2)(HOH) - k_{14}(UO_2F_2); \qquad (23)$$

$$d(UO_3)/dt = k_9(UO_2F_2)(HOH) - k_{15}(UO_3), \qquad (24)$$

where k_4, k_5, k_9, k_{14}, and k_{15} are the specific rates of the appropriate chemical reactions.

By means of these equations (21-24) we estimated the specific rates of appropriate reactions and phase transitions and determined conditions for the most efficient plasma conversion of UF_6 in the gaseous phase.

Figure 2 shows how the condensation period of UO_2F_2 changes during heating. When the temperature of the reaction mixture is near the boiling point of UO_2F_2, the condensation period increases dramatically. Here **t** is the time at which the partial pressure $P(UO_2F_2)$, reduces from $3.4*10^4\ Pa$ to $3.4*10^2\ Pa$. The proper temperature regime (near 1600K) was maintained by addition of H_2O that resulted in promotion of reactions 4, 5, and 9 and suppression of reaction 14. In this case, the primary product of the UF_6 -conversion will be UO_3, which may be easily condensed on completion of the gaseous reactions mentioned above.

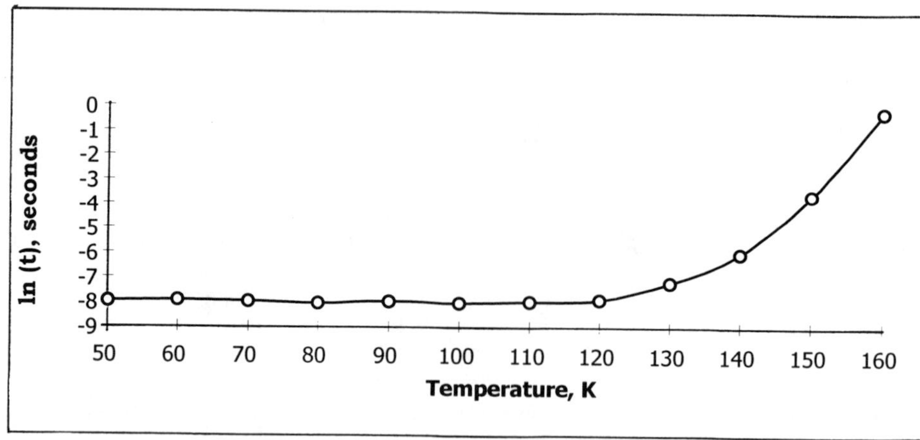

Figure 2. Dependence of the condensation time of UO_2F_2 on the plasma temperature

PRINCIPAL TECHNICAL SCHEME

The overall process may be subdivided into five stages:

1. Generation of the (H-OH)-plasma flow;
2. Interaction of the gaseous UF_6 with the (H-OH)-plasma flow , resulting in formation of two principal products, U_3O_8 and HF, and the by-product O_2;
3. Separation of the solid and gaseous adducts;
4. Condensation of the gaseous hydrogen fluoride;
5. Distillation and purification of saturated HF-solution.

The principal uniqueness of this plasmachemical reaction is the utilization of the H-OH plasma as one of the active components which makes possible the conversion of uranium hexafluoride using a stoichiometric quantity of water vapor, i.e. a molar proportion of $H_2O/UF_6 = 3$. As a result, UF_6 can be transformed into the thermodynamically stable and nonpyrophoric compound, U_3O_8, and into hydrofluoric acid that may be utilized for the electrolytic production of pure gaseous fluorine.

It is necessary to emphasize another advantage of the approach described above - the absence of any harmful by-products (the only by-product is gaseous oxygen).

The designed technological scheme (see Fig.3) for reprocessing of the waste uranium hexafluoride starts with evaporation of UF_6 from storage tanks, followed the conversion of UF_6 to uranium oxides and hydrofluoric acid. The distillation process for HF (saturated solution) includes a standard fractionation method usually applied for the production of anhydrous HF.

The basic products of this process are: hydrofluoric acid (97-98% HF), containing less than 5 mg/l of dissolved uranium compounds, and the residue - azeotropic HF solution. The first product may be efficiently refined and concentrated up to unhydrous HF using a distillation column. The U-contaminated residue may be used as an additional source for the H-O-H plasma.

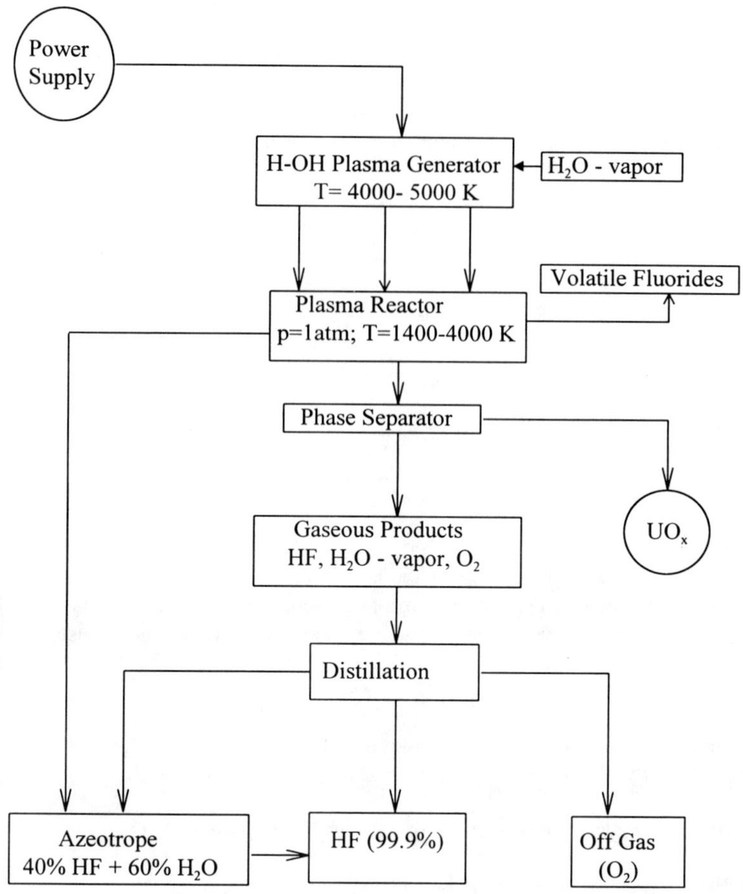

Figure 3. General scheme of the UF_6 plasma conversion

TECHNICAL APPROACH AND METHODOLOGY

The high temperature conversion of volatile rare and less-common metal halides to powdered oxides was used before the plasma chemical technique was developed. For example, a gas-flame method is known to produce uranium oxyfluorides from UF_6 using a hydrogen-air mixture. However, the gas-flame technique had no large-scale application because of some substantial disadvantages:

- as a rule, by-products of this process are diluted acids (e.g. HF and HCl) which have very limited application;
- the granularity of the oxide powders produced cannot be regulated properly;
- there is no reliable methodology to model and design an appropriate reactor, as a result the parameters of the equipment used do not meet requirements for product quality;

- insufficient productivity of the gas-flame apparatus due to the low enthalpy of the gaseous torch in comparison with chemical plasma;
- high explosion hazard of the gas-flame systems.

The principal scheme for plasma chemical reprocessing of UF_6 with the production of UO_x and hydrogen fluoride was developed following appropriate thermodynamic calculations and laboratory experiments.

The heart of the converter is the plasma reactor supplied by a water-vapor plasma generator. The water-vapor plasma generator consists of an arc heater with a water-cooled anode of up to 200 kW.

<u>Basic technological characteristics of the plasma chemical reactor:</u>

- productivity- UF_6, kg per hour   **150;**
- H_2O/UF_6 - mole ratio   **3.3**
- electric power of plasma generator, kW   **up to 200;**
- consumption of energy, MW*h per kg of UF_6 **1.3-1.4**
- conversion rate   **99%**

<u>Quality of final UO_x:</u>

- bulk density, g/cm^3 **4.5-4.7**
- specific surface area of UO_x -powder, m^2/g **0.037-0.138**
- composition (%): U- 84.6; F<0.02; Ni<0.001; W < 0.0001; Cr< 0.001; Cu< 0.0005.

Total consumption of energy may be decreased by means of a higher efficiency plasma generator and power supply and by increasing reactor volume - due to reduction of heat losses.

CONCLUSIONS

1. The uranium hexafluoride can be completely converted to U_3O_8 and HF using a plasma flow with a stoichiometric quantity of water vapor;

2. The limiting stage of the process is the conversion of gaseous UF_6 to gaseous U_3O_8;

3. The lowest acceptable temperature to maintain the conversion is 1600 K;

4. The process is practically waste free. Anhydrous HF (98%) and its azeotrope (40% HF + 60% H_2O may be produced from saturated (90-93%) HF-solution using additional distillation;

5. The proposed technique is efficient, easily controlled and free of any harmful by-products.

REFERENCES

1. V.S. Ivanov, V.S. Astavin, Yu.N. Tumanov et al. <u>Thermodynamics and kinetics of uranium fluorides in (Ar-H$_2$O) - plasma</u>. *Proceeding of the 3-rd Conference on Uranium Chemistry (USSR)*. Moscow: Nauka, p. 90 (1985)

2. Yu.N. Tumanov. <u>Electrothermical processes in modern chemical industry and metallurgy (Rus.)</u> Moscow: Energoizdat (1981)

DEPLETED URANIUM OXIDES AND SILICATES AS SPENT NUCLEAR FUEL WASTE PACKAGE FILL MATERIALS

C. W. FORSBERG
Oak Ridge National Laboratory, Oak Ridge, TN 37831, forsbergcw@ornl.gov

ABSTRACT

A new repository waste package (WP) concept for spent nuclear fuel (SNF) is being investigated. The WP uses depleted uranium (DU) to improve performance and reduce the uncertainties of geological disposal of SNF. The WP would be loaded with SNF. Void spaces would then be filled with DU (~0.2 wt % ^{235}U) dioxide (UO_2) or DU silicate-glass beads.

Fission products and actinides can not escape the SNF UO_2 crystals until the UO_2 dissolves or is transformed into other chemical species. After WP failure, the DU fill material slows dissolution by three mechanisms: (1) saturation of WP groundwater with DU and suppression of SNF dissolution, (2) maintenance of chemically reducing conditions in the WP that minimize SNF solubility by sacrificial oxidation of DU from the +4 valence state, and (3) evolution of DU to lower-density hydrated uranium silicates. The fill expansion minimizes water flow in the degraded WP. The DU also isotopically exchanges with SNF uranium as the SNF degrades to reduce long-term nuclear-criticality concerns.

INTRODUCTION

The use of DU as a WP fill material has been investigated. A synthesis of laboratory data on oxides and silicates, hot-cell tests on SNF, geological field data, and other data provide the technical basis for the concept.

MECHANICAL AND CHEMICAL DESCRIPTION

SNF repository WPs are loaded with SNF and then the void space is filled with DU dioxide and/or DU silicate-glass beads, which are sufficiently small (<1 mm) such as to fill coolant channels in the SNF assemblies. The DU enrichment levels are as low as 0.2 wt % ^{235}U. DU is added to the package to reduce the total fissile concentration in the waste package below 1 wt % of the heavy metal. The technology is applicable to light-water reactor SNF and potentially applicable to high-enriched-uranium SNF.

Filling has been demonstrated [1] with steel shot on light water reactor (LWR) dummy fuel elements. Canada [2] has conducted extensive tests to fill empty spaces with small particles within WPs containing simulated Canadian Deuterium Uranium SNF. The Canadian WP is a thin-walled, titanium container in which the SNF and fill material must support the WP against external hydrostatic pressure after burial. Canadian tests investigated different particle sizes, different mixtures of particle sizes, alternative fill materials, vibratory filling with different vibration frequencies and amplitudes, and other factors. Tests included successful full-scale WP hydrostatic tests (10 MPa, 150 °C) of the filled WP.

Representative designs of WPs with DU fill are shown in Table I. These designs cover the expected range of designs from a relatively low-uranium content silicate glass to UO_2. Calculations [3] using DU silicate fills show that the total WP weight does not significantly change for self-shielded WPs. The fill adds weight but reduces the WP wall thickness needed for radiation shielding.

TABLE I. Alternative WP Designs* with Depleted Uranium

Property	Fill Material	
	Low-Uranium Glass**	UO$_2$
SNF (MTIHM***)	9.96	9.96
Solid bead density (g/cm^3)	4.1	10.96
DU (wt % of fill)	25.	88.
DU mass (t)	3.43	32.3
Bead mass (t)	13.7	36.7
Assay (wt %) of DU ^{235}U	0.2	0.2
Equivalent ^{235}U assay (wt %) of SNF	1.6	1.6
Equivalent ^{235}U assay (wt %) of WP	1.24	0.53

*Design basis was the U.S. Multi-Purpose Canister WP system for 21 PWR SNF assemblies [4]. The MPC has an internal volume of 7.9 m^3; the SNF baskets have a solid volume of 1.1 m^3; and the 21 SNF assemblies have a solid volume of 1.6 m^3. A total of 5.2 m^3 of the 7.9 m^3 of internal volume can be filled with beads.
**Multiple formulations exist for uranium silicate glass [3]
***Metric tons initial heavy metals

REPOSITORY BENEFITS

The expected repository failure mode is radionuclide migration to the open environment by (1) water leaching of SNF, (2) dissolution of radionuclides and generation of colloids, and (3) transport of those radionuclides in dissolved or colloidal forms to the open environment. Two mechanisms reduce radionuclide release rates: (1) minimize the water flow through the WP and (2) reduce the dissolution of radionuclides into groundwater.

The use of DU fill slows each of these mechanisms for radionuclide release. The details of the mechanisms are somewhat different for oxidizing versus reducing groundwater. The choice of DU dioxide or DU silicate-glass depends upon specific SNF type and groundwater conditions.

Reduction of Repository Long-Term Radionuclide Release Rate Under Oxidizing Groundwater Conditions (Yucca Mountain)

Maintenance of Chemically Reducing Conditions To Minimize SNF UO$_2$ Solubility

Most fission products and actinides within SNF are incorporated into UO$_2$ spent fuel pellets. Upon WP failure and after entry of groundwater into the WP, these fission products and actinides can not escape the UO$_2$ crystals until the UO$_2$ dissolves or is transformed into other chemical species.

The solubility of uranium in reducing groundwater is very low - about 1 ppb [5]. This solubility is about four orders of magnitude less than the solubility of uranium under oxidizing conditions. Radionuclide releases are extremely limited under chemically reducing conditions [5] because the UO$_2$ does not dissolve and release the radionuclides incorporated within its structure. Chemically-reducing conditions also minimize the solubility and transport of several other long-lived radionuclides (e.g., neptunium and technetium) and reduce the formation and transport of radionuclides as colloids.

With the use of UO$_2$ or uranium silicates with uranium in the +4 valence state, the DU fill ensures chemically reducing conditions within the WP for an extended time independent of external groundwater chemistry. After a WP failure, any oxygen in the groundwater would first encounter the DU before it encounters the SNF. The DU in the +4 chemical state would be oxidized to the +6 chemical state through a series of oxidation steps, thus removing the oxygen from the groundwater.

The DU fill minimizes generation of oxidizing acids that degrade the WP and accelerate SNF dissolution. Water and air in radiation fields generate corrosive acids (nitric acid etc.). WPs are dried and filled with inert gases before sealing to minimize this problem. After WP failure, DU fill reduces acid generation by two mechanisms. The high-density fill reduces internal WP radiation fields by a factor of 2 to 3. The available WP air or water volume for acid generation is reduced by displacement with DU fill. Each mechanism reduces acid generation proportionally.

Saturation of groundwater with DU.

Surrounding the SNF with DU saturates any groundwater within the degraded WP with uranium. This saturation slows the SNF dissolution process. DU-saturated groundwater can not dissolve additional SNF uranium.

Creation of Mass Transfer Barriers to Radionuclide Transport

In oxidizing groundwater, after a loss of chemically reducing conditions in the WP, the addition of uranium silicates may also reduce SNF transformations and dissolution - but by other mechanisms. Because oxidizing groundwater usually contains silicates, the long-term evolution of a WP initially filled with DU dioxide or DU silicate in many ways will be similar. The groundwater can provide silicates to a WP containing only oxides.

Solubility limits. Uranium silicates are less soluble in most groundwaters than are most other uranium compounds. This low solubility reduces the quantity of uranium that can dissolve in a unit of groundwater. The slower dissolution of uranium silicates compared to uranium oxides in oxidizing groundwater implies that longer times will occur before the uranium in the WP is dissolved and transported away from the WP. Low rates of dissolution of uranium imply slow formation of flow channels from uranium dissolution through the SNF with slower groundwater flow.

However, the solution is saturated with respect to uranium silicates, not uranium oxides. The availability of silicates can accelerate initial transformations of uranium oxides to less soluble silicates in the SNF and bring about the release of soluble radionuclides to the groundwater during the transformation. The details of the chemical transformation from uranium oxides that are somewhat soluble in oxidizing groundwater to much less soluble uranium silicates determine which actinides and fission products are better isolated after the transformation as compared to before the transformation.

Mass transfer barriers. A DU-silicate-saturated solution can form insoluble, protective layers of uranium silicates over the UO_2 crystals within the SNF. This phenomenon is expected to slow uranium alteration and dissolution after initial dissolution of surfaces and has been observed in natural uranium ore deposits under oxidizing groundwater conditions [6,7]. Any surface silicates that are formed are stable because the groundwater is saturated in uranium silicates that are thermodynamically stable in oxidizing groundwater.

Protection Against Variable Groundwater Chemistry

The use of DU, particularly UO_2, provides a mechanism to counteract any unexpected changes in groundwater that may accelerate degradation of the SNF. Groundwater chemistry can change because of climatic changes and human activities (e.g., irrigation, groundwater pumping, liquid waste injection, etc). The DU UO_2 acts as a sacrificial chemical to absorb changes in groundwater chemistry and delay their effects on the SNF UO_2.

Minimization of Groundwater Flow in WP

After WP failure, DU fill lowers the hydraulic conductivity within the degraded WP and thus ensures low groundwater flow near the SNF. Water flow is the primary mechanism for the transport of radionuclides to the open environment. The release of many radionuclides is controlled by solubility limits. Other radionuclides may be transported by colloids. The slower the flow of groundwater, the slower the transport of radionuclides. The goal is to create a WP where water flow and radionuclide migration is by diffusion only. Diffusion is a very slow process.

During oxidizing conditions, UO_2 will oxidize to higher valence-state uranium oxides which have lower densities [8]. In oxidizing groundwater, the DU oxide will evolve to lower-density, hydrated uranium oxides. Silicates and hydrated uranium oxides further evolve to hydrated uranium silicates. These changes have been observed in the laboratory and in uranium ore deposits [9,10]. This volume expansion fills void space and thus minimizes water flow. Figure 1 shows this evolutionary sequence as it has been observed in natural uranium ore bodies with oxidizing groundwater [11]. Because of the high surface area of the beads, they will expand and choke off water flow before the SNF uranium evolves to hydrated uranium silicates.

Expansion of the fill will be controlled by two mechanisms. The selection of a specific DU compound determines the maximum possible expansion. Kinetic factors will control actual expansion. After WP failure, DU swelling will begin to occur but swelling slows ingress of water and gas into the WP. The evolution of the DU to more stable forms must be designed to stop water movement while avoiding excess expansion.

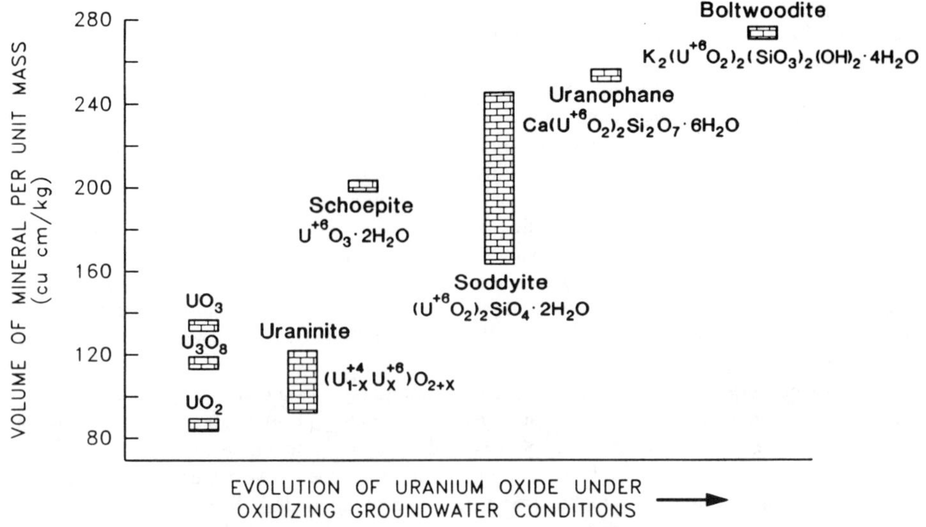

Fig. 1. Specific volume of selected uranium minerals versus evolution over time in oxidizing groundwater.

<u>Reduction of Repository Long-Term Radionuclide Release Rate Under Reducing Groundwater Conditions</u>

With chemically-reducing groundwater, UO_2 fill is preferred. DU dioxide helps maintain within the WP strong chemically reducing conditions that minimize the oxidation and dissolution of SNF UO_2. The DU saturates the local WP groundwater with uranium; this saturation effect further depresses dissolution of SNF UO_2.

<u>Reduced Potential For Repository Nuclear Criticality</u>

As a fill material DU reduces the potential for repository nuclear criticality events. Long-term, low-power, repository nuclear-criticality events are a major concern because they generate heat that (1) accelerates degradation of WPs and (2) accelerates the water movement that may transport radionuclides to the environment.

Criticality is prevented in a repository by neutron absorbers and geometric spacing of fissile materials. Neutron absorbers include [238]U, boron, gadolinium, and other materials. Neutron absorbers (except [238]U) leach from WPs and travel at rates different from the SNF uranium through the geologic media because of the different chemistries of the neutron absorbers in groundwater. Uranium groundwater transport and redeposition (a mechanism that creates uranium ore bodies) creates the potential for criticality events if the fissile concentration in the repository SNF is sufficiently high. That the fissile content of LWR SNF is sufficient to cause nuclear criticality can be demonstrated by comparing the expected fissile concentrations in the repository (1.47 wt %) with fissile concentrations at shutdown (1.3 wt %) in naturally occurring reactors [12] in the Oklo mining district in Gabon, Africa. Theoretical analysis yields similar conclusions [13].

Whether nuclear criticality will, in fact, occur in a specific repository depends upon the long-term evolution of the chemical conditions. The addition of DU to a SNF WP lowers the uranium enrichment in each WP and, consequently, in the repository as a whole below 1 wt %. DU is the only neutron absorber that will not separate from SNF uranium over time. Nuclear criticality can not occur with such low uranium enrichment levels in a geological environment [13].

In LWR SNF, half or more of the fissile material is ^{239}Pu. The foregoing analysis is based on the assumption that plutonium remains with the uranium until the major plutonium isotope (^{239}Pu) decays to ^{235}U and can be isotopically diluted by the DU. This is ensured if the rate of plutonium decay to uranium is faster than the rate of dissolution and transport of uranium within the repository. The primary plutonium isotope, ^{239}Pu, has a half-life of 24,000 years (i.e., the decay rate is 3×10^{-5}/year).

Performance assessments indicate that plutonium migration is slow in most geological environments. Uranium migration is extremely slow in reducing groundwater; thus, the uranium will remain near the plutonium until the plutonium decays to ^{235}U and is isotopically mixed in the groundwater. In oxidizing groundwater, uranium is more mobile; however, uranium silicates are several orders of magnitude less mobile than many other compounds. Most groundwater contains silicates; thus, formation of uranium silicates from uranium oxides would be expected to slow uranium transport sufficiently to allow plutonium decay to uranium and isotopic dilution of the SNF uranium from plutonium with the DU. Higher assurance of criticality control may be achieved by including within the WP system some DU as uranium silicates or use of excess DU in the WP.

DU can be added exterior to the WP for criticality control; however, this has several disadvantages. It eliminates the chemical benefits of DU within the WP. Furthermore, it adds uncertainties in terms of nuclear criticality control. Water can channel through a WP without fill material, transport the SNF uranium, and bypass DU exterior to the WP. Last, DU exterior to the WP does not address large-WP internal-criticality issues.

BASIS OF DESIGN

The central problem of repository design is how to predict long-term performance. Because the kinetic dissolution mechanisms of SNF degradation may be temperature dependent, results from accelerated testing of materials are difficult to interpret. The use of natural materials in WPs can minimize these uncertainties.

Uranium dioxide is currently evolving into hydrated uranium silicate minerals in multiple ore deposits with conditions similar to those of the proposed Yucca Mountain repository. Geological data on the performance of UO_2 over millions of years can be combined with laboratory data to provide estimates of WP performance. The use of natural WP fill materials reduces the limitations of using short-term laboratory experiments to predict repository performance for very-long times.

MANAGEMENT OF DEPLETED URANIUM

For LWR fuel production, natural uranium with a ^{235}U content of 0.71% is separated into an enriched-uranium fraction with the by-product production of 4 to 6 tons of DU per ton of enriched uranium. Currently, there are only limited uses for this material. In the United States, about 400,000 t of DU is in storage with various ^{235}U assays. About a third of the inventory has assay levels below 0.21 wt % ^{235}U.

If this material is declared a waste, existing regulatory requirements indicate that some type of geological disposal will be required [14,15]. Performance assessments [16] also indicate that large quantities of uranium are probably not acceptable for shallow land disposal. The uranium must be converted to a stable form such as an oxide for disposal. Use of DU within SNF WPs eliminates the separate, not insignificant, cost of disposal of this material. The use of DU as a beneficial material to improve WP performance reduces legal and institutional constraints in the management of DU.

CONCLUSIONS

DU inside the WP creates the potential for order-of-magnitude improvements in repository performance, provides higher assurance of WP performance based on uranium geochemistry, and provides a method to reduce the potential for nuclear criticality in a repository. As a new concept, significant uncertainties exist.

REFERENCES

1. J. A. Cogar, Waste Package Filler Material Testing Report, BBA000000-01717-2500-00008REV00 (U.S. Department of Energy, Las Vegas, Nevada, 1996).

2. J. L. Crosthwaite, The Performance Assessment and Ranking of Container Design Options for the Canadian Nuclear Fuel Waste Management Program, TR-500, COG-93-410 (Atomic Energy of Canada Limited Research Co.,1994).

3. C. W. Forsberg, et. al., DUSCOBS - A Depleted-Uranium Silicate Backfill For Transport, Storage, and Disposal of Spent Nuclear Fuel, ORNL/TM-13045 (Oak Ridge National Laboratory, Oak Ridge, TN., 1995).

4. Office of Civilian Radioactive Waste Management, Multi-Purpose Canister System Evaluation: A Systems Engineering Approach (U. S. Department of Energy, Washington, D.C., 1994).

5. J. A. T. Smellie, F. Karlsson, and B. Grundfelt., Proc. GEOVAL'94: Validation Through Model Testing (Nuclear Energy Agency, Organization for Economic Co-operation and Development, Paris, France, 1995), p.363-385.

6. P. A. Finn, J. C. Hoh, J. K. Bates, and S. F. Wolf, Proc. Topical Meeting: DOE Spent Nuclear Fuel - Challenges and Initiatives, Salt Lake City, Utah, December 13-16, 1994 (American Nuclear Society, La Grange Park, Illinois, 1994).

7. W. M. Murphy, Technology Today (June 1992).

8. D. K. Smith, B. E. Schettz, C. A. F. Anderson, and K. L. Smith, Uranium, 1, (1982), p 79.

9. E. C. Pearcy, J. D. Prikryl, W. M. Murphy, and B. W. Leslie, Applied Geochemistry, **9**, (1994), p. 713-732.

10. W. M. Murphy and E. C. Pearcy, (Mat Res. Soc. Symp. Proc., **257**, Pittsburgh, PA, 1992), p. 521-527.

11. W. M. Murphy, Radwaste Mag., **2** (6), (November 1995), p. 44.

12. International Atomic Energy Agency, <u>Natural Fission Reactors, Proc. of a Mtg. of the Technical Committee on Natural Fission Reactors, Paris France, December 19–21, 1977</u> (Vienna, Austria, 1978).

13. S. R. Naudet, <u>Natural Fission Reactors, Proc. of a Meeting of the Technical Committee on Natural Fission Reactors, Paris, France, December 19–21, 1977</u>, IAEA-TC-119/22 (International Atomic Energy Agency, Vienna, Austria,1978), p. 589-599.

14. U.S. Nuclear Regulatory Commission, <u>Safety Evaluation Report for the Claiborne Enrichment Center, Homer, Louisiana</u>, Docket No. 70-3070, NUREG-1491 (Washington, D.C., January 1994).

15. J. W. N. Hickey, <u>Letter from J. Hickey, Chief Fuel Cycle Safety Branch, U.S. Nuclear Regulatory Commission To Louisiana Energy Services, L. P.</u>, Docket No. 70-3070 (U.S. Nuclear Regulatory Commission, Washington, D.C., September 22, 1992).

16. Martin Marietta Energy Systems, Inc.; EG&G Idaho, Inc.: and Westinghouse Savannah River Co., *Radiological Performance Assessment for the E-Area Vaults Disposal Facility*, WSRC-RP-94-218 (Oak Ridge National laboratory, Grand Junction, CO., 1994).

STUDY ON THE FORMATION OF LANTHANUM PHOSPHATE FOR HIGHLY EFFICIENT INCORPORATION OF ACTINIDES FROM RADIOACTIVE LIQUID WASTE

Michitaka Sasoh, Shinya Miyamoto, Masumitu Toyohara, Mikio Wada*
Nuclear Engineering Laboratory, Toshiba Corporation.
* Isogo Engineering Center, Toshiba Corporation

Abstract

We have studied a co-precipitation method to remove actinides from radioactive liquid waste. Lanthanum phosphate was selected as a co-precipitation material because actinides and lanthanides are among the homology series in the periodic table. In this report, the conditions, under which the lanthanum compound is precipitated was looked at, as well as, ways to widen the pH range in which the precipitation occurrs in sodium nitrate solution. The properties of the resulting precipitates were also investigated. Further, we investigated the incorporation ratio of americium(decontamination factor), one of the actinides, into the precipitate from sodium nitrate solution. Lanthanum phosphate was found to be more effective compared to ferric compounds as the co-precipitation material to remove americium. The incorporation conditions for strontium using lanthanum phosphate were also investigated. The removal of strontium, by this method, was more effective when the pH is above 7.0.

Introduction

The actinides and lanthanides are among the homology series in the periodic table. Making use of similarities of chemical properties for actinides and lanthanides, a separation method of actinides from solution has been demonstrated in which precipitates of lanthanum compounds are produced [1],[2]. Actinides are incorporated into the precipitant with high efficiency even if the concentration actinide has not reached its solubility. Miyata et al. is reported separation method entails adding lanthanum nitrate solution to a solution containing mainly phosphoric acid; this also recovers the actinides with good efficiency[2]. However in a solution containing no fluoride ions nor phosphate ions, the lanthanum precipitate ($La(OH)_3$) appear in alkaline solution. Consequently, in order to precipitate the lanthanum compound, it is difficult to control the pH range of the solution without fluoride ions or phosphate ion. Focusing on the phosphate ion, we studied the conditions under which the lanthanum compound is precipitated and looked at ways to widen the pH range within which precipitation would occurr in sodium nitrate solution. We also investigated the properties of the precipitate produced. Further, we investigated the incorporation ratio of americium (decontamination factor), one of the actinides, into the

Mat. Res. Soc. Symp. Proc. Vol. 465 © 1997 Materials Research Society

precipitate from sodium nitrate solution. We also investigated the incorporation conditions for strontium. We think this method is applied to low and intermediate level waste from reprocessing plant.

Experimental

^{241}Am and ^{85}Sr were used as tracers. These were supplied by Amersham International plc.. The chemicals used were of A.R. grade.

To measure the lanthanum and phosphate concentration in the precipitates, the generated precipitates were dissolved in nitric acid solution. The atomic absorption spectrometry and absorptiometry were used to determine these species in the solution. The crystal structure of precipitate was identified by X-ray diffraction method.

Figure 1 shows the experimental procedure to remove of americium and strontium by co-precipitation using lanthanum phosphate. The sample of liquid phase was taken for alpha and gamma counting. Alpha-ray and gamma-ray were detected by liquid scintillation counting (for ^{241}Am) and NaI scintillation counter (for ^{85}Sr). We used 3.5M NaNO$_3$ solution in experiment from NaNO$_3$ concentration of the low and intermediate level waste from reprocessing plant[3].

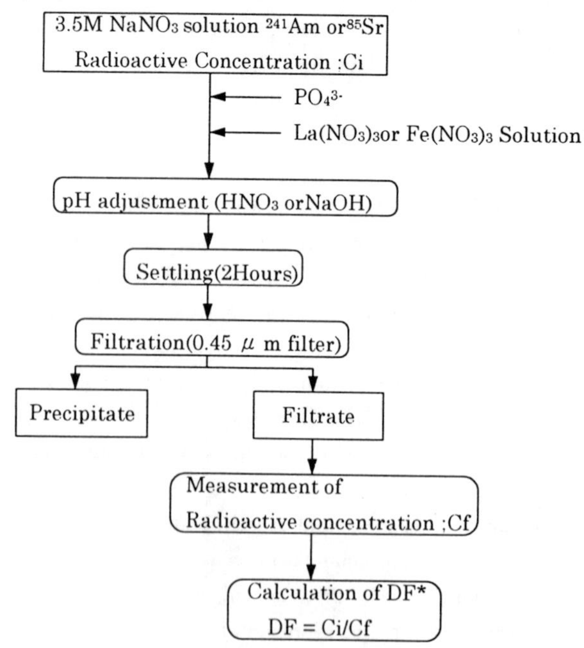

Figure 1. Flowchart of Co-precipitation Procedure

Result and Discussion

Study on Formation and Precipitation of Lanthanum Compounds

The precipitation ratio of lanthanum was investigated in a sodium nitrate solution with pH as the parameter. When there was no phosphate ions in the solution, lanthanum precipitated only if the pH value was higher than 10 (Figure 2). However when phosphate ions were present in the solution, the precipitation ratio achieved almost 100% within pH values between 2 and 12.

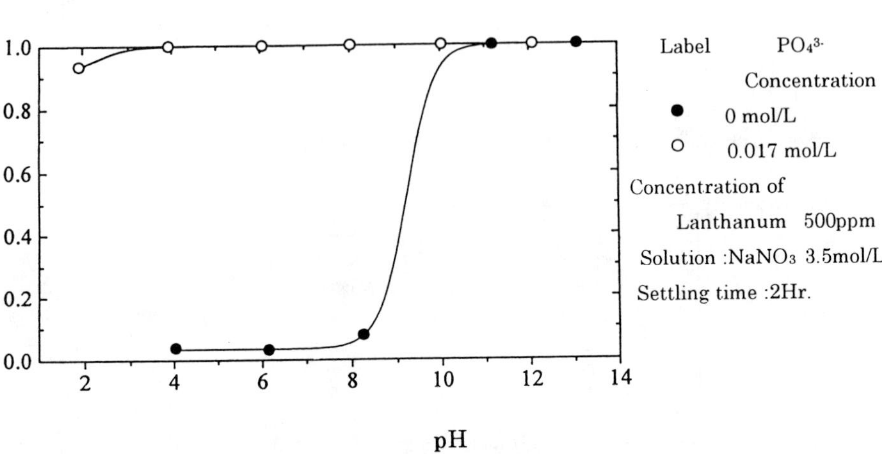

Figure 2   The precipitation ratio of Lanthanum

The mole ratio of phosphate and lanthanum in the precipitate was measured. To compare the case of using ferric compound as the precipitant, the mole ratio of phosphate and iron in the precipitate was measured under the same condition. The mole ratio of phosphate and iron in the precipitate was changed considerably by the pH value. However the mole ratio of phosphate and lanthanum in the precipitate proved to be 1 : 1.2, and it was found that this ratio gradually rose as the pH increased (Figure 3). As the solution increased in phosphate ion, phosphate and lanthanum existed in the precipitate.

From the X-ray diffraction patterns of the precipitate of lanthanum compound, we determined that LaPO$_4$ and La$_4$(P$_2$O$_7$)$_3$ were formed in the precipitate(Figure 4).

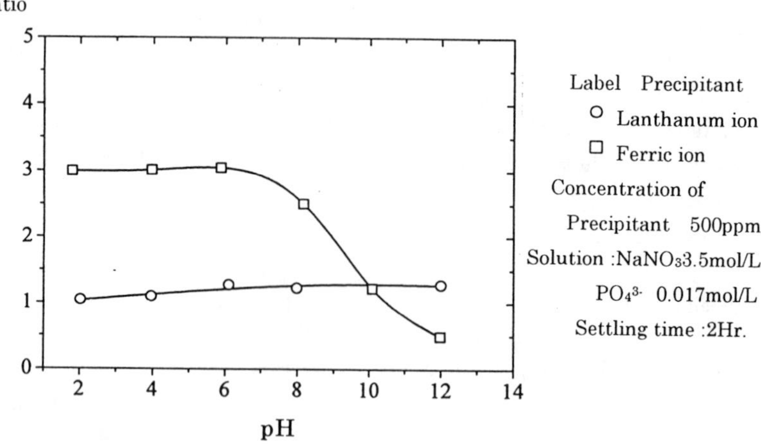

Figure 3 The mole ratio of PO_4^{3-}/Metallic ion in precipitation

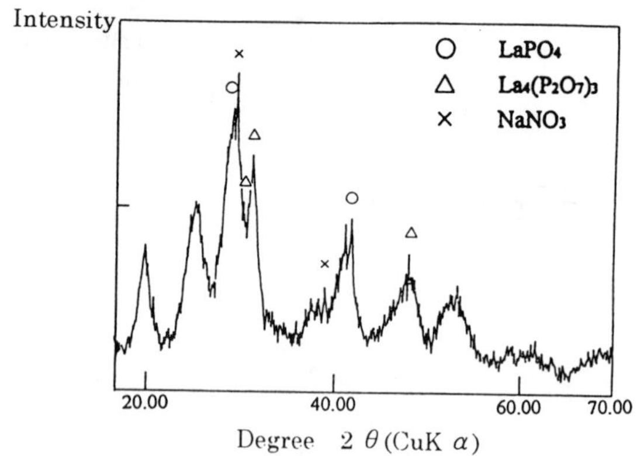

Figure 4   X-ray diffraction pattern of Lanthanum phosphate at pH2.0

Incorporation of Americium into the Precipitant

Figure 5 shows the decontamination factor of americium as a function of pH. The precipitation ratio of lanthanum phosphate (Figure 2) and the americium decontamination factor were quite similar over the pH range. This means that americium is also incorporated into the precipitate if lanthanum is precipitated.

The decontamination factor of americium was investigated as a function of increased lanthanum concentration. The comparison was made with a ferric compound.

(Figure 6). In the case of Fe^{3+}, DF of Am depended on ferric concentration. Cross et. al. is reported the Am removal as the function of Fe^{3+} concentration[4]. Same tendency was occurrd in their experiment, too.This demonstrated that the addition of lanthanum at 5ppm results in effective incorporation of americium into the precipitate while 10ppm of iron fails to incorporate more than a tiny amount.

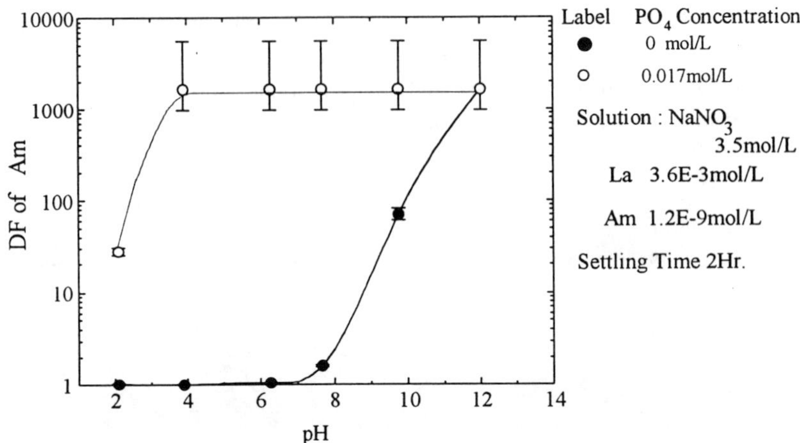

Figure 5 Removal of Amercium by Lanthanum

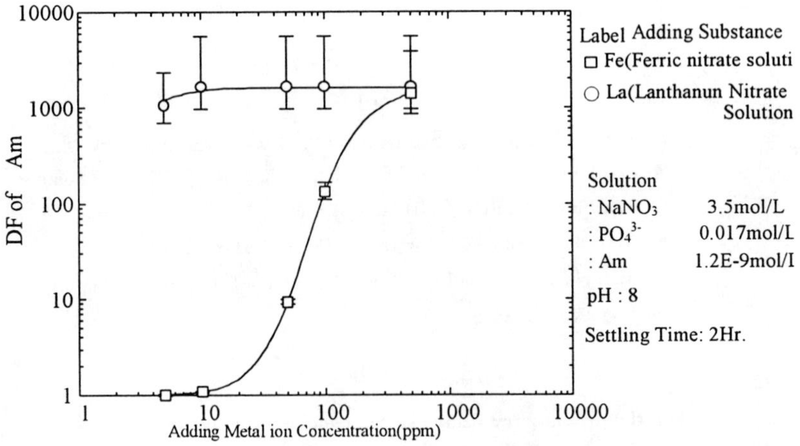

Figure 6 Removal of Americium by Lanthanum and Iron

<u>Incorporation of Strontium into the Precipitant</u>

The decontamination factor of strontium was also examined. Using the previous experiment, the strontium concentration was found to be below the solubility limit. Figure 7 shows the decontamination factor of strontium as a function of pH. The lanthanum phosphate produced a better incorporation of strontium into the precipitate than the ferric compound, especially when the pH is around 7.0. The decontamination factor with lanthanum as the precipitant was much higher than when iron was the precipitant. This means that the co-precipitation method using lanthanum phosphate has superior performance in the removal of strontium, which is commonly contained in liquid radioactive waste along with the actinides. This is assumed to be attributable to the similar ionic radii of strontium and lanthanum.

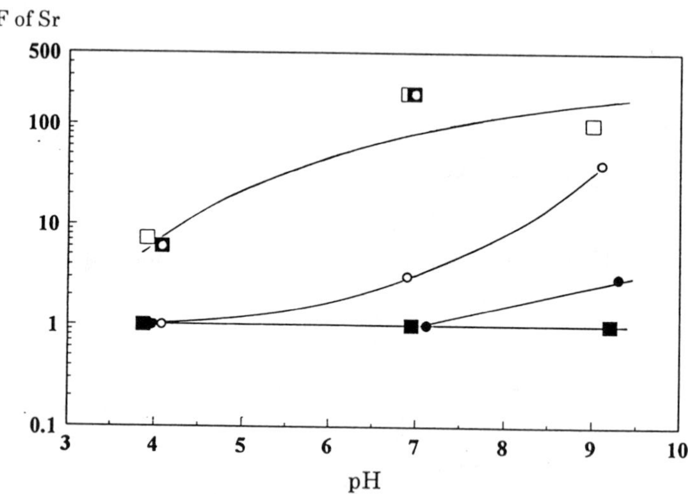

<u>Figure 7   Removal of Strontium by Lanthanum and Iron</u>

■ Lanthanum- PO_4^{3-} 0mol/L   ◨ Lanthanum- PO_4^{3-} 0.0085mol/L
□ Lanthanum- PO_4^{3-} 0.017mol/L

● Iron - PO_4^{3-} 0mol/L   ○ Iron- PO_4^{3-} 0.017mol/L

Solution $NaNO_3$: 3.5mol/L   Concentration of Precipitant : 500ppm

Settling Time : 2Hr.

Conclusion

Based on the preliminary study, we can conclude the following:
· To widen the pH range within which lanthanum precipitation occurred in solution,

adding phosphate ion is effective.

· According to the ratio of La^{3+} vs. PO_4^{3-} in the precipitation, it is sufficient adding quantity of PO_4^{3-} is 1.2 times of that La^{3+}.

· Co-precipitation by lanthanum phosphate is more effective than the method of using ferric compound. The adjustment of pH (above 7.0) enable to remove americium and strontium at once.

References

1)Hirayama,F., et al., Nuclear Instruments and Method in Physics Research, 223 ,188(1984)

2)Miyta,K., et al., Proceeding of the 1989 Joint International Waste Management Conference, Kyoto Vol.I (189),pp.157-162

3)Toyohara,M., et al., The Third International Conference on Nuclear Fuel Reprocessing and Waste Management 'RECOD '91' , pp.417-421(1991)

4)Cross,J.,E., Hooper,E.,W., Waste Management '89 (Tuscon),Vol.2, pp.667-673(1989)

Part VIII

Radiation Effects in Ceramics, Glasses, and Nuclear Waste Materials

MOLECULAR DYNAMICS SIMULATION OF THE TRAJECTORY
OF A RECOIL NUCLEUS IN A SIMPLIFIED NUCLEAR GLASS

J.M. DELAYE* and D. GHALEB**

*Commissariat à l'Energie Atomique (CEA) Centre d'Etudes de Saclay,
 F–91191 Gif-sur-Yvette Cedex, France
**Commissariat à l'Energie Atomique (CEA) Centre d'Etudes de la Vallée du Rhône, BP 171,
 F–30207 Bagnols-sur-Cèze Cedex, France

ABSTRACT

In a simplified (SiO_2, B_2O_3, Na_2O, Al_2O_3, ZrO_2) glass, corresponding to the basic matrix for the French nuclear waste containment glass, the authors developed a molecular dynamics simulation of the atom displacement cascades resulting from α disintegration of the actinides. After simulating the secondary cascades resulting from the first atoms displaced by a collision with a recoil nucleus, an actinide (U) was added to the model and the cascade produced by accelerating this type of atom was explicitly investigated, notably to observe the morphological evolution of the cascades and the resulting changes in the glass structure.

INTRODUCTION

Borosilicate glass has been selected in France for containment of high-level radioactive waste produced by reprocessing spent nuclear fuel. After fabrication and surface interim storage, the glass packages intended for disposal in a geological repository must be capable of withstanding both environmental corrosion and radioactive decay of the nuclides contained in the glass, for periods of 10^4 to 10^5 years.

Particularly, radiation effects can influence the corrosion by modifying the structure. It can be creation of defects (O_2 or He bubbles), changes in the diffusion rates, modification of the exchange surface (evolution of crack density) ...

Alpha decay of long-lived radioactive waste nuclides (actinides) in the glass matrix constitutes the primary source of atom displacements [1]. The basic phenomenon in the irradiation damage process, a displacement cascade lasting a few picoseconds, was analyzed using molecular dynamics to assess its ballistic effects on the glass structure. The molecular dynamics approach, which allows simulation at atomic scale, is ideal for analysis of displacement cascades in very minute time and distance frames.

Our approach to the problem involved two steps. Born-Mayer-Huggins potentials were first used to simulate the basic French nuclear glass matrix (SiO_2 + B_2O_3 + Na_2O + Al_2O_3 + ZrO_2)[2,3]; displacement cascades were then simulated within these structures and the effects of accelerating an atom in the matrix were analyzed[4].

Earlier published work discussed the effects of the simulated acceleration of an oxygen atom to energies not exceeding 1.5 keV – well below the energy of a recoil nucleus (up to 100 keV), but sufficient to obtain interesting results. When cascades involved atoms with a kinetic energy exceeding 1 keV, the disordered zone produced by the collision sequence was characterized by a peculiar morphology, divided into several elementary zones each surrounded by displaced Na atoms. Moreover, the disordered zones tended to be depolymerized, with lower mean coordination numbers for certain elements (notably B and O) after the passage of the projectile. A third characteristic observed was the large number of collective displacements that occurred during the ballistic process.

Using a parallel processing architecture we extended the investigation to higher energy levels and included heavy environmental nuclei with the same mass as uranium.

Mat. Res. Soc. Symp. Proc. Vol. 465 © 1997 Materials Research Society

METHOD

The modeled structure comprised 82 944 atoms in an invariant rectangular prismatic cell measuring $L \times l \times l$, where $L = 159.5$ Å and $l = 79.75$ Å. Periodic conditions were applied to obtain an infinite system. The structural composition was as follows :

$$63.8\%SiO_2+17.0\%B_2O_3+13.4\%Na_2O+4.0\%Al_2O_3+1.8\%ZrO_2.$$

i.e. the main elements of the actual nuclear glass were included in comparable proportions.

The potentials used to simulate the atomic interactions are pair potentials (Born-Mayer-Huggins potentials) and three-body terms. The formulations of these potentials and the values of the adjustable parameters are described in reference [3].

Two heavy atoms (with an atomic mass equal to that of uranium) were inserted in the structure.

The U-O interactions are simulated by a pair potential coming from the work of Lindan & al. [5]. The interactions between U^{4+} and the other cations are purely coulombic.

Initially, an amorphous system of 82 944 atoms was prepared at room temperature after liquid quenching. The liquid was relaxed at 6000 K for 13 600 time steps of 10^{-15} s before quenching at a rate of 5×10^{14} K·s^{-1} to 300 K, then relaxed at 300 K for 5000 time steps of 10^{-15} s to obtain the initial configuration in which the cascades were simulated. The cascade was initiated by accelerating one of the heavy atoms with energies of 4 keV or 6 keV [7], and following the modifications sustained by the glass matrix: displacements, coordination numbers and energies.

A time step Δt of 10^{-17} s was initially selected, then Δt was gradually increased to 10^{-15} s as the displacement cascade progressed. An outer layer 3.5 Å thick was maintained at room temperature by controlling the atomic velocities to dissipate the excess thermal agitation propagated during the collision sequence : the thermal wave was absorbed when it reached the outer layer. We have shown [6] that the control of the outer layer has little influence when the cascade volume is much smaller than the simulation cell volume.

DESCRIPTION OF 4 keV AND 6 keV CASCADES

<u>Displacement Cascade Morphology</u>

The first interesting aspect of the 4 keV and 6 keV cascades concerns the morphology of the damage zones. An atom was considered "displaced" if the distance between its initial and final positions exceeded 1 Å. **Figure 1** shows the displaced atoms at the end of the 4 keV cascade; in this calculation the path of the accelerated atom was relatively straight.

The damage zone can be clearly seen to be broken down into several elementary zones in **Figure 1**. Three zones may be distinguished: a very dense region around the projectile path, a second less dense region extending from the first one and located around the final position of the projectile, and a third less dense region visible at the right of the figure. The third damage zone corresponds to a secondary cascade: in a collision with the accelerated particle, an oxygen atom received considerable kinetic energy, initiating another cascade. The distance covered by the initial accelerated particle was 82.6 Å.

Figure 2 is the corresponding plot for the 6 keV cascade. The damage zone created by the 6 keV cascade exhibited a different morphology, with a single large damage zone corresponding in shape to the path of the projectile. The damage zone became narrower as the particle progressed, indicating a direct correlation between the width of the damage zone and the particle energy. The distance covered by the initial accelerated particle in this cascade was 106.76 Å.

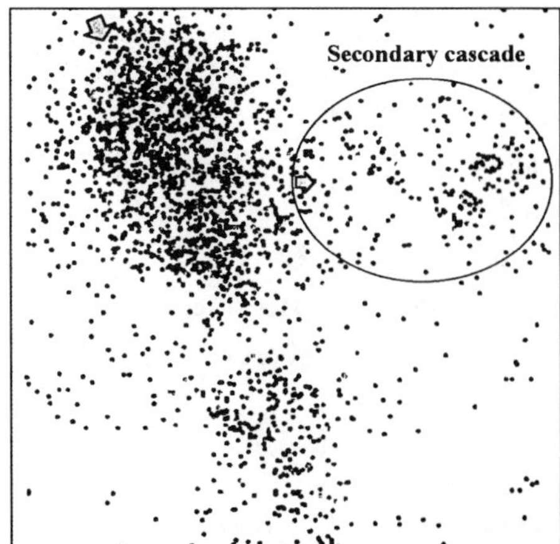

Figure 1. Morphology of 4 keV displacement cascade (only atoms displaced by more than 1 Å are shown).

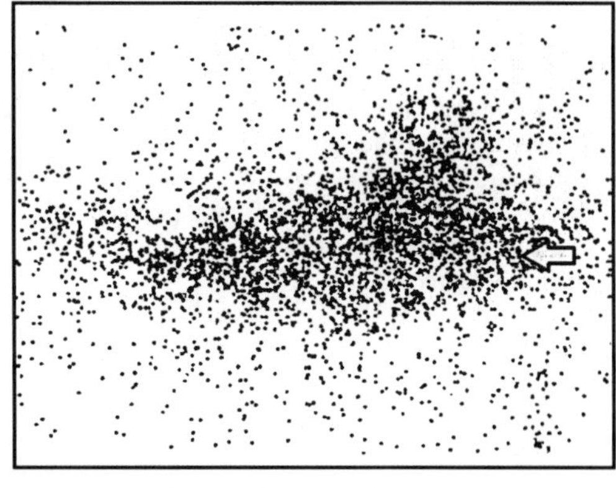

Figure 2. Morphology of 6 keV displacement cascade (only atoms displaced by more than 1 Å are shown)

Figure 3 shows the morphology of the 6 keV cascade damage zone, but with a "displaced atom" cutoff distance of 2.5 Å. The zonal breakdown that was not visible in **Figure 2** reappears in this representation, i.e. the separation between highly perturbed and slightly perturbed regions is again observed, as for the 4 keV cascade.

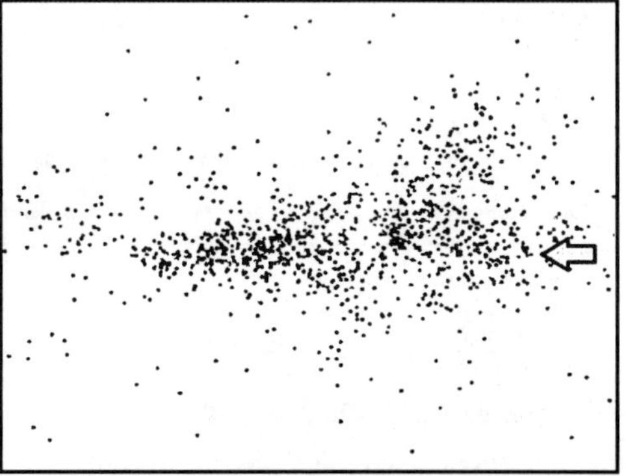

Figure 3. Morphology of 6 keV displacement cascade (only atoms displaced by more than 2.5 Å are shown).

Analysis of Displacements

The distribution of atom displacements for each element (**Figure 4**) shows that the broadest distributions were those of the sodium and oxygen atoms; the Si atom distribution was narrower, comparable to the distributions observed for the B and Al atoms. **Figure 5** shows the number (N) of atoms displaced by more than 1 Å versus the cascade energy (E) for three types of systems containing 5184 atoms (losanges), 41 972 atoms (crosses) and 82 472 atoms (squares). The number of displacements clearly increased with the incident energy, but not in a linear manner.

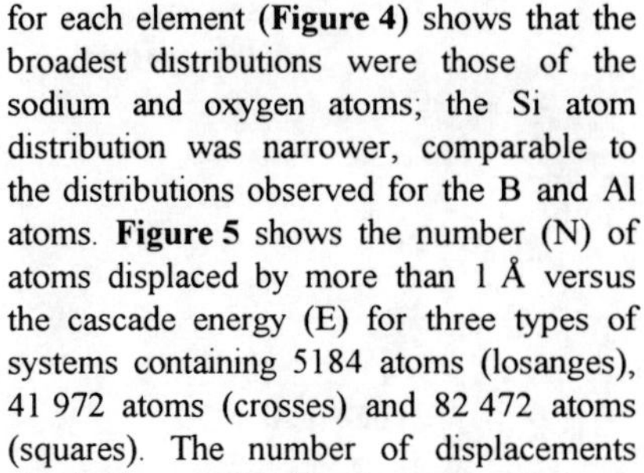

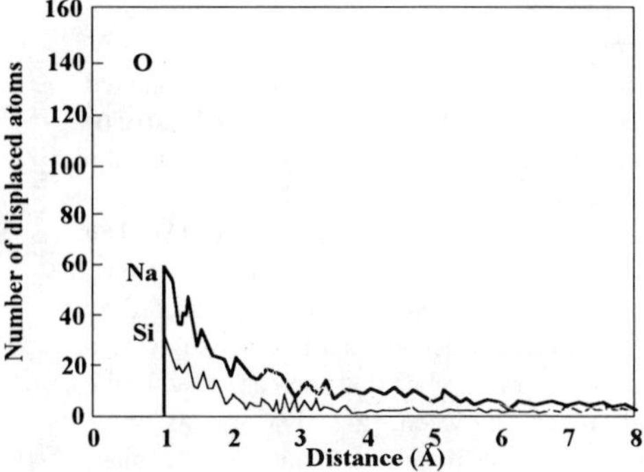

Figure 4. Distribution of atom displacements for various species.

The slope of the $N = f(E)$ curve increases with E (this phenomenon is discussed at the end of this article). Numerous collective displacements occurred during low-energy calculations with the 5184-atom system[4]. We pursued this analysis by examining changes in the local environment.

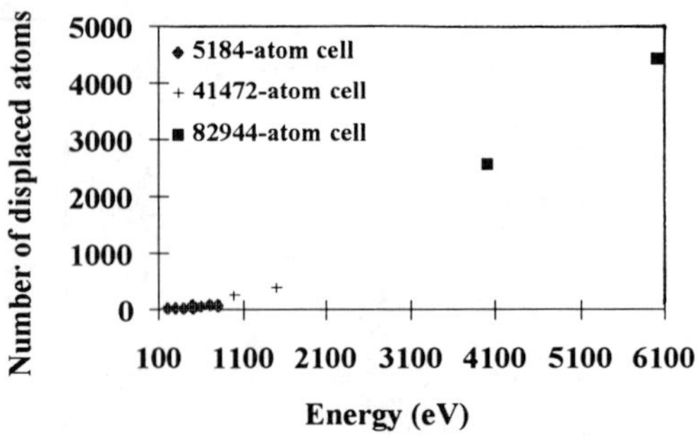

Figure 5. Number of displacements versus cascade energy (symbols correspond to calculations performed in cells of variable dimensions).

Table I lists the number of displaced Si atoms with changed local environments, itemized according to the number of changed first-neighbor atoms (from 0 to 4).No more than one neighbor was changed for about 72% of the displaced atoms, compared with only 28% for which two or more neighbors were different. The atoms with no more than one changed first neighbor may logically be considered to have taken part in a collective displacement. This was the case for a large majority of the displaced atoms, suggesting that atom displacements are mainly collective. This interesting phenomenon, and its repercussions, are discussed in greater detail below.

Table I Number of modified neighbors around displaced Si atoms

Number of modified neighbors	0	1	2	3	4
Displaced Si atoms (%)	33.1	39.3	17.6	6.7	3.3

Depolymerization and Annealing

Significant intermediate depolymerization was observed during the collision sequences for both the 4 keV and 6 keV cascades. "Depolymerization" refers to a gradual decrease observed in the mean coordination numbers of the Si, B and O atoms. Numerous Si(4) → Si(3), B(4) → B(3), O(3) → O(2) and O(2) → O(1) conversions occurred during the early stages of the collision sequence (by A(n) we meant an atom A with a coordination number equal to n).

Figure 6 shows the progressive depolymerization versus time for the 4 keV and 6 keV cascades. The calculated depolymerization (Y-axis in **Figure 6**) corresponds to the overall drop in the coordination number: $\Sigma N(i \to i\text{-}1) - \Sigma N(i \to i\text{+}1)$, where N = Si, O, B or Al. The first sum corresponds to the number of atoms for which the coordination number diminished by 1, and the second sum to those with a coordination number that increased by 1; the difference between the two sums thus represents the depolymerization of the glass matrix.

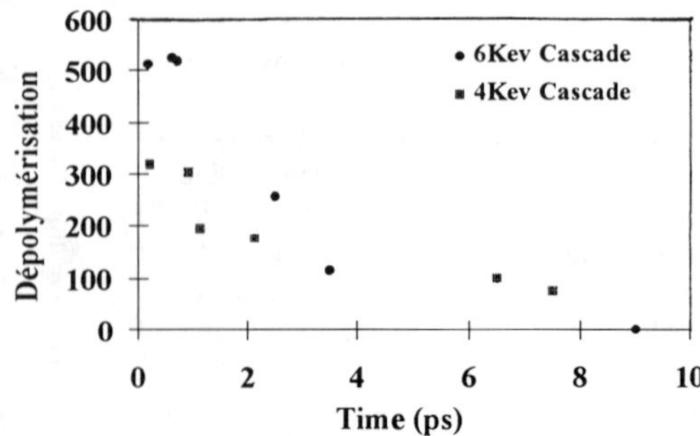

Figure 6. Depolymerization versus time (refer to text for a discussion of depolymerization).

A depolymerization peak is clearly visible at about 1 picosecond; a repolymerization process then began, and the residual depolymerization at the end of the cascade was considerably lower. This depolymerization/repolymerization process is directly related to the temperature map during the collision sequence. The higher the simulated cascade energy, the higher the temperature peak inside the damage zone, but also the more violent the collision sequence. The intermediate depolymerization level thus increases with the cascade energy, as shown in **Figure 6**. Repolymerization is also more effective at higher energies, however, as the intermediate temperature peak is higher. This observation highlights a difference in the time scales between the ballistic propagation of the cascade and the propagation of the thermal wave that accompanies the collision sequence. The multiplication of ballistic collisions is damaging and creates numerous defects, but it precedes the thermal wave, which then tends to restore the damage caused by the collisions.

Behaviour of Sodium Atoms

Another phenomenon previously observed during lower energy cascades is the role of sodium atoms. When cascades were initiated in 5184-atom systems[4], the sodium atom displacements occurred later than for the other species, and their final positions tended to surround the damage zone. The same phenomenon was again observed during these higher energy cascades.

Figure 7 shows the number of displaced atoms as a function of time; the Na atoms are represented separately from the other atoms to reveal the approximately 0.1 ps time lag in the Na displacement curve.

Moreover, the extent of the zone containing the displaced Na atoms encloses the damage zone, as shown in **Figure 8**. This figure was plotted after first identifying the center of gravity of the primary disturbed zone (the zone with the highest density) during the 4 keV cascade, then determining the distribution of distances between the displaced atoms and the center of gravity, while discriminating between Na atoms and all other atoms. For greater clarity, the distributions shown in the figure have been smoothed and the Na distribution amplified. The Na atom distribution is clearly broader than for the other atoms: the displaced Na atom volume encompasses the damage zone.

Furthermore, the damage zone morphology plots (**Figure 1** and **Figure 2**) show a number of isolated displaced atoms situated far from the projectile path. These were always Na atoms.

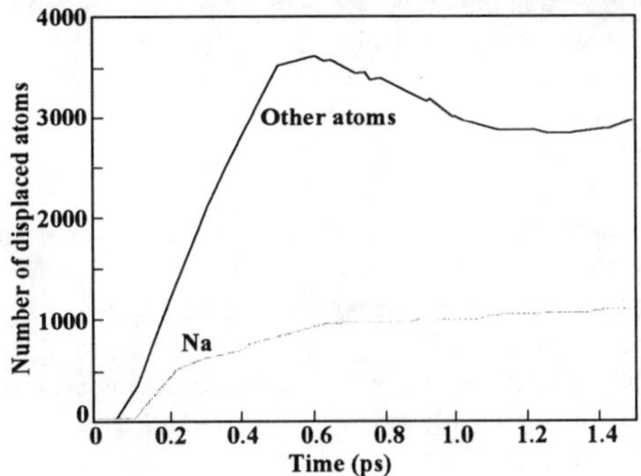

Figure 7. Number of displacements versus time; Na atoms plotted separately.

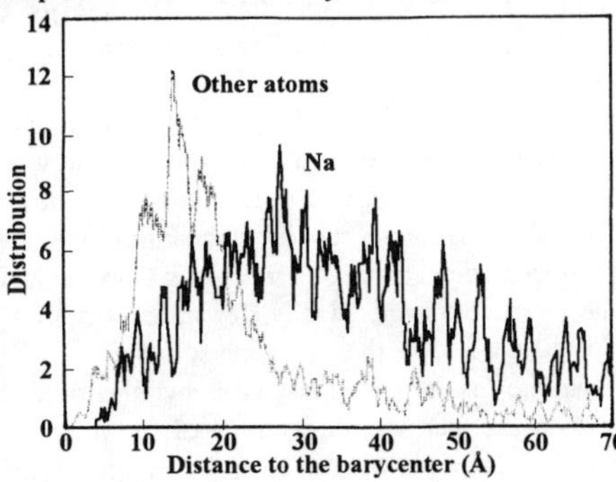

Figure 8. Distribution of distances between displaced atoms and center of gravity of the damage zone (first zone damaged during the 4 keV cascade); Na atoms are plotted separately

<u>Stored Energy</u>

We also analyzed the evolution of the stored energy: 492 eV for the 4 keV cascade and 710 eV for the 6 keV cascades. Expressed in terms of unit mass, these values are 31 $J \cdot g^{-1}$ for the 4 keV cascade (corresponding to a dose of 0.032 dpa) and are 45 $J \cdot g^{-1}$ for the 6 keV cascade (corresponding to a dose of 0.05 dpa). The available experimental stored energy values[1] (about 150 $J \cdot g^{-1}$ for a dose of 10^{18} α disintegrations per gram i.e. approximately 0.5 dpa) are not to far from the simulated results considering the differences between simulations and experiments.

DISCUSSION

The first point requiring clarification concerns the validity of the results, considering that the potentials used here – like all empirical potentials – are highly simplified representations of the actual atomic interactions. Nevertheless, they correctly account for the glass structures, distinguishing between network formers with a high local order and network modifiers with a low local order. This distinction is very important, and is responsible for the behavior of the atoms observed.

Experimental observations [10] have shown that most (about 80%) of the projectile energy is dissipated by kinetic energy transfer to other atoms through collisions. Therefore, although electron excitation processes have not yet been incorporated in the simulations, it seems reasonable to consider the calculated phenomena as physically representative and indicative of the behavior of network formers and modifiers.

A more troubling point is the inability of these potentials to stabilize oxide structural defects, i.e. the E' centers (a silicon atom surrounded by three oxygen atoms and an oxygen vacancy) and the O–O bonds that are liable to appear during the collision sequence. The structure at the heart of the cascade could be highly enriched in such defects, but this eventuality cannot be predicted by molecular dynamics simulations.

The first result requiring more thorough discussion concerns the observed zonal patterns in the damage zone. This phenomenon has already been observed in metallic alloys [8] at energies around 50keV. Two complementary explanations may be advanced to account for this phenomenon. One is statistical, i.e. along its path the projectile has a certain probability of sustaining a frontal collision with a glass matrix atom, depending on the collision cross section that increases as the kinetic energy of the projectile decreases [9]. The most severely damaged zones would then be the product of frontal collisions. The other possible explanation assumes the existence of zones offering little resistance to a high-energy atom, and denser zones in which the number of collisions is greater. Work is now in progress to characterize the possible zone in which the projectile encounters few obstacles.

It was also observed that the subsequent annealing effect was greater for high-energy cascades. Annealing is directly related to the amplitude of the thermal peak at the center of the damage zone. The number of defects created increases with the incident energy, but the annealing rate also increases with the incident energy. Finally, for an energy of 6 keV, the rate of annealing is sufficient to restore the defects created by the collision sequence.

In the case of an actual glass, this phenomenon will clearly depend on the defect formation energies and on the thermal conductivity of the material. There is probably an optimum irradiation energy at which the number of created and stabilized defects reaches a maximum.

In areas where the thermal wave is not sufficient to restore the glassy structure, a final depolymerization can subsist. If the relaxation time for these depolymerized areas is long, they could accumulate in the glassy structure, and could modify on a larger time scale the corrosion of the glass network.

Moreover, we have seen that, despite a complete repolymerization in the case of the 6keV cascade, a certain amount of energy is stored in the structure. It means that some structural rearrangements occurred. In reference [6], we have seen that for low energy cascades, an increase of the angular disorder appears. We can attribute the stored energy observed here to this phenomenon. And here again, if the relaxation time of these structural rearrangements is long enough, a certain amount of disorder could accumulate and influence the corrosion rate of the glass.

Sodium atoms are displaced with some delay compared with the other atoms, and their final positions tend to be situated around the periphery rather than at the heart of the damage zone. This raises two questions that will be investigated in the next simulations. First, it will be interesting to see whether the damage zone volume depends on the Na_2O concentration. Second, this seems to indicate that there are two types of displacements in the glass matrix: direct displacements following collisions between two atoms, and displacements induced by changes in the electric field as seen by certain atoms. Na atoms are highly mobile; it is thus logical to assume that they easily readjust their positions according to modifications in the electric field and thereby facilitate network relaxation, contrary to the other atoms, which are more likely to be displaced by ballistic collisions. The peripheral region occupied by the Na atoms surrounding the damage zone could thus tend to screen the electric field disturbances in the remainder of the glass matrix.

It is also clear that the displaced Na atoms situated far from the projectile path were displaced by modification of long-range $(1/r)$ Coulomb interactions capable of inducing displacements at considerable distances. Moreover, both types of displacements can account for the nonlinear curve observed in **Figure 5**. The number of displacements following readjustments to changes in the electric field may not be a linear function of the projectile energy. The more extensive the damage zone, the greater the number of atoms affected by a change in the electric field. Another explanation for this nonlinearity may involve the method used to control the outer layer, which limits the number of displacements to the higher energy values. Nevertheless, the highly nonlinear aspect of the curve suggests that a physical interpretation of the phenomenon is required.

Finally, the existence of large-scale collective displacements during a cascade must be taken into account, notably in the code based on approximations of binary collisions [10]. Such codes predict fewer displacements than are observed by molecular dynamics, because of the high migration energies required for individual displacements of network formers. The migration energies for collective movements are lower on the whole, and the total number of displacements is greater.

CONCLUSION

Displacement cascades at energies of 4 keV and 6 keV were simulated by molecular dynamics in 5-component oxide glass compositions including the major constituents of nuclear containment glass. Atoms with the same mass as uranium were incorporated in the simulation cells to reproduce the effect of accelerating a heavy atom in the glass matrix.

Analysis of the displacement cascades shows that the damage zones may be subdivided into highly disturbed and slightly disturbed zones. A depolymerization peak occurs during the displacement cascade; the higher the cascade energy, the more effective the damage annealing.

Network forming and network modifying atoms exhibit different behavior. Displaced Na atoms are found in a region more extensive than the zone containing the other displace atoms, indicating that the periphery of the damage zone is surrounded by displaced sodium atoms. This phenomenon could shield the remainder of the glass matrix from the electric field disturbances produced by the damage zone.

ACKNOWLEDGMENTS

The authors are grateful to Dr. Y. Limoge for an interesting critical reading of the paper and to Dr. Doan for his contribution to the computational code.

The authors wish to express their gratitude to COGEMA for its financial support in this study.

REFERENCES

1. W.J. Weber, Nucl. Instr. & Meth. in Phys. Rev. **B32**, p.471 (1988).
2. J.M. Delaye, D. Ghaleb, J. Non-Cryst. Sol., **195**, p.239 (1996).
3. J.M. Delaye, D. Ghaleb, Mat. Sc. & Ing. **B37**, p.232 (1996).
4. J.M. Delaye, D. Ghaleb, Proc. of 3rd International Conference on Computer Simulation of Radiation Effects in Solids COSIRES'96, Guildford U.K., 22-26 July 1996 (to be published).
5. P.J.D. Lindan, M.J. Gillan, J. Phys. Condens. Mat., **3**, p.3929 (1991).
6. J.M. Delaye, D. Ghaleb, J. Nucl. Mat. (in press).
7. N.V. Doan, F. Rossi, Sol. St. Phen., **30&31**, 75 (1993).
8. K.L. Merkle, in <u>Radiation Damages in Metals</u>, edited by M.L. Petersen, S.D. Harkners, A.S.M., 1975, p.58-94.
9. W. Jäger, J. Microsc. Spectrosc. Electron., **6**, p.437, (1981).
10. M.T. Robinson, J. Nucl. Mat., **216**, p.1 (1994).

EFFECT OF RADIATION ON TOPOPAH SPRING TUFF MECHANICAL PROPERTIES

P.A. BERGE, S.C. BLAIR
Lawrence Livermore National Lab., Livermore, CA 94550, berge1@llnl.gov, blair5@llnl.gov

ABSTRACT

The effect of radiation on the mechanical properties of Topopah Spring tuff was investigated by performing uniaxial compressive tests on irradiated and control samples of the tuff from the potential repository horizon at Yucca Mountain. Test results are presented, including stress-strain curves and peak strength and Young's modulus values. The results from this preliminary study show that for uncracked samples of Topopah Spring tuff, exposure to gamma radiation had no discernible effect on the unconfined compressive (peak) strength or the Young's modulus. However, results for samples that contained partially healed subvertical cracks indicate that exposure to radiation may reduce the strength and Young's modulus significantly. This is attributed to weakening of the cementing materials in the cracks and fractures of the samples that were irradiated. These results are preliminary, and additional studies are warranted to evaluate whether radiation weakens cementing materials in welded tuff.

INTRODUCTION

We present results of a suite of uniaxial compressive tests conducted to provide laboratory data to determine how radiation affects the compressive strength of Topopah Spring tuff, which is the host rock type for the proposed geologic repository at Yucca Mountain, NV. Both repository design and performance assessment require information about the effects of radiation on mechanical properties of rock in the near field of a repository. Until now, data describing the effects of radiation on tuff from the potential repository horizon have been unavailable.

We made precise measurements of mechanical properties of rock in uniaxial compression for irradiated and non-irradiated samples of Topopah Spring tuff. Details of the sample preparation, irradiation, and testing equipment and methods can be found elsewhere [1]. To evaluate the effects of radiation, we used a gamma-irradiation method similar to that used in previous investigations of the effects of radiation on Climax granite [2]. For our tests, identical procedures were used for preparing and mechanical testing of all samples, except that some samples were exposed to gamma radiation. Results for the irradiated and non-irradiated samples were then compared. The results are presented in the form of stress-strain curves and tabulated strength and modulus values.

EXPERIMENT

Core samples were machined from a piece of Topopah Spring tuff material that was broken from an outcrop during blasting to excavate a test area at Fran Ridge, Nevada Test Site. Samples for the experiments were prepared as right circular cylinders, 7.6 cm long and 2.5 cm in diameter. In order to have control samples and irradiated samples that could be compared, it was necessary to pair the cores that had similar appearance and that had similar cracks and vugs. All samples were described in detail, including number and location of visible cracks and vugs [1]. We formed 19 pairs from the 39 usable core samples. The sample chamber for the irradiation pool was large enough to hold 15 samples. For each of 15 pairs, we flipped a coin to determine which sample would be irradiated and which would be a control sample. The other nine samples also were included in the study as additional control samples.

One statistical method for analyzing the effect of radiation is known as blocking. A block is a unit of sample material within which the variation of some attribute is less than its variation between blocks. Treatment comparisons are then made within blocks rather than across blocks. The different blocks can be viewed as independent replications of the comparison. The block size in our experiments is two, so the method is also known as the method of matched pairs. For each

Mat. Res. Soc. Symp. Proc. Vol. 465 © 1997 Materials Research Society

pair, one sample is exposed to a massive dose of gamma radiation, while the second sample acts as a control. Any radiation effect is detected by comparing the measured parameter between the members of a pair.

Fifteen samples were subjected to a 9.5-MGy (0.9-Grad) dose of gamma irradiation from a ^{60}Co source over a 47-day period, at the LLNL Standards and Calibrations Laboratory. The remaining samples were held as controls. When the radiation exposure was completed, we found that the 15 irradiated cores had developed a distinctive gray color. This finding was not surprising because rock with high quartz content develops color centers when displaced electrons and holes are trapped by impurities and crystal defects. (In most cases, the color change can be reversed by putting the sample under an ultraviolet light source.)

We used a test apparatus equipped with a 50-ton hydraulic loading ram to perform unconfined compression tests on the 39 cylindrical core samples of tuff. Each sample was loaded in uniaxial compression (at a constant strain rate of 10^{-5} s^{-1}) until it failed.

An automatic data acquisition system was used to record the load cell input and output voltages and displacement transducer output voltages, for the uniaxial compressive tests. These data were used to calculate axial stress applied to each sample and the resulting displacements and strains. Stress-strain curves for all 15 pairs of irradiated and control samples were plotted [1].

RESULTS AND DISCUSSION

Typical stress-strain plots are shown in Figures 1 and 2. The solid lines represent the control samples and the heavy dashed lines represent the irradiated samples. Peak strength and Young's modulus values were determined for all the samples in the 15 pairs (Table I).

The irradiated samples had a mean peak strength of 139 ± 73 MPa, whereas the 15 corresponding control samples had a mean peak strength of approximately 154 ± 36 MPa. These values of peak strength are consistent with those reported elsewhere for welded tuff [3]. The large amount of scatter in the values was expected and is generally attributed to the heterogeneity in the form of cracks and vugs present in the rock. Average Young's modulus values found for the 15 irradiated and 15 control samples were 23 ± 5 GPa and 25 ± 3 GPa, respectively.

Stress-strain curves show that most of the samples behaved in a linear elastic manner up to the point of brittle failure, and that the Young's modulus for matched cores was similar (e.g., Figure 1, top). Stress-strain curves for some samples show nonlinear behavior at stress levels below the peak stress (e.g., Figure 1, bottom). For many of the pairs, the behavior for each of the samples was quite similar, but for some of the pairs, dissimilar behavior was observed for the two samples (e.g., Figure 2). To further evaluate this behavior, we divided the 15 pairs into two groups, termed the homogeneous and heterogeneous groups, based on stress-strain behavior for the pair. The heterogeneous group (pairs d, f, i, j, l, n) contained the pairs having one sample that failed at a very low stress level (e.g., below 50 MPa) compared to the other sample (typically above 100 MPa). Nearly all of the samples in the homogeneous group (pairs a, b, c, e, g, h, k, m, o) exhibited catastrophic brittle failure (e.g., Figure 1). Both irradiated and non-irradiated samples in this group had a higher mean peak strength than that determined for the total dataset, i.e. 185 ± 49 MPa for the irradiated and 169 ± 24 MPa for the control samples. Values for Young's modulus for the irradiated and non-irradiated samples in the homogeneous group were identical with similar standard deviations, i.e. 26 ± 2 GPa. For the welded tuff samples in the homogeneous group, radiation has little effect on the mechanical behavior in compression.

For the pairs in the heterogeneous group, we found a significant difference between the mean peak strength observed for the irradiated and non-irradiated samples (e.g., Figure 2). The irradiated samples had a mean strength of approximately 70 ± 38 MPa, and the non-irradiated had a mean strength of approximately 131 ± 37 MPa. Among the pairs in the heterogeneous group, the average value of Young's modulus for the irradiated samples, 19 ± 6 GPa, was also significantly lower than that for the non-irradiated samples, 23 ± 3 GPa. Preliminary examination of the core descriptions [1] for these samples indicated that, for many of the pairs in the heterogeneous group, both samples in the pair contained preexisting vertical or subvertical cracks. Also, for the irradiated samples, the failure occurred along one of these preexisting cracks. For the non-irradiated samples in this group, failure occurred more frequently by catastrophic or explosive fracture. A possible explanation of these results is that exposure to radiation weakened the cementing material in the pre-existing fractures. This tuff commonly contains carbonate cementing materials, but because of cost these samples were not analyzed chemically to determine the composition of the cementing materials. It is possible that the fracture orientation and characteristics of the samples in each pair are not identical. However, our preliminary results (summarized in Table II) show that irradiated samples with pre-existing fractures were weakened by the irradiation, when compared to the control samples.

One possible mechanism that could weaken a carbonate cementing material in a high radiation field is degradation of the carbonate by nitric acid formed by irradiation of moist air. This hypothesis could be evaluated by performing another similar set of experiments and changing the experimental procedure so that the irradiation vessel would be flooded with an inert gas such as argon instead of using air. A second possible mechanism is alteration of some of the hydrated minerals in the cementing material through radiolysis of the waters of crystallization. The alteration would weaken the cementing material and thus degrade the compressive strength.

Additional studies are warranted to evaluate whether radiation does weaken cementing materials in welded tuff. Careful analysis of such factors as geochemistry of tuff and cementing materials, temperature, dose rate, and irradiation geometry are beyond the scope of this study, and must be considered to establish whether radiation weakens the cementing materials in the tuff. However, if this is a real phenomenon, it has significant implications for the behavior of rock in the near-field region of the proposed nuclear waste repository, and should be taken into account when the shielding for the waste packages is designed. If shielding were to be inadequate, the radiation field would be expected to affect only rock exposed on the surface of excavated drifts and to penetrate only a few centimeters into the rock. However, the rock in this region would also experience the highest temperatures and stresses in a repository, and possibly high humidity. Weakening of fracture-filling materials may cause unanticipated spalling, which may change the amount and nature of rock fragments that would come in contact with the waste containers. In addition, changes in fracture properties, such as fracture shear strength, compressibility, and permeability could also occur. Changes in these properties would affect the thermomechanical and thermohydrological behavior of the rock in the near-field region. In particular, changes in the shear strength of cementing material in fractures would enhance stress gradients that would occur within the rock mass and may affect rock mass behavior in unanticipated ways, including movement of rock blocks along fractures.

CONCLUSIONS

We tested 15 pairs of Topopah Spring tuff samples and found that the irradiated samples had a mean peak strength of 139 ± 73 MPa, whereas the corresponding control samples had a mean peak strength of approximately 154 ± 36 MPa (see Tables I and II). These peak strength values are typical for the tuff [3], and the scatter is expected because different tuff cores contain varying amounts of vugs and cracks. Average Young's modulus values found for the 15 irradiated and 15 control samples were 23 ± 5 GPa and 25 ± 3 GPa, respectively. Our preliminary results (Table II) indicate that for intact samples of Topopah Spring tuff, exposure to gamma radiation had no discernible effect on the unconfined compressive (peak) strength or the Young's modulus. However, results for samples that contained partially healed, preexisting vertical or subvertical cracks (pairs in the heterogeneous group) indicate that radiation may cause significant degradation of the strength and Young's modulus. These results are preliminary, and additional studies are warranted to evaluate whether radiation weakens cementing materials in welded tuff.

Table I. Results from the uniaxial tests on each sample in pairs a–o.

Pair	Sample ID No.	Peak strength (MPa)	Young's modulus (GPa)	Rad. (Y/N)	Comments
a	1	170.5	25.4	N	Chipping at 122, 158, and 166 MPa.
	43	123.0	23.3	Y	Vertical crack formed through a preexisting vug near 100 MPa.
b	4	135.6	24.6	N	A sliver cracked off at about 87 MPa.
	8	221.2	27.1	Y	No chipping or cracking before failure.
c	7	191.1	27.8	N	No cracking or chipping before failure.
	28	213.4	25.9	Y	No cracking or chipping before failure.
d	9	58.9	17.9	N	Failure along preexisting fracture.
	10	150.5	21.2	Y	Preexisting crack opened at about 122 MPa.
e	11	192.6	27.5	N	Ram pressure increased somewhat slower than in other tests.
	16	151.9	26.1	Y	No cracking or chipping before failure.
f	12	164.5	25.0	N	No cracking or chipping before failure.
	37	63.0	23.9	Y	See comments in [1].
g	14	201.6	26.7	N	Chipping at about 193 MPa.
	22	236.0	26.4	Y	No cracking or chipping before failure.
h	20	158.3	27.1	N	Cracking and chipping around 148 MPa.
	6	204.0	26.8	Y	A chip fell off at 160 MPa.
i	25	149.0	25.4	N	Chipping at about 131 MPa.
	2	47.9	6.8	Y	Failure along preexisting fracture.
j	26	128.4	24.2	N	Chipping at about 122 MPa.
	40	76.3	19.0	Y	Chipping along a preexisting crack.
k	30	132.4	23.1	N	Large chips fell off; the test was paused and restarted.
	3	96.0	23.3	Y	Chipping at about 87 MPa.
l	31	116.8	22.7	N	Some cracking.
	39	31.9	21.0	Y	Failure along a large preexisting fracture.
m	33	168.3	23.2	N	No chipping or cracking before failure.
	32	213.0	26.6	Y	Cracking and chipping at about 193 MPa.
n	41	166.6	25.2	N	Chipping at about 140 MPa.
	35	51.3	22.5	Y	Failure due to preexisting crack.
o	44	174.1	27.7	N	No cracking or chipping before failure.
	15	205.8	27.3	Y	Chipping at about 193 MPa.

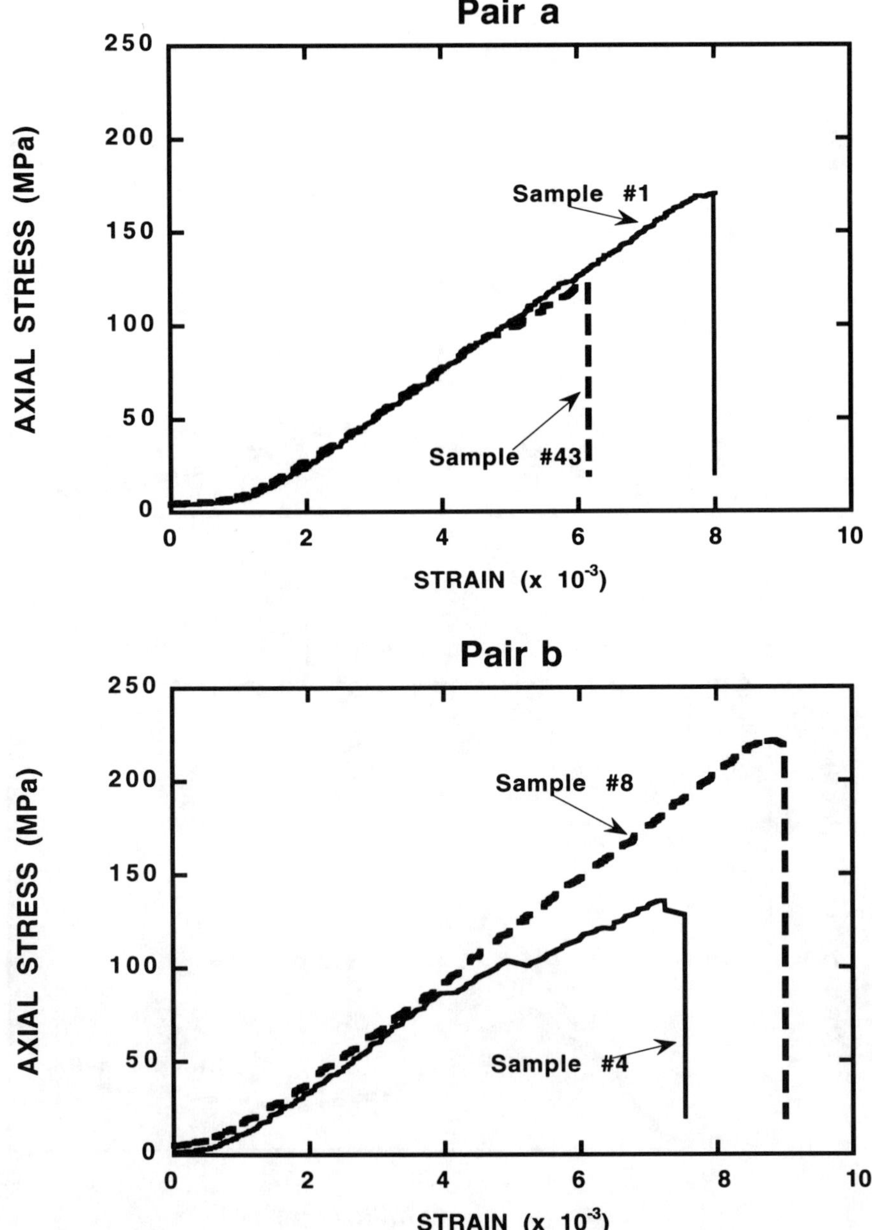

Figure 1. Stress-strain curves for four cores from homogeneous group. In pair a (top), sample #43 was irradiated and sample #1 was the control. In pair b (bottom), sample #8 was irradiated and #4 was the control.

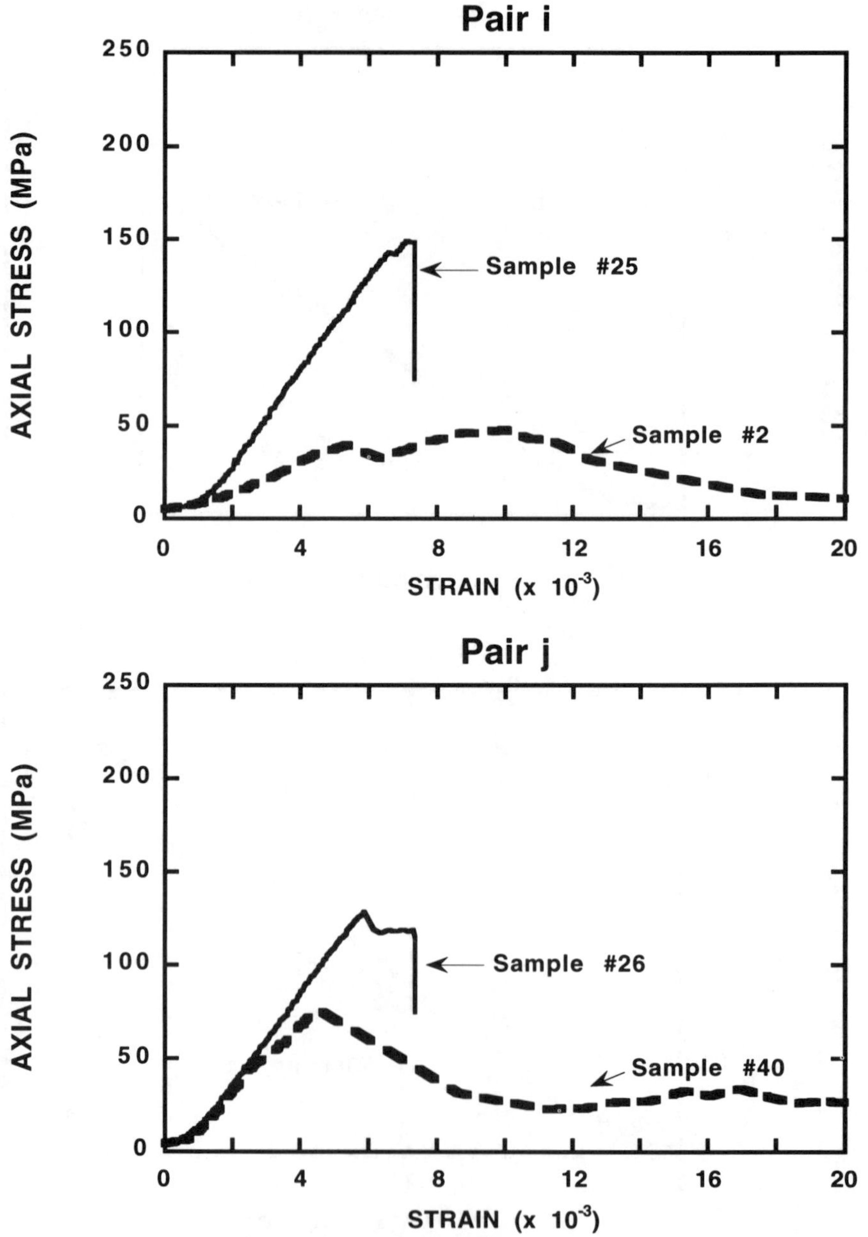

Figure 2. Stress-strain curves for four cores from heterogeneous group. In pair i (top), sample #2 was irradiated and sample #25 was the control. In pair j (bottom), sample #40 was irradiated and #26 was the control.

Table II. Average values for peak strength and Young's modulus for various groups of samples.

Samples	Rad. (Y/N)	Average peak strength (MPa)	Standard deviation	Average Young's modulus (GPa)	Standard deviation
Pairs a–o	Y	139.0	73.4	23.1	5.2
(15 pairs)	N	153.9	36.1	24.9	2.6
Homogeneous group	Y	184.9	48.9	25.9	1.5
(9 pairs)	N	169.4	24.3	25.9	1.9
Heterogeneous group	Y	70.2	38.4	19.1	5.7
(6 pairs)	N	130.7	36.8	23.4	2.6

ACKNOWLEDGMENTS

The authors thank J. M. Kelly, O. Pine, and R. Pletcher for assistance in performing the laboratory experiments, S. Trettenero and R. Pletcher for preparing the test apparatus, R. Pletcher for sample preparation, K. Keller for the automated data acquisition system, and O. Pine for data reduction. This work was performed under the auspices of the U. S. Department of Energy by the Lawrence Livermore National Laboratory under contract W-7405-Eng-48 and was supported specifically by the Yucca Mountain Site Characterization Project, LLNL.

REFERENCES

1. S.C. Blair, J. M. Kelly, O. Pine, R. Pletcher, and P. A. Berge, *Effect of Radiation on the Mechanical Properties of Topopah Spring Tuff*, LLNL report UCRL-ID-122899 (Lawrence Livermore National Laboratory, Livermore, CA, 1996), 29 pp.

2. W.B. Durham, J. M. Beiriger, M. Axelrod, and S. Trettenero, *The effect of gamma irradiation on the strength and elasticity of Climax Stock and Westerly granites,* Nuclear and Chemical Waste Management **6**, 159–168 (1986).

3. R. H. Price, K. G. Nimick, and J. A. Zirzow, *Uniaxial and Triaxial Compression Test Series on Topopah Spring Tuff,* SNL report SAND82-1723 (Sandia National Laboratory, Albuquerque, NM, 1982).

TRANSMISSION ELECTRON MICROSCOPY STUDY OF α-DECAY DAMAGE IN AESCHYNITE AND BRITHOLITE

W.L. Gong[1,2], L.M. Wang[1], R.C. Ewing[1], L.F. Chen[1], and W. Lutze[2]
1, Department of Earth and Planetary Sciences, University of New Mexico, Albuquerque, NM 87131, wgong@unm.edu; 2, Center for Radioactive Waste Management, University of New Mexico, Albuquerque, NM 87131

ABSTRACT

The aeschynite structure-type $(Ce,Nd,La,Th,U,Ca)(Nb,Ti)_2O_6$, and the rare-earth silicate apatite structure-type with the formula $(Ce,La,Nd,Ca,Th)_{10}(SiO_4,PO_4)_6(O,F,OH)_2$ are important rare-earth and actinide host phases for high-level nuclear waste. Natural phases of these structure-types have calculated alpha-decay doses up to $\sim10^{17}$ α-events/mg which have accumulated over hundreds of millions of years. Transmission electron microscopy has been used to study the microstructure of α-decay damage in aeschynite and britholite. Electron diffraction analysis of natural aeschynite revealed that minerals originally crystalline gradually lost their crystallinity with increasing alpha-decay doses. Helium bubbles were found in the aeschynite which have accumulated up to $\sim2\times10^{16}$ α-events/mg. These bubbles may nucleate within collision cascades during α-decay damage. Electron irradiation has an enhanced rare-gas migration and the formation of larger bubbles. High-resolution electron microscopy (HRTEM) revealed that amorphization during accumulation of α-decay damage was from alpha-recoil nuclei collision cascades, in both the aeschynite and britholite.

INTRODUCTION

Aeschynite (AB_2O_6) and britholite $(Ce,Ca)_{10}(SiO_4,PO_4)_6(O,OH,F)_2$ a natural silicate apatite phase, are important rare-earth minerals occurring in the Baiyun Obo, Nb-REE-Fe complex ore deposit, China. Because uranium and thorium are contained at the 0-25 weight percentage level in these natural phases, they are excellent candidates for the study of radiation damage resulting from α-decay events, receiving doses up to 10^{17} α-events/mg. Aeschynite and euxenite structure-types have been proposed as host phases for rare-earths and actinides in high-level nuclear wastes [1]. Rare-earth silicate phases of the apatite structure-type, $Ca_2RE_8(SiO_4)_6O_2$ (RE=La, Ce, Nd, etc.) to which britholite is analogous, have been extensively studied as actinide host-phases in high-level nuclear waste forms [2-4]. Actinides in nuclear waste have also been observed to preferentially partition into orthosilicate apatites in partially devitrified waste glass [2, 5]. Natural apatite are often used for geologic age-dating, because uranium can occupy the large 6h site. Due to the chemical durability of orthosilicate and natural apatites, the apatite structure-type has been suggested as possibly host phases for the immobilization and disposal of actinides, particularly plutonium [6].

These natural phases provide specimens which cover the full range of α-decay doses over which the radiation-induced transition from the crystalline-to-aperiodic state occurs. These doses are comparable to those reached in actinide doping experiments. In addition, these natural phases have accumulated their damage over geologic periods of time (10^6 to 10^9 years) during which annealing may have occurred. Thus, the naturally occurring phases may be used as structure-type analogues in order to understand the long-term radiation damage effects in actinide-bearing phases of crystalline ceramic nuclear waste forms. In this study, we present the results of a transmission electron microscopy study of α-decay damaged, natural aeschynite and the rare-earth silicate apatite, britholite.

Detailed studies of the effects of alpha-decay damage in $Ca_2Nd_8(SiO_4)_6O_2$ doped with ^{244}Cm have been previously completed [5, 7, 8]. Irradiation-induced amorphization in this material occurs directly within the collision cascade of the heavy recoil nuclei (Pu) emitted during α-decay of Cm, and the fully amorphous state is reached at a dose equivalent to 0.3 dpa. Recently, studies of the effect of temperature and ion mass on amorphization in $Ca_2La_8(SiO_4)_6O_2$ have been

Mat. Res. Soc. Symp. Proc. Vol. 465 © 1997 Materials Research Society

systematically reported [9-11]. In this study, we report direct observations of individually resolved collision cascades in lattice images of britholite, which are compared to those revealed in ion-irradiation induced amorphization in synthetic rare-earth apatite, $Ca_2Nd_8(SiO_4)_6O_2$ [11].

STRUCTURE DESCRIPTION OF AESCHYNITE AND BRITHOLITE

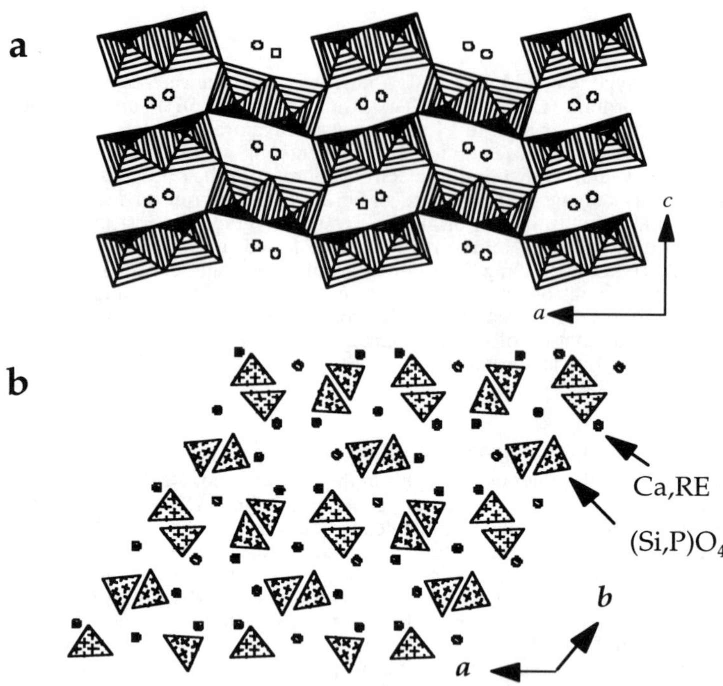

Fig. 1. The polyhedral representation of aeschynite and rare-earth silicate apatite
a) aeschynite; b) rare-earth silicate apatite

The complex Nb-Ta-Ti oxides of the type formula AB_2O_6 (where A = RE, Y, Th, U, Ca, Pb, Fe^{2+}, Mn; B = Nb, Ta, Ti, Fe^{3+}, Al) are characterized by isomorphism and dimorphism. Two distinct isomorphous series are known: the aeschynite series, *Pnma*, and the euxenite series, *Pncn* [12, 13]. The aeschynite structure-type occur in granites, pegmatites, alkaline and carbonatite complexes, and hydrothermal veins. The chemical compositions of naturally occurring aeschynite and euxenite structure-types from worldwide localities indicate that $(Th,U)O_2$ correlates well to the TiO_2, and the $(Th,U)O_2$ content is as high as 30 wt. % [13]. This suggests that aeschynite structure-types are solid-solutions of $RENbTiO_6$-$ThTi_2O_6$-UTi_2O_6-$CaNb_2O_6$ end-members. Charge balance is maintained through coupled atomic substitutions, such as $(Th,U)^{4+}+Ti^{4+} = Nb^{5+} + RE^{3+}$. The basic structural features of the aeschynite series are a "puckered" hexagonal packing of oxygen, whose three-dimensional network contains very large holes. Nb, Ti, and Ta occupy the octahedral, B-sites of the framework, while the A-site cations such as REE, Y, Th, U, and Ca, occupy the large cavities (Fig.1a). The projection of the crystal structure along the b-axis shows zigzag strips of the octahedral double chains of octahedra. The B-site cations form a basic BO_6 octahedron parallel to the b-axis, and the large holes are centered on the 2_1 screw axes. Two of the large A-site cations lie in the tunnel formed by the holes in a distorted 8-coordinated polyhedron [14]. Thus, a variety of large cations can be accommodated in the tunnels of connected holes in the aeschynite structure-type.

Britholite, $(Ce,La,Nd,Ca)_{10}(SiO_4,PO_4)_6(O,OH,F)_2$, is a natural rare-earth silicate apatite. Uranium and thorium enter the apatite structure by the atomic replacement: $Ca^{2+} + (Th,U)^{4+} = 2RE^{3+}$. In a previous study [15] utilizing X-ray and neutron powder diffraction techniques, the hexagonal crystal structure of $Ca_2Nd_8(SiO_4)_6O_2$ has been shown to belong to F63/m space group and to be isostructural with natural apatite, $Ca_{10}(PO_4)_6(F,OH)_2$. This orthosilicate apatite has a very open structure in which the SiO_4 tetragonal monomers are isolated from one another and then linked by the various metal cations (Fig. 1b). Two Ca ions and two Nd ions occupy the 8-fold cation sites (4f), with the six remaining Nd ions occupying the 7-fold sites (6h).

SAMPLES AND EXPERIMENTAL PROCEDURES

The aeschynite and britholite in this study were collected from the Baiyun Obo, Nb-REE-Fe complex ore deposits, China [16, 17]. Aeschynites are one of the most common rare-earth minerals in the Baiyun Obo ore deposit. The varieties of aeschynite are very abundant, including Ce- and Nd-members of aeschynite and niobo-aeschynite. The aeschynite used in this study is niobo-aeschynite-(Ce) with a general formula of $(Ce,Nd,La,Sm,Pr,Th,U,Ca)(Nb,Ti)_2O_6$. The britholite is britholite-(Ce) with a general formula of $(Ce,La,Nd,Pr,Ca,Mg,Th)_{10}(SiO_4, PO_4)_6(O,OH,F)_2$. The britholite consists of 65 mol. % $Ca_2RE_8[SiO_4]_6O_2$ and 35 mol. % $Ca_{10}(PO_4)_6(F,OH)_2$. The Equivalent Uranium (EU) which equals U+Th/3 µg/g ranged $0.1\text{-}15\times10^3$ for the minerals.

The TEM samples for aeschynite were prepared from thin foils by Ar ion-milling which was finished at 3.5 keV and 11°. The TEM samples for britholite were prepared by dispersing the powder in acetone and then depositing on holey-carbon filmed copper grid substrates. TEM was performed using a JEM 2000FX microscope operated at 200 keV. The α-recoil collisional cascade images were recorded by high-resolution TEM using and JEM 2010 microscope having a point-to-point resolution of 0.194 nm operated at 200 keV. Natural samples commonly display chemical zoning and heterogeneity. The concentrations of uranium and thorium were measured on areas corresponding to those where selected area diffraction patterns and HRTEM images were taken in order to correlate microstructural observations with α-decay dose.

In this study, the α-decay dose is given in units of alpha-decay events per mg (α/mg). The dose calculation is similar to that used by Holland and Gottfried [18], and includes two terms each for the decay of ^{238}U and ^{232}Th, respectively:

$$D = (8N_{238}\lambda_{238} + 6N_{232}\lambda_{232})t \qquad (1)$$

where D is dose, N_{238} and N_{232} are the present numbers of atoms/mg of ^{238}U and ^{232}Th, respectively, λ_{238} and λ_{232} are half-decay periods of ^{238}U and ^{232}Th, respectively, and t is the age of the samples. The geologic ages for aeschynite and britholite are 270 Ma [16]. Thus, the alpha-decay doses can be calculated by application of the equation 1. The displacement per atom (dpa) values were calculated by assuming 1500 atomic displacements per alpha-decay event:

$$dpa = 1500 \, (D \times M)/N_f N_a \qquad (2)$$

where M equals the molecular weight in mg units, N_f is the number of atoms per formula unit, and N_a is Avogradro's number.

RESULTS AND DISCUSSION

The progressive amorphization process by α-decay damage in natural aeschynite was evidenced by changes in the electron diffraction patterns (Fig. 2). Generally, at a dose less than 4×10^{15} α-events/mg, no apparent damage was observed. At a dose of $4\text{-}12\times10^{15}$ α-events/mg, a diffusive diffraction halo appeared, which was superimposed over the diffraction spots caused by the remaining crystalline matrix. At doses >12×10^{15} α-events/mg, complete amorphization was reached. The α-decay dose for amorphization of aeschynite is close to that of minerals in the pyrochlore group, ~10^{16} α-events/mg [19]. For a comparison, the [244]Cm doped $Gd_2Ti_2O_7$ with the pyrochlore structure-type becomes full amorphous at a dose of 3.1×10^{15} α-events/mg [20]. $CmAlO_3$ with the perovskite structure-type becomes amorphous after a dose of 1.4×10^{15} α-events/mg which was achieved in 7.5 days from the decay of [244]Cm [21]. A HRTEM image for aeschynite receiving ~8×10^{15} α-events/mg shows that collisional cascade overlap may be responsible for amorphization (Fig. 3). The apparent size of the collisional cascades in aeschynite averages ~5 nm in diameter, as compared to 3.8 nm for titanite which has received 7.5×10^{14} α-events/mg [22].

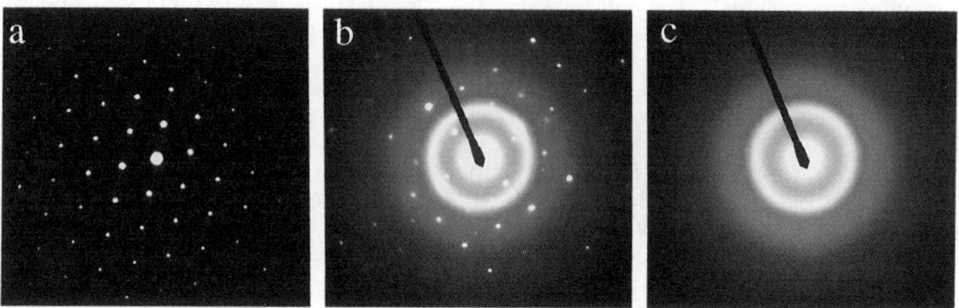

Fig. 2. Electron diffraction patterns of aeschynite at different accumulated α-decay doses: a) < 4×10^{15}; b) 4-12×10^{15}; and c) >12×10^{15} α-events/mg.

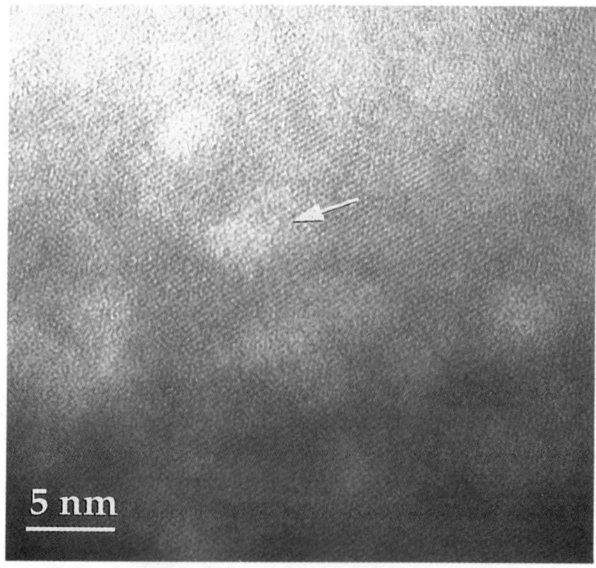

Fig. 3. HRTEM image showing collision cascades in aeschynite (~8×10^{15} α-events/mg).

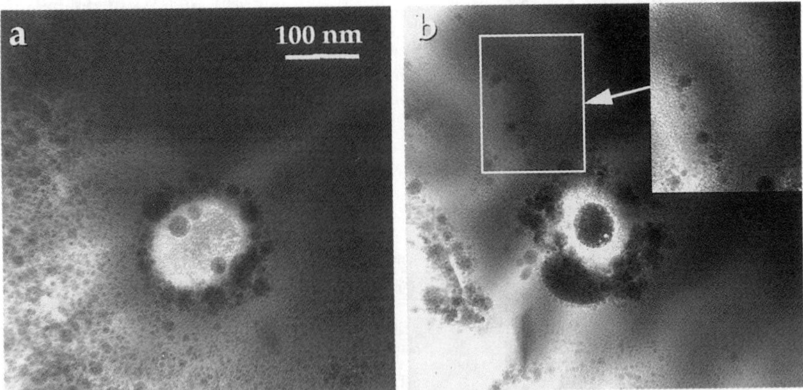

Fig. 4. Bubble formation in amorphous aeschynite after receiving a dose of 18×10^{15} α-events/mg. The bubbles may contain He which diffuses due to the electron-beam irradaition.

Alpha-decay damage in minerals is caused by two separate but simultaneous processes associated with the alpha-decay events [23]: (1) An alpha-particle (~4.5 MeV) with a range of 10,000 nm dissipates most of its energy by ionization; however, at low velocities near the end of its track, the α-particle displaces several hundred atoms, creating Frenkel defect pairs. (2) The alpha-recoil atom (~0.09 MeV) with a range of 10 to 20 nm produces several thousand atomic displacements which appear as collisional cascades in lattice images.

Helium, which results from the capture of two electrons by an α-particles, must be accommodated interstitially or substitutionally, and the He diffuses to irradiation induced defects [24]. Fig. 4 shows spherical bubbles in the amorphous matrix. These bubbles migrate and coalesce under continued electron irradiation (200 keV electrons at 300 K). The spherical bubbles (5-50 nm in diameter) most probably contain helium produced by the α-decay process. H_2O species in the samples were extremely low, as determined by the infrared spectrometry [13]. The bubble formation was not observed in partly amorphized aeschynite samples. The helium atoms are probably traped in defects produced by irradiation in the crystalline matrix, and they may diffuse along the tunnels of the crystalline aeschynite structure. The focused electron-beam enhances the diffusion process of helium atoms, thus forming larger bubbles in the amorphous matrix. A similar phenomenon has been observed in Xe and Kr ion-implanted muscovite at 25 K and room temperature [25]. Visible cavities ranging from 5-50 nm in diameter, filled with fluid rare gas, formed in the mica irradiated by 120 keV Kr ions up to a dose of 10^{16} ions/cm^2. Inagaki et al. [26] observed bubble formation in ^{238}Pu- and ^{244}Cm-doped glasses as a result of α-decay. When higher helium concentrations (up to 10^4 ppm) are introduced by (n, α) reactions, helium bubble formation in nuclear waste glasses was observed at ambient irradiation temperature (<100 °C) by Dé et al. [27] and at a slightly elevated irradiation temperature (<230 °C) by Sato et al. [28]. Thus, the bubble formation in the metamict aeschynite may be indicative to helium accumulation and bubble formation in nuclear waste glasses.

The atomic arrangement on the (001) plane of britholite is shown in Fig. 1b. A HRTEM image generated from a computer simulation of a 4 nm thick foil with a -20 nm defocus. The simulation shows that most of the atom columns on this plane can be resolved with HRTEM with a 0.19 nm point-to-point resolution (Fig. 5). The HRTEM image of britholite matches very well with the computer simulated image (small insert with defocus, Δf = -25 nm; sample thickness, t = 3 nm). The circular white dots represent oxygen columns and the six dark dots represent rare-earth and calcium atoms, assuming rare-earths and calcium are disordered among the 6h and 4f sites. In the sample receiving a dose of ~5×10^{15} α-events/mg (correspondingly 0.44 dpa), brighter spots of 5 nm and more in diameter were observed in the thin region of the britholite grain (Fig. 6). Although the lattice fringes were not disrupted within some of the brighter spots, they are believed to be the images of small amorphous domains created by the collisional cascades of α-recoil nuclei in a crystalline matrix. Image simulation of structures containing various amount of amorphous volumes has displayed similar contrast and has indicated that when the amorphous volume fraction is less than 35 %, the lattice fringes may not be disrupted or the amorphous domains may not be detected at all [29, 30]. A damage domain where lattice fringes are completely disrupted is also observed as indicated by the arrow in Fig. 6. The average size of these amorphous domains is ~5 nm in diameter, very close to that of collision cascades after irradiation by 1.5 MeV Kr ions up to 0.5 ions/nm^2 at 300 K [11].

Fig. 5. HRTEM plan view image of a (001) britholite foil which matches well with the computer simulated images (inserts).

The cascade microstructure in this natural rare-earth silicate apatite, britholite is consistent with that of ion beam irradiation of $Ca_2La_8(SiO_4)_6O_2$ [11]. The results of experimental HRTEM and the image processing on ion beam irradiated $Ca_2Nd_8(SiO_4)_6O_2$ suggest that amorphization occurs directly in the displacement cascades. The amorphous domain is not a column through the entire sample thickness like those amorphous tracks induced by GeV ions, but rather they are spherical in dimension [11]. In addition, earlier work on $Ca_2Nd_8(SiO_4)_6O_2$ concluded that amorphization occurs by direct impact collision for 95 keV ^{240}Pu [8]. Britholite which has received an α-event dose equivalent to 0.4 displays only individual cascade damage (Fig. 6); whereas, the fully amorphous state is reached at a dose equivalent to 0.3 dpa in the Cm-doped $Ca_2Nd_8(SiO_4)_6O_2$ at room temperature. The critical amorphization dose of $Ca_2La_8(SiO_4)_6O_2$ due to a 1.5 MeV Kr$^+$ irradiation is 0.3-0.5 dpa at room temperature [9, 10]. Recently, Wang and Weber [11] experimentally confirmed that the size of surviving collision cascades decreases with the increasing temperature due to increased dynamic recovery of the crystalline phase at higher temperatures through epitaxial regrowth from the crystalline-amorphous boundary for each cascade. In natural samples of great age, there is annealing of α-decay damage [18]. The annealing may occur at low temperatures and involves the removal of isolated Frenkel defect pairs caused by α-particle and epitaxial recrystallization of amorphous volumes caused by the collision cascades of the recoil nuclei. Thus, for a natural rare-earth silicate apatite of ~270 Ma, a greater α-decay dose is necessary to reach a similar microstructural effect to that observed for ion-beam irradiated or Cm-doped rare-earth silicate apatites.

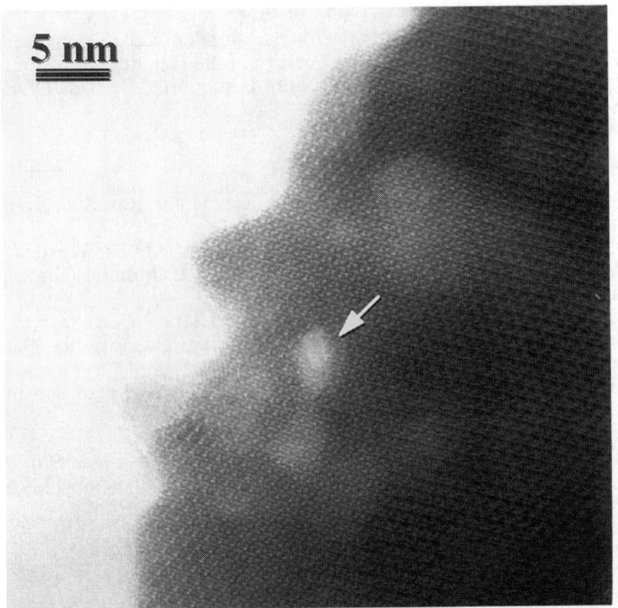

Fig. 6. HRTEM image of a (001) britholite foil after accumulationof 5×10^{15} α-events/mg.
The bright domains in the crystalline region are believed to be caused by the survivingcollisional
   cascades.

CONCLUSIONS

TEM and HRTEM have been used to study α-decay damage in naturally occurring aeschynite and the rare-earth silicate apatite, britholite, which both have received doses up to ~10^{17} α-events/mg (~9 dpa). The results lead to the following conclusions:

(1) Aeschynite and britholite gradually lose their crystallinity due to the accumulation of α-decay damage. They become amorphous at a dose of approximately 1.2×10^{16} α-events/mg (~1 dpa).

(2) Helium bubbles possibly occur in the aeschynite samples which are completely amorphous and have received α-decay doses >2×10^{16} α-events/mg. The focused electron beam has an enhancing effect on the diffusion of helium gas retained in the amorphous matrix, and this leads to the formation of bubbles.

(3) Amorphization (metamictization) occurs directly within the collisional cascades caused by the α-recoil nuclei. Complete amorphization occurs due to the overlap of these individual cascades. Annealing at low temperature over several hundreds of million years may lead to a decrease in the size of the residual amorphous domains generated by each collision cascade and to an increase of α-decay dose required for amorphization in natural samples.

(4) Microstructural effects of α-decay damage in natural samples is an important consideration in the evaluation of long-term radiation effects in crystalline, ceramic, nuclear waste forms.

ACKNOWLEDGMENT

The authors gratefully acknowledge the help of Professors C.J. Guo, Y.X. Wang, Y.Z. Qiu, P.S. Zhang, K.J. Tao, and W.X. Hong at Chinese Academy of Sciences. The TEM was performed at the Electron Microbeam Analysis Facility of the Department of Earth and Planetary Sciences of the University of New Mexico. This work is supported by BES/DOE Grant No. DE-FG03-93ER45498.

REFERENCES

1. W.L. Gong, R.C. Ewing, L.M. Wang, and H.S. Xie, Mater. Res. Soc. Symp. Proc. **412**, (1996).
2. W.J. Weber, P.R. Turcotte, L.R. Bunnell, F.P. Roberts, J.H. Jr. Westsik, in *Ceramics in Nuclear Waste Management*, edited by T.D. Chikalla and J.E. Mendel (Springfield, Virginia, National Technical Information Service, 1979), pp. 544.
3. G.J. McCarthy, J.G. Pepin, D.E. Pfoertsch, D.G. Clarke, *Ceramics in Nuclear Waste Management*, edited by T.D. Chikalla and J.E. Mendel (Springfield, Virginia, National Technical Information Service, 1979), pp. 315.
4. P.R. Turcotte, J.W. Wald, F.P. Roberts, J.M. Rusin, and W. Lutze, J. Amer. Ceram. Soc. **65**, 589 (1982).
5. W.J. Weber, J. Amer. Ceram. Soc. 65, 544 (1982).
6. R.C. Ewing, W.J. Weber, and W. Lutze, in *Disposal on Weapon Plutonium*, edited by E.R. Merz and C.E. Walter (The Netherlands, Kluwer Academic Publishers, 1996), pp. 65.
7. W.J. Weber, Radiat. Eff. **77**, 295 (1983).
8. W.J. Weber, J. Amer. Ceram. Soc. **76**, 1729 (1993).
9. W.J. Weber and L.M. Wang, Mater. Res. Soc. Symp. Proc. **279**, 523 (1993).
10. W.J. Weber and L.M. Wang, Nucl. Instr. Meth. Phys. Res. **B91**, 63 (1994).
11. L.M. Wang and W.J. Weber, Philos. Mag. A (in press).
12 R.C. Ewing and B.C. Chakoumakos, Mineralogical Association of Canada Short Course Handbook, Vol.**8** (1982).
13. W.L. Gong, Ph.D dissertation, Chinese Academy of Sciences, 1988.
14. G. Giuseppeetti and C. Tadini, N. Jb. Miner. Mh **20**, 301 (1990).
15. J.A. Fahey, W.J. Weber, F.J. Rotella, J. Solid State Chem. 60, 145 (1985).
16. Institute of Geochemistry, Chinese Academy of Sciences, *Geochemistry of the Baiyun Obo Nb-REE-Fe Ore Deposits* (Science Press, Beijing 1988).
17. P.S. Zhang, Z.M. Yang, K.J. Tao, and X.M. Yang, *Mineralogy and Geology of Rare Earths in China* (Science Press, Beijing, 1995).
18. D.D. Hogarth and D. Gottfried, Acta Cryst. **8**, 291 (1955).
19. G.R. Lumpkin and R.C. Ewing, Phys. Chem. Minerals **16**, 2 (1988).
20. W.J. Weber, J.W. Wald, and Hj. Matzke, J. Nucl. Mater. 138, 196 (1986).
21. W.C. Mosley, J. Am. Ceram. Soc. **54**, 474 (1971).
22. R.K. Eby and R.C. Ewing, Mater. Res. Soc. Symp. Proc. **183**, 297 (1990).
23. R.C. Ewing, B.C. Chakoumakos, G.R. Lumpkin, T. Murakami, R.B. Greegor, and F.W. Lytle, Nucl. Instr. Meth. Phys. Res. **B32**, 487 (1988).
24. R.C. Ewing, W.J. Weber, and F.W. Clinard, Jr., Prog. Nucl. Energy **29**, 63 (1995).
25. G.A. Hishmeh, L. Cartz, F. Desage, C. Templier, and J.C. Desoyer, J. Mater. Res. **9**, 3095 (1994).
26. Y. Inagaki, H. Furuya, K. Idemitsu, T. Banba, S. Matsumoto, and S. Muraoka, Mater. Res. Soc. Symp. Proc. **257**, 199 (1992).
27. A.K. Dé, B. Luckscheiter, W. Lutze, G. Malow, and E. Schiewer, Ceram. Bull. **55**, 500 (1976).
28. S. Sato,H. Furuya, T. Kozaka, Y. Inagaki, and T. Tamai, J. Nucl. Mater. **152**, 265 (1988).
29. M.L. Miller and R.C. Ewing, Ultramicroscopy **48**, 203 (1993).
30. L.M. Wang, M.L. Miller, R.C. Ewing, Ultrsmicroscopy **51**, 339 (1993).

FORMATION OF PARAMAGNETIC DEFECTS IN OXIDE GLASSES DURING THE BOMBARDMENT OF THEIR SURFACE WITH CHARGED PARTICLES

L.D. BOGOMOLOVA*, S.V. STEFANOVSKY**, Y.G. TEPLYAKOV**, S.A. DMITRIEV**
*Institute of Nuclear Physics, Moscow State University, Leninskiye gory, Moscow, Russia
*SIA "Radon", 7th Rostovskii per., 2/14, Moscow 119121, Russia

ABSTRACT

Heavy-ion bombardment of a glass surface is a conventional laboratory technique for producing damage of interest for radioactive waste encapsulation. At energy of order 100 keV such a bombardment simulates the damage produced by α-recoil nuclei and fission fragments resulting from the nuclear decay. The damage region is 100-500 nm depending on conditions of the bombardment.

In the present work some results of EPR study of point defects formed in silicate, borate, borosilicate, phosphate and other oxide glasses irradiated with different charge particles (C, N, O, Ar. Mn, Cu, Pb) at energy E=150 keV and large total fluence of ions (up to 10^{17} cm^{-2}) are reported. Electron paramagnetic resonance (EPR) is a very sensitive technique which gives an information on the structure of point defects and their content. It is shown that in some cases (for example, in borate glasses) the oxygen hole centers similar to ones observed in γ-irradiated glasses are formed after ion bombardment. However, in the majority of cases new defects which are not typical of γ-irradiated oxide glasses were found. They were large molecular oxygen ions (O_2^-, O_3^-, O_4^-) located in the cavities formed under ion bombardment in the near surface layer of glass. It should be noted that the relative content of these defects is of the order of several tens per 1000 incident ions. This content decreases with increasing fluence and atomic mass of incident ions. It indicates indirectly that point defects are clustered when the damage of the near surface layer becomes strong. The formation of gaseous oxygen is possible in cavities of the damage surface layer.

It was found that some elements (for example C, N and transition metals) form chemical compounds with oxygen. The migration of alkali ions promotes the formation of such compounds since the chemical compounds were detected by means EPR in glasses rich in alkali oxides.

INTRODUCTION

As has been noted in Ref.[1], the bombardment of glass surface by heavy charged particles is a convenient laboratory technique for producing damage in glasses which are of interest for radioactive waste encapsulation. When the energy of particles is of order of hundreds keV this technique permits to simulate the damages produced by α-recoil nuclei resulting from radioactive decay process [1]. The charged particles lose their energy in near-surface layer of glass due to atomic collisions and ionization processes. Traversing the glass one particle produces a cascade of secondary energetic atoms. The formation of point defects (vacancies and interstitials), diffusion of atoms and recombination processes occur during ion bombardment.

Mat. Res. Soc. Symp. Proc. Vol. 465 ©1997 Materials Research Society

In the present work we used electron paramagnetic resonance (EPR) for the study of the stable paramagnetic defects induced by ion bombardment in oxide glasses. EPR is very sensitive method which specifically yields information about the nature and density of radiation defects.

We studied multicomponent borosilicate, borophosphate and aluminophosphate glasses which are interest for immobilization of radioactive waste. In addition, we investigated more simple silicate, borate and phosphate glasses which can be considered as models for understanding the nature of defects formed in complex glasses. Glasses were irradiated by light (N^+, O^+, C^+), middle (Ar^+) and heavy (Mn^+, Cu^+, Pb^+) ions at different energies and doses.

EXPERIMENTAL

We studied more than 100 implanted samples of 13 compositions in 9 vitreous oxide systems [2,3]. The molar compositions of some glasses mentioned in the present report are given in Table I The glasses were melted in an electric furnace in air at temperature depending on glass composition. Then they were cast and annealed. Polished plates were prepared. These plates were irradiated with different ions at energies (E) ranged between 80 and 380 keV to nominal doses from $D=3\times10^{15}$ to $D= 1\times10^{17}$ cm^{-2}. After implantation the slabs were crushed and used for EPR experiments which were performed using a modified RE-1306 spectrometer (Soviet model) operating at X-band frequency. The procedure of evaluation of the number of defects by means of EPR is described in details in Refs. [2-5].

Table I. Glass compositions (in mol. %)

Nos.	Name of glass	SiO$_2$	P$_2$O$_5$	B$_2$O$_3$	Al$_2$O$_3$	Na$_2$O	MgO	CaO
1	S-2	70	-	10	-	20	-	-
2	S-3	50	-	5	5	20	5	15
3	P-1	-	36	-	20	44	-	-
4	P-2	-	50	-	-	-	-	50
5	B-1	-	-	65	5	30	-	-

The EPR lines from ion-implanted glasses were analyzed by comparison with computer generated simulations. To calculate EPR spectra we used an IBM PC compatible with Intel 486-4DX-100 processor and Microsoft Excel 7.0 spreadsheet. The details of calculations are given in the same Refs [3,4].

RESULTS AND DISCUSSION

Figure 1 shows the EPR spectrum of a phosphate sample P-1 implanted with Ar^+ at $D_1 = 3 \times 10^{15}$ cm^{-2}. It can be seen that the spectrum contains an A- shoulder and S-line. Similar spectra were observed for some borosilicate S-2 and S-3 glasses. The isochronal annealing data for the S-signal and the A-shoulder [3] allowed to assume that they belong to distinct centers. Computer simulation of these spectra gives that they are superpositions of the isotropic Lorentzian line (labeled S) with g=2.0025±0.0005 and ΔH=0.6 mT and broad anisotropic spectrum (A$_1$) with g_z=2.017±0.001; g_y=2.010±0.001 and g $_x$ =2.0020±0.005.

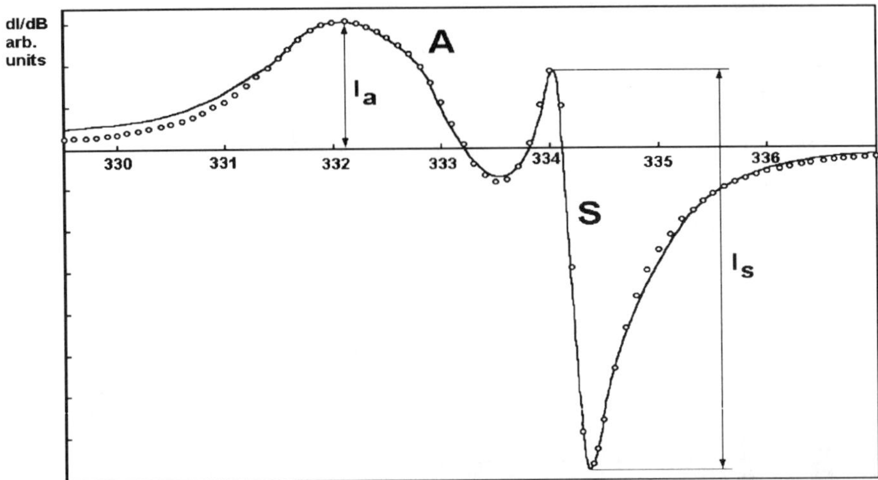

Figure 1. Experimental (solid curve) and calculated (circles) EPR spectra of glass P-1 implanted with Ar^+ ions. $D=3 \cdot 10^{15}$ cm^{-2}, E=150 keV, T=300 K, P=30 mW, A=0.06 mT. For calculated spectrum $\beta=S/A_1=0.1$.

It is found that the concentration of S-centers increases with increasing fluence of incident particles very slowly, and as follows from Figure 2 the number of defects per 1000 incident ions decreases with an increasing dose. This result can be interpreted in terms of generation and annihilation of defects [6] when the number of defects decreases due to the overlapping of collision cascades at high doses.

Based on the fact that the intensity of S-spectrum is several times greater in O^+ implanted glasses than in the samples implanted with other ions we assign this to a defect involving oxygen. The analysis of data presented in Refs [7,8] shows that a rather narrow almost symmetric S- line with $g \sim g_e$, where g_e is g-factor of free electron, can be attributed to large molecular oxygen O_4^- or $O_2 \cdot O^-$ ions. At the same time we cannot exclude that this line belongs to carbon-oxygen centers [8] which exhibit the similar signals. The additional experiments with C^+-implanted silicate and phosphate glasses pointed out that CO_2^+ ions can be responsible for the line of S-type. It should be noted that high near-surface content of carbon was detected by XPS in ion-implanted glasses which exhibited S-signal. Carbon impurity can penetrate into surface the layer of the substrate from oil pump during the implantation process.

Both groups of species are related to aggregation of oxygen in some regions of the implanted layer. Such regions can be formed as a result of sodium depletion which occurs near the surface of implanted substrate [9]. It is necessary to emphasize that that S-signals were found for the glasses containing Na (P-1, S-2, S-3). However, in the S-3 sample the broad anisotropic spectra dominate except for O^+ -implanted ones.

The spectrum of the ion-implanted borosilicate N^+-implanted S-3 samples is shown in Figure 3. The presence of two low-field peaks can be seen. The relative intensities of these peaks and the positions of their minimum were various for different samples indicating that the spectrum

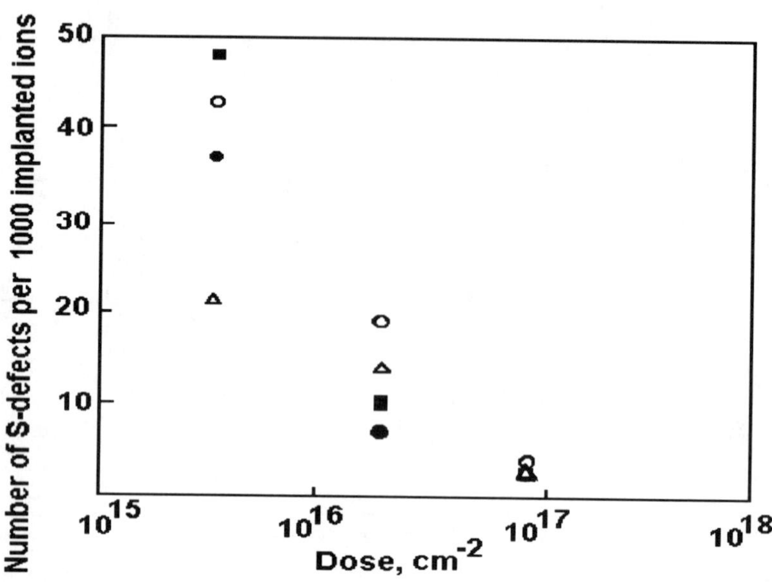

Figure 2. Effect of implantation dose of O^+ and N^+ ions on the number of S-defects per 1000 implanted ions in the glasses S-2 and P-1.
● - S-2 implanted with O^+, △ - S-2 implanted with N^+, ■ - P-1 implanted with O^+, ○ - P-1 implanted with N^+.

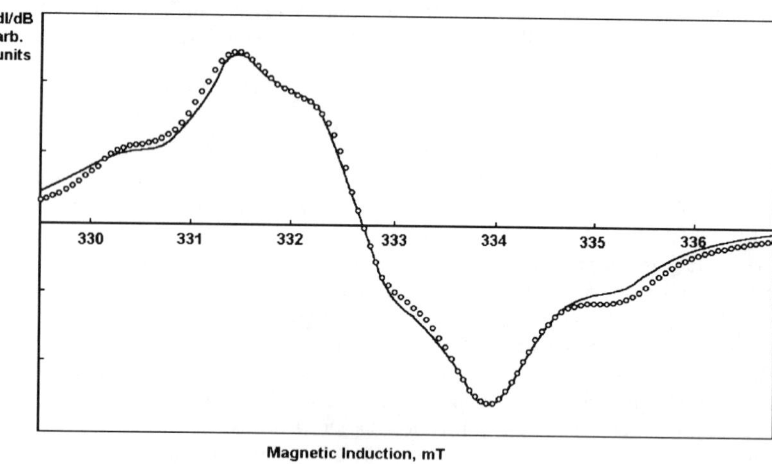

Figure 3. EPR spectrum of the sample S-3 implanted with Ar^+ ions. D=3 10^{15} cm^{-2}, E=150 keV. Solid curve shows experimental spectrum. Circles shows computed spectrum prepared from spectrum A_1 and A_2 at ratio A_1/A_2=0.7.

is a superposition of at least two anisotropic spectra. Computer simulation of this spectrum shows that it contains the above mentioned spectrum A_1 and anisotropic spectrum A_2 with $g_z=2.023\pm0.001$; $g_y= 2.004\pm0.015$ and $g_x =1.993\pm0.0015$. The spectra of A-type were observed for all the implanted glasses studied up to now [2-5]. Commonly the spectra of implanted samples contain from one to three item spectra A_i. Their principal g-values vary from 2.0165 to 2.058 for g_z ; from 2.004 to 2.011 for g_y and from 1.993 to 2.0028 for g_x. Since only oxygen is constituent common to all the studied glasses the spectra of A-type can be attributed to oxygen-related states. The absence of the resolved hyperfine structure (HFS) in the A- spectra of glasses containing only cations with non-zero nuclear spin indicates the weak bond between oxygen species and a glass network. The relationship between principal g-components permits us to attribute the spectra of A-type to molecular O_2^- ions [7]. The structure of ion-implanted layers contains cavities, tunnels and channels of various shapes and sizes. The oxygen atoms displaced from their sites due to collision processes induced by incident particles can form the O_2^- ions located in these voids.

Figure 4 shows that the number of defects responsible for the spectra of A-type in phosphate glasses decreases with increasing atomic mass. Similar dependence was found for all the glass systems. The concentration of defects in C^+ -implanted samples is 10 or 15 times larger than in Pb^+-implanted glasses. This result together with SIMS data indicates that the projected range of incident particles plays the important role for the content of O_2^- defects.

The number of O_2^- ions evaluated from EPR spectra is maximum for borate glasses and decreases in order borate>silicate>borosilicate~phosphate>alumophosphate glasses. For example

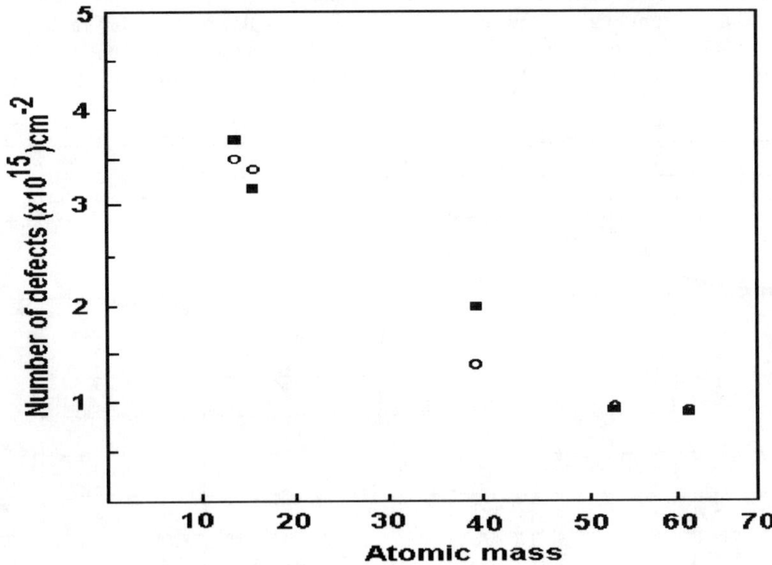

Figure 4. Effect of atomic mass of implanted ions on the number of A-type defects formed in phosphate glasses.
■ - P-1,   O - P-2.

the ratio (R) of the content of O_2^- ions to the number of incident N^+ ions at dose $D{\sim}10^{16}$ cm^{-2} decreases from ~1 to 0.1 in this direction. The weak decrease in R value occurs with increasing concentration of modifiers in all the system. The high concentration of O_2^- ions in borate glasses in comparison with other glasses can be associated with strong sputtering of light boron atoms after collision with more heavy particles [10] and , as result, with the appearance of excess oxygens.

In Figure 5 the EPR spectrum of B-1 sample is presented. Several lines separated by ~1mT are well resolved in low-field and central parts of this spectrum. The additional signal (F) is seen at magnetic field H=340 mT. It should be noted that in P-1,P-2 and S-2 samples implanted with N^+ ions doublet separated by about 13 mT was observed. The position of high-field line of this doublet coincides with one of F-line of the B-1 glass spectrum whereas the low-field line is at magnetic field where the peak of other spectrum of B-1 glass is located. Therefore, we assume that the F-line belongs to doublet observed for other glasses and labeled below as F-doublet. The appearance of this doublet in EPR spectra of glasses of distinct systems containing no common constituent excepting for oxygen indicates that it belongs to identical defects associated with common implant, i.e. with nitrogen. Computer simulation of EPR spectra of N^+-implanted samples exhibited F-doublets were performed under assumption of the presence of HFS due to the interaction of unpaired electron with ^{14}N nucleus having spin I=1 (m_I =0±1). The best-fit HFS spectrum is characterized by g-values : g_1 =1.996; g_2 =g_3 =2.002 and HFS constants (in mT): A_1=4.2; A_2=7.0; A_3=6.5.These parameters resemble ones reported for NO_2 molecules in N^+-implanted silica glass [11] and in other compounds [8].Therefore we attributed this spectrum to NO_2 molecules. The ratio of NO_2 content to the number of implanted N atoms is highest in borate glasses (~5%) and decreases to 0.04% in S-2 sample.

As follows from Figure 5 the spectrum of the B-1 samples contains several lines in low-field and central parts. Similar lines were observed in the spectra of other borate glasses. We assumed that they associated with HFS due to interaction with ^{11}B nucleus (I =3/2). Computer

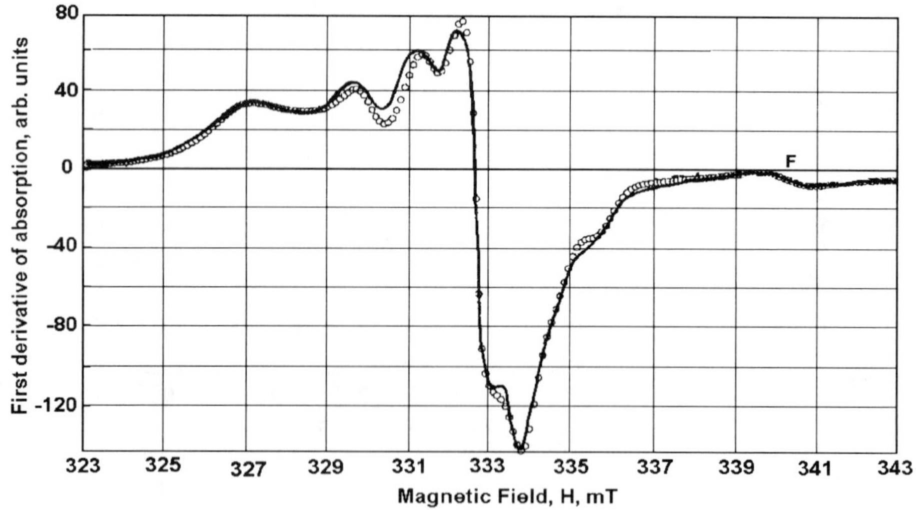

Figure 5. EPR spectrum of the sample B-1.

simulation shows that the HFS spectrum with $g_z=2.035\pm0.005$, $g_y=2.0118\pm0.0005$, $g_x=2.0020\pm0.0005$ and HFS constants (in 10^{-4} cm^{-1}): $A_z=9\pm2$, $A_y=13.5\pm1$, $A_x=12.5\pm1$ contributes to the spectrum of the B-1 glass. The shape of this HFS spectrum and its parameters are similar to ones of well-known spectrum "five-lines-plus-a shoulder" [12] observed for γ-irradiated low-alkali borate glasses and assigned to oxygen hole centers (OHC). Therefore, we assume that OHC can be formed in ion-implanted oxide glasses. The number of these defects is small ($\sim 10^{14}$ cm^{-2}).

In implanted phase separated borosilicate glasses, an E'- center located in a silica phase was found. In oxide glasses implanted with Mn and Cu [13] the EPR spectra of Mn^{2+} and Cu^{2+} was detected, respectively. The chemical compounds (for example, MnO) were found at high fluences of transition metals.

CONCLUSION

1. The analysis of EPR data for more than 100 implanted samples in glassy oxide systems shows that the molecular O_2^- ions are formed in all the systems. These ions are located in voids of near-surface layers of glass. At very high fluences of incident particles ($>10^{18}$ cm^{-2}) EPR signal of these ions disappears, presumably, due to a formation of non-paramagnetic clusters or gaseous oxygen. The highest concentration of the O_2^- molecules is observed in borate glasses containing boron having mass smaller than one of incident ions and recoil atoms.
2. In ion implanted glasses containing Na the larger oxygen ions are formed. In samples containing carbon impurity CO_2^+ ions are perhaps formed.
3. Molecules NO_2 are formed in the samples implanted with nitrogen. Their concentration is small in comparison with number of implanted nitrogen. The rest of N is present in glasses as gaseous N_2 [11] or reacts chemically with Si, P and B [11,13].

REFERENCES

1. G.W.Arnold, Radiat. Eff., **98**, p. 55 (1986).
2. L.D. Bogomolova, V.A. Jachkin, S.A. Prushinsky, S.A. Dmitriev, S.V. Stefanovsky, Yu.G. Teplyakov, F. Caccavale, E. Cattaruzza, R. Bertonello, F. Trivillin, J. Non-Cryst. Solids, in press
3. L.D. Bogomolova, V.A. Jachkin, S.A. Prushinsky, S.V. Stefanovsky, Yu.G. Teplyakov, J. Non-Cryst Solids, in press
4. L.D. Bogomolova, Yu.G. Teplyakov, V.A. Jachkin, V.L. Bogdanov, V.D. Khalilev, F. Caccavale, Lo Russo, Opt. Materials, **5**, p.311 (1996).
5. L.D. Bogomolova, Y.G. Teplyakov, S.V. Stefanovsky, S.A. Dmitriev, in <u>Proc. Fifth International Conference on Radioactive Waste Management and Environmental Remediation ICEM '95</u> (Berlin, Germany. 1995), pp.. 409-411
6. R.A.Devine and A.Golansky, J. Appl. Phys. **54**, p.3833 (1983).
7. M.Che and A.J.Tench, in Advances in Catalysis **32**, p.1 (1983).
8. Landolt-Bornstein. <u>Numerical Data and Functional Relationships in Science and Technology. New Series. Magnetic Properties of Free Radicals. Group II.</u> (Vol. 9 Part A. Ed. by H.Fisher and K.-H.Hellwege, Springer-Verlag, Berlin-Heidelberg-N.Y., 1977), p. 81-119.
9. G. Battaglin, A. Bascoletto, G. Della Mea, G.D. Marchi, P. Mazzoldi, A. Miotello, B. Tiveron, Radiat. Eff. 98, p.101 (1986).

10. Wang Chenguy, Wang Bo, in Proc. of XV Intern. Congress on Glass, (Leningrad, Ed. by O.V. Mazurin, Leningrad, "Nauka", 1989),V.1b,p. 263
11. D.L.Griscom, J. Non-Crystal. Solids 13, p.251 (1973/1974).
12. H.Hosono, Y.Abe, K.Oyoshi and S.Tanaka, Phys. Rev. B 43, p.1196 (1992).
13. L.D. Bogomolova, Yu.G. Teplyakov, F. Cacccavale, J. Non-Crystal. Solids, 194, p.291 (1996).

Part IX

Waste Package Materials

NUMERICAL PREDICTIONS OF DRY OXIDATION OF IRON AND LOW-CARBON STEEL AT MODERATELY ELEVATED TEMPERATURES

G. A. HENSHALL
Lawrence Livermore National Laboratory, L-355, Livermore, CA 94551.

ABSTRACT

Wrought and cast low-carbon steel are candidate materials for the thick (e.g. 10 cm) outer barrier of nuclear waste packages being considered for use in the potential geological repository at Yucca Mountain. Dry oxidation is one potential degradation mode for these materials at the moderately elevated temperatures expected at the container surface, e.g. 323–533 K (50–260 °C). Therefore, numerical predictions of dry oxidation damage have been made based on experimental data for iron and low-carbon steel and the theory of parabolic oxidation. A numerical approach employing the Forward Euler method has been implemented to integrate the parabolic rate law for arbitrary, complex temperature histories. Assuming growth of a defect-free, adherent oxide, the surface penetration of a low-carbon steel barrier following 5000 years of exposure to a severe, but repository-relevant, temperature history is predicted to be only about 0.127 mm, less than 0.13% of the expected container thickness of 10 cm. Allowing the oxide to spall upon reaching a critical thickness increases the predicted metal penetration values, but degradation is still computed to be negligible. Based on these physically-based model calculations, dry oxidation is not expected to significantly degrade the performance of thick, corrosion allowance barriers constructed of low-carbon steel.

INTRODUCTION

Wrought and cast low-carbon steel are candidate materials for the thick (e.g. 10 cm) outer barrier of nuclear waste packages (WPs) being considered for use in the potential geological repository at Yucca Mountain. Dry oxidation is one possible degradation mode for these materials at the moderately elevated temperatures expected at the container surface, e.g. 323–533 K (50–260 °C). Due to the long times for which such barriers must remain intact (e.g. 5000 years), mathematical models are required with which to extrapolate the available short-time experimental data for the purpose of predicting container performance.

Stahl and McCoy [1] recently presented the following empirical equation to predict the penetration rate of a carbon steel barrier due to high temperature oxidation:

$$p = 1.787 \times 10^5 \, t^{\,0.33} \exp(-6870/T) \tag{1}$$

where p is the penetration depth in μm, t is the exposure time in years, and T is temperature in Kelvins. Equation (1) predicts a negligible penetration of less than 0.02 μm in 1000 years at the boiling point of water (approximately 368 K at Yucca Mountain) [2]. Equation (1) was also written in a differential form by McCoy [2] so that variable-temperature histories could be simulated. As in the derivation of Eq. (1), the activation energy was assumed to be independent of temperature in this analysis.

A somewhat different approach to the problem of predicting dry oxidation was taken in the present study. This approach was based on the theory of parabolic oxidation for iron and low-carbon steel, and employed a numerical integration technique to solve the resulting differential equation for arbitrary thermal histories. A simple method for estimating the effects of cracking or spalling of the oxide was also included.

Mat. Res. Soc. Symp. Proc. Vol. 465 © 1997 Materials Research Society

DEVELOPMENT OF THE MODEL

A variety of physical theories exist to describe the oxidation of metals at constant temperature [3]. In iron, a logarithmic relationship exists between oxide thickness and exposure time at low temperatures (i.e. near room temperature) and for oxide films less than about 0.1 μm thick [4]. For oxide thicknesses of practical importance to WP degradation, however, both theoretical treatments and the available experimental data show that the dry oxidation of iron and low-carbon steels at moderately elevated temperatures ($293 < T \leq 843$ K) is diffusion controlled, and therefore obeys a parabolic rate law if the temperature is constant [3-9].

In general, the parabolic oxidation of iron and low-carbon steel produces multi-layered scales due to the variable oxidation state iron. In order of proximity to the metal surface these are: FeO (wüstite), Fe_3O_4 (magnetite), and Fe_2O_3 (hematite). However, at temperatures below 843 K (570 °C), FeO is thermodynamically unstable, so only Fe_3O_4 and Fe_2O_3 form [3]. At these moderately elevated temperatures, the Fe_3O_4 layer is found to be much thicker than the Fe_2O_3 layer, at least for scales up to about 20 μm thick [5]. Therefore, it was assumed in the present study that the oxide scale consists of only Fe_3O_4 at the temperatures of interest. This assumption significantly simplifies the model while providing predictions that are reasonably accurate, as shown in the next section.

In addition to the assumptions of parabolic growth (under isothermal conditions) and the presence of only magnetite, several other assumptions were made in developing the model. First, the Fe_3O_4 oxide was assumed to be 100% dense, consistent with observations for layers up to about 20 μm thick [4,5,9]. Second, Arrhenius behavior with a constant activation energy over the relevant temperature range was assumed. As discussed later, this appears to be a reasonable approximation of the behavior for low-carbon steel oxidized in the 323–533 K (50–260 °C) temperature regime. Third, it was assumed that oxidation occurs uniformly across the surface of the metal. Finally, the oxide was assumed to be adherent to the metal surface; a specific exception to this assumption is described later.

Mathematically, the parabolic increase in the thickness of a single-species oxide with increasing exposure time can be expressed [3]:

$$\mathrm{d}x/\mathrm{d}t = k'_\mathrm{p}/x \tag{2}$$

where x is the oxide thickness, t is exposure time, and k'_p is the parabolic rate constant. The temperature dependence of the oxidation rate is governed by the Arrhenius behavior of k'_p:

$$k'_\mathrm{p} = k_\mathrm{p} \exp(-Q/RT) \tag{3}$$

where k_p is a constant, Q is the activation energy for the oxidation process, R is the gas constant, and T is absolute temperature. Literature values of k_p and Q for iron and low-carbon steel are given in Table 1. The values of k_p vary considerably, at least partly because the measurements were made using a variety of materials, surface treatments, and environments. In the parabolic regime and at the relatively low temperatures of 523–736 K (250–463 °C), Q appears to be constant at about 104 kJ/mol (25 kcal/mol). This value is consistent with that for grain-boundary diffusion of Fe ions in Fe_3O_4 [4]. Table 1 shows that lower values of Q have been measured at the higher temperatures of 673–853 K (400–580 °C). The influence of a possible change in Q at temperatures below 523 K, which are most relevant to WP performance, is discussed in the next section.

Equations (2) and (3) can be easily integrated analytically for simple (e.g. constant) temperature histories. However, to perform the integration for arbitrary, complex thermal histories, such as those expected in the repository, a numerical approach employing the Forward Euler method [10] was implemented. Combining Eqs. (2) and (3) and writing the result as a difference equation gives:

$$\Delta x = \{(1/x)\,[k_\mathrm{p} \exp(-Q/RT)]\}\Delta t \tag{4}$$

where Δx is the increase in oxide thickness during the time step Δt. Writing Δx as

$$\Delta x = x_i - x_{i-1}, \tag{5}$$

where the subscript "i" denotes the current time step, Eq. (4) can be written

$$x_i - x_{i-1} - \{(1/x_i)\,[k_p\,\exp(-Q/RT_i)]\}\Delta t = 0. \tag{6}$$

This non-linear equation in the single unknown, x_i, can be solved using any standard method for solving non-linear equations in one variable. A bisection technique [10] was chosen because it is simple and robust. Also note that Eqs. (4)–(6) represent an implicit scheme for solving the differential equation (2). Therefore, while this method is unconditionally stable, the computed oxide thickness will depend on the time-step size. For this reason, convergence of the solution was confirmed by decreasing Δt until the final x values varied by less than a specified tolerance, typically 0.1%, for a five-fold decrease in Δt.

For predicting the degradation of WPs, it is the metal penetration caused by oxidation, p, not the oxide thickness itself, that needs to be calculated. This can be accomplished by again assuming a fully dense Fe_3O_4 oxide forming on pure iron, and computing p from the oxide thickness using:

$$p = \left[(\rho_o/M_o)(M_M/\rho_M)N_M\right] x \tag{7}$$

where ρ_o and ρ_M are the density of the oxide and the metal, M_o and M_M are the molecular or atomic weight of the oxide and the metal, and N_M is the number of moles of metal atoms per mole of oxide (3 for Fe_3O_4). For simplicity, ρ_o and ρ_M were assumed to equal their room temperature values in the numerical calculations presented in the next section.

It is unrealistic to assume that a 10-cm thick metal barrier will completely oxidize without cracking or spalling of the oxide layer. Such defects in the oxide decrease its protectiveness and increase the oxidation rate because fresh metal is exposed or a fast diffusion path for oxygen is created. Therefore, a conservative approach was taken to estimate the effects of mechanical defects in the scale. It was assumed that mechanical stresses build up in the oxide as its thickness increases, and that a critical thickness exists at which these stresses exceed the fracture strength of the oxide. Once this critical thickness, x_c, is reached, the oxide is assumed to completely spall away, exposing the bare metal surface. At this point, the parabolic oxidation process begins as if no oxide had ever been present; i.e. $x = 0$. Of course, the metal penetration does not return to zero when $x = 0$, but continues to accumulate. In this case, the accumulated penetration, P, is computed from Eq. (8):

$$P = \sum_{j=1}^{n} p_j \tag{8}$$

where n is the number of spall events or "cycles" and p_j is computed from Eq. (7) for each cycle. While such a model is certainly over simplified, it provides a conservative way in which to treat defects in the scale. The use of such a conservative approach is justified later.

Table 1. Oxidation of Iron and Carbon Steel at Moderately Elevated Temperatures.

$k_p{}^a$ (cm²/s x 10⁵)	Q (kJ/mol)	T (K)	$x_{max}{}^b$ (µm)	Regime[c]	Alloy	Ref.
———	25.1	473 – 623	0.05	L	0.2–0.8% C	4
———	30.5	473 – 623	0.1	L	0.2–0.8% C	4
0.7 – 1.7[d]	104	523 – 623	1.4	P	0.2–0.8% C	4
0.02[e]	104	541 – 736	0.4	P	0.04% C	6
0.03[e]	102	541 – 666	2	P	0.04% C	6
2.5[f]	34 / 106[g]	673 – 923	22	P	0.02% C	5
0.02[h]	72.4	673 – 823	10.5	P	Pure Fe	7,8
4.3	———	773	15	P	0 – 0.99% C	9

Notes:

[a] Representative or maximum values selected from the data provided

[b] Maximum oxide thickness observed, based on a fully dense layer of Fe_3O_4

[c] L = logarithmic, P = parabolic

[d] Measured at 573 K

[e] Measured at 569 K

[f] Measured at 723 K

[g] Q=106 kJ/mol for $670 \leq T \leq 723$ K; Q=34 kJ/mol for $723 \leq T \leq 853$ K

[h] Measured at 673 K

NUMERICAL RESULTS

The validity of the assumptions used in developing the model, other than those associated with spalling, was tested by simulating literature data for the oxidation of iron and low-carbon steel at constant temperature. Figure 1 presents such data for both pure iron and for a low-carbon steel. In both cases, the increase in oxide thickness with increasing exposure time is nearly parabolic over the range studied. These results support the hypothesis that the dry oxidation of these metals is diffusion controlled and can be modeled using Eq. (2). As shown in Fig. 1(a), the simulated oxidation rate is consistent with the data of Caplan and Cohen [5] for iron up to an oxide thickness of about 6 µm. The data of Boggs and Kachik [9] for 0.99% C steel are simulated with nearly the same accuracy up to $x = 16$ µm, as shown in Fig. 1(b). The accuracy of the simulations shown in Fig. 1 suggests that the assumptions used to develop the model are reasonable for oxide thicknesses up to at least about 20 µm.

With the accuracy of the model confirmed for simple, constant-temperature conditions, predictions of oxidation were then made for complex, repository-relevant thermal histories. First, calculations were made using the WP surface temperatures calculated by Buscheck et al. [11] for a 27.3 kg U/m² thermal loading. The computed oxide thickness and metal penetration values were very small, e.g. $x = 20$ µm for 5000 years of exposure. Then, to increase the conservatism of the predictions, the temperatures calculated by Buscheck et al. were increased by 100 K for all exposure times. Figure 2(a) compares this "extreme" temperature history to a polynomial approximation that was used for the oxidation calculations. Based on the experimental data given in Table 1, values of $k_p = 4.3$ x 10^{-5} cm²/s and $Q = 104$ kJ/mol were used in the oxidation predictions. The use of $k_p = 4.3$ x 10^{-5} cm²/s is conservative since this was the largest value indicated by the literature data. Figure 2(b) shows the resulting oxide thickness and metal penetration predictions for the case of a defect-free, adherent oxide. Note that the oxidation rate is not parabolic because the temperature is not constant. The metal penetration following 5000 years of exposure to this severe temperature history is predicted to be only about 0.127 mm, less than 0.13% of the expected container thickness of 10 cm. Using an empirical approach and assuming an adherent oxide, Stahl and McCoy [1] also predicted that very little metal penetration of carbon steel would occur due to dry oxidation under repository-relevant conditions.

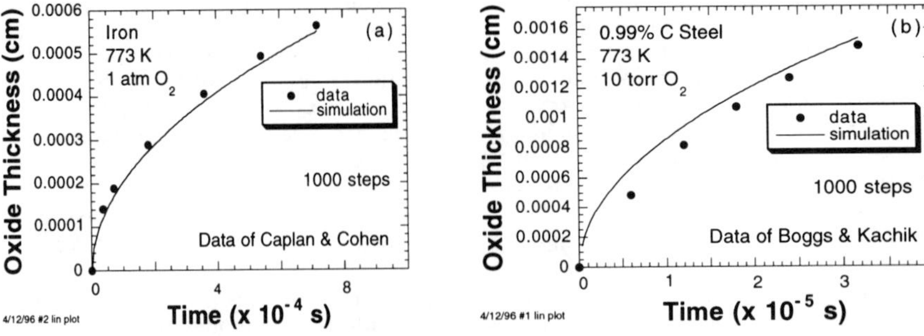

Fig. 1   Model simulations of isothermal oxide growth are compared with data for (a) iron oxidized at 773 K in 1 atmosphere of O_2 [5] and (b) 0.99% C steel oxidized at 773 K in a 10 torr O_2 atmosphere [9].

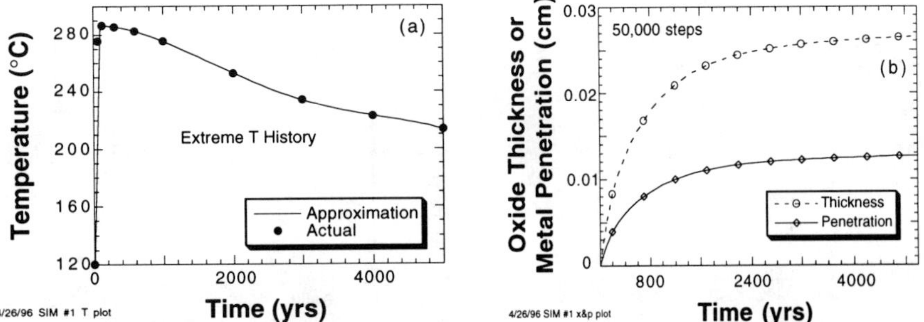

Fig. 2. (a) The "extreme" temperature history (solid circles) is compared to a polynomial approximation. (b) Corresponding model predictions of the oxide layer thickness and metal surface penetration histories assuming a fully dense, adherent oxide.

The data in Table 1 support the use of $Q = 104$ kJ/mol for oxidation of iron and carbon steel at temperatures between 523 and 723 K (250–450 °C). Since this value corresponds to that for grain-boundary diffusion of Fe ions in Fe_3O_4, it is likely that it will not decrease at the lower temperatures expected on WP surfaces, 323–533 K (50–260 °C). Nevertheless, to evaluate the sensitivity of the predicted metal penetration on Q, the effect of a decreased activation energy at low temperatures was computed. The lowest value of Q cited in the literature for parabolic oxidation at $T \leq 723$ K is 72.4 kJ/mol [7], Table 1. Using this value together with $k_p = 4.3 \times 10^{-5}$ cm²/s [9], the calculations using the thermal history shown in Fig. 2(a) were repeated. The conservatism of this calculation must be emphasized: (1) the thermal history assumes temperatures 100 K higher than expected, (2) the activation energy equals the lowest value cited

671

in the literature, which probably relates to much higher temperatures (Table 1), and (3) the pre-exponential constant, k_p, is the largest that can be expected based on the available data. Despite this conservatism, the model predicts that $p = 4.6$ mm, or only 4.6% of the expected container thickness, following 5000 years of exposure. Of course, it would be prudent to experimentally measure Q in the temperature range of interest to confirm that some change in mechanism does not result in a Q value significantly lower than that used here.

Defects in the oxide, such as cracking or spalling, would increase the metal penetration rate above that computed assuming a defect-free, adherent oxide layer. Using the approach described previously to account for such defects and the temperature history shown in Fig. 2(a), predictions of the surface penetration into iron or low-carbon steel were made. Values of the critical oxide thickness for spalling, x_c, were chosen to be either $x_c = 0.02$ mm or $x_c = 0.1$ mm. The former roughly corresponds to the maximum oxide thickness cited in literature studies [4-9] of fully dense, adherent oxide growth on iron and steel at the relevant temperatures (see Table 1). Since significant cracking or spalling was never observed in these studies, this value provides a reasonable lower limit on x_c. The second value, $x_c = 0.1$ mm, was chosen arbitrarily to assess the influence of this parameter on the predictions. The values of $k_p = 4.3$ x 10^{-5} cm^2/s and $Q = 104$ kJ/mol again were used in these calculations.

As shown in Fig. 3, the predicted metal penetration depths are larger than for the case of defect-free oxide growth, Fig. 2(b), but are still small. Therefore, even though the method used to treat defects in the oxide is very conservative, the model predicts that dry oxidation damage is minimal. For $x_c = 0.02$ mm, Fig. 3(a) shows that P is predicted to be approximately 1.7 mm, or about 1.7% of the expected container thickness, following 5000 years of exposure to the extreme temperature history. Increasing x_c to 0.1 mm decreases the predicted value of P to approximately 0.35 mm. In this case, the discontinuous rate of metal penetration caused by discrete spall events is clearly apparent.

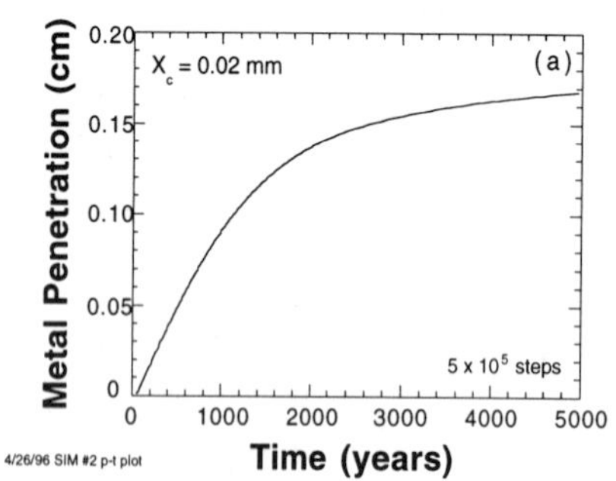

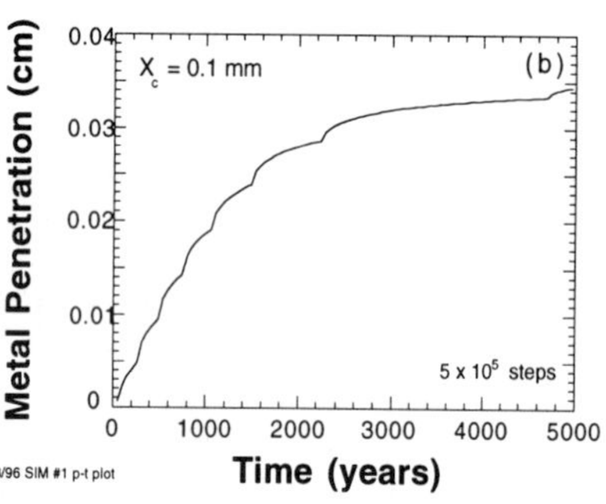

Fig. 3. Predictions of the time-dependent penetration of an iron or low-carbon steel surface subjected to the extreme temperature history shown in Fig. 2(a) and periodic spalling of the oxide at thicknesses of: (a) $x_c = 0.02$ mm and (b) $x_c = 0.1$ mm.

CONCLUSIONS

Based on calculations using the physically-based model presented here, dry oxidation is not expected to significantly degrade the performance of thick, corrosion allowance barriers constructed of low-carbon steel. This conclusion remains valid when a very conservative thermal history is combined with conservative values of the activation energy and pre-exponential factor in the oxidation rate equation, or with a conservative model of the effects of oxide damage. However, it is suggested that the activation energy and pre-exponential constants used in these calculations be confirmed experimentally at the expected waste package surface temperatures of 323–533 K (50–260 °C).   Further studies should also be conducted to consider internal oxidation and the effects of water vapor in the atmosphere.

ACKNOWLEDGMENTS

This work was performed under the auspices of the U. S. DOE under contract no. W-7405-ENG-48 at Lawrence Livermore National Laboratory. The support of the Yucca Mountain Site Characterization Project is gratefully acknowledged.

REFERENCES

1. D. Stahl and J. K. McCoy, "Impact of Thermal Loading on Waste Package Material Performance," in *Scientific Basis for Nuclear Waste Management XVIII*, Materials Research Society, Pittsburgh, PA, pp. 671-678 (1995).

2. J. K. McCoy, "Updated Report on RIP/YMIM Analysis of Designs," Report prepared for the U.S. Dept. of Energy, Yucca Mountain Site Characteriztion by TRW Environmental Safety Systems, Inc., Report BBA000000-01717-5705-00002 REV 02 (1995).

3. N. Birks and G. H. Meier, *Introduction to High Temperature Oxidation of Metals*, Edward Arnold, London (1983).

4. R. B. Runk and H. J. Kim, *Oxidation of Metals* **2**, 285 (1970).

5. D. Caplan and M. Cohen, *Corr. Sci.* **6**, 321 (1966).

6. E. J. Caule, K. H. Buob, and M. Cohen, *J. Electrochem. Soc.* **108**, 829 (1961).

7. O. Kubaschewski and B. E. Hopkins, *Oxidation of Metals and Alloys*, Academic Press, London, p. 108-110 (1962).

8. M. H. Davies, M. T. Simnad and C. E. Birchenall, *J. Metals* **3**, 889 (1951).

9. W. E. Boggs and R. H. Kachik, *J. Electrochem. Soc.* **116**, 424 (1969).

10. G. E. Forsythe, M. A. Malcolm and C. B. Moler, *Computer Methods for Mathematical Computations*, Prentice-Hall, Englewood Cliffs, New Jersey (1977).

11. T. A. Buscheck, J. J. Natao and S. F. Saterlie, in *High Level Radioactive Waste Management: Proceedings of the Fifth International Conference*, ANS, La Grange Park, IL and ASCE, New York, pp. 592-610 (1994).

THE CORROSION BEHAVIOR OF IRON AND ALUMINUM UNDER WASTE DISPOSAL CONDITIONS

R. FUJISAWA*, T. CHO*, K. SUGAHARA*, Y. TAKIZAWA*, Y. HORIKAWA**,
T. SHIOMI** and M. HIRONAGA***
* Mitsubishi Materials Co., Central Research Institute, Saitama, Japan
** Kansai Electric Power Co.,Inc., Osaka, Japan
*** Central Research Institute of Electric Power Industry, Chiba, Japan

ABSTRACT

The generation of hydrogen gas from metallic waste in corrosive disposal environment is an important issue for the safety analysis of low-level radioactive waste disposal facilities in Japan. In particular iron and aluminum are the possibly important elements regarding the gas generation. However, the corrosion behavior of these metals has not been sufficiently investigated under the highly alkaline non-oxidizing disposal conditions yet.

We studied the corrosion behavior of iron and aluminum under simulated disposal environments. The quantity of hydrogen gas generated from iron was measured in a closed cell under highly alkaline non-oxidizing conditions. The observed corrosion rate of iron in the initial period of immersion was 4 nm/year at 15 °C, 20 nm/year at 30 °C, and 200 nm/year at 45 °C. The activation energy was found to be 100 kJ/mol from Arrhenius plotting of the above corrosion rates.

The corrosion behavior of aluminum was studied under an environment simulating conditions in which aluminum was solidified with mortar. In the initial period aluminum corroded rapidly with a corrosion rate of 20 mm/year. However, the corrosion rate decreased with time, and after 1,000 hours the rate reached 0.001 to 0.01 mm/year.

Thus we obtained data on hydrogen gas generation from iron and aluminum under the disposal environment relevant to the safety analysis of low-level radioactive disposal facilities in Japan.

INTRODUCTION

In Japan the low-level radioactive solid wastes including metals are planned to be buried in the repository after solidified with mortar in containers. In the wastes there will be a large quantity of iron having a very low corrosion rates and a small quantity of aluminum having a high corrosion rate under non-oxidizing highly alkaline conditions in the repository. The hydrogen gas generation is an important issue from the viewpoint of accumulation of inner pressure of the disposal facilities. Therefore it is necessary to evaluate both the gas generation rate and gas permeability through the disposal facilities to analyze the safety of the disposal facilities. However, the corrosion behavior of these metals has not been sufficiently investigated yet. With respect to iron, although some measurements have been reported by NAGRA [1] and SKB [2], corrosion tests under conditions of Japanese repository are necessary. With respect to aluminum, it is known that it corrodes at a high rate under alkaline conditions but the corrosion rate is reduced due to film formation when it is embedded in a cement structure [3]. Since the corrosion data characteristic to the disposal environment must be used for the safety analysis, we conducted the corrosion test for aluminum in pulverized mortar and solidified mortar.

A part of the results has been reported already [4].

Mat. Res. Soc. Symp. Proc. Vol. 465 © 1997 Materials Research Society

<u>CORROSION TEST METHOD</u>

(a) Specimens

Carbon steel wires (JIS SWRS82A) with the size of ϕ 0.4 x 100 mm were used as iron specimens. The carbon steel was composed of Fe(98.5%), C (0.81%), Si (0.19%), and Mn (0.46%). The specimens were rinsed with 5% sulfuric acid and then with deionized water. After that they were degreased with acetone, and dried. A bundle of 1280 carbon steel wire specimens (total surface area of 0.16 m^2) were placed in a cell.

Aluminum specimens of commercial pure aluminum (JIS A 1070) at the size of 25 x 25 x 3 mm (surface area of 17 cm^2) were polished with up to #1200 emery grade and degreased with acetone, and dried.

Mortar was prepared by mixing an ordinary Portland cement, water, and sand at a ratio of 1 : 0.35 : 1.

(b) Test procedure

Fig. 1 shows a glass cell apparatus used for the tests on iron. In a glovebox, the test solutions were repeatedly degassed under vacuum and replaced with high-purity nitrogen gas so as to reduce dissolved oxygen in the solution to 0.06 mg/l or less. Three kinds of test solutions, i.e. mortar-equilibrated water (pH = 12.6) simulating the early stage of the disposal environment, aqueous solution saturated with calcium hydroxide (pH = 12.8), and synthetic groundwater simulating the disposal environment in which the influence of mortar has been reduced; were employed. The pH value of the synthetic groundwater was adjusted to 8 by adding bentonite powder. Table 1 shows the chemical composition of the mortar-equilibrated water and the synthetic groundwater.

After placing the carbon steel wires into the glass cell, the upper portion of the cell was sealed, then the test solution (205 ml) was introduced to the cell through an empty manometer tube. After that mercury was put into the manometer. The quantity of generated hydrogen gas was obtained the pressure change in the glass cell measured from the manometer [1, 5]. The temperature of glass cell was maintained at 15, 30, or 45 °C using a thermostat.

The test was repeated three times under the same condition.

Fig. 2 shows a glass container used for the tests on aluminum. Pulverized mortar (grain size of 1 to 3 mm) was placed in an inner glass cell with holes and two aluminum specimens were inserted near the center of the pulverized mortar medium. The inner glass cell had numerous holes (diameter of 1 mm) through which the surrounding solution was let go into the inner cell freely to ensure the contact with the specimens. After sealing the bottom portion of the container, 350 ml of mortar-equilibrated water was poured into the cell. The container was maintained at 15 °C using a thermostat.

The corrosion rate was estimated from the quantity of hydrogen gas generated due to corrosion. During the initial period of corrosion test, the corrosion rate was high, and hydrogen gas was collected over water in a meascylinder. After the initial period the generation rate of hydrogen gas was so low that quantity could not be measured by the above method. The concentration of hydrogen gas accumulated in the glass container over a specified time-period was measured by gas chromatography for later periods of the test.

The second corrosion test was conducted in which 100 ml of each test solution was replaced with a flesh solution every 4 weeks.

The test was repeated three times under the same condition.

In the third test, the corrosion rate of aluminum was measured in an aqueous solution which was saturated with mortar without mortar-grain coexistence to study the mechanism of film formation on aluminum.

In the fourth test, using a similar test apparatus mentioned the above, the quantity of hydrogen gas generated from three aluminum plates (25 x 25 x 3 mm) which was solidified with mortar was measured to compare with the tests using mortar grains.

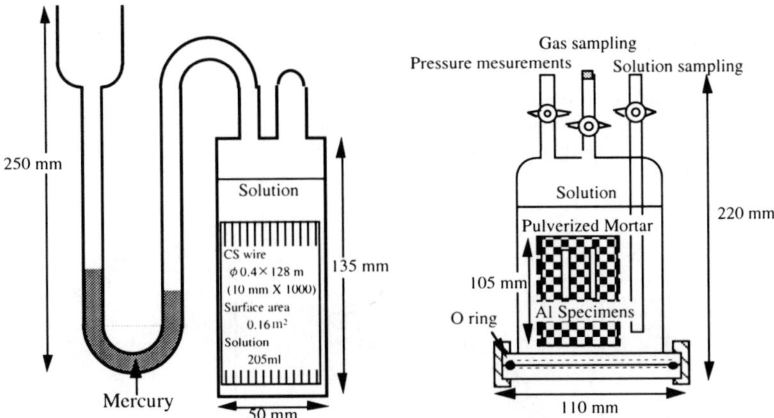

Fig.1 Glass cell for iron corrosion tests   Fig. 2 Glass cell for aluminum corrosion test

Table 1.   Chemical composition of the test solutions

Composition	Mortar-equilibrated wate (ppm)	Synthetic groudwater (ppm)
Mg^{+2}	< 0.1	2
Ca^{+2}	700	3
Na^+	29	10
K^+	40	—
SO_4^{-2}	2.9	4
Cl^-	0.4	15
Al	0.4	—
Fe(total)	-	3
SiO_2 (total)	1.1	8

RESULTS AND DISCUSSION

(a) Iron

Fig. 3 shows the quantity of hydrogen gas (mole) generated form the iron in mortar-equilibrated water at 15, 30, and 45 °C. At 15 °C no hydrogen gas was detected during an initial period of 2,000 hours. After the initial period, the quantity of hydrogen gas increased at an almost constant rate with time. The corrosion rate was estimated to be 4 nm/year obtained from the gradient of the curve. At 30 °C hydrogen gas was not generated during an initial period of 200 hours. After this initial period, corrosion proceeded at an almost constant rate for 3,800 hours with the corrosion rate of 20 nm/year and then the rate started decreasing. At 45 $^{\circ}$C hydrogen gas was generated immediately with a corrosion rate of 200 nm/year and then the corrosion rate became lower after 1,000 hours. The corrosion rates were estimated on the basis of uniform corrosion according to the following reaction:

$$3Fe + 4H_2O \rightarrow Fe_3O_4 + 4H_2. \qquad (1)$$

Fig. 4 shows the corrosion rates of the above initial periods versus reciprocal of temperature. The corrosion rate of iron under this test condition shows temperature dependency with the apparent activation energy of 100 kJ/mol. The above results can be compared with those of NAGRA [1], that is, 2 to 25 nm/year in highly alkaline cement water under a non-oxidizing atmosphere at 21 $\pm$ 3 $^{\circ}$C. Our data is compatible with that of NAGRA [1].The activation energy of corrosion of carbon steel was reported by Posey [4] to be 18 kJ/mol at pH 2 and 33 kJ/mol at pH 5 to 7 in 4M NaCl solutions. Our data is compatible with that of Posey[6].

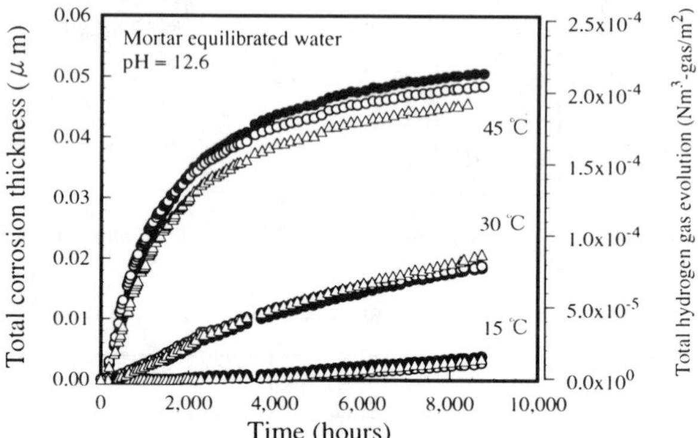

Fig. 3   Total corrosion thickness and hydrogen gas evolution at different temperatures in mortar-equilibrated water.

(experiments were repeated three times under each condition.)

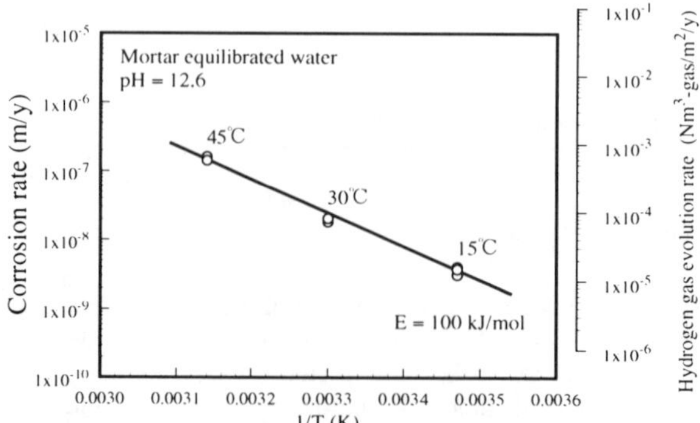

Fig. 4   Effect of the 1/T(K) on corrosion rate and hydrogen gas evolution rate for iron in mortar-equilibrated water.

The Eh value of the test solution which was measured in a glovebox after the corrosion test was - 644 mV vs. NHE at pH 12.08. Therefore it is considered that the results shown in Fig. 3 were obtained under a sufficiently non-oxidizing atmosphere. The surface of the specimens tested at 45 °C were observed immediately after test. The specimens were covered with uniform black films. Auger electron spectroscopic analysis revealed that the film was 1 μ m thick and composed of Si, Fe, Ca, and O.

Since the surfaces of the iron specimens were covered with the above-mentioned thick films, the reduction of corrosion rate shown in Fig. 3 were attributed to this film formation.

For the test with the aqueous solution saturated with Ca(OH)$_2$ solution, iron exhibited almost the same corrosion behavior as that shown in the aqueous solution which were coexisted with mortar.

Fig. 5 shows the quantity of hydrogen gas generated from iron in a synthetic groundwater of pH 8. It is understood from Fig. 5 that hydrogen gas was generated at a constant rate. From the slope of the curve, the corrosion rate is estimated to be 90 nm/year. The difference in corrosion behavior between in a highly alkaline solution and in the synthetic groundwater may be explained as follows: iron does not form the surface oxide film in the synthetic groundwater because the pH of the solution is lower than 9.5 which value is required for iron to be in a passive state. However, a stable surface oxide film at a pH ranging from 12.5 to 13 is formed[1, 7].

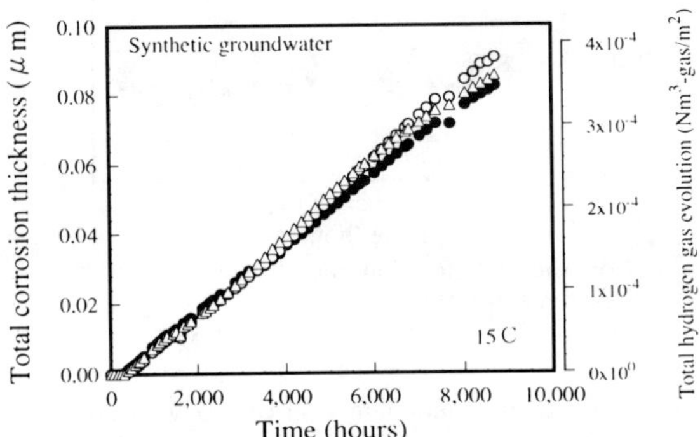

Fig. 5 Total corrosion thickness and hydrogen gas evolution in
synthetic groundwater at 15 °C.
(experiments were repeated three times under each condition.)

(b) Aluminum

Fig. 6 shows the corrosion rate of aluminum in mortar grains. It was understood from Fig. 6 that during an initial period of immersion, the corrosion rate was 20 mm/year and that after 400 hours it was reduced to 0.1 mm/year. And after 1,000 hours the rate reached a constant value ranging from 0.001 to 0.01 mm/year. The corrosion rates were estimated on the basis of uniform corrosion according to the following reaction:

$$Al + 3\ H_2O \rightarrow Al(OH)_3 + 3/2\ H_2. \qquad (2)$$

Similar results were obtained from the second corrosion test in which the solution were changed every 4 weeks. It was indicated that the change of the solution has no influence on the corrosion behavior.

On the surface of the aluminum specimens placed in the pulverized mortar formed films composed of the corrosion products and the pulverized mortar. EPMA was employed for the analysis of the films. Near the interface with the test solution Ca, Al and O was observed and near the interface with the metallic bulk specimens Al and O was observed and no Ca was detected. Aluminum were uniformly corroded under this film.

Furthermore, since the changing solutions had almost no influence on the corrosion rate, it is considered that the films would not readily dissolve even when the solution properties near the specimen surface altered due to the changing solutions. Therefore, it was considered that after film formation, the corrosion rate depended on the dissolving rate of the films. The films were identified as gibbsite from X-ray diffraction.

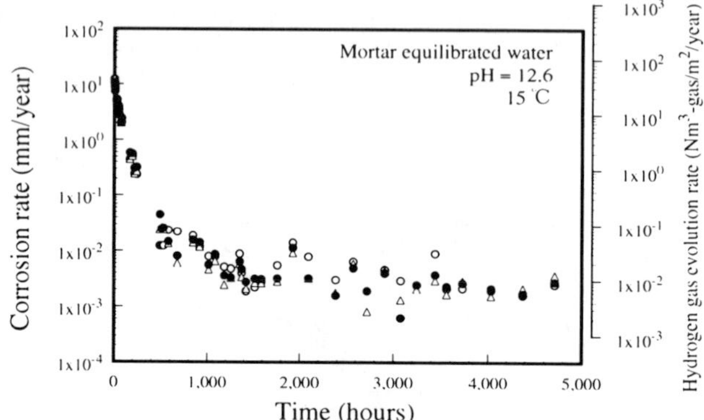

Fig. 6 Corrosion rate for aluminum in mortar- equilibrated water versus time.
(experiments were repeated three times for each condition)

In the third corrosion test, the aluminum plates were vertically immersed into mortar-equilibrated water without mortar grains. Fig. 7 shows the results of this test. Reduction in the corrosion rate, which is shown in Fig. 6, is not observed in Fig. 7. This difference is explained as follows; corrosion products formed on the specimens during the initial period could not stay on the vertical surfaces.

Meanwhile, horizontally immersed aluminum plates had an Al-Ca complex salt deposited on the surface thereof and gibbsite formation near the interface the metallic surface was confirmed by X-ray diffraction.

From the above, it was found that under conditions in which the Al-Ca complex salt was allowed to stay on the specimen surface, for example, by placing the specimens in mortar grains, the complex salt contributed to forming films which suppressed corrosion. The corrosion suppression effect on aluminum by corrosion products has been reported by Walton [3]. And the corrosion resistance of aluminum is determined essentially by the behavior of its layer of oxide and the corrosion rate of this oxide is affected by pH[8]. Therefore the reason why the corrosion rate of aluminum plates placed in mortar grains decrease is regarded as follows: corrosion products restrict movement of hydroxide ion at interfaces between aluminum plates and corrosion products. With the changing the pH, the corrosion rate of aluminum is limited by the dissolving rate of the films and becomes constant, as in shown in Fig. 6.

Fig.8 shows the results of the fourth test with results of the first test. Each plot indicates a mean value obtained from three experiments repeated under the same conditions. From the results it is confirmed that in the initial period the corrosion rates of aluminum specimens solidified with mortar are a tenth of those of aluminum specimens in pulverized mortar. However, after 1,000 hours, the difference becomes smaller and the specimens of both experiments show almost the same corrosion rate. Therefore, the tests in which the aluminum specimens are placed in pulverized mortar can simulate the conditions in which aluminum waste are solidified with mortar in disposal condition.

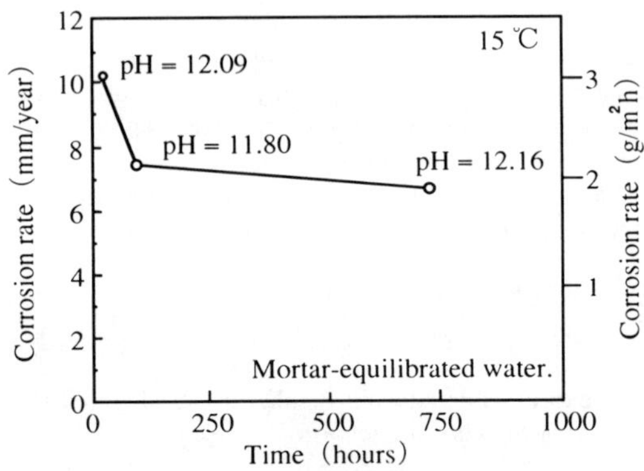

Fig. 7   Corrosion rate for aluminum with free surface in a mortar-equilibrated water without crushed mortar.

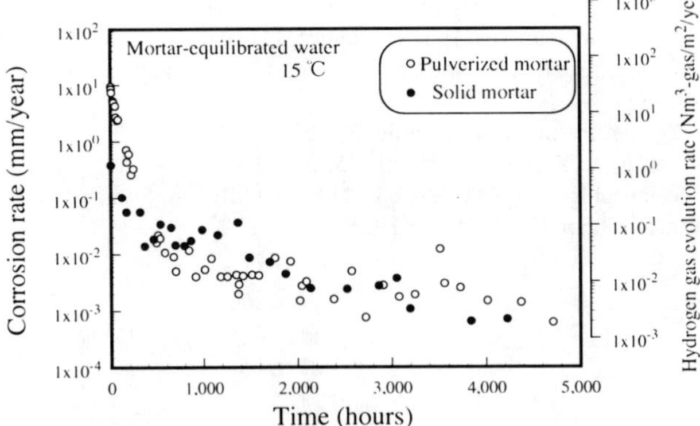

Fig. 8   Corrosion rate for aluminum solidified with mortar compared with that of aluminum in the pulverized mortar.

CONCLUSIONS
(1) Iron

The corrosion behavior of iron was investigated by measuring the quantity of hydrogen gas generated under highly alkaline non-oxidizing conditions simulating the disposal environments.

a) The corrosion rate of iron in an initial corrosion period was 4 nm/year at 15 °C, 20 nm/year at 30 °C, and 200 nm/year at 45 °C. The apparent activation energy was approximately 100 kJ/mol.

b) After an experiment conducted at 45 °C, the specimens were covered with uniform film to which the reduction in the corrosion rate was attributed.

c) The reduction in corrosion rate was not observed in the synthetic groundwater of pH 8 and the corrosion rate was 90 nm/year.

(2) Aluminum

The corrosion behavior of aluminum was investigated by placing aluminum in pulverized mortar grains to simulate disposal conditions in which aluminum is solidified with mortar.

a) The corrosion rate of aluminum in an initial immersion period was 10 to 20 mm/year and then reached a constant value of 0.01 to 0.001 mm/year after 1,000 hours.

b) The reduction in the corrosion rate was attributed to deposition of corrosion products on the aluminum surface, in particular formation of gibbsite.

Thus we obtained data on hydrogen gas generation from iron and aluminum under the disposal environment relevant to the safety analysis of low-level radioactive disposal facilities in Japan.

ACKNOWLEDGMENTS

The authors wish to express their sincere thanks to Dr. Shigeo Tsujikawa, Professor of The University of Tokyo for his kind suggestion.

REFERENCES

1. P. Kreis, Technical Report NTB 91-21, NAGRA, Winterthur, Switzerland, 1991
2. JNFL Low-Level waste repository ; SKB experience form SFR facility (draft 1993)
3. C.T. Walton, F.L. McGeary and E.T. Englehart, Corrosion, 13, 807 (1957)
4. "Advances in Cement-Based Solidification Materials for Streamlined Disposal" (in Japan), CRIEPI, in press
5. G.Schikorr, Zeitschrift fur Electrochemie, 35, 62 (1929)
6. F. A. Posey and A. A. Palko : Corrosion, 35, 38 (1979)
7. R. Grauer, Technical Report NTB 88-02E, NAGRA, Baden, Switzerland, 1988
8. H.T.S Britton, J. Chem. Soc., 21, 20(1925)

DEVELOPMENT OF MATHEMATICAL MODELS FOR LONG-TERM PREDICTION OF CORROSION BEHAVIOUR OF CARBON STEEL OVERPACKS FOR RADIOACTIVE WASTE DISPOSAL

A. R. HOCH, A. HONDA*, F. M. PORTER, S. M. SHARLAND AND N. TANIGUCHI*
AEA Technology, 424.4 Harwell, Oxfordshire, United Kingdom, fiona.porter@aeat.co.uk
* PNC, Power Reactor and Nuclear Fuel Development Corporation Tokai Works, Ibaraki, Japan

ABSTRACT

Mathematical models to enable long-term prediction of the corrosion behaviour of carbon steel overpacks for radioactive waste have been developed. An existing model of the growth of pits, implemented in the CAMLE software, has been extended and used to investigate the sensitivity of the predictions to input parameters, including cathodic reaction kinetics and the relative position of the anode and cathode. Predictions have also been made of the aeration period of the repository, during which localised corrosion is possible.

1. INTRODUCTION

Plans for the disposal of High-Level Radioactive Waste (HLW) in Japan involve burial in a deep underground, multi-barrier repository. The waste will be placed in metal overpacks which will be surrounded by a bentonite backfill. Carbon steel is one of the materials being considered for the overpacks. The overpacks may be required to remain intact for several hundred years, to provide a significant contribution to the repository performance. It is likely that the main factor limiting their performance on this timescale will be corrosion and it is anticipated that the required lifetime can be attained by constructing the containers with an adequate metal thickness to allow for corrosion. To estimate this thickness, mathematical modelling of the corrosion behaviour has been carried out, based on the key physical and chemical processes. Validation of such models is made by comparing the short-term predictions with the results of laboratory experiments that are designed to simulate repository conditions.

Carbon steel can be subject to both uniform and localised corrosion, depending on the environment. Pitting corrosion can result in the relatively rapid corrosion at isolated locations on the metal. A mathematical model of pit growth in metals has been developed with the aim of providing a long-term predictive capability. The model is implemented in the computer program CAMLE (Corrosion and Migration in Localised Environments)[1] and includes a representation of the chemical, electrochemical and transport processes that control pit growth rates. The model development has been supported by an experimental programme to provide input for the model, and data on pit growth rates under different conditions[2].

The aim of this study is to develop the model by including a limitation on the supply of cathodic charge, to better reflect conditions inside a repository, and to test the sensitivity of the predictions to input parameters. In section 2 the CAMLE conceptual model is described. Section 3 describes the experimental determination of oxygen reduction kinetics. In Section 4, describes a study of the sensitivity of pit growth rates to the model parameters, including cathodic reaction kinetics and the relative position of the anode and cathode, together with a comparison with experimental data.

Mat. Res. Soc. Symp. Proc. Vol. 465 © 1997 Materials Research Society

2. CONCEPTUAL MODEL OF PIT GROWTH

A mathematical model of pit growth, implemented in the CAMLE code, is used to predict the solution chemistry and electrochemistry (and hence metal penetration rates) within a corroding cavity. It is necessary to model the solution chemistry within the cavity, the electrochemical-reaction rates and their dependence on parameters such as the electrostatic potential, the migration of ions under concentration and potential gradients, the effect of the changing dimensions of the pit as propagation proceeds and the effect of the blocking of the cavity with solid corrosion product. Due to the complexity of the system, certain approximations are made in the mathematical representation of these processes. The assumptions are:

i)   The electrolyte is static.
ii) Cavity propagation is slow compared with ionic-migration rates, and therefore both the moving boundary and any induced electrolyte motion can be ignored.
iii) Migration processes are considered in two dimensions only.
iv) Dilute-solution theory may be used throughout.
v)   Any time before pit initiation is neglected - the pit grows from time zero.
vi) The metal surface outside the pit is covered by a passive oxide film. Either there is sufficient generation of cathodic charge on the outer surface to support the localised corrosion and the metal is held at a fixed potential, or the potential is determined by the supply of cathodic charge, as described by cathodic reactions over the metal surface.
vii) A simplified geometry can be used to represent the pit. This is modelled as a plane, and it is assumed that the pit is cylindrically symmetric. The model can alternatively represent the cavity by a rectangular, parallel sided slot geometry, for crevice corrosion. The model domain can be extended to include the region immediately outside the pit, which is assumed to contain solid corrosion product. This is illustrated schematically in Figure 1.
viii)It is assumed that the pit solution is in equilibrium with specified solid phases in the pit.
ix) It is assumed that the cavity solution is anaerobic and cathodic reactions within the pit are neglected. These occur on the bulk metal surface. (The model is currently being extended to give the option of including cathodic reactions within the pit.)

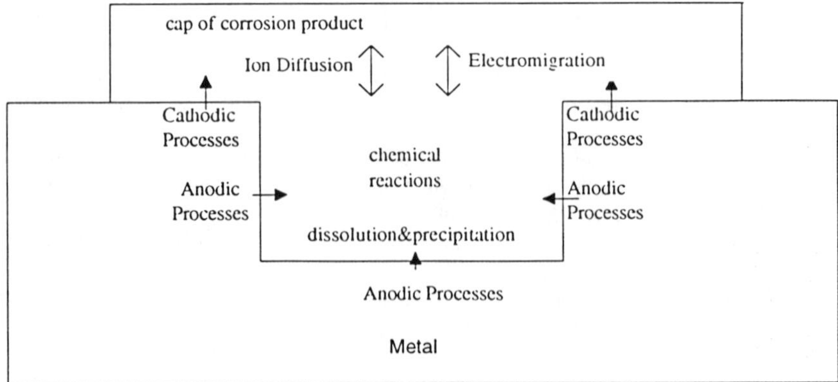

Figure 1 Schematic illustration of processes included in CAMLE model of growth of localised corrosion cavity

The mathematical equations describing mass transport of aqueous chemical species in a dilute electrolytic solution, the distribution of electrostatic potential and chemical reactions in solution are given in reference 1. The region modelled may include solely the cavity, or may extend beyond this, over the metal surface, in which case it is assumed that the region external to the pit includes a cap of corrosion product. The boundary conditions of the model are that the concentrations of the species are fixed at the interface of the corrosion product cap and the bulk medium beyond and remain constant and the fluxes of species either produced or consumed in the electrode processes are proportional to the corresponding electrochemical reaction rates on the metal surface in the modelled regions. The fluxes of the other species at the walls are zero.

CAMLE uses a generalised reaction scheme. The chemistry is simplified by considering a small number of chemical species and a limited number of reactions. The chemical reactions are much faster than the diffusion processes, so the detailed kinetics of several parallel reaction schemes are collected into a single scheme: it is the overall equilibrium constant which matters.

In previous versions of the model[1,2] one electrochemical reaction is included within the pit, for iron-dissolution:

$$Fe \rightarrow Fe^{2+} + 2e^-$$

(1)

It was assumed that the metal potential remained fixed, (although the potential of the solution varies throughout the region) and that there was sufficient generation of cathodic charge to support localised corrosion. Experiments have shown that the maximum pit depth for specimens of carbon steel in bentonite increases with increasing oxygen concentration in the system[8], although inside the pit itself it is largely anaerobic, as the oxygen has been consumed more rapidly than diffusion replaces it. In a repository situation, pit growth may be limited by the supply of cathodic charge. In the model, the metal potential is determined from a charge balance condition on charge flowing between the anode and the cathode of the corrosion cell:

$$\int_{anode} i_A \; dx + \int_{cathode} i_C \; dx = 0$$

(2)

Where i_A is the anodic current and i_C the cathodic current.

The model has now been extended to include a representation of the cathodic reaction over the bulk surface. The dominant cathodic reaction is the reduction of oxygen:

$$O_2 + 4e^- + 2H_2O \rightarrow 4OH^-$$

(3)

The kinetics of the anodic and cathodic reactions are important inputs to the model.

There are three timescales associated with the growth of a pit. On the shortest timescale, the chemical reactions take place, and on an intermediate timescale, the mass transport of chemical species in the cavity occurs. The wall of the cavity moves on a much longer timescale, over which the chemical reactions are in equilibrium and the concentrations of the chemical species have reached their steady-state values. In predicting the growth of a corroding pit, the approach adopted is therefore to solve a succession of steady-state systems with varying pit depths.

Various combinations of the steady-state mass-balance equations are added and subtracted in order to remove the chemical-reaction terms, following the approach of Turnbull[3]. The resulting equations for the migration of the ionic species, combined with the chemical-equilibrium equations and the charge-neutrality equation, are solved numerically by CAMLE, using the finite-element subroutine library TGSL[4].

The model predicts the spatial distribution of both the aqueous chemical species and the electrostatic potential in the pit. The corrosion-current density on the wall of the pit is used to calculate the instantaneous rate of growth of the pit, which is assumed to maintain its original shape. Equations for the time evolution of the depth and the width of the pit are determined using values of the current density that have been averaged over the base and over the side - wall of the pit respectively.

3. EXPERIMENTAL DETERMINATION OF OXYGEN REDUCTION KINETICS

Experiments were carried out previously to determine the kinetics of the metal-dissolution reaction[2]. The kinetics of oxygen reduction are relatively poorly known, by comparison with those of the anodic process. It is assumed that the oxygen-reduction kinetics take the form[5-7]:

$$i = K \, [O_2] \, \exp \, (-\beta F \, (E-E_o)/RT) \tag{4}$$

where $[O_2]$ is the concentration of oxygen at the metal surface, K and β constants, T temperature in K, F Faraday's constant, and R the gas constant. ϕ_M is the electrical potential of the corroding metal relative to some standard electrode in the bulk solution, and $\phi(x)$ is the potential drop in the solution in the cavity: the difference between the potential at the boundary with the bulk solution and the potential at position $\phi(x)$ just outside the electrical double layer on the metal surface. The potential driving the corrosion reaction is then given by $E = \phi_M - \phi(x)$.

The literature data available seem to span a wide range[5-7]. Experiments have been carried out to measure the kinetics of the oxygen reduction reaction for carbon steel, under a range of conditions. The chemical composition, as a mass percentage of carbon steel, was 0.12%C, 0.15%Si, 0.64%Mn, 0.020%P, 0.004%S. Specimens were wet-polished using #800 SiC paper and covered by silicon resin, with an exposed area of 10 by 10 mm remaining. The air or the gas containing fixed O_2 concentrations was purged into the test solution during the measurement. In initial measurements, the pH 10 solutions contained either 600 or 6000 ppm of CO_3^{2-} and HCO_3^-, added as Na_2CO_3 and $NaHCO_3$. The cathodic polarization curves were measured at 80°C, for three different scan rates: 100 mV/min, 50 mV/min and 10 mV/min. Also, measurements were made in the presence of different levels of oxygen; 50% O_2 and 50% N_2 and for 5%O_2 and 95% N_2. The data fitted to these results is shown in Table I.

Table I

(a) 5 % O_2 concentration, $[O_2]$ of 0.021 mol/m^3

	Carbonate Conc. 600 ppm			Carbonate Conc. 6000 ppm		
Scan Rate (mV/min)	10	50	100	10	50	100
$K[O_2]$ (A/m^2)	1.25×10^{-3}	1.17×10^{-3}	3.14×10^{-3}	3.25×10^{-5}	4.60×10^{-4}	3.98×10^{-3}
K (A m/mol)	5.84×10^{-2}	5.47×10^{-2}	1.46×10^{-1}	1.52×10^{-3}	2.15×10^{-2}	1.86×10^{-1}
β	0.32	0.40	0.35	0.59	0.50	0.33

(b) 20 % O_2 concentration, $[O_2]$ of 0.086 mol/m^3

	Carbonate Conc. 600 ppm			Carbonate Conc. 6000 ppm		
Scan Rate (mV/min)	10	50	100	10	50	100
$K[O_2]$ (A/m^2)	2.34×10^{-3}	9.26×10^{-4}	2.45×10^{-3}	1.96×10^{-4}	8.04×10^{-4}	5.19×10^{-4}
K (A m/mol)	2.73×10^{-2}	1.08×10^{-2}	2.86×10^{-2}	2.29×10^{-3}	9.39×10^{-3}	6.06×10^{-3}
β	0.39	0.44	0.39	0.72	0.61	0.66

(c) 50 % O_2 concentration, $[O_2]$ of 0.214 mol/m^3

	Carbonate Conc. 600 ppm			Carbonate Conc. 6000 ppm		
Scan Rate (mV/min)	10	50	100	10	50	100
$K[O_2]$ (A/m^2)	6.44×10^{-4}	9.16×10^{-4}	1.25×10^{-3}	3.07×10^{-5}	5.11×10^{-3}	1.13×10^{-2}
K (A m/mol)	3.01×10^{-3}	4.28×10^{-3}	5.84×10^{-3}	1.43×10^{-3}	2.38×10^{-2}	5.28×10^{-2}
β	0.44	0.50	0.48	0.93	0.51	0.42

4. SENSITIVITY OF PIT GROWTH RATES TO INPUT PARAMETERS

Tests have been carried out to test the sensitivity of the model to input parameters such as changes in diffusion coefficients, cathodic kinetics scheme and the relative position of the anode and cathode.

<u>Base case parameters</u>

In the calculations described here, the parameters are shown in Table II:

Table II
Parameters for CAMLE Simulations

Chemical Parameters: External and initial conditions	High Cl⁻	Low Cl⁻
Solids	$FeCl_2/Fe_3O_4$	Fe_3O_4
pH	5.064	5.6073
Cl^-,M	4.66	2.82×10^{-2}
$FeCl^+$,M	4.33	2.15×10^{-3}
Fe^{2+},M	0.158	1.3×10^{-2}
Na^+,M	1.1×10^{-2}	1.3×10^{-4}
Model geometry		
Cavity shape	cylindrical	
Extent of modelled region	pit and cap of corrosion product.	
Initial pit depth and width, mm	0.2	
Mass Transport Parameters:		
Ionic diffusion coefficient of H^+, $m^2 s^{-1}$	9.3×10^{-10}	
Ionic diffusion coefficient of OH^-, $m^2 s^{-1}$	5.3×10^{-10}	
Ionic diffusion coefficients of other species, $m^2 s^{-1}$	1×10^{-10}	
Temperature, °C	80	

The chemical reactions used are:

$$H^+ + OH^- \Leftrightarrow H_2O$$
$$Fe^{2+} + H_2O \Leftrightarrow FeOH^+ + H^+$$
$$3Fe^{2+} + 4H_2O \Leftrightarrow Fe_3O_4 + 6H^+ + H_2$$
$$Fe^{2+} + Cl^- \Leftrightarrow FeCl^+$$

and for situations involving high concentrations of chloride in the bulk solution:

$$FeCl^+ + Cl^- \Leftrightarrow FeCl_2$$

Seven aqueous chemical species Fe^{2+}, $FeOH^+$, Na^+, Cl^-, $FeCl^+$, H^+ and OH^- were included in the model. The base case simulation assumed the high chloride concentration case, and a fixed anode inside the pit and a cathode outside. The actual cathode radius is not known, but it was arbitrarily assumed that the cathode extended to 4 times the pit radius. Initial calculations used the cathodic kinetics of reference 5, and the anodic kinetics determined in an earlier study[2].

Time evolving simulations of pit growth in 30 days were carried out. The metal potential was determined by a balance between the anodic current and cathodic current. The predicted pit dimensions at 30 days are shown in Table III. The base case parameters result in relatively low pit growth rates.

Table III
Pit Growth Predictions At 30 Days

Cathode Radius (x pit half width)	$D=o(10^{-10})m^2s^{-1}$ Depth (mm)	$D=o(10^{-10})m^2s^{-1}$ Width (mm)	$D=o(10^{-9})m^2s^{-1}$ Depth (mm)	$D=o(10^{-9})m^2s^{-1}$ Width (mm)
4 (base case)	0.213	0.226	0.302	0.403
5	0.220	0.241	0.350	0.501
10	0.277	0.356	0.647	1.10

<u>Sensitivity to diffusion coefficients</u>

Calculations were performed with a higher value of the effective diffusion coefficient of order 10^{-9} m^2s^{-1}. These are presented in Table III and show more rapid pit growth.

<u>Measured cathodic reaction kinetics</u>

Simulations were carried out using the measured cathodic kinetics data for 20% O_2 and 50% O_2 for different carbonate concentrations and scan rates. The results are shown in Table IV, with a comparison with a calculation using the kinetics in reference 5, where K is -4.9×10^{-3}A dm/mole and β 0.5. For these cases, there is relatively little sensitivity to the cathodic kinetics.

Table IV
Effect of different cathodic reaction kinetics

Cathode Reaction Kinetics	Predicted Depth (mm)	Predicted Width (mm)
K-4.9$\times10^{-3}$A dm/mole and β 0.5K [5]	0.302	0.403
6000 ppm, 10mV/min [O_2] 0.086 mol/m^3	0.320	0.441
600 ppm, 100mV/min [O_2] 0.086 mol/m^3	0.317	0.434
600 ppm,:[O_2] of 0.214 mol/m^3	0.309	0.419

<u>Comparison with experimental data</u>

Experiments were carried out to measure pit growth rates under simple immersion conditions, in which the potential was not controlled artificially. The corrosion potential was measured in a similar system, varying with time between -350 to -400 mV[8]. CAMLE was used to predict pit depths for these tests, using the parameters for low chloride concentration in Table II. The results are shown in Table V. This compares potentiostatic results with those obtained assuming a limitation on cathodic charge. For the base case parameters, the results show a considerable reduction in predicted pit depth, to below the experimental value of 0.67 mm. Two additional calculations are presented, with increased diffusion coefficient and increased cathodic radius. These show an improvement toward the experimental value and suggest that realistic predictions can be obtained within a reasonable parameter range.

Table V
Comparison with Experimental Data, measured value = 0.67 mm

Cathodic reaction	Depth (mm) $D=o(10^{-10})m^2s^{-1}$ r=4	Depth (mm) $D=o(10^{-9})m^2s^{-1}$ r=4	Depth (mm) $D=o(10^{-9})m^2s^{-1}$ r=10
None, Potential -0.35 V	1.62	4.47	
6000 ppm, 10mV/min	0.21	0.31	0.7

r = cathode radius/pit radius

<u>Extension of Anode onto Metal Surface</u>

When examining pitting experiments, it was observed that there was additional crevice corrosion between a corrosion product cap, outside the corrosion pit and the outer metal surface. This may have arisen due to differential mass-transport properties between the corrosion cap and the environment, which has lead to the creation of an occluded area on the metal. The CAMLE model has been used to examine the effect of extension of the anode onto the metal surface on the pit growth rates. The model is illustrated in Figure 2. Two sets of calculations were performed: in the first the anode was located within the pit and in the second, the anode is arbitrarily extended beyond the pit mouth to a distance of one pit radius. Steady state calculations were performed using the low chloride set of parameters in Table I and cathodic reaction kinetics from reference 5, where it is assumed that K is 4.9 10^{-3} A dm/mole and β is 0.5.

The predicted current on the pit wall, is compared in Figure 3, where it can be seen that smaller currents are obtained at the base of the pit (hence pit growth rates are reduced) when the anode extends onto the metal surface. This has the effect of reducing predicted pit growth rates.

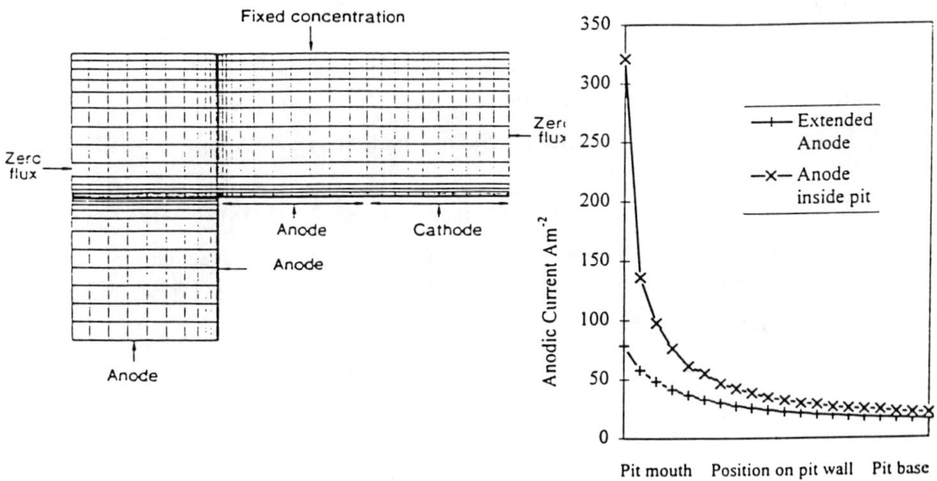

Figure 2 Finite element model

Figure 3   Predicted steady state anodic current on the pit wall

5. CONCLUSIONS

Sensitivity tests have been carried out to assess the dependence of pit growth rates on input parameters. The pit growth rate has been found to show a strong dependence on assumed cathode size and to be proportional to the diffusion coefficient, over the range tested. In these tests there was little difference in the results obtained with different kinetic reaction schemes to describe the cathodic reduction of oxygen.

Model predictions have been compared with experimental measurements of pit growth rates under simple immersion conditions. For the base case parameters, the model predicted pit depths below the experimental data. Additional calculations with increased diffusion coefficient and cathodic radius showed an improvement toward the experimental value and suggest that realistic predictions can be obtained within a reasonable parameter range.

CAMLE has been used to predict the extension of the anode onto the surrounding metal surface. This has the effect of reducing predicted pit growth rates.

6. ACKNOWLEDGEMENTS

This work was funded by the Power Reactor and Nuclear Fuel Development Corporation (PNC), Tokyo, Japan, and their permission to publish these results is gratefully acknowledged.

7. REFERENCES

1   S. M. Sharland, C. P. Jackson and A. J. Diver, Corros. Sci. **29**(9), 1149 (1989).

2   A. Hoch, A. Honda, F. M. Porter, S. M. Sharland and N. Taniguchi, A modelling and experimental study for long-term prediction of localised corrosion in carbon steel overpacks for high-level radioactive waste, (MRS Conference, Kyoto, Japan, October 1994).

3   A Turnbull and J G N Thomas, J. Electrochem. Soc. **129**, 1412 (1982).

4   C P Jackson, 'The TGSL Finite-Element Subroutine Library', AEA Report AERE-R 10713 (1982).

5   J C Walton, Nuclear and Chemical Waste Management, **8**, 143, (1988).

6   C C Naish, N J M Wilkins, A Harker and R F A Carney, 'Concrete Inspection : Interpretation of Potential and Resistivity Measurements', AEA Report, AERE- R 13429 (1989)

7   A Turnbull, Br. Corros. J., **15**, 162-171, (1980).

8   M. Nodaka, H. Ishikawa, Y. Yusa and N. Sasaki, 'Corrosion Studies on Candidate Overpack Materials', ASME/JSME/AESJ Joint International Waste Management Conference Proc. p589-593 (1989).

Part X

Radionuclide Solubility and Speciation

SPECTROSCOPIC INVESTIGATION OF ACTINIDE SPECIATION IN CONCENTRATED CHLORIDE SOLUTION

W. Runde, M.P. Neu, S.D. Conradson, D.L. Clark, P.D. Palmer, S.D. Reilly, B.L. Scott, C.D. Tait.
Environmental and Structural Actinide Chemistry, Los Alamos National Laboratory, Los Alamos, NM 87545, USA.

ABSTRACT–Actinide solubilities in highly concentrated chloride solutions are about one order of magnitude higher than in similar inert electrolyte ($NaClO_4$) solutions. This increased solubility is due to interactions between actinide and chloride ions. Contradictory results exist regarding the interaction mechanism between actinide and chloride ions. Specifically, both inner-sphere complex formation and ion pair association have been implicated in the interpretation of spectrophotometric and extraction data. To address this controversy, we investigated the interaction between actinide ions in the (III), (IV), (V) and (VI) oxidation states and chloride ions using a multi-method approach. Spectroscopic techniques (TRLFS, Raman, UV-Vis absorption, EXAFS) were used to distinguish between changes in the inner coordination sphere of the actinide ion and effects of ion pairing. X-ray absorption spectroscopy and single crystal X-ray diffraction were used to determine structural details of the actinide chloro complexes formed in solution and solid states.

INTRODUCTION

The proposed disposal of nuclear waste in geological salt formations, e. g. the Waste Isolation Pilot Plant (USA) and the Gorleben site (Germany), raises a fundamental question: To what degree actinides will be solubilized and mobilized upon interaction with chloride ions? Numerous studies have been performed to investigate the formation and stability of actinide chloro complexes; however, only a few actinide chloro compounds formed in aqueous solution have been structurally characterized. Stability constants have been determined using various methods, such as solvent extraction, spectrophotometry, potentiometry, and ion exchange [1, 2, 3, and literature therein]. Despite the large number of investigations, the interaction mechanism between actinides and chloride ions is still under discussion including both inner-sphere complexation and electrostatic ion pair association (outer-sphere complexation). Extraction and ion exchange methods cannot distinguish between ion associates and complex formation which both affect the activity of the actinide ions in solution. The objective of this study is to evaluate spectroscopically the interaction of actinides in the oxidation states +III, +IV, +V, and +VI with chloride ions and to elucidate the formation of inner-sphere actinide chloro complexes in aqueous solution.

Mat. Res. Soc. Symp. Proc. Vol. 465 © 1997 Materials Research Society

EXPERIMENTAL

Am(III) stock solution was obtained by dissolving AmO_2 in 9M HCl and precipitating Am(III) as pink $Am(OH)_3$. The hydroxide was then dissolved in 0.1M $HClO_4$ and aliquots were used for the spectrophotometric investigations.

Np(IV) was prepared from a 1mM Np(V) solution in 0.1M $HClO_4$ by electrochemical reduction. The solution was characterized spectroscopically to verify the absence of Np(V) and its absorption band at 980 nm and the presence of the characteristic Np(IV) bands, i. e. 723 and 959 nm. Aliquots of this stock solution were added to guanidinium carbonate and potassium chloride solutions, and the Np(IV) absorption spectra were collected immediately.

Pure solutions of Pu(VI) were prepared by heating a Pu, concentrated perchloric acid solution to near dryness, and diluting the residue to desired volume. The solution was assayed for Pu(VI) by use of the absorption band at 830 nm (ε = 525 cm^{-1} M^{-1}), and to assure the absence of Pu in lower oxidation states. Sample solutions were prepared by adding aliquots of this stock to solutions with different NaCl or LiCl concentrations.

Single crystals of $[K(18\text{-}Crown\text{-}6)]_2[AnO_2Cl_4]$ (An = U, Pu) were obtained by adding excess 18-Crown-6 (1,4,7,10,13,16-hexaoxacyclooctadecane) to a 5M HCl/KCl solution with 0.1M in actinyl. The crystals formed within a few days at 4° C and were characterized using single crystal X-ray diffraction and diffuse reflectance spectroscopy.

RESULTS AND DISCUSSION

Trivalent and Tetravalent Actinides

Chloride complexation of trivalent actinides has been widely studied [1,3,4]. It is well known that mono- and dichloro complexes of actinides, $AnCl^{2+}$ and $AnCl_2^+$, form in aqueous solution, and stability constants have been derived using various techniques. The transuranium tri- and tetra chloro complexes have not been reported to form in aqueous solution. Anionic americium(III) chloro complexes, $AmCl_4^-$ and $AmCl_6^{3-}$, have been proposed to be stable only in non-aqueous media or as the dominant Am(III) species formed on ion-exchange columns [3]. Since the chloro complexes are relatively weak (log β_1 = -0.1±0.1, I = 1 [1, 3]) high concentrations of chloride are required to replace water molecules in the inner-coordination sphere to form $AmCl_n^{3-n}$ complexes. The exchange of water in the outer-sphere, and the formation of ion pairs does not affect the most intense Am(III) absorption band at 503.2 nm. The absorption spectra of Am(III) in 5M NaCl and 5M $NaClO_4$ show no significant difference in both the peak location and the peak shape. However, using the recommended stability constants of the Am(III)/Cl- system for species distribution calculation (log β_1 = 0.5 (I=0.1), 0.1 (I=0.5), -0.1 (I=1), -0.15 (I=4); log β_2 = -0.5 (I = 1), -0.7 (I=4) [1]), only about 12% of the total amount of americium in 5M NaCl solution should exist as the free Am^{3+} ion. In addition, the fluorescence band of Am^{3+} at 687 nm (excitation wavelength 503.2 nm) is not influenced by the addition of

NaCl up to 5 M. We used X-ray absorption spectroscopy to study the change in the Am(III) hydration sphere. In Figure 1 the EXAFS Fourier transforms of Am(III) in 5M NaCl is compared with that of the Am^{3+} aquo ion in 1 M $HClO_4$. The most intense peaks in both FT spectra correspond to the O atoms of the hydration water in the inner-coordination sphere. With addition of NaCl the bond distance $Am-OH_2$ in the first coordination shell decreases slightly. This effect is a first hint on a change in the hydration sphere of Am(III). Whether this change is due to a replacement by hydration water in the inner shell or an effect relating to an outer-sphere interaction of chloride ions remains unknown. However, the EXAFS data can not confirm 90% chloride complexation as calculated using published extraction data. These results agree with the spectrophotometrically determined stability constants of $AmCl^{2+}$ and $AmCl_2^+$ reported by Marcus [4], which are about 2 orders of magnitude lower than those determined by phase extraction [5].

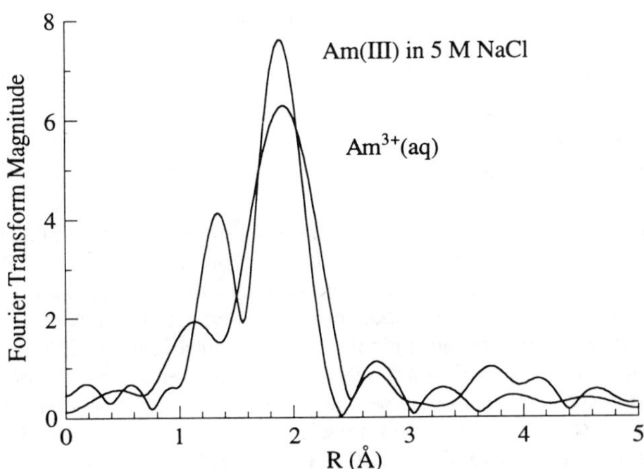

Figure 1. Fourier transform moduli of EXAFS data for Am(III) in 1 M $HClO_4$ and 5M NaCl.

Similar to Am(III), the spectroscopic evidence for chloride complexation of tetravalent actinides is limited. Figure 2 shows the absorption spectra of Np(IV) in aqueous solution with different potential ligands. Addition of 5M KCl does not affect the main absorption bands of Np(IV) with absorption maxima at 723 and 959 nm. In contrast, the absorption spectra of Np(IV) show obvious changes due to addition of carbonate and the formation of strong Np(IV) carbonate complexes.

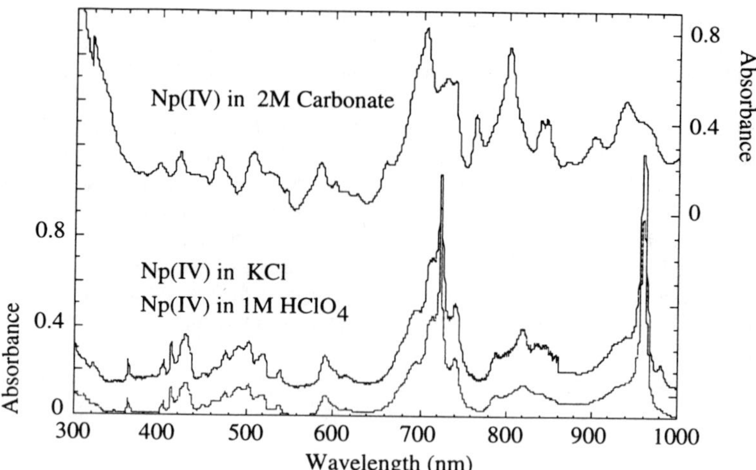

Figure 2. Absorption spectra of Np(IV) in 2M carbonate media, in 5M KCl and in 1M HClO$_4$.

Pentavalent Actinides

Literature data on chloride complexation of pentavalent actinides are limited. The formation and stability of the mono- and dichloro complexes of Np(V), NpO$_2$Cl and NpO$_2$Cl$_2$, have been reported [1]. However, the results are contradictory depending on the method used for investigation: spectroscopic investigations show no significant change in the absorbance spectra up to chloride concentrations of 2M chloride, whereas organic/aqueous extraction data show significant differences in the Np(V) separation from NaCl and NaClO$_4$ solutions with similar concentrations. To date, no formation constants of An(V) chloro complexes are recommended in recent reviews [1]. Solubility studies have shown an increase of solubility at [CO$_3^{2-}$] < 10^{-3} mol kg^{-1} by about one order of magnitude in NaCl compared to that in inert electrolyte solutions (NaClO$_4$) at ionic strength I = 5M [7]. With increasing carbonate concentration, the solubility difference decreases and the Np(V) concentration in solution at [CO$_3^{2-}$] > 10^{-3} mol kg^{-1} is approximately equal in NaCl and NaClO$_4$ solutions. The solubility difference (Figure 3) is reflected by the concentration ratio of the Np(V) species in 5M NaCl and 5M NaClO$_4$ as calculated from the solubility of NaNpO$_2$(CO$_3$)·nH$_2$O in 5M NaCl and 5M NaClO$_4$ [7]:

$$\log \frac{[NpO_2^+]^{(NaCl)}}{[NpO_2^+]^{(NaClO_4)}} = 0.4 \qquad \log \frac{[NpO_2(CO_3)^-]^{(NaCl)}}{[NpO_2(CO_3)^-]^{(NaClO_4)}} = 1.2$$

$$\log \frac{[NpO_2(CO_3)_2^{3-}]^{(NaCl)}}{[NpO_2(CO_3)_2^{3-}]^{(NaClO_4)}} = 0.7 \qquad \log \frac{[NpO_2(CO_3)_3^{5-}]^{(NaCl)}}{[NpO_2(CO_3)_3^{5-}]^{(NaClO_4)}} = -0.3$$

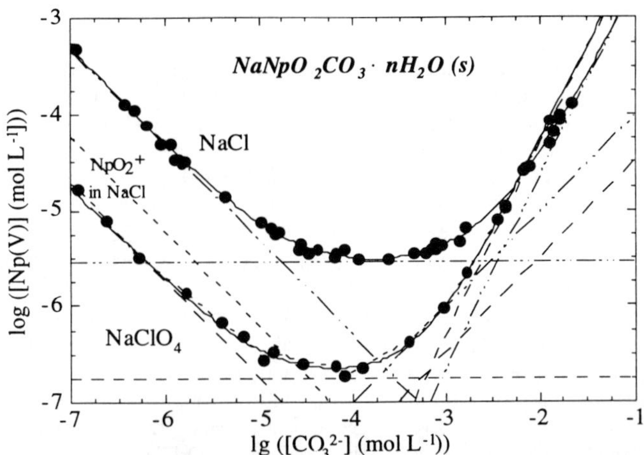

Figure 3. Comparison of the Np(V) species concentrations in 5M NaCl (upper curve) and 5M NaClO$_4$ (lower curve) solution [7]. The slopes of the calculated Np(V) species concentrations are -1 for NpO$_2^+$, 0 for NpO$_2$(CO$_3$)$^-$, +1 for NpO$_2$(CO$_3$)$_2^{3-}$, and +2 for NpO$_2$(CO$_3$)$_3^{5-}$.

Under the latter conditions the Np(V) triscarbonato complex is the predominant solution species. The neptunyl ion is coordinated bidentately with three carbonate ligands in a hexagonal bipyramidal fashion [8]. This geometry does not allow any further inclusion of chloride in the inner coordination sphere around the neptunyl ion. An exchange of carbonate ligands with chloride ions can be excluded due to the strong complexation of carbonate and about six orders of magnitude higher stability constants. With decreasing carbonate complexation water molecules have been determined in the inner-coordination sphere: one in the biscarbonato complex, and two in the monocarbonato complex [8]. These water molecules may be replaced by chloride ions to form mixed Np(V) chlorocarbonato complexes. In the range [CO$_3^{2-}$] < 10^{-5} mol kg^{-1} the neptunyl ion, NpO$_2^+$, is the dominant solution species and chloride interaction raises the solubility by about 1.4 orders of magnitude. By taking into account the formation constants derived from solvent extraction including both ion pairing and potential complex formation, the solubility of Np(V) is still about 0.3 units higher than in NaClO$_4$. This could be due to different crystallinity of the solid phases used in the solubility experiments. At [CO$_3^{2-}$] between 10^{-5} and 10^{-3} mol kg^{-1} the monocarbonato complex, NpO$_2$CO$_3^-$, determines the Np(V) concentration in solution. The difference in Np(V) solubility is observed to be about 1.2 orders of magnitude. This difference is close to that observed in the neptunyl ion stability range at [CO$_3^{2-}$] < 10^{-5} mol kg^{-1} where chloride complexation is proposed fromextraction data. These differences in experimental solubility and calculated species concentrations in NaCl and NaClO$_4$ suggest a different chloride interaction mechanism of the different Np(V) species.

Hexavalent Actinides

The mono- and dichloro complex of uranyl, plutonyl and neptunyl have been reported in the literature and few data exist for $UO_2Cl_3^-$ [1, 2] In Figure 4 the absorption spectra of uranyl ions in NaCl solutions at $[H^+] = 0.1$ m are shown. The addition of NaCl results in a significant change of the U(VI) absorption spectra. The main absorption peak of UO_2^{2+} at 413.3 nm is shifted to 423.4 nm, and in addition two absorption bands appear at 458.2 and 473.8 nm. An isosbestic point at 392.4 nm at [NaCl] < 1.5 m indicates the existence of only two species, UO_2^{2+} and UO_2Cl^+. At higher NaCl concentrations the absence of an isosbestic point in the absorption spectra indicates the existence of more than two uranyl species in solution, and further chloride complexation of the uranyl ion. A more significant change is observed in the Pu(VI) absorbance spectra for the coordination of plutonyl with chloride. Figure 5 shows the absorption spectra of Pu(VI) in LiCl at 1 M $[H^+]$. The appearance of three new absorption bands centered at 837, 843, and 849 nm are consistent with the formation of $PuO_2Cl_n^{2-n}$ for n = 1-3. A shoulder at about 857 nm is observed only at very high LiCl concentrations due to the formation of the tetrachloro complex, $PuO_2Cl_4^{2-}$. The limiting chloro species, $AnO_2Cl_4^{2-}$ (An = U, Pu), were isolated after addition of 18-Crown-6. Here we report the first Pu(VI) chloride complex isolated from aqueous solution. The Pu compound crystallized in the monoclinic space group, P2(1)/c, with a = 10.113 Å, b = 9.973 Å, c = 19.548 Å, and β = 93.78° for 2 molecular

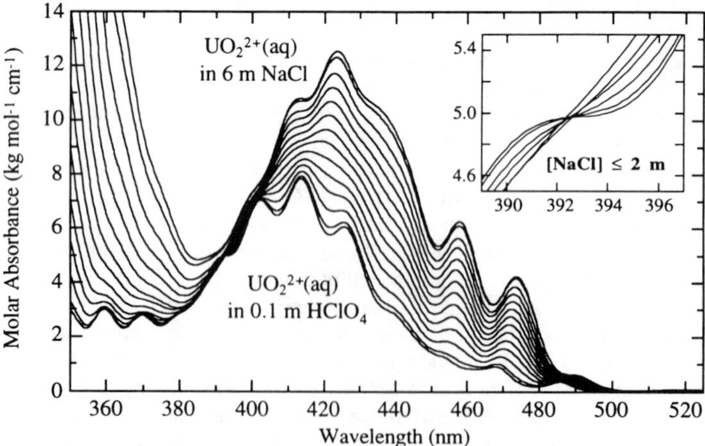

Figure 4. UV-Vis absorption spectra of U(VI) in 0.1 M $HClO_4$ and in NaCl solutions ($[H^+]$ = 0.1M).

units. The plutonium atom is coordinated in a tetragonal bipyramidal geometry with two trans plutonium oxygen atoms, and four chlorides in the equatorial plane. Two potassium cations interact with two crown ethers and with two cis-chloro ligands forming a sandwich complex with the $PuO_2Cl_4^{2-}$ anion. For the analogous U(VI) compound we obtained a U=O bond distance of 1.768(3) Å. The Pu–Cl (2.66(1) Å) and U–Cl (2.6771(14) and 2.6740(12) Å) bond distances are similar to those reported by Rogers et al. [9] for $[(H_2O)_5(18\text{-}Crown\text{-}6)][UO_2Cl_4]$.

The NIR characteristics provide information about the number of chloride ligands in the innner coordination sphere of Pu(VI). To investigate the effect of the chloride complexation on bond distances and stabilities, we initiated an investigation into the structure of $PuO_2Cl_n^{2-n}$ using Extended X-ray Absorption Fine Structure (EXAFS). X-ray absorption measurements were performed at the plutonium LIII edge for Pu(VI) in $HClO_4$ and solutions with various chloride concentrations. Figure 6 shows a comparison of the Fourier Transforms of the experimental data for Pu(VI) in 1M $HClO_4$ and in solutions with different chloride concentrations. The components used to fit the experimental data are plotted with negative FT amplitudes. The Fourier Transform spectra show two significant well-resolved peaks. The peak at 1.75 Å corresponds to two trans plutonium oxygen atoms, while the second and third shell contains the contributions from water oxygen and chloride ligands in the inner coordination sphere. For the PuO_2^{2+} aquo ion the assignment of the shell at 2.45 Å corresponding to five water molecules in the

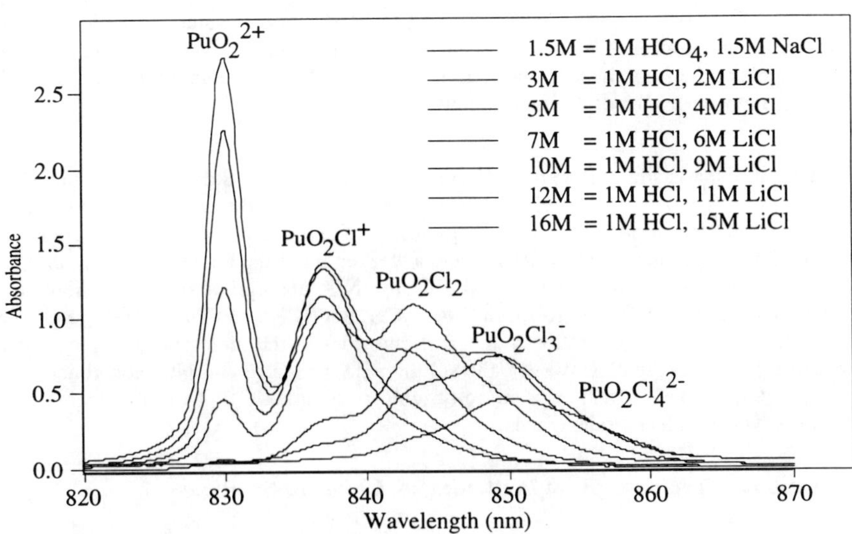

Figure 5. NIR absorption spectra of Pu(VI) in acidic LiCl solutions.

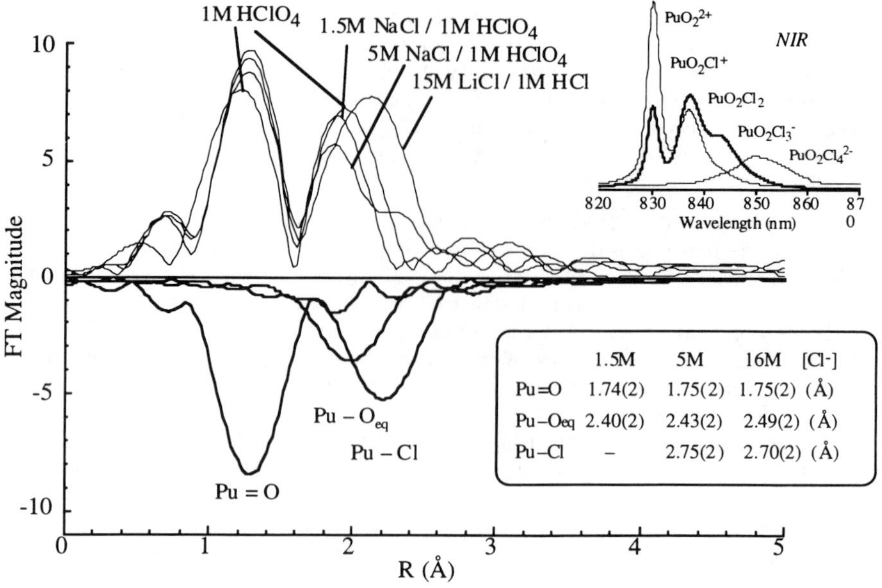

	1.5M	5M	16M [Cl-]
Pu=O	1.74(2)	1.75(2)	1.75(2) (Å)
Pu–Oeq	2.40(2)	2.43(2)	2.49(2) (Å)
Pu–Cl	–	2.75(2)	2.70(2) (Å)

Figure 6. Fourier transforms for Pu(VI) in 1M $HClO_4$ and in solutions with different chloride concentrations. Plotted with negative amplitudes are the three main wave components of the fit of the data obtained in 15M LiCl/1M HCl, representing the plutonyl, equatorial oxygen and chloride contributions.

equatorial plane is straightforward. Curve fitting of the EXAFS data of Pu(VI) in 16M chloride reveals an additional shell, at 2.70 Å, corresponding to the contribution of chloride ligands in the inner coordination sphere. Inclusion of chloride induces a small increase of the Pu=O bond distance of about 0.01 Å and a weakening of the Pu–O_{eq} bond distance by about 0.06 Å. Since in 15M LiCl/1M HCl the NIR absorption spectrum shows the existence of three Pu(VI) chloro species, PuO_2Cl_2, $PuO_2Cl_3^-$, and $PuO_2Cl_4^{2-}$, we cannot determine the exact coordination number of chlorides in the inner coordination sphere around the plutonyl ion. However, the solution EXAFS data and NIR absorbance are consistent with the formation of plutonyl chloro complexes and the replacement of hydration water with chloride ligands.

Vibrational Spectroscopy of Actinides in Chloride Solutions

The Raman frequencies for the actinyl-stretching frequencies have been reported to be very similar across the series: U(VI) 870 cm⁻¹, Np(VI) 854 cm⁻¹, Pu(VI) 833 cm⁻¹ [11].

We used this technique to obtain information on the chloride concentrations for Am(III), Np(V) and U(VI). We did not observe any bands for the proposed Am-Cl bond; only frequencies corresponding to impurity ClO_4^- were measured (Figure 7). The neptunyl stretching frequency in 5M NaCl was measured to be 767 cm^{-1}, which is equal to that reported for the neptunyl(VI) ion. In contrast, the uranyl stretching frequency depends on the chloride concentration and shifts from 885 in 1M $HClO_4$ for the free UO_2^{2+} ion to 865 cm^{-1} and 851 cm^{-1} in 3M NaCl and conc. HCl, respectively. These changes in An=O stretching frequencies, where An = Np(V) and U(VI), agree with those reported: relatively large shift in the uranyl frequency upon carbonate complexation (812 cm^{-1} for $UO_2(CO_3)_3^{4-}$ [10]), and small difference in the neptunyl frequencies (756 cm^{-1} for $NpO_2(CO_3)_3^{5-}$ [10]). In fact, identical frequencies have been reported for $NpO_2(CO_3)_2^{3-}$ and $NpO_2(CO_3)_3^{5-}$.

Two actinyl–chloride Raman active vibrations were observed in the region 200 to 400 cm^{-1}. Figure 8 shows the Raman spectrum of [K(18-Crown-6)]$_2$[PuO$_2$Cl$_4$] in comparison of that of a 0.75M KCl/18-Crown-6 solution. The frequencies at 258 and 307 cm^{-1} can be assigned to the plutonyl chloride complex, while the signal at 283 cm^{-1} corresponds to the organic ligand. The most intense peak at 307 cm^{-1} corresponds to the symmetric A_{2g} mode where all bonds lengthen and contracting in unison. The weaker

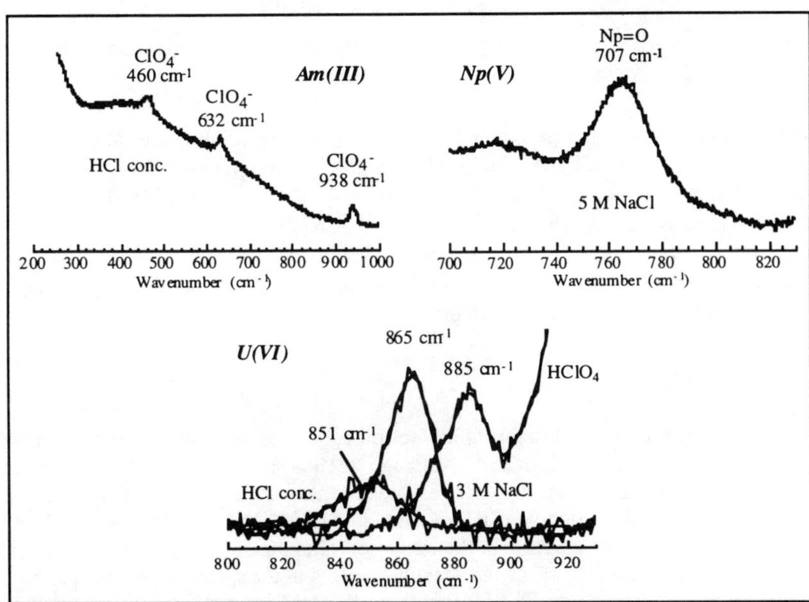

Figure 7. Raman spectra of Am(III), Np(V), and U(VI) in chloride solutions.

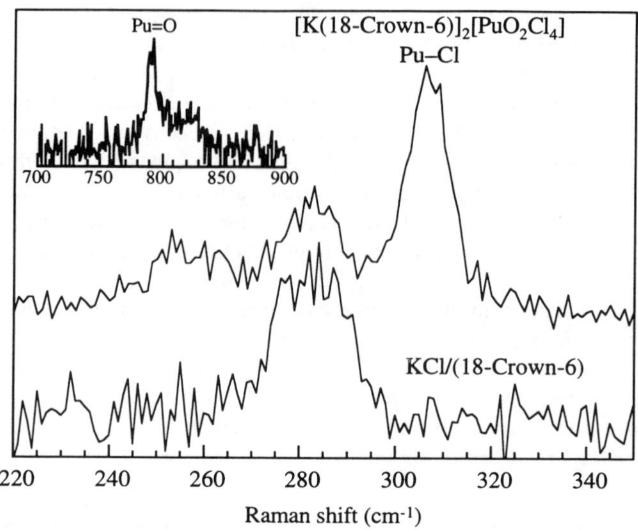

Figure 8. Raman spectrum of [K(18-Crown-6)][PuO$_2$Cl$_4$] in comparison with that of a 0.75M KCl/(18-Crown-6) solution.

signal at about 258 cm^{-1} originates from the B_{2g} mode (in-out-in-out at adjacent M-Cl bonds). These observations are in good agreement with reported Raman frequencies for square planar D_{4h} molecules: [Au(Cl)$_4$]$^-$: B_{2g} at 324, A_{1g} at 347 cm^{-1}; [Pd(Cl)$_4$]$^{2-}$: B_{2g} at 275, A_{1g} at 303 cm^{-1}; [Pt(Cl)$_4$]$^{2-}$: B_{2g} at 312, A_{1g} at 330 cm^{-1} [12]. Since the Pu-Cl A_{1g} mode (at about 306 cm^{-1}) and the plutonyl vibration (at 791 cm^{-1}) are well-separated in energy, only very little coupling between these modes should occur and hence little Fermi resonance artificially separate their frequencies further.

CONCLUSIONS

The formation and stability of actinide chloro complexes have been widely determined using extraction and exchange techniques. These techniques cannot differentiate between ion activity changes due to inner-sphere complexation or ion pairing effects. Whereas spectroscopic (TRLFS, Raman, UV-Vis–NIR absorption) results cannot prove the association of chloride in the inner-coordination sphere of Am(III), Np(IV) and Np(V), hexavalent actinides show substantial changes in the spectroscopic data upon inner-sphere chloro complexation. The formation of the chloro complexes AnO$_2$Cl$_n{}^{2-n}$ with n =1-4 has been established. The limiting species in solution have been found to be the tetrachloro complex, $AnO_2Cl_4^{2-}$, which forms only at high chloride concentrations (> 15M). Further experiments have to be performed to establish the formation of chloro complexes with tri-, tetra-, and pentavalent actinides at chloride concentrations relevant for natural aquifer systems (up to 5M $[Cl^-]$).

ACKNOWLEDGEMENTS

We are grateful the Office of Basic Energy Sciences (OBES), and 94-1 Core Technology Program for Plutonium Residue Stabilization, US DOE.

LITERATURE

[1]   J. Fuger, I.L. Khodakovsky, E.I. Sergeyeva, V.A. Medvedev, J.D. Navratil, <u>The Chemical Thermodynamics of Actinide Elements and Compounds. Part 12. The Actinide Aqueous Inorganic Compexes</u>. IAEA, Vienna, 1992.

[2]   I. Grenthe, J. Fuger, R.J.M. Donigs, R.J. Lemire, A.B. Muller, C. Nguyen-Trung, H. Wanner, <u>Chemical Thermodynamics of Uranium, Vol. 1</u>. Elsevier Science Publishing Company Inc., North-Holland, 1992.

[3]   R.J. Silva, G. Bidoglio, M.H. Rand, P.B. Robouch, H. Wanner, I. Puigdomenech, <u>Chemical Thermodynamics of Americium, Vol. 2</u>. Elsevier Science Publishing Company Inc., North-Holland, 1995.

[4]   Y. Marcus, M. Shiloh, Israel J. Chem. **7**, 31 (1969).

[5]   I. Grenthe, Acat Chem. Scand. **16**, 2300 (1962).

[6]   M. Bansal, L. Sommer, Z. Phys. Chem. **226**, 309 (1964).

[7]   W. Runde, M.P. Neu, D.L. Cark, Geochim. Cosmochim. Acta **60**(12), 2065-2073, (1996).

[8]   D.L. Clark, S.D. Conradson, S.A. Ekberg, N.J. Hess, M.P. Neu, P.D. Palmer, W. Runde, C.D. Tait, J. Am. Chem. Soc. **118**, 2089-2090, (1996).

[9]   R.D. Rogers, A.H. Bond, W.G. Hipple, A.N. Rollins, R.F. Henry, Inorg. Chem. **30**, 2671, (1991).

[10] D.L. Clark, S.D. Conradson, S.A. Ekberg, N.J. Hess, D.R. Janecky, M.P. Neu, P.D. Palmer, C.D. Tait, New J. Chem. **20**, 211-220, (1996).

[11] L. Maya, G.M. Begun, J. Inorg. Nucl. Chem. **43**, 2827, (1981).

[12] K. Nakamoto, <u>Infrared and Raman Spectra of Inorganic and Coordination Compounds</u>. 4th edition, Wiley-Interscience, New York, 1986, pp. 141-145.

ASSESSING THE EFFECT OF DISSOLVED ORGANIC LIGANDS ON MINERAL DISSOLUTION RATES: AN EXAMPLE FROM CALCITE DISSOLUTION

Teri DeMaio and D. E. Grandstaff
Department of Geology, Temple University, Philadelphia, PA 19122 U.S.A.

ABSTRACT

Experiments suggest that dissolved organic ligands may primarily modify mineral dissolution rates by three mechanisms: (1) metal-ligand (M-L) complex formation in solution, which increases the degree of undersaturation, (2) formation of surface M-L complexes that attack the surface, and (3) formation of surface complexes which passivate or "protect" the surface. Mechanisms (1) and (2) increase the dissolution rate and the third decreases it compared with organic-free solutions.The types and importance of these mechanisms may be assessed from plots of dissolution rate versus degree of undersaturation.

To illustrate this technique, calcite, a common repository cementing and vein-filling mineral, was dissolved at pH 7.8 and 22°C in Na-Ca-HCO_3-Cl solutions with low concentrations of three organic ligands. Low citrate concentrations (50 μM) increased the dissolution rate consistent with mechanism (1). Oxalate decreased the rate, consistent with mechanism (3). Low phthalate concentration (< 50 μM) decreased calcite dissolution rates; however, higher concentrations increased the dissolution rates, which became faster than in inorganic solutions. Thus, phthalate exhibits both mechanisms (2) and (3) at different concentrations. In such cases linear extrapolations of dissolution rates from high organic ligand concentrations may not be valid.

INTRODUCTION

Performance of a high-level nuclear waste repository will be affected by many factors; among these are rates of waste-form and repository host mineral dissolution. The former will affect radionuclide release rates, the latter will affect repository water chemistry, the rate and nature of secondary minerals formed, and rates of radionuclide retention in secondary minerals. Previous studies have shown that many variables may affect the rate of mineral and waste form dissolution including: temperature, solution pH and composition, degree of mineral undersaturation, and mineral surface area, structural state, and composition. Recent results have also shown that microorganisms [1] and dissolved organic ligands can alter the rate of mineral and waste-form dissolution [2-6]. Organic ligands will undoubtedly be present in repository waters, either from natural sources or from waste-form processing or repository operation. These ligands may affect dissolution rates. Some studies have shown that some organic ligands may cause increases in mineral dissolution rates [5,6], whereas other studies have shown rate decreases [2,3]. However, little has been done to systematically organize these observations or to evaluate mechanisms.

This paper presents a method for evaluating some mechanisms by which organic ligands may alter dissolution rates. To illustrate this method, the effects of low concentrations of three organic ligands (citrate, oxalate, and phthalate) on the calcite dissolution rate were investigated. Calcite is a common vein-filling and cementing mineral at the Nevada repository site. Because calcite is a common rock-

Mat. Res. Soc. Symp. Proc. Vol. 465 © 1997 Materials Research Society

forming mineral, calcite dissolution kinetics has been the subject of a number of studies [*e.g.*, 7-10]. However, only a limited number of studies have investigated the effect of trace elements or organic species on calcite dissolution rates [2, 3]. Citrate, oxalate, and phthalate are well-characterized organic ligands which are present in low concentrations in many natural waters. In addition, they may be used as analogs for the more poorly characterized materials, such as the humic and fulvic acids, which comprise the bulk of natural dissolved organic matter. The organic concentrations investigated in this study are generally similar to those in natural waters, and much lower than in most previous dissolution studies.

MECHANISMS BY WHICH ORGANIC LIGANDS MAY MODIFY DISSOLUTION RATES

Organic ligands may modify the rate of mineral dissolution by at least one of three mechanisms:

(1) The ligand may form complexes with constituent ions, such as Ca^{2+}, in solution, decreasing their activity, and thereby decreasing the ion activity product (IAP) of the mineral. This formation of complexes will increase the degree of undersaturation, and increase the rate of dissolution.

(2) The ligand may complex with mineral surface sites. The complexed site may be destabilized relative to an uncomplexed site. This destabilization will increase the rate of dissolution [6].

(3) The ligand may complex with surface dissolution sites. The presence of kinks, because of their higher energy states, are favored locations for the adsorption of species from solution. These adsorbed ligands may bind or immobilize the kinks, thereby decreasing the rate of dissolution. The complexed site may be "protected" from attack by the solution and the rate of mineral dissolution decreased. In one respect these complexed sites could be viewed as, or modeled as, a decrease in the effective surface area of the mineral.

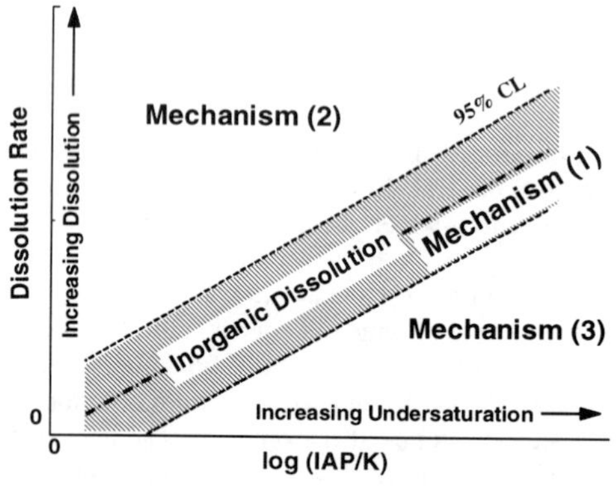

The importance and effect of these mechanisms may be evaluated from plots of mineral dissolution rate *vs.* degree of undersaturation (IAP/K) as shown in Figure 1. The rate of mineral dissolution in inorganic solutions is shown by the diagonal line and confidence band (shaded area) in Figure 1.

In mechanism (1), the change in the dissolution rate is due only to changes in IAP/K. That is, by complexing with Ca^{2+} ions, for example, the ligands remove free Ca^{2+} from solution, decreasing the activity. This will increase the degree of undersaturation and the dissolution rate. However, the rate of dissolution in the presence of such organic ligands will be the same as that predicted from the degree of saturation

Figure 1. The rate of dissolution increases with increasing undersaturation [log(IAP/K)]. 95% of data generating the regression line will plot within the 95% Confidence Limits (CL). For discussion of Mechanisms (1), (2), and (3) see text.

706

and should plot along the inorganic dissolution rate line (Figure 1).

Mechanism (2) may also increase the dissolution rate; however, because the increase is due to surface complexes, the rate of dissolution will be greater than that predicated from the inorganic (or non-complexing) rate model and will plot above the trend line in Figure 1. The chemical reaction that takes place at the water-mineral interface in calcite involves a weakening of the Ca^{2+} and CO_3^{2-} bonds at the surface so these ions can be detached more easily and rapidly. That is, the adsorption of a ligand to a Ca^{2+} site destabilizes the metal-anion bonds. This destabilization results in the metal being released into solution from the mineral surface [5].

Mechanism (3) will cause a decrease in dissolution rate which is not related to saturation state and thus the rate data will plot below the trend in Figure 1. This slower rate occurs because the higher energy sites in the mineral lattice are effectively "covered" with an organic coating.

To illustrate this technique, calcite was dissolved in the presence of low concentrations of three organic ligands.

METHODS

Calcite used was optical-grade Iceland spar, from the 250-125 μm size fraction, having a BET surface area of 1,890 cm^2 gm^{-1} and a geometric surface area of about 150 cm^2 gm^{-1}. Fresh calcite was used in each set of experiments. Experiments were conducted in continuously-stirred flow-through reactor systems [7] maintained at 22°C. Input Na-Ca-HCO_3-Cl solutions were prepared from reagent grade, $CaCl_2$, $NaHCO_3$, and Na- or K-salts of citrate, oxalate, or phthalate, and having ionic strengths between 0.001 and 0.02. All input HCO_3 concentrations were approximately 30 ppm. The pH of input solution was maintained at 7.8 by titration with either HCl or NaOH and solutions were in equilibrium with atmospheric CO_2. Output solutions had higher pH, averaging about 7.85. Input Ca concentrations were varied between zero and 20 ppm to investigate a range of undersaturation ($-2 <$ log (IAP/K) < 0). Dissolved species concentrations were measured by Atomic Absorption, titration, and specific ion electrodes. Individual experiments varied from one to three weeks and were continued until steady-state rates were obtained. Reaction rates were determined from the difference in Ca^{2+} ion between the input and reacted solutions. Precision of rates measured averaged approx. ± 10%. Solution speciation and saturation indices (IAP/K_{cal}), which included organic interactions, were calculated using standard computer algorithms and association constants from Ball and Nordstrom [11] and Sillén and Martell [12]. Calculated activity values were confirmed for some experiments using a calcium specific ion electrode.

RESULTS AND DISCUSSION

The rate of calcite dissolution in Na-Ca-HCO_3-Cl solutions at pH 7.8 as a function of undersaturation is shown in Figure 2. The rate of dissolution is zero at calcite saturation and increases linearly with respect to log undersaturation (proportional to ΔG of the reaction), within the range investigated. The intercept at a dissolution rate of zero shown in Figure 2 is consistent (within ca. ± 0.1) with log K_{cal} = -8.48 (25°C) and other equilibrium constants from Ball and Nordstrom [11].

The rate of dissolution (R) is given by:

$$R = K_1 \, A \, \log (IAP/K_{cal}) \tag{1}$$

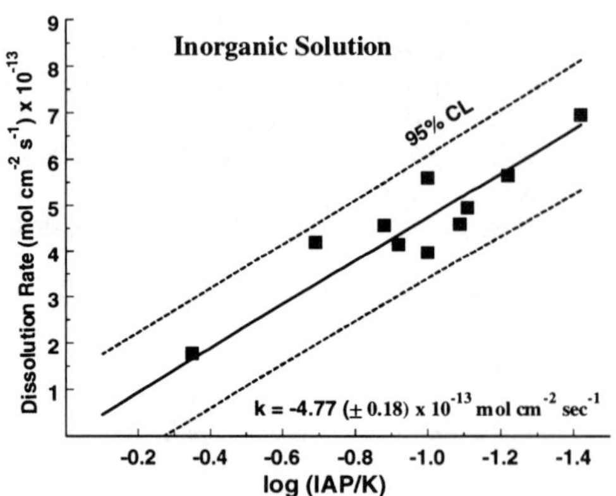

Figure 2. Calcite dissolution in Na-Ca-HCO$_3$-Cl solutions at 22°C as a function of undersaturation (log(IAP/K)).

where K_1 is the rate constant, A the calcite BET specific surface area (cm^2 gm^{-1}), IAP is the ion activity product, and K_{cal} the calcite equilibrium constant. Also shown in Figure 2 is the 95% confidence interval for the data. The regression value of K_1 is -4.77 (± 0.18) * 10^{-13} mol cm^{-2} sec^{-1}. This is similar to previous results, such as those of Chou *et al.* [7], when allowance is made for the difference between geometric [7] and BET surface areas (this study).

The data shown in Figures 3, 4, 5, and 6 show that for all organic ligand concentrations investigated the rate of dissolution is proportional to the log of undersaturation index (log (IAP/K)). Thus the form of equation 1 is applicable throughout all three data regions.

Calcite Dissolution in Citrate and Oxalate

Figure 3 shows the calcite dissolution rate in 50 μM citrate (C$_6$H$_6$O$_7^{3-}$) ligand (DOC = 3.6 ppm). Also included for reference is the calcite dissolution rate prior to addition of citrate.

The rate of calcite dissolution appeared to increase upon addition of citrate; however, the data points plot on or slightly below the line established by the inorganic solution. The points do plot within the 95% confidence interval for the inorganic data, suggesting that, at this concentration, most, if not all, of the effect of citrate on calcite dissolution is through complexing with dissolved calcium (Mechanism 1).

Figure 3 also shows the dissolution rate in 50 and 100 μM oxalate (C$_2$O$_4^{2-}$, 1.2 and 2.4 ppm DOC). In contrast to citrate, the oxalate data plot well below the line for the inorganic solution. Thus, the dissolution rate in oxalate solutions is slower than in the Na-Ca-HCO$_3$-Cl solutions.

These data suggest that the oxalate ligand is complexing with surface dissolution sites. Because of this adsorption of oxalate ligand, the rate of dissolution is considerably reduced from the dissolution rates found in

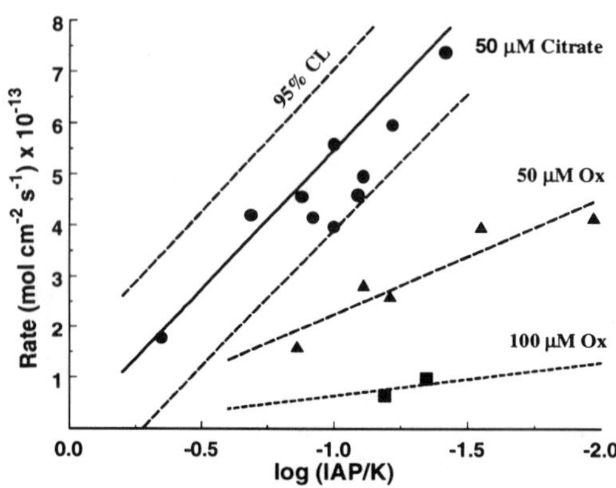

Figure 3. Calcite dissolution rate in inorganic solutions (marked by 95% CL) and in citrate (●) and oxalate-bearing (▲ and ■) solutions.

purely inorganic solutions. Higher oxalate concentrations were not investigated because of possible precipitation of calcium oxalate.

Calcite Dissolution in Phthalate

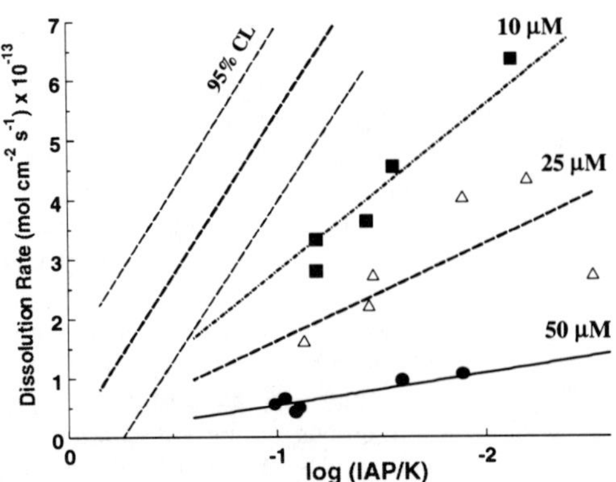

Figure 4. Calcite dissolution in inorganic solutions (marked by the 95% CL) and in solutions containing 10, 25, and 50 µM phthalate.

Figures 4 and 5 show the dissolution rate for calcite in the presence of phthalate ($C_8H_4O_4^{2-}$) ligand. As the ligand concentration increased from 0 to 50 µM, the dissolution rate decreased (Figure 4). At 50 µM phthalate, calcite dissolution rates were less than 10% of the rate in the inorganic solution. Note that the intercepts of the data for all four different concentrations pass through zero at or near log (IAP/K) = 0. (Saturation calculations include effects of organic complexing.) Therefore, it appears that for both oxalate and phthalate, the presence of organic ligand does not greatly alter the relationship of dissolution rate to saturation, *i.e.*, that the rate becomes zero at log(IAP/K) = 0. However, at concentrations > 50 µM (Figure 4) the dissolution rate increased as the concentration of phthalate increased, until, at 5000 µM, it was greater than in inorganic solutions. Data in figures 4 and 5 were obtained, with overlap and replication, in three separate experiments using different samples of calcite. Thus, the variations observed are not likely to result from irreversible changes to the calcite surfaces during the experiments.

A summary of rate data as a function of phthalate concentration at log (IAP/K) = -1.25 undersaturation is shown in Figure 6. The rate of dissolution decreases at phthalate concentrations of 50 µM and less, while at higher concentrations (100 to 5000 µM phthalate) the rate of dissolution increases (more or less) linearly with organic ligand concentration.

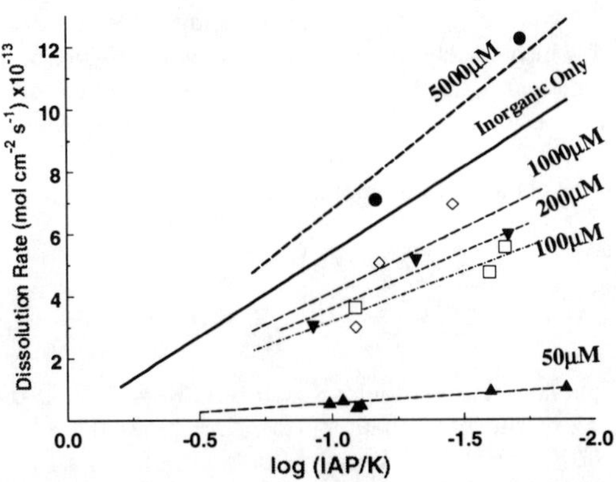

Figure 5. Calcite dissolution rates in inorganic Na-Ca-HCO₃-Cl solutions and identical solutions containing higher concentrations (50 to 5,000 µM) of phthalate.

This behavior may be modeled using two curves (R_1 and R_2 in Figure 6). The term, 1 - R_1, represents the decrease in rate observed at low concentrations. R_2 represents the increase in rate observed at higher phthalate concentrations (> 50 μM). R, the overall rate of dissolution, may be represented as:

$$R = K_1 A \log(IAP/K)(1 - R_1 + R_2) \qquad (2)$$

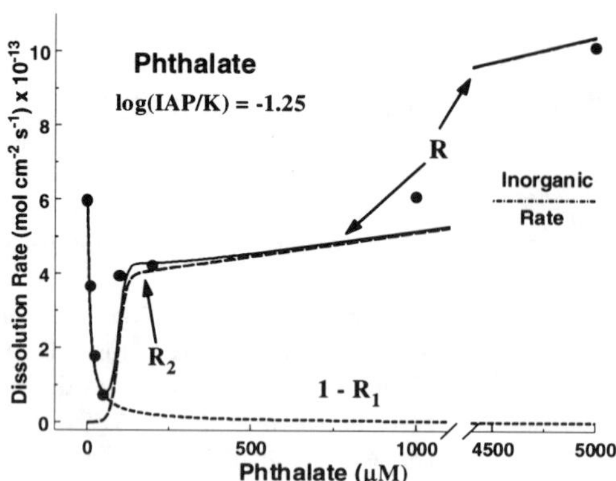

Figure 6. Summary of calcite dissolution in phthalate-bearing solutions at log(IAP/K) = -1.25. R is the rate of dissolution from equation (3). For R_1 and R_2 see explanation in text. At low concentrations phthalate decreases the dissolution rate; however, at higher concentrations the rate becomes greater than in inorganic solutions. Note X axis break between 1100 and 4400 μM.

If the initial decrease in dissolution rate results from "protection" of surface dissolution sites (kinks or steps) to which organic ligands are adsorbed, then the rate decrease may be modeled as resulting from a decrease in available mineral surface area (surface dissolution sites) due to adsorption of ligand onto the mineral surface [3]. The degree of surface coverage (R_1 or Θ) may be related to the dissolved concentration by an adsorption isotherm, such as the Langmuir adsorption isotherm.

Although the dissolution rate decreased by almost 90% between 0 and 50 μM, the data show that between 50 and 100 μM the dissolution rate increased rapidly (Figure 6). The dissolution rate continued to increase at higher phthalate concentrations until it became greater than in inorganic solutions. The rate increase at higher ligand concentrations may be modeled mathematically using a logistic growth curve. A general logistic growth curve (R_2), which closely fits the data, is shown in Figure 6. The logistic growth curve probably has no direct physical significance. The increase in calcite dissolution rate at higher phthalate concentrations may result from changes in phthalate complexing and coordination on the calcite surface [13].

Thus, the rate of calcite dissolution may be obtained by summing the two curves:

$$R = K_1 A \log\left(\frac{IAP}{K_{cal}}\right)\left(1 - \Theta_{max}\frac{K_2 C}{(1 + K_2 C)} + \frac{K_3 + K_4 C}{(1 + \exp(-K_5(C - \beta)))}\right) \qquad (3)$$

Adjustable parameters were estimated using non-linear curve fitting. Results are: $K_1 = 4.77 (\pm 0.18)$ x 10^{-13} mol cm^{-2} sec^{-1} {obtained from inorganic experiments, equation (1)} , $K_2 = 135{,}000 \pm 8{,}000$ lit mol^{-1}, $K_3 = 0.74 \pm 0.05$, $K_4 = 163 \pm 16$ lit mol^{-1}, $K_5 = 0.09$ x 10^6 lit mol^{-1}, $\beta = 8.53 (\pm 0.05)$ x 10^{-5} mol lit^{-1}, with C = phthalate concentration (mol lit^{-1}). K_2 and Θ_{max} describe the Langmuir isotherm (R_1 in equation (2)). Θ_{max}, the maximum amount of site coverage, was set to 1.0 to simplify curve fitting. K_3, K_4, K_5, and β describe the shape of the general logistic growth curve {R_2 in equation (2)}.

The oxalate data were regressed for K_2 in the R_1 term. The data have a multiple $R^2 = 0.92$ and $K_2 = 26{,}400 \pm 2{,}600$ mol lit^{-1} (with $\Theta_{max} = 1$). The difference in K_2 values for oxalate and phthalate is probably related to the difference in adsorption of these two species on the surface of the calcite.

Association Constants

The decrease in dissolution rate did not correlate with the value of the Ca-ligand association constant. In spite of its larger association constant (log $K_{Ca\text{-}cit}$ = 4.68, [12]), citrate had only a negligible effect on the dissolution rate. Conversely, phthalate, which has the smallest association constant (log $K_{Ca\text{-}phth}$ = 2.43, [12]) has the greatest effect on the dissolution rate. This may indicate that steric factors are more important in governing the interaction between ligands and mineral surfaces.

The effects of citrate, oxalate and phthalate appear to be reversible. That is, if the organic ligand is removed from the input solution, the calcite dissolution rate eventually returns to the initial inorganic value. However, this return may take a week or longer; apparently desorption of the organic ligand or removal of any "protecting" Ca-L layers takes a protracted period of time. Thomas and Longo [3] found that many of the adsorbates they studied were also reversible; these included long chain amines, alcohols, and branched and aromatic carboxlyates. It is conceivable that some organic compounds could bind irreversibly to the surface.

SCANNING ELECTRON MICROSCOPY

Scanning electron microscopy also suggests that organic ligands affect the dissolution mechanism. Surface morphology of calcite grains dissolved in inorganic solutions is different from those dissolved in oxalate or phthalate, suggesting that organic ligands do alter the calcite dissolution mechanism. Calcite grains dissolved in inorganic solutions show formation of typical V-shaped or triangular etch pits and dissolution by recession of ledges, steps, and cleavage planes [14]. Dissolution in phthalate produced few etch pits, but tooth-like projecting features, ranging from 0.1 to 0.7 μm in height, formed in large numbers on some crystal surfaces.

CONCLUSIONS

Dissolution rates of repository minerals, and release rates of various radionuclide and other species, may be significantly affected by organic ligands present in repository groundwaters. However, there has been no general method to assess the mechanism and degree of any organic effect. This paper presents one method by which the nature and magnitude of the effect may be assessed; plots of dissolution rate *versus* degree of undersaturation (IAP/K). This method combines thermodynamic (saturation state) and kinetic factors. Thus, it is best suited for thermodynamically well-characterized phases or phases for which baseline dissolution values may be determined. It is also best suited for near-equilibrium conditions. Solution transport may become a rate-determining factor far from equilibrium; however, due to slow rates of groundwater movement, repository fluids are likely to be fairly near equilibrium and thus this method of assessment should be appropriate.

Organic ligands may affect mineral dissolution rates through at least three different mechanisms. Low concentrations of different organic ligands (oxalate, citrate, and phthalate), concentrations similar to TOC concentrations in natural waters, affect calcite dissolution by all three mechanisms. Low citrate concentrations (50 μM) increased dissolution rates, primarily by formation of Ca-citrate complexes in solution which decreased the saturation index. Oxalate decreased the rate of calcite dissolution by formation of surface complexes which protect or passivate the surface. Formation of surface complexes with phthalate decreased the rate of calcite dissolution at low concentrations but increased the rate at at higher concentrations (> 1000 µM). Glass dissolution rates may also show similar trends, decreasing at low citrate concentrations (< 1 mM) and increasing at higher concentrations [4]. Most previous experiments have only investigated high organic ligand concentrations, concentrations much high than are found in most natural environments. Previous workers might have missed decreases caused by organic complexing because they used high organic concentrations. Previous workers have extrapolated effects from high concentrations, more or less linearly, to lower environmental values. Such extrapolations may not always be valid.

Dissolution rates measured in the laboratory are often much greater than those estimated for natural conditions. For example, based on the rate constants given above, at 1 order of magnitude undersaturation [log(IAP/K) = -1.0], a calcite crystal 1 cm^3 in size would dissolve within about 300 years. This is much faster than geological experience would suggest. Based on most previous work, the presence of dissolved organic matter appeared generally to increase the difference between laboratory and natural environments. Although many factors may affect dissolution rates, it is possible that low concentrations of organic ligands may significantly decrease the dissolution rate of calcite and other minerals. This partially may resolve the discrepancies between natural and laboratory dissolution rates. Thus leach rates of repository waste-forms and minerals measured in the laboratory may not be valid unless the effects of organic complexing and coating are taken into account.

REFERENCES

1. R. D. Rogers, M. A. Hamilton, R. H. Veeh, and J. W. McConnell in *Scientific Basis for Nuclear Waste Management XIX*, edited by W. M. Murphy and D. A. Knecht (Materials Research Society, Pittsburgh, 1996), pp. 475-482.
2. K.E. Chave, Science **148**, 1723 (1965).
3. M. M. Thomas and J. M. Longo, in *Water-Rock Interaction V* (Orkustofnun, Reykjavik, 1986), pp. 573-576.
4. H. Teng and D. E. Grandstaff, in *Scientific Basis for Nuclear Waste Management XIX*, edited by W. M. Murphy and D. A. Knecht (Materials Research Society, Pittsburgh, 1996), pp. 249-256.
5. S. A. Welch and W. J. Ullman, in *Water-Rock Interaction*, edited by Y. K. Kharaka and A. S. Maest (Balkema, Rotterdam, 1992), pp. 127-132.
6. D. E. Grandstaff, in *Rates of Chemical Weathering of Rocks and Minerals*, edited by S. M. Colman and D. P. Dethier (Academic Press, New York, 1986), pp. 41-60.
7. L. Chou, R. M. Garrels, and R. Wollast, Chemical Geology **78**, 269 (1986).
8. R. A. Berner and J. W. Morse, Am. Jour. Science **274**, 108, (1974).
9. E. L. Sjöberg and D. T. Rickard, Geochim. Cosmochim. Acta **48**, 485 (1984).
10. L. N. Plummer and T. M. L. Wigley, Geology **40**, 191 (1976).
11. J. W. Ball and D. K. Norstrom, Users manual for WATEQ4F, with revised thermodynamic data base and test cases for calculating speciation of major, trace, and redox elements in natural waters. *U.S. Geol. Survey Open-File Report, 91-183* (1991).
12. L. G. Sillén and A. E. Martell, *Stability Constants* (The Chemical Society, London, 1964).
13. W. Stumm and J. J. Morgan, *Aquatic Chemistry* (Wiley Interscience, New York, 1996)
14. I. N. MacInnis and S. L. Brantley, Geology **56**, 1113 (1992).

RETROGRADE SOLUBILITIES OF SOURCE TERM PHASES

WILLIAM M. MURPHY
Center for Nuclear Waste Regulatory Analyses, Southwest Research Institute,
6220 Culebra Road, San Antonio, TX 78228-0510 USA

ABSTRACT

Natural analog, experimental, and thermodynamic studies indicate that the properties of secondary uranyl minerals are likely to control the source term for U and other radioelements incorporated in these phases in the proposed Yucca Mountain repository. Thermodynamic calculations using data from the 1992 NEA data base indicate an increase in the equilibrium constant for schoepite dissolution from $10^{3.1}$ to $10^{4.8}$ with decreasing temperature from $100°$ to $25°C$, i.e., retrograde solubility. Enthalpies for mineral transformation reactions that consume protons and release cations are typically negative, suggesting that solubilities of other uranyl phases such as uranophane increase more than that of schoepite with decreasing temperature. Solubilities of mineral phases associated with spent nuclear fuel will be initially relatively low under the elevated repository temperature regime. As the temperature of the repository decreases due to radioactive decay and heat dissipation, source term mineral solubilities increase. The rate of release of U and other species is controlled by a series of processes: transport of oxidants and flux of water; oxidative dissolution of spent fuel; uranyl mineral precipitation; uranyl mineral dissolution or transformation; and radionuclide transport. Decreasing diffusion and reaction rates and increasing uranyl mineral solubilities with decreasing temperature may lead to a change with time from solubility to transport or reaction rate as a source term controlling mechanism. Preservation of large quantities of uranyl minerals formed by oxidation of uraninite and radiometric ages of secondary uranophane at the natural analog site at Peña Blanca indicate that oxidative alteration of uraninite was fast relative to transport of U away from the deposit. The successive formation of schoepite and uranophane in natural settings where uraninite has been oxidized may represent a paragenesis reflecting increasing temperature or increasing incorporation of environmental components. In contrast, diminishing temperature conditions in a repository source area could lead to the reverse sequence of mineral formation.

INTRODUCTION

It is widely recognized from theoretical, experimental, and natural analog studies related to the oxidizing environment at Yucca Mountain that secondary minerals, particularly those containing uranyl, are likely to play an important role in controlling the source term for U and much of the inventory of nuclear waste, e.g., [1], [2]. Furthermore, after a few tens of years following emplacement, the continuously decreasing temperature of the source area at the proposed repository will affect differentially the properties of secondary phases and the processes that lead to radionuclide release. Nevertheless, little recognition of the role of secondary uranyl minerals or the relative temperature dependence of solubilities and reaction rates is acknowledged in source term analyses for performance assessments. This article examines thermodynamic data for likely secondary uranyl phases to support realistic predictions of their role in the evolution of the source term and ultimately the performance of the proposed repository. In particular, the temperature dependence of uranyl mineral solubilities is examined and implications are developed for their role as source term phases and for the application of natural analog data to predictions of repository performance.

Mat. Res. Soc. Symp. Proc. Vol. 465 © 1997 Materials Research Society

SOLUBILITY DATA

Limited empirical thermodynamic data exist for uranyl minerals particularly above 25°C. Properties for the uranyl hydrate, with the nominal composition of schoepite[1] ($UO_3 \cdot 2H_2O$), are recommended in the recently published, comprehensive review of chemical thermodynamics of U by Grenthe et al. [3]. These data are derived from calorimetric studies, primarily by Tasker et al. [4]. Attempts have been made to determine equilibrium solubilities of uranyl minerals using synthetic and natural samples [5], [6], [7], [8]. However, principal problems in these solubility experiments and data interpretations are: reaction stoichiometries were not firmly ascertained; chemical formulas used in thermodynamic interpretations were inconsistent with analytical data; phases other than those of primary interest were present as contaminants or precipitates and demonstrably affected some results; and solubility reversals were not attempted. Reversed solubility measurements of schoepite of unknown hydration at 90°C have been determined [8], but complete solution compositions were unspecified and no thermodynamic interpretation was reported.

Dissolution studies for soddyite [5], [7] provide the most likely basis for retrieval of reliable thermodynamic data for a uranyl silicate mineral. However, it was assumed in these studies that the soddyite conformed to its nominal ideal stoichiometry. In one study [5] excess Si detected in the reactant material and deficit silica in solution were assumed to be a consequence of the initial presence and experimental precipitation of amorphous silica, respectively. For the soddyite dissolution reaction

$$(UO_2)_2 SiO_4 \cdot 2H_2O + 4H^+ \rightleftharpoons 2UO_2^{2+} + SiO_{2(aq)} + 4H_2O \tag{1}$$

the corresponding equilibrium constant is defined by

$$K_{sod} = \frac{[UO_2^{2+}]^2 \, [SiO_{2(aq)}] \, [H_2O]^4}{[(UO_2)_2SiO_4 \cdot 2H_2O] \, [H^+]^4} \tag{2}$$

where square brackets represent thermodynamic activities corresponding to standard states of a one molal solution referenced to infinite dilution for aqueous species and pure minerals and H_2O. Thermodynamic data reported for reaction (1) are $\log K_{sod} = 5.74 \pm 0.21$ at 30°C, and the corresponding standard free energy of formation of soddyite from the elements, $\Delta G_{f,sod}^\circ = -3{,}658 \pm 4.8 \; kJ \cdot mole^{-1}$ [5]. (Reported uncertainties are based on uncertainties in standard state properties for the aqueous species and H_2O and propagated standard deviations of analytical results.) However, equilibrium was approached only from undersaturation (i.e., the reaction was not reversed), so the solubility must be regarded as a lower limit. These results for soddyite solubility are supported by a dissolution solubility study of synthetic $(UO_2)_2SiO_4 \cdot 2H_2O$ at 25°C [7]. At pH 3, releases of UO_2^{2+} and $SiO_{2(aq)}$ were generally in proportions corresponding to soddyite stoichiometry. However, mass balance relations indicate that silica (probably amorphous) began to precipitate prior to reaching soddyite solubility. The equilibrium constant deduced from the "steady state" solution compositions

---

[1]Structural and compositional relations between schoepite, metaschoepite, and dehydrated schoepite have been detailed recently [9]. The phase corresponding to $UO_3 \cdot 2H_2O$ was classified as metaschoepite, and schoepite was characterized as having a higher degree of hydration. Reported calorimetric data [3] are for the crystalline uranyl hydrate, $UO_3 \cdot 2H_2O$, which is called schoepite in this article.

in the study is log K_{sod} = 6.03$\pm$0.45, where the reported uncertainty is based on analytical uncertainties in pH and dissolved species concentrations. Although uncertainties are too large to confirm a trend in solubility with temperature between 25° and 30°C, the data from these two dissolution studies suggest retrograde solubility of soddyite (i.e., decreasing solubility with increasing temperature).

TEMPERATURE DEPENDENCE OF URANYL MINERAL SOLUBILITY

Reported thermodynamic data [3] permit evaluation of the solubility of schoepite as a function of temperature. Dissolution of schoepite is expressed as

$$UO_3 \bullet 2H_2O + 2H^+ \rightleftharpoons UO_2^{2+} + 3H_2O . \tag{3}$$

The equilibrium solubility of schoepite is expressed by the mass action relation for reaction (3) as

$$K = \frac{[UO_2^{2+}][H_2O]^3}{[UO_3 \bullet 2H_2O][H^+]^2} \tag{4}$$

where K stands for the equilibrium constant for reaction (3). The temperature dependence of K is given by the van't Hoff relation

$$\frac{d\ln K}{dT} = \frac{\Delta H^\circ}{RT^2} \tag{5}$$

where T denotes temperature in kelvins, ΔH° stands for the standard state molar enthalpy of reaction, and R represents the ideal gas constant. The standard enthalpy of reaction as a function of temperature is given as the integral of the standard heat capacity of reaction (ΔC_p°)

$$\Delta H^\circ_{(T)} - \Delta H^\circ_{(T^\circ)} = \int_{T^\circ}^{T} \Delta C_p^\circ dT \tag{6}$$

where $\Delta H^\circ_{(T^\circ)}$ stands for the standard enthalpy of reaction at the reference temperature (T°). $\Delta H^\circ_{298.15K}$ = −50.39 $\pm$ 2.27 kJ $\cdot$ mole^{-1} for reaction (3) [3]. The heat capacity of reaction (3) is given by

$$\Delta C_p^\circ = 3C_{p,H_2O}^\circ + C_{p,UO_2^{2+}}^\circ - C_{p,schoepite}^\circ - 2C_{p,H^+}^\circ \tag{7}$$

where $C_{p,i}^\circ$ represents the standard molar heat capacity of species i. Using heat capacity functions and values from [3] for UO_2^{2+} (extrapolated above 328 K) and schoepite, and a constant heat capacity for H_2O of 75.48 J $\cdot$ K$^{-1}\cdot$ mole^{-1}, which is accurate within 1 percent between 25° and 100°C [10], the heat capacity of reaction (3) calculated using Eq. (7) is given by

$$\Delta C_p^\circ = 492.71\,(J \cdot K^{-1} \cdot mole^{-1}) - 1.1668\,(J \cdot K^{-2} \cdot mole^{-1})\,T - \frac{5308\,(J \cdot mole^{-1})}{T-190\,(K)} \tag{8}$$

$$= A - BT - \frac{C}{T-D}$$

Using Eqs. (6) and (8), the integral of Eq. (5) yields the equilibrium solubility constant of schoepite as a function of temperature, which can be expressed as

$$\ln K - \ln K^\circ = \int_{T^\circ}^{T} \frac{\Delta H^\circ_{(T)}}{RT^2}\,dT$$

$$= \int_{T^\circ}^{T} \left\{ \frac{A(T-T^\circ)}{RT^2} - \frac{B\,[T^2-(T^\circ)^2]}{2RT^2} - \frac{C}{RT^2}\ln\left[\frac{T-D}{T^\circ-D}\right] + \frac{\Delta H^\circ_{(T^\circ)}}{RT^2} \right\}\,dT \tag{9}$$

where K° denotes the equilibrium constant at T°. Using standard state free energies of formation of reactants and products in reaction (3) taken from [3] gives

$$K_{298.15} = \exp\left[\frac{-\Delta G^\circ_{(T^\circ)}}{RT}\right] = 6.50\times10^4 \tag{10}$$

where $\Delta G_{(T^\circ)}$ stands for the standard state Gibbs free energy of reaction (3) at 298.15 K. Solving the definite integral in Eq. (9) yields

$$\ln K - \ln K^\circ = \frac{A}{R}\ln\frac{T}{T^\circ} + \left[\frac{\Delta H^\circ_{(T^\circ)} - AT^\circ}{R} + \frac{B(T^\circ)^2}{2R}\right]\left[\frac{1}{T^\circ}-\frac{1}{T}\right]$$

$$-\frac{B(T-T^\circ)}{2R} - \frac{C}{DR}\left[\frac{T-D}{T}\ln\left[\frac{T-D}{T^\circ-D}\right] + \ln\frac{T^\circ}{T}\right] \tag{11}$$

The dependence of equilibrium solubility on temperature is illustrated in Figure 1, which shows the natural logarithm of the equilibrium constant calculated using Eq. (11) plotted as a function of inverse temperature. The inclusion of heat capacity and its temperature dependence in the integral affects the predicted solubility constant curve by a small amount between 25° and 100°C as indicated by the minimal curvature.

It is evident from Figure 1 that the equilibrium solubility constant of schoepite increases with decreasing temperature from $10^{3.1}$ to $10^{4.8}$ between 100° and 25°C. This retrograde solubility is further illustrated by comparing the aqueous activity of UO_2^{2+} at different temperatures. At pH 8, $[UO_2^{2+}]$ at equilibrium with schoepite equals 6.48×10^{-12} at 25°C and 0.17×10^{-12} at 100°C. Total dissolved U concentrations would be much greater if U speciation is dominated by carbonate or other complexes as anticipated in the Yucca Mountain system. In a progressively cooling repository source term environment, the solubility of schoepite would progressively increase.

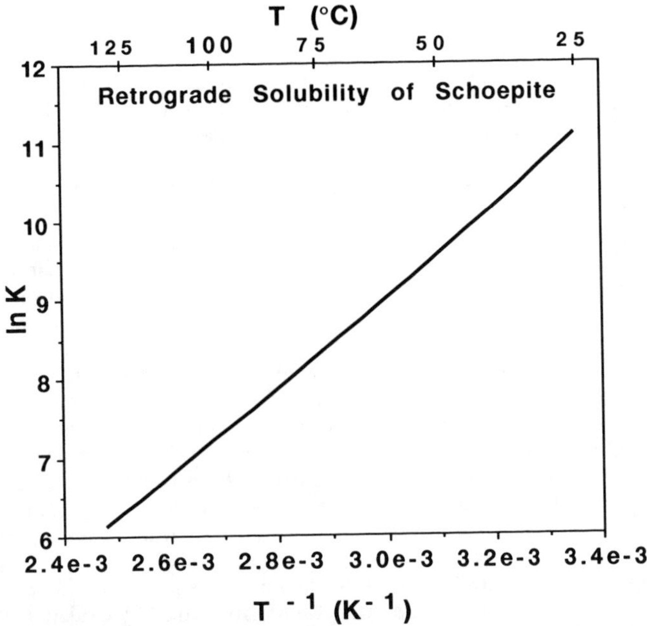

Figure 1. Natural logarithm of the equilibrium constant for the dissolution of schoepite (reaction 3) as a function of the inverse of the absolute temperature calculated using Eq. (11). Temperature in °C is given at the top of the figure.

Equilibrium between uranophane and schoepite can be expressed as

$$Ca(UO_2)_2[SiO_3(OH)]_2 \cdot 5H_2O + 2H^+ \rightleftharpoons$$
$$2UO_3 \cdot 2H_2O + Ca^{2+} + 2SiO_{2\,(quartz/amorphous\ silica)} + 3H_2O \qquad (12)$$

Reliable enthalpy and Gibbs free energy data are presently unavailable to perform a solubility analysis for uranophane. However, enthalpies for mineral transformation reactions that consume protons and release cations are typically negative. If the standard enthalpy of formation of uranophane is assumed to be close to that of schoepite plus quartz or amorphous silica and water, the enthalpy of reaction (12) would be dominated by the enthalpy of the Ca^{2+} ion, and its value would be negative. For these conditions, with decreasing temperature in the near field, schoepite plus a silica phase will be favored thermodynamically to form at the expense of uranophane despite the retrograde solubility of schoepite, which follows from the van't Hoff relation applied to reaction (12). These arguments suggest that the solubility of uranophane is also retrograde with temperature, even more strongly than schoepite solubility.

A mechanism that augments the retrograde solubility of uranyl minerals in general is the increasing stability of dominant aqueous uranyl carbonate complexes relative to uranyl plus bicarbonate with decreasing temperature between 200° and 25°C [11].

DISCUSSION

Interactions between water and uranyl minerals are likely to control the source term for U and other radioactive waste species at the proposed repository at Yucca Mountain. Solid solution has been observed in several natural uranyl silicates [6], Th has been noted as an impurity in uranophane [12], and Cs is largely included in secondary phases in experimental oxidative alteration of spent nuclear fuel [2]. The rate of release of U and other species will be controlled by a series of processes in which slow steps tend to control the overall rate. These processes include transport of oxidants and flux of water, oxidative dissolution of spent fuel, mineral precipitation (and coprecipitation), mineral dissolution or transformation, and radionuclide transport. If reaction and transport rates are sufficiently fast and solubilities are low, the system will approach equilibrium at the source; solubilities of secondary uranyl minerals limit concentrations of U and other waste species sequestered in these secondary phases; and the rate of dissolution will be controlled by the product of water flux and solubility. Such conditions would be favored by elevated temperatures relatively early in repository evolution. As solubilities increase with decreasing temperature, diffusion and reaction (e.g., oxidation and dissolution) rates decrease with decreasing temperature. For example oxidative dissolution of spent fuel in oxidizing Na carbonate/bicarbonate solutions follows Arrhenius behavior with an activation energy of 30 kJ/mole [13]. Increasing uranyl mineral solubilities and decreasing diffusion and reaction rates with decreasing temperature may lead to a change with time from solubility to transport, oxidation rate, or dissolution rate as the source term controlling mechanism in the cooling repository source area.

Preservation of large quantities of uranyl minerals formed by oxidation of uraninite and radiometric ages of secondary uranophane at the natural analog site at Peña Blanca [14] indicate that oxidative alteration was fast relative to transport of U away from the deposit [15]. Under oxidizing conditions in nature, uraninite alters to secondary uranyl minerals which tend to form in a sequence, such as uraninite $\rightarrow$ schoepite $\rightarrow$ soddyite $\rightarrow$ uranophane, characterized by increasing incorporation of cations from the environment, e.g., [14]. However, the successive formation of schoepite, soddyite, and uranophane in natural settings may represent a paragenesis reflecting increasing temperature. In contrast, equilibrium solubility under conditions of decreasing temperature in a closed system, as described in relation to reaction (12), would lead to the reverse sequence. However, open system behavior is typical of alteration of rocks in the unsaturated zone at Yucca Mountain, where altered (zeolitized) rocks are commonly enriched in Ca and Mg relative to the primary bulk rock composition [16]. It is likely that environmental chemical conditions will also have a strong effect on formation of secondary uranyl minerals at Yucca Mountain.

Spent nuclear fuel oxidizes to U_4O_9 in moist air. At temperatures below 200°C this phase is apparently metastable, and complete oxidation in moist air to a U^{6+} phase does not occur over periods of time in excess of 4 yr under laboratory conditions [17]. However, in the presence of synthetic groundwater at 95°C, unirradiated UO_2 alters to secondary uranyl minerals including schoepite, soddyite, and uranophane [18]. Secondary products are also observed in long term spent fuel leaching tests designed to mimic conditions at a hypothetical Yucca Mountain repository [2]. In these tests it was concluded that a minimum of 70 to 90 percent of released gap and grain boundary ^{137}Cs was incorporated in these secondary phases. Secondary products have been observed in numerous other spent fuel and UO_2 alteration studies under oxidizing conditions (e.g., [19], [20], [21]). Experimental studies (as well as natural analog studies) indicate rapid formation of secondary phases relative to transport in systems relevant to Yucca Mountain. Liquid water is clearly necessary for formation of

Liquid water is also required for transport of waste species out of the engineered barrier system (except for potential gas phase species such as ^{14}C). It is logical that spent nuclear fuel in the proposed repository at Yucca Mountain will react with liquid water to form secondary uranyl minerals prior to aqueous phase releases from the engineered barrier system by groundwater flow or aqueous diffusion. With a sufficient increase in the activity of aqueous Ca, uranophane may form rather than schoepite or soddyite despite diminishing temperature. Also, Fe derived from corrosion of waste container materials could affect the composition of secondary uranyl minerals in the repository near field. Uranyl is commonly found associated with ferric Fe in nature, e.g., at the Peña Blanca natural analog site [15].

CONCLUSIONS

The realistic evolution of the source area and near field environment at the proposed high-level nuclear waste repository at Yucca Mountain includes transformation of spent nuclear fuel to uranyl minerals, which would be the source term phases for uranium and much of the inventory of radioelements. Thermodynamic analysis indicates that uranyl minerals have retrograde aqueous solubilities. As the repository near field cools, solubilities of these source term phases will increase. This effect will be augmented by the increasing thermodynamic stability of dominant aqueous uranyl carbonate complexes with decreasing temperature. Retrograde solubility of source term phases, coupled with decreasing rates of transport and reaction with decreasing temperature, may lead to a change in the dominant source term mechanism from solubility to transport, oxidation, or dissolution rate in a cooling repository environment. Typical uranyl mineral paragenesis in natural analog systems indicate increasing incorporation of environmental components. However, this paragenesis is also consistent with prograde oxidation of primary uraninite. Open system interaction of the oxidation products of spent fuel with environmental components may stabilize more complex uranyl minerals despite stronger retrograde solubility characteristics.

ACKNOWLEDGMENTS

This article documents work performed by the Center for Nuclear Waste Regulatory Analyses (CNWRA) for the Nuclear Regulatory Commission (NRC) under Contract No. NRC-02-93-005. R.T. Pabalan, W.C. Patrick, C. Degueldre, B.W. Leslie, and two anonymous reviewers provided valuable suggestions on the text. The activities reported here were performed on behalf of the NRC Office of Nuclear Regulatory Research, Division of Regulatory Applications, and the NRC Office of Nuclear Material Safety and Safeguards, Division of Waste Management. The article is an independent product of the CNWRA and does not necessarily reflect the views or regulatory position of the NRC.

REFERENCES

1. W.M. Murphy, and E.C. Pearcy, in Scientific Basis for Nuclear Waste Management XV, edited by C.G. Sombret (Mater. Res. Soc. Proc. 257, Pittsburgh, PA, 1992) pp. 521–527.

2. P.A. Finn, J.C. Hoh, S.F. Wolf, S.A. Slater, and J.K. Bates, in Scientific Basis for Nuclear Waste Management XIX, edited by W.M. Murphy and D.A. Knecht (Mater. Res. Soc. Proc. 412, Pittsburgh, PA, 1996) pp. 75–81; Radiochim. Acta 74 pp. 65–71.

3. I. Grenthe, J. Fuger, R.J.M. Konings, R.J. Lemire, A.B. Muller, C. Nguyen-Trung, and H. Wanner, Chemical Thermodynamics of Uranium, edited by H. Wanner, and I. Forest (Elsevier Science Publishers B.V., Amsterdam, North-Holland, 1992).

4. I.R. Tasker, P.A.G. O'Hare, B.M. Johnson, G.K. Johnson, E.H.P. Cordfunke, Can. J. Chem. **66** (1988) pp. 620–625.

5. S.N. Nguyen, R.J. Silva, H.C. Weed, and J.E. Andrews, Jr., J. Chem. Thermo. **24** (1992) pp. 259–276.

6. I. Casas, J. Bruno, E. Cera, R.J. Finch, and R.C. Ewing, in Kinetic and Thermodynamic Studies of Uranium Minerals Assessment of the Long-Term Evolution of Spent Nuclear Fuel, (Swedish Nuclear Fuel and Waste Management Company, Technical Report 94-16, Stockholm, Sweden, 1994) p. 73.

7. H. Moll, G. Geipel, W. Matz, G. Bernhard, and H. Nitsche, Radiochim. Acta **74** pp. 3-7 (1996).

8. H.D. Holland and L.H. Brush, in Proceedings of the Conference on High-Level Radioactive Solid Waste Forms, edited by L.A. Casey, Nuclear Regulatory Commission, NUREG/CP-005, 1979, p. 597–615.

9. R.J. Finch, F.C. Hawthorne, and R.C. Ewing, in Scientific Basis for Nuclear Waste Management XIX, edited by W.M. Murphy and D.A. Knecht (Mater. Res. Soc. Proc. 412, Pittsburgh, PA, 1996) pp. 361–368.

10. R.C. Weast, CRC Handbook of Chemistry and Physics, edited by M.J. Astle, and W.H. Beyer, (CRC Press Inc., Boca Raton, Florida, 1988).

11. Based on the EQ3/6 data base data0.com.R2, Lawrence Livermore National Laboratory (1995).

12. C. Frondel, Systematic Mineralogy of Uranium and Thorium, (U.S. Geological Survey Bulletin 1064. Washington, DC: U.S. Government Printing Office, 1958).

13. W.J. Gray, H.R. Leider, and S.A. Steward, J. Nuc. Mater. **190** (1992) pp. 46–52.

14. E.C. Pearcy, J.D. Prikryl, W.M. Murphy, and B.W. Leslie, App. Geochem. **9** (1994) pp. 713–732.

15. D.A. Pickett, J.D. Prikryl, W.M. Murphy, and E.C. Pearcy, Submitted to App. Geochem. (1996).

16. D.E. Broxton, D.L. Bish, and R.G. Warren, Clay Clay Min. **35** (1987) pp. 89–110.

17. R.E. Einziger, L.E. Thomas, H.C. Buchanan, and R.B. Stout, J. Nuc. Mater. **190** (1992) pp. 53–60.

18. D.J. Wronkiewicz, J.K. Bates, T.J. Gerding, E. Veleckis, and B.S. Tani, J. Nuc. Mater. **190** (1992) pp. 107–127.

19. C.N. Wilson, in Scientific Basis for Nuclear Waste Management XIV, edited by T.A. Abrajano, Jr., (Mater. Res. Soc. Proc 212, Pittsburgh, PA, 1991) pp. 197–204.

20. Ch. Cachoir, M.J. Guittet, J.-P. Gallien, and P. Trocellier, Radiochim. Acta **74** (1996) pp. 59–63.

21. R. Taylor, R.J. Lemire, and D.D. Wood, in High Level Radioactive Waste Management, (Amer. Nuc. Soc., La Grange Park, IL, 1992) pp. 1,442-1,448.

THE EFFECTS OF THE CHEMICAL AND RADIOLYTIC DEGRADATION OF ASPHALT ON PLUTONIUM SOLUBILITY

B.F. GREENFIELD*, M. ITO**, M.W. SPINDLER*, S.J. WILLIAMS* and M. YUI**
*AEA Technology plc, 220.32 Harwell, Didcot, Oxfordshire, OX11 0RA, United Kingdom
**Power Reactor and Nuclear Fuel Development Corporation (PNC), Tokai works, Ibaraki, Japan

ABSTRACT

The influence of alkaline degradation or radiolytic degradation of asphalt on plutonium solubility has been investigated. Asphalt has been contacted with water, sodium hydroxide solution or concrete leachate at 80°C for periods of up to approximately 2 years. Sodium nitrate was also present in some of the experiments. Plutonium solubilities were measured at pH 12 in the leachates and found to be less than 10^{-8} mol dm^{-3} for most degradations. Relatively low levels of Total Organic Carbon were measured in the leachates. Alpha radiolysis of asphalt in the presence of concrete and water has also been studied. Samples of asphalt were encapsulated in concrete after coating with the ^{238}PuO$_2$, crushed and leached at room temperature. The solubility of plutonium was measured in samples of the leachates after approximately 90 days and 180 days had elapsed. The results showed that the solubility of plutonium in the α-radiolysis leachates remained low and was in the range 2×10^{-11} to 8×10^{-9} mol dm^{-3}. A consideration of these results, and data published elsewhere, suggests that chemical and radiolytic attack on asphalt or bitumen under anaerobic, alkaline conditions typical of a deep cementitious repository is unlikely to generate complexants for plutonium which are effective at high pH. Any enhancement of plutonium solubility is likely to be less significant than that arising from the degradation of some other organic materials.

INTRODUCTION

Many elements have isotopes of importance in assessments of the post-closure performance of a radioactive waste repository. Measurements of the solubility and sorption of these elements under repository near-field conditions, and of their sorption under far-field conditions, are important input data in post-closure safety assessments of radioactive waste disposal. It is therefore necessary to consider any processes that may affect solubility or sorption.

The degradation of organic material within a radioactive waste repository can result in the formation of water-soluble complexants which may alter the retention behaviour of some elements. For example, the solubilities of some elements in the near-field porewater can be increased by complexation with organic ligands [1]. These complexes can also reduce the sorption of elements to near-field components (e.g. cements and concretes) and geological materials (the host rock) [2]. Investigations of the potential effects of disposed organic materials on radionuclide behaviour increase the understanding of repository performance.

Asphalt or bitumen has been employed in several countries, including Japan, as an encapsulation matrix for radioactive wastes. Therefore, any effects that this organic material, or its degradation products, may have on repository performance (for example the release of radionuclides in groundwater) is of interest.

Asphalt can potentially be degraded by chemical [3], thermal [4], radiolytic [5] and microbiological processes [6]. However, chemical processes involve mainly oxidation

Mat. Res. Soc. Symp. Proc. Vol. 465 © 1997 Materials Research Society

mechanisms, and carboxylic acids and carbonates appear to be the main degradation products. Chemical oxidation is likely to be insignificant under anaerobic repository conditions. Although the radiolytic degradation of bitumen and bitumenised waste forms has been studied extensively [5], this has mainly been with regard to gas formation and related modifications to physical properties (swelling, bubble formation, etc.). Van Loon and Kopajtic [7] have investigated the aqueous soluble degradation products from the gamma radiolysis of bitumen by irradiating bitumen in water at pH 12.5 with an external ^{60}Co source. Oxalate and formate were identified as the main degradation products and complexation studies showed that the effects of bitumen γ-irradiation degradation products on radionuclide speciation would not be significant in a cementitious environment. Alpha-radiolysis may offer an alternative degradation mechanism.

In this work the influence of any chemical or α-radiolytic degradation of asphalt on plutonium solubility has been investigated under conditions which simulate those typical of a repository vault backfilled with a cementitious material [8]. Some of the wastes encapsulated in asphalt in Japan may contain significant amounts of sodium nitrate and this was included in the degradation experiments.

EXPERIMENTAL DETAILS

Chemical Degradation

Asphalt (10g) was cut from a solid block supplied by PNC and contacted with 200 cm^3 of water, 0.1 mol dm^{-3} sodium hydroxide solution or concrete leachate in the absence of oxygen for periods of up to approximately 2 years. 3 mol dm^{-3} or 0.03 mol dm^{-3} sodium nitrate solutions were used in some of the experiments. Concrete (90g of cured and crushed 4.8:1 sand/ordinary Portland cement (OPC)) was present in those experiments with asphalt which contained concrete leachate or sodium nitrate solution. The degradation experiments were heated at 80°C in an oven purged with Ar/4%H$_2$; this was installed in a positive-pressure glove box purged with nitrogen. The bitumen became fluid when heated to this temperature and globules floated to the top of the aqueous phase. The samples were agitated by hand occasionally. Control experiments with concrete leachate (or sodium nitrate solution) but no asphalt were also set up. Most of the degradation experiments were carried out for up to 270 - 280 days; a few continued to approximately 750 days.

Radiolytic Degradation

Alpha radiolysis of asphalt in the presence of concrete and water was also studied. Samples of asphalt were encapsulated in concrete after coating with the oxide of a high specific activity α-emitter, in this case ^{238}Pu. The use of a high specific activity α-emitter allows an accelerated dose rate and may thus simulate, on a short timescale, many years of accumulated dose compared to a typical waste. The asphalt was fragmented by cooling in liquid nitrogen for about 10 minutes and, whilst cold, mixed with the oxide (0.19g of plutonium oxide (~80% ^{238}PuO$_2$) to 10g of asphalt). When the asphalt had warmed to near room temperature and become sticky (this aided adherence between the plutonium oxide and the asphalt) dry cement/sand (14g OPC : 67.6g) mixture was added, mixed with minimum stirring and allowed to stand for 2 days. 10 cm^3 of water were then added to each sample, mixed and allowed to cure for 30 days at room temperature. The concrete blocks were then broken up and the fragments contacted with either de-ionised water or 3 mol dm^{-3} sodium nitrate solution at room temperature in order to accumulate an α-dose in an aqueous environment. Surface coating the asphalt with oxide, rather than mixing the oxide with melted asphalt and allowing to cool, allows access of water to irradiated areas. These experiments were undertaken at room temperature in a nitrogen-atmosphere glovebox for 180 days.

<u>Sampling Procedures and Plutonium Solubility Determination</u>

The chemical degradation experiments were sampled after cooling to room temperature. A minimum of duplicate experiments were sampled for all the degradations containing asphalt. Samples were filtered through a 0.45μm filter (Millipore Millex HV, polyvinylidene fluoride membrane) to remove particulate material and some retained for Total Organic Carbon (TOC) analysis. 10 cm^3 of the remaining filtrate were filtered through a 30,000 nominal molecular weight cut-off (NMWCO) filter (Millipore TTK, polysulphone membrane) and used for plutonium solubility experiments. Samples taken after 737 or 767 days were only filtered through 0.45μm filters prior to plutonium solubility experiments. The pH was adjusted, if necessary, to 12.0 - 12.2 with 1 mol dm^{-3} NaOH solution as a common basis for comparison of plutonium solubilities.

Duplicate α-irradiated experiments were sampled by taking 10 cm^3 of leachate and filtering through a 0.45μm filter. 1 cm^3 of this filtrate was filtered through a 30,000 NMWCO filter into excess nitric acid to allow the plutonium concentration in the leachate from the dissolution of ^{238}PuO$_2$ present during the irradiation to be determined. The remaining portion of the 0.45μm filtrate was retained for plutonium solubility measurements.

Plutonium solubility determinations were carried out from oversaturation by the addition of excess Pu(IV) in nitric acid. A 5 mm^3 aliquot of plutonium nitrate solution (^{238}Pu at 2×10^{-2} mol dm^{-3} in 1 mol dm^{-3} HNO$_3$) was added to the asphalt leachate while the solution was stirred with a magnetic follower. The excess plutonium precipitated and, after stirring for 48 hours, a 1 cm^3 aliquot of the solution was filtered through a 30,000 NMWCO filter into an excess of nitric acid. Coulometric analysis of the plutonium nitrate solution used for these experiments gave the oxidation state distribution as 99.86% Pu(IV) and 0.14% Pu(VI).

The filters used for plutonium determinations were pre-conditioned using a procedure based on that outlined by Rai and co-workers [9]. The filter was washed with de-ionised water followed by 1 cm^3 of water at the pH of the experiment. A 1 cm^3 volume of solution from the experiment was then filtered to saturate any possible sorption sites on the filter and discarded. Samples for plutonium determination were then filtered into nitric acid to prevent sorption onto the walls of the collection vessel.

The plutonium concentrations in the 30,000 NMWCO filtrates were assayed by gross α-counting of prepared counting trays using standard scintillation counting techniques (Harwell Instruments Scintillation Counter Type 1749A).

RESULTS

<u>Chemical Degradation</u>

The chemical degradation experiments were sampled over a period of 10 - 280 days, with a few experiments being continued to a sampling at 737 or 767 days. The concentrations of Total Organic Carbon in the leachates after filtration through 0.45μm filters are given in Table I.

TABLE I

Total Organic Carbon Concentrations in 0.45 μm Filtrates of Leachates from the Chemical Degradation of Asphalt after approximately 10, 100, 200, 300 and 700 days contact at 80°C

	TOC [a,b] in 0.45 μm Filtrate / $\mu g\ cm^{-3}$				
Composition	10 days	101 days	193 days	280 days	700 days
concrete + H_2O (no asphalt)	6	11	12	13	-
	-	-	-	10	-
asphalt + concrete + H_2O	19	14	16	18	-
	9	17	15	17	-
	NS	-	-	14	23
	-	-	-	-	9
asphalt + H_2O	3	7	9	14	-
	3	5	9	10	-
	-	-	-	7	-
asphalt + 10^{-1} mol dm^{-3} NaOH solution	38	27	39	27	-
	52	25	39	26	-
	-	-	-	15	12
	-	-	-	-	16
	10 days	96 days	185 days	270 days	700 days
0.03 mol dm^{-3} NaNO$_3$ solution (no asphalt)	20	-	-	5	-
3 mol dm^{-3} NaNO$_3$ solution (no asphalt)	-	-	-	5	-
asphalt + concrete + 0.03 mol dm^{-3} NaNO$_3$ solution	10	14	22	16	-
	10	14	18	17	-
	-	-	-	12	-
asphalt + concrete + 3 mol dm^{-3} NaNO$_3$ solution	32	28	31	27	-
	22	23	23	26	-
	-	-	-	21	24
	-	-	-	-	30

(a)   TOC = Total Organic Carbon, all values = ±10%.
(b)   Multiple entries in a given box are from individual experiments under the same conditions.
-       This experiment was not sampled at this time.

Although there is some scatter in the TOC results shown in the Table, the concentrations are in the range 3 to 52 μg cm^{-3}. Filtrates from experiments which contain asphalt give similar

724

TOC concentrations to experiments which contain no asphalt. It is possible that leaching of organic material from the filters contributes to the TOC observed. However, a trend of increasing TOC concentration with time is not observed for any of the systems studied. This tends to suggest that any organic carbon in solution is unlikely to be due to a continuing asphalt degradation process. No trends due to the presence of nitrate or concrete can be observed.

The plutonium solubilities measured in the leachates at pH 12 are given in Table II. The solubilities were obtained in leachates sampled from individual degradation experiments, e.g. values for asphalt + H_2O at 280 days are from three degradations under these conditions.

TABLE II

Plutonium Solubilities Determined in Leachates from the Chemical Degradation of Asphalt after approximately 10, 100, 200, 300 and 750 days contact at 80°C

	Plutonium solubility [a, b, c] in leachates / mol dm^{-3}				
Composition	10 days	101 days	193 days	280 days	767 days
concrete + H_2O	7×10^{-10}	7×10^{-10}	7×10^{-10}	7×10^{-10}	-
	-	-	-	7×10^{-10}	-
asphalt + concrete + H_2O	4×10^{-9}	5×10^{-9}	5×10^{-9}	5×10^{-9}	-
	4×10^{-9}	5×10^{-9}	5×10^{-9}	5×10^{-9}	-
	-	-	-	5×10^{-9}	3×10^{-11}
	-	-	-	-	3×10^{-10}
asphalt + H_2O	2×10^{-9}	3×10^{-9}	3×10^{-9}	3×10^{-9}	-
	3×10^{-9}	3×10^{-9}	3×10^{-9}	3×10^{-9}	-
	-	-	-	3×10^{-9}	-
asphalt + NaOH solution	4×10^{-9}	5×10^{-9}	5×10^{-9}	5×10^{-9}	-
	4×10^{-9}	5×10^{-9}	5×10^{-9}	5×10^{-9}	-
	-	-	-	5×10^{-9}	5×10^{-6}
	-	-	-	-	6×10^{-6}
	10 days	96 days	185 days	270 days	737 days
0.03 mol dm^{-3} NaNO$_3$ solution	4×10^{-10}	-	-	7×10^{-10}	-
3 mol dm^{-3} NaNO$_3$ solution	5×10^{-10}	7×10^{-10}	7×10^{-10}	7×10^{-10}	-
asphalt + concrete + 0.03 mol dm^{-3} NaNO$_3$ solution	7×10^{-10}	1×10^{-9}	1×10^{-9}	1×10^{-9}	-
	8×10^{-10}	9×10^{-10}	1×10^{-9}	1×10^{-9}	-
	-	-	-	1×10^{-9}	-
asphalt + concrete + 3 mol dm^{-3} NaNO$_3$ solution	9×10^{-10}	1×10^{-9}	1×10^{-9}	1×10^{-9}	-
	1×10^{-9}	1×10^{-9}	1×10^{-9}	1×10^{-9}	-
	-	-	-	1×10^{-9}	3×10^{-10}
	-	-	-	-	1×10^{-10}

(a)   pH adjusted to 12.0 - 12.2 as necessary before solubility determination.
(b)   Plutonium solubilities determined at room temperature.
(c)   Multiple entries in a given box are from individual experiments under the same conditions.
-    This experiment was not sampled at this time.

In the leachates from the experiments which do not contain asphalt the plutonium solubility is in the range 4×10^{-10} to 7×10^{-10} mol dm^{-3} and there are no observable trends with elapsed time or the presence of sodium nitrate.

In the experiments which contain asphalt but no nitrate, all the plutonium solubilities are similar in leachates sampled in the period 10 - 280 days and are in the range 2×10^{-9} mol dm^{-3} to 5×10^{-9} mol dm^{-3}.

Two of these sets of experiments (asphalt + concrete + H_2O and asphalt + NaOH solution) were continued until 767 days had elapsed. The plutonium solubilities measured in the leachates from the experiments containing asphalt and concrete had decreased to 3×10^{-11} mol dm^{-3} and 3×10^{-10} mol dm^{-3}. However, the plutonium solubilities in the leachates from the experiments in which asphalt was degraded in contact with sodium hydroxide solution (i.e. no concrete was present) increased to 5×10^{-6} and 6×10^{-6} mol dm^{-3} compared to 5×10^{-9} mol dm^{-3} measured after 280 days. It was not clear why the plutonium solubility in the sodium hydroxide leachates should have increased. The solubility determinations in the asphalt/sodium hydroxide leachates were re-sampled after being allowed to stand for 22 days to determine whether the elevated plutonium concentrations were stable. These re-determinations were similar at 2×10^{-6} and 5×10^{-6} mol dm^{-3}. For comparison, the solubility of plutonium in sodium hydroxide solution at pH 12 was also measured and was 4×10^{-9} to 5×10^{-9} mol dm^{-3}. Although this is higher than 7×10^{-10} mol dm^{-3} measured in water equilibrated with concrete, it is 3 orders of magnitude lower than measured in the asphalt/sodium hydroxide leachates.

Table I shows that there is no increase in the TOC concentrations from the asphalt/sodium hydroxide degradations which can be correlated with the increase in plutonium solubility. The values of 12 and 16 μg cm^{-3} measured after 767 days degradation compare with the range of 15 to 27 μg cm^{-3} measured after 280 days. The increase in plutonium solubility cannot therefore be related to an increase in the levels of organic carbon in solution. Visual inspection of the glass vessels used for the degradations in the presence of sodium hydroxide showed evidence of attack. It is possible that dissolution of silicate from the glass and the formation of silicate polymers in solution may have occurred in the sodium hydroxide experiments. Plutonium may interact with these silicate species resulting in the elevated plutonium solubilities observed.

For the experiments which contain nitrate and asphalt, the plutonium solubilities are similar in leachates sampled in the period 10 - 270 days and are $\leq 1 \times 10^{-9}$ mol dm^{-3}. One set of experiments (asphalt + concrete + 3 mol dm^{-3} sodium nitrate solution) were continued until 737 days had elapsed. The plutonium solubilities measured in the leachates had fallen to 1×10^{-10} mol dm^{-3} and 3×10^{-10} mol dm^{-3}.

<u>Radiolytic Degradation</u>

The α-irradiated experiments were sampled after 90 days and 180 days contact with the aqueous solutions. These times are assumed to be equivalent to between 10^4 - 10^5 years and 10^4 - 10^6 years respectively for the real waste, based solely on the increased α-activity. The pH values of all the leachates were in the range 12.0 - 12.2. The concentrations of plutonium due to the dissolution of plutonium oxide during irradiation, and the plutonium solubilities measured in the leachates by oversaturating the solution, are shown in Table III. The upper part of the table shows the concentrations of plutonium in the leachates from the dissolution of the $^{238}PuO_2$

present during irradiation. The lower part of each table shows the plutonium solubilities measured after the addition of excess ^{238}Pu as a small aliquot of acid solution.

After 90 days the plutonium concentrations from the dissolution of the oxide were 1×10^{-10} and 1×10^{-11} mol dm^{-3} in the leachates from the experiments containing concrete-equilibrated water. In the presence of 3 mol dm^{-3} NaNO$_3$ the leached plutonium concentration in both experiments was 6×10^{-11} mol dm^{-3}. After the addition and precipitation of excess plutonium the plutonium solubilities were determined as 2×10^{-11} and 4×10^{-11} mol dm^{-3} in the concrete-equilibrated water whereas in the leachates containing sodium nitrate they were 2×10^{-10} and 3×10^{-10} mol dm^{-3}.

After 180 days of radiolytic degradation the results were similar to those obtained after 90 days. In the leachates from the experiments containing no nitrate the leached plutonium concentrations were 9×10^{-12} and 6×10^{-10} mol dm^{-3}. In the presence of sodium nitrate the plutonium concentrations were 1×10^{-11} and 2×10^{-11} mol dm^{-3}. The corresponding plutonium solubilities were 4×10^{-11} and 1×10^{-10} mol dm^{-3} for the experiments in concrete-equilibrated water and 9×10^{-10} and 8×10^{-9} mol dm^{-3} for those in sodium nitrate solution.

TABLE III
Plutonium Concentrations in Leachates
from Pu-238 Coated Asphalt Samples after 90 and 180 Days contact

Experiment Composition	Plutonium Concentration[a] in Leachates	
	90 days	180 days
	Leached plutonium[b] / mol dm^{-3}	
Asphalt + Concrete + Water	1×10^{-10}	9×10^{-12}
	1×10^{-11}	6×10^{-10}
Asphalt + Concrete + 3 mol dm^{-3} NaNO$_3$	6×10^{-11}	1×10^{-11}
	6×10^{-11}	2×10^{-11}
	Plutonium Solubility[c] / mol dm^{-3}	
Asphalt + Concrete + Water	4×10^{-11}	4×10^{-11}
	2×10^{-11}	1×10^{-10}
Asphalt + Concrete + 3 mol dm^{-3} NaNO$_3$	3×10^{-10}	9×10^{-10}
	2×10^{-10}	8×10^{-9}

(a)   Multiple entries in a given box are from individual experiments under the same conditions.

(b)   These values are the results of assays of the plutonium concentration in the leachates from the dissolution of the 238-plutonium oxide present as the alpha-irradiation source.

(c)   Portions of the leachates were filtered through 0.45μm filters and then oversaturated by an addition of a small volume of solution of plutonium nitrate in 1 mol dm^{-3} nitric acid. The solutions were then filtered through a 30,000 NMWCO ultrafilter, to give the plutonium solubilities in the leachate solutions.

Comparison of the results show that the plutonium concentrations in the leachates from the dissolution of the plutonium oxide are in the range 9×10^{-12} mol dm^{-3} to 6×10^{-10} mol dm^{-3}. From these results there is no trend with water chemistry or time of leaching. The solubilities of plutonium are in the range 2×10^{-11} mol dm^{-3} to 8×10^{-9} mol dm^{-3}. The only trend suggested by these results is that the plutonium solubility may be higher in the α-irradiated leachates containing sodium nitrate although this is a small set of data on which to draw a conclusion.

It was observed that the solutions in the alpha radiolysis experiments containing sodium nitrate had become brown. The UV/visible absorption spectra of samples of the leachates taken after 120 days contain a large peak at approximately 390 nm in the α-irradiated nitrate experiments which is not present in the experiments which do not contain nitrate. The UV/visible absorption spectrum of nitrite in acid solution peaks at 383 nm.

CONCLUSIONS

A consideration of these results and data published elsewhere [7] suggests that chemical and radiolytic attack on asphalt or bitumen under anaerobic, alkaline conditions typical of a deep cementitious repository is unlikely to generate complexants for plutonium which are effective at high pH. Plutonium solubilities in leachates from the chemical or α-radiolytic degradation of asphalt in the presence of concrete were in the range 2×10^{-11} to 8×10^{-9} mol dm^{-3}. These are lower than plutonium solubilities measured in leachates from the degradation of some other organic materials such as cellulose tissues [1,2]. Although high plutonium solubilities of about 5×10^{-6} mol dm^{-3} were measured in asphalt and sodium hydroxide leachates after 767 days this may be due to attack on the glass vessel rather than degradation of the asphalt.

ACKNOWLEDGEMENTS
This work was funded by the Power Reactor and Nuclear Fuel Development Corporation (PNC), Tokyo, Japan, and their permission to publish these results is gratefully acknowledged.

REFERENCES

1. B.F. Greenfield, A.D. Moreton, M.W. Spindler, S.J. Williams and D.R. Woodwark, in Scientific Basis for Nuclear Waste Management XV, edited by C.G. Sombret, (Mater. Res. Soc. Proc. **257** 1992) pp. 299-306.

2. B.F. Greenfield, C.M. Linklater, A.D. Moreton, M.W. Spindler and S.J. Williams, in Actinide Processing: Methods and Materials, edited by B. Mishra and W.A. Averill (TMS, 1994) pp. 289-303.

3. e.g. S.E. Moschopedis and J.G. Speight, Fuel **50**, 211 (1971).

4. e.g. I. Lorinc, Nature **212**, 1459 (1966).

5. e.g. H. Duschner, W. Schorr and K. Starke, Radiochimica Acta **24**, 133 (1977).

6. e.g. N. Ait-Langomazino, R. Sellier, G. Jouquet and M. Trescinski, Experientia **47**, 533 (1991).

7. L.R. Van Loon and Z. Kopajtic, Radiochimica Acta **54**, 193 (1991).

8. A.V. Chambers, S.J. Williams and S.J. Wisbey, UK Nirex Ltd Science Report S/95/011, 1995.

9. D. Rai, J.A. Schramke, D.A. Moore and G.L. McVay, Nuclear Technology **75**, 350 (1986).

Characterization of Th Carbonate Solutions Using XAS and Implications for Thermodynamic Modeling

N.J. Hess*, A.R. Felmy*, D. Rai*, and S.D. Conradson**
*Pacific Northwest National Laboratory, Richland WA 99352
** Los Alamos National Laboratory, Los Alamos NM 89545

ABSTRACT

The chemical behavior of actinide elements in tank solutions, in soil, and in groundwater is dependent upon the chemical species that form when aqueous solutions come in contact with the actinide compounds. In particular the chemical speciation of the reduced actinide oxidation states (III and IV) are important, for example, to DOE waste tank processing and, more generally, to nuclear waste disposal issues. Predicting the solubility of the actinides in these solutions requires identification of the strong aqueous complexes, such as carbonates and organic chelating agents, that can form in aqueous solution.

Previous speciation work has often relied on indirect techniques such as potentiometric titrations or solubility measurements. Recent XAS experiments determine directly the speciation of the Th carbonato species of seven solutions under a range of carbonate concentrations and pH conditions. The presence of the pentacarbonato complex is confirmed and the complex's stability at low carbonate concentrations is determined. These experimental results support a proposed thermodynamic model that describes the solubility of Th(IV) hydrous oxide in the aqueous Na^+-HCO_3^--CO_3^{2-}-OH^--ClO_4^--H_2O system extending to high concentrations at 25°C. This model is relatively simple in that only two aqueous species are included $Th(OH)_3CO_3^-$ and $Th(CO_3)_5^{6-}$.

INTRODUCTION

The large volumes of high level waste currently stored in tanks at DOE sites contain actinides both in the sludge and supernate components of the tanks. The chemical partitioning of actinide elements between tank solutions and sludge and in the soil and groundwater is dependent upon the chemical species in the aqueous solutions that come in contact with the actinide compounds. In particular, the chemical speciation of the reduced actinide oxidation states (III and IV) is of particular importance to DOE waste tank processing and nuclear waste disposal issues. As examples, the presence of reducing agents, such as nitrite and H_2, in the high level waste tank solutions at Hanford can stabilize actinides, for example Pu and U, in lower oxidation states. In order to predict the solubility of the actinide elements in these solutions, knowledge of the strong actinide aqueous complexes, such as carbonate and organic chelating agents, is required. In a similar fashion, the presence of iron metal and organic compounds in the Waste Isolation Pilot Plant wastes can reduce the actinide elements present in these wastes dramatically altering the elemental solubilities and affecting the performance assessment of a repository. The prediction of the solubility of these compounds requires knowledge of the aqueous complexes that can form particularly with strong complexing agents such as carbonate. As a result, the aqueous complexation of tetravalent actinide species, including species of Th(IV), U(IV) , and Np(IV) with carbonate has been the subject of intense research efforts [1-4]. Although all of these studies clearly demonstrate the potential importance of carbonate complexation, definitive evidence for the species in solution is often lacking.

The speciation of the actinide complexes determined by potentiometric titrations or solubility measurements involves the comparison of the predicted slopes of the solubility or titration curves

729

of hypothesized complexes to the experimental data (curve-fitting). The species that give the best overall fit are then assumed to be present and predictive thermodynamic models are then developed from these data. Although the aqueous speciation in these solutions is often examined by standard spectroscopic methods such as UV-VIS or Raman it is often difficult to unambiguously determine the aqueous speciation. This is especially true for the actinide elements in these solutions since solid phase analogues for comparing the spectra are frequently unavailable, multiple species can form in solution and complicate the observed spectra, and certain valences (eg. Th(IV)) do not exhibit the appropriate absorption spectra. A preferred approach is to employ a more direct method of determining the speciation of actinides in solution, such as measuring the x-ray absorption spectrum. This approach was successfully demonstrated by Clark et al [2] in determining the complexation of Np(V) carbonato solutions over a range of pH and concentration conditions.

EXPERIMENT

The details of the solution preparation have been presented previously [5]. X-ray absorption measurements of the Th carbonate solutions listed in Table I were conducted at the Stanford Synchrotron Radiation Laboratory on endstation 4-2 under dedicated operating conditions (3.0 GeV and 40 to 90 mA current). Spectra were collected at the thorium L_{III} edge in transmission and fluorescence modes simultaneously to photoelectron wavevector of 13 Å^{-1}. While the transmission was measured using standard ionization chamber detectors, the fluorescence signal was collected using a 13-element Ge detector. Energy calibration was based on either assigning the first inflection point in the absorption edge of a zirconium metal foil to 17999.35 eV or to a thorium metal foil at 16300 eV. Normalization of the absorption spectrum was accomplished by fitting polynomials through the pre- and post-edge regions, setting the value of the extrapolated pre-edge to zero at E_0, and setting the difference between the extrapolations of the pre- and post-edge polynomials to unity at E_0.

Table I. Sample solution conditions

sample	solution conditions	Th concentration
Th15	2M Na_2CO_3, 0.10M NaOH	1.0 mM
Th33	2M Na_2CO_3, 0.50M NaOH	1.0 mM
Th36	2M Na_2CO_3, 1.00M NaOH	1.0 mM
Th76	0.25M $NaHCO_3$, pH 9.3	7.0 mM
Th74	0.10M $NaHCO_3$, pH 9.36	0.9 mM
Th71	0.10M $NaHCO_3$, pH 9.39	0.11 mM
Th70	0.07M $NaHCO_3$, pH 9.43	0.002 mM

EXAFS were extracted by fitting a polynomial spline function through the post-edge region and normalizing the difference between this approximation to the solitary atom EXAFS and the actual data with the absorption decrease given by the McMaster tables [6]. Fourier transforms were taken over photoelectron wavevector ranges based on signal-to-noise ratios for each sample. Nodes in EXAFS were selected as endpoints to the transform range and a Gaussian window with a 1 sigma width of 0.5 Å-1 was used to dampen the EXAFS oscillations at the transform range endpoints. The phase shift has not been removed from the Fourier transforms; as a result the peaks in the transform moduli correspond to contributions from shells that are 0.2 to 0.5 Å shorter than the actual absorber-scatter distances. Fourier transforms of selected samples are shown in Figure 1.

At high carbonate concentrations the Fourier transforms reveal two clearly resolved peaks centered at 1.9 Å and 3.6 Å. The first peak at 1.9 Å corresponds to the first coordination shell around the thorium cation and consists of oxygen atoms. Using the guanidinium pentacarbonato-thorate tetrahydrate structure [7] as a qualitative model, the identity of the atoms in the second shell can be assigned to distal oxygens. Note that the coordination shell due to the five carbons is not resolved from the first shell oxygens. As illustrated in Figure 2, it's existence can only be determined through quantitative analysis using the multiple scattering code, FEFF6.01 [8,9]. At lower carbonate concentrations, in addition to a lower signal to noise ratio, the amplitude of both the first oxygen and distal oxygen coordination shells is reduced.

FEFF6.01 was used to calculate the phase and amplitude for the individual scattering paths to the equatorial oxygen, the carbon, and the distal oxygens, denoted Th-O, Th--C, and $Th--O_d$, respectively. The 180° scattering angle formed by $Th--C-O_d$ generates large multiple scattering amplitudes at the same effective distance as the $Th--O_d$ path and these are included in the parameterization of the $Th--O_d$ path. The parameterized scattering paths were used to curvefit the EXAFS over the photoelectron wavevector region 2.9 to 12.5 Å^{-1}, except for sample Th71 which was fit over a shorter photoelectron wavevector region 2.9 to 10.0 Å^{-1}. Fits to the total EXAFS are given in Figure 3.

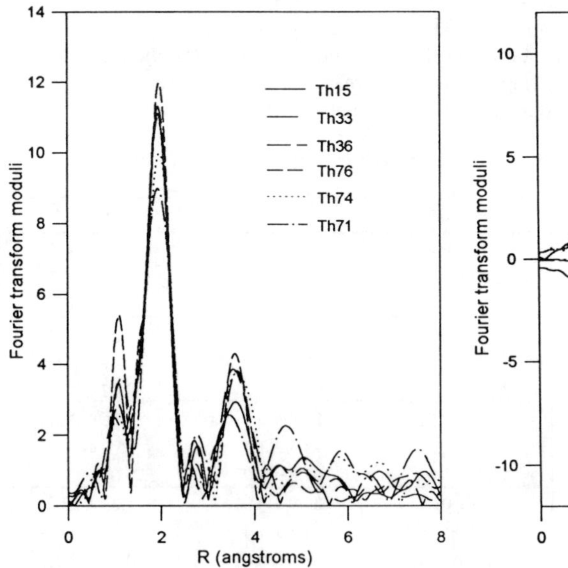

Figure 1. Fourier transforms of selected Th solutions.

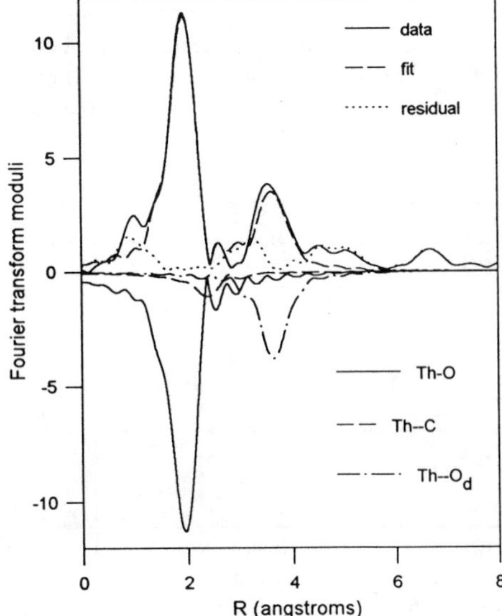

Figure 2. Comparison of the fit to the data for Th33.

RESULTS

At high carbonate concentrations, the thorium cation is coordinated by 10 oxygens at 2.49 Å.

A carbon shell is identified at 3.00 Å and a distal oxygen shell at 4.27 Å. The calculated scattering distances for Th--C and Th--O_d are a few hundredths of an Å longer than that reported for the solid structure [6]; however are in good agreement with other actinide bidentate carbonato complexes such as the Np(V)mono-, bis-, tri- carbonato species [2]. Determination of the number of carbonate groups involved in the complexation is hindered because (1) the carbon shell resides within the Th-O shell as shown in Figure 2, (2) photoelectron scattering off of low z elements is intrinsically weak, and (3) the EXAFS originating from second and third coordination shells are dampened from increased disorder contributions, and thus leads to increased uncertainty in the metrical parameters. These effects are evident in the greater uncertainty in the metrical parameters for the Th--C and Th--O_d given in Table II. For these reasons, efforts to determine the number of carbonate groups were focused on the distal oxygen because, although it is more distant, it is well isolated from other contributions. Filtered curvefits of the back Fourier transform of the Th--O_d shell were performed over the photoelectron region of 3.10 to 4.20 Å^{-1}. The resulting parmeters were fixed during the subsequent fit to the total EXAFS. The metrical parameters for filtered curvefit are given in Table II. Using this approach it is clear that five carbonate groups are involved in the solution complexation of thorium in bidentate fashion.

Preliminary detailed analysis of the Th--O_d shell indicates while there is no change in the total number of carbonate groups is observed, there may be a change from pure bidentate carbonate complexation to a mixture of bidentate and monodentate complexation as the carbonate concentration decreases below 0.25M $NaHCO_3$. FEFF6.01 models of the monodentate configuration were constructed and used successfully to fit the total EXAFS for samples Th76 and Th74. In the monodentate configuration the Th-O-C bond angle is assumed to be 180° and results in a large multiple-scattering contribution at the Th--C bond distance. In the monodentate configuration, the resulting Th--O_d scattering distance is increased from 4.25 Å to 4.70 Å. For both samples the best fit consists of 4 bidentate and 1monodentate carbonate groups.

Table II. Metrical parameters determined from curvefits to the EXAFS.

		Th - O[1]	Th -- C[2]	Th -- O_d[3]
Th15	distance	2.49 ± 0.02	3.00 ± 0.02	4.26 ± 0.02
	number	9.9 ± 2.6	3.4 ± 1.1	5.1 ± 1.57
$r^2 = 0.9897$	sigma	0.08 ± 0.01	0.08 ± 0.01	0.08 ± 0.01
Th33	distance	2.50 ± 0.02	2.98 ± 0.02	4.26 ± 0.02
	number	11.1 ± 2.8	3.1 ± 1.0	5.2 ± 1.7
$r^2 = 0.9568$	sigma	0.09 ± 0.01	0.10 ± 0.01	0.07 ± 0.01
Th36	distance	2.50 ± 0.02	2.99 ± 0.02	4.26 ± 0.02
	number	10.6 ± 2.7	3.4 ± 1.1	5.0 ± 1.1
$r^2 = 0.9395$	sigma	0.09 ± 0.01	0.08 ± 0.01	0.07 ± 0.01
Th76	distance	2.50 ± 0.02	2.99 ± 0.01	4.25 ± 0.02
	number	11.1 ± 3.0	4.3 ± 1.4	3.6 ± 0.8
$r^2 = 1.3270$	sigma	0.09 ± 0.01	0.09 ± 0.01	0.04 ± 0.02
Th74	distance	2.50 ± 0.02	2.98 ± 0.02	4.25 ± 0.02
	number	10.2 ± 2.7	3.1 ± 1.0	5.2 ± 1.5
$r^2 = 1.1447$	sigma	0.09 ± 0.01	0.09 ± 0.01	0.07 ± 0.01
Th71	distance	2.50 ± 0.02	2.98 ± 0.08	4.23 ± 0.02
	number	8.7 ± 2.5	1.8 ± 0.6	2.5 ± 0.8
$r^2 = 1.9259$	sigma	0.08 ± 0.02	0.05 ± 0.04	0.00 ± 0.02

1. ΔE_0 for Th-O = -6.5 eV; 2. ΔE_0 for Th--C = -2.4 eV; 3. ΔE_0 for Th--O_d = -2.6 eV

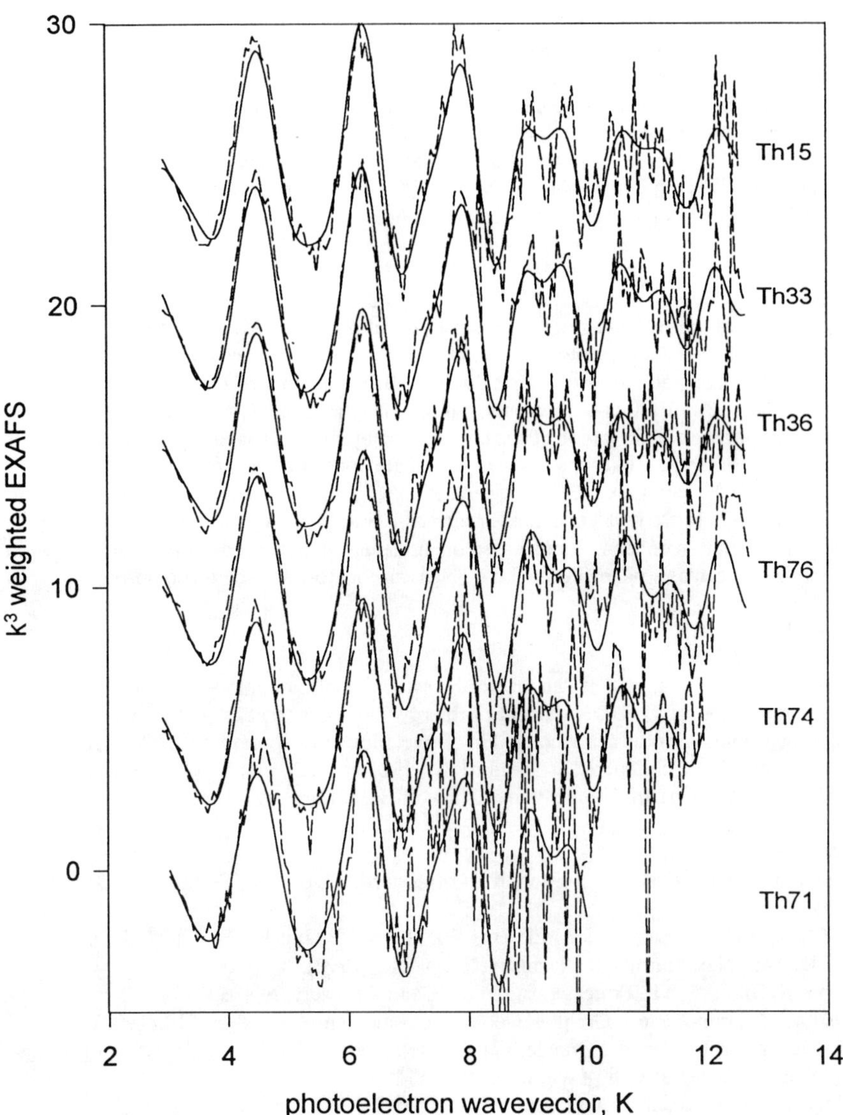

Figure 3. Comparision of the fits to the data in k-space.

At carbonate concentrations below 0.10 M NaHCO$_3$ the thorium speciation changes dramatically. The distance to the Th-O shell decreases from 2.50 Å to 2.46 Å and the number oxygens decreases from ten to eight. Additionally, there is a reduction in the number of distal oxygens at from five to three, indicating only three bidentate carbonate groups are involved in the complexation of Th at these pH and carbonate concentration conditions. Several models of bidentate and monodentate carbonate coordination were evaluated, however attempts to quantify the bidentate-monodentate configuration were frustrated due to the low signal-to-noise ratio of this low concentration sample. The data suggest that the three carbonate groups are bound in bidentate fashion to the Th ion accounting for six of the eight equatorial oxygens. The remaining two equatorial oxygens could represent hydroxyl groups.

CONCLUSIONS

Complexation of tetravalent actinide species with carbonate species has been the subject of intense interest. The results presented here support earlier hypothesized speciation models in which Th exists as a pentacarbonato complex at high carbonate concentrations and is consistant with speciation models for other tetravalent actinides in highly concentrated carbonate solutions [9]. At low carbonate concentration, Th is coordinated by three carbonate groups and an unknown number of hydroxyl groups at lower carbonate concentrations forming a $ThO_{2-x}(CO_3)_3(OH)_x^{(6-x)-}$ complex. Other researchers have proposed analogous mixed carbonate-hyrdoxide complexes for tetravalent actinides such as U(IV) [10], Np(IV) [10] and Pu(IV) [11] at lower carbonate solution concentrations based on modeling of titration data, however this work presents the first spectroscopic evidence for a Th triscarbonato-dihydroxide complex.

ACKNOWLEDGEMENTS

The research was conducted at Pacific Northwest National Laboratory, operated by Battelle Memorial Institute for the U.S. Department of Energy under Contract DE-AC06-RLO 1830, and was funded by Power Reactor and Nuclear Fuel Development Corporation (PNC) of Japan. XAFS experiments were conducted at Stanford Synchrotron Radiation Laboratory which is operated by the U.S. Department of Energy, Office of Basic Energy Sciences.

References
1. D. Rai, A.R. Felmy, D.A. Moore, and M.J. Mason, Mat. Res. Soc. Symp. Proc. **353**, p. 1143 (1995).
2. D.L. Clark, S.D. Conradson, S.A. Ekberg, N.J. Hess, D.R. Janeky, M.P. Neu, P.D. Palmer, and C.D. Tait, New Journal of Chemistry, **20**, p. 211 (1996).
3. A. Joao, S. Bigot, and F. Fromage, Bull. Soc. Chim. France **1**, p. 42 (1987).
4. E. Osthols, J. Bruno, and I. Grenthe, Geochim Cosmochim Acta **58**, p. 613 (1994).
5. W.H. McMaster, N. Kerr del Grande, J.H. Mallett, and J.H. Hubbell, <u>Calculation of X-Ray Cross Sections</u>, Univ of Calif, Livermore, 1969, 350p.
6. S. Voliotis and A. Rimsky, Acta Cryst. **B31**, p. 2612 (1975).
7. J.J. Rehr, S.I. Zabinsky, and R.C. Albers, Phys. Rev. Let. **69**, p. 3397 (1992).
8. S.I. Zabinsky, J.J. Rehr, A. Ankudinov, R.C. Albers, and M.J. Eller, Phys. Rev. B. **52**, p. 2995 (1995).
9. D.L. Clark, D.E. Hobart, and M.P. Neu, Chemical Reviews **95**, p. 25 (1995).
10. M.I. Pratopo, H. Moriyama, and K. Higahi, Radiochim. Acta. **51**, p. 27 (1990).
11. T. Yamaguchi, Y. Sakamoto, T. Ohnuki, In *Migration '93*;Charleston SC, (1993).

THE EFFECT OF MINERAL VARIABILITY ON THE SOLUBILITY OF SOME ACTINIDES: AN UNCERTAINTY ANALYSIS

Christian Ekberg, Allan T. Emrén and Anders Samuelsson
Department of Nuclear Chemistry, Chalmers University of Technology
S-412 96 Göteborg

ABSTRACT

The use of computer simulations in the performance assessment for a repository for spent nuclear fuel, are in many cases the only method to get information on how the rock-repository system will work. One important factor is the solubility of the elements released if the repository is breached. This solubility may be determined experimentally or simulated. If it is simulated, several factors such as thermodynamical uncertainties will affect the reliability of the results. If these uncertainties are assumed to be small, the composition of the water used in the calculations may play a major part in the uncertainties in solubility. The water composition, in turn, is either determined experimentally or calculated through water-rock interactions. Thus, if the mineral composition of the rock is known, it is possible to foresee the water composition. However, in most cases a determination of the rock composition is made from drilling cores and is thus quite uncertain. Therefore, if solubility calculations are to be based on water properties calculated from rock-water interactions another uncertainty is introduced. This paper is focused on uncertainty and sensitivity analysis of rock-water interaction simulations and the uncertainties thus obtained are propagated through a program making uncertainty and sensitivity analysis of the solubility calculations. In both cases the latin hypercube sampling technique have been used.
The results show that the solubilities are in most cases log normal distributed while the different elements in the simulated groundwater in some cases diverge significantly from such a distribution. The numerical results are comforting in that the uncertainty intervals of the solubilities are rather small, i.e. up to 30%.

INTRODUCTION

The governmental licensing of a repository for spent nuclear fuel must be based on a judgement of its allowability from several aspects, the most important being safety and radiation protection. The necessary judgement is based on an assessment of the performance of the repository in different time frames ranging from the present situation to a time 10^6 years in the future. Further, in connection with a safety analysis of a repository, there are many disciplines to take into account, e.g. hydrology, geology, rock mechanics and geochemistry. Since the repository is supposed to function a long time without maintenance, many uncertainties arise in each discipline concerned. A characterisation of the chemical environment is of importance for the prediction of canister corrosion as well as for the prediction of the transport of any released radionuclides from a failed canister. One of the important factors determining the migration speed of the released radionuclides are the solubility of concerned solid phases. Such solubilities may be determined experimentally or calculated. If calculated, the results are subject to uncertainties in such experimental input data as stability constants, and water composition.

In many cases it is interesting to know how the system will behave in the future or when the environment is changed, e.g. the land covered with ice. Further, it is not always possible to perform measurements. For those cases computer simulations together with expert judgement may be the only source of information.

It is a common problem when trying to simulate reality that the models become very complicated and that it is necessary to have input data that are not easily available. In some cases the input data have to be guesses, or attempts must be made to determine them under difficult experimental conditions. In any case the input data will be encumbered with uncertainties of varying magnitude. To run the computer program with such data and pretend that the results reflect reality is clearly unacceptable for the strategy of a performance assessment; a thorough sensitivity / uncertainty (s/u) analysis should instead be made. Giving the uncertain parameters as intervals with a probability distribution produces the result as an interval with a given confidence level and thus make it possible to select the worst case (if desired).

With the complexity of the codes used, it has become difficult to ascertain the correctness of the results. Obviously impossible answers are usually discovered, but an erroneous result obtained because of

Mat. Res. Soc. Symp. Proc. Vol. 465 © 1997 Materials Research Society

the incorrectness of some input data may be difficult to pinpoint. In such cases, an uncertainty / sensitivity analysis is an important tool in the understanding of the system concerned. The effect of input data uncertainties may be studied in such an analysis, and the common question of how important a parameter is in the simulated system may be answered, thus making the effort of obtaining reliable data less time consuming and cheaper[1][2][3].

This paper presents results from s/u calculations of a water-rock interaction program, CRACKER [4], and of solubility calculations made by PHREEQE [5] assuming that the element concentrations in the water are uncertain.

THE PROBLEM

Since granitic rocks in most cases are strongly heterogeneous on all scales, characterisation from sampled drill cores gives an uncertain description of the local mineralogy. Thus, uncertainty analyses are important tools to describe variability in rock-water interactions and thereby caused groundwater properties. The water composition influences solubilities, which are important factors for limiting the migration of radionuclides into the environment of a failed repository. One should note that variability in rock - groundwater chemistry is merely one source of uncertainties in solubility calculations. Other sources are assumptions in conceptual models and uncertainties in thermodynamical data [6] [7].

COMPUTER PROGRAMS

Two computer programs are primary used in the calculations presented in this paper, MINVAR and UNCCON [8]. Both may be seen as consisting of a simulation part and a statistical sampling part. A short description is given below.

MINVAR

The simulation part consists of the CRACKER program [4] in wich the fracture is modelled as two parallel planes acros which minerals are randomly ditrsibuted with abundances according to the mineral composition of the rock. Water flows through the fracture and is allowed to equilibrate locally. Dissolved material diffuses through the water. In this way, even mineral sets large enough to violate the phase rule may be treated.

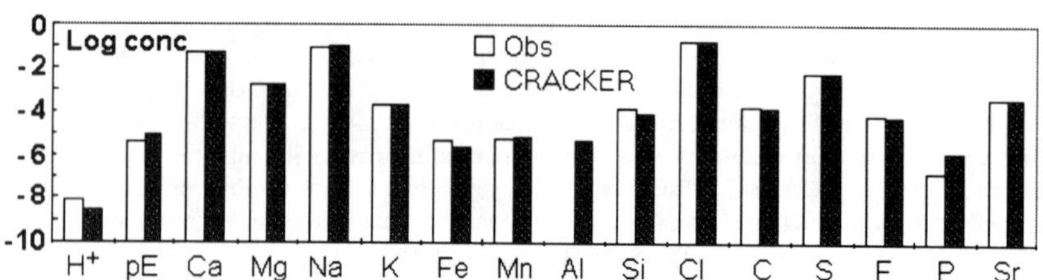

Figure 1, Log element concentrations observed at Äspö and simulated with the CRACKER program.

It has been found that the CRACKER program successfully simulates groundwater composition. An example is shown in Figure.1.

As indicated earlier one of the input data to CRACKER is the mineral abundances, which are rather uncertain in a geologic system. To MINVAR these percentages are given as intervals rather than as fixed values. The samples are then taken from these intervals with the well-known Latin Hypercube sampling (LHS) technique [9]. Further it may be noted that the abundances are assumed to be uniformly distributed.

From the MINVAR calculation it is possible to obtain e.g. the mean concentration, a 95% confidence interval for the concentration and the cumulative distribution function (cdf) of the concentrations. The pH and pe are also included in the calculations.

UNCCON

The simulation motor in this case is the well known thermodynamical equilibrium code PHREEQE [5], in which one of the input data is the water composition. In UNCCON it is possible to give the element concentrations as intervals rather than fixed values. An input matrix is then generated using the LHS technique. In order to obtain the solubility corresponding to each input set PHREEQE is run once for each input set. The solubilities of the selected solid phase is then collected, and the whole sample is analysed with common statistical tools. Examples of output are the density function for the solubility and some common statistics, as for MINVAR.

RESULTS

As two different programs have been used, the results are divided into two parts, one consisting of MINVAR results and the other consisting of UNCCON results. However, both cases have naturally common features. The data for both cases are taken from the Äspö site in the southeast of Sweden.

MINVAR

The MINVAR calculations were based on a sampled groundwater with a composition according to Table 1.

Table 1. Composition of reference water from the Äspö site, borehole KAS 02, level 530-535m [9,10]. Concentrations in mM.

Ca	47.2		Cl	181		[a]U	5.45E-07
Mg	1.73		HCO_3^-	0.164		Sr	0.399
Na	91.3		S_{tot}	5.83		Li	0.144
K	0.207		SO_4^{2-}	5.83		N_{tot}	1.85E-03
[a]Fe	4.37E-03		S^{2-}	4.68E-03		NH_4	1.67E-03
Mn	5.28E-03		F	7.90E-02		pH	8.1
[a]Al	1.00E-03		Br	0.501		[b]pe	-4.37
Si	0.146		P	1.61E-04		Temp (°C)	15

[a] The analysis did not contain any value for this species. The concentration has been estimated on the basis of other Äspö groundwater samples.

[b] The pe value was adjusted from the measured value (-5.42) with regard to additions of Al, Fe, U, and the equilibrium between SO_4^{2-}/S^{2-} in the solution.

The water is then virtually contacted with the mineral set, shown in Table 2, within a CRACKER fracture that is 20 cells wide and 100 cells long. These numbers are not generally good ones, but in this case they are sufficient.

Table 2, Estimated mineral composition at Äspö together with assumed uncertainty intervals [11].

Mineral	Abundance, (%)
Chlorite	40±10
Calcite	25±10
Epidote	20±10
Fluorite	10±5
Hematite	5±5
Quartz	2±2
Illite	2±2
Montmorillionite	2±2
Pyrite	2±2

The resulting water is given with largest and smallest element concentrations according to Table 3. A 95% confidence interval is also shown, but it is assumed there that the concentrations follow a normal distribution. Whether or not this assumption is correct may either be visually or statistically ascertained.

The solid line in Figure 2 is the cumulative distribution function P(x), for a normal distribution with the calculated mean logarithm of the concentration and standard deviation, according to Equations 1 and 2.

$$P(x) = \frac{1}{\sqrt{2\pi}} \int_{-\infty}^{x} e^{-t^2/2} dt \tag{1}$$

were P(x) is the probability that a randomly selected value will be smaller than x and t is an integration variable. In order not to evaluate this integral, and instead use a table of values, x must be N(0,1), i.e. x will assume values according to Equation 2.

$$x = \frac{y - \mu}{\sigma} \tag{2}$$

where y is the logarithm of the concentration values, i.e. the x-axes in Figure 2, and μ and σ are the mean of the logarithm of the concentration and the standard deviation, respectively.

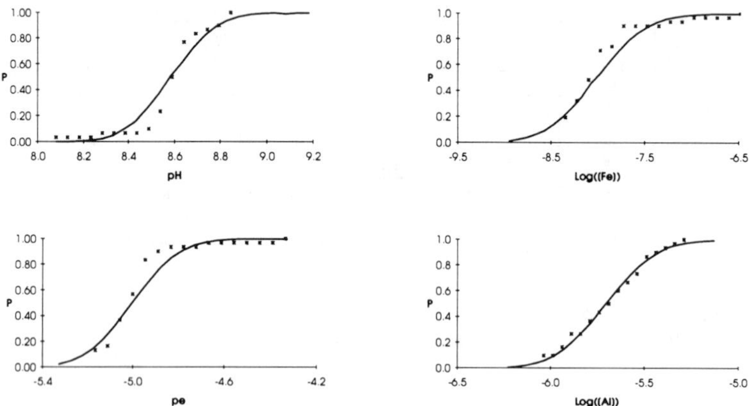

Figure 2, Cumulative distribution functions for some important elements,
x are the sampled results and the solid line is the theoretical cdf

The hypothesis that the concentrations are log-normally distributed is rejected at 95 % confidence in all cases. However, aluminium may be close enough to normality that such an assumption induces a smaller error than the method itself, i.e. sample size and thermodynamical errors. Using the assumption of normality it is possible to make a confidence interval for the concentration, pH and pe. Table 3 shows the largest and smallest values obtained together with a 95% confidence interval. In the case of aluminium a normal distribution is assumed, but a confidence interval based on Chebychevs inequality is used for the others [12].

Table 3, Minimum and maximum concentration together with
95% confidence interval for the mean concentrations.

Parameter	Min. value	Max. value	Confidence interval for the mean log(conc.)(95%)
pH	8.0	8.8	8.6±0.99
pe	-5.2	-4.3	-5.0±1.07
Al	8.2E-07	5.2E-06	-5.70±0.41
Fe	3.4E-09	1.0E-07	-8.01±2.75
Si	6.1E-05	1.4E-04	-4.09±0.74
C	3.2E-05	1.2E-04	-4.35±0.75

The reasons for choosing to write the confidence interval in term of logarithms rather than linear values are that the results are closest to a log-normal distribution and the confidence interval stretches over several orders of magnitude. The logarithm is the only method to make a symmetric interval. Obviously the results shown in Table 3 do not agree with the sampled Äspö water. This should be the case if the rock selected was in fact similar to the one at Äspö and the database and model used for the calculation clearly represented reality. However, this shows how difficult it is to predict the mineral composition in a rock by sampling from boreholes.

UNCCON

As mentioned earlier the UNCCON program calculates the uncertainty interval for the solubility of a selected solid phase owing to uncertainties in element concentration in the water used. Further, the selected phase is assumed to be the solubility limiting one.

The UNCCON program was run for two different cases, first for the water given in Table 1 together with the experimental and statistical uncertainties [13][14], and second for the water produced by the MINVAR calculations as described in Table 3. The elements not given in Table 3 are assumed stable with respect to the selected rock and are thus taken from Table 1. The deviation from normality in the MINVAR results does not affect the UNCCON calculations because, in order to be somewhat conservative, the distributions in that program are assumed to be uniform with the obtained minimum and maximum values as endpoints for the interval.

The calculations were performed for some actinide salts deemed important as solubility limiting phases, and the numerical results are shown in Table 4. However, it must be noted that this paper will not stress the subject of which phase is the solubility limiting one, but only indicate the uncertainties associated with this type of solubility calculation.

Table 4, Minimum and maximum solubility together with 95%
confidence interval for the mean concentrations.

| Solid phase | Sampled case | | | MINVAR case | |
	min ; max solubility	dist	confidence interval (95%)	min ; max solubility	dist
$Pu(OH)_2CO_3$	2.615E-07 ; 2.769E-07	log norm	2.695E-07 +/- 7.59E-09	1.076E-09 ; 1.427E-09	log uniform
$Pu(OH)_4$	1.896E-06 ; 2.155E-06	log norm	2.027E-06 +/- 1.30E-07	3.456E-06 ; 4.283E-06	log uniform
Schoepite	7.490E-03 ; 9.343E-03	log norm	8.448E-03 +/- 1.20E-03	1.029E-04 ; 1.168E-04	log uniform
$UO_2(am)$	3.019E-07 ; 3.047E-07	log norm	3.031E-07 +/- 1.55E-09	3.337E-07 ; 3.343E-07	log uniform
$Am_2(CO_3)_3$	1.475E-06 ; 1.555E-06	log norm	1.516E-06 +/- 5.01E-08	3.044E-07 ; 3.287E-07	log uniform

No confidence intervals are given for the uniformly distributed cases since such an interval would not provide any relevant information. As mentioned earlier, the fit of a distribution to the sampled data points may be evaluated either by statistical tests, e.g. the chi square test, or by visual examination. In the case of a chi square test it is important to remember that such a test will reject the hypothesis that the sampled points follow the assumed distribution if the deviations lie in the "tails" of the distribution. Therefore, for practical purposes, a visual examination may suffice. The distributions given in Table 4 are the results of visual examination since the chi square test rejected similarity between sampled values and the assumed distribution in all cases. Some of the cases, including both sampled values and assumed distributions, are shown in Figures 3 and 4.

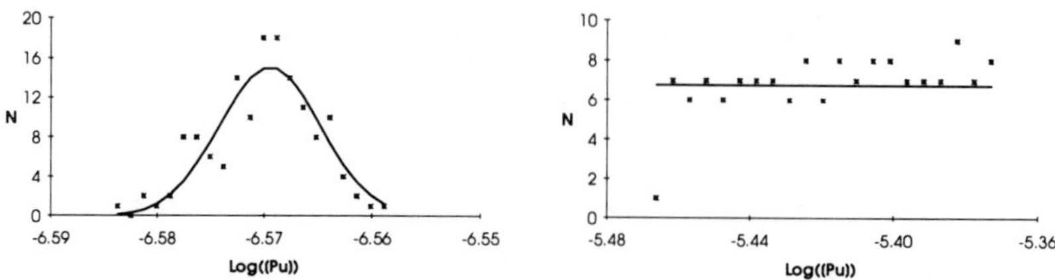

Figure 3, Density functions for the solubility of Pu(OH)$_4$, in the sampled water (right) and the MINVAR water (left); x are the sampled results and the solid line is the theoretical df.

In the case of a uniform distribution, the solid line in the above figures, represents the mean value and, in the case of a log normal distribution, the curve is calculated according to Equation 3 with mean and standard deviation obtained from the sample.

$$N = \frac{1}{\sigma\sqrt{2\pi}} e^{\frac{x-\mu}{\sigma}} \qquad\qquad (3)$$

where N is the number of samples at x, and μ and σ are the mean and standard deviation, respectively.

The reason for uniformity in the MINVAR cases is probably the fact that only a few elements are given with an uncertainty interval and the others are fixed. According to the central limit theorem a sum of uniformly distributed variables will be normally distributed. Therefore, if there are only a few variables with a uniform distribution, their distribution will be reflected in the result. It is also noticable that the solubility for the same solid phase is different in the two cases. This is due to the fact that the concentrations for some elements differ by more than one order of magnitude in the two cases. Such uncertainty can not be assigned to either determination errors or simulation errors, but must rather be concidered as solubility calculations for two different waters.

At this point it is important to remember that the uncertainties in solubility calculations arising from uncertainties in thermodynamical data is much larger than the ones presented here. In Table 5 these uncertainties for some solid phases are listed. The calculations was made by first determing with constants were important, by a sensitivity analysis, and then these constants were given an uncertainty interval of one log unit, e.g. the logatithm of the stability constant for $PuCO_3^{2+}$ was varied within the interval [18.6;19.6].

Table 5, some uncertainty results originating from uncertainties in thermodynamical data [6]

Solid Phase	Mean solubility (M)[a]	Confidence interval for the mean (95%)[a]	Min;Max solubility (M)
$ThO_2(c)$	2.57E-15	2.57E-15±1.14E-16	9.38E-16 ; 7.12E-15
$UO_3\cdot2H_2O$	4.15E-05	4.15E-05±7.24E-06	4.54E-06 ; 9.65E-04
$UO_2(c)$	3.85E-12	3.85E-12±2.97E-13	6.90E-13 ; 2.10E-11
$PuO_2(c)$	1.72E-08	1.72E-08±3.54E-10	9.56E-09 ; 3.27E-08

a The values are calculated on logarithmic scale and then converted to linearity.

CONCLUSION

In the case of a failed repository the solubility of the released radionuclides is important in the calculation of the migration speed. It is thus interesting to know whether calculated solubilities are reliable, with respect to uncertainties in input data.

Modelling such phenomena as water-rock interactions is in many respects a difficult task, not only in the building of the computer models, but in the search for acceptable input data as well. If a model is deemed to mirror reality reasonably well, it is of great interest to try to calculate the effect of uncertainties in input data. The present paper describes such an approach, in which the CRACKER code is used together with experimental data from the Äspö site in Sweden. The results show that the mineral uncertainties used resulted in approximately ten percent uncertainty in the calculated element concentrations. One should realise, however, that the low variability is partly a result of the method selected for simulating the water sampling process. This process is similar to the sampling of water from a borehole, which causes waters from different parts of fractures to be mixed and thus to cancel out local variations in the composition.

It is also obvious that the rock (determined from drilling cores) used in the simulations does not represent a rock which is in steady state with the sampled water. In the present work, the variability in the rock composition was considered genereous. One should realise, however, that the local variability might be much greater than assumed in the present simulations.

It should not be taken for granted that the small variabilities found in the examples here will be valid for other cases in which the behaviour of a future repository is simulated, i.e. other water and rock composition. As is discussed elsewhere [6][7][15], the uncertainties in calculated solubilities are heavily influenced by factors such as uncertainties in thermodynamical data and the approaches that have been used to calculate the solubilities. Compared with these, the effect of uncertainties in water composition, whether they originate from experimental errors or are results of a modelling attempt, are negligible.

ACKNOWLEGEMENT

The authors gratefully acknowlege Susanne Börjesson for her help during the writing of this paper. This work was funded by the Swedish Nuclear Power Inspectorate (SKI).

REFERENCES

1. Helton, J. C., Uncertainty and sensitivity analysis techniques for use in performance assessment for radioactive waste disposal, Reliability Engineering and System Safety, 42, p. 327-367, 1993.

2. Iman, R. L., Helton, J. C., Comparison of Uncertainty and Sensitivity Analysis for Computer Models., Rep. NUREG/CR-3904, Sandia National Laboratories, Albuquerque, NM, 1985.

3. Kleijnen, J. P. C., Sensitivity Analysis, Uncertainty Analysis, and Validation: a Survey of Statistical Techniques and Case Studies. Proceedings of the Symposium on Theory and Applications of Sensitivity Analysis of Model Output in Computer Simulations (SAMO), Belgirate, Italy, September, 1995.

4. Emrén, A. T. , CRACKER - A program coupling chemistry and transport Version 92-11, SKI TR 93:6, The Swedish Nuclear Power Inspectorate (SKI), S-106 58 Stockholm, Sweden, 1992.

5. Parkhurst D. L., Thorstenson D. C., Plummer N. L., PHREEQE - A computer program for geochemical calculations., Water-Resources Investigations 80-96, U. S. Geological Survey, 1990.

6. Ekberg, C., Skarnemark, G., Emrén, A. T. and Lundén, I., Uncertainty and sensitivity analysis of solubility calculations at elevated temperatures, Scientific Basis for Nuclear Waste Management XIX, Materials Research Society Symposium proceedings, 1995.

7. Ekberg, C., Lundén, I, Uncertainty analysis for some actinides under groundwater conditions.,to be published in the Journal of Statistical Computation and Simulation, 1996.

8. Ekberg, C., Emrén, A. T. and Samuelsson A., <u>Tools for Uncertainty Analyses of Solubility Calculations in Geological Systems</u>, In preparation, 1997.

9. Smellie, J.,Laaksoharju, M., <u>The Äspö Hard Rock Laboratory: final evaluation of the hydrochemical pre-investigations in relation to existing geologic and hydraulic conditions</u>, Swedish Nuclear Fuel and Waste Management co. (SKB), TR 92-31 Stockholm, 1992.

10. Nilsson, A-C., <u>Groundwater chemistry monitoring at Äspö during 1991</u>, Swedish Nuclear Fuel and Waste Management Co. (SKB), progress report, 1992.

11. Tullborg, E-L., personal communication, TERRALOGICA AB, Gråbo, 1995.

12. C. Ekberg, <u>SENVAR: a Code for Handling Chemical Uncertainties in Solubility Calculations</u>, SKI report 95:18, Swedish Nuclear Power Inspectorate (SKI), Stockholm, 1995.

13. Samuelsson, A., Diploma thesis, Department of Nuclear Chemistry, Chalmers University of Technology, S 41296, Göteborg, Sweden, in preparation.

14. Nilsson, A-C., Personal communication, Dept. Inorganic Chemisty, Royal Technical Institute, 1995.

15. Arthur R., Emrén A.T., Glynn P. and McMurry J., <u>The modeler's influence on results for calculated Plutonium(IV) Hydroxide Solubility in Hypothetical Environments at the Äspö Hard-Rock Laboratory</u>, SKI Report in preparation, SKI, Stockholm, 1996.

COMPLEXATION OF TRANSURANIC IONS BY HUMIC SUBSTANCES: APPLICATION OF LABORATORY RESULTS TO THE NATURAL SYSTEM

KEN CZERWINSKI and JAE-IL KIM*
Nuclear Engineering Department, Massachusetts Institute of Technology, Bldg. 24-210, 77 Massachusetts Ave., Cambridge, MA 02139-4308, USA
*Forschungszentrum Karlsruhe, Institut fuer Entsorgungstechnik, 76021 Karlsruhe, Germany

ABSTRACT

Environmental investigations show transuranic ions sorb to humic substances. The resulting species are often mobile and are expected to be important vectors in the migration of transuranic ions in natural systems. However, theses environmental studies yield no quantitative data useful for modeling. Laboratory complexation experiments with transuranic ions and humic substances generate thermodynamic data required for complexation modeling. The data presented in this work are based on the metal ion charge neutralization model, which is briefly described. When a consistent complexation model is used, similar results are obtained from different experimental conditions, techniques, and laboratories. Trivalent transuranic ions (Cm(III), Am(III)) have been extensively studied with respect to pH, ionic strength, origin of humic acid, and mixed species formation. The complexation of Np(V) has been examined over a large pH and metal ion concentration range with different humic acids. Some data does exist on the complexation of plutonium with humic acid, however further work is needed. Calculations on the Gorleben aquifer system using the thermodynamic data are presented. Critical information lacking from the thermodynamic database is identified.

INTRODUCTION

Predicting the environmental behavior of transuranic ions in aquifer systems is important for the safety assurance of proposed high level nuclear waste repositories and assessing contaminated groundwaters. To model the behavior of transuranic ions, reliable thermodynamic data to describe the complexation with environmentally relevant ligands is required. A critical component of groundwater systems is humic substances; mainly humic and fulvic acids. The acids are polyelectrolytes with high complexation affinities [1-4]. In the Gorleben aquifer systems, humic substances vary in concentration from 0.03 mg/L to 100 mg/L [5] with a proton exchange capacity (PEC) between 1.2 to 11.2 meq/g [6]. These exchanging groups are primarily carboxylic and phenolic acids [7]. The ability of humic acids to complex metal ions, even in the presence of competing natural inorganic ligands, is well documented [8 -11]. This paper discusses the complexation of transuranic ions by humic acid substances. Environmental observations are presented. The metal ion charge neutralization model is briefly described and applied to the Gorleben aquifer system. The utility of thermodynamic data from laboratory experiments for describing environmentally relevant transuranic species is examined.

ENVIRONMENTAL OBSERVATIONS

In the environment, humic substances can complex Pu [12-,20], trivalent transuranic ions (Am(III) and Cm(III)) [12-14,16,18,21,22] and Np [12,18] in soil and aquatic systems. In a few cases the resulting species are at times mobile and represent an important migration vector [12,16,18]. Interaction of transuranic ions with the biosphere seems to occur through humate complexes. Uptake of transuranic ions bound to humic substances by plants has been observed [18].

From the environmental observations a number of complexation trends can be discerned. Some studies indicate the actinide ions are primarily complexed by smaller fractions of humic substance in the 1000-3000 molecular weight range [12,14]. However, this should not be taken as complexation only occurs through smaller fractions. Differences between the complexation of Pu and the trivalent actinide ions tend to be small, with Pu bound in a slightly higher percentage at times [12-14,16]. Mixed complexes have been seen. In the environment, ternary systems of Pu,

Mat. Res. Soc. Symp. Proc. Vol. 465 © 1997 Materials Research Society

humic substances, and inorganic colloids can exist [17,20]. Generally, less is known about the behavior of Np. Observations show the ratio of Pu to Np bound to humic substance can vary [12,18]. This fact suggests the environmental conditions effect the actinide ion oxidation state, which in turn changes the complexation behavior.

LABORATORY EXPERIMENTS

Complexation Model

The modeling approach used in this work is founded on the idea of metal ion charge neutralization upon complexation to humic acid functional groups [23]. The complexation constants are derived by introducing the operational humic acid concentration and loading capacity. The loading capacity is the mole fraction of the maximum available complexing sites of humic acid and changes with ionic strength, pH, origin of humic acid and metal ion. The resulting complexation constants are invariant with respect to experimental conditions.

In a metal ion charge neutralization process the metal ion occupies the number of proton exchange sites equal to its charge (z^+). A group of complexing sites needed to neutralize a metal ion is considered as a one humic acid complexation unit. The reaction is described as:

$$M^{z+} + HA(z) <\!-\!> MHA(z) \tag{1}$$

Even though humic acid is heterogeneous in structure a metal cation of higher charge can attract anionic sites for charge neutralization. Based on Eq. (1), the complexation constant is:

$$\beta = \frac{[MHA(z)]}{[M^{z+}]_f [HA(z)]_f} \tag{2}$$

where $[MHA(z)]$ is the concentration of metal ion humate complex, $[M^{z+}]_f$ the free metal ion concentration, and $[HA(z)]_f$ the free humic acid concentration. All concentrations under consideration are defined in mol/L. The terms $[MHA(z)]$ and $[M^{z+}]_f$ are quantified by various speciation methods. For the metal ion charge neutralization model, $[HA(z)]_f$ is calculated by inclusion of a loading capacity determined under given experimental conditions. The value for $[HA(z)]_f$ is different from $[HA(z)]_t$ introduced in a given experiment. Both quantities are defined below.

Operational humic acid concentration

The molecular weight of humic acid is difficult to define due to its irregular and heterogeneous nature. However, using the known proton exchange capacity (PEC), the operational concentration of humic acid is defined as:

$$[HA(z)]_t = \frac{(HA)(PEC)}{z} \tag{3}$$

where (HA) is the concentration of humic acid in g/L, PEC in eq/g and z the charge of the complexing metal ion. The PEC is found through pH titration of a known quantity of humic acid from pH 3.5 to pH 10.5. From Eq. (3), $[HA(z)]_t$ can be determined in mol/L and is equivalent to the concentration of a metal ion with charge z that can be neutralized by a given amount of humic acid. For complexation to trivalent actinide ions, Pu(IV), Np(V), the humic acid is written as HA(III), HA(IV), and HA(I) respectively. Mixed species can also be considered. In this case, z is taken as the overall charge of the actinide ion and the secondary complexing ligand. As an example, in the case of $CmOH^{2+}$ complexation, humic acid is rewritten as HA(II).

<u>Loading capacity and free humic acid concentration</u>

Humate complexation studies are often performed in conditions where the metal ion does not undergo overlapping reactions such as hydrolysis or carbonate complexation. These conditions are reached by selecting a suitable pH range; pH $\leq$ 6 for M^{3+} [8,24-252627-28,29,30] pH $\leq$ 4 for $MO2^{2+}$ [30-313233], and pH $\leq$ 8 for $MO2^{+}$ [30,34,35]. In the pH range of each metal ion, the total proton exchanging sites concentration, as defined by $[HA(z)]_t$, is not fully available for complexation. At a given pH in the acidic or near neutral range, the proton exchange sites are partially dissociated. Steric hindrance within cross linkages of the humic acid restricts accessible functional sites for a given metal ion. Therefore, the complexation strength of each metal ion towards humic acid differs.

As a consequence, the loading of a metal ion onto the proton exchange sites of humic acid is dependent on the pH, ionic strength, charge of the metal ion and the origin of humic acid [7,23-29]. A quantity, defined as a loading capacity (LC) of humic acid for a particular metal ion, can be expressed as:

$$LC = \frac{[MHA(z)]_m}{[HA(z)]_t} \tag{4}$$

where $[MHA(z)]_m$ represents the maximal M^{z+} concentration for complexation with functional sites of a given humic acid. The value for $[MHA(z)]_m$ varies under the experimental conditions. From LC, $[HA(z)]_f$ in Eq. (2) can be calculated as:

$$[HA(z)]_f = [HA(z)]_t LC - [MHA(z)] \tag{5}$$

A combination of Eqs. (2) and (5) results in:

$$\beta = \frac{[MHA(z)]}{[M^{z+}]_f \left([HA(z)]_t LC - [MHA(z)]\right)} \tag{6}$$

The values for $[MHA(z)]$ and $[M^{z+}]$ are determined by speciation experiments, and $[HA(z)]_t$ is given for the experiment. Loading capacity is determined by introducing a factor F

$$F = \frac{[M^{z+}]_f [HA(z)]_t}{[MHA(z)]} \tag{7}$$

A combination of Eqs. (6) and (7) gives:

$$[M^{z+}]_f = LC \times F - \frac{1}{\beta} \tag{8}$$

Plotting $[M^{z+}]_f$ against F at constant conditions gives a linear correlation with the slope corresponding to the LC value. The consequence of LC is seen by Fig. 1. The value of $[HA(z)]_t$ is defined in mol/L by Eq. (3), $[HA(z)]_t LC$ is the effective concentration of humic acid available for complexation. The $[HA(z)]_f$ value is found from Eq. (5). The humic acid amount depicted by $[HA(z)]_t(1-LC)$ is unavailable for complexation under the given experimental conditions.

The degree of ionization (α) is measured by pH titration of humic acid in the absence of complexing metal ions and varies with pH at constant ionic strength for a given humic acid. The degree of ionization at a certain pH can vary amongst different humic acids. Due to steric hindrance from the cross linkage of humic acid and charge repulsion within polyelectrolyte structures, LC is always smaller than α at the same pH and ionic strength as seen from current experimental data. Both values tend towards 1 at higher pH where humic acid is structurally unimpeded by the dissociation of all the proton exchanging sites.

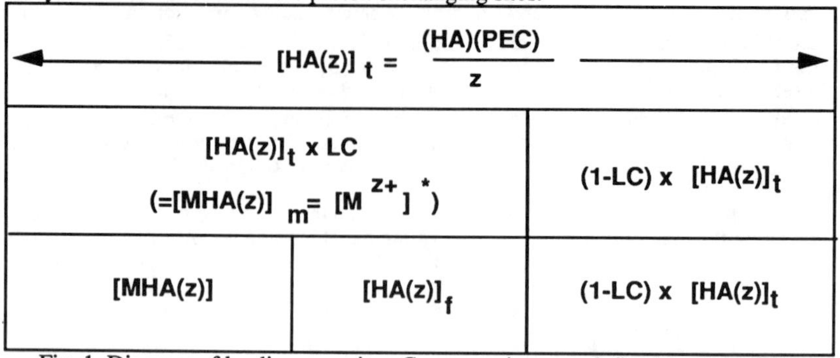

<table>
<tr><td colspan="3" align="center">← $[HA(z)]_t = \dfrac{(HA)(PEC)}{z}$ →</td></tr>
<tr><td colspan="2" align="center">$[HA(z)]_t \times LC$

$(=[MHA(z)]_m = [M^{z+}]^{*})$</td><td align="center">$(1\text{-}LC) \times [HA(z)]_t$</td></tr>
<tr><td align="center">$[MHA(z)]$</td><td align="center">$[HA(z)]_f$</td><td align="center">$(1\text{-}LC) \times [HA(z)]_t$</td></tr>
</table>

Fig. 1 Diagram of loading capacity. Concentrations are defined by Eqs. (3-5).

<u>Complexation studies</u>

As mentioned earlier, most experiments on the complexation of transuranic ions with humic acid are conducted in conditions which exclude carbonate complexation and hydrolysis. An expansive range of experimental methods has been used; UV-Visible spectroscopy [24-28,32,35] time resolved laser fluorescence spectroscopy (TRLFS) [26,28,32,36,37], laser induced photoacoustic spectroscopy [27], ultrafiltration [27,28,32,37], and ion exchange [37]. Of the transuranic ions, trivalent Cm and Am have been extensively examined with respect to pH, metal ion concentration, humic acids of different origin, ionic strength, and formation of mixed complexes [24,26-28,32,38-39,40]41424344]. At 0.1 M ionic strength, the resulting complexation constants of Cm(III) with Gohy-573 humic acid are given below [44]. In addition to the complexation of the aquo ion, mixed hydroxide and carbonate species are also considered.

Species	$\log\beta$
CmHA(III)	6.23 ± 0.11
CmOHHA(II)	12.9 ± 0.2
Cm(OH)$_2$HA(I)	16.8 ± 0.3
CmCO$_3$HA(I)	12.3 ± 0.3

The complexation of Np(V) has been examined over a large pH range with different humic acids [35,45,46]. A pH dependent reduction of Np(V) to Np(IV) is observed. The evaluated complexation constant with Gohy-573 humic acid at 0.1 M ionic strength is [35]:

$$\log \beta = 3.66 \pm 0.02 \text{ for } NpO_2HA(I)$$

The interaction of plutonium with humic acid has been studied [47-,48,49], but further work is required to yield useful thermodynamic data, especially for the tetravalent oxidation state. Reduction of penta- and hexavalent plutonium to the tetravalent oxidation state is mediated by humic acid.

CALCULATIONS

The above complexation constants are used to calculate the amount of transuranic metal ions bound to humic substances from different Gorleben groundwaters. The Gorleben site is a proposed high level nuclear waste repository. Physical and chemical parameters of selected Gorleben groundwaters are shown in Table I. The deep groundwaters are reducing and the assumed oxidation states of the transuranic ions are Np(IV), Pu(IV), and Cm(III) (Figure 2).

Table I. Physical and chemical parameters of some Gorleben groundwaters

Groundwater	Depth (m)	Ionic Strength (mol/L)	Eh (mV)	pH	[HA(I)] (µmol/L)	[CO_3^{2-}] (µmol/L)	logP$_{CO2}$ (atm)
Gohy-532	65-68	0.009	26±60	8.8	268	175	-3.76
Gohy-573	134-137	0.02	11±30	7.9	989	35.2	-2.65
Gohy-1271	80-82	0.03	75±63	8.3	433	76.1	-3.12
Gohy-2211	83-85	0.05	35±4	8.3	979	104	-2.98
Gohy-2227	128-130	0.04	25±70	7.8	742	28.0	-2.55

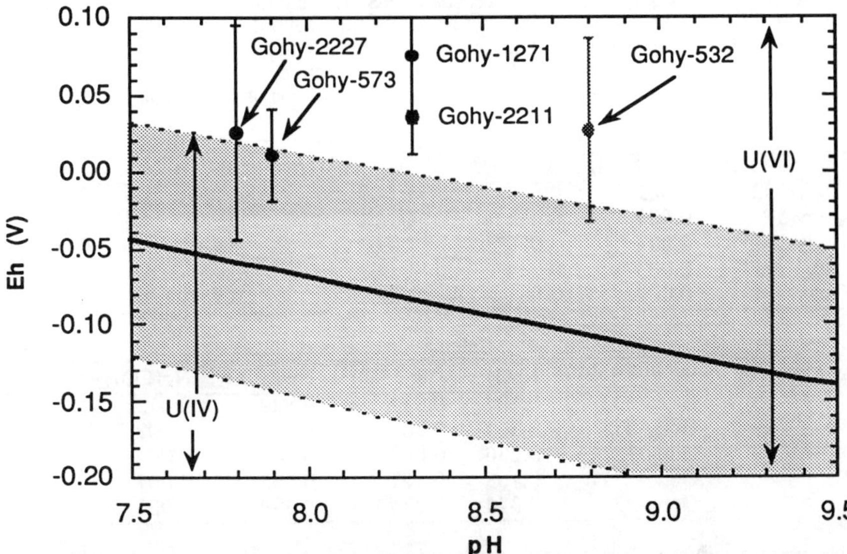

Figure 2. Eh-pH diagram in the region of selected Gorleben groundwaters. The shaded area represents the error range of the redox potential for U(VI)/U(IV) [50]. The oxidation states of the transuranic ions are Np(IV), Pu(IV), Am(III), and Cm(III).

Since data on the tetravalent actinides is lacking, calculations will be done for Np(V). The calculations for Np(V) are only an exercise and are not intended as an estimation of Np(IV) behavior. For the calculations, the transuranic metal ion concentration is taken as 1 nmol/L. Using the available data for humate, hydroxide, and carbonate complexation, calculations can be performed on Np(V) and the trivalent transuranic ions. The hydrolysis and carbonate constants for Np(V) [34,51],Cm(III) [52], and Am(III) [9] are given below.

Species	logß	Species	logß
NpO_2OH	2.44	$CmOH^{2+}$	6.67
$NpO_2(OH)_2^-$	4.10	$Cm(OH)_2^+$	12.06
$NpO_2CO_3^-$	4.58	$AmCO_3^+$	6.33
$NpO_2(CO_3)_2^{3-}$	6.60	$Am(CO_3)_2^-$	9.54
$NpO_2(CO_3)_3^{5-}$	7.09		

For considering the humate complexation, the calculations utilize the metal ion charge neutralization model and include the loading capacity. In the pH region of the groundwaters, the loading capacity for Cm(III) is taken as unity while the value for Np(V) varies with the pH according to:

$$LC = 0.0761pH - 0.415 \tag{9}$$

The results from the calculations are given for trivalent actinides (Table II) and Np(V) (Table III). Only those species with a mole fraction greater than 0.001 are included. For the trivalent actinides, the dominant form is the mixed humate carbonate. The results for the trivalent transuranic ions can be compared to field measurements. Examination of Gohy-2227 groundwater shows trivalent metal ions (REE, Cr, Fe) are sorbed to humic colloids with a relative concentration of 0.873±0.097 [53], which is in good agreement with the calculations. With exclusion of mixed species, one would calculate carbonate alone to be the important complexing ligand, resulting in a flawed supposition when compared to environmental observations of suitable homologs.

Table II. Calculation results for trivalent transuranic ions

Groundwater	AnHA(III)	AnOHHA(II)	An(OH)$_2$HA(I)	AnCO$_3$HA(I)	AnCO$_3$
Gohy-532	0.001	0.078	0.009	0.905	0.004
Gohy-573	0.008	0.051	0.001	0.940	0.001
Gohy-1271	0.003	0.059	0.002	0.933	0.002
Gohy-2211	0.003	0.044	0.002	0.951	0.001
Gohy-2227	0.009	0.051	0.001	0.938	0.001

Table III. Calculation results for Np(V)

Groundwater	NpO$_2^+{}_f$	NpO$_2$HA(I)	NpO$_2$CO$_3^-$	Np(CO$_3$)$_2^{3-}$
Gohy-532	0.124	0.039	0.822	0.015
Gohy-573	0.314	0.264	0.420	0.002
Gohy-1271	0.230	0.098	0.666	0.005
Gohy-2211	0.168	0.162	0.663	0.007
Gohy-2227	0.374	0.226	0.399	0.001

Neptunium carbonate is calculated to be dominate under the groundwater conditions. However, the free and humic bound species are also calculated to be present. The amount of each species is primarily effected by the CO_2 partial pressure. In the examined groundwaters with the lowest CO_2 partial pressure (Gohy-2227 and Gohy-573), the amounts of free NpO_2^+, $NpCO_3$, and $NpO_2HA(I)$ are roughly equal. Reduction of Np(V) to the tetravalent oxidation state should profoundly increase the amount of Np bound to humic acid.

Using the literature plutonium carbonate [54] and hydrolysis [55] constants, estimates for the plutonium humate complex can be made. In Gohy-2227 groundwater, the relative concentration of tetravalent metal ions (Zr, Hf, Th) sorbed on humic colloids is 0.887±0.10 [53]. The tetravalent metal ions are suitable plutonium homologs in this instance. From the literature constants, $Pu(OH)_4$ is expected to be the predominant species at the groundwater pH in the absence of humate complexation. Plutonium could be sorbed on humic substance as $Pu(OH)_4$, or as a mixed species $Pu(OH)_3HA(I)$. Assuming the mixed plutonium species $Pu(OH)_3HA(I)$ would be dominate with a relative concentration of 0.90, an estimate of the stability constant is:

$$log\beta = 56.4 \text{ for } Pu(OH)_3HA(I)$$

Further examinations on the complexation of plutonium, especially the tetravalent oxidation state, are required for evaluating relevant stability constants.

CONCLUSIONS

Humic substances have been shown to complex transuranic species in the environment, act as transport vectors, and need to be considered as important complexing agents. The metal ion charge neutralization model thermodynamically describes the complexation of actinide ions to humic acid. The resulting stability constants are invariant under different experimental conditions, can be used to explain the formation of mixed constants, and are useful in geochemical calculations. For the transuranic ions, there exists data for Cm(III), Am(III), and Np(V), mostly in low ionic strength solutions. Good agreement is found for trivalent ions between field observations and calculations using laboratory results. Additional field studies should be performed when possible to supplement and improve our knowledge base. Further laboratory experiments are needed to evaluate stability constants for tetravalent ions. Additionally, more data is required for higher ionic strength conditions, near neutral and higher pH, and mixed complexes.

REFERENCES

1. Carlsen, L.: The Role of Organics in the Migration of Radionuclides in the Geosphere. CEC Report EUR 12024 (1989).
2. Nash, K.L and Choppin, G.R. Inorg. Nucl. Chem. **42**, 1045 (1980).
3. Ibarra, V., Osacar, T., and Gavilan, M. An. Quim. **77**, 224 (1981).
4. Li, W.C., Victor, D.M., and Chakrabarti, C.L.: Anal. Chem. **52**, 520 (1980)
5. Kim, J.I., and Buckau, G: Huminstoffuntersuchungen an Gorleben Grundwässern, RCM 01590, TU München (1990).
6. Schnitzer, M.: Recent Findings on the Characterization of Humic Substances Extracted from Soils from Widely Differing Climatic Zones. in: Proc. Symposium on Soil Organic Matter Studies, International Atomic Energy Agency, Vienna, 117 (1977).
7. Kim, J.I., Buckau, G., Klenze, R., Rhee, D.S., and Wimmer, H.: Characterization and Complexation of Humic Acid. RCM 01090, TU München (1990).
8. Stadler, S. and Kim, J.I. Radiochim. Acta **44/45**, 39 (1988).
9. Meinrath, G., and Kim, J.I. Radiochim. Acta **52/53**, 29 (1991).
10. Kim, J.I., Buckau, G., and Klenze, R.: Natural Colloids and Generation of Actinide Pseudocolloids in Groundwater, in: Natural Analogues in Radioactive Waste Disposal. (B. Come and N. Chapman eds.), Graham + Trottman, London (1987).
11. Kim, J.I., Buckau, G., Baumgärtner, F., Moon, Ch., and Lux, D. Mat. Res. Soc. Symp. Proc. **26**, 31 (1984).
12. Cooper, E.L., Hass, M.K., and Mattie, J.F. Appl. Radiat. Isot. **46**, 1159 (1995).
13. Bunzl, K., Flessa, H., Kracke, W., and Schimmack, W. Environ. Sci. Technol. **29**, 2513 (1995).
14. Agapkina, G.I., Tikhomirov, F.A., Shcheglov, A.I., Kracke, W., and Bunzl, K. J. Environ. Radioactivity **29**, 257 (1995).
15. Graham, M.C., MacKenzie, A.B., McDonald, P., and Cook, G.T.: Actinide Interaction with Humic Substances. 5th International Conference on the Chemistry and Migration Behaviour of Actinides and Fission Products in the Geosphere. St. Malo, France 1995.
16. Marley, N.A., Gaffney, J.S., Orlandini, K.A., and Cunningham, M.M. Sci. Technol. **27**, 2456 (1993).
17. Goryachenkova, T.A., Pavlotskayas, F.I., and Myasoedov, B.F. J. Radioanal. Nucl. Chem. Art. **147**, 153 (1991).
18. Pavlotskayas, F.I., Goryachenkova, T.A., and Myasoedov, B.F. J. Radioanal. Nucl. Chem. Art. **147**, 133 (1991).
19. Livens, F.R., Baxter, M.S., and Allen, S.E. Soil Sci. **144**, 24 (1987).
20. Shen, G.T., Sholkovitz, E.R., and Mann, D.R. Earth Plan. Sci. Lett. **64**, 437 (1983)
21. McCarthy, J.F., Sanford, W.E., and Czerwinski, K.R.: Natural Organic Matter Transports Transuranic Radionuclides in Shallow Groundwater. International Humic Substances Society Meeting, Poland 1996
22. McCarthy, J.F., Marsh, J.D., and Tipping, E.: Mobilization of Actinides from Disposal Trenches by Natural Organic Matter. 5th International Conference on the Chemistry and

MIgration Behaviour of Actinides and Fission Products in the Geosphere. St. Malo, France 1995.

23. Kim, J.I. and Czerwinski, K.R. Radiochim. Acta **73**, 5 (1996).

24. Kim, J.I., Rhee, D.S., and Buckau, G. Radiochim. Acta **52/53**, 49 (1991).

25. Moulin, V., Robouch, P., Vitorge, P., Allard, B. Inorg. Chim. Acta **140**, 303 (1987).

26. Buckau, G., Kim, J.I., Klenze, R., Rhee, D.S., and Wimmer, H. Radiochim. Acta **57**,105 (1992).

27. Kim, J.I., Buckau, G., Bryant, E., and Klenze, R. Radiochim. Acta **48**, 145 (1989).

28. Kim, J.I., Rhee, D.S., Wimmer, H. Buckau, G., and Klenze, R. Radiochim. Acta. **62**, 35 (1993).

29. Wimmer, H., Klenze, R. and Kim, J.I. Radiochim. Acta **56**, 79 (1992).

30. Choppin, G.R. Radiochim. Acta **32**, 43 (1983).

31. Edelstein, N. Bucher, J. Silva, R. Nitsche, H.: Thermodynamic Properties of Chemical Species in Nuclear Waste, Report LBL-14325, Lawrence Berkeley Laboratory, 1983.

32. Czerwinski, K.R., Kim, J.I., Rhee, D.S., and Buckau, G. Radiochim. Acta, **72**, 179 (1996).

33. Grenthe, I., Lagerman, B. Acta Chem. Scand. **45**, 122 (1991).

34. Neck, V., Kim, J.I., and Kanellakopulos, B. Radiochim. Acta **56**, 25 (1992).

35. Kim, J.I., and Sekine, T. Radiochim. Acta. **55**, 187 (1991).

36. Kim, J.I, Wimmer, H., and Klenze, R. Radiochim. Acta **54**, 35 (1991).

37. Czerwinski, K.R., Buckau, G., Scherbaum, F, and Kim, J.I. Radiochim. Acta. **65**, 111 (1994).

38. Torres, R.A., and Choppin, G.R. Radiochim. Acta **35**, 143 (1984).

39. Bertha, E., and Choppin, G.R. J. Inorg. Nucl. Chem. **40**, 655 (1978).

40. Yamamoto, M., and Sakanoue, M. J. Radiat. Res. **23**, 261 (1982).

41. Moulin, V. Robouch, P., and Vitorge, P. Inorg. Chim. Acta **140**, 303 (1987).

42. Moulin, V., Tits, J., Moulin, C., Decambox, P., Mauchien, P. and de Ruty, O. Radiochim. Acta **58/59**, 121 (1992).

43. Panak, P.: Untersuchung von Intramolekular Energietransferprozessen in Cm(III) und Tb(III)-Komplexen mit Organischen Liganden mit Hilfe der Zeit-Aufgelösten Laserfluoreszenzspektroskopie, Dissertation, Institut für Radiochemie, TU München 1996.

44. Panak, P., Klenze, R., and Kim, J.I. Radiochim. Acta **74**, 141 (1996).

45. Rao, L. and Choppin, G.R. Radiochim. Acta **69**, 87 (1995).

46. Marquardt, C., Herrmann, G., and Trautmann, N. Radiochim. Acta.73, 119 (1996).

47. Choppin, G.R. J. Radioanal. Nucl. Chem. **147**, 109 (1991).

48. Mahajan, G.R., Rao, V.K., and Natarajan, P.R. J. Radioanal. Nucl. Chem. **137**, 219 (1989).

49. Shchebetkovskii, N. and Bochkov, A.A.: Interaction of Plutonium with Natural Humic Substances. Radiokhim. **17**, 952 (1975).

50. Allard, B. Kipatsi, H., and Liljenzin, J.O. J. Inorg. Nucl. Chem. **42**, 1015 (1980).

51. Kim, J.I., Klenze, R., Neck., V., Sekine, T., and Kanellakopulos, B.: Hydrolyse, Carbonat- und Humat-Komplexierung von Np(V). Institut fuer Radiochemie der TU Muenchen, RCM 01091 (1991).

52. Wimmer, H., Klenze, R. and Kim, J.I. Radiochim. Acta **56**, 79 (1992).

53. Zeh, P., Kim, J.I., and Buckau, G.: Aquatic Colloids Composed of Humic Substances. in Binding Models Concerning Natural Organics in Performance Assessment, Proceedings, OECD NEA, pg. 81 (1995).

54. Nitsche, H. and Silva, R. Radiochim. Acta **72**, 65 (1996).

55. Baes, C.F. and Mesmer, R.E.: The Hydrolysis of Cations Wiley-Interscience, New York 1976, p.188.

SOLUBILITY OF Sn(IV) OXIDE IN DILUTE NaClO₄ SOLUTION AT AMBIENT TEMPERATURE

Takayuki AMAYA[1], Tamotsu CHIBA[2], Kazunori SUZUKI[1], Chie ODA[3], Hideki YOSHIKAWA[3], Mikazu YUI[3]
[1] JGC, Nuclear Research Center, 2205, Naritacho Oaraimachi. Ibaraki Pref., 311-13, Japan
[2] JGC, No.2 Project Div. Landmark tower 2-2-1-1, Minato mirai, Yokohama 220-81, Japan
[3] PNC, Muramatsu, Tokai-mura, Naka-gun, Ibaraki Pref., 319-11, Japan

ABSTRACT

The solubility of Sn(IV) oxide was determined in a dilute $NaClO_4$ solution with pH 2 through 12 at ambient temperature. Both oversaturation and undersaturation experiments were carried out in an inert gas glovebox where the concentration of the oxygen and carbon dioxide were less than 1 ppm. The solubility of Sn(IV) oxide was 3×10^{-8} mol/l at neutral pH, and increased at pH >7.5. Equilibrium constants of soluble reactions were calculated from the experimental data, using curve fitting method.

$$SnO_2(am) + 2H_2O \rightleftharpoons Sn(OH)_4^0 \qquad : \log[K^0_{am}] = -7.46 \quad (am: amorphous)$$
$$Sn(OH)_4^0 + H_2O \rightleftharpoons Sn(OH)_5^- + H^+ \qquad : \log[K^0_{11}] = -7.65$$
$$Sn(OH)_4^0 + 2H_2O \rightleftharpoons Sn(OH)_6^{2-} + 2H^+ \qquad : \log[K^0_{12}] = -17.31$$

The study suggests that the solubility of Sn(IV) oxide would be higher than that provisionally used in current safety assessments of HLW disposal sites.

INTRODUCTION

Solubility is an important item for study of the migration behaviour of radionuclides. As ^{126}Sn (β nuclide, fission yield = 0.06%) in high level radioactive waste (HLW) is one of the important radionuclides because of its long half life of 1×10^5 year, the migration behavior of ^{126}Sn in the geologic media should be well understood [1]. But, in the current safety assessments , the solubility values of Sn oxide varied from 10^{-11} mol/l to 10^{-5} mol/l were provisionally used [2] because only a few experimental data are available. The chemical species of Sn are changeable with pH, and the valence charge strongly controls the chemical behavior of Sn itself at a specific pH condition. Therefore, the relation between the solubility and the chemical species is an important item to understand the migration behavior of Sn.

In this study, the solubility of Sn was measured by both oversaturation tests using soluble Sn(IV) and undersaturation tests using SnO_2 crystalline solid in a dilute $NaClO_4$ solution at neutral and weakly alkaline pH ranges at ambient temperature. The dominant species of Sn in the solution were estimated from the experimental data.

EXPERIMENTAL

The experiments were carried out in an inert gas glovebox, in which the concentrations of oxygen and carbon dioxide were less than 1 ppm. The equilibrium of solubility was approached under both oversaturation and undersaturation conditions. Pure water was degassed for 12 hours before it was used in the experiment. The NaOH solution without CO_2 was prepared using sodium metal and pure water in the inert gas glovebox. All reagents used in this test were special grade or nearly same grade.

Mat. Res. Soc. Symp. Proc. Vol. 465 © 1997 Materials Research Society

<u>Oversaturation Experiment</u>

The standard solution of ^{113}Sn (LMRI) and SnCl$_4$ (Kanto chemical company) were used to prepare several solutions with specific radioactivity of 1.2×10^7 Bq/mol through 1.6×10^{10} Bq/mol, and chemical concentration of 1×10^{-4} mol/l and 1.5×10^{-2} mol/l.

The experimental conditions are shown in Table I. Sn(IV) chloride stock solution labeled by ^{113}Sn was spiked in the 50 ml of 0.1M NaClO$_4$ solution in teflon container, and then pH was adjusted by the addition of NaOH solution and finally stored at 25℃. The solution was agitated twice a week during the test period. In order to determine the equilibrium of the solubility, the preliminary solubility experiment was carried out. One month and six months were selected for the test period. After the test period, a small aliquot of the solution was filtered through 10,000 MW ultrafilter. Prior to the filtration, the filter was twice immersed in the same solution by filtration in order to avoid any adsorption on the filter. pH and Eh were measured by pH meter (Horiba B-112) and Eh meter (Horiba D-14). The Eh measurement was carried out after the electrode was set in the solution for 30 minutes. Another experiments using stable Sn were carried out at the same conditions. X-ray diffraction analysis was carried out for the precipitates using X-ray spectrometer (Rigaku geiger flex RAD-IIA).

Table I Experimental condition (1) -oversaturation test-	
Solution : 0.1M NaClO$_4$ Test period and pH	: one month pH 2-12
	: six month pH 6-12

^{113}Sn is an electron capture nuclide and does not release strong γ radiation. Therefore, ^{113}Sn can be determined by the detection of the daughter nuclide, ^{113m}In (Er=0.391 MeV), which was analyzed by the well type NaI scintillation analyzer (Aloka ARC-300). The specific radioactivity (SA) was corrected in each calculation because of the short life time of ^{113}Sn (half life 115.1 day). After the radiation equilibrium of ^{113}Sn -^{113m}In was reached, the concentration of soluble Sn was calculated by the following equation.

$$Conc(Sn)\ [mol/l] = \frac{1,000 \times Conc(^{113}Sn)}{SA}$$

where,   Conc(Sn):chemical concentration of Sn [mol/l],   Conc(^{113}Sn):radioactive concentration of ^{113}Sn [Bq/ml], SA:specific radioactivity [Bq/mol] .

<u>Undersaturation test</u>

A purchased SnO$_2$ crystalline (Rare Metallic Inc.) was purified by immersing and rinsing in pure water for one month, to remove impurity ions such as SO$_4{}^{2-}$ and Cl$^-$. The test condition was shown in Table II. 500 mg of purified SnO$_2$ solid was added in 50 ml of 0.01M NaClO$_4$ solution at a fixed pH value at 25℃. After the immersion period of one month and three months, the solutions were filtered through 10,000 MW ultrafilter and the concentration of Sn was measured by ICP-MS (YOKOKAWA PMS-200). The precipitate was also analyzed by XDA method.

Table II Experimental condition (2) -undersaturation test	
Solution : 0.01M NaClO$_4$ Test period and pH : three month pH 2-8	
Solid phase : SnO$_2$ (Crystalline) with particle size <74 μ m	

RESULTS AND DISCUSSION

<u>Immersion period</u>

At first, the adequate immersion time of the solubility was experimentally evaluated. The concentrations of soluble Sn with immersion times were shown in Fig. 1. In the case of oversaturation, the concentration at pH 10 could be constant after one month, although pH of each test solution was slightly different from one other. In the case of undersaturation, the concentration at pH 8 was constant at an immersion period of both one month and three months.

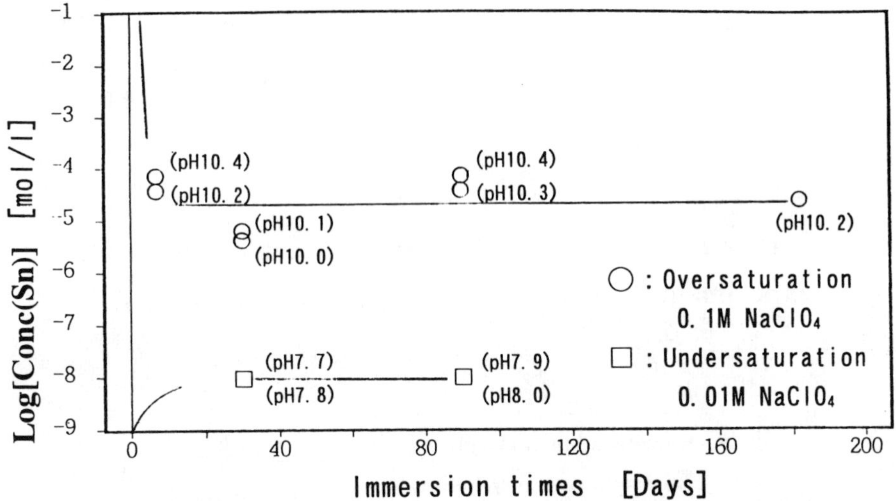

Fig.1 Cocentration of Sn with immersion times

<u>Soluble behavior of Sn</u>

The typical XRD patterns of the precipitates in the oversaturation were shown in Fig.2. It was clear that the precipitate was amorphous, because no peak appeared. The results of the solubility experiments are shown in Table III and Table IV. The concentration of Sn was measured by NaI scintillation analyzer for Table III and ICP-MS for Table IV. Fig.3 shows solubilities of both crystalline and amorphous Sn oxide with pH values. At neutral pH ranges, the solubility was different from each other. The average solubility of amorphous was about 3×10^{-8} mol/l, and it was three times higher than that of crystalline (9×10^{-9} mol/l).   This observation indicates that the different solid states contributed to the difference of the solubility.

The solubility increased at higher than pH 7.5, and the slope of the logarithm of concentration of Sn against pH values was 1 to 2 at pH 7.5 to 12, as shown in Fig.2.   This suggests the formation of Sn-complex containing OH^- ion. Therefore, the concentration of OH⁻ ion is an important parameter in the solubility of Sn. Fig. 4 shows the Eh-pH diagram of Sn [3]. The condition of Eh-pH in this study were remarked in this figure. It is indicated that the precipitate of Sn oxide corresponds to the oxide of Sn(IV) under the experimental conditions.

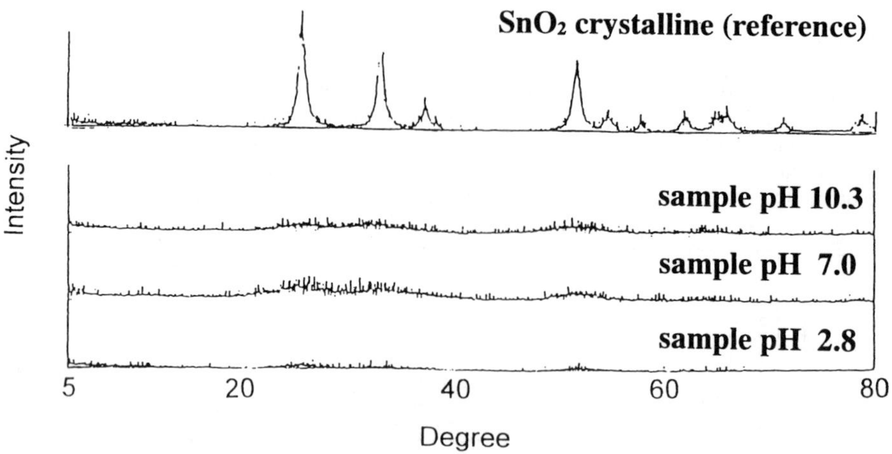

Fig.2 X-ray diffraction pattern of precipitate in oversasuration experiment for one month (0.1M NaClO₄ solution)

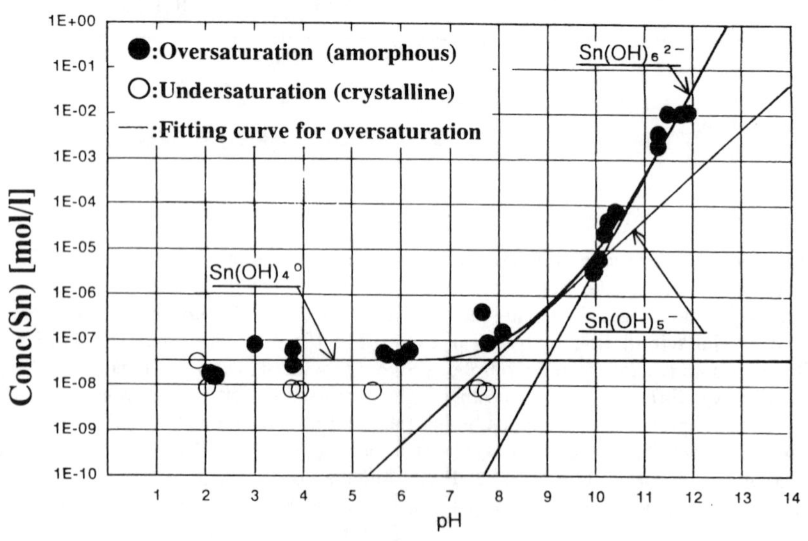

Fig.3 The concentration of Sn with pH in dilute NaClO₄ solution at 25 ℃

Table III The result of the solubility experiment of Sn (IV) - oversaturation -

No	pH	Eh	Concentration [mol/l]	period	No	pH	Eh	Concentration [mol/l]	Period
1	2.1	374	(1.8±1.3)E-8	one month	16	8.1	270	(1.36±0.02)E-7	six month
2	2.1	371	(1.5±1.3)E-8	one month	17	10.4	232	(6.90±0.04)E-5	six month
3	3.0	338	(7.9±1.3)E-8	one month	18	10.3	242	(3.60±0.02)E-5	six month
4	3.0	335	(8.4±1.3)E-8	one month	19	10.2	300	(2.35±0.02)E-5	six month
5	3.8	324	(6.2±1.3)E-8	one month	20	10.2	306	(2.52±0.02)E-5	six month
6	3.8	307	(2.9±1.3)E-8	one month	21	10.3	275	(5.10±0.04)E-5	six month
7	5.7	240	(5.6±1.2)E-8	one month	22	10.3	295	(3.18±0.03)E-5	six month
8	6.2	241	(5.9±1.3)E-8	one month	23	11.3	243	(3.71±0.05)E-3	six month
9	7.8	226	(9.1±1.3)E-8	one month	24	11.3	225	(1.94±0.03)E-3	six month
10	7.7	227	(4.7±1.3)E-7	one month	25	11.5	213	(1.02±0.01)E-2	six month
11	10.0	210	(4.1±0.2)E-6	one month	26	11.7	219	(0.91±0.01)E-2	six month
12	10.1	199	(6.1±0.3)E-6	one month	27	11.8	186	(1.02±0.01)E-2	six month
13	5.8	268	(4.7±0.2)E-8	six month	28	11.9	179	(1.10±0.02)E-2	six month
14	6.0	273	(4.1±0.1)E-8	six month					
15	8.1	266	(1.50±0.02)E-7	six month					

Initial concentration of Sn = 1E-4 mol/l (No.1 to No.22) and 1.5E-2 mol/l (No.23 to No.28)
Temp. = 25 ℃,   Solution : 0.1M·NaClO$_4$,   [Eh] = mV$_{SHE}$

Table IV The result of the solubility experiment of Sn (IV) - undersaturation -

No	pH	Eh	concentration [mol/l]	period	No	pH	Eh	concentration [mol/l]	period
1	2.0	391	9E-9	three month	5	5.5	345	7E-9	three month
2	1.9	396	3E-8	three month	6	7.8	292	8E-9	three month
3	3.8	376	9E-9	three month	7	7.6	272	1E-8	three month
4	3.9	385	9E-9	three month					

Initial concentration of Sn =500 mg/50 ml (No.1-No.8)
Temp. = 25 ℃,   Solution : 0.01M·NalCO$_4$,   [Eh] = mV$_{SHE}$

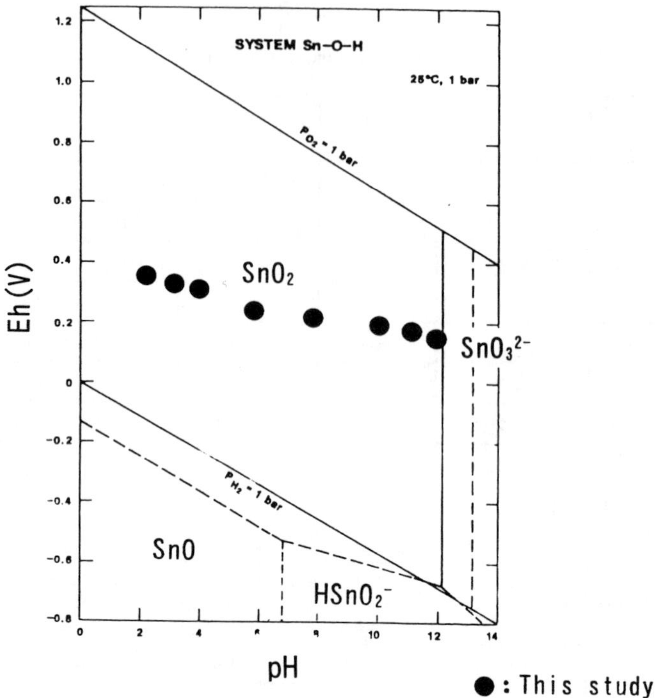

Fig.4 Eh-pH diagram of Sn[3]

<u>Chemical equilibrium of solubility</u>

Based on the experimental results, the equilibrium reactions for solubility of Sn(IV) were estimated as follows.

$$SnO_2(am) + 2H_2O = Sn(OH)_4^0 \qquad : \log[K^0{}_{am}] \qquad (1)$$

$$Sn(OH)_4^0 + H_2O = Sn(OH)_5^- + H^+ \qquad : \log[K^0{}_{11}] \qquad (2)$$

$$Sn(OH)_4^0 + 2H_2O = Sn(OH)_6^{2-} + 2H^+ \qquad : \log[K^0{}_{12}] \qquad (3)$$

where

$$K^0{}_{am} = \gamma_4[Sn(OH)_4^0]$$

$$K^0{}_{11} = \frac{\gamma_5[Sn(OH)_5^-]\,\gamma_{H^+}[H^+]}{\gamma_4[Sn(OH)_4^0]} \qquad\qquad K^0{}_{12} = \frac{\gamma_6[Sn(OH)_6^{2-}]\,\gamma_{H^+}{}^2[H^+]^2}{\gamma_4[Sn(OH)_4^0]}$$

γ_i : activity coefficient of component i,   am: amorphous

Ion activity coefficient (γ) of the $Sn(OH)_5^-$ and $Sn(OH)_6^{2-}$ was given by Davies equation (4), and γ of the $Sn(OH)_4^0$ was assumed to be 1 as the ionic strength was less than 0.1 [4].

$$\log(\gamma_i) = -0.509 z_i^2 \left[\frac{\sqrt{I}}{1+\sqrt{I}} - 0.3\, I \right] \tag{4}$$

where z_i is charge of species i, and I is ionic strength.

The total concentration of Sn, Sn_{total}, was described as follows;

$$Sn_{total} = [Sn(OH)_4^0] + [Sn(OH)_5^-] + [Sn(OH)_6^{2-}]$$

$$= \frac{K^0_{am}}{\gamma_4} + \frac{K^0_{am}K^0_{11}}{\gamma_5 \gamma_{H^+}[H^+]} + \frac{K^0_{am}K^0_{12}}{\gamma_6 \gamma_{H^+}^2[H^+]^2} \tag{5}$$

These data were analyzed by the curve fitting calculation using least square method, and the result is shown in Fig.3. The equilibrium constants of each reaction were summarized in Table V.

Table V   Equilibrium constant of the reaction   (temp.=25°C)

$SnO_2(am) + 2H_2O \rightleftharpoons Sn(OH)_4^0$	$: \log[K^0_{am}] = -7.46$
$Sn(OH)_4^0 + H_2O \rightleftharpoons Sn(OH)_5^- + H^+$	$: \log[K^0_{11}] = -7.65$
$Sn(OH)_4^0 + 2H_2O \rightleftharpoons Sn(OH)_6^{2-} + 2H^+$	$: \log[K^0_{12}] = -17.31$

Adsorption to the container

Another study done by authors indicates that the Sn adsorption on the container was negligible at pH 2 to 10 when the soluble Sn concentration was higher than 1×10^{-4} mol/l. Then, there is no influence of adsorption of Sn on the container in the oversaturation test.

Comparison with published data

Table VI shows the solubility data of Sn provisionally used in HLW safety assessments in several countries [2]. Solubilities varied 6 order of magnitude from 10^{-11} mol/l (SKI-90) to 10^{-5} mol/l (Kristallin-I). The solubility values of other studies except Kristalline-I were lower than that of this study. It should be noted that these values were probably derived from simulation code, and discrepancy in the values would be due to the different thermodynamic data for SnO_2 and/or the aqueous species of Sn [2]. Therefore it is clearly important that experimental solubility studies at each chemical conditions of disposal sites can be carried out to correctly evaluate HLW safety disposal system.

Table VI   Solubility data of Sn provisionally used in several safety reports [2]   (mol/l)

	Kristallin-I	PG'85	TVO-92	PNC/H3	SKB-91	SKI-90	This Study	
pH[a]	9	9	8~10	7~10	7~8	7~8	8	10
Sn conc	1E-5	1E-9	3E-8	2E-8	3E-8	1E-11	1E-7	1E-5

a: values evaluated by authors, based on several references.

CONCLUSIONS

The solubility of the Sn(IV) was experimentally determined at dilute $NaClO_4$ solution at pH 2 through 12 at ambient temperature. The conclusions are summarized.

(1) The soluble concentration of Sn(IV) at neutral pH is constant value of 3×10^{-8} mol/l, and increases at higher than pH 7.5.

(2) The equilibrium constants were calculated from the experimental data.

REFERENCES

1. PNC: Research and Development on geological disposal of high level radioactive waste : First progress report. PNC Tech. Rep. TN 1410 93-081, Tokyo 1992, ～pp. 4-95.
2. I. G. McKinley and D. Savage in <u>Fourth International Conference on the Chemistry and Migration Behaviour of Actinides and Fission Products</u> in the Geosphere, (Charleston SC USA, December 12-17, 1993) pp. 657-665 (1994)
3. D. G. Brookins, <u>Eh-pH Diagrams for Geochemistry</u> (Springer-Verlag, Berlin 1988)
4. V. L. Snoeyink, et al., <u>WATER CHEMISTRY</u> (1980)

PLUTONIUM SOLUBILITY AND SPECIATION TO BE APPLIED TO THE SEPARATION OF HYDROTHERMAL WASTE TREATMENT EFFLUENT

M.P. NEU, S.D. REILLY, W.H. RUNDE
Chemical Science and Technology Division, Mail Stop G739, Los Alamos National Laboratory, Los Alamos, NM 87545, mneu@lanl.gov

ABSTRACT

One of the most complex problems concerning nuclear waste management and the restoration of plutonium production sites is the treatment and disposition of mixed and TRU wastes. Hydrothermal oxidation, which has been shown to be effective in oxidizing a wide variety of organic material to CO_2, water, salts and other nonhazardous oxides, is a promising new technology for the treatment and volume reduction of actinide-containing waste. Information on the speciation and solubility of plutonium under process effluent conditions will facilitate the development of separation techniques for removing it from the treated solutions. Such a strongly oxidizing environment will generate plutonium(VI); and upon the destruction of organics, hydrothermal reactor solutions will contain carbonate. We are investigating the solubility and speciation of the plutonium(VI) carbonate system as a function of ionic strength (0.1 to 5.0 M). Formation constants for the tris- and biscarbonato complexes of plutonium(IV) were determined to be, $\log \beta_{130} = 17.7$ and $\log \beta_{120} = 13.6$, respectively, by spectrophotometry. These formation constants indicate that $PuO_2CO_{3(aq)}$ is the plutonium (VI) carbonate solution species with the largest relevant stability range. We prepared and characterized the corresponding solid using XRD, EXAFS, and diffuse reflectance, and initiated solubility experiments in 0.1, 0.2, 0.5, 1, 2, and 5 M NaCl at $22\pm1°C$ under 100% CO_2. Data collected thus far yield the solubility products, $\log K_{sp}$ mol^2/kg^2 = -12.9 (0.1 m NaCl), -12.4 (0.2 m NaCl), -12.5 (0.5 m NaCl), -12.3 (1 m NaCl), -12.2 (2 m NaCl), -12.3 (5 m NaCl).

INTRODUCTION

Operations at D.O.E. facilities have created a large legacy of combustible wastes that are contaminated with transuranic compounds, other radioactive elements, and strong oxidizers such as nitrates. Hydrothermal oxidation is one of several technologies being studied for the destruction of residues and wastes.[1] Oxidation of the organic and reduction of the nitrate components of waste will mitigate safety hazards, reduce waste volume, and facilitate separation of radioactive elements.[2] For aqueous/organic mixtures, pure organic liquids, or contaminated combustible solids such as ion exchange resins, plastic filters, cellulosic rags, etc., hydrothermal processing removes >99.999% of the organic and nitrate components.[2] In this process organic material is pressurized, then combined and mixed with pressurized hydrogen peroxide. The reaction mixture is fed into a high temperature, high pressure reactor (on the order of 450-550°C and 5000-7000 psig) and allowed to react for 20 to 60 seconds. The primary products of mixed waste destruction by this process will be carbon dioxide, water, salts, and oxides. The dissolved CO_2 will be available to form carbonato complexes with metal ions in the effluent, including Pu and RCRA metals. The process is very oxidizing and effluent metals will likely be present as aquo or complexed ions of the highest oxidation states. Understanding the speciation and kinetic behavior of actinides during and after hydrothermal treatment is crucial for the design of post-processing

Mat. Res. Soc. Symp. Proc. Vol. 465 © 1997 Materials Research Society

effluent separations, and ultimately for the deployment of this technology for nuclear waste reduction and plutonium stabilization.

Depending upon the pH, salts present, and total Pu(VI) and carbonate concentrations in the process effluent, a number of solution and solid state species may be present (Table 1).[3] We have begun investigating the structure, solubility, and stability of each of these species. Based upon the total carbon content in anticipated waste, process conditions, and experimental detection limits, we investigated Pu(VI) carbonate chemistry within ionic strengths from 0.1 to 2 m and Pu concentrations of 10^{-2} to 10^{-6} m. In order to obtain thermodynamic data which may be applied to waste isolation and actinide source term studies we expanded the ionic strength range under investigation to 5.6 m.

Table 1. Pu(VI) Carbonate Solution and Solid State Species Potentially Present in Actinide Hydrothermal Waste Processing Effluents.[*]

Pu(VI) to Carbonate ratio	1:1	1:2	1:3
Solution	$PuO_2(CO_3)$	$PuO_2(CO_3)_2^{2-}$ or $[(PuO_2)_3(CO_3)_6]^{6-}$	$PuO_2(CO_3)_3^{4-}$
Solid	$PuO_2(CO_3)$	$M_2PuO_2(CO_3)_2$, $M_6[(PuO_2)_3(CO_3)_6]$, $M'PuO_2(CO_3)_2$, or $M'_3[(PuO_2)_3(CO_3)_6]$	$M_4PuO_2(CO_3)_3$ $M'_2PuO_2(CO_3)_3$

[*]Strongly pH dependent. M = Li^+, Na^+, K^+, Cs^+, etc., M' = Mg^{2+}, Ca^{2+}, Sr^{2+}, Ba^{2+}, etc.

EXPERIMENTAL

Plutonium(VI) stock solutions were prepared by dissolving ^{239}Pu metal in 7 M $HClO_4$ and fuming aliquots of this solution to near dryness with concentrated $HClO_4$, diluting with H_2O, and determining the concentration and verifying the oxidation state purity using liquid scintillation counting (LSC) and absorbance spectrophotometry. The isotopic composition of the material was determined using alpha spectroscopy, proportional counting, and LSC.

Plutonyl carbonate, PuO_2CO_3 was prepared by bubbling CO_2 through a stirred acidic stock solution for 3 to 5 days, washing the resulting precipitate with distilled deionized water, redissolving, and repeating precipitation. Ozone was also bubbled through the suspension for the final 2 days to re-oxidize any plutonium reduced via radiolysis The resulting pale tan solid was characterized using powder X-ray diffraction (Inel, CPS-120), extended X-ray absorbance fine structure (EXAFS) spectroscopy, and diffuse reflectance spectroscopy.[4]

Spectrophotometric titrations are performed using a fiber optic spectrophotometer (Guided Wave, Model 260) and a standard glass electrochemical cell fitted with a teflon cap (Brinkman). Combination pH electrodes (Orion, Ross) used in titration and solubility experiments were filled with 3 M $NaClO_4$ or NaCl, corresponding to the sample media.

Solubility experiments were performed using approximately 50 mg of the PuO_2CO_3 and 5 mL of NaCl solution contained in loosely closed vessels with 100% CO_2 bubbling through. Solution pH were adjusted by addition of NaOH/NaCl or HCl/NaCl. The electrode was calibrated in pH units against buffers. Solution p[H] was calculated using p[H] = pH_{obs} - ΔpH, where ΔpH values are experimentally determined using analytical solutions prepared and measured for each ionic strength. Solution $[CO_3^{2-}]$ was calculated using log $[CO_3^{2-}]$ = log K^* + log P_{CO2} + 2pH, where K* is the equilibrium constant for the formation of CO_3^{2-} from CO_2 gas in equilibrium in NaCl solutions which was calculated using NONLIN.[5] Samples removed from the solubility experimental solutions for analysis were filtered using 250 nm pore syringe filters (Millipore). LSC data are obtained for 25μL aliquots of filtrate using a Packard, Tricarb 2500 instrument.

RESULTS AND DISCUSSION

We collected three sets of spectrophotometric titration data and a subset of one of these pH dependent absorption spectra is shown in Figure 1. Based upon a preliminary fit of the titration data we have calculated formation constants for the tris- and biscarbonato complexes to be, log β_{130} = 17.7 and log β_{120} = 13.6, respectively. These values are in excellent agreement with those calculated from the solubility of plutonium carbonate in 3.5 M $NaClO_4$ and S.I.T. ion interaction parameters, β_{130} = 17.4 and log β_{120} = 13.4.[6] Full details on the spectrophotometric titration experiments and data analysis will be reported elsewhere. It is noteworthy that the bis- and triscarbonato plutonyl species are much less stable than the analogous uranyl species, β_{130} = 21.8 and log β_{360} = 48.6.[7] The species distribution diagram, Figure 2, produced using these constants and that for $UO_2CO_{3(aq)}$ [7], shows the $PuO_2CO_{3(aq)}$ species has quite a large stability range; thus, our initial focus is on the solubility of the corresponding solid, PuO_2CO_3.

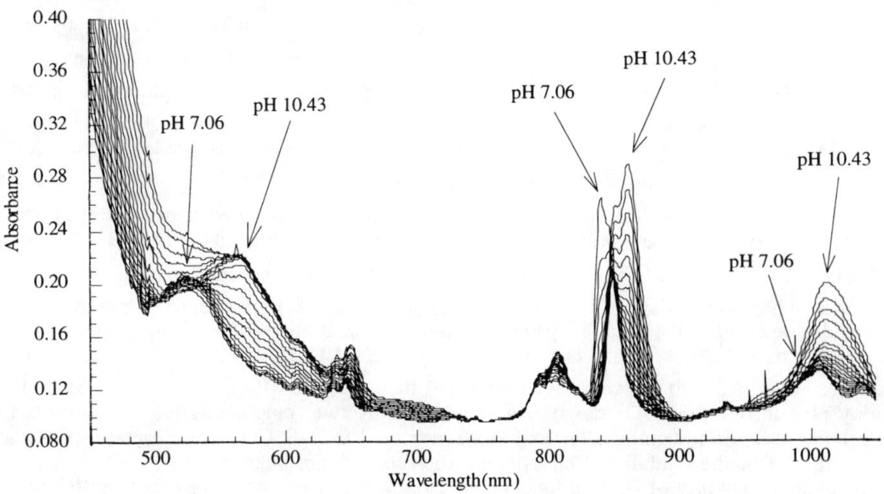

Figure 1. A subset of absorption spectra (20 of 35) used to determine Pu(VI) carbonate complex formation constants. Experimental conditions were: $[PuO_2^{2+}]$ = 4.45 mM, $[CO_3^{2-}]$ = 12.0 mM, ionic strength = 0.1 M $NaClO_4$, 25°C.

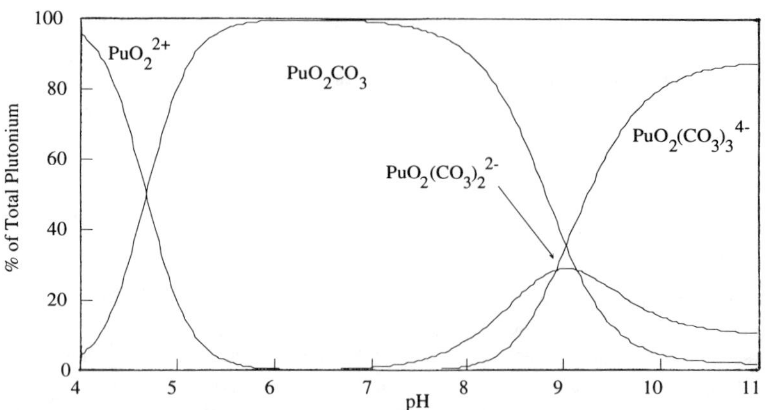

Figure 2. A species distribution of plutonyl carbonate calculated using the formation constants which we determined by spectrophotometric titrations, $\beta_{110} = 9.68$, from the literature[7], under the conditions: $[Pu]_{total} = 0.1$ mM, $[CO_3^{2-}]_{total} = 0.1$ mM, ionic strength $= 0.1$ M $NaClO_4$, 25°C.

We have prepared and characterized the plutonyl carbonate, PuO_2CO_3, and verified that it is isostructural with UO_2CO_3 based upon powder XRD and EXAFS analysis. The 2θ positions and peak intensities agree well with those reported by Navtril[8] and what we have measured for UO_2CO_3;[9] and the EXAFS spectra PuO_2CO_3 of UO_2CO_3 contain all the same features with nearly identical intensities.[4] We determined the solubility product of PuO_2CO_3 to be log $K_{sp} = -12.9$ in 0.1 m NaCl. This value may be compared with the previously reported values of log $K_{sp} = -13.98$ $\pm$ 0.12 in 0.1 $NaClO_4$,   log $K_{sp} = -13.5 \pm 0.3$ in 3.5 m $NaClO_4$[10], and 12.8 in 1 M $(NH_4)_2CO_3$.[11] To precisely determine the solubility product and associated error we must collect more data; however, it compares well with the solubility product determined for UO_2CO_3, log K_{sp} $= -13.2$ in 0.1 m NaCl[9] and the solubility product calculated for PuO_2CO_3, log $K_{sp} = -13.4$ at 0.1 m NaClO4 based upon data collected at higher ionic strength and ion interaction parameters.[10] We have also determined the solubility product at higher concentrations of NaCl and found log K_{sp} $= -12.4$ in 0.2 m NaCl, -12.5 in 0.5 m NaCl, -12.3 in 1 m NaCl, -12.2 in 2 m NaCl, -12.3 in 5 m NaCl. The value at 2.1 m compares well with that measured for UO_2CO_3, -12.5.[9] Since the solubility product for PuO_2CO_3 has been reported at only two ionic strengths, 0.1 and 3.5 m $NaClO_4$, and both differ significantly from those measured for UO_2CO_3 in the same media, the effect of chloride on the solubility of plutonium carbonates remains unknown.

The overall solubility of Pu as a function of carbonate concentration is lower than that of U (Figure 3b and reference 9). We believe that this results from a substantial difference in the solution speciation; most notably, the highly charged, trimeric uranyl species, $[(UO_2)_3(CO_3)_6]^{6-}$, is approximately 3 orders of magnitude more stable than the corresponding bis carbonate plutonyl species, $PuO_2(CO_3)_2^{2-}$. Based upon NMR, EXAFS, and the spectrophotometric and solubility studies, we conclude that either a trimeric plutonyl species, $[(PuO_2)_3(CO_3)_6]^{6-}$, does not form, or it is a very minor species.[4]

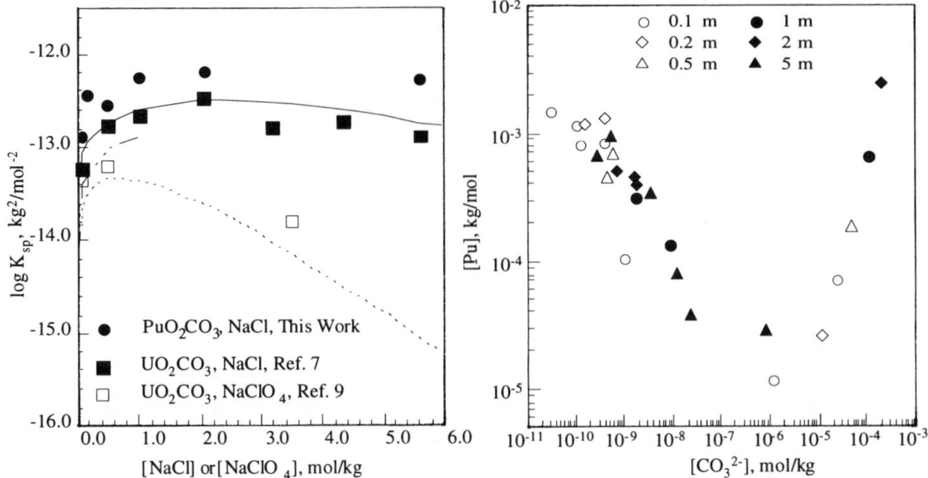

Figure 3. Solubility of PuO$_2$CO$_3$ in NaCl a. The log of the solubility product vs electrolyte concentration. Lines shown are calculated solubility products; that for plutonyl carbonate in NaClO$_4$ is from reference 6; that for uranyl carbonate in NaClO$_4$ is from reference 7; and that for for uranyl carbonate in NaCl is from reference 9; b. Overall solubility of Pu vs free carbonate concentration.

Plutonium(VI) undergoes some interesting chemistry at low NaCl concentrations. From the first day in the more dilute NaCl solubility experiments optical absorbance spectra of the filtered solution showed an absorbance band at 569 nm, indicative of Pu(V). We also noticed a color change from tan to bright green at the surface of the solid. Since the solution was maintained at pH 4 and Pu(V) does not form carbonato and hydroxo complexes at that pH (and Pu(V) carbonato and hydroxo solids are generally white or pale pink) Pu(V) solids were not responsible for the change. The solubility experiment was continued for 4 months and the new phase was allowed to age. At that time a portion of the solid was removed for XRD, diffuse reflectance, and solution spectral analysis. The powder XRD pattern matches that of the initial solid and has no new peaks, although the peaks corresponding to PuO$_2$CO$_3$ are broader and the diffraction is weaker. The diffuse reflectance spectrum of the solid is similar to that of PuO$_2$CO$_3$, but contained additional bands at 620, 690, and 738 nm (Figure 4). Addition of 1 M HClO$_4$ to a portion of the solid resulted in only partial dissolution. The material that dissolved was determined to be a Pu(VI) solid since the measured absorption spectrum of acid solution contained characteristic Pu(VI) absorption maxima at 830 and 615 nm, and no bands corresponding to other species. The fraction which did not dissolve remained bright green. Since the insoluble fraction, is bright green, not readily soluble in acid, has 3 absorbance bands in the visible between 600 and 750, and did not diffract x-rays, we conclude that the Pu(V) present in solution disproportionated to form Pu(VI) and polymeric Pu(IV) hydroxide. We note that Pu(VI) in low ionic strength carbonate solutions is reduced to Pu(V) within hours to days, while concentrated NaCl appears to stabilize the higher oxidation state, presumably via chloro complexation and the radiolytic production of hypochlorite.

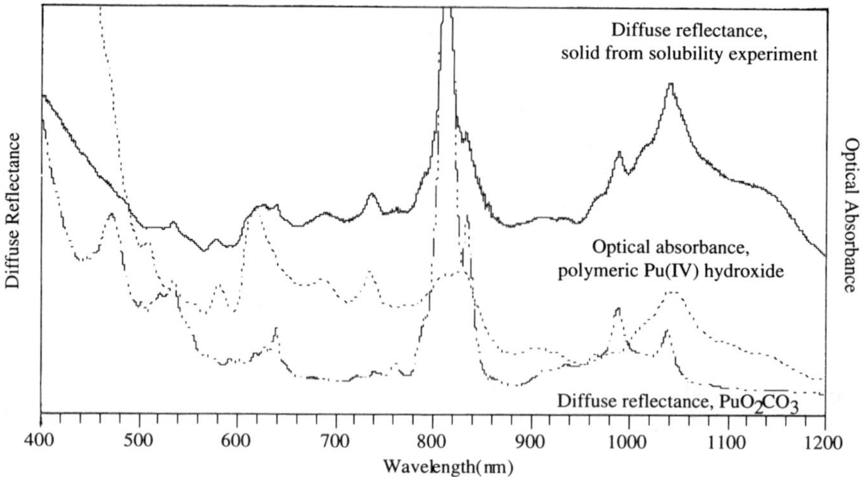

Figure 4. Optical absorbance of colloidal Pu(IV) hydroxide (dashed) and diffuse reflectance of PuO$_2$CO$_3$ (solid-dashed) and of solid from solubility experiment of PuO$_2$CO$_3$ in 0.1 m NaCl, 100% CO$_2$, maintained at pH 4 for over 4 months (solid).

ACKNOWLEDGMENTS

This work was supported by Los Alamos National Lab Laboratory Directed Research and Development and The D.O.E. Plutonium Residue Stabilization Program. We gratefully acknowledge Laura Worl, Steve Buelow, and John Berg for helpful discussions on hydrothermal processing.

REFERENCES

[1] a. M. Modell, in Standard Handbook of Hazardous Waste Treatment and Disposal edited by H.M. Freeman, (McGraw Hill, New York, 1989) pp. 8.153-8.168. b. J.W. Tester, H.R. Holgate, F.J. Armellini, P.A. Webley, W.R. Killilea, G.T. Hong, H.E. Barer, in Supercritical Water Oxidation Technology: Process Development and Fundamental Research," ACS Symp. Ser. **518** 35 (1993).

[2] J.M. Robinson, B.R. Foy, P.C. Dell'Orco, G. Anderson, F. Archuleta, J. Atencio, D. Breshears, R. Brewer, H. Eaton, R. McFarland, R. McInroy, T. Reynolds, M. Seillo, E. Wilmanns, S.J. Buelow, in Technology and Programs for Radioactive Waste Management and Environmental Restoration, Vol. 1, (Waste Management Symposia, Tucson, AZ, March 1993) p. 709-716.

[3] D.L. Clark, D.E. Hobart, M.P. Neu, Chem. Rev. **95**, 25 (1995) and references therein.

[4] Details to be published in a forthcoming paper.

[5]   NONLIN 94.05.13b developed by A.R. Felmy, Pacific Northwest National Laboratory; a chemical equilibrium program based on the Gibbs free energy minimization procedure by Harvie, *et al.* Geochim. Cosmochim. Acta **48**, 723 (1984).

[6]   P. Robouch, P. Vitorge, Inorg. Chim. Acta **140**, 239 (1987).

[7]   I. Grenthe, J. Fuger, R.J.M. Donigs, R.J. Lemire, A.B. Muller, C. Nguyen-Trung, H. Wanner, <u>Chemical Thermodynamics of Uranium</u>; Elsevier Science Publishing Company, Inc.: North-Holland, 1992; Vol.1.

[8]   J.D. Navratil, H.L. Bramlet, J. Inorg. Nucl. Chem. **35**, 157 (1973).

[9]   W.H. Runde, Ch. Lierse, B. Eichhorn, to be published in Geochim. Cosmochim. Acta.

[10]   I. Pahalidis, W. Runde, J.I. Kim, Radiochim. Acta **61**, 141 (1993).

[11]   A.D. Gel'man, A.I. Moskvin, V.P. Zaitseva, Radiochim. Acta **4**, 154 (1962).

Part XI

Radionuclide Sorption

THE USE OF LABORATORY ADSORPTION DATA AND MODELS
TO PREDICT RADIONUCLIDE RELEASES FROM A GEOLOGICAL REPOSITORY:
A BRIEF HISTORY

DONALD LANGMUIR[1], U.S. Nuclear Waste Technical Review Board
1100 Wilson Boulevard, Suite 910, Arlington, VA 22209

ABSTRACT

Radionuclide (RN) adsorption has long been recognized as important to assure the isolation of nuclear wastes in a geological repository [1]. Laboratory measured RN adsorption data have generally been expressed as distribution coefficient (K_d) values or adsorption isotherms. The proper application of these models is to site conditions nearly identical to those used in the laboratory adsorption experiments. This has required that multiple K_d's and isotherms be determined in a wide range of experiments designed to bracket expected repository conditions.

The surface complexation (SC) adsorption models were introduced in the late 1970's. The best known of these models incorporate electrical double layer (EDL) theory [2]. Their use requires that the water chemistry and surface properties of adsorbing rocks and minerals be fully characterized. Adsorption is then studied as reactions involving specific aqueous RN species (often complexes) and specific surface sites. Because the SC models are relatively mechanistic, they may allow extrapolation of adsorption results to repository conditions that lie outside the limited experimental range used to parameterize a given model. Turner [3] has shown that the diffuse layer model (the simplest SC model) fits a wide range of RN adsorption data as well as the more complex models. Others have suggested ways to generalize and estimate SC model parameters for a variety of minerals, rocks and engineered materials (cf. [4,5,6,7,8,9,10,11,12]. Degueldre and Werlni [12] and Degueldre et al. [13] have proposed a simplified SC model for RN adsorption that avoids EDL theory, in which the adsorption of RN species is estimated from linear free energy relationships.

It is appropriate to ask how accurately RN adsorption behavior must be known or understood for total system performance analysis (TSPA). In most geological settings now being considered for repository development globally, it may suffice to select bounding K_d values for the different rock types (cf. [14,15]). Use of the SC models to describe RN adsorption can provide us with increased confidence that minimum K_d's and the distribution of K_d values we might propose for TSPA are in fact conservative.

DISTRIBUTION COEFFICIENT AND ISOTHERM ADSORPTION MODELS

Early adsorption measurements involving RN's and rocks and minerals from potential repository formations were used to obtain K_d or adsorption isotherm parameters [1]. Such approaches simply curve-fit the adsorption data and provide little insight into controls on adsorption. Accordingly, they have been termed empirical. Most such measurements were done in batch tests using crushed and sized rock or mineral samples. The amount of RN adsorbed on a weight of sorbent was calculated from the total added minus that in solution

---

[1]Mailing address: 129 S. Eldridge Way, Golden, CO 80401

Mat. Res. Soc. Symp. Proc. Vol. 465 © 1997 Materials Research Society

at the end of the experiment. Only occasionally were pH or Eh controlled, or solution compositions reported in detail.

Common problems with such experiments were that the experimental solutions and sorbents often differed from those of concern, that the sorbents were improperly prepared and not preequilibrated with the solutions, and that the high concentrations of RN's used often exceeded the solubility of likely RN precipitates, making the experimental results difficult to interpret. Consequently, much of the early adsorption data and models are of qualitative value only.

The typical appearance of the adsorption isotherm for a RN is shown in Fig. 1. For RN concentrations in the linear range (if all else in the water-rock system remains constant), a K_d (linear sorption) model may be used to describe the data. At higher RN concentrations, the isotherm becomes non-linear, and Freundlich and Langmuir isotherm models may fit the data. At even higher concentrations precipitation of a RN solid may occur, RN(aq) is then fixed, and no sorption model applies. Alternatively, at high RN concentrations all available sorption sites may be occupied. The RN adsorbed is then fixed and independent of the RN(aq) concentration. For example, measurements of adsorption by zeolitic tuffs [17] indicate that Np(V) adsorption is linear for aqueous Np(V) concentrations between 1×10^{-7} and 3×10^{-5} M. U(VI) adsorption by the same solids which is slightly greater, is non-linear and is better fit with Freundlich or Langmuir isotherms in nearly the same concentration range (8×10^{-8} to 1×10^{-4} M).

An isotherm for U(VI) adsorption by hydrous ferric oxide (HFO) is shown in Fig. 2. The linear portion is the tangent to the curve at zero concentration. In this case adsorption is truly linear for U(VI) (aq) only below about 1 μg/L (4×10^{-9} M). Above 5 μg/L (2×10^{-8} M) the curve may be approximated with a second linear isotherm, but with a K_d value that is about 1/3rd as large as the first. However, the curve is better fit using a Freundlich or Langmuir isotherm equation. This shows the danger of assuming a single K_d value to describe U(VI) adsorption. In fact, as shown in Fig. 3, if pH and CO_2 pressure are allowed to vary, K_d values for U(VI) adsorption by HFO may range from near 0 to 3.4×10^6 ml/g.

In early studies, RN removal from solution by all processes was often described by an apparent distribution coefficient. For example, Taylor [19] (see [11]) report apparent K_d values for Th(IV),

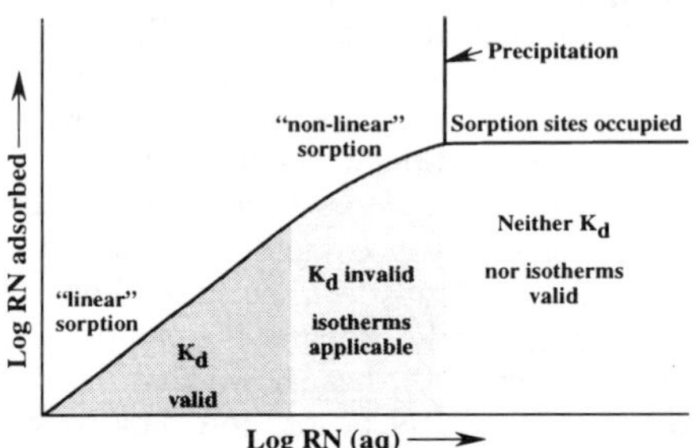

Fig. 1. Schematic plot of typical adsorption isotherm for a radionuclide (RN)[16].

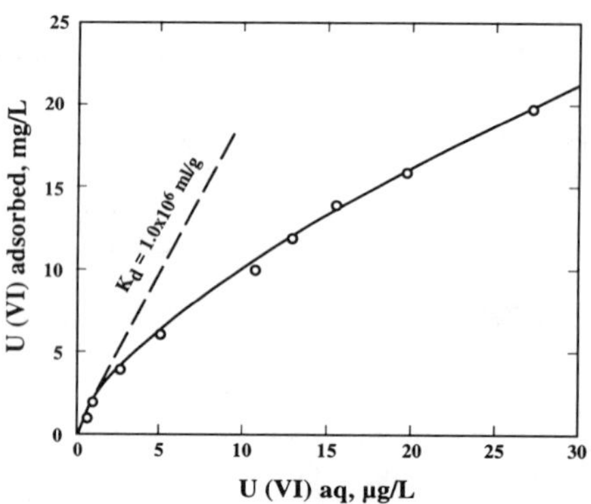

Fig. 2. Isotherm for adsorption of U(VI) by hydrous ferric oxide (HFO) at pH = 7.23, $P_{CO_2} = 0$ and 25°C. In the experiments 51.3 mg of HFO (as Fe) was suspended in 50 ml of a 0.01 M KCI solution. Added ΣU(VI) was increased from 1 to 20 mg/L [11].

Pb(II), U(VI), Pa(V) and Ra(II) computed from their loss from solution during neutralization of an acid uranium mill tailings solution with $CaCO_3$. Removal processes include adsorption, precipitation and coprecipitation in Fe(III), Al(III) and Mn(IV) oxyhydroxide and sulfate precipitates. From pH 2 to 7, apparent K_d values increase by about 10^2 times for Ra, to 10^4 times for Th, and tell us nothing about the specific removal processes involved.

SURFACE COMPLEXATION ADSORPTION MODELS

In the 1970's, a number of researchers developed surface complexation (SC) adsorption models (cf. [2,20,21,22]). Among the SC models that incorporate

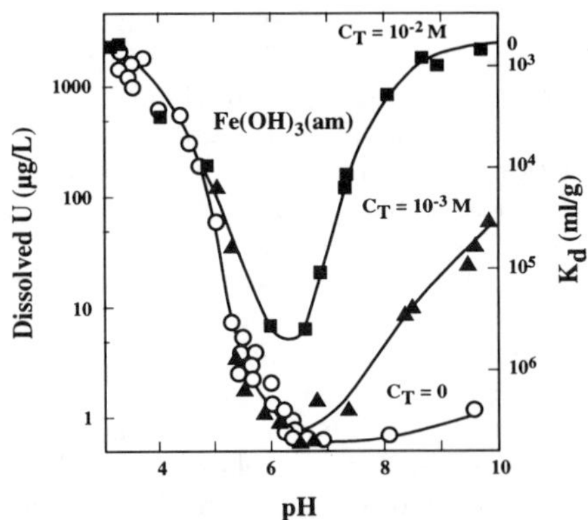

Fig. 3. The effects of changes in pH and total carbonate (C_T) on the extent of U(VI) adsorption by a 1 g/L suspension of HFO in 0.1 M NaNO$_3$ solutions at 25°C, with ΣU(VI) = 10^{-5} M [18].

electrical double layer (EDL) theory (SC-EDL models) are the constant capacitance (CC), diffuse layer (DL) and triple layer (TL) models. These three models are found in the geochemical computer model MINTEQA2 [23], and in the computer program HYDRAQL [24]. The DL model is also included in the geochemical code PHREEQC [25].

The SC-EDL adsorption models treat sorbent surfaces as electrically charged, with acidic (+), neutral, and basic (-) surface sites. Surface charge and potential are consequences of chemical reactions involving the surface sites or functional groups. A surface complex is formed when a RN species is adsorbed onto a surface site. Reactions that form surface complexes can be described with mass-action equations, which are corrected for electrostatic effects using EDL theory. For example, the desorption of a surface proton from the charged surface site, SOH_2^+, may be described by the reaction

$$SOH_2^+ = SOH + H_s^+ \tag{1}$$

for which the so-called intrinsic (thermodynamic) constant expression is

$$K_{a1}^{int} = \frac{(SOH)\,(H_s^+)}{(SOH_2^+)} \tag{2}$$

According to EDL theory, the surface concentration of protons, H_s^+, which cannot be measured directly, is related to aqueous H^+ through a Boltzmann expression

$$(H_s^+) = (H^+)\exp(-\psi_o F/RT) \tag{3}$$

where ψ_o is the electrical potential at the surface. A similar Boltzmann expression, $(X_s^z) = (X^z)\exp(-\psi_o F/RT)^z$, may be used to relate the surface concentration of an adsorbed RN species, X_s^z, to its aqueous concentration, X^z, where z is the charge of the RN species.

SC model data requirements include a complete speciation calculation for the solution at adsorption equilibrium to determine ionic strength and the concentrations and activities of all species that may be involved in RN complexing, or adsorption or competition for sorption sites. Adsorption is studied at RN concentrations below those shown or computed to result in the precipitation of RN solids. Sorbent solids to be used in laboratory experiments should be carefully sized and characterized [22]. The concentration of sorbing surface sites must be determined. This involves the measurement or estimation of surface area and surface charge density for the sorbents. Sorption sites may be defined as strong or weak sites [11,23,26]. Except in the DL model, one or two surface capacitances must also be assumed for each sorbent.

Laboratory adsorption measurements are used to derive the intrinsic equilibrium constants (K^{int} values) for surface adsorption-desorption reactions. The K^{int} values for H^+ adsorption-desorption are obtained first, from acid/base titrations in the presence and absence of the sorbing solids. Once these constants are known, adsorption experiments involving electrolyte ions (for the TL model only), and involving the RN and potentially competing ions are performed to obtain K^{int} values for these species. The detailed procedures used in such experiments and their interpretation are described elsewhere [2,27,28,29,30].

THE SC MODELS AND TSPA

Westall [31] has questioned the merit of using the SC models to predict RN adsorption in complex natural systems when isotherm models may be adequately reliable for TSPA purposes. However, use of the SC models can clarify our understanding of adsorption mechanisms and controls for a particular RN. They may also allow us to predict RN adsorption for conditions outside those used in the laboratory adsorption experiments. The SC models have seen limited use in TSPA in part because of the lack of appropriate experimental RN adsorption data. To an extent this reflects the substantial time and skill required to measure RN adsorption and to derive SC model parameters. The following discussion examines recent efforts to simplify these models and to estimate SC model parameters, making the models more readily available for TSPA.

The Diffuse Layer Model

Turner [32,33,34] has shown that in most cases, the DL, CC and TL models can all fit the same empirical RN adsorption data (see also [21]). Usually, the same single surface reaction provides a good fit of the data for all three models. Turner [34] points out that since all the models match the data about equally well, the simplest (the DL model) is probably the best choice to model RN adsorption in TSPA.

Estimation of SC Model Parameters for Adsorption by Minerals

Numerous authors have noted that the strength of adsorption of a species by oxyhydroxide solids is proportional to its tendency to form hydroxyl complexes in solution. Thus, for example, plots of $\log K_M^{int}$ for a particular SC model versus $\log \,^*K_{11}$ for dissolved cations, M^z, often approach a strait line (cf. [11,35]), where K_M^{int} and $^*K_{11}$ are intrinsic and equilibrium constants for the respective surface and aqueous reactions

$$M_s^z + SOH = H_s^+ + SOM^{z-1} \tag{4}$$

$$M^z + H_2O = H^+ + MOH^{z-1} \tag{5}$$

Such plots, which are called linear free energy relationships (LFER's), have chiefly been employed to estimate K^{int} values for relatively simple metal cations (cf. [36,37]). Their use to predict K^{int} values for actinide species has been limited (cf. [38]).

Degueldre et al. [13] and Degueldre [39,40] have used LFER's to estimate pH-dependent RN cation adsorption by AlOH, FeOH and SiOH surface sites from the stepwise hydrolysis constants of RN cations. Their approach, which avoids EDL theory, leads to reasonable predictions of K_d values for a wide variety of radionuclides at different pH and Eh values and surface site concentrations.

Sverjensky and Sahai [10] have shown that K^{int} values for H^+ adsorption-desorption by oxyhydroxides and aluminosilicates can be predicted from the dielectric constants and average Pauling bond strengths of sorbent minerals. Sverjensky [5] used a similar approach to predict K^{int} values for adsorption of metal cations. LFER plots and Sverjensky's relationships may be useful for estimating the SC model parameters for adsorption of simple RN species. They are less likely to facilitate the estimation of SC parameters for the adsorption of multivalent actinide cations, most of which have a more complex solution chemistry.

A number of authors have noted the similarity in adsorption behavior of actinide (An) cations of the same charge, such as those that share the general formulas AnO_2^+, AnO_2^{2+}, An^{3+} and An^{4+} (cf. [11]). Silva and Nitsche [41] in fact have suggested respective average K_d values of 1, 5, 50 and 500, for adsorption of these type species by 12 different minerals and 4 rock types. Because of this similarity in behavior, Puigdomenech and Bergstrom [42] assumed identical K^{int} values for adsorption onto goethite of NpO_2^+ and PuO_2^+, of UO_2^{2+}, NpO_2^{2+} and PuO_2^{2+}, and of U^{4+}, Np^{4+} and Pu^{4+}. For the conditions modeled, their computed adsorption of Np by goethite as a function of Eh and pH gave K_d values that ranged from near 1 (for NpO_2^+) at low pH and high Eh, to as high as 10^6 (for Np^{4+}) in anaerobic groundwaters at high pH.

The Linear Adsorptivity Model

The geological formations in repositories are likely to contain multiple sorbent phases. In some repository groundwaters, Fe(III), Al(III) or Mn(IV) oxyhydroxide coatings on mineral and fracture surfaces may dominate RN adsorption (cf. [4,42]). The geochemical computer code MINTEQA2 [23] is capable of modeling simultaneous adsorption by multiple sorbents, each having strong or weak adsorption sites. The program adapts the linear adsorptivity model (LAM), in which the total adsorption of a species is assumed the arithmetic sum of its adsorption by individual phases present. The LAM has been found reliable in systems that do not contain interacting colloidal sorbents [11,43].

Radionuclide Adsorption by Complex Materials

In a few cases, RN adsorption by specific aluminosilicate minerals has been measured and SC modeled (cf. [34,44,45,46]. More often, particularly in recent years, scientists within the nuclear waste community have used SC models and a LAM approach to estimate RN adsorption by complex materials. Some of this work is summarized in Table 1. As shown, it has often been assumed that sorbing aluminosilicates are mixtures of simple oxides present in amounts that are defined by the bulk composition of the mineral, rock or engineered material. Total RN adsorption is then computed using SC model parameters for the oxides, and assuming the solid is a mechanical mixture of its oxides.

As an indication of the reliability of such estimates, Baston et al. [7,8] have compared retardation coefficients (R_d values) estimated using this approach to R_d values measured in laboratory experiments for the adsorption of Am(III), Pu(V) and U(IV) by sorbents that include bentonite, tuff, granodiorite, sandstone, mortar and concrete. Assuming rapid and reversible adsorption, the retardation coefficient, R_d, is related to K_d through the expression

$$R_d = \frac{\overline{v}}{\overline{v}_{RN}} = 1 + \left(\frac{\rho_b}{n}\right)K_d \quad (6)$$

[54], where $\overline{v}$ and $\overline{v}_{RN}$, and X and X_{RN} are the average velocities and average migration distances of the groundwater and of the dissolved RN, and ρ_b and n are the bulk density of the geomedia and its porosity as a volume fraction, respectively. A plot comparing the measured and estimated R_d values from Baston et al. [7,8] is shown in Fig. 4, which indicates that these values generally agree within 0.5 to 1 log units.

A study of Bradbury and Baeyens [4] provides a further opportunity to compare measured RN adsorption to RN adsorption estimated using a simplified SC model. Many authors have noted that goethite, hematite and gibbsite have similar adsorption behavior. This similarity led Bradbury and Baeyens (1993) to propose a simplified DL model for Np(V) adsorption by Fe and Al oxyhydroxides and Al-silicate minerals. Their simplifying assumptions included:

1. All adsorption is onto AlOH and FeOH surface sites which are of equivalent energy. All sites are therefore

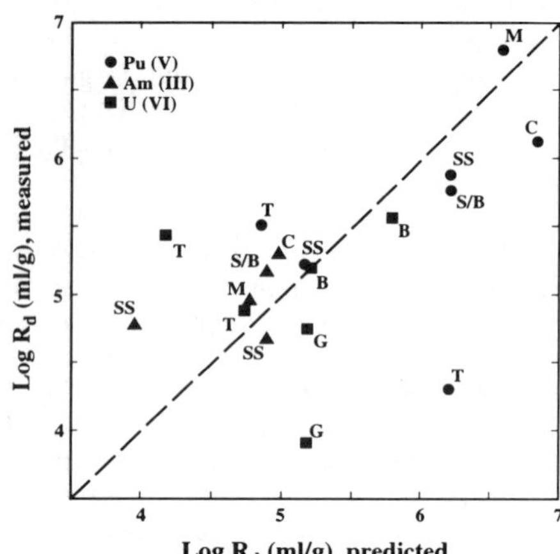

Fig. 4. Laboratory measured and TL model-predicted R_d values for adsorption of Pu(V), Am(III) and U(VI) [7,8]. Letters indicate which sorbents were used. B is bentonite, C cement, G granodiorite, M mortar, S/B sand/bentonite mixture, SS sandstone, and T tuff.

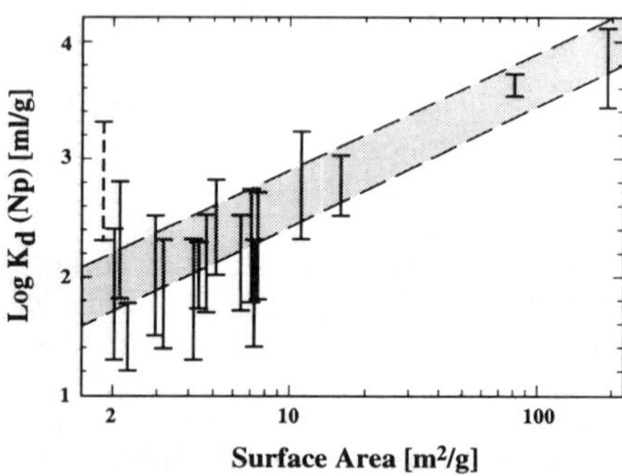

Fig. 5. The shaded area indicates the range K_d values for Np(V) adsorption predicted by Bradbury and Baeyens [4] using a simplified DL model. Vertical bars are measured ranges of K_d values for individual minerals at pH 8.2 as reported by Allard [50].

Table 1. Some simplified surface complexation (SC) electrical double layer (EDL) model applications to radionuclide adsorption by complex sorbents. See also [39,40].

Sorbent	Radionuclide	SC model(s)	Conditions and simplifying model assumptions (if any)	Reference(s)
biotite	Np(V)	DL, CC, TL	Biotite assumed a 1:3 mixture of 0.5 α-Al_2O_3 and quartz (α-SiO_2)	Nakayama & Sakamoto [47], Turner [34]
kaolinite	Np(V)	DL, CC, TL	Kaolinite assumed a 1:2 mixture of 0.5 α-Al_2O_3 and α-SiO_2	Kohler et al. [48], Turner, [34]
kaolinite	U(VI)	DL, CC, TL	Kaolinite assumed a 1:2 mixture of 0.5 α-Al_2O_3 and α-SiO_2	Payne et al., [49], Turner, [34]
Fe(III) & Al oxyhydroxides, & Al silicates	Np(V)	DL	Assumed all sorption onto AlOH and FeOH sites, which are of equivalent energy. All sites are therefore considered FeOH sites. Ca^{2+} competes with Np(V) for adsorption sites.	Bradbury & Baeyens,[4]. Rock and mineral adsorption data from Allard [50]
CaSi cement (concrete & mortar)	Pu(IV), Pu(V), Am(III)	DL	Ca^{2+} and $CaOH^+$ compete with Pu and Am for SiOH sites. pH 8-12.	Baston et al., [7]
granodiorite	U(IV), U(VI), Tc(IV), Tc(VII)	TL	Sorbent sites assumed a 1:1:0.5 mixture of FeOH, AlOH, and SiOH groups.	Baston et al., [8]
bentonite & tuff	U(IV), U(VI), Tc(IV), Tc(VII)	TL	Goethite (α-FeOOH) assumed the only important sorbent. Its amount computed from rock Fe content. Near-neutral pH.	Baston et al., [8]
goethite	Cs(I), Ra(II), Np, Pu, U,	TL & Ion Exchange	Goethite and a clay assumed the only important sorbents. K^{int} values for Np, Pu and U onto α-FeOOH assumed equal for species with same oxidation states (e.g. equal for NpO_2^+ and PuO_2^+). Eh considered.	Puigdomenech & Bergstrom, [42]
cementitious materials	Cs^+, I^-, Sn(IV), Pu(IV), U(IV), Np(IV), Am(III)	DL	Cement surface model assumes SiOH and CaOH surface sites dominate, and that the CaOH/SiOH site ratio increases from 0 to 2.5 between pH 10 and 12.5.	Heath et al., [57]
smectite (Fe(III)-biedellite) montmorillonite	U(VI)	TL	Ion exchange assumed on interlayer SiOH and AlOH sites. Edge sites assumed mixture of gibbsite [α-$Al(OH)_3$] and α-SiO_2. U(VI) competes with Na^+ and Ca^{2+} for surface sites.	Turner et al., [52] (See also McKinley et al. [53])

assumed to be FeOH sites.

2. Total surface site density on all minerals is fixed as 3.8 μmol/m^2 as for hydrous ferric oxide (HFO)[22].

3. BET N$_2$-gas adsorption measured surface areas are the mineral surface areas available for adsorption.

4. Only strong surface (SOH) sites adsorb Np(V). These are 2.5% of total surface sites as assumed for HFO [26].

5. DL model parameters for HFO adsorption of competing H$^+$ and Ca^{2+} are obtained from Dzombak and Morel [26].

6. The K^{int} value for Np(V) adsorption by FeOH sites is found by optimal fitting of the SC model to the data for Np(V) adsorption by HFO [55], assuming the adsorption reaction

$$SOH + NpO_2^+ = SO-NpO_2 + H_s^+ \tag{7}$$

Based on N$_2$-BET measured surface areas of individual minerals, Bradbury and Baeyens [4] next predicted Np(V) adsorption by these minerals. Their model calculations indicated that the Np(V) adsorbed onto kaolinite, for example, occupied about 10^3 to 10^5 times fewer sites than were occupied by adsorbed H$^+$, and 10^5 to 10^6 times fewer sites than occupied by adsorbed Ca^{2+}. Such behavior can be expected in general, when H$^+$ and major ions compete with trace concentrations of RN species for adsorption sites on minerals.

In Fig. 5, the adsorption of Np(V) by a variety of minerals at pH 8.2, predicted by Bradbury and Baeyens [4] with their simplified DL model, is compared to measured Np(V) adsorption by the same minerals as reported by Allard [50]. The plot shows that measured K$_d$ values are generally lower than predicted for mineral surface areas below 10 m^2/g, but that the two are in good agreement for higher surface areas.

What K$_d$ Value Is Large Enough to Assure Waste Isolation?

A TSPA analysis should decide how large the K$_d$ value for a given RN must be to provide adequate confidence that the RN will not become a health hazard at the accessible environment. To examine the effect of K$_d$ on RN transport, it is useful to consider what R$_d$ values correspond to different K$_d$ values. The bulk density (ρ_b) of silicate rocks is about 2.65 g/cm^3. If fractured, crystalline igneous and metamorphic rocks have fractional porosities as high as n = 0.1, although their porosities are generally lower. Assuming n$\le$0.1 leads to $\rho_b/n \ge (2.65-0.27)/0.1 = 24$. Substituting into Eq. 6 gives

$$R_d \ge 1 + 24K_d \tag{8}$$

A K$_d$ of 100 ml/g thus corresponds to an R$_d$ of about 2400. If a groundwater moves 1 m a year, then after 10,000 y the groundwater will have moved 10 km, and the RN about 4 m. For the same conditions, if the K$_d$ value for a RN exceeds 10^4 ml/g, the RN moves less than 4 cm in 10,000 y, obviously a trivial distance. In many of the national programs, measurements of RN adsorption are being made to accurately determine K$_d$ values that are known to exceed 10^4 ml/g at dissolved RN concentrations below 10^{-8} to 10^{-10} M. It is appropriate to ask why such work is necessary. In any case its relevance should be decided through a TSPA analysis of the applicable repository site.

The most soluble forms of the 4+ valent actinides (usually their amorphous oxyhydroxides) have solubilities of about 10^{-7} to 10^{-9} M or less in potential repository groundwaters of low Eh [11]. Adsorption studies to evaluate controls on lower RN concentrations than these may be of questionable value.

CONCLUDING REMARKS

We've come a long way since the 1970's in our understanding of fundamental controls on the adsorption of most RN's by minerals and rocks, and how to model the adsorption. This has resulted from a number of developments:

(1) Improved thermodynamic data bases for the RN's (there's still a long way to go). This permits us to more reliably speciate the RN's in solution using geochemical computer models.

(2) Introduction of the surface complexation (SC) models, and the assembling of a sizeable data base for RN adsorption by individual minerals using these models.

(3) Standardization of more scientifically sound experimental measurements of adsorption. An understanding of controls on adsorption, which is required by the SC models, has led to the use of better experimental protocols, generally, including in experiments to obtain data for K_d and the adsorption isotherms. This has included careful selection, preparation and characterization of sorbent phases, and improved design, performance and reporting of experimental results. Researchers must still be careful to avoid adsorption measurements when sorbent phases are dissolving or precipitating incongruently [52,56,57]. Adsorption measurements should also be made reversibly and under equilibrium conditions. This requires understanding the complexities of adsorption-desorption reaction rates [22,58-62].

(4) Development of approximation and estimation methods for obtaining SC model parameters to simplify use of the models for TSPA. This effort continues and has expanded in recent years.

(5) Use of the SC models to compute K_d values and to increase confidence in modeling results obtained with the K_d and isotherm models. Distribution coefficients are still the preferred adsorption parameters for TSPA [14,63]. The limitations and applications of the K_d values being measured and reported today, are better understood than ever before.

(6) Improved understanding of the role in RN adsorption and transport, of mineral and rock properties and of the hydrogeology of proposed repository sites. In the final analysis, the application of adsorption data and models to TSPA requires an understanding of the nature of groundwater flow and transport. Recent studies suggest useful relationships between K_d and rock properties such as the reactive surface area in pores and fractures, and rock porosity, permeability and tortuosity [54,57,64-68].

We will always have a limited knowledge of groundwater flow paths and travel times, and of the reactive surface areas of sorbing rocks between the repository and the accessible environment. Given these inevitable uncertainties, a TSPA analysis may show that rather large uncertainties in our assumed K_d values are unimportant by comparison [15].

REFERENCES

1. L.L Ames and D. Rai, *Radionuclide Interactions with Soil and Rock Media*, vol. 1, Report EPA 520/6-78-007, Battelle Pacific Northwest Laboratories, Richland, WA.
2. J.A. Davis, R.O James, J.O. Leckie, J. Colloid Interface Sci. **63**, 480 (1978).
3. D.R. Turner, *A Uniform Approach to Surface Complexation Modeling of Radionuclide Sorption*, Report CNWRA **95-001** (Center for Nuclear Waste Regulatory Analysis, San Antonio, TX, 1995).

4. M.H. Bradbury, B. Baeyens, J. Colloid Interface Sci. **158**, 364 (1993).
5. D.A. Sverjensky, Nature **364**, 776 (1993).
6. D.A. Sverjensky, Geochim. Cosmochim. Acta **58(14)**, 3123 (1994).
7. G.M.N. Baston, J.A. Berry, M. Brownsword, T.G. Heath, C.J. Tweed, S.J. Williams, in *Scientific Basis for Nuclear Waste Management XVIII*, eds. T. Murakami & R.C. Ewing (Mat. Res. Soc. Symp. Proc. **353**, Pittsburgh, PA, 1995), p. 957.
8. G.M.N. Baston, J.A. Berry, M. Brownsword, M.M. Cowper, T.G. Heath, C.J. Tweed, in *Scientific Basis for Nuclear Waste Management XVIII*, eds. T. Murakami & R.C. Ewing (Mat. Res. Soc. Symp. Proc. **353**, Pittsburgh, PA, 1995), p. 989.
9. T. Fujita, M. Tsukamoto, T. Ohe, S. Nakayama, Y. Sakamoto, in *Scientific Basis for Nuclear Waste Management XVIII*, eds. T. Murakami & R. C. Ewing (Mat. Res. Soc. Symp. Proc. **353**, Pittsburgh, PA, 1995), p. 965.
10. D.A. Sverjensky, N. Sahai, Geochim. Cosmochim. Acta **60(20)**, 3773, (1996).
11. D. Langmuir, *Aqueous Environmental Geochemistry*, (Prentice-Hall, Upper Saddle River, NJ, 1997).
12. C. Degueldre, B. Wernli, J. Envir. Radioactivity **20**, 151 (1993).
13. C. Degueldre, H.J. Ulrich, H. Silby, Radiochimica Acta **65**, 173 (1994).
14. A. Meijer, in *Proc. DOE/Yucca Mountain Site Characterization Project Radionuclide Adsorption*, Rept. **LA-12325-C** (Workshop, Los Alamos Natl. Lab, Los Alamos, NM, 1992), p. 9.
15. OECD, *The Status of Near-Field Modeling. Proc. Technical Workshop, Cadarache, France* (Organisation for Econ. Cooperation & Devel., Paris, 1993).
16. I.G. McKinley, W.R. Alexander, J. Contaminant Hydrol. **13**, 249 (1993).
17. I. Triay, C.R. Cotter, S.M. Kraus, M.H. Huddleston, S.J. Chipera, D.L. Bish, *Radionuclide Sorption in Yucca Mountain Tuffs with J-13 Well Water: Neptunium, Uranium and Plutonium.* Report **LA-12956-MS** (Los Alamos National Laboratory, Los Alamos, NM, 1996)
18. C-K. Hsi, D. Langmuir, Geochim. Cosmochim. Acta **49(9)**, 1931 (1985).
19. M.J. Taylor, *Practical Study Requirements Groundwater and Seepage Uranium Mill Waste Disposal Systems.* Written testimony for Royal Commission of Inquiry Health and Envir. Protection, Vancouver, BC, (1979).
20. R.O. James, T.W. Healy, J. Colloid Interface Sci. **40(1)**, 65 (1972).
21. J. Westall, H. Hohl, Adv. Colloid Interface Sci. **12(4)**, 265 (1980).
22. J.A. Davis, D.B. Kent, in *Mineral-Water Interface Geochemistry.* eds. M.F. Hochella and A.F. White. *Reviews in Mineralogy* **23** (Min. Soc. Am.) p. 177.
23. J.D. Allison, D.S. Brown, K.J. Novo-Gradac. *MINTEQA2, A Geochemical Assessment Data Base and Test Cases for Environmental Systems.* Version **3.0**, Users Manual. Report EPA/600/3-91/-21 (U.S. Envir. Protection Agency, Athens, GA, 1991).
24. C. Papelis, K.F. Hayes, J.O. Leckie, *HYDRAQL: A Program for the Computation of Chemical Equilibrium Composition of Aqueous Batch Systems Including Surface-Complexation Modeling of Ion Adsorption at the Oxide/Solution Interface,* Tech. Rept. **306** (Dept. of Civil Eng., Stanford Univ., 1988).
25. D.L. Parkhurst, *Users Guide to PHREEQC-A Computer Program for Speciation, Reaction-Path, Advective-Transport, and Inverse Geochemical Calculations.* U.S. Geol. Survey Water Resources Inv. Rept. 95-4227 (1995).
26. D.A. Dzombak, F.M.M. Morel, *Surface Complexation Modeling. Hydrous Ferric Oxide* (Wiley-Interscience, New York, 1990).
27. R.O. James, G.A. Parks, Surface Colloid Sci. **12**, 119 (1982).

28. D.A. Dzombak, F.M.M. Morel, J. Hydraulic Eng. **113**, 430 (1987).
29. W. Stumm, editor, *Aquatic Surface Chemistry. Chemical Processes at the Particle-Water Interface* (John Wiley & Sons, New York, NY, 1987).
30. W. Stumm, *Chemistry of the Solid-Water Interface* (Wiley-Interscience, New York, NY, 1992).
31. J.C. Westall in *Scientific Basis for Nuclear Waste Management XVIII*, eds. T. Murakami and R.C. Ewing (Mater. Res. Soc. Symp. Proc. **353**, Pittsburgh, PA, 1995), p. 937.
32. D.R. Turner, *Sorption Modeling for High-Level Waste Performance Assessment: A Literature Review*. Rept. CNWRA **91-011** (Center for Nuclear Waste Regulatory Anal., San Antonio, TX, 1991).
33. D.R. Turner, *Mechanistic Approaches to Radionuclide Sorption Modeling*. Rept. CNWRA-**93-019** (Center for Nuclear Waste Regulatory Anal., San Antonio, TX, 1993).
34. D.R. Turner, *A Uniform Approach to Surface Complexation Modeling of Radionuclide Sorption*. Rept. CNWRA **95-001** (Center for Nuclear Waste Regulatory Anal., San Antonio, TX, 1995).
35. P.W. Schindler, W. Stumm, in *Aquatic Surface Chemistry*, editor W. Stumm (Wiley-Interscience, New York, NY, 1987), p. 83.
36. L. Balistrieri, P.G. Brewer, J.W. Murray, Deep-Sea Research **28A**, 101 (1981).
37. R.M. Smith, E.A. Jenne, *Compilation, Evaluation and Prediction of Triple-Layer Model Constants for Ions on Fe(III) and Mn(IV) Hydrous Oxides*. Rept. **PNL-6754** (Battelle Pacific Northwest Laboratory, Richland, WA, 1988)
38. T. Fujita, M. Tsukamoto, T. Ohe, S. Nakayama, Y. Sakamoto, in *Scientific Basis for Nuclear Waste Management XVIII*, eds. T. Murakami & R. C. Ewing (Mat. Res. Soc. Symp. Proc. **353**, Pittsburgh, PA, 1995), p. 965.
39. C. Degueldre, J. Envir. Radioactivity **29(1)**, 75 (1995).
40. C. Degueldre, J. Envir. Radioactivity **34(2)**, 211 (1997).
41. R.J. Silva, H. Nitsche, Radiochim. Acta (in press) (1996).
42. I. Puigdomenech, U. Bergstrom, Nuclear Safety **36(1)**, 142 (1995).
43. M.D. Siegel, V.S. Tripathi, M.G. Rao, D.B. Ward, in *Proc. 7th Intl. Water-Rock Interaction*, eds. Y.K. Kharaka and A.S. Maest (A.A. Balkema, Rotterdam, 1992) p. 63.
44. A.C. Riese, *Adsorption of Radium and Thorium onto Quartz and Kaolinite: A Comparison of Solution/Surface Equilibria Models* PhD thesis T-2625 (Colorado School of Mines, Golden, CO, 1982).
45. R.F. Vochten, L. Van Haverbeke, F. Goovaerts, J. Chem. Soc. Farad. Trans. **86(4)**, 4095. (1990).
46. R.T. Pabalan, P.M. Prikryl, T.B. Dietrich, in *Scientific Basis for Nuclear Waste Management XVI*. eds C. Interrante and R. Pabalan (Mater. Res. Soc. Symp. Proc. **294**, Pittsburgh, PA, 1993) p. 777.
47. S. Nakayama, Y. Sakamoto, Radiochim. Acta **52/53**, 153 (1991).
48. M. Kohler, E. Wieland, J.O. Leckie, in *Proc. 7th Intl. Symp. on Water-Rock Interaction*, vol. 1: *Low Temperature Environments*. eds. Y.K. Kharaka and A.S. Maest (A.A. Balkema, Rotterdam, 1992), p. 51.
49. T.E. Payne, K. Sekine, J.A. Davis, T. D. Waite in *Alligtor Rivers Analogue Project Annual Rept. 1990-1991*. ed. P. Duerden (Australian Nuclear Sci. & Technol. Org., 1992), p. 57.
50. B. Allard, *Sorption of Actinides in Granite Rock*, Rept. KBS **82-21** (Dept. of Nuclear Chem., Chalmers Univ. of Technol., Goteborg, Sweden, 1992) p. 61.
51. T.G. Heath, D.J. Ilett, C.J. Tweed, in *Scientific Basis for Nuclear Waste Management XIX*. eds. W.M. Murphy and D.A. Knecht (Mater. Res. Soc. Symp. Proc. **412**, Pittsburgh, PA,

1996) p. 443.

52. G.D. Turner, J.M. Zachara, J.P. McKinley, S.C. Smith, Geochim. Cosmochim. Acta **60(18)**, 3399 (1996).

53. J.P. McKinley, J.M. Zachara, S.C. Smith, G.D. Turner, Clays & Clay Minerals **43**, 586 (1995).

54. R.A. Freeze, J.A. Cherry, *Groundwater* (Prentice-Hall, Englewood Cliffs, NJ, 1979).

55. D.C. Girvin, L.L. Ames, A.P. Schwab, J.E. McGarrah, J. Colloid Interface Sci. **141(1)**, 67 (1991).

56. W.M. Murphy, R.T. Pabalan, J.D. Prikryl, C.J. Goulet, Am. J. Sci. **296**, 128 (1996).

57. I.R. Triay, C.R. Cotter, M.H. Huddleston, D.E. Leonard, S.C. Weaver, S.J. Chipera, D.L. Bish, A. Meijer, J.A. Canepa, *Batch Sorption Results for Neptunium Transport through Yucca Mountain Tuffs*. Rept. LA-12961-MS (Los Alamos Natl. Laboratory, Los Alamos, NM).

58. E.A. Jenne, in *Metal Speciation and Contamination of Aquatic Sediments*. ed. H.E. Allen (Ann Arbor Press, Ann Arbor, MI, 1995), p. 81.

59. P.R. Grossl, D.L. Sparks, C.C. Ainsworth, Envir. Sci. & Technol. **28**, 1422 (1994).

60. K.E. Bencala, A.P. Jackman, V.C. Kennedy, R.J. Avanzino, G.W. Zellweger, Water Resources Res. **19(3)**, 725 (1983).

61. K. Skagius, I. Neretnieks, Water Resources Research **24(1)**, 75 (1988).

62. I.R.Triay, A.C. Furlano, S.C. Weaver, S.J. Chipera, D.L. Bish, *Comparison of Neptunium Sorption Results Using Batch and Column Techniques*. Rept. LA-12958-MS (Los Alamos Natl. Laboratory, Los Alamos, NM).

63. R.A. Andrews et al., *Total system Performance Assessment-1995: An Evaluation of the Potential Yucca Mountain Repository* (Civilian Radioactive Waste Management System Management and Operating Contractor, Las Vegas, NV, 1995).

64. F.P. Bertetti, R.T. Pabalan, D.R. Turner, M.G. Almendarez, in *Scientific Basis for Nuclear Waste Management XIX*, eds. W.M. Murphy and D.A. Knecht (Mater. Res. Soc. Symp. Proc. **412**, Pittsburgh, PA, 1996) p. 631.

65. A.F. White, M.L. Peterson in *Chemical Modeling of Aqueous Systems II*, eds. D.C. Melchior and R.L. Bassett. Am. Chem. Soc. Symp. Ser. **416**, (1990), p. 461.

66. R.B. Wanty, C.A. Rice, D. Langmuir, P. Briggs, E.P. Lawrence in *Scientific Basis for Nuclear Waste Management XIV*, edited by T. Abrajamo, Jr. and L.H. Johnson (Mater. Res. Soc. Symp. Proc., **212**, Pittsburgh, PA, 1991), p. 695.

67. R.W. Smith, A.L. Schafer, A.F.B. Tompson, in *Scientific Basis for Nuclear Waste Management XIX*, eds. W.M. Murphy & D.A. Knecht (Mater. Res. Soc. Symp. Proc. **412**, Pittsburgh, PA, 1996), p. 693.

68. A.F.B. Tompson, Water Resources Res. **29(11)**, 3709 (1993).

INFLUENCE OF CARBONATE IONS ON EUROPIUM SORPTION ONTO IRON-OXIDES

T. FUJITA and M. TSUKAMOTO
Central Research Institute of Electric Power Industry (CRIEPI)
2-11-1, Iwado-Kita, Komae, Tokyo 201, JAPAN

ABSTRACT

The sorption behavior of europium onto two iron-oxides, goethite and magnetite, focused on the effect of carbonate species concentration was investigated in 0.01M $NaClO_4$ solution by the batch method over a pH range of 4 to 11. It was found that sorption of europium was enhanced in the higher concentration solution of carbonate species for both materials. The zeta-potential of goethite in the solutions including sodium bicarbonate was also measured to estimate the effect of sorbed carbonate species. Our experimental results and model calculation suggest that the production of surface carboxylic group $FeOCOO^-$ by the sorption of carbonate species lowered the zeta-potential and enhanced europium sorption.

INTRODUCTION

Most groundwaters have fairly high concentrations of carbonate species (in this paper the term carbonate species includes carbonate ion, bicarbonate ion, and carbonic acid) which is supplied by the geochemical reactions such as dissolution reaction of calcite and/or decay of organic matter[1,2]. The carbonate species which is often the dominant solute in groundwaters may affect sorption behavior of metal ions released from radioactive waste due to competition for sorption sites and change of surface properties by carbonate species sorption. While sorption of metal ions released from radioactive waste has been measured in electrolytes such as NaCl, $NaNO_3$, and $NaClO_4$ usually because these studies are focused on ionic strength for application of surface comlexation modeling[3]. The estimation of influence of major solutes in groundwater such as Ca^{2+}, Mg^{2+}, SO_4^{2-}, and CO_3^{2-} is, therefore, still an important issue as these ions bind specifically to surface and compete for sorption sites with metal ions. The sorption behavior of carbonate species has been almost neglected with the exception of several works[4,5]. Geen et al.[6] showed that carbonate species prevent chromate anion, CrO_4^{2-}, from sorbing onto goethite due to sorption site coverage by sorbed carbonate species. The influence for cation sorption has, however, not been investigated in detail.

In the present work, the sorption behavior of europium onto two iron-oxides, goethite and magnetite, is investigated to elucidate the influence of carbonate species concentration by the batch method. Europium may serve as an analogue for trivalent lanthanides and actinides which are present in radioactive waste and forms highly coordinated carbonate complexes. Iron-oxides are well investigated as a sorbent for transition element ions in many research spheres. Goethite and magnetite are used as the potential corrosion products of iron-based overpack material which are expected to be effective sorbents for some radioisotopes released from radioactive waste. The zeta-potential of the iron-oxides in the solutions including carbonate species was also measured to estimate the sorbed carbonate species effect. The zeta-potential of a sorbent is itself strongly influenced by sorption of cations and anions. Conversely, the sorption behavior of ions should be strongly influenced by the zeta-potential of the sorbent[7]. Therefore the effect of carbonate species sorption on europium sorption onto goethite and magnetite can be estimated by measurements of zeta-potential qualitatively. Finally, preliminary analysis using a surface complexation model is applied to interpret our experimantal results.

Mat. Res. Soc. Symp. Proc. Vol. 465 © 1997 Materials Research Society

MATERIALS & METHODS

Materials

Two iron-oxides, magnetite and goethite, were used as sorbents. Analytical grade magnetite powder was obtained from Wako Pure Chemical Industries Ltd. Goethite was prepared using the method described by Atkinson et al[8]. It was stored in deionized water at least 60 days for crystal growth, and used after washing three times with deionized water. The BET surface areas of magnetite and goethite were 4.8m^2/g and 36m^2/g, respectively. The grain sizes of these materials were not measured.

Europium sorption experiment

Sorption of europium onto these materials was examined as a function of pH by batchwise procedure. The experiments were carried out at room temperature (ca. 25°C) in polypropylene bottles under aerobic conditions.

The procedure of sorption experiment is shown in Fig.1. The 0.005g of solid sample was contacted with the 30ml of the europium solution including Eu-152 tracer in an ionic medium of 0.01M NaClO$_4$. A small fraction of sodium bicarbonate solution was added to the suspensions to attain a concentration of 5x10^{-3}M initially, for investigation of the effect of carbonate species concentration.

The pH was adjusted by sodium hydroxide and perchloric acid. After an equilibration of 3days, the solid and liquid phase were separated with 0.45μm-pore size filter. The value of pH was measured by glass electrode, the concentration of europium by gamma-ray spectrometry, and the total carbonate concentration by IR analyzer, in the filtrates. The filtrates were separated again with Millipore ultrafilter membrane (10,000MWCO) to investigate the effect of colloidal particles on sorption behavior of europium. It was negligible in this case. The sorption onto the wall of polypropylene bottles was also investigated by blank tests. A negligible percent of europium sorption onto the wall was found at the pH11.

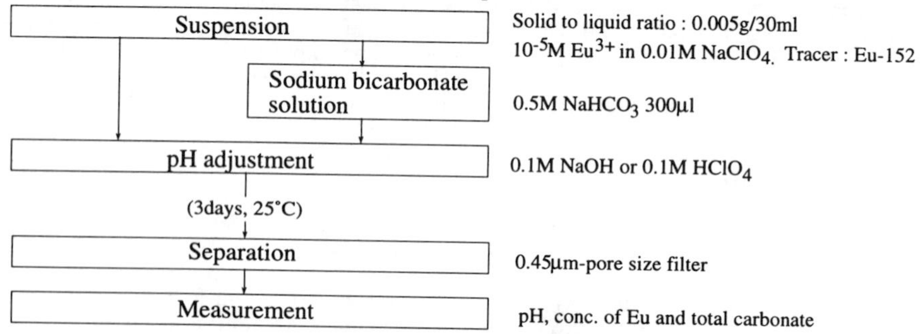

Fig.1 Flow diagram for the sorption experiment procedure.

Zeta-potential measurement

The pH dependence of iron-oxide zeta-potential was investigated in 0.01M NaClO$_4$, 0.01M NaClO$_4$+carbonate species(the initial concentration of sodium bicarbonate is 5x10^{-3}M), and 0.01M NaClO$_4$+10^{-5} mol l^{-1} europium to estimate the effect of carbonate species sorption and europium sorption. The concentration of these solutions is identical to the solution used in sorption experiments. The solid to liquid ratio was also the same. The zeta-potential was determined by measuring the electrophoretic mobility of these particles in the solutions described above with a LASER-assisted microscopic electrophoresis analyzer (System 3000, PEN KEM, USA) at 25°C. The suspension was dispersed for 1day by an ultrasonic bath to avoid particle condensation before the zeta-potential measurement.

RESULTS AND DISCUSSION

Total carbonate concentration

The total carbonate concentration of the suspension for sorption experiments is indicated as a function of final pH in Fig.2 for the solutions in which sodium bicarbonate solution is added or not added. It was almost saturated according to the partial pressure of carbon dioxide in the air at pH<7, while at pH>7, a constant concentration of total carbonate of ca. 3×10^{-4}M was found for the solution in which sodium bicarbonate solution was not added. For the solution in which sodium bicarbonate solution was added, it remained 5×10^{-3}M at pH>8 as initially established. While at pH<8, a considerable amount of carbonate escaped from the solution depending on the pH value of the solution. We postulated the total carbonate concentration profiles as indicated in Fig.2 as solid lines when we use this data for calculations of chemical equilibrium.

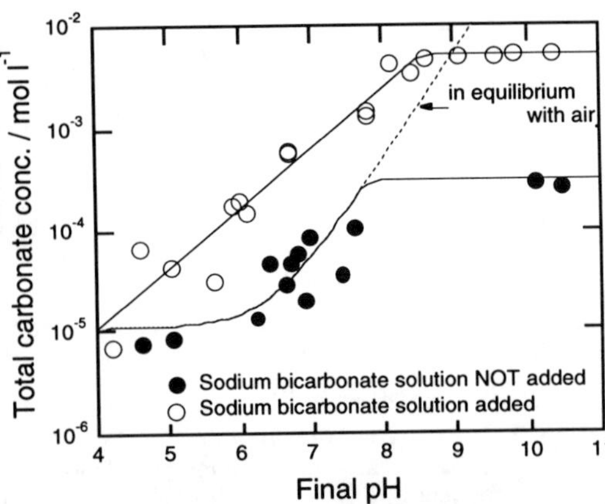

Fig.2 Total carbonate concentrations as a function of finalpH of sorption tests. The value as indicated by solid lines are used for chemical equilibrium calculations.

Chemical speciation of europium

The chemical species of europium in the solutions without suspended particles were identified by using MINEQL+[9] that uses the chemical equilibrium program, MINEQL[10] as its numerical engine, combined with the thermodynamic data[11] summarized in Table I. The calculated results are illustrated for cases when sodium bicarbonate solution is added or not added in Fig.3.

The major chemical species of europium is Eu^{3+} up to pH7, and $EuCO_3$ beyond pH8 when sodium bicarbonate solution is not added as shown in Fig.2(a). Precipitation of $Eu(OH)_3$(s) may occur in the high alkaline pH region beyond pH9.3. In cases when sodium bicarbonate solution is added, the major chemical species of europium is still Eu^{3+} up to pH7 as indicated in Fig.2(b). Highly coordinated carbonate complexes of europium such as $Eu(CO_3)_2^-$, $Eu(CO_3)_4^{5-}$ became dominant in the alkaline pH region as pH increases. In this case, the precipitation of $Eu(OH)_3$(s) may occur in high alkaline pH region beyond pH10.3 similarly as described above.

Table I Equilibrium reactions and their constants

Reaction No.	Equilibrium reaction			log β
1	$Eu^{3+} + H_2O$	$\rightleftarrows$	$EuOH^{2+} + H^+$	-7.8
2	$Eu^{3+} + CO_3^{2-}$	$\rightleftarrows$	$EuCO_3^+$	8.06
3	$Eu^{3+} + 2CO_3^{2-}$	$\rightleftarrows$	$Eu(CO_3)_2^-$	12.91
4	$Eu^{3+} + 3CO_3^{2-}$	$\rightleftarrows$	$Eu(CO_3)_3^{3-}$	12.40
5	$Eu^{3+} + 4CO_3^{2-}$	$\rightleftarrows$	$Eu(CO_3)_4^{5-}$	14.30
6	$Eu^{3+} + 3H_2O$	$\rightleftarrows$	$Eu(OH)_3 + 3H^+$	17.50
7	$2H^+ + CO_3^{2-}$	$\rightleftarrows$	H_2CO_3	16.68
8	$H^+ + CO_3^{2-}$	$\rightleftarrows$	HCO_3^-	10.33
9	$Na^+ + CO_3^{2-}$	$\rightleftarrows$	$NaCO_3^-$	1.27
10	$Na^+ + H^+ + CO_3^{2-}$	$\rightleftarrows$	$NaHCO_3$	10.08

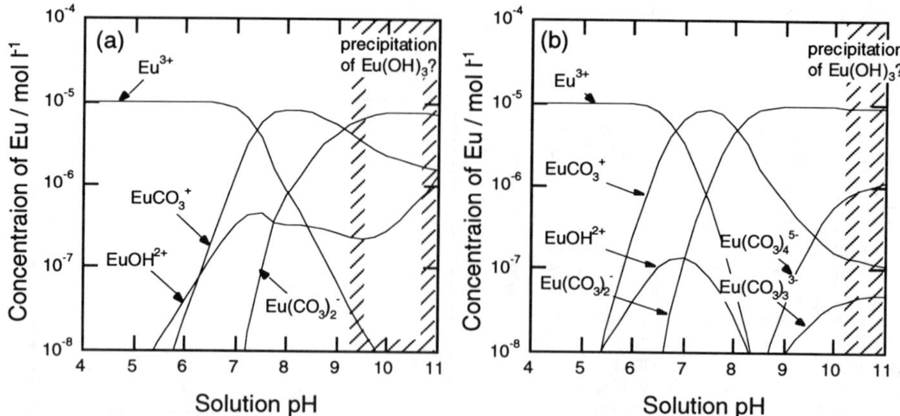

Fig.3 Concentration of europium species in the 0.01M $NaClO_4$ solution in case the sodium bicarbonate solution is added(a) or not added(b). Total carbonate concentrations used for this calculation is indicated as solid lines in Fig.2.

<u>Europium sorption experiments</u>

The percentage of sorbed europium onto goethite and magnetite is shown in Fig.4(a) and Fig.4(b) respectively as a function of final pH. The amount of sorbed europium onto these materials increased sharply with the increase of solution pH as reported for other cations[3]. When the sodium bicarbonate solution is added, the influence of carbonate species concentration on the sorption of europium is recognized. It is found that sorption of europium onto both magnetite and goethite is enhanced in the solution where sodium bicarbonate solution is added. Consequently the sorption edge (the pH of 50% sorption of europium) for goethite is loweredfrom 6.6 under equilibration condition with laboratory air to 6.0 when sodium bicarbonate solution is added. For magnetite, the shift of the sorption edge is observed similarly. It may noticeable that

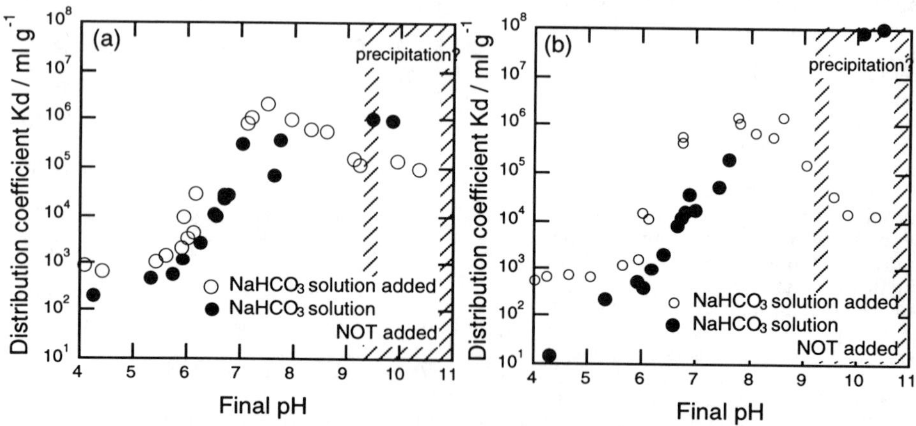

Fig.4 The pH dependence of the distribution coefficient Kd of europium sorption onto (a) goethite and (b) magnetite in case sodium bicarbonate solution is added or not added.

sorbed europium doesn't decrease to zero in case sodium bicarbonate solution is added for either materials. Data beyond pH9.3 was not evaluated because the precipitation of $Eu(OH)_3(s)$ may occur in high alkaline pH region beyond pH9.3 as indicated in Fig.3

The ionic strength of the solution including sodium bicarbonate is almost equal to that of the other solution in the pH region (<pH7.5) in which europium sorption is enhanced. Therefore, there is no effect of ionic strength change of the solutions on the sorption behavior.

<u>Zeta-potential change by carbonate species sorption</u>

As indicated in previous part, it is found that the europium sorption onto two iron-oxides was enhanced relative to the total carbonate concentration. It is thought that, in part, the cause of this effect is the change of europium speciation by total carbonate concentration in the liquid phase. While the surface properties of the solid phase should be changed by sorption of carbonate species. Estimation of sorbed carbonate species influence may be possible by measurements of zeta-potential of sorbents. The zeta-potential of metal oxide particle is itself strongly influenced by sorption of cations and anions such as carbonate species. Conversely, the sorption behavior of ions is strongly influenced by the zeta-potential of metal oxide particle.

The zeta-potential of goethite in 0.01M $NaClO_4$ and in 0.01M $NaClO_4$ containing sodium bicarbonate is shown in Fig.5(a) as a function of pH. The total carbonate concentrations for these solutions are almost similar to the values measured at sorption tests respectively. The zeta-potential of magnetite cannot be determined exactly because the particles of magnetite tend to adhere to the wall of electrophoresis capillary tube and electrodes.

The zeta-potential of goethite in 0.01M $NaClO_4$ decreased with the increase of solution pH. This is caused by the proton dissociation reaction of the surface hydroxyl group, which can be written as

$$\equiv Fe - OH_2^+ \longleftrightarrow \equiv Fe - OH + H^+$$
$$\equiv Fe - OH \longleftrightarrow \equiv Fe - O^- + H^+$$

where $\equiv Fe-$ represents a metal ion at the goethite surface, and H^+ signifies a proton released to the bulk solution. Therefore the equilibrium of these reactions depends on the bulk solution pH. The result of zeta-potential of goethite in 0.01M $NaClO_4$ reflects the equilibrium

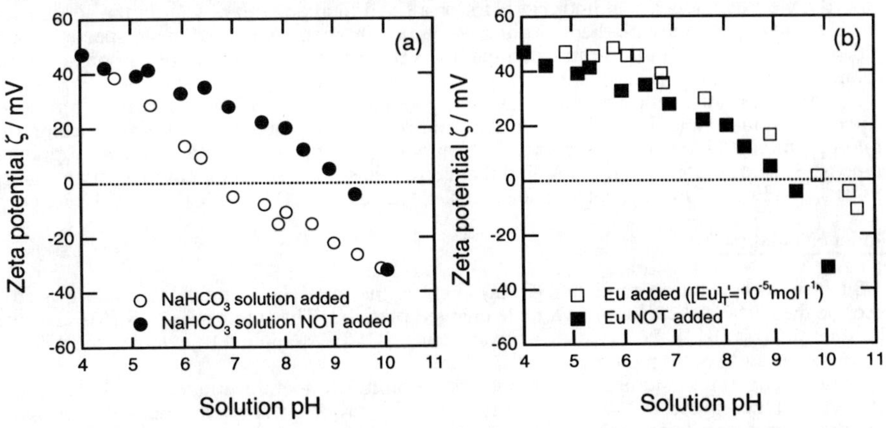

Fig.5 Zeta-potentials of goethite as a function of pH.
   (a): Sodium bicarbonate added or not added in the solution of 0.01M $NaClO_4$
   (b): Europium added or not added in the solution of 0.01M $NaClO_4$

of these reactions. The value of isoelectric point (IEP; the pH value of zero point of zeta-potential) in the solution was 9.1 as shown in Fig.5(a). The PZNPC (point of zero net protonic charge[12]) of this goethite measured by potentiometric titration under argon gas atmosphere was 9.0, which is almost the same as reported value[6,12]. When there is no specific adsorption reaction onto goethite except for a protonation reaction of surface hydroxyl group is negligible, the value of IEP is equal to the value of PZNPC. Therefore it is thought that the specific adsorption of carbonate species onto goethite in 0.01M $NaClO_4$ is not notable although the solution includes carbonate species as indicated in Fig.2. When sodium bicarbonate solution is added, the zeta-potential decreased by 5 to 30mV over wide pH range. The IEP value in this solution lowered from 9.1 under equilibration condition with laboratory air to 7.0. Generally, the value of zeta-potential of metal oxide particles is affected by the ionic strength of the suspension. Hence, the effect of ionic strength may be not negligible in alkaline pH regions when total carbonate concentration is relatively high. But there is no difference between ionic strengths of both solutions at pH7 when europium sorption is enhanced. Therefore the zeta-potential change is thought to be affected by only carbonate species sorption. Therefore our results indicate that carbonate anions were specifically sorbed on these goethites over wide pH range.

The amount of specifically sorbed ions can be roughly estimated from zeta-potential change by the following equation when the Gouy-Chapmann electrical double layer model is postulated.

$$\sigma = (8n\varepsilon_0\varepsilon_r kT)^{1/2} \sinh(ze\psi_\delta / 2kT) \tag{1}$$

where, σ; surface charge density, n; density of ions; ε_0 and ε_r; permittivity of vacuum and relative permittivity of water respectively, k; Boltzmann constant, T; absolute temperature, z; charge number of ion, e; elementary charge, ψ_δ; potential of Stern layer. As it is assumed usually that zeta-potential is very close to the potential of Stern layer, ψ_δ, the surface charge density, σ, that is the amount of specifically sorbed ions, can be estimated from the above equation. The estimate by this equation doesn't have good precision generally, but is convenient for simplicity of zeta-potential measurement. According to this approach, the difference of sorption amount of carbonate species onto goethite suggested a maximum around pH7 at which the difference of two zeta-potential curves showed a maximum in Fig.5(a). Sorption of carbonate species onto metal oxide particles was investigated by measuring headspace P_{CO_2} in equilibrium with solid suspension[6]. The authors found that sorption of carbonate species onto goethite showed a well developed maximum at pH6 for a fixed total amount of CO_2 in the suspension. This may be concerned with the europium sorption enhancement by carbonate species around pH6, although it is difficult to estimate quantitatively as the concentration of carbonate species was not constant in our experiment.

Fig.5(b) shows the zeta-potential change of goethite when europium is sorbed as a function of pH. The zeta-potential of goethite in the solution including europium (initial concentration of europium is 10^{-5}M) is increased slightly over wide pH range compared with that of goethite in the absence of europium. It is implied that europium sorbed as cation complex such as Eu^{3+}, $EuOH^{2+}$, and $EuCO_3^+$ mainly, not as anion complex such as $Eu(CO_3)_2^-$, $Eu(CO_3)_3^{3-}$, and $Eu(CO_3)_4^{5-}$.

Model calculations

Surface complexation models (SCMs) such as the triple layer model are being used to describe the surface chemistry of variable charged particles. The summaries of SCMs are given elsewhere in details[3]. Preliminary analysis using SCM was applied to evaluate the effect of carbonate species sorption on europium sorption.

Infrared absorption studies of CO_2 adsorption on moist goethite suggest that coordination interaction takes place at the oxide surface[12]. If carbonate species bind as inner-sphere complexes of the goethite surface, it is reasonable to postulate surface complex stoichiometries as described below:

$$\equiv Fe-OH + H^+ + CO_3^{2-} \leftrightarrow \equiv Fe-OCOO^- + H_2O$$
$$\equiv Fe-OH + 2H^+ + CO_3^{2-} \leftrightarrow \equiv Fe-OCOOH + H_2O.$$

The two reactions amount to replacing an Fe-bound surface hydroxyl with a more acidic carboxylic group and would, therefore, tend to lower the IEP and PZC of the suspension. Our zeta-potential results which the zeta-potential of goethite decreased in higher concentration solution of carbonate species supports formation of surface carboxylic group. Geen et al. [6] determined the model parameters for triple layer model using stoichiometries of the type Fe-OCOOH and Fe-OCOO$^-$ for surface bound carbonate species by calculating their potentimetric titration data under controlled CO_2 conditions using the least squares optimization program FITEQL[14]. In addition to this, they showed that carbonate ions prevent chromate anion, CrO_4^{2-}, from sorbing onto goethite owing to sorption site coverage by sorbed carbonate species as described in the above equations. The production of carboxylic group of type FeOCOO$^-$ may act as a sorption sites for cations such as europium. We applied the surface complexation model using a triple layer model to interpret our sorption results for preliminary analysis.

The set of surface complexation reactions and the corresponding complexation constants for our model calculations are summarized in Table II. The chemical reactions and their equilibrium constants in liquid phase are the same as the data summarized in Table I. The surface complexes combined with anion complexes of europium such as $Eu(CO_3)_2^-$, $Eu(CO_3)_3^{3-}$, and $Eu(CO_3)_4^{5-}$ were not considered because europium sorbed as cation such as Eu^{3+}, $EuOH^{2+}$, and

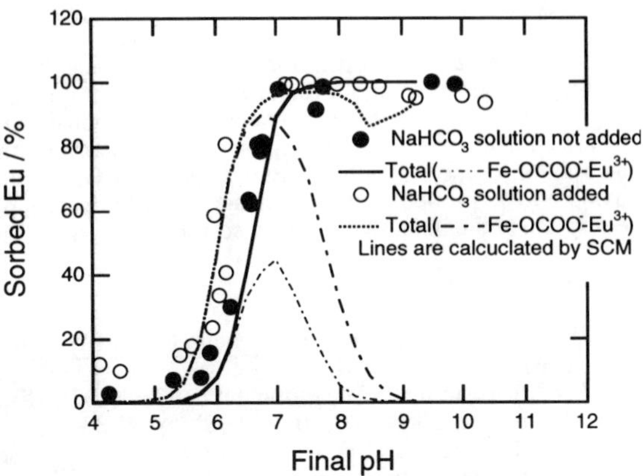

Fig.6 Simulation of europium sorption on goethite in 0.01M $NaClO_4$ and in 0.01M $NaClO_4$ + $NaHCO_3$. The concentration of carbonate species are indicated in Fig.2 as solid lines.

Table II Surface complex reactions and the corresponding complexation constants.

$KFe\text{-}OH_2^+$	$= [Fe\text{-}OH_2^+]\exp(F\psi_0/RT)/[Fe\text{-}OH]\gamma_1[H^+]$	$=10^{7.90*1}$
$KFe\text{-}O^-$	$= [Fe\text{-}O^-]\gamma_1[H^+]\exp(-F\psi_0/RT)/[Fe\text{-}OH]$	$=10^{-9.90*1}$
$KFe\text{-}OH_2^+\text{-}ClO_4^-$	$= [Fe\text{-}OH_2^+\text{-}ClO_4^-]\exp(F(\psi_\beta-\psi_0)/RT)/[Fe\text{-}OH]\gamma_1[H^+]\gamma_1[ClO_4^-]$	$=10^{8.89*1}$
$KFe\text{-}O^-\text{-}Na^+$	$= [Fe\text{-}O^-\text{-}Na^+]\gamma_1[H^+]\exp(F(\psi_0-\psi_\beta)/RT)/[Fe\text{-}OH]\gamma_1[Na^+]$	$=10^{-8.76*1}$
$KFe\text{-}OCOO^-$	$= [Fe\text{-}OCOO^-]\exp(-F\psi_0/RT)/[Fe\text{-}OH]\gamma_1[H^+]\gamma_2[CO3_2^-]$	$=10^{12.45*1}$
$KFe\text{-}O^-\text{-}Eu^{3+}$	$= [Fe\text{-}O^-\text{-}Eu^{3+}]\gamma1[H^+]\exp(2F\psi_0/RT)/[Fe\text{-}OH]\gamma_3[Eu^{3+}]$	$=10^{5.0}$
$KFe\text{-}O^-\text{-}EuOH^{2+}$	$= [Fe\text{-}O^-\text{-}EuOH^{2+}](\gamma1[H^+])^2\exp(F\psi_0/RT)/[Fe\text{-}OH]\gamma_3[Eu^{3+}]$	$=10^{-3.0}$
$KFe\text{-}OCOO^-\text{-}Eu^{3+}$	$= [Fe\text{-}OCOO^-\text{-}Eu^{3+}]\exp(2F\psi_0/RT)/[Fe\text{-}OH]\gamma1[H^+]\gamma2[CO_3^{2-}]\gamma_3[Eu^{3+}]$	$=10^{28.0}$

*1 Geen et al.[6]

γ_1, γ_2, and γ_3 ; activity coefficient for single, double, and triple charged ions

F; Fraday's constant, ψ_0, ψ_β ; surface potentials at the 0 and β planes, respectively.

Specific surface area 36m^2/g, Site density 2.3 site/nm^2

$EuCO_3^+$ mainly as discussed from the results of zeta-potential measurement. It is thought that metal cations such as Co^{2+}, Ni^{2+}, Cd^{2+}, and Cu^{2+} sorb on surface hydroxyl groups as bare ions (i.e. M^{C+}) or as its hydroxide ions usually[3]. But our sorption data couldn't be simulated by SCM calculations considered the sorption reactions on surface hydroxyl group conbined with Eu^{3+} and $EuOH^{2+}$ because the chemical equilibrium of these reactions were not affected by the concentration of carbonate species. When the europium sorption on surface carboxylic group $(Fe\text{-}OCOO^-\text{-}Eu^{3+})$ was considered in addition to above reactions, the sorption data was reproduced as described in Fig.6. This may suggest that the surface carboxylic group, $FeOCOO^-$, produced by carbonate species sorption acts as a sorption site for cations such as Eu^{3+}, $EuOH^{2+}$ and enhanced europium sorption. It should be noted that such an approach to sorption modeling is essentially based on curve fitting and therefore, it must be verified whether a set of chosen constants for surface complexes represents actual cases.

SUMMARY

It is observed that sorption of europium was enhanced in higher concentration solutions of carbonate species for both materials, goethite and magnetite. The zeta-potential measurements suggest that the surface property change by carbonate species sorption on solid phases contribute to the enhancement of europium sorption and europium sorbs as a cation complex mainly, not as a anion complex such as $Eu(CO_3)_2^-$, $Eu(CO_3)_3^{3-}$, and $Eu(CO_3)_4^{5-}$. The preliminary analysis using SCM implied that the surface carboxylic group, $FeOCOO^-$ acts as a sorption site for europium and enhances europium sorption.

REFERENCES

1. W.Stumm and J.J. Morgan, Aquatic Chemistry, 3rd ed. (John Willey and Sons, New York, 1996), p.149.
2. PNC, Research and Development on Geological Disposal of High-Level Radioactive Waste, PNC report, TN1410 93-059(1992), pp.62-83.
3. D. A. Dzombak and F. M. M. Morel, Surface complexation modeling, (John Wiley & Sons, New York, 1990).
4. C. P. Schulthess and J. F. McCarthy, Soil Sci. Soc. Amer. J., **54**, 688(1990).
5. J. M. Zachara, D. C. Girvin, R. L. Schmidt and C. T. Resch, Environ. Sci. Tech., **21**, 589(1987).
6. A. V. Geen, A. P. Robertson, and J. O. Leckie, Geochim. et Cosmochim. Acta, **58**, 2073 (1994).
7. H. Meier, E. Zimmerhackl, G. Zeitler and P.Menge, in Scientific Basis for Nuclear Waste Management XVIII, Edited by T. Murakami and R. C. Ewing (Mat. Res. Soc. Proc.353, Pittsburgh, PA, 1994) pp.1061-1068.
8. F. J. Atkinson, A. M. Posner, and J. P. Quirk, J. Phys. Chem., **71**, 550(1967).
9. W. D. Schecher and D. C. McAvoy, Computers, Enviroment and Urban Systems, **16**, 65(1992).
10. J. C. Westall, J. L.Zachary, and F. M. M. Morel, MINEQL:A Computer Program for the Calculation of the Chemical Equilibrium Composition of Aqueous Systems, (Report 86-01, Dept. of Chem., Oregon State Univ., Corvallis, Oregon, 1986).
11. J. E. Cross and F. T. Ewart, Radiochim. Acta, **52/53**, 421(1991).
12. D. G. Lumsdon and L. J. Evans, J. Colloid and Interface Sci., **164**, 119(1992).
13. W. A. Zeltner and M. A. Anderson, Langmuir, **4**, 469(1988)
14. J. C. Westall, FITEQL VERSION 2.0:A Computer Program for Determining of Chemical Equilibrium Constants for Experimental Data, (Report 82-01, Dept. of Chem., Oregon State Univ., Corvallis, Oregon, 1982).

THE EFFECT OF SPECIFIC SURFACE AREA ON RADIONUCLIDE SORPTION ON CRUSHED CRYSTALLINE ROCK

P. HÖLTTÄ[*], M. SIITARI-KAUPPI[*], P. HUIKURI[*], A. LINDBERG[**] and
A. HAUTOJÄRVI[***]

[*] University of Helsinki, Department of Chemistry, Laboratory of Radiochemistry, P.O.Box 55, FIN-00014 University of Helsinki, Finland.

[**] Geological Survey of Finland, Kivimiehentie 1, FIN-02150 Espoo, Finland.

[***] VTT Energy, Nuclear Energy, P.O.Box 1604, FIN-02044 VTT, Finland.

ABSTRACT

The sorption of sodium (^{22}Na), calcium (^{45}Ca) and strontium (^{85}Sr) was studied on mica gneiss, unaltered, moderately altered and strongly altered tonalite samples taken from hole SY-KR7 drilled in the Syyry area in Sievi, Western Finland. The crushed rock samples were sieved into six fractions from 71 μm to 1250 μm. A proportional mineral composition for the different fractions were estimated by X-ray diffraction. The specific fraction surface areas were determined by the BET nitrogen adsorption method. The fractal method was applied to characterize rocks and to describe quantitatively surface irregularity. The mass distribution ratio values for each fraction were determined using the static batch method. The sorption of tracers onto different minerals was observed using rock thin sections. K_d-values calculated from thin section K_a-values and K_d-values obtained from batch experiments were in good agreement. Mass distribution ratios for different size fractions are given, and the effect of the specific surface area is discussed. Owing to larger specific surface areas considerably higher sorption on smaller fractions was found for altered tonalites.

INTRODUCTION

The Finnish high-level nuclear waste disposal concept is based on a multi-barrier arrangement in crystalline rock, which is the natural barrier isolating radionuclides from the biosphere. Radionuclides possibly released from an underground repository are transported mainly by the groundwater flow in fractures. The migration of dissolved radionuclides is retarded both by interaction with the fracture surfaces and fracture filling materials, and by diffusion into the fissures of the rock. To be able to predict the transport and radionuclide retardation in rock fractures and rock matrices, it is essential to understand the different physical and chemical phenomena involved, as well as the retardation properties of unfractured rock.

In transport models, radionuclide retardation has been taken into account by the K_d concept, in which a retardation factor is used to apply the distribution ratio to radionuclide transport. Determination of the retardation factor by the means of fracture flow experiments is a direct approach to application of the effects of sorption in radionuclide transport models. It may, however, be difficult to relate the parameters measured by the static method for crushed rock to a heterogeneous intact rock, owing to differences in trace mineral compositions and in the availabilities of minerals to the aqueous phase. The results of static and dynamic experiments can be compared using hydrologically homogeneous crushed rock columns. Different approaches for measuring the interaction between radionuclides and rock matrix are needed in order to test the compatibility of transport models and retardation experiments.

Mat. Res. Soc. Symp. Proc. Vol. 465 © 1997 Materials Research Society

This study is part of the research project in which different laboratory scale methods have been introduced in order to study radionuclide migration. The transport and retardation behavior of non-sorbing radionuclides and sodium in rock fracture and crushed rock columns has been studied earlier as well as some batch K_d values were determined for mica gneiss, unaltered and strongly altered tonalite [1, 2, 3]. These results showed stronger sodium sorption on altered tonalite than expected. In this work sorption of sodium, calcium and strontium was studied in the same materials including moderately altered tonalite. These experiments aim to explain sorption behavior of sodium and examine the effects of mineral composition and particle size on radionuclide retardation.

EXPERIMENTAL

The rock samples representing different rock features and porosities were obtained from the hole SY-KR7 drilled in the Syyry area in Sievi, Western Finland. Syyry was one of the five sites selected for preliminary investigations with regard to the final disposal of spent nuclear fuel in Finland by Teollisuuden Voima Oy. Drill hole SY-KR7 [4] intersects a medium-grained slightly schistose tonalite zone containing a few fine-grained and homogeneous mica gneiss inclusions. A quantitative mineral composition for the rock samples was analyzed by point counting method from the thin sections. The main minerals in unaltered tonalite sample (280 m) were plagioclase (53.2 %), quartz (19.6 %), biotite (14.0%) and hornblende (10.0 %). The accessory minerals were chlorite (1.2 %), sphene (0.2 %), muscovite/sericite (1.4 %) and calcite (0.4 %). Apatite, epidote, opaques and analcime were optically observed. Calcite filling was observed in some pores between mineral grains. In strongly altered tonalite (108 m) plagioclase has been altered to sericite and partly to kaolinite, hornblende to biotite and chlorite. In alteration process plenty of fine grained opaque was formed. In moderately altered tonalite (201 m) mineral grains were cemented by analcime. Grain size in mica gneiss (87 m) was very small ($\leq$0.1-0.3 mm) with clear foliation. The main minerals are quartz, plagioclase and biotite. The accessories found were chlorite, calcite, muscovite/sericite, sphene, apatite, epidote and opaques. Fissures filled with calcite cement and surrounded by alteration zones were observed. In the alteration zone plagioclase was strongly altered to sericite and biotite totally to chlorite. There is also found a younger fissure cutting the first one. The alteration is similar but it contains also opaque minerals.

The crushed rock samples were sieved into the six fractions from 71 μm to 1250 μm. The specific surface areas for each fraction given in Table I were determined by the BET nitrogen adsorption method. Highest specific surface areas were determined for the smallest fractions. A qualitative mineral composition for the different fractions were estimated by X-ray diffraction method, which gives a rough understanding how minerals are separated by crushing and sieving into the fractions. In all fractions the main mineral was plagioclase, which was sericitized in altered rocks. In altered tonalites biotite was chloritized. Some chlorite was found also in mica gneiss and unaltered tonalite samples which had quite similar mineral composition except amphibole only in unaltered tonalite. In the largest fraction of strongly altered tonalite (1250 μm) proportional mineral composition was 46% plagioclace, 25 % prehnite and 11 % chlorite, 7 % quartz, 6 % laumontite and 5 % muscovite. The smallest fraction (71μm) contained 34 % plagioclase, 28 % chlorite, 13 % laumontite, 12 % prehnite, quartz and muscovite approx the same as above. Moderately altered tonalite contained one half of plagioclase and one third of zeolite analcime in each fraction. Proportional composition of other minerals in the smallest fraction was 13 % chlorite and 9 % mica. In the largest fraction the composition was 7 % chlorite and 9 % mica.

Table I. Specific Surface Area ($m^2 \cdot g^{-1}$), Particle Diameter (μm) and Fractal Dimension D for Fractions of Mica Gneiss and Tonalites.

Particle diameter (μm)	Specific surface area ($m^2 \cdot g^{-1}$)					
	Strongly altered tonalite	Mica gneiss	Unaltered tonalite	Moderately altered tonalite	Reference Quartz	Reference Goodland
2605	-----	-----	-----	-----	-----	11.7
1300 - 1410	-----	-----	-----	-----	0.00723	-----
>1250	0.41	0.09	0.08	0.61	-----	-----
850- 1250	0.48	0.18	0.12	0.71	0.01074	-----
651	-----	-----	-----	-----	-----	11.54
300 - 850	0.52	0.16	0.15	1.05	0.01765	12.23
200 - 300	0.43	0.21	0.18	1.07	0.02368	-----
150 - 200	0.89	0.29	0.28	1.55	0.03147	12.10
71 - 150	1.25	0.69	0.77	2.5	0.0547	12.63
D = Fractal dimension	*2.62*	*2.43*	*2.32*	*2.54*	*2.21*	*2.91*

Table II. The Composition of the Synthetic Granitic Groundwater [5].

Ion	(M)	($mg \cdot l^{-1}$)
HCO_3^-	$2.01 \cdot 10^{-3}$	123
H_4SiO_4	$2.05 \cdot 10^{-4}$	12
SO_4^{2-}	$1.00 \cdot 10^{-4}$	9.6
Cl^-	$1.97 \cdot 10^{-3}$	70
Ca^{2+}	$4.47 \cdot 10^{-4}$	18
Mg^{2+}	$1.77 \cdot 10^{-4}$	4.3
K^+	$1.00 \cdot 10^{-4}$	3.9
Na^+	$2.83 \cdot 10^{-3}$	65

pH=8.2
I=0.0008 mol·l⁻¹

The batch method was used to determine the mass distribution ratios of sodium (^{22}Na), calcium (^{45}Ca) and strontium (^{85}Sr) on the crushed rock fractions. The rock samples were equilibrated with the synthetic granitic groundwater (Table II) for 20-30 days by changing the water 4-6 times. Sodium, potassium, magnesium and calcium concentrations were followed by atomic adsorption spectroscopy determinations. Spiked solutions containing ^{22}Na, ^{45}Ca or ^{85}Sr came into contact with crushed rock in polypropylene centrifuge tubes which were shaken gently. The initial strontium concentration was 10^{-5} M. Experiments were carried out for mass/liquid ratios of 1:10. After 7 and 21 day contact times groundwater and crushed rock was separated by centrifugation and the samples were washed five times. The mass distribution ratio, K_d, was determined in batch experiments as the ratio of the radionuclide concentration in the solid phase to the aqueous concentration.

The sorption of tracers onto different minerals was estimated from autoradiographs taken from the spiked thin sections thickness of 30 μm. The thin sections were equilibrated with the synthetic granitic groundwater for 7 days by changing the water 3 times. After 7 days contact time washed thin sections were exposed on a β film (Hyper film βmax) and on X-ray film (Kodak X-OMAT MA). ^{22}Na and ^{45}Ca had suitable beta energies for rough autoradiographic measurement, eg. the range of isotopes was below the size of mineral grains. The surface based distribution coefficient, K_a, was calculated using geometric sorption area determined by digital image processing of autoradiographs. K_d-values were calculated from K_a-values using thin section thickness of 30 μm and densities of 2.4 g·cm^{-3} for altered rock type and 2.8 g·cm^{-3} for unaltered rock type.

RESULTS AND DISCUSSION

A fractal method [6] was applied to characterize rocks and to describe quantitatively surface irregularity. Determination of surface fractal dimension (D) is based on an assumption that in BET nitrogen specific surface area determination the apparent monolayer value of absorbent (n, moles·g^{-1}) depend on the change in average radius (R) of the spheroidal particle:

$$n \propto R^{D-3} \qquad (1)$$

Equation (1) may be expressed in terms of apparent surface areas (A, m^2·g^{-1}):

$$A = Nn\sigma \propto R^{D-3} \qquad (2)$$

where σ is the effective cross sectional area of the adsorbate and N is Avogadro's constant.

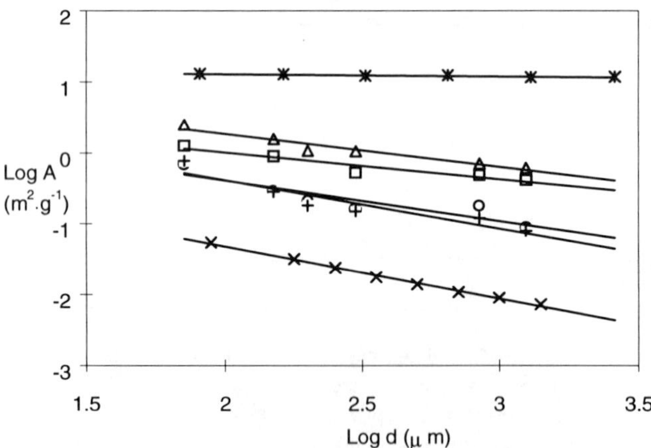

Figure 1.   Surface Areas as a Function of the Minimum Particle Diameter d for Sieved Fractions of Strongly Altered (square), Moderately Altered (Δ), Unaltered Tonalite (+) and Mica Gneiss (o). Ref. 1 (*) is Carbonate Rock (Goodland) and Ref. 2 (x) Quartz [6].

In practice the same adsorbate is used for all sieved fractions. Surface areas are presented as a function of minimum particle diameter (d) for sieved fractions of each rock (Figure 1). If the specific surface follows a fractal model distribution then log A/log d plot gives a straigth line, and D is obtained from the slope [6]. Obtained fractal dimension values for strongly altered, moderately altered, unaltered tonalite mica gneiss are given in Table I with reference values for carbonate rock (Goodland) and for quartz. D values in the range of $2 \leq D < 3$ have been reported [6]. Higher surface irregularity is characterized by higher fractal dimension.

Slight sorption of sodium, less than 10 % was obtained for unaltered tonalite and mica gneiss. Sodium sorption was very strong on moderately altered and strongly altered tonalite. After one week contact time sodium sorption on unaltered rocks was higher than after three weeks. The effect of contact time was lower on altered rocks. Mass distribution ratio, K_d, values of 0.0001-0.003 $m^3 \cdot kg^{-1}$ for unaltered rocks and 0.03-0.6 $m^3 \cdot kg^{-1}$ for altered rocks were determined.

Owing to very slight sorption on mica gneiss and unaltered tonalite no effect of fraction size was observed. Sodium sorption on moderately altered and strongly altered tonalite as a function of fraction size after one week contact time is shown in Figure 2. Owing to larger specific surface areas considerably higher sorption on smaller fractions was found for altered tonalites. Mineral distribution was quite even between different fractions except chlorite and mica minerals (biotite and muscovite) which were slightly enriched into the smallest size fraction.

On the basis of the autoradiographs taken from the spiked thin sections sodium was sorbed onto sericitized and saussuritized plagioclase grains and chloritized biotite in strongly altered tonalite. Sorption onto laumontite, prehnite and calcite was very slight. Main reason for higher sodium sorption on moderately altered tonalite was zeolite analcime. In thin sections analcime was observed around unaltered plagioclase grains. Sodium sorption onto mica gneiss and unaltered tonalite was slight and only some biotite and chlorite grains were observed in the autoradiographs.

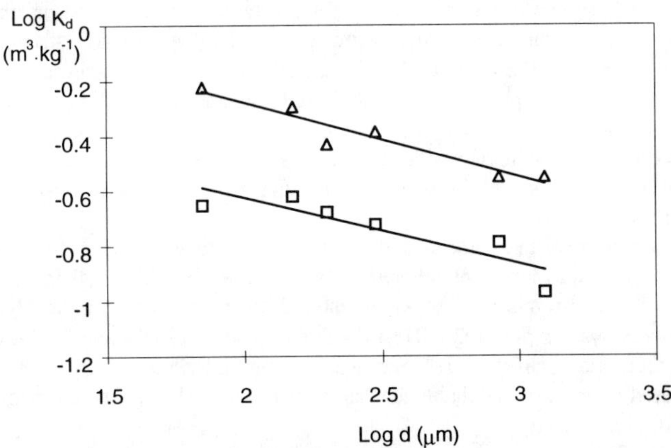

Figure 2.   Sodium Sorption on Strongly Altered (square) and Moderately Altered Tonalite (Δ) as a Function of Minimum Particle Diameter d (μm).

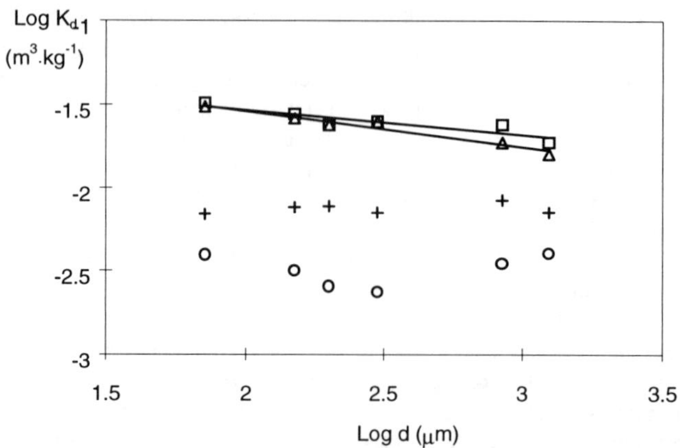

Figure 3.   Calcium Sorption on Mica Gneiss (o), Unaltered tonalite (+), Moderately Altered Tonalite (Δ) and Strongly Altered Tonalite (square) as a Function of Minimum Particle Diameter d (μm).

Calcium sorption on mica gneiss and unaltered tonalite was slightly higher than sodium sorption. Sorption on strongly altered and moderately altered tonalite was considerably lower. Mass distribution ratio, K_d, values of 0.002-0.009 m^3·kg^{-1} for unaltered rocks and 0.02-0.04 m^3·kg^{-1} for altered rocks were determined. After three weeks contact time, only a slight increase was found in sorption compared to values after one week probably owing to diffusion into the grains. Difference in the K_d values is partly result of 3-7 times higher specific surface areas for altered than for unaltered rocks (Table I). Calcium sorption as a function of grain size after one week contact time is shown in Figure 3. No grain size effect was observed on mica gneiss and unaltered tonalite. For strongly altered tonalite and moderately altered tonalite difference in sorption was 10 % between the smallest (71 μm) and largest (1250 μm) grain size.

The mineral specific sorption was not obvious for calcium according to autoradiographs of thin sections. Calcium was observed in fine grained opaques and controlled by altered minerals in strongly altered and moderately altered tonalites. In mica gneiss and unaltered tonalite calcium was observed in some biotite grains and in a few slightly sericitized plagioclase grains.

Strontium sorption on mica gneiss and unaltered tonalite was similar to calcium sorption. Sorption on strongly and moderately altered tonalite was slightly higher than calsium sorption. Compared to sodium strontium sorption on altered tonalites was considerably lower. Mass distribution ratio, K_d-values of 0.002-0.015 m^3·kg^{-1} for unaltered rocks and 0.02-0.05 m^3·kg^{-1} for altered rocks were determined. After one week strontium sorption on both unaltered and moderately altered rocks was slightly higher than after three weeks but the effect of contact time, however, was not remarkable.

Strontium sorption on rocks as a function of fraction size after one week contact time is shown in Figure 4. Owing to quite slight sorption on unaltered rocks no effect of fraction size was observed. For moderately altered tonalite no remarkable difference was obtained due to fraction size. Fair sorption dependence on surface area was found for strongly altered tonalite owing to largest specific surface areas of the smallest fractions. Stronger sorption on small fractions is explained by larger specific surface areas and possibly by changeable calcium and magnesium containing minerals, which were separated mostly into the smallest size fractions.

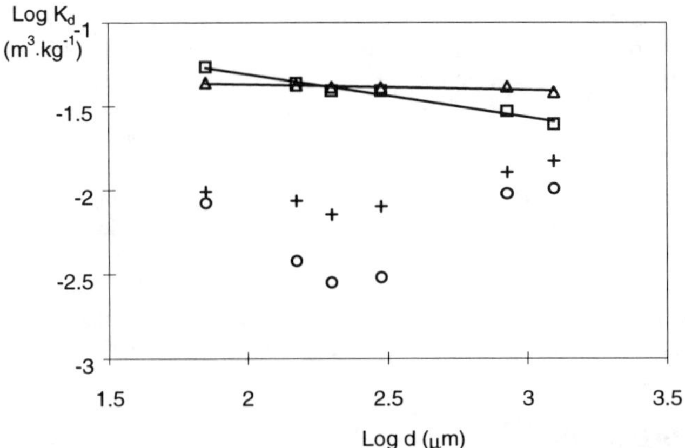

Figure 4.   Strontium Sorption on Mica Gneiss (o), Unaltered Tonalite (+), Moderately Altered Tonalite (Δ) and Strongly Altered Tonalite (square) as a Function of Minimum Particle Diameter d (μm).

Table III.   Comparison of K_d ($m^3 \cdot kg^{-1}$) Values of Sodium, Calcium and Strontium Obtained from Batch to Values Evaluated from Thin Section Experiments. Batch Experiment Values are from Fractions 15-22 μm and One Week Contact Time.

	Strongly altered tonalite	Mica gneiss	Unaltered tonalite	Moderately altered tonalite
^{22}Na				
Batch	0.21	0.0014	0.0018	0.37
Thin section	0.15	0.0064	0.0061	0.33
^{45}Ca				
Batch	0.024	0.003	0.008	0.024
Thin section	0.029	0.002	0.004	0.017
^{85}Sr				
Batch	0.04	0.003	0.007	0.04
Thin section	0.03	0.003	0.007	0.02

The surface based distribution coefficient, K_a, was evaluated from values from batch K_d-values on the basis of specific surface area. Evaluated K_a-values and values from thin section experiments are not comparable. Owing to porous thin sections geometric surface used in calculations was much smaller than real specific surface area resulting in higher values from thin section experiments. K_d-values calculated from thin section K_a-values are compared to K_d-values obtained from batch experiments (Table III). Obtained good agreement indicates that the specific surface area usable in thin section experiments was equal to that in batch experiments.

CONCLUSIONS

Sodium, calcium and strontium sorption on unaltered tonalite and mica gneiss was slight and no difference due to fraction size was observed. Owing to only weak physical interactions it is not reasonable to expect any specific surface area dependence. Higher sodium and strontium sorption on altered tonalites was obtained. Owing to larger specific surface areas considerably higher sorption on smaller fractions was found for altered tonalites. High sodium sorption on altered tonalites was explained by the composition of sodium sorptive alteration minerals, especially analcime in moderately altered tonalite. This work has been our first attempt to take account the effect of fraction size and surface area on radionuclide sorption. More experiments with stronger sorptive radionuclides and wider fraction size distribution are needed. Evaluated K_a-values and values from thin section experiments were not comparable. K_d-values calculated from thin section K_a-values and K_d-values obtained from batch experiments were in good agreement. Thin section method was found useful in order to estimate radionuclide sorption onto different minerals.

ACKNOWLEDGMENTS

This study is part of the publicly financed nuclear waste management research programme in Finland.

REFERENCES

1. P. Hölttä, M. Hakanen, A. Hautojärvi, J. Timonen, K. Väätäinen, J. Contam. Hydrol. <u>21</u>, (1996) 165-173.
2. P. Hölttä, M. Hakanen, M. Siitari-Kauppi, A. Hautojärvi in <u>Scientific Basis for Nuclear Waste Management XVIII</u>, (Mater. Res. Soc. Proc. 353, Pittsburgh, 1995) 355-362.
3. P. Hölttä, M. Siitari-Kauppi, M. Hakanen, T. Huitti, A. Hautojärvi, A. Lindberg, in print J. Contam. Hydrol. (1997).
4. A. Lindberg and M. Paananen, Teollisuuden Voima Oy, Site Investigations Work Report 92-34, Helsinki, 1992 (in Finnish).
5. B. Allard and J. Beall, J. Environm. Sci. Health <u>6</u>, 507 (1979).
6. Avnir, D. Farin, P. Pfeifer, J. Coll. Imterf. Sci. <u>103</u>, No 1 (1985) 112-123.

URANYL SORPTION ONTO ALUMINA

ANNA-MARIA M. JACOBSSON* AND ROBERT S. RUNDBERG **
*Department of Nuclear Chemistry, Chalmers University of Technology, 412 96
Gothenburg, Sweden
** Los Alamos National Laboratory, MS-J514, Los Alamos, NM 87545, USA

ABSTRACT

The mechanism for the adsorption of uranyl onto alumina from aqueous solution was studied experimentally and the data were modeled using a triple layer surface complexation model. The experiments were carried out at low uranium concentrations ($9x10^{-11}$ - $5x10^{-8}$ M) in a CO_2 free environment at varying electrolyte concentrations (0.01 - 1 M) and pH (4.5 - 12). The first and second acid dissociation constants, pK_{a1} and pK_{a2}, of the alumina surface were determined from potentiometric titrations to be 7.2 ± 0.6 and 11.2 ± 0.4, respectively. The adsorption of uranium was found to be independent of the electrolyte concentration. We therefore conclude that the uranium binds as an inner sphere complex. The results were modeled using the code FITEQL. Two reactions of uranium with the surface were needed to fit the data, one forming a uranyl complex with a single surface hydroxyl and the other forming a bridged or bidentate complex reacting with two surface hydroxyls of the alumina. There was no evidence from these experiments of site heterogeneity. The constants used for the reactions were based in part on predictions made utilizing the Hard Soft Acid Base, HSAB, theory, relating the surface complexation constants to the hydrolysis of the sorbing metal ion and the acid dissociation constants of the mineral oxide surface.

INTRODUCTION

Naturally distributed uranium is found in much larger proportions in fracture filling materials, especially in the clay minerals, than in granite [1]. Clay minerals are made up of alumina and iron silicates, and are believed to have several different mechanisms for retarding the mobility of uranium. To be able to predict whether the uranium will sorb or not under the various possible conditions in the fractures, it is of interest to find what specie(s) will sorb and what type of complex it will form with the various groups.

The sorption on a given surface will be affected by the pH, electrolyte concentration, and the concentration of complexing ligands. The complexes with the surfaces can be inferred from the stoichiometry of sorption. Kinniburgh [2] showed that the surface complexation constants of alkaline earths and transition metals are proportional to the first hydrolysis constants of the metals. Dzombak and Morel [3] recognized this proportionality to be a consequence of the Hard Soft Acid Base theory [4]. They further demonstrated the proportionality to the first hydrolysis constant for the surface complexation of transition metal ions onto hydrous ferric oxide. Schindler [5] found that for the bidentate attachment of metal ions the surface complexation constant is proportional to the aqueous second

Mat. Res. Soc. Symp. Proc. Vol. 465 © 1997 Materials Research Society

hydrolysis constant. A complete Hard Soft Acid Base theory will allow prediction of surface complexation constants on the basis aqueous hydrolysis constants, which are easier to obtain.

Uranium(VI) was chosen for study because it is a moderately soft Lewis acid and it is relatvely soluble. Both the uranium and the metal oxide surface will hydrolyze to different extents at different pH depending on their relative acidity. The surface acid dissociation constants for the mineral oxides are determined by potentiometric titration. The surface acid dissociation constants reported for alumina range from 5.7 to 8.5 for pK_{a1} and 8.7 to 11.5 for pK_{a2} [6].

Uranium forms a number of different complexes in solution. The carbonate complexes are already among the dominating species at pH 6 and polynuclear complexes are significant at uranium concentrations of 10^{-4} M [7]. Del Nero [8] studied the uranium sorption at two different uranium concentrations [10^{-6} and 10^{-4} M] and found a dependence on the uranium concentration. EXAFS of U(VI) sorption onto ferrihydrites indicate that bidentate complexation is the dominating mechanism for the complexation with the surface [7]. It could not be determined from these studies whether polynuclear complexes did form with the surface or not.

In this work we have studied both the surface acid dissociation reactions of the alumina and the sorption behavior of uranium under varying pH(5-11), ionic strengths (0.01-1 M), low uranium concentrations (9×10^{-11}-5×10^{-8} M) and solid to solution ratios (0.5 and 10 $g{\cdot}l^{-1}$) in an argon atmosphere, minimizing the influence of complexing carbonate.

EXPERIMENTAL

<u>Reagents and Reagent preparations</u>

All chemicals used were reagent grade. A 50 wt % NaOH was made CO_2 free by addition of barium nitrate, and diluted as needed. The concentration was determined by Gran titration[9].
Alumina: The alumina was a neutral aluminum oxide powder from JT Baker; its surface area was determined with multipoint BET to 142.9 m^2/g. The alumina was not pretreated.
U(VI) solution: The ^{232}U solution was provided by LANL (produced by neutron activation of ^{231}Pa). The ^{232}U solution was separated from its daughters by means of an anionic exchange before the sorption experiments. A >99.92% pure solution of natural uranium from JT Baker was used for sorption at higher concentrations.

<u>Procedure</u>

Titrations
All titrations were carried out in a temperature controlled glass beaker at $25^{\circ}C$ under argon atmosphere. The automatic titration unit consisted of a Hamilton injection dispenser, a Keithley 617 Electrometer and a LABVIEW program for measuring the

potential and controlling the titrations (size and time interval for the additions), developed by Rundberg[10]. Prior to all titrations the solution and solid were mixed, a specific amount of acid added and the solution was purged by argon for at least 1h to remove CO_2. For each ionic strength a Gran calibration titration without the solid was made in the same manner. The pH was measured with a combination glass electrode with a Ross reference sleeve junction for NaCl titrations. By using a double junction reference electrode with the same outer solution as the one titrated, the influence of diffusion was minimized. Additions were made in intervals of 15 min. for the calibration titrations and 2 hr. for the surface titrations. It was estimated that equilibrium was reached to at least 98 % for these time intervals. One back titration was made for 1 M $NaNO_3$.

Sorption of U(VI)
The sorption of the uranium onto the alumina was calculated from measurement of the ^{232}U activity of the aqueous phase at varying conditions (pH, ionic strength, electrolytes, uranium concentrations and solid to solution ratios). Two different methods were used: batch sorption and continuous sorption with samples taken from the same solution at different pH. The uranium concentrations in the aqueous solutions were measured by liquid scintillation counting.

The batch sorption was performed in polycarbonate Oak Ridge centrifuge tubes (NALGENE) at room temperature. After two days of equilibrium time the liquid was separated from the solid by two consecutive centrifugations. The pH was measured in the original tube after the sample was taken. All but centrifugation was performed in a CO_2 free glove box under Ar atmosphere. The conditions were: 0.01 and 0.1 M NaCl, 0.1 and 1 M $NaNO_3$, 5×10^{-8} - 9×10^{-11} M uranium, pH 5-12 for 0.01 g solid in 20 ml solution. The estimated dissolution of alumina at the maximum pH is less than 5 percent [11].

Two sets of continuous sorption measurements were made, 0.01 M $NaNO_3$, 4.2×10^{-10} M uranium and 0.1 M NaCl, 2.8×10^{-9} M uranium at $25^{\circ}C$, 10 g solid$\cdot$liter^{-1}. Before the solid was added the original uranium concentration in the aqueous phase was measured. The samples were withdrawn by means of vacuum tubes (VACUTAINER) allowing at least two hours of equilibrium time since the last addition of acid or base. The samples were centrifuged at 5000 rpm for half an hour, part of the liquid fraction was withdrawn and centrifuged for 1 hour at 15000 rpm.

RESULTS AND DISCUSSION

The results and discussion are divided into surface titrations and uranium sorption.

<u>Surface titrations, determination of alumina surface acidity constants</u>

The intrinsic equilibrium constants for the surface acidity reactions were determined to $pK_{a1}^{int} = 7.2 \pm 0.6$, $pK_{a2}^{int} = 11.2 \pm 0.4$ from the titrations by extrapolating the apparent equilibrium constants, K_{a1}^{app} and K_{a2}^{app}, to zero inner surface charge (Figure 1).

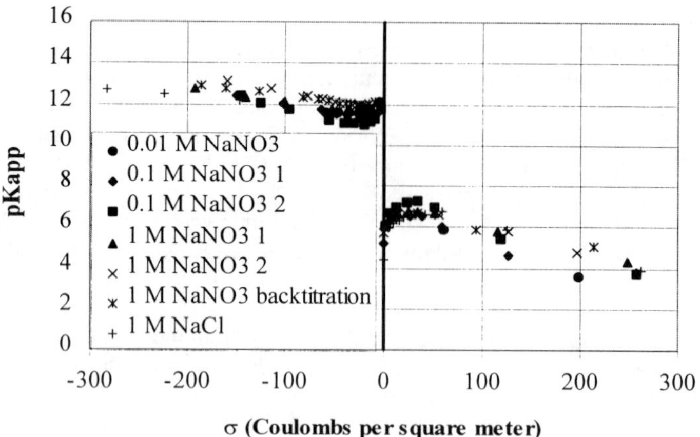

Figure 1   The negative log of the apparent equilibrium constants as a function of surface charge.

The data close to zero inner surface charge were not used in the extrapolation as these deviate. These data deviate because the errors in hydrogen ion mass balance are greatest near the point of zero charge. In addition , a small degree of aggregation of the particles could diminish the number of available hydroxyl sites, slow the rate of equilibration, and cause a suspension error in the pH determination. The apparent equilibrium constants were calculated from the bulk concentrations as:

$$K_{a1}^{app} = \frac{[\equiv SOH]\left\{H_{bulk}^{+}\right\}}{\left[\equiv SOH_{2}^{+}\right]} = \frac{\left(\left[\equiv SOH_{tot}\right]-\Delta[H^{+}]\right)\bullet\left\{H_{bulk}^{+}\right\}\bullet(1+K_{A-}\bullet\{A^{-}\})}{\Delta[H^{+}]}$$

$$K_{a2}^{app} = \frac{\left[\equiv SO^{-}\right]\left\{H_{bulk}^{+}\right\}}{[\equiv SOH]} = \frac{-\Delta[H^{+}]\bullet\left\{H_{bulk}^{+}\right\}}{(1+K_{Na+}\bullet\{Na^{+}\})\bullet([\equiv SOH_{tot}]+\Delta[H^{+}])}$$

where $\Delta[H^{+}]$ is the difference between the measured bulk hydrogen ion concentration and added amount of H^{+}, symbols for molecular species contained in curly brackets, {}, signify the activity of the molecular species, K_{A-} and K_{Na+} are the constants for the outer sphere complexation of the anion and respective cation with the surface, and SOH_{tot} is the total concentration of surface sites calculated from the site density, surface area and concentration of solid. All the constants are given in table 1. The surface charge, σ_{0}, was calculated from:

$$\sigma_0 = \frac{\left(\left[\equiv SOH_2^+\right]+\left[\equiv SOH_2^+Cl^-\right]-\left(\left[\equiv SO^-\right]+\left[\equiv SO^-Na^+\right]\right)\right)\bullet F}{SA} = \frac{\Delta[H^+]\bullet F}{SA} \quad (C\bullet m^{-2})$$

where F is Faradays constant, SA is the surface area and $\equiv SO$ denotes the surface species. No hysterisis was observed in the backtitration indicating that the surface had not been altered during the titrations and that any aggregation would be reversible.

<u>Sorption of U(VI) onto Al$_2$O$_3$</u>

The sorption is expressed as the distribution coefficient, K_d, calculated as:

$$K_d = \text{amount sorbed onto solid/amount in solution} = \frac{\left([U]_0 - [U]\right)}{[U]}\left(\frac{V}{M}\right) \quad (ml\bullet g^{-1})$$

where $[U]_0$ is the original uranium concentration, $[U]$ is the concentration in the liquid phase after contact with the solid, and M/V is the mass to volume relationship $(g\bullet ml^{-1})$. Figure 2 shows how the K_d varies with the pH.

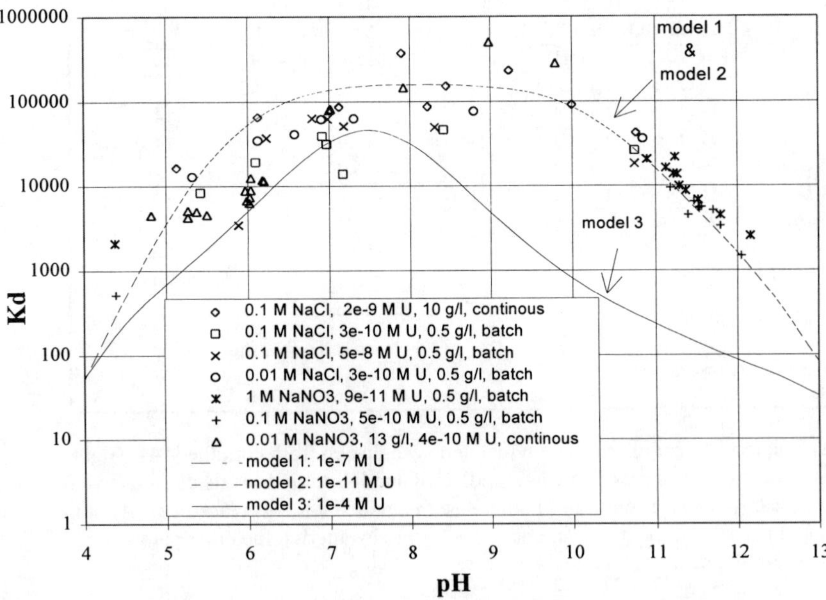

Figure 2   **Kd as a function of pH. The reactions and the constants used for the simulations are summarized in table 1. (Note that models 1 and 2 overlap)**

The fraction sorbed increases with pH until pH 7, reaches a maximum and decreases linearly from pH 10. There is no significant difference observed in the sorption behavior for the different ionic strengths, electrolytes, experimental methods and uranium concentrations. The independence of the sorption on the ionic strength indicates an inner sphere complexation with the alumina surface sites [12]. The experimental independence on uranium concentration indicate that there are no site heterogeneities.

The sorption was modeled using a three layer model with the code FITEQL[13]. The reactions used and the constants are summarized in table 1.

Table 1. Reactions and constants

Reaction	Constant		Ref
Acid dissociation of alumina surface:			
$SOH + H^+ \Leftrightarrow SOH_2^+$	$pK_{a1}^{int} = 7.2$		this
$SOH \Leftrightarrow SO^- + H^+$	$pK_{a2}^{int} = 11.2$		this
$SO^- + A^- \Leftrightarrow SO^-A^-$	$K_{A-}^{int} = 0$		
$SO^- + Na^+ \Leftrightarrow SO^-Na^+$	$K_{Na+}^{int} = 15$		15
Aqueous uranium speciation	original	adjusted	
$UO_2^{2+} + H_2O \Leftrightarrow UO_2(OH)^+ + H^+$	-4.9	-4.9	14
$UO_2^{2+} + 2 H_2O \Leftrightarrow UO_2(OH)_2 + 2H^+$	-9.7	-11.4	14
$UO_2^{2+} + 3 H_2O \Leftrightarrow UO_2(OH)_3^- + 3H^+$	-18.3	-21.5	14
$UO_2^{2+} + 4 H_2O \Leftrightarrow UO_2(OH)_4^{2-} + 4H^+$	-31.8	-34.5	14
$2 UO_2^{2+} + H_2O \Leftrightarrow (UO_2)_2(OH)^{3+} + H^+$	-2.4	-2.4	14
$2 UO_2^{2+} + 2 H_2O \Leftrightarrow (UO_2)_2(OH)_2^{2+} + 2H^+$	-5.02	-5.02	14
$3 UO_2^{2+} + 4 H_2O \Leftrightarrow (UO_2)_3(OH)_4^{2+} + 4H^+$	-10.7	-10.7	14
$3 UO_2^{2+} + 5 H_2O \Leftrightarrow (UO_2)_3(OH)_5^+ + 5H^+$	-14.05	-14.05	14
$3 UO_2^{2+} + 7 H_2O \Leftrightarrow (UO_2)_3(OH)_7^- + 7H^+$	-28.9	-28.9	14
$4 UO_2^{2+} + 7 H_2O \Leftrightarrow (UO_2)_4(OH)_7^+ + 7H^+$	-19.8	-19.8	14
Surface complexation			
$SOH + UO_2^{2+} \Leftrightarrow SOUO_2^+ + H^+$	-3.1		this
$2 SOH + UO_2^{2+} \Leftrightarrow (SO)_2UO_2 + 2H^+$	3.4		this

constants used:

Inner layer capacitance:	$1.1\ C/m^2$	Site density:	10 sites / nm^2
Outer layer capacitance	$0.2\ C/m^2$	pKw:	13.7
Surface area (BET):	$142.9\ m^2/g$	I:	0.1 M

The aqueous uranium speciation reactions were taken from Grenthe [14]. Slight modifications of the estimated second, third and fourth hydrolysis constants were necessary, (Table 1, column 2) otherwise the modeled K_d decreased already at pH 7 and not at pH 10. This effect could not be achieved by altering the other parameters within justifiable ranges.

As can be seen in Figure 2 the sorption of uranium is independent of the uranium concentration (or within experimental error) in the range $9x10^{-11}$ to $5x10^{-8}$ M. The decrease in K_d at pH 10 is at a higher pH than observed by Del Nero [8] (pH ~7-8) sorbing U onto hydrargilite at higher uranium concentrations (10^{-4} and 10^{-6} M). A uranium concentration dependence was observed by Waite [7] for uranium sorption onto ferrihydrite. The sorption started to decrease at pH 7 for 10^{-4} M U and at pH 8 for 10^{-6} M, and this shift in the sorption was explained by a decrease in the average free energy by increasing surface coverage. However, modeling the uranium sorption with the altered hydrolysis constants for three different uranium concentrations, 10^{-9}, 10^{-7} and 10^{-4} M gave us the same effect. The model did not predict any difference for the two lowest concentrations (Figure 2 models 1 and 2) but a rather large shift towards decreasing the K_d for the higher uranium concentration at the same pH (Figure 2 model 3). This effect was only due to the formation of polynuclear complexes in the solution. If there were polynuclear complexes formed with the surface the fraction sorbed would increase with increasing uranium concentrations. Therefore it seems unlikely that polynuclear complexes form with the surface.

A good fit of the modeled uranium sorption was obtained using two reactions, a monodentate and a bidentate inner sphere. The stoichiometry for a bidentate inner sphere uranyl complex is identical to that of the monodentate attachment of a hydrolyzed uranyl ion (as suggested by DeGueldre et al for Americium [16]). It is therefore impossible to differentiate between these structure on the basis of sorption measurements. The bidentate mechanism was chosen because there is spectroscopic evidence supporting the bidentate attachment of divalent cations onto alumina [17]. Modeling using only the bidentate complexation gave equally good fit in the pH range studied but would not be able to predict the sorption at lower pH. EXAFS of U onto ferrihydrites indicate that a bidentate complex forms. Since the pK_{a2}s for the alumina and Fe oxides are so close (11.3 and 11.1 [18], respectively) the same surface complexation model should be applicable. The good fit, the EXAFS and the pK_{a2}s indicate that the bidentate binding is likely to be the dominating surface reaction.

CONCLUSIONS

Uranium forms primarily an inner sphere bidentate complex with the alumina surface, with a maximum sorption at pH 8-9.

There are no low density high energy sites affecting the surface complexation of uranium at low concentrations (9×10^{-11} - 5×10^{-8} M).

The same surface complexation model should be applicable to iron oxides because the pK_{a2}s are so close. The observed sorption is found to be in close agreement with that observed for uranium onto goethite[19] when the surface area differences are taken into account. The observation of Bradbury and Baeyens[20] that the surface complexation of neptunyl on many aluminosilcate minerals and rocks can be modeled as iron oxide surfaces may be a consequence of this property of alumina.

The dependence of the surface complexation constant on the pK_{a2} of the metal oxide needs to be further explored to develop a quantitative model. The bidentate surface complex will have a different dependence on pK_{a2} from the monodentate complex. Uranyl surface complexation needs to be studied on more acidic metal oxides, such as silicon dioxide. Further spectroscopic studies of the structure of the uranyl surface complex are needed to complement the sorption experiments.

ACKNOWLEDGMENTS

This work was funded by the Swedish Nuclear Fuel and Waste Management Company, SKB. This work was partially funded by the Yucca Mountain Project of the U.S. Department of Energy.

REFERENCES

1. D. Cui, Thesis: Sorption Processes and Solubilities of Radionuclides in Deep Granite Fracture Systems, Royal Institute of Technology, Stockholm, Sweden, (1996).
2. D. G. Kinniburgh, M. L. Jackson, and S. K. Syers, Soil Sci. Soc. Am. J., 40 796-799, (1976).
3. D. A. Dzombak and F. M. M. Morel, Surface Complexation Modeling: Hydrous Ferric Oxide, *John Wiley and Sons*, New York (1990).
4. R. G. Pearson, J. Chem. Ed. **45**(9) 581-587 (1968).
5. P. W. Schindler, Österreichische Chemie-Zeitschrift., **86**(6), 141-147 (1985).
6. M. A. Anderson and A. J. Rubin eds., Adsorption of Inorganics at Solid-Liquid Interfaces, pp. 203, *Ann Arbor Science Publishers*, Ann Arbor (1981).
7. T. D Waite, J. A Davis, T. E. Payne, G. A. Waychunas, N. Xu, Geochim. Cosmochim. Acta **58**, 5465-5478 (1994)
8. M. Del Nero, T. Miura, G. Bontems, G. Duplâtre, A. Clément, extended abstract, 4th International conference Nuclear and Radiochemistry (1996).
9. G. Gran, Acta Chemica Scandinavica **4**, 559-577 (1950).
10. R. S. Rundberg, unpublished work.
11. C. F. Baes Jr. and R. E. Mesmer, The Hydrolysis of Cations, *Krieger Publishing Company*, Malabar Florida (1986).
12. K. F. Hayes, A. L. Roe, G. E. Jr. Brown, K. O. Hodgson, J. O. Leckie, and G. A. Parks, Science **238**, 783-786 (1987).
13. J. Westall, Report 82-01 Dept. of Chemistry, Oregan State University, Corvallis Oregon (1982).
14. I. Grenthe *et al.* Chemical Thermodynamics of Uranium, *Elsevier*, (1992).
15. R. S. Rundberg, Y. Albinsson, K. Vannerberg, Radiochim. Acta **66/67**, 333-339 (1994).
16. C. Degueldre, H. J. Ulrich, and H. Silby, Radiochim. Acta **65**, 173-179 (1994).
17. H. Motschi, Adsorption Sci. and Technol. **2**, 39-54 (1985).
18. K. F. Hayes, C. Papelis, and J. O. Leckie, J. Colloid Interface Sci., **125**(2) 717-726 (1988).
19. C-K. D. Hsi and D. Langmuir, Geochim. Cosmochim. Acta **49**, 1931-1941 (1985).
20. M. H. Bradbury and B. Baeyens, J. Colloid Interface Sci., **128** 364-371 (1993).

THE EFFECT OF TEMPERATURE ON THE SORPTION OF TECHNETIUM, URANIUM, NEPTUNIUM AND CURIUM ON BENTONITE, TUFF AND GRANODIORITE

G M N BASTON, J A BERRY, M BROWNSWORD, T G HEATH, D J ILETT, C J TWEED AND M YUI*
AEA Technology plc, 220 Harwell, Didcot, Oxfordshire, UK
*Power Reactor and Nuclear Fuel Development Corporation, Tokai Works, Ibaraki, Japan

ABSTRACT

A study of the sorption of the radioelements: technetium; uranium; neptunium; and curium onto geological materials has been carried out as part of the PNC programme to increase confidence in the performance assessment for a high-level radioactive waste repository in Japan. Batch sorption experiments have been performed in order to study the sorption of the radioelements onto bentonite, tuff and granodiorite from equilibrated de-ionised water under strongly-reducing conditions at both room temperature and at 60°C.

Mathematical modelling using the geochemical speciation program HARPHRQ in conjunction with the HATCHES database has been undertaken in order to interpret the experimental results.

INTRODUCTION

Much work has been carried out world-wide on the sorption of technetium, uranium and neptunium on geological materials [e.g. 1 - 4 and references therein]. Comparatively little has been undertaken with curium [3], or with any of these radioelements at elevated temperature under reducing conditions. The research described here is part of the Power Reactor and Nuclear Fuel Development Corporation (PNC) programme to increase confidence in the performance assessment for a generic study of Japanese high-level radioactive waste (HLW) disposal.

Measurements have been made of the sorption of technetium, uranium, neptunium and curium onto bentonite (Kunigel V1 [5]), which is a potential backfill material in the repository, and onto two rocks, tuff and granodiorite. The study has involved the investigation of sorption behaviour at two temperatures and is an extension of work reported previously [6].

The room temperature batch sorption data have been interpreted using a pragmatic surface complexation model. Mathematical modelling using the geochemical speciation program HARPHRQ version 1.4 [7] and the HATCHES database version 7 [8] has been used to aid the interpretation of the effect of temperature.

EXPERIMENTAL

Batch sorption experiments were carried out at both room temperature (21 $\pm$ 3°C) and at 60$\pm$ 1°C in polypropylene centrifuge tubes in a nitrogen-atmosphere glove-box with an oxygen level of less than 1 ppm. The high-temperature experiments were undertaken in a shaker-incubator. Conditions were chosen to simulate, as closely as possible, those anticipated to be relevant to a Japanese HLW repository.

All solutions were de-oxygenated prior to use. Samples of the crushed rocks were mixed with the filtered radioelement solution under appropriate redox conditions (Eh). Strongly-reducing conditions were achieved using sodium hydrosulphite at an initial concentration of 2×10^{-3} M. The Eh value was monitored every 7-14 days, and kept at a low value by the addition of further hydrosulphite. The pH of the solutions was maintained at the value of the rock-equilibrated de-ionised water. In the higher temperature experiments, Eh and pH were measured at 60°C.

The experiments involving technetium and uranium were carried out in triplicate. In these cases, one of the three tubes was sampled to monitor the approach to steady-state. The distribution ratios (R_D values) reported here were obtained from the remaining two tubes. Curium and neptunium experiments were carried out in duplicate.

Mat. Res. Soc. Symp. Proc. Vol. 465 © 1997 Materials Research Society

Analysis of the radioelements was by α-spectrometry for uranium, neptunium and curium, (following electrodeposition of the solution samples) and by γ-spectrometry for technetium (Canberra S100). The concentration of neptunium, however, fell below the limit of detection ($\sim$0.5 mBq cm^{-3}) of the α-spectrometer, but R_D values were obtained from single experiments carried out on a tenfold larger scale. In all cases, the contents of the tubes were analysed after four months. All tubes were gently agitated continuously. Three phase separation techniques were employed: centrifugation at 1100g; filtration through a 0.45μm filter ('Millex HV'), and filtration through a 10000MWCO filter ('Millipore TGC') in sequence. Filters were washed and pre-conditioned prior to use [9]. Filtrations were carried out at the appropriate temperature.

The tubes were checked to confirm that sorption was onto rock rather than onto the vessel walls. A liquid:solid ratio of 5:1 was used for tuff and granodiorite, and 20:1 for bentonite. These were the lowest values at which phase separation could be conveniently carried out.

The geological materials were supplied by PNC. The tuff and granodiorite were crushed in an automatic mortar and pestle and sieved (<250μm) in a nitrogen-atmosphere glove-box. The bentonite [5] was used as received. Solutions were prepared by equilibrating crushed samples of each rock with de-ionised water for one month. Each resulting solution was analysed, and synthetic equilibrated solutions were prepared accordingly [6].

Technetium-95m (Los Alamos National Laboratory, USA), received as pertechnetate, and uranium-233, (received as uranium(VI), from the Actinide Services Department, AEA Technology plc, Harwell), were purified and reduced to Tc(IV) and U(IV) as described previously [6]. Neptunium-237 and curium-244, (also obtained from Actinide Services Department, AEA Technology plc, Harwell as calibrated stock solutions) were similarly added to sodium hydrosulphite solution before addition to the synthetic equilibrated de-ionised water.

Initial concentrations used for the batch experiments were $\sim$8 x 10^{-11}M for technetium, $\sim$1 x 10^{-7}M for uranium, $\sim$6 x 10^{-9}M for neptunium and $\sim$1 x 10^{-10}M for curium.

RESULTS AND DISCUSSION

The distribution ratios (R_D values) were determined for each of the three phase separation techniques, but only the values calculated from the filtered solutions are included in the Tables. Results for technetium, uranium, neptunium and curium are presented in Tables I to IV

TABLE I: TECHNETIUM SORPTION ONTO GEOLOGICAL MATERIALS

Rock	Temperature °C	Final pH	Final Eh/mV	R_D(cm^{3}g^{-1}) 0.45 μm	10000 MWCO	Model
Bentonite	21 ± 3	10.0	-460	1.4x10^3±0.1x10^3	1.5x10^3±0.1x10^3	4.5x10^3
		10.1	-450	1.6x10^3±0.1x10^3	1.7x10^3±0.1x10^3	
	60 ± 1	9.4	-390	1.3x10^4±0.1x10^4	1.5x10^4±0.1x10^4	1.6x10^4
		9.4	-390	8.0x10^3±0.4x10^3	2.1x10^4±0.2x10^4	
Tuff	21 ± 3	9.4	-400	1.2x10^4±0.2x10^4	1.8x10^4±0.4x10^4	3.3x10^3
		9.4	-400	1.2x10^4±0.2x10^4	1.1x10^4±0.2x10^4	
	60 ± 1	Not Measured				
Grano-diorite	21 ± 3	10.2	-330	3.5x10^4±0.7x10^4	3.8x10^4±0.9x10^4	4.7x10^3
		10.1	-340	2.8x10^4±0.5x10^4	3.3x10^4±0.6x10^4	
	60 ± 1	Not Measured				

TABLE II: URANIUM SORPTION ONTO GEOLOGICAL MATERIALS

Rock	Temperature °C	Final pH	Final Eh/mV	$R_D(cm^3g^{-1})$		
				0.45 µm	10000 MWCO	Model
Bentonite	21 ± 3	10.1	-450	$3.9x10^4 \pm 1.0x10^4$	$2.8x10^4 \pm 0.7x10^4$	$2.2x10^5$
		10.1	-470	$1.7x10^4 \pm 0.5x10^4$	$1.8x10^4 \pm 0.5x10^4$	
	60 ± 1	9.4	-400	$2.7x10^4 \pm 0.7x10^4$	$5.2x10^4 \pm 1.8x10^4$	$1.4x10^5$
		9.4	-410	$3.3x10^4 \pm 1.0x10^4$	$3.8x10^4 \pm 1.2x10^4$	
Tuff	21 ± 3	9.5	-410	$1.9x10^4 \pm 0.3x10^4$	$5.6x10^4 \pm 1.0x10^4$	$2.0x10^5$
		9.5	-420	$5.6x10^4 \pm 1.5x10^4$	$5.6x10^4 \pm 0.9x10^4$	
	60 ± 1	Not Measured				
Grano-diorite	21 ± 3	10.1	-450	$6.3x10^3 \pm 1.6x10^3$	$2.8x10^4 \pm 0.6x10^4$	$1.7x10^5$
		10.1	-450	$4.7x10^3 \pm 1.6x10^3$	$6.3x10^3 \pm 2.0x10^3$	
	60 ± 1	Not Measured				

respectively. Errors shown for the R_D values are 2σ, and are based on counting statistics only.

Sorption data for technetium and uranium at room temperature were obtained earlier [6], at two liquid:solid ratios for each material, but only R_D values at higher liquid:solid ratios (50:1 for tuff and granodiorite, and 100:1 for bentonite) were presented then [6]. The corresponding values for the lower ratios (5:1 and 20:1) are shown in Tables I and II for direct comparison with the results from the present study of the sorption of technetium and uranium at 60°C and neptunium and curium at both temperatures.

For technetium at 21°C, sorption was strong for all three geological materials, R_D values being in the range 1.5×10^3 to 3.8×10^4 cm^3 g^{-1} after 10000MWCO filtration. In the case of technetium at 60°C, only sorption onto bentonite was studied. R_D values were a factor of approximately ten higher at the elevated temperature.

TABLE III: NEPTUNIUM SORPTION ONTO GEOLOGICAL MATERIALS

Rock	Temperature °C	Final pH	Final Eh/mV	$R_D(cm^3g^{-1})$		
				0.45 µm	10000 MWCO	Model
Bentonite	21 ± 3	10.4	-440	$4.5x10^3 \pm 1.3x10^3$	$5.7x10^3 \pm 2.0x10^3$	$2.8x10^3$
	60 ± 1	9.4	-320	$4.7x10^3 \pm 1.4x10^3$	$4.0x10^3 \pm 1.1x10^3$	$5.3x10^3$
Tuff	21 ± 3	9.4	-310	$8.1x10^2 \pm 1.7x10^2$	$6.8x10^2 \pm 1.5x10^2$	$2.8x10^3$
	60 ± 1	9.2	-330	$6.4x10^2 \pm 1.3x10^2$	$2.1x10^2 \pm 0.4x10^2$	$3.9x10^3$
Grano-diorite	21 ± 3	10.1	-460	$1.3x10^3 \pm 0.3x10^3$	$3.2x10^3 \pm 1.1x10^3$	$1.5x10^3$
	60 ± 1	9.6	-470	$7.9x10^2 \pm 1.9x10^2$	$1.0x10^3 \pm 0.3x10^3$	$2.9x10^3$

R_D values could only be obtained from the single large-scale experiments.

TABLE IV: CURIUM SORPTION ONTO GEOLOGICAL MATERIALS

Rock	Temperature °C	Final pH	Final Eh/mV	R_D(cm^{3}g^{-1})	
				0.45 µm	10000 MWCO
Bentonite	21 ± 3	10.1	-400	$3.5\times10^4\pm0.5\times10^4$	$6.5\times10^4\pm1.2\times10^4$
		10.1	-410	$5.2\times10^4\pm0.8\times10^4$	$2.1\times10^5\pm0.4\times10^5$
	60 ± 1	9.4	-350	$2.3\times10^4\pm0.4\times10^4$	$6.9\times10^4\pm1.6\times10^4$
		9.4	-350	$2.1\times10^5\pm0.4\times10^5$	$5.1\times10^5\pm1.5\times10^5$
Tuff	21 ± 3	9.4	-430	$3.0\times10^4\pm0.5\times10^4$	$4.5\times10^4\pm1.0\times10^4$
		9.4	-440	$2.9\times10^4\pm0.5\times10^4$	$1.1\times10^5\pm0.3\times10^5$
	60 ± 1	9.2	-370	$6.4\times10^4\pm1.2\times10^4$	$1.3\times10^5\pm0.3\times10^5$
		9.2	-370	$7.1\times10^4\pm1.2\times10^4$	$2.1\times10^5\pm0.5\times10^5$
Grano-diorite	21 ± 3	10.1	-430	$1.1\times10^4\pm0.2\times10^4$	$1.2\times10^4\pm0.2\times10^4$
		10.1	-440	$7.3\times10^3\pm1.0\times10^3$	$1.2\times10^4\pm0.2\times10^4$
	60 ± 1	9.6	-390	$3.2\times10^3\pm0.5\times10^3$	$4.0\times10^4\pm0.4\times10^4$
		9.6	-390	$2.1\times10^3\pm0.3\times10^3$	$2.3\times10^4\pm0.5\times10^4$

Sorption of uranium was also strong, R_D values at 21°C being in the range 6.3 x 10^3 to 5.6 x 10^4 cm^3 g^{-1} after 10000MWCO filtration. At 60°C, sorption onto bentonite was slightly stronger (as in the case of technetium, sorption onto tuff and granodiorite was not measured at the higher temperature).

At room temperature, neptunium was quite strongly sorbed onto bentonite and granodiorite (R_D values were in the range 3.2 x 10^3 to 5.7 x 10^3 cm^3 g^{-1} following 10000MWCO filtration), but less strongly sorbed onto tuff (R_D = 6.8 x 10^2 cm^3 g^{-1}). At 60°C, sorption of neptunium on all three geological materials was generally slightly weaker than at the lower temperature, R_D values being up to a factor of approximately three lower.

Curium was more strongly sorbed than any of the other three radioelements. At room temperature, R_D values for the three geological materials were in the range 1.2 x 10^4 to 2.1 x 10^5 cm^3 g^{-1} after 10000MWCO filtration. At the higher temperature, sorption was slightly stronger, R_D values being up to a factor of approximately three higher. Differences between R_D values following the two different filtrations were more pronounced in the case of curium, than with the other radioelements, particularly in the case of granodiorite at 60°C. This suggests the possible formation of colloids or particulate matter in this instance.

<u>Prediction of Radionuclide Speciation</u>

Before modelling the sorption processes, it is important to consider the predicted speciation of radionuclides under the range of experimental conditions studied. For technetium, uranium and neptunium, the speciation was predicted using the HARPHRQ [7] program, in conjunction with the HATCHES thermodynamic database [8]. This database contains values of the equilibrium constants for the various chemical equilibria at 25°C and the corresponding standard molar enthalpy change for the equilibrium. The latter quantity has been used to derive the equilibrium constants at 60°C. For uranium and neptunium, enthalpy data were available in the HATCHES database. In the case of technetium, however, no enthalpy data were present and a zero value for each technetium species was assumed. The effect of this assumption is to make the equilibrium constants for the formation of technetium species independent of temperature. For curium, no sorption modelling has been carried out due to uncertainties in the thermodynamic data.

At 25°C and 60°C, the technetium speciation under strongly-reducing conditions is predicted to be dominated by the neutral $TcO(OH)_2$ species. Similarly for uranium and neptunium, the neutral species $U(OH)_4$ and $Np(OH)_4$ respectively dominate the speciation at both temperatures, under strongly-reducing conditions. The observed drop in pH at 60°C (Tables I to IV), is consistent with the increased dissociation of water at this temperature.

<u>Interpretation of Results using a Pragmatic Surface Complexation Model</u>

The room temperature batch sorption data have been interpreted using a relatively simple surface complexation model reported by Degueldre et al. [10]. This model neglects electrostatic effects but provides a useful method for the prediction of sorption equilibrium constants for the hydrolysis products of an element. These predicted values are based on correlations derived between the equilibrium constant for the stepwise formation of the nth hydrolysis product and that for the sorption of the $(n-1)^{th}$ hydrolysis product. Such correlations are given for both alumina (γ-Al_2O_3) and goethite (FeOOH). The model details are the same as those described in reference [10] except that sorption of carbonate complexes is neglected and the model has been applied at the experimental liquid:solid ratio and measured pH value. The thermodynamic data [11] used in the original paper have been supplemented with data from the HATCHES database. The site concentration reported in reference [10] has been used (6.25×10^{-5} mol g^{-1}) although this could be treated as a fitting parameter due to uncertainty in its value.

Comparison of the experimental R_D data with the model predictions for bentonite is given in Table V. The FeOOH surface model provides better agreement with the observed data. The agreement between the observed and predicted trends in R_D with element suggests that the assumption of a goethite-type surface is reasonable. The agreement between the predicted and observed values may reflect that the value chosen for the concentration of sites was reasonable.

<u>Interpretation of Results using a Goethite-based Triple-Layer Model</u>

Thermodynamic modelling using the HARPHRQ program has been performed to aid the interpreation of the effect of temperature on sorption. For uranium and technetium sorption onto bentonite and tuff, an existing detailed triple-layer model has been applied based on an iron hydroxide (goethite) surface phase [6]. This may be present as a coating on smectite clay [12] which is a major component of both these minerals. In the case of granodiorite, there is only a small proportion of iron-bearing minerals [13]. However, earlier modelling work [6] showed that

Table V: Comparison of Observed Sorption with that Predicted from the pragmatic Model [10]

Element	Mean Experimental R_D (cm^{3}g^{-1})	γ-Al_2O_3 Model R_D (cm^{3}g^{-1})		FeOOH Model R_D (cm^{3}g^{-1})	
		data from [11]	HATCHES	data from [11]	HATCHES
Tc(IV)	1.6×10^3	7.2×10^4	7.2×10^4	1.1×10^3	1.2×10^3
U(IV)	2.3×10^4	-	-	-	-
Np(IV)	4.9×10^3	9.0×10^4	7.5×10^4	2.0×10^4	2.0×10^3
Cm(III)	1.4×10^5	-	1.0×10^5	-	6.9×10^4

a better fit to the range of sorption data could be obtained by assuming sorption onto $\equiv$FeOH groups or $\equiv$AlOH groups rather than $\equiv$SiOH groups. Furthermore, biotite, at 4.2% of the granodiorite, contains significant amounts of iron and surface analysis results confirm this as the major sorbing phase [6]. For these reasons, the goethite-based model was also applied to granodiorite. The core model used for these calculations was developed in other studies [14,15].

In the present work, the model has been extended to aid the interpretation of data at elevated temperatures (60°C) and at lower liquid:solid ratios (20:1 for bentonite and 5:1 for tuff and granodiorite) than reported previously [6]. Some model refinement has been carried out and the full details of the revised model used are summarised in Table VI. The sorption modelling methodology involved three steps, as follows.

Firstly, the model was parameterised based on fitting batch sorption data determined at room temperature and high liquid:solid ratios (100:1 for bentonite, 50:1 for tuff and granodiorite) [6]. The concentration of sites was used as a fitting parameter whereas the site density was fixed at a constant value for all the systems. The specific surface area was constant for each rock. For each radionuclide, the dominant aqueous species was assumed to be the important sorbing species.

Secondly, the model was applied to predict the R_D value at the lower liquid:solid ratios at 25°C. The predicted values are given in Tables I to III. In the case of neptunium, no data at the higher liquid:solid values were available, and the corresponding sorption equilibrium constant given in Table VI was obtained by fitting the data presented in Table III.

Finally, the model was applied to predict the sorption at 60°C and at the lower liquid:solid

TABLE VI: TRIPLE LAYER SORPTION PARAMETERS ASSUMED IN THE GOETHITE SORPTION MODEL FOR SORPTION ONTO BENTONITE, TUFF AND GRANODIORITE

Reactions	LogK
Surface: $SOH + H_{(s)}^+ = SOH_2^+$ $SOH = SO^- + H_{(s)}^+$	 5.57 -9.52
Tc: $SOH + TcO(OH)_{2(s)} + H_{(s)}^+ = SOH_2TcO(OH)_2^+$	 13.6
U(IV): $SOH + U(OH)_{4(s)} = SOHU(OH)_4$ $SOH + U(OH)_{4(s)} + Ca_{(s)}^{2+} + H_2O = SOCaOH.U(OH)_4 + 2H_{(s)}^+$	 6.5 -8.0
Np(IV): $SOH + Np(OH)_{4(s)} = SOHNp(OH)_4$	 5.15

Subscript (s) denotes the surface activity of the solute. The surface activity is related to the bulk solution activity through the expression: $a_{(s)} = a_b \exp(-ze\Psi/KT)$ where z is the charge of the species, e is the electronic charge, Ψ the potential at the surface, K the Boltzmann constant and T the absolute temperature.

Other parameters used were:

Site density of α–FeOOH		4.25×10^{-6} mol m^{-2}
Triple layer parameters:	C_1	140 μF cm^{-2}
	C_2	20 μF cm^{-2}
Concentration of solid		50g dm^{-3} (bentonite)
		200g dm^{-3} (tuff)
		200g dm^{-3} (granodiorite)
Concentration of sites		2.88×10^{-2} mol dm^{-3} (bentonite)
		2.31×10^{-1} mol dm^{-3} (tuff)
		1.16×10^{-1} mol dm^{-3} (granodiorite)

Data for the sorption of groundwater elements were taken from references 16 and 17.

ratios. In carrying out these predictions further assumptions have been made. Although enthalpydata were available for the formation of species involving the major groundwater elements and uranium and neptunium, no enthalpy data were available for technetium or for the surface equilibria. For technetium, the equilibrium constant values were assumed to be independent of temperature; the $TcO(OH)_2$ species was effectively assumed to be the dominant species at both temperatures. Similarly for the sorption equilibria, the equilibrium constant for the sorption of the major aqueous species onto the surface was assumed to be independent of temperature. Measured and predicted R_D values may therefore be compared to determine whether the observed changes in sorption at the elevated temperature can be explained simply by the effect of temperature on pH and the aqueous speciation. A large discrepancy between the observed and predicted sorption will suggest that the temperature dependence of one or more sorption equilibria is important.

For technetium sorption onto bentonite, increased sorption at 60°C is observed and predicted, although the predicted increase is slightly smaller. This predicted increase is due to the pH dependence of the technetium sorption equilibrium.

CONCLUSIONS

The results from batch sorption experiments show that the sorption of technetium, uranium, neptunium and curium from equilibrated de-ionised water onto bentonite, tuff and granodiorite is generally strong under strongly-reducing conditions and is not greatly affected by increased temperature.

The observed sorption behaviour has been interpreted using a surface complexation model. For the uranium and neptunium systems studied, both the observed and predicted R_D values show only a small variation with temperature, suggesting that there are no important temperature-dependent surface effects. Where a greater temperature dependence was observed, in experiments with technetium on bentonite, the results are in reasonable agreement with the model prediction.

ACKNOWLEDGEMENTS

This work was funded by the Power Reactor and Nuclear Fuel Development Corporation (PNC), Tokyo, Japan, and their permission to publish these results is gratefully acknowledged.

REFERENCES

1.   T. Amaya, W. Kobayashi and K. Suzuki in Scientific Basis for Nuclear Waste Management XVIII, edited by T. Murakami and R.C. Ewing (Mater. Res. Soc. Proc. **353**, Pittsburgh, PA, 1995) pp 1005-1012.

2.   F.P. Bertetti, R.T. Pabalan, D.R. Turner and M.G. Almendarez in Scientific Basis for Nuclear Waste Management XIX, edited by W.M. Murphy and D.A. Knecht (Mater. Res. Soc. Proc. **412**, Pittsburgh, PA, 1996) pp 631-638.

3.   J.A. Berry. A Review of Sorption of Radionuclides Under the Near- and Far-Field Conditions of an Underground Radioactive Waste Repository. Parts I - III. UK DoE Report DOE/HMIP/RR/92/061 (1992).

4.   M.J. Stenhouse. Sorption Databases for Crystalline, Marl and Bentonite for Performance Assessment. Nagra Technical Report NTB 93-06 (1995).

5. H. Sato, T. Ashida, Y. Kohara and M. Yui in <u>Scientific Basis for Nuclear Waste Management XVI</u>, edited by C.G. Interrante and R.T. Pabalan (Mater. Res. Soc. Proc. **294**, Pittsburgh, PA, 1993) pp 403-408.

6. G.M.N. Baston, J.A. Berry, M. Brownsword, M.M. Cowper, T.G. Heath and C.J. Tweed in <u>Scientific Basis for Nuclear Waste Management XVIII</u>, edited by T. Murakami and R.C.Ewing (Mater. Res. Soc. Proc. **353**, Pittsburgh, PA, 1995) pp 989-996.

7. A. Haworth, T.G. Heath and C.J. Tweed, HARPHRQ: A Computer Program for Geochemical Modelling, UK Nirex Ltd. Report NSS/R380, (1995).

8. J.E. Cross and F.T. Ewart. Radiochimica Acta, **52/53**, 421 (1991).

9. D. Rai. Radiochimica Acta, **34**, 97 (1984).

10. C. Degueldre, H.J. Ulrich and H. Silby. Radiochimica Acta, **65**, 173-179 (1994).

11. J. Pearson, U. Berner and W. Hummel. NAGRA Thermochemical Data Base II: Supplemental Data, NAGRA, Wettingen, report NTB 91-18 (1992).

12. J.J.W. Higgo. Progress in Nuclear Energy, **19**, 173 (1987).

13. H. Sato and T. Shibutani, PNC, Private Communication, 19 February 1993.

14. K.A. Bond, J.E. Cross and F.T. Ewart. Thermodynamic Modelling of Uranium(VI) Sorption onto London Clay. UK Nirex Ltd. Report NSS/R207 (1990).

15. K.A. Bond and C.J. Tweed. Geochemical Modelling of the Sorption of Tetravalent Radioelements. UK Nirex Ltd. Report NSS/R227 (1992).

16. L.S. Balistrieri and J.W. Murray. Am.J.Sci, **281**, 788 (1981).

17. A.L. Sanchez, J.W. Murray and T.H. Sibley, Geochim et Cosmochim Acta, **49**, 2297 (1984).

Characterization of Electroactive Cs Ion-Exchange Materials using XAS

N.J. Hess, J.H. Sukamto, S.D. Rassat, R.T. Hallen, R.J. Orth, M.A. Lilga, and W.E. Lawrence
Pacific Northwest National Laboratory, Richland WA 99352

ABSTRACT

Various ion exchange materials have been proposed for the removal of Cs from high level waste streams produced during the reprocessing of fuel rods. Cs can be released from loaded traditional exchange resins by elution and then the resin can be reused. However large quantities of secondary wastes are generated. Another class of "single use" exchangers is directly incorporated in the loaded state into a solid waste form (e.g. borosilicate glass logs). A third alternative is electroactive ion-exchange materials, where the uptake and elution of Cs are controlled by an applied potential. This approach has several advantages over traditional reusable ion-exchange resins including much reduced secondary waste, higher Cs selectivity, and higher durability.

XAS experiments were conducted at the Fe K-edge and Cs L_{III}-edge on a series of electrochemically produced nickel ferrocyanide films to determine the effects of deposition conditions and subsequent alkali exchange on structural and chemical aspects of the films. The deposition conditions include methods described in the literature and PNNL proprietary procedures. Although the performance and the durability of the films do vary with processing conditions, Fe K-edge EXAFS results indicate that all deposition conditions result in the formation of the cubic phase. Initial results from Cs L_{III}-edge EXAFS analysis suggest that the Cs ion is present as a hydrated species.

INTRODUCTION

Several sites within the Department of Energy's weapons complex have numerous underground storage tanks containing radioactive high-level waste (HLW) from nuclear fuel processing activities. Separation and concentration of the radionuclides would drastically reduce the volume of HLW and facilitate the permanent disposal of the treated tank waste in a vitrified glass matrix. Soluble radionuclides are removed from liquid waste streams using ion exchange separation techniques. Various ion-exchange materials have been proposed for the removal of cesium from high level waste streams. Cesium can be released from loaded traditional organic and inorganic ion-exchange resins by elution and then the resin can be reused. However, large quantities of secondary wastes are generated due to the numerous process steps necessary for cesium elution. These steps include acid elution, exchanger water rinse, and caustic sodium loading of the exchanger. In addition, organic ion-exchangers can lose approximately 3% of their exchange capacity per use cycle and as a result need to be replaced, and subsequently disposed of, after 20 to 30 cycles.   Another class of "single use" exchangers is directly incorporated as loaded ion-exchange materials into a solid waste form such as borosilicate glass logs by melting. However, a third alternative employs electroactive ion-exchange materials. In these new materials the uptake and elution of cesium is controlled by an applied electric potential. This approach has several advantages over traditional reusable ion-exchange resins including reduced secondary waste generation, higher cesium selectivity, and higher durability.

Currently, the most promising electroactive ion-exchange materials for cesium separations consist of nickel ferrocyanide films grown electrochemically on nickel substrates. Fundamental work by Bocarsly [1-5] documented the electrochemical processing of nickel ferrocyanide films in

Mat. Res. Soc. Symp. Proc. Vol. 465 © 1997 Materials Research Society

addition to other anionic metallocyanide complexes and the charge-transfer behavior of these films under oxidizing and reducing conditions. The high Cs selectivity of nickel-ferrocyanides has been reviewed by Loos-Neskovic [6-7]. The combined technologies of ion exchange and electrochemistry have been demonstrated previously [8]. PNNL researchers have modified the ion exchange process by controlling the ion exchange properties of the film by modulating the potential of the film directly instead of regulating the interfacial pH through localized acid and base generation by water hydrolysis. The resulting films show a selectivity for Cs in the presence of high Na concentrations and have a lifetime approximately 10 times that of traditional organic ion exchange materials. However, the primary advantage of electroactive ion exchange materials is that secondary waste is minimized as compared to traditional organic ion exchange processing. This advantage arises because several process steps are eliminated by controlling the ion exchange properties of the film by an applied potential instead of the altering interfacial pH by acid elution and caustic sodium loading.

Much of the previous work on electroactive ion exchange processes has focused on the engineering and applications aspects of the processing. As a result there is limited understanding of the behavior of these films at the molecular level. For example, while it has been recognized that there is a relationship between the crystalline structure of these films and their chemical reactivity[9], the details of this relationship are not known. In addition, a relationship between film deposition conditions has also been observed to impact the ion exchange capacity of the film as well as film durability, yet it is not known whether film deposition conditions affect the crystalline structure of the film or the chemical composition. Most significantly, the coordination environment and chemical speciation of the exchange ions in the film are also unknown. This could strongly impact the selectivity of ion exchangers in chemically complicated waste streams.

Transition metal ferrocyanides can form in a variety of symmetries including hexagonal and cubic. However, the cubic structure, where the metal atoms are linked by cyanide units, is commonly formed when nickel is anodized in the presence of $Fe(CN)^{6-}$ [10]. Iron and nickel atoms alternate at the corners of the cube and the cyano groups located along the cube edges are oriented so that all the irons are coordinated by eight carbon atoms and all the nickel atoms are coordinated by eight nitrogen atoms. The center of the cube is occupied by an alkali metal cation. The occupancy of the alkali site reflects the redox potential of the film, and a net diffusion of cations in and out of the film is required for redox reactions to occur. Because the alkali metal site is quite large in the cubic structure (the $Fe\text{--}M^+$ atomic distance is about 3.2 Å and the $CN\text{--}M^+$ distance is 2.9 Å) there is much speculation as to whether the alkali metal is hydrated as it enters and leaves the film [1,2,10]. A hexagonal modification, $K_2Zn_3[Fe(CN)_6]_2 \cdot n\mathrm{H2O}$ [11], has basically the same structure as the cubic analog; however it contains zeolitic water as well as an alkali metal cation at the center of the cube.

EXPERIMENT

<u>Film Preparation</u>

The experimental apparatus used to deposit and characterize the films includes a PAR 273A potentiostat/galvanostat, and an electrochemical cell. All samples mounted on the electrochemical cell had a circular exposed area with a diameter of 1.9 cm. All equipment is computer controlled via a GPIB card using LabView software.

A 99.98 % pure nickel substrate (Goodfellow) was used for all the data presented below. All chemicals were A.C.S. reagent grade, and all solutions were prepared with 18.2 MΩ-cm water. Prior to each film deposition, the nickel substrate surface was abraded using a 600-grit sandpaper and thoroughly rinsed. Three different deposition procedures were used. One procedure was similar to that reported by Bocarsly et al., where the nickel surface was exposed to a solution of 5 mM $K_3Fe(CN)_6$ and 0.1 M KNO_3, and a 1.0 V(SCE) potential was applied to the nickel electrode for 300 seconds. This procedure is designated as the literature procedure. The other two procedures were PNNL proprietary methods and designated as PNNL-1 and PNNL-2. The characteristics of most of the films were determined from their cyclic voltammograms in 1.0 M $NaNO_3$ solution. The applied potential was cycled from 0.25 to 0.8 to -0.1 and back to 0.25 V(SCE) at 50 mV/sec.

Atomic Force Spectroscopy

A Digital Instruments Nanoscope III atomic force microscope was used to probe the morphology of the electrochemically deposited nickel ferrocyanide films. A standard square pyramidal Si_3N_4 tip was used and all images were acquired in 'constant force' or 'constant deflection' mode. The piezoelectric scanner was chosen to view the large scale morphologic features on the order of a few microns. Measurements of grain diameter were based on a vertical relief measurement as opposed to the measurement of grain diameter in the plane of the image because of the greater vertical sensitivity of the AFM.

Raman Spectroscopy

Raman vibrational spectra of the nickel ferrocyanide films was excited with 20 mW of 488.0 nm radiation from an argon ion laser. The laser excitation was focused on to the film using an optical microprobe and a 40X objective. The inelastically scattered light was collected in backscattering geometry through the optical microprobe and dispersed by a 0.85 meter triple-spectrometer on to liquid nitrogen cooled Charge Coupled Device detector. Both the cyanide stretching region, from 2000 to 2250 cm^{-1}, and O-H stretching region, from 3200 to 3800 cm^{-1} of the vibrational spectrum were examined and subsequently curve-fit.

X-Ray Absorption Spectroscopy

XANES/XAFS experiments were conducted at the Fe K-edge and Cs L_{III}-edge on a series of nickel ferrocyanide films produced by the three different deposition conditions. These experiments were performed at Stanford Synchrotron Radiation Laboratory on sidestation 2-3 under dedicated operating conditions (3.0 GeV and 40 to 90 mA current). In addition to the films, two powder samples, $Na_2NiFe(CN)_6$ and $Cs_2NiFe(CN)_6$, were analyzed as standards. Spectra were collected by measuring the Auger electrons emitted as the absorbing atom relaxes to its ground state using an electron yield detector to a photoelectron wavevector of 13 $Å^{-1}$. Energy calibration was based on assigning the first inflection point in the absorption edge of an iron metal foil to 7111.3 eV. Normalization of the absorption spectrum was accomplished by fitting polynomials through the pre- and post-edge regions, setting the value of the extrapolated pre-edge to zero at E_0, and setting the difference between the extrapolations of the pre- and post-edge polynomials to unity at E_0. EXAFS were extracted by fitting a polynomial spline function through the post-edge region and normalizing the difference between this approximation to the solitary atom EXAFS and the actual data with the absorption decrease given by the McMaster tables [12]. Fourier transforms were taken over photoelectron wavevector region from 2.5 to 12.5 $Å^{-1}$. Nodes in EXAFS were selected as endpoints to the transform range and a Gaussian window with a 1 sigma width of 0.5 $Å^{-1}$ was used to dampen the EXAFS oscillations at the transform range endpoints. The phase shift has not been removed from the Fourier transforms; as a result the peaks in the transform moduli correspond to contributions from shells that are 0.2 to 0.5 Å shorter than the actual absorber-scatterer distances.

RESULTS

Cyclic voltammetry indicated that films deposited using the PNNL deposition techniques showed an initial 40% increase in capacity for Na ion exchange than films deposited using literature techniques. Also, the PNNL deposited films displayed greater resistance to degradation losing only 25% of their initial capacity after 2000 uptake/elution cycles compared to a 50 % decrease for films deposited using literature techniques.

AFM images of the films deposited following PNNL procedures show complex, large-scale microstructure that is on the order of a few to 10 microns in diameter. The films are not continuous across the substrate and vary in thickness from bare substrate to several 100 nanometers thick on the micron length scale. The film itself consists of grains that are tens of nanometers in diameter.

Raman spectroscopy of the ferrocyanide films in the cyanide stretching region revealed distinctive vibration spectra that can be associated with each of the alkali metals. The Raman spectra of Na-rich films display a broad, low intensity mode centered at 2185 cm^{-1}. The Raman spectra of K-rich films consist of three, broad low intensity modes at 2105 cm^{-1}, 2148 cm^{-1}, and 2177 cm^{-1}. Cs-rich films display two intense modes at 2105 cm^{-1}, 2142 cm^{-1}. However, for K-rich films that are subsequently ion exchanged with Cs, the frequencies of the Raman modes diagnostic for Cs are shifted to higher Raman shift by 5 cm^{-1}. Only films deposited using the literature deposition technique displayed vibrational modes in the O-H stretching region.

The energy position of the Fe K-edge absorption edge shifts to higher energy as the Fe oxidation state increases. A comparison of the edge positions of the prepared films to Fe^{2+} and Fe^{3+} cyanide standards indicates that the Fe in all the films is trivalent despite even though some of the films were deliberately left under reducing conditions.

FEFF6.01 [13,14]was used to calculate the phase and amplitude for the individual scattering paths to the carbon, the nitrogen, the alkali metal, and the distal nickel atoms, and are denoted Fe-C, Fe--N, and Fe--M, and Fe--Ni respectively. The 180° scattering angle formed by the Fe-C-N-Ni linkage generates large multiple scattering amplitudes at the same effective distance as the Fe--N and Fe--Ni paths and the multiple scattering contributions are included in the parameterization of these paths. In addition, a large multiple scattering contribution exists at twice the Fe-C bond distance. This path was also included in the fits. The parameterized scattering paths were used to curvefit the EXAFS over the photoelectron wavevector region 2.5 to 12.5 $Å^{-1}$

Comparisons of the FEFF6.01 calculations of the Fe K-edge EXAFS spectra for the cubic and hexagonal structural modifications reveal only subtle changes in bond distances for the first two coordination shells. Analysis of our initial EXAFS measurements, shown in Table I, yield bond distances that are intermediate to the hexagonal and cubic structures. We anticipate that experimental measurement of ferrocyanide compounds in cubic, hexagonal and monoclinic structure will aide in determining the variation in molecular structure of the deposited films. The measured EXAFS amplitudes of films deposited using procedures in the literature typically had lower amplitudes than the PNNL procedures perhaps suggesting greater static disorder. Fits to the EXAFS for four films representative of the three deposition techniques are shown in Figure 1. The Fourier transform of the individual scattering paths used to fit a film deposited using PNNL1 deposition procedure is shown in Figure 2.

Initial Cs L_{III}-edge EXAFS spectra have a very low signal-to-noise ratio due to the low concentration of cesium in the films. However a comparison of a FEFF6.01 Cs L_{III}-edge EXAFS calculation for Cs residing in the center of the cube to the measure EXAFS film with the highest concentration suggest that at least some of the Cs ions are present as hydrated species. The alkali metal site in the ferrocyanide structure is large enough to accomodate a hydrated Cs ion. The

Table I. EXAFS fitting results over the k range from 1.7 to 11.0 Å^{-1}.

path →		Fe - C	Fe -- O	Fe -- N	Fe -- M	Fe -- Ni
NaNiFeCN	d	1.87 ± 0.02		3.09 ± 0.02	4.30 ± 0.02	5.12 ± 0.02
standard	n	5.9 ± 1.5		5.3 ± 1.8	4.9 ± 1.0	5.8 ± 1.3
	s	0.06 ± 0.02		0.06 ± 0.01	0.10 ± 0.02	0.10 ± 0.02
CsNiFeCN	d	1.87 ± 0.02		3.08 ± 0.021	4.52 ± 0.02	5.08 ± 0.02
standard	n	7.0 ± 1.8		5.0 ± 1.3	8.4 ± 2.2	6.4 ± 1.3
	s	0.07 ± 0.01		0.06 ± 0.01	0.10 ± 0.02	0.10 ± 0.02
Literature	d	1.92 ± 0.02	2.68 ± 0.02	3.09 ± 0.02	4.38 ± 0.02	5.17 ± 0.02
K IX film	n	3.8 ± 0.9	3.9 ± 1.2	6.1 ± 1.6	2.1 ± 0.8	2.6 ± 0.5
	s	0.02 ± 0.01	0.00 ± 0.01	0.06 ± 0.01	0.01 ± 0.02	0.08 ± 0.02
PNNL1	d	1.94 ± 0.01		3.04 ± 0.02	4.37 ± 0.01	5.10 ± 0.02
K IX film	n	3.6 ± 1.0		7.8 ± 2.2	1.7 ± 0.5	1.3 ± 0.4
	s	0.00 ± 0.00		0.08 ± 0.01	0.00 ± 0.02	0.11 ± 0.01
PNNL2	d	1.95 ± 0.02		3.06 ± 0.02	4.39 ± 0.02	5.12 ± 0.02
K IX film	n	4.3 ± 1.1		8.7 ± 2.6	2.1 ± 0.7	2.7 ± 0.8
	s	0.00 ± 0.0		0.09 ± 0.02	0.00 ± 0.0	0.09 ± 0.01
PNNL2	d	1.95 ± 0.02		3.06 ± 0.02	4.40 ± 0.01	5.14 ± 0.02
Cs IX film	n	3.8 ± 1.0		8.3 ± 0.8	1.2 ± 0.4	3.4 ± 0.7
	s	0.00 ± 0.0		0.04 ± 0.02	0.00 ± 0.0	0.07 ± 0.01

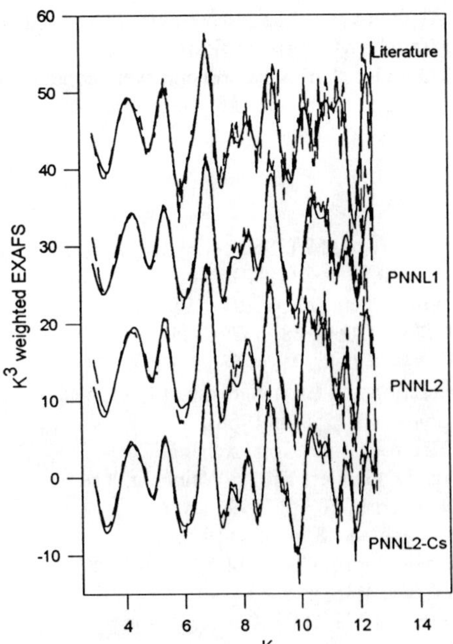

Figure 1. Comparison of the k^3 weighted EXAFS (dashed line) to the fit (solid line) for four films representative of the three deposition techniques.

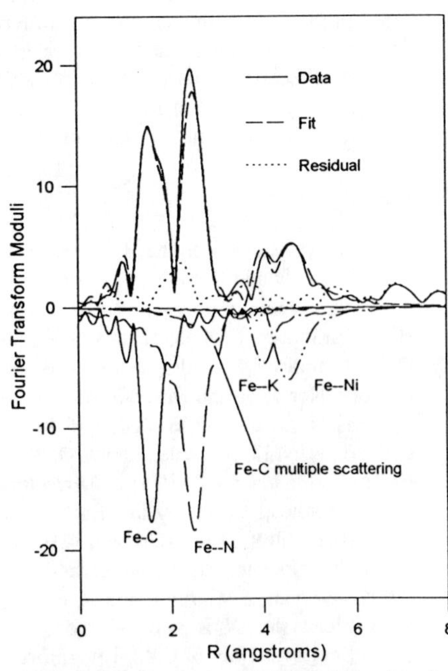

Figure 2. Fourier transform of the EXAFS and the fit for film PNNL1. The individual scattering paths are shown with negative amplitudes.

speciation of cesium in these films is critically important in developing predictive models of film performance in complex chemical environments. Future EXAFS experiments will be optimized to improve the data quality acquired at the Cs L_{III}-edge. EXAFS analysis is one the few tools that can provide speciation information at the low concentration levels that cesium exists in these films. EXAFS measurements of several cesium standards where cesium occupies large zeolitic like sites will be used to help interpret our data at the Cs L_{III}-edge. However it is possible that low intensity Cs signal results from dampening of the EXAFS due to the thermal motion of the Cs atom in the large zeolitic site; therefor satisfactory data can only be attained at low temperature.

CONCLUSIONS

X-ray absorption spectroscopy, in conjunction with Raman spectroscopy, has been useful in determining the structural and chemical dependence of nickel ferrocyanide films on deposition technique and suggests a possible correlation with electrochemical performance. Raman studies suggest that films prepared using literature deposition techniques may contain hydroxides. The presence of hydroxide may result in lower ion exchange capacity and reduced durability.

ACKNOWLEDGEMENTS

The authors are thankful Dr. Steve Conradson, Los Alamos National Laboratory, and Dr. Farrel Lytle, The EXAFS Company, for their the assistance and comments in XAFS data collection and analysis, and to Dr. A.S. Lea, Pacific Northwest National Laboratory, for his assistance in the collection and interpretation of the AFM images. The research was conducted at Pacific Northwest National Laboratory, operated by Battelle Memorial Institute for the U.S. Department of Energy under Contract DE-AC06-RLO 1830. XAFS experiments were conducted at Stanford Synchrotron Radiation Laboratory which is operated by the U.S. Department of Energy, Office of Basic Energy Sciences.

REFERENCES

1. A.B. Bocarsly and S. Sinha, J. Electroanal. Chem. **137**, p. 157 (1982).
2. A.B. Bocarsly and S. Sinha, J. Electroanal. Chem. **140**, p. 167 (1982).
3. S. Sinha, B.D. Humphrey, and A.B. Bocarsly, Inorg. Chem. **23**, p. 203 (1984).
4. B.D. Humphrey, S. Sinha, and A.B. Bocarsly, J. Phys. Chem. **88**, p. 736 (1984).
5. B.D. Humphrey, S. Sinha, and A.B. Bocarsly, J. Phys. Chem. **91**, p. 586 (1987).
6. C. Loos-Neskovis and M. Fedoroff, Solvent Extraction and Exchange 7, p. 131 (1989).
7. C. Loos-Neskovis and M. Fedoroff, Reactive Polymers 7, p. 173 (1988).
8. C.P. Jones, M.D. Newville, and A.D. Turner, "Electrochemical Ion Exchange" in *Electrochemistry for a Cleaner Environment*, eds. D. Genders and N. Weinberg, The Electrosynthesis Company Inc, East Amherst, New York, 207 p. 1992.
9. C.H. Luangdilok, D.J. Arent, and A.B. Bocarsly, Langmuir **8**, p. 650 (1992).
10. L.F. Schneemeyer, S.E. Spengler, and D.W. Murphy, Inorg. Chem. **24**, p. 3044 (1985).
11. P.P. Gravereau, E. Garnier, and A. Hardy, Acta Cryst. **B35**, p. 2843 (1979).
12. W.H. McMaster, N. Kerr del Grande, J.H. Mallett, and J.H. Hubbell, <u>Calculation of X-Ray Cross Sections</u>, Univ of Calif, Livermore, 1969, 350p.
13. J.J. Rehr, S.I. Zabinsky, and R.C. Albers, Phys. Rev. Let. **69**, p. 3397 (1992).
14. Zabinsky, J.J. Rehr, A. Ankudinov, R.C. Albers, and M.J. Eller, Phys. Rev. B. **52**, p. 2995 (1995).

ADSORPTION OF Sn(IV) ON GOETHITE IN 0.01M NaCl SOLUTION AT AMBIENT TEMPERATURE

Takayuki AMAYA [1], Kazunori SUZUKI [1], Chie ODA [2], Hideki YOSHIKAWA [2], Mikazu YUI [2]
[1]JGC Nuclear Research Center, 2205, Naritacho Oaraimachi, Ibaraki Pref., 311-13, Japan
[2]PNC, Muramatsu, Tokai-mura, Naka-gun, Ibaraki Pref., 319-11, Japan

ABSTRACT

The adsorption of soluble Sn(IV) species on goethite (FeOOH), a naturally occurring mineral, was investigated in 0.01M NaCl solution at ambient temperature. The sequential spike method was used to obtain the distribution coefficient (Kd). Kd values of higher than 1×10^7 ml/g at neutral pH, and 3×10^5 ml/g at pH 11 were obtained. It is observed that Kd values decrease with increasing pH. In the desorption experiment, the sequential extraction method was applied. The extractants such as 0.1M NaOH, Tamm's acid oxalate and Coffin's reagent were used. The result indicates that the adsorption consists of both reversible and irreversible reactions, and that the most of fractions in adsorption of Sn on goethite correspond to the irreversible adsorption or a stable fixation. Moreover, it is suggested that the reversible adsorption converts to the irreversible adsorption with time.

INTRODUCTION

The solubility and the adsorption behavior of radionuclides in the geological media are important items in the migration study of high level radioactive waste (HLW) disposal system. ^{126}Sn (β nuclide, fission yield = 0.06%) in HLW is one of the important radionuclides because of its long half life of 1×10^5 year [1]. Our previous study experimentally indicated that the solubility of Sn(IV) in a solution is a higher value [2] than that provisionally used in the current performance assessment [3], and also that goethite (FeOOH), a naturally occurring mineral, has a strong adsorption ability of Sn. However, the adsorption behavior of Sn on minerals and rocks is not enough studied. Therefore, it is necessary to confirm whether Sn can be retarded in a HLW disposal system. To obtain the distribution coefficient accurately, a new procedure, sequential spike method, was applied to the batch type test. This procedure was useful on the conditions where there is low solubility and strong adsorption of the nuclide. Regarding the adsorption, it is also important to know whether the adsorption consists of a reversible and/or an irreversible reaction. The sequential extraction method was applied to the desorption test to study the adsorption mechanism of Sn(IV) on goethite.

EXPERIMENTAL

The experiments were carried out in an inert gas glovebox, in which the concentrations of oxygen and carbon dioxide were less than 1 ppm. Pure water was degassed 12 hours before it was used in the experiment. NaOH solution without CO_2 was prepared using sodium metal and pure water in the inert gas glovebox. All reagents used in this test were special grade or nearly same grade.

Mat. Res. Soc. Symp. Proc. Vol. 465 © 1997 Materials Research Society

<u>Adsorption experiment</u>

The standard solutions of [113]Sn (LMRI) and $SnCl_4$ (Kanto Chemical Company) were used to prepare the Sn stock solutions with specific radioactivity of 2.2×10^{13} Bq/mol and the chemical concentration of 7.3×10^{-6} mol/l. The experimental conditions are shown in Table I. The experiments are carried out in duplicate.

<u>Table I   Experimental conditions of adsorption test</u>

test type   : sequential spike method   solid phase : goethite (grain size<1 μm)
liquid phase : 0.01 mol/l·NaCl (pH 6,9,11 temp.=25℃)
immersion period : 300 days   solid / liquid ratio = 1 g / 40,000 ml

10 mg of Goethite (FeOOH:Kanto Chemical Company) was immersed in 400 ml solution at pH 6, 9 and 11 solution, and then Sn(IV) chloride stock solution labeled by [113]Sn was spiked in the solution. The initial concentration of Sn (IV) in the solution was 7×10^{-9} mol/l which is less than solubility of Sn at these pH values [2]. After one week, a small aliquot of the solution was filtered through 10,000 MW ultrafilter and the radioactivity of Sn in 1 ml sample was measured. If it was less than the detectable limit, the Sn stock solution was spiked again. The same procedure continued until the adsorption achieved the equilibrium state. For Sn species to be always soluble, the concentration of Sn(IV) to be spiked was controlled correctly. This procedure, named the sequential spike method, was applied to all batch type tests in this study in order to accurately obtain distribution coefficient, Kd values. All filters were washed and immersed in the same solution prior to use, in order to avoid any adsorption on the filter. The pH and Eh were measured by pH meter (Horiba B-112) and   Eh meter (Horiba D-14), respectively. The Eh measurement was carried out after the electrode was set the solution for 30 minutes. The concentration of Sn was calculated by equation (1) using the measured radioactivity after the radiation equilibrium of [113]Sn-[113m]In was reached. The specific radioactivity (SA) in each calculation was corrected because of the short life time of [113] Sn (half life 115.1 day). Kd value was calculated by the equation (2).

$$C_{Sn} = \frac{1,000 \times C_{(^{113}Sn)}}{SA} \qquad (1)$$

where C_{Sn} is the chemical concentration of Sn [mol/l], $C_{(113Sn)}$ is the radioactive concentration of [113]Sn [Bq/ml], SA is the specific radioactivity of Sn [Bq/mol].

$$Kd = \frac{C_{solid}}{C_{liquid,end}} = \frac{\sum [R_n - C_{liquid,n} \times V_n]}{W \times C_{liquid,end}} \qquad (2)$$

where C_{solid} is the concentration of Sn on goethite after test [mol/g], $C_{liquid,end}$ is the concentration of Sn in the solution after test [mol/l], Rn is the quantity of Sn by spiking [mol], $C_{liquid,n}$ is the concentration of Sn in the solution just before (n+1) spike [mol/l], Vn is the volume of the solution just before (n+1) spike [l], W is the weight of goethite [g].

<u>Desorption experiment</u>

In order to carry out the desorption test, goethite adsorbed Sn(IV) was prepared under the experimental conditions shown in Table II. 100 mg of goethite was added to 50 ml of 0.01M NaCl in teflon container at pH 10. The concentration of Sn was less than the solubility. After one month, the solid phase was filtered and dried.

Table II   Preparation of goethite adsorbed Sn

Test type	: one time spike
Liquid phase	: 0.01 mol/l NaCl temp.=25°C
Immersion period :	S1: 30 days pH 10
	S2: 139 days pH 6,7,8,9,10

The experimental conditions are shown in Table III. At first, 100 mg of goethite obtained in this preparation was added in the 2 ml centrifuge tube, and then the initial γ radioactivity of ^{113m}In in the tube ($A_{initial}$) was measured directly by the NaI(Tl) scintillation analyzer (Aloka 300). After the measurement, 2 ml of the extractant was added in the tube containing the goethite at the solid : liquid ratio of 1 g : 20 ml. The tube was vibrated by the shaker (Iwaki tube mixer TWIN 3-28) for a fixed time. The mixture in the tube was centrifuged at 10,000 G for 30 minutes to separate the goethite from the extractant. To remove the residual extractant, the goethite was rinsed by 0.01M NaCl after each extraction. Then, γ radioactivity of ^{113m}In on the goethite in the tube (A_{ext}) was measured. Three kinds of extractant, 0.1M NaOH solution, Tamm's acid oxalate and Coffin's reagent were step by step used for one goethite sample. The rate of residual Sn on goethite in each extraction was calculated by the equation (3).

Table III   Extractants for desorption experiment

step	extractant	compositions	mixing time	remarks
1st	0.1M NaOH	0.1M NaOH	1 hr$\times$3 times	(for S1 sample)
			2 hr$\times$1 time	(for S2 sample)
2nd	Tamm's	0.2M Ammonium oxalate	2 hr$\times$3 times	(for S1 sample)
	acid oxalate	0.2M Oxalic acid	4 hr$\times$1 time	(for S2 sample)
		pH3 in dark		
3rd	Coffin's	0.175M Sodium citrate	3 hr$\times$3 times	(for S1 sample)
	reagent	0.025M Citric acid	6 hr$\times$1 time	(for S2 sample)
		5% Na$_2$S$_2$O$_4$		

liquid/solid ratio = 20(ml/g) temp.=25°C

$$Rate\ of\ Sn\ (\%) = \frac{100 \times A_{ext}}{A_{initial}} \qquad (3)$$

where A_{ext} [Bq/ml]   : concentration of ^{113}Sn on goethite after each extraction
        $A_{initial}$ [Bq/ml] : concentration of ^{113}Sn on goethite before the desorption experiment

RESULTS AND DISCUSSION

<u>Adsorption reaction</u>

The results of the adsorption experiment at pH 6, 9 and 11 were shown in Fig.1, Fig.2 and Fig.3, respectively. The concentrations of Sn at pH 6, 9 were less than detectable limit during the experimental period of more than 12 times of spiking. At pH 11, the equilibrium of the adsorption reaction would reach at the second spiking. The detectable limit increases with immersion times because of short half life of ^{113}Sn.

Kd value was calculated at the test period of 300 days. Kd values of higher than 1×10^7 ml/g at neutral pH, and 3×10^5 ml/g at pH 11 were obtained, and it was also observed that Kd values decrease with increasing pH. Because PZC of goethite is 8.4 [5], it is reasonably suggested that the positively charged sites on the surface of goethite, $FeOH_2^+$ will be progressively disappeared to form neutral and also negative charged sites, FeOH, FeO^- at higher pH values. In addition, the chemical valence state of Sn in this study would be tetravalent from the Eh-pH diagram of Sn [6] as shown in Fig. 4. Another study done by authors indicated that dominant species of Sn(IV) in the solution were $Sn(OH)_4^0$ at pH neutral, $Sn(OH)_5^-$ and $Sn(OH)_6^{2-}$ at the alkaline pH range [2]. Therefore, pH dependance of the adsorption is controlled by relationship between charge of goethite surface charge and chemical species of Sn(IV). The behaviour of this adsorption can be explained by surface complexation model.

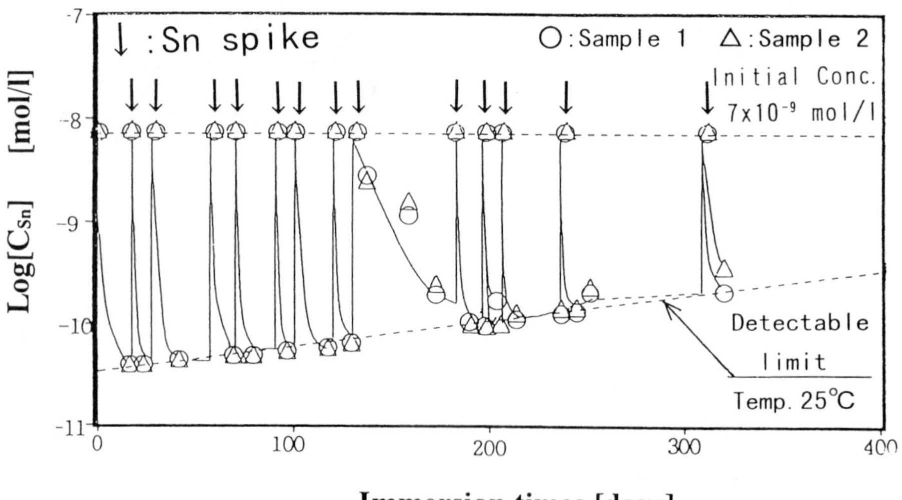

Immersion times [days]

Fig.1   The adsorption of Sn on goethite by sequential spike method at pH 6

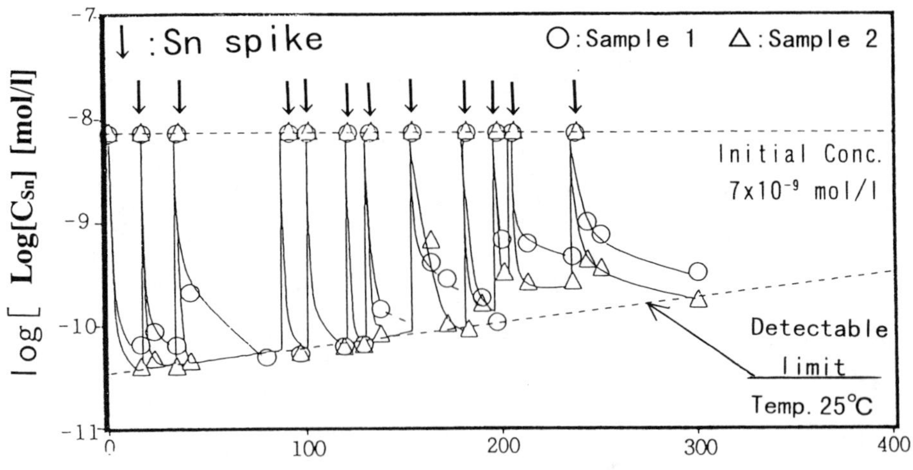

Fig.2   The adsorption of Sn on goethite by sequential spike method at pH 9

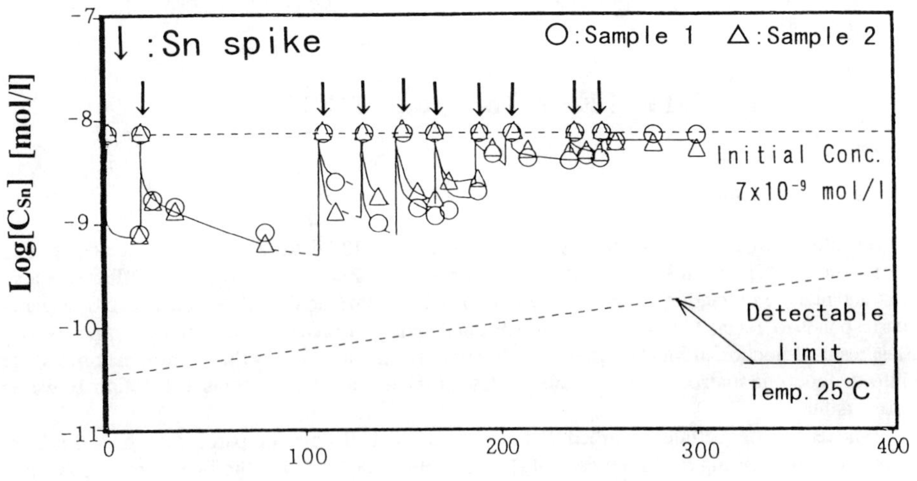

Fig.3   The adsorption of Sn on goethite by sequential spike method at pH 11

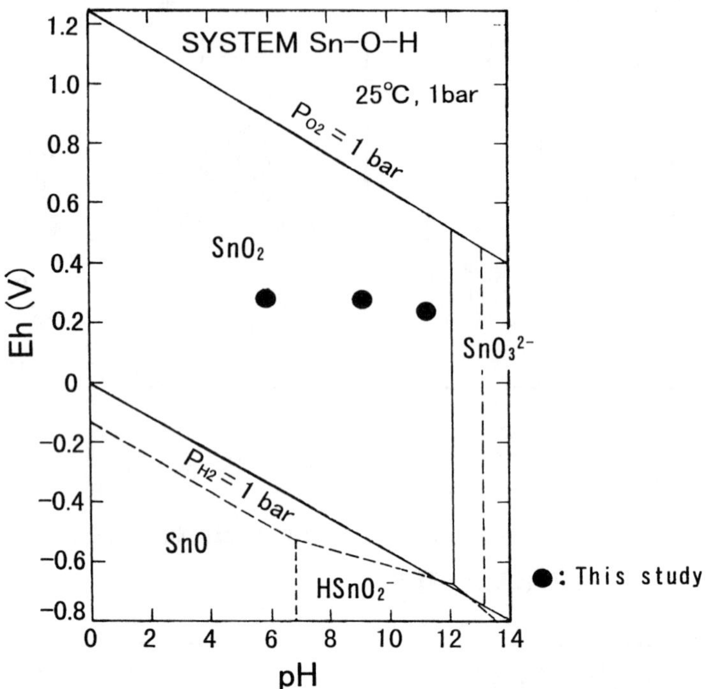

Fig.4 Eh-pH diagram of Sn[6]

<u>Adsorption mechanism</u>

Goethite samples adsorbed Sn at pH 6,7,8,9 and 10 for 139 days were used in the desorption experiments. The results of the extraction are shown in Fig.5. Regarding the effect of each extractant, the extraction reaction by 0.1M NaOH, Tamm's acid oxalate and Coffin's reagent corresponds to reversible, weakly irreversible and strongly irreversible adsorption, respectively. The residual fraction of Sn (IV) after all extraction treatments, means that Sn was incorporated into the goethite matrix and fixed stably. Most of the adsorption reactions at pH 6 to 10 were irreversible.

The results of goethite adsorbed Sn for 30 days at pH 10 were compared with those for 139 days as shown in Fig.6. In the case of the adsorption for 30 days, the rate of reversible and weakly irreversible fractions were higher than those for 139 days. It is suggested that the reversible adsorption or weakly irreversible adsorption gradually converts to the strongly irreversible adsorption, and also the stable fixation. The experimental results indicate the irreversible adsorption contributes to adsorption reaction of Sn(IV) on goethite.

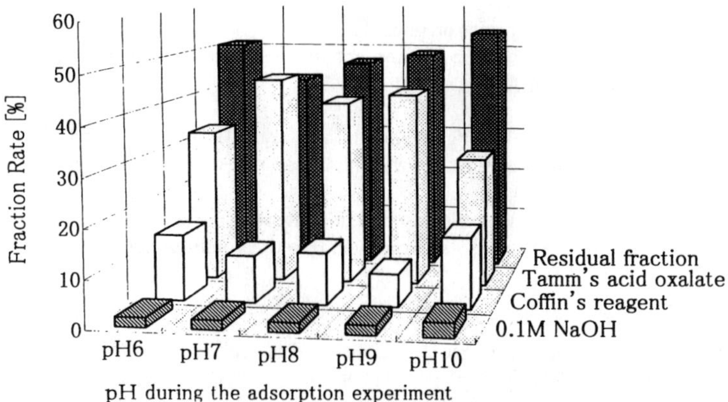

Fig.5   The fraction of Sn in the desorption experiment
using sequential extraction method

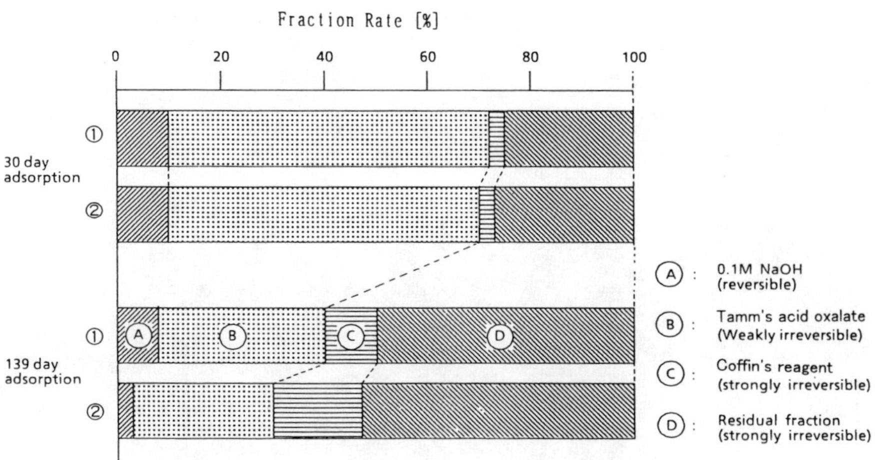

**Fig.6   The desorption fraction of adsorbed Sn on goethite
for 30 days and 139 days at pH 10**

CONCLUSIONS

The adsorption of soluble Sn(IV) species on goethite (FeOOH), a naturally occurring mineral, was investigated in 0.01M NaCl solution at ambient temperature by both adsorption and desorption experiment. The conclusions are summarized.

(1) Kd value was higher than 1×10^7 ml/g at a neutral pH, and 3×10^5 ml/g at pH 11.
(2) The contribution of irreversible adsorption was more than that of reversible in the adsorption system of Sn on goethite.
(3) Both reversible and weakly irreversible adsorption convert to the strongly irreversible adsorption with time, and Sn was also incorporated into the goethite matrix .

REFERENCES

1. PNC: Research and Development on geological disposal of high level radioactive waste : First progress report. PNC Tech. Rep. TN 1410 93-081, Tokyo 1992, ～pp. 4-95.
2. T. AMAYA, T. Chiba, K. SUZUKI, C. ODA, H. YOSHIKAWA and M. YUI in Scientific Basis for Nuclear Waste Management XX (Mater. Res. Soc. 1996) This volume.
3. I. G. McKinley and D. Savage in Fourth International Conference on the Chemistry and Migration Behaviour of Actinides and Fission Products in the Geosphere, (Charleston SC USA, December 12-17, 1993) pp. 657-665 (1994)
4. A .J. Plater, M. Ivanovich and R. E. Dugdale Applied Geochemistry, vol. 7, pp.101-110, (1992)
5. R. S. Rundberg, Y. Albinsson and K. Vannnerberg, Radiochimica Acta **66/67**, 333-319 (1994)
6. D. G. Brookins, Eh-pH Diagrams for Geochemistry (Springer-Verlag, Berlin 1988)

SORPTION PROCESSES OF UO_2^{2+} IONS ONTO ZIRCONIUM AND THORIUM PHOSPHATE SURFACES : THERMODYNAMICAL AND STRUCTURAL ASPECTS

R. Drot[*], E. Simoni[*], R. Cavellec[*], V. Pradin[*], Ch. Denauwer[**], JJ. Ehrhardt[***]

[*] Institut de Physique Nucléaire, Groupe de Radiochimie, Bât. 100, 91406 Orsay Cedex.

[**] CEA, DCC/DRDD, service d'études et de modélisation des procédés ValRho, BP 171,30207 Bagnols/cèze, au LURE, université Paris-Sud, 91405 Orsay Cedex.

[***] LCPE-CNRS-UHP, 405 rue de Vandœuvre, 54600 Villers-lès-Nancy.

ABSTRACT

Most of the studies on sorption processes take into account only a thermodynamic approach, and only few studies have paid attention to the structural aspect (i.e. structure of the postulated surface complex). Experiments were conducted to study UO_2^{2+} sorption onto zirconium and thorium phosphate surfaces. Moreover, in order to identify the nature of the sorption sites and the structure of the surface complex, various spectroscopic techniques such as EXAFS, XPS and laser induced fluorescence have been used. Two different types of sorption sites on the thorium phosphate but only one on the zirconium phosphate have been detected.

INTRODUCTION

One of the most important processes affecting the safety during the storage of nuclear waste in the underground disposal, is the migration through the geosphere. The radionuclides potentially released from the nuclear waste, particularly actinides such as U, Np, Pu, Am and Cm, may sorb onto non-transportable mineral surfaces which may enhance retardation. However, a quantitative understanding is rather complicated by the dependence of sorption processes on various geochemical parameters such as pH and ionic strength of the aqueous phase, radionuclide concentration, surface area and sorption site density of the mineral substrates. In this context, a new thorium phosphate Th4(PO4)4P2O7 has already been studied as a possible candidate for long-term storage because of its low solubility. Therefore, the study of the sorption equilibria of actinides onto such phosphate surfaces is of particular interest.

Sorption phenomena were carried out mainly on oxide [1,2] and hydrogenophosphates [3] but, to our knowledge, no study was made on phosphate surfaces, other than apatite. These adsorption processes can be modeled as chemical complexation reactions between the metal ion and the surface ligands or ion exchange. But this thermodynamical approach does not lead to accurate information in regards to the nature and the structure of the sorbed complex. It seems thus very interesting to investigate not only the sorption isotherms but to relate structure of surface complex to surface coordination models as well [4,5]. The thorium phosphate considered in this work presents two types of phosphate groups (orthophosphate and diphosphate) [6]. In order to understand the surface complexation, we have also investigated, for comparison, the zirconium diphosphate (ZrP2O7). The UO_2^{2+} ion was chosen as a model of hexavalent actinides. The thermodynamic study was realized in batch experiments and EXAFS, XPS and optical spectroscopy have been used as direct molecular probes to identify the sorbed species.

EXPERIMENTAL PROCEDURE

Both phosphates have been synthetized as described in the literature [6,7]. The zirconium phosphate is cubic, while the thorium phosphate is orthrhombic (space group Pcam, Z = 2), and can be describe from layers, parallel to (010), formed by both PO_4 and P_2O_7 groups

Mat. Res. Soc. Symp. Proc. Vol. 465 © 1997 Materials Research Society

where each Th atom,which links these layers, is connected to one P_2O_7 group and to five PO_4 groups. They were ground and washed several times with distilled water. The powder is dried at room temperature and ground again. The X-ray powder diffraction patterns,before and after this treatment, are identical. The average grains size, measured by a Malvern Instruments Mastersizer, is about 10 μm. The external surfaces areas were determined using a N_2-BET isotherm measured with a Coulter analyser SA 3100. The surfaces were prepared by heating at 150 °C under vacuum for 6 hours. Several analysis, on ZrP_2O_7 and $Th_4(PO_4)_4P_2O_7$, lead to the following values : 5.4 +/- 0.1 m^2/g and 1.2 +/- 0.1 m^2/g respectively.

The U^{VI} stock solutions were prepared by dissolution of the solid $UO_2(NO_3)_2,6H_2O$ in a acidic solution (HNO_3 1M). The cation concentration is determined by a spectrophotometric method (complexation by arsenazo I).

Batch experiments were conducted at room temperature by reacting weighed amount of the solid phase with a volume of cation solution in polypropylene tubes. The initial pH is adjusted in the range of 2.0 to 7.0 by addition of HNO_3 and KOH. The ionic strength is controlled by addition of KNO_3 in the solution. At first, for the hydratation step, a mass M of the solid is stirring for 24 hours with water containing KNO_3 (0.5 M) at the desired pH : 15 hours are neccessary to reach equilibrium. Then, after centrifugation, part of the liquid is removed and replaced by an equal volume of the solution containing the cation (at the same pH). After 24 hours, the solution is removed after centrifugation (3500 rpm during 30 min) and the solid is dried at room temperature. Kinetics experiments indicate that 20 hours are sufficient to reach the sorption equilibrium. The amount of sorbed cation onto the mineral phase is determined from the difference between the initial and final cation concentrations in solution. The obtained dried solids are then used for the spectroscopic measurements.

EXPERIMENTAL RESULTS

I. Figure 1 shows the percentage of the sorbed U^{VI} concentration as a function of pH for the zirconium phosphate surface. As expected, the curve exhibits the usual shape; U sorption is at a maximum at pH around 3. Figure 2 shows the sorption rate Γ (moles number of sorbed U^{VI} / mass of solid) respectively on $Th_4P_6O_{23}$ and ZrP_2O_7 versus the logarithm of the final U^{VI} concentration in the solution. As usual, the rate Γ increases with the U^{VI} concentration. Moreover, the percentages of sorption are nearly identical for both surfaces.

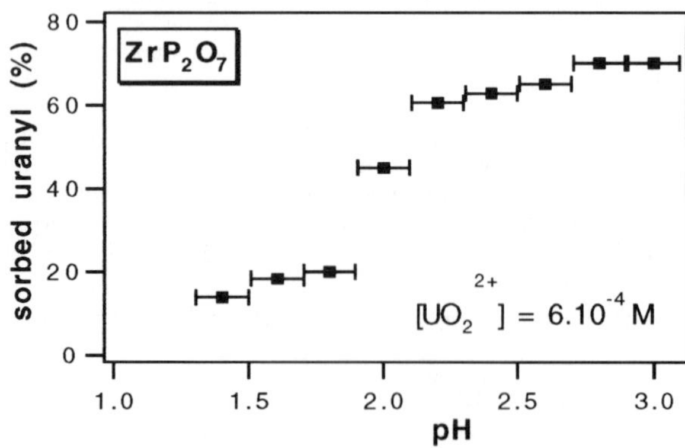

Fig. 1 : U^{VI} sorbed percentage versus pH

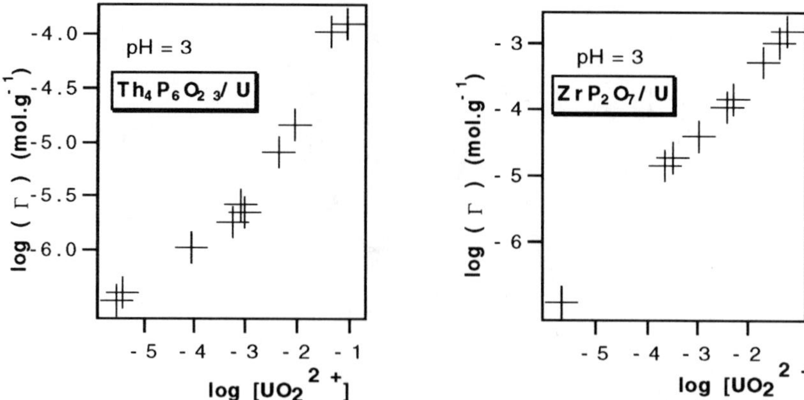

Fig. 2 : sorption rate function of U^{VI} concentration

Under the conditions of experiments, the total exchange capacity is around 2 mequiv./g for the ZrP_2O_7 and around 0.2 mequiv./g for $Th_4P_6O_{23}$.

II. In order to examine the chemical nature of surface species, photoelectron spectroscopy (XPS), laser induced fluorescence (LIF) and Extended X-ray Absorption Fine Structure (EXAFS) at the L$_{III}$-edge of U have been performed for U ions sorbed onto ZrP_2O_7, and $Th_4(PO_4)_4P_2O_7$ substrates. For these experiments, the sorbed solid phase is washed with distilled water at the sorption experiment pH and dried at room temperature.

The binding energies observed in the XPS spectra (fig. 3) corresponding to U^{VI} ions sorbed onto the zirconium and thorium phosphate surfaces, are compared to those obtained for reference samples such as $UUO_2(PO_4)_2$ and $UO_2(NO_3)_2,6H_2O$. This first comparison indicates that the sorbed cation seems to be surrounded by phosphate and not by nitrate ions. Furthermore, we have observed only one binding energy value for U sorbed onto ZrP_2O_7 surfaces while the U sorbed onto thorium phosphate exhibits two different values. Moreover, one of these two binding energies has the same value as the single one detected for the zirconium matrix while the second value is close to that one found for the $UUO_2(PO_4)_2$ compound (Table I). That could be interpreted as follows : one binding energy corresponds to an orthophosphate environment, while the other corresponds to a diphosphate environment.

Table I : Binding energy of U 4f$_{5/2}$ orbitals (eV)

	U4f$_{5/2}$
$UO_2(NO_3)_2,6H_2O$	396.8
$UUO_2(PO_4)_2$	391.5
U^{VI} / ZrP$_2$O$_7$	393.8
U^{VI} / Th$_4$P$_6$O$_{23}$	393.4 and 391.9

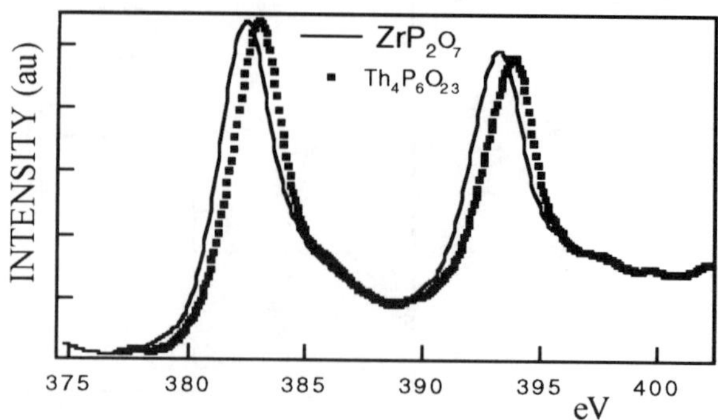

Fig. 3 : XPS spectra of U^{VI} sorbed on ZrP2O7 and Th4P6O23

The first optical experiments, at room temperature, have been carried out on UO_2^{2+} ions sorbed onto zirconium and thorium phosphates. The two corresponding spectra (not presented in this paper) are different which seems to indicate that the environment of U^{VI} onto the ziconium surface is different than onto the thorium surface. In order to check this result, emission decay times were performed. The results are summarized on the Table II.

We observe two different lifetime values for the cation sorbed onto the thorium phosphate but only one corresponding to the sorption onto zirconium phosphate.

Moreover, one of these two values (76 μs) is roughly the same as the one obtained for the zirconium diphosphate (73 μs).

Therefore, the obtained lifetime values and emission spectra indicate clearly that there are two different phosphate environments for U^{VI}/thorium phosphate and only one for the zirconium case. These first results are in agreement with those obtained by XPS measurements.

Table II : U^{VI} emission decays

U^{VI} sorbed on pH = 3, C(U) = 4.10⁻⁴ M	decay (μs) +/- 8 μs
ZrP2O7	73
Th4P6O23	76 and 310

EXAFS experiments were carried out on L_III edge of U^{VI} sorbed onto the zirconium diphosphate surface. These measurements are not possible with thorium substrate due to the large absorption of the thorium itself. X-ray absorption spectra were collected at room temperature in the French synchrotron facility (LURE, Orsay) operating at 1.8 GeV and 200-300 mA, over the range energy 17-18 keV. The incident beam was monochromatized with a Si (311) double crystal. Due to the small amount of uranium sorbed onto the surface, fifteen to twenty scans were collected for each sample. At first, in order to follow some change in the coordination environment of the sorbed cation with different intial conditions (pH of the solution, uranium concentration) we have recorded absorption spectra with several samples which have been prepared with different concentrations of uranium in solution (around 10⁻² M) and at different pH values. The EXAFS oscillations, all extracted identically, are presented in Figure 4. We observe clearly that there is no change with the uranium concentration while differences between the two spectra seem to indicate that the pH has some influence on the sorbed species. Data analysis, in order to determine the structural environment of the sorbed U^{VI}, is still under investigation.

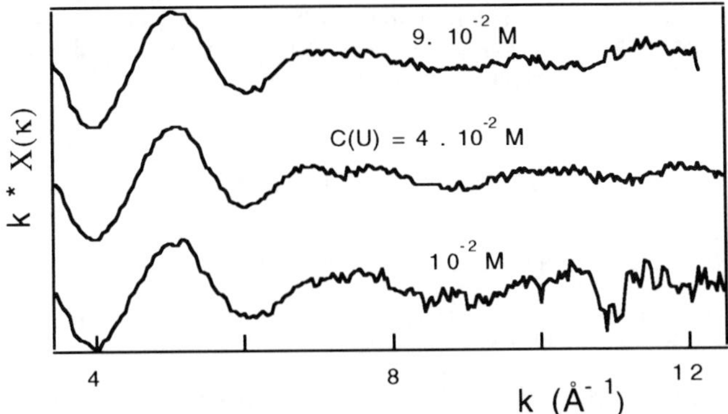

Fig.4 a : EXAFS spectra versus U^{VI} concentration

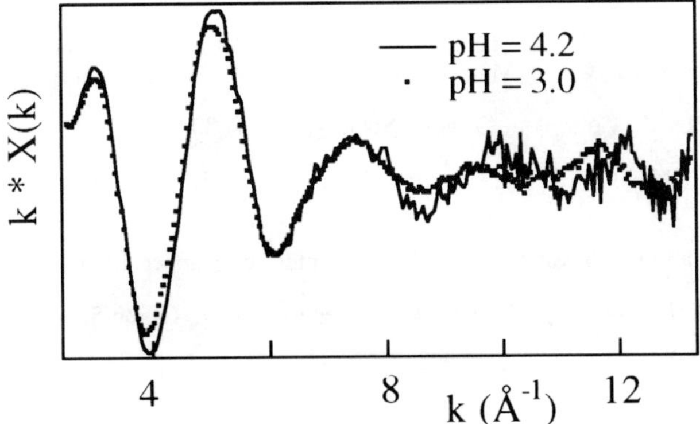

Fig.4 b : EXAFS spectra versus pH solution

Conclusion

Sorption isotherms of U^{VI} on zirconium and thorium phosphates, have been realized by batch experiments. The average rates of uranium sorption are roughly identical for both surfaces. The maximum of sorption is at around pH = 3 for ZrP_2O_7; the isotherm function of pH for $Th_4P_6O_{23}$ is under investigation. In order to take into account the pH dependence of the sorption, the electrostatic surface complexation model [8,9] will be investigated using the FITEQL code [10].

On the other hand, XPS and LIF experiments seem to indicate that there are two types of sorption sites on the thorium phosphate according to the formula : $Th_4(PO_4)_4P_2O_7$, and only one site on the ZrP_2O_7 substrate. Moreover, the preliminary EXAFS investigations show a pH dependence in the structural environment of the sorbed U^{VI} onto ZrP_2O_7 surface. With regards to these results, the same spectroscopic studies of the sorption process on the $Zr_2O(PO_4)_2$ compound are expected to confirm the presence of the two different sorption sites on the thorium phosphate material. Then, the thermodynamical results and their interpretation, in regards with the uranyl speciation in solution, will be compared to those obtained with spectroscopic methods and this comparison will probably lead to a good understanding of the sorption mechanism of U^{VI} on phosphate surfaces.

References

1. Prikryl J. D., Pabalan R. T., Turner D. R., Leslie B.W., Radiochimica Acta, 66 (1994) 291.

2. Jaffrezic-Renault N., Poirier-Andrade H., Trang D. H., J. Chroma.201 (1980) 187

3. Song Yinjie, Zhang Hui, Yang Qiaoling, Zhao Aimin, J. radioanal. Nuclear Chem., Articles, Vol. 198, No. 2 (1995) 375.

4. Manceau A., Charlet L., Boisset M. C., Didier B., Spadini L., Appl. Clay Sci., 7 (1992) 201

5. Bidoglio G., Gibson P. N., E. Haltier, N. Omenetto, M. Lipponen, Radiochimica Acta, 58/59 (1992) 191.

6. Benard P., Brandel V., Dacheux N., Jaulmes S., Launay S., Lindecker C., Genet M., Louer D., Quarton M., Chem. Mater., Vol. 8, No. 1 (1996) 181.

7. Alamo J., Roy R., Com. Am. Ceramic Society, (1984) 80

8. Hayes K., Redden G., Ela W., Leckie J., J. Colloid and Interface Science, Vol. 142, No 2, (1991) 448

9. Westall J., Hohl H., Advances in Colloid and Interface Science 12 (1980) 265

10. Herbelin A. L., Westall J., Report 96-01, Depart. Chem., Orgon State Univ. Corvallis, Or 97331. 1996

Part XII

Radionuclide Transport

GROUNDWATER COLLOID PROPERTIES AND THEIR POTENTIAL INFLUENCE ON RADIONUCLIDE TRANSPORT

C. DEGUELDRE
Paul Scherrer Institute, LWV & PE, 5232 Villigen, Switzerland.
Also, Centre des Sciences Naturelles de l'Environnement, University of Geneva.

ABSTRACT

Colloid facilitated transport is still an issue in radioactive waste management. Sophisticated phenomenological transport models are available, but progress is required to fully understand mechanisms and parameters. This study lead in this direction. Investigation of the marl groundwater colloids in steady-state conditions at Wellenberg, shows that their concentration is independent of the water flow rate. Their generation is caused by the re-suspension of the rock clay fraction only. The re-suspension process is presently being studied under flow transient conditions. Extension of our measurements to other safety relevant systems as well as a literature survey show that the colloid concentration under steady-state conditions is correlated to the concentration of alkali elements and earth alkali elements. The higher their respective concentration, the fewer colloids are occurring. Considerations on colloid contamination models are also included. The paper emphasises various colloid transport mechanisms including colloid generation from the irreversibly contaminated aquifer.

INTRODUCTION

Groundwater colloids are being studied because they may play a role in contaminant transport [1,2,3]. Trace elements sorbed or associated on these particles may be transported over large distances in an aquifer. In the framework of safety assessment studies of radioactive waste repositories, it is necessary to quantify the occurrence or concentration of colloids, their stability and transport properties under deep groundwater conditions as well as their contamination or generation with safety relevant nuclide species.

To be effective, studies on groundwater colloids require artefact-free sampling of suspended mobile material in groundwater and accurate characterisation of this material [2]. The water chemistry must be fully understood and, for colloid generation study, the host rock mineralogy must be compared with the results of colloid phase analysis. This type of work has been performed over the last few years at well SB6 in Wellenberg, Switzerland [4].

The study of colloid transport also requires determination of the attachment-detachment properties of natural colloids [5,6,7]. Because the flow rate is the only free parameter in this field study (well system), it should be the only parameter that can be varied to study the attachment-detachment properties of the natural colloids in the investigated aquifer. The colloid concentration has therefore to be quantified under steady-state conditions and under different out-flow rates. The research work must also be extended to study of colloid concentrations (breakthrough) during out-flow transients.

Mat. Res. Soc. Symp. Proc. Vol. 465 © 1997 Materials Research Society

To extend the study to other systems, collaboration with other groups was required. The goal of these collaborations is to produce accurate colloid data from a variety of aquifers. A link between the colloid concentration and the hydrogeochemical properties of the system is at least necessary to justify on a non-site-specific basis why the colloid concentration is low enough in a selected aquifer.

Finally, the interaction of nuclide species with the colloids remain to be quantified. The irreversible sorption or association of radionuclides on colloids must be understood in the broad concept of colloid generation.

WELLENBERG GROUNDWATER COLLOIDS, A SITE SPECIFIC STUDY

The Swiss National Co-operative for the Disposal of Radioactive Waste (NAGRA) is investigating a potential site for a low-level radioactive waste repository in the Wellenberg Area of central Switzerland, a few kilometres south of Lake Lucerne. The host rock is a strongly deformed, slightly metamorphosed marl of early Cretaceous age [4]. The study included the construction of seven boreholes with depths from 430-1870 m. The well selected (SB6) is the only one accessible from which artesian water flows from the marl formation without contact with cement. The water sampling in this study was carried out with a PVC line to a double packer. The marl groundwater is of a Na-HCO_3 type. It is reducing with a redox potential of -310 mV vs SHE at a pH of 8.5.

Characterisation of the Wellenberg marl groundwater colloids was carried out, with emphasis on their properties under steady-state flow conditions. Characterization tests were carried out coupling micro- or ultra-filtration with further analysis for example by SEM, atom force microscopy, or ICP spectroscopy of the colloid phase dissolved in nitric acid solution. The size distribution and concentration of the colloids were routinely measured utilising a single particle unit (Horiba monitor) for sizes of 100, 200 and 500 nm. In addition, for quantitative analysis, tests were performed combining filtration and quantitative X-ray diffractometry (XRD) and IR-spectroscopy. The groundwater colloids were identified as clay particles.

The groundwater colloid size distributions are reported in Figure 1. That obtained using a single particle counting unit utilising a Horiba (colloid illumination with an Ar cylindrical laser beam) are compared with that provided by the Particle Measuring Systems PMS unit (illumination with a helium planar laser beam) from Los Alamos National Laboratory (Nuclear Chemistry Division). The size distributions for the range 100 - 200 nm are consistent for the two units. For size smaller than 100 nm, the concentration detected with the Horiba unit is larger (more conservative) than for the PMS spectrometer. This is due to the differences in both illumination modes and detection geometries. From the particle size distribution a concentration of the order of 30 ng·ml^{-1} is calculated for the size range 100-1000 nm and a density of 2 g·cm^{-3} for the clay colloids. This is in agreement with weights calculated from the chemical results found for this size range considering Si as SiO_2, Al as Al_2O_3 and Fe as Fe_2O_3 (as oxide composing chemically the mineralogical phase of the clay colloids). This result was obtained, combining cascade micro-filtration and ICP-AES for the major elements associated with the colloid phase.

The analysis of trace elements in the colloid phase was also carried out by coupling filtration and ICP-MS. Only cesium results can be considered artefact free i.e. other concentration results are below the detection limit (ppt) for the blank. The cesium concentration of the water was 1.45 ± 0.05 ng·ml^{-1}. Cesium associated to the colloids was found to be 0.50 ± 0.05 and 0.12 ± 0.08 pg·ml^{-1} respectively for size ranges of 1000-10000 and 100-1000 nm. Taking into account the mass concentrations of 0.25 and 0.05 µg·ml^{-1}, for these respective cut-off's, one can calculate Rp values of 1.4 and 1.6 10^3 ml·g^{-1} respectively. These Rp values are comparable with the values obtained for the caesium sorption on colloids from the Grimsel Test Site [4].

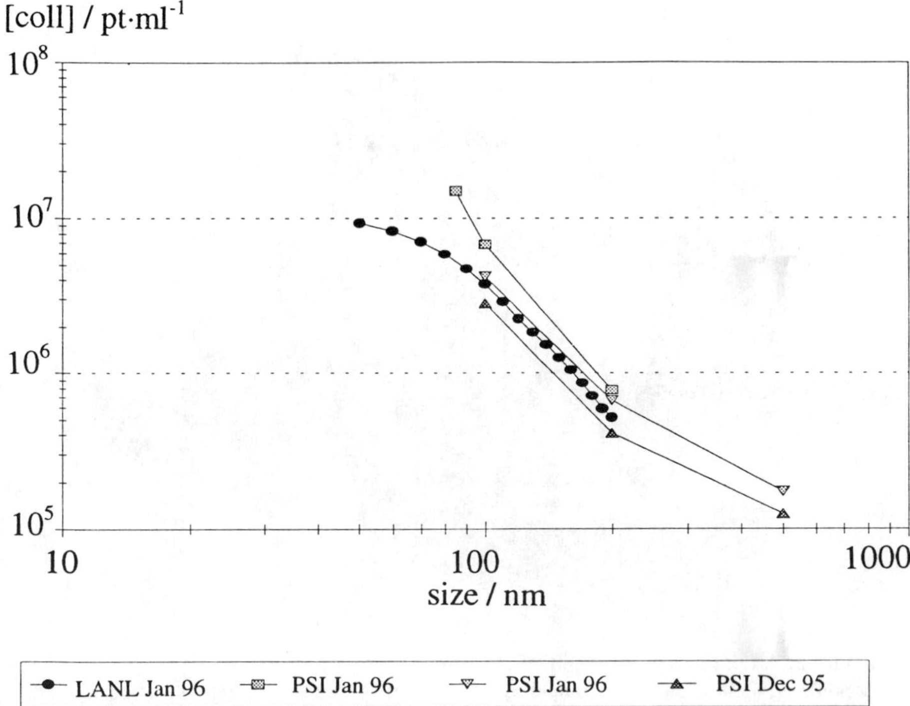

Figure 1: Comparison of single particle size distribution of the marl groundwater colloids from Wellenberg. Conditions: Single particle by counting using an Horiba PLCA-311 unit (**PSI**) and PMS spectrometer (**LANL**), sampling in early Dec. 95, measurement comparison in mid Dec. 95 and mid Jan. 96, using normal or high sensitivity mode, note the colloid phase is stable over a one month storage.

Further colloid characterisations were carried out by filtration and quantitative X-ray diffraction spectrometry (XRD). Colloid cakes were filtered out and air dried prior to analysis. In addition, glycolation was performed to identify smectite particles. The quantitative XRD allowed 3 classes of colloid mineral to identified i.e. smectite, illite and chlorite and indicated their relative concentrations in the colloid phase. The colloid cake analysis was also performed by IR-spectroscopy. On the colloid phase spectrum, clay phases are also identified as well as organic components. Other potential componants such as oxide which could be present as amorphous colloids are not found on the IR spectrum. Although this spectrum should also allow determination of the organic concentration in the colloid phase, only semi-quantitative results are possible i.e. a few per cent. XRD and IR-spectrometry results are given in Table 1. The morphology of these particles is visualised on the atom force micrograph as shown in Figure 2. Only flat particles are identified which are currently ascribable to clay colloids.

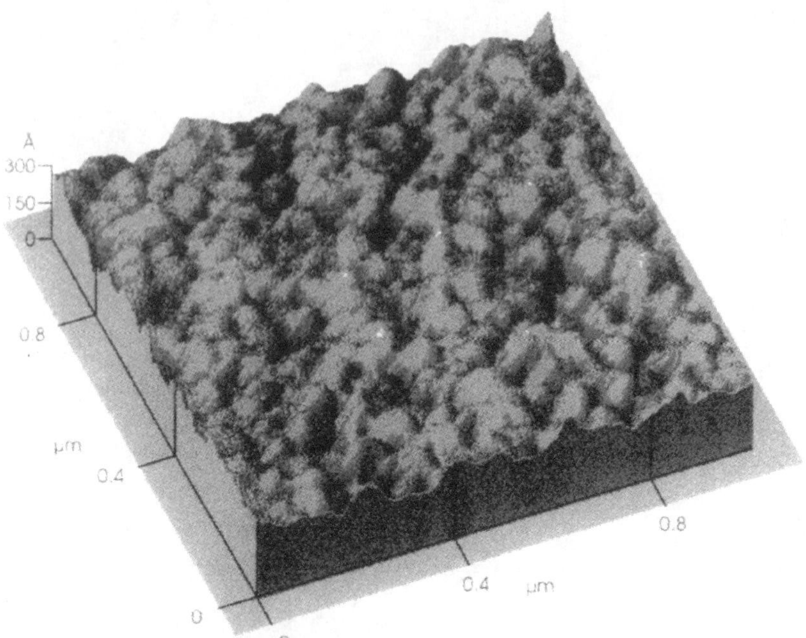

Figure 2: Atom force micrography of marl groundwater colloids.
Conditions: Colloid sample obtained by filtration of 0.2 ml SB6 groundwater on a 3 nm pore size membrane (Amicon XM50). AFM Autoprobe CP, Park Scientific Instruments. Note: The observed colloids have apparent sizes (widths) of the order of 30 (detection limit) to about 100 nm. Their height is of the order of 5 to 15 nm. These dimensions are typical for clay colloids. It must also be noted that the AFM picture is less stable than the SEM picture, because both suffer from the instability of the membrane during observation. Micrography with courtesy H. Van Swygenhoven.

Table 1 compares the composition of the colloid phase with the average composition of the marl [4]. The generation of the groundwater colloid phase may be interpreted as follow: the carbonates (calcite, dolomite and ankerite) are absent or present at a negligible concentration from the artefact-free groundwater colloid phase because they are either soluble (under-saturated) or should crystallise onto the solid phase (saturated). Quartz particles are relatively large and are not found in the colloid phase; suspended particles of quartz are only occasionally collected. Clay particles, with concentrations in the rock of about 10% illite, 10% smectite, 10% chlorite, are present at about 33% for each component in the colloid phase. This result on colloid generation in a marl system is important. Since colloids are generated from the aquifer, contaminated colloids resulting from the detachment of host rock particles sorbed by nuclides may play a significant role. Meanwhile, long-term monitoring demonstrated that colloid concentrations were constant with constant flow rate and independent of this rate (Figure 3). This result is important, it reflects the concept of a background colloid concentration corresponding to a steady state between colloid generation and colloid attachment on the host rock.

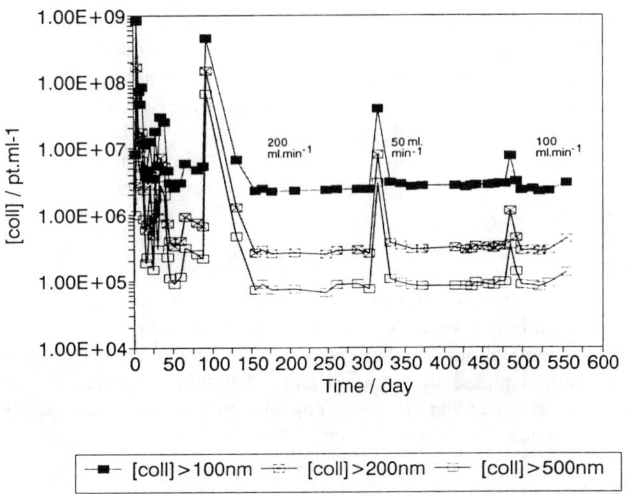

Figure 3: Colloid concentration as a function of discharge. Conditions: the gray area's correspond to non-steady state period, white area's correspond to steady state periods.

Component	Marl w%	Colloids w%
Calcite	45 ± 21	<2
Dolomite/ankerite	8 ± 5	-
Quartz	14 ± 5	<5
Feldspars	<1	-
Illite	12 ± 6	33 ± 5
Illite/Smectite	11 ± 5	30 ± 5
Chlorite	8 ± 4	30 ± 5
Kaolinite	<1	<5
Pyrite	1.2±0.8	-
Organic C	0.6±0.3	5 ± 5

Table 1: Comparison of mineralogical components of the rock and the groundwater colloid phases. Colloid data from XRD and IR-spectroscopy and for all sizes.

COLLOID STUDIES UNDER TRANSIENT CONDITIONS

Because the colloid concentrations were found to be independent of water flow rate under steady-state conditions (Figure 3), colloid generation has been studied during flow transient tests, the only way to observe colloid concentration changes. Under laminar flow conditions, only transients (physical e.g. water flow changes, or, chemical e.g. concentration changes) induce colloid concentration changes [9].

During the long-term survey, preliminary tests showed that the colloid concentration increased considerably during non-planned transients (for example due to large flow-rate changes because of valve clogging). Such colloid peaks were previously observed by Vilks et al. [10] and by Laaksoharju et al. [11], although no effort was undertaken to quantify the phenomena. The goal of the transient studies is to evaluate colloid attachment - detachment parameters from the colloid breakthrough curve.

<u>Conceptual background.</u>

The flow rate change in the aquifer induces a three step process.
1- the propagation of the flow rate change in the aquifer,
2- activation of the colloid generation by detachment from the aquifer,
3- transmission of the new colloid population through the aquifer to the bore hole.

The three processes are depicted in Figure 4 and treated in a separated way in the following sections. The effect of the flow-rate changes on colloids in the borehole and sampling system must be quantitatively eliminated.

Propagation of the flow rate change

Because the activation is transferred by a flow rate change in the aquifer, it starts at the borehole surface and is conducted through the aquifer. In a simplified model, the aquifer is a homogeneous water bearing zone of thickness h and porosity ε. The system is described radially starting at the borehole/aquifer interface, which is at a distance r_1 from the axis of the borehole. The flow rate v at a distance r ($>r_1$) is simply given by:

$$v = \frac{Q}{2 \cdot \pi \cdot r \cdot h \cdot \varepsilon}$$

where Q is the volumetric flow rate at the well head. The flow rate variation is then given by:

$$\frac{\delta v}{\delta t} = \frac{1}{2 \cdot \pi \cdot r \cdot h \cdot \varepsilon} \cdot \frac{\delta Q}{\delta t}$$

This simplified model could be completed by taking into account the hydrogeological properties of the aquifer. However, colloids only pass through large pathways and not through the nanoscopic porous structure of the aquifer.

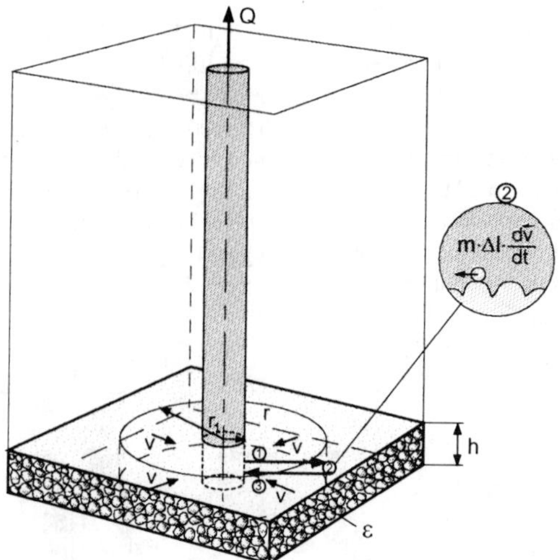

Figure 4: Schematic presentation of the 3 processes related to the transient study.
1- propagation of the flow rate change
2- activation of colloid generation
3- transmission of the new colloid population

Activation of the colloid generation

No colloid concentration increase was observed with stationary flows ($\delta v \cdot \delta t^{-1} = 0$) when increasing stepwise the flow rate, however, increases are produced by significant positive changes of $\delta v \cdot \delta t^{-1}$. Only a part of the energy $m \cdot \Delta l \cdot \delta v \cdot \delta t^{-1}$ is transferred to the colloid, Δl denotes the average size of the host rock particle from which the colloid of mass m is freed; a minimum of energy is required for detachment.

Consequently, the idea that activation of the colloid generation is coupled to the flow rate change was suggested. The energy transferred from the water flow change to detach the colloids must be more than the minimum detachment energy required to free the colloid from a host rock particle. This was observed on the well SB6 when comparing the effect of a strong and a weak transient (Figure 5).

Transmission of the new colloid population

Once the new colloids are generated at the distance r from the borehole axis, they are transported to the well by advection in the aquifer. The time t required for water to flow from r to the bore hole is approximately given by:

$$t = \frac{\pi \cdot (r^2 - r_1^2) \cdot h \cdot \varepsilon}{Q}$$

The new colloids generated by the variation in flow arrive after this time (plus the transit time from in situ to the well head) and are then detectable. The concentration after break through from the activation place (r) to the borehole interface (r_1) is affected by filtration in the aquifer.

<u>Results</u>

For the SB6 well geometry (55 mm tubing internal diameter, 203 mm borehole diameter) and a packer interval from 356.3 to 365.3 m, the volumes of line and double packer interval are 8.67 10^5 and 2.70 10^5 cm^3 respectively. For a 300 ml·min^{-1} flow rate, transit times of 48 h in the line and 15 h (maximum) in the inter-packer section are required. Results from the transients obtained when full opening the well valve after 2 weeks shut down showed that a colloid concentration peak appeared after 2.5 days delay. This delay corresponds to the transit of the colloid front from the aquifer to the well head for an average flow rate of 300 ml·min^{-1} (suggesting no retention in the tubing). Colloid generation by detachment from the aquifer is then observed yielding a colloid concentration peak (Figure 5a).

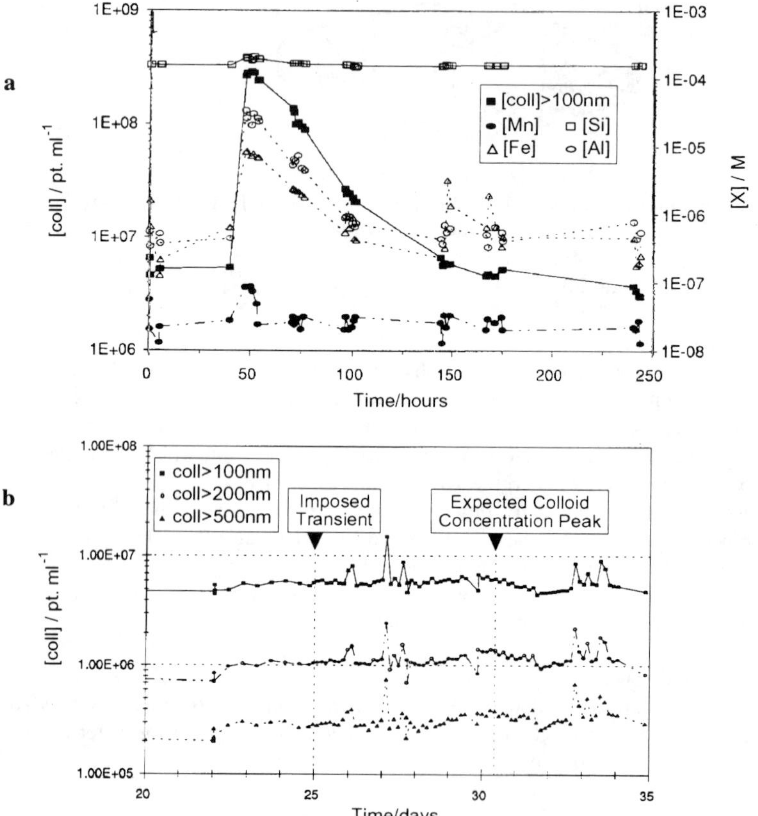

Figure 5: Colloid concentration break through curve during transient experiments. Conditions: **a** natural transient obtained with strong flow variation (from 0 to a maximum of 600 ml·min^{-1} in about 10 s followed by a decrease to about 150 ml·min^{-1}), with flow variations noted in the graph and **b** transient obtained with flow variation: 100 to 150 ml·min^{-1} in about 30 s.

This test is fully reproducible (duplicate experiment). When analysing the size distribution of the colloids in the concentration peak, it was striking to find a distribution containing more large particles than under the steady state conditions. Analysis of the non-filtered groundwater for the colloid elements (Al, Si, Fe) revealed that the element concentration peak is sharper than the colloid number concentration peak. This confirms that the flow rate change generates larger particles than small ones (Figure 5a). Because the system to model is relatively complex under natural flow conditions, it was decided to repeat the experiment under controlled flow conditions. A separation / flow-control / sampling unit was built. The unit was installed at the borehole head and the first controlled transient results are reported in Figure 5b. It must be noted that a minimum flow rate change is necessary to achieve the colloid detachment.

COLLOID PROPERTIES, A NON-SITE SPECIFIC APPROACH

<u>Colloid occurence</u>

The extention of the study to other systems relies on collaborative work with other groups. Utilisation of accurate colloid data from a variety of aquifers (e.g. Triay et al. [12] and Pedersen et al. [13]) is needed to derive general laws about the occurrence of groundwater colloids e.g. [9,14,15]. Artefact-free colloid concentrations are likely to be linked to the water chemistry and, more especially, to the concentration of alkali elements and in a cumulative way to the concentration of alkaline earth elements. This was verified for many data from systems with low organic concentrations and under steady-state flow conditions (Fig. 6). However, when the dissolved organic carbon concentration is large, colloid populations are stabilised as associated organic/inorganic colloids.

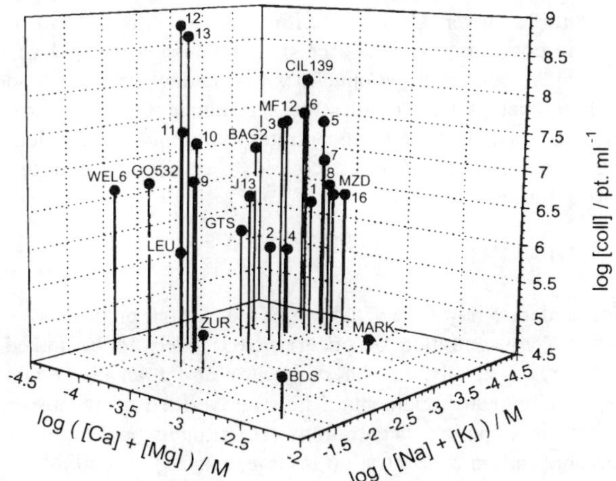

Figure 6: Colloid concentration as a function of alkali and alkaline earth element concentrations in relevant aquifers. Conditions: [coll]/pt·ml⁻¹ for size larger than 100 nm. No 1 to 16, GTS, ZUR, LEU, BDS, MZD from [14,15], J13 from [12], CIL Cigare Lake from [10], GO Gorleben from [16], BAG Bagombé from [13], MF Morro do Ferro from [17].

From these studies, a consistent picture is emerging today. Artefact-free colloid concentration or occurrence is likely to be linked to the water chemistry and, more especially, to the concentration of alkali elements and to the concentration of alkaline earth elements (see [14] and [15]). A literature reseach [18] confirms that the critical coagulation concentration of colloids (e.g. Al_2O_3) with NaCl or KCl is around $5 \ 10^{-2}$ M and that it drops to $5 \ 10^{-4}$ M for $CaCl_2$. The situation is different, however, when the dissolved organic carbon concentration is large, when the colloid populations are stabilised as associated organic and inorganic colloids.

<u>Contaminant transport by colloids</u>

Colloid contamination by safety relevant radionuclides can result from two main processes: contamination of natural colloids free in the groundwater, or generation of contaminated particles from the host rock followed by their resuspension.

For the first contamination mechanism, nuclide ions sorb onto groundwater colloids by ion exchange or by surface complexation and a sorption model such as described by Bradebury and Baeyens may be applied [19]. In addition, nuclide complexes may also sorb onto the colloids by surface complexation and a model such as developed by Degueldre et al. [20] may be adapted for a given colloid population. These two first models state that the sorption takes place by reversible processes and with distribution coefficients smaller than 10^7 ml·g^{-1}.

However, for systems where the contaminant plume may not be filtered, e.g. by a bentonite near field, radionuclide polymers or radiocolloids (e.g. Newton et al. [21]) may contaminate the aquifer. These species are able to associate or attach on the groundwater colloids reversibly or irreversibly. In this case, the distribution ratios derived from the attachment of the species may be larger than 10^7 ml·g^{-1}.

For nuclides sorbing irreversibly into the host rock (encrustation or secondary phase generation) particle detachment is a source of contaminated colloid generation. These contaminated colloids contain all radionuclides that cannot be freed or desorbed in the groundwater. These contaminated entities may be transported within the aquifer with a retention which is a function of their concentration in the natural aquifer. Therefore, the work on sampling and characterisation of colloids is of direct relevance to their role as contaminant carrier through the aquifer.

CONCLUDING REMARKS

From studies under steady-state situation, a consistent picture is emerging today. Artefact-free colloid concentration or occurrence is likely to be linked to the water chemistry and, more especially, to the concentration of alkali elements and to the concentration of alkaline earth elements. This was verified using numerous data from groundwaters with low organic concentrations. The situation is different, however, when the dissolved organic carbon concentration is large, causing the colloid populations to be stabilised as associated organic and inorganic colloids.

The impacts of transients such as flow rate or chemical changes, of natural origin for example glaciation/deglaciation, neotectonic, or, of anthropogenic origin such as well drilling or tunnel excavation, generate changes in the hydrogeology of the system and may induce colloid concentration increases. These colloid concentration enhancements are however local and restricted in time period. Nature trends always to restore the original background concentration.

Colloid contaminant transport is also linked to the colloid generation process. In addition, the studies on sorption behaviour on colloids and colloid re-suspension require more efforts to fully understand and quantitatively describe the transport process.

ACKNOWLEDGEMENTS

A. Laube (LES) carefully performed carefully the experimental work. Mr. Gerodetti B. from LWV is thanked for constructing and installing the sampling unit in Wellenberg. A. Michel from LANL/INC is acknowledged for determining the colloid size distribution with their PMS spectrometer. J. Hadermann (Head of LES) is thanked for his interest and support for this work. A. Scholtis from NAGRA is acknowledged for his help and useful discussions. NAGRA partially funds the research work.

REFRENCES

1. C. Degueldre. Colloid properties in granitic groundwater systems, with emphasis on the impact on safety assessment of a radioactive waste repository. Mat. Res. Soc. Symp. Proc. 294 (1993) 817-823.

2. J. McCarthy, C. Degueldre. Sampling and characterisation of colloids in groundwater for studying their role in the subsurface transport of contaminants. In characterisation of environmental particles. J. Buffle and H.P. van Leeuwen Eds. IUPAC Environmental Analytical Chemistry Series Vol. II Chapt. 6, 247-315.

3. J. McCarthy, J. Zachara. Subsurface transport of contaminants. Environ. Sci. Technol. 23 (1989) 496-503.

4. NAGRA: Geologische Grundlagen und Datensatz zur Beurteilung der langzeitsicherheit des Endlagers für schwach- und mittelaktive Abfälle am Standort Wellenberg. NAGRA technischer Bericht NTB 93-28, Wettingen, Switzerland (1993).

5. J. Ryan, M. Elimelech. Colloid mobilization and transport in groundwater. Colloids and Surfaces 107 (1996) 1-56.

6. P. Smith, C. Degueldre. Colloid facilitated transport of radionuclides through fractured media. J. Contam. Hydrol. 13 (1993) 143-166.

7. L. McDowell-Boyer, J. Hunt J., N. Sitar. Particle transport through porous media. Water Resour. Res. 22 (1986) 1901-1921.

8. P. Vilks, C. Degueldre. Sorption behaviour of ^{85}Sr, ^{131}I and ^{137}Cs on colloids and suspended particles from the Grimsel Test Site, Switzerland. Appl. Geochem. 6 (1991) 553-563.

9. M. Laaksoharju, C. Degueldre, C. Skarman. (1995) Studies of colloids and their importance for repository performance assessment. SKB Technical report 95-24,

Stockholm, Sweden.

10. P. Vilks, J. Cramer, D. Baschinski, D. Doern, H. Miller. Studies of colloids and suspended particles, Cigar lake uranium deposit, Saskatchewan, Canada. Appl. Geochem. 8 (1993) 605-616.

11. M. Laaksoharju, U. Vuorinen, M. Snellman, B. Allard, C. Petterson, J. Helenius, H. Hinkkanen. Colloids or artefacts? A TVO/SKB co-operation project in Olkiluoto, Finland. Report YJT-94-01, TVO, Helsinki, Finland (1994).

12. I. Triay, C. Degueldre, A. Wistrom, C. Cotter, W. Lemos. Progress report on colloid facilitated transport at Yucca Montain. LANL report LA-12959-MS, May 1996, Los Alamos, NM.

13. K. Pedersen, J. Arlinger, R. Bruetsch, C. Degueldre, L. Hallbeck, M. Laaksoharju, T. Lutz, C. Petterson. Bacteria, colloids and organic carbon in groundwater at the Bangombé site in the Oklo area. SKB Technical Report 96-01, 1996, Stockholm, Sweden, Also in EC report EUR 16704 EN, Bruxelles, CEC, 1996.

14. C. Degueldre, H.R. Pfeiffer, R. Alexander, B. Wernli, R. Bruetsch. Colloid properties in granitic groundwater systems I: sampling and characterisation. Appl. Geochem. 11 (1996) 677-695.

15. C. Degueldre, R. Grauer, A. Laube, A. Oess, H. Silby. Colloid properties in granitic groundwater systems II: stability and transport study. Appl. Geochem. 11 (1996) 697-710.

16. J. Kim, Zeh P., Delakovitz . Chemical interactions of actinide ions with groundwater colloids in Gorleben aquifer systems. Radiochim. Acta 58/59, (1992) 147-154.

17. N. Miekeley, H. Countinho de Jesus, C. Porto da Silveira, C. Degueldre. Chemical and physical characterization of suspended particles and colloids in waters from the Osamu Utsumi mine and Morro do Ferro analogue study sites, Poços de Caldas, Brasil. J. Geochem. Exploration, 45 (1992) 409-437.

18. S. Swanton. Modelling colloid transport in groundwater, the prediction of colloid stability and retention behaviour. Adv. Coll. Interf. Sci. 54 (1995) 129-208.

19. M. Bradbury, B. Baeyens. A quantitative mechanistic description of Ni, Zn and Ca sorption on Na-montmorillonite.PSI Bericht Nr 95 10,11 &12, Jul. 1995 Paul Scherrer Institute, Villigen, Switzerland.

20. C. Degueldre, H.J. Ulrich, H. Silby. Sorption of [241]Am onto montmorillonite, illite, and hematite colloids. Radiochim. Acta 65 (1994) 173-179.

21. T. Newton, D. Hobart, P. Palmer. The formation of Pu(IV)-colloid by alpha-reduction of Pu(V) or Pu(VI) in aqueous solutions. Radiochim. Acta 39 (1986) 139-147.

INVESTIGATION OF THE INFLUENCE OF HETEROGENEOUS POROSITY ON MATRIX DIFFUSION: A NOVEL APPROACH USING ADAPTIVE TREE-MULTIGRID TECHNIQUE AND REAL POROSITY DATA

P. Simbierowicz and M. Olin
VTT Chemical Technology, P.O.Box 1404 (Otakaari 3A, ESPOO), FIN-02044 VTT, Finland

ABSTRACT

Last year we developed a two-dimensional deterministic heterogeneous matrix diffusion model, which is capable of utilising porosity information originating from real drill-core samples. The results of numerical infiltration experiment we had performed with the model displayed substantial spatial variations in the penetration depth. Because it is practically impossible to verify experimentally those two-dimensional penetration profiles we had computed, this time we decided to try modelling of measured leaching curves. Unfortunately we have not succeeded in acquiring such curves for the exact same samples, which we have used in numerical leaching experiments. Nevertheless it can be seen, that the shape of leaching curves computed with the heterogeneous model is clearly closer to the shape of measured curves, than the shape of curves provided by the standard model. These differences can be utilised as a basis for an approximate numerical method of assessing the geometric factor, which has traditionally been a purely empirical parameter.

The results of the new numerical experiments agree with our older results from last year: the heterogeneity of the rock matrix has highly significant impact on the diffusion. However, when interpreting the results, one must not neglect numerous limitations of the model, and hence, one should not attempt to overgeneralise the conclusions.

INTRODUCTION

Matrix diffusion, advection and chemical reactions are the three most important mechanisms responsible for radionuclide migration in porous media. Thorough understanding of matrix diffusion is crucial for proper assessment of the risks originating from nuclear waste disposal in the ground. Matrix diffusion is typically extremely slow, and big time lag is a good thing in conjunction with the spread of nuclear waste, but on the other hand it makes direct experimental investigation of the diffusion phenomenon very time-consuming. There is another intriguing circumstance related to experimental leaching curves: apparently it is difficult or even impossible to numerically simulate them with the standard Fickian diffusion laws under the assumption of homogeneous diffusivity. In this study we shall develop a two-dimensional heterogeneous diffusion model, which will utilise actual porosity data obtained from several real drill-core samples. Unfortunately at the moment we do not have experimental leaching curves from the exact same samples, so for the time-being direct comparison with measured data will not be feasible, nevertheless an attempt at qualitative simulation of the behaviour of real leaching curves can and will be made. We shall also discuss the possibility of using our model for numerical assessment of the geometric factor of the diffusion coefficient, which conventionally has been a purely empirical parameter. In the model exclusively geometric structure of the pores will be considered, moreover only geometric details greater than 0.08 mm will be accounted for. Even though microstructure and mineralogical composition of the porous media will be left out altogether, the amount of geometric data our model must handle will be enormous. In order to deal with such amount of data and actually solve some useful diffusion problems on the data, we shall apply a special numerical method called tree-multigrid technique. This method will allow nearly optimal

Mat. Res. Soc. Symp. Proc. Vol. 465 © 1997 Materials Research Society

utilisation of the geometric data, as well as fast repetitive solution of generalised Poisson-equations resulting from the Fick's law.

MODELLING EXPERIMENTAL LEACHING CURVES

SOME BASICS OF THE DIFFUSION THEORY

In our model we concentrate on chemically and electrically inactive non-sorbing species. Furthermore we assume, that its flux obeys the following equation:

$$\vec{J} = -\vec{\vec{D}} \cdot \vec{\nabla} c \tag{1}$$

where c means the total concentration of the investigated species, $\vec{\vec{D}}$ stands for the diffusion tensor and $\vec{J}$ is the flux of the species. The above equation is Fick's first diffusion law, which has been generalised to handle multi-dimensional, anisotropic and heterogeneous media. Combining it with the equation of continuity:

$$\frac{\partial c}{\partial t} + \vec{\nabla} \cdot \vec{J} = 0, \tag{2}$$

we acquire generalised version of Fick's second law of diffusion:

$$\frac{\partial c}{\partial t} = \vec{\nabla} \cdot \vec{\vec{D}} \cdot \vec{\nabla} c . \tag{3}$$

When modelling matrix diffusion, the anisotropy and heterogeneity of the $\vec{\vec{D}}$ tensor are generally neglected, and for non-sorbing species the so-called pore diffusion constant D_p is used instead, which leads to the standard model:

$$\frac{\partial c}{\partial t} = D_p \Delta c . \tag{4}$$

Although the above equation looks quite simple, it is just the final result in a chain of rather complicated numerous reductions and simplifications; for detailed discussion of matrix diffusion refer to Neretnieks [1] or Olin [2].

STANDARD MODEL AND DEAD-END POROSITY MODEL

The standard diffusion model cannot simulate measured leaching curves correctly (see Fig. 1 below). One attempt to rectify this problem is the so-called dead-end porosity model (see e.g. [3]), which does indeed provide good agreement with measurements, but it requires several unknown parameters to be calibrated afterwards from the measured leaching curve.

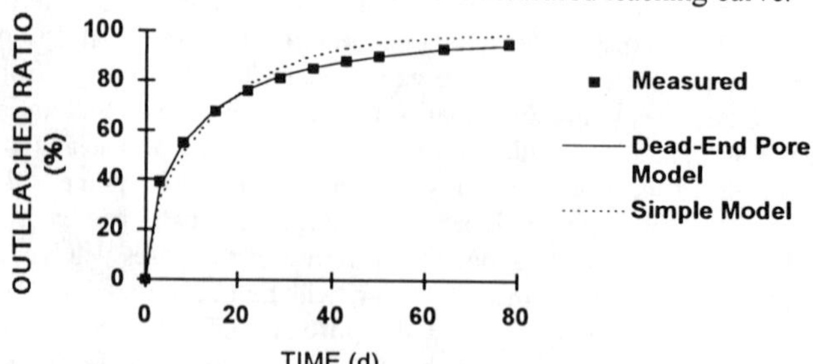

Figure 1. Behaviour of measured leaching curves vs. standard and dead-end pore diffusion model. Courtesy of Mr Valkiainen [4].

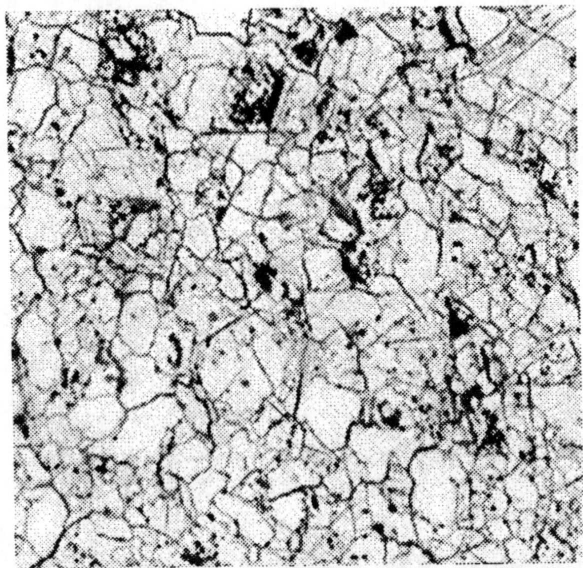

Figure 2. Intensity bitmap obtained from a real drill-core sample.

Hellmuth and Siitari-Kauppi [5] have described a technique in which a rock sample is impregnated with ^{14}C-polymethylmethacrylate (PMMA), and the sample then cured. An autoradiograph can then be made of a section of the sample, and the autoradiograph scanned to obtain a two-dimensional porosity map of the section in bitmap form (see Fig. 2). This gives us an opportunity to develop a deterministic heterogeneous porosity and diffusion model, as opposed to widely used homogeneous or stochastic approach.

GENERATING QUADTREE GRID FROM A REAL SAMPLE

In order to utilise the bitmap information in the development of deterministic heterogeneous diffusion model, the method of Gáspár [6, 7, 8] and co-workers is used to build a quadtree grid based on the intensity data. In this method, a square 'root cell' is chosen to contain a typical area of the bitmap, and then all the bitmap pixels contained within the root cell are scanned for maximum and minimum intensity value. If the difference between the maximal and minimal intensity exceeds some prescribed limit parameter, then the root cell is divided into four congruent child cells. Each of the children is then scanned separately for their corresponding extremes, and those violating the uniformity criterion, i.e., exceeding the aforementioned limit parameter, become subdivided further. This recursive subdivision process continues until all originating grid cells contain sufficiently uniform intensity values, or until prescribed maximal allowed tree depth is reached. In this way we acquire a non-equidistant grid consisting of many different size squares (see Fig. 3), which can be very fine at some spots and simultaneously rather coarse elsewhere, depending on the distribution of the intensity in the bitmap. Since starting at some tree level not all, but only part of the cells become subdivided, the overall number of grid cells is most often substantially smaller than that of an equidistant grid of the same resolution.

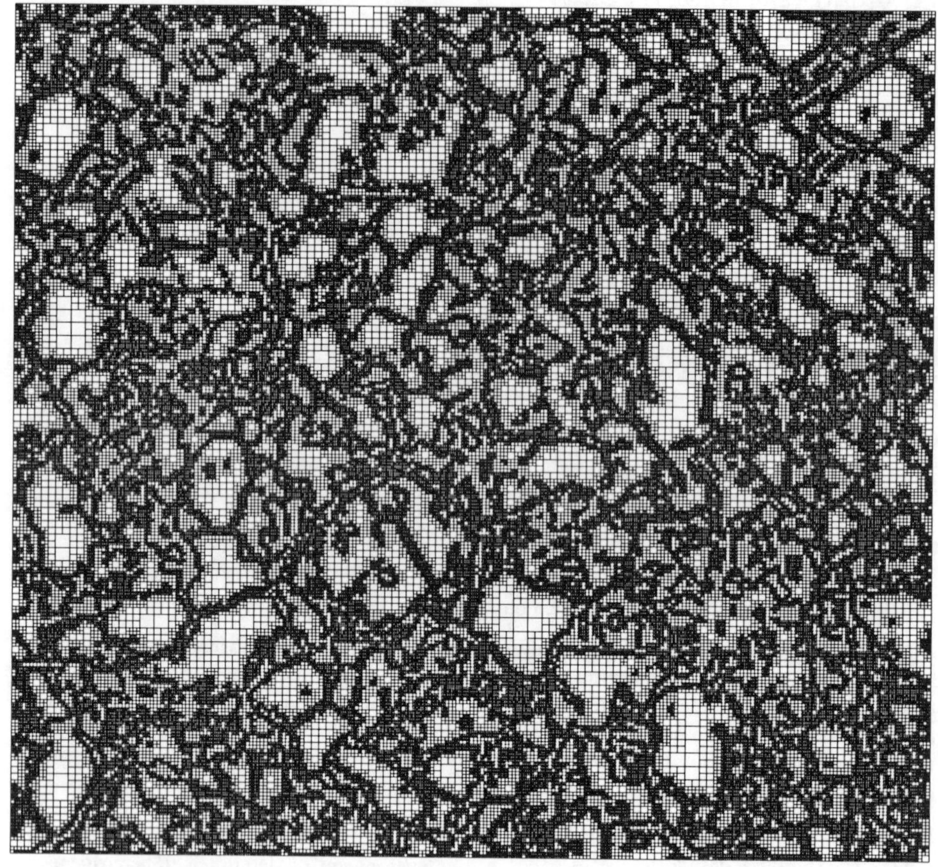

Figure 3. Ten-level quadtree generated from the bitmap shown in Fig. 2.

After we have generated the grid structure, we convert the bitmap intensities to porosity values with the aid of a conversion table, and then we assign to each leaf cell (i.e. not subdivided cell) the average of all porosity values contained in the cell. For very deep trees it may happen, that not all leaf cells get porosity value in this process, not to mention the non-leaf cells, which certainly do not have any porosity values yet. In order to remedy this situation, we first traverse the tree up to the root assigning to each non-leaf cell the average of its children's porosity values, and then back down to leaves, prolonging father's porosity to each cell without one already assigned.

It is our intention to solve later diffusion equation on the generated quadtree grid, so we also must assign diffusivity parameters to each leaf cell. Before we finally do it, we make two simplifying assumptions: neglect the cross term of the otherwise still anisotropic diffusivity tensor and presume that at the leaves both diagonal diffusivity terms are equal and proportional to the porosity, the proportionality factor being constant. The diffusivity parameters are given at cell faces, and for non-leaf cells they are computed according to the following averaging schemes:

$$DX_{i,j}^{l-1} = \frac{DX_{2i,2j}^{l} + DX_{2i,2j+1}^{l}}{2} \quad \text{and} \quad DY_{i,j}^{l-1} = \frac{DY_{2i,2j}^{l} + DY_{2i+1,2j}^{l}}{2}. \tag{5}$$

In the next section it will become evident, why it is not sufficient to assign diffusivity (and porosity) values to leaf cells only, why we must compute them for all levels.

THE TREE-MULTIGRID TECHNIQUE

Fick's second diffusion law is spatially a purely elliptic equation, and therefore relatively easy to solve numerically. The most efficient method for solving elliptic boundary value problems is called multigrid, which utilises simultaneously several grids of different resolution. Transferring information back and forth between the grids accelerates iterative solution of elliptic equations substantially: the computational work of this method is directly proportional to the first power of the number of unknowns. For more detailed discussion of multigrid methods refer to Hackbusch [9].

The multigrid method requires several hierarchic grids of different resolution to work. The quadtree grid generated previously offers such grid system automatically. For example, cutting off all the finest leaf cells produces a grid that is coarser than the former one. Also the restriction and prolongation procedures are given automatically by exploiting the father-child relationships of the tree. It is therefore easy to define a multigrid iteration on a tree-structured grid system. It is also very reasonable, since the multigrid minimises the computational work per grid cell, while the tree-structured grid helps to keep the overall number of cells moderate. This combination of tree-structure with the multigrid method is called tree-multigrid technique. The grid generation procedure is also very fast, which ensures the adaptability of the technique.

HETEROGENEOUS VS. HOMOGENEOUS DIFFUSION MODEL

Now that we have all the necessary tools available, we can proceed to compare our new heterogeneous matrix diffusion model to the standard homogeneous one. To do that in a meaningful way, we must generate two grids which exhibit exactly the same quadtree structure, but differ in diffusivity (and porosity) distribution, the first one being heterogeneous and the second one homogeneous. All the other differences must be eliminated. The simplest and safest method to achieve this is to convert the original bitmap to an abstract two-state intensity data. For the heterogeneous model one state means permeable and the other state means impermeable, for the homogeneous model correctly weighted average of the two permeabilities must be used. Now we can perform numerical leaching experiment for both models, keeping the boundary and initial conditions as well as all the other computation parameters exactly the same. We have done it for the sample of Figure 2 and the results of the two numerical experiments are shown in Figure 4. We have also performed comparable numerical experiments on a few different samples and those results were quite similar to the ones presented here (i.e., there were similar differences between the heterogeneous and the homogeneous models). As we have already mentioned before, experimental data for the very same samples is not at our hand, so direct comparison with measurements is not possible. Still it would seem as if the heterogeneous model gave qualitatively better agreement with the behaviour of actual leaching curves. This is most probably due to the fact, that in the heterogeneous model there are actually many different diffusion paths simultaneously present in the system. One would expect that to be closer to the reality, than the single diffusion path of the standard model. But even if this reasoning should be considered speculation, our numerical experiments still do point out at least one thing: there is a distinct difference between the heterogeneous and the standard model, or, in another words, heterogeneity of porosity (and diffusivity) has clear impact on matrix diffusion.

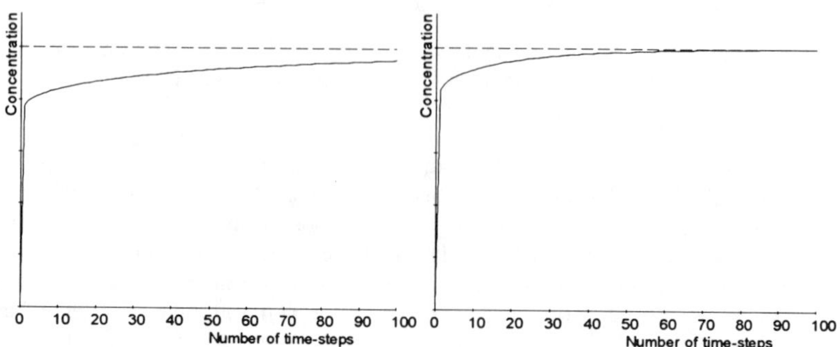

Figure 4. Total outleached concentration for heterogeneous (left) and standard model (right).

SIMULATION OF THE GEOMETRIC FACTOR

Our heterogeneous matrix diffusion model is capable of providing much more information than what is needed for simulating experimental leaching curves, which hide all spatial dependencies showing only the total outleached concentration. Based on the leaching curves only, there is no way to tell, whether the leaching speed is everywhere uniform or not. But this sort of information is quite easily generated by the heterogeneous model in the form of two-dimensional time-dependent leaching profiles. We are not going to show such leaching profiles in this paper, but we have performed numerical infiltration experiment in another work of ours (Simbierowicz & Olin [10]), according to which significant spatial differences in infiltration speed (and hence in leaching speed) are indeed possible. If, despite the availability of the new heterogeneous approach to the matrix diffusion, one should insist on applying the standard model instead, the heterogeneous model can still be used for numerical assessment of the so-called geometric factor. This factor, G, is commonly utilised to relate the pore diffusion constant D_p (Eq. 4) to the water diffusion constant D_w:

$$D_p = GD_w \tag{6}$$

Detailed description of G and associated parameters can be found in Olin [2]; for our purposes it is sufficient to know that the geometric factor is supposed to account for the geometrical properties of porous media, especially the prolonged path of the diffusion in the pores.

Now let us perform the numerical leaching experiments shown in Figure 4 once again, this time simulating the heterogeneous case for much longer period. The results can be seen in Figure 5, which differs from Figure 4 in the time-scale of the heterogeneous simulation. It has been shrunken by the factor of ¼, and if the two curves were ideally similar, this would be precisely the value of the geometric factor. However, the curves are not similar, especially at the beginning there are visible differences, although they could be attributed to the inaccuracies of the numerical calculation. Because the curves are not similar, there is no single way to fit them onto each other. If we stress the asymptotic behaviour, then the factor ¼ is about right, if we look at the beginning, when the most of the concentration is leached out, then a greater value is more appropriate. All in all, although not perfect and clearly needs improvement, this might be considered a numerical alternative for assessing the value of the geometric factor, as opposed to the traditional experimental way.

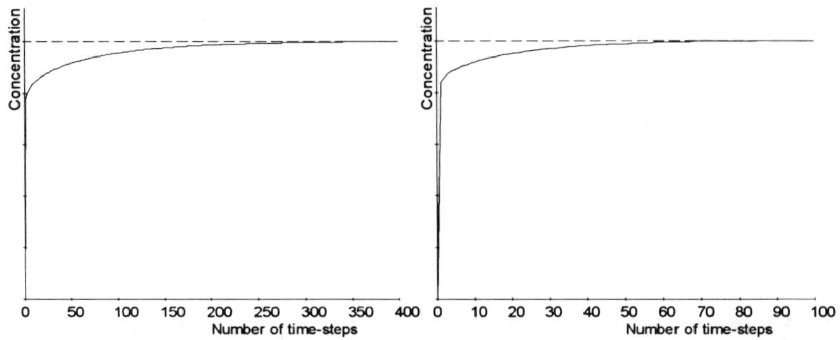

Figure 5. Total outleached concentration for heterogeneous (left) and standard model (right).

CONCLUSIONS

The results acquired from the numerical simulations are potentially highly significant, but they require careful interpretation. We must not jump into conclusions, because our model has several limitations. First of all, we only performed numerical and not physical experiments, and therefore comparisons with measurements were omitted. Secondly, we only had a few samples available, and instead, the simulation should have been performed for dozens of samples. Thirdly, the transport model is extremely simplistic taking into account only plain diffusion, which additionally is itself also greatly simplified. Lastly, neither the simulated concentrations nor the time-scale bears any absolute physical meaning and should only be considered relatively.

Noting all the aforementioned limitations and simplifications, we can now focus on the actual results of the numerical experiments. If they are valid, their message is, in our opinion, very important: heterogeneous structure of porous media has major impact on matrix diffusion. Of course, the investigated bitmaps come from relatively small drill-core samples, so our model can only tell us about small scale effects. However, we are convinced, that analogous effects do take place also in large scale, even though they are invisible from the small drill-core samples. Our numerical experiments do not in any way invalidate the valuable results obtained from the safety analysis under the homogeneous media assumption. However, our results do point out, that this assumption might be questionable, and in any case requires further investigations.

FURTHER DEVELOPMENT

Our model requires more extensive applications and especially verifications with measurements, though it is not yet certain, how to accomplish this. The main reason for the uncertainty is the fact that the model is two-dimensional, while real leaching experiments are three-dimensional. The model depends very much on the correct geometrical information, therefore we do not know if we can ever achieve sufficiently good agreement with three-dimensional measurements. If we should ever get a hold on real three-dimensional bitmap describing the entire interior of a sample and not just one slice of it, we could easily extend our model to three dimensions. We could also improve our diffusion equation by allowing cross-term diffusivities as well as adding advection, sorption and chemical reactions.

ACKNOWLEDGEMENTS

The authors thank the Finnish Centre for Radiation and Nuclear Safety for the financing of this work and especially Dr Karl-Heinz Hellmuth from the same institute for many important comments concerning the actual work.

REFERENCES

1. I. Neretnieks, J. Geosphys. Res. **85** (B8), 4379-4397, (1980).
2. M. Olin, PhD thesis, VTT Publications 175, (1994).
3. S.J. Hemingway, M.H. Bradbury & D.A. Lever, Report AERE-R 10691, (1983).
4. J. Lehikoinen, A. Muurinen, M. Olin, K. Uusheimo & M. Valkiainen, VTT Research Notes 1394, (1992). 28 pp.
5. K-H. Hellmuth & M. Siitari-Kauppi, Report STUK-B-VALO 63, (1990).
6. C. Gáspár, J. Józsa 6 P. Simbierowicz, in First International Conference on Water Pollution, edited by L.C. Wrobel & C.A. Brebbia, (Elsevier, 1991). p. 299.
7. C. Gáspár & P. Simbierowicz, in HYDROCOMP '92 Conference edited by J. Gayer, O. Starosolszky & C. Maksimovic (1992). p. 131.
8. C, Gáspár & P. Simbierowicz, in IX. International Conference on Computational Methods in Water Resources, edited by T.F. Russell, R.E. Ewing, C.A. Brebbia, W.G. Gray & G.F. Pinder (Elsevier, 1992). p. 555.
9. W. Hackbusch, Multi-Grid Methods and Applications. (Springer-Verlag, Heidelberg 1985). 377 pp.
10. P. Simbierowicz & M. Olin, Porosity and pore-structure as relevant parameters of matrix diffusion. VTT internal report (Espoo, 1995). 8 pp.

A NOTE ON RADIONUCLIDE TRANSPORT BY GAS BUBBLES

IVARS NERETNIEKS AND MÄRTA-LENA ERNSTSON
Department of Chemical Engineering and Technology, Royal Institute of Technology
S-100 44 Stockholm, SWEDEN

ABSTRACT

In a repository for spent nuclear fuel, gas generated by corrosion of the iron in the canister may form small bubbles that will escape and rise to the ground surface. Colloidal particles may attach to the surface of the bubbles and be carried by them. If the colloids are supplied by the montmorillonite clay of the buffer material surrounding the canister, the clay can be carried away. Nuclides sorbed in the clay can be carried with the bubbles. We have estimated the carrying capacity of the gas of the clay particles and the escape rate of nuclides carried by the gas bubbles. The latter is also compared to the escape rate by the conventional escape mechanisms from the near field. We have further estimated the detachment of the nuclides from the clay and their sorption onto the fracture surfaces of the rock as well as their uptake by diffusion into the rock matrix along the bubble transport paths.

The present paper is speculative and uses some hypothetical assumptions. Although the processes that are modelled are known to exist there is not enough known of several of them to quantify them accurately. The carrying capacity of the gas used in the calculations is an upper bound and probably very much exaggerated. Even so, the consequences are minor for the release of radionuclides.

BACKGROUND AND INTRODUCTION

Gas may form by corrosion of the iron in the canisters. The montmorillonite clay surrounding the canisters consists of very small particles of colloidal size. One concern is that the buffer may be carried away or at least be depleted so much that it loses its swelling capacity and increases its hydraulic conductivity. Another concern is that the colloids may have sorbed nuclides which may be quickly carried by the bubbles to the ground surface. We have tried to address these issues in a simplified way by making exploratory calculations based on a set of assumptions that may be more or less reasonable but at least were made to be conservative.

APPROACH TO THE PROBLEM

It was recently found that particles with positive as well as negative surface charge and hydrophilic as well as hydrophobic particles sorb on gas water interfaces [1]. The particles will not desorb even when the bubbles are washed with colloid free water.

Our main assumptions are that 1) All gas escaping from the canister will form small bubbles and their surfaces will be entirely covered with a monolayer of colloidal particles. 2) The particles are attached irreversibly and can migrate with the bubbles. 3) All gas escapes as bubbles and reaches the ground surface. 4) The colloidal particles emanate from the montmorillonite clay in the backfill. 5) The clay has sorbed the radionuclides to the equilibrium concentration with the water in the canister.

RATE OF GAS GENERATION AND ESCAPE

The gas is generated by corrosion of the iron in the canisters. The rate of gas generation is taken to be 16 litres/a at the ambient pressure of 50 bar at the depth of 500 m [2]. The gas passes through a multitude of small paths in the clay. We assume that as the gas reaches the water in the fractures in the rock it forms small bubbles that immediately are fully covered by a monolayer of bentonite particles. The bubble size is estimated to be about 0.2 to 1 mm in diameter if formed in bulk water [3].

For 0.2 mm diameter bubbles fully coated with 0.5 μm diameter colloid particles, about 0.02 kg colloids per litre of gas can be attached and transported. By the 16 l/a gas 0.4 kg/a clay can

Mat. Res. Soc. Symp. Proc. Vol. 465 © 1997 Materials Research Society

be transported. There is about 18 tonnes of bentonite in one deposition hole. The nearly 5 tonnes of iron in the canister can generate the 16 l/a of hydrogen gas at 50 atm during 2 400 years. At most a few percent of the bentonite could be carried away by the gas under these quite conservative assumptions.

NUCLIDE TRANSPORT BY THE COLLOIDS

The clay colloids may carry sorbed nuclides. With our near field model [4] we calculate the concentration of each nuclide in the clay at the mouth of the fracture where the gas escapes as well as the release rate of the nuclides to the water in the fractures. Data and conditions from SKB-91 [5] are used. The clay at the fracture mouth irreversibly sorbs the radionuclides to equilibrium concentration. The gas bubbles carry the colloids that have the sorbed nuclides.

The effective diffusion coefficients and the distribution coefficients in bentonite used in the calculations, and the resulting concentrations in pore water for different radionuclides, are given in Table 1.

The amount of nuclides on the colloidal particles transported by the gas bubbles is the mass flowrate of colloids into the fracture times the concentration of radionuclides on the colloids, $\dot{m}_c\, q_i$, and is given in Table 2. This transport has to be compared with the release of radionuclides by diffusion through the barriers. Table 3 gives the release of radionuclides at peak release calculated from the non-stationary transport model. The time for the peak release is also shown. For some radionuclides the time to reach steady state is much longer than the time for the gas to escape from the canister.

It is found that Sr-90 would escape at a rate of 0.06 GBq/a with the colloids. Similar values are found for Cs-137. This applies for one damaged canister. For comparison it can be noted that this is 10-100 times more than is released to the flowing water in the near field by the conventional diffusion processes to the groundwater.

The actinides Pu, Np, U and Th will not reach their peak concentrations during the few thousand years the gas is generated. The actinide concentration in the clay at the mouth of the fracture in the rock will be many orders of magnitude less than the later peak concentration.

Table 1:   Effective diffusion coefficients in bentonite, distribution coefficients between colloids and water and concentrations of radionuclides in pore water at the mouth of the fracture.

Nuclide	D_e m^2/year	K_d m^3/kg	c_i moles/m^3	c_i GBq/m^3
Cs-135	$7.884 \cdot 10^{-1}$	0.050	$3.07 \cdot 10^{-4}$	$1.38 \cdot 10^{-3}$
Cs-137	$7.884 \cdot 10^{-1}$	0.050	$1.28 \cdot 10^{-5}$	5.62
Sr-90	$7.884 \cdot 10^{-1}$	0,010	$2.88 \cdot 10^{-5}$	13.2
Pu-239	$3.154 \cdot 10^{-3}$	3.00	$9.26 \cdot 10^{-10}$	$5.08 \cdot 10^{-7}$
U-238	$3.154 \cdot 10^{-3}$	3.00	$1.12 \cdot 10^{-6}$	$3.31 \cdot 10^{-9}$
U-235	$3.154 \cdot 10^{-3}$	3.00	$1.25 \cdot 10^{-8}$	$2.35 \cdot 10^{-10}$
U-234	$3.154 \cdot 10^{-3}$	3.00	$6.13 \cdot 10^{-11}$	$3.31 \cdot 10^{-9}$
Np-237	$3.154 \cdot 10^{-3}$	3.00	$1.65 \cdot 10^{-8}$	$1.02 \cdot 10^{-7}$
Th-230	$3.154 \cdot 10^{-3}$	3.00	$3.16 \cdot 10^{-11}$	$5.23 \cdot 10^{-9}$
Ra-226	$7.884 \cdot 10^{-1}$	0.010	$1.34 \cdot 10^{-8}$	$1.11 \cdot 10^{-4}$
Pa-231	$7.884 \cdot 10^{-1}$	3.00	$4.08 \cdot 10^{-9}$	$1.65 \cdot 10^{-6}$
Am-241	$3.154 \cdot 10^{-3}$	3.00	$6.76 \cdot 10^{-12}$	$2.07 \cdot 10^{-7}$

Table 2: Concentrations of radionucides on the colloids and transport of the radionuclides by the colloids on gas bubbles.

Nuclide	q_i moles/kg colloids	q_i GBq/kg colloids	$\dot{m}_c\,q_i$ moles/year	$\dot{m}_c\,q_i$ GBq/year
Cs-135	$1.53 \cdot 10^{-5}$	$6.88 \cdot 10^{-5}$	$6.6 \cdot 10^{-6}$	$3.0 \cdot 10^{-5}$
Cs-137	$6.39 \cdot 10^{-7}$	$2.81 \cdot 10^{-1}$	$2.8 \cdot 10^{-7}$	$1.2 \cdot 10^{-1}$
Sr-90	$2.88 \cdot 10^{-7}$	$1.32 \cdot 10^{-1}$	$1.3 \cdot 10^{-7}$	$6.0 \cdot 10^{-2}$
Pu-239	$2.78 \cdot 10^{-9}$	$1.52 \cdot 10^{-6}$	$1.2 \cdot 10^{-9}$	$7.0 \cdot 10^{-7}$
U-238	$3.35 \cdot 10^{-6}$	$9.93 \cdot 10^{-9}$	$1.5 \cdot 10^{-6}$	$4.3 \cdot 10^{-9}$
U-235	$3.74 \cdot 10^{-8}$	$7.04 \cdot 10^{-10}$	$1.6 \cdot 10^{-8}$	$3.0 \cdot 10^{-10}$
U-234	$1.84 \cdot 10^{-10}$	$9.93 \cdot 10^{-9}$	$8.0 \cdot 10^{-11}$	$4.3 \cdot 10^{-9}$
Np-237	$4.96 \cdot 10^{-8}$	$3.07 \cdot 10^{-7}$	$2.2 \cdot 10^{-8}$	$1.3 \cdot 10^{-7}$
Th-230	$9.49 \cdot 10^{-11}$	$1.57 \cdot 10^{-8}$	$4.1 \cdot 10^{-11}$	$7.0 \cdot 10^{-9}$
Ra-226	$1.34 \cdot 10^{-10}$	$1.11 \cdot 10^{-6}$	$6.0 \cdot 10^{-11}$	$4.8 \cdot 10^{-7}$
Pa-231	$1.23 \cdot 10^{-8}$	$4.94 \cdot 10^{-6}$	$5.0 \cdot 10^{-9}$	$2.2 \cdot 10^{-6}$
Am-241	$2.03 \cdot 10^{-11}$	$6.20 \cdot 10^{-7}$	$9.0 \cdot 10^{-12}$	$2.7 \cdot 10^{-7}$

Table 3: Transport of radionuclides by diffusion at peak release and time to reach the peak.

Nuclide	Time years	$\dot{m}_c\,q_i$ moles/year	$\dot{m}_c\,q_i$ GBq/year
Cs-135	$5.0 \cdot 10^3$	$7.67 \cdot 10^{-8}$	$3.44 \cdot 10^{-7}$
Cs-137	$3.8 \cdot 10^1$	$3.20 \cdot 10^{-9}$	$1.41 \cdot 10^{-3}$
Sr-90	$2.8 \cdot 10^1$	$7.19 \cdot 10^{-9}$	$3.30 \cdot 10^{-3}$
Pu-239	$2.1 \cdot 10^5$	$2.22 \cdot 10^{-13}$	$1.22 \cdot 10^{-10}$
U-238	$4.5 \cdot 10^8$	$2.68 \cdot 10^{-10}$	$7.94 \cdot 10^{-13}$
U-235	$1.7 \cdot 10^8$	$2.99 \cdot 10^{-12}$	$8.85 \cdot 10^{-15}$
U-234	$4.0 \cdot 10^8$	$1.47 \cdot 10^{-14}$	$4.35 \cdot 10^{-17}$
Np-237	$1.3 \cdot 10^7$	$3.97 \cdot 10^{-12}$	$2.46 \cdot 10^{-11}$
Th-230	$4.0 \cdot 10^8$	$7.59 \cdot 10^{-15}$	$1.52 \cdot 10^{-6}$
Ra-226	$4.0 \cdot 10^8$	$3.35 \cdot 10^{-12}$	$2.77 \cdot 10^{-8}$
Pa-231	$3.6 \cdot 10^5$	$1.02 \cdot 10^{-12}$	$4.12 \cdot 10^{-10}$
Am-241	$1.0 \cdot 10^3$	$1.62 \cdot 10^{-15}$	$4.95 \cdot 10^{-11}$

However, if the peak concentration were to be reached the calculations show that Pu-239 and Np-237 are released about 10 000 times more by the colloids than what goes into the water in the near field by diffusional release.

It should be kept in mind that even with these assumptions the release rate is on the order of 1000 Bq/a down to much less than one Bq/a for the other actinides.

There is thus a theoretical possibility for releasing some short lived nuclides at a higher rate by the colloid transport than by ordinary diffusion from the near field.

RETARDATION MECHANISMS ALONG THE GAS FLOW PATH

The nuclides dissolved in the water leaving the near field will be retarded by sorption on the rock. The same mechanisms may be operating for the nuclides sorbed on the particles, provided they can be desorbed from the particles.

When the bubbles with the attached colloidal particles rise, they will enter water with a lower nuclide concentration. The nuclides on the colloids may desorb into the water and may then sorb on the fracture surfaces. In a porous rock matrix the nuclides may diffuse into the rock matrix and sorb on the inner surfaces.

We have made a simple model to describe the migration of the bubbles in a fracture and the nuclide transfer from the bubbles to the rock. It is based an important assumption, namely that the nuclides are reversibly sorbed on the colloids. This is known to be true for cesium and strontium sorption on bentonite.

DERIVATION OF A MODEL FOR NUCLIDE TRANSPORT WITH COLLOIDS ON BUBBLES MOVING IN A FRACTURE.

Figure 1 shows a vertical fracture in which the bubbles rise. The nuclides that are reversibly sorbed on the colloids are dissolved into the water, sorbed on the fracture wall and diffuse into the rock matrix where they are sorbed.

In the model a flowrate of M kg/s of colloids are transported by the bubbles into a fracture with a width W and aperture δ.

A differential mass balance of the nuclide accounting for in- and outflow, accumulation on the colloids on the bubbles and uptake into the rock matrix by diffusion and sorption on the micropore surfaces gives the following expression.

$$\frac{\partial c}{\partial t}(1 + K_c c_c) + \frac{M}{\delta W}\frac{\partial q}{\partial z} = \frac{2D_e}{\delta}\frac{\partial c_p}{\partial x}\Big|_{at\ x=0} - \lambda c(1 + K_c c_c) \qquad (1)$$

It is assumed that the sorption on the colloids as well as on the rock is reversible and fast. This implies that the water adjacent to the colloids on the bubbles is in local equilibrium with that on the colloids.

$$q = K_c c \qquad (2)$$

Introducing Equation (2) into (1) gives

$$\frac{\partial c}{\partial t}(1 + K_c c_c) + \frac{M K_c}{\delta W}\frac{\partial c}{\partial z} = \frac{2D_e}{\delta}\frac{\partial c_p}{\partial x}\Big|_{at\ x=0} - \lambda c(1 + K_c c_c) \qquad (3)$$

which also can be written

$$\frac{\partial c}{\partial t} + \frac{M K_c}{\delta W(1 + K_c c_c)}\frac{\partial c}{\partial z} = \frac{2D_e}{\delta(1 + K_c c_c)}\frac{\partial c_p}{\partial x}\Big|_{at\ x=0} - \lambda c \qquad (4)$$

The differential mass balance for the rock matrix is

$$\frac{\partial c_p}{\partial t} = \frac{D_e}{K_d \rho_p} \frac{\partial^2 c_p}{\partial x^2} - \lambda c_p \tag{5}$$

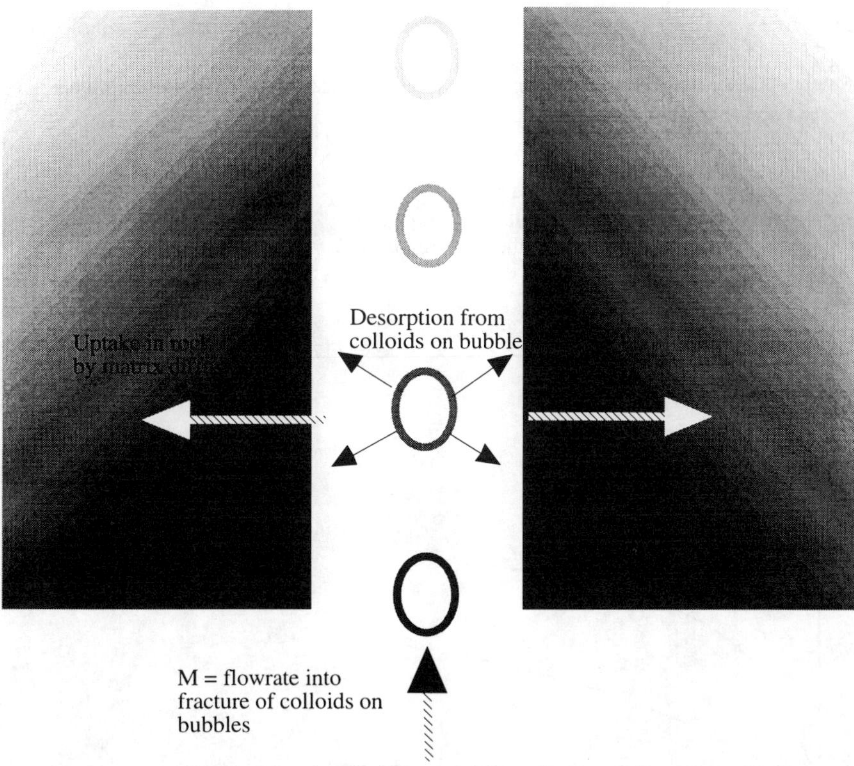

Figure 1. Bubbles rising through a vertical fracture. The nuclides on the colloids on the bubble surfaces are desorbed, migrate to the fracture surface and diffuse into the rock matrix.

With the initial condition that there are no nuclides in the system initially, boundary conditions that the flux of nuclides into the fracture is constant, and that the concentration at the fracture surface is equal to that in the water, the solution is found in [6]. For simplification we neglect the gas volume in the fracture. The error introduced is negligible in this context. The outflow concentration c from the fracture for $t > t_{bubble}$ then is obtained from

$$\frac{c}{c_o} = e^{-\lambda t}\, \mathrm{Erfc}\!\left(\left(\frac{D_e K_d \rho_p}{t - t_{bubble}}\right)^{1/2} \frac{(1 + K_c c_c)}{K_c}\, \frac{z_o W}{M}\right) \tag{6}$$

It is seen that the ratio of colloid mass inflow M to the flow wetted surface $z_o W$ in the fracture plays a dominating role and that the fracture aperture does not influence the results.

The larger the value of the argument in the error function is the smaller will be the concentration at the outlet of the fracture. This means that the longer the travel pathway and the wider the fracture the more will be sorbed into the rock. Obviously the more colloids that go into the fracture and the higher their sorption capacity is the higher the effluent concentration will be. It may be noted that when the bubbles rise quickly the term $(1 + K_c c_c)$ is near 1. When they rise very slowly the term $\dfrac{(1 + K_c c_c)}{K_c} \approx c_c$ for large K_c's and in this case the presence of the colloids in the fracture actually help decrease the breakthrough of the nuclides.

The inlet concentration c_o is the concentration in the water in equilibrium with the incoming colloids. The nuclide outflow rate is $M \cdot c \cdot K_c$.

For illustration the following data is used in a sample calculation.

D_e	Effective diffusivity in rock matrix, 10^{-13} m²/s
$K_c = K_d$	Distribution coefficients for colloids and rock, 0.1, 1 and 5 m³/kg
M	Mass flowrate of colloids into the fracture, 0.4 kg/a
W	Fracture width, 1 m
z_o	Flowpath length, 500 m
ρ_c	Density of rock matrix, 2 700 kg/m³

The rise time of the bubble t_{bubble} is assumed to be fast [7] and the nuclide is long lived.

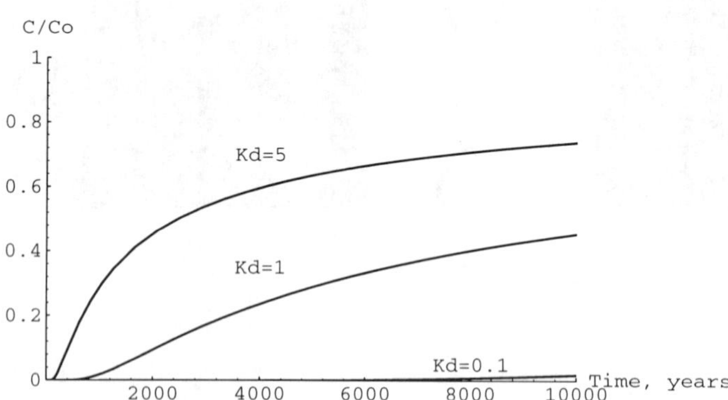

Figure 2. Breakthrough curves for nuclides carried by colloids.

Figure 2 shows breakthrough curves for three different sorption coefficients, 0.1, 1 and 5 m³/kg. This range covers most important sorbing nuclides. It is seen from the figure that there is some retardation in the breakthrough but eventually the nuclides will emerge. Cs-137 and Sr-90 which have sorption coefficients below 0.1 and half-lives of about 30 years would decay to insignificance during the travel.

The data for this example were chosen such that it may represent a possible case but obviously data can be selected within a wide range.

Using this model it is found that the actinides will not desorb from the colloids or decay much during their travel. Sr-90 and Cs-137 will desorb from the colloids, sorb on the fracture surfaces, diffuse into the rock matrix, be retarded and decay to practically insignificance (to less than 10^{-10} of the inflow concentration) as the bubbles pass up through the fractures. Ra-226 will also decay by many orders of magnitude.

Considering uncertainties in data and that it is not sufficiently shown that the sorption of actinides on the colloids is reversible in this situation it is prudent to assume that the released actinides will be transported by the bubbles to the ground surface and that no large decay can be expected.

DISCUSSION

In the present analysis we have tried to make the assumptions needed as realistic as possible but on the conservative side. Due to the very large uncertainties the possible range of values often vary over several orders of magnitude.

The bubbles are larger than the fracture apertures and will have to squeeze through the fractures helped by new bubbles pushing them on. We think that the surface to volume ration of the gas in the system will be considerably less than that of 0.2 mm bubbles. Also it seems to be a very conservative assumption that the bubble surfaces are fully covered with particles. Furthermore the particle size used 0.5 μm may also be too large because the montmorillonite clay particles are often much smaller. The real colloid escape rate should be considerably smaller than the values presented here.

The retardation of the nuclides in the far field is much less than for those nuclides which are dissolved in the water and move with it. However, the stream tube principle for the water movement makes it possible for practically all flowing water traversing the repository to end up in a very small region at the surface e.g. in one well. In that case many leaking canisters might contaminate the same well. The escape by the bubbles, on the other hand, will not collect at the same point at the ground surface. The gas will come up in independent (assumed vertical) paths and thus be spread out over a large area.

Other uncertainties are that the base case used for the near field calculations also has a large variability and it is entirely conceivable that more than ten times higher concentrations can be found in the clay. A special problem is that as corrosion progresses, the damage in the canister grows and the release of nuclides to the clay increases.

The assumption that the particles are sorbed irreversibly on the bubbles has been observed but this applies to transport over distances of centimetres and in glass conduits. It is conceivable that in real fractures over long distances the colloids are detached and stay in the fractures at least to some degree.

There is a need for further experiments to assess the actual carrying capacity of the gas bubbles under realistic repository conditions. Such experiments are being discussed.

CONCLUSIONS

With our present understanding of colloid transport it cannot be excluded that the gas can carry noticeable amounts of clay particles. These in turn can carry nuclides. The amount of gas generated cannot carry away more that a few percent of the clay in a deposition hole. For some nuclides the clay could initially carry up to ten thousand times more nuclides than what can be released to the near field water by the conventional diffusion processes.

Even if the actinides were reversibly sorbed on the clay particles they will not desorb sufficiently during the colloid transport to the ground surface to be significantly retarded to allow decay. The maximum escape rate is in the order of 10^4 Bq/a for the actinides if the concentration on the clay has reached peak values. This will not occur during the few thousand years that the gas evolution takes place. The concentration on the clay carried by the bubbles will be several orders of magnitude less than that which can be reached at steady state.

For Cs-137 and Sr-90 the release rate is less than a few thousand Bq/a due to colloid transport. These nuclides will desorb from the colloids and sorb on the rock. They will then decay to insignificance during their transport to the ground surface. Cs-135 would escape at a rate about 100 times larger than by the conventional diffusion processes but will decay by several orders of magnitude during the transport to the ground surface. Ra-226 will escape at a rate about 20 times higher that by the conventional escape processes but will decay by many orders of magnitude during the transport in the fracture.

ACKNOWLEDGEMENTS

We gratefully acknowledge the stimulus and financial support from the Swedish Nuclear Fuel and Waste Management Co. SKB.

NOTATION

c	Concentration of radionuclides in water in fracture, mol/m^3
c_c	Concentration of colloids in water, kg/m^3
c_p	Concentration of radionuclides in matrix pore water, mol/m^3
c_0	Concentration of radionuclides at fracture inlet, mol/m^3
D_e	Effective diffusion coefficient, m^2/s
K_c	Distribution coefficient between colloids and water, m^3/kg
K_d	Distribution coefficient between rock and water, m^3/kg
$\dot{m}_c$	Mass flowrate of colloids on bubbles, kg/s
M	Mass flowrate of colloids into the fracture, kg/s
q_i	Concentration of radionuclides on colloids, moles/kg or Bq/kg
t	Time since bubbles started entering the fracture, s
t_{bubble}	Travel time of bubble in fracture, s
W	Fracture width, m
x	Distance into rock matrix from fracture, m
z	Distance along fracture, m
z_0	Flowpath length, m
δ	Fracture aperture, m
λ	Decay constant, s^{-1}

REFERENCES

1.   Wan J., Wilson J.L. Visualization of the role of the gas-water interface on the fate and transport of colloids in porous media. Water Resources Res. Vol. 30, No 1, p 11-23, 1994.
2.   Wikramaranta R.S., Goodfield M., Rodwell W.R., Nash P.J. Agg P.J. A preliminary assessment of gas migration from the Copper/Steel Canister. SKB Technical Report 93-31, Stockholm, November 1993.
3.   Perry R.H., Green D. Perry´s Chemical engineers handbook, 6:th Ed., McGraw Hill, 1984.
4.   Romero L., Moreno L., Neretnieks I. Fast multiple path model to calculate radionuclide release from near field of a repository, Nuclear Technology 112, p 99-107, 1995
5.   SKB 91, Final disposal of spent nuclear fuel. Importance of the bedrock for safety, Swedish Nuclear Fuel and Management Co. SKB TR 92-20, May 1992.
6.   Neretnieks I. Diffusion in the rock matrix: An important factor in radionuclide retardation? J. Geophys. Res. 85, p4379-4397, 1980
7.   Coulson J.M., Richardson J.F. Chemical Engineering, Vol 2, 4:th Ed., Pergamon, 1993.

INVESTIGATION OF ROCK SAMPLES USING X-RAY-MICROCOMPUTER-TOMOGRAPHY BEFORE AND AFTER MERCURY INTRUSION POROSIMETRY

P. Klobes[*], H. Riesemeier[*], K. Meyer[*], J. Goebbels[*], M. Siitari-Kauppi[**], K-H. Hellmuth[***]

[*]Federal Institute for Materials Research and Testing (BAM), Unter den Eichen 87 ,
D-12200 Berlin, Germany
[**]Laboratory of Radiochemistry, University of Helsinki, PO Box 55, FIN-00014 University of
Helsinki, Finland
[***]Finnish Centre for Radiation and Nuclear Safety, PO Box 14, FIN-00881 Helsinki, Finland

ABSTRACT

A new method for the physical characterization of rock matrices for use in site investigations of nuclear waste repositories has been developed. The method can provide information needed in the assessment of the performance of the geosphere working as a natural barrier retarding the migration of radionuclides by diffusion into the rock matrix. Most conventional methods for the physical characterization of rocks give only bulk information. The combination of mercury porosimetry and computer tomography can give 3-D data on mineral-specific porosity distributions with additional pore size information. Additionally, limits for mineral-specific internal surface areas can be estimated, which is essential for the assessment of water-rock interaction and reactive interaction with radionuclides (sorption). Results of measurements on granitic rock (granodiorite) from the Baltic shield are discussed and integrated with results by complementary methods.

INTRODUCTION

For the assessment of migration processes of radionuclides in fractured crystalline rock information on the complex microstructure and the related transport properties of the geomatrix is needed. Besides pure physical transport processes such as Darcy flow and diffusion, in particular, chemical interactions of groundwater and transported components with mineral surfaces play a major role as migration/retardation processes. Therefore information on mineral-specific pore size distributions and internal surface areas are needed. Macropores and the connectivity of the pore network determine the transport of fluids under a pressure gradient. Meso- and micropores determine internal surface areas which govern e.g. sorption.

Methods for the characterization of physical rock properties generally give only bulk data (porosity, pore size distribution, diffusivity, flow conductivity, internal surface area) or half-quantitative data (e.g. impregnation with fluorescent resins). Mercury intrusion porosimetry of compact samples of dense rock cannot give realistic information on the pore system due to i.a. limitations of the theoretical pore models used in the interpretations of the results [1]. By taking advantage of the hysteresis observed in large (as compared to the grain size), intact, dense rock samples X-ray computer tomography before and after mercury intrusion porosimetry can overcome the principal difficulties of that method and give information on properties of transport pathways and specific mineral phases [2]. Quantitative, mineral-specific porosities and spatial porosity distributions can be obtained also by impregnation with carbon-14-polymethylmethacrylate (C-14-PMMA) [3], but the method does not give pore size information, however. Selection of appropriate mercury intrusion pressure steps with subsequent application of X-ray com-

Mat. Res. Soc. Symp. Proc. Vol. 465 © 1997 Materials Research Society

puter tomography allows the characterization of 3-D pore systems between preselected limits of pore apertures and the estimation of limits for mineral-specific surface areas.

In a previous work [2] a rock type was chosen which was supposed to have ideal properties to develop and show the potential of the combined methods: (i) large grains of potassium feldspar with porosity nearly exclusively in tiny fissures and fissure fillings and (ii) heavily altered, fine-grained phases with porosities about one order of magnitude higher. In this work a pervasively altered porphyritic rock was selected where also the feldspar showed high porosity which was distributed over the whole area of the grains [4]. The purpose was to use the method to distinguish various porous phases by the nature of their pore structure and to give more quantitative information on mineral-specific pore-size distributions and internal surface areas.

EXPERIMENTAL

A crystalline rock sample consisting of mainly distinct, altered mineral phases was selected. The rock was porphyritic granodiorite, a plutonic rock of Proterozoic age with large potassium feldspar phenocrysts from an investigation site for a deep nuclear waste repository in Finland (core Ki7, Konginkangas (Central Finland), sample position: 122 m core length). The rock in general was composed of about 25-30% quartz, 25-30% potassium feldspar, 25% plagioclase and 10% biotite. Due to the maximum sample size of about 25 x 30 mm which fits into the dilatometer of the mercury porosimeter the measured sample not does exactly represent this composition. Alteration is visible as significant colour change, in particular, in plagioclase (light green) and potassium feldspar (reddish brown) as well as staining of fissure fillings in both phases by hematite. Biotite is altered to chlorite.

The mercury porosimeter used was a Porosimeter 2000 by Fisons Instruments. The use of a special dilatometer chamber for large samples up to about 30 mm diameter limited the maximum pressure which could be applied to 200 MPa (2000 bars). The pressure steps of different experiments were 7.25, 19.9 and 196.6 MPa corresponding to pore openings of about 207, 75 and 7.6 nm, respectively. Because of pronounced hysteresis of the extrusion curves a tomographic investigation of the mercury injected rock sample indicating the distribution of trapped mercury inside the sample could be achieved. The hysteresis curves were measured at every pressure step and pore size distributions derived.

Contrast cone beam tomography (CT) was used which can give significantly higher resolution than medical scanners which produce in general only single slices with medium spatial resolution (0.40 mm). The sample was rotated in the X-ray cone beam and up to 1024 slices were measured at one time using an area detector (Fig. 1). By geometric magnification the spatial resolution could be improved by a factor of about twenty compared with medical scanners. The cone beam tomograph developed at the BAM uses a microfocal X-ray tube (200 kV, 0.5 mA, 5 mm Cu filter, beam diameter below 10 μm) and an image intensifier combined with a CCD-camera as detector (volume element: 65x65x65 μm) [4]. Porosities were measured within rectangular frames 1 to 5 mm wide.

The delay between mercury intrusion and tomography was couple of hours due to transport of the samples between the laboratories. The tomographic measurement itself took usually about 1-1.5 hours. The samples should be measured as fast as possible because slow extrusion of small amounts of mercury could be observed over time periods of days. Small droplets of mercury appeared mainly on the surface of potassium feldspar and some minor altered inclusions therein. In contrast, in the rapakivi of the previous work delayed mercury extrusion was more intensive and concentrated on biotite alteration products.

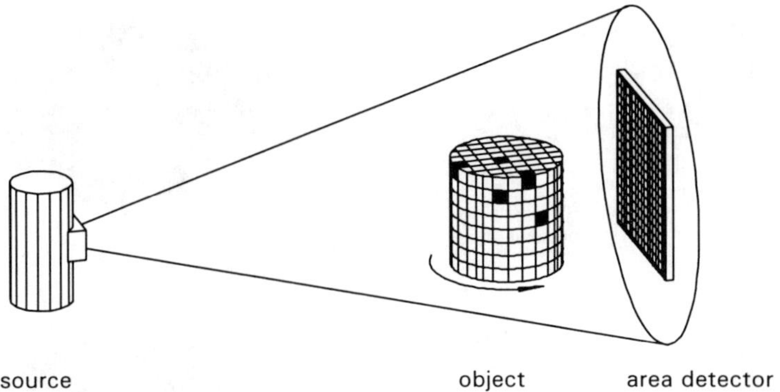

Fig. 1. Scheme of cone beam tomography arrangement

RESULTS AND DISCUSSION

The hysteresis between mercury intrusion and extrusion was very pronounced at all intrusion pressures and comparable to findings on altered rapakivi granite in a previous work [2] which gives the basis for the application of tomographic measurement of entrapped mercury distribution in the rock matrix (Fig. 2). Within the range covered by the instrument no decrease of the curve was observed. The extrusion range below 0.1 MPa could not be measured.

The intrusion patterns at various mercury pressures show distinct differences demonstrated by exemplary 2-D CT-pictures (Fig. 3). The main mineral components in that plane are shown in Fig. 4. At low pressures mercury intrudes only into single fissures in potassium feldspar which are filled with hematite-rich material and into spot-like regions in biotite alteration products (mostly chlorite). At rising mercury pressures the areas of impregnated regions in chlorite phases increase steadily. In plagioclase mercury intrusion takes place only at the highest pressure and in potassium feldspar a network of small fissures appears.

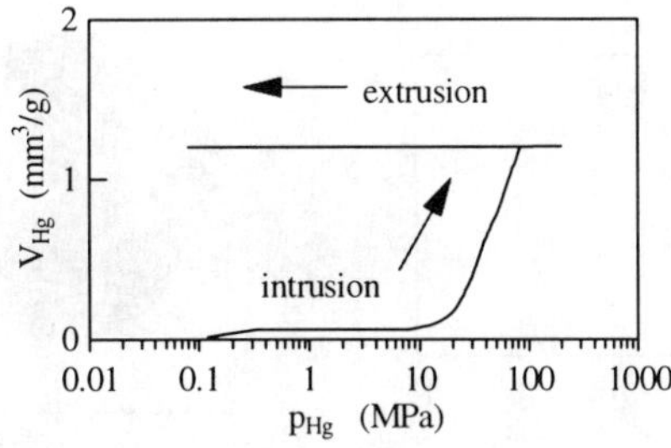

Fig. 2. Hysteresis between mercury intrusion and extrusion; sample: porphyritic granodiorite (Ki7/122), maximum pressure 197 MPa.

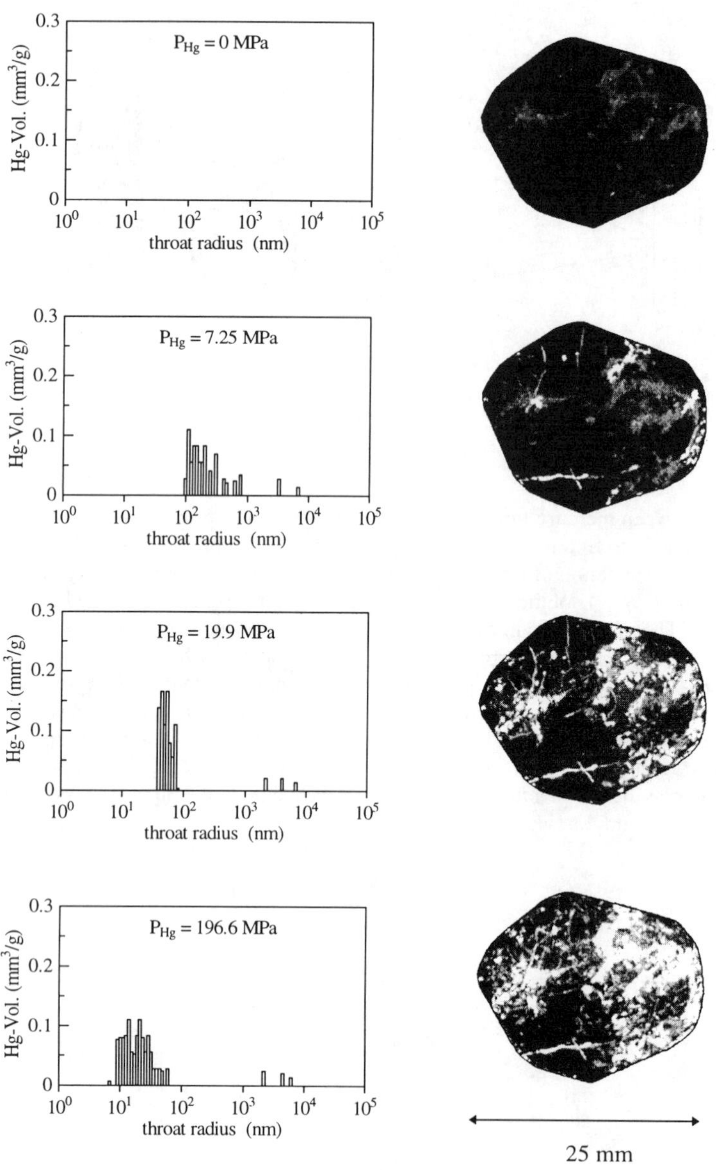

Fig. 3. Pore(throat) size distributions (left) in porphyritic granodiorite (Ki7/122) measured by mercury porosimetry at different intrusion pressures and corresponding spatial distributions of trapped mercury measured by X-ray microcomputer tomography and displayed in 2-D planes at identical location in the specimen (right). Mercury pressures: 0, 7.25, 19.9, 196.6 MPa.

866

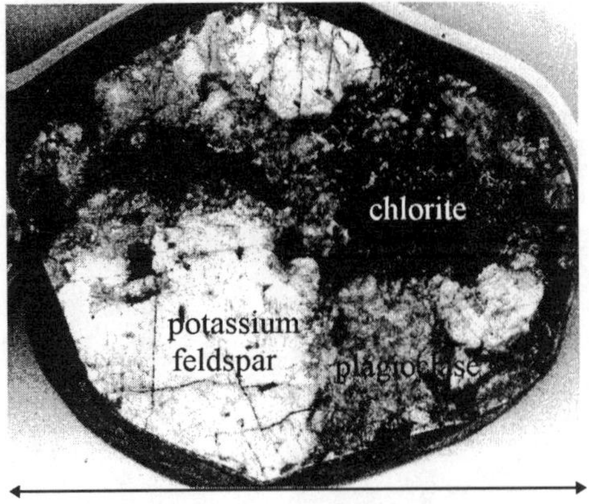

Fig. 4. Image of rock section prepared at identical location as tomographic images in Figure 3 showing the mineral composition at that plane.

When compared with a C-14-PMMA autoradiograph of an equivalent sample (Fig. 5: thin rock section and corresponding autoradiograph) which gives the distribution of porosity accessible by a wetting fluid the following observations can be made. At the highest mercury intrusion pressure the porosity pattern seems to be very similar in dark phases (chlorite etc.). In plagioclase the spot-like regions observed in the autoradiograph are not resolved. A marked difference can be recognized in potassium feldspar (largest grains in Fig. 5) where the patchy, medium-level porosity pattern obviously caused by formation of sericite is not intruded by mercury.

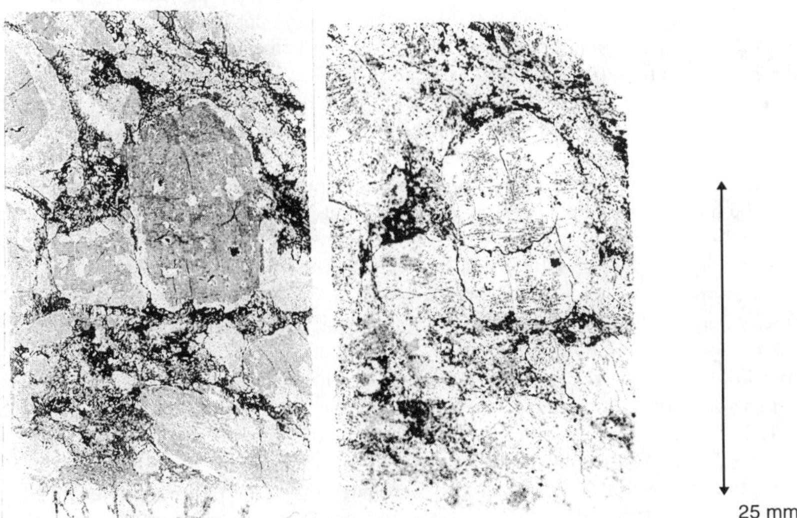

Fig. 5. Rock section (left) and C-14-PMMA autoradiograph (right) showing the spatial porosity distribution in an exemplary 2-D plane in a sample of porphyritic granodiorite Ki7/122.

Table I. Physical properties of granodiorite Ki7/122 and rapakivi granite Y24/54 [1, 5]

Quality (unit)	Method	Porphyr. granodiorite Ki7/122	Rapakivi granite Y24/54
Porosity (vol.%)	Water saturation	1.3-1.7	1.6-2.3
	Hg porosimetry	0.7	2.7
	N_2 adsorption	0.25	--
Surface area (m^2/g)	Hg porosimetry	0.13	1.0
	N_2 adsorption	0.38	0.62
	Kr adsorption	0.08	0.22
Hydraulic conductivity (m^2)	He flow	5.3×10^{-18}	1.2×10^{-18}
Diffusivity, effective (m^2/s)	HTO (average)	2×10^{-13}	7×10^{-13}
	Cl⁻ (average)	0.7×10^{-13}	2×10^{-13}

In Table I information available from other methods on physical properties of the porphyritic granodiorite Ki7/122 is summarized in comparison to rapakivi granite Y24/54 [1] for the purpose to show discrepancies and make further interpretations using results from this work. One has to note that considerable scatter is introduced due to the natural heterogeneity of the rock material, because various methods could not be applied to the same specimen. The porosities measured gravimetrically by saturation of the same samples with monomer MMA were by a factor of 0.9-0.8 and 0.9, respectively lower than those determined with water which seems to indicate a larger fraction of small pores not filled by MMA in sample Ki7/122 [3]. Low pore apertures can be inferred, too, from the low mercury intrusion porosity and the nitrogen adsorption results showing a steep pore volume increase at pore diameters near 2 nm [1]. Mercury porosimetry pore size distributions of both rock types are rather similar, although large pores are more common in the granodiorite and small pores below 100 nm in the rapakivi granite. Mercury extrusion hysteresis is more pronounced in granodiorite. Structural features such as cracks and fissures with large apertures and high connectivity, on the other hand, seem to be the reason for the high hydraulic conductivity measured in the granodiorite. Delayed extrusion of mercury was observed mainly on the surface of potassium feldspar.

The internal surface area in rapakivi granite derived from mercury intrusion measurements [1] appears to be high as compared with that of granodiorite, although mercury porosimetry seems to overestimate the surface area in the case of rapakivi granite as indicated by gas adsorption data. Gas adsorption did not deliver consistent results for sample Ki7/122. Microporosity might be a reason for the observed discrepancy (adsorption of water; kinetic effects, etc. [1]).

The effective diffusivity of the granodiorite is somewhat higher than that of the rapakivi granite, both being at comparatively high levels [1]. The anion exclusion effect of chloride ions is observable, but low in both matrices in comparison to other rock types, probably due to the obviously open structures dominating through-diffusion [5]. The apparent diffusivity of granodiorite Ki7/122 (4.4×10^{-11} m^2/s) was found to be significantly lower than that of fresh granodiorite from core Ki7 [5]. In through-diffusion experiments retardation of diffusing species in heterogeneous matrices is difficult to quantify by measurement of the time-lag. In the non-stationary phase of He diffusion experiments the tailing can be analyzed (more pronounced in case of sample Ki7/122 when compared with Y24/54). In case of non-stationary out-diffusion of MMA diffusion curves obtained from both rock types were composed of more than one component. The reason for this might be low-aperture pore space in altered phases which is not participating significantly in through-diffusion, and is more slowly filled/emptied during the non-stationary phase. Recently, influences of the heterogeneity of the rock matrix on diffusion have been verified by theoretical modelling based on real lateral porosity distributions [6].

The conclusions made hitherto are supported by direct observations: by scanning electron microscopy (SEM) no pores were observed in potassium feldspar in granodiorite Ki7/122, and impregnation by fluorescent resins not impregnating nm-pores made only a few large hematite-stained fissure fillings visible. In the fine-grained, dark, heavily altered minerals of both rock types regions as well as fissures and cracks were impregnated by the resin. Additionally, microfissures and grain boundary pores were detected by SEM in both rock types, indicating good accessibility of these potentially sorbing phases.

The combined methods mercury porosimetry and computer tomography give information on the internal structure not available by other methods which, on the other hand, support suggestions based on results of other methods. The coarse-grained potassium feldspar in both rock types shows peculiarities: mercury does neither intrude large seritized regoins in that mineral in sample Ki7/122 nor the network of tiny fissures in sample Y24/54. Estimations of mineral-specific porosities and internal surfaces in larger grains of a few main minerals based on these measurements are given in Table II. The corresponding porosity results from C-14-PMMA autoradiographs of equivalent samples are consistent within the order of magnitude although the summing up the pore size ranges from the CT measurements gave a slightly lower value, 0.47% against about 1% for potassium feldspar and a slightly higher value, 3.6%, against about 2.6% for chlorite. The bulk values obtained by mercury porosimetry were 0.81% for granodiorite, Ki7/122 and 1.8% for rapakivi granite Y24/54. In the altered, dark phases a significant fraction of the porosity is found in the smallest pore size range covered by the mercury intrusion pressure used. The quantitative treatment of the fine-grained regions needs a further development of the image handling and analysis routines.

Table II. Mineral-specific porosities within three different pore size ranges measured by X-ray microcomputer tomography after mercury intrusion at different pressures and derived average (median) and maximum mineral-specific internal surfaces in porphyritic granodiorite Ki7/122.

	Pore diameter range (nm)		
	15000 - 207	207 - 75	75 - 7.6
Porosity (vol.%)			
K-feldspar	0.17	0.1	0.20
Plagioclase	0.53	0.44	0.78
Chlorite	0.36	1.48	1.73
Surface area (av./max.) (m^2/g)			
K-feldspar	0.0019 / 0.012	0.013 / 0.020	0.086 / 0.39
Plagioclase	0.0058 / 0.038	0.061 / 0.087	0.33 / 1.53
Chlorite	0.0039 / 0.026	0.21 / 0.29	0.73 / 3.4

The calculation of surface areas in the granodiorite, a rough estimation based on the assumption of capillary pores which gives an upper limit, was made using average (median; 1355, 106, 35 nm) and minimum (207, 75, 7.6 nm) pore diameters, respectively. In the lower pore size ranges calculation of the median is feasible while at large pore sizes it seems arbitrary, due to single peaks scattered over a wide range. The significant contribution of the smallest pores is as expected and the predominance in chlorite and to some extent plagioclase is evident (Table II).

CONCLUSIONS

Alteration of rocks by hydrothermal or weathering processes leads to significant structural changes in the rock matrix such as in porosity, pore size as well as size and accessibility of internal surfaces finding expression in changes in diffusive and sorptive properties of the rock. Mercury porosimetry combined with computer tomography can provide information allowing conclusions on potential retarding properties of rocks. In the altered granodiorite Ki7/122 low-aperture pores mainly in altered phases provide reactive surfaces which are accessible through conductive pathways.

ACKNOWLEDGEMENTS

This work has been carried out as part of the Publicly Administrated Nuclear Waste Management Research Programme in Finland. The authors are grateful to B. Illerhaus (BAM, Berlin) for handling and preparing the tomographical images.

REFERENCES

[1] K-H. Hellmuth, P. Klobes, K. Meyer, B. Röhl-Kuhn, M. Siitari-Kauppi, J. Hartikainen, K. Hartikainen, J. Timonen, Z. Geol. Wiss. **23 (5/6),** p. 691-706 (1995).

[2] P. Klobes, H. Riesemeier, K. Meyer, J. Goebbels, K-H. Hellmuth, Fresenius J. Anal. Chem. (1997) (in print).

[3] K-H. Hellmuth, S. Lukkarinen, M. Siitari-Kauppi, Isotopenpraxis Environ. Health Stud. **30,** p. 47-60 (1994).

[4] A. Lindberg, K-H. Hellmuth, M. Siitari-Kauppi, J. Suksi, in Water-Rock Interaction - WRI-7, edited by Y.K. Kharaka, A.S. Maest, 1992, Balkema, Rotterdam, p. 573-576.

[5] M. Siitari-Kauppi, K-H. Hellmuth, A. Lindberg, T. Huitti, Radiochimica Acta **66/67**, p. 409-414 (1994).

[6] P. Simbierowicz, M. Olin, (submitted; this symposium).

MATRIX DIFFUSION OF SOME ALKALI- AND ALKALINE EARTH-METALS IN GRANITIC ROCK

H. Johansson, J. Byegård, G. Skarnemark and M. Skålberg
Department of Nuclear Chemistry, Chalmers University of Technology, S-412 96 Göteborg, Sweden.

ABSTRACT

Static through-diffusion experiments were performed to study the diffusion of alkali- and alkaline earth-metals in fine-grained granite and medium-grained Äspö-diorite. Tritiated water was used as an inert reference tracer. Radionuclides of the alkali- and alkaline earth-metals (mono- and divalent elements which are not influenced by hydrolysis in the pH-range studied) were used as tracers, *i.e.* $^{22}Na^+$, $^{45}Ca^{2+}$ and $^{85}Sr^{2+}$. The effective diffusivity and the rock capacity factor were calculated by fitting the breakthrough curve to the one-dimensional solution of the diffusion equation. Sorption coefficients, K_d, that were derived from the rock capacity factor (diffusion experiments) were compared with K_d determined in batch experiments using crushed material of different size fractions.

The results show that the tracers were retarded in the same order as was expected from the measured batch K_d. Furthermore, the largest size fraction was the most representative when comparing batch K_d with K_d evaluated from the diffusion experiments. The observed effective diffusivities tended to decrease with increasing cell lengths, indicating that the "transport" porosity decreases with increasing sample lengths used in the diffusion experiments.

INTRODUCTION

In the safety assessment of a repository for spent nuclear fuel, the surrounding rock will act as a barrier for the retardation of released radionuclides. The retardation process in waterconducting fractured rock is believed to be due to sorption of radionuclides on the fracture walls combined with diffusion into the fracture coatings and further into micropores in the rock (1). In order to perform a correct safety assessment, it is important to quantify and understand these rock-groundwater interaction processes both in the laboratory scale and in the field scale. Important parameters in order to understand the retardation of radionuclides, are parameters such as the flow wetted surface and the sorption capacity of the fracture walls, etc. These parameters can be determined in *in-situ* experiments if the sorption properties of the tracer are well understood. Byegård *et al.* (2) proposed tracers to be used for in-situ field experiments and studied the sorption properties of these elements.

In the present study, the diffusion behaviour of the proposed tracers on the laboratory scale is presented. Diffusion and sorption of both sorbing and non-sorbing radioactive tracers is studied by using a through-diffusion experimental technique. With this method the effective diffusivity, D_e, and the distribution coefficient, K_d, can be evaluated. Earlier studies have indicated that batch experiments using crushed material may overestimate the sorption distribution coefficient, K_d, compared to intact rock material (3). Crushing of the material creates new surfaces and may increase the accessibility for pores that may not be representative for the intact rock. The rock discs used in the diffusion experiments are still somewhat disturbed, compared to the intact rock, due to the pressure release when the core is drilled out of the bedrock and by treatment methods (cutting and grinding). The results obtained for sorption measured with crushed material and through diffusion experiments are compared.

Mat. Res. Soc. Symp. Proc. Vol. 465 © 1997 Materials Research Society

EXPERIMENTAL

Materials

Drill cores ($\varnothing$ 35.1 mm) of about 1800 Ma old fine-grained granite and Äspö-diorite were obtained from the Äspö Hard Rock Laboratory (HRL) situated near Oskarshamn on the south-east coast of Sweden. The Äspö-diorite is a medium-grained, porphyritic monzo/granodiorite. It contains large K-feldspar phenocrysts of 1-2 cm in size. The porosity was determined with the water saturation method to be 0.55±0.1%. The samples obtained were slightly foliated and the dominating minerals were determined to be plagioclase (45%), biotite (15%), K-feldspar (15%), quartz (14%) and epidote (6%).

The fine-grained granite is a homogeneous fine-grained alkali-granite. The samples were foliated in about 45 degrees angle to the core axis. The main minerals were K-feldspar (38%), quartz (31%) and plagioclase (23%) and it contained small amounts of chlorite (3%) and muscovite (2.5%). The porosity was determined to be 0.25±0.1%.

Geological material for the batch experiments was collected from the Äspö HRL. It was crushed and dry sieved into 6 different size fractions (0.045-0.090, 0.090-0.25, 0.25-0.5, 0.5-1, 1-2 and 2-4 mm).

The tracers used were tritiated water, H^3HO (HTO), the alkali metal, $^{22}Na^+$, and the alkaline earth metals, $^{45}Ca^{2+}$ and $^{85}Sr^{2+}$. The radioactive tracers were injected in concentrations that gave a negligible increase of the natural elemental concentrations in the groundwater.

A synthetic groundwater representative for the site was used. The composition of the synthetic groundwater is typical for a depth of 400 m at the Äspö HRL, see Table I.

In order to obtain non-oxic conditions, all experiments were performed in glove boxes with N_2-atmosphere. All experiments were conducted at ambient temperature (20-25 $^\circ$C).

Through-diffusion experiments

A cylindrical rock disc from a drill core was cast in an epoxy resin (Araldite F with triethylenetetraamin as a hardener, Ciba Geigy). After grinding the end surfaces, the cylinder was put in vacuum to remove gas trapped in pores and then saturated with the synthetic groundwater for at least one month. The rock disc was mounted between two water containers, see Fig. 1, sealed with o-rings on both sides. The volume of the containers were 20 ml. One side of the diffusion cell was spiked with a sorbing tracer and HTO as an inert reference. The concentration increase on the other side of the diffusion cell was monitored under stagnant groundwater conditions. Samples of 2 ml and 10 µl were collected in the measurement container and the injection container, respectively. The removed sample volume was replaced with synthetic groundwater.

Three different lengths of the rock discs were used; 1,2 and 4 cm. Each side of the diffusion cell container was closed by a screw cap that was ventilated through a hole connected to a hose. The screw caps were closed by putting the hose on a thin pin. This was done in order to minimize pressure build-up when closing the reservoirs after sampling.

Radioactivity of HTO, $^{22}Na^+$ and $^{45}Ca^{2+}$ was measured with liquid scintillation counting (LKB Wallac, Rackbeta 1219) and $^{85}Sr^{2+}$ was measured with a NaI-detector (Intertechnique, CG 4000).

Batch sorption experiments

The experimental technique used in the batch sorption experiments has been previously described by Byegård et al.(2), with the changes that the contact time was two weeks and that the desorption agent was synthetic groundwater. K_d, the sorption distribution coefficient between the solid and the liquid phase, was obtained in two ways: 1) by measuring the concentration decrease in the solution, 2) by measuring the activity which was desorbed into the solution after contacting with new groundwater and the activity remaining in the solution before rinsing.

The radioactive tracers were measured with a NaI-detector (Intertechnique, CG 4000), except for $^{45}Ca^{2+}$ which was measured by liquid scintillation counting (LKB Wallac, Rackbeta 1219). Tritiated water, HTO, was not used in the batch experiments.

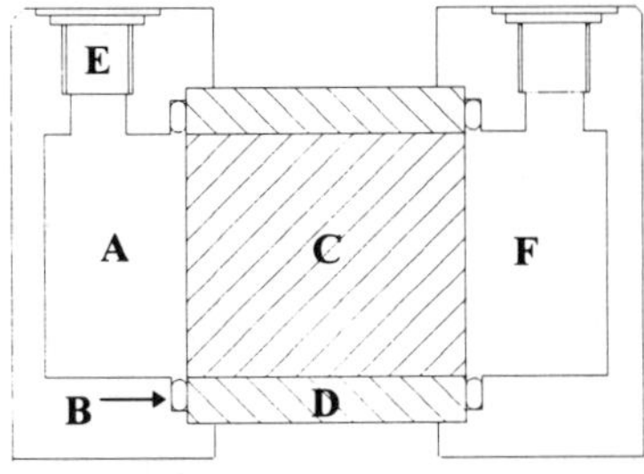

Fig. 1. Schematic drawing of the diffusion cell. A injection container, B o-ring, C rock cylinder, D epoxy resin, E sampling hole, F measurement container.

Table I. Composition of the synthetic groundwater used in the experiment.

Element	(mg/l)	Element	(mg/l)
Li^+	1.3	Cl^-	8350
Na^+	2400	SO_4^{2-}	170
K^+	10	SiO_4^{4-}	16
Rb^+	0.026	HCO_3^-	17
Cs^+	0.001		
Mg^{2+}	50		
Ca^{2+}	2540	pH	7.5
Sr^{2+}	35	Ionic	0.25
Ba^{2+}	0.14	strength	

THROUGH-DIFFUSION THEORY

The accumulated scaled ratio of concentration, C_r, of the diffusing species which has passed through the porous, homogeneous rock disc (initially at zero concentration) at time t, can be obtained by solving the diffusion equation for the boundary conditions: constant inlet concentration C_1 from $t = 0$, and outlet concentration C_2 ($C_2 \ll C_1$) at a distance l. The result is, (4)

$$C_r = \frac{C_2 V_2}{C_1 Al} = \frac{D_e t}{l^2} - \frac{\alpha}{6} - \frac{2\alpha}{\pi^2} \sum_{n=1}^{\infty} \frac{(-1)^n}{n^2} \exp\left\{ -\frac{D_e n^2 \pi^2 t}{l^2 \alpha} \right\} \tag{1}$$

where C_r is the scaled ratio of the concentration in the measuring cell, C_2, relative to that of the injection cell, C_1. A is the geometrical surface area of the sample which has thickness l and V_2 is the volume of the measuring container. $D_e = D_p \varepsilon^+$ is the effective diffusivity where D_p is the pore diffusivity, ε^+ is the "transport" porosity (the pores that are connected and contributes to the transport of the dissolved species from one side to the other), $\alpha = \varepsilon_t + K_d \rho$ is the rock capacity factor where ε_t is the total porosity (the sum of the transport porosity and the storage porosity), K_d is the sorption distribution coefficient and ρ is the density of the rock sample. The theory has previously been described in detail (5).

Equation (1) is only valid if: D_e is constant (independent of C and t), the concentration in the injection cell, C_1, is constant with time, the concentration in the measuring cell, C_2, is negligible compared to C_1 for all times, no bulk flow (convection) occurs and if the rock sample is homogeneous. These assumptions seems to be reasonable, except for the assumption of homogeneous media which is not obvious, because the K-feldspar grains in the Äspö-diorite are of the same size as the thickness of the samples, furthermore the fine-grained granite is brittle containing microfissures and is foliated.

The effective diffusivity, D_e, and the rock capacity factor, α, were calculated by fitting the data to eq. (1). A steepest gradient method was used to minimize the error sum of squares according to the least square method. The errors of the parameters were estimated assuming that the errors were normal distributed.

The total porosity used in the calculations was determined by the rock capacity factor for the non-sorbing tritiated water, for which $\alpha = \varepsilon_t$. K_d for the sorbing tracers could then be derived from the rock capacity factor for each sorbing tracer.

RESULTS

Batch sorption experiments

The K_d obtained in the batch experiments increase in the order $Na^+ < Ca^{2+} < Sr^{2+}$ for the fine-grained granite and $Na^+ < Ca^{2+} \approx Sr^{2+}$ for the Äspö-diorite, see Fig.2. K_a, the surface related sorption coefficient, was calculated by dividing K_d with the BET-surface area (m^2/kg) which was measured for all particle sizes.

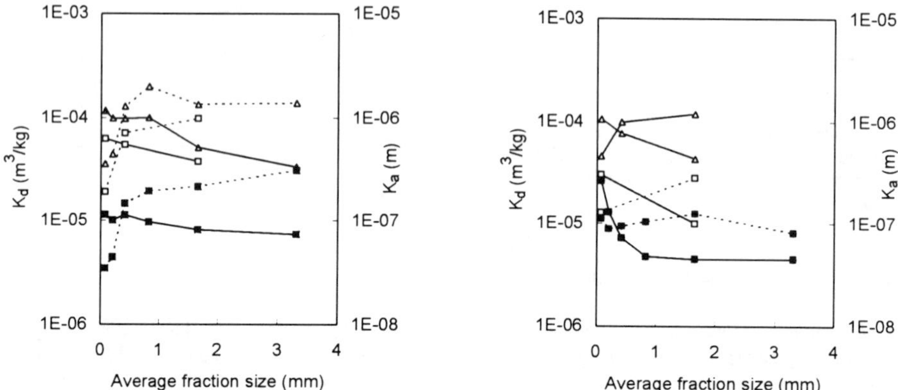

Fig. 2. K_d (left axis, straight lines) and K_a (right axis, dotted lines) vs fraction size for Äspö-diorite (left) and Fine-grained granite (right). K_d for Sr^{2+} ($\triangle$) was evaluated from measurements of the aqueous phase. K_d for Na^+ (■) and Ca^{2+} (□) were evaluated from measurements of the aqueous phase and the desorption solution. The contact time was 14 days and the desorption time 14-32 days. The K_a values corresponds to the K_d values in the diagram. $K_a = K_d$ / specific surface area. The experimental errors were ±25%.

The K_d varied with particle size, the smallest particle size gave the largest K_d. The variations were approximately one order of magnitude between the largest particle size (2-4 mm) and the smallest particle size (0.045-0.090 mm). This can probably be explained by the larger surface areas in the smaller size fractions. K_a varied less with particle size than K_d, or even tended to show increased sorption with increasing particle size, which indicates that there is a large influence of the surface area on sorption capacity.

Diffusion experiments

The measurements of HTO showed that there was a large deviation in D_e and α between the different diffusion cells, even within the same rock type, the same length and with rock discs taken from the same drill core. In Fig. 3a and b all HTO data are plotted together for the two rock types. The plots show a tendency that an increase in D_e is associated with an increase in α for both materials. Since D_e is dependent on ε^+, the "transport" porosity, and α is the total porosity for HTO, it seems that both the storage and the transport porosity varies between the samples.

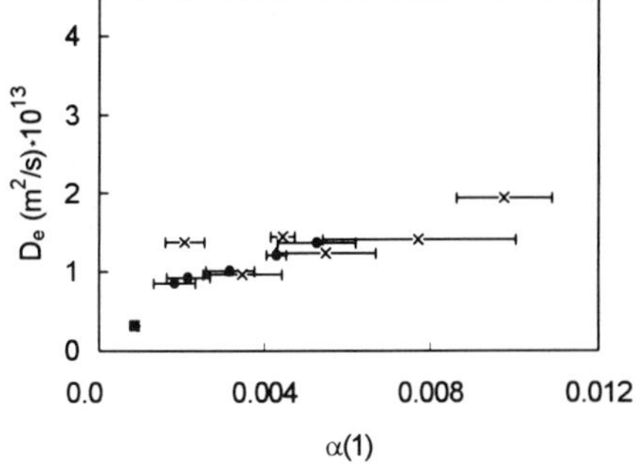 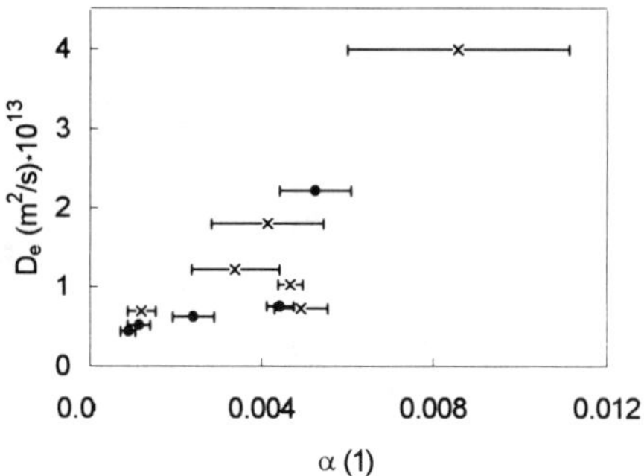

Fig. 3. D_e vs α for HTO in Äspö-diorite (left) and fine-grained granite (right) for 1 cm (×), 2 cm (•), and 4 cm (■) cells. Error bars represent 1σ. The error bars in D_e are of the same size as the height of the dots. The large deviations in α for the 1 cm cells are due to experimental problems with contamination of HTO in the glove box in combination with the short break through time of HTO in the 1 cm cells.

The spread in data, see Fig.3., is larger for the fine-grained granite, which is probably due to the fact that it is brittle and contains even visible micro-fissures in some samples. There is a possibility for a decrease in D_e and α in the 4 cm cells containing Äspö-diorite due to the fact that the 1-2 cm sized K-feldspar grains may give directly connecting grain boundary pores in the 1 and 2 cm cells but not in the 4 cm cells. The fine-grained granite is foliated with veins of chlorite which may give directly connective diffusion pathways in the shorter cells but not in the 4 cm cells. However, more data would be needed to show this.

There is a tendency towards decreasing D_e and α with increasing cell lengths, see Fig.4 and Table II. Increased D_e with decreasing cell lengths, indicates that the "transport" porosity is larger in the shorter samples.

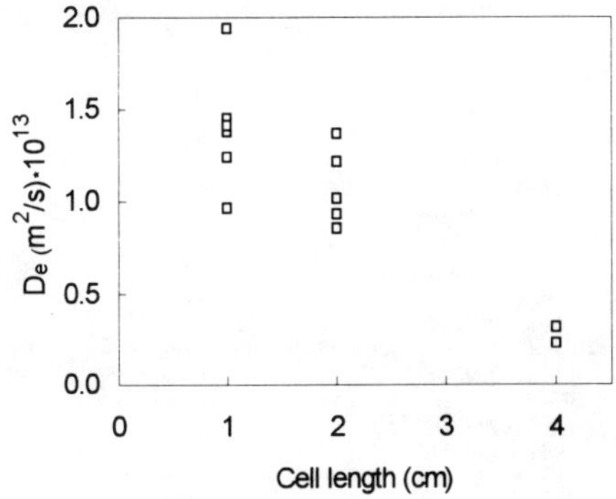

Fig.4. D_e vs cell length for HTO diffusion in Äspö-diorite. One of the two 4 cm samples is calculated from Na^+ diffusion (cell 7) assuming that the formation factor for HTO is equal to the formation factor for Na^+. The error bars are of the same size as the square dots.

Some typical break-through curves for HTO (cell 3), Na^+ (cell 3), Ca^{2+} (cell 15) and Sr^{2+} (cell 10) are presented in Fig.5. Ca^{2+} and Sr^{2+} are more retarded than Na^+ as expected from the K_d-values obtained in the batch experiments. The K_d that have been evaluated from the diffusion experiments are presented together with the diffusion data in Table II.

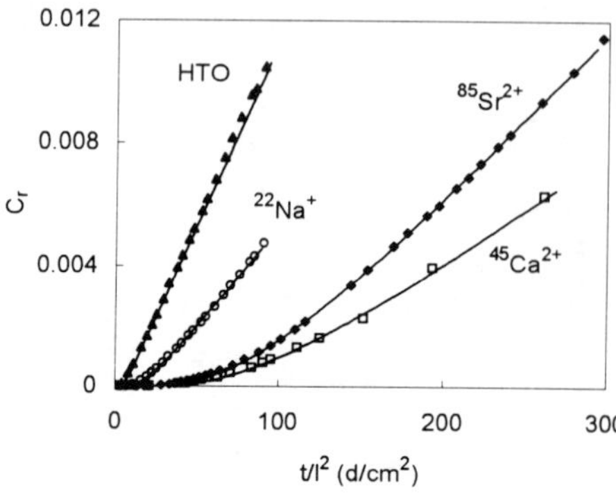

Fig.5. Measured C_r vs scaled diffusion time for HTO, Na^+, Ca^{2+} and Sr^{2+} in 1 cm cells of Äspö-diorite. Drawn lines are the fit of eq. 1.

Table II. Evaluated effective diffusivities, D_e, rock capacity factors, α, sorption coefficients, K_d and rock formation factors, F, for Äspö-diorite (Äd) and fine-grained granite (Fgg). The errors are given as 1σ.
1) The North American definition of the formation factor is the reciprocal of the European definition.

Cell nr	Nuclide	Rock type	Cell size (cm)	D_e (m^2/s)	α (1)	K_d (m^3/kg)	$F_{(HTO)}$ 1) (1)	$F_{(sorb)}$ 1) (1)	$F_{(sorb)}/F_{(HTO)}$ (1)
1	HTO	Fgg	1	1.0±0.01E-13	4.7±0.3E-03		4.3E-05		
	Na-22			6.4±0.01E-14	1.1±0.02E-02	2.6±0.2E-06		4.8E-05	1.13
2	HTO	Fgg	2	7.5±0.12E-14	4.4±0.3E-03		3.1E-05		
	Na-22			6.0±0.1E-14	1.3±0.01E-02	3.2±0.2E-06		4.5E-05	1.44
3	HTO	Äd	1	1.4±0.01E-13	4.4±0.3E-03		6.0E-05		
	Na-22			7.8±0.1E-14	8.5±0.2E-03	1.5±0.2E-06		5.9E-05	0.98
4	HTO	Äd	2	1.2±0.01E-13	4.3±0.2E-03		5.1E-05		
	Na-22			7.3±0.1E-14	1.1±0.02E-02	2.3±0.2E-06		5.5E-05	1.08
5	HTO	Äd	4	3.2±0.12E-14	8.6±0.8E-04		1.3E-05		
	Na-22			1.3±0.1E-14	2.1±0.01E-03	4.6±0.3E-07		9.9E-06	0.75
6	Na-22	Fgg	4	6.2±0.1E-14	7.0±0.1E-03			4.7E-05	
7	Na-22	Äd	4	1.6±0.1E-14	2.4±0.03E-03			1.2E-05	
8	HTO	Fgg	1	1.2±0.03E-13	3.4±1.0E-03		5.1E-05		
	Sr-85			3.5±0.1E-14	2.5±0.02E-02	8.2±0.5E-06		4.5E-05	0.88
9	HTO	Fgg	2	6.2±0.35E-14	2.4±0.5E-03		2.6E-05		
	Sr-85			7.9±0.3E-15	4.4±0.05E-03	7.5±2.0E-07		9.9E-06	0.39
10	HTO	Äd	1	1.9±0.03E-13	9.8±1.1E-03		8.1E-05		
	Sr-85			6.3±0.1E-14	2.9±0.03E-02	6.9±0.5E-06		7.9E-05	0.98
11	HTO	Äd	2	1.4±0.07E-13	5.3±0.9E-03		5.7E-05		
	Sr-85			2.8±0.1E-14	1.2±0.01E-02	2.5±0.4E-06		3.6E-05	0.63
14	HTO	Fgg	1	4.0±0.08E-13	8.6±2.6E-03		1.7E-04		
	Ca-45			1.4±0.03E-13	1.8±0.1E-02	3.6±1.5E-06		1.8E-04	1.06
15	HTO	Äd	1	1.2±0.03E-13	5.5±1.2E-03		5.2E-05		
	Ca-45			4.3±0.1E-14	2.1±0.1E-02	5.5±0.8E-06		5.4E-05	1.05
16	HTO	Fgg	2	2.2±0.07E-13	5.3±0.8E-03		9.2E-05		
	Ca-45			7.3±0.12E-14	1.4±0.08E-02	3.5±E0.6-06		9.3E-05	1.00
17	HTO	Äd	2	8.5±0.48E-14	1.9±0.5E-03		3.6E-05		
	Ca-45			3.2±0.01E-14	1.5±0.08E-02	4.9±0.5E-06		4.1E-05	1.15
18	HTO	Fgg	1	6.8±0.12E-14	1.2±0.3E-03		2.8E-05		
19	HTO	Äd	1	9.6±0.18E-14	3.5±1.0E-03		4.0E-05		
20	HTO	Fgg	2	4.3±0.21E-14	8.9±1.8E-04		1.8E-05		
21	HTO	Äd	2	9.3±0.56E-14	2.2±0.5E-03		3.9E-05		
22	HTO	Fgg	1	1.8±0.03E-13	4.1±1.3E-03		7.5E-05		
23	HTO	Äd	1	1.4±0.01E-13	2.1±0.5E-03		5.7E-05		
24	HTO	Fgg	2	5.1±0.25E-14	1.1±0.3E-03		2.1E-05		
25	HTO	Äd	2	1.0±0.05E-13	3.2±0.6E-03		4.2E-05		
26	HTO	Fgg	1	7.2±0.12E-14	4.9±0.6E-03		3.0E-05		
27	HTO	Äd	1	1.4±0.05E-13	7.7±2.3E-03		5.9E-05		

The D_e (slopes of the concentration curves) tend to increase in the order $Ca^{2+} \approx Sr^{2+} < Na^+ < HTO$. It is very important, when comparing the different sorptive tracers, to relate D_e to the D_e of HTO in each diffusion cell due to the variations in porosity between the cells. The difference in D_e between the different tracers could simply be an effect of the difference in bulk diffusivities, D_w, (the diffusion coefficient of the tracer in water) between the tracers. Instead the formation factor, F, calculated from both HTO and the sorbing tracer in each diffusion cell should be compared. The formation factor (properties of the material such as transport porosity, ε^+, tortuosity, τ, and pore constrictivity, σ) relates the effective diffusivity, D_e, to the water bulk phase diffusivity, D_w:

$$D_e = F \cdot D_w \quad ; \quad F = \frac{\varepsilon^+ \cdot \sigma}{\tau^2} \qquad (2)$$

The formation factors, see Table II, for HTO and the accompanying sorbing tracer in the same diffusion cell, are generally of the same size, ($F_{(sorb)}/F_{(HTO)} \approx 1$), which indicates that the difference in D_e between the sorbing tracers is only due to the difference in D_w and not due to the variation in material properties. The bulk diffusion coefficients used in the calculations ($D_w HTO=2.4E-9$, $D_w Na^+=1.33E-9$, $D_w Ca^{2+}=0.79E-9$ and $D_w Sr^{2+}=0.79E-9$) were calculated from ionic conductivities at infinite dilution (6).

Comparison of K_d-values

The K_d-values evaluated from the diffusion experiments correspond best to the batch K_d-values obtained with the largest size fractions (2-4 mm). This indicates that the large particle size is the most representative for estimating K_d for bulk rock. Possibly, this is due to the fact that the largest size fraction consists of whole mineral grains. This is not the case for the smaller size fractions where the mineral grains have been divided during the sample preparation. An attempt was made to distinguish the sorption onto outer surfaces from inner surfaces, which only can be reached by diffusion. Assuming that the particles are spherical, the K_d can be expressed as the sum of the inner $K_{d,i}$ and the outer K_a times the geometrical surface area:

$$K_d = K_{d,i} + K_a \cdot \frac{6}{\rho <d_p>} \qquad (3)$$

where $<d_p>$ is the average particle diameter. A plot of K_d vs $1/<d_p>$ would then give $K_{d,i}$ as the intercept with the y-axis. K_d for Na^+ and Sr^{2+} are plotted in fig 6.

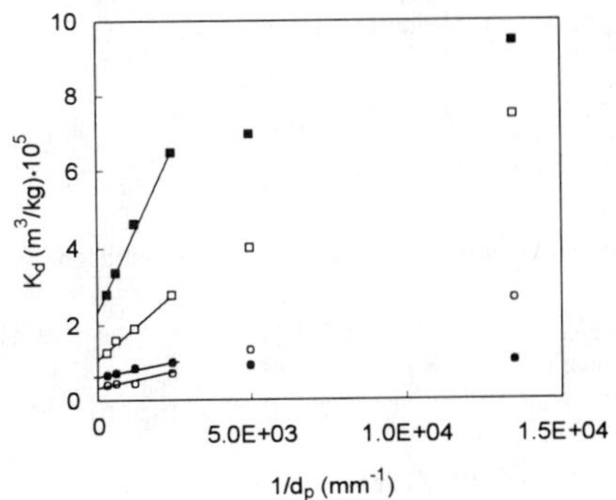

Fig.6 K_d for Na^+ (● Äspö-diorite, o fine-grained granite) and Sr^{2+} (■ Äspö-diorite, □ fine-grained granite) vs the inverse of the mean particle diameter of Äspö-diorite and fine-grained granite The lines are the fit of Eq.(4) to the four largest size fractions.

Only the four largest size fractions were fitted since those size fractions consist of polymineralic grains. When estimating K_d which is valid for the bulk rock from batch experiments, the $K_{d,i}$ value should be most correct to use since the largest size fractions consist of only about 10% outer surface. With this method the $K_{d,i}$ can be estimated without measuring the surfaces of the size fractions. A summary of the batch K_d for the largest size fraction, the evaluated $K_{d,i}$ and the average diffusion K_d are presented in table III. The diffusion K_d for both the Äspö-diorite and the fine-grained granite tend to increase in the same order as observed in the batch experiments. It should also be noted that the $K_{d,i}$ gives an overestimation compared to the diffusion K_d by a factor of 2-6.

Table III. K_d from the 2-4 mm fraction, extrapolated $K_{d,i}$ and average diffusion K_d for Na^+, Ca^{2+} and Sr^{2+}.

Tracer	Rock type	K_d, largest size fraction (m^3/kg)	$K_{d,i}$ (m^3/kg)	K_d, diff. exp (m^3/kg)
Na^+	Äd	6.6±0.6E-6	6.3±0.2E-6	1.4±0.6E-6
	Fgg	4.0±0.5E-6	3.3±0.4E-6	2.9±0.4E-6
Ca^{2+}	Äd	3.8±1E-5	3.2±1E-5	5.2±1.3E-6
	Fgg	1±0.3E-5	9±0.5E-6	3.6±1.5E-6
Sr^{2+}	Äd	3.4±0.8E-5	2.8±1.3E-5	4.7±0.8E-6
	Fgg	1.3±0.5E-5	1.1±0.2E-5	4.5±0.6E-6

CONCLUSIONS

Prediction of K_d, for weakly, ion exchangeable sorbing tracers in laboratory diffusion experiments, based on laboratory batch experiments, may overestimate the matrix diffusion related K_d if small size fractions are used. The comparisons show that large size fractions are the most representative for estimating the K_d in bulk rock samples possibly due to that these fractions consists of polymineralic grains.

The observed differencies in D_e between HTO and Na^+, Ca^{2+} or Sr^{2+} were related to differences in water diffusivity, D_w, and no influences of the material properties were observed on the D_e.

There are tendencies of decreasing D_e and α with increasing cell lengths, indicating that the transport porosity (and the storage porosity) decreases with increasing cell lengths. It is probable that the shorter samples overestimates the D_e and α since the length of the samples are of the same size as the heterogeneity of the samples.

ACKNOWLEDGEMENTS

This work was funded by SKB, the Swedish Nuclear Fuel and Waste Management Company.

REFERENCES

(1)   I. Neretnieks, J. Geophys. Res., **85**, 4379, (1980)

(2)   J. Byegård, G. Skarnemark and M. Skålberg, Mat. Res. Soc. Symp. Proc. **353**, Pittsburgh, PA 1995, pp. 1077-1084

(3)   Bradbury, M.H. and Stephen, I.G. in <u>Scientific Basis for Nuclear Waste Management IX</u>, edited by L.O. Werme, (Mat. Res. Soc. Symp. Proc. **50**, Pittsburgh, PA, 1985) pp. 81-90.

(4)   Crank, J., The Mathematics of Diffusion, 2nd ed., Oxford University Press, NewYork, pp. 50-51, (1975)

(5)   Skagius, K., Neretnieks, I., Water Resources Res., **22:3**, 389-398, (1986)

(6)   D. E. Gray, *American Institute of Physics Handbook*, McGraw-Hill , New York (1972)

BEHAVIOUR OF PLUTONIUM INTERACTING WITH BENTONITE AND SULFATE-REDUCING ANAEROBIC BACTERIA

A. KUDO[*#], J. ZHENG[*], I. CAYER[*], Y. FUJIKAWA[**], H. ASANO[***], K. ARAI[***]
H. YOSHIKAWA[****], M. Ito[****]
*National Research Council of Canada, Ottawa, Ontario, Canada K1A 0R6
**Research Reactor Institute, Kyoto University, Kumatori, Osaka-fu, Japan 590
***Ishikawajima-Harima Heavy Industries Co. Ltd., Isogo-ku, Yokohama, Japan 235
****Power Reactor and Nuclear Fuel Development Co., Tokai-mura, Ibaraki, Japan 319
Presently Research Reactor Institute, Kyoto University, Kumatori, Osaka-fu, Japan 590

ABSTRACT

The interactions between sulfate reducing anaerobic bacteria and plutonium, with or without bentonite present, were investigated using distribution coefficients {Kd (ml/g)} as an index of the radionuclide behaviour. Plutonium Kds for living bacteria varied within a large range, from 1,804 to 112,952, depending on the pH, while the Kds ranged from 1,180 to 5,931 for dead bacteria. In general, living bacteria had higher plutonium Kds than dead bacteria. Furthermore, the higher Kd values of 39,677 to 106,915 for living bacteria were obtained for a pH range between 6.83 and 8.25, while no visible pH effect was observed for dead bacteria. These Kd values were obtained using tracers for both ^{236}Pu and ^{239}Pu, which can check the experimental procedures and mass balance.

Another comparison was conducted for plutonium Kd values of mixtures of living bacteria with bentonite and sterilized bacteria with bentonite. The range of Kd values for the non-sterilized bacteria with bentonite were 1,194 to 83,648 while Kd values for the sterilized bacteria with bentonite were from 624 to 17,236. Again, the Kd values for the living bacteria with bentonite were higher than those of sterilized bacteria with bentonite. In other words, the presence of living anaerobic bacteria with bentonite increased, by roughly 50 times, the Kd values of ^{239}Pu when compared to the mixture of dead bacteria with bentonite. The plutonium Kd values for bentonite alone, both non-sterilized and sterilized, were within a constant range of around 10,000 even though some of the data are not yet available.

The bentonite used for this experiment was a product of Japan and the sulfate reducing anaerobic bacteria was previously used for the treatment of a pulp and paper wastewater. The results indicate that the effects of anaerobic bacteria within the engineered barrier system (in this case bentonite) will play a significant role in the behaviour of plutonium in geologic repositories.

INTRODUCTION

Many countries have been developing technology for the disposal of high level radioactive wastes in deep underground geologic formations. Based on Japan's "Long -term program for development and utilization of nuclear energy", Japanese high level radioactive wastes produced by reprocessing plants will be disposed of in a geologic repository. And in the engineered barrier system, bentonite will be used to minimize the movement of radionuclides by various forces, such as groundwater [1-3]. The repositories will be secured for a period of more than 10,000 years. During this period groundwater will penetrate the engineered barrier system and anaerobic bacteria can grow within the deep underground environment, interacting with long half-life radionuclides such as ^{239}Pu [4-9].

It is well understood that the underground disposal of radioactive wastes requires an understanding of the environment in which the waste will be stored. The facility and technique

Mat. Res. Soc. Symp. Proc. Vol. 465 © 1997 Materials Research Society

used should be designed and manufactured to meet the requirements of ensuring and maintaining a safe operation for future generations. Engineering barrier systems, used to prevent the movement of radionuclides, have long been investigated by many researchers[10-12]. Facilities and techniques, for the same purpose, have also been studied vigorously, however, few papers can be found which discuss the effects of sulfate reducing anaerobic bacteria in geologic repositories of high level radioactive wastes.

This paper reports preliminary experimental results of an important transuranium radionuclide, ^{239}Pu, interacting with bentonite, which would be used as an engineering barrier system at the disposal site. It is assumed that the bentonite would provide anaerobic conditions over centuries because of a lack of oxygen, thus allowing anaerobic bacteria to flourish within the bentonite. This investigation explores the possibility that bentonite would be contaminated by anaerobic bacteria, especially sulfate reducing bacteria, which could play an important role in the behaviour of ^{239}Pu at the disposal sites for centuries to come.

EXPERIMENT

The bentonite used in this experiment originated in a Japanese mine and was not treated before use. The bentonite was used as a dilute mixture with distilled deionized water at a concentration of 0.1% (dry weight). There was some quartz in the bentonite, which was not suspended in the dilute mixture. In order to produce a homogenous bentonite-water mixture, the mixture was well shaken before use. The x-ray diffraction analysis, x-ray photo spectroscopy analysis and other analyses indicated that the main components of the bentonite were quartz and montmorillonite and minor components such as beidellite.

The anaerobic sulfate reducing bacteria used in this experiment were originally used for the treatment of a pulp and paper wastewater, and were in granular form. The gases produced during the treatment of the wastewater were H_2S, CH_4, CO_2 and H_2. The bacteria granules were homogenized, and diluted into distilled deionized water after expelling dissolved oxygen by pure N_2 gas. The concentration of the anaerobic bacteria was adjusted to 0.1% (dry weight). In order to compare the Kd (ml/g) values of the living and dead anaerobic bacteria, an autoclave was used for 20 minutes at 120°C to sterilize the mixtures containing sulfate reducing anaerobic bacteria. Each of six different combinations in distilled water; sterilized and non-sterilized anaerobic bacteria, bentonite, and a mixture of anaerobic bacteria and bentonite, were mixed with an exact amount of ^{239}Pu. The reaction period between ^{239}Pu and the distilled water mixtures, after adjusting pH, was 4 hours.

To obtain the distribution coefficients for ^{239}Pu, membrane filters with a pore size of 0.20 μm were used to separate ^{239}Pu in distilled water mixtures into solid and liquid phases. The mixtures containing bacteria, bentonite or both were filtered and the glass container was rinsed with 2 ml of distilled deionized water and filtered. These filtered solution were treated as the liquid phase. The solid phase was the contents of the actual membrane filter and the remaining solids on the reaction vessel wall and other glassware. These walls were initially rinsed with 3 ml of (3+2) HNO_3, then 2 ml of (3+2) HNO_3 and the ^{239}Pu contained in the rinsed solution was classified as solid phase ^{239}Pu.

The statistical error for the distribution coefficients was less than 5% with a confidence level of 90%. The range of the errors for each value indicated in Figures 1, 2 and 3 is not visible because the scale of the distribution coefficients were expressed as a logarithmic scale.

The analysis of ^{236}Pu and ^{239}Pu was conducted after long chemical extraction, concentration, purification and electro-plating to a polished stainless steel disc, following an analytical method developed by, and combined with, methods used in the USA, Germany, and Japan[13-18]. The electro-plated plutonium samples were analyzed using alpha spectrometry. The certified internal standard, ^{236}Pu, was purchased from Harwell, United Kingdom.

RESULTS

Figure 1 shows the Kd of ^{239}Pu interacting with non-sterilized and sterilized sulfate reducing anaerobic bacteria for pHs ranging from 1.74 to about 13.0. The Kd of non-sterilized bacteria was high as opposed to the Kd of sterilized bacteria, which was low. Apparently, there were some active reaction sites which increased Kd values for the living bacteria, over much of this pH range while there was not much change in the Kd for sterilized bacteria over this pH range, Fig. 1. The coefficients for non-sterilized bacteria sharply increased to 81,853 at pH 5.86 and to more than 100,000 around the neutral pH area, although the coefficients stayed within a range of a few thousand for sterilized bacteria. The effect of the living bacteria on Kd is maximized in the neutral pH area and decreases towards the two pH extremes, where there is no significant difference in distribution coefficients between sterilized and non-sterilized bacteria. In an extremely harsh environment, pH below 3.01 or above 12.0, however, there was no significant difference in the Kd of ^{239}Pu between non-sterilized and sterilized anaerobic bacteria. The Kd ranged from 1,804 to 5,266 in the low pH region to about 8,000 in the high pH region, for both non-sterilized and sterilized anaerobic bacteria, Fig. 1.

There is a decrease in the Kd at pH 6.83 for the non-sterilized bacteria, and a significant increase at pH 6.83 and 7.35, for sterilized bacteria. At present, these phenomena can not be explained. Further investigations are warranted to pinpoint the cause of this interesting phenomenon.

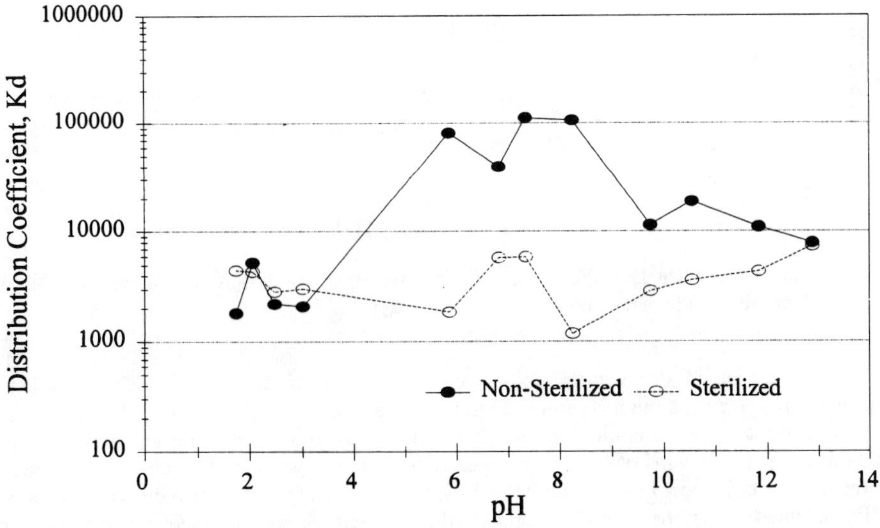

Fig. 1 Changes in ^{239}Pu Distribution Coefficient with pH for an Anaerobic Bacteria (0.1%) Suspended in Distilled Water.

The effects of the living (non-sterilized) anaerobic bacteria on the distribution coefficients [239]Pu were further investigated for a mixture of anaerobic bacteria and bentonite, Fig. 2. For the distilled water mixture, the addition of bentonite does not really affect the sorption capacity of the bacteria for [239]Pu. The Kd of the mixture of non-sterilized bacteria and bentonite was higher than the Kd of sterilized bacteria and bentonite. The Kd values for the non-sterilized mixture ranged from 51,091 to 83,648 for a pH range of 5.53 to 8.62, while the Kds for the sterilized mixture ranged from 1,101 to 10,386. The distribution coefficient for the non-sterilized mixture was roughly 50 times higher than that of the sterilized mixture for this pH range, which was a comfortable environment for the anaerobic bacteria. These [239]Pu Kd values behaved similarly to those for the anaerobic bacteria mixtures alone.

The addition of bentonite to the bacteria did not change the Kd values in a harsh environment, i.e., at a pH lower than 2.26 and higher than 12.77. Therefore, when the environment became extremely acidic or alkaline, the distribution coefficient for both the non-sterilized and sterilized solution reached a similar value. This is further evidence that there is a definite biological contribution for the distribution coefficient of [239]Pu in the mixtures studied.

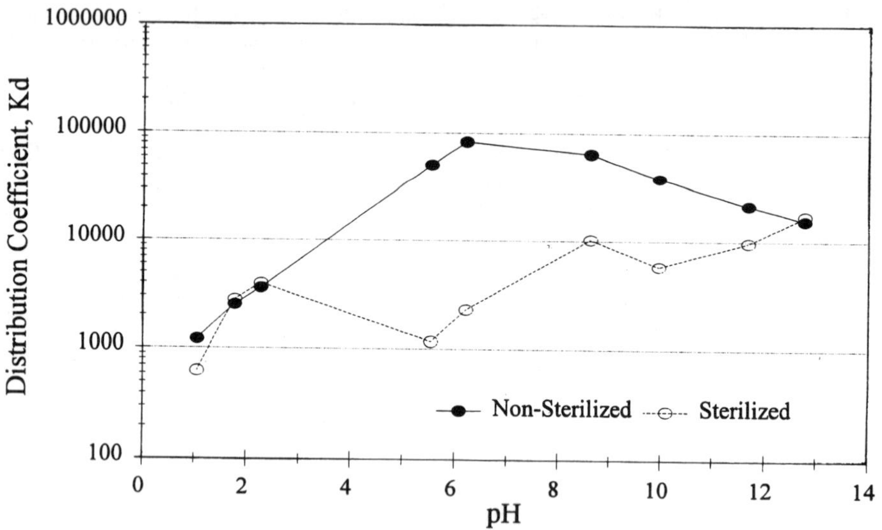

Fig. 2 Changes in [239]Pu Distribution Coefficient with pH for Anaerobic Bacteria (0.1%) and Bentonite (0.1%) Suspended in Distilled Water.

The effects of sterilization (autoclave) of bentonite on the Kd of [239]Pu was studied by changing the pH, even though complete data are not yet available. The Kd of [239]Pu for both non-sterilized and sterilized bentonite were within a constant range of a few thousand, around a Kd of 10,000, except at a low pH of 1.02. The values ranged from several thousand to twenty thousand at the highest pH range, Fig. 3. In other words, the difference in the distribution coefficients of [239]Pu obtained for the non-sterilized and sterilized anaerobic bacteria were not caused by the sterilization of the bentonite but rather by the anaerobic bacteria, either living or dead.

882

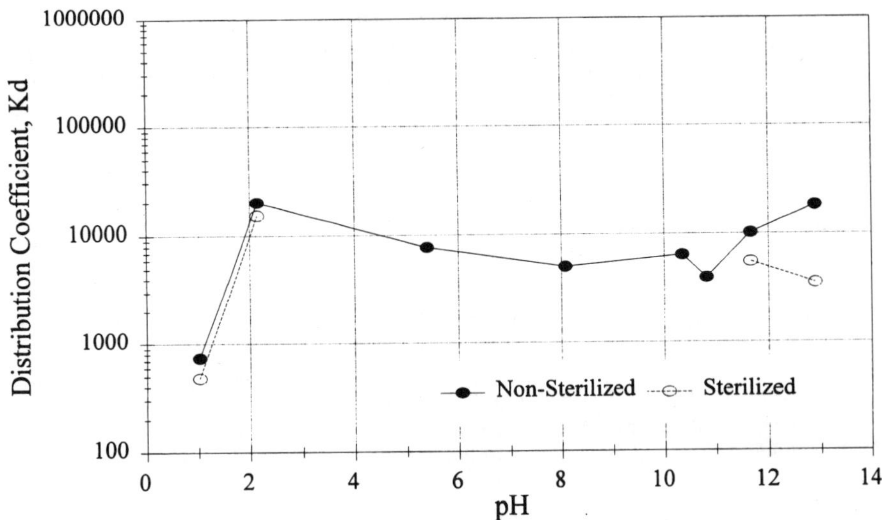

Fig. 3 Changes in [239]Pu Distribution Coefficient with pH for Bentonite (0.1%) Suspended in Distilled Water.

It seems that almost any type of bacteria could be present at the man-made geologic disposal site because the processes used in the bentonite back filling will definitely introduce all types of bacteria. However, bacteria may not be able to penetrate through the swelled compacted bentonite[19], thus the movement of plutonium would be greatly reduced because of the large Kd obtained for living bacteria by this investigation. The effects of the bacteria, including enhancing the mobility of radionuclides , should be investigated further to establish a safe geologic disposal technology for high level radioactive wastes.

CONCLUSIONS

Based on the preliminary experimental results reported in this paper, the following conclusions were obtained. The living bacteria played a considerable role in interacting with plutonium. The Kd values for the living anaerobic bacteria ranged from 39,677 to 106,915 between neutral pH values of 6.83 and 8.25, while the values for sterilized bacteria ranged from 1,180 to 5,931 in this pH range, Fig.1.

For the mixture of bentonite and bacteria, the plutonium Kd values ranged from 51,091 to 83,648 for the living bacteria while they were between 1,101 and 10,386 for the sterilized bacteria, for relatively comfortable pH values between 5.53 and 8.62, Fig. 2. The interaction of plutonium with bentonite was not affected by sterilization even though data were not fully available to support this conclusion, Fig. 3.

ACKNOWLEDGEMENTS

The authors thank Drs. D.C. Santry, D.L. Singleton and T. Kauri and Mrs. P. Zhang and other staff of the National Research Council of Canada for their contribution. A part of this investigation was financially supported by the Ministry of Foreign Affairs (JSTF) of Canada.

REFERENCES

1. H. Kimura, J. Nuclear Science and Technology, **32** (5), pp. 439-449 (1995).
2. M. Tsukamoto, Radiochimica Acta, **66/67**, pp. 397-403 (1994).
3. Y. Sakamoto, Radioactive Waste Management and the Nuclear Fuel Cycle, **15** (1), pp. 13-25 (1994).
4. G.S. Strobel, Measurements of the Resistive and Capacitive Properties of a Bentonite Clay/ Sand Mixture, (AECL-10668, COG-92-195, Geotechnical Science and Engineering Branch, Whiteshell Laboratories, Manitoba, Canada, 1994) pp. 1-23.
5. H.D. Sharma and D.W. Oscarson, Diffusion of Plutonium in Mixtures of Bentonite and Sand at pH 3, (AECL-10435, Geotechnical Science and Engineering Branch, Whiteshell Laboratories, Manitoba, Canada, 1991) pp. 1-15.
6. B. Torstenfelt, Chemistry and Mobility of Radionuclides in Geologic Environments, (Chalmers University of Technology, Göteborg, Sweden, 1983).
7. S. Fukunaga, H. Yoshikawa, F. Fujiki and H. Asano. Experimental Investigation on the Active Range of Sulfate-Reducing Bacteria for Geological Disposal (Mater. Res. Soc. Proc. **353**, 1995) pp. 173-180.
8. R.E. Wildung and T.R. Garland, Applied and Environmental Microbiology, **43** (2), 418-423 (1982).
9. T. Kauri, T. Kauri, D.C. Santry, A. Kudo and D.J. Kushner. Environmental Toxicology and Water Quality, **6** (1) pp. 109-112 (1991).
10. Z. Jishu and X. Deying. J. Nuclear and Radiochemistry, **14** (1), pp. 49-57 (1992).
11. B. Torstenfelt, Radiochimica Acta, **44/45**, pp. 111-117 (1988).
12. Y. Albinsson, Radiochimica Acta, **52/53**, pp. 283-286 (1991).
13. Y. Mahara, A. Kudo, T. Kauri, D.C. Santry and S. Miyahara, Health Physics, **54**, 107-111 (1988).
14. A. Kudo, Y. Mahara, T. Kauri, and D.C. Santry, Water Science and Technology, **23**, 291-300 (1991).
15. A. Kudo, Y. Mahara, D.C. Santry, S. Miyahara and J-P. Garrec, J. Environmental Radioactivity, **14**, 305-316 (1991).
16. A. Kudo, T. Suzuki, D.C. Santry, Y. Mahara, S. Miyahara, M. Sugahara, J. Zheng and J-P. Garrec, J. Environmental Radioactivity, **21**, 55-63 (1993).
17. A. Kudo, Y. Mahara, D.C. Santry, T. Suzuki, S. Miyahara, M. Sugahara, J. Zheng and J-P. Garrec, J. Applied Radiation and Isotopes, **46**, 1089-1098 (1995).
18. A. Kudo, R.M. Koerner, D.A. Fisher, J. Bourgeois, D.C. Santry, Y. Mahara and M. Sugahara, in Ice Core Studies of Global Biogeochemical Cycles, edited by R.J. Delmas (Springer-Verlag Berlin Heidelberg, Germany, 1995) pp. 417-427.
19. S. Kurosawa, M. Yui and H. Yoshikawa, Experimental Study of Colloidal Filtration by Compacted Bentonite (Mater. Res. Soc. Proc. **in Ppress** , 1997).

MONTE CARLO SIMULATION OF RADIOACTIVE CONTAMINANT TRANSPORT IN FRACTURED GEOLOGIC MEDIA: DISORDER AND LONG-RANGE CORRELATIONS

Sumit Mukhopadhyay and John H. Cushman
Center for Applied Mathematics, Purdue University, West Lafayette, Indiana 47907-1395.

Abstract

The geologic media near Yucca mountain site consist of fractured welded tuffs along with less fractured unwelded tuff. Numerical simulation of flow and transport in such media poses a number of challenging problems, due mainly to the heterogeneities and disorder in the media. In addition, because of different dominant transport mechanisms in different regions of the media, investigations need to be carried out at different time-scales. "Time-marching" will pose a considerable problem in analyzing such multi-scale transient problems. We develop a field-scale network model of fractures and study transport of radionuclides through geologic media as a function of disorder and correlated fracture-permeabilities.

Introduction

The possible leakage and dissolution into groundwater of radioactive materials from the proposed high-level waste repository near Yucca mountain has necessitated a thorough investigation of transport in fractured media. It has been argued that the geology at Yucca mountain would retard the movement of the radioactive materials, minimizing the risk to the environment. However, the geologic strata near Yucca mountain contain an irregular network of fractures [1, 2]. This fracture network provides a fast route for the contaminants to move. In addition, it has been observed that the geologic media are heterogeneous and highly disordered. The heterogeneity interacts with the flow pattern, and subsequently with the spread of the contaminants, in a complex fashion. In a disordered geologic medium, such as the one near the proposed repository, the wide variation in morphology (such as the orientation of the fractures and their lack of connectivity) causes the contaminants to follow paths of greatly varying lengths. This, compounded by the effect of local pressure gradients caused by local variation in hydraulic properties (such as shape, size, and hydraulic conductivity of an individual fracture), makes for a chaotic local velocity field [3, 4]. Thus the contaminant particles, which are carried across the medium by this underlying flow field, experience a broad variation of residence times, *i.e.*, the time over which an individual particle resides inside the medium changes widely, resulting in a very complex phenomenon.

The modeling effort is further complicated by additional factors. Geologic media show strong correlations in their hydraulic properties, and only limited data are available on such correlations. Also, due to the presence of both the fractures (providing relatively faster transport by advection and dispersion) and a porous rock matrix (transport is mainly by slow dispersion), two distinct time-scales for transport operate [5]. Any model of transport in fractured media has to capture both these short-term and long-term behaviors of the system. This point can never be overemphasized, particularly when we are considering transport of radioactive contaminants with a wide range of half-lives.

Our goal here is threefold: Firstly, we want to investigate the influence of the various factors (retardation by adsorption on the fracture surface and in the matrix; radioactive decay;

885

dispersion) for a given geology (permeability and size distribution; background flow-field) on transport. Secondly, we study the effect of the structure itself on the overall transport. We are specifically interested in investigating the influence of long-range correlations in the hydraulic properties on the transport processes. Thirdly we study the effect of disorder in the connectivity of the fractures and its impact on the overall transport of contaminants.

We model the disorder in the fracture network by utilizing some concepts of percolation. There are many reasons for investigating percolation in the context of contaminant transport by groundwater. The single most important characteristic of a percolation system is that it has a critical point, the percolation threshold p_c, above which the system permits transport across the system. Below the critical point, no transport is possible. On or near p_c, the network is very poorly connected, and its topology can greatly affect the residence time distribution - owing to the existence of the many slow paths that the particles can take. This essentially captures the physics of the problem. Thus, if we represent the transport process through a distribution of residence times, the structure of the distribution and its various moments can yield information on both the morphology of the medium and the mixing process [6]. In addition, recent results [7, 8] indicate that at *large length scales* (of the order of a kilometer or more), the fractured network of heterogeneous rocks is fractal, and is similar to the sample-spanning percolation cluster at p_c. Near the percolation threshold, the percolation system itself behaves like a fractal object at all length scales [9].

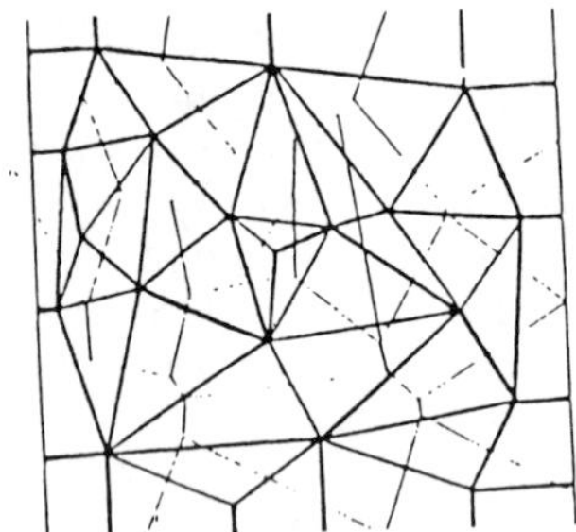

Figure 1: A Voronoi network.

Figure 2: A Square network close to percolation threshold.

Fracture networks embedded in a geologic medium have a random connectivity, *i.e.*, the number of fractures that a particular fracture connects to is a distributed quantity, and it varies from point to point. We can generate such a network of fractures using a complicated algorithm. An example will be the Voronoi network, which is shown in Fig. 1. However, previous work [10] has shown that, as long as the connectivity of a regular network is equal to the average connectivity of an irregular network, their transport properties will be similar. This allows us to model a fracture network with a regular structure, and then introduce disorder through some simple algorithm, such as percolation. A comparison of the Voronoi network and a square network close to p_c (see Fig. 2) reveals that the later can indeed represent highly disordered geologic media.

In classical random percolation [11], one blocks the fractures at random. Physically, this implies that even the highest permeable fracture may not be available to flow. This, however, goes against the real-life situation, where the lowest permeable fractures are the least likely to permit flow. Thus, if any fracture is to be blocked to flow, it should be the least permeable one. We adopt this approach here, where the fractures are blocked (assigned zero permeability) progressively, starting with the least permeable bonds. This is a new kind of percolation, and assists in explaining various physical phenomena which cannot be explained with regular percolation [12, 13, 14].

Our objective here is to present a field-scale network model of flow and transport of radioactive contaminants in a fractured network embedded in a rock matrix. Network models have been widely used for investigating flow and transport in porous media (for a review see [9, 15]). Contaminant transport in the discrete fracture network has been analyzed by many researchers with a particle tracking technique (see [4] and references therein). The effect of percolation disorder on transport was studied by Sahimi and Imdakm [3] and Koplik *et al.* [6]. These authors were concerned mainly with the convective-diffusive transport at the molecular level. However, we intend to move beyond the molecular-level simulations, and employ the methodology of network simulation in investigating the transport of contaminants in a field-scale medium.

Temporal moments

It is assumed that the transport process in an individual fracture ij (connecting node i to node j) of the network is governed by a CDE [16]. The description of the transport processes at the network level is then developed by assuming perfect mixing at every node of the network. The transport of contaminant through a single fracture with one-dimensional flow is written as

$$\frac{\partial c_{ij}}{\partial t} + \frac{V_{ij}}{R_{ij}} \frac{\partial c_{ij}}{\partial z} - \frac{D_{ij}}{R_{ij}} \frac{\partial^2 c_{ij}}{\partial z^2} + \lambda c_{ij} = 0 \quad . \tag{1}$$

where z is the coordinate along the fracture axis, t is the time, R is a retardation coefficient [17], λ is the radioactive decay constant, and the subscript ij refers to quantities pertaining to the channel between node i and node j, and the direction of flow is assumed to be from i to j. The hydrodynamic dispersion coefficient D is expressed as $D_{ij} = \alpha_L V_{ij} + D^*$ [18], where α_L is the dispersivity, and D^* is the molecular diffusion coefficient in water. Without loss of generality we will henceforth neglect D^*, which accounts for transport primarily on the molecular level.

Using this equation for the transport of the contaminant through the fracture ij, assuming an initial condition of $c(0,t) = 0$, we can write an expression for the mass influx of contaminant through fracture ij in Laplace-transform space,

$$\bar{m}_{ij} \mid_{z=0} = \frac{V_{ij}}{2} \left[\beta_{ij} \coth \phi_{ij} + 1 \right] \bar{C}_i - \frac{V_{ij}}{2} \left[\frac{\beta_{ij} e^{-\nu_{ij} l_{ij}}}{\sinh \phi_{ij}} \right] \bar{C}_j = \bar{Q}_{1_{ij}} \bar{C}_i - \bar{Q}_{2_{ij}} \bar{C}_j \quad , \tag{2}$$

where $\nu_{ij} = V_{ij}/(2D_{ij})$, $\beta_{ij} = \sqrt{1 + \frac{4R_{ij}D_{ij}S}{V_{ij}^2}}$, $\phi_{ij} = \nu_{ij} l_{ij} \beta_{ij}$, and $S = s + \lambda$. The symbols with an overbar refer to quantities in Laplace space where s is dual of time. We need to mention here that s is in general a complex number. This will render both β_{ij} and ϕ_{ij} complex numbers. A similar expression can be developed in Laplace space for the flux of contaminant leaving the channel ij,

$$\bar{m}_{ij}\mid_{z=l_{ij}} = -\frac{V_{ij}}{2}\left[\beta_{ij}\coth\phi_{ij} - 1\right]\bar{C}_j + \frac{V_{ij}}{2}\left[\frac{\beta_{ij}e^{\nu_{ij}l_{ij}}}{\sinh\phi_{ij}}\right]\bar{C}_i = \bar{Q}_{3_{,j}}\bar{C}_i - \bar{Q}_{4_{,j}}\bar{C}_j \quad . \tag{3}$$

Note that $\bar{C}_i(s)$ and $\bar{C}_j(s)$ are not yet known, and depend on the imposed boundary conditions.

Considering the entire network of fractures, and assuming perfect mixing at the nodes, we can write $\sum_{i,j}\bar{m}_{ij} = 0$, where the summation is taken over all the available neighboring fractures of a particular node. For the particular case, where a pulse of tracer is injected at the inlet of the network and a vanishing concentration gradient is maintained at the outlet, it can be easily shown that the *first-passage probability density function* (FPPDF) [19] $M(t)$ is just the flux leaving the network, *i.e.*, in Laplace space $\bar{M}(s) = \sum_j \bar{m}_{Oj}$, where the summation is over all the nodes residing on the outlet plane of the network. The various transit moments are then easily calculated by using the following equations.

$$\langle t^n \rangle = \int_0^t t^n M(t)dt = (-1)^n \frac{d\bar{M}(s)}{ds}\mid_{s=0} \tag{4}$$

Simulation strategy

The hydraulic conductivity field

We start our simulation by assigning hydraulic conductances g_{ij}s, and widths b_{ij}s to each fracture of a two-dimensional square network. Experimental evidence [20] has confirmed that the permeability and the channel-aperture of a fracture network often follow a fractional Brownian motion (fBm), and that they exhibit fractal characteristics. The fBm is best described in terms of a Hurst's exponent H. A distribution is positively correlated for $H > 0.5$, and negatively correlated for $H < 0.5$. $H = 0.5$ represents an ordinary Brownian motion. We investigate three different cases: A network whose all the fractures have the same g_{ij} and b_{ij}, a network where uniformly random values are assigned to these quantities, and a network where these quantities follow a fBm. While the uniform random distribution was generated using a pseudo-random number generator, DURAND, the fBm distribution was generated using a spectral synthesis scheme [15] that employs a discrete fast Fourier transform in two dimensions.

Initially the network is perfectly connected (we will call this a full network). We next introduce disorder in the network by blocking a certain fraction of the flow-paths by arranging the permeabilities in an ascending order and then blocking those paths that have low permeabilities, *i.e.*, the path with the smallest permeability is selected first and a zero permeability is assigned to it. This process is then carried out until a pre-assigned fraction of the fractures has zero permeability.

The background flow-field

We maintain some equipotential at the extreme left edge of the network and a lower equipotential at the extreme right of the network. The top and bottom edges are no-flow boundaries. By drawing a parallel with current flow in a network of resistors, we can assume that volumetric flow rate at each node is conserved ($\sum_{i,j} Q_{ij}$), which yields

$$\sum_{i,j} \frac{k_{ij}a_{ij}}{\mu l_{ij}}(P_i - P_j) = 0 \quad , \tag{5}$$

where the symbols have their usual meanings. The summation is over all the *available* channels of the network. The set, Eqn 5, is solved by using a biconjugate gradient method, yielding the pressures at each node of the network. The fluid velocities in each fracture are then calculated as $V_{ij} = -\frac{k_{ij}}{\mu l_{ij}}(P_i - P_j)$. Depending upon the pressures at the nodes, V_{ij}s can be either positive or negative.

The mass balance

The direction of V_{ij} determines the direction of contaminant movement. A flux entering a fracture is assumed to be positive, and a flux leaving the fracture is treated as negative. If V_{ij} is zero, no flux (of contaminants) enters that fracture. The set of linear equations is solved as above for the nodal concentrations $\bar{C}_i$ in the Laplace space. The sum of the flux leaving the network is then calculated, and $\bar{M}(s)$ is constructed.

The moments in real time

The conversion to real time is performed using Crump's[21] algorithm, along with a partial sums algorithm [22] to accelerate convergence. Because this inversion algorithm requires knowledge of the value of the transformed variable for different values of the Laplace parameter $s = s_k$, $k = 0, 1, 2...K$, the system of equations must be solved anew for each s_k. We have used $K = 50$, *i.e.* the set of equations above is solved for fifty different s_k values for each configuration of the network. This, though computationally demanding, makes our results free from any unexpected convergence problem.

Monte Carlo simulations

We have described the steps that constitute one realization of our simulation. Next we change the initial seed number, generate another configuration of the fracture-width and permeability distribution, and go through the entire computation to complete another realization. We perform calculations over a large number of such realizations, and the final result is the arithmetic average over all realizations. All simulations are carried out in a two-dimensional square network of size $L \times L$ nodes, where L lies between 4 and 128. Each fracture-channel is of length 10 meters, and the mean width of the channels is 50×10^{-6} meters. We also assume that the mean permeability of each channel is 500 md. A pressure gradient of $0.1172psi/m$ is maintained across the network. Simulations are carried out for various values of the dispersivity ($\alpha_L = 0.1$, 0.5, 1.0, 1.5, and 2.0 meters).

Results and Discussions

First we investigate the influence of long-range correlations on the transport process. To stress their importance, we consider a full network. In Fig. 3 we show the vertically averaged concentration of a radioactive contaminant with a half-life period of 10 years. The left-hand side of the network is maintained at a fixed concentration, and the concentrations shown are at 1500 days. We observe that the long-range correlations strongly affect the overall transport process and the shape of the contaminant front. Whereas transport in the presence of a fBm distribution lags that with a homogeneous property distribution, it moves ahead of that with uniformly random property distribution. This can be explained as follows. Recall H determines the extent of correlation. A large value of H represents a strong positive correlation, $H = 1$ being the compact homogeneous distribution, and $H = 0.5$ being the Gaussian distribution. These figures (having an underlying hydraulic structure with $H = 0.6$) show a transport behavior somewhere in between the homogeneous and Gaussian distribution. We expect with increase in H the concentration profile will move more towards that of the homogeneous distribution.

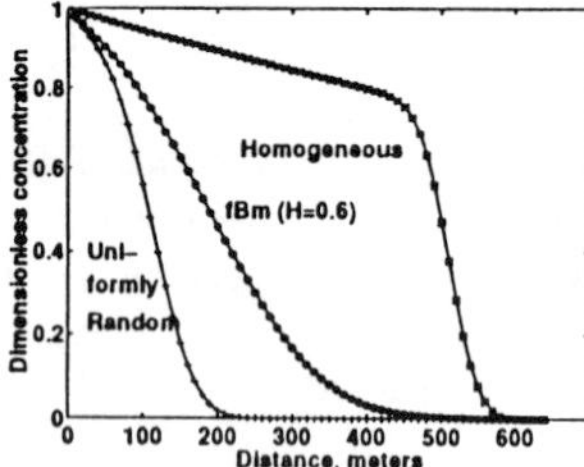

Figure 3: Effect of correlation: Small half-life.

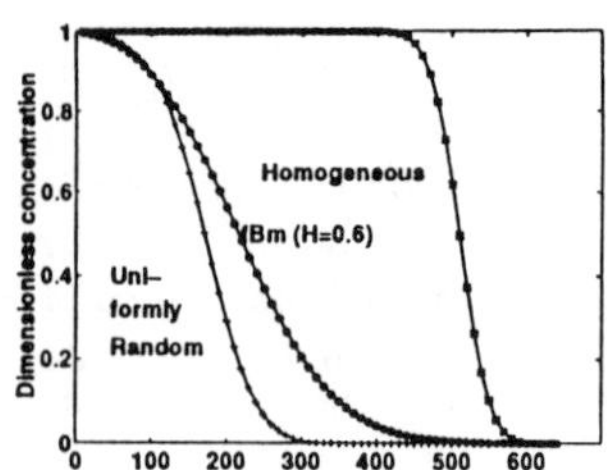

Figure 4: Effect of correlation: Large half-life

All parameters in Fig. 4 are as in Fig. 3, except here we choose a very *long* half-life. Observe that at any point, and at any time, for a given hydraulic field, the longer the half-life, the higher is the concentration in the network. Adsorption on the fracture surface also influences the propagation of the radioactive contaminants. This is shown in Fig. 5, which is similar to Fig. 3, except that here we introduce a strong retardation $(R = 5)$ due to adsorption on the fracture surface. Also in contrast to Figs. 3 and 4, the concentrations are shown at 3000 days. The larger retardation slows the propagation of the contaminants, and reduces the influence of long-range correlations. This is even more pronounced at shorter times (not shown). Due to the strong retardation, the contaminants move at a slower rate and hence do not experience much of the heterogeneity until later times.

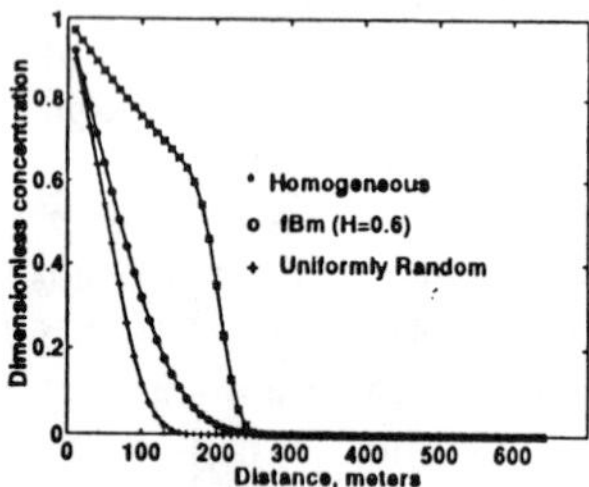

Figure 5: Effect of correlation: Retardation and decay.

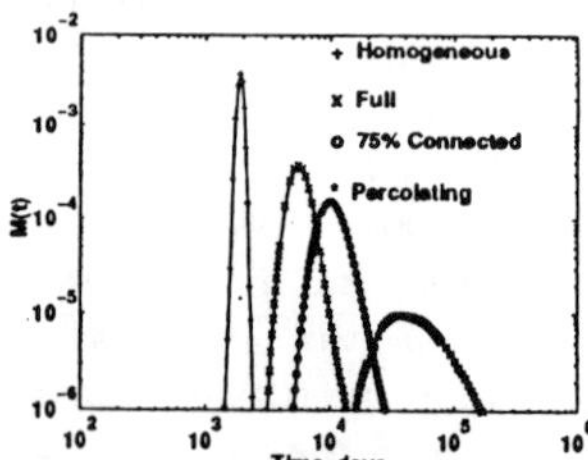

Figure 6: FPPDF without retardation and decay.

Having stressed the importance of long-range correlations on the fate of radioactive contaminants, we next analyze the effect of disorder on the transport process. In Fig. 6 we show the FPPDF for a square network of size 64 × 64 nodes. We see that in going from the homogeneous, to the full, to the 75% connected, to the system near p_c, the FPPDF becomes wider and wider. While the breakthrough in a homogeneous network happens earlier and the contaminant pulse moves out of the network over a relatively shorter period of time, it takes a longer time to breakthrough and move out completely in a percolating system. This is expected as the system near p_c has disorder and heterogeneities at many scales, thus slowing of the particles. These results pertain to the case where the injected pulse has an infinite half-life period and no adsorption on the fracture surface is considered. The FPPDFs for the above configurations, where the contaminants have a half-life period of 10 years, and when there is a retardation factor of 5, are shown in Fig. 7.

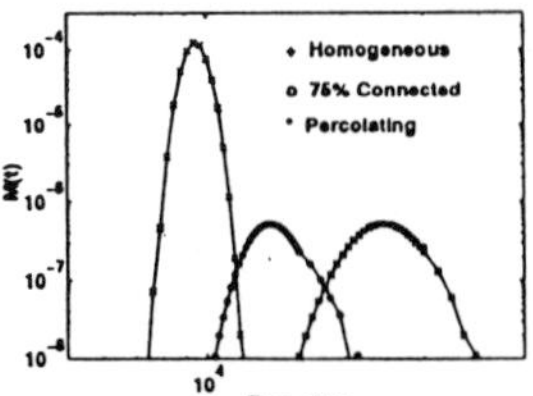

Figure 7: FPPDF with retardation and decay.

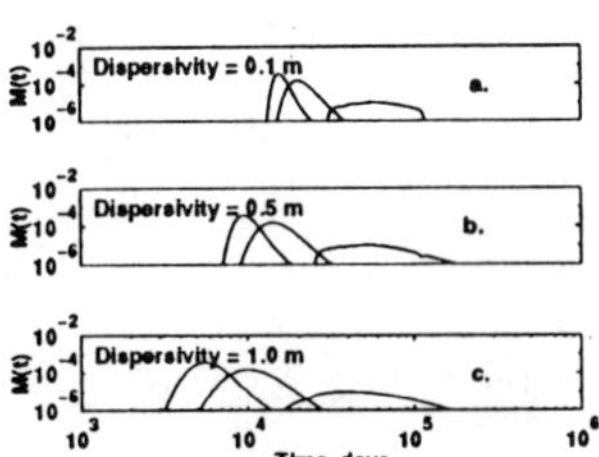

Figure 8: FPPDF for different dispersivity α_L in the fractures.

In Figs. 8a-8c we show the dependence of the FPPDF on dispersivity. The three plots correspond to $\alpha_L = 0.1$, 0.5, and 1.0 meters, respectively from top to bottom. The curves in each of these plots represent the three configurations (from left to right): a network with homogeneous properties, a full network, and a network close to p_c. All networks have 64×64 nodes. An increase in the value of α_L in an individual fracture results in a wider FPPDF. Observe that the first appearance of radioactive contaminant at the outlet happens earlier in a network with a higher dispersivity. In addition, observe that for a given α_L, the shape of the FPPDF flattens considerably as p_c is approached.

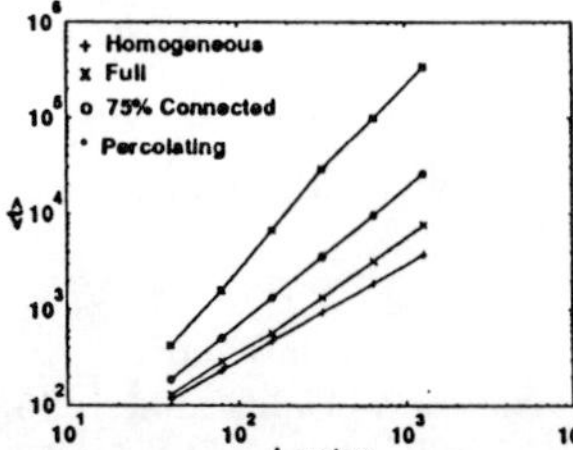

Figure 9: The first moment of the FPPDF as a function of the network size.

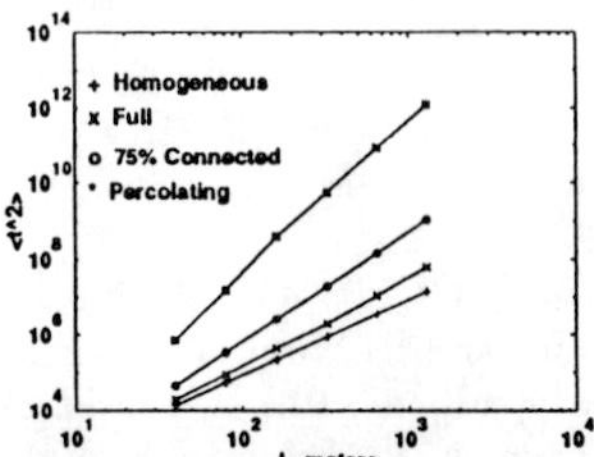

Figure 10: The second moment of the FPPDF as a function of the network size.

In Fig. 9 we plot the dependence of the first moment on the size of the network. The radioactive particles have a half-life of 10 years and the retardation factor is 5. We start with a network of some linear size L. We perform the Monte Carlo simulations as described above, and calculate the first moment. We next change the size of the network, and perform the same calculations. We show the first moment for four different configurations. It can be easily shown that the first moment in a homogeneous network changes linearly with size. Our simulation results agree with this prediction. A network near p_c shows a stronger dependence on size than a fully connected or partially connected network. This indicates that disorder strongly affects the residence time distribution, and the contaminant particles, on an average, reside longer in a network near p_c. Observe also that for smaller networks near p_c, the first moment does not show a power-law type dependence on size. This may be the result of the finite size of the network. Recall that the percolation threshold itself is a function of size, and a finite-size correction is in order for smaller networks. In Figure 10, we show the second moments as a function of network size for the above configurations. As in the case of the first moment, the second moment also depends strongly on size, especially near p_c.

Acknowledgments

This research has been supported by the U.S. Army Corps of Engineers Waterways Experiment Station under contract DACA 39-95-K-0056.

References

[1] P. Montazer and W. E. Wilson, Report **84-4345**, U. S. Geological Survey (1984).

[2] D. T. Hoxie, in: *GEOVAL-1990*, Stockholm, Sweden (1990).

[3] M. Sahimi and A. O. Imdakm, *J. Phys. A: Math. Gen.* **21**, 3833-3870 (1988).

[4] B. Berkowitz and Braester, C., *Water Resour. Res.* **27**, 3159-3164 (1991).

[5] E. A. Sudicky and R. G. McLaren, *Water Resour. Res.* **28**, 499-514 (1992).

[6] J. Koplik, S. Redner, D. Wilkinson, *Phy. Rev. A* **37**, 2619-2636 (1988).

[7] C. C. Barton and P. A. Hsieh, *Guidebook T385*, AGU, Las Vegas, Nevada (1989).

[8] M. Sahimi, M. C. Robertson, and C. G. Sammis, *Phys. Rev. Lett.* **70**, 2186-2189 (1993).

[9] M. Sahimi, *Rev. Mod. Phys.* **65**, 1393 (1993).

[10] G. R. Jerauld, J. C. Hatfield, L. E. Scriven, and H. T. Davis, *J. Phys. C* **17**, 1519-1529 (1984).

[11] D. Stauffer and A. Aharony, *Introduction to Percolation Theory*, 2nd edition, Taylor and Francis, London (1992).

[12] Prakash, S., Havlin, S., Schwartz, M., and Stanley, H. E., *Phys. Rev. A* **46**, R1724 (1992).

[13] M. Sahimi, *AIChEJ* **41**, 229-240, 1995b.

[14] M. Sahimi and S. Mukhopadhyay, *Phys. Rev. E*, 1995 (submitted).

[15] S. Mukhopadhyay, Ph. D. thesis, University of Southern California, 1995.

[16] D. H. Tang, E. O. Frind, and E. A. Sudicky, *Water Resour. Res.* **17**, 555-564 (1981).

[17] R. A. Freeze and J. A. Cherry, *Groundwater*, Prentice-Hall, Englewood Cliffs, N. J., 1979.

[18] J. Bear, *Dynamics of Fluids in Porous Media*, Elsevier, New York (1972).

[19] M. Sahimi, B. D. Hughes, L. E. Scriven, and H. T. Davis, *Chemical Engg. Sci.* **41**, 2103-2122 (1986).

[20] T. A. Hewett, paper SPE 15386 presented at the 1986 SPE Annual Technical conference and Exhibition, New Orleans (1986).

[21] K. S. Crump, *J. Assoc. Comput. Mach.* **23**, 89-96 (1976).

[22] J. R. MacDonald, *J. Appl. Phys.* **10** 3034-3041 (1964).

Effect of Dry Density on Activation Energy for Diffusion of Strontium in Compacted Sodium Montmorillonite

Tamotsu KOZAKI, Hiroki SATO, Atsushi FUJISHIMA, Nobuhiko SAITO, Seichi SATO, Hiroshi OHASHI

Division of Quantum Energy Engineering, Graduate School of Engineering, Hokkaido University, Sapporo, 060, Japan, kozaki@hune.hokudai.ac.jp

ABSTRACT

For performance assessments of geological disposal of high-level radioactive waste, activation energies for the diffusion of strontium ions and the basal spacings of compacted sodium montmorillonite in the water-saturated state were determined.

Basal spacings determined by XRD indicated changes in the interlamellar space from a three-water layer hydrate state to a two-water layer hydrate state as the dry density of the montmorillonite increased from 1.0 to 1.8 Mg m^{-3}. Activation energies from 17.3 to 30.8 kJ mol^{-1} for the apparent diffusion coefficients of strontium ions were obtained. The lower activation energies than for diffusion of strontium ions in free water were determined for montmorillonite specimens of lower dry density (1.2 Mg m^{-3} and below), while the higher activation energies were at higher dry densities (1.4 Mg m^{-3} and above).

These findings cannot be explained by changes in only the geometric parameters, which the pore water diffusion model is based upon. Possible explanations for the dry density dependence of the activation energy are the changes of the temperature dependence of the distribution coefficients and/or of the diffusion process with increasing dry density.

INTRODUCTION

Compacted bentonite is a promising candidate material for engineered barriers in geological disposal of high-level radioactive waste[1,2]. For performance assessment of geological disposal, it is necessary to study the migration behavior of radioactive nuclides in compacted bentonite. A pore water diffusion model, in which nuclides are considered to diffuse through pore water in the compacted bentonite, has been widely adopted to explain the migration behavior of the nuclides[3-5]. For highly sorbing cations, however, the cations migrate more rapidly than predicted by the pore water diffusion model and to explain this the contribution of surface diffusion has been proposed by

Mat. Res. Soc. Symp. Proc. Vol. 465 © 1997 Materials Research Society

several investigators[6-9].

The activation energy for diffusion of nuclides in compacted bentonite is an important parameter to clarify the diffusion process. In previous studies, we have determined the activation energies for the apparent diffusion of sodium[10] and cesium ions[11] in compacted sodium montmorillonite. It was found that the activation energies depended on the dry density of montmorillonite and were different from those in free water. In the present study, the apparent diffusion coefficients and their activation energies were determined for strontium ions in compacted sodium montmorillonite. The diffusion process is discussed with the results comparing the basal spacing of the montmorillonite obtained by X-ray diffraction.

EXPERIMENTAL

The sodium montmorillonite used in the present study was Kunipia-F, a product of Kunimine Industries Co. Ltd. The montmorillonite content of the Kunipia-F is 99%. The chemical composition of the Kunipia-F is given elsewhere[4]. Homoionic montmorillonite was prepared from the Kunipia-F by dispersing it in 1M NaCl solution for 24 h. The dispersion was repeated three times, with settling and exchanging of the supernatant with new NaCl solution. Excess salt was removed with a dialysis tube in distilled water, until no chloride ions were detected with $AgNO_3$ solution. The homoionized sodium montmorillonite was mortared and sieved to separate the fraction from 75 to 150 μm, after drying in an oven at 378 K.

One-dimensional non-steady diffusion experiments were carried out using ^{85}Sr as a radiotracer at temperatures from 278 to 323 K, after the homoionized montmorillonite was compacted to a prescribed dry density from 1.0 to 1.8 Mg m^{-3} in an acrylic resin cell and saturated with distilled water. The apparent diffusion coefficients were obtained from the concentration profiles of ^{85}Sr in compacted sodium montmorillonite. The radioactive tracer, ^{85}Sr, was obtained from the Japan Radioisotope Association (JRIA). The other chemicals used in the present study were reagent grade, and obtained from Kanto Chemical Co. Inc. The detailed procedures of the homoionization, the water saturation, and the diffusion experiments were described in earlier papers[10,11].

The basal spacing of the water-saturated montmorillonite specimens was determined from the X-ray diffraction (XRD) profiles measured in the range 2θ from 3 to 8 degrees with the Cu-Kα line. The compacted montmorillonite specimen in an acrylic resin tube, which was just removed from a water saturation cell, was placed in a sample holder together with the tube. The measuring time, including the time for taking the cell apart, was within approximately 5 minutes to avoid changes in the water content of the specimen.

The water content of the montmorillonite specimen (the weight of the water in the montmorillonite specimen $\times$ 100% / the weight of the water-saturated montmorillonite specimen) was determined by weighing the specimens before and after drying at 378 K for 24h.

RESULTS AND DISCUSSION

Figure 1 shows the X-ray diffraction profiles of water-saturated montmorillonite specimens with different dry densities. The diffraction peak around 1.88 nm, corresponding to the three-water layer hydrate[12], was observed at lower dry densities (1.0 and 1.2 Mg m^{-3}), and the diffraction peak around 1.56 nm, corresponding to the two-water layer hydrate, at higher dry densities (1.6 and 1.8 Mg m^{-3}). Both peaks were observed at the dry density of 1.4 Mg m^{-3}. The basal spacings obtained from the profiles are listed in Table 1, together with the values of the water content of the montmorillonite specimens. Average basal spacings were calculated from water contents in the same way reported by Madsen and Kahr[13] by assuming that the specific surface area of the montmorillonite specimen is 800 m^2 g^{-1} and that all water in the specimen is related to the mont-morillonite surfaces. The calculated values of the average basal spacing are also listed in Table 1. The average basal spacing was almost the same as the measured one at higher dry densities, but it was larger than that measured at lower dry densities. This suggests that there are large spaces such as pores in compacted bentonite only at lower dry densities.

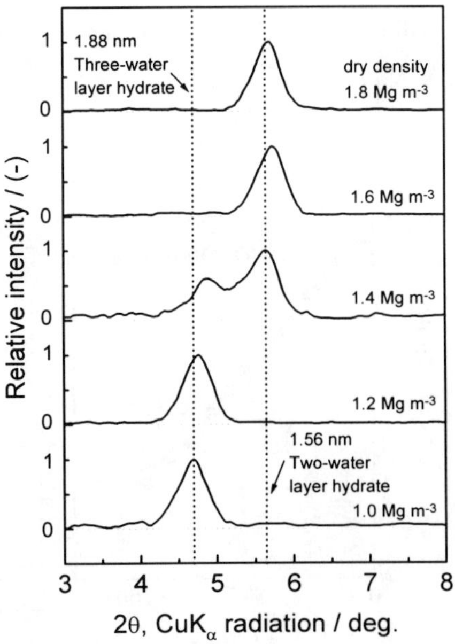

A typical concentration profile of ^{85}Sr in compacted sodium montmorillonite in a diffusion experiment is shown in Fig. 2. Assuming that ^{85}Sr is an infinitely thin source, the distribution of ^{85}Sr can be

2θ, CuK$_\alpha$ radiation / deg.

Fig.1 X-ray diffraction profiles for water-saturated Na-montmorillonite with different dry density

Table 1 Basal spacing and water content of water-saturated Na-montmorillonite
with different dry density

Dry density (Mg m^{-3})	Basal spacing (nm)		Water content (%)	Average basal spacing(nm)
1.0	1.87	-	38.1	2.54
1.2	1.85	-	31.9	2.17
1.4	1.80	1.57	27.2	1.93
1.6	-	1.54	22.6	1.73
1.8	-	1.53	19.3	1.60

expressed by the equation[14]:

$$C(x,t) = \frac{M}{2\sqrt{\pi D_a t}} \exp\left[-\frac{x^2}{4 D_a t}\right] \tag{1}$$

where D_a(m^2 s^{-1}) is the apparent diffusion coefficient of strontium ions in the compacted montmorillonite, C(counts m^{-3}) is the concentration of diffusing [85]Sr, t(s) is the diffusion period, x(m) is the distance from the montmorillonite surface on which [85]Sr was spiked, and M(counts m^{-2}) is the total quantity of the [85]Sr source per unit area. The apparent diffusion coefficients were determined by the least-squares fitting. The fitting line is in good agreement with the measured values, as seen in Fig. 2.

Figure 3 shows the dry density dependence of the apparent diffusion coefficients of strontium ions at 298 K by the present authors and elsewhere[4,5,7]. The values of the diffusion coefficients of strontium ions at 298 K are very close to those reported by others.

The activation energy was determined from the temperature dependence of the

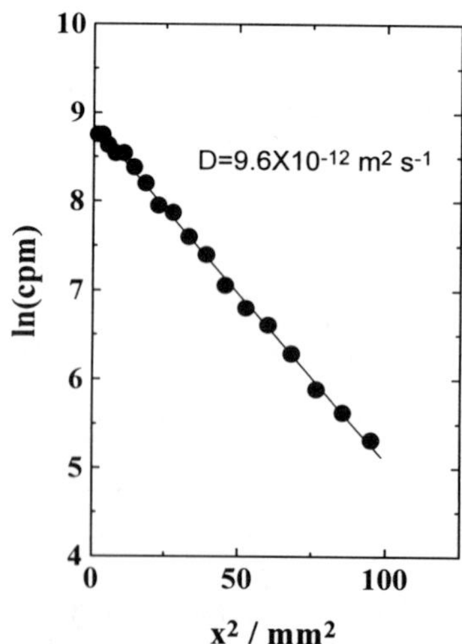

Fig.2 Concentration profile of [85]Sr in water-saturated montmorillonite compacted at dry density of 1.2 Mg m^{-3} at 298 K; diffusion period, 8.0 d

apparent diffusion coefficients of strontium for the compacted montmorillonite specimens in the water-saturated state. The Arrhenius plots of the apparent diffusion coefficients are shown in Fig. 4. The slope of the line of the least-squares fit yields activation energies for montmorillonite specimens with different dry densities. The activation energies obtained in this study are shown in Fig. 5 as a function of the dry density of the montmorillonite specimens, together with those of sodium[10] and cesium ions[11]. The activation energies obtained for the montmorillonite specimens with dry densities of 1.2 Mg m^{-3} and below were found to be lower than that for diffusion of strontium ions in free water, 20.1 kJ mol^{-1} [15], while higher activation energies were obtained for the montmorillonite specimens with dry densities of 1.4 Mg m^{-3} and above. It is possible that the lower activation energies were obtained for montmorillonite specimens with higher water contents and a basal spacing of approximately 1.88 nm (the three-water layer hydrate), and that the higher activation energies of the montmorillonite specimens with lower water content had the basal spacings of approximately 1.56 nm(two-water layer hydrate). A similar dependence of the activation energy has been found for sodium[10] and cesium ions[11]. These findings cannot be explained just by changes in the geometric parameters, which the pore water diffusion model is

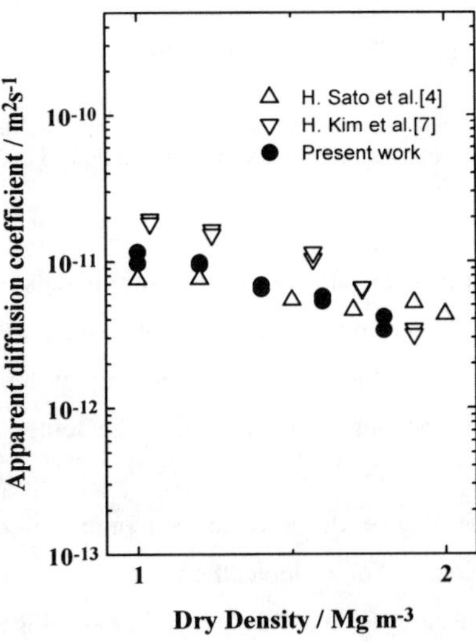

<table>
<tr><td>

Fig.3 Apparent diffusion coefficients of Sr as a function of dry density at 298 K by the present authors, ●; by H.Kim, et al.[7],▽; and at room temperature by H. Sato, et al.[4],△

</td><td>

Fig.4 Temperature dependence of apparent diffusion coefficients of Sr in compacted Na-montmorillonite with dry densities of 1.0 Mg m^{-3},●; 1.2 Mg m^{-3},▲; 1.4 Mg m^{-3},■; 1.6 Mg m^{-3},◆; 1.8 Mg m^{-3},▼

</td></tr>
</table>

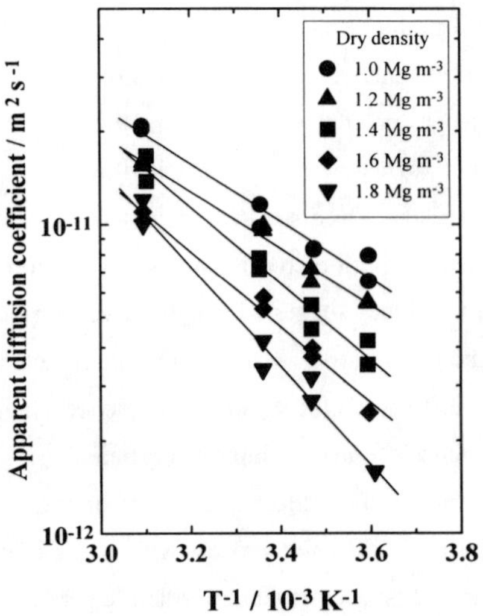

based on.

A possible explanation for the dry density dependence of the activation energy is the change in the temperature dependence of the distribution coefficients of strontium ions on the montmorillonite. According to the pore water diffusion model, high distribution coefficients reduce the apparent diffusion by a retardation factor. If the temperature dependence of the distribution coefficient increases with increasing dry density of montmorillonite, i.e. the distribution coefficient becomes higher at low temperatures and lower at high temperatures as the dry density of the montmorillonite increases, the result will be a higher activation energy, as the activation energies were determined from the apparent diffusion coefficients in the present study. Oscarson *et al*. reported that the distribution coefficients of cesium ions on compacted clay decreased with increasing dry density, and they suggested that the lower sorption can be attributed to small and occluded pores that Cs cannot enter[16]. If so, the change in activation energy cannot be explained only by the pore water diffusion model, since geometric changes in the pore will not affect the temperature dependence of the distribution coefficient.

Another possible explanation for the dry density dependence of the activation energy is changes in the diffusion process with increasing dry density. For example, the external surfaces and interlamellar spaces of montmorillonite particles can be considered alternative diffusion paths in the compacted montmorillonite. Madsen, *et al*. pointed out the strong dependence of the apparent diffusion coefficients of ions in compacted bentonite on the water content of the specimens, and suggested a contribution to the diffusion process of a three-water layer hydrate on the external surfaces of the montmorillonite [13]. Torikai, *et al*. have studied thermodynamic properties of water

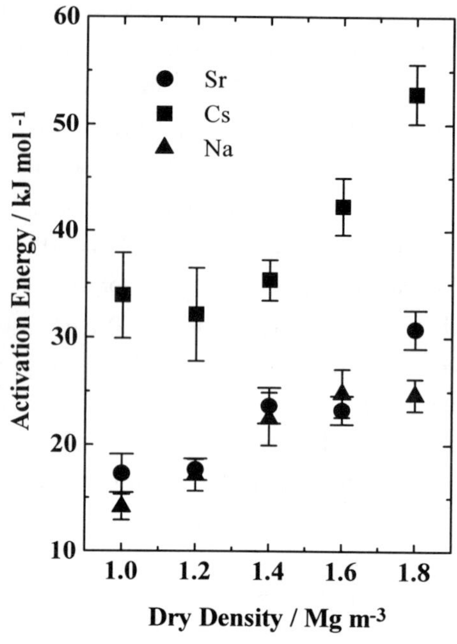

Fig.5 Activation energies for diffusion of Sr in compacted montmorillonite as a function of dry density

in compacted sodium montmorillonite (Kunipia-F)[17]. According to their work, the activity of water in compacted montmorillonite decreased with increasing dry density of montmorillonite, and the three-water layer hydrate disappeared at the dry density of 1.76 Mg m^{-3}, although it was observed at the dry density of 1.2 Mg m^{-3}. If strontium ions diffuse in the interlamellar spaces and/or on the external surfaces of montmorillonite particles, the change in the number of the hydrate water-layers will affect the diffusion process and a difference in activation energy should be observed. In addition, the correlation between changes in the basal spacings and the activation energy seems very possible, if the interlamellar spaces are the main diffusion path.

CONCLUSIONS

Basal spacings of the water-saturated montmorillonite were measured by XRD. A three-water layer hydrate was observed at lower dry densities (1.0 and 1.2 Mg m^{-3}) and a two-water layer hydrate at higher dry densities (1.6 and 1.8 Mg m^{-3}); both were observed at the dry density of 1.4 Mg m^{-3}.

Activation energies from 17.3 to 30.8 kJ mol^{-1} were obtained from the temperature dependence of the apparent diffusion coefficients of strontium ions. Lower activation energies than that for diffusion of strontium ions in free water were found for montmorillonite specimens at lower dry densities (1.2 Mg m^{-3} and below), while the higher activation energies were at higher dry densities (1.4 Mg m^{-3} and above).

These findings cannot be explained by changes in only the geometric parameters, which the pore water diffusion model is based upon. Possible explanations for the dry density dependence of the activation energy are changes in the temperature dependence of the distribution coefficients of strontium ions on the montmorillonite and/or of the diffusion process with increasing dry density.

ACKNOWLEDGMENTS

This work has been performed in part by using facilities of the Central Institute of Isotope Science, Hokkaido University. The financial support was provided by the Ministry of Education, Science, Sports and Culture, Japan (a Grant-in-Aid for Encouragement of Young Scientists, No. 08780462) and by the TEPCO Research Foundation.

REFERENCES

1. Power Reactor and Nuclear Fuel Development Corporation, PNC TN1410 93-059, 1992.

2. F. Bucher and M. Müller-Vonmoos, Appl. Clay Sci. **4**, 157 (1989).

3. B. Torstenfelt, SKB Tech. Rep. 86-14, 1986.

4. H. Sato, T. Ashida, Y. Kohara, M. Yui and N. Sasaki, J. Nucl. Sci. Technol. **29**, 873-882 (1992).

5. D.W. Oscarson, Clays & Clay Miner. **42**, 534-543(1994).

6. I. Neretnieks, Nucl. Technol. **71**, 458-470 (1985).

7. H. Kim, T. Suk, S. Park and C. Lee, Waste Manag. **13**, 303-308 (1993).

8. A. Muurinen, P. Penttila-Hiltunen and J. Rantanen in <u>Diffusion Mechanisms of Strontium and Cesium in Compacted Sodium Bentonite</u>, edited by J. K. Bates and W. B. Seefeldt (Mater. Res. Soc. Proc. **84**, Pittsburgh, PA, 1987) pp. 803-812.

9. J. O. Lee, W. J. Cho, P. S. Hahn and K. J. Lee, Ann. Nucl. Energy **23**, 727-738(1996).

10. T. Kozaki, A. Fujishima, S. Sato and H. Ohashi, Bull. Fac. Eng. Hokkaido Univ. **175**, 87-95(1995) (in Japanese).

11. T. Kozaki, H. Sato, A. Fujishima, S. Sato and H. Ohashi, J. Nucl. Sci. Technol. **33**, 522-524(1996).

12. T. Watanabe and T. Sato, Clay Sci. **7**, 129-138 (1988).

13. F. T. Madsen and G. Kahr, Proc. of the 1993 International Conference on Nucl. Waste Manage. and Environmental Remediation, **1**, 239-246(1993).

14. J. Crank, <u>The Mathematics of Diffusion</u>, 2nd ed. (Clarendon Press, Oxford, 1975), pp. 11-21.

15. R. A. Robinson and R. H. Stokes, <u>Electrolyte Solutions</u>, 2nd ed. (Academic Press, New York, 1959).

16. D. W. Oscarson, H. B. Hume and F. King, Clays & Clay Miner. **42**, 731-736(1994).

17. Y. Torikai, S. Sato and H. Ohashi, Nucl. Technol., **71**, 458-470 (1985).

CHLORINE-36 INVESTIGATIONS OF GROUNDWATER INFILTRATION IN THE EXPLORATORY STUDIES FACILITY AT YUCCA MOUNTAIN, NEVADA

Schön S. Levy, June T. Fabryka-Martin, Paul R. Dixon, Beiling Liu, H.J. Turin, and Andrew V. Wolfsberg, Los Alamos National Laboratory, Los Alamos, New Mexico 87545

ABSTRACT

Chlorine-36, including the natural cosmogenic component and the component produced during atmospheric nuclear testing in the 1950's and 1960's (bomb pulse), is being used as an isotopic tracer for groundwater infiltration studies at Yucca Mountain, a potential nuclear waste repository. Rock samples have been collected systematically in the Exploratory Studies Facility (ESF), and samples were also collected from fractures, faults, and breccia zones. Isotopic ratios indicative of bomb-pulse components in the water ($^{36}Cl/Cl$ values $> 1250 \times 10^{-15}$), signifying less than 40-yr travel times from the surface, have been detected at a few locations within the Topopah Spring Tuff, the candidate host rock for the repository. The specific features associated with the high $^{36}Cl/Cl$ values are predominantly cooling joints and syngenetic breccias, but most of the sites are in the general vicinity of faults. The non-bomb pulse samples have $^{36}Cl/Cl$ values interpreted to indicate groundwater travel times of at least a few thousand to possibly several hundred thousand years. Preliminary numerical solute-travel experiments using the FEHM (Finite Element Heat and Mass transfer) code demonstrate consistency between these interpreted ages and the observed $^{36}Cl/Cl$ values but do not validate the interpretations.

INTRODUCTION

One of the key information needs for performance assessment of a potential high-level nuclear waste repository at Yucca Mountain, Nevada, is the quantification of ground-water infiltration and travel times in the unsaturated zone. Related to this need is the identification of fast fluid pathways – features such as faults and fractures through which water may largely bypass the rock matrix. These needs are being addressed by chlorine-36 isotopic studies of surface samples, drill cores and cuttings, and, most recently, samples from the Exploratory Studies Facility (ESF). The ESF is an ~8-m-diameter tunnel beneath Yucca Mountain, extending as deep as the potential repository level. The tunnel exposures allow detailed characterization of the petrologic and structural settings of sample sites. Chlorine-36 investigations are part of an extensive hydrologic testing program in the ESF. The isotopic studies are fully integrated with mineralogic and textural analysis of transmissive features and with numerical modeling of groundwater infiltration.

PRINCIPLES OF CHLORINE-36 ISOTOPIC STUDIES

The interpretation of chlorine isotopic data is simple in principle and complex in application. The use of ^{36}Cl as a tracer and as a means of dating ground water is based on the production of this cosmogenic isotope in the atmosphere and its incorporation into infiltrating water. Once isolated from the atmosphere, the ^{36}Cl content of the ground water gradually decreases as a result of radioactive decay. Chloride extracted from the accessible pore spaces of subsurface samples is representative of the water that moved through the rock. Water ages, representing estimates of ground-water travel time from the surface to the underground sample sites, may be calculated from the decay constant for ^{36}Cl, the sample $^{36}Cl/Cl$, the initial atmospheric $^{36}Cl/Cl$, and the secular equilibrium $^{36}Cl/Cl$ resulting from *in situ* production by the subsurface neutron flux (from

Mat. Res. Soc. Symp. Proc. Vol. 465 © 1997 Materials Research Society

the decay of uranium and thorium isotopes) according to the standard equation of Bentley et al. [1].

The most important uncertainties affecting interpretation of the isotopic data are temporal variations in the production and deposition rates of cosmogenic ^{36}Cl relative to the deposition of stable chloride on the ground surface. The production rate of cosmogenic ^{36}Cl is controlled by variations in the earth's magnetic field strength, and the deposition of the cosmogenic isotope is influenced by climatic factors.

The deposition of stable chloride on the ground surface, which affects the ^{36}Cl/Cl value independent of variations in cosmogenic ^{36}Cl production, is largely a function of climate. The compound effect of climatic change includes varying eolian deposition of salt derived from salt flats and dry lake beds and changing patterns and frequency of storms from the Pacific, carrying salt of marine origin. A theoretical reconstruction of ^{36}Cl production rates for the last million years and a ~40 ky record of ^{36}Cl/Cl variations from regional packrat middens, dated by the ^{14}C method, both suggest that ^{36}Cl/Cl ratios were higher throughout most of the Pleistocene than during the last few thousand years [2].

Waters entering the subsurface during the last 40 years contain high concentrations of ^{36}Cl relative to natural background values. The elevated values are traceable primarily to global fallout from more than 70 above-ground nuclear tests conducted between 1952 and 1958 [3]. This input provides a fortuitous tracer to identify infiltrating water of very recent origin. ESF samples with ^{36}Cl/Cl values indicating a component of bomb-pulse ^{36}Cl are the basis for identifying fast pathways in the subsurface.

Chlorine-36 of cosmogenic origin is also produced in surficial materials and is a potential complicating factor in the interpretation of isotopic data from subsurface samples. At Yucca Mountain, Ca-rich soil calcretes are accumulation sites for cosmogenic ^{36}Cl. Our working hypothesis, subject to refinement, is that the release of ^{36}Cl from calcretes by dissolution is too small and too slow to significantly affect the isotopic signature of infiltrating water. Other potential sources of extraneous ^{36}Cl are discussed in [4].

COLLECTION AND ANALYSIS OF SAMPLES

The silicic tuffs encountered in the ESF include three main stratigraphic units: the mostly welded Tiva Canyon Tuff, the nonwelded Paintbrush Tuff, and the mostly welded Topopah Spring Tuff, in order of increasing age and depth. The PTn hydrologic unit, used in our modeling efforts, consists of the Paintbrush Tuff plus immediately overlying and underlying nonwelded rocks of the Tiva Canyon and Topopah Spring Tuffs. Geologic samples were selected to provide a systematic representation of the bedrock and to include likely transmissive features such as faults, fractures, and breccia zones. Bedrock samples were collected at 200-m intervals along the tunnel. A large variety of features was sampled to test for the existence of fast fluid pathways. ESF sample locations are measured inward from the north portal. For example, a location 115 m inward is designated as Station (abbreviated Sta.) 1+15. Depths from the ground surface to the ESF range from about 40 m at Sta. 2 to about 300 m at Sta. 34.

Chloride was extracted from one- to five-kg rock samples by leaching in an equal mass of deionized water for 48 hours. The Cl of interest is on the outer surfaces of particles or fractures. Poorly cohesive material was leached without further comminution, but other samples were crushed to 1- to 2-cm size fragments prior to leaching. An aliquot of the leachate was analyzed for Cl and Br by ion chromatography to estimate the contribution of Cl from construction water traced with LiBr. The remaining leachate was decanted, acidified to promote settling of particulates, and filtered. Known quantities of isotopically pure ^{35}Cl were added to samples with low Cl concentrations. Silver nitrate was added to the leachates to precipitate silver chloride, AgCl. The AgCl was purified of S by multiple cycles of dissolution in ammonium hydroxide, addition of barium nitrate to precipitate barium sulfate, followed by reprecipitation of the AgCl with nitric acid. The purified AgCl precipitates were sent to the Purdue Rare Isotope Measurement (PRIME) Laboratory for Cl isotopic analysis by accelerator mass spectrometry.

The factors and uncertainties that must be taken into account are described in [2, 4]. Isotopic compositions of samples were corrected for contamination by construction water, which has been isotopically characterized and labeled with lithium bromide to achieve a known Br/Cl ratio. Water ages, representing estimates of maximum ground-water travel time from the surface to the underground sample sites, were calculated from the corrected data assuming a maximum initial $^{36}Cl/Cl$ value of 1250×10^{-15} according to the standard equation of [1]. The complete data set of isotopically analyzed samples is contained in [4]. Table I lists the corrected $^{36}Cl/Cl$ values and calculated ages for all mineralogically characterized samples.

RESULTS AND INTERPRETATIONS

In the complete data set [4], fourteen samples (13%) have corrected $^{36}Cl/Cl$ ratios less than 500×10^{-15}, whereas 60 samples (57%) have values in the range of 500×10^{-15} to1250×10^{-15}. Thirty-one samples (30%) have ratios above 1250×10^{-15}. Except for two systematic (bedrock) samples, the samples with ratios above 1250×10^{-15} are from faults, fractures, or breccia zones. Our working hypothesis is that samples with $^{36}Cl/Cl$ ratios above 1250×10^{-15} contain a component of bomb-pulse ^{36}Cl and therefore indicate the presence of water less than 40 years old. Data from packrat middens and theoretical calculations of past cosmogenic ^{36}Cl production rates suggest that cosmogenic input was higher in the past but possibly not high enough to produce infiltration-water $^{36}Cl/Cl$ ratios above 1250×10^{-15}.

Most bomb-pulse samples are from the vicinities of fault zones. Only a few bomb-pulse samples are directly from fault traces; these include several breccia samples from the Bow Ridge fault zone (E008, E011) and a breccia from the Sundance fault (E175). Most Topopah Spring samples with bomb-pulse values are associated with syngenetic features, mainly cooling joints and breccias, that formed while the pyroclastic deposit was cooling and do not extend into overlying units. This result implies that paths of rapid movement of water from the surface must be more complex than individual throughgoing fault traces. In some locations, fault traces locally follow pre-existing syngenetic features. For example, the Sundance fault follows a syngenetic breccia zone at the ESF level.

The calculation of water ages/travel times from the $^{36}Cl/Cl$ data is conceptual model-dependent. For ESF samples with $^{36}Cl/Cl$ less than 1250×10^{-15}, four alternative interpretations were developed as possible explanations of the observed values:

Interpretation 1: Modern water. Travel times for water in the unsaturated zone are assumed to be sufficiently fast that an initial ratio of 500×10^{-15} (present cosmogenic background value) can be assumed. Chloride in the samples represents a mixture of waters traveling along various flow paths and for different periods of time. Samples with $^{36}Cl/Cl$ ratios above ~500×10^{-15} contain at least a small component of bomb-pulse water, and therefore almost all samples from the ESF received infiltration during the last 40 years.

TABLE I
Structural Settings, ^{36}Cl/Cl Values, Water Ages, and Secondary Mineralogy of ESF Sample Sites

Sample[1]	Station	Sampled feature[2]	Corrected ^{36}Cl/Cl ×10^{-15}	Water age (upper limit, ky)[3]	Calcite	Opal[4]	Clay/Mord.[5]	Clay/Mord. 2 or more[6]	Feldspar±Cr. Silica±Fe-Ti oxides[7]	Transported Particulates[8]	Mn minerals	Other mineral(s)[9]
E001	1+98	fault breccia	518±17[10]	396	-	-	●	-	●	●	●	-
E008	1+99.8	fault breccia	2132±139	●	●	-	●	-	●	●	●	-
E010	1+99.8	fault breccia	722±51	245	-	-	●	-	?	●	●	-
E011	1+99.8	fault breccia	2424±160	●	●	●	?	?	?	●	●	-
E007	2+3	bedrock	519±14	394	-	-	-	-	-	-	-	-
E073	5+04	fracture	463±16	446	-	-	●	-	-	-	●	-
E028	12+44	cooling jts.	2636±90	●	-	-	-	-	●	●	-	●
E029	13+00	bedrock	632±24	305	-	-	-	-	●	-	-	-
E030	13+67	cooling jts.	1613±93	●	-	-	-	-	●	-	-	-
E031	14+00	shear zone	2352±213	●	-	-	-	-	●	-	-	-
E033	14+41	fault breccia	853±44	170	●	●	-	-	●	-	-	-
E035	15+05	fracture	619±43	314	●	●	●	-	●	-	●	-
E036	16+12	cooling jt.	385±35	532	●	●	-	-	●	-	-	●
E037	16+19	fracture	975±36	110	-	-	●	-	-	●	-	-
E038	17+00	bedrock	632±123	305	-	-	-	-	●	-	●	-
E040	18+96	broken rock	1642±57	●	●	-	●	-	●	-	-	-
E041	19+00	bedrock	755±26	225	-	-	-	-	●	-	●	-
E042	19+31	breccia	3044±119	●	●	-	-	-	-	-	-	-
E044	19+42	breccia	2298±74	●	●	-	-	-	●	-	-	-
E045	21+00	bedrock	791±32	204	-	-	-	-	●	-	-	-
E046	22+71	fractures	865±37	164	●	-	-	-	●	-	-	-
E047	23+00	bedrock	648±33	294	-	-	-	-	-	-	-	-
E050	24+40	fault breccia	2582±98	●	●	-	-	-	●	-	-	-
E020	24+68	fracture	804±55	197	●	●	-	-	●	-	-	-
E052	26+79	shear zone	2034±66	●	●	-	-	-	●	-	-	-
E141	29+00	bedrock	924±51	134	-	-	-	-	●	-	-	-
E142	29+21	fracture	581±25	343	●	-	-	-	-	-	-	-
E144	29+73	cooling jt.	816±32	190	●	-	-	-	-	-	-	-
E149	31+64	cooling jt.	633±33	304	●	-	●	-	-	-	●	-
E150	33+00	fr. bedrock	1340±56	●	●	-	-	-	-	-	●	-
E152	34+28	fr. bedrock	4064±321	●	●	●	-	-	-	-	●	-
E153	34+32	cooling jts.	3305±205	●	●	●	-	-	-	●	●	-

TABLE I (cont.)

Structural Settings, ^{36}Cl/Cl Values, Water Ages, and Secondary Mineralogy of ESF Sample Sites

Sample[1]	Station	Sampled feature[2]	Corrected ^{36}Cl/Cl ×10^{-15}	Water age (upper limit, ky)[3]	Calcite	Opal[4]	Clay/Mord.[5]	Clay/Mord. 2 or more[6]	Feldspar±Cr. Silica±Fe-Ti oxides[7]	Transported particulates[8]	Mn minerals	Other mineral(s)[9]
E154	34+71	cooling jts.	3769±114	(•)	•	-	-	•	-	•	•	-
E155	35+00	bedrock	988±56	105	-	-	•	-	•	•	•	-
E157	35+03	cooling jts.	1339±83	(•)	•	-	-	-	-	-	•	-
E158	35+08	cooling jts.	2579±172	(•)	•	-	-	•	-	-	•	-
E160	35+45	cooling jts.	3533±199	•	(•)[11]	-	-	•	•	•	•	-
E161	35+58	cooling jt.	2141±78	(•)	•	-	-	•	•	-	•	-
E175	35+93	fault breccia	2833±234	(•)	(•)[11]	-	-	•	•	•	•	-

[1]Samples were divided into separate splits for isotopic and mineralogic analysis. Mineralogic data were also recorded for the sample sites In cases where more than one split of a sample was measured for chlorine isotopic ratios, the value reported in this table is the highest value obtained.

[2]Abbreviations: cooling jt. = cooling joint; fr. bedrock = fractured bedrock.

[3]The black dot symbol in this column denotes samples containing a component of "bomb-pulse" chlorine inferred to be less than 40 years old. Parentheses around the symbol indicate that bomb-pulse values were measured in a sample or sub-sample closely associated with the mineralogically characterized sample.

[4]As used here, opal is transparent, colorless to light-colored, and typically fluoresces yellow-green in short-wave UV light. X-ray diffraction analysis of selected samples indicates opal-A.

[5]This category includes clay and/or mordenite.

[6]An entry in this column indicates the presence of two or more distinct deposits of different colors.

[7]This category includes minerals inferred to be of early to late syngenetic origin. Reported occurrences in this category are limited to minerals in growth position on the rock surfaces. Cr. Silica = crystalline silica, including quartz, chalcedony, cristobalite, tridymite, opal-CT.

[8]This category includes physically transported particulates, mostly silt- and sand-size material. Deposits of clay-size material are not included here.

[9]This category includes fluorite and unidentified minerals.

[10]1-σ standard deviation.

[11]Calcite was not present in the aliquot for mineralogic characterization but was observed in fractures at the collection site (E160) or in fractures adjacent to the fault (E175).

Interpretation 2: Holocene to late Pleistocene water, with travel times <50 ky. The travel times for most waters are likely to have been more than 40 years. If ground-water travel times to the ESF level were less than 50 ky, changes in the initial ^{36}Cl/Cl ratios due to radioactive decay would be negligible. The range of ^{36}Cl/Cl values below the bomb-pulse threshold would reflect temporal variations in the cosmogenic input function and mixing of waters with different input ratios. Samples with ratios between 500×10^{-15} and 1200×10^{-15} would have lower age limits (because of the no-decay assumption) of about 0 to 15 ky, based on reconstructed cosmogenic input values and packrat midden data for this time period [4].

Interpretation 3: Mid- to late Pleistocene water, with travel times >50 ky. The travel times for most water at the ESF level, other than bomb-pulse samples, may be as much as several hundred thousand years. Initial ^{36}Cl/Cl ratios for most of the infiltrating water were much higher than present-day values, and the initial ratios of older samples have been reduced due to radioactive decay. Upper limits for water travel time may be calculated assuming a maximum initial ^{36}Cl/Cl value of 1250×10^{-15} (see Table I). Calculated upper age limits for non-bomb-pulse samples in the complete data set [4] average about 280 ky and range up to 670 ky.

Interpretation 4: Water ages/travel times indeterminate. This alternative is included to account for factors that have not yet been fully evaluated. One such factor is the contribution of ^{36}Cl from cosmic irradiation of soil calcite. This source is unlikely to have a significant effect on the measured values of subsurface samples [4].

Independent data are required to help assess which of these interpretations are most applicable. As shown below, Interpretations 2 and 3 are most compatible with modeling results. They also imply maximum water ages comparable to U-series ages of calcites and opals from the ESF [5].

FLOW AND TRANSPORT MODELING

A series of numerical modeling experiments using the FEHM (Finite Element Heat and Mass Transfer) code [6] was run to assess the extent to which the various interpretations of ^{36}Cl data are compatible with current understanding of the hydrogeologic framework of Yucca Mountain and with current estimates of hydrologic parameters. The goal of the modeling study was to simulate chloride transport in the unsaturated zone at Yucca Mountain with a specific emphasis on interpretation and analysis of ^{36}Cl/Cl ratios measured in the ESF. Details of the simulations and input parameters are given in [4].

The conceptual model for ^{36}Cl transport is that it enters the unsaturated zone with infiltrating water and migrates downward either in the matrix or fractures of the variably welded tuff units. The distribution of ^{36}Cl between fractures and matrix depends on the flux rate as well as the unit in which the flow occurs. Higher flow rates increase the potential for sustained fracture flow in all units. The welded units have lower matrix permeabilities than the nonwelded units and hence fracture flow is more readily generated and sustained in those units. The nonwelded units have higher matrix permeabilities, making fracture flow less likely. The bulk permeability of fractures in nonwelded units is low due to lower fracture density, and the potential for water to flow from fractures into the matrix is higher in nonwelded units due to higher moisture gradients in that direction than in welded units.

The nonwelded Paintbrush Tuff hydrologic unit (PTn) is expected to decrease the velocity of downward moving water because most, if not all, flow occurs in the matrix. However, because bomb-pulse ^{36}Cl values have been measured in ESF samples below the PTn, there must be a mechanism for fast flow through the unit. Fast flow could occur if (1) the matrix is bypassed by a fraction of flow moving quickly through fractures in the PTn, or (2) the matrix saturation of the PTn is locally high, allowing fracture flow to occur.

Simulations of ^{36}Cl/Cl values in the unsaturated zone indicate that infiltration rates between 1 and 5 mm/yr are more consistent with the measured values in ESF samples than infiltration rates on the order of 0.1 mm/yr. Using current hydrologic material parameters, simulations using infiltration rates of 0.1 mm/yr yield ^{36}Cl/Cl values significantly less than 500×10^{-15} throughout the ESF level. Infiltration rates of 1 and 5 mm/yr lead to simulated ^{36}Cl/Cl ratios between 500×10^{-15} and 800×10^{-15}, consistent with measured values of most non-bomb-pulse samples. A cumulative breakthrough curve for Sta. 35+00 at a 1 mm/yr infiltration rate and unenhanced fracture permeability shows travel times on the order of 10^4 to 10^5 yr for most water (fracture and matrix) arriving at the ESF.

The simulations support the hypothesis that bomb-pulse ^{36}Cl/Cl ratios in the ESF reflect fast pathways associated with faults or other features of increased fracture permeability in the PTn. For simulations performed with PTn bulk fracture permeability increased above "base-case" values and infiltration rates of 1 to 5 mm/yr, small amounts of water containing bomb-pulse ^{36}Cl arrived at the ESF level within the required 40 years. At an infiltration rate of 0.1 mm/yr, bomb-pulse arrivals could not be simulated at the ESF level for any of the cases examined.

Using preliminary estimates of the spatially distributed infiltration flux [7] ranging from 0 to 30 mm/yr and averaging about 4 mm/yr, three-dimensional site-scale simulations projected a distribution of ^{36}Cl/Cl values consistent with ESF measurements. Under areas predicted to have no infiltration, simulated values are very low. However, some lateral flow brings younger water to units below the Tiva Canyon Tuff in some of those zones.

MINERALOGIC ASSOCIATIONS

Mineralogic studies of samples collected for isotopic analysis document the geochemical context of fast fluid pathways. Studies of void-filling minerals in surficial pedogenic deposits and in the subsurface [e.g., 5, 8] suggest that the most common secondary minerals deposited during the last few hundred thousand years are calcite, opal, and clays. Mineralogic data for samples in this study show that calcite is usually present at sample sites that have received infiltration during the last 40 years (Table I), with 15 of 19 (79%) "bomb-pulse" values from samples containing calcite. By comparison, calcite is present in only 8 of the 20 (40%) samples with ages greater than 40 years, or 8 out of 12 (67%) if bedrock samples are excluded.

Amorphous opal (opal-A) is much less common in the samples than calcite and is associated with only one bomb-pulse sample from the Bow Ridge fault zone (E011). This finding is significant because U-series geochronology studies in the ESF, intended like the ^{36}Cl studies to identify temporal patterns of infiltration, have concentrated on the dating of opal associated with calcite [5]. However, the mineralogic data for the ^{36}Cl study suggest a possibility that opal deposition is not associated with pathways of very recent infiltration or perhaps that pathways of relatively rapid infiltration are distinctive geochemical and hydrologic environments not conducive to opal deposition. Additional research will establish the compatibility of fast-path infiltration interpretations with conceptual models of infiltration based on data from calcite-opal assemblages.

CONCLUSIONS

The data and modeling results in this study are compatible with a conceptual model of flow in which bomb-pulse ^{36}Cl/Cl values occur under zones of low alluvial cover, indicating that water readily enters the fractured Tiva Canyon Tuff and is transported into the underlying Paintbrush Tuff which acts as a barrier to downward movement. In the vicinity of faults or fracture zones, water rapidly traverses the Paintbrush Tuff by a combination of fracture and matrix flow and enters the Topopah Spring Tuff where downward flow occurs along fractures. The Topopah Spring Tuff is sufficiently fractured that flow is not restricted to fault traces. Infiltrating water that moves downward by routes other than fast pathways requires at least a few thousand years to reach the ESF level. Mineralogic data suggest that some geochemical distinctions exist between fast pathways and pathways without enhanced fault/fracture permeability.

ACKNOWLEDGMENTS

The comments of J. William Carey and two anonymous reviewers are greatly appreciated. This work was supported and managed by the U.S. Department of Energy, Yucca Mountain Site Characterization Office.

REFERENCES

1. H.W. Bentley, F.M. Phillips, and S.N. Davis in Handbook of Environmental Isotope Geochemistry, Vol. IIB, edited by J.C. Fontes and P. Fritz, Elsevier, Amsterdam, 1986, 427-480.

2. J.T. Fabryka-Martin, H.R. Turin, D. Brenner, P.R. Dixon, B. Liu, J. Musgrave, and A.V. Wolfsberg, Summary Report of Chlorine-36 Studies, Los Alamos National Laboratory YMP Milestone Report 3782M (in press).

3. S. Glasstone (ed.), The Effects of Nuclear Weapons, Revised Edition, U.S. Atomic Energy Commission, Washington D.C., 1962, 730 pp.

4. J. Fabryka-Martin, A.V. Wolfsberg, P.R. Dixon, S. Levy, J. Musgrave, and H.J. Turin, Summary Report of Chlorine-36 Studies: Systematic Sampling for Chlorine-36 in the Exploratory Studies Facility, Los Alamos National Laboratory, YMP Level 3 Milestone Report 3783M (in press).

5. J.B. Paces, L.A. Neymark, B.D. Marshall, J.F. Whelan, and Z.E. Peterman, Ages and Origins of Subsurface Secondary Minerals in the Exploratory Studies Facility (ESF), U.S. Geological Survey - Yucca Mountain Project Branch 1996 Milestone Report 3GQH450M (in press).

6. G.A. Zyvoloski, Z.V. Dash, and S. Kelkar, FEHMN 1.0: Finite Element Heat and Mass Transfer Code, Los Alamos National Laboratory Report LA-12062-MS, Rev. 1 (1992).

7. A.L. Flint, J.A. Hevesi, and L.E. Flint, Conceptual and Numerical Model of Infiltration for the Yucca Mountain Area, Nevada, U.S. Geological Survey Water Resources Investigation Report (in press).

8. D.T. Vaniman, S.J. Chipera, and D.L. Bish, Los Alamos National Laboratory Report LA-13096-MS (1995).

DIFFUSION OF TECHNETIUM IN COMPACTED BENTONITES IN THE REDUCING CONDITION WITH CORROSION PRODUCTS OF IRON

Yuji Kuroda, K.Idemitsu, H.Furuya, Y.Inagaki and T.Arima
Department of Nuclear Engineering, Kyushu University, Fukuoka 812-81, JAPAN.

ABSTRACT

In the vicinity of a high-level waste repository, corrosion of carbon steel overpacks will create a reducing environment. Reducing conditions are expected to retard the migration of redox-sensitive radionuclides such as technetium.

The apparent diffusion coefficients of technetium were measured in compacted bentonites (Kunigel V1® and Kunipia F®, JAPAN) in contact with carbon steel and its corrosion products under reducing conditions or without carbon steel under oxidizing conditions for comparison. The apparent diffusion coefficients measured were 10^{-10} to 10^{-11} m^2/s under oxidizing conditions and 10^{-12} to 10^{-13} m^2/s under reducing conditions. There were significant effects of redox condition, dry density (0.2 to 2.3 g / cm^3) and montmorillonite content (50 % for Kunigel V1 or 100 % for Kunipia F) on the apparent diffusion coefficients. Montmorillonite density could be a good index to explain density dependence of the diffusion coefficients under both reducing and oxidizing conditions.

INTRODUCTION

Compacted bentonite and carbon steel are being considered as a candidate buffer and overpack material for high-level waste disposal in Japan. In the real repository, a carbon steel overpack will be corroded by consuming oxygen trapped in the repository after closure. This will create a reducing environment in the vicinity of the repository. Reducing conditions are expected to retard the migration of redox-sensitive elements such as uranium [1, 2] and technetium [3].

The objective of this study is to investigate the effects of redox conditions, dry density of bentonite and the composition of bentonite on the diffusion of technetium in compacted bentonite. The apparent diffusion coefficients of technetium were measured in two types of compacted bentonites (Kunigel V1® and Kunipia F®, JAPAN) in contact with carbon steel and in others without carbon steel. On the basis of the experimental results, the diffusion mechanism of technetium in bentonite is discussed.

EXPERIMENTAL

Two types of sodium bentonite were used in this experiment ; Kunigel V1® and Kunipia F®, JAPAN. Kunigel V1 and Kunipia F contain approximately 50 % and 100 % of montmorillonite, respectively. The mineral composition of Kunigel V1 is shown in Table I. The chemical composition of the bentonites is shown in Table II. Fig. 1. shows a schematic of the experimental apparatus. Bentonite powder was compacted into a 10 mm diameter, 10 mm high specimen holder. The dry density of the compacted bentonite was in the range of 0.2 to 2.3 g/cm^3. The compacted bentonite was inserted in an acrylic resin column, which was then submerged in deionized water at reduced pressure for one month to saturate the clay.

Mat. Res. Soc. Symp. Proc. Vol. 465 © 1997 Materials Research Society

<table>
<tr><td colspan="4">Table I. Mineral Composition of 'Kunigel V1 ®'</td></tr>
<tr><td>Mineral</td><td colspan="3">weight %</td></tr>
<tr><td>Montmorillonite</td><td>46</td><td>to</td><td>49</td></tr>
<tr><td>Quartz</td><td>0.5</td><td>to</td><td>0.7</td></tr>
<tr><td>Chalcedony</td><td>37</td><td>to</td><td>38</td></tr>
<tr><td>Plagioclase</td><td>2.7</td><td>to</td><td>5.5</td></tr>
<tr><td>Calcite</td><td>2.1</td><td>to</td><td>2.6</td></tr>
<tr><td>Dolomite</td><td>2.0</td><td>to</td><td>2.8</td></tr>
<tr><td>Analcime</td><td>3.0</td><td>to</td><td>3.5</td></tr>
<tr><td>Pyrite</td><td>0.5</td><td>to</td><td>0.7</td></tr>
</table>

Table II. Chemical composition of bentonites used in this study (wt%)

Component	Kunigel V1 ®	Kunipia F ®
SiO_2	70.7	58.36
Al_2O_3	13.8	20.36
Fe_2O_3	1.49	1.34
FeO	0.62	0.51
TiO_2	0.20	0.13
MnO	0.22	<0.01
Na_2O	2.56	2.93
K_2O	0.33	0.09
CaO	2.30	0.42
MgO	2.26	2.97
P_2O_3	0.05	<0.01
CO_2	2.20	-
S	0.29	-

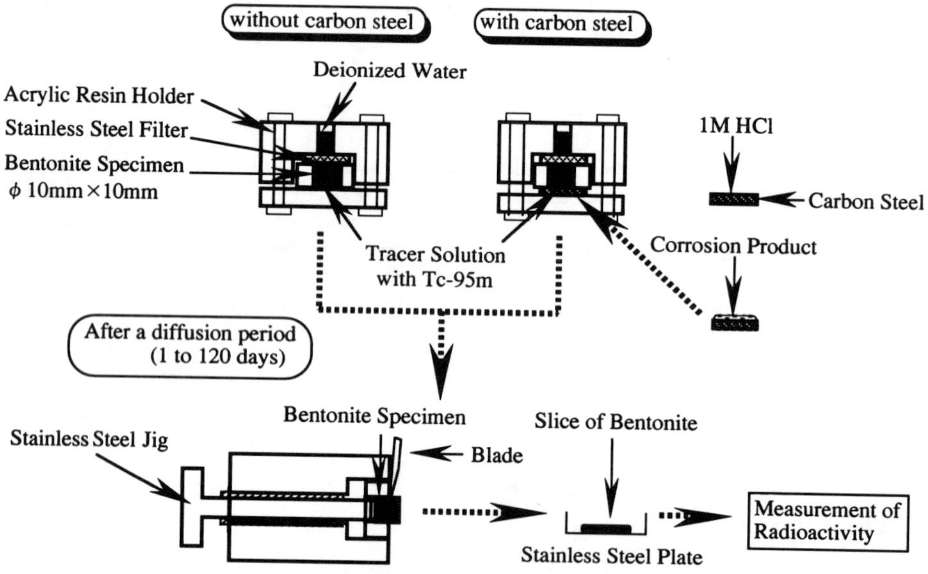

Fig. 1. Schematic of experimental apparatus

The carbon steel used in this experiment was SS401 and the specimen diameter and thickness were 12 mm and 2 mm, respectively. One side of the carbon steel specimen was corroded with 100 mL of 1M HCl for two days at room temperature. In the experiments under reducing conditions the partly corroded carbon steel was contacted with a water-saturated bentonite for 90 days and the corrosion products of iron were diffused into the bentonite. Then 10 μ L of technetium solution containing ca. 10^{-14} g of Tc-95m was put on the surface between the carbon steel and the bentonite. After a diffusion period of 1 to 120 days, the column was disassembled. Then the bentonite was pushed out and was sliced in increments of 0.2 to 2 mm. The amount of Tc-95m in each slice was measured with a Ge detector.   Same experiments were carried out without the carbon steel and corrosion products.

RESULTS

<u>Observation of the bentonite specimens</u>

A lot of corrosion products were observed at the interface between the bentonite specimen and the carbon steel. The color of the corrosion product was dark-green and turned red on exposure to air in several minutes. This means that the carbon steel maintained a reducing condition in the bentonite during contact.

<u>Determination of apparent diffusion coefficients</u>

Fig.2. shows penetration profiles of technetium in Kunigel V1 with and without carbon steel.

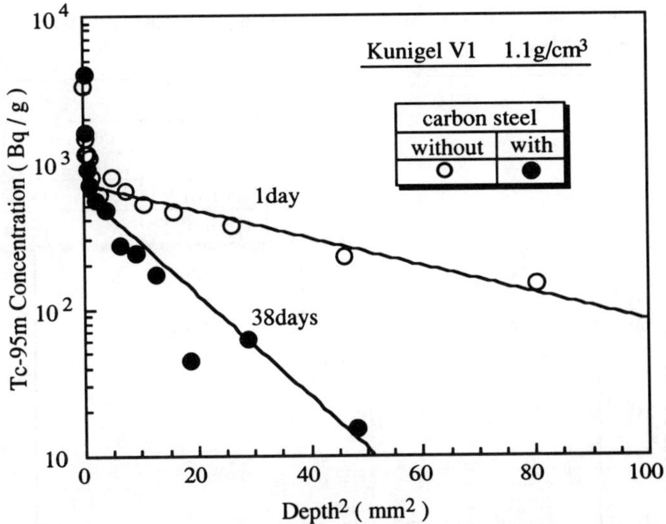

Fig. 2. Technetium penetration profiles in Kunigel V1 with a dry density of 1.1 g / cm^3

The apparent diffusion coefficients of Tc-95m were determined from its penetration profiles. When it is assumed that the diffusion coefficients are independent of the position, one-dimensional non-steady state diffusion is described by Fick's second law

$$\frac{\partial C}{\partial t} = Da \frac{\partial^2 C}{\partial x^2} \qquad (1)$$

where D_a is the apparent diffusion coefficient. In this case, it is subject to the following initial and boundary conditions:

$$t = 0 , \; x > 0 , \; C = 0 , \tag{2}$$
$$t \to 0 , \; x = 0 , \; C \to \infty \qquad t > 0 , \; x \to \infty , \; C \to 0 , \tag{3}$$
$$\int_0^\infty C \, dx = M(\text{constant}) , \tag{4}$$

where M is the total amount of diffusing substance. Eq.(1) is solved under these conditions, and the value of C is given by

$$C = \frac{M}{\sqrt{\pi D_a t}} \, exp\left(-\frac{1}{4 D_a t} x^2 \right) . \tag{5}$$

The value of D_a is calculated by plotting lnC versus x^2 to obtain a straight line of slope $(4 \, D_a \, t)^{-1}$ [4]. It seems appropriate to apply this analytical model, because the concentration of Tc-95m introduced into bentonite (ca. 1.3×10^{-11} M) was much lower than the solubility (ca. $10^{-9} \sim 10^{-10}$ M under reducing condition [5]).

The apparent diffusion coefficients are shown in Table III. and they are plotted as a function of the dry density of bentonites as shown in Fig. 3. It can be seen that there are significant effects of dry density and redox conditions on the apparent diffusion coefficients in this study. The apparent diffusion coefficients decreased with increasing dry density of the bentonite under redox conditions. The presence of carbon steel decreased the apparent diffusion coefficients two orders of magnitude. Thus it is expected that technetium could be kept at the reduced state during diffusion period in the case with carbon steel since the apparent diffusion coefficients of ferrous ion is on the order of 10^{-12} m^2/s [6].

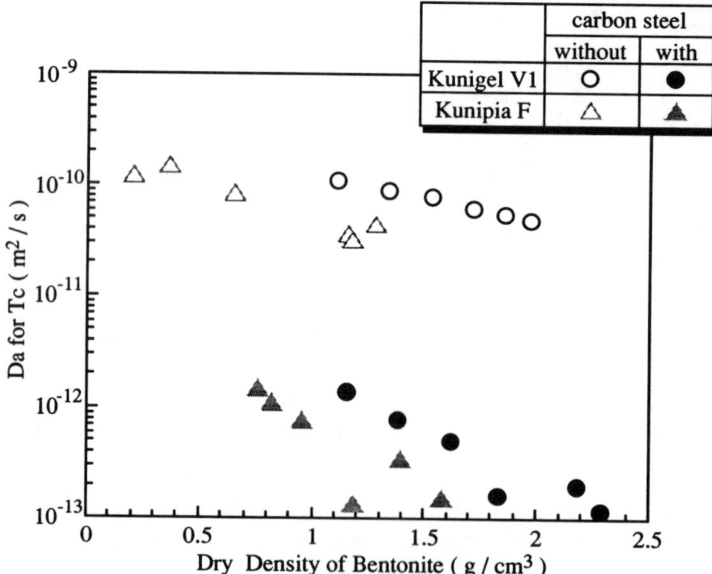

Fig. 3. Dry density dependence of apparent diffusion coefficients of technetium

Table III. Apparent technetium diffusion coefficients at different densities of Kunigel V1 and Kunipia F under oxidizing and reducing conditions

< Kunigel V1 >

Ox.	Dry Density (g/cm^3)	1.11	1.34	1.53	1.72	1.86	1.97
	Da ($\times 10^{-11}$ m^2/s)	11	8.9	7.9	6.1	5.5	4.9
Red.	Dry Density (g/cm^3)	1.15	1.38	1.62	1.83	2.18	2.29
	Da ($\times 10^{-13}$ m^2/s)	14	8.0	5.1	1.6	2.0	1.2

< Kunipia F >

Ox.	Dry Density (g/cm^3)	0.20	0.36	0.65	1.15	1.17	1.28
	Da ($\times 10^{-11}$ m^2/s)	12	15	8.6	3.6	3.2	4.5
Red.	Dry Density (g/cm^3)	0.76	0.82	0.95	1.18	1.39	1.58
	Da ($\times 10^{-13}$ m^2/s)	15	11	7.8	1.4	3.5	1.5

DISCUSSION

Normalization for montmorillonite density

The major clay mineral in bentonite is montmorillonite. It contains water layers in which the diffusion of species might occur. The apparent technetium diffusion coefficients for the two types of bentonite (Kunigel V1 and Kunipia F) exhibit a trend, as shown in Fig. 4., when plotted as a function of normalized montmorillonite density in bentonite.

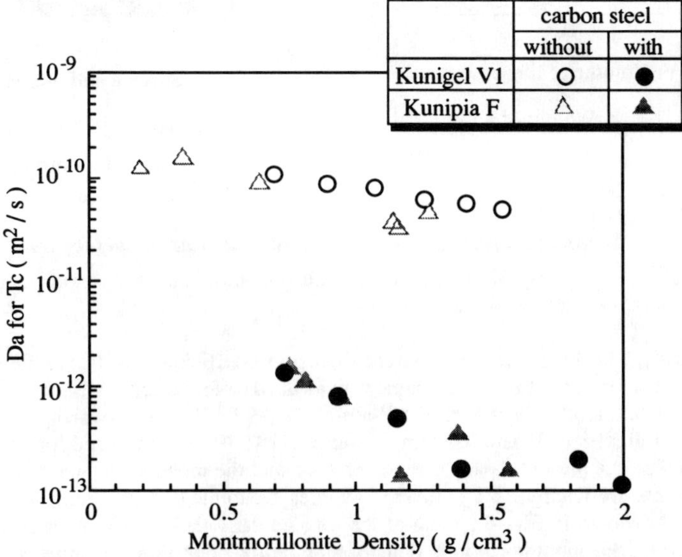

Fig. 4. Montmorillonite density dependence of apparent diffusion coefficients of technetium

Here montmorillonite density is defined as the density of montmorillonite itself after swelling and expressed according to the following equation,

$$\rho_{mont} = \frac{f\rho}{1 - \frac{(1-f)\rho}{\rho_{others}}} = \frac{\text{weight of montmorillonite in unit volume}}{\text{unit volume} - \text{volume of other minerals in unit volume}}$$

Where ρ_{mont}, ρ_{others}, ρ and f are montmorillonite density, true density of other minerals ($2.7\mathrm{g}/\mathrm{cm}^3$), dry density and fraction of montmorillonite in bentonite (50% for Kunigel V1, 100% for Kunipia F), respectively. For example, montmorillonite density of Kunipia F equals its dry density because f equals 1. The apparent diffusion coefficients lie in nearly the same position in the figure in spite of different montmorillonite contents between the two types of bentonites. The apparent diffusion coefficients of technetium could be estimated as a function of montmorillonite density. This suggests that technetium could be assumed to diffuse only in the swelled montmorillonite portion of the bentonite as shown in Fig. 5.

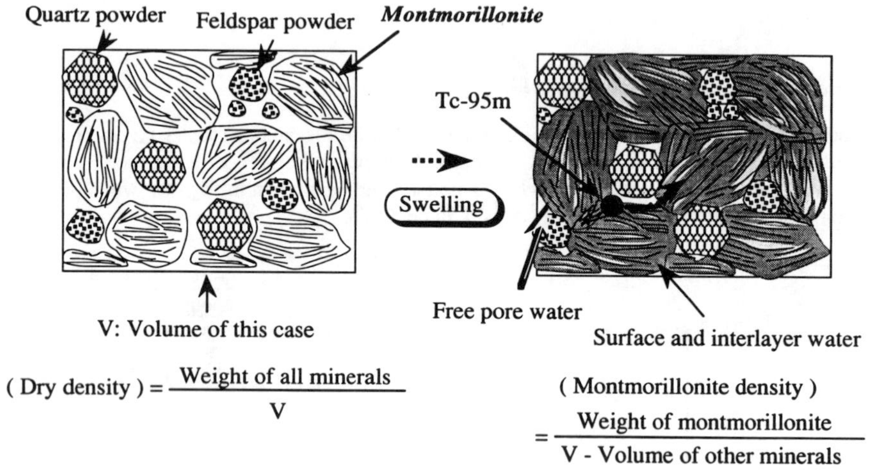

$$(\text{Dry density}) = \frac{\text{Weight of all minerals}}{V}$$

$$(\text{Montmorillonite density}) = \frac{\text{Weight of montmorillonite}}{V - \text{Volume of other minerals}}$$

Technetium can diffuse in free pore water and/or surface water.

Fig. 5. Schematic of swelled montmorillonite

Consequently, the data on the apparent diffusion coefficients of technetium in different bentonites also can be normalized in the same way as the data for Kunigel V1 and Kunipia F. Fig. 6. shows the data for other bentonites (H.Sato et al.,1992 [7], H.G.Sawatsky et al.,1991 [8], B.Torstenfelt et al.,1985 [3] and Y.Albinsson et al.,1991 [9]) normalized for montmorillonite density as well as the present data. The bentonites (and the montmorillonite contents) used in their experiments are Kunigel V1 (50%), Avonlea bentonite (80%) and MX-80 (75%), respectively. As shown in Fig. 6., most of the data similarly lie in nearly the same position, for the some normalized montmorillonite density and redox condition, in spite of the different bentonites used. This suggests that, for the diffusion of technetium in different types of bentonite, montmorillonite density could be a good index to explain the density dependence of the diffusion coefficients.

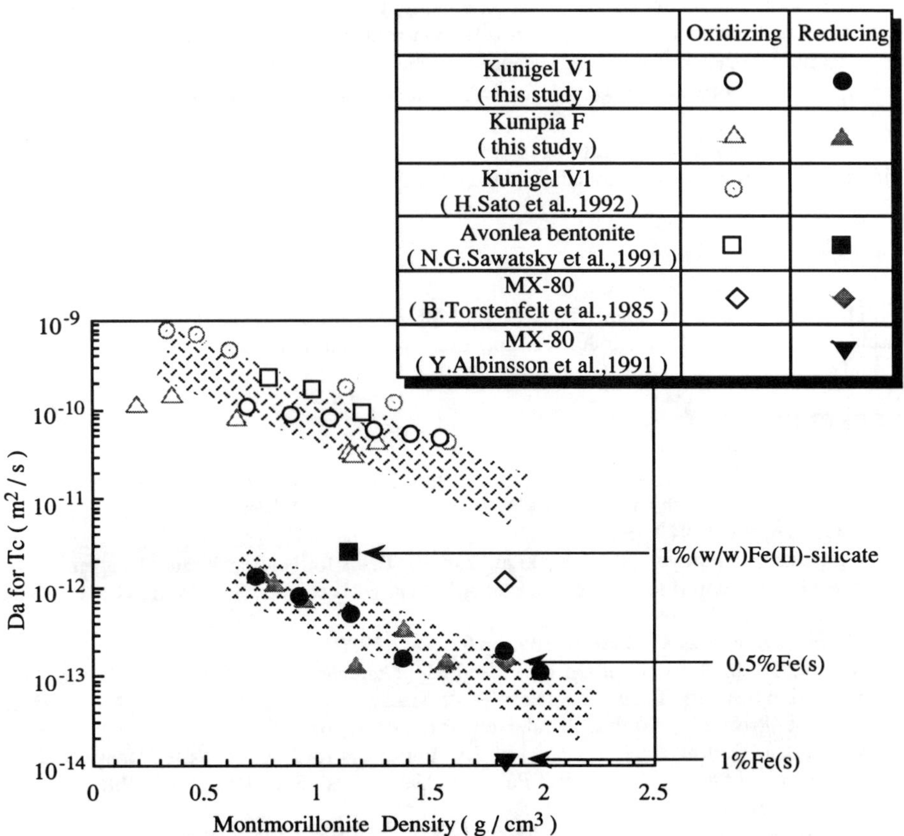

Fig. 6. The data for other bentonites normalized for montmorillonite density

Change in apparent diffusion coefficients with redox condition

The oxidation state of technetium can vary with redox condition. The dominant species of technetium is TcO_4^- in oxidizing conditions and $TcO(OH)_{2-x}^{x+}$ (TcO^{2+}, $TcO(OH)^+$ or $TcO(OH)_2$, depending on pH) in reducing conditions [10]. The change in the apparent diffusion coefficients with the redox condition as shown in Fig. 4. results from the change in the species caused by corrosion products diffused in bentonite. It is likely that $TcO(OH)_{2-x}^{x+}$ was adsorbed on negatively charged montmorillonite surface, therefore, diffusion of technetium was delayed in reducing conditions. On the other hand, it seems that TcO_4^- diffused in free pore water primarily since TcO_4^- and negatively charged montmorillonite repel each other in oxidizing conditions.

CONCLUSION

There were significant effects of redox condition, dry density (0.2 to 2.3 g / cm^3) and montmorillonite content on the apparent diffusion coefficient of technetium in bentonite. The apparent diffusion coefficients of technetium under reducing conditions (1.1×10^{-10} to 3.2×10^{-11} m^2/s) were two orders of magnitude lower than those under oxidizing conditions (1.4×10^{-12} to 1.2×10^{-13} m^2/s).

Montmorillonite density could be a good index to explain the density dependence of the diffusion coefficients under both reducing and oxidizing conditions.

ACKNOWLEDGMENT

The authors wish to thank Mr.K.Miwa, Mr.E.Asano, Mr.T.Kanno and Ms.Y.Yamanaka (Ishikawajima-Harima Heavy Industries Co., Ltd.) for their technical support.

REFERENCES

1. K.Idemitsu, H.Furuya and Y.Inagaki in Scientific Basis for Nuclear Waste Management XVII, edited by A.Barkatt and Richard A.Van Konynenburg (Mater. Res. Soc. Proc. 412, Pittsburgh, PA, 1994), pp.939-945.
2. K.Idemitsu, H.Furuya and Y.Inagaki in Scientific Basis for Nuclear Waste Management XIX, edited by William M.Murphy and Dieter A.Knecht (Mater. Res. Soc. Proc. 412, Pittsburgh, PA, 1996), pp.683-690.
3. B.Torstenfelt in Radiochimica Acta 39, pp.97-104 (1986).
4. J.Crank, The Mathematics of Diffusion, 2nd ed. (Clarendon Press, Oxford, 1975), pp.11-13.
5. B.Allard in Scientific Basis for Nuclear Waste Management VII, edited by G.L.McVay (Mater. Res. Soc. Proc. 263, North Holland, New York, 1984), pp.219-226.
6. K.Idemitsu, H.Furuya and Y.Inagaki in Scientific Basis for Nuclear Waste Management XVI, edited by C.G.Interrante and R.T.Pabalan (Mater. Res. Soc. Proc. 294, Pittsburgh, PA, 1993), pp.467-474.
7. H.Sato, T.Ashida, Y.Kohara and M.Yui in Scientific Basis for Nuclear Waste Management XVI, edited by C.G.Interrante and R.T.Pabalan (Mater. Res. Soc. Proc. 294, Pittsburgh, PA, 1993), pp.403-408.
8. N.G.Sawatsky and D.W.Oscarson in Soil Sci. Soc. Am. J., VOL 55, pp.1261-1267.
9. Y.Albinsson, et al. in Radiochimica Acta 52 / 53, pp.283-286 (1991).
10. B.Skytte JENSEN, Migration Phenomena of Radionuclides into the Geosphere (Harwood Academic Publishers, New York, NY, 1982), pp A44-A50.

ESTIMATION OF THE RADIONUCLIDE TRANSPORT BY APPLYING THE MEAN, THE STANDARD DEVIATION AND THE SKEWNESS OF PERMEABILITY

Y. NIIBORI*, O.TOCHIYAMA* and T.CHIDA**
*Department of Quantum Science and Energy Engr.,Tohoku University, Sendai 980-77 Japan
**Department of Geoscience and Technology Engr.,Tohoku University, Sendai 980-77 Japan

ABSTRACT

A new method for estimating the mass transport by using the stochastic values (the arithmetic mean, the standard deviation and the skewness) of permeability is presented. Generally, detail of permeability distribution cannot be obtained except for moments of the distribution. Also, measurement results of permeability for the rock matrix including cracks or fast flowpaths do not always follow the log-normal distribution frequently applied. In such a situation, we must evaluate the characteristic permeabilities for the whole or some regions of the disposal site including the accessible environment.

The authors have investigated the characteristic permeability on the basis of some probability density functions of permeability, applying the Monte Carlo method and FEM. It was found that its value does not depend on type of probability density function of permeability, but on the arithmetic mean, the standard deviation and the skewness of permeability [1].

This paper describes the use of the stochastic values of permeability for estimating the rate of radioactivity release to the accessible environment, applying the advection-dispersion model to two-dimensional, heterogeneous media. When a discrete probability density function (referred to as 'the Bernoulli trials') and the lognormal distribution have common values for the arithmetic mean, the standard deviation and the skewness of permeability, the calculated transport rates (described as the pseudo impulse responses) show good agreements for Peclet number around 10 and the dimensionless standard deviation around 1. Further, it is found that the transport rates apparently depends not only on the arithmetic mean and the standard deviation, but also on the skewness of permeability. When the value of skewness dose not follow the lognormal distribution which has only two independent parameters (the mean and the standard deviation), we can replicate the three moments estimated from an observed distribution of permeability, by using the Bernoulli trials having three independent parameters.

INTRODUCTION

Some useful models describing the transport of the radioactive nuclides in the near/far-field have been designed (e.g., [2-7]). These models consider the spatial variability of permeability, assuming a probability density function. However, there are not always sufficient experimental data to define the function. While the detail of the permeability distribution is generally not available [8,9], the available data at least give us the moments such as the arithmetic mean, the standard deviation and the skewness. The purpose of this paper is to connect such stochastic values with estimation of the rate of radioactivity release to the accessible environment. Applying the simple advection-dispersion model to two-dimensional region having various permeability distribution, this paper discusses, in particular, the relations between the skewness of permeability and the average transport rates.

PROBLEM CONSIDERED

Probability Density Functions

Figure 1 shows two examples of the permeability distribution in two-dimensional medium. These examples follow the Bernoulli trials described by

917

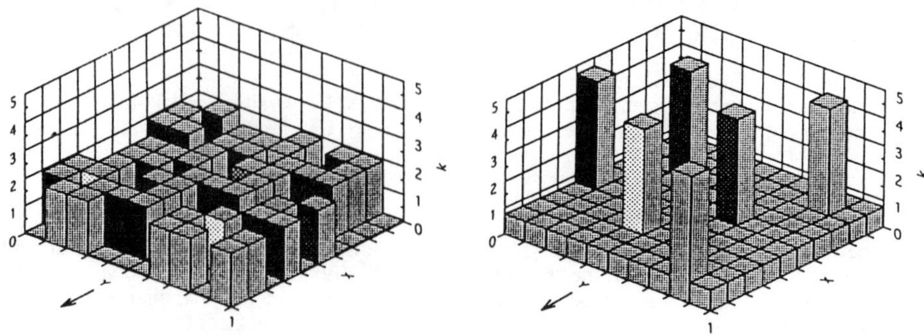

(a)K_A=1, σ=0.947 and S_K=0
(i.e., K_S=0.053, K_L=1.947 and f=0.5)

(b)K_A=1, σ=0.947 and S_K=3.633
(i.e., K_S=0.757, K_L=4.682 and f=0.938)

Figure 1 Examples of the permeability distribution on the basis of the Bernoulli trials.

$$F_{Ber}(K) = f\delta(K - K_S) + (1 - f)\delta(K - K_L) , \qquad (1)$$

where F_{Ber} is the probability density function for the value of K in each cell of a rectangular mesh, K is dimensionless permeability value defined by $K=k/k_A$ (k:permeability (m²), k_A:the arithmetic mean (m²) of k), and δ is the delta function. As shown in Eq.(1), F_{Ber} depends on three values, i.e., K_S (= k_S/k_A), K_L(= k_L/k_A) and f, where k_S and k_L are two values of permeability (m²) in $k_S \leq k_L$, and f is the fraction of K values taking the value K_S (thus the fraction taking the value K_L is $1-f$). These values are determined by the other three values, i.e., the arithmetic mean , the standard deviation and the skewness of permeability. About the skewness, its value, S_K, is defined by $S_K=\mu_3/\sigma^3$, where μ_3 is the central moment of third-order, and σ is the standard deviation, respectively given by

$$\mu_3 = \int_0^\infty F(K)(K - 1)^3 dK , \qquad (2)$$

$$\sigma^2 = \int_0^\infty F(K)(K - 1)^2 dK , \qquad (3)$$

where $F(K)$ is a probability density function of the dimensionless permeability, K. In this paper, the arithmetic mean value of K, i.e., K_A is always set at 1. Using the pseudo random-numbers and the ratio, f, we can set K_S or K_L at each cell. In Figure 1, the two-dimensional region (the ranges $[0,1]$ of X and Y) is divided into regular square cells of 10 by 10 in the X- and the Y-directions. The distribution of Fig.1(a) is on the basis of K_A=1, σ=0.947, and S_K=0, i.e. , K_S=0.053, K_L=1.947, and f=0.5. On the other hand, Fig.1(b) is the distribution of K_A=1, σ=0.947, and S_K=3.633, i.e., K_S=0.757, K_L=4.682 and f=0.938. This value of S_K equals the skewness of the lognormal distribution with K_A=1 and σ=0.947 (then, the standard deviation of ln(K) is 0.8)[1].

The lognormal distribution is described by the following equation:

$$F_{LN}(K) = \frac{1}{\sqrt{2\pi}\sigma^* K} exp\{-\frac{\{ln(K) - \mu\}^2}{2\sigma^{*2}}\} , \qquad (4)$$

where μ and σ^* are the arithmetic mean of ln(K) and its standard deviation, respectively. Since

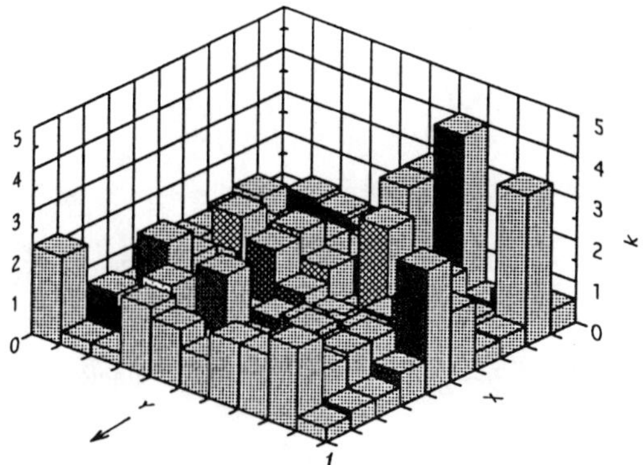

Figure 2 Example of the permeability distribution on the basis of the lognormal distribution. with $K_A=1$ and $\sigma=0.947$ (Then $S_K=3.633$).

$K_A=\exp(\mu+\sigma^{*2}/2)$ and $\sigma^2=\exp(2\mu+\sigma^{*2})\{\exp(\sigma^{*2})-1\}$, i.e., $\sigma^2=K_A^2\{\exp(\sigma^{*2})-1\}$, the standard deviation σ of K is

$$\sigma^2 = exp(\sigma^{*2}) - 1 \quad , \tag{5}$$

when $K_A=1$.(Note that $\mu \neq 0$.)

Figure 2 is the example of the permeability distribution based on the lognormal distribution with $K_A=1$ and $\sigma=0.947$ ($\sigma^* = 0.8$, then $S_k=3.633$). This values of σ is in the range of realistic values [3]. As shown in Figs.1 and 2, these permeability distributions are obviously different, while the density functions have the common values of the mean and the standard deviation, respectively.

Apparently we are hypothesizing a spatial medium with a distribution of permeability having a arithmetic mean that can be estimated from a finite sample. Thus, if we divide each observation in the sample by the mean, we get dimensionless permeability values K, from which we can estimate the standard deviation σ and the skewness S_K. Then we construct a model of the medium with a rectangular mesh and assume that permeability is constant inside each cell. By setting the three values K_S, K_L and f, the arithmetic mean, the standard deviation and the skewness can be replicated.

<u>Fundamental Equations and Boundary Conditions</u>

This paper applies the simple advection-dispersion model to the two-dimensional regions with each permeability distribution. Some models describing the decay, the retardation and the diffusion to the rock matrix around the flow paths have been proposed (e.g.,[2-7]), these models need the measurement data not only for the spatial variability of permeability, but also the practical values of parameters for the retardation of the radionuclides and its diffusion to the rock matrix. However, the actual values depend on disposal site and these data are limited.

In this paper, the effects of the retardation and the decay on the radionuclide transport are neglected, in order to get a general understanding for the spatial variability of permeability. Then, the dimensionless, fundamental equations applied are:

the continuity equation,

$$\frac{\partial U}{\partial X} + \frac{\partial W}{\partial Y} = 0 \quad , \tag{6}$$

Darcy's law,

$$U = -K\frac{\partial P}{\partial X}, \qquad W = -K\frac{\partial P}{\partial Y}, \tag{7}$$

and the mass balance equation,

$$\frac{\partial C}{\partial T} = \frac{1}{P_e}\left(\frac{\partial^2 C}{\partial X^2} + \frac{\partial^2 C}{\partial Y^2}\right) - U\frac{\partial C}{\partial X} - W\frac{\partial C}{\partial Y}. \tag{8}$$

In the equations above, the dimensionless variables and parameter are defined by

$$X = \frac{x}{x_1}, \qquad Y = \frac{y}{x_1}, \qquad T = \frac{t}{t^*}, \qquad t^* = \frac{x_1 \epsilon}{u^*},$$

$$K = \frac{k}{k_A}, \qquad C = \frac{c}{c^*}, \qquad U = \frac{u}{u^*}, \qquad W = \frac{w}{u^*},$$

$$P = \frac{k_A(p - p_1)}{u^* \mu_f x_1}, \qquad P_e = \frac{x_1 u^*}{D_e}, \qquad u^* = \frac{k_A(p_0 - p_1)}{\mu_f x_1},$$

where x_1 is the characteristic length (m) defining x and y regions of interest, t is time (s), ϵ is porosity, k is the permeability (m^2), u^* is the Darcy flow velocity (m/s), whose value is calculated by assuming k to be uniform (the homogeneous medium), c is the concentration (kg/m^3), c^* is the injection concentration (kg/m^3) , u and w are Darcy flow velocities (m) in x- and y-directions, respectively, p is pressure (Pa), p_0 and p_1 are the pressures at $x=x_0$ and $x=x_1$, respectively, μ_f is fluid viscosity (Pa s), and D_e is the dispersion coefficient (m^2/s).

The boundary conditions of dimensionless pressure are $P=1$ at $X=0$, $P=0$ at $X=1$ and $\partial P / \partial Y = 0$ at $Y=0$ and $Y=1$. The injection of the radionuclides into this flow system, i.e., the boundary conditions of the concentration at $X=0$ is described by $C=1$ in $0 \leq T \leq T_{in}$, and $C=0$ in $T_{in} < T$, in use of the closed vessel boundary condition [10,11]. The other boundaries of C are assumed to be $\partial C / \partial X=0$ at $X=1$ and $\partial C / \partial Y=0$ at $Y=0$ and $Y=1$.

To solve the fundamental equations, at the first stage, the finite difference method is applied to the system of Eqs.(6) and (7) and the boundary conditions about the dimensionless pressure, P. Next, by substituting the numerical solution of P into Eq.(7), the dimensionless velocities, U and W, are obtained. Based on those values, the dimensionless concentration, C, is solved numerically in Eq.(8). In the numerical implementations, each cell with a permeability is further divided into 5 by 5. About T_{in}, its value must be less than 0.05 (that is, the length of the injection time has to be shorter than 5% of the characteristic residence time, t^*) to get an impulse response for a δ -function input signal [11].(Hereinafter, referred to as 'pseudo impulse response'.)

RESULTS

Figure 3 display some examples of the pseudo impulse responses, $\overline{C}_{out}$, for each distribution. Here, the permeability distributions of Trial 1, 2 and 3 are equivalent to Fig.1(a) in the arithmetic mean, the standard deviation and the skewness. And, the Trial 4 and 5 are equivalent to the distribution of Fig.1(b). The normalized response $\overline{C}_{out}$ is evaluated by

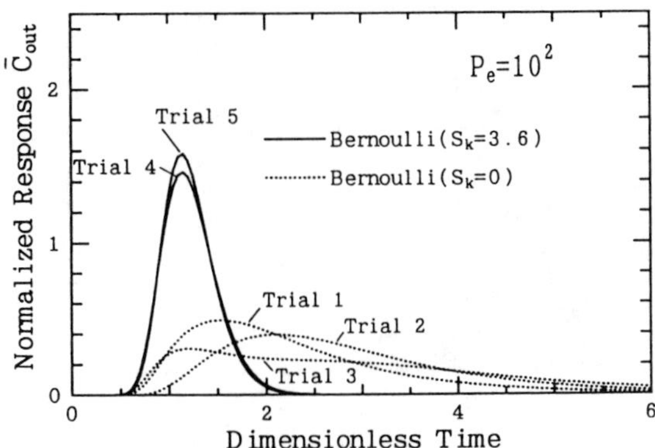

Figure 3 The pseudo-impulse responses for $P_e=10^2$, based on the Bernoulli trials of $K_A=1$ and $\sigma=0.947$ (the solid lines: the cases for $S_K=3.633$,the dotted lines: for $S_K=0$).

$$\overline{C}_{out} = \frac{\int_0^1 C_{X=1}U_{X=1}dY}{\int_0^\infty (C_{in}\int_0^1 U_{X=0}dY)dT} \ , \tag{9}$$

where C_{in} is the dimensionless injection concentration at $X=0$ and the denominator in the right side of Eq.(9) is the amount injected. From the mass balance, we can derive

$$\int_0^\infty \overline{C}_{out}dT = 1 \ . \tag{10}$$

The responses for Trail 2 and Trial 3 were the most different cases in arrival time of the peak through the trials of ten times. In these calculations, P_e , i.e., the Peclet number, is set at 10^2. Figure 4 is the responses for $P_e=10$. The calculated results in Figs.3 and 4 show that the responses almost agree when the permeability distributions have the same values for K_A, σ and S_K, respectively.

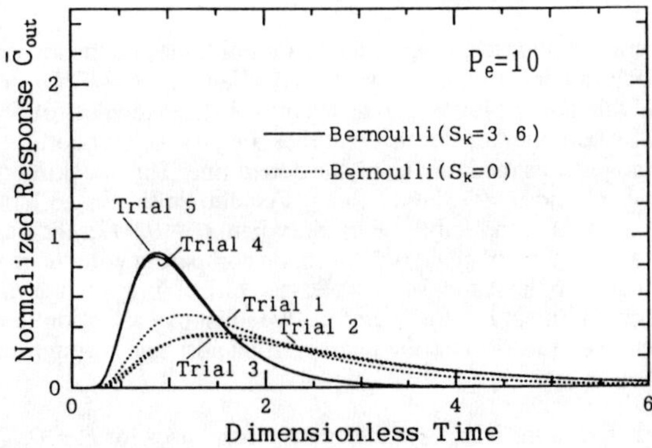

Figure 4 The pseudo-impulse responses for $P_e=10$,based on the Bernoulli trials of $K_A=1$ and $\sigma=0.947$ (the solid lines are the cases for $S_K=3.633$,the dotted lines are for $S_K=0$).

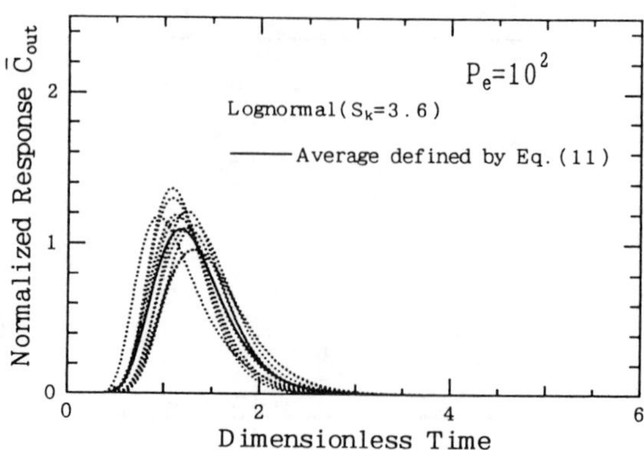

Figure 5 The pseudo-impulse responses for $P_e=10^2$, based on the lognormal distribution with $K_A=1$ and $\sigma=0.947$ (the dotted line displays each $\bar{C}_{out}$ obtained from the computations of ten times and the solid line is the average normalized response defined by Eq.(11)).

To confirm this tendency, further, we apply the permeability distribution based on the lognormal distribution, as shown in Fig.2. Figure 5 displays the responses of the trials of ten times, when $K_A=1, \sigma=0.947$ (then, $\sigma^*=0.8, S_k=3.633$). These responses yield an average response, i.e.,

$$\bar{\bar{C}}_{out} = \frac{\sum\limits_{i=1}^{n} \bar{C}_{out,i}}{\sum\limits_{i=1}^{n} \int_0^\infty \bar{C}_{out,i}dT}, \qquad (\sum\limits_{i=1}^{n} \int_0^\infty \bar{C}_{out,i}dT = n) \qquad (11)$$

where n is the trail number of computations of $\bar{C}_{out}$, and the subscript i is the counter. From Eq.(11) , we can confirm

$$\int_0^\infty \bar{\bar{C}}_{out,i}dT = 1 \quad. \qquad (12)$$

Therefore $\bar{\bar{C}}_{out}$ is also the normalized response. Similarly, we can get an average response for the Bernoulli trials where P_e, K_A, σ and S_k are fixed, respectively.

Figure 6 shows the average responses, $\bar{\bar{C}}_{out}$, for the Bernoulli trials and the lognormal distribution, when $P_e=10^2$. These probability density functions have $K_A=1$ and $\sigma=0.947$. Of them, the Bernoulli trail with $S_k=3.633$ is equivalent to the lognormal distribution in the skewness of permeability. These responses were obtained by computations of the ten times, i.e., $n=10$. In addition, Fig.6 shows the response through the homogeneous medium, using the dotted line. This breakthrough curve denotes the case with $P_e=10^2$, $K_A=1$ and $\sigma=0$ (then, $S_k=0$) . As shown in Fig. 6, the dependency of skewness on the responses is remarkable. Further, when $\sigma=0.947$, the degree of mixing is apparently large and the arrival time of peak is delayed, on comparing with the responses of $\sigma=0$. This tendency appears clearly in the response for $S_k=0$ (the broken line) rather than $S_k=3.633$ (the solid line and the dotted chain line in Fig.6). Therefore, we must pay attention to the values of not only K_A and σ, but also S_k , to estimate the average transport rate through the heterogeneous media.

In the same way as Fig.6, Figure 7 displays the average responses for $P_e=10$. The solid line and the dotted chain line show good agreement, although the permeability distributions based on the Bernoulli trails and the lognormal distribution are quite different, as shown in Fig.1(b) and Fig.2. On the other hand, the average response with $K_A=1, \sigma=0.947$ and $S_k=0$ (the broken line in Fig.7) is

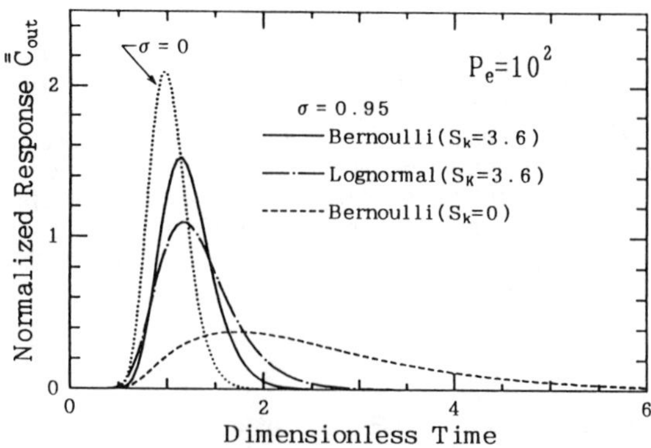

Figure 6 The average normalized responses, $\bar{\bar{C}}_{out}$, for $P_e=10^2$ (the solid line : the average response based on the Bernoulli trial of $K_A=1$, $\sigma=0.947$ and $S_K=3.633$, the dotted chain line: on the lognormal distribution of $K_A=1$ and $\sigma=0.947$ (then, $S_K=3.633$), the broken line: on the Bernoulli trials of $K_A=1$, $\sigma=0.947$ and $S_K=0$, and the dotted line: the pseudo impulse response of $K_A=1$, and $\sigma=S_K=0$).

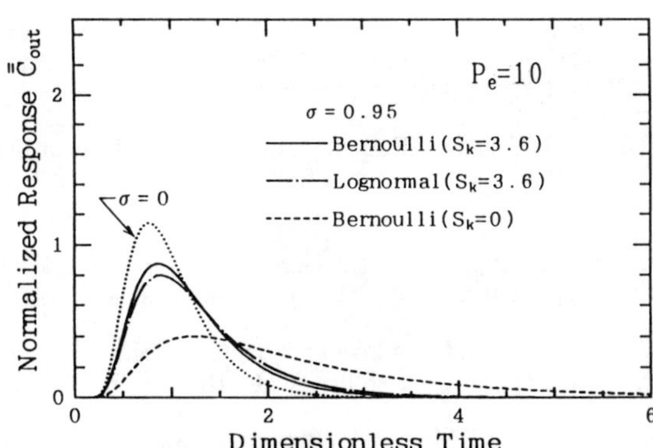

Figure 7 The average normalized responses, $\bar{\bar{C}}_{out}$, for $P_e=10$ (the solid line : the average response based on the Bernoulli trial of $K_A=1$, $\sigma=0.947$ and $S_K=3.633$, the dotted chain line :on the lognormal distribution of $K_A=1$ and $\sigma=0.947$ (then, $S_K=3.633$), the broken line: on the Bernoulli trials of $K_A=1$, $\sigma=0.947$ and $S_K=0$, and the dotted line: the pseudo impulse response of $K_A=1$, and $\sigma=S_K=0$).

clearly different from those with $K_A=1$, $\sigma=0.947$ and $S_K=3.633$. Further, we can confirm from the broken lines in Figs.6 and 7 that the height of peak for $P_e=10$ almost agree with that for $P_e=10^2$, though the arrival time of peak for $P_e=10$ is faster.

CONCLUSIONS

This paper applied the most conservative model for estimating the release rate of the radionuclides. When the permeability distributions had the common values in the arithmetic mean, the standard deviation and the skewness of permeability, respectively, the calculated results demonstrated good agreements of the transport rates (described as the pseudo impulse responses) for Peclet number around *10* and the dimensionless standard deviation around *1*. Also, the responses apparently depended on the value of the skewness. Therefore, we must estimate not only the arithmetic mean and the standard deviation, but also the skewness from the experimental data, before assuming a specific probability density function as a distribution of permeability. When the value of skewness dose not follow the lognormal distribution expressed by the two independent parameters (the arithmetic mean and the standard deviation), we can use the Bernoulli trials to replicate the three moments, which in turn will give us a more reliable estimate for the release rate. In addition, we confirmed through this approach that the degree of mixing is apparently large, as the standard deviation is large and the skewness is almost zero.

REFERENCES

1. Niibori,Y. and Chida,T.:The Use of Standard Deviation and Skewness for Estimating   Apparent Permeability in Two-Dimensional, Heterogeneous Medium, Transport in Porous Media, **15**,1-14 (1994)

2. Gylling,B.,Moreno,L. and Neretnieks,I.:A Channel-Network-Model for Radionuclide Transport in Fractured Rock-Testing Against Field Data, in *Scientific Basis for Nuclear Waste Management XVIII*, edited by T.Murakami and R.C.Ewing (Mater.Res.Soc.Proc., **353**, Pittsburgh, PA, 1995), pp.395-402.

3. Chesnut,D.A.:Groundwater Flux, Travel Time, and Radionuclide Transport, in *Scientific Basis for Nuclear Waste Management XVIII*, edited by T.Murakami and R.C.Ewing(Mater.Res. Soc. Proc., **353**, Pittsburgh, PA, 1995), pp.463-470.

4. Nordqvist,A.W.,Tsang,Y.W.,Tsang,C.F.,Deverstop,B. and Andersson,J.:A variable aperture fracture network model for flow and transport in fractured rocks, Water Resour. Res., **28**, 1703-1713 (1992).

5. Dverstorp,B.,Andersson,J. and Nordqvist,W.:Discrete Fracture Network Interpretation of Field Tracer Migration in Sparsely Fractured Rock, Water Resour. Res., **28**, 2327-2343 (1992)

6. Moreno,L.,Tsang,C.F.,Tsang,Y., and Neretnieks,I.:Some anomalous features of flow and solute transport arising from fracture aperture variability, Water Resour. Res., **26**, 2377-2391 (1990)

7. Neretnieks,I.:Diffusion in the Rock Matrix: An Important Factor in Radionuclide Retardation?, Water Resour. Res., **85**, 4379-4397 (1980)

8. Grindrod,P.,McEwen,Robinson,P. and Savege,D.:Chapter 5 The far-field, in *The Scientific and Regulatory Basis for the Geological Disposal of Radioactive Waste*, edited by D.Savege (JOHN WILEY & SONS, Chichester,1995), pp.119-183.

9. Chapman,N.A., and McKinley,I.G.:*The Geological Disposal of Nuclear Waste*, (JOHN WILEY & SONS,Chichester,1987), pp.108-132.

10. Levenspoel,O.:Chemical Reaction Engineering, 2nd Edn., Wiley, New York, pp.275-278 (1972)

11. Niibori,Y.,Ogura,H. and Chida,T.:Identification of Geothermal Reservoir Structure Analyzing Tracer Responses Using The Two-Fractured-Layer Model, Geothermics, **24**, 49-60 (1995)

NEAR-FIELD CHEMICAL COMPOSITION OF POREWATERS IN A NEAR-SURFACE LOW-LEVEL RADIOACTIVE WASTE VAULT

F. Caron*, M. K. Haas*, G. Manni** and J. Torok***
Atomic Energy of Canada Limited, *Environmental Research Branch, and ***Waste technology,
Chalk River Laboratories, Chalk River, ON, K0J 1J0, Caronf@aecl.ca.
**CANDU Operations, Sheridan Park 2285 Speakman Dr. Mississauga, ON, L5K 1B2

ABSTRACT

A long-term waste degradation experiment has been performed with actual low-level radioactive wastes (LLRW) at the Chalk River Laboratories (CRL), to support the licensing and modelling efforts for near-surface disposal. The wastes consist of paper, mop heads, paper towels, used clothing, etc. The wastes were compacted into bales and sealed into separate steel containers, which were connected to leachate collection systems for sampling. The leachates collected had a composition typical of landfill leachates. The major inorganic ions were Na, Ca, Cl, and Fe, and the ionic strength was ~0.05 M. The relative distribution of inorganic ions in the leachates was remarkably similar between bales. Volatile fatty acids (VFA) were the major species of dissolved organic carbon (DOC; total DOC up to 7000 mg/L). A typical composition of leachates is proposed, which can be used in geochemical and source term modelling.

INTRODUCTION

Low-level radioactive wastes (LLRW) intended for near-surface disposal at Chalk River Laboratories (CRL), consist mostly of compacted bales (mop heads, used clothing, paper wipes, gloves, etc.) generated on-site, from universities, and research centres across Canada. The wastes contain various radionuclides (e.g., ^{3}H, ^{14}C, ^{60}Co, ^{90}Sr, ^{137}Cs, $^{239+240}Pu$ [1,2]) present in low but detectable levels. This type of waste constitutes most of the LLRW volume (>90%). Upon emplacement and contact with water, the release of salts from the wastes and the microbial degradation of the organic matter present in the wastes could impact on the safety of the facility. Some potential impacts are: gaseous transport of volatile nuclides and other hazardous waste components; a change in nuclide speciation due to the presence of dissolved organic carbon (DOC) originating from microbial degradation; and competition for sorption sites due to high salt contents, resulting in higher nuclide mobility in geological media; and a decrease in the long-term stability of the engineered structures due to the presence of these salts, etc.

As a first step to quantifying these impacts, a long-term degradation experiment was performed with LLRW at CRL [1]. Typical wastes, collected in the active area at CRL, were compacted into bales (dimensions 56 × 56 × 100 cm) and sealed in separate carbon steel containers. Each container was connected to a separate leachate collection system and tubing to allow gas and liquid sampling. Gases were sampled for their pressure and composition, and the liquid leachates were sampled for major ions, selected metals and nuclides, and DOC. The main objective of this work is to determine a source term of electrolytes contributing to the ionic strength of the solutions leached from the bales, to be used eventually for geochemical modelling.

Mat. Res. Soc. Symp. Proc. Vol. 465 © 1997 Materials Research Society

EXPERIMENTAL

The bale degradation experiment has been described in detail elsewhere [1,2]. A total of 11 bale containers and sampling lines were initially set up. The bales were allowed to degrade as-received for approximately one year, to obtain a gas generation term [3]. After this period, water was added (Jan. 1992) to 8 of these containers, and excess water was allowed to drain by gravity into a container located outside each box. Four of these bales were maintained in wet mode (called "wet bales" henceforth), simulating wastes located in the vadose zone. In this situation, some infiltrating water would be retained by the bale and the excess water would percolate through the bale. All the excess water from the wet bales (~10 L) was located in the overflow container. The outlet tubing was modified to flood the bottom half of the four remaining bales (called "flooded bales" hernceforth), to simulate wastes buried under saturated conditions. The leachates collected in each of the eight overflow containers were recirculated daily to promote degradation and leaching. The levels of leachate in the overflow containers were maintained relatively constant by periodic addition of make-up water to replace the sampled liquids.

The bale containers were sampled at 5 regular intervals for approximately 16 months, starting approximately 30 d after initial addition of water, and on an intermittent basis for the following 3 years. The results presented in this work are based mostly on the initial 16-month sampling period. Each sampling episode consisted of collecting gas and leachate samples, usually starting with gas collection. The gas pressure was recorded, and a sample was taken for analysis by gas chromatography (GC) [1]. Leachates were collected under an argon atmosphere in a glove box. An aliquot was acidified (5% v/v) with nitric acid upon collection for cation analysis by ICP-MS and for volatile fatty acid (VFA) analysis by GC [4]. An aliquot of the unacidified leachate collected for anion analysis was passed through a small bed of cation exchange resin upon collection, to remove cationic species that could precipitate in the sample. Analysis was done by liquid chromatography. Finally, DOC was analyzed on a Dohrmann total carbon analyzer.

RESULTS AND DISCUSSION

Dissolved electrolytes

The concentrations of the major dissolved inorganic cations and anions present in the leachates, for all the containers and all 5 sampling events, are shown graphically in Figure 1. Each one of the triangles represent the relative distribution of the major species as a function of Fe, Ca+Mg, Na+K for the cations, and NO_3, SO_4 and Cl for the anions. The diagram shows, for example, that the anions are made up almost exclusively of the Cl ion, whereas for the cations, the Ca+Mg constitute almost always ~25% of the species, Na+K is dominant (~100% of the cations for a few sampling events for Bale #1), and Fe is variable. The diamond-shaped field between the two triangles is used to show the composition with respect to both anions and cations. The Na^+ ion constituted ~75% of the combined $Na^+ + K^+$ in most cases.

It is remarkable that, despite the diversity of sources of wastes, and the number of "typical" garbage bags (~40 /bale) that were sampled, the composition of inorganic electrolytes is consistently dominated by the Na^+ and Ca^{2+} cations and by the Cl^- anion. Also, relatively high concentrations of Fe were observed in some of the leachates. It is possible that much of the Fe was introduced as an experimental artifact by corrosion of the containers that were used to hold the waste bales. These amounts of Fe, however, are not necessarily atypical of LLRW because

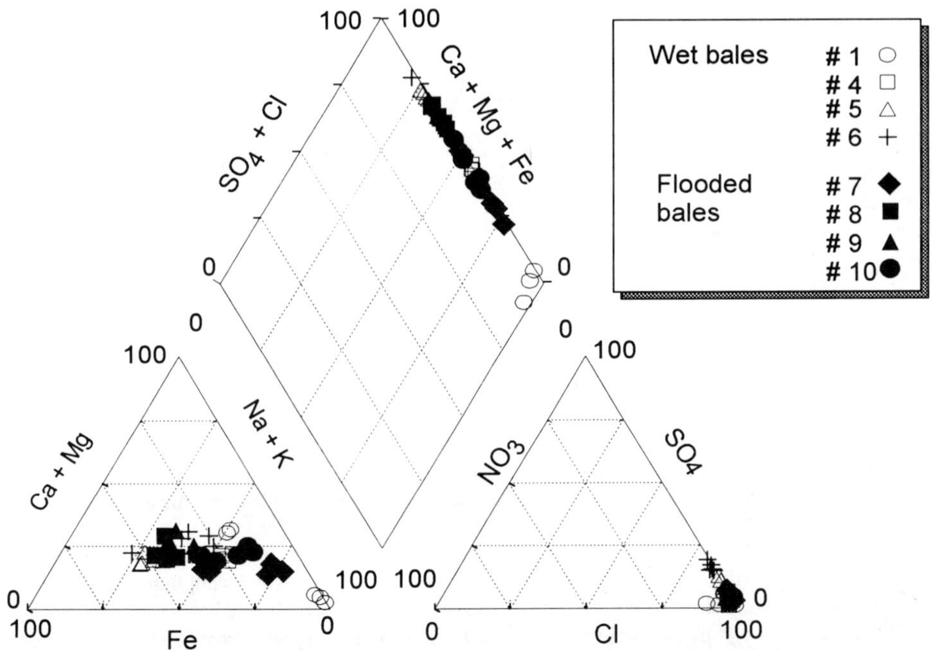

Figure 1: Ternary diagrams showing the distribution of the major inorganic ions in the leachates.

steel drums are often used as containers. The Fe concentrations observed in this experiment (~28-2600 mg/L; median ~500 mg/L) were comparable to the range of ~0.2-1400 mg/L found in another LLRW trench leachate [5], and some landfills (up to ~1500 mg/L; [6]). Moreover, Fe concentrations were highly variable from one bale to another (e.g., ~28-450 mg/L for bale 1, ~1200-2600 mg/L for bale 5), which suggests that the wastes themselves, as opposed to the bale containers, could be the main source of dissolved Fe in the leachates.

Transient organic compounds such as acetate and other VFA, although not shown in Figure 1, also contributed significantly to the ion balance of the leachates. The geochemical significance of these compounds is discussed in a separate section.

<u>Leaching behaviour</u>

There were no consistent trends with time in the amounts of ions leached per unit waste (A, mmole/kg waste), normalized to the maximum amount leached for the ion in each bale (A_{max}) for most inorganic ions in each bale. Some large fluctuations were noted, such as for Cl^- and Na^+ (Figure 2). Similar behaviour was observed for the other salts (Ca, Mg, etc.). Nevertheless, Figure 2 shows that with the exception of bales 4 and 10 (for Cl^-), and bales 1 and 4 (for Na^+), A/A_{max} was nearly constant with time, in the high fractional range. If these amounts represent steady-state ionic concentrations for each bale for the duration of the experiment, then the amount leached per unit of waste in a bale can be summarized by the information displayed in Table I, under "Amount of ions leached". This amount is tentatively called the "fast release fraction", and it represents the amount of easily leachable constituents per unit waste, for the time period shown.

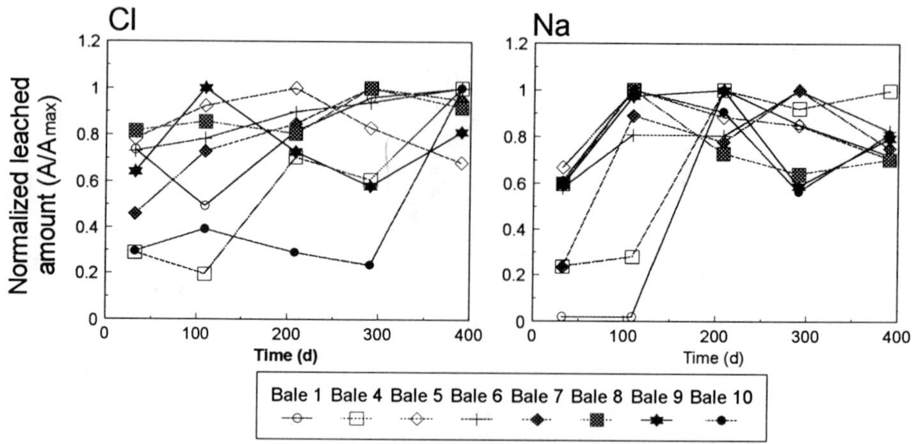

Figure 2: Normalized concentration of Na^+ and Cl^- ions vs time for all bales (the amounts leached (mmol/kg waste) were normalized to the maximum amount leached for each bale, A_{max}).

Although the variations in leachate between and within bales are much too large to apply statistical analysis, the median of the population for each ion (8 bales × 5 sampling events = 40 values) is reasonably close to the average calculated. We interpret the median as being a measure of the centre of the population of sampling events, if each sampling was treated as an independent event. The 1σ around the average is indicated in Table I to show that the variability with respect to the major constituents (Na, K, Ca, Mg, Fe, Cl, acetate & VFA, DOC) is less than 100%, which is high but reasonable given the diverse origins of these wastes. Minor constituents (metals, SO_4^{2-}, NO_3^-, etc.) exhibited large fluctuations. A discussion of their behaviour is not useful in the current context, since their concentrations are low relative to the major constituents. This also holds for the radionuclides.

Estimates of porewater concentrations in engineered vaults

If one considers a generic near-surface LLRW disposal vault with an impermeable cover or similar engineered cover to limit water movement, and if the vault contained only baled wastes, the fast released fraction of ions ("average", Table I) could be used as a source term to calculate ionic compositions of the porewaters. Since the "fast released fraction" was obtained during a relatively short period, the ionic composition could be representative of the early post-closure period of the vault. In the absence of groundwater advection, the dissolved ions from the "fast release fraction" would migrate into the rest of the vault primarily by diffusion in the saturated or unsaturated medium. The porewater concentrations have been calculated for four hydraulic conditions, assuming hydrostatic equilibrium in a vault (Table I, Scenarios 1-4): 1) saturated bale (base case), 2) flooded vault, 3) vault after cover failure, without a cover, or under humid conditions, and 4) dry vault, under arid or dry initial conditions. The calculations are based on the average leached fraction (mmole compound/kg waste), for each element or compound listed in Table I, a bale density of 570 kg/m^3 [1,2], and a bale porosity of 0.579 (n = 8 [1, 2]). In scenarios 2, 3, and 4, it is assumed that the bales are with an equal volume of a loamy sand or a sand/clay mixture (typical porosity ~0.35 [7,8]). It was also assumed that aqueous reactions with the sandy backfill do not contribute signficantly to the ionic loadings, which is a reasonable assumption for the siliceous sand at CRL.

The porewater compositions calculated for the four scenarios have ionic strengths ranging from 0.027 mol·kg^{-1} (flooded vault) to 0.165 mol·kg^{-1} ("dry" vault). The composition reported for Scenario #4 is expected to be similar to the near-surface disposal concepts at CRL, where the engineered facility would isolate the wastes from water infiltration. Given various assumptions in the calculations, however, the ionic strength of porewater for Scenario 4 could vary considerably. For example, the water content in the bales (0.08 m^3/m^3) was estimated from the initial conditions of the bales as-received. If the predicted water content in the vault was somewhat higher (e.g., 0.100 m^3/m^3 instead of 0.076 m^3/m^3), the resulting vault porewater would have a lower ionic strength (~0.125 mol·kg^{-1} instead of 0.165 mol·kg^{-1}). Conversely, if the water content of the sandy backfill was only 0.036 m^3/m^3, as has been suggested [9], then the total vault water content

Table I: Normalized quantities of leached ions per unit of waste, and porewater composition given water content scenarios.

Species (nominal)	Amount of ions leached* (millimole comp./kg of waste)			Scenarios for water content (mmole/kg solution)			
	Median	Average	Rel. std dev.	#1	#2	#3	#4
Na+**	8.68	9.63	69%	9.5	5.9	13.6	36.1
K+	1.92	2.64	76%	2.6	1.6	3.7	9.9
Ca++	3.86	3.92	57%	3.9	2.4	5.6	14.7
Mg++	1.14	1.36	65%	1.3	0.83	1.9	5.1
Fe++	5.61	7.73	78%	7.6	4.7	11.0	29.0
Mn++	0.065	0.069	74%	0.07	0.04	0.10	0.26
Al+++	0.029	0.060	198%	0.06	0.04	0.09	0.22
Zn++	0.041	0.215	151%	0.21	0.13	0.30	0.81
Cu++	0.003	0.005	130%	0.005	0.003	0.007	0.019
SO4--	0.047	0.132	130%	0.13	0.08	0.19	0.49
Cl-	3.71	6.50	85%	6.4	3.99	9.2	24.4
NO3-	0.005	0.069	289%	0.07	0.04	0.10	0.26
Acetic acid	14.5	23.5	119%	23.2	14.5	33.4	88.2
Total VFA (mmol C/kg)***	81.3	81.7	84%	80.4	50.2	116	306
Total DOC (mmol C/kg)‡	92.6	92.6	50%	90.8	56.6	131	346
Ionic strength (mol/kg)	0.044			0.043	0.027	0.062	0.165
pH (all scenarios)	7-8 (range observed)						
Water content (m^3/m^3)				0.579	0.464	0.201	0.076

*n = 40 (8 bales × 5 sampling episodes) unless otherwise specified.
**n = 37 for the average (1 sampling episode and the subsequent 2 were not included for Bale 1.
***n = 24 (8 bales × 3 sampling episodes), because of analytical problems.
‡n = 32 (4 sampling episodes), also because of analytical problems.

Scenarios #:
1. Flooded bale. Water content = 0.579 m^3/m^3.
2. Flooded vault. Water content = 0.464 m^3/m^3.
3. Field capacity, which approximates a vault with at least a partial cover failure. Values of field capacity: 0.290 m^3/m^3 (wastes), 0.113 m^3/m^3 (sand) [8], total = 0.2 m^3/m^3.
4. Dry vault: water content = 0.072 m^3/m^3 ("Initial values" for sand [8]); wastes: 0.08 m^3/m^3 (estimate for bales; unpublished), total = 0.076.

would be 0.058 m^3/m^3, and the resulting ionic strength of the vault porewater would be 0.210 $mol \cdot kg^{-1}$. Weak ionic interaction effects become increasingly important at high ionic strength, which potentially could limit the use of geochemical modelling to predict the speciation of porewater and its saturation state with respect to various solids. However, all of the ionic strengths calculated in these variations are low enough (< 0.50 $mol \cdot kg^{-1}$) that geochemical modelling is still feasible, provided that ion interaction effects are accounted for by one of several ion activity correction methods, such as the Davies equation or the WATEQ Debye-Huckel equation [10].

Transients and other specific cases: NO_3^-, SO_4^{2-}, VFA, DOC

In all of the leachates, the NO_3^- and SO_4^{2-} concentrations were low (near or below the detection limit of 0.5 mg/L for NO_3^-). Particularly, the SO_4^{2-} data were at the low end (<1-250 mg SO_4^{2-}/L; median ~4.4 mg SO_4^{2-}/L) of a range expected in landfills (55-456 mg/L [11]). We believe that by the time that water was added to the bales to mark the beginning of the wet degradation stage, most of the sulfate ions had already been consumed by microbial activity to produce H_2S, and/or organic S species. During the compaction process, we have observed that several of the bags contained water, in addition to the solid wastes. This water content would have sufficed to sustain the growth of microbial populations within the bales. By the same process, some of the nitrate and nitrogen gas would have been converted to NH_4^+, which was detected in concentrations of about 50-250 mg/L at later stages of this experiment [2].

Concentrations of VFA and DOC varied considerably (~170-7000 mg C/L). VFA were not abundant in the flooded bales, which produced mostly methane gas instead, as expected, because of the high water content [6]. VFA were most abundant in the wet bales. The VFA result from the degradation of structural polysaccharides (paper, cellulose, etc.) that make up a large part of the matrix ([12]). The VFA are transient in landfills [6], and they are not likely to be part of a long-term source term of material coming from the bales. Chromatographic separations of leachates from the bales also revealed the presence of humic/fulvic-like compounds [13]. Such compounds are likely to be the dominant type of DOC produced upon ageing in landfills [14;15;16].

Effects of possible disposal system conditions

Under the conditions of the experiment, all of the bales were in a reducing environment (~0 to -0.4 V; unpublished). The "wet bales" were mostly in acetogenesis or acid-producing mode, whereas the "flooded bales" were in methanogenesis mode [6]. In real disposal or burial situations, there is considerable debate whether conditions will be reducing or oxidizing. It is likely that reducing conditions will prevail in the bales if an engineering facility is flooded or below the water table, but the question is not resolved for an engineered facility in the vadose zone. If oxygen availability is restricted by a low porosity medium surrounding the wastes, conditions could eventually become reducing. Any water that would eventually be absorbed from the surroundings would restrict oxygen access within a bale, and this may create pockets where reducing conditions could prevail.

Regardless of the oxidation potential of the system, most of the inorganic elements listed in Table I are not strongly redox-sensitive, and so source terms would be relatively unaffected, except for Fe. An analysis of the leachate compositions using the geochemical modelling code MINTEQA2 [10] suggests that, if the Fe content in the leachates was allowed to oxidize, virtually all of the Fe would have precipitated as Fe oxide phases.

Fe concentrations also may be affected by the degradation of organic compounds over time in a disposal system. Relatively high concentrations of VFA were observed in wet bales, but these compounds are transient. Because VFA constitute an important part of the ion charge balance, their degradation (stochiometrically: to CO_2 and CH_4) has to be accompanied by the removal of cationic species. Indications in landfills of various ages suggest that VFA removal is accompanied by a decrease in Fe concentration [6].

CONCLUSIONS

Despite the heterogeneous content of compacted wastes in the bales, the leachates from this LLRW degradation experiment had a relatively constant chemical composition, dominated by the Na-Ca-Cl ions. Considerable variations in concentration were observed but were nevertheless not as diverse as might have been expected given the expected variability of the wastes contained in the bales, and their many sources. If the analyzed leachates were used to constitute the porewater composition in a wet or dry engineered vault, the ionic strength of the vault porewater is estimated to be in the range of ~0.1 - 0.2 mole/L. This is well within the limit for the application of a simple method, such as the Davies equation, to correct for ion interaction effects as part of modelling speciation and solubilities in the vault porewaters. Even in the case of a dry, near-surface disposal vault in which the expected water content differs by ±25-30% from the scenario modelled in this study, the resulting changes in ionic strength would still be well within the limit of application for such activity coefficient correction methods.

ACKNOWLEDGEMENTS

The authors wish to thank J. McMurry (AECL, Whiteshell Laboratories) for providing useful suggestions for and improvements of this manuscript. We also thank D. Spagnolo and A.I. Miller (Chemical Engineering, CRL) who have provided us with a suitable location to perform part of this experiment. C.M. Bertrand (AECL, Chalk River) has provided editorial comments on the manuscript.

REFERENCES

1. M. K. Haas, G. Manni, F. Caron, J. Torok, Description and analytical methods used in Low-level waste leaching experiment. Atomic Energy of Canada Limited, Chalk River Laboratories. Report No.TR-662, 1996.*

2. F. Caron, J. Torok, M.K. Haas, G. Manni, These Proceedings, 1996.

3. J. Torok and M. K. Haas, in Gas Generation and Release From Radioactive Waste Repositories, workshop held Sept. 23-26, 1991, Aix-en-Provence, France. (NEA, OECD, Paris, France, 1992), pp. 175-187.

4. G. Manni and F. Caron, J. Chromatogr **690**, 237 (1995).

5. R. Dayal, R. F. Pietrzak, and J. H. Clinton, Nucl. Technol. **72**, 158 (1986).

6. H.-J. Ehrig, Waste Manag. Res. **1**, 53 (1983).

7. B. L. Woods, <u>Physical properties of backfill materials selected for use in a low-level waste repository</u>. Atomic Energy of Canada Limited, Chalk River Laboratories Report No. TR-259, 1986.*

8. D. B. Stephens and L. M. Coons, Ground Water Monit. Remed. **14**, 101 (1994).

9. M. I. Sheppard, D. H. Thibault, and J. H. Rowat, <u>Predictions of average moisture content and capillary rise of the candidate sand and clinoptilolite for the IRUS disposal concept</u>. Atomic Energy of Canada Limited, Whiteshell Laboratories, 1995.*

10. S. R. Peterson, C.J. Hostetler, W.J. Deutsch, C.E. Cowan, <u>MINTEQ user manual.</u> Pacific Northwest Laboratory Report No. NUREG/CR-4808; PNL-6106, 1987.

11. D. J. Lisk, Sci. Tot. Environ. **100**, 415 (1991).

12. F. Caron, <u>Dissolved Organic Matter in Low-Level Waste Leachates: first year of activities</u>. Atomic Energy of Canada Ltd., Chalk River Laboratories, Report No. TR-560 (COG-93-100), 1994.*

13. F. Caron, S. Elchuk, and Z. H. Walker, J. Chromatogr. **739**, 281 (1996).

14. M. Weis, G. Abbt-Braun, and F. H. Frimel, Sci. Tot. Environ. **81/82**, 343 (1989).

15. F. H. Frimmel and M. Weis, Wat. Sci. Tech. **23**, 419 (1991).

16. O. Castagnoli, L. Musmeci, E. Zavattiero, M. Chirico, Water, Air, Soil Pollut. **53**, 1 (1990).

------

*AECL unpublished report, available from SDDO, Chalk River Laboratories, Chalk River, Ont. Canada K0J 1J0.

DETAILED DESCRIPTION OF A LONG-TERM LOW-LEVEL WASTE DEGRADATION EXPERIMENT

F. Caron*, J. Torok**, M. K. Haas*, G. Manni***
Atomic Energy of Canada Limited, *Environmental Research Branch, and **Waste Technology, Chalk River Laboratories, Chalk River, ON, K0J 1J0, Caronf@aecl.ca.
***CANDU Operations, Sheridan Park 2285 Speakman Dr. Mississauga, ON, L5K 1B2

ABSTRACT

This work gives a detailed description of the important aspects of a long-term Low-Level Radioactive Waste (LLRW) degradation experiment, performed at Chalk River Laboratories (CRL). This experiment utilized actual LLRW. The wastes consist of unconditioned compacted refuse (paper, mop heads, paper towels, used clothing, etc), which represents the bulk of the waste volume intended for near-surface disposal at CRL. Waste material was collected and compacted to make a total of 11 bales for this experiment. Each bale was then placed and sealed in separate steel containers which were connected to sampling lines. After a dry monitoring period, water was added to promote leaching and decomposition of the wastes. The leachate sampled had a composition similar to landfill leachates. Some applications of this experiment, used to support the safety case of near-surface disposal, are briefly discussed in this paper, e.g., the production of colloidal material, the nature and role of dissolved organics of microbial origin, etc.

INTRODUCTION

In a first step to determine the factors affecting leaching and degradation of LLRW, a long-term degradation experiment has been performed with wastes representative of the material intended for disposal at CRL. LLRW contain radionuclides that usually constitute only small quantities of material, on a mole basis. Therefore, degradation processes taking place in the wastes are likely to control the chemical environment, which in turn affects the source term and behaviour of radionuclides. For example, leaching and degradation of the matrix will mobilize inorganic ions and produce dissolved organic carbon (DOC). The latter processes affect the speciation and the mobility of toxic metals and radionuclides contained in the wastes.

Actual LLRW were used in this experiment, which consisted of unconsolidated material (paper, mop heads, paper towels, used clothing, etc.), generated on-site in the active area laboratories at CRL, and compacted into bales. A total of 11 bales were prepared and sealed into separate carbon steel containers, equipped with gas and liquid sampling lines. The wastes were allowed to degrade as-received for a period of approximately one year in order to obtain a gas generation rate [1]. After this period, water was added to 8 bales to promote leaching and degradation, while the other 3 bales were reserved for future use. The water added to these bales was intended to simulate some of the natural long-term processes that would take place in a disposal vault if, for example, the wastes were located under the water table, or after water infiltration in a vault following the failure of an engineered cover.

The aim of this manuscript is to present a detailed description of the experiment. Applications of this experiment to support licensing cases for near-surface LLRW disposal at CRL are also discussed.

Mat. Res. Soc. Symp. Proc. Vol. 465 © 1997 Materials Research Society

EXPERIMENTAL

Waste collection and bale preparation

Individual bags of LLRW were collected at the CRL active area site from January to August 1988. Each bale consisted of ~40 bags of wastes, processed with a 50-ton compactor in two stages of ~20 bags each. After the first stage of compaction, probes (tensiometers and thermocouples) were placed on top of the compacted bags, followed by the addition and compaction of the other 20 bags. The final dimensions of these bales were ~ $0.56 \times 0.56 \times 1.0$ m, for a volume of ~0.31 m^3 (the normal bales produced at CRL have a volume of ~0.45 m^3). Each bag was assayed with a Aptec Ge detector before compaction. The inventory of gamma emitters, based on the total amount detected in individual bags, is shown in Table I. All the bags in 5 of these bales were injected with 10 mL of KI solution (100 mg/L as I⁻), which serves as a conservative (i.e., non-sorbing) tracer.

Container preparation, sealing and commissioning

Each container was constructed of 1/8" (3.2 mm) welded carbon steel sheet. Swagelok™ fittings were placed at the bottom, near the top, and through the top sheet of each container for instrumentation and sampling. Two pieces of lumber and ~5 cm of sand were placed at the bottom of each container to support the bale. The sand also acted as a filter to keep the outlet tubing from clogging. After placement of the bales in their respective containers, the thermocouple leads and the tensiometer connection lines (previously inserted during compaction) were directed through the Swagelok™ fitting located on the top sheet. Finally, the container top was welded, and the fitting routing the tensiometer and thermocouple leads was sealed with epoxy cement and/or silicone sealant.

After these steps, each container was pressurized to ~5 psi (34 kPa) with air to test for leaks. Any leak was re-welded and re-sealed with epoxy and/or silicone cement, as appropriate. This was repeated until the pressure inside each container remained at ~5 psi (34 kPa) for 1 to 2 d. The containers were then vented to atmospheric pressure prior to starting the actual experiments. The containers remained sealed thereafter.

A peristaltic pump and a 10 L leachate collection flask were connected to the bale containers using copper tubing (Figure 1) to constitute independent bale container leaching and sampling stations. Valves and fittings were placed appropriately on the lines to facilitate sampling. The

Table I: Inventory of gamma emittors in the 8 bales used in this experiment, based on individual bag counting.

Bale #	Radionuclide (MBq/bale)						
	Co-60	Cs-137	Cs-134	Ce-144	Zr-95	Nb-95	Zn-65
1	0.29	0.51	0.01	0.19	0.62	1.3	0.003
4	2.3	0.84	0.22	1.7	0.68	1.7	0.16
5	1.6	0.72	0.07	1.9	0.48	1.1	1.9
6	1.8	77.8	0.16	0.85	0.42	0.90	0.14
7	7.2	13.3	3.7	25.1	0.99	2.2	
8	6.7	0.35	0.05	0.50	0.28	0.83	0.007
9	2.0	1.2	0.05	0.72	1.31	2.1	0.36
10	12.0	4.9	1.4	14.1	0.30		

"gas pressure and equalization line" was necessary to equalize the pressure between the overflow container and the steel container containing the bale, which allowed leachate to flow by gravity in a subsequent part of this waste degradation experiment.

The bale porosity was calculated during the leak testing operation. The principle was based on the differences in pressure before and after venting. The dead volumes (between the walls of the container and the bale, the tubing, etc.) are known, and the pressures within each container before and after venting were measured. The pressure difference was attributed to the pore volumes. The calculation was done using the ideal gas law relationship:

$$\frac{V_1}{V_2} = \frac{P_2}{P_1}$$

(1)

Where: V_1 = Bale container void volume (dead volume + pore volume)
P_1 = Absolute pressure of gas in container (Atmospheric + ~5 psi (34 KPa))
V_2 = Gas volume at P_2 (= V_1 + V_v (vented gas volume), at ambient atmospheric conditions)
P_2 = Final pressure in the container (slightly above atmospheric).

This approach to measure the bale porosity was based on the assumption that there was a good pneumatic conductivity between the bale pores. The measurements were made within a relatively short period of time to minimize barometric and temperature variations, and also to ensure that gas generation from the degrading bale was negligible during this period. The porosities measured at the start of the experiment and the bale weights are shown in Table II. These numbers represent the bale characteristics, "as-received". The bales contain a small amount

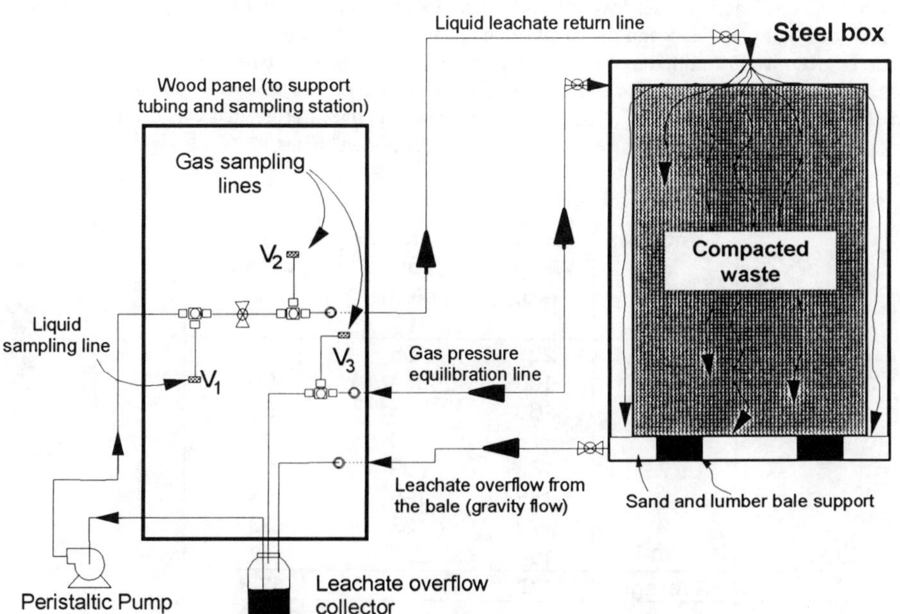

Figure 1: Schematic of a bale container leaching and sampling station.

of water, as revealed by water seeping out of the bales during compaction. A crude estimate of the original water content was calculated, based on a pressurization and venting operation done at the end of the monitoring period. Additions and removal of water for sampling (see later) were recorded in our data log, and using the approach described above to determine the final pore volumes, the original water content was back-calculated. The value of the original water content was estimated to ~8% (by weight). We did not correct the bale weights for their water content, because of the uncertainty associated with this value.

<u>Monitoring period for the dry bales ("as-received")</u>

The bales were initially monitored "as-received" for about one year starting April 1989, to simulate the conditions of degradation prior to the roof failure of a vault [1]. The water already present in the bales was sufficient to promote microbial degradation and gas generation. No water was added, and only the container pore gases were sampled and analyzed during this period. The results showed that the gas composition started to change within a few months after sealing. Oxygen was quickly consumed, and CO_2 and CH_4 gases were generated, indicating microbial degradation. After this gas sampling period (ending ~June 1990), the containers were sampled sporadically for an additional ~18 months (~January 1992).

<u>Wet monitoring period</u>

After the "as-received" monitoring, the experimental set-up was modified to simulate conditions in a vault where water ingress had occurred. Water was added through the leachate return line (Figure 1) until water trickled by gravity from the bottom outlet of the container, to fill most of the 10 L leachate overflow collector. From 10 to 36 L of water were added to each of the first four bales (#1, 4, 5 and 6). Water remained in these bales by capillary forces, and all of the excess free water drained into the leachate collector. These four bales are henceforth referred to as the "wet" bales. The liquid outlet tubing was modified on the other four containers (#7, 8, 9 and 10), to flood the lower half of each bale. The amounts of water needed was from 60 to 210 L. These four bales are henceforth referred to as the "flooded" bales.

Table II: Bale characteristics*

Container #	Bale Weight (kg)	Container void volume (L)	Bale porosity %
1	221	156	50
4	180	204	66
5	163	215	69
6	176	130	42
7	182	209	67
8	153	195	63
9	172	103	33
10	166	227	73
Average	**177**	-	**57.9**

*Bale volume = 0.31 m^3; dead volume = 0.144 m^3

Approximately 10 L of leachate was kept in each overflow container for the flooded bales as well. This level was maintained in all overflow containers by periodic addition of make-up water to individual containers to replace leachate that had been sampled or absorbed by the bales.

The leachates were recirculated daily using the pump connecting the container to the top of each bale. Recirculation was carried out over a ~30-45 min period each day, at a rate of ~0.35-0.4 L/min, for a total of ~10-18 L/d per bale. This corresponded to ~4-5.4 cm/day of rainfall infiltration on each bale. Additions of fresh water, when needed (e.g., when no or little water was observed in an overflow container), were done at least 6 days prior to sampling. This gave us a good assurance that the freshly added water was completely mixed and conditions inside each bale container had time to return to "steady-state". The fresh water used in the early part of the experiment was supplied from the building steam condensate, whereas regular service water was used at late stages of the program (1994-1996). The source of water did not matter, because the bale decomposition processes controlled the chemistry of the water.

Gas and liquid sampling were performed regularly for approximately 16 months, from January 1992 to May 1993. The most complete set of results was obtained during that period. An inventory of liquid added and removed was kept for each bale container. The gas pressures were taken at least 2-3 times/month to obtain gas generation data. Excess gases were vented off to prevent overpressurization.

After this intensive monitoring period, the four wet bales were moved to a different location in the fall of 1994 and they were sampled intermittently until May 1996. The flooded bales were not monitored.

<u>Sampling</u>

Sampling consisted of collecting gas and leachate samples, usually starting with gas collection. The order of sampling (i.e., sampling gas or leachate first) did not affect the results. The sampling procedures for the wet sampling period are summarized below. The detailed procedures and most of the results obtained in the 16-month sampling period are listed in the primary reference [2]. A detailed log of the material that was added and removed from each container was kept for mass balance purposes.

Since the conditions in the bale containers were reducing (measured values ~0 to -0.4 V), a glove box, flushed with argon gas to displace atmospheric oxygen, was built to accommodate the sampling flasks and equipment for the collection of all leachate samples. A piece of flexible tubing was attached from the liquid sampling line (V_1 in Figure 1) to the glove box. The pump was turned on to withdraw leachate from the leachate overflow container. The pH and conductivity were measured immediately upon sampling. An aliquot of this solution was immediately acidified with nitric acid for cation, DOC and volatile fatty acid (VFA) analysis. The original unacidified solution was prepared for anion analysis by passing triplicate aliquots of 10 mL of leachate through a 1-mL cartridge of cation exchange resin to remove cations that could precipitate. Cations were analyzed using ICP-AES, anions using a Dionex HPLC, VFA were analyzed by gas chromatography [3], and DOC was analyzed on a Dohrman DC-80 carbon analyzer. The ^{90}Sr was determined by Cerencov counting of its ^{90}Y daughter, after a "grow in" period of 5 d. The gamma emittors (Cs isotopes, ^{60}Co) were determined using a high purity Ge detector.

RESULTS AND DISCUSSION

Upon the first addition of fresh water (January 1992), the water that came down the overflow collector was already quite dark brown, coffee-like, with a pungent smell of decomposition. This water was likely present originally in the bale, and it was displaced by the addition of fresh water. This observation, in addition to gas production from microbial degradation, further confirms that leaching and decomposition were already taking place even in the unsaturated bales (as-received).

One example of leachate analysis for container #6 is shown in Table III. These results were typical of the leachate compositions obtained throughout this program [2,4], and more interestingly, this composition was observed after only one month of monitoring in the wet mode. This ionic composition was typical of landfill leachates (see Table III). VFA made up two thirds or more of the DOC in the "wet" bales, and [5]the relative abundance of the various VFA was remarkably similar between sampling dates for the same bale, despite significant fluctuations in DOC [2,5]. The similarity between the composition of these leachates and landfill leachates is important, because of the apparent differences between the types of wastes: compostable material such as food and yard wastes are disposed of in municipal landfills [6], whereas these types of wastes are excluded from LLRW. This suggests that, despite the apparent source of the waste material, LLRW will undergo similar degradation steps compared to a landfill. Microbial degradation produce high amounts of VFA in young landfills, and this process controls, at least in part, the degradation rates of the carbon material in the LLRW [5,7]. The pieces of wood supporting the bale in the containers, would produce similar degradation compounds. Chromatographic separations of the leachate DOC confirmed that 70-93% of the DOC consisted of hydrophilic compounds (VFA are predominantly hydrophilic), and that the hydrophobic compounds exhibited some capacity to complex Co [8]. The "flooded" bales produced mostly CH_4 and CO_2 instead of VFA ([2]; results not shown). This is consistent with what is found in flooded landfills [9].

The amounts of radionuclides (^{60}Co, [134]Cs, ^{137}Cs) found in the leachates represent only a small fraction of the total inventory (inventory not available for ^{90}Sr). If the amounts reported in Table III are used to calculate a fast-release fraction (defined as the ratio of the total amount present in leachate to the total inventory; see [4]), this fraction is, respectively, 0.013%, 0.06% and 0.002% for ^{60}Co, [134]Cs and [137]Cs. This fraction is small compared to that of the major ions (~5-10% for Na, Ca, Mg, K, Fe [10]).

It is also possible that the containers, made of carbon steel, have introduced an artefact in the amount of iron leached. In our experiment, the amount of Fe leached (defined above) was not preferentially enriched compared to the other major constituents [10]. The Fe concentration reported in Table III is not atypical of LLRW trench leachate (~0.2-1400 mg/L; [11]) and some landfills (up to ~1500 mg/L; [6]). Moreover, steel drums are often used as containers in LLRW, so leaching of Fe is expected as well.

The levels and species of ions in the leachates also constitute a potential source term of ions that could affect the groundwater quality in the vicinity of a disposal site. The waste degradation experiment provided a range of porewater compositions [4], which could be used in geochemical and performance assessment modelling.

CONCLUSIONS

This low-level waste degradation experiment is unique, partly because it was performed with real wastes in a closed system. This allowed us to have a reasonable mass balance of active and inactive material. The composition of the leachates is typical of municipal landfill leachates. This is important, because landfills contain compostable material that is usually not present in LLRW, yet VFA are being produced, which are related to microbial decomposition.

Aqueous species (nominal)	Concentrations (mg/L)	
	This study, container #6 day #32	Range observed in landfills (from [12], Table 6)
Na^+	400	43-2500
K^+	215	20-650
Ca^{++}	389	165-1150
Mg^{++}	68	12-480
Fe^{++}	720	0.09-380
Mn^{++}	8.8	0.32-26.5
Al^{+++}	43	-
Zn^{++}	92	<0.05-0.95
Cu^{++}	0.3	<0.01-0.15
SO_4^{2-}	250	55-456
Cl^-	404	70-2777
NO_3^- (as N)	<D.L.*	<0.2-4.9
NH_4^+ (as N)	50-250**	5-730
I^-	32	-
PO_4^{3-}	<D.L.*	<0.02-3.4
Acetic Acid (as C)	1200-2500***	-
Propionic Acid (")	155	-
Isobutyric Acid (")	62	-
Butyric Acid (")	380	-
Isovaleric Acid (")	12	-
Valeric Acid (")	66	-
Isocaproic Acid (")	245	-
Total VFA (mg C/L)	2120-3420***	-
COD†	8700-14200‡	66-11600
DOC (mg C/L)	3700	21-4400
pH	4.83	6.2-7.4
Conductivity (mS/cm)	5.43	-
Gases (Volume %)		
CH_4	3.9	
CO_2	27.5	
O_2	0.3	
N_2	68.3	
Radionuclides		
^{137}Cs (filtered, Bq/L)	4400	-
^{134}Cs (filtered, Bq/L)	280	-
^{60}Co (filtered, Bq/L)	700	-
^{90}Sr (unfilt., Bq/Kg)	110	-

*<D.L.: below detection limit; D.L. was ~0.5 mg/L for NO_3^- and PO_4^{3-}.

**Ammonia unavailable for BBLSS#6; the range reported was obtained for two other bales at a different date.

***Acetic acid unavailable at that date because of analytical problems; range reported for this bale at other dates.

†Chemical Oxygen Demand

‡Using the relationship between VFA and COD [6] (COD (mg/L) = 111*VFA (meq/L) + 1148).

Some radionuclides were found in the leachates, which corresponded to approximately 0.002 to 0.013% of their total inventories. This quantity was small compared to other elements (Na, K, etc.). The experiment should be run for some additional time with analyses customized for radionuclides to determine an appropriate source term of nuclides from this type of LLRW.

ACKNOWLEDGEMENTS

The authors have sincerely appreciated the comments made by B.L. Woods (Waste Technology, AECL, Chalk River) on this manuscript, and the help of D. Spagnolo and A.I. Miller (Chemical Engineering Branch, AECL, Chalk River) for providing us with a suitable location to perform part of this experiment. This manuscript has also benefited from the editorial contribution of C.M. Bertrand (AECL, Chalk River).

REFERENCES

1. J. Torok and M. K. Haas, in <u>Gas Generation and Release From Radioactive Waste Repositories</u>, *workshop held Sept. 23-26, 1991, Aix-en-Provence, France.* (NEA, OECD, Paris, France, 1992), pp. 175-187.

2. M. K. Haas, G. Manni, F. Caron, J. Torok, Description and analytical methods used in Low-level waste leaching experiment. Atomic Energy of Canada Limited, Chalk River Laboratories. Report No.TR-662, 1996.*

3. G. Manni and F. Caron, J. Chromatogr **690**, 237 (1995).

4. F. Caron, M.K. Haas, G. Manni and J. Torok, Near-field chemical composition of porewaters in a near-surface Low-Level Radioactive Waste vault, These Proceedings, 1996.

5. F. Caron, Dissolved Organic Matter in Low-Level Waste Leachates: first year of activities. Atomic Energy of Canada Ltd., Chalk River Laboratories, Report No. TR-560 (COG-93-100), 1994.*

6. H.-J. Ehrig, Waste Manag. Res. **1**, 53 (1983).

5. D. J. Lisk, Sci. Tot. Environ. **100**, 415 (1991).

7. F. Caron, in:.Waste Management'96, held in Tucson, AZ, Feb 25-29, 1996, Edited by R.G. Post and M.E. Wacks. WM Symposia Inc., Tucson, AZ, 1996; ibid, Atomic Energy of Canada Ltd. Report No RC-M-20, 1996*.

8. F. Caron, S. Elchuk, and Z. H. Walker, J. Chromatogr. **739**, 281 (1996).

9. J. F. Rees, J. Chem. Tech. Biotechnol. **30**, 161 (1980).

10. P. Vilks, F. Caron, and M. K. Haas, Appl. Geochem. *Accepted* (1996).

11. R. Dayal, R. F. Pietrzak, and J. H. Clinton, Nucl. Technol. **72**, 158 (1986).

12. D. J. Lisk, Sci. Tot. Environ. **100**, 415 (1991).

---

*AECL unpublished report, available from SDDO, Chalk River Laboratories, Chalk River, Ont. Canada K0J 1J0.

ENUMERATION OF MICROBIAL POPULATIONS IN RADIOACTIVE ENVIRONMENTS BY EPIFLUORESCENCE MICROSCOPY

M.E. PANSOY-HJELVIK*, B.A. STRIETELMEIER*, M.T. PAFFETT*, S.M. KITTEN*, P.A. LEONARD*, M. DUNN**, J.B. GILLOW**, C.J. DODGE**, R. VILLARREAL*, I.R. TRIAY*, and A.J. FRANCIS**. *Los Alamos National Laboratory, CST-7 MS J514, Los Alamos, NM 87544, meph@lanl.gov. **Brookhaven National Laboratory, PO 5000, Bldg. 318, Upton, NY 11973

ABSTRACT

Epifluorescence microscopy was utilized to enumerate halophilic bacterial populations in two studies involving inoculated, actual radioactive waste/brine mixtures and pure brine solutions. The studies include an initial set of experiments designed to elucidate potential transformations of actinide-containing wastes under salt-repository conditions, including microbially mediated changes.

The first study included periodic enumeration of bacterial populations of a mixed inoculum initially added to a collection of test containers. The contents of the test containers are the different types of actual radioactive waste that could potentially be stored in nuclear waste repositories in a salt environment. The transuranic waste was generated from materials used in actinide laboratory research. The results show that cell numbers decreased with time. Sorption of the bacteria to solid surfaces in the test system is discussed as a possible mechanism for the decrease in cell numbers.

The second study was designed to determine radiological and/or chemical effects of ^{239}Pu, ^{243}Am, ^{237}Np, ^{232}Th and ^{238}U on the growth of pure and mixed anaerobic, denitrifying bacterial cultures in brine media. Pu, Am, and Np isotopes at concentrations of $\leq 1 \times 10^{-5}$ M , $\leq 5 \times 10^{-6}$ M and $\leq 5 \times 10^{-4}$ M respectively, and Th and U isotopes at concentrations of $\leq 4 \times 10^{-3}$ M were tested in these media. The results indicate that high actinide concentrations affected both the bacterial growth rate and morphology. However, relatively minor effects from Am were observed at all tested concentrations with the pure culture.

INTRODUCTION

Epifluorescence microscopy has been used to count bacterial populations in two studies related to the disposal of transuranic (TRU) radioactive waste [1,2]. The motivation for the microscopy studies was to determine the relevant microbially-influenced system variables which could be crucial with respect to transuranic waste repositories. Bacteria present at potential TRU waste disposal sites can play roles in, for example, 1) actinide retardation or transport [2], 2) biological transformation of organic or inorganic materials in radioactive waste containers [3], and 3) gas generation due to bacterial metabolic activity [4]. Prior to addressing any of these variables, it is first necessary to confirm the presence of the bacteria and to quantify their populations.

In the first study, bacterial total numbers were determined in tests of actual low-level transuranic waste under high-ionic strength environmental conditions. Tests are being conducted at Los Alamos National Laboratory on 54 different test containers which hold miscellaneous laboratory waste items used in actinide research and processes such as gloves, cellulosic materials, plastics, etc. [5]. Several of the test containers have added solidified inorganic process sludge, pyrochemical salts, and various cementitious materials [5]. Fifteen large (55-gal sized) drums and 39 small (2-3 liter sized) titanium test containers hold the varying types of material. A mixed bacterial inoculum prepared by investigators at Brookhaven National Laboratory, consisting of muck pile salt, hypersaline lake brine and sediment slurry, was added to the test containers. Additional experiments performed at BNL verified the presence of various types of micro-organisms in the inoculum. Synthetic brines similar to brines that could be found at potential salt-based radioactive waste repositories were also added to the containers [5]. The microbial enumeration is part of a much larger study which examines the physical and chemical characteristics of the test containers with time [5].

941

Mat. Res. Soc. Symp. Proc. Vol. 465 © 1997 Materials Research Society

In the second study, pure and mixed anaerobic, denitrifying bacterial cultures isolated by BNL from a salt environment were used [4]. Actinides of varying concentrations were added to the growth media prior to inoculation with the bacterial cultures [1,2]. Cell numbers were measured as a function of incubation time in the growth cycle of each culture and as a function of actinide concentration. From monitoring temporal data, growth curves of the individual cultures were generated, and radiological or chemical effects of the actinides to the cultures were observed as trends in the curves. These types of toxicity effects are important factors underlying actinide retardation or transport by bacteria. This study included identifying the effects to the two cultures when exposed to ^{239}Pu, ^{243}Am, ^{237}Np, ^{232}Th and ^{238}U isotopes and employed turbidimetric, liquid scintillation and pH measurements, in addition to epifluorescence microscopy [1,2]. We report herein only bacterial cell counts from studies of the pure culture with Pu and Am.

Epifluorescence microscopy offers several advantages when working with radioactive or hazardous samples [6,7]. Small volumes of the samples are used (typically 0.01 to 0.50 ml) and as a result, there is minimal radioactive waste generation and minimal radiological exposure to the analyst. Additionally, analysis time is rapid. These features are in contrast to the larger sample volumes and ancillary materials that end up as waste, the much longer growth and analysis time required with plate count methods, and the increased potential for exposure to personnel during analysis.

EXPERIMENTAL

Epifluorescence microscopy [8], when used with the appropriate fluorochrome, allows for the visualization of the smallest bacterial cells as well as the ability to differentiate them from non-microbial debris. The Zeiss Axioscope and the specific objective used in the studies gives a 1000-fold overall magnification. Accurate field of view determinations were made using either a NIST traceable micrometer or from measurements of transmitted light through TEM calibration grids.

In the data presented here, the specific fluorochromes used to render the bacteria fluorescent include deoxyribonucleic acid (DNA)-specific stain 4',6-diamidino-2-phenylindole dihydrochloride (DAPI) [9]. This stain has an absorbance maximum at 340 nm and a fluorescence maximum at 488 nm. The dye fluoresces when bound to adenine-thymine (A-T) rich regions of DNA, and the fluorescence appears white-blue to blue. This particular fluorochrome is useful when dealing with environmental samples because it also binds to phosphate-rich regions of non-bacterial organic matter with fluorescence emissions appearing yellow [9]. Thus, in the study of bacterial populations of the complex mixtures comprising the test radioactive waste containers, bacteria are readily distinguished from other organic matter. Because both viable and non-viable cells contain DNA, DAPI staining provides total bacterial counts.

The slides are prepared by first obtaining a small volume of the sample. In order to provide accurate measurements of cell numbers in time-dependent experiments, the cells must be fixed immediately after the sample is taken to prevent further growth of cells. Since the cells will be added to deionized water for staining by DAPI, fixation is also required to prevent cell lysis due to drastic changes in ionic strength. Chemical fixation with formaldehyde causes cross-linking of the polymeric layers of the cell wall, which increases rigidity and stability of the cell, preventing lysis. Approximately 10-40 µl of a 1 mg/ml DAPI solution is added to an aqueous suspension of the bacteria. The sample is placed in the dark for 5-7 minutes. The sample is vacuum-filtered onto a 0.20 µm poresize black Poretics membrane. Filtration is performed quantitatively and in a manner such that the sample material is filtered uniformly across the membrane. The membrane is mounted on a microslide and a coverslip placed over the membrane is sealed into place with silicon sealant. Since the membrane contains radioactive or hazardous material, the sealed coverslip provides containment of any contamination which may be present.

Bacteria concentration is determined by the following equation [10]:

cells/ml = [(avg.cells/field) x (fields/filter) x (dilution factor)]/[sample vol. (ml)]

-(avg.cells/field) is the average of bacterial counts from 10-30 different fields of view of the emission image.

-(fields/filter) is an experimentally determined factor which is unique to the filtration apparatus and the microscope objective used. In our studies, this factor is 25,430 fields/filter.
-(dilution factor) must be taken into account if the total sample volume is first diluted, and only a portion of the diluted sample volume is used.
-(sample volume) is the total sample volume used in ml.

RESULTS AND DISCUSSION

The focus of this paper is to demonstrate the utility of epifluorescence microscopy for enumerating microbial populations in the presence of radioactive materials. The results presented from the two studies are shown only as examples of the technique being used successfully in radioactive systems [1-2,6-7].

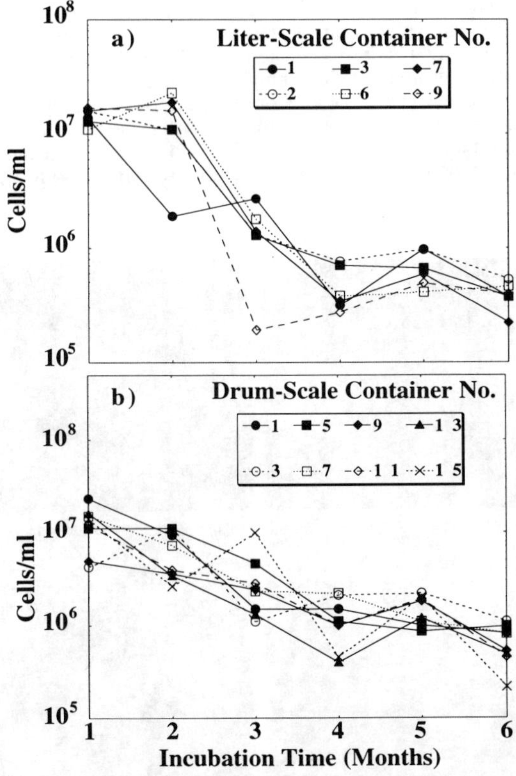

Figure 1.   Cell counts measured in a) small (2-3 liter-sized) and b) large (55 gal-sized) test containers versus incubation. Data points are derived from analysis of one slide, 30-50 fields counted per slide. Error bars not shown so as not to complicate the graphs. Error ranged from 15-30% for the data shown and was derived from taking an average of the number cells counted per field.

In the first study, epifluorescence microscopy with DAPI staining was used to enumerate total bacterial populations in test containers with actual low-level radioactive waste [5]. Figure 1 shows the results of total cell counts versus time in selected a) small (2-3 liter-sized) titanium test containers and selected b) large (55 gal-sized) drum test containers. The waste in the small test containers is comprised specifically of ground-up solidified waste such as crucibles, etc., whereas the waste in the large test containers is made up of heterogeneous waste such as tygon tubing, lab coats, etc. As mentioned previously, inorganic sludge, pyrochemical salts and cementitious materials were also added to several of the containers. All containers were inoculated with the mixed inoculum. The requirement to work within the guidelines of this experiment in its entirety [5], the hazardous nature of the test container contents, and the economics involved made it impossible to prepare an uninoculated control. However, the main point of the study was to determine bacterial populations within the brine solution with time, hence the lack of blanks or controls would not bias in any way the bacteria cell numbers.

The results show that the total bacterial cell numbers decrease with time which may be a result of cell lysis and death. The decrease may be attributed, however, to cell aggregation or sorption to solid surfaces.

The waste material and/or sludge in the containers settles out of the brine to the bottom of the container. For the data shown in Figure 1, the samples analyzed are obtained only from the test container brine solution, so any bacteria associated with the sludge are not detected. It is plausible that the bacteria are removed from the fluid column through aggregation and settling or sorption to test container surfaces. For any viable bacteria present, sorption to sludge material is feasible since this material will provide a source of carbon, nitrogen, sulfur and phosphorus, essential nutrients for bacterial survival. In addition, many bacteria grow and survive more successfully when sorbed to surfaces [11-13]. All of the mentioned nutrients are present in the solid waste, sludge and cementitious materials [5].

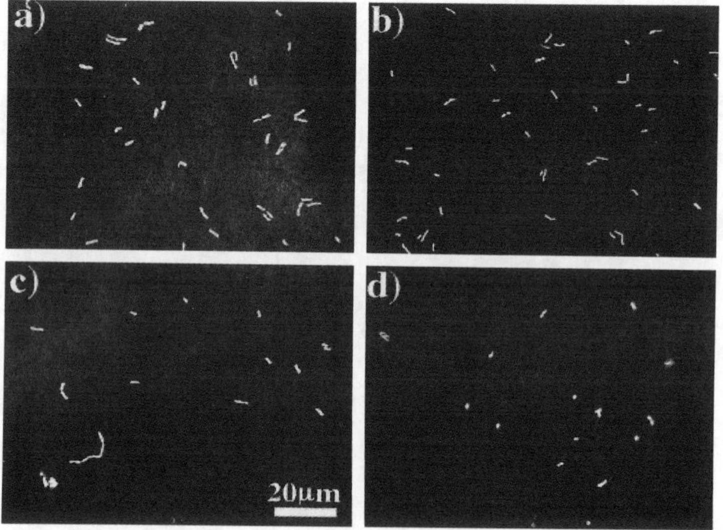

Figure 2. Epifluorescence micrographs of the pure culture at day 1 in its growth cycle. Growth medium contains a) [Pu]=0; b) [Pu]=1x10-7; c) [Pu]=1x10-6; and d) [Pu]=1x10-5. Cells in panel d) are more coccoid in appearance in comparison to rod-shaped bacteria shown in panel a). Changes to cellular morphology may be a result of cytotoxic effects of Pu on the bacteria.

To better ascertain the importance of microbial processes in the test containers, research has recently been initiated to determine the viability of the bacteria within the brine solution. Additional studies are aimed at determining the viability and types of bacteria attached to the sludge at the bottom of the test containers. Bacterial viability is implicated since several of the test containers have evolved gases that can be associated specifically with anaerobic denitrifying, or fermenting bacterial metabolic processes. As a result, the additional studies focus on identifying these types of bacteria, and the results will be presented in future reports.

In the second study reported here, total bacterial counts from both a mixed and a pure culture were obtained in order to construct growth curves (cell numbers versus incubation time) of the cultures exposed to various concentrations of actinides [1]. Any appreciable radiological or chemical effects from the actinides appear as variations in the growth curves. Figure 2 is a fluorescence micrograph showing four different fields of view of the pure culture at day 1 in the incubation period as a function of Pu concentration. Plutonium concentrations are as listed in the figure caption. The micrographs show that for the same volume of sample analyzed, there is a decrease in cell numbers, which indicates that growth is increasingly affected as the Pu concentration is increased [1].

Through the images obtained, changes to cellular morphology can be observed and represent another piece of evidence regarding the effect of Pu on the cells. In this study we observed that the morphology of the cells changed such that the initial rods became shorter and more coccoid in appearance with increasing Pu concentration. The changes to cell morphology can be observed readily by comparing the cells in Figure 2a) ([Pu]=0) with those shown in Figure 2d) ([Pu]=1 x 10^{-5} M). The overall bacterial size, from random sampling observations, was reduced in specific cases.

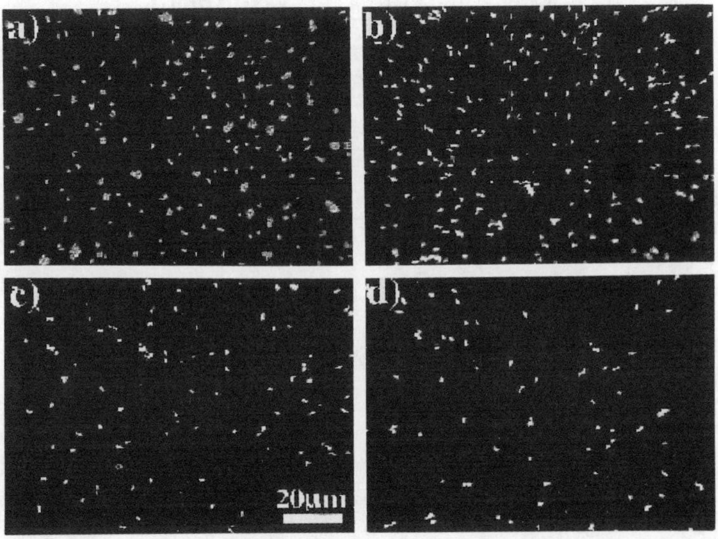

Figure 3. Epifluorescence micrographs of the pure culture at day 7 in its growth cycle. Pu concentrations are as described in Figure 2.

Figure 3 is another fluorescence micrograph showing the same pure culture at day 7 in the incubation period, again as a function of Pu concentration. The decrease in cell numbers with increasing Pu concentration is clearly evident.

These results are shown graphically in Figure 4a) and are in excellent agreement with turbidimetric measurements of the growth medium with time [1]. The agreement between the microscopy and turbidity measurements additionally insures that differences in cell counts at day 1 and day 7 are not a result of the variation in cell counts at day 0 [1]. Also shown in Figure 4 are growth curves of the pure culture with different concentrations of Am.

The inset table in each graph shows the radiological activity of the actinides as a function of their soluble concentration in the growth medium. Appropriate control experiments were performed to measure the association (i.e., sorption, precipitation, etc.) of the actinides with the cells [1]. These experiments involved separating the bacteria from the growth medium by 0.20 μm filtration and performing liquid scintillation counting (LSC) measurements of α-activity in the growth medium before and after filtration [1]. The association of the actinides with the cells could then be determined by difference calculations.

Figure 4 indicates that the presence of Pu indeed has an effect on the bacterial growth cycle as a function of Pu concentration; however, the presence of Am has minimal effect at all

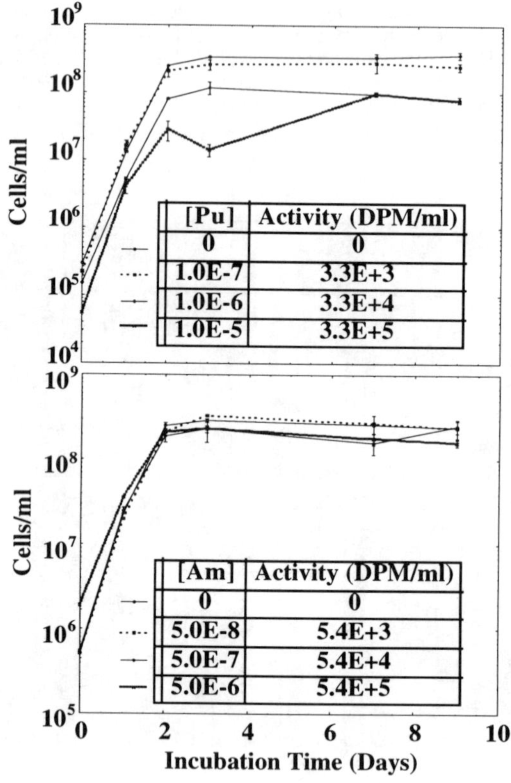

[Pu]	Activity (DPM/ml)
0	0
1.0E-7	3.3E+3
1.0E-6	3.3E+4
1.0E-5	3.3E+5

[Am]	Activity (DPM/ml)
0	0
5.0E-8	5.4E+3
5.0E-7	5.4E+4
5.0E-6	5.4E+5

Figure 4. Growth curves of pure culture exposed to different concentrations of a) Pu and b) Am. The data presented is the mean of analysis of six sample slides, 5-30 fields counted per slide. Error shown is one Standard Deviation.

concentrations tested. For the same relative concentration of actinide, Am has approximately 60% more α-radiological activity than Pu in these experiments. This suggests that the observed effects on bacterial growth in the Pu experiments is a result of a chemical effect versus a radiological effect. However, more experiments must be performed to confirm the mechanism of this effect.

SUMMARY

Epifluorescence microscopy with the fluorescent dye DAPI has been successfully demonstrated to be a useful tool for enumeration of bacterial populations in radioactive environments. The technique offers several advantages over other microbiology methods for determining bacterial numbers in complex, radioactive systems. Total bacterial (viable and non-viable) populations can be enumerated using a single sample. Only small volumes of radioactive or hazardous materials are required, resulting in minimal contaminated waste generation and radiological exposure to the analyst. Also, rapid analysis time is possible and results are available in a minimum period of time, since no incubation is required. The technique, coupled with other complementary studies, allows assessment of the role of bacteria in radioactive waste repository environments with respect to actinide retardation or transport mechanisms, biological interactions with organic materials, and gas generation due to bacterial metabolism.

ACKNOWLEDGEMENT

The studies described in this report were performed under the auspices of the U.S. Department of Energy.

REFERENCES

1. B.A. Strietelmeier, J.B. Gillow, C.J. Dodge, M.E. Pansoy-Hjelvik, S.M. Kitten, P.A. Leonard, I.R. Triay, A.J. Francis and H.W. Papenguth, in <u>Radionuclide Speciation in Real Systems</u>, edited by Reed, Clark and Rao (Am. Chem. Soc. Proc., Orlando, FL, 1996), in press.

2. J.B. Gillow, B.A. Strietelmeier, C.J. Dodge, K. Mantione, M. Dunn, A.J. Francis, M.E. Pansoy-Hjelvik, S.M. Kitten, I.R. Triay and H.W. Papenguth, in <u>Radionuclide Speciation in Real Systems</u>, edited by Reed, Clark and Rao (Am. Chem. Soc. Proc., Orlando, FL, 1996), in press;

3. A.J. Francis, Experentia **46**, 840 (1990).

4. A.J. Francis and J.B. Gillow, Sandia National Laboratory Report No. SAND93-7036 (1994).

5. a) R. Villarreal and M.L.F. Phillips, Los Alamos National Laboratory Test Plan No. CLS1-STP-SOP5-012/0 (1993). b) M.L.F. Phillips and M.A. Molecke, Sandia National Laboratory Report No. SAND91-2111 (1993).

6. J.B. Gillow and A.J. Francis, Brookhaven National Laboratory Report No. BNL-45756 (1990).

7. A.J. Francis, G. Joshi-Tope, J.B. Gillow and C.J. Dodge, Brookhaven National Laboratory Report No. BNL -49737 (1994).

8. H.G. Kapitza, <u>Microscopy from the Very Beginning</u>, edited by S. Lichtenberg (Carl Zeiss, Oberkochen, 1994), p. 23.

9. R.L. Kepner, Jr. and J.R. Pratt, Microbiol. Rev. **58**, 603 (1994).

10. <u>Standard Methods for the Examination of Water and Wastewater</u>, edited by A.E. Greenberg, L.S. Clesceri, A.D. Eaton (American Public Health Association, Washington, 1992), pp. 9-39 - 9-40.

11. Y. Bar-Or, Experientia **46**, 823 (1990).

12. M.P. Dawson, B.A. Humphrey and K.C. Marshall, Curr. Microbiol. **6**, 195 (1981).

13. S. Kjelleberg and M. Hermansson, Appl. Environ. Microbiol. **48**, 497 (1984).

Part XIII

Repository Backfill

DIRECT CHARACTERIZATION OF TRANSPORT PARAMETERS IN NEAR-FIELD AND ENGINEERED BACKFILL/INVERT MATERIALS

J. L. CONCA*, B. A. ROBINSON**, I. R. TRIAY** and G. Y. BUSSOD**
*UFA Ventures, Inc., 2000 Logston Blvd, Richland WA 99352   http://ufa.owt.com
**Box 1663, Los Alamos National Laboratory, Los Alamos, NM 87545

ABSTRACT

Performance assessment of a potential repository at Yucca Mountain includes flow and transport modeling of the unsaturated zone as the critical predictive component, and involves a series of model calculations that provide predictions of the migration of important radionuclides in the inventory to the water table. The modeling requires the relevant properties of both the natural environment and the engineered systems. The Unsaturated Flow Apparatus, UFA, was used to directly measure the unsaturated and saturated transport properties of whole rock tuff cores and candidate barrier materials to provide real input parameters to the models. These properties included hydraulic conductivity, matric potential, air permeability, and diffusion coefficient, all of which are strong functions of the volumetric water content. Results show that for all recharges above 0.1 mm/yr, fractures will be partially saturated and conducting at that recharge rate. Whenever the thermal conditions relax enough to allow rewetting of the host rock, there will be dripping from the drift ceiling. Unsaturated transport of colloids through fractured cores of Topopah Spring and Prow Pass tuffs was also investigated and found to depend primarily upon colloid charge, and not size, for the rock cores investigated.

INTRODUCTION

Yucca Mountain, Nevada, is being studied as a potential site for the high level nuclear waste repository. Site characterization activities include evaluating the hydrologic and transport properties of the unsaturated zone where the potential repository may be located, and the saturated zone along the likely pathways to the accessible environment. The unsaturated zone consists of alternating layers of welded and nonwelded tuffs that are tilted, uplifted, fractured, and faulted. Infiltration rates into the bedrock are thought to be low because of the low precipitation in the region and large subsequent evapotranspiration rates. The small amount of recharge or infiltration that makes it into the vadose zone becomes the primary carrier for radionuclides exiting the repository. For radionuclides to reach the accessible environment, the waste canisters must be breached and radionuclides must become mobilized by dissolution. Radionuclides must then travel through the near field environment of backfill and invert material (if used) and rock altered by thermally induced geochemical reactions, through the unsaturated rock beneath the potential repository, and subsequently through the saturated zone groundwater flow system. In addition to the hydrologic processes that obviously are important to radionuclide migration, transport processes such as chemical sorption and physical retardation will control the ultimate movement of radionuclides in the geologic environment. Robinson and co-workers[1] examined the transport of radionuclides in the unsaturated zone at Yucca Mountain, concluding that at infiltration rates less than 0.1 mm/yr, the unsaturated zone alone will likely provide sufficient transport times to limit the dose at the accessible environment under any realistic regulatory scenario. At higher infiltration rates the most mobile radionuclides begin to reach the water table in less than one million years, but the unsaturated zone still provides a barrier to the accessible environment. The primary goal of the present study is to provide actual transport property measurements on real samples, and to re-examine the conclusions of Robinson[1] for key radionuclides.

Figure 1 is a schematic of the drift and near-field environment showing possible geometries. The drift may or may not be filled with backfill either as a one- or two-component fill, may or may not have invert layers with retarding or criticality barrier materials, and may have a large air-space at the top even if backfilled. Although a final design has not been chosen, possible candidate backfill/invert materials were suggested and were examined in this study using the Unsaturated

Mat. Res. Soc. Symp. Proc. Vol. 465 © 1997 Materials Research Society

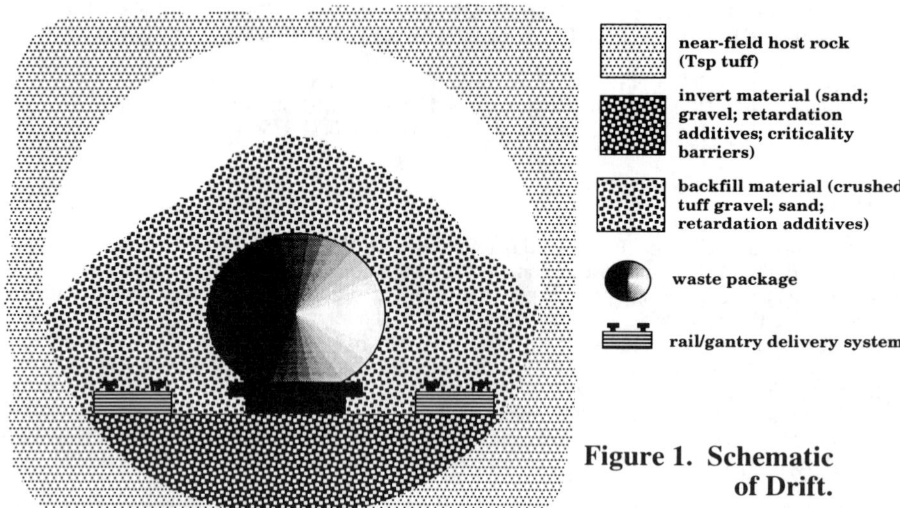

Figure 1. Schematic of Drift.

Unsaturated Flow Apparatus (UFA) to measure hydraulic conductivity, matric potential, air permeability, and diffusion coefficient as a function of volumetric water content over a range of conditions equivalent to all infiltration or recharge rates possible at Yucca Mountain.

EXPERIMENTAL METHODS

The UFA achieves hydraulic steady state in a few hours for most geologic materials, even those with very low water or fluid contents.[2] There are specific advantages to using acceleration as the fluid driving force. It is a whole-body force similar to gravity and, so, acts simultaneously over the entire system and independently of other driving forces, e.g., gravity or matric potential. Although the centrifugal force is usually thought of as an inertial effect in the non-rotating frame, it is a real force in the rotating frame of reference of the sample. Centrifugation has been used to investigate pressure potential relationships in soils for over 60 years, but the use of centrifugation to measure steady-state hydraulic conductivity on various porous media has only recently been demonstrated, for soils by Nimmo[3] and for sediments and rocks by Conca .[2,4,5]

The UFA instrument consists of an ultracentrifuge with a constant, ultralow flow pump that provides fluid to the sample surface through a rotating seal assembly and microdispersal system.[5] The ultracentrifuge can reach accelerations of up to 20,000 g (soils are generally run only up to 1,000 g), temperatures can be adjusted from -20° to 150°C, and constant flow rates can be reduced to 0.001 ml/h. Effluent from the sample is collected in a transparent, volumetrically-calibrated chamber at the bottom of the sample assembly. Using a strobe light, an observer can check the chamber while the sample is being centrifuged. Materials can be run in the UFA as recomposited samples, or as *in situ* samples subcored directly into the UFA chamber. Cores of whole rock, ceramics, grouts, and other solids are cast in an appropriate epoxy sleeve for use in the UFA.

The UFA Method is effective because it allows the operator to set the variables in Darcy's Law. Darcy's Law states that the fluid flux equals the hydraulic conductivity times the fluid driving force. Under an acceleration, in which water is driven by both the matric potential gradient and the centrifugal force per unit volume, Darcy's Law is given by[3]

$$q = -K(\psi) \, [d\psi/dr - \rho\omega^2 r] \qquad (1)$$

where q is the flux density into the sample, K is the hydraulic conductivity, ψ is the matric potential, $d\psi/dr$ is the matric gradient, $\rho\omega^2 r$ is the centrifugal force per unit volume, r is the radius from the axis of rotation, ρ is the fluid density, and ω is the rotation speed in radians per second. Hydraulic conductivity is a function of either the matric potential or the volumetric water content. Above speeds of about 300 rpm, provided that sufficient flux density exists, the matric potential is much less than the acceleration, $d\psi/dr \ll \rho\omega^2 r$. Therefore, Darcy's Law is given by $q = -K(\psi)\ [-\rho\omega^2 r]$ under these conditions. Rearranging the equation and expressing hydraulic conductivity as a function of volumetric water content, θ, Darcy's Law becomes

$$K(\theta) = q/\rho\omega^2 r \tag{2}$$

As an example, an alluvial silt from the Yucca Mountain site accelerated to 2500 rpm with a flow rate of 0.01 ml/h achieved hydraulic steady state in 10 hours at a target volumetric water content of 18% and an unsaturated hydraulic conductivity of 4×10^{-10} cm/s. Appropriate values of rotation speed and flow rate into the sample are chosen to obtain desired values of flux density, water content, and hydraulic conductivity within the sample. This method provides hydraulic conductivity to within ±8% at a volumetric water content known to within ±2%.[6] Once steady-state ia achieved, other methods are used to measure related properties at that water content, e.g., air permeability, electrical conductivity, retardation factor for a specific contaminant species.

The UFA is a flux-controlled system and is open at either end. The pressures at the top and bottom of the sample are not fixed at any particular value. The matric potential at all points throughout the sample is allowed to attain whatever magnitude is in equilibrium with the choice of flux and rotation speed. Because the driving force is an acceleration, the bottom of the sample does not need to saturate in order for water to leave the sample, removing the requirement of traditional columns that $\psi = 0$ at the sample bottom. Although the acceleration can be used to calculate an equivalent pressure for comparison with the matric potential, the acceleration is not a pressure, but is a whole-body force acting at all points simultaneously within the system, whereas the matric potential is a surface force acting as a function of distance and is slow to respond to changes in water content. Redistribution of water and attainment of hydraulic steady-state occurs rapidly, within hours, in response to the imposed acceleration and flux. The water content within the sample also attains the steady-state value at every point, and $d\psi/dr \to 0$ throughout the sample. If the sample is homogeneous, then the water content will be uniform throughout the sample and is easily measured.[2,3] If the sample is heterogeneous, such as a silt lense in a coarse sand or a fractured rock core of lithic tuff, then each component reaches its own steady-state water content, but $d\psi/dr$ still approaches zero throughout the sample.

EXPERIMENTAL RESULTS

Hydraulic conductivity, K, matric potential, ψ (also referred to as water potential, moisture potential, or soil suction), and air permeability, k_a, were experimentally determined as a function of the volumetric water content, θ, on the whole rock cores and on the candidate engineered barrier materials: fractured and unfractured cores of Topopah Spring tuff (Tsp) and other Paintbrush tuff units, fractured Prow Pass tuff (Tcp), 3/8-inch crushed Tsp tuff gravel (Tsp gravel), 3/8-inch crushed basalt gravel (basalt gravel), Yucca Mountain Split Wash alluvium, fine and coarse clean quartz sands, and North Carolina apatite, a phosphate ore-rock. These samples represent a broad range of materials at the site and materials that could be used as backfill or invert materials. The range of volumetric water contents investigated cover the range of conditions that would exist under all infiltration or recharge rates possible at Yucca Mountain. However, the conditions interior to the drift in the absence of advection are still uncertain. Experiments are underway that will address steady-state non-advective conditions in the drift. Also, experiments are underway that will assess the effects of the thermal drying period on the hydrologic properties of whole rock Topopah Spring tuff.

Using the new ASTM Standard Test Method for the UFA in conjunction with other techniques[7] each material was characterized in only 11 days. The diffusion coefficient, D(θ), was determined from the empirically-derived unsaturated aqueous diffusion curve for all porous media which was measured using the UFA Method.[8] The following several figures display these data in various ways. Graphs of all the transport properties for each material can be assembled or graphs can be made of a single property for several materials for comparison. Examples are shown in Figures 2 and 3 for the tuff whole rock and gravel, which are composed of the same material but in very different forms. Figure 2 shows the $K(\theta)$, $\psi(\theta)$, and $k_a(\theta)$ for just the fractured Tsp whole rock core with an expanded x-axis to show the large changes in behavior with small changes in water content. Hydraulic conductivity within this whole rock core is dominated by a single small fracture with a mean aperture of 20 μm. At saturation, when the fracture is filled, the volumetric water content is 15%, the hydraulic conductivity is 10^{-7} cm/s, the matric potential is effectively zero, the vapor diffusivity is effectively zero, and the aqueous diffusion coefficient is 8×10^{-7} cm^2/s.[8] A slight desaturation to 14% drops the hydraulic conductivity to 10^{-10} cm/s, raises the matric potential to over 2 bars, increases vapor diffusivity to 3×10^{-4} cm/s (a coefficient of air permeability of 3×10^{-8} cm^2), and leaves the aqueous diffusion coefficient almost unchanged at 6×10^{-7} cm^2/s. After the fracture desaturates, $K(\theta)$ is almost a vertical line. At matric potentials above 2 bar, $\psi(\theta)$ becomes a steadily decreasing function up to 47 bars as the finer and finer pores desaturate. The tuff gravel (Figure 3) exhibits very different behavior. At saturation, the volumetric water content is 56%, the hydraulic conductivity is 10.4 cm/s, the matric potential is effectively zero, the vapor diffusivity is effectively zero, and the aqueous diffusion coefficient is

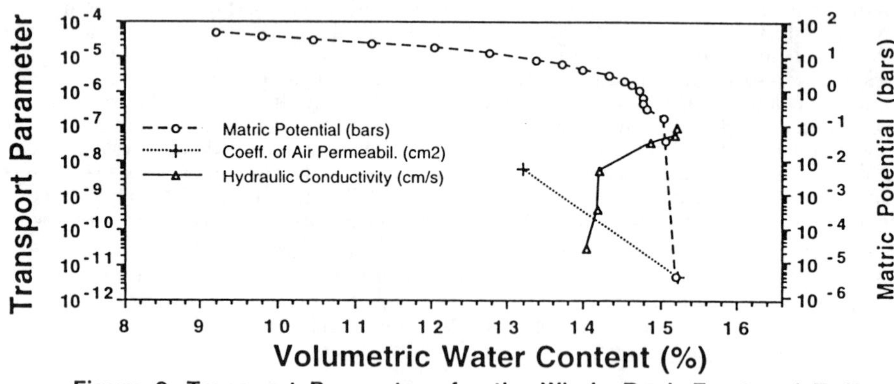

Figure 2. Transport Parameters for the Whole Rock Fractured Tuff.

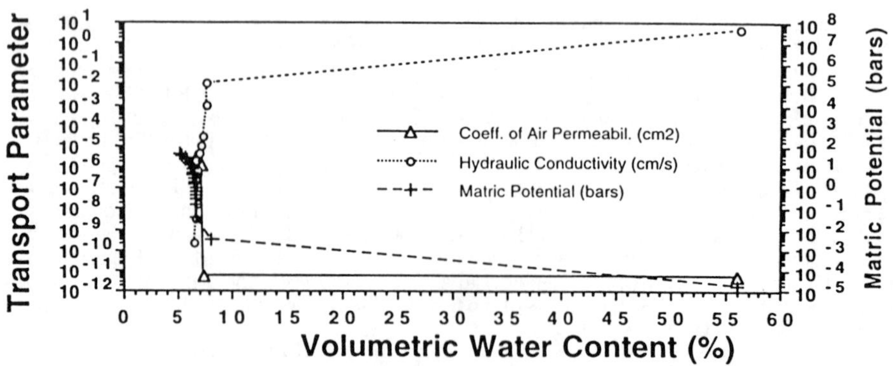

Figure 3. Transport Parameters for the Candidate Tuff Gravel.

10^{-5} cm^2/s. However, gravel cannot saturate at any flux under Yucca Mountain, and any degree of desaturation causes the following: 1) the volumetric water content drops to 7% (that portion of the total porosity internal to the gravel particles themselves), 2) the hydraulic conductivity falls to $< 10^{-1}$ cm/s (in fact, whatever the recharge rate is at that point), 3) $\psi(\theta)$ becomes a vertical line at intermediate pressures, but at high pressure the curve turns over to mimic the whole rock curve as the internal pores of the tuff gravel particles begin to desaturate, 4) the vapor diffusivity increases to 0.1 cm/s ($k_a = 10^{-5}$ cm^2), and 5) the aqueous diffusion coefficient falls to 10^{-7} cm^2/s. The basalt gravel has similar behavior except that the volumetric water content is about 4% lower at any point, because the basalt particles have no internal porosity. It is uncertain what effect the tuff gravel internal porosity has on the rewetting phase of the repository, but percolation box experiments indicate that the tuff will not rewet to 7% in the absence of advection.[9]

Figure 4 shows K(θ) for several whole rock tuff cores. Most of these samples have a low saturated hydraulic conductivity, even the fractured Tsp core with an aperture of 20 μm. Only the Tcp tuff core had a high saturated hydraulic conductivity of about 10^{-3} cm/s. This core had a large-aperture fracture of about 200 μm. The drop in hydraulic conductivity from 10^{-3} cm/s to 10^{-7} cm/s, with a drop in water content of only 1%, represents the drainage of this one fracture. In the field, the behavior of whole rock with significant fracture apertures will be similar in that their water contents and transport behaviors will depend strongly on how close the infiltration rate is to their range of fracture conductivities. This means that at an infiltration rate greater than 0.2 mm/yr, the matrix will be saturated and conducting at a rate of about 0.2 mm/y and a 200 μm fracture would be partially saturated and conducting water and any dissolved radionuclides and that infiltration rate. But at an infiltration rate less than 0.2 mm/yr, the fracture would be completely desaturated and non-advective, and the matrix will be partially saturated and conducting water and any dissolved radionuclides and that infiltration rate. Figures 5 and 6 show the air permeability and matric potential results for the sand-sized materials investigated. The Split Wash alluvium never desaturated enough to allow air to migrate, even at water contents in steady-state with a recharge of 10^{-10} cm/s (0.003 cm/year). The other materials exhibited the common step function associated with air permeability, i.e., as soon as the largest pores desaturated enough to allow vapor transport, the coefficient of air permeability jumped to near the intrinsic permeability and did not change very much with further desaturation. The data became noisy for the two sands probably as a result of random geometric changes in water films and pendular water elements at point contacts in these two well-sorted materials. With the data obtainable using the UFA, comparisons can be made among various materials for any of these properties (Figure 7).

A preliminary experiment in unsaturated transport of colloids in fractured tuff was also performed. The mobility of radionuclides may be enhanced by colloidal material, e.g., weathered rock matrix, or silica and iron colloids from the waste form and canister.[10,11] Experimental studies on colloid transport through soil under partially saturated conditions indicate that colloids can migrate even under relatively dry conditions, although migration rates exhibit strong dependence upon colloid size and character as the system desaturates.[12,13]

We investigated colloid transport in the two fractured rock cores of tuff, described above, using the UFA under isothermal, steady-state flow conditions with J13 wellwater at a hydraulic conductivity of 4.1 x 10^{-8} cm/s, equivalent to a recharge of 1 cm/year at Yucca Mountain. The solutions contained two different colloids, a 150 nm diameter colloid of positively-charged Fluorescent Amidine Polystyrene Latex and a 340 nm colloid of negatively–charged Fluorescent CML Polystyrene Latex. The solution run through the fractures consisted of a 5 ppm total colloid concentration; 2.5 ppm of the 150 nm (+) colloid and 2.5 ppm of the 340 nm (–) colloid. The effluents were then analyzed for colloid size and quantity using fluorescent spectroscopy.[14] Charged colloids were run as analogs to various true polymerized radiocolloids and radionuclide-bearing pseudocolloids, e.g., ^{239}Pu, ^{237}Np, ^{243}Am, and ^{247}Cm from high level waste, or ^{137}Cs, ^{90}Sr, and ^{60}Co from low level waste. Under most geologic conditions, transport of radionuclides by pseudocolloids should dominate over transport of radionuclides by true radiocolloids. The results for the tuff cores are given in Table 1 and show that colloids do migrate through fractures in the tuff, even small fractures. The larger negatively-charged 340 nm (-) colloid broke through both fractures immediately, $R_f \approx 1$, while the smaller positively-charged 150 nm (+) colloid did not break through either core over the duration of this experiment, $R_f > 100$. Therefore, in this

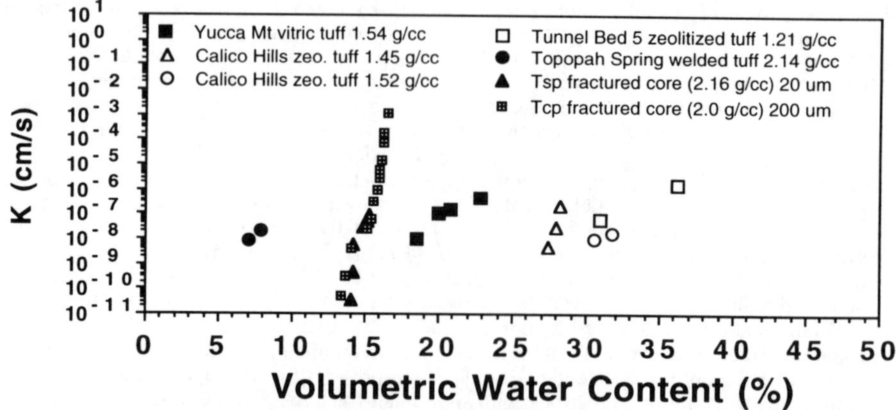

Figure 4. Unsaturated Hydraulic Conductivities for Whole Rock Cores of Different Tuffs from Yucca Mountain.

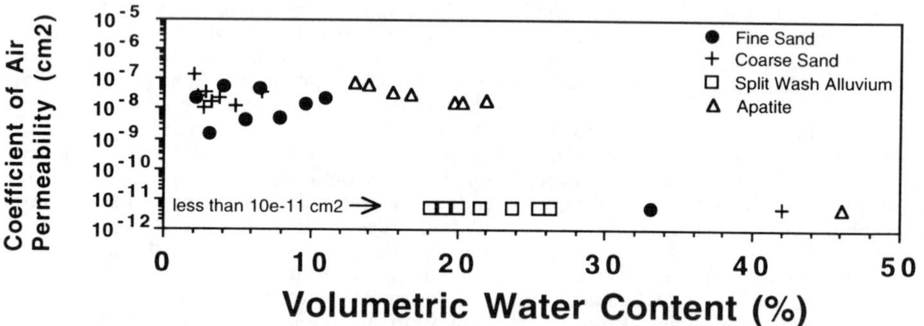

Figure 5. Air Permeability Curves for the Sandy Materials

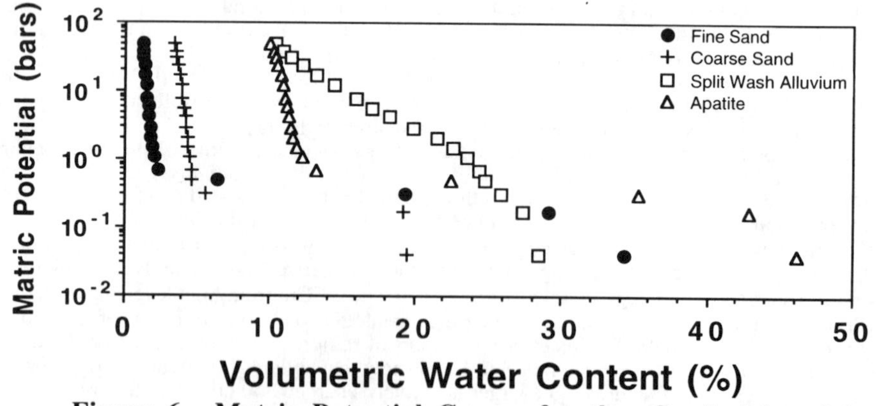

Figure 6. Matric Potential Curves for the Sandy Materials

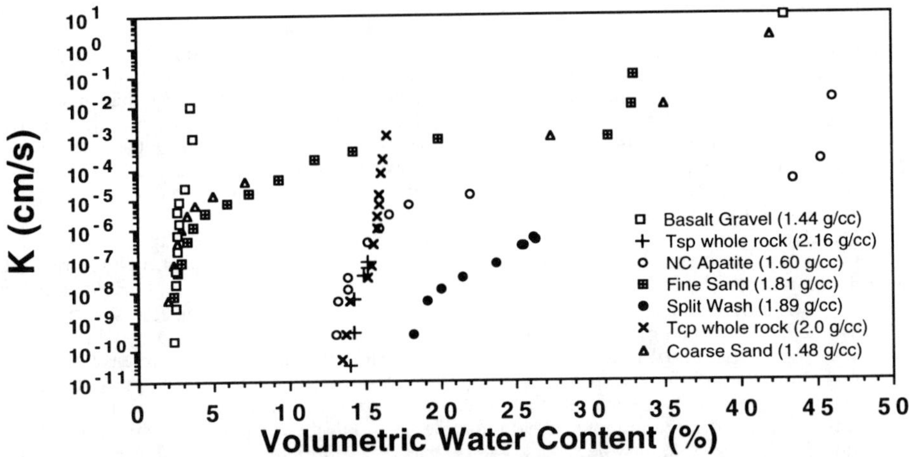

Figure 7. Hydraulic Conductivity Curves for Selected Materials Directly Measured in Three Days.

situation, colloid charge controlled transport through these cores and not colloid size. This work is continuing, and will investigate the effects of fracture size, rock type, colloid characteristics and degree of saturation on the transport of colloids.

Table 1. Preliminary Retardation Factor Experiments using Two Colloids in Fractured Whole Rock Cores of Prow Pass (Tcp) and Topopah Spring (Tsp) Tuff Determined Under Variably Saturated Conditions.

Rock Type and Fracture Aperture	Colloid Type	Retardation Factor
Tsp devitrified tuff 20 um fracture	340 nm (-)	≈ 1
Tsp devitrified tuff 20 um fracture	150 nm (+)	> 100
Tcp zeolitized tuff 200 um fracture	340 nm (-)	≈ 1
Tcp zeolitized tuff 200 um fracture	150 nm (+)	> 100

REVIEW OF MODELING RESULTS

This section reviews the modeling results of Robinson and co-workers[15] which used the above UFA properties as input parameters for the near-field performance and radionuclide transport calculations. The details, assumptions and a complete discussion of the modeling are provided in the unsaturated zone study by Robinson and co-workers.[15] The results given here only highlight some of the effects of having a gravel backfill versus no backfill, and are extremely shortened from their full development. The purpose of this part of the modeling was to assess different design concepts (backfill versus no backfill, sorbing invert versus no sorption), and to provide abstractions for source terms for the unsaturated zone site scale transport calculations.[15]

The physics of the hydrology at the interfaces of the rock and air space is not properly captured in a porous medium model. Therefore, bounding cases were simulated. A model result in which no water percolates through the drift was considered one end-member limiting case. At the other extreme, the drift and backfill provided no resistance to percolating fluid. This end-member case was simulated by assuming that the emplacement drift (air gap and backfill) had hydrologic properties identical to those of the host rock. Examination of these two extremes allows assessment of the impact of this uncertainty on the radionuclide source term from the near field. In the present study, simulations which assume the measured properties of the crushed tuff gravel apply in the drift are called "Gravel Properties" simulations, while the runs that set the properties equal to those of the host rock are called "Host Rock Properties" simulations.

The release rates and concentrations for a specific radionuclide are dependent on many processes and parameter values, including infiltration rate, solubility, waste form dissolution rate, and the nature of the hydrology near the emplacement drift. These are detailed in Robinson et al. (1997).[15] To study the influence of backfill and invert material on performance, it is instructive to examine the transport in a generic way before presenting results for specific radionuclides. We do this by taking a single infiltration rate (1 mm/y) and examining the predicted release rate under solubility-limited conditions. The "Gravel Properties" hydrologic model will be presented in this development. Before considering the effect of a limited supply of radionuclide, we present results based on an assumed large amount of radionuclide, as would be the case for a sparingly soluble radionuclide. Figure 8 shows the release rate versus time up to 1 million years, from the waste package and from the bottom of the model domain for a solubility of 10^{-3} moles/l (probably an upper bound on the solubility of ^{237}Np at Yucca Mountain[16]). The release curve for the bottom of the model is assumed to be an appropriate source term for site-scale unsaturated zone calculations, since this drift-scale model is intended to capture the detail that is occurring within a single source node of a site scale simulation. The difference between the release rate from the package and the release rate from the model at early times (less than about 1 ky) represents the finite time required for radionuclides to migrate from the waste package to the bottom of the model. If releases are presumed to occur for times much greater than 1 ky, then this transient can be ignored in favor of a simpler constant source rate. The impact of an invert material consisting of apatite (using a K_d value of 500 cc/g, the nominal value for ^{237}Np) is to reduce the source rate somewhat at any given time. Eventually, the source rate approaches that of the case assuming no sorption, but this does not occur until near the end of the one million year simulation. For the more soluble radionuclides, the time for complete release is much less than one million years. So the apatite will probably result in a performance benefit for a nuclide such as ^{237}Np. The rather modest performance benefit of only about a factor of 2 at 1 ky for this simulation is due to the fact that radionuclides have a diffusion pathway laterally to the edge of the drift in which it does not encounter invert. Radionuclides traveling downward do in fact exhibit significant retardation, but a substantial fraction of the released material could bypass the sorbing invert if the diffusion coefficient in the gravel is high enough. Because water content alone determines the aqueous diffusion coefficient, it is critical to know the water content in the backfill.

It has generally been assumed that under conditions in which no advective flux travels through the drift, the release rates are vanishingly small. The present study shows this to be a nonconservative result of assuming a low value for the volumetric water content, rather than the value of about 7% as measured in this study (Figure 3). This value results in a diffusion coefficient of 1.6×10^{-11} m^2/s, large enough to result in significant mass transport by molecular diffusion.[8] It could be argued that in the presence of even a small amount of decay heat, the backfill would remain much drier than is predicted in these isothermal calculations, with a resulting delay in the time at which molecular diffusion through the backfill would become significant. Also, the value of 7% is an advective value measured in the UFA. In the absence of advection, the value will be lower.[9] In these simulations, we chose to take the very conservative assumption of 7% volumetric water content. In the future, nonisothermal and nonadvective conditions should be simulated. In the near-term, examination of LLNL's nonisothermal simulations would indicate the approximate time at which the backfill rebounds to a large enough water content for diffusion to become important, i.e., greater than about 3% volumetric water content.[8]

A comparison of the release rates from the near field environment for the two different hydrologic model assumptions is shown in Figure 9. Advective flux through the drift results in a more rapid release of radionuclides to the bottom boundary of the model, as expected, but the magnitude of this difference is not as pronounced as previously assumed. At times on the order of 1 ky, the release rate has approached a steady state value after the initial transient representing the time required for diffusion through the backfill and transport to the bottom of the model. The actual magnitude of the curves in Figure 9 are directly proportional to the radionuclide solubility.

Figure 10 shows the effective concentration versus time exiting the drift-scale model domain for the two hydrologic models. The concentrations are normalized with respect to the solubility of the radionuclide, so that the value is actually a dilution factor (the concentration is diluted by this factor from the solubility-limited value assumed to occur immediately adjacent to the waste package). For the "Gravel Properties" case, the diffusional limitations through the backfill are sufficiently great to lower the effective concentration leaving the near field by an order of magnitude below the concentration at the waste package wall. The dilution effect is less pronounced for the "Host Rock Properties" case because flowing fluid comes in direct contact with the waste package. However, there is still a tendency for some dilution by fluid that does not contact the package. Sorption in the invert reduces the effective concentration somewhat in the time frame of interest, but the effectiveness of sorption is limited if it is assumed to occur only in the invert.

We now examine the impact of infiltration rate on the release of radionuclide to the far field for the two hydrologic models. For the "Host Rock Properties" hydrologic model, water travels immediately adjacent to the waste package, and the steady state rate at which radionuclides are leached from the near field is approximately proportional to the infiltration rate, as shown in Figure 11. At higher infiltration rates the approach to steady state occurs more rapidly than for lower fluxes. For the "Gravel Properties" case (Figure 12), there is a significant diffusion limitation through the backfill that limits the rate of release at any infiltration rate so that the steady state release rate is less dependent on infiltration rate. The effective concentration plots for these cases are shown in Figure 13 ("Host Rock Properties" model) and Figure 14 ("Gravel Properties" model). The "Gravel Properties" case, with diffusion limitations in the backfill, shows a larger dilution effect with increasing infiltration rate. In other words, the impact of increased infiltration rate is to *lower* the effective concentration leaving the near field, rather than increasing the release rate, even though the total cumulative amount of radionuclide leached is greater.

CONCLUSIONS

The UFA was used to directly measure unsaturated and saturated hydraulic conductivity, matric potential, air permeability, and diffusion coefficient of whole rock tuff cores and candidate barrier materials to provide real input parameters to the models. Results show that for all infiltration rates greater than 0.2 mm/yr, the matrix in welded tuff will be saturated and conducting at a rate of about 0.2 mm/y and any fractures will be partially saturated and conducting water and any dissolved radionuclides and that infiltration rate. When thermal conditions relax enough to allow rewetting of the near-field host rock, there will be dripping from the drift ceiling. Unsaturated transport of colloids through fractured cores of Topopah Spring and Prow Pass tuffs was also investigated and found to depend primarily upon colloid charge, and not size, for the rock cores investigated.

Using these results, predictive modeling[15] concluded that: 1) in the absence of advection, diffusion through any backfill will be significant as long as the steady-state volumetric water content is above about 3%; 2) for release functions that are effectively a short pulse, rather than a step change of long duration, the sorption coefficient has a strong effect on the predicted peak dose, and 3) the exact nature of the near-field radionuclide source term is not important to the prediction of peak dose. If the time over which the majority of the radionuclide is released is short compared to the transit time through the unsaturated zone, then the performance will be insensitive to the exact details of waste package failure scenarios. Although not discussed in this paper, distribution coefficients were determined from batch tests, and unsaturated retardation factors were measured using the UFA for selected radionuclides.[16,17]

ACKNOWLEDGMENTS

The authors gratefully acknowledge Andrew Wolfsberg, Hari Viswanathan, Carl Gable, George Zyvoloski and Jake Turin. This work was supported by the Yucca Mountain Project.

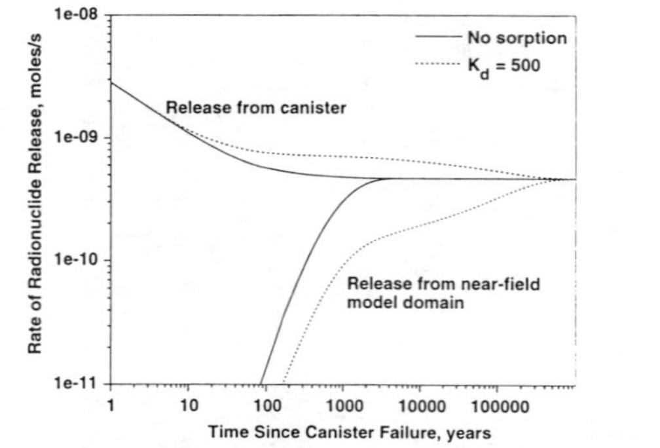

Figure 8. Release rate versus time from the canister and from the bottom of the near-field model domain. Infiltration rate: 1 mm/y, "Gravel Properties" model.

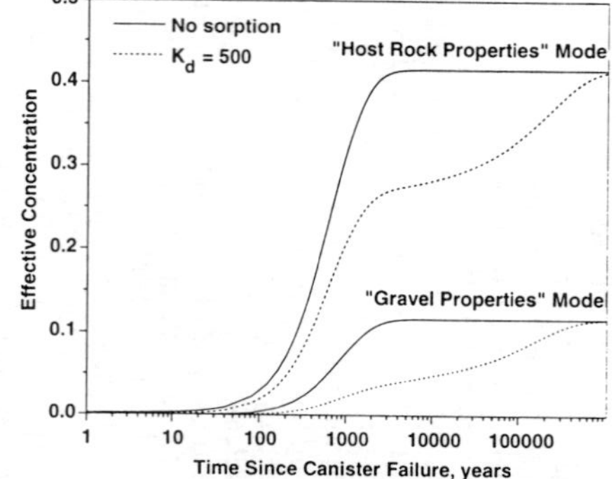

Figure 10. Effective concentration versus time exiting the near-field model domain. Infiltration rate: 1 mm/y.

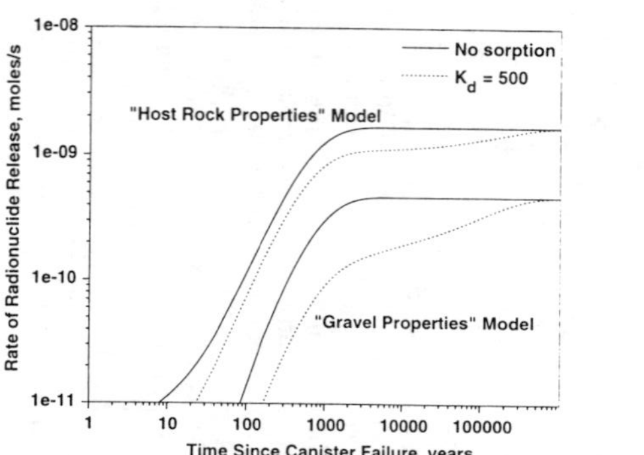

Figure 9. Release rate versus time from the bottom of the near-field model domain comparing the "Gravel Properties" and "Host Rock Properties" models. Infiltration rate: 1 mm/y, model.

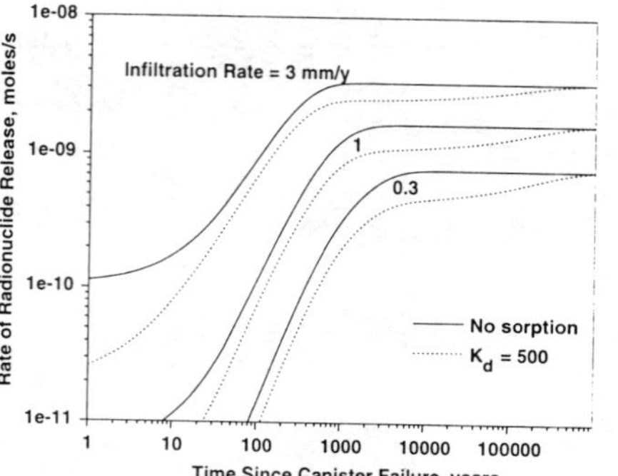

Figure 11. Release rate versus time from the bottom of the near-field model domain for different infiltration rates. "Host Rock Properties" model.

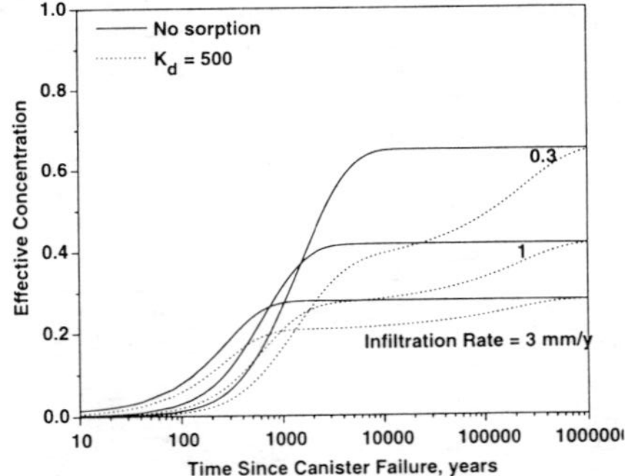

Figure 12. Release rate versus time from the bottom of the near-field model domain for different infiltration rates. "Gravel Properties" model.

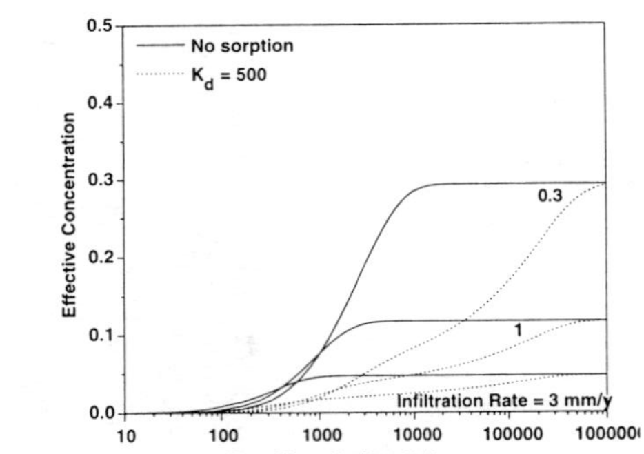

Figure 13. Effective concentration versus time from the bottom of the near-field model domain for different infiltration rates. "Host Rock Properties" model.

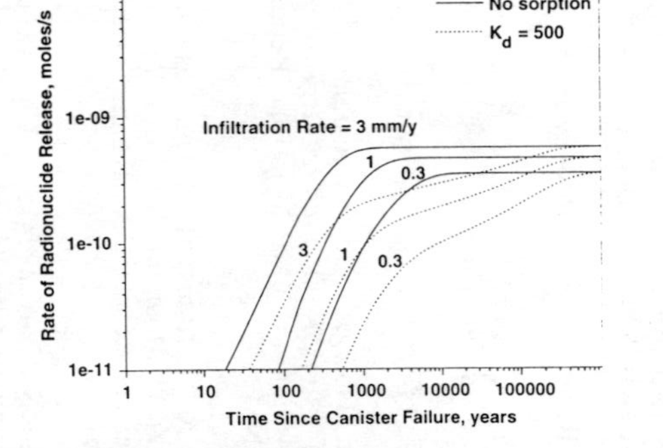

Figure 14. Effective concentration versus time from the bottom of the near-field model domain for different infiltration rates. "Gravel Properties" model.

REFERENCES

1. B. A. Robinson, A. V. Wolfsberg, G. A. Zyvoloski, and C. W. Gable, *Modeling of Flow, Radionuclide Migration, and Environmental Isotope Distributions at Yucca Mountain*, YMP Milestone Report 3468, Los Alamos National Laboratory, Los Alamos, NM, 1996.

2. J. L. Conca and J. V. Wright, *Applied Hydrogeology* **1**, p. 5 (1992).

3. J. R. Nimmo, Rubin, J., and Hammermeister, D. P. *Water Resour. Res.* **23**, p. 124 (1987).

4. J. L. Conca and J. V. Wright, *Water Resources Research* **26**, p. 1055 (1990).

5. J. V. Wright, J. L. Conca, and X. Chen, *Hydrostratigraphy and Recharge Distributions from Direct Measurements of Hydraulic Conductivity Using the UFA™ Method*, Technical Report PNL-9424, Pacific Northwest Laboratory, Richland, WA, 1994, 150 p.

6. J. R. Nimmo and K. C. Akstin, *Soil Science Society of America Journal* **52**, p. 303 (1988).

7. J. L. Conca, *Measurement of Unsaturated Hydraulic Conductivity and Chemical Transport in Yucca Mountain Tuff*, Technical Report LA-12596-MS, Los Alamos National Laboratory, Los Alamos, NM, 1993, 28 p.

8. J. L. Conca, M. J. Apted and R. C. Arthur in *Scientific Basis for Nuclear Waste Management XVI*, edited by C. G. Interrante and R. T. Pabalan (Materials Research Society Symposium Proceedings 294, Pittsburgh, PA 1993), p. 395-402.

9. J. L. Conca in *Proceedings of the First International High-Level Radioactive Waste Management Conference* (American Nuclear Society Symposium Proceedings, La Grange Park, IL 1990), **1**, p. 394-401.

10. D. I. Kaplan, P. M. Bertsch, and D. C. Adriano, *Ground Water* **33**, p.708 (1995).

11. J. Buffle and H. P. van Leeuwen in *Environmental Analytical and Physical Chemistry Series; Environmental Particles,* Lewis Publishers, Boca Raton, 1993, **2**, pp. 247-315.

12. J. Wan and J. L. Wilson, *Water Resources Research* **30**, p. 857 (1994).

13. M. A. McGraw, *The Effect of Colloid Size, Colloid Hydrophobicity and Matrix Saturation on Colloid Transport in The Subsurface, PhD.* Dissertation, Univ. of California, Berkeley, 1996.

14. J. R. Lackowicz, *Principles of Fluorescence Spectroscopy*, Plenum Press, New York, 1983.

15. B. A. Robinson, A. V. Wolfsberg, H. S. Viswanathan, C. W. Gable, G. A. Zyvoloski and H. J. Turin *Modeling of Flow, Radionuclide Migration, and Environmental Isotope Distributions at Yucca Mountain*, YMP Milestone Report 3672, Los Alamos National Laboratory, Los Alamos, NM, 1997.

16. I. R. Triay, C. R. Cotter, S. M. Kraus, M. H. Huddleston, S. J. Chipera, and D. L. Bish, *Radionuclide Sorption in Yucca Mountain Tuffs with J13 Well Water: Neptunium, Uranium, and Plutonium*, YMP Milestone Report 3338, Technical Report LA-12956-MS, Los Alamos National Laboratory, Los Alamos, NM, 1996.

17. I. R. Triay, A. C. Furlano, S. C. Weaver, S. J. Chipera, and D. L. Bish, *Comparison of Neptunium Sorption Results Using Batch and Column Techniques*, YMP Milestone Report 3041, Technical Report LA-12958-MS, Los Alamos National Lab, Los Alamos, NM, 1996.

EXPERIMENTAL STUDY OF COLLOID FILTRATION BY COMPACTED BENTONITE

Susumu KUROSAWA*, Mikazu YUI** and Hideki YOSHIKAWA**

*Naka Energy Research Center, Mitsubishi Materials Corporation, 1002-14 Mukouyama, Naka-machi, Naka-gun, Ibaraki-ken, 311-01, Japan
**Power Reactor and Nuclear Fuel Development Corporation, 4-33 Muramatsu, Tokai-mura, Naka-gun, Ibaraki-ken, 319-11, Japan

ABSTRACT

In this study, the transport behavior of colloids through compacted bentonite and sand-bentonite mixtures was investigated. Colloidal gold was used to simulate mobile colloids because it was well characterized and its dispersivity was controlled. The bentonite used was a sodium bentonite. The sand-bentonite mixtures were prepared by mixing up to 50wt.% silica sand with the bentonite. The bentonite and the sand-bentonite mixtures were compacted to dry densities of 1000 and 1800 kg/m^3 and then saturated with distilled water. The sand-bentonite mixture was also saturated with synthetic sea-water. Column experiments were performed to investigate colloidal transport. Further, colloidal particles stabilities in high ionic strength water such as bentonite porewater or saline groundwater were interpreted based on the repulsion potential from the double layer force and the attraction potential from the van der Waals force.

The results indicated that the colloidal particles were effectively filtered by both the compacted bentonite and the sand-bentonite mixtures. This study indicated that the effect of colloids on radionuclide transport in compacted bentonite is negligible for the safety assessment of high level radioactive waste (HLW) disposal.

INTRODUCTION

The behavior of colloids in environments relevant to safety assessment of high-level radioactive waste disposal is still uncertain. The compacted bentonite surrounding the waste in the PNC concept is considered to behave as a filter which traps colloids because of its microstructure [1]. This microstructure depends on the degree of compaction and the swelling properties of bentonite. Experimental studies have not yet been performed to identify the correlation between colloidal transport and the bentonite microstructure.

This paper presents experimental and theoretical studies performed to clarify the transport behavior of colloids. The transport media examined were compacted bentonite and sand-bentonite mixtures at different densities after saturation with distilled water and synthetic sea-water, respectively.

EXPERIMENTAL

Procedures

The bentonite used was a sodium bentonite, Kunigel-V1(Kunimine Industries Co. Ltd., Japan). The sand-bentonite mixture was prepared by mixing silica sand with the bentonite. A sand-bentonite mixture is one of the candidate buffer materials for the Japanese reference design for HLW disposal. The silica sand used was 0.92 - 1.18 mm in diameter and was mixed at ratios of 30, 40 and 50 wt.% with the bentonite. The bentonite and the sand-bentonite mixture were packed into columns of 50mm diameter and 5mm height at dry densities of 1000 and 1800 kg/m^3 and saturated with distilled water. The sand-bentonite mixture was also saturated with synthetic sea-water (0.5M NaCl solution). Tracer solutions with the colloidal gold were prepared with both distilled water and 0.5M NaCl solution.

Mat. Res. Soc. Symp. Proc. Vol. 465 © 1997 Materials Research Society

Column experiments were carried out to investigate colloidal transport behavior. Figure 1 shows the experimental equipment. Ultrafilters (Millipore Corp.) with effective 10,000 molecular weight cutoffs (MWCO), equivalent to approximately 1.5nm pore sizes [2], were used to separate the colloids from the effluent solutions. The concentrations of gold in the filtered and unfiltered effluents were measured by inductively coupled plasma-mass spectrometry (ICP-MS) with a detection limit of 5.1×10^{-11}M. The difference in the gold concentration between filtered and unfiltered effluents directly corresponds to the concentration of the colloid passing through the compacted bentonite and the sand-bentonite mixture. Colloids in the bentonite samples were observed using transmission electron microscopy (TEM) and electron probe micro-analysis (EPMA) at the end of the column experiment.

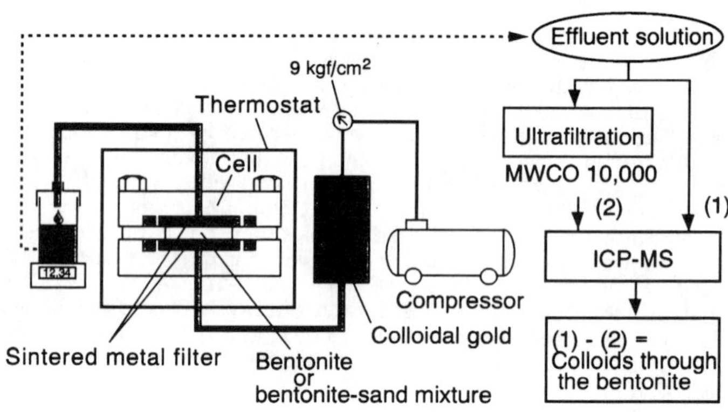

Figure 1 Schematic view of colloidal transport test equipment

Characteristic of Colloidal Gold

Colloidal gold was prepared by reduction of chloroauric acid ($HAuCl_4 \cdot 4H_2O$) solution with sodium citrate. To maintain colloidal stability, a surface-active agent was added before the reduction according to the method of Ogino et al. [3]. The composition of colloidal gold prepared is shown in Table I.

The stability of gold colloids is affected by the major ions in the bentonite porewater. The concentrations of such major ions, which were leached from the bentonite into distilled water, are shown in Table II. The major ionic concentration (Na^+, SO_4^{2-} and HCO_3^-) was adjusted in the colloidal gold solution to be consistent with measurements of equilibrated bentonite-distilled water at a liquid/solid ratio of 0.3ml/g.

Table I Composition of colloidal gold

concentration of gold colloids	98ppm
pH	6.6
surface potential	-6.9mV
surface-active agent	
i) ethanol	10.0wt.%
ii) polyoxyethylene hydrogenated castor oil	0.5wt.%

Table II Major ionic concentration leached from Kunigel-V1   (mol/l)

liquid/solid (ml/g)	Na^+	SO_4^{2-}	HCO_3^-
0.3	3.78×10^{-1}	1.97×10^{-1}	3.40×10^{-3}
1.0	7.39×10^{-2}	3.33×10^{-2}	5.57×10^{-3}

Figure 2 shows photographs of colloids by TEM after the adjustment of major ionic concentration. Slight aggregations were found in the equilibrated bentonite-synthetic sea-water mixture, but this was not considered to be significant because the composition of colloidal gold was not changed. Figure 3 shows the aggregation between gold colloids and bentonite colloids. Since the authors have confirmed that the chemical and physical properties of bentonite change very little following addition of a surface-active agent [4] and that colloidal gold is difficult to sorb on the bentonite, the colloidal gold was considered an appropriate material to examine colloidal transport behavior through the bentonite.

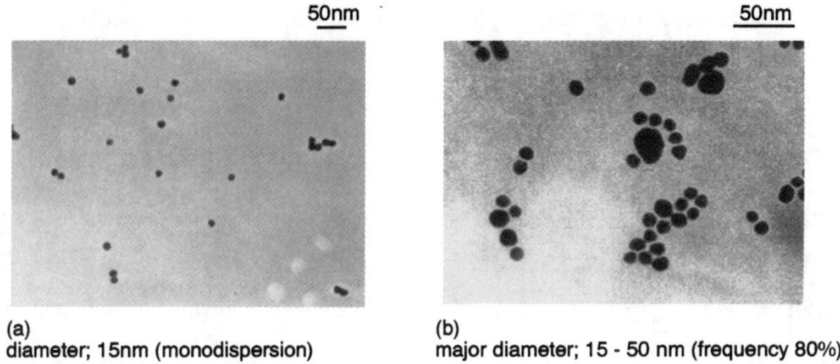

Figure 2   TEM image of colloidal gold dispersion, (a) after the adjustment of major
ionic concentration to simulate the equilibrated bentonite-distilled water,
(b) after further addition of 0.5M NaCl

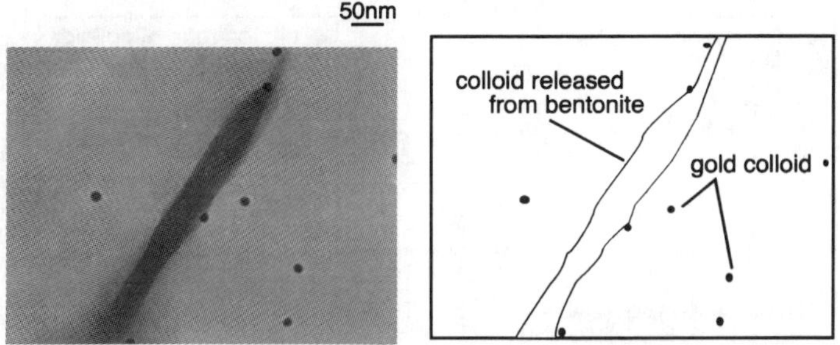

Figure 3   TEM image of aggregation between gold colloids and bentonite colloid

RESULTS AND DISCUSSION

<u>Colloidal Transport through Bentonite Saturated with Distilled Water</u>

Figure 4 shows the typical effluent volume of the solution and the elution ratio of colloids passing through compacted bentonite and sand-bentonite mixtures at a dry density $1000kg/m^3$. The concentrations of gold (Au) in filtered effluents were generally the same as in unfiltered effluents. This means that no measurable colloidal gold transport occurs through the compacted bentonite and sand-bentonite mixtures with a sand content less than 40wt.%. However, a small amount of gold colloid was found to pass through the sand-bentonite mixture with a sand content of 50wt.%. This finding was interpreted as follows.

Water uptake by dry compacted bentonite causes swelling of montmorillonite platelets which, under confined conditions, constricts available pores and builds up a swelling pressure. The crystal layer separation within platelets, and many of the pores between platelets, are of sub-nm dimensions [5] and clearly inaccessible to gold colloids 15nm in diameter. There are indications of connected networks of larger sized pores (e.g. measured anion diffusivities) and the tendency towards larger dimension pores will increase with increasing sand content. The measurements indicate that, for sand contents up to at least 40wt.%, colloids are efficiently filtered out by the constrictions encountered along the flow path. For 50wt.% sand content, however, some colloid breakthrough is observed but there is, even here, an indication of significant colloid retention due to filtration or sorption effects.

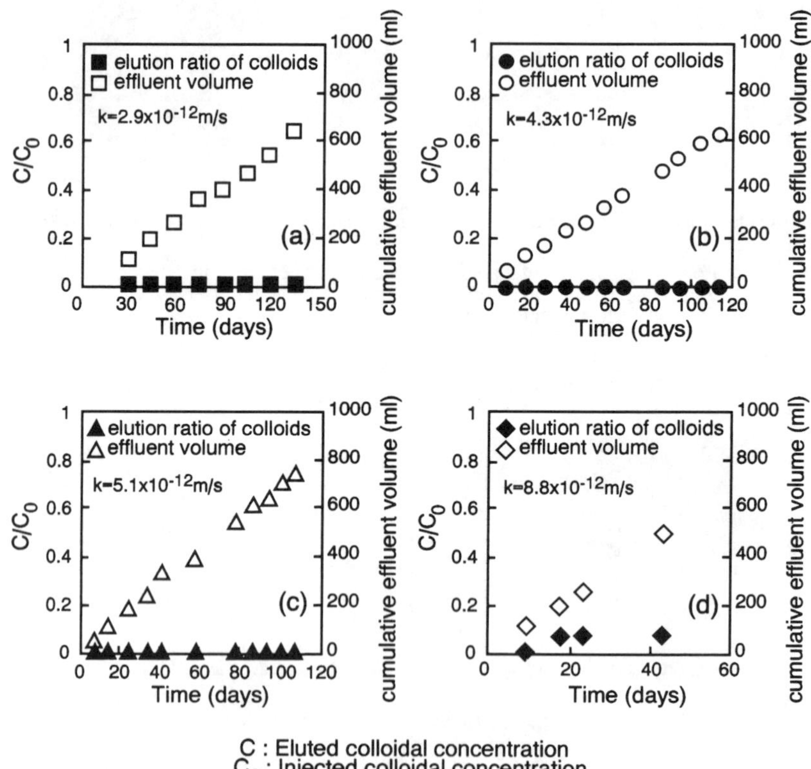

Figure 4   Results of colloidal transport test through compacted bentonite and sand-bentonite mixtures, saturated with distilled water at dry density 1000kg/m^3, k the hydraulic conductivity, (a) without sand, (b) with 30wt.% sand, (c) with 40wt.% sand and (d) with 50wt.% sand

The pore structure also depends on the extent of compaction. This is clearly illustrated by the results for the 50wt.% sand-bentonite mixture at a higher dry density of 1800kg/m^3 (Figure 5).

In this case, colloids are efficiently filtered out by the bentonite sample. The breakthrough results are supported by TEM studies of solid phase samples after the flow experiment (Figure 6). Although some larger pores exist, they do not appear to be well connected and hence colloids are immobilized in the flow path constrictions.

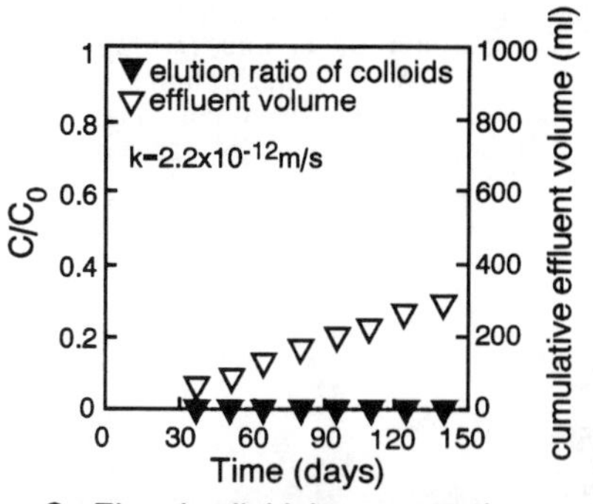

Figure 5

Result of colloidal transport test through the compacted 50wt.% sand-bentonite mixture, saturated with distilled water at dry density 1800kg/m^3, k the hydraulic conductivity

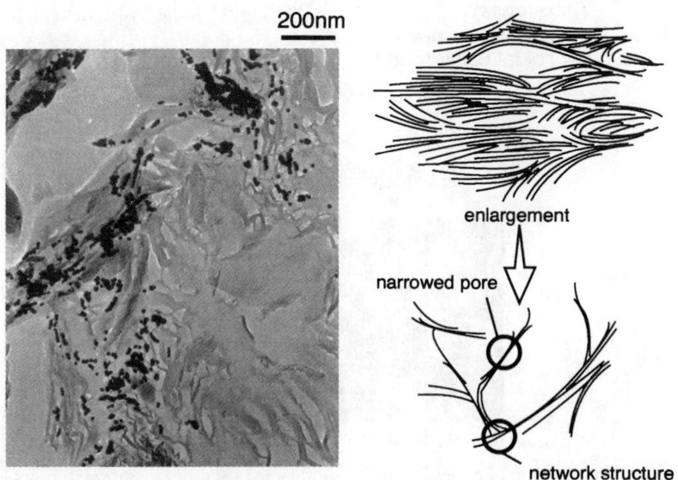

Figure 6 TEM image of colloids filtration and the pattern of the microstructure in the bentonite block

Colloidal Transport through Bentonite Saturated with Synthetic Sea Water

The microstructure of the compacted bentonite also depends on the porewater chemistry. For example, the swelling pressure of compacted bentonite decreases with increasing salinity and hence larger pores might be expected [6]. Thus experiments were performed to examine colloidal transport through 50wt.% sand-bentonite mixture saturated with 0.5M NaCl solution at a dry density of 1800kg/m^3. The 0.5M NaCl was added to the solution containing gold colloids. The major colloids were found to be 15 - 50 nm in diameter.

Figure 7 shows the typical effluent volume for solution containing gold colloids and the elution ratio of colloids passing through the compacted sand-bentonite mixture. The water flux through the column was double that for the experiments using the compacted sand-bentonite mixture saturated with distilled water. Colloids were not detected passing through the compacted sand-bentonite mixture. Figure 8 shows a photograph of colloids within the compacted sand-bentonite mixture, which was obtained by EPMA after the column experiment. The concentration of Au was found to be rich on the surface of the bentonite. This result indicates that the colloids were readily filtered on the surface layer of the compacted sand-bentonite mixture.

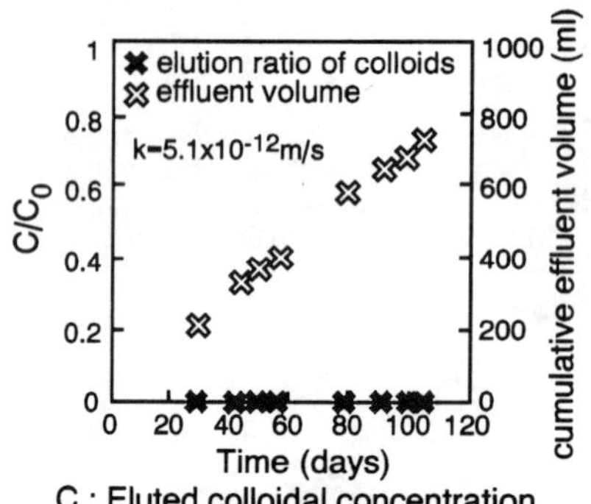

Figure 7

Result of colloidal transport test through the compacted 50wt.% sand-bentonite mixture, saturated with synthetic sea-water at dry density 1800kg/m^3, k the hydraulic conductivity

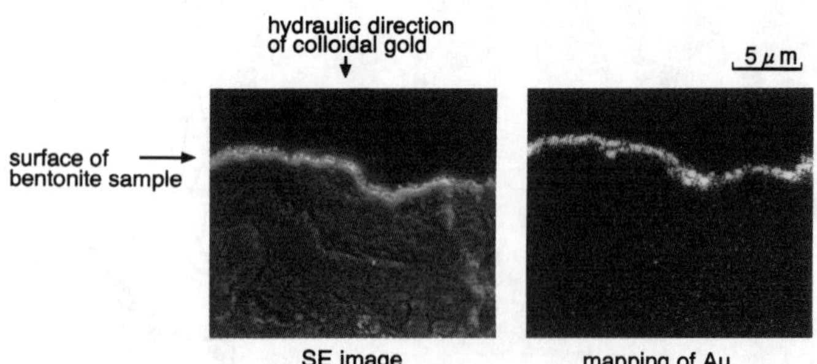

Figure 8 Distribution of colloidal gold in compacted sand-bentonite mixture by EPMA

Colloidal Stability

The stability of colloidal particles can easily be determined from the curve of total energy of an interaction between particles [7]. This curve is derived by addition of the repulsion potential (V_R) from the double layer force and the attraction potential (V_T) from the van der Waals force.

$$V = V_R + V_T \tag{1}$$

The repulsion potential between spherical particles from the double layer force is

$$V_R = \frac{\varepsilon a^2 \psi_0^2 (1+a)}{H+2a} \ln\left[1 + \frac{a}{H+2a}\exp(-\kappa H)\right] \qquad (2)$$

where $\kappa^2 = (8\pi nz^2 e^2)/\varepsilon kT$, κ is the concentration of electrolytes ($1/\kappa$ the thickness of electrical double layer), n the concentration in the bulk solution, z the ionic valency, e the elementary electric charge, ε the dielectric constant of fluid, k the Boltzmann constant, T the absolute temperature, a the half diameter of a colloidal particle, Ψ_0 the surface potential of a colloidal particle and H the distance between the colloidal particles.

The attraction potential from the van der Waals force between spherical particles has been described as

$$V_T = -\frac{A}{6}\left(\frac{2}{S^2-4} + \frac{2}{S^2} + \ln\frac{S^2-4}{S^2}\right) \qquad (3)$$

where $S=(H/a)+2a$, and A is the Hamaker constant. It is expected that stability will be reached when maximum total energy is larger than $15kT$.

For example, Figure 9 illustrates the curves of total energy between colloidal particles as a function of concentration of electrolyte κ. The curve of total energy moves into the negative range with increasing κ (decreasing $1/\kappa$), i.e. the colloids will be unstable in the range of repulsion potential which decreases with increasing electrolyte concentration. Colloidal particle stabilities in the porewater of compacted bentonite were estimated based on the repulsion potential from the double layer force and the attraction potential from the van der Waals force. However, as the porewater composition is known from previous studies, the values in Table II (liquid/solid ratio, 1.0ml/g) were used to calculate the concentration of electrolytes κ. The calculated κ value is approximately $10^9 \mathrm{m}^{-1}$ and indicates that colloidal particles will be unstable and flocculated according to Figure 9. In the case where compacted bentonite is saturated with synthetic sea-water, particles will be more unstable and flocculated.

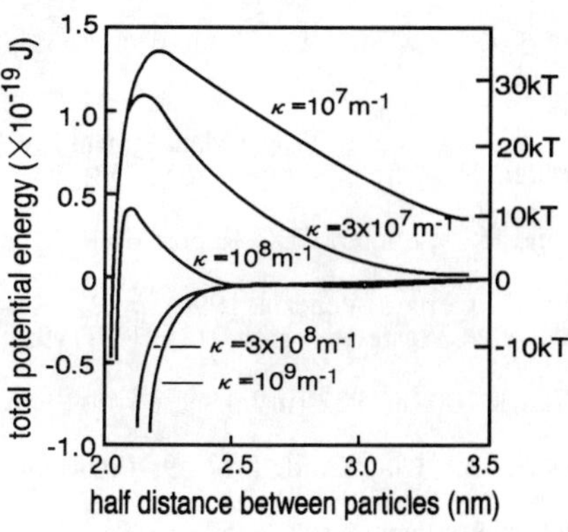

Figure 9

Influence of the concentration of electrolyte on the total potential energy of interaction of between spherical particles, a=10^{-7}m; T=298K; A=10^{-19}J; Ψ_0=25.6mV

In this work, it was found that the colloids which are larger than 15nm in diameter were filtered by the pore structure of the highly compacted bentonite and sand-bentonite mixture. On the other hand, the filtration of colloids smaller than 15nm has not yet been confirmed, because obtaining colloids smaller than 15nm is practically difficult. However, the colloids will be readily flocculate in high ionic strength water such as bentonite porewater or saline groundwater based on the repulsion potential from the double layer force and the attraction potential from the van der Waals force. Therefore, colloids smaller than those in this experiment would probably not be present. And colloids can easily be filtered by the compacted bentonite with sand content less than 40wt.% and at dry density more than 1000kg/m^3.

CONCLUSIONS

The transport behavior of colloids through compacted bentonite and sand-bentonite mixture was studied by column experiments. The colloid used was colloidal gold with 15nm (in distilled water) or 15 - 50 nm (in sea-water) in diameter. The conclusions of this study are as follows:
1)   The compacted bentonite and sand-bentonite mixture containing up to 40wt.% silica sand, which were saturated with distilled water at dry density 1000kg/m^3 were able to effectively filter out the colloidal gold. The compacted sand-bentonite mixture containing 50wt.% silica sand, which was saturated with distilled water at dry density 1000kg/m^3, was not able to completely filter out the colloids. The compacted sand-bentonite mixture, which was saturated with both distilled water and synthetic sea-water at dry density 1800kg/m^3 was able to filter out the colloidal gold.
2)   Based on the repulsion potential from the double layer force and the attraction potential from the van der Waals force, the colloidal particles will be flocculated in high ionic strength water, for example bentonite porewater or saline groundwater. Therefore, colloids smaller than those in this experiment are probably uncommon. This indicates that colloids can easily be filtered by microstructure of the compacted bentonite with sand content less than 40wt.% and at dry density more than 1000kg/m^3. Accordingly, the effects of colloid on radionuclide transport in the bentonite buffer are negligible for safety assessment of high level radioactive waste disposal.

ACKNOWLEDGEMENT

The colloidal gold used in this experimental work was provided by from Kanebo Ltd., Japan. The authors are grateful to Mr. Ogino K., Mr. Ohta M. and Mr. Matsui J., Kanebo, Ltd., for their support in this work.

REFERENCES

1.H. Sato, M. Yui and H. Yoshikawa, in Scientific Basis for Nuclear Waste Management XVIII, edited by Takahashi M. and Rodney C. E. (Mater. Res. Soc. Symp. Proc. 353, Kyoto, 1994), pp269-276.
2.J. I. Kim, G. Buckau, R. Klenze, D. S. Rhee and H. Wimmer, CEC Report EUR 13181, 41(1991).
3.K. Ogino, M. Ohta and J. Matsui, 18th Int. I.F.S.C.C. Cong., (Venezia, 1994), pp.272-284.
4.S. Kurosawa, H. Yoshikawa and M. Yui, Radioactive Waste Research 1(2), 177(1995)(in Japanese).
5.M. Nakano, Y. Amemiya and K. Fujii, Trans. JSIDRE 100, 8(1982)(in Japanese with English abstract, figures and tables).
6.H. Ishikawa, K. Amemiya, Y. Yusa and N. Sasaki, Int. Clay Conf. Proc. 9, (Strasbourg, 1989).
7.H. R. Kruyt, Colloid Science, 1st ed. (Elsevier Publishers, New York, 1952), p.271.

EXPERIMENTAL STUDY ON SELF-HEALING OF BENTONITE / SAND MIXTURES AND ITS IMPACT ON HYDRAULIC PERMEABILITY

Toshiyuki Hokari*, Mitsunobu Okihara*, Takashi Ishii* and Keiji Kojima**
*Shimizu Corporation, Power & Energy Project Div., 1-2-3 Shibaura, Minato-ku, Tokyo 105-07, Japan
**University of Tokyo, Dept. of Geosystem Engineering, 7-3-1 Hongo, Bunkyo-ku, Tokyo 113, Japan

ABSTRACT

The effects of gas permeation on the hydraulic permeability of bentonite/sand mixtures were studied experimentally by conducting permeability tests with various fluids and by microscopic observations. In these tests, hydraulic permeabilities were measured before and after helium gas was applied to permeate the mixtures.

Although gas formed preferential migration paths through the mixtures, the imperviousness of the bentonite/sand mixtures to the tested fluids never deteriorated because the paths became filled by swelling bentonite on re-saturation.

INTRODUCTION

Bentonite/sand mixtures have impervious characteristics owing to the swelling capacity of montmorillonite. These mixtures are expected to be used as a natural impervious barrier in facilities for the disposal of radioactive wastes, and are planned to be emplaced around the wastes themselves or the disposal facilities[1]. It is assumed that several kinds of gases could be generated inside the facilities because of chemical reaction of the wastes with groundwater[2] and that the gases could pass through the bentonite/sand mixtures after some period of storage inside the mixtures.

Komine[3] reported a mechanism accounting for the hydraulic properties of bentonite/sand mixtures. He argued that the mixtures have an impervious capacity because voids between aggregates are filled with bentonite gel which swells with water. Pusch[4] proposed a generalised microstructural model of the bentonite/sand mixtures, and reported that, among various factors, the densities of the bentonite gels could have the greatest effect on hydraulic properties, swelling pressure and gas migration. As a property related to gas migration, he concentrated on the gas pressure (referred to as the critical pressure) above which gas started to penetrate into the mixtures, and concluded that when there was a low bentonite content in the mixtures, the critical pressure was nearly equal to the swelling pressure. Pusch[5] conducted experiments in which a cyclic process of saturation and application of gas pressure to bentonite/sand mixtures was studied. He explained that reproducible measures of the critical pressure could be attributed to self-healing of the bentonite/sand mixtures.

We assumed that the self-healing of the bentonite/sand mixtures would cause no changes in imperviousness of the mixtures by gas migration. We conducted experiments designed to assess the validity of this assumption and discuss the results in this report.

EXPERIMENT

Changes in permeability were examined by performing hydraulic tests after and before gas permeation through bentonite/sand mixtures saturated with water. A hydrophobic fluid was used in some permeability tests after gas breakthrough in order to investigate the effects of affinities between the bentonite and the fluids on self-healing .

In addition to hydraulic permeability tests, microscopic observations were made after and before gas permeation in the mixtures in order to observe self-healing from a structural point of view.

Mat. Res. Soc. Symp. Proc. Vol. 465 © 1997 Materials Research Society

<u>Apparatus</u>

Figure 1 shows a schematic view of the cylindrical apparatus used for hydraulic permeability and gas permeation tests. The moulds have a size of 100 mm in diameter and 50 mm in height. The moulds are made of acrylic so as to be able to observe specimens.

Microscopic observations were made of specimens removed from the moulds. The observation apparatus is a VH-5910 Monitor microscope made by KEYENCE Ltd., which allows continuous observation by recording on video tape and obtaining photographs by hard-copying the screen. The magnification could be changed by attachment of lenses, i.e. x 50, 100, 200, 500 and 1000.

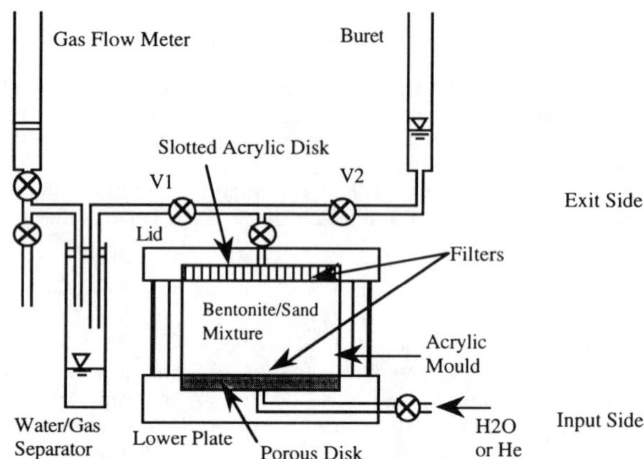

Figure 1. Schematic View of the Cylindrical Apparatus for Hydraulic & Gas Test

<u>Test material description</u>

Table I gives the characteristics of interest and compaction test results for bentonite/sand mixtures used for the permeability tests.

Table I. Description of the Tested Bentonite/Sand Mixtures

Specimen Characteristics				Compaction test results[4]	
$B/(B^{1)}+S^{2)})$ (%)	Fines content[3] (%)	Dry density (kg/m³)	Water content (%)	Maximum dry density (kg/m³)	Optimum Water content (%)
20	9	1 682	16.2	1 732	16.2
	1	1 774	14.0	1 785	14.0
17.5	1	1 714	14.1	1 794	14.1
		1 734	14.1		
		1 818	14.1		
15	8	1 670[5]	15.8[5]	1 762	15.8
		1 667	15.8		
		1 675	15.8		
	1	1 780	12.6	1 799	14.6
		1 806	14.6		
		1 817	14.6		
		1 812	14.6		

1) B : dry weight of bentonite
2) S : dry weight of sand
3) Fines content : weight content of smaller particles than sands (0.074mm)
4) Compaction test performed in conformity to JIS A 1210
5) specimen re-saturated with kerosene

The bentonite used was commercial Na-bentonite, Kunigeru-V1[6] made by Kunimine Industries Co., Ltd. in Japan. The aggregates used were terrace deposit sands from the Diluvium Epoch in the Quaternary Period. To allow for the variation in fines content [weight content of particles smaller than sand particles (less than 0.074mm)] among them, we used two representative aggregates with different fines content.

As shown in Table I, one specimen that was re-saturated with a hydrophobic fluid had a bentonite content of 15%, a dry density of 1 670 kg/m^3, a water content of 15.8 % and a fines content of 8%. Commercially available kerosene was used as the hydrophobic fluid.

<u>Experimental procedure</u>

Uniaxial hydraulic and gas tests were conducted as follows.
(1) Preparation of Specimens
Sand, bentonite and water were measured so that specimens would have the specified dry density and water content. Then, those materials were compacted inside moulds as prescribed in JIS A 1210[7].
(2) Saturation
Water was injected at a pressure of approximate 20 kPa at the bottom of the moulds. Water flow rates at the exit were measured with burets every day; it is judged that steady flow rates should represent saturation of the specimens.
(3) Hydraulic Test
Initially, the valve V1 was shut and V2 opened. While constant water pressure was applied to the inlet side with gas cylinders, the water flow rate was measured at the exit.
The water flow rate was calculated from the differences in water levels inside the burets over timed intervals.
The test was performed at three different pressure, i.e. 10, 20 and 30 kPa.
(4) Preparation for Gas Migration Test
After removing the lid, lower plate, porous disks and filters from the mould, a slotted acrylic disk was attached to the top of the mould. The lower porous disk was dried, reattached and filled with helium gas. Water in the inlet tube was displaced with the gas and the upper part of the slotted acrylic disk was filled with water.
(5) Gas Breakthrough Test
Initially, a gas pressure of 10 kPa is applied to the specimen. When no additional water was expelled by the gas over a fixed period (1 day), the gas pressure was increased to 20 kPa. In the same manner, the gas pressure was increased in 10 kPa steps until a pressure is reached above which water started to be expelled by the gas. This pressure is referred to as the threshold pressure. The water volume expelled was measured by changes in the level inside the buret.
Subsequently the pressure was increased by 20 kPa at the end of each fixed period (1 day). The water volume expelled was measured and gas breakthrough monitored. The breakthrough was confirmed by observation of the upper part of the specimen, and gas pressure and time were recorded at the breakthrough.
(6) Gas Permeability Test
With the valve V2 shut and V1 opened, the gas flow rate was measured at the breakthrough gas pressure, by measuring the change in water level inside a double tube buret at the outlet.
Gas flow rates were measured at various gas pressures.
(7) Re-saturation and Hydraulic Permeability Test
Water was injected at a pressure of approximate 20 kPa at the bottom of the mould. Water flow rates at the exit were measured with burets, and it is judged that steady flow rates should represent saturation of the specimens. Hydraulic permeability was calculated from the measured flow rate and the pressure .

Figure 2 shows a flow chart of the experimental procedure.

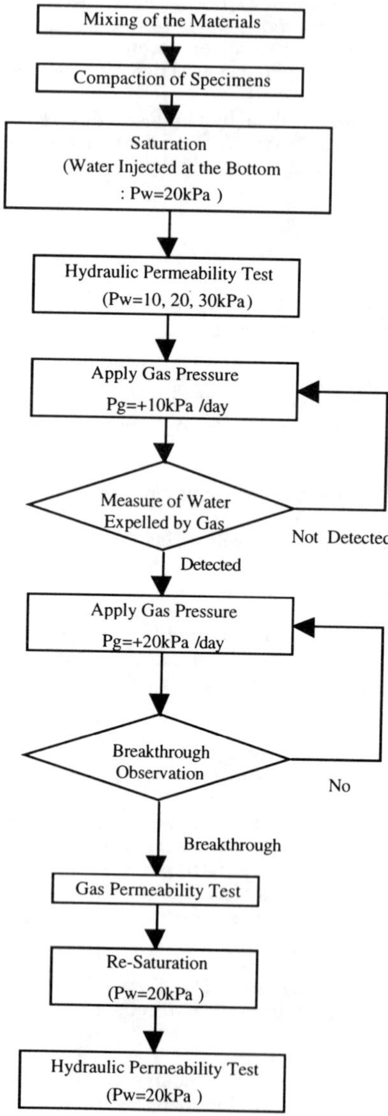

Figure 2 A Flow Chart of the Experimental Procedure

Microscopic observations were made on the surface of the specimens as follows:
(1) Compaction of the specimens

Compacted specimens were unsaturated by water to conform to the specified dry density and water content. This observation was made on specimens in which the bentonite had not yet swollen
(2) Saturation completed
The aggregate voids are supposed to be filled with bentonite gel swelling with water. The observation was made of the aggregate voids under saturation.
(3) Gas Breakthrough
Comparison with the surface under saturation provides information on gas migration
(4) Re-Saturation
Observation was made of the series of changes occurring when, following gas breakthrough, the specimen was supplied with water.

RESULTS

Calculation of Permeability

As this experiment used various fluids, it is convenient to express hydraulic and gas migration properties of the bentonite/sand mixtures as permeability. The permeability is calculated from the formula for Darcy linear flow under steady conditions[8]. While formula (1) is applied to an incompressible liquid phase, formula (2) is used for a compressible gas phase.

$$K = \frac{\mu L Q}{A(P_1 - P_2)} \qquad (1)$$

$$K = \frac{2\mu L Q_b P_b}{A(P_1^2 - P_2^2)} \qquad (2)$$

where K is permeability(m^2); μ, viscosity of fluids(Pa s); L, length of the specimen(m); Q, flow rate(m^3/s); A, sectional area of the specimen(m^2); P, absolute pressure(Pa); subscript b, measurement condition; subscript 1 and 2, inlet and outlet conditions, respectively.
Here, as no back pressure is applied to the output gas, the gas flow rate is measured under atmospheric conditions.

<u>Properties of Fluids</u>

The experiments use three kinds of fluids. Table II shows the viscosity and density of each fluid.

Table II. Physical Properties of the Fluids Used

Fluid	Density (kg/m^3)	Viscosity (Pa s)	Tempreture (C°)
Water	998.20[1]	1.002E-3[1]	20
Helium	0.1663[2]	19.6E-6[1]	20
Kerosene	789.2[3]	1.72E-3[4]	25

1) Reference to Chronological Science Tables[9]
2) Modified value from the C.S.T.[9] by Charles Law
3) Reference to Specifications of Idemitsu Kohsan Co.,
   Approximation at a temperature of 15 C°
4) Calculated with density (3) and kinematic viscosity measured in
   Uberhode type viscometer

<u>Permeability</u>

Table III shows the permeability values obtained in the hydraulic and gas tests based upon the formulae given above and the fluid properties in the previous section.

Table III. Results of Hydraulic and Gas Tests

Specimen Parameters				Permeability (m^2)		
B/(B+S) (%)	Fines content (%)	Dry density (kg/m^3)	Water content (%)	Hydraulic test before gas applied	Gas test	Hydraulic test after gas applied
20	9	1 682	16.2	1.61E-18	2.01E-16	7.96E-19
	1	1 774	14.0	5.53E-19	8.27E-15	5.23E-19
17.5		1 714	14.1	1.14E-18	3.26E-15	9.03E-19
	1	1 734	14.1	1.11E-18	2.95E-15	9.92E-19
		1 818	14.1	4.65E-19	1.45E-14	2.32E-19
15	8(1)	1 670	15.8	2.60E-18	4.80E-17	2.90E-17
	8	1 667	15.8	1.57E-18	1.58E-15	1.96E-18
	8	1 675	15.8	1.50E-18	3.63E-16	1.61E-18
	1	1 780	12.6	4.60E-19	3.41E-15	7.17E-19
	1	1 806	14.6	4.92E-19	8.36E-15	1.64E-20
	1	1 817	14.6	8.49E-19	6.31E-16	7.68E-19
	1	1 812	14.6	7.07E-19	3.23E-17	6.65E-19

(1) Kerosene used in re-saturation process

<u>Observation Results with Microscope</u>

Microscopic pictures are shown in Figure 3 to Figure 6.
Figure 3 represents a specimen of bentonite/sand mixture prepared with compaction. It shows incomplete occupation of the voids with bentonite gel before saturation.
Figure 4 represents a specimen saturated with water to which gas is about to be applied. It is found that the aggregate voids are filled with the swelling bentonite gel.

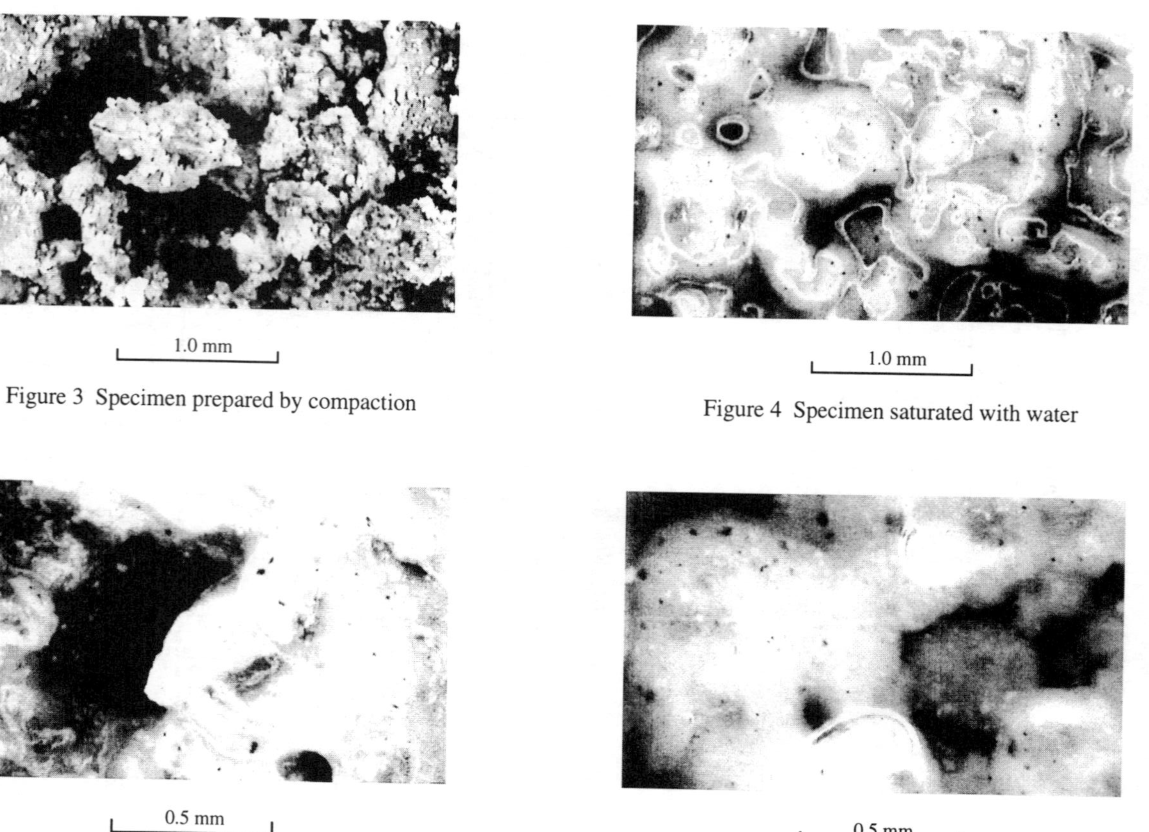

Figure 3 Specimen prepared by compaction

Figure 4 Specimen saturated with water

Figure 5 Effect of gas breakthrough

Figure 6 Re-saturation of the specimen

Figure 5 represents a specimen in which gas has broken through. While most of the voids are found to be occupied with the bentonite gel, there are a few voids which are empty. Considering the gas breakthrough, the void holes could be paths through which gas migrates in the mixtures. It is estimated that the permeability of gas would be larger than that of water because gas migrates through the formed paths in a pipe flow after gas breakthrough.

Finally the effect of re-saturation was observed as water was supplied to the paths formed by gas breakthrough. It was observed that the gas paths were occupied with bentonite gel expanding in the supplied water. Figure 6 represents a final stage at which bentonite gel re-fills the voids.

The above behaviour is considered to be self-healing as bentonite gel re-occupies the voids of the gas paths on re-saturation.

DISCUSSION

Microscopic observations show that re-saturation lets bentonite gel block gas paths and re-fill voids. The behaviour of the gel suggest a high possibility of self-healing in the bentonite/sand mixtures.

This section discusses self-healing to restore the impervious property required for the bentonite/sand mixtures, using the results of hydraulic and gas permeability tests with three kinds of fluid: water, helium and kerosene.

Figures 7, 8 and 9 show the results of the gas permeability and hydraulic permeability tests before and after breakthrough. They represent the relationship between permeabilities obtained from the tests with three kinds of fluids. Judging from the microscopic observations and test results, self-healing can occur as follows:

(1) Before Gas Breakthrough

Bentonite/sand mixtures saturated with water have aggregate voids filled with pore water and swelling bentonite(bentonite gel). As a result, the mixtures have an impervious nature because water migrates through the bentonite gel in a uniform flow but not in isolated dominant paths. Actually, the specimens in this experiment show this imperviousness in a permeability of less than $3 \times 10^{-18} m^2$ (corresponding to 3×10^{-11} m/s).

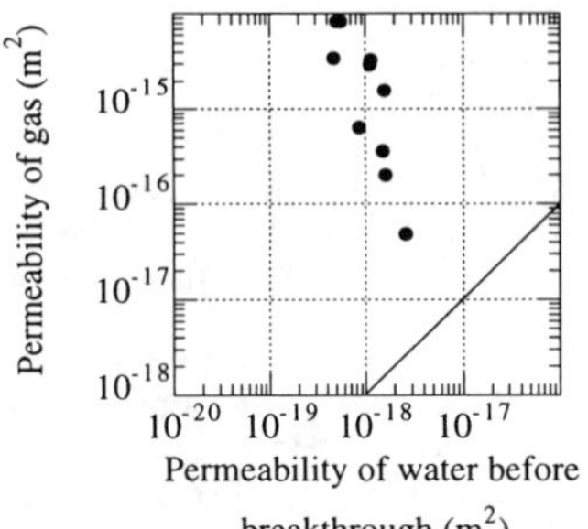

Figure 7 Comparison between gas permeability test and hydraulic test before breakthrough

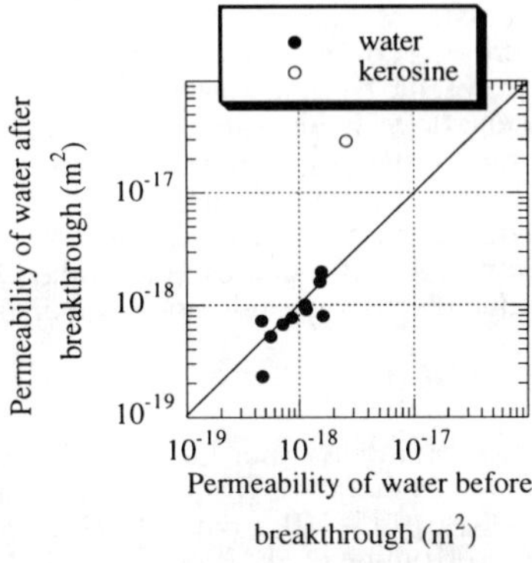

Figure 8 Comparison between permeability from hydraulic test before/after breakthrough

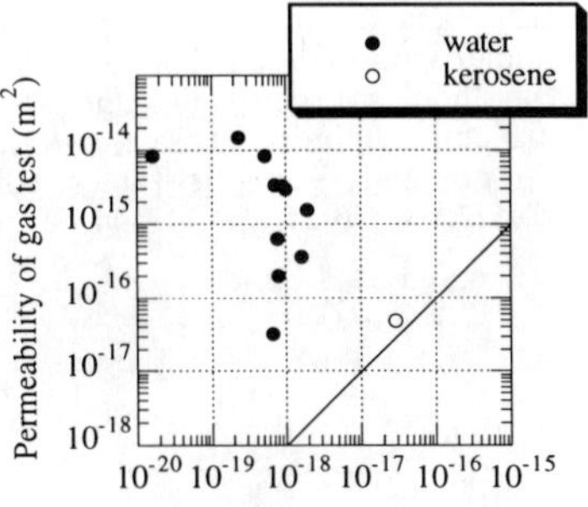

Figure 9 Comparison between gas permeability test and hydraulic permeability test after breakthrough

(2) Gas Breakthrough

Water was drained from the specimens by application of gas pressure. It is supposed that during the transient period of gas penetration, gas pressure shrinks and dehydrates bentonite gel to form gas paths. Preferential paths are formed because gas penetration is concentrated in the voids filled with gel whose swelling capacity is weakest. This is confirmed by microscopic observation.

Figure 7 shows that the permeability measured by the gas tests is larger than that from the hydraulic tests by one to four orders of magnitude. It is inferred that while gas flows through preferential paths in a pipe flow, water migrates as a uniform flow, not in a dominant flow path.

(3) After Gas Breakthrough

According to Figure 8, except for kerosene, the permeability measured by the hydraulic test after gas breakthrough is almost equal to that before breakthrough. So it is found that gas permeation causes no increase in the hydraulic permeability. This can be explained as follows.

In the hydraulic test after gas breakthrough, the addition of water induces re-swelling of the bentonite gel which had been shrunk and dehydrated by the gas pressure. After the swelling gel blocks the gas paths again, the mixtures, finally, resume the initial condition in which the voids are filled with the gel. This leads to no change in permeability.

Figure 9 also shows that permeability to kerosene is larger than that from the initial hydraulic test. The reason is that hydrophobic kerosene does not enable bentonite to swell, which results in no blocking of gas paths.

Kerosene could flow through the bentonite gel and the gas paths created by gas breakthrough. The kerosene concentrates in the latter which are more conductive than the former. Because kerosene flows by the same pipe flow as gas, permeability to kerosene is larger than that from the hydraulic water permeability test. This is supported by the fact that permeability to kerosene is equal to that of gas in Figure 9.

Judging from the microscopic observations and the comparison between permeabilities with different fluids, it is confirmed that the imperviousness of bentonite /sand mixtures to water flow is reproducible after gas migration, which is due to self-healing by expansion of bentonite gel.

REFERENCES

1. Hideo Shimoda et al., J. Atomic Energy Soc. Jpn., Vol.33, No.11(1991)
2. S. Voinis et al., in Gas Generation and Release - Proceedings. of a Workshop by NEA with ANDRA, p111-119, France(1992)
3. Hideo Komine et al., Proc. Sym. Soil Mech., No.40, VI-5(1995)
4. Pusch et al., SKB technical report 90-43, 1990
5. Pusch et al., SFR Arebetsrapport 87-06, 1987
6. Masakazu Itoh et al., J. Atomic Energy Soc. Jpn., Vol.36, No.11(1994)
7. Test method for soil, compaction using a rammer - JIS A 1210, published by Japanese Standards Association, Tokyo, 1990
8. Handbook of Petroleum Industry, edited by Soc. Jpn. Petroleum Engineers(Soc. Jpn. Petroleum Engineers, Tokyo, 1983), p.504
9. Chronological Scientific Tables, edited by National Astronomical Observatory (Maruzen Co., Ltd.,Tokyo, 1992), pp. 442-443 & pp. 446-448

COMPARATIVE TESTS FOR EVALUATING PERMEABILITY CHANGES OF A COMPACTED BENTONITE/SAND MIXTURE DURING SHEAR

T. ESAKI*, M. ZHANG* AND Y. MITANI**
*Institute of Environmental Systems, Faculty of Engineering, Kyushu University, Fukuoka 812-81, JAPAN
**SHIMIZU Corporation, Shibaura 1-2-3, Minato-ku, Tokyo 105-07, JAPAN

ABSTRACT

A compacted mixture of Kunigeru VI bentonite and D-sand is being considered for use as an engineered barrier in low-level radioactive waste disposal facilities in Japan. An important issue is the maintenance of the retardation property of the mixture during shear that might be induced in such barriers by earthquakes and/or gradual tectonic deformations occurring over the design life of the facility. To investigate this issue, comparative tests on a bentonite-sand mixture and kaolin-sand mixture were conducted by means of a recently-developed coupled shear and permeability testing apparatus under temperature controlled condition. In addition to permeability, the specific storage of bentonite-sand specimen during shear is also systematically evaluated with the new analytical theory for the constant flow permeability test. The present study reveals that 1) temperature control is preferred for measuring the permeability of extremely-low permeability materials with the constant-flow pump method; 2) both the permeability and specific storage of the mixture of Kunigeru VI bentonite and D-sand (bentonite to sand weight ratio = 15:85) were not significantly influenced by shear strains up to 3% whereas the permeability of the kaolin-sand mixture increased almost linearly with the increment of shear strain; 3) the swelling of bentonite in the mixture under low confining stress decreases both the permeability and specific storage of bentonite-sand mixture. This feature of bentonite helps maintain the long-term retardation property of the mixture; and 4) the constant flow permeability test method, with the newly derived theoretical analysis, promises to become a very effective means of investigating, rapidly and systematically, the permeability and specific storage of extremely-low permeability materials with relatively-low (i.e. close to in situ) hydraulic gradients. This is necessary for the design and long-term performance assessment of engineered barriers.

INTRODUCTION

Permeability, a porosity-related property of porous materials, greatly affects the retardation properties of engineered barriers. The first requirement of an engineered barrier is low permeability. Shear may cause dilatancy of over-consolidated soils and thus increase their permeability. During the service life of radioactive waste disposal facilities, shear strains may be induced in engineered barriers by earthquakes as well as other kinds of geologic events and may consequently alter the retardation property of the barriers.

This study verifies the constancy of the retardation property of a mixture of Kunigeru VI bentonite and D-sand during shear. The mixture is being considered for use in Low-Level Radioactive Waste disposal facilities in Japan. The verification tests were conducted by means of a recently-developed coupled shear/permeability testing apparatus which includes temperature control. Tests on kaolin-sand mixture were also conducted for a comparison. This paper presents systematic and detailed information on the comparative tests. In addition, this paper briefly introduces a new theoretical analysis of the constant flow permeability test that was used to obtain the permeability and specific storage of low permeability (lower than 10^{-10} m/s) barrier materials. This method is rapid, effective, and eliminates experimental errors associated with measurements of low-permeability with relatively low hydraulic gradients, that is gradients as close as possible to those occurring in situ.

EXPERIMENTAL

Testing apparatus

A coupled shear and permeability testing apparatus, capable of investigating the permeability in the

Mat. Res. Soc. Symp. Proc. Vol. 465 © 1997 Materials Research Society

direction parallel to shear planes, has recently been developed by the authors[1]. The permeability in such a direction is assumed to be the maximum and thus provides useful information for the safe design of barriers. The apparatus mainly consists of a system for shear and measuring specimen volume change, a flow pump system for extremely-low permeability measurement, and a data acquisition and monitoring system (Fig. 1). Shear strain was generated by applying different pressures within and outside the thick-walled hollow cylindrical specimen in the triaxial cell. The specially designed flow pump system could generate a specified low and constant flow rate ranging from 2.6×10^{-6} to 1.2×10^{-5} ml/s. The computer-based data acquisition and monitoring system display the plots and the current digital values of all the physical parameters measured.

For testing the compacted kaolin-sand mixture with relatively high permeability, a constant-head water supply apparatus, generally used in geotechnical laboratories, was used instead of the flow pump system. In this case, the flow rate was measured by means of an electronic balance.

Furthermore, to eliminate the previously reported influence of temperature variation on permeability measurement[1], the coupled shear and permeability tests were performed under temperature controlled condition.

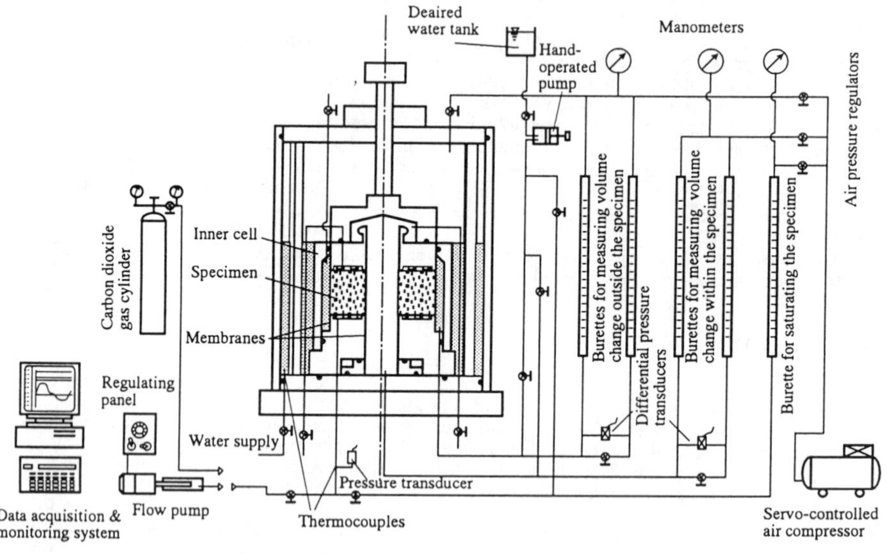

Fig. 1 The coupled shear and permeability testing systems

<u>Testing materials</u>

Bentonite is a natural material which may vary considerably from one source to another in regards to mineral composition, chemical state and grain-size distribution. The bentonite used in the present study is Kunigeru VI bentonite made by the Kunimine Industry Co. It is a high-swell bentonite containing sodium montmorillonite. The D-sand is also a typical kind of sand in Japan with stable physicochemical properties. The kaolin, with little or no swelling potential, and a so-called standard sand used for the comparative test were made by the Katayama Chemical Industry Co and the Toyoura Industry Co, respectively. Physical properties of the materials used in the present study are provided in Table I.

<u>Specimen preparation</u>

Both bentonite-sand and kaolin-sand mixtures were prepared with a weight ratio (either bentonite or kaolin to sand) of 15:85. To determine the optimum water contents for preparing the specimens, the Standard Proctor compaction tests for both mixtures were conducted in advance. The maximum dry density and the optimum water content of the mixture of Kunigeru VI bentonite and D-sand were 1.77 Mg/m^3 and 13.8%, respectively. The maximum dry density and the optimum water content of the mixture of kaolin and Toyoura standard sand were 1.78 Mg/m^3 and 11.5%, respectively. The thick-walled hollow cylindrical specimens were then compacted in a specialized mould using a fan-shaped rammer with the same compaction energy per unit volume as that of the Standard Proctor compaction. Specifically, a thick-walled hollow cylindrical specimen was prepared by compacting the mixture in three equal layers in an annular mould (ϕ 4 cm × ϕ10 cm × H10 cm), by dropping a 1.25 kg hammer through a distance of 37.5 cm, each layer being subjected to 21 blows. The compacted sample was finally cut and trimmed to be 4 cm high by means of a wire saw and an apparatus for pushing the sample out from the mould.

Table I Physical properties of bentonite, kaolin and sands

Trade name	Kunigeru VI bentonite	Trade name	D-sand
Density (Mg/m^3)	2.79	Density (Mg/m^3)	2.67
Liquid limit (%)	473.9	Max. particle size (mm)	4.75
Plastic limit (%)	26.6	Uniformity coefficient	1.57
Plastic index	447.3	Coarse-grained particle (%)	89.0
Shrinkage limit (%)	10.6	Fine-grained particle (%)	7.0
Activity	6.9	Silt (%)	2.0
Plastic ratio	16.8	Fine gravel (%)	2.0
Trade name	Katayama kaolin	Trade name	Toyoura standard sand
Density (Mg/m^3)	2.70	Density (Mg/m^3)	2.65
Liquid limit (%)	51.6	Mean particle size (mm)	0.17
Plastic limit (%)	28.0	Uniformity coefficient	2.0
Plastic index	23.6	Max. void ratio	0.977
		Min. void ratio	0.597

<u>Testing procedures and conditions</u>

The procedures for the coupled shear and permeability test on the thick-walled hollow cylindrical specimen and the cautions that should be exercised during the test have been described in detail by the authors[1]. Here the procedures are briefly summarized:
1) Set the specimen onto the testing apparatus.
2) Apply relatively low confining pressures (external and internal pressures) to the specimen.
3) Replace the air in the specimen with carbon dioxide (for a rapid saturation of bentonite-sand mixture).
4) Saturate the specimen with deaired water.
5) Apply specified high confining pressures to the specimen for consolidation.
6) Decrease the internal pressure step by step and measure the permeability of the sheared specimen on each of shear steps.
In addition, when measuring the flow rate in an open system with the electronic balance during the constant-head permeability test on kaolin-sand mixture, the evaporation effect was calibrated and considered in flow rate measurements.
The specified values of confining pressures on the specimen of compacted bentonite-sand mixture and the specimen of compacted kaolin-sand mixture are listed in Table II in the next subsection.

<u>Experimental results</u>

The volume changes measured within and outside the hollow cylindrical specimen during shear are listed in Table II. The corresponding radial displacements of the inside and outside walls of the specimen can be calculated from these volume changes by assuming that the radial deformations along the vertical direction of specimen are uniform (see also Table II). These displacements are obtained in order to evaluate the shear strains generated in the specimen, as described later in this paper.

Table II The volume changes measured within and outside the specimen, and the calculated radial displacements of the inside and outside wall of the specimen

Specimen	External pressure (kPa)	Internal pressure (kPa)	V_i (ml)	u_i (mm)	V_o (ml)	u_o (mm)
Bentonite-sand mixture	500	500	0	0	0	0
		100	2.10	0.42	2.07	0.17
		50	3.17	0.64	3.01	0.24
		30	3.94	0.80	3.67	0.29
Kaolin-sand mixture	600	600	0	0	0	0
		200	2.07	0.42	2.05	0.16
		100	3.31	0.67	3.27	0.26
		50	5.42	1.11	5.38	0.43

V_i and V_o are the volume changes measured within and outside the specimen

u_i and u_o are the calculated radial displacements of inside and outside wall of specimen

Fig. 2 shows typical curves obtained during the constant flow permeability test for bentonite-sand mixture. **They are provided for an infusion state, transient decay state and withdrawing state.** Although the previous study[1] reported a sensitivity of the measurements to circumference temperature variation, the head difference versus time curve is very smooth because the temperature was controlled to be almost constant in this study.

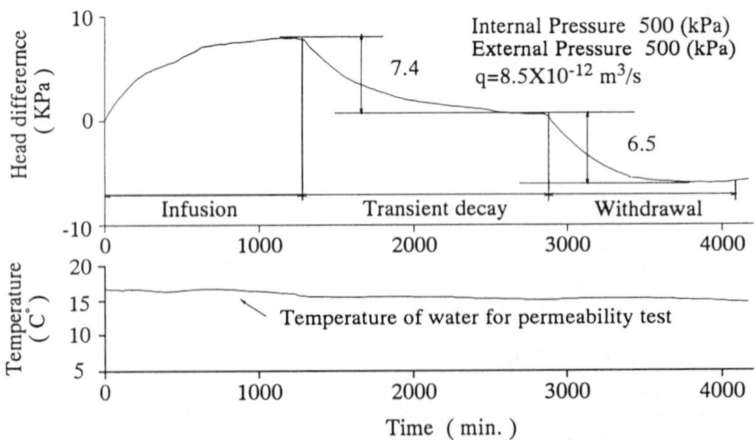

Fig. 2 The time-dependent variation of the induced head-difference and temperature for bentonite D-sand mixture in temperature controlled condition.

Flow rate was measured during the constant-head permeability test on the kaolin-sand mixture. During the test, the hydraulic gradient was maintained constant at 15. The flow rates measured were 0.0123, 0.0140, 0.0177 and 0.0211 ml/s under the condition of internal pressures being 600, 200, 100 and 50 kPa, respectively. As can be seen from Table II, external pressure during each of shear steps was fixed at the specified value of 600 kPa.

THEORETICAL ANALYSIS

Shear strain

The maximum shear strains γ_{max} generated in the specimen were evaluated through the radial displacements of the inside and outside walls of the hollow cylindrical specimen (see Table II) using the following equation:

$$\dot{\gamma}_{max} = -\frac{u_o + u_i}{R_o + R_i} + \frac{u_o - u_i}{R_o - R_i} \tag{1}$$

in which, R_o and R_i are external and internal radii of the thick-walled hollow cylindrical specimen. The first term and second term express the average radial strain (the maximum strain) and the average tangential strain (the minimum strain) generated in the specimen[2]. Therefore, the maximum shear strain was defined as the difference between the maximum and minimum strains[3]. The shear strains generated in the specimen for both mixtures under various differential pressures are evaluated in Table III.

Table III Evaluated shear strains in the specimen under various differential pressures

Specimen	Bentonite-sand mixture				Kaolin-sand mixture			
External pressure (kPa)	500				600			
Internal pressure (kPa)	500	100	50	30	600	200	100	50
Shear strain (%)	0	1.7	2.6	3.3	0	1.7	2.7	4.4

Permeability and specific storage

Generally, the permeability of a specimen during permeability testing is calculated using Darcy's law for flow through saturated porous media:

$$Q = -K \cdot A \cdot h \tag{2}$$

where:
Q = the volumetric flow rate, L^3/T (for a constant flow permeability test, this is the controlled flow rate of flow pump);
h = the hydraulic gradient across the specimen (for a constant flow permeability test, this is the stabilized head difference, as shown in Fig. 2, divided by the length of specimen.);
K = permeability, L/T;
A = cross-sectional area of specimen, L^2.

Recently, an exact theoretical analysis of the flow pump permeability test has also been developed by the authors and expressed as follows.

For the infusion state[4, 5]:

$$H = \frac{qL}{AK} \left\{ \frac{z}{L} - 2 \sum_{n=0}^{\infty} \frac{\exp\left(-\frac{K}{S_s}\beta_n^2 t\right) \sin\left(\beta_n z\right)}{L\delta\beta_n \cos\left(\beta_n L\right)\left[L\left(\beta_n^2 + \frac{1}{\delta^2}\right) + \frac{1}{\delta}\right]} \right\} \tag{3}$$

For the transient decay state following the steady infusion state[5]:

$$H = 2 \cdot \Delta H_s \cdot \sum_{n=0}^{\infty} \frac{exp\left(-\dfrac{K}{S_s}\beta_n^2 t\right) sin\left(\beta_n z\right)}{L\delta\beta_n cos\left(\beta_n L\right)\left[L\left(\beta_n^2 + \dfrac{1}{\delta^2}\right) + \dfrac{1}{\delta}\right]} \tag{4}$$

in which $\delta = C_e /(AS_s)$ and, β_n are the roots of the following equation

$$tan\left(\beta L\right) = \frac{1}{\beta\delta} \tag{5}$$

Where
   H = the hydraulic head across the specimen, L;
   S_s = specific storage, L^{-1};·
   q = constant flow rate of flow pump, L^3/T;
   L = the length of the specimen, L;
   t = time, T;
   z = vertical distance along the specimen, L;
   C_e = storage capacity of the flow pump system, L^2;

   ΔH_s = steady-state differential head across the specimen, L.

Evaluation of the unknown parameters K, S_s and C_e of bentonite-sand mixture from the measurements of the flow pump permeability test were performed by minimizing the following error function. It represents a least squares reduction of the discrepancy between the M head differences $H(L, t_i)^*$ measured at times t_i and the corresponding data $H(L, t_i)$ obtained by the theoretical analysis. The Variable Alternating Method coupled with the Golden Section Method was used for the minimization process[4, 5]. Calculated values for parameters K, S_s and C_e are given in Table IV.

$$\varepsilon = \left\{\sum_{i=1}^{M}\left[H\left(L, t_i\right)_{(K, S_g, C_e)} - H\left(L, t_i\right)^*\right]^2\right\}^{1/2} \tag{6}$$

In addition, permeabilities of the kaolin-sand mixture were calculated from the measurements of constant-head permeability test using the Darcy's law and also listed in Table IV.

Table IV Parameters evaluated from the permeability test at each of the shear steps

| Specimen | | | Bentonite-sand mixture | | | | Kaolin-sand mixture | | | |
| Ext. pressure (kPa) | | | 500 | | | | 600 | | | |
Int. pressure (kPa)			500	100	50	30	600	200	100	50
Infusion	Early time	K (m/s)$\times 10^{-11}$	5.38	6.57	4.39	3.54				
		S_s (1/m)$\times 10^{-3}$	2.78	3.29	2.97	1.03				
		C_e (m^3/kPa)$\times 10^{-9}$	1.37	2.89	4.65	19.9				
	Whole time	K (m/s)$\times 10^{-11}$	5.99	6.60	4.32	3.82				
		S_s (1/m)$\times 10^{-3}$	2.39	3.29	3.09	1.03				
		C_e (m^3/kPa)$\times 10^{-9}$	1.48	2.75	4.22	18.6				
Transient decay		K (m/s)$\times 10^{-11}$	5.46	5.73	3.69	3.45				
		S_s (1/m)$\times 10^{-3}$	2.97	2.99	3.14	1.06				
		C_e (m^3/kPa)$\times 10^{-9}$	1.12	3.06	4.74	19.8				
Average		K (m/s)$\times 10^{-11}$	5.61	6.30	4.13	3.60				
		S_s (1/m)$\times 10^{-3}$	2.71	3.19	3.07	1.04				
		C_e (m^3/kPa)$\times 10^{-9}$	1.32	2.90	4.54	19.4				
Steady state	Infusion	K (m/s)$\times 10^{-11}$	7.01	5.89	5.54	4.84	12400	14100	17900	21300
	Withdrawal	K (m/s)$\times 10^{-11}$	7.93	5.60	5.15	5.15				

Average is the mean of the values of Early time, Whole time and Transient decay

<u>Impact of shear strain on the permeability</u>

For comparison, the relationships between the shear strain and permeability for the bentonite-sand mixture (with swell potential) and the kaolin-sand mixture (without swell potential) are both plotted in Fig. 3.

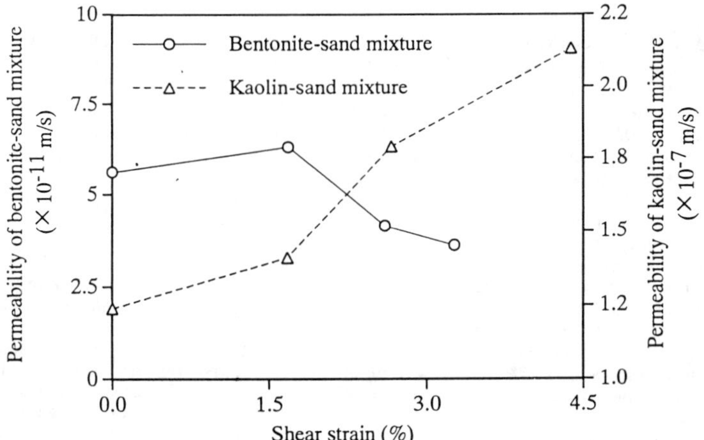

Fig. 3 The relationship between the shear strain and permeability
for bentonite-sand and kaolin-sand mixtures

DISCUSSION AND OBSERVATIONS

From the Table IV, the following observations can be made:

(1) The permeabilities evaluated for all conditions with the new theoretical analysis of the constant flow permeability test are relatively lower than those obtained from steady condition evaluation using Darcy's law. This can be explained by the fact that steady flow in the specimen was not completely established at the end of the infusion or withdrawing process. The actual head differences across the specimen at the steady state would be a little larger than those illustrated in the Fig. 2. However, the time required for establishing the steady flow state is longer when testing specimens with lower permeability and/or higher specific storage. Also, determination of the steady state during the permeability test is sometimes difficult without observation over a relatively long time interval. Therefore, the new method is suggested for evaluating low-permeability.

(2) The permeabilities calculated from the early time measurements, complete time measurements and transient decay measurements are almost the same. Therefore, it is possible to shorten the permeability test time because the early time evaluation showed almost the same values of parameters as those obtained from the complete time evaluation.

(3) The storage capacity C_e of the flow pump system increased step by step during the series of shear and permeability tests. This maybe due to the increase of the ratio of air-rich water in the flow pump system which was withdrawn from the specimen during the downward flow process. The phenomenon is well recognized by the new method for constant flow permeability test. Therefore, the new method obtains not only the permeability but also the storage capacity of specimen, eliminating the experimental errors.

(4) The permeability and specific storage of the specimen remained almost unchanged or showed only little increase at the first shear step (i.e. when the internal pressure was decreased from 500 kPa to 100 kPa.) This little increase in permeability might be due to shear-induced dilatancy and the consequent increase in the void ratio of specimen. However, in the next two shear steps, both the permeability and specific storage of specimen decreased a small amount owing to the time and confining pressure related swelling effect of bentonite in the mixture. This trend continues even when larger shear strains were generated in the specimen.

In addition, Fig. 3 clearly shows that the permeability of bentonite-sand mixture was not significantly influenced by shear strain, whereas the permeability of kaolin-sand mixture increased almost linearly with the increment of shear strain.

CONCLUSIONS

To investigate the retardation property of a mixture of Kunigeru VI bentonite and D-sand, which is being considered for use in Low-Level Radioactive Waste disposal facilities in Japan, a series of laboratory tests were performed under temperature controlled conditions with the recently developed coupled shear/permeability test apparatus. The permeability and specific storage of the mixture were evaluated by means of a new theoretical analysis of the flow pump permeability test. In addition, comparative tests on a kaolin-sand mixture were also presented. Conclusions drawn from the study are briefly summarized as follows: ·

1) Temperature control is preferable for the permeability measurement on extremely-low permeability material with the constant-flow pump method.

2) The permeability of the mixture of Kunigeru VI bentonite and D-sand, with a bentonite to sand weight ratio of 15:85, varied in the range of about 3.6×10^{-11} m/s to 6.3×10^{-11} m/s. The flow pump method could measure the low permeabilities rapidly in several tens of hours with relatively low hydraulic gradients. Therefore, the constant-flow method is a very effective means for evaluating the extremely-low permeability of barrier materials.

3) The permeability of the mixture of Kunigeru VI bentonite and D-sand was not significantly influenced by shear strains up to 3%. Also, swelling of bentonite at low confining stress could decrease both the permeability and specific storage of the bentonite-sand mixture. Therefore, the permeability of bentonite-sand mixture could be kept at a low value even when shear deformation was generated in the mixture. This partially supports that bentonite-sand mixtures can be effectively used as barrier materials.

4) The permeability of the kaolin-sand mixture increased almost linearly with the increment of shear strain because the kaolin has no or little swelling potential. This is a collateral evidence supporting the effectiveness of bentonite with swelling potential as a barrier material.

5) The new analysis for the flow pump permeability test method permitted the evaluation of both the permeability and specific storage of extremely-low permeability materials from the early time measurements of the permeability test. Therefore, this approach shows promise in rapidly obtaining the hydraulic parameters of extremely low permeability material under relatively-low hydraulic gradients, as close as possible to the in situ conditions. However, determination of a proper duration time for completing the flow pump permeability test requires additional study.

REFERENCES

1. M. Zhang and T. Esaki, et al., in <u>Scientific Basis for Nuclear Waste Management XVIII</u>, edited by T. Murakami and R. C. Ewing, (Mater. Res. Soc. Proc. **353**, Pittsburgh, 1995), pp. 261-268.
2. A. S. Saada, in <u>Advanced Triaxial Testing of Soil and Rock</u>, edited by R. T. Donaghe, R. C. Chaney and M. L. Silver, (ASTM, **STP 977**, Philadelphia, 1988), pp. 766-795 .
3. W. F. Chen and E. Mizuno, <u>Nonlinear Analysis in Soil Mechanics: Theory and Implementation</u>, (Elsevier Science Publishers, New York, 1990), p. 661.
4. T. Esaki and M. Zhang, et al., Geotechnical Testing Journal **19** (3), 241-246 (1996).
5. M. Zhang, PhD thesis, Kyushu University, 1996.

THE CHANGE IN BIOAVAILABILITY OF ORGANIC MATTER ASSOCIATED WITH CLAY-BASED BUFFER MATERIALS AS A RESULT OF HEAT AND RADIATION TREATMENT

S. Stroes-Gascoyne, S.A. Haveman and P. Vilks. AECL, Whiteshell Laboratories, Pinawa, Manitoba, Canada, R0E 1L0.

ABSTRACT

Compacted clay-based buffer surrounds corrosion-resistant waste containers in the Canadian concept for nuclear fuel waste disposal. Clays naturally contain small quantities of organic matter that may be resistant to bacterial degradation. The containers with highly radioactive material would subject the surrounding buffer to both heat (max. 95°C) and radiation (max. 52 Gy/h). Both could potentially break down complex organic material to smaller, more bioavailable compounds. This could stimulate microbial growth and possibly affect gas production, microbially-influenced corrosion or radionuclide migration. Experiments were carried out in which buffer (50 wt.% Na-bentonite, 50 wt.% sand) was heated at 60 and 90°C for periods of 2, 4 and 6 weeks, in some cases followed by irradiation to 25 kGy. Unheated buffer was also irradiated to 25 and 50 kGy at different moisture contents. The treated materials were subsequently suspended in distilled water, shaken for 24 h and centrifuged to remove the solids. The 0.22 µm filter-sterilized leachates were inoculated with equal volumes of fresh groundwater and incubated at room temperature for 10 d to determine the increase in total and viable bacteria compared to a groundwater control. Results indicated that leachates from buffer subjected to heat, radiation or combinations of these, had a stimulating effect on both total and viable cell counts in groundwater, compared to unamended groundwater controls. This stimulating effect was generally most pronounced for viable counts and could be larger than two orders of magnitude. Leachates from untreated buffer material also stimulated the growth of groundwater bacteria, but to a lesser extent than leachates from heat-and radiation-treated buffer material. The effects of heat and radiation on nutrient availability in clay-based sealing materials (at relevant clay/moisture ratios) should, therefore, be taken into account when attempting to quantify the effects of microbial activity on vault performance.

INTRODUCTION

AECL has developed a concept for the permanent disposal of nuclear fuel waste in plutonic rock of the Canadian Shield [1]. The concept involves the disposal of nuclear fuel waste in an engineered excavation (vault), at a depth of 500 to 1000 m in plutonic rock. Fuel wastes would be isolated in corrosion-resistant metal (Ti or Cu) containers and emplaced in disposal rooms or in boreholes in the floor of disposal rooms, surrounded by a compacted clay-based buffer (consisting of 50 wt.% Na-bentonite and 50 wt.% silica sand). The rooms would be backfilled with a mixture of 75 wt.% crushed and graded host rock and 25 wt.% glacial lake clay. Clays naturally contain small quantities of organic matter that may be resistant to bacterial degradation. The containers with highly radioactive material would subject the surrounding buffer to both heat (max. 95°C) and radiation (max. 52 Gy/h) [2]. Both could potentially break down complex organic material to smaller, more bioavailable compounds. This could stimulate microbial growth which then could affect processes such as gas production, microbially-influenced corrosion and radionuclide migration in a vault environment.

Mat. Res. Soc. Symp. Proc. Vol. 465 © 1997 Materials Research Society

Organic matter can persist in soils for centuries, even millenia, and only a fraction may be available for microbial use (e.g., [3], [4]). The mechanisms of organic matter preservation in soils are not well understood, although both chemical and physical processes have been implicated. Skjemstad et al. [5] discuss two different mechanisms that appear to be important in the long-term protection of soil organic matter against microbial decomposition. Physical incorporation into microaggregates that are able to withstand physically disruptive forces probably occurs to some extent in all soil types. The chemical composition of this physically-protected organic material appears to be very similar to that of the organic matter in the soil in general [5]. The second mechanism involves the formation of charcoal or 'charred' organic matter which is determined as organic carbon in chemical analysis but is highly resistant to microbial and probably chemical decomposition, and has a highly aromatic chemistry. Its biological and physico-chemical properties, and distribution in soils are largely unknown at present [5].

Much of the soil organic matter that is recalcitrant may be humic material or related compounds. Irradiation of humic matter may cause changes in its structure and form water soluble products [6, 7]. Organic polymeric materials exposed to large doses of ionizing radiation may produce extensive molecular rearrangements which may bring about either the degradation or the polymerization of constituent molecules [8]. In many instances, both phenomena may occur simultaneously within the same material. Senesi et al. [8] irradiated a humic and fulvic acid as a solid with electrons and gamma-rays, using doses ranging from 8 to 930 kGy. In addition, the humic materials were gamma-irradiated in aqueous solutions at different pH values. Irradiation caused only minute changes, which were limited mainly to decreases in particle size with increasing radiation dose when humic and fulvic acid was exposed as a solid. When exposed in solution, irradiation at neutral and acid pH tended to bring about aggregation (formation of larger particles), whereas irradiation at high pH favoured dispersion (formation of smaller particles) [8]. Groneman [9] concluded that gamma irradiation of sludges increased considerably the total organic carbon (TOC) in the aqueous phase. Yamazaki et al. [10] studied treatment of landfill leachates by gamma-irradiation in order to degrade TOC, and found a distinct reduction of TOC with increasing radiation time. Aromatic hydrocarbons are more stable in the presence of ionizing radiation than aliphatic hydrocarbons, and it may be that the aliphatic side chains are cleaved by irradiation [11, 12]. Dahl et al. [11] investigated the effects of irradiation from uranium decay on extractable organic matter in the Alum Shales of Sweden. The ratio of aromatic to saturated hydrocarbons increased with increasing uranium concentration, presumably because aromatic hydrocarbons are more stable in the presence of radiation than saturated hydrocarbons.

Fukunaga et al. [13] showed that heat increased the amount of organic material eluting from clay in a groundwater suspension. The variation of dissolved organic carbon (DOC) in 10% (w/v) bentonite suspensions in water was measured during incubation at 30, 50 and 70°C for 5 days and more than 5, 6 and 8 mg/L, respectively, of dissolved organic carbon was eluted from the bentonite. Bentonite also affected the viable count of aerobic heterotrophs when it was suspended (2 and 10% (w/v)) in groundwater and incubated at 30°C with shaking for 14 d, compared to unamended groundwater. The viable count for the 2% and 10% (w/v) bentonite suspensions were about 2 and 3 times higher than for the unamended groundwater [13].

The work described in this paper investigates if, and how much, heat and radiation effects on organics in buffer may enhance microbial growth in granitic groundwater.

MATERIALS AND METHODS

The buffer used in this study was a mixture of two batches of buffer prepared at AECL's

Underground Research Laboratory (URL), for use in large scale engineering studies. Both of these mixtures contained 50 wt.% Avonlea Na bentonite, 50 wt.% silica sand, and Fracture Zone 2 (FZ2) groundwater from the URL [14]. The moisture content ranged from 12 to 13 wt. % at the start of the experiments. The buffer was heated in an oven at 60 or 90°C for periods of 0 (control), 2, 4 and 6 weeks. Some of the material was subsequently irradiated in a Gammacell 220 Research Irradiator to 25 kGy to determine combined heat and radiation effects. Unheated buffer material was irradiated to 25 kGy (at moisture contents of 0, 13 and 18%), and to 50 kGy (at a moisture content of 12%). After heat and/or radiation treatment, 500 g of buffer was suspended in 1500 ml sterile distilled deionized water. The suspensions were shaken for 24 h and subsequently centrifuged to remove the solids, followed by sterilization with a 0.22-μm filter. The leachates were analyzed for DOC by high temperature (875°C) platinum catalyzed oxidation using an Ionics Model 1555 carbon analyzer. Twenty-five milliliters of each sterile leachate were inoculated with an equal volume of fresh FZ2 groundwater. The groundwater/leachate mixtures were incubated, with shaking, at room temperature in the dark for 10 to 11 d under aerobic conditions. The leachate mixtures and corresponding, unamended groundwater controls were sampled several times during incubation. The samples were analyzed for total cell counts by staining with acridine orange (AO) followed by microscopic counting [15]. Viable cell counts were obtained by pour plating serial dilutions on R2A medium [16] followed by incubation at 30°C for 7 d, after which the colonies were counted.

RESULTS AND DISCUSSION

Table 1 shows the DOC content of the leachates, as well as the maximum total and viable cell counts during the incubation period in the leachate-groundwater mixtures and in the corresponding groundwater controls.

Constant levels of DOC were obtained after 4 weeks of heat treatment, and the concentrations were significantly higher in the leachate from the buffer heated at 90°C. All buffer treatments increased the concentration of DOC in the leachates, compared to untreated buffer, except a radiation dose of 50 kGy at 12% moisture. Heating at 90°C increased DOC levels the most, in some cases almost threefold, compared to the DOC levels in leachate from untreated buffer. Radiation at 0% moisture for both untreated and heat-treated buffer (60 and 90°C, 6 weeks) also increased DOC concentrations considerably, by as much as a factor of two compared to leachate from untreated buffer. However, radiation of unheated buffer at higher moisture contents caused only slight increases in DOC in the leachates compared to leachate from untreated buffer. These results suggest that on one hand radiation appears to liberate DOC in buffer (i.e., the results obtained at 0% moisture), but that on the other, radiation may cause the organics to polymerize and become more resistant to desorption, or that radiolysis effects in moist samples break down liberated DOC at almost the same rate of desorption (i.e., the results at 13 and 18% moisture).

Figure 1 shows the maximum total and viable counts obtained during incubation for the inoculations with leachates from buffer treated at 60 and 90°C for 0 (untreated), 2, 4 and 6 weeks, compared to the corresponding groundwater controls. Almost invariably the presence of a buffer leachate increased both the total and viable cell counts, compared to the unamended groundwater controls. The increase ranged from about half to almost two orders of magnitude, and was generally slightly larger for the higher temperature. This implies that the presence of buffer leachates can stimulate the growth of groundwater bacteria by as much as two orders of magnitude, compared to unamended groundwater. However, when comparing the effects of leachates from heat-treated buffer to those of leachate from untreated buffer on groundwater bacteria, the effects were smaller. Heat treatment increased the maximum total cell count by

TABLE I DISSOLVED ORGANIC CARBON (DOC) AND MAXIMUM TOTAL AND VIABLE CELL COUNTS IN BUFFER LEACHATES AND GROUNDWATER CONTROLS

Buffer Treatment	DOC (mg/L)	Maximum Total Cell Count in Leachate and Groundwater (bacteria/mL)	Maximum Total Cell Count in Groundwater Control (bacteria/mL)	Maximum Viable Cell Count in Leachate and Groundwater (CFU/mL)	Maximum Viable Cell Count in Groundwater Control (CFU/mL)
Untreated	7.3 ± 0.5	$(1.5 \pm 0.3) \times 10^6$	$(2.8 \pm 1.4) \times 10^5$	$(2.9 \pm 0.8) \times 10^6$	$(8.0 \pm 7.8) \times 10^5$
60°C, 2 weeks	9.8 ± 1.5	$(2.9 \pm 1.2) \times 10^6$	$(3.9 \pm 1.3) \times 10^5$	$(3.0 \pm 1.6) \times 10^5$	$(3.5 \pm 3.0) \times 10^5$
90°C, 2 weeks	15.8 ± 1.1	$(1.3 \pm 0.2) \times 10^7$	$(3.9 \pm 1.3) \times 10^5$	$(5.0 \pm 2.1) \times 10^6$	$(3.5 \pm 3.0) \times 10^5$
60°C, 4 weeks	12.2 ± 0.5	$(7.7 \pm 0.9) \times 10^6$	$(6.5 \pm 0.6) \times 10^5$	$(3.1 \pm 0.7) \times 10^6$	$(5.9 \pm 6.0) \times 10^4$
90°C, 4 weeks	19.6 ± 2.1	$(1.2 \pm 0.2) \times 10^7$	$(6.5 \pm 0.6) \times 10^5$	$(3.5 \pm 2.8) \times 10^6$	$(5.9 \pm 6.0) \times 10^4$
60°C, 6 weeks	12.6 ± 0.5	$(5.8 \pm 1.0) \times 10^6$	$(6.0 \pm 1.8) \times 10^5$	$(3.2 \pm 0.5) \times 10^6$	$(1.8 \pm 0.7) \times 10^5$
90°C, 6 weeks	17.9 ± 0.5	$(1.7 \pm 0.4) \times 10^7$	$(6.0 \pm 1.8) \times 10^5$	$(1.0 \pm 0.4) \times 10^7$	$(1.8 \pm 0.7) \times 10^5$
60°C, 6 weeks, 25 kGy	15.4 ± 0.5	$(1.7 \pm 0.2) \times 10^7$	$(1.3 \pm 0.5) \times 10^6$	$(2.5 \pm 0.1) \times 10^7$	$(1.1 \pm 1.1) \times 10^5$
90°C, 6 weeks, 25 kGy	17.2 ± 0.5	$(2.1 \pm 0.2) \times 10^7$	$(1.3 \pm 0.5) \times 10^6$	$(2.6 \pm 0.3) \times 10^7$	$(1.1 \pm 1.1) \times 10^5$
25 kGy, 0% moisture	14.5 ± 0.5	$(4.8 \pm 2.4) \times 10^5$	$(1.3 \pm 0.5) \times 10^6$	$(8.0 \pm 9.6) \times 10^3$	$(1.1 \pm 1.1) \times 10^5$
25 kGy, 13% moisture	9.2 ± 0.5	$(4.1 \pm 1.9) \times 10^6$	$(5.9 \pm 3.1) \times 10^5$	$(1.9 \pm 0.6) \times 10^6$	$(3.7 \pm 0.2) \times 10^4$
25 kGy, 18% moisture	8.5 ± 0.5	$(5.8 \pm 0.7) \times 10^6$	$(1.3 \pm 0.5) \times 10^6$	$(1.9 \pm 0.2) \times 10^6$	$(1.1 \pm 1.1) \times 10^5$
50 kGy, 12% moisture	7.4 ± 0.5	$(1.9 \pm 0.5) \times 10^7$	$(1.3 \pm 0.5) \times 10^6$	$(1.3 \pm 0.3) \times 10^6$	$(1.1 \pm 1.1) \times 10^5$

about half to one order of magnitude, but the maximum viable cell count either decreased (one case), did not change or increased by only a factor of three. These results suggest that the effects of heated buffer leachates may increase the growth of groundwater bacteria by at most an order of magnitude when compared to unheated buffer leachate.

Figure 2 shows results for the experiments with leachates from heated and irradiated buffer. Maximum total cell counts increased only slightly as a result of the combined effects of heat and radiation, compared to the effects of heat alone (factor of three), or about one order of magnitude compared to untreated buffer. Maximum viable cell counts increased by at most about an order

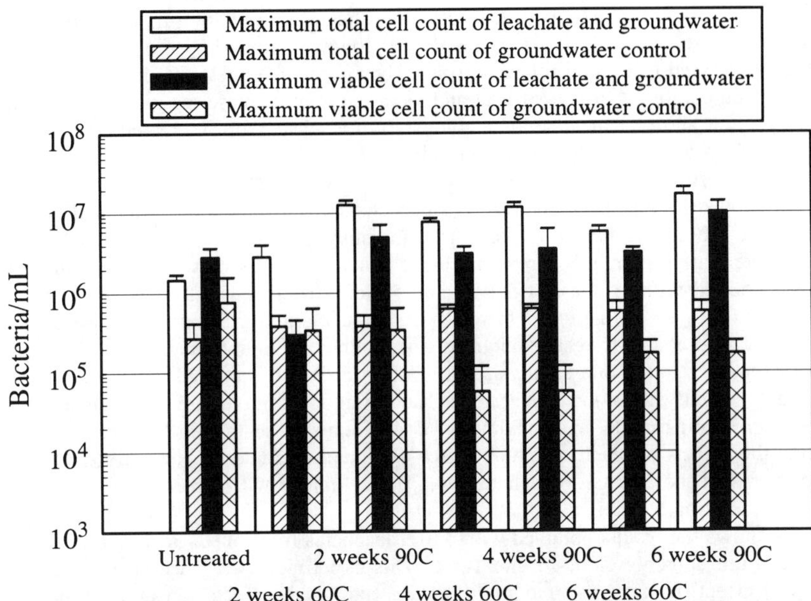

Figure 1 : Maximum total and viable cell counts in buffer leachate/ groundwater solutions as a function of heat treatment

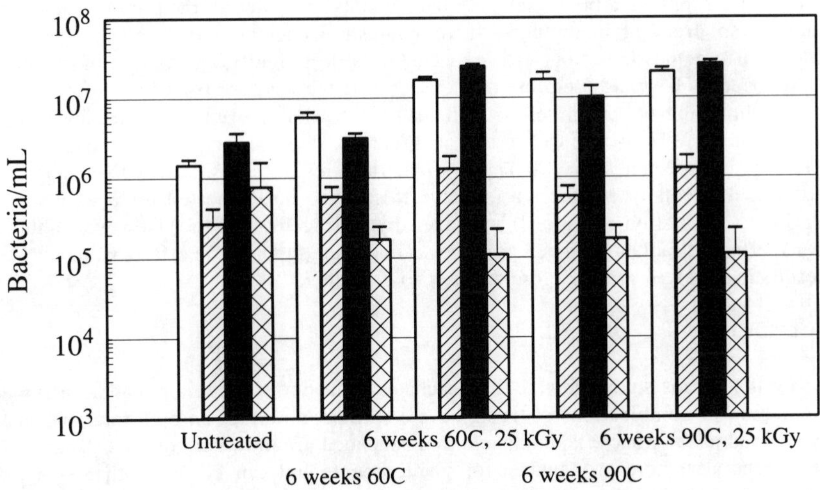

Figure 2 : Maximum total and viable cell counts in buffer leachate/ groundwater solutions as a function of heat and radiation treatment

of magnitude, compared to untreated or heat-treated-only buffer leachates. However, maximum total cell counts in experiments with leachates from heat- and radiation-treated buffer increased by as much as thirtyfold and maximum viable cell counts increased by more than two orders of magnitude compared to groundwater controls. These results suggest that leachates from buffer subjected to the combined effects of heat and radiation may increase the growth of groundwater bacteria by about an order of magnitude compared to leachate from untreated buffer. Compared to unamended groundwater, the stimulating effects from the presence of leachates from heated and irradiated buffer may be more than two orders of magnitude.

Figure 3 shows the results of the effects of leachates from unheated, irradiated buffer. The radiation dose was either 25 or 50 kGy, at a moisture content of 12 to 13% (the effects of moisture content are shown in Figure 4, discussed below). Results show no significant difference in results obtained with leachates from 25 and 50 kGy-irradiated buffer, for maximum viable counts, and only a small increase (factor of four) in maximum total counts with increasing radiation dose. Compared to results obtained with leachate from untreated buffer, irradiation had no significant effect on viable counts and at most an order of magnitude effect on total counts. Compared to groundwater controls, these leachates stimulated maximum total cell counts by less than an order of magnitude, and maximum viable counts by at most 1.5 orders of magnitude. Therefore, the effects of leachates from buffer irradiated at moderate moisture contents appear to be smaller than the effects of leachates from buffer treated with either heat or a combination of heat and radiation.

Figure 4 shows the results obtained with buffer leachates from buffer material irradiated to 25 kGy at moisture contents of 0, 13 and 18%. This experiment was carried out because water content will affect the extent of radiation damage (radiolysis effects) on organic material. The leachates from the buffers irradiated to 25 kGy at moisture contents of 13% and 18% stimulated both total and viable cell counts by about half to 1.5 orders of magnitude, respectively, compared to groundwater controls. Results showed that the leachate from the buffer irradiated at 0% moisture suppressed both total and viable counts compared to the groundwater control by as much as an order of magnitude. At zero moisture content, radiation damage and hence breakdown of organics in the buffer is expected to be less significant because of the absence of radiolysis effects. It would therefore be expected that the leachate from this buffer would have less of a stimulating effect on cell growth than the leachates from buffer irradiated at higher moisture contents. However, the results in Figure 4 show an actual suppression of growth. Reasons for this are not clear. It appears that radiation in a dry state either suppresses the liberation of nutrients or produces a toxicity effect. The buffer heated for 6 weeks at 60 and 90°C prior to irradiation (Figure 2) was also dry during irradiation. However, although the additional effect of radiation on cell growth was not large, suppression effects were not observed. The results in Figure 4 are therefore difficult to explain.

CONCLUSIONS

The results of this study indicate that leachates from buffer materials that were subjected to heat, radiation or combinations of these, have a stimulating effect on both total and viable cell counts of groundwater bacteria, compared to unamended groundwater controls. This stimulating effect is generally more pronounced for viable counts and can be larger than two orders of magnitude. In a vault environment, compacted buffer will be subjected to heat and radiation, and water will be in short supply. Therefore, 'dilute' leachates as were used in the laboratory studies described here would be much smaller in volume but likely more concentrated in a vault. Such concentrated leachates could potentially enhance bacterial growth almost anywhere in a vault, at

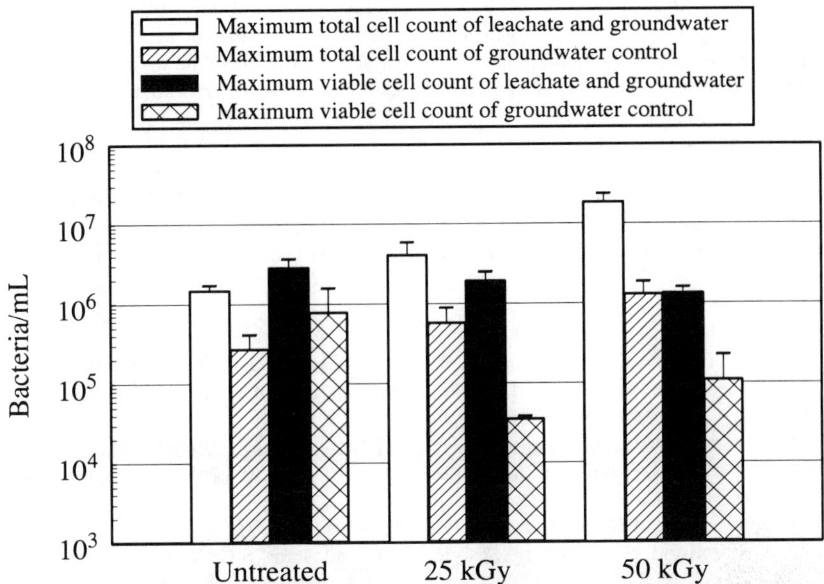

Figure 3 : Maximum total and viable cell counts in buffer leachate/ groundwater solutions as a function of irradiation dose

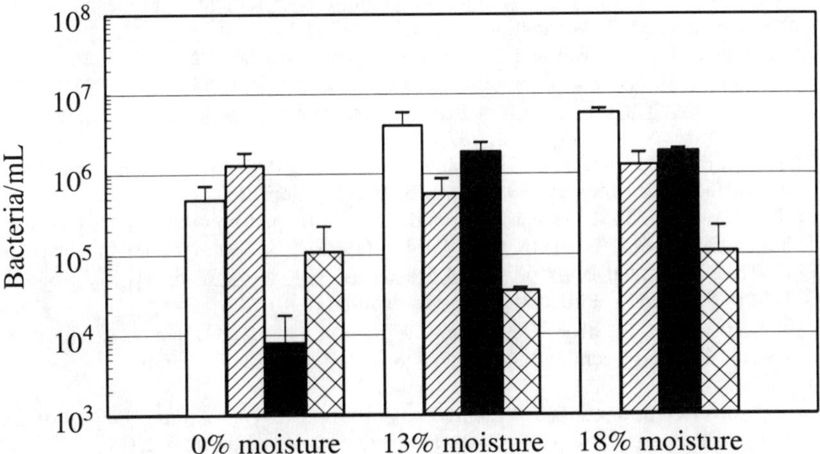

Figure 4 : Maximum total and viable cell counts in buffer leachate/ groundwater solutions as a function of buffer moisture content during 25 kGy irradiation

least from a nutrient-availability point of view. Any stimulation of bacterial growth and activity could potentially affect processes such as gas production, microbially-influenced corrosion and radionuclide migration. The effects of heat and radiation on nutrient availability in clay-based sealing materials (at relevant clay/moisture ratios) should, therefore, be taken into account when attempting to quantify the effects of microbial activity on vault performance.

ACKNOWLEDGEMENTS

This work was funded by Ontario Hydro. Technical assistance from D. Bachinski, C.J. Hamon, M. Goulard, R. Comba, L.K. Krupa, E.L. Beebe and R.L. Clarke is gratefully acknowledged, as well as useful comments by L.M. Lucht and D.W. Oscarson.

REFERENCES

1.   Atomic Energy of Canada Limited Report AECL-10711, COG-93-1 (1994).

2.   S. Stroes-Gascoyne and J.M. West, Atomic Energy of Canada Limited Report, AECL-10808, COG-93-54 (1994).

3.   G. Stotzky, In <u>Interactions of soil minerals with natural organics and microbes,</u> Edited by P.M. Huang and M. Schnitzer, (1986), pp. 305-428. Soil Science Society of America, Inc., Madison, WI.

4.   D.B. Nedwell, FEMS Microbiol. Ecol. **45**, 47-52 (1987).

5.   J.O. Skjemstad, P. Clarke, J.A. Raylor, J.M. Oades and S.G. McClure, Aust. J. Soil Res. **34**, 251-271 (1996).

6.   K. Bunzle and W. Schimmack, Radiat. Environ. Biophys. **27**, 165-176 (1988).

7.   N.L. Korenevich, Tr. VNIITP **53**, 141-148 (1984).

8.   N. Senesi, Y. Chen and M. Schnitzer, Fuel **56**, 171-176 (1977).

9.   A.G. Groneman, In <u>Proc. Radiation for Pollution Abatement</u>. European Society of Nuclear Methods in Agriculture, Munich, 107 (1976).

10.   M. Yamazaki, T. Sawai and T. Sawai, Radiat. Phys. Chem. **20**, 329-332 (1982).

11.   J. Dahl, R. Hallberg and I.R. Kaplan, Org. Geochem. **12**, 559-571 (1988).

12.   F.A. Bovey, <u>The effects of ionizing radiation on natural and synthetic polymers.</u> Wiley Interscience, New York, 1958, 287 pp.

13.   S. Fukunaga, K. Fujiki and H. Asano, Abstract for 1993 International Symposium on Subsurface Microbiology (ISSM-93), Bath U.K. Sept. 19-24,1993.

14.   S.A. Haveman, S. Stroes-Gascoyne and C.J. Hamon, Atomic Energy Of Canada Limited Technical Record, TR-654*, COG-94-488 (1995).

15.   ASTM., In <u>Annual Book of ASTM Standards, Vol. 11.02 Water (II)</u>, 632-635. Method D4455-85. ASTM, Philadelphia, Pennsylvania (1990).

16.   S.A. Haveman, S. Stroes-Gascoyne, C.J. Hamon and T.L. Delaney, Atomic Energy of Canada Limited Technical Record, TR-677*, COG-95-017 (1995b).

* Unrestricted, unpublished report available from SDDO, Atomic Energy of Canada Limited Research Company, Chalk River, Ontario, Canada, K0J 1J0.

EXPERIMENTAL STUDY OF GAS PERMEABILITIES AND BREAKTHROUGH PRESSURES IN CLAYS

K.TANAI[*], T.KANNO[*], C.GALLÉ[**]
[*]Geological Isolation Technology Section, PNC Tokai Works, Tokai-mura, Ibaraki, JAPAN.
[**]Département d'Entreposage et de Stockage des Déchets, CEA Saclay, 91191 Gif-sur-yvette, FRANCE.

ABSTRACT

In this study, gas migration experiments in unsaturated and saturated states were carried out to clarify the fundamental gas migration characteristics in compacted bentonite to be used for the geological disposal of high-level radioactive waste. In unsaturated experiments, the gas permeability for Japanese bentonite (Kunigel V1) as a function of degree of saturation was measured to examine the applicability of conventional two-phase flow models to compacted bentonite. The intrinsic permeability obtained in this study was about five orders of magnitude larger than that obtained in water permeation tests with the same density. The difference seems to originate from the change of pore structure due to the swelling phenomenon of the bentonite. Since these effects have not been evaluated quantitatively yet, various relative gas permeability functions of conventional two-phase flow models were applied as a first approximation.

Saturated experiments designed to simulate the gas migration phenomenon in a repository for the waste were carried out to obtain relationship between breakthrough and swelling pressures using Kunigel V1 and French Fo-Ca clay in saturation state. The reproducibility of the breakthrough pressure was also examined for Kunigel V1 bentonite. The breakthrough pressure was almost the same as swelling pressure irrespective of the type of clay. As to the reproducibility of breakthrough pressure, it was observed that first and second breakthrough pressures were almost the same for Kunigel V1 specimens with the dry densities of 1.7 and 1.8 g/cm^3.

INTRODUCTION

Carbon steel is a candidate for the overpack materials of geological disposal of high-level waste in Japan. The corrosion of the carbon steel overpack in aqueous solution under anoxic conditions will be accompanied by the generation of hydrogen gas, which may affect hydrological and mechanical conditions of the bentonite buffer [1]. To evaluate the consequences of gas generation on radioactive waste repository in deep underground, it is necessary to develop a gas migration model for bentonite buffer material based on the information obtained from experiments. In this study, unsaturated and saturated experiments were carried out to clarify the fundamental characteristics of gas migration in bentonite.

EXPERIMENTAL APPARATUS

The schematic set-up of apparatus for unsaturated and saturated experiments is shown in Figure 1. The apparatus comprises four parts : (1)gas supply unit, (2)vessel, (3)soap bubble flow meters and (4)data acquisition system. The inlet pressure of gas is measured by a pressure transducer with a precision of 0.2 %. The gas flow rate is measured by soap bubble flow meters. The vessel is composed of a cylinder, a bottom flange and a piston. The cylinder is 50 mm in inner diameter and contains a bentonite specimen with the length up to 50 mm. The bottom flange and the piston both contain a filter (diameter 30 mm, thickness 5 mm pore-size 5 μm) made from stainless steel powder. For the saturated experiment, the swelling pressures of specimens are measured by a load cell.

UNSATURATED EXPERIMENT

Prior to the saturated experiment which simulates the gas migration phenomenon in a repository, experiments in unsaturated state were performed to examine the applicability of

Mat. Res. Soc. Symp. Proc. Vol. 465 © 1997 Materials Research Society

conventional two-phase flow models to compacted bentonite. The gas permeability for Japanese bentonite as a function of degree of saturation was measured.

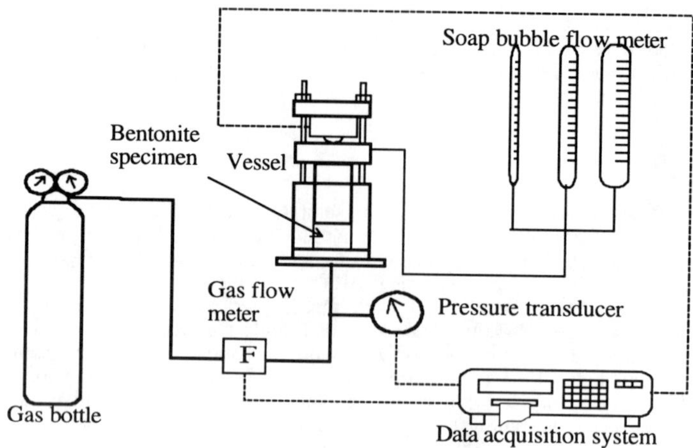

Figure 1 Experimental apparatus

Materials

Kunigel V1 made by Kunimine Industry Co.,Ltd. was used as the bentonite material. It consists mostly of montmorillonite (49 %) and quartz (39 %). The other important minerals are plagioclase (5 %), calcite (2 %) and dolomite (3 %). For a dry density of 1.8 g/cm^3 under the distilled water conditions, its swelling pressure is approximately 4.0 MPa and saturated hydraulic conductivity is of the order of 10^{-13} m/s [1].

Procedures

To determine the gas permeability as a function of degree of saturation, the experiments were carried out on bentonite specimens at different degrees of saturation with dry densities of 1.6, 1.7 and 1.8 g/cm^3. These specimens were made by uniaxial compaction of bentonite powder in the test vessel. The specimen is 50 mm in diameter and 10 mm thick.
The following methods were used to produce bentonite powder with different water contents :
    (1) 0%   ; dried in an oven at 110 °C for 24 h,
    (2) w < 9% ; in equilibrium with the atmosphere,
    (3) 9% < w ; controlled by spraying a fine mist over a thin layer of bentonite powder.
The pressure gradient over the bentonite specimen was controlled only by changing the inlet pressure, while the outlet pressure was always kept at atmospheric pressure. Argon gas was used in this experiment. The gas flow rate was measured at the outlet with soap bubble flow meters. The gas permeability of each bentonite specimen was determined at three different inlet pressures (P = 0.025, 0.05 and 0.1MPa).

Results and Discussion

The gas permeability Kg was calculated by the following equation [2].

$$K_g = 2Q\mu LP / (A(P_i^2 - P_o^2)) \tag{1}$$

where ;

Kg = gas permeability (m^2)
Q = volumetric flow rate (m^3/s)
μ = the viscosity of gas (Pa·s)
L = the specimen thickness (m)
P = pressure at which the flow rate is measured (Pa)
A = cross sectional area of specimen (m^2)
P_i = absolute injection pressure (Pa)
P_o = absolute exit pressure (Pa)

At a saturation of 0% (S_l=0), the measured gas permeability can be taken as the intrinsic permeability which does not depend on the kind of fluid for common non-expansive porous media. The intrinsic permeability of Kunigel V1 measured by water permeation tests in saturated condition [3] is shown in Figure 2.

However, it turned out that the intrinsic permeabilities obtained in this study were larger by about five orders of magnitude than those obtained in water permeation tests with the same dry densities. The difference seems to originate from the change of pore structure by the swelling phenomenon of the bentonite. Recent studies performed by CEA teams on compacted Fo-Ca clay revealed that there are important modifications of clay structure during hydration/dehydration cycle tests. In particular, mercury intrusion tests show the presence of an important pore access family, around 1 μm, for Fo-Ca clay under hydrated conditions - 30 % water content - which disappear for lower hydration - 10 % water content - [4]. Therefore, special attention should be paid to the effects of swelling on structure change when applying conventional two-phase flow models including intrinsic or relative permeability concepts to compacted bentonite.

However, because these effects have not been evaluated quantitatively, relative gas permeability functions of conventional two-phase flow models were evaluated as a first approximation. The measurements cover the range between 0 % and about 80 % saturation. The variation of relative gas permeability with saturation is shown in Figure 3, taking the permeability at S_l=0 as the intrinsic permeability. The relative gas permeability of the Kunigel V1 approaches 0 at a saturation of about 72 %, which means that at this saturation no continuous gas channels remain available and the gas breakthrough pressure becomes > 0.

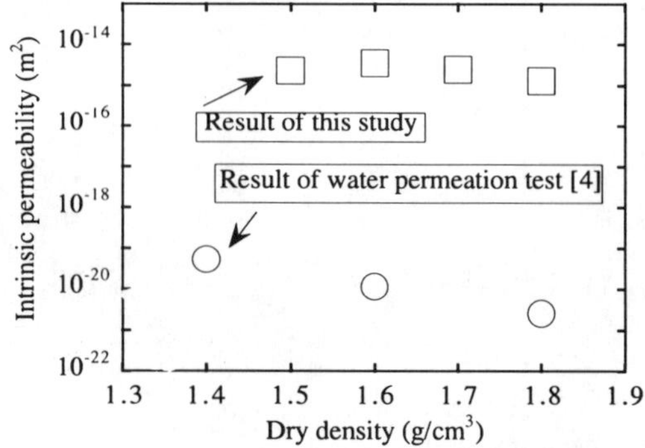

Figure 2 Measurement results of intrinsic permeability

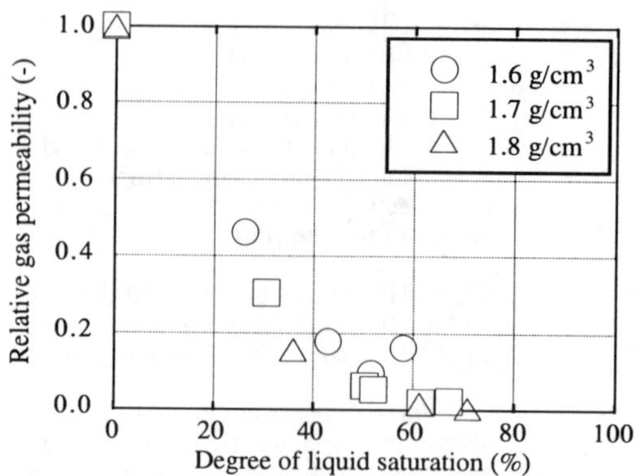

Figure 3 Measurement results of relative gas permeability

Next, applicability of the relative permeability function models to this test result is examined. For non-expansive porous media, some analytical function models were proposed to describe the relationship between the relative permeability and the degree of liquid saturation. The following three models were selected as candidates for fitting the test results:

Corey model [5] ;

$$K_{rg} = (1-S^*)^2(1-[S^*]^2) \tag{2}$$

$$S^* = (S_l - S_{lr}) / (1 - S_{lr} - S_{gr}) \tag{3}$$

Fatt and Klikoff model [6] ;

$$K_{rg} = (1-S^*)^3 \tag{4}$$

$$S^* = (S_l - S_{lr}) / (1 - S_{lr}) \tag{5}$$

Sandia model [7] ;

$$K_{rg} = (1 - K_{rl}) \tag{6}$$

$$K_{rl} = S^{*1/2}(1 - (1 - S^{*1/\lambda})^\lambda)^2 \tag{7}$$

$$S^* = (S_l - S_{lr}) / (1 - S_{lr} - S_{gr}) \tag{8}$$

where,

K_{rg} = relative permeability of gas
K_{rl} = relative permeability of liquid
S_l = liquid saturation
S_{lr} = residual liquid saturation
S_{gr} = residual gas saturation
λ = pore-size distribution

For the residual liquid saturation and pore-size distribution parameter for the compacted bentonite, the following values were used based on results of laboratory experiments [8] with the same bentonite.

Residual liquid saturation (S_{lr}) ; 0.09,

Pore-size distribution parameter (λ) ; 0.52.

The residual gas saturation is judged to be $S_{gr} = 0.28$ from the results shown in Figure 3. Figure 4 shows relative gas permeability function curves obtained by substituting above values for the parameters in the models. Among the three models, the Corey model agrees best with the experimental results. Fatt and Klikoff model also shows good agreement. Therefore, further detailed investigations are required to determine a model giving the best fitting curve for the compacted bentonite.

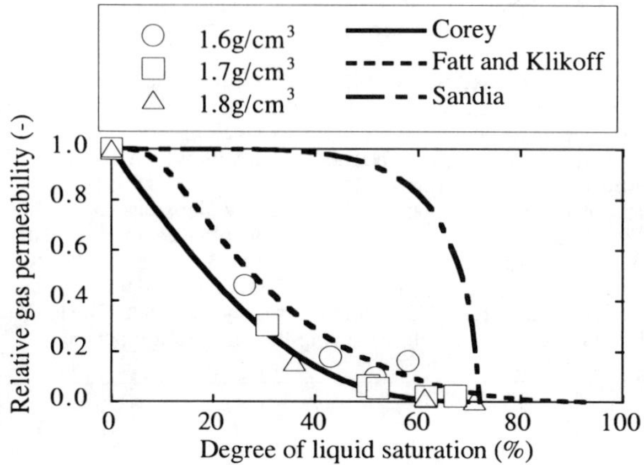

Figure 4 Measured relative gas permeability with fitting curves

SATURATED EXPERIMENT

Saturated experiments were carried out to obtain breakthrough pressure and relationship between breakthrough and swelling pressures of Kunigel V1 and French Fo-Ca clay in saturation state. In this experiment, the breakthrough pressure, gas permeability and swelling pressure were measured.

Materials

Kunigel V1 and French clay (Fo-Ca) [9] were used as experimental materials. Fo-Ca clay is sedimentary clay from the Vexin region located west of the Parisian basin in France. It consists mostly (80 %) of interstratified kaolinite and beidelite. The other important minerals are kaolinite (4%), quartz (6%) and geothite (6%). The average specific gravity of the clay particle is 2.67. The swelling pressure and the hydraulic conductivity of the clay with a dry density of 1.75 g/cm³ are approximately 13 MPa and 5×10^{-14} m/s, respectively.

Procedures

Clay powder was placed in the test vessel and compacted uniaxially to predetermined dry densities. Water was supplied from the lower side of bentonite specimen by water head. The swelling pressure was measured by a load cell. Hydrogen gas was injected from the lower end of the bentonite specimen and injection pressure was increased stepwise up to the pressure at which breakthrough occurred. The initial gas injection pressure was 0.5 MPa and the period of each step was 120 h. The increment of the injection pressure between each step was 0.5 MPa.

The reproducibility of the breakthrough pressure was examined for specimens with densities of 1.7 and 1.8 g/cm^3. In these cases, after the occurrence of the first breakthrough, the gas injection was suspended for approximately 336 h to make the specimens resaturated, then gas injection was restarted to measure the second breakthrough pressure.

Results and Discussion

The measured results of gas permeability and breakthrough pressure as function of dry densities of Kunigel V1 (1.6, 1.7 and 1.8 g/cm^3) and Fo-Ca clay (1.6 g/cm^3) are shown in Figure 5. Breakthrough pressure increased for increasing dry density for Kunigel V1.
The magnitude of gas permeabilities obtained are $10^{-18} \sim 10^{-21}$ m^2 for Kunigel V1 and 10^{-18} m^2 for Fo-Ca clay, respectively.

The relationship between breakthrough and swelling pressures of Kunigel V1 and Fo-Ca clay, as well as MX-80 bentonite measured by Pusch et al. [10], are shown in Figure 6. The breakthrough pressure seems to be almost the same as swelling pressure irrespective of the type of clay. This result suggests that there are some mechanisms by which swelling property of bentonite prescribes the breakthrough phenomenon.

For two Kunigel V1 bentonite specimens with dry densities of 1.7 and 1.8 g/cm^3, the reproducibility of the gas breakthrough pressure were measured. The first and second breakthrough pressures measured are shown in Table 1. As to the reproducibility of the breakthrough pressure, it was observed that first and second breakthrough pressures were almost the same for the specimens, which suggests that gas pathways created during the first gas-injection period were closed due to bentonite swelling during the resaturation period.

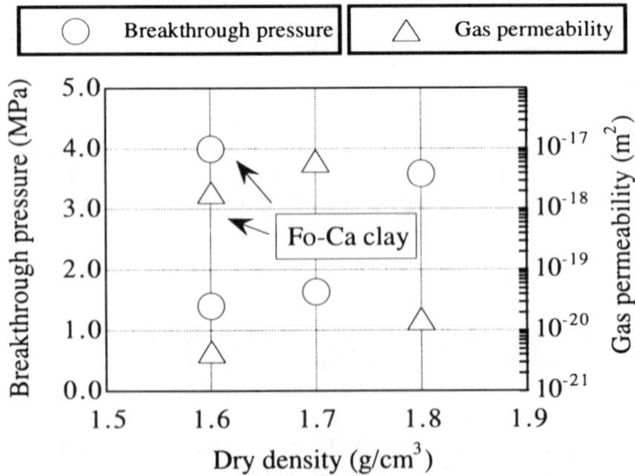

Figure 5 Measurement results of gas permeability and breakthrough pressure

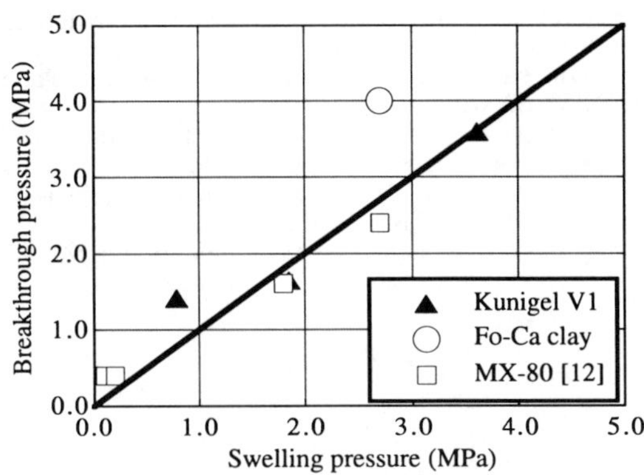

Figure 6 Relationship between swelling pressure and breakthrough pressure

Table 1 Repeatability of the breakthrough pressure

Dry density	Breakthrough pressure	
	First step	Second step
1.7 g/cm^3	1.64 MPa	1.70 MPa
1.8 g/cm^3	3.59 MPa	3.15 MPa

CONCLUSIONS

The major conclusions obtained in this study are as follows :
(1) The intrinsic permeabilities for the compacted bentonite measured in this study at S_l=0% with argon gas was about five orders of magnitude larger than those obtained by water permeation tests at S_l=100% with the same bentonite.
(2) Calculated results of the relative permeability for unsaturated bentonite as a function of the degree of liquid saturation with the Corey model were in good agreement with measured results.
(3) The breakthrough pressure of hydrogen gas for saturated clays was almost the same with the swelling pressure irrespective of the type of clay.
(4) No significant difference was observed between the first and second breakthrough pressures.

REFERENCES

[1] PNC, Research and development on geological disposal of high level radioactive waste, First progress report, PNC TN1410 93-059, 1992.
[2] Handbook of Petroleum Industry (Society of Japan petroleum Engineers, 1983, in Japanese).
[3] T.Kanno, S.Takeuchi and H.Suzuki, Temperature dependence of hydraulic conductivity of compacted bentonite, PNC TN 1100 94-003, 1994.
[4] L.Michaux and J.M.Loubignac, Etude microstructurale du matériau Fo-Ca, Evolution de la microstructure de l'argile Fo-Ca 7 compactée lors d'un cycle d'hydration/dehydration à 25 °C, CEA NT-SESD/95.37,1995.
[5] A.T.Corey, Mechanics of immiscible fluids in porous media, Water resources publications, Littleton, Colorado, 1986.

[6] I.Fatt and W.A.Klikoff, Effect of fractional wattability on multiphase flow through porous media, AEA Transactions, vol.216, p.246, 1959.

[7] S.W.Webb, Sensitivity studies for gas release from the waste isolation pilot plant (WIPP), Proceeding of OECD/NEA workshop, pp.309-326, Paris, 1992.

[8] S.Takeuchi and K.Hara, Two-phase water movement in buffer material for geological isolation of high-level radioactive waste, PNC TN 8410 93-302.1994, (in Japanese).

[9] C.Gallé and K.Tanai, Gas transport in engineered barrier for nuclear waste disposal "H2 migration experiments in Fo-Ca clay", 1996.

[10] R.Pusch, L.Ranhagen and K.Nillson, Gas migration through MX-80 bentonite, Nagra TR 85-36, 1985 .

GAS MIGRATION IN MX80 BUFFER BENTONITE

S.T. HORSEMAN *, J.F. HARRINGTON * AND P. SELLIN **
* Fluid Processes Group, British Geological Survey, Nottingham, NG12 5GG, UK
** Svensk Kärbränslehantering AB (SKB), PO Box 5864, S102 48 Stockholm, Sweden.

ABSTRACT

Controlled flow-rate gas injection experiments have been performed on pre-compacted samples of KBS-3 specification Mx80[1] buffer bentonite using helium as a safe replacement for hydrogen. By simultaneously applying a confining pressure and backpressure, specimens were isotropically-consolidated and fully water-saturated under pre-determined effective stress conditions, before injecting gas using a syringe pump. Ingoing and outgoing gas fluxes were monitored. All tests exhibited a conspicuous threshold pressure for breakthrough, somewhat larger than the sum of the swelling pressure and the backpressure. All tests showed a post-peak negative transient leading to steady-state gas flow. Using a stepped history of flow rate, the flow law was shown to be nonlinear. With the injection pump stationary (i.e. zero applied flow rate), gas pressure declined with time to a finite value. When gas flow was reestablished, the threshold value for gas breakthrough was found to be significantly lower than in virgin clay. There is strong evidence to suggest that the capillary pressure for the penetration of interparticle pore space of buffer bentonite is of such a magnitude that normal two-phase flow is impossible. Gas entry and breakthrough is therefore accompanied by the development of microcracks which propagate through the clay from gas source to sink. The experiments suggest that these pathways open under high gas pressure conditions and partially close if gas pressure falls, providing a possible explanation of the nonlinearity of the flow law.

INTRODUCTION

In the Swedish KBS-3 repository concept, copper/steel canisters containing spent nuclear fuel will be placed in large diameter disposal boreholes drilled into the floor of the repository tunnels. The space around each canister will be filled with pre-compacted bentonite blocks. Over time, the bentonite blocks will draw in the surrounding groundwater and will swell, closing up any construction gaps. Because of the important buffering effect of the bentonite on the local water chemistry, this barrier is usually referred to as the bentonite buffer. The copper/steel waste canisters are expected to have a very substantial life in the repository environment. However, for the purposes of performance assessment, one must consider the possible impact of penetration of the canisters by water.

Assuming that available oxygen has been flushed out of the repository or consumed in chemical reactions, corrosion of the steel inner of each canister under anoxic conditions will lead to the formation of hydrogen. Radioactive decay of the waste and the radiolysis of water will produce some additional gas. Depending on the gas production rate and the rate of diffusion of gas molecules in the pores of the bentonite, it is possible that the solubility limit of the gas will be exceeded. The gas would then accumulate in the void-space of each canister until its pressure becomes sufficiently large for it to enter the bentonite. Since penetration of the KBS-3 canister is a prerequisite for the development of a discrete gas phase in the buffer, the timing of gas movement in the clay might coincide with the timing of radionuclide release into the buffer porewater. The possibility of an interaction between gas and radionuclide migration therefore emerges as an important issue in performance assessment.

The problem of gas migration in a compact clay is of sufficient complexity that it demands the development of a fully-quantified process model. A programme of work is presently underway

---

[1] Mx80 is a trade name of Volclay Ltd., a subsidiary of the American Colloid Company. The material is fine-grained sodium bentonite from Wyoming containing aound 90% montmorillonite.

Mat. Res. Soc. Symp. Proc. Vol. 465 © 1997 Materials Research Society

for a consortium of waste management companies[2] (GAMBIT Consortium) to develop a model of gas transport in buffer bentonite. Gas transport data from a number of laboratories will be pooled.

In the specification of experiments, it was considered that the gas migration phase of repository evolution would post-date both the resaturation and the thermal phases, prompting the requirement that the clay be tested in an initially water-saturated condition at normal laboratory temperature. An option to test at higher temperatures was retained.

EXPERIMENTS

The BGS controlled flow-rate apparatus is shown in Figure 1 and is described in detail in Sen et al. (1996). The specimen is subject to an isotropic confining (boundary) stress, σ, which is held constant. The specimen is equilibrated with water at a fixed pressure. The thermodynamic equilibrium condition is characterised by

$$\sigma = \sigma' + p_{wo} = \Pi_{sw} + p_{wo} \tag{1}$$

where σ' is the conventional (Terzaghi) effective stress, p_{wo} is the backpressure of the water and Π_{sw} is the swelling pressure of the clay. For a high swelling clay such as bentonite, the effective stress and the swelling pressure are directly equivalent (Horseman et al., 1996).

The 4.9 x 4.9 cm cylindrical clay specimen is sandwiched between two tapered end-caps, each with a sintered stainless steel porous disc, and jacketed in a thin-walled (0.2 mm) copper sheath to exclude confining fluid and prevent diffusional losses of gas. Tapered locking rings ensure a very gas-tight seal. The combination of sizes guarantees an interference fit between the clay and the copper. The injection end-cap has a central inflow duct and a circular groove cut into its load-bearing surface and linked to an outflow duct, allowing the gas to sweep radially through the porous disc during preliminary flushing operations.

The volumetric flow-rate of the injected fluid and the pressure of the downstream fluid are controlled using a pair of ISCO-500, Series D, syringe pumps, operating from a single digital control unit. A pressure transducer monitors the outgoing pressure and provides a feedback signal to the microprocessor when the pump is set in pressure control mode. Piston motion gives a direct measure of flow into the pump. The pump can also be set to constant pumping rate mode, whereby the syringe piston is advanced at constant velocity. Given the possibility that gas might leak past the seal of the syringe, we have opted to pump water, not gas, and develop a constant helium flow-rate by displacing the gas from a pre-charged vessel. This also ensures that the helium is water-saturated and cannot cause dessication of the specimens.

The ISCO pump controller has an RS232 serial port, which allows volume, flow-rate and pressure data from the pumps to be transmitted to the equivalent port of a 32-bit personal computer. A programme written in QBASIC prompts the pump controller to transmit the data to the computer at preset time intervals. Typical acquisition rate is one scan per 15 minutes. Confining pressure is monitored using a pressure transmitter linked to a Druck DPI-203 digital pressure indicator. The apparatus is assembled in an air-conditioned chamber at $20 \pm 0.3°C$.

Test Material and Experimental Procedures

Four bentonite blocks were manufactured by Clay Technology AB (Lund, Sweden) by rapidly compacting Volclay Mx80 bentonite powder in a mould under a one-dimensionally applied stress. The blocks were divided into two batches, identified as MS for "medium swelling" and HS for "high swelling". The average dry density of the MS-blocks was 1.558 Mg.m^{-3}, with a corresponding value for the HS-blocks of 1.638 Mg.m^{-3}.

---

[2] The consortium is led by Svensk Kärbränslehantering AB (SKB). Model development will be undertaken by AEA Technology PLC, Winfrith, Dorset, UK.

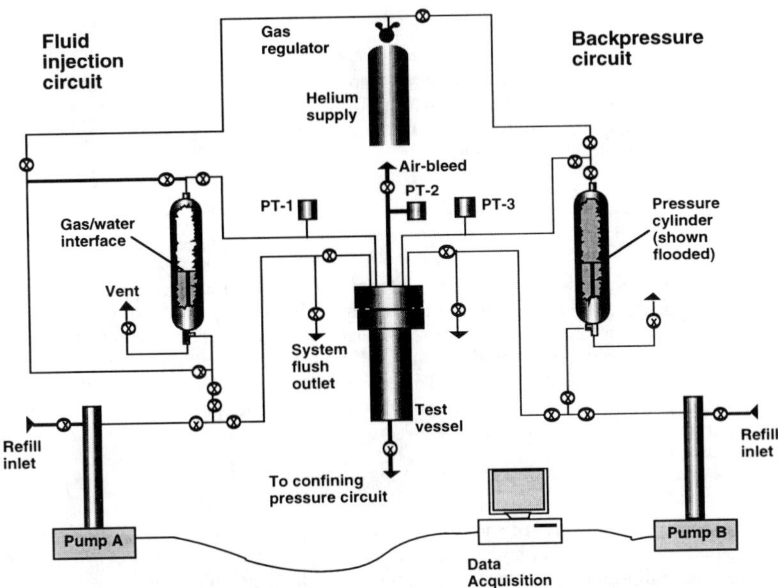

Figure 1 - Schematic of the BGS controlled flow-rate gas migration apparatus.

The blocks were cut into rectangular sticks using a bandsaw. The cylindrical specimens were prepared with the aid of a 4.90 cm I.D. steel ring-former with a sharpened leading edge. The clay was carefully pared away around the periphery, allowing the former to move downward under a light, manually-applied pressure. The upper and lower surfaces were trimmed and finished-off by a scraping action, leaving the ends perfectly flat and parallel. The clay was extruded from the former into the copper sheath. The porous discs were saturated in Purite (distilled water) before making up the specimen assembly.

Specimens were fully saturated ($S_w > 99\%$) and equilibrated under confining stress with a backpressure of 1.01 MPa, applied at both ends. Net flows were monitored to establish the point of equilibration. Helium was then admitted into the upper part of the gas-water interface cylinder and the tubing and porous disc were gas-flushed. The initial gas pressure in the injection system was set using a regulator. After closing the valve to the regulator, the injection pump was set to constant flow-rate mode. Depending on the specific objectives of each test, the rate was then changed in step-wise manner in order to investigate the sensitivity of the gas permeability to the flow rate.

DATA REDUCTION MODEL

An analytical model has been developed to describe post-breakthrough gas migration. Although based on a number of rather restrictive assumptions, the model provides a clear explanation of the test responses, a straightforward means of quantifying the transport parameters, and a reasonable fit to certain subsets of the available data. The governing differential equation of the testing apparatus is

$$V_{gi} \frac{dp_{gi}}{dt} + p_{gi} \frac{dV_{gi}}{dt} = B\left(p_{go}^2 - p_{gi}^2\right) \tag{2}$$

where p_{gi} and V_{gi} are the pressure and volume of gas in the upstream system and B is transport parameter which depends on the gas flux equation for its specification. Assuming continuum gas transport properties[3], the post-breakthrough volumetric flux of gas, Q_{st} ($m^3.s^{-1}$), at standard temperature and pressure (STP) is given by

$$Q_{st} = \frac{v_{mst} \, k_g \, A_s}{2 \, RT \, \eta_g \, L_s} \left(p_{gi}^2 - p_{go}^2\right) \tag{3}$$

where A_s and L_s are the cross-sectional area and length of the specimen, v_{mst} ($= 0.024$ m^3) is the molar volume of the gas at STP, k_g (m^2) is the gas permeability, R (8.314 $J.mol^{-1}.K^{-1}$) is the gas constant, T (K) is the temperature, η_g (Pa.s) is the viscosity of the gas, p_{gi} (Pa) is the pressure of the gas just inside the specimen at the upstream end, and p_{go} (Pa) is the pressure of the gas just inside the specimen at the downstream end. The downstream gas pressure, p_{go}, cannot be directly measured in the experiments but should be related to the backpressure of the water at the downstream end, p_{wo}, by the relationship

$$p_{go} = p_{wo} + p_{co} \tag{4}$$

where p_{co} is the "apparent value" of the capillary pressure. Analytical solution[4] of the governing differential equation demands that both k_g and p_{co} be taken as constant (Sen et al., 1996). The upstream pressure build-up before gas entry into the bentonite is given by

$$p_{gi} = p_{gi0}\left(\frac{V_{gi0}}{V_{gi0} - Ct}\right) \qquad \text{(before gas entry)} \tag{5}$$

where V_{gi0} and p_{gi0} are the initial volume and pressure of gas in the injection system, C ($m^3.s^{-1}$) is the volumetric pumping rate, and t is the elapsed time from the start of pumping.

All specimens exhibit a conspicuous pressure transient after a change in pumping rate. Evolution of the upstream gas pressure from the initial condition $p_{gi} = p_{gi0}$ and $V_{gi} = V_{gi0}$ at $t' = 0$ is described by

$$p_{gi} = \frac{\left[E\left(p_{gi0} - D\right) - D\left(p_{gi0} - E\right)\left(\dfrac{V_{gi0}}{V_{gi0} - Ct'}\right)^{G}\right]}{\left[\left(p_{gi0} - D\right) - \left(p_{gi0} - E\right)\left(\dfrac{V_{gi0}}{V_{gi0} - Ct'}\right)^{G}\right]} \qquad (C = \text{const.}) \tag{6}$$

where t' is the elapsed time from the change in the pumping rate and

$$D = \frac{C}{2B} + \sqrt{\left(\frac{C}{2B}\right)^2 + p_{go}^2} \qquad E = \frac{C}{2B} - \sqrt{\left(\frac{C}{2B}\right)^2 + p_{go}^2} \tag{7}$$

Exponent G is given by

---

[3] Since the flow law can be re-written in terms of capillary flow along discrete pathways, the assumption of continuum gas flow properties is not actually necessary for application of the model.

[4] Numerical solution of the governing differential equation would make these assumptions unnecessary.

$$G = \frac{B(D - E)}{C} \tag{8}$$

Provided that pumping is maintained at a constant rate, the upstream gas pressure p_{gi} evolves to a steady-state value given by

$$p_{gi} = \frac{C}{2B} + \sqrt{\left(\frac{C}{2B}\right)^2 + p_{go}^2} \qquad \text{(steady-state; } C = \text{const.)} \tag{9}$$

The upstream gas pressure at steady-state should therefore be independent of test path followed to achieve the steady-state condition and, within a single test, gas pressure at steady-state should, in theory, be a repeatable quantity. If, some time after gas breakthrough, the syringe pump is stopped, the upstream gas pressure decays with time. We refer to this as the "shut-in" condition. Solution of the differential equation for this particular case gives

$$p_{gi} = p_{go}\left[\frac{(p_{gi0} + p_{go})\exp\{Ht'\} + (p_{gi0} - p_{go})}{(p_{gi0} + p_{go})\exp\{Ht'\} - (p_{gi0} - p_{go})}\right] \qquad \text{(shut-in; } C = 0) \tag{10}$$

where

$$H = \frac{2B\, p_{go}}{V_{gi0}} \tag{11}$$

The upstream gas pressure, p_{gi}, therefore asymptotically approaches the downstream gas pressure, p_{go}, and after infinite elapsed time:

$$p_{gi} = p_{go} = (p_{co} + p_{wo}) \qquad (t' = \infty) \tag{12}$$

Since p_{wo} is held constant in the experiments, determination of the upstream gas pressure at the asymptote during a shut-in provides a direct method of quantifying parameter p_{co}. The decay time of the negative transient is determined by parameter H which depends on the initial volume and pressure of the gas in the injection system and on the gas permeability of the clay. Large gas volumes in the upstream system lead to very lengthy experimental transients.

TYPICAL RESULTS

Figure 2 shows the first stage of experimental history Mx80-4 on HS-batch clay, where confining stress and backpressure were 16.00 and 1.01 MPa, respectively. An initial pumping rate of 375 μL.hr^{-1} was used to raise the upstream pressure. The initial response is quantified by (5). Gas breakthrough occurs at an excess gas pressure ($p_{gi} - p_{wo}$) of 15.19 MPa, which is fractionally larger than the swelling pressure of the clay:

$$p_{gi} - p_{wo} > \sigma - p_{wo} = \Pi_{sw} \qquad \text{(at breakthrough)} \tag{13}$$

The peak response, at 15.29 MPa, is probably indicative of some sort of fracturing process and is followed by a spontaneous and very well-defined negative transient which approaches an asymptote of 14.22 MPa. The gas injection pump was then stopped. Excess pressure follows a second negative pressure transient which is quantified by (10). The extrapolated asymptote of the curve is the apparent capillary pressure, p_{co}, with a value around 12.5 MPa.

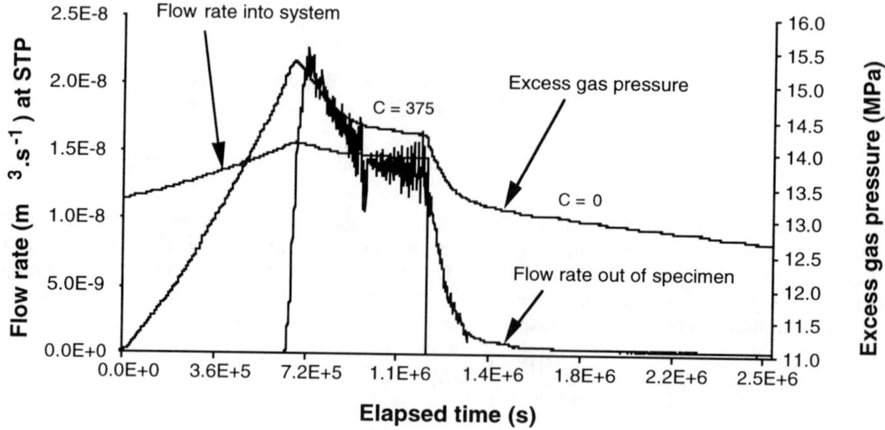

Figure 2 - Excess gas pressure (p_{gi} - p_{wo}) and volumetric flow rates (STP) into the testing system and out of the specimen plotted against elapsed time. The material is HS-batch clay with an average dry density of 1.638 $Mg.m^{-3}$ and a swelling pressure close to 15 MPa. The peak pressure response is probably indicative of the propagation of microcracks.

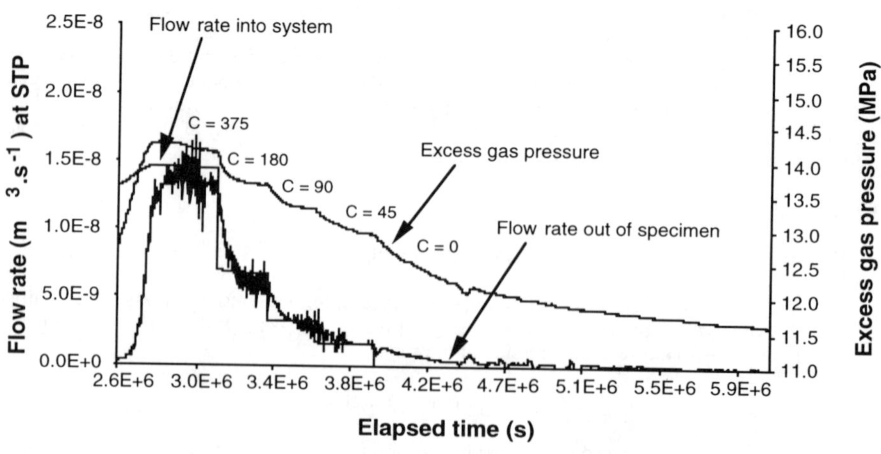

Figure 3 - Excess gas pressure (p_{gi} - p_{wo}) and volumetric flow rates (STP) into the testing system and out of the specimen plotted against elapsed time for the same specimen as in Figure 2, but retested after the period of shut-in. The peak is substantially lower than for the virgin clay suggesting that microcracks formed in the first cycle were not fully closed at the onset of the second testing cycle.

Gas injection was reinstated at a pumping rate of 375 μL.hr^{-1}. Figure 3 shows no well-defined breakthrough event and gas flow through of the specimen commences at an excess pressure substantially lower than that of the virgin specimen. Differential pressure climbs to a rather inconspicuous secondary peak at an excess pressure of 14.25 MPa. This is 1.04 MPa lower than the primary peak. The post-peak negative transient is also rather poorly defined, with the steady-state asymptote around 14.11 MPa, which is a little lower than before. The possible explanation for the lack of a distinctive breakthrough event and the substantial reduction in the peak excess pressure is that the gas pathways in the bentonite failed to reseal completely during the preceding shut-in stage. This might be explained by the presence of residual gas bubbles left along the routes of gas migration.

The specimen was then subjected to a descending history of pumping rates (180, 90, 45 and 0 μL.hr^{-1}). Each stage gives the negative transient response characterised by (6). As a rough rule of thumb, halving the gas flow rate at each stage leads to a differential pressure history with more or less equal decrements in the steady-state excess gas pressure. The gas flow law is very clearly nonlinear (i.e. non-Darcian) over this history, leading to a variable gas permeability. The very substantial duration of the shut-in response is a direct consequence of this underlying nonlinearity in the flow law. After 24 days of shut-in, the excess gas pressure falls to 11.60 MPa, well below the previous value of around 12.5 MPa.

Gas permeability is calculated using (3), (4) and (12) and plotted (Figure 4) against net mean stress, σ", which is the difference between the confining stress, σ, acting on the specimen and the average internal gas pressure (taken as $(p_{gi} + p_{go})/2$). This quantity should be indicative of the tendency of the gas pathways to dilate or close with changes in their internal pressure. The history starts at the gas breakthrough point, passes through the primary peak and evolves to the first steady-state. The increase in gas permeability along the initial section of this line is indicative of decreasing resistance to flow, resulting from the dilation of pathways. Maximum permeability occurs some time after the peak pressure.

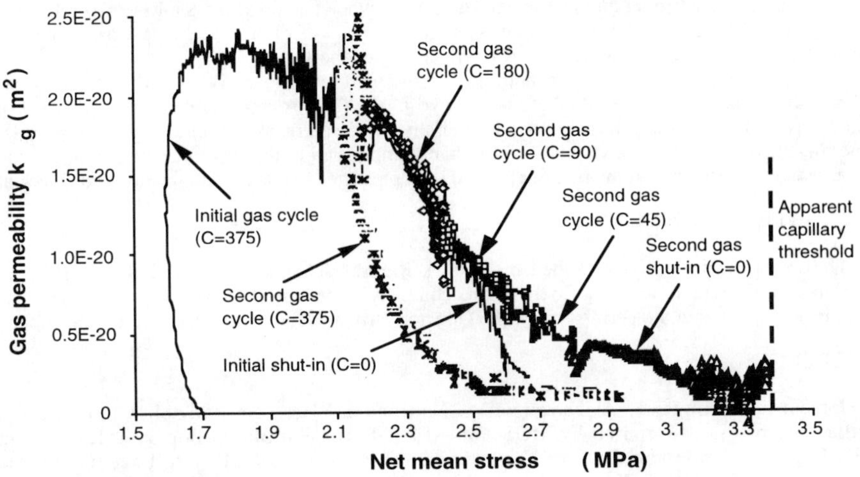

Figure 4 - Gas permeability plotted against net mean stress, based on the experimental histories depicted in Figures 2 and 3. The plot is indicative of pathway dilatancy after the breakthrough event and partial pathway closure during periods of declining gas pressure.

The shut-in transient is characterised by a very substantial fall in permeability, although this stabilises at around 1.5×10^{-21} m^2 in the latter part of the transient. The zero flow-rate condition is never reached. It seems probable that the cross-sectional area of each of the gas pathways decreases throughout the period of shut-in. Gas flow continues throughout the repressurisation period. The maximum permeability occurs at the steady-state condition and is almost identical to the steady-state value of the first cycle. The stepped history is marked by a substantial decline in permeability, but stabilises at around 2.0×10^{-21} m^2. It seems probable that, for repeated cycles of increasing and decreasing flow rate, the k_g versus σ" relationship settles into a unique, sigmoidal-shaped, hysteresis loop.

CONCLUSIONS

Passage of a gas phase through the initially water-saturated buffer clay is only possible if the gas pressure slightly exceeds the sum of the external water pressure and the swelling pressure. Bentonite which is pre-compacted to a high dry density will have a large swelling pressure. Such bentonite will therefore display a commensurately large gas entry pressure.

All experiments on virgin clay showed a peak in gas pressure, followed by a spontaneous negative transient. The peak response is probably indicative of crack propagation, suggesting that gas moves through an interconnected network of microcracks which are formed by tensile rupture of the clay under high applied gas pressure. Microcrack dilation and closure provide a possible explanation of the nonlinearity of the flow law revealed by multi-stage testing.

With the injection pump stationary (i.e. zero applied flow rate), gas pressure declined with time to some finite value. The excess gas pressure (p_{gi} - p_{wo}) in this shut-in condition is the apparent capillary pressure, p_{co}, of the clay. No gas flow has ever been detected at an excess gas pressure lower than this value.

When gas flow was reestablished in a specimen, the threshold pressure for gas breakthrough was found to be significantly lower than in the virgin clay. This suggests that the gas pathways do not close up completely over an experimental time-scale when gas pressure is allowed to fall to the lower threshold value. The sealing of gas migration pathways emerges as an important consideration in assessing the long-term performance of the buffer.

The capillary pressure for the penetration of interparticle pore space of buffer bentonite is likely to be of such a magnitude that normal two-phase flow is impossible. In the absence of dilated gas pathways (i.e. microcracks), the initially water-saturated bentonite is probably totally impermeable to gas. Hydrodynamics are probably implicated in the gas migration problem and future research will focus on the role of elevated pore pressures in crack initiation and propagation.

ACKNOWLEDGEMENTS

The experimental study is funded by Svensk Kärbränslehantering AB (SKB), Stockholm, Sweden, and the researchers wish to thank this company for its continuing financial support. The paper is published with the permission of the Director, British Geological Survey (NERC).

REFERENCES

S. Horseman, J. Higgo, J. Alexander and J. Harrington. Water, gas and solute movement in argillaceous media. Prepared by BGS for NEA SEDE Working Group on Measurement and Physical Understanding of Groundwater Flow Through Argillaceous Media. Rept. No. CC-96/1, OECD Nuclear Energy Agency, Paris, 1996, 290 pp, 26 figs.

M. Sen, S. Horseman and J. Harrington. Further studies on the movement of gas and water in an overconsolidated clay. EU MEGAS Project Phase 2. Fluid Processes Group, British Geological Survey, Technical Rept. No. WE/96/8, 1996.

DETERMINATION OF POREWATER CHEMISTRY IN COMPACTED BENTONITE

J. Lehikoinen[*], A. Muurinen[*], A. Melamed[**], and P. Pitkänen[**]
[*]VTT Chemical Technology, P.O. Box 1404, FIN-02044 VTT, Finland
[**]VTT Communities and Infrastructure, P.O. Box 19041, FIN-02044 VTT, Finland

ABSTRACT

Laboratory experiments were performed to study the interaction between groundwater and compacted sodium bentonite (Volclay MX-80). The solutions used were the fresh and saline groundwater simulants. The experiments were carried out in aerobic and anaerobic conditions at elevated temperature. Of main interest in the present study were the chemical changes in the reacting solution, bentonite porewater, and bentonite itself. The results for major cations display a principal difference between the interactions with fresh and saline solutions, while the differences between aerobic and anaerobic conditions within each solution case seem to be minor. The experimental results for the bentonite-water equilibria were interpreted in terms of a multi-site surface complexation model and the computer program HYDRAQL. The apparent diffusivities for sodium and sulphate in bentonite samples sandwiched between two filter-plates were also determined.

INTRODUCTION

The safety of the disposal of spent nuclear fuel is ultimately based on the multiple barrier concept in which the waste is surrounded by several redundant engineered and natural barriers. Compacted sodium bentonite, Volclay MX-80, has been selected as the reference buffer material in Finnish nuclear waste research. The objective of the present contribution was to investigate the interaction of sodium bentonite with fresh and saline groundwater simulants at elevated temperature in anaerobic and aerobic conditions. The main interests of the study were the chemical and mineralogical changes in the pore and bulk solutions, and in bentonite. An attempt to model the bentonite-water equilibrium as well as the diffusion of sodium and sulphate in these systems was also made.

EXPERIMENTAL

Bentonite was compacted in copper tubes equipped with steel sinters at each end and the tubes were immersed in experimental solutions (Fig. 1, Table I) for varying times. The diameter of bentonite in the tubes was 24 mm, the length 44 mm and the dry density 1.5 g/cm^3. The experiments were performed in a water bath at a temperature of 68-70 °C. The bentonite-to-water ratio was fixed at 337 g/L. The parameters varied in the experiments were the chemical composition of the solution, and the oxygen content in the system, i.e., either aerobic or anaerobic.

The samples were pre-saturated with deionized water by a vacuum saturation technique for two weeks and then moved to the experimental solution. The samples to be studied in anaerobic conditions were handled in a glove box. All the solutions used for these experiments were made anaerobic by nitrogen bubbling. In the anaerobic experiments, bottles with a copper cylinder and experimental solution were closed in a hermetic autoclave, which was then transferred into the water bath outside the glove box. The aerobic sample bottles were handled in ambient conditions and taken into the water bath without any shield.

Mat. Res. Soc. Symp. Proc. Vol. 465 © 1997 Materials Research Society

After different time intervals the bottles were removed from the water bath
and the copper tube with bentonite was taken out of the bottle. In the case of
the anaerobic samples, this procedure was carried out in the glove box, while
the aerobic samples were handled in ambient atmosphere.

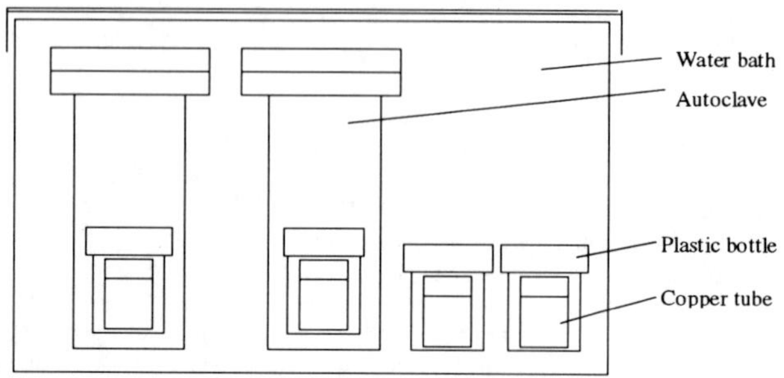

Fig. 1. Experimental setup for the groundwater-bentonite interaction test.

Table I. Initial compositions of equilibrating solutions (mmol/L).

Solute	Fresh water		Saline water	
Na	2.35		240.6	
K	0.10		0.00	
Ca	0.44		116.8	
Mg	0.16		0.00	
Cl	1.55		474.2	
SO_4	0.12		0.01	
	Anaerobic	Aerobic	Anaerobic	Aerobic
HCO_3	1.77	2.08	0.01	0.01
pH	9.21	8.38	7.72	6.32

The bentonite was pushed out of the tube and 6 successive slices, each 3 mm
thick, were cut from one end of the bentonite cylinder. The slices were dried
and analyzed in order to find any concentration profiles and mineralogical
alterations. A 10 mm-thick slice cut from the other end of the bentonite
cylinder was used for porewater studies. The porewater samples were obtained by
a compression technique [1].

RESULTS

Equilibration

The chemical equilibria with respect to the two solutions are summarized in
Table II (abbreviation "B"). The temporal progress for the attainment of these
equilibria in anaerobic conditions is presented in Figs. 2 and 3. The
concentration plots of individual cations show a prominent difference between
the interactions with fresh and saline solutions, while the differences between
aerobic and anaerobic conditions within each solution case seem to be
insignificant. The major processes with the fresh water were the diffusion of
sodium, potassium, sulphate, bicarbonate and chloride from bentonite to
solution, and the diffusion of calcium and magnesium from solution into
bentonite. The soluble non-clay fractions in MX-80 serve as the main sources for
the above anions [1]. The pH's of the solutions increased during the experiment,

the highest values being obtained in aerobic conditions. The concentration changes in the solutions were fairly pronounced for many ions with the fresh water, while the interaction in bentonite in both aerobic and anaerobic conditions resulted in a marked time-dependent concentration decrease for sodium but only minor changes for magnesium, calcium and potassium.

Table II. Chemical composition after 10 months of equilibration (mmol/L). B = bulk (solution in the plastic bottle - see Fig. 1), P = pore and M = model.

Solute	Soln. type	Fresh water Anaerobic	Aerobic	Saline water Anaerobic	Aerobic
Na	B	38.7	42.1	336.1	331.0
	P	7.48	10.1	125.8	131.8
	M	40.4	39.0	274.0	274.0
K	B	0.29	0.34	2.24	1.92
	P	ND	0.23	1.22	0.43
	M	0.45	0.43	1.53	1.53
Ca	B	0.07	0.04	32.2	20.5
	P	ND	ND	4.57	1.60
	M	0.32	0.29	103.0	103.0
Mg	B	0.07	0.04	8.68	4.32
	P	ND	ND	7.53	2.63
	M	0.06	0.05	3.55	3.55
Cl	B	2.48	2.99	392.4	350.2
	P	2.65	1.27	211.7	186.5
	M	2.99	2.99	476.0	476.0
HCO_3	B	10.8	7.01	2.22	0.50
	P	3.57	(17.0)	14.0	11.6
	M	4.42	1.99	0.25	0.15
SO_4	B	12.9	15.1	11.3	10.3
	P	1.22	0.87	6.26	6.71
	M	15.0	15.1	5.14	5.21
pH	B	8.51	9.51	7.43	8.53
	P	8.35	9.50	9.49	8.66
	M	8.13	8.53	7.10	7.34

ND - not detected

The concentration changes with dependence on distance from the bentonite-water interface were negligible except for some notable difference with respect to calcium and sodium between the first and second bentonite slice. In general, the results from 10 months of interaction represent somewhat stronger concentration changes compared to those found in an earlier study of bentonite interaction with fresh water in ambient conditions [2].

The major processes in the experiments with saline water were the diffusion of the sodium, magnesium, sulphate and bicarbonate from bentonite to the solution, and the diffusion of calcium from the solution into bentonite. The highest pH's as well as significant consumption of chloride were observed in aerobic conditions. The main mineralogical change in bentonite was the alteration from Na- to Ca-rich form. Concentration changes of calcium and sodium in bentonite with dependence on distance from the bentonite-water interface were notable in both aerobic and anaerobic conditions already after 14 days of interaction. It seems that the most significant differences between the results of interaction with the two solution cases arise to the largest extent from the differences in water chemistry.

The initial concentration of calcium in the saline water (Table I) seems to account for its strong concentration increase in the first slices of bentonite already after 14 days. However, for longer interaction periods it appears that the concentration gradients level out when the system approaches equilibrium.

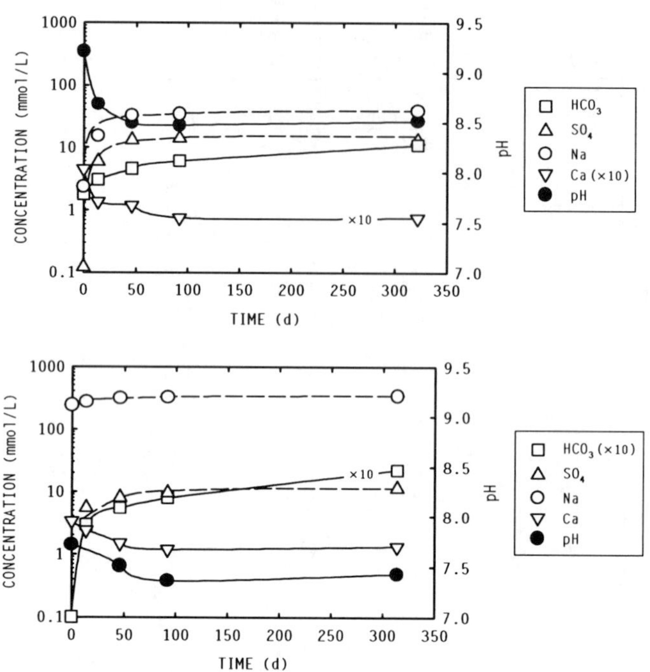

Fig. 2. Concentration in the bulk solution in anaerobic conditions. Upper: fresh water, lower: saline water. Broken line: diffusion model prediction.

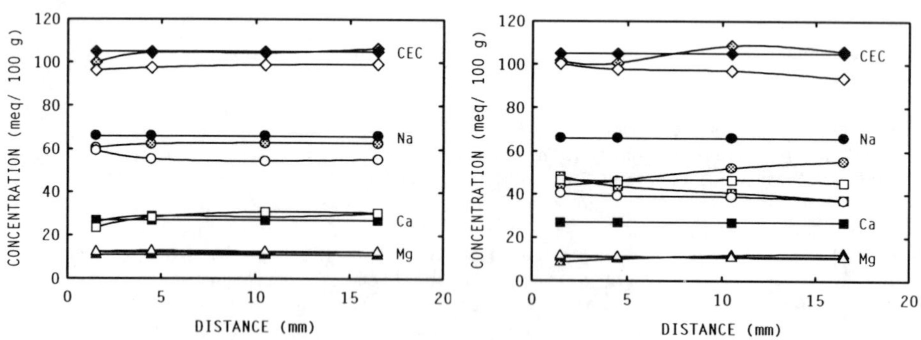

Fig. 3. Exchangeable cations in bentonite in anaerobic conditions as a function of distance from the bentonite-water interface. Left: fresh water, right: saline water. Symbol: full - 0 d, gray - 14 d, open - ca. 10 months.

<u>Squeezing of porewater</u>

The extraction of porewater by gradually increasing compaction [1] was employed to make comparisons between the compositions of the bulk aqueous phase and that of porewater from the bentonite samples reacted for 10 months (Table II, abbreviation "P"). The results, which are in line with earlier compaction studies (e.g. [3]), indicate a clear exclusion for nearly all ions. Exclusion is strongest for divalent sulphate, in qualitative agreement with the electric double-layer theory. The pH's of equilibrating solutions and porewaters seem in most cases to be fairly equal. Only in the saline anaerobic solution was there a marked discrepancy between the pH of porewater and that of the bulk.

<u>Equilibrium modelling</u>

The surfaces of smectites carry at least two types of reactive sites. Cationic substitutions within the tetrahedral layers give rise to the permanent negative structural charge and consequent cation-exchange capacity. In addition to these "layer sites", crystallite edges contain exposed broken-bond hydroxylated aluminol (AlOH) and silanol (SiOH) sites, which exhibit acid/base behaviour and co-ordinative properties similar to their component oxides [4,5].

The calculations for the bentonite-water equilibria were performed using this multi-site surface complexation approach and the program HYDRAQL [6]. Cation-exchange reactions were modelled in half-reaction format [7]. The constant capacitance model (CCM) was applied to the edge-site complexation and involved only proton equilibria. The selected properties of the sorbent and the surface chemical equilibria on MX-80 are summarized in Tables III and IV. We have assumed that the variation of binding constants in Table IV with temperature is a second-order effect. The basis of the bentonite model is given elsewhere [8].

Table III. Properties of MX-80.

Property	Value
Cation-exchange capacity, CEC	0.810 meq/g (this study)
Tetrahedral site density, SiOH	6.39 mmol/m^2 [9]
Octahedral site density, AlOH	5.32 mmol/m^2 [9]
Capacitance, C	1.06 F/m^2 [10]
Edge surface area, S	7.0 m^2/g† [5]
Exchangeable Na	80.6 % (this study)
Exchangeable K	1.4 % (this study)
Exchangeable Ca	12.9 % (this study)
Exchangeable Mg	5.0 % (this study)
Carbonate (as $CaCO_3$)‡	1.4 w% [11]
Quartz‡	10 w% [11]
NaCl‡	0.025 w% (this study)
$CaSO_4$‡	0.57 w% (this study)

† Corrected for the montmorillonite content of MX-80, ~75 w% [11].
‡ Impurity.

The amounts of NaCl and $CaSO_4$ impurities were adjusted to meet their experimental bulk equilibrium values in the fresh aerobic water. The oxidation of pyrite, FeS_2, was not considered in our calculations since, by inspection of Table II, no considerable differences between the amounts of SO_4 released to the bulk solutions in anaerobic and aerobic conditions can be observed. It was assumed that the anaerobic and aerobic equilibria are reached according to a closed and open system respectively.

Table IV. Equilibrium constants.

Surface reaction	log K
$X^- + Na^+ = XNa$ †	20.00 (by def.)
$X^- + H^+ = XH$	20.10 [6]
$X^- + K^+ = XK$	20.26 [6]
$2X^- + Ca^{2+} = X_2Ca$	40.24‡ [12]
$2X^- + Mg^{2+} = X_2Mg$	40.20‡ [12]
$SiOH = SiO^- + H^+$	-6.53 [13]
$AlOH + H^+ = AlOH_2^+$	7.2 [13]
$AlOH = AlO^- + H^+$	-9.5 [13]

† X represents one equivalent of cation exchanger.
‡ Conforms to the mole fraction convention.

For the fresh water, the agreement of the modelled (Table II, abbreviation "M") and bulk data was satisfactory except for pH, calcium and bicarbonate. However, the deviation of modelling results from experimental data is substantially more pronounced in the saline water. Typical for both aerobic experiments was the pH difference of about one unit between the model calculations and experiments. In the anaerobic conditions, the pH is controlled by the $CaCO_3$-CO_2-H_2O system, while the influence of bentonite through cation exchange and surface protonation is comparatively small [14].

The calculated equilibrium concentrations for major exchangeable cations are compared with those from experiments carried out in anaerobic conditions in Table V. The agreement is adequate for the fresh water, whereas for the saline water major quantitative differences appear. It is noteworthy here that, according to model calculations, the experimentally observed alteration of MX-80 to a calcium-dominated form did not take place in the saline water.

Table V. Exchangeable cations in MX-80 in anaerobic conditions: model predictions vs. experimental results at various distances from the bentonite-water interface (percentage of the CEC).

Cation	Solution	Model	0-3 mm	3-6 mm	9-12 mm	15-18 mm
Na	Fresh	62.0	61.5	56.7	55.1	55.8
	Saline	64.1	40.2	40.0	40.0	39.6
Ca	Fresh	31.7	24.5	29.0	31.2	30.7
	Saline	34.1	46.6	47.2	48.1	48.3
Mg	Fresh	5.1	12.8	13.1	12.6	12.4
	Saline	1.2	11.5	11.4	11.0	11.1

<u>Diffusion</u>

The governing equation for the 1D diffusive transport is

$$\frac{\partial c}{\partial t} = D_a \frac{\partial^2 c}{\partial x^2} \qquad (1)$$

where c is the (total) concentration, and D_a the apparent diffusivity to conform to the conventional K_d-model based on the linear adsorption isotherm [15]. Here, we seek a solution for diffusion between a stirred solution of limited volume and a composite medium consisting of a cylindrical clay plug sandwiched between two filter-plates [16,17]. The fitted quantity was the D_a of Na and SO_4 in bentonite while, otherwise, parameters determined from the equilibrium state were applied [1]. The estimated diffusivities of Na and SO_4 in the filter-plates were 2.5×10^{-9} m^2/s and 2.0×10^{-9} m^2/s respectively. The governing equations were solved with the Laplace finite difference technique, which differs from its better-known analog [18] in that the spatial discretization is realized with finite differences instead of the Galerkin finite element technique. The results from fitting to the bulk data in Fig. 2 are depicted in Table VI and Fig. 2. The diffusivities in Table VI are subject to the presumption that no solubility constraints are imposed upon either sodium or sulphate.

Table VI. Apparent diffusivities of Na and SO_4 in anaerobic and aerobic conditions ($\times 10^{11}$ m^2/s).

Ion	Fresh water	Saline water
Na	1.1-1.2	2.9-3.0
SO_4	6.7-11.4	4.8-5.0

An attempt was also made to simultaneously fit the diffusion model to the bulk and adsorbed profile for sodium. This is, in fact, an absolute prerequisite for obtaining diffusivities with a sound basis. A reasonable fit was possible only by assuming an unrealistically high diffusion resistance for the filter-plates. Due to the inability of the present K_d-model to reproduce both profiles, the diffusivities of Na in Table VI are to be considered somewhat dubious. One may thus conclude that the linear K_d-model is not a valid desciptor of the diffusion of sodium here. The situation is obviously different for non-sorbing solutes (such as SO_4) because then no K_d is involved.

DISCUSSION

Although the trend of the cation-exchange process and the behaviour of major ions in reacted bentonite samples and bulk water can be identified, it should be taken into consideration that the results from the chemical analyses discussed above are preliminary and represent only comparatively short interaction time periods.

The copper cylinders also participated in the chemical reactions. This was most obvious in the saline, aerobic experiments, where a bluish precipitate formed in the solution. This was characterized by XRD as ~50 % $CuCl_2 \cdot 3Cu(OH)_2$. Another major component was CuO, as well as traces of $Cu_3Cl_4(OH)_2$. The formation of copper precipitates explains the substantial consumption of chloride, particularly in the saline aerobic solution (Table II).

The deviation of modelling results from the experimental data is ascribable to various factors; uncertainties in the inventory and dissolution pathways of the soluble impurities [7], effect of temperature [1], chemical composition and compaction on the binding constants in Table IV, and the effect of compaction on the accessibility of surface sites, to name but a few. With regard to compaction, our earlier study with a bentonite-to-water ratio of 79 g/L [19] demonstrated the influence of increasing sample density on the equilibria attained. The onset of this effect was at a density of about 1 g/cm^3. It is likely that the influence of density is even more enhanced upon the present system with a higher bentonite-to-water ratio.

Particularly in the aerobic conditions the actual pCO_2 is not known. It is probably intermediate between that of open air and a closed system. The CO_2 content of the atmosphere has implications in particular on the pH and (bi)carbonate concentration in these systems.

On the grounds of what has been presented, it is obvious that additional laboratory studies as well as further elaboration of both equilibrium and diffusion models are required. The relevant aspects to be looked at in future research include a CO_2-controlled atmosphere, the role of accessory non-clay minerals, the coupling of geochemistry and solute transport, and the relationships between bulk water and porewater compositions.

ACKNOWLEDGMENTS

This work was financially supported by the Ministry of Trade and Industry of Finland (KTM) and has been carried out as a part of the publicly administrated nuclear waste management programme. The authors are grateful to Mrs. Kirsti Helosuo and Mrs. Jaana Rantanen for performing the water analyses, and Dr. Antti Vuorinen from the Department of Geology, University of Helsinki, Finland, for analysing the exchangeable cations of bentonite samples.

REFERENCES

1.   A. Muurinen and J. Lehikoinen (to be published).
2.   A. Melamed, P. Pitkänen, M. Olin, A. Muurinen and M. Snellman, in _Scientific Basis for Nuclear Waste Management XV_, edited by C.G. Sombret (Mater. Res. Soc. Proc. 257, Pittsburgh, PA, 1992) pp. 557-566.
3.   M.S. Rosenbaum, Clays and Clay Minerals, __24__, 118 (1976).
4.   G. Sposito, _The Surface Chemistry of Soils_ (Oxford University Press, New York, 1984), p. 234.
5.   J.M. Zachara and J.P. McKinley, Aquatic Sci., __55__, 250 (1993).
6.   C. Papelis, K.F. Hayes and J.O. Leckie, Dept. Civil Engrg., Stanford Univ., CA, Technical Report No. 306 (1988), p. 130.
7.   P. Flecther and G. Sposito, Clay Minerals, __24__, 375 (1989).
8.   E. Wieland, H. Wanner, Y. Albinsson, P. Wersin and O. Karnland, SKB 94-26, 1994.
9.   G.N. White and L.W. Zelazny, Clays and Clay Minerals, __36__, 141 (1988).
10. S. Goldberg, in _Emerging Technologies in Hazardous Waste Management III_, edited by D.W. Tedder and F.G. Pohland (Am. Chem. Soc. Symp. Ser. 518, 1993) pp. 278-307.
11. M. Müller-Vonmoos and G. Kahr, NTB 83-12, 1983.
12. G. Sposito, C. Jouany, K.M. Holtzclaw and C.S. LeVesque, Soil Sci. Soc. Am. J., __47__, 1081 (1983).
13. P.W. Schindler and G. Sposito, in _Interactions at the Soil Colloid - Soil Solution Interface_, edited by G.H. Bolt, M.F. De Boodt, M.H.B. Hayes and M.B. McBride (Kluwer Academic Publishers, 1991) pp. 115-145.
14. M. Ochs, H. Wanner, C. Oda and M. Yui, presented at the Fifth International Conference on Chemistry and Migration behaviour of Actinides and Fission Products in the Geosphere, Saint-Malo, France, 1995.
15. A. Muurinen, VTT Publications 168, 1994.
16. M.J. Put, Radioactive Waste Management Nuclear Fuel Cycle, __16__, 69 (1991).
17. J. Crank, _The Mathematics of Diffusion_, 1st ed. (Clarendon Press, Oxford, 1957), p. 347.
18. E.A. Sudicky, Geoderma, __46__, 209 (1990).
19. A. Muurinen, H. Aalto, T. Carlsson, J. Lehikoinen, A. Melamed, M. Olin and P. Salonen, VTT Research Notes 1714, 1995.

EXPERIMENTAL STUDY ON SCALE EFFECTS OF BENTONITE / SAND MIXTURES ON GAS MIGRATION PROPERTIES

Toshiyuki Hokari*, Mitsunobu Okihara*, Takashi Ishii*, Takao Ishii** and Hiroyuki Ikuse***
*Shimizu Corporation, 1-2-3 Shibaura, Minato-ku, Tokyo 105-07, Japan
**Mitsubishi Materials Corporation, 1-3-25 Koishikawa, Bunkyo-ku, Tokyo 112, Japan
***Radioactive Waste Management Center, 2-8-10 Toranomon, Minato-ku, Tokyo 105, Japan

ABSTRACT

Gas migration properties of bentonite/sand mixtures for a large-scale structure were estimated on the basis of laboratory test results obtained with small test specimens. Water and gas migration tests were carried out to clarify scale effects on migration properties of the mixtures.

INTRODUCTION

Bentonite/sand mixtures are impermeable owing to the content of montmorillonite which has a high swelling capacity. These mixtures are expected to be used to form impervious barriers in radioactive waste disposal facilities [1]. It is believed that several kinds of gases could be generated inside the facility because of chemical reaction of the wastes with groundwater[2]. In this event, the generated gases are predicted to pass through the bentonite/sand mixtures after first accumulating inside the mixtures.

Pusch proposes a generalised microstructural model[3] to provide a reasonable explanation of properties of bentonite/sand mixtures; for example, hydraulic conductivity, swelling pressure, gas migration and so on. He emphasises that the density of suspensions (referred to as bentonite gel) of pore water and swelling bentonite in aggregate voids has the greatest effect of all factors in these properties. The property of gas migration, to which he attaches most importance, is the pressure (referred to as critical pressure) above which gas starts to penetrate the bentonite/sand mixtures. It is reported that the critical pressure is nearly equal to the swelling pressure in mixtures of low bentonite content.

Bentonite/sand mixtures are expected to be homogeneous in density and composition because they would be prepared using compaction machinery to a fixed density, after being well mixed at a controlled water content. Thus it is realistic that data on fundamental properties should be obtained in laboratory experiments. When predicting the behaviour of gas migration in full-scale facilities on the basis of properties determined from laboratory data, it is essential that scale effects and variations in preparation and handling should be taken into account. For the purpose of understanding scale effects on gas migration, Hokari et al. [4] reported gas migration tests with specimens of different sizes. It is hypothesised that gas migration through bentonite/sand mixtures is dependent on the gas path length and the area to which gas pressure is applied. This report discusses effects of vertical and lateral scale on the gas migration properties of bentonite/sand mixtures on the basis of the data obtained by Hokari et al.[4].

EXPERIMENT

Apparatus

A schematic view of the apparatus for hydraulic and gas tests is shown in Figure 1. Because it takes a long time to saturate specimens above 200 mm in height, the apparatus for these samples was designed for water injection and drainage at the side wall.

Mat. Res. Soc. Symp. Proc. Vol. 465 © 1997 Materials Research Society

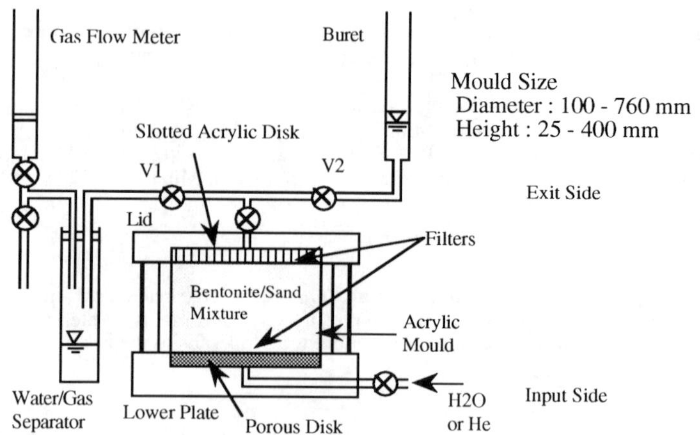

Figure 1 A Schematic View of Hydraulic & Gas Test

Specimen conditions

Specimen sizes are varied for hydraulic and gas tests of bentonite/sand mixtures. The experiments are grouped into two sets according to the vertical and lateral size. Set (A) is aimed at the investigation of height scale effects [25, 50, 100, 200 and 400 mm in height (diameter is fixed at 100 mm)]. Set (B) is aimed at the investigation of diameter scale effects [100, 200, 300 and 760 mm in diameter (height is fixed at 200 mm)]. The data for the specimen of height 200 mm in Set (A) is used to provide data for a specimen of diameter 100 mm in Set (B).

The bentonite used is commercially available Na-bentonite, Kunigeru V1[5], made by Kunimine Industries Co., Ltd. in Japan. The aggregates used are terrace deposit sands. The fine content vary little between batches of sands. Representative specimen conditions are as follows; bentonite content 15%; dry density 1 799 kg/m^3 (the maximum dry density), and water content 14.6 % (the optimum water content).

Procedure

(1) Preparation of Specimens
Sand, bentonite and water are weighed to ensure that specimens conform to a fixed dry density and fixed water content. Then, those materials are compacted inside moulds in the way prescribed in JIS A 1210[6].
(2) Saturation
Water is injected at the bottom of the moulds at a pressure of approximately 20 kPa. Water flow rates at the exit are measured with burets. Saturation is assumed once steady state flow is reached. The specimens above 200 mm in height are subjected to accelerated saturation by water injection and drainage at the side wall in the direction of spiral climb from the bottom to the top.
(3) Hydraulic Test
While a constant water pressure is applied at the inlet of the specimen by means of gas cylinders, the water expelled is measured at the outlet.
A measure of the water flow rate is obtained from differences in water levels inside the burets over selected time intervals.
The tests are conducted at 3 different values of the water pressure. If these pressures are too low to produce detectable water flow rates for a particular specimen, the pressure is fixed at the lowest value at which flow is observed.
(4) Preparation for Gas Migration Test

After removing a lid, a lower plate, porous disks and filters from the mould, a slotted acrylic disk is installed at the upper side. The lower porous disk is dried, re-settled and filled with helium gas. Water in the lower tube is displaced with the gas. The upper part above the slotted acrylic disk is filled with water.

Because the moulds above 200 mm in height consist of several moulds of 100 mm height, only the lid is removed. The lower porous disk is drained with being attached to the mould.

(5) Gas Breakthrough Test

While a gas pressure of 10 kPa is applied to the specimen, water expelled by gas is measured at the outlet. When no water is detected over a fixed period of time, the gas pressure is increased to 20 kPa. In the same manner, the gas pressure is increased in steps of 10 kPa until a pressure measure (referred to as threshold gas pressure) is obtained above which water starts to be expelled by gas. The water volume expelled is measured by changes in the level inside the buret.

After gas entry, the gas pressure is increased in steps of 20 kPa over a fixed period of time, and the water volume expelled is measured until gas breakthrough is observed. The breakthrough is confirmed by observation of the upper part of the specimen. The gas pressure and time are recorded at the breakthrough.

It is supposed that the volume of water expelled is dependent upon the hydraulic properties of the specimen. The time interval for changing the pressure was fixed for each specimen by consideration of the measurement precision and the ease of performing the experiment, as the follows: (A) 100 mmϕ x 25mmH 1day, 50mmH 1day, 100mmH 2day, 200mmH 4day and 400mmH 4day; (B) 200 mmH x 200mmϕ 2day, 300mmϕ 2day and 760mmϕ 4day.

(6) Gas Permeability Test

The gas flow rate is measured at the breakthrough pressure. It is measured by changes in water levels inside a double tube buret at the outlet.

The gas flow rates are measured at various gas pressures.

RESULT

A value of hydraulic conductivity is calculated from the slope of a linear regression for 3 data points for the flow rate and pressure in accordance with Darcy's Law for steady linear flow. A value for gas conductivity is calculated using the corresponding formula for steady linear flow of a compressible fluid[7].

Figures 2 - 5 show respective relationships between hydraulic and gas migration properties of bentonite / sand mixtures and specimen scales. Each Figure consists of relationships between the property, height and diameter. The results are analysed below.

(1) Hydraulic conductivity

Figure 2 shows the dependence of hydraulic conductivity on the heights or the diameters of the specimens. The hydraulic conductivity is on the order of 10^{-11} m/s regardless of variations in specimen heights or diameters. The conductivity is judged to have no significant relationship to specimen scale.

(2) Threshold pressure

The time interval for changing the pressure which was initially determined as described in the previous section was judged to be inadequate to measure the threshold pressure of Set (A). Another experiment was conducted with a longer time interval. The details are discussed in the following paragraph.

Figure 3 shows the effects on threshold pressure of the heights or the diameters of the specimens. Threshold pressure is the gas pressure above which gas starts to penetrate into water saturated specimens. The lower the threshold pressure is, the easier gas can migrate through the specimen. In these experiments, the threshold pressure is determined as the initial gas pressure at which drained water is detected at the upper face of the specimens. Measured threshold pressures show no significant relationship with specimen height, but exhibit a gradual trend of a decrease with specimen diameter.

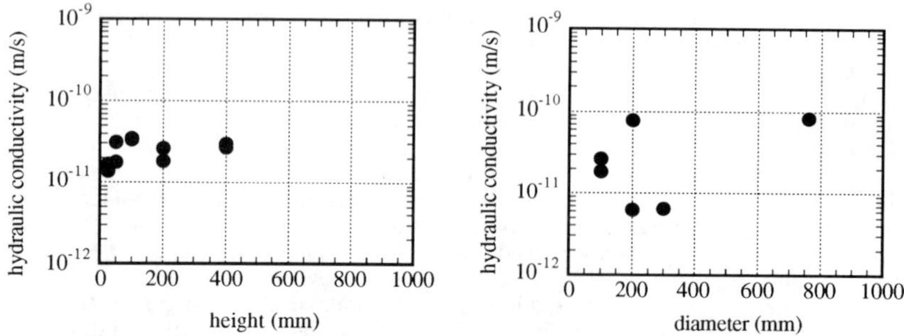

Figure 2 Relationship between hydraulic conductivity and scale
(left : height effects ; and right : diameter effects)

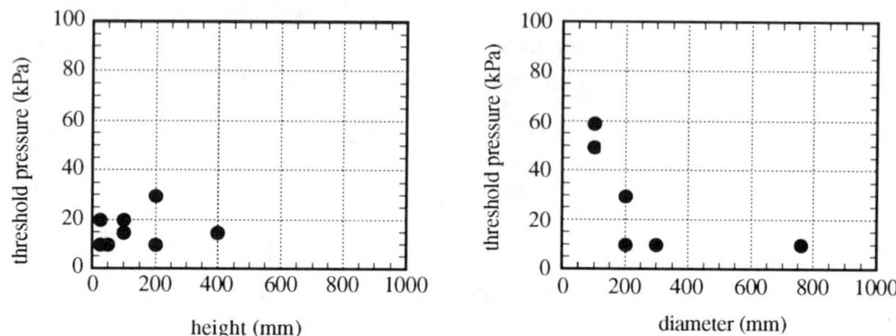

Figure 3 Relationship between threshold pressure and scale
(left : height effects ; and right : diameter effects)

(3) Effective gas porosity

Figure 4 shows the effect on the effective gas porosity of the heights or the diameters of the specimens. Here the effective gas porosity is defined as the ratio of the total volume of drained water prior to breakthrough, when gas paths are formed through the specimen, to the specimen volume. A smaller value of the effective gas porosity means swifter gas migration through the bentonite/sand mixture. Measures of effective gas porosity show a decreasing trend with both height and diameter of the specimens.

(4) Gas conductivity

Figure 5 shows the effects on gas conductivity of the heights or the diameters of the specimens. Judging from the way that the measured gas conductivity is distributed around 10^{-11} m/s with a large variance, there is not a significant correlation between gas conductivity and specimen size.

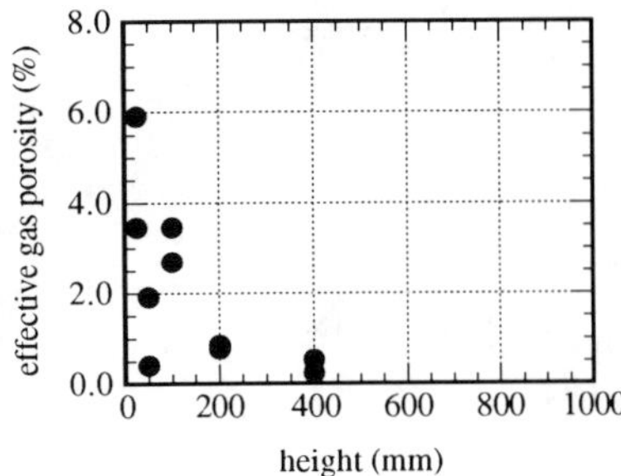
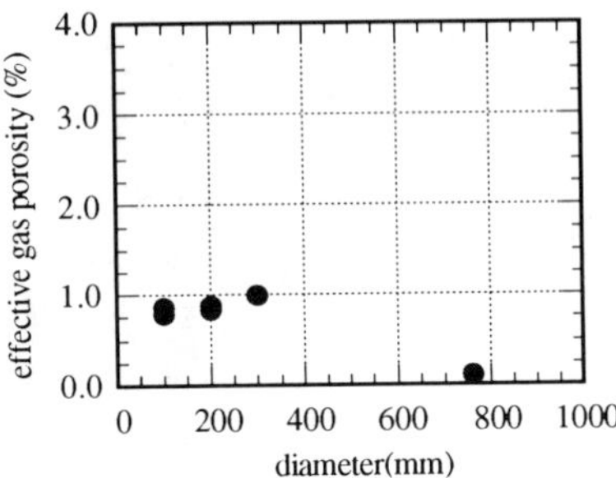

Figure 4 Relationship between effective gas porosity and scale
(left : height effects ; and right : diameter effects)

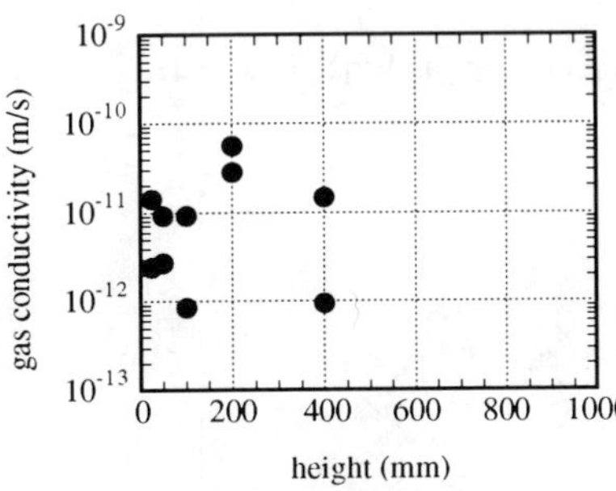
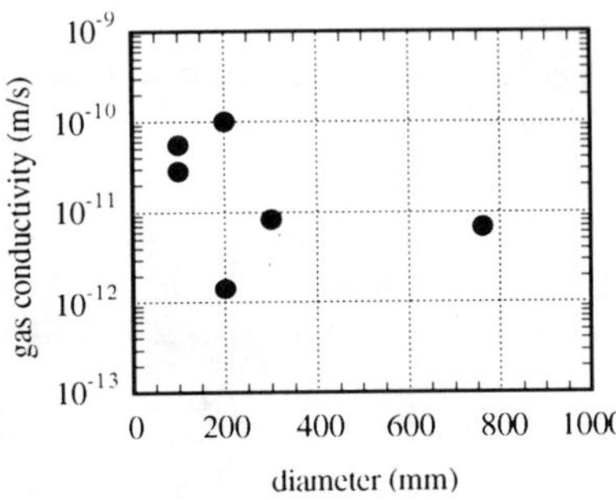

Figure 5 Relationship between gas conductivity and scale
(left : height effects ; and right : diameter effects)

DISCUSSION

The experimental results show that hydraulic conductivity and gas conductivity of bentonite/sand mixtures have no significant relationship with scale, while there are correlations with scale for threshold pressure and effective gas porosity.

It is supposed that when gas is applied to water saturated bentonite/sand mixtures whose aggregate voids are filled with bentonite gel, gas would penetrate in the manner of fingers, compacting and dehydrating parts of the bentonite gel to form preferential paths. In one of these experiments, it is confirmed that only water was expelled by gas from a specimen because bentonite was not detected in drained liquid. Figure 6 represents a concept of gas path formation at a microscopic scale. Hokari et al.[8] identified both entrances and exits of gas paths by means of microscopic observation on both end surfaces of specimens with which gas migration tests were conducted. It is not until gas paths develop a 3 dimensional network to form an initial path traversing the soil that breakthrough is observed.

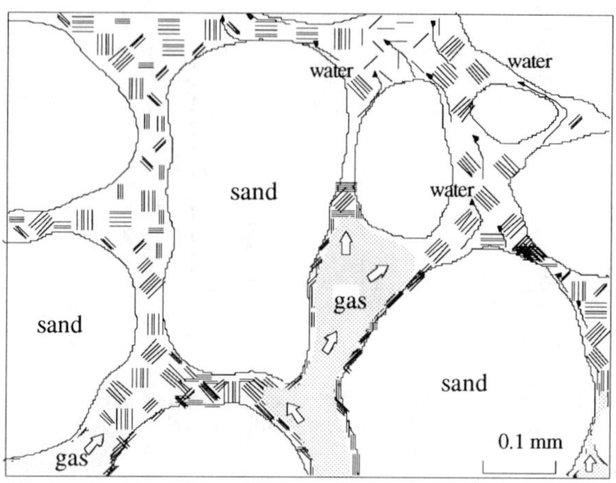

Gas penetrates the voids filled with thinner bentonite gel and the void water is expelled to develop gas paths.

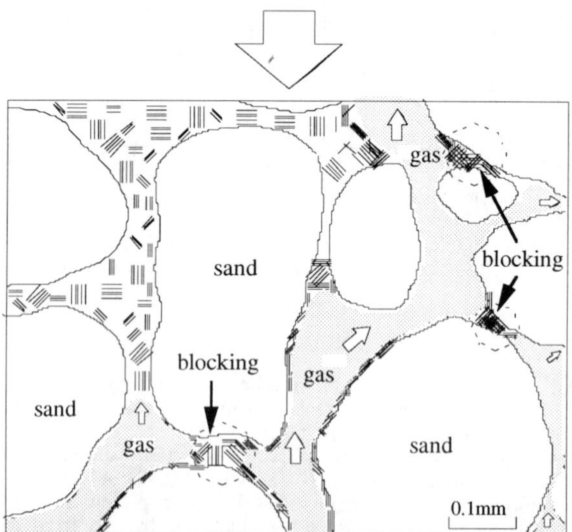

Bentonite gel is compressed with removal of the void water. Some voids are blocked by denser bentonite gel.
Gas, which develops a 3 dimensional network of paths, pierces through specimens.

Figure 6 Concept of gas migration in bentonite/sand mixtures
(Montmorillonite in bentonite is illustrated as stacks. These consist of multilayers of approximate
0.1 µm wide, whose thickness is dependent on water intake into the interlayers. Not to scale.)

The results can be studied on the basis of the previous concept of gas migration as follows.

The threshold, at which the gas pressure starts to cause water movement in aggregate voids, is presumed to be governed by the hydraulic resistance to void water and the friction between aggregates and liquid/gas interface in voids. When specimens vary in diameter, the threshold pressure is mainly dependent on the friction [corespondent to capillary pressure] because they have the same hydraulic resistance that consists of hydraulic conductivity and pressure gradient. The results show hydraulic conductivity has no significant relationship to scale. It is predicted that the threshold pressure would be lower in specimens of larger diameter because the relative abundance of conductive voids increases in a wider specimen with a larger contact area of gas, void water and aggregates.

When specimens vary in height, the hydraulic resistance is not equal because the pressure gradient is dependent on the specimen height while the conductivity is nearly equal. Another experiment concerning the effect of vertical scale on threshold pressure was conducted under the same condition of hydraulic resistance [pressure gradient] by using the following time intervals for increasing the pressure: 100 mmϕ x 25 mmH 1 day, 50 mmH 2 day, 100 mmH 4 day, 200 mmH 8 day and 400 mmH 16 day. The threshold pressure would be predicted not to be dependent upon changes in height because these do not result in changes in the contact area. The above theory is consistent with Figure 3.

It is also supposed that a traversing gas path would be more swiftly generated in specimens of larger diameter for the same reason as above. The effective gas porosity is measured at breakthrough when the initial traversing gas path is formed, which implies a decrease in the measured effective gas porosity for the larger diameter specimens. The number of traversing gas paths would possibly be decreased in taller specimens because their lateral development is restricted by side boundaries. Consequently effective gas porosity shows a downward trend with both diameter and height.

This experiment applied simple uniaxial moulds made of acrylic with various scales. In case of gas leakage between the specimen and the mould, its side wall was observed on application of gas pressure in every test so that it was found that any gas leakage did not occur on the interface of the specimen and the mould.

Breakthrough, which includes the onset of gas penetration into the bentonite/sand mixtures, and the preferential development of gas paths across the specimens, results from microscopic displacement of liquid by gas in the voids. The breakthrough should be an unstable phenomenon. Since tests on each sample size involved only one or two specimens, some effects of instability in the data could not be avoided. However, the experimental data confirmed that gas migrated more swiftly in the larger specimens, that is, threshold pressure showed a downward trend with diameter and effective gas porosity tended to decrease with both height and diameter. Bentonite/sand mixtures in a full-scale facility are planned to be 2 m in thickness and more than 30 m in width, which means that the diameter should be more than 10 times greater than the height. It is predicted that gas migration behaviour would show a trend of far swifter breakthrough in the full scale situation than in the laboratory specimens.

CONCLUSION

In order to estimate scale effects on gas migration through bentonite/sand mixtures, experiments are performed with specimens of different height and diameter. Results are as follows;
(1) It is not found that scale affects hydraulic conductivity.
(2) Threshold pressure shows no correlation with height, while it displays a gradual trend of a decrease with diameter
(3) Effective gas porosity shows a downward trend with both height and diameter .
(4) It is not found that gas conductivity is affected by scale.

It takes long periods of time to perform a series of gas migration experiments with bentonite/sand mixtures, as such experiments involve saturation of the specimen, measurement of hydraulic conductivity, breakthrough by gas application and measurement of gas conductivity. It is realistic to conduct numbers such experiments with smaller size specimens in order to estimate an unstable phenomenon such as gas breakthrough. In consideration of the above effects of scale, it is conservative to estimate gas migration behaviour of bentonite/sand mixtures in the full scale facility on the basis of properties fixed from experimental data for specimens of several 10 mm in size. It is concluded that gas migration behaviour would show a trend of swifter breakthrough in the full scale situation than in the laboratory specimens.

REFERENCES

1. Hideo Shimoda et al., J. Atomic. Energy Soc. Jpn., Vol.33, No.11 (1991)
2. S.Voinis et al., in Gas Generation and Release - Proc. of a Workshop organised by NEA in co-operation with ADNDRA , pp111-119, France(1992)
3. Pusch et al., SKB technical report 90-43, SKB(1990)
4. Toshiyuki Hokari et al., J. Radioactive Waste Research, Vol.3, No.2 (1996), printing.
5. Masakazu Itoh et al., J. Atomic Energy Soc. Jpn., Vol.36, No.11(1994)
6. Test method for soil, compaction using a rammer - JIS A 1210, published by Japanese Standards Association, Tokyo, 1990
7. Handbook of Petroleum Industry, edited by Soc. Jpn. Petroleum Engineers (Soc. Jpn. Petroleum Engineers, Tokyo, 1983), 504p,
8. Toshiyuki Hokari et al. in Scientific Basis for Nuclear Waste Management XX, edited by W.J. Gray and I.R.Triay (Mater. Res. Soc. Proc. , Pittsburgh, PA, 1996), printing

Part XIV

Performance Assessment

NEAR-FIELD THERMAL–HYDROLOGICAL BEHAVIOR FOR ALTERNATIVE REPOSITORY DESIGNS AT YUCCA MOUNTAIN

THOMAS A. BUSCHECK,* JOHN J. NITAO,* AND LAWRENCE D. RAMSPOTT**
*Geological Sciences and Environmental Technologies Division, LLNL, **TRW (all at Lawrence Livermore National Laboratory, L-206, P.O. Box 808, Livermore, CA 94551)

ABSTRACT

Three-dimensional calculations that explicitly represent a realistic mixture of waste packages (WPs) are used to analyze decay-heat-driven thermal–hydrological behavior around emplacement drifts in a potential high-level waste facility at Yucca Mountain. Calculations, using the NUFT code, compare two fundamentally different ways that WPs can be arranged in the repository, with a focus on temperature, relative humidity, and liquid-phase flux on WPs. These quantities strongly affect WP integrity and the mobilization and release of radionuclides from WPs. Point-load spacing, which places the WPs roughly equidistant from each other, thermally isolates WPs from each other, causing large variability in temperature, relative humidity, and liquid-phase flux along the drifts. Line-load spacing, which places WPs nearly end to end in widely spaced drifts, results in more locally intensive and uniform heating along the drifts, causing hotter, drier, and more uniform conditions. A larger and more persistent reduction in relative humidity on WPs occurs if the drifts are backfilled with a low-thermal-conductivity granular material with hydrologic properties that minimize moisture wicking.

INTRODUCTION

The U.S. Department of Energy is investigating the feasibility of disposing of radioactive wastes, including spent nuclear fuel (SNF) from electrical utilities and wastes stored at federal facilities, in the unsaturated zone (UZ) at Yucca Mountain, Nevada. To be feasible, the waste isolation system must limit the release and transport of radionuclides to the accessible environment. A key concern is how water contacts a WP, which affects WP integrity and the mobilization and release of radionuclides from WPs. The ambient relative humidity in the UZ is high ($RH \approx 99\%$) and is therefore corrosive for most candidate WP materials. If RH were reduced enough, WP corrosion rates would be minimal [1]. Two primary modes by which water may contact a WP are (1) condensation of water vapor that forms a liquid film on the WP and (2) liquid-phase flow. The critical factor for the first mode is RH on the WP surface. For the second mode, liquid-phase contact may arise in three ways:

Drift seepage: Advective liquid-phase flow of water that enters the drift (and flows through the backfill if present) as a result of ambient percolation or decay-heat-driven condensate flow. This can include episodic nonequilibrium fracture flow or a steady seep.

Wicking: Diffusive moisture transport driven by matric potential gradients (resulting from capillary and surface forces, and osmotic effects). Because this process occurs both in the gas phase (as vapor diffusion) and the liquid phase (as imbibition), it does not require a continuous liquid phase.

Cold-trap effect: Axial vapor flow and condensation within the drift is driven by axial variations in temperature T and partial pressure of water vapor P_v along the drift. Water vapor is transported from areas of higher to lower T and P_v where it condenses, causing RH to locally increase (up to 100%). Large condensation rates arise in cooler areas if the following conditions are met: (1) RH is high in the adjacent rock, (2) WP heat output varies substantially from WP to WP, and (3) WPs are thermally isolated from one another.

Much of the debate over various thermal loading designs has focused on the areal mass loading (AML, expressed in metric tons of uranium per acre, MTU/acre). Of the many competing factors (e.g., operational constraints, costs, material responses) to consider in evaluating alternative thermal designs, perhaps most important is the role that decay heat plays in determining the thermal–hydrological (T–H) conditions in and around the emplacement drifts and, in particular, RH and liquid-phase flux on WPs. Decay heat influences the distribution of liquid saturation and liquid-phase flux in the near field and altered zone, affecting radionuclide transport. Decay heat also drives coupled thermal–hydrological–geomechanical–geochemical (T–H–M–C) processes that permanently alter flow and transport properties. These effects may continue to affect radionuclide transport well after decay heat has stopped mobilizing gas- and liquid-phase flow.

Waste packages can be arranged in the repository at Yucca Mountain in two fundamentally different ways. The first approach (the point-load design) attempts to evenly distribute the WP decay heat over the repository area by placing the WPs with roughly the same axial and lateral spacing between WPs, as is done in the advanced conceptual design (ACD) [2]. The second approach (the line-load design) lineally concentrates the WP decay heat by placing the WPs nearly end to end along the drifts. For a given AML, this maximizes the spacing between drifts and minimizes the total length of drifts. We analyzed the reference AML of 83.4 MTU/acre [2] with and without engineered backfill emplaced at the time of repository closure (100 yr); however, other AMLs have also been considered [3]. Decay-heat-driven T–H behavior is markedly different for the point and line-load designs.

Three-dimensional calculations are required to adequately represent drift-scale T–H behavior. Models based on the NUFT code [4] were developed to carry out such calculations [3]. A major factor to be considered in the selection of a repository design is that WP heat generation varies substantially in the WP inventory. Whereas defense high-level waste (DHLW) generates a negligible amount of heat, spent nuclear fuel (SNF) initially can generate as much as 18 kW per WP. The influence of WP heat output variability on T, RH, and liquid-phase flux experienced by WPs is markedly different for the point- and line-load designs. A key motivation for the point-load design is to limit peak drift-wall and WP temperatures regardless of the WP sequence along the drifts. A certain degree of thermal management of the waste stream (i.e., controlling WP sequencing) may be required for line-load spacing to sufficiently limit peak temperatures.

NUMERICAL MODELS AND ASSUMPTIONS

Model calculations were conducted with the NUFT code [4], which was developed at Lawrence Livermore National Laboratory to simulate the coupled transport of water, vapor, air, and heat in fractured porous media. We included all major hydrostratigraphic units in the UZ [5] and assumed they are horizontal and of uniform thickness. The initial and boundary conditions are the same as those used in past studies [3,6]. We assumed a bulk permeability k_b of 280 millidarcy, and an initial vertical liquid saturation profile based on a percolation flux of 0.3 mm/yr. However, a wide range of other conditions (including percolation fluxes ranging from 0 to 5 mm/yr) have also been considered [3,6]. The atmospheric RH is assumed to be 100%, so the model allows no loss of moisture by vapor diffusion to the atmosphere. Because actual (desert) RH is much less than 100%, the model underrepresents this loss.

The models include six major WP types (Fig. 1), resulting in a WP inventory that is representative of that assumed for the ACD [2]. The four SNF WP types are: (1) 40-yr-old 12-PWR WPs, (2) 26-yr-old 44-BWR WPs, (3) 26-yr-old 21-PWR WPs, and (4) 10-yr-old 21-PWR WPs. The two types of DHLW are Hanford site and Savannah River site DHLW WPs. The ACD assumes that 27% of the WPs contain DHLW, which is similar to the assumption made in this study that 30% of the WPs contain DHLW. The WP inventory described in Ref. 2 has an average heat output at the time of emplacement of 7.18 and 1.5 kW for the SNF and DHLW WPs, respectively. In our models, the SNF and DHLW averages are 7.84 and 2.4 kW. The SNF and DHLW WPs are assumed to have lengths of 5.68 and 3.68 m, respectively; all WPs are assumed to have a diameter of 1.8 m. The drift diameter is assumed to be 5.5 m, which is similar to the ACD assumption of 5 m [2]. We also considered a 5-m-diameter drift and found little difference in T–H behavior between 5- and 5.5-m-diameter drifts. For thermal radiation in the drift, a WP emissivity of 0.8 is assumed; a WP emissivity of 0.3 has also been considered [3]. For the backfill cases, a granular backfill with hydrological properties similar to sand was assumed; crushed rock from the welded Topopah Spring tuff (TSw2) unit, which is where the potential repository horizon occurs, has also been considered [3]. Backfill thermal conductivities K_{th} of 0.3 and 0.6 W/m°C were considered. The invert is assumed to consist of a granular fill with the same properties as the backfill. For the line-load backfill cases, it is assumed that measures are taken to prevent backfill from filling the small gap separating WPs. For the ACD backfill cases, backfill fills the space separating WPs.

The models are applicable to areas of the repository far enough away from the edge to remain unaffected by edge-cooling effects. This allows the use of periodic no-mass/heat-flow boundaries that correspond to (1) the drift centerline, (2) the pillar centerline, (3) the plane transverse to the middle of the Hanford site DHLW WP, and (4) the plane transverse to the middle of the 10-yr-old 21-PWR WP. The ACD assumes a drift spacing of 22.5 m and places WPs on the basis of a constant lineal mass loading (LML) of 0.46 MTU/m along the drift (Fig. 1). For

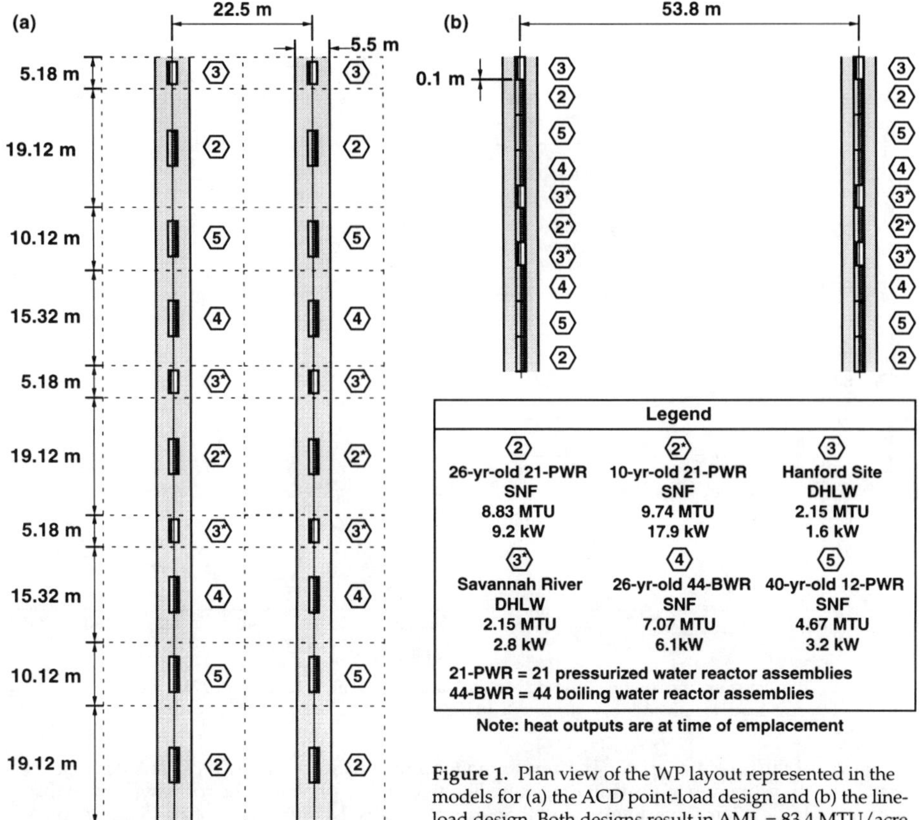

Figure 1. Plan view of the WP layout represented in the models for (a) the ACD point-load design and (b) the line-load design. Both designs result in AML = 83.4 MTU/acre.

the line-load design, we assumed a 0.1-m gap between WPs, an LML of 1.11 MTU/m and a drift spacing of 53.8 m (Fig. 1). The respective designs have linear power loadings (LPL) of 0.5 and 1.2 kW/m. Assuming the WP inventory described in Ref. 2, a 5.335-m length for all SNF WPs (which is a more recent assumption), and a 0.1-m gap between WPs results in an LPL of 1.13 kW/m, similar to the value used in this study.

DISCUSSION OF MODEL RESULTS

A key question is whether liquid-phase flux, arising from ambient percolation flux and decay-heat-mobilized condensate flux, may enter emplacement drifts or, at the very least, prevent dryout (and *RH* reduction) from occurring in the adjacent rock. For an 83.4-MTU/acre repository, the overall magnitude of the condensate flux depends on AML, not on LML [3]; consequently, the point-load and line-load 83.4-MTU/acre designs mobilize the same overall magnitude of condensate flux. Moreover, liquid-phase flux in the vicinity of an 83.4-MTU/acre repository is dominated by condensate flux for 2000–5000 yr [3]. Superheated conditions and a reduction in *RH* in the repository rock will occur if the local heat flux q_H is enough to evaporate the local incoming liquid-phase mass flux q_{liq} as expressed by the following relation:

$$q_H > q_{liq}\rho_{liq}h_{fg}, \tag{1}$$

where ρ_{liq} is the mass density of water and h_{fg} is the specific latent heat of vaporization. If q_{liq} is too large, q_H is insufficient to generate superheated conditions. Near the drift, q_H is proportional to LML; further out in the rock, it is proportional to AML after the thermal fields have coalesced.

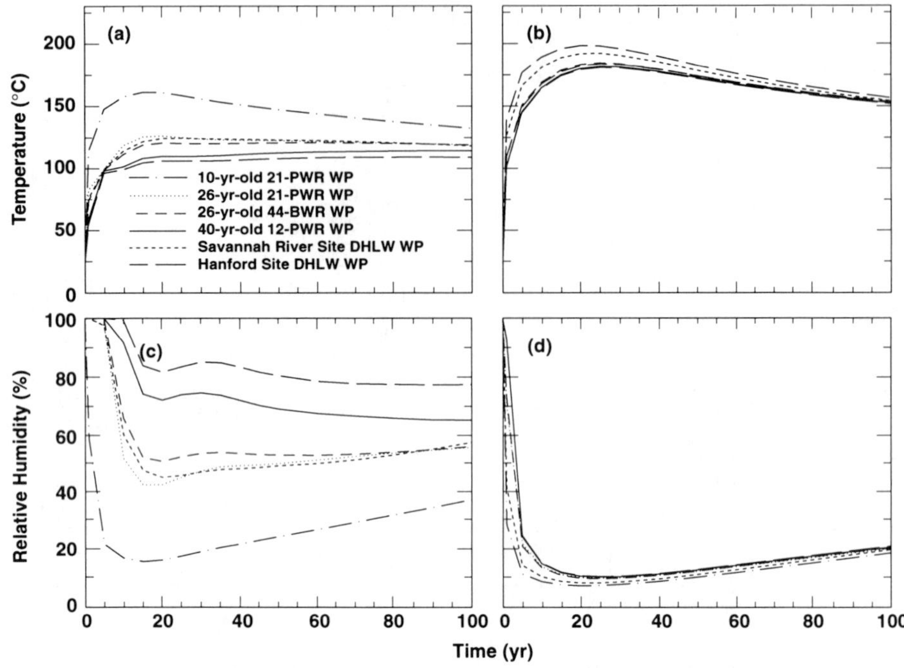

Figure 2. Temperature and relative humidity in rock at upper drift wall at various WP locations for (a,c) the ACD point-load design and (b,d) the line-load design.

Spatial variability in either q_H or q_{liq} can result in local regions where q_{liq} will prevail. Because q_H increases with proximity to the drift, it is more likely for q_{liq} to prevail in the rock away from the drifts (e.g., at the pillar centerline), rather than near the drifts. Near the drifts, q_H will be 2.4 times greater (on average) for the line-load design than for the point-load design. Moreover, the q_H at the pillar centerline will be less for the line-load design than for the point-load design, increasing the likelihood that condensate will shed through the pillar of the line-load design. The point-load design results in spatial variability in both q_H and q_{liq} along the drift. Cooler intervals of the point-load drifts are more likely to be overwhelmed by q_{liq} as a result of condensate being displaced from hotter to cooler regions and because q_H is less there; the net result is higher RH and heat pipes above the cooler intervals of the point-load drift, which increase the local cooling rate and the likelihood of seepage into the drift.

The marked difference in drift-scale T–H behavior between the point-load and line-load designs for no backfill is evident in Fig. 2. The axial WP spacing in the point-load design is large enough to thermally isolate the WPs from each other; peak drift-wall temperatures range from 107 to 158°C, and peak WP temperatures range from 112 to 186°C (Table I). The close axial WP spacing in the line-load design results in efficient WP-to-WP thermal-radiative heat transfer, so that, in spite of their different heating histories, all WPs experience similar (and more beneficial) T and RH conditions. For the no-backfill cases, the coolest and most humid WP in the line-load design always has a lower RH than the hottest and least humid WP in the ACD point-load design (Fig. 3 and Table I). The line-load design results in: (1) more locally intensive (and uniform) rock dryout around the drifts, (2) more effective condensate shedding between the drifts, and (3) less condensate buildup above the drifts.

The last two rows of Table I list the time required to rewet to $RH = 65$ and 90%. These RH thresholds are important because corrosion studies of the candidate WP materials indicate that the critical RH for significant atmospheric corrosion is 65% if a hygroscopic salt is present on the WP, or

No backfill		
Repository design	ACD point load	Line load
T_{peak} ($t < 100$ yr)	112–186 °C	184–214 °C
RH ($t = 100$ yr)	26–71%	14–19%
RH ($t = 2000$ yr)	85–94%	71–74%
RH ($t = 10,000$ yr)	93–99.9%	94–98%
$t(RH = 65\%)$	25–660 yr	1540–1710 yr
$t(RH = 90\%)$	1540–2390 yr	4260–4790 yr

Table I. Temperature and relative humidity RH ranges on WPs for AML = 83.4 MTU/acre. The last two rows give the time to rewet the RH thresholds for the most humid and least humid WP, respectively.

Backfill (K_{th} = 0.6 W/m °C)			Backfill (K_{th} = 0.3 W/m °C)		
Repository design	ACD point load	Line load	ACD point load	Line load
T_{peak} ($t > 100$ yr)	119–344 °C	245–268 °C	132–537 °C	344–362 °C
RH ($t = 120$ yr)	1–55%	2–3%	0.5–38%	1–1%
RH ($t = 2000$ yr)	43–96%	48–52%	24–95%	34–38%
RH ($t = 10,000$ yr)	62–99.7%	72–80%	41–99.6%	57–64%
$t(RH = 65\%)$	170–12,250 yr	3550–4770 yr	200–27,820 yr	10,600–15,960 yr
$t(RH = 90\%)$	1340–58,960 yr	26,200–41,030 yr	1440–104,880 yr	53,330–71,610 yr

90% if the WP surface is free of salt [7]. Whether liquid water can enter the drift and reach (and evaporate on) a WP, leaving a salt buildup, is of critical importance to WP integrity. For the no-backfill cases, the line-load design significantly extends the time required to reach these RH thresholds (relative to the point-load design). When a low-K_{th} backfill (that limits wicking) is used, the time to reach these RH thresholds is further increased for all WPs in the line-load design, while for the point-load design, it is only increased for the SNF WPs (Fig. 4); consequently, DHLW WPs do not experience the benefit of RH reduction in the point-load design. Low-K_{th} (e.g., 0.6 W/m°C) backfill accentuates thermal isolation in the point-load design, causing large variability in peak WP temperatures (119–344°C), while for the line-load design, peak WP temperatures only range from 245 to 268°C. For the backfill cases, the line-load design substantially decreases the maximum peak WP temperatures (relative to the point-load design).

The additional RH reduction for the backfill cases arises from the large temperature difference (ΔT_{drift}) between the WP and drift wall [6]. This effect (called the "drift-ΔRH effect") occurs in addition to RH reduction resulting from rock dryout. Assuming uniform P_v in the drift, RH on the WP is given by

$$RH_{wp} = RH_{dw} \frac{P_{sat}(T_{dw})}{P_{sat}(T_{wp})}, \tag{2}$$

where RH_{dw} is RH at the drift wall, T_{dw} and T_{wp} are the drift-wall and WP temperatures. For decreasing backfill K_{th}, ΔT_{drift} increases, resulting in a larger and more persistent reduction in RH. The influence of thermal design on WP integrity is pronounced. For example, the time to reach RH = 65% is only 25–660 yr for the ACD no-backfill case, while it is 10,600–15,960 yr for the line-load backfill case with K_{th} = 0.3 W/m°C (Table I). The time to reach RH = 90% is only 1540–2390 yr for the ACD no-backfill case, while it is 53,330–71,610 yr for the line-load backfill case.

The possibility of hydrothermal alteration of the vitric nonwelded Paintbrush tuff (PTn) unit that overlies the repository is an important issue. If it occurs, it could significantly reduce the ability of the PTn to imbibe (and thereby attenuate) fast fracture flow. These changes may also

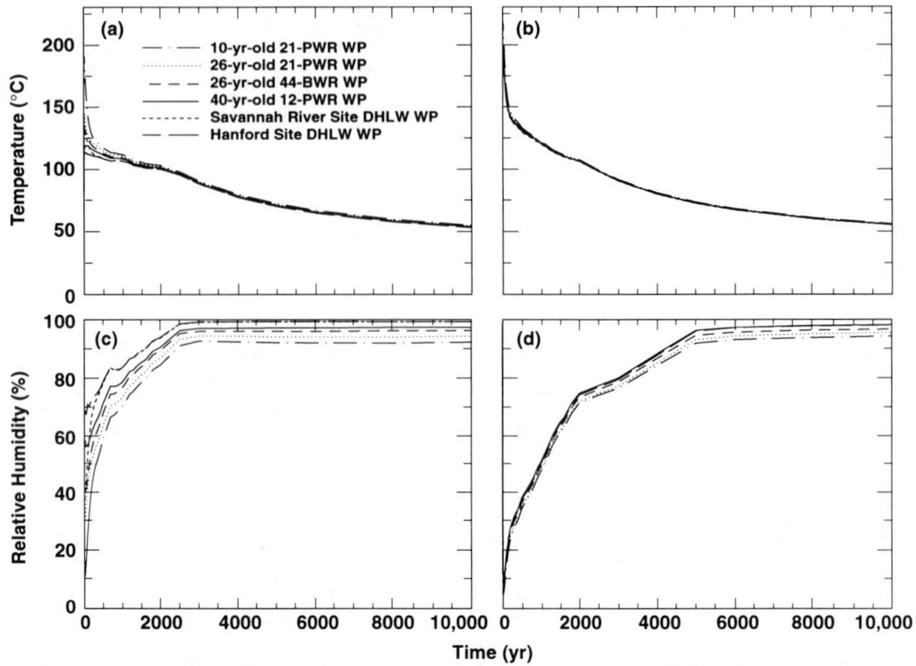

Figure 3. Temperature and relative humidity on upper WP surface for no backfill. Curves are plotted for (a,c) the ACD point-load design and (b,d) the line-load design.

increase the percolation flux at the repository. The 83.4-MTU/acre point-load and line-load designs drive markedly different changes in T and liquid saturation S_{liq} in the PTn, which are the result of how effectively condensate sheds between the drifts. More efficient condensate shedding between the line-load drifts results in less condensate buildup above the repository than the point-load design. The vertical extent of the upper heat-pipe zone depends on the amount of condensate available (above the repository) to reflux. For the point-load design, the maximum vertical extent of the upper heat-pipe zone is 100 m closer to the ground surface than in the line-load design (Fig. 5a). Figure 5b shows how the larger condensate buildup above the ACD repository inundates the matrix (raising S_{liq} to 99%) in the PTn and causes the peak temperature rise in the PTn to be 30°C higher than in the line-load design. The condensate buildup above the ACD repository causes S_{liq} in the PTn to be greater than ambient for thousands of years. The S_{liq} increase, together with the temperature rise, may be enough to cause hydrothermal alteration that reduces the ability of the PTn to imbibe fast fracture flow. The upper extent of the condensate zone overlying the line-load repository remains well below the PTn; consequently, S_{liq} in the PTn never exceeds ambient. Greater condensate buildup above the ACD point-load repository also causes the peak temperature rise near the ground surface to be almost twice that for the line-load design.

CONCLUSIONS

A three-dimensional model representing a realistic mixture of WPs was used to compare T–H behavior for two different approaches to arranging the WPs in a potential repository at Yucca Mountain. The point-load approach used in the ACD places WPs equidistant from each other to spread the WP decay heat, thereby limiting peak drift-wall and WP temperatures. The line-load approach lineally concentrates the WP decay heat by placing WPs nearly end to end in widely spaced drifts. A key consideration in evaluating alternative WP layouts is the markedly

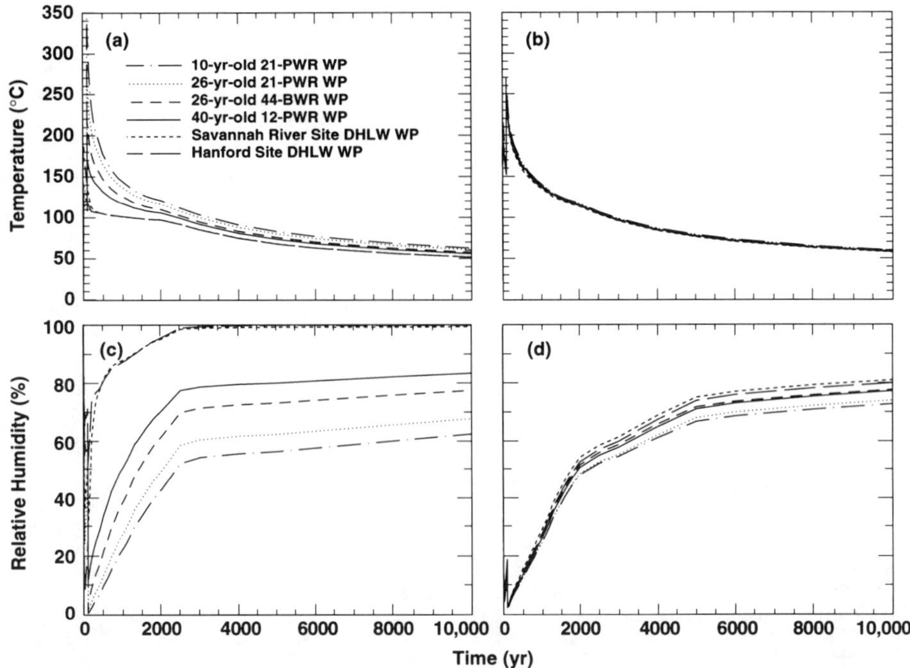

Figure 4. Temperature and relative humidity on upper WP surface for backfill emplaced at 100 yr with K_{th} = 0.6 W/m°C. Curves are plotted for (a,c) the ACD point-load design and (b,d) the line-load design.

different T–H behavior that occurs in and around drifts, including: (1) how condensate returns to the repository, (2) how much heat is available near the drifts to vaporize incoming liquid water, and (3) the spatial variability in heat flux and condensate flux along the drifts. The line-load design causes the pillar to be a more preferential pathway for condensate drainage than the point-load design. Moreover, the line-load design has 2.4 times the local heat flux near the drifts (on average) to vaporize incoming liquid water than the point-load design. Consequently, it is more difficult for condensate to drain into line-load drifts than point-load drifts; the coolest and most humid WP in the line-load design is always less humid than the hottest and least humid WP in the point-load design. The point-load design thermally isolates WPs from each other, causing large variability in T, RH, and condensate flux along the drifts. The close axial WP spacing in the line-load design allows efficient WP-to-WP heat transfer that homogenizes the heat flux distribution along the drift. When a low-K_{th} backfill is used, this heat-flux homogenization greatly limits peak WP temperatures compared to the point-load design. A larger and more persistent reduction in RH on WPs occurs when the drifts are backfilled with a low-K_{th} granular material that limits wicking.

The line-load design provides the following advantages (compared to the ACD): (1) reduction (up to 60%) in the required length (and number) of emplacement drifts with a corresponding cost reduction, (2) large reduction in backfill volume, (3) narrower range of T and RH for which natural and engineered materials must be tested, (4) less spatially variable drift-scale T–H behavior to be accounted for in per-formance analyses, (5) all WPs (including DHLW) experience the benefit of RH reduction, (6) decreased probability of condensate flow entering the drifts, (7) decreased tendency for decay-heat-driven temperature and liquid saturation increase (and any resulting hydrothermal alteration) in the PTn unit, and (8) a reduction in peak temperature rise near the ground surface. The cost associated with any WP sequencing that may be necessary to implement the line-load design is worthwhile given the important benefits of this approach.

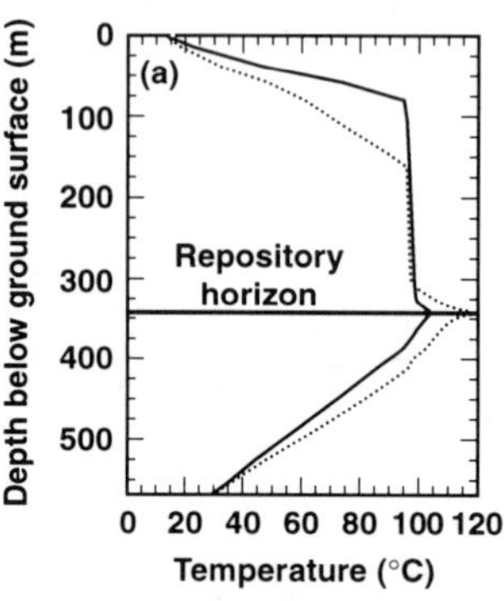
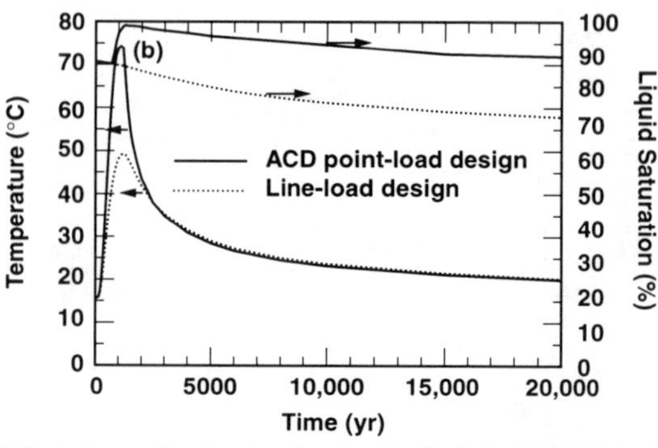

Figure 5. Vertical temperature profile (a) at 1000 yr along a line intersecting the Hanford site DHLW WP for the ACD point-load design and the line-load design. Also plotted (b) are liquid saturation and temperature histories in the PTn unit 295 m above the repository horizon.

ACKNOWLEDGMENTS

We acknowledge the review of Jim Blink, the editorial assistance of Robert Kirvel, and the graphical support of Rick Wooten and Dan Fletcher. This work was performed under the auspices of the U.S. Department of Energy by the Lawrence Livermore National Laboratory under contract W-7405-ENG-48 and was supported specifically by the Yucca Mountain Site Characterization Project at LLNL.

REFERENCES

1.   Stahl, D., J.K. McCoy, and R.D. McCright, "Impact of Thermal Loading on Waste Package Material Performance," *Proceedings Material Research Society XVIII Symposium on the Scientific Basis for Nuclear Waste Management*, Material Research Society, Pittsburgh, PA, Oct. 23–27 (1994).

2.   CRWMS Management and Operating Contractor, "Mined Geologic Disposal System Advanced Conceptual Design Report," Vol. 2, B00000000-01717-5705-00027, Rev. 00 (1996).

3.   Buscheck, T.A. in Wilder, D.G. (editor), *Near-Field and Altered-Zone Environment Report*, Volume II, Chapter 1: "Hydrothermal Modeling," UCRL-JC-124998, Lawrence Livermore National Laboratory, Livermore, CA (1996).

4.   Nitao, J.J., "The NUFT Code for Modeling Nonisothermal, Multiphase, Multicomponent Flow and Transport in Porous Media," *EOS*, American Geophysical Union, Vol. 74, no. 3, pg. 3 (1992).

5.   Peters, R.R., E.A. Klavetter, I.J. Hall, S.C. Blair, P.R. Hellers, and G.W. Gee, *Fracture and Matrix Hydrologic Characteristics of Tuffaceous Materials from Yucca Mountain, Nye County, Nevada*, SAND84-1471, Sandia National Laboratories, Albuquerque, NM (1984).

6.   Buscheck, T.A., J.J. Nitao, and L.D. Ramspott "Localized Dryout: An Approach for Managing the Thermal–Hydrological Effects of Decay Heat at Yucca Mountain," *Proceedings Materials Research Society XIX International Symposium on the Scientific Basis for Nuclear Waste Management*, Materials Research Society, Pittsburgh, PA, Nov. 27–Dec. 1, 1995. Also UCRL-JC-121232, Lawrence Livermore National Laboratory, Livermore, CA (1995).

7.   Jones, D.A., *Principles and Prevention of Corrosion* (Macmillan Publishing Company, New York, (1992).

TRANSPORT FROM THE CANISTER TO THE BIOSPHERE: USING AN INTEGRATED NEAR-AND FAR-FIELD MODEL

B. GYLLING, L. ROMERO, L. MORENO and I. NERETNIEKS
Department of Chemical Engineering and Technology, Royal Institute of Technology, Stockholm, Sweden

ABSTRACT

A coupled model concept which may be used for performance assessment of a nuclear repository is presented. The tool is developed by integration of two models, one near field and one far field model. A compartment model, NUCTRAN, is used to calculate the near field release from a damaged canister. The far field transport through fractured rock is simulated by using CHAN3D, based on a three-dimensional stochastic channel network concept. The near field release depends on the local hydraulic properties of the far field. The transport in the far field in turn depends on where the damaged canister(s) is located. The very large heterogeneities in the rock mass makes it necessary to study both the near field release properties and the location of release at the same time. In order to demonstrate the capabilities of the coupled model concept it is applied on a hypothetical repository located at the Hard Rock Laboratory in Äspö, Sweden. Two main items were studied; the location of a damaged canister in relation to fracture zones and the barrier function of the host rock. In the study of the near field rock as a transport barrier the effect of different tunnel excavation methods which may influence the damage level of the rock around the tunnel was addressed.

INTRODUCTION

The Swedish repository for spent nuclear fuel is planned to be located at large depth in crystalline rock. In the repository, radionuclides escaping from a damaged canister will spread into the backfill material surrounding the canister and then migrate through different pathways into the mobile water in the rock fractures. The transport through the near field, occurring mainly by diffusion in the bentonite surrounding the canister and the sand-bentonite mixture in the tunnel, is modelled by using NUCTRAN [1,2]. The far field transport of the escaping radionuclides is then simulated by using CHAN3D based on a Channel Network model [3]. In addition to advection in the channels, the model can account for diffusion into the rock matrix and sorption within the matrix, which are the by far most important mechanisms retarding the nuclides.

The aim of this paper is to describe the coupling of these two models and how they may be used to calculate the release from a repository by an integrated procedure. The coupling of the two models makes it possible to simulate release and transport over the entire distance from a damage in a canister, into and through the backfill by various simultaneous paths, through the geosphere to the biosphere and still consider the detailed scale related to the near field, and also model the most important mechanisms in the far field. The concept may provide a useful tool in the performance assessment of a nuclear waste repository. To demonstrate the concept the integrated models are applied on a hypothetical repository layout [4]. Among the addressed items, the influence of the position of the initially damaged canisters and the importance of the near field rock as a barrier for release of escaping nuclides were studied.

THE NEAR AND FAR FIELD MODELS

The Near Field model

In the repository the canisters will be deposited in vertical boreholes below the floor of the tunnels. The deposition holes are backfilled with compacted bentonite and the tunnels with a mixture of sand and bentonite. Some canisters may be initially defective or later damaged by

Mat. Res. Soc. Symp. Proc. Vol. 465 © 1997 Materials Research Society

mechanical or corrosion processes. Nuclides will then dissolve in the intruding water and leak out of the canisters. The clay and the rock matrix have very low hydraulic conductivities. Since there is no water flow the nuclide transport through them is only by diffusion. Advection is restricted to the fractures in the rock surrounding the repository. Nuclides escaping from the canisters may diffuse into fractures intersecting the deposition holes and/or diffuse into the altered rock in the disturbed zone around the deposition tunnels. Figure 1 shows four possible pathways: path Q_1 directly to the water in a fracture intersecting the repository hole, path Q_2 up to the disturbed zone around the tunnel, path Q_3 into the tunnel backfill and further to a fracture (zone) intersecting the tunnel and path Q_4 through the rock to a nearby fracture or fracture zone. The release of nuclides from the canisters is calculated by using a compartment model.

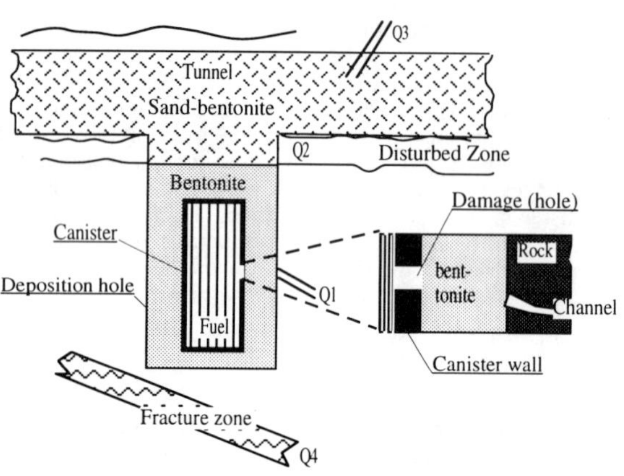

Figure 1.   Schematic view of the used repository design, showing the small hole in the canister and the location of the various escape routes.

Radionuclides leaking from a damaged canister diffuse into the backfill material surrounding the canister and then migrate through different pathways into water-bearing fractures in the rock surrounding the repository. In the model, the processes governing the nuclide transport in the near field repository are dissolution, diffusion, advection, and sorption.

For the purpose of numerical simulations the model is formulated in terms of integrated finite differences [5], where the concept of "compartments" is introduced to define the discretization of the system. Average properties are associated over these compartments with representative nodal points within the compartment. From the theoretical point of view the compartments may have any shape, but are of the same material. The compartments are defined by their diffusion length(s), contact area(s) to other compartments and their volume. These dimensions together with the transport properties such as diffusion and sorption will define the capacity and the diffusion resistance(s) characterizing the compartments. The compartments can be arranged in a three-dimensional structure. For the fuel dissolution two approaches may be considered: a) the concentration of a nuclide in the canister is controlled by its own solubility and b) congruent dissolution where the release rate of the nuclides is determined by the dissolution rate of the spent fuel matrix.

The model uses a rather straightforward discretization process into compartments considering the geometry of the system and the materials. To avoid a fine discretization at regions with very narrow pathways, analytical or semi-analytical solutions are introduced in these zones. This kind of approach is used, for example, when nuclides are released through the small hole at the wall of the damaged canister into the bentonite. This concept is also used when the nuclides migrate from the bentonite surrounding the canister into a fracture intersecting the deposition hole. Finally, at the boundaries of the repository, the diffusive transport into the flowing water in the fractures is accounted for by using the notion of the equivalent flow rate which is a fictitious flow rate of water that carries with it a concentration equal to that at the contact interface. The magnitude of the equivalent flow rate is assessed using the boundary layer theory [1]. The data needed for NUCTRAN are the canister inventory, nuclide half-lives, diffusion coefficients for nuclides in bentonite, sand-bentonite mixtures and rock, and properties for the materials involved. The output from NUCTRAN is the transient release through the different paths. Details about the model formulation and the analytical or semi-analytical expressions used may be found in [1] and [2].

<u>The Far Field model</u>

In the far field the nuclides escaping from the near field are transported by the flowing water through fractures in the rock mass. In our concept the water flows through channels mainly located in fractures and fracture zones. The fractures intersect in the rock and they form a three-dimensional network of interconnected channels. Solute transport through the far field is calculated by using the code CHAN3D based on a channel network model [3]. The model may accommodate the transport of interacting and hypothetically non-interacting solutes. For interacting solutes diffusion into the rock matrix and sorption on the interior of the matrix are considered. These are very important processes when assessing retardation of radionuclides that may escape from repositories for nuclear waste.

Fluid flow and solute transport through the far field are calculated by using the code CHAN3D based on the Channel Network model [3]. For visualization purposes, the network is depicted as a rectangular grid. The channel length is not explicitly used in the model, it is implicitly included in the conductance and volume of the channels and the flow-wetted surface. For these reasons, the exact form of the grid is not significant. To generate the grid a hydraulic conductance is assigned to each channel of the network. This is the only entity needed to calculate the flow, if the pressure field is known. The conductance is defined as the ratio between the flow in a channel and the pressure difference between its ends. If the residence time for noninteracting solutes is to be calculated, then the volume of the channels is needed as well. If sorption onto the fracture surface or diffusion into the matrix will be included in the model, the surface area of the flow-wetted surface must also be included. Then some properties of the rock are needed, such as rock matrix porosity, diffusivity, and sorption capacity for sorbing species.

The conductances of the channels are assumed to be log-normally distributed and not correlated in space. For tracer transport calculations, owing to lack of data, it is assumed that the flow in the channel may be calculated by using the cubic law, this means that the volume of the channel is proportional to the third root of the conductance. This is important for noninteracting species but not for those that sorb and/or diffuse in the rock matrix. The residence times of the latter are determined by matrix diffusion and sorption effects.

The dominating water flow paths are determined by the major fracture zones. These fracture zones are responsible for most of the water flow. In the model, we have assumed that the fracture zones may be described by two parallel planes delimiting the zone with high permeability. Channels present between these two limiting planes have higher conductances than channels within the rock mass. Conductances are obtained from a distribution with a mean value that matches the experimentally estimated transmissivity of the fracture zone.

The solute transport is simulated by using a particle-following technique [6,7]. Many particles are introduced, one at a time, into the known flow field at one or more locations. Particles arriving at an intersection are distributed in the outlet channels with a probability proportional to their flow rates. This is equivalent to assuming total mixing at the intersections. Each individual particle is followed through the network. The residence time for noninteracting tracers in a given channel is determined by the flow through the channel and by its volume. For sorbing solutes matrix diffusion and sorption are included in addition to the advection. The residence time of an individual particle along the whole path is the sum of residence times in every channel that the particle has traversed. The residence time distribution is then obtained from the residence times of a multitude of individual particle runs. Hydrodynamic dispersion in individual channels is considered to be negligible. The dispersion in the system is caused by the uneven flow distribution due to the heterogeneity of the rock and the interaction mechanisms.

THE INTEGRATION OF THE MODELS

To calculate the release from the near field, the number of canisters initially damaged and their locations are determined by a random process. The data needed for NUCTRAN are the canister design and information about the nuclide properties. Data for CHAN3D may be obtained from hydraulic tests and field observations [8]. For transport calculations with CHAN3D information about the interactions between the rock and the nuclides are also needed. Sorption and matrix diffusion data may be obtained from laboratory measurements. The magnitude of the flow-wetted surface and hydraulic conductivity distribution are obtained from borehole measurements [8]. The hydraulic model is solved using the corresponding boundary conditions. The water flow rates obtained in these calculations are then used to determine the advective transport through the channel network. Once the water flow rate is determined by using the flow part of the model (CHAN3D-flow), the release from the near field may be addressed. Flow rates in the disturbed zone and in the fracture/fractures which intersect the deposition hole are needed to calculate the release from the near field. Once the release from the near field has been determined, the transport part of the model (CHAN3D-transport) is used to calculate the transport of radionuclides through the far field. The model integration is shown schematically in Figure 2.

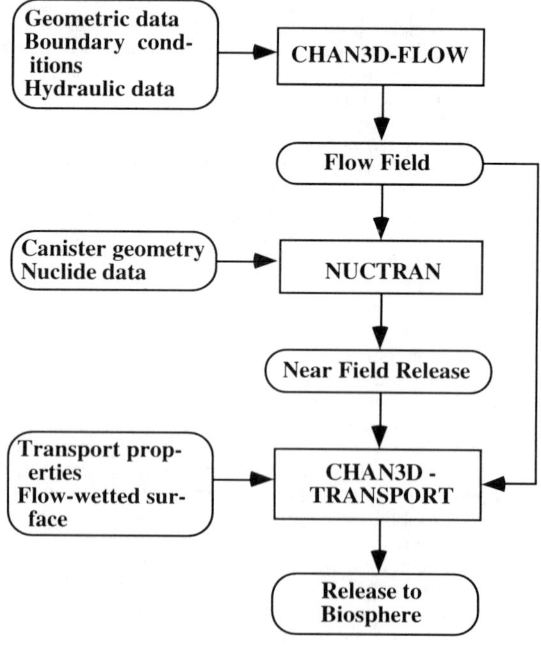

Figure 2.   Flow-sheet over the model integration.

APPLICATION OF THE INTEGRATED CONCEPT

As the location of the repository for nuclear waste is not yet decided, there is no detailed repository layout and for this reason a hypothetical repository located at the Äspö HRL site [4] is used in this scenario study. In principle, the model repository consists of a few central tunnels and from these tunnels several horizontal deposition tunnels are excavated, where the deposition holes will be drilled. Due to the fracture zone geometry the hypothetical repository is made up of three sections. We have chosen to study one of these sections.

<u>The Scenario</u>

The dominating water flow paths are determined by the major fracture zones which then are responsible for most of the water flow. The most important conductive fracture zones at Äspö HRL that have been used and the repository location are shown in Figure 3. In addition to the fracture zones the access and deposition tunnels may also act as water conducting features. The boundary conditions are taken from a regional study [9] calculated by using the code NAMMU. In the studied repository section, 25 deposition tunnels are located, which together may contain about 2 100 canister. A distance between deposition holes (canister) of 6 m and a distance of 25 m between deposition tunnels are used. The deposition tunnels are located at a depth of 450 m below the sea level. It is assumed that the probability that a canister is initially damaged is 0.1% and that the initial, 1 mm^2, hole in the canister will increase in size after 10 000 years.

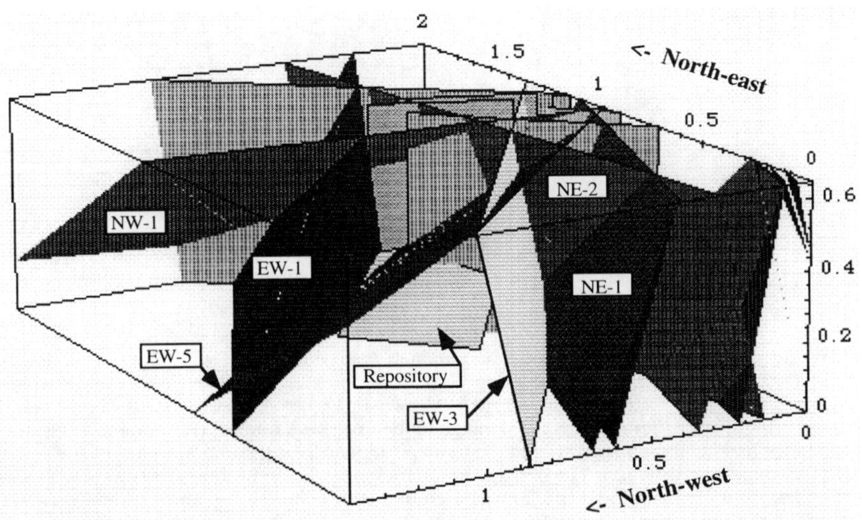

Figure 3.   A section of a hypothetical repository located in the rock mass between the fracture zones EW-1, EW-3, EW-5, NE-1, NE-2 and NW-1.

RESULTS

In the calculations, the radionuclides C-14, I-129, Cs-135, Cs-137, U-238, Pu-240, and Ra-226 were chosen for our study. Realizations of the channel network were generated and the flow field was calculated. Then three different realizations were made for transport calculations. The number of damaged canisters in each run depends on the probability for an initial damage. The expected value for each run is 2.1 damaged canisters.

The near field release is studied at a fracture intersecting the deposition hole and in the disturbed zone around the tunnel. The release from these paths are then used as input to the far field model and the breakthrough to the biosphere is recorded. This is shown in Figure 4 using the total release in the near field and the corresponding far field release of Cs-135 for two different runs as an example. In run R1 the damaged canisters are located in favourable locations and in run R2 one canister is near a fracture zone. It was found for the last case that the retardation in the far field was small. It has been found that the release from a certain run depends mainly on the locations of the damaged canisters.

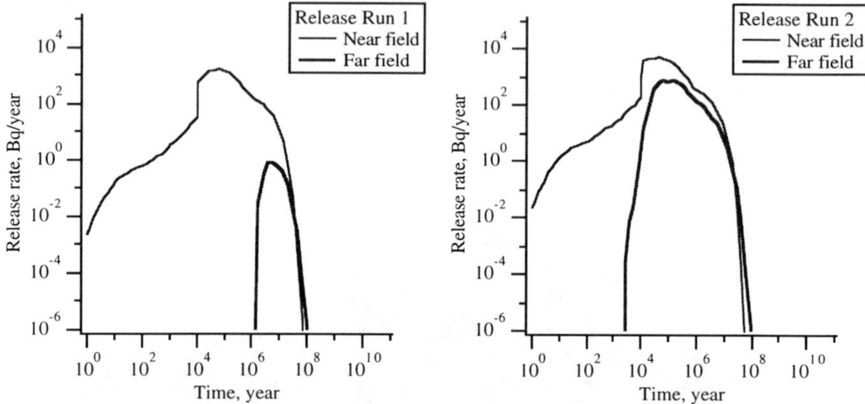

Figure 4.   Near field and far field release of Cs-135 from the runs R1 and R2. In run R2 there is a damaged canister located in a region strongly connected to a fracture zone.

The near field rock as a transport barrier and the influences of different tunnel excavating techniques were also studied. Again, using Cs-135 as an example the release to the biosphere for three levels of damage to the disturbed zone is shown in Figure 5. The far field release results are obtained using three different levels of damage to the disturbed zone. It was assumed that the mean conductance for the altered rock around the tunnel is higher than the conductance of the rock mass due to the tunnel construction. The damage to the rock may depend on the construction method. In the simulations a factor of 3, 10 and 30 was used to increase the conductance in the disturbed zone from the rock mass value. Canister 7 which was used in this case was located in a region connected to a fracture zone.

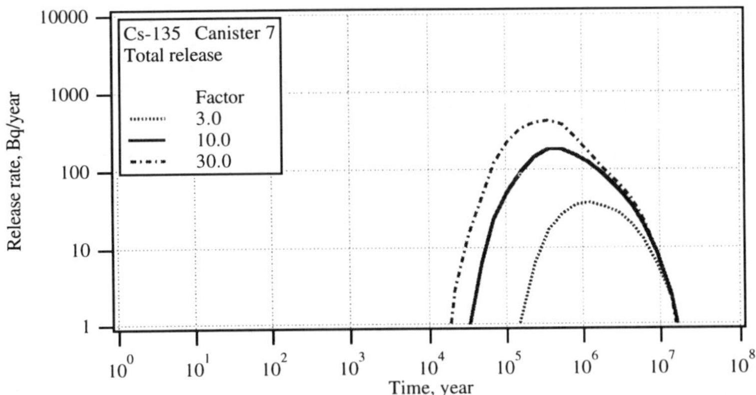

Figure 5.   The release of Cs-135 to the biosphere for three different mean conductance values of the disturbed zone.

CONCLUSIONS

In the performance assessment of a planned repository the concept using a near field model integrated with a far field model may be very useful. Using this concept the detailed calculations of the near field may be used in addition to the far field modelling of the most important mechanisms. The coupled models can be used as an efficient tool to study the impact of various factors including the location of the damaged canister(s) and the importance of the host rock as a barrier against release of potentially escaping nuclides. Here, the effect of the damage level to the rock around the tunnel due to different tunnel excavation methods was studied.

From the results it may be seen that the location of the canister with an initial damage is of vital importance. The damage level of the disturbed zone have some influence on the release to the biosphere for the nuclides C-14, Cs-135 and Ra-226, whereas for I-129 and U-238 it had no large effect.

In the near field calculations it is assumed that the damage is very small initially but grows abruptly to a large hole after 10 000 years. In these calculations 5% of the Cesium is assumed to have migrated out of the uranium oxide matrix and is available for direct dissolution and transport. The slow build up of the Cs-135 concentration initially is due to the large transport resistance through the initially small damage. The rate of transport increases considerably when the damage has grown after 10 000 years. Then the rate of transport is controlled by the diffusion into the flowing water in the release paths. The transport through the far field is considerably retarded. When the Cs-135 is released in a region connected to a fracture zone the retardation is considerably smaller.

REFERENCES

1. L. Romero, L. Moreno, and I. Neretnieks, Fast multiple path model to calculate radionuclide release from the near filed of a repository, Nuclear Technology, Vol. 112,1 pp 89-98, (1995).

2. L. Romero, L. Moreno, and I. Neretnieks, The fast multiple path model "Nuctran" - Calculating the radionuclide release from a repository, Nuclear Technology, Vol. 112,1 pp 99-107, (1995).

3. L. Moreno and I. Neretnieks, Fluid flow and solute transport in a network of channels, J. of Contaminant Hydrology, 14, pp 163-192, (1993).

4. R. Munier and H. Sandstedt, SR-95: Hypothetical layout using site data from Äspö, Swedish Nuclear Fuel and Waste Management Co., SKB Progress Report 94-53, Stockholm, (1994).

5. T. N. Narasimhan, and P. A. Witherspoon, An integrated finite difference method for analyzing fluid flow in porous media, Water Resour. Res., 12, 57, (1976).

6. P.C. Robinson, Connectivity, flow and transport in network models of fractured media, Ph. D. Thesis, St. Catherine's College, Oxford University, Ref. TP 1072, (May 1984).

7. L. Moreno, Y.W. Tsang, C.F. Tsang, F.V. Hale, and I. Neretnieks. Flow and tracer transport in a single fracture. A stochastic model and its relation to some field observations, Water Resour. Res., 24, pp 2033-3048, (1988).

8. L. Moreno, B. Gylling, and I. Neretnieks. Solute transport in fractured media - the important mechanisms for performance assessment, In print, J. of Contaminant Hydrology, (1996).

9. A. Boghammar, Kemakta Konsult AB, Sweden, (private communication).

APPLICATION OF INTERACTION MATRICES IN PERFORMANCE ASSESSMENT - THE FUEL MATRIX AS AN EXAMPLE

PATRIK SELLIN*, JORDI BRUNO**, ESTER CERA**
*Swedish Nuclear Fuel and Waste Management Company (SKB), Box 5864, S-102 40 Stockholm, Sweden, Patrik.Sellin@skb.se
**QuantiSci, Parc Tecnològic del Vallés, 08290 Cerdanyola, Spain

ABSTRACT

In a safety analysis of a repository for high level nuclear waste, it is of primary importance to identify and document all processes that are acting on the repository system. An interaction matrix methodology have been applied to the spent fuel subsystem. The purpose of this application is to identify, structure and rank the Process System and to discuss how the identified processes can be treated.

INTRODUCTION

A repository for high-level waste consists of a chain of barriers. The purpose of the barriers is to isolate the waste from the biosphere and to retard and delay any release of radioactivity from the repository. The purpose of performance assessment/safety analysis is to investigate if the repository system fulfils the given criteria for all realistic futures.

FEP AND PROCESS SYSTEM

A difficult question, that has to be answered in a safety analysis, is if today's knowledge is sufficient to describe the performance of the repository. Have all possible future events been considered? Have all the properties and processes in the repository system been taken into account? To answer these questions, features, events and processes (FEP) acting on the repository have to be identified. The FEP that describe the repository performance and development over time is separated from FEP describing external events. The first is defined as the Process System. There is a need for a systematic and transparent description of the properties and processes and their links in the Process System. However, this will never give the full answer to the questions above, the Process System has to be updated, reviewed and revised all the time to include opinions from other scientists and results from new research.

INTERACTION MATRIX METHODOLOGY

Several different methodologies, to systemise and to visualise all the FEP that can influence a repository for radioactive waste in the future, have been investigated [1] and compared. The "Rock Engineering Systems" [2] approach, using interaction matrices, was found to be one suitable tool for the purpose. The idea with interaction matrices was originally developed for rock engineering problems. It has been slightly modified to suit the purposes of repository performance assessment.

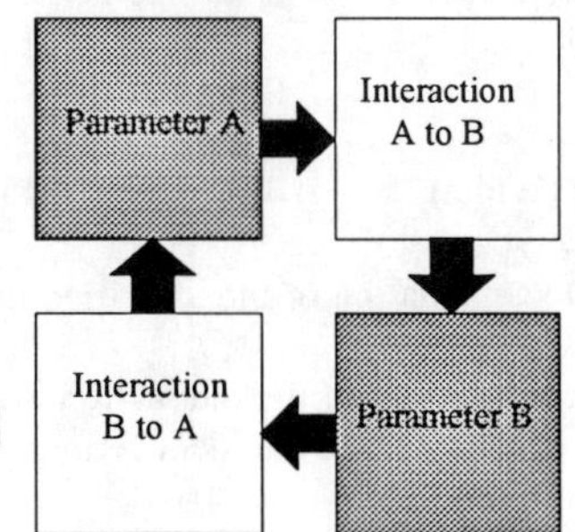

Figure 1. Principle of the interaction matrix.

Mat. Res. Soc. Symp. Proc. Vol. 465 © 1997 Materials Research Society

The basic device used in this approach is the interaction matrix, in which the main variables or parameters of the system are identified and listed along the leading diagonal in a square matrix. The interactions between the parameters are listed in the off-diagonal terms. This is illustrated in Figure 1 together with the clockwise convention for the influence direction.

The application consists of the following steps (Figure 2):

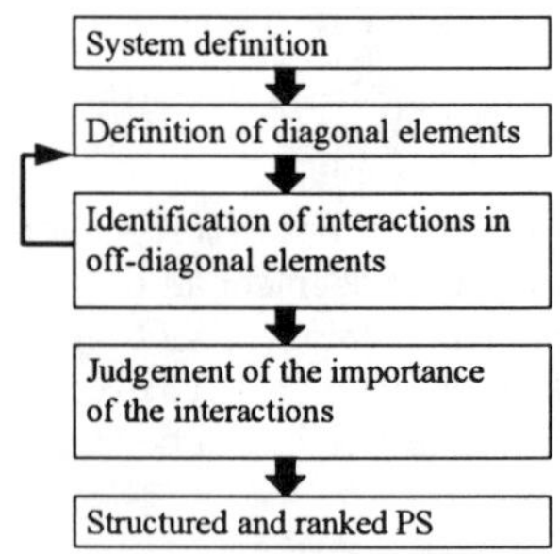

Figure 2. Procedure for the development of a structured and ranked Process System using an Interaction Matrix

- Definition of the purpose of the study and boundaries of the system;

- Selection and definition of the Diagonal Elements. These are usually, but not necessarily, physical parts of the system;

- Construction of a square matrix with the Diagonal Elements on the leading diagonal;

- Identification of clockwise binary interactions between the Diagonal Elements and placing them in the off-diagonal squares. This also gives feedback to the assignment of properties to the diagonal elements;

- Determination of the importance of the binary interactions on a scale where:

    3 (Red)   is an important interaction, a part of the performance assessment which can influence other parts of the Process System;

    2 (Yellow)   is an interaction, which has to be considered in the performance assessment and has an uncertain, but probably limited influence on other parts of the Process System, or important under very special circumstances;

    1 (Green)   is an interaction, which does not have to be considered in performance assessment and has a negligible influence on other parts of the Process System;

    0 (White)   is no identified interaction;

- Documentation of the identified interactions.

The construction of the interaction matrices has two main purposes: to show that no FEP have been overlooked and to identify important interactions and interaction pathways.

In 1995 a template for future safety assessments was presented (SR-95) [3]. The purpose of the template was to create a continuity in safety reporting and a uniformity between reports. The SR 95 was also intended to facilitate the preparation and review of safety reports. It was presented as a synopsis for future safety reports and to further clarify the structure of the template, the proposed outline was exemplified with chapter text.

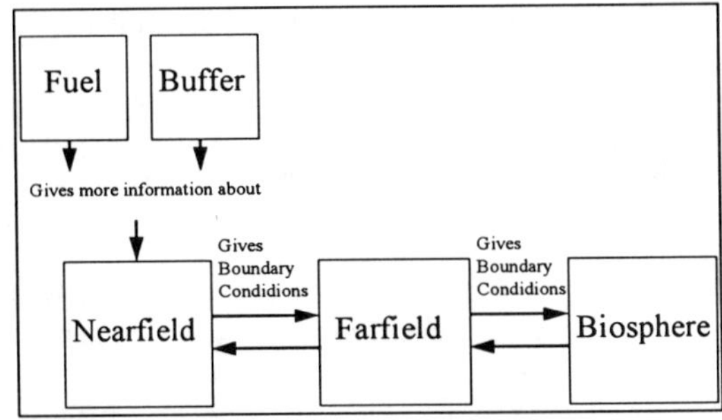

Figure 3. The different interaction matrices and their coupling to each other.

In the SR 95 assessment the repository system has been divided into three main matrices (Near Field, Far Field and Biosphere) and two supporting matrices for the near field matrix (Fuel and Buffer) as seen in Figure 3. The reason for this division is that the repository system is very complex and using one single matrix would greatly reduce the transparency of the matrix. The fuel matrix will be discussed in more detail.

THE FUEL MATRIX

The spent fuel together with the inner near field has been studied separately [4]. All processes found in this matrix should also be found in the Near Field matrix. However, the Fuel matrix has more detail and more comprehensive process descriptions.

The matrix is based on a KBS-3 type repository with a copper/steel canister, it assumes that the canister is damaged and water is allowed to enter, the repository is closed, backfilled and water saturated. The external boundary condition is given by the canister/buffer boundary.

Nine Diagonal Elements were selected and were assigned properties:

> UO_2 *matrix (1,1)* The UO_2 matrix is defined by the matrix itself, the element includes the physical and chemical properties of the matrix, but does not, by definition, include the radionuclides. The assigned properties are: confinement, volume, chemical composition, structure and availability.

> *Matrix bound elements (2,2)* Matrix bound elements include the radionuclides that are <u>initially</u> contained inside the spent fuel matrix. However, the radionuclides are <u>not</u> given a defined location during the analysis of the system. (The separation of radionuclides and radionuclide source is helpful in this application, since it is focused on binary interaction). The assigned properties are: location, release, chemical composition and quantity.

> *"Segregated" elements (3,3)* "Segregated" elements include the radionuclides that are <u>initially</u> located outside the spent fuel matrix. This includes both fission products in gaps and grain boundaries in the fuel as well as activation products in cladding and structural parts. However, the radionuclides are <u>not</u> given a defined location during the analysis of the matrix (see above). The assigned properties are: location, source, chemical composition and quantity.

> *Radiation (4,4),* This element includes all the radiation given off by the different radionuclides. The assigned properties are: source, type (α, β, γ) and intensity

Temperature (5,5), The temperature of the system. The assigned properties are: energy and temperature.

Zircaloy and other metal parts (6,6) The cladding and structural parts of the fuel elements. The assigned properties are: integrity, location and chemical composition.

Water (7,7) This element includes the water that is present inside a defective canister. The assigned properties are: saturation degree, chemical composition and physical properties.

Canister materials (8,8) The canister materials as well as corrosion products. Hydrogen gas from corrosion is included here. The assigned properties are: Chemical composition, structure, material and gas quantity.

Bentonite (9,9) This element is the outer boundary for the studied sub-system. It includes the buffer outside the canister. The assigned properties are: chemical composition, physical properties and heat sink.

The matrix comprises 72 off-diagonal interaction boxes. 64 binary interactions were identified in 49 boxes. These interactions have been ranked according to their importance and documented into a database. Some examples of different types are:

3 (Red)	Interaction (2,4). Matrix bound elements on Radiation. Matrix bound radioelements are the source of radiation. i.e. the dose rate can be calculated from the radionuclide inventory;
2 (Yellow)	Interaction (2,9). Matrix bound elements on bentonite. Released radionuclides may contaminate the buffer material. Radionuclides are released in trace concentrations and should not affect the properties of the bentonite. However, there is a possibility that the bentonite pores may be blocked by precipitated radionuclides;
1 (Green)	Interaction (7,6). Water on Zircaloy cladding. Water in the canister will corrode the cladding. Can be neglected because the rate is very low and the Zircaloy is not considered to be a barrier;
0 (White)	Interaction (1,9) UO_2 matrix on Canister materials. There is no <u>direct</u> contact between the fuel matrix and the canister (Any interaction goes through water)

All the interactions are documented in a database. For each interaction, a description of the process and a justification of the ranking is recorded, together with the names of the group of people that did the ranking.

MATRIX USE IN PERFORMANCE ASSESSMENT

The matrix has two purposes in performance assessment: completeness and basis for modelling, but can also be used as a guide for future research.

<u>Completeness</u>

It is necessary to ensure that all FEP have been found and documented. Many of the interactions ranked as "green" are usually neglected in a safety analysis document, but can be found in the documentation to the matrix. It also simplifies the revision, since it is easy to introduce "new" processes into the matrix. It is also important to document the elements where no interaction are found, since reviewers may find processes affecting this particular

interaction and there has to be a justification to why the element was considered to be
"white".

<u>Modelling</u>
 All "red" interactions have be treated in a PA/SA exercise. This can be done in different
ways, some of the interactions can be considered as prerequisite for the assessment, some are
handled by conservative assumptions and some are modelled with mathematical models. It is
also important to discuss how the "red" interactions are treated in models and assumptions.
Some examples of this is illustrated in Table 1.

DISCUSSION

FEP databases are dynamic bodies. They can never be considered to be completed,
independently of the methodology of generation. They have to be updated, reviewed and
revised all the time. Expert teams looking at the matrix from a new perspective may introduce
interactions that was previously overlooked. The ranking of the processes may also change due
to new information from ongoing research. The matrix and database can also be used for
comparison with other FEP databases produced with different methodologies by other groups.
This has been done for the this matrix [4] and the SITE-94 database [5] and the result indicated
that the processes identified were basically the same even though two different methodologies
were used. The few differences identified can be explained, to a large extent, by the
differences in system definition.

Table 1. Some examples of treatment of "red" interactions

Interaction in Matrix	Assumption	Model used	Alternative model or assumption
1.2 Confinement	Radionuclides are bound in the fuel matrix and evenly distributed	Fuel dissolution model	Uneven distribution, Instant release fractions
2.4, 3.4 Source term for radiation	Radionuclides cause radiation	Inventory calculation	Different computer codes
7.1 Alteration and dissolution of the UO$_2$ matrix	The alteration rate is dependent on radiolysis	Fuel dissolution model	The UO$_2$ matrix is solubility limited
7.2, 7.3 Precipitation of secondary phases	Secondary radioelement phases precipitate when the equilibrium concentration is reached	Near field transport code	Alternative solubility limiting phases or groundwater compositions

CONCLUSION

This study has shown that the Interaction Matrix methodology is feasible to use both for the structuring of the Process System and for visualisation of the identified processes. The documentation system increases the transparency of the system description and makes it possible to trace back the judgements made during the construction of the matrix. This will facilitate review work and future revisions as well as consistent treatment of different issuses in the system. Furthermore, it aslo constitutes a valuable tool to focus the R&D work requiered into the critical interactions which have a less satisfactory treatment.

ACKNOWLEDGEMENTS

The participation of Jordi Bruno and Ester Cera in this work was fully financed by the Swedish Fuel and Waste Management Co.

REFERENCES

1.   T. Eng, J. Hudson, O. Stephansson, K. Skagius, M. Wiborg, SKB Technical Report 94-28, Stockholm 1994

2.   J. Hudson, Rock Engineering Systems: Theory and Practice. Ellis Horwood, Chichester, UK

3.   SKB Technical Report 96-05, Stockholm 1995

4.   J. Bruno, E. Cera, M Stenhouse, P Sellin, SKB Technical Report (in progress)

5.   N. Chapman, J. Andersson, P. Robinson, K. Skagius, C-O. Kene, M. Wiborg and S. Wingefors, SKI Technical Report 95:26, 1995

SOURCE-TERM ANALYSIS FOR HANFORD LOW-ACTIVITY TANK WASTE USING THE REACTION-TRANSPORT CODE AREST-CT

Y. Chen, B. P. McGrail, D. W. Engel
Pacific Northwest National Laboratory,[a] PO Box 999, K6-81, Richland, WA, 99352, USA.

ABSTRACT

A general, integrated performance assessment code, AREST-CT, was used to analyze the influence of various factors on the release rates of radionuclides from a proposed facility for disposal of low-activity tank wastes. The code couples various process models together based on the framework of reaction-transport theory. The disposal facility was modeled as a 1-D column surrounded by soil. A borosilicate waste glass, LD6-5412 was the waste form considered in the analysis. Included in the simulations were 38 aqueous species, 14 minerals, 21 equilibrium reactions, and 16 kinetic reactions. Dissolution rate of the glass and the release rates of Tc, Pu, U, Np, I, Se under different conditions were calculated for 50,000 years. The simulations revealed that 1) open exchange between the atmosphere and pore-water within the vault significantly improves the performance; 2) an ion-exchange reaction between the glass and aqueous phase increases the release rates significantly; and 3) at the hydrogeologic conditions under consideration, variation of the pore-water velocity has little effect on the release rate of radionuclides. These results provide a scientific basis for formulation of waste forms and engineering design of the disposal facility. Reaction-transport modeling can provide information on the long-term performance of disposal systems that are not obtainable from laboratory experiments alone or by conventional decoupled process models.

INTRODUCTION

The long-term performance of disposal facilities for nuclear and hazardous waste is determined by many factors, including the formulation of waste forms, the design of the engineered-barrier system, and the surrounding geological environment. One of the key issues in performance assessment is to analyze the influence of those factors on the release rate of radionuclides or hazardous materials from the disposal facilities. The analysis is usually achieved by simulations based on process models, such as waste glass dissolution models, spent fuel dissolution models, container corrosion models, adsorption models, solubility models, and groundwater transport models.

However, these individual process models do not account for interactions among processes that can strongly affect each other. The interactions always exist because these processes share common components: the host rock, pore water, and gases. The movement of pore water, gaseous species, and the transfer of energy links the processes together. Ignoring such interactions could yield inaccurate model calculations or even qualitatively incorrect results.

---

[a]Operated for the U.S. Department of Energy by Battelle Memorial Institute under Contract No. DE-AC06-76RLO 1830.

Mat. Res. Soc. Symp. Proc. Vol. 465 © 1997 Materials Research Society

Realizing the deficiency of conventional decoupled models, we have been developing such an integrated computer model, AREST-CT (**A**nalyzer of **R**adionuclid**E** **S**ource-**T**erm with **C**hemical **T**ransport), for several years [1,2,3,4]. The model is based upon the framework of reaction-transport theory [5,6,7]. With support from the Hanford Low-Activity Tank Waste (LAW) Disposal and Yucca Mountain Projects, it was first released in 1995 [8]. In this paper, we present its first application—the source-term analysis for Hanford LAW, which is a critical part of the two performance assessments scheduled for the Hanford LAW disposal project.

AREST-CT CODE

The underlying mathematics in AREST-CT are a set of coupled, nonlinear, partial differential equations that describe the time change rate of the solute concentrations of pore water in a porous medium, and the alteration of waste forms, packaging materials, backfill, and host rocks. The detailed mathematics are discussed elsewhere [2,3,4,5,8]. In its current version, the following processes are accounted for:

- kinetic dissolution of waste forms
- kinetic dissolution of host rocks
- kinetic precipitation and dissolution of secondary phases
- aqueous equilibrium speciation
- gas-aqueous equilibria
- redox reactions
- advection
- diffusion
- dispersion
- nucleation thresholds for precipitation of new solids.

Other physical and chemical features in the code are:

- 1-D or 2-D simulation domains
- general interface to take user specified chemistry
- non-isothermal or isothermal chemistry
- ionic strength correction according to modified B-dot equation [9]
- spatially varying distribution of solid phases
- effective reaction surface depending on texture of solids.

The primary output of AREST-CT consists of:
- concentrations of aqueous species as a function of time and space
- release flux of aqueous species and chemical components
- pH and Eh changes
- radii, surface areas, and volume fractions of solids as a function of time and space
- dissolution/precipitation rates of solids
- porosity and permeability changes.

SIMULATION CONDITIONS

The remediation of underground storage tanks at Hanford will generate high-level and low-activity wastes. The low-activity waste stream is proposed to be vitrified and disposed of on-site.

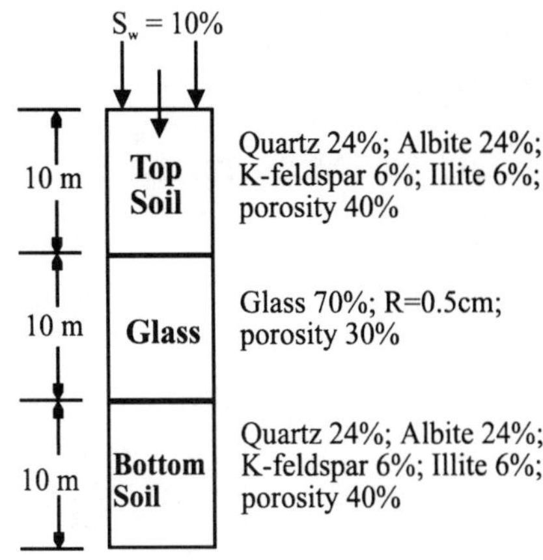

Figure 1. Simulation Domain and Conditions

The disposal facility will be located 10 meters below the ground surface and will be 10 meters in height.

With the existence of the proposed hydraulic barrier, the average pore-water velocity in the disposal facility is 1 cm/year=3.17×10^{-10} m/s, with a water saturation of 10% [10]. The average temperature is about 14°C. As shown in Figure 1, the surrounding soil contains quartz (24% by volume), albite(high) (24% by volume), K-feldspar (6%), and illite (6%), with a porosity of 40% [11].

The waste glass, LD6-5412, with a formula for the major glass components of $Al_{0.1535}B_{0.094}Ca_{0.046}K_{0.0202}Na_{0.4208}Si_{0.6068}O_{1.8514}$, is considered in our simulations. It is assumed that the waste glass will occupy 70 vol.% of the waste vault and that it is spherical in shape with a radius of 0.5 cm. Flow-through experiments show that two reaction mechanisms are involved in the glass dissolution process [12]. One is the matrix dissolution reaction:

$$
\begin{aligned}
Glass + 0.5OH^- &+ 0.45H_2O = \\
&0.0202K^+ + 0.4208Na^+ + 0.046Ca^+ + 0.094BO_2^- \\
&+ 0.154Al(OH)_3(aq) + 0.2712HSiO_3^- + 0.3348H_2SiO_4^{2-} \\
&+ 2.28 \times 10^{-8}SeO_3^{2-} + 1.62 \times 10^{-6}TcO_4^- + 3.83 \times 10^{-8}IO_3^- \\
&+ 2.72 \times 10^{-5}U(OH)_4(aq) + 1.93 \times 10^{-9}Pu(OH)_5^- \\
&+ 2.73 \times 10^{-9}NpO_2(OH)(aq)
\end{aligned}
\tag{1}
$$

with a rate law as:

$$
R = kAS_w a_{OH^-}^{0.5} \left(1 - \frac{Q}{K^{eq}}\right)
\tag{2}
$$

where R is the dissolution rate, k is the rate constant, A the surface area, S_w the water saturation index, Q the activity product of glass matrix components, and K^{eq} the equilibrium constant. The other is an ion-exchange reaction

$$
Glass\cdots 0.4208Na + 0.4208H^+ = 0.4208Na^+ + Glass\cdots 0.4208H.
\tag{3}
$$

The measured rate constant for Reaction (1) at 14°C is 3.03×10^{-14} mol/m²·s, for Reaction (3) it is 1.74×10^{-11} mol/m²·s. Radionuclide release is assumed to occur only via Reaction (1) and the release of all radionuclides is congruent. The stoichiometric coefficients for the radioactive elements given in Reaction (1) were calculated from the total waste inventory by assuming that 210,000 m³ of glass with a uniform composition will be produced [13].

Included in the simulations were 38 aqueous species and 14 solid species plus glass. The selection of species is based on speciation calculations and observations from glass testing experiments. Besides the aforementioned two glass dissolution reactions, 21 equilibrium reactions and 14 kinetic reactions are included in our simulations as listed in Table I.

Table I. List of Other Reactions Considered in the Simulations

1.	$H_2O = H^+ + OH^-$
2.	$CO_2(aq) + H_2O = H^+ + HCO_3^-$
3.	$HCO_3^- = H^+ + CO_3^{2-}$
4.	$Al(OH)_4^- + H^+ = Al(OH)_3(aq) + H_2O$
5.	$H_2SiO_4^{2-} + 2H^+ = SiO_2(aq) + 2H_2O$
6.	$HSiO_3^- + H^+ = SiO_2(aq) + H_2O$
7.	$BO_2^- + H^+ + H_2O = B(OH)_3(aq)$
8.	$CaCO_3(aq) + H^+ + Ca^{2+} + HCO_3^-$
9.	$CaHCO_3^+ = Ca^{2+} + HCO_3^-$
10.	$CaOH^+ + H^+ = Ca^{2+} + H_2O$
11.	$(UO_2)_2CO_3(OH)_3^- + 4H^+ = 2UO_2^{2+} + HCO_3^- + 3H_2O$
12.	$UO_2(CO_3)_3^{4-} + 3H^+ = UO_2^{2+} + 3HCO_3^-$
13.	$UO_2(CO_3)_2^{2-} + 2H^+ = UO_2^{2+} + 2HCO_3^-$
14.	$UO_2(OH)_2(aq) + 2H^+ = UO_2^{2+} + 2H_2O$
15.	$UO_2(OH)_3^- + 3H^+ = UO_2^{2+} + 3H_2O$
16.	$UO_2CO_3(aq) + H^+ = HCO_3^- + UO_2^{2+}$
17.	$UO_2OH^+ + H^+ = UO_2^{2+} + H_2O$
18.	$U(OH)_4(aq) + 0.5O_2(aq) + 2H^+ = UO_2^{2+} + 3H_2O$
19.	$NpO_2OH(aq) = OH^- + NpO_2^+$
20.	$NpO_2(CO_3)^- = NpO_2^+ + CO_3^{2-}$
21.	$HSeO_3^- = H^+ + SeO_3^{2-}$
22.	$Quartz + OH^- = HSiO_3^-$
23.	$Calcite = Ca^{2+} + CO_3^{2-}$
24.	$Albite(high) + 2OH^- = Na^+ + Al(OH)_3(aq) + 3HSiO_3^-$
25.	$K\text{-feldspar} + 2OH^- = K^+ + Al(OH)_3(aq) + 3HSiO_3^-$
26.	$Illite + 2.4OH^- = 0.6K^+ + 0.25Mg^{2+} + 2.3Al(OH)_3(aq) + 3.5HSiO_3^-$
27.	$Analcime + 0.96H^+ = 0.96Na^+ + 0.96Al(OH)_3(aq) + 2.04SiO_2(aq) + 5.04H_2O$
28.	$Chalcedony = SiO_2(aq)$
29.	$NaAlSi2O6{\cdot}6H_2O + OH^- = Na^+ + Al(OH)_3(aq) + 2HSiO_3^- + 4H_2O$
30.	$Phillipsite + 3H^+ = K^+ + Ca^{2+} + 3Al(OH)_3(aq) + 5SiO_2(aq) + 3H_2O$
31.	$Tobermorite + 10H^+ = 5Ca^{2+} + 6SiO_2(aq) + 10.5H_2O$
32.	$Gobbinsite + 6H^+ = 4Na^+ + Ca^{2+} + 6Al(OH)_3(aq) + 10SiO_2(aq) + 6H_2O$
33.	$Haiweeite + 6H^+ = Ca^{2+} + 6SiO_2(aq) + 2UO_2^{2+} + 8H_2O$
34.	$Schoepite + 2H^+ = UO_2^{2+} + 3H_2O$
35.	$PuO_2(s) + 3H_2O = H^+ + Pu(OH)_5^-$

* 1-21 are equilibrium reactions, 22-35 are kinetic reactions .

RESULTS AND DISCUSSIONS

The simulation with the conditions described above is the references case (called Case A hereafter). Results for Case A are presented in Figures 2, 3, and 4. Figure 2 shows the calculated pH profiles in the vault as a function of time and space. The peak pH values occur at the center of the vault. As time advances, pH continues to rise. Figure 3 shows that the dissolution rate of the glass decreases slightly with time. The largest dissolution rates appear at the interfaces between the vault and the surrounding soil at t=20,000 years because diffusive mass transfer lowers the concentration of Si and other glass components in the aqueous phase there.

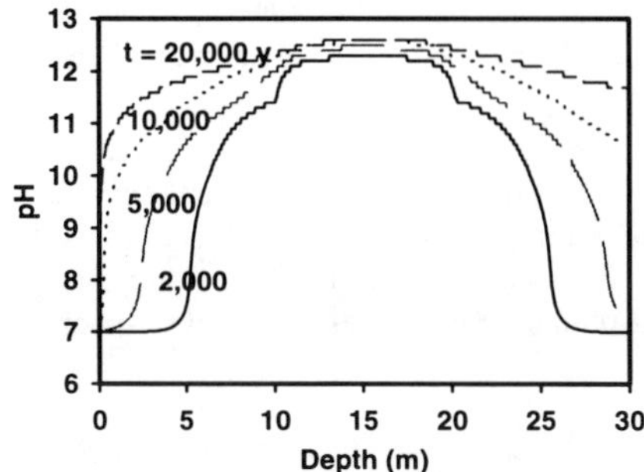

Figure 2. Computed pH Profiles for Case A

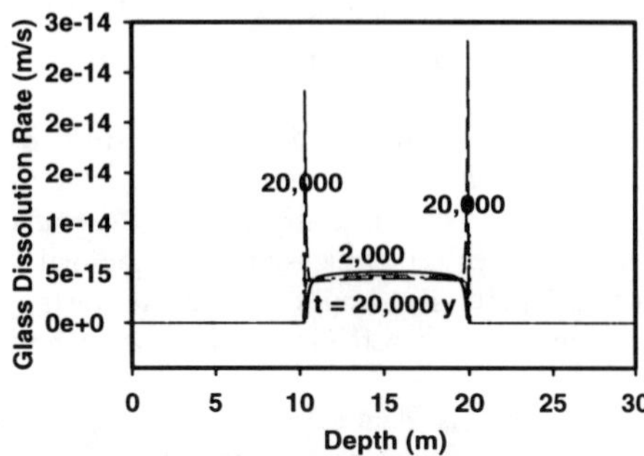

Figure 3. Dissolution Rate of Glass for Case A

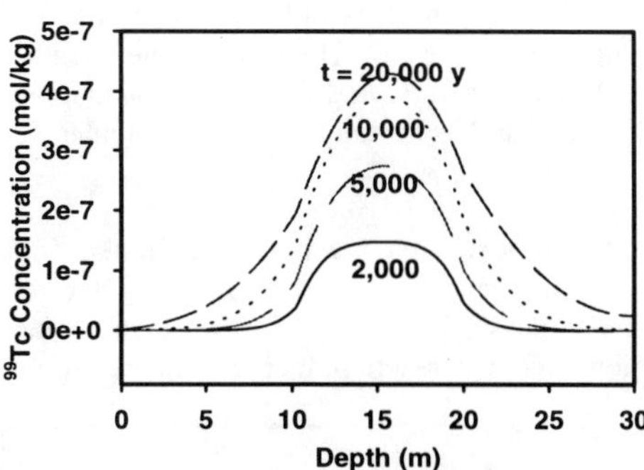

Figure 4. ^{99}Tc Concentration for Case A

Figure 4 shows time-dependent profiles of the ^{99}Tc concentration in the aqueous phase. As the dissolution rate of the glass slows with time, the increase in the ^{99}Tc concentration also slows down. The peak value appears at the center of the vault at about 4×10^{-7} mol/kg.

It is worthwhile to note that these profiles are strongly symmetric with respect to the center of the vault. This is because when the pore-water velocity is as low as 3.17×10^{-10} m/s, transport is dominated by diffusion. The diffusive flux of ^{99}Tc at the depth of 20 meters comprises more than 95% of the total flux.

Additional simulations were done to analyze the effects of various factors on the release rates. Only one parameter at a time was varied from those of Case A.

Case B: Case B was designed to investigate the effect of gas-water equilibrium on release rates. Equilibrium with atmospheric CO_2 was assumed so that one more reaction is added to the reaction network described for Case A :

$$CO_2(g) = CO_2(aq) \qquad \log_{10} pCO_2(g) = -3.5 \qquad (4)$$

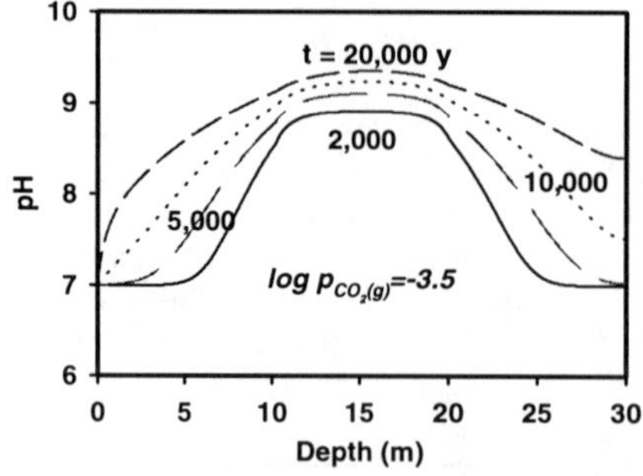

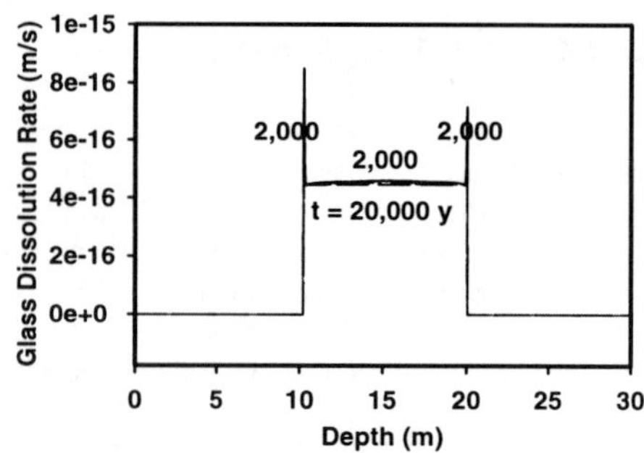

Figure 5. Calculated pH Profiles in Case B.　　**Figure. 6.** Dissolution Rate of Glass in Case B.

Figures 5 and 6 show the simulation results for Case B. The peak pH value is about 3 units lower than that for Case A. Consequently, the overall glass dissolution rate is about one order of magnitude lower than in Case A. Near the boundaries of the vault, it is two orders of magnitude lower. As a result, the release rates of radionuclides from the vault are reduced significantly, as shown in Figure 7. It implies that if the waste vault remains freely open to the atmosphere, the performance of the waste glass can be improved significantly.

Case C: Case C is identical to Case A, except that the ion-exchange reaction (Reaction (3)) has been deleted from the reaction network .

Figure 8 presents the calculated dissolution rate of the glass for Case C. The glass dissolution rate (ranging from 10^{-17} - 10^{-18} m/s) in the vault interior is two to three orders of magnitude lower than that of Case A and one to two orders lower than that of Case B. However, at the boundaries of the vault, the dissolution rates are as high as that of Case A, and two orders of magnitude higher than that of Case B. As mentioned above, the high dissolution rates at the boundaries are caused by the diffusive transport.

The differences in Case A and Case C reveal that the ion-exchange reaction has a dramatic effect on the performance of the glass. With the ion-exchange reaction included, the computed pH in the vault rises to over 12. At this high pH, LD6-5412 glass is less stable. If new glasses could be formulated that either eliminate or minimize the ion-exchange reaction, its performance would be

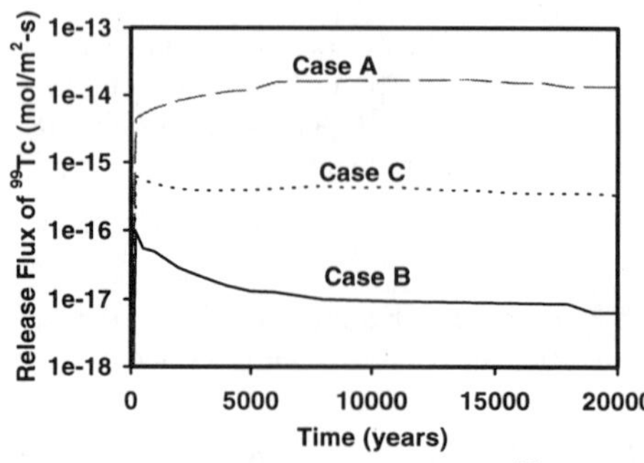

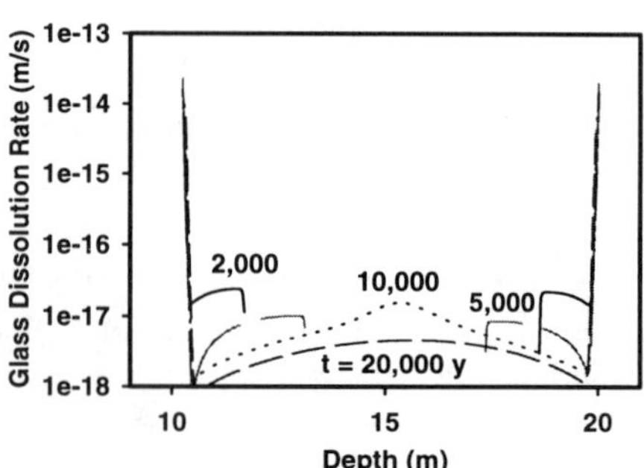

Figure 7. Calculated Release Rates of ^{99}Tc from the Bottom of the Vault for Cases A, B, and C.　　**Figure 8.** Dissolution Rate of Case C, Which Does Not Include the Ion-Exchange Reaction.

significantly improved. It should be noted that laboratory experiments with LD6-5412 glass have only shown the Na ion-exchange reaction to be of minor significance in increasing the early-time pH. Only through the computer simulations, was the long-term effect on the performance of the glass revealed.

The calculated release fluxes of ^{99}Tc from the bottom of the vault for Cases A, B, and C (Figure 7) show that Case A has the highest release flux, because it has the highest overall glass dissolution rate. However, lower overall dissolution rate does not necessarily guarantee a lower ^{99}Tc release rate. For example, although the overall dissolution rate in Case C is one to two orders of magnitude lower than Case B, the release rate from the vault is higher than in Case B. This is caused by the peak dissolution rates in Case C at the boundaries of the vault. In this case, the contaminant release rates are determined more by the corrosion behavior of wastes near the vault boundaries, which is controlled by local geochemical conditions, and less by the corrosion behavior of wastes in the interior of the vault. The detailed complex dependence of release rates on the local geochemical conditions can only be assessed through coupled process models.

Case D: Case D was designed to investigate the effect of variation in pore-water velocity. Two simulations, one (Case D-1) with the pore-water velocity 10 times faster (3.17×10^{-9} m/s = 10 cm/y) than in Case A, and the other (Case D-2) with the velocity 10 times slower (3.17×10^{-11} m/s = 0.1 cm/y) than in Case A were performed.

Release rates of ^{99}Tc from the bottom of the vault in Cases D-1 and D-2 are presented in Figure 9, along with that of Case A. A pore-water velocity 10 times higher than Case A causes the peak release rate to increases only 3 times. On the other hand, the peak release rate of Case D-2 is only about 20% lower than Case A, though its pore-water velocity is 10 times lower. The sub-linear effect of pore-water velocity variation on release rates occurs because diffusion is the dominant transport process at pore-water velocities as low as 3.17×10^{-10} m/s.

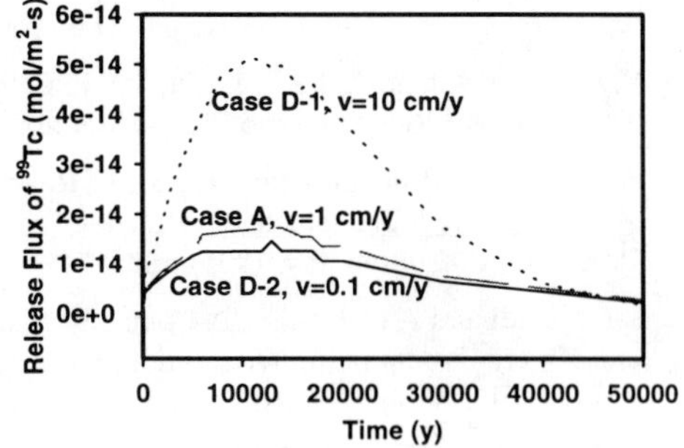

Figure 9. Calculated Release Rates of ^{99}Tc from the Bottom of the Vault for Case A, D-1 and D-2

The results show that the proposed hydraulic barrier would effectively reduce the release rates of contaminants through advection. Further reducing the pore-water velocity from the average value of 3.17×10^{-10} m/s does not appear to improve the performance of the disposal facility very much.

CONCLUSIONS

Analyzing the long-term performance of the disposal facility using the coupled reaction-transport model reveals that: 1) open exchange between the atmosphere and pore-water within the waste vault significantly improves the performance of the disposal system by lowering the pH and overall dissolution rate of the glass; 2) ion-exchange between the glass and aqueous phase was found to increase the release rates of contaminants significantly; thus, by formulating waste glass that minimize ion-exchange, reduction in the release rates of radionuclides can be realized; and

3) the effects of pore-water velocity variations on the release rate of radionuclides is not strong at the hydrologic conditions under consideration.

Reaction-transport modeling provides new information on the long-term performance of disposal systems that are not obtainable from laboratory experiments alone or by conventional de-coupled process models. The proper application of reaction-transport models provides a scientific basis for the engineering design of the waste forms and the disposal facility.

REFERENCES

1. B.P. McGrail, and D.W. Engel, in Scientific Basis for Nuclear Waste Management XVI, edited by C.G. Interrante and R.T. Pabalan (Mater. Res. Soc. Proc. **294**, Pittsburgh, PA, 1993) pp. 215-224.

2. B.P. McGrail, C.I. Steefel, J.A. Fort, D.W. Engel, S.B. Yabusaki, in Scientific Basis for Nuclear Waste Management XVIII, edited by T. Murakami and R. C. Ewing, (Mater. Res. Soc. Proc. **353**, Pittsburgh, PA, 1995) pp. 551-557.

3. Y. Chen, B.P. McGrail, and D.W. Engel, in Proceedings of International Conference on Deep Geological Disposal of Radioactive Waste, edited by M.M. Ohta and K. Nuttall, (Can. Nucl. Soc.) September 16-19, 1996, Winnipeg, Manitoba, Canada.

4. Y. Chen, B.P. McGrail, and D.W. Engel, The Development of a General, Coupled Reaction-Transport Code: AREST-CT (in preparation).

5. Y. Chen, W. Chen, A. Park, J. Mu, and P. Ortoleva, A General, Coupled Model for Solute Transport and Reactions and Porous Medium Alteration (in preparation)

6. P.C. Lichtner, Principles and Practice of Reactive Transport Modeling, in Scientific Basis for Nuclear Waste Management XVIII, edited by T. Murakami and R. C. Ewing, (Mater. Res. Soc. Proc. **353**, Pittsburgh, PA, 1995) pp. 117-130.

7. C.I. Steefel and A.C. Lasaga, A Coupled Model for Transport of Multiple Chemical Species and Kinetic Precipitation/Dissolution Reactions with Application to Reactive Flow in Single Phase Hydrothermal Systems, American Journal of Science, Vol. 294, 1994, pp529-592.

8. Y. Chen, D.W. Engel, B.P. McGrail, and K.S. Lessor, Report No. PNL-10692, Pacific Northwest Laboratory, Richland, Washington, USA, 1995.

9. T.J. Wolery, Report No. UCRL-MA-110662 PT III, Lawrence Livermore National Laboratory, Livermore, California, USA, 1992.

10. N.W. Kline, Report No. WHC-SD-WM-RPT-189, Westinghouse Hanford Company, Richland, Washington, USA, 1995.

11. R.J. Serne, J.L. Conca, V.L. LeGore, K.J. Cantrell, C.W. Lindenmeier, J.A. Campbell, J.E. Amonette, and M.I. Wood, Report No. PNL-8889 Vol.1, Pacific Northwest Laboratory, Richland, Washington, USA, 1993.

12. B.P. McGrail, W.L. Ebert, A.J. Babel, and D.K. Peeler, Measurement of Kinetic Rate Law Parameters on a Na-Ca-Al Borosilicate Glass for Low-Activity Waste, submitted to J. Nuc. Mat., 1996.

13. F.A. Schmittroth, Report No. WHC-SD-WM-RPT-164, Westinghouse Hanford Company, Richland, Washington, USA, 1995.

Development of a New Mechanistic Low-Level Waste Source Term Model

Man-Sung Yim*
*Department of Nuclear Engineering, North Carolina State University, Raleigh, NC 27695

ABSTRACT

A new mechanistic low-level waste source term model was developed. Key features of this effort are: use of cumulative probability functions to describe the failures of waste containers; capability to describe diffusion controlled release which is dependent on the conditions of the waste package surroundings; consideration of the effects of gas generation on the source term, and; use of a source inventory characterization routine to provide a built-in capability for defining the distributions of radionuclides in various waste forms and streams. The model is capable of describing the diffusion of radionuclides in waste forms with the use of concentration and flux continuity boundary conditions. Release of radionuclides from various waste streams is modeled by the combination of diffusion, dissolution, and surface release. Failures of waste containers are portrayed by the use of probabilistic failure functions based on Weibull and lognormal distributions. The model for radionuclide release from waste packages is coupled with the near-field transport model to describe the effects of migration and dispersion within the disposal unit. Characteristics of the new model were evaluated through sensitivity analysis and compared with existing source term codes.

INTRODUCTION

Source term analysis is one of the key steps in the performance assessment of nuclear waste disposal facilities. Present consensus in low-level waste (LLW) site performance assessment is that the currently predicted source term is either too uncertain or conservative. The uncertainty or conservatism in LLW source term analysis is partly attributed to the inability of computer models to mechanistically describe the release of radionuclides from various waste forms. This is in part due to the use of the following assumptions: (1) Instantaneous complete failure of waste containers at the time of failure (the single time-to-failure approach); (2) Release of radionuclides controlled by diffusion is independent of the surrounding conditions of the waste packages (zero concentration boundary condition), and; (3) Effects of gas generation on source term are negligible.

The objective of this study is to develop a new mechanistic source term model to test these assumptions. The model needs the capability of describing distributed failures of waste containers, diffusion controlled release which is dependent on the conditions of the waste package surroundings, and the effects of gas generation on the source term. Knowing that the distributions of radionuclides in various waste forms and streams are important in source term analysis, it is also proposed that the new model be combined with a source inventory characterization routine which is based on the actual waste manifest information from the waste generators. Most of present source term models require the description of radionuclide inventory distributions as input.

MODELING APPROACHES

Failure of Waste Containers

The presently used single time-to-failure (STF) approach is not compatible with the stochastic nature of waste container failure processes. Probabilistic modeling of the failure of waste containers is an alternative. Carbon steel containers (both drums and liners) which are widely used in waste disposal could undergo general corrosion and pitting corrosion under the conditions expected in disposal facilities and a database is available for the support of more elaborate modeling. In this study, failure of carbon steel containers from both general and

Mat. Res. Soc. Symp. Proc. Vol. 465 © 1997 Materials Research Society

pitting corrosion is described by the use of three parameter Weibull distribution [1]. These three parameters include the mean container lifetime, the threshold container lifetime, and the Weibull slope (failure rate at the mean container lifetime). Based on the information compiled by Sullivan and Suen [2] for carbon steel in subsurface environment, the rate of carbon steel container penetration is obtained. Using this rate, the fraction of carbon-steel containers that has failed up to time (t), was derived:

$$W_{csd}(t) = 1 - \exp[-(\frac{t}{3})^{0.9}] + W_{csd}^{init} \quad \text{for carbon steel drums} \quad Eq.\ (1)$$

$$W_{csl}(t) = 1 - \exp[-(\frac{t-5}{16})^{0.5}] + W_{csl}^{init} \quad \text{for carbon steel liners} \quad Eq.\ (2)$$

where,
W^{init} = *initial failed fraction*; $0 < W(t) < 1$
t = *time*

The mean time to failure is estimated as the time required for complete penetration of the container thickness. The thickness of carbon steel drums and liners assumed was 0.127 cm (50 mil) and 0.3 cm, respectively. The threshold failure time of 0 and 5 years is used for carbon steel drums and liners, respectively. In this derivation, it is assumed that the rate of general corrosion is 5.7×10^{-3} cm/yr and the pitting corrosion takes place under neutral pH with good aeration in the facility[2].

Figure 1. Cumulative Failure Fractions of Waste Containers with Weibull/ Lognormal Distributions

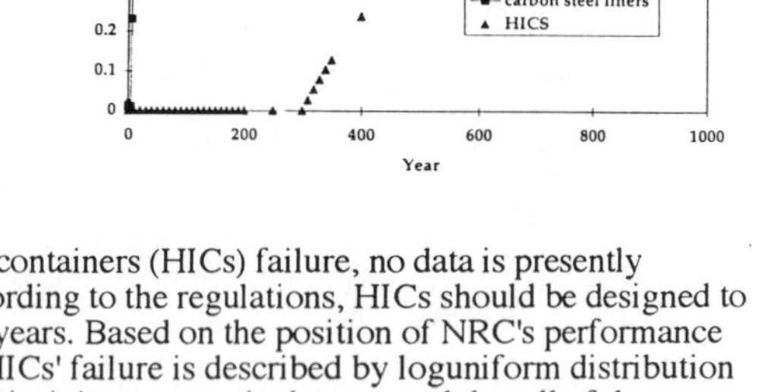

Regarding the modeling of high integrity containers (HICs) failure, no data is presently available to support any predictions. According to the regulations, HICs should be designed to maintain their structural stability for 300 years. Based on the position of NRC's performance assessment group [3], it is assumed that HICs' failure is described by loguniform distribution between 300 and 1000 years. In the analysis, it is conservatively assumed that all of the activity become available for release once the container fails. Figure 1 depicts the cumulative failure function of waste containers used in this study.

<u>Release of Radionuclides from Waste Forms</u>

As currently practiced, LLWs are disposed in a wide variety of materials and forms, including metals, resins, filters, mixed trash, cement, or sorbent media. This has a considerable impact on the manner and degree to which the accompanying radionuclides are retained in the waste form.

Release of radionuclides from various waste forms results from the contact of water and can be described by the combination of three major release mechanisms, i.e., surface rinse, diffusion, and dissolution [4]. Surface rinse applies when the waste is an absorbing material and radionuclides are retained at the surface. Diffusion controlled release is appropriate to describe the movement of radionuclides in the pore water of solidified waste forms or when radionuclides are retained in the waste through ion-exchange processes. Dissolution describes the release of immobile radionuclides bound to the waste form, such as the activation products in metals. The solubility of radionuclides needs to be considered in all of these processes.

In this study, the surface rinse release is modeled based on the equilibrium partitioning of radionuclides between the solid and the liquid phase. The partitioned activity in the liquid phase is released by the exchange of percolating water in the waste packages at a rate given by the total water inflow rate divided by the total volume of water available within failed waste packages. This is represented by,

$$Q_{rinse}(t) = \frac{S_r}{(\theta + K_d\rho)\cdot d_{waste}}(1 - \frac{C(t)}{C_{sat}})\cdot p(t) \quad Eq.(3)$$

where $Q_{rinse}(t)$ is the release rate by surface rinse, S_r is the source activity available for surface rinse, $C(t)$ is the solution concentration, C_{sat} is the solubility limit, θ is the mositure content, K_d is the distribution coefficient, ρ is the density of waste, d_{waste} is the depth of waste region, and $p(t)$ is the percolation flux. For the description of diffusion-controlled release, most of the existing models assume zero nuclide concentration at the waste form surface boundary to facilitate the derivation of analytical solutions. This boundary condition leads to release rates that are conservative and independent of the surrounding conditions. Effects of surrounding conditions (backfills) on the diffusive release of radionuclides have been investigated by Kim et al. [5]. Kim's work relied on the numerical inverse Laplace transform technique to calculate the changes in the release rate due to the variations in the diffusion coefficient in the backfill media. No analytical solution was given to describe the effects of backfill on the diffusion from waste.

A two region cylindrical model to describe the backfill-dependent radionuclide diffusion can be written as

$$\frac{\partial C_1(r,t)}{\partial t} = \frac{D_w}{R_w r}\frac{\partial}{\partial r}(r\frac{\partial C_1(r,t)}{\partial r}) - \lambda C_1(r,t), \ 0 < r < r_w, t > 0 \quad Eq.(4)$$

$$\frac{\partial C_2(r,t)}{\partial t} = \frac{D_s}{R_s r}\frac{\partial}{\partial r}(r\frac{\partial C_2(r,t)}{\partial r}) - \lambda C_2(r,t), \ r \geq r_w, t > 0 \quad Eq.(5)$$

where,

$C_1(r,t) =$ *concentration of radionuclide in pore water within the waste form*

$C_2(r,t) =$ *concentration of radionuclide in pore water of the surrounding medium*

$D_w =$ *diffusion coefficient of radionuclide in pore water within the waste form*

$D_s =$ *diffusion coefficient of radionuclide in pore water of the surrounding medium*

$R_w =$ *retardation coefficient of radionuclide within the waste form*

$R_s =$ *retardation coefficient of radionuclide in the surrounding medium*

$\lambda =$ *decay constant*

$r_w =$ *radius of the waste form*

The initial and boundary conditions for these equations are [5]:
$$C_1(r, t = 0) = C_0, \quad 0 \leq r \leq r_w$$
$$C_2(r, t = 0) = 0, \quad r > r_w$$
$$C_1(r_w, t) = C_2(r_w, t), \quad t > 0$$
$$-\varepsilon_1 D_w \frac{\partial C_1(r_w, t)}{\partial r} = -\varepsilon_2 D_s \frac{\partial C_2(r_w, t)}{\partial r}, \quad t > 0$$

$$\frac{\partial C_1(0,t)}{\partial r} = 0, \quad t > 0$$

$$C_2(\infty,t) = 0, \quad t \geq 0$$

where,

$\varepsilon_{1,2}$ = *porosity of waste / backfill*

C_0 = *initial nuclide concentration in pore water within the waste form* $= \dfrac{S_{diff}}{\pi r_w^2 H(1 + K_d\rho/\theta)}$

S_{diff} = *source activity available for diffusion*

H = *height of the waste form*

Solutions of these equations can be sought by the Laplace transform approach. The rate of radionuclide release from the waste form in the Laplace transformed domain can be given by[1],

$$Q_{diff}(s) = \int dS \cdot \left(-\varepsilon_1 D_w \frac{d\overline{C}_1(r,s)}{dr}\right)\Bigg|_{r=r_w} = 2\pi r_w H F_c\left(-\varepsilon_1 D_w \frac{d\overline{C}_1(r_w,s)}{dr}\right)$$

$$= 2\pi r_w H F_c \varepsilon_1 \sqrt{D_1}\, C_0 \frac{1}{\sqrt{s}} \frac{1}{\left[\dfrac{I_0(\sqrt{sr_w^2/D_1}}{I_1(\sqrt{sr_w^2/D_1}} + \alpha\, \dfrac{K_0(\sqrt{sr_w^2/D_2}}{K_1(\sqrt{sr_w^2/D_2}}\right]} \qquad Eq.\,(6)^2$$

where,

$Q_{diff}(s)$ = *total diffusive release rate in the Laplace transformed domain* (s)

F_c = *correction factor from an infinite to a finite cylinder* $= 1 + r_w/H$

$D_1 = D_w / R_w$ = *effective diffusion coefficient within the waste form*

$D_2 = D_s / R_s$ = *effective diffusion coefficient in the surrounding medium*

$$\alpha = \frac{\varepsilon_1 R_w \sqrt{D_1}}{\varepsilon_2 R_s \sqrt{D_2}}$$

Due to the complexity in the inversion of the Laplace transform, it is difficult to obtain the exact solution for the entire time domain. However, for the period of early release[3] which is important in performance assessment an asymptotic solution can be derived. The solution is:

$$Q_{diff}(t) = \frac{2\pi r_w H F_c \varepsilon_1 D_1^{1/2} C_0}{(1+\alpha)} \left[\frac{\alpha D_2^{1/2} - D_1^{1/2}}{2(1+\alpha)r_w} + \frac{t^{-1/2}}{\sqrt{\pi}} - \frac{D_1 + 3\alpha(D_1 + D_2) + \alpha^2 D_2 + 4\alpha D_1^{1/2}D_2^{1/2}}{4(1+\alpha)^2 r_w^2}\frac{t^{1/2}}{\sqrt{\pi}}\right.$$
$$\left. + \frac{3(D_1^{3/2} + \alpha D_1^{1/2}D_2 - \alpha D_1 D_2^{1/2} - \alpha^2 D_2^{3/2})}{8(1+\alpha)^2 r_w^3}t\right] \cdot e^{-\lambda t} \qquad Eq.\,(7)$$

The third main process, dissolution, is modeled assuming that radionuclides are released congruently and that the rate of release is constant. The equation used for the total release rate is [4]:

$$Q_{dis}(t) = \frac{v_{dis}S_{dis}SA_{waste}^{dis}}{V_{waste}^{dis}}\left(1 - \frac{C(t)}{C_{sat}}\right) \qquad Eq.\,(8)$$

---

[1] In this equation, radioactive decay is ignored and is described in the final solution.

[2] I_0, I_1, K_0, K_1 are the modified Bessel functions of the first and second kind.

[3] In performance assessment, the peak dose usually comes from the initial release of mobile radionuclides.

where,
$Q_{dis}(t) = $ *release rate by dissolution*
$v_{dis} = $ *dissolution velocity*
$S_{dis} = $ *source activity available for dissolution*
$SA_{waste}^{dis} = $ *surface area of waste*

Gas Generation and Release from Waste Forms

Effects of gas generation are currently not taken into account in source term modeling. However, gas generation could affect the source term of ^{14}C, ^{3}H, and other volatile nuclides [6]. Many of the waste materials that contain ^{14}C are susceptible to microbial attack and can release the activity through the generation of CO_2 and methane. A major portion of the ^{3}H inventory can become incorporated into the gas-phase through anaerobic corrosion of metals. Once gases are generated, they migrate to void spaces in waste containers and can be released by the pumping effect of atmospheric pressure changes, if there is a breach in the container. This pumping effect is found to be capable of releasing all of the generated gases in a facility to the atmosphere [7]. HICs allow free release of gases, without the failure of the container, through the vent designed to relieve pressure buildup.

Detailed modeling of microbial gas generation involves the description of hydrolysis of organics, growth kinetics of various biomass groups, and acetate utilization [7]. Acetate utilization by bacteria is an essential part of gas generation modeling during the early period of disposal. For the long-term history of a facility, hydrolysis of materials could become rate-limiting. What becomes the rate limiting step depends on the level of biological activity in the waste. In this study, a first-order kinetics approach is used based on this rate-limiting step approach to consider the effects of gas generation on the source term. Appropriate rate constants for the rate limiting step are evaluated elsewhere [7] and utilized in the current model. The rate of gas generation (or release) is represented by:

$$Q^{gas}(t) = \sum_{i=1}^{n} S_{bio}^{i}[1 - \exp(-k_i(T) \cdot t)] \quad Eq.\,(9)$$

where,
$Q^{gas}(t) = $ *gas generation rate*
$S_{bio}^{i} = $ *source activity available for biodegradation*

$$k_i(T) = k_i(T_o) \cdot \exp[-E_A(\frac{1}{RT} - \frac{1}{RT_o})]$$

$\qquad = $ *time constant for rate – limiting step in gas generation*
$T_o = $ *reference temperature* (K)
$T = $ *system temperature* (K)
$E_A = $ *Arrhenius activation energy* $(kcal\,/\,mol)$
$R = $ *gas constant*
$i = $ *index for organic waste according to biodegradation rates*

NEAR FIELD TRANSPORT

After radionuclides are released from waste packages, they will migrate through the backfill within the disposal unit. This process is governed by advection, dispersion, and the sorption characteristics of the backfill. If sorption in the backfill become significant, near-field transport can have a major impact on the source term. A one-dimensional transport equation is used for the near-field transport:

$$\frac{\partial C(x,t)}{\partial t} = -\frac{v_D}{R\theta}\frac{\partial C(x,t)}{\partial x} + \frac{D}{R}\frac{\partial^2 C(x,t)}{\partial x^2} - \lambda C + \frac{q(x,t)}{R\theta} \quad Eq.\,(10)$$

where,
v_D = *Darcy velocity*
$q(x,t)$ = *source concentration* (*based on the predicted release from waste packages*)

This equation is solved by the finite difference method to consider the space-dependent nature of parameters with boundary conditions of zero concentration at the top of the facility and zero flux gradient at the bottom boundary of the facility.

SOURCE TERM CALCULATIONS

To analyze the characteristics of the new model, various calculations were made. As a test case, ^{14}C is chosen in this study considering its importance in LLW performance assessment. A hypothetical waste facility was set up as an engineered concrete vault with a cover system where Class A waste containers are placed in the upper layers and Class B/C containers are placed at the bottom. For a base case, it was assume that the vault is backfilled with soil.

Infiltration of water into waste disposal facilities is dependent on the performance of the engineered barriers. In this study, infiltration is conservatively assumed to be the saturated hydraulic conductivity of each component of the barrier system [8] depending on the time-dependent scenario of engineered barrier failures. When engineered barriers are completely failed, natural water infiltration at the site is used. Natural infiltration at the site was assumed to be 0.1 m/yr. Parameters used in the analysis are listed in Table 1. These input parameters are either based on the test case analysis of NRC [8] or conservative estimates based on available literature.

Table 1. Parameters used in the Source Term Analysis
K_d of ^{14}C: 0.01 (in waste); 0.01 (in soil); 10 (in concrete)
D_{eff} of ^{14}C: 10^{-10} cm^2/sec (in waste); 10^{-12} cm^2/sec (in grout backfill); 10^{-6} cm^2/sec (in water)
Solubility of ^{14}C: 1.7×10^{-6} g/ml
Saturated hydraulic conductivity: 10^{-9} cm/s (of concrete); 10^{-7} cm/s (of the cover system)
Time when the barrier begins to fail: 75 yr (for concrete vault); 775 yr (for the cover system)
Time when the barrier is considered fully failed: 125 yr (for concrete vault); 825 yr (for the cover system)

Figure 2 represents the predicted diffusive ^{14}C release from the use of Eq. (7). Results are presented for both the case of single-time-to-failure (STF) and distributed failure (DF). For the STF case, the assumed time of failure was: 0 years for carbon steel containers; 5 years for carbon steel liners, and; 300 years for HICs. Results are also compared with the diffusion model of DUST [9]. Results show that using STF will have a significant impact on the peak release. Comparison with DUST shows that DUST results, which are based on a zero concentration boundary condition, are conservative. Use of an asymptotic solution (Eq.(7)) is expected to be appropriate for a period of several hundred years [5]. The backfill-dependent diffusive release of ^{14}C is represented in Figure 3 where the release rates are compared for different cases (i.e., different backfill materials). When cement grout backfill is used, the peak release was reduced by a factor of up to 7 due to the lower diffusion coefficient ($D_{backfill} <$ D_{waste}). If it is assumed that the waste is placed in water, the release rate is much higher ($D_{backfill} >> D_{waste}$).

Figure 4 shows the impact of using the DF approach on the surface rinse release. Results show that the surface rinse is mainly governed by water infiltration. However, the results also varied significantly between the DF and the STF case. The results also show that the present model is much more conservative than the PRESTO model [10]. The difference between the present model and PRESTO arises mainly from the difference in how the activity available for surface rinse is defined. In PRESTO, only the activity in a saturated condition was defined as the available inventory. This issue of the effects of unsaturated condition on surface rinse needs to be further investigated.

Figure 5 shows the effects of gas generation on the ^{14}C source term. The results are strongly dependent on the temperature of the facility and how the release of ^{14}C from biodegradable materials is modeled. The results show that the predicted failure of carbon steel liners has significant effect on the source term since the vault longevity does not exceed the longevity of carbon steel liners in the assumed scenario. In this prediction, it was assumed that the release of ^{14}C from biodegradable materials (predominantly ion-exchange resins) is described by either diffusion or surface rinse. If the surface rinse controls the release, the effect of gas generation is somewhat greater. Overall the effect does not control the source term but is significant for the release from Class B/C wastes. Gas generation is also dependent on the water infiltration, initial water content in the waste and the rate of biodegradation.

Figure 6 shows the prediction of source term for different near-field transport characteristics. The results confirm that the source term is reduced when near-field transport is described compared to the single-mixing cell case. When the backfill is not highly sorptive, the source term is slightly dispersed with time delay. When the backfill is highly sorptive, the predicted source term is much lower than that of the single mixing cell.

Figure 2. Comparisons of Predictions of Diffusive Release

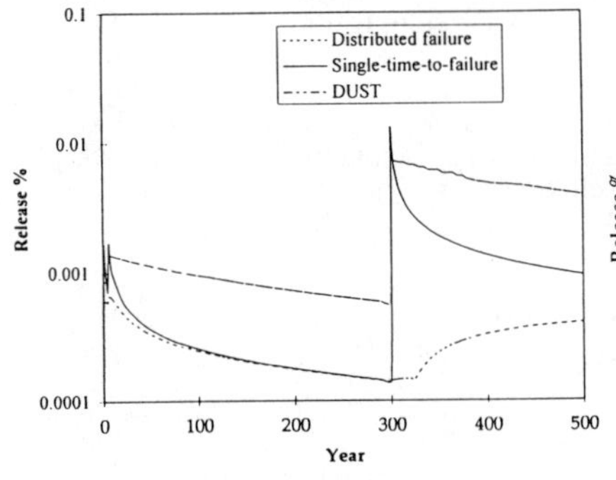

Figure 3. Effects of Backfills on Diffusive Release.

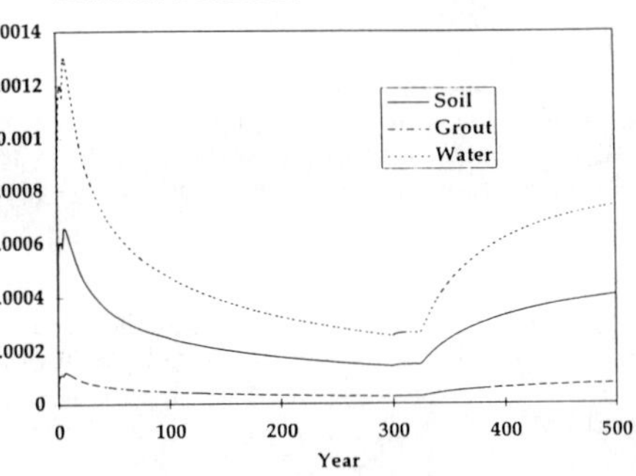

Figure 4. Comparisons of Surface Rinse Predictions.

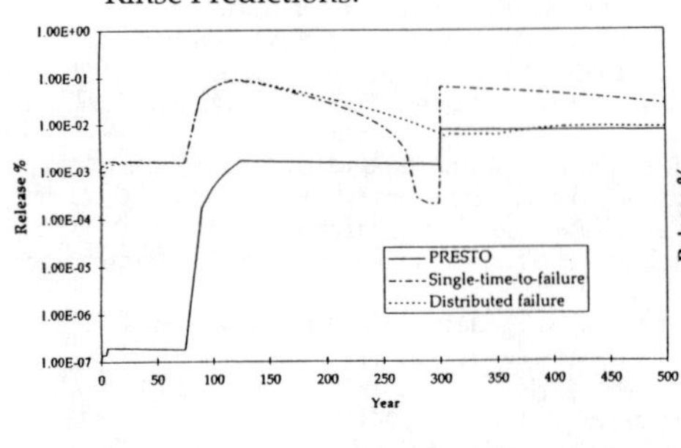

Figure 5. Predicted Effects of Gas Generation on Source Term (Source Term: Surface Rinse + Diffusion + Dissolution)

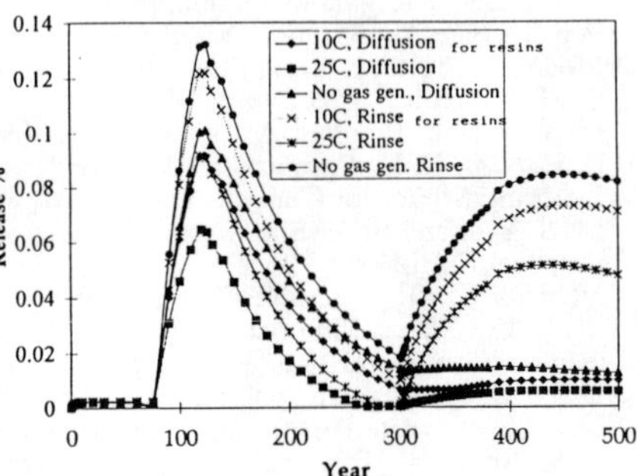

CONCLUSIONS

A new mechanistic LLW source term model was developed. Key innovations of this modeling effort compared with prior work include: use of cumulative failure functions to describe the failure of waste containers; capability to describe diffusion controlled release which is dependent on the conditions of the waste package surroundings; consideration of the effects of gas generation on the source term, and; use of a source inventory characterization routine to provide built-in capability for defining the distributions of radionuclides in various waste forms and streams. The following observations are made in the study:

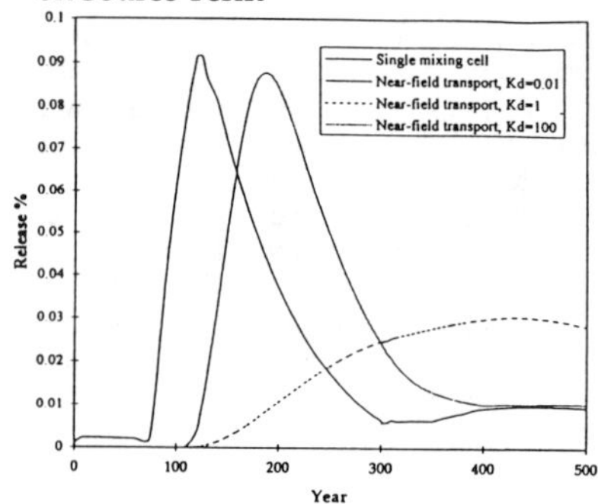

Figure 6. Effects of Near-Field Transport on Source Term

(1) Use of the distributed approach for waste container failure has a significant impact on the peak dose in performance assessment; (2) Describing backfill-dependent diffusion is important when a material with a low diffusion coefficient (or high sorption) is used as backfill; (3) Effects of unsaturated conditions on surface rinse release need to be further studied; (4) Effects of gas generation on the source term depend on water infiltration into the facility and how the release from biodegradable material is modeled. Gas release was not a controlling factor in the present simulations, and; (5) When a highly sorptive material is used as backfill, the source term is significantly reduced by near-field transport.

ACKNOWLEDGMENTS

The author gratefully acknowledges the support of Carol Hornibrook of Electric Power Research Institute for this work and the comments of Scott A. Simonson of MIT.

REFERENCES

1. D. B. Bullen, "A Model for Container Performance in An Unsaturated Repository," Nuclear Technology, Vol. 113, pp. 29-45, 1996.
2. T. M. Sullivan and S. J. Suen, "Low-Level Waste Shallow Land Disposal Source Term Model: Data Input Guides," NUREG/CR-5387, U. S. Nuclear Regulatory Commission, 1989.
3. U.S. Nuclear Regulatory Commission, "Source Term Implementation for Initial Performance Assessment Calculation," Technical Review of Performance Assessment Modeling Strategy, Rockville, MD, July 12-13, 1993.
4. T. M. Sullivan, "Selection of Models to Calculate the LLW Source Term," NUREG/CR-5773, U. S. Nuclear Regulatory Commission, 1991.
5. C. L. Kim, C. H. Cho, and H. J. Choi, "Equilibrium Concentration and Diffusion Controlled Leaching Model for Cement-Based Waste Forms," Waste Management, Vol.15, no. 5/6, 1995.
6. M.-S. Yim and S. A. Simonson, "Generation and Release of Radioactive Gases in LLW Disposal Facilities," The 23rd DOE/NRC Nuclear Air Cleaning and Treatment Conference, NUREG/CP-0141, CONF-940738, 1995.
7. M.-S. Yim, S. A. Simonson, and T. M. Sullivan, "Investigation of C-14 Release from a Generic Low-Level Waste Disposal Facility," Nuclear Technology, Vol. 114, no. 2, 1996.
8. U.S. Nuclear Regulatory Commission, "Overview of NRC Staff LLWPA Modeling Efforts," Low-Level Waste Performance Assessment Workshop, Rockville, MD, November 17, 1994.
9. T. M. Sullivan, "Disposal Unit Source Term (DUST) Data Input Guide," NUREG/CR-6041, U. S. Nuclear Regulatory Commission, 1993.
10. V. Rogers and C. Hung, "PRESTO-EPA-CPG: A Low-Level Radioactive Waste Environmental Transport and Risk Assessment Code, Documentation and User's Manual," EPA 520/1-87-026, U. S. Environmental Protection Agency, 1987.

A MATHEMATICAL MODEL FOR ADVECTIVE RELEASE FROM PERFORATED WASTE PACKAGE CONTAINER UNDER DRIPPING WATER

Joon H. Lee
INTERA, Inc., CRWMS M&O, 1180 Town Center Drive, Las Vegas, NV 89134, USA

ABSTRACT

A defense-in-depth engineered barrier system (EBS) is employed in the current design concept for the potential high-level nuclear waste repository at Yucca Mountain, Nevada, USA. Simplifying the geometry of the cylindrical waste container into the equivalent spherical configuration, and incorporating detailed analysis of the mechanics of water flow around the waste container surface, a mathematical model is developed for advective release from a "failed" (or perforated) waste container under dripping water. It is shown that the advective release rates are controlled by diffusion through the perforations in the waste container, and affected *insignificantly* by the dripping flow rate for the flow rate range considered. The release rates depend strongly on the number of perforations (or pit penetrations) in the waste container. The insensitivity of the release rate to the dripping flow rate is explained by the fact that radionuclide is released from the container surface to the boundary layer of the water film which contacts the container surface and is relatively stagnant. Also, since a laminar flow around the waste container surface is assumed in the model development, radionuclides transport across the water layers in the film by diffusion only. Additionally, the insignificant effect of the flow rate is contributed by the "short" penetration depth of radionuclide into the water film assumed in the model development. The analyses show that the number of perforations, the size of perforation, the container wall thickness, and the geometry (i.e., radius) of the waste container are important parameters that control the advective release rate. It is emphasized that the "failed" (or perforated) waste package container can still perform as a potentially important barrier to radionuclide release.

INTRODUCTION

A defense-in-depth engineered barrier system (EBS) is employed in the current design concept for the potential high-level nuclear waste repository at Yucca Mountain, Nevada, USA. The primary component of the EBS is a two-layer waste package container, which will be designed to achieve substantially complete containment of the waste and gradual release after it is breached [1]. Except for a small fraction of the waste containers that may fail prematurely due to materials and/or manufacturing defects, the waste containers are likely to fail by localized corrosion, especially by pitting corrosion [1,2]. Mathematical models were developed for diffusive mass transfer of radionuclides from a "failed" (or perforated) waste container [3]. Previous analyses showed that advective release from the waste packages under dripping fractures is the dominant EBS release process [4]. This paper discusses a mathematical model developed for advective release from a perforated waste container under dripping water by incorporating detailed analyses of fluid mechanics of water flowing around the waste container surface. The mathematical model discussed in this paper is developed specifically for the source-term calculation in the Repository Integration Program (RIP) [5], which was used in the recent iteration of total system performance assessment-1995 (TSPA-1995) of the potential repository [1].

DESCRIPTION OF THE CONCEPTUAL MODEL

The current design concept for the potential repository describes a cylindrical waste package container placed horizontally on the invert in the emplacement drift. In this paper, the waste (either

Mat. Res. Soc. Symp. Proc. Vol. 465 © 1997 Materials Research Society

spent nuclear fuel or vitrified high-level waste) is assumed to be distributed uniformly inside the waste container. The waste containers are assumed to fail by pitting corrosion. [Failure of waste container is defined as the first pit penetration through the container wall.] After a failure of waste container, the waste inside the container is assumed to be exposed immediately to the near-field environment, making it available for alteration/dissolution. Then, radionuclides may be released and transported through the failed waste package.

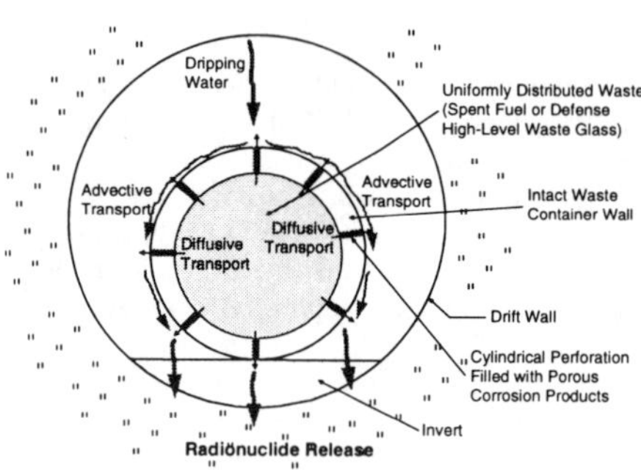

Fig. 1 Schematic of the conceptual model for advective release from waste package with multiple perforations.

To simplify the model development, the cylindrical waste package container is approximated by an equivalent spherical configuration. Fig. 1 shows a schematic of the conceptual model for advective release of radionuclide from a perforated waste container under dripping water. The perforations are assumed to be filled with fine, gel-like porous corrosion products [6-8], which may keep dripping groundwater from flowing through the waste package and thus from directly contacting the waste form. Thus, dripping groundwater is assumed to be diverted around the waste package container. The waste form inside the perforated container is assumed to undergo degradation under thin water film that forms on the waste form surface, and the radionuclides that were released are assumed to be transported through the perforations (filled with porous corrosion products) by diffusion to the outside surface of the waste container. Once outside, these radionuclides are released from the waste container by advection. The diffusive mass transfer rate of radionuclide from the perforated waste container is restricted by the number of perforations on the container (i.e., by the area available for diffusion). This conceptual model accounts for potential performance of the "partially" failed (or perforated) waste container as a barrier to radionuclide release. More details of the conceptual model are given elsewhere [1,3,4].

DIFFUSIVE MASS TRANSFER FROM PERFORATED WASTE CONTAINER

Steady-state diffusion is assumed in the perforations filled with porous corrosion products in the waste container. The perforation is assumed to be cylindrical in shape of length l and radius r. The steady-state diffusive mass transfer rate ($\dot{M}_{diff}$) from a waste container with N uniformly distributed perforations is expressed as follows [3]

$$\dot{M}_{diff} = \frac{C_0 - C_2'}{\dfrac{1}{N\pi r^2}\left[\dfrac{\pi r}{4}\left(\dfrac{1}{D_0\varepsilon_0} + \dfrac{1}{D_2\varepsilon_2}\right) + \dfrac{l}{D_1\varepsilon_1}\right]} \tag{1}$$

C_0 is the concentration of radionuclide at the surface of waste form, and C_2' is the concentration of radionuclide at the interface between the perforation and the boundary layer of the water film. ε_i and D_i are the porosity and diffusion coefficient, respectively, in the region i ($i = 0, 1,$ and 2): Region 0 is the opening (or boundary) of the perforation from the waste form side; Region 1 is the cylindrical portion of the perforation which is filled with porous corrosion products; and Region 2 is the exit of the perforation separating the perforation from the boundary layer of the water film, which contacts the waste container surface. Details of the model derivation is given in Ref. [3]. Since the diffusive release model (and advective release model developed in the next section) is to be implemented into the RIP code [5], and since the code has a module to calculate the radionuclide decay, the decay term is not considered in the model development. The diffusion coefficient in the porous corrosion product is calculated as a function of the percent volumetric water content, Φ, which was developed using the data from Ref. [9].

$$\log D = -8.255 + 1.898 \log \Phi \tag{2}$$

D is the diffusion coefficient (cm^2/sec). The bulk water diffusion coefficient (10^{-5} cm^2/sec) is assumed in Regions 0 and 2 (D_0 and D_2), and the porosities in the regions (ε_0 and ε_2) are set to 1. The porosity of the corrosion product is assumed to be 0.4.

ADVECTIVE MASS TRANSFER FROM PERFORATED WASTE CONTAINER UNDER DRIPPING WATER

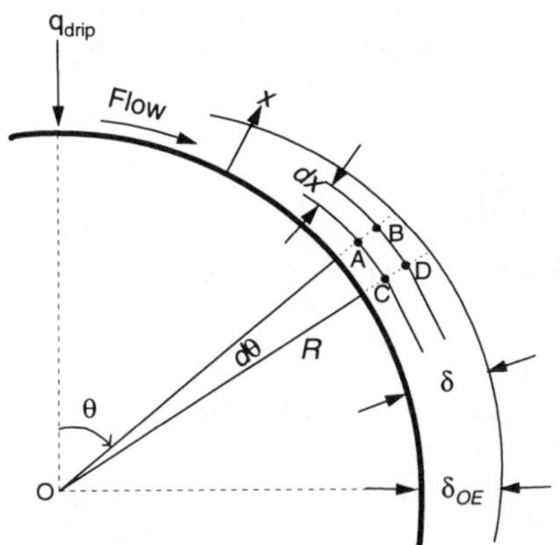

Fig. 2 Schematic of laminar flow of water over a solid sphere viewed at the vertical section through the sphere center (not to scale).

As discussed in the conceptual model, advective release of radionuclide from a perforated waste container under dripping water is assumed to involve diffusive transport of radionuclide through the perforations and "forced-convective" release from the container by the water flowing around the container. This involves diffusion of radionuclide from the container surface into a moving water stream in which the velocity field is known. A schematic of the water flow around the "spherical" waste container at the vertical section through the sphere center is shown in Fig. 2 (not drawn to scale). Since the waste container has a surface area of about 37 m^2, and since relatively low dripping flux (10^{-5} to 10 mm/yr) are expected in the emplacement drift [1], a laminar flow is assumed over the entire container surface. It is assumed that along a radial line (for example, OA), the water velocity (v) is in the plane of the diagram and is nearly perpendicular to the radial line. In the figure, x is radial distance from the sphere surface, R is the radius of the sphere, δ is the water film thickness, δ_{OE} is the water film thickness at the equator ($\theta = \pi/2$), and q_{drip} is the volumetric flow rate of water dripping onto the container.

It is also assumed in the model development that 1) no evaporation from the water film occurs; 2) flow field around the sphere is not affected by the diffusion; and 3) since release (or diffusion) of radionuclide from the waste container into the flowing water film can occur only where a pit (assumed to have a constant surface area of 1-mm^2 or a radius of 0.56-mm [10]) has penetrated through the container wall, the contact time between the laminar water flow and the perforation surface is "short". The last assumption is equivalent to assuming a small penetration depth by the diffusing radionuclide into the water film.

Writing a solute balance on a differential ring of the water film having a perimeter $2\pi R \sin\theta$ (at constant angle θ) and a radial thickness dx, which is indicated in Fig. 3 by the differential element ABCD (bounded by two streamlines AC and BD and by two radial lines AB and CD), gives

$$vC(2\pi R \sin\theta)dx - v\left[C + \left(\frac{\partial C}{\partial\theta}\right)_\psi d\theta\right](2\pi R \sin\theta)dx =$$

$$(2\pi R \sin\theta)(R\,d\theta)D\left(\frac{\partial C}{\partial x}\right)_\theta - (2\pi R \sin\theta)(R\,d\theta)D\left[\left(\frac{\partial C}{\partial x}\right)_\theta + \frac{\partial}{\partial x}\left(\frac{\partial C}{\partial x}\right)_\theta dx\right] \tag{3}$$

where v is the tangential water velocity at x, C is the radionuclide concentration in the water film, and D is the diffusion coefficient in the water film. In deriving Eqn. (3), x and δ are neglected in comparison with R; also, no convection across the streamlines is assumed, and diffusion across the radial lines is neglected. After summing up the terms, Eqn. (3) becomes

$$v\left(\frac{\partial C}{\partial\theta}\right)_\psi = RD\left(\frac{\partial^2 C}{\partial x^2}\right)_\theta \tag{4}$$

The partial differential on the left is taken at constant volumetric water flow (ψ) along a streamline between any two values of x. It is easily shown that the flow (ψ) at any θ is a function of x/δ ($= p$) only [11]. Thus, by introducing the parameter p, Eqn. (4) is rewritten as follows

$$v\left(\frac{\partial C}{\partial\theta}\right)_p = \frac{RD}{\delta^2}\left(\frac{\partial^2 C}{\partial p^2}\right)_\theta \tag{5}$$

Expressions for v, q_{drip}, and δ are obtained by equating the force of gravity to the resisting shear force on the differential element ABCD as follows

$$v = \frac{\rho g \delta^2 \sin\theta}{\mu}\left[\frac{x}{\delta} - \frac{1}{2}\left(\frac{x}{\delta}\right)^2\right] = 2v_s\left[\frac{x}{\delta} - \frac{1}{2}\left(\frac{x}{\delta}\right)^2\right] \tag{6}$$

$$q_{drip} = \frac{2\pi R \rho g \delta^3 \sin^2\theta}{3\mu} \tag{7}$$

$$\delta = \left(\frac{3\mu q_{drip}}{2\pi R \rho g \sin^2\theta}\right)^{1/3} = \delta_{OE}\,\sin^{-2/3}\theta \tag{8}$$

v_s is the surface velocity at $x = \delta$, g is the acceleration due to gravity, and ρ and μ are respectively the density and viscosity of water in the film. As mentioned previously (Fig. 2), a small penetration depth of radionuclide into the water film is assumed, thus $x \ll \delta$, and $p \gg \frac{1}{2}p^2$ in Eqn. (6). This gives

$$p - \frac{1}{2}p^2 \approx p \tag{9}$$

Then, Eqn. (5) is re-written as follows

$$\frac{2v_s\delta^2}{R}\,p\left(\frac{\partial C}{\partial\theta}\right)_p = D\left(\frac{\partial^2 C}{\partial p^2}\right)_\theta \tag{10}$$

Eqn. (10) is further simplified by introducing the following relationship [11].

$$\frac{d\theta}{d\phi} = \frac{2v_s\delta^2}{R} \tag{11}$$

Thus, Eqn. (10) becomes

$$p\left(\frac{\partial C}{\partial \phi}\right)_p = D\left(\frac{\partial^2 C}{\partial p^2}\right)_\phi \tag{12}$$

This is the final form of the desired differential expression. Eqn. (12) can be solved with the following boundary conditions

$$B.C.\ 1 \qquad C = C_2 \quad at \quad p = 0 \quad (or \quad x = 0) \tag{13a}$$

$$B.C.\ 2 \qquad \frac{\partial C}{\partial p} = 0 \quad at \quad p = 1 \quad (or \quad x = \delta) \tag{13b}$$

$$B.C.\ 3 \qquad C = 0 \quad at \quad \phi = 0 \quad (or \quad \theta = 0) \tag{13c}$$

Now seek a solution to Eqn. (12) in a form of

$$\frac{C}{C_2} = f(\eta) \tag{14}$$

where η is a dimensionless parameter defined as

$$\eta = p\left(\frac{1}{9D\phi}\right)^{1/3} \tag{15}$$

This solution technique was applied to a similar problem [12, p. 551]. Utilizing Eqns. (14) and (15), Eqn. (12) is reduced to

$$\frac{d^2 f}{d\eta^2} + 3\eta^2 \frac{df}{d\eta} = 0 \tag{16}$$

Solution to Eqn. (16) is found with the following boundary condition that is derived from Eqn. (13a).

$$f = 1\ (or\ C = C_2) \qquad at \quad \eta = 0\ (or\ p = 0) \tag{17}$$

The resulting expression for the solution to Eqn. (12) is

$$C = f C_2 = \frac{C_2}{\Gamma\left(\frac{4}{3}\right)} \int_\eta^\infty e^{-\eta^3}\, d\eta \tag{18}$$

where $\Gamma(n)$ is the gamma function. The advective release rate over the waste container "sphere" from the top to a point of withdrawal at $\theta = \theta_2$ is obtained as follows

$$\dot{M}_{adv} = \int_0^{\theta_2} \left(\frac{N\pi r^2}{4\pi R^2}\right)(-2\pi R^2 \sin\theta)D\left(\frac{\partial C}{\partial x}\right)_{x=0} d\theta \tag{19}$$

With the aid of Eqns. (6) to (8) and (11), Eqn. (19) becomes

$$\dot{M}_{adv} = -3D\,q_{drip}\left(\frac{Nr^2}{4R^2}\right)\int_0^{\phi_2}\left(\frac{\partial C}{\partial p}\right)_{p=0} d\phi \tag{20}$$

Evaluating the integration in Eqn. (20) gives

$$\dot{M}_{adv} = \frac{9D\,q_{drip}}{2\Gamma\!\left(\dfrac{4}{3}\right)}\left(\frac{Nr^2}{4R^2}\right)\left(\frac{1}{9D}\right)^{1/3}\phi_2^{2/3}\,C_2 \tag{21}$$

The advective release rate from the entire waste container surface is obtained by substitution into Eqn. (21) of ϕ_2 that is determined by integrating Eqn. (11) from $\theta = 0$ to π. Integrating Eqn. (11) and using Eqns. (6) to (8),

$$\phi_2 = \frac{R}{2}\int_0^\pi \frac{1}{v_s\,\delta^2}\,d\theta = \frac{\mu R}{\rho g\,\delta_{OE}^4}\int_0^\pi \sin^{5/3}\theta\;d\theta = \frac{1.68\,\mu R}{\rho g}\left(\frac{2\pi R\rho g}{3\mu\,q_{drip}}\right)^{4/3} \tag{22}$$

The value of the integration (i.e., 1.68) is taken from Ref. [11]. After substitution of ϕ_2 into Eqn. (21) and re-arranging the resulting equation, the equation for $\dot{M}_{adv}$ is obtained as follows

$$\dot{M}_{adv} = Q\,q_{drip}^{1/9}\,N\,C_2 \tag{23}$$

where

$$Q = 0.85591\,r^2 D^{2/3}\left(\frac{2\pi}{3}\right)^{8/9}\left(\frac{\rho g}{\mu}\right)^{2/9} R^{-4/9} \tag{24}$$

The value (i.e., 0.893) of the gamma function in Eqn. (21) is taken from Ref. [13, p. 349]. C_2 in Eqn. (23) can be equated with $C_2{}'$ in Eqn. (1). Thus, equating $\dot{M}_{adv}$ in Eqn. (23) with $\dot{M}_{diff}$ in Eqn. (1) and solving for C_2, an equation for the steady-state advective mass transfer rate from perforated waste container under a constant dripping flow rate is expressed as

$$\dot{M}_{adv} = \frac{Q\,q_{drip}^{1/9}\,N\,C_0}{\dfrac{1}{\pi r^2}\left(\dfrac{\pi r}{2D_0\varepsilon_0} + \dfrac{l}{\varepsilon_1 D_1}\right)Q\,q_{drip}^{1/9} + 1} \tag{25}$$

Eqn. (25) shows that a power of $1/9$ raised to q_{drip} diminishes the effect of dripping flow rate on the advective release rate. It is also shown that the area of perforation (r), the number of perforations (N), and the container wall thickness (l) are important parameters controlling the advective release rate. Eqn. (25) is valid for the volumetric dripping rates in which the water film flow is laminar (i.e., Reynolds number less than 2,100) [12].

ADVECTIVE RELEASE RATE FROM PERFORATED WASTE CONTAINER

Advective release rates from a perforated waste container are calculated with Eqn. (25) for a range of perforation numbers (10^3 to 10^6 perforations) [1,2] and volumetric dripping rates up to 1.0 m³/yr for which the water film flow is laminar (i.e., Reynolds number less than 2,100). The range of the perforation numbers used in this paper are from the pit density (10 pits/cm²) of the waste package used in related studies [1, 2, 14]. In the calculations, the radius of perforation is assumed 0.56-mm, and the volumetric water content of the corrosion products in the perforations is assumed 20%. The density of water assumed is 1.0 g/cm^3, the viscosity 1.0 centipoise (0.01 g/cm·sec), and the acceleration of gravity 980 cm/sec^2. The results are shown in Figs. 3-A and 3-B for a container wall thickness of 12-cm and 2-cm, respectively. The radius of the waste container sphere is 1.721-m for the case of 12-cm thick container wall (Fig. 3-A) and 1.610-m for the case of 2-cm thick container wall (Fig. 3-B).

As shown in the figures, the advective release rates are controlled by diffusion through the perforations in the waste container, and not affected by the dripping flow rates for the flow rate range considered in this analysis. The release rates depend strongly on the perforation numbers. When the container wall thickness is reduced from 12-cm to 2-cm, the release rates increase by a factor of about 5. The insensitivity of the release rate to the dripping flow rate is indicated from the *1/9* power raised to the dripping flow rate in the model (Eqn. (25)). This is explained by the fact that radionuclide is released from the container surface to the boundary layer of the water film, which contacts the container surface and is relatively stagnant due to the viscosity of the water. Since a laminar flow is assumed in the model development, there is no water flow across the water layers in the flowing water film, and mixing of radionuclide within the water film is minor. In this model, radionuclides transport across the water layers by diffusion only (the concentration gradients being established in the direction from the (nearly stagnant) boundary layer contacting the container surface to those away from the surface). In addition, the insignificant effect of the flow rate is caused by a "short" penetration depth of radionuclide into the water film assumed in the model development (i.e., Eqn. (9)).

Additionally, alternative conceptual models and assumptions for the advective release from a perforated waste package under dripping water should be evaluated to investigate the sensitivity of the waste package release behavior.

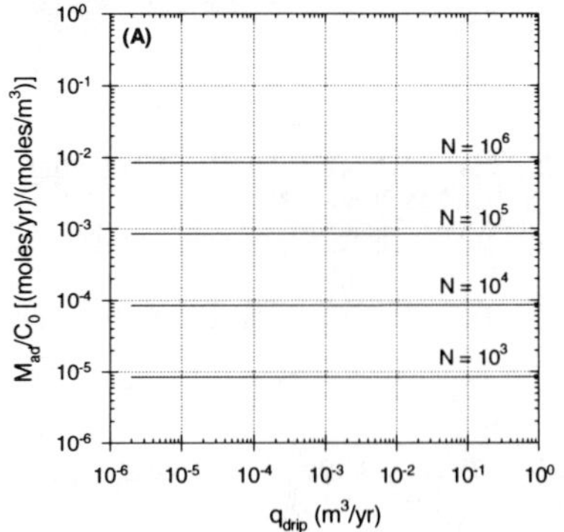

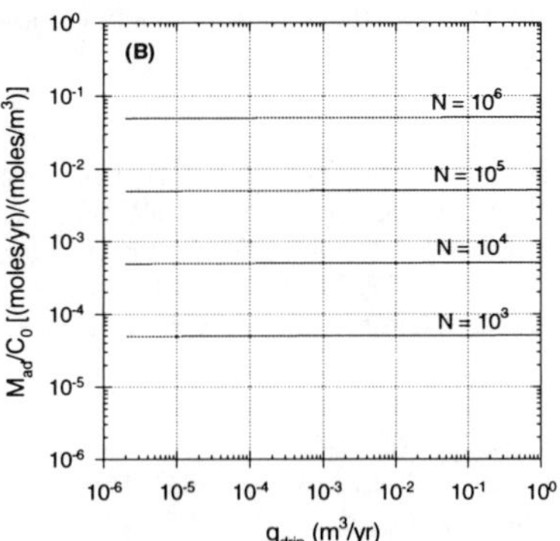

Fig 3. Advective release rates from a waste container with multiple perforations for the waste container wall thickness of (A) 12 cm and (B) 2 cm. *N* is the number of perforations.

SUMMARY AND CONCLUSION

A defense-in-depth EBS is employed in the current design concept for the potential high-level nuclear waste repository at Yucca Mountain, Nevada, USA. The primary component of the EBS is a multi-barrier waste package container, which will be designed to achieve substantially complete containment of the waste and controlled release after it is breached. Simplifying the geometry of the cylindrical waste container into the equivalent spherical configuration, and incorporating detailed analysis of the mechanics of water flow around the waste container surface, a mathematical model is developed for the advective release rate from a "failed" (or perforated) waste container under dripping water. Using the model, advective release behaviors from the perforated waste container are studied.

It is shown that the advective release rates are controlled by diffusion through the perforations in the waste container, and affected *insignificantly* by the dripping flow rate for the flow rate range considered. The release rates depend strongly on the perforation numbers. When the container wall thickness is reduced from 12-cm to 2-cm, the release rates increase by a factor of about 5. The insensitivity of the release rate to the dripping flow rate is explained by the fact that radionuclide is released from the container surface to the boundary layer of the water film, which contacts the container surface and is relatively stagnant due to the viscosity of water. Since a laminar flow is assumed in the model development, there is no water flow across the water layers in the flowing water film, and mixing of radionuclide within the water film is minor. In this model, radionuclides transport across the water layers by diffusion only. Additionally, the insignificant effect of the flow rate is contributed by the "short" penetration depth of radionuclide into the water film, assumed in the model development.

The analyses show that the number of perforations, the size of perforation, the container wall thickness, and the geometry (i.e., radius) of the waste container are important parameters that control the advective release rate. It is emphasized that the "failed" (or perforated) waste package container can still perform as a potentially important barrier to radionuclide release. Finally, alternative conceptual models and assumptions for the advective release from a perforated waste package under dripping water should be evaluated to investigate the sensitivity of the waste package release behavior.

ACKNOWLEDGMENTS

This work was funded by the US DOE Yucca Mountain Site Characterization Office under Contract #DE-AC01-91RW00134 to TRW Environmental Safety Systems, Inc. The author would like to thank two anonymous reviewers for their valuable comments.

REFERENCES

1. CRWMS M&O, Total System Performance Assessment-1995: An Evaluation of the Potential Yucca Mountain Repository," B00000000-01717-00136, Rev. 01, Las Vegas, NV, 1995.
2. J.H. Lee, J.E. Atkins, and R.W. Andrews, Scientific Basis for Nuclear Waste Management XIX, Mat. Res. Soc. Proc. 412, W.M. Murphy and D.A. Knecht (eds.), p. 603 (1996).
3. J.H. Lee, P.L. Chambré, and R.W. Andrews, Proc. 1996 Int'l Conf. Deep Geological Disposal of Radioactive Waste, p. 5:61 (1996).
4. J.H. Lee, J.E. Atkins, J.A. McNeish, and V. Vallikat, Proc. 1996 Int'l Conf. Deep Geological Disposal of Radioactive Waste, p. 8:65 (1996).
5. Golder Associates, RIP Performance Assessment and Strategy Evaluation Model: Theory Manual and User's Guide, Version 3.20, Golder Associates, Redmont, WA, 1994.
6. W.H. Vernon, Trans. Electro. Soc., **64**, p. 31 (1933).
7. A. Raman, and S. Nasrazadani, Corrosion, **46**, p. 601 (1990).
8. D. Fyfe, Corrosion, 3rd ed., L.L. Shreir, et al (eds.), Vol. 1, Butterworth-Heinemann, p. 2:31 (1994).
9. J.L. Conca, and J. Wright, Appl. Hydro., **1**, p.5 (1992).
10. G.A. Henshall, Modeling Pitting Degradation of Corrosion Resistant Alloys, UCRL-ID-125300, Lawrence Livermore National Laboratory, Livermore, CA, 1996.
11. J.F. Davidson, and E.J. Cullen, Trans. Inst. Chem. Engrs., **35**, p. 51 (1957).
12. R.B. Bird, W.E. Stewart, and E.N. Lightfoot, Transport Phenomena, John Wiley & Sons, 1960.
13. W.H. Beyer, CRC Standard Mathematical Tables, 28th Ed., CRC Press, 1987.
14. J.H. Lee, J.E. Atkins, and B. Dunlap, this volume.

INCORPORATION OF "CORROSION-TIME" AND EFFECTS OF CORROSION-PRODUCT SPALLING IN WASTE PACKAGE DEGRADATION SIMULATION IN THE POTENTIAL REPOSITORY AT YUCCA MOUNTAIN

Joon H. Lee, Joel E. Atkins, and Bryan Dunlap
INTERA, Inc., CRWMS M&O, 1180 Town Center Drive, Las Vegas, NV 89134, USA

ABSTRACT

A two-layer waste package (carbon steel outer barrier and Alloy 825 inner barrier) is specified to dispose of high-level nuclear waste at the potential repository at Yucca Mountain. A set of improvements and more realism have been added to a stochastic waste-package degradation model which was developed for a recent total system performance assessment of the potential repository [1]. The waste-package surface is divided into "patches" to better represent the general corrosion of the carbon-steel outer barrier. The "corrosion-time" concept is developed to represent the corrosion of the carbon-steel outer barrier in changing exposure conditions with time such as those expected in the potential repository. With the patches approach and the corrosion-time concept implemented into the waste-package degradation model, sensitivity of the waste package degradation (failure and pitting degradation) to different threshold spalling thicknesses of the corrosion products from the carbon-steel outer barrier is analyzed. The results show that the waste-package pitting degradation is sensitive to the corrosion-products spalling thickness of the carbon-steel outer barrier. A greater pitting degradation of the waste packages is predicted with a smaller spalling thickness. Further understanding of the corrosion-products spalling in different exposure conditions (i.e., water chemistry, water contact mode, etc.) and its effects on carbon steel corrosion is needed to enhance the confidence in the waste-package performance modeling in the potential repository.

INTRODUCTION

In the current design concept for the potential high-level nuclear waste repository at Yucca Mountain, Nevada, USA, a two-layer waste package container is employed to provide substantially complete containment of the waste and controlled release of radionuclide after it is breached [2]. Except for a small fraction (less than 1%) of the waste packages that are expected to fail prematurely due to materials and/or manufacturing defects [2], the waste packages are likely to fail by localized corrosion, especially pitting corrosion [1,3]. A stochastic waste package degradation model was developed as part of the recent iteration of total system performance assessment (TSPA-1995) of the potential repository [1,3]. The waste-package degradation model incorporates general and pitting corrosion models for the carbon-steel outer barrier [4] and a pitting corrosion model of the Alloy 825 inner barrier developed from an expert elicitation [5]. The model provides as output the time history of waste package failure (defined as the first pit penetration through the waste package) and subsequent pitting degradation of the waste packages. Currently, the waste-package degradation model is evolving to include improvements on the existing corrosion models and additional important corrosion processes such as crevice corrosion and stress corrosion cracking, and to incorporate more realism in the long-term waste-package degradation simulation. The data and information for these further developments of the waste-package degradation model are being developed from the on-going, site-specific testing programs [6]. This paper discusses recent improvements on the carbon steel outer-barrier corrosion modeling and their effects on the waste-package lifetime and subsequent pitting degradation.

Mat. Res. Soc. Symp. Proc. Vol. 465 © 1997 Materials Research Society

THE PATCHES APPROACH FOR THE CARBON STEEL OUTER-BARRIER CORROSION MODELING

In the TSPA-1995 version [1], the general corrosion depth of the carbon-steel outer barrier was sampled independently at the location of every pit. Both the humid-air and aqueous general-corrosion models are based on data from the literature [4], which were derived from the weight loss measurements of the sample coupons. Thus the general-corrosion depth at a given exposure time is an average over the entire exposed surface of the sample coupons. Most sample coupons were prepared following ASTM standard procedures [7]. The standards recommends a sample coupon size of 10 by 15-cm, which gives the surface area of the sample coupon of about 310 cm^2.

To reflect more adequately the general corrosion data from the literature, the waste package container surface area (about 37 m^2) is divided by the sample coupon surface area. A portion of the waste package surface which is equivalent to the surface area of a single sample coupon is referred to as a "patch." This process results in a total of 1,200 patches over the waste-package surface. A uniform general-corrosion depth is assumed within each patch, and the uniform general-corrosion depth is multiplied by (normally distributed) pitting factors to calculate the pit depth distribution within the patch [4]. This "patching" concept is particularly useful to incorporate other relevant corrosion processes, as their models become available, into the waste-package degradation model. Such corrosion processes that are considered important for the potential repository are the galvanic protection of the inner barrier by the carbon-steel outer barrier (e.g., ratio of exposed surface areas between the two metal layers and/or thickness reduction of the outer barrier) [8,9], effects of localized variations of the chemistry of the water contacting the waste container, locally enhanced corrosion within a waste package due to localized dripping and salt-scale buildup, etc [6].

THE "CORROSION-TIME" CONCEPT FOR THE CARBON STEEL OUTER-BARRIER CORROSION MODELING

The "corrosion-time" concept is developed to represent adequately the carbon-steel outer barrier corrosion under changing exposure conditions such as those expected in the potential repository. In the TSPA-1995 version [1,3], the models for general corrosion of the outer barrier in humid-air and aqueous conditions were developed using the literature data which were measured in "relatively" constant exposure conditions [4]. The general-corrosion model and the corrosion rate may be expressed simply as follow [1,4]:

$$D_g = f_{EC}\, t^{\,b} \tag{1}$$

$$\frac{dD_g}{dt} = f_{EC}\, b\, t^{\,b-1} \tag{2}$$

where D_g is the general corrosion depth (its unit depending on the choice of the units of other terms in the model), f_{EC} includes a constant term and the exposure conditions (i.e. temperature and, for humid-air corrosion, relative humidity), t is the (real) time elapsed after the initiation of corrosion, and b is a constant having a value between 0.0 and 1.0. The time term raised to a power of b (between 0 and 1) reflects a decrease in corrosion rate with time. This is due mostly to the accumulation of corrosion products, providing a protective layer as corrosion proceeds. As the corrosion product layer becomes thicker, the corrosion rate is further reduced because the movement of reaction species (both reactants and products) is further restricted. Eqns. (1) and (2) are applicable to constant exposure conditions. As an approximation in using Eqn. (1) in the TSPA-1995 version, the time steps were divided in such a way that the exposure conditions (i.e., f_{EC}) were relatively constant (within ±2%) in each time step.

<u>Implementation of the Corrosion-Time Concept</u>

Since the near-field environment in the emplacement drift of the potential repository changes with time, f_{EC} also changes with time. Therefore, a direct use of Eqn. (1) does not properly capture the effects of the changing exposure conditions. One possible way to adequately represent the effects of changing exposure conditions is to utilize the concept of "corrosion-time" (t_c) which entails adjusting the value of t (the real-time term in Eqn. (1)) for changing exposure conditions. At the beginning of each time step in which the exposure condition is relatively constant (within ±2%), the corrosion-time for the current time step is determined based on *the corrosion depth in the previous time step* and *the exposure conditions in the current time step*. This may be expressed as follows

$$t_{c,i} = \left(\frac{D_{g,i-1}}{f_{EC,i}} \right)^{1/b} \tag{3}$$

where $t_{c,i}$ is the corrosion-time at the beginning of the current time step i, $D_{g,\,i-1}$ is the general corrosion depth at the end of the previous time step i-1, $f_{EC,\,i}$ is the exposure condition for the current time step i, and b is constant. At the beginning of each time step, the time ($t_{c,\,i}$) that is required to corrode the corrosion depth ($D_{g,\,i-1}$) in the exposure conditions ($f_{EC,\,i}$) of the current time step is determined by Eqn. (3). Then, the estimated corrosion-time ($t_{c,\,i}$) is used for the time term in Eqn. (1) to calculate the general corrosion depth during the current time step. The use of $t_{c,\,i}$ in Eqn. (1) would properly capture the effects of the changing exposure conditions in the outer barrier general corrosion modeling.

EFFECTS OF CORROSION-PRODUCTS SPALLING OF THE CARBON STEEL OUTER-BARRIER ON WASTE PACKAGE DEGRADATION

Corrosion products of carbon steel are generally characterized as fine, gel-like particles that have a greater specific volume than the uncorroded metal [10-12]. If the metal under corrosion is confined, these corrosion products generate an expansive force. As the corrosion of the carbon steel barrier (10-cm thick) proceeds, the corrosion products formed previously would be pushed away from the surface of the bare carbon steel due to the expansive action generated by the corrosion products forming at the bare metal surface. Once the corrosion-products layer reaches a "threshold" thickness, the corrosion products at the outer edge of the layer would start to flake off the waste package. This process is referred to in this paper as "spalling", and the threshold corrosion-products thickness is referred to as the "spalling thickness" (l_{st}). From this stage, the corrosion-products layer would maintain a relatively constant thickness with continued spalling at a rate that corresponds to the rate of formation of new corrosion products at the surface of the bare carbon steel. There is some indication that different types of corrosion-products morphology (i.e., adherent/non-adherent, continuous/discontinuous, dense/cracked/porous, etc.) may result from different exposure conditions (temperature, humidity and chemistry) and water contact modes [13,14]. These factors and others such as chemical and physical conditions of carbon steel may contribute to variability in the spalling thicknesses.

<u>Implementation of the Corrosion-Products Spalling Effects</u>

It is assumed that spalling of corrosion products does not occur until the general corrosion depth of the carbon-steel outer barrier reaches some threshold spalling thickness, l_{st}. Once it reaches that threshold thickness, then a constant thickness of the corrosion-product layer remains on the waste package surface. Thus, once the general corrosion depth has reached l_{st}, general corrosion of the outer barrier continues at a rate that is determined by the exposure condition of each time step and the (constant) spalling thickness. The corrosion-time ($t_{c,\,st,\,i}$) for the current time step i is calculated from the constant spalling thickness in the exposure condition ($f_{EC,\,i}$) of the current time step as follows

$$t_{c,st,i} = \left(\frac{l_{st}}{f_{EC,i}} \right)^{\frac{1}{b}} \tag{4}$$

In the waste-package degradation modeling, once the general corrosion depth of the outer barrier reaches the spalling thickness, the corrosion-time estimated with Eqn. (4) is used for the time term in Eqn. (2). Thus, after spalling thickness is reached, the corrosion rate is dependent on the exposure condition (f_{EC}) only (or independent of time, t).

TEMPERATURE AND RELATIVE HUMIDITY AT THE WASTE PACKAGE SURFACE

The waste-package surface temperatures and relative humidities (RH) are calculated using the thermal-hydrological computer code NUFT (Nonisothermal Unsaturated-Saturated Flow and Transport) [15]. A two-dimensional grid is used for the thermal-hydrological simulation that reflects the center-in-drift waste-package emplacement. Symmetrical repository drift spacing and waste package spacing are assumed, thus the simulation reflects the conditions that an average waste package is sufficiently far away from the repository edge. The simulation is conducted for a thermal loading of 83 MTU per acre with no-backfill. Since the no-backfill scenario is analyzed, radiative heat transport connections are established between the waste package and the drift wall with an assumed waste-package surface emissivity of 0.3. A ground surface infiltration flux of 0.3 mm/year is specified and a static water table assumed. The material property data used in the simulation are derived from Ref. [16]. The simulation is performed for up to 100,000 years. Further details of the thermal-hydrological simulation are given elsewhere [17].

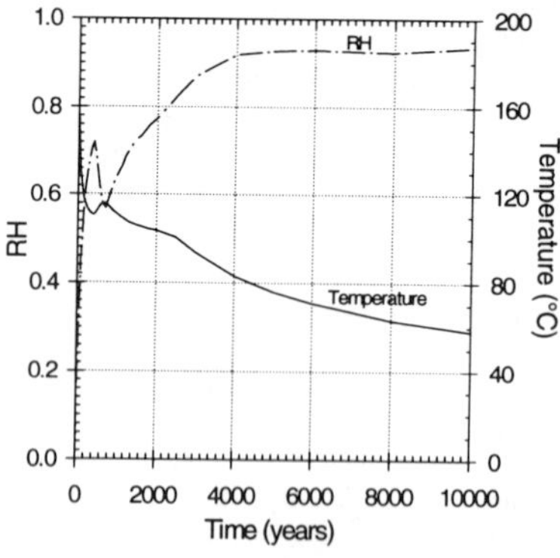

Fig. 1 Temperature and RH profiles at the waste package surface for the case of 83 MTU/acre and no-backfill.

The simulation results for the first 10,000 years are shown in Fig. 1. The waste-package surface temperature peaks to about 155°C, then gradually cools down with time. It reduces to 100°C at about 2,600 years and to about 60°C at 10,000 years. Relative humidity at the waste package surface increases sharply after the peak waste package surface temperature. It reaches about 72% RH, then drops sharply to about 56% RH. The initial spike of the relative humidity and the corresponding hump in temperature are caused by the development and decay of a porous media thermal flow phenomena referred to as a "heat-pipe" effect [17]. After the breakdown of the "heat-pipe", the relative humidity increases steadily and reaches about 84% at 2,600 years (at the corresponding time, the waste package surface temperature cools to 100°C) and about 92% at about 4,000 years. Then the relative humidity gradually levels off approaching the ambient relative humidity (about 98%).

WASTE PACKAGE DEGRADATION SIMULATION RESULTS

The temperature and relative humidity profiles at the waste-package surface (developed in the previous section) are input as a lookup table to the waste-package degradation model. A total of 400 waste packages are considered in each simulation. A pit density of 10 pits/cm^2 is assumed for the carbon-steel outer barrier [18,19], and the same pit density is assumed for the corrosion-resistant inner barrier (preliminary short-term test results indicate a higher pit density [20]). This gives 3,100 pits per patch for a total of 3.72 million pits per waste package. It is assumed that corrosion of waste package initiates once the surface temperature drops to 100°C.

In the potential repository, about 12,000 waste packages will be spread over the repository area. The exposure conditions (and the corrosion rates) would vary across the repository (this is referred to here as waste-package to waste-package variability). Also, since a waste package has a relatively large surface area (about 37 m^2), the corrosion rate would vary over the waste package surface (this is referred to here as pit to pit variability within a single waste package). Currently, information on the degree of the variability among waste packages and among pits within a single waste package is not available. As a way to explore the impact of such as-yet unquantified variability in the repository exposure conditions, half of the assumed variability in the outer barrier general-corrosion (humid-air and aqueous) models [3] is used for the variability between the patches within a single waste package, and the other half for the variability among waste packages. Following the similar approach, half of the assumed variability in the inner-barrier pitting model [2] is used for the variability between waste packages; a quarter of the assumed variability is used for the variability between patches on the same waste package; and the rest (i.e., the remaining quarter) of the assumed variability in the inner-barrier pitting model is used for the variability between pits in the same patch. Details of other assumptions used in the current simulations are found elsewhere [1,2].

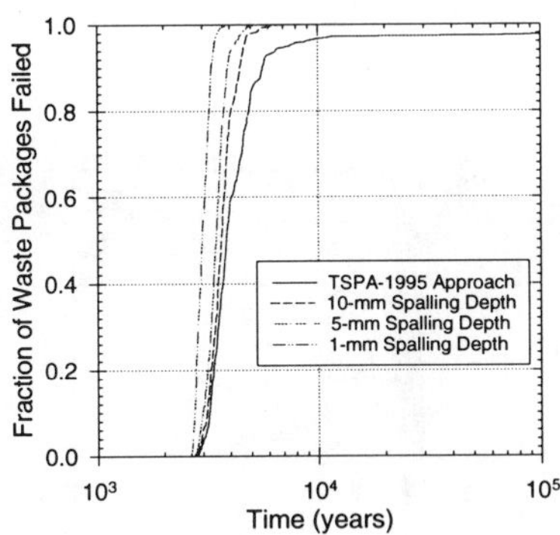

Fig. 2 Waste-package failure histories for the cases considered in this paper.

Fig. 2 shows the simulation results for the waste-package failure history. The failure histories with the patches and the corrosion-time approach with different spalling thicknesses (i.e., 10-mm, 5-mm and 1-mm) are compared with the failure history with the TSPA-1995 approach. In the TSPA-1995 approach, general corrosion depth of the outer barrier is calculated at the location of every pit using the real-time (instead of the corrosion-time); and spalling of the corrosion products of the outer barrier is not considered. No significant differences in the initiation time of waste package failure are predicted with the TSPA-1995 approach and the current approach (with the different spalling thicknesses). For all the cases, waste packages are predicted to begin to fail from about 2,600 to 2,800 years. However, the 'empirical' cumulative density functions (CDFs) of the failure history become "steeper" with a smaller spalling thickness. For example, with the TSPA-1995 approach, waste packages begin to fail at about 2,800 years, and after a rapid rise in the failure distribution, the waste-package failure rate spreads over a very long period. With the current approach and the 10-mm threshold spalling thickness, the failure begins at about 2,700 years, and all the waste packages fail by about 7,000 years. With the 1-mm spalling thickness and with the current approach, the failure begins at about 2,600 years, and all the waste packages fail by 4,000 years. The results show that the waste-package failure distribution is affected considerably by the corrosion-products spalling thickness of the carbon-steel outer barrier.

The waste-package degradation model also provides as output the pitting degradation history of waste packages, which gives a pit-penetrated area of the waste container as a function of time, through which radionuclides may release (either by diffusion, advection, or a combination of both) [1,21-23]. Pit penetration histories of 25 representative waste packages are shown in Fig. 3-A for the TSPA-1995 approach, and in Figs. 3-B to 3-D for the current approach with 10-mm, 5-mm, and 1-mm spalling thickness, respectively. With the TSPA-1995 approach (Fig. 3-A), the waste package with the most severe pitting degradation is predicted to have about 1% of the total pits (i.e., 40,000 pits out of 4 million pits) penetrating the container by 100,000 years. Most of the waste packages have much less pit penetrations (ranging from 5×10^{-2} to 10^{-6} of the total of 4 million pits) for the 100,000-year simulation period.

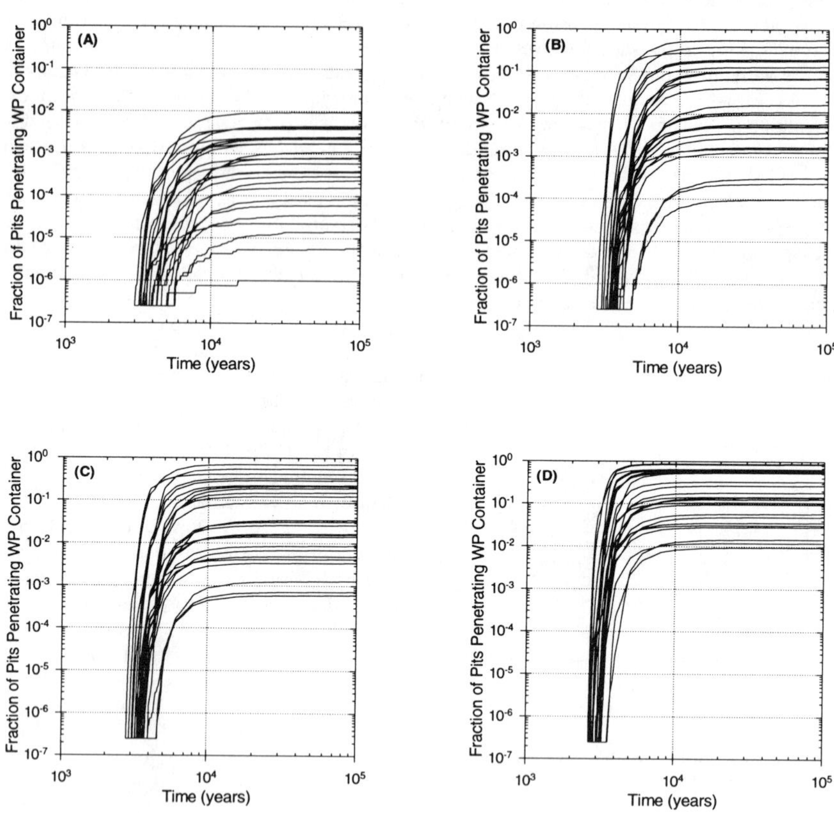

Fig. 3 Pitting profiles of 25 representative waste packages for (A) the TSPA-1995 approach, and for the current approach with the spalling thickness of (B) 10-mm, (C) 5-mm, and (D) 1-mm.

As shown in Figs. 3-B to 3-D, with the current approach, a more severe pitting degradation is predicted with smaller spalling thicknesses. With the 10-mm spalling thickness (Fig. 3-B), for the most severely pitted waste package, about 60% of the total number of pits penetrate the container by 100,000 years, and for the least pitted, about 0.1% penetrate the container. With the 1-mm spalling thickness, by 100,000 years, about 90% of the total number of pits penetrate the most pitted waste package, and it is about 1% for the least pitted waste package. It is shown that the waste-package pitting degradation is highly sensitive to the corrosion-products spalling thickness of the carbon-steel outer barrier. More severe pitting degradation of waste package is predicted with smaller spalling thicknesses, which would cause a greater release of radionuclides from the waste package. Further information on the corrosion-products spalling and its effects on the carbon steel corrosion is needed to reduce the uncertainty in the waste-package degradation modeling.

One of the major uncertainties in the current sensitivity study of waste-package pitting degradation to the spalling thickness is the uncertainty in the results produced from the inner-barrier pitting model (which is also highly dependent on temperature, another uncertain parameter) employed in the waste-package simulation model [1,2]. Another major uncertainty is the galvanic protection of the inner barrier by the carbon-steel outer barrier. It was demonstrated that, with a preliminary galvanic-protection model that is based on an expert elicitation [9], the life-time of a waste package is greatly enhanced, and that the failure rate and the pitting degradation rate are significantly reduced [8].

SUMMARY AND CONCLUSION

A concept of "patches" (i.e., dividing the waste-package surface into small patches) is developed to represent the general corrosion of the carbon-steel outer barrier more closely with the data from the literature. This "patches" concept is one useful way to incorporate other corrosion and relevant processes important to waste package degradation, as their models become available, into the waste-package degradation model. Such processes that are considered important for the potential repository are the galvanic protection of the inner barrier by the carbon-steel outer barrier, localized variations in the chemistry of the water contacting the waste package, locally enhanced corrosion within a waste package due to localized dripping and salt-scale buildup, etc [6].

The corrosion-time concept is developed to better represent the corrosion of the carbon-steel outer barrier under the changing exposure conditions expected in the potential repository. In general, simulation results with the corrosion-time concept predict greater pit penetrations than using the real-time. This is expected in light of the evolution of the exposure conditions (or near-field environments) into more corrosive (i.e., more humid) conditions with time in the potential repository. The effects of the evolving exposure conditions on the waste package performance (i.e. failure and pitting degradation) are better represented with the corrosion-time concept.

With the patches approach and the corrosion-time concept implemented into the waste-package degradation model, the results of a sensitivity study of waste package degradation (failure and pitting degradation) to the threshold spalling thickness of the corrosion products from the carbon-steel outer barrier are presented. It is shown that the waste-package pitting degradation is highly sensitive to the corrosion-products spalling thickness of the carbon-steel outer barrier. Enhanced pitting degradation of waste packages is predicted with a smaller spalling thickness, which could cause a greater radionuclide release from the waste package. Further information on the corrosion-products spalling in different exposure conditions (i.e., chemistry, water contact mode, etc.) and its effect on the carbon steel corrosion is needed to enhance the confidence in the waste-package performance modeling in the potential repository.

One of the major uncertainties in the waste-package degradation simulation results discussed in this paper is the uncertainty of the inner-barrier pitting model (and its high temperature-dependency) [3,5]. Another major uncertainty is the galvanic protection of the inner barrier by the carbon-steel outer barrier, which could significantly reduce the pitting degradation of the inner barrier.

ACKNOWLEDGMENTS

This work was funded by the US DOE Yucca Mountain Site Characterization Office under Contract #DE-AC01-91RW00134 to TRW Environmental Safety Systems, Inc. The authors would like to thank two anonymous reviewers and Bryan Bullard of INTERA, Inc., for their valuable comments.

REFERENCES

1. CRWMS M&O, <u>Total System Performance Assessment-1995: An Evaluation of the Potential Yucca Mountain Repository</u>, B00000000-01717-2200-00136, Rev. 01, Las Vegas, Nevada, 1995.
2. CRWMS M&O, <u>Controlled Design Assumption Document</u>, B00000000-01717-4600-00032, Rev. 04, Las Vegas, Nevada, 1996.
3. J.H. Lee, J.E.Atkins, and R.W. Andrews, <u>Scientific Basis for Nuclear Waste Management XIX</u>, Mat. Res. Soc. Proc., **412**, W.M. Murphy and D.A. Knecht (eds.), p. 603 (1996).
4. J.H. Lee, J.E. Atkins, and R.W. Andrews, <u>Scientific Basis for Nuclear Waste Management XIX</u>, Mat. Res. Soc. Proc., **412**, W.M. Murphy and D.A. Knecht (eds.), p. 571 (1996).
5. R.W. Andrews, T.F. Dale, and J.A. McNeish, <u>Total System Performance Assessment-1993: An Evaluation of the Potential Yucca Mountain Repository</u>, B00000000-01717-2200-00099-Rev. 01, CRWMS M&O, Las Vegas, Nevada, 1994.
6. R.D. McCright, <u>Engineered Materials Characterization Report for the Yucca Mountain Site Characterization Project, Vol. 3: Corrosion Data and Modeling</u>, Rev. 1, UCRL-ID-119564, Lawrence Livermore National Laboratory, Livermore, CA, 1997.
7. <u>Standard Practice for Conducting Atmospheric Corrosion Tests on Metals</u>, ASTM G50-76, Annual Book of ASTM Standards, Vol. 03.02, ASTM, Philadelphia, PA, 1992.
8. J.E. Atkins, J.H. Lee, and R.W. Andrews, <u>Proc. 7th Int'l Conf. High Level Radioactive Waste Management</u>, p. 459 (1996).
9. R.D. McCright, Galvanized Effects in Multi-Barrier Waste Package Containers, personal communication to J.H. Lee, July 12, 1995.
10. W.H. Vernon, Trans. Electro. Soc., **64**, p. 31 (1933).
11. A. Raman, and S. Nasrazadani, Corrosion, **46**, p. 601 (1990).
12. D. Fyfe, <u>Corrosion</u>, 3rd ed., L.L. Shreir, et al (eds.), Vol. 1, Butterworth-Heinemann, p. 2:31 (1994).
13. G.E. Gdowski and J.C. Estill, <u>Scientific Basis for Nuclear Waste Management XIX</u>, Mat. Res. Soc. Proc. 412, W.M. Murphy and D.A. Knecht (eds.), p. 533 (1996).
14. J.C. Estill and G.E. Gdowski, <u>Proc. 7th Int'l Conf. High Level Radioactive Waste Management</u>, p. 457 (1996).
15. J.J. Nitao, <u>User's Manual for the USNT Module of the NUFT Code, Version 1.0</u>, Lawrence Livermore National Laboratory, Livermore, CA, 1996.
16. US DOE, <u>Yucca Mountain Project Reference Information Base</u>, YMP/CC-0002, V. 04.002, Las Vegas, NV, 1990.
17. CRWMS M&O, <u>Drift-Scale Thermo-Hydrology Analysis/Effects Report</u>, BA00000000-01717-2200-0083, Las Vegas, NV, 1996.
18. J.E. Strutt, J.R. Nicholls, and B. Barbier, Corr. Sci., **25**, p. 305 (1985).
19. G.P. Marsh, and K.J. Taylor, Corr. Sci., **28**, p. 289 (1988).
20. G.A. Henshall, <u>Modeling Pitting Degradation of Corrosion Resistant Alloys</u>, UCRL-ID-125300, Lawrence Livermore National Laboratory, Livermore, CA, 1996.
21. J.H. Lee, J.E. Atkins, J.A. McNeish, and V. Vallikat, <u>Proc. 1996 Int'l Conf. Deep Geological Disposal of Radioactive Waste</u>, p. 8-65 (1996).
22. J.H. Lee, P.L. Chambré, and R.W. Andrews, <u>Proc. 1996 Int'l Conf. Deep Geological Disposal of Radioactive Waste</u>, p. 5-61 (1996).
23. J.H. Lee, this volume.

NUCLEAR REPOSITORY PERFORMANCE ASSESSMENT: INSIGHTS INTO CRITICAL MODELS AND PARAMETERS AFFECTING PROJECTED FUTURE DOSES

R.K. MCGUIRE *, J.H. KESSLER **, J.A. VLASITY *
*Risk Engineering, Inc., 4155 Darley Avenue, Suite A, Boulder, CO 80303
**Electric Power Research Institute, P.O. Box 10412, Palo Alto, CA 94303

ABSTRACT

The Phase 3 Total System Performance Assessment (TSPA) sponsored by the Electric Power Research Institute (EPRI) has led to new insights into critical models and parameters affecting estimated doses to humans from a potential repository of high-level radioactive wastes at Yucca Mountain, Nevada. The Phase 3 model has been extended to encompass time-varying climate and infiltration, detailed modeling of the source term and hydrology, and detailed specification of possible interaction between percolating ground water and waste containers. The model estimates doses to a time of one million years.

The three key radionuclides contributing to estimated total doses are Tc-99, I-129, and Np-237. Five other nuclides contributing to dose in lesser (but significant) amounts are U-233, Th-229, Pa-231, Ac-227, and Se-79. These results are consistent with other TSPAs.

From sensitivity studies, the most critical models and parameters are as follows. Infiltration and percolation assumptions, including the amount of lateral diversion of infiltration water, are important and need verification with site data and/or more detailed modeling. Parameters of the unsaturated zone (UZ) and saturated zone (SZ) determine dilution and delay of concentrations and peak doses downstream. The fraction of containers that become wet are not critical in our model, but this lack of sensitivity reflects our coupling of the fraction with a model of focused flow past the containers; an different model might indicate higher sensitivity. Also, the degree of coupling between fracture and matrix flow is important in affecting the times of peak doses but not their magnitudes.

Other critical design assumptions that could lead to reduced and/or delayed doses are a more robust container design, a capillary barrier around each container, the dilution during hydrologic transport from the repository to a potential agricultural community downstream, and the characteristics of an "average" individual in that community who might receive a dose.

INTRODUCTION

Current estimates of repository performance are made with the Integrated Multiple Assumptions and Release Calculations (IMARC) software, developed over a period of years under EPRI sponsorship. For input we use models, parameters, and assumptions developed by consultants to EPRI; these models, parameters, and assumptions are documented in detail elsewhere [1,2] and are the best current interpretations by these consultants.

The purpose of total system performance assessment (TSPA) is to provide a sound basis for decision making. The main "added value" delivered by a systems-level model such as IMARC as compared to detailed subsystem models is the ability to capture interactions and correlations among different parts of the system. Thus when we abstracted a subsystem for IMARC, we aimed to capture as many of these correlations as possible rather than to maximize the accuracy of approximation to more detailed subsystems. Another important point of system-level models is their ability to capture uncertainties about physical processes and mechanisms.

Mat. Res. Soc. Symp. Proc. Vol. 465 © 1997 Materials Research Society

The purpose of performing calculations with the current IMARC Phase 3 software is to explore the various models, parameters, and assumptions affecting estimated repository performance and to draw conclusions about which of those are robust, which are unimportant, and which have substantial uncertainties that lead to uncertainties in repository behavior. Through this process we can identify the most critical engineering and scientific areas that need further investigation or quantitative characterization of their uncertainties.

This paper presents a summary of results; more detailed documentation is contained in [1, 2 and 3].

ADVANCES IN TSPA

The EPRI Phase 3 TSPA has incorporated several significant advances over previous versions [4,5]. The IMARC software has been extended to estimate individual dose rates up to one million years in the future. This is necessary because peak dose rates generally occur between a few tens of thousands of years and one million years, so a code that calculates doses for only ten thousand years is not capable of estimating peak doses when they occur at longer times.

In the area of future climate conditions, we model three climate assumptions, as follows. (1) A minor greenhouse effect where greenhouse conditions persist for 3,000 yrs, followed by changes in climate up to a full glacial maximum precipitation at 50,000 yrs, then a resumption of cyclicity in climate and precipitation between glacial and interglacial conditions. (2) A moderate greenhouse effect where greenhouse conditions persist for 10,000 yrs, followed by changes in climate up to 2/3 of the full glacial maximum precipitation at 50,000 yrs, then a resumption of cyclicity in climate and precipitation between glacial and interglacial conditions. (3) A permanent greenhouse climate under which greenhouse conditions persist for 10,000 yrs, followed by a permanent interglacial conditions. These provide major differences among future climates as they affect PA calculations; within each climate assumption, one sequence of climate changes is used. Net infiltration results calculated from detailed models of precipitation and the effects of soils, vegetation, and hydrologic units were used to define infiltration values for source term and hydrologic models.

It is assumed that 80% of the infiltrating water is diverted laterally above the repository to locations of fast flow, and that during placement of containers, none will be located in those regions of diverted water. The remaining 20% of water is focused from the entire repository area onto the fractions of repository that become wet, and these wet areas do result in water contacting containers. A "focused-flow factor" quantifies the factor by which flux is increased by this focusing effect, and values of 3 and 15 quantify our uncertainty on this factor. The focused-flow factors are applied so that the total volume of water infiltrating the repository is maintained.

The waste container failure model uses Weibull distributions to represent failure of the containers due to corrosion. The mechanisms considered were general corrosion, localized corrosion (crevice and pitting corrosion), stress corrosion cracking, degradation due to metastable microstructure, embrittlement due to hydride formation, and microbiologically influenced corrosion. Galvanic protection was not considered.

New estimates of hydrothermal system behavior and associated uncertainties for use in IMARC are made. The estimates are in the form of a five-dimensional matrix whose axes are as follows. (1) Alternative repository thermal loadings (83 or 25 MTU/acre). (2) Flow regimes (strong or weak coupling between fractures and matrix). (3) Dominant heat transfer mechanism (conduction, convection plus conduction, or heat pipe). (4) Waste package temperature histories (hot=above boiling, warm=limited to boiling, or cool=below boiling). (5) Modes of water contact with the waste package (dry, wet drip, episodic flow, and reflux following the heat pulse).

Values in the matrix are the fractions of the waste in the repository at different temperatures and exposed to different modes of water contact. These values depend on design choices (in particular, the thermal loading), on properties of the rock (which will control the mode of heat transfer), and on which conceptual model of fracture-matrix coupling in unsaturated fractured rock is more accurate. For each alternative thermal loading, heat transfer mode, and conceptual model, the fractions of the repository that are wet take on different values. It should be understood that values assigned to these fractions represent judgmental estimates rather than the results of calculations.

In addition to the basic model for focused flow on the fraction of repository that is wet, "recirculation factors" are also included during the thermal period (first 50,000 years). These judgmentally determined factors account for the water that continually evaporates, condenses, and refluxes past the containers during this time period.

The new source term model used in the EPRI Phase 3 TSPA is a compartment model that begins by assuming all waste form surfaces are wetted immediately after their container fails. Inventories are based on a hybrid of 25-yr old, 30,000 MWd/MTU boiling water reactor spent nuclear fuel and 25-yr old, 40,000 MWd/MTU pressurized water reactor spent nuclear fuel (this assumption is made in [6]). Advection and diffusion between the following compartments can be modeled: waste form; corrosion products found in the corroded sections of the canister; gravel backfill below and sometimes above the container; concrete invert (both concrete matrix and fractures); and rock matrix and fractures immediately surrounding the drift.

The far-field flow and transport model used in Phase 3 provides a more realistic portrayal of hydrologic processes that affect the flow of water and radionuclides than had been used previously [4,5]. The most important concern in developing the new code was to provide a capability to model "fast fracture flow" when it may occur. Fast fracture flow is a condition where the water in the fractures moves faster than the associated zone of wetting within the matrix. The new hydrology code within IMARC also adds considerably more power and flexibility in the treatment of processes and parameters. The unsaturated zone is modeled as multiple one-dimensional columns that calculate flow and transport numerically accounting for changing infiltration and different temperature histories within regions of the repository. The flow and transport is modeled in the saturated zone with a 3-dimensional calculation that accounts for changing vertical infiltration with time and for horizontal flow through the saturated zone. Because the new code is numerically based, it can incorporate complex climate-related changes in infiltration at the surface, various patterns of layering, the in-growth of daughter products, and a rigorous treatment of flow and transport in the saturated zone.

Factors to convert elemental radionuclide concentrations to human doses for this study are determined using a biosphere model [2]. Both the drinking water exposure pathway only and all exposure pathways are considered. It is assumed that water is continuously extracted from a deep well and used for spray irrigation of crops, watering of livestock, and domestic supply (including drinking water). Irrigated crops are assumed to become contaminated directly by interception of spray irrigation water or through root uptake and soil splash. Contamination in such crops are assumed to be in equilibrium with concentrations in well water and surface soil.

RESULTS OF CALCULATIONS

Figure 1 shows expected dose rates for the drinking water exposure pathway ("water only") vs. time for the base-case calculations. These are expected results calculated over uncertainties in climate/infiltration, solubility/alteration time, fracture-matrix coupling/flow-focusing factor, retardation, and dominant thermal process. Calculated doses at a location 5 km downstream from

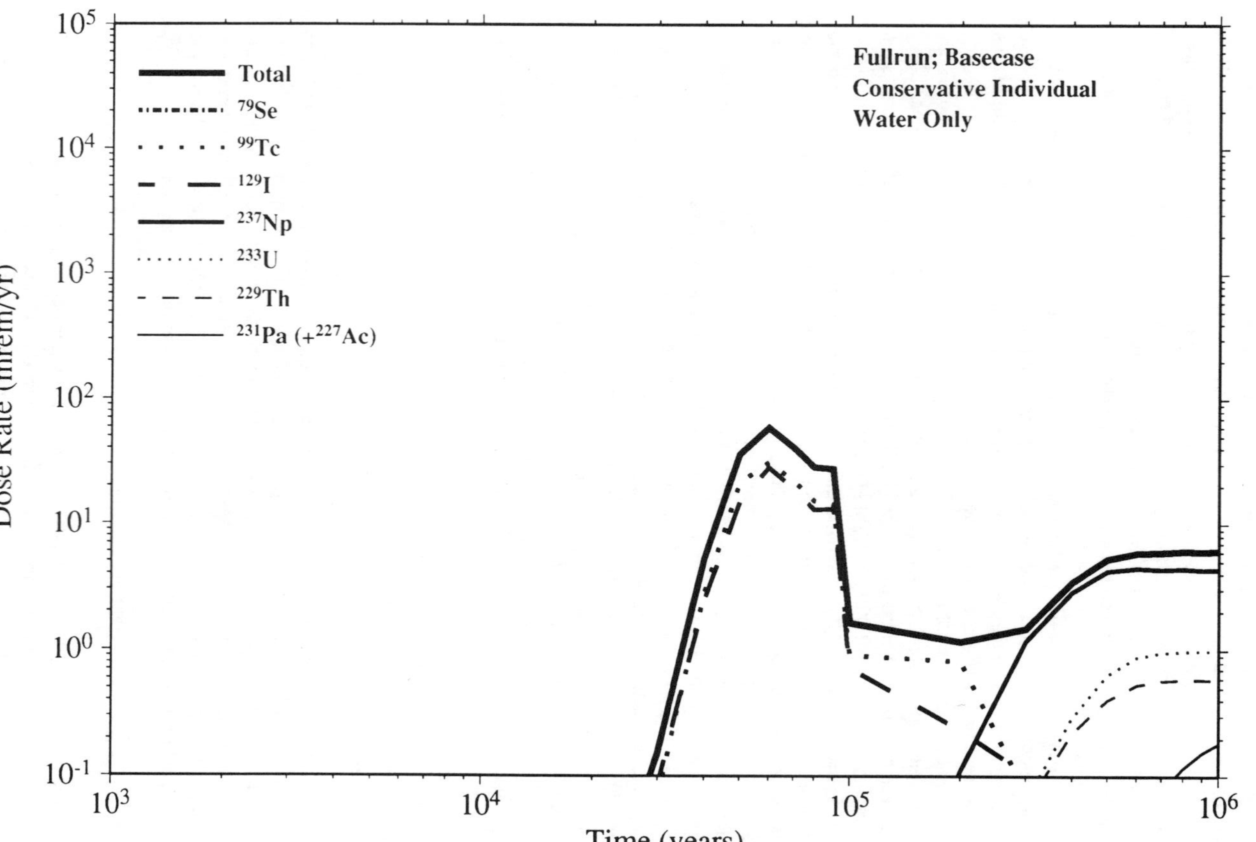

Figure 1. Annual dose vs. time from highest concentration at the accessible environment.

the repository (the "accessible environment") are shown for seven radionuclides: Se-79, Tc-99, I-129, Np-237, U-233, Th-229, and Pa-231 (which includes the dose from Ac-227). Fifteen other nuclides were also investigated but they led to doses lower than these seven. Figure 1 shows the contribution for each nuclide and indicates that Tc-99 and I-129 give the highest doses and that these doses occurs within 100,000 years but well after 10,000 years. These elements are not retarded and are transported most quickly to the accessible environment. At times past 300,000 years, Np-237 gives the highest dose. Lower doses within one million years come from Pa-231 (which includes Ac-227), Th-229, and U-233. Se-79 is included in these calculations, although it does not show up in the drinking water exposure pathway dose calculations, because it contributes to total dose from other biosphere pathways.

Detailed sensitivity studies on the input assumptions indicate that the conceptual model of the thermal-hydrologic process is important. The fraction of the repository that becomes wet affects concentrations and doses, but this sensitivity is offset in our model by flow-focusing factors (larger wet areas imply less focusing of percolated ground water, and vice versa). Solubility limits are not critical for Tc-99 and I-129, which are alteration-rate limited. The minor- and moderate-greenhouse climates indicate similar doses, whereas the permanent greenhouse condition shows somewhat lower doses over a longer period. Fracture-matrix coupling affects the wet fraction of the repository and the degree of focused flow within those fractions; slow fracture flow implies later releases than does fast fracture flow. Retardation only affects nuclides (notably Np-237) that are sorbed; peak doses are delayed but not reduced. Finally, the dominant thermal process (conduction, convection plus conduction, or heat pipe) has only a slight effect on the time and magnitude of the calculated peak dose.

Figure 2 indicates expected dose vs. time for our base case calculated for all potential dose pathways. In addition to drinking water, these include human ingestion of meat and vegetable products raised or grown with contaminated ground water. These use general (not Yucca Mountain-specific) biosphere models and account for the dilution in ground water when flowing from the repository to a point 25 km downstream to a postulated community in Amargosa Valley. The dose calculations indicate that average doses to an individual in the local population are three to five orders of magnitude lower than to a "conservative" individual at the "accessible environment" (5km downstream from the repository) who is assumed to consume all of his or her water and food from a contaminated ground water supply.

CONCLUSIONS

The EPRI Phase 3 TSPA calculations indicate that very low doses will occur over the next 10,000 to 20,000 years to a conservative individual in the future from a potential repository at Yucca Mountain. Doses at longer times are caused primarily by Tc-99, I-129, and Np-237. The first two cause peak doses at 50,000 to 90,000 years, the last cause peak doses at times greater than 300,000 years. These results are consistent with those from other TSPAs [6,7].

New insights gained from this application are that the calculations depend heavily on conceptual models of climate/infiltration and of the interaction between the fraction of the repository wet and focusing of hydrologic flow. The importance of solubility of radioelements (for solubility-limited nuclides) is confirmed from previous studies. Less critical is the thermal transfer mechanism (conduction, convection plus conduction, or heat pipe) and the waste alteration time (for alteration-rate limited nuclides). Several design changes could potentially result in reduced and/or delayed doses, including a more robust container design (e.g. constructed of titanium and Alloy C-22) and a capillary barrier placed around the waste. The latter is a low-permeability backfill placed to keep advective flow from reaching the waste packages.

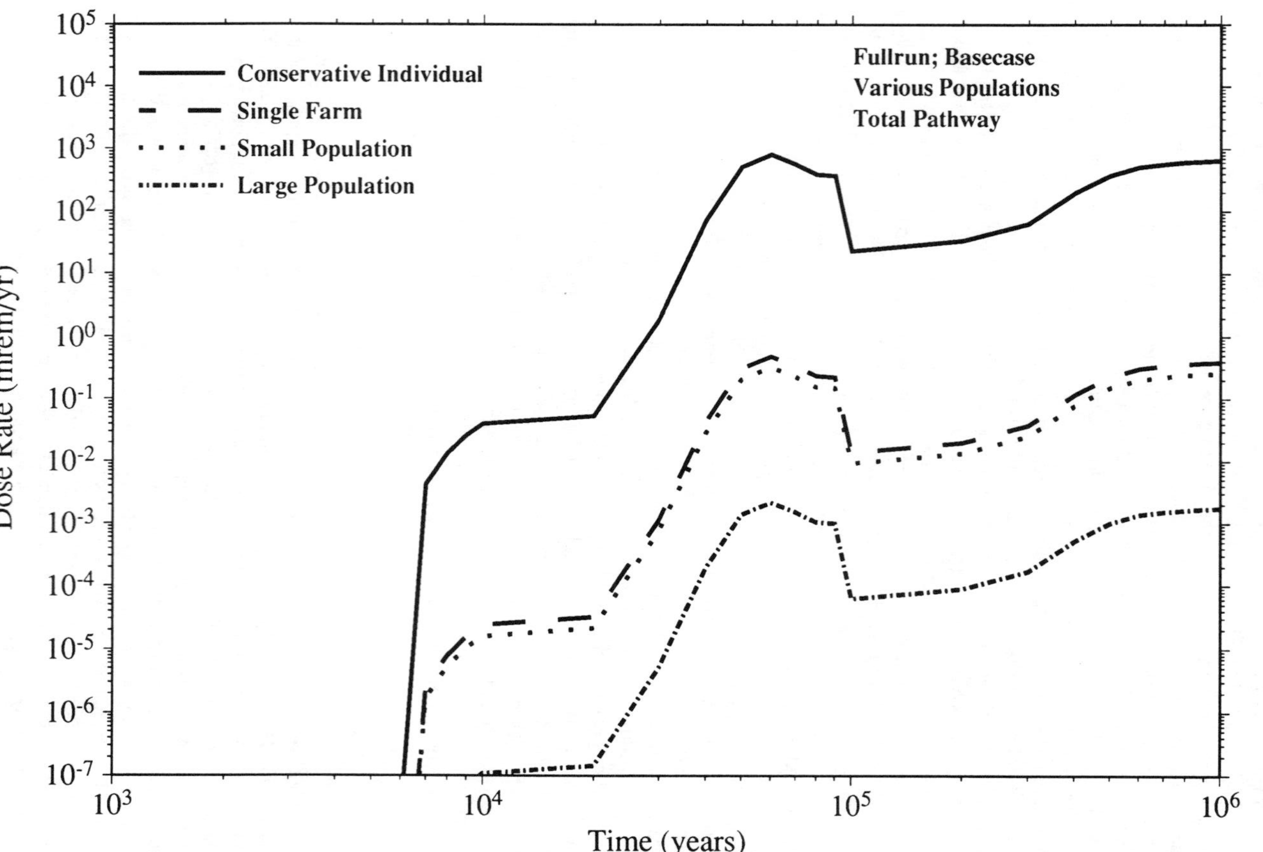

Figure 2. Total annual dose vs. time for various population scenarios, 25 km downstream, compared to a conservative individual at 5 km.

REFERENCES

1. Electric Power Research Institute, *Yucca Mountain total system performance assessment, Phase 3*, EPRI TR-107191, December 1996.

2. Electric Power Research Institute, *Biosphere modeling and dose assessment for Yucca Mountain*, EPRI TR-107190, Palo Alto, CA, December 1996.

3. Electric Power Research Institute, *Analysis and confirmation of robust performance for the flow-diversion barrier system at Yucca Mountain*, EPRI TR-107189, November 1996.

4. Risk Engineering, Inc., *Demonstration of a risk-based approach to high-level waste repository evaluation*, Elec. Power Res. Inst., Rept. NP-7057, Palo Alto, CA, Oct. 1990.

5. Risk Engineering, Inc., *Demonstration of a risk-based approach to high-level waste repository evaluation: phase 2*, Elec. Power Res. Inst., Rept. TR-1000384, May 1992.

6. Sandia National Laboratories, *Total system performance assessment for Yucca Mountain - SNL second iteration (TSPA-1993)*, SAND93-2675, 2 Vol, Albuquerque, NM, 1994.

7. TRW, *Total system performance assessment - 1995: an evaluation of the potential Yucca Mountain repository*, US Dept. of Energy Rept. B00000000-01717-2200-00136, Rev. 1, Las Vegas, NV, Nov. 1995.

BOUNDING ANALYSIS FOR SOLUBILITY

T. Ohi *, K. Nakajima **

*Power Reactor and Nuclear Fuel Development Corporation, Tokai-mura, Ibaraki-ken,
 JAPAN 319-11, ooitakao@tokai.pnc.go.jp.
**Nuclear Energy System Incorporation, Tokai-mura, Ibaraki-ken, JAPAN 319-11.

ABSTRACT

In performance assessment of geological disposal systems for High Level Radioactive
Waste (HLW), the change of environment over the long-term must be considered. Therefore, it is
necessary to consider a wide range of parameters concerned with radionuclides migration,
especially the dependence of solubility on geochemical environment. In this study, assuming that
the release rate of the nuclides from buffer material is limited by inventory ultimately, the
relationship between the initial inventory and the solubility that produces a solubility-invariant
maximum release rate from the buffer is examined by using a simple steady-state analytical
solution without decay. As the result, the threshold of "effective" solubility in the performance
assessment of the geological disposal systems for HLW is obtained as a function of initial
inventory, distribution coefficient (Kd), diffusion coefficient, and thickness, porosity and density
of the buffer. Also, the threshold of "effective" steady dissolution rate corresponding to the
threshold of "effective" solubility is obtained.

INTRODUCTION

Uncertainties included in the associated scenarios, models, and data make it difficult to
exactly evaluate performance of geological disposal systems for HLW. In order to more
comprehensively evaluate the performance of the systems, it is required that performance
measures for various conditions considering these uncertainties be analyzed. To analyze these
efficiently, it is necessary to understand the relationship between the performance measures and
the parameters concerned with radionuclides migration, and to determine the relative importance
of the parameters. Until now, analytical solutions for various mass-transfer analyses have been
derived, and the relationship between the performance measures and the parameters has been
shown analytically [1], [2]. In addition, the relative importance of the parameters for the
performance measures has been shown by bounding analysis [3], [4].

For the performance assessment of the geological disposal systems for HLW over the
long-term, the change of environment depending on geological events such as volcanism,
faulting / seismic and uplift / erosion, and on climate change must be considered. Therefore, it is
necessary to consider a wide range of the parameters concerned with the nuclides migration,
especially the dependence of solubility on geochemical environment.

Mat. Res. Soc. Symp. Proc. Vol. 465 © 1997 Materials Research Society

In the previous analyses of release rate from buffer material imposed by solubility limits, the calculations show a proportional increase in the release rate from the buffer with increasing up to some limiting, "effective" solubility value, and that the release rate from the buffer for solubility values greater than this threshold value is equal to the release rate for this threshold value [5]. These results indicate that the release rate from the buffer is limited by inventory ultimately. In this study, the relationship between the initial inventory and the solubility that produces a solubility-invariant maximum release rate from the buffer is examined by using a simple steady-state analytical solution without decay.

DERIVATION OF THE THEORY AND ANALYSIS

Transport Model

The model used in this study is based on one-dimensional cylindrical geometry with diffusive transport, reversible and instantaneous sorption, and without decay. In this model, solubility limit and zero-concentration boundary conditions are imposed at the waste glass / buffer material interface and at the buffer / surrounding rock interface, respectively.

In the analyses, the transport for a single radionuclide is analyzed without considering the effects of precipitation and isotope ratio relating to decay series [6], in order to individually evaluate release rate from the buffer of each nuclide. This means that shared solubility for each nuclide is equal to elemental solubility. The dissolution of the nuclides from the glass is assumed to start at 1000 year after waste emplacement. Also, initial inventory is defined as the maximum value of the inventory in closed-system waste form calculated by ORIGEN 2 after the start of the nuclides dissolution.

Mathematical Formulation

The nuclide migration through the buffer is formulated as below.

$$\frac{\partial C}{\partial t} = \frac{De}{\varepsilon R} \left(\frac{\partial^2 C}{\partial r^2} + \frac{1}{r} \frac{\partial C}{\partial r} \right) \tag{1}$$

where,

C	: concentration of radionuclide in porewater [gm^{-3}]
De	: effective diffusion coefficient [$m^2 y^{-1}$]
ε	: porosity of buffer material [-]
R	: retardation coefficient [-] [$R = 1 + (1 - \varepsilon) \rho Kd / \varepsilon$]
ρ	: density of buffer material [kgm^{-3}]
Kd	: distribution coefficient [$m^3 kg^{-1}$]
t	: time [y]
r	: distance from the center of waste glass [m].

Imposing the solubility limit and the zero-concentration boundary conditions and not considering the decay, the analytical solution for the concentration in a steady-state is as follows.

$$C = \frac{\ln(r/b)}{\ln(a/b)} C_0 \tag{2}$$

where,
C_0 : solubility limit [gm^{-3}]
a : radius of the inner surface of buffer material [m]
b : radius of the outer surface of buffer material [m].

The analytical solution for the release rate from the buffer in a steady-state is as follows.

$$\Phi = - \frac{2\pi H D e}{\ln(a/b)} C_0 \tag{3}$$

where,
Φ : release rate from buffer material [gy^{-1}]
H : length of buffer material [m].

Relationship between Initial Inventory and Solubility Limit

In a steady-state, the total amount of the nuclide in the buffer is given by integrating equation (2) over the buffer with sorption included, as follows.

$$W_b = (\varepsilon + (1-\varepsilon)\,\rho Kd)\pi H\,C_0 \left(\frac{a^2 - b^2}{\ln(a/b)^2} - a^2 \right) \tag{4}$$

where,
W_b : total amount of radionuclide in buffer material at a steady-state [g].

The W_b may be represented by the difference between the total amount of the nuclide dissolved from the glass (W_g) until the time to reach a steady-state and the total amount of the nuclide released from the buffer (W_o) until the time to reach a steady-state, as follows.

$$W_b = W_g - W_o \tag{5}$$

In the case that the W_g equals to initial inventory, the W_b is maximized. Thus, the solubility limit necessary for the nuclide transport to reach a steady-state in buffer is maximized. If it is assumed that the W_o can be ignored, the relationship between the initial inventory and the maximum value of the solubility limit that will limits steady-state release is given as follows.

$$I_o = (\varepsilon + (1-\varepsilon)\,\rho Kd)\pi H\,C_{max} \left(\frac{a^2 - b^2}{\ln(a/b)^2} - a^2 \right) \tag{6}$$

where,

I_o : initial inventory [g]

C_{max} : maximum value of solubility limit necessary for the nuclide transport to reach a steady-state in buffer [gm^{-3}].

The C_{max} calculated by equation (6) is conservatively greater than the actual maximum value of the solubility limit because of neglect of W_o. In this study, the C_{max} is approximately regarded as the solubility limit which maximizes the release rate from the buffer. The validity of this approximation is evaluated from the comparison with numerical solution, stated in later of this paper. Therefore, the C_{max} is referred to as the threshold of solubility (C_T) for the release rate of the nuclides.

Precipitation relating to in-growth of daughter nuclide in the buffer depends on the solubility of the daughter and gives lower release rate than the release rate with no limit of the solubility in buffer [6]. Therefore, the C_T value for the daughter must be the solubility in which no precipitation of the daughter in the buffer occurs. In the analysis considering the decay-ingrowth and using the initial inventory at start time of the nuclides dissolution, the total amount of the daughter in buffer never exceeds the maximum value of the inventory in closed-system waste form. This is true, even if the in-growth is considered in the buffer. This means that the C_T value for the daughter calculated by the maximum value of the inventory make it difficult for the precipitation of the daughter to occur. Therefore, the C_T value given by these approach can be approximately regarded as the threshold of "effective" solubility, including the effects of the in-growth of the nuclides.

The C_T value for a element can be calculated by defining the summation over the maximum value of the inventory of all isotopes included in such element as the initial inventory. The validity of this approach is explained as follows. The conservative condition of the C_T value for a element is that the shared solubility of the nuclide, which is given by multiplying the C_T value for the element by the isotope ratio depending on a spatial and time, is greater than the C_T value for the amount of the nuclide in a spatial and time. This relation is represented as follows.

$$K * (A_{max} + B_{max}) * \frac{A}{A + B} \geq K * A \tag{7}$$

where,

K : constant $[\, 1/K = (\varepsilon + (1 - \varepsilon) \, \rho \, Kd) \pi H ((a^2 - b^2) / \ln(a/b)^2 - a^2) \,]$

A_{max} : maximum value of the inventory of a isotope "a" in closed-system waste form [g]

B_{max} : summation over the maximum value of the inventory of all isotopes except "a" [g]

A : amount of the isotope "a" in a spatial and time $[A_{max} \geq A]$

B : total amount of all isotopes except "a" in the spatial and time $[B_{max} \geq B]$.

This relation should always hold.

Relationship between the C_T Value and Steady Dissolution Rate

In a steady-state, the steady dissolution rate of radionuclides from the glass equals the steady release rate from the buffer. Therefore, the threshold of "effective" steady dissolution rate corresponding to the threshold of "effective" solubility is given by equation (3), as follows.

$$\Phi_T = -\frac{2\pi H D e}{\ln(a/b)} C_T \tag{8}$$

where,

Φ_T : the threshold of "effective" steady dissolution rate corresponding to C_T [gy^{-1}].

The threshold of "effective" steady leaching rate of the glass is equal to Φ_T divided by the mass fraction of the nuclide in the glass. The steady leaching rate of the glass is given by product of surface area of the glass including large uncertainties and experimental reaction rate of the glass (g/m^2/y). Therefore, using the threshold of "effective" steady leaching rate of the glass and experimental reaction rate of glass, the threshold of "effective" surface area of the glass is estimated.

Also, the steady dissolution rate of the nuclide corresponding to the input value of solubility used in analysis is given by equation (3). The comparison of this value with the threshold of "effective" steady dissolution rate of nuclide indicates which factors limit dissolution of radionuclide from the glass, when the input values such as solubility, steady dissolution rate and initial inventory, which are obtained by experiment or calculation, are given for the analysis.

Data Used in Analysis

The radionuclides and the parameters used in H3 report [5] are used in this analysis. The radionuclides, the initial inventory and the parameters used in this analysis are shown in Table I and II, respectively.

RESULTS AND DISCUSSION

In order to confirm the validity of the approximation, in Figure 1, the threshold of "effective" solubilities (C_T) for Pu-239 calculated by equation (6) are compared with relationships between maximum release rate and solubility, which are calculated by a numerical release model used in H3 report [5]. In these results, the relationship for a single nuclide calculated in this study with decay and the exhaustion of the inventory (lines indicated by each Kd value), and the relationship to the Kd value of 10 (m^3/Kg) calculated in H3 report with the decay-ingrowth and the exhaustion of the inventory (the bars for the solubilities used in H3 report and the length of them) are included. Also, steady leaching rate used in H3 report is included in Figure 1.

Table I Nuclides Used in this Analysis and Initial Inventory

Nuclides	Half life (y)	Initial inventory(g)		Nuclides	Half life (y)	Initial inventory(g)	
		Nuclides	elements			Nuclides	elements
FP				U -233	1.59E+05	5.26E+01	
Se-79	6.50E+04	7.99E+00	7.80E+01	U -234	2.45E+05	2.37E+00	2.20E+03
Zr-93	1.53E+06	1.00E+03	5.64E+03	U -235	7.04E+08	1.58E+02	
Tc-99	2.13E+05	1.06E+03	1.06E+03	U -236	2.34E+07	4.98E+01	
Pd-107	6.50E+06	2.75E+02	1.76E+03	U -238	4.47E+09	1.94E+03	
Sn-126	1.00E+05	3.58E+01	1.18E+02	Np-237	2.14E+06	8.89E+02	8.89E+02
Cs-135	2.30E+06	4.96E+02	2.05E+03	Pu-239	2.41E+04	8.35E+01	1.25E+02
ACTINIDE				Pu-240	6.54E+03	3.76E+01	
Th-230	7.70E+04	4.40E-01	5.14E+01	Am-241	4.32E+02	5.54E+01	1.42E+02
Th-232	1.41E+10	4.85E+01		Am-243	7.38E+03	8.71E+01	

Table II Parameters Used in this Analysis

a	radius of the inner surface of buffer material [m]		0.52
b	radius of the outer surface of buffer material [m]		1.5
H	length of buffer material [m]		2.47
De	effective diffusion coefficient $[m^2 y^{-1}]$		9.47E-03
ε	porosity of buffer material [-]		0.33
ρ	density of buffer material $[kgm^{-3}]$		2.70E+03
WG	weight of glass [Kg]		4.12E+02

Table III Threshold of "Effective" Solubility for Elements (C_T ; gm^{-3})

	Distribution coefficient (m3/Kg)						
Elements	0.00E+00	1.00E-03	1.00E-02	1.00E-01	1.00E+00	1.00E+01	1.00E+02
Se	4.588E+01	7.078E+00	8.220E-01	8.354E-02	8.354E-02	8.354E-02	8.354E-02
Zr	3.317E+03	5.118E+02	5.943E+01	6.041E+00	6.051E-01	6.052E-02	6.052E-02
Tc	6.235E+02	9.619E+01	1.117E+01	1.135E+00	1.137E-01	1.137E-01	1.137E-01
Pd	1.035E+03	1.597E+02	1.855E+01	1.885E+00	1.888E-01	1.888E-02	1.888E-02
Sn	6.941E+01	1.071E+01	1.243E+00	1.264E-01	1.264E-01	1.264E-01	1.264E-01
Cs	1.206E+03	1.860E+02	2.160E+01	2.196E+00	2.199E-01	2.200E-02	2.200E-02
Th	3.023E+01	4.664E+00	5.416E-01	5.505E-02	5.505E-02	5.505E-02	5.505E-02
U	1.294E+03	1.996E+02	2.318E+01	2.356E+00	2.360E-01	2.360E-01	2.360E-01
Np	5.229E+02	8.067E+01	9.368E+00	9.522E-01	9.537E-02	9.539E-03	9.539E-03
Pu	7.353E+01	1.134E+01	1.317E+00	1.317E+00	1.317E+00	1.317E+00	1.317E+00
Am	8.353E+01	1.289E+01	1.289E+01	1.289E+01	1.289E+01	1.289E+01	1.289E+01

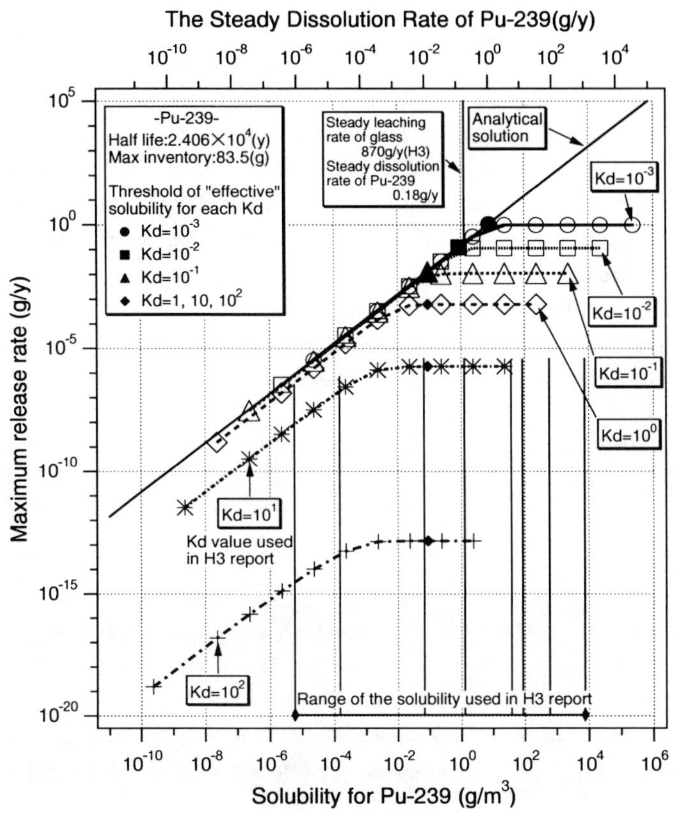

Figure1   Comparison with Numerical Solution

The values for Kd in which the effect of decay is significant are replaced by a value for Kd in which the effect of the decay is not significant.

In the comparison between the numerical results calculated in H3 report and the numerical results calculated in this study without in-growth, the significant differences for relatively low solubility are attributed to the effects of precipitation of Pu-239, and slightly differences for relatively high solubility are attributed to in-growth of Pu-239 in buffer with no precipitation.

Figure 1 shows that the results obtained from a simple steady-state analytical solution without decay can represent the threshold of "effective" solubility. The validity of the "effective" solubility approximation is evident from figure 1. The threshold of "effective" solubilities for elements to the variation of Kd are given in Table III.

CONCLUSIONS

The relationship between the initial inventory and the solubility that produces a solubility-invariant maximum release rate from the buffer is derived. From this relationship, the threshold of "effective" solubility (C_T) in the performance assessment of the geological disposal systems for HLW is obtained as a function of initial inventory, distribution coefficient, diffusion coefficient, and thickness, porosity and density of the buffer material by the equation (6). Also, the threshold of "effective" steady dissolution rates corresponding to C_T and the steady dissolution rates corresponding to the solubility used in analysis are identified.

Such simple analysis helps to bound and integrate the range of the parameters that impact on the nuclides migration. It also helps to guide the effective data acquisition and the analysis in the performance assessment over which the change of environment must be considered.

ACKNOWLEDGMENT

The authors would like to thank M. J. Apted for kindly advice in editing this paper.

REFERENCES

1. I. Neretnieks, Diffusivities of Some Constituents in Compacted wet Bentonite Clay and the Impact on Radionuclide Migration in the Buffer, Nuclear Technology, 71 pp. 458-470, (1985).
2. M. J. Apted, D. P. Hodgkinson, Logic Over Rhetoric : The Role of Performance Analysis in Guiding Near-Field Data Collection, Proceedings Nuclear Waste Packaging FOCUS `91, pp. 24-31, (1991).
3. S. J. Zavoshy, P. L. Chambre`, and T. H. Pigford, Mass Transfer in a Geologic Environment, Scientific Basis for Nuclear Waste Management VIII, Materials Research Society, Pittsburgh, PA, 44, pp. 311-322, (1985).
4. T. H. Pigford, and P. L. Chambre`, Near- Field Mass Transfer in Geologic Disposal Systems A Review, Scientific Basis for Nuclear Waste Management XI, Materials Research Society, Pittsburgh, PA, 112, pp. 125-141, (1988).
5. PNC, Research and Development on Geological Disposal of High-Level Radioactive Waste, PNC TN 1410 93-059, (1993).
6. T. Ohi, K. Miyahara, M. Naito, H. Umeki, Effects of Transport Model Alternatives Incorporating Precipitation on The Performance of Engineered Barriers, High Level Radioactive Waste Management Proceedings of the Seventh Annual International Conference, pp. 274-275, (1996).

MAXIMUM INDIVIDUAL DOSE AND VICINITY-AVERAGE DOSE
FOR A GEOLOGIC REPOSITORY

T. H. PIGFORD, E. D. ZWAHLEN
Nuclear Engineering Department, University of California, Berkeley, CA 94720,
tpigford@nas.edu

ABSTRACT

Recent proposals for a new U.S. standard for high-level waste disposal would limit the
average dose to individuals in the vicinity surrounding a geologic repository. This would be a
new approach to protecting the public from environmental releases of radioactivity.
Heretofore, criteria adopted for geologic disposal have limited the reasonable maximum
exposure to a future hypothetical individual. Here we present quantitative analyses of the
relation between maximum exposure and vicinity-average exposure, resulting from future
human use of ground water contaminated by radioactive releases from a repository.

Estimating the vicinity-average exposure would require postulates and guesses of location
and habits of future people. Exposure probabilities postulated by others show that proposed
dose limit to the vicinity-average individual would be a far more lenient standard than the
traditional dose limit to reasonably maximally exposed individuals. The proposed vicinity-
average dose limit would allow far greater concentrations of contaminants in ground water
than would be allowed by normal standards of ground water protection. A safety standard that
limits vicinity-average exposure should also include limits on maximum exposure.

INTRODUCTION

Industry and Congress have proposed a new standard for protecting public health from the
Yucca Mountain repository for radioactive waste. The standard would limit the average
radiation dose to a future population in the general vicinity, instead of limiting the maximum
dose to an individual. The standard would allow the <u>vicinity-average</u> dose to be about tenfold
higher than what is now adopted in this country and abroad as a limit for the <u>maximum</u> dose.
Calculational methods proposed by an advocate of the new standard predict doses to
individuals that could exceed internationally accepted levels by factors of several thousand and
more. Concentrations of radioactivity in ground water in the vicinity could exceed safe
concentrations by factors of thousands. Safety problems of a repository with a high maximum
individual dose could be obscured by low calculated values of the vicinity-average dose.

Instead, the standard for Yucca Mountain should limit the annual radiation dose to the
reasonable maximally exposed individual. For estimating compliance with radiation
protection standards for geologic disposal, national and international radiation protection
agencies and bodies have long calculated reasonable maximum exposures for future
subsistence farmers, who drink contaminated ground water and obtain a substantial portion of
their food from crops irrigated by that ground water. If the reasonable maximum dose
estimate is within acceptable limits, the doses to others, who by definition should receive
lower doses, will also be acceptable [see Appendix]. This is accepted international practice
for protection of public from disposal of radioactive waste in geologic repositories.
Subsistence farming is not a rare event. Family farms are a way of life for many residents in
the Amargosa Valley, who utilize ground water from an aquifer that flows under Yucca
Mountain [1].

The Yucca Mountain project needs a standard that is stringent enough to build confidence
in the face of legal and political challenges. At present no scientific bases exist to support a
policy less stringent than that now used in the U.S. and in other countries. Policy makers
must reject pressures for short-term expediency and economy lest, by enacting policy that
compromises scientific validity and credibility, they undermine public confidence and end
needed nuclear research and application.

International practice protects the public from nuclear radiation by ensuring that the

Mat. Res. Soc. Symp. Proc. Vol. 465 © 1997 Materials Research Society

reasonable maximum radiation dose received by a member of the public is less than a specified limiting dose [2 through 19]. However, recent proposals [20,21,22,23,24] would adopt a new less stringent standard for protecting public health from geologic disposal of radioactive waste. The proposed standard would be much more lenient than dose limits now in use in this country and abroad [25,26,27]. Here we consider the proposals to calculate an average dose to a future (and unknowable) population in the general vicinity and to allow that average dose as large or larger than what we now limit for the reasonable maximum exposure to individuals. In so doing, the many people exposed to above-average doses would be unprotected. Also, the proposed "vicinity-average dose" standard has ill-defined notions of a future population "in the vicinity". It would invite manipulation, such as increasing the "vicinity" size, to obtain lower calculated radiation doses.

People living outside the vicinity can also be exposed to radioactivity from the geologic repository, by consuming contaminated food grown in the vicinity near Yucca Mountain and by importing water extracted from wells in the vicinity. For example, a 1994 proposal by Amargosa Resources Inc. would purchase rights to ground water in the Amargosa Valley, to supply more water to Las Vegas [28] By protecting the subsistence farmer who could use the most contaminated water near Yucca Mountain, individuals outside the vicinity would receive lower individual doses and would also be protected.

THE EFFECT OF THE PROPOSED VICINITY-AVERAGE DOSE LIMIT ON CONCENTRATION OF RADIOACTIVITY IN GROUND WATER

The traditional subsistence-farmer calculation of dose, together with dose limits in the range of 10 to 100 mrem/year [20], effectively limit the concentration of key radionuclides in ground water near a geologic repository. Over two decades ago the U.S. Nuclear Regulatory Commission adopted 10 mrem/year to the public as the design criterion that controls radiological emissions from nuclear power plants [29]. It is appropriate to adopt 10 mrem/year as the U.S. design criterion for a geologic waste repository [18], tenfold lower than the 100 mrem/year limit now proposed by industry and Congress for geologic repositories. Both the proposed higher dose limit of 100 mrem/year and the new proposed lenient method of calculating doses to compare to that limit will allow higher concentrations of contaminants in ground water.

Focusing only on the vicinity-average dose violates the long-established principle in protecting public health from radiation. The International Commission on Radiological Protection emphasizes that calculations should be made of the maximum individual dose [1]. The individual receiving that dose should be protected, so that all other individuals will be protected. Here the person receiving that calculated maximum dose is referred to as the "reference subsistence farmer".

PROBABILITY OF A WELL INTERSECTING CONTAMINATED GROUND WATER

Calculating vicinity-average doses would require knowing the spatial distribution of contaminants in ground water and the locations and habits of future people [22,30]. To illustrate, we show in Figure 1 our calculation of the steady-state concentration field of a long-lived radionuclide, e.g. I-129, in ground water that flows in the saturated zone beneath and beyond Yucca Mountain. The concentration isopleths are normalized to the concentration in ground water near the projected edge of the repository. This assumes a repository 2 miles in breadth, a uniform porous medium, and a constant pore velocity of 1 m/yr. Normalized concentrations are plotted as functions of distance in the direction of ground-water flow and distances transverse to flow. Defining the "vicinity" as a circle surrounding the repository, and assuming a distance of 35 miles to the edge of the "vicinity", the vicinity-average

---

1   In the modern terminology of EPA. we speak of "the reasonable maximum exposure", where exposure is measured by dose.

normalized concentration is about 0.018.

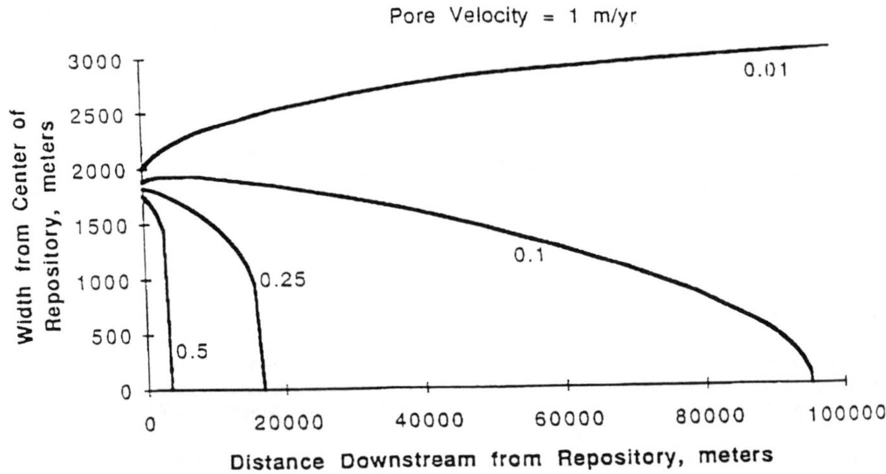

Figure 1. Normalized Concentration Field in Ground Water
Downstream From a Repository

The ratio of maximum concentration to the vicinity-average concentration, as calculated from data in Figure 1, is close to that estimated by the Electric Power Research Institute (EPRI) [22] on the basis of a more simplified transport model. Investigations of the regional flow of ground water in the region beyond Yucca Mountain are not yet sufficient for more precise estimates of pore velocities and of dispersion coefficients. If the pore velocities are lower than assumed here, and if the transverse dispersion coefficients are higher, the concentration ratio would be expected to be smaller than estimated here.

To illustrate the effect of the regional concentration field on radiation doses from using ground water, we now adopt EPRI's assumption [22] that future people are uniformly distributed over the entire circular vicinity surrounding the repository, excluding the area occupied by the repository. If all such people are subsistence farmers, and assuming a linear relation between concentration and dose, the ratio of vicinity-average dose to maximum dose is also equal to 0.018. EPRI interprets this dose ratio as the probability that any person living within the circular vicinity will drill a well that will intersect the most contaminated water [2]. Use of such probability estimates that depend in part on location and habits of future people has recently been endorsed by some [20,22,30,31], even though there is no scientific basis for predicting the locations and habits of future populations [20].

The probability estimated above is the "location probability". Additional probabilities that consider possible actions of future people to protect themselves have been proposed [20,22,30,31,33] and will be considered later.

If the allowable vicinity-average dose is 100 mrem/year, as proposed by industry and Congress [22,23,24], and for a location probability as low as 0.018, the dose to a subsistence farmer who uses contaminated ground water extracted from near the repository could be as

---

[2]   EPRI's calculated probability would be even lower if some of the people in the vicinity obtain their food and water from uncontaminated sources.

high as 5.6 rem/year. At this unacceptable dose level there would be a chance of roughly 1 in 8 for the maximally exposed person to suffer a lifetime cancer fatality.

Similarly, if an outer-zone radius of 100 miles is assumed, a dose ratio of 0.0063 would result. For a 100 mrem/year limiting average dose, the corresponding maximum individual dose would be 15 rem/year!

EARLY TIME CORRECTIONS

The dose ratios calculated above represent steady-state calculations, assuming that the contaminant plume has progressed to the assumed outer boundary of the vicinity. However, contaminated ground water is expected to move slowly towards the environment beyond Yucca Mountain. During early times, when the underground contaminant front has propagated only a short distance beyond the edge of the repository footprint, fewer wells will intersect contaminated ground water. As compared to the steady-state calculations, there will be less chance that any well in the general vicinity will produce contaminated water for farming than later, when the contaminant plume extends to the outer boundary. The location probability will be lower, as will the ratio of average dose to maximum dose. For example, for a plume length of only 3 miles, repository breadth of 2 miles, and vicinity outer radius of 35 miles, the ratio of vicinity-average dose to maximum dose would be 0.0016. If a dose of 100 mrem/year is allowed for the vicinity-average individual, a farmer using water extracted from the contaminant plume could receive an individual dose as high as 62 rem/year!

FURTHER REDUCTION IN AVERAGE DOSE BY POSTULATED HABITS OF FUTURE PEOPLE

The above calculations are based on assumptions of the location of future people. However, recent proposals [20,22,30,31,33] postulate several additional probabilities that could further reduce the calculated average doses below the dose calculated for a subsistence farmer. For example, a person might reside in the area for only part of his lifetime. The fraction of total population who are subsistence farmers could be quite small. Some of the water and food consumed could be derived from noncontaminated sources, such as bottled water. Wells penetrating the contaminated plume might not extract contaminated water. The presence of contamination in ground water could be detected, resulting in postulated decontamination of the water or greater use of water from other sources. These "habit probabilities" are guesses and postulates that the ratio of vicinity-average dose to the maximum dose will be lower than estimated above on the basis of location of future people. A low habit probability does not mean that the reference subsistence farmer would not exist. It means that there might be fewer subsistence farmers relative to all other individuals who receive lower doses. If ICRP recommendations are to be followed, the dose to the reference subsistence farmer will still need to be included in the dose calculations.

These habit probabilities are not derived on scientific grounds. Some [22,30] observe that the limited quantity of water in the vicinity of Yucca Mountain makes it likely that the population that depends solely on groundwater for its water needs will be limited in size. One of the habit probabilities reflects the assumed technology for detecting radioactivity and clean-up of the well water before use [20,22,30,33], claimed to be more effective for a future population with advanced technology [3]. For a small population, more appropriate for the

---

3   It is arguable whether the possibility that future humans will detect radioactivity in water and remove it before using the water should be included in performance assessment. A goal of geologic disposal is to dispose of the waste sufficiently carefully so that future humans need not take action to protect themselves. Society now has sensitive means of detecting contaminants in soil and ground water. There are many instances in present society wherein contamination is known but remediation is difficult or impracticable. It is not reasonable to presume constant monitoring of ground water for tens and hundreds of millennia, let alone

Yucca Mountain vicinity, EPRI [22] postulates net habit probabilities of 0.11 assuming present technology and 0.0038 for advanced technology. Each of these values would multiply the calculated probability of intersecting the plume, as calculated above, to result in even lower vicinity-average dose relative to the subsistence-farmer dose. Thus, the corresponding ratios of average dose to maximum dose and the maximum dose that could result if the allowed average dose is 100 mrem/year (0.1 rem/year) are listed in Table I.

Table I. Subsistence-Farmer Doses that Could Result From 100 mrem/year
Vicinity-Average Dose

(Assumes 35 mile outer radius of population zone, uniform distribution of population)

	Small population, current technology	Small population, advanced technology
Habit probability *	0.11	0.0038
Location probability	0.018	0.018
Calculated ratio of vicinity-average dose to subsistence-farmer dose **	2×10^{-3}	7×10^{-5}
Calculated ratio of subsistence-farmer dose to vicinity-average dose	5×10^{2}	1.4×10^{4}
Subsistence-farmer dose for 0.1 rem/year average dose, rem/year	50	1,400

*   Probability guesses proposed in [22,30].

**   The dose ratio is equal to the product of the location probability and the habit probability. The subsistence farmer dose is for a subsistence farmer who uses contaminated ground water from a well near the repository.

The annual doses of 50 rem and 1,400 rem that could theoretically be received by the reference subsistence farmer are of uncertain reliability, because of the many guesses postulated by those who suggested the probability values. Further, for doses in the range of hundreds of rem and above, radiation effects would become acute, including prompt life threatening, rather than the latent cancer and genetic effects assumed implicitly when calculating the lower doses (ca. 10 mrem/year) that could be reasonably considered for licensing.

EPRI also estimates habit and location probablities for a large population in the vicinity surrounding Yucca Mountain. EPRI suggests that a large population would exert a larger force to detect and mitigate contaminated ground water. Assuming advanced technology for detecting and mitiating contamination in ground water, EPRI's probabilities result in a ratio of vcinity-average dose to subsistence-farmer dose of 2×10^{-8}. For a vicinity-average dose limit of 0.1 rem/year, the corresponding subsistence-farmer dose would be 5 million rem/year. The corresponding concentrations of contaminants in ground water would be millions of times

---

continued technical and financial capability to perform such testing far into the future.

higher than normally considered safe [4].

The postulated calculational model with guesses at habit probabilities can lead to even greater estimated doses to the reference subsistence farmer due to early-time corrections, as discussed earlier.

CONCLUSION CONCERNING VICINITY-AVERAGE ALLOWABLE DOSE

These calculations illustrate that determining compliance on the basis of a vicinity average dose, allowed to reach a level of 100 mrem/year, could expose some individuals to doses that would not be tolerable in any reasonable analysis of public-health protection. Public health is better protected by limiting the reasonable maximum individual dose, as calculated for the reference subsistence farmer. Limiting that dose is consistent with the recommendation of the National Council on Radiation Protection and Measurements, the Nuclear Regulatory Commission's general public protection standard, the recommendation by the International Commission for Radiological Protection, and the current international consensus on health protection for geologic disposal [2 through 19].

GROUND-WATER PROTECTION

These calculations have much significance for ground-water protection. The probabilities suggested by EPRI [22,30] and the TYMS committee [20,31] all serve to allow greater concentrations of contaminants in ground water, by large factors. In its repromulgation of 40 CFR 191 the U.S. EPA added the requirement that a geologic repository should meet requirements imposed by the Safe Drinking Water Act. If those requirements are to apply to the proposed Yucca Mountain repository, the ground water would have to meet maximum contaminant levels (MCL's). There is now some uncertainty as to what MCL's would be appropriate. Whatever levels are set, there is the danger of exceeding those levels by adopting the more lenient vicinity-average dose limit now proposed by industry and Congress, as discussed above.

The traditional method of calculating MCL's is to assume that a person drinks the contaminated water as his only source of drinking water [18]. The assumption is akin to that of the subsistence farmer, but doses due to drinking contaminated water are typically more than tenfold less than doses from eating food grown in contaminated water. If the proposed vicinity-average dose were used, with a dose limit of 100 mrem/year, and assuming the probabilities adopted for Table 1, drinking the contaminated ground water could result in annual doses about tenfold lower than those estimated in Table I for the subsistence farmer. Even without invoking EPRI's "habit probabilities", the allowable doses from drinking contaminated ground water could still be far above any tolerable levels for drinking water. Therefore, ground-water protection could easily be compromised by the proposed lenient vicinity-average dose standard.

RELAXING THE PERFORMANCE REQUIREMENTS FOR A GEOLOGIC REPOSITORY

These calculations also illustrate how adopting probabilities [20,22,30,31,33] of future human activities and locations could relax the performance requirements for a geologic repository. For example, EPRI and the National Research Council's TYMS Committee proposed such probabilistic analyses after learning of calculations by the Yucca Mountain project [2] that the maximum calculated dose to the reference subsistence farmer was as high

---

4   The calculations do not consider self-limiting effects on ground water concentrations, such as solubility.

as 30 rem/year [5]. The postulated probabilities in Table I, proposed for a small population, would reduce the vicinity-average dose estimates to 60 mrem/year (assuming current technology for water purification) and to 2 mrem/year (assuming advanced technology for water purification). The subsistence-farmer dose of 30 rem/year (30,000 mrem/year) is 3,000 times greater than the allowable individual dose of about 10 mrem/year now adopted in the U.S. and abroad. However, the far more lenient limit of 100 mrem/year proposed [22,23,24] for the vicinity-average dose would not be exceeded. For those who would allow the low vicinity-average dose to obscure the high individual dose, the repository would be said to be safe!

If the reference subsistence farmer dose is 30 rem/year, as quoted above from the 1994 Yucca Mountain report, EPRI's large-population estimate would lead to a vicinity-average dose would be only 0.00075 mrem/year (7.5×10^{-7} rem/year). Although the reference subsistence farmer dose greatly exceeds any reasonable limit for an individual dose, the calculated vicinity-average dose would be far below the 100 mrem/year level proposed [21,22,23] for public health protection.

If compliance with health protection standards were to be based on the very low vicinity-average dose calculated above, the geologic repository would seem to be better located in the midst of a metropolitan desert community, such as Las Vegas! A contributor to this bizarre conclusion is expectation by some [22,30,32] that the force and technology of a large population would result in likely detection and mitigation of contaminated well water. Also, the aquifer under Yucca Mountain could supply contaminated water to only a limited number of people. The additional people would have to obtain water and food from uncontaminated sources, would receive low doses, and would reduce the vicinity-average dose. If a vicinity-average dose limit were adopted and if the much higher dose to the reasonable maximally exposed individual were ignored, calculated population dilution by a metropolitan area like Las Vegas could seem to make Yucca Mountain compliance easy, even if high individual doses could occur. Whether the repository would be safe would be obscured by the faulty logic of a vicinity-average dose limit.

THE NATIONAL RESEARCH COUNCIL'S TYMS REPORT

The National Research Council's TYMS report [20] proposes to develop probabilities that could lower the calculated average dose to a hypothetical critical group of individuals in the vicinity. The critical group would include the individual receiving maximum dose and all others whose doses are within tenfold of the maximum [6]. The TYMS committee's proposal was developed after the committee reviewed the high individual doses calculated by the Yucca Mountain project [2] and after it reviewed EPRI's probabilistic approach described above. There have been strong dissents from the TYMS committee's proposed probabilistic critical group on the grounds that there is no scientific basis for predicting probabilistic distributions of habits of future people, that it would be an unjustifiably lenient approach to public health protection, and that it would do irreparable damage to the Yucca Mountain project [25,26,27].

One cannot yet estimate how low the calculated individual dose would be using the TYMS approach, because the TYMS report does not present example estimates of the probabilities that it advocates, nor does it explain how such probabilities would be obtained. It does acknowledge that there is no scientific basis for estimating such probabilities.

However, the TYMS report does indicate that the authors may have been thinking of the same kinds

---

5 The Yucca Mountain project acknowledges that the calculated dose is much too high and would be unacceptable. It believes that the high dose is largely a result of extreme conservatism in the choice of parameters and in modeling release of radioactivity and its transport through the geosphere. It has underway several analyses and experimental programs that are expected by the project to provide a technical basis for reducing the calculated dose.

6 Because of mathematical errors in Appendix C of the TYMS report [20,25,32], these two ICRP criteria for the critical group could not both be fulfilled by the method endorsed by the TYMS committee.

of probabilities that were described to the committee by EPRI. Appendix C of the TYMS report speaks of calculating the probability that future people will be present over the contaminated plume of ground water. This is akin to EPRI's location probability described above. Also, the TYMS report and a member of the TYMS committee [32] speak of the benefits of a future society monitoring ground-water quality and either treating or avoiding use of contaminated sources. This is a key feature of EPRI's habit probability, particularly for an assumed future large population with advanced technology.

Basing performance assessment on such conjured probabilities and adopting a vicinity-average dose limit that requires estimates of such probabilities are not the ingredients of a standard that would build confidence by the scientific community and the public. The above calculations illustrate that a standard that would limit the reasonable maximum individual dose to a subsistence farmer is a far more reliable means of ensuring public health protection from geologic disposal.

REFERENCES

1. C. A. Johnson, "National Research Council Report 'Technical Bases for Yucca Mountain Standards': A State of Nevada View," Proceedings International Conference on **High-Level Waste Management**, Las Vegas, April 1996.

2. R. W. Andrews, T. F. Dale, and J. A. McNeish, "Total System Performance Assessment -- An Evaluation of the Potential Yucca Mountain Repository," Yucca Mountain Site Characterization Project, INTERA, Inc., Las Vegas, Nevada, 1994.

3. I. M. Barraclough, S. F. Mobbs, and J. R. Cooper, "Radiological Protection Objectives for the Land-Based Disposal of Solid Radioactive Wastes," Documents of the National Radiation Protection Board (NRPB), 3, No. 3, 1992.

4. D. Charles, and G. M. Smith, "Project 90 Conversion of Releases From the Geosphere to Estimates of Individual Doses to Man," Swedish Nuclear Regulatory Commission, SKI Technical Report 91:14, 1991.

5. P. A. Davis, R. Zach, M. E. Stephens, B. D. Amiro, G. A. Bird, J. A. K. Reid, M. T. Sheppard, S. C. Sheppard, M. Stephenson, "The Disposal of Canada's Nuclear Fuel Waste: The Biosphere Model, BIOTRAC, for Postclosure Assessment," AECL-10720, 1992.

6. W. T. Farris, B. A. Napier, T. A. Ikenberry, J. C. Simpson, D. B. Shipler, "Atmospheric Pathway Report, 1944-1992," Hanford Environmental Dose Reconstruction Project, Pacific Northwest Laboratories, PNWD-2228 HEDR, 1994.

7. W. T. Farris, B. A. Napier, T. A. Ikenberry, J. C. Simpson, D. B. Shipler, "Columbia River Pathway Dosimetry Report," 1944-1992," Hanford Environmental Dose Reconstruction Project, Pacific Northwest Laboratories, PNWD-2227 HEDR, 1994.

8. C. D. Leigh, *et al.*, "User's Guide for GENII-S: A Code for Statistical and Deterministic Simulation of Radiation Doses to Humans from Radionuclides in the Environment," Sandia National Laboratories, SAND-91-0561, 1993.

9. T. McCartin, R. Codell, R. Neel, W. Ford, R. Wescott, J. Bradbury, B. Sagar, J. Walton, "Models for Source Term, Flow, Transport and Dose Assessment in NRC's Iterative Performance Assessment, Phase 2," Proc. International Conference on High Level Radioactive Waste Management, Las Vegas, NV, 1994.

10. B. A. Napier, R. A. Peloquin, D. L. Streng, J.V. and J.V. Ramschell, "GENII: The Hanford Environmental Radiation Dosimetry Software System," Richland, Washington, Pacific Northwest Laboratory, PNL-6584, 1988.

11. R. B. Neel, Dose Assessment Module", in <u>NRC Iterative Performance Assessment Phase 2: Development of Capabilities for Review of A Performance Assessment for a High-Level Waste Repository,</u>" R. G. Wescott, M. P. Lee, T. J. McCartin, N. A. Eisenberg, and R. B. Baca, eds., U. S. Nuclear Regulatory Commission, NUREG-1464, 1995.

12. R. P. Rechard, Ed., "Performance Assessment of the Disposal in Unsaturated Tuff of Spent Nuclear Fuel and High-Level Waste Owned by U.S. Department of Energy, Vol. 1: Methodology and Results," Sandia National Laboratories, SAND94-2563/2, March 1995.

13, Switzerland National Cooperative for the Storage of Radioactive Waste, "Nuclear Waste Management in Switzerland: Feasibility Studies and Safety Analyses," Project Report NGB 85-09, June 1985.

14. Switzerland National Cooperative for the Storage of Radioactive Waste, "Kristallin - I: Safety Assessment Report," Technical Report 93-22E, February 1994.

15. UK Department of the Environment, "Review of Radioactive Waste Management Policy: Preliminary Conclusions," August, 1994.

16. F. van Dorp, NAGRA Project Manager for Biosphere Models, Switzerland, Private Communication, 1994.

17. T. Vieno, A. Hautojarvi, L. Koskinen, and H. Nordman, "TVO-92 Safety Analysis of Spent Fuel Disposal," Report YJT-92-33E, Technical Research Centre of Finland, Nuclear Engineering Laboratory, Helsinki, 1992.

18. T. H. Pigford, J. 0. Blomeke, T. L. Brekke, G. A. Cowan, W. E. Falconer, N. J. Grant, J. R. Johnson, J. M. Matuszek, R. R. Parizek, R. L. Pigford, D. E. White, "A study of the Isolation System for Geologic Disposal of Radioactive Wastes,", National Academy Press, Washington, D.C., 1983.

19. International Atomic Energy Agency, "Critical Groups and Biospheres in the Context of Radioactive Waste Disposal," Position Paper Produced for the Working Group on Principles and Criteria for Radioactive Waste Disposal, July 1995.

20. R. W. Fri, *et al.,* "Technical Bases for Yucca Mountain Standards," National Academy Press, Washington, D.C. 1995.

21. U.S. Congress, Proposed Legislation: HR-1020 (Upton, 1995)), S-167 (Johnston, 1995), S-1271 (Craig, 1996).

22. Electric Power Research Institute (EPRI), "A Proposed Public Health and Safety Standard for Yucca Mountain: Presentation and Supporting Analysis," Report EPRI-TR10412, pp. 3-26 - 3-27, April 1994.

23. Nuclear Energy Institute, "NEI Comments on Behalf of the Nuclear Industry on Selected Findings and Recommendations of the NAS Report, *Technical Bases for Yucca Mountain Standards,*" October 1995.

24. J. H. Kessler, "Initial EPRI Reaction to the NAS Yucca Mountain Standards Recommendations, "**Proceedings International Conference on High-Level Waste Management**, Las Vegas, April 1996.

25. T. H. Pigford, "Personal Supplementary Statement," Appendix E in **Technical Bases for Yucca Mountain Standards**, National Academy Press, Washington, D.C. 1995.

26. T. H. Pigford, "The Yucca Mountain Standard: Proposals for Leniency,"**Proc.
Materials Research Society: Scientific Basis for Radioactive Waste Management**, October 1995.

27. T. H. Pigford, "The Yucca Mountain Standard: How Lenient Should it Be?," **Proceedings
International Conference on High-Level Waste Management**, Las Vegas, April 1996.

28. M. Hynes, **Las Vegas Review Journal**, May 12, 1994.

29. U.S. Nuclear Regulatory Commission, Code of Federal Regulations, Chapter 10, Part 50,
Appendix I, "Numerical Guides for Design Objectives and Limiting Conditions for
Operation to Meet the Criterion 'As Low as is Reasonably Achievable' for Light-Water-
Cooled Nuclear Power Reactor Effluents, May, 1975.

30. R. E. Wilems, "Illustrative Probabilistic Biosphere Model For Yucca Mountain
Individual Risk Calculations," Proceedings Waste Management **1994**, R. G. Foss and M.
E. Macks, ed., University of Arizona, 1994.

31. F. M. Phillips, Presentation of "Technical Basis of Yucca Mountain Standards" to Advisory
Committee on Nuclear Waste, Nuclear Regulatory Commission, June 1996.

32. T. H. Pigford, "Invalidity of the Probabilistic Exposure Scenario Proposed by the
National Research Council's TYMS Committee," Report UCB-NE-9523, Rev. 1, May 1996.

33. C. G. Whipple, Presentation of "Comments Regarding the NAS Report on Yucca Mountain
Standards," International Conference on High-Level Waste Management, Las Vegas,
April 1996.

APPENDIX: VARIOUS INTERPRETATIONS OF "MAXIMUM INDIVIDUAL DOSE"

The dose to the "reference subsistence farmer" should not be confused with a dose to the
"hypothetical maximally exposed individual", as used in many studies. The "reference subsistence
farmer" would be the individual who receives the highest dose, among all those individuals considered
in calculating radiation doses. However, there could be some individuals who could receive higher
doses, such as individuals with unusual sensitivity to radiation or with unusual diets. It has been the
policy or practice in the international community to calculate protection to future individuals whose
diets and sensitivity to radiation are typical of present-day people in the vicinity.

Further, the reference subsistence-farmer doses calculated in performance assessment of geologic
repositories are not the maximum doses that could be received even by a subsistence farmer. As
explained elsewhere [25], it is the practice to express uncertainties in geosphere parameters as
probabilistic distributions of those parameters. The doses calculated are the expected values of the
resulting probabilistic distribution of doses, not the doses at the high-dose end of the distribution. The
highest dose of that distribution is referred to by EPA as the Theoretical Upper Bound Estimate
(TUBE). It is calculated by assuming most unfavorable and conservative values of each parameter that
affects the dose calculation. This extremely conservative deterministic calculation of the TUBE is not
the calculated subsistence-farmer dose referred to herein. Here "dose" is understood to be calculated
as the mean of the probabilistic distribution of doses. The probabilistic distribution of doses is to be
calculated based on scientifically based distributions of parameters that affect those calculations.

Thus, there are several examples of calculation of doses to a "hypothetical maximally exposed
individual" that are far more conservative and extreme than the subsistence-farmer calculation that is
international practice. In the current language of the U.S. EPA, the dose calculated to the reference
subsistence farmer could be better construed as the calculation of the "reasonable maximum dose".

THERMAL/HYDROLOGICAL MODELING OF THE RADIOACTIVE SCRAP AND WASTE FACILITY (RSWF) WITH THE TOUGH2 (TRANSPORT OF UNSATURATED GROUNDWATER AND HEAT) CODE

D. W. ESH and R. W. BENEDICT
Argonne National Laboratory-West, PO Box 2528, Idaho Falls, ID 83403

ABSTRACT

Thermal/hydrological modeling of the Radioactive Scrap and Waste Facility (RSWF) has been completed with the TOUGH2 (Transport of Unsaturated Groundwater and Heat) Code.[1] The RSWF will be utilized as an interim storage facility for ceramic and metallic waste forms developed from the electrometallurgical treatment of spent nuclear fuel. The RSWF is an array of 1,350 carbon steel liners located at grade near Argonne National Laboratory-West on the Idaho National Engineering Laboratory (INEL).

The primary driving force for this modeling research was to assess thermal capacity limits for RSWF liners so that heat generating materials can be safely stored. Maximum wasteform temperatures will be governed by both the amount of heat the system can dissipate and the orientation and characteristics of the wasteform. The focus of this report is on the amount of heat the interim storage system can safely dissipate. The effect of the temporal variation of soil moisture on the performance of the RSWF is assessed. The facility was analyzed to determine the maximum allowable thermal loading of the RSWF liners.

INTRODUCTION

Argonne National Laboratory (ANL) is actively pursuing a treatment process for the conditioning of Department of Energy spent nuclear fuel. A demonstration is currently being performed on Experimental Breeder Reactor-II (EBR-II) sodium-bonded, metallic spent nuclear fuel. The process involves the electrorefining of spent nuclear fuel, and the subsequent production of ceramic and metallic waste forms. The metallic waste form will have a relatively low thermal output. However, the ceramic waste form may have a specific thermal loading that is up to an order of magnitude larger than typical Defense High-Level Waste (DHLW).[2] The high thermal loading of the ceramic waste form limits interim storage and final disposal options.

One possible candidate for an interim storage facility is the Radioactive Scrap and Waste Facility (RSWF) located at ANL-West on the Idaho National Engineering Laboratory's (INEL) 2300 km^2 site. Argonne West occupies about 810 acres of the INEL. The RSWF is a 2D array of 1350 carbon steel liners located near Argonne National Laboratory's other facilities at the INEL. The design approval by the Atomic Energy Commission for the RSWF was given in 1964. In June 1965 the first waste was emplaced.

There are different types of liners at the RSWF, but the standard liner is cylindrical with a 0.4064 m (16 in) diameter, 0.00635 m (0.25 in) thickness, and a 3.760 m (12.33 ft) length. A 0.762 m (30 in) thick concrete plug is utilized to provide shielding. The top 0.10 m of each liner is above the land surface. The liners have a 1.83 m pitch in each row and the rows are 3.66 m apart. The current maximum allowable heat load per liner is 300 W. Each liner

Mat. Res. Soc. Symp. Proc. Vol. 465 © 1997 Materials Research Society

is cathodically protected to prevent corrosion of the carbon steel. The objective of this study is to determine a safe design capacity (thermal load) for the RSWF's carbon steel liners.

MODELING

Two computer codes, TOUGH2 and HEATING7.2i, have been utilized to examine liner capacity issues of the RSWF. TOUGH2 is a well documented code that was developed to numerically simulate nonisothermal flows of multicomponent, multiphase fluids in porous and fractured media.[3]-[4] TOUGH2 can be applied to a variety of problems, including unsaturated zones, groundwater aquifers, and geothermal reservoirs. TOUGH2 solves mass and energy balance equations that describe fluid and heat flow in general multiphase multicomponent systems. HEATING7.2i is a widely used general-purpose heat transfer program.[5] With TOUGH2, the general methodology utilized in this study is to solve for the steady-state moisture distribution, under isothermal conditions, using a time-averaged infiltration rate and then use the steady-state moisture distribution as the initial conditions in the coupled thermal/hydrological simulation.

The factors considered in this study will include: (1) radioactive decay of the waste, (2) the heat flux distribution of the source on the liner surface, (3) drying of the facility soil and consequent decrease in thermal conductivity of the soil, and (4) liner geometry and liner spacing. By assuming a uniform liner spacing or in the case of the analysis of a single liner, semi-infinite conditions apply and the three-dimensional nature of the problem becomes pseudo two-dimensional. The heat and mass transfer problem can be modeled with a two-dimensional cross section in rz coordinates (if land surface area is preserved, i.e. for a 3.66 m liner spacing the equivalent cylindrical radius is given by $R=(3.66^2/\pi)^{0.5}=2.065$ m). Initially, the boundary conditions are taken as follows: (1) the top surface (land) is constant temperature, (2) the bottom surface (water table at 100 m) is constant temperature, (3) the side boundaries (r=0, r=R, where R is a problem variable) are symmetry boundaries with a null valued heat flux, and (4) the region of the waste can is taken as a temporally varying heat flux boundary in which the flux sums to the total heat generation in the waste can. Figure 1 is a representation of the problem which is modeled in this study. The general features of the modeled domain are given and the boundary conditions are shown.

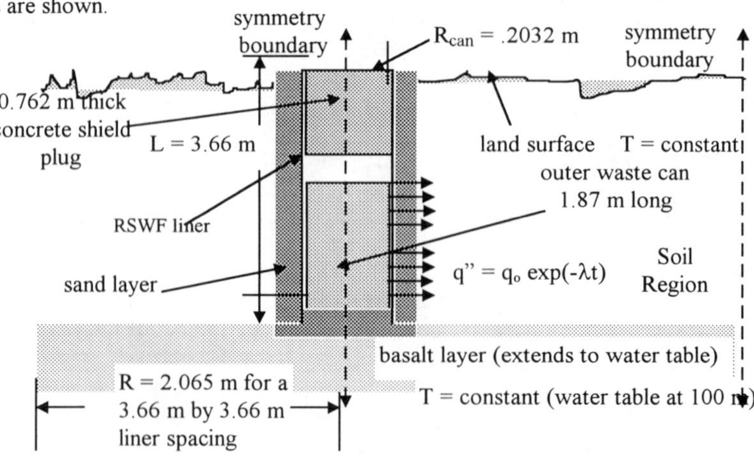

Figure 1. Representation of problem geometry and boundary conditions.

One of the most difficult tasks in most modeling projects is obtaining accurate input data, this project is not an exception. The variation of soil thermal conductivity with moisture content has been studied by others for various soils, but the variation has not been measured for RSWF soils. In addition, the soil moisture characteristic curve has not been measured at ANL-West. Table I is a summary of the thermal data that was utilized in this modeling study. The thermal data was obtained from a literature search on the properties of soils.[6]-[7] Table II is a summary of the hydrological data that was utilized in this project. Most of the data in Table II was obtained from technical reports on INEL hydrological studies.[7]-[8] The characteristics of ANL-West soils are expected to vary from those of other INEL soils, however until detailed

Table I.   Summary of thermal data utilized in the modeling study

Material	Density (kg/m^3)	k_t (W/m K)	C_p (J/kg K)
Sand	1400-2000	1.0-2.5	800-1000
Soil	1400-2000	1.0-2.5	800-1000
Basalt	1700-2300	1.6	800-1000
Steel	7854	60.5	430

measurements are completed, the data obtained in the literature are the best available. Very little information is available on the heterogeneity of the soils and basalt at ANL-West. Therefore, properties were assumed to be homogeneous. Future work plans include detailed measurement of each estimated variable so that a firmer estimate of liner capacity can be determined. Some experimental measurements have been completed and a more in-depth experimental study is anticipated. The data in the last five columns of Table II are the variables utilized in the option (IRP=7) in TOUGH2 for the capillary pressure functions and relative permeability functions.[9]

Table II.   Summary of hydrological data

Material	Permeability (m^2)	Porosity (m^3/m^3)	λ	S_{lr} (m^3/m^3)	S_{ls} (m^3/m^3)	P_o (Pa)	P_{max} (Pa)
Sand	1.10E-11	0.3	0.72	0.20	1.00	3.18E+03	4.90E+04
Soil	2.28E-12	0.51	0.33	0.33	1.00	4.07E+03	1.96E+05
Basalt	8.33E-12	0.23	0.32	0.065	1.00	2.58E+03	4.90E+04

The capillary pressure and relative permeability functions are used to describe the movement of moisture in the vadose zone. For a complete description see ref. 9. The relative permeability functions for the liquid and gas phases are given by

$$k_{rl} = \begin{cases} \sqrt{S^*}\left\{1-\left(1-\left[S^*\right]^{1/\lambda}\right)^{\lambda}\right\}^2 & \text{if } S_l < S_{ls} \\ 1 \text{ if } S_l \geq S_{ls} \end{cases} \qquad \text{where } S^* = \frac{S_l - S_{lr}}{S_{ls} - S_{lr}}$$

$$k_{rg} = 1 - k_{rl}$$

S_l is the liquid saturation, S_{lr} is the residual liquid saturation, and S_{ls} is the saturated liquid saturation. The capillary pressure function is given by

$$P_{cap} = \begin{cases} 0 \ (\text{if } S_l \geq S_{ls}) \\[2ex] -P_o\left\{\left[S^*\right]^{-1/\lambda} - 1\right\}^{1-\lambda} \\[2ex] -P_{max} \ \left(\text{if } P_o\left\{\left[S^*\right]^{-1/\lambda} - 1\right\}^{1-\lambda} \geq P_{max}\right) \end{cases} \qquad \text{where } S^* = \frac{S_l - S_{lr}}{S_{ls} - S_{lr}}$$

and P_{max} is the maximum capillary pressure. Many other relationships such as a linear relationship, Milly's function, or Leverett's function could be utilized to describe the capillary pressure function.[10]-[11] In addition, many functions exist which could be used to describe the relative permeability functions.[9]

EXPERIMENTAL MEASUREMENTS

Experimental measurements were completed in the past in which a 240 watt heat source was placed within an RSWF liner.[12] Temperature measurements were taken with four sets of thermocouples located at radial distances of 0, 0.152, 0.406, and 0.711 m (0, 6, 16, and 28 in.) from the liner surface. The measured temperature profiles can be compared to modeled results.

Soil moisture measurements were taken in July 1996. Six soil cores were taken with a hand auger instrument. The moisture content was then estimated from *Standard Methods*.[13] Soil moisture was found to be quite variable at the RSWF, as was expected. A discussion of soil moisture variations and the observations to date follow in the next section. A more detailed study of soil moisture is planned with a neutron probe instrument.

RESULTS AND ANALYSIS

As mentioned previously, the initial step in the modeling study was to simulate the steady-state moisture distribution, under isothermal conditions, assuming a time-averaged infiltration rate on the land surface. Historical measurements suggest that average rainfall is 0.203 m/yr at ANL-West. In the arid climate, evapotranspiration is expected to greatly reduce the amount of infiltration. However, measurements suggest that in disturbed areas of the INEL and because rainfall tends to be very episodic, infiltration can be as much as 95% of rainfall.[7] For this reason, infiltration was considered to be a problem variable. The range of infiltration used in these simulations was 0.0 to 0.193 m/yr. Figure 2a is the simulated and measured steady-state soil moisture distribution, at a distance of one meter from an RSWF liner, for an infiltration rate of 0.193 m/yr. The experimental measurements suggest that moisture content decreases, with increasing depth, as the basalt layer is approached. It is expected that the moisture content should increase above the basalt layer, if the basalt layer is less permeable than the soil. An agreement was achieved between the simulated and measured distributions by assigning a low capillary pressure function and low residual liquid saturation to the basalt. It is expected that fractures in the basalt

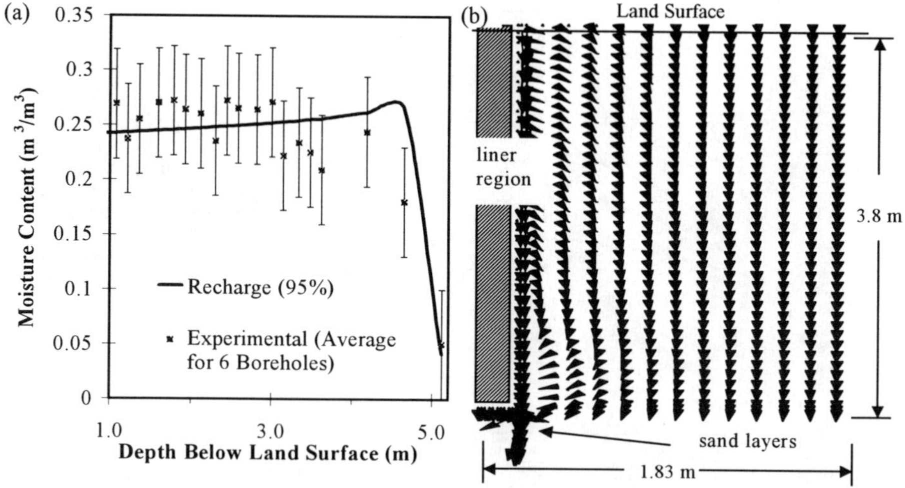

Figure 2. (a) Simulated steady-state moisture distribution for an infiltration rate of 0.193 m/yr. (b) Steady-state Darcy velocity vector field for an infiltration rate of 0.193 m/yr. The sand layer near the liner has much larger liquid phase velocities and the direction of transport is essentially vertical. The length of time of the steady-state simulation was 10^7 years. Only the top 3.8 m of the Darcy velocity vector field is shown.

will result in a large effective permeability for the basalt unit and it will drain rather easily. Figure 2b is the simulated Darcy velocity vector field at steady-state for an infiltration rate of 0.193 m/yr. The magnitude of the Darcy vectors are much larger in the sand layers, as expected. The reason for the observed pattern in the Darcy velocity vector field is unknown at this time.

A comparison of the simulated and measured temperature responses of an RSWF liner to a 240 watt heat load is given in Figure 3. Since only a single liner was loaded for the experimental measurements, the distance for the radial adiabatic boundary condition was determined by running simulations with an increasing radial dimension until the maximum liner temperature no longer changed. A soil thermal conductivity of 1.5 W/m K and a heat loss through the top of the liner to the land surface of 10% gave the best fit to the experimental data. The reason for adding a heat loss term (a reduction in the total heat source) was to account for the transfer of heat through the concrete shield plug to the land surface. The heat flux distribution which was applied to the surface of the liner was calculated with a HEATING7.2i simulation which considered the geometry of the source and the radiant exchange of energy within a liner.

Simulations predict that the transfer of heat from a liner through the soil is dominated by conduction for a thermal loading of less than 500 watts. Figures 4a to 4d are examinations of maximum predicted liner temperatures as a function of liner loading for some of the different problem variables. These simulations were possible because of the successful agreement between the simulated and measured temperature responses of an RSWF liner, given in Figure 3. As expected, a number of variables affect RSWF liner capacity including: material properties, liner geometry, array geometry, geometry of the heat source, half-life of the heat source, and soil

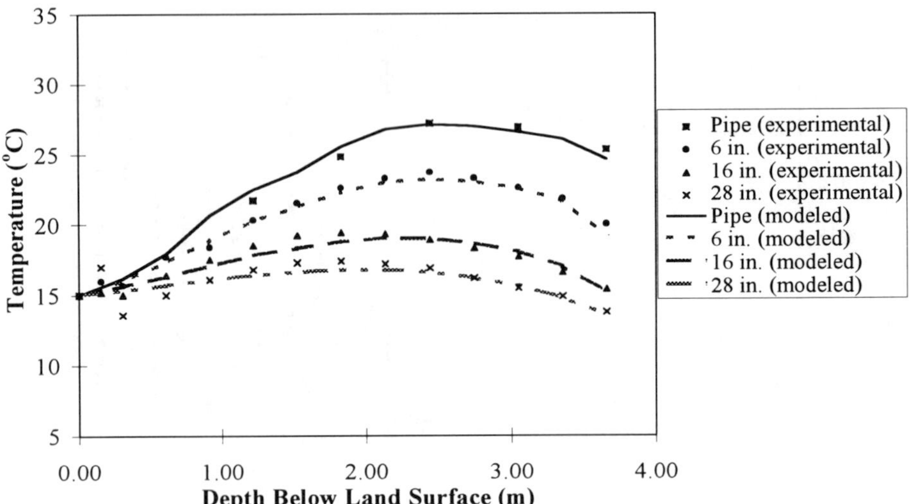

Figure 3. Simulated and measured temperature distributions for an RSWF liner loaded with a 240 watt heat source. The best fit of the experimental data was achieved with a soil thermal conductivity of 1.5 W/m K and a heat loss through the top of a liner of 10 %.

moisture variations/dry-out. Soil moisture variations and soil dry-out were found to have a minimal effect on liner capacity for simulations in which the soil temperature remained below 90°C. Liquid saturation changed by as much as 30% in localized areas near the liner surface, but this change was not large enough or long enough in duration to create substantial perturbations of the maximum observed temperatures.

Maximum liner temperatures were estimated as a function of a number of problem variables. Figure 4a shows the simulated difference between maximum soil temperatures and land surface boundary temperatures for a range of thermal outputs per liner and a range of soil thermal conductivity values. Both TOUGH2 and HEATING7.2i were utilized to simulate the maximum temperature difference curves. The agreement between these codes was excellent, generally to within 1%, when the maximum temperatures of the soil were maintained below the boiling point of water. If thermal output of RSWF liners is maintained below 550 W, then the effect of soil moisture should be unimportant. However, if soil temperatures are increased above 90°C, no experimental information currently exists on the performance of the system. A future study of soil moisture at the RSWF with a neutron probe will be utilized to examine the effect of high liner temperatures, i.e. liner thermal outputs for which the soil temperature is above 90°C. The simulations to estimate the relationship between maximum soil temperatures and liner thermal output utilized a conservative heat flux distribution. Therefore, RSWF liner capacity will most likely be larger than the 550 W shown in figure 4a for a soil thermal conductivity of 1.5 W/m K (to avoid soil moisture loss). Based on the calculations completed for this project, liner thermal capacity can be safely increased to 550 W from 300 W, with a 600 to 650 W limit likely when a less conservative heat flux distribution on the inner liner surface is utilized.

A 600 W heat source and a heat flux distribution over the inner surface of the waste can ,not the total liner surface, was utilized for the calculations of 4b and 4c. Therefore the

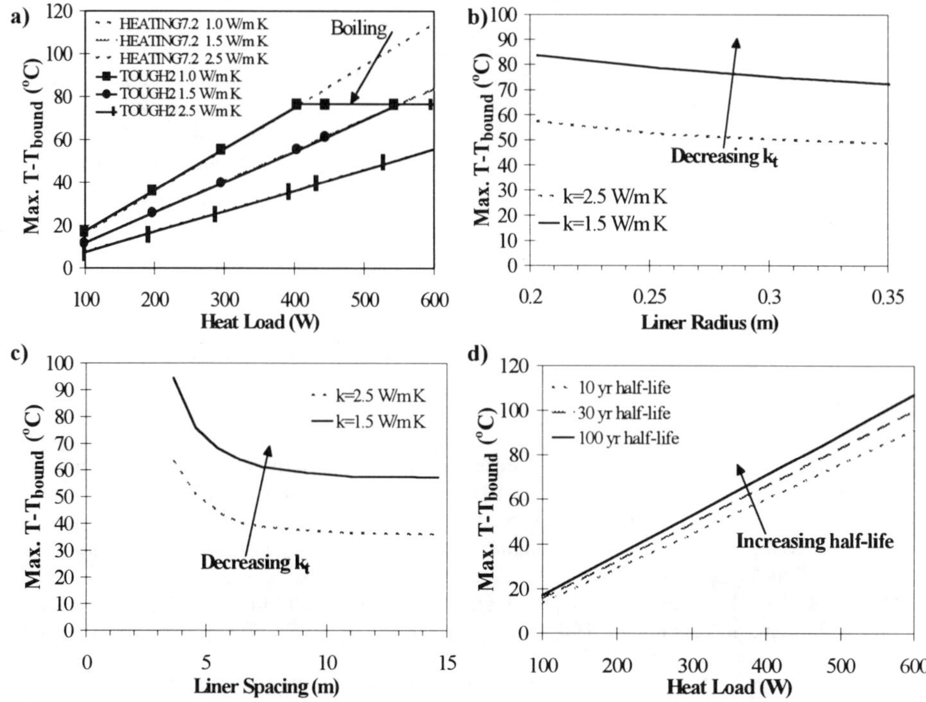

Figure 4. Maximum predicted soil temperatures as a function of: (a) soil thermal conductivity and liner heat load, (b) liner radius, (c) liner spacing, and (d) composite half life of the waste. The effects of soil moisture phase change is not included in (b), (c), or (d).

temperatures shown in 4b and 4c are larger in magnitude than those expected for a less conservative heat flux distribution. Figure 4c is a plot of the dependence of maximum temperature differences on liner spacing and soil thermal conductivity. When the distance between liners approaches 5 m (16.5 ft) the maximum observed soil temperatures are substantially less than for smaller liner spacing. The decrease in maximum temperatures (Figure 4b) with increasing liner radius is much less pronounced, although an extra 10 to 20% larger heat source can be safely stored in a liner with a 0.30 m radius compared to a 0.20 m radius liner. The effect of the half-life of the waste on maximum observed temperatures is not extremely important for this system and the range of half-lives under consideration.

CONCLUSIONS

RSWF liner capacity is not a simple function of one variable. RSWF liner capacity can be safely increased to 550 W from 300 W. A limit in the 600 to 650 W range is possible if a less conservative heat flux distribution is utilized. Also, increasing the liner radius and liner spacing are suitable methods of increasing the individual liner capacity, although an overall reduction in the capacity of the field would result. The system is dominated by heat conduction but soil moisture loss will become increasingly important at higher liner loading. Soil thermal conductivity, liner geometry, and soil moisture content are all important problem variables. Future work includes making temporal soil moisture measurements with a neutron probe instrument. A liner will be loaded with up to an 800 watt heat source and the temperature and moisture responses will be measured. The collected data from the aforementioned and other experiments should provide the necessary information to estimate the response of RSWF soils to large heat loads. More parameter measurements will be completed to reduce the uncertainty in the problem analysis.

ACKNOWLEDGEMENTS

This work was funded by US DOE under contract No. W-31-109-ENG-38 to Argonne National Laboratory. The authors gratefully acknowledge T. Zahn and R. Clarksean for the information they provided about the facility and previous experimental measurements.

REFERENCES

1. Pruess, K., "TOUGH2--A General Purpose Numerical Simulator for Multiphase Fluid and Heat Flow," *LBL-29400*, Lawrence Berkeley Laboratory, Berkeley, CA (1991).
2. Notz, K. J., "Characteristics of Potential Repository Wastes," Department of Energy, Washington, DC, *Proceedings International Conference for High-Level Radioactive Waste Management*, Las Vegas, NV, April 1990.
3. Moridis, G. and K. Pruess, "Flow and Transport Simulations Using T2CG1, A Package of Conjugate Gradient Solvers for the TOUGH2 Family of Codes," *LBL-36235*, Lawrence Berkeley Laboratory, Berkeley, CA (1995).
4. Moridis, G. and K. Pruess, "TOUGH Simulations of Updegraff's Set of Fluid and Heat Flow Problems," *LBL-32611*, Lawrence Berkeley Laboratory, Berkeley, CA (1992).
5. Childs, K. W., "HEATING 7.2 User's Manual," *ORNL/TM-12262*, Oak Ridge National Laboratory, Oak Ridge, TN (1993).
6. Farouki, O. T., Thermal Properties of Soils, Series on Rock and Soil Mechanics Vol. 11, Trans Tech Publications, Federal Republic of Germany, 1986.
7. Martian, P. and S. O. Magnuson, "A Simulation Study of Infiltration Into Surficial Sediments at the Subsurface Disposal Area, Idaho National Engineering Laboratory," *EGG-WM-11250*, EG&G Idaho, Idaho Falls, ID (1994).
8. Magnuson, S. O. and A. J. Sundrop, "A Modeling Study of Contaminant Transport Resulting From Flooding of Pit 9 at the Radioactive Waste Management Complex, Idaho National Engineering Laboratory," *EGG-EEL-10498*, EG&G Idaho, Idaho Falls, ID (1992).
9. Pruess, K., "TOUGH User's Guide," *LBL-20700*, Lawrence Berkeley Laboratory, Berkeley, CA (1987).
10. Milly, P. C. D., "Moisture and Heat Transport in Hysteretic, Inhomogeneous Porous Media: A Matric-Head Based Formulation and a Numerical Model," Water Resources Research, **18**, 3, 489-498, 1982.
11. Leverett, M. C., "Capillary Behavior in Porous Solids," AIME Trans., **142**, 152, 1941.
12. Clarksean, R. L., Personal Communication, Argonne National Laboratory, Idaho Falls, ID (1994).
13. "Standard Test Method for Determination of Water (Moisture) Content of Soil By Direct Heating Method," ASTM *D 4959-89*, American Society for Testing and Materials, Philadelphia, PA (1989).

BOX MODEL OF RADIONUCLIDE DISPERSION AND RADIATION RISK ESTIMATION FOR POPULATION IN CASE OF RADIOACTIVITY RELEASE FROM NUCLEAR SUBMARINE #601 DUMPED IN THE KARA SEA

E.I.YEFIMOV, D.V.PANKRATOV, S.V.IGNATIEV
Institute of Physics and Power Engineering, Obninsk 249020, Russia
Phone: (084) 399-8703, Fax: (095) 230-2326, E-mail: yefimov@ippe.rssi.ru

ABSTRACT

A relatively simple model is suggested for radionuclide dispersion and assessment of additional population dose and radiation risk for radionuclide release from the nuclear submarine (NS) #601 with Pb-Bi coolant that was dumped in the Kara Sea.

INTRODUCTION

When ships with nuclear reactors or nuclear materials aboard suffer shipwreck or in the case of burial or dumping of radioactive wastes, atmospheric fallout, etc. radionuclides may be released and spread in the sea, contaminating the sea water and the sea bottom.

When a nuclear submarine is dumped this spread of activity may occur, due to gradual core destruction by corrosion over many years. Of obvious interest is also fast release of the total activity caused by unlikely but real events, like an earthquake or military activities in the burial area.

The main radiological consequences of sea contamination are that local residents are exposed to additional radiation due to this contamination.

The task of predicting the consequences of shipwrecks of ships having nuclear materials aboard is a vital issue from the viewpoint of determining sources of radiation for population and most probable doses.

This task requires a step-by-step consideration and mathematical description of complex space-and time processes (evaluation of the rate of structure disintegration and radionuclide release into the sea; spatial transfer and radionuclide diffusion around the radioactivity source).

The objective of this paper is to develop a mathematical model of radionuclide dispersion and to assess the population dose and radiation risk for radionuclide release from the NS #601, with Pb-Bi coolant that was dumped in the Kara Sea.

MATHEMATICAL MODEL

Estimation of additional population exposure doses

The problem of population exposure caused by radioactive waste dumped in sea water has been considered by IAEA experts [1,2].

Ingestion of contaminated seafood (hydrobiota) proved to be the main path way of population exposure. Let us evaluate individual exposure dose in Sv per year after time t from the beginning of radionuclide release into sea water:

$$D_o(t) = \sum_j \alpha_j g_j \sum_i K_{ij}^H d_i \overline{a_i}(t) \tag{1}$$

Mat. Res. Soc. Symp. Proc. Vol. 465 © 1997 Materials Research Society

where:

g_j - total consumption of the j-th form of seafood, t/year;

α_j - edible fraction of j-th form of seafood;

K_{ij}^H - concentration coefficient of i-th nuclide in j-th sea food, m³/t;

d_i - effective equivalent dose for ingestion of i-th nuclide, Sv/Bq;

$\overline{a_i}(t)$ - average sea water concentration of the i-th nuclide, Bq/ m³.

There are some difficulties in determining these parameters. Considerable differences are observed in the values of K_{ij}^H [1,2,3]. The recommended data from Reference [1] were used in our calculations. There is no trustworthy data on seafood use by coastal populations [4,5].

Box model for radionuclide concentration calculation

The essence of the box model lies in the division of the sea water into multiple volumes, i.e. boxes, for which box-to-box water exchange coefficients are determined.

The main assumption of the box model is prompt activity equilibration within each box volume. This assumes that the processes of water intermixing are so intensive that the total box intermixing time is much less than the radionuclide half-lives.

One can consider a simple one-box model. Let us assume that a well mixed water volume V contains a radionuclide source, and some amount Q(t) of the i-th nuclide is released into that volume at time t.

The specific activity a(t) of the i-th nuclide changes as follows

$$\frac{da(t)}{dt} = \frac{Q(t)}{V} - \left(\lambda + \lambda_o + \lambda_{oc}\right)a(t) \tag{2}$$

where

$\lambda, \lambda_o, \lambda_{oc}$ are constants for radionuclide exit from the water volume due to decay, exchange and bottom deposition, respectively, s⁻¹.

The solution of (2) is

$$a(t) = e^{-\lambda_{ef}t}\left[a_o + \frac{1}{V}\int_0^t Q(t')e^{\lambda_{ef}t'}dt'\right] \tag{3}$$

where $\lambda_{ef} = \lambda + \lambda_o + \lambda_{oc}$, and a_o is initial concentration of the i-th, nuclide, in Bq/m³.

In general, the concentration of the i-th nuclide during time interval (t, t+T_o) is

$$\overline{a(t)} = \frac{1}{T_o}\int_t^{t+T_o} a(t')dt' \tag{4}$$

Supposing that radionuclide release goes on uniformly (Q(t') is a constant) during time T from the time t_o, when that source appeared on the sea bottom, one can obtain

$$\overline{a(t)} = \frac{Q_o e^{-\lambda t}}{(\lambda_o + \lambda_{oc})T\,T_o V}\left[\frac{1 - e^{-\lambda T_o}}{\lambda} + \frac{1 - e^{-\lambda_{ef}T_o}}{\lambda_{ef}}e^{-(\lambda_o + \lambda_{oc})(t - t_o)}\right] \tag{5}$$

for $t_o < t \leq t_o + T$

and

$$\overline{a(t)} = \frac{Q_o e^{-\lambda t}\, 1 - e^{-\lambda_{ef} T_o}}{(\lambda_o + \lambda_{oc})\, T\; T_o V\; \lambda_{ef}} \left[e^{-(\lambda_o + \lambda_{oc})(t - T - t_o)}\, e^{-(\lambda_o + \lambda_{oc})(t - t_o)} \right] \qquad (6)$$

$$\text{for } t < T + t_o$$

where Q_0 is the total amount of the i-th nuclide. As in the case of equation (1), there are some difficulties for estimation of the model parameters, which were calculated using data from References [5,6, and 7].

ESTIMATION OF MAXIMUM ALLOWABLE RADIONUCLIDE CONCENTRATION

In accordance with the IAEA recommendation, the value of 1 μSv/year was chosen as the maximum allowable dose D_0 in equation (1). Two seafood rations were considered:
1). 300 grams of fish and 100 gram of mollusks, crawfish and seaweed per day, and
2). 300 grams of fish per day.

Table I. Calculated radionuclide concentration in sea water equivalent to a population exposure dose of no more than 1 μSv/year for the seafood ingestion path way, Bq/m^3.

Nuclide	Allowable concentration for drinking water	Allowable concentration for sea water		Nuclide	Allowable concentration for drinking water	Allowable concentration for sea water	
		Ration -1	Ration-2			Ration-1	Ration-2
Co-60	1.3+06	172	660	Bi-210m		18	10
Ni-59	7.4+06	6.3+04	8.5+04	Th-228	4.81+03	69.4	77.2
Ni-63	1.04+06	2.3+04	3.1+04	Th-229		6.94	7.72
Se-79		400	335	Th-230	8.14+02	46.3	51.4
Sr-90	1.5+06	5.0+04	5.8+04	Pa-231	1.22+04	3.1	3.2
Zr-93		8.0+03	5.52+04	U-232	1.66+04	2.27+03	6.44+03
Nb-94		4.63+03	1.1+05	U-239	8.14+04	2.27+03	6.44+03
Tc-99	5.9+06	2.56+04	4.4+05	U-234	8.14+04	2.30+03	6.82+03
Pd-107	4.44+06	6.67+05	4.17+05	U-235	8.54+04	2.40+03	6.82+03
Cd-113m		18.3	110	U-236	8.51+04	2.44+03	6.92+03
Sb-125	3.66+06	1.8+04	1.6+04	U-238	2.18+04	2.44+03	6.92+03
I-129	7.0+03	2.91+03	2.08+04	Pu-238	9.25+04	10	230
Cs-134	3.2+05	3.3+03	2.3+03	Pu-239	8.1+04	8.6	195
Cs-135	4.1+06	3.5+04	2.4+04	Pu-240	8.1+04	8.6	195
Cs-137	5.6+05	4.8+03	3.5+03	Pu-241	4.1+06	43	980
Pm-147	8.5+06	1.1+04	3.8+04	Np-237	5.6+04	190	1.8+03
Sm-151	1.4+07	2.9+04	1.0+05	Am-241	7.0+04	1.6	153
Eu-152	2.8+06	1.45+03	6.0+03	Am-242m	8.14+04	1.7	161
Eu-154	8.5+05	940	3.9+03	Am-243	8.14+04	1.6	156
Eu-155	7.4+06	6.3+03	2.6+04	Cm-242	4.44+05	53.9	5.15+03
Bi-207	2.3+06	6.0+03	3.5+03	Cm-243	9.25+04	2.5	238
Bi-208		4.5+03	2.6+03	Cm-244	1.33+05	3.23	309

Under these conditions the concentrations a_o were calculated and compared with allowable concentrations for drinking water. The results are presented in Table I.

One can see that with the exception of Sr-90 and I-129 the calculated allowable concentrations for sea water are two to four orders of magnitude less than for drinking water. This fact indicates the conservative nature of the approach proposed.

CONSEQUENCES AND RADIATION RISK

NS #601 with lead-bismuth coolant was dumped in 1981 in the Kara Sea (Stepovoy fjord) at a ~50 m depth with the reactor cores. Prior to dumping special coatings and chemical additives were used to improve corrosion resistance. At present there are no signs that would indicate radioactivity release into the environment [9]. However , due to increased in public attention, the analysis of possible pathways for radioactive release to sea water is still vital.

Radionuclide activities in the dumped NS #601.

Based on the actual reactor operating history, the fission product and actinide compositions as well as the activation product composition in metal structures and coolant

Table II. Major long-lived nuclide activity in NS #601, Bq.

Nuclide	Half-lives, years	Calendar years				
		1981	1994	2050	2100	2200
Ni-60	5.27	5.06+14	9.23+13	6.05+10	8.5+08	1.74+02
Ni-59	7.5+04	1.56+12	1.56+12	1.56+12	1.56+12	1.56+12
Ni-63	100.1	1.52+14	1.39+14	9.37+13	6.6+13	6.3+13
Se-79	6.5+03	8.4+08	8.4+08	8.4+08	8.4+08	8.4+08
Sr-90	28.6	1.58+14	1.15+14	2.97+13	8.86+12	7.86+11
Zr-93	1.53+06	4.5+09	4.5+09	4.5+09	4.5+09	4.5+09
Te-99	2.13+05	3.0+10	3.0+10	3.0+10	3.0+10	3.0+10
Pd-107	6.5+06	2.62+07	2.62+07	2.62+07	2.62+07	2.62+07
Cd-113m	13.7	2.3+10	1.19+10	7.02+08	5.6+07	3.55+05
I-129	1.57+07	6.5+07	6.5+07	6.5+07	6.5+07	6.5+07
Cs-135	3.0+06	3.2+09	3.2+09	3.2+09	3.2+09	3.2+09
Cs-137	30.17	1.7+14	1.26+14	3.49+13	1.10+13	1.11+12
Sm-151	90	4.1+12	3.71+12	2.42+12	1.64+12	7.59+11
Eu-152	13.3	1.2+14	6.08+13	3.29+12	2.44+11	1.33+09
Eu-154	8.8	3.6+13	1.3+13	1.58+11	3.07+09	1.17+06
Eu-155	4.96	5.9+11	9.6+10	3.84+07	3.55+04	
Bi-207	32.2	2.2+10	1.69+10	6.68+09	2.27+09	2.64+08
Bi-208	3.68+05	6.3+09	6.3+09	6.3+09	6.3+09	6.3+09
Bi-210m	3.0+06	3.36+09	3.36+09	3.36+09	3.36+09	3.36+09
Pu-238	87.7	9.45+09	8.56+09	5.37+09	4.14+09	2.26+09
Pu-239	24119	3.39+11	3.39+11	3.38+11	3.38+11	3.37+11
Pu-240	6563	6.5+09	6.5+09	6.45+09	6.42+09	6.37+09
Pu-241	14.35	5.1+10	2.76+10	1.74+09	1.54+08	1.23+06
Am-241	432.7	8.37+08	1.14+09	1.64+09	1.55+09	1.34+09

have been calculated [10,11]. The results obtained are presented in Table II.

<u>Individual exposure dose estimation.</u>

Using the approach developed above, the individual population doses D from complete radioactivity release were estimated. The following conditions were taken into account:
- the radioactivity release is assumed to being in 2050 and continues for 50 years;
- all nuclides are intermixed in the whole volume of the Kara Sea ($V \approx 10^5$ km^3) [5];
- λ_0=286 year^{-1};
- deposition rate m_s=5·10^{-12} t/m^2·s
The annual exposure dose was calculated for ration-2. The concentration coefficients K_d are taken from Reference [1] and the λ_{oc} are calculated[8]. The result of calculations are shown in Table III.

Table III. Radiation parameters and population exposure doses for complete activity release into the Kara Sea over 50 years.

Nuclide	K_d, m^3/t	λ_{oc}, year^{-1}	a, Bq/m^3	D, nSv/year	Nuclide	K_d, m^3/t	λ_{oc}, year^{-1}	a, Bq/m^3	D, nSv/year
Ni-60	3+06	0.246	5.1-06	7.7-03	Sm-151	1+06	0.218	2.06-04	2.06-03
Ni-59	1+06	0.218	1.33-04	1.56-03	Eu-152	4+06	0.25	2.76-04	0.046
Ni-63	1+06	0.218	7.97-03	0.257	Eu-154	4+06	0.25	1.33-05	0.0034
Se-79	1000	1.31-03	7.75-08	2.3-04	Eu-155	4+06	0.25	3.22-09	1.2-07
Sr-90	200	2.62-04	2.74-03	0.0472	Bi-207	500	6.55-04	2.07-07	5.9-05
Zr-93	5+05	0.187	3.86-07	6.9-06	Bi-208	500	6.55-04	5.75-07	2.21-04
Tc-99	100	1.31-04	2.75-06	6.25-06	Bi-210m	500	6.55-04	3.07-07	0.0307
Pd-107	5000	6.55-03	2.39-09	5.7-09	Pu-238	1+05	0.087	4.7-07	0.02
Cd-113m	5000	6.55-03	4.39-08	4.0-04	Pu-239	1+05	0.087	3.0-05	0.154
I-129	200	2.62-04	6.0-09	2.9-07	Pu-240	1+05	0.087	5.65-07	2.9-03
Cs-135	2000	2.62-03	3.0-07	1.25-05	Pu-241	1+05	0.087	1.52-08	1.55-05
Cs-137	2000	2.62-03	4.4-03	1.257	Am-241	2+06	0.238	1.39-07	9.1-04

One can see that the total exposure dose is ~2 µSv/year or ~10^{-4} % of the average natural background. More than 60% of the added exposure dose is due to Cs-137.

<u>Collective dose and radiation risk estimation.</u>

The expected collective dose in men-Sv can be estimated

$$D_K^{exp} = \sum_j \alpha_j G_j \sum_i K_{ij}^H d_i a_{t_i} \qquad (7)$$

where $a_{t_i} = \int_0^\infty a_i(t)dt$ is the time integral of i-th nuclide concentration in sea water.

In the calculations λ_o was taken as zero, i.e. the model assumes a closed sea without water exchange with other neighboring seas. The annual fish take from the Kara Sea is ~1000 t/year [5], so the resting annual collective dose from the beginning of radioactivity release for the scenario considered is ~10^{-5} men-Sv. This dose is distributed among ~5000 people and it is assumed that ~300 grams fish per day are ingested.

Following the recommendations of the ICRP [12] we considered this exposure to be constant.
Then the individual risk due to radioactive release from NS #601 can be estimated as $r = D_k\,R/N$
$\approx 10^{-10}$ year^{-1}.   This value is approximately 5 orders of magnitude less than natural background
[13].

Potential radiotoxicity and some limit estimations.

As one can see, the individual and collective additional exposure doses for the scenario
considered are not significant in comparison with natural radioactive background. In this
connection another scenario was analyzed and the total nuclide radiotoxicity versus time was
calculated.

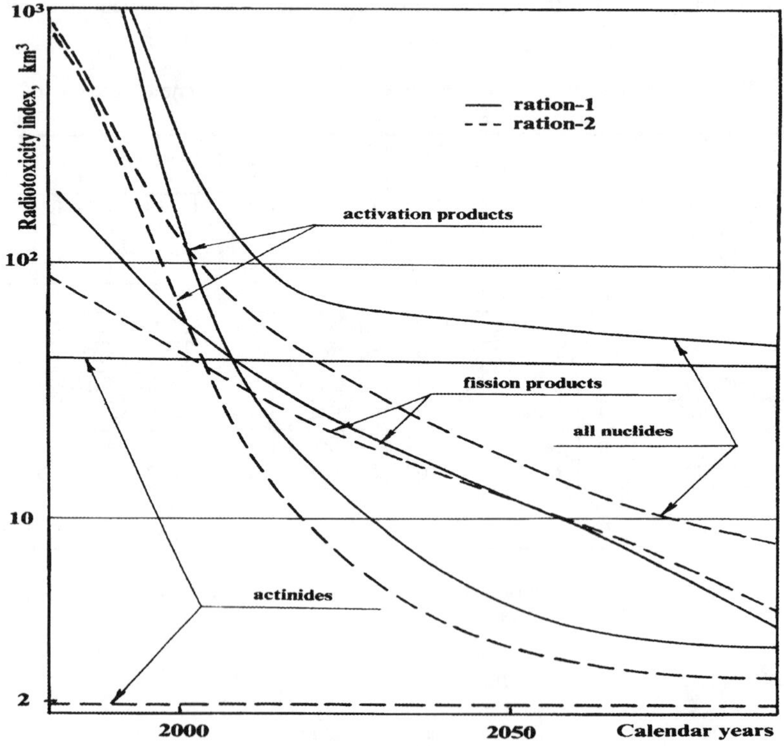

Fig. 1 Nuclide radiotoxicity index for NS #601

In Fig.1 the calculated radiotoxicity for all nuclides for rations 1 and 2 are presented. Indeed,
for the Kara Sea conditions, ration 2 is more realistic and the associated radiotoxicity index in
1995 is ~200 km^2. In the case of a hypothetical prompt complete release of all the radioactivity,
the sea water will be contaminated in range of ~35 km around NS #601, i.e. the contamination
will be local.

If the released activity is intermixed through the whole of the Kara Sea volume the relative concentration in the sea C_r would be ~$2 \cdot 10^3$ and the population additional exposure dose due to eating contaminated fish would be ~2 µSv per year or ~0.1% of natural background. Accordingly, collective dose will be ~ 0.01 men-Sv, which corresponds to additional individual radiation risk for the population group considered of ~10^{-7} year-1 .

CONCLUSION

A relatively simple method of calculating the average radionuclide concentration in sea water and doses of additional exposure is suggested for the Kara Sea conditions. The radionuclide composition aboard the dumped submarine #601 is determined more precisely as well as the total radioactivity, which at present amounts to 16,500 Ci (1995). The latter is expected to drop to 4, 600 Ci by 2050. Two scenarios of radioactivity release are considered. For the first scenario with fast release of the total activity occurring in 1995, it is shown that the individual additional exposure dose would be ~ 2 µSv per year, or ~ 0.1 % of the natural background. Radiation risk in this case equals ~10^{-7} per year with background risk being ~10^{-5} per year. For the second scenario with a total activity release occurring over~ 50 years starting in ~2050, these values are ~1,000 times less. Thus, radiological consequences are insignificant in both cases.

REFERENCES

1.   Definitions and Recommendations for the Convention on Prevention of Contamination by Waste Discharge and Other Materials, IAEA Document of Safety Series #78, Vienna, 1988.
2. The Oceanographic and Radiological Basis for Definition of High-Level Wastes Unsuitable for Dumping in Sea, IAEA Document of Safety Series #66, Vienna, 1984.
3. UNO Scientific Committee 1982 Annual Report on Atomic Radiation Effects, Appendix F, Irradiation Exposure Due to Nuclear Energy Production, V. 1, NY, 1982, p. 575.
4. Katkov A.E., Introduction to Regional Sea Radioecology, Moscow, Energoatomizdat, 1985, p. 33 (in Russian).
5. Benchmark Scenarios to Modeling of Radiological Impacts of Radioactive Wastes Dumping in the Arctic Seas, IAEA, Vienna, 1994.
6. UNO Scientific Committee 1982 Annual Report on Atomic Radiation Effects, Appendix A, Dose Estimation Models, V. 1, NY, 1982, p. 160.
7. Barents Sea, Big Soviet Encyclopedia, 3rd Edition, V. 2., 1970, p. 629 (in Russian).
8. Yefimov E.I., Pankratov D.V., Zakcharova S.M., Models of Radionuclide Dispersion in the Sea and Some Estimations of Radiological Consequence of Radiotoxicity Release from NS #601 Dumped in the Kara Sea, IPPE Report #8974, 1994 (in Russian).
9. The Joint Norwegian - Russian Expedition to the Dumped Sites for Radioactive Waste in the Abrosimov Bay and the Stepovoy Bay, August - September 1994, Report from Expedition, 1994.
10. Kolobashkin V.M., Radiation Parameters of Irradiated Nuclear Fuel. Moscow, Atomizdat, 1983, p. 51 (in Russian).
11. Kochetkov A.L., The Code for Calculation of Isotope Composition of Fast Reactor Fuel, IPPE Report #2654, 1980 (in Russian).
12. Radiation Protection, ICRP Publication #60, Pergamon Press, 1991, p.118(in Russian).
13. Vorobyev E.I., Kovalev E.E., Radiation Safety of the Aircraft Crew, Moscow, Energoatomizdat, 1983, p. 27 (in Russian).

AN ANALYSIS OF BIAS IN GROUNDWATER MODELLING DUE TO THE INTERPRETATION OF SITE CHARACTERIZATION DATA

K. J. Clark*, T. Ikeda**, M.D.Impey*, T. McEwen***, M. White***.
* QuantiSci Ltd , Chiltern House, 45 Station Rd, Henley-on-Thames, Oxon, RG9 1AT, UK, kclark@quantisci.co.uk
** JGC Corporation, Yokohama, Japan.
*** QuantiSci Ltd , Melton Mowbray, Leics, UK, tmcewen@quantisci.co.uk

ABSTRACT

Bias is a difference between model and reality. Bias can be introduced at any stage of the modelling process during a site characterisation or performance assessment programme. It is desirable to understand such bias so as to be able to optimally design and interpret a site characterisation programme. The objective of this study was to examine the source and effect of bias due to the assumptions modellers have to make because reality cannot be fully characterised in the prediction of groundwater fluxes. A well-defined synthetic "reality" was therefore constructed for this study. A limited subset of these data were independently interpreted and used to compute groundwater fluxes across specified boundaries in a cross section. The modelling results were compared to the "true" solutions derived using the full dataset. This study clarified and identified the large number of assumptions and judgements which have to be made when modelling a limited site characterisation dataset. It is concluded that bias is introduced at each modelling stage, and that it is not necessarily detectable by the modellers even if multiple runs with varied parameter values are undertaken.

INTRODUCTION

Mathematical models of groundwater flow around potential repository sites are an integral part of site characterisation programmes. When constructing a mathematical model of groundwater flow at a potential repository site, a modelling team has to interpret and interpolate the necessarily sparse data from a site characterization programme in order to provide all the data required by a model. Within this paper, we use the term "bias" to label the difference between our model and reality. Bias can be introduced by the choice of conceptual models, by the interpretation of the data and by the interpolation methods used. Such modeller-induced bias, together with uncertainty or error in the measured data, lead to errors in model predictions.

The primary objective of this paper is to study the source and effect of modeller-induced bias due to the interpretation of site characterisation data. However, on a real site the "true" solution can never be known completely and so it is difficult to determine the extent or cause of bias. For this study a high resolution synthetic dataset was constructed to act as our reality. This dataset consisted of a vertical cross section through a site with a simple geology in which a hypothetical repository might be built. This allowed a study to take place of the causes and effects of the biases inevitable in such interpretive modelling. Such an approach was used by Cole [1], and has recently been applied to groundwater flow by several authors, for example [2],[3],[4].

When planning a characterisation programme, decisions must be made such as where and how many boreholes to drill, how frequently to make measurements and what quantities to measure, all within time and cost restraints. Each of these decisions effects the data available for subsequent computer modelling. It is hoped that this study, and future iterations thereof, will contribute to the design of more effective site characterisation programmes.

Within this paper we first briefly describe the methodology. We then describe the construction of the synthetic datset and the interpretation of the subset of borehole data, before presenting and comparing the results of calculations. We then present an analysis of the sources of bias, and their significance, before drawing our final conclusions.

Mat. Res. Soc. Symp. Proc. Vol. 465 © 1997 Materials Research Society

METHODOLOGY

Two teams were involved in this study. The Site Team generated a high resolution heterogeneous permeability field for a 2-D vertical cross section through a potential disposal site. They then defined performance measures, and used their full dataset to calculate these performance measures. These were the so-called "true" solution. The Assessment Team interpreted a subset of the Site Team's conductivity and head data, and calculated the same performance measures, without any knowledge of the "true" solution. The Site Team then carried out the final comparisons of the two set of results.

The Site

The site consists of three sedimentary and one sedimentary-volcanic formation deposited on a basin margin and thickening from right to left (East to West). A siltstone and a sandstone overlie a mudstone (the repository host formation), which in turn overlies the tuffs. One planar extensional fault displaces all the formations. The cross -section is shown in Figure 1.

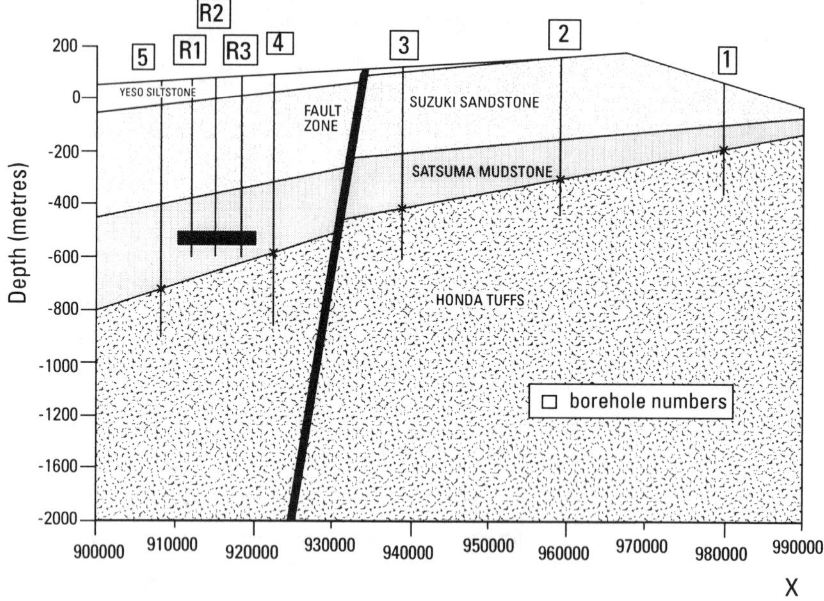

Figure 1: The geological cross-section of the site, showing the layered sedimentary and volcanic environment, the extensional fault, and the vertical boreholes along which data was supplied to the Assessment Team. A potential repository is shown in black within the low permeability Satsuma Mudstone.

Statistical distributions and fractal scaling laws were defined for the hydraulic conductivity fields within each formation. Table 1 gives the data sources, although the values used in this exercise were based only loosely on these sources. Lateral variations in hydraulic properties were derived assuming a deposition environment typical of a basin margin in an active tectonic setting, such as the Neogene Green Tuff area of Japan. A fractal model was used to generate the hydraulic conductivity field (Figure 2). A computer code, AZURE [5], was then used to calculate flows and head distributions based on the full dataset.

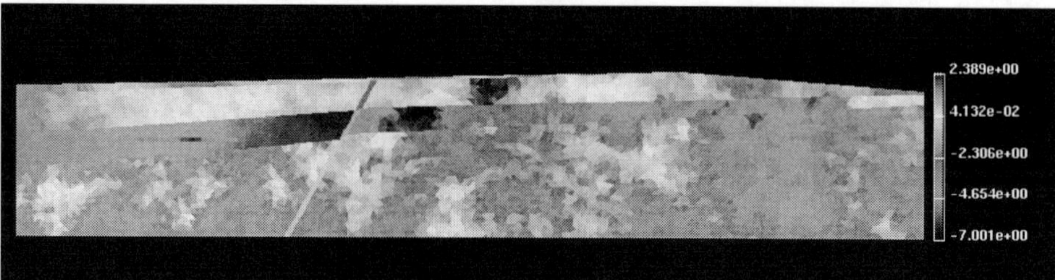

Figure 2: The hydraulic conductivity field generated by the Site Team. Note the fault zone which provides a partial seal within Suzuki Sandstone, but a more transmissive pathway within the Satsuma Mudstone. Scale gives log transmissivity values (m^2y^{-1}).

Table 1: Data sources for hydraulic conductivities for synthetic cross section

Unit	Data Source
Yeso Siltstone	St Bees sandstone, Sellafield, UK [3]
Suzuki Sandstone	St Bees sandstone, Sellafield, UK [3]
Satsuma Mudstone	Opalinus clay, northern Switzerland [4]
Honda Tuffs	Borrowdale Volcanic Group, UK (modified) [3]

The subset of the full dataset given to the independent Assessment Team consisted of hydraulic conductivities, lithological information and hydraulic heads at 20m intervals down the 8 hypothetical boreholes (Figure 1). The presence of the fault zone was not mentioned to the Assessment Team. In addition realistic errors were introduced into the supplied subset of borehole hydraulic head data. The required performance measures were:

- Flux of groundwater across the left, right and upper boundaries of repository zone,

- Flux of groundwater across the upper surface of model to the left of borehole 4,

- Flux of groundwater across the left-hand boundary of model.

The Interpretation

The dip of the formation boundaries was observed to change between the eastern three boreholes and the rest. It was decided to interpret this dip change as the existence of an extensional fault, dipping to the west. A standard dip of 60% was assumed since no other information was available.

The given descriptions of the rock types were compared and statistical tests carried out on the variation in hydraulic conductivity within and between boreholes. From these investigations it was decided that interpolation of the hydraulic conductivities should be done separately for each lithological unit, and should take account of lateral variation. Various interpolation methods were considered. Issues considered included the degree of anisotropy (i.e. bedding) within a unit, the existence of larger scale trends in mean hydraulic conductivity values, and the scale of heterogeneity within a unit. Plots of hydraulic conductivity versus lateral distance for each unit (e.g. Figure 3), variograms to test for fractal scaling laws and statistical tests of the inter-unit differences in conductivities were each used in deciding suitable interpolations.

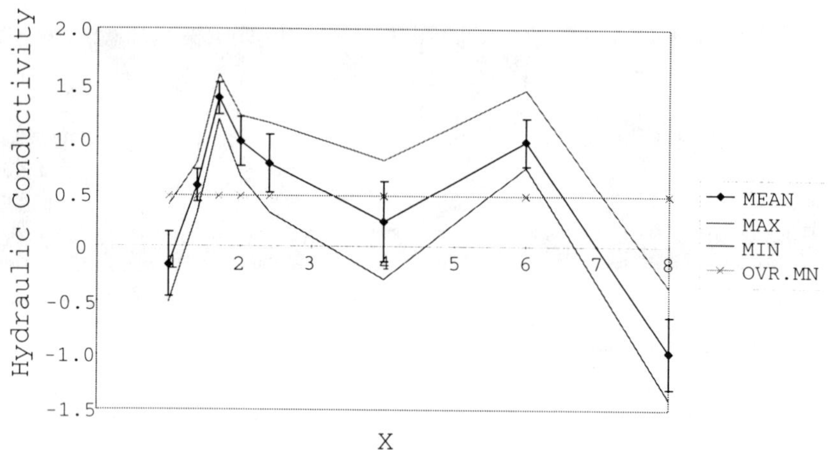

Figure 3: Plot of the hydraulic conductivity versus distance x for the Suzuki Sandstone. Data from each borehole is plotted along the x axis at its position along the cross section. The vertical axis values are log conductivity (my^{-1}). Mean, maximum and minimum values are plotted (solid lines). The horizontal line shows the overall mean of all the data from all the boreholes. The error bars at each borehole mean values show one standard deviation either side of the mean.

Two alternative methods were used for interpolation. In the first, termed "Layer Mean", conductivity values were assumed to vary linearly parallel to the bedding plane, to be uniform perpendicular to bedding plane, and to interpolate the mean value of conductivity for that layer at each borehole. In the second method, a fractal scaling law was fitted to all borehole data within a unit, and the hydraulic conductivity field was given by a fractal generated with the fitted dimension which interpolated all the borehole values. Figure 4 shows two of the conductivity fields thus generated.

No information was available to the Assessment Team regarding fault transmissivities or head or flow boundary conditions. It is known that faults can either seal a unit or provide a more transmissive pathway. High conductivity, low conductivity and unchanged conductivity alternatives were therefore all modelled. There was an order of magnitude difference in conductivity between each alternative. Two sets of boundary conditions were both modelled: a default case with boundary heads set equal to surface level, and an alternative case in which the borehole heads were extrapolated to give the right and left hand boundaries. In all cases, the lower boundary was assumed to be no-flow.

Two phases of calculations were carried out by the Assessment Team using the QuantiSci-JGC code AZURE [1]. Twelve cases were run in the first phase to cover the two interpolation options, two boundary condition options and the three fault conductivity options. Having examined the results of these first twelve runs (see Table 2 below), four additional runs were made with modified versions of the boundary conditions. After this second round of calculations, the Assessment Team drew its conclusions about the biases within their results. Finally, the Site Team compared their results with the Assessment Team results, and considered the Assessment Teams's decision processes to analyse the sources and effects of bias.

RESULTS

The performance measure values computed by the Assessment Team are summarised in Table 2 together with the performance measures calculated by the Site Team using the "true" conductivity field and boundary conditions.

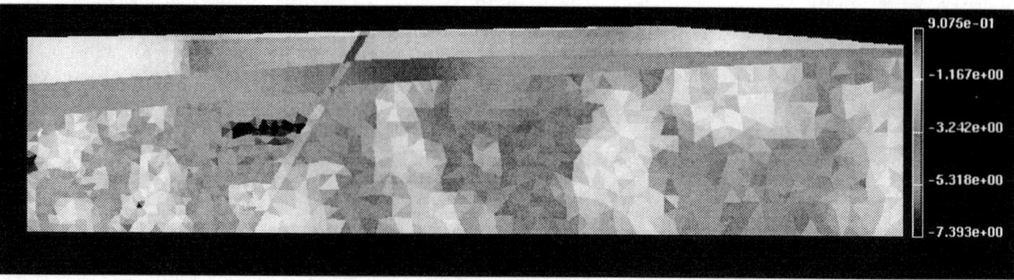

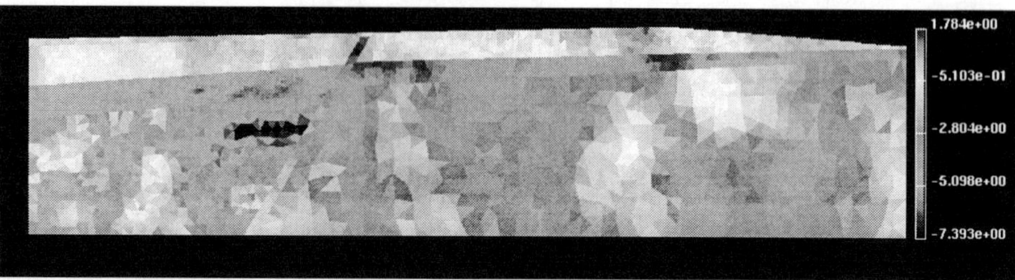

Figure 4: The hydraulic conductivity fields generated by the Assessment Team for the Layer (above) and the Fractal (below) interpolation options. Note that a fractal interpolation was used in both cases for the Honda Tuff. These examples show the high fault conductivity option. The scale gives log transmissivity values (m^2y^{-1})

The Assessment Team's results, were first considered without reference to the Site Team's values. The most striking feature of these results was the sensistivity of the predicted fluxes across the upper and left-hand boundary, including a reversal in flow direction, to the choice of boundary condition. These fluxes were dominated by flow through the upper high conductivity sedimentary layers. The impact on the upper and left-hand boundary fluxes of the assumptions about the fault conductivity and interpolation method was small compared with the impact of the boundary conditions. The choice of interpolation method made a factor of two difference, and the choice of fault conductivity a few percent difference. The latter bias was seen to involve a small but systematic increase of flux with fault conductivity.

The predicted fluxes through the repository zone, which were three orders of magnitude smaller, depended most strongly upon the interpolation method used, that is upon the local conductivity distribution.

Four additional runs using modified boundary conditions were carried by the Assessment Team to investigate the effect of the assumed upper and left-hand boundary conditions. These calculations confirmed the initial findings on flux sensitivities, and the Assessment Team requested further data on the boundary conditions should another iteration of the exercise be undertaken.

The distribution of pressure heads from each run was plotted and interpreted. The effect of the boundary conditions were very clear, and the impact of the fault visible. Comparison of the predicted heads along the borehole positions and heads from the given borehole data suggested that further investigation and modelling of the fault zone and the eastern lower layers should also be undertaken.

When the performance measures calculated by the Assessment Team and the Site Team (Table 2) are compared, the variability and lack of agreement between the performance measures from different model runs is confirmed. We now dicuss possible sources for this bias, by consideration of the assumptions at each step of the modelling.

Table 2: Performance measures from assessment team calculations, and from site team calculations of "true" solution: positive fluxes (m^3y^{-1}) are outward across the boundary

Case definition				Perfromance measures				
Case No	Boundary condition	Interpo-lation method	Fault cond-uctivity	Left hand boundary	Partial upper boundary	Left boundary repository	Right boundary repository	Upper boundary repository
1	Default	Fractal	Low	16.6	11.4	7.3×10^{-3}	3.6×10^{-5}	-0.048
2	Default	Fractal	Equal	17.0	12.4	7.5×10^{-3}	3.7×10^{-5}	-0.050
3	Default	Fractal	High	17.2	13.0	7.6×10^{-3}	3.6×10^{-5}	-0.50
4	Default	Layer	Low	8.52	4.49	1.6×10^{-3}	-7.2×10^{-5}	-0.047
5	Default	Layer	Equal	8.78	5.59	1.7×10^{-3}	-7.3×10^{-5}	-0.049
6	Default	Layer	High	8.91	6.13	1.7×10^{-3}	-7.4×10^{-5}	-0.051
7	Extrap'd	Fractal	Low	-1.03	-3.9×10^{-4}	7.6×10^{-3}	7.0×10^{-5}	-0.088
8	Extrap'd	Fractal	Equal	-1.04	-3.9×10^{-4}	7.6×10^{-3}	7.1×10^{-5}	-0.088
9	Extrap'd	Fractal	High	-1.05	-4.0×10^{-4}	7.6×10^{-3}	6.9×10^{-5}	-0.087
10	Extrap'd	Layer	Low	-1.22	-1.5×10^{-3}	-1.1×10^{-3}	7.0×10^{-5}	-0.099
11	Extrap'd	Layer	Equal	-1.24	-1.6×10^{-3}	-1.1×10^{-3}	-6.8×10^{-5}	-0.099
12	Extrap'd	Layer	High	-1.24	-1.6×10^{-3}	-1.1×10^{-3}	-6.7×10^{-5}	-0.099
13	Modified	Fractal	Equal	23.0	8.1×10^{-3}	8.3×10^{-3}	4.3×10^{-5}	-0.057
14	Modified	Layer	Equal	14.0	1.3×10^{-2}	2.2×10^{-3}	-8.7×10^{-5}	-0.068
15	Modified	Fractal	Equal	8.87	3.1×10^{-3}	1.1×10^{-2}	8.8×10^{-5}	-0.11
16	Modified	Layer	Equal	5.3	4.7×10^{-3}	-1.1×10^{-4}	-1.1×10^{-4}	-0.13
Site Team				3.18	1.38×10^{-1}	6.33×10^{-3}	-2.30×10^{-4}	-0.0806

ANALYSIS OF SOURCES OF BIAS

The simple cross-section and "standard" dip-angle used for the fault meant that the Assessment Team's interpretation of the lithology of the model site matched the reality created by the Site Team very closely. Therefore it was concluded that little bias had been introduced by the geological interpretation. Further studies in more complex lithologies will be required for the issue of bias introduced by geological interpretion to be tackled.

Relatively litttle bias appeared to have been introduced during the interpretation of the borehole conductivity data. Statistical tests successfully demonstrated that data from different lithological units could not have been samples from data with the same mean. This is of generic interest because it established that quantitative links could be made between lithological core logs and conductivity data. The statistics of the borehole data were therefore calculated separately for each lithological unit. Broadly speaking, the means and fractal dimensions computed by the Assessment Team matched those used by the Site Team in generating the real site, though for two of the units the mean values calculated were up to an order of magnitude in error.

The assumptions and decisions made during the interpolation of the conductivity data onto the cross section did introduce several systematic biases. The method of interpolation systematically biased the calculated fluxes, though the size and sign of the effect depended upon the location of the flux (performance measure) and the boundary conditions. The key role of the fault became obvious only when details of the "true" solution were available and the variation in fault:unit conductivity ratio with rock unit was revealed. The use of a relatively narrow fault and low fault:unit conductivity ratio somewhat obscured the bias within the Assessment Team's results. The Assessment Team did not consider the possibility of the fault:unit conductivity ratio varying between units. This bias in the fault conductivity appears to be a major cause of the differences between the Site Team's and Assessment Team's performance measures.

It is of generic interest that the existence of very local low conductivity regions can strongly bias the flow through the relatively high permeability regions, as in the true solution for the Suzuki Sandstone, whilst the existence of a high conductivity fault through a low conductivity region (the Satsuma Mudstone) has little impact. This reinforces the truism that robust modelling can be undertaken in the low conductivity near-field, where uncertainties can be rather limited, whilst there are greater biases and uncertainties in modelling flow in the higher conductivity far-field.

The final set of assumptions and decisons in the Assessment Team's modelling concerned the setting of the flow boundary conditions. Bias was introduced in this stage, as clearly shown by the variability in predicted fluxes. The Assessment Team introduced bias by changing the boundary conditions because of the assumption that the boundary conditions must be more complex than the initial simple assumptions. It is interesting to note that these default boundary conditions assumed initially by the Assessment Team were actually those assumed by the Site Team. This "convergence by ignorance", or the mutual use of the simplest assumptions, may prevent some biases being revealed in an exercise of this type.

The distribution of the supplied borehole data contributed to the bias in the flow boundary conditions. The wider spacing of the eastern boreholes increased the uncertainty in the permeability interpolation to the east of the fault. Boundary conditions for the regions of interest were not known. The lack of data along or through the fault also increased bias, since the modellers had to make arbitrary assumptions, based on generic knowledge, about the fault. In each of these areas, additional data would be requested in the next iteration of the exercise. This would correspond to additional drilling or sampling in a real site characterisation programme. It is anticipated that this will demonstrate how the results of initial model calculations can guide the more effective planning of subsequent phases of a site characterisation programme.

CONCLUSIONS

This study clarified and identified the large number of assumptions and judgements which have to be made by an Assessment Team faced with a limited site characterisation dataset. The use of a synthetic dataset as our "reality" removed any uncertainties as to what was reality, and allowed us to define reality at any scale required.

Bias is introduced at every step of the modelling process. It is not necessarily detectable by the modeller, even after multiple model runs. However in the areas of most concern, the low conductivity regions srrounding a possible repository site, the effects of these biases can be relatively small.

The main sources of bias found were in the interpolation of the boreholes data, and in the assumptions made about missing data such as the boundary conditions. The actual intepretation of the available data in terms of identifiying lithological units and in statistically characterising hydraulic conductivities agreed well with "reality" (the full synthetic dataset).

This study has given valuable insight into biases due to modelling assumptions possible within a real site characterization study. Further iterations of this study with the aim of studying how or if modeller bias can be reduced with the phased supply of additional data (as might take place within a real site characterisation programme) are proposed.

REFERENCES

1. C R Cole *et al* (1985). Understanding, testing and development of stochastic approaches to hydrologic flow and transport through the use of the multigrid method and synthetic datasets. In *Preprints of the Symposium on the Stcohastic Approach to Subsurface Flow Greco 35 Hydrogeologie Ecole des Mines de Paris, Fontainbleau*, pp 364-378. Ecole des Mines de Paris, Fontainbleau.

2. A J Desbarats and R M Srivastava (1991) Geostatistical characterisation of groundwater flow parameters in a simulated aquifer. *EOS* 73(43) pp200

3. T D Sheibe and D L Freyberg (1995). Use of sedimentological information for geometric simulation of natural porous media structure. *Water Resources Research* 31 (12) pp 3259-3270

4. M J Williams, M D Impey, C Sellar. AZURE User Guide. QuantiSci report IM4138-3 Version 1.0 May 1996.

5. Nirex, *The geology and hydrogeology of the Sellafield area. Volume 3: The Hydrogeology*, Nirex Report 524, 1993.

6. Nagra, *Sediment Study. Intermediate report 1988. Disposal options for long lived radioactive waste in Swiss sedimentary formations: Executive Summary*, Nagra Technical Report 88-25E, 1989.

THERMAL ANALYSIS OF OPTIONS FOR SPENT FUEL DISPOSAL IN SWITZERLAND

TETSUO SASAKI*, KENICHI ANDO*, HIDEKI KAWAMURA*, JÜRG W. SCHNEIDER**, AND IAN G. McKINLEY**
*Design Department No. 2, Civil Engineering Technical Division, OBAYASHI CORPORATION, 2-2-9 Hongo, Bunkyo-ku, Tokyo 113, Japan.
**NAGRA, National Cooperative for the Disposal of Radioactive Waste, Hardstrasse 73, CH-5430 Wettingen, Switzerland.

ABSTRACT

In parallel to studies of disposal of vitrified high-level waste from reprocessing, projects have been initiated to examine options for direct disposal of spent fuel in Switzerland. The basic concept involves in-tunnel emplacement of encapsulated spent fuel in a deep repository which is backfilled with compacted bentonite. Two possible host rocks are considered - crystalline basement and Opalinus Clay. This paper reports the results of a thermal analysis which was carried out to evaluate constraints on repository layout set by the desire to limit temperatures experienced by the bentonite backfill.

INTRODUCTION

The Swiss concept for disposal of high-level radioactive waste involves a deep mined repository in either the crystalline basement or in the Opalinus Clay sediment formation. The tunnels with waste packages - containing either spent fuel (SF) for direct disposal or vitrified high-level waste (HLW) from reprocessing - will be backfilled with highly compacted bentonite. Extensive performance assessments for HLW disposal in the crystalline basement (Project Gewähl 1985 [1], Kristallin-I [2]) and in sediments (Zwischenbericht 1988 [3]) have illustrated that this massive engineered barrier system (EBS) performs very effectively - minimizing the requirements on the geological barrier. The thick compacted bentonite backfill, in particular, has important roles as a plastic mechanical buffer, as an extremely low hydraulic conductivity barrier to advective water flow and as a colloid filter. To ensure that these key properties are retained, it is considered desirable that the outer annulus of bentonite does not experience temperatures above 100°C.

Thermal analysis associated with the original study of HLW in the crystalline basement [4] supported the feasibility of the design adopted - 3.7 m diameter tunnels separated by a pitch of 40 m and a canister emplacement pitch of 5 m [1]. This reference design was based on the assumption of the presence of large blocks of relatively undisturbed crystalline rock in the siting region. Syntheses of regional geological investigations indicate that the size of blocks of rock between large shear zones are a critical issue and hence the Kristallin-I assessment included an analysis of the thermal consequences of decreasing tunnel spacing [5]. The work reported in this paper leads on from this work for the case of direct disposal of spent fuel - in both the crystalline basement and the Opalinus Clay.

This study evaluates the radiogenic thermal transients for times up to 25,000 years after emplacement. The parameters which are varied include:
- emplacement tunnel spacing,
- canister emplacement pitch,
- thermal properties of bentonite,

Mat. Res. Soc. Symp. Proc. Vol. 465 © 1997 Materials Research Society

- presence or absence of air gaps in the engineered barrier system,
- heat production profile of spent fuel,
- emplacement tunnel diameter, and
- host rock type.

MODELING AND THERMAL PROPERTIES

<u>Analysis code and model</u>

As for Kristallin-I, temperature calculations in this study are performed using the ADINA-T (Automatic Dynamic Incremental Nonlinear Analysis of Temperatures) finite element analysis code [6] with a 3-dimensional mesh. The solution methodology is based on the finite element formulation of the non-steady and non-linear heat flow problem.

The EBS Components are spent fuel, canister, and bentonite. Two types of host rock are considered, crystalline basement (KRI) and Opalinus Clay (OPA). The canister for spent fuel is taken to be a version of the Advanced Cold Process (ACP) canister developed by TVO [7]. This canister is a composite of steel surrounded by copper. It contains a steel rack to hold the spent fuel and void spaces are infilled with quartz sand. The emplacement tunnels are of circular cross section. Canisters are emplaced horizontally, and a single-level repository layout variant is analyzed. The repository reference depth is 1200 m below ground level in KRI or 600 m deep in OPA. Key differences between KRI and OPA are thermal properties, initial temperature and temperature gradient.

Although air gaps within the EBS are not considered in the base case, the following air gaps are considered in parametric studies:
- between canister and bentonite with a width of 1 cm, and
- between bentonite and host rock with a width of 1 cm.

The schematic illustration of canisters and emplacement tunnel is shown in Figure 1 and the boundary conditions of the analysis model are shown in Figure 2.

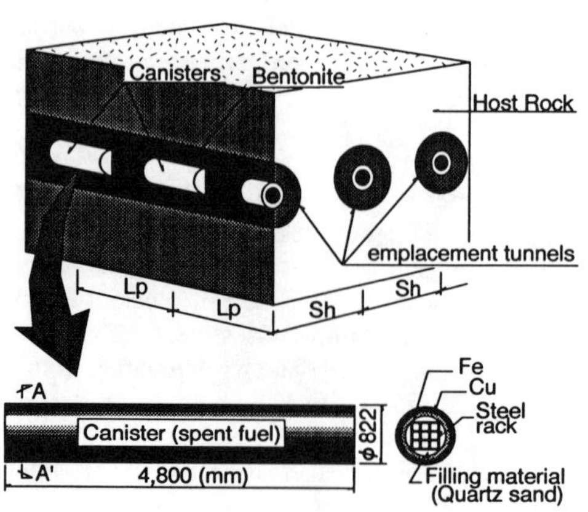

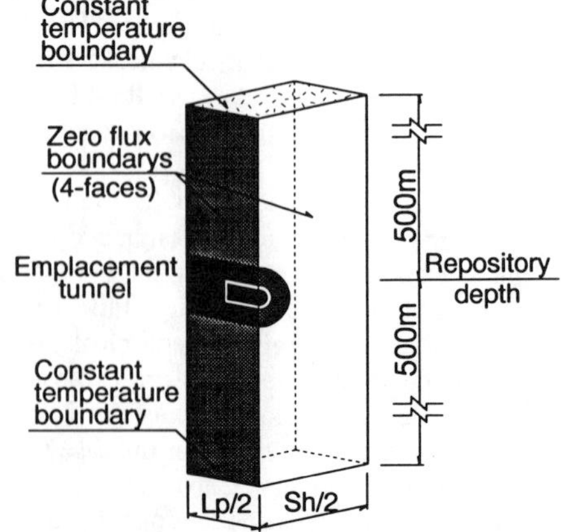

Figure 1 Canisters and emplacement tunnels

Figure 2 Three-dimensional model representation and boundary conditions (after NTB 94-13)

<u>Thermal conditions and properties</u>

Initial temperatures and thermal properties of the components of the EBS and host rock (crystalline basement) are taken from NTB 85-54 [4] and NTB 88-25 E [8]. Initial temperatures of spent fuel and canister are 65°C, and that of bentonite is 35°C. The host rocks KRI and OPA have initial temperature gradients, 0.03°C/m and 0.0412°C/m and initial temperatures of 55°C (at the repository depth of 1200 m) and 35°C (at 600 m) respectively. The thermal properties, thermal conductivity and heat capacity are shown in Table 1.

Table 1   Thermal properties and initial temperatures of the different materials considered

Material	Thermal Conductivity (W/mK)	Heat Capacity (MJ/m³K)	Initial Temperature (°C)
Spent fuel	6.1	3.19	65
Steel rack	52	3.05	65
Quartz sand	52	1.26	65
Carbon steel	52	3.05	65
Copper	380	3.44	65
Air	3.17E-02	9.56E-04	65
Bentonite: Type A	0.7	1.1	35
Type B	1.7	2.1	35
Host Rock*[1]:KRI	2.5	2.3	55 (at 1200 m) *[2]
OPA	2.2	2.0	35 (at 600 m) *[3]

*1) Host Rock Type: KRI (crystalline rock), OPA (Opalinus Clay)
*2) Temperature gradient of KRI is 0.03 °C /m.
*3) Temperature gradient of OPA is 0.0412 °C /m.

<u>Heat production</u>

Heat productions of two generic types of packages of spent fuel shown in Figure 3 are used for thermal calculations. Type-1 is the base heat production type taken from TVO-92 [9]. The decay heat is normalized to yield 1000W at 40 years after unloading. The reason for this normalization is that the detailed characteristics of the waste package (e.g. fuel burn-up, mixture of fuel of different types) have not yet been specified. Heat production type-2 is set 1.5 times higher than type-1 corresponding to a higher heat-loading. The start time for the thermal calculations is at emplacement, 40 years after fuel unloading.

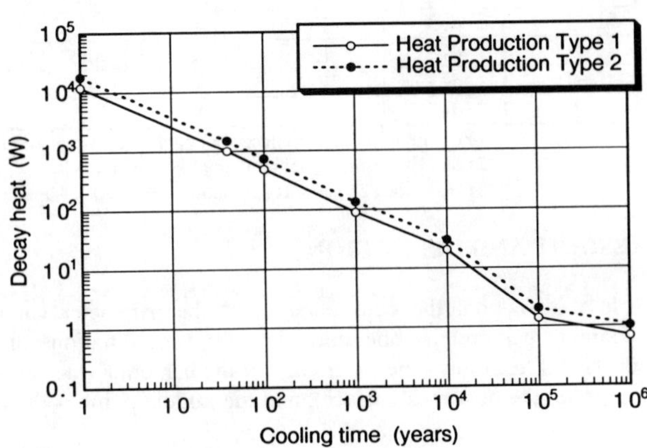

Figure 3   Heat production considered in the calculations

PARAMETERS AND CALCULATIONAL CASES

Parameters for this study are as follows :
(a) Horizontal spacing of emplacement tunnels : $Sh = 10, 20, 40$ (m)
(b) Longitudinal emplacement pitch of canisters : $Lp = d+3, d+5, d+7$ (m)
     (d : canister length, d+3 : corresponds to the reference case of vitrified HLW)
(c) Bentonite type with respect to thermal property : Type A, Type B (see Table 1)
(d) Presence of air gaps :   yes, no
(e) Heat production profile :   Type 1, Type 2 (see Figure 3)
(f) Emplacement tunnel diameter : $Dt = 3.7, 2.4$ (m)
(g) Host rock type :   KRI, OPA

Cases and parameters used are summarized in Table 2. In total, 15 cases are analyzed.

Table 2   Calculational cases

Calculational Case (No.)	Lp[*1] (m)	Sh[*2] (m)	Bentonite (type)	Air Gaps (yes/no)	Heat Production (type)	Tunnel Diameter (m)	Host Rock[*3] (type)
1	d+3	20	A	no	1	3.7	KRI
2	d+5	20	A	no	1	3.7	KRI
3	d+7	20	A	no	1	3.7	KRI
4	d+3	10	A	no	1	3.7	KRI
7	d+3	40	A	no	1	3.7	KRI
10	d+3	20	B	no	1	3.7	KRI
11	d+3	20	A	yes	1	3.7	KRI
1B	d+3	20	A	no	2	3.7	KRI
1C	d+3	20	A	no	1	2.4	KRI
1D	d+3	20	A	no	1	2.4	OPA
1E	d+3	20	A	no	2	2.4	OPA
7B	d+3	40	A	no	2	3.7	KRI
7C	d+3	40	A	no	1	2.4	KRI
7D	d+3	40	A	no	1	2.4	OPA
7E	d+3	40	A	no	2	2.4	OPA

*1) Lp: Longitudinal emplacement pitch of canisters, d: Length of spent fuel canisters,
*2) Sh : Horizontal spacing of emplacement tunnels
*3) Host Rock Type: KRI (crystalline rock), OPA (Opalinus Clay)

RESULTS AND DISCUSSION

In this section, the consequences of the parameter variations are evaluated in terms of maximum bentonite temperatures and the time durations of bentonite temperature exceeding 100°C at 3 locations - the inner edge of the bentonite (the canister surface), the outer edge of the bentonite (the host rock surface), and the middle point of the bentonite (half way between these locations).

The maximum bentonite temperatures and the time durations of bentonite temperature exceeding 100°C at the 3 locations are summarized in Figure 4 and in Figure 5 respectively to compare the results of the different cases.

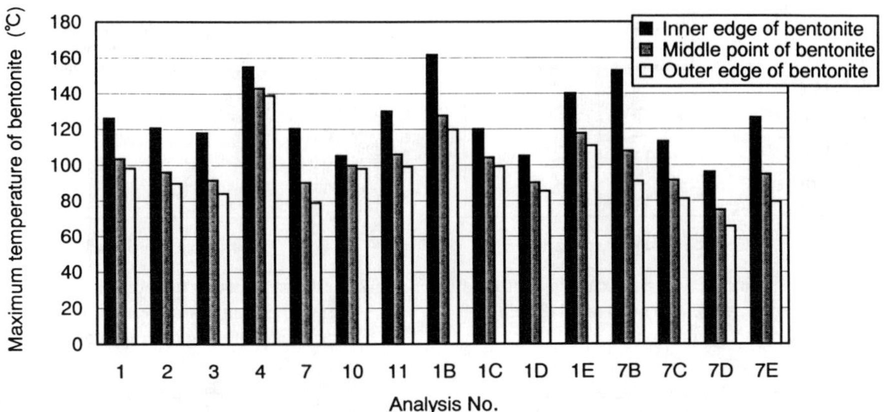

Figure 4   Maximum bentonite temperatures at 3 locations, for the cases considered in the sensitivity analysis

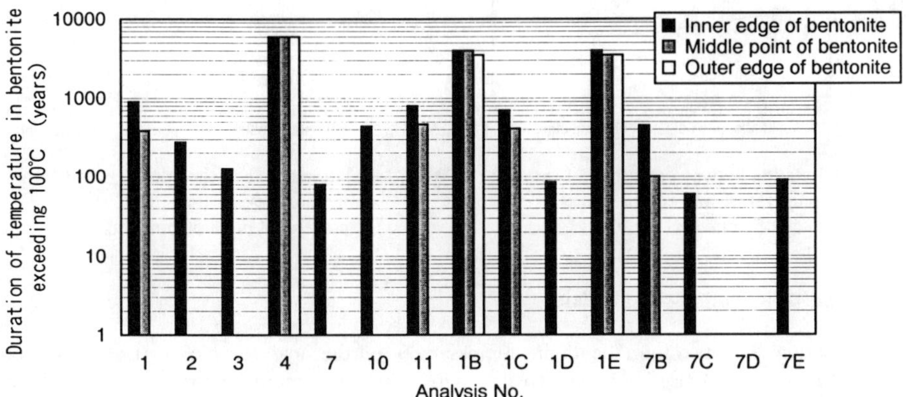

Figure 5   Time durations of maximum bentonite temperature exceeding 100°C at 3 locations for the cases considered in the sensitivity analysis.

Effect of emplacement tunnel separation

Comparing the maximum bentonite temperatures of the analysis No. 1 (Sh=20m), No. 4 (Sh=10m), and No. 7 (Sh=40m), shown in Figure 6, the maximum bentonite temperature decreases with increasing the horizontal tunnel spacing Sh from 10 m to 20 m and/or 40 m. There is a dramatic decrease between Sh=10m and Sh=20m, whereas the decrease between Sh=20m and Sh=40m is much less pronounced. This tendency is similar to the result for vitrified waste in the Kristallin-I study [2]. Likewise, the duration of bentonite temperatures exceeding 100°C decreases with increasing Sh and it decreases dramatically between Sh=10m and Sh=20m, compared to between Sh=20m and Sh=40m, as shown in Figure 7. These results lead to the recommendation that the horizontal emplacement tunnel spacing should not be less than about 20 m to ensure that thermal interactions between the tunnels are acceptable.

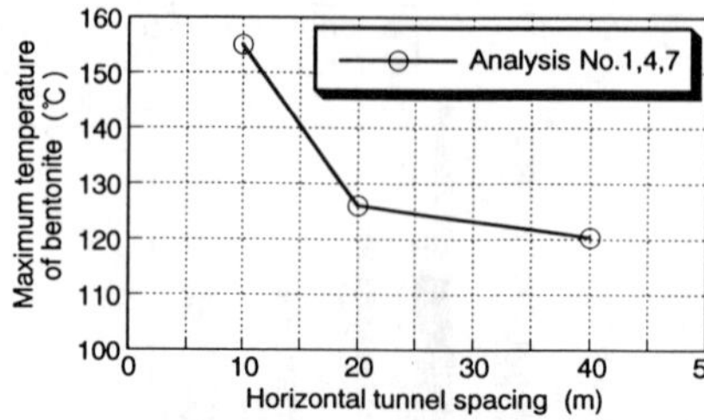

Figure 6 Relation between maximum bentonite temperature and horizontal spacing of emplacement tunnel.

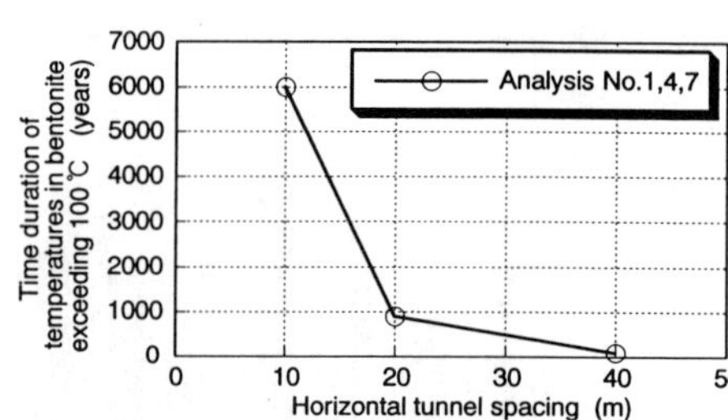

Figure 7 Relation between time duration of bentonite temperature exceeding 100°C and horizontal spacing of emplacement tunnel.

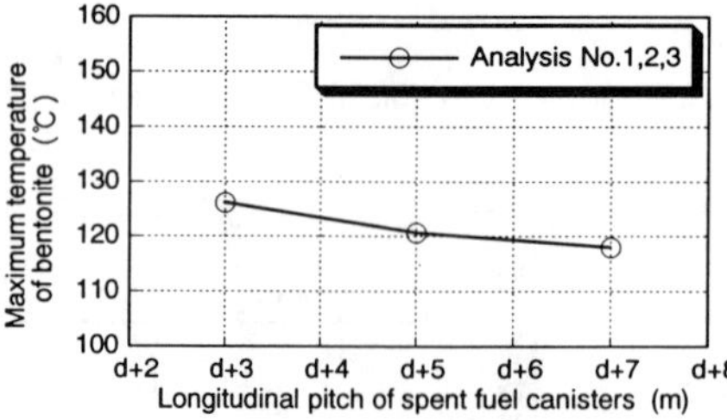

Figure 8 Relation between maximum bentonite temperature and emplacement pitch of canisters.

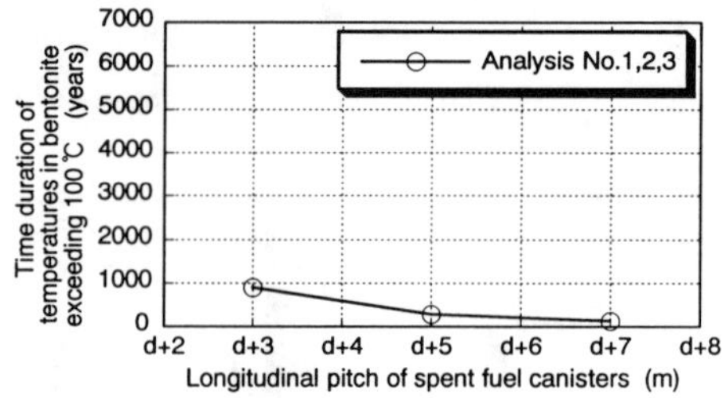

Figure 9 Relation between time duration of bentonite temperature exceeding 100°C and emplacement pitch of canisters.

Effect of canister emplacement pitch

Comparing the maximum bentonite temperatures of the analyses No. 1 (Lp=d+3), No. 2 (Lp=d+5), and No. 3 (Lp=d+7), shown in Figure 8, the maximum bentonite temperature with increasing Lp decreases from d+3 (m) to d+7 (m). As long as the canister emplacement pitch is within this range, the effect of canister emplacement pitch is much smaller than that of emplacement tunnel separation. The tendency for the duration of bentonite temperature exceeding 100°C is similar, as can be seen in Figure 9.

Effect of bentonite thermal properties

From Figure 4, comparing the maximum bentonite temperatures of analyses No. 1 (bentonite type-A; thermal conductivity of 0.7W/mK and heat capacity of 1.1MJ/m^{3}K) and No. 10 (bentonite type-B; thermal conductivity of 1.7W/mK and heat capacity of 2.1MJ/m^{3}K), the difference between the two temperatures at the inner edge of the bentonite is approximately 20°C, while the difference is within 5°C at the middle point and outer edge of the bentonite. From Figure 5, comparing the duration of the bentonite temperature exceeding 100°C of analyses No. 1 and No. 10, it can be said that, at the inner edge, assuming type-B bentonite reduces this duration by more than 400 years compared to type-A; also, for type-B bentonite temperatures at the middle point or outer edge never exceed 100°C. Thus, it can be inferred that the length of the time interval above 100°C is more sensitive to bentonite thermal properties than the maximum bentonite temperature.

<u>Effect of presence or absence of air gaps in the engineered barrier system</u>

Comparing the maximum bentonite temperatures of analyses No. 1 (no air gaps) and No. 11 (with air gaps) in Figure 4 and the duration of bentonite temperatures exceeding 100°C in Figure 5, it can be inferred that air gaps do not strongly affect either the maximum bentonite temperature or the time duration of bentonite temperatures exceeding 100°C, given the conditions in this study.

<u>Effect of heat production type of spent fuel</u>

In Figure 4 and Figure 5 comparing analyses No. 1 (heat production type-1) and No. 1B (heat production type-2; 1.5 times higher than type-1 at all times) for Sh=20m and No. 7 (type-1) and No. 7B (type-2) for Sh=40m in crystalline rock, it can be said that the maximum bentonite temperature for heat production type-2 is higher than that for type-1 and that the duration of bentonite temperatures exceeding 100°C for heat production type-2 is longer than that for type-1 for Sh=20m and Sh=40m in crystalline rock. Although both the maximum bentonite temperature and the duration of bentonite temperature exceeding 100°C are sensitive to heat production, the duration is more sensitive than the maximum temperature.

In Opalinus Clay, although both the maximum bentonite temperature and the duration of bentonite temperature exceeding 100°C are sensitive to heat production, the duration is more sensitive than the temperature, comparing analyses No. 1D (type-1) and No. 1E (type-2) for Sh=20m and No. 7D (type-1) and No. 7E (type-2) for Sh=40m .

<u>Effect of emplacement tunnel diameter</u>

From Figure 4, comparing analyses No. 1 (emplacement tunnel diameter: Dt=3.7m) and No. 1C (Dt=2.4m) for Sh=20m and No. 7 (Dt=3.7m) and No. 7C (Dt=2.4m) for Sh=40m, it can be said that the case of Dt=2.4m has lower maximum bentonite temperatures at the inner edge of the bentonite than the case of Dt=3.7m, but that there is almost no difference for the maximum temperatures at the outer edge of the bentonite. From Figure 5, comparing analyses No. 1 and No. 1C for Sh=20m and No. 7 and No. 7C for Sh=40m, it can likewise be said that the case Dt=2.4m has a shorter duration of bentonite temperature exceeding 100°C at the inner edge of the bentonite than the case Dt=3.7m but that there is rarely any difference in the duration at the outer edge of the bentonite.

<u>Effect of host rock type</u>

Comparing analyses No. 1C (KRI, Dt=2.4m) and No. 1D (OPA, Dt=2.4m) for Sh=20m in Figure 4 and Figure 5 and No. 7C (KRI, Dt=2.4m) and No. 7D (OPA, Dt=2.4m) for Sh=40m, it can be said that the maximum bentonite temperature and the duration of bentonite temperature exceeding 100°C are sensitive to the type of host rock and its thermal properties and initial temperatures. The maximum bentonite temperatures of No. 1D and No. 7D are 10°C to 20°C lower than those of No. 1C and No. 7C respectively. The duration of bentonite temperature exceeding 100°C of No. 1D and No. 7D is approximately one tenth or less of that of No. 1C and No. 7C respectively.

CONCLUSIONS

In this study we considered the thermal features of disposal of spent fuel compared to that of vitrified waste and the effects of the following parameters on maximum temperature of bentonite and on the duration of bentonite temperature exceeding 100°C:
(1) Horizontal spacing of emplacement tunnels,
(2) Pitch of spent fuel canisters,
(3) Bentonite thermal properties,
(4) Presence or absence of air gaps,
(5) Heat production type,
(6) Emplacement tunnel diameter, and
(7) Host rock type.

The results indicate that a temperature ceiling of 100°C - which simplifies the performance assessment - is rather restrictive. For particular combinations of parameters, indeed, temperatures greater than 100°C may persist in the bentonite for several thousand years. Nevertheless, various layouts can be defined in which an outer annulus of bentonite remains below 100°C at all times. The restrictions on layout involved - together with the present trend towards higher burn-up fuel (with correspondingly higher thermal output) - suggests, however, that it will be valuable to re-assess the expected degradation of performance (if any) of bentonite subjected to temperatures in the range 100 - 150°C for periods of a few decades to a few centuries.

ACKNOWLEDGEMENT

This study has been carried out jointly by the Obayashi Corporation, Japan and NAGRA, Switzerland under the agreement between them in the field of radioactive waste management, specifying collaborative modeling studies on HLW and TRU disposal concepts.

REFERENCES

1. Nagra, <u>Project Gewähr 1985 - Nuclear waste management in Switzerland - Feasibility studies and safety analyses</u>, Project Gewähr Report series, NGB 01-09, Nagra, 1985.
2. Nagra, <u>Kristallin-I - Safety Assessment Report</u>, Nagra Technical Report series, NTB 93-22, Nagra, 1994.
3. Nagra, <u>Sediment Study - Disposal options for long lived radioactive waste in Swiss sedimentary formations - Executive summary</u>; Nagra Technical Report series, NTB 88-54, Nagra, 1989.
4. R. J. Hopkirk, W. H. Wagner, <u>Thermal loading in the near field of repositories for high and intermediate level nuclear waste</u>, Nagra Technical Report series, NTB 85-54, Nagra, 1986.
5. Obayashi Corporation, <u>Preliminary calculation of temperature distributions around an emplacement tunnel</u>, Kristallin-I repository layout study, Nagra-Obayashi Joint Study, Nagra Technical Report series, NTB 94-13, Nagra, 1996.
6. W. D. Rolph and K. J. Bathe, <u>An efficient algorithm for analysis of nonlinear heat transfer with phase changes</u>, International Journal for Numerical Methods in Engineering, Vol. 18, pp. 119-134, 1982.
7. H. Raiko and J. Salo, <u>The design analysis of ACP-canister for nuclear waste disposal</u>, Teollisuuden Voima Oy, YJT 92-05, Nuclear Waste Commission of Finnish Power Companies, 1992.
8. Nagra, <u>Sediment Study - Disposal options for long lived radioactive waste in Swiss sedimentary formations - Executive summary</u>; Nagra Technical Report series, NTB 88-25 E, Nagra, 1989.
9. T. Vieno, A. Haukojärvi, L. Koskinen, H. Nordman, <u>TVO-92 Safety analysis of spent fuel disposal</u>, Teollisuuden Voima Oy, Report YJT 92-33 E, Nuclear Waste Commission of Finnish Power Companies, 1992.

THE WIPP PRE-DISPOSAL PHASE EXPERIMENTAL PROGRAM

M. L. MATTHEWS
Office of Regulatory Compliance, U.S. Department of Energy, Carlsbad Area Office, P.O. Box 3090, Carlsbad, NM 88221-3090, matthem@wipp.carlsbad.nm.us

ABSTRACT

The WIPP is a transuranic waste disposal project of the U.S. Department of Energy and administered by the Carlsbad Area Office. The WIPP facility is a mined repository located near Carlsbad, New Mexico. It was developed and constructed 658 meters below the surface to demonstrate the safe, permanent isolation of transuranic radioactive wastes in a bedded salt formation. After the site selection phase and construction phase were completed, DOE embarked upon a pre-disposal experimental program phase. There are four major WIPP experimental program areas: shaft seals and rock mechanics; disposal room interactions; fluid flow and transport; and direct releases due to human intrusion.

 The Pre-Disposal Phase Experimental Program provided the necessary information to support the WIPP performance assessment process. The results of the program have been instrumental in demonstrating that the WIPP meets the regulatory requirements.

INTRODUCTION

The Waste Isolation Pilot Plant (WIPP) is a project of the US Department of Energy (DOE) and is administered through the Carlsbad Area Office (CAO). The WIPP facility is a repository mined in a bedded salt formation at a depth of 658 meters (Figure 1). It has been developed to safely and permanently isolate transuranic (TRU) radioactive wastes in a deep geological formation. The WIPP is located approximately 40 kilometers east of Carlsbad, New Mexico (Figure 2).

In October 1996, DOE submitted a Compliance Certification Application (CCA) to the US Environmental Protection Agency (EPA) in accordance with the requirements of Title 40 of the Code of Federal Regulations (40 CFR) Parts 191 and 194 [2,3]. The DOE plans to be in disposal operations in November 1997 following receipt of certification by the EPA. The disposal phase is expected to last for 35 years, and will include re-certification activities every five years.

The certification of WIPP relies mostly on the demonstration of the long-term safety of the repository through the performance assessment calculations. The Pre-Disposal Phase Experimental Program has provided the necessary information to support the WIPP performance assessment process. The results have been instrumental in demonstrating compliance with the regulatory requirements.

WIPP BACKGROUND

One of the fundamental reasons for selecting bedded salt deposits as the location for a geological repository for TRU waste is that the salt will creep and close around the waste, thereby isolating it. Beginning with siting studies in the 1970s, scientific programs have been conducted to obtain

Mat. Res. Soc. Symp. Proc. Vol. 465 © 1997 Materials Research Society

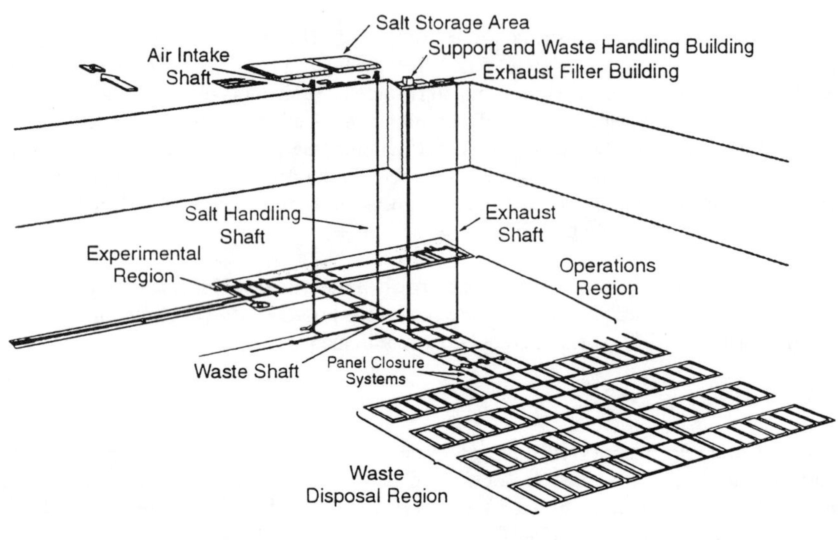

Figure 1. Spatial View of the WIPP Facility

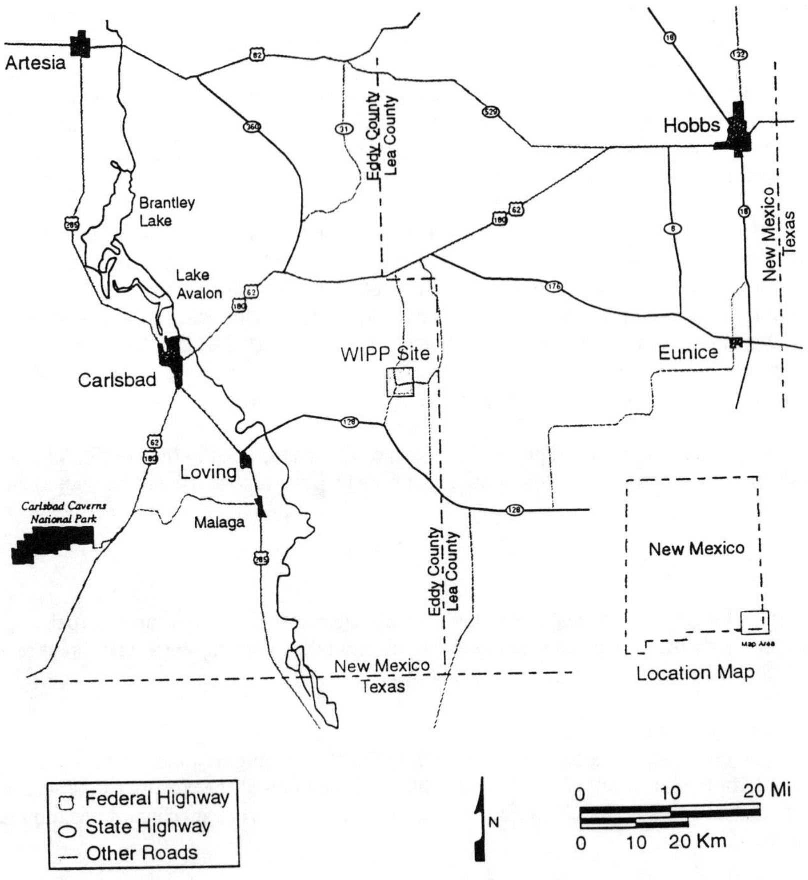

Figure 2. WIPP Site Location in Southeastern New Mexico

information about the WIPP site. The WIPP siting phase ended in 1983 with the completion of the Site and Preliminary Design Validation investigations. This underground excavation study resulted in the conclusion that the site met the previously established siting criteria.

The WIPP site-construction phase began in 1983. The initial underground excavations and the underground experiments were implemented between 1983 and 1990. The experiments and other activities conducted during the 1980s were undertaken primarily to: (1) develop a understanding of natural phenomena, as deemed prudent by WIPP Project scientists and/or scientists on the WIPP Panel of the National Academy of Sciences (NAS) and the Environmental Evaluation Group (EEG); and (2) satisfy negotiated agreements with the State of New Mexico. In the late1980s, the WIPP Project began developing and refining the tools needed to assess the long-term performance of the disposal system, including the performance assessment process.

Since 1990, the pre-disposal phase has been used to develop data and perform the analyses needed to demonstrate compliance with the regulations. The performance assessment process is the DOE's method for demonstrating the long-term containment of the waste, supported by information from the experimental program. The results of the performance assessment process are a critical element of the Compliance Certification Application (CCA) [1]. The US Environmental Protection Agency (EPA) will use the CCA as the basis for determining WIPP's compliance with the disposal system performance regulations.

During the 35 years of the disposal phase, TRU waste will be emplaced in the WIPP. This will then be followed by a decommissioning phase (during which the repository will be sealed) and a post-decommissioning phase.

EXPERIMENTAL ACTIVITIES

The pre-disposal phase experimental program was organized into four major areas: rock mechanics and shaft seals, disposal room interactions, fluid flow and transport, and direct releases that specifically address human intrusion disruptions.

Rock Mechanics and Seal System
A primary attribute of a salt repository is the ability of salt to creep over time, thereby healing damages caused by excavations. To allow quantitative prediction of the closure of the waste rooms and creep around the repository access shafts, an extensive rock mechanics program has been carried out.

The WIPP rock mechanics program included small-scale measurements of the creep and fracture properties of WIPP rock salt, development of detailed models for creep and fracture of the salt, and room-scale verification of the predictive capabilities of these models [4]. The data provided information on processes including salt creep closure of underground openings, development of a Disturbed Rock Zone (DRZ) around underground openings, and healing of the DRZ by creep closure around rigid inclusions. The DRZ, which is present around the repository, can provide a potential pathway for migration of brine containing radionuclides. The results of the rock mechanics program have provided DOE with sufficient understanding of the processes to incorporate them into the performance assessment models and calculations.

Shaft seals play a fundamental role in isolating the waste. Effective seals will serve two functions: (1) preventing brine containing radionuclides and/or hazardous constituents from migrating up the four shafts to the overlying units; and (2) preventing brine from the overlying units from migrating down the shafts. Key properties affecting shaft seal performance include the permeability of the seal materials and characteristics of the DRZ around each shaft. The shaft seals program has included measurements of component properties and performance, shaft seal design, and shaft behavior characterization (including measurements of creep closure and DRZ development around the shaft). The current seal system design includes several materials (compacted salt, salt-saturated concrete, clay, and asphalt) [5]. These materials are incorporated into multiple independent seal components that are designed to be emplaced using available construction methods and equipment and to meet the system performance requirements. The redundant components are designed to provide additional assurance of seal performance. The seal system is incorporated into the performance assessment models and calculations, and appropriate assumptions are used to address uncertainties.

<u>Disposal Room Interactions</u>
One of the key elements in the performance assessment calculations is the understanding of the possible physical and chemical conditions in the disposal rooms that will determine the potential releases of radionuclides. The processes inside the disposal room are very complex. They include creep closure of the room, consolidation of the waste, and chemical interactions with brine that may flow into the room from the surrounding formation. These processes are all closely coupled and can significantly influence one another.

Mobilization of the actinides in brine has been characterized through the actinide chemistry program. The actinide source term directly affects radionuclides releases for scenarios that involve the transport of brine to overlying geological units or the surface.

Laboratory tests have been conducted to establish solubility for the possible oxidation states of important actinides at a variety of possible chemical conditions in the repository [6]. Also, experiments were carried out to determine the quantities of actinides that can be mobilized in colloidal forms in WIPP brines [6]. The models and parameters used in the performance assessment calculations are based on the extensive set of experimental measurements that have been made in the laboratory chemistry program. As a result, the CAO has sufficient understanding of the chemical processes, and has identified the TRU waste components and characteristics of the waste that are important to the models and calculations of disposal system performance. Conservatism has been incorporated into the performance assessment models as needed to account for uncertainties in the processes.

Gas may be generated in the closed repository through two primary processes: (1) anoxic corrosion of metals in the disposal areas (the resulting gas is primarily hydrogen), and (2) anaerobic microbial degradation of cellulosic materials in the waste (the gas generated is primarily carbon dioxide). The gas can have both mechanical and chemical effects. Mechanically, the gas may pressurize the repository, thus retarding closure, reducing brine inflow, and perhaps fracturing the nearby brittle anhydrite layers. Carbon dioxide will dissolve in the brine; this will affect the pH, and thus the solubility of actinide species.

In order to provide additional performance assurance, the DOE will use magnesium oxide backfill in the disposal rooms to modify the chemical conditions in the repository and control the actinide source term. The magnesium oxide will react with the carbon dioxide gas formed by microbial degradation; the carbonate formed will buffer the pH of the brine present in the repository, thereby controlling actinide solubility. The chemical reactions which provide the control are well understood at the laboratory scale. The current model includes conservative assumptions to bound the uncertainties. The specifications for packaging and placement of the backfill have been developed [1].

Fluid Flow and Transport
Since the late 1980s, it has became clear that the most possible groundwater pathway for radionuclides to be released from the repository is through a future intrusion borehole into the aquifer system (Culebra) overlying the repository. It was also found that the Castile formation that underlies the repository may contain pressurized brine reservoirs. In order to characterize the processes that determine repository performance, it is therefore necessary to understand the presence and flow of brines in Salado, in the overlying Culebra, and in the Castile. Brines entering the repository support gas-generating reactions and provide a medium for mobilizing actinides in the waste. The brines may come from the Salado (in the undisturbed system), or from the Castile (in the case of human intrusion into an area of the Castile containing pressurized brine), while the Culebra forms a potential release pathway under certain conditions.

A series of measurements around the repository has provided data on the hydrologic properties of Salado halite and anhydrite units near the repository level; these properties (with their spatial variability) have been incorporated into the performance assessment models and parameters.

The Culebra is a potential off-site pathway for groundwater transport in the case of human intrusion. Brine containing dissolved and colloidal actinides may flow through a human-intrusion borehole to the Culebra. The human-intrusion borehole intercepting a Castile brine reservoir may also provide a pathway for brine from the Castile to flow through the repository, and up to the Culebra.

The Culebra exhibits complex flow properties with significant spatial variability. There is considerable variability in the structure and size of porous features through which flow (and potential radionuclide transport) occurs, and the transmissivity varies over six orders of magnitude in the area around the WIPP site. A higher-transmissivity zone, located in the southeastern area of the site, could provide the preferred path for radionuclide transport in the Culebra. An extensive hydrologic test program was carried out to characterize the Culebra flow properties. Extensive hydraulic and conservative-tracer testing was conducted at the H-19 hydropad, near the north end of the Culebra high-transmissivity zone. Additional testing was also conducted at the H-11 hydropad near the southern site boundary. The most significant output from these tests was confirmation of the role of matrix diffusion in physical transport within the Culebra [7]. The test results have been used to develop the Culebra conceptual model and the parameter values which incorporate the spatial variability for the performance assessment calculations.

In addition to the field studies, extensive tests have been conducted to establish retardation factors for the important actinides. Many of these tests were batch tests that determine the sorption coefficients of actinides on crushed Culebra rocks in controlled laboratory conditions. In addition, laboratory evaluation was also conducted using intact cores from the Culebra. During these tests, chemical retardation of actinides were determined by flowing through the core columns. Using the results from both batch and core column tests, reasonable yet conservative chemical retardation values were used in the compliance calculations to account for the uncertainty in these processes.

<u>System Response to Human-Initiated Activities</u>
The natural phenomena and processes described in the previous sections are considered when assessing the performance of: (1) the undisturbed system, and (2) the system when disturbed through human intrusion or other human-initiated activities.

In the disturbed-performance scenarios, direct releases can contribute significantly to the total releases of radionuclides. Direct releases include material brought to the surface during the drilling process: the drill cuttings themselves; material eroded from the borehole wall by circulating drill fluid; material carried by depressurizing gas; and material contained in releases of pressured brine. Calculations and experimental measurements have addressed heterogeneity and uncertainty in the properties of the consolidated wastes and disposal room, and have provided conservative bounds on the processes involved in direct releases.

The presence and characteristics of brine reservoirs in the Castile Formation are important to system response to human intrusion, as the pressurized brine may provide a driving force for both direct releases and long-term releases of brine carrying dissolved and colloidal radionuclides through the intrusion boreholes. The DOE's models of actinide solubility and colloid stability in Castile brine include conservatism as needed to address uncertainties in the models and variability in composition of repository brines.

Indirect releases may also occur as a result of human-initiated activities as radionuclides migrate to the accessible environment using pathways and conditions created by the human intrusion boreholes. Many of the properties of the Culebra related to physical transport and chemical retardation are important in predicting the magnitude and time of indirect releases. In addition, the properties of an intrusion borehole will change over time as plugs placed after abandonment degrade. The borehole and plug properties that are incorporated into the performance assessment models are based on current borehole plugging practices in the Delaware Basin in New Mexico. As the plugs degrade, the borehole permeability will increase, affecting fluid flow. The DOE has a good understanding of the process, and the models include conservatism to account for the variability in current plugging practices.

The federal government is committed to preventing adverse activities at the WIPP site for as long as it has jurisdiction at the site. To mitigate the potential of human intrusion into the repository, the DOE will incorporate passive institutional controls to reduce the probability of inadvertent human intrusion in the performance assessment calculations. The current design of the passive controls system includes multiple and redundant elements (berms with multiple sets of markers, and records) to provide performance assurance [1].

CONCLUSION

The Pre-Disposal Experimental Program has been an integral part of demonstrating that the WIPP meets the regulatory requirements. The Program provided the necessary data and rationale to support the WIPP performance assessment process. The information obtained from the four main experimental areas of the Program has enabled the DOE to submit a compliance application which meets the EPA standards.

REFERENCES

1. DOE (U.S. Dept. Of Energy), 1996: Title 40 CFR Part 191 Compliance Certification Application for the Waste Isolation Pilot Plant: DOE/CAO-2184; Carlsbad Area Office, Carlsbad, New Mexico.

2. EPA (U.S. Environmental Protection Agency), 1985: 40 CFR Part 191: Environmental Standards for the Management and Disposal of Spent Nuclear Fuel, High-Level and Transuranic Radioactive Waste: Final Rule; *Federal Register*, Vol. 50, No. 182, pp. 38066-38089, September 19, 1985. Office of Radiation and Air, Washington, DC.

3. EPA (U.S. Environmental Protection Agency), 1996: 40 CFR Part 194: Criteria for the Certification and Re-Certification of the Waste Isolation Pilot Plant's Compliance with the 40 CFR Part 191 Disposal Regulations; *Federal Register*, 61, pp. 5223-5245, February 9, 1996. Office of Air and Radiation, Washington, DC.

4. DOE (U.S. Department of Energy), 1996: Title 40 CFR Part 191 Compliance Certification Application for the Waste Isolation Pilot Plant; DOE/CAO-2184; Carlsbad Area Office, Carlsbad, New Mexico.

5. Repository Isolation Systems Department 6121; 1996: Waste Isolation Pilot Plant: Shaft Sealing System, Compliance Submittal Design Report; SAND96-1326, Sandia National Laboratories, Albuquerque, New Mexico.

6. DOE (U.S. Department of Energy), 1996: Title 40 CFR Part 191 Compliance Certification Application for the Waste Isolation Pilot Plant; Appendix SOTERM; DOE/CAO-2184; Carlsbad Area Office, Carlsbad, New Mexico.

7. Meigs, L.C., R.L. Beauheim, and J.T. McCord, 1996: Design, Modeling, and Current Interpretations of the H-19 and H-11 Tracer Tests at the WIPP Site; presented at the (NEA/OECD) GEOTRAP Workshop on Field-Tracer Transport Experiments, Cologne, Germany, August 28-30, 1996.

Compiled by I. R. TRIAY* and M. J. APTED**
*Los Alamos National Laboratory, Los Alamos, NM
**QuantiSci, Inc.

The role of performance assessment was discussed by a group of panelists and the participants of the 20th MRS symposium on the "Scientific Basis for Nuclear Waste Management." Panel members were Professor Thomas Pigford, Dr. Alan Cooper, and Dr. Patrik Sellin; Dr. Michael J. Apted served as moderator. For discussion purposes, "performance assessment" (PA) was defined as the *analysis of the release of radionuclides from a repository system of barriers to the accessible environment.*

As stated by some of the panelists, performance assessment:

- provides the major tool for the dialog between the implementor and the regulator;
- assesses the degree of safety of a repository with respect to national and international safety criteria;
- identifies the components and processes of the system that are responsible for safety;
- identifies the key uncertainties; and
- serves as a tool for scientific management that can be used to make decisions regarding design and research priorities.

In summary, as stated by one of the panelists, PA is a process for showing that a safe system is being designed and developed for the containment of radionuclides over the long term, which provides a technically defensible basis for decisions that must be made.

Members of the audience expressed concern regarding the role of PA and its implementation. Some of the concerns and proposed solutions to address these concerns are paraphrased here:

1) Concern: The role of PA for prioritization of research must be pursued cautiously because lack of data could result in analysis erroneously assigning a low priority to an important part or process of the system. PA can be used to help decision makers prioritize research but prioritization can't be done in the absence of scientific judgment from the larger technical community.

Proposed Solution: PA must actively and continually integrate new technical research from the advancing state of knowledge in the various areas relevant to radioactive waste isolation. Given that assumptions must be used in all models, technical experts are obligated to carefully study the importance or effect of such assumptions. PA experts and research scientists should team together to explore and identify weak spots in all PA models. Decisions on the prioritization of research and the abstraction of PA models necessary for making a licensing case should be fundamentally sound and based on knowledge, not ignorance or wishful thinking.

Mat. Res. Soc. Symp. Proc. Vol. 465 © 1997 Materials Research Society

2) Concern: Scientific investigations and performance assessment calculations are not always well-integrated. During a licensing hearing, lack of consensus among groups responsible for performance assessment and data collection could lead to failure of the entire process.

Proposed Solution: Performance assessment should be built and maintained by continually integrating all relevant scientific disciplines.

3) Concern: PA analyses may result in favoring either the engineered barriers or the natural barrier. In some cases this may be due to the lack of knowledge concerning these barrier subsystems.

Proposed Solution: An approach based on multiple barriers for long-term isolation of radioactive wastes is essential. The barriers play different and complementary roles in assurance of both technical and public acceptability for waste disposal.

4) Concern: The PA models are not always calibrated, and there are examples of conflicting results obtained using PA models.

Proposed Solution: Field studies, laboratory tests, and natural analogues should be used to develop confidence in PA models. In addition, PA should be done more often by independent groups working in a parallel, "double-blind" manner, so that eventual comparison of results can be used to evaluate the impacts of alternative conceptual models, treatments of uncertainties, and simplifying assumptions on the robustness of the safety assessment.

5) Concern: The PA models often do not represent or propagate uncertainties correctly in their assessment calculations.

Proposed Solution: Some investigators evaluating different aspects of the disposal system use probabilistic techniques to reflect the uncertainty of the important parameters. In addition, PA modelers should address conceptual model uncertainties. Decisions for licensing should be made on the basis of compiling and including all of the various sources of uncertainty in the calculated results.

6) Concern: Public acceptance of long-term disposal of radioactive wastes rests, in large part, on acceptance by the broad, non-waste technical community. More needs to be done to obtain acceptance from both the non-waste technical community and the general public.

Proposed Solution: Following the example of reactor safety and licensing, experts should first answer the challenges of the broad, non-waste technical community as a key part to ensuring public acceptance.

Consensus points of the discussion were:
  - *integration between research scientists and performance assessment experts is essential; otherwise, it is likely we will fail in the licensing of any repository; and*
  - *public acceptance must be aggressively pursued.*

Part XV

Natural Analogues

THE HYRKKÖLÄ NATIVE COPPER MINERALIZATION: A NATURAL ANALOGUE FOR COPPER CANISTERS

NURIA MARCOS PEREA

Helsinki University of Technology, Engineering Geology & Geophysics Lab, FIN-02150 Espoo, FINLAND

ABSTRACT

The Hyrkkölä U-Cu mineralization is located in south-western Finland, near the Palmottu analog site, in crystalline, metamorphic bedrock. The age of the mineralization is estimated to be between 1.8 and 1.7 Ga. The existence of native copper and copper sulfides in open fractures in the near-surface zone allows us to study the native copper corrosion process in conditions analogous to a nuclear fuel waste repository.

From the study of mineral assemblages or paragenesis, it appears that the formation of copper sulfide (djurleite, $Cu_{1.934}S$) after native copper (Cu^0) under anoxic (reducing) conditions is enhanced by the availability of dissolved hydrogen sulfide (HS^-) in the groundwater circulating in open fractures in the near-surface zone. The minimum concentration of HS^- in the groundwater is estimated to be of the order of 10^{-5} M ($\sim 10^{-4}$ g/l) and the minimum pH value not lower than about 7.8 as indicated by the presence of calcite crystals in the same fracture.

The present study is the first one performed on occurrences of native copper in reducing, neutral to slightly alkaline groundwaters. Thus, the data obtained is of most relevance in improving models of anoxic corrosion of copper canisters.

INTRODUCTION

In Finland and Sweden, disposed of spent nuclear fuel is planned to be in a repository to be constructed at a depth of about 300 - 800 m in crystalline bedrock. The nuclear waste will be isolated from the biosphere by multiple technical and natural barriers. In assessing the performance of the system, the corrosion of a copper canister (technical barrier) is considered.

The native copper mineralization at Hyrkkölä is related to the main uranium mineralization in granite pegmatites which occur as veins (0.05 - 5m thick) in quartz-feldspar gneisses and amphibolites. Petrological and mineralogical studies have demonstrated that this mineralization has many geological features comparable to sites being considered for nuclear waste disposal in Finland.

The evaluation of this native copper mineralization as a natural analogue for copper canister corrosion behavior under repository conditions is based on the study of the mineral assemblage native copper - copper sulfide occurring within open fractures in the near-surface zone. The occurrence of uranyl compounds in these fractures also allows consideration of the sorption properties of the engineered barrier material (metallic copper) and its corrosion product (copper sulfide).

A mineral assemblage at a given location may have been fixed by an externally controlled medium such as a moving fluid or pervasive atmosphere [1]. Therefore a detailed study of the fracture mineral assemblages may reveal the geochemical conditions (redox conditions, pH) in which they formed and the chemistry of the groundwaters in which they have been preserved. The hydrogeochemical conditions (water-rock interactions) that control the stability of the native copper and related fracture minerals are evaluated here.

Mat. Res. Soc. Symp. Proc. Vol. 465 © 1997 Materials Research Society

MATERIAL AND METHODS

The fracture mineralogy assemblage of native copper-copper sulfide is present in an open fracture zone at about 8.3 m depth. So far, this fracture zone is the only one where, among other minerals, native copper and copper sulfides are known to be present. Gummites are also visible to the eye in these open fractures.

Fracture surfaces, polished sections and polished thin sections were examined using a stereomicroscope and polarizing microscope; α-particle autoradiography was used to determine the occurrences of radionuclides and their spatial relations to other fracture minerals. Spatial relation and quantitative and qualitative analyses of the native copper and copper sulfide in fractures were obtained using an electron probe microanalyzer (EPMA).

RESULTS

Copper sulfide composition

Table I shows the results of microprobe analyses from random points in copper sulfide grains. The calculated average of the Cu/S ratio is 1.934 and the standard deviation $\sigma_{n-1} = 0.033$. This result indicates that the copper sulfide phase is djurleite, $Cu_{1.934}S$, the upper thermal stability of which varies between 72 ± 2 and 93 ± 2 °C [2]. It has been found that djurleite is the solid phase precipitated at low temperatures under alkaline pH (pH ~ 8)[3] and reducing conditions (Eh < 0).

Table I. Chemical composition (weight %) of copper sulfide grains.

Cu	Fe	Ca	S	U	O	Total
78.85	0.03	0.01	20.68	0.13	0.37	100.07
78.90	0.02	0.00	20.66	0.18	0.28	100.04
79.27	0.02	0.02	20.13	0.10	0.34	99.88
79.34	0.01	0.02	20.60	0.00	0.27	100.24
79.42	0.00	0.01	20.29	0.00	0.34	100.07
79.18	0.04	0.02	20.68	0.00	0.21	100.13
79.85	0.01	0.01	20.75	0.00	0.47	100.09
77.38	0.04	0.02	20.75	0.00	0.93	99.12

Native copper composition

The data of microprobe analyses of native copper (Table II) was normalized to 100%. Thus, the maximum Cu content obtained is 99.39%. These results are affected by the background "noise" produced by the surrounding silicate minerals, and the small grain size of the analysed copper. The true composition of native copper is estimated to be about 99.99% Cu.

Table II. Chemical composition (weight %) of metallic copper.

Cu	Fe	Ca	S	U	O	Total
99.33	0.05	0.01	0.00	0.00	0.60	100
99.39	0.03	0.02	0.02	0.08	0.45	100
97.46	0.09	0.02	0.00	0.06	2.37	100
97.36	0.08	0.04	0.02	0.00	2.50	100

<u>Fracture surface mineralogy and uranium distribution</u>

The X-ray spectrum in Figure 1b shows that the main mineral at the fracture surface is a silicate, rich in aluminum and magnesium, most likely a clay mineral of the montmorillonite group (smectites). Calcite (Fig. 1a) is also a common mineral on the fracture surface. The lighter zones in the backscattered secondary electron (BSE) image (Fig. 1a) correspond to a concentration of heavy materials, mostly metals. Figure 2b shows the X-ray spectrum at the center of a heavy mineral phase (Fig. 2a). The relatively high peaks of sulfur and copper are an indication that copper sulfide is underlying the U-rich silicate and that the maximum thickness of the U-rich clay layer is about 0.5 μm (depth from which X-rays emerge).

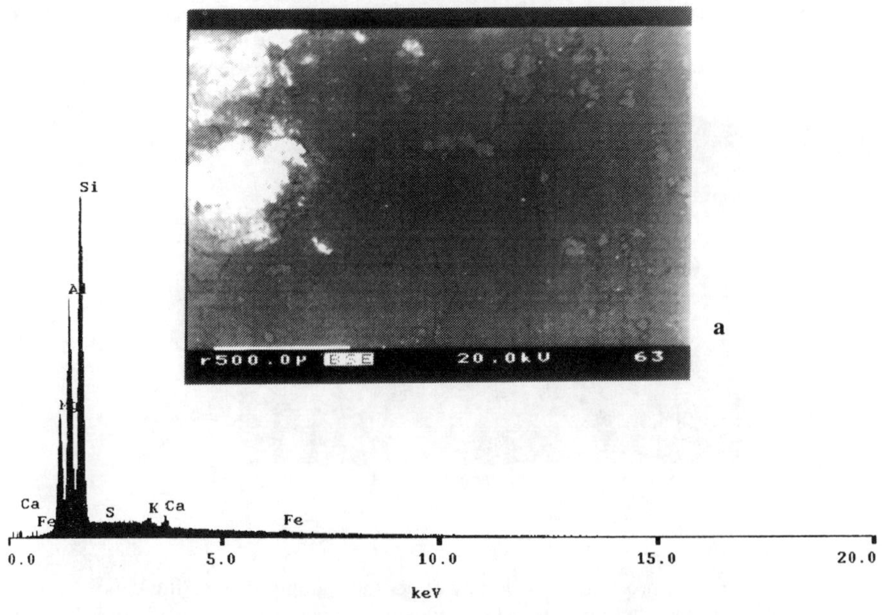

FIG. 1. a)Fracture surface. BSE image. b) X-ray spectrum at the center of the image.

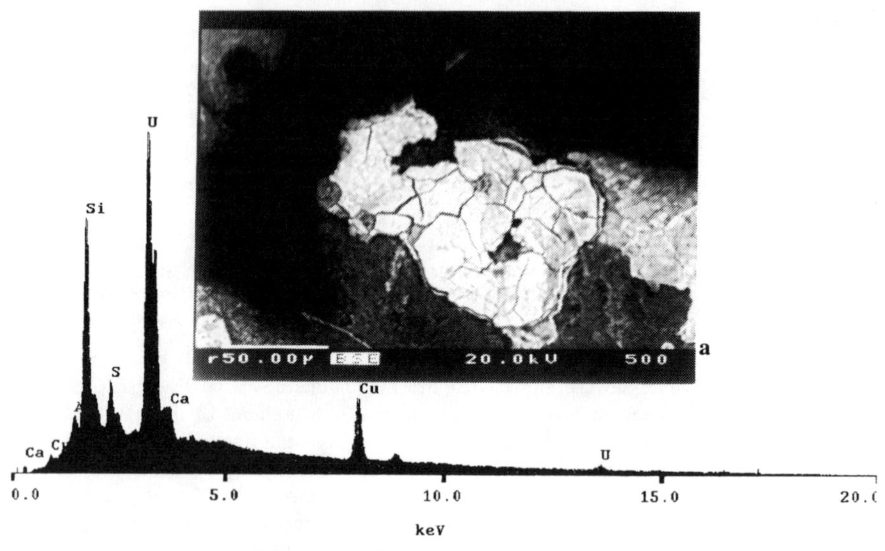

FIG. 2. a) A heavy mineral phase, BSE image. b) X-ray spectrum at the center of the image.

Figure 3 shows the BSE image of a copper sulfide grain. The X-ray spectrum at the center of the grain showed a small peak of uranium, indicating that uranium is not significantly adsorbed in the bulk of copper sulfide grains, but adsorbed around the grain boundaries. This was also shown by the results of α-particle autoradiography.

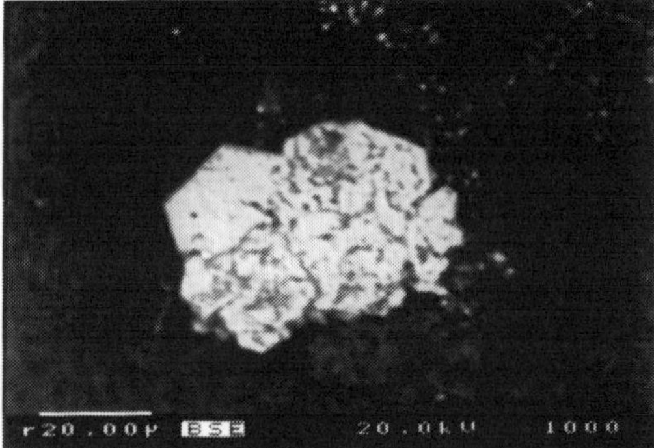

FIG. 3. Grain of copper sulfide, BSE image.

DISCUSSION

Of the various corrosion processes which may affect the stability of metallic copper in deep bedrock [4], sulfidization, a process in which metallic copper reacts with dissolved sulfide forming a solid sulfide phase, is considered the most important one at Hyrkkölä. The secondary sulfidization of native copper in fractures is dependent on the availability of hydrogen sulfide (HS^-) in the groundwater. Water acts as an electron acceptor (oxidant) in the corrosion reaction (3), a summation of (1) and (2):

$$1.934Cu^0 + HS^- = Cu_{1.934}S + 2e^- + H^+ \qquad (1)$$

$$\underline{2H^+ + 2e^- = H_2(aq)} \qquad (2)$$

$$1.934\ Cu^0 + HS^- + H^+ = Cu_{1.934}S + H_2(aq) \qquad (3)$$

Djurleite $Cu_{1.934}S$ is a mineral of low temperature supergene origin formed by sulfidization of native copper. It is a mineral phase of the system copper-sulfur. The phase boundaries of this system have been represented in pE - pH diagrams (Figs. 4 and 5; Eh = 59.155pE at 25 °C and at 5 °C Eh = 55.164pE). The occurrence of calcite crystals at the fracture surface indicates a minimum pH value of about 7.8 (25 °C; [5]) for circulating groundwaters in the fractures.

Figure 4a shows that at 25 °C, and a pH ~ 7.8 the minimum concentration of HS^- for the stability of djurleite is of the order of 10^{-5} M ~ 10^{-4} g/l. At the same values of temperature and pH, and a $[HS^-] = 10^{-6}$ M, the stability limits of djurleite lie outside the range to which water can act as an oxidant (Fig. 5a). At lower temperatures (Figs. 4b and 5b) and the same pH, the stability limits of djurleite lie within the range at which water can act as an oxidant even at concentrations of HS^- of the order of 10^{-6} M ~ 10^{-5} g/l.

CONCLUSIONS

The presence of native copper and copper sulfide within open fractures in the near-surface (8.3 m depth) implies the existence of reducing conditions, which are also predicted to prevail between 300 and 800 m depth in granitic bedrock. This finding allowed the study of native copper corrosion process in anoxic conditions.

It has been estimated that the formation of copper sulfide (djurleite, $Cu_{1.934}S$) after native copper (Cu^0) under anoxic (reducing) conditions is enhanced by the availability of dissolved HS^- in continuously or intermittently circulating groundwaters. The minimum quantity of dissolved HS^- has been calculated to be of the order of 10^{-4} to 10^{-5} g/l (Eh of about -350 to -248 mV), at temperatures of 25 °C and 5 °C respectively, with a pH of about 7.8. These values are not exceptional in granitic groundwaters, such as [6], [7].

Secondary uranium compounds were observed to be significantly adsorbed at the djurleite grain boundaries. Sorption was not significant in the bulk of copper sulfide grains nor in the bulk or boundaries of native copper grains.

The scale of the processes considered is still unclear. Further hydrogeochemical studies would provide a unique opportunity to determine the scale of the phenomena and to test the thermodynamic models (and their associated databases) used for copper corrosion analyses in performance assessment.

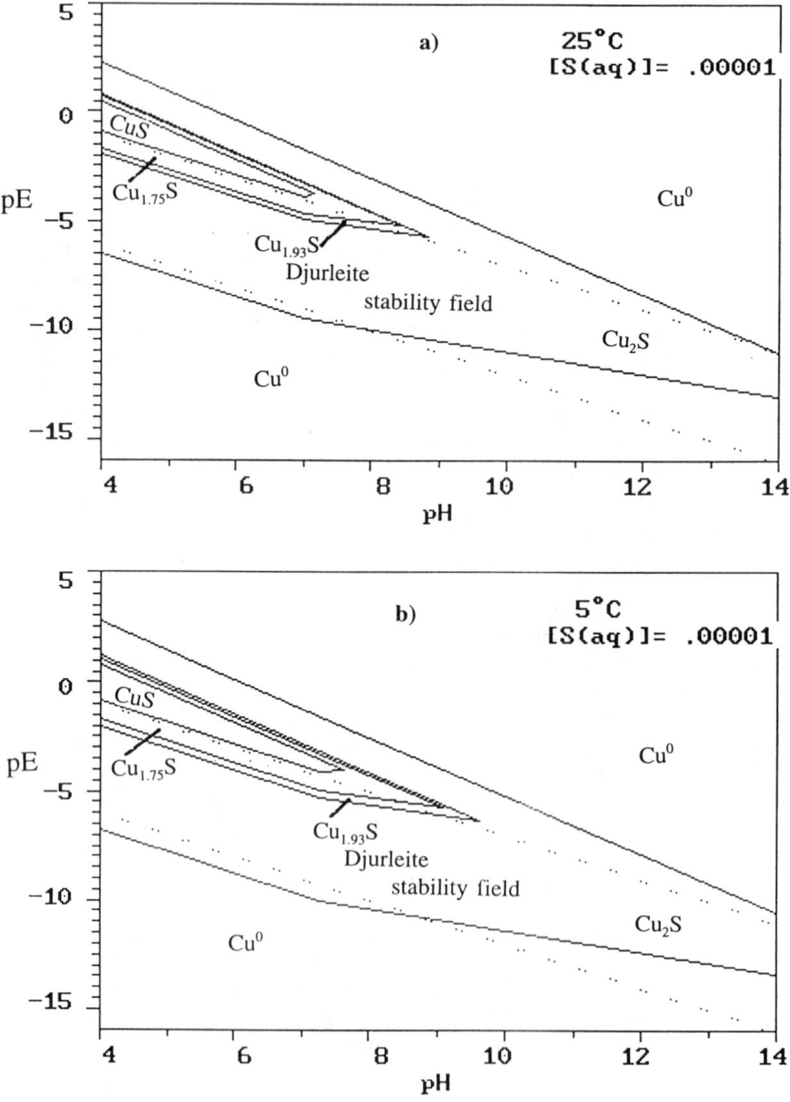

FIG. 4. Stability fields of metallic copper and copper sulfides. a) T = 25 °C, b) T = 5 °C. Activity of dissolved sulfur [S] = 10^{-5} M (either as HS^- or $SO_4^=$). Dotted lines indicate the pE-range at which water can act as an oxidant, upper limit ($[H_2] = 10^{-9}$) corresponds to the estimated lowest concentration of H_2(aq), lower limit corresponds to the H_2(aq) activity of 1.

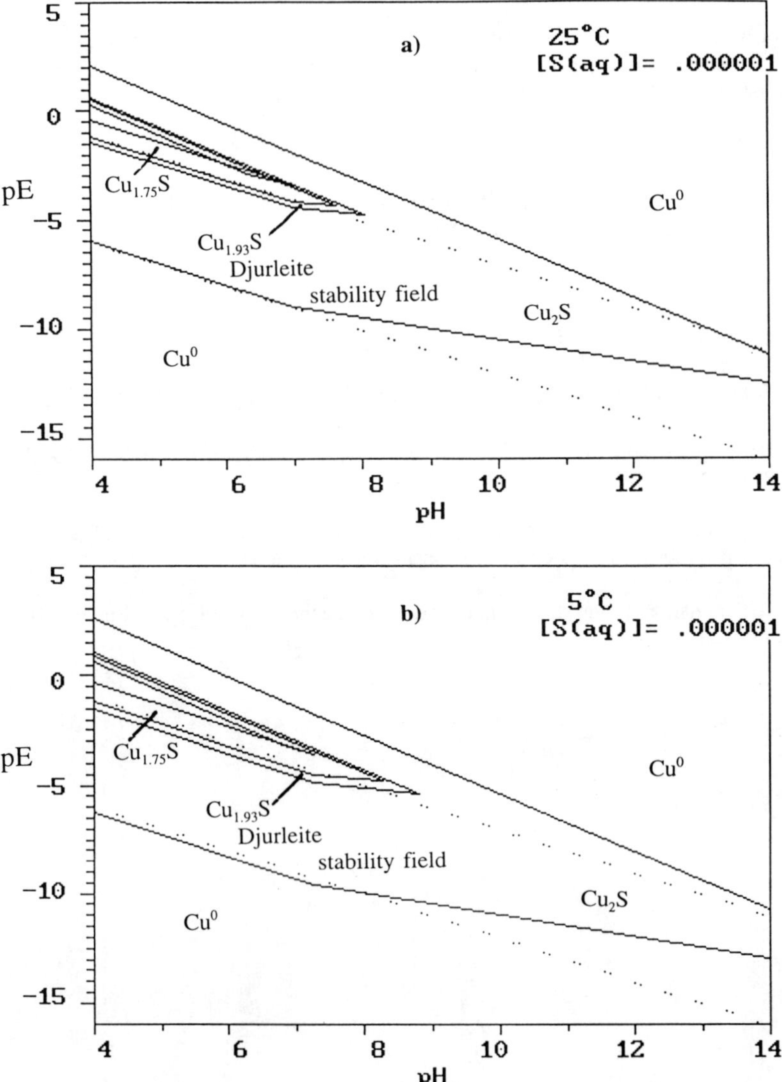

FIG. 5. Stability fields of metallic copper and copper sulfides. a) T = 25 °C, b) T = 5 °C. Activity of dissolved sulfur [S] = 10^{-6} M (either as HS^- or $SO_4^=$). Dotted lines indicate the pE-range at which water can act as an oxidant, upper limit ($[H_2]$ = 10^{-9}) corresponds to the estimated lowest concentration of H_2(aq), lower limit corresponds to the H_2(aq) activity of 1.

ACKNOWLEDGEMENTS

The author is very grateful to Lassi Pakkanen for performing the SEM analysis, to Lasse Ahonen and Jalle Tammenmaa for performing the algorithm for Figures 4 and 5, and Juhani Aav for technical assistance. This work was supported by Posiva Oy.

REFERENCES

1.   D.S. Korzhinskii, Physicochemical Basis of the Analysis of the Paragenesis of Minerals, New York: Consultants Bureau (1959).

2.   R.W. Potter, Econ. Geol.72, p. 1524 - 1542, (1977).

3.   R.A. Snellgrove and L.H. Barnes, in Trans. Am. Geophys. Union, v. 55., p. 484, (1974).

4.   B. Mason, Principles of Geochemistry, John Wiley & Sons, Inc. New York, (1966).

5.   L. Ahonen, Report **YJT-95-19**, (1995), (Helsinki, Nuclear Waste Commission of Finnish Power Companies).

6.   P. Pitkänen, M. Snellman, and H. Leino-Forsman, Report **YJT-92-30**, (1992), (Helsinki, Nuclear Waste Commission of Finnish Power Companies).

7.   R. Blomqvist, R. Pilviö, L. Ahonen, and T. Ruskeeniemi, Geol. Surv. Finland, Rep. **YST-78** (1992), p. 32 - 67.

MECHANISMS FOR THE FORMATION OF A PERCHED WATER ZONE IN FRACTURED TUFF: A NATURAL ANALOGUE STUDY

E. G. WOODHOUSE*, R. L. BASSETT*
*Department of Hydrology and Water Resources, P.O. Box 210011, University of Arizona, Tucson, AZ 85721-0011, betsy@hwr.arizona.edu

ABSTRACT

Perched water zones have been identified in the fractured, welded tuff in the semi-arid to arid environments of Yucca Mountain, Nevada and near Superior, Arizona. An understanding of the formation of such zones is necessary in order to predict where future perched water might form at Yucca Mountain, the proposed site of a high-level nuclear waste repository. The formation or growth of a perched zone near a repository is one of the factors to be considered in the risk assessment of the Yucca Mountain site.

The Apache Leap Research Site near Superior, Arizona is a natural analog to the Yucca Mountain site in terms of geology, hydrology, and climate. Information used to study possible mechanisms for the formation of the perched zone included data regarding isotopic and geochemical properties of the waters in and above the perched water zone; measured hydrologic parameters of the perched zone; geophysical and measured parameters of the tuff; megascopic and microscopic observations of the tuff, including mineralogical, alteration, and structural features; and the lateral and vertical extent of perched water in the region.

Aquifer test, geophysical, geochemical, and radioisotopic data show that fractures are the means by which water is recharging the perched zone. The reduced hydraulic conductivity of the formation in the perched zone appears to result from both a severe reduction in matrix porosity and permeability caused by welding, devitrification, and vapor phase crystallization; and by an increase in fracture filling which restricts the pathways for flow.

INTRODUCTION

Yucca Mountain in southwestern Nevada is being investigated by the U.S. Department of Energy as a potential site for a high-level nuclear waste repository. The site was chosen in part because it has a thick unsaturated section of welded and zeolitized tuff, on the order of several hundred meters. This unsaturated zone should impede a significant flux of water from the land surface down to a deeply-buried repository, and retard transport from the repository to the zone of permanent saturation several hundred meters below, thus minimizing the likelihood of corrosion or spent fuel dissolution problems if any release of radionuclides were to occur from the repository. Locally saturated water conditions have been identified at Yucca Mountain [1]. The existing zones have the potential to expand, or new zones may form in the future, due possibly to heterogeneities in the geologic material and/or to intense meteorological events on a short time scale, near-field effects of a hot repository, or climatic changes on a larger scale. If such a perched water body formed above the repository, it would have the potential to channel water via fracture flow down to the repository; if perched water formed at or beneath the repository, it could provide a more rapid pathway of radionuclide transport to the accessible environment. Therefore, an understanding of the mechanisms of formation of such perched zones is necessary to properly evaluate the suitability of the site for high level waste disposal.

The Apache Leap Research Site (ALRS) near Superior, Arizona (Figure 1) is composed of fractured welded tuff with climatic and topographic conditions similar to Yucca Mountain, making it a highly suitable natural analog for an independent characterization of hydrological processes in such material. Furthermore, a perched water body covering a region of at least 16 km^2 [2] has been identified at the ALRS. The objective of this research was to study and describe the physical and hydrologic characteristics of the Apache Leap Tuff, and then to compare those characteristics against

Mat. Res. Soc. Symp. Proc. Vol. 465 ©1997 Materials Research Society

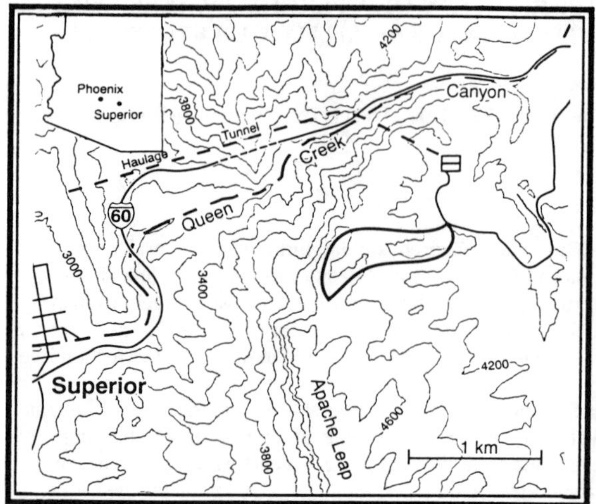

Figure 1. Location of the Apache Leap Research Site.

possible mechanisms for the formation of perched water in order to develop a model for perching at the ALRS.

The Apache Leap Tuff is a mid-Miocene ash-flow tuff approximately 20 my old. The present deposit has a maximum exposed thickness of 600 m and covers about 260 km^2. Peterson [3] concluded that the Apache Leap Tuff represents a single cooling unit composed of many (several tens) flows deposited in rapid succession. Identification of individual flow units has not been made; instead, the deposit has been divided into five units based on the character of the groundmass [3]. From the bottom, these units are the basal tuff, the basal vitrophyre, the brown unit, the gray unit, and the white unit. The upper three units were differentiated on the basis of the color of their fresh surfaces and are primarily mapping units, although microscopic distinction in their alteration characteristics has also been made. The perched aquifer under investigation occurs in the gray unit; therefore, characterization of the properties of the tuff focused on the gray and white units, in the upper 200 m of the deposit.

CHARACTERIZATION OF THE APACHE LEAP TUFF

Sources of information for the characterization of the physical and hydrologic properties of the Apache Leap Tuff included a detailed geologic analysis of the entire deposit by Peterson [3], recent or in-progress investigations by others at the ALRS [2,4,5,6,7], and new information obtained primarily from two boreholes which were installed at the ALRS during this investigation. A deep slant borehole (DSB) extends 200 m down at an approximate 45° angle, and a vertical observation borehole (VOB), collared upgradient and 15 m from the top of the DSB, extends down 170 m. Both boreholes intercept perched water, and terminate just below that zone. Continuous core was collected from the entire length of the DSB and was used for a variety of physical measurements and descriptions, including mineralogy and alteration characteristics, density, moisture content, hydraulic conductivity, geochemistry, and degree of fracturing, weathering, welding, and devitrification. These characteristics were examined through analysis of hand specimens, thin sections, x-ray diffraction results, and scanning electron microscope techniques, and were compared and correlated with the results of other investigations. Hydrologic characteristics were determined from isotopic and geochemical investigations of pore, fracture, and perched zone water [5], geophysical surveys of both the DSB and VOB [7], and aquifer tests performed on the VOB.

Physical Properties

The upper 200 m section of the Apache Leap Tuff at the ALRS is characterized by densely to partially welded tuff that has been substantially altered by devitrification and vapor phase crystallization. The white unit is a very light, mottled, brownish-pink color, and is speckled with oxidized micas and weathered iron-oxide minerals. It contains randomly oriented, relatively equidimensional pumice fragments. In contrast, the gray unit is darker, and contains pumice fragments which have been distinctly flattened, imparting a visible fabric to the rock. The transition from white to gray units occurs gradually, over approximately 10 m.

Two of the 45 2.5-cm x 3.5-cm blocks of the DSB core from which thin sections were cut illustrate the macroscopic textural changes that occur in the gray unit with depth to the perched water zone. Figure 2 shows the gray unit at depths of 36 m and 148 m. The core at 36 m appears relatively uniformly heterogeneous, with white pumice fragments and some larger quartz and feldspar grains scattered randomly throughout the pinkish-brown matrix. A faint linearity is seen running diagonally from the lower left to upper right sides of the photograph. The porosity occurs in regions generally associated with the pumice fragments, but is scattered throughout the rock. In contrast, the core from 148 m shows a smaller number of larger pumice fragments, with regions of comparatively pumice-free matrix-plus-phenocrysts; the orientation has become more distinct, with the top at the left side of the photograph. Porosity is significantly reduced, and is only observed in a few of the larger pumice fragments. This downward trend parallels the borehole geophysical grain density and porosity data [7]. This evidence suggests that the weight of compaction may have completely flattened out some of the pumice fragments as well as much of the matrix, resulting in a visible reduction in porosity that is substantiated by matrix hydraulic conductivity meausrements reduced from on the order of 100 mm/yr in the white unit down to 0.001 mm/yr in the gray unit at the perched zone [4].

Seven blocks of core were vacuum-impregnated at a pressure of 1500 Pa with blue epoxy prior to cutting the thin sections. Figure 3 shows representations of the porosity seen in thin sections; in these images, the portions that were dyed blue (pore spaces) are shown black, and all solid rock material is white. These figures further illustrate the changes in porosity with depth. The top image in Figure 3 is from the white zone, at a depth of 23 m. Pore spaces are shown scattered thoroughly throughout the matrix, implying considerable connectivity between them. The apparent vertical lineation that can be seen is an artifact of the imaging process. In the section from 36 m, from the upper part of the gray zone, the pattern of porosity has changed considerably (Figure 3, left). It is concentrated primarily in small but discrete regions generally associated with pumice fragments; it also fills small fractures in the phenocrysts. Small, isolated regions containing virtually no porosity are starting to appear at this depth. At 148 m depth, in the perched water zone, Figure 3 (right) shows even further reduction in the porosity and large impermeable regions. The pumice-related porosity zones correlate to the thin white regions shown in the macroscopic view of this core in Figure 2.

The mineralogy of the Apache Leap Tuff is quite consistent through the upper 200 m. Phenocrysts comprise 35-45% of the rock and are dominantly plagioclase, with lesser amounts of quartz, biotite, sanidine, and magnetite [3]. Lithic fragments comprise about 2-4% of the rock, and pumice fragments make up about 25% of the rock and contain approximately the same proportion and type of phenocrysts as the matrix. The matrix consists predominantly of cryptocrystalline cristobalite and feldspar, although some pumice fragments may contain tridymite as a result of vapor phase crystallization [3]. X-ray diffraction analysis revealed a small but consistent presence of clay minerals, dominantly smectite, throughout the entire 200-m thickness. Scanning electron microscope and thin section analyses showed the smectite filling approximately 8% of the fractures and 4% of the pore spaces [8]; in general it is not a significant presence.

The Apache Leap Tuff is cut by many high-angle fractures, as well as by fractures parallel to the depositional surface of the rock. Fractures appear to become more abundant towards the bottom of both the DSB and VOB, although the presence of a fault is suspected [7] near the bottom of the DSB. Below the depth of the perched water, fracture density was observed to decrease in two other boreholes in the Apache Leap Tuff [6,8]. In the quantification and analysis of the DSB fracture characteristics, thin fractures completely filled with silica minerals, most likely cristobalite, were observed in many samples from the gray unit, and were most abundant towards the bottom of the borehole. A noticeable increase in fracture-filling silica minerals was observed in the gray and brown units in the two other boreholes in the tuff [6,8], supporting evidence of increased fracture filling with depth below the perched zone. Fracture filling by silica minerals is expected to be enhanced deeper in the unit where

Figure 2. Cut sections of DSB core from depths of 36 m (left) and 148 m (right).

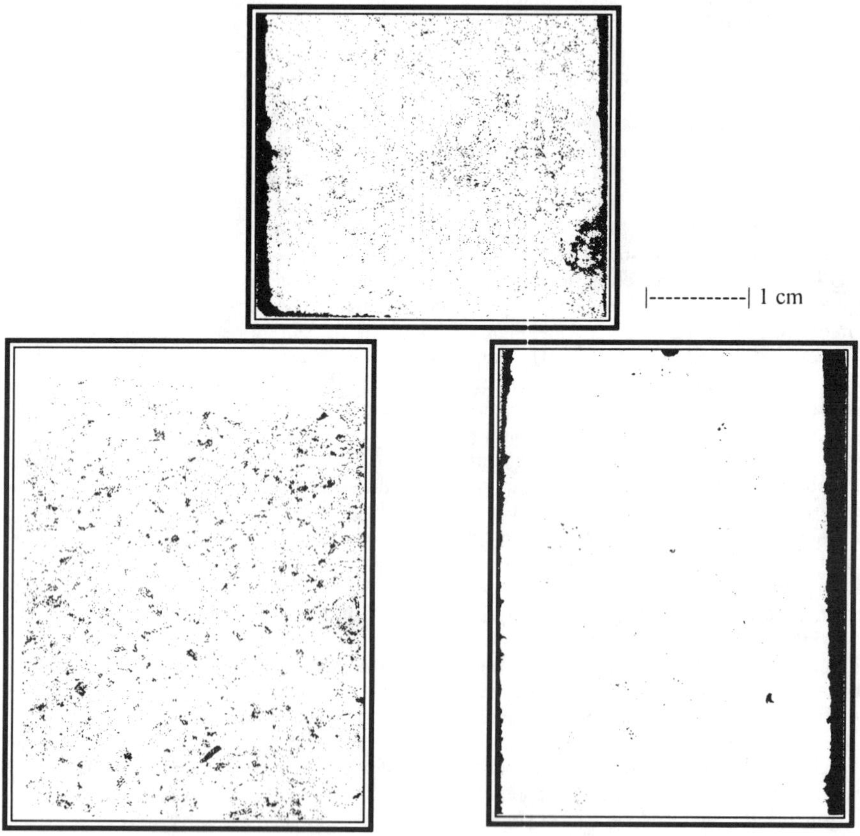

Figure 3. Porosity distribution representations of DSB core from 23 m (top), 36 m (left) and 148 m (right). Pore spaces were made black; all matrix and minerals are white.

sustained high temperatures were likely to exist, enhancing the solubility of silica.

Hardin [4] measured matrix hydraulic conductivity at intervals along the DSB. He determined that matrix hydraulic conductivity declines sharply with depth to values as low as 0.001 mm/yr in the region of the perched water zone. At a depth of 20 to 30 m from the surface, hydraulic conductivity drops from on the order of 10 to 140 mm/yr down to 2 mm/yr or less, suggesting that below this depth, the matrix is virtually impermeable.

Hardin et al. [7] measured wet bulk density, volumetric moisture content, porosity, and apparent saturation of sections of core from the DSB. Wet bulk density values for the white unit ranged from 2.2 to 2.4 g/cm^3 with depth, whereas the gray unit values were more consistent, at 2.4 to 2.5 g/cm^3, reflecting the change in characteristics of the two units. Volumetric moisture content measurements showed an increased moisture content in the upper 61 m of the section, which would correspond to higher porosity in the white unit. Deeper in the section, below 91 m, the scatter of the data increases, suggesting a change in formation conditions [7]. Porosity and apparent initial saturation measurements were made for the gray unit only; the porosity of that unit varies from 5 to 8%, and nearly all the core had an initial saturation of at least 80%, with some sections appearing to be completely saturated.

<u>Hydrologic Properties</u>

Davidson [5] compared ^{14}C activities from water sampled directly from the perched aquifer, and from pore waters from core within that saturated zone, and concluded that perhaps as much as half of the water in the aquifer beneath the DSB may be derived from nearby fractures. He found that the ^{14}C activity of the pore water obtained from core in the saturated zone was much higher than from the pumped aquifer, and proposed that the water from fractures enters a mixing zone, resulting in a higher ^{14}C activity in the localized region where fractures intercept the perched zone. In contrast, the pumped perched water represents an older, averaged sample from a larger area.

Davidson [5] used chemical data from surface water runoff and DSB pore waters, combined with additional analyses from Bassett et al. [2], to model the perched aquifer water. The solutions from several models consistently indicated that the aquifer water is composed dominantly of surface water runoff, with at most a 10% contribution of pore water that has reacted with the rock matrix. Davidson's geochemical model results substantiate the interpretation of rapid fracture flow as the primary mechanism for aquifer recharge.

Hardin et al. [7] collected nuclear, electrical induction, and borehole television logs from the DSB and electrical induction logs from the VOB. Distinct responses in the neutron, density, and resistivity logs from the DSB were observed at small features which are likely to be high-moisture fractures. Many of these features could be correlated to fractures observed on the borehole television log. Several rather broad peaks may be indicative of water movement in the fractures, and of higher moisture in the rock matrix adjacent to fractures [7].

Both an pumping and a slug test were performed on the DSB to measure properties of the perched aquifer. To perform the pumping test, the aquifer was pumped for 18.8 hours; the VOB was equipped as an observation well, but no response was ever recorded. To perform the slug test, the aquifer was allowed to completely recover from pumping, then a slug of water was added to the DSB; the response was again observed only in the DSB. The pumping data and the slug test data were analyzed by three different methods each, for comparison purposes [8]. The results of the comparison showed close agreement in most cases, and indicated that the Apache Leap perched aquifer is more accurately portrayed by a discrete fracture model than by a dual porosity system. The drawdown data were fit to Gringarten and Ramey's [9] discrete fracture model; the fit indicated that for a measured matrix hydraulic conductivity of 2.5 x 10^{-7} m/day in the perched zone [4], the fracture hydraulic conductivity is 238 m/day, or nine orders of magnitude greater. These results indicate that water movement in the ALRS perched aquifer system is clearly fracture-dominated, and that matrix flow is insignificant.

MECHANISMS FOR THE FORMATION OF PERCHED WATER

In order for the formation of a perched water zone to be understood, two processes must be addressed. First, water must be able to move down through the formation by some pathway. Second, the hydraulic conductivity of the formation must be reduced so that the movement of water is restricted, creating the saturated zone. Whatever mechanism allows water transport from the surface must be altered such that it becomes less effective at the perched zone. Water must reach the perched zone by flow through fractures and/or the matrix, therefore models for reduction of permeability must be tested on both the fractures and matrix. The different characteristics of the tuff described in the previous section were analyzed in view of which factors appear to affect the processes of water transport and pathway restriction, in order to develop a model for the formation of perched water.

The model for perching at Apache Leap was developed by considering eight different models for perching against the physical and hydrological characteristics of the Apache Leap Tuff that have been identified. These potential models included heterogeneities in the boundary between individual flow units, cooling zone boundaries, a hydrologic trap caused by faulting, variations in fracture density, connectivity, or fill characteristics, changes in matrix characteristics, and weathering or chemical alteration [8]. These models were each rejected or incorporated into the Apache Leap model based on the characteristics of the tuff that have been described.

Two characteristics of the Apache Leap Tuff appear to exert the strongest influence on the formation of the perched water zone. First, the hydrological evidence provided by geochemistry, geophysics, and aquifer test analyses all indicate that fracture flow is responsible for the transport of water from the surface to the perched zone. Second, very low matrix porosity and permeability appear to be one component responsible for the lack of, or severely reduced, transport of water through the perched zone, as indicated by the lithologic and petrographic data, alteration characteristics, and porosity and permeability data. A change in the characteristics of the fractures at the perched zone is also necessary, however, in order to reduce fracture flow. Models considered for fracture flow restriction included disconnected fracture sets above and below the perched zone, reduced fracture density below the perched zone, and increased fracture filling below the perched zone. Data were not available to support the presence of disconnected fracture sets; a reduction in fracture density below the perched zone was observed, however, as discussed previously. In addition, the fracture fill model is supported by the observation of increased fracture filling by silica minerals in the region below the perched water in three different boreholes [6,7,8].

Additional mechanisms that could theoretically cause reduced hydraulic conductivity in fractured tuff were also considered, and rejected on the basis of the physical evidence. Vertical heterogeneities caused by changes in the composition of individual flows could create differences in permeability, however no individual flows have been identified, and the mineralogy shows no noticeable variations with depth. Syngenetic boundaries between cooling zones, such as the vitrophyre/welded zone contact, are likely to have differing hydrologic properties. This mechanism has been identified as one means for perched water at Yucca Mountain [1], but at the ALRS, the perched zone occurs entirely within the gray unit. A previous high stand of water could have created a chemically altered zone, such that the product of feldspar weathering created a clay-rich, reduced permeability zone; however the x-ray diffraction and thin-section analyses showed no concentrations of iron hydroxides, aluminum hydroxides, weathered phenocrysts or residual oxyhydroxides of iron and aluminum.
Faults could create pathways for selective weathering, or they could offset stratigraphic units such that perched water "traps" are created; although faults are relatively abundant in the ALRS region, their occurrence was not consistently associated with the location of the perched zone.

A combination model is therefore proposed to explain the formation of the perched water table at the ALRS; the model features changes in the matrix property with depth, fractures as the primary recharge mechanism, and less abundant, but more frequently filled fractures beneath the perched water zone, as shown in Figure 4. In addition to fitting the characterizations made in the region of the DSB, VOB, and additional boreholes [6,8], the model must be applicable over a broad region, as perched

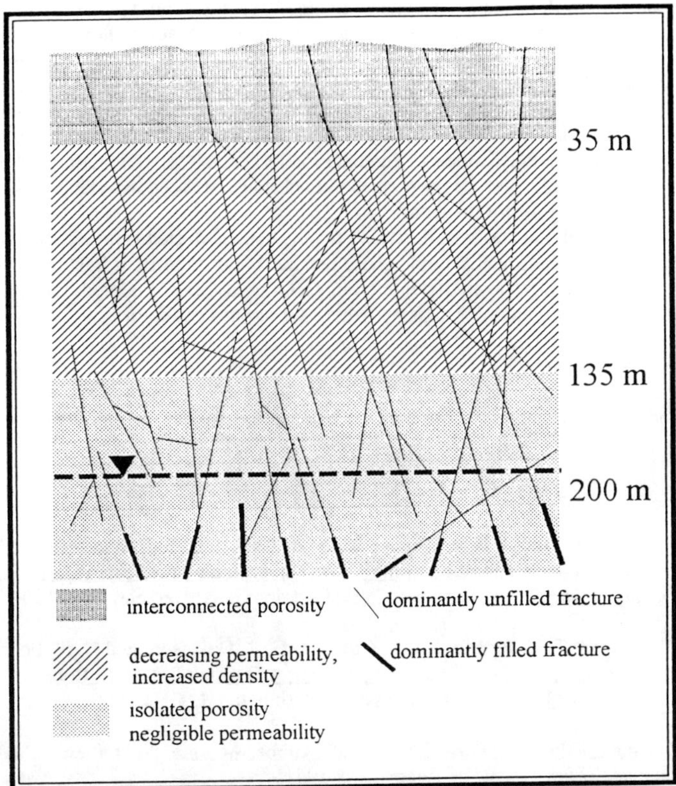

Figure 4. Combination model for perched water at the Apache Leap Research Site.

water has been identified over at least a 16 km² area of Apache Leap Tuff [2]. The mechanisms that have been proposed would appear to fit a regional model. Peterson [3] documented the consistency of petrology, mineralogy, and fracture pattern over much of the entire Apache Leap Tuff deposit, and showed that the deposit is a single cooling unit. This information suggests that the matrix properties would in turn be likely to be consistent over the region. Furthermore, the zone of greatest welding and heat would also be expected to be continuous over approximately the same depth, allowing for topographic variations, such that fracture filling would occur in a similar pattern over the region. Thus, the proposed combination model appears to fit for the Apache Leap region as well as the ALRS in particular.

CONCLUSIONS

The evaluation of the physical and hydrologic characteristics of the Apache Leap Tuff against possible mechanisms for perching resulted in the elimination of some of the models, and the acceptance of others into a combination model that fits the conditions of Apache Leap. Matrix properties and fracture characteristics were determined to be the most significant factors contributing to the formation of the perched zone. The porosity in the matrix is generally associated with pumice fragments; the pumice fragments become more flattened and isolated with depth, such that both porosity and permeability are reduced to very low values in the perched water region. Fracture flow is the primary mechanism by which water is transported from the surface to the perched zone, and an increase in the amount of fractures that are filled, typically by silica mineralization, beneath the perched zone suggests a reduction in the secondary permeability at the perched zone as well. Other factors which could contribute to perched zone formation include a reduction in fracture density or disconnected fracture sets; data were not available to test those models. Perching models calling for weathering and chemical alteration by a former high stand of water, or for permeability changes between individual flow units, were found not to be supported by any of the data.

While the combination model that has been described fits the data collected from Apache Leap, it is not the only combination of mechanisms that could cause perched water to form in fractured, welded tuff. This portion of the research has shown that numerous factors must be considered, ranging from microscopic to megascopic in scale, in order to assess where such zones could form.

ACKNOWLEDGMENTS

The authors acknowledge research support from the U.S. Nuclear Regulatory Commission (contract no. NRC-04-95-041).

REFERENCES

1. Y. S. Wu, G. Chen, and G. S. Bodvarsson, Perched water analysis, in G. S. Bodvarsson and T. M. Bandurraga, eds., <u>Development and Calibration of the Three-Dimensional Site-Scale Unsaturated Zone Model of Yucca Mountain, Nevada</u>, ch.7, Lawrence Berkeley Laboratory, Berkeley, CA, 1996.

2. R. L. Bassett, S. P. Neuman, T. C. Rasmussen, A. Guzman, G. R. Davidson, and C. F. Lohrstorfer, <u>Validation Studies for Assessing Unsaturated Flow and Transport Through Fractured Rock</u>, NUREG/CR-6203, U. S. Nuclear Regulatory Commission, Washington, D.C., 1994.

3. D. W. Peterson, <u>Dacitic ash-flow sheet near Superior and Globe, Arizona</u>, Ph.D. thesis, Stanford University; U.S. Geological Survey Open-File Report 130, 1961.

4. E. L. Hardin, Ph.D. dissertation, University of Arizona, Tucson, 1996.

5. G. R. Davidson, R. L. Bassett, E. L. Hardin, and D. L. Thompson, Geochemical evidence of preferential flow of water through fractures in unsaturated tuff, Apache Leap, Arizona, <u>Applied Geochemistry</u>, accepted for publication.

6. Sample Management Facility, Geologic summary of borehole USW UZP-4 phase 1e prototype drilling, Apache Leap, Arizona, prepared for U.S. Department of Energy Yucca Mountain Project, 1990.

7. E. L. Hardin, R. L. Bassett and M. T. Murrell, $^{234}U/^{238}U$ fractionation in vadose zone pore waters of the Apache Leap Tuff, <u>Waste Management '96: Proceedings on Waste Management</u>, Tucson, AZ, 1996, CD-ROM.

8. E. G. Woodhouse, Ph.D. dissertation, University of Arizona, Tucson, in preparation.

9. A. C. Gringarten and H. J. Ramey, Jr., The use of source and Green's functions in solving unsteady-flow problems in reservoirs, Soc. Pet. Eng. J., 1973, pp. 285-296.

RECONCILIATION OF EXPERIMENTAL AND MODELLING CONCEPTS IN A NATURAL ANALOGUE OF RADIONUCLIDE MIGRATION

J. Suksi*, K. Rasilainen**
* University of Helsinki, Laboratory of Radiochemistry, Finland
**VTT Energy, Espoo, Finland

ABSTRACT

One of the major problems in testing radionuclide transport models with a natural analogue is the potential discrepancy between experimental and modelling concepts. In the matrix diffusion studies at Palmottu the measured U-series concentration profiles and mathematical simulations showed disagreement, suggesting discrepancy in respective concepts of radionuclide attachment on the rock pores. Here, the discrepancy was approached by studying uranium fixation in more detail within the most loosely-bound fraction.

INTRODUCTION

Model validation is usually based on comparing measured and simulated behaviour of a system. In our efforts to validate matrix diffusion model with the help of natural analogues, we have observed a considerable discrepancy between the modelled and observed behaviour of natural radioactive decay chain members. The measured concentrations appear much higher than predictions based on measured K_d values of the same samples [1]. Since the concentrations were obtained with a technique assumedly specific for mobile phases, there must be a gap between the respective experimental and modelling radionuclide fixation concepts.

The discrepancy stems from the fact that in simulations we use a simple well-defined fixation mechanism, but to obtain the experimental reference we use a technique whose specificity is operationally defined. The problem is genuine, because the rigour of any model validation depends directly on the correctness of the experimental reference. The question about the experimental reference is by no means limited to our case, but applies, of course, to all natural analogues used for model testing purposes (see discussions in [2]). Attempts to separate sorbed U phases have been reported in studies at Alligator Rivers [3] and Palmottu [4,5].

In the present work we report on a systematic study initiated to reconcile the experimental results and modelling concepts. Cases in which the problem appeared, and others in which it did not, will be briefly described. The harmonization strategy aims at improving the selectivity of the chemical extraction technique. To start the experimental fine tuning, we study the distribution of U in loosely-bound phases.

THE PROBLEM

In earlier studies at <u>Palmottu</u> we measured U-series profiles in the rock matrix away from a water-carrying fracture, in which the concentrations were three orders of magnitude higher than predicted by the K_d values based on batch experiments of the *same* rock samples. This was unexpected, because a batch experiment, with its sample crushing procedure, yields a maximum value for K_d: the specific surface area in the crushed sample can be much higher than in the intact rock matrix. Sensitivity studies showed that this inventory can not be mobile, because the strong concentration gradients observed would have levelled out. Studies of radioactive disequilibrium within the profile indicated a very old, possibly steady-state system [1].

Mat. Res. Soc. Symp. Proc. Vol. 465 ©1997 Materials Research Society

In a <u>boulder</u> sample found in glacial till at Hämeenlinna, also studied to validate the matrix diffusion model, no discrepancy between the measured and simulated results was observed [6]. This was surprising, because in this case a chemically more aggressive extraction method was used; on the other hand, batch experimental procedures, yielding the K_d value, were identical. The boulder sample has been studied in more detail by the U-series disequilibrium method and the U accumulation appears unambiquously recent [7].

Comparing these two cases, young systems appear clearly easier to interpret, because, for some reason, there is no discrepancy between experimental and simulated concentration profiles in the rock matrix. This indicates fundamental differences in the mode of radionuclide fixation between young and old systems.

COMPARISON OF EXPERIMENTAL AND MODELLING FIXATION CONCEPTS

We have utilized <u>selective extraction</u> to obtain the experimental reference for the simulations. The most loosely-bound radionuclides were assumed to represent the mobile portion of the total radionuclide inventory for the purpose of the diffusion model. This assumption seems now partly erroneous, because the extracted inventory appears to include also other phases, in addition to the adsorbed one.

The <u>K_d concept</u> used in migration modelling represents fast reversible adsorption on the mineral surfaces; furthermore, K_d is assumed to be independent of the concentration of the tracer. The last assumption is apparently valid for low concentrations. Ion exchange and physical adsorption are examples of sorption mechanisms that can be described with K_d; as such, K_d does not give any direct information about the sorption mechanisms, only of the final partition of radionuclides between solid and water phases. The partition is therefore the result of all adsorption mechanisms. Mechanisms like precipitation and mineralisation are clearly excluded from the K_d concept. Equilibrium is assumed to prevail in the batch experiments, normally conducted to measure K_d's. Considering the short-term (days) batch experiments, the instantaneous sorption equilibrium assumption in the model is well in line with the K_d value obtained from batch experiments.

<u>Reversibility of sorption</u> is one of the fundamental assumptions in the model. Below we study what happens to the attached radionuclide inventory if sorption is not assumed to be 100 % reversible. The partly irreversible sorption concept appears useful in illustrating the fundamental differences between K_d -based modelling and experimental extraction procedures.

We assume first that equilibrium sorption adequately describes the distribution of radionuclides between the aqueous and mineral phases. We assume further that a fraction λ' of the sorbed radionuclides is irreversibly removed into a geochemical sink of some kind per time unit. Because of sorption equilibrium, this removed inventory must be immediately replaced by an equal amount from the water phase. In the following we assume that water phase is an inexhaustible source of nuclides.

The following mass balances in Fig. 1 can be established:

$$\frac{dC_1}{dt} = -\lambda' C_1 + C_1^+, \qquad C_1(t=0) = C_1^0 \tag{1}$$

$$\frac{dC_2}{dt} = \lambda' C_1, \qquad C_2(t=0) = C_2^0 = 0 \tag{2}$$

where C_1 is the adsorbed concentration of the tracer (Bq/kg), λ' is the release rate of the tracer (1/a), C_1^+ is the inventory added into the adsorbed inventory from groundwater to preserve the sorption equilibrium (Bq/kg/a), and C_2 is the concentration of the tracer in the irreversible geochemical sink (Bq/kg). The second initial condition equation states that the geochemical sink starts from zero. Assuming steady-state for the adsorbed inventory C_1, in line with the sorption equilibrium, it can be directly seen that $C_1(t)=C_1^0$. Now the second equation can be solved easily:

$$C_2(t) = \lambda' \, C_1 t = \lambda' \, C_1^0 t \qquad (3)$$

The total solid-phase inventory of the tracer, summing adsorbed and irreversible sink inventories, is:

$$C_{tot} = C_1 + C_2 = C_1^0 + \lambda' \, C_1^0 t = C_1^0(1 + \lambda' \, t) \qquad (4)$$

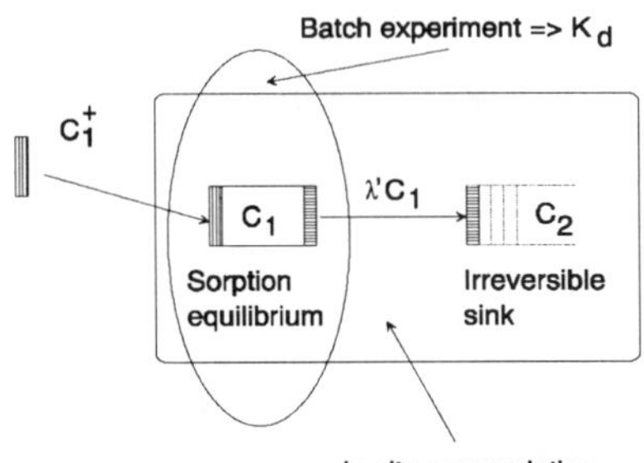

Fig. 1 Schematic concept of partly irreversible sorption. The quantity accumulated into the irreversible geochemical sink is replaced instantaneously by an equal amount from the aqueous phase.

Equation (4) has interesting consequences in long-term natural analogues. Assume e.g. that $\lambda'=1\bullet10^{-2} \Rightarrow$ after $1\bullet10^6$ a the original inventory has increased to 10^4 -fold. The release fraction used, actually, states that sorption is 99 % reversible on one year perspective. Assuming further that $\lambda'=1\bullet10^{-4}$ (sorption is 99.99 % reversible) would make the original sorbed inventory still 100-fold after $1\bullet10^6$ a. It is, therefore, obvious that very small fractions of the sorbed inventory, if accumulated gradually into a co-located irreversible geochemical sink, will eventually overgrow the adsorbed inventory. However, only the adsorbed inventory being in sorption equilibrium is continuously communicating with the respective inventory in the groundwater.

It appears that conceptually K_d represents C_1, whereas the experimental reference obtained by extraction is between C_1 and C_{tot}. It seems probable that short-term batch experiments have difficulties in detecting small mass losses due to other sources of error, and therefore, K_d values cannot include slow irreversible accumulation of material. It also appears that even a slightly over-aggressive chemical extraction can easily lump the adsorbed and sink fractions in to one "measured" inventory.

RECONCILIATION STRATEGY

In this work we try to harmonise the experimental and modelling concepts by redefining the chemical extraction concept to correspond to K_d, the modelling concept. Essentially, we match three partly different but interlinked elements: the long-term in situ accumulation of radionuclides on the rock material, the sequential extractions, and the batch experiments, Fig. 2. The following discussion is based on an extension of the partly irreversible sorption concept.

Long-term in situ accumulation of radionuclides is assumed to initially consist of a set of fast sorption mechanisms (M_1, M_2, and M_3 in Fig. 2). Each of these mechanisms may have an irreversible geochemical sink of some kind, see Fig. 1. The total in situ inventory of radionuclides in the rock material is, therefore, composed of a component in sorption equilibrium (i.e. the adsorbed component) and an irreversible component (i.e. the slow irreversible geochemical sinks).

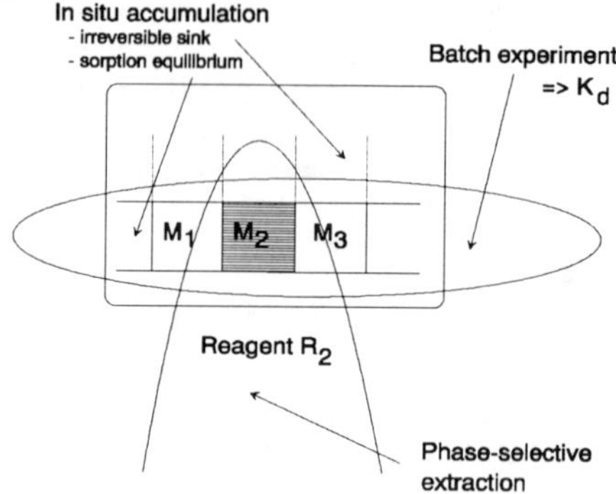

Fig. 2 Harmonisation of extraction and K_d concepts.

Short-term sequential extractions are based on using phase-selective reagents to obtain the desired phase; in Fig. 2 we assume that reagent R_2 is selective for the sorption mechanism M_2. The extraction, however, is not 100 % selective, and, therefore, parts of other adsorbed inventories, due to other fast sorption mechanisms, can be extracted; in addition, part(s) of the irreversible sinks, especially the one linked to M_2, may also be extracted. Obviously, the extraction time is an important factor linked to the extent of unwanted dissolution.

Batch experiments are short-term experiments to measure K_d values. Basically, they do not give any direct information about the sorption mechanisms involved, only of the final distribution of radioactivity between rock and water. This distribution is a function of all available fast sorption mechanisms in the system (i.e., M_1, M_2, and M_3 in Fig. 2). Because of its short duration, a batch experiment does not record any slow sinks, only the fast sorption.

Referring to Fig. 2, the ultimate aim of our fine-tuning of extraction techniques can be described as making the intersection of the three linked elements to cover the set of available fast sorption mechanisms in situ (i.e. M_1, M_2, and M_3 in Fig. 2). In extraction terms, this requires a reagent that is equally selective for all sorption mechanisms M_1, M_2, and M_3.

BATCH EXPERIMENTS AND CHEMICAL EXTRACTIONS

The radionuclide inventory fixed in the rock matrix in situ may contain a whole spectrum of phases from very loosely-held fractions to more permanently bound material, all readily dissolvable. In this study we are basically interested in the adsorbed phase within the mobile inventory, and, therefore, we applied batch experiments, traditionally used in sorption studies. Standard approach was used in which K_d's are determined from sorption and desorption stages, thus obtaining information of the reversibility of sorption. The idea is that the adsorbed natural U, if present, behaves in the batch experiment in same way as the tracer U added to the system. The phase-selective extraction was used to remove the adsorbed inventory from the rock. Tracer U is necessary, because with it we can be sure of having only fast U fixation.

The rock <u>samples</u> used in our matrix diffusion studies were also used in this study (Table 1). Two of the samples were divided into two parts: altered rock close to the fracture (a), and slightly altered matrix further away from the fracture (ua). Concentrations of loosely-bound U varied from some to hundreds of ppm. For samples R357/168 and R346/103 corresponding rock and water pairs were available. Both these waters exhibited oxidizing conditions (Eh ~ +140 mV) with the same pH 7.8. U concentrations in the groundwaters were around $3 \bullet 10^{-8}$ M.

Table 1. Sample information. Samples were taken around water-carrying fractures in the vicinity of U deposit. Loosely-bound U is defined as U extractable in 1M ammonium acetate (pH 4.8). The current work examines uranium distribution in more detail within this phase.

Granitic rock samples	Sample code used hereafter	Distance from fracture (mm)	Total U (ppm)	Loosely-bound U (% of tot)
R357/168 (168 m) very slightly altered	A	5 - 10	~ 50	~ 10
R346/103 (101 m) intensely altered (a)	B1	0 - 3	~ 400	~ 80
slightly altered (ua)	B2	17 - 20	~ 130	~ 50
R325/211 (50 m) intensely altered (a)	C1	0 - 3	~1500	~ 90
slightly altered (ua)	C2	4 - 8	~1000	~ 60

Rock samples were gently crushed to particle size approximating pore opening, in order to study radionuclide inventories on pore surfaces. Crushed rock (0.6-2.0 g) and dried ^{236}U tracer on a Teflon plate was placed in 20 ml glass vessels, and groundwater was added. The solid/solution ratio of the batches ranged from 1/40 to 1/10 g/ml. The system was then allowed to equilibrate for 30 days (sorption). After the equilibration, the solutions were removed, and additional 20 ml of groundwater was added for an other 30 days' equilibration period (desorption). After the equilibrations the amount of adsorbed ^{236}U was quantified. Eh and pH of each solution was measured and the concentrations of natural U and its isotopic composition were determined by α-spectrometry.

The purpose of the <u>extraction</u> here is to remove adsorbed ^{236}U and the similarly adsorbed natural U component. In our earlier matrix diffusion studies (see Table 1) the extracted U most probably included, in addition to physically adsorbed and ion-exchange phases, also other phases (possibly specific U compounds, dissolved during the longer extraction time, see [8,3]). To avoid unwanted dissolution here, the reagent aggressivity was considerably decreased by replacing the ammonium acetate used earlier by a less aggressive 0.5 M CsCl solution prepared in groundawaters. The use of high cationic concentration for desorbing U species is based on competition of sorption sites in the new equilibrium.

RESULTS AND DISCUSSION

<u>The equilibration of tracer U and desorption of natural U</u> were studied first. Experimental conditions and the release of natural U in equilibrations are presented in Table 2. A striking change, from oxidising to reducing conditions was observed during both the sorption and desorption experiments. Furthermore, during equilibrations a considerable U release was observed: concentrations were orders of magnitude higher than in the original groundwater. Both observations suggest that the inventories released were not in equilibrium with the groundwater used; this is true at least for sample R325/211, because corresponding groundwater could not be used. It also raises a question about the stability of U phases during the long

storage of drill cores. The amounts of ^{236}U adsorbed were small (9-22% of the added amount). The calculated K_d-values were, 0.002 - 0.009 m^3/kg in sorption, and 0.007 - 0.142 m^3/kg in desorption equilibration.

Table 2. Experimental conditions and results of the two batch experiments. Equilibration time in both experiments was 30 days. Initial conditions were those of the used groundwaters: pH = 7.8, Eh = +140 and U concentration ~3•10^{-8} mol/l. Counting errors in the activity ratios due to counting statistics are of the order of 10%.

Sample code	Sorption equilibration					Desorption equilibration				
	pH	Eh (mV)	U (mol/l)	$\frac{^{234}U}{^{238}U}$	$\frac{^{238}U}{^{236}U}$	pH	Eh (mV)	U (mol/l)	$\frac{^{234}U}{^{238}U}$	$\frac{^{238}U}{^{236}U}$
A	9.3	-100	5.1•10^{-7}	1.3	0.01	8.0	-193	1.2•10^{-7}	1.9	0.03
B1	8.2	-220	7.1•10^{-6}	0.73	0.13	8.1	-203	1.3•10^{-6}	.85	0.51
B2	8.2	-160	4.8•10^{-6}	1.1	0.10	8.7	-184	7.0•10^{-7}	1.4	0.27
C1	8.5	-130	1.8•10^{-5}	0.74	0.37	8.5	-206	5.5•10^{-6}	0.81	0.94
C2	8.1	-150	5.1•10^{-5}	0.91	0.96	7.9	-205	1.1•10^{-5}	.03	2.3

The different ^{234}U/^{238}U activity ratios (consistently smaller in the sorption equilibration) suggest the release of different natural U inventories. The lower relative amount of ^{236}U in the desorption equilibration appears to support this view.

Incidentally, the U concentrations in the water phase after the equilibrations (sorption equilibrium) can be utilised in assessing the U concentrations in pore water. This information is valuable, because pore water analyses are difficult to do, due to the low porosity of Finnish rocks; for this reason, matrix diffusion simulations usually use 0-concentration as the initial pore-water composition for the pore water.

Extraction of adsorbed ^{236}U. Only part of the adsorbed ^{236}U tracer could be removed (Fig. 3). Extraction yield increases quickly to a steady state value, indicating depletion of the fast sinks. The yield varies among rock samples, suggesting variably effective fast sinks, see Fig. 2. The CsCl reagent clearly appears too weak to remove all fixed U species. However, compared to the products of ammonium acetate extraction, the leachates obtained here are significantly closer to the adsorbed component. Studies with modified reagent composition are under way.

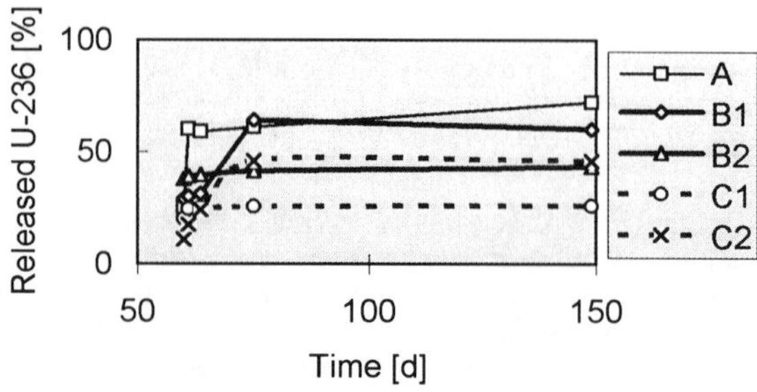

Fig.3. Cumulative release of adsorbed ^{236}U tracer from crushed rock during the CsCl extraction. Desorption equilibration (leftmost points in the graph) represents "extraction" with water.

Extraction of "sorbed" natural U. Tracer ^{236}U was actually used to examine the sorbed natural U inventory. Theoretically, both inventories should behave in the same manner, if their fixing mechanisms are identical, and, consequently, their release ratio during the extraction should be constant. The actual release ratio during extraction can be seen in Fig. 4.

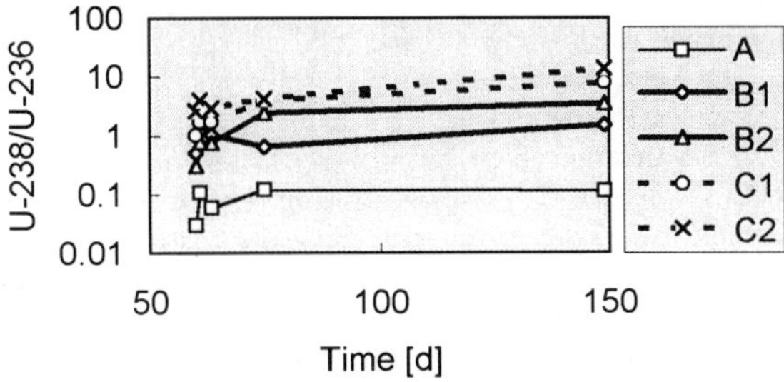

Fig.4. ^{238}U/^{236}U activity ratios in the leachate during the CsCl extraction. Desorption equilibration (leftmost points in the graph) represents "extraction" with water.

At the beginning of the extraction no marked change in the activity ratios was seen, indicating identical release mechanisms for both U inventories. With time, however, the relative release of natural U increases, suggesting other release mechanism for natural U, or in other words, different natural U sinks. Of course, the original concentration of natural U was also higher. The oscillating ^{234}U/^{238}U activity ratios in equilibrations and extractions (Fig. 5) suggest heterogeneity in the natural U inventories, and, the preferential release of ^{234}U can be seen clearly. The combined results from the tracer and natural U extractions indicate that we have at least touched the respective adsorbed components. The considerable variations in uranium isotopic composition, from equilibration experiments to CsCl extraction, however, indicate that the inventories are extremely sensitive to experimental conditions.

The experimental observations above certainly warrant more detailed discussion, e.g. concerning U fixation mechanisms, but in this intermediate paper we limit ourselves to the main trends most relevant to concept harmonisation.

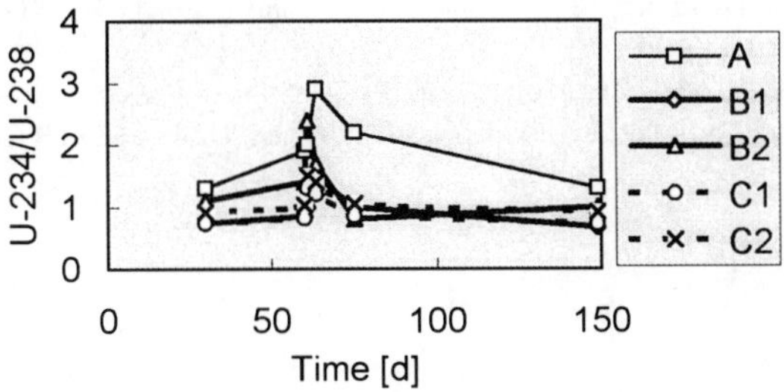

Fig. 5. ^{234}U/^{238}U activity ratios in the leachate during the CsCl extraction (starts from the third point). Activity ratios at the end of sorption and desorption equilibrations are included (first two points, respectively).

CONCLUSIONS

Our earlier attempts to validate a matrix diffusion model with natural analogues indicated conceptual discrepancy between the modelling result and the experimental reference. In this study we try to tune the technique, used to obtain the experimental reference, so that it corresponds the modelling concept. The tuning was based on the idea of dividing the in situ radionuclide inventory into a reversible (i.e., adsorbed) and an irreversibly bound component.

We did batch experiments with tracer U, and mild salt extractions to study the adsorbed natural U inventory. The idea was that tracer and natural U should behave similarly if the respective fixation mechanisms on rock phase were similar. The results obtained indicate that we have succeeded at least to touch this phase, but in order to remove all adsorbed U species more specific reagent than CsCl is needed. The reagent development work is under way.

Considering the technical difficulty to quantify only the adsorbed inventory from the rock samples for all decay chain members, it is noteworthy that the discrepancy problem may be circumvented by studying geologically young systems. Recent accumulations make it easier to obtain an experimental reference in line with the modelling concept, simply because of the smaller number of possible U sinks. Furthermore, recent accumulations are easier to interpret, because boundary conditions may be quantified more reliably, see [6].

ACKNOWLEDGEMENTS

This study is part of the Publicly Administrated Nuclear Waste Management Research Programme in Finland.

REFERENCES

1. Rasilainen, K., Suksi, J., Hakanen, M., and Olin, M.: Mat. Res. Soc. Symp. Proc. Vol. **353** (1995) 1169-1177.
2. McKinley, I.G., Alexander, W.R..: J. Contaminant Hydrology **13** (1993) 249-259.
3. Yanase, N., Nightingale, T., Payne, T., and Duerden, P.: Radiochimica Acta 52/53 (1991) 387-393.
4. Suksi, J., Ruskeeniemi, T., Lindberg, A., and Jaakkola, T.: Radiochimica Acta 52/53 (1991) 367-372.
5. Suksi, J., Ruskeeniemi, T., and Saarinen, L.: J. Contaminant Hydrology **21** (1996) 47-58.
6. Rasilainen, K., Suksi, Hellmuth, K.-H., Lindberg, A., and Kulmala, S.: Mat. Res. Soc. Symp. Proc. Vol. 412 (1996) 855-862.
7. Rasilainen, K., Suksi, J.: Mat. Res. Soc. Symp. Proc., this volume.
8. Bolle, J.N., Martin, H., Sondag, F., and Cardoso Fonseca, E.: Uranium 4 (1988) 327-340.

ON THE ACCURACY OF URANIUM-SERIES DATING: A COMPARISON WITH A KNOWN MATRIX DIFFUSION CASE

K. Rasilainen[1], J. Suksi[2]

[1] VTT Energy, P.O. Box 1604, FIN-02044 VTT, Finland
[2] University of Helsinki, P.O. Box 55, FIN-00014 HELSINKI UNIVERSITY, Finland

ABSTRACT

The accuracy of uranium-series dating was studied quantitatively using the natural long-lived decay chains 4n+2 and 4n+3. An unusually well-bounded matrix diffusion case was used as the reference for USD simulations. The simulations applied the traditional closed and open system models, as well as detailed multistage models that incorporated the known accumulation history of the decay chain members. The results show clearly the improved accuracy and consistency in dating with detailed mass flow information.

INTRODUCTION

Uranium-series disequilibria (USD) have traditionally been used for dating purposes, see [1]. The popularity of USD is due to, e.g., the ubiquity of uranium in nature, the technical observability of radionuclides, and the wide time span that can be covered with the different daughter/parent activity ratios. Qualitative interpretation of the activity ratios is also straightforward: radioactive disequilibria always indicate mass-exchange during the time period it takes for a specific daughter/parent pair to reach radioactive equilibrium. Quantitative interpretation requires, however, knowledge of mass flows of the radionuclides into or out of the system, and these data are hard to obtain. Therefore, most USD dating studies have been limited to closed systems with fewer input data needs than open or periodically open systems.

Earlier we studied a rock sample that contains unusual pulse-like concentration distributions of radionuclides belonging to the natural decay chains 4n+2 and 4n+3 [2]. Preliminary USD dating, applying an open system model, unambiguously indicated a young system. Similar time information could be independently obtained from the known glacial history of the site, and from matrix diffusion simulations using site-specific input data (see Table I). The rock sample, a boulder, was studied as a natural analogue of matrix diffusion, and therefore offers an unusually well-bounded basis for detailed USD dating studies.

The current study investigates the extent to which detailed mass diffusion information improves the accuracy of USD dating. Due to the well-defined earlier matrix diffusion study, we can fully concentrate on the of USD dating features. Thus, we can also study how accurately the USD method of net mass flows approximates the known diffusion process.

REFERENCE FOR USD SIMULATIONS

The accuracy of USD dating is difficult to study in complex mass flow cases, because natural systems are usually poorly constrained. For this reason, we use as reference an earlier study of postglacial matrix diffusion of two decay chains: 4n+2 (U-238 - U-234 - Th-230 - Ra-226) and 4n+3 (U-235 - Pa-231), see [2]. The modelled diffusion history consisted of fast in-diffusion (a short-term discharge of uranium-rich waters as a withdrawing ice margin passed the site), fast

Mat. Res. Soc. Symp. Proc. Vol. 465 © 1997 Materials Research Society

partial out-diffusion (short stage below Yoldia sea level), and a long period of isolation with only chain decay (the Yoldia sea level decreased, exposing the site and the boulder), see Fig. 1.

Table I. Available time information of the rock sample [2].

Method	Time (years)
Uranium-series disequilibria	
Open system (max. age)	30 000 y
Glacial history	
Ice margin at the site	10 000 y BP
Sample under Yoldia sea	50 - 100 y
Matrix diffusion simulation	
In-diffusion of nuclides	100 y
Out-diffusion of nuclides	50 y
Isolated stage	10 000 y

In this exercise, we are essentially studying whether the earlier matrix diffusion case can also be explained by the USD model. As a reference we take the earlier modelled matrix diffusion case: we know the mass flow history within the sample exactly, and can, therefore, investigate the accuracy of the USD simulations free of experimental uncertainties. In other words, the reference case is an idealised self-contained system that we have perfect control over; furthermore, this idealised system appears to be very close to the actual world, see [2].

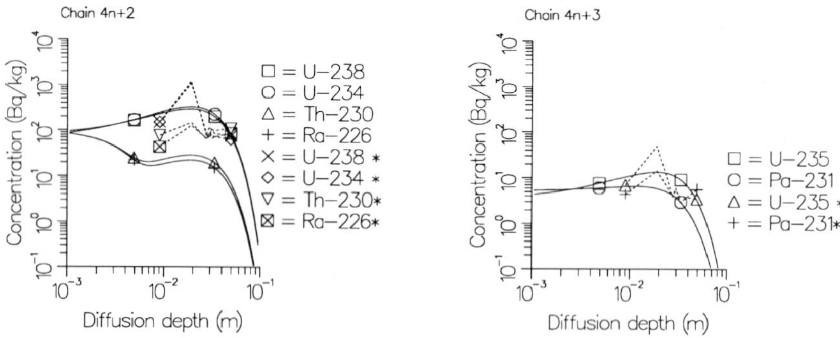

Fig. 1 The matrix diffusion case. The diffusion history behind the simulated final concentration profiles is used to derive successive mass flow episodes for USD dating. The measured concentration distributions, denoted with asterisks (*), are shown for comparison.

USD simulations concentrate on daughter/parent activity ratios, and, thus, the essential input parameter in simulations is the ratio of the respective accumulation rates. Therefore, we apply reference inventories (Bqm/kg) and the corresponding reference mass flows below; the actual inventory (Bq) can be obtained by simply converting the sample unit length (m) into the corresponding mass (kg). The modelled diffusion history is approximated by single-event closed and open system models, and by more detailed multistage system models. In order to incorporate the known in- and out-diffusion steps in the multistage system models, the respective accumulation/release rates must be quantified. In this case we derive the mass flow rates from the modelled matrix diffusion case. The overall mass flow of radionuclides into the system is assumed to have occurred in three stages, in line with the diffusion history. The accumulation rate in the in-diffusion stage is calculated from:

$$C_i^{in} = \int_0^\infty C_i\,(z,t_{sim})dz \Rightarrow R_i^+(in) = C_i^{in}\,/\,t_{in}\,, \qquad t_{sim} = t_{in} \tag{1}$$

where C_i^{in} is the reference inventory of nuclide i (Bqm/kg) after an in-diffusion stage lasting t_{in} (a), $C_i\,(z,t_{sim})$ is the modelled concentration distribution of nuclide i (Bq/kg) after the in-diffusion stage, z is the distance from fracture surface into the rock matrix (m), t_{sim} is the simulation time (a), and $R_i^+(in)$ is the constant reference accumulation rate of nuclide i during the in-diffusion stage (Bqm/kg/a). The release rate in the out-diffusion stage is calculated from:

$$C_i^{out} = \int_0^\infty C_i\,(z,t_{sim})dz \Rightarrow R_i^+(out) = \left(C_i^{in} - C_i^{out}\right)/\,t_{out}\,, \quad t_{sim} = t_{in} + t_{out} \tag{2}$$

where C_i^{out} is the reference inventory of nuclide i (Bqm/kg) after the out-diffusion stage lasting t_{out} (a), and $R_i^+(out)$ is the constant reference release rate of nuclide i (Bqm/kg/a) during the out-diffusion stage. The third stage in the multistage system model was a simple closed system period lasting t_{iso} (a).

The successive accumulation/release rates in the USD simulations were derived from the earlier modelled matrix diffusion case. The derivation was based on the assumption that in this short period of time, the system is dominated by diffusion. In order to make sure that this actually is the case, we also simulated the diffusion of all radionuclides as stable isotopes, with otherwise identical input data as for the original matrix diffusion case; the simulations were done using the code FTRANS. With the help of this purely diffusive case we could confirm that for the long-lived uranium isotopes there was no difference whatsoever between the original and purely diffusive case, as expected. The largest difference in accumulation rates was for Th-230 (4 %), due to its low mobility, which makes even small chain decay contributions notable. The purely diffusive simulation proved that the potential error in using the original matrix diffusion case to derive mass flow rates is acceptably small.

The reference activity ratios for the USD simulations were calculated as ratios of the respective reference inventories at the end of the matrix diffusion simulations (see Table II). By so doing we can specifically study the accuracy of the USD simulations in this system. For comparison, Table II also shows the experimental results of the previous study [2]. The experimental results appear to be very close to the modelled data except for Pa-231/U-235. This exception is not unexpected, however, because the analytical technique to measure Pa-231 is quite demanding. In addition, U-235 concentrations were not measured, but were based on the standard assumption of constant mass ratio with U-238. All in all, the two reference cases are actually surprisingly close to each other, considering that we had measured K_d values only for uranium in the matrix diffusion case [2].

Table II. Reference case for the USD simulations. The modelled activity ratios are ratios of the respective reference inventories at the end of the simulation time ($t = t_{sim}$). The measured activity ratios are shown for comparison only.

Reference case	U-234/U-238	Th-230/U-234	Ra-226/Th-230	Pa-231/U-235
Modelled	1.11	0.11	0.84	0.45
Measured	1.14 ± 0.14	0.12 ± 0.02	0.86 ± 0.11	0.29 ± 0.07

USD SIMULATIONS

Models

USD dating simplifies the time- and space-dependent diffusion process into a time-dependent decay chain evolution, and the diffusive fluxes of radionuclides are converted into averaged net mass flows. In other words, the evolution is simulated as a function of time at one point in space. In the following, we simulate radionuclide inventories within the rock matrix as whole (i.e. using the integrated concentration distributions from above); in taking the integral, the explicit spatial information is inevitably lost.

The models below are written for an arbitrary nuclide i, and each represents a group of coupled ordinary differential equations (one per nuclide) that must be solved simultaneously. The <u>closed</u> system assumes instantaneous accumulation of all uranium at t = 0:

$$\frac{dC_i}{dt} = -\lambda_i C_i + \lambda_i C_{i-1}, \qquad 0 < t \le t_{sim} \tag{3}$$

where C_i is the concentration of nuclide i (Bq/kg), C_{i-1} is the concentration of the parent nuclide i-1 (Bq/kg), and λ_i is the decay constant of nuclide i (1/a). The <u>open</u> system assumes continuous accumulation of uranium beginning at t = 0:

$$\frac{dC_i}{dt} = -\lambda_i C_i + \lambda_i C_{i-1} + R_i^+, \qquad 0 < t \le t_{sim} \tag{4}$$

where R_i^+ is the constant accumulation rate of nuclide i (Bq/kg/a), $R_i^+ > 0$. The <u>multistage</u> system assumes successive periods of continuous accumulation, release, and chain decay:

$$\frac{dC_i}{dt} = -\lambda_i C_i + \lambda_i C_{i-1} + R_i^+, \qquad 0 < t \le t_{in}$$

$$\frac{dC_i}{dt} = -\lambda_i C_i + \lambda_i C_{i-1} - R_i^+, \qquad t_{in} < t \le t_{in} + t_{out} \tag{5}$$

$$\frac{dC_i}{dt} = -\lambda_i C_i + \lambda_i C_{i-1}, \qquad t_{in} + t_{out} < t \le t_{in} + t_{out} + t_{iso}$$

The initial condition for all models above can be expressed as:

$$C_i(t = 0) = C_i^0 \tag{6}$$

In practice, the USD simulations were done using the code URSE, which can combine successive closed and open system mass flow episodes into a continuous scenario [3]. Technically this is done by using the output of one episode as the initial condition in the next one.

Scenarios

We simulated the evolution of the decay chains 4n+2 and 4n+3 assuming four different mass diffusion scenarios. The traditional <u>closed</u> and <u>open</u> system models were complemented by two <u>multistage</u> system models in an effort to utilise the modelled mass flow history of the matrix diffusion case. The first multistage system included successive accumulation/release episodes only for uranium, and the second for all radionuclides. The last scenario, therefore, represents the most detailed description of the mass flows in this study.

The scenarios were made comparable by deriving the respective input data directly from the reference case, in particular the U-234/U-238 activity ratio in the accumulating uranium was set equal for all scenarios [R_2^+(in)/R_1^+(in), see Eq. (1)]. The closed and open system cases can be simulated without detailed knowledge of the mass flow history. Our previous studies with closed and open systems that were comparable in this sense, indicated that the two systems yield the minimum and maximum system ages, respectively [4]. In this work these single-event systems were used to obtain a first order age bracket for the system in an effort to mimic the (usual) situation in which the mass flow history is not known in detail. The multistage cases represent stepwise improvements in the data base, due to increasingly detailed description of the mass flow history of the radionuclides.

RESULTS AND DISCUSSION

The results of the USD simulations are shown in Figs. 2 - 4. The scenario-specific results comprise U-234/U-238, Th-230/U-234, Ra-226/Th-230, and Pa-231/U-235 activity ratios as a function of time. In the figures, the reference value for each activity ratio is projected on the time axis via the respective simulated curve. The projections are denoted with dashed lines; the exact scenario-specific system ages indicated by different activity ratios are shown in Table III.

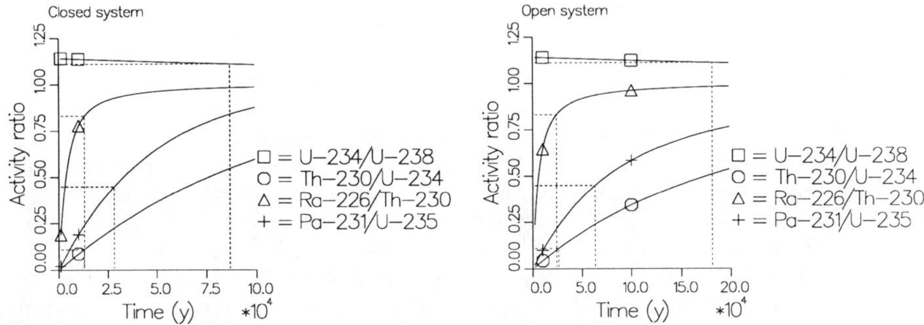

Fig. 2 USD simulation results for the closed and open system models. Reference activity ratios (dashed lines) are projected on the time axis via the respective simulated curves.

From the closed system results in Fig. 2 it can be seen that the two decay chains yield different system ages, and that within the 4n+2 chain different activity ratios yield different system ages. The U-234/U-238 activity ratio curve is very flat, and the system age indicated by the U-234/U-238 ratio is clearly higher than those given by the other activity ratios. The Ra-226/Th-230 system age is, actually, outside its valid time range, see e.g. [1, 3]; the flattening slope of the Ra-226/Th-230 curve at later times in Fig. 2 illustrates the time range point. Against this background, the good coherence between the Ra-226/Th-230 and the Th-230/U-234 system ages is quite interesting. The U-234/U-238 and Ra-226/Th-230 activity ratio curves are included in this study, however, because we specifically aim to compare the system ages from different activity ratios.

The open system results are similar to the closed system results, except for a systematic increase in all system ages by a factor of two. The reason to this increase is obvious, because an open system, with continuous accumulation of nuclides, can be conceptually described by successively accumulating closed systems, and, naturally, the later accumulated isolated system also reaches radioactive equilibrium later. The successive addition of nuclides increases the overall time it takes for the whole (open) system to reach radioactive equilibrium, and thereby also the projected system ages.

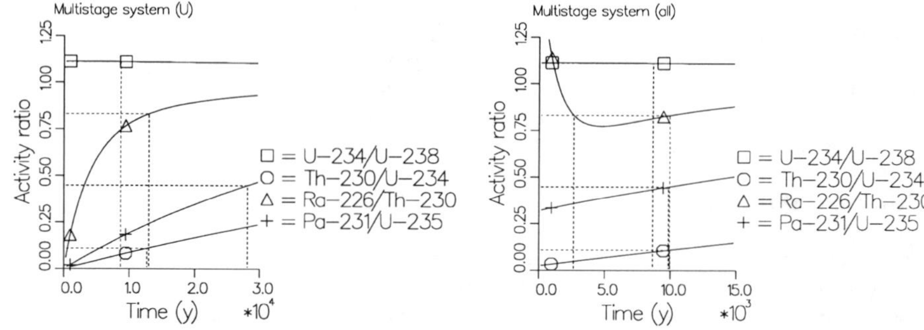

Fig. 3 USD simulation results for the multistage systems. The curves on the left represent uranium accumulation in three stages. The curves on the right represent accumulation of all radionuclides in three stages. The reference case activity ratios (dashed lines) are projected on the time axis via the respective simulated curves.

The multistage system results yield clearly younger system ages, and also a more consistent set of system ages from different activity ratios. The results based on multistage accumulation of uranium only, still yield an inconsistent picture between the two decay chains. In contrast, in the case of multistage accumulation of all radionuclides very good coherence is obtained. A new problem appears, however, in that the Ra-226/Th-230 system age becomes ambiguous. This open system feature is due to the relative mobilities and radioactive half lives of Ra-226 and Th-230. In the in-diffusion period, the more mobile Ra-226 accumulates at a higher rate than Th-230, and, therefore, the Ra-226/Th-230 activity ratio at early times is determined by the respective mass flows. The accumulated "unsupported" Ra-226 will, however, decay according to its radioactive half-life, and this causes the Ra-226/Th-230 activity ratio to decrease and ultimately reach a level that is supported by the Th-230 inventory, and in between there is a local minimum for the activity

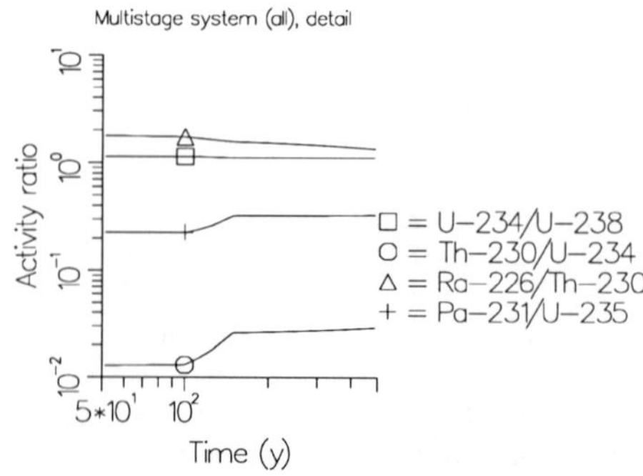

Fig. 4 USD simulation results for the multistage system model. A detailed view of the successive changes from continuous accumulation (t < 100 y) through continuous release (100 $y \leq t \leq 150$ y) to a closed system (t > 150 y).

ratio. The results also show that only when the mass flow of Pa-231 is described accurately, is the Pa-231/U-235 system age in agreement with the other system ages.

The successive episodes of radionuclide movement into and out of the system took place in so short a time period, compared with the duration of the period of isolation, that they are not apparent in the linear scale curves of Figs. 2 and 3. Therefore, Fig. 4 shows a detailed picture of the most detailed mass flow description of Fig. 3 in logarithmic scale. It can be seen that, due to the low mobility of Th-230, the Th-230/U-234 activity ratio is highly sensitive to the changes in the accumulation rate of U-234. In comparison, the Pa-231/U-235 activity ratio is not as sensitive to changes in the accumulation rate of the parent U-235, because the daughter Pa-231 is also mobile. The U-234/U-238 activity ratio is, of course, constant during the time period shown in the figure. The gradual decrease of the Ra-226/Th-230 activity ratio can also be seen.

An overall comparison of the system ages obtained for the different scenarios is shown in Table III. It is noteworthy that while the closed and open system models, which can be used without detailed knowledge of mass flows, showed scattering around the real system age, the most detailed multistage system modelling approximated the real age very accurately. This increased our confidence in the validity and accuracy of the USD technique, provided the detailed mass flow information can be obtained.

Table III. Simulated system ages (y) obtained for the different scenarios, from different activity ratios. The real system age (i.e. simulation time for the matrix diffusion reference case) was 10 150 y [2].

Scenario	U-234/U-238	Th-230/U-234	Ra-226/Th-230	Pa-231/U-235
Closed system	86 600	12 700	13 000	28 100
Open system	180 900	26 000	24 000	63 200
Multistage system (U)	8 700	12 700	13 000	28 100
Multistage system (all)	8 700	10 000	2 600, 10 000	9 900

It can be seen in Table III that the USD system ages usually are higher than the real system age. Only with the most detailed mass flow description do the system ages converge on the real age. The fact that the projected system ages are not exactly the same as the real age is not surprising, due to the coarse time intervals used in the USD simulations (standard discretisation), and also due to the simple linear interpolation technique used to obtain the projected system ages. The good convergence is, however, disturbed by the ambiguity of the Ra-226/Th-230 system age. Therefore, the Ra-226/Th-230 activity ratio does not appear reliable for dating purposes in cases where Ra-226 has been mobile. The U-234/U-238 system age, although in line with other system ages in this study, is based on a very flat simulated curve, which in practice makes it too sensitive for dating recent, e.g. postglacial, systems like this one. The Th-230/U-234 and Pa-231/U-235 activity ratios appear to be the most useful for dating application; because they represent two independent decay chains, their mutual agreement is quite valuable. In this study the errors in the system ages (relative differences from the real age) based on the Th-230/U-234 and Pa-231/U-235 activity ratios were -1 % and -2 %, respectively. These errors, actually, define the accuracy of the USD *method* (i.e. no errors assumed in the input) in this diffusive system.

The conclusion made earlier about the possibility of obtaining limiting values for the system age by simulating closed and open system models, see [4], appears reasonable for the case in which there is only accumulation of uranium in the system. The essential thing is, of course, that the sooner the final uranium mass is accumulated in the system, the sooner the decay chain system reaches radioactive equilibrium, and, therefore, the shorter the system age obtained. In this sense, closed and open system models are extremes between which periodically opening systems must fall. But, as seen in this study, when other elements beside uranium have been mobile, the closed-open age bracket gives too high a system age.

CONCLUSIONS

The study showed a clear improvement in USD dating accuracy and consistency as mass flow history becomes available to use in the analysis. This knowledge may be relevant in cases in which one must decide whether it is worth the effort to try to obtain the actual mass flow history of the system under study. The techniques used to bracket the system age by closed and open system simulations proved feasible to the cases in which they were applied, but when there are mobile elements other than uranium in the system, this age bracket yields erroneously high system ages.

The Th-230/U-234 and Pa-231/U-235 activity ratios appeared to be the most useful for dating purposes; the Ra-226/Th-230 activity ratio proved ambiguous in cases with mobile Ra-226, and simulated U-234/U-238 vs. time curves appeared too flat for practical dating of recent systems, because small changes in the activity ratio would mean huge changes in the respective system age.

The errors in the system ages derived from Th-230/U-234 and Pa-231/U-235 activity ratios were 1 % and 2 %, respectively. This accuracy, however, required a detailed knowledge of the mobility of all nuclides in the system, and, therefore, it can be considered as the accuracy of the USD method for this diffusive system. In general, the accuracy of the USD method depends, of course, on how accurately the real time- and space-dependent radionuclide migration events in the system can be reconstructed with the net mass flow approach of the USD method.

The simple integration technique used here to convert diffusive mass fluxes into net mass flows appeared to work well for all radionuclides, and for the long-lived uranium isotopes the technique was accurate. The main reason for this is the short accumulation period, during which there was not much time for change in inventory from chain decay.

ACKNOWLEDGMENTS

This study is part of the Publicly Administrated Nuclear Waste Management Research Programme in Finland.

REFERENCES

[1]   Ivanovich, M. In: Ivanovich, M. & Harmon, R.S (eds.) Uranium series disequilibrium: Applications to environmental problems, Oxford, 1982, Clarendon Press, pp. 56-78.

[2]   Rasilainen, K., Suksi, J., Hellmuth, K.-H., Lindberg, A. & Kulmala, S. In: Murphy, W.M. & Knecht, D.A. (eds.) Scientific Basis for Nuclear Waste Management XIX. (Mat. Res. Soc. Symp. Proc. Vol. 412, Pittsburgh, PA, 1996), pp. 855-862.

[3]   Rasilainen, K. & Suksi, J. On the quantitative interpretation of uranium series disequilibria (in Finnish). VTT Research Notes 1404, 1992, 63 p. + app. 14 p.

[4]   Finch, R.J., Suksi, J., Rasilainen, K. & Ewing, R.C. In: Murphy, W.M. & Knecht, D.A. (eds.) Scientific Basis for Nuclear Waste Management XIX. (Mat. Res. Soc. Symp. Proc. Vol. 412, Pittsburgh, PA, 1996), pp. 823-830.

THERMODYNAMIC STABILITIES OF U(VI) MINERALS: ESTIMATED AND OBSERVED RELATIONSHIPS

ROBERT J. FINCH

University of Manitoba, Winnipeg, Manitoba, R3T 2N2, Canada
Present address: Argonne National Laboratory, 9700 South Cass Ave., Argonne, IL 60439

ABSTRACT

Gibbs free energies of formation (ΔG°_f) for several structurally related U(VI) minerals are estimated by summing the Gibbs energy contributions from component oxides. The estimated ΔG°_f values are used to construct activity-activity (stability) diagrams, and the predicted stability fields are compared with observed mineral occurrences and reaction pathways. With some exceptions, natural occurrences agree well with the mineral stability fields estimated for the systems SiO_2-CaO-UO_3-H_2O and CO_2-CaO-UO_3-H_2O, providing confidence in the estimated thermodynamic values. Activity-activity diagrams are sensitive to small differences in ΔG°_f values, and mineral compositions must be known accurately, including structurally bound H_2O. The estimated ΔG°_f values are not considered reliable for a few minerals for two major reasons: (1) the structures of the minerals in question are not closely similar to those used to estimate the ΔG_f^* values of the component oxides, and/or (2) the minerals in question are exceptionally fine grained, leading to large surface energies that increase the effective mineral solubilities.

INTRODUCTION

The thermodynamic stabilities of uranium(VI) minerals are of interest for understanding the role of these minerals in controlling uranium concentrations in oxidizing ground waters associated with uranium ore bodies, uranium mining and mill tailings and geological repositories for nuclear waste. Unfortunately, few thermodynamic data are available for these minerals, and the large number of U(VI) minerals makes the experimental determination of thermochemical data daunting. However, if the compositions of these minerals are accurately known, the Gibbs free energies of formation may be calculated using the method of Tardy and Garrels [1], in which the estimated ΔG°_f value for each mineral is the arithmetic sum of the constituent oxide contributions.

Using this method provides a straightforward way to estimate ΔG°_f values; however the confidence in the estimates is uncertain without an independent check on the estimated values. We can use the estimated ΔG°_f values to construct activity-activity diagrams (also called "stability" diagrams) and compare the estimated stability fields with reaction pathways and mineral relationships observed in Nature. Thus natural occurrences are used to support (or refute) the estimated ΔG°_f values.

Two systems are analyzed in detail: SiO_2-UO_3-CaO-H_2O and CO_2-UO_3-CaO-H_2O. The SiO_2- and CO_2-bearing system are important in Nature, and the uranyl silicates and uranyl carbonates have been identified as potentially important U(VI) alteration phases of U-bearing waste forms [2,3]. The mobility of U(VI) in carbonate ground waters is potentially high, as uranyl-carbonate complexes are generally quite stable in solution [4]. The uranyl carbonate

Mat. Res. Soc. Symp. Proc. Vol. 465 © 1997 Materials Research Society

minerals tend to be soluble, and the geochemical conditions under which these minerals precipitate is of interest for understanding the geochemical controls on U transport [5]. Dissolved silica is an important constituent of natural ground waters, where it occurs predominantly as orthosilicic acid, H_4SiO_4, below approximately pH = 9 [6]. The concentration of H_4SiO_4 in most ground waters is limited by the precipitation of amorphous silica, which occurs where $\{H_4SiO_4\}$ = $10^{-2.7}$ mol·liter^{-1} [6]. The Ca-containing systems are considered because Ca is virtually ubiquitous in near-surface ground waters, and many Ca-bearing uranyl minerals are known, representing a variety of near-surface environments. Fortunately, thermodynamic data are available for several of these minerals. The uranyl phosphates are also important in natural waters [7,8,9]; however, no uranyl phosphates have been identified from experimental studies of U-bearing waste forms.

CALCULATED FREE ENERGY VALUES

Tardy and Garrels [1] describe a method for estimating the Gibbs free energies of formation for oxides, hydroxides and layer silicates, including clay minerals [1,10,11]. The method is straightforward: Given the ΔG_f° values for stoichiometrically simple oxides and hydroxides, one may estimate the contribution of the constituent oxides to the total ΔG_f° value of each mineral; that is, the ΔG_f^* of each oxide *in the mineral structure*. The ΔG_f° value of the mineral is the arithmetic sum of the oxide contributions:

$$\Delta G_f^\circ = \Sigma \Delta G_f^*{}_i .$$

In this study, the oxide components in the structures of the uranyl minerals are indicated with an asterisk: UO_3^*, CaO^*, SiO_2^*, CO_2^* and H_2O^*.

The ΔG_f^* values of UO_3^* and H_2O^* are calculated using the data for the UO_3 oxy-hydroxides, including (meta)schoepite and dehydrated schoepite [12] {The phase synthesized by O'Hare et al. [12] was probably metaschoepite, based on the temperature of synthesis (~50 °C) as well as the composition (see [13]). For this reason, the stoichiometry for metaschoepite, $UO_3 \cdot 2H_2O$ is used in the calculations in this paper.} The experimentally determined ΔG_f° values for several UO_3 oxy-hydroxides [12], when plotted as a function of total water content, define a straight line. A least-squares fit to these data is used to estimate the ΔG_f^* contribution by H_2O^* and UO_3^* in the UO_3 oxy-hydroxides. The slope of the least-squares-fit line is equal to ΔG_f^* for structurally bound water (i.e., H_2O^*, without regard to whether it occurs in the structure as OH$^-$ or H_2O), and the ΔG_f^* value for anhydrous UO_3^* is given by the intercept at $H_2O = 0$. The derived ΔG_f^* values for H_2O^* and UO_3^* are given in Table I. The value for SiO_2^* is calculated by subtracting the ΔG_f^* values for H_2O^* and UO_3^* from ΔG_f° for soddyite [14].

Table I. Calculated ΔG_f^* values for component oxides (kJ·mol^{-1})

UO_3^*	-1153.5
H_2O^*	-242.6
CaO^*	-716.4
SiO_2^*	-865.9
CO_2^*	-404.6

$$\frac{2UO_3^* + SiO_2^* + 2H_2O^* - 2UO_3^* - 2H_2O^*}{SiO_2^*} \qquad \frac{\Delta G_f^\circ{}_{soddyite} - 2(\Delta G_f UO_3^*) - 2(\Delta G_f H_2O^*)}{\Delta G_f SiO_2^*}$$

The value of ΔG_f^* for CaO* is then calculated from uranophane, $Ca(UO_2)_2(SiO_3OH)_2(H_2O)_5$ [14].

$$\frac{\begin{array}{ll} CaO^* + 2UO_3^* + 2SiO_2^* + 6H_2O^* & \Delta G_f^\circ{}_{uranophane} \\ -2UO_3^* - 2SiO_2^* - 6H_2O^* & -2(\Delta G_f UO_3^*) - 2(\Delta G_f SiO_2^*) - 6(\Delta G_f H_2O^*) \end{array}}{\begin{array}{ll} CaO^* & \quad\quad\quad\quad \Delta G_f CaO^* \end{array}}$$

Two uranyl carbonates (Table II) are used to determine $\Delta G_f CO_3^*$, using the experimental solubilities of rutherfordine [15] and liebigite [5], respectively. Using the previously determined ΔG_f^* values for H_2O^* and UO_3^* (Table I) gives $\Delta G_f CO_3^*$:

$$\frac{\begin{array}{ll} UO_3^* + CO_2^* & \Delta G_f^\circ{}_{rutherfordine} \\ -UO_3^* & -\Delta G_f UO_3^* \end{array}}{\begin{array}{ll} CO_2^* & \quad\quad\quad \Delta G_f CO_2^* \end{array}}$$

Table II. Gibbs free energy values used to construct Figs. 1 and 2 (kJ·mol^{-1})

MINERAL	FORMULA	ΔG_f°	M/C[†]	REF
schoepite	$[(UO_2)_8O_2(OH)_{12}](H_2O)_{12}$	-14045.8	C	[16]
metaschoepite	$[(UO_2)_8O_2(OH)_{12}](H_2O)_{10}$	-13092.1	M	[12]
DS*	$[(UO_2)O_{0.25 - 0.1}(OH)_{1.5 - 1.8}]$	-1362.3	M	[12]
becquerelite	$Ca[(UO_2)_6O_4(OH)_6](H_2O)_8$	-10305.8	C	[17]
"Ca-protasite"	$Ca_2[(UO_2)_6O_6(OH)_4](H_2O)_8$	-10779.6	C	
metacalciouranoite	$Ca_3[(UO_2)_6O_8(OH)_2](H_2O)_6$	-10768.3	C	
calciouranoite	$Ca_3[(UO_2)_6O_8(OH)_2](H_2O)_9$	-11496.1	C	
soddyite	$(UO_2)_2SiO_4(H_2O)_2$	-3658.0	M	[14]
swamboite	$(UO_2)[(UO_2)(SiO_3OH)]_6(H_2O)_{30}$	-21092.4	C	[17]
uranosilite	$(UO_2)(SiO_4)_7$	(-7214.5)	C	[17]
uranophane	$Ca[(UO_2)(SiO_3OH)]_2(H_2O)_5$	-6210.6	M	[14]
haiweeite	$Ca[(UO_2)_2(Si_2O_5)_3](H_2O)_5$	(-9431.4)	C	[18]
ursilite	$Ca_4[(UO_2)_4(Si_2O_5)_5(OH)_6](H_2O)_{15}$	(-20504.6)	C	[18]
rutherfordine	UO_2CO_3	-1563.1	M	[15]
joliotite	$UO_2CO_3(H_2O)_{1-2}$	-2043.3	C	[17]
sharpite	$Ca(UO_2)_6(CO_3)_5(OH)_4(H_2O)_3$	-11601.1	C	[17]
fontanite	$Ca(UO_2)_3(CO_3)_4(H_2O)_3$	-6523.1	C	[19]
urancalcarite	$Ca(UO_2)_2(CO_3)(OH)_6(H_2O)_3$	-6037.0	C	[17]
zellerite	$Ca(UO_2)(CO_3)_2(H_2O)_3$	(-3892.1)	C	[17]
liebigite	$Ca_2(UO_2)(CO_3)_3(H_2O)_{10}$	(-6226.0)	M	[5]

[†]M/C indicates measured or calculated values. * DS = dehydrated schoepite; the ΔG°_f value cited corresponds to $UO_3 \cdot 0.85H_2O$ [12]. Metaschoepite (= $UO_3 \cdot 2H_2O$) is used for all reactions; the structural formula is inferred [16]. References refer to reported ΔG°_f values where available; for phases with calculated ΔG°_f values, references refer to the formula cited. Values in parentheses are not considered reliable (see text).

The value of $\Delta G_f CaO^*$ determined from liebigite is the same as that determined using uranophane, reflecting the similar structural roles of Ca in the structures of uranophane and liebigite. In fact, the structural role of Ca is similar in most of the uranyl minerals for which crystal structures are known [20]. This also suggests a similar free energy contribution by CO_2^* in the structures of these two uranyl carbonates.

The values for these five hypothetical oxides are used to estimate ΔG_f° for various uranyl minerals by adding the contributions from the constituent oxides in their stoichiometric proportions. The $\Delta G f^\circ$ values thus calculated are listed in Table II, along with the structural formulas of the minerals considered in this study. The $\Delta G f^\circ$ values for all dissolved species, as well as calcite and CO_2, used for the construction of the activity-activity diagrams are taken from Grenthe et al. [4].

RESULTS

Activity-activity diagrams illustrate the stabilty fields occupied by solid phases in equilibrium with an aqueous phase. The underlying assumption of these diagrams is that the concentration of dissolved U is sufficient to stabilize the solids. This is also implicit in the way that the equilibrium reactions are written: all U is conserved in the solids.

System CO_2-CaO- UO_3-H_2O

Calculated using the ΔG°_f values from Table II, Fig. 1 illustrates the stability fields for several uranyl carbonate minerals. The concentration of Ca in many ground waters is controlled by equilibrium with calcite, $CaCO_3$, and dissolved $CO_{2(g)}$:

$$CaCO_3 + 2H+ \Leftrightarrow Ca^{2+} + H_2O + CO_{2(g)} .$$

This equilibrium is indicated as a diagonal dotted line in Fig. 1. The composition of water in equilibrium with calcite will plot on this line. Also shown in Fig. 1 is the partial pressure of CO_2 for waters open to the atmosphere (horizontal dotted line: $P_{CO_2} = 10^{-3.5}$ atm). The composition of a ground water, open to the atmosphere and in equilibrium with calcite, will plot at the intersection of the two dotted lines in Fig. 1. Notably, this falls within the stability field of becquerelite, the most common uranyl oxide hydrate mineral in Nature [21].

The calcite equilibrium line passes through the fields for becquerelite, schoepite and rutherfordine, the most common minerals among those represented in Fig. 1, and rutherfordine is by far the most common uranyl carbonate [21]. The other carbonates, fontanite, sharpite and urancalcarite are relatively rare. These occur where P_{CO_2} values reach quite high values, usually where calcite equilibrium does not control ground water chemistries. In fact, of the three uranyl carbonates that lie above the calcite line in Fig. 1, sharpite and urancalcarite are rare and occur in deposits where the predominant carbonate mineral in the host rocks is magnesian calcite or dolomite [22]. Fontanite, which coexists with becquerelite and uranophane, occurs in pelitic silts and shales [19].

Two uranyl carbonates are not shown in Fig. 1: liebigite and zellerite (Table II). The stability fields calculated for these two minerals replace most of the area shown in Fig. 1, contrary to observation. Liebigite and zellerite are commonly found as efflorescences on mine walls, surface outcrops, and elsewhere that evaporation is high. These minerals tend to be extremely fine grained, and this contributes to the surface free energy of the minerals. It is

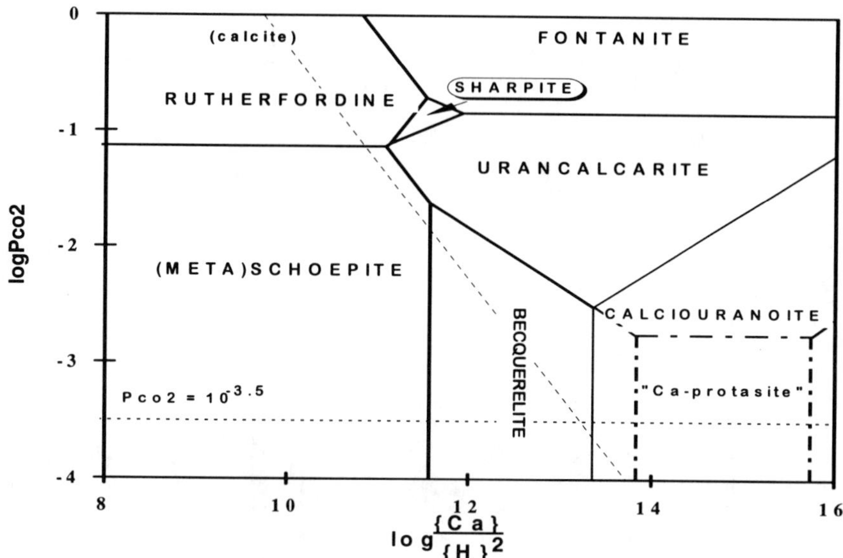

Figure 1. Activity-activity diagram for the system CO_2-CaO-UO_3-H_2O, calculated using estimated $\Delta G°_f$ values from the mineral formulas in Table II (empirical $\Delta G°_f$ values used for metaschoepite and rutherfordine). Lines dashed where metastable. Diagonal dotted line is calcite equilibrium (see text). Pco_2 value for water open to the atmosphere is shown by horizontal dotted line ($Pco_2 = 10^{-3.5}$).

unlikely that significant structural strain exists in the structure of liebigite, as well-formed crystals are known, however, some degree of supersaturation may be required to precipitate liebigite under most geochemical conditions. Similar arguments can be made for the lack of agreement between the estimated stability range of zellerite and known natural occurrences. On the other hand, neither mineral bears a close structural similarity to the other carbonates, which are all based on structural sheets. The structure of liebigite consists of isolated clusters of $(UO_2)(CO_3)_3$ groups; the structure of zellerite is unknown.

System SiO_2-CaO- UO_3-H_2O

Figure 2 illustrates the stability fields of minerals common in Si-bearing ground waters. The most common Pb-free U(VI) minerals are schoepite, becquerelite, soddyite, uranophane and rutherfordine. Schoepite forms early in the alteration paragenesis of uraninite oxidation products. Though common, schoepite is not usually abundant at most oxidized uranium deposits [7,21]. This is because schoepite is commonly replaced by uranyl silicates and carbonates, especially uranophane, soddyite and rutherfordine [7,21,23,24,25]. Although the direct replacement of schoepite by becquerelite is not readily confirmed in Nature, this reaction has been reported from several experimental studies [26,27,28]. The stability field shown for schoepite probably represents a maximum. Schoepite commonly forms fine-grained masses, effectively increasing schoepite solubility. Schoepite dehydrates spontaneously, becoming polycrystalline [23,29]. The lower hydrates also have higher solubilities in water at ~25 °C [12].

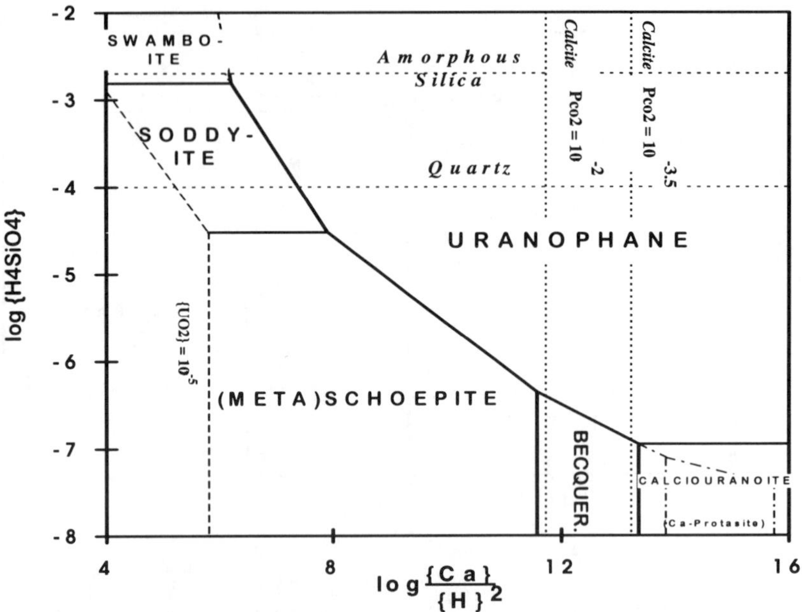

Figure 2. Activity-activity diagram for the system SiO$_2$-CaO-UO$_3$-H$_2$O, calculated using estimated ΔG°_f values from the mineral formulas in Table II (empirical ΔG°_f values are used for metaschoepite, rutherfordine, uranophane and soddyite). Lines dashed where metastable. Horizontal lines are equilibria with silica; vertical dotted lines represent calcite equilibrium for P_{CO_2} values for water open to the atmosphere ($P_{CO_2} = 10^{-3.5}$) and for a typical water-saturated soil ($P_{CO_2} \cong 10^{-2}$) [6]. BECQUER. = becquerelite.

Uranophane is the most common U(VI) mineral in Nature [30]. This observation is consistentwith the large field of stability indicated in Fig. 2, as well as the fact that uranophane is the stable U(VI) mineral in contact with ground waters whose compositions are controlled by calcite and silica equilibria [31]. Soddyite is another common mineral in oxidized U deposits, where it replaces schoepite and, less commonly, replaces uranophane (due to interaction with dilute meteoric waters [24]).

The rare mineral swamboite commonly occurs with soddyite and uranophane [22], suggesting a genetic relationship. According to Fig. 2, swamboite forms near the upper limit for {H$_4$SiO$_4$} in natural waters. The extremely rare mineral, calciouranoite defines a stability field in Si-poor waters at high values of pH and/or dissolved Ca. This is consistent with reports of the occurrences for calciouranoite and metacalciouranoite [18]. The stability field for calciouranoite represents a maximum, and assumes coarsely crystalline material. In fact, calciouranoite is always cryptocrystalline, and the small grain size will contribute to the surface free energy, increasing its solubility and decreasing its field of stability. The small grain size of calciouranoite is probably due to strain caused by structural limitations on the ability of the uranyl oxide hydrate sheets to accommodate a large amount of interlayer Ca atoms [24]; the same effect may explain the apparent lack of the hypothetical mineral, "Ca-protasite" in Nature (it is known only as a synthetic phase). In practice, becquerelite may well replace most of these fields, as suggested by the formation of synthetic becquerelite in Ca-rich solutions [26,27,28].

The stability field for becquerelite is approximately bounded by the $\{Ca\}/\{H\}^2$ values defined by a range of P_{CO_2} values common for waters in equilibrium with calcite, from water open to the air to water-saturated soils ($10^{-3.5} \leq P_{CO_2} \leq 10^{-2}$) [6]. Thus, although the stability field for becquerelite appears relatively small compared to that of schoepite in Fig. 2, it falls within the expected range of compositions for most Si-poor near-surface ground waters. This is why becquerelite is more abundant than schoepite at most oxidized U deposits [21].

Not shown in Fig. 2 are the stability fields estimated for the three uranyl silicates, haiweeite, uranosilite and ursilite (Table II). The stability fields of these three minerals, when estimated using the ΔG_f^* values in Table I, replace the entire field of uranophane illustrated in Fig. 1, contrary to observation. Haiweeite, uranosilite and ursilite always form as fine-grained masses, and surface-free energy must contribute to the increased solubilities of these minerals. However, though the structures of these minerals remain unknown, they are not structurally related to uranophane or soddyite. Haiweeite and ursilite are probably structurally related to weeksite [18]. The structure of the rare mineral uranosilite is also unknown, and is probably unique among the uranyl silicates. All three minerals are probably based on a framework of SiO_4 tetrahedra, rather than isolated SiO_4 tetrahedra as in uranophane and soddyite. Therefore, the estimate of $\Delta G_f SiO_2^*$ derived from soddyite is not appropriate for estimating the contribution of silica to the Gibbs free energy of these minerals.

SUMMARY AND CONCLUSIONS

The uranyl minerals whose stability fields are illustrated in Figures 1 and 2 have several features in common: (1) There is little compositional variation observed for uranyl minerals formed during weathering (~20 - 40 °C); these minerals tend to be well defined compositionally [24]; (2) The uranyl minerals represented in Figures 1 and 2 commonly occur as well-formed crystals; estimates of ΔG_f° values for poorly crystalline or very fine-grained minerals tend to be inconsistent with the natural occurrences; (3) the structures of these phases are related. Most are sheet structures, based on polymerized U polyhedra [20,32,33]. The unique role of U^{6+} in most of these minerals suggests that the ΔG_f^* contributions of UO_3^* to ΔG_f° may be similar for a variety of U(VI) minerals.

The method described here for estimating the Gibbs free energies of several structurally related uranyl phases is a potentially powerful tool for helping us better understand the geochemical role of this large family of minerals. With the aid of activity-activity diagrams, the natural occurrences can be used to place constraints on reasonable estimates of ΔG_f° values. However, this method is not a panacea, and it cannot replace accurate thermodynamic determinations. Especially important is the fact that estimates of ΔG_f° values using constituent oxides in minerals of one structure type are not appropriate for estimating the ΔG_f° values of structurally distinct phases. It is also important to accurately know the compositions of the minerals in question, as even small variations in structurally bound H_2O can significantly affect estimated ΔG_f° values (see e.g., [12]). This method is best used for estimating ΔG_f° values of minerals for which structural studies have confirmed the compositions. Mineral solubilities can also be strongly influenced by grain size, and most fine-grained mineral will not precipitate under conditions expected (compare with clay minerals). Another drawback to this method is the inability to distinguish among polymorphs. The most important polymorphs in the systems described here are alpha-uranophane and beta-uranophane. The stability field shown in Fig. 2 represents alpha-uranophane, but beta-uranophane is important in several oxidized U deposits [21,25]. Given these caveats, this method appears useful, especially with the large number of U(V) minerals that can influence U transport in oxidizing ground waters.

ACKNOWLEDGMENTS

This work was completed at the University of Manitoba during a postdoctoral fellowship from the National Science and Engineering Research Council (Canada). Financial support for the early stages of this work were provided by the Swedish Nuclear Fuel and Waste Management Co. (SKB). Travel funds were generously provided by Argonne National Laboratory.

REFERENCES

1. Y. Tardy and R.M. Garrels, Geochim. Cosmochim. Acta **41**, 1051 (1976).
2. R.J. Finch and R.C. Ewing, SKB Technical Report **91-15** (SKB, Stockholm) 1991.
3. D.J. Wronkiewicz, J.K. Bates, T.J. Gerding, E. Veleckis and B.S. Tani, J. Nucl. Mater. **190**, 107 (1992).
4. I. Grenthe et al. *Thermodynamics of Uranium*. North-Holland (NY, Amsterdam) (1992) 656 p.
5. A.K. Alwan and P.A. Williams, Mineral. Mag. **43**, 665 (1980).
6. W. Stumm and J.J. Morgan, *Aqueous Chemistry, 2nd Edition*, Wiley-Interscience (NY, 1981) 780 p.
7. R.J. Finch and R.C. Ewing, J. Nucl. Mater. **190**, 133 (1992).
8. A. Sandino and J. Bruno, Geochim. Cosmochim. Acta **56**, 4135 (1992).
9. T. Murakami, T. Ohnuki, H. Isobe and T. Sato, Amer. Mineral. (submitted).
10. Y. Tardy and R.M. Garrels, Geochim. Cosmochim. Acta **42**, 87 (1977).
11. Y. Tardy and J. Duplay, Geochim. Cosmochim. Acta **56**, 3007 (1992).
12. P.A.G. O'Hare, B.M. Lewis and S.N. Nguyen, J. Chem. Thermodynamics **20**, 1287 (1988).
13. R.J. Finch, F.C. Hawthorne and R.C. Ewing in *Scientific Basis for Nuclear Waste Management XIX*, edited by W.M. Murphy and D.A. Knecht (Mater. Res. Soc. Proc. **412**, Pittsburgh, PA, 1996) pp. 361-368.
14. S.N. Nguyen, R.J. Silva, H.C. Weed and J.E. Andrews, J. Chem. Thermodyn. **24**, 359 (1992).
15. E.I. Sergeyeva, A.A. Nikitin, I.L. Khodakovkiy and G.B Naumov, Geokhimiya **11**, 1340 (1972) [English translation in: Geochem. Internat. **9**, 900 (1972)].
16. R.J. Finch, M.A. Cooper, F.C. Hawthorne and R.C. Ewing, Can. Mineral. **34**, 1071 (1996).
17. E.H. Nickel and M.C. Nichols, *Mineral Reference Manual*. Van Norstrand Reinhold (NY) 1992, 250 p.
18. J. Cejka, and Z. Urbanec, *Secondary Uranium Minerals*. Academia, Czechoslovak Academy of Sciences (Prague, 1990) 93 p.
19. M. Deliens and P. Piret, Eur. J. Mineral. **4**, 1271 (1992).
20. P.C. Burns, M.L. Miller, R.C. Ewing, Can. Mineral. **34**, 845 (1996).
21. C. Frondel, *Systematic mineralogy of uranium and thorium*. Geol. Soc. Amer. Bull. **1064** (1958) 399 p.
22. M. Deliens, P. Piret and G. Comblain, *Les Minéraux Secondaires d'Uranium du Zaire*. Musée Royal de l'Afrique Centrale (Tervuren, 1981) 113 p.
23. R.J. Finch, M.L. Miller and R.C. Ewing, Radiochimica Acta **58/59**, 433 (1992).
24. R.J. Finch, Ph.D. Thesis, University of New Mexico, 1994.
25. E.C. Pearcy, J.D. Prikryl, W.M. Murphy and B.W. Leslie, Appl. Geochem. **9**, 713 (1994).
26. A. Sandino and B. Grambow, Radiochim. Acta **66/67**, 37 (1994).
27. R. Vochten and L. Van Havebeke, Mineral. Petrol. **43**, 65 (1990).
28. A.G. Sowder, S.B. Clark and R.A. Fjeld, Radiochim. Acta **74**, 45 (1996).
29. C.L. Christ and J.R. Clark, Amer. Mineral. **45**, 1026 (1960).
30. D.K. Smith, in: *Uranium Geochemistry, Mineralogy, Geology, Exploration and Resources*, edited by B. DeVivo, F. Ippolito, G. Capaldi & P.R. Simpson. Instit. Min. Metall., London (1984). pp.43-88.
31. D. Langmuir, Geochim. Cosmochim. Acta **42**, 547 (1978).
32. H.T. Evans, Science **141**, 154 (1963).
33. M.L. Miller, R.J.Finch, P.C.Burns, and R.C. Ewing, J. Mater. Res. **11**, 3048 (1996).

THE CRYSTAL STRUCTURE OF IANTHINITE, A MIXED-VALENCE URANIUM OXIDE HYDRATE

Peter C. Burns[1], Robert J. Finch[2], Frank C. Hawthorne[3], Mark L. Miller[4] and Rodney C. Ewing[4]

[1]Department of Geology, University of Illinois at Urbana-Champaign, 245 Natural History Building, 1301 West Greet Street, Urbana, IL 61801, U.S.A.

[2]Argonne National Laboratory, Chemical Technology Division, 9700 South Cass Avenue, Argonne, IL 60439-4837, U.S.A.

[3]Department of Geological Sciences, University of Manitoba, Winnipeg, Manitoba R3T 2N2, Canada.

[4]Department of Earth and Planetary Sciences, University of New Mexico, Albuquerque, NM 87131-1116, U.S.A.

ABSTRACT

Ianthinite, $[U^{4+}_2(UO_2)_4O_6(OH)_4(H_2O)_4](H_2O)_5$, is the only known uranyl oxide hydrate mineral that contains U^{4+}, and it has been proposed that ianthinite may be an important Pu^{4+}-bearing phase during the oxidative dissolution of spent nuclear fuel. The crystal structure of ianthinite, orthorhombic, a 7.178(2), b 11.473(2), c 30.39(1) Å, V 2502.7 Å^3, $Z = 4$, space group $P2_1cn$, has been solved by direct methods and refined by least-squares methods to an R index of 9.7 % and a wR index of 12.6 % using 888 unique observed $[\,|F| \geq 5\sigma\,|F|\,]$ reflections. The structure contains both U^{6+} and U^{4+}. The U^{6+} cations are present as roughly linear $(U^{6+}O_2)^{2+}$ uranyl ions (Ur) that are in turn coordinated by five O^{2-} and OH^- located at the equatorial positions of pentagonal bipyramids. The U^{4+} cations are coordinated by O^{2-}, OH^- and H_2O in a distorted octahedral arrangement. The $Ur\phi_5$ and $U^{4+}\phi_6$ (ϕ: O^{2-}, OH^-, H_2O) polyhedra link by sharing edges to form two symmetrically distinct sheets at $z \approx 0.0$ and $z \approx 0.25$ that are parallel to (001). The sheets have the β-U_3O_8 sheet anion-topology. There are five symmetrically distinct H_2O groups located at $z \approx 0.125$ between the sheets of $U\phi_n$ polyhedra, and the sheets of $U\phi_n$ polyhedra are linked together only by hydrogen bonding to the intersheet H_2O groups. The crystal-chemical requirements of U^{4+} and Pu^{4+} are very similar, indicating that extensive $Pu^{4+} \leftrightarrow U^{4+}$ substitution can occur within the sheets of $U\phi_n$ polyhedra in the structure of ianthinite.

INTRODUCTION

Ianthinite is the only known uranyl oxide hydrate mineral that contains U^{4+}. It has been identified from numerous uraninite-bearing deposits [1,2,3,4,5,6], and it may play a key role in the paragenesis of the complex assemblage of uranyl minerals that form where uraninite has been exposed to oxidizing meteoric water [1,7]. Due to its rarity, its instability in the presence of oxygen, and the difficulty in obtaining pure specimens, the crystal structure of ianthinite has resisted solution.

Recently, there has been renewed interest in the paragenesis and structure of uranyl oxide hydrate minerals, particularly schoepite, $[(UO_2)_8O_2(OH)_{12}](H_2O)_{12}$, ianthinite and becquerelite, $Ca[(UO_2)_3O_2(OH)_3]_2(H_2O)_8$, as they not only occur as the secondary products of alteration of uraninite under oxidizing conditions [1,5,8] but are also prominent alteration phases in laboratory experiments on UO_2 and spent nuclear fuel subjected to oxidative dissolution [9,10,11,12,13,14]. Details of the occurrence of uranyl oxide hydrate minerals are an important test of the

1193

extrapolation of the results of short-term experiments to periods relevant to nuclear-waste disposal [15]. Moreover, they provide important constraints on models used to predict the long-term behavior of spent nuclear fuel [16]. Finch and Ewing [17] proposed that ianthinite may be an important Pu^{4+}-bearing host phase formed during the oxidative dissolution of spent nuclear fuel, and ianthinite may control Pu concentrations in U-saturated solutions under oxidizing conditions (cf., [10]). However, the details of the crystal structure are required to assess the likelihood of Pu^{4+} incorporation.

The conditions of formation of ianthinite are not well characterized. Ianthinite may form due to incomplete oxidation of uraninite, although the growth of synthetic ianthinite requires partial reduction of U^{6+} in solution [6,18]. The range of stability for ianthinite in natural waters has been estimated by Taylor et al. [19], who calculated a limited stability field in U-saturated aqueous solutions at 25 °C: $PH_2O > 10^{-2.5}$, $10^{-50} < PO_2 < 10^{-30}$ (atm). The stability field of ianthinite decreases with increasing temperature and vanishes near 100 °C, where it is replaced by α-U_3O_8 and β-$UO_2(OH)_2$ [19]; however, Taylor et al. [20] showed evidence for the formation of synthetic ianthinite during the oxidation of UO_2 at 200 °C caused by the depletion of oxygen in sealed reaction vessels. Ianthinite commonly lines cavities within corrosion rinds on uraninite where oxygen may become depleted during uraninite oxidation [17]. The coexistence of ianthinite with uranyl carbonates may reflect anoxic or moderately reducing conditions coincident with elevated Pco_2 values caused by biologic respiration or decomposition [17].

EXPERIMENTAL

The ianthinite crystal studied was extracted from sample CSM 91.62 that was collected from the Shinkolobwe (Kasolo) mine in Shaba, southern Zaire, and obtained by us from the Colorado School of Mines Geology Museum. The sample consists of a coarsely crystalline matrix of intergrown schoepite, becquerelite, vandendriesscheite and ianthinite, in contact with altered uraninite. The specimen also contains veins of fine-grained soddyite and uranophane. The crystal used was adjacent to uraninite and surrounded by granular schoepite or metaschoepite, much of which may have formed due to the oxidation of pre-existing ianthinite, as suggested by dark inclusions within the yellow crystals of schoepite.

A thin cleavage fragment with dimensions 0.21 x 0.19 x 0.03 mm was removed from the sample and examined optically and by X-ray precession photography. Optical examination of the crystal showed no evidence for schoepite intergrowths as described by Frondel and Cuttita [21] and Schoep and Stradiot [22]; however, schoepite or metaschoepite is evident in $\{0kl\}$ precession photographs as faint diffuse diffraction spots parallel to c^*. The precession photographs confirmed orthorhombic symmetry. The crystal was mounted on a Nicolet $R3m$ automated four-circle diffractometer equipped with a graphite monochrometer and Mo-$K\alpha$ X-radiation. Forty-five diffraction maxima were centered, and the unit-cell dimensions: a 7.178(2), b 11.473(3), c 30.39(1) were refined by least-squares methods.

Data were collected using the ω scan-mode with a variable scan-rate proportional to the peak intensity; minimum and maximum scan-speeds were 4.0 and 29.3° 2θ/min, respectively. A total of 4147 reflections were measured over the range $4° \leq 2\theta \leq 60°$, with index ranges $0 \leq h \leq 10$, $0 \leq k \leq 17$, $0 \leq l \leq 43$. Two standard reflections were measured after every 58 reflections. One hundred and two reflections were discarded due to peak asymmetry and one due to bad backgrounds, leaving 4043 data. An empirical absorption-correction was applied, based on 36 psi-scans of each of 11 diffraction maxima at least every 5° from 7.6 to 51.5° 2θ. The crystal was modeled as a $\{001\}$ plate, and 539 reflections with a plate-glancing angle of less than 6° were discarded. The value of 6° was arrived at by using a range of glancing angles and determining where the effect of varying the glancing angle produced no improvement in the refinement. The absorption correction reduced R(azimuthal) from 27.9% to 5.7%. Reflections were corrected for drift, Lorentz, polarization and background effects.

<u>Diffraction-peak overlap</u>

Peak overlap was a significant problem for the data due to the relatively large unit-cell volume (2502 Å^3) and long c axis (30.34 Å) of ianthinite. Data exhibiting peak overlap were evident from asymmetric background counts; only the worst of these were removed by the data-reduction program. Peak overlap commonly results in (erroneously) negative peak intensities due to background enhancement; however, not all peaks affected by overlap displayed negative intensities. Because the relatively large mosaic spread in the crystal contributes to peak broadening, we did not attempt to reduce peak overlap by narrowing the scan ranges; instead, peak intensities with asymmetric backgrounds were discarded following the data collection. All data for which left and right background (raw) counts differed by more than one standard deviation of the integrated peak intensity (I_o) were deleted from the data set. Thus 359 reflections were removed.

STRUCTURE SOLUTION AND REFINEMENT

Scattering curves for neutral atoms, together with anomalous-dispersion corrections, were taken from Cromer and Mann [23] and Cromer and Liberman [24], respectively. The Siemens SHELXTL PLUS (PC version) system of programs was used throughout this study.

A recognizable partial-structure model was obtained in the space group $P2_1cn$ using direct methods. The model included U positions as well as a few anion positions. Successive cycles of least-squares refinement followed by the calculation of difference-Fourier maps gradually revealed additional anion positions. Difference-Fourier maps calculated parallel to (001) at $z = 0.125$ revealed five peaks attributable to interlayer H_2O groups. One of these peaks [$H_2O(9)$] was split into three roughly equal maxima separated by ~1.8 Å; an O atom with occupancy factor of 0.33 was placed on each of these peaks. The least-squares matrix was ill-conditioned because of the pseudo-symmetric arrangement of the U positions and the relatively poor parameter-to-data ratio; significant damping was necessary to obtain convergence. The uranyl ion U^{6+}-O bond-lengths were constrained to be ~1.8 Å by adding extra weighted observational equations to the least-squares matrix. Full-matrix least-squares refinement of the model, which included positional parameters for all atoms, anisotropic-displacement parameters for the U positions, and overall isotropic-displacement parameters for the anion positions, converged to a final R of 9.7% and a wR of 12.6% for 888 observed reflections ($|F| \geq 5\sigma|F|$), although the displacement parameters for some of the atomic positions were poorly behaved. The high R factors and poor displacement-factors reflect problems with the data which include peak overlap, weak intensities, possible crystal inclusions, and an imperfect absorption correction. Bond lengths are listed in Table I.

DESCRIPTION OF THE STRUCTURE

<u>Uranium coordination</u>

There are six symmetrically unique U positions in the structure of ianthinite. Where present in crystal structures, the U^{6+} cation is usually part of a nearly linear $(U^{6+}O_2)^{2+}$ uranyl ion [25] (designated Ur) with U^{6+}-O bond-lengths of ~1.8 Å. The uranyl ion occurs in structures

Table I
Bond lengths (Å) for ianthinite

$U(1)$-$O(1)$	1.81(5)	$U(4)$-$O(11)$	1.96(5)
$U(1)$-$O(2)$	1.82(5)	$U(4)$-$O(14)$	2.11(5)
$U(1)$-$O(9)$	2.32(5)	$U(4)$-$H_2O(2)$	2.36(9)
$U(1)$-$OH(2)$	2.19(6)	$U(4)$-$H_2O(3)$	2.37(10)
$U(1)$-$O(9)$a	2.45(4)	$U(4)$-$O(12)$f	2.00(5)
$U(1)$-$OH(1)$b	2.60(6)	$U(4)$-$O(13)$f	2.07(5)
$U(1)$-$OH(1)$	3.01(6)	$<U(4)$-$\phi>$	2.14
$<U(1)$-$\phi_{eq}>$	2.51		
		$U(5)$-$O(5)$	1.80(5)
$U(2)$-$O(9)$	2.05(5)	$U(5)$-$O(6)$	1.78(5)
$U(2)$-$O(10)$	2.35(5)	$U(5)$-$O(11)$g	2.50(6)
$U(2)$-$OH(1)$	2.18(7)	$U(5)$-$O(12)$h	2.36(6)
$U(2)$-$OH(3)$	2.24(6)	$U(5)$-$O(13)$i	2.35(4)
$U(2)$-$H_2O(1)$	2.54(11)	$U(5)$-$O(14)$i	2.27(5)
$U(2)$-$H_2O(4)$	2.36(11)	$U(5)$-$OH(4)$g	2.19(6)
$<U(2)$-$\phi>$	2.29	$<U(5)$-$\phi_{eq}>$	2.33
$U(3)$-$O(3)$	1.78(5)	$U(6)$-$O(7)$	1.78(5)
$U(3)$-$O(4)$	1.80(5)	$U(6)$-$O(8)$	1.80(5)
$U(3)$-$O(10)$	2.20(6)	$U(6)$-$OH(4)$	2.54(6)
$U(3)$-$OH(3)$	2.92(7)	$U(6)$-$O(11)$j	2.45(5)
$U(3)$-$O(10)$c	2.34(4)	$U(6)$-$O(12)$k	2.40(5)
$U(3)$-$OH(2)$d	2.52(6)	$U(6)$-$O(13)$l	2.45(6)
$U(3)$-$OH(3)$e	2.48(6)	$U(6)$-$O(14)$m	2.46(6)
$<U(3)$-$\phi_{eq}>$	2.49	$<U(6)$-$\phi_{eq}>$	2.46

a = x+½, $\overline{y},\overline{z}$; b = x-½, $\overline{y},z$; c = x+½, $\overline{y}$ -1,$\overline{z}$; d = x, y-1, z; e = x-½, $\overline{y}$ -1,$\overline{z}$; f = x-1, y, z; g = x, y+1, z; h = x-1, y-1, z; i = x-½, y+1½, ½-z; j = x-½, y+½, ½-z; k = x+½, y+½, ½-z; l = x+1, y+1, z; m = x, y+1, z.

coordinated by four, five or six anions, such that the coordinating anions and the U^{6+} cation are approximately coplanar, and the oxygen atoms that are part of the uranyl ions are the apices of square, pentagonal or hexagonal bipyramids. Four of the six U positions in the structure of ianthinite contain a uranyl ion and thus correspond to U^{6+}: $U(1)$, $U(3)$, $U(5)$ and $U(6)$. In each case, the uranyl ion is coordinated by five equatorial anions in a roughly planar arrangement, resulting in pentagonal bipyramids (Fig. 1a). The $Ur\phi_5$ polyhedra (ϕ: O^{2-}, OH^-, H_2O) have a uranyl ion U^{6+}-O bond length of ~1.8 Å, as imposed by constraints during the refinement of the structure. The O-U^{6+}-O bond angles of the uranyl ions, which were not constrained during the refinement, are $U(1)$: 162(2); $U(3)$: 176(2); $U(5)$: 177(2) and $U(6)$: 166(2)°. The $U(1)$ and $U(3)$ uranyl ions are each coordinated by two O atoms and three OH groups, and the $<U$-$\phi_{eq}>$ bond lengths are 2.51 and 2.49 Å, respectively, values which are in the range of $<U$-$\phi_{eq}>$ observed for $Ur\phi_5$ polyhedra in well-refined structures. The $U(5)$ and $U(6)$ uranyl ions are each coordinated by four O atoms and a single OH group; the $<U$-$\phi_{eq}>$ bond lengths are 2.33 and 2.46 Å, respectively, values that are typical for $Ur\phi_5$ polyhedra.

The $U(2)$ and $U(4)$ sites are each coordinated by four anions and two H_2O groups in a distorted octahedral arrangement (Fig. 1b). The $U(2)O_2(OH)_2(H_2O)_2$ and $U(4)O_4(H_2O)_2$ polyhedra have $<U$-$\phi>$ bond-lengths of 2.29 and 2.14 Å, respectively. These polyhedra do not contain a uranyl ion, suggesting that these polyhedra are probably occupied by U^{4+}. The expected $^{[6]}U^{4+}$-O bond-length from sums of effective ionic-radii is 2.25 Å ($^{[6]}U^{4+}$ = 0.89, $^{[3]}O^{2-}$ = 1.36 Å [26]) whereas the $^{[6]}U^{6+}$-O bond length is 2.09 Å ($^{[6]}U^{6+}$ = 0.73 [26]). The $<U(2)$-$\phi>$ bond-length of 2.29 Å is consistent with complete occupancy of the site by U^{4+}. The $<U(4)$-$\phi>$ bond length of 2.14 Å, however, is somewhat shorter than expected for an octahedron that contains only U^{4+}.

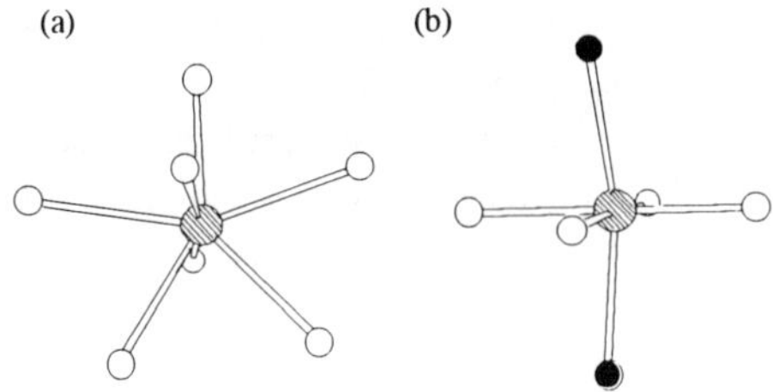

Figure 1. Uranium atom coordination geometries in the structure of ianthinite. U: circles shaded with parallel lines; O and OH: open circles; H_2O: solid circles.

Bond-valence analysis

A bond-valence analysis, calculated using the parameters for U^{6+} and U^{4+} from Brese and O'Keeffe [27], permits the identification of O^{2-}, OH^- and H_2O anions. The analysis gives cation bond-valence sums consistent with occupancy of the $U(1)$ (6.01 *vu*), $U(3)$ (6.22 *vu*), $U(5)$ (6.86 *vu*) and $U(6)$ (6.06 *vu*) sites by U^{6+}, and the $U(2)$ (4.05 *vu*) site by U^{4+}. However, the sum of 6.02 *vu* obtained for the $U(4)$ site using the bond-valence parameters for U^{4+} is much higher than expected. Calculation of this sum using the parameter for U^{6+} gives 5.44 *vu*. The higher-than-expected bond-valence sum at the $U(4)$ position may be attributable to a mixture of U^{4+} and U^{6+}, or even possibly to some U^{5+} at the site. Inclusion of some U^{5+} or U^{6+} at the $U(4)$ site could be locally charge-compensated by deprotonation of the H_2O groups that are bonded to $U(4)$. It is also plausible that partial substitution of another chemical species for U^{4+} at the $U(4)$ site causes the observed discrepancy of bond-valence sums.

Structural formula of ianthinite

The determination of the exact formula of ianthinite from the present study is difficult owing to the low quality of the data (due to crystal problems) and the resulting imprecise bond-lengths, the lack of a suitable bond-valence parameter for U^{6+} [28], and the possibility of partial oxidation of U^{4+} to U^{6+}. The $U(1)$, $U(3)$, $U(5)$ and $U(6)$ sites are all occupied by U^{6+}, as shown by the presence of a uranyl ion. The $U(2)$ and $U(4)$ polyhedral geometries are generally consistent with occupancy by U^{4+}. There are 22 distinct anion positions within the two sheets; bond-valence arguments indicate that there are 14 O atoms, 4 OH groups and 4 H_2O groups bonded within or directly to the two sheets. In addition, there are 5 H_2O groups between the sheets of $U\phi_n$ polyhedra. Thus, as each site is on a general position, the structure indicates that the formula of ianthinite is $[U^{4+}{}_2(UO_2)_4O_6(OH)_4(H_2O)_4](H_2O)_5$, where the square brackets enclose sheet constituents. The ratio of U^{4+} to U^{6+} for this formula is greater than that proposed by Guillemin and Protas [4], although the total H_2O content is in reasonable agreement. The formula gives a calculated density of 5.00 g/cm^3, which agrees reasonably well with the measured density of 5.16 ± 0.05 g/cm^3 [4].

Connectivity of the structure

A projection of the structure of ianthinite along [100] (Fig. 2) shows $U\phi_n$ polyhedra connected to form sheets parallel to (001). The only constituents that occur between the sheets of $U\phi_n$ polyhedra are H_2O groups. Thus, linkage between the sheets is by hydrogen bonds only. The data is not of sufficient quality to allow resolution of the H atom positions, as is usually the case for uranium-rich phases.

There are two symmetrically distinct sheets of $U\phi_n$ polyhedra, one at $z \approx 0$ and the other at $z \approx 0.25$. Although symmetrically distinct, these sheets are topologically identical; the sheet at $z \approx 0.25$ is shown in Figure 3. Each sheet contains three unique U positions; two are associated with

$Ur\phi_5$ polyhedra and one with a $U^{4+}\phi_6$ polyhedron. The sheets of $U\phi_n$ polyhedra result from sharing of corners and edges between $Ur\phi_5$ and $U^{4+}\phi_6$ polyhedra. $Ur\phi_5$ polyhedra share edges, forming chains parallel to x that are one $Ur\phi_5$ polyhedra wide; these chains are in turn linked together by the sharing of corners between $Ur\phi_5$ polyhedra of adjacent chains and by the sharing of edges with $U^{4+}\phi_6$ polyhedra located between the chains (Fig. 3).

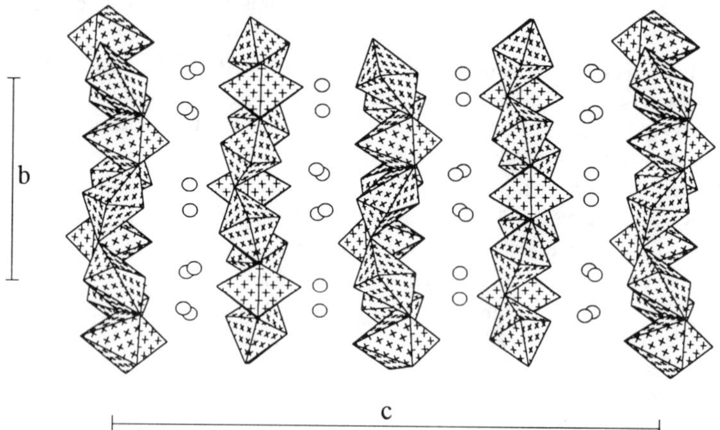

Figure 2. The structure of ianthinite projected along [100]. The $U\phi_n$ polyhedra are shaded with crosses and the H_2O groups of the interlayer are shown as open circles.

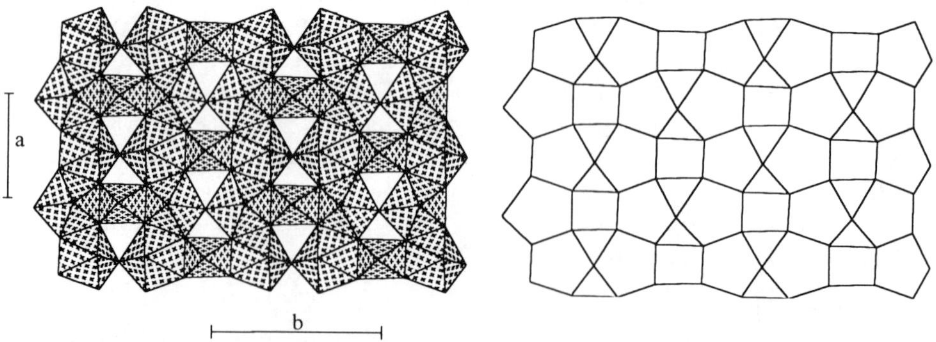

Figure 3. The sheets of $U\phi_n$ polyhedra in the structure of ianthinite. Polyhedral representation of the sheet at $z \approx 0.25$ (left) and the corresponding anion topology (right).

Representation of the sheet of $U\phi_n$ polyhedra (Fig. 3) with its corresponding sheet anion-topology following the procedure of Burns et al. [29] results in the β-U_3O_8 anion-topology (Fig. 3). The sheets in the structures of β-U_3O_8 [30] and billietite [31], $Ba[(UO_2)_3O_2(OH)_3]_2(H_2O)_4$, also have this anion topology. The sheets in the structure of β-U_3O_8 are linked by sharing O atoms between adjacent sheets, and there are no interlayer constituents. The structure of billietite contains two symmetrically distinct sheets of $U\phi_n$ polyhedra; one has the α-U_3O_8 anion-topology [29]; whereas, the other is more similar to the β-U_3O_8 anion-topology [29]. However, if the $U(2)$-$O(5)$ bond-length of 3.02 Å is included in the latter sheet, it is better described as having the α-U_3O_8 anion-topology [29]. In the structure of billietite, the sheets of $U\phi_n$ polyhedra are linked by interlayer Ba cations and H_2O groups.

Schoepite is chemically similar to ianthinite, but it contains no U^{4+}. In air, ianthinite oxidizes to form schoepite [22]. The structure of schoepite was recently solved by Finch et al. [32]; their work showed that the structure contains sheets of $Ur\phi_5$ polyhedra that are connected through hydrogen bonding to interlayer H_2O groups. The same sheet of $Ur\phi_5$ polyhedra also occurs in the structure of fourmarierite [33], $Pb[(UO_2)_4O_3(OH)_4](H_2O)_4$, and the topological arrangement of anions is referred to as the fourmarierite anion-topology [29]. Note that the structure of ianthinite contains both $Ur\phi_5$ pentagonal bipyramids and distorted $U^{4+}\phi_6$ octahedra; whereas, the structure of schoepite contains only $Ur\phi_5$ pentagonal bipyramids.

INCORPORATION OF Pu^{4+} DURING THE ALTERATION OF SPENT NUCLEAR FUEL

The structure of ianthinite is of considerable interest because it contains both U^{4+} and U^{6+}, and it, together with schoepite, is likely to form early during the oxidative dissolution of UO_2 in spent nuclear fuel. As such, Finch and Ewing [17] and Burns et al. [34] have proposed that the structure may be an important Pu^{4+}-bearing phase, and that it may possibly control Pu^{4+} concentration in U-saturated oxidizing solutions. The structure determination presented here provides the basis to assess the possibility of Pu^{4+} incorporation into the structure of ianthinite.

The crystal-chemical requirements of Pu^{4+} are very similar to those of U^{4+}. Pu^{4+} is known to occur in coordination polyhedra with coordination numbers ranging from six to eight. The U^{4+} in the structure of ianthinite is in octahedral coordination, as is Pu^{4+} in the structure of $BaPu^{4+}O_3$ [35]. Bond lengths for $^{[6]}Pu^{4+}$-O and $^{[6]}U^{4+}$-O from sums of effective ionic-radii [26] are 2.22 and 2.25 Å, respectively. Thus, there is no structural limitation for the amount of Pu^{4+} $\leftrightarrow$ U^{4+} substitution that may occur in ianthinite, and a Pu-analogue of ianthinite with composition $[Pu^{4+}_2(UO_2)_4O_6(OH)_4(H_2O)_4](H_2O)_5$ could conceivably be stable. Ianthinite is likely to be an important Pu^{4+}-bearing phase during the oxidative dissolution of the UO_2 in spent nuclear fuel.

ACKNOWLEDGMENTS

This work was supported by the Natural Sciences and Engineering Research Council of Canada in the form of Operating, Major Equipment and Infrastructure Grants to FCH and by a Post-Doctoral Fellowship to PCB. This work was also supported by the Office of Basic Energy Sciences (Grant No. DE-FG03-95ER14540).

REFERENCES

[1] E.C. Pearcy, J.D. Prikryl, W.M. Murphy and B.W. Leslie, Appl. Geochem. **9** (1994) 713.
[2] M. Deliens, P. Piret, and G. Comblain, Les Minéraux Secondaires d'Uranium du Zaïre (Musée Royal de l'Afrique Centrale, Tervuren, Belgium, 1981) 99 pp.
[3] M. Deliens, Bull. Soc. Franç. Minéral. Cristallogr. **100** (1977) 32.
[4] C. Guillemin and J. Protas, Bull. Soc. Franç. Minéral. Cristallogr. **82** (1959) 80.

[5] C. Frondel, Systematic Mineralogy of Uranium and Thorium. U.S. Geological Society Bulletin **1064** (1958) 400 pp.

[6] C. Bignand, Bull. Soc. Franç. Minéral. Cristallogr. **78** (1955) 1.

[7] R.J. Finch and R.C. Ewing, Alteration of uraninite in an oxidizing environment and its relevance to the disposal of spent nuclear fuel. SKB Technical Report 91-15 (1991) (SKB, Stockholm).

[8] R.J. Finch and R.C. Ewing, J. Nucl. Mater. **190** (1992) 133.

[9] L.H. Johnson and L.O. Werme, Mater. Res. Soc. Bull. **XIX**(12) (1994) 24.

[10] R.S. Forsyth and L.O. Werme, J. Nucl. Mater. **190** (1992) 3.

[11] D.J. Wronkiewicz, J.K. Bates, T.J. Gerding, E. Veleckis and B.S. Tani, J. Nucl. Mater. **190** (1992) 107.

[12] S. Stroes-Gascoyne, L.J. Johnson, P.A. Beeley, and D.M. Sellinger, In: Scientific Basis for Nuclear Waste Management IX (L.O Werme, ed.), Materials Research Society Proceedings Volume **50** (Materials Research Society, Pittsburgh, 1985) 317.

[13] R. Wang and J.B. Katayama, Nucl. Chem. Waste Management **3** (1982) 83.

[14] T. Wadsten, J. Nucl. Mater. **64** (1977) 315.

[15] R.C. Ewing, In: Scientific Basis for Nuclear Waste Management XVI (C.G. Interante & R.T. Pabalan, eds.). Materials Research Society Proceedings, Volume **294** (Materials Research Society, Pittsburgh, 1993) 559.

[16] J. Bruno, I. Casas, E. Cera, R.C. Ewing, R.J. Finch and W.O. Werme, In: Scientific Basis for Nuclear Waste Management XVIII (T. Murakami & R.C. Ewing, eds.). Materials Research Society Proceedings, Volume **353** (Materials Research Society, Pittsburgh, 1995) 633.

[17] R.J. Finch and R.C. Ewing, In: Scientific Basis for Nuclear Waste Management XVI (edited by A. Barkatt and R.A. von Konynenburg,), Materials Research Society Proceedings Volume **333** (Materials Research Society, Pittsburgh, 1994) 625.

[18] E.H.P. Cordfunke, G. Prins and P. van Vlaanderen, J. Inorg. Nucl. Chem. **30** (1968) 1745.

[19] P. Taylor, R.J. Lemire and D.D. Wood, In: Proceedings of the Third International Conference on High-Level Radioactive Waste Management, Las Vegas, Nevada (American Nuclear Society, La Grange, IL, and American Society of Civil Engineers, New York, 1992) 1442.

[20] P. Taylor, D.D. Wood, D.G. Owen and G.-I. Park, J. Nucl. Mater. **183** (1991) 105.

[21] J.W. Frondel and F. Cuttita, Am. Mineral. **39** (1953) 1018.

[22] A. Schoep and S. Stradiot, Am. Mineral. **32** (1947) 344.

[23] D.T. Cromer and J.B. Mann, Acta Crystallogr., **A24** (1968) 321.

[24] D.T. Cromer and D. Liberman, J. Chem. Phys. **53** (1970) 1891.

[25] H.T. Evans, Jr., Science, **141** (1963) 154.

[26] R.D. Shannon, Acta Crystallogr. **A32** (1976) 751.

[27] N.E. Brese and M. O'Keeffe, Acta Crystallogr. **B47** (1991) 192.

[28] In prep.

[29] P.C. Burns, M.L. Miller and R.C. Ewing, Can. Mineral. **34** (1996) 845.

[30] B.O. Loopstra, Acta Crystallogr. **B26** (1970) 656.

[31] M.K. Pagoaga, D.E. Appleman and J.M. Stewart, Am. Mineral. **72** (1987) 1230.

[32] R.J. Finch, M.A. Cooper, F.C. Hawthorne and R.C. Ewing, Can. Mineral. **34** (1996) (in press)

[33] P. Piret, Bull. Minéral. **108** (1985) 659.

[34] P.C. Burns, R.C. Ewing and M.L. Miller, J. Nucl. Mater. (1996) (accepted).

[35] G.G. Christoph, A.C. Larson, P.G. Eller, I.D. Purson, J.D. Zahrt, R.A. Penneman and G.H. Rinehart, Acta Crystallogr. **B44** (1988) 575

Uraninite-Water Interactions in an Oxidizing Environment

M. Fayek*, T.K. Kyser**, R.C. Ewing*, M.L. Miller*
*Department of Earth and Planetary Sciences, University of New Mexico, Albuquerque, NM, USA 87131,
fayek@unm.edu, rewing@unm.edu, mmiller@unm.edu
**Department of Geological Sciences, Queen's University, Kingston, Ontario, Canada, K7L 3N6,
kyser@Geol.QueensU.ca

ABSTRACT

Exceptionally low $\delta\ ^{18}O$ values of primary uraninite and pitchblende (i.e. -32 per mil to -19.5 per mil) from Proterozoic unconformity-type uranium deposits in Saskatchewan, Canada, in conjunction with theoretical uraninite-water oxygen isotope fractionation factors suggest that primary uranium mineralization is not in oxygen isotopic equilibrium with clays and silicates. The low $\delta\ ^{18}O$ values have been interpreted to have resulted from the recrystallization of primary uranium mineralization in the presence of modern meteoric fluids having low $\delta\ ^{18}O$ values of ca. -18 per mil. The absence of apparent alteration in many of the uraninite and pitchblende samples requires that the uranium minerals exchange oxygen isotopes with fluids, with only minor disturbances to their original chemical compositions and textures. However, experiments on the interaction between water and natural uraninites, from these deposits, and detailed electron micro-probe analyses of natural uraninite and pitchblende indicate that, in the presence of water, old uraninite rapidly alters to curite ($Pb_2U_5O_{17}\ 4H_2O$). The hydration of uraninite to curite releases uranium and calcium into solution and becquerelite ($Ca(UO_2)_6O_4(OH)_6H_2O$) is precipitated. In the presence of Si-saturated waters, uranium silicate minerals, soddyite ($(UO_2)_2(SiO_4)2H_2O$) and kasolite ($Pb(UO_2)SiO_4H_2O$) are precipitated in addition to curite and schoepite ($(UO_2)_8O_2(OH)_{12}(H_2O)_{12}$). The mineral paragenesis observed in these experiments is similar to sequences observed in oxidized zones in uranium deposits and UO_2-water experiments. Therefore, it is unlikely that natural uraninite and pitchblende can simply exchange oxygen with an oxidizing fluid without a concomitant change in phase chemistry or structure, nor will oxidation of uraninite lead to the formation of U_3O_7 as predicted by theoretical calculations used in natural analogue studies for the disposal of high level nuclear waste.

INTRODUCTION

Uranium deposits have been used as natural analogues in order to understand the potential migration of uranium and other radionuclides around a spent fuel repository because uraninite and UO_2 (the major constituent of spent fuel) are chemically and structurally similar [1, 2]. In this regard, high grade, unconformity-type uranium deposits have been studied extensively, although the extent and process of alteration of uraninite from these deposits is poorly understood.

Isotope fractionation in solids depends on the nature of the bonds between the isotopic species of an element and the nearest atom in the structure. Bonds with small force constants, involving atoms with high masses, produce low vibrational frequencies and tend preferentially to incorporate light isotopes [3]. Exceptionally low $\delta\ ^{18}O$ values of primary uraninite and pitchblende (i.e. -32 per mil to -15 per mil) from unconformity-type uranium deposits in Saskatchewan and high $\delta\ ^{18}O$ values of coexisting silicate and clay minerals (i.e. 10 per mil in quartz) seem to confirm this mass effect phenomenon [4, 5, 6, 7]. However, theoretical uraninite-water fractionation factors proposed by Hattori and Halas [3] and Zheng [8] indicate that primary uraninite and pitchblende in the Athabasca Basin would have been in equilibrium with fluids that had a $\delta\ ^{18}O$ value of less than -11 per mil at ca. 200 °C, the temperature at which the deposits formed [9, 10]. The dominant fluids that equilibrated with silicate minerals in unconformity-type uranium deposits were saline fluids having $\delta\ ^{18}O$ values of 4±2 per mil [6, 10]. Based on these theoretical fractionation factors, Kotzer and Kyser [6, 11] interpreted the low $\delta\ ^{18}O$ values of primary uranium mineralization from unconformity-type uranium deposits to reflect complete recrystallization of uraninite and pitchblende in the

Mat. Res. Soc. Symp. Proc. Vol. 465 © 1997 Materials Research Society

presence of relatively recent meteoric waters having low $\delta\ ^{18}O$ values of ca. -18 per mil. They also proposed that the lack of petrographically observable alteration in many of the uraninite and pitchblende samples requires that exchange of oxygen isotopes with meteoric waters resulted in only minor disturbances to the chemical compositions and original textures of the uraninite.

In more recent studies, Fayek and Kyser [7] have shown that uraninite and pitchblende from unconformity-type uranium deposits in Saskatchewan are variably and subtly altered to coffinite, $U(SiO_4)_{1-x}(OH)_{4x}$, and CaU-oxide hydrate minerals such as becquerelite, $Ca(UO_2)_6O_4(OH)_6H_2O$. This type of alteration normally occurs on a scale that is detectable only by quantitative electron microprobe analyses and back-scattered electron imaging of samples. Kotzer and Kyser [6] and Fayek and Kyser [7] have shown that a positive correlation exists between $\delta\ ^{18}O$ values and increasing Ca and Si contents of seemingly unaltered uraninite and pitchblende samples.

Despite the absence of higher oxides, such as U_3O_7, in natural samples [12], theoretical models used in many natural analogue studies report that the oxidation of uraninite in the presence of a fluid results in the formation of U_3O_7 and eventually UO_3 [13, 14, 15]. They concluded that the presence of U_3O_7 along with UO_2 and U_4O_9 is indicative of reducing conditions, and that the dissolution rate of the uranium oxide does not become significant until it is oxidized beyond the U_3O_7 [13]. Therefore, the presence of U_3O_7 is a rate limiting phase in the dissolution of uraninite. The detailed characterization of the secondary actinide-bearing phases that form during the alteration of uraninite under oxidizing conditions is required to understand the chemistry of these phases and therefore provide a base-line context for the interpretation of uranium mobilization under repository conditions. The purpose of this study is to document the effects of uraninite-water interaction in an oxidizing environment on the phase composition and structure of uraninite and determine whether uranium minerals can exchange oxygen isotopes with fluids with only minor disturbances to their chemical compositions and original textures.

EXPERIMENTAL AND ANALYTICAL TECHNIQUES

Three sets of experiments on the UO_2 and uraninite interaction with water were conducted. The first set of experiments were conducted at 100 °C and consisted of 300-500 mg of natural uraninite powder and 20 g of distilled water. The second set of experiments were conducted at 150 °C and consisted of two charges that contained 300-500 mg of uraninite powder and 20 g of silica-saturated distilled water with a pH of 4 and 6, to simulate the chemical composition of ground waters percolating through the quartz-rich sandstones which are associated with the unconformity-type uranium deposits in the Athabasca Basin. The third set of experiments also conducted at 150 °C, consisted of two charges that contained 500 mg of synthetic UO_2 and 20 g of silica saturated distilled water with a pH of 4 and 6, to simulate the interaction between spent nuclear fuel and oxidizing fluids. The water for the second and third set of experiments was prepared by equilibrating 20 g of water with 25 mg of crushed quartz and powdered silicic acid for two weeks at 150 °C.

The uraninite powder used in the experiments was made by crushing a sample of high-grade uraninite from the McArthur River unconformity-type uranium deposit located along the eastern margin of the Athabasca Basin in northern Saskatchewan, Canada. The 200-230 mesh size fraction of uraninite was separated using conventional sieving techniques, and the purity of the uraninite and synthetic UO_2 samples was verified by X-ray diffraction and scanning electron microscopy (SEM). The constituents were placed in 25 ml teflon containers and metal bombs, and then held at a constant temperature. Uraninite and UO_2 from each charge was sampled periodically over several months. Reaction products were examined using a Siemens D5000 x-ray diffractometer and a Hitachi S-800 scanning electron microscope (SEM) at an operating voltage of 20 keV.

Chemical compositions of the uranium oxides were determined by wavelength dispersive spectroscopy (WDS) using a fully automated Jeol JXA-8600 X-ray microanalyzer at an operating voltage of 20 keV, a beam diameter of $2\mu m$, and counting times of 40 s per element. Synthetic UO_2, ThO_2, ZrO_2, $CaMO_4$, InAs, and minerals (orthoclase for K, Al, and Si; apatite for P; galena for Pb and S; rutile for Ti;

spessartine for Mn; hematite for Fe; chalcopyrite for Cu; and Y-garnet for Y) were used as analytical standards. Detection limits of the elements were on the order of 0.1 wt percent. Data reduction for the various elements was done by taking into account the matrix corrections between standards and samples and the analytical parameters (i.e., take-off angle and operating voltage) of the electron microprobe. Chemical ages of uranium minerals were calculated based on their U, Pb, and Th content using the formula $t=(Pb*10^{10} yr)/(1.612U+4.95Th)$ [16]. The chemical ages from unreacted uraninite grains were relatively consistent, ranging from 1502 (Ma) to 1599 (Ma)

RESULTS

Natural uraninite and water at 100 °C

Uraninite grains prior to reaction were relatively unaltered (Fig. 1a) and chemically homogenous (Fig. 1b). These grains reacted with oxidizing distilled water to produce crystals of curite, $Pb_2U_5O_{17} 4H_2O$ (Fig. 1c), and minor amounts (<10%) of becquerelite, $Ca(UO_2)_6O_4(OH)_6H_2O$ (Fig. 1c). Continued interaction produced a reaction front which progressively moved inwards, eventually resulting in the complete replacement of uraninite by curite and becquerelite (Fig. 1d).

Newly formed curite and becquerelite in the experiments are orange in color and are similar in appearance to uranyl minerals described by Fayek and Kyser [7], in unconformity-type uranium deposits in the Athabasca basin. Electron microprobe analyses of residual uraninite cores and newly formed curite and becquerelite, from the experiments, indicate that during the reaction process, U and Pb are remobilized and incorporated into the newly formed phases. Uraninite cores continually lose Pb during the reaction process so that their U-Pb chemical ages decrease with time (Fig. 2a). In contrast, curite continually incorporates Pb with time so that the U-Pb chemical ages of curite increase with time (Fig. 2b). Becquerelite shows a positive correlation between U content and time and has very low Pb contents, resulting in very young chemical ages (Fig. 2c).

Natural uraninite-silica-water at 150 °C

Uraninite reacted with silica-saturated oxidizing water to produce crystals of curite, $Pb_2U_5O_{17}$ $4H_2O$ (Fig. 3a), and minor amounts (<10%) of schoepite, UO_32H_2O (Fig. 3b). Continued interaction resulted in the precipitation of fine-grained soddyite, $(UO_2)_2(SiO_4)2H_2O$ (Fig. 3c), and minor amounts (<10%) of kasolite, $Pb(UO_2)SiO_4H_2O$ (Fig 3c). The pH of the water affected the rate of reaction of uraninite and therefore the relative abundances of the minerals precipitated. Reaction products after 115 days at 150 °C, from charges with water that had a pH of 6, consisted mainly of curite (~80%), only minor amounts of soddyite (~20%), and no kasolite. However, reaction products from charges with water that had a pH of 4 consisted of curite (~50%), soddyite (~30%), kasolite (<10%) and schoepite (<10%).

Synthetic UO₂-silica-water at 150 °C

Grains of the synthetic UO_2 starting material have a platy surficial texture and range in size from 3 to 6 μm (Fig. 3d). Reaction of these grains with silica-saturated water produced soddyite (Fig. 3e). As in experiments using natural uraninite, the reaction rate of UO_2 was influenced by the pH of the water. Reaction products after 115 days of reaction at 150 °C from water that had a pH of 6 consisted mainly of UO_2 (~70%) and soddyite (~30%). However, reaction products from charges with water that had a pH of 4 consisted mainly of soddyite (~70%) (Fig. 3e) and only minor amounts of UO_2 (~30%) (Fig. 3f).

Published studies of the dissolution, oxidation, and secondary phase formation during the interaction between UO_2-and silica-saturated oxidizing fluids have shown that shortly after the onset of UO_2 dissolution, uranyl-oxide hydrates precipitate followed by uranyl-silicate minerals such as soddyite and eventually uranyl alkali silicates [17]. These results are similar to the observations from this study.

Figure 1.　　　(a) SEM image of unreacted uraninite grains: (b) BSE image of unreacted uraninite grains (Ur) showing an area analyzed for a chemical U-Pb age: (c) SEM image of a cluster of uraninite grains (Ur) that have reacted with water for 15 days at 100 °C and are held together by curite (Cur) and becquerelite (Bec): (d) BSE image of a cluster of uraninite grains (Ur) that have reacted with water for 15 days at 100 °C showing a reaction front that consists of curite (Cur). The grains are held together by becquerelite (Bec). Also shown are the chemical U-Pb ages.

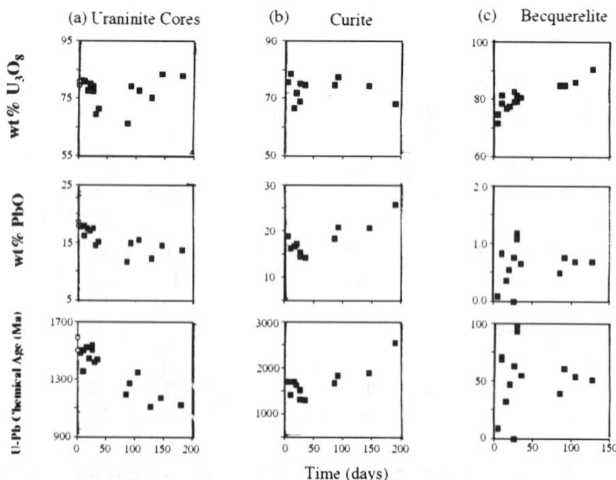

Figure 2.　　　Variations in U₃O₈, PbO, and chemical U-Pb ages with time for (a) uraninite cores: (b) curite: and (c) becquerelite from the reaction between natural uraninite and water at 100 °C. Open symbols are values for unreacted natural uraninite.

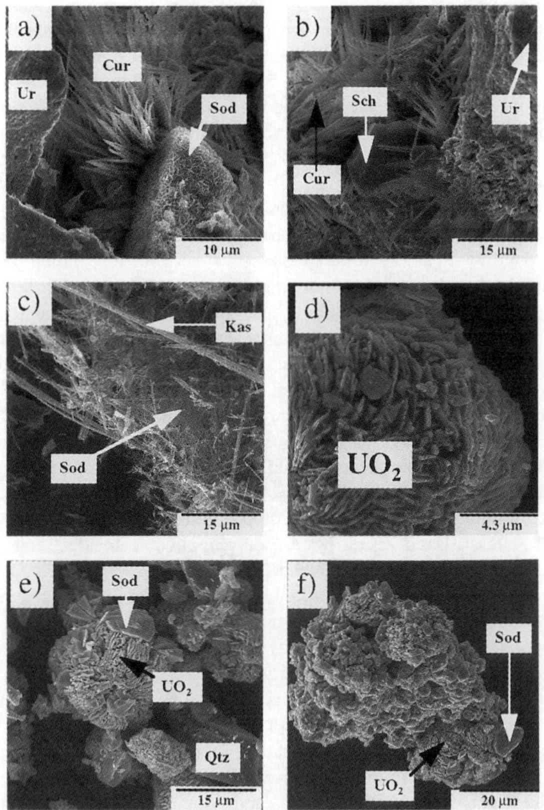

Figure 3.   (a) SEM image of a cluster of uraninite grains (Ur) that have reacted with silica-saturated water with a pH of 4 for 20 days at 150 °C. Grains have partially altered to curite (Cur) and fine-grained soddyite (Sod): (b) SEM image of a cluster of uraninite grains (Ur) that have reacted with silica-saturated water with a pH of 4 for 30 days at 150 °C. Grains have altered to curite (Cur) and schoepite (Sch): (c) SEM image of a uraninite grain (Ur) that has reacted with silica-saturated water with a pH of 4 for 115 days at 150 °C. The grain has completely altered to fine-grained soddyite (Sod) and Kasolite (Kas): (d) SEM image of an unreacted grain of synthetic UO_2: (e) SEM image of a cluster of grains of UO_2 that have reacted with silica-saturated water with a pH of 6 for 115 days at 150 °C. Grains have partially altered to soddyite (Sod): (f) SEM image of a cluster of grains of UO_2 that have reacted with silica-saturated water with a pH of 4 for 115 days at 150 °C. Grains have almost completely altered to soddyite (Sod).

DISCUSSION

Uraninite and pitchblende are easily oxidized and extremely soluble in oxidizing fluids [18]. Therefore, they vary widely in composition from $UO_{2.07}$ to $UO_{2.25}$, and are susceptible to alteration and oxidation by meteoric fluids. Oxidation of uraninite likely occurs stepwise by an oxygen interstitial mechanism, affecting only parts of the sample at a time [19, 20]. Compounds with a composition richer in oxygen than $UO_{2.37}$ have crystal structures that can differ significantly from the fluorite structure of pure uraninite. In addition, natural uraninite and pitchblende are complex minerals that consist of both U^{6+} and U^{4+} and minor amounts of Ca, Si, REEs and Th [6, 7, 20, 21, 22]. Trace elements (e.g. REE) have a great effect on the distribution of oxygen in the uraninite structure. The fluorite-type structure representing uraninite has sites approximately the size of the oxygen ion at (1/2, 0, 0) and (1/2, 1/2, 1/2) positions which are filled with oxygen atoms in the end-member UO_3. In natural uraninite, this is rarely the case, and these sites are only partially filled [20].

The major uncertainty in calculating fractionation factors is estimating frequency shifts caused by isotopic substitution, particularly for lattice vibrations required in calculating reduced partition coefficients for minerals [23]. Although Kieffer [24] formulated a set of rules for calculating these frequency shifts for simple silicate minerals, such a data set does not exist for oxide minerals with heavy elements, especially minerals containing elements, such as uranium, that exist in multiple oxidation states. Therefore, calculating vibrational frequencies for uraninite requires several assumptions (e.g. that isotopic substitution does not affect the crystal field and thus does not cause frequency shifts). In addition, the substitution of one metal ion by another (i.e. U^{4+} by U^{6+}) will affect the vibrational frequencies of the lattice nodes [25]. The theoretical uraninite-water fractionation factors of Hattori and Halas [3] and Zheng [8], likely lead to erroneous results because of the assumptions concerning errors on force constants, which translate to significant errors in the vibrational frequencies used in calculating reduced partition coefficients. Their calculations are based on end member uraninite crystal structures, which do not adequately represent the more complex natural uraninite and pitchblende compositions. Therefore, the theoretical fractionation factors of Hattori and Halas [3] and Zheng [8] are only applicable to ideal end-member structures and cannot be applied to natural uraninite and pitchblende.

The experimental interactions between uraninite and water indicate that uraninite alters rapidly to uranyl-oxide and eventually to uranyl-silicate minerals, similar to those minerals associated with natural uraninite and pitchblende samples from unconformity-type uranium deposits in Saskatchewan. Therefore, the simple exchange of oxygen isotopes between natural uraninite and fluids may occur only to a limited extent without uraninite changing its chemical composition or altering to new phases. Primary uranium minerals from the Saskatchewan deposits with $\delta\ ^{18}O$ values of ca. -28 per mil likely have retained, in large part, their original oxygen isotopic composition rather than having their oxygen isotopic composition completely overprinted by low-temperature ground water. This is geologically reasonable because of the high grade nature of the unconformity-type uranium deposits in Saskatchewan, where most of the uranium mineralization (~90 percent) consists of uraninite and pitchblende which are reduced uranium-bearing mineral phases. Large scale interaction between uraninite or pitchblende and oxidizing groundwater would likely have extensively altered uraninite and pitchblende to uranyl-oxide and uranyl-silicate minerals, which are only present in unconformity-type uranium deposits as minor phases.

Recent studies of uraninite from unconformity-type uranium deposits have suggested that U_3O_7 is present in these ores [13, 14, 15]. They also suggested that the dissolution rate of uranium oxide does not become significant until it is oxidized beyond U_3O_7 [13]. Therefore, the presence of UO_2, U_4O_9, and U_3O_7, and the absence of U_3O_8 or higher oxides is indicative of reducing conditions, and U_3O_7 is a rate limiting phase in the dissolution of uraninite [13]. However, this study clearly shows that in the presence of oxidizing fluids uraninite initially alters to uranyl-oxide hydrate minerals and eventually to uranyl-silicate minerals, when sufficient silica is present. In addition, Janeczek et al. [12] analyzed the data from the studies mentioned above, as well as conducted their own study, in an effort to identify U_3O_7 in natural samples. They concluded that there was no evidence of U_3O_7 and the reported occurrences of U_3O_7 were likely mixtures of isometric uraninites of slightly different compositions.

The rapid alteration of uraninite and pitchblende in the presence of an oxidizing fluid must be considered: in the disposal of spent nuclear fuel in underground vaults where the UO_2 may come into contact with oxidizing ground waters and in storing uranium ores on the surface (e.g., uranium mine mill tailings) where the ore is in direct contact with highly oxidizing meteoric waters. Although alteration of uraninite and pitchblende may be detected in hand specimen, the alteration can occur on a micro-scale, making detection more difficult.

CONCLUSIONS

Anomalously low $\delta^{18}O$ values (-32 per mil to -19.5 per mil) of high reflectance primary uranium mineralization from unconformity-type uranium deposits in Saskatchewan have been interpreted to have resulted from the relatively recent interaction between uranium minerals and low-temperature meteoric fluids [4, 5, 6, 10]. The high reflectivities and lack of conspicuous alteration (i.e. high SiO_2 and CaO) of primary uranium minerals are interpreted to indicate that uranium minerals can exchange oxygen isotopes with fluids resulting in only minor disturbances to their chemical composition and texture [6]. However, uraninite-water experiments show that uraninite, in the presence of an oxidizing fluid, readily alters to uranyl-oxide and uranyl silicate minerals such as curite, schoepite, becquerelite, and soddyite, which are similar to uranyl minerals found in natural uraninite samples, but not to U_3O_7. Electron microprobe analyses and paragenetic studies of natural uraninite and pitchblende samples [7] indicate that, in addition to the formation of uranyl-oxide minerals and coffinite, alteration of uraninite causes an increase in Ca and Si contents of uraninites and oxidizes uranium. The alteration may occur on a micro-scale which is difficult to detect; however, it increases the $\delta^{18}O$ values of uraninites. Uraninite likely cannot exchange oxygen with an oxidizing fluid without changing its chemical composition or altering to a new phase.

In calculating reduced partition coefficients, theoretical uraninite-water fractionation factors are based on the fluorite structure, which is isostructural with stoichiometric UO_2, because of the paucity of thermodynamic data available for more complex uraninite structures. However, stoichiometric UO_2 rarely exists in nature because some of the uranium is readily oxidized to U^{6+}. Consequently natural uraninite samples are complex mixtures of both U^{6+} and U^{4+}, with minor amounts of Ca, Si, and REEs. Uranium compounds that deviate from stoichiometric UO_2 have crystal structures that differ from the fluorite structure. Therefore, theoretical uraninite-water fractionation factors cannot be appropriately applied to natural uraninite samples. Based on the uncertainty of these theoretical fractionation factors in conjunction with the results of the experiments in this study and lack of alteration in natural uraninite, primary uraninite minerals likely have retained their original oxygen isotopic compositions.

ACKNOWLEDGMENTS

The Natural Sciences and Engineering Research Council of Canada (NSERC) supported this work via operating and equipment grants to T.K. Kyser and a Post-Doctoral Fellowship to M. Fayek. This work was also partially supported by the Basic Energy Science DOE grant DE-FG03-95ER14540. The Authors also wish to thank Drs. Audunn Ludviksson, Leo Archer and Vlad Sopuck for their help in collecting the data for this manuscript.

REFERENCES

1.   R.J. Finch and R.C. Ewing, J. Nucl. Mater., **190**, p. 133-156 (1992).

2.   J. Janeczek, R.C. Ewing, V.M. Oversby and L.O. Werme, J. Nucl. Mater., in press (1996).

3.   K. Hattori and S. Halas, Geochim. Cosmochim. Acta, **46**, p. 1863-1868 (1982).

4.   H.R. Hoekstra and J.J. Katz, U.S. Geological Survey Professional Paper **300**, p. 543-547 (1956).

5.   K. Hattori, K. Muehlenbachs, and D. Morton, Geol. Soc. Amer.-Geol. Soc. Can., Pgm. w Abst., **10**, A417 (1978).

6.   T.G. Kotzer and T.K. Kyser, Am. Mineral., **78**, p. 1262-1274 (1993).

7.   M. Fayek and T.K. Kyser, accepted, Can. Min (1996).

8.   Y. Zheng, Geochim. Cosmochim. Acta., **55**, p. 2299-2307 (1991).

9.   M.R. Wilson and T.K. Kyser, Econ. Geol., **82**, p. 1450-1557 (1987).

10.   T.G. Kotzer and T.K. Kyser, Sask. Energy and Mines, Sask. Geol. Surv., Misc. Rep. **90-4**, p. 153-157 (1990).

11.   T.G. Kotzer and T.K. Kyser, Chem. Geol., **120**, p. 45-89 (1995).

12.   J. Janeczek, R.C. Ewing and L.E. Thomas, J. Nucl. Mater., **207**, p. 177-191 (1993).

13.   J.J. Cramer and J.A.T. Smellie, AECL Rep. **10851** (1994).

14.   J.J. Cramer, AECL Rep. **11204** (1995).

15.   J.A.T.Smellie and F. Karlsson, SKB Tech. Rep. **TR 96-08**, (1996).

16.   J.F.W. Bowles, Chem. Geol., **83**, p. 47-53 (1990).

17.   D.J. Wronkiewicz, J.K. Bates, T.J. Gerding, E. Veleckis and B.S. Tani, J. Nucl. Mater., **190**, p. 107-127 (1992).

18.   D.E. Grandstaff, Econ. Geol., **71**, p. 1493-1506 (1976).

19.   F. Gronvold, J. Inorg. Nucl. Chem., **1**, p. 357-370 (1955).

20.   R.M. Berman, Journal of the Mineralogical Society of America, **42**, p. 705-731 (1957).

21.   L. Powers and M. Stauffer, Can. J. Earth Sci., **25**, p 1945-1954 (1985).

22.   J. Janeczek and R.C. Ewing, R.C., J. Nucl. Mater., **190**, p. 128-132 (1992).

23.   J.R. O' Neil, <u>Stable Isotopes in High Temperature Geological Processes</u>, eds. J. W. Valley, H.P. Taylor, J.R. O' Neil (Reviews in Mineralogy, 16, Mineral. Soc. of Amer, 1986), p. 1-37.

24.   S.W. Kieffer, Rev. Geophys. Space Phys., **30-4**, p. 827-849 (1982).

25.   R.H. Becker and R.N. Clayton, Geochim. Cosmochim. Acta, **40**, p. 1151-1165 (1976).

URANINITE: A 2 GA SPENT NUCLEAR FUEL FROM THE NATURAL FISSION REACTOR AT BANGOMBÉ IN GABON, WEST AFRICA

K.A. Jensen[*], R.C. Ewing[**], F. Gauthier-Lafaye[***]

[*] Department of Earth Sciences, Aarhus University, DK-8000 Aarhus C, keld@geo.aau.dk

[**] Department of Earth and Planetary Sciences, University of New Mexico, Albuquerque, NM 87131

[***] Centre National de la Recherche Scientifique, Centre de Geochemie de la Surface, 67084 Strasbourg Cedex, France

ABSTRACT

Uraninites from the Bangombé natural fission reactor (RZB) and "normal" uranium-ore occur as fine veins in the sandstone host-rock as well as altered, broken, and slightly displaced grains in an illitic matrix, and in nodules and veins of solid bitumen. Inclusions of galena, (Y,Gd)-rich phosphates, a Pb-oxide and a Ti-oxide? were observed. Uraninites just below RZB were partially altered to a uranyl-sulfate. Three generations of uraninite were identified based on their PbO-contents of 8-11.06 wt%, 6 wt% (the largest population), and a younger generation with 3 wt%. The high Pb-uraninites may be the precursor to the low Pb-uraninites. Diffusional loss of Pb is indicated by the presence of a Pb-oxide at the interface to the uraninites. The behaviour of the metallic fission products, incompatible with the uraninite structure, may mimic the behaviour of Pb in these uraninites. The averaged impurity-content ranges from 4.29 to 6.89 wt%, and consists mainly of SiO_2, TiO_2, ZrO_2, FeO, CaO, Al_2O_3 and P_2O_5. The averaged content of Y_2O_3 and the Ln's is less than 0.78 wt% and there is a scattered positive correlation with P_2O_5. The content of Y + Ln's is generally highest in the uraninites from RZB. Uraninite hydration and the formation of "uranopelite/zippeite" have caused complete loss of Y and the Ln's. These elements seems also to be partially lost by weak phosphatian coffinitization. The analytical results indicate that Y and the Ln's, which are high yield fission products, may be released from uraninite during alteration in the presence of P.

INTRODUCTION

Uraninite (UO_{2+x}) from the natural fission reactors and "normal" ore at Oklo, Okelobondo and Bangombé in Gabon, West Africa, is studied as a natural analogue to the UO_2 in spent nuclear fuel [1,2,3,4,5]. A detailed study of the mineralogy of the Bangombé ore-deposit has recently been initiated. In this paper we present a preliminary petrographic and chemical description of uraninite from the Bangombé ore body. Our final goal is to obtain a better understanding of the long-term physical and chemical stability of uraninite in the geological environment.

The Bangombé uranium-ore deposit is located ~ 25 km W-NW of Franceville in SE-Gabon, and includes a high grade ore-body (~50 wt% U) in which natural fission reactions started approximately 1.97 Ga ago [1,2,6]. The ore body is situated in an anticlinal fold at the top of the FA-sandstone, the basal formation of the Lower Proterozoic sedimentary Franceville Series [1]. The amounts of fissiogenic Nd and Sm indicate that the maximum burnup of U-235 in RZB (**R**eactor **Z**one **B**angombé) was approximately 1.25 %, and the neutron flux was up to $0.488 \ 10^{21}$ n/cm^2 [6,7].

RZB has been verified in three drill-cores (BA145, BA145bis, and BAX03). In BAX03, the reactor core is approximately 5 cm thick (the maximum observed thickness in RZB) and located 11.80 to 11.85 m below the surface. This is approximately 1.1 m above the normal

Mat. Res. Soc. Symp. Proc. Vol. 465 © 1997 Materials Research Society

water table in the drill hole [8]. In BAX03, the reactor core is overlain by a 0.3 m thick clay mantle which was produced by hydrothermal alteration of the surrounding rocks during criticality [2,9]. The clay mantle is covered by a 2.2 m thick layer of black shales (rich in organic material) followed by green pelites which both belong to the so-called FB-formation of the Franceville series [3,6]. In BA-145 and BA-145bis, RZB is very thin and located at 11.60 m depth. The uraninites in the RZB typically lie in an illitic groundmass. However, uraninite in the reactor samples from BA145 and BA145bis are also strongly associated with organic matter occurring as solid bitumen. Immediately beneath RZB, the FA-sandstone is highly silicified and contains nodules and veins of uraninite-bearing solid bitumen.

Since criticality ceased (after ~ 300,000 years of reactor operation [10]), the sediments in the Franceville basin have undergone a late diagenesis [11]. Subsequently, a 950-750 Ma regional thermal event associated with dolerite dyke emplacement caused local remobilization of U and a loss of Pb from most of the uraninites in the ore-bodies [5,12,13]. This resulted in abundant galena formation. At Oklo, a later remobilization event (330 Ma) has been suggested by Nagy et al. [14]. A similar age (375 Ma) was obtained by Gancarz [15], who argued that this was an artifact of volume diffusion of Pb. Today, the Bangombé uranium-deposit is exposed to supergene alteration. This involves kaolinitization of the phyllosilicates (illite and chlorite) and strong Fe-enrichment at about 9 m depth [6].

ANALYTICAL METHODS

The uraninite texture, alteration and mineral chemistry were studied in seven samples from four drill cores (BAX02, BAX03, BA-145 and BA-145ter; see Table 1). Sample 705 is from RZB; 854B and 859B are also from RZB or very close by. Sample 801 is from just below RZB in BAX03, while the samples 783, 786 and 790 are from BAX02 where criticality has not been demonstrated.

The textures were described by optical microscopy. The mineral chemistry was determined by electron microprobe analyses completed with a JEOL 733 Electron Microprobe at the University of New Mexico, and a JEOL JXA-8600 Superprobe at the Aarhus University, Denmark. The electron microprobes were operated with an acceleration voltage of 20 kV and a beam current of 20 nA. All elements were analysed with Wave Dispersion Spectrometry, except O, which was calculated by stochiometry.

Table 1: Information about the samples used for the textural and chemical study of uraninite. Abbreviations: Ass. = associated, w/ = with, SB = solid bitumen.

Sample #	Drill-core	Depth (m)	Uraninite association
UNM783	BAX02	10.10-10.20	uraninite ass. w/ SB-nodules
UNM786	BAX02	11.45	uraninite ass. w/ SB-nodules and a thin uraninite vein
UNM790	BAX02	13.35	uraninite ass. w/ SB-nodules and a SB-vein
UNM801	BAX03	12.20-12.30	altered uraninite veins and grains in the sandstone
UNM705	BA145	11.00-11.80	uraninite grains in an illite matrix + in SB nodules
UNM854B	BA-145ter	11.60	uraninite ass. w/ SB-nodules and a SB-vein
UNM859B	BA-145	11.60	uraninite ass. w/ SB-nodules and in phyllosilicate veins

Nineteen elements (U, Th, Pb, Si, Ti, Zr, Ni, Co, Fe, Ca, Mg, Al, Sr, Y, La, Ce, Gd, and S) were analysed simultaneously with the JEOL 733. The standards included UO_2 (U); ThO_2 (Th); galena (Pb, S); kyanite (Si, Al); rutile (Ti); zirconia (Zr); metallic Ni and Co; chromite (Fe), scheelite (Ca); olivine (Mg); $SrBaNb_4O_{10}$ (Sr); and phosphates for P, Y, La, Ce and Gd.

The elements were analysed for a maximum of 40 sec or until 30,000 counts were reached on the peak position. The background positions were analysed for a maximum of 20 sec or until 15,000 counts. Results were accepted which had a standard deviation of 3 sigma or less on the counts. The JEOL 733 was operated with the Oxford Link system and data reduction was performed by ZAF-corrections.

On the JXA-8600, twenty elements were analysed in two sessions. Ni, Co and Sr from the setup on the JEOL-733 were not analysed. Instead, the content of Pr, Nd, Sm, and Eu were analysed in addition to the previously mentioned elements. The elements were calibrated against UO_2 (**U**); $ThSiO_4$ (Th); PbS (**Pb**); albite (Si, **Al**); $CaTiO_3$ (**Ca**, Ti); olivine (Mg); troilite (S), and phosphates for **P**, <u>Y</u>, <u>La</u>, **Ce**, <u>Pr</u>, <u>Nd</u>, <u>Sm</u>, <u>Eu</u>, and <u>Gd</u>. The underlined elements were analysed in the second session, while those in bold were analysed in both the first and second sessions to ensure a better matrix correction. The elements were analysed for up to 60 sec or until the standard deviation reached a value of 1%; the minimum time of measurement was, however, set to 5 sec. The background intensity was counted for 30 sec on each position. The results were accepted by exceeding the minimum detection limit. The JXA-8600 was operated with the Tracor Northern Software and data reduction was performed by PRZ-corrections.

Peak-overlaps between the Ln's (lanthanides), and between $Nd_{L\alpha}$ and $Pb_{L\alpha2}$ were controlled by measurements on the standards. The peak-overlaps between the Ln's are neglible at the measured concentrations. The Nd_2O_3-content was reduced by [-(0.37/96)*wt%PbO].

RESULTS

In the "normal" uranium-ore outside RZB, the uraninites typically occur as up to 0.05 mm large inclusions in veins and nodules of solid bitumen (figure 1a). Uraninite outside RZB does, however, also appear in thin fracture veins and as single grains emplaced in quartz dissolution cavities in the sandstone close to the reactor (figure 1a and 2a,b).

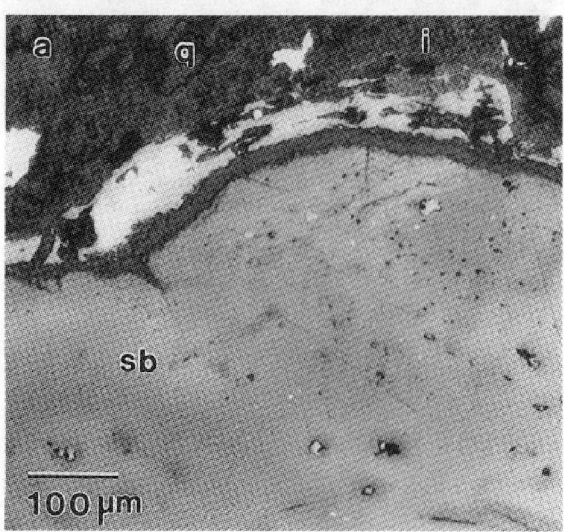
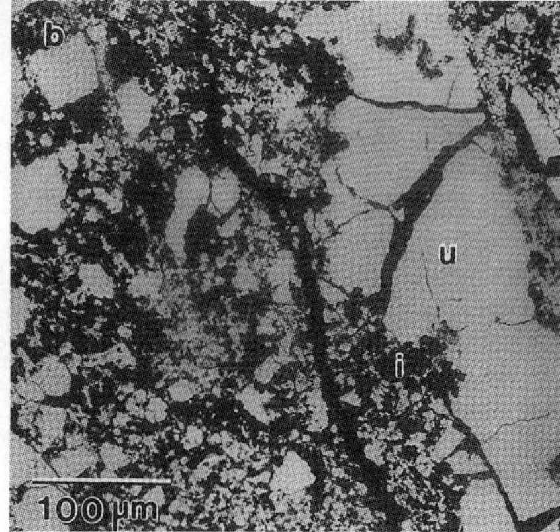

Figure 1: Microtextures observed in the samples from Bangombé U-ore body. **a)** Reflected light micrograph from 859B: Uraninite inclusions in a nodule of solid bitumen (sb) and a uraninite vein at the interface between the solid bitumen and the altered sandstone (q=quartz and i=illite). The grains with high reflectivity are galena. **b)** Backscattered electron image of 705 showing the corroded and brecciated uraninites (u) from the reactor core. The matrix mainly consists of illite (i).

In the reactor core (705), the uraninites lie in a groundmass of illite and minor amounts of Mg-rich chlorite partially impregnated with Fe_2O_3 (figure 1b). The grains are fractured and have a corroded rim. The grain size varies from a few microns to approximately 0.5 mm. Some of the fractured grains are slightly displaced. The uraninites in the other samples (854B and 859B) from or close to the reactor core mainly occur in association with solid bitumen. The sandstone in which these nodules and veins are emplaced is much more fractured as compared with the sandstone further from the RZB. These fractures are filled with uraninite, sulfides (galena, chalcopyrite and/or pyrite) and phyllosilicates. Uraninite also occurs as coatings on the solid bitumen.

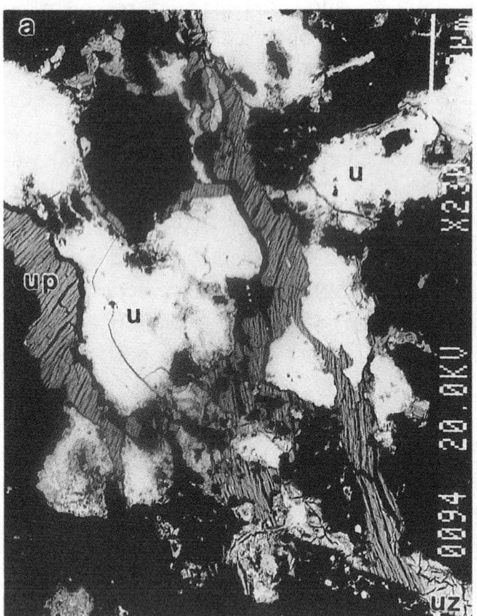
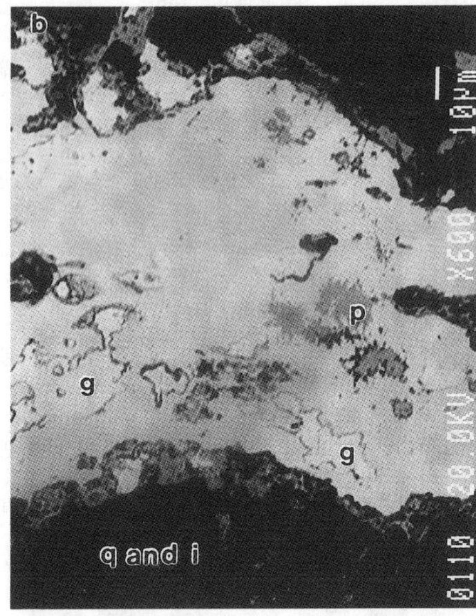

Figure 2: Backscattered electron images showing the microtextures observed in samples 801 (a) and 859B (b) from the Bangombé uranium ore-deposit. **a)** A vein filled with uranopelite/zippeite (uz) which has been cut by a later vein filled with a uranyl phosphate (up). Note the laminated structure of the uranyl phosphate. Uranopelite/zippeite also occurs at the rim of the corroded uraninites (u). The dark matrix consists of quartz. **b)** The image shows part of a uraninite vein in 859B. The uraninite vein contains inclusions of galena (g) and a Pb-oxide (minium Pb_3O_4?) (p). The Pb-rich phase is also observed at the interface between the uraninite vein and the altered sandstone consisting of quartz (q) and illite (i).

The uraninites trapped in the solid bitumen often display a spherical anhedral morphology. Euhedral uraninites are seldom observed. The uraninite inclusions are sometimes broken and may be displaced by only a few tens of μm's. The majority of the uraninite inclusions appear corroded, as observed in 705. The uraninite inclusions are also commonly rimmed by a TiO_2-phase (rutile?). The solid bitumen generally appears as a compact and homogeneous solid. However, some of the solid bitumen veins have regions showing flow structures and regions where up to 0.25 mm large individual "drops" can be recognized.

Three samples (783, 801, 859B) from the FA-sandstone contain uraninite veins with a maximum thickness of approximately 0.5 mm. The uraninites in 801 have clearly been subjected to hydrous alteration, as some of them are rimmed by a yellow-greenish alteration product. Microprobe analyses show the presence of U and S, and the phase may be either uranopelite ($[UO_2])_6SO_4(OH)_{10}].12(H_2O)$ or zippeite ($[UO_2])_3(SO_4)_2(OH)_2].8(H_2O)$. In 801 uranopelite/zip-peite was also observed in a vein which is cut by a later ~75 μm wide fracture filled with a uranyl-phosphate (figure 2a). Recent alteration was observed in 790, where parts of a solid bitumen nodule, adjacent to an illite vein, contained UO_2, but did not contain any Pb.

The uraninites normally contain small galena inclusions which are sometimes several μm large (figure 1a and 2b). Microprobe studies also indicate the presence of hydrous phosphate inclusions, rich in Y and Ln (see Y-phos; table 2). High Ti-contents in a few microprobe analyses indicate that Ti-oxide may also occur as inclusions in the uraninites. In 859B, several Pb-rich patches (minium: Pb_3O_4?) were observed in addition to abundant galena inclusions in a uraninite vein (figure 2b). The Pb-oxide was also observed at the interface to several of the uraninite veins in this sample.

A total of 113 sulfur free (<0.05 wt%) uraninite compositions were analysed. The uraninites in 801 were very beam-sensitive, and the UO_2-content decreased by almost 4 wt%, when analysed as the last unknown. The same effect was observed for PbO. This may be due to partial hydroxylation of U. Representative analyses and average compositions for the uraninites are given in Table 2.

The uraninites primarily consist of UO_2 and PbO with up to almost 7 wt% impurities (Table 2). The averaged UO_2-content is quite constant and varies between 85.05 wt% and 87.78 wt%. The averaged PbO-content varies between 3.05 wt% (801) and 7.74 wt% (859B). However, some grains have higher PbO-contents (figure 3a). A plot of UO_2 vs. PbO shows that the analyses cluster around three compositions (figure 3a). One cluster is located at 88 wt% UO_2: 6 wt% PbO and another at 87 wt% UO_2: 3 wt% PbO. A third group contain 8 to 9 wt% PbO, and a single grain has a PbO-content of 11.06 wt%.

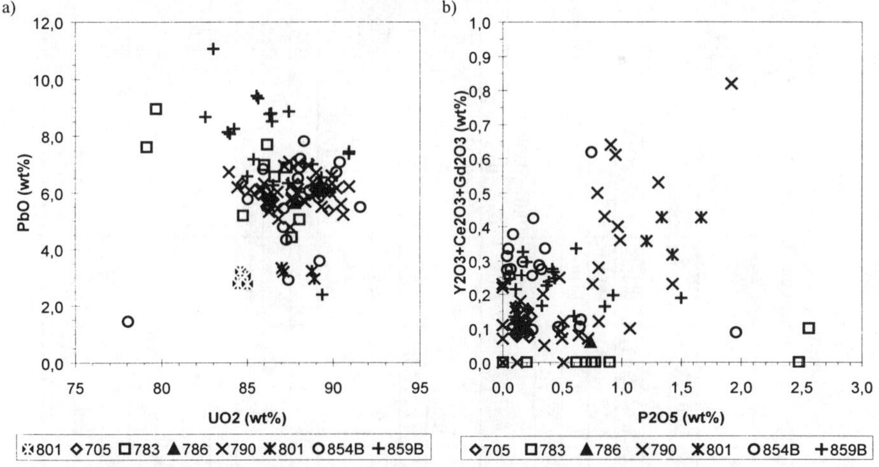

Figure 3: **a)** Plot of the UO_2 vs. PbO in the unaltered uraninites (Table 1) from the Bangombé U-ore body. **b)** Plot of P_2O_5 vs. $Y_2O_3+Ce_2O_3+Gd_2O_3$ for all the unaltered uraninites.

Table 2: Uraninite analyses in wt.% from the U-ore deposit at Bangombé. The last two analyses are a phosphate inclusion (Y-phos) in a uraninite grain and a completely hydroxylized uraninite grain (Uran$_{OH}$). Sample number 705 is from RZB. 854B and 859B are also from RZB or very close by. The other samples represent the "normal" U-ore body. Abbrevations: n is number of grains analysed. - not analysed element. < the average content (if present) and the standard deviation for the population (shown in brackets) is below 0.01.

Oxide	783 (n=9)	786 (n=1)	790 (n=31)	801 (n=9)	854B (n=19)	859B (n=21)	705 (n=23)	783 Y-phos	786 Uran$_{OH}$
UO_2	85.05 (3.15)	87.78	85.19 (1.73)	86.16 (1.73)	87.04 (2.70)	86.29 (2.44)	87.36 (1.72)	58.94	77.16
ThO_2	<	<	<	0.12 (0.03)	0.01 (0.02)	<	0.09 (0.03)	<	<
SiO_2	0.94 (0.65)	0.40	0.89 (0.21)	0.14 (0.06)	0.83 (0.28)	0.80 (0.27)	0.98 (0.18)	5.31	0.27
TiO_2	2.23 (1.46)	0.86	1.26 (0.70)	0.36 (0.13)	2.32 (2.11)	0.79 (0.65)	0.25 (0.07)	1.21	<
ZrO_2	<	0.14	0.06 (0.10)	0.16 (0.06)	1.01 (0.49)	0.08 (0.05)	0.28 (0.22)	<	<
PbO	6.59 (1.38)	5.63	5.85 (0.59)	3.05 (0.20)	5.41 (1.58)	7.74 (1.67)	5.99 (0.31)	0.86	<
SrO	<	<	0.03 (0.04)	-	-	-	-	<	<
NiO	0.01 (0.01)	0.03	0.03 (0.02)	-	-	-	-	<	<
CoO	<	<	0.01 (0.02)	-	-	-	-	<	<
FeO	0.77 (0.16)	0.44	0.58 (0.26)	1.11 (0.10)	0.63 (0.09)	0.81 (0.30)	0.62 (0.11)	0.12	0.07
CaO	0.52 (0.36)	1.62	1.46 (0.37)	0.49 (0.13)	1.12 (0.37)	1.07 (0.36)	1.11 (0.13)	1.71	<
MgO	0.02 (0.01)	0.02	0.08 (0.05)	0.04 (0.01)	0.03 (0.01)	0.03 (0.01)	0.38 (0.10)	<	<
Al_2O_3	0.47 (0.42)	0.19	0.22 (0.13)	0.27 (0.04)	0.11 (0.07)	0.08 (0.03)	0.12 (0.05)	0.65	0.01
P_2O_5	1.00 (0.86)	0.74	0.60 (0.47)	1.38 (0.12)	0.39 (0.43)	0.41 (0.34)	0.18 (0.05)	8.95	<
Y_2O_3	<	<	0.11 (0.18)	0.01 (0.01)	0.07 (0.05)	0.03 (0.03)	0.01 (0.02)	2.61	<
La_2O_3	-	-	-	0.07 (0.02)	0.02 (0.02)	0.02 (0.02)	0.02 (0.02)	-	-
Ce_2O_3	0.01 (0.03)	0.06	0.12 (0.08)	0.30 (0.03)	0.17 (0.06)	0.16 (0.04)	0.09 (0.02)	0.09	<
Pr_2O_3	-	-	-	0.05 (0.04)	0.01 (0.03)	0.01 (0.01)	0.01 (0.02)	-	-
Nd_2O_3	-	-	-	0.27 (0.04)	0.09 (0.05)	0.12 (0.05)	0.09 (0.03)	-	-
Sm_2O_3	-	-	-	0.04 (0.02)	0.02 (0.03)	0.05 (0.04)	0.02 (0.03)	-	-
Eu_2O_3	-	-	-	<	<	<	<	-	-
Gd_2O_3	<	<	<	0.04 (0.03)	0.05 (0.09)	0.03 (0.03)	0.01 (0.02)	0.76	<
S	0.01 (0.02)	<	< (0.01)	<	0.01 (0.01)	0.01 (0.01)	0.04 (0.02)	<	<
Total	97.62 (1.55)	97.91	96.49 (2.00)	94.06 (2.03)	99.34 (1.71)	98.52 (1.97)	97.64 (1.79)	81.21	77.51

DISCUSSION

The general textural features of the uraninites observed are consistent with previous observations [3,16,17]. However, some of the new observations are the presence of the phosphate inclusions, the Pb-oxide associated with the uraninite veins just below RZB, the uranyl phases, and the locally hydrous altered uraninites in the sandstone close to RZB.

The morphology of the individual uraninite grains in the ore-body suggests that most of the grains have been subjected to corrosion at some time during their geological history. However, the homogeneity of the uraninites in 705 indicates that this ancient corrosion had negligble effect on the present day chemistry of the residual grains. The presence of corroded uraninite inclusions and broken grains in the solid bitumen suggest that either (1) corrosion of these uraninites occurred before they were trapped by the organic matter, and/or (2) the solid bitumen no longer protects the uraninites against hydrous alteration at Bangombé. This needs to be further clarified.

The chemistry of the uraninites is complex and they contain a wide variety of impurities which add up to 7 oxide wt% in addition to the PbO-content. Based on ionic radii, the majority of the most abundant impurities (Si, Ti, Zr, Pb, Al, and P) are not expected to substitute for U in uraninite. However, experimental studies have shown that approximately 0.2 wt% ZrO_2 can substitute for synthetic UO_2 at 1473°K and the amount increases at higher temperatures (T>1573°K) [18]. Only the ZrO_2-content in the uraninites in sample 705 from RZB, which formed at much lower temperatures, exceed this experimental value (table 2).

The PbO-content in the uraninites indicates the presence of three generations of uraninite: (1) The high PbO-generation (8-11 wt% PbO); (2) the 6 wt% PbO generation, which has been observed before [5]; and (3) the 3 wt% PbO generation, which was observed in fracture veins and nodules of solid bitumen with flow structures. The events related to generations 2 and 3 are presently unknown, but texturally generation 3 seems to have been formed by remobilisation of U and organic matter. This is further supported by the distinct difference in the composition of the minor elements as compared with the other grains (higher contents of Th, Y and Ln's are most significant). The high Pb-uraninites in generation 1 have similar PbO-contents (~10 wt%) as those which recrystallized during the regional heating 950-750 Ma ago (see [5]). At least the uraninites with intermedier PbO-contents may have been produced by diffusional loss of Pb from the high Pb-uraninites. This is further indicated by the presence of the Pb-oxide which occurs in association with the uraninite veins in 859B (figure 2b).

Lead is not of concern with respect to the disposal of spent nuclear fuel. However, this behaviour of Pb in uraninite may resemble the behaviour of the metallic fission products (e.g. Mo, Tc, Ru, Rh, and Pd) in spent nuclear fuel, as previously suggested by Janeczek and Ewing [5]. In that respect, figure 2b becomes quite important since it shows that such elements may be liberated from the UO_2-matrix to the surrounding environment where the distance of migration will depend on the local physico-chemical conditions.

Yttrium and the Ln's are among the high yield fission products of U-235 and are expected to appear in high concentrations in the samples from the RZB (705, 859, and 854B). Experimental studies have shown that these elements are easily incorporated into synthetic UO_2 with more than 48 mole% at temperatures above 1273°K [18]. Phosphorus may also be produced during fission reactions by neutron capture by Si. A positive correlation between P and Y+Ln is then expected as shown in figure 3, but the $(Y_2O_3+Ce_2O_3+Gd_2O_3)/P_2O_5$-ratio shows that P is more dominant in the uraninites from outside RZB. This suggests that the primary ore-forming fluids had a relatively high concentration of P rather than P having been formed during the fission reactions. The heat released during the fission reactions and/or the

DISCUSSION

The general textural features of the uraninites observed are consistent with previous observations [3,16,17]. However, some of the new observations are the presence of the phosphate inclusions, the Pb-oxide associated with the uraninite veins just below RZB, the uranyl phases, and the locally hydrous altered uraninites in the sandstone close to RZB.

The morphology of the individual uraninite grains in the ore-body suggests that most of the grains have been subjected to corrosion at some time during their geological history. However, the homogeneity of the uraninites in 705 indicates that this ancient corrosion had negligble effect on the present day chemistry of the residual grains. The presence of corroded uraninite inclusions and broken grains in the solid bitumen suggest that either (1) corrosion of these uraninites occurred before they were trapped by the organic matter, and/or (2) the solid bitumen no longer protects the uraninites against hydrous alteration at Bangombé. This needs to be further clarified.

The chemistry of the uraninites is complex and they contain a wide variety of impurities which add up to 7 oxide wt% in addition to the PbO-content. Based on ionic radii, the majority of the most abundant impurities (Si, Ti, Zr, Pb, Al, and P) are not expected to substitute for U in uraninite. However, experimental studies have shown that approximately 0.2 wt% ZrO_2 can substitute for synthetic UO_2 at 1473°K and the amount increases at higher temperatures (T>1573°K) [18]. Only the ZrO_2-content in the uraninites in sample 705 from RZB, which formed at much lower temperatures, exceed this experimental value (table 2).

The PbO-content in the uraninites indicates the presence of three generations of uraninite: (1) The high PbO-generation (8-11 wt% PbO); (2) the 6 wt% PbO generation, which has been observed before [5]; and (3) the 3 wt% PbO generation, which was observed in fracture veins and nodules of solid bitumen with flow structures. The events related to generations 2 and 3 are presently unknown, but texturally generation 3 seems to have been formed by remobilisation of U and organic matter. This is further supported by the distinct difference in the composition of the minor elements as compared with the other grains (higher contents of Th, Y and Ln's are most significant). The high Pb-uraninites in generation 1 have similar PbO-contents (~10 wt%) as those which recrystallized during the regional heating 950-750 Ma ago (see [5]). At least the uraninites with intermedier PbO-contents may have been produced by diffusional loss of Pb from the high Pb-uraninites. This is further indicated by the presence of the Pb-oxide which occurs in association with the uraninite veins in 859B (figure 2b).

Lead is not of concern with respect to the disposal of spent nuclear fuel. However, this behaviour of Pb in uraninite may resemble the behaviour of the metallic fission products (e.g. Mo, Tc, Ru, Rh, and Pd) in spent nuclear fuel, as previously suggested by Janeczek and Ewing [5]. In that respect, figure 2b then becomes quite important since it shows that such elements may be liberated from the UO_2-matrix to the surrounding environment where the distance of migration will depend on the local physico-chemical conditions.

Yttrium and the Ln's are among the high yield fission products of U-235 and are expected to appear in high concentrations in the samples from the RZB (705, 859, and 854B). Experimental studies have shown that these elements are easily incorporated into synthetic UO_2 with more than 48 mole% at temperatures above 1273°K [18]. Phosphorus may also be produced during fission reactions by neutron capture by Si. A positive correlation between P and Y+Ln is then expected as shown in figure 3, but the $(Y_2O_3+Ce_2O_3+Gd_2O_3)/P_2O_5$-ratio shows that P is more dominant in the uraninites from outside RZB. This suggests that the primary ore-forming fluids had a relatively high concentration of P rather than P having been formed during the fission reactions. The heat released during the fission reactions and/or the regional heating 950-750 Ma ago may have caused aggregation of the impurities and formation of the phosphate inclusions. Apatite inclusions with fissiogenic Ln have been reported from reactor zone 10 at Oklo [19]. The hydrothermal influence during the fission reactions, diagenesis and subsequent regional heating may have caused additional loss of P from RZB. This release of P and combination with actinides and Ln's could explain the formation of crandallite group minerals with fissiogenic Ln's which are observed in the surrounding rocks at Bangombé [7].

Coffinitization has previously been described as a mechanism for the alteration of uraninite under reducing conditions [e.g. 20]. Even though coffinite is a well known alteration product, the elemental behaviour of the impurities in uraninite during coffinitzation is almost unknown. This investigation and previous studies [20] indicate that Pb is readily lost from uraninite by this process. However, loss of Pb may also be caused by diffusion, as the low Pb-content is observed in the outer 10 μm of the vein (figure 3). The behaviour of Y and the Ln's is unclear at this stage of alteration. However, the Ce and Nd contents seem to decrease at the rim of the vein. Loss of Y and Ln's during coffinitization is unexpected, but similar behaviour was observed in reactor zone 10 at Oklo [21].

Hydrous alteration results in the formation of uranyl phases such as schoepite, uranopelite, zippeite and uranyl orthophosphate. The hydrous alteration phases observed here are possibly uranopelite or zippeite and hydroxylized uraninite/schoepite (Table 2). Microprobe studies of these grains indicated no presence of Y and Ln. Hence these elements do not appear to substitute for U in the uranyl phases and therefore seem to be lost during the hydrous alteration.

CONCLUSIONS

- The uraninites from the Bangombé ore deposit are generally corroded and partially altered. They occur in an illite matrix (RZB), as veins and grains in the sandstone, and as inclusions in veins and nodules of solid bitumen. The uraninites contain inclusions of Y- and Ln-rich phosphates, a Ti-oxide, a Pb-oxide, and galena.
- Three generations of uraninites with 8 to 11.06 wt%, 6 wt%, and 3 wt% PbO are observed.
- The impurities in the uraninites mainly consist of SiO_2, TiO_2, ZrO_2, FeO, CaO, Al_2O_3 and P_2O_5 and reach a maximum average concentration of 6.89 wt%, in addition to the content of PbO. The average content of Y_2O_3 and the Ln's is less than or equal to 0.78 wt%.
- Loss of Pb seems to have occurred by diffusion from some uraninites and resulted in the formation of a Pb-oxide at the rim of the vein in sample 859B. The behaviour of metallic fission products in spent nuclear fuel may resemble the behaviour of Pb in these uraninites.
- The uraninites from or close to RZB have a higher content of Y and Ln's relative to P as compared with the normal uranium ore. Y+Ln has a positive correlation with P. Hence, P may have been an important constituent of the primary ore-forming fluids. Phosphorus may also have played a role in actinide and REE-migration from the RZB.
- Coffinitization is very limited, but this process is related to Pb-release from the uraninites and may also be related to the release of Y and Ln's in the presence of P.
- The uraninites have locally been subjected to hydrous alteration. In some cases this has resulted in the formation of uranyl phases and loss of Y and the Ln's.

ACKNOWLEDGEMENTS

We thank the Commissariat á l'Energie Atomique, France, for providing the well-documented samples; Janusz Janeczek for discussions of the analytical results; and Sidsel Grundvig and

Mike Spilde for their assistance during the microprobe sessions. The work was supported by the Faculty of Science, Aarhus University, Denmark and the *Svensk Kärnbränslehantering AB* (SKB), Stockholm, Sweden. The work was performed in cooperation with the Oklo Working Group organized by the European Union and the Commissariat á l'Energie Atomique, France. The electron microprobe in the Department of Earth and Planetary Sciences, University of New Mexico was supported by NSF, NASA, DOE/BES, and the State of New Mexico. The electron microprobe at the Department of Earth Sciences, Aarhus University was supported by *Statens Naturvidenskabelige Forskningsråd, Carlsbergfondet*, and *Aarhus Universitetsfond.*

REFERENCES

[1] F. Gauthier-Lafaye and F. Weber, Econ. Geol., **84**, 2267-2285 (1989)

[2] F. Gauthier-Lafaye, F. Weber, and H. Ohmoto, Econ. Geol., **84**, 2286-2295 (1989)

[3] F. Gauthier-Lafaye, Rapport final; Volume 2: Les Reacteurs de Fission et les Systemes Geochemiques; 2éme partie: Geologie des réacteurs, études des épontes et transferts anciens, Commissariat a l'Energie Atomique. Institution de Protection et de Sûreté Nucléaire, 105-118 (1995).

[4] P.-L., Blanc, Final report: Volume 1: Acquierements of the natural analogy programme. Commissariat á l'Energie Atomique, Institut de Protection et de Sûrete Nucléaire, 135 pp. (1995)

[5] J. Janeczek and R.C. Ewing, Geochim. Cosmochim. Acta, **59**, 1917-1931 (1995)

[6] R. Bros, F. Gauthier-Lafaye, P. Larque, and P. Stille, Mineralogy and isotope geochemistry of Bangombé reaction zone - Migrations of U, Th and fission products. Internal report for SKB-CNRS "Oklo - Analogues Naturels", Centre National e la Recherche Scientifique, Centre de Geochemie de la Surface, 1, rue Blessig, 67084 Strasborg Cedex, France, 14 p. (1993)

[7] J. Janeczek and R.C. Ewing, Am. Min., **81**, 1263-1269 (1996)

[8] P. Toulhaut, J.P. Gallien, D. Louvat, V. Moulin, P. l'Henoret, R. Guerin, E. Ledoux, I. Gurban, J.A.T. Smellie, and A. Winberg, Fourth International Conference on the Chemistry and Migration Behaviour of Actinides and Fission Products in the Geosphere. Charleston, SC USA, December 12-17, 1993), 383-390.

[9] P.O. Eberly, R.C. Ewing, J. Janeczek, and A. Furlano. Radiochim. Acta, **74**, 271-275 (1996).

[10] Ph. Holliger, Terme Source: Caracterisation Isotopique Parametres Nucleaires et Modelisation. Commissariat a l'Energie Atomique, Institut de Protection et de Sûreté Nucléaire, p. 40 (1994)

[11] M.G. Bonhomme, F. Gauthier-Lafaye, and F. Weber, Precambrian Research, **18**, 87-102 (1982)

[12] F. Gauthier-Lafaye, P. Holliger, and P.-L. Blanc, Geochim. Cosmochim. Acta, in press (1996)

[13] P.O. Eberly, J. Janeczek, and R.C. Ewing, Radiochim. Acta, **66/67**, 455-461 (1994)

[14] B. Nagy, F. Gauthier-Lafaye, Ph. Holliger, D.W. Davis, D.J. Mossman, J.S. Leventhal, M.J. Rigali, and J. Parnell, Nature, **354**, 472-475 (1991)

[15] A.J. Gancarz, Natural Fission Reactors, (Proc. Tech. Comm. Meet., Paris, December 1977, International Atomic Energy Agency: Vienna), 513-520 (1978)

[16] J. Janeczek, and R.C. Ewing, In Nuclear Science and Technology - Report EUR 16098 edited by H. von Marevic (Oklo Working Group - Proceedings of the third joint EC-CEA progress meeting held in Brussels on 11 and 12 October 1993, 181-211 (1995)

[17] B. Nagy, In Nuclear Science and Technology - Report EUR 16098 edited by H. von Marevic (Oklo Working Group - Proceedings of the third joint EC-CEA progress meeting held in Brussels on 11 and 12 October 1993, 169-179 (1995).

[18] H. Kleykamp, J. of Nucl. Mat., **206**, 82-86 (1993)

[19] H. Hidaka, K. Takahashi, and Ph. Holliger, Radiochim. Acta, **66/67**, 463-468 (1994).

[20] J. Janeczek, and R.C. Ewing, J. Nucl. Mat., **190**, p. 157 (1992).

[21] R.C. Ewing and J. Janeczek, In Nuclear Science and Technology - Report EUR 14877EN edited by H. von Marevic (Oklo Working Group - Proceedings of the second joint CEC-CEA progress meeting held in Brussels on 6 and 7 April 1992, 177-188 (1993).

MINERALOGICAL AND MICROTEXTURAL CHARACTERIZATION OF "GEL-ZIRCON" FROM THE MANIBAY URANIUM MINE, KAZAKHSTAN

K.B. HELEAN[1], B.E. BURAKOV[2], E.B. ANDERSON[2], E.E. STRYKANOVA[2], S.V. USHAKOV[2] AND R.C. EWING[1]
[1]Department of Earth & Planetary Sciences, University of New Mexico, Albuquerque, NM 87131
[2]V.G. Khlopin Radium Institute, St. Petersburg, Russia

ABSTRACT

"Gel-zircon", an unusual Zr-silicate phase from the Manibay uranium mine, northern Kazakhstan, was studied using X-ray diffraction (XRD), electron microprobe energy dispersive X-ray spectroscopy (EDS) and high resolution transmission electron microscopy (HRTEM). XRD results indicate that gel-zircon is mostly amorphous and occurs with numerous impurity phases. Microprobe EDS results indicate a UO_2 content up to 9.14 wt. %. HRTEM images revealed that the microtexture of gel-zircon consists of nanocrystallites of zircon, 2-10 nm in size, in a dominantly amorphous matrix. Despite the U-Pb age of 420±25 my and the lack of significant crystallinity, the gel-zircon is an apparently chemically durable phase. Leaching of uranium ores which contain gel-zircon as the major U-bearing phase is impossible using existing uranium plant technologies. The alpha-decay dose, 2.64 displacements per atom (dpa), corresponding to the age of gel-zircon is much higher than that (0.5 dpa) required to cause metamictization of crystalline zircon. However, the morphology of gel-zircon which occurs as veins up to 5 mm thick and tens of mm long does not indicate initial crystallinity. Initially crystalline natural zircons often preserve their crystal morphology after metamictization. This amorphous phase is analogous to the highly damaged state characteristic of zircon proposed as a waste form for the disposition of excess weapons plutonium.

INTRODUCTION

Zircon ($ZrSiO_4$, *I4_1/amd*) has long been used for U-Th-Pb dating owing to its ability to accommodate uranium and thorium in substitution for zirconium. These same properties (i.e., incorporation of actinides and high chemical and physical durability) are the basis for the proposal that zircon be used as a waste form for actinides (e.g., excess weapons plutonium) [1,2,3,4,5]. Natural zircons commonly contain up to 5,000 ppm U and up to 2,000 ppm Th [6]. The maximum concentrations that can be incorporated are significantly higher. This is shown by zircons crystallized from the Chernobyl "lava" which were found to contain 12.9 wt.% U [7]. Even small concentrations of actinides expose zircons of great age to α-recoil event doses sufficient to render them metamict [8]. Numerous studies have characterized radiation effects on both metamict [9,10,11] and ion beam irradiated zircons [11,12,13,14]. Complete amorphization of natural zircons was observed at 0.5 dpa [14]. Natural zircons are particularly useful in evaluating the chemical durability of proposed waste forms under proposed repository conditions.

This paper reports the results of high resolution transmission electron microscopy, electron microprobe and X-ray diffraction analysis of an unusual gel-zircon from the Manibay uranium mine, northern Kazakhstan. This phase is similar in composition to natural zircon, but

Mat. Res. Soc. Symp. Proc. Vol. 465 © 1997 Materials Research Society

lacks significant crystallinity. Microtextures were examined to determine if the uranium is incorporated in the gel-zircon or if it resides in impurity crystallites.

GEOLOGIC OCCURRENCE

Gel-zircon is the main U-bearing phase in the P-Zr-U ore found at the Manibay uranium mine, Kazakhstan. The country rock consists of diorite and quartz diorite interbedded with their extrusive counterparts, porphyritic andesite and tuff. The P-Zr-U ore is associated with alkaline metasomatites which are located in contact aureoles proximal to an aplite dike system. Gel-zircon is found in association with coffinite ($USiO_2 \bullet nH_2O$), pitchblende (UO_2), and apatite $Ca_5(PO_4)_3(F,Cl,OH)$.

EXPERIMENTAL TECHNIQUES

Mineral phases in the host rock were identified by powder X-ray diffractometry using a Scintag diffractometer operated at 20kV using Cu Kα radiation (λ-1.5418 A). The samples were crushed, suspended in acetone and allowed to settle onto a low background quartz plate. X-ray diffraction   analysis of the gel-zircon phase was conducted at the V.G. Khlopin Radium Institute.

Quantitative energy-dispersive X-ray spectroscopy (EDS) analyses were obtained using a fully automated JEOL JXA-733 electron microprobe operated at 15kV, a focused beam diameter of 1 μm and a counting time of 60 s per element. Elements were analyzed against the following standards: uraninite (U), zirconolite (Zr), albite (Na, Al), diopside (Mg, Si, Ca), rutile (Ti), apatite (P), galena (S), orthoclase (K), chromite (Fe). Back scattered electron imaging was used to examine the element partitioning between phases in the gel-zircon sample.

Microtextural characterization of the gel-zircon by high resolution transmission electron microscopy (HRTEM) was made using a JEOL 2010FX transmission electron microscope operated at 200kV. TEM samples were prepared as crushed grains on holey carbon films supported by copper grids. Energy-dispersive X-ray spectroscopy (EDS) was completed using a JEOL 2000FX transmission electron microscope operated at 200kV. The point-to-point resolution of the JEOL 2010FX is 0.19 nm; for the 2000FX, 0.25 nm.

RESULTS

X-ray Diffraction Analysis

X-ray diffraction analysis indicated that the host rock of the uranium ore is quartz rich and contains abundant low-albite ($NaAlSi_3O_8$). Ninety percent of the peaks in this pattern with relative intensities greater than 1% are accounted for by low-quartz (JCPDS card #33-1161) and low-albite (JCPDS card #20-554). The X-ray diffraction pattern of the gel-zircon sample shows broad peaks which indicates that the sample is largly amorphous (Fig. 1). These broad peaks correspond to the reported *d*-spacings and relative intensities of the zircon structure type (JCPDS card # 6-266).

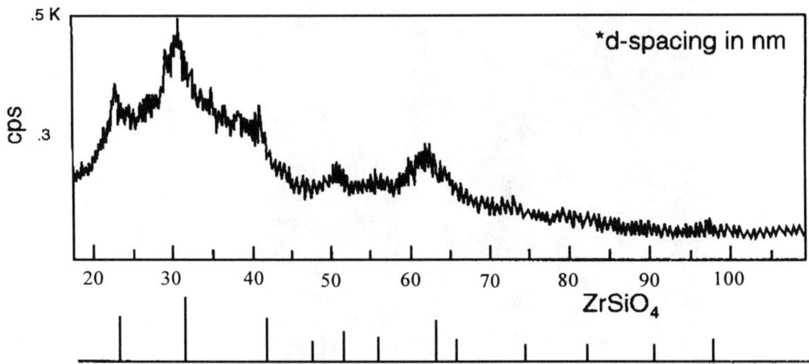

Fig. 1. X-ray diffraction pattern for gel-zircon plots *d*-spacing (nm) vs. Counts per second (cps). The breadth of the peaks indicates that gel-zircon is largely amorphous. These broad peaks correspond to the diffraction maxima of the zircon structure type. For comparison, the zircon pattern is shown below the gel-zircon diffraction pattern.

Electron Microprobe Analysis

Elemental maps were made of the gel-zircon (Fig. 2). Gel-zircon occurs as a fine-grained matrix with small inclusions, 2 to 10s of μm, of albite and pyrite (FeS$_2$). The Zr is enriched along the edges of fractures. Uranium overlaps with Zr and shows a slight concentration along the lower edge of the sample. Concentrations of Si correlate with Al and Na in albite, however aluminum is distributed more extensively than Na. Si, Na, and Al overlap in the albite phase. A minor concentration of Ca suggests a possible calcium carbonate phase, and Fe and S correspond to pyrite. Copper shows some overlap with Fe and S, and Mg concentrations overlap with Fe. A Ti phase occurs as small grains (1 to 2 μm). Oxygen is homogeneously distributed.

Quantitative energy dispersive X-ray spectroscopy was used to measure the chemical constituents of the gel-zircon phase. The results of twelve analyses are given in Table I. Totals for all twelve analyses are below 100 wt.% suggesting the presence of water. The range of the totals is from 81.71 wt.% to 91.65 wt.% with a mean of 87.50 wt.%. An analysis of a standard natural zircon was made with the same instrumental parameters as used with the gel-zircon analyses (Table I). The natural zircon analysis totals to 99.22%, where SiO$_2$ and ZrO$_2$ contents total to 32.3 wt.% and 66.5 wt.%, respectively. The total oxide wt.% is low for this standard because Hf was not included in the analysis. The gel-zircon sample showed UO$_2$ values ranging from 3.86 wt.% oxide to 9.14 wt.% with a mean content of 6.09 wt.%. Both the mean SiO$_2$ (25.73 wt.%) and ZrO$_2$ (35.12 wt.%) components are low compared to the normal composition of natural crystalline zircon.

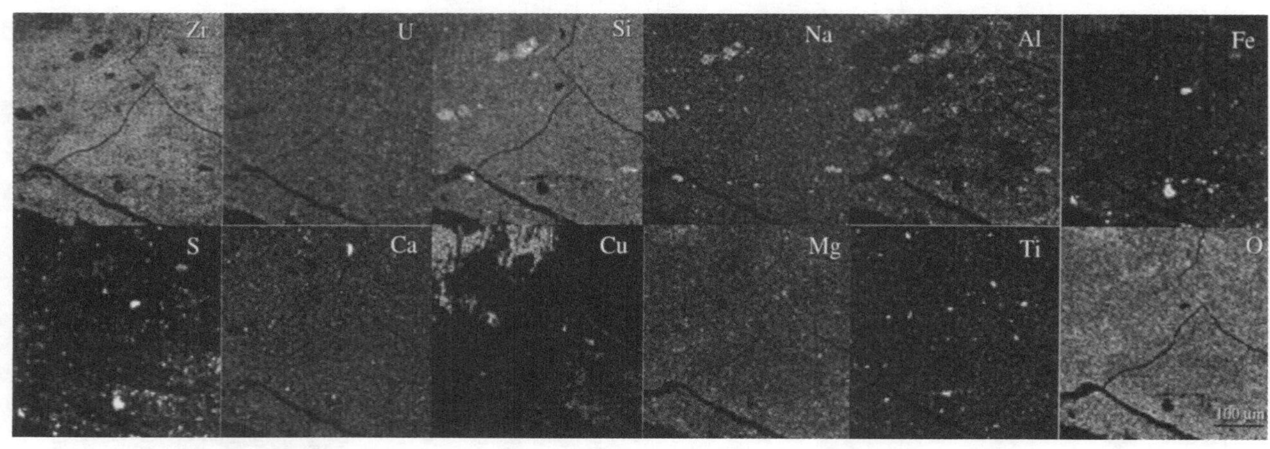

Fig. 2. Elemental maps of the constituents found in gel-zircon. Element enrichments are indicated as light areas. Zr is ubiquitous. U shows a slight concentration along grain boundaries. Si, Na, Al overlap in the albite phase. Fe and S correspond to pyrite. Minor Ca concentrations indicate the possibility of calcium carbonate. Cu shows some overlap with Fe and S as a Cu-sulphide phase. Mg concentrations overlap Fe. Ti occurs as small grains. Oxygen is homogeneously distributed across the sample.

Table I. Quantitative EDS analyses of gel-zircon (wt.%). The last column shows an analysis of a standard natural zircon. Error is indicated in parentheses and the symbol <d.l. indicates an element below the detection limit.

	#1	#2	#3	#4	#5	#6	#7	#8	#9	#10	#11	#12	$ZrSiO_4$
K_2O	0.276	0.285	0.328	<d.l.	0.381	0.233	0.522	<d.l.	<d.l.	1..357	<d.l.	<d.l.	-
	(.067)	(.067)	(.064)	(.071)	(.075)	(.062)	(.073)	(.067)	(0.64)	(0.77)	(0.67)	(0.71)	
CaO	2.031	2.121	2.825	1.862	2.462	2.076	2.298	3.305	2.261	2.089	3.109	2.643	-
	(.067)	(.069)	(.072)	(.068)	(.072)	(.068)	(.071)	(.076)	(.070)	(.069)	(.075)	(.073)	
Na_2O	11.165	1.329	4.834	3.583	4.690	5.484	1.104	0.987	<d.l.	<d.l.	<d.l.	<d.l.	-
	(.086)	(.060)	(.068)	(.061)	(.070)	(.077)	(.055)	(.057)	(.050)	(.049)	(.050)	(.066)	
UO_2	6.168	5.762	4.047	7.594	9.140	4.265	6.975	5.095	3.858	7.167	5.722	7.285	-
	(.290)	(.294)	(.279)	(.311)	(.326)	(.275)	(.311)	(.293)	(.286)	(.318)	(.296)	(.310)	
ZrO_2	28.825	32.510	40.268	41.110	32.105	25.704	35.968	41.973	48.299	25.483	42.054	27.140	66.94
	(.323)	(.345)	(.365)	(.363)	(.340)	(.320)	(.352)	(.376)	(.389)	(.315)	(.373)	(.326)	
FeO	5.179	6.373	2.831	1.325	4.276	9.187	3.318	3.0469	2.613	6.265	3.818	6.781	-
	(.171)	(.185)	(.148)	(.133)	(.163)	(.205)	(.157)	(.156)	(.153)	(.182)	(.161)	(.189)	
MgO	1.539	2.828	0.992	0.537	1.626	4.185	1.521	1.349	0.488	2.613	1.690	2.362	-
	(.055)	(.055)	(.050)	(.044)	(.053)	(.064)	(.049)	(.048)	(.043)	(.051)	(.047)	(.051)	
TiO_2	0.921	0.966	0.810	0.967	0.875	0.960	0.885	0.808	1.097	0.325	0.721	0.597	-
	(.078)	(.080)	(.079)	(.083)	(.079)	(.077)	(.080)	(.082)	(.084)	(.077)	(.082)	(.080)	
Al_2O_3	5.471	7.190	4.604	2.050	6.195	10.337	6.542	3.331	2.325	9.935	4.209	5.853	-
	(.057)	(.062)	(.055)	(.046)	(.060)	(.071)	(.059)	(.052)	(.047)	(.067)	(053)	(.058)	
SiO_2	24.116	27.395	25.496	22.278	27.829	28.390	28.934	23.431	24.653	28.996	24.042	23.200	32.29
	(.091)	(.096)	(.092)	(.086)	(.096)	(.100)	(.097)	(.089)	(.089)	(.097)	(.089)	(.089)	
P_2O_5	<d.l.	1.111	0.806	<d.l.	<d.l.	<d.l.	<d.l.	1.010	<d.l.	0.956	2.117	0.888	-
	(.152)	(.163)	(.172)	(.170)	(.160)	(.151)	(.167)	(.176)	(.184)	(.149)	(.176)	(.152)	
SO_3	1.114	1.487	1.263	1.504	1.869	0.829	0.765	4.127	<d.l.	0.795	0.669	4.965	-
	(.071)	(.078)	(.092)	(.081)	(.077)	(.072)	(.075)	(.094)	(.082)	(.070)	(.080)	(.086)	
Total	86.804	89.358	89.106	82.809	91.447	91.649	88.832	88.462	85.595	85.983	88.211	81.714	99.22
	(1.508)	(1.554)	(1.536)	(1.517)	(2.435)	(2.118)	(1.546)	(1.566)	(1.541)	(1.521)	(1.549)	(1.551)	

<u>Transmission electron microscopy</u>

HRTEM micrographs show that gel-zircon is primarily amorphous with nanocrystallites (Fig. 3). These crystallites range in size from 2 to 10 nm and show no preferred orientation. Selected area electron diffraction (SAED) patterns with the beam centered on the nanocrystallites in the upper left corner of Fig. 4 show distinct rings which were indexed and found to correspond to the d-spacings reported for zircon (Fig. 3 inset).

Textural variation on the scale of a few nanometers corresponds to chemical zonation of the same order (Fig 4). Electron diffraction patterns corresponding to region (A) indicate this region is amorphous (Fig. 4 inset A), however region (B) is nanocrystalline (Fig. 4 inset B). This diffraction ring pattern was measured, indexed and compared to the calculated d-spacings of the zircon structure type. The measured d-spacings correspond to the calculated values for the zircon structure within 7%. The diffraction pattern corresponding to (C) shows faint rings indicating nanocrystallinity (Fig. 4 inset C). EDS analyses collected from all three regions indicate the presence of Si, Zr, and U. Region (A) has a higher iron content relative to regions (B) and (C); region (B) contains little iron, and region (C) is Zr-rich compared with (A) and (B).

Brookite (TiO_2) grains measuring up to 50 nm were found to be completely crystalline. Qualitative EDS analysis of the Ti-rich areas indicates the presence of impurity elements including Ca, Si, Zr, Fe and minor Pb. EDS results also show these Ti nanocrystallites contain little uranium relative to the Zr-rich regions in gel-zircon. The uranium signal for the Ti-rich areas may originate from the surrounding Zr-rich regions which contain the highest U-concentration.

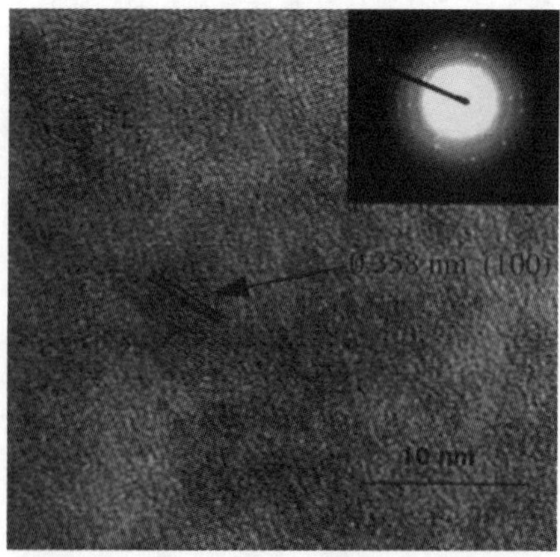

Fig. 3. HRTEM micrograph showing nanocrystallites in gel-zircon. The lattice fringes of the nanocrystallites correspond to the zircon structure-type.

K-Ar isochron ages of silicates from the metasomatic host rocks are similar to the U-Pb ages of the uranium minerals. The U-Pb age of the Manibay uranium ore deposit, 420±25 my, was determined from analyses of pitchblende (UO_2) and coffinite ($USiO_4 \cdot nH_2O$) (V.G. Khlopin Radium Institute, unpublished). An α-decay dose of 2.64 displacements per atom was calculated from the mean UO_2 content (6.09 wt.%) and the U-Pb age. This is well beyond the dose (0.5 dpa) required for the metamictization of crystalline zircon [14]. The presence of nanocrystallites is, therefore, attributed to thermal recrystallization. The morphology of gel-zircon, which occurs as irregular veins up to 5 mm thick and tens of mm long, does not, however, indicate any previous crystal form. Initially crystalline natural zircons usually preserve their crystal morphology after metamictization. Gel-zircon is assumed, therefore, to have precipitated as a gel. Despite its essentially amorphous structure, gel-zircon is resistant to leaching during the uranium extraction process, and this makes gel-zircon bearing uranium ores at the Manibay mine useless (V.G. Khlopin Radium Institute, unpublished data).

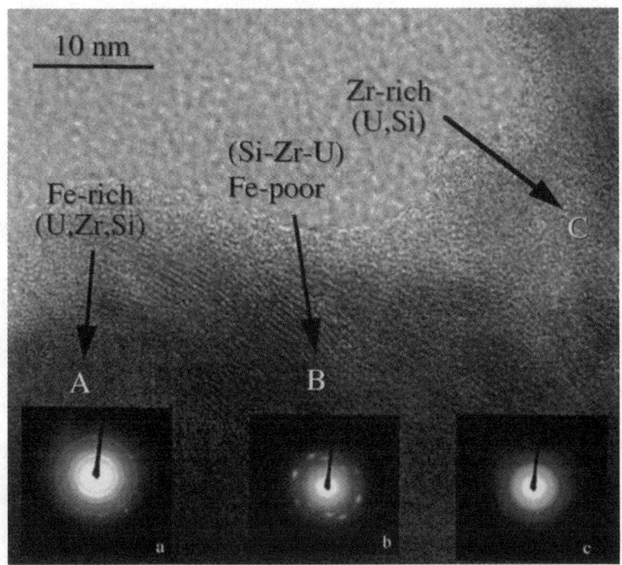

Fig. 4. HRTEM micrograph of gel-zircon showing textural variation on the scale of a few nm. Electron diffraction patterns corresponding to the regions marked A, B, and C (insets a, b, and c, respectively) show varying degrees of crystallinity associated with differences in chemical composition.

CONCLUSIONS

Gel-zircon contains high concentrations of impurity elements and has a mixed nanocrystalline and amorphous state. Microprobe EDS analyses were consistently below 100%

suggesting gel-zircon is a hydrous phase. Elemental maps indicate that gel-zircon has been subjected to alteration along fractures and, yet, retains up to 9 wt.% UO_2. The lack of any crystal form suggests that gel-zircon precipitated as a gel. Subsequent recrystallization resulted in the final nanocrystalline microtexture. Gel-zircon is chemically durable to such an extent that leaching uranium from ores containing gel-zircon is impossible using existing uranium plant technologies. Gel-zircon is structurally analogous to the highly damaged amorphous state of metamict zircon.

ACKNOWLEDGMENTS

This work was supported by a travel grant from NATO, Los Alamos National Laboratory, and the Research Experience for Undergraduates program of the NSF.

REFERENCES

1. B.E. Burakov, Safe Waste, 2, 19 (1993).
2. E.B. Anderson and B.E. Burakov, Safe Waste, 2, 29 (1993).
3. R.C. Ewing and W. Lutze, Journal of Materials Research, 10(2), 243 (1995).
4. R.C. Ewing, W.J. Weber, and W. Lutze in Disposal of Weapon Plutonium, edited by E.R. Merz and C.E. Walter (NATO Workshop, St. Petersburg, Russia, 1995) pp. 65-83.
5. B.E Burakov, E.B. Anderson, B. Ya. Galkin, V.A. Starchenko, and V.G. Vassiliev in Disposal of Weapon Plutonium, edited by E.R. Merz and C.E. Walter (NATO Workshop, St. Petersburg, Russia, 1995) pp. 85-89.
6. L.H. Ahrens, R.D. Cherry, and A.J. Erlank, Geochimica et Cosmochimica Acta, 31, 2379 (1967).
7. E.B. Anderson, B.E. Burakov, and E.M. Pazukhin, Radiochimica Acta 60, 149 (1993).
8. R.C. Ewing, Nucl. Instr. and Methods in Phys. Res. B91, 22 (1994).
9. T. Murakami, B.C. Chakoumakos, R.C. Ewing, G.R. Lumpkin, and W.J. Weber, American Mineralogist 76, 1510 (1991).
10. B.C. Chakoumakos, T. Murakami, G.R. Lumpkin, and R.C. Ewing, Science 236, 1556 (1987).
11. H.D. Holland and D. Gottfried, Acta Crystallographica 8, 291 (1955).
12. L.M. Wang, R.K. Eby, J. Janeczek, and R.C. Ewing, Nucl. Instr. and Methods in Phys. Res. B59/60, 395 (1991).
13. L.M. Wang, R.C. Ewing, W.J. Weber, and R.K. Eby, Mat. Res. Soc. Symp. Proc. 279, 451 (1993).
14. W.J. Weber, R.C. Ewing, and L.M. Wang, Journal of Materials Research, 9(3), 688 (1994).

Part XVI

Excess Plutonium Dispositioning

DEVELOPMENT AND TESTING OF A GLASS WASTE FORM
FOR THE IMMOBILIZATION OF PLUTONIUM

D. B. CHAMBERLAIN,[1] J. M. HANCHAR,[1] J. W. EMERY,[1] J. C. HOH,[1] S. F. WOLF,[1]
R. J. FINCH,[1] J. K. BATES,[1] A. J. G. ELLISON,[2] and D. B. DINGWELL[3]
[1]Argonne National Laboratory, 9700 S. Cass Ave., Argonne, IL 60439. [2]Corning Inc., Corning,
NY, 14831, USA. [3]Bayerisches Geoinstitut, Universität Bayreuth, 95440 Bayreuth, Germany.

ABSTRACT

Two alkali-tin-silicate (ATS) glasses have been prepared at Argonne National Laboratory
(ANL) as part of our ongoing research in radioactive waste glass development. These glasses
dissolved 5% and approximately 7% Pu. Early corrosion test results indicate that Pu-bearing
ATS glass is extremely durable. The initial goal in this project concerned equally both the
solubility of Pu and the durability of the ATS glasses; however, our primary emphasis has
changed recently to maximizing the loading of Pu in the glass. ATS-based glasses, using Th(VI)
and Ce(III) as surrogates for Pu(IV), are now being investigated to increase the solubility of Pu
without substantially sacrificing the durability of the current ATS formulations. The solution
data from various corrosion tests on the original Pu-containing ATS glasses are also presented.

INTRODUCTION

The United States has declared about 50 metric tons of weapons-grade Pu surplus to
national security needs. The President has directed that this Pu be placed in a form that provides
a high degree of proliferation resistance in which the surplus Pu is both unattractive and
inaccessible for use by others [1]. Three alternatives are being evaluated for the disposal of this
material: (1) use of the Pu as a fuel source for commercial reactors; (2) immobilization, where Pu
is fixed in a glass or ceramic matrix that also contains or is surrounded by highly radioactive
material; and (3) deep bore hole, where Pu is emplaced at depths of several kilometers. The
immobilization alternative is being directed by the staff at Lawrence Livermore National
Laboratory (LLNL). The staff at ANL are assisting by developing a glass for the immobilization
of Pu and in the corrosion testing of glass and ceramic material prepared both at ANL and at
other DOE laboratories. As part of this program, we have developed an ATS glass into which
5-7 wt% Pu has been dissolved. The ATS glass was engineered to accommodate high Pu
loading and to be durable under conditions likely to accelerate glass reactions in the geological
environment during long-term storage.

ATS DEVELOPMENT

Background

The ATS glass was originally designed to be used in a new melter (greenfield) or adjunct
melter facility (located next to the Defense Waste Processing Facility, DWPF). With this in
mind, three objectives were set for developing this glass. First, keep the pH of the solution in
contact with the glass as low as possible. In solutions with a pH greater than 11, the glass will
react more rapidly because of the increased solubility of Si; therefore, minimizing the
concentrations of ionized silica species such as $H_3SiO_4^-$ in solution will improve durability.
Second, reduce the potential for alteration-phase (i.e., clays) formation by reducing or
eliminating Al, Ca, Mg, and Fe from the glass. In addition, keep the total silica concentration in
the glass as low as possible to help reduce the potential for zeolites or other alumino-silicate
phase formation and the potential for increasing the glass corrosion rate. Third, demonstrate a
Pu solubility of 10 wt% in the glass. Using this rationale, an ATS glass was prepared and tested
as described by Bates [2].

Mat. Res. Soc. Symp. Proc. Vol. 465 © 1997 Materials Research Society

This glass, referred to as "P10", was developed using principles of glass chemistry and the knowledge of the glass reaction-progress pathway [3]. Although this glass met the first two objectives described above, the glass contained both dissolved (~7 wt%) and some residual, undissolved PuO_2 (~3 wt%). Some of the undissolved PuO_2 was dispersed throughout the glass, while some settled to the bottom of the crucible during fusion.

After the development and preparation of this initial ATS glass, the intended use was changed from a greenfield or adjunct melter to a "can-in-can" concept. For can-in-can, the Pu-containing glass would be placed into small cans which would then be located inside a larger DWPF canister. For this disposal method, the need for high glass durability and for the presence of ^{137}Cs as a terrorist deterrent became less important; increasing the Pu solubility became a higher priority. Additionally, based upon repository criticality calculations, the Pu-to-Gd mole ratio was increased to 1:1. With this opportunity to improve the glass composition, a more systematic approach was used to gain a better understanding of the ATS glass composition and the interactions between the glass components.

Based upon this approach, a second ATS glass containing 5 wt% Pu was prepared. The Pu content was set at 5 wt% for this glass for two reasons. Based upon initial criticality concerns, the Pu:Gd ratio in the initial ATS glass was set at 2.7:1 (on an atomic % basis). Based upon current criticality calculations, a Pu:Gd ratio of 1:1 is required. Because Gd is expected to compete for Pu sites in the glass, the Pu content was decreased to 5 wt%. Second, a glass without undissolved PuO_2 was needed for corrosion testing. This new formulation, called "P5", successfully dissolved the Pu. The composition of the initial ATS glass (P10) and the nominal composition of this new P5 ATS glass are reported in Table I.

Table I. Composition of Pu ATS Glasses Subjected to MCC-1 and PCT.

Compound	Glass Compositions[a] (element wt%)	
	ATS-P10	ATS-P5
Al	1.38	1.19
B	3.94	2.76
Cs	0.47	0.85
K	3.32	6.41
Li	1.58	1.82
Na	6.68	6.79
Si	20.0	21.2
Sn	1.42	1.74
Ti	1.32	1.24
Zr	3.70	3.66
Zn	0	2.46[b]
Pu	10.3[c]	5.00
Gd	2.60	3.29

[a]Values in table for ATS-P10 are a bulk sample analysis. Values for ATS-P5 are the nominal composition of the feed.

[b]Because of Zn's classification as a hazardous metal, it will be eliminated from future ATS formulations.

[c]Glass contains approximately 2.8 wt% undissolved Pu as PuO_2 (see reference 2 for more details).

<u>Use of Pu Surrogates</u>

Because of the difficulty of working with Pu-containing materials, either Ce^{3+} (ionic radius: 0.1143 nm) or Th^{4+} (ionic radius: 0.105 nm) are commonly used as chemical surrogates for Pu^{4+} (ionic radius, 0.096 nm). Thorium appears to be an ideal substitute because of its ionic radius and charge. However, our experiments show that Th is approximately two times more soluble than Pu in the ATS glass. This may be explained by the disparate electronic structures of Th^{4+} (with no $5f$ valence electrons) and Pu^{4+} (which has $5f$ electrons). This has led some investigators to use Ce^{3+} as a surrogate for Pu^{4+}, despite charge and radius differences, because Ce has an electronic structure more similar to that of Pu (i.e., $4f$ electrons). The electronic structures affect the structural roles of these elements in the glass, information that is not currently known. Furthermore, uncertainties remain as to the oxidation state of Pu (which may also occur as Pu^{3+}, Pu^{5+}, and Pu^{6+}) and Ce (which may also be 4+).

We chose to use Th^{4+} (as ThO_2) as our primary surrogate for Pu because, based on the chemistry of Th, we know that it exists solely as a 4+ cation, whereas Ce may exist as a mixture of 3+ and 4+. In a study by Hanchar [4], which utilized a highly oxidizing environment to synthesize zircons doped with rare earth elements, Ce was found to occur as Ce^{3+}. Therefore, it is likely that most, if not all, of the Ce in ATS is in the 3+ oxidation state. Knowledge of the exact extent to which this occurs in ATS glass awaits further work.

<u>Elemental Roles in ATS Glass</u>

The ATS glass currently includes the following component oxides: Li_2O, Na_2O, K_2O, Cs_2O, ZnO, B_2O_3, Al_2O_3, SiO_2, TiO_2, SnO_2, ZrO_2, and Gd_2O_3 (see Table I). In the ATS glass, B, Al, Si, and possibly Zr are "network formers". The Ti, Al, Zr, Th, and Zn (if two-fold coordinated by oxygen) have intermediate structural roles depending on their concentrations. The Sn, Th, Li, Zn, Na, K, and Cs act as "network modifiers" [5]. The exact role an element plays in the ATS glass depends in part on the concentration of each element and on the interactions between the element of interest and the other elements in the glass [6].

Elements are added to the ATS glass for the following reasons: Li(I) increases glass durability (because of its strong bonding with oxygen [7]) and, along with the other alkalis and Al, increases the solubility of the 4+ cations in the glass [7]. In addition to increasing the solubility of the 4+ cations, the alkali elements (with Li having the greatest effect [7]) tend to decrease the viscosity of the melt. Also, both Na(I) and K(I) can reduce glass durability. The ^{137}Cs, replaced by a small amount of nonradioactive Cs in our tests, has little impact on melt viscosity [7].

Zinc(II) increases glass durability and lowers the melting point. Boron(III) lowers the melt viscosity and improves the ease with which the glass components react to form a homogeneous melt [7]. Aluminum(III), in conjunction with the alkalis, stabilizes the 4+ cations [7,8] and is present in the ATS glass at a low concentration (~1.2wt%). Both Ti(IV) and Sn(IV) improve glass durability; SnO_2 significantly decreases the melt viscosity. The benefit of adding Sn to increase durability outweighs the disadvantage caused by decreased melt viscosity. The ZrO_2 increases durability owing to the strong Zr-O bond [6]. The ZrO_2 also increases the viscosity of the melt, although not to the extent that Ti or Th do in similar amounts. Gadolinium(III) is added as a neutron absorber at a 1:1 atomic percent basis with Pu.

<u>Glass Development Rationale for Increasing Pu Loading</u>

As noted above, the primary objective for the continued development of the ATS glass formulation is to increase the Pu content of the glass. To decide how to increase the Pu loading, information on the structural environment of the elements in the glass at their respective concentrations would be useful. Important questions are: (1) what is the most suitable surrogate for Pu in ATS glass, and (2) how do the structural environments of Th, Ce, and Pu change as a function of concentration? Our preliminary results indicate that Th^{4+} is approximately 1.8 times as soluble in ATS glass as Pu. This is in general agreement with the work by Veal et al. [8], who determined that Th was 2.5 times as soluble as Pu. Since their glass composition was a simple alkali silicate glass, the difference is not too surprising. Knowledge of structural environments may help us to explain observed solubility differences among these elements. The structural environments of other elements in the ATS glass are also uncertain, and how these components may affect the solubilities of Th and Pu in the glass is of interest. Thus, a long-term goal in the development of Pu-containing ATS glass is to obtain information on the structural environments of the elements in these glasses using atomic-absorbance methods, such as X-ray absorption spectroscopy (XAS).

In addition to removing Zn, we will attempt to increase the solubility of the surrogate elements Th and Ce, as well as Pu, by taking advantage of the alkali/aluminum ratio and its effect on the solubility of 4+ cations. Studies by Watson [9] and Ellison and Hess [10] demonstrated the importance of this ratio on the solubility of 4+ cations in silicate melts.

Until spectroscopic data are available, however, we are proceeding in the following manner. The 4+ elements originally added to increase the durability are being systematically removed in an attempt to increase the Pu solubility, with the hope that removing those 4+ elements that have structural roles similar to Pu(IV) will free these atomic sites for Pu. We believe that Th(IV), Ce(III and possibly IV or a mixture of oxidation states) or Pu(IV) can serve a dual role: that of fulfilling the intended task of increased loading as well as maintaining (or increasing) the glass durability.

The consequences of removing the 4+ cations Zr, Sn, and Ti to increase the Pu loading will be evaluated by measuring both viscosity and changes in Ce, Th, and Pu solubilities. The viscosity of the melt must be at least 2-10 Pa•s (20-100 poise) at 1200°C [3], so a value of 2 Pa•s (20 poise) is being used as a benchmark to which the glasses will be compared. Our preliminary viscosity measurements (Table II) on the ATS base glass (i.e., no Pu, Th, Ce, or Gd) indicate that the viscosity at 1200°C is 3.1 Pa•s (30.6 poise), a factor of approximately 1.5 times greater than required.

ATS GLASS TESTING

Corrosion testing of the ATS glass (samples P10 and P5) includes the following tests. These tests are done in deionized water (DIW) to provide a consistent, comparable set of data to demonstrate whether the waste form may undergo catastrophic reaction under test conditions that limit the reaction of the waste form with water.

(1) MCC-1. A 3-day test in DIW at 90°C and a surface area-to-volume ratio (S/V) of 10 m^{-1}, which provides an estimate of the forward rate of waste form reaction. This information bounds the rate of the waste-form reaction under a specific test condition but does not provide useful information on the release of Pu or the neutron absorber Gd.

(2) PCT-B. A variable time test in DIW at 90°C and a S/V of 20,000 m^{-1}, which allows evaluation of the waste form reaction at high S/V ratios and static conditions known to increase the concentration of glass components in the leachate (DIW) and to accelerate the onset of alteration phase formation. This information provides an indication of initial Pu and Gd release to solution and also can be used to infer the mechanism(s) of waste-form reaction.

Table II. ATS Base Glass Viscosity Data

Temperature (°C)	Temperature 10000/T(K)	Viscosity (Poise)	Viscosity ($\log_{10}$ Pa•s)
1200	6.789	30.6	0.486
1150	7.027	47.6	0.678
1100	7.283	77.8	0.891
1050	7.558	133.6	1.126
1000	7.855	244.1	1.387
950	8.177	487.6	1.688
900	8.525	1093	2.039

(3)   PCT-A. A 7 day test in DIW at 90°C and a S/V of 2,000 m^{-1}, which provides comparative information (batch to batch) required by the Waste Acceptance Product Specifications used for DWPF glass.

(4)   Vapor Hydration Test (VHT). A 7-56 day test at 200°C provides qualitative information on the tendency of a waste form to form alteration phases [3], and allows us to evaluate the distribution of Pu and Gd within the alteration phases, if applicable.

Normalized release values were calculated for the MCC-1 and product consistency tests (PCT) based upon the analyzed (P10) and as-prepared (P5) glass compositions (Table 1), the geometric surface area, and the solution-analyses from the terminated tests. The equation used to calculate normalized releases is

$$NL(i) = \frac{\left(C_i - C_i^\circ\right)}{\left(\left(\dfrac{S}{V}\right) \bullet f_i\right)} \tag{1}$$

where NL(i) is the normalized release in g/m^2, C_i is the concentration of element i in the leachant in g/m^3, C_i° is the concentration of element i in the initial leachant solution in g/m^3, S is the surface area of the glass in m^2, V is the volume of the glass in m^3, and f_i is the weight fraction of element i in the glass.

MCC-1 Results

Three-day MCC-1 tests were completed in triplicate for the P5 ATS glass. Based on a comparison of various test methods performed earlier [11], these tests give an estimate of the forward rate of glass reaction. The precision of short term MCC-1 is generally less than that obtained for tests run at higher S/V for longer times, and the concentrations of the least-soluble elements are known with correspondingly low precision. Thus the measured releases of B, Si, and the alkalis are considered reliable and are used to estimate the forward reaction rate. The release of Pu and Gd are not used to estimate the release of these elements under repository conditions. Three-day MCC-1 tests were not completed for P10 because, at the time of testing, the test matrix was not designed to obtain this information.

The glass monoliths were polished using 240-grit abrasive paper, then placed in Teflon vessels with DIW. Solutions were collected at-temperature after 3 days at 90°C and were analyzed using inductively coupled plasma-mass spectrometry (ICP-MS); the pH of each solution was measured within 5 minutes of termination of the experiment. Specific details of the experimental method used to conduct MCC-1 tests are reported elsewhere [12]. Because Teflon vessels were used, acid strips of the vessel walls were not performed. The results are reported in Table III. All reported concentrations were blank-corrected by subtracting the average concentrations obtained from 24 blanks from the test solution results[1]. The data for each sample for each glass component are in good agreement; the scatter is typical of this type of test.

PCT Results

Samples of both P10 and P5 glasses were subjected to PCT-A and PCT-B. The 7-day PCT-A were completed in duplicate with 1 gram of glass [13]; sufficient glass required for triplicate testing was not available. These tests were completed in stainless-steel vessels. Acid rinses were performed and analyzed using ICP-MS. Normalized releases, based upon the sum of both solutions, are reported in Table IV.

---

[1] In addition to conducting tests with glass monoliths (MCC-1) or powders (PCT), a blank test is completed for the same time period and temperature as the glass samples. The solutions from the blank tests are analyzed, and a running average is calculated.

Table III. Results from Triplicate MCC-1 Testing of the P5 ATS Glass

Element	Normalized Release (g/m^2)			
	Sample P5M3A1	Sample P5M3B1	Sample P5M3C1	Average
Li	4.4	4.4	3.9	4.2
B	3.6	2.7	3.1	3.2
Na	6.5	6.4	6.5	6.5
Al	3.6	0.7	2.3	2.2
Si	4.2	2.2	3.5	3.3
K	3.9	3.2	3.5	3.5
Ti	<MDL[a]	<MDL	<MDL	<MDL
Zn	0.45	1.6	0.49	0.8
Zr	0.01	0.03	0.007	0.02
Sn	0.02	0.03	0.005	0.02
Cs	4.6	2.8	3.5	3.6
Gd	0.03	0.05	0.01	0.03
Pu	0.06	0.10	0.05	0.07
pH	9.2	7.4	8.4	8.3

[a]<MDL = less than minimum detection limit.

Table IV. Normalized PCT-A Releases from of the ATS Glasses P10 and P5

Element	Normalized Release (g/m^2)[a]			
	P10 ATS Glass		P5 ATS Glass	
	Sample A	Sample B	Sample A	Sample B[b]
Al	0.007	0.008	0.014	0.13
B	1.6	1.6	0.98	1.3
Cs	0.23	0.23	0.20	0.33
K	0.66	0.71	0.66	0.94
Li	1.6	1.7	1.2	1.6
Na	1.1	1.2	0.98	1.3
Si	0.38	0.40	0.53	0.55
Sn	0.017	0.017	0.024	0.092
Ti	0.001	0.0006	0.009	0.088
Zr	0.0006	0.0008	0.006	0.064
Zn	0	0	0.007	0.12
Pu	0.003	0.003	0.015	0.15
Gd	0.001	0.012	0.018	0.21
pH	10.2	10.2	11.1	11.0

[a]Tests performed at 2,000 m^{-1} and 90°C, and (NL)$_i$ values are calculated using the total elemental contribution released from the glass. Concentrations were blank-corrected.

[b]The elemental concentrations for this sample are suspect because of possible contamination of the acid-strip solution.

Results from PCT-B were obtained using the same procedures as PCT-A; a single sample was used for each period. These tests were also performed in stainless-steel vessels, which were acid stripped after termination of the tests. Results are reported in Table V for both P10 and P5. For the P5 ATS glass, only the 3-and 28-day tests have been terminated and analyzed to date. The Si and B data reported in Table V are plotted in Fig. 1.

Table V. Normalized Release Data for PCT-B Tests for the Pu ATS Glasses P10 and P5

| | Normalized Release (g/m^2)[a] | | | | | | |
| | P10 ATS Glass | | | | | P5 ATS Glass | |
Component	3 day	28 day	98 day	182 day	364 day	3 day	28 day
Al	0.001	0.002	<MDL[b]	<MDL	0.004	0.001	0.005
B	0.35	0.59	0.99	1.2	1.4	0.26	0.48
Cs	0.037	0.065	0.076	0.079	0.20	0.045	0.058
K	0.12	0.21	0.24	0.28	0.40	0.16	0.22
Li	0.37	0.80	1.0	1.1	1.4	0.35	0.56
Na	0.25	0.76	0.69	0.76	0.92	0.32	0.57
Si	0.057	0.054	0.086	0.072	0.088	0.14	0.26
Sn	0.002	0.001	0.0004	0.0007	0.035	0.002	0.0008
Ti	0.001	0.003	0.0007	<MDL	0.010	0.002	0.003
Zr	0.0001	<MDL	0.0002	0.0005	0.024	0.0005	0.001
Zn	na[c]	na	na	na	na	0.002	0.002
Pu	0.0009	0.002	0.0014	0.001	0.002	0.002	0.002
Gd	0.003	0.007	0.005	0.004	0.008	0.003	0.003
pH	10.4	10.3	10.5	10.7	10.7	11.2	11.3

[a]Tests performed at 20,000 m^{-1} and 90°C, and (NL)$_i$ values are calculated using the total elemental contribution released from the glass. All concentrations were blank-corrected.

[b]<MDL = less than the minimum detection limit.

[c]na = not analyzed; zinc was not included in the P10 formulation.

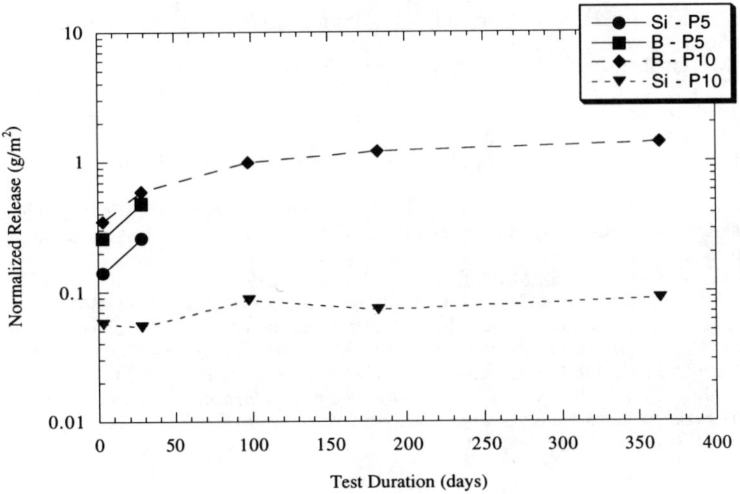

Fig. 1. Plot of Si and B Calculated Normalized Release for PCT-B

CONCLUSIONS

A systematic approach is being used to examine the solubility of Pu in an ATS glass, with the aim of increasing Pu loading while maintaining glass durability. Two ATS glasses containing Pu have been prepared and are currently being examined by a variety of standard leach tests. Also, several glasses containing the Pu surrogates Th and Ce have been prepared. The solubility of Th in ATS glass was determined to be approximately two times that of Pu, which is consistent with a previous study. The viscosity of the ATS base glass was measured to be 3.06 Pa•s (30.6 Poise) at 1200°C, which is a factor approximately 1.5 times the minimum required for a Pu waste glass. Normalized releases for several elements were determined from 3-day MCC-1 testing of the P5 ATS glass. The Si- and B-based normalized releases are 3.3 g/m^2 and 3.2 g/m^2, respectively. Results from the PCT-B testing of the initial ATS glass (P10) have been compiled through 364 days. For the revised composition (P5), 3-and 28-day tests have been completed. Results appear to be somewhat similar for the two glasses; however, for the P5 glass, the pH is much higher as is the Si release. We caution that long-term data are still needed to evaluate the release of Pu and Gd from the glass. In addition, how these elements are distributed between the solution and the acid strip are relevant to the potential segregation of each element in a repository environment.

ACKNOWLEDGMENTS

This work was supported by the U.S. Department of Energy, Office of Materials Disposition, under contract W-31-109-ENG-38.

REFERENCES

1. President Clinton's March 1, 1995 Address to the Nixon for Peace Freedom Policy Conference.
2. J. K. Bates, A. J .G. Ellison, J. W. Emery, and J. C. Hoh, Mater. Res. Soc. Symp. Proc. 412, pp. 57-64 (1996).
3. A. J. G. Ellison, J. J. Mazer, and W. L. Ebert, *Effect of Glass Composition on Waste Form Durability: A Critical Review*, Argonne National Laboratory Report ANL-94/28 (1994).
4. W. D. Kingery, H. K. Bowen and D. R. Uhlmann, *Introduction to Ceramics*, John Wiley & Sons, New York (1976).
5. J. M. Hanchar, Ph.D. Thesis, Rensselaer Polytechnic Institute (1996).
6. F. Farges, G. E. Brown, Jr., A. Navrotsky, H. Gan, and J. J. Rehr, "Coordination Chemistry of Ti (IV) in Silicate Glasses and Melts: II. Glasses at Ambient Temperature and Pressure," Geochim. Cosmochim. Acta 60, 3039 (1996).
7. M. B. Volf, *Chemical Approach to Glass. Glass Science and Technology, 7*, Elsevier, New York (1984).
8. B. W. Veal, J. N. Mundy, and D. J. Lam, in *Handbook on the Physics and Chemistry of the Actinides*, edited by A.J. Freeman and G.H. Lander, Elsevier, Amsterdam, pp. 271-309 (1987).
9. E. B. Watson, Contrib. Mineral. Petrol. 70, 407 (1979).
10. A. J. G. Ellison and P. C. Hess, Contrib. Mineral. Petrol. 94, 343 (1986).
11. W. L. Ebert, Argonne National Laboratory, personal communication (1996).
12. *Test Method for Static Leaching of Monolithic Waste Forms for Disposal of Radioactive Waste*, ASTM Standard C1220-92, ASTM, Philadelphia PA (1992).
13. *Standard Test Methods for Determining Chemical Durability of Nuclear Waste Glasses: The Product Consistency Test (PCT)*, ASTM Standard C1285-94, ASTM, Philadelphia PA (1994).

Proliferation Resistance of Borosilicate Glass
as a Host Form for Weapons-Grade Plutonium

G. S. CEREFICE*, K. W. WENZEL
 Department of Nuclear Engineering, Massachusetts Institute of Technology
*Bldg. NW12-312, 138 Albany St., Cambridge, MA. 02139; cerefice@mit.edu (Primary contact)

ABSTRACT

To examine the proliferation resistance of borosilicate glass, a process to extract and recover a plutonium analog (thorium) from borosilicate glass was developed and examined. The glass matrix examined was a modified standard frit consisting of the ARM-1 frit (with simulated fission products) loaded with 2 wt. % thorium (as an analog for plutonium) and 2 wt. % each of three rare earth elements (Gd, Sm, Eu), which were added for criticality control and to possibly increase the proliferation resistance of the glass matrix. The plutonium analog was extracted from the crushed glass with a nitric acid dissolution process, and subsequently decontaminated using a solvent extraction process. The acid dissolution process was able to extract 88.4 ± 6.8 % of the plutonium surrogate from the glass host form. The bench top solvent extraction process was 30.2 ± 10.9 % efficient in recovering the plutonium analog as a purified product. Overall, this process was able to extract 26.7 ± 9.9 % of the plutonium analog from the glass as a purified product. To quantify the proliferation resistance of borosilicate glass as a host form for weapons-grade plutonium, MCNP was used to determine the compressed critical mass of a plutonium alloy with the same composition as the product of the extraction process. For the average product composition, the compressed critical mass was 4.7 kg of material. On average, one compressed critical mass could be recovered from 613 kg of borosilicate glass (2 wt. % Pu loading).

INTRODUCTION

In 1994, the National Academy of Sciences' Committee on International Security and Arms Control recommended, among other things, that the United States and Russia pursue long-term disposition options for surplus weapons plutonium that "result in a form from which the plutonium would be as difficult to recover for weapons use as the larger and growing quantity of plutonium in commercial spent fuel" (the spent fuel standard).[1] The primary goal of this work is to determine whether the vitrification option would meet the spent fuel standard by examining the recovery of weapons-grade plutonium (WGPu) from a borosilicate glass host form and comparing the recovery process to the PUREX process for recovering plutonium from spent fuel.

To examine the recovery of WGPu from a borosilicate glass host form, the Crushed Glass-Acid Dissolution (CA) process was developed and then used to recover thorium from a borosilicate glass matrix in order to determine the effectiveness of the recovery process. This paper will describe the CA process, the results of the thorium extraction experiments, and then will compare the CA process for recovering WGPu from borosilicate glass to the PUREX process for recovering plutonium from spent fuel to determine by analogy whether the borosilicate glass host form will meet the spent fuel standard.[2]

Mat. Res. Soc. Symp. Proc. Vol. 465 ©1997 Materials Research Society

EXPERIMENT

Methodology

The borosilicate glass examined in this work incorporated simulated high-level waste (HLW), a plutonium surrogate (thorium), and additional criticality control elements (rare earths) as part of the glass matrix. Thorium was used as the plutonium surrogate in these experiments due to its similar behavior in a TBP solvent extraction decontamination process[3], which was used to decontaminate the recovered thorium as part of the CA recovery process.

The Crushed Glass - Acid Dissolution process (CA process) was formulated after some initial scoping experiments as a process for recovering and decontaminating the thorium from the borosilicate glass. The CA process was then tested to determine the extraction (Eq. 1) and recovery efficiencies (Eq. 2), as well as the contamination factors (Eq. 3) and decontamination factors (Eq. 4) obtained for the elements of interest, defined below. For these definitions, the term Feed solution refers to the starting material for a given process and Product solution refers to the final product of a given process.

$$e_{Th} = \frac{(\text{Mass of Th in Product Solution})}{(\text{Mass of Th in Glass})} \qquad (1)$$

$$r_{Th} = \frac{(\text{Mass of Th in Product Solution})}{(\text{Mass of Th in Feed Solution})} \qquad (2)$$

$$CF_i = \frac{(\text{Concentration of Element } i \text{ in Solution})}{(\text{Concentration of Th in Solution})} \qquad (3)$$

$$DF_i = \frac{(\text{Contamination Factor in Feed Solution})}{(\text{Contamination Factor in Product Solution})} \qquad (4)$$

For this work, the key elements examined were the added rare earths (europium, samarium, and gadolinium), some of the simulated fission products (neodymium, and cerium), the glass former elements (sodium, silicon, and boron), and the thorium. The concentration of these elements in solution was determined using ICP flame spectroscopy. The results from these experiments were then used along with the Monte Carlo Neutron Photon (MCNP) transport code to determine the compressed critical mass of the product.

Borosilicate Glass Host Form

The borosilicate glass host form examined in this experiment was based on the Advanced Reference Material frit (ARM-1), with the addition of thorium and select rare earth elements (europium, gadolinium, and samarium). ARM-1 is a borosilicate glass that contains simulated fission products, representative of HLW. Thorium was added as an analog for plutonium, and the rare earth elements were added for criticality control as well as increased proliferation resistance. The addition of 2% by mass thorium oxide (ThO_2) and 6% by mass rare earth oxides (2% Eu_2O_3, 2% Sm_2O_3, and 2% Gd_2O_3) to the ARM-1 glass caused no visible damage to the glass (i.e. gross devitrification), which was confirmed by SEM-EDAX. The nominal mass fractions and contamination factors for the elements of interest in the borosilicate glass used in this work is given in table 1.[4] A complete description of the glass composition can be found in Sylvester (Ref. 4).

<u>**Table 1: Nominal Glass Composition**</u>[4]
Table 1: This table lists the mass fraction of the elements of interest for the borosilicate glass examined in this work. The component mass fraction (fraction) is given in terms of the mass of oxide per gram of glass. The contamination factors, relative to thorium, are also provided.

element	fraction	CF$_{Th}$
ThO_2	2.0 %	1.000
CeO_2	1.4 %	0.642
Eu_2O_3	2.0 %	0.983
Gd_2O_3	2.0 %	0.989
Sm_2O_3	2.0 %	0.977
Nd_2O_3	5.5 %	2.67
SiO_2	42.8 %	11.35
Na_2O	8.9 %	3.75
B_2O_3	10.4 %	1.835
Other Oxides	23.0 %	N/A

<u>Crushed Glass - Acid Dissolution Recovery Process</u>

The process for the recovery of WGPu from borosilicate glass can be broken down into three phases: Extraction, Decontamination, and Reduction to metal. The Crushed Glass - Acid Dissolution process is designed to accomplish the first two of these stages. The final stage (reduction to metal) can be accomplished by a variety of established techniques.[5] The first two stages of the recovery process are an acid dissolution of the crushed glass followed by a single contact solvent extraction process.

<u>Extraction Stage</u>

The glass sample was crushed using a ceramic mortar and pestle to pass through a 60 mesh filter (particle size less than 250 microns). Then the crushed glass was put in a beaker, and a 3.9M solution of nitric acid (approximately 17.6 wt. % HNO3) was added in a 30:1 mass ratio (acid:glass) and stirred and covered with a watch glass before being placed in the oven. The samples were then heated in an oven for 2 hours at 90°C. Every 30 minutes, the samples were removed from the oven for stirring. After 2 hours, the samples were weighed, and then the liquid solution was decanted off using a disposable pipette, leaving the undissolved glass in the beaker. A small portion of this solution was taken, filtered, and diluted for ICP analysis to determine the effectiveness of the extraction stage of the recovery process.

<u>Decontamination Stage</u>

The remaining sample was then added to a glass bottle (shaker) and mixed with an equal volume of pre-equilibrated 30 vol. % TBP in kerosene. (The pre-equilibrated organic solution was prepared by mixing the TBP + kerosene mixture with an equal volume of 3M nitric acid) The sample was then shaken for 1 minute and allowed to settle. The organic phase was recovered with a disposable pipette and transferred to a clean shaker. Next, an equal volume of 3M nitric acid was added to the organic solution. The mixture was then shaken for 1 minute and allowed to settle before the organic solution was again recovered and transferred to a clean shaker with a disposable pipette. Finally, the organic solution was then mixed with an equal volume of DI water and shaken for 1 minute. After settling, the organic phase was removed, leaving only the aqueous

solution. The aqueous phase was then acidified to approximately 3M HNO3 by the addition of
concentrated nitric acid. This sample was then filtered and diluted for ICP analysis.

Critical Mass Determination
        The compressed critical mass of the final product solution was determined using the
Monte Carlo Neutron Photon (MCNP) transport code. We assumed a homogeneous, bare sphere
geometry, and used the cross sections provided by MCNP for the average fission products in
place of the cross sections for the cerium and neodymium in the metal product. Since the
reduction to metal was not performed, it was assumed for these calculations that the reduction of
the plutonium did not provide any further decontamination. To examine the relative effect of the
contaminants on the compressed critical mass (and to simplify the analysis), it was assumed that
the compressed density of the recovered plutonium metal was 39 g/cm^3. The compressed critical
mass was determined iteratively by assuming some radius for the sphere and then using MCNP to
determine k_{eff}. The estimate of the critical radius was then adjusted until a k_{eff} value of 1.0 ± 0.001
was reached.

RESULTS

        Using the CA process, thorium was extracted and purified from borosilicate glass to
determine the thorium extraction efficiency, the thorium recovery efficiency, the contamination
factors, and the decontamination factors for the CA process. Glass produced in 2 batches was
used, and 5 samples were prepared using each glass. The extraction and decontamination of the
thorium from the borosilicate glass was performed in two stages: the extraction stage (I) and the
decontamination stage (II).
        The average contamination factors for the product solution of the extraction process
(stage I) are shown in table 2. The extraction efficiency (e_{Th}) for the first stage of the CA was
determined to be 88.4 ± 6.8 %. Note that contamination factors after the first stage are essentially
the same as in the borosilicate glass, with the exception of the relative concentration of silicon
(relative to thorium), which has been decreased by a factor of more than 80.
        The recovery efficiency (r_{Th}) for the second stage of the CA was determined to be $30.2 \pm$
10.9 %.The average contamination factors for the product solution from the decontamination
process (stage II) along with the average decontamination factors obtained with the extraction
process, are also shown in table 2, along with the highest and lowest decontamination factors
observed in this experiment. The extreme range of the observed decontamination factors is
indicative of the crudeness of the decontamination stage.
        These data were then used as the composition of the product plutonium metal recovered
from the borosilicate glass. MCNP was then used to determine the compressed critical mass of
the product plutonium metal. The compressed critical mass was determined for the average
product composition, as well as for the 'best' and 'worst' products obtained experimentally to
determine if the CA process would be sufficient to recover weapons-usable plutonium from
borosilicate glass. Also examined was a case in which the experimental decontamination process
was assumed to be replaced by a PUREX-type decontamination process with a decontamination
factor of 10^4 and a recovery efficiency of 95%[6]. Using the observed (or assumed in the case of
PUREX) extraction and recovery, the amount of 2 wt. % WGPu borosilicate glass required to
recover 1 compressed critical mass (without recovering plutonium from the "waste" streams of
the CA process) was determined for each of these cases. The results are shown in table 3.

Table 2: Decontamination Stage Results
Table 2: This table contains the average contamination factors (relative to thorium) in the 'product' nitric acid solution of the extraction (stage I) and decontamination (stage II) stages. Also included are the decontamination factors obtained with the decontamination process (relative to the extracted nitric acid feed solution), along with highest and lowest observed decontamination factors.

element	Stage I CF_{Th}	CF_{Th}	Stage II DF_{Th}	High / Low
Ce	0.65	0.023	28.3	55.2 / 13.8
Eu	1.06	0.073	14.5	35.2 / 8.9
Gd	1.02	0.064	15.9	35.1 / 9.6
Sm	1.14	0.074	15.4	34.2 / 9.3
Nd	2.53	0.104	24.3	68.1 / 12.6
Si	0.14	0.013	10.7	1500 / 2.4
Na	4.14	0.051	81.2	1600 / 24.5
B	2.13	0.038	56.0	570 / 19.0

Table 3: Compressed Critical Mass Calculation Results
Table 3: This table contains the compressed critical mass (CCM) results determined for the listed cases using MCNP. The amount of WGPu-bearing borosilicate glass (2 wt. % loading WGPu) required to recover 1 CCM is also provided.

Case Examined	CCM	Glass Required
Average	4.7 kg	613.4 kg
"Best" Case	3.3 kg	527.4 kg
"Worst" Case	8.2 kg	821.8 kg
PUREX Decontamination	2.4 kg	144.8 kg

CONCLUSIONS

From the analysis of the CA process, it is apparent that plutonium could be recovered from a borosilicate glass matrix. Acid dissolution of the crushed glass provides a straightforward technique for extracting the plutonium into a nitric acid solution, which can then be purified via a number of established chemical processes or could be used as the feed to a PUREX decontamination process.

Pilot Plant Scale Recovery of WGPu

A pilot-plant scale version of the CA process for recovering WGPu from borosilicate glass would require the following stages: crushing and grinding, dissolution, decontamination, reduction, and possibly some pre-processing before decontamination. The crushing and grinding stage could be accomplished with a ball mill and possibly some mechanical cutting process to chop the glass logs into smaller pieces for the mill. After crushing the glass, the glass powder could be transferred to a reaction vessel or mixer-settler where it would be contacted with the acid solution to extract the plutonium. The plutonium-laden solution could then be sent into an existing PUREX decontamination process, possibly after some pre-treatment. Once the extracted solution enters the

PUREX process, the recovery of the plutonium should be no different from that for spent fuel.

<u>Borosilicate Glass and the Spent Fuel Standard</u>

The only significant differences between the recovery of plutonium from borosilicate glass via the CA process and the recovery of plutonium from spent nuclear fuel are in the initial extraction of the plutonium into solution and in the possible need for further processing of the feed solution from the CA process before it enters the PUREX process. The initial stage of the CA process involves crushing and dissolving the borosilicate glass, which should not be particularly easier than decladding the spent fuel and dissolving the fuel in nitric acid to extract the plutonium into solution. This feed solution differs from the typical PUREX process feed solution primarily in that it will contain some of the glass former elements (boron, silicon, and sodium) which may or may not cause a problem in the PUREX process. Any additional processes to purify the nitrate solution, however, will result in a more complicated recovery process.

From the similarities between the CA process for recovering plutonium from borosilicate glass and the PUREX process for recovering plutonium from spent fuel, it can be said that the borosilicate glass host form that contains both fission products (a radiation barrier) and the WGPu as an integral part of the host matrix should meet the National Academy of Science's <u>spent fuel standard</u>, provided that the isotopic dilution of the WGPu is not considered to be part of the <u>spent fuel standard</u>.

ACKNOWLEDGEMENTS

The authors gratefully acknowledge the support of the W. Alton Jones Foundation and Northeast Utilities for their support of this work. The authors would also like to acknowledge the support and contribution of Dr. Scott Simonson and Kory Budlong Sylvester to this project.

REFERENCES

[1] National Academy of Sciences, Committee on International Security and Arms Control. Management and Disposition of Excess Weapons Plutonium. (National Academy Press, Washington, D.C., 1994) p. 2.

[2] A more complete description of the details of this work is given in Cerefice, Gary S. Proliferation Resistance of Borosilicate Glass as a Host Form for Weapons-Grade Plutonium. (Masters Thesis, MIT. June 1996)

[3] Marcus, Y and A.S. Kertes. Ion Exchange and Solvent Extraction of Metal Complexes. (Wiley-Interscience, New York, 1969) p. 741.

[4] The description, composition, and fabrication of the borosilicate glass samples used for these experiments is described in Sylvester, Kory W. B. A Strategy for Weapons-Grade Plutonium Disposition. (Masters Thesis, MIT. September, 1995), pp 47-58 and p. 88.

[5] Benedict, M., Pigford, T., and Hans Wolfgang Levi. Nuclear Chemical Engineering, (McGraw-Hill, Inc., New York, 1981), pp 442-447.

[6] A value of $2*10^4$ is a typical decontamination factor for the first stage of a 2 stage commercial PUREX process. Katz, J.J., Seaborg, G.T., and L.R. Morss. The Chemistry of the Actinide Elements, Second Edition, Vol. 1, (Chapman and Hall Publishing, 1986). p. 534.

COMPOSITION EFFECTS ON VISCOSITY AND CHEMICAL DURABILITY OF SIMULATED PLUTONIUM RESIDUE GLASSES

S. A. BULKLEY[†] and J. D. VIENNA[‡]
† New York State College of Ceramics, Alfred University, Alfred, New York 14802
‡ Pacific Northwest National Laboratory, Richland, Washington 99352

ABSTRACT

Vitrification is currently being explored as an option for plutonium (Pu) scrap and residue stabilization because of its ability to convert plutonium-bearing materials into a safe, durable, and accountable waste form with reduced need for safeguarding. To develop this glass, the effect of composition on key glass properties (viscosity and chemical durability) were measured. A one-component-at-a-time change test matrix with 36 glasses was designed. The viscosity is reported as the temperature at 5 Pa·s (T_5), and the durability as the normalized release of boron and sodium (r_B and r_{Na}) from MCC-1 and PCT static leach tests and the negative logarithm of fractional aluminum and lithium release ($-\ln[f_{Al}]$ and $-\ln[f_{Li}]$) from TCLP. The effects of Al_2O_3, B_2O_3, CaO, CeO_2, Gd_2O_3, K_2O, Li_2O, MgO, Na_2O, PbO, SiO_2, SnO, TiO_2, and ZrO_2 were found to be linear for T_5 and non-linear for chemical durability. T_5 is increased most by the addition of SiO_2 followed by Al_2O_3, and decreased most by Li_2O followed by Na_2O. MCC-1 $\ln[r_B]$ and $\ln[r_{Na}]$ are increased most by the addition of Li_2O followed by MgO and decreased most by Al_2O_3 followed by ZrO_2. PCT $\ln[r_B]$ and $\ln[r_{Na}]$ are increased most by the addition of MgO followed by B_2O_3 and decreased most by Al_2O_3 followed by SiO_2. The relative effect of components on r_B and r_{Na} are similar for both PCT and MCC-1. TCLP $-\ln[f_{Al}]$ and $-\ln[f_{Li}]$ are increased most by the addition of SiO_2 followed by SnO and decreased most by Li_2O.

INTRODUCTION

Vitrification is an attractive treatment option for meeting stabilization (94-1) and final disposal needs for many plutonium (Pu) bearing materials at Rocky Flats Environmental Technology Site (RFETS), Hanford, and other Department of Energy (DOE) sites. Vitrification is a demonstrated technology[1] able to convert Pu residues into a durable[2] and accountable[3] waste form with reduced need for safeguarding.[4] The potential waste streams are low in Pu. For instance, the average Pu loading in the 106 metric tons (MT) of residues at RFETS is 2.7 wt%, and that for Hanford's 3.7 MT of lean scraps is 5 wt%. The remaining 95 wt% of these materials varies widely in composition and is generally not well characterized. The goal of this study is to observe how variations in these waste compositions will affect key glass properties.

For vitrification to be successful, a glass that is processable, durable, and tolerant to changing or uncertain feed composition must be developed. The uncertainty and variation of feed composition is a challenging aspect to glass formulation, since waste loading is low (currently dictated by transportation requirements to be ~200 g Pu in a 208 L drum (55 gal)). Changes in waste composition will affect glass processing through viscosity, electrical conductivity, and phase behavior and glass performance through leach resistance and phase behavior. The effects of composition on these properties have been extensively studied for high-level waste (HLW) glasses,[5-10] low-level waste glasses,[11,12] and industrial glasses.[13,14] Generally, models capable of predicting the properties of glasses outside of fixed (narrow) glass composition regions do not exist. Thus, the approach taken in this study is to measure the properties of glass as a function of changing composition and later to interpolate the properties to glasses within the measured region using physically defensible property / composition relationships. This paper reports the preliminary property data for simulated Pu residue glasses. The properties measured are glass viscosity and glass leach resistance using the Materials Characterization Center (MCC) leach test 1, the Product Consistency Test (PCT), and the Toxicity Characteristic Leaching Procedure (TCLP).

Mat. Res. Soc. Symp. Proc. Vol. 465 © 1997 Materials Research Society

EXPERIMENTAL APPROACH

To measure component effects on glass viscosity and leach resistance a centroid or baseline glass composition was developed. This glass, designed using existing, property / composition relationships of glasses that melt at 1150°C and have high durability (normalized PCT release < 2.5 g/m^2), is roughly central in the composition region studied. Fourteen key waste components expected of having a strong influence on glass properties were varied one-at-a-time from the baseline composition. The one-component-at-a-time type design is used to cover a broad composition region with few glasses and permits the direct measurement of component effects on measured properties.[6] Components are varied to the extent initially expected in Pu residue glasses with CeO$_2$ used as a PuO$_2$ surrogate. Table I lists the glass compositions studied.

Table I. Glass compositions in as batched wt%.

	Al$_2$O$_3$	B$_2$O$_3$	CaO	CeO$_2$	Gd$_2$O$_3$	K$_2$O	Li$_2$O	MgO	Na$_2$O	PbO	SiO$_2$	SnO	TiO$_2$	ZrO$_2$	Others[*]
BL	**5.0**	**6.0**	**4.9**	**4.5**	**4.5**	**3.0**	**4.0**	**2.0**	**10.0**	**1.0**	**45.0**	**0.0**	**1.0**	**3.0**	**6.1**
Al-1	**0.0**	6.3	5.2	4.7	4.7	3.2	4.2	2.1	10.5	1.1	47.4	0.0	1.1	3.2	6.4
Al-2	**10.0**	5.7	4.6	4.3	4.3	2.8	3.8	1.9	9.5	0.9	42.6	0.0	0.9	2.8	5.8
B-1	5.3	**0.0**	5.2	4.8	4.8	3.2	4.3	2.1	10.6	1.1	47.9	0.0	1.1	3.2	6.5
B-2	5.1	**4.0**	5.0	4.6	4.6	3.1	4.1	2.0	10.2	1.0	46.0	0.0	1.0	3.1	6.2
B-3	4.7	**12.0**	4.6	4.2	4.2	2.8	3.7	1.9	9.4	0.9	42.1	0.0	0.9	2.8	5.7
Ca-1	5.3	6.3	**0.0**	4.7	4.7	3.2	4.2	2.1	10.5	1.1	47.3	0.0	1.1	3.2	6.4
Ca-2	5.2	6.2	**2.0**	4.6	4.6	3.1	4.1	2.1	10.3	1.0	46.4	0.0	1.0	3.1	6.3
Ca-3	4.8	5.8	**8.0**	4.4	4.4	2.9	3.9	1.9	9.7	1.0	43.5	0.0	1.0	2.9	5.9
Ce-1	5.2	6.3	5.1	**0.0**	4.7	3.1	4.2	2.1	10.5	1.0	47.1	0.0	1.0	3.1	6.4
Ce-2	5.1	6.1	5.0	**2.5**	4.6	3.1	4.1	2.0	10.2	1.0	45.9	0.0	1.0	3.1	6.2
Ce-3	4.9	5.9	4.8	**6.5**	4.4	2.9	3.9	2.0	9.8	1.0	44.1	0.0	1.0	2.9	6.0
Gd-1	5.2	6.3	5.1	4.7	**0.0**	3.1	4.2	2.1	10.5	1.0	47.1	0.0	1.0	3.1	6.4
Gd-2	5.1	6.1	5.0	4.6	**2.5**	3.1	4.1	2.0	10.2	1.0	45.9	0.0	1.0	3.1	6.2
Gd-3	4.9	5.9	4.8	4.4	**6.5**	2.9	3.9	2.0	9.8	1.0	44.1	0.0	1.0	2.9	6.0
Gd-4	4.8	5.8	4.7	4.3	**8.0**	2.9	3.9	1.9	9.6	1.0	43.4	0.0	1.0	2.9	5.9
K-1	5.2	6.2	5.1	4.6	4.6	**0.0**	4.1	2.1	10.3	1.0	46.4	0.0	1.0	3.1	6.3
K-2	4.8	5.8	4.7	4.4	4.4	**6.0**	3.9	1.9	9.7	1.0	43.6	0.0	1.0	2.9	5.9
Li-1	5.2	6.3	5.1	4.7	4.7	3.1	**0.0**	2.1	10.4	1.0	46.9	0.0	1.0	3.1	6.4
Li-2	5.1	6.2	5.0	4.6	4.6	3.1	**1.5**	2.1	10.3	1.0	46.2	0.0	1.0	3.1	6.3
Li-3	5.0	6.0	4.9	4.5	4.5	3.0	**4.5**	2.0	9.9	1.0	44.8	0.0	1.0	3.0	6.1
Mg-1	4.8	5.8	4.8	4.4	4.4	2.9	3.9	**5.0**	9.7	1.0	43.6	0.0	1.0	2.9	5.9
Mg-2	4.7	5.6	4.6	4.2	4.2	2.8	3.8	**8.0**	9.4	0.9	42.2	0.0	0.9	2.8	5.7
Na-1	5.6	6.7	5.4	5.0	5.0	3.3	4.4	2.2	**0.0**	1.1	50.0	0.0	1.1	3.3	6.8
Na-2	5.1	6.1	5.0	4.6	4.6	3.1	4.1	2.0	**8.0**	1.0	46.0	0.0	1.0	3.1	6.2
Pb-1	5.1	6.1	4.9	4.5	4.5	3.0	4.0	2.0	10.1	**0.0**	45.5	0.0	1.0	3.0	6.2
Pb-2	4.8	5.8	4.7	4.3	4.3	2.9	3.8	1.9	9.6	**5.0**	43.2	0.0	1.0	2.9	5.9
Pb-3	4.5	5.5	4.5	4.1	4.1	2.7	3.6	1.8	9.1	**10.0**	40.9	0.0	0.9	2.7	5.5
Si-1	5.5	6.5	5.3	4.9	4.9	3.3	4.4	2.2	10.9	1.1	**40.0**	0.0	1.1	3.3	6.7
Si-2	4.1	4.9	4.0	3.7	3.7	2.5	3.3	1.6	8.2	0.8	**55.0**	0.0	0.8	2.5	5.0
Sn-1	4.9	5.9	4.8	4.4	4.4	2.9	3.9	2.0	9.8	1.0	44.1	**2.0**	1.0	2.9	6.0
Sn-2	4.8	5.8	4.7	4.3	4.3	2.9	3.8	1.9	9.6	1.0	43.2	**4.0**	1.0	2.9	5.9
Ti-1	5.1	6.1	4.9	4.5	4.5	3.0	4.0	2.0	10.1	1.0	45.5	0.0	**0.0**	3.0	6.2
Ti-2	4.8	5.8	4.7	4.3	4.3	2.9	3.8	1.9	9.6	1.0	43.2	0.0	**5.0**	2.9	5.9
Zr-1	5.2	6.2	5.1	4.6	4.6	3.1	4.1	2.1	10.3	1.0	46.4	0.0	1.0	**0.0**	6.3
Zr-2	4.9	5.9	4.8	4.4	4.4	2.9	3.9	2.0	9.8	1.0	44.1	0.0	1.0	**5.0**	6.0

[*]Others—8.2wt% Cl, 6.6 Cr$_2$O$_3$, 1.6 F, 26.2 Fe$_2$O$_3$, 10.5 MoO$_3$, 13.1 Nd$_2$O$_3$, 3.6 BaO, 8.2 NiO, 0.7 CuO, and 4.9 P$_2$O$_5$

Each glass was melted in a Pt-10% Rh crucible for 1 h in 350 g batches, ground in a tungsten carbide mill for 12 min and remelted for 1 h. A test bar of approximately $75 \times 13 \times 13$ mm was poured from the final melt and was annealed at 515°C for 2 h. The remaining glass was quenched to room temperature on a steel plate. A monolith was cut from the annealed bar and was tested by MCC-1 for 3 d in DIW at 90°C with a glass surface-area-to-water-volume ratio (S/V) of 10 m^{-1}. PCT was performed on quenched glass for 7 d in DIW at 90°C with a S/V of 2000 m^{-1}, and TCLP was performed on quenched glass in acetic acid at pH 4.7 for 18 h at 25°C with a S/V of 45 m^{-1}. The solutions were analyzed using inductively coupled plasma atomic emission spectroscopy (ICP/AES). PCT and MCC-1 releases were normalized according to procedures previously reported.[5] TCLP releases were normalized to mass of element released per mass of element in glass in g/g. The viscosity/temperature profile of each glass was measured between approximately 2 and 20 Pa·s with a calibrated spindle viscometer.

RESULTS AND DISCUSSION

The results are listed in Table II. The viscosity data are reported as the temperature at which the melt viscosity is 5 Pa·s (T_5), interpolated from measured data using the Arrhenius equation. Normalized boron and sodium releases (r_B and r_{Na}) are reported for both MCC-1 and PCT, while the negative logarithms of lithium and aluminum fractional releases ($-\ln[f_{Li}]$ and $-\ln[f_{Al}]$) are listed for TCLP.

Table II. Measured values for T_5 (°C), MCC-1 and PCT r_B and r_{Na} (g/m^2), and TCLP $-\ln[f_{Li}]$ and $-\ln[f_{Al}]$

glass	T_5	MCC-1		PCT		TCLP		glass	T_5	MCC-1		PCT		TCLP	
		r_B	r_{Na}	r_B	r_{Na}	$-\ln[f_{Al}]$	$-\ln[f_{Li}]$			r_B	r_{Na}	r_B	r_{Na}	$-\ln[f_{Al}]$	$-\ln[f_{Li}]$
BL	1,120	5.1	5.2	0.5	0.6	7.4	7.7	Li-1	1,343	2.9	2.9	0.3	0.3	8.6	
Al-1	1,075	17.9	17.6	6.1	5.0		7.4	Li-2	1,248	4.2	4.1	0.3	0.4	8.4	8.3
Al-2	1,187	4.2	4.3	0.2	0.3	8.2	8.1	Li-3	1,106	5.7	5.7	0.6	0.6	7.4	7.5
B-1	1,209		3.9		0.6	8.9	8.3	Mg-1	1,111	6.5	5.6	1.3	1.1		
B-2	1,154	4.5	4.5	0.3	0.3	8.4	8.2	Mg-2	1,103	8.0	7.8	4.0	3.3		
B-3	1,065	8.1	8.3	1.8	1.5	7.5	7.2	Na-1	1,282	2.7		0.1		8.4	8.7
Ca-1	1,188	3.0	3.0	0.2	0.4	8.4	8.4	Na-2	1,159	4.7	5.3	0.4	0.5	8.0	8.1
Ca-2	1,158	4.4	4.4	0.4	0.5	8.2	8.2	Pb-1	1,128	6.4	6.2	0.5	0.6	8.2	8.0
Ca-3	1,103	6.2	6.1	0.6	0.7	7.1	7.4	Pb-2	1,103	5.3	5.2	0.5	0.5	8.1	7.8
Ce-1	1,124	5.5	5.5	0.6	0.6	7.3	7.4	Pb-3	1,096	3.7	3.6	0.5	0.5		
Ce-2	1,129	5.4	5.1	0.6	0.6	7.9	7.8	Si-1	1,039	6.4	6.3	0.9	0.9	7.8	7.4
Ce-3	1,118	5.4	5.2	0.6	0.6	8.0	7.9	Si-2	1,314	2.8	2.9	0.3	0.3	8.9	8.8
Gd-1	1,124	6.0	5.7	0.3	0.4			Sn-1	1,136	5.3	5.2	0.5	0.6	8.1	7.9
Gd-2	1,132	5.4	5.3	0.5	0.5			Sn-2	1,139	4.4	4.2	0.4	0.5	8.3	8.0
Gd-3	1,124	5.4	5.2	0.6	0.6			Ti-1	1,130	5.0	5.0	0.5	0.6	7.5	7.6
Gd-4	1,154	5.5	5.3	0.6	0.6			Ti-2	1,098	4.9	5.0	0.5	0.5	7.9	8.0
K-1	1,138	5.0	4.7	0.5	0.5	8.6	8.1	Zr-1	1,116	6.5	6.4	0.6	0.6	7.4	7.5
K-2	1,107	6.3	5.8	0.6	0.6	8.0	7.6	Zr-2	1,131	4.6	4.6	0.5	0.5	7.8	7.9

Table III shows each component's effect on the properties. Component effect is defined as the first-derivative of the property with respect to component addition and can be directly measured as the slope of the property vs. component addition curve.[6]

Table III. Component effects for measured properties

	T_5	MCC-1 $\ln[r_B]$	MCC-1 $\ln[r_{Na}]$	PCT $\ln[r_B]$	PCT $\ln[r_{Na}]$	TCLP $-\ln[f_{Al}]$	TCLP $-\ln[f_{Li}]$
Al_2O_3	1,120	-14.46	-14.07	-32.96	-28.39	7.18	6.23
B_2O_3	-1,197	7.32	6.56	20.79	19.94	-20.09	-9.66
CaO	-1,073	5.70	5.57	7.12	6.38	-17.82	-12.07
CeO_2	-92	-0.32	-0.72	-0.81	-0.74	8.67	6.63
Gd_2O_3	379	-1.02	-0.91	4.12	1.58		
K_2O	-517	3.64	3.55	4.03	4.73	-14.97	-9.31
Li_2O	-5,285	13.78	13.95	18.73	15.75	-27.31	-23.59
MgO	-283	7.48	6.69	35.40	29.78		
Na_2O	-1,596	6.54	2.67	12.81	12.23	-11.46	-8.97
PbO	-479	-4.37	-4.54	-1.07	-0.63	-2.74	-2.48
SiO_2	1,849	-5.57	-5.29	-7.79	-6.67	19.95	9.46
SnO	475	-3.64	-5.46	-2.36	-3.92	16.31	8.83
TiO_2	-614	-0.46	-0.29	-1.82	-3.02	11.30	8.06
ZrO_2	287	-6.95	-6.83	-4.73	-4.72	9.76	7.44

The T_5 values range from 1,039°C for Si-1 (40% SiO_2) to 1,343°C for Li-1 (0% Li_2O). T_5 is increased most by the addition of SiO_2 (1,849°C) followed by Al_2O_3 (1,120°C) and decreased most by Li_2O (-5,285°C) followed by Na_2O (-1,596). Table IV compares the effect of some of the components on T_5 from this study with those from a composition variation study (CVS) baseline glass.[7] There is reasonably good agreement ($\pm$ 30% for all components except ZrO_2) in component effects on the T_5 of the two glasses. This suggests that pronounced interactive effects are unlikely.

Table IV. Component effects on T_5 measured by one-component-at-a-time change

	SiO_2	B_2O_3	Na_2O	Li_2O	Al_2O_3	ZrO_2
This Study (°C)	1,849	-1,197	-1,598	-5,285	1,120	287
CVS[7] (°C)	1,916	-1,546	-1,964	-5,620	1,100	164
Difference (%)	4	29	23	6	-2	-43

Figure 1 shows T_5 vs. changing mass fraction of each component. This plot shows no dramatic single component curvature. The linearity of viscosity with composition over a narrow composition region, reported elsewhere,[5,7] is supported by these data. It is assumed that T_5 data can be interpolated linearly within the composition region between the measured data.

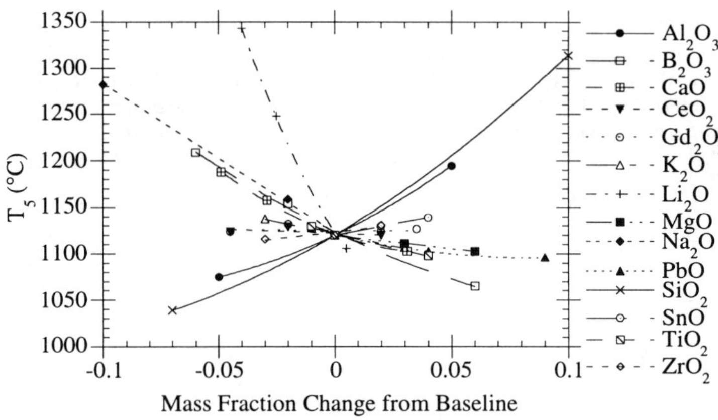

Figure 1. Effect of component mass fraction change on T_5

The MCC-1 r_B values range from 2.70 g/m² for Na-1 (0% Na₂O) to 17.87 g/m² for Al-1 (0% Al₂O₃). MCC-1 ln[r_B] is increased most by the addition of Li₂O (13.78) followed by MgO (7.48) and decreased most by Al₂O₃ (-14.46) followed by ZrO₂ (-6.95). The MCC-1 r_{Na} values range from 2.88 g/m² for Si-2 (55% SiO₂) to 17.61 g/m² for Al-1 (0% Al₂O₃). MCC-1 ln[r_{Na}] is increased most by the addition of Li₂O (13.95) followed by MgO (6.69) and decreased most by Al₂O₃ (-14.07) followed by ZrO₂ (-6.83). The effect of components on r_B and r_{Na} are similar. Figure 2 shows the effect of changing mass fraction of each component on ln[r_{Na}].

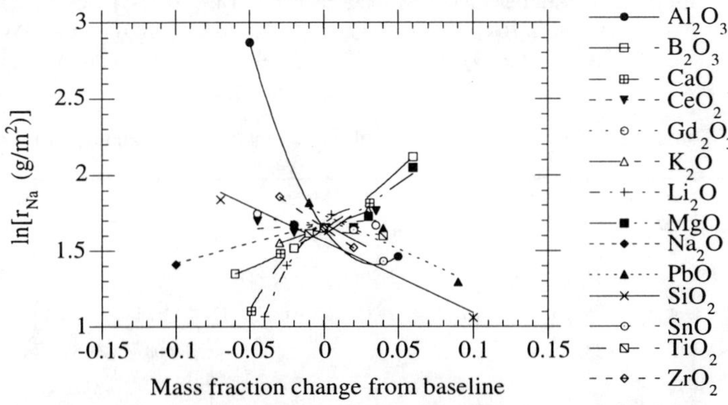

Figure 2. Effect of component mass fraction change on MCC-1 ln[r_{Na}]

Component effects on MCC-1 ln[r_{Na}] and ln[r_B] are not well described with a linear (or first-order) models such as free energy of hydration or first-order empirical forms. Curvature is seen in Figure 2 for Al₂O₃, B₂O₃, and Li₂O, and was reported for other components.[5] Interactive effects of components on MCC-1 normalized release have also been reported.[5] Second- or higher-order models better approximate MCC-1 release in glasses. Interpolations can be made only for compositions near to those for which measured data are available.

The PCT r_B values range from 0.14 g/m² for Na-1 (0% Na₂O) to 6.10 g/m² for Al-1 (0%

Al_2O_3). PCT $\ln[r_B]$ is increased most by the addition of MgO (35.40) followed by B_2O_3 (20.79) and decreased most by Al_2O_3 (-32.96) followed by SiO_2 (-7.79). The PCT r_{Na} values range from 0.29 g/m^2 for Al-2 (10% Al_2O_3) to 4.96 g/m^2 for Al-1 (0% Al_2O_3). PCT $\ln[r_{Na}]$ is increased most by the addition of MgO (29.78) followed by B_2O_3 (19.94) and decreased most by Al_2O_3 (-28.39) followed by SiO_2 (-6.67). The effect of components on r_B and r_{Na} are similar. Figure 3 shows the effect of changing mass fraction of each component on $\ln[r_B]$ as an example.

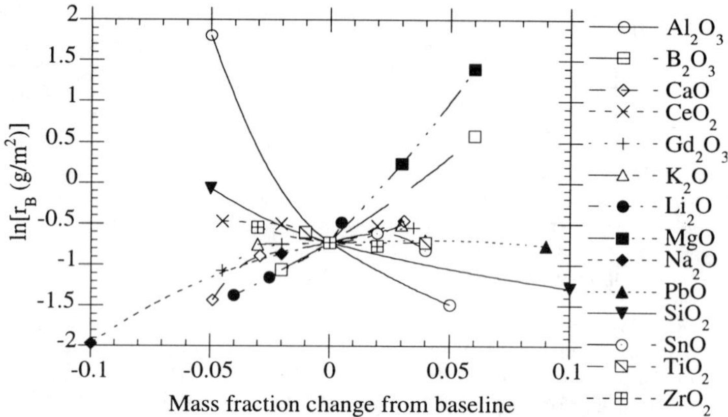

Figure 3. Effect of component mass fraction change on PCT $\ln[r_B]$

Table V provides a summary of the component effects on $\ln[r_{Na}]$ from similar, one-component-at-a-time change studies including the effects of Al_2O_3, B_2O_3, Li_2O, Na_2O, and SiO_2 on the CVS3-1 baseline;[6] Al_2O_3, B_2O_3, CaO, Li_2O, MgO, Na_2O, and SiO_2 on the HW39-4 baseline;[5] Al_2O_3, SiO_2, TiO_2, and ZrO_2 on the WV205 baseline;[8] and TiO_2 on the HTB651 baseline[9] glasses.

Table V. Component effects on PCT $\ln[r_{Na}]$ measured by one-component-at-a-time change

Baseline	Al_2O_3	B_2O_3	CaO	Li_2O	MgO	Na_2O	SiO_2	TiO_2	ZrO_2
This Study	-28.39	19.94	6.38	15.75	29.78	12.23	-6.67	-3.02	-4.72
CVS3-1	-17.34	-9.48		12.20		27.23	-4.45		-7.79
HW39-4	-42.95	10.47	16.33	42.00	42.10	32.52	-11.75		
WV205	-14.23						-608	-21.89	-8.85
HTB651								-5.03	

As for MCC-1, the effects of composition on PCT $\ln[r_{Na}]$ and $\ln[r_B]$ are nonlinear. Figure 3 shows the single component curvature. The differences in component effects from baseline to baseline composition shown in Table V suggest that interactive effects are present. Non-linearity in logarithm of normalized PCT release has been reported elsewhere.[6] These data confirm the need for second- or higher-order model forms for the interpolation of PCT data.

The TCLP $-\ln[f_{Al}]$ values range from 7.14 for Ca-3 (8% CaO) to 8.91 for B-1 (0% B_2O_3). TCLP $-\ln[f_{Al}]$ is increased most by the addition of SiO_2 (19.95) followed by SnO (16.31) and decreased most by Li_2O (-27.31) followed by B_2O_3 (-20.09). The TCLP $-\ln[f_{Li}]$ values range from 7.23 for B-3 (12% B_2O_3) to 8.75 for Si-2 (55% SiO_2). TCLP $-\ln[f_{Li}]$ is increased most by the addition of SiO_2 (9.46) followed by SnO (8.83) and decreased most by Li_2O (-23.59) followed by CaO (-12.07). The effect of components on $-ln[f_{Al}]$ and $-ln[f_{Li}]$ are similar. Figure 4 shows the effect of changing mass fraction of each component on $-ln[f_{Li}]$ as an example.

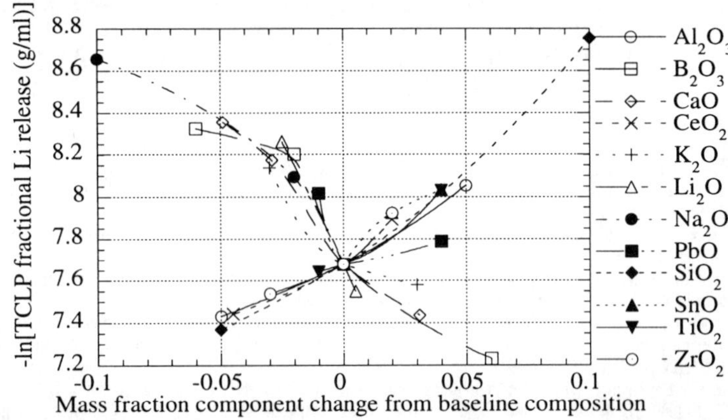

Figure 4. Effect of component mass fraction change on TCLP $-ln[f_{Li}]$

The effect of composition on TCLP release has long been difficult to model.[15] However, the curves in Figure 4 suggest that first- and second-order polynomials may be capable of describing composition effects on the negative logarithm of fractional release by TCLP. This one component-at-a-time change data is only preliminary. Further studies are planned to statistically vary many components-at-a-time to investigate relationship between composition and TCLP release.

SUMMARY AND CONCLUSIONS

The effects of composition on glass properties important to vitrification of plutonium residues(viscosity and chemical durability) were measured based on a one-component-at-a-time change experimental design. The effects of Al_2O_3, B_2O_3, CaO, CeO_2, Gd_2O_3, K_2O, Li_2O, MgO, Na_2O, PbO, SiO_2, SnO, TiO_2, and ZrO_2, taken at the composition of the baseline simulated Pu residue glass, on T_5, $ln[r_{Na}]$ and $ln[r_B]$ for MCC-1 and PCT, and $-ln[f_{Al}]$ and $-ln[f_{Li}]$ for TCLP are listed in Table III.

The temperature at a viscosity of 5 Pa·s (T_5) was shown to change in a nearly linear fashion with composition. T_5 is increased most by the addition of SiO_2 followed by Al_2O_3 and decreased most by Li_2O followed by Na_2O. The chemical durability has been characterized using three tests: 1) MCC-1 leaching of a glass monolith for 3 d in DIW at 90°C with a glass surface-area-to-water-volume ratio (S/V) of 10 m[-1], 2) PCT leaching of crushed glass for 7 d in DIW at 90°C with a S/V of 2000 m[-1], and 3) TCLP leaching of chunked glass in acetic acid at pH 4.7 for 18 h at 25°C with a S/V of 45 m[-1]. A non-linear relationship between logarithm of normalized elemental release and composition was found for MCC-1 and PCT. The relationship between composition and the negative logarithm of TCLP fractional release appears to be describable by first- and second-order polynomials, based on these preliminary data.

Since glass durability was shown here and elsewhere[5,6,11] to be a nonlinear function of composition, interpolations of glass durability measurements can be made only for compositions near to those for which measured data are available. These data are thus used to guide expansion of the experimental database for Pu scrap and residue glass durability currently underway at PNNL. The linearity of viscosity with composition allow the T_5 data to be useful in formulating Pu scrap and residue glasses for further testing.

ACKNOWLEDGEMENTS

The authors would like to thank M. J. Schweiger and D. E. Smith for laboratory assistance, H. Li and P. Hrma for helpful conversations and suggestions, and M. L. Gong and R. J. Barbera for solution analyses. SAB acknowledges the U. S. Department of Energy's (DOE) Science and Engineering Research Semester program for partial funding. The Pacific Northwest National Laboratory is operated for the DOE by Battelle under contract DE-AC06-76RLO 1830.

REFERENCES

1. K. R. Kormanyos, J. C. Marra, and T. J. Overcamp, "Vitrification of Simulated Radioactive Rocky Flats Plutonium Containing Waste Ash with a Stir-Melter System," in Ceram. Trans. in press (1997).

2. W. J. Lutze and R. C. Ewing, Radioactive Waste Forms for the Future, North-Holland Publishing, New York, NY (1988).

3. C. W. Forsberg, What Is Plutonium Stabilization, and What Is Safe Storage of Plutonium?, ORNL/M-4322, Oak Ridge National Laboratory, Oak Ridge, TN (1995).

4. Memorandum, Edward J. McCallum, Director, Office of Safeguards and Security, US Department of Energy, Subject: Additional Attractiveness Level E Criteria for Special Nuclear Materials (SNM), July 22, 1996.

5. P. Hrma, G. F. Piepel, M. J. Schweiger, D. E. Smith, D. S. Kim, P. E. Redgate, J. D. Vienna, C. A. LoPresti, D. B. Simpson, D. K. Peeler, and M. H. Langowski, Property / Composition Relationships for Hanford High-Level Waste Glasses Melting at 1150 °C, PNL-10359, vol. 1 and 2, Pacific Northwest National Laboratory, Richland, WA (1994).

6. J. D. Vienna, P. Hrma, M. J. Schweiger, and M. H. Langowski, "Compositional Dependence of Elemental Release from HLW Glasses by the Product Consistency Test: a One Component-at-a-Time Study," in Ceram. Trans. in press (1997).

7. J. D. Vienna, P. Hrma, D. S. Kim, M. J. Schweiger, and D. E. Smith, "Compositional Dependence of Viscosity, Electrical Conductivity, and Liquidus Temperature of Multicomponent Borosilicate Waste Glass," in Ceram. Trans. in press (1997).

8. X. D. Feng, I. L. Pegg, A. Barkatt, P. B. Mecedo, S. J. Cucinell, and S. Lai, "Correlation Between Composition Effects on Glass Durability and the Structural Role of the Constituent Oxides," Nuclear Technology, vol. 85, 334-45 (1989).

9. H. Li, M. H. Langowski, P. Hrma, M. J. Schweiger, J. D. Vienna, and D. E. Smith, Minor Component Study for Simulated High-Level Nuclear Waste Glasses, PNNL-10996, Pacific Northwest National Laboratory, Richland, WA (1995).

10. J. C. Cunnane, High-Level Waste Borosilicate Glass: A Compendium of Corrosion Characteristics, DOE-EM-0177, U.S. DOE, Office of Waste Management, Washington D.C. (1994).

11. D. S. Kim, P. Hrma, S. E. Palmer, D. E. Smith, and M. J. Schweiger, "Effect of B_2O_3, CaO, and Al_2O_3 on the Chemical Durability of Silicate Glasses for Hanford Low-Level Waste Immobilization," Ceram. Trans., vol. 61, 531-8 (1995).

12. H. Li, J. G. Darab, P. A. Smith, X. D. Feng, and D. K. Peeler, "Chemical Durability of Low-Level Simulated Nuclear Waste Glasses with High-Concentrations of Minor Components," in Nuclear Materials Management, vol. XXIV, INMM, Palm Desert, CA (1995).

13. B. C. Bunker, "Waste Glass Leaching: Chemistry and Kinetics," in Scientific Basis for Nuclear Waste Management X, Materials Research Society, Pittsburgh, PA (1987).

14. D. E. Clark and B. Z. Zoitos, Corrosion of Glass, Ceramics, and Ceramic Superconductors, Noyes Publications, Park Ridge, NJ (1992).

15. J. L. Resce, T. J. Overcamp, C. A. Cicero, and D. F. Bickford, The Effect of Compositional Parameters on the TCLP and PCT Durability of Environmental Glasses, WSRC-MS-95-0294, Westinghouse Savannah River Co., Aiken, SC (1995).

GLASS-CERAMIC WASTE FORMS FOR IMMOBILIZING PLUTONIUM

T. P. O'Holleran, S. G. Johnson, S. M. Frank, M. K. Meyer, M. Noy, and E. L. Wood
Argonne National Laboratory-West
P.O. Box 2528
Idaho Falls, ID 83403-2528

D. A. Knecht, K. Vinjamuri and B. A. Staples
Lockheed Martin Idaho Technologies Co.
P.O. Box 1625
Idaho Falls, ID 83415-5218

ABSTRACT

Results are reported on several new glass and glass-ceramic waste formulations for plutonium disposition. The approach proposed involves employing existing calcined high level waste (HLW) present at the Idaho Chemical Processing Plant (ICPP) as an additive to: 1) aid in the formation of a durable waste form and 2) decrease the attractiveness level of the plutonium from a proliferation viewpoint. The plutonium, PuO_2, loadings employed were 15 wt% (glass) and 17 wt% (glass-ceramic). Results in the form of x-ray diffraction patterns, microstructure and durability tests are presented on cerium surrogate and plutonium loaded waste forms using simulated calcined HLW and demonstrate that durable phases, zirconia and zirconolite, contain essentially all the plutonium.

INTRODUCTION

In its report on plutonium dispositioning the National Academy of Sciences recommended two primary options for long-term plutonium disposition: 1) Burning the material in nuclear reactors as a mixed oxide (MOX) fuel or 2) Dissolving the material in a silicate glass along with enough fission products to protect the material with a lethal radiation field [1]. Dissolving the material in a silicate glass along with fission products is sometimes referred to as meeting the "spent fuel standard." This approach puts weapons-grade material into a form very similar, from a nuclear non-proliferation standpoint, to the plutonium in spent nuclear fuel. The intense radiation field generated by the fission products greatly complicates the process required to recover the material for weapons production, making it unlikely that any but the most technologically advanced countries,which presumably already have nuclear weapons technology, would be able to recover the material. The major drawback to this approach is that the protection afforded by the radiation field from the fission products decays relatively quickly. As the radiation field diminishes, the process required to dissolve the glass and separate the plutonium becomes less daunting. The objective of this project is to show the feasibility of preparing plutonium for final dispositioning in a safe and proliferation-resistant form by disbursing it in a durable, multi-phase silicate-based matrix spiked with fission products from HLW stored at the ICPP.

EXPERIMENTAL

Technical Approach

The present work builds upon previous experience developing waste forms to immobilize HLW currently stored at the ICPP. Both amorphous [2] and crystalline [3] waste forms have been developed. Amorphous waste forms, e.g., borosilicate glasses, offer the advantage of simplified processing conditions, but higher waste loadings and more durable host phases are usually obtained with crystalline waste forms. In this case, the limited solubility of plutonium in silicate

Mat. Res. Soc. Symp. Proc. Vol. 465 © 1997 Materials Research Society

glasses is particularly problematic. Devitrification was selected as a potential route to combine the desirable features of both types of waste form while minimizing the disadvantages. Devitrification is accomplished by heating a glass above its crystallization temperature, and nucleating agents can sometimes be used to control the process. In order to produce an acceptable material for plutonium immobilization by devitrification of a borosilicate glass, it is desirable to partition both the plutonium and neutron absorbers, added for criticality control, into the same phase(s). Furthermore, sufficient quantities of the plutonium and neutron absorber host phase(s) must be formed to result in an acceptable plutonium loading, 10-25 wt.% of the bulk material.

The project used two closely related but distinct approaches to the issue of producing a devitrified waste product with maximum Pu content. The first approach focused on lower temperatures easily achievable in "conventional" Joule-heated HLW melters. The second approach increased the simulated HLW content to a loading of 50-60 wt% and utilized higher melt temperatures, 1450 °C.

For the first approach the starting materials, i.e., surrogate calcine, glass frit and/or other reagent additives, CeO_2 or PuO_2, and Sm_2O_3 (neutron absorber), were mixed as dry powders, and melted together in alumina crucibles. Heat treatments to devitrify the glasses were performed in two ways: 1) by slow cooling the melt to the temperature selected to start the heat treatment schedule, and 2) by pouring the molten glass into a graphite mold to form small ($\approx$1cm x 1cm) cylindrical "buttons" that were re-heated for devitrification. The first method more closely resembles processing conditions that would be encountered in a full-scale process. Because the material bonds to in the alumina crucible, samples for characterization and testing were obtained by core drilling. The second method also gives some insight into the nature of the melt because of the rapid cooling rates that result from pouring the glass buttons.

Initial devitrification experiments used a glass additive (Frit 127) that was developed for immobilizing zirconia calcine[2]. The composition of Frit 127 is shown in Table I. The frit-to-calcine ratio was fixed at 7/3 (by weight), because this ratio is known to result in a durable waste form for HLW. Cerium oxide (or plutonium oxide) and samarium oxide were added to the frit/calcine mixture to produce the starting materials for these experiments. The three formulations tested are shown in Table IV. These formulations were melted in air at 1050°C for 4 hours in an electric furnace. For devitrification by slow cooling, the furnace was cooled to 700°C, followed by a 144 hr hold. Following the hold, the furnace was switched off and allowed to cool. Quenched buttons were prepared by pouring the molten material into a 1 cm by 1 cm cylindrical graphite mold. Quenched buttons were also prepared from melts at 1150°C and 1350°C. The buttons were devitrified by heating to 700°C at 2°C/min.

Additional formulations were developed that, upon devitrification, were expected to produce durable crystalline phases similar to those observed in glass-ceramic materials developed to immobilize, calcine (see Table V). These formulations were melted at 1200°C for 5 hours in an electric furnace. Devitrification was accomplished by cooling the furnace to 550°C and varying the hold time from 12 hr to 14 days.

The second approach consisted of a glass ceramic formulated with the following constituents: zirconia calcine (52 wt%), Frit 127 (21 wt%), PuO_2 (17 wt%), Sm_2O_3 (6 wt%), and Ti (4 wt%). The higher zirconia calcine content dictates a higher melt temperature. The glass ceramic was formed by heating this mixture in air at 1450 °C for a total of 4 hours to allow for mixing, and then the product was brought to 500 °C by lowering the temperature at a rate of 350 °C/hr. The product was then held at 500 °C for 4 hr to allow for devitrification.

Samples of the formulations produced in alumina crucibles were obtained by core drilling. Quench-cooled samples were analyzed directly. Materials were characterized by Scanning Electron Microscopy (SEM) with Energy Dispersive X-ray Spectroscopy (EDS), and X-ray Powder Diffraction (XRD). Durability was measured using the 7-day Product Consistency Test (PCT) [4]. Elemental analyses were performed on the leachate solutions using Inductively Coupled Plasma Atomic Emission Spectroscopy.

RESULTS AND DISCUSSION

The results of the two approaches used in this project will be presented separately here so as to be clear; however, the conclusions drawn from these experiments are general and will be presented together.

Low Temperature Glass Approach

The first devitrification test used a 7/3 mixture (by weight) of Frit 127 to surrogate calcine, with cerium and samarium admixed to form CE1 in Table II. Figure 1 is a backscattered electron SEM image of devitrified CE1 glass. Four distinct phases were identified in backscattered electron images. EDS analysis was used to estimate the elemental composition of these phases, with XRD providing crystallographic information. Phase 1, labeled zirconia in Figure 1, produces an XRD pattern similar to that of cubic zirconia, but cerium is heavily substituted for zirconium ($\approx$50 atom%). The zirconia phase also contains 7-8 atom% samarium. The sodium-calcium aluminosilicate phase tentatively identified as anorthite in Figure 1 also contains 4-5 atom% cerium and samarium. The unidentified phase labeled 2 in Figure 1 is a calcium silicate phase with 7-8 atom% fluorine, and 4-5 atom% zirconium, but with no detectable cerium or samarium. Cerium and samarium concentrations in the residual glassy matrix range from undetectable to $\approx$0.5 atom%.

When quenched as buttons from 1050 °C and 1150 °C as described earlier, the CE1 composition in Table IV exhibited residual cerium oxide in an otherwise homogeneous glassy matrix. When quenched from 1350 °C, CE1 produced a homogeneous glass. Figure 2 is a backscattered electron SEM image of the CE1 glass quenched from 1350 °C. The bright particle near the center of the image is a dust particle high in chromium, and the diagonal feature is a scratch. Even at a magnification of 10,000X this glass appeared homogeneous, suggesting that the 9 wt% cerium oxide dissolved completely at 1350 °C. These buttons produced crystalline phases similar to those in the slow-cooled material when re-heated to 700 °C for 144 hours, except that the cerium- and samarium-rich cubic zirconia phase was conspicuously absent. The cerium and samarium content in the anorthite-like phase was somewhat higher, and some remnants of cerium oxide remained.

When the glasses of composition shown in Table III were devitrified as described earlier, calcium aluminosilicate and zirconia phases similar in composition to the anorthite-like and zirconia phases found in the devitrified CE1 and CE2 glasses, as well as fluorite, were produced. Cerium and samarium were detected in both phases (except in glasses 647.1, 648.1, and 649.1 where no samarium was present). Partitioning of the cerium and samarium was generally good, however, in some samples pure Sm_2O_3 formed during devitrification.

Seven-day PCT leach tests were performed on selected cerium-loaded devitrified samples, and the leachates analyzed as described earlier (see Table IV.). All these materials performed well

Fig. 1 Backscattered Electron SEM Image of Devitrified CE1 (see Table IV). Cerium is heavily substituted in the bright zirconia phase (1), which also contains a high concentration of samarium. The dendritic phase (4) tentatively identified as anorthite contains smaller amounts of cerium and samarium. Only trace amounts of cerium and samarium were detected in the residual glassy phase (3). Phase (2) is an unidentified calcium silicate phase also containing zirconium and fluoride.

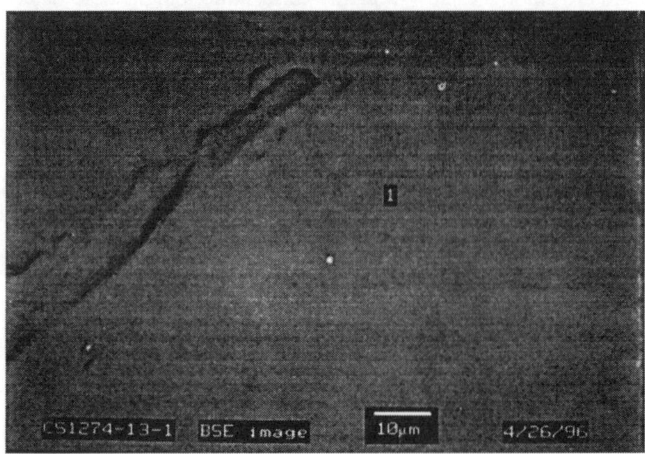

Figure 2: Backscattered electron SEM image of quenched CE1 glass containing 9 wt% CeO_2 (15 wt% PuO_2 equivalent). The bright spot near the center of the image is a dust particle high in chromium, and the diagonal feature is a scratch in the sample surface. No atomic number variations were evident in this sample up to a magnification of 10,000X.

in terms of overall mass loss compared to results considered acceptable for HLW glasses. Concentrations of plutonium surrogate cerium and samarium in the leachates were of particular interest, because of concerns that, in a repository, leaching could separate plutonium from the neutron poison and concentrate the plutonium, producing a critical configuration. The "<" symbols in Table IV indicate that both cerium and samarium were below detectability limits in the leachates.

Materials containing plutonium were prepared in order to verify that cerium was performing as a good surrogate for plutonium in these experiments. In order to minimize sample preparation requirements, the first plutonium-containing material (PU1 in Table II) was prepared as a quenched button and devitrified by re-heating at 700 C for 144 hours. The devitrified material was virtually identical to the quenched and devitrified CE1 material, except that no plutonium or samarium could be detected in the residual glassy matrix.

Table I. Compositions (in wt%) of Frit 127,and the surrogate calcine used in these experiments.

Frit 127		Surrogate Zirconia Calcine			
Component	w/w%	Components	w/w%	Components	w/w%
SiO_2	70.3	Al_2O_3	14.4	CaF_2	34.1
Na_2O	12.8	B_2O_3	0.3	Fe_2O_3	0.2
Li_2O	6.2	CaO	25.9	K_2O	<0.1
B_2O_3	8.5	CdO	<0.1	MnO_2	<0.1
CuO	2.1	Cr_2O_3	0.7	Na_2O	0.3
		Cs_2O	0.3	NiO	0.1
		HgO	<0.1	SrO	0.8
				ZrO_2	22.8

Table II. Starting mixtures for initial devitrification tests (in wt %). For cerium-loaded samples the equivalent weight percent plutonium is shown in parentheses.

Glass ID	Frit 127	Surrogate Calcine	CeO_2	Sm_2O_3	PuO_2
CE1	61	26	9 (15)	4	--
CE2	54	23	15 (25)	8	--
PU1	56	24	--	5	15

Table III. Formulation for immobilizing plutonium based on previous experience with glass-ceramic waste forms developed for immobilizing HLW. Values are wt %, with the equivalent plutonium oxide loading shown in parentheses.

Calcine	SiO_2	Na_2O	B_2O_3	Al_2O_3	CeO_2	Sm_2O_3
27.9-30	27.9-46.6	10.8-19.6	5.9-9.8	0-1.4	2.5 (3.5)-12.7 (20)	0-12.8

Table IV. Normalized release rates from 7-day PCT (g/m^2 day). Please see Table II for CE1 composition. Please see Table III for "New Formulation Glasses" composition.

Sample	Si	Na	Al	B	Ce	Sm
CE1	3E-2	7E-2	1.9E-1	3.3E-1	< 2.6E-4	< 2.2E-4
New Formulation Glasses	0.03-1.0E-2	4-15E-2	1-2.4E-1	1-8E-2	< 2.9E-3	< 1.4E-3

<u>High Temperature Glass-Ceramic Approach</u>

A series of CeO_2 surrogates was produced to verify that the desired crystalline phases would form and that Ce and Sm would be immobilized within durable hosts. The CeO_2 concentrations varied between 7 and 16 wt % which mimic PuO_2 concentrations of 10-25 wt%. The SEM and XRD results supported the hypothesis that CeO_2 at these levels was soluble in the resulting multi-phase product. A PuO_2-loaded formulation containing 17 wt % PuO_2 was then produced and characterized.

The SEM analysis was performed on an axially split core sample. A total of 11 phases in addition to the glassy matrix were observed. Two phases were found to contain essentially all the Pu. The observed stoichiometries for these two phases are contained in Table V below.

Table V. Observed stoichiometries for two Pu containing phases from SEM* (wt %).

Phase	Al	Si	Ca	Ti	Zr	Sm	Pu
1	-	-	3-6	<1	20-44	5-8	23-52
2	2.8	-	8.1	16.3	27.3	11.1	5.7

* Balance is assumed to be oxygen. Light elements (O, F) are not observable with SEM. Quantification was achieved using standardless method.

Phase 1 is thought to be a cubic zirconia (fluorite) type structure with phase 2 being a zirconolite type structure. These Pu containing phases are featured in Fig. 4. Phase 1 has variable Pu-content, 23-52 wt %, with calcium and samarium also present, 3-8 wt %, and a spherical morphology. Phase 2 contains approximately 6 wt % Pu but is of fixed composition with a higher samarium content of 11 wt %.

 The XRD analysis of the powdered product revealed a very complex pattern with a multitude of phases present. We should first note that PuO_2 was not found in the pattern search thus indicating that the feed material did go into solution. The most prominent match was the fluorite structure.   Other phases were also present: zirconium oxide (cubic ZrO_2), zirconolites ($CaZrTi_2O_7$, $(ZrCaTi)O_2$), aluminum silicates and sodium and/or calcium silicates ($CaAl_2SiO_4$, $Ca_3Al_2(SiO_4)_3$). An example of a typical powder pattern is shown in Fig. 3.

A PCT type test, SA/V = 2000 m^{-1}, T = 90 °C, t = 9 days, was performed on a crushed, sieved sample with the -100 to +200 mesh size fraction used in the leach process. The results of the PCT test are presented in Table VI. The quantities reported here are for an abbreviated version of the conventional PCT method, where pH, conductivity and normalized release rate (NRR) of Pu are determined [5]. Values for EA glass are also included for comparison. Each sample and standard was run in triplicate.

Table VI. PCT test results for Pu loaded glass-ceramic and EA glass.

Sample	pH of leachate	NRR Pu (g/m^2 day)	Conductivity (µS/cm)
Pu Glass Ceramic	8.8	1.2×10^{-3}	160
EA glass	11.8	N/A	6500

The normalized release rate for Pu compares favorably with that obtained by other ceramic phase waste forms that contain actinides [6]. It is important that all the detectable plutonium in the glass ceramic is found in two distinct phases, the cubic zirconia and zirconolite, which also are hosts for samarium, the neutron absorber. These phases are known to be very durable hosts with

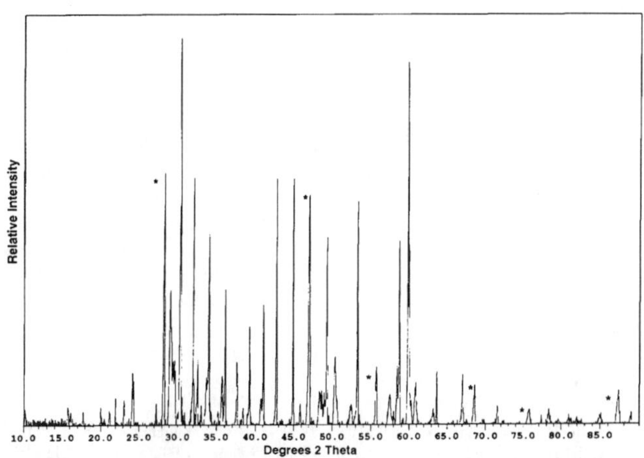

Fig. 3 XRD pattern of the 17 wt% plutonium glass ceramic waste form with the prominent lines for the fluorite phase indicated with an asterisk to the left of the peak.

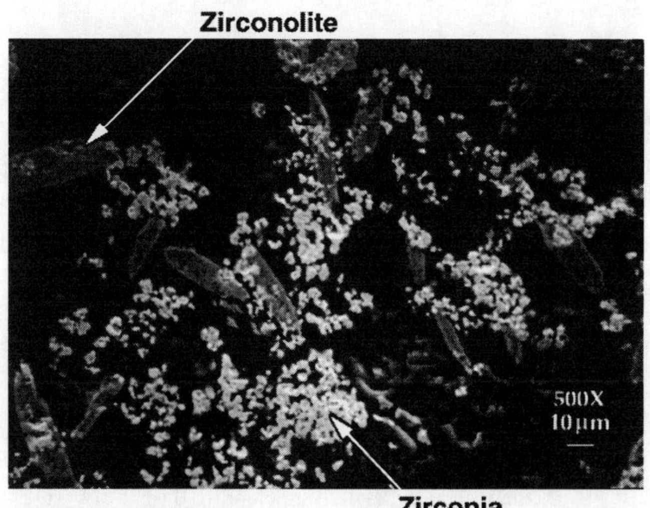

Fig. 4 SEM micrograph of the 17 wt % plutonium glass ceramic waste form with the zirconia and zirconolite phases highlighted. Please see Table V for further detail.

extensive geological evidence supporting their longevity [6,7]. These phases are also considered desirable for other waste product formulations [8].

CONCLUSIONS

A variety of candidate plutonium dispositioning materials have been prepared using surrogate calcined HLW. Experiments with plutonium have verified the choice of cerium as a surrogate for plutonium. The waste product formulations presented here exhibit desirable characteristics for a plutonium disposition option. Chiefly they: 1) display a high capacity for plutonium (15-25 wt%), 2) they incorporate the neutron poison into the same phases as the Pu, 3) they are durable as demonstrated by the PCT test, 4) they incorporate existing HLW to discourage recovery of the plutonium and 5) they are feasible to produce using several different methods detailed here. The moderate processing conditions required to prepare these materials would allow conventional processes and equipment used to immobilize HLW to be adapted for this mission with only minor modifications.

ACKNOWLEGMENTS

The Idaho National Engineering Laboratory is operated for the U. S. Department of Energy by Lockheed Martin Idaho Technologies Co. under contract DE-AC07-99ID13223. B. Scholes, L. Torres, R. Hague, J. Waddell, B. Boyle and M. Hankins are gratefully acknowledged for their contributions to this work. Argonne National Laboratory is operated for the U.S. Department of Energy by the University of Chicago. This work was supported by the U.S. Department of Energy, Reactor Systems, Development and Technology, under contract W-31-109-ENG-38. T. Moschetti is gratefully acknowledged for her contributions in the cerium oxide surrogate work.

REFERENCES

1) "Management and Disposition of Excess Weapons Plutonium," Committee on International Security and Arms Control, National Academy of Sciences, National Academy Press, 1994.

2) Staples, B. A., H. S. Cole and D. A. Pavlica, in Scientific Basis for Waste Management II, Ed. V. G. Brookins, (Mater. Res. Soc. Proc. 2, 1979), p. 125.

3) Vinjamuri, K., in Ceramic Transactions, (45), Ed. D. Bickford, S. Bates, V. Jain and G. Smith, (1994), p.3.

4) ASTM, C1285-94, "Standard Test Methods for Determining Chemical Durability of Nuclear Waste Glasses: The Product Consistency Test."

5) Gougar, M. L. D., D. D. Siemer and B. E. Scheetz, in Scientific Basis for Waste Management XIX, Ed. W. M. Murphy and D. A. Knecht (Mater. Res. Soc. Proc. 412, 1995), p.395.

6) Vance, E. R., B. D. Begg and C. J. Ball, in Scientific Basis for Waste Management XVIII, Ed. T. Murakami and R. C. Ewing (Mater. Res. Soc. Proc. 353, 1994), p. 767.

7) Anderson, E. B., B. Burakov, B. Galkin, V. Stracherko and V. Vasilev, in Proceedings of the Plutonium Stabilization and Immobilization Workshop, Washington, D.C., U. S. DOE, (1995).

8) Swenson, D., T. G. Nieh and J. H. Fournelle, in Scientific Basis for Waste Management XIX, Ed. W. M. Murphy and D. A. Knecht (Mater. Res. Soc. Proc. 412, 1995), p. 337.

CHARACTERIZATION OF A PLUTONIUM-BEARING ZIRCONOLITE-RICH SYNROC

E. C. BUCK, B. EBBINGHAUS,* A. J . BAKEL, and J. K. BATES
Argonne National Laboratory, Argonne, IL 60439
*Lawrence Livermore National Laboratory, Livermore, CA 94550

ABSTRACT

A titanate-based ceramic waste form, rich in phases structurally related to zirconolite $(CaZrTi_2O_7)$, is being developed as a possible method for immobilizing excess plutonium from dismantled nuclear weapons. As part of this program, Lawrence Livermore National Laboratory (LLNL) produced several ceramics that were then characterized at Argonne National Laboratory (ANL). The plutonium-loaded ceramic was found to contain a Pu-Gd zirconolite phase but also contained plutonium titanates, Gd-polymignyte, and a series of other phases. In addition, much of the Pu was remained as PuO_{2-x}. The Pu oxidation state in the zirconolite was determined to be mainly Pu^{4+}, although some Pu^{3+} was believed to be present.

INTRODUCTION

Titanium minerals such as rutile, ilmenite $(FeTiO_3)$, and arizonite (Fe_2TiO_5) are extremely insoluble in water, and therefore varieties of titanate-based mineral phases (or Synrocs) have been studied for some time as possible nuclear waste storage materials [1-4]. Evidence from the laboratory testing and from the study of natural analogues of the constituent phases of Synroc suggests that these materials are extremely corrosion resistant [1-3]. Recently, zirconolite-rich ceramics have been considered for the disposal of actinide-bearing waste streams, in particular excess Pu from dismantled nuclear weapons [5]. However, the heterogeneous nature of Synroc waste forms may make it difficult to predict the long-term behavior of these types of materials.

In this paper we report on the production and microstructural characterization of a preliminary Fissile Materials Disposition (MD) ceramic formulation which has been designed for high-Pu loadings. The purpose of this study is to obtain data to support the choice of a waste form for Pu immobilization. The data presented in this paper are only preliminary, as the MD ceramic composition has not, as yet, been finalized.

EXPERIMENTAL PROCEDURES

Preparation of the Pu-Loaded Ceramic

The plutonium oxide (PuO_{2-x}) used to prepare the Pu-loaded ceramic at LLNL was made from a low temperature $(<200°C)$ oxidation of $PuH_{2.7}$ in a 4% O_2/Ar atmosphere. The hydride was formed at about 100°C from a weapons Pu pit. The PuO_2 was split into two batches: half was calcined at 450°C for four hours (low fired) and half was calcined in air at 1000°C for four hours (high fired). The PuO_2 particles ranged in size from 20 μm to about 100 μm, with an average at around 65 μm.

The ceramic material was prepared by two different methods. In one series of samples the precursor material was made from precipitation of alkoxides, with the Gd_2O_3 (the low energy neutron absorber) and PuO_2 added as powders in a wet slurry. This material was then dried and calcined in air at 750°C for about 4 hours. In another series, the precursor material which was ground, dried, pulverized, and then calcined in air at 600°C for 1 hour, consisted of $Zr(NO_3)_4 \cdot 5H_2O$, TiO_2, Gd_2O_3, CaO, $Al(OH)_3$, and BaO. The product was dry blended with PuO_2 in a mortar and pestle. The blended powder was cold pressed at 1055 kg/cm^2 in a 1.27 cm stainless steel die to produce green pellets. These pellets were sintered at 1325°C for 1-4 hours. The target compositions for the Pu-loaded ceramic materials are presented in Table I.

1259

Mat. Res. Soc. Symp. Proc. Vol. 465 © 1997 Materials Research Society

Table I. Target Composition of Pu-Loaded Ceramic Material
(values reported in wt%)

PuO$_x$ Calcine Conditions Prep. Route	Low-Fired, Hot-Pressed Alkoxides	High-Fired, Hot-Pressed Alkoxide	High-Fired, Cold-Pressed Oxides
CaO	7.3	6.5	8.1
ZrO$_2$	16.1	14.4	17.7
TiO$_2$	38.6	34.4	40.4
Al$_2$O$_3$	10.9	9.7	4.7
BaO	3.0	2.7	3.4
PuO$_2$	14.4	19.3	16.9
Gd$_2$O$_3$	9.6	12.9	8.7

Methods for Characterization

Scanning electron microscopy (SEM) analyses were performed at ANL on an ISI microscope with a backscattered electron detector. Representative particles of the material, around 5-30 μm in diameter, were selected with the aid of an optical microscope and embedded into epoxy blocks. Thin sections of the MD ceramic were produced with an ultramicrotome from the blocks; these thin sections were suitable for transmission electron microscopy (TEM). Phase characterizations were performed with a JEOL 2000 FXII TEM, operating at 200 kV, which was equipped with an X-ray energy dispersive spectrometer (EDS) and parallel electron energy loss spectrometer. Several zone axis patterns were used to determine the structure of each phase encountered. The camera lengths for electron diffraction were calibrated with a polycrystalline aluminum sample. Compositional analysis was achieved with TEM/EDS while remaining off the Bragg angle to avoid electron beam channeling.

Leach Test Procedure

Static leach tests were carried out in accordance with the PCT-A procedure [6]. Crushed ceramic (-100+200 mesh) was prepared and examined with SEM to verify the absence of fines. Approximately 1 g of the crushed ceramic and 10 g of deionized water (ASTM-1) were placed in a 22 mL stainless steel (Type 304) Parr vessel. The surface-area-to-solution-volume (S/V) ratio of the PCT-A test is assumed to be around 2000 m^{-1} for a material with a density of 2.7 g/cm^3. Since the ceramic used in these tests has a higher density (4.7 g/cm^3), the actual S/V of these tests is significantly less than 2000 m^{-1}. The vessels were sealed with a Teflon gasket and placed in a 90°C ($\pm$ 2°C) oven for 7 days along with an experimental blank test. At the end of the test period, the leachates were analyzed for pH and cation concentrations and the reacted material was examined with SEM. The insides of the vessels were soaked in 2% HNO$_3$ for 8 hours, and the "acid strip" solution was collected and analyzed for cations. The cation concentrations from the leachates and the "acid strip" solutions were summed to calculate normalized mass losses. Previously determined background values were subtracted from the cation concentrations.

Vapor Hydration Testing

Sample discs were prepared by cutting ~1 x 10 mm wafers from the MD ceramic and polishing the discs to a 600 grit finish. The wafers were suspended via Pt-Rh wires inside a Parr vessel. Sufficient deionized water was added to saturate the air inside the vessel once it had been sealed and at the set temperature. Diagrams of the apparatus have been presented elsewhere [7]. The tests were conducted at 200°C for up to 91 days. After test termination the vessel was examined with SEM and TEM.

MICROSTRUCTURAL CHARACTERIZATION

Ceramic waste forms designed for Pu immobilization are composed of at least 80% zirconolite and pyrochlore phases [5]. Zirconolite has the highest durability of the all the phases in Synroc and is capable of accepting into its structure a variety of ions. The grain size of the ceramic was generally <1 μm, which meant that we were unable to distinguish individual phases with the SEM (see Fig. 1). Therefore, phase characterizations were performed with the analytical TEM. Although the major titanate-based phases are well known, waste loadings can induce the formation of unique phases, which may not be visible with X-ray diffraction (XRD). In the MD ceramic, the high concentration of Pu and Gd may also induce the formation of different phases. Based on SEM examination, it was clear that the ceramic contained a substantial amount of undissolved PuO_2 (see Fig. 1). This result was expected, due to the large PuO_2 particle size used. The material that was characterized was a mixture of the compositions described in Table I.

Zirconolite and Related Structures

The major phase observed in the ceramic was a Pu- and Gd-bearing zirconolite phase, containing nearly equal amounts of Pu and Gd in the structure. Gadolinium-rich phases, identified as being closely related to polymignyte, were also found in the ceramic. With electron energy loss spectroscopy (EELS), we were able to demonstrate the presence of a trace amount of Ce, possibly an impurity in the added Gd, in the zirconolite phase (see Figs. 2a-c).

Zirconolite does not always incorporate radionuclides by isomorphic substitution. Rather, distinct phases form as a consequence of particular elements present in the waste [8]. Radioactive elements are not distributed uniformly throughout the material but concentrated in extended defects [7]. It is most constructive to consider the zirconolite and related structures in terms of layered hexagonal tungsten bronze (HTB) motifs [8]. The repeating unit can be observed in high-resolution images of zirconolite (see Fig. 2d).

The diffraction patterns from these phases can be interpreted in terms of the stacking of the HTB which allows easy comparison of the various zirconolite polytypes. Using procedures developed by White for HTB arrangements [8], it is possible to interpret the electron diffraction patterns of zirconolite polytypes without the need for high resolution imaging. These methods were used to understand the nature of the zirconolite phases formed in the Pu-ceramic formulation. The diffraction pattern of a zirconolite polytype is shown in Fig. 3a.

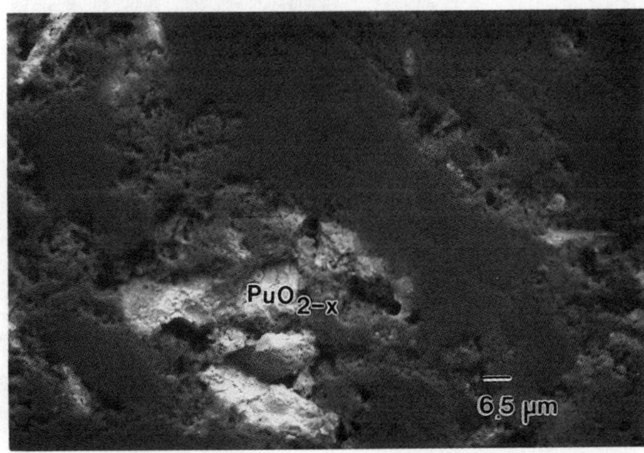

Fig. 1. Backscattered SEM Image of the Surface of the Pu-Ceramic which was Formed with High-Fired PuO_{2-x} and Hot Pressed at 1275°C for 1 hr. The material is porous and shows the presence of undissolved plutonium oxide (bright contrast). The porous regions may contribute to increased levels of release.

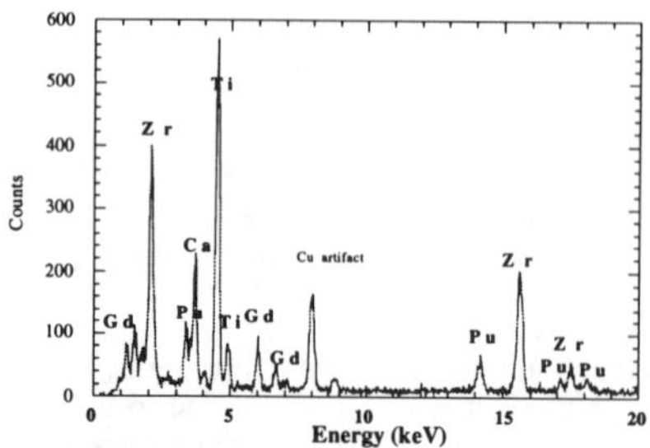

Fig. 2a. EDS Analysis of Pu-Gd Zirconolite.

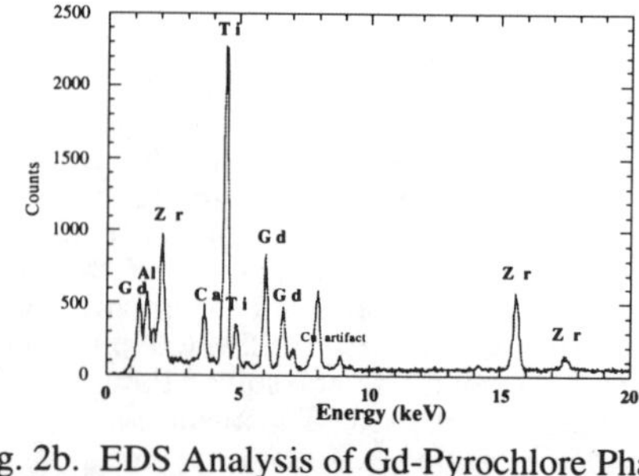

Fig. 2b. EDS Analysis of Gd-Pyrochlore Phase.

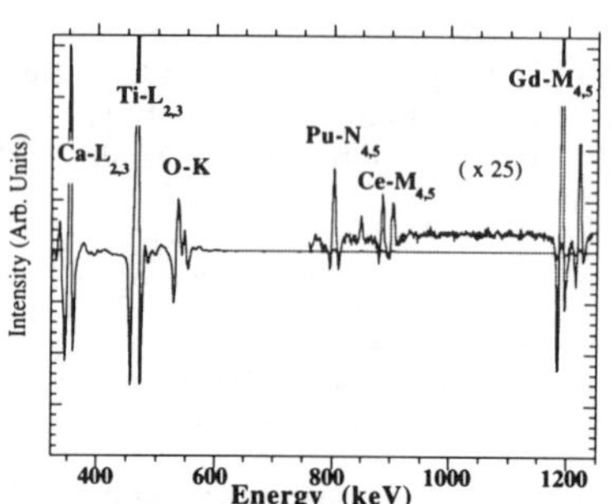

Fig. 2c. TEM/EELS of Zirconolite Phase Showing the Presence of a Trace Amount of Ce. The spectrum collected is the second derivative which helps to reveal some of the finer features of the spectrum.

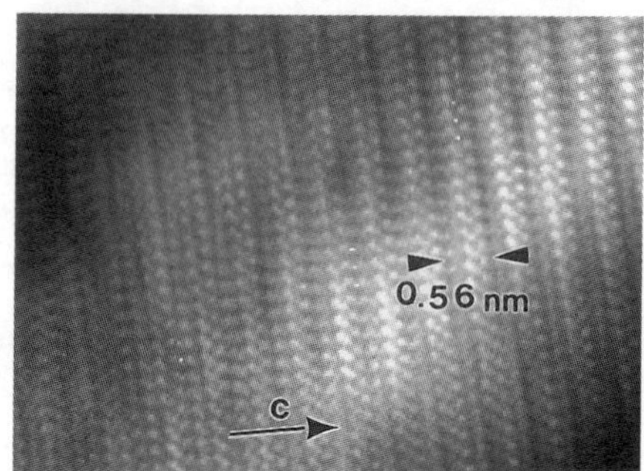

Fig. 2d. Fourier Filtered Multibeam Image of Zirconolite Phase. As described by White [7], the stacking of HTB layers results in bimodular repeat with $c^* = \sim 1.11$ nm.

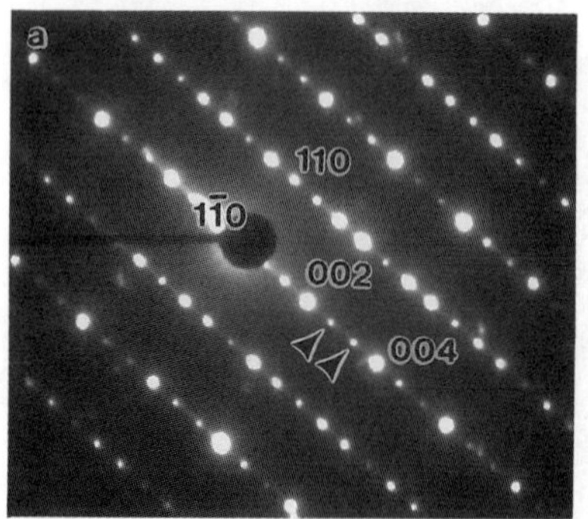

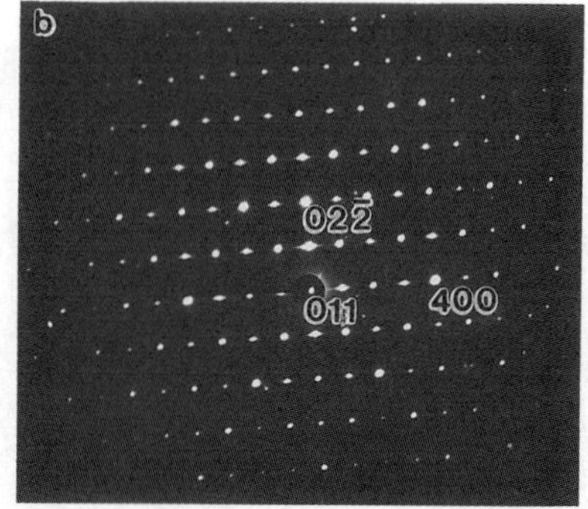

Fig. 3. Electron Diffraction Pattern Taken along $\mathbf{B}[110]_c$ of a Zirconolite Polytype. Structure (a), estimated to be $Ca_{0.75}Gd_{0.2}Pu_{0.4}Zr_{0.75}Ti_{1.5}Al_{0.5}O_7$, exhibits superlattice reflections along the [001] direction (white arrows). The pattern has forbidden reflections for $00l$ where l is odd. A longer exposure was required to reveal the $\mathbf{c}/3$ spots. (b) $\mathbf{B}[011]$ zone axis pattern from a polymignyte phase.

Trivalent rare earths and transuranics (e.g., Gd) may also form zirconolite-related structures of general formula $M_2Ti_2O_7$. Aluminum was found in several phases in minor amounts. Aluminum was found in more significant amounts in a titanium oxide phase and as aluminum oxide. It has been reported that aluminum (Al^{3+}) can readily exchange for Ti^{4+} in titanates with added charge compensating cations [8].

Determination of the Pu Oxidation State in Zirconolite

The oxidation state of the Pu in the zirconolite phase was determined with EELS to be in a reduced state with a technique described by Fortner and Buck [9]. The value obtained from the zirconolite phase indicated that the Pu was probably a mixture of Pu^{4+} and Pu^{3+}. The Pu-$N_{4,5}$ absorption edges exhibit sharp 'white line' transitions; these $4d_{3/2} \Rightarrow 5f_{5/2}$ (N_4) and $4d_{5/2} \Rightarrow 5f_{7/2}$ (N_5) transitions were used to determine the chemical state of the Pu in the zirconolite. The 'white lines' of Pu, Gd, and Ce can be seen in Fig. 2c. The ratio of N_4/N_5 absorption edges for a range of other actinide-bearing phases, as well as other Pu phases have been found to decrease with increasing $5f$-orbital occupancy. In the plot shown in Fig. 4, the alkali tin silicate (ATS) glass was determined by optical methods [10] to contain Pu^{4+}. The Pu oxide was assumed to be in the +3.5 state, based on electron diffraction analysis of this phase; this suggested that it was Pu_4O_7, which is sometimes referred to as cubic Pu_2O_3 [11].

The oxidation state of the Ce present in the zirconolite in trace quantities (< 50 ppm) was determined with EELS to be +3.5 using the Ce-M_4/M_5 absorption edge ratio [9]. The Pu appears to be more oxidized than the Ce in the zirconolite. Fairly oxidizing conditions are required to move Ce into +4 state; therefore, it is expected that the Pu will have a higher average oxidation state than the Ce. However, Ce is still an effective analogue for Pu in zirconolite, as it can substitute into the same sites as the Pu.

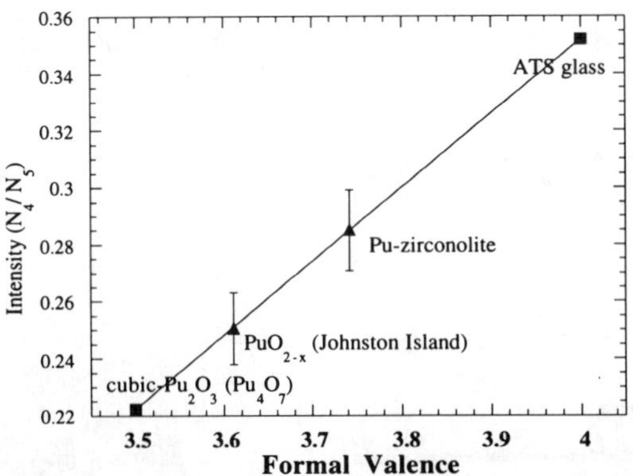

Fig. 4. The Pu Valence in the MD Ceramic was Determined using the Ratio of the Intensity of the N_4 to N_5 White Lines. The reference data points used were the ATS glass for the Pu^{4+} [10] and the reduced plutonium oxide present in the ceramic waste form. The N_4/N_5 ratio from a partially reduced plutonium oxide obtained from a contaminated soil from Johnston Island is also shown for comparison [14]. Although this procedure apparently allows determination of the Pu oxidation state, the $N_{4,5}$ edges are complicated by multiple scattering, multiplet effects, and interactions between the bound states and the continuum. This method, therefore, requires further investigation before it can be used routinely for determination of the Pu oxidation state.

<u>Separation of Gd and Pu in the MD Ceramic</u>

The formulation of the MD ceramic was designed to incorporate both Gd and Pu in the zirconolite-based phases [5]. However, several phases were identified, apart from the occurrence of undissolved Pu oxide, where Gd and Pu were not associated in the same phase (see Fig. 5a). Electron diffraction analysis of a Gd-bearing calcium titanate phase suggested that it was a cubic perovskite. The introduction of Gd into this structure may have induced stabilization of this symmetry which is otherwise orthorhombic. Perovskite is known to be the least durable of the ceramic phases [3]. Brannerite phases (UTi_2O_6 and $ThTi_2O_6$) [12] have also been observed in some formulations [13]. Therefore we might expect similar Pu-bearing titanate phases to occur in the MD ceramic. Indeed, a plutonium titanate phase was identified (see Fig. 5b).

TESTING

Solution results from PCT-A are useful for comparing the reactivities of various waste forms. One of the goals of the MD program is to compare the reactivity of the Pu-ceramic to the reactivity of Pu-glasses. However, this comparison is difficult because the major components of glass (boron and silicon) are not present in the ceramic and the major components of Pu-ceramic (calcium, zirconium and titanium) are not present as major components in the Pu-glasses. In this section we will present normalized loss data for Al, Ba, Ca, Ti, and Zr. We propose that zirconium is released exclusively from zirconolite, while Ti, Al, Ba, and Ca may be released from multiple phases.

Normalized mass loss, NL(i), values (Table II) were calculated based on the mass of each element in the leachate and "acid strip" solutions, the composition of the ceramic, and the assumed S/V (2000 m^{-1}).

The particularly low values of NL(Zr) indicate that zirconolite is dissolving at a very low rate in these tests. The similar values of NL(Al) and NL(Ca) indicate that some phase containing both elements is dissolving at a significantly higher rate than the zirconolite. The relatively high NL(Ba) value shows that some relatively soluble barium phase is present in the ceramic. The NL(Ba) gives an indication of the dissolution rate of the most soluble component, and the NL(Zr) gives an indication of the solubility of the zirconolite in the ceramic.

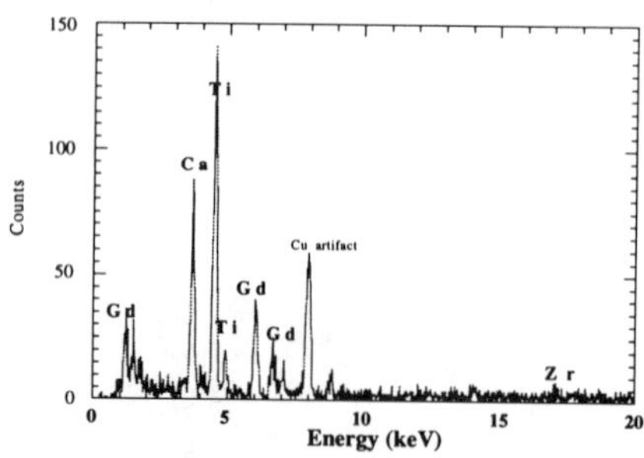

Fig. 5a. Compositional Analysis of Perovskite-Type Particle.

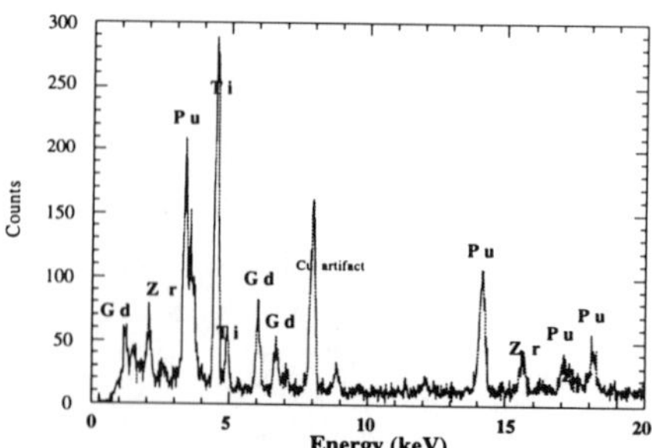

Fig. 5b. Compositional Analysis of a Pu-Titanate Phase.

Table II. Normalized Mass Loss Values for Major Components from Triplicate PCT-A with the Pu-Doped Ceramic

Element	Normalized Mass Loss (g/m^2)[a]	σ Normalized Mass Loss
Al	0.0048	0.0008
Ba	0.092	0.005
Ca	0.0059	0.0006
Ti	$<7.8\times10^{-7}$ [b]	—
Zr	$<1.5\times10^{-5}$ [b]	—

[a]Average of triplicate PCT-A.

[b]The normalized mass loss values are derived from limits of quantification and are therefore maximum values.

EXAMINATION OF CORRODED MD CERAMIC

The MD ceramic, reacted under vapor conditions for 35 days at 200°C, exhibited some surface alteration. In Figs. 6a,b the TEM images of the corroded surface of the ceramic show the formation of a 100-200 nm thick layer of altered material, including a region which has become enriched in Pu.   The ceramic appears to have been in contact with an iron-bearing material, either from the test vessel or from contaminant iron in the ceramic.

The identification of Pu-rich regions on the outer corrosion rind in Fig. 6a suggests that dissolution of the ceramic has occurred, whereas in Fig. 6b the iron-bearing material has precipitated on the zirconolite and there is little evidence of zirconolite dissolution.

CONCLUSIONS

The characterization of the MD ceramic will continue as additional samples are received from LLNL. Knowledge of the phase distribution in the ceramic waste form is critical for understanding the results from the corrosion tests on this material. We have shown that the ceramic contains about 10-20 vol% of phases other than zirconolite or plutonium oxide, although the formulation conditions have not, as yet, being optimized for the MD ceramic. In the case of any titanate-based heterogeneous ceramic waste form, the selection of a marker element for

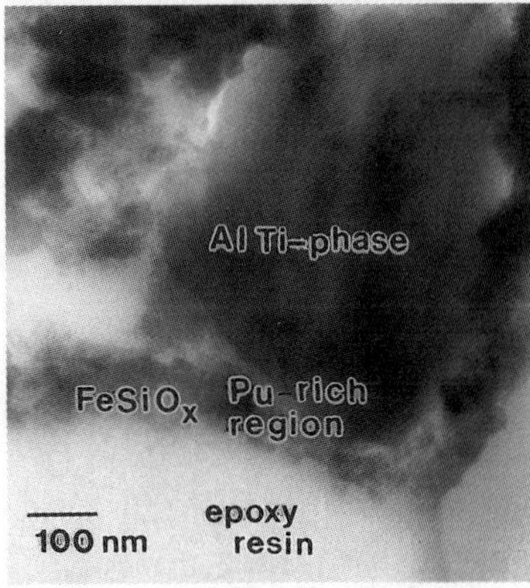

Fig. 6a. Micrographs of Al-Bearing Titanate Phase with a Surface Alteration Phase.

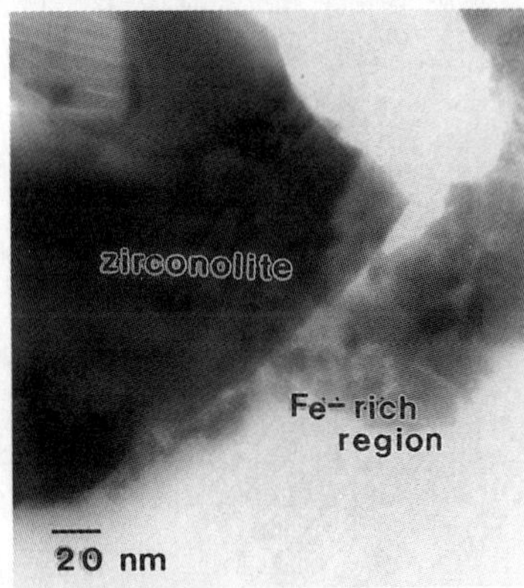

Fig. 6b. Zirconolite Crystal with Precipitated Material on Surface.

monitoring dissolution will be difficult, unless the phase distribution in the particular waste form is known. As we are interested in the corrosion of the Pu-bearing phase, which is structurally and compositionally related to zirconolite ($CaZrTi_2O_7$), we could choose Ca or Zr as a marker element. However, no single element is unique to any one phase; indeed, even plutonium is present in a number of different phases. The use of Ca to monitor the corrosion rate of the ceramic may not provide the correct data on the dissolution term. In addition, Gd and Pu do not always appear to follow each other into the different phases in the MD ceramic, and this may impact criticality assessments. In some cases, Gd is concentrated and in others Pu is concentrated. Lastly, a significant amount of the Pu is still present as an oxide, which suggests that finer PuO_2 should be used or a more reactive form such as $Pu(NO_3)_4$ be used. It is evident that a detailed TEM study will be required of any formulated ceramic to interpret corrosion testing results.

ACKNOWLEDGMENTS

This work was supported by the U.S. Department of Energy, Office of Materials Disposition, under contract W-31-109-ENG-38. Assistance in handling of the Pu ceramic was provided by J. Emery for testing, and analysis with ICP-MS was provided by S. F. Wolf and M. T. Surchik. Samples for TEM analysis were prepared by T. DiSanto.

REFERENCES

1. V. M. Oversby and W. E. Ringwood, "Leaching Studies on Synroc at 95° and 200°C," Rad. Waste Mgmt. 2, 223-237 (1982).

2. G. R. Lumpkin and R. C. Ewing, "Geochemical Alteration of Pyrochlore Group Minerals: Pyrochlore Subgroup," Am. Miner. 80, 732-743 (1995).

3. P. J. McGlinn, K. P. Hart, E. H. Loi, and E. R. Vance, "pH Dependence of the Aqueous Dissolution Rates of Perovskite and Zirconolite at 90°C," Mater. Res. Soc. Symp. Proc. 353, 847-854 (1995).

4. R. A. Van Konynenburg and M. W. Guinan, *Plutonium Doping of Synroc-D*, Lawrence Livermore National Laboratory Report UCRL-53425 (1983).

5. E. R. Vance, C. J. Ball, R. A. Day, K. L. Smith, M. G. Blackford, B. D. Begg, and P. Angel, "Actinide and Rare Earth Incorporation into Zirconolite," J. Alloys Comp. 213/214, 406-409 (1994).

6. *Standard Test Methods for Determining Chemical Durability of Nuclear Waste Glasses: The Product Consistency Test (PCT)*, ASTM Standard C1285-94, ASTM, Philadelphia PA (1994).

7. D. J. Wronkiewicz, L. M. Wang, J. K. Bates, and B. S. Tani, "Effects of Radiation on Glass Alteration in a Steam Environment," 294, 183-190 (1993).

8. T. J. White, "The Microstructure of Synthetic Zirconolite, Zirkelite, and Related Phases," Am. Miner. 69, 1156-1172 (1984).

9. J. A. Fortner and E. C. Buck, "The Chemistry of the Light Rare-Earth Elements as determined by Electron Energy Loss Spectroscopy," Appl. Phys. Lett. 68, 3817-3819 (1996).

10. L. Nuñez and J. A. Fortner, Argonne National Laboratory, private communication (1996).

11. R. B. Roof, *X-ray Diffraction Data for Plutonium Compounds*, Los Alamos National Laboratory Report LA-11619 (1989)

12. R. Ruh and A. D. Wadsley, "The Crystal Structure of $ThTi_2O_6$ (Brannerite)," Acta Cryst. 21, 974-978 (1966).

13. E. R. Vance, P. J. Angel, B. D. Begg, and R. A. Day, "Zirconolite-Rich Ceramics for High-Level Actinide Wastes," Mater. Res. Soc. Symp. Proc. 333, 293-298 (1994).

14. S. F. Wolf, N. R. Brown, J. A. Fortner, E. C. Buck, N. L. Dietz, and J. K. Bates, Environ. Sci. Technol. 31 (1997).

CHARACTERIZATION AND LEACHING BEHAVIOR
OF PLUTONIUM-BEARING SYNROC-C

KATHERINE L. SMITH, GREGORY R. LUMPKIN, MARK G. BLACKFORD,
MICHELLE HAMBLEY, R. ARTHUR DAY, KAYE P. HART and ADAM JOSTSONS

Materials Division, Australian Nuclear Science and Technology Organisation, Private Mail Bag
1, Menai, NSW 2234, Australia.

ABSTRACT

Synroc-C containing 10 wt% simulated PW-4b-D HLW including 0.62 wt% ^{239}Pu was subjected to MCC-1 type leach tests at 70°C in deionised water, silicate and carbonate leachates for 53 d and deionised water for 2472 d. The normalised total (i.e. unfiltered leachate + vessel wall) Pu leach rates in deionised water, silicate and carbonate leachates for periods up to 53 d were found to be of the order of 10^{-5}, 10^{-4} and 10^{-4} g m^{-2} d^{-1} respectively. After 2472 d, the differential, normalised, Pu leach rate in deionised water dropped to ~5 x 10^{-6} (total) and ~5 x 10^{-8} (solution - after filtration through a 1000NMW filter) g m^{-2} d^{-1}. SEM and AEM were used to characterise our starting material and investigate the secondary phases on the surfaces of leached Synroc-C discs. Calculated and measured normalised Pu leach rates are compared and the partitioning of Pu between zirconolite and perovskite is discussed.

INTRODUCTION

Synroc-C and zirconolite-rich Synroc have been proposed as media for the disposition of excess weapons plutonium [1]. Synroc-C is a polyphase titanate ceramic primarily containing hollandite, zirconolite, perovskite, and rutile-Magnéli phases with minor hibonite, loveringite, calcium aluminium titanate (CAT) and intermetallic phases. It was originally developed for the immobilisation of high level waste (HLW) from the reprocessing of spent nuclear power reactor fuel [2]. However it is an obvious choice for Pu disposition because i) it can immobilise up to 12 wt% PuO$_2$ [3], ii) it can be fabricated with internal radiation barriers (a ^{137}Cs spike or a complete suite of fission products), iii) it can accommodate large amounts of Gd$_2$O$_3$, Sm$_2$O$_3$ or HfO$_2$ (neutron absorbers) to limit the risks of nuclear criticality [1] and iv) it is highly durable in aqueous fluids. For example, when Synroc-C containing 10 wt% simulated Purex waste was leached in deionised water at 150°C for a total of 336 d, the average normalised leach rates of Cs, Mo and Sr asymptotically tend towards ~10^{-4} g m^{-2} d^{-1} [4].

In this paper, the results of leach tests on Synroc-C containing 0.62 wt% Pu and 10 wt% simulated waste are reported.

METHODOLOGY

Synroc-C containing 10 wt% simulated PW-4b-D HLW including 0.62 wt% ^{239}Pu was prepared by the alkoxide method and hot pressing in a graphite die. Polished discs were cut from the bulk sample and leached in triplicate at 70°C in deionised water and in silicate and carbonate solutions. Pu analysis of leachants was carried out using α-spectrometry on unfiltered solutions, filtrates from 0.45 μm and 1000 nominal molecular weight (NMW) filters and on the concentrated nitric acid solution used to strip the activity off the walls of the leaching vessel [6]. The results reported here are the median of triplicate measurements. For the purposes of this paper, 'total' is used to designate unfiltered Pu activity plus that on the vessel walls. During the 53 d leach tests, the samples were leached for one 25 d period then for one 28 d period. During the 2472 d leach tests in deionised water, the leachant was initially replaced every 7 days, then the leach duration was lengthened to 28 d and subsequently to 120 d and longer as the elemental levels in the leachants decreased to the limits of detectability.

After the leach tests, the surfaces of the discs were investigated using scanning electron microscopy (SEM) and analytical transmission electron microscopy (AEM). SEM investigations

Mat. Res. Soc. Symp. Proc. Vol. 465 © 1997 Materials Research Society

were performed with a JEOL 6400 operated at 15 kV in standard secondary electron and backscattered electron modes. AEM samples were made by the replication technique [7] that extracts loosely bound secondary phases onto a replica carbon film. AEM analyses were performed using a JEOL 2000FX TEM operated at 200 kV fitted with a Tracor Northern Si(Li) energy dispersive x-ray spectrometer (EDS) with a Be window and a Link ISIS microanalyser. EDS spectra were analysed with the ISIS software package, "TEMQuant". Empirically determined correction factors were used for Al, Ca, Ti, Fe, Sr, Y, Zr, Ce and Gd. Theoretical correction factors calculated by the TEMQuant software were used for Mo and Nd. Because of the difficulty in obtaining Pu standards, the empirically determined correction factor for U was used to estimate a correction factor for Pu.

RESULTS

<u>Leach data</u>

After 53 d, the normalised, differential, total (i.e. unfiltered leachate + vessel wall) Pu leach rates from Pu-doped Synroc-C leached in deionised water, silicate and carbonate solutions were of the order of 10^{-4}, 10^{-5} and 10^{-5} g m^{-2} d^{-1} respectively.

Figure 1 shows the total and solution, normalised, differential ^{239}Pu leach rates from Synroc in deionised water at 70°C. Both the total and solution leach rates slowly decrease with time, and after a leaching time of 2472 d, they are about 10^{-6} (total) and 10^{-8} (solution) g m^{-2} d^{-1}. Furthermore, the activity of ^{239}Pu detected on the vessel walls remains constant even for leach periods of up to 420 d, suggesting that the long-term release of Pu is not dominated by the solubility of Pu in deionised water but that it is probably controlled by the Synroc matrix.

<u>SEM and AEM of unleached Synroc-C</u>

The microstructure of Pu-doped Synroc-C is typical of Synroc-C in general (e.g. [7]). It is a dense matrix composed of hollandite, zirconolite, perovskite, and rutile-Magnéli phases with minor hibonite, loveringite, CAT and intermetallic phases. All phases except the intermetallics have grain sizes of 0.2-1.0 μm. The intermetallics have grain sizes of 0.025-1.0 μm.

The composition (oxide wt%) of the major and some of the minor primary phases are given in Table I. The results reported here represent the average of 5 to 15 analyses from different grains of each phase. Two distinct populations of perovskite were found: population 1 incorporated significant amounts of rare earth elements and plutonium; population 2 contained no observable amounts of rare earth elements or plutonium. The presence of two perovskites is probably due to poor mixing prior to hot-pressing. It is unlikely to occur in future samples, as improved homogenisation procedures have supplanted the tray drying method used in the preparation of this sample.

Structural formulae for zirconolite, both perovskites, and hollandite were calculated from the data in Table I and are shown below. They include an estimate of the amount of Ti^{3+} obtained from charge balancing considerations.

zirconolite

$$(Ca_{.87}Ce_{.02}Nd_{.04}Gd_{.02}Pu_{.01}Y_{.01}Zr_{.04})_{1.01}(Zr_{.65}Ti_{.35})_{1.00}$$
$$(Al_{.15}Fe_{.01}Ti^{3+}_{.03}Ti^{4+}_{1.81})_{2.00}O_{7.00}$$

perovskite #1

$$(Ca_{.83}Sr_{.02}Ce_{.05}Nd_{.05}Gd_{.01}Pu_{.01})_{0.97}(Al_{.02}Zr_{.01}Ti^{3+}_{.15}Ti^{4+}_{.82})_{1.00}O_{3.00}$$

perovskite #2

$$(Ca_{.82}Sr_{.01})_{0.83}(Al_{.01}Zr_{.01}Ti^{4+}_{.97})_{0.99}O_{3.00}$$

hollandite

$$(Ba_{.78}Cs_{.11}Sr_{.02}Ca_{.09})_{1.00}[(Ti^{3+}_{1.16}Al_{.70}Fe_{.02})(Ti^{4+}_{6.03}Zr_{.07}Mo_{.02})]_{8.00}O_{16.00}$$

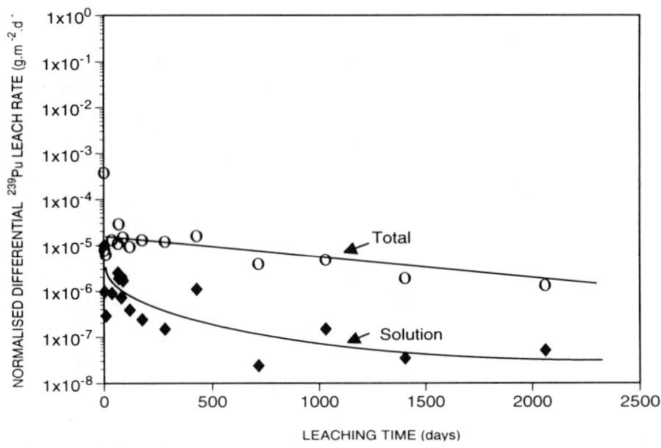

Figure 1. Total and solution normalised, differential ^{239}Pu leach rates measured at 70°C in deionised water.

The partitioning coefficient, $D^{Z/P}$, is defined as the (wt% metal oxide in zirconolite) / (wt% metal oxide in perovskite) [8]. From the data in the table the $D^{Z/P}$ of Pu is 0.3.

Following the procedure detailed in Lumpkin et al. [8], we found that the relative amounts of zirconolite, hollandite, perovskite #1, perovskite #2, and rutile in the Pu10 Synroc are 29%, 35%, 11%, 6%, and 15%, respectively. Based on this calculated phase distribution and the Pu_2O_3 content of zirconolite and perovskite #1 in this Synroc, the ratio of the total mass of Pu in zirconolite to that in perovskite is 3:4.

Table I. Average AEM analyses of primary phases in unleached Pu-doped Synroc (ion beam thinned section prepared from sample Np10).

oxide formula	zirconolite (oxide wt%)	perovskite 1 (oxide wt%)	perovskite 2 (oxide wt%)	hollandite (oxide wt%)	alumina (oxide wt%)	CAT (oxide wt%)	rutile (oxide wt%)
Al_2O_3	2.16	0.85	0.60	4.49	95.79	12.87	0.98
CaO	14.97	31.49	36.16	0.55	0.34	5.26	0.38
TiO_2	53.63	53.09	60.76	75.59	3.09	77.63	96.14
Cr_2O_3	—	—	—	—	0.08	—	—
Fe_2O_3	0.24	0.14	0.19	0.18	0.10	0.56	0.13
SrO	—	1.56	1.11	0.33	—	0.44	—
Y_2O_3	0.41	—	—	—	—	—	—
ZrO_2	24.93	0.85	1.19	1.11	0.60	1.87	2.37
MoO_2	—	—	—	0.41	—	—	—
Cs_2O	—	—	—	2.23	—	—	—
BaO	—	—	—	15.08	—	—	—
Ce_2O_3	0.76	5.17	—	—	—	0.73	—
Nd_2O_3	1.72	4.75	—	—	—	0.65	—
Gd_2O_3	0.72	0.48	—	—	—	—	—
Pu_2O_3	0.46	1.62	—	—	—	—	—

<u>SEM and AEM of leached Synroc-C</u>

SEM showed that Pu-doped Synroc-C suffers only slight alteration after being leached for 53 d in deionised water, silicate and carbonate leachants at 70°C (Figure 2a). No preferential leaching of individual phases was observed, and secondary phases were generally only associated with polishing scratches or surface roughness. AEM of replicas taken from Synroc-C leached in deionised water for 53 d showed that ~95% of the secondary material is anatase. The remaining material appears to be Fe-rich oxide/hydroxide spheres (diam. ~0.1 μm). It was not possible to estimate secondary phase percentages from replicas of samples leached in carbonate and silicate leachants, because there was so little material on them.

After leaching for 2472 d in deionised water, the surface of Pu-doped Synroc-C was largely covered with secondary material (Figure 2b & c). This agrees with the findings of previous long-term (> 2000 d) leaching studies of Synroc [9]. There appears to be no correlation between the underlying primary phases and the presence or absence of the layer of secondary material, which suggests that handling caused some abrasion of the layer after leaching. SEM of a cross-section of a disc leached for 2472 d did not clearly show the layer of secondary material because of preferential erosion of the layer during polishing, which resulted in the edge of the sample having a curved profile. However, it did reveal that the thickness of the layer is << 1μm.

AEM of replicas taken from a disc leached for 2472 d showed that ~70% of the secondary material was crystalline Fe-Nb-Ti-Cr-Al oxide/hydroxide and that the rest was anatase. It is likely that the Fe, Nb and Cr found in secondary phases result from the corrosion of the stainless steel specimen supports used in the leaching measurements to keep the sample off the bottom of the vessel.

No Pu was observed in any of the secondary phases examined. However it is possible that they contain minor Pu (less than the AEM detection limit, ~0.2 wt%) and/or that Pu is adsorbed in very thin layers (<< 1 μm) onto the surfaces of the secondary phases.

As expected, SEM of the surfaces of leached discs after AEM replicas had been taken showed no obvious differences from the surfaces of discs before replicas were taken. This shows that the replica technique only collects material from the external surface of the layer of secondary phases. Consequently, sampling may be biased.

DISCUSSION

Recent investigations [10] have shown that surficial perovskite in Synroc-C containing 20 wt% simulated HLW corrodes to a depth of ~250 nm after 365 d in MCC-1 type tests in deionised water at 150°C. From this data and the composition of perovskite in our Synroc, one can calculate that the normalised total Pu leach rate should be ~3 x 10^{-3} g m^{-2} d^{-1}, prior to the precipitation of aqueous Pu species. This rate is two orders of magnitude larger than the measured total Pu leach rate after 365 d (~8 x 10^{-5} g m^{-2} d^{-1}, see Figure 1), which indicates that dissolved Pu species rapidly plate out back onto the primary Synroc phases from which the Pu originally dissolved and/or onto the secondary phases intimately associated with the primary Synroc phases (e.g. anatase in the case of perovskite). This suggestion is supported by Matzke et al.'s [11] data. They found that the actinide leach rate from Np-doped Synroc-B was much smaller than the bulk leach rate and showed (using α-spectrometry) that the actinides become enriched at the surface after leaching.

The Synroc-C sample in this study has the same composition and was fabricated under the same conditions as that investigated by Blackford et al. [12]. However, the value of D$^{Z/P}$ and ratio of the amount of Pu in zirconolite to that in perovskite are different (0.3 & 3:4 and 1.6 & 5:1 respectively). There are two reasons for these differences: i) we used improved EDS data analysis procedures ([13]; Table II) and ii) sampling limitations (AEM samples only a small area of the whole sample, and the compositions of Synroc major phases show minor variation on the 50 μm scale).

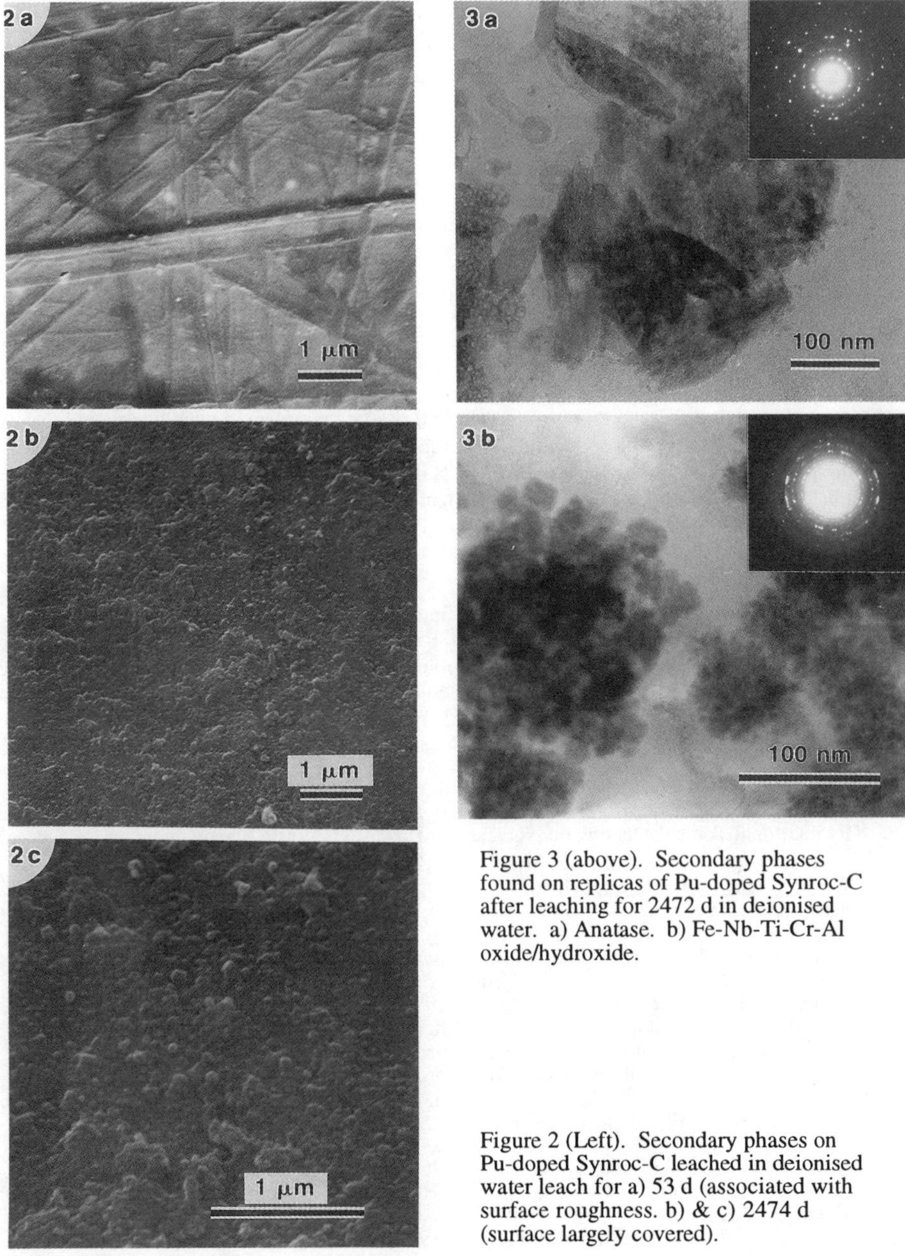

Figure 3 (above). Secondary phases found on replicas of Pu-doped Synroc-C after leaching for 2472 d in deionised water. a) Anatase. b) Fe-Nb-Ti-Cr-Al oxide/hydroxide.

Figure 2 (Left). Secondary phases on Pu-doped Synroc-C leached in deionised water leach for a) 53 d (associated with surface roughness. b) & c) 2474 d (surface largely covered).

Table II. Comparsion of data collected and analysis method

	This study	Blackford et al. (1991)
$D^{Z/P}$	0.3	1.6
Ratio of the amount of Pu in zirconolite to that in perovskite.	3:4	5:1
Pu content of phases calculated using:	1. Experimental U k-factor (because of the difficulty in obtaining Pu standards) 2. Pu-M peak because it contains more counts than the Pu-L peak 3. Improved deconvolution procedures	1. Theoretical U k-factor (because the analysis software would not permit actinides other than U to be analysed) 2. Pu-L peak (because of the unreliability of theoretical M line k-factors) 3. Region of interest overlap separation.

$D^{Z/P}$ (and consequently the Pu ratio) quoted in this study conform to what one would expect from other studies. Begg et al. [14] found that the valence of Pu in zirconolite hot-pressed in graphite dies is predominantly 3+, which suggests that the Pu in our Synroc has a valence of 3+. Lumpkin et al. [8] showed that $D^{Z/P}$ varies systematically with ionic radius and also with valence. Although they did not measure the partitioning of Pu^{3+}, their data suggest that the $D^{Z/P}$ of Pu^{3+} will be somewhere between 0.1 and 1.0. Our value for $D^{Z/P}$ is 0.3.

CONCLUDING REMARKS

This study showed that Pu-doped Synroc-C and its constituent phases are very durable (the normalised, differential, total Pu leach rate from Pu-doped Synroc-C after 2472 d was found to be ~5 x 10^{-5} g m^{-2} d^{-1}) and suggested that dissolved Pu species rapidly plate out back onto the primary Synroc phases from which the Pu originally dissolved and/or secondary phases on the surface of leached Synroc. This latter suggestion will be investigated using alpha recoil techniques.

REFERENCES

1. A. Jostsons, E.R. Vance, R.A. Day, K.P. Hart and M.W.A. Stewart (1996), Spectrum '96, International Topical Meeting on Nuclear and Hazardous Waste Management.
2. A.E. Ringwood, S.E. Kesson, K.D. Reeve, D.M. Levins and E.J. Ramm, in Radioactive Waste Forms for the Future, edited by W. Lutze and R.C. Ewing (North-Holland, Amsterdam, 1988), p. 233.
3. E.R. Vance (1994), MRS Bull. 19, 28-32.
4. K.L. Smith, K.P. Hart, G.R. Lumpkin, P.J. McGlinn, J. Bartlett, P. Lam and M.G. Blackford (1991), Mat. Res. Soc. Symp. Proc., 212, 167-174.
5. K.P. Hart, W.E. Glassley and P.J. McGlinn (1992), Radiochimica Acta, 58/59, 33-35.
6. C.W. Sill and R.L. Williams (1981), Anal. Chem., 53, 412-415.
7. G.R. Lumpkin, K.L. Smith and M.G. Blackford (1991), J. Mater. Res. **6**, 2218-2233.
8. G.R. Lumpkin, K.L. Smith and M.G. Blackford (1995), J. Nuc. Mater., 224, 31-42.
9. K.L. Smith, M.G. Blackford, G.R. Lumpkin, K.P. Hart and B.J. Robinson (1996), Mat. Res. Soc. Symp. Proc. 412, 313-319.
10. K.L. Smith, M. Colella, G.T. Thorogood, M.G. Blackford, E. Loi, K.P. Hart, K.E. Prince, G.R. Lumpkin and A. Jostsons (1996), Proc. Mat. Res. Soc., 465.
11. Hj. Matzke, E. Toscano, C.T. Walker and A.G. Solomah (1988), Advanced Ceramic Materials, 3, 285-288.
12. M.G.Blackford, K.L. Smith and K.P. Hart (1992), Mat. Res. Soc. Symp. Proc., 257, 243-248.
13. M.G. Blackford (1996), Improved EDS. ANSTO Mat. Div. internal report, pp. 12.
14. B.D. Begg, E.R. Vance, S.D. Conradson and M. Hambley (1996), Proc. Mat. Res. Soc., 465.

UNDERGROUND CRITICALITY IN GEOLOGIC DISPOSAL[1]

J-S CHOI *, T. H. PIGFORD **
*Lawrence Livermore National Laboratory, Livermore, CA 94550
**Nuclear Engineering Department, University of California, Berkeley, CA 94720

ABSTRACT

Disposition of weapons fissile material as once-through MOx fuel in reactors or immobilization in waste glass would result in end products requiring geologic disposal. Similar considerations apply to some spent fuels in a geologic repository. The criticality potential of these end products in a geologic setting is an important consideration for the final disposal of these materials. In this document we present calculations that help define parameters and configurations for criticality, including those that, if the criticality is achieved, could lead to autocatalytic power transients. It is not the purpose of this paper to define scenarios that could reasonably lead to criticality or to establish the probabilities that critical configurations could occur. We do suggest some parameters that should be considered by waste package designers and repository analysts to ensure that criticality cannot occur.

INTRODUCTION

Several kinds of radioactive wastes now destined for geologic disposal can theoretically reach nuclear criticality in a geologic medium. These are surplus weapons-grade plutonium from dismantled nuclear weapons, spent nuclear fuel from commercial light-water reactors, and DOE-owned spent nuclear fuel from naval vessels, research reactors, and production reactors. One option for disposing of weapons-grade plutonium involves incorporating the plutonium into glass, along with sufficient fission products, so that the resulting waste will be as self protecting as spent fuel when placed in a geologic repository.

Another option is to mix the weapons-grade plutonium with natural uranium, to form mixed-oxide fuel for light-water reactors. The discharged mixed-oxide fuel, containing a few percent of fissile U-235 and Pu-239, would be placed in a geologic repository. About 50 tons of weapons-grade plutonium might be disposed of in one of these two ways. Spent fuel from commercial reactors contains ~1.5 wt% of fissile uranium and plutonium. About 63,000 tons of commercial spent fuel, containing about 950 tons of fissile material are now destined for geologic disposal in the proposed Yucca Mountain repository. DOE-owned spent fuel contains a wide range of fissile compositions, including some highly enriched U-235. Amounts of Pu-239 and U-235 in typical waste containers planned for geologic disposal in the U.S. are listed in Table 1.

Table 1. Amounts of Fissile Material (Pu-239, U-235) in Waste Containers:

Waste Type:		Amount (kg)	
	wt%	small container[a]	MPC[b]
Commercial spent fuel			
UO$_2$ spent fuel	1.5	20	143
Surplus weapons-grade plutonium			
MOx spent fuel	~4	54	378
Glass waste	~5	84	336
Ceramic waste	~12	80	320
DOE-owned spent fuel			
N-reactor	1.4	74[c]	495[d]
ATR	30.4	94	664
Shippingport	60.9	44	262
Graphite	3.1	1.5	10

a. Small container holds 3 UO$_2$ or MOx spent PWR assemblies, or 1 glass or ceramic waste form.

b. Multi-Purpose Container (MPC) holds 21 UO$_2$ or MOx spent PWR assemblies, or 4 glass or ceramic waste forms.

c. Fissile content in small container designed for light-weight truck transport.

d. Fissile content in MPC designed for rail-transport.

---

[1] This is an independent study by the authors. It is not associated with Lawrence Livermore National Laboratory.

Mat. Res. Soc. Symp. Proc. Vol. 465 © 1997 Materials Research Society

Possible mechanisms of reaching criticality in a geologic repository, after the groundwater has leaked into a failed waste container, include (1) preferential leaching of neutron absorbers from a failed waste container, and (2) preferential leaching and deposition of plutonium or fissile uranium in surrounding rock.

Bowman and Venneri [1] have pointed out that fissile material mixed with varying compositions of water and silica can undergo a nuclear chain reaction. Some configurations can become autocatalytically supercritical resulting in considerable energy release, terminated finally by disassembly.

Some reviews [2] rejected Bowman and Venneri's warning as implausible because of low probabilities of scenarios that could lead to such configurations. Sanchez, et al.[3] reported possible supercritical conditions in systems of Pu-SiO$_2$-H$_2$O and Pu-tuff-H$_2$O but concluded that the probability of forming such combinations is extremely low.

Kastenberg, et al.[4] studied the potential for autocatalytic criticality of plutonium or highly enriched uranium in the proposed Yucca Mountain geologic repository, focusing on the possibility of accumulating a critical assembly of migrating fissile material in the far field. They concluded that plutonium or uranium could theoretically become supercritical, although the critical mass required would be in hundreds of kilograms. The requirement of such incredibly large critical masses for an underground criticality, apparently the result of the use of a one-dimensional slab model in their analysis, again led to the conclusion that underground criticality was unlikely given the hydrology, geology and geochemistry of the Yucca Mountain site.

More technical analysis is needed to understand the potential for underground autocatalytic criticality. Here we present our calculations of possible conditions for criticality in a container holding fissile material in various different chemical forms, including conditions for autocatalytic criticality. We discuss possible criteria for designing waste solids for the disposition of weapons fissile material so that long-term dissolution of neutron absorbers or of fissile material would not cause criticality.

These results for simplified geometries help define conditions for more detailed analysis of practical systems. Potential for underground criticality is also studied for other fissile waste forms, including spent fuel from commercial reactors, research reactors, and naval reactors [5], all of which may be destined for geologic disposal. The calculations are made using the Monte-Carlo neutron-transport MCNP code [6].

ANALYSIS

Pu-239, SiO$_2$, H$_2$O System

The effective multiplication factors k_{eff} for homogeneous systems involving Pu-239, SiO$_2$ and H$_2$O, reflected by an infinite reflector of silica, are shown in Figures 1, 2, and 3 for mole ratios of SiO$_2$ to H$_2$O of 0, 0.67, and 2.70, respectively. The silica reflector is assumed to contain 15 volume percent

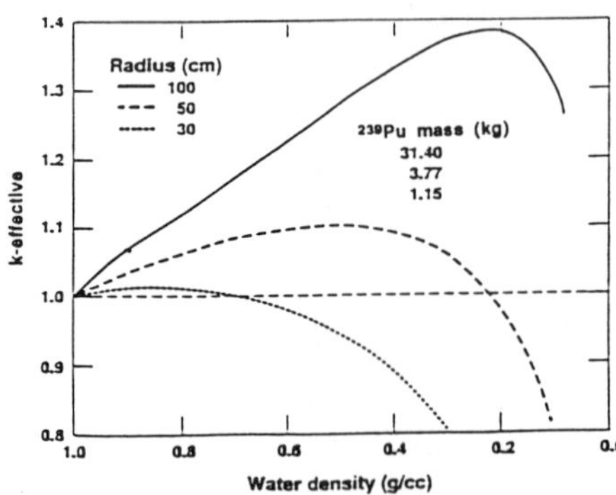

Figure 1   Homogeneous Spherical System of ^{239}Pu and H$_2$O

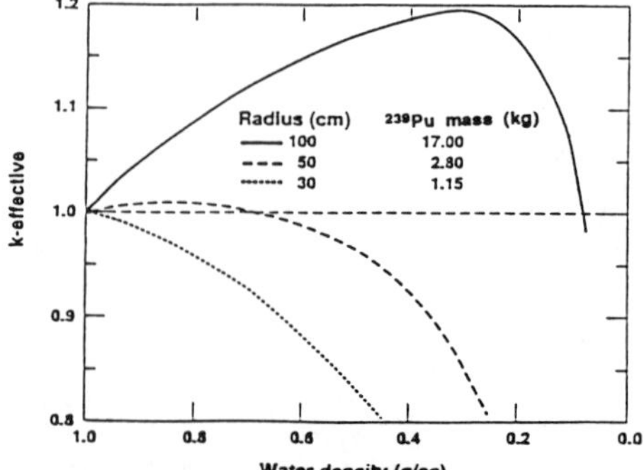

Figure 2   Homogeneous Spherical System of ^{239}Pu, SiO$_2$ and H$_2$O (mole ratio of SiO$_2$ to H$_2$O of 0.67)

water. These results for an idealized system help define parameter space for more detailed evaluation of the possibility of criticality in practical systems. The water is assumed to be contained in cracks and voids, which are assumed to be initially saturated with water at 20 °C. For a given assumed spherical radius,the amount of Pu-239 is adjusted to make the system initially just critical. We then assume that fission heating brings the temperature to the boiling point of water, and vaporization reduces the average density of water in the pores and cracks. If there is sufficiently large ratio of initial water to Pu-239, the system is initially over-moderated and k_{eff} increases as water density decreases. This describes the conditions for instability and autocatalytic behavior. Eventually a maximum k_{eff} is reached, when the remaining amount of water is small enough for transition into under-moderated conditions. The system is then self stabilizing.

Larger system volumes lead to greater potential for autocatalytic criticality. The greater neutron leakage for small systems tends to greater stability. In Figure 4 we compare our calculated results with those of others[1,3].

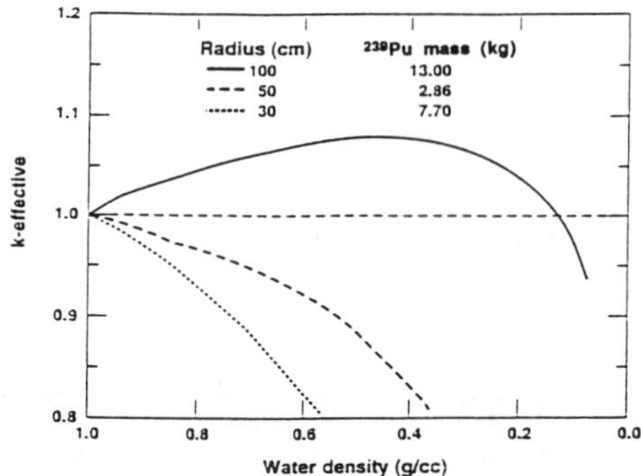

Figure 3   Homogeneous Spherical System of ^{239}Pu, SiO$_2$ and H$_2$O (mole ratio of SiO$_2$ to H$_2$O of 2.70)

Figure 4   Comparison of k_{eff} Results for Systems of ^{239}Pu, SiO$_2$ and H$_2$O

If a real system with such assumed configuration were to become critical and generate heat, spatial variation in neutron flux could result in spatially nonuniform changes in water density, instead of the spatially uniform density assumed here. Similarly, if fission heating from criticality is sufficiently rapid, the extent of vaporization and reduction in water density will be affected by the permeability for flow of water vapor. These are some of the details that would have to be calculated in assessing the autocatalytic response of a realistic system. An example is given later.

Although we do not attempt to define mechanisms and scenarios for reaching these critical configurations, one can easily imagine a mechanism by which an over-moderated autocatalytic critical condition could occur without first reaching under-moderated criticality. A system of fissile material in highly degraded silica might initially contain no water, but calculations could indicate the need of sufficient neutron absorber so that criticality would not occur if water were later to leak into the cracks and pores. Thus, with sufficiently volume of cracks and pores, the system could become over-moderated, even though never critical. If the neutron absorber were to preferentially dissolve into the water and to transport from the system more rapidly than the fissile material, autocatalytic criticality could eventually occur. If the dissolved neutron absorber and fissile material were to have the same diffusive transport properties, a simple criterion for safety would be for the ratio of solubility to inventory of the neutron absorber to be less than that of the fissile material.

Figure 5 shows the calculated critical mass of Pu-239 in spherical volumes of various radii, surrounded by SiO$_2$, with the same assumptions as described previously. For each mole ratio, the critical mass is calculated as a function of the effective density of water in the pores. The curves proceed to lower water densities, implying that some liquid water may have been converted to steam by fission heating. Autocatalytic criticality is indicated where a curve shows lower critical masses for lower water densities, because boiling the water would require less fissile mass for criticality. Where the curve shows higher critical masses for lower water densities the system is self regulating, because fission heating increase the required critical mass and would thereby cause the system to become subcritical.

Autocatalytic criticality occurs more readily for large-volume systems. For smaller systems neutron leakage becomes more important. Lower water density increases neutron leakage and adds self-regulating feedback.

The lowest critical mass occurs for the curves of lower void fraction and larger volume, showing that there is an optimum composition and size for lowest critical mass. If plutonium were dissolved in the water instead of being a solid, boiling the plutonium-water solution could cause the egress of some plutonium and water, and the curves would depart significantly from those in Figure 5.

Yucca Mountain tuff is not pure silica. A typical material composition of tuff is shown in Table 2.

Table 2.   Composition of Yucca Mountain tuff

Compound	wt%	Compound	wt%
SiO_2	70.3	Al_2O_3	15.6
K_2O	6.6	Na_2O	4.1
CaO	1.1	Fe_2O_3	1.1
MgO	0.6	FeO	0.6

Criticality conditions for systems with pure silica and with tuff are shown in Figure 6. Both tuff and pure silica can lead to conditions of autocatalytic criticality. It also indicates that the tuff system yields a lower k_{eff} than the pure silica system, due mainly to neutron absorption by ingredients other than SiO_2 in tuff.

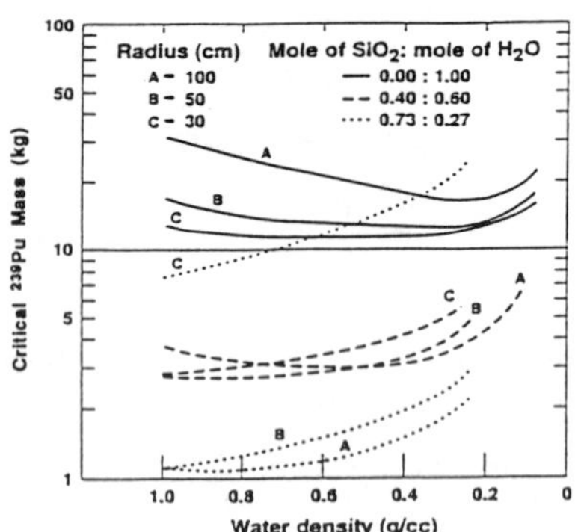

Figure 5   Spherical Critical Mass of ^{239}Pu, SiO_2 and H_2O

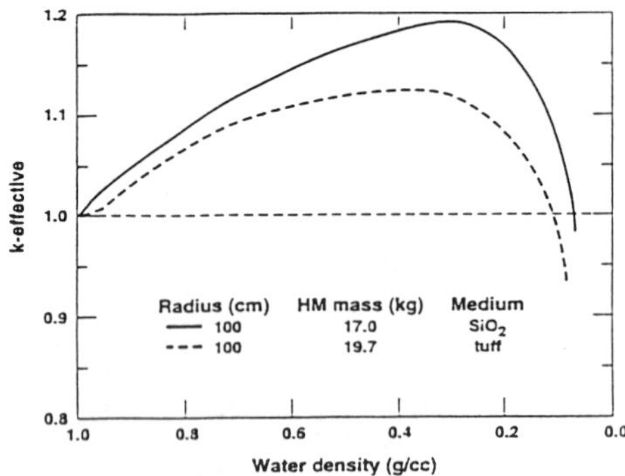

Figure 6   Homogeneous Spherical System of ^{239}Pu, tuff (or SiO_2) and H_2O (mole ratio of SiO_2 to H_2O of 0.67)

Pu-239, U-238, H$_2$O System

Figure 7 shows the criticality conditions presented in terms of k_{eff} for reflected spherical homogeneous mixtures of Pu-239 (5 wt% of heavy metal), U-238, and water, roughly simulating fresh MOx fuel assemblies in water. Other neutron absorbers are neglected. The maxima of curves for radii of 45.7 cm and 30.0 cm occur when enough water has been boiled away that further loss of water reduces k_{eff} because of less neutron moderation and increased leakage. The transition from autocatalytic to self-regulating criticality occurs at a higher water density than in Figure 5 because of the self-regulating effect of the absorption resonances of U-238.

Figure 8 shows the k_{eff} factor for reflected spherical homogenous mixtures of 10-year-old spent MOx fuel and water. Fission products (i.e., Sm-149, Gd-155, Rh-103 etc.) are included in the fuel mixtures. These neutron-absorbing isotopes significantly reduce the criticality potential of the spent fuel-water mixture. To achieve the initial criticality, a system with radius of 45.7 cm would require almost twice the mass, as compared to the fresh MOx fuel of Figure 7. The transition from autocatalytic to self-regulating criticality also occurs at a much higher water density than that in Figure 7. For a radius of 30 cm, the system requires more than twice the mass as that in Figure 7, and it does not become autocatalytically critical.

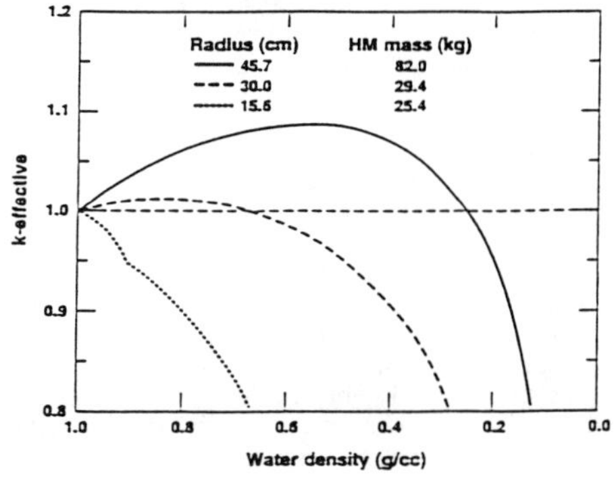

Figure 7   Homogeneous Spherical System of ^{239}Pu (5 wt%), ^{238}U and H$_2$O

Figure 8   Homogeneous Spherical System of Spent MOx Fuel and H$_2$O

U-235, U-238, H$_2$O System

Figure 9 shows similar reactivity curves of k_{eff} for homogeneous spherical mixtures of solid enriched uranium (5 wt% U-235) and water. More U-235 is required for criticality, and there is less tendency to autocatalytic criticality than for the Pu-239 system of Figure 7.

Figure 10 shows k_{eff} curves for homogeneous spherical mixtures of 10-year old spent UO$_2$ fuel in water. Fission products (i.e., Sm-149, Gd-155, Rh-103 etc.) are included in the spent-fuel containing about 1.5% of fissile content. These neutron-absorbing isotopes significantly reduce the criticality potential of the spent fuel-water mixture. As for the spent MOx fuel and water system, more spent UO$_2$ fuel mass is required for criticality, even for a larger system volume (radius of 50 cm instead of 45.7 cm), and the transition to self-regulating criticality occurs at a higher water density than that in Figure 9.

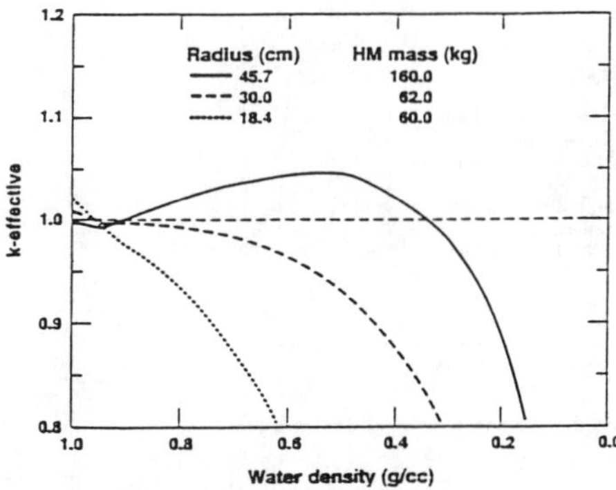
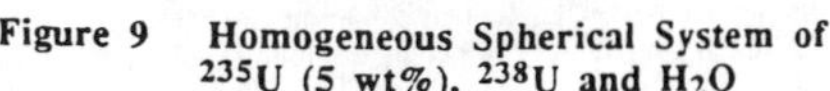
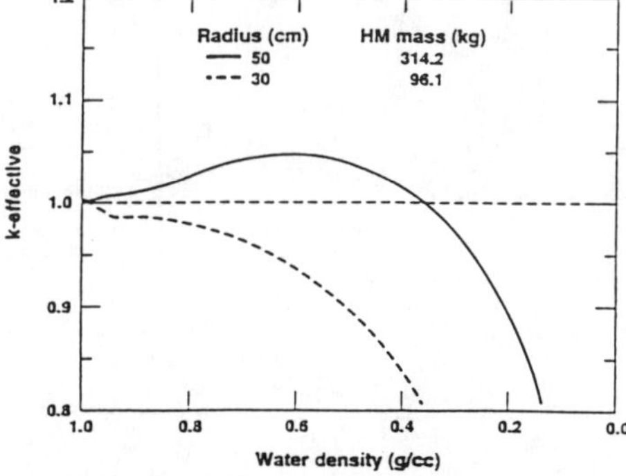

Figure 9   Homogeneous Spherical System of ^{235}U (5 wt%), ^{238}U and H$_2$O

Figure 10   Homogeneous Spherical System of Spent UO$_2$ Fuel and H$_2$O

U-235,238 Spent Fuel in a Waste Canister

The k_{eff} curve for low-enrichment uranium fuel discharged from a pressurized light-water reactor (PWR) is shown in Figure 11, for a burnup of 16 GWd/Mg, selected to illustrate partially irradiated fuel. It also shows the k_{eff} curve for MOx spent fuel discharged with a burnup of 42.6 GWd/Mg. Fresh UO_2 fuel with enrichment of 4.5% U-235 is included for comparison. The k_{eff} results are presented as a function of the mole ratio of water and heavy metal in the fuel assemblies, simulating changes in spacing of fuel pieces that might hypothetically occur in a large degraded spent-fuel canister, and assuming no mitigation from structural materials and from extra low-solubility neutron absorbers. The assumed canister contains 21 fuel assemblies. All three of the fuel assemblies considered here can become critical if the canister becomes flooded, as shown for the condition for mole ratio of 7.1 which represents a water density of 1 g/cc. Additional neutron absorbers must be added for safety. If the system does become critical, fission heat can boil water and decrease the mole ratio of water to heavy metal. At constant fuel-rod spacing, reactivity decreases and the system is self regulating.

However, the waste container and the fuel pellets and assemblies could become degraded over long time period of underground storage. If there is some chance that fuel pellets and corrosion products could be so rearranged such that the mole ratio of water to heavy metal becomes much larger, corresponding perhaps to an effectively larger fuel-rod spacing, then a region of over-moderation could be reached, as shown in Figure 11. Criticality could still be prevented by sufficient neutron absorber. However, if that absorber were to preferentially leach from the container, autocatalytic criticality could occur. For spent UO_2 fuel, the transition to over-moderation and potential autocatalytic criticality occurs at mole ratio of about 15. It is important that the waste container design not allow enough initial voids for water-filled criticality even with dispersed fuel and that it preclude enough water for autocatalytic criticality. Criteria for selecting neutron absorbers to avoid preferential leaching and criticality were discussed earlier.

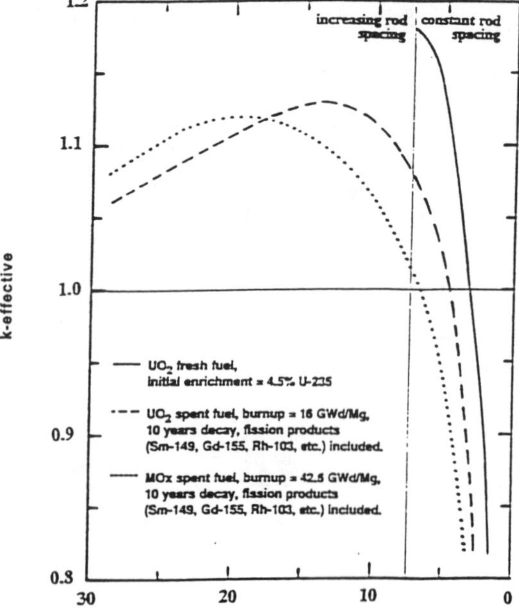

Figure 11   keff of LWR fuel assemblies in a multiple purpose canister (MPC) containing 21 PWR assemblies (Westinghouse 17x17), 0.12 wt% boron, and void filled with water

<u>Pu-SiO$_2$-H$_2$O System with Heating</u>

Previous figures have assumed water boiling at constant temperature. However, if the system is autocatalytic, appreciable fission heating can occur. Figure 12 presents a recent calculation by Bowman and Venneri [1], who have kindly agreed to its use in the present paper. The three-dimensional surface shows k$_{eff}$ as affected by the system temperature and by the relative amount of water in the Pu-H$_2$O-SiO$_2$ mixture. In the left corner, at boiling temperature and water-saturated porosity of 35%, the system is slightly supercritical. If water can easily escape when it boils, the effective volume of liquid water in the pores (calculated at normal boiling temperature) decreases and the reactivity increases. The pathway would be similar to the trace on the coordinate plane at a constant low temperature.

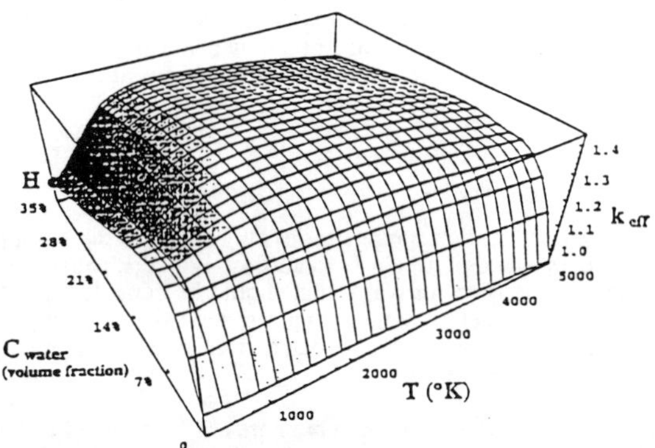

Figure 12   Criticality for Pu-SiO$_2$-H$_2$O System
(from C. D. Bowman and F. Venneri, for Sphere of 200 cm diameter and SiO2 reflected)

However, if the fission heating is sufficiently rapid, appreciate vapor pressure of water may be required to force the rapid escape of water. The kinetic properties of reactivity feedback and fluid dynamics can generate a new pathway along the three-dimensional surface. High temperatures and pressures are included in the calculation, sufficiently high for the water to become a supercritical fluid. The equation of state of silica at high temperatures and pressures was also included in the calculation. The pathway along this three-dimensional surface would depend also on the neutron lifetime and the permeability of the medium. If the pores that contain water are of sufficiently small diameter, the permeability of the porous solid can be quite low. The system will heat further until the vapor pressure of water is sufficient to force water from the pores. In the limit of very low permeabilities, the pathway will lie on a path of nearly constant water density. Very large pressures could be generated.

Along any of the pathways that involve heating, the reactivity first increases because the average thermal energy of the neutrons increases, and the neutron thermal energy spectrum overlaps more of the large low-lying 0.3-eV absorption resonance of Pu-239, thereby increasing the effective fission cross section of Pu-239. This temperature effect is essentially absent for U-235 systems. For MOx fuel Doppler broadening of the U-238 absorption resonances will reduce the net autocatalytic effect from Pu-239.

<u>Neutron Absorbers and Uranium Diluents to Avoid Criticality</u>

Simply adding neutron poison to avoid criticality may not be a sufficient solution. It is important to select an absorber with low enough solubility and high enough inventory to preclude preferential dissolution, as discussed earlier. The use of gadolinium as neutron poison may also be problematic due to its significant decrease in neutron absorption at elevated neutron energy, above 10 eV. Neutron absorbers such as erbium, with low-energy resonance absorption to compensate for the 0.3 eV Pu-239 resonance, should be considered.

Another solution may involve the use of depleted uranium, in UO_2, as filler material in the waste container. The depleted uranium could mix with U-235, as the uranium degraded and leached out from the fuel assemblies, and dilute the overall enrichment level required for criticality.

<u>Criticality from U-233, a decay daughter of Np-237</u>

The amount of U-233 in the spent fuel from low-enriched-uranium LWRs, originally very small, increases as its parent Np-237, a radionuclide with half-life of 2.1 million years, decays. U-233 generated from decay of Np-237 in the fuel matrix could increase the overall enrichment of the fuel matrix. During this time Pu-239 in spent fuel will have decayed to U-235. If depleted uranium is used as filler material, the overall uranium enrichment level by U-233 and U-235 may still be below the level required for criticality. However, if neptunium leaches away from the fuel matrix faster than uranium, as may occur in an oxidizing environment, it would decay into U-233 along the migration path and away from the depleted uranium diluent. This is one of the mechanisms by which fissile material could accumulate in the geologic medium, a subject for further studies of possible criticality.

CONCLUSIONS

We conclude that

• Designers of waste packages and geologic repositories should avoid the compositions and volumes that could give rise to criticality, and particularly to autocatalytic criticality, such as those presented herein for idealized systems. The types of waste that should be considered with regard to potential criticality include spent fuel, MOx fuel, plutonium in glass, and various forms of DOE spent fuel. The present calculations help define the needs for more detailed treatment of geometry and composition in regions where criticality can potentially occur.

• Criticality and autocatalytic conditions in waste containers can be avoided by careful design, such as low-release neutron absorbers or uranium diluents (e.g., erbium, depleted-uranium, or massive iron), and by limiting the amount of void that could become filled with water.

• There is a theoretical potential for appreciable energy release from autocatalytic criticality. The extent and consequences are not yet resolved. Hence, additional technical analysis is required.

REFERENCES

1. C. D. Bowman and F. Venneri. "Autocatalytic Criticality from Plutonium and Other Fissile Material," Report LA-UR-94-4022, Los Alamos National Laboratory, 1995, also published in *Science & Global Security*, Vol. 5, pp.279-302, 1996.

2. "Comments on the draft paper -Underground supercriticality from plutonium and other fissile material,- written by C. D. Bowman and F. Venneri (LANL)," compiled by R. A. Van Konynenburg, Lawrence Livermore National Laboratory, *Science & Global Security*, Vol. 5, pp. 303-322, 1996.

3. R. Sanchez, W. Myers, D. Hayes, R. Kimpland, P. Jaegers, R. Paternoster, S. Rojas, R. Anderson and W. Stratton, "Criticality Characteristics of Mixtures of Plutonium, Silicon Dioxide, Nevada Tuff, and water," Report LA-UR-95-21130, Los AlAmos National Laboratory, 1995.

4. W. E. Kastenberg, et al., "Mechanisms for autocatalytic criticality of fissile materials in geologic repositories," UCB-NE-4214, Department of Nuclear Engineering, University of California, Berkeley, June 1996.

5. R. P. Rechard, et al., "Performance Assessment of the Direct Disposal in Unsaturated Tuff of Spent Nuclear Fuel and High-Level Waste Owned by U.S. Department of Energy," SAND94-2563, Sandia National Laboratory, March 1995.

6. "MCNP - A General Monte Carlo Code for Neutron and Photon Transport," Report LA-7396-M, Rev.2, Los Alamos National Laboratory, 1991.

ACTINIDE SOLUBILITY IN LANTHANIDE BOROSILICATE GLASS FOR POSSIBLE IMMOBILIZATION AND DISPOSITION

T.F. MEAKER[*], D.K. PEELER, J.C. MARRA, J.M. PAREIZS, W.G. RAMSEY
Savannah River Laboratory, Savannah River Site, Aiken, SC, 29808

ABSTRACT

Immobilization by vitrification is one potential disposition option for a portion of the United States' excess plutonium inventory. Research has been performed to determine the glass forming region of a frit, plutonium and rare earth system. The frit contains mainly oxides of aluminum, silicon and boron; small amounts of ZrO_2 and SrO are also included. The rare earth elements provide a flux to the glass during processing. The rare earths are also added as neutron absorbers (to prohibit criticality) in the final vitreous product.

This report will show the compositional region, using Th as a Pu surrogate, that should be targeted in future studies for maximum Pu solubility in the lanthanide borosilicate glass system. Durability data and process variables will also be provided.

INTRODUCTION

In support of the Fissile Material Disposition Program, SRTC is evaluating a lanthanide borosilicate (LaBS) glass composition for possible plutonium immobilization. This glass has been shown to accommodate up to 10 weight percent Pu along with rare earth elements (REE)[1], which may be used as neutron absorbers (Gd). The durability of Pu-loaded LaBS glasses has been shown to be ~50x better than HLW glasses produced at the DWPF[2]. Several different LaBS glass compositions have been processed with Th (a Pu surrogate) and Pu as glass formulation efforts strive to optimize the system. Initially, the LaBS glass composition contained PbO and BaO (for Am/Cm Demonstration Project) [2-3]. More recent compositions have removed the lead and barium, resulting in a glass composition without RCRA listed materials[4]. This composition is represented in this study as 'B-24-17' and represents 15 elemental weight percent Th. This document summarizes the work performed to develop a compositional envelope around this composition with ThO_2 used as a plutonium surrogate in the LaBS glass without lead or barium. The compositional region showing maximum ThO_2 solubility will be a basis for future tests with PuO_2 in the glass.

EXPERIMENTAL

Glass compositions were batched in platinum crucibles with reagent chemicals in the oxide form with the exception of boron. The boron content was added as boric acid. Glasses were melted at 1475° C for 4-6 hours with manual stirring after ~2 hours at temperature. The ramp rate was 560° C/hr. As the melts were removed from the furnace, the crucibles were placed in a pan of water to 'shock' the glass from the crucible surface which allowed for total recovery of the processed glass. The compositional envelope was defined by varying the percentages of:
        (1) Frit (Table I),
        (2) Rare Earth (lanthanide) oxides, and
        (3) ThO_2 (PuO_2 surrogate).
The base frit composition (renormalized components of B-24-17 without ThO_2 and rare earth oxides) did not change, but was used in decreasing amounts as the rare earth oxides and ThO_2 content increased.

Mat. Res. Soc. Symp. Proc. Vol. 465 © 1997 Materials Research Society

Table 1. Base frit composition for the LaBS glass. Includes all components except lanthanides and actinides.

Oxide	Weight Percent
SiO_2	44.0
B_2O_3	17.7
Al_2O_3	32.5
SrO	3.8
ZrO_2	2.0

The rare earth contributors were lanthanum and neodymium with gadolinium, used as the neutron absorber. As the actinide content increased, the gadolinium content increased, maintaining a 1:1 atomic ratio with thorium. The amount of La and Nd decreased to compensate for the increase in Gd.

In order to bound the processing region, initial ratios of the three components were determined by previous experience with this system (both successful and unsuccessful melts). First, as homogeneous glasses were produced, new compositions with systematic increases in lanthanide and/or actinide loadings were attempted. Second, an equilateral triangle was used by placing two corners on successful melts, the third corner of the triangle determined a new compositon to be processed. Finally, when an outline of a processing region was apparent, compositions near the boundary (between the homogeneous and non-homogeneous glass region) were tested to better determine the boundary line.

Glass formulations were evaluated for homogeneity by XRD, SEM and TEM (for amorphous phase separation). Three homogeneous glasses from different regions of the processing range were analyzed by sodium peroxide fusion followed by ICPES for chemical composition. The 7-day Product Consistency Test (PCT) for durability[5] was performed on these glasses to determine if the durability of the glasses changed significantly with respect to different lanthanide and/or thorium loadings.

RESULTS AND DISCUSSION

The ternary diagram (Figure 1) shows the homogeneous glass forming region for the base glass, rare earth oxide (lanthanide content) and thorium oxide system at 1475° C. Table II shows the batch compositions and processing results of all glasses shown in Figure 1.

At high frit loadings (>75 weight percent), the glass was not processable due to the frit being highly refractive. Over 94 percent of the frit is composed of SiO_2, B_2O_3 and Al_2O_3. The phase diagram for these three components indicates that the melting point for this composition is ~1650° C[6]. Crystalline species in this part of the ternary diagram were found to be aluminum borate. All rare earth oxides and thorium oxide were incorporated into a glassy phase. Crystalline species or undissolved oxides containing these elements were not detected by XRD. As the frit drops below 75 percent, the fluxing affects of the rare earth elements effectively lowers the melting to allow processing in a bushing melter.

At high rare earth oxide loadings (>65 weight percent oxide) the solubility limit of rare earth oxides (only, no thorium oxide) was exceeded. XRD detected strontium-rare earth-silicates. As thorium oxide was added, the combined solubility limit of rare earths and thorium decreases. The highest combined solubility (Th and REE) reached 65 weight percent (10 percent ThO_2, 55 percent Ln_2O_3). At the highest ThO_2 loading (25 weight percent oxide), the combined solubility (Th and REE) was only 50 weight percent. The indication of solubility limits being exceeded was the formation of a thorium oxide layer on the bottom of the melt pool and/or crystalline strontium-neodymium silicates throughout the melt.

Table II. Batched compositons of glasses that were tested for durability in weight percent oxide.

Glass Id	Base Frit	REE	ThO_2	Homogeneous?
B-24-17	59	24	17	YES
Base	100	0	0	$AlBO_3$
B-5-0	95	5	0	$AlBO_3$
B-5-10	85	5	10	$AlBO_3$
B-15-0	85	15	0	$AlBO_3$
B-25-0	75	25	0	YES
B-45-0	55	45	0	YES
B-55-0	45	55	0	YES
B-65-0	35	65	0	YES
B-75-0	25	75	0	$Sr-Nd-SiO_4$
B-70-0	30	70	0	$Sr-Nd-SiO_4$
B-15-10	75	15	10	$ThSiO_4$
B-20-5	75	20	5	$ALBO_3$
B-25-10 (PCT)	65	25	10	YES
B-35-10	55	35	10	YES
B-45-10	45	45	10	YES
B-55-10 (PCT)	35	55	10	YES
B-65-10	25	65	10	$Sr-Nd-SiO_4$[†]
B-60-10	30	60	10	$Sr-Nd-SiO_4$[†]
B-20-20	60	20	20	YES
B-25-20	55	25	20	YES
B-30-20	50	30	20	YES
B-45-20	35	45	20	ThO_2
B-40-20	40	40	20	ThO_2
B-35-20	45	35	20	YES
B-20-25	55	20	25	YES
B-25-25 (PCT)	50	25	25	YES
B-30-25	45	30	25	ThO_2
B-20-30	50	20	30	A.P.S.[††]
B-25-30	45	25	30	ThO_2
B-35-30	35	35	30	ThO_2
B-15-30	55	15	30	A.P.S.[††]
B-15-25	60	15	25	A.P.S.[††]
B-15-20	65	15	20	A.P.S.[††]
B-17.5-22.5	60	17.5	22.5	YES
B-20-10	70	20	10	YES
B-15-15	70	15	15	YES
B-20-15	65	20	15	YES
B-45-15	30	45	15	YES
B-50-15	35	50	15	ThO_2

[†] Also contained a ThO_2 layer on the bottom of the melt pool.
[††] Amorphous Phase Separation.

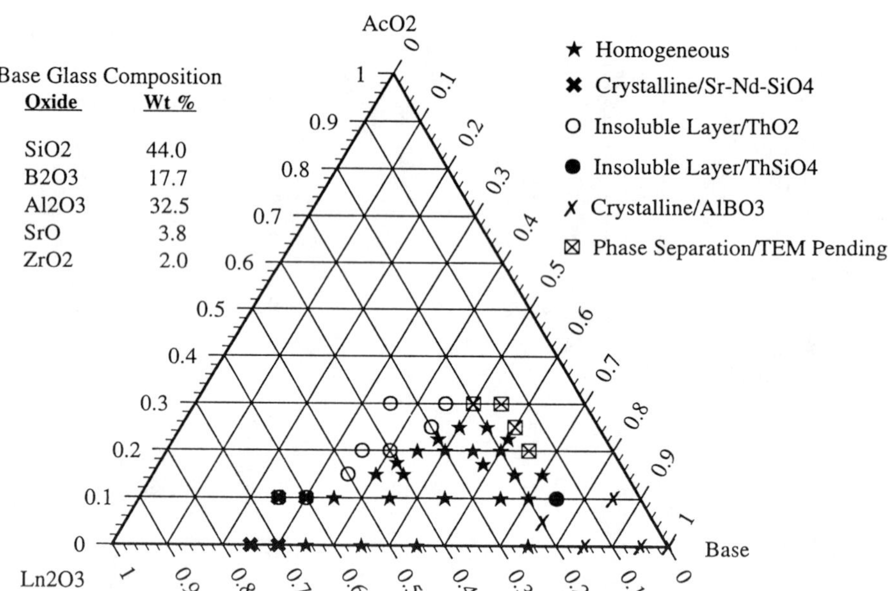

Figure 1. Ternary diagram of LaBS glass compositions with ThO$_2$ loadings.

Glass compositions were attempted with 30 weight percent thorium oxide in addition to 15 and 20 weight percent Ln$_2$O$_3$. These glasses had an opalescent appearence indicative of amorphous phase separation. XRD and SEM could not identify any crystalline structures. TEM/EDS detected the presence of silicon-rich droplets 20 to 30 nm in diameter. Figure 2 is TEM micrograph of the amorphous phase separation of the B-15-30 glass. The microstructure of the amorphous phase separated glass is quite stable. No beam damage was observed in the glass as a result of prolonged beam exposure (~5 minutes) in the B-15-20 glass.

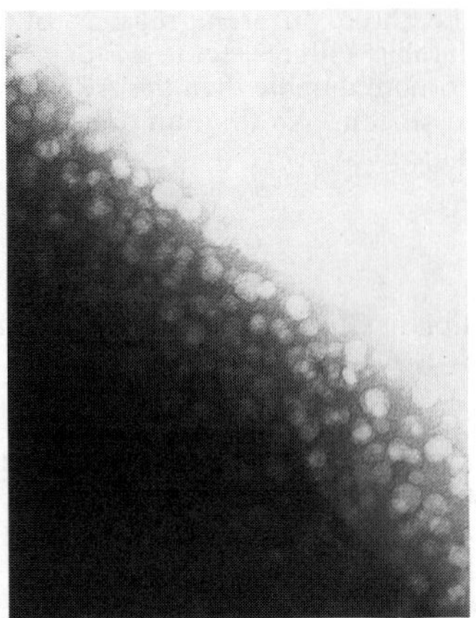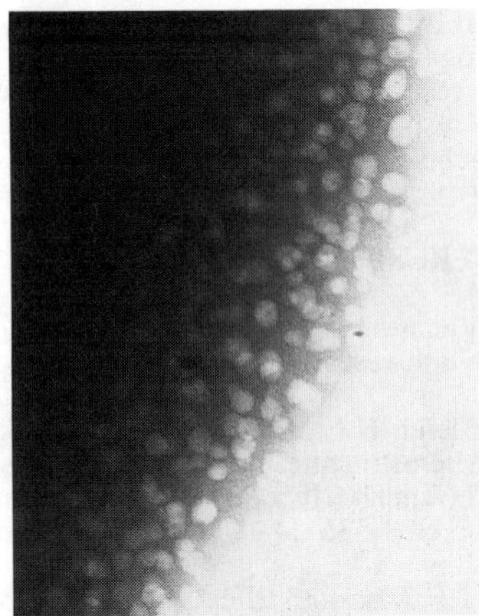

Figure 2. TEM micrograph of a phase separated glass (B-15-30) showing Si-rich droplets that are 20-30 nm in diameter.

Durability

Three glasses were analyzed and tested for durability. The glasses that were tested are indicated in Table II by "PCT" in the Glass Id column. These three glasses were chosen to compare the durability from three different regions of the homogeneous glass processing region. They were also selected to compare durabilities between two glasses with similar ThO_2 concentrations but different total Ln_2O_3 concentrations and two glasses with similar Ln_2O_3 concentrations but different ThO_2 loadings. The durability, based on normalized releases of boron and silicon, of the three glasses were compared to the Approved Reference Material - 1 (ARM) glass[7]. All three glasses tested were found to be ~1000x more durable than HLW glass with respect to normalized release rates of silicon and boron. No appreciable differences were noted in the durabilities of the three glasses tested. Release of Th and rare earth elements (neutron absorbers) were found to be below detectable limits.

CONCLUSIONS

A ternary diagram showing the homogeneous glass processing region of a base frit, rare earth oxide and thorium oxide has been developed at 1475° C. Thorium oxide was used as a plutonium surrogate. All ThO_2 glasses processed included a 1:1 mole ratio of Th to Gd. Gadolinium is added to the glass as a neutron absorber. Forty individual compositions were melted at 1475° C for 4 to 6 hours with periodic stirring. Two glasses (B-20-25 and B-25-25) were processed with a ThO_2 loading of 25 weight percent (oxide) without amorphous phase separation or crystalline species. These were processed with 55 weight percent frit, 20 weight percent rare earth oxides and 50 percent frit, 25 percent rare earth oxides.

Crystalline species that formed outside of the homogeneous glass processing region due to solubility limits or insufficient processing temperature were identified. Amorphous phase separation was detected and examined by transmission electron microscopy (TEM) at high ThO_2 loadings (25 to 30 weight percent oxide). The base frit was able to dissolve up to 65 weight percent rare earth oxides without thorium oxide.

Durability tests showed all three glasses from the three different regions of the homogeneous glass processing region were equally durable with respect to a 7-day PCT. PCT results show that the glasses tested were ~1000x more durable than the ARM glass with respect to normalized release rates of boron and silicon. No thorium or rare earth elements (used as neutron absorbers)was leached from the glass.

REFERENCES

1. Vienna, J.D., et. al., "Plutonium Dioxide Dissolution in Glass," PNNL-11346, Pacific Northwest National Laboratory, Richland, WA, (September 1996).

2. Bibler, N.E., Ramsey, W.G., Meaker, T.F. and Pareizs, J.M., "Durabilities and Microstructures of Radioactive Glasses for Immobilization of Excess Actinides at the Savannah River Site", Materials Research Society Symposia Proceedings, Vol. 233, pp. 333-336, Materials Research Society, Pittsburgh, PA (1995).

3. T.F. Meaker, "Effects of Periodic Manual Stirring and Uranium Addition on Surrogate Plutonium Glass Processing," WSRC-TR-96-0321, Westinghouse Savannah River Company, Aiken, SC, (September, 1996)

4. T.F. Meaker, "Compositional Development of a Lanthanide Borosilicate Glass for Plutonium Immobilization Without Listed RCRA Materials (Lead and Barium) (U)," WSRC-TR-96-0322, Westinghous Savannah River Company, Aiken, SC, (September 1996).

5. C.M. Jantzen, N.E. Bibler, D.C. Beam, and W.G. Ramsey, "Nuclear Waste Glass Product Consistency Test (PCT) - Version 7.0 (U)," USDOE Report WSRC-TR-90-539, Rev. 3, Westinghouse Savannah River Co., Aiken, SC (June 1994).

6. Figure No. 763, Phase Diagrams for Ceramists, American Ceramic Society, (1996).

7. Mellinger, G.B., and Daniel, J.L., "Approved Reference and Testing Materials for Use in Nuclear Waste Management Research and Deveoolopment Programs," U.S. DOE Report PNL-4955-2, Materials Characterization Center, Batelle Pacific Northwest Laboratory, Richland, WA, (December 1984).

Part XVII

Chernobyl-Related Waste
Disposal Issues

THE 4TH UNIT OF THE CHERNOBYL NPP: ITS PRESENT STATE AND PROBLEMS RELATED TO ITS TRANSFORMATION INTO AN ECOLOGICALLY SAFE SYSTEM

E. B. ANDERSON*, A. A. BOROVOY**, E. M. PAZUKHIN*
* V. G. Khlopin Radium Institute, St. Petersburg, Russia
** I. V. Kurchatov Institute of Atomic Energy, Moscow, Russia

ABSTRACT

The following problems related to the destroyed 4th Unit of the Chernobyl Nuclear Power Plant (NPP) are considered.

Monitoring of the unit's current state, including:

1) Monitoring of the atmospheric release of radioactive elements in the form of aerosols. In this case the problem is the unquantified area of cracks and holes that have formed in the "Sarcophagus" (the "Shelter") since 1986. In addition, the method used to evaluate the release, in which an uncontrolled escape of radioactive contamination is possible, raises objections;

2) Monitoring of the aquatic medium both inside the "Shelter" and on the industrial site. At present, the quantity of water in the 4th Unit is about 3000 m^3, which in fact is low- and intermediate-level radioactive waste. This quantity is not constant and essentially varies depending on the season, amount of atmospheric precipitation, the intensity of dust suppression, etc. In this case the problem is that the paths of water migration inside the unit are not known, so that an uncontrolled penetration of water into the ground, via the damaged foundation slab or the unit's walls, is quite possible;

3) Monitoring of the state of lava-like fuel-containing masses (LFCM) and the fuel. The problem is that the exact location of the fuel in the 4th Unit, the amount of fuel in major aggregates of LFCM, and the nuclear safety of these aggregates are all unknown to date.

The arrangements for the stabilization of the 4th Unit first of all include the reinforcement of the structural construction.

The transformation of the "Shelter" into an ecologically safe system is subject to many questions still unresolved. In particular:

- Whether to create a common "Shelter-2" for the 3rd and 4th Units or to merely encase the destroyed 4th Unit in concrete (the project "Cube");
- Whether to peace a plant for conditioning and reprocessing of radioactive waste under the roof of the "Shelter-2", or to build this plant separately;
- Where to create a temporary repository for containers of radioactive waste from the ChNPP 4th Unit;
- How to solve the problem of radioactive waste permanent storage in the territory of the Ukraine with proper observation of all nuclear and radiation safety requirements

INTRODUCTION

At present there are many problems with the destroyed 4th Unit of the Chernobyl NPP, the most important among which are the following:

- Monitoring of the unit's current state;
- Arrangements for the stabilization of the 4th Unit, preventing possible unfavorable impacts to the environment [1–3];
- Transformation of the "Shelter" ("Sarcophagus") into an ecologically safe system.

Mat. Res. Soc. Symp. Proc. Vol. 465 © 1997 Materials Research Society

THE MONITORING OF THE "SARCOPHAGUS'S" CURRENT STATE

Monitoring of the "Shelter's" current state includes:
- Monitoring of the state of lava-like fuel-containing masses (LFCM) and nuclear fuel;
- Monitoring of the aquatic medium both inside the "Shelter" and on the industrial site;
- Monitoring of the atmospheric release of radioactive elements in the form of aerosols.

The state of LFCM

Researchers working at the ChNPP 4th Unit first came across a highly radioactive material resembling black glass in the autumn of 1986. In one of the under-reactor locations (217/2) a giant congealed "drop" (about 1.5 m in diameter at the base) was found, that was later named "the elephant foot". In the course of further investigations the LFCM were found in many other locations under the reactor. Their composition was determined to include a considerable portions of the uranium and radionuclides that had been in the reactor core before the accident. Therefore the LFCM became the subject of an intensive study that had several purposes:
- To determine the degree of nuclear criticality hazard of the large accumulation of the lavas;
- To determine the radiation hazard of the lavas and their degradation products;
- To use the information about the composition and structure of the lavas for reconstructing the scenario of the course of the accident's active stage.

It is established that the process of formation of the lava took place in the under-reactor Room 305/2 (mark 9.00 m), more precisely, in its south-eastern part. During that process a part of the reactor base slab became molten. With the melt's mass increasing, it spread into the floor of Room 305/2, into the lower locations (the steam-distribution corridor, the bubbler pool), while simultaneously spreading in the horizontal direction through a breach in the wall between Rooms 305/2 and 304/3.

The main problems concerned with the LFCM are the following:
- Up to now the exact quantity of fuel in the LFCM and the final location of fuel accumulations in the 4th Unit have not been determined. By different estimates, there are from 20 to 90 tons of fuel (as uranium) in the under-reactor locations [4–6]. In addition, one may assume that a considerable amount of the fuel is in the Central Hall under the filling material that was deposited by helicopters.
- By now the nuclear safety of LFCM aggregates has decreased for at least two reasons:

First, the fuel's specific heat-generation has decreased, which increases the possibility of water penetration into the accumulated LFCM. Thus, we assume that at the time of the accident the total heat-generation was $\sim 3 \cdot 10^2$ kW/ton of uranium and the temperature inside the accumulated lava in the first years exceeded the boiling temperature of water. At present the specific heat-generation is estimated at only $3.5 \cdot 10^{-1}$ kW/ton of uranium, and the temperature inside the lava accumulations does not exceed 40 °C.

Second, the LFCM are a metastable system that is destroyed over time, as a roughly annealed glass. Therefore, if at the time of formation the LFCM were monolithic with a strength not inferior to basalt, then they now are porous masses permeated with cracks which also considerably facilitates the penetration of water.

For these two reasons, the effective neutron multiplication factor in the LFCM–water is increased, decreasing the nuclear safety of the Chernobyl lava accumulations. An illustration of this may be the incident that happened in Rooms 304/3 in the summer of 1990 when the monitoring system "Finish" indicated a sharp increase of the neutron background (approximately by 2 orders of magnitude). An analysis of that event does not exclude the beginning of a self-maintaining chain reaction (SCR).

<u>Monitoring of the aquatic medium both inside the "Shelter" and in the industrial site</u>

The ChNPP accident presented a number of problems requiring timely solution and long-term monitoring. One of the topical ecological problems is the solution of the issue of radionuclide contamination of surface and underground waters that are the only source of domestic-drinking water-supply of a considerable part of the Ukraine. Taking into account the intercommunication of ground waters with pressure waters as well as the discharge into the Pripet river with a subsequent penetration to the Dnieper river, they should be the primary object of permanent monitoring.

As a result of the accident at the ChNPP 4th Unit, according to certain estimates, up to 0.5% of the total nuclear fuel charge of the reactor and other elements of the core were ejected onto the industrial site territory. The decontamination work, carried out on the industrial site in 1986, resulted in the removal of the major ionizing radiation sources. Nevertheless, significant quantity of the fuel (by preliminary assessments, from 500 to 1000 kg) remains on the industrial site under a cover of stones and sand, layers of concrete and asphalt. This remaining portion of the fuel creates a real threat of environmental contamination over time, because the protective layers of stones, sand and concrete cannot reliably isolate the fuel-containing masses from the physico-chemical impact of natural and man-made factors. Both may result in transfer of radionuclides from the fuel matrix into a water-soluble state, which considerably increases the probability of radionuclide penetration into ground waters.

The second possible source of ground water contamination on the industrial site of the "Shelter" may be accumulations of water inside the 4th Unit proper. By preliminary estimates the total quantity of water in the "Shelter" premises is about 3,000 m^3, about 2,000 m^3 concentrated in the Machine Hall. Of this volume only about 1,000 m^3 are monitored.

The presence of considerable masses of radioactive waters and the quite possible non-leak-tightness of the 4th Unit's basement does not exclude contact of the unit's contaminated water with ground water. It should be also taken into account that water penetrating into the "Shelter" is a potential source of the increase of not only radioecological, but also nuclear criticality hazard. This is due to the following reasons:

- By contacting accumulated LFCM, water as already mentioned, increases the effictive neutron multiplication factor in the lava-water system;
- By dissolving fissile materials, water promotes an uncontrolled displacement of these materials and their accumulation in unknown places;
- Water washes out neutron-absorbing additives from areas of nuclear fuel accumulation;
- Water destroys LFCM and promotes uncontrolled displacement of radionuclides inside the facility and transport of radionuclides to ground waters;
- Water increases the rate of formation of LFCM fine particulates and transport to the atmosphere;
- Water speeds the chemical and mechanical destruction of LFCM and the structural destruction of the "Sarcophagus".

Besides these direct impacts on the state of the "Shelter", water exerts an indirect negative influence, disturbing the normal operation of diagnostic systems and hindering the course of investigations aimed at increasing the safety.

Therefore at present the most important task for the safety of the "Shelter" is to undertake measures for decreasing the quantity of water penetrating inside, and to organize permanent monitoring of the radionuclide composition and the presence of dissolved fissile material in the water.

The sources of water penetration into the 4th Unit may be divided into natural and man-made. The natural sources of water include atmospheric precipitation and a condensate arising from the temperature difference between the "Shelter" and the surrounding air. To obtain the upper values of the quantity of atmospheric precipitation one may use the StateComhydromet data: in the north of the Kiev region the average annual precipitation is 500–600 mm. Then the annual maximum quantity of water contacting the roof structures and in principle capable of penetrating into the interior premises of the "Shelter" may be assessed as 6,000 m^3. It should be noted that the penetration of water into the unit is subject to seasonal variations, with the maximum occurring during the spring melting of ice and snow.

The man-made water problem arises in conducting the boring work, dust suppression, decontamination, and as a result of emergency leaks. The volume of these waters does not exceed 200 m^3 annually.

At present the largest volumes of water go through:
- The Central Hall;
- The southern and the northern separator drum;
- The Cascade Wall.

It is evident that water inside the unit will interact with soluble components of the filling materials, for example trisodium phosphate. This is why there are high concentrations of sodium and phosphate ions in the unit's waters. Infiltrating through the "Sarcophagus" filling materials, water interacts with structural concrete, as well as with LFCM, destroying the matrices and leaching out soluble radionuclides. As a result the unit's water is essentially an alkali-carbonate solution in which compounds of cesium and uranium are very soluble, but in which the neutron-absorbing additive gadolinium is practically insoluble because of hydroxide formation.

It follows that at present at the ChNPP 4th Unit favorable conditions have arisen for the dissolution and redistribution of uranium isotopes. Under these conditions the gadolinium added in dust suppression as a neutron-absorbing additive most probably precipitates in the form of carbonate or hydroxide in the upper filling layer in the Central Hall and does not get to the LFCM accumulations in the under-reactor premises.

The composition of the unit's waters is presented in Tables 1 and 2.

Table 1

The radionuclide composition of water from various premises of the "Shelter"

Volume activity, Bq/l				Uranium µg/l
Cs-137	Cs-134	Sr-90	Σ Pu	
$1.75 \cdot 10^3 - 2.1 \cdot 10^8$	$1.52 \cdot 10^2 - 9.60 \cdot 10^6$	$1.2 \cdot 10^2 - 6.0 \cdot 10^6$	$2 \cdot 10^{-1} - 3.4 \cdot 10^3$	$1.1 - 2.3 \cdot 10^4$

Table 2

The radionuclide composition of ground waters from various boreholes at the "Shelter" site

Volume activity, Bq/l				Uranium µg/l
Cs-137	Cs-134	Sr-90	Σ Pu	
$0 - 5 \cdot 10^3$	$2 - 814$	$1.9 - 5.8 \cdot 10^3$	$0.05 - 5$	background – 4.3

<u>Monitoring of the atmospheric release of radioactive elements in the form of aerosols</u>

In 1986 the ecological impact of the Chernobyl catastrophe was determined chiefly by the transport of radioactive fission products and nuclear fuel particles by air currents. Now, one of the main sources of potential radioecological hazard from the destroyed 4th Unit of the ChNPP still remains the process of airborne radionuclide migration to the environment. In spite of measures which were taken, the "Shelter" is not a hermetic structure at present, as a result air currents can transport radioactive aerosols to the environment. Such radionuclide transport was quite often observed before 1988 and was related to work conducted at the damaged reactor or to specific meteorological conditions. Both radioactive aerosols formed at the time of the accident and remaining as dust inside the "Shelter" (by some estimates, this quantity is about 10 tons) and those formed in the course of the physico-chemical destruction of LFCM, including pure fuel, can be transported from the destroyed unit.

The destruction of the LFCM occurs under the influence of both natural factors (through radiation fields, humidity, temperature variations, etc.) and man-made factors (through carrying out of any work at the "Sarcophagus"). The radiobiological hazard of the aerosols is related to the presence in them of highly toxic fission product and transuranium elements, in particular, strontium-90 and isotopes of plutonium.

The most chemically active aerosols were formed as a result of evaporation and subsequent condensation of highly volatile isotopes of cesium, antimony, ruthenium and other radionuclides. This type of particles, to a considerable extent, determines the level of surface contamination of the interior of the "Shelter" with the greatest contribution introduced at present by the isotopes Cs-137 and Sb-125. From recent data, the total release of cesium-137 in the course of the Chernobyl accident was not less than $7.4 \cdot 10^{16}$ Bq. As the temperature of boiling and condensation of cesium compounds is about 1300 °C, most condensed on relatively cold surfaces inside the "Shelter".

To assess of the release of radioactive aerosols from the "Shelter", the results of measurements of the contamination of adsorbing plates are used at present. For this purpose special plate-holders are installed directly over selected engineered hatches on the "Sarcophagus's" roof. The plates are impregnated with venyl copolymer based compounds with special collecting additives to retain the aerosols and sharply reduce the breakthrough in the filter.

In this case the main problem is the unquantified area of cracks and holes formed in the "Sarcophagus" since 1985 and the resulting uncontrolled release of radioactive contamination.

To determine the concentration of radioactive aerosols at the industrial site of the "Shelter" (at distances of 60–100 m from it) four calibrated intake filters were installed: north, north-west, west and south of the "Shelter". The filter exposure time was 10–15 days.

The data for transport of radionuclides from the "Shelter" are presented in Table 3; the data on the concentrations of radioactive aerosols in the air of the industrial site, in Table 4.

Table 3

The transport of radionuclides from the "Shelter" for the period 1990–1996, mCi/km^2

Ce-144	Pu	Rh-106	Cs-137	Cs-134
$3 \cdot 10^{-2}$ – $1.3 \cdot 10^{4}$	$5 \cdot 10^{-2}$ – $2.2 \cdot 10^{2}$	$1.2 \cdot 10^{-1}$ – $7.7 \cdot 10^{2}$	5.3 – $3.8 \cdot 10^{3}$	$1.7 \cdot 10^{-1}$ – $1.4 \cdot 10^{4}$

Table 4
The concentration of aerosols in the air of the industrial site of the "Shelter" for the period 1992–1996, Ci/l

Ce-144	Rh-106	Cs-134	Cs-137	Pu + Am	Eu-154	Sb-125
$5.8 \cdot 10^{-19}$ – $2.5 \cdot 10^{-16}$	$1.0 \cdot 10^{-18}$ – $3.0 \cdot 10^{-16}$	$8.0 \cdot 10^{-19}$ – $2.2 \cdot 10^{-16}$	$2.8 \cdot 10^{-17}$ – $3.8 \cdot 10^{-15}$	$7.7 \cdot 10^{-19}$ – $1.1 \cdot 10^{-16}$	$5.8 \cdot 10^{-19}$ – $8.7 \cdot 10^{-18}$	$8.0 \cdot 10^{-19}$ – $5.1 \cdot 10^{-18}$

ARRANGEMENTS FOR THE STABILIZATION OF THE 4TH UNIT

The most important action for stabilization of the 4th Unit is the reinforcement of the constructed structures, especially the "Mammoth" and B-1, B-2 beams. For 10 years (since 1986) structures which during construction have been fixed only by friction, without application of screw fasteners or concreting, have to a considerable extent, lost their supporting structural ability. Warping of the beams threatens to collapse the unit's roof and result in a large-scale radiation accident with release of a considerable amount of aerosols. Therefore efforts are being made for additional reinforcement and stabilization of the weakened structures. Such stabilization actions should include inslallation of powerful dust-suppressing systems around the destroyed unit. These systems should be capable of shooting, within fractions of a second, many tons of water mist to collect and contain radioactive aerosols during a possible catastrophic collapse of the structure.

THE TRANSFORMATION OF THE "SHELTER" INTO AN ECOLOGICALLY SAFE SYSTEM

Because of many unresolved problems, a coherent concept for the transformation of the "Shelter" into an ecologically safe system does not yet exist.

Among the problems are the following:

- Whether to create a "Shelter-2" only for the 4th Unit alone or to make a joint structure, simultaneously including the 3rd and the 4th Units in it;
- Whether it is possible to realize the cheapest option: concreting the 4th Unit, without building a protective dome over it;
- What is to be done with radioactive wastes resulting from the possible dismantling of the 4th Unit if the "Shelter-2" project is realized: where to build, and whether to build at all, a plant for radioactive waste conditioning and reprocessing. There are no suitable geologic horizons for underground disposal in the Ukraine, and Russia hardly will agree to accept wastes from the Chernobyl NPP 4th Unit.

The same questions are true for a temporary repository for containers of radioactive waste from the ChNPP 4th Unit.

In conclusion it should be emphasized, that problems concerned with the final elimination of the consequences of the ChNPP accident require an immediate solution and can be solved only by joint efforts of the world community counties.

REFERENCES

1. At. Energ. **61**, p. 301 (1986) (in Russian).
2. Nuclear News **29**, p. 59 (1986).
3. Publication series on safety No. 75-INSAG-1, Proceedings of Conference on Studies of Causes and Consequences of Chernobyl Accident. IAEA, STI/PUB/740, ISB N-92-0-423088-6, Wien (1988).
4. A. A. Borovoy, Inside and outside the "Sarcophagus". Preprint KE IAE, Chernobyl (1990).
5. A. A. Borovoy, B. Ya. Galkin, A. P. Krinitsyn, V. M. Markushev, E. M. Pazukhin, A. N. Kheruvimov, K. P. Checherov, New-formed products of the interaction of fuel with structural materials of the 4th Unit of the Chernobyl NPP, Radiochem. **32**, p.103-113 (1990). (in Russian).
6. A. N. Kiselev, A. Yu. Nenagliadov, A. I. Surin, K. P. Checherov, A study of fuel containing masses on the ChNPP 4th Unit (by results of investigations of 1986–1991). Preprint IAE-5533/3, Moscow (1992). (in Russian).

THE BEHAVIOR OF NUCLEAR FUEL IN FIRST DAYS OF THE CHERNOBYL ACCIDENT

B.E. BURAKOV*, E.B. ANDERSON*, S.I. SHABALEV*, E.E. STRYKANOVA*, S.V. USHAKOV*, M. TROTABAS**, J-Y. BLANC**, P. WINTER**, J. DUCO***
*V.G. Khlopin Radium Institute, 28, 2nd Murinsky ave., 194021, St.Petersburg, Russia,
**Center D'Etudes de Saclay, 91191 Gif-Sur, Yvette CEDEX, France
***IPSN, 60-68 ave. du General Leclerc, BP6-92265 Fontenay-Aux-Roses CEDEX, France

ABSTRACT

Various types of Chernobyl fuel containing masses named black "lava", brown "lava", porous "ceramic" and "hot" particles that formed during first days of the accident at the Chernobyl Nuclear Power Plant 4th Unit were studied by methods of optical and electron microscopy, microprobe and x-ray diffraction. Data about their chemical, phase and radionuclide composition are summarized. The products of interaction between fuel, zircaloy and concrete, produced under experiments in laboratory were examined for comparison with samples of Chernobyl "lava" and "hot" particles. The behavior of nuclear fuel in first days of the Chernobyl accident was a three-stage process. The first stage occurred before the moment of the Chernobyl explosion and was exceptionally short-lasting, perhaps, less than a few seconds. It was characterized by reaching a high temperature, ≥2600 °C, in the epicenter of accident and formation of a Zr–U–O melt in a local part of the core, which is estimated to be not more than 30% of whole core volume. The second stage lasted for about 6 days since the explosion, during which there was interaction between uranium products of the destroyed reactor: UOx, UOx with Zr, Zr-U-O, with the environment and silicate structural materials of the 4th Unit. The third stage, after 6 days involved the process of final formation of the radioactive silicate melt or Chernobyl "lava" at one of the sections of the destroyed 4th Unit. During this stage the melt's lamination occurred, followed by a break-through of the "lava" reservoir on the 11th day of the accident and penetration of the "lava" into space under the reactor.

INTRODUCTION

The accident at the Chernobyl Nuclear Power Plant (ChNPP) 4th Unit on 26 April 1986 was accompanied by the destruction of the reactor core, a fire, and the release to the environment of a tremendous amount of solid and gaseous radioactive products. The activity of release is estimated to be for Kr-85: $3.3 \cdot 10^{16}$ Bq, Xe-133: $1.7 \cdot 10^{18}$ Bq and all other nuclides $1 \cdot 10^{18}$ - $2 \cdot 10^{18}$ Bq [1-3]. As a result of the accident, a part of nuclear fuel of the 4th Unit was dispersed by the explosion, and as dust or "hot" particles settled on the surface of soil and house roofs over a distance of hundreds kilometers from the ChNPP, as well as contaminating premises of the forth Unit [4-14]. The quantity of the dispersed fuel released to the environment outside the boundaries of the ChNPP industrial site is estimated as 3.5% of the whole fuel volume (190 metric tons of UO_2) of the 4th Unit [1,3,7]; however, this figure may be an underestimate. The amount of fuel dust in spaces of the "Sarcophagus" or the object "Shelter", built over the ruins of the 4th Unit, perhaps, is considerably greater [7], but this is difficult to estimate correctly.

Another part of the 4th Unit's nuclear fuel, remained after explosion inside the ruins of the reactor, reacted with silicate construction materials: sand, concrete, serpentenite and formed "lava-like" fuel-containing masses or Chernobyl "lava". The "lava" spilled and hardened under

Mat. Res. Soc. Symp. Proc. Vol. 465 © 1997 Materials Research Society

the reactor [7,15–17]. A black and brown "lava" may be distinguished, as well as "porous ceramics" a product of brown "lava" coming into contact with water. The volume of visually studied "lava" exceed 190 m^3, the quantity of fuel bound with them , in the dissolving form and solid inclusions, accounted, presumably, from 11 to 33% of the whole fuel volume of 4th Unit [17].

Considerable amount of nuclear fuel, which is hard to estimate, retained its main physico-chemical characteristics during the course of the accident. This part of the fuel includes fuel rods and their fragments scattered by the explosion over the 4th Unit ruins, as well as that which remained in broken off technological channels of the reactor's upper slab [7,16]. Thus, the Chernobyl accident caused a structural and chemical transformation of a considerable part, not less than 30%, of the 4th Unit's nuclear fuel. The behavior of this fuel in the course of the accident was determined not so much by its mechanical destruction as by the formation of various uranium containing compounds [11,14,19–22]. This affected the dynamics of radionuclide release and features of the radioactive contamination of the surrounding territories.

In this paper, we consider the Chernobyl accident as a three-stage process of destruction and transformation of uranium fuel. Each stage was distinguished by comparing the results of our study of the different uranium-containing products of Chernobyl "lava" and "hot" particles with the dynamics of radionuclide release in the period from 26 April until 6 May 1986 [3,7]. The paper is based on materials investigated at the V. G. Khlopin Radium Institute, carried out at Chernobyl and in St. Petersburg in the years 1990–1994. Also used are results of studies of a series of Chernobyl samples by the Saclay Nuclear Center (France) performed in 1992–93 within the scope of collaboration between the Institut de Protection et de Surete Nucleaire, Departement de Mecanique et de Technologie (France) and the Radium Institute (Russia).

SAMPLES

Without doubt, before the moment of explosion of ChNPP 4th Unit, a considerable part of nuclear fuel had already experienced overheating. For a limited part of the fuel of the damaged reactor calculations suggest temperatures of 3000 °C [2] and in some cases above 6000 °C [23].

Thus, the "hot" particles ejected during the explosion of the 4th Unit, and not undergoing a subsequent interaction with structural materials, must contain information about the initial stage of the accident [14]. These small solid particles from a few to tens of microns in size can record the temperature and other conditions that occurred inside reactor before the moment of explosion, and reflect these conditions in their chemical and radionuclide compositions of phases. After the destruction of the Chernobyl reactor an active transmission of radionuclides bound to the "hot" particles continued for about 10-11 days [3,7]. However, the chemistry and phases formed in these particles was influenced by later processes, such as burning of graphite [8,10].

To identify "hot" particles, whose matrices reflect the temperature and chemical changes of the fuel before the moment of explosion, samples of soil were obtained along the axis of the Western Plume over the distance of 12 km from ChNPP. This territory is characterized by the maximum levels of radioactive contamination (more than 40 Ci/km^2 for Cs-137) caused by the earliest fallout of "hot" particles after the destruction of ChNPP 4th Unit. Biological investigations of radiation damage to the pine-trees revealed unique peculiarities of the process of formation of the Western Plume. It was determined that the radioactive cloud in the immediate vicinity of the reactor was suprisingly moving as a narrow band 1-1.5 km in width piercing the forest to the west. Along the first 5 kilometers cloud moved into relatively thick forest and a further 5–6 kilometers under this cover without significant spreading . [24]. Along the whole length of that portion at distances up to 12 km from the ChNPP, very large fuel particles were found up to hundreds of microns in size. The transport of a large amount of comparatively heavy particles was probably the result of the explosion, rather than subsequent events of the accident.

After the destruction of the reactor core a part of nuclear fuel interacted with silicates used as structural materials of the reactor, thus forming a silicate "lava" [7,15–17]. Visual studies of hardened Chernobyl "lava", hardened in premises of the "Sarcophagus" or "Shelter", revealed three main flows with a common source [17]. In our investigations the sample collection was determined by: accessibility of samples under conditions of extremely high radiation fields, reaching more than 1000 R/hour in some places; visual differences of the "lava" by color and porosity; vertical and horizontal location of sampling areas. A simplified scheme of disposition of the Chernobyl "lava" by height levels with an indication of sample location is shown in Fig. 1.

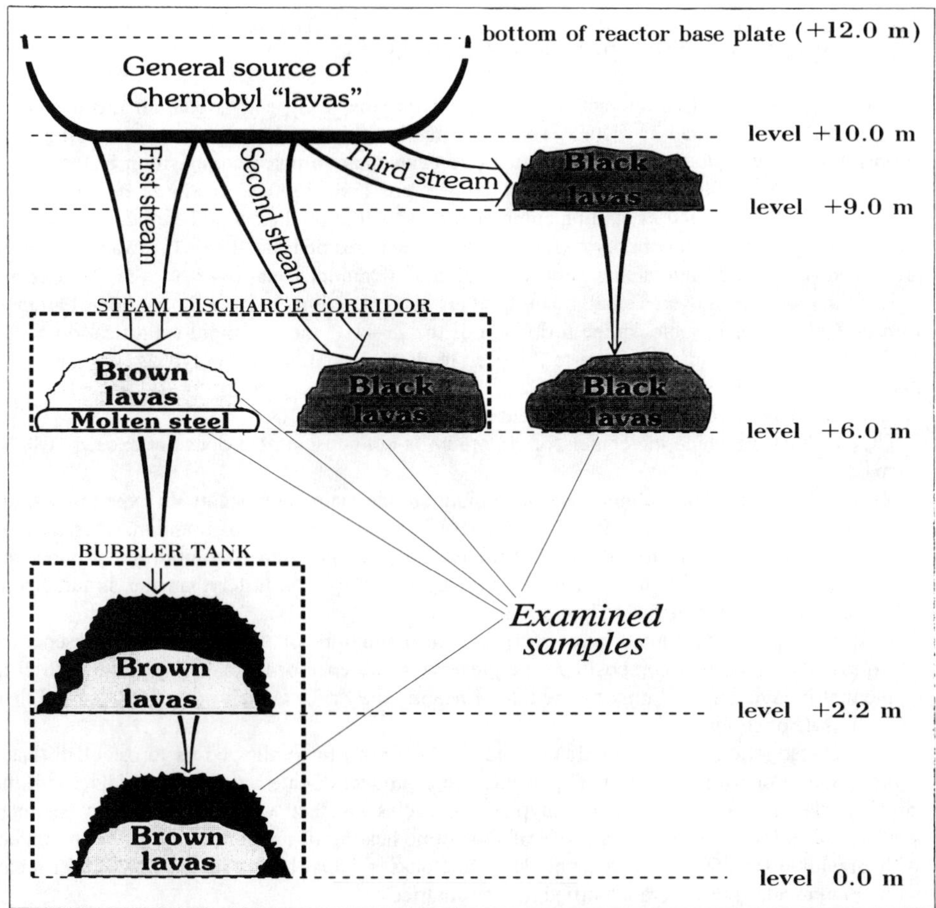

Fig. 1. Height distribution of various types of Chernobyl "lavas" by height levels with indication of sampling places

In general, the samples we were able to collect are not representative or numerous enough. Most representative samples are of the brown "lava" from the steam discharge corridor where the "lava" layer thickness did not exceed 15 cm. It was considerably more difficult to obtain samples of black "lava" from thicker formations, therefore samples were studied from the subsurface sections of "lava" streams. For a more representative and detailed study of mineral inclusions found in the matrix of the "lava", large pieces of fuel-containing masses, tens of cubic centimeters in volume were dissolved in cold hydrofluoric acid. The insoluble residue, obtained as a result of this procedure, consisted of grains of various crystalline and amorphous phases. However, the separation of the inclusions was not complete. A certain portion of them, for example, small particles of uranium oxides, dissolved simultaneously with the glass-like matrix of the "lava".

THE FIRST STAGE: NUCLEAR FUEL INTERACTION WITH THE ZIRCALOY CLADDING PRIOR TO THE CHERNOBYL EXPLOSION

The process of active interaction of uranium dioxide, UO_2, and zirconium metal at high temperatures was studied in detail. The solidification of the Zr–U–O melt, occurring at a temperature of 1900 °C, led to the formation of an eutectic two-phase composition UO_2/α-Zr(O) [25]. The behavior of the Zr–U–O melt at a temperature 2500–2600 °C to a large extent depends on the ratio of the UO_2 and Zr starting components, as well as on the rate of heating.

Experiments under laboratory conditions at temperatures of 1900, 2600 °C in vacuum and in argon showed that the interaction between UO_2 and zirconium metal α-Zr or α-Zr(O) does not proceed in presence of even a small quantity of air in the reaction chamber. Formation of an inert film of ZrO_2 completely blocks the formation of the Zr–U–O melt. A rapid solidification of the Zr–U–O melt heated up in vacuum to 2600 °C leads to formation of a polyphase ceramic. The matrix of the ceramic consists of:

1. UO_x that may be considered as unreacted relicts of the initial UO_2.
2. UO_x with Zr, in which the content of zirconium is not constant, but does not exceed 10% by mass.
3. Zr–U–O, in which both uranium and zirconium are the main components. In experiments, the zirconium concentration in the Zr–U–O phase is higher than that of uranium. This may be caused by the fact that the diffusion of uranium and oxygen into zirconium metal under the non-equilibrium conditions of rapid heating up to 2600 °C is faster than the simultaneous process of zirconium diffusion into uranium oxide.
4. Zirconium metal, presumably, α-Zr(O) with an admixture of uranium. In the presence of excess UO_2 in starting composition, the presence or absence of the α-Zr(O) phase with U in the matrix is determined only by the rate of heating the Zr–U–O melt up to 2600 °C and then by the rate of cooling.

An investigation of "hot" particles from the Western Plume allowed us to establish that a considerable proportion of the "hot" particles are a product of interaction of fuel UO_x with the zircaloy cladding [11,14]. Some polyphase particles in fact were analogues to samples synthesized in the laboratory as a result of short-time heating of uranium oxide pellet in contact with zircaloy up to 2600 °C in vacuum. The same phases: UOx, UOx with Zr, Zr-U-O, Zr metal with uranium admixture were identified in their matrices.

Results included: *i*) identification of Zr-U-O and UOx with Zr phases in the matrices of "hot" particles; *ii*) experimental synthesis of the same phases in association with UOx and Zr metal with uranium admixture under the non-equlibrium conditions of rapid heating up to

2600 °C and only in the non-air media, are the basis of our proposal that the initial stage of the Chernobyl accident was short-lasting, not more than a few seconds, was characterized by a temperature ≥2600 °C at the epicenter prior to the explosion. In Fig. 2 is shown a physico-chemical transformation of fuel before the moment of explosion on the level of a fuel rod fragment at the accident epicenter.

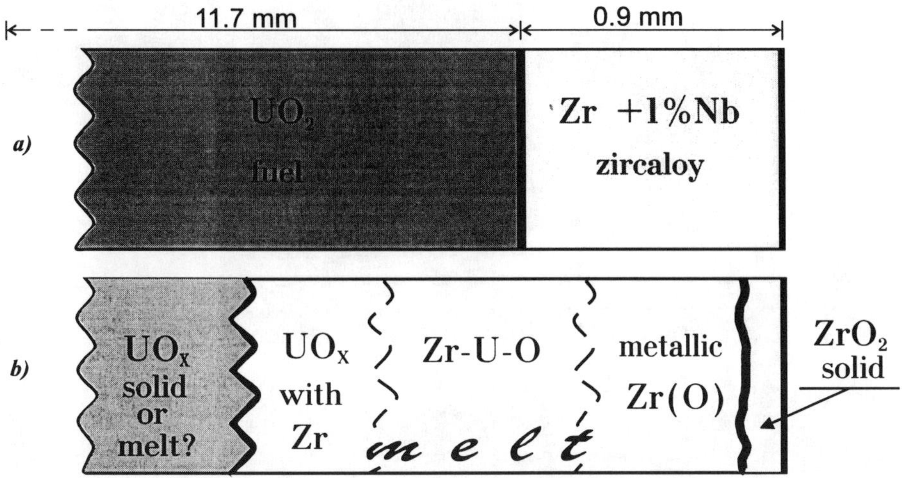

Fig. 2. A section view of a fuel rod fragment: a) initial structure;
b) before the explosion in the accident epicenter.

THE SECOND STAGE: SIX DAYS AFTER THE EXPLOSION

The explosion of the ChNPP 4th Unit completely destroyed the reactor core. The scorching fuel in the form of uranium oxide, as well as products of UO$_x$ with Zr and Zr–U–O, escaped to and interacted with the environment. Overheating of the fuel materials caused a release of radionuclides that had evolved from the uranium matrix. Some of them, for example Ru-106, formed separate types of "hot" particles, having a matrix of pure Ru-106 metal [4,5]. In our investigations [11,14] we did not find these particles. However, there was noticed that "hot" particles, containing in their matrix metallic inclusions Fe-Cr-Ni, were characterized by enrichment in Ru-106.

The release of different forms of radionuclides from fuel products at the 4th Unit ruins was going on during all 10 days after the explosion; however, in the first 6 days its intensity was continuously decreasing from $4.5 \cdot 10^{17}$ to $1 \cdot 10^{17}$ Bq/day (error ±50%) [3,7]. Such a prolonged and active radionuclide release was caused mainly by residual heat-release of the destroyed fuel. Although one should not exclude an influence of a fire at the center for the very first days after the explosion [7,8,10].

The melting of structural materials and steel began in places of fuel conglomeration. During this stage of the accident only the initial formation of the Chernobyl "lava" source occurred. In our opinion, this represented a non-uniform mixture of "cold" structural and "hot" fuel materials with the melt in the zone of direct contact. This stage ended on 2 May 1986, when separate sources of silicate melt began to consolidate in the common source of Chernobyl "lava" (Fig. 3). In general, the process of heat absorption due to destruction and melting of concrete, steel and sand, prevailed over the process of its release from separate local sources. Thus, there was a gradual decrease of radionuclide release in the period between 26 April untill 2 May 1986.

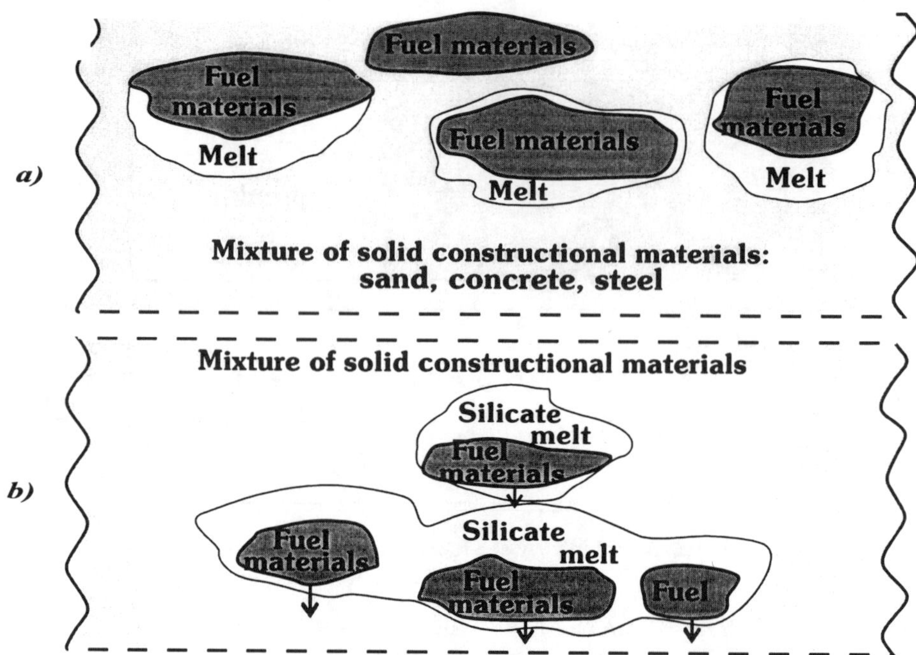

Fig. 3. General scheme of interaction between fuel in the form UO_x, UO_x with Zr and Zr-U-O, with structural materials in the period of the 2nd stage of the Accident: a) after the Chernobyl explosion on 26 April 1986, and b) on 2 May 1986.

THE THIRD STAGE: FINAL FORMATION OF THE CHERNOBYL "LAVA" SOURCE AND ITS BREAK-THROUGH

After 2 May, 1986, an increase of the radionuclide release was noted over the ChNPP 4th Unit's ruins. For several days the release continuously increased up to $3 \cdot 10^{17}$ Bq/day (error $\pm$ 50%) and on 6 May, 1986, it dropped abruptly to $0.2\text{-}0.6 \cdot 10^{16}$ Bq/day [3,7].

During the period from 2 till 6 May 1986, we consider this as the time of the most vigorous physico-chemical transformation of the ChNPP nuclear fuel. We propose that on 2-3 May a united center of radioactive melt formed due to confluence of previously isolated sources, and by 5–6 May the temperature homogenation of the melt and its stratification into three layers was completed.

The main peculiarities of the formation of the Chernobyl "lava" center and the chemical transformation of fuel materials include:

• the uranium phases UO_x, UO_x with Zr partly dissolved in the silicate melt;

- a considerable part of the Zr–U–O phase came into chemical contact with the silicate melt, which caused the crystallisation of high-uranium zircon $(Zr,U)SiO_4$ [20,21];
- a redistribution of the Ru-106 occurred, that selectively concentrated Ru-106 in drops of molten steel;
- under the action of gravity the molten metal and solid inclusions of the uranium phases began to gather in the lower part of the melt center.

In order to understand the peculiarities of the process of the "lava" formation, we used the results of studies of various solid inclusions in their matrices, for example, zircon $(Zr,U)SiO_4$. It was shown, that inclusions of high-uranium zircon $(Zr,U)SiO_4$ are present in all the studied types of Chernobyl "lava", including the "porous ceramics". It is evident that crystallization of zircon, $(Zr,U)SiO_4$, had occurred before the moment of spreading-out of "lava" streams, because "porous ceramics" is a product of the practically instantaneous solidification of brown "lava" as it came into contact with water. It may be also stated, that the temperature of the silicate melt in the Chernobyl "lava" center did not exceed 1660-1700 °C [21] – the limit of zircon's thermal stability [26,27]. An attempt was made to indirectly assess the time required for the crystallisation of the high-uranium zircon $(Zr,U)SiO_4$ in the "lava", admitting a certain analogy to results for the synthesis of crystalline zircon by the method of crystallisation from a melt under laboratory conditions [29-31]. Calculations showed that the formation of high-uranium zircon $(Zr,U)SiO_4$ with crystals reaching a size of several hundreds microns requires approximately three days.

Thus, by 6 May 1986 the site where the Chernobyl "lava" originated was, presumably, represented by a radioactive melt separated into three layers:

1. Lower layer: metallic Fe-Cr-Ni close to the composition of structural stainless steel. The given layer did not contain dissolved uranium admixtures or solid inclusions of uranium phases. The radioactivity was caused by radionuclide ^{106}Ru and, partly, ^{60}Co and ^{125}Sb.
2. Middle layer: silicate with dissolved admixtures of U, Zr, Al, K, Ca, as well as with numerous dispersed solid inclusions of UO_X, UOx with Zr, the phase Zr–U-O, crystals of high-uranium zircon $(Zr,U)SiO_4$.
3. Upper layer: silicate, similar in composition to the middle layer, but with a lesser amount of uranium phase solid inclusions.

For several days up to 6 May, 1986, something prevented the spreading out of Chernobyl "lava". We assume that in the zone of direct contact of the melt with concrete a stable thermoinsulating crust have formed, thus preventing a melting-through of the concrete constructions.

A sharp decrease of the radionuclide release on 6 May 1986 was probably caused by the spreading-out and cooling of the Chernobyl "lava". The three independent flows of "lava" left the common underreactor space, level + 10.0 meters (Fig. 1). The primary source of the melt, however, was much higher, because the "lava" melted through the south-eastern part of the reactor base slab (+12.0 meters). A small "leak" deposit of "lava" 0.5 meters in diameter and 1.2 meters in height, resembling a stalagmite remained at the surface of the reactor base slab in its north-eastern part [17].

A comparative analysis of the results from study of materials in the "lava" and of visual observations directly in the ruined 4th Unit shows that the primary site of formation of Chernobyl "lava" was outside the reactor shaft. In the course of spreading-out the composition of the silicate melt varied. Admixtures of magnesium appeared in it when the "lava" melted through the serpentinite-filled reactor base slab. Probably, the melting of the base occurred quickly. This caused the different magnesium contents in the matrices of brown and black "lava" (Table 1).

Table 1. The chemical composition of the matrices from EDS analysis. (wt.%).
The number of analyses: porous ceramic - 55, brown ceramic - 75, black ceramic - 79.

Types of "lava"	Fe	Na	Si	Al	Mg	K	Ca	Zr	U
porous ceramic	0.2	0.5	35.2	3.8	4.5	2.3	7.5	4.0	2.9
brown "lava"	0.2	0.6	36.6	4.0	4.4	2.3	7.2	2.9	2.0
black "lava"	0.3 to 6.7 *	0.4	37.2	3.8	1.3 to 3.2 *	2.7	8.2	3.7	3.2

* - contents in black "lava" from vertical stream (13 analyses)

The general succession of events concluding this stage of the accident were:
- On 5–6 May 1986 a break-through of the Chernobyl "lava" center occurred. The melt entered the reactor shaft and melted through the reactor base slab in the south-eastern part. The melt's spreading was not uniform, but comparatively rapid. In general, this initial stratification remained.
- A molten metal melt and a part of the silicate melt saturated with uranium phases as solid inclusions was the first to reach the + 10.0 meter level and form a vertical flow of brown "lava" (Fig. 1).
- The upper part of the silicate melt reached the + 10.0 meter level in the last turn following metal melt and brown "lava" and formed horizontal and vertical flows of black "lava" (Fig.1).

SOLIDIFICATION OF CHERNOBYL "LAVA"

Different types of "lava" solidified at different rates. This affected both the chemical composition of the matrix and peculiarities of the morphology, structure and chemical composition of uranium phases (Tables 1,2). The black "lava" solidified slowly, therefore no inclusions of the phase Zr–U–O was found in their matrices. Crystals of high-uranium zircon $(Zr,U)SiO_4$ in this type of "lava" sometimes contain inclusions of a few microns in size of monoclinic zirconium dioxide ZrO_2 with a U admixture up to 6-7 wt%. These inclusions may be considered as a result of crystallization of relicts of the Zr–U–O phase in course of zircon formation: $Zr–U–O + SiO_2 \rightarrow (Zr,U)SiO_4 (zircon) + ZrO_2$ with U.

In comparison with the brown "lava", the black "lava" are sharply depleted in uranium phase inclusions. However, the concentration of uranium dissolved in the glass-like matrix of the black "lava" is 1.5 times higher, than in the silicate matrix of the brown "lava" (Table 1). This was caused by the dissolution of the most dispersed solid inclusions of the UO_x and UO_x with Zr, during the black "lava" spreading-out, as well as by the slow cooling of this type of "lava".

The brown "lava" matrix has a large number of solid inclusions of uranium phases. It is in these solid inclusions of the UO_x with Zr the main part of uranium present in the brown "lava". These inclusions of own brown color are the reason for the coloring of this type of "lava". The morphology of the UO_x + Zr phase inclusions is very diverse. There are particles with a fused morphology, cubic and cubo-octahedric crystals, as well as dendrites. These inclusions of UO_x with Zr have, presumably, different origins:
1. Particles with fused morphology: They are relicts of solidified drops of the initial melt of UO_x with Zr;
2. Crystalline inclusions: They are products of the recrystallization of the UO_x with Zr melt;

3. Fine inclusions of 1-3 microns in size inside crystals of high-uranium zircon: They are the products of reaction between the Zr–U–O phase and silicate melt: Zr–U–O + SiO_2 → $(Zr,U)SiO_4$ (zircon) + UO_x with Zr;

4. Dendrite-like particles: They are products of recrystallization of part of the uranium initially dissolved in silicate melt in the course of the "lava" cooling.

Results of X-ray diffraction analysis (XRD) show that the structure of the main part of inclusions of the phase UO_x with Zr corresponds to the U_4O_9. Inclusions of the Zr–U–O phase found in the brown "lava" and the "porous ceramics" are nearly amorphous to X-ray diffraction analysis with individual diffraction peaks of monoclinic zirconia phase, ZrO_2. This indirectly confirms the rapid solidification of this type of "lava". It is important to note that own inclusions of ZrO_2 are not typical for the matrices of Chernobyl "lava". We have found single grains of ZrO_2 only in "porous ceramic". The assumption of a possible dissolution of ZrO_2 in the silicate melt has not been corroborated experimentally. A simulation of the process of "lava" formation, carried out under laboratory conditions, showed that at 1600 °C the penetration of fragments of oxidized zircaloy cans by the silicate melt was not accompanied by dissolution of ZrO_2 by the melt. The zirconium admixture in the silicate melt was obtained by interaction between an artificially synthesized Zr–U–O composition and concrete at 1600°C.

Table 2. The composition of different types of uranium-zirconium-bearing
oxide phases in the Chernobyl "lava" (wt. %).

Types of inclusions	Types of "lava"	U	Zr	Interpretation
The inclusion of irregular morphology in the matrix	black	78.3	2.7	UO_x with Zr
The inclusion of irregular morphology inside the crystal of high-uranium zircon	black	84.9	2.5	UO_x with Zr
Isolated crystal of cubic form in the matrix	black	83.3	3.0	UO_x with Zr
Stellate crystal in the matrix	brown	82.6	4.7	UO_x with Zr
The aggregate of inclusions of irregular morphology in the matrix	Porous ceramic	80.3 18.3	7.1 65.5	UO_x with Zr Zr-U-O
The aggregate of inclusions of irregular morphology in the matrix	brown	80.4 18.1	6.1 66.0	UO_x with Zr Zr-U-O
The inclusion of irregular morphology inside the crystal of high-uranium zircon	black	5.8	75.0	ZrO_x with U
The porous inclusion with microblock structure in the matrix	brown	86.0	-	UO_{2-x}

CONCLUSIONS

The investigations described provide an understanding of the behaviour of nuclear fuel in the first days of the Chernobyl accident. An integrated scheme of physico-chemical transformation of the starting uranium fuel is presented (Fig. 4). As a result of the accident the formation of new solid uranium phases took place: UOx with Zr, Zr-U-O, uranium-containing glass, $(Zr,U)SiO_4$ - zircon, ZrO_2 with U and Zr metall with uranium admixture.

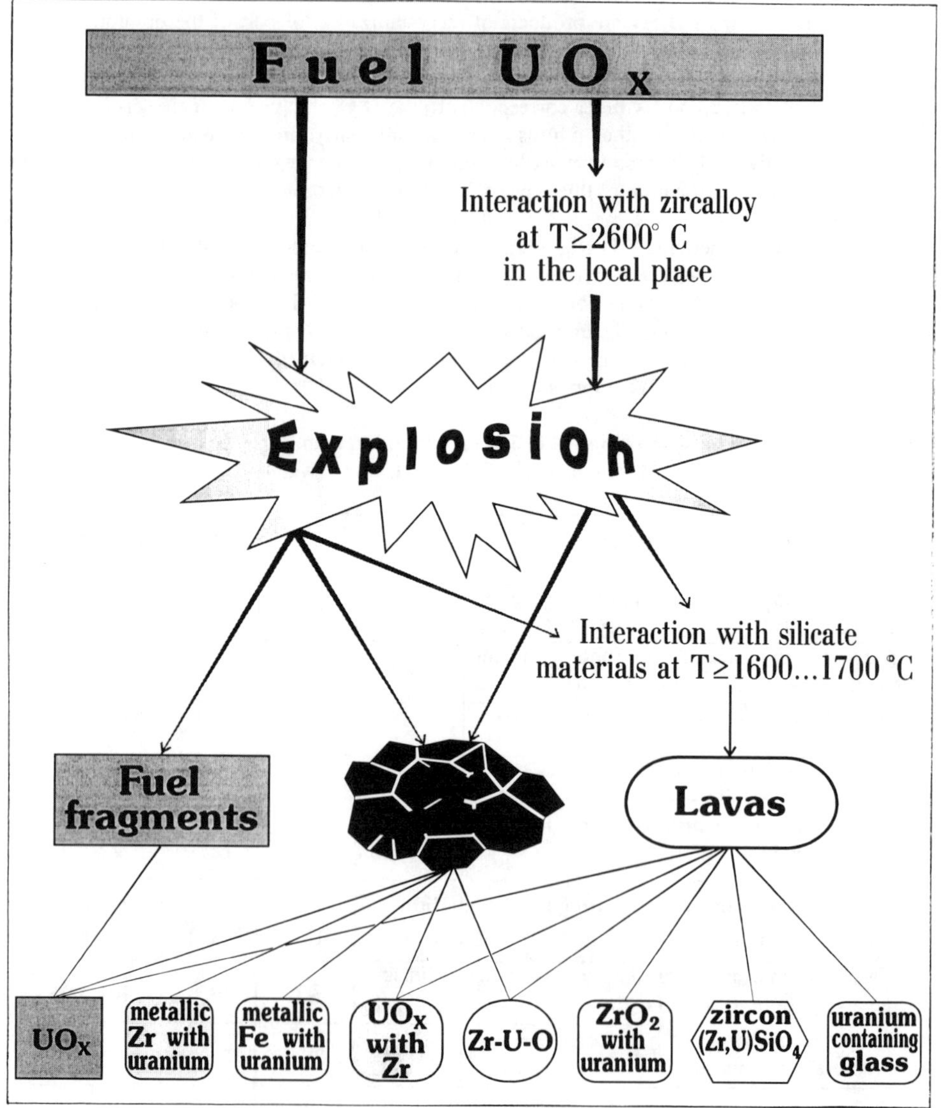

Fig. 4. Physico-chemical transformation of nuclear fuel during Chernobyl accident.

In our opinion, the most important conclusion is that an essential chemical change of the fuel took place even before the moment of Chernobyl explosion. The transformation of the fuel began in the volume of nuclear reactor at the level "UO_2 – Zircaloy". It allows us to use our results to simulate of major accidents on various types of nuclear reactors. It is important to compare our results with the data of Three Mile Island accident.

The main unanswered question concerns the amount of fuel involved in the interaction with zircaloy cladding at the early stage of the accident. The value of instantaneous energy release during the accident, based on determining the activity ratio of Xe-133 and Xe-133m in air, allows us to assess the proportion of overheated fuel as being 0.01–0.1% of its total mass [31]. More precise assessments, however, may be obtained by determining the proportion of "hot" particles with zirconium-containing matrices in the form UO_x with Zr, Zr–U–O, out of the total amount of fuel particles.

Note, that the high-temperature event in the fuel prior to the explosion was quite short, less than first seconds, because most of the "hot" particles with matrices of UO_x with Zr and Zr–U–O retained their essential compositions of light-volatile radionuclides, for example, Cs-134 and Cs-137. An experimental simulation of the dynamics of radionuclides release from the Zr–U–O melt may be used to evaluate the duration of the accident's initial phase. Thus, our work is not complete. In some points it contradicts the already published versions [32] of the Chernobyl accident scenario, proposing the possibility of interaction between fuel and zircaloy cladding after explosion. Therefore, these investigations must continue, as they are an essential part of safety assessment of nuclear reactors and their behavior during melt-down events.

ACKNOWLEDGMENTS

The authors are very grateful to R.C. Ewing and W. Lutze of the University of New Mexico for help in the preparation of this review and to N. Rutchkovski of IPSN and A. P. Zavragzin of Radium Institute for the translation of this article into English. The collaboration between University of New Mexico and V.G. Khlopin Radium Institute is supported by the NATO International Scientific Exchange Program.

REFERENCES

1. At. Energ. **61**, p. 301 (1986) (in Russian).
2. Nuclear News **29**, p. 59 (1986).
3. Publication series on safety N 75-INSAG-1, Proceedings of Conference on Studies of Causes and Consequences of Chernobyl Accident. IAEA, STI/PUB/740, ISB N-92-0-423088-6, Wien (1988).
4. L. Devell, H. Tovedal, U. Bergström, A. Appelgren, J. Chyssler, L. Andersson, Nature **321**, (1986) p. 192
5. P. Schubert, U. Behrend, Radiochimica Acta **41**, p. 149-155 (1987)
6. S.A. Bogatov, A.A. Borovoy, Yu.V. Dubasov, V.V. Lomonosov About forms of the fuel release in the accident at the Chernobyl NPP. Preprint IAE-4952, Moscow (1989) (in Russian).
7. A.A. Borovoy Inside and outside the "Sarcofagus". Preprint KE IAE, Chernobyl (1990). (in Russian).
8. S.A. Bogatov, A.A. Borovoy, V.I. Dvoretskiy, L.A. Elesin, N.V. Isaev, V.V. Lomonosov, L.I. Lebedeva, I.V. Matveev, V.N. Nikitin, L.A. Obukhova, I.A. Semin, Yu.N. Simirskiy,

An investigation of the most radiologically hazardous nuclides in various forms of the fuel release of the Chernobyl accident. Preprint IAE-5022/3, Moscow (1990). (in Russian).

9. E. Piasecki, P. Jaracz, S. Mirowski, J. of Radioanalytical and Nuclear Chemistry, Articles **141**, 2, p. 221-259 (1990)

10. S.A. Bogatov, A.A. Borovoy Preprint IAE-5344/3, Moscow (1991). (in Russian).

11. E.B. Anderson, B.E. Burakov, E.M. Pazukhin, Radiokhimia **5** (1992) p. 139 (in Russian).

12. L.M. Khitrov, V.O. Cherkezian, O.V. Rumiantsev, Geochemistry **7** (1993), p. 963. (in Russian).

13. E.P. Petriaev, S.L. Leynova, G.A. Sokolik, E.M. Danil'chenko, V.V. Duksina, Geochemistry 7, (1993), p. 930 (in Russian).

14. B.E. Burakov, E.B. Anderson, B.Ya. Galkin, E.M. Pazukhin, S.I. Shabalev, Radiochimica Acta **65**, p. 199-202 (1994).

15. A.A. Borovoy, B.Ya. Galkin, A.P. Krinitsyn, V.M. Markushev, E.M. Pazukhin, A.N. Kheruvimov, K.P. Checherov, New-formed products of the interaction of fuel with structural materials of the 4th unit of the Chernobyl NPP, Radiochem. **32**, p.103-113 (1990).

16. A.A. Borovoy, G.D. Ibraimov, S.S. Ogorodnik, V.D. Popov, K.P. Checherov The state of the ChNPP 4th unit and nuclear fuel being in it (by results of investigations of 1988–1989). Preprint KE IAE. Chernobyl (1990). (in Russian).

17. A.N. Kiselev, A.Yu. Nenagliadov, A.I. Surin, K.P. Checherov, Preprint IAE-5533/3, Moscow (1992). (in Russian).

18. A.S. Krivokhatskiy, V.G. Savonenkov Classification of radioactive man-made products from debris of 4th Unit of the Chernobyl nuclear power plant. Preprint RI-216, Ts'NII atominform, (1990). (in Russian).

19. E.B. Anderson, B.E. Burakov, E.M. Pazukhin Did the fuel of the 4th Unit of Chernobyl NPP melt? Radiochem. 5 (1992). (in Russian).

20. E.B. Anderson, B.E. Burakov, E.M. Pazukhin High uranium zircon (Zr,U)SiO4 from "Chernobyl lavas". Radiochimica Acta 60, p. 149 (1993).

21. B.E. Burakov in <u>Proceedings of international conference SAFEWASTE'93</u> **2**, (13-18/06/1993, Avignon, (FRANCE)) p. 19-28.

22. M.Trotabas, J-Y. Blanc, B. Burakov, E. Anderson, J. Duco Examination of Chernobyl samples, impact on the accident scenario understanding. Report IPSN/93/02 (1993), DMT/92/309, RI -1-63/92 (1993).

23. Report of the US Department of Energy's Team analyses of the Chernobyl atomic energy station accident sequence. DOE/NE-0076, DE 87 003614, Washington (1986).

24. N.I. Gol'tsova, N.I. Abaturov in <u>Eurochernobyl-2, Materials of the international conference.</u> (Kiev, 1992), p. 155 (in Russian).

25. A. Skokan, Proc. on Thermal Nuclear Reactor Safety, KfK-3880/2, p. 1035-1041. (1984)

26. Y. Kanno, J. Mater. Sci., 24 (1989) p. 2415

27. P. Pena, S. de Aza, J. of Mater. Sci., 19 (1984) p. 135-142.

28. M.A. O'Donoghue J. of Gemology 3, vol. XV, , (1976) p. 119-124.

29. B.M. Wanklyn, J. Cryst. Growth 37, 1 (1977) p. 51-56

30. <u>Synthesis of minerals.</u> v. 2, Moscow: "Nedra", 1987. (in Russian).

31. S.A. Pakhomov, K.S. Krivochatskiy, I.L. Sokolov, Radiochem. **6**, (1991), p.125. (in Russian)

32. E.M. Pazukhin, Radiochem. **36**, iss. 2, (1994), p. 97–142. (in Russian).

SECONDARY URANIUM MINERALS
ON THE SURFACE OF CHERNOBYL "LAVA"

B.E.BURAKOV*, E.E.STRYKANOVA*, E.B.ANDERSON*
* V.G.Khlopin Radium Institute, 28, 2nd Murinsky ave., 194021, St.Petersburg, Russia

ABSTRACT

The formation of uranium minerals is still continuing in Chernobyl Unit No. 4. Yellow products of alteration that stain the surface of Chernobyl "lava" have been examined by SEM and X-ray diffraction methods. Secondary minerals of uranium identified are: $UO_4 \cdot 4H_2O$ studtite; $UO_3 \cdot 2H_2O$ epiianthinite; $UO_2 \cdot CO_3$ rutherfordine; also, $Na_4(UO_2)(CO_3)_3$ was identified together with the sodium carbonate phases $Na_3H(CO_3)_2 \cdot 2H_2O$ and $Na_2CO_3 \cdot H_2O$. These minerals formed due to the interaction between fuel-containing masses or "lava", water and air. The matrices of the "lava" do not contain significant amounts of sodium. The source of sodium may be water that has penetrated into the "Sarcophagus". All identified secondary minerals of uranium are highly unstable, and their continued formation can seriously endanger the radiological situation of the 4th Unit.

INTRODUCTION

The accident at the fourth unit of Chernobyl Nuclear Power Plant on 26 April 1986 resulted in the destruction of the reactor core [1]. The interaction of heated fuel with structural materials led to the formation of a highly active silicate melt or Chernobyl "lava", which flowed under reactor compartments and solidified [2]. It is supposed that significant amounts (up to 23-56 t [3]) of the 4th unit nuclear fuel are present in the "lava". Uranium is present in the "lava" in two main forms:

1) dissolved uniformly in a glass-like matrix of the "lava" and
2) in the form of solid inclusions of uranium phases UO_x, UO_x with Zr, Zr-U-O, $(Zr, U)SiO_4$ etc. [4, 5].

The total uranium content varies with "lava" type. For example, in brown "lava", which is localized in a steam discharge corridor, the uranium content reaches 10 wt %, whereas black "lava" of so-called "elephant foot" contains about 5 wt % uranium. It is important to note that most of the uranium in the brown "lava" occurs as inclusions; and in contrast, the uranium in the black "lava" is predominantly dissolved in the matrix.

The silicate matrix of the "lava" is a vitreous material containing silicon (35-38 wt %) and the following elements (in wt %): Al (3.8-4.0), Ca (7.2-8.2), Mg (1.3-4.5), K (2.3-2.7), Zr (2.9-4.0), U (2.0-3.2), Fe (0.2-6.7).

The "lavas" are still heated by radioactive decay, and even in open areas not covered by concrete, the "lava" temperature is somewhat higher than that of the environment.

There is no doubt that the "lava" is not a chemically stable material. In spite of the erection of a "Shelter" or "Sarcophagus" over the 4th unit ruins, the "lava' is not isolated from surrounding environment. Even rain water can penetrate inside the "Shelter" through gaps in the roof [6].

In 1990 in the steam discharge corridor so-called "yellow stains" were discovered on the surface of black "lava" , those were not observed before in any type of "lava". These new formations attracted our attention as a probable product of "weathering" of the Chernobyl "lava".

Mat. Res. Soc. Symp. Proc. Vol. 465 © 1997 Materials Research Society

COMPOSITION OF SECONDARY URANIUM MINERALS

The SEM (scanning electron microscopy) study of the yellow products of alteration has shown that they are aggregates of acicular crystals (Fig. 1).

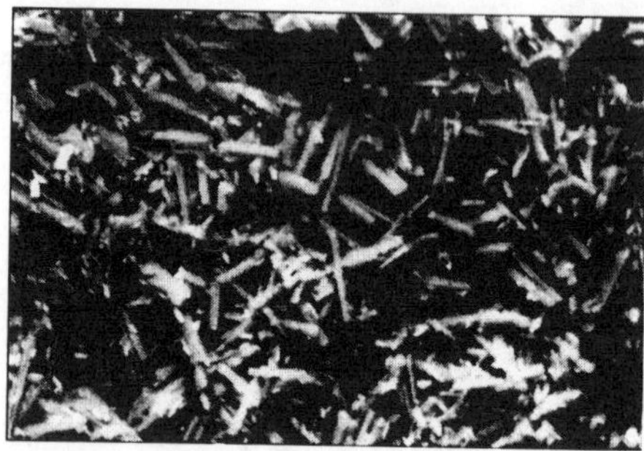

Fig. 1. SEM image of secondary uranium minerals on the surface of Chernobyl "lava".

30 μm

Chemical analysis of the yellow alteration products shows that uranium is present as the only heavy element; its total content is about 13 wt %. In some cases, the EDS spectrum shows an intense peak that corresponds to sodium. X-ray diffraction analysis showed that the yellow alteration products are polyphase (Fig. 2). The crystalline material contains the following phases (wt %, error ±40 rel. %):

- $Na_3H(CO_3)_2 \cdot 2H_2O$	~35
- $UO_3 \cdot 2H_2O$ - epiianthinite	~25
- $Na_4(UO_2)(CO_3)_3$	~12
- $Na_2CO_3 \cdot H_2O$	~15;

two other phases were also found:
- $UO_4 \cdot 4H_2O$ - studtite
- UO_2CO_3 - rutherfordine.

CONCLUSION

The results obtained allow us to conclude that:
1. At present the Chernobyl "lava" interacts actively with its environment, and secondary uranium minerals are forming on its surface.
2. One of principal causes of secondary uranium mineral formation is the contact of the "lava" surface with sodium-containing aqueous alkaline solutions whose origin is unknown.
3. Uranium phases formed as a result of alteration of the "lava" are unstable, and some dissolve in water. In the future, this process may seriously effect the radiation conditions and situation inside the "Shelter".

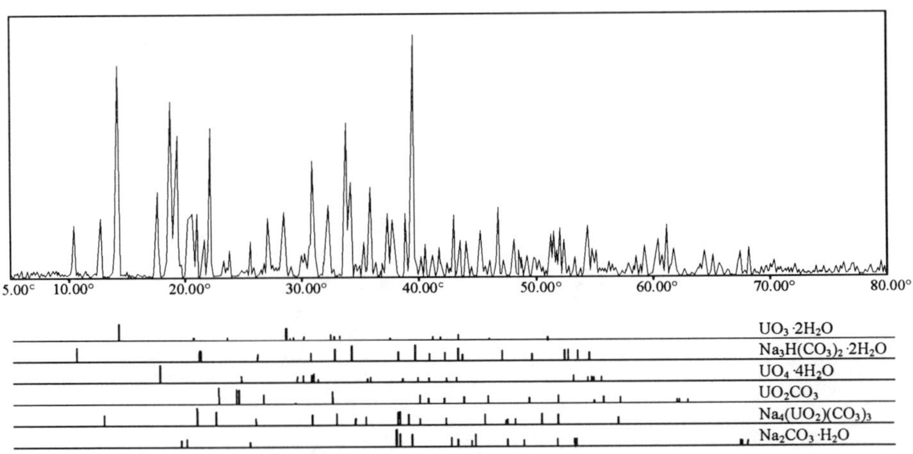

Fig. 2. X-ray diffractogram of "yellow stains" on the surface of Chernobyl "lava".

REFERENCES

1. Information about the accident at the Chernobyl Electric Power Station and its consequences: Prepared for IAEA. At. Energ. **61**, 301 (1986) (in Russian).
2. Borovoy A.A., Galkin B.Ya., Krinitsyn A.P., Markushev V.M., Pazukhin E.M., Kheruvimov A.N., Checherov K.P. New-formed products of the interaction of fuel with structural materials of the 4th Unit of the Chernobyl NPP. Radiochem. **32** (1990). (in Russian).
3. Kiselev A.N., Nenagliadov A.Yu., Surin A.I., Checherov K.P. A study of fuel-containing masses on the ChNPP 4th unit, (by results of investigations of 1986-1991). Preprint IAE-5533/3, Moscow (1992). (in Russian).
4. Anderson E.B., Burakov B.E., Pazukhin E.M. High uranium zircon $(Zr,U)SiO_4$ from Chernobyl lavas. Radiochimica Acta **60**, 149 (1993).
5. Burakov B.E., Anderson E.B., Galkin B.Ya., Pazukhin E.M., Shabalev S.I. Study of Chernobyl "hot" particles and fuel containing masses: implication for reconstructing the initial phase of the accident. Radiochimica Acta **65**, 199-202 (1994).
6. Borovoy A.A. Inside and outside the "Sarcofagus". Preprint KE IAE, Chernobyl (1990). (in Russian).

INTERACTION OF UO$_2$ AND ZIRCALOY DURING THE CHERNOBYL ACCIDENT

S.V. USHAKOV, B.E. BURAKOV, S.I. SHABALEV, E.B. ANDERSON
V.G. Khlopin Radium Institute, 28, 2-nd Murinsky ave., 194021, St.Petersburg, Russia,

ABSTRACT

A summary of the results collected during the studies of the products of a chemical interaction between uranium oxide fuel and Zircaloy cladding in the Chernobyl accident is presented in this paper. The reaction products are mainly Zr-U-containing phases with different U/Zr ratio and are described on the basis of electron microprobe and X-ray diffraction (XRD) analyses. The Zr-U-bearing phases were discovered among the inclusions in different types of Chernobyl fuel-containing masses ("lava") inside the destroyed 4th Unit and in hot particles collected up to 12 km from the 4th Unit along the West Plume. A correlation of data on the chemical composition and phase interrelations obtained in investigated samples with a phase diagram of Zr(O) - UO$_2$ shows, that a temperature >1900 °C was reached in a part of the core before the explosion. The detection of hot particle with segregated morphology points out that liquid immiscibility existed between U-rich and Zr-rich melts. This and other observations indicate that the core temperature locally was above 2400-2600°C.

INTRODUCTION

The Chernobyl accident probably is the worst fuel damage accident in the history of nuclear power. Based on the accident's chronology, the time interval did not exceed one minute [1] between the assumed beginning of the core's temperature rise above the normal, a 100-fold leap of the reactor's power and the subsequent explosions. The explosions destroyed and unsealed the reactor core; a part of its material was dispersed and thrown out beyond the boundaries of the reactor unit (hot particles) [1]. A part of the remaining material later reacted with the structural materials with a formation of a melt ("lava") [3]. Thus, the temperature and time conditions of fuel interaction with the Zircaloy cladding in the Chernobyl accident considerably differed from both the simulated experiments under Power Cooling Mismatch (PCM) conditions [2], and the conditions at the TMI-2 accident [5]. Thus, an analysis of the processes that took place is necessary for both a reconstruction of temperatures reached during the Chernobyl accident and a deeper understanding of the kinetics of uranium dioxide-zirconium reactions.

SAMPLE CHARACTERIZATION

The Zr-U-bearing phases were discovered among the inclusions in different types of Chernobyl fuel-containing masses ("lava") inside the destroyed 4th Unit and in hot particles collected up to 12 km from the 4th Unit along the West Plume. The separation of the hot particles from the soil samples was performed in heavy liquids and using radioactivity for select hot particles from the concentrates. The products of fuel-Zircaloy interaction observed as the inclusions in "lava" were examined as observed in the matrices of the "lava" as well as in separated samples made by dissolving the "lava" in HF acid. The appearance of Zr-U-containing hot particles and results of XRD analysis of the Zr-U-containing inclusions in the "lava" indicate that they are products of a rapid congealing of the heterogeneous Zr-U-O melt.

Mat. Res. Soc. Symp. Proc. Vol. 465 © 1997 Materials Research Society

<u>Zr-U-bearing phases observed in hot particles</u>

As a result of these investigations, concentrates of hot particles were separated from 11 soil samples at distances 1-12 km from the Chernobyl Nuclear Power Plant along the axis of the West Plume. For most samples the radioactivity level was used as a criterion of particles selection from the separated heavy fraction.

In the concentrates obtained the Zr-U-containing particles accounted for 10 to 45% of the total amount of hot particles. The remaining part was represented by fuel particles with block morphology characteristic of irradiated fuel. The size of the examined particles (from 10 to 200 μm) and their distance from the 4th unit allow the conclusion that the given particles were thrown out as a result of the explosions and, hence, did not undergo possible changes connected with interaction with structural materials after the accident. For detailed investigations using scanning electron microscopy and quantitative energy-dispersion analysis, Zr-U-containing particles with the most complex phase composition were selected.

The quantitative analysis of the hot particles and the Zr-U-containing inclusions in the "lava" was done on a microprobe "Camscan" with a use of an EDS extension block Link-10000. The content of oxygen was calculated by difference. The results of the quantitative analysis are presented in Table I of several hot particles that are the most informative for revealing the conditions of interaction between the fuel and the cladding. For phases with significant variations in the chemical composition, there are the maximum and the minimum values of the U/Zr atom ratio in Tables I and II.

Table I. The chemical composition of Zr-U-containing phases in some hot particles.

Hot particles	Zr wt%	U wt%	O wt%	U/Zr atom ratio	Phases
B-1	33.0	46.7	20.3	0.49-0.62	Zr-U-O
	22.1	59.8	18.1	1.04-1.03	U-Zr-O
S-1	-	89.8	10.2	-	UO_2
	5.2	79.5	15.3	5.69-6.15	$(UZr)O_{2-x}$
	10.1	72.7	17.2	2.76	U-Zr-O
RL-3	50.9	28.8	20.3	0.22	Zr-U-O
RL-2	17.3	73.7	9.1	1.58-1.69	U-Zr-O
	21.0	66.3	12.6	1.17-1.25	U-Zr-O

By the composition and morphology one can separate three Zr-U-containing phases in the hot particles: Zr-U-O, U-Zr-O* and $(U,Zr)O_{2-x}$. The morphology of the Zr-U-O and U-Zr-O phases indicates, that they are rapidly congeals melts (Fig. 1). In association with the above phases in the hot particles there were found drop-like inclusions of metallic phases, similar in composition to the steel of the spacing lattices of fuel assemblies (Fig. 1C). In a few samples there was found a particle consisting of ZrO_2 in contact with the phase U-Zr-O and a particle with the matrix Zr(O) containing inclusions with the metallic zirconium-uranium phase [4].

---

* In the previous papers, such as in reference 4 phases U-Zr-O and $(U,Zr)O_{2-x}$ were not distinguished and named as UOx with Zr.

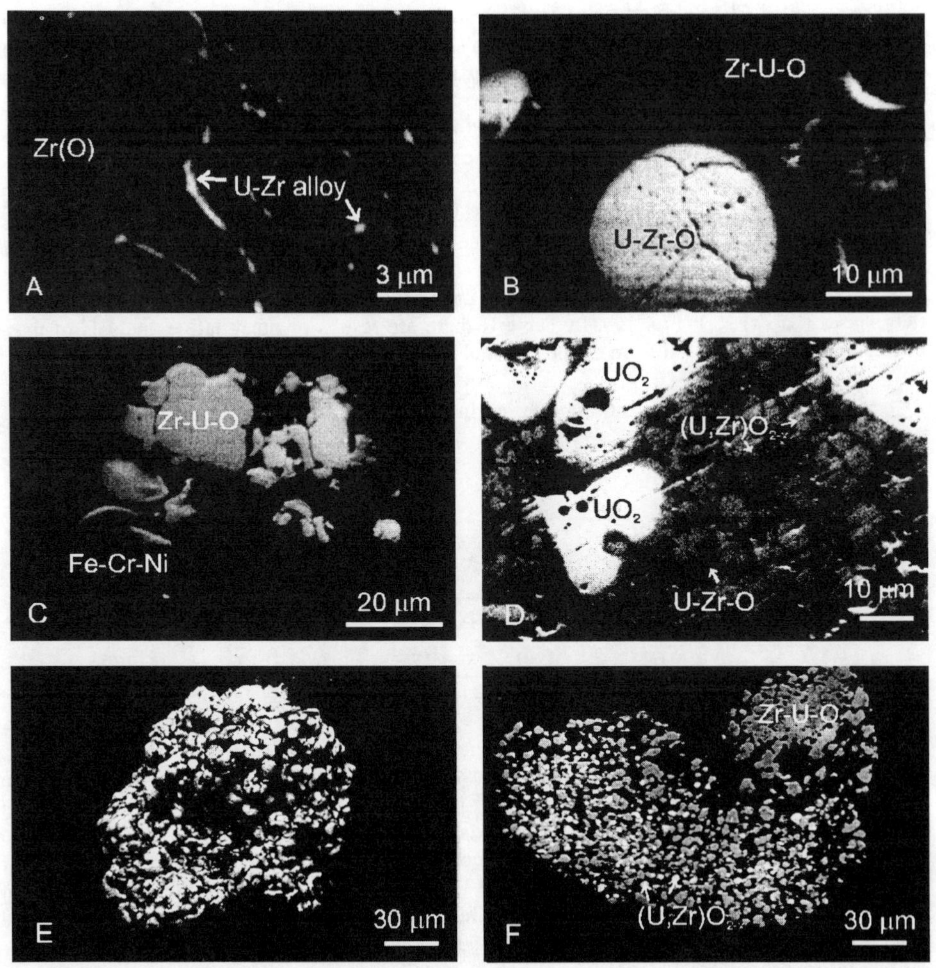

Fig. 1. SEM backscattered images of Zr-U-containing hot paricles and inclusions in Chernobyl "lava" A) particle 2-8 ; B) particle B-1 ; C) particle RL-3; D) particle S-1; E),F) inclusions in the brown "lava", separated from concentrate.

<u>Products of UO$_2$ and Zircaloy interaction in Chernobyl "lava".</u>

The so called Chernobyl "lava" is a congealed silicate melt [3] with numerous inclusions of the Zr-U-containing phases, particles of molten steel, high-uranium zircon [3,4] etc. By the color and structure three varieties of fuel-containing masses are distinguished: a black and a brown "lava" and a porous ceramic.

To study the inclusions in various types of fuel containing masses, concentrates were studied obtained at the Radium Institute by means of a partial dissolution of "lava" pieces in hydrofluoric acid. The Zr-U-containing inclusions account for about 20% in concentrates of porous and brown varieties of the lava. Mainly they represent cluster globular aggregates up to 200 μm in size (Fig. 1E). By the results of X-ray phase analysis these inclusions are amorphous in the main mass and contain some amount of a monoclinic modification of zirconium dioxide and Zr_3O_{1-x}. Usually their matrix consist of a zirconium-rich phase (Zr-U-O) in which there are inclusions of uranium-rich phases $((U,Zr)O_{2-x})$ (Fig. 1F). In Table II there are results of quantitative analysis of these phases in individual inclusions in the porous and brown varieties of the fuel-containing masses.

Table II. The composition of phases in aggregates found as inclusions in "lava".

Type of "lava"	Zr wt%	U wt%	O wt%	U/Zr atom ratio	Phases
Brown "lava"	6.2	78.8	15.0	3.84-6.33	$(U,Zr)O_{2-x}$
	65.7	18.3	15.9	0.10-0.11	Zr-U-O
Porous ceramic	7.1	80.4	12.5	4.06-4.69	$(U,Zr)O_{2-x}$
	62.9	19.7	17.3	0.11-0.13	Zr-U-O

RESULTS AND DISCUSSION

Hoffman, conducting out-of-pile experiments showed that the process of a partial reduction of UO$_2$ by Zircaloy reveals itself beginning from 1100 °C in the formation of the following succession of reaction layers: $[UO_2+U] \rightarrow [\alpha\text{-}Zr(O)+U,Zr] \rightarrow U,Zr$ alloy $\rightarrow \alpha\text{-}Zr(O)$ [2]. In studies of our samples, such structures were not found. However, we discovered a particle of Zr(O) with inclusions of veins and globules of U,Zr alloy (Fig. 1A) [4], of which suggests that the given process could take place. Hence, during the Chernobyl accident the occurrence of the first portion of liquid phases was possibly at 1400°C as a result of melting of U,Zr alloy.

During investigations of TMI-2 core debris samples [5], it was shown, that the U/Zr atom ratio can serve as an indicator of melt temperatures, since the oxygen concentrations may change due to a subsequent steam oxidation. Calculations of oxygen content carried out by us by difference from EDS analysis data also introduces a significant error into its determination. Taking into account these circumstances, we used the U/Zr atom ratio in the analyzed phases for a correlation of the obtained data with the phase diagram Zr(O)-UO$_2$ [6] (Fig. 2). It should be noted, that this diagram only approximates to the Zr-U-O melt condition in the Chernobyl accident, since: i) the equilibrium of the melt may be unattainable; ii) the oxygen content may considerably vary due to the processes of oxidation of the Zircaloy by steam and a possible presence of metallic uranium. iii) an impurity of niobium (1 wt%) in the Zircaloy composition may cause a shift of equilibrium fields, as well as an impurity of iron from the steel of spacing lattices.

The U/Zr ratio in the Zr-U-containing phases in the porous and brown varieties of Chernobyl "lava" indicates the temperature of the initial Zr-U-O melt as 2000-2200 °C. The U/Zr ratio in the phases of hot particles gives a temperature range 2200-2760 °C. Criteria for evaluation of temperature by the $Zr(O)-UO_2$ phase diagram and the temperature ranges for specific particles are discussed below.

In Fig. 2 the points A-A` and B-B` represent the composition of the Zr-U-O and $(U,Zr)O_{2-x}$ phases in aggregates from the brown "lava" and from the porous ceramics, respectively. The phases Zr-U-O in the investigated inclusions (A and B) are close by the composition to the eutectic point of this system, that corresponds to a temperature 1900 °C. However, the compositions of phases $(U,Zr)O_{2-x}$ (A` and B`) coexisting with them testify that a temperature of 2000-2200 °C might have been reached by the melt that was the origin of these particles.

The composition of coexisting Zr-U-containing phases in some hot particles point to higher temperatures. The points C-C` in Fig. 2 correspond to the composition Zr-U-O and U-Zr-O phases in the particle b-1. The U-Zr-O phase in the particle b-1 is in the form of an ideal round inclusion in the Zr-U-O phase and has with it a common system of cracks (Fig. 1B). This is a proof of a liquid immiscibility between them in the initial melt. In the system $Zr(O)-UO_2$ the liquid immiscibility between Zr-rich and U-rich melts is presumed beginning from 2400 °C [6].

Unfortunately, this high-temperature region of the given system is poorly studied experimentally, which does not allow to determine more precisely the temperature range of the melt. This particle can also serve as an "experimental" proof of the existence of liquid immiscibility in the U-Zr-O system.

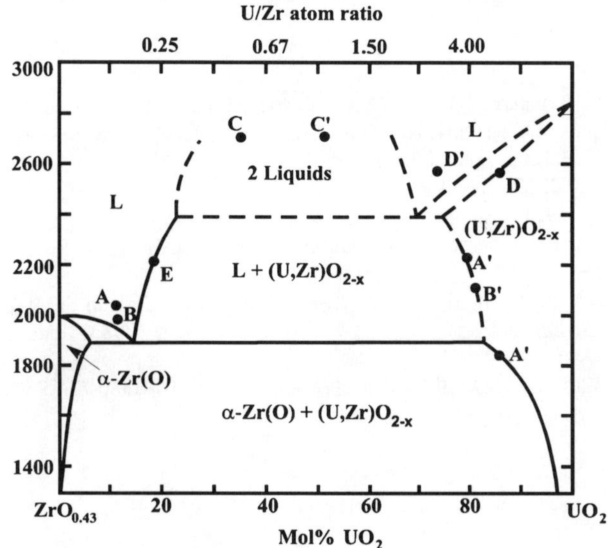

Fig. 2. $Zr(O)-UO_2$ pseudo-binary phase diagram [6]. See text for labelling.

Also one may attribute to the high-temperature region the particle S-1 (Fig. 1D) in which three coexisting phases are observed: UO_2, $(U,Zr)O_{2-x}$ and U-Zr-O. The morphology of the UO_2 inclusion does not allow one to determine if it underwent melting. The composition of the phases $(U,Zr)O_{2-x}$ and U-Zr-O (in Fig. 2 the points D-D`, respectively) also indicates a temperature > 2400 °C. Therefore, for this particle, the temperature range may be determined as 2400-2760 °C.

The particles RL-2 and RL-3 represent congealed melts U-Zr-O and Zr-U-O, respectively. From the U/Zr atom ratio, the lower temperature limit of the RL-3 particle's formation may be determined as 2200 °C (point E). Obviously, at a temperature >2400 °C the formation of mainly U-Zr-O or Zr-U-O melt determined by the mass ratios UO_2/Zr. A more complete interpretation of obtained data is possible after an experimental determination of the high-temperature part of the ternary diagram Zr-U-O.

CONCLUSIONS

The investigations of the Zr-U-containing phases from the Chernobyl "lava" and hot particles allows us to draw the following conclusions about the UO_2-Zircaloy interaction in the Chernobyl accident.

1. Before the explosion, in a part of the core the temperature exceeded 2000 °C, which caused an active interaction of the UO_2 fuel with the Zircaloy cladding, with formation of a Zr-U-O melt and phases of the compositions $(U,Zr)O_{2-x}$.

2. In local areas the temperature exceeded 2400-2600 °C, which caused melting with formation of two immiscible melts: U-Zr-O and Zr-U-O.

3. All the products of interaction of the fuel with the cladding underwent a rapid congelation.

REFERENCES

1. Publication series on safety N 75-INSAG-1, Proceedings of Conference on Studies of Causes and Consequences of Chernobyl Accident. IAEA, STI/PUB/740, Wien (1988).
2. P. Hofmann., C. Pollitis, J. Nucl. Mater. **87**, 375 (1979).
3. A.A. Borovoy, B.Ya. Galkin, A.P. Krinitsyn, V.M. Markushev, E.M. Pazukhin, A.N. Kheruvimov, K.P. Checherov, Radiochem. **32**, 103 (1990).
4. B.E. Burakov., E.B. Anderson, B.Ya. Galkin, E.M. Pazukhin, S.I. Shabalev, Radiochimica Acta **65**, 199 (1994).
5. D.W. Akers, E.R. Carlson, B.A. Cook, S.A. Ploger, J.O. Carlson, TMI-2 core debris grab samples examination and analysis Part 1, GEND-INF-075, EG&G Idaho, Inc. Idaho Falls, Idaho 83415, (Sept.1986) p. 73
6. A. Skokan, in Proc. on Thermal Nuclear Reactor Safety, KfK-3880/2, (1984) pp.1035-1041

PROPERTIES AND GENESIS OF HOT PARTICLES FROM THE CHERNOBYL REACTOR ACCIDENT

Peter Schubert-Bischoff[1], Werner Lutze[2], and Boris E. Burakov[3]
1, Hahn-Meitner -Institut Berlin, Glienicker Strasse 100, 14109 Berlin, Germany;
2, University of New Mexico, Department of Chemical and Nuclear Engineering,
Albuquerque, NM 87131, USA, 3, Khlopin Radium Institute, St. Petersburg, Russia

ABSTRACT

On April 25, 1986, the nuclear reactor Unit 4 (RBMK) at Chernobyl, Ukraine, exploded. Besides molecular species, the fallout contained particles of relatively high specific activity (hot particles) with a wide range of chemical compositions. The composition of a hot particle bears information about its genesis. Particle sizes ranged from a few to 100s of micrometers. Data on a hot particle, found in Berlin, Germany, is presented and discussed in context with earlier measurements on other particles to understand their genesis. The chemical composition was determined by electron probe micro analysis. Our particles are either reactor fuel (one) or fission product alloys (nine). The alloys were formed during normal reactor operation. Strongly varying concentrations of Fe and Ni suggest that at least some of our particles reacted with molten structural material of the reactor. The particles were mobilized by fuel oxidation or fuel dust generation during the accident. The fission product composition can only be explained if we assume that the alloys remained in the solid state in the course of the accident. Some particles may have been ejected during the explosion, others later while the reactor was burning. Activities (^{103}Ru and ^{106}Ru, originally up to 160,000 Bq) of our ten year old particles were re-measured but were no longer detectable. No long-lived γ-emitters were found. The ^{99}Tc activity was calculated and found to only 1Bq. The γ -spectrum of the fuel particle still shows ^{137}Cs (1 Bq) and ^{60}Co (<1 Bq).

INTRODUCTION

Ten years ago, on April 25, 1986, the nuclear reactor Unit 4 (RBMK) at Chernobyl, Ukraine, exploded. The containment and the reactor building were destroyed. An enormous amount of gaseous and solid radioactive materials, corresponding to 10^{18} Bq to 10^{19} Bq, were released including 1.7×10^{18} of ^{133}Xe and 3.3×10^{16} Bq of ^{85}Kr [1]. Burning graphite carried nuclides higher than 1,000 m into the air followed by wide-range atmospheric distribution over Europe.

Besides molecular species, the fallout contained particles of relatively high specific activity (hot particles). Their size ranged from a few to 100 micrometers. As soon as the accident became known, plant, soil, and water samples were collected in Western European countries to determine the extent of regional fall-out. As an example, vegetables coming from Eastern European countries into then West-Berlin, Germany, were sampled and counted day and night at the border and returned or dumped if certain activity levels were exceeded. The authors 1 and 2 were part of the counting teams at the border.

The first hot particles were found on aerosol filters in Sweden on the third day after the accident [2]. Soon thereafter hot particles were found in various countries in soils, plants and on air filters. There are various types of hot particles. Anderson et al. [3] classified hot particles found in the Chernobyl area by their chemical composition. They identified three types of particles: (1) UO_x fuel, (2) UO_x fuel with Zr, Zr-U-O and Fe-Cr-Ni alloy , and (3) a mixture of graphite, ZrO_x , steel with radionuclide adsorbed.

Broda [4] analyzed 65 particles found in Poland for their radionuclide composition and discussed the possibility of metal formation by reduction of volatile oxides, for example RuO_x, by the burning reactor graphite. Balashazy et al. [5] found hot particles on an air conditioning filter in Budapest and calculated a concentration of 5×10^{-5} particles per m^3 of air. Osuch et al [6] compared

Mat. Res. Soc. Symp. Proc. Vol. 465 © 1997 Materials Research Society

the radionuclide composition of 206 hot particles from northeastern Poland with that of the general fallout and discussed the mechanism of depletion of Ru and Cs in the hot particles. Saari et al. [7] distinguish between three different types of particles, collected from pine needles in Finland, according to their radionuclide composition as measured by γ-spectroscopy. Uranium was the main constituent of the particles and most of them contained ^{141}Ce, ^{144}Ce, ^{95}Zr, and ^{95}Nb (type 1). Some particles contained these nuclides and additionally ^{103}Ru and ^{106}Ru (type 2). The ^{134}Cs, ^{137}Cs content was low. The third type contained only ^{103}Ru and ^{106}Ru. The resemblance of type 1 and 2 composition with the reactor fuel composition was pointed out. Mandjoukov et al. [8] conducted α, β, γ-activity measurements and determined the elemental composition of their hot particles, which were identified as similar to the Chernobyl reactor fuel. Sandalls et al. [9] performed an evaluation of hot particles in terms of migration in soils and plant uptake and conclude that these may be important factors in land reclamation. They suggested that hot particle formation should also be taken into consideration in nuclear reactor contingency planning. Kerekas et al. [10] report on two types of particles found on the ground surface in Sweden and calculated a particle density of 0.03-0.07 particles per m^2 for the scanned area of 470 m^2. Their particles were similar in composition to the ones reported in this paper: reactor fuel and metallic particles, respectively. Their hot particle activity was also comparable with ours and was on the order of tens of KBq, whereas many of the other authors quoted here reported lower particle activities. Tcherkezian et al. [11] collected hot particles, mostly resembling the reactor fuel composition, within the 30 km zone around the destroyed reactor and reported that these particles contributed not less than 65% of the total radioactivity on the ground. Kashparov et al. [12] modeled the formation of hot particles experimentally by heating unaffected nuclear fuel from the Chernobyl reactor and measuring the thermal destruction resulting from oxidation and fracturing. They investigated the sequence of selective evaporation of fission products and the re-condensation and formation of hot particles of mostly Ru on steel surfaces. Burin et al. [13] concluded from their analyses of hot particles collected in Bulgaria that there are two types of hot particles, reactor fuel and particles that originated from chemical and thermal interaction of the fuel with other reactor and building materials.

This paper was written as a contribution to the commemoration of the tenth anniversary of the reactor accident at Chernobyl in the former USSR. Part of the technical information presented here has been published by Schubert and Behrend [15] nine years ago. In the meantime additional measurements were conducted and new insight into the genesis of hot particles was gained based on work conducted by our Russian co-author.

EXPERIMENTAL

Ten particles were found, nine in the Masurian lake region in northeastern Poland and one in Berlin, Germany. The particles were detected in the field with the help of a Geiger Mueller (GM) counter. In the laboratory, they were separated from mud, always using activity as the indicator of the presence of the tiny particles during processing. All manipulations of the particles including their transfer to a sample holder were conducted using a GM counter and an optical microscope.

The morphology of the samples was investigated and their sizes determined by scanning electron microscopy. The samples were analyzed quantitatively for their radionuclide and chemical composition. Radioactivity measurements were conducted in 1986 and in 1996. A Ge/Li solid state detector and a multi-channel analyzer were used to measure γ-emitting radionuclides. α and β emitters were not analyzed. Count rates were converted to activities and calculated for the day of the reactor accident. The chemical composition of the hot particles was determined by nondestructive methods using wavelength- and energy-dispersive electron probe microanalysis (EPMA, EDX).

One particle was analyzed for oxygen. To distinguish between an oxidized surface layer and an oxide, the particle was polished and etched to remove the surface layer. Oxygen was then detected qualitatively by EPMA, measuring the K$_\alpha$X-ray emission.

RESULTS

Size and morphology of the hot particles

Sizes of the particles ranged between 4 µm and 14 µm. Nine particles had a spherulitic shape, one (HS6, table 1) was more fragile than the others upon sample preparation. The morphology of the hot particles HS1, HS6 and HS10 are shown in figure 1a, c, and e, respectively. Figs. 1a and 1c are adapted from [15]. Particle HS10 is the one found in Berlin and has not been studied before. Results on HS1 and HS6 have been reported in reference [15]. As can be seen in figure 1 particles HS1 and HS10 have a spherulitic shape, HS6 does not. Particle HS10 is almost a perfect sphere. The properties of particle HS10 will be compared with HS1 and HS6 in terms of activity and chemical composition.

Radioactivity of hot particles

The results of the γ radioactivity measurements of hot particle from Poland and Berlin, Germany, are compiled in table 1.

Table 1: γ activities of hot particles

Sample No.	^{103}Ru/KBq		^{106}Ru/KBq		^{99}Tc/Bq**	^{137}Cs/Bq	^{60}Co/Bq
	1986	1996	1986	1996	1986/96	1986/96	1986/96
HS1 (Pol)*	139.00	<DL	28.00	<DL	≤1	-	-
HS2 (Pol)*	4.99	<DL	1.05	<DL	≤1	-	-
HS3 (Pol)*	28.10	<DL	4.96	<DL	≤1	-	-
HS4 (Pol)*	23.20	<DL	4.54	<DL	≤1	-	-
HS5 (Pol)*	13.00	<DL	2.79	<DL	≤1	-	-
HS6 (Pol)*	46.80	<DL	5.81	<DL	-	≈1	≈1/ <1
HS7 (Pol)*	34.40	<DL	5.45	<DL	≤1	-	-
HS8 (Pol)*	41.70	<DL	8.73	<DL	≤1	-	-
HS9 (Pol)*	32.10	<DL	7.27	<DL	≤1	-	-
HS10 (Ber)	3.30	<DL	0.69	<DL	≤1	-	-

Values for 1986 marked (*) were taken from the previous work [15]. <DL values indicate that the samples were re-measured in 1996 and that the radioactivity of ^{103}Ru and ^{106}Ru have decreased below the detection limit (DL) of our instrument. The measurement of the hot particles in 1996 revealed that no long-lived γ-emitters are contained in the respective particles that might have gone undetected in the presence of the Ru isotopes. Values marked (**) indicate that an upper limit of the ^{99}Tc activity was calculated in 1996. The hot particles, except HS6, can be considered as practically non-radioactive. They do no longer contribute to the contamination of land surfaces affected by the Chernobyl reactor fallout.

Chemical composition of hot particles

EDX spectra of the particles HS1, HS6 and HS10 are shown in figure 1b,d, and f. As seen in figure 1b, HS1 consists of Ru, Fe, Ni, Mo, Tc, and Rh. Particle HS6, figure 1d, contains mainly U and Fe. Oxygen was also detected. Analysis of the whole particle by X-ray mapping [15] revealed that elements found in HS1 and HS10 were also present but as small spots within the U,Fe matrix. The composition of HS10 is similar to particle HS1. It consists of Ru, Ni, Tc, Rh, and Pd. There are distinct differences between HS1 and HS10. Whereas HS10 does not contain Fe and little Mo, HS1 does not contain Pd. However, the presence of Pd is not unique. Particle HS4 from Poland [15], for example, does contain Pd. There were even particles almost free of Fe and Ni. Hence particles of different compositions were deposited in the same place. Particle HS1 was analyzed quantitatively by EPMA and so were some of the other particles, except HS6 [15]. The sum of the relative compositions of different particles varied between 92 and 98 weight percent,

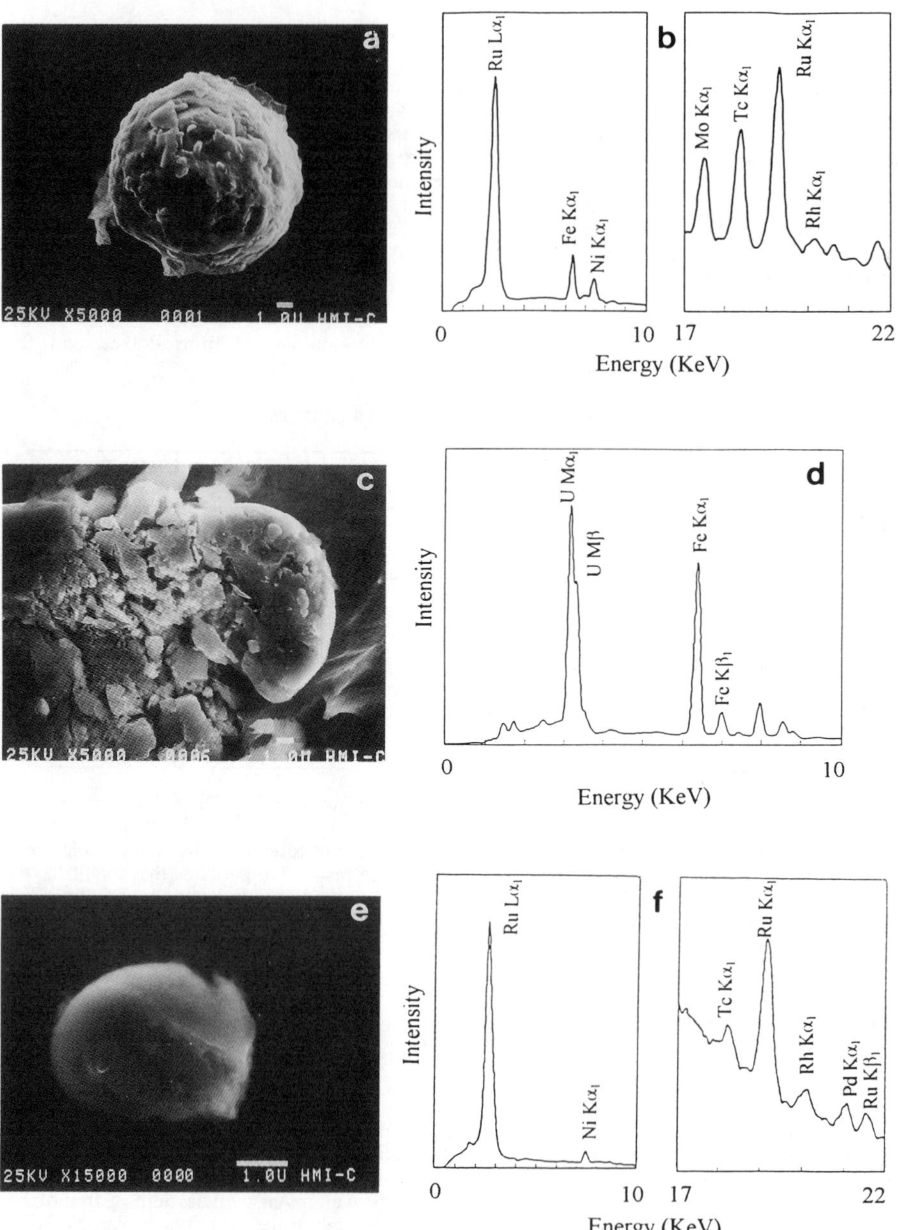

Fig. 1: Hot Particle HS1, HS6 and HS10 shown in a), c), e), respectively.
EDX spectra are shown in b), d) and f).

suggesting that these particles are alloys of the respective metals, rather than oxides. Direct oxygen measurements confirmed the metallic nature of the respective particles. Particle HS10 was analyzed using the same procedure as described in reference [15]. The relative concentrations of the metals in HS10 add up to 94.1 weight percent, suggesting that HS10 is a metallic particle as well. The quantitative compositions of the hot particles HS1 and HS10 are given in table 2.

Particle HS6 could not be analyzed quantitatively, but the qualitative chemical composition and the presence of some ^{137}Cs suggests that this particle originates from the reactor fuel. This explains its lower mechanical durability compared with the alloys. A high Fe and Ni content was found in this particle.

Fuel particles contributed considerably to the contamination of the ground surface within the 30 km perimeter of the reactor [11] and will continue to do so. Fuel particles found by Tcherkezian et al. [11] were larger than HS6 and contained higher levels of activity other than that from Ru isotopes.

Table 2: Chemical composition of the hot particles HS1 and HS10.

element	particle HS1 wt%	particle HS10 wt%
Fe	9.8	-
Ni	7.1	2.9
Mo	9.1	2.1
Tc	17.7	6.9
Ru	42.2	47.1
Rh	8.1	16.1
Pd	3.7	17.5
Σ	97.6	94.1

(*) from reference [15]

Table 2 shows that both particles contain technetium. This element must be present as ^{99}Tc ($E_\beta = 0.3$ MeV, $T_{1/2} = 2.1 \times 10^5$ y). However, the small amount of Tc in one particle provides a β-activity of less than 1 Bq that is of no concern. The size of the particle investigated in this work and the sizes reported in [15] are compatible with the sizes of metallic particles typically found in spent reactor fuel [14]. It was speculated in [15] that the metallic particles were entrained with the stream of material released from the reactor into the atmosphere. However, the presence of Ni and Fe in the hot particles is not typical of those found in spent fuel. This suggests that the metallic particles may have reacted with other reactor components prior to their release.

DISCUSSION

Most of our particles found in Poland and the one in Berlin are alloys. One particle from Poland is reactor fuel. The different chemical compositions of the particles may reveal information about details of the accident. To elucidate the behavior of the nuclear fuel shortly before and in the first days after the accident, Burakov et al. [16] collected samples of the exploded reactor in the vicinity and directly from solidified lava within the reactor building. Laboratory experiments were conducted by Burakov et al. [16] to study the interaction between the fuel (UO_2) and its cladding (Zr). The authors assumed that the accident can be characterized by three phases. Phase one lasted probably only seconds and ended in an explosion after extremely high temperatures were reached in the superheated reactor fuel. The second phase encompassed the first six days after the explosion. The third phase began after six days and ended on the eleventh day with a breakthrough of lava into rooms beneath the reactor core.

During phase one, parts of the fuel may have reached temperatures above 6000°C [17]. Burakov et al. [18] assumed that the material ejected upon explosion and shortly thereafter carry information about what was going on seconds before the explosion. Samples from this phase were collected on the ground along the main wind direction up to 12 km to the west of the reactor (the western plume). This was the most contaminated area at all (^{137}Cs $> 1.5 \times 10^{12}$ Bq/km^2).

Typical sizes of the hot particles exceeded 100µm. They consisted of UO_x and Zr-U-O phases, indicating a chemical reaction between the fuel and its zirconium cladding at a temperature above 2600°C. Metallic inclusion of Fe-Cr-Ni-U-Zr were also detected, suggesting that melting of structural material had occurred before the explosion. Particles of this size are unlikely to be transported very far and were most likely ejected by the explosion. Our particles from northeastern Poland and Berlin are much smaller, 4 - 14 µm. They do not contain Zr and therefore did not participate in reactions with the fuel cladding.

Emission of hot particles continued for about 11 days. At least 3.5% of the mass of the 190 tons of the UO_2 fuel were dispersed in the atmosphere and then deposited. In phase two, the fuel and its reaction products with Zr came into contact with other components such as steels and concrete. Melting processes supported homogenization and accelerated chemical reactions. This process led to the formation of lava and continued until May, 2nd. Burning graphite enhanced the chimney effect and enhanced further release of hot particles. The release of activity decreased slightly between the 26th of April and the 2nd of May from 4.5×10^{17} to 1×10^{17} Bq/day [1].

Phase three of the accident covers the period between May, 2nd and May, 6th. On May, 2nd, the release of activity increased again. The values reached 3×10^{17} Bq/day [19] by May, 6th, when the release dropped by an order of magnitude. Burakov et al. [16] assumed that isolated streams of lava formed a larger pool leading to an increase in temperature and finally to an escape and distribution of the lava pool through a hole in the ceiling. Cooling of the lava could explain the sudden drop in the release of activity.

The spectrum of hot particles emitted during phase two and three is expected to be different from those released during the explosion. Smaller particles are more likely to be carried by hot gases into the atmosphere and transported over larger distances. We assume that our particles were emitted during all three phases. But during phase two and three these particles were emitted preferentially if not exclusively. The genesis of our particles is not easy to explain. The fuel particle HS6 appears to be a mixture of Fe, Ni or their oxides and UO_x suggesting that the fuel was in contact with burning or molten steel, prior to its escape from the reactor. The missing Cr could have been oxidized [19]. A quantitative analysis of this particle was not possible, but the content of Fe and Ni is substantial and not just an impurity in the fuel. Small spots ($< 5µm$) within the UO_x matrix consist of the fission product alloy typical of spent reactor fuel. Their presence is not surprising. Precipitates of this kind form in the reactor fuel during burn-up under normal operating conditions. The absence of zirconium in particle HS6 can be explained by insufficient heating ($< 2500°C$). Particles HS1 and HS10 have a qualitative composition close to that of the small spots found in particle HS6. HS1 and HS10 are much larger (10 - 15 µm) but are of the same origin as the spots in HS6. The particles could have been separated mechanically from the fuel, for example, during the explosion when fuel dust was formed. Oxidation of the fuel has also been suggested as a possibility to generate small particles [20].

Analytical results reported here and in [15], suggest that there are different ways for hot particles to escape from the destroyed reactor during phases two and three. The high concentrations of Fe (9.8%) and Ni (7.1%) in HS1, table 2, and the very low concentrations of Fe (0.2%) and Ni (0.2%) in HS7 and HS9 [15], can have entirely different reasons. (1) The fuel in the Chernobyl reactor may have been fairly inhomogeneous with respect to impurities. Trace amounts of steel were detected in UO_2 fuel [19]. Metallic Fe is a known grain-boundary impurity in UO_2 fuel and was detected in the fission product alloys [21]. Hence, different particles of fission product alloys could have different Fe and Ni concentrations. (2) Alternatively , impurities could have been low in the fuel. In this case, HS7 and HS9 represent the typical composition of the fission product alloys in the fuel. HS1 could have been in contact with steel at a temperature high enough for Fe and Ni to diffuse into this particle. Upon ejection and transport by hot gases, attached steel would oxidize after escape from the reducing environment, whereas the more noble metal alloy would not. To our knowledge, steel particles without fission products were not found.

The very high concentration of Ru (42 - 47%, table 2), comparable with values reported for spent fuel, supports the assumption that these particles were solid at any time. Otherwise, considerable oxidation and volatility losses of Ru should have occurred. The absence of Fe in HS10 is surprising but was encountered only in this particle. We exclude, that our multi-element particles formed by reduction of volatile oxides and condensation. Formation by condensation may have occurred with particles that consist of pure or almost pure Ru [4].

Independent of the process of escape from the destroyed reactor, the hot particles exhibit high concentrations of Ru and Tc. Both elements are more or less volatile depending on their oxidation state. We conclude that all our particles were emitted during all phases of the accident after the explosion. Their chemical composition provides some, but not enough, information to describe their genesis completely.

REFERENCES

1   Publication Series on Safety N 75-INSAG-1, Studies of Causes and Consequences of the Chernobyl Accident, Conf. Proceedings, IAEA, STI/PUB/740. ISB N-92-0-423088-6, Vienna (1988)
2. L. Devell et al., Nature 321, 192 (1986)
3. E. B. Anderson, B. E. Burakov, E. M. Pazukhin, Radiochimia 5, 139-144 (1992)
4. R. Broda, in report No. 1342/b, Institute of Nucl. Physics Krakow, Krakow, Poland (1986)
5. I. Balashazy, Radiation Protection Dosimetry 22(4), 263-267 (1988)
6. S. Osuch et al., Health Physics 57(5), 707-716, (1989)
7. H. Saari et al., Health Physics 57(6), 975-984 (1989)
8. I. G. Mandjoukov, K. Burin, B. Mandjoukova, and E. I. Vapirev, Radiation Protection Dosimetry 40(4), 235-244 (1992)
9, F. J. Sandalls, M. G. Segal, N. Victorova, J. Environm. Radioactivity 18(1), 5-22 (1993)
10. A. Kerekas, R. Falk, and J. Suomela, EUR-13574, Vol.1, 211-222 (1991)
11. V. Tcherkezian, V. Shkinev, L. Khitrov, and G. Kolesov, Journal of Environmental Radioactivity 22(2), 127-139 (1994)
12. V. A. Kashparov et al., Radiochemistry 36, 98-104 (1994)
13. K. Burin, Ts. Tsacheva, I Mandjoukov., and B. Mandjoukova, INIS-mf-14777, Conf. in Gyulechitsa Bulgaria, Sept. 6-10 (1993) - abstract only
14. J. I. Bramman, R. M. Sharpe, D. Thom, and G. Jates, J. Nucl. Mat. 25, 201 (1968)
15 .P. Schubert and U. Behrend, Radiochimica Acta 41,149-155 (1987)
16. B. E. Burakov et al., The Behavior of Nuclear Fuel in the First Days of the Chernobyl Accident, this proceedings volume
17. Report of the US Department of Energy's Team Analyses of the Chernobyl Atomic Energy Station Accident Sequence, DOE/NE-0076, DE 87003614, Washington 1986
18. B. E. Burakov et al., Radiochimica Acta 65, 199-202 (1994)
19. H. Kleykamp, KfK report No. 2696 (1979), Kernforschungszentrum Karlsruhe, Germany
20. V. A. Kashparov et al. Nuclear Technology 114(2), 246-253 (1996)
21. I. J. Hastings, H. D. Rose, and J. J. Baird, J. Nucl. Mat. 61, 229-231 (1976)

STABILITY OF MINERAL MATTER IN AQUEOUS MEDIA OF THE CHERNOBYL UNIT-4 SHELTER: THERMODYNAMIC EVALUATION

V.A.SINITSYN[*,***], D.A.KULIK[**], M.S.KHODORIVSKI[***], V.A.KUREPIN,[*]
A.Y.ABRAMIS[***], I.L.KOLYABINA[***], N.A.SHURPACH[*]
[*]Institute of Geochemistry, Mineralogy & Ore Formation, NAS Ukraine, 252180 Kyiv, Ukraine;
[**]State Scientific Centre for Environmental Radiogeochemistry, 252180 Kyiv, Ukraine;
[***]R&D Centre "META", Minchornobyl Ukraine, 255620 Chornobyl, Ukraine;

ABSTRACT

A special geochemical environment exists within the Shelter ("Sarcophagus") erected in 1986 over the destroyed Unit-4 of Chernobyl nuclear power plant (NPP). Based upon the available *in situ* and compositional data, thermodynamic models of solid-aqueous interactions were developed to clarify the leaching behaviour of various materials within the Shelter. The "Selektor-A" code, based on a convex programming approach to Gibbs free energy minimization, was used for the calculations. A built-in flexible hybrid thermodynamic database for the system Na-K-Ca-Mg-Cl-S-N-H-O-Si-P-Fe-Al-Sr-Cs was extended with the critically selected and matched parameters for aqueous species and solid phases in the U-Zr-Si-O-H subsystem, secondary U-minerals, mineral phases of fully hydrated Portland cements and U-bearing zircons. Modeling results show that the "Shelter waters" can selectively leach a significant quantity of U and Si from the fuel-containing masses, while Zr, Fe, Ca, Mg and some other components are rather insoluble. Serpentinite, assemblages of fully-hydrated phases of Portland cements, and oxidation products of steel structural elements are estimated to be sufficiently stable in the aqueous environment of the Shelter. Our calculations also define some feasible pathways for secondary mineral formation from evaporation of Shelter water solutions and interactions between these waters with the mineral matter inside the Shelter.

INTRODUCTION

The problem of the nuclear hazard of residual fuel-containing masses (FCM) from the destroyed Unit 4 of the Chernobyl nuclear power plant (NPP) has been investigated since 1986 [1,2]. But systematic research on chemical processes inside the Shelter were conducted only during the last few years [3]. The possibility for these processes to occur was caused by the chemical heterogeneity of the Interior of Chernobyl Shelter (ICS), by the presence of moisture in it, and by the existence of gas exchange between the environment and the ICS. Clearly, hydrogeochemical studies within the Chernobyl Shelter are crucial to both nuclear hazard and environmental impact assessment, as well as to solution of the practical problems of transforming the object into an environmentally clean state. The specific conditions inside the *Sarcophagus* (mainly, very high levels of radioactivity) impose serious restrictions upon the *in situ* and experimental investigations. Therefore, quantitative modeling is the most important method for studying the chemical interactions within the ICS. Aqueous speciation of major chemical components and radionuclides in Sarcophagus waters, as well as the degree of saturation of the water with respect to some solids and atmospheric gases, has been modeled recently [4]. The results of this modeling clarified important peculiarities of chemical state of the water and demonstrated that these waters should be regarded as chemically reactive with respect to the FCM and some other materials. The results of the thermodynamic modeling of water-solid materials interaction within the Chernobyl Unit-4 Shelter are presented in this work.

Mat. Res. Soc. Symp. Proc. Vol. 465 © 1997 Materials Research Society

ICS SYSTEM DESCRIPTION

The ICS is a complex heterogeneous system formed as a result of the accident processes (April-May, 1986), the activities conducted to mitigate the accident after-effects (May-December, 1986), and the activities required to control the Shelter state and spontaneous physicochemical processes in the ICS (1986 to present time).

ICS construction materials are poorly studied. Thus, for the major components, serpentinite (basement material in the destroyed reactor), and the concrete and steel structural elements of the reactor and building, the data on composition and physical parameters is practically unknown.

Fuel-containing masses (FCM) have been studied in more detail. In previous work [1,2], FCM were classified into three types: (1) unaltered nuclear fuel in the UO_2 form; (2) dispersed fuel in the form of "hot" particles up to 10-40 μm size. The composition of the "hot" particles varies depending on the percentage of UO_2 and different components of the structural materials (Fe, Zr, Si etc.) up to the pure UO_2; (3) Lava-type FCM, the matrix of which is a silicate glass (>65 wt.% of SiO_2), containing K, Ca, Mg, Al, U, Zr impurities with no more than 3-4% of each element. In the lava, matrix there are numerous inclusions of crystalline phases, such as uranium oxides containing Zr impurity, crystals of high uranium zircon (HUZ), and Zr-U-O phases with undefined composition, crystal structure and oxidation state of the metals. The HUZ uranium zircon contains up to 10 - 11% of uranium [2].

Secondary uranium phases. In previous work [2,6], the following newly formed uranium minerals are described: $Na_4(UO_2)(CO_3)_3 \cdot nH_2O$, $UO_3 \cdot 2H_2O$ and UO_2CO_3.

Waters (aqueous solutions). Abramis et al. [5] have performed a detailed study of the Shelter water composition. The waters are found to be mainly of chloride-bicarbonate, potassium-sodium, alkaline (pH 7.6 to 10.6) type, mildly reducing (*in situ* determined values of Eh are -83 to -112 mV) with total dissolved solids (TDS) 1 to 13 g/L. The analyzed samples were classified into two groups. In the waters of the group (I), bicarbonate and carbonate anions and alkali metal cations dominate, TDS varies between 2 to 13 g/L, and pH varies correspondingly between 9.0 and 10.6. The waters of group (II) are usually less mineralized, less alkaline (pH<9), and higher in chloride. The high values of the oxidant demand (permanganate addition) point to a high amount of the dissolved organics, though their specific composition is still not known. The compositions of three typical water samples used in the modeling calculations are given in the following table:

Sample #	Group id.	pH	Na^+ mg/L	K^+ mg/L	Ca^{+2} mg/L	Mg^{+2} mg/L	Fe mg/L	U mg/L
13	II	7.61	862	46	24.4	12.7	<0.1	1.7
20	I	9.72	273	526	5.2	9.5	<0.1	1.4
21	I	10.58	5346		43.5	21.4	<0.1	7.7

Sample #	Si mg/L	Cl^- mg/L	SO_4^{-2} mg/L	CO_3^{-2} mg/L	HCO_3^- mg/L	PO_4^{-3} mg/L	TDS mg/L	Oxidant demand, mg/L O_2
13	12.5	839	130	<5	321	46.2	2281	48
20	8	138	200	200	687	4.8	2045	31
21	n.d	963	880	1814	4300	8.4	13380	n.d

(n.d = not determined)

Atmosphere. The gas composition of the atmosphere inside the Shelter was not studied. However, it can be supposed that it results from the interaction of the outside air with the waters and solids in the ICS. In particular, it was suggested [5] that CO_2 is absorbed from the air by the alkaline water solutions, while O_2 is consumed by the redox reactions occurring in the ICS.

Humidity and temperature conditions of the ICS are strongly dependent on weather and climate conditions, because there is no complete isolation from the air environment.

THERMODYNAMIC DESCRIPTION OF CONSTITUENTS

Aqueous solution is modeled using a comprehensive set of aqueous species for the ion-association model, including dissolved gases and solutes for the system Na-K-Ca-Mg-Si-P-Fe-Al-Sr-Cs-U-Zr-C-Cl-S-N-H-O [4]. Thermodymamic values were taken from previous work [9-13,18]. Standard values of partial molal Gibbs free energy (g^o_{298}) for zirconium aqueous species (hydroxide, chloride and sulphate complexes) were calculated using published association constants [14], and $g^o_{298}(Zr^{+4})$ = -521745 J/mol [15]. The $g^o_{298}(Zr(OH)_3(HCO_3)_2^-)$ = -2462703 J/mol and $g^o_{298}(Zr(OH)_3(HCO_3)^o)$ = -1871306 J/mol were derived from the experimental data [16], and $g^o_{298}(Zr(OH)_5^-)$ = -1643210 J/mol - from experimental data [17]. The activity coefficients of the individual aqueous species were calculated with the Debye-Hueckel equation in third approximation, with the common third parameter set to 0.064. The bulk chemical composition of the aqueous phase was included in accordance with the analytical data measured for the typical water samples (see Table).

Gas atmosphere is represented by an ideal mixture of ideal gases (CH_4, CO, CO_2, H_2S, NH_3, NO_2, SO_2, SO_3, H_2, N_2, O_2) which can cover a wide range of redox conditions. Thermodynamic properties of the gases were taken from reference 18.

Minerals were included in the calculations of the "solid-water-atmosphere" equilibria as single phases (kaolinite Kl, gibbsite, quartz, brucite, amorphous SiO_2 [18], serpentine $Serp$, hydrogoethite Hgt, hydromagnetite, mackinawite, whitlockite, disordered dolomite Dol, [19-21]) and the ideal solid solution phases (Ca,Sr)CO$_3$ Arg-Str, and (Ca,Sr)SO$_4$ [18].

Salts in the subsystem Na-K-Ca-Mg-CO$_2$-SO$_3$-P$_2$O$_5$-H$_2$O were taken into consideration for the calculation of water evaporation. The thermodynamic properties of the salts were taken from references 11,18,20,21.

Secondary uranium minerals are represented by single phases (Na_2UO_4, $Na_2U_2O_7$, Na_3UO_4, $NaUO_3$, β-$UO_2(OH)_2$, $(UO_2)_3(PO_4)_2(H_2O)_6$, UO_2CO_3 [13]) and solid solutions: autunite (Aut) $(H_2,K_2,Na_2,Cs_2,Ca,Mg,Sr,Fe)(UO_2)_2(PO_4)_2$ [11,21] and $(Na,Cs)_4UO_2(CO_3)_3$ [13,21].

Solids in the subsystem U-Zr-Si-O-H. HUZ phase was modeled by the non-ideal (U,Zr)SiO$_4$ solid solution. The thermodynamic properties of the solution end-members, as well as of ZrO_2, UO_2, UO_3, were taken from references 11 and 18. The non-ideal behavior of the (U,Zr)SiO$_4$ solid solution is described by a subregular model with the miscibility parameters (W_{Zr} = 24.4 kJ/mol and W_U = 21.9 kJ/mol) estimated using the theory of isomorphic miscibility [22] and the available data on uranium and other actinides incorporated in zircon. The values of apparent Gibbs energy of formation $\Delta G^o_{f,298}(Zr_{0.95}U_{0.05}SiO_4)$ = -1917234 J/mol and $\Delta G^o_{f,298}(Zr_{0.9}U_{0.1}SiO_4)$ = -1914722 J/mol, calculated on this basis, were used to model the solubility of metastable HUZ ("Chernobylite" [1]). The values of $\Delta G^o_{f,298}(Zr(OH)_{4,cr})$ = -1483210 J/mol and $\Delta G^o_{f,298}(ZrO(OH)_{2,cr})$ = -1260868 J/mol were derived from the experimental solubility data [23] and [16], respectively.

Mineral phases of fully hydrated Portland cements are represented by: ettringite $Ettr$ - $Ca_6Al_2S_3O_{12}(OH)_{12}$, hydrotalcite Htc - $Mg_4Al_2O_7(H_2O)_{10}$, portlandite $Port$ - $Ca(OH)_{2,cr}$, solid solutions of amorphous calcium silicate hydrogel (CSH1, CHS2 phases) and "hydrogarnet" solid solution Hgr. The values of the ion activity product for the dissolution of $Ettr$ and Htc derived from experimental data [23] have been used for calculations of $\Delta G^{o}_{f,298}$ values for these single-component phases. $\Delta G^{o}_{f,298}(Port)$ was taken from reference 18. CSH hydrogel was modeled by three phase associations. For Ca/Si ratios from 0 to 1 and 1 to 1.7, CSH is represented by the two-component ideal solid solutions, respectively, $SiO_{2,amorph}$ - CaH_2SiO_4 (CSH1) and CaH_2SiO_4 - $Ca_{1.7}H_{3.4}SiO_{5.4}(H_2O)_4$ (CSH2). At Ca/Si > 1.7, the composition of CSH is described by two single-component phases, $Ca_{1.7}H_{3.4}SiO_{5.4}(H_2O)_4$ and $Ca(OH)_{2,cr}$. It is important that the Gibbs energy minimization method (see below) can explicitly calculate composition of any solid solution at the equilibrium with aqueous electrolyte, gas and other solids in the system. Thus, we avoid the determination of solubility products of CSH phase as functions of Ca/Si ratio (as it is done, e.g., by D.Bennett et al. [23]). The values of ΔG^{o}_{f} for the end-members of CHS1 and CHS2 solutions were calculated from the ion activity products for the dissolution of $SiO_{2,amorph}$, CaH_2SiO_4 [23] and $Ca_{1.7}H_{3.4}SiO_{5.4}(H_2O)_4$ [24]. The consistency between the calculated and experimental solubility of CSH hydrogel at Ca/Si ratios from 0 to 2.5 was achieved by including the $Ca(OH)_2^{o}$ complex to the aqueous electrolyte model. Estimated value of $g^{o}_{298}(Ca(OH)_2^{o})$ is -886419 J/mol.

The model of the *Hgr* solid solution is represented by three components: $Ca_3Al_2O_6(H_2O)_6$, $Ca_3Fe_2O_6(H_2O)_6$ and $Ca_3Al_2Si_3O_{12}$, their thermodynamic properties were taken from references 23,24 and 25, respectively.

MODEL CALCULATIONS

<u>Gibbs energy minimization and "Selektor-A" code.</u> The calculations were conducted using the *Selektor-A* code (ver. 3.0.243) based on a convex programming approach to Gibbs energy minimization [7,8]. The code finds the explicit (meta)stable phase-component composition x (including aqueous speciation, gas and solid-solution compositions); values of chemical potentials of stoichiometry units u; pH, Eh and pP(gases) directly from the input temperature T, pressure P, bulk chemical composition of the whole system b, apparent standard-state Gibbs energies for all species g^{θ}_{TP}, and set of parameters for the non-ideality models θ (if needed). One aqueous electrolyte, one gas mixture, and any number of crystalline and dispersed (multicomponent) solid phases (also with surface species) can be included in the system. Calculation of the dual solution values u leads to the Karpov' stability criteria f_{α} [7], allowing the stability of each component (phase) to be tested, and the consistency of input thermodynamical data to be improved.

<u>Interaction of U-Zr-Si-O phases with aqueous solution.</u> The results of modeling metastable dissolution of the uranium oxides in the Shelter waters at different redox conditions are described elsewhere [4]. The interactions of the Shelter waters and inclusions of the crystalline U-Zr-Si phases in the lava matrix of FCM (at ambient *TP)* were modeled in the following sequence:
(1) Calculation of the stable composition of "aqueous-air" system for excess air;
(2) Titration of the aqueous solution by CO_2 to fit measured pH values in order to obtain a consistent speciation for the real water at oxic conditions;
(3) Determination of the metastable solubility of HUZ in the real water at varying redox conditions.

At stage (3), the total composition of the modeled system was defined by 1 mole of $Zr_{0.9}U_{0.1}SiO_4$ and 1 kg of aqueous solution with a composition, as determined at stage (2), without the measured amount of dissolved uranium. The equilibrium calculations were performed for titration of the "solids-aqueous" system by O_2 or H_2 to reach *pe* values in the -6 ÷ +13 range. At the model conditions (25°C, 1 bar), the mole fraction of $USiO_4$ in $(U,Zr)SiO_4$ phase, calculated using the accepted solid solution model, was less than 10^{-4}. Therefore, $Zr_{0.9}U_{0.1}SiO_4$, being a metastable phase, should decompose to practically pure $ZrSiO_4$. We suppose that the HUZ does

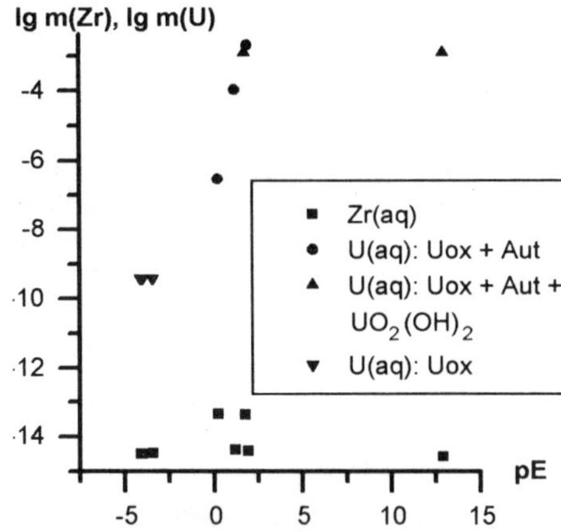

Fig. 1. Metastable solubility of $Zr_{0.95}U_{0.05}SiO_4$ and stable products of decomposition of $Zr_{0.9}U_{0.1}SiO_4$ (uranium oxide -*Uox*, Na-autunite - *Aut* and $UO_2(OH)_2$) versus *pe* in the Shelter water sample #13 (mol/kg H_2O). $SiO_{2,amorph}$ and $Zr_{0.95}U_{0.05}SiO_4$ phases are present at each point.

not decompose right away to a pure $ZrSiO_4$, but dissolves incongruently, producing a metastable $(Zr,U)SiO_4$ phase, like $Zr_{0.95}U_{0.05}SiO_4$, having a lower uranium content. To model this pathway, $ZrSiO_4$ was exluded from the list of possible equilibrium phases. The solids included were UO_3, UO_2, secondary uranium minerals, $SiO_{2,amorph}$, Zr oxide and hydroxide. The results of the model calculation of metastable HUZ solubility are shown in Fig.1.

Concrete-aqueous solution interaction. Since the real composition of concrete in the ICS is unknown, the following composition of Portland cements was used for the model (wt.%): CaO - 65, SiO_2 -23, Al_2O_3 -7, Fe_2O_3, - 3, MgO - 1.5, SO_3 - 0.5. For the calculations, solution compositions corresponding to real Shelter waters, were equilibrated with air. The modeling was performed for different solid/water ratios ranging from 10^{-7} to 1. At each point, the equilibrium phase assemblage, quantities of phases and compositions of solutions were determined (Fig.2).

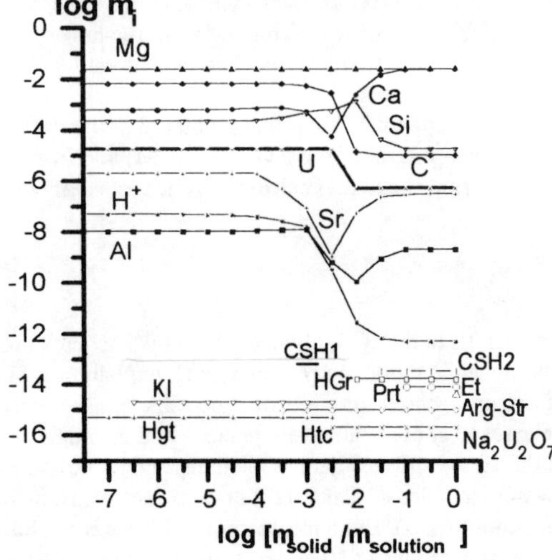

Fig.2 Modeled equilibrium total dissolved element concentrations (mol/kgH_2O) and phase assemblages formed as a result of interaction of Portland cement with the Shelter water sample #13 as function of the solid/water mass ratio ξ.
Solid phases: *CSH1* and *CHS2* - solid solutions phases of amorphous calcium silicate hydrogel; *Hgr* - "hydrogarnet" solid solution, *Arg-Str* - solid solution phases $(Ca,Sr)CO_3$, *Et* - ettringite, *Htc* - hydrotalcite, *Prt* - portlandite; *Kl* - kaolinite, *Hgt* - hydrogoethite.

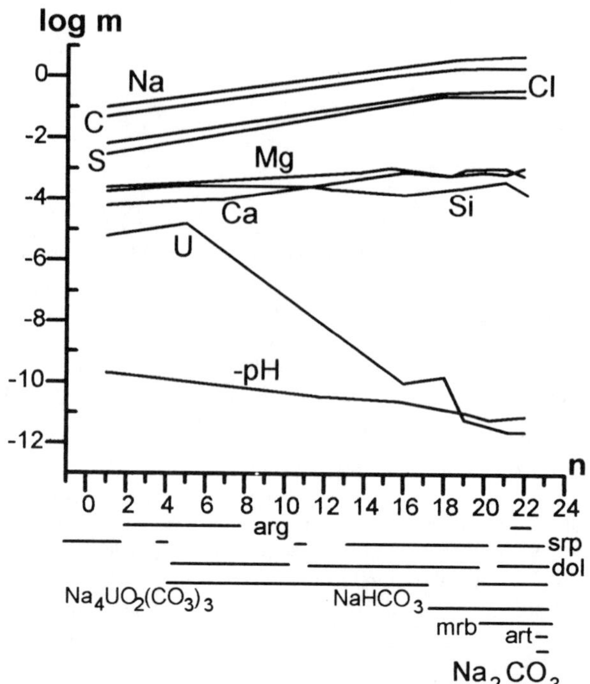

Fig. 3 The evolution of the total dissolved element concentrations and assemblages of the precipitated minerals for an evaporation path of sample #20 water. Each step (n) of "sequential reactor" algorithm corresponds to 25% loss of H_2O from the bulk composition.

Solid phases: *Arg* - aragonite, *srp* - serpentinite, *dol* - dolomite, *mrb* - mirabilite, *art* - artinite.

The evaporation pathways were modeled using a "sequential reactors" algorithm to follow the sequence of precipitation of secondary minerals and changes of the dissolved U content of residual brines. The initial compositions were real Shelter water samples, with excess dry air. After the first calculation, ¼ of the current amount of H_2O-solvent was extracted from the bulk   composition to simulate the next "reactor" (step of evaporation) and so on. The typical modeled evaporation path (Shelter water sample #20) is shown on Fig.3.

Stability of other mineral constituents such as serpentinite (basement material in the destroyed reactor), $SiO_{2,amorph}$ (main component of silicate glass matrix of the lava-type FCM) and iron oxides (oxidation products of steel) were estimated from the model calculations described above, as well as in reference 4.

RESULTS AND DISCUSSION

The results of research on chemical interactions in the ICS, obtained from hydrochemical studies of the Shelter waters [3,5] and thermodynamic modeling [4, this work], explain many observations as well as outline major regularities permitting to predict some possible consequences.

1. Two groups of water composition, classified above and in references 3 and 5, reflect the degree of water solution transformations in the ICS. It seems that the alkaline, highly bicarbonate mineralized waters of the group I ("meta" water; samples #20 and #21) are products of transformation of the group II water ("fresh" water; sample #13). Evidently, sample #13 water is undersaturated in Ca and Mg carbonates, while the water samples #20 and #21 appear to be supersaturated with Ca and Mg carbonates and Mg silicate [4].

2. Calculations [4; this work] show that analytical concentrations of Si_{aq} correspond to the stability of amorphous SiO_2 (sample #13) or Mg silicate (samples #20, #21), but exceed the solubility of quartz. Measured Fe_{aq} concentrations (10^{-4} to 10^{-6} molal) correspond to the theoretical solubility at the *in situ* Eh (suboxic), while in more oxic (stable hydrogoethite) or anoxic (Fe sulphide) conditions, the calculated Fe_{aq} is much lower.

3. The solid UO_2 phase controls U solubility at *pe* < 0; at higher *pe*, at oxic conditions, uranium oxide solubility in the waters can reach values up to 10^{-1} molal (ca. 20 g/L), although measured concentrations of U_{aq} (see Table) are still much lower than the limiting solubilities of both UO_3 and UO_2CO_3 [4]. However, measured U_{aq} seems to be controlled by the Na-autunite $Na_2(UO_2)_2(PO_4)_2$ (sample #13) or $Na_4(UO_2)_2(CO_3)_3$ phase (sample #21) , which points out the significance of secondary minerals in the geochemical system of the ICS.

4. HUZ interaction with both "fresh" (sample #13, Fig.1) and "meta" water in a *pe* range from -6 to +13 results in leaching of uranium from this metastable phase. One can see in Fig.1 that the Zr_{aq} concentration is considerably less than the U_{aq} concentration, even under reducing conditions. It is evident that the phase assemblage of the HUZ decomposition products should be dependent at the degree of U leaching. For high leaching (for example, at oxidizing conditions), HUZ decomposition can trigger precipitation of secondary U-minerals (Fig.1). It also should be noted that during HUZ decomposition Zr oxide phases are not formed, owing to the high concentrations of Si_{aq} in the Shelter water.

5. Interaction of concrete with a highly bicarbonate Shelter water leads to the carbonatization of high calcium cement phases, accompanied with a decrease in dissolved carbonate. As a consequence, secondary uranium minerals may precipitate, and Sr_{aq} concentration may drop as a result of $(Ca,Sr)CO_3$ formation (Fig.2). In Fig.2, it is also shown that Mg_{aq} concentration in the Shelter water is close to the concrete saturation values. Si_{aq}, Al_{aq} exceed, while Ca_{aq} does not approach the concrete saturation values. Evidently, this can lead to Ca leaching via portlandite dissolution, though the aluminosilicate phases of the concrete are sufficiently stable in this environment.

6. Along the modeled evaporation path, both soda-type water (sample #20, Fig.3) and "fresh" chloride-enriched water lose Ca and Mg and quickly evolve to highly mineralized soda-type water similar to sample #21. At pH$\approx$10.5 and TDS from 5 to10 g/L, precipitation of $Na_4(UO_2)_2(CO_3)_3$ phase occurs, which can be regarded as a very important mechanism of uranium redistribution within the Shelter. It can effectively decrease contents of U_{aq} down to 10^{-8}- 10^{-9} mol/L as evaporation proceeds for the initially soda-type waters.

Thus, these waters must be regarded as chemically reactive with respect to FCM and other materials. Future changes in redox state or migration pathways may cause leaching of large quantities of uranium and other radionuclides from the FCM and their precipitation inside the *Sarcophagus* or escape from it, could create a potential environmental hazard.

ACKNOWLEDGEMENTS

The work was supported in part by the International Atomic Energy Agency (IAEA) via the Technical Cooperation project UKR/9/010. Authors would like to express their gratitude to two MRS reviewers who helped very much to improve the lucidity of the English presentation.

REFERENCES

1. B.Burakov, E.Galkin, E.Pazukhin et al., Radiochim. Acta **65**, 1994.
2. A.Kiselev, A.Nenaglyadov, A.Surin et al., Preprint IAE, 5533/3, Moscow, 1992 (in Russian).
3. M.Khodorivski, V.Sinitsyn, D.Kulik et al., NRC4 Conf. Extended Abstracts. (Eds: F.David, J.Krupa), St.Malo, 1996.
4. D.Kulik, V.Sinitsyn, M.Khodorivski et al., NRC4 Conf. Extended Abstracts. (Eds: F.David, J.Krupa), St.Malo, 1996.
5. A.Abramis, M.Khodorivski, V.Sinitsyn et al., in Problems of Chernobyl Exclusion Zone 5, Naukova Dumka Publishers, Kyiv, 1996 (in Russian).
6. A.Vishnevskii, I.Kuzmina, V.Tkach and V.Tokarevskii, in Problems of Chernobyl Exclusion Zone N 3, Naukova Dumka Publishers, Kyiv, 1996 (in Russian).
7. I. Karpov, Computer-aided physical-chemical modeling in geochemistry (Nauka Publishers, Novosibirsk,1981) (in Russian).
8. D.Kulik, in Water-Rock Interaction 8, edited by Y. Kharaka and O.Chudaev (Balkema, Rotterdam, 1995).
9. J.Johnson, E.Oelkers and H.Helgeson, Comput. Geosci. **18**, 1992.
10. R.Lemire and P.Tremaine, J. Chem. Eng. Data **25**, 1980.
11. D.Nordstrom, L.Plummer, D. Langmuir et al., in Chemical modeling in aqueous systems II (ACS Symp. Series 416, 1990).
12. J.Fuger, I.Khodakovsky, E.Sergeyeva et al., The actinide inorganic complexes, Part 12 (IAEA, Vienna., 1992).
13. I.Grenthe, J.Fuger, R.J.M.Konings et al, Chemical thermodynamics of uranium (North-Holland, Amsterdam et al. 1992).
14. S.Aja, S.Wood and A.Williams-Jones, Applied Geochem., **10**, 1995.
15. V.Vasiliev, L.Kochergina and A. Lytkin, J. Inorg. Chem. (11), 1974 (in Russian).
16. A.Samchuk and E.Dorofey, Geokhimiya (2), 1983 (in Russian).
17. S.Sheka and T. Pevzner, J. Inorg. Chem (5), 1960 (in Russian)
18. R.Robie and B.Hemingway, U.S.Geol.Surv.Bull. 2141, 1995.
19. Yu.Melnik, Genesis of Precambrian Iron Formations (Naukova Dumka Publishers, Kyiv, 1986) (in Russian).
20. Y.Kharaka, W.Gunter, P.Aggarval et al., U.S.Geol.Surv.Water-res.Invest.Rep.88-42, Menlo Park,CA 1988.
21. J.Ball and D.Nordstrom, U.S.Geol.Surv.Open-File Rep. 91-183, Menlo Park,CA,1992.
22. V.Urusov, Theory of isomorphous miscibility (Moscow, Nauka Publishers, 1977) (in Russian).
23. D.Bennett, D.Read, M.Atkins, and F.Glasser, Jour. Nucl. Materials, **190**, 1992.
24. J.Lee, D.Roy, B.Mann and D.Stahl, Mat.Res.Soc.Symp.Proc.,**353**,1995.
25. T.Holland and R. Powell. Metamorphic Geol., **8**(1), 1990.

MODELING THE MIGRATION OF ^{90}Sr AND ^{137}Cs IN THE SUBSURFACE FROM THE CHERNOBYL NPP UNIT-4 SHELTER

S.L.KIVVA *, O.B.STELYA **, N.I.PROSKURA ***
*EM Department, Institute of Mathematical Machines&System Problems, 42, Glushkova pr., Kiev, 252187, Ukraine, slk@dem.ipmms.kiev.ua
**Kiev University, 64, Vladimirskaya st., Kiev, 252601, Ukraine
***Administration of Exclusion Zone, 14, Sovetskaya st., Chernobyl, 255620, Ukraine

ABSTRACT

The Shelter constructed above the destroyed Unit-4 of the Chernobyl NPP contains 20 MCi of nuclear fuel. More than 1000 m^3 of intermediate-level radioactive water is disposed in its basement. The purpose of this work was to simulate migration of the radionuclides ^{90}Sr and ^{137}Cs from the Shelter into the subsurface environment to evaluate their migration rate and migration paths. A mathematical model accounting for the coupled transport of water and radionuclides in variably saturated media was used. The lack of suitable experimental and field studies excluded the possibility of complete validating the model. Results of the simulations will be useful for future field studies.

INTRODUCTION

A Shelter constructed above the destroyed Unit-4 at the end of 1986 contains from 135 to 180 tons of nuclear fuel with an activity of up to 20,000,000 Ci [1]. Within the Shelter the largest part of the nuclear fuel from the completely destroyed reactor core is contained in fuel-containing masses(FCM), and the rest in fuel assemblies and their fragments. The FCM can be represented as lava-like materials with a glass silicate matrix containing many inclusions of various uranium compounds, finely dispersed high-level active fuel dust, and aerosols. Experimental studies of the FCM show that long-term interaction with water and the atmosphere within the Shelter leads to (1) a decrease in fuel particle size; (2) degradation of the surface of the lava-like materials with formation of dust fuel on the surface; (3) appearance of new uranium compounds including soluble compounds on the lava surface; (4) leaching of radionuclides from the FCM; (5) and an abrupt increase in migration rate of radionuclides.

According to previous research [2-3] the Shelter contains more than 3,000 m^3 of intermediate-level and low-level radioactive water. More than 1,000 m^3 of intermediate-level radioactive water is disposed in the basement of the Unit-4. It is also reasonable to consider such major water sources within the Shelter as: precipitation, water used for dust suppression, moisture condensation onto building structures, and water leakage from the service lines of the 3rd unit of the Chernobyl NPP. Most precipitation onto the roof penetrates through cracks to the interior of the Shelter. The volume of this precipitation is approximately 1,000 m^3 per year. For dust suppression, 15 m^3 of water are sprayed twice a month within the Shelter. Upon movement through the FCM and interaction with FCM, the water is contaminated by radionuclides. The contaminated water penetrates to the subsurface environment. The total activity of the intermediate-level radioactive water ranges from 2×10^{12} to 9×10^{13} Bq/L. Cesium isotopes constitute the main portion of the water activity; the cesium activity varies from 1.6×10^5 to 5.5×10^7 Bq/L. The total ^{90}Sr activity in the water was determined to be 3.6×10^3 to 1.1×10^6 Bq/L.

The Chernobyl NPP is located on the right bank of the Pripyat River. The NPP site is a flat

Mat. Res. Soc. Symp. Proc. Vol. 465 © 1997 Materials Research Society

river bench. The aquifer system is composed of fine, medium and sometimes coarse alluvial sands with interlayers of sand loam and loam. Medium sands prevail. According to field studies the hydraulic conductivity of the alluvial sands ranges from 6.0 to 15.0 m day^{-1}. These alluvial sands lie above 25 to 40 m of clay marls of the Kiev Eocene suite. The marls have very low permeability and act as an aquitard. The infiltration is estimated between 80 mm and 200 mm per year [1,4].

The purpose of this work was to simulate, under a variety of conditions, migration of ^{90}Sr and ^{137}Cs through geologic media due to water leakage from the Shelter.

MODEL DESCRIPTION

Neglecting hysteresis and temperature gradients, movement of moisture through a saturated-unsaturated porous medium is described by

$$\frac{\partial \theta}{\partial t} = div(k_r \frac{k \rho g}{\mu} \, grad(h)) + f, \tag{1}$$

where t is time; h is the hydraulic head; θ is the volumetric water content; k_r is the relative permeability $(0 \leq k_r \leq 1)$; k is the intrinsic permeability; μ is the dynamic viscosity of soil water; ρ is the fluid density; g is the acceleration of gravity; and f represents sources and sinks of water in the system.

The initial conditions for (1) consist of initial values of hydraulic head, specified by:

$$h(0,x) = \varphi(x), \tag{2}$$

where x is the spatial coordinate vector and φ is a prescribed function.

Boundary conditions may be Dirichlet

$$h(t,x) = g_1^h(t,x) \qquad \text{on } \Gamma_1, \tag{3}$$

or Neuman

$$\left[k_r \frac{k \rho g}{\mu} grad(h(t,x)) \right] n = g_2^h(t,x) \qquad \text{on } \Gamma_2, \tag{4}$$

where g_1^h is the prescribed hydraulic head at the boundary Γ_1; n is the unit vector normal to Γ_2, direction outward; and g_2^h is the prescribed outward fluid flux normal to Γ_2. Γ_1 and Γ_2 comprise the entire boundary of the domain. Usually Γ_1 is the part of boundary at which surface-water bodies are intercepted by aquifers; Γ_2 are the boundaries across which fluid passes at a specific rate (impervious boundaries, areas of infiltration or evapotranspiration, points of injection and withdrawal, etc.).

The leaching of radionuclides from the fuel particles is modeled by a first-order equation

$$\frac{\partial c_p}{\partial t} = -(\alpha_p + \lambda)c_p, \tag{5}$$

$$c_p(0,x) = c_p^0(x), \tag{6}$$

where c_p is the radionuclide activity in the fuel particles per soil solid volume, α_p is the first-order constant of radionuclide leaching from fuel particles, λ is the radionuclide decay constant, c_p^0 is the initial radionuclide activity in the fuel particles.

Let c_l represent the volumetric radionuclide activity in the aqueous phase, c_s the volumetric exchangeable sorbed activity in solid phase, and c_f the volumetric activity which is fixed in the mineral lattice. Assuming that the ion-exchange reaction transferring activity between aqueous

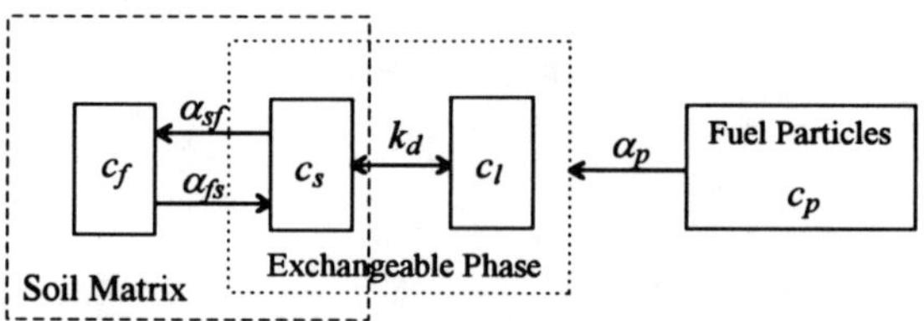

Fig. 1. Schematic representation of the kinetic sorption model.

and solid exchangeable phases has a sufficiently short timescale, we will consider this reaction as instantaneous with a distribution coefficient k_d. Transfers of activity to and from the fixed phase are modeled by first-order rate constants α_{sf} and α_{fs}, respectively, as shown in Fig.1.

Ignoring chemical diffusion in the solid phase, the equation for advective-dispersive transport of radionuclide activity in the exchangeable phase in variably saturated media may be written in the form

$$\frac{\partial}{\partial t}(\theta c_l + (1-n_T)c_s) = div(\theta D\, grad(c_l) - Vc_l) -$$

$$\lambda(\theta c_l + (1-n_T)c_s) + (1-n_T)(\alpha_p c_p + \alpha_{fs}c_f - \alpha_{sf}c_s), \qquad (7)$$

subject to the following initial and boundary conditions:

$$c(0,x) = c^0(x), \qquad (8)$$
$$B_c(c) = g_c(t,x), \qquad (9)$$

where $c = \theta c_l + (1-n_T)c_s$, n_T is the porosity, B_c is the boundary operator for radionuclide activity, c^0 is the initial radionuclide activity in the exchangeable phase, g_c is prescribed radionuclide activity in the exchangeable phase at the boundaries, and c_s is related to c_l by

$$c_s = k_d c_l \frac{\rho_s}{\rho},$$

where ρ_s is the soil solid density.

D is the dispersion tensor, defined by

$$\theta D = \alpha_T |V|\delta + (\alpha_L - \alpha_T)VV/|V| + D_0\theta\tau\delta, \qquad (10)$$

where α_T is the transverse dispersivity; δ is the Kronecker delta tensor; $|V|$ is the magnitude of the Darcy velocity V; α_L is the longitudinal dispersivity; D_0 is the molecular diffusion coefficient; and τ is the tortuosity.

The transport of activity in the fixed phase is described by

$$\frac{\partial c_f}{\partial t} = \alpha_{sf}c_s - \alpha_{fs}c_f - \lambda c_f, \qquad (11)$$

$$c_f(0,x) = c_f^0(x), \qquad (12)$$

where c_f^0 is the initial activity in the fixed phase.

RESULTS

Numerical simulations of the ^{90}Sr and ^{137}Cs migration from the Shelter into and through the subsurface environment were conducted for a two-dimensional vertical cross section selected along the streamlines of groundwater flow. These streamlines were obtained from field

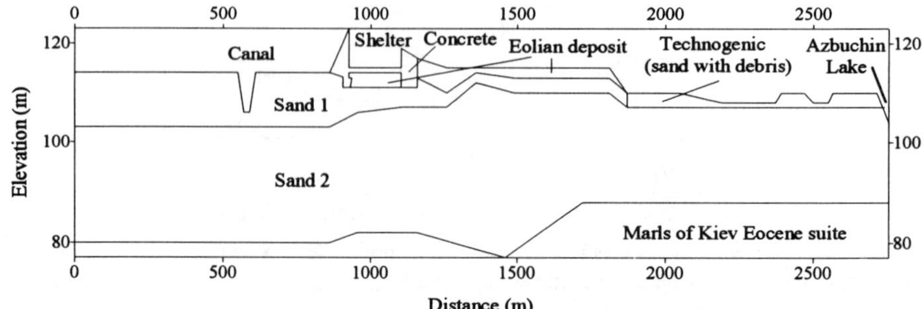

Fig. 2. Schematic representation of geologic section.

observations of groundwater level distribution in the vicinity of the Chernobyl NPP. The geology of this cross section is shown in Fig.2.

The MSTS code [5,6] adapted to implement the model described above was used for the numerical simulations. The flow domain was discretized by a non–uniform rectangle mesh, nodes of which spaced at 5 to 60 m intervals in the horizontal direction, and 0.2 to 3 m in the vertical direction. Time steps of 30 days were used. Values for the hydraulic and species properties used in the simulations are given in Table I [1,4,7]. According to previous simulations the results are very sensitive to the distribution coefficient, the first-rate constants α_{sf} and α_{fs}, the hydraulic conductivity and so their values were taken to obtain upper (worst) estimation of groundwater contamination. Parameters α_{sf} and α_{fs} vanish for the case of ^{90}Sr migration in the subsurface. The coefficient of radionuclide leaching from fuel particles was 3.2×10^{-4} day^{-1} [1].

TABLE I.   Values of Parameters Used in Numerical Simulations.

Soil or Rock	K	n_T	ρ_s	α_L	α_T	k_d^{Sr}	k_d^{Cs}	α_{sf}^{Cs}	α_{fs}^{Cs}
	m day^{-1}		gm cm^{-3}	m	m	L kg^{-1}	L kg^{-1}	day^{-1}	day^{-1}
Eolian deposit	13.	0.316	2.34	1.2	0.1	1.0	5.2	0.001	0.00005
Sand 1	9.	0.35	2.46	1.2	0.1	1.0	5.2	0.001	0.00005
Sand 2	10.	0.342	2.43	1.2	0.1	1.0	5.2	0.001	0.00005
Concrete	0.001	0.06	2.66	1.	0.1	0.5	5.2	0.001	0.00005
Technogenic	11.	0.36	2.5	1.2	0.1	1.0	5.2	0.001	0.00005

Boundary conditions for the liquid phase were the specified values of hydraulic head at the left and right boundaries, the Shelter basement, boundaries on which surface water bodies are intercepted by the aquifer, and infiltration and evapotranspiration rates at the ground surface. The

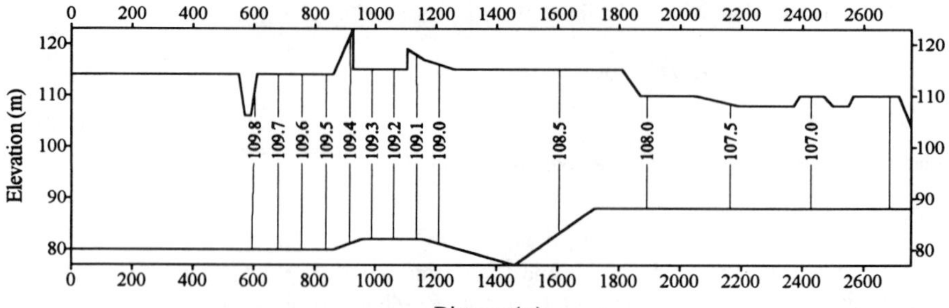

Fig. 3. Initial hydraulic head distribution.

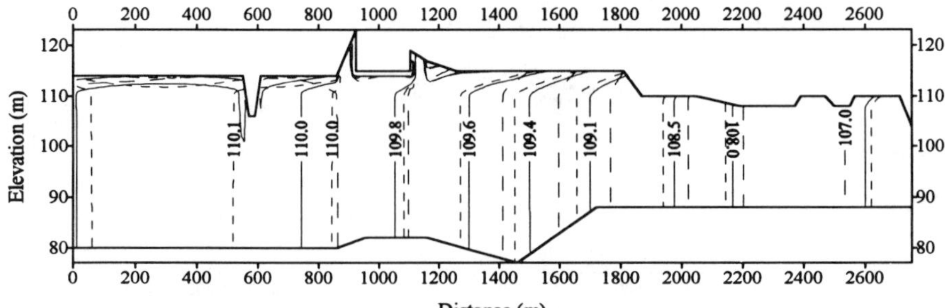

Fig. 4. Distribution of hydraulic head in: (——) summer; (- - -) autumn; (– –) spring. Case of a
120-mm infiltration rate and a 10-cm water depth in the basement.

marls were considered as impervious. Seasonal fluctuations of the water level in the canal varied
between 109.5 m and 110.2 m; in Azbuchin Lake, between 106.0 m and 106.8 m; and for the
hydraulic head at the left boundary, between 109.5 m and 110.0 m [4]. These seasonal
fluctuations of water level in surface water bodies and the variability over time of precipitation
and evapotranspiration were taken into account in the simulations. The initial condition for the
liquid phase is illustrated in Fig. 3.

A zero radionuclide flux condition was employed at all boundaries, excluding the Shelter
basement and boundaries that contact surface water bodies. The ^{90}Sr activity of water from the
basement was 2.3×10^6 Bq/L, and 2,800 Bq/L in Azbuchin Lake [4]. Similarly, ^{137}Cs activity in the
basement water and Azbuchin Lake were 7.1×10^7 and 84 Bq/L, respectively. From 1986 to 1995,
the radionuclide activity of canal water was assigned the annual average activity of water samples
from Cooling pond. The radionuclide activity of canal water after 1995 was specified as the
radionuclide activity of canal water in 1995. For the area surrounding the Chernobyl NPP, land
surface contamination by ^{90}Sr and ^{137}Cs was 600 and 1,000 Ci per sq.km, respectively. Just after
the accident, exchangeable ^{90}Sr was 5% of the total ^{90}Sr activity and exchangeable ^{137}Cs was 10%
of the total ^{137}Cs activity. The subsurface was assumed to be uncontaminated before the accident.
The date of the Chernobyl accident was taken as the start date for the simulations.

There is no reliable data on water level in the Shelter basement and infiltration rate for the
region surrounding the Chernobyl NPP. Therefore, in order to estimate migration rates of
radionuclides from the Shelter through the subsurface, numerical simulations were conducted for
various values of the infiltration rate and water depth in the basement. The effects of different

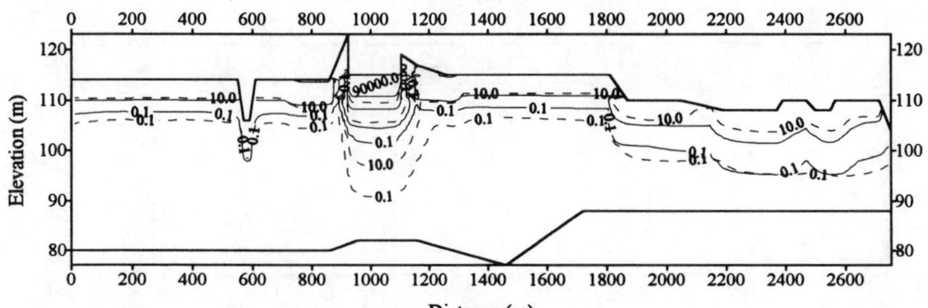

Fig. 5. Activity of ^{90}Sr (Bq/L) in the aqueous phase in (——) 1995 and (- - -) 2045. Case of a
120-mm infiltration rate and a 10-cm water depth in the basement.

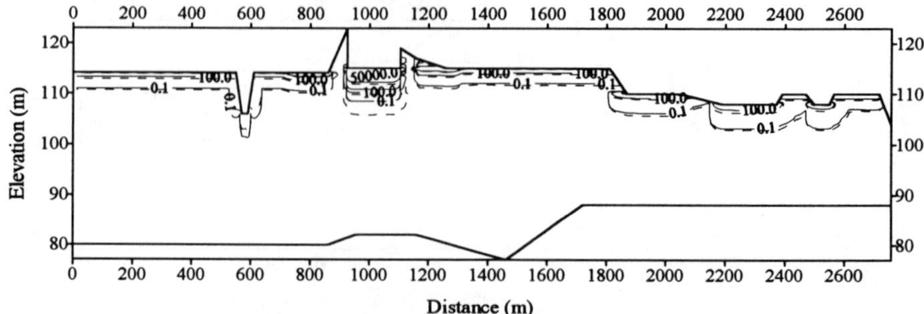

Fig. 6. Activity of ^{137}Cs (Bq/L) in the aqueous phase in (——) 1995 and (- - -) 2045. Case of a 120-mm infiltration rate and a 10-cm water depth in the basement.

water depths in the basement and infiltration rates on radionuclide migration from the Shelter are compared to a base case of radionuclide migration for a 120-mm infiltration rate per year and a 10-cm water depth. According to this case, Fig. 4-6 shows the model-generated seasonal fluctuation of the hydraulic head and radionuclide activity distribution for 1995 and 2045. Activity-depth profiles of ^{90}Sr and ^{137}Cs in the aqueous phase below the Shelter for 1995, 2020, and 2045 are illustrated in Fig. 7.

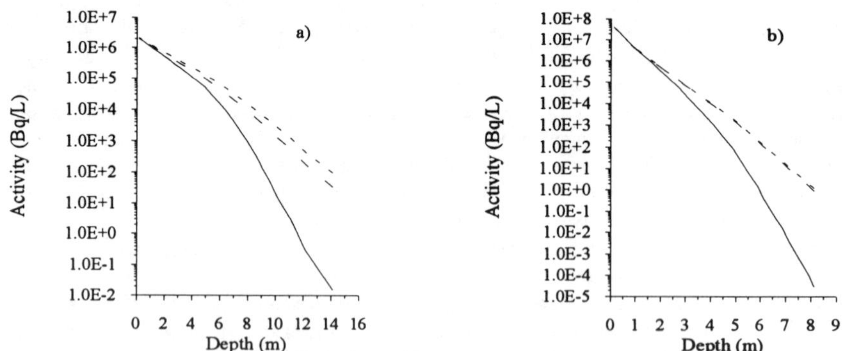

Fig. 7. Activity-depth profiles of a) ^{90}Sr and b) ^{137}Cs in the aqueous phase below the Shelter in (——) 1995, (– – –) 2020, and (- - -) 2045. Case of a 120-mm infiltration rate and a 10-cm water depth in the basement.

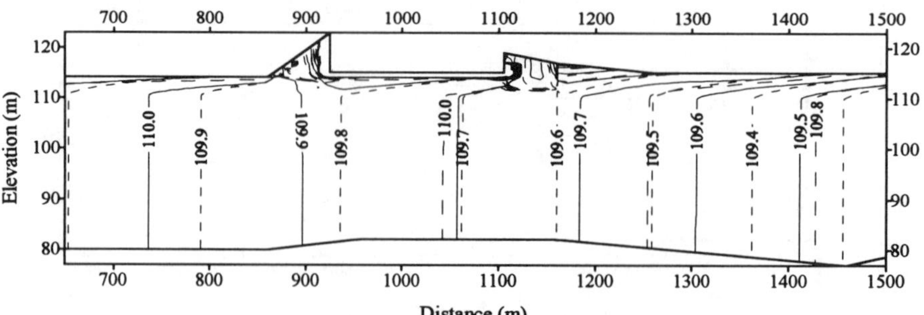

Fig. 8. Summer distribution of hydraulic head for a 10-cm water depth in the basement and infiltration rates of (——) 120 mm, (- - -) 80 mm, and (– –) 140 mm per year.

The influence of infiltration rate on hydraulic head distribution and migration of ^{90}Sr from the Shelter through the subsurface is shown in Fig. 8-9. The distribution of hydraulic head and ^{90}Sr activity were obtained for infiltration rates of 80 mm, 120 mm, and 140 mm per year, respectively. The level of water in the basement was 115.1 m. Analysis of the results indicates that the lateral velocity of the contamination front below the Shelter decreases with increasing annual infiltration rate. The effect of infiltration on radionuclide migration below the Shelter at depth is negligible.

In order to determine the effect of water depth in the basement on the ^{90}Sr migration from the

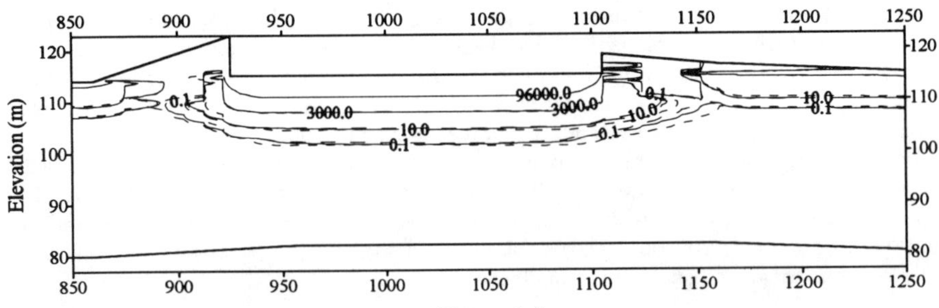

Fig. 9. Activity of ^{90}Sr (Bq/L) in the aqueous phase in 1995 for a 10-cm water depth in the basement and infiltration rates of (——) 120 mm, (- - -) 80 mm, and (– –) 140 mm per year.

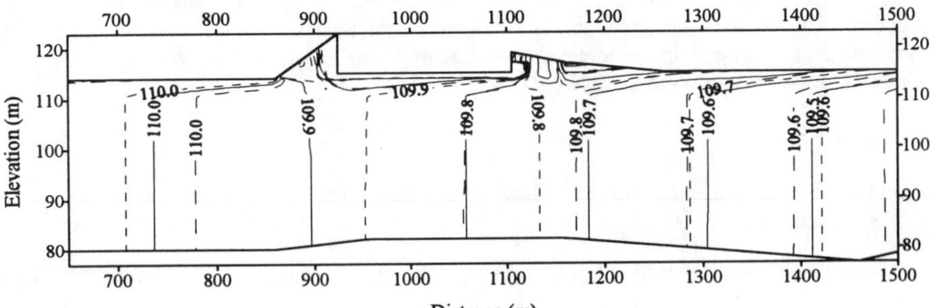

Fig. 10. Summer distribution of hydraulic head for a 120-mm infiltration rate and water depths in the basement of (——) 10 cm, (- - -) 2 cm, and (– –) 50 cm.

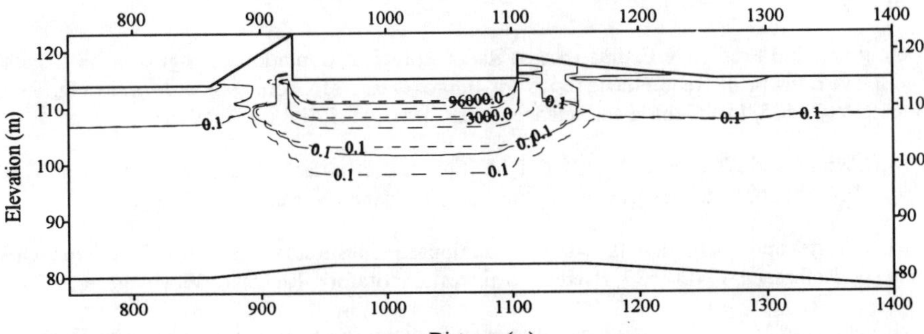

Fig. 11. Activity of ^{90}Sr (Bq/L) in the aqueous phase in 1995 for a 120-mm infiltration rate and water depths in the basement of (——) 10 cm, (- - -) 2 cm, and (– –) 50 cm.

Shelter through the subsurface environment, numerical solutions were obtained for depths of water in the basement of 0.02 m, 0.1 m, and 0.5 m. The infiltration rate was 120 mm per year. Fig. 10-11 shows the summer distribution of hydraulic head and activity of ^{90}Sr in the aqueous phase in 1995. These results indicate that the migration rate of ^{90}Sr below the Shelter increases with increasing water depth in the basement.

CONCLUSIONS

This work has concentrated on modeling ^{90}Sr and ^{137}Cs migration from the Shelter into and through the subsurface environment. A two-dimensional mathematical model accounting for the coupled transport of water and radionuclides in variably saturated porous media was used. The lack of any field studies of the subsurface environmental contamination below the Shelter, reliable data about water location, and its depth within the basement excluded complete validation of the model. However, simulation results presented in this report illustrate the basic effects of infiltration rate and basement water depth on migration rate of radionuclides from the Shelter. This migration rate increases with increasing basement water depth and decreasing infiltration rate. Simulation results will be useful for future field studies on the subsurface contamination assessment due to water leakage from the Shelter.

ACKNOWLEDGMENTS

The authors are grateful to Dr. S.P.Djepo and Dr. A.S.Skalsky from the Institute of Geological Sciences of Ukrainian National Academy of Sciences for their assistance in obtaining field and experimental data needed for modeling. This research was supported in part by IAEA via the Technical Cooperation project UKR/9/010.

REFERENCES

1. <u>Chernobyl Accident</u>, edited by V.G.Baryakhtar, Naukova dumka, Kiev, 1995, pp.170-259, 376-378.(in Russian)

2. Observation Results of Water Location within the Shelter, Report No. 2849, ISTC Ukrytie, Chernobyl, 1993, pp.23-87.(in Russian)

3. Water within the Shelter, Report No. 09-013/22, ISTC Ukrytie, Chernobyl, 1993, pp.13-31.(in Russian)

4. On-going and Predictive Estimations of Radiogeological Condition Changes of the Exclusion Zone. Development of Recommendations for Improvement of Groundwater Monitoring, Report No. 0195V022415, Inst.Geological Sciences, Kiev, 1995, pp.17-38.(in Russian)

5. M.D.White and W.E.Nichols, MSTS: Multiphase Subsurface Transport Simulator Theory Manual, PNL-8636, Pacific Northwest Laboratory, Richland, Washington, 1992.

6. M.D.White and W.E.Nichols, MSTS: Multiphase Subsurface Transport Simulator User's Guide and Reference, PNL-8637, Pacific Northwest Laboratory, Richland, Washington, 1993.

7. Groundwater Monitoring and Field Studies of Radionuclides Migration in Sites of Their Local and Regional Accumulation within Exclusion Zone, Report No. 0194V030392, Inst.Geological Sciences, Kiev, 1994, pp.90-127.(in Russian)

GENERAL CLASSIFICATION OF "HOT" PARTICLES FROM THE NEAREST CHERNOBYL CONTAMINATED AREAS

S. I. SHABALEV, B. E. BURAKOV, E. B. ANDERSON
V. G. Khlopin Radium Institute, 28, 2nd-Murinskiy Ave., St. Petersburg, 194021, RUSSIA
Phone/Fax: +7+812 346-1129, E-mail: stas@riand.spb.su

ABSTRACT

The morphology and composition both chemical and radionuclide of the main types of the solid-phase "hot" particles formed following the accident on the Chernobyl NPP have been studied by SEM, electron microprobe and gamma-spectrometry methods. Differences in many isotopes including: ^{106}Ru, ^{134}Cs, ^{137}Cs dependent upon the hot particle matrix chemical composition was observed. The classification of hot particles based upon the chemical composition of their matrices has been done. It includes three main types: 1) fuel particles with UO_X matrix; 2) fuel-constructional particles with Zr-U-O matrix, 3) hot particles with metallic inclusions of Fe-Cr-Ni. Moreover, there are more rare types of hot particles with silicate or metal matrices. It was shown that only metallic inclusions of Fe-Cr-Ni are concentrators of ^{106}Ru, which caused this nuclides assimilation in the molten stainless steel during the initial stages of the accident. Soils contamination of non-radioactive lead oxide particles in the Chernobyl NPP region were noticed. It was supposed that part of metallic lead, dropped from helicopters into burning reactor during first days of accident, was evaporated and oxidized accompanying solid oxide particles formation.

INTRODUCTION

On 26th April 1986 the active zone of the 4th Unit of the Chernobyl Nuclear Power Plant (NPP) reactor was destroyed following an accident. Dispersed radioactive products were thrown out over large distances around the Chernobyl NPP and have contaminated considerable areas of Ukraine, Belorussia, Russia, Poland and other countries [1–3]. In the nearest contaminated area the most dangerous nuclide composition is connected with dispersed solid-phase particles, called "hot particles".

The V. G. Khlopin Radium Institute (St. Petersburg, Russia) has been carrying out a detailed examination of hot particles since 1990. Special attention has been given to the investigation of individual hot particles while simultaneously compiling results on a computer database. The main goals of this work have been:

- scenario reconstruction of the initial stage of the Chernobyl accident;
- creation of computer programs (codes) about the interaction between nuclear fuel, zircaloy rods and constructional materials and also products of this interaction with the environment.

Our classification of hot particles is based upon the chemical composition of their matrices and includes three main types:

- fuel particles with UO_X matrix;
- fuel-constructional particles with Zr-U-O matrix;
- hot particles with metallic inclusions of Fe-Cr-Ni.

Certainly this classification is not exhaustive. For example multi-component particles have been found consisting of two primary phases: UO_X and Zr-U-O; Zr-U-O and ZrO_2; metallic Zr and Zr-U-O and so on [4]. Also it is difficult to classify single particle matrices of UO_X with

Mat. Res. Soc. Symp. Proc. Vol. 465 © 1997 Materials Research Society

insignificant impurities of Zr. However the application of this classification for the general appreciation of the system "chemical composition of matrices – radionuclide composition" is thought to be most suitable. Therefore in contrast to our previous papers [4, 5], we combine here all Zr-U-containing phases into the cathegory "Zr-U-O".

GENERAL CLASSIFICATION

<u>Fuel hot particles with UO_X-matrix</u>

This is the basic type of hot particles caused by the mechanical destruction and dispersion of nuclear fuel, UO_X. In the Western Plume at a distance not more than 12 km from Chernobyl NPP, fuel particles were composed of 60 to 80% by volume of the hot particles with sizes more than 10 μm.

Hot particle sizes vary in broad limits. As a rule particles with sizes from 10–20 μm to 60–80 μm are due to individual microblocks of UO_X (Fig. 1a). Larger particles with sizes up to 200 μm are more usually aggregates of microblocks (Fig. 1b, c). Radionuclide compositions of hot particles-fragments of unmodified fuel with different burnups are shown in Table I (particle ##1–5).

Fig. 1. Fuel hot particles with UO_X-matrix – the main type of hot particles in the nearest area. Backscattering electron image:
a) sole microblocks of unmodified nuclear fuel of different burn-up levels;
b) aggregate of microblocks;
c) polished section of an aggregate;
d) a polished section of a fragment of Chernobyl NPP nuclear fuel.

Table I. Data of gamma-spectrum analysis of some typical hot particles from the nearest Chernobyl contaminated areas.

#	distance to the West from the ChNPP, km	type of the matrix	Gamma-activity, Bq									
			on the 26th April 1986								on the measure day	
			^{144}Ce	^{134}Cs	^{137}Cs	^{106}Ru	^{125}Sb	^{154}Eu	^{155}Eu	^{60}Co	^{241}Am	date
1	1.5 (HP-4)	UO_X	3225±128	180±10	362±1	1359±38	26.4±4.1	9.34±1.54	6.61±0.66	—	1.6±0.4	6.03.91
2	1.5 (HP-5)	UO_X	7352±553	504±35	912±3	3596±162	61.5±11.3	17.1±3.4	16.2±1.6	—	4.2±0.7	27.02.91
3	5.5 (S-2)	UO_X	43587±2517	67±11	2318±13	5206±420	106±15	16.2±3.7	45±6	—	—	3.06.91
4	8 (3-37)	UO_X	13152±931	97±6	625±13	2867±533	36.2±4.3	1.89±0.23	11.10±0.75	—	1.35±0.20	11.04.92
5	8 (3-15)	UO_X	1928±193	48.0±4.8	138±2	637±75	8.59±1.31	1.01±0.08	2.36±0.17	—	0.56±0.06	21.05.92
6	8 (3-18)	UO_X	15778±1027	584±39	146±1	6479±227	98±8	17.4±2.7	19.4±3.0	—	5.13±1.55	4.06.91
7	8 (3-21)	UO_X	1844±173	4.90±1.27	9.67±0.09	173±48	—	2.1±1.2	2.59±0.26	—	—	10.07.91
8	8 (3-8)	UO_X	6598±254	12.60±0.52	30.3±0.2	1566±38	—	10.8±2.9	9.1±1.1	—	1.57±0.27	9.07.91
9	1.5 (HP-1)	UO_X	1857±145	—	—	426±93	3.88±0.23	1.76±0.09	3.07±0.10	—	1.75±0.10	9.12.90
10	1.5 (HP-2)	UO_X	4383±206	167±11	296±3	2021±238	27.5±5.8	5.52±1.45	8.15±0.63	—	2.83±0.63	9.12.90
11	5.5 (2-20)	Zr-U-O	41917±4352	98.1±13.2	240±3	6176±758	161±38	21.1±4.2	43.0±11.2	—	—	27.08.91
12	8 (3-7)	Zr-U-O	12980±1142	90.7±10.9	241±18	2339±398	—	13.2±1.8	—	—	—	30.10.91
13	12 (BGR1-1)	Zr-U-O	119017±5595	1046±119	2904±52	21718±123	563±60	213±22	—	—	—	9.10.91
14	5.5 (S-1)	Zr-U-O + Fe-Cr-Ni	343500±11600	1984±228	4284±73	145070±6870	2054±110	790±94	671±55	138±11	116±5	4.06.91
15	8 (3-27)	—«—	6991±140	12.3±2.2	18.0±1.1	2856±357	9.14±2.74	4.83±0.91	8.42±0.55	3.03±0.69	3.22±0.10	5.06.91
16	8 (3-9)	FeO_X+U(O?)	42.6±1.9	15.1±3.4	21.4±0.5	245±26	50.6±16.2	—	—	1.51±0.04	—	10.06.91
17	5.5 (2-7)	alumosilicate matrix	14594±1469	316±29	716±1	5372±256	98.9±9.5	29.2±6.8	25.9±2.3	—	5.05±1.05	28.05.91
18	5.5 (2-2)	serpentinite matrix + Zr-U-O	681±16	9.42±1.66	23.6±0.8	993±96	23.7±1.8	0.75±0.15	2.12±0.17	—	—	29.05.91

More rarely, fuel particles with irregular morphology have been discovered (Fig. 2), which are also distinguished by their unusual radionuclide compositions (Table I, particles 6–10). Unlike hot particles ##1–5 which evidently displayed the microblock structures particle #9 is characterized by a fine-grained surface relief (Fig. 2c) and by the complete absence of ^{134}Cs and ^{137}Cs in the gamma-spectrum. Particles ##7–8 have a molten morphology (Fig. 2b) and are typified by an absence of ^{125}Sb and by a significant decrease in the concentration of isotopes ^{134}Cs and ^{137}Cs. For particle #6 with irregular morphology, an anomalous ratio of ^{134}Cs/^{137}Cs was found.

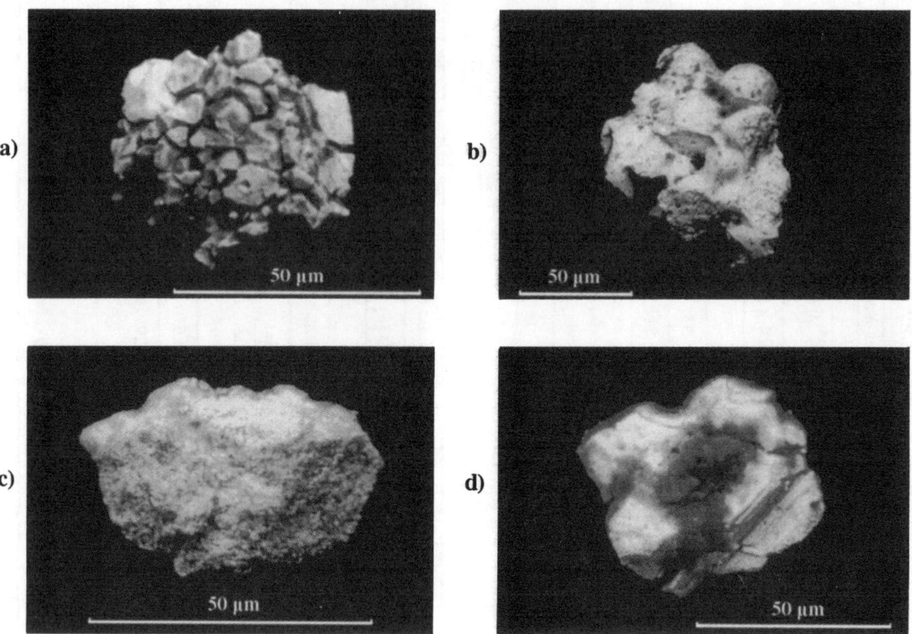

Fig. 2. Backscattering electron image of hot particles illustrating different forms of fuel's modification:
a) disturbed aggregate of microblocks (radionuclide composition of analogous particle is presented in Table I, particle #6);
b) hot particles with melted surface (analogous particles are in Lines 7 and 8 in Table I);
c) hot particle with fine-grained structure. This particle lost cesium completely (Table I, particle #9);
d) hot particle with tabular structure (particle #10 in Table I).

Fuel-constructional hot particles with a Zr-U-O matrix

This type of hot particle is composed of 20 to 40% by volume of particles in the Western Plume at a distance less than 12 km from ChNPP. They are the result of the interaction between nuclear fuel and zircaloy tubes which was accompanied by the formation of the Zr-U-O melt. As a rule, fuel-constructional particles are characterized by a molten or irregular morphology (Fig. 3a, b, c) and polyphase structure (Fig. 3d). Usually gamma-spectra analysis reveals a reduction in the main gamma-nuclides relative to ^{144}Ce (Table I, particles #11–13). There are hot particles with Zr-U-O matrix and a fragment of unmodified fuel (Fig. 3f). On the another hand we have identified a particle with zircaloy matrix and thin veins of uranium (Fig. 1A in the paper [5]).

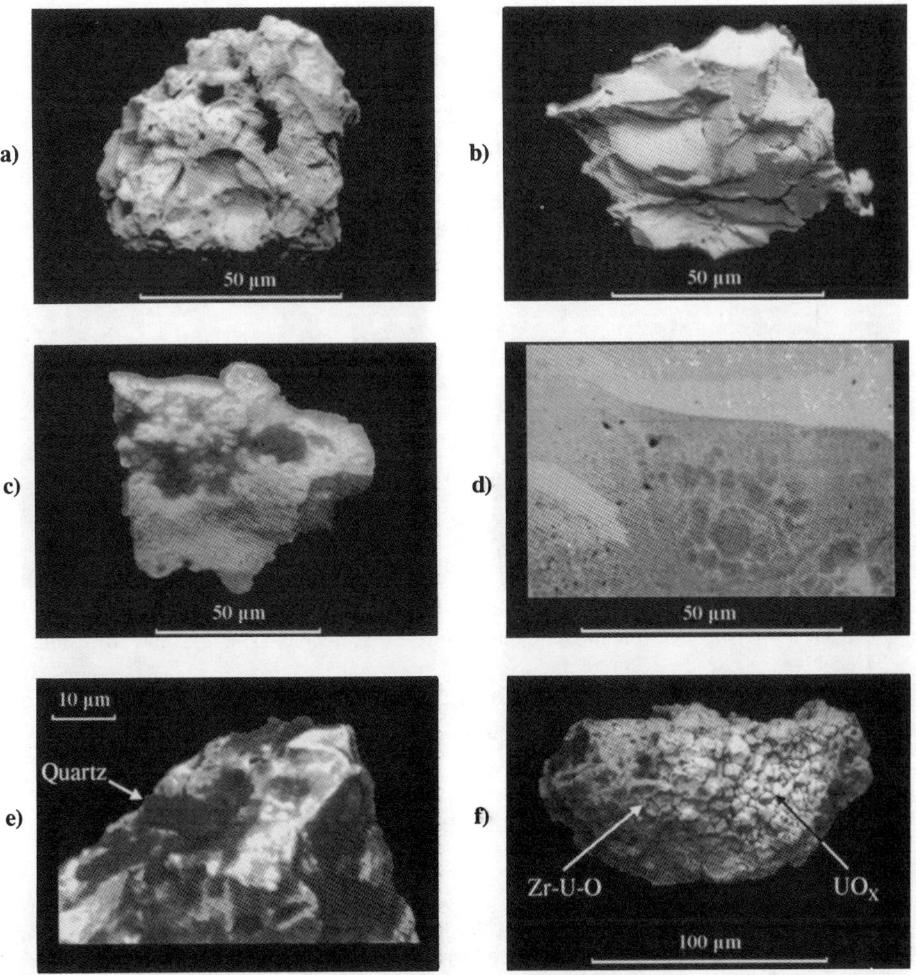

Fig. 3. Backscattering electron image of hot particles with Zr-U-O matrix:

a, b, c) typical forms of hot particles with Zr-U-O matrix. Radionuclide composition of hot particle shown in photo (a) is presented in Line 12 of Table I;

d) a fragment of polished section of hot particle with Zr-U-O matrix. Different shades of gray color correspond to different phases of Zr-U-O system;

e) a small crystal of quartz – an inclusion in Zr-U-O matrix;

f) "boundary" hot particles. The right half of the particle is an aggregate of unmodified fuel microblocks, the left one is the Zr-U-O matrix.

It was experimentally proven that particles with Zr-U-O matrices are significantly more stable in acids than fuel hot particles.

Hot particles with metallic inclusions of Fe-Cr-Ni

This type of hot particle is not so broadly distributed in the most contaminated areas of the Western Plume. As a rule in these areas, we find some fuel-construction particles with a Zr-U-O matrix containing metallic inclusions of Fe-Cr-Ni (Fig. 4). So the location of these particles in isolation is not completely true. However only metallic inclusions of Fe-Cr-Ni have concentrators of [106]Ru (Table I, particle #14), which caused this nuclide's assimilation in the molten stainless steel during the initial stages of the accident. Perhaps in the more distant areas, for example in Poland, particles of Fe-Cr-Ni form the specific type of radioactive contamination named "Ruthenium stains" [3].

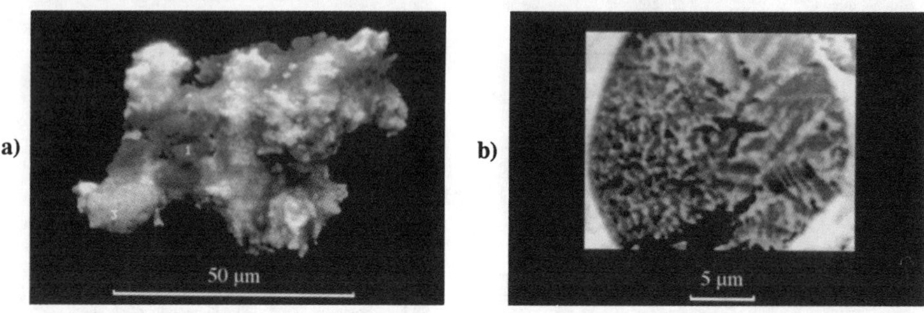

Fig. 4. Back scattering electron image of hot particles with Zr-U-O matrix and inclusions of melted steel:

a) a typical shape of such type hot particles. White color is Zr-U-O matrix, gray inclusions are Fe-Cr-Ni (melted stainless steel). Radionuclide composition of this particle (#15) is given in Table I;

b) a polished section of poliphase steel inclusion in a Zr-U-O matrix (particle #14 in Table I). Different shades of gray color correspond to different phases of U–Fe-Cr-Ni system.

Rare types of hot particles

Other types of particles are less common. We found a few hot particles with a silicate matrix (Fig. 5a, b; particles ##17, 18 in Table I), particles with iron oxide (Fig. 5c, particle #16 in Table I) or brass matrix. As a rule their activity is caused by the inclusion of very small Zr-U-O particles in fragments of structural materials outside of the reactor core.

Soil contamination of lead oxide particles in the area nearest to the ChNPP

During the first days of the accident, sand, dolomite, boron carbide, metallic lead and other materials were dropped from helicopters into the core of the 4-th Unit of the Chernobyl NPP in order to stop the fire, decrease the radionuclide release and reduce the possibility of criticality [2]. At the present time the question about the efficiency of this action is open because a significant part of that material did not get into the core reactor site and was subsequently found in the Central Hall of the 4th Unit.

More than once the possibility of lead contamination in the soil nearest to the ChNPP due to lead evaporation has been discussed. However detailed analysis of the nearest area soils have not been carried out. In our investigations we did not try to define the lead contamination level although lead oxide particles were found together with hot particles in the soil samples at a distance of 1–1.5 km to the West of the ChNPP (Fig. 5d). Some of these particles are aggregates of crystalline shaped grains and reach 60–80 μm in size.

We suppose that part of the lead was evaporated and oxidized as it was thrown on the burning reactor and then it fell on the soil surface forming lead oxide particles in the area nearest to the ChNPP. At present it is impossible to estimate quantitatively the level of lead contamination without additional examinations.

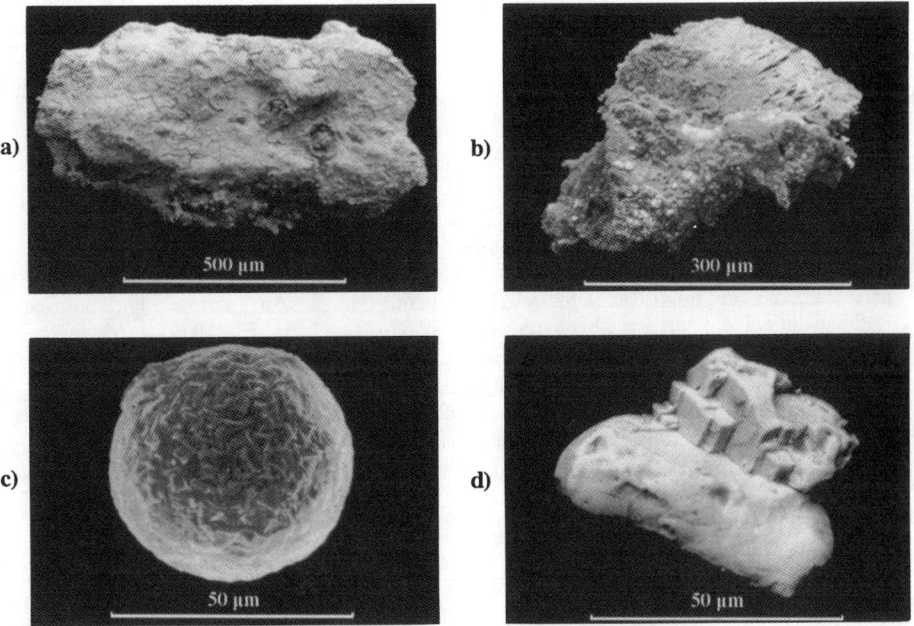

Fig. 5. Rare types of hot particles and a typical lead oxide particle. Photos (a), (b), (d) are backscattering electron image, photo (c) is secondary electron image:

a) hot particle with melted Ca-Fe-alumosilicate matrix with traces of Zr on its surface (particle #17 in Table I);
b) hot particle with serpentinite (?) matrix (gray) with inclusions of Zr-U-O (white) (particle #18 in Table I);
c) iron oxide spherule with admixture of uranium (UO$_X$?) (particle #15 in Table I);
d) typical lead oxide particle.

CONCLUSIONS

Investigations have shown that individual hot particles are a unique source of information about the process of the physico-chemical modification of the nuclear fuel and construction materials resulting from the Chernobyl accident.

The radionuclide composition of hot particles is not homogeneous and it depends on the chemical composition of their matrices. Particles differ significantly in their stability to the environment, which must be taken into consideration in making a prognosis of radionuclide migration in the contaminated areas.

ACKNOWLEDGMENTS

Presentation of this paper and the participation of the authors in this symposium were financially supported by Material Research Society and International Science Foundation / Logovaz.

REFERENCES

1. Publications series on safety No. 75-INSAG-1, Proceedings of Conference on Studies of Causes and Consequences of Chernobyl Accident, IAEA, Vienna, 1988, STI / PUB / 740, ISBN-92-0-423088-6.
2. A. A. Borovoy, Inside and outside "Sarcophagus", Chernobyl: CE IAE, 1990 (in Russian).
3. P. Jaracz, E. Piasecki, S. Mirowski, Z. Wilhelmi, Analysis of gamma-radioactivity of "hot particles" released after the Chernobyl accident. Part 2. An Interpretation, Journal of Radioanalytical and Nuclear Chemistry, Vol. 141, No. 2, p. 243–259, (1990).
4. B. E. Burakov, E. B. Anderson, B. Ya. Galkin, E. M. Pazukhin and S. I. Shabalev, Study of Chernobyl "hot" particles and fuel containing masses: implications for reconstructing the Initial Phase of the Accident, Radiochimica Acta 65, 199–202 (1994).
5. S. V. Ushakov, E. B. Anderson, B. E. Burakov, S. I. Shabalev Interaction of UO_2 and zircaloy during the Chernobyl accident, (published in these Proceedings).

AUTHOR INDEX

Kulik, D.A., 1327
Kurepin, V.A., 1327
Kuroda, Yuji, 909
Kurosawa, Susumu, 963
Kushnikov, V.V., 55
Kyser, T.K., 1201

Lacombe, J., 355
Langmuir, Donald, 769
Lawrence, W.E., 813
Lee, Joon H., 1067, 1075
Lehikoinen, J., 1011
Leonard, P.A., 941
Leturcq, G., 355
Levy, Schön S., 901
Lewis, M.A., 433
Li, H., 261, 277
Lierzer, J.R., 33
Lilga, M.A., 813
Lindberg, A., 789
Lindenmeier, C.W., 253
Liu, Beiling, 901
Liu, J., 277
Loi, Elaine, 349
Lumpkin, Gregory R., 349, 1267
Luo, J.S., 157
Lutze, Werner, 649, 1319
Lytle, F.W., 229

Macedo, Pedro B., 139
Maddrell, Ewan, 481
Maeda, T., 213
Mamakin, Igor V., 595
Manni, G., 925, 933
Marinin, D.V., 601
Marra, J.C., 105, 1281
Marra, S.L., 181
Martin, P.F., 253
Matsuda, M., 221
Matsumoto, S., 213
Matthews, M.L., 1141
Matyunin, Yu.I., 55
Mazer, J.J., 157
McConnell, J.L., 503, 511
McEwen, T., 1125
McGrail, B.P., 253, 1051
McGuire, R.K., 1083
McKinley, Ian G., 1133
Meaker, T.F., 95, 105, 1281
Melamed, A., 1011
Mertz, Carol J., 165
Mesko, M.G., 105
Messi de Bernasconi, N.B., 205
Meyer, K., 863
Meyer, M.K., 1251
Mika, M., 71
Miller, Mark L., 581, 1193, 1201
Mitani, Y., 979
Miyamoto, Shinya, 623
Mokhov, A.V., 363, 401
Molera, M., 535
Mones, E.T., 557

Monteith, J.E., 123
Moreno, L., 1037
Mukhopadhyay, Sumit, 885
Muller, Isabelle S., 139
Murphy, William M., 713
Muurinen, A., 1011

Naitoh, T., 221
Nakajima, K., 1091
Namekawa, T., 221
Nardova, Anna K., 595
Nelson, L.O., 317
Neretnieks, Ivars, 573, 855, 1037
Neu, M.P., 693, 759
Niibori, Y., 917
Nikonov, B.S., 363, 401, 417
Nishi, T., 221
Nitao, John J., 1029
Noshita, K., 221
Noy, M., 1251
Nuñez, L., 237

Oda, Chie, 751, 819
Ohashi, Hiroshi, 893
Ohi, T., 1091
O'Holleran, T.P., 1251
Ojovan, M.I., 117, 245
Ojovan, N.V., 245
Okihara, Mitsunobu, 971, 1019
Olin, M., 847
Ollila, K., 549
Omelianenko, B.I., 363, 371, 401
Orth, R.J., 813

Paffett, M.T., 941
Palmer, P.D., 693
Pankratov, D.V., 1117
Pansoy-Hjelvik, M.E., 941
Pareizs, J.M., 95, 111, 1281
Pazukhin, E.M., 1289
Pedersen, J., 287
Peeler, D.K., 105, 1281
Pegg, Ian L., 139
Perea, Nuria Marcos, 1153
Pérez, I., 565
Perić, Aleksandar, 311
Petit, J.C., 295
Pigford, T.H., 1099, 1273
Pitkänen, P., 1011
Plodinec, M.J., 181
Porter, F.M., 683
Porth, R.J., 511
Pradin, V., 827
Prastalo, S., 205
Prince, Kathryn, 349
Proskura, N.I., 1335

Quillin, K., 287

Rai, D., 729
Ramsey, W.G., 95, 105, 111, 1281
Ramspott, Lawrence D., 1029

SUBJECT INDEX